TEACHER WRAPAROUND EDITION

Chemistry

CONCEPTS AND APPLICATIONS

GLENCOE

McGraw-Hill

New York, New York   Columbus, Ohio   Mission Hills, California   Peoria, Illinois

A GLENCOE PROGRAM

Chemistry:
Concepts and Applications

Student Edition
Teacher Wraparound Edition
Laboratory Manual Student Edition
Laboratory Manual Teacher Edition
Study Guide Student Edition
Study Guide Teacher Edition
ChemLab and MiniLab Worksheets
Section Focus Transparency Package
Basic Concepts Transparency Package
Problem Solving Transparency Package
Mastering Concepts in Chemistry Software
Consumer Chemistry
Chapter Review and Assessment
Applying Scientific Methods in Chemistry

Tech Prep Applications
Chemistry and Industry
Critical Thinking/Problem Solving
Lesson Plans
Spanish Resources
CD-ROM Multimedia System
Videodisc Program
Problems and Solutions Manual
Computer Test Bank
Supplemental Practice Problems
MindJogger Videoquizzes
Chapter Summaries, English/Spanish
Audiocassettes

AUTHORS
CHERYL WISTROM
JOHN PHILLIPS
VICTOR STROZAK

CONTRIBUTING WRITERS
NICHOLAS HAINEN
RICHARD G. SMITH

SAFETY CONSULTANT
DOUGLAS MANDT

Glencoe/McGraw-Hill
A Division of The McGraw-Hill Companies

ISBN 0-02-827453-9

Send all inquiries to
Glencoe/McGraw-Hill
936 Eastwind Drive,
Westerville, Ohio 43081

Printed in the United States of America
2 3 4 5 6 7 8 9 0 - QPK/MC - 03 02 01 00 99 98 97 96

CONTENTS
Teacher Guide

CONTENTS IN BRIEF

Student Edition

A chemistry program for all your students

Chemistry: Concepts and Applications

Whether your students are in a tech-prep curriculum or bound for college, *Chemistry: Concepts and Applications* will help more of them succeed in chemistry than ever before. Chemistry may be the most challenging course your students take in high school, and traditional chemistry texts often make it difficult for students to succeed. They overly burden students with quantitative exercises and fail to explain abstract concepts in terms they can understand. Glencoe's new *Chemistry: Concepts and Applications* combines a variety of creative approaches to guide your students successfully through the submicroscopic world of chemistry.

✓ Focus on developing core concepts

First and foremost, *Chemistry: Concepts and Applications* focuses on developing the core concepts of chemistry. From the first chapter to the final chapter, the text gradually and incrementally expands students' understanding of the fundamental structure and properties of matter. Students learn the basics of chemistry and use the periodic table as an information tool. Then students explore bonding, kinetic theory, and the nature of matter in solution. Finally, students extend their knowledge to special areas of chemistry such as biochemistry, electrochemistry, and nuclear chemistry.

✓ Unique presentation for greater understanding

While you'll find the topics that organize the text familiar, the presentation of these topics is unique. The instructional objective on every page is to make the abstract more real. Each lesson begins with the macroscopic world of the student, with the chemistry observable in everyday life. Analogies, student-friendly narrative, and a profusion of illustrations relate the observable to the submicroscopic level of electrons and molecular structures. As each lesson unfolds, still more examples of chemistry in everyday life and industry are introduced.

Use Glencoe's interactive CD-ROM and videodisc programs to enhance the learning and presentation of chemistry.

✓ The right amount of quantitative chemistry

In *Chemistry: Concepts and Applications*, you'll find just the right emphasis on the quantitative aspects of chemistry because understanding chemistry, not solving mathematical problems, is the primary goal of the program. Students are not asked to do more exercises than are necessary to reinforce a concept. Plus, the text gives students the problem-solving tools they need, including carefully modeled problem solutions.

*If your students need more practice, select additional problems from the **Supplemental Practice Problems** book.*

$8e^-$ $2e^-$ $1e^-$

✓ Integrated hands-on experiences

Full-period *ChemLabs* and shorter *MiniLabs* are integrated within each chapter where they best support the concepts being presented. *ChemLabs* include a variety of large-scale and small-scale laboratory investigations. Because *MiniLabs* require minimal effort to set up and take down, they offer an ideal way to integrate hands-on experiences when less than a class period is available. A separate Laboratory Manual offers many other laboratory options that will suit your requirements.

Consumer Chemistry and Tech Prep Applications books offer even more hands-on activities.

✓ Connections to the world outside the classroom.

No other program shows your students the relevance of chemistry to the world outside the classroom in as many ways as does *Chemistry: Concepts and Applications*. *Everyday Chemistry, How It Works, Chemistry and Society*, and *People in Chemistry* are just some of the features that appear throughout the text to reveal the diverse ways in which chemistry affects their lives.

Chemistry and Industry and Consumer Chemistry books provide additional real-world connections to chemistry.

✓ Teaching support and classroom resources for every situation

From the comprehensive and easy-to-find-everything *Teacher Wraparound Edition* to more than a score of ancillary components, *Chemistry: Concepts and Applications* gives you manageable options for all types of students and for all teaching situations. For example, you'll find the strategies and resources for tech-prep students especially useful. The *Section Focus Transparencies* give you a successful technique to engage students in each new section of the text.

✓ Technology choices that fit your requirements

From videoquizzes to interactive CD-ROM, *Chemistry: Concepts and Applications* offers you a variety of technology-based materials to fit your instructional needs and your budget. The CD-ROM provides individual and small groups of students engaging opportunities to manipulate animated chemical models and to conduct experiments in a computer-simulated lab environment. Videodiscs allow you to show costly or time-consuming demonstrations to the entire class. Test generators, audiocassettes, and videoquizzes are also available.

*See pages 14T-23T for more information about the wealth of print and technology resources available with **Chemistry: Concepts and Applications**.*

Scope and Sequence

Chemistry: Concepts and Applications is a chemistry program for the majority of high school students who want to learn chemistry. There are numerous chemistry programs available that are designed for the college preparatory student who will go on to take college chemistry. However, with increased emphasis on science education for all students, modern textbooks must also serve students who need to learn about the nature, composition, and changes of matter that they encounter in everyday life. Many of these students will go to college in nontechnical areas. Many will enter post-secondary training in practical fields.

The majority of these students should understand that chemistry is a quantitative science but do not need to practice hundreds of mathematical computations. Likewise, they are unlikely to need extended abstract theory in chemistry. In Chemistry: Concepts and Applications, Glencoe provides you with a conceptual chemistry program that includes just the right amount of theory and quantitative ideas to enable students to understand the applications of chemistry.

SCOPE AND SEQUENCE

Fundamentals Chemistry: Concepts and Applications is divided into three blocks of chapters. The first six chapters constitute a survey of the most basic ideas of chemistry with an emphasis on reaction chemistry. So that students can explain what takes place in chemical reactions, a sound atomic model is developed very early (Chapter 2) and the underlying principles of bonding are presented (Chapter 4). Emphasis is placed on the use of the periodic table as a practical source of chemical information from the beginning of the course, including Chapters 1 and 2, even before it is taken up as a chapter topic in Chapter 3.

Chapter 1	Chemistry: The Science of Matter
Chapter 2	Matter Is Made up of Atoms
Chapter 3	Introduction to the Periodic Table
Chapter 4	Formation of Compounds
Chapter 5	Types of Compounds
Chapter 6	Chemical Reactions and Equations

In-Depth Explorations The second block of chapters, 7 through 15, expands upon the basic ideas presented in the first block. More details of electron structure are presented and then the full potential of the periodic table as a predictor of chemical properties is revealed. Using this as a foundation, the whole spectrum of bonding is presented. The interaction of energy and the structure of matter are emphasized in chapters dealing with kinetic theory, changes in states, and the behavior of gases (Chapters 10 and 11). Quantitative relationships in reactions are introduced in Chapter 12, then water, water solutions, acids, bases, and acid-base reactions are covered in an integrated series of chapters (13-15).

Chapter 7	Completing the Model of the Atom
Chapter 8	Periodic Properties of the Elements
Chapter 9	Chemical Bonding
Chapter 10	The Kinetic Theory of Matter
Chapter 11	Behavior of Gases
Chapter 12	Chemical Quantities
Chapter 13	Water and Its Solutions
Chapter 14	Acids, Bases, and pH
Chapter 15	Acids and Bases React

Extensions The final block of chapters, 16 through 21, take up several topics that may be considered extensions of basic chemistry. At this point, teachers may want to jump to topics that are of special interest or importance to classes. The topics covered can be divided into three categories: redox and electrochemistry, organic and biochemistry, and energy and nuclear chemistry.

Chapter 16	Oxidation-Reduction Reactions
Chapter 17	Electrochemistry
Chapter 18	Organic Chemistry
Chapter 19	The Chemistry of Life
Chapter 20	Chemical Reactions and Energy
Chapter 21	Nuclear Chemistry

Meeting National Science Standards

The *National Science Education Standards*, recently published by the National Research Council and representing the contribution of thousands of educators and scientists, offers a comprehensive vision of a scientifically literate society. The Standards describe not only what students should know, but also offer guidelines for science teaching and assessment. If you are using, or plan to use, the standards to guide changes in your science curriculum, you can be assured that *Chemistry: Concepts and Applications* aligns exceedingly well with *The National Science Education Standards*.

CONTENT STANDARDS

Chemistry: Concepts and Applications does an exceptional job helping students understand and perform the science content standards. Below are just a few of the ways Glencoe's program supports the content standards.

Unifying Concepts and Processes in Science By using *Chemistry: Concepts and Applications*, a teacher will be able to meet the national standards for unifying concepts and processes. In every chapter, themes that run throughout the textbook help students see the big picture of chemistry. Themes including energy, conservation, equilibrium, and macro to submicroscopic interpretation of chemical reactions will help students build upon their knowledge, step by step, and see relationships to the world around them.

Science as Inquiry *Chemistry: Concepts and Applications* reflects the real nature of science on every page. The *ChemLabs* and *MiniLabs* in every chapter give students frequent opportunities to learn and apply a variety of skills and science processes. Real-world applications and open-ended activities in the CD-ROM, Tech Prep Applications, and Consumer Chemistry offer a wide range of options to address this standard.

Physical Science (Standards for Grades 9-12) *Chemistry: Concepts and Applications* will provide your students with an in-depth understanding of the physical science standards.

Science and Technology *Chemistry: Concepts and Applications* uses numerous examples of how chemistry is important in technological development in the world. Not only are there special features that deal with *Chemistry and Society* and *Chemistry and Technology*, but throughout the text are references to how chemistry is used in industry and everyday life.

Science in Personal and Social Perspectives *Chemistry: Concepts and Applications* is designed to involve more students in the excitement of everyday chemistry. Core concepts in chemistry are clearly demonstrated in a logical developmental sequence with exciting visuals and applications that will maintain student interest in chemistry throughout their lives. The narrative is student friendly, using many everyday associations to keep the reader engaged. Analogies and familiar processes are used throughout the text to enhance students' understanding of core concepts.

History and Nature of Science *Chemistry: Concepts and Applications* presents historical perspectives on the development of chemistry concepts and principles. Students develop a clear understanding that chemistry is an ongoing human enterprise.

ASSESSMENT STANDARDS

The assessment standards are supported by many of the components that make up the *Chemistry: Concepts and Applications* program. The *Teacher Wraparound Edition* and *Teacher Classroom Resources* provide many opportunities for assessing students' understanding of important concepts and ability to perform a wide range of skills. Ideas for portfolios, performance activities, written reports, and other assessment activities accompany every lesson. For more details about assessment ideas and resources, see pages 20T-21T and 40T-44T. Rubrics and performance task assessment lists can be found in *Glencoe's Professional Series* booklet *Performance Assessment in the Science Classroom*.

TEACHING STANDARDS

Chemistry: Concepts and Applications helps you make the most of every instructional moment. It offers more strategies and a wider variety of effective strategies than any other program for presenting the content, processes, and skills of chemistry.

Unifying Themes in Chemistry

In *Chemistry: Concepts and Applications Student Edition*, themes provide a framework for integrating chemistry core concepts. Themes in chemistry provide the big ideas that link the core concepts. Although there are several possible themes around which to unify the study of chemistry, we have chosen five: Energy, Macro to Submicroscopic, Conservation, Systems and Interactions, and Equilibrium and Change.

Energy Energy is the ability of an object to change itself or its surroundings—the capacity to do work. Nearly all chemical and physical changes either take in or give up energy. The theme of energy goes hand in hand with the theme of equilibrium and change because the availability of energy can determine whether or not a change will occur and to what degree it occurs.

Macro to Submicroscopic A central idea in chemistry is that the physical and chemical properties of matter are a consequence of the structure and arrangement of matter's basic particles—atoms, molecules, and ions. We observe the behavior of matter on a macroscopic scale and then attempt to explain that behavior by using models of matter at the sub-microscopic level, that is, at the level of individual particles and their interactions.

Conservation The theme of conservation is a basic one in chemistry. Matter is conserved in all chemical and physical changes. Energy is conserved because it changes from one form to another without loss or gain. Even in nuclear reactions where some matter is converted to energy, a broader conservation law holds, stating that the sum of all matter and energy is constant. Through these laws, the principles of chemistry extend further into environmental, economic, and social contexts as students realize that nothing can really be thrown away and that energy released in industrial processes must always affect the surroundings in some way.

Systems and Interactions A system can be as small as a nucleus' atom with its electrons or as large as the stars that make up a galaxy. By defining the boundaries of the system, one can study the interactions among its parts. The interactions may be the force of attraction between the positively charged nucleus and its negatively charged electrons or they may be the way two molecules come together in a reaction.

Equilibrium and Change A system that is stable is constant. Often, this stability is the result of a system being in equilibrium. If a system is not stable, it will undergo change. Changes in an unstable system may be characterized as trends as in electron configurations, cycles as in the actions of enzymes, or irregular as in radioactive decay.

THEME DEVELOPMENT • Each chapter of *Chemistry: Concepts and Applications* incorporates at least one of these five themes

Chapters	Energy	Macro to Submicro	Conservation	Systems and Interactions	Equilibrium and Change
1 Chemistry: The Science of Matter		✓	✓		
2 Matter Is Made up of Atoms		✓	✓		
3 Introduction to the Periodic Table		✓			
4 Formation of Compounds	✓	✓			
5 Types of Compounds		✓			
6 Chemical Reactions and Equations				✓	✓
7 Completing the Model of the Atom		✓			
8 Periodic Properties of the Elements		✓			
9 Chemical Bonding		✓			
10 The Kinetic Theory of Matter	✓	✓			
11 Behavior of Gases		✓		✓	
12 Chemical Quantities		✓	✓		
13 Water and Its Solutions		✓			
14 Acids, Bases, and pH				✓	✓
15 Acids and Bases React					✓
16 Oxidation-Reduction Reactions		✓		✓	
17 Electrochemistry	✓			✓	
18 Organic Chemistry		✓			
19 The Chemistry of Life	✓			✓	
20 Chemical Reactions and Energy	✓		✓		
21 Nuclear Chemistry	✓		✓		

Developing and Applying Thinking Processes

Science is not just a collection of facts for students to memorize. Rather, it is a process of applying observations and intuitions to situations and problems, formulating hypotheses, and drawing conclusions. This interaction of the thinking process with the content of science is the core of science and should be the focus of science study. *Chemistry: Concepts and Applications* encourages the interaction between science as a process and thinking skills by offering many hands-on activities that provide excellent opportunities for students to expand their comprehension of important core concepts while practicing the basic science process skills and applying their thinking skills.

BASIC SCIENCE PROCESS SKILLS

Observing The most basic process is observing. Through observation—seeing, hearing, touching, smelling, tasting—the student begins to acquire information about an object or event. Observation allows a student to gather information regarding size, shape, texture, or quantity of an object or event.

Measuring is a way of quantifying information in order to classify or order it by magnitude, such as area, length, volume, or mass. Measuring can involve the use of instruments and the necessary skills to manipulate them.

Using Numbers Numbers are used to quantify information, including variables and measurements. Quantified information is useful for making comparisons in tables or graphs and for classifying data or objects.

Classifying One of the simplest ways to organize information gathered through observation is by classifying. Classifying involves the sorting of objects according to similarities or differences.

THINKING SKILLS

Predicting involves suggesting future events based on observations and inferences about current events. Reliable predictions are based on making accurate observations and measurements.

Recognizing Cause and Effect involves observing actions or events and making logical inferences about why they occur. Recognizing cause and effect can lead to further investigation to isolate a specific cause of a particular event.

Interpreting Data involves synthesizing information to make generalizations about the problem under study and apply those generalizations to new problems. Interpreting data may, therefore, involve many of the other process skills, such as predicting, inferring, classifying, and using numbers.

Sequencing involves arranging objects or events in a particular order and may imply a hierarchy or a chronology. Developing a sequence involves the skill of identifying relationships among objects or events. A sequence may be in the form of a numbered list or a series of objects or ideas with directional arrows.

Comparing and Contrasting Comparing is a way of identifying similarities among objects or events, while contrasting identifies their differences.

Inferring Inferences are logical conclusions based on observations and are made after careful evaluation of all the available facts or data. Inferences are a means of explaining or interpreting observations and are based on making judgments.

Using Space/Time Relationships This process skill involves describing the spatial relationships of objects and how those relationships change with time. For example, a student may be required to describe the motion, direction, or shape of an object.

Formulating Models A model is a way to concretely represent abstract ideas or relationships. Models can also be used to predict the outcome of future events or relationships. Models may be expressed physically in three dimensions: verbally, mathematically, or diagrammatically.

Communicating information is an important part of science. Once all the information is gathered, it is necessary to organize the observations so that the findings can be considered and shared by others. Information can be presented in tables, charts, a variety of graphs, or models.

CHAPTERS

Basic Process Skills	1	2	3	4	5	6	7	8	9	10	11	12	13	14	15	16	17	18	19	20	21
Observing	✓	✓	✓	✓	✓	✓	✓	✓	✓	✓	✓	✓	✓	✓	✓	✓	✓	✓	✓	✓	✓
Measuring	✓	✓			✓		✓	✓		✓	✓	✓	✓	✓	✓	✓		✓	✓		✓
Using Numbers	✓	✓			✓		✓	✓				✓			✓				✓	✓	✓
Classifying	✓	✓	✓	✓	✓				✓				✓	✓		✓	✓		✓		✓
Thinking Skills																					
Predicting				✓			✓		✓	✓	✓		✓	✓			✓				
Recognizing Cause and Effect				✓	✓	✓		✓			✓	✓	✓					✓	✓		
Interpreting Data		✓			✓	✓	✓		✓	✓	✓	✓		✓	✓	✓	✓	✓	✓		✓
Sequencing				✓			✓		✓		✓					✓					
Comparing and Contrasting				✓	✓		✓	✓			✓						✓	✓		✓	
Inferring	✓	✓	✓	✓		✓	✓	✓	✓		✓		✓								✓
Using Space/Time Relationships			✓			✓			✓		✓			✓							
Formulating Models			✓			✓			✓			✓									✓
Communicating	✓	✓	✓	✓	✓	✓	✓	✓	✓	✓	✓	✓	✓	✓	✓	✓	✓	✓	✓	✓	✓

Single-Class Scheduling and Block Scheduling

Chemistry: Concepts and Applications provides a complete selection of core concepts that are presented in a way to meet the needs of all your students. As the teacher, you are in the best position to design a chemistry course that meets the needs of your individual students and classes. To assist you in planning your course for long-range and daily teaching, the following Course Planning Guide is provided.

If you prefer to teach chemistry in a different sequence, *Chemistry: Concepts and Applications* is adaptable to your needs. Following the Course Planning Guide, you will find tables of Alternate Sequences showing how you can tailor the program to your preferences.

Chemistry: Concepts and Applications may be used in a full-year, two-semester program that is comprised of 180 periods of approximately 45 minutes each. This type of schedule is represented in the table under the heading of Single-Class Scheduling. This table also outlines a plan under the heading Block Scheduling for schools that use a block scheduling system. With block scheduling, it is assumed that the course will be taught for 90 class periods of approximately 90 minutes each.

Please remember that the planning guide is provided as an aid in planning the best course for your students. Use this guide in relation to your particular situation, curriculum requirements, and ability levels of your students.

Course Planning Guide for *Chemistry: Concepts and Applications*

COURSE PLANNING GUIDE

Course Planning Guide for *Chemistry: Concepts and Applications*

Chapter/Section	Single-Class Scheduling (180 days)	Block Scheduling (90 days)
17 Electrochemistry	**6**	**3**
17.1 Electrolysis: Chemistry from Electricity	3	1 1/2
17.2 Galvanic Cells: Electricity from Chemistry	2	1
Chapter Review	1	1/2
18 Organic Chemistry	**10**	**5**
18.1 Hydrocarbons	5	2 1/2
18.2 Substituted Hydrocarbons	2	1
18.3 Plastics and Other Polymers	2	1
Chapter Review	1	1/2
19 The Chemistry of Life	**5**	**2 1/2**
19.1 Molecules of Life	3	1 1/2
19.2 Reactions of Life	1	1/2
Chapter Review	1	1/2
20 Chemical Reactions and Energy	**12**	**6**
20.1 Energy Changes in Chemical Reactions	4	2
20.2 Measuring Energy Changes	4	2
20.3 Photosynthesis	3	1 1/2
Chapter Review	1	1/2
21 Nuclear Chemistry	**12**	**6**
21.1 Types of Radioactivity	4	2
21.2 Nuclear Reactions and Energy	4	2
21.3 Nuclear Tools	3	1 1/2
Chapter Review	1	1/2
TOTAL	**(180 days)**	**(90 days)**

Resources For All Your Needs

In addition to the wide array of instructional options provided in the student and teacher editions, *Chemistry: Concepts and Applications* offers an extensive list of support materials and program resources. Some of these materials offer alternative ways of presenting your chemistry program, others provide tools for reinforcing core concepts and evaluating student learning, and still others will help you extend and enrich your course. You won't have time to use them all, but the ones you use will help you make the best use of the time you have.

Hands-On Learning

Laboratory Manual, Student and Teacher Editions

If you want more hands-on options, the **Laboratory Manual** offers you at least two more labs for each chapter. There is a selection of both traditional and small-scale labs throughout. A complete list of chemicals and equipment is provided for convenient ordering of supplies, and basic lab and chemical safety guidelines are included.

ChemLab and MiniLab Worksheets

Each of the ChemLabs and MiniLabs in the student text is also available in copymaster form in *ChemLab and MiniLab Worksheets.* As they proceed with a ChemLab or MiniLab, have students fill in their observations and data on the expanded data tables provided in these worksheets.

Tech Prep Applications

With these *Tech Prep Applications* worksheets, your students will learn more about careers in chemistry and how everyday problems can be solved by using their knowledge of chemistry. They can be the problem-solving manager of the Sherlock Chemical Company, or they can follow the step-by-step procedures in carrying out a simple hands-on activity that demonstrates the applications of chemistry in society.

PROGRAM RESOURCES

Review & Reinforcement

Study Guide, Student and Teacher Editions

Use section-by-section masters from the *Study Guide* book to reinforce the core concepts presented in the student text. These worksheets are ideal for students of average and below-average ability. A consumable student edition of the *Study Guide* is also available.

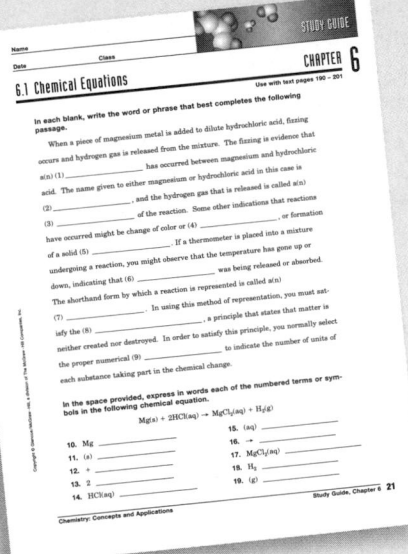

Supplemental Practice Problems

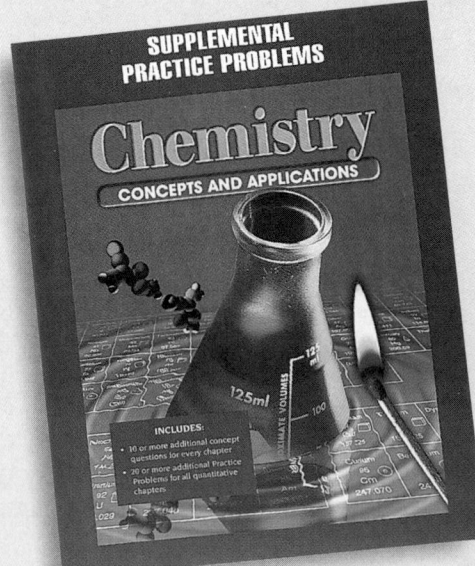

For further review and reinforcement, the *Supplemental Practice Problems* booklet provides you with ten or more questions for each chapter. More practice problems are provided for those chapters that include quantitative chemical concepts.

Enrichment & Application

Critical Thinking/Problem Solving

Challenge students to apply their critical thinking and problem-solving abilities with the *Critical Thinking/Problem Solving* book. It is especially suitable for average and above-average ability students.

Consumer Chemistry

Show students how chemistry affects their everyday lives by using worksheets from *Consumer Chemistry*. There is one for every chapter.

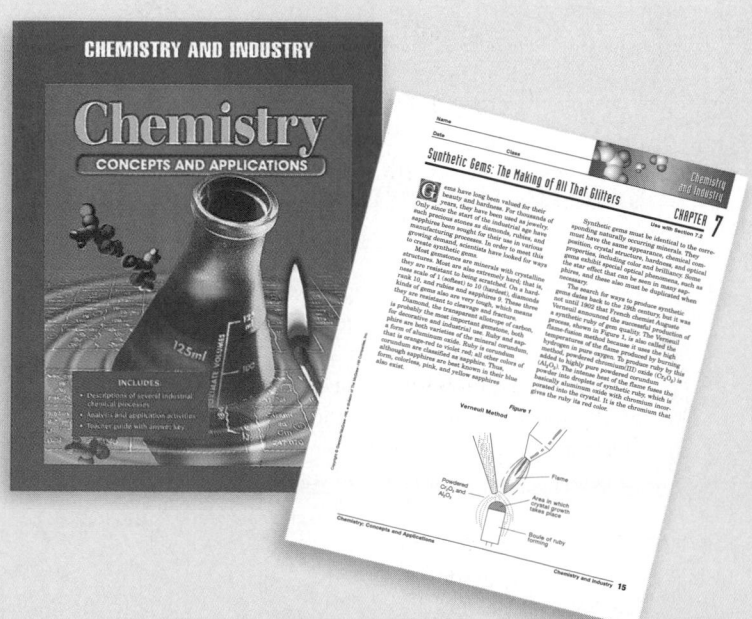

Chemistry and Industry

Students will learn about important commercial chemical processes as they use the selections from *Chemistry and Industry.*

Teaching Aids

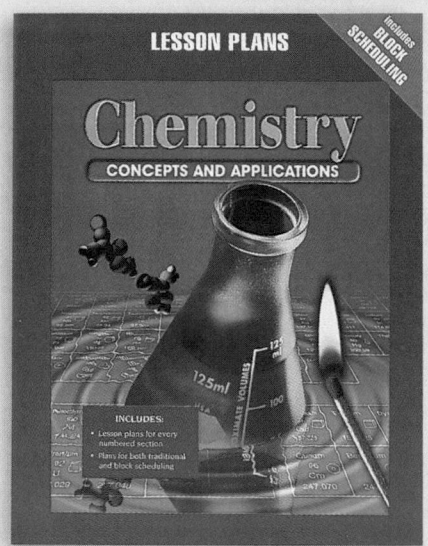

Lesson Plans, including Block Scheduling

Chapter-by-chapter **Lesson Plans** will help you organize your lessons more efficiently, either with traditional single-period scheduling or block scheduling.

Problems and Solutions Manual

For your convenience, the **Problems and Solutions Manual** contains all solutions and answers to every Section Review and Chapter Review question, and for every Practice Problem from each chapter.

Spanish Resources

Help your Spanish-speaking students get more out of your science lessons by reproducing pages from the **Spanish Resources** book. In addition to the complete English/Spanish glossary, the book contains translations of all objectives, key terms, and summaries for each chapter of the student text.

Three Transparency Packages

Enhance your presentation of science concepts with one or all of the three color transparency packages.

Section Focus Transparencies and Masters

The *Section Focus Transparency Package* includes more than 40 full-color transparencies, each of which introduces one section of the student text. The package contains a booklet of copymasters of each transparency.

Basic Concepts Transparencies, Masters, and Worksheets

The *Basic Concepts Transparency Package* includes two or more full-color transparencies per chapter designed to teach basic chemistry concepts. A transparency master and worksheet accompany each transparency.

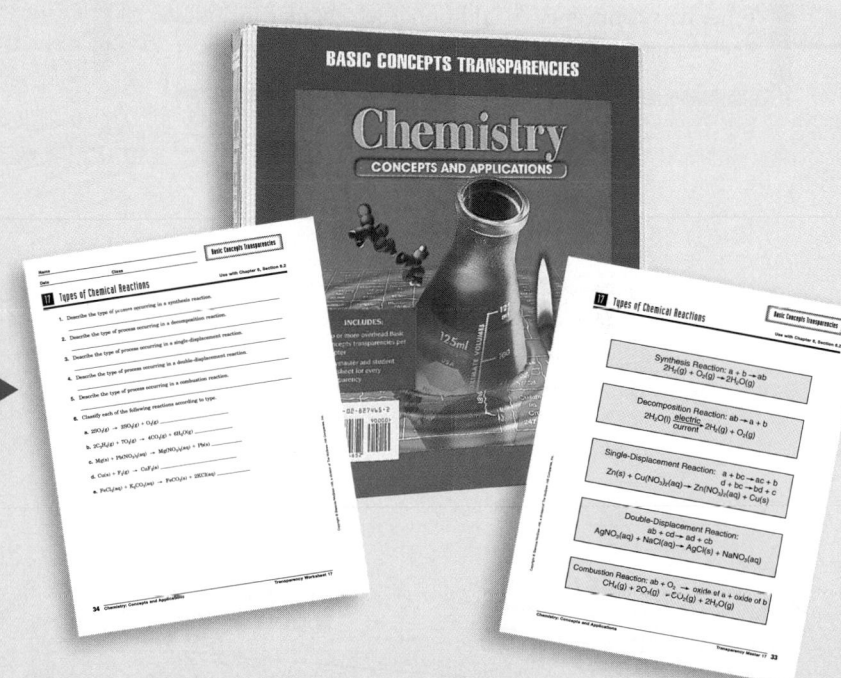

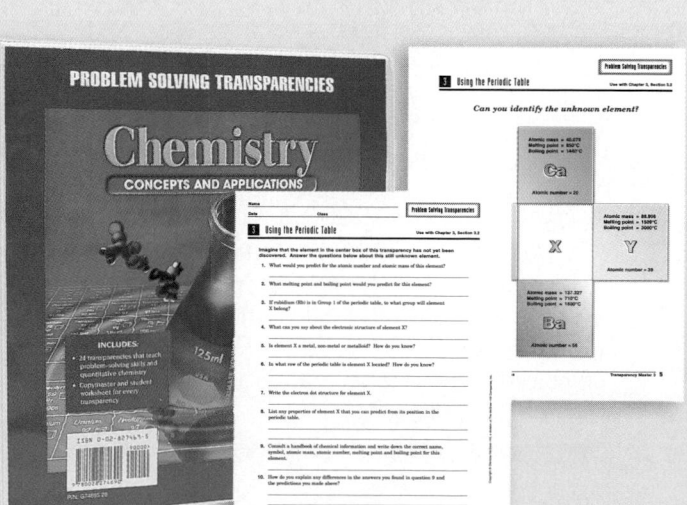

Problem Solving Transparencies, Masters, and Worksheets

The *Problem Solving Transparency Package* includes at least one transparency per chapter designed to help students solve problems, write formulas, or balance equations. A transparency master and worksheet accompany each transparency.

Assessment

Chapter Review and Assessment

CHAPTER REVIEW and ASSESSMENT

Chemistry
CONCEPTS AND APPLICATIONS

INCLUDES:
- Six-page assessment for each chapter
- Vocabulary review
- Concept review
- Skill assessment
- Performance assessment
- Answer pages

Assess student learning and performance through a variety of questioning strategies and formats given in *Chapter Review and Assessment.* For each chapter, a one-page Vocabulary Review precedes five pages of test items that cover activity procedures and analysis as well as core chemistry concepts. In addition, each chapter includes a performance assessment and corresponding rubrics.

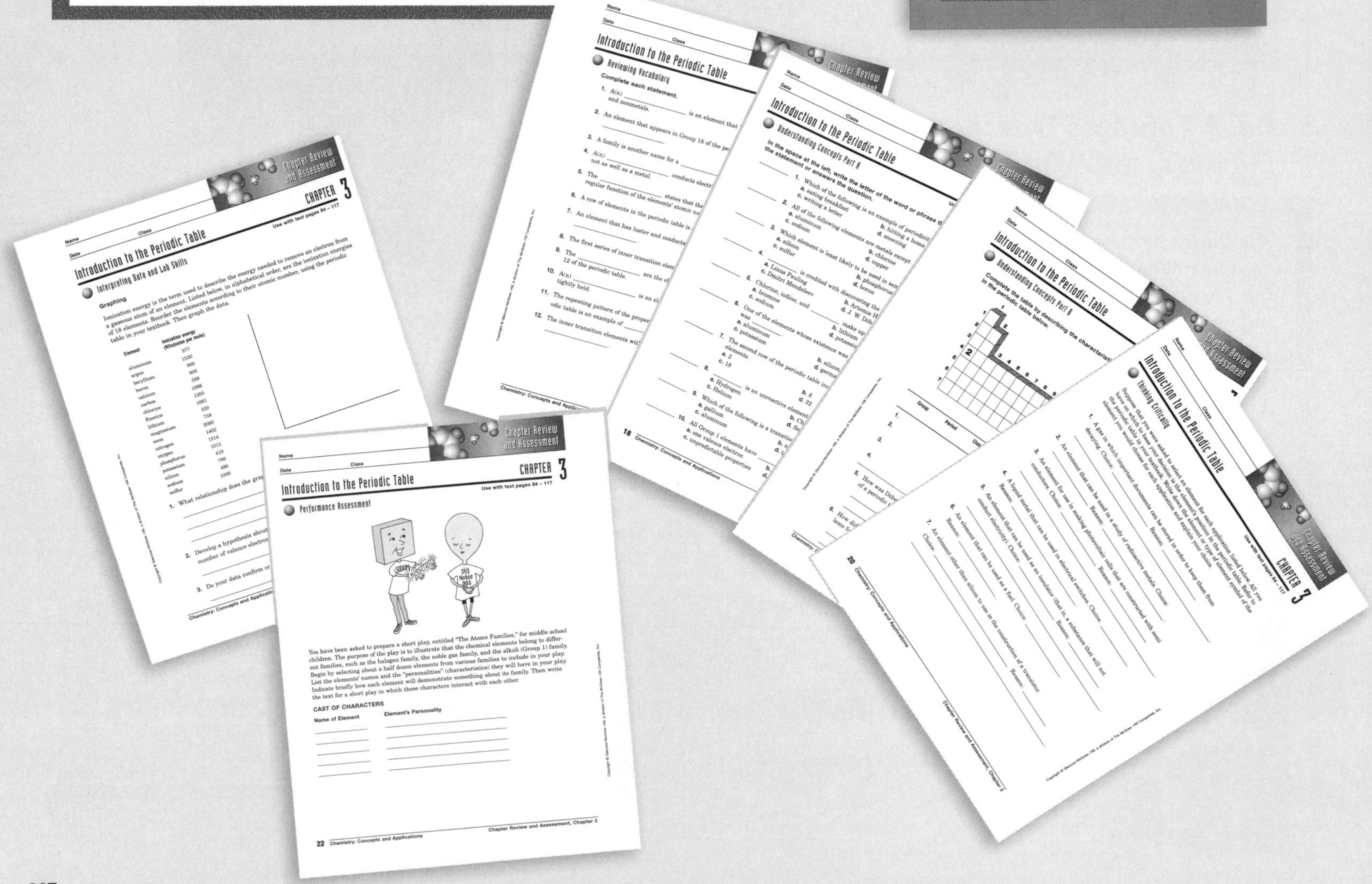

Computer Test Bank
(DOS and Macintosh)

Design and create your own test instruments from our *Computer Test Bank* for DOS and Macintosh. Select your own test items by objective from two different levels of difficulty, or write and edit your own.

Performance Assessment
in the Science Classroom

Use the performance task assessment lists and rubrics in *Performance Assessment in the Science Classroom* to assess students' work.

Alternate Assessment
in the Science Classroom

Alternate Assessment in the Science Classroom provides additional background and examples of performance assessment.

Glencoe Technology

Glencoe's *Chemistry: Concepts and Applications* offers a broad array of new technology to enhance your students' chemistry experience.

CD-ROM Multimedia System

Help your students visualize submicroscopic behavior of atoms and bonding by showing them three-dimensional animations. Involve them in interactive explorations to practice basic chemistry skills. Have them conduct lab experiments on the computer without using a gram of chemicals. The *Chemistry: Concepts and Applications CD-ROM Multimedia System* will excite all your students by providing them with their own chapter-by-chapter, personalized learning guide.

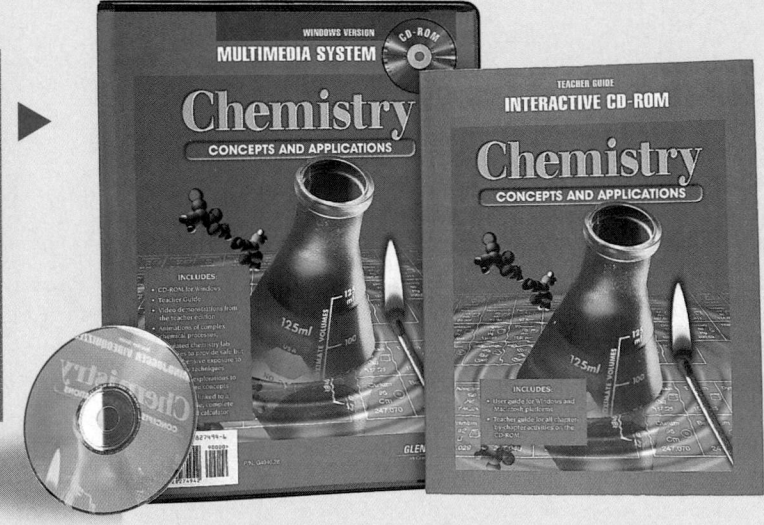

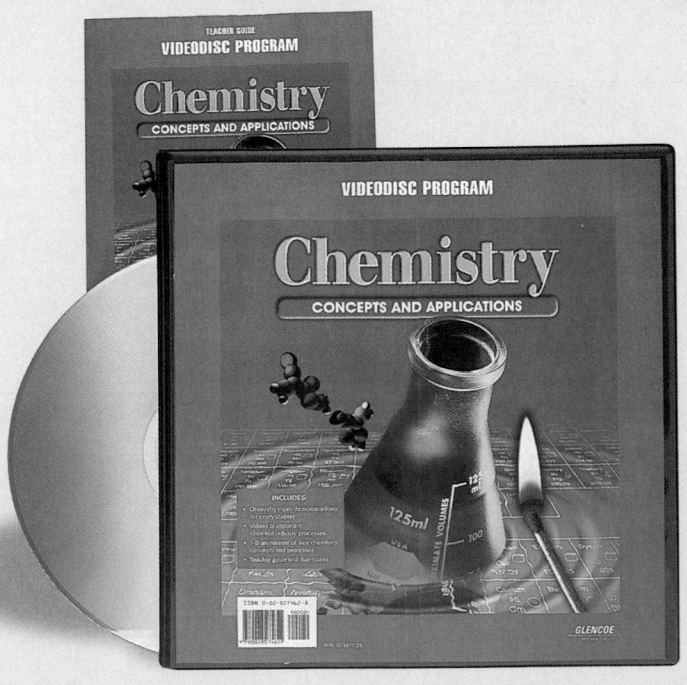

Videodisc Program

Provide your students with visual explanations of major chemistry concepts by showing them the wide variety of videos on *Chemistry: Concepts and Applications Videodisc Program.* Chapter-by-chapter demonstrations, animations, and video clips help you bring chemistry to life.

Mastering Concepts in Chemistry

For whole-class presentations or individual student learning, this 15-unit software program guides your students step by step through all the core concepts and processes of chemistry. The accompanying Study Guide reinforces each software lesson.

MindJogger Videoquizzes

The interactive quiz-show format of *Chemistry: Concepts and Applications MindJogger Videoquizzes* provides fun for your students while reviewing core concepts for every chapter.

Chapter Summaries, English and Spanish Audiocassettes

Each chapter of *Chemistry: Concepts and Applications* is summarized in English or Spanish for you and your limited-English-proficient students on *Chapter Summaries, English/ Spanish Audiocassettes.*

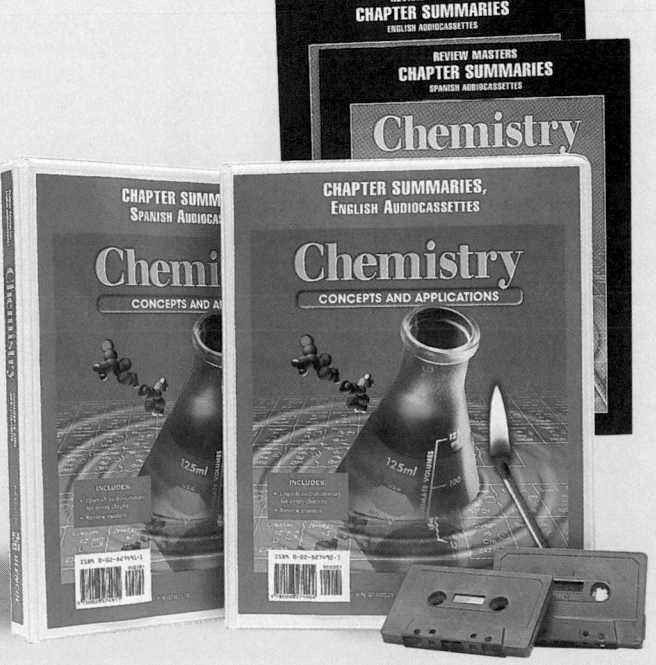

Using Your Teacher Wraparound Edition

Your *Teacher Wraparound Edition* is designed to make teaching *Chemistry: Concepts and Applications* as easy as possible. All relevant teacher information is presented in wraparound margins at point of use.

CHAPTER ORGANIZER

Each chapter is preceded by a *Chapter Organizer*, a complete planning guide that lists the objectives, text features, laboratory activities, and demonstrations for each numbered section. Also correlated to each section are all the components of the *Teacher Classroom Resources*.

Assessment Resources available for the chapter are also given in the *Chapter Organizer*. With all this information at your fingertips, you can quickly plan and organize each day's lessons.

Laboratory Planning To aid you with laboratory planning, a table of Activity Materials needed for the Chapter's *MiniLabs* and *ChemLab* is provided.

Key to Teaching Strategies describes all the icons used in the wraparound margins that identify different teaching strategies.

Glencoe Technology lists software, videodisc, and videotape materials for the chapter that are available exclusively from Glencoe.

Tech Prep identifies Glencoe resources and teaching strategies that are especially appropriate for your Tech Prep students.

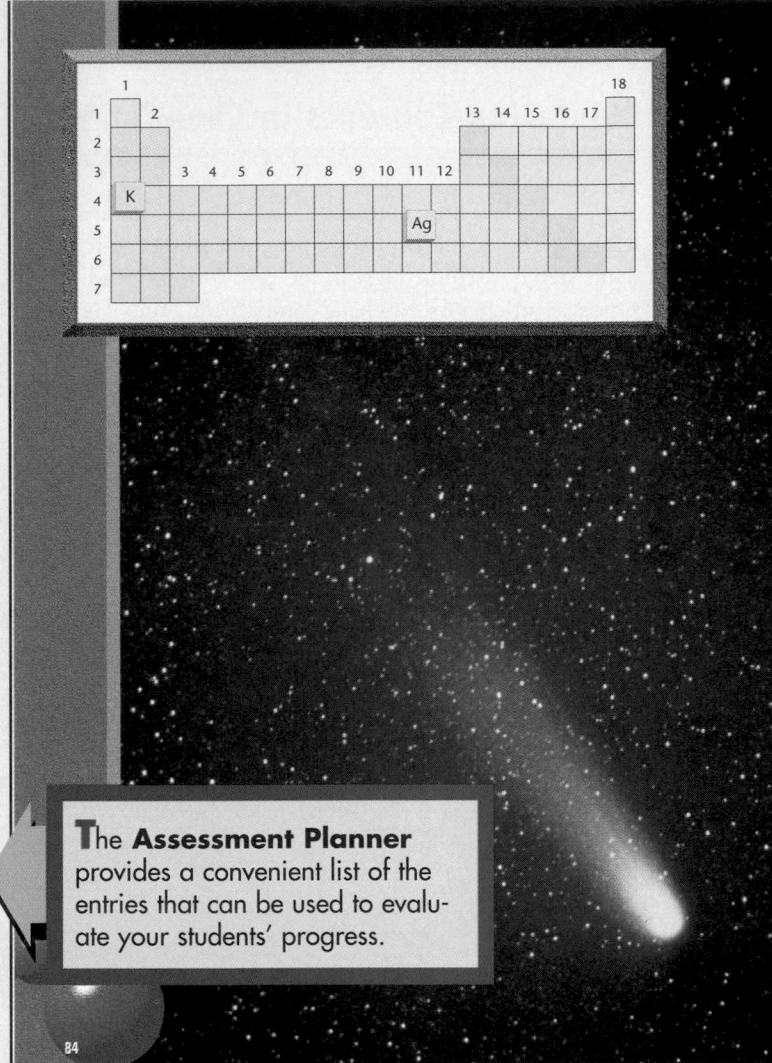

CHAPTER

3 Introduction to the Periodic Table

Chapter Overview
Students begin to learn about trends in the chemical and physical properties of elements. The concept of periodicity and the historical development of a classification system based on atomic mass are used to build a foundation for understanding the modern periodic table. Trends in properties and behavior are related to repeating patterns in the number of valence electrons.

Theme Connection
Macro to Submicro This chapter emphasizes macroscopic observations of periodic properties of the elements and submicroscopic explanations for these observations.

Assessment Planner

Choose assessment strategies from the following pages to evaluate the progress of your students.
Assess, pp. 94, 98, 99, 112
MiniLabs, pp. 89, 97
ChemLab, p. 100
Portfolio, pp. 89, 94, 96, 98, 113, 114
Chapter Review, p. 114

The **Assessment Planner** provides a convenient list of the entries that can be used to evaluate your students' progress.

Discovery Demo 3.1
Activity of Alkali Metals
Purpose: To demonstrate that chemical reactivity follows a predictable pattern related to an element's position on the periodic table.
Materials: overhead projector, explosion shield, three 600-mL beakers, 300 mL of water, 10 drops phenolphthalein indicator, clear kitchen plastic wrap, small cubes of Li, Na, and K, approximately 2 mm on

a side, 10 cm 3 10 cm wire screen
Safety Precautions:

Wear safety goggles and an apron. Use an explosion shield.
Disposal: Code C.1 Check the pH, neutralize with acetic acid or dilute HCl to a pH of 5 to 8, then flush down the drain with copious amounts of water.

Procedure: Cover the stage and lower lens of an overhead projector with clear plastic wrap. Place a 600-mL beaker containing 100 mL of water on the projector. **CAUTION:** *Place an explosion shield around the projector.* Turn on the projector and darken the room. Drop a small piece of lithium into the water. **CAUTION:** *Quickly cover the beaker with a wire screen to prevent the metal from flying out of the*

The **Discovery Demo** provides suggestions for an inquiry approach to starting the chapter.

CHAPTER 3

Introduction to the Periodic Table

In 1986, Halley's comet came streaking into Earth's neighborhood—a guest but not an unexpected one. Last seen 76 years earlier, people were waiting for this collection of ice and rock to visit again. When it comes around once more in 2062, some young people like you may be waiting to see Halley's comet again. Astronomers can predict the appearance of known comets because they show up regularly, just as the full moon appears in the sky every month. Both the moon and the comet illustrate that in the natural world, patterns of behavior often repeat in predictable ways. The moon repeats its cycle every month; Halley's comet takes 76 years.

Are there any repeating patterns in the behavior of the elements? Can you make any predictions about what they might do? Elements have different properties. Oxygen and nitrogen are colorless, whereas mercury and bromine are bright-colored liquids. Some solid elements melt at temperatures over 1000°C, and others melt in your hand. Silver won't react with water, but, as you can see below, potassium reacts violently with water. Examples like these make it seem as if there couldn't be any way of making predictions about the behavior of elements. But when elements are organized in the periodic table, repeating patterns do emerge.

Chemistry Around You

Like the appearance of Halley's comet every 76 years, the properties of the elements repeat in a regular way when they're arranged in the periodic table. When you understand the table's arrangement, you'll be able to explain why silver and potassium don't act in the same way. The repeating patterns in the periodic table are the key to differences as well as similarities in behavior and properties of all the elements.

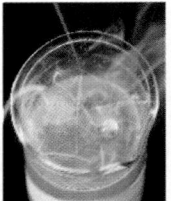
Potassium in water

Silver in water

85

Concept Check

Review the following concepts before studying this chapter.
Chapter 1: physical and chemical properties of elements
Chapter 2: relationship of atomic number to number of protons and electrons in an atom

Chapter Preview

3.1 Development of the Periodic Table

3.2 Using the Periodic Table

Laboratory Activities
ChemLab
The Periodic Table of the Elements

MiniLabs
1. Predicting the Properties of Mystery Elements
2. Trends in Reactivity Within Groups

---

beaker. In separate beakers, repeat the procedure using Na and K. The reactions of Na and K are similar to that of Li.

$$2Li + 2H_2O \rightarrow 2LiOH + H_2$$

Results: As students observe the metals skimming across the water's surface, they will relate the speed to the metal's activity. Li is relatively slow (less active), Na is fast (active), and K bursts into flame upon hitting the water (very active).

Analysis: Ask these questions.
1. Which metal reacts the fastest with water? *K*
2. How does the element's position in the column on the periodic table relate to its reactivity? *Li is first and least reactive; K is third and most reactive.*
3. Which metal has its outer-level electron farthest from the nucleus? *K*

Introducing the Chapter

Discuss with students the difference between elements and compounds, and ask them to give examples of objects made of elements. Point out that many gases in the atmosphere are also elements. Note similarities and differences in the properties of some elements and point out that the periodic table will help them predict properties and behaviors.

Using the Illustrations Just as there is predictability in the appearance of Halley's comet, there is predictability in the properties of the elements. Elements that are widely separated on the periodic table have different chemical properties.

Revealing Misconceptions

Students may think that all metals have the same properties and behavior. Emphasize that there is a gradual change in the properties of elements as the number of valence electrons changes from one to eight.

GLENCOE TECHNOLOGY

 **Videodisc**

Chemistry: Concepts and Applications
Activity of Alkali Metals
Disc 1, Side 1, Ch. 10

Show this video of the Discovery Demo for this chapter to introduce the concept of trends in chemical properties within a group of elements in the periodic table. **L1** **LEP**

CD-Rom

Chemistry: Concepts and Applications
Activity of Alkali Metals
Have students use this video to reinforce the concept of trends in reactivity within a group in the periodic table.
L1 **LEP**

85

Glencoe Technology Videodisc Program and **CD-ROM Multimedia System** references remind you that many of the demonstrations can also be viewed on these two technology resources.

WRAPAROUND MARGINS

The bulk of the teacher information is conveniently placed in the margin next to the student page. At the beginning of each chapter, as shown here, the **Chapter Overview** previews the main ideas presented in the chapter. A **Theme Connection** points out how the relevant themes of the text are incorporated into the chapter.

Using the Illustrations suggests a way to begin teaching the chapter by using a visual link between chemistry concepts and real-world situations.

Revealing Misconceptions suggests strategies for eliciting and clearing up student misconceptions about specific ideas in the chapter.

THE FOUR-STEP TEACHING CYCLE

Each numbered section begins with *PREPARE*, which includes **Key Concepts**. In many sections, **Planning Ahead** assists you in advance preparation of **ChemLab**, **MiniLab**, or **Demonstration** materials. The **PREPARE** section is followed by a comprehensive, four-part teaching cycle designed to help you introduce, teach, assess, and close each text section.

1 FOCUS

The first step to teaching any set of concepts or skills is to focus the students' attention. Each focus section suggests using a **Focus Transparency** from the **Section Focus Transparency Package** to focus the class on the key topic of the text section.

TEACHING STRATEGIES

2 Teach

The primary aim of the *Chemistry: Concepts and Applications Teacher Wraparound Edition* is to give you the tools to teach chemistry concepts and problem-solving skills to your students. Most of these tools are presented under **TEACH**. Many teaching suggestions are provided under a series of clearly defined headings, such as *Using an Analogy, Discussion, Concept Development, Reinforcement, Tying to Previous Knowledge,* and *Extension.*

Visual Learning The extensive use of visuals to teach concepts is further enhanced by additional *Visual Learning* teaching strategies. *Assessment* strategies throughout the section provide ideas for monitoring your students' progress.

Content Background offers in-depth background information either to aid your understanding of the material or to enrich the lesson for your more capable students. *Correcting Misconceptions* and *Correcting Errors* point out possible problems students may have so you can correct them.

Glencoe Technology and *Transparency* references tell you where and how you can integrate these resources into your lessons. *Applying Chemistry* entries offer additional applications of the chapter material to the everyday world.

Special Features You are also provided with teaching strategies and answers needed to teach each of the special features of the student edition that include *ChemLab, MiniLab, Chemistry and Society, Chemistry and Technology, Everyday Chemistry, Connections, People in Chemistry,* and *How it Works.* Finally, many *Quick Demos* are provided for your use in helping students visualize chemistry concepts.

SECTION 3.2

Applying Chemistry

Miniaturization of Electronic Devices The importance of semiconductor devices in modern electronics, in particular the importance of miniaturization, may become clearer to students if they can see the circuits that run the ordinary computer. Arrange with the person in charge of the school's computers for your class to visit the computer room at a time when a computer can be opened to reveal the circuit boards. Students are surprised to find that the heart of the computer is small compared to the amount of empty space. If the circuit boards can be taken out, students can get a good look at the complexity of the circuits contained in a small amount of space.

3 ASSESS

Check for Understanding

- Check for understanding of the section material by having students answer the Section Review questions.
- In an oral exercise, select elements from the periodic table and ask students to give this information: period number; group number; family name, if applicable; and classification as metal, nonmetal, or metalloid. **L1**

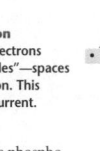

WORD ORIGIN

semiconductor:
semi (L) half
conductus (L) to
escort or guide
Semiconductors do not conduct electricity as well as metals do.

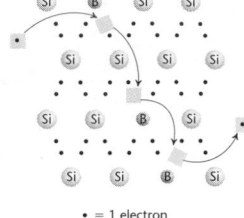

Figure 3.17
Silicon Doped with Boron
In boron-doped silicon, electrons move into and out of "holes"—spaces that are lacking an electron. This movement is an electric current.

• = 1 electron

A semiconductor such as phosphorus-doped silicon is called an *n*-type semiconductor because extra electrons (negatively charged) are present in the crystal structure.

Silicon can also be doped with an element such as boron that has three valence electrons. Boron has fewer valence electrons than silicon, so when boron is added to a crystal of silicon, a shortage of electrons results. "Holes" are created in the silicon crystal structure. These are locations where there should be an electron but there is not because boron has one less electron than silicon. The movement of electrons into and out of these holes produces an electric current. Boron-doped silicon is an example of a *p*-type semiconductor because the holes act as if they are positive charges moving throughout the crystal. **Figure 3.17** illustrates boron-doped silicon.

Diodes

Many semiconductor devices are made by combining *n*- and *p*-type semiconductors to form a diode. When a battery is connected to this combination of semiconductors, the electrical current flows only in one direction, from the negative terminal to the positive terminal, just as it would if the battery terminals were connected by a copper wire.

Transistors like those shown in **Figure 3.18** are the key components in electrical circuits in devices such as computers, calculators, hearing aids, and televisions. They are used to amplify (increase the strength of) electrical signals. Transistors are exceedingly small, and their compact size and efficient operation have allowed the miniaturization of many electronic devices, such as laptop computers, heart pacemakers, and hearing aids. Transistors may be constructed by placing a *p*-type semiconductor between two *n*-type semiconductors, called an *npn*-junction, or by placing an *n*-type semiconductor between two *p*-type semiconductors, called a *pnp*-junction.

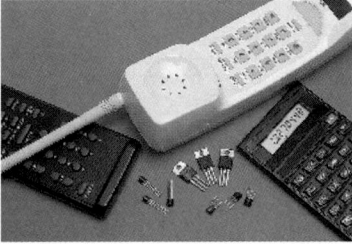

Figure 3.18
Importance of Transistors
Transistors are used to increase the strength of electrical signals in such devices as television remote controls, telephones, and hearing aids.

112   Chapter 3 Introduction to the Periodic Table

Chemistry Journal

Household Elements
Ask each student to choose ten common household items or materials and, insofar as possible, list the elements that are the primary components of the items or materials. The elements

112

Ample suggestions also are provided for **Meeting Individual Needs** of students with diverse learning styles and abilities. **Practice** gives the answers to all practice problems in the student edition and offers additional practice problems for independent practice by your students. Finally, the *Teacher Wraparound Edition* provides between four and seven **Demonstrations** for each chapter. Each demonstration includes analysis questions and an assessment strategy.

Also under **TEACH**, special features can be found at the bottom of many pages. **Integrating the Sciences** and **Across the Curriculum** provide cross-curricular connections with other sciences and nonscience disciplines. Many **Chemistry Journal** suggestions offer writing opportunities for your students. At least one **Cultural Diversity** entry in each chapter looks at the contributions of different people or cultures to chemistry.

Transistors, diodes, and other semiconductor devices are incorporated onto thin slices of silicon to form integrated circuits. Integrated circuits like the one shown in **Figure 3.19** may contain hundreds of thousands of devices on a slice of silicon called a chip. The small size of a chip—only a few millimeters in width—has made possible the amazing growth of computer technology.

Connecting Ideas

You already know that elements combine chemically to form compounds. The ways that elements combine depend entirely on their valence electrons. With your knowledge of the periodic table, you will be able to explain why elements combine and predict what compounds they will form.

Figure 3.19
Silicon and the Computer Revolution
Miniaturization of integrated circuits on silicon chips (top) allowed the size of computers to shrink from room-size (left) to laptop (right).

SECTION REVIEW

Understanding Concepts

1. Where are metals usually found on the periodic table? Where are nonmetals found? The metalloids?
2. How does the arrangement of elements in periods reflect electron structure?
3. What are the major differences in the physical properties of metals, nonmetals, and metalloids?

Thinking Critically

4. **Observing and Inferring** What can you deduce from the periodic table about the properties of the element barium?

Applying Chemistry

5. **Semiconductors** Germanium has the same type of structure and semiconducting properties as silicon. What type of semiconductor would you expect arsenic-doped germanium to be?

3.2 *Using the Periodic Table* 113

SECTION 3.2

Reteach

To find areas of weakness in understanding or misconceptions, the following is a class exercise providing opportunities for discussion and clarification. Write on the chalkboard two lists of words and phrases. One list might be *conductivity, group, nonmetal, Mendeleev, energy level, silver, pool of electrons, liquid, lanthanides, sodium, alkaline earth metal, Group 18, Halley's comet, solid, metalloid*, and many more. Another list might be *inner transition element, family, state of matter, semiconductor, metal, alkali metal, mercury, bromine, noble gas, Group 2, periodic law, Group 1, magnesium, transition element, actinide, gas, doping elements, period*, and many more. Ask students to match words and phrases in one column with words and phrases in the other. There should be multiple connections for many of the entries. Encourage students to find as many connections as possible, but require that they justify or explain the connections. **L1** **COOP LEARN**

Extension

Ask each student to write an essay about the benefits and/or adverse effects of efforts to synthesize new chemical elements in nuclear experiments. These essays could be put into student portfolios. **L2** **P**

TECH PREP

SECTION REVIEW

1. Metals are on the left side and in the center; nonmetals are in the upper-right corner; metalloids form a border between the metals and nonmetals.
2. The period number of an element is equal to the number of the energy level that is occupied by the valence electrons.
3. Metals are lustrous, are easily deformed and drawn into wires, and are good conductors of heat and electricity. Nonmetals are not lustrous, may be hard or soft, are usually brittle, and are poor conductors of heat and electricity. Metalloids look like metals but have physical properties that are between those of metals and nonmetals.
4. Barium is a metal in Group 2 with two valence electrons in the sixth energy level. It has all the properties of a typical metal.
5. Arsenic has five valence electrons, and germanium has four. Therefore, arsenic-doped germanium would be an *n*-type semiconductor.

---

4 CLOSE

Extension

Ask students to predict the appearance of the periodic table if 120 elements existed. What would it look like if 140 elements existed? If 168 elements existed? **L3**

113

---

3 ASSESS

To assist you in monitoring students' progress and adjusting your teaching accordingly, a **Check for Understanding** idea is provided for you in each section. For students who are having trouble, a **Reteach** suggestion immediately follows, suggesting a way to teach important concepts differently to adjust to students' individual learning styles. For those students needing enrichment, **Extension** ideas are also provided.

4 CLOSE

Closing the lesson is the last, but one of the most important, steps in teaching the section because it helps students relate and synthesize the ideas presented in the section. The *Teacher Wraparound Edition* gives you a variety of ways to provide effective closure to a section.

And There's More!

As you review the *Chemistry: Concepts and Applications, Teacher Wraparound Edition*, you will discover that it accommodates an amazing range of teaching and learning styles and student ability levels. By giving you convenient access to an enormous number of teaching options and classroom resources, the *Teacher Wraparound Edition* offers the flexibility you need to satisfy the ordinary and extraordinary demands of your chemistry classes.

Glencoe Technology For Your Chemistry Classroom

GLENCOE'S CD-ROM MULTIMEDIA SYSTEM

Students can better understand complex chemical processes and structures when they see them in action. Turn your students into interactive explorers and excited learners.

Students will be able to:

- Make choices and change variables in Explorations to challenge their critical thinking skills.
- Practice lab skills and learn core chemistry concepts with the Interactive Experiments in a virtual lab setting.
- Watch 3-D animations illustrate chemical processes at the submicroscopic level.
- See full-color videos of chemical processes in industry and in the lab.
- Hear key terms pronounced.
- Operate an easy-to-use, intuitive navigation system with active file-folder tabs for instant access to text sections, glossary, activities, and much more.

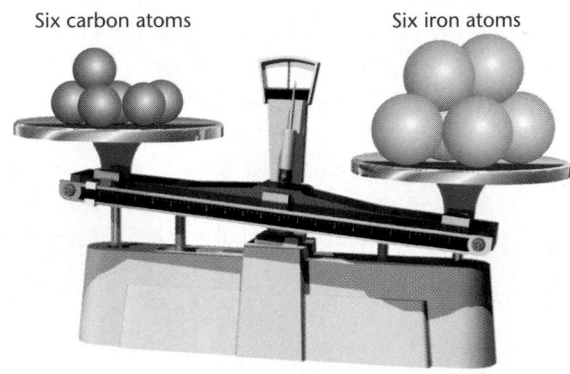

Six carbon atoms   Six iron atoms

DESIGNED FOR LEARNING

With this flexible interactive multimedia system, any part of the chemistry curriculum can be introduced, enriched, or remediated at any time during the course. During one lesson, students can review concepts with interactive activities, view videos or animations, experiment in the virtual lab, review key terms and their definitions, keep a journal, answer quiz questions, and calculate answers to practice problems.

VERSATILITY IN THE CLASSROOM

Use the flexible Chemistry **CD-ROM Multimedia System** to suit the special needs of your classroom.

Whole-Class Instruction Connect your computer to a large monitor or LCD panel so that you can integrate animation and video into your whole-class presentations.

Individual/Small-Group Instruction Students may work individually or in small groups to perform the activities, view the chemistry in their everyday lives, or answer the quiz questions.

Media Center/Computer Lab Because this program is so self-contained and easy-to-use, a set of CD-ROMs can be placed in a computer lab or media center for students to access at any time with no need for special instructions.

Pb S

Galena PbS

MASTERING CONCEPTS IN CHEMISTRY

All your students can master core concepts of chemistry with *Mastering Concepts in Chemistry* software, available for IBM-compatible computers with Microsoft's Windows 3.1, or better. Chapter-by-chapter correlations to *Chemistry: Concepts and Applications* are given in the *Teacher Wraparound Edition*. This unique teaching tool provides ways for you to enhance your classroom instruction with colorful, animated, and interactive presentations. There are two versions of the software: one for your classroom presentation and another designed for use by individual students.

Students who need reinforcement or remediation will find *Mastering Concepts in Chemistry* an exciting way to learn. The step-by-step presentations allow students to review presentations at their own pace. In order for all your students to succeed in chemistry, they must master the core concepts. These concepts are presented in 15 logical units, and each presentation is supplied with frequent review questions for quick assessment of student progress. An accompanying *Student Study Guide* includes printed scripts of the teaching screens.

GLENCOE'S VIDEODISC PROGRAMS

Glencoe provides two videodisc programs that are correlated to *Chemistry: Concepts and Applications*. Bar codes to both programs are displayed in the *Teacher Wraparound Edition*, right where they are needed to enhance your instructional lesson plan.

The *Chemistry* videodisc included in the *Science and Technology Videodisc Series* contains full-motion video reports that cover a broad spectrum of topics relating to research in the field. The video reports are derived from the Mr. Wizard television program and are ideal for illustrating scientific methods, laboratory techniques, and careers in chemistry.

Chemistry: Concepts and Applications Videodisc Program was developed along with the textbook. It provides exciting opportunities to see chemistry come alive from the page. For example, as students read about the five different types of chemical reactions, they will be able to view them instantly on the screen.

For you, the teacher, we have saved you time and money by filming many of the demonstrations that you would like your students to experience. If you have a limited budget, strict safety rules, and not enough time to perform the demonstrations yourself, you will find this program invaluable. Many of the submicroscopic processes of chemistry have been animated for both the CD-ROM program and the Videodisc program. As an instructional advantage, while your students manipulate the CD-ROM animations frame-by-frame on their computer screens, you can show the animation on your full-size screen at the front of the classroom.

Glencoe Technology For Your Chemistry Laboratory

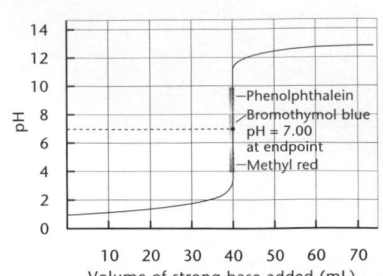

LAB PARTNER

Integrate mathematical practices into your chemistry laboratory by providing your students with **Lab Partner**, a software program available for IBM or Apple that will help them analyze and graph their experimental data. The program is self guided and allows students to enter their data as they are collected. **Lab Partner** includes a variety of mathematical functions that provide instant averaging and plotting of data on graphs. Students have the opportunity to manipulate their data to achieve the best fit curves. Up to eight variables can be plotted at one time. The program is another valuable tool for groups of students to work together cooperatively.

Lab Partner makes use of many standard mathematical functions, including square root, natural logarithm, parentheses used in algebraic logic, ten to the nth power, factorial, sine, cosine, and tangent. The program also has incorporated many useful constants including Avogadro's constant, molar volume, charge and mass of electrons, the universal gas constant, and pi. Students of all learning abilities will find this software program help them gain a greater understanding of chemical relationships.

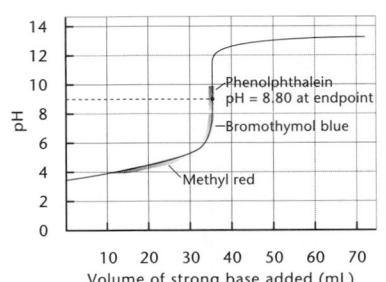

MICROCOMPUTER-BASED LABS

Many of the laboratory activities in **Chemistry: Concepts and Applications** can be conducted using probes and computer software. **Quantum Technology** has adapted a number of *Chemlabs* from the Glencoe program to their Leap System Hardware and Software. Using the customized lab guide, Leap Probes, and a DOS or Macintosh computer, groups of students can quickly record and graph a variety of measurements. Call **Quantum Technology** for information about the Leap System and the Lab Pacs available for **Chemistry: Concepts and Applications.**

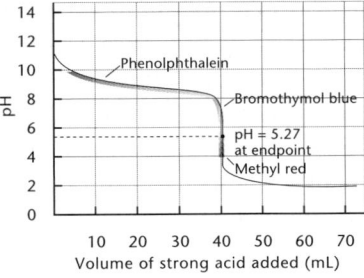

Quantum Technology
30153 Arena Drive
Evergreen, CO 80439
Phone: (303) 674-9651
Fax: (303) 674-6763
Internet: leapsys @ aol.com

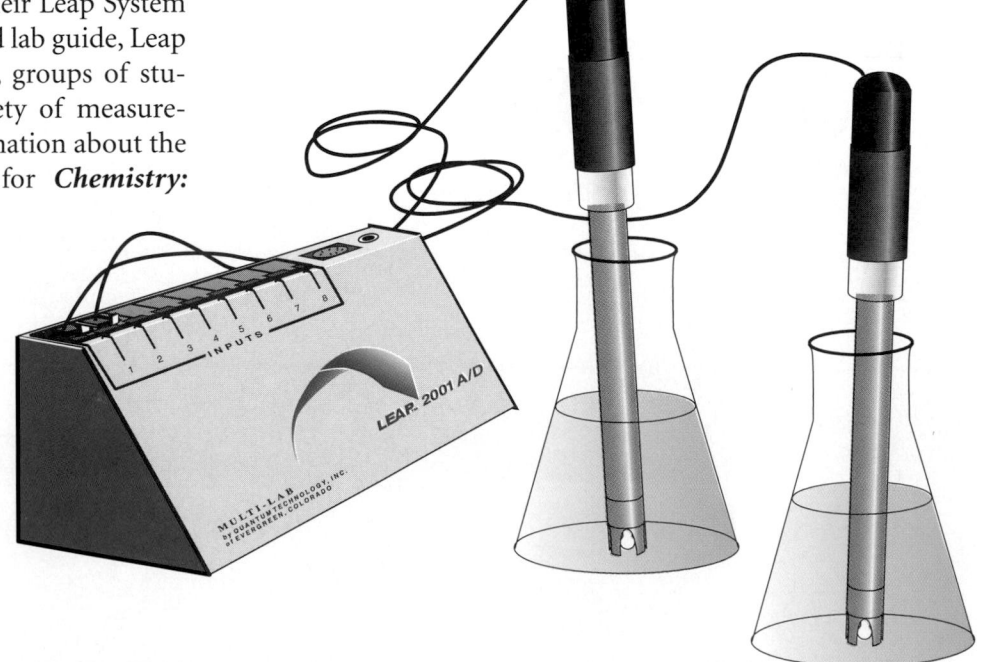

The Internet in Your Classroom

*inter*NET CONNECTION

ACCESSING THE INTERNET

The Internet is the world's largest computer network. It is a network of networks linked by high-speed data lines and wireless systems that freely exchange information. It provides access to individuals, corporations, educational institutions, government departments, and other groups.

To access the Internet, you need a computer with communication software, a modem, and a phone line. A modem is a device that enables computer data to travel to another computer via phone lines. Many computers come equipped with an internal modem. You will need a microprocessor of class 486 or greater, at least 8 megabytes of random access memory, and a modem that runs at 9600 baud or faster.

To access the Internet, you must first set up an account with an Internet service provider (server). Although these commercial providers can be expensive, depending on how many of their services you use, it is also possible in some cities to access the Internet without charge through university or local library systems.

Internet for Teachers After you have access to the Internet, you can use it to enhance your teaching in a variety of ways. You will be able to obtain numerous lessons, activities, project ideas, and labs on any topic you choose. Learn the locations of useful teacher contacts and place them into your favorite-places files for easy access at a later time. When you find good ideas that you would like to use, simply download, print out, and hand out to your students. If you have students with special needs or wish to incorporate cultural diversity into an existing lesson, you can find help on the Internet.

The Internet can provide you with easy access to the most up-to-date information on any topic you may wish to research. It's like having your own personal university library. There are thousands of free sources for graphics, videos, animations, and shared software. By using the electronic mail (e-mail) feature of your server, you can exchange ideas with teachers worldwide, consult with research chemists, and order products. You can look for or set up your own bulletin board to share information daily on special topics, such as new ideas for chemistry demonstrations or how to use cooperative groups in the chemistry classroom.

Internet for Students Students will also benefit from using the Internet. Students who become familiar now with the technology of computers and the new method of communicating by asking questions and searching for answers on the Internet will be better prepared for the future in business, technology, education, and leisure. Students could use the Internet to compare and contrast lab data with students in the next town or across the globe. Many servers provide a staff of teachers that are available to advise students about homework, or they could e-mail peers for advice on term papers. Students soon learn how to express themselves unambiguously when they send written messages to scientists, teachers, or friends over the Internet.

As students use the Internet for information to enhance their lab reports and other classroom activities, they develop their search-and-retrieval strategies, which require careful analysis, evaluation, and application. An added benefit of communicating over the Internet is that student differences in personal appearance, gender, culture, and type of disability are not apparent. Students learn to value expertise without bias in this culturally diverse setting.

The World Wide Web

To access a wider source of multimedia information, you and your students can access the World Wide Web (WWW) by using a web browser. Many servers now provide the use of a web browser. The first page you see when you enter a WWW site is the home page. From the home page, you can obtain complete texts and graphics from books, clips from movies, graphical replicas of art from museums, or items from a shopping mall.

You can browse through books, magazines, and scientific periodicals. You can view pictures, maps, paintings, movies with CD-quality sound, or software games. Go to the home pages of scientific-supply companies and order your chemicals, equipment, books, or posters. You and your students can even create your own home page on the WWW to explain to other classrooms around the country your topics of study for the week, share data collected from classroom experiments, or ask other classes for suggestions for projects.

SOME USEFUL FREE WEB SITES

AskERIC The Q & A Service
http://ericir.syr.edu
This is an ongoing project that is building a digital library for educational information. You use it as you would a library.

Chemistry–Resources from ACS
gopher://acsinfo.acs.org
This gopher site is sponsored by the American Chemical Society, and it gives Internet chemistry resources.

National Renewable Energy Laboratory
http://www.nrel.gov/
The U.S. Department of Energy Lab offers information about renewable energy, including energy data, resource maps, publications, and job opportunities.

U.S. Department of Education
http://www.ed.gov/
This site provides information about various programs in the Department of Education.

Webcrawler
http://webcrawler. com
This is a search engine with useful links to high school chemistry home pages with opportunities for sharing ideas and resources.

WebElements
http://www.cchem.berkeley.edu/Table/index.html
This server provides you with a database for the periodic table on-line.

World Wide Web Virtual Library: Chemistry
http://www.chem.ucla.edu
This is a great starting point for access to many resources for chemistry education, including demos, free materials, conferences, and education information.

Yahoo
http://www.bham.wednet.edu
Another search engine with an extensive inventory of Web sites. Type in the key words *Chemistry Teacher* and it will list many useful resources for general chemistry.

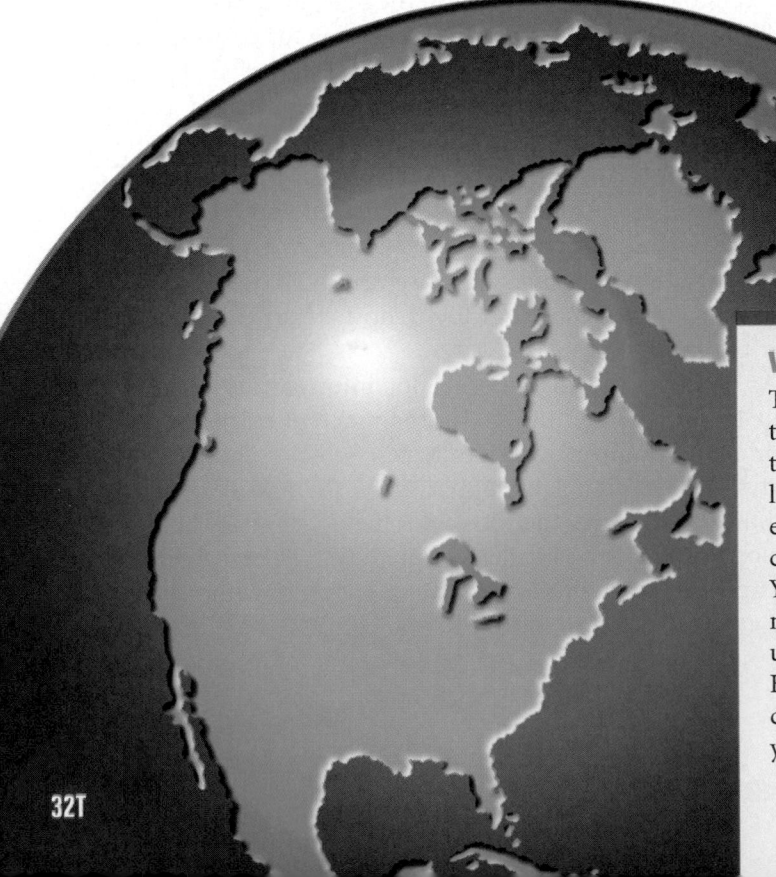

Words of Caution

The sites referenced in the *Teacher Wraparound Edition* are not under the control of Glencoe, and therefore, Glencoe can make no representation concerning the content of these sites. Extreme care has been taken to list only reputable links by using educational and government sites whenever possible. Internet searches have been used that return only sites that contained no content apparently intended for mature audiences.

You might wish to caution your users that any scientific information they may read on the Internet has not been reviewed and authenticated by the usual peer-review system practiced in textbook and journal publishing. However, with this caution in mind, you will soon discover the use of the computer in the classroom to be an exciting and rewarding addition to your lesson plans.

Meeting Individual Needs

Each student brings his or her own unique set of abilities, perceptions, and needs into the classroom. It is important that the teacher try to make the classroom environment as receptive to these differences as possible and to ensure a good learning environment for all students.

It is important to recognize that individual learning styles are different and that learning style does not reflect a student's ability level. Whereas some students learn primarily through visual or auditory senses, others are kinesthetic learners and do best if they have hands-on exploratory interaction with materials. Some students work best alone and others learn best in groups. Some students seek first to understand the big picture in order to deal with specifics, whereas others need to understand the details first in order to construct the whole concept.

In an effort to provide all students with a positive science experience, *Chemistry: Concepts and Applications* offers a variety of ways for students to interact with materials so that they can utilize their preferred method of learning the concepts. The variety of approaches also provides opportunities for students to become familiar with other ways to learn.

ABILITY LEVELS

Each activity is placed into one of three ability levels that provide a range to accommodate all student levels of learning. *Chemistry: Concepts and Applications Teacher Wraparound Edition* designates the three levels of activities as follows:

L1 activities are basic activities designed to be within the ability range of all students. These activities reinforce the concepts presented on the student pages where they are found.

L2 activities are application activities designed for students who have mastered the basic concepts presented. These activities give students an opportunity for practical application of the concepts.

L3 activities are challenging activities designed for the students who have mastered the basic concepts and are ready to expand upon the perspectives presented in the text.

LIMITED ENGLISH PROFICIENCY

In providing for the student with limited English proficiency, the focus needs to be on overcoming a language barrier. Once again, it is important not to confuse ability in speaking/reading English with academic ability or intelligence. In general, the best method for dealing with students with **LEP**, variations in learning styles, and different ability levels is to provide all students with a variety of ways to learn, apply, and be assessed on the concepts. Look for the symbol **LEP** in the teacher margin for specific strategies for students with limited English proficiency.

The chart on pages 34T-35T gives tips you may find useful in structuring the learning environment in your classroom to meet students' special needs. In addition, *Meeting Individual Needs* strategies are provided in the bottom margins of every chapter of this *Teacher Wraparound Edition.*

Meeting Individual Needs

	DESCRIPTION	SOURCES OF HELP/INFORMATION
Learning Disabled	All learning-disabled students have an academic problem in one or more areas, such as academic learning, language, perception, social-emotional adjustment, memory, or attention.	*Journal of Learning Disabilities* *Learning Disability Quarterly*
Behaviorally Disordered	Children with behavior disorders deviate from standards or expectations of behavior and impair the functioning of others and themselves. These children may also be gifted or learning disabled.	*Exceptional Children* *Journal of Special Education*
Physically Challenged	Children who are physically disabled fall into two categories–those with orthopedic impairments and those with other health impairments. Orthopedically impaired children have the use of one or more limbs severely restricted, so the use of wheelchairs, crutches, or braces may be necessary. Children with other health impairments may require the use of respirators or other medical equipment.	Batshaw, M.L. and M.Y. Perset. *Children with Handicaps: A Medical Primer.* Hale, G. (Ed.). *The Source Book for the Disabled.* New York: Holt, Rinehart & Winston, 1982. *Teaching Exceptional Children*
Visually Impaired	Children who are visually disabled have partial or total loss of sight. Individuals with visual impairments are not significantly different from their sighted peers in ability range or personality. However, blindness may affect cognitive, motor, and social development, especially if early intervention is lacking.	*Journal of Visual Impairment and Blindness* *Education of Visually Handicapped* American Foundation for the Blind
Hearing Impaired	Children who are hearing impaired have partial or total loss of hearing. Individuals with hearing impairments are not significantly different from their hearing peers in ability range or personality. However, the chronic condition of deafness may affect cognitive, motor, and social development if early intervention is lacking. Speech development also is often affected.	*American Annals of the Deaf* *Journal of Speech and Hearing Research* *Sign Language Studies*
Limited English Proficiency	Multicultural and/or bilingual children often speak English as a second language or not at all. The customs and behavior of people in the majority culture may be confusing for some of these students. Cultural values may inhibit some of these students from full participation.	*Teaching English as a Second Language Reporter* R.L. Jones (Ed.). *Mainstreaming and the Minority Child.* Reston, VA: Council for Exceptional Children, 1976
Gifted	Although no formal definition exists, these students can be described as having above-average ability, task commitment, and creativity. Gifted students rank in the top five percent of their class. They usually finish work more quickly than other students and are capable of divergent thinking.	*Journal for the Education of the Gifted* *Gifted Child Quarterly* *Gifted Creative/Talented*

TIPS FOR INSTRUCTION

With careful planning, the needs of all students can be met in the science classroom.

1. Provide support and structure; clearly specify rules, assignments, and duties.
2. Practice skills frequently. Use games and drills to help maintain student interest.
3. Allow students to record answers on tape and allow extra time to complete tests and assignments.
4. Provide outlines or tape lecture material.
5. Pair students with peer helpers, and provide class time for pair interaction.

1. Provide a clearly structured environment with regard to scheduling, rules, room arrangements, and safety.
2. Clearly outline objectives and how you will help students obtain objectives.
3. Reinforce appropriate behavior and model it for students.
4. Do not expect immediate success. Instead, work for long-term improvement.
5. Balance individual needs with group requirements.

1. Openly discuss with the student any uncertainties you have about when to offer aid.
2. Ask parents or therapists and students what special devices or procedures are needed, and if any special safety precautions need to be taken.
3. Allow physically disabled students to do everything their peers do, including participating in field trips, special events, and projects.
4. Help nondisabled students and adults understand physically disabled students.

1. Help the student become independent. Modify assignments as needed.
2. Teach classmates how to serve as guides.
3. Limit unnecessary noise in the classroom.
4. Provide tactile models whenever possible.
5. Describe people and events as they occur in the classroom.
6. Provide taped lectures and reading assignments.
7. Team the student with a sighted peer for laboratory work.

1. Seat students where they can see your lip movements easily, and avoid visual distractions.
2. Avoid standing with your back to the window or light source.
3. Using an overhead projector allows you to maintain eye contact while writing.
4. Seat students where they can see speakers.
5. Write all assignments on the board, or hand out written instructions.
6. If the student has a manual interpreter, allow both student and interpreter to select the most favorable seating arrangements.

1. Remember, students' ability to speak English does not reflect their academic ability.
2. Try to incorporate the student's cultural experience into your instruction. The help of a bilingual aide may be effective.
3. Include information about different cultures in your curriculum to help students' self-image. Avoid cultural stereotypes.
4. Encourage students to share their cultures in the classroom.

1. Make arrangements for students to take selected subjects early and to work on independent projects.
2. Let students express themselves in art forms such as drawing, creative writing, or acting.
3. Make public services available through a catalog of resources, such as agencies providing free and inexpensive materials, community services and programs, and people in the community with specific expertise.
4. Ask "what if" questions to develop high-level thinking skills. Establish an environment safe for risk taking.
5. Emphasize concepts, theories, ideas, relationships, and generalizations.

The **Student Edition** and **Teacher Wraparound Edition** contain a variety of strategies for addressing the individual learning styles of students:
- Both small-scale and traditional *ChemLabs*
- Hands-on *MiniLabs*
- *Visual Learning* strategies
- *Cross-Curricular* strategies
- *Extension* activities
- *Demonstrations* and *Quick Demos*
- *Cooperative Learning* activities

The following Glencoe products will provide you with additional means to help you tailor your instruction to meet the individual needs of your students:
- **Study Guide**
- **Basic Concepts Transparencies**
- **Section Focus Transparencies**
- **Problem Solving Transparencies**
- **MindJogger Videoquizzes**
- **English/Spanish Audiocassettes**
- **CD-ROM Multimedia System**
- **Critical Thinking/Problem Solving**
- **Cooperative Learning in the Science Classroom**
- **Videodisc Program**
- **Mastering Concepts in Chemistry** software

Cooperative Learning

What Is Cooperative Learning?

In cooperative learning, students work together in small groups to learn academic material and develop interpersonal skills. Group members each understand that they are responsible for accomplishing an assigned group task, as well as for personally learning the material. Cooperative learning fosters academic, personal, and social success for all students.

Research has shown that cooperative learning results in:
- ◆ development of positive attitudes toward science and toward school.
- ◆ lower drop-out rates for at-risk students.
- ◆ building respect for others regardless of race, ethnic origin, or sex.
- ◆ increased sensitivity to and tolerance of diverse perspectives.
- ◆ improved capability for problem solving in the sciences.

ESTABLISHING A COOPERATIVE CLASSROOM

Cooperative groups in the high school usually contain from two to five students. Heterogeneous groups that represent a mixture of abilities, genders, and ethnicity expose students to ideas different from their own and help them learn to work with different people.

Initially, cooperative learning groups should work together for only a day or two. After the students are more experienced, they can work with a group for longer periods of time. It is important to keep groups together long enough for each group to experience success and to change groups often enough that students have the opportunity to work with a variety of students.

Students must understand that they are each responsible for all group members learning the material. This is achieved by using interpersonal skills that include disagreeing constructively, encouraging other group members to participate, checking for understanding, and reaching a group consensus. Before beginning, discuss the basic rules for effective cooperative learning: (1) listen while others are speaking, (2) respect other people and their ideas, (3) stay on task, and (4) be responsible for your own actions.

USING COOPERATIVE LEARNING STRATEGIES

The *Teacher Wraparound Edition* uses the symbol **COOP LEARN** at the end of activities and teaching ideas where cooperative learning strategies might be useful. For additional help, refer again to these pages of background information on cooperative learning and to *Glencoe's Professional Series* booklet *Cooperative Learning in the Science Classroom*.

During cooperative learning activities, monitor the functioning of groups. Praise group cooperation and good use of interpersonal skills. When students are having trouble with the task, clarify the assignment, reteach, or provide background as needed. Answer questions only when no students in the group have possible answers.

ACCESSING COOPERATIVE LEARNING

At the close of the lesson, have groups share their products or summarize the assignment. You can evaluate group performance during a lesson by asking questions of group members picked at random or by having each group take a quiz together. You might have all students write papers and then choose one at random from each group to grade. Assess individual learning by your traditional methods.

Cultural Diversity

American classrooms reflect the rich and diverse cultural heritage of the American people. Students come from different ethnic backgrounds and different cultural experiences into a common classroom that must assist all of them in learning. The diversity itself is an important focus of the learning experience.

Diversity can be repressed, creating a hostile environment; ignored, creating an indifferent environment; or appreciated, creating a receptive and productive environment. Responding to diversity and approaching it as a part of every curriculum is challenging to a teacher, experienced or not. The goal of science is understanding. The goal of multicultural education is to promote the understanding of how people from different cultures approach and solve the basic problems all humans have in living and learning.

CULTURAL DIVERSITY IN SCIENCE

Every culture has utilized science to explore fundamental questions, meet challenges, and address human needs. The growth of knowledge in science has advanced along culturally diverse roots. No single culture has a monopoly on the development of scientific knowledge. The history of science is rich with the contributions and accomplishments of men and women from diverse cultural backgrounds. A brief look at some of the milestones in science reveals this diversity. Cultural groups on all continents have devised methods for measuring time and space. For example, the Mayan civilization, which flourished in what is now southern Mexico, Guatemala, and Belize, developed sophisticated and accurate mathematical systems and calendars. All cultures have a history of uses of local plants in folk medicine that has led to our modern understanding of the healing powers of plant chemicals. Today, individuals from diverse cultural and ethnic backgrounds continue to make significant contributions to science and society. The *People in Chemistry* feature in the Student Edition of **Chemistry: Concepts and Applications** highlights some of these individuals as possible role models for the next generation of culturally diverse scientists.

The scientific enterprise is a human enterprise that utilizes science processes (observing and inferring, classifying, and so on) to discover, explore, and explain the environment. Consequently, students should understand that all people, regardless of their cultural background or ethnic origin, have made and continue to make contributions to science. In addition to the individuals highlighted in the **Student Edition**, the *Cultural Diversity* features in the **Teacher Wraparound Edition** provide information about people and groups not commonly covered in science texts. The intent is to build awareness and appreciation for the global community in which we all live.

GOALS OF MULTICULTURAL EDUCATION

A teacher can integrate multicultural education into the curriculum by supporting its four main goals.
1. Promote the strength and value of cultural diversity.
2. Promote human rights and respect for those who are different from oneself.
3. Promote social justice and equal opportunity for all people.
4. Promote equity in the distribution of power among groups.

These goals are accomplished when students see others like themselves in positive settings in their textbooks. They also are accomplished when students realize that major contributions in science are the result of the work of hundreds of individuals from diverse backgrounds. The philosophy and design of **Chemistry: Concepts and Applications** attempts to help you with these goals by showing students of diverse ethnic origins taking part in the scientific endeavor.

Additional information on multicultural education can be found in the following references.

Atwater, Mary, et al. *Multicultural Education: Inclusion of All.* Athens, Georgia: University of Georgia Press, 1994.

Banks, James A. & Cherry A. McGee Banks. *Multicultural Education: Issues and Perspectives.* Boston: Allyn and Bacon, 1989.

Selin, Helaine. *Science Across Cultures: An Annotated Bibliography of Books on Non-Western Science, Technology and Medicine.* New York: Garland, 1992.

Tech Prep Education

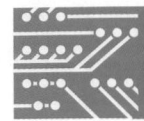

What Is Tech Prep?

Tech Prep is a rigorous and focused program of study that aims to create a workforce in the United States that is technically literate. It is designed to prepare students enrolled in a general curriculum for the demands of further education or for employment by providing them with essential academic and technical foundations, along with problem-solving, group-process, and lifelong-learning skills. These goals can be achieved by integrating vocational study with higher-level academic study.

What Are the Characteristics of the Tech Prep Curriculum?

In 1990, Congress passed the Carl D. Perkins Vocational and Applied Technology Act to set aside funds for the development and administration of Tech Prep programs. The criteria outlined in Title III of the Perkins Act specify that Tech Prep programs take place during the last two years of high school, followed by two years of post-secondary occupational education, and that this education culminate in a certificate or associate degree. The Secretary's Commission on Achieving Necessary Skills (SCANS), an arm of the U.S. Department of Labor, published a report in June 1991 that outlined several competencies that characterize successful workers. The Tech Prep curriculum seeks to address these competencies, which include:

- the ability to use resources productively.
- the ability to use interpersonal skills effectively, including fostering teamwork, teaching others, serving customers, leading, negotiating, and working well with individuals from culturally diverse backgrounds.
- the ability to acquire, evaluate, interpret, and communicate data and information.
- the ability to understand social, organizational, and technological systems.
- the ability to apply technology to specific tasks.

What is the role of the community in Tech Prep education?

The key to implementing a Tech Prep course of study is a partnership among high schools, post-secondary educational institutions, businesses, industry, and labor. Teachers and schools can enlist the support of their communities in a number of ways to enhance tech prep education in the science classroom. They include corporate partnerships, inviting scientists into the classroom, enlisting the aid of industrial tutors, and encouraging students to become involved in community service projects.

How does *Chemistry: Concepts and Applications* address Tech Prep issues?

Chemistry: Concepts and Applications helps you develop scientific and technological literacy in your students through a variety of performance-based activities that emphasized problem solving, critical thinking skills, and teamwork. In the **Student Edition**, many applications of chemistry are used throughout the text and in the features to illustrate the relevance of chemistry to everyday life and the world of work. *ChemLabs* and *MiniLabs* provide opportunities for practical applications of chemistry concepts. The *Practice Problems* given wherever there is a quantitative aspect of chemistry provide practice in quantitative applications and the use of calculators. Projects listed in the *Chapter Review Authentic Assessment* provide extra activities that will give the student practical experience with problem solving and that can be used as science fair projects or as extensions for the interested student. In the **Teacher Wraparound Edition**, a unique Tech Prep icon highlights the wide variety of features, activities, and strategies that support your Tech Prep curriculum. *Everyday Chemistry, How It Works, Chemistry and Technology, Chemistry and Society*, and *People in Chemistry* are just some of the features of the **Student Edition** that offer strong Tech Prep connections. The **Teacher Classroom Resources** includes a **Tech Prep Applications** booklet that gives additional activities and career ideas designed to fulfill the goals of the tech prep education.

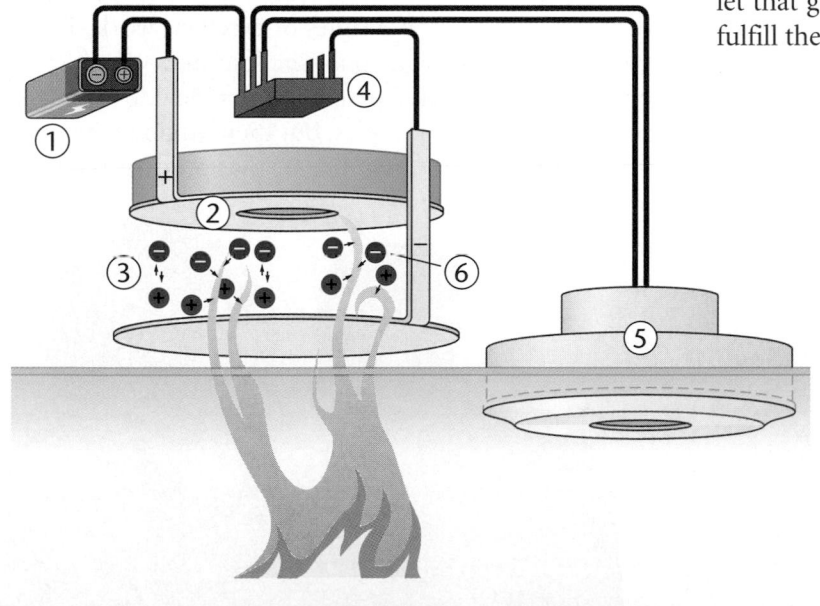

Assessment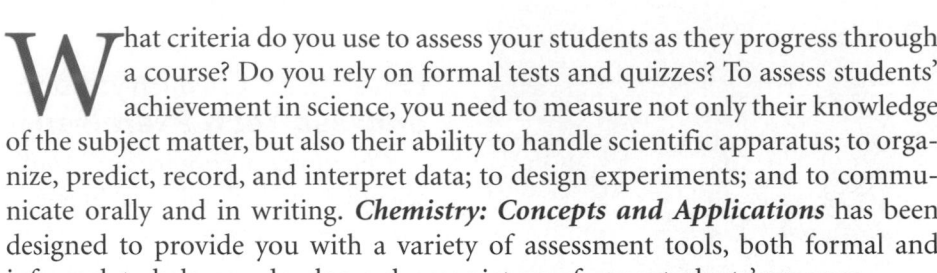

What criteria do you use to assess your students as they progress through a course? Do you rely on formal tests and quizzes? To assess students' achievement in science, you need to measure not only their knowledge of the subject matter, but also their ability to handle scientific apparatus; to organize, predict, record, and interpret data; to design experiments; and to communicate orally and in writing. *Chemistry: Concepts and Applications* has been designed to provide you with a variety of assessment tools, both formal and informal, to help you develop a clearer picture of your students' progress.

PERFORMANCE ASSESSMENT

Performance assessments are becoming more common in today's schools. Science curricula are being revised to prepare students to cope with change and with futures that will depend on their abilities to think, learn, and solve problems. Although learning core concepts will always be important in the chemistry curriculum, the concepts alone are no longer sufficient in a student's scientific education.

Defining Performance Assessment

Performance assessment is based on judging the quality of a student's response to a performance task. A performance task is constructed to require the use of important concepts with supporting information, work habits important to science, and one or more of the elements of scientific literacy. The performance task attempts to put the student in a real-world context so that the class learning can be put to authentic uses later on.

Performance Assessment Is a Learning Activity

Performance assessment is designed to improve the student. Whereas a traditional test is designed to take a snapshot of what the student knows, the performance task involves the student in work that makes the learning more meaningful and builds on the student's knowledge and skills. As a student is engaged in a performance task with performance assessment lists and examples of excellent work, both learning and assessment are occurring.

Performance Task Assessment Lists

A performance task can be assessed according to many different criteria. These criteria, which may include observing, data collecting, or modeling, are then divided up into a number of well-defined categories that are assigned points. These categories can then be scored separately by both teacher and student. The teacher scores not only the quality of the product, but also the quality of the student's self-assessment. Examples of performance assessment lists can be found in the *Glencoe Professional Development Series* booklet: ***Performance Assessment in the Science Classroom***.

In addition to using the performance task assessment list to help guide his or her work and then to self-assess it, the student needs to study examples of excellent work. These examples could come from published sources, or they could be other students' work. These examples include the same type of product for different topics, but they would all be rated excellent using the assessment list for that product type.

Assessing Student Work with Rubrics

A rubric is a set of descriptions of the quality of a process and/or a product. The set of descriptions includes a continuum of quality from excellent to poor. Rubrics for various types of assessment products can be found in *Performance Assessment in the Science Classroom.*

When to Use Performance Task Assessment Lists and Rubrics

Performance task assessment lists are used for all or most performance tasks. If a grade is also necessary, it can be derived from the total points earned on the assessment list. Rubrics are used much less often. Their best use is to help students periodically assess the overall quality of their work. After making a series of products, the student is asked how he or she is doing overall on one of these types of products. With reference to the standards of quality set for that product at that grade level, the student can decide where on the continuum of quality his or her work fits. The student is asked to assign a rubric score and explain why the score was chosen.

Because the student has used a performance task assessment list to examine the elements of each of the products, he or she can justify the rubric score. Experience with performance task assessment lists comes before use of a rubric.

ASSESSMENT AND GRADING

Assessment is giving the learner feedback on the individual elements of his or her performance or product. Therefore, assessment provides specific information on strengths and weaknesses, and allows the student to set targets for improvement. Grading is an act to evaluate the overall quality of the performance or product according to some norms of quality. From the performance task assessment list, the student can evaluate the quality of the pieces. The points on an assessment list can be totaled and an overall grade awarded. The grade alone is not very helpful to the student, but it does describe overall quality of the performance or product. The information from both the assessment list and the grade should be reported to the student and parents.

PERFORMANCE ASSESSMENT IN *CHEMISTRY: CONCEPTS AND APPLICATIONS*

As a part of the *Chapter Review* at the end of every chapter in **Chemistry: Concepts and Applications**, the *Authentic Assessment* page provides yet another source of alternative assessment opportunities. It offers a wide range of ideas for assessing student performance in applying thinking and problem-solving skills, in communicating ideas and concepts, and in conducting long-term projects.

A variety of skill performance tasks are provided in the **Chapter Review and Assessment** book that accompanies the **Chemistry: Concepts and Applications** program. The lists can be used to support *MiniLabs*, *ChemLabs*, and the analysis questions that accompany *Chemistry and Technology*, *Chemistry and Society*, *Everyday Chemistry*, *How it Works*, and the *Literature, Art, History, Physics, Earth Science, Biology*, and *Health Connection* features. The assessment lists and rubrics found in **Performance Assessment in the Science Classroom** can also be used to assess the activities in the *Authentic Assessment* of each *Chapter Review*. Glencoe's **Alternate Assessment in the Science Classroom** provides additional background and examples of performance assessment.

The **MindJogger Videoquizzes** offer videos that provide a fun way for your students to review chapter concepts. You can extend the use of the videoquizzes by implementing them in a testing situation. Questions are at three difficulty levels: basic, intermediate, and advanced.

ASSESSING COOPERATIVE GROUPS

Recent research has shown that a cooperative learning environment improves student learning outcomes for students of all ability levels. **Chemistry: Concepts and Applications** provides many opportunities for cooperative learning and, as a result, many opportuni-

ties to observe group work processes and products. **Cooperative Learning in the Science Classroom** provides strategies and resources for implementing and evaluating group activities. In cooperative group assessment, all members of the group contribute to the work process and the products it produces. For example, if a mixed-ability, four-member laboratory work group conducts an activity, you can use a rating scale or checklist to assess the quality of both group interactions and work skills. An example, along with information about evaluating cooperative work, is provided in the booklet **Alternate Assessment in the Science Classroom**.

Journals and Portfolios

STUDENT JOURNALS

A student journal is intended to help the student organize his or her thinking. It is not a lecture or laboratory notebook. It is a place for students to make their thinking explicit in drawings and writing. It is the place to explore what makes science fun and what makes it hard.

PORTFOLIOS

The portfolio should help the student see the big picture of how he or she is performing in gaining knowledge and skills and how effective his or her work habits are. The portfolio is a way for students to see how the individual performance tasks done during the year fit into a pattern that reveals the overall quality of their learning. The process of assembling the portfolio should be both integrative (of process and content) and reflective. The performance portfolio is not a complete collection of all worksheets and other assignments for a grading period. At its best, the portfolio should include integrated performance products that show growth in concept attainment and skill development.

The Portfolio Criteria for Success

To be successful, a science portfolio should:
- improve the student's performance in science.
- promote the student's skills of self-assessment and goal setting.
- promote a sense of ownership and pride of accomplishment in the student.
- be a reasonable amount of work for the teacher.
- be highly valued by the teacher receiving the portfolio the following year.
- be useful in parent conferences.

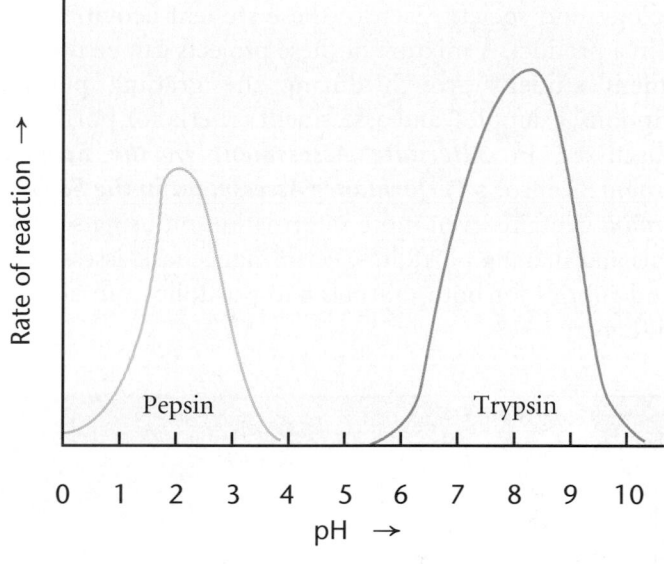

The Portfolio: Its Contents

Include in a student's portfolio evidence of his or her growth in the following five categories.
1. Range of thinking and creativity
2. Use of scientific methods
3. Inventions and models
4. Connections between science and other subjects
5. Readings in science

For each of the five categories, students should make an index that describes what they have selected for their portfolios and why. From their portfolio selections, have students pick items that show the quality of their work habits. Finally, have each student write a letter to next year's science teacher introducing himself or herself to the teacher and explaining how the work in this section shows what the student can do.

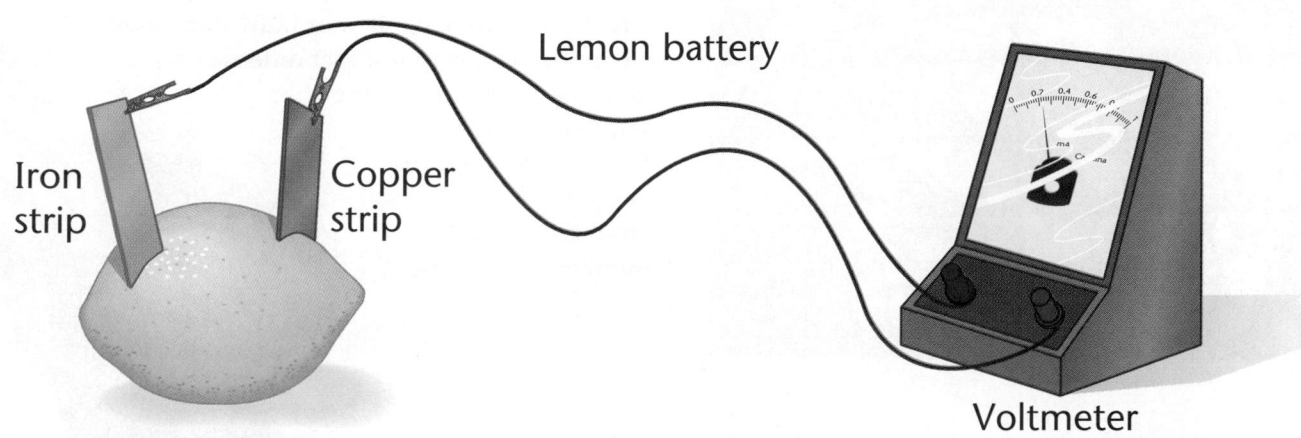

TEACHING STRATEGIES

OPPORTUNITIES FOR USING SCIENCE JOURNALS AND PORTFOLIOS

Chemistry: Concepts and Applications presents many opportunities for portfolio development. Each chapter in the *Student Edition* contains *Projects, ChemLabs, MiniLabs,* and *Connections* with other sciences, other curricular subjects, technology, and society. Each of these student activities will result in a product. A mixture of these projects can be used to document student growth during the grading period. Descriptions, examples, and assessment criteria for portfolios are discussed in *Alternate Assessment in the Science Classroom.* Glencoe's *Performance Assessment in the Science Classroom* contains even more information on using science journals and making portfolios. Performance task assessment lists and rubrics for both journals and portfolios can also be found there.

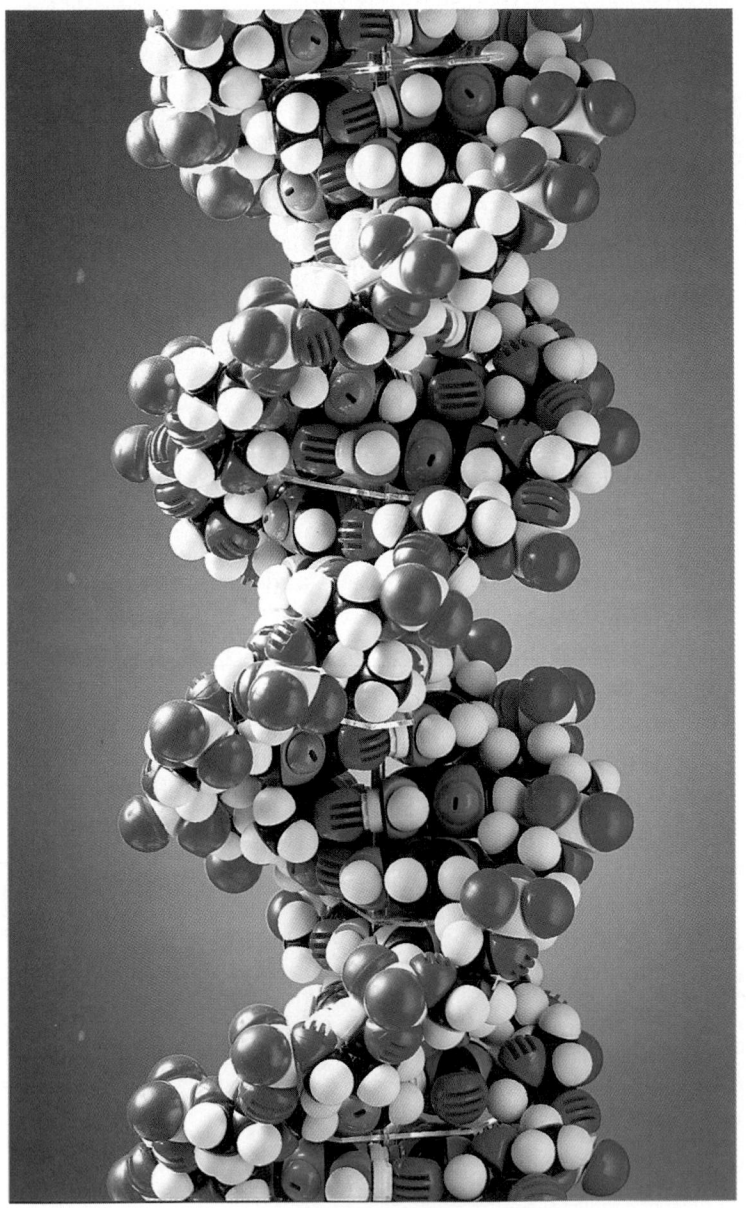

CONTENT ASSESSMENT

Although new and exciting performance skill assessments are emerging, paper-and-pencil tests are still a mainstay of student evaluation. Students must learn to conceptualize scientific processes, and prepare for traditional content assessments. Presently, and in the foreseeable future, students will be required to pass written tests to exit high school and to enter college, trade schools, and other training programs.

Chemistry: Concepts and Applications contains numerous strategies and formative checkpoints for evaluating student progress toward mastery of chemistry concepts. Throughout the chapters in the *Student Edition*, each numbered section has a set of five *Section Review* questions. This spaced review process helps build learning bridges that allow all students to confidently progress from one lesson to the next.

For formal review, *Chemistry: Concepts and Applications* presents a three- to five-page *Chapter Review* at the end of each chapter. By evaluating the student responses to this comprehensive review, which also includes *Cumulative Review* questions, you can determine whether any substantial reteaching is needed. Throughout the text, wherever *Sample Problems* are presented, they are always followed by *Practice Problems.* Additional practice problem sets are placed wherever students would benefit by practicing problem-solving skills. For further assessment, extra practice problems are provided in the *Teacher Wraparound Edition* and in the *Supplemental Practice Problems* booklet of the *Teacher Classroom Resources.*

For the formal content assessment, a six-page chapter test is provided for each chapter in the *Chapter Review and Assessment* book. If your individual assessment plan requires a test that differs from the chapter test in the *Teacher Classroom Resource Package,* customized tests can be easily produced using the *Computer Test Bank.*

Laboratory Safety

Outlined here are some considerations on laboratory safety that are intended primarily for teachers and administrators. Safety awareness must begin with the principal, be supervised by the department head, and be important to the individual teacher.

Principals and supervisors should be familiar with the guidelines for laboratory safety and provide continual supervision to ensure compliance with those guidelines. Teachers have the ultimate responsibility for enforcing safety standards in the laboratory. Planning is essential to laboratory safety, and must include what to do in an emergency, as well as how to prevent accidents.

Numerous books and pamphlets are available on laboratory safety, with detailed instructions on planning and preventing accidents. However, much of what they present can be summarized in the phrase "Be prepared." Know the rules and what common violations occur. Know where emergency equipment is stored and how to use it. Practice good laboratory housekeeping by observing the following guidelines.

Guidelines

1. Store chemicals properly.
    *   Segregate chemicals by reaction types.
    *   Label all chemical containers; include purchase date, special precautions, and expiration date.
    *   When outdated, discard chemicals according to appropriate disposal methods.
2. Prohibit eating, drinking, and smoking in the laboratory.
3. Refuse to tolerate any behavior other than a serious, businesslike approach to laboratory work.
4. Require that eye and clothing protection be worn at all times in the laboratory.
5. Expect proper use of equipment.
    *   Do not leave equipment unattended.
    *   Shield systems under pressure or vacuum.
    *   Never use open flames when a flammable solvent is in use in the same room. Use open flames only when necessary; substitute hot plates whenever possible.
    *   Instruct students in the proper handling of glass tubing.
    *   Instruct students in the proper use of pipets, cylinders, and balances.
    *   See that laboratory benches do not become catchalls for books, jackets, or other personal items.
6. Dispose of wastes correctly.

Local and EPA Regulations Waste disposal is subject to various federal and state regulations. The agencies charged with administering disposal of wastes are the U.S. Environmental Protection Agency (EPA) and its equivalent state agency. The regulations promulgated by the EPA are found in the Code of Federal Regulations (CFR), Title 40. Because some wastes may have to be disposed of away from the school, transportation of hazardous materials becomes a problem. Regulations concerning the movement of hazardous materials are established by the Department of Transportation (DOT) in CFR, Title 49. Both CFR Title 40 and Title 49 are published annually. Changes are published as they occur in the Federal Register.

Practical Guides A practical guide to waste disposal for laboratories is *Prudent Practices for Disposal of Chemicals from Laboratories*, published by the National Research Council, Washington: National Academy Press, 1983. Other resources on waste disposal and other aspects of laboratory management include the following: the chemical catalog published by Flinn Scientific Co., PO Box 219, Batavia, IL 60510, (800)452-1261; a handbook, *Less Is Better*, and two pamphlets, *Hazardous Waste Management* and *Chemical Risk: A Primer*, available free from the Office of Federal Regulatory Programs, ACS Department of Government Relations and Science Policy, 1155 16th St. NW, Washington, DC 20036; and the ACS Chemical Health and Safety Referral Service at the same address, which provides referrals to literature, films, educational courses, or organizations that can provide safety information, (202) 872-4511.

Safety Symbols

These safety symbols are used throughout the text to alert you and your students to any potential safety problems while carrying out *MiniLabs*, *ChemLabs*, *Quick Demos*, or *Demonstrations*.

Disposal Alert
This symbol appears when care must be taken to dispose of materials properly.

Biological Hazard
This symbol appears when there is danger involving bacteria, fungi, or protists.

Open Flame Alert
This symbol appears when use of an open flame could cause a fire or an explosion.

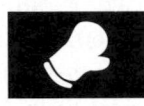

Thermal Safety
This symbol appears as a reminder to use caution when handling hot objects.

Sharp Object Safety
This symbol appears when a danger of cuts or punctures caused by the use of sharp objects exists.

Fume Safety
This symbol appears when chemicals or chemical reactions could cause dangerous fumes.

Electrical Safety
This symbol appears when care should be taken when using electrical equipment.

Plant Safety
This symbol appears when poisonous plants or plants with thorns are handled.

Animal Safety
This symbol appears whenever live animals are studied and the safety of the animals and the students must be ensured.

Radioactive Safety
This symbol appears when radioactive materials are used.

Clothing Protection Safety
This symbol appears when substances used could stain or burn clothing.

Fire Safety
This symbol appears when care should be taken around open flames.

Explosion Safety
This symbol appears when the misuse of chemicals could cause an explosion.

Eye Safety
This symbol appears when a danger to the eyes exists. Safety goggles should be worn when this symbol appears.

Poison Safety
This symbol appears when poisonous substances are used.

Chemical Safety
This symbol appears when chemicals used can cause burns or are poisonous if absorbed through the skin.

Chemical Storage & Disposal

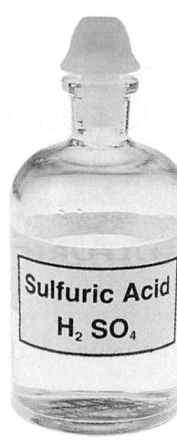

Sulfuric Acid
H₂ SO₄

Positive Ions	Negative Ions
aluminum	borate
ammonium	bromide
bismuth	carbonate
calcium	chloride
copper	hydrogen sulfate
hydrogen	dycroxide
iron	iodide
lithium	nitrate
magnesium	phosphate
potassium	sulfate
sodium	sulfite
strontium	tetraborate
tin	thiocyanate
titanium	
zinc	

Reducing chemical waste A common problem that teachers face is that of chemical waste disposal. One way to reduce waste is to conduct experiments using small amounts of chemicals and smaller sized equipment. Many of the **ChemLabs** in the *Student Edition* and experiments in the Laboratory Manual employ small-scale equipment and techniques. Another solution to excess chemical waste is to carry out demonstrations.

Several complications can make life difficult for the chemistry teacher. The regulations by OSHA, EPA, and DOT are numerous and complex. In addition, some teachers may have to work with the prohibition, by a principal or governing board, of students working with any chemical substance that may be considered even slightly hazardous. In those circumstances, the teacher could perform an experiment as a demonstration with students acting as observers, note takers, and data recorders. Even if the demonstration is carried out behind a safety shield, the students will at least have experienced important chemical reactions. *Chemistry: Concepts and Applications Teacher Wraparound Edition* offers you an abundant resource for demonstration ideas that can be used with every lesson.

Liquid waste disposal A frequent question from chemistry teachers is about which liquids may safely be disposed of in the sanitary drains. First, the teacher must be certain that the sewer flows to a wastewater-treatment plant and not to a stream or other natural waterway. Second, any substance from a laboratory should be flushed with at least 100 times its own volume of tap water. Third, the disposal suggestions given on the next page should be checked with local authorities because local regulations are often more stringent than federal requirements. The National Research Council book *Prudent Practices for Disposal of Chemicals from Laboratories* lists many substances that can be disposed of in the sanitary drain.

Some positive and negative ions from the NRC lists are given in the table shown here. Please note that it is important for both the positive and negative ion of a salt to be listed in order for its drain disposal to be considered safe. Also note that although hydrogen and hydroxide ions are listed, acids and bases should be neutralized before disposal. A good rule of thumb is that nothing of pH less than 3 or greater than 8 should be discarded without neutralizing it first.

Disposal of organic waste Of the organic compounds most often found in high school laboratories, the following can be disposed of in the drain: methanol, ethanol, propanols, butanols, pentanols, ethylene glycol, glycerol, sugars, formaldehyde, formic acid, acetic acid, oxalic acid, sodium and potassium salts of carboxylic acids, esters with fewer than five carbon atoms, and acetone. More extensive lists can be found in *Prudent Practices for Disposal of Chemicals from Laboratories*.

Substances that do not fall into one of the disposal categories described here can be treated in one of three of the following ways.

1. Treat waste chemically to convert it to a form that is drain-disposable. A good example is iodate ion, which is too strong an oxidizing agent to be disposed of untreated. However, it is readily reduced to iodide (disposable) by acidified sodium hydrogen sulfite. Many procedures for processing laboratory waste to a discardable form can be found in *Prudent Practices*.

2. Recycle waste. A good example is the recovery of valuable metals such as mercury and silver. Solvents can be recycled through distillation.

3. If waste cannot be recycled or processed to disposable form, then it must be packed and shipped by a Department of Transportation-approved shipper to a landfill designated to receive chemical and hazardous waste.

Numerous processes are described in *Prudent Practices* for reducing both the bulk and hazard of wastes. An example is the reduction of chromate and dichromate waste solutions to chromium(III) solutions, which are then made basic. The chromium(III) oxide precipitate is filtered off, dried, and crushed, resulting in significant savings. As hazardous wastes are accumulated awaiting shipment, it is important to observe the proper storage procedures for separating incompatible substances.

Disposal Procedures

Local, state, and federal laws regulate the proper disposal of chemicals. Before using the disposal instructions given here, you should confirm them with local regulators so that you do not violate your local or state laws. No representation, warranty, or guarantee is made by Glencoe Publishing Company or the authors as to the completeness or accuracy of these disposal procedures.

The proper disposal of chemicals should be done only by the teacher, who must be wearing a laboratory apron, goggles, and rubber gloves. The disposal procedures should be carried out in an operating fume hood. Even teachers should never be alone in the laboratory when disposing of chemicals. They should have the proper type of fire extinguisher and a telephone or intercom nearby.

The disposal procedures listed are for relatively small amounts of wastes. If a larger amount is produced, the stoichiometry of the chemical reactions in the disposal procedure will have to calculated.

Disposal advice given in the **Teacher Wraparound Edition** is keyed to the following lettered disposal procedures.

Disposal A

These materials can be packaged and sealed in separate plastic containers and buried in a landfill approved for chemical and hazardous waste disposal. Contact your local regulatory agency before assuming it is acceptable to place these materials in your school's trash dumpster. Do not mix chemicals; place each one in its own container. Place the containers in a cardboard box and separate with vermiculite. Seal the closed box before disposing of it.

Disposal B

Decant the water layer into a separate beaker. Then, discard the water layer down the drain. Label the other chemical and save it for future activities, following proper storage directions.

Disposal C

Rinse the chemical down the drain with at least a 100-fold increase in the volume of water. Rinse only one chemical down the drain at a time using plenty of water to dilute it. Do not mix chemicals in the sink or drain. Dissolve any solids in a beaker before rinsing them down the drain. Do not rinse these chemicals down the drain if your school's drains go into a septic system. This disposal procedure is intended for schools whose sanitary sewers go to a wastewater-treatment plant. Due to local soil and water conditions, local regulations may prohibit the disposal of these chemicals in this manner. Check with the local government regulatory agency before using this procedure.

Disposal D

While being stirred, the acidic solution should be slowly added to a large beaker of cold water. Prepare a $1M$ $Na_2CO_3 \cdot 10H_2O$ solution by dissolving 62 g in 500 mL of water. Slowly add the $1M$ Na_2CO_3 solution to the diluted acid. Carbon dioxide gas will be evolved. When there is no more evolution of gas as more Na_2CO_3 solution is added, the solution can be tested with pH paper to verify that it is neutral. Rinse the neutralized solution down the drain with a large volume of water.

Disposal E

Filter the solution through filter paper. Open the filter paper and allow it to dry. Place the dried solid and filter paper in a plastic container surrounded by vermiculite in a box. Seal the box and dispose of it in a landfill approved to receive chemical and hazardous waste. Dilute the filtrate by adding it slowly, while stirring, to a large beaker of cold water. Verify that it is neutral by using pH paper. Rinse the neutral solution down the drain with a large volume of water.

Disposal F

Place the substance in a plastic container and seal it. Completely label the container and, following proper storage directions, save it for future activities.

Disposal G

For the iodine: add 18 g of sodium thiosulfate to the solution. Stir while warming the solution to 50°C. After the iodine is consumed, check with pH paper and add enough $3M$ NaOH solution (12 g NaOH in 100 mL of water) to neutralize the solution.

For the manganese or cobalt: add 6 g of sodium sulfide. After one hour of stirring, neutralize the solution with $3M$ NaOH. Verify with pH paper that the solution is neutral. Filter and place the MnS solid in a plastic bottle for disposal in an approved landfill. Treat the filtrate for excess sulfide by adding 12 g of iron(III) chloride with constant stirring. Filter the precipitate, FE_2S_3. Place the solid in a plastic container, and seal it for disposal in an approved landfill. Rinse the neutralized solution down the drain with a large volume of water.

Disposal H

Use an operating fume hood to disperse the small volume of gas produced.

Disposal I

Slowly add the substance to a large beaker of cold water while stirring. Then in an ice bath, slowly add $6M$ HCl until the solution is neutralized. Verify by using pH paper. Rinse the neutralized solution down the drain using a large volume of water.

Small-Scale Chemistry Techniques

Chemistry teaching at the high school level is presently facing several problems at once. Concern for student safety, environmental questions, cost of materials, and the necessity of adhering to a prescribed curriculum have all worked to make the laboratory program difficult to carry out. However, the lab is still the most tangible, best-remembered, and most visible aspect of chemistry instruction. Without direct observation of chemical phenomena and the hands-on manipulation of the substances referred to in the classroom and the textbook, chemistry can become no more than a survey of chemical theory with little real meaning to students, especially students who are visual learners.

Chemistry: Concepts and Applications provides small-scale chemistry techniques both in the textbook in the form of *ChemLabs* and in the **Laboratory Manual**. When you have limited time and a limited budget, small-scale chemistry provides a safe, inexpensive, and time-efficient means to conduct a laboratory program.

Laboratory activities and experiments that use micro amounts of chemicals provide a way to involve students in observation and manipulation of some substances that might otherwise be regarded as hazardous. Because drastically reduced amounts of chemicals are used, safety is greatly increased. Likewise, the expense of running a laboratory chemistry program is cut. Because of the savings in both time and expense with small-scale chemistry, students are afforded the opportunity to do much more experimental chemistry than was possible with the more traditional lab techniques. As with traditional chemistry, all processes should be conducted with safety goggles and protective clothing.

SMALL-SCALE DISPOSAL

Disposal J

This applies only to solutions used in small-scale chemistry procedures. Collect all chemical solutions, precipitates, and rinse solutions in a polyethylene dishpan or similar container devoted to that purpose. Retain the solutions until the end of the period or day. In the fume hood, set up a hot plate with a 1-L or 2-L beaker. Pour the collected solutions into the beaker. Turn the hot plate on low and allow the beaker to heat with the hood running and the hood door closed. The liquids and volatiles in the mixture will evaporate, leaving dried chemicals. Allow the beaker to cool. Continue to add solutions and waste until the beaker is two-thirds full. Treat the waste as heavy metal waste. Dispose of the beaker and its dry contents in an approved manner.

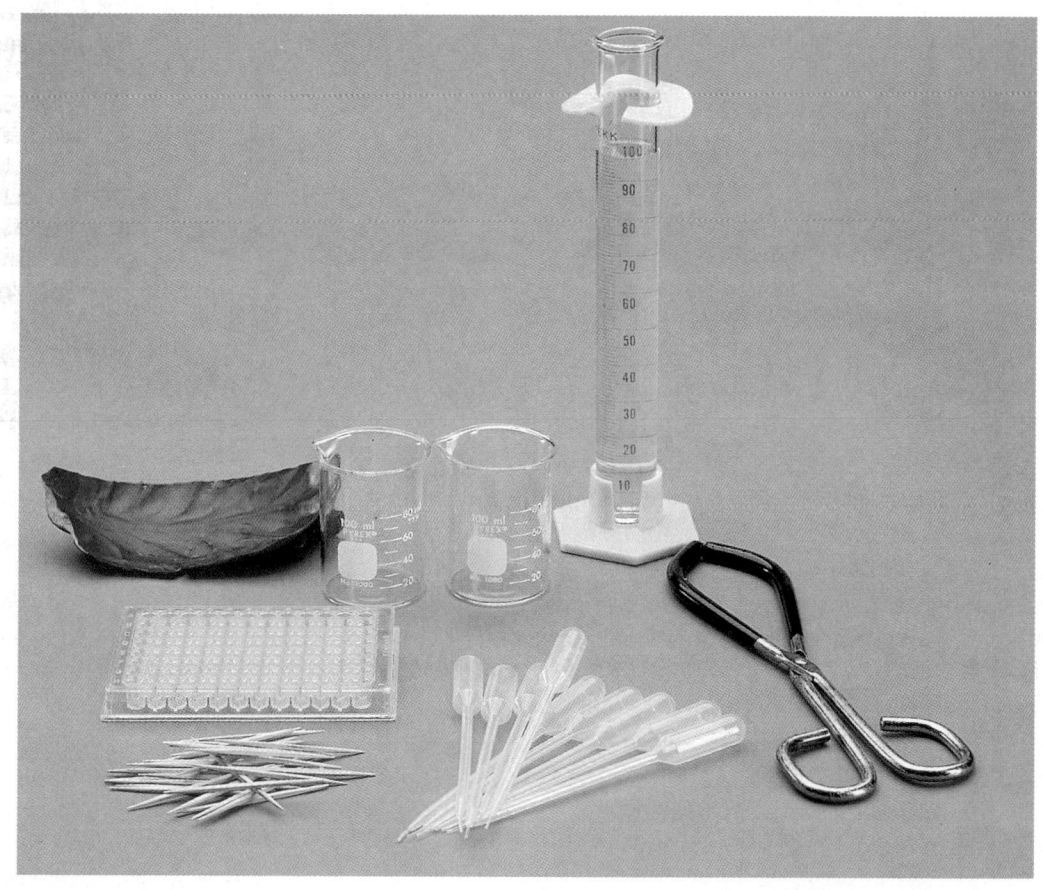

Small-Scale Equipment

Small-scale chemistry uses only two basic tools:

The microplate and the plastic pipet

THE MICROPLATE

The microplate is a plastic tray containing several shallow wells arranged in lettered rows and numbered columns. These wells are used in the same way that test tubes and beakers are used. There are three types of microplates, as shown here. A 24-well microplate has four rows and six columns of wells. A 96-well microplate has eight rows and 12 columns of smaller wells. There is also a combination microplate, which consists of four rows of 12 wells and two rows of six wells. Most of the activities in *Chemistry: Concepts and Applications* that use microplates are carried out with either 24-well or 96-well microplates.

PLASTIC PIPETS

Small-scale chemistry uses two main types of plastic pipets: the thin-stem and microtip pipets shown here. These pipets are used to deliver chemicals and to collect products, including gases. The pipets are

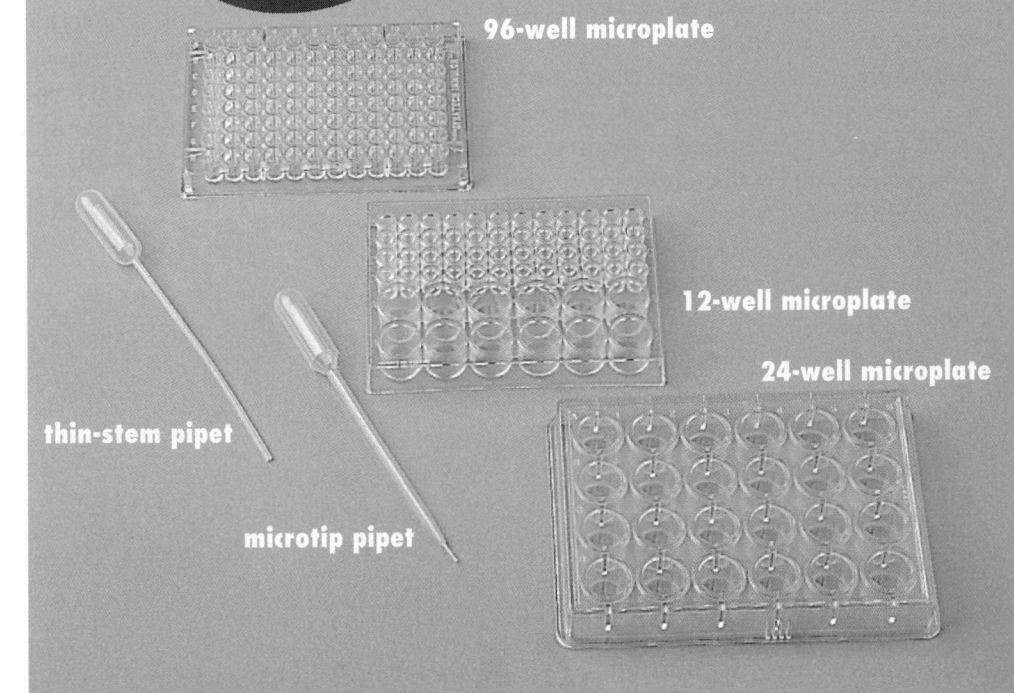

96-well microplate

12-well microplate

24-well microplate

thin-stem pipet

microtip pipet

made of polyethylene, which makes them soft and flexible. The microtip pipet is used when it is necessary to deliver very small droplets of liquid to a reaction. The pipets may be modified in a variety of ways to serve diverse purposes. For example, the thin-stem pipets can be used as chemical scoops, gas generators, and reaction vessels. The long, flexible tube of a thin-stem pipet makes a convenient delivery tube. Some small-scale chemistry techniques are based on the fact that the tube of a thin-stem pipet will slide into the tube of a microtip pipet that has had its tip cut off. This arrangement produces a typical generator-collector setup.

Either type of pipet can be reused simply by rinsing the stem and bulb with water. The inside plastic surface of the pipet is non-wetting and therefore does not hold water or solutions the way glass does. This means that, with proper technique, the entire contents of the pipet can be dispensed, leaving none of the solution behind.

Chemicals and Equipment

These easy-to-use tables of materials can help you prepare for your chemistry classes for the year. All quantities are for one lab setup of each *Demonstration*, *MiniLab*, or *ChemLab* for the entire course. Before placing your order for supplies, determine how many classes you will be teaching and how many students you expect in each class. The amounts listed for *Demonstrations* are for one performance of the demonstration. Therefore, if you have seven classes, then you will need to multiply the quantities of materials for *Demonstrations* by seven. However, if you have ten groups of students in each of your seven classes, multiply the quantities of materials for *ChemLabs* and *MiniLabs* by 70 to arrive at your total course requirements.

The standard list of equipment is made up of a set of equipment that is generally recommended for each lab bench station in the chemistry laboratory. For all lab activities in this program, it as assumed that your classroom is equipped with these items for each setup of a *Demonstration*, *MiniLab*, or *ChemLab*. Additional equipment required for the course is listed under **Nonconsumables**. The listed amounts of **Chemicals** and **Other Consumables** for *Demonstrations*, *MiniLabs*, and *ChemLabs* are sufficient for one demonstration or one lab setup per student or group of students.

STANDARD EQUIPMENT LIST (for each station)

apron, 1 per student
goggles, 1 pair per student
beakers, 5: two 400-mL, two 250-mL, one 100-mL
beaker tongs
Bunsen burner and tubing
clay triangle
crucible and cover
crucible tongs
droppers, 2
Erlenmeyer Flask, 125-mL
Erlenmeyer Flask, 250-mL
evaporating dish
forceps
funnel

graduated cylinder, 10-mL
graduated cylinder, 50-mL or 100-mL
microplate, 24-well
microplate, 96-well
scissors
spatula, stainless steel
stirring rods, 2
test-tube holder
test-tube rack
test tubes, 6 large
test tubes, 6 small
wash bottle
watch glass
wire gauze

CLASSROOM EQUIPMENT (for general use)

balance
beakers, assorted large 600-mL, 800-mL, 1-L, 2-L
clamps, assorted including buret clamps
conductivity tester (see directions on page 60T)
dishpan, plastic
Erlenmeyer flasks, 50-mL, 500-mL, 1-L
hot plate
iron rings, assorted

iron tripod
lighter for burner
mortar and pestle
ring stands, 2
rubber or tygon tubing
rubber stoppers, assorted
thermometer, -10°C to 150°C
thermometer clamp

NON-CONSUMABLES

ITEM	ChemLabs	miniLABS	Demonstrations
aquarium			1
battery, 9-V	1	1	1 used
bottle, 250-mL polyethylene or polypropylene			1
bottle, 6-gal transparent plastic			1
bottle, 500-mL widemouth with stoppers			3
burner, Meker and tubing			1
cafeteria tray			1
calculator	1		
can opener			1
cans of food			12+
capillary tube			4
ceramic plate, white			1
clamp, double-post screw	1		
cloth, silk			1
cloth, wool (or piece of fur)			1
clothespin, spring-type			1 box
cloud chamber			1
conductivity apparatus, lightbulb			1
copper tubing, small diameter.			45 cm
Crooke's tube (CRT)			1
cylinder, 25-mL graduated		1	
dart		1	
deflagrating spoon			1
deionizing cartridge		1	
diffraction grating		1	
dish, plastic	1		
dish, shallow metal	1		
dropper			3
electroscope			1
Erlenmeyer flask, 1-L or 2-L			1
explosion shield			1
file, triangular	1		
flashlight bulb, for two AAA batteries	1		
flashlight with batteries	1		
forceps	1		
funnel, powder			1
Geiger counter with meter			1
glass plates			4
glass tubing, 10-mm diameter			35 cm
glass tubing, 6-mm diameter			35 cm
glove, thermal	1	1	
graduated cylinder, 100-mL			1
graduated cylinder, 250-mL			1
hammer, small	1	1	1
Hoffman apparatus			1
hot plate/magnetic stirrer			1
iron sheeting			10 cm × 10 cm
jar, 1-pint canning with ring lid			
jar, large chromatography, with lid	1		
jar, with screw-on lid		1	24

NON-CONSUMABLES

ITEM	ChemLabs	miniLABS	Demonstrations
knife		1	1
lead foil			10 cm × 10 cm
lightbulb, incandescent		1	1
magnet, permanent bar			1
magnet, small	1	1	
magnet, teflon-coated white			1
magnetic stirrer			1
marbles, glass			12
marking pen, permanent	1	1	1
meterstick		1	1
microscope		1	
microscope slide, glass	1	1	1
molecular model kit		1	1
overhead projector			1
paper clip, small			1 box
pencil, glass-marking	1		
penny, post-1982	60		
penny, pre-1982	60		
periodic table, notebook size			1
petri dish		1	4
pipet, 10-mL	1		
plastic film, dark			1 piece
plastic lid, clear			1
pliers			1 pair
pliers, needle-nose			1 pair
power supply, 6- to 12-V DC			1
power supply, high voltage DC		1	1
radiation source, needle-mounted alpha			1
radio, 9-V transistor			1
rod, metal			50 cm length
rod, rubber or plastic			1
rolling pin		1	
ruler, metric	1	1	1
screwdriver, small			1
spectroscope, student			24
spectrum tube, helium		1	
spectrum tube, hydrogen		1	
spectrum tube, mercury		1	
spectrum tube, neon		1	
stirring rod			1
thermometer, -10°C to 150°C	1		2
thumbtack		64 assorted colors	1
timer, second	1	1	1
tubing, soft glass	24 cm		50 cm
vacuum flask, 500-mL with stopper			1
vegetable shredder			1
voltmeter, DC with 2-V or 3-V scale	1		
watch glass			2
wire leads with alligator clips	2		2
wire screen, window	10 cm × 10 cm		

CHEMICALS

ITEM	ChemLabs	miniLABS	Demonstrations
1,6-diaminohexane			1.1 g
acetic acid, glacial	145 mL	4 mL	1 mL
acetone	6 mL		325 mL
acetylsalicylic acid (aspirin)	2 g		1 tablet
aluminum foil	10 cm^2		100 cm^2
aluminum strip		1	
aluminum sulfate octadecahydrate		0.5 g	
ammonium chloride	10 g	10 g	
ammonium hydroxide, 14.8M		10 mL	10 mL
ammonium meta-vanadate			10 g
ammonium thiocynate			15 g
baking powder	3 g		
barium chloride dihydrate	2 g	20 g	1.5 g
barium hydroxide octahydrate			33 g
barium nitrate	1.3 g		
barium sulfate			1 g
boric acid	0.1 g		
bromine			23 mL
bromothymol blue			0.1 g
bromothymol blue reagent solution		1 mL	
calcium carbide			20 g
calcium carbonate (marble) chips	10	3	
calcium chloride, granular		30 g	10 g
calcium chloride, powdered			10 g
calcium hydroxide	13 g		
calcium nitrate tetrahydrate	1 g		
calcium oxide			1 g
calcium sulfate			1 g
calcium turnings	6 g		2 g
carbon	10 g		5 g
charcoal, powdered	1 g		
chlorine gas	20 mL		
chlorine water		2 mL	
cobalt(II) chloride hexahydrate		1 g	2 g
cobalt chloride test paper			1 strip
copper(II) chloride	5 g		6.8 g
copper(II) oxide	1 g		
copper(II) sulfate pentahydrate	12 g	3 g	31 g
copper(III) nitrate trihydrate	0.3 g		1 g
copper foil	5 cm^2	1 cm^2	20 cm^2
copper, granular	1.6 g		1 g
copper turnings	2 g		
copper wire, insulated 14-gauge	20 cm		23 cm
cyclohexane			5 mL
cyclohexene			5 mL
dextrose			17 g
dichloromethane			30 mL
dry ice		40 g	800 g
ethanol		55 mL	525 mL
ferric ammonium sulfate dodecahydrate			2 g

ITEM

ITEM	ChemLabs	miniLABS	Demonstrations
ferrous ammonium sulfate hexahydrate			2 g
ferrous sulfate heptahydrate			2 g
formic acid	.6 mL		
fructose	.2 g		
germanium			5 g
glass wool	.3 g		
glycerol			10 mL
hexane		5 mL	30 mL
hydrochloric acid, 12M	.25 mL	10 mL	160 mL
hydrogen gas, lecture cylinder with regulator			1
hydrogen peroxide, 3%			120 mL
hydrogen peroxide, 6%			100 mL
iodine	.12 g.		5 g
iron(II) chloride tetrahydrate	.1 g		
iron(III) chloride hexahydrate	.1 g		
iron filings			2 g
iron, powdered		25 g	
isopropanol, 91%		100 mL	
kerosene		5 mL	100 mL
lead	.5 g		5 g
lead acetate test paper			1 strip
lead(II) nitrate	.0.4 g		
lead strip		1	
lithium			2 g
lithium chloride		20 g	
luminol			0.2 g
lycopodium powder			2 g
magnesium chloride hexahydrate	.0.3 g		
magnesium ribbon	.3 g		5 g
magnesium sulfate heptahydrate (Epsom salt)	.12 g.		1 g
manganese dioxide			7 g
methanol		3 mL	5 mL
methyl orange reagent solution		1 mL	
methylene blue			1 crystal
methylene blue reagent solution		1 mL	1 mL
nitric acid, 15M	.5 mL		50 mL
nitrogen gas			20 mL
oxygen gas			20 mL
paraffin wax	.2 g		100 mL
phenolphthalein indicator reagent solution	.1 mL	3 mL	2 mL
phosphorus, red	.10 g		
polyvinyl alcohol		20 mL	80 g
potassium			2 g
potassium chloride		20 g	
potassium dichromate		0.3 g	
potassium hexacyanoferrate(II)			0.5 g
potassium hexacyanoferrate(III)		0.3 g.	0.5 g
potassium hydroxide			3.4 g
potassium iodate		0.1 g	
potassium iodide	.0.1 g		8 g

CHEMICALS

CHEMICALS

ITEM	ChemLabs	miniLABS	Demonstrations
potassium nitrate		0.1 g	
potassium permanganate			0.7 g
potassium thiocyanate			0.5 g
salicylic acid		1 g	
sebacoyl chloride			1 mL
selenium	.5 g		
silicon	.10 g.		5 g
silver nitrate		1 g.	13 g
sodium			6 g
sodium acetate		0.1 g	4 g
sodium borate(tetra) decahydrate (borax)	.11 g.	0.5 g	12 g
sodium bromide	.0.1 g		
sodium carbonate	.3.1 g.	0.1 g	1 g
sodium chloride (table salt)	.42 g.	1 g	135 g
sodium hydrogen carbonate (baking soda)	.3 g	5 g	235 g
sodium hydrogen sulfite			1.5 g
sodium hydroxide	.20 g.	2 g.	58 g
sodium iodide	.0.3 g		
sodium oxalate	.2 g		
sodium silicate (water glass)			25 mL
sodium sulfate decahydrate	.0.5 g		140 g
sodium sulfite		4 g	
sodium thiosulfate			1 g
starch, soluble		1 g	3 g
stearic acid	.20 g		
strontium chloride hexahydrate	.2 g.	20 g	
strontium nitrate	.1 g		
sucrose (table sugar)	.2 g.	10 g	50 g + 1 cube
sulfur, powdered	.11 g.		2 g
sulfur, roll			22 g
sulfuric acid, 18M	.12 mL	3 mL	35 mL
tannic acid			0.5 g
thermite mixture			50 g
thermite starter			5 g
tin	.10 g.		5 g
trichlorotrifluoromethane (TTE)			80 mL
universal indicator reagent solution			10 mL
urea	.2 g		
vinegar	.1 mL each of 6 brands	100 mL	540 mL
vinegar, white	.12 mL		
zinc chloride		3 g	
zinc, dust	.1 g.		1 g
zinc, granular	.0.25 g.	1 g	
zinc–mercury(II) chloride amalgam			63 g
zinc, mossy		3 pieces	
zinc strip		2.	1

OTHER CONSUMABLES

ITEM	ChemLabs	miniLABS	Demonstrations
ammonia soln., household		5 mL	
antacid		1 tablet each of 4 types	3 tablets
apple			1
bag, plastic large		1	
bag, plastic sealable		11	
bag, plastic small	1	1	4
balloon, large			10
balloon, small			12
blotter paper			2 pieces
bone, small chicken uncooked		1	
bottle, colorless glass soda			1
bottle, 2-L soft-drink with cap			1
bottle, small clear capped			4
bubblegum, sugared			5 pieces
can, metal	1		1
can, soft-drink	1		1
candles, large birthday	2		6
candy, assorted colors	6 pieces		
canola oil		20 mL	
carbon paper		1 sheet	
card, white index		1	
cardboard	40 cm × 40 cm		
carrot	1		1
cereal, fortified dry		50 g	
chalk			1 piece
cheesecloth	20 cm x 20 cm piece		20 cm × 20 cm
chlorine bleach		40 mL	20 mL
cloth towel			1
coffee filter		4	
cooking oil			1 spray can
cork			1
cotton ball			6
cotton swab		2	
craft stick	1	1	
cup, 3-oz. plastic			100
cup, clear plastic 6"	1	5	
cup, large Styrofoam			1
cup, paper		2	1
detergent, liquid dish			26 mL
diaper, disposable		2	
egg white		1	
fabric samples, 7 types	6 squares, 0.5 cm × 0.5 cm each		
filter paper, qualitative	1 strip / 2 circles (11-cm diameter)		1 circle / 1 piece, 10 cm × 10 cm
flashcube			1
flour		10 g	20 g
food coloring, assorted colors		1 mL each	1 mL each
food coloring, yellow	1 mL		
gelatin, clear unflavored		1 pkg	
glue			10 g
gumdrop			144
ink marker, water soluble		8 colors	1
lemon		1	
lemon juice	10 mL		10 mL
lifesaver, wintergreen			1
lighter fluid		2 mL	

OTHER CONSUMABLES

ITEM	ChemLabs	miniLABS	Demonstrations
litmus paper, blue	.4 strips	3 strips	
litmus paper, red	.2 strips.		2 strips
manila file folder			4
marshmallow, small	.2		
matches, book	.4	2.	7
milk			5 mL
modeling clay			50 g
nail, aluminum		1	
nail, galvanized		1	
nail, iron		2.	1
nail, painted iron		1	
orange juice			50 mL
paper, black construction		1 (8 1/2" × 11" sheet)	2 (8 1/2" × 11" sheets)
paper clip, large	1	1	
paper, graph		1	
paper towels			10
paper, assorted colors		8 (8 1/2" × 11" sheets)	
peanut oil		20 mL	
peanuts			30 g
pecan	1		
pencil lead		2.	4
pencil, wood		2.	1
penny, dull/darkened			10
penny, pre-1982	1		
pickle juice			5 mL
popcorn kernels			30
poster board		1 m^2	
potato chip			1
radon detection kit		1	
red cabbage	1 leaf		
rock salt			10 g
roofing shingle			1
rubber band			8
sand			20 lbs
shampoo, baby			5 mL
silicone caulking			1 tube
soap, laundry	.0.5 g		
soap, liquid			5 mL
steel wool, fine	1 pad	1 pad	1 pad
straw, soda		24	4
string			6 m
tape, masking	15 cm	40 cm.	30 cm
tape, transparent waterproof	.5 cm	5 cm	
thumbtacks, assorted colors		64	
toothpaste			2 g
toothpick	1 box		1 box
trash bag, plastic			1
tubing, dialysis 25-mm flat width	15 cm		
vanilla flavoring			2 mL
vegetable shortening, solid			25 g
vermiculite		30 g	
wax		10 g	
wheat germ		5 g	
wood splint		6	11
wrap, clear plastic	.400 m^2	400 cm^2	1 cm^2
yeast, dry		1 pkg	1 pkg

Suppliers

Chem Scientific, Inc.

67 Chapel Street
Newton, MA 02158
Toll free: (888) 527-5827
Fax: (617) 527-5827

Chem Scientific can supply you with all the chemicals and equipment, including small-scale supplies, required to do all the *ChemLabs*, *MiniLabs*, and *Demonstrations* in Glencoe's program. Request their customized catalog and price list and ask about the special discounts they are offering to users of *Glencoe's* **Chemistry: Concepts and Applications.**

Chemical and Equipment Suppliers

Aldrich Chemical Co., Inc.
PO Box 2060
Milwaukee, WI 53201
(800) 558-9160

Arbor Scientific
PO Box 2750
Ann Arbor, MI 48106-2750
(800) 367-6695
arborsci@bizserve.com

Carolina Biological Supply Co.
2700 York Road
Burlington, NC 27215
(800) 334-5551

Central Scientific Co.
11222 Melrose Avenue
Franklin Park, IL 60131
(800) 262-3626
cencophys@aol.com

Fisher Scientific Co.
1600 W. Glenlake
Itasca, IL 60143
(800) 766-7000

Flinn Scientific Co.
PO Box 219
Batavia, IL 60510
(800) 452-1261
flinnsci@aol.com

Frey Scientific
100 Paragon Parkway
Mansfield, OH 44903
(800) 225-3739

Kemtec Educational Corp.
9889 Crescent Drive
West Chester, OH 45069
(800) 733-0266

McKilligan Supply Corp.
435 Main Street
Johnson City, NY 13790
(800) 477-6512

Nasco
901 Janesville Avenue
Fort Atkinson, WI 53538
(800) 558-9595

Pasco Scientific
10101 Foothills Blvd.
PO Box 619011
Roseville, CA 95678-9011
(800) 772-8700
sales@pasco.com

Quantum Technology
30153 Arena Drive
Evergreen, CO 80439
(303) 674-9651
leapsys@aol.com

Sargent-Welch Scientific Co.
PO Box 5229
Buffalo Grove, IL 60089
(800) 727-4368
sarwel@sargentwelch.com
http://www.sargentwelch.com

Ward's Natural Science Establishment, Inc.
PO Box 92912
Rochester, NY 14692
(800) 962-2660

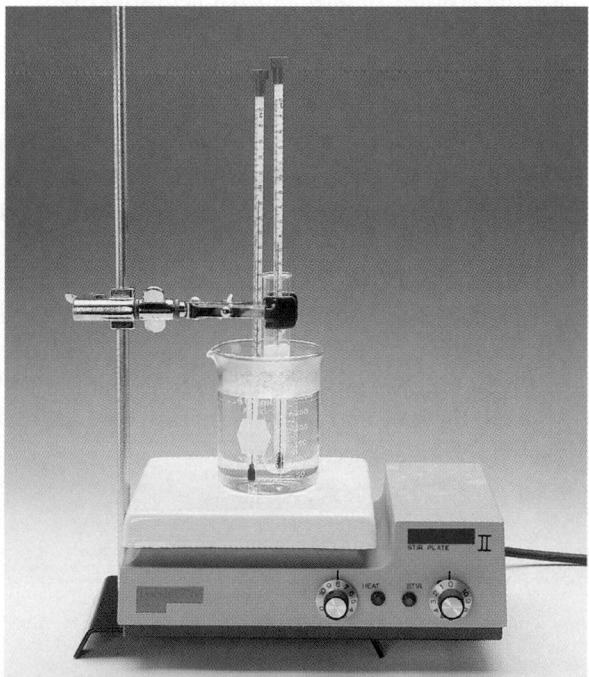

Company names, mailing addresses, phone numbers, e-mail addresses, and World Wide Web sites were accurate at the time of publication and are subject to change.

Instructions for Making a Conductivity Tester

The use of a simple conductivity tester is required for several of the labs in *Chemistry: Concepts and Applications*. It is easy to construct a conductivity tester from a few simple items easily obtained from your local electronics or hardware store.

For each conductivity tester, you will need the following items.

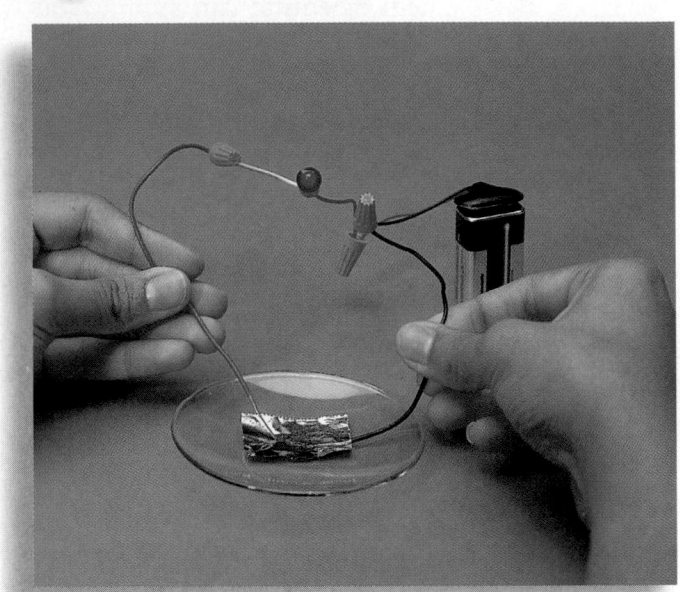

Radio Shack Part Numbers

- 1 9-V battery . . . . .23-583 [one per package]
- 1 9-V battery clip . . . . .270-325 [five per package]
- 1 LED lamp assembly with a resistor . . . . .276-011 [one per package]
- 3 very small wire nuts . . . . .64-3057 [16 per package]
- 2 lengths of solid wire, 20-gauge . . . . .278-1222 [75 ft. per package]

ASSEMBLY PROCEDURE

1. Carefully strip about 1/2 in. of the insulation from each of the wires connected to the battery clip. Be careful not to cut through the slender copper wire inside.
2. Twist the strands of wire in each lead together.
3. Repeat this process with the leads of the LED lamp assembly if they are not already stripped and twisted together.
4. Cut two pieces of solid wire and strip about 1/2 in. of the insulation from one end of each piece.
5. Look at the instructions that came in the packet with the LED lamp assembly. There should be some indication of which wire is the cathode. Place the bare part of the cathode wire and the bare part of the black wire from the battery clip together side by side, and twist them together just enough to hold them together.
6. Fold back the free end of this junction to double up the wire and to secure the nut.
7. Take one of the small wire nuts and twist it into the bare end of the wires. Be careful not to twist the wires too much. Over twisting could break the wires.
8. Connect the other wire from the LED to one of the pieces of solid wire by carefully twisting them together.
9. Fold the wires and attach the wire nut as before.
10. Repeat steps 8 and 9 with the remaining wire from the battery clip and the solid wire.
11. Strip insulation from the free ends of the solid wires to a length sufficient to pass through the item being tested in the individual lab.
12. Test the conductivity tester by attaching the battery to the clip and touching the free ends of the wires together. The LED should light up. If it does not, you may have a broken wire, a faulty LED, or a wiring error. In rare cases, the wire colors on the battery clips or LEDs may be incorrect.

Audiovisuals

Chapter 1
Software
Lake Study for Windows. JCE: Software.
Doing Chemistry. American Chemical Society.
Videodisc
The World of Chemistry, Selected Demonstrations and Animations I and II. JCE: Software.
The Periodic Table Videodisc. JCE: Software.
Chemistry at Work Image Database for Chemistry. Videodiscovery.

Chapter 2
Software
Spectral Lines Experiment. AR201. Project Seraphim.
Electron Arrangement. PC 2401. Project Seraphim
Atomic Spectroscopy, for IBM. JCE: Software.
Atomic Spectra-Graph, for Macintosh. JCE: Software.
Chemistry Explorer. The atom. Logal Software, Inc.

Chapter 3
Software
KC? Discoverer: Exploring the Properties of the Chemical Elements. JCE: Software.
Videodisc
Periodic Table Videodisc Database. Falcon Software, Inc.
The Periodic Table Videodisc. JCE: Software.

Chapter 4
Videodisc
The Periodic Table Videodisc. JCE: Software.
Cosmic Chemistry. The Living Textbook, Physical Sciences, Side 3 *Chemical Changes*. Optical Data Corporation.
CD-ROM
The Periodic Table CD-ROM. JCE: Software.

Chapter 5
Software
Introduction to General Chemistry. Disk 2, *Inorganic Nomenclature*. Falcon Software, Inc.
Videodisc
ChemDemos. JCE: Software.

Chapter 6
Software
An Equilibrium Simulation, AP603, IB601. Project Seraphim.
Video
Chemical Equilibrium, Concepts in Science. TVOntario.
Videodisc
ChemDemos. JCE Software.
CD-ROM
Exploring Chemistry IV. Lesson 1 *Chemical Reactions*. Falcon Software, Inc.

Chapter 7
Software
Orbitals and Electrons. Falcon Software, Inc.
Atomic Structure. PC2401, Electron Arrangement. Project Seraphim.
Electron Cloud Diagrams. PC2403. Project Seraphim.
Videodisc
The World of Chemistry: Selected Demonstrations and Animations I and II. JCE: Software.

Chapter 8
Software
KC? Discoverer with Knowledgeable Counselor. JCE: Software.
KC? Discoverer: Exploring the Properties of the Elements. JCE: Software.
Videodisc
The Periodic Table Videodisc. JCE: Software.
The World of Chemistry: Selected Demonstrations and Animations I and II. JCE: Software.

Chapter 9
Software
Element Search. MC902. Project Seraphim.
Videodisc
The Periodic Table Videodisc. JCE: Software.
CD-ROM
The Periodic Table CD-ROM. JCE: Software.

Chapter 10
Software
Concentrated Chemical Concepts. Trinity Software.
Concepts in General Chemistry. Trinity Software.
Videodisc
The World of Chemistry: Selected Demonstrations and Animations I and II. JCE: Software.

REFERENCES

Chapter 11

Software

Chemistry Explorer. Gas Laws. Logal Software, Inc.

CD-ROM

Exploring Chemistry IV. Lesson 7 *Gases.* Falcon Software, Inc.

Videodisc

ChemDemos. JCE: Software.

Chapter 12

Software

Introduction to General Chemistry. Section II, *Molecules.* Projected Learning Programs, Inc.

Moles in Space. MC304. Project Seraphim.

Chapter 13

Software

Introduction to General Chemistry. SectionIV, *Solutions.* Projected Learning Programs, Inc.

Chapter 14

Software

Abs Game. MC501. Project Seraphim.

Videodisc

The Periodic Table Videodisc. JCE: Software.

CD-ROM

The Periodic Table CD-ROM. JCE: Software.

Chapter 15

Software

Lake Study, for DOS, Macintosh, and Windows. JCE: Software.

Videodisc

The Periodic Table Videodisc. JCE: Software.

ChemDemos. JCE: Software.

CD-ROM

The Periodic Table CD-ROM. JCE: Software.

Exploring Chemistry IV. Lesson 4 *Acids and Bases.* Falcon Software, Inc.

Chapter 16

Software

Redox Game. AP 306. Project Seraphim.

Chemistry Explorer. Electrochemistry. Logal Software, Inc.

Introduction to General Chemistry. Section III, *Chemical Reactions.* Projected Learning Programs, Inc.

Videodisc

The World of Chemistry: Selected Demonstrations and Animations I and II. JCE: Software.

Chapter 17

Software:

ElectrodeP, AP 604. Project Seraphim.

Redox Game, AP306, PC3101. Project Seraphim.

Video

Electrochemistry, Chemistry Explorer Series, Logal Software, Inc.

Videodisc

Cosmic Chemistry. The Living Textbook, Physical Sciences, Side 5 *Redox and Electrochemistry.* Optical Data Corporation.

Chapter 18

Software

Biochemistry Series, for Apple. Projected Learning Programs, Inc.

CD-ROM

Comprehensive Chemistry. Organic Chemistry Laboratory. Falcon Software, Inc.

Biochemistry Series. Projected Learning Programs, Inc.

Videodisc

Demonstrations in Organic Chemistry. Gary Trammell. JCE: Software.

Chapter 19

CD-ROM

Biochemistry Series. Projected Learning Programs, Inc.

Videodisc

Demonstrations in Organic Chemistry. Gary Trammell. JCE: Software.

Chapter 20

Software

Chemistry Explorer. Chemical Kinetics. Logal Software, Inc.

CD-ROM

Introductory Chemistry. Chemical Reactions. Falcon Software, Inc.

Videodisc

ChemDemos. JCE: Software.

Chapter 21

Software

Nuke 4. AP1001. Project Seraphim.

Decay. AP1001. Project Seraphim.

Video

Eminent Chemists Videotape Series. Glenn T. Seaborg. American Chemical Society.

Videodisc

Cosmic Chemistry. The Living Textbook, Physical Sciences, Side 1 *Nuclear Chemistry.* Optical Data Corporation.

Audiovisual Suppliers

Agency for Instructional Technology (AIT)
Box A
Bloomington, IN 47402-0120
(800) 457-4509

Ambrose Video Publishing, Inc.
Exclusive Distributor of Time-Life Videos
28 West 44th Street,
Suite 2100
New York, NY 10036
(800) 526-4663

American Chemical Society
Educational Distribution Office
PO Box 2537
Kearneysville, WV 25430
(800) 209-0423

American Chemical Society
1155 16th Street NW
Washington, DC 20036
(800) 227-5558
software@acs.org

Barr Media Group
12801 Schabarum Avenue
Irwindale, CA 91706
(800) 234-7878

Bergwall Productions, Inc.
540 Baltimore Pike
PO Box 2400
Chadds Ford, PA 19317
(800) 645-3565

Carolina Biological Supply Co.
2700 York Road
Burlington, NC 27215
(800) 334-5551

Clearvue, Inc.
6465 N. Avondale Avenue
Chicago, IL 60631
(800) 253-2788

Coronet/MTI Film and Video
4350 Equity Drive
Columbus, OH 43228
(800) 777-2400

Cross Educational Software
PO Box 1536
508 E. Kentucky Avenue
Ruston, LA 71270
(800) 768-1969
markcross@aol.com

Educational Images, Ltd.
PO Box 3456, West Side
Elmira, NY 14905
(800) 527-4264

Educational Materials and Equipment Co. (EME)
PO Box 2805
Danbury, CT 06813-2805
(800) 848-2050
emecorp@aol.com

Encyclopedia Britannica Educational Corp.
310 South Michigan Avenue
Chicago, IL 60604
(800) 323-1229

Falcon Software, Inc.
PO Box 200
Wentworth, NH 03282
(603) 764-5788

Hawkhill Associates
125 East Gilman Street
PO Box 1029
Madison, WI 53701
(800) 422-4295

IBM Educational Systems
Department PC
4111 Northside Parkway
Atlanta, GA 30327
(800) 426-9920
ibmk12@vnet.ibm.com
http://www.solutions.ibm

Intellimation
PO Box 1922
Santa Barbara, CA 93116-1922
(800) 346-8355

International Film Bureau
Suite 450
332 South Michigan Avenue
Chicago, IL 60604-4382
(800) 432-2241

J and S Software
PO Box 1276
Port Washington, NY 11050
(800) 767-8696
gradebook@aol.com

JCE: Software
University of Wisconsin-Madison
Department of Chemistry
1101 University Avenue
Madison, WI 53706-1396
(608) 262-5153
jcesoft@macc.wisc.edu

Logal Software, Inc.
PO Box 1499
East Arlington, MA 02174
(800) LOGAL-US (564-2587)
http://www.logal.om

Mindscape, Inc.
Educational Division
88 Rowland Way
Novato, CA 94945
(800) 866-5967

Minnesota Educational Computing Corp. (MECC)
6160 Summit Drive
Minneapolis, MN 55430
(800) 685-6322
http://www.mecc.com/

Optical Data Corp.
30 Technology Drive
Warren, NJ 07059
(800) 248-8478

Projected Learning Programs, Inc.
PO Box 3008
Paradise, CA 95967-3008
(800) 248-0757

Project Seraphim
University of Wisconsin-Madison
Department of Chemistry
1101 University Avenue
Madison, WI 53706
(608) 263-2837
jwmoore@macc.wisc.edu

Queue, Inc.
338 Commerce Drive
Fairfield, CT 06432
(800) 232-2224
queueinc@aol.com

Rock Ware
2221 East Street
Suite 101
Golden, CO 80401
(800) 775-6745
rockware@rockware.com

TVOntario
1140 Kildare Farm Road
Suite 308
Cary, NC 27511
(800) 331-9566
u.s.sales@tvo.org

Videodiscovery
1700 Westlake Ave. North
Suite 600
Seattle, WA 98109-3012
(800) 548-3472
http:// www.videodiscovery.com/vdyweb

Wm. K. Bradford Publishing Co.
16 Craig Road
Acton, MA 01720
(800) 421-2009
wkb@wkbradford.com
http://www.wkbradford.com

Company names, mailing addresses, phone numbers, e-mail addresses, and World Wide Web sites were accurate at the time of publication and are subject to change.

REFERENCES

Readings...

for the Teacher

TEACHING METHODS

Anholt, Robert R.H. *Dazzle 'em with Style: The Oral Art of Scientific Presentation.* New York: W.H. Freeman, 1994.

Hart, Diane. *Authentic Assessment: A Handbook for Educators.* Menlo Park, CA: Addison-Wesley, 1994.

Johnson, David W., Roger T. Johnson, and Edythe Johnson Holubec. *Cooperation in the Classroom.* Edina, MN: Interaction Book Company, 1991.

Lunn, George and Eric B. Sansone. *Destruction of Hazardous Chemicals in the Laboratory,* 2nd ed. New York: Wiley, 1994.

Mitchell, Sharon and Frederick Juergens. *Laboratory Solutions for the Science Classroom.* Batavia, IL: Flinn Scientific, Inc., 1991.

National Research Council. *National Science Education Standards.* Washington, DC: National Academy Press, 1996.

Penick, John. "Where's the Science?" *The Science Teacher,* May 1991, pp. 27-29.

CHEMISTRY CONTENT

Anderson, G.M. *Thermodynamics of Natural Systems.* New York: Wiley, 1996.

Balibar, Francoise. *The Science of Crystals.* New York: McGraw-Hill, 1993.

Bernstein, Jeremy. *Cranks, Quarks, and the Cosmos: Writings on Science.* New York: BasicBooks, 1993.

Braben, Donald. *To Be a Scientist.* New York: Oxford University Press, 1994.

Brock, William H. *The Norton History of Chemistry.* New York: Norton, 1993.

Gordon, J.E. *The Science and Structure of Materials.* New York: Scientific American Library, 1988.

Hoffman, Roald and Vivian Torrence. *Chemistry Imagined: Reflections on Science.* Washington, DC: Smithsonian Institution Press, 1993.

Jacques, Jean. *The Molecule and Its Double.* New York: McGraw-Hill, 1993.

Knight, David. Ideas in Chemistry: *A History of the Science.* New Brunswick, NJ: Rutgers University Press, 1992.

Laidler, Keith J. *The World of Physical Chemistry.* New York: Oxford University Press, 1993.

Montgomery, Scott L. *Minds for the Making: The Role of Science in American Education,* 1750-1990. New York: Guilford Press, 1994.

Moody, Christopher J. and Gordon H. Whitham. *Reactive Intermediates.* New York: Oxford University Press, 1992.

Rodricks, Joseph V. *Calculated Risks: The Toxicity and Human Health Risks of Chemicals in Our Environment.* New York: Cambridge University Press, 1994.

Solomon, Sally. "Qualitative Analysis of Eleven Household Compounds," *Journal of Chemical Education,* April 1991, pp. 328-329.

Taubes, Gary. *Bad Science: The Short Life and Weird Times of Cold Fusion.* New York: Random House, 1993.

for the Student

Asimov, Isaac. *Atom: Journey Across the Subatomic Cosmos.* New York: Dutton, 1991.

Binkerhoff, Richard F. *One-Minute Readings: Issues in Science, Technology, and Society.* Menlo Park, CA: Addison-Wesley, 1992.

Bombaugh, Ruth. *Science Fair Success.* Hillside, NJ: Enslow Publishers, 1990.

Chemical Education for Public Understanding Program. *Toxic Waste: A Teaching Simulation.* New York: Addison-Wesley, 1993.

Churchill, E. Richard. *Amazing Science Experiments with Everyday Materials.* New York: Sterling Publishing Co., Inc., 1991.

Fisher, Leonard Everett. *Marie Curie.* New York: Macmillan, 1994.

Friedhoffer, Robert. *Matter and Energy.* New York: Watts, 1993.

Gardner, Robert. *Experimenting with Water.* New York: Watts, 1993.

Gardner, Robert. *Famous Experiments You Can Do.* New York: Watts, 1990.

Gardner, Robert. *Robert Gardner's Challenging Science Experiments.* New York: Watts, 1993.

Gardner, Robert. *Science Projects About Chemistry.* Hillside, NJ: Enslow, 1994.

Gardner, Robert and Eric Kemer. *Making and Using Scientific Models.* New York: Watts, 1993.

Grand, Gail L. *Student Science Opportunities: Your Guide to Over 300 Exciting National Programs, Competitions, Internships, and Scholarships.* New York: Wiley, John and Sons, Inc., 1994.

Gray, Harry B., John D. Simon, and William C. Trogler. *Braving the Elements.* Sausalito, CA: University Science Books, 1995.

Gutnik, Martin J. *Experiments That Explore: Acid Rain.* Brookfield, CT: Millbrook Press, 1992.

Hanson, Robert M. *Molecular Origami: Precision Scale Models from Paper.* Sausalito, CA: University Science Books, 1995.

Haven, Kendall. *Marvels of Science: 50 Fascinating 5-Minute Reads.* Englewood, CO: Libraries Unlimited, 1994.

Hazen, Robert M. *The New Alchemists: Breaking Through the Barriers of High Pressure.* New York: Times Books, 1993.

Hoffman, Roald. *The Same and Not the Same.* New York: Columbia University Press, 1995.

Hudson, John. *The History of Chemistry.* New York: Chapman and Hall, 1992.

Mebane, Robert C. and Thomas R. Rybolt. *Plastics and Polymers.* New York: Twenty-First Century Books, 1995.

Mebane, Robert C. and Thomas R. Rybolt. *Salts and Solids.* New York: Twenty-First Century Books, 1995.

Mercer, Ian. *Crystals.* Cambridge, MA: Harvard University Press, 1990.

Morgan, Sally and Adrian. *Materials.* New York: Facts on File, 1994.

Newmark, Ann. *Chemistry.* New York: Dorling Kindersley, 1993.

Newton, David E. *Consumer Chemistry Projects for Young Scientists.* New York: Watts, 1991.

Nye, Bill. *Bill Nye the Science Guy's Big Blast of Science.* New York: Addison-Wesley, 1993.

Snyder, Carl H. *The Extraordinary Chemistry of Ordinary Things,* 2nd ed. New York: Wiley, 1995.

Chemistry

CONCEPTS AND APPLICATIONS

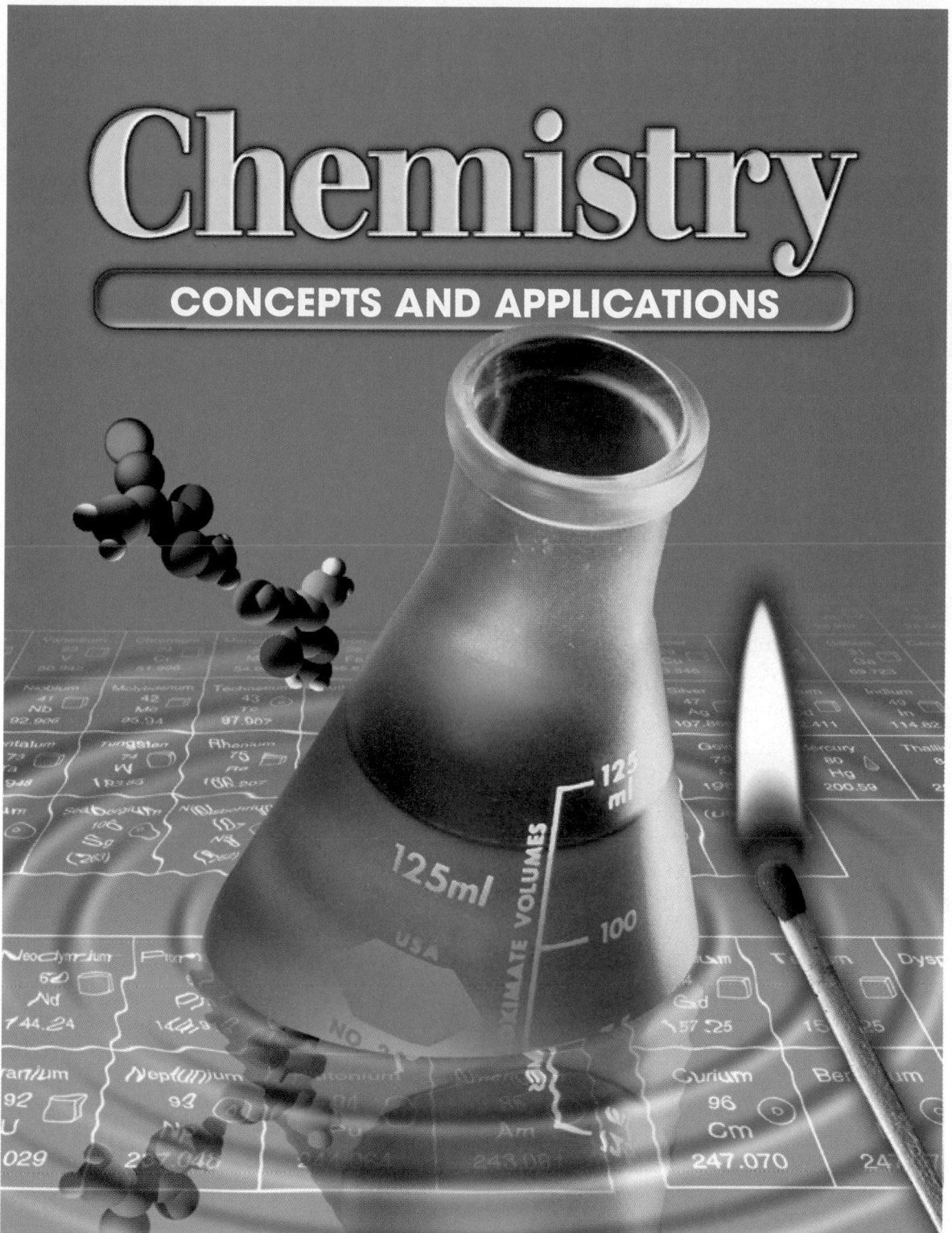

GLENCOE

McGraw-Hill

New York, New York   Columbus, Ohio   Mission Hills, California   Peoria, Illinois

Chemistry
Concepts and Applications

Student Edition

Teacher Wraparound Edition

Teacher Classroom Resources:

Laboratory Manual SE and TE
Study Guide SE and TE
Section Focus Transparency Package
Problem Solving Transparency Package
Basic Concepts Transparency Package
Transparency Masters and Worksheets
ChemLab and MiniLab Worksheets
Chapter Review and Assessment
Critical Thinking/Problem Solving
Consumer Chemistry
Tech Prep Applications
Lesson Plans
Spanish Resources
Problems and Solutions Manual
Chemistry and Industry
Applying Scientific Methods in Chemistry
Supplemental Practice Problems

Technology Resources

Computer Test Bank: For DOS and Macintosh
MindJogger Videoquizzes
CD-ROM Multimedia System: Windows and
  Macintosh versions
Videodisc Program
Chapter Summaries, English and Spanish
  Audiocassettes
Mastering Concepts in Chemistry software

The Glencoe Science Professional Development Series

Cooperative Learning in the Science Classroom
Alternate Assessment in the Science Classroom
Lab and Safety Skills in the Science Classroom

Glencoe/McGraw-Hill

A Division of The **McGraw·Hill** *Companies*

Send all inquiries to:
Glencoe/McGraw-Hill
936 Eastwind Drive
Westerville, Ohio 43081

ISBN 0-02-827452-0
Printed in the United States of America.

2 3 4 5 6 7 8 9 QPH/MC 03 02 01 00 99 98 97 96

AUTHORS

John S. Phillips is a chemistry teacher at Forest Ridge School in Bellevue, Washington. He has been teaching chemistry at the high school and college levels for 16 years. Dr. Phillips has coordinated and led programs and workshops for teachers from kindergarten through college levels that encourage and support creative science teaching. He earned a B.A. degree in chemistry at Western Maryland College and a Ph.D. in chemistry from Purdue University. He is a member of the American Chemical Society, National Science Teachers Association, and Sigma Xi.

Victor S. Strozak is a professor of chemistry at New York City Technical College of the City University of New York. He has been teaching chemistry at the high school and college levels for 34 years. He earned his B.S. in Chemistry from St. John's University and his M.S. in Chemistry and Ph.D. in Science Education from New York University. He is a member of the American Chemical Society, the Two Year College Chemistry Conference, and the ACS Division of Chemical Education. He has participated in grants administration for the National Science Foundation, concentrating on systemic change and improvements in science education.

Cheryl Wistrom is an associate professor of chemistry at Saint Joseph's College in Rensselaer, Indiana. She has taught chemistry and chemical education at the college level for six years. She earned her B.S. degree at Northern Michigan University and her M.S. and Ph.D. at the University of Michigan, where she carried out research on gene expression during aging of human cells. She has participated in summer institutes for educators at Pennsylvania State University and Miami University of Ohio. She is a member of the National Science Teachers Association and the American Chemical Society.

Contributing Writers

Helen Frensch, M.A.
Santa Barbara, CA

Nicholas Hainen, M.A.
Former Chemistry Teacher
Worthington High School
Worthington, OH

Zoe A. Godby Lightfoot, M.S.
Former Chemistry Teacher
Carbondale Community High School
Marion, IL

Mark V. Lorson, Ph.D.
Chemistry Teacher
Jonathan Alder High School
Plain City, OH

Robert Roth, M.S.
Pittsburgh, PA

Richard G. Smith, M.A.T.
Chemistry Teacher
Bexley High School
Bexley, OH

Patricia West
Oakland, CA

Safety Consultant

Douglas K. Mandt, M.S.
Science Education Consultant
Sumner, WA

High School Reviewers

Jon L. Allan, M.S.
University High School
Spokane, WA

William Allen, M.Ed.
Stevens Point Area Senior High School
Stevens Point, WI

Eddie Anderson
Oak Ridge High School
Oak Ridge, TN

Susan H. Brierley
Garfield High School
Seattle, WA

Robert A. Cooper, M.Ed
Pennsbury High School
Fairless Hills, PA

Sharon Doerr
Oswego High School
Oswego, NY

Jeffrey L. Engel, M.Ed., Ed.S.
Madison County High School
Danielsville, GA

Richard A. Garst
Ironwood High School
Glendale, AZ

Jo Marie Hansen
Twin Falls High School
Twin Falls, ID

Cynthia Harrison, M.S.A.
Parkway South High School
Manchester, MO

Vince Howard, M.Ed
Kentridge High School
Kent, WA

Stephen Hudson
Mission High School
San Francisco, CA

Michael Krein, M.S.
Coordinator of Chemistry
Stamford High School
Stamford, CT

**Sister John Ann Proach,
O.S.F., M.A., M.S.**
Science Curriculum Chairperson
Archdiocese of Philadelphia
Bishop McDevitt High School
Wyncote, PA

Eva M. Rambo, M.A.T.
Bloomington South High School
Bloomington, IN

Nancy Schulman, M.S.
Manalapan High School
Manalapan, NJ

Tim Watts, M.Ed.
Assistant Principal
Warren County Middle School
Front Royal, VA

Consultants

Larry B. Anderson, Ph.D.
Associate Professor
The Ohio State University
Columbus, OH

Ildiko V. Boer, M.A.
Assistant Professor
County College of Morris
Randolph, NJ

Marcia C. Bonneau, M.S.
Lecturer
State University of New York
Cortland, NY

James H. Burness, Ph.D.
Associate Professor
Penn State University
York, PA

Larry Cai
Graduate Teaching Associate
The Ohio State University
Columbus, OH

Sheila Cancella, Ph.D.
Department Chair, Science &
Engineering
Raritan Valley Community College
Somerville, NJ

James Cordray, M.S.
Berwyn, IL

Jeff Hoyle, Ph.D.
Associate Professor
Nova Scotia Agricultural College
Truro, Nova Scotia
Canada

Teresa Anne McCowen, M.S.
Senior Lecturer
Butler University
Indianapolis, IN

Lorraine Rellick, Ph.D.
Assistant Professor
Capital University
Columbus, OH

Marie C. Sherman, M.S.
Chemistry Teacher
Ursuline Academy
St. Louis, MO

Charles M. Wynn, Ph.D.
Chemistry Professor
Eastern Connecticut State University
Willimantic, CT

Chemistry: Concepts and Applications
CONTENTS IN BRIEF

CONTENTS

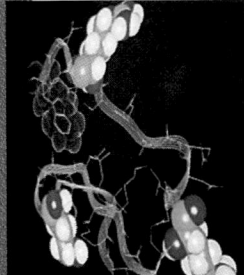

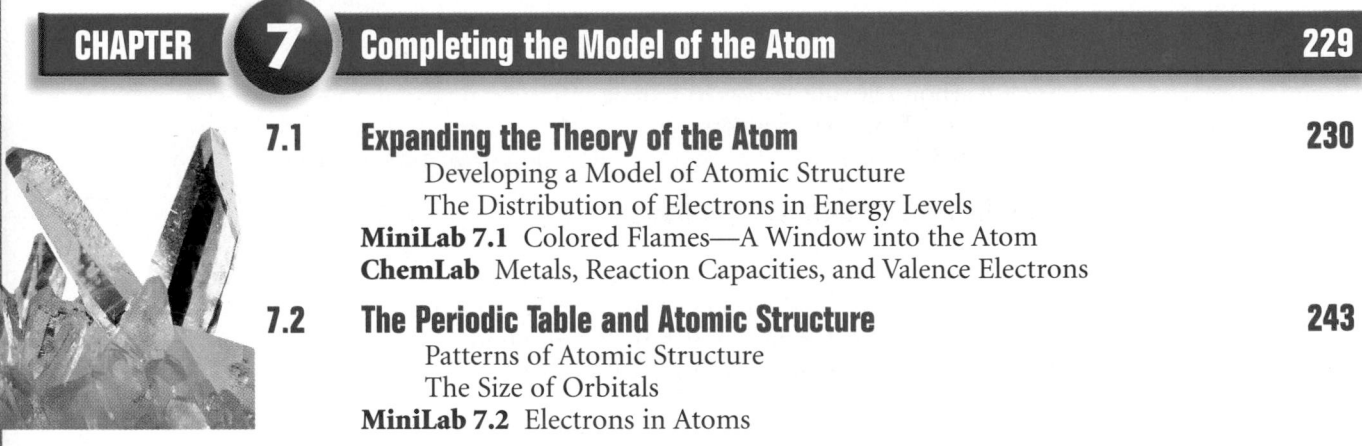

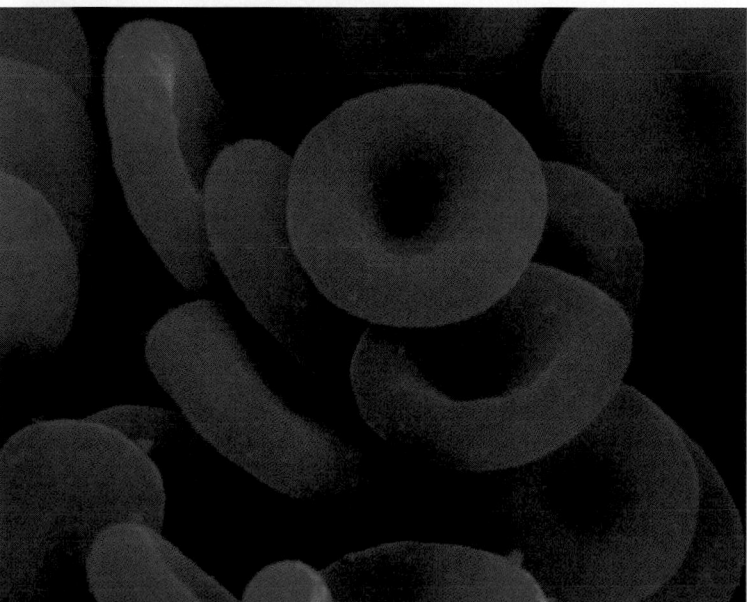

APPENDICES

Everyday Chemistry

How does your microwave work? Why do you hiccup? What makes a ruby red? These and other everyday questions are answered here as you explore their underlying chemistry.

ChemLabs

ChemLabs offer you the opportunity to discover how elements form compounds, why matter changes but is not created, what gives candies their colors, and many other chemical wonders. Develop your lab skills and become a practicing chemist.

miniLABS

With these short and easy activities, discover for yourself how chemistry is simple and exciting. Just as a picture is worth a thousand words, a MiniLab can take the place of many hours of study.

How it Works

Look inside a tire gauge, an electric lightbulb, or a rechargeable battery to find out what makes them work. How does chemistry help you understand these simple everyday items?

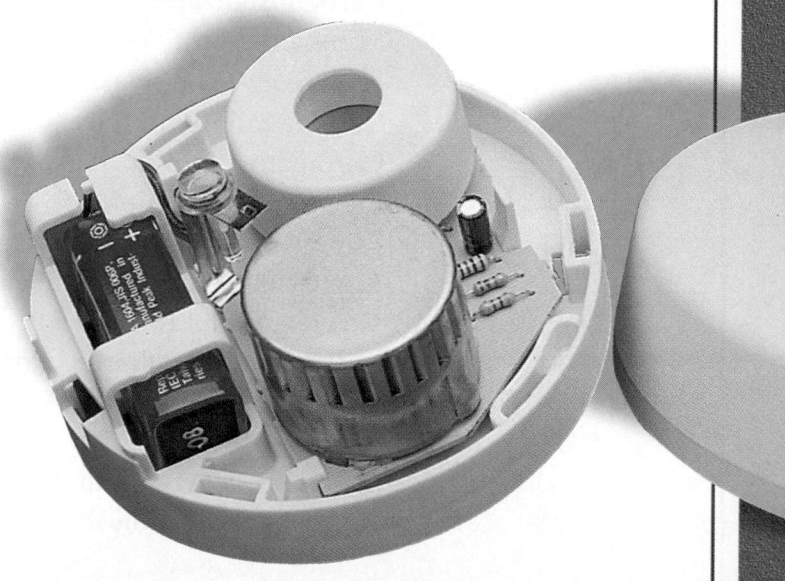

Chemistry and SOCIETY

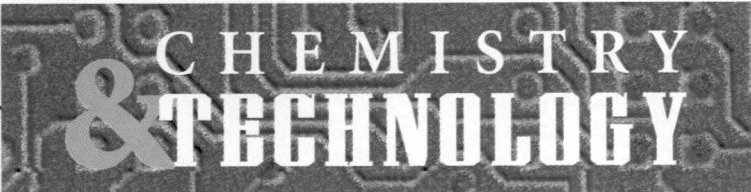

How has chemistry helped solve some of the world's problems, such as finding cures for rare diseases, cleaning up pollution, and clothing the world's ever-increasing population? Take this opportunity to learn how chemistry is used to benefit society.

CHEMISTRY & TECHNOLOGY

What new advances have been developed from the application of chemical principles? Explore how metals can retain memory, where to look for alternative energy sources, and how microscopes can now be used to see atoms in these amazing new stories of science in progress.

PEOPLE in CHEMISTRY

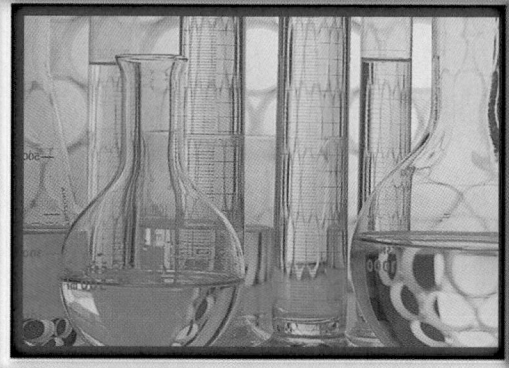

How do people use chemistry in their daily work?

What excites people about chemistry, and how did they ever get interested in the first place? Read these interviews to find out who or what inspired these people to become successful in their work.

xxi

Cross-Curricular Connections

It may not have occurred to you that chemistry is an intergral part of all your courses, not just the sciences. Learn in these features how chemistry is connected to literature, art, and history, as well as to the sciences of physics, biology, health, and earth science.

LITERATURE

ART

HISTORY

PHYSICS

BIOLOGY

HEALTH

EARTH SCIENCE

Chapter 1 Organizer

Section	Objectives	Activities/Features
1.1 **The Puzzle of Matter** (6 days)	1. **Classify** matter according to its composition. 2. **Distinguish** among elements, compounds, homogeneous mixtures, and heterogeneous mixtures. 3. **Relate** the properties of matter to its structure.	**ChemLab 1:** pp. 8-9 **People in Chemistry:** pp. 12-13 **ChemLab 2:** pp. 16-17 **Everyday Chemistry:** p. 19 **MiniLab 1:** p. 21 **MiniLab 2:** p. 22 **MiniLab 3:** p. 25 **Literature Connection:** p. 26 **MiniLab 4:** p. 30 **Chemistry and Society:** p. 32 **Discovery Demo:** 1.1, pp. 2-3 **Demonstrations:** 1.2, pp. 6-7; 1.3, pp. 28-29
1.2 **Properties and Changes of Matter** (6 days)	4. **Distinguish** between physical and chemical properties. 5. **Contrast** chemical and physical changes. 6. **Apply** the law of conservation of matter to chemical changes.	**ChemLab 3:** pp. 38-39 **MiniLab 5:** p. 40 **Demonstrations:** 1.4, pp. 36-37; 1.5, pp. 40-41; 1.6, pp. 42-43

Activity Materials

ChemLab 1 (pages 8-9) candles, matches, shallow metal dish, 25 mL of limewater solution, 250-mL beaker, 500-mL Erlenmeyer flask, solid rubber stopper to fit flask, wire gauze square, tongs

ChemLab 2 (pages 16-17) 96-well microplate, 9 test tubes, spatula, 3 thin-stemmed pipets, masking tape, marking pen

ChemLab 3 (pages 38-39) 5 pennies of varying mint dates, balance sensitive to 0.01 g, 50-mL graduated cylinder

MiniLab 1 (page 21) 100-mL graduated cylinder, food coloring, dropper, thermometer, ethanol, stirring rod

MiniLab 2 (page 22) water-soluble ink markers, coffee filter, 6 cm high clear plastic cup, clear plastic wrap, pencil

MiniLab 3 (page 25) pre-1982 penny, 1 g granular zinc, 20 mL 1M $ZnCl_2$ solution, 150-mL beaker, evaporating dish, steel wool, tongs, hot plate, Bunsen burner

MiniLab 4 (page 30) 250 mL dry fortified cereal, 250-mL beaker, small plastic bag, rolling pin, small magnet

MiniLab 5 (page 40) 20 mL polyvinyl alcohol, 3 mL borax solution, food coloring (optional), craft stick or tongue depressor, plastic cup, small self-sealing plastic bag

Demonstrations For a complete list of materials for the demonstrations in this chapter, see pages 2, 6, 28, 36, 40, and 42.

KEY TO TEACHING STRATEGIES

The following designations will help you decide which activities are appropriate for your students.

L1 Level 1 activities should be within the ability range of all students.

L2 Level 2 activities should be within the ability range of the average to above-average student.

L3 Level 3 activities are designed for the ability range of above-average students.

LEP LEP activities should be within the ability range of Limited English Proficiency students.

COOP LEARN Cooperative Learning activities are designed for small group work.

P These strategies represent student products that can be placed into a best-work portfolio.

Chemistry: The Science of Matter

Teacher Classroom Resources

STUDENT MASTERS

Study Guide: pp. 1-2
Critical Thinking/Problem Solving: p. 1
Laboratory Manual: 1.1 Determining Boiling Point, pp. 1-4
ChemLab and MiniLab Worksheets: pp. 1-6, 10-13

TEACHING AIDS

Section Focus Transparency 1
Section Focus Transparency Master: p. 1
Basic Concepts Transparency 1, 2
Basic Concepts Transparency Masters: pp. 1-4
Problem Solving Transparency 1
Problem Solving Transparency Master: pp. 1-2
Lesson Plans: p. 1

Study Guide: pp. 3-4
Laboratory Manual: 1.2 The Eight Solution Problem pp. 5-10
ChemLab and MiniLab Worksheets: pp. 7-9, 14

Section Focus Transparency 2
Section Focus Transparency Master: p. 2
Basic Concepts Transparency 3
Basic Concepts Transparency Master: pp. 5-6
Lesson Plans: p. 2

ASSESSMENT RESOURCES

Chapter Review and Assessment: pp. 5-10
Alternate Assessment in the Science Classroom
Computer Test Bank: Chapter 1
Problems and Solutions Manual: Chapter 1
Supplemental Practice Problems: Chapter 1
Performance Assessment in the Science Classroom

CHAPTER RESOURCES

Applying Scientific Methods in Chemistry: pp. 15-18
Spanish Resources: Chapter 1
English/Spanish Audiocassettes: Chapter 1
Cooperative Learning in the Science Classroom
Lab and Safety Skills in the Science Classroom

GLENCOE TECHNOLOGY

Videotape
MindJogger Videoquizzes, Chapter 1
Videodisc
Chemistry: Concepts and Applications,
Demonstrations, Videos, and Animations
Videodisc and Bar Code Teacher Guide
Science and Technology Videodisc Series: Disc 2
Chemistry *Shock Impact Gun, Fire Resistant Clothing*
CD-ROM Multimedia System
Chemistry: Concepts and Applications, Same as
for Videodisc, plus Interactive Exploration and Major
Simulation, *The Periodic Table*

TECH PREP

The following Glencoe resources provide opportunities for integrating science and technology.

Student Edition: People in Chemistry, pp. 12-13; Chemistry and Society, p. 32

Teacher Wraparound Edition: Extension, p. 18

Teacher Classroom Resources: Tech Prep Applications, pp. 1-2; Chemistry and Industry, pp. 1-4; Consumer Chemistry, pp. 1-2

CHAPTER 1

Chemistry: The Science of Matter

Chapter Overview

The study of chemistry is compared with the solving of a puzzle. Matter is viewed as puzzle pieces connected through composition, structure, and behavior. Classification is introduced as a first step in simplifying the study of matter. Models are used to connect the macroscopic and the submicroscopic views of matter. Chemical language and symbols are introduced.

Theme Connection

Macro to Submicro/Conservation
Although atoms and molecules are not introduced in this chapter, students' prior knowledge of these familiar terms is used to establish the link between *macroscopic* observations and *submicroscopic* explanations. The theme of *conservation* appears as the law of conservation of mass.

✓ Assessment Planner

Choose assessment strategies from the following pages to evaluate the progress of your students.
Assess, pp. 20, 33, 43
MiniLabs, pp. 21, 22, 25, 30, 40
ChemLabs, pp. 8, 16, 38
Portfolio, pp. 20, 33, 45
Chapter Review, p. 45

GLENCOE TECHNOLOGY

⊙ CD-Rom

Chemistry: Concepts and Applications
The Magic of Chemistry
Have students use this video of the Discovery Demo to introduce the concept of chemical change. L1 LEP

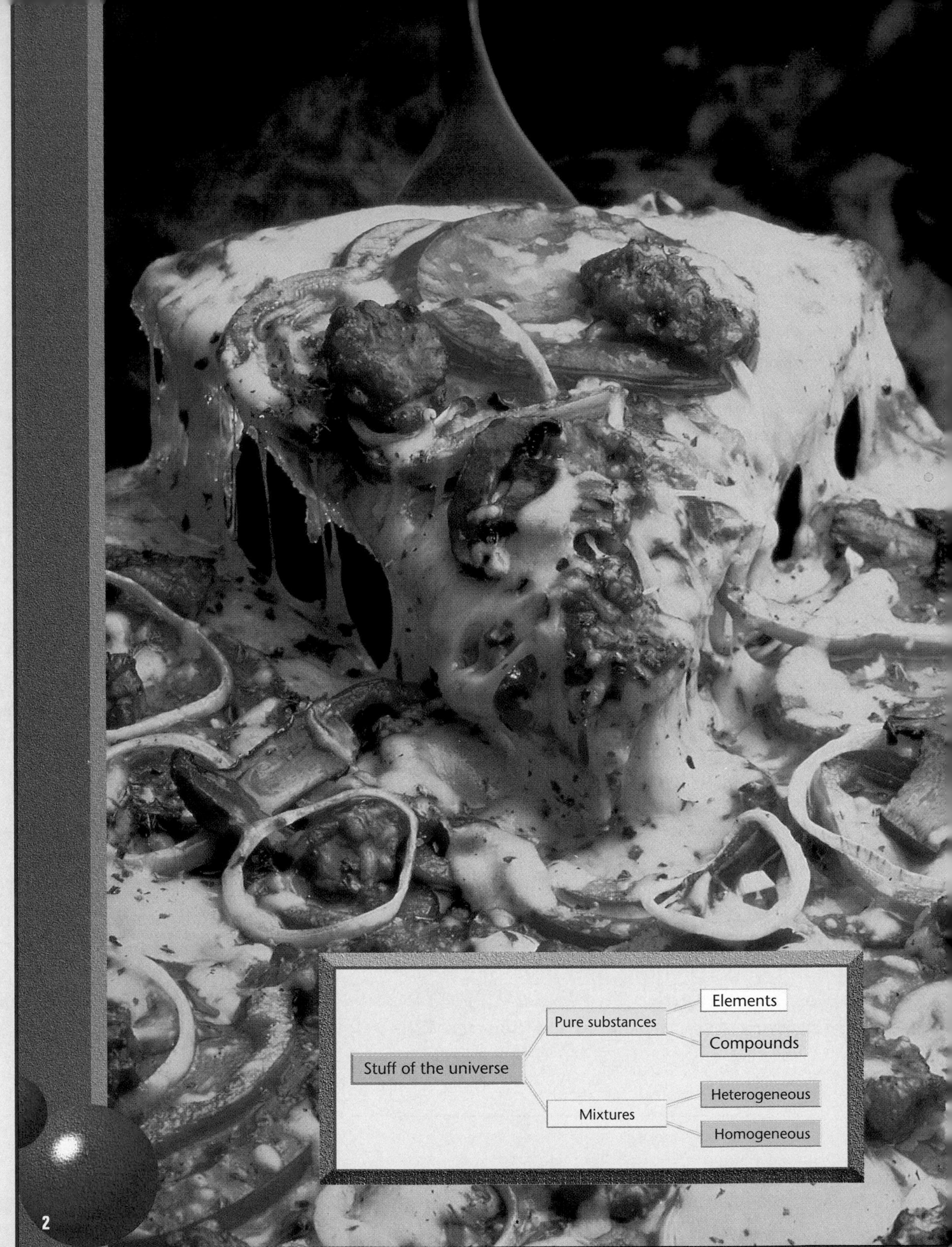

2

		Elements
Pure substances		
		Compounds
Stuff of the universe		
		Heterogeneous
Mixtures		
		Homogeneous

2

Discovery Demo 1.1

The Magic of Chemistry

Purpose: To show that new substances can be prepared to replace materials that are in short supply.
Materials: 0.05 g $KMnO_4$, 1 g $NaHSO_3$, 1 g $BaCl_2 \cdot 2H_2O$, three 400-mL beakers, 2 small test tubes
Safety Precautions: 🧤 🥽 ☠️
Disposal: Code E

Procedure: Dissolve three or four small crystals of $KMnO_4$ in 250 mL of water in a beaker. In a test tube, dissolve 1 g of $NaHSO_3$ in 1 mL of water. In a second test tube, dissolve 1 g $BaCl_2 \cdot 2H_2O$ in 1 mL of water. **CAUTION:** *The solutions are toxic.* Before the demo, place the $NaHSO_3$ solution in a beaker labeled *1* and the $BaCl_2$ solution in a beaker labeled *2*. Show students the "wine" ($KMnO_4$ solution). Pour

CHAPTER

1

Chemistry: The Science of Matter

Take a look at what's around you. You don't have to look far to see an incredible assortment of stuff—everyday, ordinary stuff like this book, a piece of paper, a pencil, a can of soda, puffy white clouds, a pair of jeans, an ice-cream cone, a photograph, a dog, and dog hair. And maybe there's a pizza—that wonderful combination of tomato sauce, spices, cheese, and your favorite toppings all carefully layered on a bed of dough and baked in a hot oven until it's just right.

But what is all this stuff? There's paper in the book and wood in the pencil. The can is aluminum and the jeans are denim. There's sauce and cheese and crust in the pizza, but what are paper and wood and denim? What are sauce and cheese and crust made of? Why is one piece of stuff different from another? One way that chemists begin to answer these questions is by classifying matter into categories according to its composition.

Chemistry Around You

There's a tremendous variety of different stuff in the universe, and yet everything is the same in one way. It's all matter—from a supernova (large exploding star) in a galaxy 30 million light-years away to a piece of sausage on your pizza. How is a supernova different from a fresh-baked pizza? That's not too tough—you can't call on the phone and order a supernova! But how is a supernova like a pizza? That's a question that chemistry answers. Chemistry is the science of asking and answering questions about stuff. Why does the cheese melt and a star explode? What makes the crust crisp and the star shine? Chemistry is the what, how, and why of stuff.

 Chapter Preview

1.1 The Puzzle of Matter

1.2 Properties and Changes of Matter

Laboratory Activities

ChemLabs
Observation of a Candle
Kitchen Chemicals
The Composition of Pennies

MiniLabs
1. 50 mL + 50 mL = ?
2. Paper Chromatography of Inks
3. Copper to Gold: The Alchemists' Dream
4. Waiter, what's this stuff doing in my cereal?
5. It's a Liquid, It's a Solid . . . It's Slime

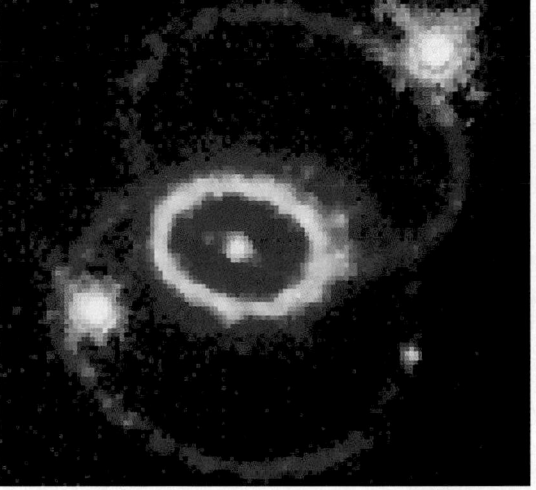

3

---

the "wine" into beaker *1*. Pour the "water" that results into beaker *2*.
Results: The "wine" turns into "water," and the "water" is then changed into "milk." Discussing the chemistry of this demo will be of little value at this time. Explain that a chemist learns how to change one substance into another substance having different properties.

Analysis: Ask these questions.
1. Have chemists produced substitutes for cotton and wool? *yes, synthetic fibers*
2. Which requires less land area: a chemical plant that produces synthetic fibers or farms where sheep graze and cotton grows? *chemical plant*
3. As Earth's population increases, will land be used to grow food or produce cotton and wool? *grow food*

Introducing the Chapter

Distribute large handfuls of puzzle pieces to groups of two or three students. Do not show students the puzzle picture. Have the groups work on assembling their puzzle pieces. As the work proceeds, encourage groups to share pieces with other groups. After a time, stop the activity and have students discuss strategies and problems. Students will identify classification as a strategy. Problems may include not knowing what the puzzle looks like and not having all the pieces. Discuss how this exercise relates to the study of chemistry. `L1` `LEP` `COOP LEARN`

Using the Illustrations The photos show the range of stuff that composes the universe. Because chemistry is the study of stuff, chemistry connects objects as seemingly different as a pizza and a supernova.

Revealing Misconceptions

Structural formulas and models are used in this chapter so that students will become familiar with them. They do not need to be discussed beyond the fact that they are the symbols and language of chemistry and will be explained and used later.

 GLENCOE TECHNOLOGY

 **Videodisc**

Chemistry: Concepts and Applications
The Magic of Chemistry
Disc 1, Side 1, Ch. 1

Show this video of the Discovery Demo to introduce the concept of chemical change. `L1` `LEP`

3

PREPARE

Key Concepts

Chemistry and matter are defined and the behavior of matter is related to its composition and structure. Pure substances and mixtures are differentiated, and the two types of mixtures are defined. Elements and compounds are defined as forms of matter. The function of models is explored.

Planning Ahead

For the MiniLabs Obtain water-soluble markers in a variety of colors for MiniLab 2.

1 FOCUS

Focus Transparency

Display the **Section Focus Transparency** for Section 1.1 to introduce two views of matter. [L1] [LEP]

2 TEACH

Correcting Misconceptions

Students may think that scientists already know everything about matter—that for them, all the puzzle pieces are in place. Point out that in science, it is impossible to know when the puzzle of matter is completely and correctly understood.

Quick Demo

What's the difference?
Set out containers of liquids and solids of varying forms and different colors. Ask students to list properties that might be used to distinguish the types of matter in the containers and objects. [L1]

The Puzzle of Matter

SECTION PREVIEW

Objectives

Classify matter according to its composition.
Distinguish among elements, compounds, homogeneous mixtures, and heterogeneous mixtures.
Relate the properties of matter to its structure.

Key Terms

chemistry
matter
mass
property
scientific model
qualitative
quantitative
substance
mixture
physical change
physical property
solution
alloy
solute
solvent
aqueous solution
element
compound
formula

Have you ever worked a jigsaw puzzle? Imagine that you've just been hired by the Acme Puzzle Company to design the world's most difficult jigsaw puzzle. You'd probably want your puzzle to have a lot of pieces—thousands, or even tens of thousands. That way, no single piece would give away much information about the whole picture. Finally, you might pack the pieces in a plain box with no picture. That would be a tough puzzle.

A Picture of Matter

When you begin a study of chemistry, you start to work on a challenging puzzle—the puzzle of matter. Every chunk of matter is a puzzle piece. Your puzzle box is the universe, and the box contains many different kinds of pieces. Your job, like that of the puzzle solver at the top of the page, is to figure out how to connect all the different pieces. Keep in mind that, as with an ordinary jigsaw puzzle, your goal is not only to connect the pieces but also to see the complete picture that emerges.

Composition, Structure, and Behavior

Chemistry is the science that investigates and explains the structure and properties of matter. Matter is the *stuff* that's all around you: the metal and plastic of a telephone, the paper and ink of a book, the glass and liquid of a bottle of soda, the air you breathe, and the materials that make up your body. A more formal definition of **matter** is anything that takes up space and has mass. **Mass** is the measure of the amount of matter that an object contains. On Earth, mass is usually equated with weight. **Figure 1.1** compares two chunks of matter—one with a large mass and one with a small mass. On pages 785-791 of the *Skill Handbook* in the back of this book, you'll find a description of the metric system and the units used in chemistry. What isn't matter? The heat and light from a lamp are not matter; neither are thoughts, ideas, radio waves, or magnetic fields.

The structure of matter refers to its composition—what matter is made of—as well as how matter is organized. The **properties** of matter describe the characteristics and behavior of matter, including the changes that matter undergoes. **Figure 1.2** compares some different kinds of matter in terms of composition and behavior.

4   Chapter 1   Chemistry: The Science of Matter

Program Resources

Study Guide, pp. 1-2 [L1]
Critical Thinking/Problem Solving, p. 1 [L2]
Laboratory Manual, 1.1, pp. 1-4 [L1]
Section Focus Transparency 1 and **Master,** p. 1 [L1] [LEP]
Basic Concepts Transparencies 1, 2 and **Masters,** pp. 1-4 [L1] [LEP]
Problem Solving Transparency 1 and **Master,** pp. 1-2 [L1] [LEP]

ChemLab and MiniLab Worksheets, pp. 1-6, 10-13 [L1]
Consumer Chemistry, pp. 1-2 [L1]
Tech Prep Applications, pp. 1-2 [L1]
Applying Scientific Methods in Chemistry, p. 15 [L1]
Chemistry and Industry, pp. 1-4 [L1]

Figure 1.1
Comparison of Masses
Mass is a measure of the amount of matter in an object.

The bus below has a large mass (about 14 000 kg or 15 tons), whereas the strand of hair at the right has a small mass (approximately 0.000 001 kg).

◀ The soda in the container has a mass of 1 kg. The kilogram is the unit of mass in the metric system.

Figure 1.2
Comparing Composition and Behavior
Salt and water have different compositions, so it isn't surprising that they have different properties. Salt is made up of the elements sodium and chlorine, while water is made up of hydrogen and oxygen. You couldn't wash your hair with salt, just as you wouldn't sprinkle water on your popcorn. ▼

◀ Aspirin and sucrose (table sugar) are both composed of carbon, hydrogen, and oxygen, but you wouldn't use aspirin to sweeten your cereal or take a spoonful of sugar for a headache. Even though aspirin and sugar contain the same elements, differences in their structures determine their individual behaviors.

Aspartame and saccharin are substances with different compositions but similar tastes. Saccharin is made of carbon, hydrogen, nitrogen, oxygen, sodium, and sulfur. Aspartame also contains carbon, hydrogen, nitrogen, and oxygen, but it contains no sodium or sulfur. The way in which the components of aspartame and saccharin are organized must be a major factor in causing both of them to have a sweet taste. ▶

1.1 The Puzzle of Matter **5**

Chemistry Journal

A Chemical Dictionary
Point out to students that they will encounter many new terms in their study of chemistry. For easy reference and review, have them keep a running list of new terms and definitions in their journals. L1

Discussion
Ask students to name some chemicals that they know from past experience. *Students are likely to name chemicals they have encountered in other science courses.* Point out that all matter is composed of chemicals, and add some well-known substances to the student list. Then ask students to name something that is not a chemical. If students name common objects or substances, point out that they are composed of matter and, therefore, are chemicals. L1

Content Background
Mass Is Defined as Inertia In physics, mass is defined as the measure of the inertia of an object. Inertia is the resistance of an object to a change in its state of motion.

Discussion
Display a container of an alcohol and a wooden object. Tell students that the two substances have something in common and ask what it might be. *Both alcohol and wood are composed of the same elements—carbon, hydrogen, and oxygen—but in different proportions and with different internal structures.* Point out that structure, as well as composition, influences the properties of a substance. L1

Visual Learning
Figure 1.2: Display a model of the sodium chloride crystal lattice and a molecular model of water. Point out that the formulas for the two substances, NaCl and H_2O, do not adequately describe and differentiate the structures. One consists of repeating sodium and chlorine units, and the other consists of discrete H_2O units.

5

Figure 1.3
Some Properties of Iron
Many properties of iron are easy to observe.

It is strong, yet can be flattened and stretched. ◀

◀ It turns to liquid at a high temperature.

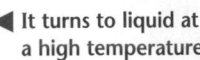

◀ It is a gray, shiny solid.

It doesn't dissolve in water, but it rusts when exposed to air and water. ▼

It conducts electricity and is attracted by a magnet. ▶

You can determine some properties of a particular chunk of matter just by examining or manipulating it. What color is it? Is it a solid, liquid, or gas? If it's a solid, is it soft or hard? Does it burn? Does it dissolve in water? Does something happen when you mix it with another kind of matter? You determine all of these properties by examining and manipu-lating a chunk of stuff, as shown in **Figure 1.3**.

Although you can find out a lot about a piece of material just by looking at it and doing simple tests, you can't tell what something is made of only by looking at it, no matter whether it is a spoonful of sugar or a rock picked up by a robot on the surface of Mars. Measurements usually must be made or chemical changes observed. **Figure 1.4** shows that it is relatively easy to determine, by a simple experiment, that sugar is composed of carbon, hydro-gen, and oxygen. Most matter does not reveal its composition so easily.

Figure 1.4

The Composition of Sugar
When concentrated sulfuric acid is added to sucrose, an interesting reaction occurs. The sugar breaks down and forms water (composed of hydrogen and oxygen). The water is released as steam (center), and black carbon is left behind (right).

6   Chapter 1   Chemistry: The Science of Matter

Figure 1.5
The Visible Results of Structure
When you look at the World Trade Center towers, you see their size and shape. These properties are the result of the building's structure, which is hidden from view under the exterior skin of the building. You cannot see the organization of the steel beams and bolts that hold the building together, or the system of pipes, wires, and ventilation ducts that thread through the building.

Tying to Previous Knowledge

Show students two or more models of water molecules to represent a sample of water. Ask them to explain what happens when water boils at 100°C. *The H₂O molecules separate from one another and go off as steam.* Ask what happens when water decomposes at 4000°C. *The H₂O molecules break apart into hydrogen and oxygen atoms.* [L1]

Reinforcement

Students are familiar with large numbers, but to make sure they understand the magnitude of 1 trillion years, ask them approximately how long Earth has existed. *probably between 3 and 12 billion years.* Write the numbers *1 billion* and *1 trillion* on the chalkboard and calculate how many times larger a trillion is than a billion. *1000 times* Ask how many times larger 100 quintillion is than 1 trillion. *100 million times* [L1]

Examining Matter: The Macroscopic View of Matter

Observations of the composition and behavior of matter are based on a macroscopic view. Matter that is large enough to be seen is called macroscopic, so all of your observations in chemistry, and everywhere else, start from this perspective. The macroscopic world is the one you touch, feel, smell, taste, and see. The properties of iron shown in **Figure 1.3** are seen from a macroscopic perspective. But if you want to describe and understand the structure of iron, you must use a different perspective—one that allows you to see what can't be seen. What you can see of the New York City World Trade Center, shown in **Figure 1.5,** is similar to a macroscopic perspective. The actual structure of the building is hidden from view.

In the same way, the appearance and properties of a piece of matter are the result of its structure. Although you may get hints of the actual structure from a macroscopic view, you must go to a submicroscopic perspective to understand how the hidden structure of matter influences its behavior.

The Submicroscopic View of Matter

The submicroscopic view gives you a glimpse into the world of atoms. It is a world so small that you cannot see it even with the most powerful microscope, hence the term *sub*microscopic. You learned in earlier science courses that matter is made up of atoms. Compared with the macroscopic world, atoms are so small that if the period at the end of this sentence were made up of carbon atoms, it would be composed of more than 100 000 000 000 000 000 000 (100 quintillion) carbon atoms. If you could count all of those atoms at a rate of three per second, it would take you a trillion years to finish. Fortunately, in chemistry, you will spend your time becoming acquainted with atoms rather than counting them.

GLENCOE TECHNOLOGY

Videodisc

Chemistry: Concepts and Applications
See the Videodisc Program Teacher Guide to access illustrations and other material available for this chapter.

---

ice cube into the beaker of water. Repeat the question and place an ice cube into the beaker of ethanol. Let students discuss among themselves any discrepancy between their predictions and observations. [L1] **COOP LEARN**
Results: The chalk does not have magnetic properties, but the white, Teflon-coated stir bar sticks to any iron or steel object. The ice cube floats in water but sinks in ethanol.

Analysis: Ask these questions.
1. Why does the piece of "chalk" stick to the metal? *It must not be chalk because it is magnetic.*
2. Why does the ice cube float in water? *Ice is less dense than water.*
3. Why does the ice cube sink in the second beaker of "water?" *The beaker must contain a liquid other than water, which is less dense than ice.*

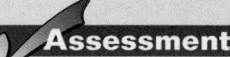

Knowledge: Have students explain how the liquids in the two beakers can be distinguished by listing their observable physical properties. *Both are clear, colorless liquids. The density of one is less than the density of ice; the density of the other is greater than the density of ice.* [L1]

1. Observation of a Candle

Time Allotment: One laboratory period.

Objectives

Review objectives with students before they begin the ChemLab.

Process Skills

Observing, inferring, interpreting data, communicating, classifying

Safety Precautions

Remind students to wear safety goggles and aprons and to keep all combustible materials and clothing away from flames.

PREPARATION

Limewater, a saturated calcium hydroxide solution, should be prepared in advance in a large bottle so that the clear supernatant solution may be poured from the bottle without disturbing any undissolved solid.

Alternative Materials

A piece of cardboard covered with aluminum foil could be used instead of the metal pan for procedure steps 1 through 5.

PROCEDURE

- Require that students' notebooks contain the following items before they begin: Objectives, Safety Precautions, Outline or Flow Chart of the Procedure, Data and Observations Table.
- Discuss your expectations regarding laboratory procedures and behavior prior to the first lab. If students know what is expected of them, there will be less inappropriate behavior.
- To make the water level more observable in procedure steps 6 and 7, add food coloring to the water.

Observation of a Candle

You have seen candles burn, perhaps on a birthday cake. But you probably have never considered the burning of a candle from a chemist's point of view. Michael Faraday, a 19th-century chemist, found much to observe as a candle burns. He wrote a book and gave talks on the subject. In this ChemLab, you will investigate the burning of a candle and the products of combustion.

Problem

What are the requirements for and characteristics of a candle flame? What are the products of the combustion of the candle?

Objectives

- **Observe** a candle flame and perform several tests.
- **Interpret** observations and the results of the tests.

PREPARATION

Materials

large birthday candles
matches
shallow metal dish
25 mL of limewater solution
250-mL beaker
500-mL Erlenmeyer flask
solid rubber stopper to fit the flask
wire gauze square
tongs

Safety Precautions

Wear an apron and goggles. Keep all combustible materials, including clothing, away from the match and candle flames. Do not allow the limewater to splash into your eyes. If it does, immediately rinse your eyes for 15 minutes and notify your teacher.

PROCEDURE

1. Light a candle and allow a drop or two of liquid wax to fall into the center of the pan. Press the candle upright onto the melted wax before it can solidify. If the candle burns too low during the following procedures, repeat this step with a new candle.

2. Observe the flame of the burning candle for a few minutes. Try to observe what is burning and where the burning takes place. Observe the different regions of the flame. Make at least eight observations, and record them in a data table like the one shown.

3. Light a second candle and hold the flame about 2 cm to 4 cm to the side of the first candle flame. Gently blow out the first candle flame, then quickly move the flame of the second candle into the smoke from the first flame. Record your observations.

4. Relight the standing candle. With tongs, hold the wire gauze over the flame, perpendicular to the candle. Slowly lower the gauze onto the flame. Do not allow the gauze to touch the candle wax. If the flame goes out, quickly move the wire gauze off to the side. Record your observations.

5. Fill the 250-mL beaker with cold tap water, dry the outside of the beaker, and hold it about 3 cm to 5 cm above the candle flame. Record your observations.

ChemLab

6. Pour tap water into the pan or dish to a depth of about 1 cm.

7. Quickly lower an Erlenmeyer flask over the candle so that the mouth of the flask is below the surface of the water. Allow the flask to remain in place for approximately one minute. Record your observations.

8. Lift the flask out of the water, turn it upright, and add about 25 mL of limewater. Stopper the flask and swirl the solution for approximately one minute. Record your observations. If the solution becomes cloudy or chalky, calcium carbonate was formed, indicating the presence of carbon dioxide in the flask.

ANALYZE AND CONCLUDE

1. **Classifying** Which changes that you noted in step 2 were physical? Which were chemical?

2. **Making Inferences** Do your results in step 3 indicate that the candle wax burns as a solid, a liquid, or a vapor? Explain.

3. **Interpreting Data** One requirement for combustion is the presence of fuel. Interpret your results from steps 4 and 7 to determine the other requirements.

4. **Interpreting Data** Based upon your analysis of the observations from steps 5 and 8, what are two products of the combustion of the candle?

APPLY AND ASSESS

1. Sir Humphry Davy invented a safety lamp for miners in which a flame was surrounded by a wire gauze cylinder. Can you explain the reason why the lamp was constructed in this way?

2. What change in water level occurred in procedure 7? Propose an explanation for this change.

DATA AND OBSERVATIONS

Procedure step	Observations
2	
3	
4	
5	
7	
8	

ANALYZE AND CONCLUDE

1. The melting and vaporizing of the candle wax were physical changes, and the burning of the wax was a chemical change.

2. The candle wax burns as a vapor. The vapor was reignited by the second candle flame even though the second flame did not touch the wick of the first candle.

3. The results from step 4 indicate that a certain minimum temperature is required for combustion, and the results from step 7 indicate that air is required.

4. from step 5, water vapor; from step 8, carbon dioxide

APPLY AND ASSESS

1. The wire gauze conducted heat away from the flame and vapors, lowering the temperature sufficiently so that other mine gases would not ignite.

2. The water level rose. Oxygen from the air inside the flask was consumed by combustion. When the combustion ceased, the temperature of the gases in the flask decreased, thereby decreasing the volume of the gases.

✓ Assessment

Oral: Give students this formula for candle wax, $C_{25}H_{52}$, and ask them how the carbon dioxide and water products were formed. *Oxygen from the air combined with carbon to form carbon dioxide and with hydrogen to form water.* L1

DATA AND OBSERVATIONS

Procedure Step	Observations
2	Candle wax melts, vapor burns. The flame is blue at the bottom, yellow at the top. The flame appears to be hollow at the bottom.
3	The candle flame relights even though the second flame did not touch the wick.
4	The flame does not burn above the wire gauze. The candle relights when the gauze is removed.
5	Water condenses on the beaker.
7	The flame goes out, and the water rises inside the flask.
8	The limewater solution becomes cloudy.

Using an Analogy

Refer students to **Figure 1.7** and ask them what a football team, a school committee, a biology class, the U.S. Congress, and a church choir have in common. *They are all composed of people.* Point out that these groups have different functions even though they are all composed of people. In the same way, compounds such as aspirin and sucrose behave differently even though they are composed of the same kinds of atoms. L1

GLENCOE TECHNOLOGY

 **Videodisc**

Science and Technology Videodisc Series
Fire Resistant Clothing
Disc 2: Chemistry
Side 1, Ch. 7

Show this video to tie into the concept of examining materials by studying their properties. L1 LEP

A probe like this one moves up and down in response to the position of carbon atoms on the surface. A computer converts the probe's motion into a bumpy-looking image. ▶

This image of a crystal of graphite, a form of carbon, is obtained by scanning the surface with a fine, sensitive probe. ▼

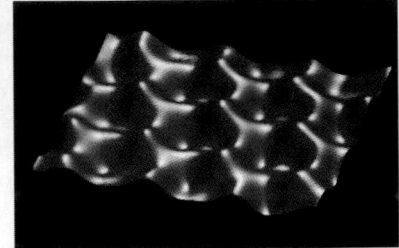

Figure 1.6
A Submicroscopic View of Matter
Although a scanning tunneling microscope provides a glimpse of the submicroscopic world, it does not truly produce pictures of atoms, at least not as we think of pictures in the macroscopic world.

Although individual atoms cannot be seen, the recently developed scanning tunneling microscope (STM), shown in **Figure 1.6,** is capable of producing images on a computer screen that show the locations of individual atoms. The STM can even be used to move individual atoms around on a surface.

Using Models in Chemistry

In your study of chemistry, you will use both macroscopic and submicroscopic perspectives. For example, sucrose and aspirin are both composed of carbon, hydrogen, and oxygen atoms, but they have different behaviors and functions. These differences must come about because of differences in the submicroscopic arrangement of their atoms. **Figure 1.7** shows models that reveal these submicroscopic differences.

Figure 1.7
Comparing the Structures of Aspirin and Sucrose
The different submicroscopic arrangements of the atoms in aspirin (left) and sucrose (right) cause the differences in their behavior. Don't worry about not understanding the complete meaning of these structures. Each ball represents an atom, and each bar between atoms represents a chemical connection between the atoms.

○ hydrogen
● carbon
● oxygen

Chemistry Journal

Rationalizing Beliefs

Ask students to make a list in their journals of things they have not personally seen but that they believe to exist because of indirect evidence. Have students write an analogy of their belief in each thing to the belief of scientists in the existence of atoms. L1

Models Connect the Macroscopic and Submicroscopic Views

The drawings in **Figure 1.7** are tools that allow you to study what you cannot actually see—in this case, the arrangement of atoms. These drawings represent one type of model used in science. Sometimes, a model is something you can see and manipulate. These are the kinds of models you are already familiar with—model cars and airplanes or perhaps an architect's scale model of a proposed building. **Figure 1.8** shows that models are important tools in many fields. Models are used, tested, and revised constantly through new experiments. A model of the submicroscopic structure of a piece of matter must be able to explain the observed macroscopic behavior of that matter and predict behavior that has yet to be observed.

The model of aspirin (acetylsalicylic acid) in **Figure 1.8** is an example of a scientific model. A **scientific model** is a thinking device that helps you understand and explain macroscopic observations. Scientific models are built on experimentation. In Greece, a model of matter based on atoms was discussed about 2500 years ago, but this model was not a scientific model because it was never supported by experiments. It took until the 1800s before a scientific model of matter was proposed. This atomic model was developed and verified by experiments. It has withstood 200 years of prediction and experimentation with only slight modifications.

◀ A model of a prospective new airliner is made and studied before an airplane is actually built.

Figure 1.8
Using Models
Models help you see and understand structure, whether you're creating a new airplane or designing a drug that you hope will be an effective treatment for AIDS.

The model on the right is another, more informative way to represent the arrangement of atoms in the compound aspirin. ▼

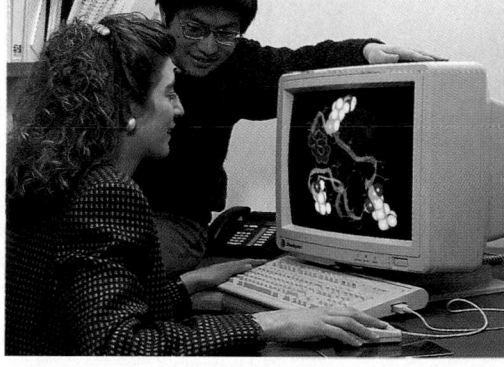

Chemists use computers to create models of new drugs in the search for new treatments for disease. ▶

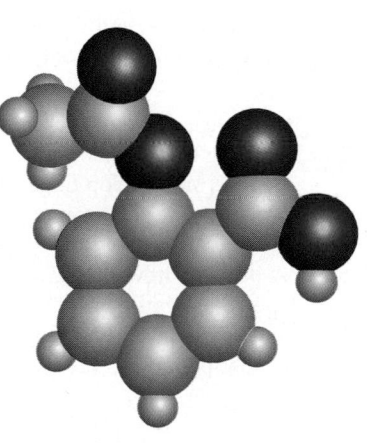

Concept Development

Explain that once a set of hypotheses agrees with all observations, the set of hypotheses is assembled into a theory or model. Therefore, a theory or model is a set of tested hypotheses that provides an overall explanation of a natural phenomenon.

Using an Analogy

To illustrate the tentative nature of theories and models, describe this scenario. Primitive people have just discovered fire, and because they have observed only sticks and logs burning, they have come to believe that only cylindrical objects burn. Then, when they see grass or material of some other shape burning, they must revise their model.

Concept Development

Point out that science is both a body of facts and a procedure for processing and understanding these facts. Explain that the procedure is often referred to as the scientific method, and generally consists of these steps: (1) making observations, (2) forming hypotheses, and (3) performing experiments to prove or disprove hypotheses. Ask students whether they think that this procedure always gives the correct result. *no* Explain that the scientific method is an ongoing process. L1

Meet Dr. John Thornton, Forensic Scientist

Teaching Strategies

- Have students use dictionaries to determine the derivation and meaning of the term *forensic*. The word is derived from the Latin word *forum*, a public debating place, and means "appropriate to court judicature." L1

- Ask students to discuss how the "quality of justice is enhanced by a full and perhaps contentious airing of all the relevant issues and the physical evidence that (forensic scientists) provide." L1

- Write the terms *biology, earth science, chemistry*, and *physics* on the chalkboard. Ask students to brainstorm specific ways that each subject could be used by a forensic scientist. Have students list their responses on the board and explain them. L1

- Tell students that besides private forensic laboratories, the federal government, the states, and many cities maintain forensic laboratories. If your local government maintains a forensic laboratory, arrange to have a speaker address the class or, if possible, have a team of students videotape a field trip to the laboratory and present it to the class. L1 LEP

COOP LEARN

Background

- In the eighth century A.D., the Chinese recognized fingerprints as substitutes for signatures.

- Fingerprints are obtained from paper by means of a color reaction between ninhydrin and amino acids found in perspiration deposited by the fingertips when the print was made.

Meet Dr. John Thornton, Forensic Scientist

A poem by William Blake contains these words: "To see a world in a grain of sand and a heaven in a wildflower." In Dr. Thornton's work, that favorite line of poetry is literally true. To discover what Dr. Thornton means, read this interview, in which he shares his thoughts about the world of physical evidence.

On the Job

 Dr. Thornton, could you tell us what you do at your lab, Forensic Analytical Specialties?

 We analyze physical evidence from the scene of a crime—hair, fiber, body fluids, bullets, paint, soil, glass, shoe impressions, fingerprints, drugs, and plant material. Any of this tangible, physical evidence can associate a person with a crime scene. That's half of it, the "whodunit"; the other half is the "howdunit."

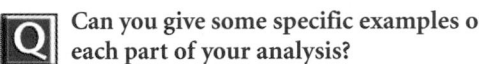 **Can you give some specific examples of each part of your analysis?**

Fingerprints are the classic means of establishing that a person was at the scene of a crime. We develop fingerprints with chemicals: ninhydrin for prints on paper, and cyanoacrylate, which is actually superglue, on other items. As for the "how," here's an example of a case. A person claims he shot a person while defending himself against strangulation. If there's a gunpowder pattern on the victim's clothing, that would tend to support the suspect's story. On the other hand, if there's no powder, indicating that the person was several feet away, that story doesn't hold up.

It must be almost impossible to avoid leaving some kind of evidence behind, right?

Yes. Frequently, the evidence that is most incriminating is so small that the perpetrator is oblivious to it, such as grains of pollen, sand, or tiny diatoms. Those would be a signature of a particular geographic location.

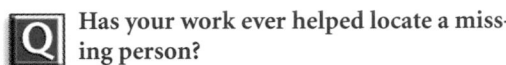 **Has your work ever helped locate a missing person?**

A few years ago, when a young woman was missing, a shoe was found near the freeway. Inside the shoe was about a thimbleful of fibers. I went through her sock drawer and found socks that matched every single type of fiber that was in the shoe. That clue indicated the direction of her disappearance, and we later found her.

Early Influences

 How did you get into this line of work?

 When I was in the seventh or eighth grade, I was just browsing through the local public library and stumbled upon a book called *Crime Investigation,* by Paul Kirk, who was probably the foremost forensic scientist at the time. From that point on, I knew what I wanted to do. What I *didn't* know was that I would later enroll in Kirk's university class. Even more improbably, I eventually took his place on the faculty after he retired.

 Were you a kid who liked things like decoder rings and puzzles?

 No, I was just a farm boy in the Central Valley. Everything there was flat, and I guess I was looking for some other horizons. I did that by reading.

Personal Insights

 What do you find most interesting about your work?

 Forensic science is really a study of how the world is put together. My work involves chemistry, physics, botany, and geology, so it's impossible to get bored.

 How do you go about tackling the problems posed by a set of clues at a crime scene?

 At any crime scene, you want to use physical evidence to tell a story. It's like looking at a tapestry from the back side: you can vaguely see that over here there's a unicorn, and over here there's a tree, and this might be a fence. It's difficult to get the clarity. I generally just sit down and think about things for a while. I let ideas wash over me from different directions. Then

I play what-if: What if this happened at the scene, then what might have happened next? I try all of these ideas on for size, then I try to winnow the scientific wheat from the chaff.

 What kind of satisfaction do you derive from your work?

 I feel that the quality of justice is enhanced by a full and perhaps contentious airing of all the relevant issues and the physical evidence that we provide.

CAREER ▶ CONNECTION

Forensic chemists are assisted by people in these lines of work:

Crime Lab Technologist (Collects and analyzes physical evidence) BA from college with crime laboratory program

Fingerprint Classifier (Files and matches fingerprints) High school followed by study at police schools

Private Investigator (Collects facts pertaining to a crime) High school and detective-training program

CAREER CONNECTION

- **Career Path** A career in forensic science would include these high school courses: biology, earth science, chemistry, and physics. Some colleges have criminal justice departments, which offer bachelor degree programs in crime laboratory work such as forensic pathology. Graduate and doctoral programs in forensic sciences are offered by universities.
- **Career Issue** Have students investigate the increasing role of DNA analysis as a forensic tool. [L1]

For More Information

- Interested students may want to read Paul Kirk's *Crime Investigation*, 2nd ed., New York: John Wiley, 1974, edited by Dr. Thornton. [L1]
- Information about careers in forensic science can be obtained from the American Academy of Forensic Sciences, Suite 203, 410 North 21st Street, PO Box 669, Colorado Springs, CO 80901-0669.

Classifying Matter

Examples of matter range from grains of rice to the stars, from a drop of water to the rocks of the Grand Canyon, from a potato chip to a computer chip. It's all the stuff around us—all part of the same puzzle. You know that all these bits of matter connect in some way, just like the pieces of a jigsaw puzzle. But, there are so many different sizes and kinds of matter. How can you begin to make sense of the puzzle of matter?

Just as you do when you work a jigsaw puzzle, you first classify the pieces before trying to make connections. You might find pieces that share common properties, such as a flat edge or the same color, and place them in separate piles.

Classification by Composition

A powerful way to classify matter is by its composition. This is the broadest type of classification. When you examine an unknown piece of stuff, you first ask, "What is it made of?" Sucrose is composed of the elements carbon, hydrogen, and oxygen. This is a **qualitative** expression of composition. A qualitative observation is one that can be made without measurement.

Figure 1.9
Measurement—A Quantitative Observation
In a medical examination, many of the doctor's observations are quantitative. Measured quantities, such as height and weight, can be compared with those of previous examinations. This boy has gained 5 pounds and grown 1 inch since his last examination.

After a qualitative analysis, the next question that you might ask is how much of each of the elements is present. For sucrose, the answer to that question is that 100 g of sucrose contains 42.1 g of carbon, 51.4 g of oxygen, and 6.5 g of hydrogen. This is a **quantitative** expression of composition. A quantitative observation is one that uses measurement. You make quantitative measurements every day when you answer such questions as What's the temperature? How long was the pass? How much do you weigh? **Figure 1.9** shows some quantitative measurements being made.

Pure substance or a mixture?

The most general way to classify matter by composition is in terms of purity. There are only two categories here. A sample of matter is either pure—made up of only one kind of matter—or it is a mixture of different kinds of matter. What does it mean to say that a chunk of matter is pure?

Cultural Diversity

The Discovery and Isolation of Radium

One of the most spectacular examples of the separation of a pure substance from a mixture is the work of Marie Sklodowska Curie and Pierre Curie when they discovered and isolated the radioactive element radium. Starting with several tons of pitchblende, a natural ore containing a mixture of uranium compounds, the couple isolated less than 0.1 g of a radium compound. The process took more than four years. For their work on radioactivity, Marie and Pierre Curie shared the 1903 Nobel Prize in Physics with Henri Becquerel. In 1911, Marie Curie was awarded the Nobel Prize in Chemistry for her work on the discovery, isolation, and purification of radium. Marie Curie is the only person to win two Nobel prizes in science.

Figure 1.10
Pure Products?
Products such as orange juice and soap are often advertised as pure. From a chemist's point of view, they are not pure but instead are complex mixtures containing many different substances.

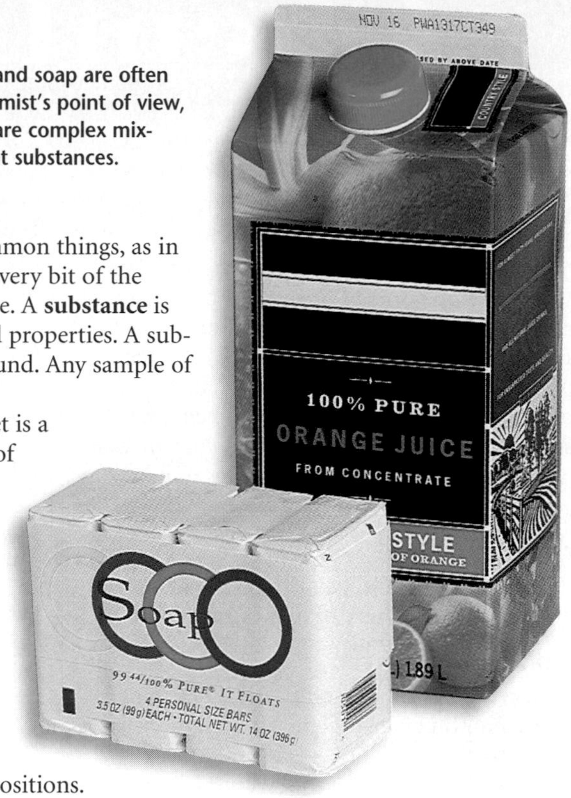

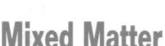

The word *pure* is often used to describe common things, as in **Figure 1.10.** In chemistry, *pure* means that every bit of the matter being examined is the same substance. A **substance** is matter with the same fixed composition and properties. A substance can be either an element or a compound. Any sample of pure matter is a substance.

If the sugar in a bag from the supermarket is a substance, then it is pure sucrose. Every bit of matter in the bag must have the same properties and the same fixed composition as every other bit. Now, consider a bag of high-quality, dry, white sand. White sand is the common name for a substance called silicon dioxide. It is white and crystalline like sugar, and, when the sand is examined, every particle has the same fixed composition (53.2 percent oxygen, 46.8 percent silicon). Both sand and sucrose are substances, but they have different properties and compositions.

Mixed Matter

Suppose you mix the pure white sand with the pure white sucrose, as in **Figure 1.11.** You may not be able to notice any difference, but if you add this mixture to your cereal or tea, it's a different story. The sweet taste of the sugar is still there, but so is the grittiness of the sand. Every particle of this stuff is not the same, so the properties are not the same throughout. Some parts taste sweet, while others are gritty and tasteless. The composition is also not fixed, but instead depends on how much sand is mixed with the sugar.

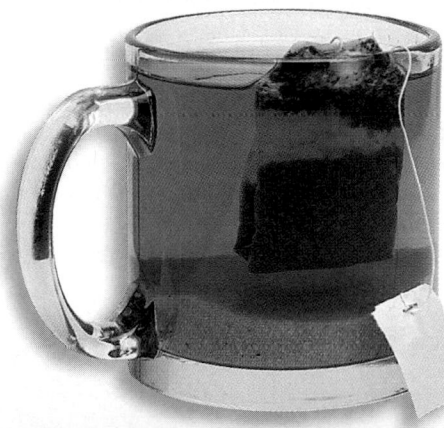

Figure 1.11
A Mixture Retains the Properties of Its Components
Pure sugar is different from a mixture of sugar and sand. Each component of a mixture retains its behavior. Sugar dissolves and sweetens the tea, but sand is insoluble. It settles to the bottom of the cup.

Discussion

Point out the uniform appearance of air and ask students whether it is a mixture. *yes, a homogeneous mixture of nitrogen, oxygen, carbon dioxide, water vapor, and other gases and solids* Ask them for evidence that the individual components retain their properties. *The oxygen in air enables us to live, though it comprises only about 20% of the air.* Ask how one might separate the components of air. *Solids could be filtered out. Gaseous components could be condensed to liquids sequentially by cooling the air.* L1

Correcting Misconceptions

Students may confuse a natural law with a theory or model. Explain that a law summarizes what happens, while a theory or model is an attempt to explain what happens.

Tying to Previous Knowledge

Ask students to recall the list of properties of water from the *Discussion* on page 6 and to determine which properties are quantitative. *freezes at 0°C; boils at 100°C* L1

Cultural Diversity

Four Elements and Five Changes

Early Greek philosophers postulated that materials were composed of four primary elements, each with two of four related properties. The elements and their respective properties are water (moist and cold), earth (cold and dry), fire (dry and hot), and air (hot and moist). All substances were made of different combinations of these elements. Properties of substances were the result of different combinations of types and proportions of the properties of the elements from which they were composed. The early Chinese philosophers viewed all materials as undergoing five fundamental stages of change—earth, wood, fire, metal, and water. All natural processes were viewed as transformations similar to the orderly transformations that take place among the five stages of change. For example, earth gives rise to wood, which evolves fire, which transforms metals, which give rise to water (dew drops condensing on metal objects). Water finally gives rise to wood as evidence of the role of water in the growth of plants.

ChemLab

2. Kitchen Chemicals

Time Allotment: One laboratory period.

Objectives

Review objectives with students before they begin the ChemLab.

Process Skills

Observing, interpreting data, inferring, communicating

Safety Precautions and Disposal

Advise students to wear safety goggles and aprons. Caution students not to taste any of the materials or test liquids. Label stockroom supplies of the kitchen chemicals.

Disposal: The solids may be disposed of in the wastebasket. The I_2-KI solution can be flushed down the drain with copious amounts of water.

PREPARATION

- Kitchen material A is table salt, B is baking soda, C is baking powder (see note below), and D is soluble starch.
- If you use commercial baking powder, be sure to test it with I_2-KI solution to be sure it does not contain starch. You may prepare your own baking powder by thoroughly mixing baking soda and citric or tartaric acid.
- Reagent liquid I is water, II is white vinegar, and III is prepared by dissolving 2 g of I_2 and 6 g of KI in 100 mL of water.
- The mixtures are prepared by thoroughly mixing the solids as follows: Mixture 1 is A and C; 2 is B and D; 3 is C and D; 4 is A, B, and D; and 5 is B, C, and D.

PROCEDURE

- Require that students' notebooks contain the following items before they begin: Objectives; Safety Precautions; Outline or Flow Chart of the Procedure; Data and Observations Table.

ChemLab 2 SMALL SCALE

Kitchen Chemicals

The chemical and physical properties of a substance make up a sort of fingerprint that characterizes the substance. In this ChemLab, you will test four unknown solids using three different liquids. The unknowns are common materials that you'd probably find in your kitchen. The results of your tests will give you the information you need to unravel the compositions of mixtures of two solids and three solids.

Problem

How can you identify a material by comparing its properties with those of known materials?

Objectives

- **Observe** the chemical and physical reactions of four common kitchen materials with three test reagents.
- **Compare and interpret** the reactions of the test reagents with five two-solid and three-solid mixtures of the common kitchen materials.
- **Infer** the composition of each of five unknown mixtures by comparing their reactions with those of the known materials.

PREPARATION

Materials

96-well microplate	3 thin-stemmed pipets
9 test tubes	masking tape
spatulas	marking pen

Safety Precautions

Wear an apron and goggles. Do not touch or taste any of the solids or liquids, even though you may believe you know their identities.

PROCEDURE

1. Label four test tubes *A*, *B*, *C*, and *D*. Label five test tubes 1 through 5.

2. In the four lettered test tubes, place about 1 g of each of the labeled samples supplied by your teacher. These are the common kitchen materials.

3. In the numbered test tubes, place about 1 g of each of the numbered samples supplied by your teacher. These are the unknown mixtures. If you use the same spatula for each material, rinse and dry it before dipping into the next solid to avoid contaminating one material with another.

4. Label three long columns of wells on the microplate *I*, *II*, and *III*. Label nine rows of wells on the microplate *A*, *B*, *C*, and *D*, and 1 through 5 as shown in the photo.

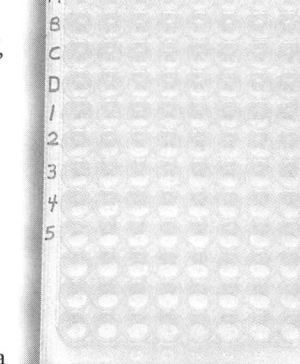

5. Place the microplate on a sheet of white paper.

6. Add a small amount of each material to the row of three wells that has the appropriate letter or number.

7. Observe and record the texture of each of the nine materials in a data table like the one shown.

- If students guess the identities of the solids, liquids, or mixtures as they perform the lab, tell them that identities will be discussed only during the post-lab discussion.

ANALYZE AND CONCLUDE

1. Solid A dissolves in all three reagents, but it does not fizz or turn blue. Solid B dissolves in reagents I and III, and fizzes with reagent II. Solid C fizzes with all three reagents. Solid D dissolves with all three reagents and produces a blue color with reagent III.

2. Mixture 2 contains solids B and D because it fizzes with reagent II only and turns blue with reagent III. Mixture 3 contains solids C and D because it fizzes with all three reagents and turns blue with reagent III. Mixture 4 contains solids A, B, and D because it does not fizz with reagents I and III.

3. Mixture 1 contains solids C and A or B because it fizzes with all three reagents but does not turn blue with reagent III. Mixture 5 contains solids C, D, and A or B because it fizzes with all three reagents and turns blue with reagent III.

ChemLab

8. Label the three pipets *I*, *II*, and *III*. From the containers of reagent liquids supplied by your teacher, draw into the bulb of each pipet the liquid corresponding to the label number on that pipet.

9. Add 3 drops of liquid *I* to each of the nine materials in column I.

10. Observe any changes that take place, and record them in the data table.

11. Repeat steps 9 and 10 using liquid *II* and then liquid *III*.

ANALYZE AND CONCLUDE

1. **Interpreting Data** What properties and reactions characterize each of the four kitchen solids?

2. **Drawing Conclusions** Can you positively identify the solids that are contained in any of the five mixtures? If so, identify the solids and explain your conclusions.

3. **Making Inferences** If you are unable to conclusively identify the solids in any of the mixtures, what are their likely identities? Explain.

APPLY AND ASSESS

1. Two of the four original solids, baking powder and baking soda, are often used in making baked goods. What characteristic probably makes them useful in baking? Which solids display this characteristic?

2. Baking powder is a mixture of two or more compounds, and it reacts with water or any other liquid that contains water. Baking soda is a single compound that reacts with acidic solutions but not with water. Which of the solids do you think is baking powder? Explain.

3. One of the solids is an organic compound you may have learned about in a biology course. It produces a characteristic color when combined with iodine. Which solid gave this reaction? What is the identity of this compound?

APPLY AND ASSESS

1. Solids B and C react to produce a gas or gases that could enable baked goods to rise.
2. Solid C is probably baking powder because it fizzes with all three liquid reagents.
3. Solid D produces the dark blue color reaction with reagent III. It is starch.

DATA AND OBSERVATIONS
See table below.

Performance: Ask students to write a formal report for the ChemLab. If this is their first experience writing a formal report, guide them through the process. L1

DATA AND OBSERVATIONS

Solid	Color	Texture	Reaction with liquid 1	Reaction with liquid 2	Reaction with liquid 3
Two-solid mixtures					
Three-solid mixtures					

A (salt)	white, clear	crystals or powder	dissolves	dissolves	dissolves
B (baking soda)	white	powder	dissolves	fizzes	dissolves
C (baking powder)	white	powder	fizzes	fizzes	fizzes
D (starch)	white	powder	dissolves	dissolves	dissolves and produces dark blue color
Two-solid mixtures 1			fizzes	fizzes	fizzes
2			dissolves	fizzes	blue
3			fizzes	fizzes	fizzes and blue
Three-solid mixtures 4			dissolves	fizzes	blue
5			fizzes	fizzes	fizzes and blue

Most of the matter you encounter every day is a mixture. A **mixture** is a combination of two or more substances in which the basic identity of each substance is not changed. In the sugar and sand mixture, the sand does not influence the properties of the sugar, and the sugar does not influence the properties of the sand. They are in contact with each other, but they do not interact with each other.

Unlike pure substances, mixtures do not have specific compositions. For example, a sand-and-sugar mixture can be any combination of sand and sugar. **Figure 1.12** shows some common mixtures.

Figure 1.12
Everyday Mixtures
This seascape shows two mixtures: seawater, a mixture of water with many salts and other soluble substances; and air, a mixture of nitrogen, oxygen, water vapor, argon, carbon dioxide, and other gases. ▶

▲ The exotic flavors of dishes from Asia and Africa arise from mixtures of spices such as cardamon, turmeric, cumin, allspice, and coriander.

▲ Solder is an alloy of tin and lead used by plumbers and electricians to seal a joint or connect one piece of metal to another.

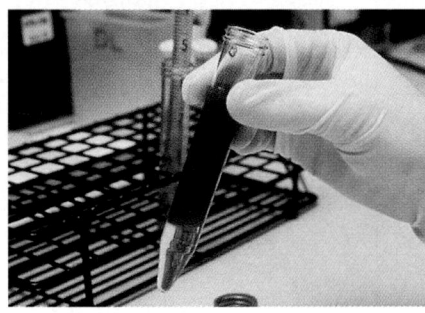

▲ Blood is a complex mixture of many substances including water, proteins, glucose, fats, amino acids, and carbon dioxide.

◀ Sand, pulverized stone, minerals, salts, and substances from decayed plants and animals make up the rich mixture called soil.

18   Chapter 1   Chemistry: The Science of Matter

Integrating THE Sciences

Physics The spectacular blue iridescent color of a *Morpho* butterfly occurs not because of the presence of a blue pigment in the butterfly's wing, but because of the well-ordered microscopic arrangement of scales on the butterfly's wings. When sunlight strikes the wings at just the right angle, the ordered scales on the wing act like a diffraction grating, reflecting blue light while canceling out the other colors of the spectrum. The ordering of scales on the butterfly's wings arises from the same process that produces soap scum on a bathtub: the slow evaporation of a water solution on a solid surface. Scientists have been able to crudely duplicate the blue coloring of the butterfly by using well-organized synthetic surfaces composed of plastics.

Everyday Chemistry

You Are What You Eat

The morning is half over and you need some quick energy. So, you eat a piece of sour-apple candy. You have just consumed four chemicals—the sugar fructose for sweet taste and energy, citric acid for tartness, methyl butanoate for apple flavor, and red dye #2 for color. *Chemical* is just another name for "substance." Whenever you eat or drink, you consume chemicals that your body needs for energy, growth, and repair.

What chemicals are found in your body? The percentages of elements that make up the human body are shown in the graph. However, the elements found in your body are not free elements; they are in the form of compounds. For example, 67 percent of your body is water, a compound of hydrogen and oxygen. The calcium and phosphorus in your bones are in a compound called hydroxyapatite, $Ca_5(PO_4)_3OH$.

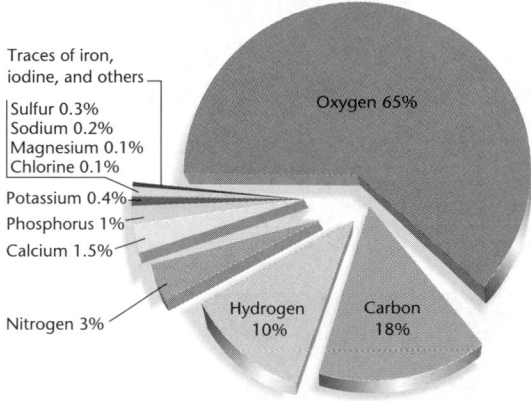

Traces of iron, iodine, and others
Sulfur 0.3%
Sodium 0.2%
Magnesium 0.1%
Chlorine 0.1%
Potassium 0.4%
Phosphorus 1%
Calcium 1.5%
Nitrogen 3%
Oxygen 65%
Hydrogen 10%
Carbon 18%

Your body is an unbelievably complex chemical factory. It continually checks to see if the right amount of each compound is present. It decomposes the food you eat and uses the new substances formed to make compounds needed for growth and repair. It breaks down other food components to obtain the energy it needs.

The chemistry of food You might start the day with a breakfast of melon, eggs, whole-wheat toast, milk, and tea. All of these foods are mixtures of many different compounds. If you decide to sweeten your tea with sugar, you will be adding a single compound—sucrose—to a mixture of water, caffeine, tannin, butyl alcohol, isoamyl alcohol, phenyl ethyl alcohol, benzyl alcohol, geraniol, hexyl alcohol, and essential oils. If you eat scrambled eggs, you are consuming a mixture of water, ovalbumin, coalbumin, ovomucoid, mucia, globulins, amino acids, lipovitellin, livetin, cholesterol, lecithin, lipids (fats), fatty acids, butyric acid, acetic acid, lutein, zeaxanthine, and vitamin A with a little sodium chloride (salt) sprinkled in. Doesn't that sound mouthwatering?

Exploring Further

1. **Analyzing** What color is burned food? What element is usually found in that color? What element do you think most foods have in common?

2. **Interpreting** Find out how organic and inorganic compounds differ. Which type makes up most of the human body? Why is the other type important to life?

Everyday Chemistry

Purpose

To show students that chemistry is a part of their lives by explaining that (1) the human body is a massive complex of reacting chemicals and (2) food supplies the body with the necessary chemicals for growth and energy.

Teaching Strategies

- Use both chemical and structural formulas to help students develop familiarity with them. They can see what elements and how many atoms of each are present.
- Introduce students to the *CRC Handbook of Chemistry and Physics* by having them look up in Section B, pages B-1 to B-40, information about the elements. L1
- Help students learn to use chemical references like the *Merck Index* by looking up the structural formula and commercial sources of caffeine. L1

Extension

Have students read the labels on food containers to see what ingredients are present. Then have them decide which ingredients are mixtures of chemicals and which pure compounds are organic or inorganic. L1

Exploring Further

1. black, carbon, carbon
2. Organic compounds contain carbon. With few exceptions—for example, carbonate compounds—inorganic compounds do not contain carbon. Water, an inorganic compound, makes up the greatest proportion of the human body by mass. Organic compounds form the structure of the body and carry out the activities of life.

The Separation of Mixtures into Pure Stuff

One characteristic of a mixture is that it can be separated into its components by physical processes. The word *physical* means that the process does not change the chemical identity of a substance. How could you separate a sand/sugar mixture into pure sand and pure sucrose? The simplest physical means would be to look at it with a microscope and separate the bits of sugar and sand with tweezers. You are right in thinking that there must be an easier way.

Separating mixtures by using physical changes is one easier way. A **physical change** is a change in matter that does not involve a change in the identity of individual substances. Examples of physical changes include boiling, freezing, melting, evaporating, dissolving, and crystallizing. Separation of a mixture by physical changes takes advantage of the different physical properties of the mixed substances. **Physical properties** are characteristics that a sample of matter exhibits without any change in its identity. Examples of the physical properties of a chunk of matter include its solubility, melting point, boiling point, color, density, electrical conductivity, and physical state (solid, liquid, or gas). Look at **Figure 1.13** to see one way to separate a mixture of sugar and sand using differences in the physical properties of the two substances.

Figure 1.13
Separating a Mixture by Physical Changes
Because sugar and sand have some different physical properties, a separation can be made by using a series of physical changes.

Water is added and the mixture is stirred. The sugar dissolves in the water, but the sand does not. ▶

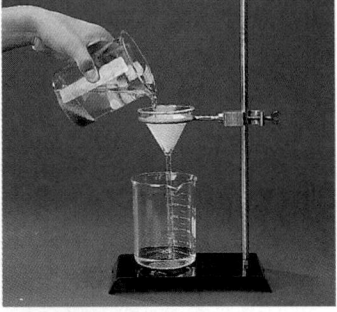
▲ The mixture is passed through a filter that traps the sand but allows the sugar solution to pass through.

The sugar solution is heated to evaporate the water. ▶

◀ When all of the water has evaporated, pure sugar remains in the beaker. The separation is possible because of the difference in solubility of sugar and sand.

50 mL + 50 mL = ?

Nothing much happens when you make a mixture of sand and sugar. Some mixtures, such as water and alcohol, can give unexpected results that might contain clues to the structure of matter.

Procedure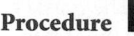

1. Wear an apron and goggles.

2. Fill a 100-mL graduated cylinder exactly to the 50.0-mL mark with water that has been tinted with food coloring. Use a dropper to adjust the volume. Insert a thermometer and read the temperature of the water. Remove the thermometer.

3. Tilt the cylinder over almost as far as you can without spilling water and begin to add ethanol very slowly so that it does not mix with the water. Raise the cylinder slowly as you add more alcohol. Add the final alcohol with a dropper to adjust the level to exactly 100.0 mL. Assume the ethanol has the same temperature as the water.

4. Use a stirring rod to mix the contents of the cylinder as rapidly as possible. Immediately insert a thermometer. Read and record the temperature of the mixture.

5. Remove the thermometer and stirring rod, taking care that all liquid drips back into the cylinder. Read and record the volume to the nearest 0.1 mL.

Analysis

1. Describe what happened to the volume when the liquids mixed. Suggest a way to account for your observation.

2. Was heat absorbed or given off as the liquids mixed? How do you know?

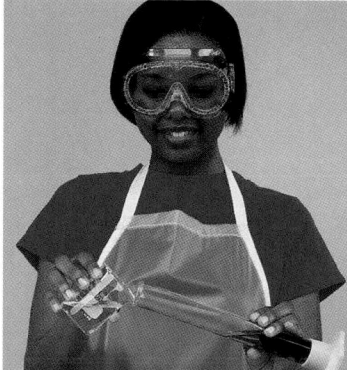

Two Kinds of Mixtures

Sometimes, when you look at a sample of matter, it's easy to tell that the sample is a mixture. This kind of mixture is called a heterogeneous mixture. The prefix *hetero* means "different." A heterogeneous mixture is one with different compositions, depending upon where you look. The components of the mixture exist as distinct regions, often called phases. Orange juice and a piece of granite, as shown in **Figure 1.14,** are examples of heterogeneous mixtures.

Figure 1.14

Heterogeneous Mixtures

If you look closely at a glass of orange juice (left), you can see pieces of solid orange pulp floating in the liquid. If you look at a granite rock (right), you can see areas of different color that indicate that the rock is composed of crystals of different substances.

1.1 **The Puzzle of Matter** 21

SECTION 1.1

miniLAB

50 mL + 50 mL = ?

Purpose: To observe evidence of interaction on the submicroscopic level.
Process Skills: Observing, measuring, inferring

Teaching Strategies

- Use only 95% denatured ethanol, available at pharmacies. Do not substitute other alcohols or use a lower concentration.
- If necessary, demonstrate how to read the volume in the graduated cylinder.
- Have all liquids and glassware equilibrated to room temperature.
- Remind students to read the thermometer while it is in the liquid.
- Demonstrate how to add alcohol to water without mixing. Emphasize slow addition.
- Stirring rods should be long enough to reach the bottom of the graduated cylinder so that the liquids can be mixed rapidly.

Expected Results

The volume of the mixture will be between 96 and 97 mL. The temperature will rise 5°C to 7°C.

Analysis

1. The volume decreased. Students will have a variety of reasons. Help them realize that submicroscopic interactions may result in the two substances being drawn closer together.
2. Heat was given off. This observation will prepare students for the introduction of exothermic processes later in the chapter.

✓ Assessment

Oral: Pose this question for students. If attractive forces between two substances in a mixture result in a smaller volume of the mixture than predicted, what would happen if two substances in a mixture repel each other? *The volume of the mixture will be greater than the sum of the individual volumes of the substances.* L1

miniLAB

Paper Chromatography of Inks

Purpose: To use chromatography to analyze the inks from several pens.

Process Skills: Observing, inferring, using space/time relationships

Teaching Strategies

• Washable marker pens of the type used for overhead projectors work well.

• With different lab groups, you could try different types of coffee filters, filter papers, or white paper towels.

Expected Results

Most water-soluble inks will separate into two or more pigments. Black and brown separate into several colors.

Analysis

1. Yes, the water and the ink components moved up the paper.

2. Yes, the marker colors separated as they moved up the paper.

3. The black marker separated into purple or blue, green, and yellow. The blue marker separated into purple and blue. The green marker separated into yellow and blue-green. The red marker separated into magenta and orange-red.

✓ Assessment

Performance: Have students perform the experiment with a dark food coloring, let the paper dry, and cut the different-colored regions apart. Ask them to place each colored piece of paper in a separate cup with a small amount of water to redissolve the dye. Then have them mix the dyes to reproduce the original color. L1

---

miniLAB 2

Paper Chromatography of Inks

The inks in marking pens are often mixtures of dyes of several basic colors. In this MiniLab, you will use the technique of chromatography to analyze the ink from several pens.

Procedure

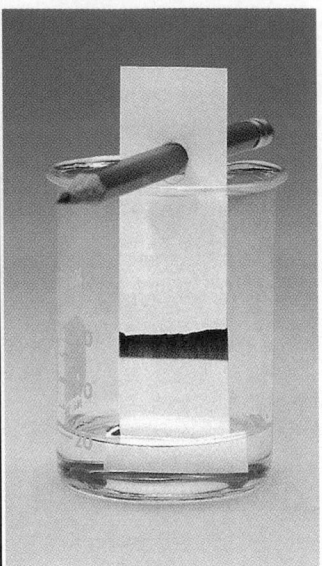

1. Wear an apron and goggles.

2. Obtain a clear plastic cup that is at least 6 cm high. From a coffee filter, cut a strip of paper about 2.5 cm wide and about 2.5 cm longer than the height of the cup.

3. Place the paper strip in the cup so that the bottom of the strip just rests on the bottom of the cup.

4. Push a pencil through the top of the paper in such a way that when the pencil rests on the top of the cup, the paper is suspended with its lower edge just touching the bottom.

5. Prepare several strips in the same way, one strip for each type of marker ink you will analyze.

6. Using one water-soluble ink marker for each strip, draw a narrow, horizontal line across the strip about 2 cm up from the bottom. If possible, include one black or brown marker.

7. Add about 1 cm of water to the cup and suspend the first strip in the water. The marker line must be above the water level when the strip is suspended in the cup.

8. Cover the top of the cup loosely with clear plastic wrap to reduce evaporation. Observe and record the effect on the marker line as the water moves up the paper.

9. Remove the strip from the cup or beaker when the water level has risen to just below the pencil. Lay the strip on a paper towel to dry.

10. Repeat procedures 6 through 9 with each of your marker strips.

Analysis

1. Capillary action is the movement of a liquid upward through small pores that exist in some materials. Did you note any evidence of capillary action in this MiniLab?

2. Does your evidence indicate that any of the marker inks were composed of more than one pigment?

3. Which colors contained the greatest number of pigments?

The separation of sand and sugar took advantage of a difference in the physical properties of the two substances. Sugar dissolves in water but sand does not. When sugar dissolves in water, the two pure substances, sugar and water, combine physically to form a mixture that has a constant composition throughout. This means that no matter where you sample the mixture, you find the same combination of sugar and water. Even with a powerful light microscope, you could not pick out a bit of pure sugar or a drop of pure water. This type of mixture is called a homogeneous mixture. The prefix *homo* means "the same." Homogeneous mixtures are the same throughout. Another name for a homogeneous mixture is **solution.** Even

Table 1.1 Some Common Alloys		
Name of Alloy	**Percent Composition by Mass**	**Uses**
Stainless steel	73-79% iron (Fe) 14-18% chromium (Cr) 7-9% nickel (Ni)	Kitchen utensils, knives, corrosion-resistant applications
Bronze	70-95% copper (Cu) 1-25% zinc (Zn) 1-18% tin (Sn)	Statues, castings
Brass	50-80% copper (Cu) 20-50% zinc (Zn)	Plating, ornaments
Sterling silver	92.5% silver (Ag) 7.5% copper (Cu)	Jewelry, tableware
14-karat gold	58% gold (Au) 14-28% silver (Ag) 14-28% copper (Cu)	Jewelry
18-karat white gold	75% gold (Au) 12.5% silver (Ag) 12.5% copper (Cu)	Jewelry
Solder (electronic)	63% tin (Sn) 37% lead (Pb)	Electrical connections

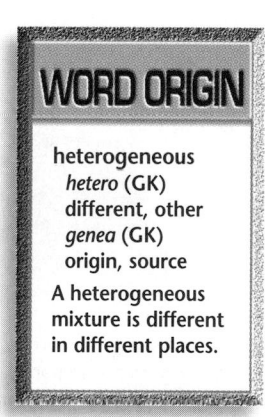

White gold alloy

though solutions may appear to be one pure substance, their compositions can vary. For example, you could make your tea very sweet by dissolving a lot of sugar in it or less sweet by dissolving only a little bit of sugar.

When you hear the word *solution,* something dissolved in water probably comes to mind. But liquid solutions do not have to contain water. Gasoline is a liquid solution of several substances, but it contains no water. Some solutions are gases. Air, for example, is a homogeneous mixture of several gases. Some solutions are solid. **Alloys** are solid solutions that contain different metals and sometimes nonmetallic substances. *Steel,* for example, is a general term for a range of homogeneous mixtures of iron and substances such as carbon, chromium, manganese, nickel, and molybdenum. Because pure gold is soft and bends easily, most gold jewelry is not made from pure gold, but rather from an alloy of gold with silver and copper. **Table 1.1** shows some common alloys and their compositions. The knife shown is stainless steel. The tableware is sterling silver, and the jewelry is white gold.

When you dissolve sugar in water, sugar is the **solute**—the substance being dissolved. The substance that dissolves the solute, in this case water, is the **solvent.** When the solvent is water, the solution is called an **aqueous solution.** Many of the solutions you encounter are aqueous solutions, for example, soda, tea, contact-lens cleaner, and other clear cleaning liquids. In addition, most of the processes of life occur in aqueous solutions.

WORD ORIGIN

heterogeneous
hetero (GK)
different, other
genea (GK)
origin, source
A heterogeneous mixture is different in different places.

Correcting Errors

Point out to students that they shouldn't casually use terms such as *matter, mixture, suspension,* and *substance* in talking or writing about chemical phenomena because these terms have specific meanings. Explain that chemistry has its own specialized vocabulary.

GLENCOE TECHNOLOGY

 CD-Rom

Chemistry: Concepts and Applications
Separate the Substances
Have students use this Interactive Experiment to practice separating mixtures using differences in physical properties. L1 LEP

Chemistry Journal

Alloys and Their Uses
Ask students to make a table in their journals with a column for each of the alloys listed in **Table 1.1.** Have them look around their homes and write, in each of the columns, the objects they find made of that alloy and the way that the object is used. Have students keep their tables open-ended and add to them whenever a new alloy is observed. L1

Correcting Misconceptions

Point out to students that, although the word *element* means "basic part or feature," elements are composed of even more fundamental particles—electrons, proton, and neutrons—and that they will learn more about them in later chapters.

Figure 1.15

Classifying Matter
The chart shows a way of classifying matter. Notice that mixtures can be either heterogeneous or homogeneous and can often be separated into pure substances by physical changes. You can also see that pure matter can be elements or compounds. You'll learn more about these classes of matter in the next section.

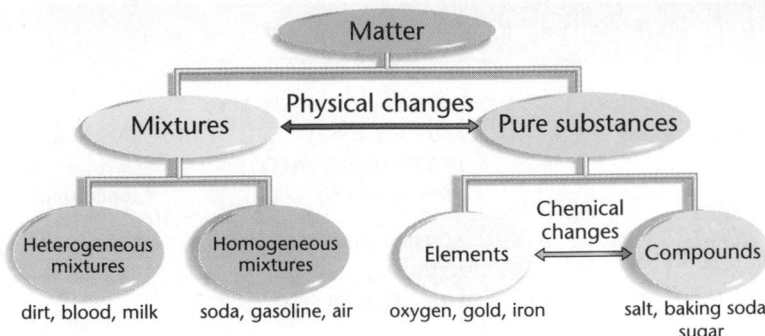

The chemistry of the real world is mostly the chemistry of mixtures. Dig a hole, buy something at a grocery store, pick an apple from a tree, or take a deep breath. The stuff you dig, buy, pick, or inhale is a mixture. However, the behavior of mixtures is based on the composition, structure, and behavior of the pure substances that compose them. **Figure 1.15** summarizes the classification of matter from a chemical point of view.

Substances: Pure Matter

All matter is composed of substances. Although a pizza and a can of paint have different compositions, both are made of some combination of pure substances, and pure substances are chemicals. You, too, are made of chemicals. This book, your desk, your clothes, the air you breathe, the water you drink, the food you eat, and all the other stuff of the universe are all made of chemicals.

Elements: The Building Blocks

If you classify an unknown piece of matter as pure, it means that the stuff is made up of only one substance. But there are two types of substances. One type of pure substance can be broken down into simpler substances. This type of substance is called a compound. Another type of substance cannot be broken down into simpler substances. Such a substance is called an **element.** Elements are the simplest form of matter. They are the building blocks from which other forms of matter are made. All the substances of the universe are either elements, compounds formed from elements, or mixtures of elements and compounds. **Figure 1.16** shows some well-known elements.

WORD ORIGIN

homogeneous
  homo (GK) same, alike
  genea (GK) origin, source
A homogeneous mixture is the same throughout.

Diamonds are one form of pure carbon. ▶

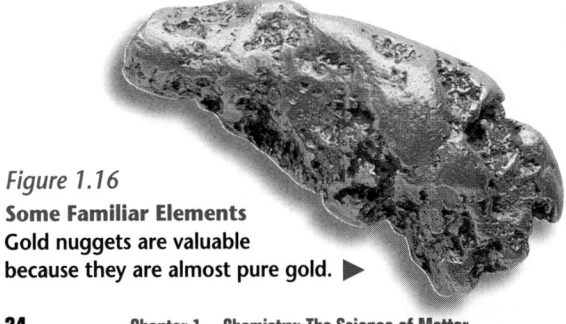

Figure 1.16

Some Familiar Elements
Gold nuggets are valuable because they are almost pure gold. ▶

Across the Curriculum

History

The major periods of human history are described by chemical substances that played important roles in the development of civilization. The earliest period of human civilization is known as the Stone Age. This was followed by the Bronze Age (bronze is an alloy of copper and tin), and then the Iron Age. In more recent history, the period of industrial revolution that started in the early 1800s is sometimes referred to as the Steam Age, and many refer to the 20th century as the Plastic Age.

Copper to Gold: The Alchemists' Dream

An alchemist was a combination of magician and metallurgist who tried unsuccessfully to convert common metals to gold. The craft flourished from ancient times until the 18th century. Alchemists were not early chemists, as some people believe, but their practical knowledge about elements and compounds contributed to the work of the earliest true chemists. Like the alchemists, you will not turn copper into gold, but by allowing the copper in a penny to react with zinc under certain conditions, you may create an interesting alloy of the two metals.

Procedure

1. Wear an apron and goggles.

2. Clean a pre-1982 penny with steel wool or a pencil eraser.

3. Place 1 g of granular zinc in an evaporating dish. Add 20 mL of $1M$ zinc chloride solution ($ZnCl_2$). Use tongs to place the penny in the dish, and put the evaporating dish on a hot plate.

4. Heat the mixture until it just starts to boil. This should take about two minutes. Carefully stir the mixture with the tongs and turn the penny. Continue to heat and stir gently until the penny becomes covered with zinc and appears gray in color. This usually takes less than a minute.

5. Use the tongs to remove the penny from the liquid. Rinse the penny in a beaker of cold tap water, then pat it dry with a paper towel.

6. Using tongs to hold the penny, gently heat it in the cooler, outer portion of a Bunsen burner flame until it changes color. Record your observations.

7. Continue heating gently for two or three seconds longer, then immediately immerse the penny in a fresh beaker of cold water.

8. After the penny has cooled for about a minute, remove it from the water and pat it dry. Record your final observations.

Analysis

1. Does the evidence indicate that you created an alloy of copper and zinc? Explain.

2. What is the probable identity of this alloy?

3. What do you think you would see if you cut the penny in two and examined the cut edge with a powerful microscope?

Mercury is the only metallic element that is a liquid at room temperature.

The element aluminum fills many needs in the modern world. ▶

interNET
CONNECTION
Information on environmental chemistry regulations and applications is available on-line from the EPA.
World Wide Web
http://www.epa.gov

---

miniLAB

Copper to Gold: The Alchemists' Dream

Purpose: To create and examine an alloy of copper and zinc.

Process Skills: Observing, inferring, communicating

Safety Precautions: Have students wear safety goggles and aprons. Use only granular zinc, not powdered zinc.

Teaching Strategies

• Hot plates should be used in the fume hood, if available; more than one evaporating dish may be heated on each hot plate.

• Don't tell students they have formed brass until the post-laboratory discussion.

Expected Results

Pennies will become gray when coated with zinc and will turn "gold" when heated gently.

Analysis

1. Yes, the distinctive color change seems to indicate the creation of an alloy.
2. brass
3. An inner layer of copper would be sandwiched inside the outer layer of brass.

✓ Assessment

Knowledge: Have students write a summary of their observations in the MiniLab and include lists of the properties of copper, zinc, and brass that they observed. **L1**

Purpose

To show how knowledge of scientific data can be used to write fiction.

Teaching Strategies

- Two related references are: B. Priess and W. Alschuler. *The Microverse.* New York: Bantam Books, 1989. This book is comprised of scientific essays followed by science fiction stories based on them. Everett, E. "Do the Write Thing." *Science Teacher,* Oct. 1994, pp. 35-37.

- Have students hold a relatively heavy, irregularly shaped object, such as a broom or a shovel, in a vertical position to experience what is meant by *center of gravity.*

Extension

After students have answered question 1, read them selected passages from Chapters 13 and 15 in Part 2 of *Twenty Thousand Leagues Under the Sea.* Using present iceberg data, decide whether Verne's descriptions of icebergs fit those in the northern or the southern hemisphere.

Connecting to Chemistry

1. North: pinnacle and dome shape; small in size, up to 500 feet; $^6/_7$ of volume is submerged
   South: flat, tabular; large in size, up to 200 miles long; $^4/_5$ of volume is submerged

2. In fresh water, less of the mass of an iceberg would be above water because the density of the iceberg is only slightly less than the density of fresh water.

Jules Verne and His Icebergs

You are underwater in a submarine in an Antarctic iceberg field. While sleeping in your bunk, you are awakened by a violent shock and thrown into the middle of your cabin. Surveying the situation, you realize that the submarine is on its side. This is what happens to P. Aronnax when he is aboard Captain Nemo's *Nautilus* in Jules Verne's science fiction novel *Twenty Thousand Leagues Under the Sea.*

How does Jules Verne explain the accident? In Verne's book, published in 1869, Captain Nemo explains that a mountain of floating ice, an iceberg, has turned over. When an iceberg is undermined at its base by warmer water or repeated shocks, its center of gravity rises, and the whole thing turns over. As the bottom of the iceberg turned upward, it struck the *Nautilus,* slid under its hull, and raised it onto an ice bed, where it lay on its side.

What concepts are involved? Density and center of gravity account for the accident. The iceberg was floating because the frozen fresh water (density 0.9 g/mL) composing it was less dense than the salt water (density 1.0 g/mL) in which it was floating. An iceberg floats in seawater because its density is less than that of the seawater. However, its density isn't a lot less than that of seawater

so it floats with only about 15 percent of its mass above the water's surface.

The center of gravity of a mass is the point at which all the weight of an object seems to be located. The higher an object's center of gravity is above its support, the less stable the object is. If something happens to the bottom of an iceberg such that the center of gravity is shifted above the waterline, the iceberg will turn over.

An iceberg's center of gravity changes when large pieces of ice break off the berg. This can be caused by water freezing in cracks, or the vibration of the waves, or thunderous vibrations resulting from the breaking of other bergs. When an iceberg becomes unstable, small movements can cause it to topple over.

Connecting to Chemistry

1. **Acquiring Information** Find out what differences exist between icebergs in the Northern Hemisphere and in the Southern Hemisphere.

2. **Applying** If an iceberg were floating in fresh water instead of seawater, would more or less of its mass be above the water's surface? Explain.

Millions of substances are known to chemists, but only 111 are elements. These 111 elements combine with each other to form all the millions of known compounds. That's why the chemical elements are often referred to as the building blocks of matter.

Of the 111 known elements, only about 90 occur naturally on Earth. The remainder are synthesized, usually in barely detectable amounts, in high-energy nuclear experiments. Less than half of the 90 naturally occurring elements are abundant enough to play a significant role in the chemistry of everyday stuff.

The fact that most stuff is composed of a relatively small number of building blocks simplifies the puzzle of matter. On the other hand, the observation that such a small number of pieces creates such a variety of compounds means that elements must connect to each other in countless ways.

Organizing the Elements

Your classroom may have a large chart labeled *The Periodic Table of the Elements* hanging on the wall. This table will become the tool you will use most in your chemistry course. A similar table is printed on pages 92-93 of this book, as well as inside the back cover. The periodic table organizes elements in a way that provides a wealth of chemical information—much more than is evident to you now. **Figure 1.17** on the following two pages is a pictorial version of the periodic table. It shows the chemical symbols for the elements and photos of samples of many of the naturally occurring elements.

The symbols of the elements are part of the language of chemistry. As **Figure 1.18** indicates, the chemical symbols are a universal shorthand that is used to make chemistry communication understandable around the world. Just as it is easier and quicker for you to write *USA* instead of *United States of America,* it is easier to write *Al* instead of the word *aluminum.* As you can see, the symbol for aluminum is taken directly from the element's name, but some elements have symbols that don't correspond to their English names. Their symbols usually correspond to their names in Latin. Some examples are shown in **Table 1.2**.

Figure 1.18
Chemical Symbols Are Universal
These chemical research articles—written in journals of different countries—show that the symbols for the chemical elements are a universal language through which chemists can communicate worldwide.

Table 1.2 Some Historic Chemical Symbols

Element	Symbol	Origin	Language
Antimony	Sb	stibium	Latin
Copper	Cu	cuprum	Latin
Gold	Au	aurum	Latin
Iron	Fe	ferrum	Latin
Lead	Pb	plumbum	Latin
Potassium	K	kalium	Latin
Silver	Ag	argentum	Latin
Sodium	Na	natrium	Latin
Tin	Sn	stannum	Latin
Tungsten	W	wolfram	German

Pewter—an alloy of tin and antimony

Reinforcement

Ask students to work in small groups to create 20 × 20 letter blocks containing as many names of the chemical elements as possible, either left to right, right to left, top to bottom, bottom to top, or diagonally. Ask them to circle and list the elements with their symbols. Award a prize to the group that achieves the greatest number of element names in their block. You may want to duplicate the best of the blocks for all students to try. L1 COOP LEARN

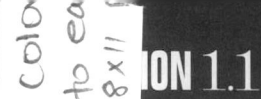

GLENCOE TECHNOLOGY

 CD-Rom

Chemistry: Concepts and Applications
The Periodic Table
Have students use this inter-active periodic table to help reinforce their knowledge of the elements in the periodic table. L1 LEP

Figure 1.17

Pictorial Periodic Table
The periodic table shown here was designed to illustrate samples of the elements. It differs somewhat from the most up-to-date table that you will find on pages 92-93. Be sure to use this latter table for reference throughout your chemistry course.

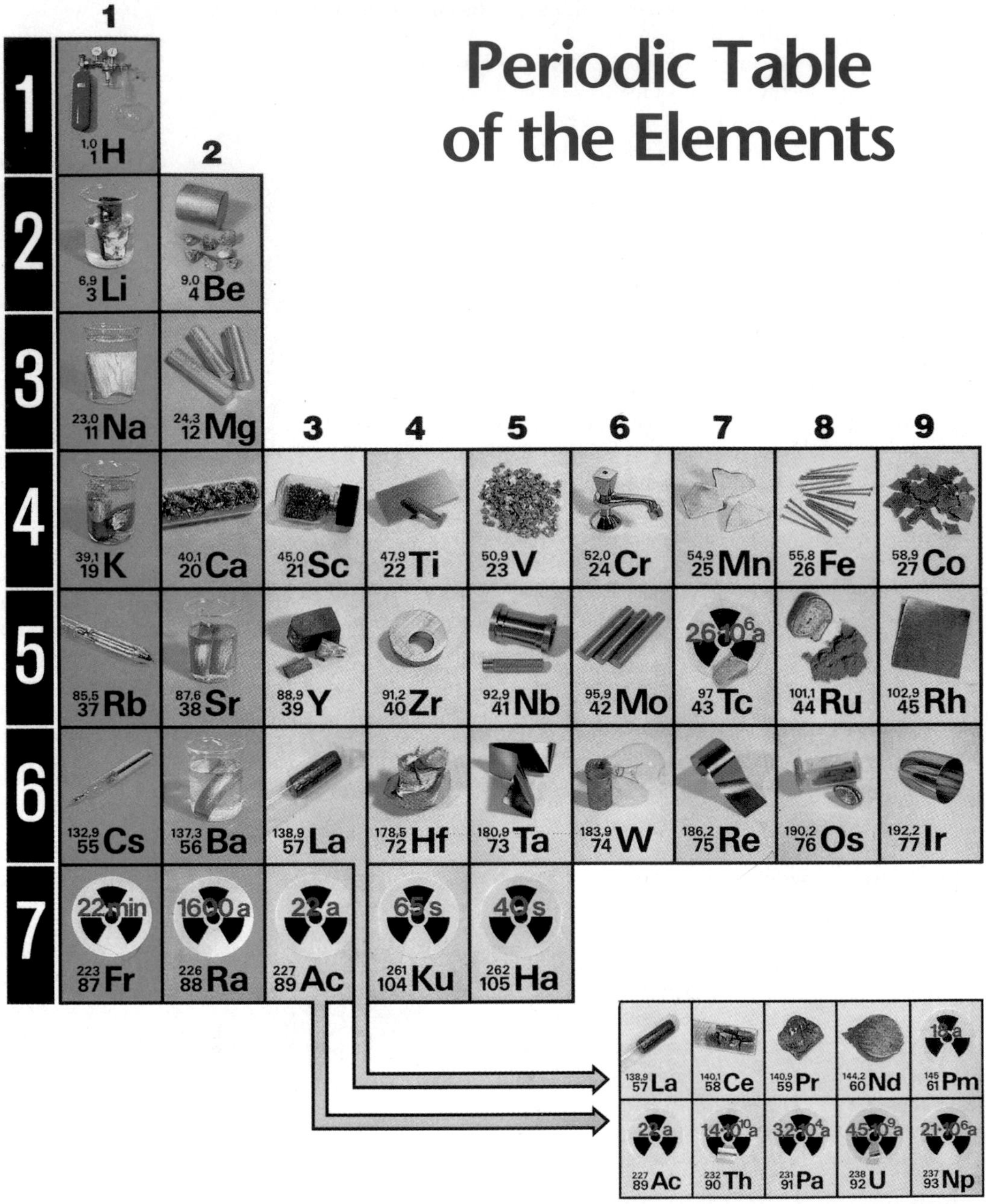

Periodic Table of the Elements

1	2	3	4	5	6	7	8	9
1 **1.0 ¹H**								
2 **6.9 ₃Li**	**9.0 ₄Be**							
3 **23.0 ₁₁Na**	**24.3 ₁₂Mg**							
4 **39.1 ₁₉K**	**40.1 ₂₀Ca**	**45.0 ₂₁Sc**	**47.9 ₂₂Ti**	**50.9 ₂₃V**	**52.0 ₂₄Cr**	**54.9 ₂₅Mn**	**55.8 ₂₆Fe**	**58.9 ₂₇Co**
5 **85.5 ₃₇Rb**	**87.6 ₃₈Sr**	**88.9 ₃₉Y**	**91.2 ₄₀Zr**	**92.9 ₄₁Nb**	**95.9 ₄₂Mo**	**97 ₄₃Tc**	**101.1 ₄₄Ru**	**102.9 ₄₅Rh**
6 **132.9 ₅₅Cs**	**137.3 ₅₆Ba**	**138.9 ₅₇La**	**178.5 ₇₂Hf**	**180.9 ₇₃Ta**	**183.9 ₇₄W**	**186.2 ₇₅Re**	**190.2 ₇₆Os**	**192.2 ₇₇Ir**
7 **223 ₈₇Fr**	**226 ₈₈Ra**	**227 ₈₉Ac**	**261 ₁₀₄Ku**	**262 ₁₀₅Ha**				

138.9 ₅₇La	**140.1 ₅₈Ce**	**140.9 ₅₉Pr**	**144.2 ₆₀Nd**	**145 ₆₁Pm**
227 ₈₉Ac	**232 ₉₀Th**	**231 ₉₁Pa**	**238 ₉₂U**	**237 ₉₃Np**

Demonstration 1.3

Light a Candle

Purpose: To generate discussion about how an element differs from a compound. This demo is useful if there is not time for the ChemLab *Observation of a Candle*.

Materials: For each team of four students: small candle, matches, white ceramic plate, 600-mL beaker, ice cube

Safety Precautions:

Disposal: Code F

Procedure: Have students work together to perform the following procedures. They should make and record all observations. Light a candle with a match. **CAUTION:** *Be aware of the risk of fire.* Place a white ceramic plate in the candle flame for a few seconds. Remove and observe the under-side of the plate. Invert the beaker over the flame and place an ice cube on top of the beaker.

Results: The solid candle melts and the liquid is drawn into the wick where it is converted into a vapor that burns in air. A black deposit appears on the white ceramic plate. The candle is slowly extinguished when covered by a beaker. Condensed water droplets appear on the inside of the beaker.

Analysis: Ask these questions.

The periodic table (partial) showing groups 10 through 18:

10	11	12	13	14	15	16	17	18
								$^{4.0}_{2}$He
			$^{10.8}_{5}$B	$^{12.0}_{6}$C	$^{14.0}_{7}$N	$^{16.0}_{8}$O	$^{19.0}_{9}$F	$^{20.2}_{10}$Ne
			$^{27.0}_{13}$Al	$^{28.1}_{14}$Si	$^{31.0}_{15}$P	$^{32.1}_{16}$S	$^{35.5}_{17}$Cl	$^{39.9}_{18}$Ar
$^{58.7}_{28}$Ni	$^{63.5}_{29}$Cu	$^{65.4}_{30}$Zn	$^{69.7}_{31}$Ga	$^{72.6}_{32}$Ge	$^{74.9}_{33}$As	$^{79.0}_{34}$Se	$^{79.9}_{35}$Br	$^{83.8}_{36}$Kr
$^{106.4}_{46}$Pd	$^{107.9}_{47}$Ag	$^{112.4}_{48}$Cd	$^{114.8}_{49}$In	$^{118.7}_{50}$Sn	$^{121.8}_{51}$Sb	$^{127.6}_{52}$Te	$^{126.9}_{53}$I	$^{131.3}_{54}$Xe
$^{195.1}_{78}$Pt	$^{197.0}_{79}$Au	$^{200.6}_{80}$Hg	$^{204.4}_{81}$Tl	$^{207.2}_{82}$Pb	$^{209.0}_{83}$Bi	$^{209}_{84}$Po	$^{210}_{85}$At	$^{222}_{86}$Rn

$^{150.4}_{62}$Sm	$^{152.0}_{63}$Eu	$^{157.3}_{64}$Gd	$^{158.9}_{65}$Tb	$^{162.5}_{66}$Dy	$^{164.9}_{67}$Ho	$^{167.3}_{68}$Er	$^{168.9}_{69}$Tm	$^{173.0}_{70}$Yb	$^{175.0}_{71}$Lu
$^{244}_{94}$Pu	$^{243}_{95}$Am	$^{247}_{96}$Cm	$^{247}_{97}$Bk	$^{251}_{98}$Cf	$^{254}_{99}$Es	$^{257}_{100}$Fm	$^{258}_{101}$Md	$^{259}_{102}$No	$^{260}_{103}$Lr

Quick Demo

Where's the iron and sulfur?

Disposal: Code A
Thoroughly mix 7.0 g of powdered reagent-grade iron with 4.0 g of powdered sulfur. Demonstrate that the iron in the mixture is attracted to a magnet through a piece of paper. Put the mixture in a large Pyrex test tube. **CAUTION:** *Wear safety goggles and perform the reaction in a fume hood or a well-ventilated room.* Heat the bottom of the test tube strongly until the reaction begins. (Be aware that the sulfur could ignite.) When the reaction is complete, break the test tube by putting it in cold water. Retrieve the product, iron(II) sulfide, and demonstrate that it is no longer magnetic. Explain that the two elements have formed a compound and that the properties of the compound are different from the properties of sulfur and iron. Ask students what differences they see in properties. *Iron sulfide is not magnetic; it is not yellow like sulfur; it does not resemble iron.* L1

---

1. Is paraffin an element or a compound? *compound*
2. Is paraffin homogeneous or heterogeneous? *homogeneous*
3. Which element do you believe is deposited on the ceramic plate? *carbon*
4. If the condensed droplets inside the beaker are water, which three elements are present in paraffin? *hydrogen, oxygen, and carbon*

✔ Assessment

Performance: Have teams of students invert different-sized beakers over burning candles, and relate the time it takes for the candle to be extinguished to the volume of air in the beaker. *The larger the volume of air, the longer the candle burns.* Have students test the droplets inside the beaker with cobalt chloride test paper. Tell them that when the blue paper turns pink, water is present. *The test paper turns pink, so the droplets are water.* L1
COOP LEARN

miniLAB

Waiter, what's this stuff doing in my cereal?

Purpose: To test a common cereal for the presence of iron.
Process Skills: Observing, interpreting data

Teaching Strategies

- Do not mention in the pre-laboratory discussion that students will be testing for iron.
- If you have a sufficient number of magnetic stirrers and stirring bars, these may be used to stir the cereal and water mixtures for about 10 minutes.
- Cow magnets, obtainable at farm-supply stores, may be used.
- Ask students to try several brands of cereals, particularly fortified ones, and to share their data with the class.

Expected Results

Small particles of iron will be found on the pencil-magnet or magnetic stirrer.

Analysis

1. iron
2. Iron is an essential element in our diets. Hemoglobin, the compound in red blood cells that enables them to carry oxygen, contains a substantial amount of iron.

✓ Assessment

Performance: Ask students to make comprehensive lists of vitamin and mineral additives, carbohydrates, fats, and calories contained in the various brands of tested cereals. Have the lists posted so that teams of students can study them and reach a conclusion about the healthiest brand. L1 COOP LEARN

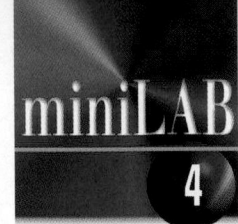

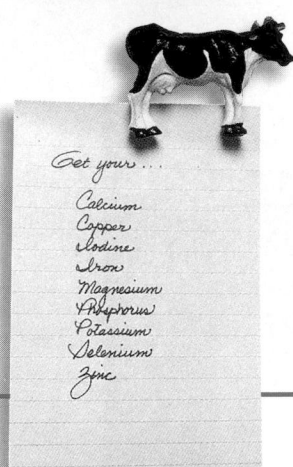

miniLAB 4

Waiter, what's this stuff doing in my cereal?

Breakfast cereals are often fortified with elements and compounds to increase their nutritional value. In this MiniLab, you will test a common cereal for the presence of one of these additives.

Procedure

1. Tape a small, strong magnet to a pencil at the eraser end.
2. Place a sample of dry, fortified, cold cereal in a plastic bag.
3. Thoroughly crush the cereal with a rolling pin or other heavy object.
4. Pour the crushed cereal into a beaker and cover it with water.
5. Stir the cereal/water mixture for about ten minutes with your pencil-magnet stirrer. Stir slowly and easily for the last minute.
6. Remove the magnet from the cereal and examine it carefully. Record your observations.

Analysis

1. The substance attracted to your magnet is a common element. What is it?
2. Why do you think that this element is added to the cereal?

Compounds Are More Than One Element

You've learned that a compound is a pure substance that can be broken down into elements. A more complete definition is that a **compound** is a chemical combination of two or more different elements joined together in a fixed proportion. For example, if you were to collect and analyze samples of the compound water from a faucet, an iceberg, a river, and a rain puddle, you would find that every sample (barring impurities) is 11.2 percent hydrogen and 88.8 percent oxygen by mass. Every compound has its own fixed composition, and that composition results in a unique set of chemical and physical properties. The properties of the compound are different from the properties of the elements that compose the compound. You can see this in **Figure 1.19.**

Figure 1.19

Two Elements Combine to Form a New Substance
The elements that make up a compound are chemically combined to form a new substance with a unique set of properties. The element silver is a solid metal. The element bromine is a poisonous red liquid. Silver bromide, a compound of silver and bromine, is a crystalline solid that is used in photographic and print paper. Silver bromide has a unique set of physical and chemical properties and a fixed composition of 57.45 percent silver and 42.55 percent bromine.

Silver

Bromine

More than 10 million compounds are known and the number keeps growing. Some common compounds are listed in **Table 1.3** on page 33. New compounds are discovered and isolated from natural chemical sources such as plants and colonies of bacteria. Compounds are also synthesized in laboratories where they are tested for a variety of uses ranging from medicine to manufacturing.

Because the supply of useful chemicals from natural sources is often limited, chemists work to synthesize these compounds in the laboratory. The effort to synthesize taxol, an anticancer compound found in the bark of the Pacific yew tree, is an example of how nature often provides the lead in compound synthesis. If taxol can be made in the laboratory, then chemical engineers will try to find a way to produce it on an industrial scale.

Formulas of Compounds

The second column in **Table 1.3** gives the chemical formulas for the compounds listed. A **formula** is a combination of the chemical symbols that show what elements make up a compound and the number of atoms of each element. For example, the chemical formula of the compound sucrose is $C_{12}H_{22}O_{11}$. The formula tells you, in a compact way, that sucrose contains carbon, hydrogen, and oxygen. It also tells you that the smallest unit of sucrose, a molecule, contains 12 carbon atoms, 22 hydrogen atoms, and 11 oxygen atoms. Formulas provide a shorthand way of describing a submicroscopic view of a compound. You probably already use formulas like H_2O and CO_2 as a way of talking about water and carbon dioxide.

In later chapters, you'll learn more about compounds and why elements combine to form compounds. You'll also learn how to determine the formulas for many compounds.

Silver bromide

1.1 The Puzzle of Matter   **31**

Discussion

 If students have difficulty understanding how more than 10 million compounds can be formed from 90 elements, ask them how many two-element compounds (containing one atom of each element) could be formed from just 20 elements. *190* Three-element combinations? *1140* Four-element combinations? *4845* L2

GLENCOE TECHNOLOGY

 Videodisc

Chemistry: Concepts and Applications
Physical and Chemical Properties of Matter
Disc 1, Side 1, Ch. 4

Show this video to help you explain the difference between physical and chemical properties. L1 **LEP**

 CD-Rom

Chemistry: Concepts and Applications
Physical and Chemical Properties of Matter
Have students use this video to help reinforce the difference between physical and chemical properties of matter.
L1 **LEP**

Transparency

Display **Basic Concepts Transparency 2** to point out the difference between elements and compounds. L1 **LEP**

Meeting Individual Needs

Gifted Ask a gifted student to work out the number of compounds containing ten elements (one atom of each element) that could be formed from 20 elements. *20! ÷ (10!10!) = 184 756*
L3

31

Chemistry and
SOCIETY

Purpose

To help students think about which drugs are best—natural herbal, natural purified, or synthetic.

Background

- The pain-killing drug in willow bark is salicin.
- When an aspirin bottle smells like vinegar, it should be thrown away because the aspirin is decomposing into salicylic acid and acetic acid.
- Scientists are not sure whether the yew got the taxol gene from the fungus or vice versa. When the taxol gene is located, another method of production might be possible. Through genetic engineering, the taxol gene might be spliced into a bacterial genome, and the genetically engineered bacteria would make taxol.

Teaching Strategies

- Make sure students know the difference between natural and synthetic chemicals.
- To initiate discussion, ask students whether they would rather take natural or synthetic vitamins and why.
- Ask students to research and prepare a written or oral report about the natural sources of curare, quinine, and heparin. L1

Analyzing the Issue

1. The bark from four yew trees is needed to treat one person, but recent forestry research indicates that the Pacific yew tree is more abundant than was once believed.

2. Both salicylic acid and aspirin have a benzene ring and a carboxyl group, but aspirin has an acetate group (CH_3COO) in place of an OH group.

Natural Versus Synthetic Chemicals

If you look carefully at the vitamin display in a drugstore, you'll see some bottles marked *all natural*. Most of these vitamins have been extracted from plant or animal sources. Are natural vitamins, drugs, and other substances better than the purified or synthetic ones produced by drug companies? Maybe this will help you decide.

Aspirin: A common synthetic drug To get rid of a headache, would you rather drink a cup of willow bark tea or take two aspirin? The active ingredients in both remedies have similar chemical structures, and both are effective against headaches. However, the chemical from the willow bark, salicylic acid, has several harmful side effects including stomach pain. So does aspirin, but it is a more effective painkiller and can be taken in lower doses. Also, willow bark contains many other chemicals. After years of research, scientists made aspirin in the laboratory from salicylic acid and acetic anhydride. It contains only one active ingredient, acetylsalicylic acid. Not only is aspirin a pure substance, but also the difference in its structure eliminates the serious stomach pain caused by salicylic acid.

Development of new drugs What happens when a chemical discovered in nature is shown to be a potential treatment or cure for a disease? Scientists use the following procedures to make safe, effective drugs: (1) isolate and purify the drug, (2) determine its composition and structure, (3) search for a way to make it synthetically, (4) look for a cheap, easy way to produce it in large quantities, and (5) try changing the structure and composition of the original compound to improve on nature's model.

Taxol: A new cancer drug Scientists discovered that taxol, a chemical found in the bark of the Pacific yew tree, reduced the size of ovarian and breast cancer tumors in 30 percent of otherwise-untreatable patients. But scientists were concerned that demand for the drug might wipe out the population of yew trees, so they started to look for new sources of the drug. Chemists Andrea and Donald Stierle found a taxol-producing fungus growing on one yew. Other scientists discovered that needles from the European yew contain a chemical similar to taxol. Chemists experimented to figure out taxol's structure. In 1994, they succeeded in producing pure taxol in the lab. You might be wondering which taxol is better—the purified natural one or the synthetic one. Actually, the chemical structure and purity of both is identical. However, being able to make taxol synthetically is a real advantage; drug companies may be able to produce it more cheaply, and they can work on modifying its structure to make it more effective.

Trunk of Pacific yew

Analyzing the Issue

1. **Acquiring Information** Find out why scientists were concerned that the use of taxol could endanger the Pacific yew and whether this is still a concern.

2. **Modeling** In what ways are the structures of salicylic acid and acetylsalicylic acid (aspirin) alike? In what ways are they different?

3. **Debating** Debate the pros and cons of using natural herbal drugs, purified natural drugs, and synthetic drugs.

3. A debate should cover these points. The natural herbal form includes many other chemicals besides the medicinal one. The purified natural form is a single chemical so experiments with it are reliable. Synthetic and purified forms are identical. Modified forms are often superior to natural forms.

Table 1.3 Some Common Compounds

Compound Name	Formula	Uses
Acetaminophen	$C_8H_9NO_2$	pain reliever
Acetic acid	$C_2H_4O_2$	tart ingredient in vinegar
Ammonia	NH_3	fertilizer, household cleaner when dissolved in water
Ascorbic acid	$C_6H_8O_6$	vitamin C
Aspartame	$C_{14}H_{18}N_2O_5$	artificial sweetener
Aspirin	$C_9H_8O_4$	pain reliever
Baking soda	$NaHCO_3$	cooking
Butane	C_4H_{10}	lighter fuel
Caffeine	$C_8H_{10}N_4O_2$	stimulant in coffee, tea, some soda
Calcium carbonate	$CaCO_3$	antacid
Carbon dioxide	CO_2	carbonating agent in soda
Ethanol	C_2H_6O	disinfectant, alcoholic beverages
Ethylene glycol	$C_2H_6O_2$	antifreeze
Hydrochloric acid	HCl	called muriatic acid, cleans mortar from brick
Magnesium hydroxide	$Mg(OH)_2$	antacid
Methane	CH_4	natural gas, fuel
Phosphoric acid	H_3PO_4	flavoring in soda
Potassium tartrate	$K_2C_4H_4O_6$	cream of tartar, cooking
Propane	C_3H_8	fuel for cooking
Salt	NaCl	flavoring
Sodium carbonate	Na_2CO_3	washing soda
Sodium hydroxide	NaOH	drain cleaner
Sucrose	$C_{12}H_{22}O_{11}$	sweetener
Sulfuric acid	H_2SO_4	battery acid
Water	H_2O	washing, cooking, cleaning

SECTION REVIEW

Understanding Concepts

1. What three characteristics of matter does chemistry deal with?

2. List two ways in which a mixture differs from a pure substance.

3. How does a compound differ from a mixture?

Thinking Critically

4. **Applying Concepts** The element oxygen is a gas that makes up about 20 percent of Earth's atmosphere. Oxygen also is the most abundant element in Earth's crust, yet Earth's crust is not a gas. Explain this apparent conflict.

Applying Chemistry

5. **Everyday Materials** Matter may be subdivided into elements, compounds, and heterogeneous and homogeneous mixtures. Describe one material found in a household that belongs in each category.

SECTION REVIEW

1. composition, structure, behavior
2. A mixture can be separated into two or more pure substances. The composition of a mixture is variable.
3. A compound has a fixed composition; a mixture has a variable composition. The components of a compound are chemically combined; the components of a mixture are physically combined.
4. Oxygen in Earth's crust occurs in compounds that are solids, for example, SiO_2.
5. element: iron; compound: sucrose; heterogeneous mixture: oil-and-vinegar salad dressing, seasoning mixtures; homogeneous mixture (solution): vinegar

---

SECTION 1.1

3 ASSESS

Check for Understanding

- Check for understanding of the section material by having students answer the Section Review questions.
- Several days before finishing Section 1.1, assign one element to each student. Have them look up information about their elements and write as many clues to the identity of the element as possible. Then, for review, have each student give clues, one by one, to another classmate until the identity of the element is determined. L1

Reteach

Cooperative Learning: Have teams of students work together to develop concept maps starting with the word *matter*. Ask each team to include as many as possible of the Key Terms listed on page 4, but all maps should include the terms *substance, mixture, element, compound, solution, solvent, solute, alloy, physical property,* and *physical change.* Emphasize finding connections between as many terms as possible. L1 **COOP LEARN**

Extension

Ask each student to bring an item or material from home and classify it for the class as a mixture, element, or compound. If it is a mixture, ask the student whether it is homogeneous or heterogeneous. L1 **LEP**

4 CLOSE

Writing About Chemistry

Ask students to write a detailed response to the comment, "Natural products are always better than synthetic products." The paper can be placed in the students' portfolios. L1 **P**

Properties and Changes of Matter

SECTION PREVIEW

Objectives
Distinguish between physical and chemical properties.
Contrast chemical and physical changes.
Apply the law of conservation of matter to chemical changes.

Key Terms
volatile
density
chemical property
chemical change
chemical reaction
law of conservation
of mass
energy
exothermic
endothermic

When the refuse truck pulls up at the landfill to empty its load, you might say, "What *is* all that stuff?" You know that it's all the throwaways of modern life—paper, glass, metal, plastic, and more—but the question still remains: "What *is* it?" Chemists want to know what every bit of stuff is. What is it made of (composition)? How are its atoms arranged (structure)? What will the stuff do (behavior)? Any characteristic that can be used to describe or identify a piece of matter is a property of the matter. In fact, each different substance has its own set of properties, just as each person has unique fingerprints. By knowing the *fingerprint* of a substance, you can identify the substance.

Identifying Matter by Its Properties

Physical properties are those that don't involve changes in composition. Many physical properties are qualitative descriptions of matter, such as: *The solution is blue; The solid is hard;* or *The liquid boils at a low temperature.* Other physical properties are quantitative, which means that they can be measured with an instrument. Examples include: *An ice cube melts at 0°C; Iron has a density of 7.86 g/mL;* or *A mass of 35.7 g of sodium chloride dissolves in 100 mL of water.*

States of Matter

All matter exists in one of three physical states: solid, liquid, or gas. A fourth state of matter, called plasma, is less familiar. You will learn about it in Chapter 10. The physical state of a substance depends on its temperature. You know that if you put liquid water into a freezer, it changes to solid water (ice), and if you heat liquid water to 100°C in a teakettle, it boils and becomes gaseous water (steam). But the physical state of a substance usually means its state at room temperature—about 20°C to 25°C. At room temperature, the physical state of water is liquid, salt is a solid, and oxygen is a gas.

Figure 1.20
The Freezing/Melting Temperature of Water
Because freezing and melting occur at the same temperature, both phases of water, solid and liquid, can be present when the temperature is exactly 0°C.

A physical property, closely related to the physical state of a substance, is the temperature at which the substance changes from one state to another. Water, for example, freezes (and melts) at 0°C. Salt (NaCl) melts (and freezes) at a much higher temperature, 804°C, and oxygen (O_2) freezes (and melts) at a much lower temperature, −218°C. The melting point and the freezing point of a substance are the same temperature, as you can see in **Figure 1.20,** which shows both water and ice at 0°C. Whether we use the word *freeze* or *melt* depends on how we usually encounter a substance. Water boils at 100°C, but it also condenses from a gas to a liquid at the same temperature. Therefore, boiling point and condensation point are the same temperature for each substance.

Changes in state are examples of physical changes because there is no change in the identity of the substance. Ice can melt back to liquid water, and steam will condense on a cool surface to liquid water. Some substances are described as **volatile,** which means that they change to a gas easily at room temperature. Alcohol and gasoline are more volatile than water. The substance naphthalene, used as mothballs, is an example of a solid substance that is volatile. You can readily smell alcohol, gasoline, and mothballs when open containers of these substances are present in a room because the liquid or solid has changed to gas and the molecules are present in the air.

Fact of the **MATTER**

When trains carrying chemicals derail or trucks loaded with chemicals overturn on highways, news stories or broadcasts are likely to describe the spilled substances as *volatile,* thinking that the word means "dangerous." Even relatively harmless substances such as sodium carbonate (Na_2CO_3) have been described in this way.

Content Background

Intra- and Interparticle Forces and Properties Forces in chemistry may be classified as *intraparticle* and *interparticle*. Intraparticle forces are the chemical bonds that hold atoms together in molecules. These are responsible for the chemical properties of a substance. Interparticle forces are the attractions or repulsions that individual particles have for each other. Interparticle forces play a major role in determining the physical properties of a substance. For example, a substance that is a solid at room temperature has stronger interparticle attractive forces than a substance that is a liquid at the same temperature. A gas at room temperature has weak interparticle forces. Melting and boiling points provide a more quantitative measure of these forces. The higher the melting point or boiling point of a substance, the stronger the interparticle forces.

Transparency

Display **Basic Concepts Transparency 3** to introduce physical and chemical changes. L1 LEP

1.2   **Properties and Changes of Matter**   **35**

Integrating THE Sciences

Chemistry Chemistry, the study of matter, plays a central role in most other sciences. An understanding of biology, particularly at the cellular level, requires an understanding of chemical processes. One example of the close connection between chemistry and physics is the diagnostic technique of magnetic resonance imaging (MRI), which results from the interaction of chemical systems with strong magnetic fields. Geology, the study of Earth's composition, structure, and origins, requires a strong knowledge of chemical reactions. A growing area of psychological study relates human behavior to brain chemistry. Psychologists have recently mapped regions of the brain where the chemical reactions responsible for the hallucinations accompanying schizophrenia arise. Such research may facilitate the development of more effective drugs to treat psychological diseases.

Discussion

Ask students whether they use or have heard the comment "You won't melt" as they go outside on a rainy day. Ask them if the comment is accurate. *No, melting is the change of a solid to a liquid at its melting point. The statement should be "You won't dissolve."* L1

Why does this water burn? 🚫 🥽 🔥 ✴️

Ten to 15 minutes before the demo, squirt about 2 mL of cigarette lighter fluid into a 500-mL Erlenmeyer flask, and swirl it to distribute the fluid evenly. Place the flask in its original carton. **CAUTION:** *Be careful that the flask is not near an open flame.* Make a point of using a "new" flask. Fill the flask to the top with water, add a pinch of salt, and light the surface with a glowing wood splint. The small amount of lighter fluid will have floated to the surface and will burn, giving off smoke and an odor. Ask students what happened, hoping that they were misled by the addition of salt. Explain your trick of putting lighter fluid in the flask before class, and point out that it is less dense than water and insoluble in or immiscible with water.

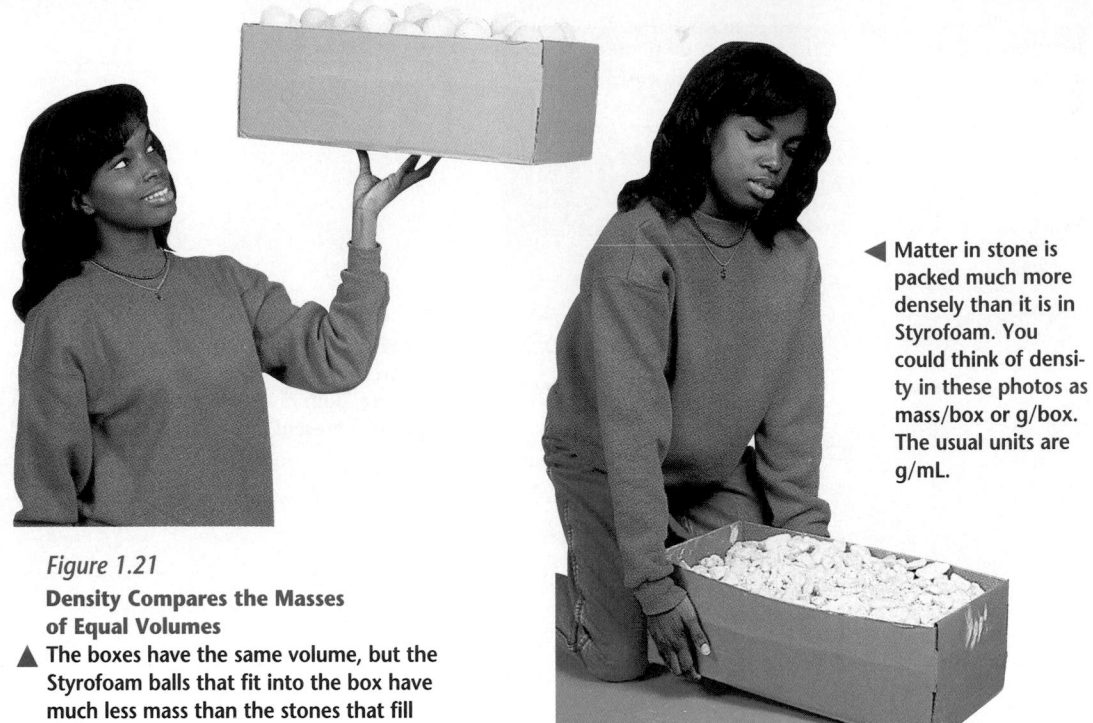

Figure 1.21
Density Compares the Masses of Equal Volumes
▲ The boxes have the same volume, but the Styrofoam balls that fit into the box have much less mass than the stones that fill the box.

◄ Matter in stone is packed much more densely than it is in Styrofoam. You could think of density in these photos as mass/box or g/box. The usual units are g/mL.

Density

Density is another physical property of matter. Consider two identical boxes, one filled with Styrofoam balls and the other filled with stones. If you were to lift both boxes, as shown in **Figure 1.21,** you might say that the box of Styrofoam balls is light while the box of stones is heavy. The Styrofoam balls occupy a certain amount of space or volume (the box), but they have little mass because of the particular structure of Styrofoam. The stones occupy the same volume (the box), but they have a larger mass because of the particular structure of stone. The structure of stone packs much more mass into a given volume than the structure of Styrofoam does. When you compare the mass of the box of Styrofoam with the mass of an identical box of stones, you are observing the different densities of the two materials. **Density** is the amount of matter (mass) contained in a unit of volume. Styrofoam has a low density or small mass per unit of volume. Stones have a large density or a large mass per unit of volume.

In science, the density of solids and liquids is usually measured in units of grams (mass) per milliliter (volume) or g/mL. **Table 1.4** gives the densities of some common materials.

Table 1.4	Densities of Some Common Materials
Material	**Density, g/mL**
Water (4.0°C)	1.000
Ice (0°C)	0.917
Helium (25°C)	0.000178
Air (25°C)	0.00185
Aluminum	2.70
Lead	11.34
Gold	19.31
Cork	0.22-0.26
Sugar	1.59
Balsa wood	0.12

Conserving Mass

Purpose: To demonstrate the law of conservation of mass.
Materials: 500-mL Erlenmeyer flask with stopper, test tube small enough to fit inside the stoppered Erlenmeyer flask, 1 g $Cu(NO_3)_2 \cdot 3H_2O$, 4 g NaOH, 10 mL H_2O, balance
Safety Precautions: 🧤 🥽 ☠️ ⚡
Disposal: Code I; the solid, black CuO will dissolve in HCl and form soluble $CuCl_2$. Dilute the solution with a large amount of water, check the pH, and adjust if necessary. Then, discharge down the drain.
Procedure: Place 1 g of $Cu(NO_3)_2 \cdot 3H_2O$ in a 500-mL Erlenmeyer flask. In a test tube, dissolve 4 g of NaOH in 10 mL of H_2O. **CAUTION:** $Cu(NO_3)_2$ *is toxic and a strong oxidizer; NaOH is caustic.* Place the test tube into the flask with the $Cu(NO_3)_2$, but the two chemicals should not mix. Stopper the flask and obtain the mass of the system. Carefully tip the stoppered flask so that the NaOH solution mixes with the $Cu(NO_3)_2$. Ask a student to feel the bottom of the flask to determine whether heat is being released by the reaction. Mass the system again.
Results: $Cu(OH)_2$ forms as a blue precipitate. The heat of the reaction may cause the

To find the density of a chunk of matter, it is necessary to measure both its mass and its volume. One technique for measuring these quantities is shown in **Figure 1.22.** It could be used for any object that is heavier than water and does not dissolve in water.

Figure 1.22
Determining Density
Here's one way to determine the density of a solid such as lead.

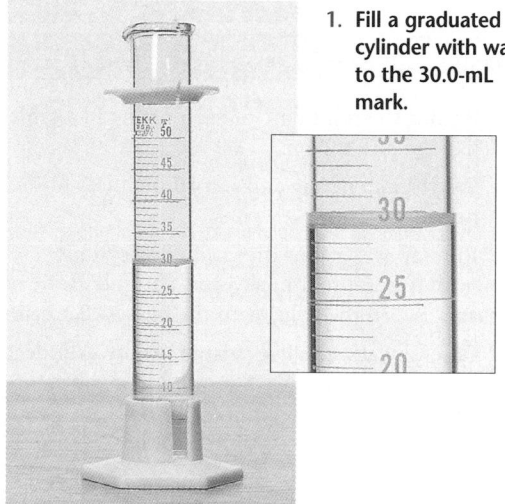

1. Fill a graduated cylinder with water to the 30.0-mL mark.

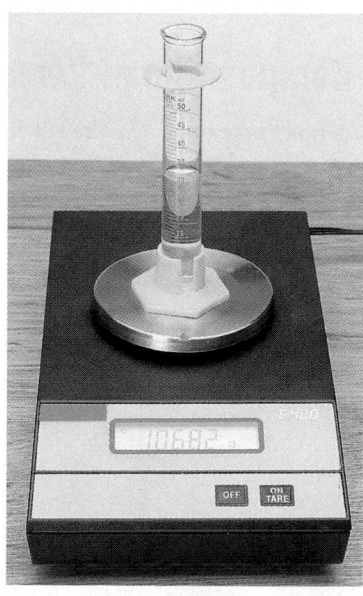

2. Weigh the cylinder with the water (106.82 g).

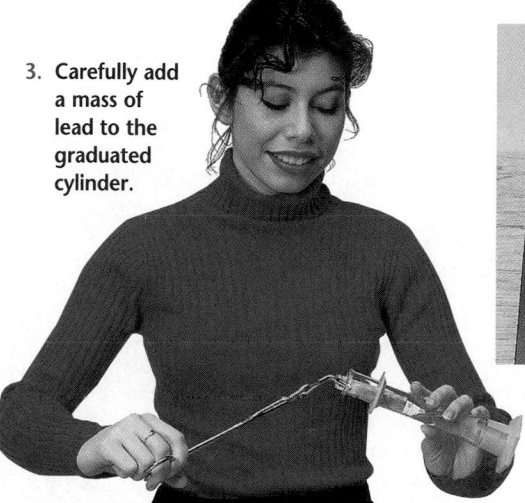

3. Carefully add a mass of lead to the graduated cylinder.

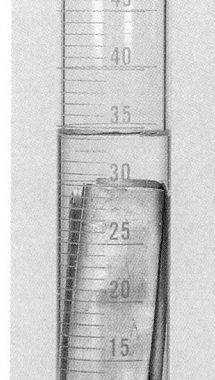

4. Reweigh the cylinder with the water and the lead (155.83 g).

5. Measure the volume of all the material in the cylinder (34.5 mL).

The mass of lead = 155.83 g − 106.82 g = 49.01 g.
The volume of lead by displacement = 34.5 mL − 30.0 mL = 4.5 mL.
Density = mass of lead/volume of lead = 49.01 g/4.5 mL = 11 g/mL.

SECTION 1.2

Quick Demo

Can raisins dance?
Put about 20 plump raisins (soak dried raisins in water for about 10 minutes) in a 100-mL graduated cylinder. Open a can of clear, colorless, carbonated beverage and immediately fill the cylinder to the top. As carbon dioxide bubbles form on the raisins, they increase the volume and decrease the density of the raisin-bubble aggregates, causing them to rise to the top of the liquid. When the aggregates reach the surface of the liquid, some of the carbon dioxide bubbles pop, decreasing the raisin densities sufficiently that they sink.

GLENCOE TECHNOLOGY

Videodisc

Science and Technology Videodisc Series
Shock Impact Gun
Disc 2: Chemistry
Side 1, Ch. 14

Show this video to illustrate how chemical change occurs.
L1 LEP

$Cu(OH)_2$ to decompose into black CuO. Students observe that a chemical reaction has occurred with no change in mass.
Analysis: Ask these questions.
1. Did the mass of the system change? *no*
2. What evidence did you observe for a chemical change? *heat, color change, formation of a precipitate*
3. What law is illustrated by the demonstration? *the law of conservation of*

mass— matter is neither created nor destroyed

Assessment

Knowledge: Ask students to contrast the energy released in this ordinary chemical reaction with the energy released in a nuclear reaction. Students may be familiar with the formula $E = mc^2$. Remind them

that changes of energy to mass and mass to energy are observable only in nuclear reactions. L2

3. The Composition of Pennies

Time Allotment: One 45- to 60-minute lab period.

Objectives

Review objectives with students before they begin the ChemLab.

Process Skills

Observing, using numbers, interpreting data, classifying

Safety Precautions

Students should wear aprons and goggles.

PREPARATION

Remind students to bring in appropriate pennies; however, have a reserve supply on hand to use, if needed.

PROCEDURE

- Require that students' note-books contain the following items before they begin: Objectives; Safety Precautions; Outline or Flow Chart of the Procedure; Data and Observations Table.
- If students have not used sensitive balances before, discuss and demonstrate their use and treatment.
- Draw a chalkboard diagram showing that the volume of a liquid in a graduated cylinder is read along a line tangent to the curved meniscus. Discuss the expected accuracy of the readings.
- Preselect pennies to provide the range of densities shown below. If you use a few 1982 pennies, students may observe that pennies minted in 1982 may be either type.

DATA AND OBSERVATIONS

26.8 mL	Volume of water + five pennies
25.0 mL	Volume of water
1.8 mL	Total volume of five pennies
0.36 mL	Average volume of one penny

ChemLab 3

The Composition of Pennies

U.S. pennies have been composed of copper and zinc since 1959, but the ratio of copper to zinc has changed over the years because of increases in the price of copper. Copper and zinc are both metallic elements and they share many physical properties, but they have different densities. Pure copper has a density of 9.0 g/mL, while pure zinc has a density of 7.1 g/mL. By measuring the density of pennies from different years, it's possible to track changes in the composition of the penny.

Problem

What are the approximate compositions of pennies having various mint dates?

Objectives

- **Measure** mass and volume and determine the density of pennies.
- **Interpret** class data to determine approximately when the composition of pennies changed.

PREPARATION

Materials

5 pennies of varying mint dates
balance sensitive to 0.01 g
50-mL graduated cylinder having 1-mL graduations

Safety Precautions

Wear an apron and goggles.

PROCEDURE

1. Record the mint date of each penny in a table like the one shown.

2. Weigh each penny and record the mass to the nearest 0.01 g.

3. Run tap water into the graduated cylinder until it is approximately one-half full. Read and record the volume to the nearest 0.1 mL.

4. Carefully add the five pennies to the cylinder so that no water splashes out. Jiggle the cylinder to dislodge any trapped air bubbles. Read and record the total volume of the water and the five pennies.

DATA AND OBSERVATIONS

Mint Date	Mass (g)	Average Volume (mL)	Density (g/mL)
1969	3.096	0.36	8.6
1976	3.173	0.36	8.8
1980	3.047	0.36	8.5
1983	2.543	0.36	7.1
1994	2.551	0.36	7.1

DATA AND OBSERVATIONS

Volume of water (mL)	
Volume of water + 5 pennies (mL)	
Volume of 5 pennies (mL)	
Average volume of a penny (mL)	

Mint Date	Mass (g)	Density (g/mL)

ANALYZE AND CONCLUDE

1. **Calculating** Subtract the initial volume of water from the volume of the water plus pennies to calculate the volume of the five pennies. Divide the volume by 5 to calculate the average volume of a penny. Record this volume in your data table.

2. **Calculating** Calculate the density of each penny by dividing its mass by the average volume. Post your results so they are available to the other members of your class.

3. **Observing and Inferring** Does your group data show any pattern that relates the densities of the pennies to the mint dates?

4. **Classifying** Classify the pennies of each year that your group examined as mostly copper (8.96 g/mL) or mostly zinc (7.13 g/mL).

5. **Making Inferences** Look at all the data from your class. In what year do you think the composition of the penny changed? Use data to support your conclusion.

APPLY AND ASSESS

1. What factors do you think might lead to error in your density measurements? Which of these factors could not be corrected by improved technique?

2. Explain how you might determine the identity of an irregularly shaped solid that is soluble in water.

3. In 1943, all pennies issued by the U.S. mint were struck from zinc-plated steel. Two of these pennies are shown in the photo. Do research to learn why steel pennies were struck in 1943. What was the purpose of the zinc coating?

4. Why would increases in the price of copper cause the mint to change the composition of pennies?

ANALYZE AND CONCLUDE

1. See Data and Observations.
2. See Data and Observations.
3. The density of the penny decreases around 1982.
4. The pennies prior to 1982 are mostly copper; those after 1982 are mostly zinc.
5. During 1982, the composition of the penny changed from mostly copper to mostly zinc. Densities prior to 1982 are slightly less than that of pure copper, and densities after 1982 are similar to the density of pure zinc. Both types were minted in 1982.

APPLY AND ASSESS

1. There are errors in measuring the mass and in reading the volume of the graduated cylinder. Coins are not uniform, even when new. Heavily circulated coins may have lost mass because of wear. The latter two factors cannot be corrected by improved technique.
2. Weigh the object to determine its mass. To determine its volume and calculate its density, submerge the object in a liquid in which the material will not dissolve.
3. Copper was in short supply in 1943 because it was being used in the manufacture of war materials for World War II. The zinc coating on the steel coins prevented the rusting of the iron.
4. If the price of the copper in a penny became greater than one cent, the coins would go out of circulation as people melted them down for the metal they contained.

✓Assessment

Performance: Have students determine the density of iron or steel objects such as large bolts or nails, and compare their results with the density of iron, 7.86 g/mL. L1

GLENCOE TECHNOLOGY

Videodisc

Chemistry: Concepts and Applications
The Law of Conservation of Mass
Disc 1, Side 1, Ch. 3

Show this quick-time animation to demonstrate that matter is neither created nor destroyed. L1 LEP

GLENCOE TECHNOLOGY

CD-Rom

Chemistry: Concepts and Applications
The Law of Conservation of Mass
Have students use this animation to reinforce the concept of conservation of matter. L1 LEP

miniLAB

It's a Liquid, It's a Solid...It's Slime

Purpose: To prepare and investigate a polymer gel.

Process Skills: Observing, interpreting data

Safety Precautions: The gel can irritate skin. Students should wear gloves when handling it.

Disposal: The slime can be disposed of as solid waste.

Teaching Strategies

- Purchase a 4% polyvinyl alcohol solution or prepare one as follows. Sprinkle 40 g of polyvinyl alcohol into 1 L of water with constant stirring. Heat the mixture on a hot plate until all the solid is dissolved and the solution is clear, approximately 45 minutes. Be careful not to scorch the viscous liquid. Allow the solution to cool, and skim off any slimy layer from the surface.
- Do not allow students to take the slime out of the laboratory.

Expected Results

The slime flows, stretches if pulled slowly, breaks if pulled quickly, and slowly flattens.

Analysis

1. Cross-linking makes the polymer more viscous. The individual strands cannot move independently.
2. carbohydrates, proteins, rubber, nylon, polyethylene, Teflon, etc.

Assessment

Oral: Ask students how cross-linking causes the polymer to gel. *It links the polymer strands, inhibiting their free movement.* L1

---

miniLAB 5

It's a Liquid, It's a Solid...It's Slime

In this MiniLab, you will work with a polymer, polyvinyl alcohol. A polymer is a large molecule that consists of a chain of smaller repeating units. By allowing polyvinyl alcohol to react with a borax solution, you will cross-link polymer molecules to form a gel not unlike the commercial green stuff you may have seen. You can then investigate some of the properties of the gel.

Procedure

1. Wear an apron, safety goggles, and disposable plastic gloves.

2. Place about 20 mL of polyvinyl alcohol solution in a plastic cup.

3. Use a craft stick to stir the solution vigorously as you add about 3 mL of borax solution. Add a drop of food coloring if you want colored slime.

4. Continue to stir the solutions together until they form a gel.

5. Remove the gel from the cup, shape it with your hands, and perform several tests on it.

Does it flow? Does it stretch or break? Can it be flattened? Record your observations.

6. Put your product in a self-sealing plastic bag, and dispose of it according to your teacher's instructions.

Analysis

1. What effect does cross-linking the polymer have on its properties? Can you explain this effect?

2. Can you name some polymers that occur in or are used to produce common materials?

Chemical Properties and Changes

Physical properties alone are not enough to describe a substance. For a complete description, you need to know about another set of properties called chemical properties. **Chemical properties** are those that can be observed only when there is a change in the composition of the substance. Chemical properties describe the ability of a substance to react with other substances or to decompose. A chemical property of iron, for example, is that it rusts at room temperature. Rusting is a chemical reaction in which iron combines with oxygen to form a new substance, iron oxide. Aluminum reacts with oxygen too, but the compound formed, aluminum oxide, coats the aluminum and protects it from further oxidation. Platinum does not react with oxygen at room temperature. Lack of reactivity is also a chemical property.

Have you ever noticed that hydrogen peroxide (H_2O_2) solution always comes in brown bottles? It's packaged in this way because in bright light, hydrogen peroxide breaks down into water and oxygen gas. The instability of a substance—its tendency to break down into different substances—is another chemical property. **Figure 1.23** illustrates other chemical properties of some substances. A chemical property always relates to a **chemical change,** the change of one or more substances into other substances. Another term for chemical change is **chemical reaction.**

Demonstration 1.5

Sulfuric Acid and You

Purpose: To introduce students to sulfuric acid and the need for eye protection in the laboratory.

Materials: 35 g sucrose (granulated sugar), 150-mL beaker, 30 × 30 cm piece of cardboard, 35 mL 18M sulfuric acid, glass stirring rod

Safety Precautions:

Disposal: Code A

Procedure: Place 35 g of sucrose in a 150-mL

beaker. Place the beaker on the cardboard. Add 35 mL of 18M sulfuric acid to the beaker. Stir briefly and wait. **CAUTION:** *H_2SO_4 is corrosive; dilutions and reactions are exothermic. Perform the demo in the fume hood or in a well-ventilated area to avoid the disagreeable fumes.* Record student observations.

Results: A column of black solid rises from the beaker. The smell of burned sugar is evident. Do not allow students to touch the solid be-

Figure 1.23
Some Chemical Properties and Changes
The chemical properties of a substance describe how that substance changes to one or more new substances. It may change by reacting with another substance or by breaking down into simpler substances.

◀ Iron combines with oxygen to form iron oxide. The reaction shown is the same reaction as the rusting of iron, but it occurs much faster in pure oxygen in the flask.

▲ When vinegar is poured on baking soda (sodium hydrogen carbonate), bubbles of carbon dioxide gas form quickly. Water and sodium acetate are also formed.

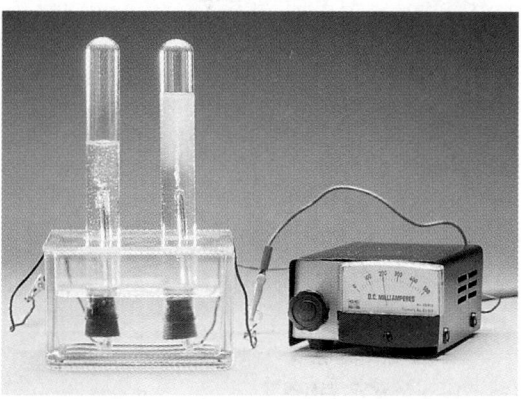

▲ The stable compound water can be decomposed into hydrogen gas (left tube) and oxygen gas (right tube) by passing an electric current through it.

▲ Hydrogen peroxide is another compound of hydrogen and oxygen. When the compound manganese dioxide is added to hydrogen peroxide, the peroxide breaks down rapidly into water and oxygen gas.

Atoms and Chemical Change

All matter is made of atoms, and any chemical change involves only a rearrangement of the atoms. Consider the chemical change shown in **Figure 1.23** in which water is broken down into hydrogen gas and oxygen gas. The reaction involves only hydrogen atoms and oxygen atoms. Whatever atoms are contained in the water that decomposes are found in the

Discussion
Tell students that when copper and nitric acid are put together, a blue substance is formed. Ask whether this is a physical or chemical property. *A chemical property; the identity of copper is changed.* Ask whether it is a physical or chemical property that copper nitrate solution is blue. *physical property* L1

Visual Learning

Figure 1.23: Ask students what they can observe in the four photos that indicates that a chemical change is occurring. *the formation of gases; the evolution of light and heat, the absorption of energy* Ask what other observations might indicate that a chemical reaction is occurring. *the formation of a precipitate, a color change* L1

GLENCOE TECHNOLOGY

📼 **Videotape**

MindJogger Videoquizzes for Chemistry
Chapter 1: Chemistry: The Science of Matter
Have students work in groups as they play the videoquiz game to review key chapter concepts. L1 **LEP** **COOP LEARN**

---

cause it is acidic. The beaker will be hot.
Analysis: Ask these questions.
1. What color change did you observe? *white to black*
2. What element is responsible for the black color? *carbon*
3. What evidence do you see that a gas was released? *white "smoke"*
4. What do you think makes up the white smoke? *water vapor and SO_2*

5. Which elements are present in sugar? *carbon, hydrogen, and oxygen*
6. Sulfuric acid is described as a dehydrating agent. What does that mean? *It removes water.*
7. How would sulfuric acid damage your skin? *It would remove the water from cells; the heat from the exothermic reaction would cause burns.*

✓ Assessment

Oral: Ask students why it is important to wear aprons, safety goggles, and gloves when handling many chemicals. *Chemicals can burn or irritate skin, splash in eyes and damage them, and stain or destroy clothing.* L1

42

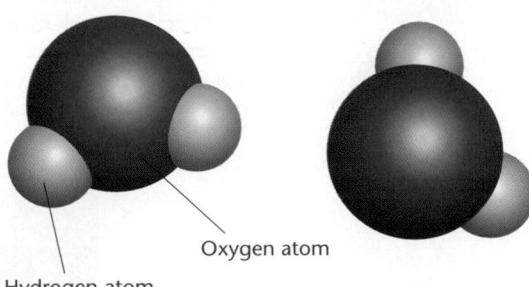

Water molecules

Figure 1.24

Conservation of Mass (Atoms)
Two water molecules contain two oxygen atoms and four hydrogen atoms. When they decompose, they form one molecule of oxygen containing two oxygen atoms and two molecules of hydrogen containing four hydrogen atoms. Because all matter consists of atoms and the number of atoms is the same before and after the chemical change, you can say that matter is conserved.

Oxygen atom
Hydrogen atom

Fact of the MATTER

Weight is different from mass. The weight of an object is a measure of the force of gravity on the object. Scientists generally say "weigh an object" when they really mean "measure the mass of an object on a balance."

hydrogen and oxygen molecules that are formed. **Figure 1.24** uses models of molecules of water, oxygen, and hydrogen to show that no atoms appear from nowhere. No atoms just disappear. This is an example of the **law of conservation of mass,** which says that in a chemical change, matter is neither created nor destroyed. It would be equally correct to call this the law of conservation of matter.

Chemical Reactions and Energy

All chemical changes also involve some sort of energy change. Energy is either taken in or given off as the chemical change takes place. **Energy** is the capacity to do work. Work is done whenever something is moved. A carpenter does work whenever he or she picks up a hammer and drives a nail. The objects being moved can also be atoms and molecules. This is the kind of work that is encountered in chemistry.

Many reactions give off energy. For example, burning wood is a chemical change in which cellulose, and other substances in the wood, combine with oxygen from the air to produce mainly carbon dioxide and water. Energy is also produced and released in the form of heat and light. Chemical reactions that give off energy are called **exothermic** reactions. A dramatic example is the rapid decomposition of nitroglycerin shown in **Figure 1.25.**

Figure 1.25

An Exothermic Reaction
The energy released in the explosion of nitroglycerin breaks and moves rock. The complex nitroglycerin molecule is converted to four gaseous products: carbon dioxide, nitrogen, oxygen, and water vapor.

Demonstration 1.6

An Exothermic Reaction

Purpose: To demonstrate activation energy and an exothermic reaction.
Materials: tube of silicone caulking, piezoelectric gas grill lighter, 250-mL polyethylene or polypropylene bottle, 5 mL methanol, cork to fit bottle, string, thumbtack, ring stand

Safety Precautions:
Disposal: Code F

Procedure: Cut a small hole on the side of the bottle near the bottom. Use silicone caulking to cement the tip of the grill lighter into the bottle. Allow the caulking to set overnight. Add 5 mL of methanol, and allow a minute for it to vaporize. Pour out any excess liquid. Use a cork (not a rubber stopper) to seal the bottle. Place a thumbtack in the cork. Tie a 1-m length of string to the thumbtack and to the neck of the bottle. Clamp the bottle to a ring stand.

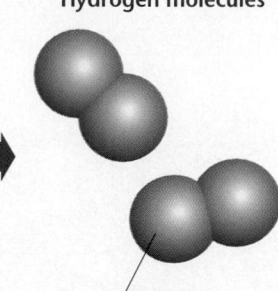

Hydrogen molecules

Oxygen molecule

+

Hydrogen atom

Oxygen atom

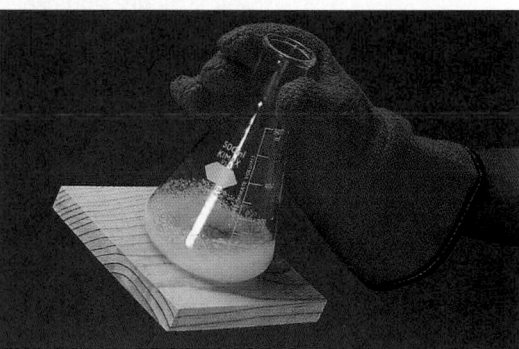

Figure 1.26
An Endothermic Reaction
When the two compounds ammonium thiocyanate and barium hydroxide octahydrate are mixed, a reaction occurs in which heat is absorbed from the surroundings. The flask becomes so cold that if a film of water is on the bottom of the flask, it will freeze the flask to a wooden block.

Some chemical changes absorb energy. Chemical reactions that absorb energy are called **endothermic** reactions. You can tell that the decomposition of water into oxygen and hydrogen is an endothermic reaction because it doesn't occur unless energy, in the form of an electric current, is passed through the water. When baking soda ($NaHCO_3$) is mixed into some kinds of cookie dough and the dough is baked in a hot oven, the baking soda absorbs energy and breaks down into carbon dioxide, water, and sodium carbonate (Na_2CO_3). The CO_2 and gaseous H_2O puff up the cookies. **Figure 1.26** shows another endothermic reaction.

Photosynthesis is probably the most important endothermic process on Earth. Photosynthesis is a series of chemical reactions that absorb light energy from the sun and produce sugars from carbon dioxide and water. Oxygen is given off as a product of the reactions. Green plants, algae, and many kinds of bacteria carry out photosynthesis. When you eat sugars and starches, you are eating the molecules formed by endothermic photosynthesis reactions, as indicated in **Figure 1.27** on the next page. Your cells break down these molecules in an exothermic process that supplies you with energy.

Quick Demo

Is energy released or absorbed? 👓

Place two 250-mL beakers each containing 150 mL of water on magnetic stirrers. Record the temperatures of the water and begin stirring. Add Epsom salt to one beaker and powdered laundry detergent to the other. After a few moments of stirring, record the temperatures and ask students to explain the results. *The dissolving of Epsom salt is endothermic, and the dissolving of laundry detergent is exothermic.* L1

3 Assess

Check for Understanding

• Check for understanding of the section material by having students answer the Section Review questions.

• ▦ Ask students to determine the following: (1) the density of regularly shaped metallic or plastic solids by massing and metric measurement, (2) the density of an irregularly shaped metallic object by massing and water displacement, and (3) the thickness of a rectangular piece of aluminum foil, given the density of aluminum, 2.70 g/mL. Some students may need assistance with procedures and calculations. L2

CAUTION: *Use an explosion shield. Do not point the cork in the direction of students.* Dim the room lights and squeeze the handle of the lighter to generate a spark. Have some students feel the outside of the bottle immediately after the explosion and report their observations to the class. Note: If you plan to use the bottle next period, flush it with air.
Results: The spark (activation energy)

causes the methanol-air mixture to react explosively, blowing off the cork.
Analysis: Ask these questions.
1. Is the reaction inside the bottle a chemical or physical change? *chemical*
2. Is the reaction endothermic or exothermic? *exothermic*
3. What evidence did you observe that supports your answer to question 2? *The bottle is warm and light is given off.*

Assessment

Oral: Ask students how the bottle resembles an internal combustion engine. *The lighter is like a spark plug. The methanol-air mixture is like the gasoline-air mixture in each cylinder. The cork is like the piston that moves and provides power to the wheels.* L1

Reteach

Discussion: Have students make up a class list of the reactions each of them observed and listed in their chemistry journals as described under *Chemistry Journal* below. Ask for evidence that each event on the list is a chemical reaction. Discuss what evidence led to the characterization of each reaction as endothermic or exothermic. L1

Extension

Ask interested students to prepare a density column in a tall cylinder using some or all of the following materials. Densities in g/mL are given: lead ball or sinker (11.3), dark karo syrup (1.4), glycerol (1.3), rubber stopper (1.2), ethylene glycol (1.1), water (1.0), piece of plastic (0.9), vegetable oil (0.9), piece of oak wood (0.9), rubbing alcohol (0.8), and cork (0.2). To prevent mixing, the liquids should be poured sequentially down a long-stemmed funnel or slowly down the side of the cylinder as it is in a tipped position. L2 LEP

4 CLOSE

Discussion

Ask students to explain how the equation for the combustion of methane, $CH_4 + 2O_2 \rightarrow CO_2 + 2H_2O$, illustrates the law of conservation of mass. *The number of atoms of each element reacting is the same as the number of atoms in the products: one atom of carbon, four atoms of hydrogen, and four atoms of oxygen. No atoms are lost or gained; therefore, mass is conserved.* L1

Figure 1.27

From Sunlight to Food Energy

When you eat vegetables from a garden such as this, you are consuming substances that plants have synthesized using energy from sunlight.

Connecting Ideas

Chemistry makes connections among the composition, structure, and behavior of matter. As you study chemistry, you will learn how the chemical and physical properties of matter are important clues to its submicroscopic structure and behavior. You'll see how knowing the structure of an atom of an element can enable you to predict the chemical behavior of that element. You'll learn how the state of a substance at room temperature provides clues to the way its atoms are arranged. You'll find out why some things dissolve in water but others do not, how metals corrode, how batteries work, why compounds containing carbon are important to life, and how a nuclear reactor works.

SECTION REVIEW

Understanding Concepts

1. What are the three states of matter?

2. Identify each of the following as either chemical or physical properties of the substance.

   a) Aluminum bends easily.
   b) Copper sulfate dissolves in water.
   c) Magnesium burns in air.
   d) Gold jewelry is unaffected by perspiration.
   e) Baking soda is a white powder.
   f) Fluorine is a highly reactive element.

3. Identify each of the following as either chemical or physical changes.

   a) A match lights when struck.
   b) Air is squeezed by a pump and forced into a tire.
   c) A lump of gold is pounded into a large, thin sheet.

   d) Baking powder bubbles and gives off CO_2 when it is moistened.
   e) A pan of water boils on the stove.
   f) Hydrogen sulfide gas causes silver to tarnish.

Thinking Critically

4. **Applying Concepts** A friend tells you that a newspaper is completely gone because it was burned up in a fire. Use the law of conservation of mass to write an explanation telling your friend what really happened to the newspaper.

Applying Chemistry

5. **Campfires** On cool evenings, a campfire warms the body and spirit. Is the burning of logs an exothermic or endothermic reaction?

Chemistry Journal

Exothermic and Endothermic

Ask students to make lists of all the chemical reactions they observe during a typical day and decide, if possible, whether each reaction is exothermic or endothermic. Ask them to include the lists in their journals. L1

SECTION REVIEW

1. solid, liquid, gas
2. **a.** physical **b.** physical **c.** chemical **d.** chemical **e.** physical **f.** chemical
3. **a.** chemical **b.** physical **c.** physical **d.** chemical **e.** physical **f.** chemical
4. When the newspaper burned, it changed into water, carbon dioxide gas, and ash. Matter was neither created nor destroyed.
5. exothermic

REVIEWING MAIN IDEAS

1.1 The Puzzle of Matter

- Chemistry deals with the composition, structure, and behavior of matter.

- Macroscopic observations of matter reflect its submicroscopic structure.

- The quantity of matter in a sample can be measured by determining its mass.

- Descriptions of matter and its behavior are called properties.

- The composition of a sample of matter determines whether it is a substance or a mixture.

- Mixtures are heterogeneous or homogeneous (solutions).

- Mixtures may be separated by using physical changes that take advantage of the physical properties of the individual substances in the mixture.

- Substances are classified as elements or compounds.

- Elements are the building blocks of all matter.

- The composition of a compound is represented by its chemical formula.

1.2 Properties and Changes of Matter

- Every substance has a unique set of physical and chemical properties.

- The states of matter are solid, liquid, and gas. The state of a substance at room temperature is a physical property.

- The temperatures at which a substance changes state are physical properties of the substance.

- The density of a sample of matter is the amount of matter (mass in grams) in a unit volume (usually a milliliter).

- The chemical properties of a substance describe the chemical changes the substance undergoes.

- Chemical changes involve substances rearranging to form different substances.

- Another name for chemical change is *chemical reaction.*

- In a chemical reaction, atoms are never created or destroyed.

Key Terms

For each of the following terms, write a sentence that shows your understanding of its meaning.

alloy	mass
aqueous solution	matter
chemical change	mixture
chemical property	physical change
chemical reaction	physical property
chemistry	property
compound	qualitative
density	quantitative
element	scientific model
endothermic	solute
energy	solution
exothermic	solvent
formula	substance
law of conservation	volatile
of mass	

✓ Assessment

Portfolio: Review the portfolio options that are provided throughout the chapter. Encourage students to select one product that demonstrates their best work for the chapter. Have students explain what they have learned and why they chose this example for placement into their portfolios. **P**

Additional portfolio options can be found in the following **Teacher Classroom Resources:**
Tech Prep Applications
Chemistry and Industry
Consumer Chemistry

- Review the Reviewing Main Ideas statements and Key Terms with your students.
- Complete solutions to Chapter Review problems can be found in the *Problems and Solutions Manual.*

UNDERSTANDING CONCEPTS

1. Chemistry is the science of matter—its composition, structure, and behavior.
2. Fe, Na, Sb, W; they were discovered a long time ago and derive their names from Latin (Fe, Na, and Sb) or German (W).
3. Only that it is higher than the melting point; melting and freezing points are the same.
4. Baking soda, salt, sodium carbonate, and sodium hydroxide contain sodium. Hydrochloric acid and salt contain chlorine.
5. It gives the number and kind of each atom in a molecule of the compound.
6. Mass is a measure of the amount of matter. The mass of an object is determined by weighing the object on a balance.
7. A property is any feature that characterizes a piece of matter. Color is a physical property.
8. gray, shiny, magnetic, conducts electricity, does not react with water, rusts
9. Water; most of the solutions we encounter are water solutions. Many life processes are carried out in aqueous solution.
10. Yes; the one with the larger mass would contain more matter.
11. matter that contains only one substance, either an element or a compound
12. Energy is the capacity to do work. Every chemical reaction involves either the absorption or the release of energy.
13. An exothermic reaction is one in which energy is released. An endothermic reaction is one in which

energy is absorbed. Burning wood is exothermic; decomposing water is endothermic.

14. A qualitative observation does not involve measurement; for example, the tree is tall. A quantitative observation involves measurement; for example, the tree is 150 feet tall.

15. A mixture is a physical combination of two or more pure substances. Each substance retains its individual properties. A mixture may have a variety of compositions. A compound is a chemical combination of two or more elements combined in a definite proportion. A compound has its own set of chemical and physical properties that are different from the properties of the elements that make up the compound.

16. a. physical **b.** chemical
c. physical **d.** chemical
e. physical

17. a. homogeneous **b.** heterogeneous **c.** homogeneous
d. heterogeneous

18. An element is a pure substance that cannot be broken down to any simpler substance. A compound is a chemical combination of two or more elements in a definite proportion. A compound can be broken down by chemical means into other elements or compounds. Iron (Fe) is an element; iron(III) oxide (Fe_2O_3) is a compound.

19. Density is the mass of a unit of volume, usually grams per milliliter. A bag of Styrofoam balls could be heavier than a bag of stones if the bag of Styrofoam balls were much larger than the bag of stones.

20. 1064°C

21. Boiling point and freezing point are physical properties of a substance that depend upon the submicroscopic characteristics of the

UNDERSTANDING CONCEPTS

1. What is chemistry?

2. List the symbols for the following elements: iron, sodium, antimony, tungsten. Explain why these symbols do not correspond to the English names of the elements.

3. If you know the melting point of a pure substance, what does it tell you about its boiling point? Its freezing point?

4. Which compounds, listed in **Table 1.3,** contain the element sodium? The element chlorine?

5. What information does the chemical formula for a compound provide about the submicroscopic structure of the compound?

6. What is mass? How is the mass of an object determined?

7. If you say that sulfur is yellow, you are stating a property of sulfur. What is meant by *property*? Is color a physical or chemical property?

8. List three properties of iron.

9. What is the solvent in an aqueous solution of salt? Why are aqueous solutions important?

10. Could two objects with the same volume have different masses? Which, if either, would contain more matter?

11. What is meant by *pure matter*?

12. What is energy? What part does it play in chemistry?

13. Distinguish between exothermic and endothermic reactions, and give an example of each.

14. How is a qualitative observation different from a quantitative observation? Give an example of each.

15. Explain how a mixture is different from a compound.

16. Are the following physical changes or chemical changes?

a) water boils
b) a match burns
c) sugar dissolves in tea
d) sodium reacts with water
e) ice cream melts

17. Classify these as heterogeneous or homogeneous mixtures.

a) salt water
b) vegetable soup
c) 14-k gold
d) concrete

18. What is an element? A compound? Give an example of each.

19. What is meant by *density*? Is there any way that a bag of Styrofoam balls could be heavier than a bag of stones?

20. Gold freezes at 1064°C. What is the melting point of gold?

21. Mercury freezes at −38.9°C; nitrogen boils at −195.8°C. How can a boiling point be lower than a freezing point?

22. Is the diagram a molecule of nitroglycerin or a model of nitroglycerin? Explain.

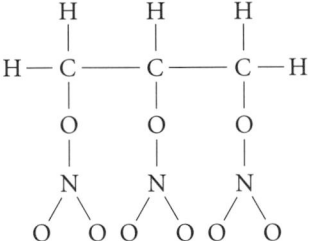

substance. Mercury and nitrogen are different substances with different states at room temperature.

22. A model; it is impossible to see a molecule. The diagram shows what has been learned through experiments about the arrangement of the atoms in the molecule.

APPLYING CONCEPTS

23. exothermic

24. The density of liquid water is greater than the density of solid water.

25. Both brands should be equally safe because they have the same chemical composition.

26. An alloy is a solid homogeneous mixture of metals and sometimes other substances. Examples are steel, sterling silver, solder, and pewter.

27. The salad dressing is a heterogeneous mixture containing a large number of substances. Vinegar, oil, pepper, and garlic are complex combinations of substances. Among the ingredients, salt is the only pure substance. Salt is a solute in the

APPLYING CONCEPTS

Everyday Chemistry

23. Is the overall processing of food in your body an exothermic or endothermic process?

Literature Connection

24. An ice cube floats in your glass of water, just as an iceberg floats in the ocean. What conclusion can you draw about the density of liquid water and the density of solid water?

Chemistry and Society

25. A drug with a well-known brand name may also be made by another company and sold at a lower price. Both drugs are specified to have the same chemical formula and are approved by the Food and Drug Administration. Would it be safer to use the well-known brand? Explain.

26. A doorknob is coated with the alloy brass. What is an alloy? Give an example of another alloy.

27. A recipe for salad dressing tells you to mix together vinegar, oil, herbs, salt, pepper, and chopped garlic. Describe the completed salad dressing in terms of substances and mixtures, both heterogeneous and homogeneous. What solutes or solvents are in the salad dressing?

28. Oxygen makes up more than 45.5 percent of Earth's crust and 65 percent of the human body. It also constitutes 20 percent of Earth's gaseous atmosphere. Explain the difference between the oxygen in the atmosphere and the oxygen in Earth's crust and the human body.

29. Some iron filings accidentally dropped into a mixture of salt crystals and water. Describe two ways that you could separate the filings.

30. Ethanol, or ethyl alcohol, melts at $-114.1°C$ and boils at $78.5°C$. What is the physical state of ethanol at room temperature?

31. Using the information in question 30, what are the freezing points and the condensation points of ethanol?

32. Is the melting of candle wax an exothermic or endothermic change?

33. The human body is like the World Trade Center in that it can be an analogy for the macroscopic and submicroscopic perspectives. Explain.

34. Which of these substances has the lowest boiling point: oxygen, water, or salt?

35. When you mix vinegar and baking soda, the mixture foams and fizzes. Suppose you weigh the vinegar and baking soda before mixing and again after they react. How will the masses compare? Explain.

36. When orange juice is advertised as *pure*, does that mean that orange juice is a pure substance? Explain.

37. List five properties of a candle. Identify them as physical or chemical properties.

38. Is a tree homogeneous or heterogeneous? Explain.

39. Identify the following as an element, compound, homogeneous mixture, or heterogeneous mixture: hydrogen, polyethylene plastic, clear apple juice, cloudy apple cider, syrup, paint, and bronze.

solvent water (the principal ingredient in vinegar).

28. The oxygen in the atmosphere is the element oxygen. The oxygen in Earth's crust is combined with other elements in compounds that make up the rocks, sand, and soil. In the human body, oxygen is also combined in chemical compounds.

29. filter the solution or use a magnet

30. liquid

31. freezing point, $-114.1°C$, and condensation point, $78.5°C$

32. endothermic

33. The human body hides the structure that underlies the body—the skeleton, muscles, tendons, and veins. These correspond to the submicroscopic perspective that cannot be seen.

34. oxygen

35. The mass after the reaction will be smaller than the mass before the reaction because one of the products is a gas and it escapes. If the gas were collected and included in the overall mass after reaction, the two masses would be equal.

36. No, orange juice is a complex mixture of substances and not a pure substance as defined in chemistry. Pure substances contain only one substance.

37. Firm, solid but soft enough to dent (physical); combines with oxygen in the air and burns (chemical); melts at the temperature of a flame (physical); the burning produces carbon (chemical); liquid candle wax is vaporized (physical). Students will have other observations.

38. Heterogeneous; a tree is a complex structure containing many substances.

39. hydrogen, element; polyethylene plastic, compound; clear apple juice, homogeneous mixture; cloudy apple cider, heterogeneous

mixture; syrup, homogeneous mixture; paint, heterogeneous mixture; bronze, homogeneous mixture

THINKING CRITICALLY

40. On the submicroscopic level, molecules of solid candle wax are separated to form a liquid, and then the molecules of the liquid are separated to form a vapor. The molecules in the vapor combine with oxygen to form carbon dioxide and water.

41. Starch does not react with water or acid to produce the bubbles needed to make cakes fluffy.

42. density = 86 g/16 mL = 5.4 g/mL

43. exothermic; physical change

44. Students may suggest that the two markers, containing pigments of the same color, be chromatographed side by side on the same paper to see whether, under the same experimental conditions, the two pigments travel the same distance.

45. It is an alloy of zinc and copper. An alloy is a homogeneous mixture.

CHAPTER 1 REVIEW

46. Most foods are compounds. Many of the compounds are large complex molecules, such as proteins, fats, starches, and sugars. Some foods contain elements such as the iron in cereals.

47. Compound; a polymer is a long-chain molecule made from a variety of elements.

48. No; many compounds burn. It would be necessary to analyze either the compounds or the products of the burning to determine whether the two compounds are the same.

49. The compound contains carbon, hydrogen, and oxygen. There are two carbon atoms and six hydrogen atoms for every oxygen atom.

● THINKING CRITICALLY

Observing and Inferring

40. **ChemLab 1** From macroscopic observations of a burning candle, what conclusions can you draw about events occurring on the submicroscopic level?

Observing and Inferring

41. **ChemLab 2** Why is starch not used in baking for the purpose of making cakes fluffy?

Measuring in SI

42. **ChemLab 3** Suppose you have a sample of an unknown mineral with a mass of 86 g. You place the sample in a graduated cylinder filled with water to the 55-mL mark. The sample sinks to the bottom of the cylinder and the water level rises to the 71-mL mark. What is the density of the sample?

Comparing and Contrasting

43. **MiniLab 1** Is the mixing of ethanol and water an exothermic or an endothermic process? Is it a chemical or physical change?

Designing an Experiment

44. **MiniLab 2** Marker inks of different colors, when separated by chromatography, may be found to contain some pigments of the same color. For example, a black marker and a blue marker might both contain a blue pigment. Design an experiment to show whether or not two pigments of the same color, found in different markers, are likely to be the same pigment.

Applying Concepts

45. **MiniLab 3** The appearance of a penny made of zinc and copper is different after heating the penny in zinc chloride solution. Is the new coating an element, a compound, or a mixture? Explain.

Observing and Inferring

46. **MiniLab 4** Are the chemicals in your food elements or compounds? Explain.

Observing and Inferring

47. **MiniLab 5** Is a polymer an element or a compound? Explain.

Observing and Inferring

48. Two solid compounds burn in the presence of oxygen. Can you conclude that they are the same compound?

Interpreting Chemical Structures

49. What does the formula C_2H_6O tell you about the substance it represents? Be as specific as possible.

● CUMULATIVE REVIEW

In Chapters 2 through 21, this heading will be followed by review questions about skills and ideas you have learned in earlier chapters.

Description	Is it chemical or physical?	Explanation
1. Brewing tea or coffee	*Physical*	*Brewing tea and coffee is an extraction where the various components of the tea or coffee dissolve in the brewing water.*
2. Heating a pot of water to cook pasta	*Physical*	*The water remains as water when heat is added.*
3. Food scraps decomposing in a compost heap	*Chemical*	*Decomposition of the food is a chemical process that breaks the large food molecules into smaller molecules.*
4. A seed germinating	*Chemical*	*The sprouting of a seed starts a series of chemical reactions that produce energy and material for plant growth.*

Skill Review

1. **Making and Using a Table** Make a table like the one shown below and fill in the columns. Identify each change or process as physical or chemical and justify your choice. When your table is complete, look at the two groups of processes. Within each group, list similarities shared by the processes. Are there differences within each group?

Writing in Chemistry

2. Look at labels on items in your home or in the grocery store. Choose one item and gather information about it. List all of its ingredients. Describe it as a mixture or pure substance. List three of its properties. For information about the substance, look it up in a chemical handbook. If it is a compound, what is its formula? What are some of its physical and chemical properties? Find out why this substance is an ingredient in the item you chose. Find a similar item, made by a different company, and determine whether the two items have the same ingredients. Summarize your findings in a report.

3. Turn to the periodic table on the inside back cover of your text, close your eyes, and point to the table. Open your eyes and find the element closest to your finger. Write a short report about that element including the following information. What is the name of your element? Where did it get its name? When and by whom was it discovered? How was it discovered? Where is it commonly found? How abundant is it? What are its uses? Is it a required mineral in the human diet?

Problem Solving

4. A miner found a nugget that had a gold color. He realized that it could be precious gold metal or iron pyrite (FeS_2), which is a compound of iron and sulfur called fool's gold. The nugget had a mass of 16.5 g and displaced 3.3 mL of water. From this information and data from a chemical handbook, find out whether the miner had found gold.

Projects

5. Get a puzzle with at least 200 pieces—the more pieces, the better. Ask four friends to solve the puzzle. Do not show your friends the picture of the finished puzzle. Borrow a video camera and record your friends as they work to solve the puzzle. If possible, edit the tape to show different stages in the solution of the puzzle. Present your videotape to your class as a model of how knowledge is gained in chemistry.

Observation	Is it chemical or physical?	Explanation
1. Brewing tea or coffee		
2. Heating a pot of water to cook pasta		
3. Food scraps decomposing in a compost heap		
4. A seed germinating		
5. Sugar dissolving in water		
6. Vinegar and oil not mixing		
7. A termite eating wood and producing methane gas		
8. The density of water increasing as salt is added		

AUTHENTIC ASSESSMENT

1. See tables below. Among the physical processes, items 1, 5, and 8 involve dissolving. Items 1 and 2 require the addition of energy. Among the chemical processes, items 3 and 7 involve decomposition.

2. Chemical handbooks such as the *Merck Index* and the *Handbook of Chemistry and Physics* will be helpful. Chemical catalogs will also be helpful.

3. Chemical handbooks, such as the *Merck Index,* and chemical catalogs will also be helpful.

4. The density of gold is 18.9 g/mL; the density of iron pyrite is 5.0 g/mL. The data give a density of 5.0 g/mL. The nugget is not gold.

Program Resources

Chapter Review and Assessment, pp. 5-10 [L1]
Alternate Assessment in the Science Classroom [L1]
Computer Test Bank, Chapter 1 [L1]
Problems and Solutions Manual, Chapter 1 [L1]
Supplemental Practice Problems, Chapter 1 [L1]
Performance Assessment in the Science Classroom [L1]

Description	Is it chemical or physical?	Explanation
5. Sugar dissolving in water	*Physical*	*Dissolving is a physical process. The chemical properties remain the same as the sugar dissolves.*
6. Vinegar and oil not mixing	*Physical*	*There is no interaction.*
7. A termite eating wood and producing methane gas	*Chemical*	*Digestion is a chemical process. The termite converts complex molecules in wood (cellulose) into simple molecules such as methane.*
8. The density of water increasing as salt is added	*Physical*	*Density is a physical property.*

Chapter 2 Organizer

Section	Objectives	Activities/Features
2.1 **Atoms and Their Structure** (5 days)	**1. Relate** historic experiments to the development of the modern model of the atom. **2. Illustrate** the modern model of an atom. **3. Interpret** the information available in an element block of the periodic table.	**ChemLab:** Conservation of Matter, pp. 56-57 **History Connection:** Politics and Chemistry—Elemental Differences, p. 58 **Chemistry and Society:** Recycling Glass, p. 60 **MiniLab 1:** A Penny for Your Isotopes, p. 63 **Discovery Demo:** 2.1 Evidence for Alpha Particles, pp. 50-51 **Demonstrations:** 2.2 Definite Proportions, pp. 54-55; 2.3 Think Very Small, pp. 58-59; 2.4 Thomson's Experiment, pp. 60-61; 2.5 A Massive Atom, pp. 66-67
2.2 **Electrons in Atoms** (5 days)	**4. Relate** the electron to modern atomic theory. **5. Compare** electron energy levels in an atom. **6. Illustrate** valence electrons by Lewis electron dot structures.	**Physics Connection:** Aurora Borealis, p. 73 **Everyday Chemistry:** Fireworks—Getting a Bang Out of Color, p. 76 **MiniLab 2:** Line Emission Spectra of Elements, p. 77 **Demonstration:** 2.6 Emission Spectra, pp. 70-71

Activity Materials

ChemLab (pages 56-57)
granular zinc, 1M HCl, 125-mL Erlenmeyer flask, 10-mL graduated cylinder, hot plate, balance, spatula, oven mitt

MiniLab 1 (page 63)
pennies (pre and post 1982), plastic bags, balance

MiniLab 2 (page 77)
lamp with incandescent bulb, hydrogen spectrum tube, neon spectrum tube, other spectrum tubes as available, spectrum-tube holder, high-voltage power supply (such as an induction coil), diffraction gratings

Demonstrations For a complete list of materials for the demonstrations in this chapter, see pages 50, 54, 58, 60, 66, and 70.

KEY TO TEACHING STRATEGIES

The following designations will help you decide which activities are appropriate for your students.

L1 Level 1 activities should be within the ability range of all students.

L2 Level 2 activities should be within the ability range of the average to above-average student.

L3 Level 3 activities are designed for the ability range of above-average students.

LEP LEP activities should be within the ability range of Limited English Proficiency students.

COOP LEARN Cooperative Learning activities are designed for small group work.

P These strategies represent student products that can be placed into a best-work portfolio.

Matter is Made up of Atoms

Teacher Classroom Resources

STUDENT MASTERS

Study Guide: pp. 5-6
Laboratory Manual: 2.1 The Thickness of Aluminum Foil, pp. 11-14
ChemLab and MiniLab Worksheets: pp. 15-18

TEACHING AIDS

Section Focus Transparency 3
Section Focus Transparency Master: p. 3
Basic Concepts Transparency 4, 5
Basic Concepts Transparency Masters: pp. 7-10
Lesson Plans: p. 3

Study Guide: pp. 7-8
Laboratory Manual: 2.2 Identifying Elements by Flame Tests, pp. 15-20
Critical Thinking/Problem Solving: p. 2
ChemLab and MiniLab Worksheets: p. 19

Section Focus Transparency 4
Section Focus Transparency Master: p. 4
Basic Concepts Transparency 6
Basic Concepts Transparency Master: pp. 11-12
Problem Solving Transparency 2
Problem Solving Transparency Master: pp. 3-4
Lesson Plans: p. 4

ASSESSMENT RESOURCES
Chapter Review and Assessment: pp. 11-16
Alternate Assessment in the Science Classroom
Computer Test Bank: Chapter 2
Performance Assessment in the Science Classroom
Problems and Solutions Manual: Chapter 2
Supplemental Practice Problems: Chapter 2

CHAPTER RESOURCES
Applying Scientific Methods in Chemistry: pp. 19-20
Spanish Resources: Chapter 2
English/Spanish Audiocassettes: Chapter 2
Cooperative Learning in the Science Classroom
Lab and Safety Skills in the Science Classroom

GLENCOE TECHNOLOGY

Software
Mastering Concepts in Chemistry: *Unit 1* Lesson 1, 2, and 3

Videotape
MindJogger Videoquizzes, Chapter 2

Videodisc
Chemistry: Concepts and Applications, Demonstrations, Videos, and Animations
Science and Technology Videodisc Series, Disc 2 Chemistry *Images of Atoms, Images of Heat*

CD-ROM Multimedia System
Chemistry: Concepts and Applications, Same as for Videodisc, plus Interactive Exploration

TECH PREP

The following Glencoe resources provide opportunities for integrating science and technology.

Student Edition: Chemistry and Society, p. 60 Everyday Chemistry, p. 76

Teacher Wraparound Edition: Chemistry Journal, p. 53; Extension, pp. 55, 73; Content Background, p. 67; Integrating the Sciences, p. 76

Teacher Classroom Resources: Consumer Chemistry, pp. 3-4

2 Matter Is Made Up of Atoms

Chapter Overview

Students learn the history of the atomic theory and are shown how experimental data led to the present-day model of the atom. The concept of energy levels in the atom is presented as a logical explanation of emission spectra.

Theme Connection

Macro to Submicro/Conservation

The *macroscopic* observations of the laws of *conservation* of matter and definite composition led to Dalton's atomic theory. The observation that matter is conserved in all chemical reactions shows that atoms themselves are conserved and therefore are only rearranged in chemical reactions.

✓ Assessment Planner

Choose assessment strategies from the following pages to evaluate the progress of your students.
Assess, pp. 67, 68, 71, 72, 78
MiniLabs, pp. 63, 77
ChemLab, p. 56
Portfolio, pp. 61, 68, 71, 80
Chapter Review, p. 80

GLENCOE TECHNOLOGY

Videodisc

Chemistry: Concepts and Applications
Evidence for Alpha Particles
Disc 1, Side 1, Ch. 5

Show this video of the Discovery Demo to show that radioactive elements decay to give off alpha particles. L1 LEP

Cloud model of atoms

Discovery Demo 2.1

Evidence for Alpha Particles

Purpose: To observe the trails of alpha particles.
Materials: cloud chamber, ethanol, dry ice, needle-mounted alpha radiation source, high-intensity light such as a slide projector, blotter paper

Safety Precautions:
Disposal: Code F; do not dispose of alpha source; keep it in its vial when not in use.

Procedure: Cloud chambers and needle-mounted alpha radiation sources may be purchased from commercial sources. You can construct a cloud chamber using a clear plastic sandwich box or other similar container. Paint the outside of the bottom of the box black. Drill a hole the size of the cork in the side of the box to mount the alpha source. Wet two pieces of blotter paper with ethanol and place them on opposite sides of the box. **CAUTION:**

Matter Is Made up of Atoms

Your family car, your lunch, and the chair you're sitting on are all made of matter. People sometimes think of matter as something solid and impenetrable, such as rock. After all, matter has mass and occupies space, so you should be able to see it. Or, should you? Can you think of a time that you have encountered invisible matter? Air is the invisible matter that keeps a hang glider aloft. You can't see air, but you know it's there. On a windy day, you can feel the force that air exerts. You take air into your body with every breath. If you look at treetops, you can see them swaying in the invisible breeze.

Chemistry Around You

Chemists have found that atoms are nearly all empty space. That may seem reasonable when you think of air, but the same is true of the atoms that make up solids and liquids. In the diagram of atoms in the box at left, almost all the mass is in the small dot in the center. If you were a mountain climber, your safety could depend on the strength of a steel piton and the rock into which it is driven. You might not want to think about the fact that the atoms of steel and rock are mostly empty space, too.

 Concept Check

Review the following concepts before studying this chapter.
Chapter 1: the difference between an element and a compound; distinguishing between chemical and physical changes; the meaning of the law of conservation of mass

 Chapter Preview

2.1 Atoms and Their Structure

2.2 Electrons in Atoms

Laboratory Activities
ChemLab
Conservation of Matter

MiniLabs
1. A Penny for Your Isotopes
2. Line Emission Spectra of Elements

Introducing the Chapter
Ask students whether they think that matter such as rocks, metal, glass, and plastic is made up of tiny solid particles or whether it's mostly empty space. Typical responses may be that these types of matter are made up of solid particles that have very little space between them. Ask students whether they can think of any types of matter that are mostly empty space. Point out that gases may fit this description of matter.

Using the Illustrations The fact that the glider is supported is evidence that air consists of invisible particles of matter. It's more difficult to imagine that rock and steel are made up of particles similar to those of air.

Revealing Misconceptions
Students may visualize atoms as having electrons that revolve around the nucleus in the same way that planets revolve around the sun. Emphasize that this model of the atom is inaccurate.

GLENCOE TECHNOLOGY

 CD-Rom

Chemistry: Concepts and Applications
Evidence for Alpha Particles
Show this video of the Discovery Demo to show that radioactive elements decay to give off alpha particles. L1 LEP

51

Alcohol is flammable; keep away from flames. Place the box on dry ice. **CAUTION:** *Dry ice will cause immediate frostbite; handle it with tongs or heavy thermal gloves.* Keep the box level for best results. Place the tip of the alpha radiation source needle in a cork. Mount the needle inside the chamber by placing the cork in the hole in the side of the box. Cover the box with its plastic lid. After a few minutes,

the air in the box will be saturated with ethanol vapor. Shine a high-intensity light beam through the box from the side. Have students look straight down into the box for best viewing of the vapor trails. You may need to warm the ends of the box using a lightbulb to increase the saturation of the air with alcohol vapor.
Results: The vapor trails will be visible.

Analysis: Ask these questions.
1. What are the particles that cause the alcohol vapor to condense in a vapor trail? *alpha particles (helium nuclei), particles consisting of 2 protons and 2 neutrons*
2. How might you show that the alpha particles have an electrical charge? *put a very strong magnet or a charged rod near the cloud chamber and observe whether the trails change direction*

PREPARE

Key Concepts

The Greek ideas about matter lead into a discussion of modern atomic theory, accompanied by the experimental evidence and contributions of such scientists as Lavoisier, Proust, Dalton, Thomson, and Rutherford.

Planning Ahead

For the ChemLab Be sure you have small granular zinc, not zinc powder.

For the MiniLab Locate a source of pennies having a variety of dates. Eliminate 1982 pennies.

1 FOCUS

Focus Transparency

Display the **Section Focus Transparency** for Section 2.1. Use it to review ideas from Chapter 1 and ask students to follow the pathway of nitrogen in the environment. L1 LEP

2 TEACH

Discussion

Hold up a piece of aluminum foil and ask students what type of material it is made of. *aluminum metal* Then cut or tear the foil in half and ask the same question. *same answer* Repeat several times, then ask students what logical question this process leads to. *"How small could I cut these pieces and still have aluminum?"* or *"If I cut this aluminum into small enough pieces, would they no longer be aluminum, but some other basic substance?"* Explain that it was perfectly logical for the Greeks to speculate about the smallest particles of matter, based simply upon their observations and ideas.

SECTION PREVIEW

Objectives
Relate historic experiments to the development of the modern model of the atom. Illustrate the modern model of an atom. Interpret the information available in an element block of the periodic table.

Key Terms
atom
atomic theory
law of definite
    proportions
hypothesis
theory
scientific law
electron
proton
isotope
neutron
nucleus
atomic number
mass number

Atoms and Their Structure

When did people first wonder about matter and its structure? Undoubtedly, early humans observed matter and questioned its behavior and composition. People in ancient Greece knew how to manipulate matter well enough to produce this beautiful vase. As years passed and methods of investigation improved, scientists from many times and cultures left written records of studies of matter. Current thoughts on the nature of matter evolved over a period of several thousand years.

Early Ideas About Matter

The current model of the composition of matter is based on hundreds of years of work that began when observers realized that different kinds of matter exist and that these kinds of matter have different properties and undergo different changes. About 2500 years ago, the Greek philosophers thought about the nature of matter and its composition. They proposed that matter was a combination of four fundamental elements—air, earth, fire, and water, as shown in **Figure 2.1.** These Greek philosophers also argued the question of whether matter could be divided endlessly into smaller and smaller pieces or whether there was an ultimate smallest particle of matter that could not be divided any further. The Greek philosophers were keen observers of nature but, unlike modern scientists, didn't test their hypotheses with experiments.

Figure 2.1
Greek Elements
The Greek philosophers thought that air, earth, fire, and water were the elements that formed all matter. Notice how the properties hot, dry, cold, and wet are associated with each element. These early ideas about the elements were not completely swept aside until the 19th century.

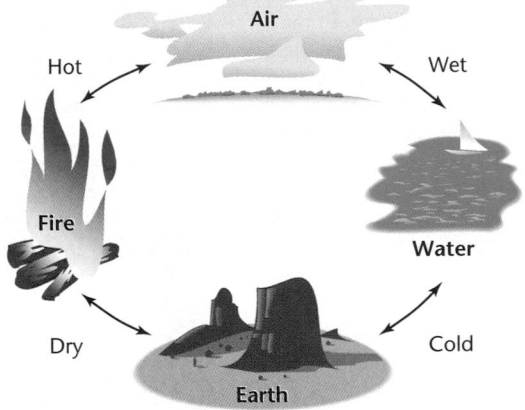

Program Resources

Study Guide, pp. 5-6 L1
Laboratory Manual, 2.1, pp. 11-14 L1
English/Spanish Audiocassettes, Chapter 2
    L1 LEP
Section Focus Transparency 3 and Master,
    p. 3 L1 LEP
Basic Concepts Transparencies 4, 5 and
    Masters, pp. 7-10 L1 LEP

Spanish Resources L1 LEP
ChemLab and MiniLab Worksheets,
    pp. 15-18 L1
Applying Scientific Methods in Chemistry,
    p. 19 L1
Consumer Chemistry, pp. 3-4 L1

Democritus, 460-370 B.C., was a philosopher who proposed that the world is made up of empty space and tiny particles called atoms. Democritus thought that **atoms** are the smallest particles of matter and that different types of atoms exist for every type of matter. The idea that matter is made up of fundamental particles called atoms is known as the **atomic theory** of matter.

Development of the Modern Atomic Theory

In 1782, a French chemist, Antoine Lavoisier (1743-1794), made measurements of chemical change in a sealed container. He observed that the mass of reactants in the container before a chemical reaction was equal to the mass of the products after the reaction. For example, in a sealed container, 2.0 g of hydrogen always reacts with 16.0 g of oxygen to give 18.0 g of water. Lavoisier concluded that when a chemical reaction occurs, matter is neither created nor destroyed but only changed. Lavoisier's conclusion became known as the *law of conservation of matter*. This is another name for the law of conservation of mass that you learned in Chapter 1. This law is illustrated in **Figure 2.2**.

Proust's Contribution

In 1799, another French chemist, Joseph Proust, observed that the composition of water is always 11 percent hydrogen and 89 percent oxygen by mass. Regardless of the source of the water, it always contains these same

Figure 2.2
Law of Conservation of Matter
When magnesium burns in air, it gives off an intense white light and some white smoke. It is difficult to measure the amounts of magnesium and oxygen involved in the reaction under these open conditions. ▼

▲ The same reaction occurs in a flashbulb, which contains fine magnesium wire and oxygen. Note the mass of the unused bulb on the left. The magnesium in the used bulb on the right has been burned, and the bulb now contains magnesium oxide. Notice that the mass did not change.

atom:
atomos (Gk) indivisible

Democritus believed that matter is made of small, indivisible particles called atoms.

Concept Development
Point out that after Democritus (about 400 B.C.) studied matter, the next 2000 years were not totally devoid of activity in chemistry; however, they were dominated by the pseudoscience of alchemy. Explain that the alchemists were obsessed with the idea of turning base metals such as iron, lead, and copper into gold. Although some believed such transformations were possible, many were simply charlatans who took advantage of human greed. However, some alchemists kept records of their discoveries of elements and other observations of matter. These records were useful to true chemists in the 18th and 19th centuries.

Extension
Ask interested students to research and report on the activities of several alchemists. Possibilities include Geber, Rhazes, Arnold of Villanova, False Geber (name unknown), Paracelsus, Libavius, van Helmont, Becher, and Stahl. L1

⚡ Quick Demo

Isn't the product heavier?

Put 50 mL of $0.1M \, K_4Fe(CN)_6$ solution in a 125-mL Erlenmeyer flask. Put about 2 mL of $0.1M$ $FeCl_3$ solution in a small test tube and stand the tube up inside the flask. Stopper and weigh the flask plus contents. Invert the flask so that the $FeCl_3$ solution spills out and reacts to produce a dark blue precipitate of $Fe_4[Fe(CN)_6]_3$. Ask students to predict how the mass has changed. *Some students will think the mass has increased because a solid formed.* Reweigh the flask plus contents to demonstrate that the mass has not changed.

Chemistry Journal

Nothing Goes Away
Ask students to keep a "trash log" for two days, entering in their journals all the garbage and trash that their families throw away and the probable ultimate destination of each type of trash. L1

TECH PREP

Concept Development

Explain that the most general version of the law of conservation of matter is the law of conservation of matter and energy, which states that the sum of the mass and energy in the universe is constant. Einstein proposed that energy and mass are interconvertible according to the equation $E = mc^2$. However, this conversion occurs only during nuclear reactions such as fission or fusion.

Discussion

Ask students why water, if it contains two atoms of hydrogen for each atom of oxygen, is only 11% hydrogen by mass. *Hydrogen atoms must be much less massive than oxygen atoms.*

Correcting Errors

Students may think that the law of definite proportions means that two elements always combine in the same proportion. Point out that this law is true for *a given compound* formed by the elements. Explain that, for example, 1.00 g of carbon may combine with either 1.33 g of oxygen to form CO or with 2.66 g of oxygen to form CO_2. Then state the law of multiple proportions: When two elements form a series of compounds, the ratios of the masses of the second element that combine with 1 g of the first element can always be reduced to small whole numbers.

Extension

Ask an interested student to research and make a class report on *phlogiston*, a hypothetical substance that a German chemist/alchemist, Georg Ernst Stahl (1660-1734), suggested flowed out of burning materials. L1

Two samples of natural FeS₂

Figure 2.3

The Law of Definite Proportions
Iron pyrite, FeS₂, looks something like gold. It fooled many gullible people and got the name *fool's gold* as a result. The composition of pyrite is 46.5 percent iron and 53.5 percent sulfur by mass. These proportions are the same for every sample of pyrite. They are independent of the origin or amount of the substance.

46.5%
Fe

53.5%
S

Composition of Iron Pyrite

percentages of hydrogen and oxygen. Proust studied many other compounds and observed that the elements that composed the compounds were always in a certain proportion by mass. This principle is now referred to as the **law of definite proportions. Figure 2.3** illustrates this law.

Dalton's Atomic Theory

John Dalton (1766-1844), an English schoolteacher and chemist, studied the results of experiments by Lavoisier, Proust, and many other scientists. He realized that an atomic theory of matter must explain the experimental evidence. For example, if matter were composed of indivisible atoms, then a chemical reaction would only rearrange those atoms, and no atoms would form or disappear. This idea would explain the law of conservation of mass. Also, if each element consisted of atoms of a specific type and mass, then a compound would always consist of a certain combination of atoms that never varied for that compound. Thus, Dalton's theory explained the law of definite proportions, as well. Dalton proposed his atomic theory of matter in 1803. Although his theory has been modified slightly to accommodate new discoveries, Dalton's theory was so insightful that it has remained essentially intact up to the present time.

The following statements are the main points of Dalton's atomic theory.
1. All matter is made up of atoms.

2. Atoms are indestructible and cannot be divided into smaller particles. (Atoms are indivisible.)

3. All atoms of one element are exactly alike, but they are different from atoms of other elements.

54   Chapter 2   Matter Is Made up of Atoms

Demonstration 2.2

 Definite Proportions

Purpose: To model the law of definite proportions.
Materials: 10 identical test tubes with matching stoppers, balance, 2 rubber bands

Safety Precautions:
Disposal: Code F
Procedure: Show students that reactant A (test tube) combines with reactant B (stopper) to form product AB (stoppered test tube). Measure and record the mass of 3 test tubes held together by a rubber band. Then measure and record the mass of three stoppers. Calculate the ratio of the two mass measurements. Repeat the above measurements and calculation using 7 test tubes and 7 stoppers.

Results: The mass ratios will be the same within experimental error. This demonstrates that the ratio of the mass of reactant A to the mass of

Dalton's atomic theory gave chemists a model of the particle nature of matter. However, it also raised new questions. If all elements are made up of atoms, why are there so many different elements? What makes one atom different from another atom? Experiments in the late 19th century began to suggest that atoms are made up of even smaller particles. Present-day chemistry explains the properties and behavior of substances in terms of three of these smaller particles. You will learn more about each of these particles in Section 2.2.

Atomic Theory, Conservation of Matter, and Recycling

Have you ever wondered what happens to the atoms in the stuff that you throw away? Where do the atoms go when waste is incinerated or buried in a landfill? As you've just learned, atoms are never created or destroyed in everyday chemical processes. So, in a sense, you can't throw anything away because there is no "away." When waste is burned or buried in a landfill, the atoms of the waste may combine with oxygen or other substances to form new compounds, but they don't go away. In natural processes, atoms are not destroyed; they are recycled. **Figure 2.4** shows how elemental nitrogen from the atmosphere is converted into compounds that are used on Earth, and then returns to the atmosphere.

In recent years, small towns, large cities, and entire states have discovered the benefits of recycling paper, plastic, aluminum, and glass. Labels on many supermarket bags, cardboard boxes, greeting cards, and other paper products say "Made from recycled paper." Scrap aluminum is easily recycled and made into new aluminum cans or other aluminum products. Have you noticed how newly paved roadways sparkle? The sparkle is a result of the addition of recycled glass to the paving material. Even ground-up tires can be added to asphalt for paving. By reusing the atoms in manufactured materials, we are imitating nature and conserving natural resources.

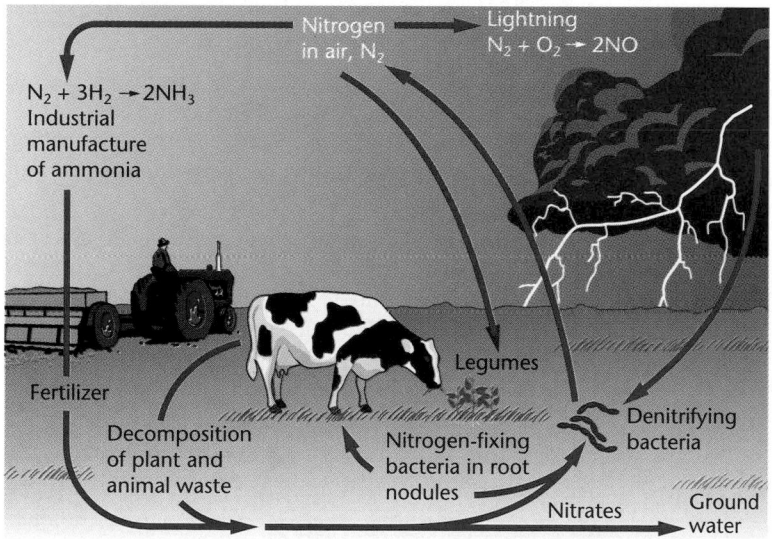

Figure 2.4

Recycling Nitrogen
Lightning, bacteria, industrial processes, and even lichens on tree branches convert nitrogen from the atmosphere into compounds. These compounds enter the plant and animal food chain. When plant and animal wastes decompose, bacteria in the soil produce free nitrogen, which returns to the atmosphere.

Reinforcement

Light a burner to demonstrate the reaction $CH_4 + 2O_2 \rightarrow CO_2 + 2H_2O$. Ask student groups to construct molecular models of both reactants and products. Ask students to count atoms of carbon, hydrogen, and oxygen for both reactants and products and to explain why the mass does not change during the reaction. *The mass does not change because both reactants and products contain the same assortment of atoms.* L1
COOP LEARN

Applying Chemistry

Do trash compactors reduce waste? Point out that although trash compactors reduce the volume of waste, the mass of the waste is not reduced and it may not degrade as readily in the environment. Also explain that disposals put organic materials into waterways, where they consume oxygen and create pollution.

Extension

 Ask interested students to investigate and make class presentations on **TECH PREP** ways in which we may compost some waste and use the product in yards and/or gardens. L1

reactant B in any sample is always the same.
Analysis: Ask these questions.
1. How do the ratios of the two samples compare? *They are the same.*
2. What would you expect to discover about the ratio of hydrogen to oxygen in samples of water molecules taken from various oceans? *The ratio would be the same.*

Assessment

Performance: Zinc reacts with iodine in the exothermic reaction $Zn(s) + I_2(s) \rightarrow ZnI_2(s)$. In a fume hood, place 0.52 g of Zn dust in an evaporating dish. Mix completely with the zinc dust 2.0 g of I_2 crystals. These are stoichiometrically correct masses for ZnI_2. Use a dropper to add a few drops of water to the mixture. **CAUTION:** *This is an exothermic reaction that produces toxic fumes.* Have students calculate the ratio of the masses of the reactants. Disposal: Code A. L1

Conservation of Matter

Time Allotment: One laboratory period.

Objectives

Review objectives with students before they begin the ChemLab.

Process Skills

Observing, measuring, using numbers, interpreting data, communicating

Safety Precautions and Disposal

Advise students to wear safety goggles and aprons and to handle hydrochloric acid and hot objects with care.

Disposal: Code A for zinc chloride. The residue in each flask may be dissolved in a small amount of water, then the solutions collected and evaporated prior to disposal.

PREPARATION

120 mL of 1*M* HCl may be prepared by adding 10 mL concentrated (12*M*) HCl to 110 mL of water with stirring. *Carry out this procedure in a fume hood and wear safety goggles and an apron.* Do not use powdered zinc, as the reaction will occur so rapidly that zinc and liquid splatter from the flask.

PROCEDURE

- Require that students' notebooks contain the following items before they begin: Objectives; Safety Precautions; Outline of the Procedure; Data and Observations Table.
- One hot plate may be used for up to four flasks as long as students clearly mark their flasks to avoid switching them.

Conservation of Matter

Hydrochloric acid reacts with metals such as zinc, magnesium, and aluminum. A colorless gas escapes, and the metal seems to disappear. What happens to the atoms of the metals in these reactions? Are they destroyed? If not, where do they go?

Problem

What happens to the atoms of a metal when they react with an acid?

Objectives

- **Infer** what happens to atoms during a chemical change.
- **Compare** experimental results to the law of conservation of matter.

PREPARATION

Materials

granular zinc	hot plate
1*M* HCl	balance
125-mL Erlenmeyer flask	spatula
10-mL graduated cylinder	oven mitt

Safety Precautions

Use care when handling hydrochloric acid and hot objects. Wear goggles and laboratory aprons. Perform this ChemLab in a well-ventilated room.

PROCEDURE

1. Weigh a clean, dry, 125-mL Erlenmeyer flask to the nearest 0.01 g. Record its mass in a data table.
2. Obtain a sample of granular zinc, and add it to your flask. Weigh the flask with the zinc in it.

Record the total mass in the data table. Subtract to find the mass of zinc. The mass of zinc must be between 0.20 g and 0.28 g. If you have too much zinc, remove some with a clean spoon or spatula until the mass is in the range described.

3. Add 10 mL of 1*M* hydrochloric acid to the flask. Swirl the contents and look for signs of a chemical change. Record your observations in the data table.
4. Set the flask on a hot plate as shown in the photo.

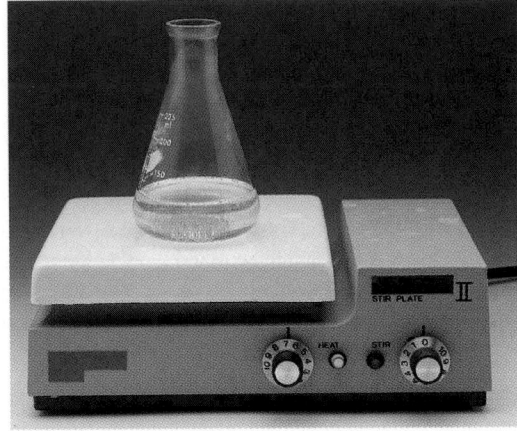

5. Set the hot plate on low to heat the flask gently. The liquid in the flask should not boil. Again look for signs that a chemical change is taking place, and record your observations in the data table.
6. Eventually, all the metallic zinc will disappear. It is important to heat the flask slowly. **CAUTION:** *Be careful not to inhale the fumes from the flask at this point.* When the liquid in the flask is almost gone, a white solid will appear. Stop heating as soon as you see this solid. Use an oven mitt to remove the flask from the hot plate The heat retained by the flask should be enough to completely evaporate the remaining liquid.

7. Let the flask cool. When all the liquid has evaporated, weigh the cool, dry flask and its contents. Record this mass in the data table. Calculate the mass of zinc chloride produced in the reaction.

ANALYZE AND CONCLUDE

1. **Interpreting Data** How did the mass of the product, zinc chloride, compare with the mass of the zinc?

2. **Interpreting Observations** How can you account for the difference in mass?

3. **Observing and Inferring** What happened to the zinc?

4. **Making Inferences** Why were the flask and its contents heated?

APPLY AND ASSESS

1. Chemists have determined that zinc chloride is 48 percent zinc. Use this information to compute the mass of zinc in your product. How does this mass compare with the mass of zinc you started with?

2. If the difference in the question above is greater than 0.04 g, how can you account for it?

3. How does this experiment support the law of conservation of matter?

4. How could you modify the procedure so that the law of conservation of matter is better

DATA AND OBSERVATIONS

Mass of empty Erlenmeyer flask	
Mass of Erlenmeyer flask with zinc sample	
Mass of zinc sample	
Mass of Erlenmeyer flask with reaction product (zinc chloride, $ZnCl_2$)	
Mass of zinc chloride produced	
What did you see when hydrochloric acid was first added to the flask?	
What did you see when you began to heat the flask?	
What did you see when all the liquid had evaporated?	

DATA AND OBSERVATIONS

Mass of empty flask	96.20 g	What did you see when hydrochloric acid was first added to the flask?	Gas bubbles formed as the acid reacted with the zinc.
Mass of flask with zinc sample	96.45 g		
Mass of zinc sample	0.25 g		
Mass of flask with reaction product (zinc chloride, $ZnCl_2$)	96.72 g	What did you see when you began to heat the flask?	The rate of bubbling increased.
Mass of zinc chloride produced	0.52 g	What did you see when all the liquid had evaporated?	A gray-white residue remained.

ANALYZE AND CONCLUDE

1. The mass of the zinc chloride is greater than the mass of the zinc.

2. The residue is a compound that contains both zinc and chlorine.

3. It has become a component of a compound.

4. to speed up the reaction between the zinc and HCl

APPLY AND ASSESS

1. Answers should be 0.48 times the mass of the product. Comparisons will vary.

2. If the product is still moist, the calculated mass will be higher than the original mass. If the product was overheated, some may have sublimed, decomposed, or spattered out of the flask. In this case, the calculated mass will be lower than the original. Other differences are probably due to weighing errors.

3. No matter was lost in the reaction. Zinc reacted with HCl, and hydrogen was evolved as a gas.

4. Students should suggest a procedure in which the masses of all products including the hydrogen and water vapor given off during the reaction are accounted for. A possibility is to cover the flask with a balloon that is included in the mass with appropriate corrections for buoyancy.

✓ Assessment

Knowledge: Ask students why the mass of the hydrochloric acid was not determined, added to the initial mass of the flask and zinc, and compared with the final mass. *The hydrochloric acid was partially water, and the water evaporated during the heating process. The hydrogen and excess HCl in the hydrochloric acid were evolved as gases. Therefore, the "lost" water, hydrogen, and HCl could not be weighed after the reaction.* L1

HISTORY CONNECTION

Politics and Chemistry—Elemental Differences

In the time of Antoine Lavoisier (1743-1794), many scientists were still trying to explain matter as combinations of the elements air, earth, fire, and water. Lavoisier's work changed the way chemistry was done, and today he is recognized as the first modern chemist. However, like other scientists of the 18th century, Lavoisier could not earn a living as a chemist, so he invested in a private firm that collected taxes for the king.

Remaking chemistry Lavoisier set out to reorganize chemistry. Lavoisier's habit of carefully weighing reactants and products in experiments led him to discover that the mass of materials before a chemical change equals the mass of the products after it, which is the basis of the law of conservation of matter. He also discovered that combustion is the result of reaction with oxygen.

International recognition Many scientists held Lavoisier in high esteem. Benjamin Franklin made a point to observe experiments by Lavoisier when he was in France soliciting support for the cause of the American Revolution. Lavoisier's experiments were also followed closely by Thomas Jefferson.

In 1774, the British chemist Joseph Priestley discussed one particular experiment with Lavoisier. Priestley explained that after heating "calx of mercury" (which we know today as mercury(II) oxide), metallic mercury remained and a gas was given off. When he placed a candle in the gas, it burned more brightly. He also found that if a mouse is placed in a closed jar with the gas, the mouse can breathe it and live. Priestley's gas was oxygen, but because he believed in an older theory of matter called the phlogiston theory, Priestley did not recognize it as an element. Lavoisier repeated Priestley's experiment and came to the history-making conclusion that air is not a simple substance but a mixture of two different gases. One of these gases, oxygen, supports combustion, promotes breathing, and rusts metals. Lavoisier gave oxygen its name.

Political price Lavoisier was not a member of the aristocracy. He belonged to the professional class from which many of the leaders of the French Revolution came. In spite of his class and the high regard for Lavoisier in the scientific world, his connection with the tax-collecting firm made him a target of suspicion. During the Reign of Terror that followed the French Revolution, Lavoisier was arrested and condemned to death in a trial that lasted less than a day. That same day, he was guillotined and his body thrown into a common grave—a victim of ignorance and mob rule.

Connecting to Chemistry
1. **Analyzing** Why was Lavoisier's role in the discovery of oxygen important even though he merely repeated Priestley's experiment?
2. **Applying** Lavoisier showed that a person uses more oxygen when working than when resting. Explain the reasoning behind Lavoisier's findings.

58

Hypotheses, Theories, and Laws

The first step to solving a problem, such as what makes up matter, is observation. Scientists use their senses to observe the behavior of matter at the macroscopic level. They then come up with a **hypothesis,** which is a testable prediction to explain their observations. For example, Lavoisier thought that matter was indestructible. He made a hypothesis that all the matter present before a chemical change would still be there after the change.

To find out whether a hypothesis is correct, it must be tested by repeated experiments. Lavoisier performed numerous careful experiments using different chemical changes. Scientists accept hypotheses that are verified by experiments and reject hypotheses that can't stand up to experimental testing.

A body of knowledge builds up as a result of these experiments. Scientists develop theories to organize their knowledge. A **theory** is an explanation based on many observations and supported by the results of many experiments. For example, Dalton's atomic theory was based on observations of matter performed again and again by many scientists. As scientists gather more information, a theory may have to be revised or replaced with another theory. In nonscientific speech and writing, the word *theory* is often used to mean an unsupported notion about something. As you can see, a scientific theory is heavily supported by evidence gained from experiments and continual observations. The methods of science just discussed are summarized in **Figure 2.5.**

A **scientific law** is simply a fact of nature that is observed so often that it becomes accepted as truth. That the sun rises in the east each morning is a law of nature because people see that it is true every day. A law can generally be used to make predictions but does not explain why something happens. In fact, theories explain laws. One part of Dalton's atomic theory explains why the law of conservation of matter is true.

Figure 2.5
The Methods of Science
Scientists make observations that lead to hypotheses. A hypothesis must be tested by multiple experiments. If the experimental results do not agree with the hypothesis, they become the observations that lead to a new hypothesis. A hypothesis that is refined and supported by many experiments becomes a theory that explains a fact or phenomenon in nature.

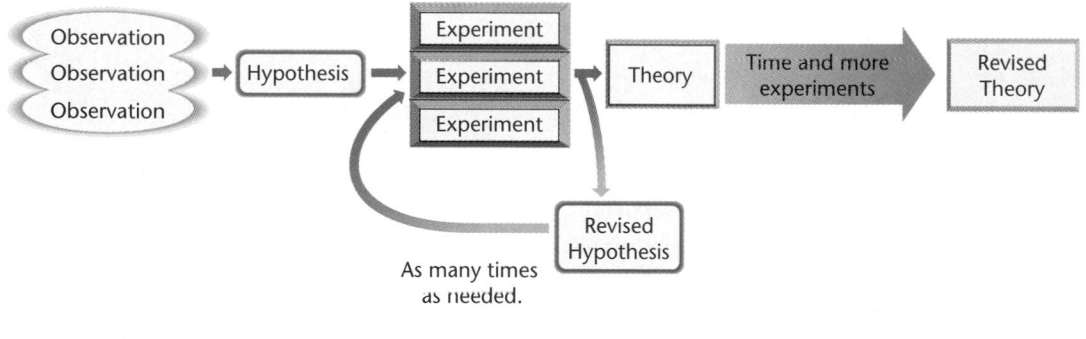

Correcting Misconceptions

Students may think that laws proceed from theories through experimentation over a long period of time. Point out that a scientific law can always be observed to be true whether or not it can be explained. For example, carefully done experiments in the late 18th century showed that no matter was lost or gained during chemical reactions. This law predated Dalton's explanation in his atomic theory.

Reinforcement

In order to help students better understand the process of creating theories and models, ask them to work in groups of three. The first student should put an object in a shoe box and tape the box closed, not allowing the remaining two students to see the object. The second student should manipulate the box for several minutes and describe the object for the third student, who should write a description or make a drawing of the object. Then the students may open the box and note the accuracy of the description or model. Explain that scientists had to use similar methods to determine the characteristics of atoms because they were not able to view atoms directly. L1
LEP   COOP LEARN

GLENCOE TECHNOLOGY

 Videodisc

Chemistry: Concepts and Applications
See the Videodisc Program Teacher Guide to access illustrations and other material available for this chapter.

---

2. Do you think helium or vanillin will leak more rapidly from equally inflated balloons? Explain your answer. *Point out to students that vanillin consists of molecules having the formula $C_8H_8O_3$, whereas helium consists of single atoms. Being much smaller, the helium atoms will leak from the balloon much faster than the vanillin molecules.*

✔Assessment

Oral: Have students discuss why helium-filled rubber balloons begin to become soft and deflated much faster than air-filled balloons. Perhaps students have noticed this difference. *Students should suggest that molecules that make up air are larger than atoms of helium.* L1

Chemistry and
SOCIETY

Purpose

Students learn how recycling relates to the law of conservation of matter.

Background

Glass that has significant levels of lead cannot be used for road construction and drainage applications. Students may think that almost any material can be used in a road because it is run over by tires and is splashed with mud. In reality, the material in a road is subject to runoff from precipitation, which eventually reaches streams and rivers. The runoff can affect wildlife and water systems.

Teaching Strategies

Mention that many wineries are switching to yellow-green bottles. This new color bottle can use only 15% green cullet, compared to 80% recycled glass allowable in all-green bottles.

Analyzing the Issue

1. Cullet is used in most glass manufactured for everyday products in which clarity and colorlessness are not important. Using cullet does not save money to the extent that recycling aluminum does. However, it reduces the cost of obtaining sand, lime, and other ingredients. In addition, the melted cullet provides a sort of solvent for the raw materials.

2. Obtaining the empty bottles should be fairly easy. Many wineries use bottles of specific shapes and colors so you would have to find a way of matching the bottles to the needs of individual wineries and still offer them at competitive prices. Students should consider other factors such as the bulkiness of empty bottles in comparison to their value.

Recycling Glass

Imagine that you arrive home thirsty from a soccer game. You reach into the refrigerator and pull out an ice-cold bottle of apple juice. In seconds, the empty bottle is all that remains. But what about the law of conservation of matter? Because of this law, the bottle lingers on—as trash. Worse yet, unless the bottle is recycled, its matter will be conserved in a landfill.

Problems of recycling glass Glass manufacturers call broken glass that can be added to new material *glass cullet.* There is no difficulty in recycling uncolored glass or amber glass because the market for these kinds of cullet is strong. Only green glass, the kind found in some bottles, presents a problem.

Solutions in recycling methods Because of the color it adds, green glass cullet often can't be mixed with uncolored or amber glass cullet. For this reason, recycling centers often sort glass by color. The sorted glass is dropped down separate chutes and processed by separate glass crushers. The end products are delivered separately to glass plants, which pay more for sorted cullet.

Finding uses for recycled glass Many areas, such as the state of Washington, are searching for new markets for recycled glass. An example of one new use is adding ten percent green glass cullet or mixed glass cullet to asphalt for roads. This seems to be working well, but engineers will continue to

study this surface for its durability. The glass particles make the roads sparkle, which is an added attraction.

Washington's Department of Trade and Economic Development has developed a list of more than 70 uses of glass cullet. These uses include manufacturing construction material such as fiberglass, foam glass, and rock wool insulation; using it in material for building roadbeds; and making pressed glass such as dinnerware and decorative glass. One other use of glass cullet is in removing pollutants in water runoff. In this extremely useful technology, glass cullet is given a static electrical charge, which makes it able to attract and hold small particles in runoff from areas such as airport runways.

Analyzing the Issue

1. **Acquiring Information** Do research to find out how new glass is manufactured. Learn whether cullet is routinely used in most glass manufacturing. Does the use of cullet reduce the cost of manufacturing glass? Does the use of cullet reduce the energy requirement for producing glass?

2. **Inferring** A company in Richmond, California, makes a business of washing wine bottles for reuse—9 million a year. If you were starting such a business, what marketing and distribution problems might you have to solve?

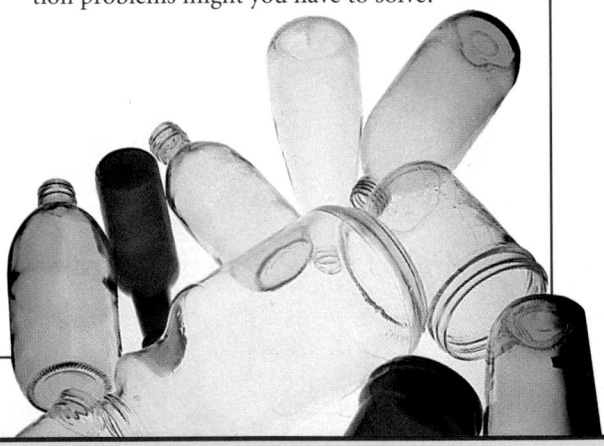

Demonstration 2.4

Thomson's Experiment

Purpose: To observe the characteristics of a cathode-ray tube.

Materials: Crooke's tube (cathode-ray tube), high-voltage DC power supply, bar magnet

Safety Precautions: 🔧 👕 👓

Disposal: Code F

Procedure: A Crooke's tube is available commercially. A physics teacher may have the high-voltage DC power supply needed to operate it.

Connect the two electrodes of the power supply to the ends of the cathode-ray tube using wire with alligator clips on each end. Turn on the power supply. **CAUTION:** *The high-voltage power supply can cause severe electric shocks.* If the tube fails to light, increase the voltage from the power supply. If the tube again fails to light, turn off the power supply and reverse the wires on the cathode-ray tube. Have students observe the electron beam on the fluorescent

The Discovery of Atomic Structure

Dalton's atomic theory was *almost* true. Dalton had assumed that atoms are the ultimate particles of matter and can't be broken up into smaller particles and that all atoms of the same element are identical. However, his theory had to be modified as new discoveries were made in the late 19th and early 20th centuries. Today, we know that atoms are made up of smaller particles and that atoms of the same element can be nearly, but not exactly, the same. In this section, you'll examine the discoveries that led to the modern atomic theory.

The Electron

Because of Dalton's atomic theory, most scientists in the 1800s believed that the atom was like a tiny solid ball that could not be broken up into parts. In 1897, a British physicist, J.J. Thomson, discovered that this solid-ball model was not accurate.

Thomson's experiments used a vacuum tube such as that shown in **Figure 2.6.** A vacuum tube has had all gases pumped out of it. At each end of the tube is a metal piece called an electrode, which is connected through the glass to a metal terminal outside the tube. These electrodes become electrically charged when they are connected to a high-voltage electrical source. When the electrodes are charged, rays travel in the tube from the negative electrode, which is the cathode, to the positive electrode, the anode. Because these rays originate at the cathode, they are called cathode rays.

Thomson found that the rays bent toward a positively charged plate and away from a negatively charged plate. He knew that objects with like charges repel each other, and objects with unlike charges attract each other. Thomson concluded that cathode rays are made up of invisible, negatively charged particles referred to as **electrons.** These electrons had to come from the matter (atoms) of the negative electrode.

Modern cathode-ray tubes are the picture tubes in TV sets and computer monitors. These tubes use a varying magnetic field to cause the electron beam to move back and forth, painting a picture on a screen that's coated with luminescent chemicals.

The pole of a magnet bends the cathode rays in a direction at a right angle to the direction of the field. ▼

Figure 2.6
Cathode-Ray Tube
When a high voltage is applied to a cathode-ray tube, cathode rays form a beam that produces a green glow on the fluorescent plate. ▶

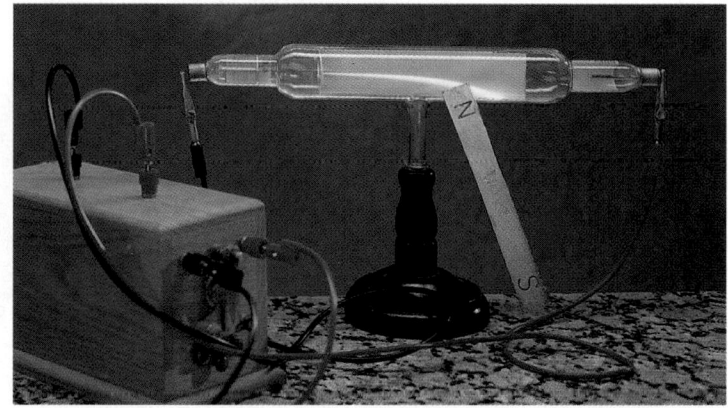

Visual Learning

Figure 2.6: Ask how Thomson was able to reason that matter also must contain positive charge, a conclusion that led him to propose his "plum pudding" model of the atom. *Because matter is electrically neutral, atoms must also be neutral and therefore contain equal amounts of positive and negative charge.*

GLENCOE TECHNOLOGY

Videodisc

Chemistry: Concepts and Applications
Thomson's Experiment
Disc 1, Side 1, Ch. 6

Show this video of the demonstration on these pages to reinforce the concept of the electron nature of cathode rays. `L1` `LEP`

CD-Rom

Chemistry: Concepts and Applications
Thomson's Experiment
Show this video of the demonstration on these pages to reinforce the concept of the electron nature of cathode rays. `L1` `LEP`

---

screen. Deflect the beam with the magnet. Reverse the magnet using the other end to show deflection in the opposite direction. **Results:** Students see evidence of an electron beam. The phosphors on the plate glow along the path of the beam. Point out that Thomson observed deflection by means of both magnets and charged plates.

Analysis: Ask these questions.
1. What properties account for the bending of the cathode ray in a magnetic field? *The ray has an electrical charge.*
2. What is the function of the electromagnets on a TV picture tube? *They deflect the electron beam that forms the picture on the face of the tube.*

✓ Assessment

Portfolio: Have students compare the effect of charged plates and magnets on electron beams and how Thomson was able to deduce the negative charge on electrons by both means. Diagrams and/or models will enhance the presentation. `L1`

Does matter contain charged particles?

Rub a glass rod with silk and touch it to an electroscope or to two pith balls that are hanging together. *Having acquired a positive charge, the leaves or balls repel each other and separate.* Charge the rubber rod by rubbing it with fur, and bring the rod close to the electroscope or the pith balls. *The leaves come back together as negative charge is forced away from the rubber rod or the pith balls are attracted to the negatively charged rubber rod.* Ask students how the glass rod acquires a positive charge and the rubber rod acquires a negative charge. *Electrons are rubbed off the glass and off the fur.*

Applying Chemistry

Fabric Softeners Explain that fabric softeners deposit molecules on fabric and that each molecule absorbs a layer of electrons. The electrons then repel each other, making the fabric soft and nonclinging.

GLENCOE TECHNOLOGY

 Software

Mastering Concepts in Chemistry
Unit 1, Lesson 1
Basic Particles
Use this lesson to reinforce the concept of the subatomic particle structure of the atom. L1
LEP

From Thomson's experiments, scientists had to conclude that atoms were not just neutral spheres, but somehow were composed of electrically charged particles. In other words, atoms were not indivisible but were composed of smaller particles, referred to as subatomic particles. Further experimentation showed that an electron has a mass equal to 1/1837 the mass of a hydrogen atom, the lightest atom.

Reason should tell you that there must be a lot more to the atom than electrons. Matter is not negatively charged, so atoms can't be negatively charged either. If atoms contained extremely light, negatively charged particles, then they must also contain positively charged particles—probably with a much greater mass than electrons. Scientists immediately worked to discover such particles.

Protons and Neutrons

In 1886, scientists discovered that a cathode-ray tube emitted rays not only from the cathode but also from the positively charged anode. These rays travel in a direction opposite to that of cathode rays. Like cathode rays, they are deflected by electrical and magnetic fields, but in directions opposite to the way cathode rays are deflected. Thomson was able to show that these rays had a positive electrical charge. Years later, scientists determined that the rays were composed of positively charged subatomic particles called **protons.** The amount of charge on an electron and on a proton is equal but opposite, but the mass of a proton is much greater than the mass of an electron. The mass of a proton was found to be only slightly less than the mass of a hydrogen atom.

At this point, it seemed that atoms were made up of equal numbers of electrons and protons. However, in 1910, Thomson discovered that neon consisted of atoms of two different masses, **Figure 2.7.** Atoms of an element that are chemically alike but differ in mass are called **isotopes** of the element. Today, chemists know that neon consists of three naturally occurring isotopes. The third was too scarce for Thomson to detect.

Because of the discovery of isotopes, scientists hypothesized that atoms contained still a third type of particle that explained these differences in mass. Calculations showed that such a particle should have a mass equal to that of a proton but no electrical charge. The existence of this neutral particle, called a **neutron,** was confirmed in the early 1930s.

Figure 2.7
Thomson's Isotopes of Neon
These diagrams represent the two isotopes of neon that Thomson observed. Both nuclei have ten protons, but the one on the left has ten neutrons and the one on the right has 12 neutrons.

Neon-20 nucleus

Neon-22 nucleus

Cultural Diversity

Properties of Materials

Islamic philosophers differed markedly from the Greek view of material properties, as described in the writings attributed to Jabir ibn Hayyan (720-815) and those of Ibn Sina Abu Ali al Husain ibn Abdula (980-1037), known as Avicenna in the Western world. To the Greeks, the property of a material was the same as the property of the individual atoms that make it up. For example, silver would appear lustrous because its individual atoms have the property of luster. Islamic philosophers attributed silver's luster to the arrangement of the atoms within it. By the Middle Ages, the Islamic alchemists and artisans had isolated so many materials that they began to classify materials by their physical and chemical properties.

A Penny for Your Isotopes

Many elements have several naturally occurring isotopes. Isotopes are atoms of the same element that are identical in every property except mass. When chemists refer to the atomic mass of one of these elements, they really mean the average of the atomic masses of the isotopes of that element. You can use pennies to represent isotopes. Determine the mass of two penny isotopes and then the average mass of a penny.

Procedure

1. Get a bag of pennies from your teacher.

2. Sort the pennies by date. Group the pre-1982 pennies and the post-1982 pennies.

3. Weigh ten pennies from each group. Record the mass to the nearest 0.01 g. Divide the total mass by ten to get the average mass for one penny in that group.

4. Count the number of pennies in each group.

5. Using the data, determine the total mass of all of the pre-1982 pennies. In the same way, calculate the total mass of the post-1982 pennies.

Analysis

1. What is the mass of all the pennies?

2. Calculate the average mass of a penny. How does this mass compare with the average masses for each group?

3. Why were you directed to weigh ten pennies of a group and divide the mass by ten to get the mass of one penny? Why not just weigh one penny from each group?

Rutherford's Gold Foil Experiment

While all of these subatomic particles were being discovered, scientists were also trying to determine how the particles of an atom were arranged. After the discovery of the electron, scientists pictured an atom as tiny particles of negative electricity embedded in a ball of positive charge. You could compare this early atomic model to a ball of chocolate-chip cookie dough (except with microscopic chocolate chips). Almost at the same time, a Japanese physicist, Hantaro Nagaoka, proposed a different model in which electrons orbited a central, positively charged nucleus—something like Saturn and its rings. Both models are shown in **Figure 2.8**.

Nagaoka's model resembled a planet with moons orbiting in a flat plane. He believed the core had a positive charge, and the negative electrons were in orbit around it. ▼

Figure 2.8
Thomson's and Nagaoka's Atomic Models
Thomson pictured the atom as consisting of electrons embedded in a ball of positive charge. ▶

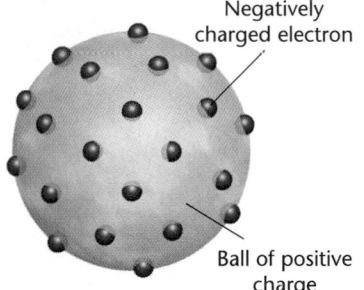

Negatively charged electron

Ball of positive charge

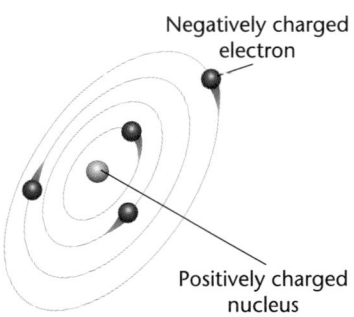

Negatively charged electron

Positively charged nucleus

2.1 **Atoms and Their Structure**   **63**

Content Background

Rutherford's Surprise You may want to describe the way in which Rutherford assigned the gold foil experiment to Ernest Marsden and Hans Geiger more or less to keep them busy. He certainly expected no remarkable result. His surprise at the result—some alpha particles deflecting at large angles or even bouncing back—was so great that he wrote of it as comparable to firing a cannon at a piece of tissue paper and having the shell bounce back.

Concept Development

Emphasize throughout the study of atomic theory that our present information and theories about the atoms are by no means absolutely correct. Emphatically dispel the idea that we now know everything about atoms.

GLENCOE TECHNOLOGY

Videodisc

Chemistry: Concepts and Applications
Rutherford's Gold Foil Experiment
Disc 1, Side 1, Ch. 7

Show this animation to reinforce the text discussion of Rutherford's discovery of the nuclear atom. `L1` `LEP`

CD-Rom

Chemistry: Concepts and Applications
Rutherford's Gold Foil Experiment
Show this animation to reinforce the text discussion of Rutherford's discovery of the nuclear atom. `L1` `LEP`

In 1909, a team of scientists led by Ernest Rutherford in England carried out the first of several important experiments that revealed an arrangement far different from the cookie-dough model of the atom. Rutherford's experimental setup is shown in **Figure 2.9.**

The experimenters set up a lead-shielded box containing radioactive polonium, which emitted a beam of positively charged subatomic particles through a small hole. Today, we know that the particles of the beam consisted of clusters containing two protons and two neutrons and are called alpha particles. The sheet of gold foil was surrounded by a screen coated with zinc sulfide, which glows when struck by the positively charged particles of the beam.

Figure 2.9
The Gold Foil Experiment
Gold is a metal that can be pounded into a foil only a few atoms thick. Rutherford's group took advantage of this property in their experiment.

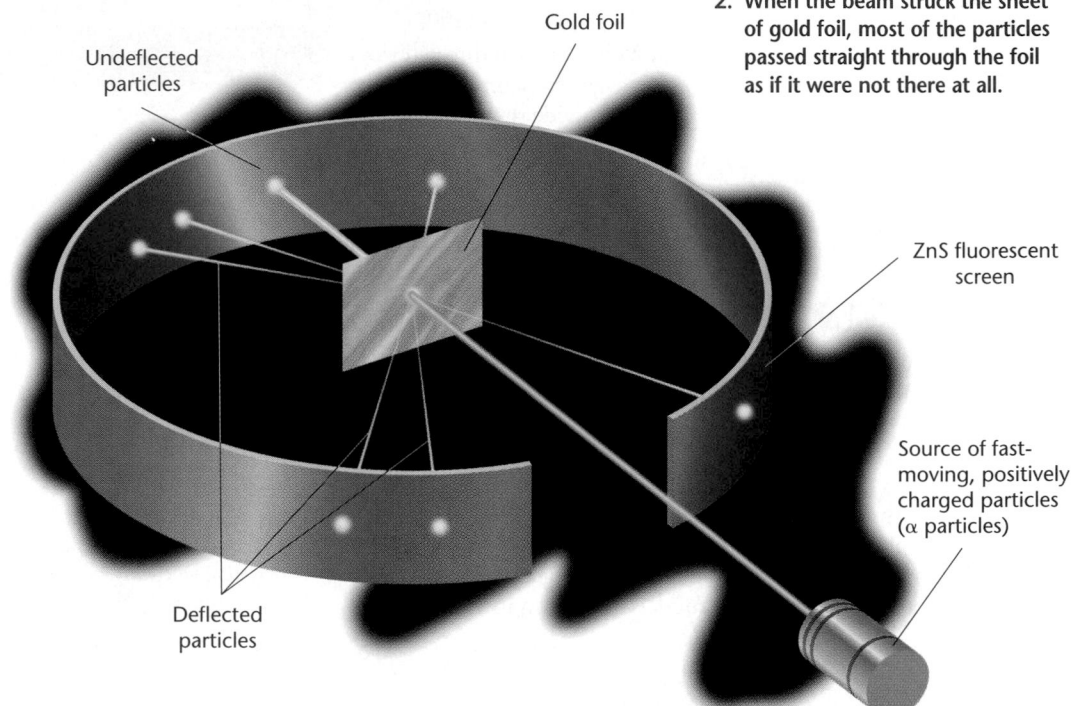

2. When the beam struck the sheet of gold foil, most of the particles passed straight through the foil as if it were not there at all.

3. However, a few of the beam's particles were deflected. Some were deflected slightly, but a few bounced almost straight back. Imagine throwing a baseball at a sheet of tissue paper and having the ball bounce back at you. Rutherford's scientists were just as surprised.

1. Rutherford's lead-shielded box contained radioactive polonium. As polonium decays, it emits helium nuclei, which consist of two protons and two neutrons. These nuclei are called alpha particles. Because they have no electrons, they are positively charged.

Meeting Individual Needs

Learning Disabled Allow students to represent atoms of helium-3 and helium-4 by using three colors of 1/4-inch-diameter self-adhesive dots to represent protons, neutrons, and electrons and sticking them to notebook paper so that protons and neutrons are in a nuclear 1/2-inch-diameter circle and the electrons are at the edge of the paper, about 8.5 inches away. Explain that if the atoms were represented to actual scale, because the diameter of an atom is about 10 000 times the diameter of the nucleus, the electrons should be shown at a distance of about 420 feet. `L1` `LEP`

The Nuclear Model of the Atom

To explain the results of the experiment, Rutherford's team proposed a new model of the atom. Because most of the particles passed through the foil, they concluded that the atom is nearly all empty space. Because so few particles were deflected, they proposed that the atom has a small, dense, positively charged central core, called a **nucleus.** The new model of the atom as pictured by Rutherford's group in 1911 is shown in **Figure 2.10.**

Imagine how this model affected the view of matter when it was announced in 1911. When people looked at a strong steel beam or block of stone, they were asked to believe that this heavy, solid matter was practically all empty space. Even so, this atomic model has proven to be reasonably accurate. As an example of the emptiness of an atom, consider the simplest atom, hydrogen, which consists of an electron and a nucleus of one proton. If hydrogen's proton were increased to the size of a golf ball, the electron would be about a mile away and the atom would be about 2 miles in diameter. To get an idea of how small an atom is, consider that there are roughly 6 500 000 000 000 000 000 000 (6.5 sextillion or 6.5×10^{21}) atoms in a drop of water. If you need more help in expressing very large and very small quantities using scientific notation, study page 795 in the *Skill Handbook*, Appendix A.

The team concluded that the nucleus is extremely small compared to the atom as a whole because so few particles were deflected. ▼

Figure 2.10
Discovery of the Nuclear Model
Remember that the beam's particles were positively charged. If the nucleus were negatively charged, the particles would attract each other. Because the beam was repelled, the nucleus must also be positively charged. ▶

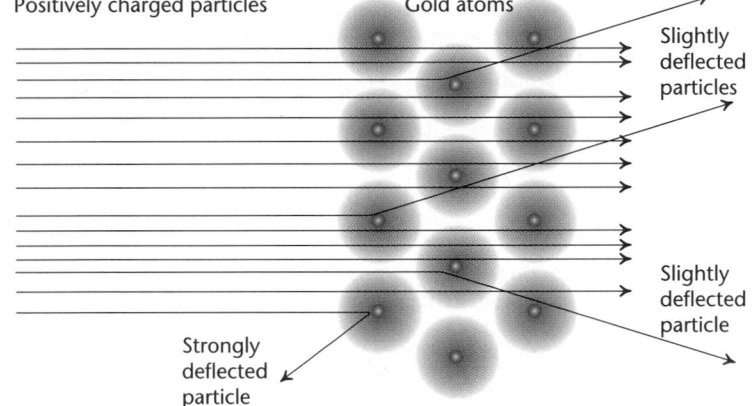

Positively charged particles   Gold atoms

Slightly deflected particles

Slightly deflected particle

Strongly deflected particle

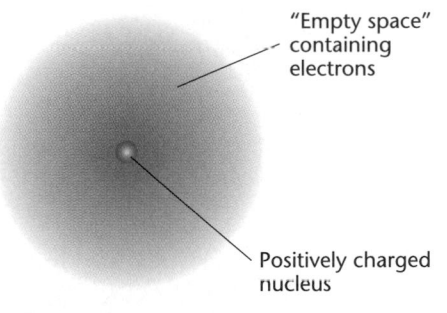

"Empty space" containing electrons

◀ Because electrons have little mass, the nucleus must contain almost all the mass of an atom. Therefore, the nucleus must be extremely dense and surrounded by empty space in which electrons were found.

Positively charged nucleus

Concept Development

To help students grasp the large amount of space within atoms and the density of the nucleus, explain that a cluster of nuclear particles the size of a pea would have a mass of about 250 million tons.

Correcting Errors

Students are often confused about the relative numbers of protons, neutrons, and electrons in atoms and may even think that, as in a carbon-12 atom, their numbers must always be equal. Explain that the number of protons determines the identity of the element, the number of electrons matches the number of protons in a neutral atom, and the number of neutrons determines the particular isotope of the element. Illustrate with a description of several atoms, negative ions, and positive ions of chlorine-35 and chlorine-37.

Transparency

 Display **Basic Concepts Transparency 4.** Use it to review Rutherford's experiment and the meaning of the results. L1 LEP

GLENCOE TECHNOLOGY

 **Videodisc**

Science and Technology Videodisc Series
Images of Atoms
Disc 2: Chemistry
Side 1, Ch. 4

Show this video to give a view of atoms on the surface of solids. L1 LEP

Across the Curriculum

History

In 1932, two British scientists used accelerated protons to disintegrate a lithium nucleus. The resulting reaction generated two helium nuclei and 17.3 million electron volts of energy. In 1934, Enrico Fermi successfully split a uranium atom in the first nuclear fission reaction. The uncontrolled fission of uranium was used in the atomic bomb, but nuclear power plants employ controlled fission. Today, the world faces the problem of what to do with the radioactive products of nuclear reactions. Proposals to dispose of nuclear waste include underground burial, use as fuel in more efficient nuclear reactors, conversion into other elements, and dumping at sea or in space. Each proposal has both political and technological drawbacks, so no solution has been reached.

Correcting Misconceptions

Correcting Misconceptions

Students often think that only certain elements have isotopes and/or that all isotopes are radioactive. Explain that most elements have naturally occurring isotopes, but not all isotopes occur naturally or are stable. Emphasize that all the isotopes of an element behave the same chemically; however, those that are unstable emit energy and/or particles when they decay and are said to be radioactive.

Extension

Ask an interested student to research the operation of a mass spectrometer and make a poster having a schematic diagram showing how a mass spectrometer can separate atoms of different masses. The student could explain the operation to the class. L1

Transparency

Display **Basic Concepts Transparency 5**. Use it to reinforce the fact that most elements are mixtures of two or more isotopes and that the average atomic mass reflects the composition of this mixture. L1 LEP

GLENCOE TECHNOLOGY

Software

Mastering Concepts in Chemistry
Unit 1, Lesson 2
Atomic Number, Atomic Mass Number, and Isotopes
Use this lesson to reinforce the concepts of nuclear structure and isotopes. L1 LEP

Atomic Numbers and Masses

Look at the periodic table hanging on the wall of your classroom or the one inside the back cover of your textbook. Notice that the elements are numbered from 1 for hydrogen, 2 for helium, through 8 for oxygen, and on to numbers above 100 for the newest elements created in the laboratory. At first glance, it may seem that these numbers are just a way of counting elements, but the numbers mean much more than that. Each number is the atomic number of that atom. The **atomic number** of an element is the number of protons in the nucleus of an atom of that element. It is the number of protons that determines the identity of an element, as well as many of its chemical and physical properties, as you will see later in the course.

Because atoms have no overall electrical charge, an atom must have as many electrons as there are protons in its nucleus. Therefore, the atomic number of an element also tells the number of electrons in a neutral atom of that element. You'll learn more about the arrangement of electrons in a later section.

Do you remember neutrons? These neutral particles are also found in the nucleus of all atoms except for the simplest isotope of hydrogen. The mass of a neutron is almost the same as the mass of a proton. The sum of the protons and neutrons in the nucleus is the **mass number** of that particular atom. Isotopes of an element have different mass numbers because they have different numbers of neutrons, but they all have the same atomic number. An isotope is identified by writing the name or symbol of the atom followed by its mass number. Recall that Thomson discovered that naturally occurring neon is a mixture of isotopes. The three neon isotopes are neon-20 (90.5 percent), Ne-21 (0.2 percent), and Ne-22 (9.3 percent). All neon isotopes have ten protons and ten electrons. Ne-20 has ten neutrons; Ne-21 has 11 neutrons; and Ne-22 has 12 neutrons. Look at the isotopes of lithium in **Figure 2.11**. The properties of the common subatomic particles are summarized in **Table 2.1**.

Figure 2.11

The Particles of an Atom
The number of protons in an atom determines what element it is. For example, all atoms that contain three protons in the nucleus are atoms of lithium.

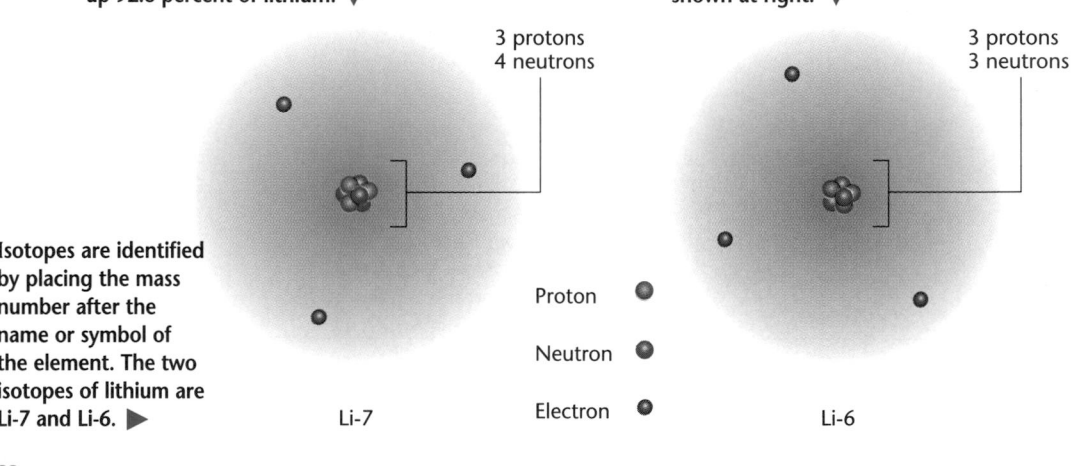

Naturally occurring lithium consists of two isotopes. The left isotope contains three protons, four neutrons, and three electrons and makes up 92.6 percent of lithium. ▼

The remaining 7.4 percent of lithium consists of atoms with three protons, three neutrons, and three electrons, shown at right. ▼

3 protons
4 neutrons

3 protons
3 neutrons

Isotopes are identified by placing the mass number after the name or symbol of the element. The two isotopes of lithium are Li-7 and Li-6. ▶

Proton
Neutron
Electron

Li-7

Li-6

Demonstration 2.5

A Massive Atom
Purpose: To demonstrate that the mass of the atom is concentrated in its nucleus.
Materials: 12 glass marbles of 2 contrasting colors if possible, modeling clay, 6 cotton balls, laboratory balance
Safety Precautions: 🧤 🥽
Disposal: Code F
Procedure: Inform students that the marbles represent protons and neutrons. Use a small amount of clay to hold the marbles together.

Weigh the "nucleus" on a laboratory balance. Inform students that the cotton balls simulate the atom's electrons. Add the cotton ball "electrons" to the clay-marble "nucleus." Weigh the model on the balance.
Results: The mass of the nucleus is not significantly different from the mass of the atom. This simulation demonstrates that nearly all the mass of an atom is in the nucleus.

Particle	Symbol	Charge	Mass Number	Mass in Grams	Mass in u
Proton	p^+	1+	1	1.67×10^{-24}	1.01
Neutron	n^0	0	1	1.67×10^{-24}	1.01
Electron	e^-	1–	0	9.11×10^{-28}	0.00055

Table 2.1 Particles of an Atom

Atomic Mass

Considering the number of atoms in just a drop of water, you can understand that atoms have extremely small masses, **Figure 2.12**. To state the mass of an atom in grams, you must deal with small numbers. **Table 2.1** shows you the mass in grams of protons, neutrons, and electrons. Notice the shorthand symbols for the particles. You will see these symbols elsewhere in the book.

As you can see, you must use some small numbers to state the mass of an atom in grams. In order to have a simpler way of comparing the masses of individual atoms, chemists have devised a different unit of mass called an atomic mass unit, which is given the symbol u. An atom of the carbon-12 isotope contains six protons and six neutrons and has a mass number of 12. Chemists have defined the carbon-12 atom as having a mass of 12 atomic mass units. Therefore, 1 u = 1/12 the mass of a carbon-12 atom. As you can see in **Table 2.1**, 1 u is approximately the mass of a single proton or neutron.

Refer again to the periodic table. Each box on the table contains several pieces of information about an element, as shown in **Figure 2.13**. The number at the bottom of each box is the average atomic mass of that element. This number is the weighted average mass of all the naturally occurring isotopes of that element. Scientists use an instrument called the mass spectrometer to determine the number and mass of isotopes of an element, as well as the abundance of each isotope. Today, scientists have measured the mass and abundance of the isotopes of all but the most unstable elements. These data have been used to calculate the average atomic masses of most elements. The isotopes of four common elements are illustrated in **Figure 2.14** on the next page.

Figure 2.12

Masses of Atoms
Recall that there are 6.5×10^{21} atoms in a drop of water. A kilogram or a gram would not be a convenient unit for objects so small.

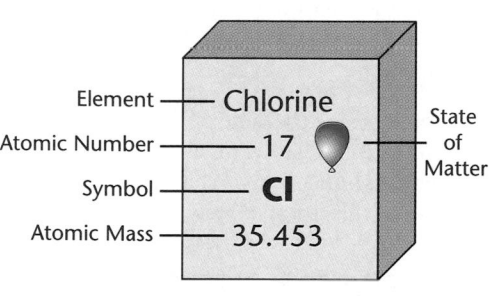

	Chlorine	
Element ———		
Atomic Number ———	17	— State of Matter
Symbol ———	**Cl**	
Atomic Mass ———	35.453	

Figure 2.13

Information in Each Block of the Periodic Table Each block of the periodic table shows the atomic number of the element, its symbol and name, its physical state at room temperature, and the average mass of its atoms.

67

 Describe a hypothetical class made up of 20 boys, each having a mass of 170 lb, and 10 girls, each having a mass of 110 lb. Ask students to calculate the average mass of a person in the class. *((20 × 170 lb) + (10 × 110 lb)) / 30 = 150 lb.* Ask them whether any single student has a mass of 150 lb. *no* Point out that average atomic masses of the elements are determined in a similar way, unless the element has only one naturally occurring form, as in the case of beryllium-9. L1

Content Background

Mass Spectrometer The mass spectrometer is an instrument that measures the charge-to-mass ratio of charged particles that pass through its detector chamber. Isotopes of an element have different masses, but each isotope acquires the same charge in the mass spectrometer. As a result of differences in the charge-to-mass ratio, each isotope arrives at a specific location in the detector where it is counted. A computer displays a chart called the mass spectrum, which shows the abundance of each isotope in the sample. The height of each peak in the mass spectrum is proportional to the percentage of that isotope in the element.

TECH PREP

✓ **Assessment**

Performance: Ask students to look through old textbooks and reference books from the library and in science classrooms to compare the accepted current atomic masses with those of years ago and to explain the reasons for differences. One reason they will discover is that atomic masses were once based on O-16 instead of C-12. L1

Analysis: Ask these questions.
1. Using the values given in **Table 2.1**, what is the mass in grams of a nucleus having 6 protons and 6 neutrons? *2.01×10^{-23} g*
2. What is the mass in grams of 6 electrons? *5.47×10^{-27} g*
3. What is the total mass in grams of the 6 protons, 6 neutrons, and 6 electrons? *2.01×10^{-23} g*
4. What percent of the total mass is the nucleus? *99.97%*

✓ **Assessment**

Knowledge: Ask students to determine which element is represented by the model that was made. *The model used for the simulation was carbon-12.* Point out to students that this atom was selected as the standard for the atomic mass scale because of its isotopic purity. L1

68

3 ASSESS

Check for Understanding

- Check for understanding of the section material by having students answer the Section Review questions.
- Give each group of three students a plastic bag containing several hundred dry beans of three different types. Ask the students to sort, count, and weigh the three types of beans to establish an average mass for each type of bean. Ask students to use the percentage abundance and average mass of each type of bean to calculate an average atomic mass for the hypothetical element beanium. **L1** **COOP LEARN**

Reteach

Discussion: Ask students to relate examples of structure and function in cells and in the human body. Then draw the analogy to scientists' quest to determine the structure of the atom in order to better understand the behavior of all matter.

Extension

Ask interested students to prepare a short description, poster, personification, role play, or model for each of the 13 key terms listed at the beginning of Section 2.1 and to make a presentation for the class. **L1**

4 CLOSE

Portfolio

Ask students to draw a time line from the year 1700 to the present and to chart the contributions of Lavoisier, Proust, Dalton, Bequerel, Thomson, Millikan, Rutherford, Chadwick, and others. Also ask students to write a short paragraph summarizing each particular discovery or theory and to include the chart and summaries in their portfolios. **L1** **P**

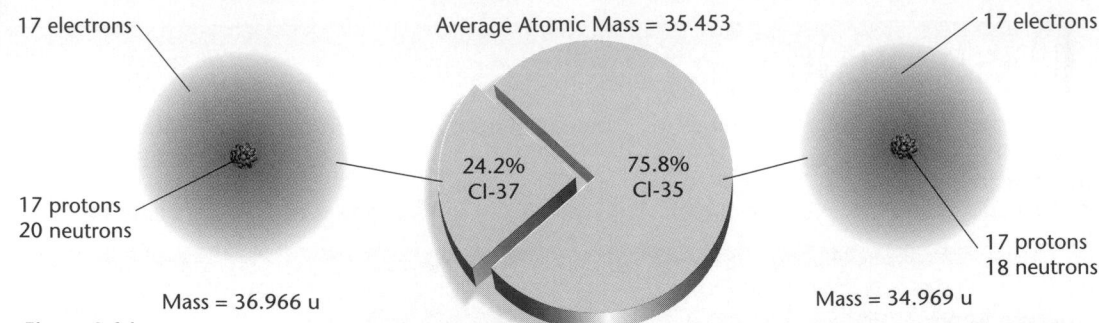

Average Atomic Mass = 35.453

17 electrons
17 protons
20 neutrons
Mass = 36.966 u

24.2% Cl-37
75.8% Cl-35

17 electrons
17 protons
18 neutrons
Mass = 34.969 u

Figure 2.14

Average Atomic Masses Chlorine-35 makes up 75.8 percent of chlorine, and chlorine-37 makes up 24.2 percent. Suppose you have 1000 atoms of chlorine. On average, 758 of those atoms would have masses of 34.969 u for a total mass of $758 \times 34.969 = 26\ 507$ u. Likewise, 242 atoms would have masses of 36.966 u for a total mass of 242×36.966 u $= 8946$ u. The total mass of the 1000 atoms is 26 507 u + 8946 u = 35 453 u, so the average mass of one chlorine atom is 35.5 u when rounded to the same number of digits as the percentages.

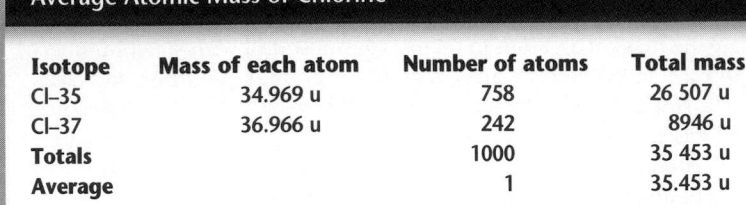

Average Atomic Mass of Chlorine

Isotope	Mass of each atom	Number of atoms	Total mass
Cl–35	34.969 u	758	26 507 u
Cl–37	36.966 u	242	8946 u
Totals		1000	35 453 u
Average		1	35.453 u

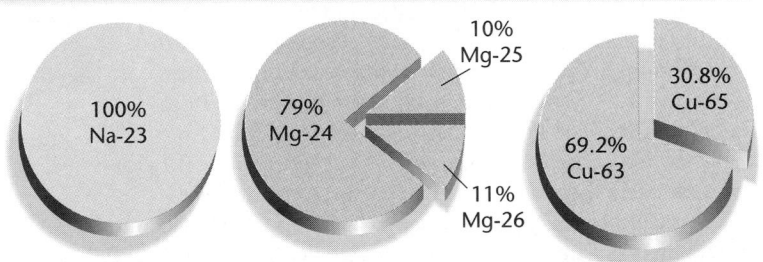

100% Na-23

79% Mg-24
10% Mg-25
11% Mg-26

30.8% Cu-65
69.2% Cu-63

▲ Several elements such as sodium consist of only one type of atom. Magnesium occurs as a mixture of three isotopes, and copper is a mixture of two.

SECTION REVIEW

Understanding Concepts

1. What conclusions about the nature of matter did scientists draw from the law of definite proportions?

2. How did scientists know that cathode rays had a negative electrical charge?

3. How do isotopes of an element differ from one another?

Thinking Critically

4. **Inferring** Why did Rutherford conclude that an atom's nucleus has a positive charge instead of a negative charge? Summarize the conclusions that Rutherford's team made about the structure of an atom.

Applying Chemistry

5. **Carbon Dating** The isotope of carbon that is used to date prehistoric fossils contains six protons and eight neutrons. What is the atomic number of this isotope? How many electrons does it have? What is its mass number?

68   Chapter 2   Matter Is Made up of Atoms

SECTION REVIEW

1. Atoms of each element are the same but are different from atoms of other elements.
2. The rays were repelled by negatively charged plates and deflected by magnets.
3. They differ in the number of neutrons in the nucleus.
4. If the nucleus were negative, the positive particles would be attracted to it instead of being deflected away. Rutherford concluded that the atom has a small, dense, positively charged nucleus, which accounts for nearly all of the atom. The volume of the atom is nearly all empty space occupied only by nearly massless electrons having negative charges.
5. The atomic number is 6; it has 6 electrons; its mass number is 14 (C-14).

Electrons in Atoms

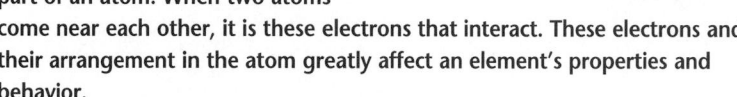

Electrons are just one of the three main kinds of subatomic particles. Why do they have special importance in chemistry? Electrons are in motion in the outer part of an atom. When two atoms come near each other, it is these electrons that interact. These electrons and their arrangement in the atom greatly affect an element's properties and behavior.

Electrons in Motion

As you learned in Section 2.1, the atom is mostly empty space, except that the space isn't entirely empty. This space is occupied by the atom's electrons. Now, look more closely at how scientists have learned about the motion and arrangement of electrons.

Electron Motion and Energy

Considering that electrons are negative and that an atom's nucleus contains positively charged protons, why aren't electrons pulled into the nucleus and held there? Scientists in the early 20th century wondered the same thing.

Niels Bohr (1885-1962), a Danish scientist who worked with Rutherford, proposed that electrons must have enough energy to keep them in constant motion around the nucleus. He compared the motion of electrons to the motion of planets orbiting the sun. Although the planets are attracted to the sun by gravitational force, they move with enough energy to remain in stable orbits around the sun. In the same way, when we launch satellites, we use rockets to give them enough energy of motion so that the satellites stay in orbit around Earth, as shown in **Figure 2.15.** In a

Figure 2.15
Motion Versus Attraction
This satellite, as it travels around Earth, is subject to balanced forces. If its energy of motion is increased, its speed increases, and it moves to an orbit farther away from Earth.

2.2 Electrons in Atoms 69

SECTION PREVIEW

Objectives
Relate the electron to modern atomic theory.
Compare electron energy levels in an atom.
Illustrate valence electrons by Lewis electron dot structures.

Key Terms
electromagnetic spectrum
emission spectrum
energy level
electron cloud
valence electron
Lewis dot diagram

PREPARE

Key Concepts
The section begins with a discussion of the Bohr model of the atom. Then, the electromagnetic spectrum is described and related to elemental emission spectra. Experimental evidence indicates that electrons within atoms have definite arrangements. Finally, Lewis dot diagrams are introduced.

Planning Ahead
For the MiniLab Locate spectrum tubes, a high-voltage transformer, and diffraction gratings. You may want to use these throughout the section to support the concepts discussed.
For the Demonstrations The glowing pickle Quick Demo on page 75 is spectacular. You should try it ahead of time in order to assess safety concerns.

1 FOCUS

Focus Transparency

Display the **Section Focus Transparency** for Section 2.2. Ask students to speculate on reasons that specific elements produce unique spectra when heated. Point out that the answer relates to electrons and that they should be able to explain this phenomenon at the end of this section. L1 LEP

2 TEACH

Discussion
Ask students whether they have any idea why chemists are concerned with the arrangements of electrons in atoms. Remind them of the space occupied by electrons compared to that occupied by the nucleus. Guide students to realize that electrons are the parts of the atom that are lost, gained, or shared when it reacts.

Program Resources

Study Guide, pp. 7-8 L1
Critical Thinking/Problem Solving, p. 2 L2
Laboratory Manual, 2.2, pp. 15-20 L1
Section Focus Transparency 4 and **Master,** p. 4 L1 LEP
Basic Concepts Transparency 6 and **Master,** pp. 11-12 L1 LEP

Problem Solving Transparency 2 and **Master,** pp. 3-4 L1 LEP
ChemLab and MiniLab Worksheets, p. 19 L1
Applying Scientific Methods in Chemistry, p. 20 L1

Correcting Misconceptions

Explain that though the Bohr model of the atom is often presented, the well-defined orbits are not correct. Point out that unlike the motion of the planets in our solar system, electron motion within an atom is irregular and an electron may, in fact, be quite close to the nucleus at times.

Reinforcement

Ask students to put a cork or table-tennis ball in a tub of water and to study its motion as waves are created at one end of the tub. They should be able to see that a disturbance at one end transfers energy to the cork or ball. This will reinforce the idea that waves transfer energy, not matter. L1 LEP

similar way, electrons have energy of motion that enables them to overcome the attraction of the positive nucleus. This energy keeps the electrons moving around the nucleus. Bohr's view of the atom, which he proposed in 1913, was called the planetary model.

When a satellite is launched into orbit, the amount of energy determines how high above Earth it will orbit. Given a little more energy, the satellite will go into a slightly higher orbit; with a little less energy, it will have a lower orbit as shown in **Figure 2.16**. However, it seemed that electrons did not behave the same way. Instead, experiments showed that electrons occupied orbits of only certain amounts of energy. Bohr's model had to explain these observations.

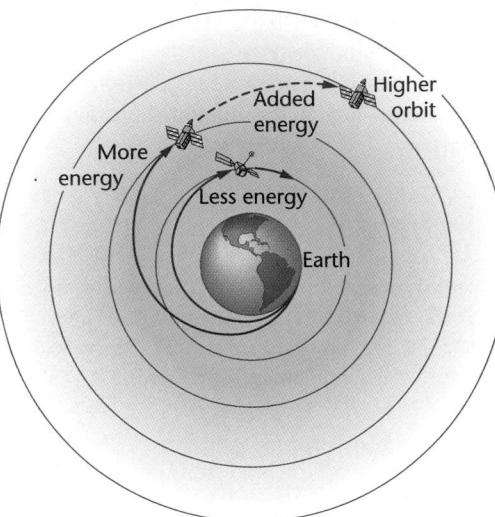

Figure 2.16

Satellites and Electrons

A satellite can orbit Earth at almost any altitude, depending on the amount of energy used to launch it. Electrons, on the other hand, can occupy orbits of only certain energies.

The Electromagnetic Spectrum

To boost a satellite into a higher orbit requires energy from a rocket motor. One way to increase the energy of an electron is to supply energy in the form of high-voltage electricity. Another way is to supply electromagnetic radiation, also called radiant energy. Radiant energy travels in the form of waves that have both electrical and magnetic properties. These electromagnetic waves can travel through empty space, as you know from the fact that radiant energy from the sun travels to Earth every day. As you may already have guessed, electromagnetic waves travel through space at the speed of light, which is approximately 300 million meters per second.

Waves Transfer Energy

If you've ever seen water waves breaking on a shoreline or heard objects in a room vibrate from the effects of loud sound waves, you already know that waves transfer energy from one place to another. Electromagnetic waves have the same characteristics as other waves, as you can see in **Figure 2.17**.

Demonstration 2.6

Emission Spectra

Purpose: To observe the bright line emission spectra of different elements.

Materials: Hydrogen and neon gas spectrum tubes, high-voltage DC power supply, student spectroscopes or handheld diffraction gratings, white light source

Safety Precautions:

Disposal: Code F

Procedure: Hydrogen and neon gas spectrum tubes are available commercially. Your school's physics teacher may have them, as well as the high-voltage DC power supply and the student spectroscopes. Have students hold the spectroscopes so that the diffraction gratings are toward their eyes and the slits are toward the white light source in a darkened room. By moving the slit slightly off the

light source, a continuous spectrum will appear to one side of the spectroscope. Repeat the procedure using a gas emission tube as the light source.

Results: Students will observe the rainbow of a continuous spectrum of the white light. Students will observe a bright line emission spectrum containing green and orange lines for neon. Often, only three of the four lines are visible for hydrogen. Try

Figure 2.17

Properties of Waves
Waves transfer energy, as you can see from the damage being done by these huge waves during a hurricane. ▼

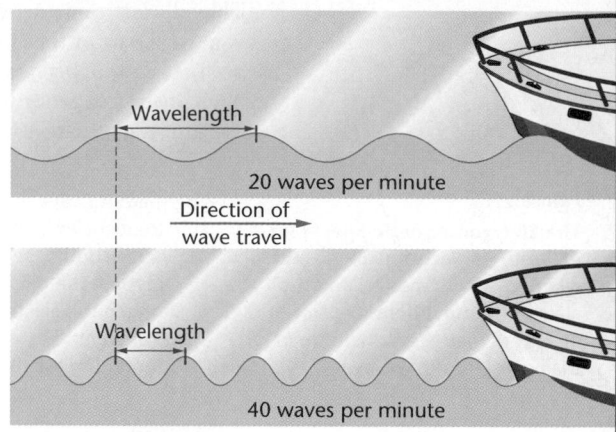

Wavelength

20 waves per minute

Direction of wave travel

Wavelength

40 waves per minute

▲ Waves are produced by something vibrating back and forth. Two properties of waves are frequency and wavelength. The number of vibrations per second is the frequency of the wave. The scientific unit for frequency is the Hertz (Hz).

▲ Wavelength is the distance between corresponding points on two consecutive waves. Because all electromagnetic waves travel at the same speed, the speed of light, wavelength is determined by frequency. A low frequency results in a long wavelength, and a high frequency results in a shorter wavelength.

Electromagnetic radiation includes radio waves that carry broadcasts to your radio and TV, microwave radiation used to heat food in a microwave oven, radiant heat used to toast bread, and the most familiar form, visible light. All of these forms of radiant energy are parts of a whole range of electromagnetic radiation called the **electromagnetic spectrum.** A portion of this spectrum is shown in **Figure 2.18.**

Figure 2.18

The Spectrum of White Light
White light is a mixture of all colors of visible light. Whenever white light passes through a prism or diffraction grating, it is broken into a range of colors called the visible light spectrum. When sunlight passes through raindrops, it is broken into the colors of the rainbow.

Assessment

Performance: Ask each student to list the frequency of his or her favorite FM and AM radio stations and to use the equation, $c = \lambda v$, in which c represents the speed of light, 3.00×10^8 m/s, the Greek letter lambda (λ) represents the wavelength, and nu (v) represents the frequency of the wave energy. *For example, a frequency of 96.3 MHz FM has a wavelength of 3.12 m; a frequency of 820 kHz AM has a wavelength of 366 m.* L2

Transparency

Display **Basic Concepts Transparency 6.** Use it to point out the regions of the spectrum and the relationship between wavelength and frequency. L1 LEP

The full electromagnetic spectrum is shown in **Figure 2.19.** Notice that only a small part of the electromagnetic spectrum is made up of visible light. Note that higher-frequency electromagnetic waves have higher energy than lower-frequency waves. This is an important fact to remember when you study the relationship of light to atomic structure.

Figure 2.19
The Electromagnetic Spectrum
All forms of electromagnetic energy interact with matter, and the ability of these different waves to penetrate matter is a measure of the energy of the waves.

A. **Gamma rays** have the highest frequencies and shortest wavelengths. Because gamma rays are the most energetic rays in the electromagnetic spectrum, they can pass through most substances.

B. **X rays** have lower frequencies than gamma rays have but are still considered to be high-energy rays. X rays pass through soft body tissue but are stopped by harder tissue such as bone.

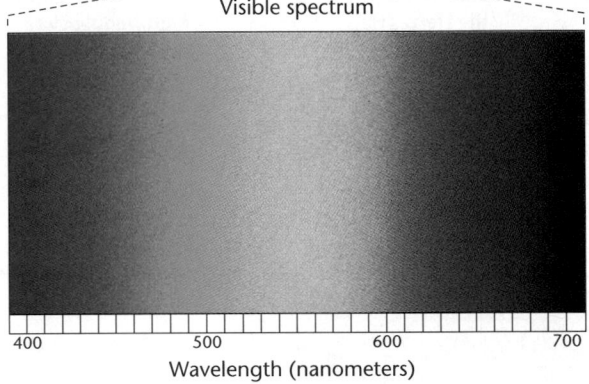

F. **Microwaves** are low-frequency, low-energy waves that are used for communications and cooking.

C. **Ultraviolet** waves are slightly more energetic than visible light waves. Ultraviolet radiation is the part of sunlight that causes sunburn. Ozone in Earth's upper atmosphere absorbs most of the sun's ultraviolet radiation.

D. **Visible light** waves are the part of the electromagnetic spectrum to which our eyes are sensitive. Our eyes and brain interpret different frequencies as different colors.

E. **Infrared** waves have less energy than visible light. Infrared radiation is given off by the human body and most other warm objects. We experience infrared rays as the radiant heat you feel near a fire or an electric heater.

G. **Radio waves** have the lowest frequencies on the electromagnetic spectrum. In the AM radio band, frequencies range from 550 kHz (kilohertz) to 1700 kHz and wavelengths from about 200 m to 600 m.

Meeting Individual Needs

Gifted Have gifted students research and describe the subatomic particles other than protons and neutrons that have been discovered. Students should delve into the idea of quarks as ultimate components of matter. L3

Aurora Borealis

In real life, you may never see colored lights such as those brightening the night sky in the photograph. You are looking at the *aurora borealis*—a fantastic light show seen only in high northern latitudes. The lights were once thought to be reflections from the polar ice fields. An aurora occurs from 100 to 1000 km above Earth.

Cause of auroras An aurora is attributed to solar wind, which is a continuous flow of electrons and protons from the sun. These high-energy, electrically charged particles become trapped by Earth's magnetic field, and they penetrate to the ionosphere. There, the particles collide with oxygen and nitrogen molecules and transfer energy to them. The energy causes electrons in these atoms and molecules to jump to higher energy levels. When the electrons return to lower energy levels, they release the absorbed energy as light.

Characteristics of the aurora When the frequencies of radiant energy released by the molecules are in the visible range, they can be seen as an aurora. Atomic oxygen, releasing energy at altitudes between 100 to 150 km, emits a whitish-green light. Molecular nitrogen gives off red light.

The aurora is most frequently seen in polar latitudes because the high-energy protons and electrons move along Earth's magnetic field lines. Because these lines emerge from Earth near the magnetic poles, it is there that the particles interact with oxygen and nitrogen to produce a fantastic display of light. Auroras also may be seen in extreme southern latitudes. These displays are the *aurora australis*.

WORD ORIGIN

aurora:
Aurora (L) the Roman goddess of dawn

The light of the aurora is a bit like the first light of dawn.

Connecting to Chemistry

1. **Applying** How does the *aurora borealis* relate to the structure of an atom?

2. **Inferring** What charactcristic of an aurora indicates that it is caused by solar winds rather than a reflection of polar ice?

Purpose
Students apply what they have learned about electron energy transitions to the aurora.

Teaching Strategies
Students may have heard of a rare occasion when an aurora sighting has occurred in the United States far from the North Pole. Explain that this happens only when the sun is very active, for example, when solar flares fling high-energy particles toward Earth at a rapid rate.

Extension

TECH PREP Ask students to investigate how neon lights and sodium-vapor lights work. Ask them to compare the operation of these lights with the way auroras are produced. Neon gas (and other gases) at low pressure is present in neon lights. When electric current passes through the gas, the atoms absorb some of the energy and electrons move to higher energy levels. When they return to lower energy levels, these electrons emit light with characteristic colors.

Connecting to Chemistry

1. The light emitted in the aurora is a result of excited electrons dropping from higher to lower energy levels. It supports the model of the atom having electrons in specific energy levels.

2. Auroras can be seen in middle latitudes and vary in color. Colors correspond to colors in the emission spectra of gases in the atmosphere.

Electrons and Light

What does the electromagnetic spectrum have to do with electrons? It's all related to energy—the energy of motion of the electron and the energy of light. Scientists passed a high-voltage electric current through hydrogen, which absorbed some of that energy. These excited hydrogen atoms returned the absorbed energy in the form of light. Passing that light through a prism revealed that the light consisted of just a few specific frequencies—not a whole range of colors as with white light, as shown in **Figure 2.20**. The spectrum of light released from excited atoms of an element is called the **emission spectrum** of that element.

Figure 2.20
Emission Spectrum of Hydrogen
When hydrogen atoms are energized by electricity, they emit light. ▼

A prism will separate the light emitted by hydrogen into four visible lines of specific frequencies (colors). Hydrogen also emits light in the non-visible ultraviolet and infrared parts of the spectrum. ▼

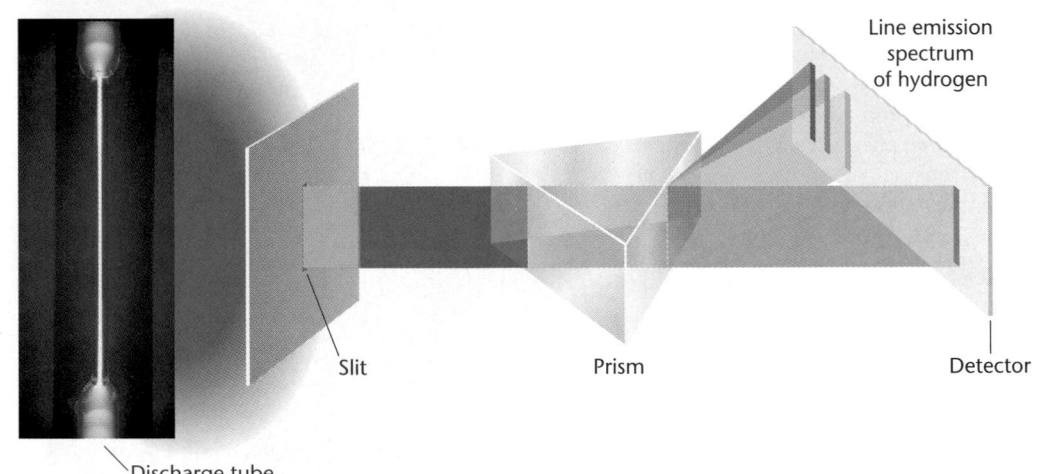

Line emission spectrum of hydrogen

Slit   Prism   Detector

Discharge tube

Evidence for Energy Levels

What could explain the fact that the emission spectrum of hydrogen consists of just a few lines? Bohr theorized that electrons absorbed energy and moved to higher energy states. Then, these excited electrons gave off that energy as light waves when they fell back to a lower energy state. But, why were only certain frequencies of light given off? To answer this question, Bohr suggested that electrons could have only certain amounts of energy. When they absorb energy, electrons absorb only the amount needed to move to a specific higher energy state. Then, when the electrons fall back to a lower energy state, they emit only certain amounts of energy, resulting in only specific colors of light. This relationship is illustrated in **Figure 2.21**.

Figure 2.21
Energy Levels in a Hydrogen Atom

This diagram shows how scientists interpret the emission spectrum. Notice that all the lines in the visible portion of the spectrum result from electrons losing energy as they fall from higher energy levels to the second energy level. The highest level represents the energy the electron has when it is completely removed from the atom.

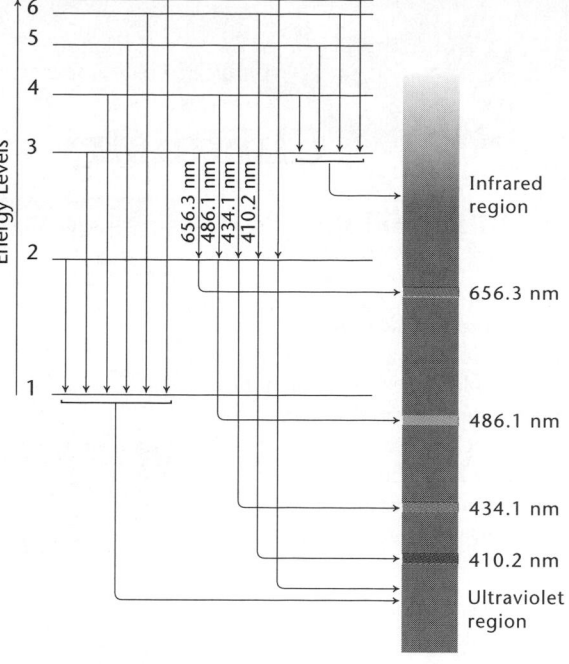

◄ Notice that drops of electrons to the lowest level emit ultraviolet frequencies, and shorter drops of electrons to the third level emit infrared frequencies.

Infrared region

656.3 nm

486.1 nm

434.1 nm

410.2 nm

Ultraviolet region

Because electrons can have only certain amounts of energy, Bohr reasoned, they can move around the nucleus only at distances that correspond to those amounts of energy. These regions of space in which electrons can move about the nucleus of an atom are called **energy levels.** Energy levels in an atom are like rungs on a ladder. You can compare the movement of electrons between energy levels to climbing up and down that ladder, as shown in **Figure 2.22.**

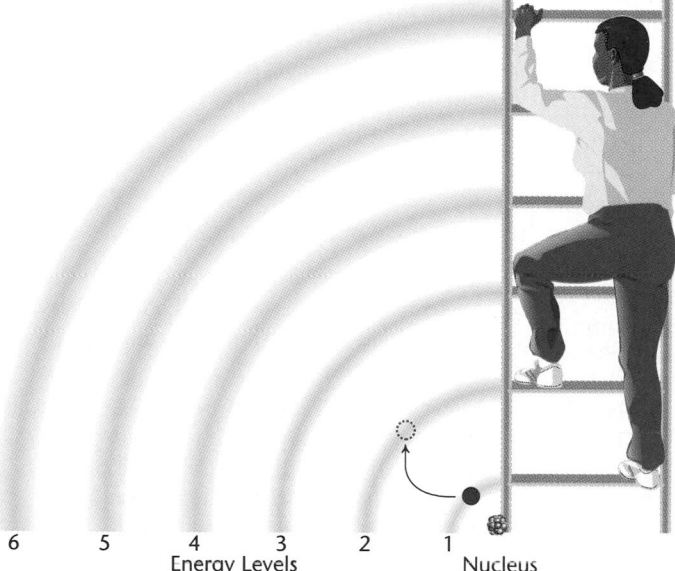

Energy Levels — 6 5 4 3 2 1 — Nucleus

Figure 2.22
Energy Levels Are Like Rungs on a Ladder

When you climb up or down a ladder, you must step on a rung. You can't step between the rungs. The same principle applies to the movement of electrons between energy levels in an atom. Like your feet on a ladder, electrons can't exist between energy levels. Electrons must absorb just certain amounts of energy to move to higher levels. The amount is determined by the energy difference between the levels. When electrons drop back to lower levels, they give off the difference in energy in the form of light.

2.2 Electrons in Atoms **75**

Quick Demo

Why does the pickle glow?

If you have a laboratory in which you can safely attach leads from 110 V AC current to two forks or nails, insert the forks into the ends of a large dill pickle. Plug the cord in, causing the sodium ions in the pickle to produce a yellow glow as they emit light at 589 nm, the bright-yellow line of the sodium emission spectrum. Support the pickle using a plastic-insulated universal clamp attached to a ring stand. **CAUTION:** *Do not allow the nails or forks to touch. Make sure everything is done before plugging in the cord. Once it is plugged in, do not touch any part of the setup or allow students to approach.*

Correcting Misconceptions

Figure 2.22 While the ladder example is useful in illustrating energies involved with electron energy level transitions, point out that energy levels within atoms are not evenly spaced like the rungs on a ladder are.

Using an Analogy

Explain that when an electron is in the lowest possible energy level within an atom, it is said to be in the ground state, just as a person on the first floor of a building might be said to be on the ground floor.

Visual Learning

Figures 2.21, 2.22: Use a ladder or shelves of a bookcase to physically move an object up and down, correlating the movement with electron energy level transitions described in **Figure 2.22.**

Meeting Individual Needs

Gifted Challenge interested students to construct models of the concentric sphere model of electron energy levels. Models should include at least two spheres and a nucleus. A cutaway model made up of concentric half-spheres would be effective. ⬜L3

Meeting Individual Needs

Gifted Ask gifted students to research and report on the dual wave-particle nature of the electron. These students may want to delve deeper into quantum theory and learn about the findings of Werner Heisenberg, Erwin Schrödinger, Louis de Broglie, Max Born, and Wolfgang Pauli. ⬜L3

Everyday Chemistry

Purpose

Students learn that light emitted in fireworks is related to energy released by excited electrons as they return to lower energy levels.

Teaching Strategies

- Have students find out how fireworks technicians obtain multiple sequential explosions of various types and colors from the same shell. *The shells are packed with several packets having different mixtures. Each has its own fuse, and these fuses burn for different times after being ignited from the main fuse.*
- Discuss how brilliant white light is produced in fireworks. Mention that it is caused by high-temperature oxidation of magnesium or aluminum by potassium perchlorate. No additional metallic salt is required.

Extension

Have students find out at what wavelength in nanometers each of the colors in the chart would appear in the electromagnetic spectrum. *red: 636 to 688 nm; yellow: 589 nm; green: 505 to 535 nm; blue: 420 to 460 nm* L2

Fireworks–Getting a Bang Out of Color

Flame Color	Color-Causing Salts
Red	Strontium salts
Yellow	Sodium salts
Green	Barium salts
Blue or Blue-Green	Copper salts

When you see a colorful display of fireworks, you probably don't think about chemistry. However, creators of fireworks displays are specialists in their craft and have to keep chemistry in mind. Packing what will provide "the rocket's red glare" for the 4th of July display is done by hand. Employees learn to follow proper precautions to avoid fires and explosions. Respect for chemical changes becomes ingrained.

Chemistry of fireworks Typical fireworks contain an oxidizer, a fuel, a binder, and a color producer. The oxidizer is the main component, making up anywhere from 38 to 64 percent of the material in fireworks. A common oxidizer is potassium perchlorate, $KClO_4$. The presence of chlorine in the oxidizer adds brightness to the colors by producing light-emitting chloride salts that make each color flame sparkle. When it oxidizes a fuel such as

aluminum or sulfur, it produces an exothermic reaction that is accompanied by noise and flashes. Aluminum or magnesium makes the flash dazzling blue-white. The loud noise comes from the rapidly expanding gases produced. Other fuels are charcoal, red gum, dextrin, and polyvinyl chloride. They not only raise the temperature but also bind the loosely assembled materials together.

Producing color A metallic salt, such as those shown in the table, is added to produce a specific color-emission spectrum. Care must be taken in selecting the ingredients so that the oxidizer is not able to react with the metallic salt during storage and cause an explosion.

Exploring Further

1. **Applying** Based on your knowledge of the electromagnetic spectrum, which of the salts in the table has an emission spectrum with the shortest wavelength?
2. **Inferring** Oxidizers in fireworks are chemicals that break down rapidly to release oxygen, which then burns the fuel. Why is it necessary to provide an oxidizer in the mixture rather than relying on oxygen in the air?

Integrating THE Sciences

Physics A new energy-efficient light uses microwaves to excite electrons in sulfur to produce intense white light similar to natural sunlight. Two or three of the new bulbs can provide the same amount of light as hundreds of ordinary floodlights. The golf-ball-sized bulbs contain sulfur in a noble gas atmosphere. There is no filament inside the bulb to burn out. The sulfur in the bulb doesn't wear out, so the life of the bulb is limited only by its microwave generator, which can last as long as 15 000 hours. At a building in Washington, DC, two of these bulbs replaced 240 lamps of 175 watts each and use only one-third the amount of electrical energy.

Line Emission Spectra of Elements

Emission spectra of elements are the result of electron transitions within atoms and provide information about the arrangements of electrons in the atoms. Observe and compare a spectrum from white light with the emission spectra of several elements.

Procedure

1. Obtain a diffraction grating from the teacher. Hold it only by the cardboard edge and avoid touching the transparent material that encloses the diffraction grating.

2. Observe the emitted light from an incandescent bulb through the grating as you hold it close to your eye. Record your observations.

3. Next, observe the light produced by the spectrum tube containing hydrogen gas and record your observations. It may be necessary for you to move to within a few feet of the spectrum tube in order to effectively observe the emission spectrum. **CAUTION:** *The spectrum tube operates at a* high voltage. *Under no circumstances should you touch the spectrum tube or any part of the transformer.*

4. Repeat procedure 3 with other spectrum tubes as your teacher designates.

Analysis

1. How do you explain why only certain colors appear in the emission spectra of the elements?

2. If each hydrogen atom contains only one electron, how are several emission spectral lines possible?

3. How do you interpret the fact that other elements emit many more spectral lines than hydrogen atoms?

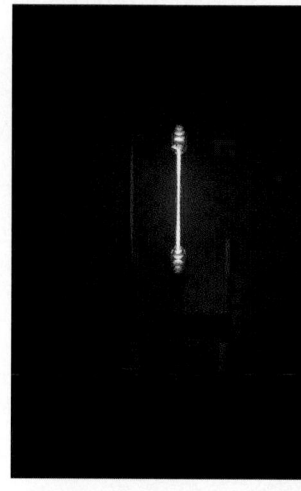

Hydrogen spectrum tube

The Electron Cloud Model

As a result of continuing research throughout the 20th century, scientists today realize that energy levels are not neat, planetlike orbits around the nucleus of an atom. Instead, they are spherical regions of space around the nucleus in which electrons are most likely to be found. Examine **Figure 2.23.** Electrons themselves take up little space but travel rapidly through the space surrounding the nucleus. These spherical regions where electrons travel may be depicted as clouds around the nucleus. The space around the nucleus of an atom where the atom's electrons are found is called the **electron cloud.**

Energy levels are concentric spherical regions of space around the nucleus. ▼

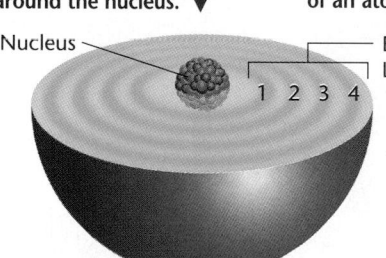

Nucleus

Figure 2.23

The Electron Cloud Model of the Atom
Pictured here is the electron cloud model of an atom.

Energy Levels

1 2 3 4

◀ The darker areas represent regions where electrons in that level are most likely to be found. Electrons are less likely to be found in the lighter regions of each level.

Analysis

1. Only certain energies are emitted by the electron drops possible in an atom of that element. A continuum of energies is emitted by an incandescent object.

2. The spectrum tube contains a large number of hydrogen atoms, though each hydrogen atom contains only one electron. The electrons in different atoms are undergoing varying energy-level transitions, some involving energy levels much higher than the first energy level.

3. Mercury and neon atoms contain many more electrons than hydrogen atoms; thus, many more electron transitions are possible.

✓ Assessment

Performance: The visible spectral lines from hydrogen represent electron transitions from energy levels 3, 4, 5, and 6 to energy level 2. Ask students to make an energy level diagram for hydrogen and to represent the visible lines on it with colored arrows. **L1**

---

miniLAB

Line Emission Spectra of Elements

Purpose: To allow students to observe, compare, and interpret the continuous spectrum of white light and the emission spectra of several elements.
Process Skills: Observing, inferring, formulating models
Safety Precautions: Be sure to set up the transformer or induction coil and spectrum tubes in such a way that neither you nor students can touch the electrodes or spectrum tubes during operation of the apparatus. Wear thermal gloves when handling the spectrum tubes because they will be hot immediately after use.

Teaching Strategies

- Be sure to instruct students on the correct way to handle and use the diffraction gratings.
- If possible, place a foil shield with a vertical slit in front of the incandescent bulb. Students will then see a rectangular spectrum similar to those they will see with the spectrum tubes.
- It may be necessary to have students move to within a few feet, especially with the hydrogen tube. At greater distances, the spectra may be dim and the lines widely separated.

Expected Results

Students should observe four spectral lines from hydrogen: violet, blue-violet, blue-green, and red. Mercury vapor produces approximately 10 visible lines across the spectrum. Neon produces approximately 22 visible lines, including many in the yellow-green, yellow, and red regions of the spectrum.

SECTION 2.2

3 ASSESS

Check for Understanding

- Check for understanding of the section material by having students answer the Section Review questions.
- Ask students to draw the Lewis dot diagrams for H, Li, Na, K, Be, Mg, Ca, B, Al, C, Si, N, P, O, S, F, Cl, He, Ne, and Ar in order for them to begin to grasp the way in which the number of valence electrons in an atom is related to the position of the element on the periodic table. L1

GLENCOE TECHNOLOGY

 Videodisc

Chemistry: Concepts and Applications
Electrons and Energy Levels
Disc 1, Side 1, Ch. 9

Show this animation to reinforce the existence of electrons in energy levels. L1 LEP

 CD-Rom

Chemistry: Concepts and Applications
Electrons and Energy Levels
Show this animation to reinforce the existence of electrons in energy levels. L1 LEP

GLENCOE TECHNOLOGY

 Videotape

MindJogger Videoquizzes for Chemistry
Chapter 2: Matter Is Made Up of Atoms
Have students work in groups as they play the videoquiz game to review key chapter concepts. L1 LEP

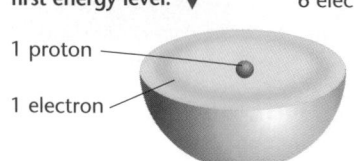

Figure 2.24
Energy Levels
A hydrogen atom has only one electron. It's in the first energy level. ▼

8 protons
8 neutrons
2 electrons
6 electrons

1 proton
1 electron

◄ An oxygen atom has eight electrons. Two of these fill the first energy level, and the remaining six are in the second energy level.

Hydrogen atom   Oxygen atom

Electrons in Energy Levels

How are electrons arranged in energy levels? Each energy level can hold a limited number of electrons. The lowest energy level is the smallest and the closest to the nucleus. This first energy level holds a maximum of two electrons. The second energy level is larger because it is farther away from the nucleus. It holds a maximum of eight electrons. The third energy level is larger still and holds a maximum of 18 electrons. **Figure 2.24** shows the energy levels and electron clouds in hydrogen and oxygen atoms.

You will learn more about the details of electron arrangement in Chapter 7. For now, it's important to learn about the electrons in the outermost energy level of an atom. The electrons in the outermost energy level are called **valence electrons.** As you can see in **Figure 2.24,** hydrogen has one valence electron and oxygen has six valence electrons. You can also use the periodic table as a tool to predict the number of valence electrons in any atom in Groups 1, 2, 13, 14, 15, 16, 17, and 18. All atoms in Group 1, like hydrogen, have one valence electron. Likewise, atoms in Group 2 have two valence electrons. Atoms in Groups 13-18 have three through eight valence electrons, respectively. You may have seen nesting dolls like the ones shown in **Figure 2.25.** These dolls serve as a visual analogy for the energy levels in an atom.

Figure 2.25
Nesting Dolls Model Energy Levels
If the nested dolls represent an atom, the largest doll is the outermost energy level. In this analogy, the largest doll also represents the valence electrons. You can continue to peel away the dolls until you reach the nucleus. ▼

▲ If you open the upper half of the doll, you will discover an identical, smaller doll inside. This doll represents the inner core of the atom, which consists of the nucleus and all electrons except the valence electrons.

78   Chapter 2   Matter Is Made up of Atoms

Chemistry Journal

Revising Models

To help students better understand the tentative nature of models, ask them to describe one or more incorrect mental models that they had at a young age and how, as they became older and wiser, they had to modify and/or replace the model(s). Ask them to place the descriptions in their journals. L1

Why do you need to know how to determine the number of outer-level electrons that are in an atom? Recall that at the beginning of this section, it was stated that when atoms come near each other, it is the electrons that interact. In fact, it is the valence electrons that interact. Therefore, many of the chemical and physical properties of an element are directly related to the number and arrangement of valence electrons.

Lewis Dot Diagrams

Because valence electrons are so important to the behavior of an atom, it is useful to represent them with symbols. A **Lewis dot diagram** illustrates valence electrons as dots (or other small symbols) around the chemical symbol of an element. Each dot represents one valence electron. In the dot diagram, the element's symbol represents the core of the atom—the nucleus plus all the inner electrons. The Lewis dot diagrams for several elements are shown in **Figure 2.26**.

Figure 2.26
Lewis Dot Diagrams
The Lewis dot diagrams for these elements illustrate how valence electrons change from element to element across a row on the periodic table.

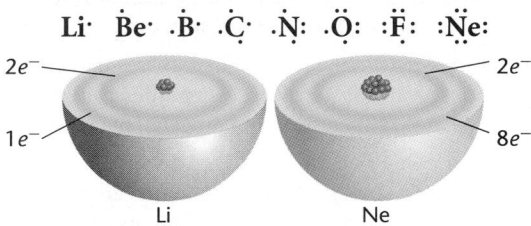

Connecting Ideas

A Lewis dot diagram is a convenient, shorthand method to represent an element and its valence electrons. You have used the periodic table already as a source of information about the symbols, names, atomic numbers, and average atomic masses of elements. In Chapter 3, you will learn that the arrangement of elements on the periodic table yields even more information about the electronic structures of atoms and how those structures can help you to predict many of the properties of elements.

SECTION REVIEW

Understanding Concepts

1. For each of the following elements, tell how many electrons are in each energy level and then write the Lewis dot symbol for each atom.

   a) argon, which has 18 electrons
   b) magnesium, which has 12 electrons
   c) nitrogen, which has seven electrons
   d) aluminum, which has 13 electrons
   e) fluorine, which has nine electrons
   f) sulfur, which has 16 electrons

2. What change occurs within an atom when it emits light?

3. How does the modern electron cloud model of the atom differ from Bohr's original planetary model of the atom?

Thinking Critically

4. **Relating Cause and Effect** Describe how scientists concluded that electrons occupy specific energy levels.

Applying Chemistry

5. **Waves** Give an everyday example of how you can tell that light waves transfer energy.

SECTION REVIEW

1. **a.** Ar: 2, 8, 8 $\ddot{A}\ddot{r}$:   **b.** Mg: 2, 8, 2 Mg:
   c. N: 2, 5 $\dot{\ddot{N}}$:   **d.** Al: 2, 8, 3 $\dot{A}\dot{l}$:
   e. F: 2, 7 $\ddot{\ddot{F}}$:   **f.** S: 2,8, 6 $\dot{\ddot{S}}$:

2. Electrons have absorbed energy and have jumped to higher energy levels. When the electrons drop from higher levels to levels of lower energy, they emit the energy difference as light waves.

3. In the electron cloud model, each energy level is a spherical region where there is a high probability of finding an electron. Bohr's model had specific orbits for each energy level.

4. When atoms absorbed energy, they returned the energy as light of only certain frequencies. Scientists concluded that if only certain energies of light were given off, then electrons must exist in only certain states of energy or energy levels.

5. Objects struck by light may increase in temperature. Another example is the electrical energy produced by photoelectric cells.

- Review the Reviewing Main Ideas statements and Key Terms with your students.
- Complete solutions to Chapter Review problems can be found in the *Problems and Solutions Manual.*

UNDERSTANDING CONCEPTS

1. They did not do experiments to test their ideas.
2. His ideas were similar to Dalton's in that he believed that atoms are ultimate, indestructible particles of matter and that different types of atoms exist. However, his ideas were based on speculation, whereas Dalton's were based on many experiments.
3. If atoms are indestructible, then matter cannot be created or destroyed.
4. He established that matter cannot be created or destroyed in a chemical reaction—the law of conservation of mass.
5. Matter is made up of atoms. Atoms are indestructible and cannot be divided into smaller particles. All atoms of one element are alike, but they are different from atoms of other elements.
6. Atoms of each element are alike in mass, but differ in mass from atoms of other elements. Therefore, a compound always having the same combination of atoms always has the same proportion of elements by mass.
7. Experiments must be repeated to be sure they yield similar results each time.
8. The rays were deflected away from a negatively charged plate and toward a positively charged plate.
9. The atomic number of Ca is 20, which indicates that a calcium nucleus has 20 protons, and therefore the atom has 20 electrons.
10. There are 12 electrons. To be neutral, an atom must have the same number of electrons as protons.

CHAPTER 2 REVIEW

REVIEWING MAIN IDEAS

2.1 Atoms and Their Structure

- Experiments in the 18th century showed that matter is not created or destroyed in ordinary chemical reactions and that a compound is always composed of the same proportion of elements.

- Dalton's atomic theory proposed that matter is made up of indestructible atoms and that the atoms of an element are alike but differ from atoms of other elements.

- Scientists make testable hypotheses based on observation. When these hypotheses are tested and refined by many experiments, they become theories.

- Experiments in the late 19th and early 20th centuries revealed that the atom has a tiny nucleus made up of protons and neutrons. Electrons move around the nucleus. The mass of the atom is concentrated in the nucleus.

- An atom's atomic number is the number of protons in its nucleus. The number of electrons in an atom equals the number of protons.

- Atoms of the same element always have the same number of protons and electrons but vary in the number of neutrons.

2.2 Electrons in Atoms

- Emission spectra led scientists to propose that electrons move around an atom's nucleus in orbits of certain energies.

- In the current model of the atom, electrons move around the nucleus in specific energy levels.

- Energy levels are spherical regions in which electrons are likely to be found. The greater the energy of the level, the farther from the nucleus the level is located.

- Electrons can absorb energy and move to a higher energy level. When they drop back to lower levels, they release specific amounts of energy in the form of electromagnetic radiation.

- Lewis dot diagrams can be used to represent the valence electrons in a given atom.

Key Terms

For each of the following terms, write a sentence that shows your understanding of its meaning.

atom	law of definite
atomic number	proportions
atomic theory	Lewis dot diagram
electromagnetic	mass number
spectrum	neutron
electron	nucleus
electron cloud	proton
emission spectrum	scientific law
energy level	theory
hypothesis	valence electron
isotope	

⬤ UNDERSTANDING CONCEPTS

1. Why could the Greek philosophers not develop a real theory about atoms?

2. Compare Democritus's ideas about the atom with those of Dalton.

3. How does Dalton's atomic theory explain the law of conservation of mass?

4. What was Lavoisier's contribution to the development of the modern atomic theory?

5. What are the major points in Dalton's atomic theory?

6. How does Dalton's atomic theory explain the law of definite proportions?

7. Why is it necessary to perform repeated experiments in order to support a hypothesis?

✓ Assessment

Portfolio: Review the portfolio options that are provided throughout the chapter. Encourage students to select one product that demonstrates their best work for the chapter. Have students explain what they have learned and why they chose this example for placement into their portfolios. **P**

Additional portfolio options can be found in the following **Teacher Classroom Resources:
Consumer Chemistry
Applying Scientific Methods in Chemistry
Laboratory Manual**

8. What evidence convinced Thomson that cathode rays consist of negatively charged particles?

9. What is the atomic number of calcium? What does that number tell you about an atom of calcium?

10. If the nucleus of an atom contains 12 protons, how many electrons are there in the neutral atom? Explain.

11. An atom of sodium has 11 protons, 11 electrons, and 12 neutrons. What is its atomic number? What is its mass number?

12. What is the number at the bottom of each box on the periodic table? What is the significance of this number?

13. The pie graph shows the abundance of the two kinds of silver atoms found in nature. The more abundant isotope has an atomic mass of a little less than 107, but the average atomic mass of silver on the periodic table is about 107.9. Explain why it is higher.

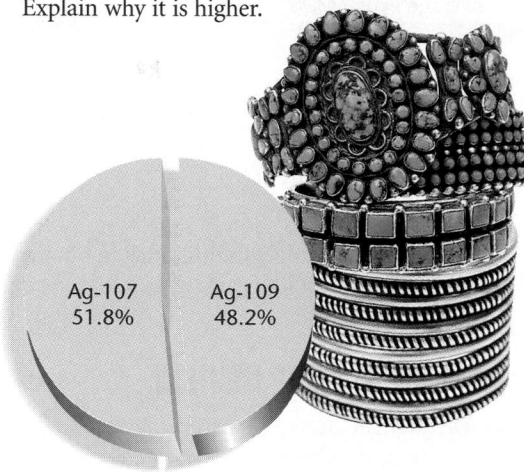

Ag-107 51.8%   Ag-109 48.2%

14. Describe the relationship among frequency, wavelength, and energy of electromagnetic waves.

15. Describe the differences among ultraviolet waves, visible light waves, and infrared waves. How are these waves alike?

16. How is the light of an emission spectrum of an element produced?

17. How did Bohr's view of energy levels differ from the way energy levels are depicted in the modern model of the atom?

18. Draw the Lewis dot diagram for silicon, which contains 14 electrons.

19. Why is the region around the nucleus of an atom referred to as an *electron cloud*?

20. What are valence electrons? Why are they important to understanding the chemistry of an element?

● APPLYING CONCEPTS

21. One atom has 20 protons and a mass number of 44. Another atom has 20 protons and a mass number of 40. What is the identity of each of these atoms? How do you account for the difference in mass numbers?

22. An atom's nucleus contains 28 protons, and the atom has a mass number of 60. How many neutrons does this atom contain? How many electrons? What is the identity of the atom?

23. Suppose you do an experiment in which you analyze the composition of aluminum chloride. You find that it consists of 20.2 percent Al and 79.8 percent Cl by mass. If you were to analyze a sample of the same compound from a different source, would you expect it to have the same or a different composition? Explain.

24. Why are the results of Rutherford's gold foil experiment more consistent with a nuclear model of the atom than with the chocolate-chip cookie dough model?

25. Give the atomic number, number of electrons, electron arrangement, and the Lewis dot structure for the elements helium, boron, chlorine, neon, and phosphorus. Use the periodic table as a source of information.

History Connection

26. Lavoisier's discovery that air was not a substance but rather a mixture of nitrogen and oxygen is described as "history-making." Why do you think this description is accurate?

22. 32 neutrons; 28 electrons; nickel
23. It should have the same composition. Atoms of aluminum have the same average mass regardless of the source. The same is true for chlorine atoms.
24. Most of the particles passed straight through the foil, indicating that atoms are mostly empty space. If the cookie-dough model had been accurate, then particles would have either all passed through undeflected or all been deflected.

25. Helium; 2; 2; 2; He ꞉

Boron; 5; 5; 2, 3; ·Ḃ·

Chlorine; 17; 17; 2, 8, 7; ꞉Cl̈꞉

Neon; 10; 10; 2, 8; ꞉N̈e꞉

Phosphorus; 15; 15; 2, 8, 5; ·P̈·

26. Until Lavoisier's experiments, many scientists still believed in the ancient Greek system of elements—air, earth, fire, and water.

11. Sodium's atomic number is 11; its mass number is 23 ($11\ p^+ + 12\ n^0$).

12. It is the average atomic mass of the element. It is the weighted average of the masses of all the naturally occurring isotopes of that element.

13. The value 107.9 represents the average of both isotopes present with their abundance taken into account.

14. As the frequency of the waves increases, the wavelength decreases. Waves of higher frequency (shorter wavelength) have higher energy.

15. They differ in frequency and wavelength, and consequently, in energy. Ultraviolet waves have the highest frequency and infrared have the lowest frequency of the three types. All are electromagnetic waves.

16. Electrons are excited to higher energy levels. When the electrons drop from their excited state, they lose discrete amounts of energy that are emitted as certain frequencies (colors) of light.

17. Bohr thought energy levels were like planetary orbits around the nucleus. The modern atomic model pictures energy levels as concentric, spherical, cloudlike regions.

18. ·Ṡi·

19. The region is nearly all empty space populated only by rapidly moving electrons, which have little mass or volume.

20. They are the electrons in an atom's outermost energy level. It is these electrons that interact when two atoms come together.

APPLYING CONCEPTS

21. Both atoms are isotopes of calcium. One atom has four more neutrons in the nucleus than the other.

27. The solar wind is composed of electrons and protons, which are particles of matter. Electromagnetic radiation is not matter.

28. The availability of oxygen decreases with altitude. At orbital distances, none is available.

29. Leaves and food waste will degrade by decay but glass will not.

THINKING CRITICALLY

30. Bohr's model showed the nucleus as a small, dense mass of particles with electrons orbiting in empty space around the nucleus.

31. They proposed that electrons can have only certain energy values and that when excited electrons drop from higher values to lower values, they emit an amount of energy equal to the difference between allowable energy values.

32. **a.** *Although atoms are composed of smaller particles, in chemical reactions,* atoms are indestructible and cannot be divided into smaller particles.
b. We know that there are naturally occurring isotopes of many elements; these isotopes have differing masses. All atoms of one element are exactly alike *except for small differences in mass,* but they are different from atoms of other elements.

33. They are isotopes of magnesium.

34. They have the same mass number but are different elements, carbon and nitrogen.

Physics Connection

27. How does solar wind differ from electromagnetic radiation from the sun?

Everyday Chemistry

28. Rocket boosters used to launch the space shuttle must carry both fuel and an oxidizer. Why would an oxidizer be needed in such rockets?

Chemistry and Society

29. Even though the law of conservation of mass applies to both, why is waste glass a more serious problem than leaves that have fallen from trees or food waste?

● THINKING CRITICALLY

Relating Cause and Effect

30. How was Bohr's model of the atom consistent with the results of Rutherford's gold foil experiment?

Observing and Inferring

31. How did scientists account for the fact that the emission spectrum of hydrogen is not continuous but consists of only a few lines of certain colors?

Applying Concepts

32. Dalton's second statement of atomic theory on page 54 says that atoms are indestructible. This is still true as far as chemical changes are concerned. However, we now know that atoms can be broken apart in atomic fission reactions, and some mass is converted to energy.
   a) Rewrite Dalton's second statement to make it true.
   b) Examine the third statement of Dalton's theory. What evidence do we have that it is untrue? Restate this part to make it fit today's knowledge.

Comparing and Contrasting

33. What is the relationship between an atom that has 12 protons, 12 neutrons, and 12 electrons and one that has 12 protons, 13 neutrons, and 12 electrons?

Comparing and Contrasting

34. What is the relationship between an atom with the atomic number 6 and mass number 14 and an atom with the atomic number 7 and mass number 14?

Interpreting Data

35. **MiniLab 1** Suppose you obtained the data in the following table for MiniLab 1. Complete the table and determine the average penny mass of the pennies in the mixture.

	Type of Penny	
	Pre-1982	Post-1982
Mass of 10	30.81 g	25.33 g
Mass of 1		
Number of Pennies in Mixture	34	55
Mass of Each Type		
Average Penny Mass		

Applying Concepts

36. **MiniLab 2** Which of the lines of the visible emission spectrum of hydrogen represents the greatest energy drop?

Observing and Inferring

37. **ChemLab** Why do you think the amount of zinc that you could use in the experiment was limited to 0.28 g?

● CUMULATIVE REVIEW

38. Distinguish among a mixture, a solution, and a compound. (Chapter 1)

39. Suppose you stir some sugar in a glass until it dissolves completely. Describe this action and its result using the terms *solvent, solute, physical change,* and *aqueous solution.* (Chapter 1)

40. Suppose you are given two identical bottles filled completely with clear liquids. Describe a way you could determine whether the liquids have different densities. You cannot open the bottles. (Chapter 1)

35.
Type of Penny	Pre-1982	Post-1982
Mass of 10	30.81 g	25.33 g
Mass of 1	*3.081 g*	*2.533 g*
Number of Pennies in Mixture	34	55
Mass of Each Type	*104.754 g*	*139.315 g*
Average Penny Mass	*2.742 g*	

Average Penny Mass is found by adding 104.754 g + 139.315 g = 244.069 g and then dividing the sum by the total number of pennies to give a weighted average of 244.069 g/89 pennies = 2.742 g/penny,

rounded back to 4 significant digits.

36. The violet line represents the greatest energy drop.

37. If there were much more zinc than this, there would not be enough acid to react with it. (A total of 0.01 mol of HCl is available to react.)

CUMULATIVE REVIEW

38. A mixture is composed of two or more substances not chemically combined. A solution is a mixture in which the substances are evenly dispersed in each other. A com-

Skill Review

1. **Relating Cause and Effect** A chemistry student decided to weigh a bag of microwave popcorn before and after popping it. Following the directions on the bag, she opened it immediately after microwaving it to let the steam escape and then weighed it. She noticed that the mass of the unpopped popcorn was 0.5 g more than the mass of the popped popcorn. Was matter destroyed when the popcorn popped? Explain your answer.

Writing in Chemistry

2. Do research to learn about the phlogiston theory. How did people who believed the phlogiston theory explain chemical changes such as burning, oxidation of metals, and the smelting of ores to obtain the pure metal? Write a paragraph in which you attempt to answer the question "What was phlogiston?" using modern scientific terms.

3. It's 1911 and a scientific breakthrough has just been announced by a team of scientists headed by Ernest Rutherford. He has just proposed a nuclear model of the atom based on research begun in 1909. You are the science reporter for a major newspaper, and your job is to write an article about this new model. Include a brief history of the atom and several drawings to illustrate your article. The article must appeal to the general reader, who may know nothing about atoms.

Problem Solving

4. A chemist recorded the following data in an experiment to determine the composition of three samples of a compound of copper and sulfur obtained from three different sources. Determine the ratio of mass of copper to mass of sulfur for each sample. How do these ratios compare? What law of chemistry does this experiment illustrate? Does the result mean that atoms of copper and sulfur occur in this compound in the same numerical ratio? Explain your answer.

Sample	Mass of Sample	Mass of Cu	Mass of S
1	5.02 g	3.35 g	1.67 g
2	10.05 g	6.71 g	3.34 g
3	99.6 g	66.4 g	33.2 g

Projects

5. Use what you have learned about electromagnetic radiation to explain the greenhouse effect. Create a poster or other display to show how the greenhouse effect works, and use labels indicating the frequencies and wavelengths of the radiant energy involved. Point out energy changes and predict what would happen if the greenhouse effect were increased or decreased by a significant amount.

6. Prepare a historical time line showing the discoveries that led to Dalton's atomic theory. Include the work of Lavoisier, Priestley, and Proust. Include also the discoveries of Becher, Boyle, Gay-Lussac, Cavendish, and others. Tell what each did and when. Point out how Dalton's theory accounted for their observations.

---

masses. In fact, the compound in this example is copper(II) sulfide, CuS.

5. Students should make clear that the greenhouse effect is a normal phenomenon that makes life on Earth possible.

6. Time lines should start with Democritus and Aristotle. You may want to have students draw conclusions about the emergence of modern chemistry during the years from (roughly) 1750 to 1810.

Program Resources

Chapter Review and Assessment, pp. 11-16 [L1]

Alternate Assessment in the Science Classroom [L1]

Computer Test Bank, Chapter 2 [L1]

Problems and Solutions Manual, Ch. 2 [L1]

Performance Assessment in the Science Classroom [L1]

Supplemental Practice Problems, Chapter 2 [L1]

---

pound consists of two or more elements chemically combined.

39. Stirring the sugar with the water results in a physical change in which the solid sugar crystal is pulled apart by water molecules to form a mixture consisting of sugar molecules (solute) evenly dispersed among water molecules (solvent), forming an aqueous solution.

40. Because the volumes are the same, the masses can be compared. If the two liquids differ enough in density, simply "hefting" the bottles will suffice.

AUTHENTIC ASSESSMENT

1. Matter was not destroyed. The apparent loss of mass can be accounted for by the loss of steam.

2. Answering the question "What was phlogiston?" will be challenging to students. For example, using a flame to release a metal from its ore was believed to add phlogiston, in a way similar to the ancient Greek idea that heating something in a flame added the fire element to it. Because the resulting metal weighed less than the ore, phlogiston must reduce the weight of a substance. Thus, believers in the phlogiston theory claimed that phlogiston had a levitating power, in other words, negative weight.

3. Students should research the members of Rutherford's team. As a side project, they could report on these scientists, particularly Geiger and Marsden.

4. The ratio by mass is approximately 2:1 copper to sulfur in each case. This result illustrates the law of definite proportion. It does not mean that atoms will occur in this same ratio because, according to Dalton's atomic theory, atoms of different elements have different

Chapter 3 Organizer

Section	Objectives	Activities/Features
3.1 **Development of the Periodic Table** (4 days)	**1. Outline** the steps in the historical development of the periodic table. **2. Predict** similarities in properties of the elements by using the periodic table.	**MiniLab 1:** Predicting the Properties of Mystery Elements, p. 89 **Discovery Demo:** 3.1 Activity of Alkali Metals, pp. 84-85 **Demonstration:** 3.2 A Search for Order, pp. 88-89
3.2 **Using the Periodic Table** (4 days)	**3. Relate** an element's valence electron structure to its position in the periodic table. **4. Use** the periodic table to classify an element as a metal, nonmetal, or metalloid. **5. Compare** the properties of metals, nonmetals, and metalloids.	**Literature Connection:** The Language of a Chemist, p. 96 **MiniLab 2:** Trends in Reactivity Within Groups, p. 97 **ChemLab:** The Periodic Table of Elements, pp. 100-101 **Chemistry and Technology:** Metals That Untwist, pp. 108-109 **Everyday Chemistry:** Metallic Money, p. 110 **Demonstrations:** 3.3 Carbon's Predictability, pp. 104-105; 3.4 Predicting a Periodic Trend, pp. 106-107

Activity Materials

ChemLab (pages 100-101)
stoppered test tubes (14); plastic dishes (6); samples of carbon, nitrogen, oxygen, magnesium, aluminum, silicon, red phosphorus, sulfur, chlorine, calcium, arsenic or selenium, tin, iodine, lead; micro-conductivity apparatus (6); 1M HCl; test tubes (6); test-tube rack; 10-mL graduated cylinder; spatula; small hammer; glass marking pencil

MiniLab 1 (page 89)
none

MiniLab 2 (page 97)
magnesium, calcium, sodium bromide solution, sodium iodide solution, chlorine water, lighter fluid, 50-mL beakers (2), beral pipets (2), test tubes (2), 10-mL graduated cylinder, forceps

Demonstrations For a complete list of materials for the demonstrations in this chapter, see pages 84, 88, 104, and 106.

KEY TO TEACHING STRATEGIES

The following designations will help you decide which activities are appropriate for your students.

L1 Level 1 activities should be within the ability range of all students.

L2 Level 2 activities should be within the ability range of the average to above-average student.

L3 Level 3 activities are designed for the ability range of above-average students.

LEP LEP activities should be within the ability range of Limited English Proficiency students.

COOP LEARN Cooperative Learning activities are designed for small group work.

P These strategies represent student products that can be placed into a best-work portfolio.

Teacher Classroom Resources

STUDENT MASTERS

Study Guide: pp. 9-10
Laboratory Manual: 3.1 An Alien Periodic Table, pp. 21-26
ChemLab and MiniLab Worksheets: p. 26

Study Guide: pp. 11-12
Laboratory Manual: 3.2 Periodicity of Halogen Properties, pp. 27-32
Critical Thinking/Problem Solving: pp. 3-4
ChemLab and MiniLab Worksheets: pp. 21-25, 27-28

TEACHING AIDS

Section Focus Transparency 5
Section Focus Transparency Master: p. 5
Basic Concepts Transparency 7, 8
Basic Concepts Transparency Master: pp. 13-16
Lesson Plans: p. 5

Section Focus Transparency 6
Section Focus Transparency Master: p. 6
Basic Concepts Transparency 8
Basic Concepts Transparency Masters: pp. 17-18
Problem Solving Transparency 3
Problem Solving Transparency Master: pp. 5-6
Lesson Plans: p. 6

ASSESSMENT RESOURCES
Chapter Review and Assessment: pp. 17-22
Alternate Assessment in the Science Classroom
Computer Test Bank: Chapter 3
Performance Assessment in the Science Classroom
Problems and Solutions Manual: Chapter 3
Supplemental Practice Problems: Chapter 3

CHAPTER RESOURCES
Applying Scientific Methods in Chemistry: p. 21
Spanish Resources: Chapter 3
English/Spanish Audiocassettes: Chapter 3
Cooperative Learning in the Science Classroom
Lab and Safety Skills in the Science Classroom

GLENCOE TECHNOLOGY

Software
Mastering Concepts in Chemistry: *Unit 3* Lesson 1 Modern Periodic Table; Lesson 2 Surveying the Table; *Unit 8* Lesson 1 Periodic Properties; Lesson 2 Typical Elements

Videotape
MindJogger Videoquizzes, Chapter 3

Videodisc
Chemistry: Concepts and Applications, Demonstrations, Videos, and Animations
Videodisc and Bar Code Teacher Guide

CD-ROM Multimedia System
Chemistry: Concepts and Applications, Same as for Videodisc, plus Interactive Exploration and Major Simulation, *The Periodic Table*

TECH PREP

The following Glencoe resources provide opportunities for integrating science and technology.

Student Edition: Chemistry and Technology, pp. 108-109; Everyday Chemistry, p. 110

Teacher Wraparound Edition: Integrating the Sciences, p. 91; Chemistry Journal, p. 112; Extension, p. 113

Teacher Classroom Resources: Tech Prep Applications, pp. 3-4; Consumer Chemistry, pp. 5-6

Chapter Overview

Students begin to learn about trends in the chemical and physical properties of elements. The concept of periodicity and the historical development of a classification system based on atomic mass are used to build a foundation for understanding the modern periodic table. Trends in properties and behavior are related to repeating patterns in the number of valence electrons.

Theme Connection

Macro to Submicro This chapter emphasizes macroscopic observations of periodic properties of the elements and submicroscopic explanations for these observations.

Assessment Planner

Choose assessment strategies from the following pages to evaluate the progress of your students.
Assess, pp. 94, 98, 99, 112
MiniLabs, pp. 89, 97
ChemLab, p. 100
Portfolio, pp. 89, 94, 96, 98, 113, 114
Chapter Review, p. 114

84

Discovery Demo 3.1

Activity of Alkali Metals

Purpose: To demonstrate that chemical reactivity follows a predictable pattern related to an element's position on the periodic table.

Materials: overhead projector, explosion shield, three 600-mL beakers, 300 mL of water, 10 drops phenolphthalein indicator, clear kitchen plastic wrap, small cubes of Li, Na, and K, approximately 2 mm on

a side, 10 cm × 10 cm wire screen

Safety Precautions:

Wear safety goggles and an apron. Use an explosion shield.

Disposal: Code C.1 Check the pH, neutralize with acetic acid or dilute HCl to a pH of 5 to 8, then flush down the drain with copious amounts of water.

Procedure: Cover the stage and lower lens of an overhead projector with clear plastic wrap. Place a 600-mL beaker containing 100 mL of water on the projector. **CAUTION:** *Place an explosion shield around the projector.* Turn on the projector and darken the room. Drop a small piece of lithium into the water. **CAUTION:** *Quickly cover the beaker with a wire screen to prevent the metal from flying out of the*

CHAPTER

3 Introduction to the Periodic Table

In 1986, Halley's comet came streaking into Earth's neighborhood—a guest but not an unexpected one. Last seen 76 years earlier, people were waiting for this collection of ice and rock to visit again. When it comes around once more in 2062, some young people like you may be waiting to see Halley's comet again. Astronomers can predict the appearance of known comets because they show up regularly, just as the full moon appears in the sky every month. Both the moon and the comet illustrate that in the natural world, patterns of behavior often repeat in predictable ways. The moon repeats its cycle every month; Halley's comet takes 76 years.

Are there any repeating patterns in the behavior of the elements? Can you make any predictions about what they might do? Elements have different properties. Oxygen and nitrogen are colorless, whereas mercury and bromine are bright-colored liquids. Some solid elements melt at temperatures over 1000°C, and others melt in your hand. Silver won't react with water, but, as you can see below, potassium reacts violently with water. Examples like these make it seem as if there couldn't be any way of making predictions about the behavior of elements. But when elements are organized in the periodic table, repeating patterns do emerge.

Concept Check

Review the following concepts before studying this chapter.
Chapter 1: physical and chemical properties of elements
Chapter 2: relationship of atomic number to number of protons and electrons in an atom

Chapter Preview

3.1 Development of the Periodic Table

3.2 Using the Periodic Table

Laboratory Activities
ChemLab
The Periodic Table of the Elements

MiniLabs
1. Predicting the Properties of Mystery Elements
2. Trends in Reactivity Within Groups

Chemistry Around You

Like the appearance of Halley's comet every 76 years, the properties of the elements repeat in a regular way when they're arranged in the periodic table. When you understand the table's arrangement, you'll be able to explain why silver and potassium don't act in the same way. The repeating patterns in the periodic table are the key to differences as well as similarities in behavior and properties of all the elements.

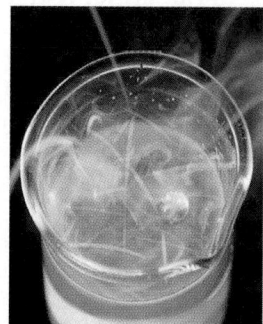
Potassium in water

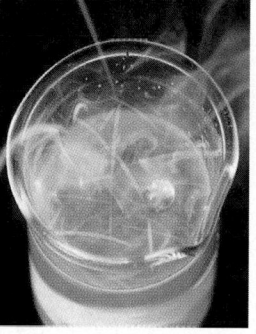
Silver in water

85

Introducing the Chapter
Discuss with students the difference between elements and compounds, and ask them to give examples of objects made of elements. Point out that many gases in the atmosphere are also elements. Note similarities and differences in the properties of some elements and point out that the periodic table will help them predict properties and behaviors.
Using the Illustrations Just as there is predictability in the appearance of Halley's comet, there is predictability in the properties of the elements. Elements that are widely separated on the periodic table have different chemical properties.

Revealing Misconceptions

Students may think that all metals have the same properties and behavior. Emphasize that there is a gradual change in the properties of elements as the number of valence electrons changes from one to eight.

GLENCOE TECHNOLOGY

 Videodisc

Chemistry: Concepts and Applications
Activity of Alkali Metals
Disc 1, Side 1, Ch. 10

Show this video of the Discovery Demo for this chapter to introduce the concept of trends in chemical properties within a group of elements in the periodic table. L1 LEP

 CD-Rom

Chemistry: Concepts and Applications
Activity of Alkali Metals
Have students use this video to reinforce the concept of trends in reactivity within a group in the periodic table.
L1 LEP

beaker. In separate beakers, repeat the procedure using Na and K. The reactions of Na and K are similar to that of Li.

$$2Li + 2H_2O \rightarrow 2LiOH + H_2$$

Results: As students observe the metals skimming across the water's surface, they will relate the speed to the metal's activity. Li is relatively slow (less active), Na is fast (active), and K bursts into flame upon hitting the water (very active).

Analysis: Ask these questions.
1. Which metal reacts the fastest with water? *K*
2. How does the element's position in the column on the periodic table relate to its reactivity? *Li is first and least reactive; K is third and most reactive.*
3. Which metal has its outer-level electron farthest from the nucleus? *K*

PREPARE

Key Concepts

Periodic characteristics and behaviors first are explained in a general sense, then related to the periodic characteristics and behaviors of the chemical elements. Through examples and analogies, the need for a systematic arrangement of the elements is demonstrated. Mendeleev's development of the periodic table is described in some detail as the basis for understanding the modern periodic table.

Planning Ahead

For the Demonstrations If you do not have a fume hood with a shatterproof door that you can use for Demonstration 3.1, an explosion shield should be constructed by connecting three 2 foot × 3 foot pieces of plexiglass with hinges. Acquire micro-conductivity testers for Demonstration 3.4.

For the ChemLab Students will need micro-conductivity testers. If you do not have them, they should be purchased, borrowed, or constructed.

1 FOCUS

Focus Transparency

Display the **Section Focus Transparency** for Section 3.1 to introduce students to the concept of repeating patterns.

L1 LEP

Development of the Periodic Table

SECTION PREVIEW

Objectives
Outline the steps in the historical development of the periodic table.
Predict similarities in properties of the elements by using the periodic table.

Key Terms
periodicity
periodic law

Changes that occur in regular, predictable cycles are natural and comfortable. The seasons change from spring to summer to fall to winter. Farmers know they can plant crops in spring and harvest them in summer or autumn. You probably look forward to summer as a time to relax, enjoy water sports, and get a break from school, but you'll be anticipating school again in the fall. Similarly, even the early chemists looked for regular, dependable changes in the properties and behavior of the elements that would allow them to make predictions. They were amassing many chemical facts. Any patterns or regularities that could organize all their data would be useful. That's why chemists began more than 100 years ago to search for an arrangement of the elements that would show patterns. It's also why the periodic table is on the walls of most chemistry labs today.

The Search for a Periodic Table

By 1860, scientists had already discovered 60 elements and determined their atomic masses. They noticed that some elements had similar properties. They gave each group of similar elements a name. Copper, silver, and gold were called the coinage metals; lithium, sodium, and potassium were known as the alkali metals; chlorine, bromine, and iodine were called the halogens. Chemists also saw differences among the groups of elements and between individual elements. They wanted to organize the elements into a system that would show similarities while acknowledging differences. It was logical to use atomic mass as the basis for these early attempts.

Program Resources

Study Guide, pp. 9-10 L1
English/Spanish Audiocassettes, Chapter 3
 L1 LEP
Laboratory Manual, 3.1, pp. 21-26 L1
Section Focus Transparency 5 and **Master,**
 p. 5 L1 LEP
Basic Concepts Transparency 7, 8 and
 Masters, pp. 13-16 L1 LEP
Spanish Resources L1 LEP

ChemLab and MiniLab Worksheets, p. 26
 L1
Tech Prep Applications, pp. 3-4 L1
Applying Scientific Methods in Chemistry,
 p. 21 L1

Döbereiner's Triads

In 1829, the German chemist J.W. Döbereiner classified some elements into groups of three, which he called triads. The elements in a triad had similar chemical properties, and their physical properties varied in an orderly way according to their atomic masses. **Table 3.1** shows the atomic mass, density, boiling point, and melting point for each of three elements in the halogen triad—chlorine, bromine, and iodine. **Figure 3.1** shows these three elements at room temperature.

Figure 3.1

Physical States and Colors of the Halogens
As atomic mass increases, the state of the elements in this triad changes. At room temperature, chlorine (left) is a gas, bromine (center) is a liquid, and iodine (right) is a solid. Colors change from greenish yellow to reddish orange to violet.

Table 3.1 shows that the atomic mass of the three elements increases from 35.5 to 79.9 to 127 from chlorine to bromine to iodine. More importantly, the atomic mass of bromine—the middle element—is 79.9, about the average of the atomic masses of chlorine and iodine:

$$\frac{35.5 + 127}{2} = 81.3$$

The fact that the atomic mass of the middle element lies about midway between the other two members of the triad is an important characteristic of triads. The table also shows that density, melting point, and boiling point all increase as atomic mass increases. The values for bromine are between those of chlorine and iodine. Iodine, with the highest atomic mass, has the highest density, boiling point, and melting point.

Table 3.1   The Halogen Triad

Element	Atomic Mass (u)	Density (g/mL)	Melting Point (°C)	Boiling Point (°C)
Chlorine	35.5	0.003 21	−101	−34
Bromine	79.9	3.12	−7	59
Iodine	127	4.93	114	185

2 TEACH

Using an Analogy

Describe the following situation, and make a drawing to illustrate it on the chalkboard. You are driving east on Kenton Ave. in an unfamiliar city and wish to find Cleveland Ave. When you intersect Main St., you turn right, drive south and intersect, in sequence, Lima Ave., Madison Ave., Nelsonville Ave., and Oberlin Ave. What should you do to reach Cleveland Ave.? *Turn around and go north on Main St. The avenues are arranged in alphabetical order, and Cleveland Ave. starts with a C.* Then, ask students if the systematic arrangements of the streets and avenues is useful. *Obviously, the systematic arrangement helps people locate the streets and avenues they wish to find.* Point out that systematic arrangements are helpful in many situations, including those encountered in chemistry. L1

Discussion

Ask students to think of situations in which the systematic arrangement of items helps in finding and using the items. *books in a library, food items in a grocery store, and listing of names in a phone book* L1

Cooperative Learning Activity

Arrange students' seating in order of something you know about them, such as: age, numerical street address, alphabetical street address, telephone number, social security number, or locker number. Ask them to talk with each other to deduce the rationale for the seating. This activity might take ten minutes at the beginning of each class period for a week. Encourage student involvement by rewarding the first student to determine the seating method. L1 COOP LEARN

What's in a name?

Borrow from your local library a compact disk, record, or tape of *An Evening Wasted with Tom Lehrer.* The lyrics for one of the songs, set to a Gilbert and Sullivan tune, are the names of the chemical elements.

Reinforcement

Make a list of puns related to the names or symbols of the elements and ask students to come up with the correct elements and to create some element puns of their own. Some examples: policeman, *copper;* well driller's chant, *boron;* fund from mother's sister, *antimony;* smelliest element, *Pu;* part of a house made of hardwood *fluorine;* foolish prisoner, *silicon.* L1

Visual Learning

Figure 3.2: Ask students to look at the modern periodic table, in which the elements are arranged according to increasing atomic number, and list pairs of elements that would be in reverse order if they were arranged by atomic mass, as Mendeleev arranged them in his table. L1

The triad in **Table 3.2** shows a relationship among the densities of three metals that is true for many triads. The density of strontium (2.60 g/mL) is near the average of the densities of calcium (1.55 g/mL) and barium (3.62 g/mL):

$$\frac{1.55 + 3.62}{2} = 2.58 \text{ g/mL}$$

Density increases with increasing atomic mass, as it does in the chlorine, bromine, and iodine triad. But in the case of the triad in **Table 3.2,** this increase in densities is more evident than it is for the halogens because the halogens exist in three different states—solid, liquid, and gas.

Table 3.2 Metal Triad

Element	Atomic Mass (u)	Density (g/mL)	Melting Point (°C)	Boiling Point (°C)
Calcium	40	1.55	842	1500
Strontium	88	2.60	777	1412
Barium	137	3.62	727	1845

Melting points for calcium, strontium, and barium show a similar trend. The average of the atomic masses of calcium and barium is about 88.5, which is close to 88, the actual atomic mass of strontium. However, the pattern of the boiling points in this triad is irregular. The irregular pattern of boiling points is typical of triads involving metals.

Döbereiner's triads were useful because they grouped elements with similar properties and revealed an orderly pattern in some of their physical and chemical properties. The concept of triads suggested that the properties of an element are related to its atomic mass.

Mendeleev's Periodic Table

The Russian chemist, Dmitri Mendeleev, was a professor of chemistry at the University of St. Petersburg when he developed a periodic table of the elements. Mendeleev was studying the properties of the elements and realized that the chemical and physical properties of the elements repeated in an orderly way when he organized the elements according to increasing atomic mass. For example, beryllium resembled magnesium, and boron resembled aluminum. Patterns of repeated properties began to appear.

In 1869, Mendeleev published a table of the elements organized by increasing atomic mass. He listed the elements in a

Figure 3.2

Mendeleev's Table of 1869

Elements in horizontal rows of Mendeleev's first table displayed similar properties. Mendeleev wrote question marks in the table in places where elements, then unknown, would eventually be placed.

A Search for Order

Purpose: To demonstrate that a system of classification can be used to organize similar items based on common properties.
Materials: A dozen or more cans of food. Include fruits, soups, and vegetables.

Safety Precautions:
Disposal: Code F
Procedure: Place cans of food randomly on the desk. Ask the class if they notice any regularities. Guide the discussion toward a classification system that might include soups, vegetables, and fruits. Within each broad classification, place the items in alphabetical or size order. Relate this demonstration to the placement of elements on the periodic table.
Results: Allow students to determine their own system of classification for the canned foods. Each class will have a similar but different result.
Analysis: Ask these questions.
1. Why is it helpful to organize food when it is brought home from the grocery store? *Organization saves time.*

Predicting the Properties of Mystery Elements

When Mendeleev arranged the elements according to their atomic masses, some elements didn't fit. He resolved this problem by predicting the existence and properties of elements that were unknown at the time. In this MiniLab, you'll predict some properties of two unknown elements.

miniLAB
1

Procedure

1. Unknown element A is in Group 14 below silicon and above tin. Unknown element B is in Group 16 below sulfur and above tellurium.

Group 13	Group 14	Group 15	Group 16	Group 17
	Si		S	
Ga	element A	As	element B	Br
	Sn		Te	

2. The following information is given for the surrounding elements and in this sequence: symbol, density in g/mL, melting point in K, atomic radius in pm. Si, 2.4, 1680, 118; As, 5.72, 1087, 121; Sn, 7.3, 505, 141; Ga, 5.89, 303, 134; S, 2.03, 392, 103; Br, 3.1, 266, 119; Te, 6.24, 723, 138.

3. Refer to the locations of these elements on the portion of the periodic table shown here, and average the values of the four surrounding elements to predict the densities, melting points, and atomic radii of elements A and B.

Analysis

1. What are the identities of elements A and B?

2. How do your predicted values for the three properties of element A compare with the actual values? Look up the actual values in a chemical handbook or obtain them from your teacher.

3. How do your predicted values for the three properties of element B compare with the actual values?

4. If your predicted values are fairly close to the actual values, how do you explain the approximate correlation?

vertical column starting with the lightest. When he reached an element that had properties similar to another element already in the column, he began a second column. In this way, elements with similar properties were placed in horizontal rows, as shown in **Figure 3.2**. Notice the question mark at atomic mass 180 and its position next to Zr = 90. Then refer to **Figure 3.3**.

Figure 3.3

Similar Elements Are Neighbors
This zircon crystal contains zirconium (Zr = 90 on Mendeleev's table). It also contains hafnium, Mendeleev's unknown element (? = 180). The chemical and physical properties of zirconium and hafnium are so similar that the two elements always occur together in nature and are difficult to separate.

3.1 Development of the Periodic Table **89**

2. What major categories were used to organize the canned foods? *soups, fruits, and vegetables*

3. Why do chemists organize the elements on the periodic table? *The organization enables them to predict the properties of similar elements.*

Assessment

Performance: Have teams of students, in a timed rotation, observe several metals and nonmetals in petri dishes. Ask them to develop a system for organizing the elements. Students should express, in writing, the reasoning they used to determine their system of classification. L1

LEP **COOP LEARN**

SECTION 3.1

Discussion

Present this information to the class. An English chemist, John Newlands, was looking for a systematic arrangement of the elements at about the same time as Mendeleev was. He noticed that when the elements were arranged according to increasing atomic mass, their properties seemed to repeat every eighth element. Newlands made an analogy of the repetition of properties to the notes on a musical scale. He called this arrangement the law of octaves. Ask if the law of octaves is still valid today. *No. The repetition occurs now every ninth element.* Ask why. *The noble gas elements were unknown at that time.*

Visual Learning

Figure 3.4: You might use these examples of periodic activities. Tell students that you mow your lawn every fifth day during the summer and you play tennis every third day. Sketch a monthly calendar on the board, marking in these periodic activities. L1

GLENCOE TECHNOLOGY

Videodisc

Chemistry: Concepts and Applications

See the Videodisc Program Teacher Guide to access illustrations and other material available for this chapter.

Mendeleev later developed an improved version of his table with the elements arranged in horizontal rows. This arrangement, shown in **Table 3.3,** was the forerunner of today's periodic table. Patterns of changing properties repeated for the elements across horizontal rows. Elements in vertical columns showed similar properties. An analogy can be made to the changes in the monthly calendar shown in **Figure 3.4.**

Table 3.3 Mendeleev's Table of 1871

Group	I	II	III	IV	V	VI	VII	VIII
Formula of Oxide	R_2O	RO	R_2O_3	RO_2	R_2O_5	RO_3	R_2O_7	RO_4
	H							
	Li	Be	B	C	N	O	F	
	Na	Mg	Al	Si	P	S	Cl	
	K	Ca	eka-	Ti	V	Cr	Mn	Fe, Co, Ni
	Cu	Zn	eka-	eka-	As	Se	Br	
	Rb	Sr	Yt	Zr	Nb	Mo	—	Ru, Rh, Pd
	Ag	Cd	In	Sn	Sb	Te	I	
	Cs	Ba	Di	Ce	—	—	—	
	—	—	—	—	—	—	—	
	—	—	Er	La	Ta	W	—	Os, Ir, Pt
	Au	Hg	Tl	Pb	Bi	—	—	
	—	—	—	Th	—	U	—	

Mendeleev's insight was a significant contribution to the development of chemistry. He showed that the properties of the elements repeat in an orderly way from row to row of the table. This repeated pattern is an example of periodicity in the properties of elements. **Periodicity** is the tendency to recur at regular intervals—like the appearance of Halley's comet every 76 years.

One of the tests of a scientific theory is the ability to use it to make successful predictions. Mendeleev correctly predicted the properties of several undiscovered elements. In order to group elements with similar properties in the same columns, Mendeleev had to leave some blank spaces in his table. He suggested that these spaces represented undiscovered elements.

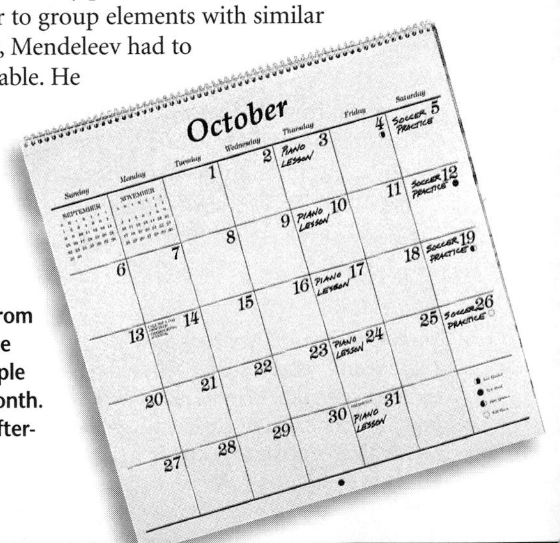

Figure 3.4

Monthly Calendar

A calendar is a table of days, with the days organized into weeks. From left to right in a given week, the days change from Sunday to Saturday and then repeat the next week. All of the days with the same name fall into vertical columns, and people generally do similar things on those days throughout the month. For example, you might have piano lessons each Thursday afternoon and soccer practice every Saturday morning.

Chemistry Journal

Periodic Table of Your Life

Ask students to list their own periodic activities in their journals. Have them divide these activities into groups, depending upon how frequently they occur. For example, birthdays happen yearly, visits to the dentist twice a year, and visits to grandparents weekly. Have students make a calendar of periodic activities and look for patterns. L1

II	III	IV
Mg	Al	Si
Ca	eka-	Ti
Zn	eka-	eka-

Figure 3.5
Prediction of Eka-Aluminum and Eka-Silicon
◀ Mendeleev left two blank spaces between zinc and arsenic and named the elements eka-aluminum and eka-silicon. *Eka* is from the Sanskrit meaning "one." In Mendeleev's earlier arrangement of elements, the unknown elements were one place from aluminum and silicon.

Figure 3.5 shows a part of Mendeleev's table in which there are two spaces for unknown elements to the right of zinc (Zn). He called these spaces eka-aluminum and eka-silicon, respectively. Based on their locations, Mendeleev predicted several of the properties of these undiscovered elements. Both elements were discovered during his lifetime. French chemists discovered eka-aluminum in 1875 and named it gallium (Ga). Eka-silicon was discovered in Germany in 1886 and was named germanium (Ge). **Table 3.4** shows the remarkable agreement between the observed properties of germanium and the properties predicted 17 years earlier by Mendeleev. The striking resemblance between Mendeleev's predicted properties and the actual properties of gallium and germanium was one of the factors that led chemists to accept his theory of periodicity among the elements and his organization of the elements into a periodic table.

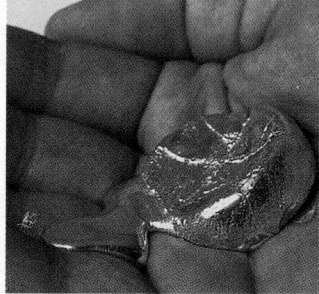

▲ Eka-aluminum was discovered in 1875 and named gallium. Gallium's melting point is so low that the metal melts from the heat of the human hand.

Table 3.4 Properties of Germanium		
Property	**Predicted (1869)**	**Actual (1886)**
Atomic mass	72 u	72.6 u
Color	dark gray	gray-white
Density	5.5 g/mL	5.32 g/mL
Melting point	very high	937°C
Formula of oxide	EsO_2*	GeO_2
Density of oxide	4.7 g/mL	4.70 g/mL
Oxide solubility in HCl	slightly dissolved by HCl	not dissolved by HCl
Formula of chloride	$EsCl_4$*	$GeCl_4$

* Es stands for eka-silicon

Mendeleev was so confident of the periodicity of the elements that he placed some elements in groups with others of similar properties even though arranging them strictly by atomic mass would have resulted in a different arrangement. An example is tellurium. The accepted atomic mass of tellurium (Te) was 128, so it should have been placed after iodine (I), which had an accepted atomic mass of 127. But the properties of tellurium logically placed it in the group with oxygen and sulfur, ahead of iodine, and the properties of iodine matched those of chlorine and bromine. Mendeleev placed tellurium with oxygen and sulfur and assumed that the atomic mass of 128 was incorrect. Mendeleev's placement of tellurium turned out to be correct even though his assumption about its atomic mass was not.

3.1 **Development of the Periodic Table** 91

SECTION 3.1

Content Background

Other Periodic Tables John Newlands, an English chemist and contemporary of Mendeleev, worked to develop a way of organizing the 62 known elements. Like Mendeleev, he listed the elements according to increasing atomic mass and, in August 1865, published an eight-column table with seven horizontal rows. Although Newlands noticed the repetition of properties when elements are arranged according to increasing atomic mass, his table lacked the versatility of Mendeleev's. However, Newlands's contribution to the development of the periodic table was recognized by the Royal Society in 1867, when it awarded him the Davy Medal.

Lothar Meyer was also a contemporary of Mendeleev and is belatedly acknowledged as an independent discoverer of the periodic law. He published a table of elements in which similar elements fell in the same column. His table clearly showed how valence changed with atomic mass. However, his table seemed to be based on the idea of natural families of elements and, although he was moving in the direction of the periodic law, he did not seem aware of it in 1864 when he published his first table in his book *Modern Chemical Theory*.

Transparency

Display **Basic Concepts Transparency 7** to introduce the periodic table. L1 LEP

Integrating THE Sciences

TECH PREP

Health/Ecology Heavy metals are dangerous to health and the environment. The U.S. Environmental Protection Agency (EPA) has listed the elements antimony, arsenic, cadmium, chromium, copper, lead, manganese, mercury, nickel, tin, vanadium, and zinc as extremely hazardous. These elements enter the body primarily through the food chain, but they can also be absorbed by direct exposure. Once ingested, heavy metals are deposited in the kidneys, liver, and brain. Have students research one or more of these elements and describe their common uses, exposure limits for humans, and health effects. L1

Discussion

Point out to students the enormous amount of information the periodic table provides about the elements. Each element has its own block with its name and symbol, atomic number, and atomic mass. Have students become familiar with the symbols that show how each element exists at room temperature (solid, liquid, or gas). Ask them to count the number of liquid elements and the number of gaseous elements. Note the symbol for the radioactive elements, and ask them to count these elements. Draw their attention to the color coding for metals, metalloids, and nonmetals. Point out the numbering of groups and periods. Give them a group number and a period number and ask them to name the element.

GLENCOE TECHNOLOGY

Software

Mastering Concepts in Chemistry
Unit 3, Lesson 1
Modern Periodic Table
Use this lesson to reinforce the concept of the periodic arrangement of the elements according to atomic number.
L1 LEP

Software

Mastering Concepts in Chemistry
Unit 3, Lesson 2
Surveying the Table
Use this lesson to help your students recognize what information is available in the periodic table. L1 LEP

PERIODIC TABLE OF THE ELEMENTS

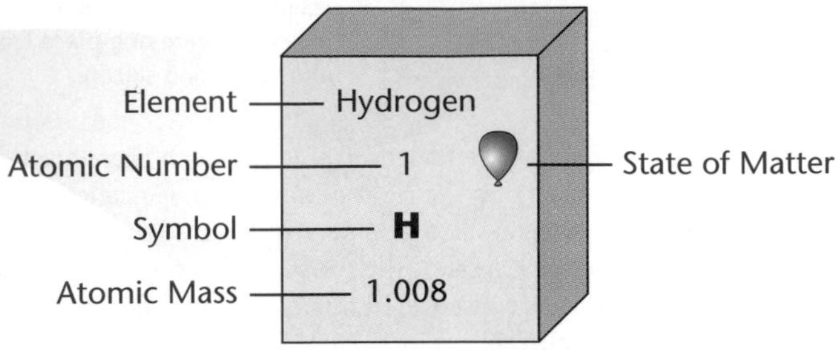

Legend:
Element — Hydrogen
Atomic Number — 1
Symbol — H
Atomic Mass — 1.008
State of Matter

Group	1	2	3	4	5	6	7	8	9
1	Hydrogen 1 H 1.008								
2	Lithium 3 Li 6.941	Beryllium 4 Be 9.012							
3	Sodium 11 Na 22.990	Magnesium 12 Mg 24.305							
4	Potassium 19 K 39.098	Calcium 20 Ca 40.078	Scandium 21 Sc 44.956	Titanium 22 Ti 47.88	Vanadium 23 V 50.942	Chromium 24 Cr 51.996	Manganese 25 Mn 54.938	Iron 26 Fe 55.847	Cobalt 27 Co 58.933
5	Rubidium 37 Rb 85.468	Strontium 38 Sr 87.62	Yttrium 39 Y 88.906	Zirconium 40 Zr 91.224	Niobium 41 Nb 92.906	Molybdenum 42 Mo 95.94	Technetium 43 Tc 97.907	Ruthenium 44 Ru 101.07	Rhodium 45 Rh 102.906
6	Cesium 55 Cs 132.905	Barium 56 Ba 137.327	Lanthanum 57 La 138.906	Hafnium 72 Hf 178.49	Tantalum 73 Ta 180.948	Tungsten 74 W 183.85	Rhenium 75 Re 186.207	Osmium 76 Os 190.2	Iridium 77 Ir 192.22
7	Francium 87 Fr 223.020	Radium 88 Ra 226.025	Actinium 89 Ac 227.028	Rutherfordium 104 Rf (261)	Hahnium 105 Ha (262)	Seaborgium 106 Sg (263)	Nielsbohrium 107 Ns (262)	Hassium 108 Hs (265)	Meitnerium 109 Mt (266)

Lanthanide Series

Cerium 58 Ce 140.115	Praseodymium 59 Pr 140.908	Neodymium 60 Nd 144.24	Promethium 61 Pm 144.913	Samarium 62 Sm 150.36	Europium 63 Eu 151.965

Actinide Series

Thorium 90 Th 232.038	Protactinium 91 Pa 231.036	Uranium 92 U 238.029	Neptunium 93 Np 237.048	Plutonium 94 Pu 244.064	Americium 95 Am 243.061

Across the Curriculum

History

In 1789, Antoine Laurent Lavoisier of France published a list of 23 known elements, some of which had been known since ancient times. However, the noble gases were not discovered until the period from 1894 to 1918. Radium, actinium, polonium, protactinium, europium, and lutetium were also discovered during this period. The remaining elements were discovered after 1923. Ask students to find out how many elements were listed in high school textbooks in the early 1900s and what events prompted the discovery of so many elements after 1894 and after 1923. *Mendeleev's table led to research to find unknown elements. Nuclear research in physics, the discovery of radioactivity, and the development of cyclotrons led to the discovery or synthesis of new elements.* L1

Legend:
- Gas
- Liquid
- Solid
- Synthetic Elements
- Metal
- Metalloid
- Nonmetal

18

Helium
2
He
4.003

13	**14**	**15**	**16**	**17**
Boron 5 B 10.811	Carbon 6 C 12.011	Nitrogen 7 N 14.007	Oxygen 8 O 15.999	Fluorine 9 F 18.998
Aluminum 13 Al 26.982	Silicon 14 Si 28.086	Phosphorus 15 P 30.974	Sulfur 16 S 32.066	Chlorine 17 Cl 35.453

(Group 18 continued: Neon 10 Ne 20.180; Argon 18 Ar 39.948)

10	**11**	**12**	**13**	**14**	**15**	**16**	**17**	**18**
Nickel 28 Ni 58.693	Copper 29 Cu 63.546	Zinc 30 Zn 65.39	Gallium 31 Ga 69.723	Germanium 32 Ge 72.61	Arsenic 33 As 74.922	Selenium 34 Se 78.96	Bromine 35 Br 79.904	Krypton 36 Kr 83.80
Palladium 46 Pd 106.42	Silver 47 Ag 107.868	Cadmium 48 Cd 112.411	Indium 49 In 114.82	Tin 50 Sn 118.710	Antimony 51 Sb 121.757	Tellurium 52 Te 127.60	Iodine 53 I 126.904	Xenon 54 Xe 131.290
Platinum 78 Pt 195.08	Gold 79 Au 196.967	Mercury 80 Hg 200.59	Thallium 81 Tl 204.383	Lead 82 Pb 207.2	Bismuth 83 Bi 208.980	Polonium 84 Po 208.982	Astatine 85 At 209.987	Radon 86 Rn 222.018
(unnamed) 110 Uun	(unnamed) 111 Uuu							

Gadolinium 64 Gd 157.25	Terbium 65 Tb 158.925	Dysprosium 66 Dy 162.50	Holmium 67 Ho 164.930	Erbium 68 Er 167.26	Thulium 69 Tm 168.934	Ytterbium 70 Yb 173.04	Lutetium 71 Lu 174.967
Curium 96 Cm 247.070	Berkelium 97 Bk 247.070	Californium 98 Cf 251.080	Einsteinium 99 Es 252.083	Fermium 100 Fm 257.095	Mendelevium 101 Md 258.099	Nobelium 102 No 259.101	Lawrencium 103 Lr 260.105

SECTION 3.1

Using an Analogy

Ask an applied arts teacher or a student in an applied arts class to cut, smooth, and number some wooden blocks in the following way: four blocks 2″ tall with the numbers 1, 6, 11, and 16 on the sides; four blocks 4″ tall with 2, 7, 12, and 17; four blocks 6″ tall with 3, 8, 13, and 18; four blocks 8″ tall with 4, 9, 14, and 19; four blocks 10″ tall with 5, 10, 15, and 20. Arrange the blocks in numerical order from the students' left to right on the demonstration table and ask students to describe the relationship between the numbers and the sizes. *The size increases gradually, then abruptly decreases in a regular or periodic manner. Each sixth block has the same size.* L1 **LEP**

GLENCOE TECHNOLOGY

CD-Rom

Chemistry: Concepts and Applications
Exploring the Periodic Table
Have students use this Interactive Exploration to increase their familiarity with the periodic table. L1 **LEP**

Transparency

Display **Basic Concepts Transparency 8** as an aid in teaching the periodic table. L1 **LEP**

*inter*NET
CONNECTION
Students can access an interactive periodic table of the elements.
World Wide Web
http://the-tech.mit.edu/~davhsu/chemicool.html

3 Assess

Check for Understanding

- Check for understanding of the section material by having students answer the Section Review questions.
- Ask students to look at the names and symbols of the elements and determine the ways in which the symbols are related to the names. Have them consult the *Handbook of Chemistry and Physics* for symbols related to Greek or Latin names. L1

Reteach

Use a transparency or slide of a jigsaw puzzle with several pieces missing. Ask students to describe the missing pieces. Explain the analogy to Mendeleev's prediction of the properties of elements that were not known to him. L1 LEP

Extension

Ask students to predict some of the properties of the yet-undiscovered element number 119. Have them describe the rationale they used in making their predictions. This exercise could be included in student portfolios. L2 P

4 CLOSE

Quick Demo

How should chemicals be organized?

Safety Precautions:

Randomly display about 30 elements and compounds and ask students whether such an arrangement facilitates finding a particular substance. Ask how the substances might be arranged, then explain your system. L1

Figure 3.6
One Use for Helium
The lightest of the elements in column 18, helium (He), is a regular at parties and celebrations. Because helium and the other noble gases are unreactive, they were not easily discovered.

The Modern Periodic Table

Döbereiner and Mendeleev both observed similarities and differences in the properties of elements and tried to relate them to atomic mass. Look at the modern periodic table shown on pages 92 and 93 and notice that as in Mendeleev's table, elements with similar chemical properties appear in the same group. For example, tellurium is in the same group as oxygen and sulfur, where Mendeleev placed it.

Mendeleev based his periodic table on 60 or so elements. At present, elements up to atomic number 111 have been discovered or synthesized. Many of the elements now known as the transition elements, lanthanides, and actinides were unknown in 1869 but today occupy the center of the table. The noble gases, such as the helium shown in **Figure 3.6,** were also unknown in Mendeleev's time, but now fill column 18 of the table.

There are several places in the modern table where an element of higher atomic mass comes before one of lower atomic mass. This is because the basis for ordering the elements in the table is atomic number, not atomic mass. The atomic number of an element is equal to the number of protons in the nucleus. Atomic number increases by one as you move from element to element across a row. Each row (except the first) begins with a metal and ends with a noble gas. In between, the properties of the elements change in an orderly progression from left to right. The pattern in properties repeats after column 18. This regular cycle illustrates periodicity in the properties of the elements. The statement that the physical and chemical properties of the elements repeat in a regular pattern when they are arranged in order of increasing atomic number is known as the **periodic law.**

SECTION REVIEW

Understanding Concepts

1. What is the modern periodic law? How does it differ from Mendeleev's periodic law?
2. What are two factors that contributed to the widespread acceptance of Mendeleev's periodic table?
3. Which of the Döbereiner triads shown are still listed in the same column of the modern periodic table?

Triad 1	Triad 2	Triad 3
Li	Mn	S
Na	Cr	Se
K	Fe	Te

Thinking Critically

4. **Using a Table** Use the periodic table to separate these 12 elements into six pairs of elements having similar properties.
Ca, K, Ga, P, Si, Rb, B, Sr, Sn, Cl, Bi, Br

Applying Chemistry

5. **Locating Elements** Use the periodic table to identify by name and symbol the elements that have the following locations.

a) Group 15, Period 3
b) Group 2, Period 6
c) Group 17, Period 2
d) Group 14, Period 5

SECTION REVIEW

1. The physical and chemical properties of the elements repeat in a regular pattern when the elements are arranged in order of increasing atomic number. Mendeleev's periodic law stated that the properties repeat when the elements are arranged in order of increasing atomic mass.
2. Two factors that led to the widespread acceptance of Mendeleev's table were its ability to predict undiscovered elements and its grouping of elements with similar chemical properties.
3. Li, Na, and K and S, Se, and Te are still in the same columns.
4. You would predict that Ca and Sr, K and Rb, Ga and B, P and Bi, Si and Sn, and Cl and Br would have similar properties.
5. a. phosphorus, P; b. barium, Ba; c. fluorine, F; d. tin, Sn

Using the Periodic Table

Have you ever tried to look up information in the tables in the sports section or the financial section of the newspaper? Or have you used a train or bus schedule to plan a trip? Tables give useful information, but they're sometimes hard to read. Some tables cram a lot of information into a small space and use many abbreviations and symbols. If you're not familiar with the codes, it's hard to obtain any useful information. But once you figure out how the data are organized and how to translate the symbols, it isn't difficult at all. It's the same with using the periodic table. Once you are familiar with the setup and symbols used in the table, you'll be able to obtain information about an element just by looking at its position in the periodic table.

Relationship of the Periodic Table to Atomic Structure

Chemists invariably have a periodic table on the walls of their offices and laboratories. It's a ready reference to a wealth of information about the elements, and it helps them think about their work, make predictions, and plan experiments based on those predictions. When you learn how to use the periodic table, you'll also find yourself referring to the periodic table to help you organize all the information you're learning about the elements.

In the modern periodic table, elements are arranged according to atomic number. Recall from Chapter 2 that the atomic number of an atom tells the number of electrons it has. If elements are ordered in the periodic table by atomic number, then they are also ordered according to the number of electrons they have. The lineup starts with hydrogen, which has one electron. Helium comes next in the first horizontal row because helium

---

SECTION PREVIEW

Objectives
Relate an element's valence electron structure to its position in the periodic table.
Use the periodic table to classify an element as a metal, nonmetal, or metalloid.
Compare the properties of metals, nonmetals, and metalloids.

Key Terms
period
group
noble gas
metal
transition element
lanthanide
actinide
nonmetal
metalloid
semiconductor

SECTION 3.2

PREPARE

Key Concepts
The structure of the periodic table is explained by noting that atoms of elements in a group have the same number of valence electrons, and atoms of elements in a period have the same number of energy levels. The properties of metals, nonmetals, and metalloids are noted, compared, and related to the number of their valence electrons. The use of metalloids in semiconductors is discussed.

1 FOCUS

Focus Transparency
Display the **Section Focus Transparency** for Section 3.2 to illustrate how a table can be made by sorting properties.
L1 LEP

2 TEACH

Concept Development
As you discuss the periodic table, it's helpful for each student to have his or her own copy. You can obtain plastic-covered, wallet-sized copies of the Fisher Periodic Chart at no charge. Call Akzo Chemicals at 1-800-227-7070 to order one for each of your students.

Purpose

To introduce the works of Primo Levi.

Teaching Strategies

- Discuss the difference between the molecular formula of benzene and its structural formula.
- Benzene is a distillate of the resin luban jawi.

Extension

Have students learn more about the work of Levi or another scientist/writer and write a report about their findings. Place the report in their portfolios. L2 P

Connecting to Chemistry

1. The ability to make accurate observations and describe them is important for writing.
2. The periodic table is a metaphor into which Levi, a chemist, could arrange the elements (events) of his life.

LITERATURE ▷ CONNECTION

The Language of a Chemist

. . . I have enjoyed looking at the world from unusual angles, inverting, so to speak, the instrumentation; examining matters of technique with the eye of a literary man, and literature with the eye of a technician.

In these words, Primo Levi describes the central paradox of his life. Born in Turin, Italy, in 1919, Primo Levi was trained as a chemist. In 1944, he was arrested as a member of the Italian Anti-Fascist resistance movement and deported to a concentration camp at Auschwitz, Poland. There, his knowledge of chemistry played a pivotal role in keeping his body as well as his spirit intact, as described in his memoirs, *Survival in Auschwitz* and *The Reawakening*. After the war, Levi continued to write until his death in Turin in April 1987.

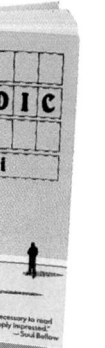

Chemist and writer In an essay, "The Language of Chemists (I)," Levi reflects on the many ways in which chemists represent reality. He traces the history of benzene from an ancient resin to the discovery of its structural formula. As he does, Levi makes the reader aware of how chemists use both words and symbols to describe a material.

Elements of a life
Each essay in his memoirs, *The Periodic Table*, carries the name of an element. Some elements recall autobiographical events; others cause the author to reflect on human nature and the natural world. "Hydrogen" is a recollection of his days as a chemistry student; Levi explores the explosive properties of the element and of youth. In his final essay "Carbon," Levi finds an element that unites all living things. He follows an atom of carbon on its journey through rocks, leaves, wine, milk, blood, and finally muscle, where the atom allows the author to end his tale.

Connecting to Chemistry

1. **Interpreting** Explain this quote by Levi: *"Chemistry is the art of separating, weighing and distinguishing: these are useful exercises also for the person who sets out to describe events or give body to his own imagination."*
2. **Inferring** Why do you think Levi chose the title *The Periodic Table?*

---

has two electrons. Lithium has three. Notice on the periodic table that lithium starts a new **period,** or horizontal row, in the table. Why does this happen? Why does the first period have only two elements? You learned in Chapter 2 that electrons in atoms occupy discrete energy levels. Only two electrons can occupy the first energy level in an atom. The third electron in lithium must be at a higher energy level. Lithium starts a new period at the far left in the table and becomes the first element in a group. A **group,** sometimes also called a family, consists of the elements in a vertical column. Groups are numbered from left to right. Lithium is the first element in Group 1 and in Period 2. Check this location on the periodic table.

Trends in Reactivity Within Groups

You've discovered that trends in the physical and chemical properties of elements occur both horizontally and vertically on the periodic table. In this MiniLab, you'll compare the reactivities of two elements in Group 2—magnesium and calcium—and three elements in Group 17—chlorine, bromine, and iodine.

Procedure

1. Wear an apron and goggles.

2. To compare the reactivities of magnesium and calcium, use forceps to drop a small piece of each element into two small beakers containing water to a depth of about 1 cm. Observe how rapidly each element reacts with water to produce bubbles of hydrogen.

3. Pour 1 mL of NaBr solution into a small test tube, and add three drops of chlorine water. Stir the solution with the tip of a micro-tip pipet. Add 1 mL of lighter fluid. Draw the entire mixture into the microtip pipet. Squeeze the pipet bulb, expelling the mixture back into the test tube. Repeat this step several times so that the liquids are thoroughly mixed. (**CAUTION:** *Use care when handling chlorine water.*)

4. Draw the liquids back into the pipet, invert it, and cover the tip with a cap made by cutting the bulb of another microtip pipet. Set the inverted pipet in a small beaker, and allow time for the upper layer to separate from the lower layer. If no color is evident in the upper layer, expel the liquids back into the test tube, and add five more drops of chlorine water. Then repeat the remainder of procedures 3 and 4 until a color is detected in the upper layer.

5. Using another test tube and microtip pipet, repeat procedures 3 and 4 with 1 mL of an NaI solution, chlorine water, and lighter fluid.

Analysis

1. Which of the two metallic elements, magnesium or calcium, is the more reactive? If this trend holds true for other groups of metallic elements, are the more reactive metals located toward the top or the bottom of the periodic table?

2. Interpret your results from procedures 3 and 4 in this way. If chlorine is more reactive than bromine, it will replace the bromine in NaBr and produce an orange color due to bromine in the upper layer. If chlorine is more reactive than iodine, it will replace iodine in NaI and produce a violet color characteristic of iodine in the upper layer. The ease with which the replacement reactions occur is an indication of the reactivities of the two elements. Given this information, how do the reactivities of chlorine, bromine, and iodine compare? If this trend holds true for other groups, are the more reactive nonmetallic elements toward the top or the bottom of the periodic table?

✔ Assessment

Knowledge: Ask students to predict which element of the following pairs is more reactive: Li or Ru, S or Se. *Ru; S* L1

---

miniLAB

Trends in Reactivity Within Families

Purpose: To compare the reactivities of two groups of elements.

Process Skills: Observing, comparing and contrasting, recognizing cause and effect, interpreting data, communicating

Safety Precautions: Students should wear aprons and safety goggles. Allow no open flames.

Teaching Strategies

- Emphasize that elements in a group have the same number of valence electrons; however, the number of energy levels increases down a group.

- **Disposal:** Allow remaining Ca and Mg to react with water in a fume hood or use Flinn disposal code 3. Allow lighter fluid mixtures to evaporate in a fume hood (Flinn disposal code 18a), then rinse the remaining solids down the drain with large amounts of water.

Expected Results

Ca is more reactive than Mg. Cl replaces I and is slightly more reactive than Br.

Analysis

1. Ca is more reactive than Mg. The metallic elements are more reactive toward the bottom of the periodic table.

2. Cl is more reactive than Br and much more reactive than I. The nonmetallic elements are more reactive toward the top of the periodic table.

Ask why, in some periodic tables, hydrogen seems to be placed in Group 1, the alkali metal family, or in Group 17, the halogen family, even though its chemical characteristics do not closely match the elements in either family. *Hydrogen has one outer-level electron so it is similar to the Group 1 elements. Like the Group 17 elements, hydrogen needs one electron to acquire the stable structure of a noble gas.* L2

Cooperative Learning Activity

Prepare a 3″ × 5″ card for three, four, or five elements in the following groups: 1, 2, 13, 14, 15, 16, 17. List this information: physical state at room temperature, color, density, electrical conductivity, formula of sodium compound (NaX, Na₂X, etc.) or formula of chlorine compound (XCl, XCl₄, etc.), and the energy level of the outermost electrons. Have each student identify one element. When all elements have been identified, have the students group themselves according to the groups of their elements and have each group research their elements and make a presentation to the class based on trends in the properties of their chemical family. The presentations could be oral, visual, rap, video, pantomime, or any combination. Look for creativity, but establish in advance the criteria upon which you will evaluate the presentations. L1 **LEP**

COOP LEARN

✓ Assessment

Portfolio: In a brief report, students should summarize their group's research in the above activity. Include the written summary in the students' portfolios. L1 **P**

Figure 3.7
Aluminum—Group 13, Period 3
This sculpture is made of aluminum, the Period 3 element in Group 13. Aluminum has two electrons in the first energy level, eight electrons in the second energy level, and three valence electrons in the third or outermost energy level.

Elements with atomic numbers 4 through 10 follow lithium and fill the second period. Each has one more electron than the element that preceded it. Neon, with atomic number 10, is at the end of the period. Eight electrons are added in Period 2 from lithium to neon, so eight electrons must be the number that can occupy the second energy level. The next element, sodium, atomic number 11, begins Period 3. Sodium's 11th electron is in the third energy level. The third period repeats the pattern of the second period. Each element has one more electron than its neighbor to the left, and these electrons are in the third energy level.

Atomic Structure of Elements Within a Period

The first period is complete with two elements, hydrogen and helium. Hydrogen has one electron in its outermost energy level, so it has one valence electron. Can you see that helium must have two valence electrons? Every period after the first starts with a Group 1 element. These elements have one electron at a higher energy level than the noble gas of the preceding period. Therefore, Group 1 elements have one valence electron. As you move from one element to the next across Periods 2 and 3, the number of valence electrons increases by one. Group 18 elements have the maximum number of eight valence electrons in their outermost energy level.

Group 18 elements are called the noble gases. Because the **noble gases** have a full complement of valence electrons, they are unreactive. The period number of an element is the same as the number of its outermost energy level, so the valence electrons of an element in the second period, for example, are in the second energy level. A Period 3 element such as aluminum, **Figure 3.7,** has its valence electrons in the third energy level.

Atomic Structure of Elements Within a Group

The number of valence electrons changes from one to eight as you move from left to right across a period; when you reach Group 18, the pattern repeats. For the main group elements, the group number is related to the number of valence electrons. The main group elements are those in Groups 1, 2, 13, 14, 15, 16, 17, and 18. For elements in Groups 1 and 2, the group number equals the number of valence electrons. For elements in Groups 13, 14, 15, 16, 17, and 18, the second digit in the group number is equal to the number of valence electrons. **Figure 3.8** illustrates the electron dot structures for the main group elements.

1	2	13	14	15	16	17	18
H·							He:
Li·	Be·	·B·	·C·	·N:	·O:	:F:	:Ne:
Na·	Mg·	·Al·	·Si·	·P:	·S:	:Cl:	:Ar:
K·	Ca·	·Ga·	·Ge·	·As:	·Se:	:Br:	:Kr:
Rb·	Sr·	·In·	·Sn·	·Sb:	·Te:	:I:	:Xe:
Cs·	Ba·	·Tl·	·Pb·	·Bi:	·Po:		:Rn:

Figure 3.8
Lewis Dot Structures for Some Elements
The number of valence electrons is the same for all members of a group. Verify the relationship between group number and the number of valence electrons.

Meeting Individual Needs

Limited English Proficiency To help students become familiar with the periodic table and how it relates to the atomic structure of the main group elements, divide the class into two teams. Have each team member make up four flash cards. One card should identify an element by its period and group number. A second card should identify an element by the number of its valence electrons and its period number. Two other cards could identify elements by their family names (alkali, alkaline earth, halogen, and noble gases) and period numbers. Have the teams challenge each other to identify the elements, setting a time limit for answering. L1 **LEP** **COOP LEARN**

Because elements in the same group have the same number of valence electrons, they have similar properties. Sodium is in Group 1 because it has one valence electron. Because the other elements in Group 1 also have one valence electron, they have similar chemical properties.

Chlorine is in Group 17 and has seven valence electrons. All the other elements in Group 17 also have seven valence electrons and, as a result, they have similar chemical properties. Throughout the periodic table, elements in the same group have similar chemical properties because they have the same number of valence electrons.

Four groups have commonly used names: the alkali metals in Group 1, the alkaline earth metals in Group 2, the halogens in Group 17, and the noble gases in Group 18. The word *halogen* is from the Greek words for "salt former" so named because the compounds that halogens form with metals are saltlike. The elements in Group 18 are called noble gases because they are much less reactive than most of the other elements.

Because the periodic table relates group and period numbers to valence electrons, it's useful in predicting atomic structure and, therefore, chemical properties. For example, oxygen, in Group 16 and Period 2, has six valence electrons (the same as the second digit in the group number), and these electrons are in the second energy level (because oxygen is in the second period). Oxygen has the same number of valence electrons as all the other elements in Group 16 and, therefore, similar chemical properties. **Figure 3.9** shows representations of the distribution of electrons in energy levels in the first three elements of Group 16.

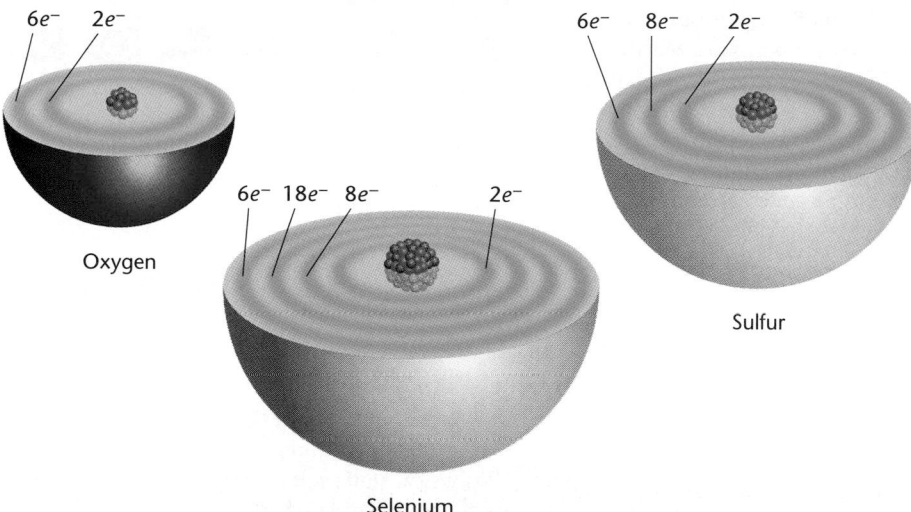

Oxygen

Selenium

Sulfur

Figure 3.9

Electrons in Energy Levels—Group 16

Oxygen, sulfur, and selenium have six electrons in their outermost energy levels as predicted by their group number. Note in the diagram of selenium that the third energy level can hold 18 e^-. You will learn more about energy levels in Chapter 7. Predict what a representation of tellurium, another Group 16 element, might look like.

WORD ORIGIN

periodic:
  periodus (GK)
  period of time
Periodic properties recur just as hours, days, and weeks repeat.

SECTION 3.2

✓ Assessment

Performance: Give blank periodic tables to students and ask them to label the periods, groups, family names, transition elements, lanthanides, and actinides. L1

Enrichment

To illustrate the international nature of chemistry, display periodic tables printed in other languages. Foreign-exchange students are a good source of these charts. It may surprise students to find that the content of the periodic table is exactly the same regardless of the language in which it is printed.

Meeting Individual Needs

Visually Impaired Attach to a board the wooden blocks created for the activity *Using an Analogy* on page 93. The blocks should be in the sequence described in the activity. Allow visually impaired students to examine the sequence of the blocks by touch in order to understand the periodic repetition of sizes as an analogy for the periodic law. L1
LEP

ChemLab

The Periodic Table of the Elements

Time Allotment: One 60-minute lab period or one and one-half 45-minute periods.

Objectives

Review objectives with students before they begin the Chem-Lab.

Process Skills

Observing, interpreting data, comparing and contrasting, classifying

Safety Precautions and Disposal

Students must wear aprons and goggles because the elements could shatter when struck with a hammer, and $1M$ HCl is harmful to eyes and clothing.
Disposal: (Code D) The HCl solution may be rinsed down the drain with large amounts of water.

PREPARATION

Concentrated HCl is $12M$. It can be diluted to $1M$ by mixing 11 parts of water with one part concentrated HCl. For example, to make 120 mL of $1M$ HCl, add 10 mL of concentrated HCl to 110 mL of water. **CAUTION:** *Add the acid to the water.*

Alternative Materials

- The lab can be done even if you do not have all of the listed elements.
- If you substitute elements, be certain that hazardous reactions will not occur.

PROCEDURE

- The first lab period should be devoted to prelab discussion and procedure steps 1 and 2. Aprons and goggles are not needed for steps 1 and 2. Steps 2 to 8 and the postlab discussion can be completed in the second lab period.
- Set up stations at various places around the lab where

100

ChemLab

The Periodic Table of the Elements

In the mid-1800s, scientists found that when all the known elements were arranged in a sequence of increasing atomic mass and according to their properties, the properties of the elements repeated in a regular, or periodic, manner. In this lab, you will investigate some representative elements in several of the vertical groups or families of the periodic table and classify the elements as metals, nonmetals, or metalloids. In general, metals are solids at room temperature. They have a metallic luster and are malleable. Metals conduct electricity, and many react with acids. On the other hand, nonmetals can be either solid, liquid, or gas at room temperature. If a nonmetal is a solid, it's likely to be brittle rather than malleable. Nonmetals do not conduct electricity and do not react with acids. Metalloids combine some of the characteristics of both metals and nonmetals.

Your experimental data will allow you to classify some elements as metals, nonmetals, or metalloids and to determine general trends in metallic and nonmetallic characteristics within the periodic table.

Problem

What is the pattern of metallic and nonmetallic properties of the elements in the periodic table?

Objectives

- **Observe** the properties of samples of the elements, including metals, nonmetals, and metalloids.
- **Classify** the elements as metals, nonmetals, or metalloids.
- **Analyze** your results to discover trends in the properties of the elements in the periodic table.

PREPARATION

Materials

stoppered test tubes containing small samples of carbon, nitrogen, oxygen, magnesium, aluminum, silicon, red phosphorus, sulfur, chlorine, calcium, selenium, tin, iodine, and lead
plastic dishes containing samples of carbon, magnesium, aluminum, silicon, sulfur, and tin
micro-conductivity apparatus
$1M$ HCl
test tubes (6)
test-tube rack
10-mL graduated cylinder
spatula
small hammer
glass marking pencil

Safety Precautions

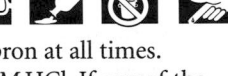

Wear goggles and an apron at all times. Be cautious when using $1M$ HCl. If any of the acid touches your skin or eyes, immediately rinse with water and notify your teacher. Never test chemicals by tasting.

PROCEDURE

1. Prepare a table like the one shown for your data and observations.

2. Observe and record the appearance of each of the elements. Your description should include physical state, color, and any other observable characteristics such as luster. Don't open any of the test tubes.

3. Remove a small sample of each of the six elements in the dishes. Place the samples on a hard surface designated by your teacher. Gently tap each of the elements with a small hammer.

DATA AND OBSERVATIONS (for some elements)

Element	Appearance	Malleable or Brittle	Electrical Conductivity	Reaction with HCl
carbon	gray or black, dull, solid	brittle	yes	no
oxygen	colorless gas			
magnesium	shiny, silver solid	malleable	yes	yes
silicon	shiny, gray solid	brittle	yes	no
sulfur	dull, yellow solid	brittle	no	no
chlorine	yellow-green gas	not tested	not tested	not tested
iodine	shiny, gray solid	not tested	not tested	not tested

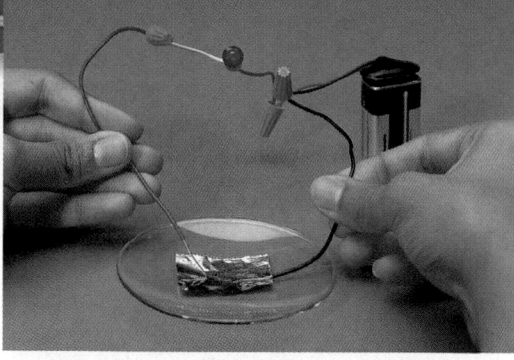

An element is malleable if it flattens when tapped. It is brittle if it shatters when tapped. Record your observations in the data table.

4. Test the conductivity of each of the six elements in the dishes by touching the electrodes of the micro-conductivity apparatus to a piece of the element. If the bulb lights, you have evidence of conductivity. Record your observations in the data table. Wash the electrodes with water, and dry between testing each element.

5. Use a graduated cylinder to measure 5 mL of water into each of the six test tubes.

6. Label each test tube with the symbol of one of the elements.

7. Using a spatula, put a small sample of each of the six elements (a 1-cm length of ribbon or 0.1-0.2 g of solid) into a test tube labeled with the symbol of the element.

8. Add approximately 5 mL of 1M HCl to each of the test tubes and observe the elements for at least one minute. Evidence of reaction is the formation of bubbles of hydrogen on the element. Record your observations in the data table.

DATA AND OBSERVATIONS

Element	Appearance	Malleable or Brittle	Electrical Conductivity	Reaction with HCl
carbon				

ANALYZE AND CONCLUDE

1. **Interpreting Data** Which elements displayed the general characteristics of metals?

2. **Interpreting Data** Which elements displayed the general characteristics of nonmetals?

3. **Interpreting Data** Which elements displayed a mixture of metallic and nonmetallic characteristics?

APPLY AND ASSESS

1. Construct an abbreviated periodic table with seven 1-inch squares across and five squares down. Label the squares across the top from left to right as Groups 1 and 2 and 13-17. Label the squares down the side as Periods 2-6. Write the appropriate atomic number and element symbol in each of the squares. Based upon your answers to the Analyze and Conclude questions, write your classification of metal, nonmetal, or metalloid for each of the elements you observed and/or tested.

2. Do the metallic characteristics of the elements across a period seem to increase from left to right or from right to left?

3. Do the metallic characteristics of the elements in a group seem to increase from top to bottom or from bottom to top?

4. The metalloids indicate the approximate border between metals and nonmetals on the periodic table. Based upon your observations, draw a dark line along this border.

ChemLab

students can observe and test the elements.

ANALYZE AND CONCLUDE

1. Mg, Al, Ca, Sn, and Pb display metallic characteristics. As, Se, and I display some metallic characteristics.

2. N, O, P, S, and Cl display nonmetallic characteristics. As, Se, and I display some nonmetallic characteristics.

3. C and Si display the characteristics of metalloids. As and Se may display metalloid characteristics.

APPLY AND ASSESS

1. See table below.
2. Metallic characteristics increase from right to left.
3. Metallic characteristics increase from top to bottom.
4. The line will go near the elements C, Si, As, Se, and I. Based upon the students' results, this line will be tentative.

✓ Assessment

Performance: Give each student a blank periodic table. Ask students to draw an arrow on their tables to show the direction of increasing metallic properties across the table and another arrow to show increasing metallic properties within a group.
L1

	Group 1	Group 2	Group 13	Group 14	Group 15	Group 16	Group 17
Period 2				6 carbon, metalloid	7 nitrogen, nonmetal	8 oxygen, nonmetal	
Period 3		12 magnesium, metal	13 aluminum, metal	14 silicon, metalloid	15 phosphor, nonmetal	16 sulfur, nonmetal	17 chlorine, nonmetal
Period 4		20 calcium, metal			33 arsenic, can't be sure	34 selenium, can't be sure	
Period 5				50 tin, metal			53 iodine, can't be sure
Period 6				82 lead, metal			

Content Background

The Newest Elements The names of elements 104 to 109 have been approved by the American Chemical Society but have not been universally accepted. The International Union of Pure and Applied Chemistry (IUPAC), which decides on nomenclature for acceptance worldwide, still considers these elements to be unnamed, as are elements 110 and 111. Like rutherfordium (Rf) and nielsbohrium (Ns), the elements hahnium (Ha), seaborgium (Sg), and meitnerium (Mt) were named to honor the physicists whose work in the field of nuclear research brought about the synthesis of these elements. They are Otto Hahn, Glenn Seaborg, and Lise Meitner, respectively. Hassium (Hs) was named for the German state of Hesse, where the element was synthesized.

Physical States and Classes of the Elements

Other practical information can be obtained from the periodic table. The arrangement of the table helps you determine the physical state of an element; whether the element is synthetic or natural; and whether the element is a metal, nonmetal, or metalloid.

Physical States of the Elements

The periodic table on pages 92 and 93 shows the states of the elements at room temperature and normal atmospheric pressure. Most of the elements are solid. Only two elements are liquids, and the gaseous elements, except for hydrogen, are located in the upper-right corner of the table.

Some elements are not found in nature but are produced artificially in particle accelerators like the one shown in **Figure 3.10.** These are known as synthetic elements. The synthetic elements, made by means of nuclear reactions, are marked on the periodic table. They include technetium, element 43, and all the elements after uranium, element 92. Although small amounts of neptunium and plutonium, elements 93 and 94, have been found in uranium ores, it is likely that they are the products of nuclear bombardment by radiation from uranium atoms.

Classifying Elements

The color coding in the periodic table on pages 92 and 93 identifies which elements are metals (blue), nonmetals (yellow), and metalloids (green). The majority of the elements are metals. They occupy the entire left side and center of the periodic table. Nonmetals occupy the upper-right-hand corner. Metalloids are located along the boundary between metals and nonmetals. Each of these classes has characteristic chemical and physical properties, so by knowing whether an element is a metal, nonmetal, or metalloid, you can make predictions about its behavior. Elements are classified as metals, metalloids, or nonmetals on the basis of their physical and chemical properties.

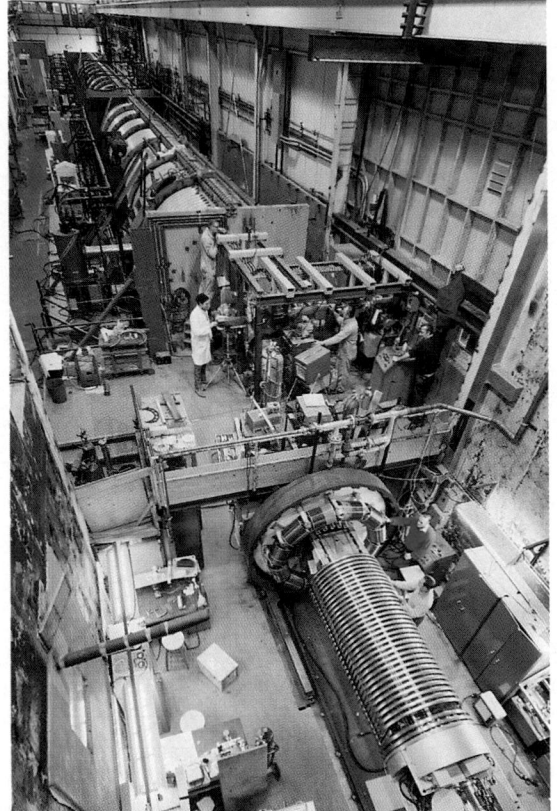

Figure 3.10

The Super HILAC Accelerator at the University of California, Berkeley
Element 106, seaborgium, was synthesized in this linear accelerator. To change a nucleus and create a new element, a high-energy collision must occur between two particles. In the case of seaborgium, highly accelerated oxygen nuclei were smashed into a target of the heavy element californium.

Across the Curriculum

Literature

The Mad Hatter in Lewis Carroll's *Alice in Wonderland* suffered from mercury poisoning because he used mercury compounds to make animal furs into felt for hats. In Carroll's time, people who made hats frequently became known as "mad hatters" because they developed serious nervous system disorders from the mercury they were exposed to every day. Mercury poisoning can be fatal. Some symptoms of mercury poisoning are memory loss, impaired vision, and insanity.

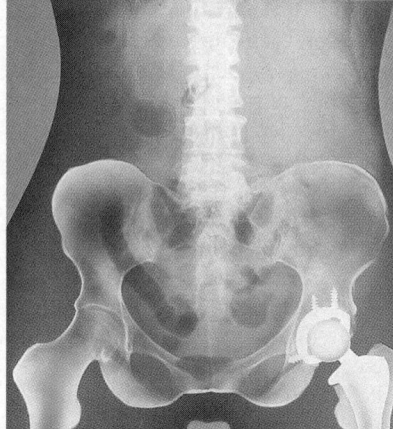

Figure 3.11
Structural Uses of Metals
Alloys of metals can be custom designed to meet a wide range of uses, such as hip replacements (left) and scaffolding for sky-scrapers (right).

Metals

Metals are almost everywhere. They make up many of the things you use every day—cars and bikes, jewelry, coins, electrical wires, household appliances, and computers. Because of their strength and durability, they're used in buildings and bridges, and for bone replacement as shown in **Figure 3.11. Metals** are elements that have luster, conduct heat and electricity, and usually bend without breaking. With the exception of tin, lead, and bismuth, metals have one, two, or three valence electrons. All metals except mercury are solids at room temperature; in fact, most have extremely high melting points.

The periodic table shows that most of the metals (coded blue) are not main group elements. A large number are located in Groups 3 through 12. Notice the elements in the fourth period beginning with scandium (Sc), atomic number 21, and ending with zinc (Zn), atomic number 30. These ten elements mark the first appearance of elements in Groups 3 through 12. From the fourth period to the bottom of the table, each period has elements in these groups. The elements in Groups 3 through 12 of the periodic table are called the **transition elements.** All transition elements are metals. Many are commonplace, including chromium (Cr), iron (Fe), nickel (Ni), copper (Cu), zinc (Zn), silver (Ag), and gold (Au). Some are less common metals but still important, such as titanium (Ti), manganese (Mn), platinum (Pt), and uranium (U). Some period 7 transition elements are synthetic and radioactive.

While the chemistry of the main group metals is highly predictable, that of the transition elements is less so. The unpredictability in the behavior and properties of the transition metals is due to the more complicated atomic structure of these elements.

SECTION 3.2

Content Background

Metals of the Ancient World The metals best known in ancient times were transition elements—gold, silver, copper, iron, mercury, and lead. They were discovered early because some of them—for example, gold and silver—often occur in Earth's crust in elemental form. Others can be separated from their ores relatively easily. The ores of transition elements are often oxides or sulfides. Iron occurs in hematite, $Fe_2O_3 \cdot nH_2O$, copper is found in chalcocite, Cu_2S, mercury is mined as cinnabar, HgS, and lead as galena, PbS.

Fact of the MATTER

A fascinating feature of chromium chemistry is the many colorful compounds it produces, such as bright yellow potassium chromate and brilliant orange potassium dichromate. These and other colored compounds are responsible for the element's name. Chromium comes from the Greek word *khroma*, which means "color." Trace amounts of chromium in otherwise-colorless mineral crystals produce the brilliant colors of rubies and emeralds.

GLENCOE TECHNOLOGY

 Software

Mastering Concepts in Chemistry
Unit 8, Lesson 2
Typical Elements
Use this lesson to reinforce information about the periodicity of elements. L1 LEP

Integrating THE Sciences

Earth Science Hydrogen and helium make up 99% of the universe. All other elements seem to have evolved from these two elements inside stars under high temperatures and pressures. Our sun is a young star with enough hydrogen fuel to last billions of years, even though it converts 400 million tons of hydrogen to helium every second. But some larger and older stars may already have fused most of their hydrogen to helium. When that happens, stars begin to collapse and enormous pressure builds up that causes another series of nuclear fusion reactions to occur. Helium fuses into carbon. Carbon then fuses into oxygen. Through the successive fusion of helium nuclei, neon, magnesium, silicon, and sulfur are built in stages by adding two protons and two neutrons at each stage. Eventually, iron is formed. Atoms, synthesized in this way, are blown away into space as star dust. All the elements on Earth were made in stars billions of years ago. The carbon and nitrogen in DNA, the calcium in bones and teeth, and the iron in hemoglobin were all made in stars. Earth is made of star dust.

Correcting Misconceptions

Students sometimes think that the inner transition elements are not really part of the periodic table. The long form of the table presented here gives you the opportunity to show that the lanthanides and actinides belong in the table. Have students follow the progression of atomic numbers following lanthanum and actinium. Then, refer them to the usual presentation of the table on pages 92 and 93 for comparison. Having seen the long form and how it relates to the more compact version, students will have a better understanding of the positions of the *f*-block elements on the table.

GLENCOE TECHNOLOGY

Software

Mastering Concepts in Chemistry
Unit 8, Lesson 1
Periodic Properties
Use this lesson to reinforce the concept of periodicity in the properties of the elements when arranged by atomic number in the periodic table.

`L1` `LEP`

Figure 3.12

Nitrogen-Containing Explosives
Dynamite and TNT (trinitrotoluene) are nitrogen-containing explosives used to blast away rock in road building or efficiently demolish unwanted buildings.

In the periodic table, two series of elements, atomic numbers 58-71 and 90-103, are placed below the main body of the table. These elements are separated from the main table because putting them in their proper position would make the table very wide, as shown in **Figure 3.13.** The elements in these two series are known as the inner transition elements. Many of these elements were unknown in Mendeleev's time, but he did know of some of them and suspected that more would be discovered.

The first series of inner transition elements is called the **lanthanides** because they follow element number 57, lanthanum. The lanthanides consist of the 14 elements from number 58 (cerium, Ce) to number 71 (lutetium, Lu). Because their natural abundance on Earth is less than 0.01 percent, the lanthanides are sometimes called the rare earth elements. All of the lanthanides have similar properties.

The second series of inner transition elements, the **actinides,** have atomic numbers ranging from 90 (thorium, Th) to 103 (lawrencium, Lr). All of the actinides are radioactive, and none beyond uranium (92) occur in nature. Like the transition elements, the chemistry of the lanthanides and actinides is unpredictable because of their complex atomic structures. What could be happening at the subatomic level to explain the properties of the inner transition elements? In Chapter 7, you'll study an expanded theory of the atom to answer this question.

Nonmetals

Although the majority of the elements in the periodic table are metals, many nonmetals are abundant in nature. The nonmetals oxygen and nitrogen make up 99 percent of Earth's atmosphere. Carbon, another nonmetal, is found in more compounds than all the other elements combined. The many compounds of carbon, nitrogen, and oxygen are important in a wide variety of applications like the one shown in **Figure 3.12.** Most **nonmetals** don't conduct electricity, are much poorer conductors of heat than metals, and are brittle when solid. Many are gases at room temperature; those that are solids lack the luster of the metals. Their melting points tend to be lower than those of metals. With the exception of carbon, nonmetals have five, six, seven, or eight valence electrons. **Table 3.5** summarizes the properties of metals and nonmetals.

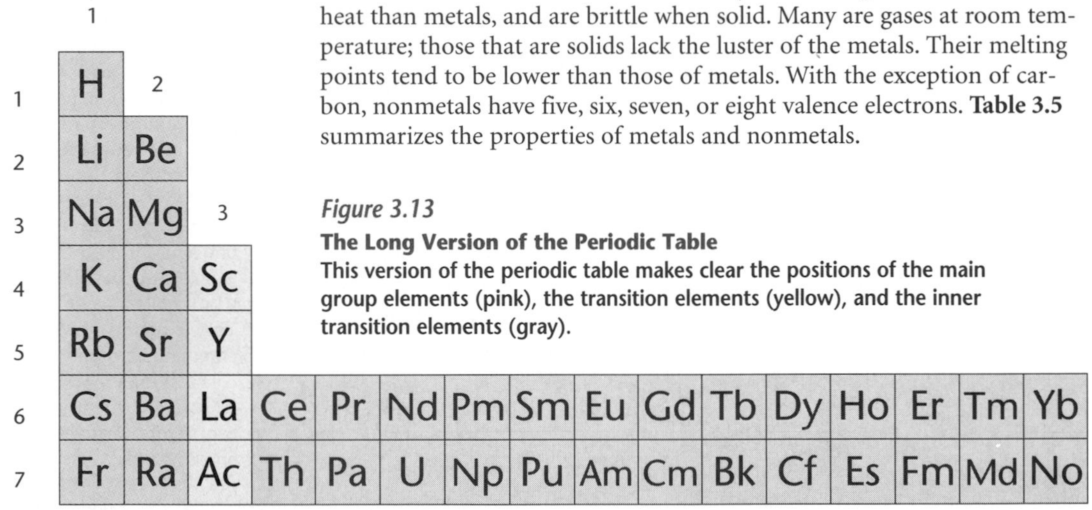

Figure 3.13

The Long Version of the Periodic Table
This version of the periodic table makes clear the positions of the main group elements (pink), the transition elements (yellow), and the inner transition elements (gray).

	1		2		3													
1	H																	
2	Li	Be																
3	Na	Mg																
4	K	Ca	Sc															
5	Rb	Sr	Y															
6	Cs	Ba	La	Ce	Pr	Nd	Pm	Sm	Eu	Gd	Tb	Dy	Ho	Er	Tm	Yb
7	Fr	Ra	Ac	Th	Pa	U	Np	Pu	Am	Cm	Bk	Cf	Es	Fm	Md	No

Demonstration 3.3

Carbon's Predictability

Purpose: To demonstrate that a gas containing carbon can be used as a fuel, but not all carbon-containing gases burn.

Materials: 100 mL of soap-water mixture, a piece of rubber tubing connected to a gas jet, wood splints, matches, 400-mL beaker, 600-mL beaker, 30 g of baking soda, 30 mL of vinegar

Safety Precautions:

Disposal: Code C

Procedure: Prepare a soap-water mixture in a 400-mL beaker, and connect a piece of rubber tubing to a gas jet. Slowly bubble the methane or propane gas into the solution until a layer of small bubbles forms on the surface. **CAUTION:** *Do not allow a large head of suds to form.* Turn off the gas

jet. Use a burning wood splint to pop the bubbles. In the 600-mL beaker, mix the 30 g of baking soda and the 30 mL of vinegar. Lower a burning splint into the 600-mL beaker containing the CO_2 gas generated by the $NaHCO_3$ and vinegar.

Results: The hydrocarbon gas in the soap bubbles is flammable, while carbon dioxide from the $NaHCO_3$ and vinegar is not. The bubbles of the hydrocarbon gas pop

Metalloids

Metalloids have some chemical and physical properties of metals and other properties of nonmetals. In the periodic table, the metalloids lie along the border between metals and nonmetals. Silicon (Si) is probably the most well-known metalloid. Some metalloids such as silicon, germanium (Ge), and arsenic (As) are semiconductors. A **semiconductor** is an element that does not conduct electricity as well as a metal, but does conduct slightly better than a nonmetal. The ability of a semiconductor to conduct an electrical current can be increased by adding a small amount of certain other elements. Silicon's semiconducting properties made the computer revolution possible.

Table 3.5 Properties of Metals and Nonmetals

Metals	Nonmetals
Bright metallic luster	Non-lustrous, various colors
Solids are easily deformed	Solids may be hard or soft, usually brittle
Good conductors of heat and electricity	Poor conductors of heat and electricity
Loosely held valence electrons	Tightly held valence electrons

Atomic Structure of Metals, Metalloids, and Nonmetals

The differences in the properties of the three classes of elements occur because of the different ways the electrons are arranged in the atoms. The number and arrangement of valence electrons and the tightness with which the valence electrons are held in an atom are important factors in determining the behavior of an element. In general, the valence electrons in a metal are loosely bound to the positive nucleus. They are free to move in the solid metal and are easily lost. This freedom of motion accounts for the ability of metals to conduct electricity. On the other hand, the valence electrons in atoms of nonmetals and metalloids are tightly held and are not easily lost. When undergoing chemical reactions, metals tend to lose valence electrons, whereas nonmetals tend to share electrons or gain electrons from other atoms.

Content Background

Silicon: The Computer-Age Metalloid
The Swedish chemist J.J. Berzelius isolated the element silicon in 1824. Silicon is found combined in almost all rocks, sand, and soil. It composes 27.7% of Earth's crust and is the second-most-abundant element after oxygen. A pure crystal of silicon is hard, dark gray, and lustrous. A 20th-century technology has developed for growing perfect crystals of silicon for use in semiconductor devices. The crystals are cut into thin slices and doped with elements from Group 13 or Group 15 to create either *n*-type or *p*-type semiconductor materials. Doping is sometimes accomplished by diffusion or by ion implantation. The diffusion process involves heating the silicon in a high-temperature oven in an atmosphere of the doping element. In the ion-implantation process, silicon is bombarded by ions of the doping element. These ions are incorporated into the top layer of the crystal structure. Doped slices, called wafers, can then be processed to become diodes, transistors, or other electronic devices.

---

and burn, producing soot. The wood splint is extinguished in the beaker filled with carbon dioxide.
Analysis: Ask these questions.
1. What element is combined with carbon in the methane or propane gas? *hydrogen*
2. What element is combined with carbon in the gas generated by reacting baking soda and vinegar? *oxygen*

3. How do the positions of the elements in questions 1 and 2 relate to the position of carbon on the periodic table? *Hydrogen is farther from carbon than oxygen is.*

Assessment

Knowledge: Ask students to describe the outer electron structure of carbon, hydrogen, and oxygen. Ask students whether they think there is a relationship between the position of the element on the periodic table, its outer electron structure, and its chemical properties. L1

Have students review the photos and descriptions of metals, nonmetals, and metalloids. Then ask them to write five clues, each of which describes an element. The clues should be specific enough that another student can guess their identity. For example, a clue might be: *It is the only liquid metal at room temperature.* Collect the clues and review them to see that all are clear and reasonable. Divide the class into two teams, and use the clues for an oral quiz/ contest between the teams. L1

COOP LEARN

GLENCOE TECHNOLOGY

 CD-Rom

Chemistry: Concepts and Applications
The Periodic Table
Have students use this interactive periodic table to enhance their appreciation and understanding of the properties of elements. L1 LEP

General Properties of Metals, Nonmetals, and Metalloids

Most properties of metals, nonmetals, and metalloids are determined by their valence electron configurations. The number of valence electrons that a metal has varies with its position in the periodic table. Valence electrons in metal atoms tend to be loosely held. Nonmetals have four or more tightly held electrons, and metalloids have three to seven valence electrons.

 Familiar Metals
Polished silverware and copper jewelry are valued for their beautiful metallic luster. The ability of copper wire to conduct electricity makes it useful in electrical circuits. Some metals can be molded into objects, like this bronze figurine. Bronze is an alloy of two metals— copper and tin.

Silver and copper

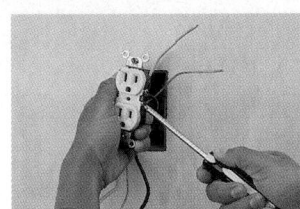
Copper wire

Bronze figurine

Many transition elements are important as structural materials. Iron is made into steel by mixing it with carbon. Sometimes, other metals are added to produce special properties. Iron mixed with manganese produces a steel hard enough to be used to make the jaws on rock-crushing machines. The combination of iron with vanadium produces a tough alloy used, among other things, to make the crankshafts in automobile engines.

Exposure to air and moisture causes the iron in steel to rust. Plating a steel surface with chromium protects it from corrosion. The appearance of some consumer products is enhanced by plating with bright, shiny chromium.

Steel rock crusher

Chrome plating

Most metals are solids at room temperature; mercury is the only metal that is a liquid at room temperature. Mercury is poisonous and should never be handled.

Mercury

Demonstration 3.4

Predicting a Periodic Trend

Purpose: To allow students to validate a prediction using the periodic table.
Materials: Notebook-sized periodic table; small pieces of carbon, silicon, germanium, tin, lead; glue; conductivity tester; heavy card stock; other small samples of elements from your stockroom. **CAUTION:** *Do not use Li, Na, K, or P.*
Safety Precautions:

Disposal: Code F
Procedure: Glue a notebook-sized periodic table to a piece of heavy card stock. Glue to the periodic table small samples of the elements that you have in your stockroom. Students will observe that most of the elements are metals. Using a conductivity tester, touch the instrument's probes to the samples glued on the table. Ask students to predict what conductivity trend

would be expected in the carbon family, then test the elements in Group 14.
Results: Students will observe the obvious difference in conductivity between the metals and the nonmetals. Students experience success in predicting the conductivity trend of the Group 14 elements.
Analysis: Ask these questions.
1. How many outer electrons are present in carbon? *four*

2 **Some Lanthanides and Actinides**
Compounds of europium and ytterbium are used in the picture tubes of color televisions. Neodymium is used in some high-power lasers.

Neodymium in a laser

Europium and ytterbium in color TVs

3 **Carbon and Some Other Nonmetals**
Coal is nearly pure carbon. It is mined, often from strip mines, and burned as a fuel. Natural gas and oil are also carbon-rich fuels. Even though their appearance and physical properties are different, graphite and diamond are both naturally occurring forms of carbon. Bromine and iodine are used in high-intensity halogen lamps. Liquid nitrogen is used to maintain low temperatures; it can freeze the moisture in air, seen here as a white cloud.

Carbon in coal

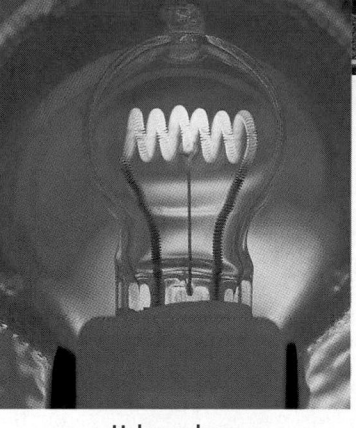

Halogen lamp

Liquid nitrogen

Graphite and diamond

4 **Metalloids**
Silicon looks like a metal, but it's brittle and doesn't conduct heat and electricity well. Its melting point, 1410°C, is close to that of many metals. Elemental silicon (left) is melted, formed into a single crystal of pure silicon, and purified (back center). The crystal is sliced into thin wafers (right), and these are used to produce electronic devices (front center).

Silicon

3.2 **Using the Periodic Table**   **107**

SECTION 3.2

Applying Chemistry

How the Elements Are Used Have students bring in photos of other uses for the metals, metalloids, and nonmetals described on these pages. Also, ask them to find photos that show the uses of other elements and their compounds. All photos should be accompanied by an identification of the element as a metal, metalloid, or nonmetal and a brief description of the use shown in the photo. Groups of students can work together to create bulletin board displays or posters. L1
COOP LEARN

Transparency

Display **Basic Concepts Transparency 9** to help students recognize elements. L1 LEP

Display **Problem Solving Transparency 3** to help students use the periodic table effectively. L1 LEP

2. Examine the electron dot diagram of carbon. .C. Does it agree with the outer electron configuration of carbon? *yes*
3. Explain the trend of increasing conductivity down the table in Group 14. *Electrons that are farther from the nucleus are more free to move. Thus, elements with large atoms are better conductors.*

✓Assessment

Knowledge: Ask students to predict the distribution of electrons in the energy levels of carbon and silicon. *Carbon has two electrons in the first energy level and four electrons in the second energy level. Silicon has two electrons in the first energy level, eight electrons in the second, and four in the third.* Point out to students that all the elements in Group 14 have four electrons in their outermost energy level, but the electrons are in energy levels that are increasingly far from the nucleus as you go down the group. Ask students what effect this has on metallic properties of the elements. *Metallic properties increase down the group.* L1

CHEMISTRY
&TECHNOLOGY

Purpose

Students will learn about the chemical properties of shape-memory alloys.

Background

One shape-memory alloy is nitinol. Its name is an acronym for *Ni*ckel *Ti*tanium *N*aval Ordnance *L*aboratory. It was developed at the Naval Ordnance Laboratory (White Oak, Maryland) by researcher William J. Buehler, who was studying alloys for use in missile nose cones. At a routine meeting, a folded strip of nitinol was repeatedly folded accordion-style and then unfolded over and over to show that it resisted fatigue. One of the participants heated the folded nitinol with a lighter, and the strip unfolded spontaneously. Besides having stamina, nitinol also had a memory!

Teaching Strategies

- Discuss the diagram showing how the phases of nitinol wire change with temperature. Have students follow the change in temperature (arrow) of the austenite phase to a temperature above the transition temperature. Explain that the wire converts to the martensite phase only after cooling below the transition temperature. Any shape assumed by the metal below the transition temperature can be reversed by heating.

- Ask students to explain the clot captor based on the discussion of transition temperatures. *The nitinol wire in the clot captor is in the martensite phase. Because the transition temperature is at body temperature, the folded captor is kept cool until it is placed in the vena cava, where it warms and unfolds.*

Metals That Untwist

Those coated-wire ties for closing garbage bags are a great invention. The plastic coating makes it easy to twist the tie without cutting your hands or slicing through the bag. The wire is easy to twist and, once twisted, stays that way. Wouldn't you be surprised if a twisted tie began to untwist spontaneously? Wires can't do that, can they?

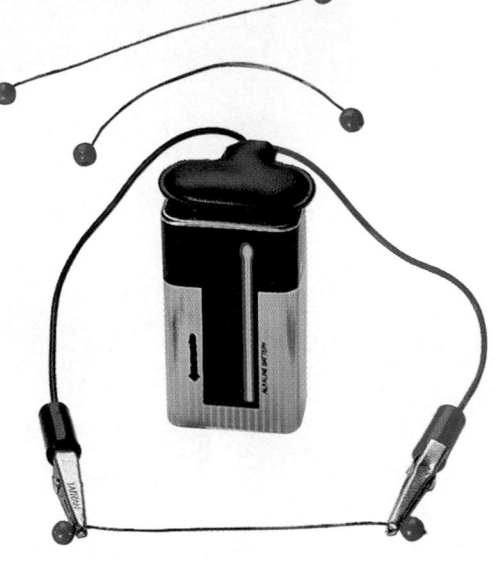

Shape-Memory Metals

Some alloys have a remarkable property. They revert to their previous shape when heated or when the stress that caused their shape is removed. These alloys are called shape-memory alloys. Above is a piece of shape-memory alloy wire that was bent and then heated by a small electric current. Notice that the wire reverted to its original shape as it was heated.

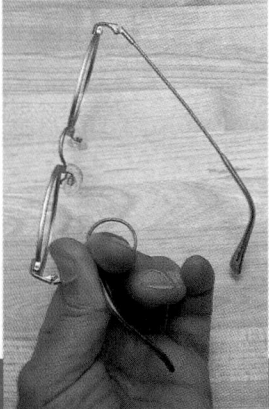

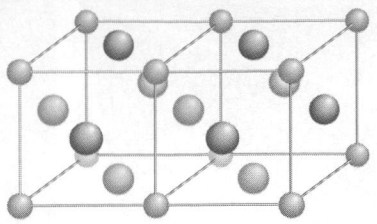

Austenite phase of nitinol

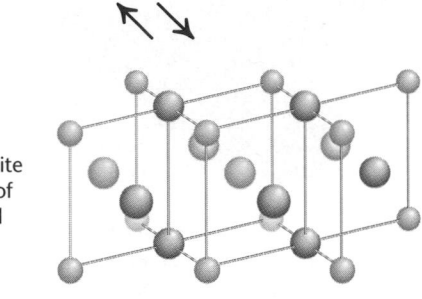

Martensite phase of nitinol

Different Solid Phases

Melting is the transition of a material from a solid to a liquid. Transitions from one phase to another can also take place within a solid. A solid can have two phases if it has two possible crystalline structures. It is the ability to undergo these changes in crystalline structure that gives shape-memory metals their properties. For example, an alloy having equal amounts of two metals may have one possible crystalline structure that is called the austenite phase. If the austenite phase is cooled under controlled conditions, the material takes on the martensite phase. The new crystal structure doesn't change the positions of the atoms throughout the material. However, the new internal organization gives the alloy new properties.

Nitinol

Nitinol is an alloy of nickel and titanium that has the austenite phase structure. Both the nickel and titanium atoms are arranged in cubes. As you can see above, each nickel atom is at the center of a cube of titanium atoms, and a titanium atom is at the center of each cube of nickel atoms. If nitinol is shaped into a straight wire (a), heated (b), and cooled past its transition temperature (c), then it takes on the martensite phase. Notice that its external shape doesn't change; it's still straight (d). But its structure in (d) allows it to be bent by an external stress in (e). Now, if the wire is heated, the stress is released and it reverts back to its initial shape in the martensite phase (f).

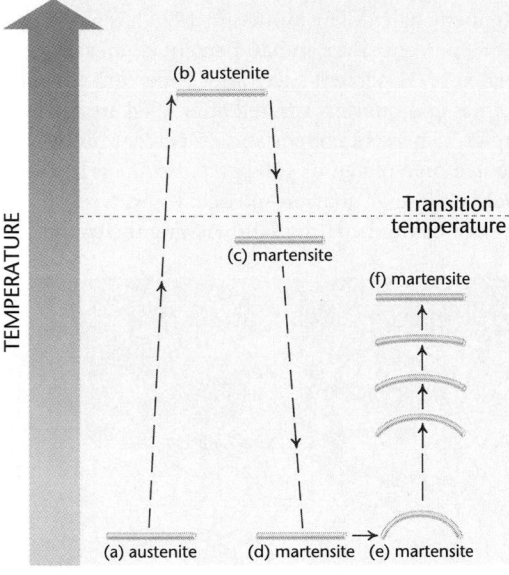

Robotic Arm

The Oaktree automatic arm contains Flexinol wires that act as muscles that can move the fingers in the robotic hand. The hand movements are so precise that

people who are both blind and deaf can feel and interpret its movements as it signs to them in American Sign Language. To do this, an optical character scanner reads texts and converts the characters into input signals, which then drive the fingers to form appropriate symbols.

Clot Captor

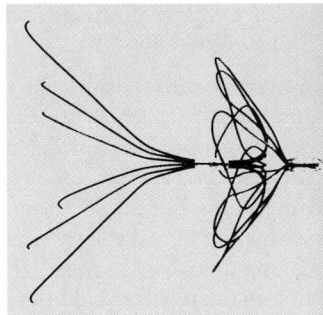

This device is used to capture blood clots in the vena cavae (main veins) before they reach the lungs. Folded inside its sheath, the clot captor is only 3 mm in diameter. But once inserted into the vein, the wire unfolds to a diameter of 28 mm. The clot captor is made of nitinol wire that has a transition temperature just below body temperature. The wire is folded and placed in a sheath. As the sheath is inserted into the vein, the wire is bathed in a cold saline solution. Slowly, the sheath is withdrawn. As the wire warms to body temperature, it unfolds into its opened, umbrella-like shape.

DISCUSSING THE TECHNOLOGY

1. **Inferring** Twisted nitinol-wire eyeglass frames unbend spontaneously at room temperature. Is the transition temperature of the frames above or below room temperature?

2. **Applying** Design a simple lever that could be raised and lowered smoothly using nitinol wires.

Visual Learning

- Have students use Styrofoam balls or gumdrops and toothpicks to make models of the austenite and martensite structures. Help them to see that both structures can be accommodated within the same array of atoms. **L1** **LEP**

- Point out that in many cases, there are several crystalline structures in the martensite phase.

Discussing the Technology

1. The transition temperature is below room temperature.

2. Accept any reasonable design in which the lever is raised and lowered by opposing sets of wires constructed so that they shorten as they're heated (i.e., are arc shaped in austenite phase and bent into straight wires at room temperature) and work like human muscles, or elongate when heated and work in a manner opposite that of human muscles.

GLENCOE TECHNOLOGY

 Videodisc

Chemistry: Concepts and Applications
Shape Memory Metals
Disc 1, Side 1, Ch. 11

Use this video to show a prosthetic arm made from Flexinol wires and the twisting and unbending of nitinol eyeglass frames. **L1** **LEP**

 CD-Rom

Chemistry: Concepts and Applications
Shape Memory Metals
Have students use this video to experience visually the properties of shape-memory metals.
L1 **LEP**

109

Everyday Chemistry

Everyday Chemistry

Purpose

Students will learn about the metals used in making historic and contemporary coins.

Teaching Strategies

- Explain that the abundance of the coinage metals in Earth's crust is small compared to the abundance of iron. Copper is 68 ppm, whereas iron is 62 000 ppm.
- Have students determine the diameter of the 1793 penny (mass: 13.478 g, thickness: 0.24 cm) and compare it to the diameter of a modern penny. L3

$$\rho = \frac{M}{V}; \quad V = \frac{M}{\rho}$$

$$\frac{\pi D^2 h}{4} = \frac{M}{P}; \quad D = \sqrt{\frac{4M}{\pi \rho h}}$$

$$D = \sqrt{\frac{4(13.478 \text{ g})}{(8.92 \text{g/mL})(3.14)(0.24 \text{cm})}}$$

$$D = 2.8 \text{ cm}$$

A modern penny is about 2.0 cm

Extension

Have students research the minting of U.S coins and write a short paper. Encourage students to find connections between the properties of the metals and the way that they are processed. L1

Exploring Further

1. chemical: not reactive; physical: malleable
2. copper, zinc, nickel
3. copper: electrical wiring; silver: photography; gold: electronics

---

Metallic Money

The concept that a certain amount of a metallic substance could stand as a measure of goods and services dates back to the ancient Greeks. They were the first to use metal coins as measures of wealth. Other civilizations used gold and silver as displays of wealth.

Coinage metals Historically, copper, silver, and gold were logical choices for use in coins. These metals are not abundant in Earth's crust and so are considered rare. Most elements occur combined in compounds, but copper, silver, and gold occur nearly pure in rocks fairly close to Earth's surface, making them easy to mine. Their lack of reactivity endows them with a kind of noble quality. People revere them because of their beauty and rarity. Finally, their properties allow them to be easily shaped, stamped, and marked for value.

Contemporary coinage metals Rarity makes metals more expensive. Gold and silver have

become so expensive that these metals have vanished from most of the coins in your pocket. Even though gold bullion is still stored at Fort Knox, Kentucky, the United States eliminated gold from its coins in 1934. Silver was eliminated in 1971. As a result, a Kennedy half-dollar minted in 1972 has a coating of 75 percent copper and 25 percent nickel on a pure copper core, called a platen. A Kennedy half-dollar minted in 1970 has a plating of 80 percent silver and 20 percent copper on a platen of 21 percent silver and 79 percent copper. Dimes and quarters minted after 1964 are made up of 75 percent copper and 25 percent nickel coated on a platen of 100 percent copper. If you hold a dime or quarter on edge, you can see the silvery-colored coating sandwiching the copper.

Cu 100%
1851

Cu 88%
Ni 12%
1859

Cu 95%
1851

Cu 95%
Sn and Zn 5%
1864

Cu 95%
Zn 5%
1962

Zn 97.5%
Cu 2.5%
1982

Exploring Further

1. **Applying** Give one chemical and one physical property of copper, silver, and gold that made them the metals of choice for use as coinage metals.

2. **Inferring** What metals make up today's U.S. coinage metals?

3. **Acquiring Information** What are some other uses of copper, silver, and gold?

Semiconductors and Their Uses

Your television, computer, handheld electronic games, and calculator are electronic devices that depend on silicon semiconductors. All have miniature electrical circuits that use silicon's properties as a semiconductor. You learned that metals generally are good conductors of electricity, nonmetals are poor conductors, and semiconductors fall in between the two extremes, but how do semiconductors work?

Electrons and Electricity

An electric current is a flow of electrons. Most metals conduct an electrical current because their valence electrons are not held tightly by the positive nucleus and are free to move. A copper wire is an example of a good conductor of electric current. **Figure 3.14** illustrates the flow of electrons in copper.

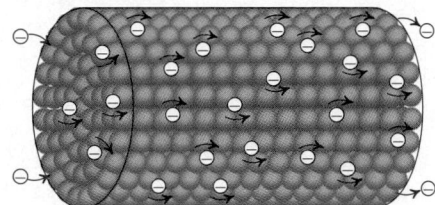

Figure 3.14

Electric Current in a Copper Wire
In a conductor such as a copper wire, valence electrons are free to move to produce an electric current.

Direction of current flow

At room temperature, pure silicon is not a good conductor of electricity. Silicon has four valence electrons, but they are held tightly between neighboring atoms in the crystal structure. You can see this clearly when you look at the structure of silicon in **Figure 3.15**.

Figure 3.15

Valence Electrons in Silicon
Valence electrons in silicon are localized between neighboring silicon atoms. These electrons hold the atoms together in the crystal, but no electrons are free to move beyond their positions. Therefore, no electrons are available to carry an electric current.

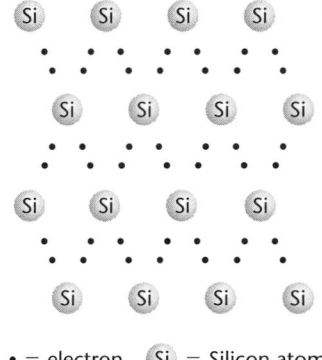

• = electron   (Si) = Silicon atom

Electrical Conduction by a Semiconductor

The electrical conductivity of a semiconductor such as silicon can be increased by a process known as doping. Doping is the addition of a small amount of another element to a crystal of a semiconductor. If a small amount of phosphorus, which has five valence electrons, is added to a crystal of silicon having only four valence electrons, each phosphorus atom provides an extra electron to the crystal structure. These extra electrons are free to move throughout the crystal to form an electric current. A phosphorus-doped silicon crystal is shown in **Figure 3.16**.

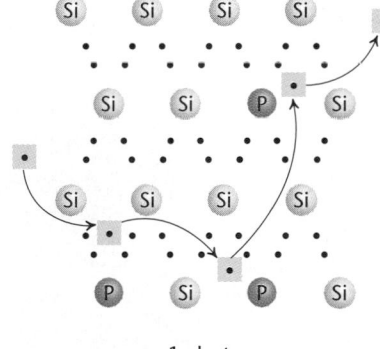

• = 1 electron

Figure 3.16

Silicon Doped with Phosphorus
In phosphorus-doped silicon, the extra electrons from the phosphorus atoms are not needed to hold the crystal together. They are free to move and carry an electric current. Arsenic and antimony are also used to dope silicon. Both have five valence electrons.

3.2 Using the Periodic Table **111**

Cultural Diversity

From A to Z: Cultural Diversity in the Elements

If the periodic table could show the people, places, and languages associated with the elements and their names, it would look like a United Nations gathering. The element *carbon* was known since early times. Its name is derived from the Latin word for charcoal. Sulfur, tin, gold, copper, iron, lead, mercury, and silver were known to all ancient civilizations. The symbols for most of these elements are based on their Latin and Greek names. But the name *sulfur* is derived from the sanskrit *sulvere*. Zinc was known in India and China before 1500. Platinum was known to the pre-Columbians of South America and was taken to Europe in 1750. Its name is derived from the Spanish *platina*. Polonium was discovered by Marie Curie and named after Poland. Vanadium was discovered in Mexico City, Mexico, by A.M. Del Rio. It is named after the Scandinavian goddess Vanadis. Tungsten was discovered in Spain and gets its name from the Swedish *tung-sten,* meaning "heavy stone." Zirconium is derived from the Arabic *zargun,* which means "gold color."

Applying Chemistry

Miniaturization of Electronic Devices The importance of semiconductor devices in modern electronics, in particular the importance of miniaturization, may become clearer to students if they can see the circuits that run the ordinary computer. Arrange with the person in charge of the school's computers for your class to visit the computer room at a time when a computer can be opened to reveal the circuit boards. Students are surprised to find that the heart of the computer is small compared to the amount of empty space. If the circuit boards can be taken out, students can get a good look at the complexity of the circuits contained in a small amount of space.

3 ASSESS

Check for Understanding

• Check for understanding of the section material by having students answer the Section Review questions.

• In an oral exercise, select elements from the periodic table and ask students to give this information: period number; group number; family name, if applicable; and classification as metal, nonmetal, or metalloid. L1

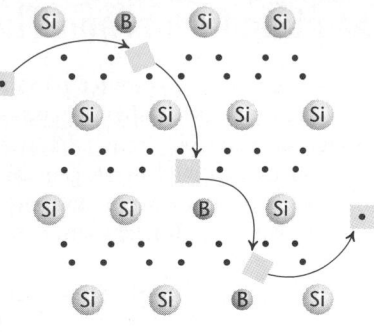

Figure 3.17
Silicon Doped with Boron
In boron-doped silicon, electrons move into and out of "holes"—spaces that are lacking an electron. This movement is an electric current.

• = 1 electron

> **WORD ORIGIN**
>
> semiconductor:
> *semi* (L) half
> *conductus* (L) to escort or guide
> Semiconductors do not conduct electricity as well as metals do.

A semiconductor such as phosphorus-doped silicon is called an *n*-type semiconductor because extra electrons (*n*egatively charged) are present in the crystal structure.

Silicon can also be doped with an element such as boron that has three valence electrons. Boron has fewer valence electrons than silicon, so when boron is added to a crystal of silicon, a shortage of electrons results. "Holes" are created in the silicon crystal structure. These are locations where there should be an electron but there is not because boron has one less electron than silicon. The movement of electrons into and out of these holes produces an electric current. Boron-doped silicon is an example of a *p*-type semiconductor because the holes act as if they are positive charges moving throughout the crystal. **Figure 3.17** illustrates boron-doped silicon.

Diodes

Many semiconductor devices are made by combining *n*- and *p*-type semiconductors to form a diode. This combination of semiconductors permits electrical current to flow in only one direction, from the negative terminal to the positive terminal.

Transistors like those shown in **Figure 3.18** are the key components in electrical circuits in devices such as computers, calculators, hearing aids, and televisions. They are used to amplify (increase the strength of) electrical signals. Transistors are exceedingly small, and their compact size and efficient operation have allowed the miniaturization of many electronic devices, such as laptop computers, heart pacemakers, and hearing aids. Transistors may be constructed by placing a *p*-type semiconductor between two *n*-type semiconductors, called an *npn*-junction, or by placing an *n*-type semiconductor between two *p*-type semiconductors, called a *pnp*-junction.

Figure 3.18
Importance of Transistors
Transistors are used to increase the strength of electrical signals in such devices as television remote controls, telephones, and hearing aids.

Chemistry Journal

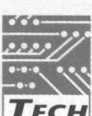

Household Elements

Ask each student to choose ten common household items or materials and, insofar as possible, list the elements that are the primary components of the items or materials. The elements should be classified as metals, nonmetals, or metalloids. Have students include the items, elements, and classifications in their journals under the heading *Common Household Elements.* L1 LEP

Transistors, diodes, and other semiconductor devices are incorporated onto thin slices of silicon to form integrated circuits. Integrated circuits like the one shown in **Figure 3.19** may contain hundreds of thousands of devices on a slice of silicon called a chip. The small size of a chip— only a few millimeters in width—has made possible the amazing growth of computer technology.

Connecting Ideas

You already know that elements combine chemically to form compounds. The ways that elements combine depend entirely on their valence electrons. With your knowledge of the periodic table, you will be able to explain why elements combine and predict what compounds they will form.

Figure 3.19

Silicon and the Computer Revolution
Miniaturization of integrated circuits on silicon chips (top) allowed the size of computers to shrink from room-size (left) to laptop (right).

SECTION REVIEW

Understanding Concepts

1. Where are metals usually found on the periodic table? Where are nonmetals found? The metalloids?
2. How does the arrangement of elements in periods reflect electron structure?
3. What are the major differences in the physical properties of metals, nonmetals, and metalloids?

Thinking Critically

4. **Observing and Inferring** What can you deduce from the periodic table about the properties of the element barium?

Applying Chemistry

5. **Semiconductors** Germanium has the same type of structure and semiconducting properties as silicon. What type of semiconductor would you expect arsenic-doped germanium to be?

SECTION REVIEW

1. Metals are on the left side and in the center; nonmetals are in the upper-right corner; metalloids form a border between the metals and nonmetals.
2. The period number of an element is equal to the number of the energy level that is occupied by the valence electrons.
3. Metals are lustrous, are easily deformed and drawn into wires, and are good conductors of heat and electricity. Nonmetals are not lustrous, may be hard or soft, are usually brittle, and are poor conductors of heat and electricity. Metalloids look like metals but have physical properties that are between those of metals and nonmetals.
4. Barium is a metal in Group 2 with two valence electrons in the sixth energy level. It has all the properties of a typical metal.
5. Arsenic has five valence electrons, and germanium has four. Therefore, arsenic-doped germanium would be an *n*-type semiconductor.

Reteach

To find areas of weakness in understanding or misconceptions, the following is a class exercise providing opportunities for discussion and clarification. Write on the chalkboard two lists of words and phrases. One list might be *conductivity, group, nonmetal, Mendeleev, energy level, silver, pool of electrons, liquid, lanthanides, sodium, alkaline earth metal, Group 18, Halley's comet, solid, metalloid,* and many more. Another list might be *inner transition element, family, state of matter, semiconductor, metal, alkali metal, mercury, bromine, noble gas, Group 2, periodic law, Group 1, magnesium, transition element, actinide, gas, doping elements, period,* and many more. Ask students to match words and phrases in one column with words and phrases in the other. There should be multiple connections for many of the entries. Encourage students to find as many connections as possible, but require that they justify or explain the connections. L1 **COOP LEARN**

Extension

 Ask each student to write an essay about the benefits and/or adverse effects of efforts to synthesize new chemical elements in nuclear experiments. These essays could be put into student portfolios. L2 P

4 CLOSE

Extension

Ask students to predict the appearance of the periodic table if 120 elements existed. What would it look like if 140 elements existed? If 168 elements existed? L3

- Review the Reviewing Main Ideas statements and Key Terms with your students.
- Complete solutions to Chapter Review problems can be found in the *Problems and Solutions Manual*.

UNDERSTANDING CONCEPTS

1. Element number 18 is in Period 3 and Group 18. It is a noble gas, and its nearest neighbors are Ne, F, Cl, Br, and Kr.
2. Mendeleev arranged the elements according to increasing atomic mass.
3. The properties of the elements repeat periodically when the elements are arranged according to increasing atomic number.
4. **a.** Group 1: Li, Na, K, Rb, Cs, and Fr
   b. Group 17: F, Cl, Br, I, and At
   c. Group 2: Be, Mg, Ca, Sr, Ba, and Ra
   d. Group 18: He, Ne, Ar, Kr, Xe, and Rn
5. **a.** Al; **b.** As; **c.** He; **d.** Br; **e.** Hg
6. **a.** 8; **b.** 7; **c.** 6; **d.** 2; **e.** 1; **f.** 5; **g.** 4; **h.** 3
7. **a.** nonmetal; **b.** nonmetal; **c.** nonmetal; **d.** metal; **e.** metal; **f.** metalloid; **g.** metal; **h.** metal
8. **a.** :C̈l· (17) **b.** ·Mg· (2) **c.** ·C̈· (14)
   d. ·B̈i· (15) **e.** :K̈r: (18) **f.** Cs· (1)
   g. :Ö· (16) **h.** ·P̈· (15)
9. Rb—atomic mass, 86.1u; K—density, 1.19 g/mL; Cs—melting point, 290 K.

APPLYING CONCEPTS

10. Hydrogen: **a.** 1, 1; **b.** nonmetal; **c.** gas; d. H·
    Lithium: **a.** 2, 1; **b.** metal; **c.** solid; d. Li·
    Nitrogen: **a.** 2, 15; **b.** nonmetal; **c.** gas; d. ·N̈·
    Fluorine: **a.** 2, 17; **b.** nonmetal; **c.** gas; d. :F̈·
    Cobalt: **a.** 4, 9; **b.** metal; **c.** solid
    Silver: **a.** 5, 11; **b.** metal; **c.** solid

REVIEWING MAIN IDEAS

3.1 Development of the Periodic Table
- In his periodic table, Mendeleev organized the elements according to increasing atomic mass and placed elements with similar properties into groups.
- The modern periodic law states that the physical and chemical properties of the elements repeat in a regular pattern when they are arranged in order of increasing atomic number.

3.2 Using the Periodic Table
- Atomic structure and the number of valence electrons an element has can be related to an element's position on the periodic table.
- Elements may be classified as metals, nonmetals, or metalloids.
- The number of valence electrons and how tightly they are held determine the chemical properties of an element.
- The conductivity of semiconductors can be increased by adding small amounts of other elements.

Key Terms
For each of the following terms, write a sentence that shows your understanding of its meaning.

actinide	nonmetal
group	period
lanthanide	periodic law
metal	periodicity
metalloid	semiconductor
noble gas	transition element

● UNDERSTANDING CONCEPTS

1. Describe element number 18 in terms of its period and group number, family name, and closest neighboring elements.

2. Mendeleev said that the properties of the elements repeat periodically when the elements are arranged in a certain way. How did Mendeleev arrange the elements?

3. The modern periodic law says that the properties of the elements repeat periodically when the elements are arranged according to a certain trend. What is that trend?

4. What is the group number of each of the following families of elements? Write the symbols for the elements in each family.
   a) alkali metals   c) alkaline earth metals
   b) halogens   d) noble gases

5. Select the symbol of the element that fits the following descriptions.
   a) the Group 13 metal in the third period
   b) the Group 15 metalloid in the fourth period
   c) the lightest of the noble gases
   d) the halogen that exists as a liquid at room temperature
   e) the only metal that is a liquid at room temperature

6. How many valence electrons are in an atom of each of the following elements?
   a) Ne   d) Sr   g) Sn
   b) Br   e) Na   h) In
   c) S   f) As

7. Classify each of the elements in question 6 as a metal, nonmetal, or metalloid.

✓ Assessment

Portfolio: Review the portfolio options that are provided throughout the chapter. Encourage students to select one product that demonstrates their best work for the chapter. Have students explain what they have learned and why they chose this example for placement into their portfolios. **P**

Additional portfolio options can be found in the following **Teacher Classroom Resources:**
Tech Prep Applications
Applying Scientific Methods in Chemistry
Consumer Chemistry

8. Write the electron dot structure for each of the following elements. What is the group number of each element?

a) Cl   e) Kr
b) Mg   f) Cs
c) C   g) O
d) Bi   h) P

9. An incomplete set of atomic mass, density, and melting point data is given for three elements in a triad. Following the patterns of Döbereiner's triads, predict a likely number for each missing value.

Element	Atomic Mass (u)	Density (g/mL)	Melting Point (K)
K	39.1		336
Rb		1.53	313
Cs	133	1.87	

● APPLYING CONCEPTS

10. The symbols of some elements are written on the periodic table shown below. Answer the following questions for each element whose symbol is shown.

H							
Li					N	F	
			Co				
				Ag		I	
				Hg			

a) What is the element's row and column number?
b) Is the element a metal or nonmetal?
c) Is the element a solid, liquid, or gas at room temperature?
d) Draw the electron dot structure for each of the main group elements.

11. Americium (Am) is an actinide that's used in smoke detectors. What property of actinides makes this element useful in a smoke detector?

12. Which elements among the following have similar chemical properties?

a) Be   c) Cs   e) Ar
b) Sr   d) F   f) I

13. Explain why electrical conductivity increases in the elements in Group 14 as atomic number increases.

14. Identify the two elements that are liquids at room temperature and pressure. Explain why one is a metal and the other is a nonmetal.

15. The noble gases share a unique chemical property. Describe this property and explain why they have it.

Chemistry and Technology

16. How is the transformation from one crystalline phase to another different from melting or boiling?

Everyday Chemistry

17. What properties of nickel and zinc make them good substitutes for copper and silver in coins?

Literature Connection

18. What two fields did Primo Levi combine in his life's work?

● THINKING CRITICALLY

Identifying Patterns

19. In the system of numbering groups of elements from 1 to 18 on the periodic table, explain the significance of the group numbers 1, 2, 13, 14, 15, 16, 17, and 18.

Making Comparisons

20. Compare the atomic structure of metals, metalloids, and nonmetals. Explain how atomic structure accounts for differences in electrical conductivity.

21. Element X is in the fourth period. Its outer energy level has three electrons. How does the number of outer-level electrons of element X compare with that of element Y, which is in the sixth period, Group 13? Write the name and symbol of each element.

21. Elements X and Y are in the same group and have the same number of valence electrons. Element X is gallium (Ga) and element Y is thallium (Tl).
22. Density increases with atomic mass. Density increases as you move from top to bottom in a group.

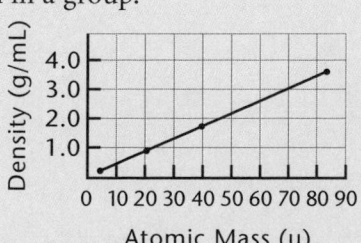

23. Metallic properties decrease from left to right in the periodic table because of increasing numbers of valence electrons.
24. scandium
25. KCl; $MgCl_2$; $AlCl_3$; $SiCl_4$

Iodine: **a.** 5, 17; **b.** nonmetal; **c.** solid; **d.** :Ï·
Mercury: **a.** 6, 12; **b.** metal; **c.** liquid

11. Americium is radioactive.
12. Be and Sr; F and I
13. Electrical conductivity increases because the outer-level electrons are farther from the nucleus and can move more freely.
14. Bromine and mercury are liquids. Mercury is a metal because it has two valence electrons that are not tightly held. Bromine is a nonmetal because it has seven valence electrons that are tightly held.
15. The noble gases are unreactive because they have a filled valence level.
16. Melting and boiling involve a change of state, whereas transformation from one crystalline phase to another occurs in the atoms.
17. Ni and Zn are corrosion resistant, malleable, and relatively abundant.
18. chemistry and literature

THINKING CRITICALLY

19. Group numbers 1 and 2 are equal to the number of valence electrons. The second digit of group numbers 13, 14, 15, 16, 17, and 18 is equal to the number of valence electrons.
20. Metals have three or fewer valence electrons; metalloids have three to six valence electrons; nonmetals have four to eight valence electrons. Metals are good conductors because their valence electrons are not tightly held and are free to move. Metalloids hold their valence electrons tightly in crystal structures, and the electrons have limited freedom to move. Nonmetals hold their valence electrons tightly, and they are not free to move.

Element	Helium	Neon	Argon	Krypton
Atomic Mass (u)	4.00	20.2	39.9	83.8
Density (g/L)	0.179	0.901	1.78	3.74

26. Element A is in the second period because it has the lower density.
27. The reactivity of metals decreases across a period from left to right.

CUMULATIVE REVIEW

28. sodium
29. Each formula has a definite ratio of whole numbers of atoms of the two elements.
30. mass number, 40; atomic number, 18
31. Argon's electrons occupy three energy levels. Level one holds two electrons, level two holds eight electrons, and level three holds eight electrons.
32. The arrangement of electrons is the same as in question 31.
33. The electromagnetic spectrum shows specific frequencies of light released by the atom. These frequencies correspond to energy differences between the levels that electrons can occupy in the atom.
34. Copper has 29 protons in its nucleus; gold has 79. No chemical reaction can change the number of protons in a nucleus.
35. Elements are pure matter composed of only one type of atom. Most matter is in the form of compounds or mixtures.

Interpreting Data

22. **MiniLab 2** Graph density versus atomic mass for the elements in the table above. Describe the relationship between atomic mass and density for the elements. Use your periodic table to locate the elements. Based on these data, what can you say about the trend in density as you move down a column of elements?

Relating Cause and Effect

23. **ChemLab** Explain how the metallic properties of the elements change as you move across a period from left to right. Why do the properties change in this manner?

Using a Table

24. **MiniLab 1** Mendeleev predicted the existence of eka-boron, which was unknown in his day. Eka-boron was located between calcium and titanium on the periodic table. What is the name of eka-boron now?

Making Predictions

25. The formulas of the chlorides of lithium, beryllium, boron, and carbon are LiCl, $BeCl_2$, BCl_3, and CCl_4, respectively. Use the periodic table to predict the formulas of the chlorides of potassium, magnesium, aluminum, and silicon.

Drawing Conclusions

26. Elements A and B are both in the same group. One element is in the second period and the other is in the fifth period. The density of element B is greater than the density of element A. Which element is in the second period? Explain your answer.

Making Inferences

27. The elements in Group 1 have one valence electron and they react vigorously with water at room temperature. The elements in Group 2 have two valence electrons and they react slowly with water at room temperature. The elements in Group 13 have three valence electrons and do not react with water at room temperature. What does this suggest about the reactivity of metals?

● CUMULATIVE REVIEW

28. Which element has the highest melting point: mercury, nitrogen, or sodium? (Chapter 1)
29. Explain how the law of definite proportions is illustrated by the formulas of the compounds given in question 25. (Chapter 2)
30. A neutral atom of argon has 22 neutrons and 18 electrons. What are the mass number and atomic number of argon? (Chapter 2)
31. Argon's 18 electrons are arranged in energy levels. How many energy levels are needed to accommodate argon's electrons and how many electrons are in each energy level? (Chapter 2)
32. Describe the arrangement of electrons in an isotope of argon that has 21 neutrons and 18 electrons. (Chapter 2)
33. What would you expect to see in the emission spectrum of argon to give evidence of the existence of energy levels? (Chapter 2)
34. Why is it impossible to change copper (Group 11, Period 4) into gold (Group 11, Period 6), as the alchemists tried to do? (Chapter 2)

The Alchemists
by Jan van der Straet

35. How do the elements differ from most of the matter you see around you? (Chapter 1)

Skill Review

1. Look at the periodic table on pages 92 and 93. Find pairs of elements that would reverse order if the table were arranged according to increasing atomic mass as Mendeleev's periodic table was.

2. Write the symbol of the element that has valence electrons that fit each description.

   a) two electrons in the third energy level
   b) seven electrons in the fourth energy level
   c) four electrons in the sixth energy level
   d) eight electrons in the fifth energy level
   e) one electron in the first energy level
   f) six electrons in the second energy level

3. Describe the relationship that exists between the number of a period and the electron structure of atoms of elements in the period. Describe the change in electron structure as you move from left to right in a period.

Writing in Chemistry

4. Mendeleev made predictions about germanium and several other elements. Three of these were gallium (Ga), scandium (Sc), and polonium (Po). Write a paragraph describing the accuracy of his predictions.

5. Consult the periodic table on pages 92 and 93 and list the elements that are naturally radioactive. Select one of these elements and write a report that describes the discovery of the element, its properties, and ways in which it is used.

6. Sodium and chlorine are never found uncombined in nature. Together, they occur as NaCl in salt mines and in ocean water. Write a report that describes how sodium and chlorine are obtained from salt recovered from the ocean or taken from salt mines.

Problem Solving

7. The density of aluminum is 2.7 g/mL and that of iron is 7.9 g/mL. Assume that manufacturers of soda cans could use the same volume of each metal to make soda cans. Compare the mass of an aluminum soda can with that of a comparable can made from iron. Explain your answer.

8. The chemical formula for zinc sulfide is ZnS. Use the periodic table to predict the formulas of the following similar compounds.

   a) zinc oxide   c) mercury sulfide
   b) cadmium oxide   d) zinc selenide

Projects

9. Do research to determine the relationship between an element's atomic number and the wavelength of an X ray emitted by the element. Include the name of the person who discovered this relationship.

10. The names of the elements are derived from a variety of sources. Some are named after planets, others after a person or place, and others after mythological characters. Find out the origin of the names of the following elements: mercury, helium, mendelevium, boron, tantalum, and francium.

11. Do research on the chemistry of chlorine, bromine, and iodine. Describe similarities that may have led Döbereiner to describe these elements as a triad.

Salt mine

Chapter 3   Review   **117**

10. Mercury is named after the Greek god Mercury. Helium is named after the sun (helios). Mendelevium is named after Mendeleev. *Boron* means "dark" because of its color—from Arabic *buraq.* Tantalum is named after the mythological character Tantalos. Francium honors the country France.

11. Chlorine, bromine, and iodine all react with metals to form salts. They react with hydrogen to form compounds of the general formula HX. The hydrogen halides dissolve in water to form acids.

AUTHENTIC ASSESSMENT

1. Cobalt and nickel; tellurium and iodine; thorium and protactinium.

2. **a.** Mg; **b.** Br; **c.** Pb; **d.** Xe; **e.** H; f. O

3. The number of the period is the number of the outermost energy level where the valence electrons are found. The number of valence electrons increases from one to eight across a period from left to right.

4. Answers will vary but should state that Mendeleev's predictions about Ga and Sc were as accurate as his predictions for Ge. His prediction about Po was not as accurate.

5. Elements that are naturally radioactive: Po, Rn, Fr, Ra, Ac, Th, Pa, and U. Reports describing discovery, properties, and uses will vary. Use a reference such as *Van Nostrand's Scientific Encyclopedia* to check accuracy.

6. Sodium is obtained by the electrolysis of molten NaCl. Chlorine is obtained by the electrolysis of aqueous NaCl.

7. The density of aluminum is 2.7 g/mL and that of iron is 7.9 g/mL. If soda cans are made from equal volumes of each metal, the mass of an iron can will be nearly three times the mass of an aluminum can because the density of iron is nearly three times the density of aluminum (7.9/2.7).

8. **a.** ZnO; **b.** CdO; **c.** HgS; **d.** ZnSe

9. The wavelength is inversely proportional to atomic number. English physicist Henry Moseley discovered this.

Chapter 4 Organizer

Section	Objectives	Activities/Features
4.1 **The Variety of Compounds** (5 days)	**1. Distinguish** the properties of compounds from those of the elements of which they are composed. **2. Compare and contrast** the properties of sodium chloride, water, and carbon dioxide. **3. Analyze** evidence to conclude that differences exist in the ways compounds form.	**MiniLab:** Evidence of a Chemical Reaction: Iron Versus Rust, p. 122 **Everyday Chemistry:** Elemental Good Health, p. 128 **Discovery Demo:** 4.1 Properties of Ethyne: A Covalent Compound, pp. 118-119 **Demonstrations:** 4.2 Chemical Synthesis, pp. 124-125; 4.3 Oxygen Gas, pp.126-127
4.2 **How Elements Form Compounds** (6 days)	**4. Model** two types of compound formation: ionic and covalent at the atomic level. **5. Demonstrate** how and why atoms achieve chemical stability by bonding. **6. Compare,** using examples, the effect of covalent and ionic bonding on the physical properties of compounds.	**MiniLab 2:** Formation of Ionic Compounds, p. 135 **ChemLab:** The Formation and Decomposition of Zinc Iodide, pp. 136-137 **History Connection:** Hydrogen's Ill-fated Lifts, p. 141 **Chemistry and Society:** The Rain Forest Pharmacy, p. 146 **Demonstrations:** 4.4 A Chemical Formula, pp. 138-139; 4.5 A Quick Thaw, pp. 140-141; 4.6 Make It Neutral, pp. 142-143

Activity Materials

ChemLab (pages 136-137)
10×150 mm test tube, test-tube rack, test-tube holder, 100-mL beaker, spatula, plastic stirring rod, zinc, iodine crystals, distilled water, 9-V battery, 20 cm insulated copper wire

MiniLab 1 (page 122)
fresh steel wool, rusty steel wool, small paper cups, 3×5 index card, magnet, plastic bag

MiniLab 2 (page 135)
paper disks (different colors), marker, corrugated cardboard, thumbtacks (different colors)

Demonstrations For a complete list of materials for the demonstrations in this chapter, see pages 118, 124, 126, 138, 140, and 142.

KEY TO TEACHING STRATEGIES

The following designations will help you decide which activities are appropriate for your students.

L1 Level 1 activities should be within the ability range of all students.

L2 Level 2 activities should be within the ability range of the average to above-average student.

L3 Level 3 activities are designed for the ability range of above-average students.

LEP LEP activities should be within the ability range of Limited English Proficiency students.

COOP LEARN Cooperative Learning activities are designed for small group work.

P These strategies represent student products that can be placed into a best-work portfolio.

Formation of Compounds

STUDENT MASTERS

Study Guide: pp. 13-14
Laboratory Manual: 4.1 Determining Melting Points, pp. 33-36
ChemLab and MiniLab Worksheets: p. 32

TEACHING AIDS

Section Focus Transparency 7
Section Focus Transparency Master: p. 7
Basic Concepts Transparency 10
Basic Concepts Transparency Master: pp. 19-20
Lesson Plans: p. 7

Study Guide: pp. 15-16
ChemLab and MiniLab Worksheets: pp. 29-31, 33
Laboratory Manual: 4.2 Distinguishing Ionic and Covalent Compounds, pp. 37-40
Critical Thinking/Problem Solving: p. 5

Section Focus Transparency 8
Section Focus Transparency Master: p. 8
Problem Solving Transparency 4
Problem Solving Transparency Master: pp. 7-8
Basic Concepts Transparency 11, 12
Basic Concepts Transparency Masters, pp. 21-24
Lesson Plans: p. 8

ASSESSMENT RESOURCES

Chapter Review and Assessment: pp. 23-28
Alternate Assessment in the Science Classroom
Computer Test Bank: Chapter 4
Performance Assessment in the Science Classroom
Problems and Solutions Manual: Chapter 4
Supplemental Practice Problems: Chapter 4

CHAPTER RESOURCES

Spanish Resources: Chapter 4
English/Spanish Audiocassettes: Chapter 4
Cooperative Learning in the Science Classroom
Lab and Safety Skills in the Science Classroom

GLENCOE TECHNOLOGY

Software
Mastering Concepts in Chemistry: *Unit 4* Lesson 2 Naming Compounds

Videotape
MindJogger Videoquizzes, Chapter 4

Videodisc
Chemistry: Concepts and Applications,
Demonstrations, Videos, and Animations Videodisc and Bar Code Teacher Guide

CD-ROM Multimedia System
Chemistry: Concepts and Applications, Same as for Videodisc, plus Interactive Exploration and Major Simulation, *The Periodic Table*

TECH PREP

The following Glencoe resources provide opportunities for integrating science and technology.

Student Edition: Minilab 1, p. 122; Everyday Chemistry, p. 128; Chemistry and Society, p. 146

Teacher Wraparound Edition: Integrating the Sciences, p. 121; Applying Chemistry, p. 124

Teacher Classroom Resources: Tech Prep Applications, pp. 5-6; Chemistry and Industry, pp. 5-8; Consumer Chemistry, pp. 7-8

4 Formation of Compounds

Chapter Overview

Macroscopic chemical and physical properties of three compounds and their component elements are introduced and compared. The formation of compounds at the submicroscopic level by transferring or sharing of electrons is described. Finally, the role of bonding in determining properties is explained.

Theme Connection

Macro to Submicro/Energy The chapter uses the *macro to submicro* link to support the model of compound formation. The role of *energetic* stability is introduced as a driving force in compound formation.

✓ Assessment Planner

Choose assessment strategies from the following pages to evaluate the progress of your students.
Assess, pp. 128, 138, 145
MiniLabs, pp. 122, 135
ChemLab, p. 136
Portfolio, pp. 121, 138, 148
Chapter Review, p. 148

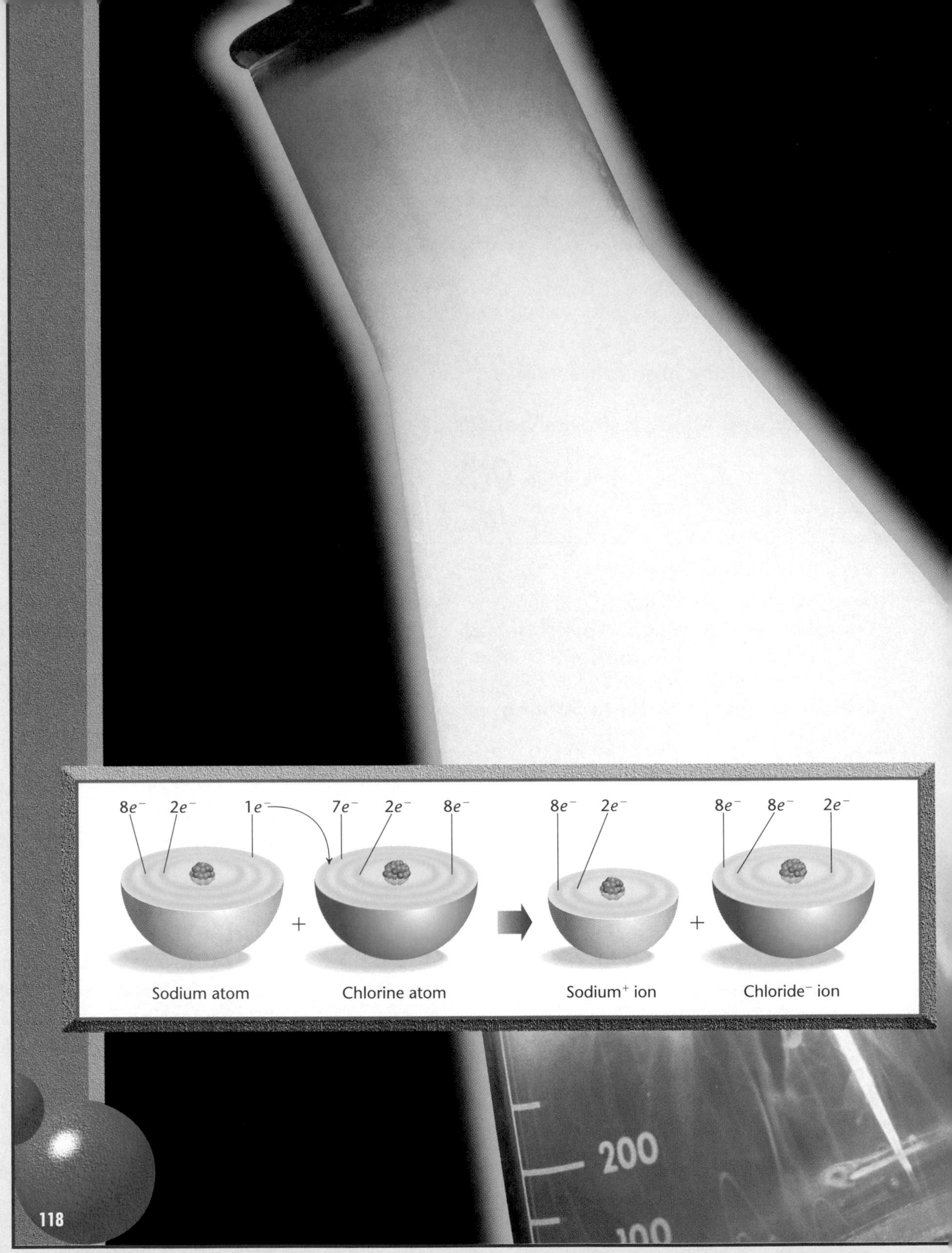

| $8e^-$ $2e^-$ | $1e^-$ | $7e^-$ $2e^-$ $8e^-$ | $8e^-$ $2e^-$ | $8e^-$ $8e^-$ $2e^-$ |

Sodium atom + Chlorine atom → Sodium⁺ ion + Chloride⁻ ion

118

200
100

Discovery Demo 4.1

Properties of Ethyne: A Covalent Compound

Purpose: To demonstrate the formation of a compound.
Materials: piece of calcium carbide, 50 mL water, 100-mL beaker, a few drops of liquid dish-washing detergent, stirring rod, wood splint, crucible tongs, laboratory burner

Safety Precautions:

Disposal: Codes C and H
Procedure: Half fill a small beaker with water. Add a few drops of liquid dish-washing detergent and stir. Use tongs to place a small piece of calcium carbide, CaC_2, in the water. Use tongs to hold a burning splint. Dim the room lights. Use

the burning splint to pop the bubbles that form. **CAUTION:** *Fire risk.* Explain to students that ethyne gas (acetylene), C_2H_2, is produced in the reaction.
Results: The chunk of calcium carbide reacts with water to form ethyne gas. This flammable gas undergoes incomplete combustion to form carbon dioxide, carbon monoxide, and water vapor. Some soot will also be present in the air. The de-

CHAPTER
4
Formation of Compounds

Suppose you went to the laboratory, took a lump of silvery-gray sodium metal, and dropped it into a flask of pale green chlorine gas. Some spectacular things would happen. You'd see a bright yellow flash and hear a bang and a sizzle, sparks would fly, and things would get hot. As the photo shows, this reaction is violent. It must be done with special equipment and experienced supervision. After the flask calmed down, you'd notice that the only thing left was some white powder. It would be obvious that some sort of reaction took place when the sodium and chlorine came into contact with each other.

The flash, bang, sparks, and powder all occur because of events that take place among atoms of sodium and chlorine. Every change in matter is caused by something happening between atoms, molecules, or ions. In other words, every macroscopic change in matter has a submicroscopic explanation. One of your tasks as a chemistry student is to observe reactions and learn how to explain them on the submicroscopic scale.

Chemistry Around You

When sodium and chlorine react, the only thing left is a white powder. Do you know what the white powder is? It's plain, everyday table salt. In this chapter, you'll learn that the elements sodium and chlorine combine by transferring electrons, while other elements combine in a different way.

 Concept Check

Review the following concepts before studying this chapter.
Chapter 1: how elements and compounds differ
Chapter 2: the structure of atoms; the arrangement of electrons in an atom

 Chapter Preview

4.1 The Variety of Compounds

4.2 How Elements Form Compounds

Laboratory Activities

ChemLab
The Formation and Decomposition of Zinc Iodide

MiniLabs
1. Evidence of a Chemical Reaction: Iron Versus Rust
2. The Formation of Ionic Compounds

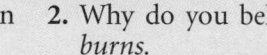

119

tergent bubbles quietly pop and burn when touched with a burning splint.
Analysis: Ask these questions.
1. Draw an electron dot diagram for the two carbon atoms and the two hydrogen atoms in acetylene, C_2H_2, that shows a triple carbon-carbon bond. *A triple-bonded structure for ethyne allows the carbon and hydrogen to obey the octet rule.* $H:C:::C:H$

2. Why do you believe this gas is reactive? *It burns.*
3. If enough oxygen were available for complete combustion, what would the products be? *Carbon dioxide and water vapor are the products of complete combustion of hydrocarbons.*

SECTION 4.1

PREPARE

Key Concepts

The physical and chemical properties of three common compounds are studied and compared with the properties of their constituent elements. It is emphasized that elements' properties change dramatically when they combine to form compounds. It is also shown that the properties of the three compounds vary widely.

Planning Ahead

For the MiniLab Obtain steel wool and magnets. Cow magnets work well. They are available at farm-supply stores that carry veterinary items. Prepare rust in advance by soaking steel wool in salt water and then allowing it to sit in a dry cup for approximately one week. Two days before the lab, gently rinse with water and allow it to dry.

For the Quick Demos You will need a cylinder of hydrogen. Investigate sources of dry ice.

1 FOCUS

Focus Transparency

Display the **Section Focus Transparency** for Section 4.1 to introduce the concept of a compound being composed of elements bonded together.

L1 LEP

SECTION 4.1

The Variety of Compounds

SECTION PREVIEW

Objectives

Distinguish the properties of compounds from those of the elements of which they are composed.

Compare and contrast the properties of sodium chloride, water, and carbon dioxide.

Analyze evidence to conclude that differences exist in the ways compounds form.

Chemistry is often like detective work. To figure out the submicroscopic structure of a compound, you first have to examine some of its macroscopic properties. Just as no two human fingerprints are the same, no two substances have exactly the same combination of physical and chemical properties. The structure of a substance at the atomic level determines its macroscopic properties. Therefore, looking at properties of compounds provides important clues about their submicroscopic structure and how they form from atoms. That is powerful detective work. You can begin this detective work by examining the properties of three familiar compounds: table salt, carbon dioxide, and water.

Salt: A Familiar Compound

What is the most popular food additive? In most kitchens, the answer is salt. It is used in cooking and at the table to enhance the flavor of food. Chemists refer to it as sodium chloride. The chemical name tells you what elements make up the compound; sodium chloride contains the elements sodium and chlorine.

Even though it's possible to make sodium chloride from its elements, salt is so abundant on Earth that it is used to manufacture the elements sodium and chlorine. Sodium chloride occurs naturally in large, solid, underground deposits throughout the world and is dissolved in the world's oceans. Salt can be obtained by mining these solid deposits, **Figure 4.1,** and by the evaporation of seawater.

Besides enhancing food's flavor, sodium chloride is an essential nutrient that plays crucial roles in living things. If you live in an area that gets snow and ice in the winter, salt is sometimes used to melt ice on roads, as shown in **Figure 4.1.**

Figure 4.1
Getting and Using Salt
Underground salt mining accounts for about 90 percent of the world's salt production. These salt deposits formed when ancient seas evaporated millions of years ago. ▼

In some regions, trucks spread salt crystals on icy roads to melt the ice. The salt lowers the melting point of ice to about 15°F. If the air temperature is below 15°F, the salt won't do much good. ▶

▲ Salt is the most common food seasoning. Some estimates say that 60 percent of the average American's sodium intake comes from salting food at the table.

Physical Properties of Salt

You already know some of the physical properties of table salt. It is a white solid at room temperature. If you look at table salt under a magnifier, you'll notice that the grains of table salt are little crystals shaped like cubes. These crystals are hard, but when you press down on them with the back of a spoon, the crystals shatter. This shattering shows that the crystals are brittle. If sodium chloride is heated to a temperature of about 800°C, it melts and forms liquid salt. Solid sodium chloride does not conduct electricity, but melted sodium chloride does. Salt also dissolves easily in water. The resulting solution is an excellent conductor of electricity, as shown in **Figure 4.2.**

Figure 4.2
Conductivity of Salt
In order to light the bulb, electric current must flow between the two electrodes. As you can see, no current flows through the dry salt crystals (left). When the salt is dissolved in water, the solution conducts electricity and the bulb lights (right). Pure water alone does not conduct electricity.

4.1 The Variety of Compounds **121**

2 TEACH

Concept Development

In this chapter, for purposes of teaching students how elements react to form compounds, elements are treated as single atoms, even though some of the elements used—such as hydrogen, oxygen, and chlorine—occur naturally as diatomic molecules.

Discussion

Ask students to describe various materials in the classroom and to suggest which elements make up each material. *Typical responses are: water—hydrogen and oxygen; wooden objects—carbon, hydrogen, and oxygen; metal objects—iron, aluminum, mixtures of iron with other elements such as carbon; glass—silicon and oxygen; air—nitrogen, oxygen, carbon, hydrogen.* Ask if any of these elements occur in pure form. *Only a few are relatively pure: nitrogen and oxygen in the air and perhaps iron and aluminum in some metallic objects.* Point out that most elements seem to combine readily with other elements to form more complex materials. Mention that this is precisely the topic of this chapter. L1

Visual Learning

Figure 4.2: Sketch on the chalkboard a diagram showing positive and negative electrodes immersed in a salt solution. Point out or demonstrate that distilled water does not conduct electricity. Ask students to speculate why dissolving salt in the water would cause the solution to conduct electricity. *Students may suggest that the solution must contain charged particles that are free to move.* L1

Integrating THE Sciences

Ecology Ask interested students to research and make class presentations about the ecological and biological effects of road salt. They might make posters showing the effects of salt on various plants or on the metallic parts of automobiles and bridges. L1 P

121

miniLAB

Evidence of a Chemical Reaction: Iron Versus Rust

TECH PREP

Purpose: To contrast the physical properties of an element with those of a compound of the element.

Process Skills: Observing, inferring, recognizing cause and effect, communicating

Disposal: Code A

Teaching Strategies

- The rusty steel wool must be dry in order for students to generate a fine powder.
- Wrap refrigerator magnets in a plastic bag to avoid getting iron filings on them.
- Place samples of fresh, dry, rusted steel wool in small paper cups.
- Begin the experiment by asking students to identify unique properties of iron.
- Have students predict whether rust will be attracted to a magnet.

Expected Results

Steel wool will be attracted to the magnet, but rust will not. Some rust will appear to be attracted to the magnet due to the presence of some unreacted iron. By tapping the card and making several passes with the magnet, students should be able to separate material into magnetic (iron) and nonmagnetic (rust) material.

Analysis

1. The magnet attracted the fresh steel wool.
2. Some of the material followed the magnet. Other particles did not move.
3. No, it is a mixture. There are at least two different substances present, each having different magnetic properties. Some of the

miniLAB 1

Evidence of a Chemical Reaction: Iron Versus Rust

Here's a chemical reaction whose result everyone has seen—the rusting of iron. Iron metal combines with oxygen in the air to form rust, Fe_2O_3, iron(III) oxide. This reaction is an example of a familiar compound forming from its component elements. One property of iron that everyone is familiar with is that it is attracted to a magnet. Attraction to a magnet is a simple property to measure—all you need is a magnet.

Procedure

1. Obtain a small wad of fresh steel wool in one small paper cup and another small wad of rusty steel wool in another small paper cup.

2. Obtain a 3" × 5" index card and a magnet wrapped in a plastic bag.

3. Test the fresh steel wool with the magnet. Record your observations.

4. Hold the rusty steel wool over the white card, and lightly rub the rusty steel wool between your thumb and forefinger. Some fine rust powder should fall onto the white card.

5. Next, hold the card up and slowly move the magnet under the card. Record your observations.

Analysis

1. What effect did the magnet have on the fresh steel wool?

2. What did you observe when the magnet was moved under the card with the rust powder?

3. Was your pile of rust powder a pure substance? How does your experimental evidence support your answer?

4. What evidence do you have that the rust is a different substance from iron itself?

Chemical Properties of Salt

The chemical properties of sodium chloride are not so useful in our detective work to determine its submicroscopic structure. Salt does not react readily with other substances. It could sit in a salt shaker for hundreds or even thousands of years and still remain salt. It does not have to be handled in any special way or be stored in a special container. Compounds with these chemical properties are referred to as stable or unreactive. You can get more clues about salt's submicroscopic structure by answering the question: How do the properties of salt compare with the properties of its component elements, sodium and chlorine?

Properties of Sodium

Sodium is a shiny, silvery-white, soft, solid element as you can see in **Figure 4.3.** From its location on the left side of the periodic table, you know that it is a metallic element. Sodium melts to form a liquid when it

material was attracted by the magnet and some was not.
4. Students may say that iron is silvery gray, rust is brown; iron is metallic and flexible whereas rust is flaky or powdery. The iron is attracted to the magnet, the brown rust is not. Other observations are possible.

✔ Assessment

Performance: Ask students to extract iron from breakfast cereals using a plastic-wrapped magnet. Have them make a slurry of cereal with water, place a plastic-wrapped magnet in the cereal, and stir for 5 minutes. They should remove the magnet and observe filings stuck to the plastic. Fortified cereals work well.

Have students report their findings. Explain that although iron is a required mineral, it is not stored by the body. **L1** **LEP**

is heated above 98°C. Sodium must be stored under oil because it reacts with oxygen and water vapor in the air. In fact, it is one of the most reactive of the common elements. When a piece of sodium is dropped into water, it reacts so violently that it catches fire and sometimes explodes. Because of its high reactivity, the free element sodium is never found in the environment. Instead, sodium is always found combined with other elements.

Properties of Chlorine

The element chlorine, shown in **Figure 4.3**, is a pale green, poisonous gas with a choking odor. Because chlorine kills living cells and is slightly soluble in water, it is an excellent disinfectant for water supplies and swimming pools. You can tell that chlorine is a nonmetal by its position in the upper-right portion of the periodic table. Chlorine gas must be cooled to −34°C before it turns to a liquid. Like sodium, it is among the most reactive of the elements and must be handled with extreme care. Chlorine is needed for many industrial processes, such as the manufacture of bleaches and plastics. Because of its industrial importance, large quantities of chlorine must be transported in railroad tank cars, tanker trucks, and river barges. If a train that is carrying chlorine derails, entire communities are evacuated until the danger of a chlorine leak passes.

Look again at the photo on the opening page of this chapter. When sodium and chlorine react to form sodium chloride, two dangerous elements combine to form a stable, safe substance that we consume every day. What could be happening in such a change? You will find the details in Section 4.2, but first look at two other common compounds whose properties are different from those of sodium chloride.

Figure 4.3

A Comparison of Sodium and Chlorine
Sodium is a metal, but it is soft enough to be cut with a knife. Where it has been freshly cut, you can see that sodium has a silvery luster that is typical of many metals. ▶

Chlorine, in the flask on the right, is a greenish, poisonous gas at room temperature. If you've ever used liquid chlorine bleach, you've probably smelled chlorine. ▶

Visual Learning

Figure 4.3: Point out the extreme difference in properties between the elements sodium (a silvery-gray, soft, highly reactive metal) and chlorine (a pale-green, poisonous gas), and the compound resulting from the reaction of the two elements, sodium chloride.

*inter*NET
CONNECTION
Students can access information on upcoming space shuttle mission payloads and archive images.
World Wide Web
http://shuttle.nasa.gov

Quick Demo

How can you extinguish a flame?

Set one or two short candles upright in a water trough. Light the candles, and set a 250-mL beaker containing a chunk of dry ice in the middle of the trough. As the dry ice sublimes, the gaseous CO_2 will spill over the edge of the beaker into the trough. Because it is denser than air, it will fill up the trough and extinguish the candles. Ask students for an explanation. *The dry ice sublimes, creating CO_2 gas, which is more dense than air and does not support combustion.*

Quick Demo

How can you make carbon dioxide?

If you have no dry ice, try the following demo. Generate CO_2 in a large beaker by reacting vinegar and baking soda. "Pour" the dense CO_2 down a trough made by folding about 50 to 75 cm of aluminum foil into a V-shape along its length. At the bottom of the trough, the CO_2 will extinguish candles as it flows over them.

Fact of the MATTER

Some people think that when we inhale, the lungs absorb all the oxygen from the inhaled air and replace it completely with CO_2. You can see from **Table 4.1** that this notion isn't true.

Carbon Dioxide: A Gas to Exhale

Carbon dioxide is a colorless gas. Take a deep breath and hold it for a few seconds. What you have inhaled is air, a colorless mixture of nitrogen and oxygen gases with small amounts of argon, water vapor, and carbon dioxide. Now, exhale. The mixture of gases that you exhale contains more than 100 times the amount of carbon dioxide that was in the air that you inhaled, as you can see in **Table 4.1.** In contrast, the quantity of oxygen is reduced by five percent. While the air was in your lungs, chemical and physical processes exchanged some of the oxygen for carbon dioxide. Carbon dioxide is an important chemical link between the plant and animal world. Green plants and other plantlike organisms take in carbon dioxide and give off oxygen during photosynthesis. Both plants and animals, including humans, use oxygen and give off carbon dioxide during respiration.

Table 4.1   Composition of Inhaled and Exhaled Air

Substance	Percent in Inhaled Air	Percent in Exhaled Air
Nitrogen	78	75
Oxygen	21	16
Argon	0.9	0.9
Carbon dioxide	0.03	4
Water vapor	variable (0 to 4)	increased

Physical Properties of Carbon Dioxide

Carbon dioxide, like sodium chloride, is also a compound, but its properties differ from those of sodium chloride. For example, salt is a solid at room temperature, but carbon dioxide is a colorless, odorless, and tasteless gas. When carbon dioxide is cooled below $-80°C$, the gas changes directly to white, solid carbon dioxide without first becoming a liquid. Because the solid form of carbon dioxide does not melt to a liquid, it is called dry ice, as shown in **Figure 4.4.** Carbon dioxide is soluble in water, as anyone who has ever opened a carbonated beverage knows. A water solution of carbon dioxide is a weak conductor of electricity. You can make carbon dioxide from its elements by burning carbon in air. Coal and charcoal are mostly carbon.

Chemical Properties of Carbon Dioxide

Like sodium chloride, carbon dioxide is relatively stable. Carbon dioxide is used in some types of fire extinguishers because it does not support burning, **Figure 4.4.** Photosynthesis is probably the most significant chemical reaction of carbon dioxide. In photosynthesis, plants use energy from the sun to combine carbon dioxide and water chemically to make simple sugars. Plants use these sugars as raw materials to make many

---

Demonstration 4.2

A Metal and a Nonmetal React
Purpose: To demonstrate that metals and nonmetals often combine to form a stable compound by releasing energy.
Materials: 1 g zinc dust, 3 g iodine crystals, a few drops water, dry test tube with stopper, evaporating dish, widemouth bottle with stopper, dropper
Safety Precautions:

Disposal: Codes G, A
Procedure: Put 1 g of zinc dust and 3 g of iodine crystals into a dry test tube. Stopper the test tube and mix the zinc and iodine by shaking. Transfer the mixture to an evaporating dish. In the fume hood, carefully add a few drops of water to the mixture using a dropper. **CAUTION:** *This demonstration should be done only in a fume hood. I_2 vapors are toxic. The dish will*

become hot. Collect the toxic, violet-colored iodine vapors in a widemouth bottle that is inverted and clamped over the evaporating dish. Immediately stopper the bottle.
Results: Violet fumes are given off from the reacting mixture as an exothermic reaction proceeds.

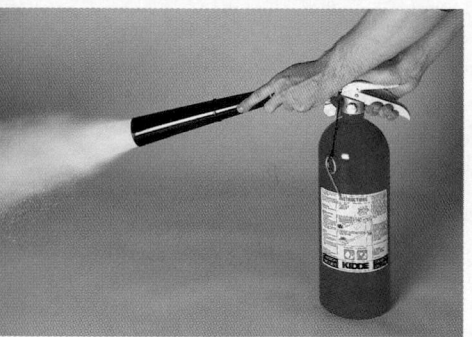

▲ Carbon dioxide does not support burning. In fact, it is often used to put out fires. Fire extinguishers like this one are filled with compressed carbon dioxide. Because CO_2 is denser than air, it displaces air and deprives the fire of a supply of oxygen.

Figure 4.4
Properties of Carbon Dioxide
 Solid carbon dioxide is called dry ice. Under ordinary conditions, it does not melt into a liquid; instead, it changes directly into a gas. The dry ice in this cylinder is immersed in water and is producing bubbles of carbon dioxide gas. The white vapors you see here are not carbon dioxide, but condensed water vapor carried along with the cold CO_2 gas. You can tell that carbon dioxide is denser than air.

other kinds of compounds, from cellulose in wood and cotton to oils such as corn oil and olive oil. Photosynthesis is only one part of a natural cycle of chemical reactions known as the carbon cycle.

The Properties of Carbon

As with sodium chloride, the properties of carbon dioxide differ from the properties of its elements. Carbon is a nonmetal and is fairly unreactive at room temperature. However, at higher temperatures, it reacts with many other elements. Charcoal is approximately 90 percent carbon. As anyone with a charcoal grill knows, carbon burns fairly easily and is an excellent source of heat, as shown in **Figure 4.5.** Carbon forms a huge variety of compounds. In fact, the majority of compounds that make up living things contain carbon. Carbon compounds are so significant that an entire branch of chemistry, called organic chemistry, is dedicated to their study.

Figure 4.5
A Chemical Property of CO_2
When you burn charcoal to cook chicken, carbon and oxygen in the air combine to produce carbon dioxide. When elements combine to form a substance that is more stable, the reaction often gives off energy in the form of heat.

4.1   **The Variety of Compounds**   **125**

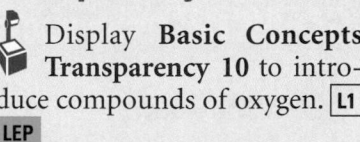

The Properties of Oxygen

Like carbon, oxygen is another nonmetal. It is a colorless, odorless, and tasteless gas that makes up about 21 percent of the air you breathe. When materials such as paper burn in air, they react with oxygen, which is why people commonly say that oxygen supports burning.

Oxygen gas becomes a liquid when it is cooled to −183°C, and it is slightly soluble in water. The gills of fishes absorb dissolved oxygen from their water environment. Oxygen is more reactive than carbon and combines with many other elements. A prime example of its reactivity is the process of rusting, in which the element iron combines with oxygen from air. Many of the compounds that make up Earth's crust contain oxygen. Oxygen is the most abundant element in Earth's crust, as shown in **Figure 4.6.**

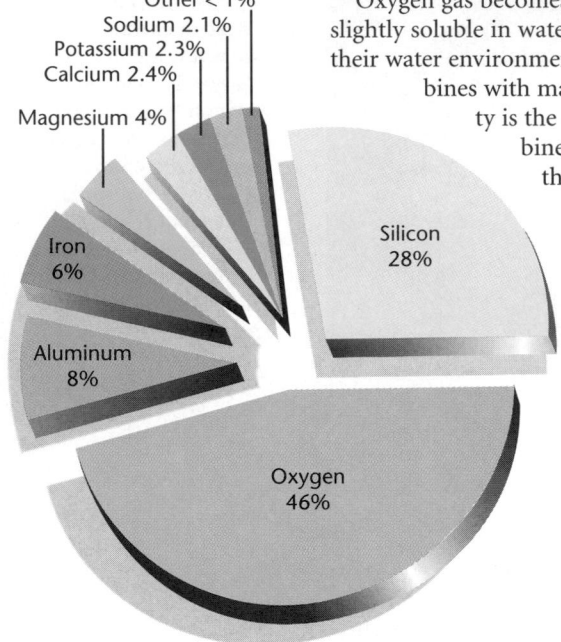

Figure 4.6

Elemental Composition of Earth's Crust
As you can see from the graph, oxygen makes up about 46 percent of Earth's crust. Nearly all of this oxygen occurs in compounds with other elements. Silicon is the next most abundant element in Earth's crust. Therefore, you won't be surprised to learn that much of the oxygen is tied up in silicon dioxide, commonly known as sand.

Water, Water Everywhere

Water is the third familiar compound you will compare in your detective work on the submicroscopic structure of compounds. The formal chemical name of water is dihydrogen monoxide, but nobody calls it that. Water covers approximately 70 percent of Earth's surface and also makes up about 70 percent of the mass of the average human body.

Physical Properties of Water

The properties of water are different from those of sodium chloride and carbon dioxide. Water is the only one of the three compounds that occurs in Earth's environment in all three states of matter, as shown in **Figure 4.7.** At sea level, liquid water boils into gaseous water (steam) at 100°C and freezes to solid water (ice) at 0°C. Pure water does not conduct electricity in any of its states. Water is also excellent at dissolving other substances. It is often called the universal solvent in recognition of this valuable property. Water plays a vital role in the transport of dissolved materials, whether the aqueous solution is flowing down a river; up the xylem in a tree; or through the veins, capillaries, and arteries of your circulatory system.

126

Chemical Properties of Water

Water is a stable compound; it doesn't break down under normal conditions and does not react with many other substances. Perhaps the most interesting property of water is its ability to act as a medium in which chemical reactions occur. Nearly all of the chemical reactions in the human body and many important reactions on Earth occur in an aqueous environment. Without water, these reactions could not occur or would occur extremely slowly. In addition, water and carbon dioxide are the starting materials for photosynthesis, the process that makes life on Earth possible. Now, compare the properties of water with those of its component elements, hydrogen and oxygen. The properties of oxygen were described on page 126.

The Properties of Hydrogen

Hydrogen is the lightest and most abundant element in the universe. Hydrogen is usually classified as a nonmetal. Like oxygen, the element hydrogen is an odorless, tasteless, and colorless gas. Hydrogen is a reactive element. Because of its reactivity, it is seldom found as a free element on Earth. Instead, it occurs in a variety of compounds, particularly water. Hydrogen reacts vigorously with many elements, including oxygen, as shown in **Figure 4.8**. This reaction forms water. The temperature of hydrogen gas must be lowered to a frigid −253°C before it turns to liquid. Hydrogen does not conduct electricity and is only slightly soluble in water.

Figure 4.8
The Reaction of Hydrogen and Oxygen
This welder is using a torch that burns hydrogen in oxygen. The heat produced is so intense that the torch can even be used underwater. The product of the reaction is water.

Visual Learning

Figure 4.7: Ask students why ice floats on liquid water and what this indicates about the structure of ice. *Ice is less dense than liquid water. The molecules in ice must be farther apart.*

Extension

Ask students to find out what causes the jet aircraft vapor trails that we see in the upper atmosphere. *The trails are condensed water vapor from the combustion of jet fuel, a mixture of compounds that contain hydrogen.* L1

Quick Demo

What are some properties of hydrogen?
Inflate a small rubber balloon with hydrogen gas from a demonstration cylinder, and tie the balloon with a string to a heavy object. Erect an explosion shield between the balloon and the class. Wearing goggles, tape a candle to the end of a meterstick, light the candle, and touch it to the balloon. A mild explosion and some flames result. Ask students what these observations indicate about hydrogen. *Hydrogen is less dense than air, and it is combustible.*

---

lower the glowing steel wool into the beaker of oxygen gas. **CAUTION:** *Sparks will fly; use an explosion shield.* Have students observe the increased reaction rate.
Results: The wood splint will burst into flame, but the O_2 gas will not burn. The increased concentration of oxygen greatly increases the rate of the reaction of the steel wool to produce iron oxides.

Analysis: Ask these questions.
1. Are the reactions with oxygen that you observed exothermic? How do you know? *Yes, the reactions gave off heat.*
2. What products are formed in the reactions of wood and iron with oxygen? H_2O and CO_2; Fe_2O_3
3. What happens to the rate of a combustion reaction as the concentration of oxygen is increased? *The rate increases.*

Assessment

Knowledge: Ask students how breathing pure oxygen before a race would affect a track runner. Point out the deleterious results of such an action. L1

Everyday Chemistry

Purpose
Students will learn about the essential nutritional minerals.

TECH PREP

Teaching Strategies
Explain to students that in nutrition, minerals are those elements that are obtained in the diet—most probably from inorganic sources. Calcium, chlorine, magnesium, phosphorus, potassium, sodium, and sulfur are classified as macrominerals because they are present in larger quantities than the other minerals.

Extension
Assign several minerals to each team of students and have them research the RDA or ESAI value, the functions in the body, and the dietary sources of each mineral. **L1** **COOP LEARN**

Exploring Further
1. Some minerals may dissolve in, and be thrown away with, the water.
2. Some minerals, such as selenium, iodine, and arsenic, are toxic in large dosages. Excess sodium intake can raise blood pressure, which increases the risk of heart disease.

3 ASSESS

Check for Understanding
- Check for understanding of the section material by having students answer the Section Review questions.
- Ask students to analyze **Figure 4.9** to describe the reaction of iron with chlorine. Students should describe how the product $FeCl_3$ differs from its elements. **L1**

Everyday Chemistry

Elemental Good Health

Do you know that 60 elements are commonly found in the human body? So far, slightly fewer than half of them have been found essential for life, although scientists think most of the others play some role in life processes. The elements currently known to be essential are listed in the table below.

Although hydrogen, carbon, oxygen, and nitrogen make up almost 96 percent of the mass of the human body, minerals are also essential for life processes.

Different roles You can easily recall the importance of some minerals. Because calcium compounds make up the hard parts of bones and teeth, calcium is needed for their growth and maintenance. Iron is an important element because it is the active part of the blood's hemoglobin molecule, which carries oxygen to the cells. Fluorine helps in the formation and maintenance of teeth and may prevent osteoporosis, which is the disintegration of bone. On the other hand, you may not know that magnesium is necessary for the functioning of nerves and muscles. So is potassium. Zinc and selenium are necessary for the activity of enzymes needed for cell division and growth and for the functioning of the immune system.

Zinc and calcium

Potassium

Selenium

Chromium

Elements Necessary for Life

Minerals	Non-minerals
F, Na, Mg, Si, P, S, Cl, K, Ca, V, Cr, Mn, Fe, Co, Ni, Cu, Zn, As, Se, Sn, I	H, C, O, N

Different amounts Maintaining the proper level of each mineral in your body is important for your health. Nutritionists have established amounts of

these elements that you should have in your daily diet. The amounts are described as the Recommended Dietary Allowance (RDA) and Estimated Safe and Adequate Dietary Intakes (ESAI). The values of the daily intakes of minerals vary greatly. For example, the RDA for calcium for teenagers is 1200 milligrams, while that of iodine is 150 micrograms (0.000 150 g). You may think an amount of 150 micrograms can't be too important, but it is crucial to the function of your thyroid gland, which helps control your metabolism and growth. If you use iodized salt, which contains a little potassium iodide, you probably are getting the proper RDA. Likewise, eating a well-balanced diet of the five food groups will maintain the proper levels of all the minerals that are elemental to good health.

Exploring Further

1. **Inferring** Why might cooking foods in boiling water reduce their mineral content?
2. **Acquiring Information** Why might consuming more than the Recommended Dietary Allowance of minerals be harmful?

Integrating THE Sciences

Biology Some students may have the mistaken impression that plants and other photosynthetic organisms only take in CO_2 and give off O_2, and never use the reverse process. In fact, green plants, animals, and most other organisms carry out respiration in which they break down food molecules, using O_2 and giving off CO_2. The difference among them is that plants make their own food molecules. Ask students to research the topic of plant respiration and trace the path of energy from the sun to the plant's life processes. **L2**

Using Clues to Make a Case

You've seen that elements combine to form compounds whose properties differ greatly from those of the elements themselves. **Figure 4.9** shows another example. On the submicroscopic level, this clue indicates that atoms of elements react chemically to form some combinations that are much different from the original atoms. Also, if atoms of elements always combined in the same way, it's likely that all compounds would be similar. However, you've just studied three compounds that have greatly differing properties. On the submicroscopic level, this clue indicates that atoms must be able to combine in different ways to form different kinds of products. With the knowledge of the structure of atoms that you learned in Chapters 2 and 3, you can now examine the different ways that atoms can combine.

Figure 4.9
Iron Reacting with Chlorine
Here is another example of two elements reacting to form a compound whose properties are different from either element. Heated steel (iron) wool reacts with chlorine gas to form iron(III) chloride, the brown cloud you see in the flask.

SECTION REVIEW

Understanding Concepts

1. Use water as an example to contrast the properties of a compound with the elements from which it is composed.

2. Classify the following substances as elements or compounds.

   a) table salt   **d)** chlorine gas
   b) water   **e)** carbon dioxide gas
   c) sulfur   **f)** dry ice

3. Give an example of a clue that indicates that atoms can combine chemically with each other in more than one way.

Thinking Critically

4. **Comparing and Contrasting** Using sodium chloride, carbon dioxide, and water as examples, what can you say about the chemical reactivity of compounds compared to the elements of which they are composed?

Applying Chemistry

5. **Using the Periodic Table** Find the elements that make up the compounds discussed in this section. Which elements are metals? Which elements are nonmetals? Compare the properties of the three compounds in terms of whether their component elements are two nonmetals or a metal and a nonmetal.

Reteach

Discussion: Ask students how they might use the different properties of salt, carbon dioxide, and water to separate the components of a carbonated, aqueous salt solution in a closed, pressurized bottle. *Open the bottle, allowing the CO_2 to escape into the air. Boil away the water, leaving the salt.*

Extension

Ask students to speculate as to why the properties of sodium chloride, carbon dioxide, and water are so different. *They consist of different elements that must be chemically combined in different ways.* L1

4 CLOSE

Discussion

As a lead-in to the next section, note the types of elements that comprise the three compounds studied in Section 4.1. Sodium chloride consists of a metal and a nonmetal. Carbon dioxide consists of two nonmetals. Water consists of a nonmetal, oxygen, and an element that is usually considered a nonmetal, hydrogen. Have students discuss how the types of elements that make up a compound have a lot to do with the characteristics of that compound. L1

SECTION REVIEW

1. Water is a liquid at room temperature; hydrogen and oxygen are gases. Hydrogen and oxygen must be cooled to very low temperatures for them to liquefy. Water is chemically less reactive than hydrogen or oxygen. Oxygen supports burning and rusting.
2. **a.** compound **b.** compound **c.** element **d.** element **e.** compound **f.** compound
3. Sodium and chlorine react to form a brittle crystalline solid; carbon and oxygen combine to form a gas.
4. In general, common compounds are less reactive than the elements of which they are composed. Water is fairly unreactive, whereas hydrogen and oxygen can react vigorously. Sodium chloride is a stable substance, whereas both sodium and chlorine are highly reactive. Carbon dioxide is an unreactive substance. Elemental carbon is somewhat unreactive, but it combines readily at high temperatures with oxygen, a highly reactive element.
5. Carbon, hydrogen, oxygen, and chlorine are all nonmetals. Sodium is the only metal. H_2O and CO_2 are each composed of two nonmetals. NaCl is composed of a metal and a nonmetal.

PREPARE

Key Concepts

The formation of ionic and co-valent compounds is related to the submicroscopic structure of the constituent elements, specifically the possible types of interactions of valence electrons that may occur in achieving stable octet configurations. Some general characteristics of ionic and covalent compounds are discussed and related to the type of bonding.

Planning Ahead

For the ChemLab Obtain several 9-V batteries, battery clips, and insulated copper wire.

1 FOCUS

Display the **Section Focus Transparency** for Section 4.2 to introduce the concept of reactive and unreactive elements. L1 LEP

How Elements Form Compounds

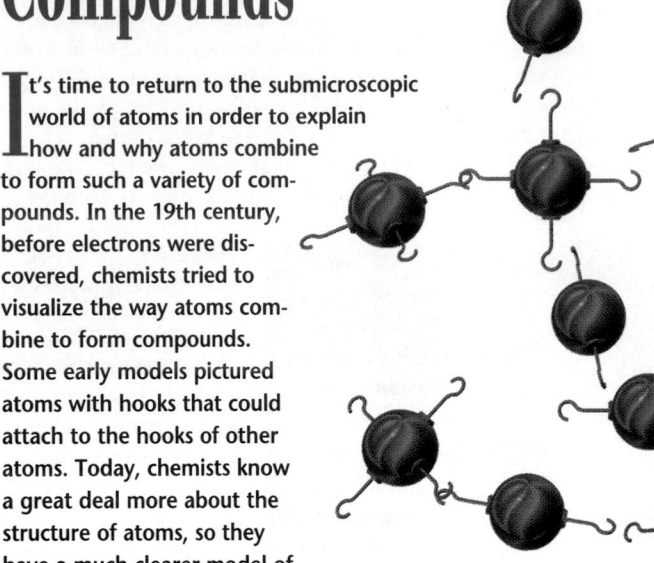

SECTION PREVIEW

Objectives
Model two types of compound formation: ionic and covalent at the atomic level.
Demonstrate how and why atoms achieve chemical stability by bonding.
Compare, using examples, the effect of covalent and ionic bonding on the physical properties of compounds.

Key Terms
octet rule
noble gas
  configuration
ion
ionic compound
ionic bond
crystal
covalent bond
covalent compound
molecule
electrolyte
interparticle force

It's time to return to the submicroscopic world of atoms in order to explain how and why atoms combine to form such a variety of compounds. In the 19th century, before electrons were discovered, chemists tried to visualize the way atoms combine to form compounds. Some early models pictured atoms with hooks that could attach to the hooks of other atoms. Today, chemists know a great deal more about the structure of atoms, so they have a much clearer model of chemical combination.

When Atoms Collide

When elements react, atoms of the elements must collide. It is what happens during that collision that determines what kind of a compound forms. How does the reaction of sodium and chlorine atoms to form salt differ from the reaction of hydrogen and oxygen atoms to form water?

When atoms collide with each other, what really comes into contact? As you learned in Chapter 2, the nucleus is tiny compared to the size of the atom's electron cloud. Also, the nucleus of an atom is buried deep in the center of the electron cloud. Therefore, it is highly unlikely that atomic nuclei would ever collide during a chemical reaction. In fact, reactions between atoms involve only their electron clouds.

When you studied the periodic table in Chapter 3, you learned that the properties of elements repeat because the pattern of the outermost (valence) electrons repeats in each period. It is this arrangement of valence electrons of an atom that is primarily responsible for the atom's chemical properties. You will not be surprised, then, to learn that it is the valence electrons of colliding atoms that interact. But what kinds of interactions between valence electrons are possible? For some additional clues to what takes place when atoms combine to form compounds, look at a group of elements with unusual properties—the noble gases.

Across the Curriculum

Etymology
Ask students to find out why the Group 18 elements are called noble gases. *In countries that have kings and queens, nobility consists of persons of high rank who are often thought of as idle and aloof. Noble gas elements also tend to do little in terms of reacting with other substances.* L1

Program Resources

Study Guide, pp. 15-16 L1
Critical Thinking/Problem Solving, p. 5 L2
Laboratory Manual, 4.2, pp. 37-40 L1
Section Focus Transparency 8 and **Master,** p. 8 L1 LEP
Basic Concepts Transparencies 11, 12 and **Masters,** pp. 21-24 L1 LEP

Problem Solving Transparency 4 and **Master,** pp. 7-8 L1 LEP
ChemLab and MiniLab Worksheets, pp. 29-31, 33 L1
Chemistry and Industry, pp. 5-8 L1
Consumer Chemistry, pp. 7-8 L1

Noble Chemical Stability

Of all of the elements, the elements of Group 18 are a curious bunch. They are notorious for their almost complete lack of chemical reactivity. In fact, they have some practical uses just because they are not reactive, as you can see in **Figure 4.10.** Despite the fact that all of these elements occur naturally in the environment, not a single *compound* of any of these elements has ever been found naturally in the environment.

This group of unreactive elements used to be called the inert gases because chemists thought these elements could never react to form compounds. However, in the 1960s, chemists were able to react fluorine, under conditions of high temperature and pressure, with krypton and with xenon to form compounds. Since that time, a few additional compounds of xenon and krypton have been produced. Still, no one has been successful at synthesizing compounds of helium, neon, and argon. Now that chemists know that these elements aren't completely inert, they are called noble gases.

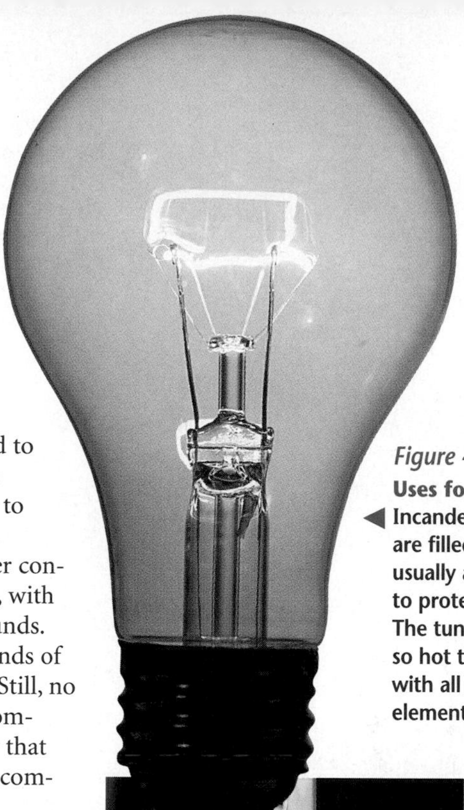

Figure 4.10

Uses for Noble Gases
◄ Incandescent lightbulbs are filled with noble gases, usually argon and krypton, to protect the filament. The tungsten filament gets so hot that it will react with all but the most inert elements.

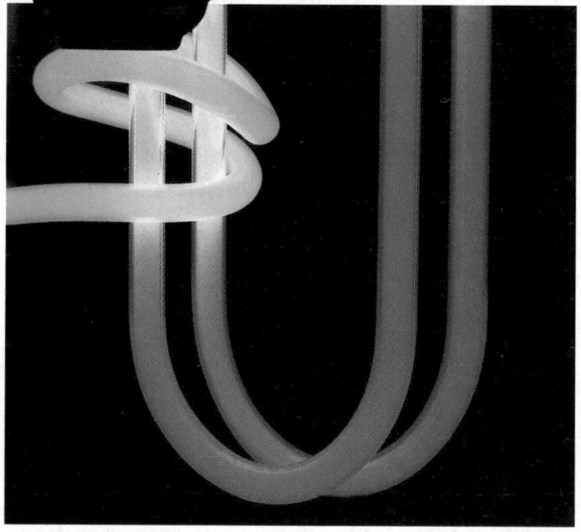

Noble gases and mixtures of noble gases are used in eye-catching light displays often referred to as neon lights. These gases give off different colors of light when a high-voltage electric current passes through them. Neon produces a bright orange color, argon produces blue, and helium gives off yellow-white. Lighting designers obtain different colors by adding mercury or other substances to the gas mixture. Sometimes, colored glass is used. ►

The Octet Rule

The lack of reactivity of noble gases indicates that atoms of these elements must be stable. Noble gases are unlike any other group of elements on the periodic table because of their extreme stability. As you know, the elements of a vertical group on the periodic table have similar arrangements of valence electrons. Each noble gas has eight valence electrons, except for helium, which has two. Because electron arrangement determines chemical properties, the electron arrangements of the noble gases must be the cause of their lack of reactivity with other elements.

2 TEACH

Content Background

Bond Formation In discussing the combining of elements at the submicro level, all elements are being treated as atomic, even though the elements hydrogen, oxygen, and chlorine used are actually diatomic molecules. This simplifies the presentation and keeps the focus on the idea that bonds form between atoms. The diatomic nature of these elements is introduced in the next chapter.

Emphasize that the valence electrons of reacting atoms are the only electrons that directly participate in chemical reactions. The inner (core) electrons are too low in energy and too close to the nucleus to play a significant role.

Using an Analogy

Ask students to picture two adjacent cities, such as Dallas and Fort Worth, Texas, growing and coming into contact. Make a comparison to two atoms coming into contact. Think of the city centers as nuclei and the suburbs as valence electrons. Then it is obvious that the suburbs (electrons) rather than the city centers (nuclei) will come into contact.

⚡ Quick Demo

How explosive is a noble gas?

Use each noble gas element— helium, neon, and argon—to fill a balloon and a small container. Demonstrate the noncombustible nature of the elements by popping each balloon with a lighted candle taped to a meterstick. Demonstrate that the elements do not support combustion by placing a burning candle in each container.

ACROSS THE CURRICULUM

Art

Gold is one of the few elements that is found in the pure state in nature because it is very unreactive. This property is probably what brought it to the attention of early civilizations. Because of its inertness, gold retains its shiny color. Artists have used the chemical and physical properties of gold for thousands of years. Because gold is the most malleable of all metals, it can be hammered into jewelry and decorative objects, as well as thin sheets called gold leaf. The sheets are so thin that 300 000 leaves make a pile only 2.5 cm thick. Artisans use gold leaf to highlight and protect their artwork.

Students may think that the atoms of all noble gas elements have completely filled outermost energy levels. This is true for helium and neon. The atoms of the remaining noble gas elements do have octets in their outer levels, but these energy levels can hold additional electrons, albeit only after the next higher level has some electrons. For example, atoms of argon have eight electrons in the third energy level, but the third level may contain up to 18 electrons.

GLENCOE TECHNOLOGY

Videodisc

Chemistry: Concepts and Applications
Ionic and Covalent Compounds
Disc 1, Side 1, Ch. 16

Use this video to introduce the formation of sodium chloride and of water from their elements. L1 LEP

CD-Rom

Chemistry: Concepts and Applications
Ionic and Covalent Compounds
Use this video to introduce the formation of sodium chloride and of water from their elements. L1 LEP

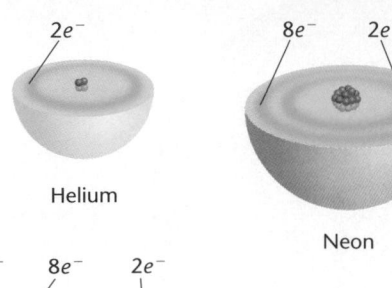
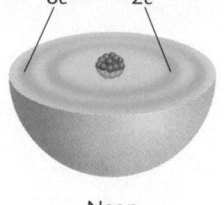

Helium

Neon

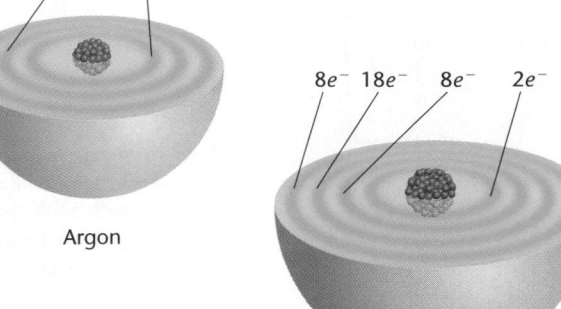

Argon

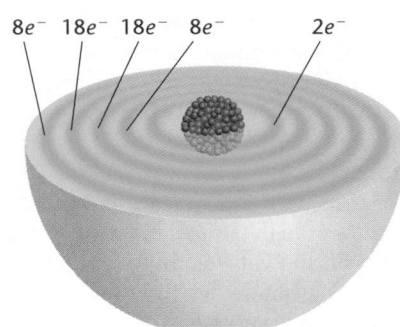

Krypton   Xenon

Figure 4.11

Electron Arrangements of Noble Gases
Note that the noble gas atoms have eight electrons in the outer energy level. This arrangement causes them to be almost completely unreactive. The lone exception to this octet arrangement is helium. The helium atom has only one energy level, which can contain only two electrons.

Fact of the MATTER

Radon, Rn, is the last member of the noble gas group. It is a radioactive element that is formed by the radioactive decay of radium. Although it is found naturally on Earth, its presence is fleeting because it decays rapidly to other elements.

What does the electron arrangement of noble gases have to do with the way other elements react? Compare the electron arrangements of the noble gases in **Figure 4.11.** Today, scientists know that atoms don't have hooks. Rather, atoms combine because they become more stable by doing so. The modern model of how atoms react to form compounds is based on the fact that the stability of a noble gas results from the arrangement of its valence electrons. This model of chemical stability is called the **octet rule.** The octet rule says that atoms can become stable by having eight electrons in their outer energy level (or two electrons in the case of some of the smallest atoms). In other words, elements become stable by achieving the same configuration of valence electrons as one of the noble gases, a **noble gas configuration.**

Ways to Achieve a Stable Outer Energy Level

If atoms collide with enough energy, their outer electrons may rearrange to achieve a stable octet of valence electrons—a noble gas configuration—and the atoms will form a compound. Remember that electrons are particles of matter, so the total number of electrons cannot change during chemical reactions. Next, think about how valence electrons might rearrange among colliding atoms so that each atom has a stable octet. There are only two possibilities to consider. The first is a transfer of valence electrons between atoms. The second possibility is a sharing of valence electrons between atoms. The reactions discussed in Section 4.1 provide good examples of both of these possibilities.

Meeting Individual Needs

Visually Impaired Students who are visually impaired will benefit from the manipulation of physical models to help them understand electron transfer or electron sharing to achieve stable octets. Office-supply stores sell small, round, colorful magnets. These magnets can serve to model electrons. Students can manipulate these on the chalkboard to represent ionic and covalent bonding. You may also want to consider expanding the modeling method used in MiniLab 2 to accomplish the same thing on a smaller scale. L1 LEP

Electrons Can Be Transferred

You know that when sodium and chlorine are mixed, a reaction occurs and a compound, sodium chloride, forms. The photograph at the beginning of this chapter showed the macroscopic view of this reaction. What can be happening at the atomic level? Begin by picturing a collision between a sodium atom and a chlorine atom. Locate these atoms on the periodic table. Sodium is in Group 1, so it has one valence electron. Chlorine is in Group 17 and has seven valence electrons.

In previous chapters, you have used Lewis dot structures to represent an atom and its valence electrons. You will now be able to use these models to show what happens when atoms combine. The electron dot structures of the atoms are shown below.

$$Na^{\cdot} \quad \cdot \ddot{\underset{..}{Cl}} :$$

How can the valence electrons of atoms rearrange to give each atom a stable configuration of valence electrons? If the one valence electron of sodium is transferred to the chlorine atom, chlorine becomes stable with an octet of electrons. Because the chlorine atom now has an extra electron, it has a negative charge. Also, because sodium lost an electron, it now has an unbalanced proton in the nucleus and therefore has a positive charge.

$$Na^{\cdot} + \cdot \ddot{\underset{..}{Cl}} : \rightarrow [Na]^+ + [: \ddot{\underset{..}{Cl}} :]^-$$

It's easy to see how chlorine has achieved a stable octet of electrons, but how does sodium become stable by losing an electron? Look at the position of sodium on the periodic table. By losing its lone valence electron, sodium will have the outer electron arrangement of neon. Sodium's stable octet consists of the eight electrons in the energy level below the level of the lost electron. **Figure 4.12** summarizes the way that sodium and chlorine react.

Figure 4.12
The Reaction of Sodium and Chlorine Atoms
The transfer of an electron from a sodium atom to a chlorine atom forms sodium and chloride ions. Examine the diagram carefully to see how this transfer gives both ions a stable octet.

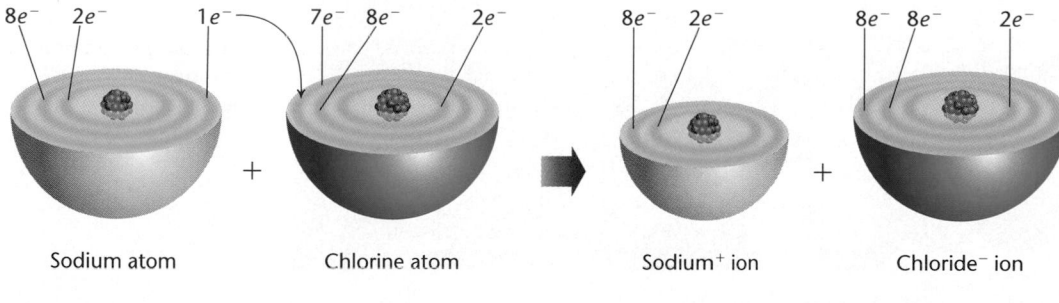

Sodium atom + Chlorine atom → Sodium⁺ ion + Chloride⁻ ion

SECTION 4.2

Correcting Misconceptions

Students may think that a chemical bond physically links atoms, just as a nail holds two pieces of wood together. This false idea is reinforced by the use of model kits in which pieces of wood, plastic, or metal springs link the atoms. Emphasize that a chemical bond is an attractive force between atoms or ions, not a physical attachment.

Transparency

Display **Basic Concepts Transparency 11** to introduce the structure of sodium chloride. L1 LEP

GLENCOE TECHNOLOGY

CD-Rom

Chemistry: Concepts and Applications
Form an Ionic Compound
Use this Interactive Experiment to have students practice making ionic compounds. L1 LEP

Fact of the MATTER

Take a deep breath. A noble gas, argon (Ar), composes approximately one percent of that breath of air you just took.

Chemistry Journal

The Perfect Team

Ask students to write a journal entry in which they consider a sodium atom to be analogous to a professional sports team that is one player over the team limit and a chlorine atom to be analogous to a team that is one player under the team limit. The transfer of one electron (player) from sodium to chlorine makes both atoms (teams) more stable. Of course, the atoms become ions by losing and gaining electrons. L1

Table 4.2   Reaction of Sodium and Chlorine

	Sodium atom Na	+	Chlorine atom Cl	→	Sodium ion Na⁺	+	Chloride ion Cl⁻
Number of protons	11		17		11		17
Number of electrons	11		17		10		18
Number of outer-level electrons	1		7		8		8

Now that each atom has an octet of outer-level electrons, they are no longer neutral atoms; they are charged particles called ions. An **ion** is an atom or group of combined atoms that has a charge because of the loss or gain of electrons. Ions always form when valence electrons rearrange by electron transfer between atoms. A compound that is composed of ions is called an **ionic compound.** The transfer of a single electron changed a reactive metal, sodium, and a poisonous gas, chlorine, into the stable and safe compound, sodium chloride. Note that only the arrangement of electrons has changed. Nothing about the atom's nucleus has changed. This result is clear when you compare the atoms and ions in **Table 4.2.**

Ions Attract Each Other

Remember that objects with opposite charges attract each other. Once they have formed, the positive sodium ion and negative chloride ion are strongly attracted to each other. The strong attractive force between ions of opposite charge is called an **ionic bond.** The force of the ionic bond holds ions together in an ionic compound.

Even the smallest visible grain of salt contains several quintillion sodium and chloride ions. Every positively charged sodium ion attracts all nearby negatively charged chloride ions and vice versa. Therefore, these ions do not arrange themselves into isolated sodium ion/chloride ion pairs. Instead, the ions organize themselves into a definite cube-shaped arrangement, as shown in **Figure 4.13.** This well-organized structure is a crystal. Solid substances are composed of crystals. A **crystal** is a regular, repeating arrangement of atoms, ions, or molecules.

Figure 4.13

Crystal Structure of NaCl
Positive sodium ions attract negative chloride ions to form a cube-shaped arrangement in sodium chloride. In this arrangement, six chloride ions surround every sodium ion, and six sodium ions surround every chloride ion. The forces holding each ion in place are ionic bonds. ▼

◄ The macroscopic result of this submicroscopic arrangement is a cube-shaped salt crystal.

Cultural Diversity

Salt of the Earth

Salt has been recognized as an important and essential substance in all cultures for thousands of years. One of the major uses of salt historically in all parts of the world was as a preservative. Salt is used to symbolize fidelity and permanence of relationships in many so- cieties. An Arab saying that "there is salt between us" indicates a lasting trust between two people. The Hebrew expression "to eat the salt of the palace" means to think highly of someone, just as does the English phrase describing a person as "the salt of the earth."

The Formation of Ionic Compounds

A sodium atom reacts by losing an electron to form a sodium 1+ ion. A chlorine atom gains one electron to form a chloride 1− ion. In this MiniLab, you will consider other combinations of atoms.

Procedure

1. Cut three paper disks about 7 cm in diameter for each of the elements: Li, S, Mg, O, Ca, N, Al, and I. Use a different color of paper for each element. Write the symbol of each element on the appropriate disks.

2. Select atoms of lithium and sulfur, and lay the circles side-by-side on a piece of corrugated cardboard.

3. Using thumbtacks of one color for lithium and another color for sulfur, place one tack for each valence electron on the disks, spacing the tacks evenly around the perimeters.

4. Transfer tacks from the metallic atoms to the nonmetallic atoms so that both elements achieve noble gas electron arrangements. Add more atoms if needed.

5. Once you have created a stable compound, write the ion symbols and charges and the formula and name for the resulting compound on the cardboard.

6. Repeat steps 2 through 5 for the remaining combinations of atoms.

Analysis

1. Why did you have to use more than one atom in some cases? Why couldn't you just take more electrons from one metal atom or add extra ones to a nonmetal atom?

2. Identify the noble gas elements that have the same electron structures as the ions produced.

The Results of Ionic Attraction

How does ionic bonding affect the macroscopic properties of a substance? The cubic shape of a salt crystal is a result of the cubic arrangement of sodium and chloride ions. Given the strong attractive force between sodium and chloride ions and the degree of organization among them, it's not surprising that sodium chloride is a solid at room temperature. All particles of matter are in constant motion. Raising the temperature of matter causes its particles to move faster. In order for a solid to melt, its temperature must be raised until the motion of the particles overcomes the attractive forces and the crystal organization breaks down. Breaking the strong crystal structure of sodium chloride requires a lot of energy, which is the reason that sodium chloride must be heated to more than 800°C before it melts.

When you press salt crystals, they don't bend or flatten out. When enough force is applied, the salt crystals suddenly shatter. This hardness and brittleness provide macroscopic evidence for the strength and rigidity of the submicroscopic structure of the salt crystal. Trying to break an ionic crystal is like trying to break a well-laid brick wall, **Figure 4.14**.

Figure 4.14

Ionic Crystals Are Strong The crystal structure of an ionic compound is a lot like a well-laid brick wall. It takes a great deal of force to break a brick wall.

135

miniLAB

Formation of Ionic Compounds

Purpose: To model the formation of stable ionic compounds by the transfer of electrons to achieve stable electron configurations.

Process Skills: Formulating models, sequencing, communicating

Teaching Strategies

Have students work in pairs or groups of three or four so that students may discuss the procedure and help each other. **L1** **COOP LEARN**

Expected Results

Li^+ and S^{2-}, Li_2S, lithium sulfide; Mg^{2+} and O^{2-}, MgO, magnesium oxide; Ca^{2+} and N^{3-}, Ca_3N_2, calcium nitride; Al^{3+} and I^-, AlI_3, aluminum iodide. Note that three atoms of the same element are required in only two combinations, Ca_3N_2 and AlI_3.

Analysis

1. Removing or adding extra electrons would not yield structures that have noble-gas configurations.

2. Li^+: He; S^{2-}: Ar; Mg^{2+}: Ne; O^{2-}: Ne; Ca^{2+}: Ar; N^{3-}: Ne; Al^{3+}: Ne; I^-: Xe

   Li^+ is an exception to the octet rule because it becomes stable by attaining a helium configuration of two electrons.

✓ Assessment

Performance: Use the same modeling method to have students create models of covalent compounds such as CH_4, NH_3, or CCl_4. **L1** **LEP**

Meeting Individual Needs

Learning Disabled Make two posters with the symbol H^+ on them and two with the symbol e^- on them. Give the H^+ posters to two persons and ask them to stand about 10 feet apart. Explaining that each H^+ represents the nucleus of a hydrogen atom, ask students if the nuclei will be attracted and bond together. *No; the + charges will repel each other.* Then give the e^- posters to two persons and ask them to stand midway between the H^+ persons. Ask students to describe the result. *Because the H^+ hydrogen nuclei are attracted to the two electrons, they are effectively held together.* Explain that a pair of electrons that effectively holds two nuclei together is called a covalent bond. **L1** **LEP**

ChemLab

ChemLab

The Formation and Decomposition of Zinc Iodide

Time Allotment: One class period.

Objectives

Review objectives with students before they begin the ChemLab.

Process Skills

Recognizing cause and effect, observing, classifying, interpreting data, comparing and contrasting

Safety Precautions and Disposal

- Use care in handling solid iodine. Solid iodine is volatile and sublimes near room temperature.
- A puff of iodine vapor may be given off when the reaction starts. If your room is poorly ventilated, this step should be done in the hood.

Disposal: Excess zinc may be collected, washed, dried, and reused. Otherwise, disposal A. Solution, disposal G.

PREPARATION

Fine-mesh granular zinc is the easiest for students to manipulate. Pieces of 22-gauge copper wire approximately 20 cm in length should be cut in advance to accommodate the entire class.

Alternative Materials

- 10 mL of a solution of iodine (10 g iodine/100 mL solution) could be used instead of solid iodine.
- Galvanized roofing nails can be used instead of granular zinc, but the reaction will be slow and may need to sit overnight.

PROCEDURE

- Placing individual containers of zinc and iodine at each station will reduce student traffic and the time required for the experiment.
- Mass measurements do not need to be precise.

The Formation and Decomposition of Zinc Iodide

Compounds are chemical combinations of elements. Many chemical reactions of elements to form compounds are spectacular but must be run under special laboratory conditions because they are dangerous. The reaction of sodium and chlorine to form sodium chloride pictured at the beginning of Chapter 4 is a good example. If elements react spontaneously to form compounds, that is a good indication that the compound state is more stable than the free element state. To break a stable compound down into its component elements, energy must be put into the compound. Electricity is often used as this energy source. The process of decomposing a compound into its component elements by electricity is called electrolysis.

Problem

Can a compound be synthesized from its elements and then decomposed back into its original elements?

Objectives

- **Compare** a compound with its component elements.
- **Observe** and monitor a chemical reaction.
- **Observe** the decomposition of the compound back to its elements.

PREPARATION

Materials
10 × 150-mm test tube
test-tube rack
test-tube holder
100-mL beaker
spatula
plastic stirring rod
zinc
iodine crystals
distilled water
9-V battery with terminal clip and leads
two 20-cm insulated copper wires stripped at least 1 cm on each end

Safety Precautions

Wear aprons and goggles.
CAUTION: *Iodine crystals are toxic and can stain the skin. Use care when using solid iodine. The reaction of zinc and iodine releases heat. Always use the test-tube holder to handle the reaction test tube.*

PROCEDURE

1. Obtain a test tube and a small beaker. Place the test tube upright in a test-tube rack.

2. Carefully add approximately 1 g of zinc dust and about 10 mL of distilled water to the test tube.

3. Carefully add about 1 g of iodine to the test tube. Record your observations in a table like the one shown.

4. Stir the contents of the test tube thoroughly with a plastic stirring rod until there is no more evidence of a reaction. Record any observations of physical or chemical changes.

5. Allow the reaction mixture to settle. Using a test-tube holder, carefully pick up the test tube and pour off the solution phase into the small beaker.

6. Add water to the beaker to bring the volume up to about 25 mL.

7. Obtain a 9-V battery with wire leads and two pieces of copper wire. Attach the copper wires

- Constant stirring speeds up the reaction and avoids the formation of "hot spots" during the reaction.
- Copper electrodes should be placed in the solution as far from each other as possible.
- Remind students that the reaction is spontaneous. This is evidence that the compound is more stable than elements. Connect to the observation that the reaction is exothermic. Note that zinc and iodine are not found as free elements in the environment.

- This lab may be done as a teacher demo. The electrolysis step can be viewed by the whole class using a petri dish as an electrolysis cell on an overhead projector. The formation of iodine at one electrode is clearly visible. The formation of Zn at the other electrode also may be seen. The electrodes can be removed, washed, and passed around the class so that each student can examine the Zn-plated electrode.

to the wire leads from the battery. Make sure that the wires are not touching each other.

8. Dip the wires into the solution and observe what takes place. Record your observations.

9. After two minutes, remove the wires from the solution and examine the wires. Again, record your observations.

ANALYZE AND CONCLUDE

1. **Observing and Inferring** What evidence was there that a chemical reaction occurred?

2. **Comparing and Contrasting** How did you know the reaction was complete?

3. **Making Inferences** What term is used to describe a reaction in which heat is given off? How can you account for the heat given off in this reaction?

4. **Checking Your Hypothesis** What evidence do you have that the compound was decomposed by electrolysis?

5. **Drawing Conclusions** Why do you think the reaction between zinc and iodine stopped?

DATA AND OBSERVATIONS

Step	Observations
3. Addition of iodine to zinc	
4. Reaction of iodine and zinc	
8. Electrolysis of solution	
9. Examination of wires	

APPLY AND ASSESS

1. What role did the water play in this reaction?

2. Do you think zinc iodide is an ionic or covalent compound? What evidence do you have to support your conclusion?

3. The formula of zinc iodide is ZnI_2. Use Lewis dot structures to show how it forms from its elements. Hint: Zinc atoms have two valence electrons.

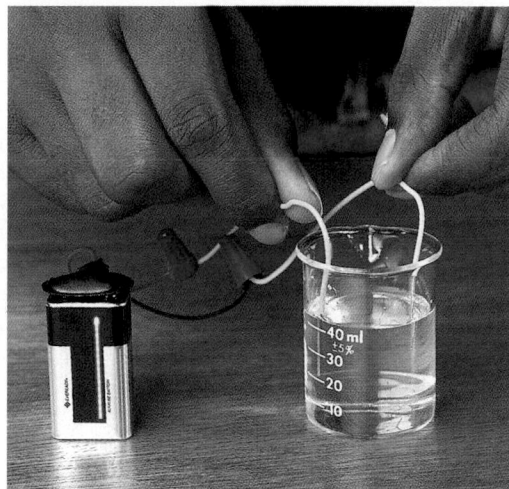

4.2 How Elements Form Compounds   **137**

ANALYZE AND CONCLUDE

1. Heat was evolved. The iodine color disappeared or almost disappeared. Students may hear a sizzling sound and see white crystals of ZnI_2.

2. The iodine appeared to be used up. The reaction also stopped generating heat and began to cool.

3. It is an exothermic reaction. Students may give a variety of reasons for the evolution of heat. When the products of a spontaneous reaction are at a more stable state than the reactants, energy is given off, usually in the form of heat.

4. Brown I_2(aq) formed at one of the copper wires. Zn metal appeared on the other copper wire.

5. All of the iodine was used up. The residue in the test tube appeared to be unreacted zinc.

APPLY AND ASSESS

1. Water acted as a reaction medium, one of the properties of water. By dissolving some of the iodine, water maximized the contact between Zn atoms and I_2 molecules.

2. Zinc iodide is an ionic compound. Zn is a metal; I_2 is a nonmetal. Metals and nonmetals usually react to form ionic compounds. The solution containing the reaction product conducts electricity.

3. $Zn\!\cdot\; + \;:\!\ddot{I}\!:\!\ddot{I}\!: \;\rightarrow\; [Zn]^{2+} +$
   $[:\!\ddot{I}\!:]^- + [:\!\ddot{I}\!:]^-$

✓ Assessment

Knowledge: Lithium burns in air to form the oxide Li_2O. Have students answer *Apply and Assess* questions 2 and 3 for that reaction. **L1**

DATA AND OBSERVATIONS

Step	Observations
3. Addition of iodine to zinc	*Iodine dissolves in water and turns dark brown.*
4. Reaction of iodine and zinc	*After about 30 seconds, the mixture gets warm. Brown color fades to clear to pale yellow. Solid Zn remains.*
8. Electrolysis of solution	*Brown substance diffuses from one electrode.*
9. Examination of wires	*Electrode from which brown liquid diffused is copper colored; other electrode is "silvery" zinc colored.*

GLENCOE TECHNOLOGY

 CD-Rom

Chemistry: Concepts and Applications
Bonding
Use this animation to help students see the movements of electrons during chemical bonding. [L1] [LEP]

 Videodisc

Chemistry: Concepts and Applications
Bonding
Disc 1, Side 1, Ch. 15

Show this video to illustrate the formation of bonds at the submicroscopic level. [L1] [LEP]

Transparency

Display **Problem Solving Transparency 4** to give students practice working with ionic and covalent bonding. [L1] [LEP]

 **Assessment**

Portfolio: The reaction between hydrogen and oxygen obviously releases significant amounts of energy. Ask students to write reports describing how this reaction might be utilized to fulfill part of our future energy needs. Ask students to include the reports in their portfolios. [L1] [P]

Representing Compounds with Formulas

Rather than writing out the name *sodium chloride* every time you refer to it, you can use a much simpler system. You can write its chemical formula, NaCl. The formula of a compound tells what elements make up the compound and how many atoms of each element are present in one unit of the compound. Water is written as H_2O. This formula means that water consists of two hydrogen atoms combined with one oxygen atom. When sodium and chlorine atoms react, they form ions, which then arrange themselves into a crystal. The ratios of the elements that make up salt crystals do not change, and so the formula NaCl does not change.

You have seen how the electrons are transferred from sodium to chlorine to form a strong crystal arrangement. There are many other ionic compounds, as you will see later in Chapter 5. Now, look at another example from Section 4.1 to see a different way that atoms can combine to achieve a stable outer level of electrons.

Electrons Can Be Shared

In Section 4.1, you learned that the reaction of hydrogen and oxygen forms water. What happens when hydrogen and oxygen atoms collide? Hydrogen has only one valence electron. Oxygen, a Group 16 element, has six valence electrons. Could these atoms achieve the stable electron configuration of a noble gas by transfer of electrons? If the oxygen atom could pick up two more valence electrons, it would have a stable octet—the noble gas configuration of neon.

What about hydrogen? Could a hydrogen lose its single valence electron? Your first inclination might be to treat hydrogen just like sodium, but be careful. If hydrogen loses an electron, it is left with no electrons, and that isn't the electron structure of a noble gas. Maybe hydrogen could gain an electron so that its electron arrangement is like that of helium. But, both atoms cannot gain electrons.

Colliding atoms transfer electrons only when one atom has a stronger attraction for valence electrons than has the other atom. In the case of

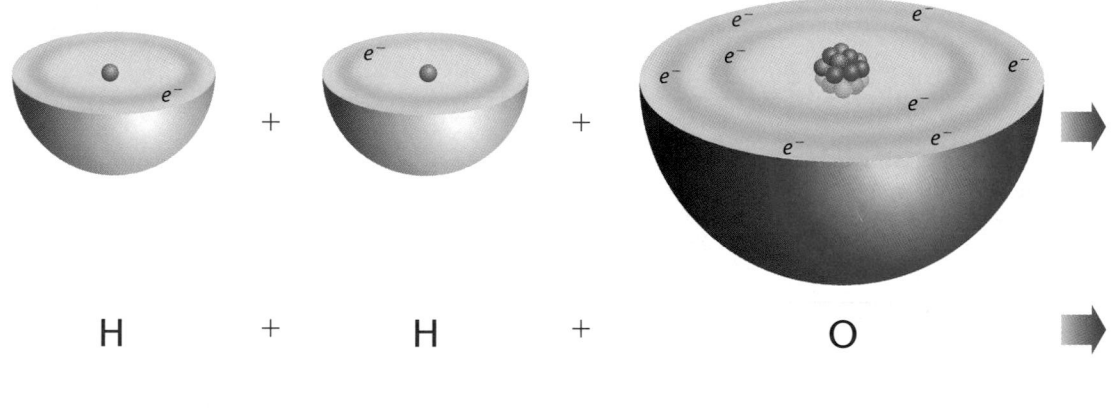

H + H + O

Demonstration 4.4

A Chemical Formula

Purpose: To demonstrate that elements combine to form compounds that can be represented by a formula.

Materials: 4 small, clear, capped bottles; small amounts of copper metal, sulfur, and copper(II) sulfate crystals

Safety Precautions:

Disposal: Code F

Procedure: In four clear, labeled, capped bottles, place some copper metal, sulfur, oxygen (air), and copper(II) sulfate crystals. Place a rubber band around the three bottles containing the Cu, S, and O_2. Discuss how these elements come together to form copper(II) sulfate, $CuSO_4$.

Results: The demonstration helps students to visualize the changes that occur in the physical

sodium and chlorine, chlorine attracts sodium's valence electron strongly, whereas sodium holds its electron weakly. Therefore, the electron moves from sodium to chlorine, forming positive and negative ions in the process. You will learn more about this process and the factors that influence it in Chapter 9. In the case of hydrogen and oxygen, neither atom attracts electrons strongly enough to take electrons from the other atom. When atoms collide with enough energy to react, but neither attracts electrons strongly enough to take electrons from the other atom, the atoms combine by sharing valence electrons.

To understand how water forms, start by looking at the electron dot structures of hydrogen and oxygen.

$$H \cdot \qquad \cdot \ddot{O} :$$

Hydrogen requires one more electron to have the same electron arrangement as helium, while oxygen requires two more electrons to have neon's arrangement. Hydrogen and oxygen can share one electron from each atom. This sharing is shown by placing two dots representing electrons between the atoms.

$$H : \ddot{O} :$$

This arrangement makes hydrogen stable by giving it two valence electrons, but it leaves oxygen with only seven valence electrons. Oxygen's octet can be completed by sharing an electron with another hydrogen atom. (This explains why water has the formula H_2O.)

$$H : \ddot{O} : \quad + \quad H \cdot \quad \rightarrow \quad H : \overset{\displaystyle ..}{\underset{\displaystyle H}{\ddot{O}}} :$$

Just as in the case of the formation of sodium chloride by ionic bonding, all the parts present before the reaction are still there after the reaction. What has changed in the combining of hydrogen and oxygen atoms? The valence electrons no longer reside in the same positions. This is what always happens in a chemical reaction; electrons rearrange. **Figure 4.15** summarizes the reaction between hydrogen and oxygen.

H₂O

Figure 4.15

Formation of Water by Electron Sharing
The stability of the atoms in a water molecule results from a cooperative arrangement in which the eight valence electrons (six from oxygen and one each from two hydrogens) are distributed among the three atoms. By sharing an electron pair with the oxygen, each hydrogen claims two electrons in its outer level. The oxygen, by sharing two electrons with two hydrogens, claims a stable octet in its outer level. By this method, each atom achieves a stable noble gas configuration.

and chemical properties of elements as they react to form compounds.
Analysis: Ask these questions.
1. List observable physical properties of the elements Cu, S, and O_2. *Answers may vary but should include the state of matter, color, hardness, and appearance. S is a hard, dull, yellow solid. Cu is a hard, red-brown solid having a luster; O_2 is a clear, colorless gas.*

2. Describe the physical appearance of the $CuSO_4$ crystals. *The crystals are small, shiny, hard, and blue in color.*

✓ Assessment

Oral: Ask students to discuss the changes in physical and chemical properties when elements combine to form compounds, and explain whether energy is involved. L1

⚡ Quick Demo

Does water always conduct electricity?
Use a low-voltage conductivity tester to show that distilled water has no (or only slight) conductivity. With the tester on, drop a small piece of dry ice into the water or blow through a straw into the water, noting the increased conductivity of the solution as some of the dissolving carbon dioxide forms carbonic acid, which then ionizes. For contrast, stir a few grains of salt into the distilled water while the electrodes of the conductivity apparatus are immersed in it.

Concept Development

Explain that a metal and a nonmetal generally form an ionic bond and an ionic compound. Nonmetals generally form covalent bonds and molecular compounds with other nonmetals. This is a good time to drive home the point that an ionic compound is composed of an assortment of independent ions that occur in a certain ratio and arrangement. Formulas such as NaCl or K_2O represent the simplest ratios of ions. On the other hand, formulas for molecular compounds such as CO_2 or C_2H_4 represent numbers of atoms bonded together in groups.

Transparency

Display **Basic Concepts Transparency 12** to show the formation of covalent compounds. L1 LEP

Figure 4.16

Comparing an Ionic and a Covalent Compound
Iron(III) chloride is a typical ionic compound. It is crystalline at room temperature, melts at a high temperature (300°C), and dissolves in water. ▼

Electron Sharing Produces Molecules

The attraction of two atoms for a shared pair of electrons is called a **covalent bond.** Notice that in a covalent bond, atoms share electrons and neither atom has an ionic charge. A compound whose atoms are held together by covalent bonds is a **covalent compound.** Water is a covalent compound. Although water is made up of hydrogen and oxygen atoms, these have combined into water molecules, each having two hydrogen atoms bonded to one oxygen atom. A **molecule** is an uncharged group of two or more atoms held together by covalent bonds. Sometimes chemists refer to covalent compounds as molecular compounds. The terms mean the same thing. **Figure 4.16** compares a compound with ionic bonds to a compound having covalent bonds.

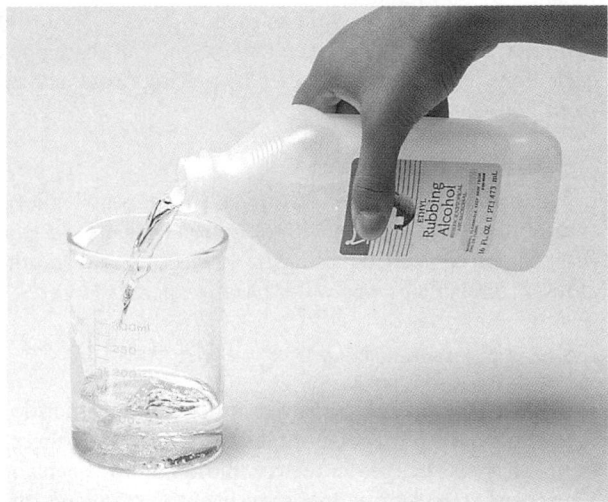

▲ Ethanol, also known as ethyl alcohol, is a typical covalent compound. It is a liquid at room temperature but evaporates readily into the air. Ethanol boils at 78°C and freezes at −114°C. Unlike many covalent compounds, ethanol dissolves in water. In fact, ethanol sold in stores as rubbing alcohol contains water.

More Than Two Electrons Can Be Shared

When charcoal burns, carbon atoms collide with oxygen to form CO_2. Carbon is in Group 14 and has four valence electrons. Oxygen is in Group 16 and has six valence electrons.

Carbon and oxygen are like hydrogen and oxygen when it comes to the question of sharing or transferring electrons. Neither atom is able to attract electrons away from the other atom. In fact, two nonmetallic elements usually achieve stability by sharing electrons to form a covalent compound. On the other hand, if the reacting atoms are a metal and a nonmetal, they are much more likely to transfer electrons and form an ionic compound.

Hydrogen's Ill-fated Lifts

Imagine the surprise of the citizens of Paris on that morning of December 1, 1783. There was Jacques Charles and his assistant gliding above their rooftops in a basket suspended from a large balloon. Charles and his assistant had filled the balloon with hydrogen and became the first humans to ride in such a lighter-than-air vehicle. By World War I, hydrogen-filled balloons were being used to carry military personnel aloft to observe troop movements.

In 1936, Germany launched the *Hindenburg*, a rigid airship originally designed to use helium for lift. Helium is slightly less buoyant than hydrogen, but it is a noble gas. Thus, it does not react with anything, whereas hydrogen burns explosively in air. However, the Germans had to continue the use of hydrogen in the airship because they had no source of helium. In the following year, the *Hindenburg* carried more than 1300 passengers in transatlantic flights. While attempting to dock in Lakehurst, New Jersey, on May 1, 1937, the airship exploded and burned when the hydrogen ignited. Thirty-six people were killed. The accident was the final chapter in the use of hydrogen in lighter-than-air vehicles.

The chemical reaction The reaction that destroyed the *Hindenburg* was the burning of hydrogen gas.

$$2H_2(g) + O_2(g) \rightarrow 2H_2O(g)$$

Once ignited, this reaction occurs spontaneously. However, the reaction can be controlled, such as in the main engine of the space shuttle.

Space shuttle fuel
The main engine of the space shuttle

booster uses liquid hydrogen and oxygen as fuel. These materials are stored in separate sections of a huge external fuel tank attached beneath the shuttle. The energy released during the reaction thrusts the shuttle into orbit.

During the launch of the space shuttle *Challenger* on January 28, 1986, the reaction became uncontrolled. One minute and 13 seconds after takeoff, the external tank and shuttle exploded, killing the shuttle's seven crew members. The accident was caused by defects in the design of O-rings that joined sections of the solid-fuel booster engines, which are attached to the sides of the shuttle.

Connecting to Chemistry

1. **Inferring** Much research has been done on developing hydrogen-fueled cars and trucks. Why would such a car be almost entirely nonpolluting? What factors do you think might affect the public's acceptance of such a vehicle?

2. **Hypothesizing** What do you think is the reason that so much energy is produced when hydrogen reacts with oxygen? Recall what happens when these atoms bond.

Challenger explosion

4.2 How Elements Form Compounds 141

HISTORY ▷ CONNECTION

Purpose
Students will learn how hydrogen combustion was involved in two dissimilar historic accidents.

Teaching Strategies
Point out that in a rigid airship, enormous gas-filled bags are encased in a metal-covered shell supported by internal metal girders. This structure gives shape to the craft. A balloon or a blimp has no such cover.

Extension
Have students research and report in their chemistry journals the reason why Germany lacked a source of helium in 1937. **L2**

Connecting to Chemistry
1. The result of hydrogen combustion is water, which is essentially nonpolluting. Cost and performance would be prime factors. Availability of fuel, ease of refueling, and hydrogen's reputation as a dangerous substance would also be important.
2. Accept reasonable answers. For example, hydrogen is reactive because it needs only one electron to achieve the stable noble-gas structure of helium.

✓ Assessment

Knowledge: Have students check the supermarket price per pound of sugar and salt. Ask them to determine which one would be a better buy as a deicing compound if the melting activity of both compounds were the same. *Salt is cheaper.* **L1**

Chemistry Journal

Fire Extinguishers
Ask students to take inventories of their homes, describing locations that are most likely to have fires and in which fire extinguishers should be kept. Have them enter the results of the surveys in their chemistry journals. If desired, you may want to describe the various types of fire extinguishers and their prescribed uses before the assignment. **L1** **LEP**

Extension

Ask students to maintain the octet rule in writing electron dot structures for acetylene, HCCH, and hydrogen cyanide, HCN. [L1]

$$H : C ::: C : H, \quad H : C ::: N :$$

Consider the reaction of carbon and oxygen to form CO_2. Look at electron dot structures for the participating atoms below.

$$: \ddot{O} . \qquad . \dot{\underset{.}{C}} . \qquad . \ddot{O} :$$

Can you arrange the 16 valence electrons from these three atoms to produce a molecule in which all three atoms have a stable configuration? You know that at least one bond must exist between the carbon and each oxygen, so start there. Here's an approach to the puzzle. Have each oxygen share an electron with carbon as in the following dot structures.

$$: \ddot{O} : \dot{\underset{.}{C}} : \ddot{O} :$$

This arrangement gives carbon six electrons and each oxygen seven, but still no atom has an octet. What else can be done? There's no law in chemistry that says atoms must bond by sharing only one pair of electrons. What happens if they share two pairs? You now have double covalent bonds, as shown in this dot structure.

$$: \ddot{O} :: C :: \ddot{O} :$$

Now, count all the electrons around each atom, including the ones that are shared. You'll see that each atom has a stable octet. Study the result of this electron sharing in **Figure 4.17.** By sharing electrons, the three atoms achieve a more stable arrangement than they had as three separate atoms. A molecule of carbon dioxide is more stable than one carbon atom and two oxygen atoms. As with water, the molecule of carbon dioxide is different from the sum of its parts. The macroscopic properties of carbon dioxide are a result of the unique properties of carbon dioxide molecules, not the properties of carbon or oxygen atoms.

Figure 4.17
Electron Sharing in CO_2
When a carbon atom and oxygen atoms react, the carbon atom shares two pairs of electrons with each oxygen. This arrangement gives all atoms a stable octet.

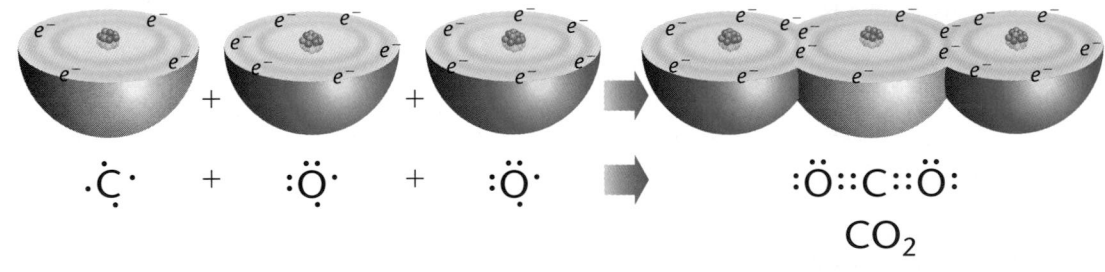

How do ionic and covalent compounds compare?

Now you can relate the submicroscopic models of the formation of NaCl, H_2O, and CO_2 to their macroscopic properties mentioned in Section 4.1. When elements combine, they form either ions or molecules. No other possibilities exist. The particles change dramatically, whether they change from sodium atoms to sodium ions or hydrogen and oxygen atoms to water molecules. This change explains why compounds have different properties from the elements that make them up.

Explaining the Properties of Ionic Compounds

The physical properties of ionic compounds are directly related to the fact that ionic compounds are composed of well-organized, tightly bound ions. These ions form a strong, three-dimensional crystal structure. This model of the submicroscopic structure explains the general observation that ionic compounds are crystalline solids at room temperature. Just like NaCl, these solids are generally hard, rough, and brittle. Ionic compounds usually have to be heated to high temperatures in order to melt them because the attractions between ions of opposite charge are strong. It takes a lot of energy to break the well-organized network of bound ions. Compare their appearance with the properties described above. Some typical ionic compounds are shown in **Figure 4.18**.

Figure 4.18
Comparing Ionic Compounds
Here are some other typical ionic compounds. Note that all are solids at room temperature and are soluble in water. Not all ionic solids are soluble in water, though.

◀ Copper(II) sulfate ($CuSO_4$) is sometimes used to treat the growth of algae in swimming pools and in water-treatment plants.

Potassium chloride (KCl) is used as a salt substitute because it has a taste similar to that of salt. ▼

◀ Sodium hydrogen carbonate ($NaHCO_3$), commonly known as baking soda, is also called sodium bicarbonate.

143

SECTION 4.2

GLENCOE TECHNOLOGY

⊙ **CD-Rom**

Chemistry: Concepts and Applications
Using Lewis Dot Structures
Have students use this Interactive Exploration to practice using Lewis dot structures. [L1]
[LEP]

⚡ **Quick Demo**

How much carbon is in a bowl of sugar?

CAUTION: *Do this demo only in a fume hood or in a well-ventilated room. Wear rubber gloves, safety goggles, and an apron when working with the concentrated sulfuric acid, which is corrosive, and when handling the black carbonaceous residue.* To illustrate that sugar contains a large amount of carbon, put about 50 g of table sugar in a 150-mL Pyrex or Kimax beaker and set it on a thick mat of paper towels. Add about 50 mL of concentrated (18*M*) sulfuric acid, and quickly stir the mixture with a glass stirring rod. Try to get the sugar completely mixed with the acid before the reaction accelerates (when the mixture has turned a distinct yellow). The color of the mixture will change from white to yellow to black as the sugar is dehydrated and burned, and the carbon residue will foam and rise from the beaker. **CAUTION:** *Do not touch the beaker immediately. It will be hot. Avoid the steam that will rise from the mixture. To dispose of the residue, thoroughly wash it with water and wrap it in paper before discarding in the trash.*

---

mastery of the concept that compounds are electrically neutral. [L1] [LEP]
Results: Model familiar compounds such as NaCl, H_2O, $AlCl_3$, NH_3, MgO, Al_2O_3, and $CaCl_2$.
Analysis: Ask these questions.
1. An ion with a 2+ charge will react with how many negative ions that have a 1– charge? *two*

2. What will be the overall charge on this compound? *0*

✓ **Assessment**

Performance: Ask teams of students to use the cards to model a particular compound whose formula you give them. [L1] [LEP] [COOP LEARN]

Concept Development

Depending on the background of students, you may want to introduce some simple concepts of kinetic theory. As energy in the form of heat is added to matter, its temperature usually rises because the particles of matter, which are in constant motion, move even faster. For matter in the solid state, the motion becomes fast enough to overcome interparticle forces, causing the organization of the solid to collapse to a liquid. With further heating, the particles move faster and faster until interparticle forces are overcome completely and the particles separate to form a gas.

Using an Analogy

Tell students this easy way to remember that interparticle attractions are *between* atoms, ions, or molecules. *Inter*scholastic sports are *between* schools, whereas *intra*mural sports are *within* a school.

Content Background

Compound Variety It is estimated that approximately 10^5 new compounds are discovered every year. These new compounds are synthesized in laboratories or are found in nature. In order for a substance to be classified as a new compound, it must have a unique set of physical and chemical properties. In addition, an elemental and structural analysis must support the fact that the compound is unique. The majority of new compounds discovered are organic compounds.

Another physical property of ionic compounds is their tendency to dissolve in water. When they dissolve in water, the solution conducts electricity, as you saw in **Figure 4.2.** Ionic compounds also conduct electricity in the liquid (melted) state. Any compound that conducts electricity when melted or dissolved in water is an **electrolyte.** Therefore, ionic compounds are electrolytes. In order to conduct electricity, ions must be free to move because they must take on or give up electrons. Ionic compounds in the solid state do not conduct electricity because the ions are locked into position. Ionic compounds become good conductors when they melt. This is evidence that the ions are less bound and free to move in the liquid state.

Explaining the Properties of Covalent Compounds

As with ionic compounds, the submicroscopic model of the formation of covalent compounds explains many of the properties of these compounds. In particular, you can use this model to explain why typical covalent compounds such as H_2O and CO_2 have properties so different from ionic compounds.

In order to explain these differences, consider the submicroscopic organization of covalent compounds. Covalent compounds are composed of molecules. As you learned in this chapter, the atoms that compose molecules are held together by strong forces—covalent bonds—that make the molecule a stable unit. The molecules themselves have no ionic charge, so the attractive forces between molecules are usually weak.

The forces between particles that make up a substance are called **interparticle forces.** These forces are illustrated in **Figure 4.19.** It is the great difference in the strength of interparticle forces in covalent compounds compared to those of ionic compounds that explains many of the differences in their physical properties.

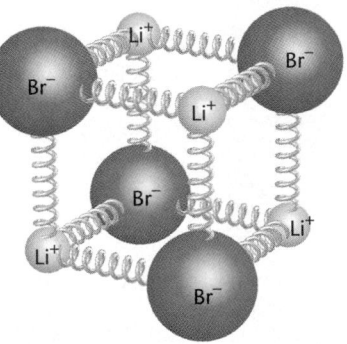

Figure 4.19
Comparing Interparticle Forces in Ionic and Covalent Compounds
In an ionic compound such as lithium bromide (LiBr), interparticle forces are strong because of the attraction between ions of opposite charges. ▶

◀ Butane (C_4H_{10}) is the fuel commonly found in plastic, disposable lighters. Butane is a covalent compound, but the molecules have no electrical charge, so the attraction between them is weak. In fact, if the butane were not held under pressure in the lighter, it would immediately boil away to a gas.

144   Chapter 4   Formation of Compounds

Gifted Ask an interested student to construct a model of an ice crystal from colored Styrofoam balls and toothpicks. Have him or her use the model to show that the intermolecular bonds are sufficiently strong to hold the water molecules apart, resulting in a lowered density. Keep the model on hand for use with Chapters 9, 10, and 13. **L2**

Physics The discovery and manipulation of electricity in the early part of the 19th century led to the discovery and isolation of many new elements by electrolysis of their compounds. Have students research and describe how metallic elements such as sodium, potassium, aluminum, and copper are obtained by electrolysis of metal-containing compounds. **L1**

Whereas all ionic compounds are solids at room temperature, many covalent compounds are liquids or gases at room temperature. Note, however, that many covalent compounds—sugar, for example—will form crystals if there is enough attractive force between molecules. In Chapter 9, you will learn why some molecules attract each other. Many of the covalent compounds that are solids at room temperature will melt at low temperatures. Examples are sugar and the compounds that make up candle wax and fat. Compare the properties of some covalent substances in **Figure 4.20.** Molecular (covalent) compounds do not conduct electricity in the pure state. Many covalent compounds, such as those in gasoline and vegetable oil, do not dissolve in water, although others such as sugar will dissolve. The solubility of covalent compounds in water varies, but in general, covalent compounds are usually less soluble in water than ionic compounds. What accounts for these differences?

Figure 4.20

Comparing Covalent Compounds
Covalent compounds are composed of molecules in which atoms are bonded by electron sharing. Because of weak interparticle forces between molecules, covalent compounds tend to be gases or liquids at room temperature. They tend to be insoluble in water, although many are extremely soluble.

Table sugar ($C_{12}H_{22}O_{11}$) is called sucrose. It is an example of a covalent compound that is a crystalline solid soluble in water. ▶

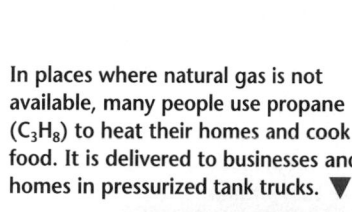

Gasoline and oil are mixtures of covalent compounds. Spilled oil does not dissolve in water, but instead floats on the water in thin layers. ▼

In places where natural gas is not available, many people use propane (C_3H_8) to heat their homes and cook food. It is delivered to businesses and homes in pressurized tank trucks. ▼

▲ Candle wax and butter are mixtures of covalent compounds. Because their molecules are larger and heavier, they form solids but melt at low temperatures.

4.2 How Elements Form Compounds   145

Integrating THE Sciences

Earth Science Limestone, a natural deposit composed primarily of calcium carbonate, $CaCO_3$, is the most widely used building stone in the United States. The strength of limestone as a building material is due to the ordered arrangement of ions in its crystal structure. Marble, another natural material, has the same composition as limestone, but it has a slightly different arrangement of ions because of the high temperature and pressure under which it crystallized. This slightly different crystal structure gives marble properties different from those of limestone.

3 ASSESS

Check for Understanding
- Check for understanding of the section material by having students answer the Section Review questions.
- Ask two students, each holding a poster with the H^+ symbol close to his or her body and a poster with the e^- symbol in an outstretched hand, to slowly walk together to illustrate the formation of the covalent bond when the electrons are midway between the two nuclei. Ask students what prevents the two hydrogen atoms from completely merging to form a helium atom. *The electrons repel each other, and the nuclei repel each other.* [L1] [LEP]

Reteach
Journal: Ask students to list in their journals some materials that they encounter daily that contain ionic compounds. [L1]

Extension
Accompany the activity described in *Check for Understanding* with a potential energy diagram, drawn as the two atoms approach and showing the minimum potential energy at the most stable bonding position.

GLENCOE TECHNOLOGY

 **Videotape**

MindJogger Videoquizzes for Chemistry
Chapter 4: Formation of Compounds
Have students work in groups as they play the videoquiz game to review key chapter concepts. [L1] [LEP] [COOP LEARN]

Purpose
Students will learn that the biodiversity of rain forests is being investigated as a source of new drugs.

Background
The evolution of the rain forests has led to myriad plants, animals, and other organisms, each occupying a niche within the system. Such niches indicate that the organism is specialized in structure, function, and behavior. The biochemistry of substances produced by these organisms is being studied by researchers in hope of isolating useful drugs from the richness of the rain forests.

Teaching Strategies
- Point out that substances derived from plants are often large, organic molecules. For example, quinine, a drug used to treat the symptoms of malaria, is derived from a rain forest plant called *Cinchona officinalis*. Quinine has the molecular formula $C_{20}H_{24}N_2O_2$.
- Have students investigate the sources and uses of these drugs: vincristine, tubocuranine, physostigmine, and reserpine. *rosy periwinkle, cancer; jungle vines, surgical anesthetic; calabar bean, eye disease; Indian snakeroot, lowers blood pressure. All are rain forest plants.*

Analyzing the Issue
1. The plant may have a toxin or substance that is biologically active and so may have some pharmaceutical use.
2. The economic potential of rain forests as areas in which to grow and harvest plants that yield pharmaceuticals would lead to their preservation.

Chemistry and
SOCIETY

The Rain Forest Pharmacy

Long ago, Samoan healers dispensed a tea brewed from the bark of a native rain forest tree, *Homalanthus nutans*, to help victims of yellow fever, a viral disease. Researchers have since identified the bark's active ingredient, prostratin. Prostratin is now being investigated for possible use as a drug for other viral diseases.

Using plants and products from plants for medicinal purposes has a long history. Such common drugs as aspirin, codeine, and quinine were originally derived from plants. However, only about 0.5 percent of all plant species have been studied intensively for chemical makeup and medicinal benefit. Because most of the world's 250 000 species of flowering plants live in rain forests, researchers are now combing these areas of dense vegetation for new substances to fight diseases.

Screening plants One method of collecting plants for drug research is to select many different species of plants from one locale and test them for possible medicinal uses. Because of the great diversity of plants, researchers hope to find an even greater diversity of chemical substances. Agencies such as the National Cancer Institute routinely use this screening method. Another method of screening plants for research is the phylogenetic survey, in which researchers study close relatives of plants already known to produce beneficial medicinal substances. Researchers hypothesize that similarities in the evolution of these plants may have produced similar properties in the biochemical substances these plants produce.

Ethnobotanical approach Interest is growing in another method of screening plants for drug research. This method is based on the work of ethnobotanists. Ethnobotanists study the medicinal uses of plants by native peoples. Ethnobotanists have learned that the sophisticated knowledge concerning the use of medicinal plants possessed by these people can be used to identify plants that may

be important for future research. The discovery of prostratin is an example of the ethnobotanical approach to screening plants for potential use in drug research.

The future Isolating and testing of substances from plants may take years. Within this time, rain forests may be significantly altered. Most rain forests are found in developing countries where the citizens want to work for better lives and grow enough food to be self-sufficient. Because of deforestation to obtain lumber as well as space for open-field agriculture, rain forest environments are disappearing. Some researchers and environmental scientists are setting up

preserves within the rain forests to maintain the forests' biodiversity. Others are attempting to grow rapidly disappearing plant species in rain forest nurseries. With these methods, scientists are trying to maintain the complex rain forest ecosystems for future research.

Analyzing the Issue

1. **Inferring** Why might drug researchers investigate a plant that has few pests as a potentially useful plant?
2. **Acquiring Information** How might research into the pharmaceutical value of rain forest substances have an impact on the present and future uses of rain forests?

Interparticle Forces Make the Difference

Interparticle forces are the key to determining the state of matter of a substance at room temperature. You already know that ions are held rigidly in the solid state by the strong forces between them. Because there are much lower interparticle forces in covalent compounds, their molecules are less tightly held to one another. Therefore, they are more likely to be gases or liquids at room temperature.

Because there are no ions in covalent compounds, you do not expect them to be electrical conductors. Ionic compounds tend to be soluble in water while molecular compounds do not. This difference is also explained by interparticle forces. Ions are attracted by water molecules, but many covalent molecules are not and, therefore, do not dissolve. Solubility in water and the nature of water solutions is a major topic in chemistry. You will learn more about solutions in Chapter 13.

Connecting Ideas

Now that you know that two major types of compounds exist, you may wonder where else you come into contact with these compounds on a day-to-day basis. In Chapter 5, you'll look at more examples of ionic and covalent compounds, including compounds more complex than the simple ones used as examples in this chapter. You'll also learn the important practical skill of naming and writing the formulas of compounds, as well as how to identify a few special categories of compounds such as acids, bases, and organic compounds.

SECTION REVIEW

Understanding Concepts

1. Explain why sodium chloride is a neutral compound even though it is made up of ions that have positive and negative charges.

2. Describe two ways in which atoms become stable by combining with each other. Compare and contrast the compounds that result from two kinds of combinations.

3. Why do you think that sodium chloride has to be heated to 800°C before melting, but candle wax will start to melt at 50°C?

Thinking Critically

4. **Making Predictions** Look at the following diagram and explain how you can determine whether the compound being formed is ionic or covalent. Do you think it is more likely to be a solid or a gas at room temperature? Explain.

$$Ca: \, + \, \cdot \ddot{B}r: \, + \, \cdot \ddot{B}r: \, \rightarrow \, [Ca]^{2+} + [:\ddot{B}r:]^- + [:\ddot{B}r:]^-$$

Applying Chemistry

5. **Using Lewis Structures** Potassium metal will react with sulfur to form an ionic compound. Use the periodic table to determine the number of valence electrons for each element. Draw a Lewis dot structure to show how they would combine to form ions. How would you write the formula for the resulting compound?

4.2 **How Elements Form Compounds** 147

SECTION 4.2

4 CLOSE

⚡ Quick Demo

Can gases form ions?
Set glass-stoppered bottles of concentrated HCl solution and concentrated NH_3 solution near each other at the front of the table. Open both bottles simultaneously, and allow them to remain open for a few moments until the fumes from each react to form a cloud of NH_4Cl. You can dramatize the reaction by darkening the room and using a halogen flashlight or similar light source directed toward the spot where the vapors will meet. Write the equation for the reaction on the chalkboard. Ask students to help you write Lewis dot structures for the reactants and the ionic products. Guide students to realize that the H^+ ion from the HCl bonds to nitrogen by using nitrogen's unshared pair of electrons. Ask them to explain why NH_4^+ has a positive ionic charge. Point out that students will learn about compounds containing more complex ions in the next chapter. **L1**

$$H:\overset{\times\times}{\underset{H}{N}}:H + H:\ddot{Cl}: \rightarrow$$

$$\left[H:\overset{H}{\underset{H}{N}}:H \right]^+ + [:\ddot{Cl}:]^-$$

SECTION REVIEW

1. Na^+ and Cl^- ions are in a 1:1 ratio; the compound is neutral.
2. An atom can become stable by gaining or losing electrons or by sharing electrons with other atoms so that each atom has a stable outer energy level.
3. There are strong forces between the ions of NaCl, so a high temperature is needed to break up the crystal. Candle wax is a mixture of covalent compounds, which consist of neutral molecules. The attraction between molecules is usually much less than that between ions; therefore, a high temperature is not needed to cause the molecules to change to a liquid state.
4. The compound $CaBr_2$ is ionic because it is formed by the transfer of electrons from Ca to Br to form one charged Ca^{2+} ion and two Br^- ions. Because of the strong attraction between ions, the compound is likely to be a solid at room temperature.
5. Potassium has one valence electron and sulfur has six. Potassium reacts with sulfur to form the compound K_2S.

$$K\cdot + K\cdot + \cdot\ddot{S}: \rightarrow [K]^+ + [K]^+ + [:\ddot{S}:]^{2-}$$

147

- Review the Reviewing Main Ideas statements and Key Terms with your students.
- Complete solutions to Chapter Review problems can be found in the *Problems and Solutions Manual*.

UNDERSTANDING CONCEPTS

1. Sodium and chlorine are highly reactive elements that react violently with each other and with other substances. At room temperature, chlorine is a poisonous, green gas, and sodium is a soft, silvery metal. In contrast, sodium chloride is a stable, unreactive, white, crystalline solid that is safe enough to use as a food seasoning.

2. CO_2 is one of the essential raw materials for photosynthesis.

3. Water is the other essential raw material for photosynthesis. In addition, water is a component of all living things. Nearly all biological processes take place in water.

4. because chemists have been able to form compounds of xenon and krypton

5. **a.** chlorine **e.** hydrogen
   b. hydrogen **f.** sodium
   c. sodium **g.** carbon
   d. oxygen

6. **a.** H_2O **e.** H_2O
   b. H_2O **f.** CO_2
   c. NaCl **g.** NaCl
   d. CO_2

7. Atoms are stable with an outer energy level of eight electrons. Helium has only one energy level, which can contain only two electrons. Therefore, it is stable with two electrons.

8. Atoms can form compounds by transferring electrons from one atom to another, resulting in positively and negatively charged ions. This process forms ionic bonds. Atoms can also react by sharing electrons. The attraction of each atom for the shared electrons is a covalent bond.

REVIEWING MAIN IDEAS

4.1 The Variety of Compounds

- When compounds form, they have properties that differ greatly from the properties of the elements of which they are made.

- Sodium, a dangerously reactive metal, reacts with chlorine, a poisonous gas, to form the stable compound sodium chloride, table salt.

- Carbon, usually found as a black solid, reacts with oxygen in the air to form carbon dioxide, an unreactive gas that produces the fizz in soda pop.

- Hydrogen, the lightest gaseous element, reacts explosively with oxygen in the air to form water, a stable compound on which all life depends.

- The properties of compounds differ widely because of differences in what happens to their constituent atoms when they form.

4.2 How Elements Form Compounds

- Atoms become stable by reacting to achieve the outer-level electron structure of a noble gas (Group 18).

- One way to achieve a stable electron structure is to transfer electrons from one atom to another, thus forming charged ions of opposite charge. The ions attract each other to form a crystal. A compound formed in this way is an ionic compound.

- Another way that atoms can achieve stability is by sharing electrons with other atoms to form molecules. A compound formed by electron sharing is a covalent compound, also called a molecular compound.

- Two atoms can share more than one pair of electrons.

- In ionic compounds, the interparticle forces are attractions between ions of opposite electrical charge. These forces are much stronger than the interparticle forces between molecules of covalent compounds.

- It is the differences in strength of interparticle forces that account for many of the differences in physical properties between ionic and covalent compounds.

Key Terms

For each of the following terms, write a sentence that shows your understanding of its meaning.

covalent bond	ionic bond
covalent compound	ionic compound
crystal	molecule
electrolyte	noble gas configuration
interparticle force	octet rule
ion	

⬤ UNDERSTANDING CONCEPTS

1. How is the compound sodium chloride different from the elements of which it is composed?

2. If you tried to breathe CO_2, you would suffocate. Why, then, is CO_2 essential to all life on Earth?

3. Why is water essential to life on Earth?

4. Why are Group 18 elements no longer called *inert*?

5. For each of the following descriptions, tell which element discussed in Section 4.1 fits best.

   a) a greenish gas
   b) a gas that burns in air
   c) a metal
   d) combines with iron to form rust
   e) the lightest element
   f) reacts violently with water
   g) forms compounds that are the subject of organic chemistry

---

✔ Assessment

Portfolio: Review the portfolio options that are provided throughout the chapter. Encourage students to select one product that demonstrates their best work for the chapter. Have students explain what they have learned and why they chose this example for placement into their portfolios. **P**

Additional portfolio options can be found in the following **Teacher Classroom Resources:**
Tech Prep Applications
Consumer Chemistry
Chemistry and Industry

6. The compounds NaCl, H_2O, and CO_2 were discussed in Section 4.1. For each of the following descriptions, tell which compound fits best.

   a) makes up 70 percent of Earth's surface
   b) important in forming solutions
   c) formed from a metal and a nonmetal
   d) a dense gas at room temperature
   e) can commonly be found in all three states of matter
   f) given off when charcoal burns
   g) composed of brittle crystals

7. Describe the electron arrangement that makes an atom stable. Why is helium stable with a different arrangement?

8. Describe two processes by which elements can combine to form stable compounds. Name the type of bonding that results from each process.

9. Compare the formation of H_2O from hydrogen and oxygen with the formation of CO_2 from carbon and oxygen. In what ways are they similar? In what ways do they differ?

10. Compare the formation of NaCl from sodium and chlorine to the formation of CO_2 from carbon and oxygen. In what ways are they similar? In what ways do they differ?

11. How is a sodium ion different from a sodium atom? From a neon atom?

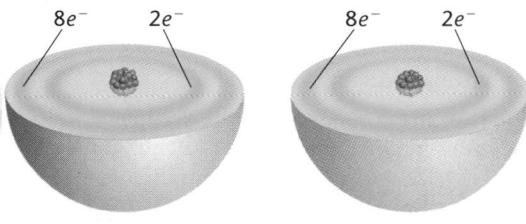

Na$^+$ ion   Ne atom

12. What are three general properties of ionic compounds? How are these properties related to the submicroscopic structure of the compounds?

13. What are three general properties of covalent compounds? How are these properties related to the submicroscopic structure of the compounds?

14. How do the basic particles of ionic and covalent compounds differ?

15. Why would you never expect to find Na_2Cl as a stable compound?

16. An unknown compound dissolves in water, but the solution does not conduct electricity. Do you think the compound is more likely to be an ionic or a covalent compound? Explain.

⬤ APPLYING CONCEPTS

17. Baking soda has the formula $NaHCO_3$. How are the properties of this compound different from its elements? Do you think it is an ionic or a covalent compound? Explain.

18. A bag of pretzels lists the sodium content of the pretzels. Considering that sodium reacts violently with water, why don't you explode when you eat these sodium-containing pretzels? What does the label really mean to a chemist?

19. Hydrazine is a compound with the formula N_2H_4. What kind of compound is hydrazine? Describe the formation of hydrazine from nitrogen and hydrogen atoms.

20. Classify each of the following compounds as ionic or covalent. Use the periodic table to decide whether an element is a metal or nonmetal.

   a) nitrogen dioxide   **d)** lithium oxide
   b) aluminum chloride   **e)** carbon disulfide
   c) zinc fluoride   **f)** hydrogen bromide

21. What does it mean to say that a chemical reaction is a rearrangement of matter?

22. Calcium chloride is an ionic compound with the formula $CaCl_2$. Benzene is a covalent compound with the formula C_6H_6. From this information, give a general description of the submicroscopic structure of each of these compounds. Describe the particles of which each compound is composed.

23. When two atoms collide, what determines whether they will react by transferring electrons or by sharing electrons?

CHAPTER REVIEW

9. Under normal circumstances, the reaction of hydrogen and oxygen gases is explosive and yields liquid water (once the product has cooled back to room temperature). The reaction of solid carbon, such as charcoal, with oxygen requires a high temperature to begin. The product (CO_2) is a gas at room temperature. Both H_2O and CO_2 are molecules held together by covalent bonds, which are formed by electron sharing.

10. See the description of CO_2 formation in the answer to question 9. NaCl is formed by a violent reaction between a metal and a nonmetal to form an ionic solid.

11. In forming a sodium ion, a sodium atom loses its single valence electron. This reaction gives the ion a net 1+ charge because it has 11 protons and only ten electrons. The Na$^+$ ion is stable because its outer energy level becomes the second energy level, which has eight electrons. This is the same electron arrangement as neon, but neon has ten protons and ten electrons.

12. Ionic compounds are usually solids at room temperature, they have high melting points, and their crystals are usually hard and brittle. These properties may be accounted for by the fact that the attraction between ions causes them to be strongly bonded and in highly organized arrangements.

13. Covalent substances tend to be gases or liquids at room temperature. Those that are solids usually melt at lower temperatures than do ionic substances. Covalent compounds consist of molecules. The atoms within each molecule may be strongly bonded by covalent bonds. However, the

molecules are neutral, and the attractions between them are generally much weaker than the attractions between oppositely charged ions.

14. Ionic compounds consist of a collection of ions of opposite charge usually arranged in well-organized crystal structures. Covalent compounds consist of molecules, which have no ionic charge.

15. Sodium atoms can achieve a stable noble-gas configuration by losing one electron. Chlorine atoms can become stable by gaining one electron. The transfer of one electron from one sodium to one chlorine atom makes both atoms stable. Therefore, they combine only in a 1:1 ratio.

16. The compound is more likely to be covalent. In solution, an ionic compound would consist of mobile ions that would conduct a current.

APPLYING CONCEPTS

17. Baking soda is a white, powdery, crystalline solid. It is safe to use in food and is sometimes ingested (perhaps unadvisedly) as an antacid. Its constituent elements consist of a reactive metal (Na), a black solid (C, at least in its most commonly observed forms), and two reactive gases (H_2 and O_2). Students should suggest that it is an ionic compound because of its physical properties and the fact that baking soda consists of a metal and nonmetals in combination.

18. The sodium listed on the label is not in the form of elemental sodium. It actually consists of sodium ions in the form of NaCl.

19. Hydrazine is a covalent compound based on the fact that it is composed of two nonmetals. Its elements would react with each other by sharing electrons.

20. **a.** covalent **d.** ionic
    b. ionic **e.** covalent
    c. ionic **f.** covalent

21. In a chemical reaction, no atoms of matter are gained or lost; they are merely rearranged to form different substances.

22. $CaCl_2$ is an ionic compound because it is composed of a metal and a nonmetal. It is composed of one Ca^{2+} ion for every two Cl^- ions in a crystal arrangement. Benzene is covalent because it contains two nonmetals. It consists of molecules.

23. whether one of the atoms has the ability to take electrons away from the other

24. A sulfide ion has the same electron structure as an argon atom.

$$\cdot \ddot{S} : \qquad \text{atom}$$

$$\left[: \ddot{S} : \right]^{2-} \qquad \text{ion}$$

25. Students should predict that it has a high melting point because it is probably an ionic compound considering that it is composed of a metal and nonmetals.

26. The presence of CO bound to iron prevents oxygen from binding. Therefore,

24. When sulfur reacts with metals, it often forms an ionic compound. Draw a Lewis dot structure of a sulfur atom. Then, draw the Lewis structure of the ion it will form. Name an element that has the same outer-level electron structure as a sulfur ion.

25. Calcium sulfate, $CaSO_4$, the main ingredient in plaster and sheets of drywall, is an ionic compound. Would you predict that this compound has a high or low melting point? How do you account for your prediction?

Everyday Chemistry

26. One of the reasons that carbon monoxide gas (CO) is poisonous is that it binds strongly to the iron atom of hemoglobin in blood. Explain how this action could cause harm to the body.

History Connection

27. Do you think a rocket powered by hydrogen and oxygen causes a great deal of atmospheric pollution? Explain.

Chemistry and Society

28. List two economic factors that contribute to the loss of rain forests.

● THINKING CRITICALLY

Applying Concepts

29. Several times in this chapter, it was stated that elements combine to form compounds. Considering what you learned in Section 4.2, what does the word *combine* mean in chemistry?

Interpreting Chemical Structures

30. **MiniLab 2** Which of these compounds could the following "tack model" represent: magnesium chloride, potassium sulfide, calcium oxide, or aluminum bromide?

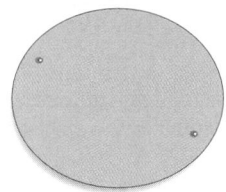 

Observing and Inferring

31. **MiniLab 1** Brass is an alloy of copper and zinc. Neither metal is magnetic. People who buy brass antiques at auctions, shows, or shops often carry a small magnet with them. What do you think they learn with this magnet?

Making Predictions

32. **ChemLab** In the electrolysis of the zinc iodide, what limits your ability to recover all of the original zinc and iodine?

Observing and Inferring

33. **ChemLab** Why does the reaction between zinc and iodine speed up when the iodine is dissolved in water?

Interpreting Data

34. **ChemLab** In the synthesis of zinc iodide, the zinc was added in excess. What does *in excess* mean? What two experimental observations did you make to confirm this?

● CUMULATIVE REVIEW

35. When carbon in charcoal burns in air to form CO_2, is the process endothermic or exothermic? How do you know? (Chapter 1)

36. Describe the particle structure of an atom of potassium. Assume an atomic mass of 39. When a potassium atom forms an ion by reacting with chlorine, how will its structure change? (Chapter 2)

37. How many energy levels are occupied by electrons in a calcium atom? A calcium ion? A bromine atom? A bromide ion? (Chapter 2)

38. What arrangement of valence electrons do carbon and other Group 14 elements have in common? (Chapter 3)

39. How does the electron arrangement of Group 18 elements relate to their chemical properties? (Chapter 3)

---

the blood cannot carry enough oxygen to supply the needs of cells, and the cells die.

27. no, because the product of the reaction is only water

28. the desire of the native population to have space to grow crops in order to be self-sufficient; the demand for types of lumber available only from rain forests

THINKING CRITICALLY

29. Atoms collide with enough energy to react. When they react, they do so by transferring electrons or by sharing electrons to achieve a stable electron configuration.

30. calcium oxide

31. Some items represented as being solid brass are in reality a thin layer of brass objects plated over iron. The magnet will be attracted to the iron core.

32. One limiting factor could be the strength of the battery but, more likely, it is the tendency of zinc and iodine to re-react.

33. When in solution, iodine molecules can contact much more of the surface area of the zinc.

34. More zinc was present than was needed to react with the iodine. All of the iodine appeared to be used up.

Skill Review

1. **Interpreting a Graph** Look at the solubility graph and answer the following questions.

   a) Which of the compounds on the graph are ionic compounds? How do you know?

   b) How many grams of KBr will dissolve in 100 g of water at 60°C?

   c) People think that salt is very soluble in water. Is that thinking accurate? Defend your answer.

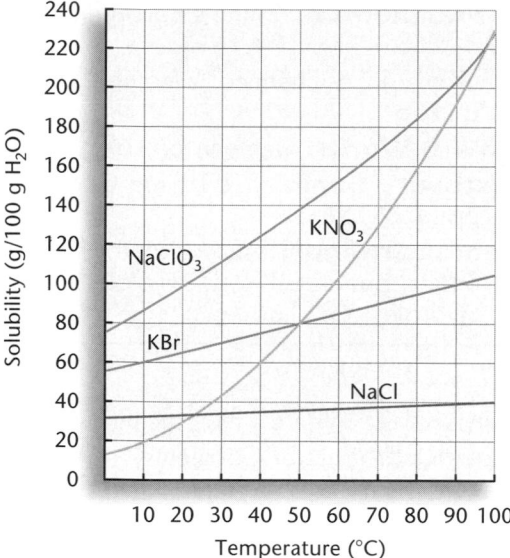

Projects

2. Choose one of the compounds in this chapter. Find a way to use models to demonstrate what happens when atoms come together to form the compound. Construct a tabletop display of 3-D models of the atoms and molecules involved and perhaps include a poster that explains what the models show. Consider showing the process by means of computer graphics. See whether you can animate your graphics to make them even more interesting. This project could be included in your portfolio.

3. Find out more about lighter-than-air craft such as blimps, dirigibles, and the like. Do not include balloons. Prepare an illustrated poster or other presentation showing two or three types of craft, how they are powered and steered, and what gases are used to fill them. Discuss how they were used in wartime and during peace.

Writing in Chemistry

4. Write an article about salt mining in the United States. Find out where the major salt deposits are located and how they came to be there. Describe what happens to the salt after it is removed from the mine. Tell how table salt is made, and find out what else it contains besides sodium chloride.

5. Write an article about the discovery of the noble gases. Find out who discovered each of them, how and when they were discovered, and where they are found on Earth. Use a chemical handbook such as the *CRC Handbook of Chemistry and Physics* to make a table of properties that includes atomic number, atomic mass, density, melting point, and boiling point. Do research to learn of practical uses for the noble gases. Point out those applications that take advantage of the lack of reactivity of the noble gases. Some compounds of the noble gases have been formed. Find out about these compounds including how and when they were produced and whether these compounds have any practical applications.

Problem Solving

6. When hydrogen and chlorine combine chemically, they form hydrogen chloride, HCl. Hydrogen chloride is a gas at room temperature; it becomes a liquid if it is cooled to −85°C. On the basis of this evidence, do you think that hydrogen chloride is ionic or covalent? Explain. Hydrogen chloride gas is extremely soluble in water. About 460 L of the gas can be dissolved in a liter of water at room temperature. In contrast, only about 1 L of CO_2 gas will dissolve in a liter of water. The water solution of HCl conducts electric current. What does this evidence tell you about the water solution of HCl? Construct a hypothesis to explain why HCl becomes a conductor of electricity when dissolved in water. Suggest a way to test your hypothesis.

Chapter 4 Review   151

CUMULATIVE REVIEW

35. The process is exothermic because heat is continuously given off after the reaction starts.

36. The atom has 19 protons, 20 neutrons, and 19 electrons. The electrons are distributed 2, 8, 8, 1 in energy levels. When reacting with chlorine, it will lose an electron and become a positively charged ion. Its electron arrangement will be 2, 8, 8.

37. Ca, four levels; Ca^{2+}, three levels; Br, four levels; Br^-, four levels

38. All have 4 electrons in the outer energy level.

39. All have eight electrons in the outer energy level except for helium. As a result, all are highly unreactive.

AUTHENTIC ASSESSMENT

1. **a.** All are ionic compounds because they are composed of metals and nonmetals. **b.** approximately 85 g **c.** NaCl is less soluble than other compounds on the graph at temperatures above about 22°C. Also, its solubility does not increase with temperature as much as the other compounds.

2. A poster can be given a 3-D treatment by using half spheres pasted onto the display board.

3. Guide students to start their research with zeppelins.

4. Have students delve into the geology of regions where salt is found and the connection between salt deposits and petroleum deposits.

5. Give this topic to a group of students, assigning one element each. Especially interesting is the discovery of helium.

6. Its properties indicate that hydrogen chloride is a covalent substance. They will learn more about this process in Chapter 13.

Program Resources

Chapter Review and Assessment, pp. 22-28
L1

Alternate Assessment in the Science Classroom L1

Computer Test Bank, Chapter 4 L1

Problems and Solutions Manual, Chapter 4
L1

Supplemental Practice Problems, Chapter 4
L1

Performance Assessment in the Science Classroom L1

Chapter 5 Organizer

Section	Objectives	Activities/Features
5.1 **Ionic Compounds** (4 days)	**1. Apply** ionic charge to writing formulas for ionic compounds. **2. Apply** formulas to name ionic compounds. **3. Interpret** the information in a chemical formula.	**Everyday Chemistry:** Hard Water, p. 160 **Art Connection:** China's Porcelain, p. 163 **MiniLab 1:** A Chemical Weather Predictor, p. 166 **How It Works:** Cement, p. 167 **Discovery Demo:** 5.1 Drying Corn, pp. 152-153 **Demonstrations:** 5.2 Variable Oxidation States, pp. 156-157; 5.3 Water in Hydrates, pp. 166-167
5.2 **Molecular Substances** (4 days)	**4. Compare** the properties of molecular and ionic substances. **5. Distinguish** among allotropes of an element. **6. Apply** formulas to name molecular compounds.	**MiniLab 2:** Where's the calcium?, p. 171 **ChemLab:** Ionic or Covalent?, pp. 172-173 **Chemistry and Technology:** Carbon Allotropes: From Soot to Diamonds, pp. 176-178 **Demonstration:** 5.4 Allotropes of Sulfur, pp. 174-175

Activity Materials

ChemLab (pages 172-173)
glass slide, grease pencil, hot plate, spatula, 4 small beakers, stirring rod, balance, conductivity tester, small graduated cylinder, thermometer, samples of salt substitute, fructose, aspirin, paraffin, urea, table salt, table sugar, Epsom salts

MiniLab 1 (page 166)
small beaker, 95% ethanol, spatula, cobalt(II) chloride, cotton swab, stirring rod, white paper, hot plate

MiniLab 2 (page 171)
uncooked chicken bone, beaker, vinegar, watch glass, forceps, paper towel, paper clip, Bunsen burner

Demonstrations For a complete list of materials for the demonstrations in this chapter, see pages 152, 156, 166, and 174.

KEY TO TEACHING STRATEGIES

The following designations will help you decide which activities are appropriate for your students.

L1 Level 1 activities should be within the ability range of all students.

L2 Level 2 activities should be within the ability range of the average to above-average student.

L3 Level 3 activities are designed for the ability range of above-average students.

LEP LEP activities should be within the ability range of Limited English Proficiency students.

COOP LEARN Cooperative Learning activities are designed for small group work.

P These strategies represent student products that can be placed into a best-work portfolio.

Types of Compounds

Teacher Classroom Resources

STUDENT MASTERS

Study Guide: pp. 17-18
Laboratory Manual: 5.1 Making Models of Compounds, pp. 41-44; 5.2 Formulas and Oxidation Numbers, pp. 45-48
ChemLab and MiniLab Worksheets: p. 38
Critical Thinking/Problem Solving: p. 6

TEACHING AIDS

Section Focus Transparency 9
Section Focus Transparency Master: p. 9
Problem Solving Transparency 5
Problem Solving Transparency Master: pp. 9-10
Basic Concepts Transparency 13, 14
Basic Concepts Transparency Masters: pp. 25-28
Lesson Plans: p. 9

Study Guide: pp. 19-20
ChemLab and MiniLab Worksheets: pp. 35-37, 39

Section Focus Transparency 10
Section Focus Transparency Master: p. 10
Basic Concepts Transparency 15
Basic Concepts Transparency Master: pp. 29-30
Lesson Plans: p. 10

ASSESSMENT RESOURCES
Chapter Review and Assessment: pp. 29-34
Alternate Assessment in the Science Classroom
Computer Test Bank: Chapter 5
Performance Assessment in the Science Classroom
Problems and Solutions Manual: Chapter 5
Supplemental Practice Problems: Chapter 5

CHAPTER RESOURCES
Applying Scientific Methods in Chemistry: p. 22
Spanish Resources Chapter 5
English/Spanish Audiocassettes: Chapter 5
Cooperative Learning in the Science Classroom
Lab and Safety Skills in the Science Classroom

GLENCOE TECHNOLOGY

Software
Mastering Concepts in Chemistry: *Unit 4* Lesson 1 Symbols and Formulas; Lesson 2 Naming Compounds
Videotape
MindJogger Videoquizzes, Chapter 5
Videodisc
Chemistry: Concepts and Applications, Demonstrations, Videos, and Animations
Science and Technology Videodisc Series: Disc 2 Chemistry *Sea Urchins and Power Plants, Uses for Super Slurper, Hole in the Ozone, Composite Materials*
CD-ROM Multimedia System
Chemistry: Concepts and Applications, same as for Videodisc, plus Interactive Exploration and Major Simulation, *The Periodic Table*

TECH PREP

The following Glencoe resources provide opportunities for integrating science and technology.

Student Edition: Everyday Chemistry, p. 160; Art Connection, p. 163; MiniLabs 1 and 2, pp. 166, 171; How It Works, p. 167; Chemistry and Technology, p. 176

Teacher Wraparound Edition: Chemistry Journal, p. 160; Applying Chemistry, p. 161

Teacher Classroom Resources: Tech Prep Applications, pp. 7-8; Chemistry and Industry, pp. 9-12; Consumer Chemistry, pp. 9-10

CHAPTER

5 Types of Compounds

Chapter Overview

Students learn how the electron configuration of an element determines the types of compounds it forms. The rules for writing formulas and names of both ionic and molecular binary compounds are presented. Polyatomic ionic compounds, transition metal compounds, hydrates, and simple hydrocarbons also are named.

Theme Connection

Macro to Submicro The theme of this chapter is how the formula and name of a compound represent the submicroscopic structure of the compound.

GLENCOE TECHNOLOGY

Videodisc

Chemistry: Concepts and Applications
See the Videodisc Program Teacher Guide to access illustrations and other material available for this chapter.

Assessment Planner

Choose assessment strategies from these pages to evaluate the progress of your students.
Assess, pp. 156, 158, 159, 162, 168, 170, 175, 180, 182
MiniLabs, pp. 166, 171
ChemLab, p. 172
Portfolio, pp. 158, 168, 169, 170, 175, 180
Chapter Review, p. 184

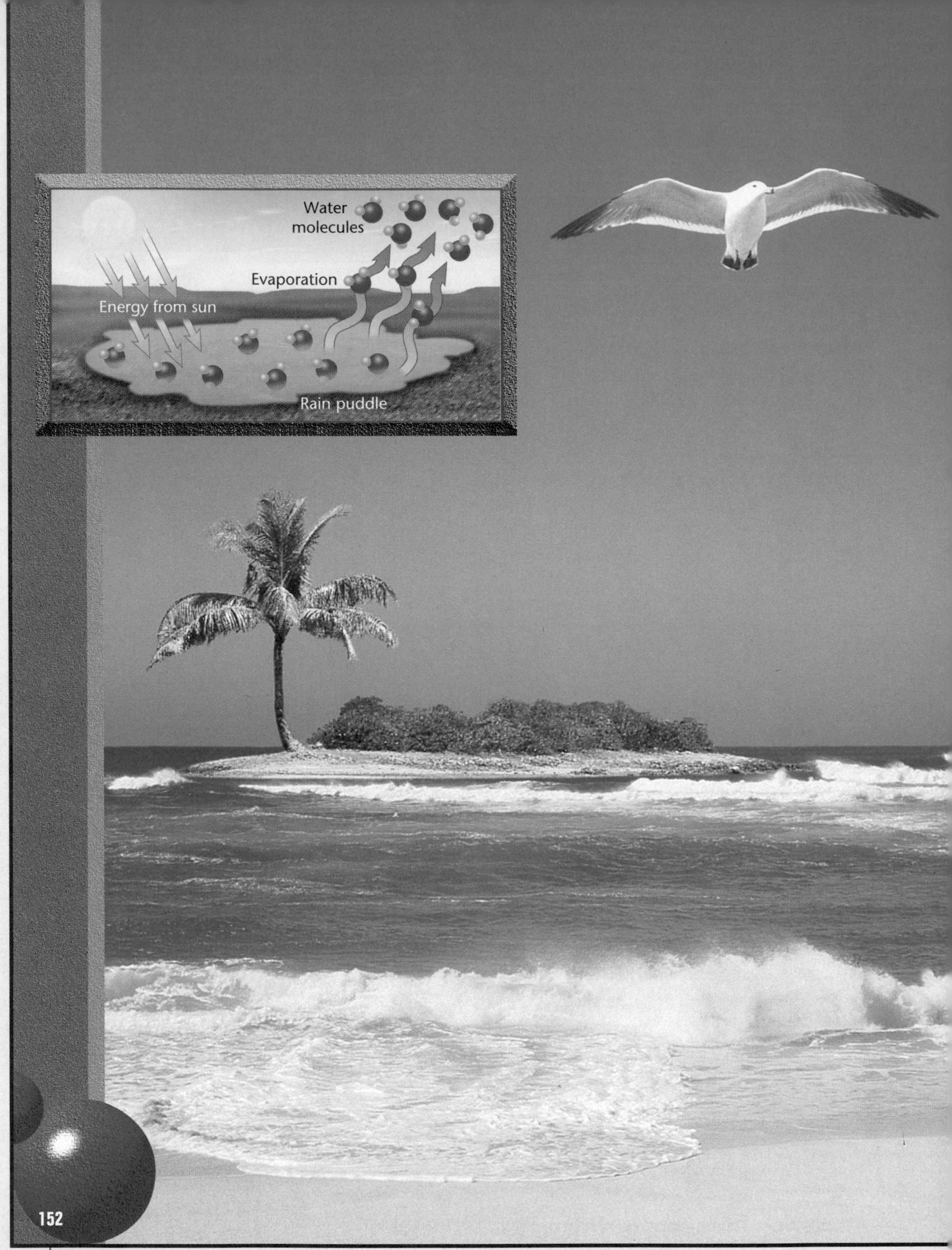
Water molecules
Evaporation
Energy from sun
Rain puddle
152

Discovery Demo 5.1

Drying Corn

Purpose: This analogy shows that a solid hydrate may lose water when heated.
Materials: 30 kernels of popcorn, spray cooking oil, 600-mL beaker, wire screen beaker cover, hot plate or burner, balance
Safety Precautions:
Disposal: Code A

Procedure: Lightly spray the inside of the beaker with cooking oil. Measure the mass of the beaker. Place the popcorn in the beaker and weigh it again. Cover the beaker with a wire screen, and heat it on a hot plate or burner until the kernels have popped. Be careful not to burn the popped kernels. Allow the beaker to cool. Remove the wire screen and weigh the beaker and its contents. **CAUTION:** *Do not allow students to eat anything prepared in the laboratory.*

Imagine being stranded on a deserted island that contains a few palm trees and sandy beaches, and is surrounded by an endless expanse of ocean. You can get food from a diet of fish and coconuts, but where will you get fresh water? There is plenty of seawater, but you can't survive on such salty water. Is there some way to separate the water from all that salt?

Which properties might allow you to separate the two compounds? Which compound is ionic? Which is covalent? Because water—like most covalent compounds—has a fairly low boiling point, it will evaporate if it gets warm enough. You have probably seen a cloud of condensed steam rising from a rain puddle on a hot summer day. Salt is an ionic compound and has a much higher boiling point than does water, so salt won't evaporate at normal, everyday temperatures.

If you can heat seawater and collect the water vapor on a cool surface, you'll have a supply of fresh water.

Chemistry Around You

Because most of Earth's water is found in the salty oceans, the salt must be removed before this valuable resource can be used for drinking or washing. The different boiling points of salt and water allow them to be separated easily. Separation can be done on a small scale, such as in portable equipment that uses the sun's energy to evaporate the water. People going on ocean voyages in small boats take along this type of solar survival still. Communities around the world use large-scale stills in desalination plants, shown here, which provide enough fresh water to meet the needs of the population.

 Concept Check

Review the following concepts before studying this chapter.
Chapter 2: arrangement of electrons in an atom
Chapter 3: periodicity of electron arrangements in atoms; importance of valence electrons
Chapter 4: formation of ionic compounds, formation of covalent compounds

Chapter Preview

5.1 Ionic Compounds

5.2 Molecular Substances

Laboratory Activities
ChemLab
Ionic or Covalent?

MiniLabs
1. A Chemical Weather Predictor
2. Where's the calcium?

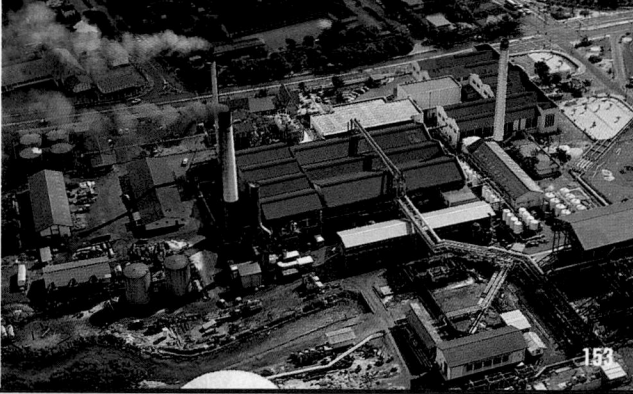

Key Concepts

The properties of ionic compounds are explained based upon the submicroscopic structure. Students then learn to write formulas and names of ionic compounds, relating the names and formulas to the ion charges.

Planning Ahead

For the MiniLab Have 95% ethanol or 91% isopropanol and cobalt(II) chloride on hand.

For the Demonstrations Purchase popcorn and spray cooking oil for the Discovery Demo.

1 FOCUS

Focus Transparency

Display the **Section Focus Transparency** for Section 5.1 to show some useful ionic compounds. [L1] [LEP]

2 TEACH

Content Background

Seawater Sodium chloride is the most abundant but not the only salt in seawater, which contains about 3.5% by weight of dissolved ions. The next four most abundant salts are magnesium chloride ($MgCl_2$), sodium sulfate (Na_2SO_4), calcium chloride ($CaCl_2$), and potassium chloride (KCl).

Ionic Compounds

SECTION PREVIEW

Objectives

Apply ionic charge to writing formulas for ionic compounds.
Apply formulas to name ionic compounds.
Interpret the information in a chemical formula.

Key Terms

binary compound
formula unit
oxidation number
polyatomic ion
hydrate
hygroscopic
deliquescent
anhydrous

Seawater contains many dissolved substances, the major one being sodium chloride. In Chapter 4, you learned that sodium chloride is an ionic compound. Another ionic compound found in seawater is magnesium chloride. Some common ionic compounds used in everyday life are potassium chloride, a salt substitute used by people avoiding sodium for health reasons; potassium iodide, added to table salt to prevent iodine deficiency; and sodium fluoride, added to many toothpastes to strengthen tooth enamel. In this section, you will learn how to use the language of chemistry to name and write the formulas of ionic compounds.

Formulas and Names of Ionic Compounds

Recall from Chapter 4 that the submicroscopic structure of ionic compounds helps explain why they share certain macroscopic properties such as high melting points, brittleness, and the ability to conduct electricity when molten or when dissolved in water. What is it about the structure of these compounds that gives them properties such as the one shown in **Figure 5.1?** The answer involves the ions of which they are made.

You have learned that ionic compounds are made up of oppositely charged ions held together strongly in well-organized units. Because of their structure, they usually are hard solids at room temperature and are difficult

Figure 5.1

Humpty Dumpty's Downfall
Eggshells are made mostly of ionic compounds such as calcium phosphate, $Ca_3(PO_4)_2$, which makes them brittle. When broken, eggshells shatter into many pieces that can't be put together again.

154   Chapter 5   Types of Compounds

Program Resources

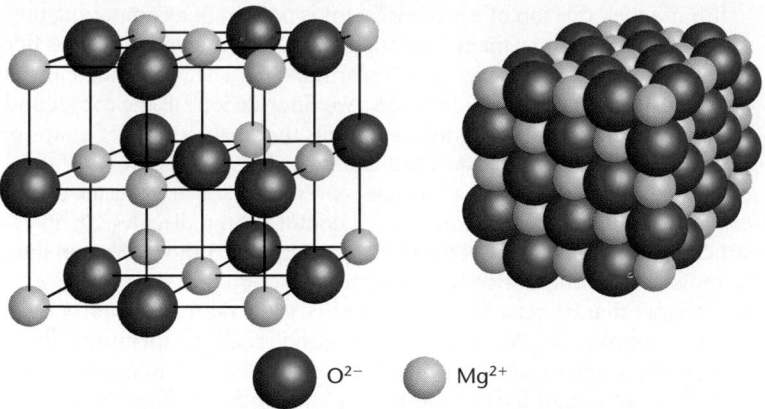

Figure 5.2
Repeating Units
The structure of magnesium oxide is a repeating pattern of magnesium and oxide ions. Each Mg^{2+} ion is surrounded by six O^{2-} ions, which, in turn, are each surrounded by six Mg^{2+} ions. The structure as a whole is neutral. In the diagram on the left, the structure is expanded so you can better see the geometric arrangement.

O^{2-} Mg^{2+}

to melt. Look at the structure of magnesium oxide in **Figure 5.2.** When ionic compounds melt or dissolve in water, their three-dimensional structure breaks apart, and the ions are released from the structure. These charged ions are now free to move and can conduct an electrical current.

Binary Ionic Compounds

Formulas are part of the language that is used to communicate information about substances. As a first step in studying this new language, you will learn how to name and write formulas for ionic compounds.

Sodium chloride (NaCl) contains only sodium and chlorine, and potassium iodide (KI) contains only potassium and iodine. Each is an example of a **binary compound,** which is a compound that contains only two elements. Binary ionic compounds can contain more than one ion of each element, as in CaF_2, but they are not composed of three or more *different* elements, as are more complex compounds.

To name a binary ionic compound, first write the name of the positively charged ion, usually a metal, and then add the name of the nonmetal or negatively charged ion, whose name has been modified to end in *-ide.* The compound formed from potassium and chlorine is called potassium chloride. Magnesium combines with oxygen to form a compound called magnesium oxide.

You are already familiar with one formula for an ionic compound—NaCl. Sodium chloride contains sodium ions that have a 1+ charge and chloride ions that have a 1− charge. You have learned that compounds are electrically neutral. This means that the sum of the charges in an ionic compound must always equal zero. Thus, one Na^+ balances one Cl^- in sodium chloride. When you write a formula, you add subscripts to the symbols for the ions until the algebraic sum of the ions' charges is zero. The smallest subscript to both ions that results in a total charge of zero is 1. However, no subscript needs to be written because it is understood that only one ion or atom of an element is present if there is no subscript. The formula NaCl indicates that sodium chloride contains sodium and chloride ions, that there is one sodium ion present for every chloride ion in the compound, and that the compound has no overall charge.

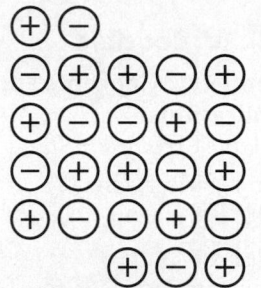

WORD ORIGIN

binary:
bini (L) two by two

Binary compounds contain two, and only two, elements.

5.1   Ionic Compounds   **155**

Assessment

Performance: Ask each student to write an explanation for the fact that calcium atoms form 2+ ions and fluorine atoms form 1− ions. *Calcium atoms have two electrons in the fourth energy level and often lose them as the atoms react. Fluorine atoms have seven electrons in the second energy level and often react by gaining one additional electron to make an octet—in this case, the stable electron arrangement of neon.* L1

GLENCOE TECHNOLOGY

Videodisc

Chemistry: Concepts and Applications
Variable Oxidation States
Disc 1, Side 1, Ch. 17

Show this video of the Demonstration on these pages to reinforce the use of Roman numerals in naming compounds of transition elements. L1 LEP

CD-Rom

Chemistry: Concepts and Applications
Variable Oxidation States
Have students use this video to help reinforce using Roman numerals in naming compounds of transition elements. L1 LEP

Fact of the MATTER

Rain falling over or near an ocean is often salty because bits of salt are picked up by the wind after bubbles of sea foam burst apart. This salt can remain suspended in the air for some time before settling. Crops raised in some coastal areas—such as the famous artichoke fields south of San Francisco—often have a distinctive and highly sought-after flavor because of the salty rainfall. Salt spread in this way also contributes to corrosion of metal objects, especially automobiles.

Figure 5.3

The Formula for Calcium Fluoride
When calcium fluoride forms from Ca and F, the two valence electrons from calcium are transferred to two fluorine atoms, leaving the Ca with a 2+ charge and each F with a 1− charge. ▶

If more than one ion of a given element is present in a compound, the subscript indicates how many ions are present. The mineral known as fluorite is calcium fluoride, which has the formula CaF_2. This formula indicates that there is one calcium ion for every two fluoride ions in the compound. In an ionic compound, a formula represents the smallest ratio of atoms or ions in the compound. In a covalent compound, the smallest unit of the compound is a molecule, so a formula represents a single molecule of a compound. However, ionic compounds do not form molecules. Their structures are repeating patterns of ions, as **Figure 5.3** shows. Should the formula of calcium fluoride be written as CaF_2, Ca_2F_4, or even Ca_3F_6? A properly written formula has the simplest possible ratio of the ions present. This simplest ratio of ions in a compound is called a **formula unit.** Each formula unit of calcium fluoride consists of one calcium ion and two fluoride ions. Each of the three ions has a stable octet configuration of electrons, and the formula unit has no overall charge. Although the sum of the ionic charges in both CaF_2 and Ca_2F_4 is zero, only CaF_2 is a correct formula. One formula unit of calcium fluoride has the formula CaF_2.

Predicting Charge on Ions

You have studied ionic compounds in which sodium becomes a positive ion with a single positive charge and calcium becomes a positive ion with two positive charges. Examine the periodic table to see if there is a way to predict the charge that different elements will have when they become ions. Which elements will lose electrons and which will gain electrons?

The noble gases each have eight electrons in their outer-energy levels. Metals have few outer-level electrons so they tend to lose them and become positive ions. Sodium must lose just one electron, becoming an Na^+ ion. Calcium must lose two electrons, becoming a Ca^{2+} ion. Most nonmetals, on the other hand, have outer-energy levels that contain four to seven electrons, so they tend to gain electrons and become negative ions. Trace the gain and loss of electrons in the example shown in **Figure 5.4.**

$$Ca: + \ddot{\underset{\times\times}{F}}\cdot + \cdot\ddot{\underset{\times\times}{F}}: \rightarrow [Ca]^{2+} + \ddot{\underset{\times\times}{F}}:^- + \ddot{\underset{\times\times}{F}}:^-$$

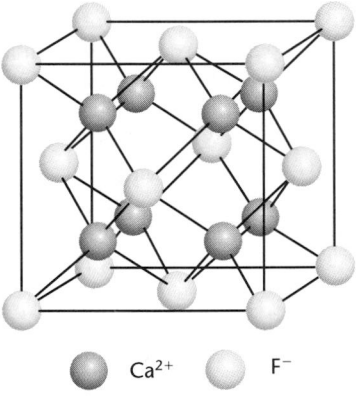

◀ An ionic bond forms between the positive Ca^{2+} ion and each negative F^- ion. Although there are many Ca^{2+} and F^- ions in a crystal of CaF_2, one formula unit of CaF_2 contains one Ca^{2+} ion and two F^- ions.

● Ca^{2+} ○ F^-

156   **Chapter 5   Types of Compounds**

Demonstration 5.2

Variable Oxidation States

Purpose: To demonstrate the importance of the use of Roman numerals in naming compounds of transition elements.
Materials: 10 test tubes, 5 droppers, 2 g $FeNH_4(SO_4)_2 \cdot 12H_2O$, 0.5 g $BaCl_2$, 0.5 g KSCN, 2 g $Fe(NH_4)_2(SO_4)_2 \cdot 6H_2O$, 0.5 g tannic acid (or strongly brewed tea), 0.5 g $NaHSO_3$, 0.5 g potassium hexacyanoferrate(II), 450 mL water

Safety Precautions: 🧤 🥽 ☠️ 🧹
Disposal: Code E
Procedure: Separately prepare solutions of iron(III) ammonium sulfate (2 g $FeNH_4(SO_4)_2 \cdot 12H_2O$ in 200 mL H_2O) and iron(II) ammonium sulfate (2 g $Fe(NH_4)_2(SO_4)_2 \cdot 6H_2O$ in 200 mL H_2O). Label five test tubes Fe^{2+} and five test tubes Fe^{3+}. Half-fill each test tube with the appropriate solution. Prepare solutions of

the following five compounds by adding 0.5 g of each compound to 10 mL H_2O: barium chloride, potassium thiocyanate, tannic acid, sodium hydrogen sulfite, and potassium hexacyanoferrate(II). **CAUTION:** *These solutions are toxic.* Use a clean dropper to add each solution to a test tube containing the Fe^{2+} ion and a tube containing the Fe^{3+} ion.

Figure 5.4
Lime Is Calcium Oxide
The compound commonly called lime is calcium oxide. It is used to make steel and cement and is added to acidic lakes and soil to neutralize the effects of acidity. ▶

Do you recognize a periodic trend in ionic charges? For the elements in the main groups of the periodic table—Groups 1, 2, and 13 through 18—group numbers can be used to predict these charges. Because all elements in a given group have the same number of electrons in their outer-energy level, they must lose or gain the same number of electrons to achieve a noble-gas electron configuration. Metals always lose electrons and nonmetals always gain electrons when they form ions. The charge on the ion is known as the **oxidation number** of the atom. The oxidation numbers for many elements in the main groups are arranged by group number in **Table 5.1.** Oxidation numbers for elements in Groups 3 through 12, the transition elements, cannot be predicted by group number.

Alumina is the common name for aluminum oxide. It is used to produce aluminum metal, to make sandpaper and other abrasives, and to separate mixtures of chemicals by a technique called chromatography. Aluminum is in Group 13, so it loses its three outer electrons to become an Al^{3+} ion; oxygen is in Group 16 and has six valence electrons, so it gains two electrons to become an O^{2-} ion.

$$\dot{A}\dot{l}\!\cdot + \ddot{:}\!\dot{O}\!\cdot \rightarrow Al^{3+} + \ddot{:}\!\ddot{O}\!\ddot{:}^{2-}$$

Notice that one of aluminum's three electrons has not been taken up by the oxygen atom. Because all the electrons must be accounted for, more than one oxygen atom must be involved in the reaction. But, oxygen cannot gain only one electron, so a second aluminum atom must be present to contribute a second electron to oxygen. In all, two Al^{3+} ions must combine with three O^{2-} ions to form Al_2O_3. Remember that the charges in the formula for aluminum oxide must add up to zero.

$$2Al^{3+} + 3O^{2-} \rightarrow Al_2O_3$$

$$\dot{A}\dot{l}\!\cdot + \dot{A}\dot{l}\!\cdot + \ddot{:}\!\dot{O}\!\cdot + \ddot{:}\!\dot{O}\!\cdot + \ddot{:}\!\dot{O}\!\cdot \rightarrow Al^{3+} + Al^{3+} + \ddot{:}\!\ddot{O}\!\ddot{:}^{2-} + \ddot{:}\!\ddot{O}\!\ddot{:}^{2-} + \ddot{:}\!\ddot{O}\!\ddot{:}^{2-}$$

$$Ca\!: \rightarrow Ca^{2+} \quad \ddot{O}\!: \rightarrow \ddot{:}\!\ddot{O}\!\ddot{:}^{2-}$$
$$Ca^{2+} + \ddot{:}\!\ddot{O}\!\ddot{:}^{2-} \rightarrow CaO$$

▲ Calcium is a metal that loses two electrons to become a Ca^{2+} ion; oxygen is a nonmetal that must gain two electrons to achieve the stable octet of the noble gas neon, so it becomes an O^{2-} ion. Because a formula unit must be neutral, one Ca^{2+} ion can combine with only one O^{2-} ion. The formula for calcium oxide is CaO.

Table 5.1 Ionic Charges of Representative Elements

Group Number	Oxidation Number	Examples
Metals		
1	1+	Li^+, Na^+, K^+
2	2+	Mg^{2+}, Ca^{2+}
13	3+	B^{3+}, Al^{3+}
Nonmetals		
15	3−	N^{3-}, P^{3-}
16	2−	O^{2-}, S^{2-}
17	1−	F^-, Cl^-, Br^-, I^-

Correcting Misconceptions

Students may refer to a formula such as CaF_2 as representing a *molecule* of calcium fluoride. Point out that the formula for an ionic compound represents only the ratio of ions present in the compound. Such formulas are often called empirical formulas or are said to represent formula units. This is different from a formula such as H_2O, which represents a molecule of water.

Correcting Errors

Students often learn to write formulas for ionic compounds by switching the absolute values of the oxidation numbers of the ions to the subscripts of the opposite ions. In calcium sulfide, for example, Ca^{2+} and S^{2-} would yield a formula of Ca_2S_2. This method is incorrect. The formula should be CaS, the simplest ratio.

GLENCOE TECHNOLOGY

☉ Software

Mastering Concepts in Chemistry
Unit 4, Lesson 1
Symbols and Formulas
Use this lesson to reinforce the use of chemical symbols when writing formulas of compounds. [L1] [LEP]

Results: The results are listed below:

Reagent	$[Fe^{2+}]$	$[Fe^{3+}]$
$BaCl_2 \cdot 2H_2O$	white	no change
KSCN	no change	deep red
tannic acid	black	blue
$NaHSO_3$	yellow	brownish-orange
$K_4Fe(CN)_6 \cdot 3H_2O$	light blue	dark blue

Ions of different oxidation states have different chemical behaviors.

Analysis: Ask these questions.
1. Do the five reagents react the same or differently with the two forms of iron? *differently*
2. Can the two different compounds that form when iron(II) and iron(III) reacted with chloride both be named iron chloride? Explain. *No, the same name would be applied to two different compounds having different properties.*

✓ Assessment

Knowledge: Ask students to become familiar with the ions listed in **Table 5.4.** Then have them use the table to name the following compounds: SnF_2 and SnF_4; Cu_2O and CuO; $NiCO_3$ and $Ni_2(CO_3)_3$. [L1]

Concept Development

Ask students to write the symbols, with Lewis electron dot notations, of the Period 3 elements. Then ask them to repeat these symbols below the original list. They should remove the valence electrons for the metallic elements because the atoms of these elements usually react by losing the valence electrons, and they should add electrons to make octets for the nonmetallic elements because the atoms of these elements usually react by gaining electrons to make the noble gas configuration. Include silicon, although it is a metalloid. Finally, ask students to write the charge or oxidation number for each of the resulting ions. *Na, 1+; Mg, 2+; Al, 3+; Si, 4+; P, 3−; S, 2−; Cl, 1−; Ar, 0.* [L1]

Assessment

Portfolio: Animals, including humans, require significant amounts of dietary calcium in order to form and maintain strong bones and teeth. Have students make a chart tracing how calcium compounds find their way into the soil and into the bodies of animals. Students can add the chart to their portfolios. [L1] [P]

Transparency

Display **Basic Concepts Transparency 14** to review oxidation numbers of groups. [L1] [LEP]

SAMPLE PROBLEM **1** **Writing a Simple Formula**

Write the formula for an ionic compound containing sodium and sulfur.

Analyze
- Sodium is in Group 1, so it has an oxidation number of 1+. Sulfur is in Group 16 and has an oxidation number of 2−.

Set Up
- Write the symbols for sodium and sulfur ions in formula form, placing the positive ion first.

$$Na^+S^{2-}$$

Solve
- The formula as written has one positive charge and two negative charges. To maintain neutrality, one more positive charge is needed to balance the 2− charge. This is accomplished by adding a second sodium ion and is indicated by placing the subscript *2* after the symbol for sodium in the formula. The correct formula is then written as Na_2S.

Check
- Check to be sure that you have not changed the charges of the ions and that the overall charge of the formula is zero.

$$2(1+) + (2-) = 0 \quad \text{The formula as written is correct.}$$

PRACTICE PROBLEMS

1. Write the formula for each of the following compounds.

   a) lithium oxide
   b) calcium bromide
   c) sodium oxide
   d) aluminum sulfide

2. Write the formula for the compound formed from each of the following pairs of elements.

   a) barium and oxygen
   b) strontium and iodine
   c) lithium and chlorine
   d) radium and chlorine

Compounds Containing Polyatomic Ions

The ions you have studied thus far have contained only one element. However, some ions contain more than one element. An ion that has two or more different elements is called a **polyatomic ion.** In a polyatomic ion, a group of atoms is covalently bonded together when the atoms share electrons. Although the individual atoms have no charge, the group as a whole has an overall charge. The formulas and names of some common polyatomic ions are shown in **Table 5.2**. Although the charge is shown to the right of the formula, it is the whole ion, rather than just the last atom listed, that is charged. **Figure 5.5** shows models of three common polyatomic ions.

158   Chapter 5   Types of Compounds

PRACTICE

Guided Practice: Guide students through the sample problem, then have them work in class on Practice Problems 1 and 2. [L1]
Independent Practice: Your homework or classroom assignment can include the additional practice problems shown below. [L1]

Answers to Practice Problems

1. a. Li_2O b. $CaBr_2$ c. Na_2O d. Al_2S_3
2. a. BaO b. SrI_2 c. $LiCl$ d. $RaCl_2$

Additional Practice Problems

1. Write the formula for each of the following compounds.
   a. cesium chloride *CsCl*
   b. calcium oxide *CaO*
   c. potassium fluoride *KF*
   d. lithium iodide *LiI*

2. List the elements that are found in each of the following compounds.
   a. magnesium sulfide *magnesium, sulfur*
   b. sodium selenide *sodium, selenium*
   c. potassium oxide *potassium, oxygen*
   d. barium bromide *barium, bromine*

Table 5.2 Common Polyatomic Ions

Name of Ion	Formula	Charge
ammonium	NH_4^+	1+
hydronium	H_3O^+	1+
hydrogen carbonate	HCO_3^-	1−
hydrogen sulfate	HSO_4^-	1−
acetate	$C_2H_3O_2^-$	1−
nitrite	NO_2^-	1−
nitrate	NO_3^-	1−
cyanide	CN^-	1−
hydroxide	OH^-	1−
dihydrogen phosphate	$H_2PO_4^-$	1−
permanganate	MnO_4^-	1−
carbonate	CO_3^{2-}	2−
sulfate	SO_4^{2-}	2−
sulfite	SO_3^{2-}	2−
oxalate	$C_2O_4^{2-}$	2−
monohydrogen phosphate	HPO_4^{2-}	2−
dichromate	$Cr_2O_7^{2-}$	2−
phosphate	PO_4^{3-}	3−

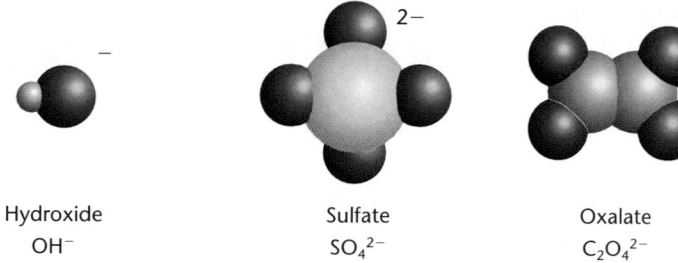

Hydroxide
OH^-

Sulfate
SO_4^{2-}

Oxalate
$C_2O_4^{2-}$

Figure 5.5
Polyatomic Ions
Polyatomic ions, such as hydroxide (left), sulfate (center), and oxalate (right), are composed of more than one atom. Electrons are shared between the atoms within the ion, forming covalent bonds, but the ion as a whole has a charge. Thus, polyatomic ions form ionic bonds with other ions to produce ionic compounds.

Ionic compounds may contain positive metal ions bonded to negative polyatomic ions, such as in NaOH; negative nonmetal ions bonded to positive polyatomic ions, such as in NH_4I; or positive polyatomic ions bonded to negative polyatomic ions, such as in NH_4NO_3. To write the formula for an ionic compound containing one or more polyatomic ions, simply treat the polyatomic ion as if it were a single-element ion by keeping it together as a unit. Remember that the sum of the positive and negative charges must equal zero.

Multiples of a polyatomic ion in a formula can be indicated by placing the entire polyatomic ion, without the charge, in parentheses. Write a subscript outside the parentheses to show the number of polyatomic ions in the compound. Never change the subscripts within the polyatomic ion. To do so would change the composition of the ion. The formula for the compound that contains one magnesium ion and two nitrate ions is $Mg(NO_3)_2$.

5.1 Ionic Compounds   159

SECTION 5.1

✔Assessment

Performance: Ask students to use Lewis electron dot symbols to depict the atoms as reactants and ions as products in the reactions between sodium and phosphorus, sodium and sulfur, sodium and chlorine, magnesium and phosphorus, magnesium and sulfur, and magnesium and chlorine. [L1]

$$3Na\cdot + \cdot\ddot{P}\colon \rightarrow 3Na^+ + \colon\ddot{P}\colon^{3-}$$
$$2Na\cdot + \cdot\ddot{S}\colon \rightarrow 2Na^+ + \colon\ddot{S}\colon^{2-}$$
$$Na\cdot + \colon\ddot{C}l\colon \rightarrow Na^+ + \colon\ddot{C}l\colon^-$$
$$3Mg\colon + 2\cdot\ddot{P}\colon \rightarrow 3Mg^{2+} + 2\colon\ddot{P}\colon^{3-}$$
$$Mg\colon + \cdot\ddot{S}\colon \rightarrow Mg^{2+} + \colon\ddot{S}\colon^{2-}$$
$$Mg\colon + 2\colon\ddot{C}l\colon \rightarrow Mg^{2+} + 2\colon\ddot{C}l\colon^-$$

Correcting Errors

Explain that it is important when writing the formula for a polyatomic ion to include the charge on the ion. SO_3^{2-}, for example, is the formula for the sulfite ion, but SO_3 is the formula for a compound. You might point out that the sulfur atoms are bonded differently in the two species and have a 4+ oxidation number in SO_3^{2-} and a 6+ oxidation number in SO_3.

Transparency

Display **Problem Solving Transparency 5** to give students practice writing formulas. [L1] [LEP]

Transparency

Display **Basic Concepts Transparency 13** showing charges of common ions. [L1] [LEP]

Integrating THE Sciences

Biology Ask whether students' fingers or toes have ever shriveled up after swimming in salt water. The shriveling is due to osmosis, which causes water to exit the cells in order to equalize the salt concentrations inside and outside the cells. Organisms that live in salty environments have special adaptations that prevent this osmosis from damaging their cells. Some bacteria can live only in concentrated salt water and are found in shallow areas of the oceans and seas where water is evaporating rapidly. These bacteria are called halophiles because halite is the mineral name for rock salt, which is large chunks of sodium chloride. These bacteria have unique cell walls that prevent water loss due to osmosis.

Purpose

TECH PREP

Students will learn how dissolved ions affect the properties of tap water.

Teaching Strategies

- Explain to students that groundwater is often hard in areas that have large deposits of limestone (calcium carbonate, $CaCO_3$) and dolomite (calcium magnesium carbonate, $CaMg(CO_3)_2$). Dissolved CO_2 in rainwater reacts with these compounds to produce soluble hydrogen carbonates of these metals, which enter the groundwater.

Extension

Ask students to design and conduct a test to determine whether the school's tap water is hard or soft. Suggest that students consider using a soap solution, tap water, and distilled water in their tests. **L1**

Exploring Further

1. The stearate ion is a polyatomic ion with a 1− charge.
2. In an ion exchanger, it takes two ions of sodium, each with a 1+ charge, to replace an ion with a 2+ charge.
3. The calcium- and magnesium-containing compounds formed with detergents are soluble in water, so they can be washed away, carrying dirt with them.

⚡ Quick Demo

Is $CaCO_3$ soluble?

Add a few drops of a $1M$ $CaCl_2$ or $Ca(NO_3)_2$ solution from a dropper to 5 mL of $1M$ Na_2CO_3 solution in a test tube to illustrate the formation of the insoluble $CaCO_3$.

Everyday Chemistry

Hard Water

The term *hard water* doesn't describe water's physical state. It describes water in which calcium, magnesium, and hydrogen carbonate ions are dissolved. It is difficult to get soap to lather in hard water.

One of the compounds in soap that helps produce lather is sodium stearate, $NaC_{18}H_{35}O_2$, which dissolves in water. In hard water, calcium ions react with the stearate ions to form calcium stearate, $Ca(C_{18}H_{35}O_2)_2$. This material is insoluble and results in the formation of soap scum. Scum is often seen as a ring around sinks or tubs. If calcium ions are removed from the water by water softeners, soap will lather and no more scum will form.

Hard water can cause serious problems. When hard water is heated, calcium carbonate is formed from the reaction of calcium ions and hydrogen carbonate ions in the water. Because calcium carbonate is not soluble in water, it forms thick scales on the interior of water heaters and water pipes. These scales often clog pipes and keep the heater from properly heating the water.

Ion Exchangers A common way of softening water—that is, reducing the number of calcium and magnesium ions—is by an ion exchanger. The ion exchanger usually contains a material, called a resin, made up of carbon, hydrogen, and sodium ions. As hard water passes through the resin, a calcium ion or magnesium ion in the water is exchanged for two sodium ions. Thus, the water leaving the ion exchanger has fewer calcium and magnesium ions but many more sodium ions. In terms of hardness, this water is softened.

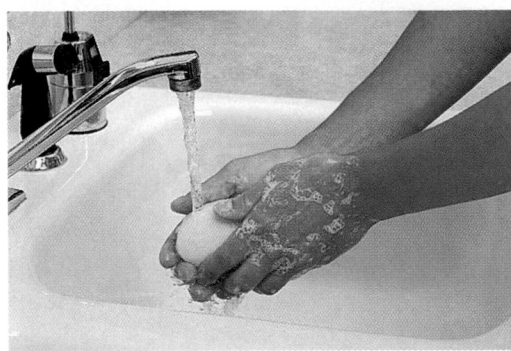

Exploring Further

1. **Interpreting** What is the charge of the stearate ion?
2. **Thinking Critically** Why do two sodium ions replace one calcium or one magnesium ion?
3. **Acquiring Information** Why are detergents more effective than soaps in hard water?

To name a compound containing a polyatomic ion, follow the same rules as used in naming binary compounds. Name the positive ion first, followed by the negative ion. However, do not change the ending of the negative polyatomic ion name. The name of the compound composed of calcium and the carbonate ion is calcium carbonate. Acids in groundwater can dissolve rocks made of calcium carbonate, such as limestone. Large, underground caverns are formed when the limestone is dissolved away slowly. Stalactites hanging from the ceiling and stalagmites rising from the ground are made when calcium carbonate precipitates from a water solution dripping through cracks in the cavern ceiling.

📓 Chemistry Journal

Water Hardness

TECH PREP

Ask students to determine what type of water they have at home. If the water is from a private well, they should determine what the hardness is from previous test results and if and how the water is treated or softened in the home. If the water is from a municipal or rural system, they should find out the source of the water and how it is treated at the water plant. If possible, they should relate the characteristics of the water at the tap to the hardness. Students should write the results of this inquiry in their chemistry journals. **L1** **LEP**

What is the formula for calcium carbonate? Calcium is in Group 2, so its ion has a 2+ charge. The carbonate ion has a 2− charge, as shown in **Table 5.2.** To form a neutral compound, one Ca^{2+} ion must combine with one CO_3^{2-} ion to give the formula $CaCO_3$.

SAMPLE PROBLEM ② **Writing a Formula Containing a Polyatomic Ion**

Write the formula for the compound that contains lithium and carbonate ions.

Analyze
- Lithium is in Group 1, so its ion has a 1+ charge. According to **Table 5.2,** the carbonate ion has a 2− charge, and its structure is CO_3^{2-}.

Set Up
- Write the symbols for lithium carbonate in formula form.

$$Li^+CO_3^{2-}$$

Solve
- Determine the correct ratio of lithium ions to carbonate ions by examining their charges. In this case, the sum of the positive and negative charges does not equal zero. Two lithium ions are needed to balance the carbonate ion. Because you cannot change the charges of the ions, you must add a subscript of *2* to Li^+. The correct formula for lithium carbonate is Li_2CO_3.

Check
- Check to be sure that the overall charge of the formula is zero.

$$2(1+) + (2-) = 0 \quad \text{The formula as written is correct.}$$

SAMPLE PROBLEM ③ **Writing a More Complex Formula**

Write the formula for the compound that contains aluminum and sulfate ions.

Analyze
- Aluminum is in Group 13 and has an oxidation number of 3+. According to **Table 5.2,** the sulfate ion has a 2− charge.

Set Up
- Write the symbols for aluminum sulfate in formula form.

$$Al^{3+}SO_4^{2-}$$

Solve
- Determine the correct ratio of aluminum ions to sulfate ions by examining their charges. In this case, the sum of the positive and negative charges does not equal zero. To achieve neutrality, you must find the least common multiple of 3 and 2. The least common multiple is 6. How many Al^{3+} ions will be needed to make a charge of 6+, and how many SO_4^{2-} ions will be needed to make a charge of 6−? It will be necessary to have two Al^{3+} ions in the compound to balance three SO_4^{2-} ions. You should add a subscript of *2* to the aluminum ion and a subscript of *3* to the sulfate ion. The entire polyatomic ion must be placed in parentheses to indicate that three sulfate ions are present. Thus, the correct formula for aluminum sulfate is $Al_2(SO_4)_3$.

Check
- Check to be sure that the overall charge of the formula is zero.

$$2(3+) + 3(2-) = 0 \quad \text{The formula as written is correct.}$$

5.1   Ionic Compounds   161

SECTION 5.1

Reinforcement

Divide students into groups of three, and give each group 33 3″ × 5″ cards. On one side of each card, they should write the name of an ion in **Tables 5.1** and **5.2.** On the other side, they should write the symbol and oxidation number. One student in the group should hold, shuffle, and extract one of the positive ion cards, while another student does the same with the negative ion cards. These two students should show both cards to the third person, who should then write the name and formula of the ionic compound that contains the two ions. Students can exchange roles after each three or four trials. [L1] [LEP] [COOP LEARN]

Applying Chemistry

 **Hard Water and Plumbing**
Ask a local plumber for small sections of some pipes that have become clogged with calcium carbonate deposits. Have students examine portions of the deposits with a hand lens or stereomicroscope. [L1] [LEP]

GLENCOE **TECHNOLOGY**

 Videodisc

Science and Technology Videodisc Series
Sea Urchins and Power Plants
Disc 2: Chemistry
Side 1, Ch. 18

Show this video to illustrate how calcium compounds in the shells of marine animals are causing problems with clogged pipes. [L1] [LEP]

Across the Curriculum

History

The element iodine was discovered by accident in seaweed by the French chemist Bernard Courtois in 1811. He was using the seaweed to extract potassium for use in making potassium nitrate explosives for Napoleon. One day, when a stronger-than-usual acid was used to clean out the extraction tanks, unusual violet vapors were produced in the tank. These vapors became a shiny solid when they condensed. Within a few years, the French chemist Joseph Gay-Lussac determined that the material was a new element, and it was given the name *iodine* after the Greek word for "violet," *iode.* Why was there iodine in the seaweed? Iodine salts in seawater become concentrated in many marine organisms.

PRACTICE PROBLEMS

3. Write the formula for the compound made from each of the following ions.

a) ammonium and sulfite ions
b) calcium and monohydrogen phosphate ions
c) ammonium and dichromate ions
d) barium and nitrate ions

4. Write the formula for each of the following compounds.

a) sodium phosphate
b) magnesium hydroxide
c) ammonium phosphate
d) potassium dichromate

Figure 5.6

Two Compounds of Iron and Sulfate
Iron forms both Fe^{2+} and Fe^{3+} ions, each of which can combine with the sulfate ion. Some people use the older spelling, *sulphate*.

Compounds of Transition Elements

In Chapter 3, you learned that the elements known as transition elements are located in Groups 3 through 12 in the periodic table. Transition elements form positive ions just as other metals do, but most transition elements can form more than one type of positive ion. In other words, transition elements can have more than one oxidation number. For example, copper can form both Cu^+ and Cu^{2+} ions, and iron can form both Fe^{2+} and Fe^{3+} ions. **Figure 5.6** shows the two compounds that iron forms with the sulfate ion. Zinc and silver are two exceptions to the variability of other transition elements; each forms one type of ion. The zinc ion is Zn^{2+} and the silver ion is Ag^+.

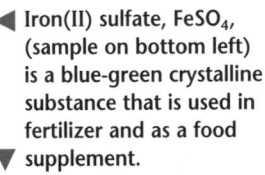

◀ Iron(II) sulfate, $FeSO_4$, (sample on bottom left) is a blue-green crystalline substance that is used in fertilizer and as a food ▼ supplement.

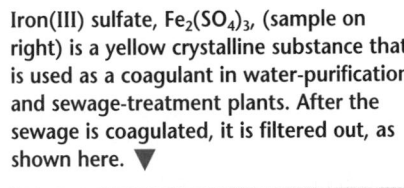

Iron(III) sulfate, $Fe_2(SO_4)_3$, (sample on right) is a yellow crystalline substance that is used as a coagulant in water-purification and sewage-treatment plants. After the sewage is coagulated, it is filtered out, as shown here. ▼

162   Chapter 5   Types of Compounds

China's Porcelain

See how proudly the dragon, an ancient symbol of Chinese culture, prances around the vase pictured here. Proud it should be, because this vase represents one of the greatest achievements of Chinese technology and art—glazed porcelain.

Clay, glaze, and fire By the third to sixth century A.D., the Chinese had invented glazed porcelain. They found that if a clay vessel, such as a bowl, is covered with a transparent glaze and then heated to a high temperature, a translucent ceramic material forms. This material is glazed porcelain. Unlike a fired-clay vessel, which remains slightly porous and opaque, the translucent vessel is sealed by a glasslike covering. By changing the chemical composition of the glaze, Chinese artisans were able to change the quality and color of the glaze. For example, when they added materials that reacted with each other to form tiny gas bubbles in the glaze, the porcelain appeared brighter because the surface of the bubbles reflected light.

Colored glazes One of the most important steps in glazing pottery was the addition of materials to the glaze to produce colored porcelains. These materials were solutions of transition element ions, such as iron, manganese, chromium, cobalt, copper, and titanium. During the firing of the glaze, these metals formed oxides. Because the metal ions in the oxides reflected only certain wavelengths of light, the glazes colored the porcelain. By varying the concentration and charges of the metal ions in the glaze, the Chinese were able to produce subtly colored porcelains. For example, cobalt produces a blue glaze, chromium a pink or green glaze depending on charge, and manganese a purple glaze. These beautiful colors have remained vivid over the course of thousands of years, and the techniques are still being used today.

Connecting to Chemistry

1. **Applying** Why are porcelain dishes superior to wooden dishes?

2. **Thinking Critically** What properties of metallic compounds make them useful as colored glazes?

ART ▷ CONNECTION

Purpose

TECH PREP

Students will learn the role of metal ions in coloring glazes of Chinese porcelains.

Teaching Strategies

- Remind students of the differences between transparent, translucent, and opaque materials. Light does not pass through opaque materials. Light passes through, and images can be seen through, transparent materials. Light passes through, but images cannot be seen through, translucent materials. Provide materials of each type for students to hold up to the light. **L1** **LEP**

- Emphasize that ceramics and glasses do not have a consistent crystalline structure throughout the solid. Therefore, they do not have a precise melting point. Their melting points and other properties can be manipulated by varying their composition.

Extension

Have students investigate the use of transition metal ions in paint pigments. **L2**

Connecting to Chemistry

1. Porcelain dishes are less porous than wooden dishes and would be easier to clean and keep sanitary.

2. Metallic salts are soluble and therefore can be applied in a liquid glaze. The electron configuration of metallic ions causes the ions to absorb almost all light except for specific wavelengths, which gives the ions color. Different colors can be produced by varying the concentrations and charges of the metal ions in the glaze.

Concept Development

Write the formulas for several compounds of the transition elements iron and manganese, such as $FeBr_2$, $FeBr_3$, MnO, Mn_2O_3, and $MnCl_2$. Ask students to name the two iron compounds (*both are iron bromide*) and the three manganese compounds (*all are manganese oxide*). Obviously, these names are not sufficient to differentiate the compounds. Use this as a lead-in to the use of Roman numerals to designate the oxidation numbers of the transition elements and other metallic elements. (Give the correct names: iron(II) bromide; iron(III) bromide; manganese(II) oxide; manganese(III) oxide; manganese(IV) oxide.)

164

Chemists must have a way to distinguish the names of compounds formed from the different ions of a transition element. They do this by using a Roman numeral to indicate the oxidation number of a transition element ion. This Roman numeral is placed in parentheses after the name of the element. No additional naming system is needed for zinc and silver compounds because their formulas are not ambiguous. **Table 5.3** shows the naming of the two different ionic compounds formed when chloride ions combine with each of the two copper ions.

Table 5.3 Compounds of Copper and Chlorine			
Copper Ion	**Chloride Ion(s)**	**Formula**	**Name**
Cu^+	Cl^-	$CuCl$	copper(I) chloride
Cu^{2+}	$2Cl^-$	$CuCl_2$	copper(II) chloride

Table 5.4 shows the chemical names of some transition element ions. When you do Practice Problems 5 and 6, you will become familiar with these names. Note in the photos accompanying the table that the different ions of a transition element often form compounds of different colors. For example, CrO is black, Cr_2O_3 is green, and CrO_3 is red. Determine the oxidation number for chromium in each of these compounds.

Table 5.4 Names of Common Ions of Selected Transition Elements		
Element	**Ion**	**Chemical Name**
Chromium	Cr^{2+}	chromium(II)
	Cr^{3+}	chromium(III)
	Cr^{6+}	chromium(VI)
Cobalt	Co^{2+}	cobalt(II)
	Co^{3+}	cobalt(III)
Copper	Cu^+	copper(I)
	Cu^{2+}	copper(II)
Gold	Au^+	gold(I)
	Au^{3+}	gold(III)
Iron	Fe^{2+}	iron(II)
	Fe^{3+}	iron(III)
Manganese	Mn^{2+}	manganese(II)
	Mn^{3+}	manganese(III)
	Mn^{7+}	manganese(VII)
Mercury	Hg^+	mercury(I)
	Hg^{2+}	mercury(II)
Nickel	Ni^{2+}	nickel(II)
	Ni^{3+}	nickel(III)
	Ni^{4+}	nickel(IV)

CrO

Cr_2O_3

CrO_3

Meeting Individual Needs

Gifted Ask interested students to determine the oxidation numbers of each of the elements in the two polyatomic ions, $Fe(CN)_6^{3-}$ and $Fe(CN)_6^{4-}$. If necessary, remind students that iron can be either 2+ or 3+. *Fe, 3+; C, 2+; N, 3–;. Fe, 2+; C, 2+; N, 3–.* Point out that the names of the polyatomic ions, hexacyanoferrate(III) or ferricyanide and hexacyanoferrate(II) or ferrocyanide, refer to the oxidation numbers of iron. **L3**

Suppose you wanted to write the formula for a compound containing a transition element. Look back at Sample Problem 1, where you learned to write the formula for a compound containing sodium and sulfur. How would you write the formula if it were iron(II) rather than sodium that combined with sulfur? Iron(II) has an oxidation number of 2+, and its ion can be written as Fe^{2+}. You know that the sulfide ion has a charge of 2− and can be written as S^{2-}. The charges balance in this case, and the formula for iron(II) sulfide is written as FeS, **Figure 5.7**.

You can write the formula for iron(III) sulfide in the same way. Just follow the steps in Sample Problem 3. The correct formula for iron(III) sulfide is Fe_2S_3. Note that the Roman numeral refers to the oxidation number of the iron and not to how many ions are in the formula.

How can you name a compound of a transition element if you are given the formula? Determining the charge of the transition element ion gives the clue needed to name the compound. In the formula $Cr(NO_3)_3$, you must determine the charge of the chromium ion in order to name the compound. Look first at the negative ion. Knowing that the nitrate ion has a charge of 1− and that there are three nitrate ions with a total charge of 3−, you can see that the chromium ion must have a charge of 3+ to maintain neutrality. Thus, this compound is named chromium(III) nitrate.

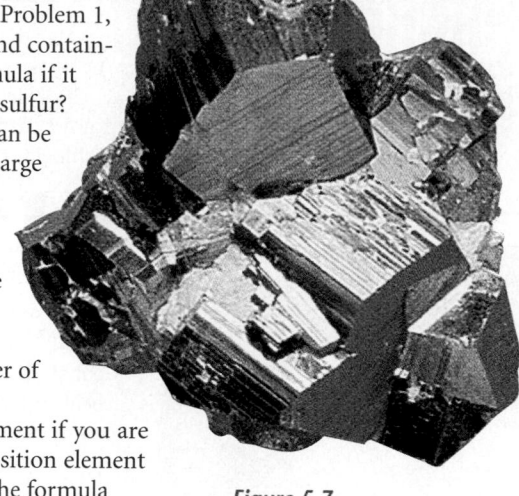

Figure 5.7
Fool's Gold
In iron disulfide, commonly called fool's gold because of its resemblance to gold, iron has an oxidation number of 2+. The complex ionic structure has the formula FeS_2.

PRACTICE PROBLEMS

5. Write the formula for the compound made from each of the following pairs of ions.

   a) copper(I) and sulfite
   b) tin(IV) and fluoride
   c) gold(III) and cyanide
   d) lead(II) and sulfide

6. Write the names of the following compounds.

   a) $Pb(NO_3)_2$
   b) Mn_2O_3
   c) $Ni(C_2H_3O_2)_2$
   d) HgF_2

Hydrates

Many ionic compounds are prepared by crystallization from a water solution, and water molecules become a part of the crystal. A compound in which there is a specific ratio of water to ionic compound is called a **hydrate.** In a hydrate, the water molecules are chemically bonded to the ionic compound.

Content Background
Old and New Nomenclature Explain that prior to the current system of using Roman numerals to differentiate the multiple positive oxidation numbers of elements, chemists used the suffixes *ous* and *ic* to represent the lower and higher oxidation numbers, respectively. The old nomenclature, which may be seen on the labels of some of the compounds in the chemical storeroom and in older textbooks, also often used the Latin names for the elements. Examples are: cuprous for Cu^+; cupric for Cu^{2+}; ferrous for Fe^{2+}; ferric for Fe^{3+}; plumbous for Pb^{2+}; plumbic for Pb^{4+}. You might wish to show students several bottles of compounds with these labels.

Discussion
Ask students this question. If you purchased a 1-pound box of Epsom salt, magnesium sulfate heptahydrate, and determined that the box contained only 12 ounces, was the company that packaged the product necessarily dishonest? *No. $MgSO_4 \cdot 7H_2O$ is about 51% water. If the product was subjected to a high temperature after packaging, some of the water of hydration may have escaped.*

PRACTICE

Guided Practice: Have students work in class on Practice Problems 5 and 6. ⌧L1⌧
Independent Practice: Your homework or classroom assignment can include the additional practice problems shown below. ⌧L1⌧

Answers to Practice Problems
5. **a.** Cu_2SO_3 **b.** SnF_4 **c.** $Au(CN)_3$ **d.** PbS
6. **a.** lead(II) nitrate **b.** manganese(III) oxide
   c. nickel(II) acetate **d.** mercury(II) fluoride

Additional Practice Problems
1. Write the formula for the compound made from each of the following pairs of ions.
   a. lead(IV) and chloride $PbCl_4$
   b. iron(III) and oxide Fe_2O_3
2. Write the formula for the compound made from each of the following pairs of ions.
   a. titanium(II) hydroxide $Ti(OH)_2$
   b. copper(II) and sulfate $CuSO_4$

miniLAB

A Chemical Weather Predictor

TECH PREP

Purpose: Students will discover that the properties of a hydrate and its anhydrous counterpart are different.

Process Skills: Observing, recognizing cause and effect, interpreting data, communicating

Teaching Strategies

- You can substitute 91% isopropanol (drugstore rubbing alcohol) for the ethanol. Check the label before purchasing it.
- Students can cut a weather-related design out of the paper they will swab. These weather indicators can be mounted on a cardboard stand and taken home to use. L1 LEP
- **Disposal:** Code A; collect solutions; evaporate, bag, and label residue for disposal.

Expected Results

The swabbed area will be blue when the air is dry and precipitation is not likely, and pink when it is humid and rain or snow is likely. Check the predictors outside.

Analysis

1. $CoCl_2$
2. $CoCl_2 \cdot 6H_2O$; cobalt(II) chloride hexahydrate
3. The test paper is a relatively reliable weather predictor.

Assessment

Portfolio: Have students videotape each other demonstrating their weather predictor and how it works. L1 LEP P

miniLAB 1

A Chemical Weather Predictor

Adding water to an anhydrous compound to form a hydrate often changes the physical properties—such as color—of the compound. Cobalt(II) chloride is such a compound. If you find that the color of the compound changes in accordance with the weather, perhaps cobalt(II) chloride can serve as a weather predictor.

Procedure

1. Place 5 mL of 95 percent ethanol in a small beaker.
2. Use a spatula to add a small amount of cobalt(II) chloride to the beaker. Stir until the compound dissolves.
3. Dip a cotton swab into the pink solution and use it to write the chemical formula of cobalt(II) chloride on a piece of white paper.
4. Dry the paper by holding it over a hot plate set on low or by putting it in a sunny location. What color is the formula now?
5. Keep your weather predictor in a convenient location, and check its color each morning and afternoon. Keep a three-week log of the time, the current weather, and the color of the treated paper.

Analysis

1. What is the formula of cobalt(II) chloride?
2. The hydrate of cobalt(II) chloride has six water molecules bonded to it. What is its formula and name?
3. From your observations, are you able to conclude that the cobalt(II) chloride test paper is a reliable weather predictor? Justify your answer.

Does your chemistry instructor often remind students to make sure that the lids on jars of chemicals are tightly closed? There is a good reason for sealing the jars tightly; some ionic compounds can easily become hydrates by absorbing water molecules from water vapor in the air. These compounds are called **hygroscopic** substances, and one example is sodium carbonate (Na_2CO_3). Substances that are so hygroscopic that they take up enough water from the air to dissolve completely and form a liquid solution are called **deliquescent**. A deliquescent substance is shown in **Figure 5.8**.

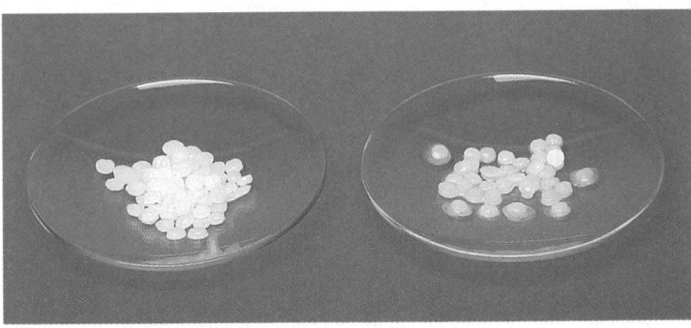

Figure 5.8

Deliquescent Substances

Sodium hydroxide (left) is an example of a deliquescent substance because it has a strong attraction for water molecules. Sodium hydroxide will absorb water molecules from the surrounding air and begin to dissolve (right). Eventually, it will absorb enough water to dissolve completely.

Demonstration 5.3

Water in Hydrates

Purpose: To demonstrate that water can be removed from a hydrate by heating and that the hydrate and its anhydrous form have different physical properties.

Materials: 30 g $CuSO_4 \cdot 5H_2O$, large test tube, laboratory burner, watch glass, 2 mL H_2O, dropper, test-tube clamp

Safety Precautions:

Disposal: Code F

Procedure: Place about 30 g of $CuSO_4 \cdot 5H_2O$ in a large test tube. Heat it over a flame until the blue color disappears. Do not overheat the hydrate as it could decompose. Allow the test tube to cool. Remove the white, anhydrous $CuSO_4$ to a watch glass, and add H_2O from a dropper.

CAUTION: *The reaction is exothermic.*

Results: Copper(II) sulfate pentahydrate loses water in an observable manner. The blue crys-

How it Works

Cement

People have been using cementing materials for thousands of years. The stones in the Egyptian pyramids are held together by a mixture of sand and the mineral compound gypsum, which is calcium sulfate dihydrate. When this dihydrate is heated, water evaporates, forming a compound with one water molecule per two calcium sulfate formula units. Today, we know this binding material as plaster of paris.

1. Cement is made from a mixture of limestone and clay. The most important minerals in clay are the combinations of aluminum, oxygen, and silicon known as aluminum silicates.

2. Before this limestone-clay mixture can be used, it must be heated. Heating drives off carbon dioxide and forms new ionic compounds. This new mixture of calcium silicates, calcium aluminates, and calcium aluminum ferrates forms in clumps called clinker.

3. The clinker is ground and mixed with small amounts of calcium sulfate. This mixture is called portland cement.

4. Cement can be used alone to form a smooth, hard surface for roads or buildings, or it can be combined with sand and gravel to form a rougher material called concrete.

5. When concrete is mixed with water, silicate compounds hydrate and form gelatinous materials called gels.

6. The hardening process takes several days. During this time, some water is removed from the gels that formed around the sand and gravel, and calcium hydroxide absorbs carbon dioxide from the air to re-form calcium carbonates. Fibers that form from the cement materials interlock and strengthen the concrete.

Thinking Critically

1. What is the formula for the calcium sulfate dihydrate that makes up gypsum?

2. Tricalcium aluminate also becomes hydrated during solidification of cement to form $Ca_3Al_2O_6 \cdot 6H_2O$. What is the name of this hydrate?

How it Works

Purpose

TECH PREP

Students learn how cement is made and why hydrates allow cement to be liquid enough to be poured and then solid and strong when hardened.

Background

During the first few hours of the solidification process, step 5, gels form around the sand and rocks in the mixture. The hardening process, step 6, can take several days and is sometimes not completed for years.

Visual Learning

• Ask students why the mixer on a cement truck must keep turning. *When the turning stops, the hardening process begins as water evaporates from hydrated minerals. Turning the wet cement keeps the water from escaping by constantly forcing the water vapors back into the mixture.*

Teaching Strategies

• Ask students why cement is stored in a dry form. *The dry cement has not solidified by forming hydrates, so it is more stable and less likely to harden.*

Thinking Critically

1. $CaSO_4 \cdot 2H_2O$
2. tricalcium aluminate hexahydrate

---

tals become pale blue and then white when heated. When the water is added to the white anhydrous form, the blue color returns with the evolution of heat.

Analysis: Ask these questions.

1. What did you observe? *A blue substance changes color when heated. A clear liquid condenses around the top of the test tube.*

2. From the demonstration, can you determine the amount of water removed from the blue substance? Explain. *No; no quantitative measurements were recorded.*

3. Why did the white substance turn blue when water was added? *The hydrate was re-formed when water bonded to the anhydrous $CuSO_4$.*

Assessment

Skill: Repeat the demonstration, but determine the mass with a balance before and after heating. Ask students to determine the mass of water that is lost upon heating and to calculate the percentage of water contained in the hydrate. L2

3 ASSESS

Check for Understanding

- Check for understanding of the section material by having students answer the Section Review questions. L1
- Ask students what each of the following symbols or formulas represents: Li, Li$^+$, O, O^{2-}, CO$_2$, 3CO$_2$, RbF, 3RbF, 3BaCl$_2 \cdot$2H$_2$O. *One lithium atom. One lithium ion. One oxygen atom. One oxygen ion. One carbon dioxide molecule. Three carbon dioxide molecules. One formula unit of rubidium fluoride, containing one rubidium ion and one fluoride ion. Three formula units of rubidium fluoride, containing three rubidium ions and three fluoride ions. Three formula units of barium chloride dihydrate, containing three barium ions, six chloride ions, and six molecules of water.* L1

Figure 5.9
Desiccants
Boxes of electronic equipment such as cameras and CD players usually contain small packets of a desiccant. The desiccant absorbs water vapor from the air, protecting the delicate metal parts against corrosion and preventing condensation of water vapor in the wiring of the equipment. Even a tiny quantity of water on a circuit board can create a short circuit. Desiccants are especially useful for packaging electronic equipment that is to be shipped overseas because ocean air in the holds of ships contains so much moisture.

Many of these compounds become hydrates by absorbing water from the air. As shown in **Figure 5.9,** compounds that form hydrates often are used as drying agents, or desiccants, because they absorb so much water from the air when they become hydrated.

To write the formula for a hydrate, write the formula for the compound and then place a dot followed by the number of water molecules per formula unit of compound. The dot in the formula represents a ratio of compound formula units to water molecules. For example, CaSO$_4 \cdot$2H$_2$O is the formula for a hydrate of calcium sulfate that contains two molecules of water for each formula unit of calcium sulfate. This hydrate is used to make portland cement and plaster of paris. To name hydrates, follow the regular name for the compound with the word *hydrate,* to which a prefix has been added to indicate the number of water molecules present. Use **Table 5.5** to find the correct prefix to use. The name of the compound with the formula CaSO$_4 \cdot$2H$_2$O is calcium sulfate dihydrate.

Heating hydrates can drive off the water. This results in the formation of an **anhydrous** compound—one in which all of the water has been removed. In some cases, an anhydrous compound may have a different color from that of its hydrate, as shown in **Figure 5.10.**

Table 5.5 Prefixes to Use in Naming Hydrates	
Molecules of Water	**Prefix**
1	mono-
2	di-
3	tri-
4	tetra-
5	penta-
6	hexa-
7	hepta-
8	octa-
9	nona-
10	deca-

Interpreting Formulas

You have learned how to write a formula to represent a formula unit of an ionic compound. Sometimes, it may be necessary to represent more than one formula unit of a compound. To do this, place a coefficient before the formula. Two formula units of NaCl are represented by 2NaCl, three formula units by 3NaCl, and so on.

Integrating THE Sciences

Earth Science Ask students to write a report about the Great Salt Lake in Utah. They should investigate how it formed, how its salinity compares with that of ocean water, what kinds of plant and animal life inhabit it, and how much salt is mined from the lake and its surroundings each year. Student reports may be put into their portfolios. L1 P

A formula summarizes how many atoms of each element are present. Each formula unit of sodium chloride contains one sodium ion and one chloride ion. How many oxygen atoms are present in $3HNO_3$? Each formula unit contains three oxygen atoms. Because there are three formula units, a total of nine atoms of oxygen are present. As another example, consider how many atoms of hydrogen are in one formula unit of ammonium sulfate. The formula for ammonium sulfate is $(NH_4)_2SO_4$. Each ammonium ion contains four atoms of hydrogen. Because two ammonium ions are present, there are eight atoms of hydrogen in a formula unit of ammonium sulfate. How many hydrogen atoms are in $3(NH_4)_2SO_4$? To find out, simply multiply the eight hydrogen atoms in one formula unit by three formula units; 24 hydrogen atoms are present.

Figure 5.10
Forming an Anhydrous Compound
When blue copper(II) sulfate pentahydrate ($CuSO_4 \cdot 5H_2O$) (left) is heated, the water is driven off (center). The anhydrous compound, $CuSO_4$ (right), is white. Hydrated copper sulfate is used as a fungicide in water reservoirs.

SECTION REVIEW

Understanding Concepts

1. Explain why ionic compounds cannot conduct electricity when they are in the solid state.

2. Write formulas for each of the following ionic compounds.

   a) manganese(II) carbonate
   b) barium iodide dihydrate
   c) aluminum oxide
   d) magnesium sulfite
   e) ammonium nitrate
   f) sodium cyanide

3. Name the ionic compound represented by each formula.

   a) Na_2SO_4
   b) CaF_2
   c) $MgBr_2 \cdot 6H_2O$
   d) Na_2CO_3
   e) $KMnO_4$
   f) $Ni(OH)_2$
   g) $NaC_2H_3O_2$

Thinking Critically

4. **Interpreting Chemical Formulas** What information does the formula $3Ni(HCO_3)_2$ tell you about the number of atoms of each element that are present?

Applying Chemistry

5. **Toothpaste Ingredients** Examine the ingredient label on a tube of toothpaste. Write formulas for as many of the chemical names listed as you can. List whether each ingredient is an ionic or a covalent compound.

Reteach

Have students make and use flash cards with the prefix for naming hydrates on one side and the number of molecules of water on the other. These prefixes, listed in **Table 5.5,** will be used later in the chapter for naming molecular compounds.
L1 LEP

Extension

Have students research a transition element of their choice. They should determine all oxidation numbers for their element, which oxidation numbers form the most stable compounds, and some typical compounds for each oxidation number. Reports or posters can be made and placed in students' portfolios. L2 P

4 CLOSE

Quick Demo

Why does NaOH gain mass?

Set an evaporating dish containing a small amount of dry NaOH on the pan of a milligram or centigram balance, noting the appearance and mass initially and after about ten minutes. Ask students to explain the difference. *The NaOH is deliquescent and has absorbed water from the air to form a solution.*

SECTION REVIEW

1. When an ionic compound is in the solid state, the charged ions that make up the compound are fixed in place and are not able to move freely. Thus, they cannot conduct an electric current.
2. **a.** $MnCO_3$ **b.** $BaI_2 \cdot 2H_2O$ **c.** Al_2O_3 **d.** $MgSO_3$ **e.** NH_4NO_3 **f.** $NaCN$
3. **a.** sodium sulfate **b.** calcium fluoride **c.** magnesium bromide hexahydrate

d. sodium carbonate **e.** potassium permanganate **f.** nickel(II) hydroxide **g.** sodium acetate
4. There are three formula units of the compound nickel(II) hydrogen carbonate present and a total of the following atoms: 3 Ni, 6 H, 6 C, and 18 O.
5. Students can classify these simple ingredients in toothpaste.

water	H_2O	covalent
sodium fluoride	NaF	ionic
trisodium phosphate	Na_3PO_4	ionic
titanium dioxide	TiO_2	ionic
hydrated silica	$SiO_2 \cdot H_2O$	covalent

Students can find more complex ingredients in a reference book.

SECTION 5.2

PREPARE

Key Concepts
The properties of ionic substances are compared with those of molecular substances. Molecular elements and allotropes are discussed and related to everyday applications. Binary molecular compounds are named, and the naming of common acids, bases, and hydrocarbons is introduced.

Planning Ahead
For the ChemLab Purchase, borrow, or construct several conductivity testers.
For the MiniLab Contact a local butcher to obtain uncooked chicken bones. They may be frozen in a plastic container and thawed prior to use.
For the Demonstrations A conductivity tester will be needed for Demonstration 5.4.

1 FOCUS

Focus Transparency
Display the **Section Focus Transparency** for Section 5.2. Use the examples of molecular substances to introduce this section. L1 LEP

2 TEACH

✓Assessment

Portfolio/Performance: Ask a few students to research water desalination techniques and usage. Each student should write a report summarizing the findings and include the report in his or her portfolio. The students should then cooperate to make a class presentation, utilizing visual aids and multimedia techniques if possible. L2 P
COOP LEARN

SECTION PREVIEW

Objectives
Compare the properties of molecular and ionic substances.
Distinguish among allotropes of an element.
Apply formulas to name molecular compounds.

Key Terms
molecular substance
distillation
molecular element
allotrope
organic compound
inorganic compound
hydrocarbon

Molecular Substances

How many compounds can you name that are liquids or gases at normal room temperature? Water, carbon dioxide, and ammonia are just a few examples. Because most ionic compounds are solids at room temperature, the odds are pretty good that any compounds you thought of are members of the other major class of compounds described in Chapter 4—the covalent compounds. However, not all of them are liquids or gases. Some covalent compounds are solids at room temperature, for example sugar, mothballs, silica (sand), and the fats that make up butter and margarine. Most of the time, it is difficult to tell whether a solid compound is ionic or covalent by visual examination alone. Compare the crystals of sugar and salt shown here.

Properties of Molecular Substances

You know that ionic compounds share many properties. The properties of a **molecular substance**—a substance that has atoms held together by covalent rather than ionic bonds—are more variable than the properties of ionic compounds. Some molecular substances, such as polyethylene plastic and the fats in butter, are soft; rubber is elastic; and diamond and quartz are hard.

Although molecular substances have varied properties, some generalities can be made to distinguish them from ionic compounds. Molecular substances usually have lower melting points, and most are not as hard as ionic compounds, **Figure 5.11.** In addition, most molecular substances are less soluble in water than ionic compounds and are not electrolytes. The properties of most ionic and molecular substances are different enough that their differences can be used to classify and separate them from one

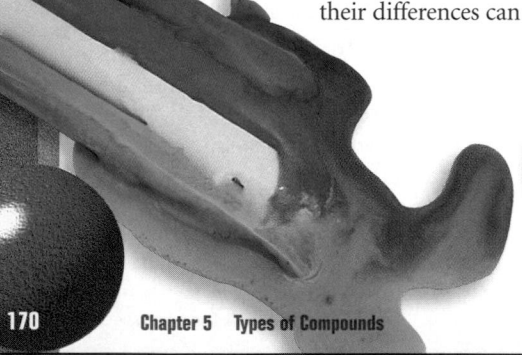

Figure 5.11
Crayons—Covalent or Ionic?
Crayons are made of covalent compounds. They are soft and are insoluble in water. If you have ever left crayons out in the sun, you know that they also have a low melting point.

Program Resources

Study Guide, pp. 19-20 L1
Section Focus Transparency 10 and **Master,** p. 10 L1 LEP
ChemLab and MiniLab Worksheets, pp. 35-37, 39 L1
Basic Concepts Transparency 15 and **Master,** pp. 29-30 L1 LEP
Consumer Chemistry, pp. 9-10 L1

Where's the calcium?

Calcium is an important part of the structure of bones and eggshells. If a bone is soaked in vinegar for several days, the structure of the bone will change. Vinegar contains acetic acid, which reacts with the calcium compounds in the bone to form calcium acetate.

Procedure

1. Pick most of the meat from a small, uncooked chicken bone.
2. Place the bone in a beaker, cover it with vinegar, and cover the beaker with a watch glass.
3. Label the beaker with your name and leave it for two days in the area indicated by your teacher.
4. Use forceps to remove the bone from the beaker, and blot it on a paper towel. Examine the bone and observe how it has changed.
5. Replace the bone in the vinegar and let it soak for two more days. Repeat step 4.
6. Straighten a paper clip. Holding the clip with forceps, dip it into the vinegar solution, then hold it in the blue flame of a Bunsen burner. If calcium is present in the vinegar, it will give an orange-red flame test.

Analysis

1. Describe the change in the properties of the bone after two days and after four days.
2. If the flame test verified the presence of calcium ions in the vinegar, what was the probable source of the calcium?
3. What do you conclude regarding the effect of ionic calcium compounds on the properties of bone? Do their properties in bone seem to correlate with those of typical ionic compounds?
4. How do the properties of the bone after soaking reflect the presence mainly of covalent compounds in the bone?

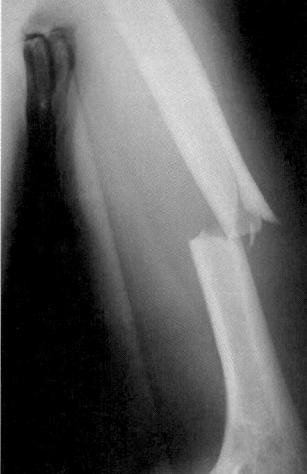

Calcium compounds make bones hard but brittle. Applying stress to brittle materials can cause them to break.

---

another. The separation of water from salt by distillation is one example that makes use of these property differences. **Distillation** is the method of separating substances in a mixture by evaporation of a liquid and subsequent condensation of its vapor. As you learned, solar stills make use of this method. A simple lab-distillation apparatus is shown in **Figure 5.12.**

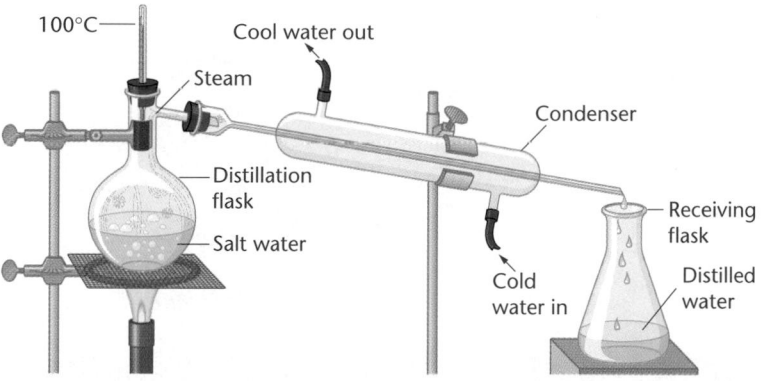

Figure 5.12

Distillation in the Lab
A soluble ionic compound such as NaCl can be separated from water using a distillation apparatus like this one. As the salt water is boiled in the distillation flask, the water turns to steam and the salt is left behind. The steam passes through a water-cooled condenser, where it condenses into pure, distilled water. The distilled water is collected in the receiving flask.

---

2. The calcium came from ionic compounds in the bone.
3. Calcium compounds, like most ionic compounds, are hard and brittle. The bone was more brittle before soaking because it contained more ionic compounds.
4. After soaking, the bone was soft and less brittle, properties that are characteristic of most covalent compounds.

✓ Assessment

Oral: Ask students what would happen to teeth that are soaked in vinegar. *Teeth would also soften as ionic calcium compounds were removed by reaction with vinegar.* L1

---

miniLAB

Where's the calcium?

TECH PREP

Purpose: Students will carry out a chemical reaction that results in the conversion of a mostly ionic material to one that is mostly covalent. They will discover that the properties of the material have changed.

Process Skills: Observing, inferring, recognizing cause and effect, predicting

Teaching Strategies

• Explain that although our bones are brittle due to ionic compounds in them, those compounds also make them strong. Without enough calcium in our diet, our bones weaken.
• After students have made their observations, explain the reaction that has taken place. You may want to use this as an introduction to Chapter 6, where reactions are explained.
• Keep the beakers covered tightly with plastic wrap to keep the water from evaporating from the vinegar and the fumes from escaping into the classroom.
• **Disposal:** Code D

Expected Results

The bone should get softer and more flexible the longer it remains in the vinegar.

Analysis

1. The longer the bone remained in the vinegar, the more soft and rubbery it became. The acetic acid in the vinegar reacted with calcium in the bone to form a soluble compound, which leached out of the bone into the vinegar. Covalent compounds in the bone remained.

ChemLab

Ionic or Covalent?

Time Allotment: One class period.

Objectives
Review objectives with students before they begin the ChemLab.

Process Skills
Observing, measuring, interpreting data, classifying, using space/time relationships, communicating

Safety Precautions and Disposal
Students should be advised to be careful when handling the hot plate and slides during and after heating. If contact with any of the samples occurs, students should rinse their hands with tap water. If thermometers contain mercury, be aware of this hazardous substance in the event of breakage. Students should wash their hands thoroughly after the lab.
Disposal: Chemicals may be flushed down the drain with water.

PREPARATION
If samples are not already a fine powder, they should be crushed with a mortar and pestle before use to permit even melting.

Alternative Materials
- Samples of pure acetylsalicylic acid, KCl, and sucrose will melt more evenly, but using commercial products such as aspirin, salt substitute, and table sugar will give the expected results.
- If small Pyrex or Kimex watch glasses are available, they may be used rather than glass slides and will be much less likely to break while being handled or heated.
- If thermometers with long metal probes are available, they may be used rather than those with glass bulbs to improve the temperature readings and reduce the likelihood of breakage.

ChemLab  SMALL SCALE

Ionic or Covalent?

Compounds can be classified by the types of bonds that hold their atoms together. Ions are held together by ionic bonds in ionic compounds; atoms are held together by covalent bonds in molecular compounds.

You cannot tell whether a compound is ionic or molecular simply by looking at a sample of it because both types of compounds can look similar. However, simple tests can be done to classify compounds by type because each type has a set of characteristic properties shared by most members. Ionic compounds are usually hard, brittle, water-soluble, have high melting points, and can conduct electricity when dissolved in water. Molecular compounds can be soft, hard, or flexible; are usually less water-soluble; have lower melting points; and cannot conduct electricity when dissolved in water.

Problem
How can you identify ionic and molecular compounds by their properties?

Objectives
- **Examine** the properties of several common substances.
- **Interpret** the property data to classify each substance as ionic or molecular.

PREPARATION

Materials
glass microscope slide
grease pencil or crayon
hot plate
spatula
4 small beakers (50- or 100-mL)
stirring rod
balance
conductivity tester

---

graduated cylinder, small
thermometer (must read up to 150°C)
1- to 2-g samples of any 4 of the following: salt substitute (KCl), fructose, aspirin, paraffin, urea, table salt, table sugar, Epsom salt

Safety Precautions
Use care when handling hot objects.

PROCEDURE

1. Use a grease pencil or crayon to draw lines dividing a glass slide into four parts. Label the parts A, B, C, and D.
2. Make a data table similar to the one shown in Data and Observations.
3. Use a spatula to place about one-tenth (about 0.1 to 0.2 g) of the first of your four substances on section A of the slide.
4. Repeat step 3 with your other three substances on sections B, C, and D. Be sure to use a clean spatula for each sample. Record in your data table which substance was put on each section.
5. Place the slide on a hot plate. Turn the heat setting to medium and begin to heat the slide.
6. Gently hold a thermometer so that the bulb just rests on the slide. Be careful not to disturb your compounds.

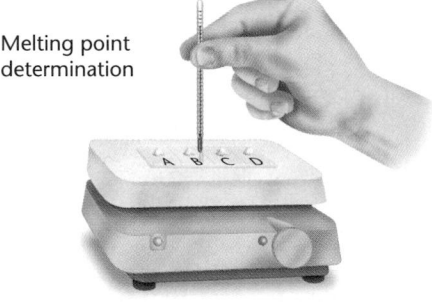
Melting point determination

PROCEDURE
- Require that students' lab notebooks contain the following items before they begin the lab: Objectives, Safety Precautions, Procedure Outline, Table for Data and Observations.
- Having the required materials available at each table or work area will reduce student traffic, the potential for spillage or breakage, and the time required for the work. All students at a table can use a single hot plate.
- Different groups can run different samples, and data can be pooled so everyone analyzes results from each sample. **L1** **COOP LEARN**

- Advise students not to press down on the glass slide with the thermometer to avoid breakage. It is sufficient to simply hold the thermometer to the slide.

ANALYZE AND CONCLUDE
1. When a substance melted, it changed from a solid to a liquid state. A substance melts when enough heat has been added to overcome the intermolecular forces holding molecules or ions together in a crystal.

7. Continue heating until the temperature reaches 135°C. Observe each section on the slide and record which substances have melted. Turn off the hot plate.

8. Label four beakers with the names of your four substances.

9. Weigh equal amounts of the four substances (1-2 g of each), and place the weighed samples in their labeled beakers.

10. Add 10 mL of distilled water to each beaker.

11. Stir each substance, using a clean stirring rod for each sample. Note on your table whether or not the sample dissolved completely.

12. Test each substance for the presence of electrolytes by using a conductivity tester. Record whether or not each acts as a conductor.

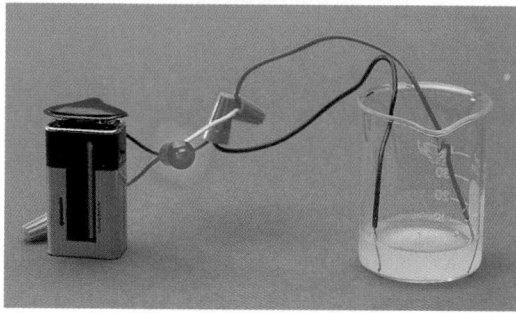

ANALYZE AND CONCLUDE

1. **Interpreting Observations** What happened to the bonds between the molecules when a substance melted?

2. **Comparing and Contrasting** Did all compounds melt at the same temperature?

3. **Classifying** Complete your data table by classifying each of the substances you tested as ionic or molecular compounds based on your observations.

APPLY AND ASSESS

1. What are the differences in properties between ionic and molecular compounds?

2. How did the melting points of the ionic compounds and the molecular compounds compare? What factors affect melting point?

3. The solutions of some molecular compounds are good conductors of electricity. Explain how this can be true when ions are required to conduct electricity.

4. Consider a mixture of sand, salt, and water. How can you make use of the differences in properties of these materials to separate them?

DATA AND OBSERVATIONS

Substance	Did it melt?	Did it dissolve in water?	Did the solution conduct electricity?	Classification
A				
B				
C				
D				

DATA AND OBSERVATIONS

Substance	Did it melt?	Did it dissolve in water?	Did the solution conduct electricity?	Classification
salt substitute (KCl)	No	Yes	Yes	Ionic
aspirin	Yes	Partially	No	Molecular
D-fructose	Yes	Yes	No	Molecular
paraffin	Yes	No	No	Molecular
Epsom salt	No	Yes	Yes	Ionic
urea	Yes	Yes	No	Molecular
table sugar	Yes	Yes	No	Molecular
table salt	No	Yes	Yes	Ionic

2. No, the melting point is a physical property that is characteristic of a given compound. In general, ionic compounds melt at much higher temperatures than covalent or molecular compounds. However, some large covalent polymers have higher melting points than many ionic substances.

3. See data table below.

APPLY AND ASSESS

1. Ionic compounds have high melting points and water solubility, are good conductors when dissolved in water, and are hard and brittle. Molecular compounds have lower melting points and water solubility, are poor conductors, and vary from soft to rubbery to hard.

2. In general, molecular compounds have lower melting points than ionic compounds. However, melting points are affected by intermolecular forces in molecular compounds. These can be strong, especially in long molecules that often have high melting points.

3. Some molecular compounds dissociate in water solution to form ions that can conduct electricity.

4. Sand is a molecular compound (mostly silicon dioxide or quartz) that does not dissolve in water, so it can be filtered out. Salt is ionic, so it has a melting point much higher than that of water, which is molecular. Water can be removed from the salt by heating the mixture to distill out the water.

✔Assessment

Knowledge: Ask students what will happen if a mixture that contains both covalent and molecular substances is tested. *It will act as an electrolyte when dissolved in water. Part of it will melt while part will remain solid.* L1

173

How does the submicroscopic structure of molecular substances contribute to their macroscopic properties? Because there are no ions, strong networks held together by the attractions of opposite charges do not form. The interparticle forces between molecules are often weak and easy to break. These weak forces explain the softness and low melting points of most molecular substances. Most molecular substances are not electrolytes because they do not easily form ions.

Molecular Elements

Molecules vary greatly in size. They can contain from just two to thousands or millions of atoms, as **Figure 5.13** shows. Most elements usually occur naturally in a combined form with another element; that is, they occur as compounds. However, in some cases, two or more atoms of the same element can bond together to form a molecule. A molecule that forms when atoms of the same element bond together is called a **molecular element.** Note that molecular elements are not compounds—they contain atoms of only one element. Why do atoms of these elements bond so readily to identical atoms? When they bond together, each atom achieves the stability of a noble-gas electron configuration.

Figure 5.13

Size of Molecules
Molecular substances can be as simple as two iodine atoms linked together as I_2 (bottom) or as complex as this protein, cytochrome *c* (top), which contains many thousands of atoms of carbon, hydrogen, oxygen, nitrogen, and sulfur linked together by covalent bonds. ▼

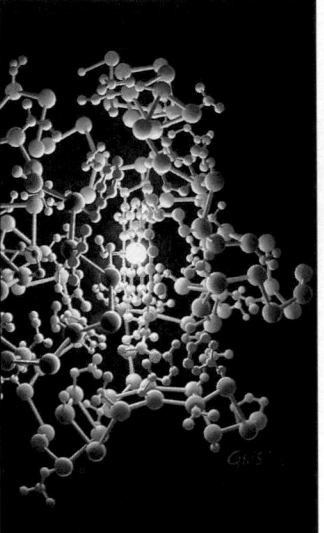

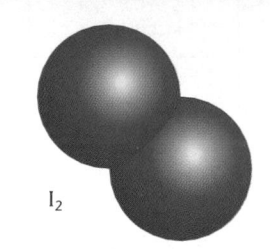

I_2

▲ Cytochrome *c* is found in all living cells that derive energy by breaking down food molecules in the presence of oxygen, and it is found in especially large quantities in hard-working muscle tissue.

Diatomic Elements

Seven nonmetal elements are found naturally as molecular elements of two identical atoms. The elements whose natural state is diatomic are hydrogen, nitrogen, oxygen, fluorine, chlorine, bromine, and iodine. Their formulas can be written as H_2, N_2, O_2, F_2, Cl_2, Br_2, and I_2, respectively. These molecules are referred to as diatomic elements. All except bromine and iodine are gases at room temperature; Br_2 is a liquid, and I_2 is a solid.

What can you learn by examining the structures of the diatomic elements? Electron dot diagrams offer clues. As an example, the chlorine atom has seven valence electrons and needs one more to achieve the configuration of the noble gas argon. If two chlorine atoms combine, they share a single pair of electrons, and each atom attains a stable octet configuration.

$$:\ddot{\underset{\cdot\cdot}{Cl}}\cdot + \cdot\ddot{\underset{\cdot\cdot}{Cl}}: \rightarrow :\ddot{\underset{\cdot\cdot}{Cl}}:\ddot{\underset{\cdot\cdot}{Cl}}:$$

Hydrogen, fluorine, bromine, and iodine molecules also are formed by the sharing of a single pair of electrons. Two oxygen atoms share two pairs of electrons to form O_2, and two nitrogen atoms share three pairs of electrons to form N_2.

$$\cdot\ddot{O}::\ddot{O}\cdot \qquad :N:::N:$$

Allotropes

Although the diatomic form of oxygen, O_2, is most common in our atmosphere, oxygen also exists as O_3—ozone. The structure of ozone is different from that of diatomic oxygen. It consists of three atoms of oxygen rather than the two atoms in diatomic oxygen.

$$\ddot{\underset{\cdot\cdot}{O}}::\ddot{O}:\ddot{\underset{\cdot\cdot}{O}}:$$

Molecules of a single element that differ in crystalline or molecular structure are called **allotropes.** The properties of allotropes are usually different even though they contain the same element. This is because structure can be more important than composition in determining properties of molecules.

Oxygen and ozone are allotropes. The oxygen we breathe is O_2; it is found in the atmosphere. Ozone also occurs naturally and is formed from diatomic oxygen by lightning or ultraviolet light. You may have smelled the sharp odor of ozone during an electrical storm as it formed in the atmosphere by the action of lightning. Small amounts of ozone also are formed in TV sets or computer monitors when an electrical discharge passes through the oxygen in the air. Have you ever smelled it when sitting close to the screen? Because ozone is harmful to living things, it is advisable that you not sit close to your TV set or computer.

Although ozone formed near the surface of Earth is an undesirable component of smog, ozone also has many uses, such as that shown in **Figure 5.14** and in purifying water. The layer of ozone found high in Earth's atmosphere is helpful because it shields living things from harmful ultraviolet radiation from the sun.

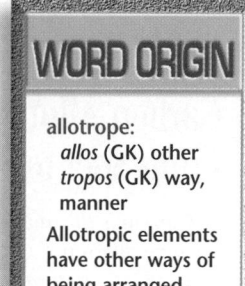
WORD ORIGIN

allotrope:
allos (GK) other
tropos (GK) way, manner
Allotropic elements have other ways of being arranged.

Figure 5.14
Uses of Ozone
Ozone produced in machines like this is used to treat clothing, carpeting, and other materials that have been damaged by smoke and soot from a fire. Hotels also use ozone machines to remove the odor of cigarette smoke from rooms. The reactive ozone oxidizes large, smelly compounds in the smoke and soot into smaller, odorless compounds.

SECTION 5.2

Concept Development
Students may have difficulty remembering the diatomic elements, which can be a problem, particularly when writing chemical equations. These may be remembered by calling them the "diatomic seven," referring both to their number and their locations in the periodic table, which form the numeral 7.

 Assessment

Performance: Supply toothpicks and small, porous Styrofoam balls or packing peanuts, and ask students to work in groups to construct models of diatomic oxygen, ozone, and the three allotropes of phosphorus. Inspect and evaluate their work. L1 LEP COOP LEARN

GLENCOE TECHNOLOGY

Videodisc
Science and Technology Videodisc Series
Hole in the Ozone
Disc 2: Chemistry
Side 1, Ch. 21

Show this video to illustrate how the ozone layer in the upper atmosphere is thinning. L1 LEP

---

tivity, and having a low melting point. The amorphous sulfur is black or dark red in color. It is elastic but slowly changes back into the orthorhombic form.
Analysis: Ask these questions.
1. What are allotropes? *They are two or more forms of an element that differ in molecular structure.*
2. Oxygen, like sulfur, is in Group 16. Give the two allotropes of oxygen. *O_2 and ozone, O_3*

 **Assessment**

Portfolio: Have students do library research to find out whether the other members of Group 16, selenium and tellurium, form allotropes. They should include the results of their research in their portfolios. L1 P

Purpose

Students will become acquainted with the structures and properties of various allotropes of elemental carbon.

Background

- Although carbon is found naturally as diamond, graphite, and some fullerenes, most carbon found on Earth is bonded to hydrogen in the form of hydrocarbons. Carbon ranks 17th (180 ppm) in abundance in Earth's crust.

- Synthetic diamonds are made industrially by subjecting graphite to pressures of 10 000 times that of the atmosphere and temperatures of 2000°C. Under these conditions, the graphite slowly changes structure and forms small diamond crystals. Because the process is costly, these diamonds are expensive.

Teaching Strategies

- Have students recall that lead pencils have various values of hardness. Ask students to explain how the hardness of a pencil is affected by the amount of clay in the "lead." Is there more or less clay in the lead of a soft-lead pencil than in that of a hard-lead pencil? *There is less clay in the lead of a soft-lead pencil. In a soft-lead pencil, the graphite is more easily rubbed off (as smudges) than in a hard-lead pencil.* L1

- To emphasize the large surface areas of activated charcoals, point out or have students calculate from its density that a crystal of sodium chloride, for example, has less than 10^{-3} m^2 of surface area per gram. L3

Carbon Allotropes: From Soot to Diamonds

Carbon is the most versatile element in forming allotropes. Organized or unorganized, atoms of carbon can take on an incredible number of arrangements, each different from the other and each forming a different allotrope. With all their diversity, these substances have one thing in common: they are made up solely of covalently bonded carbon atoms.

Graphite

The most familiar form of carbon is graphite. Mixed with a little clay and formed into a rod, it becomes the lead in a pencil. Look at the structure of graphite. As you can see, the carbon atoms are linked to each other in a continuous sheet of hexagons (six-sided figures). Note that each carbon atom connects three different hexagons. It's clear that the structure of graphite is well organized. The arrays of hexagons are arranged in layers that are loosely held together. The looseness between layers is why graphite is useful in pencil lead. As you write, the surface of the paper pulls off the loosely held layers of carbon atoms.

Graphite

Carbon Blacks

Carbon blacks make up most of the soot that collects in chimneys and becomes a fire hazard. They are formed by the incomplete burning of hydrocarbon compounds, as shown here. Each microscopic chunk of a carbon black is made of millions of jumbled chunks of layered carbon atoms. However, the layers lack the organization of graphite, giving carbon black its haphazard structure. Carbon blacks are used in the production of printing inks and rubber products.

Carbon black

Diamond

Another allotrope of elemental carbon is diamond. Besides being blinded by the brilliance of a cut diamond, you should know that diamond is the hardest natural substance. It's often used on the tips of cutting tools and drills. Can the structure of diamond explain its hardness? Look at the model of diamond. Every carbon atom is attached to four other carbon atoms which, in turn, are each attached to four more carbon atoms. Diamond is one of the most organized of all substances. In fact, every diamond is one huge molecule of carbon atoms. This organization of covalently bonded carbons throughout diamond accounts for its hardness. If you tried to write with a diamond, you'd only tear your paper because layers of carbon atoms do not slip off as they do in graphite. The organization of carbon atoms into diamond occurs under extreme pressure and temperature, often at depths of 200 km and over a long period of time. Diamonds range in age from 600 million to 3 billion years old.

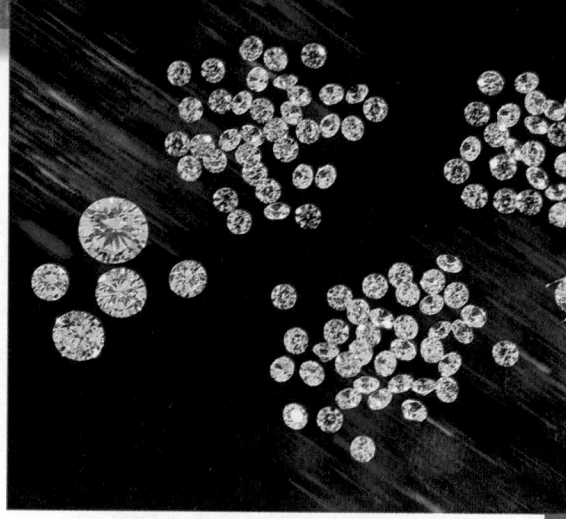

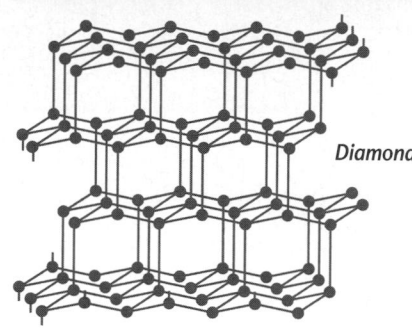

Diamond

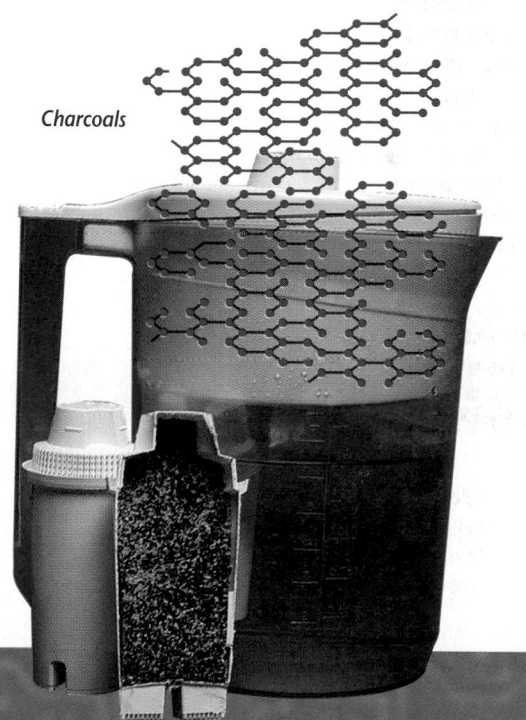

Charcoals

Charcoals

Charcoals—the kind you draw with or cook with—are another type of poorly organized carbon molecules. Charcoals are produced from the burning of organic matter. If you look closely at a chunk of charcoal, you can see that it's extremely porous. All these pores, pock marks, and holes give charcoal a large surface area. Some charcoal, called activated charcoal, has as much as 1000 m^2 of surface area per gram. This property makes activated charcoal useful in filtering water. Molecules, atoms, and ions responsible for unwanted odors and tastes in water are attracted to and held by the surface of the activated charcoal as water passes through it in this water-filtering pitcher.

5.2 Molecular Substances   177

1. In order of increasing density, buckminsterfullerene, graphite, diamond. Buckminsterfullerene appears to occupy the most volume with the smallest number of carbon atoms; therefore, its density should be the smallest. Diamond appears the most compact, so its density should be the greatest.

2. Buckminsterfullerene is a hollow, soccer-ball-shaped molecule consisting of 60 carbon atoms. Linear acetylenic carbon is a spiral string consisting of carbon atoms having molecular formulas of C_{300} to C_{500}. The entire diamond could be considered a single molecule.

3. Buckminsterfullerene is being used to produce synthetic diamonds, act as molecular reaction chambers, and form polymer coatings. Linear acetylenic carbon is being used to synthesize fullerenes and possible synthetic fuels. It also has potential use in microelectronics.

Fullerenes

This is a model of buckminsterfullerene, C_{60}, which was named after the engineer and architect Buckminster Fuller, who invented the geodesic dome shown here. Both the dome and the molecule are unusually stable. The molecule is one of a group of highly organized allotropes of carbon called fullerenes. Buckminsterfullerene was discovered in soot in 1985, and its soccer-ball shape was confirmed in 1991. Since then, other naturally occurring and artificially produced fullerenes have been identified. Fullerenes have even-numbered molecular formulas such as C_{70} and C_{78}. The molecules of some fullerenes are hollow spheres, whereas molecules of others are hollow tubes. The cagelike structures of fullerenes are very flexible. After crashing into steel plates at speeds of 7000 m/s (about 16 000 miles/hour), C_{60} molecules rebound with their original shapes intact.

Fullerenes

Linear Acetylenic Carbon

This threadlike allotrope of carbon is organized into long spirals of bonded carbon atoms. Each spiral contains 300 to 500 carbon atoms. It's produced by using a laser to zap a graphite rod in a glass container filled with argon gas. The allotrope splatters on the glass walls and is then removed. Because they conduct electricity, these carbon filaments may have uses in microelectronics. Some linear acetylenic carbons may eventually form fullerenes, whereas others form soot.

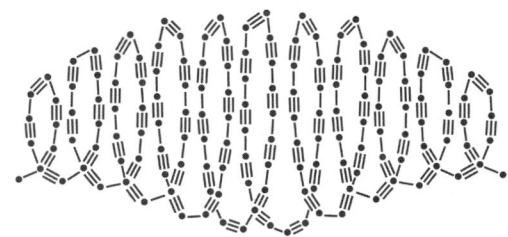

Linear acetylenic carbon

DISCUSSING THE TECHNOLOGY

1. **Applying** From their structures, predict how buckminsterfullerene, diamond, and graphite rank in increasing order of mass density. Explain.

2. **Thinking Critically** How would you describe a molecule of buckminsterfullerene? A molecule of linear acetylenic carbon? A molecule of diamond?

3. **Acquiring Information** What might be some possible uses of fullerenes and linear acetylenic carbons?

*inter*NET
CONNECTION

A student hypertext on allotropes and buckminsterfullerene research is available for use.

World Wide Web
http://fozzie.chem.wisc.edu
/curriculum_development/CurrRef
/BDGTopic/BDGtext/BDGIntro.

Phosphorus has three common allotropes: white, red, and black. All are formed from P₄ molecules that are joined in different ways, giving each allotrope a unique structure and properties, as shown in **Figure 5.15.**

Carbon has several important allotropes with different properties. Diamond is a crystal in which the atoms of carbon are held rigidly in place in a three-dimensional network. In graphite, the carbon atoms are held together closely in flat layers that can slide over each other. This property makes graphite soft and greasy-feeling and useful as a dry lubricant in locks.

Another set of carbon allotropes, the fullerenes, consist of carbon atom clusters. These molecules are unusually stable and are an exciting area of research for chemists because of their potential use as superconductors.

Black phosphorus is a semiconductor, whereas the other two forms are not. ▼

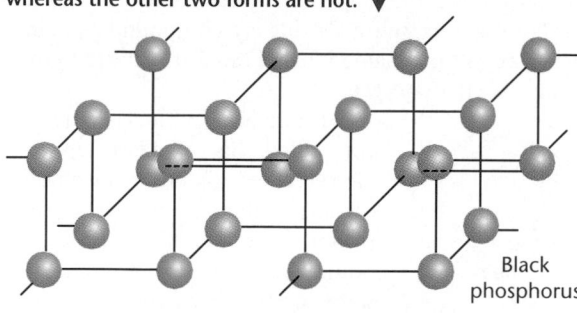

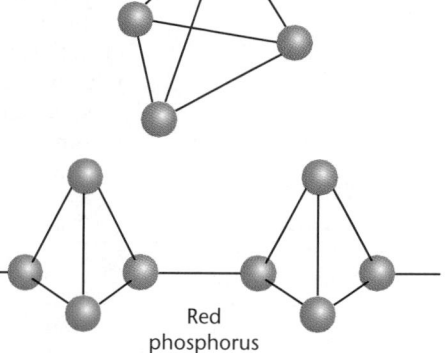

Figure 5.15
Phosphorus Allotropes
The white, red, and black allotropes of phosphorus have different properties. Note the differences in their structures.

White phosphorus

Black phosphorus

Red phosphorus

◀ White phosphorus will ignite spontaneously in air, whereas the red form won't ignite unless it contacts a flame. For these reasons, white phosphorus (left) must be stored under water, and red phosphorus (right) is used in the strike pad of safety matches.

Formulas and Names of Molecular Compounds

Molecular compounds make up a large group; millions of molecular compounds are already known, and scientists are likely to discover or create many others. How can you possibly begin to study so many compounds? Before you can study their structures and properties and learn how these properties determine their usefulness, you should be able to name the compounds and write their formulas. Fortunately, chemists have devised a naming system for molecular compounds that is based on a much smaller number of rules than there are compounds.

5.2 Molecular Substances   179

Let students browse through the Physical Constants of Inorganic Compounds and the Physical Constants of Organic Compounds sections of several handbooks of chemistry and physics. Ask them to estimate the number of known chemical compounds. *at least 10 or 11 million* Emphasize the point that formal naming systems are absolutely necessary in order to specify these compounds. L1

Assessment

Portfolio: As a great deal of practice is required to learn to name compounds and write formulas well, ask students to keep all their practice problems, worksheets, Section Review answers, and Chapter Review answers relating to these topics in their portfolios. This will facilitate studying for exams. L1 P

Transparency

Display **Basic Concepts Transparency 15** to reinforce formula writing and naming. L1 LEP

Concept Development

To decide which element to write first in the formula for a molecular binary compound, name the less electronegative element first followed by the more electronegative element with an *ide* ending.

GLENCOE TECHNOLOGY

Software

Mastering Concepts in Chemistry
Unit 4, Lesson 2
Naming Compounds
Use this lesson to reinforce the study of naming compounds.
L1 LEP

Naming Binary Inorganic Compounds

Substances are either organic or inorganic. Compounds that contain carbon, with a few exceptions, are classified as **organic compounds.** Compounds that do not contain carbon are called **inorganic compounds.** How are inorganic compounds held together? If inorganic compounds contain only two nonmetal elements, they are bonded covalently and are referred to as molecular binary compounds.

To name these compounds, write out the name of the first nonmetal and follow it by the name of the second nonmetal with its ending changed to *-ide*. How do you know which element to write first? You write first the element that is farther to the left in the periodic table, with the exceptions of a few compounds that contain hydrogen. If both elements are in the same group, name first the element that is closer to the bottom of the periodic table. For example, sulfur dioxide is a compound containing sulfur and oxygen. The sulfur is named first because it is closer to the bottom of the periodic table than oxygen is.

Figure 5.16
Carbon Disulfide
The compound represented by the formula CS_2 is named carbon disulfide because two sulfur atoms are bonded to one carbon.

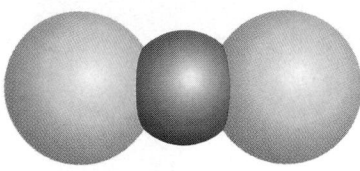

When naming a binary ionic compound, this is the last step. However, because nonmetal atoms can share different numbers of electron pairs, several different compounds can be formed from the same two nonmetal elements. Thus, an additional step is necessary to give an unambiguous name to a molecule. To name the compound correctly, add a prefix to the name of each element to indicate how many atoms of each element are present in the compound. The same prefixes that were used to indicate the number of water molecules in hydrates are used here. For example, CS_2 is named carbon disulfide, **Figure 5.16.** Refer to **Table 5.5** to review these prefixes.

A few other rules are helpful when naming molecular compounds. If only one atom of the *first* element is listed, the prefix *mono* is usually omitted. Also, if the vowel combinations *o-o* or *a-o* appear next to each other in the name, the first of the pair is omitted to simplify pronunciation. Thus, mononitrogen monoxide, NO, becomes nitrogen monoxide.

Now you are ready to practice naming molecular compounds. Several different molecules can be formed when different numbers of nitrogen and oxygen atoms combine. Look at their formulas in the first column of **Table 5.6,** and try to name them without looking at the names listed in the second column. The brown gas pictured is NO_2.

Table 5.6 Formulas and Names of Some Molecular Compounds		
Formula	**Name**	
NO	nitrogen monoxide	
NO_2	nitrogen dioxide	
N_2O	dinitrogen monoxide	
N_2O_5	dinitrogen pentoxide	

Cultural Diversity

Henna
Ask whether students know anyone who uses hair dye or shampoo with henna in it. Henna is a dye that is extracted from the henna plant of Africa, India, and the Middle East. It contains the molecular compound *lawsone,* which leaves hair with a reddish coating. Henna was a valuable trading good and has been used in many different cultures for many thousands of years. It has been detected in the hair of mummified Egyptians.

Consider the two compounds that contain carbon and oxygen. The carbon contained in wood is converted to carbon dioxide when wood burns completely. The formula for this product is CO_2. If the carbon in wood burns incompletely, the highly toxic gas carbon monoxide is formed. What is the formula for carbon monoxide?

To write the formula of a molecular compound for which you are given the name, first write the symbols of each element in the order given in the name. Then add the appropriate subscript after each element that has two or more atoms present. Remember that the prefixes in the name tell how many atoms of each element are present. For example, the compound sulfur hexafluoride contains the elements sulfur and fluorine. Because the word *sulfur* has no prefix, it is understood that there is only one sulfur atom; thus, the symbol S does not require a subscript. The prefix *hexa* tells you that six fluorine atoms are in the compound, so the subscript 6 must be added to the F. The formula for sulfur hexafluoride is SF_6. Follow the rules for writing a formula for a molecular compound as you examine the formula shown in **Figure 5.17**.

N_2O_3

dinitrogen trioxide

Figure 5.17
Formulas of Binary Molecules
The formula for dinitrogen trioxide is written N_2O_3. Analyze the name of this compound to determine how its formula is written.

PRACTICE PROBLEMS

7. Name the following molecular compounds.

   a) S_2Cl_2
   b) CS_2
   c) SO_3
   d) P_4O_{10}

8. Write the formulas for the following molecular compounds.

   a) carbon tetrachloride
   b) iodine heptafluoride
   c) dinitrogen monoxide
   d) sulfur dioxide

Common Names

A few inorganic molecular compounds have common names that all scientists use in place of formal names. Two of these compounds are water and ammonia. The chemical name for water is dihydrogen monoxide because each molecule contains two hydrogen atoms and one oxygen atom. If you wanted to get a glass of water at a restaurant, would you ask for dihydrogen monoxide? Probably not, at least not if you were really thirsty. Most people would not understand you because you used a name that even chemists never use for water. Although the formal names of both ionic and molecular compounds are simple to write once you learn the rules for the language of chemistry, there are good reasons for sometimes using common names. Which name you use will depend on your audience.

Fact of the MATTER

Vast deposits of methane trapped under high pressure in the pores of ice have been located deep under the ocean floor. This ice is as cold as the ice in your freezer, but it burns. If the methane in the ice can be harvested, it may replace Earth's dwindling supplies of fossil fuels because the deposits contain more than twice the amount of energy as in all fossil fuels combined.

5.2 Molecular Substances 181

Check for Understanding

- Check for understanding of the section material by having students answer the Section Review questions. L1
- Students may wonder how the atoms of an element are able to form varying numbers of covalent bonds. The bonding may involve varying numbers of the atom's valence electrons and in various geometric patterns. You might illustrate this by having two, then three, then four students hold hands in varying ways; for example: two students holding one hand each, two students holding both hands, three students forming a line with the middle student holding a hand with each of the other two, three students forming a circle, etc. L1 LEP

The common acids are other examples of inorganic compounds that are known by common rather than formal names. Some names of common acids and bases that you will use frequently in chemistry laboratory experiments are listed in **Table 5.7.** Although they often do not follow the rules you have been learning, they will soon become so familiar that their formulas and names will be easy to remember.

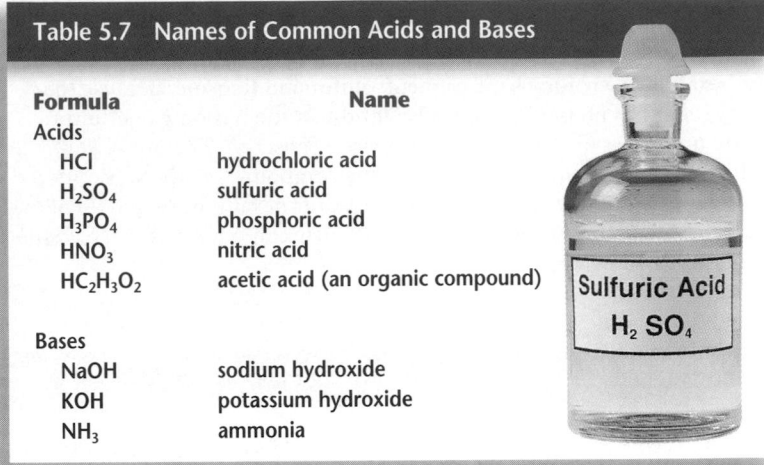

Table 5.7   Names of Common Acids and Bases

Formula	Name
Acids	
HCl	hydrochloric acid
H_2SO_4	sulfuric acid
H_3PO_4	phosphoric acid
HNO_3	nitric acid
$HC_2H_3O_2$	acetic acid (an organic compound)
Bases	
NaOH	sodium hydroxide
KOH	potassium hydroxide
NH_3	ammonia

Sulfuric Acid $H_2 SO_4$

Figure 5.18

Uses of Hydrocarbons
The structures of methane (left) and propane (right) are shown here. Count the number of carbon atoms in each. Many hydrocarbons are used for fuel. Methane is the main component in natural gas, and propane is used in gas grills.

Naming Organic Compounds

You have learned that most compounds that contain carbon are organic compounds. Organic compounds make up the largest class of molecular compounds known. This is because carbon is able to bond to other carbon atoms in rings and chains of many sizes.

Methane CH_4

Propane C_3H_8

Chemistry Journal

Ionic or Covalent?

Give students a piece of paraffin and ask how they might determine whether it is ionic or covalent. Students should write up some simple tests, perform the tests, and write their results in their chemistry journals. Tests for hardness, brittleness, and solubility in water and organic solvents can be made by bending the paraffin, scraping it with a spatula, and immersing it in water. Students should conclude that paraffin is a molecular substance. L1 LEP

The name of even the most complex organic compound is based on the name of a **hydrocarbon,** an organic compound that contains only the elements hydrogen and carbon. Hydrocarbons occur naturally in fossil fuels such as natural gas and petroleum and are used mainly as fuels and the raw materials for making other organic compounds.

A carbon atom can form four covalent bonds. In the simplest hydrocarbon, methane, a single carbon is bonded to four hydrogen atoms. Methane is the main component of the natural gas that you burn when you light a Bunsen burner. The next simplest hydrocarbon, ethane, is formed when two carbon atoms bond to each other as well as to three hydrogen atoms apiece. The formulas and names of the first ten hydrocarbon chains are shown in **Table 5.8.** Note that the names of hydrocarbons are derived from the number of carbon atoms in the molecules. Do you recognize any of these hydrocarbons? What is propane used for? **Figure 5.18** shows its structure and one common use.

Table 5.8 Hydrocarbons

Formula	Name
CH_4	methane
C_2H_6	ethane
C_3H_8	propane
C_4H_{10}	butane
C_5H_{12}	pentane
C_6H_{14}	hexane
C_7H_{16}	heptane
C_8H_{18}	octane
C_9H_{20}	nonane
$C_{10}H_{22}$	decane

Connecting Ideas

Formulas represent the known composition of real substances, but just because a formula can be written doesn't mean the compound actually exists. For example, you could easily write the formula HeP_2, but no such compound has ever been isolated. Compounds containing the noble gases helium, neon, and argon have never been found. In the next chapter, you will study the chemical changes that elements and compounds undergo and learn how to represent these changes in the language of chemistry.

SECTION REVIEW

Understanding Concepts

1. Write the formula for each of the following molecular compounds.

   a) carbon monoxide
   b) phosphorus pentachloride
   c) sulfur hexafluoride
   d) dinitrogen pentoxide
   e) iodine trichloride
   f) heptane

2. Write the name of the molecular compound represented by each formula.

   a) BF_3 d) IF_7
   b) PBr_5 e) NO
   c) C_2H_6 f) SiO_2

3. Explain what allotropes are and give two examples.

Thinking Critically

4. **Applying Concepts** Explain, in terms of electron structure, why carbon usually forms four bonds.

Applying Chemistry

5. **Tank of Gas** A tank of a substance delivered to a factory is labeled C_4H_{10}. What is the name of the substance in the tank? What is its most likely use?

SECTION REVIEW

1. **a.** CO **b.** PCl_5 **c.** SF_6 **d.** N_2O_5 **e.** ICl_3 **f.** C_7H_{16}
2. **a.** boron trifluoride **b.** phosphorus penta-bromide **c.** ethane **d.** iodine heptafluoride **e.** nitrogen monoxide **f.** silicon dioxide
3. Allotropes are different structures of molecules made up of the same element. Examples are the diamond, graphite, and fullerene allotropes of carbon; the red, white, and black allotropes of phosphorus; and the ozone and diatomic forms of oxygen.
4. Carbon is in Group 14 of the periodic table, and thus has four valence electrons. Each electron can be shared with another atom, forming bonds between carbon and four other atoms.
5. butane; fuel in disposable lighters, raw material for production of motor fuels and synthetic rubber

- Review the Reviewing Main Ideas statements and Key Terms with your students.
- Complete solutions to Chapter Review problems can be found in the *Problems and Solutions Manual.*

UNDERSTANDING CONCEPTS

1. **a.** ionic **b.** molecular **c.** molecular **d.** molecular **e.** ionic **f.** ionic
2. **a.** MnI_3 **b.** CaO **c.** AlF_3 **d.** K_2S **e.** $ZnBr_2$ **f.** PbO_2
3. **a.** calcium acetate **b.** sodium hydroxide **c.** ammonium sulfite monohydrate **d.** magnesium sulfate **e.** sodium nitrite **f.** calcium hydroxide

4.

Ionic	Molecular
electrolyte	nonelectrolyte
high melting point	low melting point
hard	soft or hard
brittle	nonbrittle

5. **a.** charge: 3+; name: iron(III) chloride
   b. charge: 2+; name: copper(II) fluoride
   c. charge: 3+; name: gold(III) bromide
   d. charge: 4+; name: tin(IV) bromide
   e. charge: 2+; name: iron(II) sulfide
   f. charge: 2+; name: lead(II) acetate
6. **a.** common **b.** common **c.** formal **d.** formal **e.** common **f.** formal
7. Ionic compounds contain both metal and nonmetal atoms, with the exception of ammonium compounds; molecular compounds contain only nonmetal atoms.
8. Na_2HPO_4 contains two sodium metal ions bonded ionically to a monohydrogen phosphate polyatomic ion by ionic bonds. The hydrogen, phosphorus, and oxygen atoms in the polyatomic ion are held together by covalent bonds.
9. One way to separate dissolved ionic compounds from water is by distillation, a process that involves heating the water until it boils, letting the water vapors

184

REVIEWING MAIN IDEAS

5.1 Ionic Compounds

- Binary ionic compounds are named by first naming the metal element and then the nonmetal element, with its ending changed to *-ide*. Subscripts are used in formulas to indicate how many atoms of each element are present in the compound.

- The position of an element in the periodic table indicates what charge its ions will have.

- Polyatomic ions have fixed charges and can combine with ions of opposite charge to form ionic compounds. These compounds are named by writing the name of the positive ion first and then the name of the negative ion.

- Most transition elements can form two or more positively charged ions. When naming a compound that contains a transition element, the oxidation number of the transition element is indicated by a Roman numeral in parentheses.

- Hydrates are ionic compounds bonded to water molecules. They are named by following the name of the compound with a prefix attached to the word *hydrate* to indicate how many water molecules are bound.

5.2 Molecular Substances

- Molecular substances have a greater variety of properties than do ionic compounds, but generally they have low melting points, low water solubility, and little or no ability to act as electrolytes.

- Seven elements occur naturally as diatomic molecules. They are hydrogen, nitrogen, oxygen, fluorine, chlorine, bromine, and iodine.

- Some elements exist in different structural forms called allotropes. Allotropes of an element have different properties.

- Binary molecular compounds are named by writing the two elements in the order they are found in the formula, changing the ending of the second element to *-ide,* and adding Greek prefixes to the element names to indicate how many atoms of each are present.

- It is important to know both the formal and common names of chemicals because both are part of the language of chemistry.

- The naming system for organic compounds is based on the names of hydrocarbons.

Key Terms

For each of the following terms, write a sentence that shows your understanding of its meaning.

allotrope	hygroscopic
anhydrous	inorganic compound
binary compound	molecular element
deliquescent	molecular substance
distillation	organic compound
formula unit	oxidation number
hydrate	polyatomic ion
hydrocarbon	

● UNDERSTANDING CONCEPTS

1. Which of the following substances are ionic and which are molecular?

   a) magnesium sulfate   **d)** ozone
   b) hexane   **e)** cesium chloride
   c) carbon monoxide   **f)** cobalt(II) chloride

2. Write the formula for the binary ionic compound that forms from each pair of elements.

   a) manganese(III) and iodine
   b) calcium and oxygen
   c) aluminum and fluorine
   d) potassium and sulfur
   e) zinc and bromine
   f) lead(IV) and oxygen

✓ Assessment

Portfolio: Review the portfolio options that are provided throughout the chapter. Encourage students to select one product that demonstrates their best work for the chapter. Have students explain what they have learned and why they chose this example for placement into their portfolios. **P**

Additional portfolio options can be found in the following **Teacher Classroom Resources:**
Tech Prep Applications
Consumer Chemistry
Chemistry and Industry

3. Write the name for each of the following compounds containing polyatomic ions.

a) $Ca(C_2H_3O_2)_2$ d) $MgSO_4$
b) $NaOH$ e) $NaNO_2$
c) $(NH_4)_2SO_3 \cdot H_2O$ f) $Ca(OH)_2$

4. Make a table comparing the properties of ionic and molecular compounds.

5. The metals in the following compounds can have various oxidation numbers. Predict the charge on each metal ion, and write the name for each compound.

a) $FeCl_3$ d) $SnBr_4$
b) CuF_2 e) FeS
c) $AuBr_3$ f) $Pb(C_2H_3O_2)_2$

6. Label the following as common names or formal names.

a) alumina
b) sulfuric acid
c) sodium iodide
d) sodium hydrogen carbonate
e) dry ice
f) potassium acetate

7. How can you tell if a compound is ionic or molecular by examining its formula?

8. How is Na_2HPO_4 a substance with two different types of bonding?

9. How can water that contains dissolved ionic compounds be separated from the ionic compounds?

10. Write the names and symbols for an ion and an element that have the same electron configuration.

11. Write formulas for a bromine atom, ion, and molecule.

12. Name the molecular compound that is represented by each of the following formulas.

a) NO d) CO
b) IBr e) SiO_2
c) N_2O_4 f) ClF_3

13. What happens to the composition of a hydrate when it is heated?

APPLYING CONCEPTS

14. In Samuel Coleridge's poem *The Rime of the Ancient Mariner,* the mariner cried the following while on his ship far from shore. "Water, water everywhere, and all the boards did shrink/Water, water everywhere, nor any drop to drink." What did he mean?

15. Predict the effect of increasing acidity of rain on the rate of formation of limestone caves.

16. Paintings discovered in a limestone cave in Lascaux, France, in 1940 were a popular tourist attraction until authorities sealed the cave to protect the paintings. Prior to the discovery of the artwork from thousands of years ago, the cave was well sealed from the atmosphere. Suggest a reason why opening the cave sped up the deterioration of the paintings.

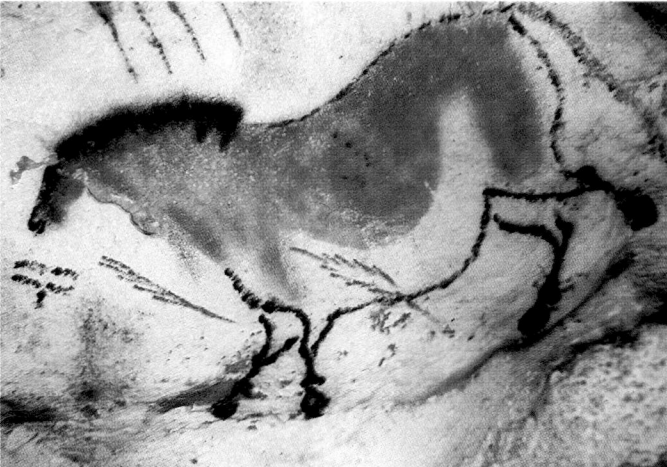

17. How could you determine quantitatively whether sodium hydroxide or calcium chloride is more deliquescent?

18. Why don't the noble gases form compounds easily?

19. How could you determine quantitatively whether the ionic compound table salt or the molecular compound table sugar is more soluble in water?

escape from the ionic impurities, which have much higher boiling points, and then condensing the water back into liquid by cooling it.

10. The sodium ion (Na^+) has the same electron configuration as neon (Ne), as does the fluoride ion (F^-); the potassium ion (K^+) has the same configuration as argon (Ar), as does the chloride ion (Cl^-).

11. bromine atom: Br; bromide ion: Br^-; bromine molecule: Br_2

12. **a.** nitrogen monoxide **b.** iodine monobromide **c.** dinitrogen tetroxide **d.** carbon monoxide **e.** silicon dioxide **f.** chlorine trifluoride

13. When a hydrate is heated, water is removed from the compound, changing the formula of the hydrate. When all the water has evaporated, an anhydrous compound remains.

APPLYING CONCEPTS

14. The ancient mariner lamented that though he was surrounded by a sea of water, it was all salt water. Our kidneys cannot filter out all of the salt, so drinking salt water leads to dehydration and death as water is pulled from other parts of the body to dilute the salt. Water was also pulled out of the cellular structure of wood in the ship, causing the wood to shrink.

15. More acidic rain will hasten the formation of caves and sinkholes in areas that contain a lot of limestone because the acids dissolve limestone as they move into the ground with the rainwater.

16. Opening the cave allowed water vapor, oxygen, and other reactive substances in the air to move inside. These substances damaged the paintings by reacting with compounds in the cave walls and in the pigment used in the paint.

17. To determine which compound is more deliquescent, equal masses of dry samples of each compound could be weighed before and after exposure to moist air. The time required for each to liquefy could also be compared.

18. The noble gases each have a full valence shell of electrons, so no electrons are available for sharing or donating to another atom and there is no room for gaining electrons from other atoms.

19. To determine whether salt or sugar is more soluble in water, the amount of each that can dissolve in equal volumes of water without precipitating out can be measured.

20. The formulas follow the pattern C_nH_{2n+2}. When n equals the number of carbon atoms, the number of hydrogen atoms is always $2n + 2$; $C_{18}H_{38}$.

21. The atoms of most elements have incomplete valence shells, so the atoms readily react with other atoms to complete the shells and attain stability.

22. Hard water contains a high concentration of ions, especially Mg^{2+} and Ca^{2+} ions. Hard water can be treated by passing it

through an ion exchange resin, in which calcium and magnesium ions are each exchanged for two sodium ions.

23. The artisan should add to the glaze a dilute solution of chromium ions of the appropriate charge.

24. Graphite consists of arrays of hexagons arranged in loosely held layers. These layers easily slide over one another, making graphite slippery.

25. Concrete is cement to which sand and gravel have been added.

THINKING CRITICALLY

26. To determine the smallest amount of water required to change the color of the anhydrous cobalt compound, water can be added to a measured amount of the compound, drop by drop, until the color changes.

27. Vinegar contains acetic acid, which reacts with calcium in the bone to form soluble calcium acetate. The calcium acetate leaches out of the bone into the vinegar solution. Ionic calcium compounds make bones hard and brittle; when they are removed, the bone is left with mostly softer, more flexible molecular compounds.

28. A warm, saturated solution of KNO_3 will probably be a better electrolyte and conduct electricity more easily. This is because the solubility of the salt increases with increasing temperature, so more ions will be dissolved in warmer water and will be available for conducting electricity.

29. The melting points generally decrease, suggesting that the strength of the bonds also decreases.

30. 2+ charge; Hg_2Cl_2

20. Examine the formulas of the simple hydrocarbons in **Table 5.8,** and devise a general rule that allows you to calculate the number of hydrogen atoms present if you are given the number of carbon atoms. Use this rule to write the formula for a simple hydrocarbon containing 18 carbon atoms.

21. Explain why most elements do not occur naturally in their pure state.

Everyday Chemistry

22. What is hard water and how is it treated?

Art Connection

23. Suppose an artisan wanted to coat a clay vessel with a faint pink glaze. What material should he or she add to a transparent glaze to achieve this result?

Chemistry and Technology

24. Use the structural organization of graphite to explain why it is a good lubricant.

How It Works

25. How does concrete differ from cement?

THINKING CRITICALLY

Designing an Experiment

26. **MiniLab 1** Design an experiment to determine the minimum amount of water required to change the color of the anhydrous cobalt compound weather predictor.

Relating Cause and Effect

27. **MiniLab 2** Why did vinegar soften the chicken bone?

Making Predictions

28. **ChemLab** Would you expect a warm or a cool saturated solution of KNO_3 in water to be a better electrolyte? Explain.

Using a Table

29. The following table lists melting points for a number of ionic compounds. Use a periodic table to help you answer the following ques-

tions. Do the melting points of the sodium and potassium compounds increase or decrease as you move down Group 17? What does this suggest about the strength of the ionic bonds between these metals and the Group 17 nonmetals?

Compound	Melting Point (°C)
NaCl	804
NaI	651
KCl	773
KBr	730
NaF	993
KI	680
NaBr	755

Applying Concepts

30. Mercury(I) is unusual in that it often forms an ion that links with another mercury(I) ion. Thus, two mercury(I) ions are linked together in a single unit. What is the charge on this double ion? Write the formula for the compound that is formed from this double ion and chlorine.

CUMULATIVE REVIEW

31. How are physical changes different from chemical changes? (Chapter 1)

32. How does the atomic number compare with the number of electrons in a neutral atom? (Chapter 2)

33. How can the periodic table be used to determine the number of valence electrons in an element? (Chapter 3)

34. You are given a white crystalline substance to identify. You attempt to determine its melting point and find that it does not melt upon heating to 600°C. The substance dissolves readily in water. Is this substance more likely to be ionic or covalent? Explain. (Chapter 4)

CUMULATIVE REVIEW

31. Physical changes involve reversible changes in matter without the formation of new kinds of matter. Chemical changes convert matter into new kinds of matter with different properties.

32. The atomic number equals the number of electrons in a neutral atom.

33. The number of valence electrons that an element has can be determined by locating the element on the periodic table. Atoms in each group have the following number of valence electrons:

Group Number	Number of Valence Electrons
1	1
2	2
13	3
14	4
15	5
16	6
17	7
18	8

34. The substance is probably ionic because it has a high melting point, dissolves readily in water, and is crystalline.

Skill Review

1. **Making and Using Graphs** Construct a graph of melting point and water solubility versus number of carbons, using the following data. What is the relationship between chain length and melting point? Can you explain why? How are chain length and water solubility related?

Number of Carbon Atoms	Melting Point (°C)	Water Solubility (g per 100 mL)
1 (methane)	−183	0.0024
2 (ethane)	−172	0.0059
3 (propane)	−188	0.012
4 (butane)	−138	0.037
5 (pentane)	−130	0.036
6 (hexane)	−95	0.0138
7 (heptane)	−91	0.0052
8 (octane)	−57	0.0015
9 (nonane)	−54	insoluble
10 (decane)	−30	insoluble

Writing in Chemistry

2. Write a set of descriptions comparing the structures of a soccer ball, a geodesic dome, and buckminsterfullerene. Is their similarity a coincidence?

Problem Solving

3. Different ions and elements that have the same electronic structure are said to be isoelectronic. Na^+ is isoelectronic with Ne; both have ten electrons, including eight valence electrons. Mg^{2+} also is isoelectronic with Na^+ and Ne. Write the symbols for two ions or elements that are isoelectronic with each of the following. Then, write a sentence for each, describing the electron configuration.

   a) Ca^{2+} c) I^-
   b) Kr   d) O^{2-}

Projects

4. Natural diamonds form slowly over many thousands of years under intense heat and pressure deep inside Earth. Small, synthetic diamonds can now be manufactured. Find out how they are made and how the size and properties of synthetic diamonds compare with those of natural diamonds.

5. To keep tropical saltwater fishes and other marine animals in an aquarium, ionic compounds must be added to the water. Use a library or local pet store as a source for finding out what combination of compounds must be added to fresh water to make synthetic seawater. If possible, set up a saltwater aquarium.

AUTHENTIC ASSESSMENT

1. Melting point generally increases with chain length, whereas water solubility decreases. As chain length increases, molecules have more area with which to form the intermolecular attractions that hold chains together in a solid. The longer the chain, the harder it is to pull molecules apart in a solid to melt it, and the harder it is to pull molecules apart to dissolve them in water. For the first four hydrocarbons, this general solubility trend is not observed. They are gases at room temperature, and vapor pressure considerations complicate the solubility.

3. **a.** K^+, Ar, Cl^- **b.** Br^-, Sr^{2+}, Rb^+ **c.** Xe, Ba^{2+}, Cs^+ **d.** F^-, Ne, Na^+, Mg^{2+}; all ions and atoms have a full outer level of electrons.

Program Resources

Chapter Review and Assessment, pp. 29-34
　L1

Alternate Assessment in the Science Classroom L1

Computer Test Bank, Chapter 5 L1

Problems and Solutions Manual, Chapter 5
　L1

Supplemental Practice Problems, Chapter 5
　L1

Performance Assessment in the Science Classroom L1

Chapter 6 Organizer

Section	Objectives	Activities/Features
6.1 **Chemical Equations** (3 days)	**1. Relate** chemical changes and macroscopic properties. **2. Demonstrate** how chemical equations describe chemical reactions. **3. Illustrate** how to balance chemical reactions by changing coefficients.	**Everyday Chemistry:** Whitening Whites, p. 194 **MiniLab 1:** Energy Change, p. 196 **How It Works:** Emergency Light Sticks, p. 197 **Discovery Demo:** 6.1, pp. 188-189 **Demonstrations:** 6.2 pp. 194-195; 6.3 pp. 198-199
6.2 **Types of Reactions** (3 days)	**4. Distinguish** among the five major types of chemical reaction. **5. Classify** a reaction as belonging to one of five major types.	**Biology Connection:** Air in Space, p. 203 **MiniLab 2:** A Simple Exchange, p. 205 **ChemLab:** Exploring Chemical Changes, pp. 206-207 **Demonstration:** 6.4 pp. 204-205
6.3 **Nature of Reactions** (4 days)	**6. Demonstrate** factors that influence the direction of a reaction. **7. Classify** factors that influence the rate of a reaction.	**People in Chemistry:** Caroline Sutliff, Plant-care Specialist, pp. 212-213 **Chemistry and Technology:** Mining the Air, pp. 216-217 **MiniLab 3:** Starch-Iodine Clock Reaction, p. 220 **Everyday Chemistry:** Stove in a Sleeve, p. 221 **Demonstrations:** 6.5 pp. 212-213; 6.6 pp. 214-215; 6.7 pp. 218-219

Activity Materials

ChemLab (pages 206-207)
125-mL flasks (4), balance, hot plate, watch glass, spatula, stirring rod, lab burner, file, new penny, 250-mL flask, ice, tongs, 100-mL graduated cylinder, large test tube and 1-hole stopper with glass tube and rubber tubing, ring stand, test-tube clamp, 0.1M $CuSO_4$, granular copper, powdered sulfur, $CaCO_3$, limewater, 6M HCl, 0.5M Na_2CO_3, 0.5M $CuCl_2$

MiniLab 1 (page 196)
powdered Fe, NaCl, resealable plastic bag, vermiculite, balance, 10-mL graduated cylinder

MiniLab 2 (page 205)
0.1M $AgNO_3$, test tube, copper wire or foil

MiniLab 3 (page 220)
wax pencil, large test tubes (5), test-tube rack, starch solution, 25-mL graduated cylinder, ice, 400-mL beaker (2), hot plate, Celsius thermometer, iodine solution, stirring rod, timer with second hand

KEY TO TEACHING STRATEGIES

The following designations will help you decide which activities are appropriate for your students.

L1 Level 1 activities should be within the ability range of all students.

L2 Level 2 activities should be within the ability range of the average to above-average student.

L3 Level 3 activities are designed for the ability range of above-average students.

LEP LEP activities should be within the ability range of Limited English Proficiency students.

COOP LEARN Cooperative Learning activities are designed for small group work.

P These strategies represent student products that can be placed into a best-work portfolio.

Demonstrations For a complete list of materials for the demonstrations in this chapter, see pages 188, 194, 198, 204, 212, 214, and 218.

Chemical Reactions and Equations

Teacher Classroom Resources

STUDENT MASTERS

Study Guide: pp. 21-22
ChemLab and MiniLab Worksheets: p. 44

Study Guide: pp. 23-24
Laboratory Manual: 6.1 Types of Chemical Reactions, pp. 49-54
ChemLab and MiniLab Worksheets: p. 45

Study Guide: pp. 25-26
Critical Thinking/Problem Solving: p. 7
Laboratory Manual: 6.2 Effects of Concentration on Chemical Equilibrium, pp. 55-60
ChemLab and MiniLab Worksheets: pp. 41-43, 46

TEACHING AIDS

Section Focus Transparency 11
Section Focus Transparency Master: p. 11
Basic Concepts Transparency 16
Basic Concepts Transparency Master: pp. 31-32
Problem Solving Transparency 6, 7
Problem Solving Transparency Masters: pp. 11-14
Lesson Plans: p. 11

Section Focus Transparency 12
Section Focus Transparency Master: p. 12
Basic Concepts Transparency 17
Basic Concepts Transparency Master: pp. 33-34
Lesson Plans: p. 12

Section Focus Transparency 13
Section Focus Transparency Master: p. 13
Basic Concepts Transparency 18
Basic Concepts Transparency Master: pp. 35-36
Lesson Plans: p. 13

ASSESSMENT RESOURCES
Chapter Review and Assessment: pp. 35-40
Computer Test Bank: Chapter 6
Performance Assessment in the Science Classroom
Supplemental Practice Problems: Chapter 6

CHAPTER RESOURCES
Applying Scientific Methods in Chemistry: p. 23
Spanish Resources: Chapter 6
English/Spanish Audiocassettes: Chapter 6
Cooperative Learning in the Science Classroom
Lab and Safety Skills in the Science Classroom

GLENCOE TECHNOLOGY

Software
Mastering Concepts in Chemistry: *Unit 7* Lessons 1, 2, 3; *Unit 3* Lesson 1
Videotape
MindJogger Videoquizzes, Chapter 6
Videodisc
Chemistry: Concepts and Applications, Demonstrations, Videos, and Animations
Science and Technology Videodisc Series Disc 2 Chemistry: *Super Grit; Grain-Dust Explosions*
CD-ROM Multimedia System
Chemistry: Concepts and Applications, Same as for Videodisc, plus Interactive Exploration and Major Simulation *The Periodic Table*

TECH PREP

The following Glencoe resources provide opportunities for integrating science and technology.

Student Edition: MiniLab 1, p. 196; How It Works, p. 197; Chemistry and Technology, pp. 216-217

Teacher Wraparound Edition: Cultural Diversity, p. 216

Teacher Classroom Resources: Tech Prep Applications, pp. 9-10; Chemistry and Industry, pp. 13-14; Consumer Chemistry, pp. 11-12

CHAPTER 6 Chemical Reactions and Equations

Chapter Overview

Students learn that chemical changes are often evidenced by physical changes. How to represent chemical changes with balanced equations is explained. The five major types of reactions are illustrated. New reactions can be classified into one of these classes by examining their equations. Students are then introduced to reversible reactions and equilibrium and factors that can influence the direction and rate of reactions.

Theme Connection

Equilibrium and Change/Conservation *Equilibrium and change* are evident in equilibria and chemical reactions. *Conservation* of matter as it applies to chemical reactions and balanced chemical equations is explained.

Assessment Planner

Choose assessment strategies from the following pages to evaluate the progress of your students.
Assess, pp. 200, 209, 223
MiniLabs, pp. 196, 205, 220
ChemLab, p. 206
Portfolio, pp. 195, 199, 201, 205, 209, 219, 223, 224
Chapter Review, p. 224

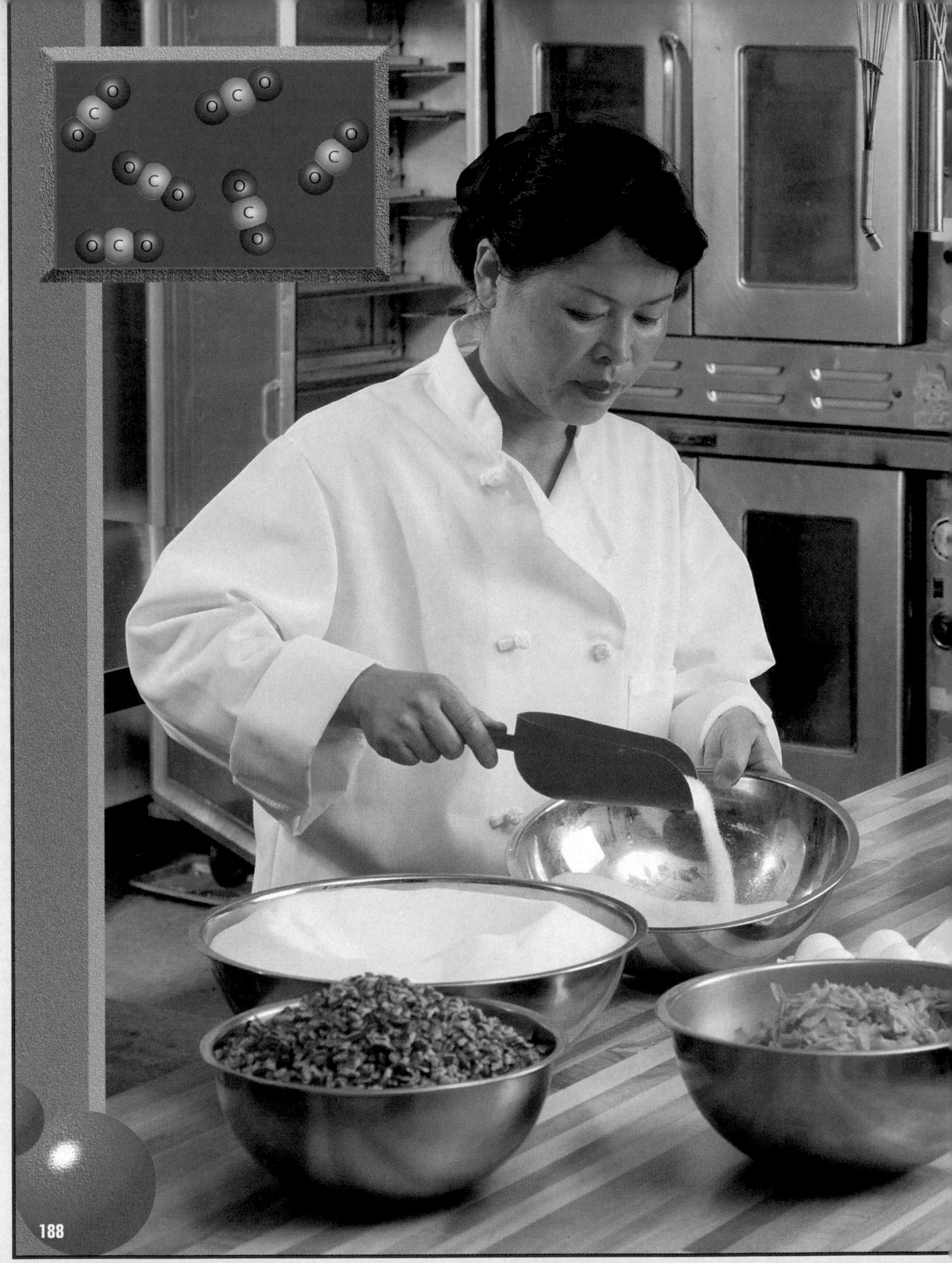

Discovery Demo 6.1

Combustion of Methane

Purpose: To illustrate the role of reactants and products in chemical reactions.

Materials: 400-mL beaker, 150-mL beaker, 25 mL of liquid detergent, 5 mL of glycerin, 5 g of sucrose, 1-m length rubber tubing, small funnel, meterstick, candle, matches, newspaper, masking tape

Safety Precautions:

Disposal: Solution may be poured down the drain.

Procedure: Prepare a soap-bubble solution by adding 160 mL of H_2O, 25 mL of liquid detergent, and 5 mL of glycerin to the 400-mL beaker. In a separate beaker, dissolve 5 g of sucrose in 60 mL of H_2O. Gently mix the two solutions. Connect rubber tubing to the gas outlet and the other end to a small funnel. Invert the funnel into the beaker of the soap mixture, then remove the funnel. Turn on the gas. Make a bubble using the gas. Darken the room. Dislodge bubbles by turning the funnel sideways and gently shaking. As the bubble rises (natural gas) or sinks (propane gas), ignite it with a burning candle taped

CHAPTER 6
Chemical Reactions and Equations

Some of the most amazing and least understood chemical changes occur right in your kitchen. Think about the many different materials used to make a cake—salt, sugar, baking powder, eggs, water, shortening, and flour. The properties of these ingredients change when the ingredients are mixed and baked. These differences all indicate that chemical changes have taken place. Some of these changes are complex, and others are simple. Baking the cake provides the heat necessary to speed up the changes so they will occur quickly.

What causes these changes to occur? In this chapter, you will learn some of the factors that determine whether a change will occur and how fast it will go if it does occur. You will also review how observing macroscopic changes in matter can indicate whether or not a chemical change has occurred. You will then use what you have learned about chemical formulas to represent chemical changes using equations.

Chemistry Around You

If you could examine what is happening at the submicroscopic level as a cake bakes, you would see specific compounds breaking down and new compounds forming. They undergo chemical changes. Chemical changes occur in all foods when they are cooked, such as when carbon dioxide bubbles form and cause a cake to rise. In this chapter, you'll learn about some of these chemical changes and how you can represent what happens during a chemical change.

 Concept Check

Review the following concepts before studying this chapter.
Chapter 1: what a compound is; the meaning of a formula
Chapter 3: what periodicity is; how periodicity applies to elements
Chapter 4: reasons why atoms combine
Chapter 5: how to write chemical formulas

 Chapter Preview

6.1 Chemical Equations

6.2 Types of Reactions

6.3 Nature of Reactions

Laboratory Activities
ChemLab
Exploring Chemical Changes

MiniLabs
1. Energy Change
2. A Simple Exchange
3. Starch-Iodine Clock Reaction

189

to the end of a meterstick. It is best to have an assistant hold the meterstick with the attached candle above or below the funnel before you shake it free. **CAUTION:** *Fire hazard. Do not do this demonstration near flammable materials.* Place newspapers on the floor to catch the wax dripping from the candle.
Results: There will be a flare-up as the bubble bursts, and the entrapped gas burns with a luminous yellow flame.

Analysis: Ask these questions.
1. The combustion you observed is a chemical reaction between what two reactants? *methane or propane and oxygen*
2. What was produced? *carbon dioxide, carbon monoxide, light, heat, water vapor, soot*
3. Did the burning fuel give off energy or absorb energy? *It gave off energy.*

SECTION 6.1

Key Concepts

Students are encouraged to associate reactions with the types of changes that are commonly observed. They then utilize the law of conservation of mass to write balanced chemical equations, which also include energy changes and physical states of reactants and products.

Planning Ahead

For the MiniLab Purchase vermiculite from a lawn and garden store.

For the Demonstrations Be sure to have a functional Hoffman electrolysis apparatus on hand for Demonstration 6.2.

1 FOCUS

Focus Transparency

 Display the **Section Focus Transparency** for Section 6.1. Relate chemical changes to the need for writing a chemical equation. L1 LEP

GLENCOE TECHNOLOGY

Videodisc

Chemistry: Concepts and Applications
See the Videodisc Program Teacher Guide to access illustrations and other material available for this chapter.

SECTION 6.1 Chemical Equations

SECTION PREVIEW

Objectives

Relate chemical changes and macroscopic properties.
Demonstrate how chemical equations describe chemical reactions.
Illustrate how to balance chemical reactions by changing coefficients.

Key Terms

reactant
product
coefficient

What do you remember about last summer's Fourth of July? Amazing bursts of color from fireworks shot over a lake? Mouthwatering aromas coming from a barbecue grill? Do the processes that result in those colors and smells have anything in common?

You learned in Chapter 1 that substances undergo both physical and chemical changes. A physical change does not change the substance itself, but a chemical change does. Did the chemicals making up the fireworks, charcoal, and barbecued food undergo chemical changes?

Recognizing Chemical Reactions

When a substance undergoes a chemical change, it takes part in a chemical reaction. After it reacts, it no longer has the same chemical identity. While it may seem amazing that a substance can undergo a change and become part of a different substance, chemical reactions occur around you all the time. Chemical reactions can be used to heat a home, power a car, manufacture fabrics for clothing, make medicines, and produce paints and dyes in your favorite colors. Reactions also provide energy for walking, running, working, and thinking.

Many important clues indicate when chemical reactions occur. None of them alone proves that such a change occurs because some physical changes involve one or more of these signs. Examine the photographs in **Figure 6.1** to see what clues to look for.

Figure 6.1

Signs of Chemical Reaction
When substances undergo chemical changes, observable differences usually occur. If you know what signs to look for, you can determine whether or not a chemical reaction has taken place.

Color changes often accompany chemical changes. If you place brownish-red iodine solution on a freshly cut potato, it reacts with white starch to produce a blue compound. ▶

Program Resources

Study Guide, pp. 21-22 L1
English/Spanish Audiocassettes, Chapter 6
L1 LEP
Consumer Chemistry, pp. 11-12 L1
Section Focus Transparency 11 and **Master,** p. 11 L1 LEP
Spanish Resources L1 LEP
ChemLab and MiniLab Worksheets, p. 44
L1

Basic Concepts Transparency 16 and **Master,** pp. 31-32 L1 LEP
Problem Solving Transparency 6, 7 and **Masters,** pp. 11-14 L1 LEP
Tech Prep Applications, pp. 9-10 L1

◀ Precipitation of a solid from a solution can result from a chemical change. Using soap with hard water produces a precipitate called soap scum because the soap reacts chemically with ions in the water.

Energy changes occur during all chemical changes. Heat or light can be absorbed or released during a chemical reaction, such as when wood or another fuel rapidly combines with oxygen during burning. ▶

◀ Odor changes can indicate that a substance has undergone a chemical change. When food spoils, odors change as a result of chemical changes within the food.

◀ Gas release sometimes occurs as a result of a chemical change. Automobile exhaust contains gases produced by the combustion of gasoline.

6.1   Chemical Equations   **191**

2 TEACH

Discussion

Ask students which of the following illustrate chemical changes: **a.** growth of a tree, **b.** melting butter, **c.** the use of food in our bodies, **d.** CO_2 escaping from a can of soda, **e.** separation of the components of crude oil by distillation, **f.** use of gasoline in an auto, **g.** a lake freezing over on a winter day. *a, c, f. The others are physical changes.* L1

Visual Learning

Figure 6.1: Ask students for other examples of the signs of chemical reaction that they may have seen. Have a student list student responses on the chalkboard. L1 LEP

Chemistry Journal

Evidence of Change

Ask students to keep a running list of chemical reactions that they use or observe. Have them include the list in their Journals at the completion of the chapter. Ask them also to state any of the macroscopic changes that accompany each reaction. L1

Figure 6.2: Lead students through the process of writing a word equation (glucose + oxygen → carbon dioxide + water) and a balanced chemical equation by using molecular models of glucose, oxygen, carbon dioxide, and water to depict the reaction, $C_6H_{12}O_6 + 6O_2 \rightarrow 6CO_2 + 6H_2O$. It is best to start with just one molecular model of each substance so students will realize the necessity of adding more molecules in order to correctly represent or balance the equation for the reaction. What are the reactants in this reaction? $C_6H_{12}O_6$, O_2 What are the products? CO_2, H_2O L1

LEP

Using an Analogy

Ask students to write a word equation for the construction of a house, a process that requires energy. *Example: energy + wood + concrete + steel + metal → a house.* Point out the inexactitude of the word equation regarding the amounts of the various building materials. L1

WORD ORIGIN

product:
pro (L) forward
ducere (L) to lead
Formation of a product helps pull a reaction forward.

Figure 6.2

Parts of a Chemical Reaction

▲ Compounds in wood chemically combine with oxygen when they burn to form water and carbon dioxide.

Energy is needed by our bodies to perform daily activities. This energy is provided when glucose combines with oxygen in cells. This reaction forms the same two substances produced when compounds in wood combine with oxygen. What are the reactants in these reactions? What are the products? ▶

Writing Chemical Equations

In order to completely understand a chemical reaction, you must be able to describe any changes that take place. Part of that description involves recognizing what substances react and what substances form. A substance that undergoes a reaction is called a **reactant.** When reactants undergo a chemical change, each new substance formed is called a **product.** For example, a familiar chemical reaction involves the reaction between iron and oxygen (the reactants) that produces rust, which is iron(III) oxide (the product). The simplest reactions involve a single reactant or a single product, but some reactions involve many reactants and many products. Examine the chemical reactions shown in **Figure 6.2.**

Complete the Description

Several possible observations help determine when a chemical reaction has taken place. But these observations don't completely describe what happens between reactants to form products. Have you ever seen what happens when baking soda and vinegar are mixed together? They react quickly, as you can tell by the bubbles that seem to explode out of the mixture, as shown in **Figure 6.3.** In describing this reaction, you could say that baking soda and vinegar turn into bubbles. But does that completely explain what happens? What are the bubbles made of? Do all of the atoms in vinegar and baking soda form bubbles? The reaction involves more than what can be determined by observation alone. Just as you can write a sentence to tell others what happened on your way to school today, chemists represent the changes taking place in a reaction by writing equations.

Across the Curriculum

Literature

In *Charlie and the Chocolate Factory* by Roald Dahl, a moral lesson is learned at the expense of a greedy boy. Read the description of his likely fate to students. Ask them to list changes that occur, classifying each as either chemical or physical, and justify the classifications. L1

He will be altered quite a bit.
He'll be quite changed from what he's been.
A hundred knives go slice, slice, slice;
We add some sugar, cream, and spice;
We boil him for a minute more,
Until we're absolutely sure

That all the greed and all the gall
Is boiled away once and for all.
This greedy brute, this louse's ear,
Is loved by people everywhere!
For who could hate or bear a grudge
Against a luscious bit of fudge?

Figure 6.3
A Sample Word Equation
Vinegar and baking soda react vigorously, forming a bubbly product. This reaction was formerly used in fire extinguishers because the bubbles produced contain carbon dioxide, which is effective in putting out fires. This reaction can be described by the following word equation.

vinegar + baking soda →
sodium acetate + water + carbon dioxide

Concept Development
Emphasize the importance of writing balanced chemical equations by mentioning that scientists in different countries may use different written and spoken languages, but all represent chemical reactions by the same symbolic language.

Correcting Errors
Strongly emphasize the importance of writing correct formulas for all reactants and products first. Once correct formulas are written, they may not be changed during the balancing part of the process. Only the coefficients may be changed. A common student error in writing equations is that of incorrectly changing formulas in order to balance equations.

Visual Learning

Figure 6.3: Point out to students that the volume of the products of this reaction is greater than the volume of the reactants. Ask them to try to explain why soda-acid extinguishers are no longer used. *Increased pressure sometimes caused the extinguisher to explode.* L1

Transparency

Display **Basic Concepts Transparency 16** to introduce students to chemical equations. L1 LEP

Word Equations

The simplest way to represent a reaction is by using words to describe all the reactants and products, with an arrow placed between them to represent change, as shown in **Figure 6.3.** As you can see in this word equation, reactants are placed to the left of the arrow, and products are placed to the right. Plus signs are used to separate reactants and also to separate products.

Vinegar and baking soda are common names. The compound in vinegar that is involved in the reaction is acetic acid, and baking soda is sodium hydrogen carbonate. These scientific names can also be used in a word equation.

acetic acid + sodium hydrogen carbonate →
sodium acetate + water + carbon dioxide

Chemical Equations

Word equations describe reactants and products, but they are long and awkward and do not adequately identify the substances involved. Word equations can be converted into chemical equations by substituting chemical formulas for the names of compounds and elements. Recall from Chapter 5 that these formulas can be written by using the oxidation numbers of the elements and the charges of the polyatomic ions. For example, the equation for the reaction of vinegar and baking soda can be written using the chemical formulas of the reactants and products.

$$HC_2H_3O_2 + NaHCO_3 \rightarrow NaC_2H_3O_2 + H_2O + CO_2$$

By examining a chemical equation, you can determine exactly what elements make up the substances that react and form.

6.1 **Chemical Equations** 193

Purpose

Students will learn some of the reactions that occur during the bleaching of laundry.

Teaching Strategies

- Point out that the color of a pigment (stain) is caused by the pigment's ability to absorb specific energies of electromagnetic radiation while reflecting others. This is often referred to as color subtraction. (Laundry stains usually are organic compounds that have chains of alternating single and double bonds. The energies between the electrons' ground and excited states are small enough to absorb specific energies from incident electromagnetic radiation while reflecting others.)

Extension

Have students research how optical bleaches are used in many synthetic fibers to maintain whiteness. L2

Exploring Further

1. $2NaClO(aq) \rightarrow 2NaCl(aq) + O_2(g)$
2. One of the products of sodium hypochlorite bleaching is hydrochloric acid. It reacts with fabrics more than water, which is one of the products of hydrogen peroxide bleaching.

Whitening Whites

Why might eating this meal be worrisome? Look closely. There isn't a paper napkin in sight—only white linen ones! Not to worry. You'll be able to remove most evidence of sloppy etiquette with a six percent aqueous solution of sodium hypochlorite.

Household bleach Perhaps the most popular type of household bleach is an aqueous solution of sodium hypochlorite, NaClO. Sodium hypochlorite is made by reacting chlorine gas with an aqueous solution of sodium hydroxide, as shown in the following equation.

$$Cl_2(g) + 2NaOH(aq) \rightarrow$$
$$NaClO(aq) + NaCl(aq) + H_2O(l)$$

In an aqueous solution, NaClO does not exist as a complete unit. It exists as sodium ions, Na^+, and hypochlorite ions, ClO^-. The ingredient responsible for the bleaching action in this type of bleach is the hypochlorite ion. Many other solid and liquid bleaches contain hydrogen peroxide, H_2O_2, instead of sodium hypochlorite. In these bleaches, the active substance in solution is the perhydroxyl ion, HOO^-. What do the ClO^- and HOO^- ions have in common?

Bleaching reactions As you can see, each of these polyatomic ions carries a single negative charge. For example, if each could react with a hydrogen ion, the following reactions would happen.

$$ClO^- + H^+ + \rightarrow HCl + O$$
$$HOO^- + H^+ + \rightarrow H_2O + O$$

Because the compounds HCl and H_2O are more stable than the hypochlorite and perhydroxyl ions,

these reactions tend to occur. The reactions result in a bleaching action because the released oxygen reacts with molecules of materials that cause the stain. The molecules of the compounds that cause color in a stain are structured in a way that gives them the physical property of producing color. In the reaction between these compounds and atomic oxygen, the compound or compounds formed have different structures. These structures do not have physical properties of producing color. So a bleach bleaches by rendering a colored compound colorless.

Exploring Further

1. **Applying** Liquid bleaches containing sodium hypochlorite are often sold in opaque, plastic containers because sunlight causes the compound to decompose to produce oxygen gas and sodium chloride. Write the balanced chemical equation for this reaction.

2. **Inferring** Why do you think bleaches containing sodium hypochlorite tend to damage finer fabrics more than bleaches containing hydrogen peroxide?

Demonstration 6.2

Electrolysis of Water

Purpose: To demonstrate to students the meaning of the coefficients used to balance a chemical equation.

Materials: Hoffman electrolysis apparatus, 6- to 9-volt power supply and wires, 140 g $Na_2SO_4 \cdot 10H_2O$, 0.1 g bromthymol blue, 16 mL 0.01M NaOH, 735 mL distilled water, 2 wood splints, 2 small test tubes, 1-L beaker

Safety Precautions:

Disposal: Code D, then pour down the drain.

Procedure: Dissolve 140 g of $Na_2SO_4 \cdot 10H_2O$ in 500 mL of water. Prepare bromthymol blue solution by mixing 0.1 g of bromthymol blue powder with 234 mL water and 16 mL of 0.01M NaOH. Add this solution to the sodium sulfate

solution to give a green color. If necessary, adjust the color of this solution with 0.1M HCl or 0.1M NaOH. Fill the Hoffman apparatus with the green sodium sulfate solution. Pass electricity through the solution to produce hydrogen and oxygen gases. Allow the apparatus to operate until an amount of gas sufficient to fill a test tube is collected. Have students note the relative amounts of the two gases pro-

It may also be important to know the physical state of each reactant and product. How can we indicate that the bubbles we see during this reaction are CO_2? Symbols in parentheses are put after formulas to indicate the state of the substance. Solids, liquids, gases, and water (aqueous) solutions are indicated by the symbols (s), (l), (g), and (aq). The following equation shows these symbols added to the equation for the reaction of vinegar and baking soda.

$$HC_2H_3O_2(aq) + NaHCO_3(s) \rightarrow NaC_2H_3O_2(aq) + H_2O(l) + CO_2(g)$$

Now the equation tells us that mixing an aqueous solution of acetic acid (vinegar) with solid sodium hydrogen carbonate (baking soda) results in the formation of an aqueous solution of sodium acetate, liquid water, and carbon dioxide gas. If you had examined this equation before you mixed vinegar and baking soda, you could have predicted that bubbles would form.

Energy and Chemical Equations

Noticeable amounts of energy are often released or absorbed during a chemical reaction. Some reactions absorb energy. If energy is absorbed, the reaction is known as an endothermic reaction. **Figure 6.4** shows an example of an endothermic reaction. For a reaction that absorbs energy, the word *energy* is sometimes written along with the reactants in the chemical equation. For example, the equation for the reaction in which water breaks down into hydrogen and oxygen gases shows that energy must be added to the reaction.

$$2H_2O(l) + energy \rightarrow 2H_2(g) + O_2(g)$$

Figure 6.4

An Endothermic Reaction

▲ The reaction of ammonium chloride and barium hydroxide octahydrate is endothermic.

If these reactants are mixed at room temperature, which is about 20°C, the temperature in the mixture drops as energy is absorbed by the reaction. ▶

Concept Development

Ask students to go back to the example of baking a cake in the chapter opener, writing a word equation for the process and specifying the reactants and products. *salt + sugar + baking powder + eggs + water + shortening + flour → cake (reactants → product)* [L1]

Discussion

Ask students whether the preceding process was exothermic or endothermic. *endothermic because energy was added to cause the reaction* Ask where energy should be included in a word equation for an endothermic reaction. *as a reactant* [L1]

GLENCOE TECHNOLOGY

Software

Mastering Concepts in Chemistry
Unit 7, Lesson 1
Chemical Equations
Use this lesson to reinforce the concept of writing chemical equations. [L1] [LEP]

duced by the electrolysis of water. Collect the gas from each arm of the apparatus into a test tube. Test the hydrogen (greater amount produced) with a burning splint, and test the oxygen with a glowing splint. **Results:** The arm that fills with more gas, H_2, turns blue. The gas pops when burned. The arm that contains oxygen turns yellow. The glowing splint bursts into flame when it contacts the oxygen

gas. Twice as much H_2 is produced as O_2.
Analysis: Ask these questions.
1. How much hydrogen gas (blue arm) was produced? *Answers will vary.*
2. How much oxygen gas (yellow arm) was produced? *half the volume given in the answer to question 1*
3. Write the balanced equation for the decomposition of water. *$2H_2O(l) + energy \rightarrow 2H_2(g) + O_2(g)$*

4. How does the experimental volume ratio of hydrogen to oxygen compare to the coefficients in a balanced equation? *They are the same, 2:1.*

✔ **Assessment**

Portfolio: Have students write a summary of this demonstration and add it to their portfolios. [L1] [P]

miniLAB

Energy Change

Purpose: Students will feel the heat given off during an exother-mic reaction.

Process Skills: Observing, in-ferring, recognizing cause and effect, communicating

Teaching Strategies

- Have students compare their homemade hot pack to a commercial hand or foot warmer. L1
- Use heavy-duty resealable bags to prevent leaks.

Expected Results

The bag should give off heat for up to an hour after the contents have been mixed well.

Analysis

1. The iron and salt take part in an exothermic reaction, and the heat given off causes the bag to feel warm.
2. commercial hot packs for warming hands, feet, and bodies; food warmers for camping trips

Assessment

Knowledge: Ask students to define *endothermic* and *exo-thermic* and explain what ob-servable physical change accompanies each type of re-action. *temperature decreases; temperature increases* L1

miniLAB

Energy Change

All chemical reactions involve an energy change. This change may be so slight that it can be detected only with sensitive instruments, or it may be quite noticeable.

Procedure

1. Place 25 g of iron powder and 1 g of NaCl in a resealable plastic bag.
2. Add 30 g of vermiculite to the bag, seal the zipper, and shake the bag to mix the contents.
3. Add 5 mL of water to the bag, reseal the zipper, and gently squeeze and shake the contents to mix them.
4. Hold the bag between your hands and note any changes in temperature.

Analysis

1. What did you observe? What type of reaction produces this kind of change?

2. Using the photo, what practical application can you think of for this reaction?

All reactions that occur in a Bunsen burner or gas grill or those used to power an automobile release energy. The How It Works feature shows another example of such a reaction. As you will recall from Chapter 1, reactions that release energy are called exothermic reactions. When writ-ing a chemical equation for a reaction that produces energy, the word *energy* is sometimes written along with the products. For example, the equation for the reaction that occurs when you light methane in a Bunsen burner shows that energy is released. Some of this energy is in the form of light.

$$CH_4(g) + 2O_2(g) \rightarrow CO_2(g) + 2H_2O(g) + energy$$

You may have noticed that the word *energy* is not always written in an equation. It is used only if it is important to know whether energy is released or absorbed. For the burning of methane, energy would be writ-ten in the equation because the release of heat is an important part of burning a fuel. Energy would also be written in the equation that describes the reaction when water is broken down into hydrogen and oxy-gen because the reaction would not occur without the addition of energy. In many reactions, such as the formation of rust, energy might be released or absorbed but it is not included in the equation because it is not impor-tant to know about the energy involved in that particular reaction.

*inter*NET CONNECTION

Students can download and use a molecular weight calculator.

Anonymous FTP
ftp://ftpserver.sfu.ca/pub/chemcai/
molecalc.sea.hqx

Emergency Light Sticks

The light given off by an emergency light stick is energy released from a chemical reaction. Light sticks are a good source of light when no electricity is available. Reactions in which light is given off are called chemiluminescent reactions. Different chemicals used in these reactions give off different colors, so light sticks come in many colors. The light from a light stick is only temporary; when the reactants are used up, light is no longer produced.

1. Light sticks are plastic rods that contain two solutions of chemicals. Enclosed inside a thin glass ampoule is one solution that is an oxidizing agent, and surrounding it in the rod is a second solution that contains a fluorescent dye.

2. When the light stick is bent, the glass ampoule is broken and the two solutions mix. The reaction begins. Energy is given off when the two solutions react.

3. The energy raises the energy level of the electrons of the dye molecules.

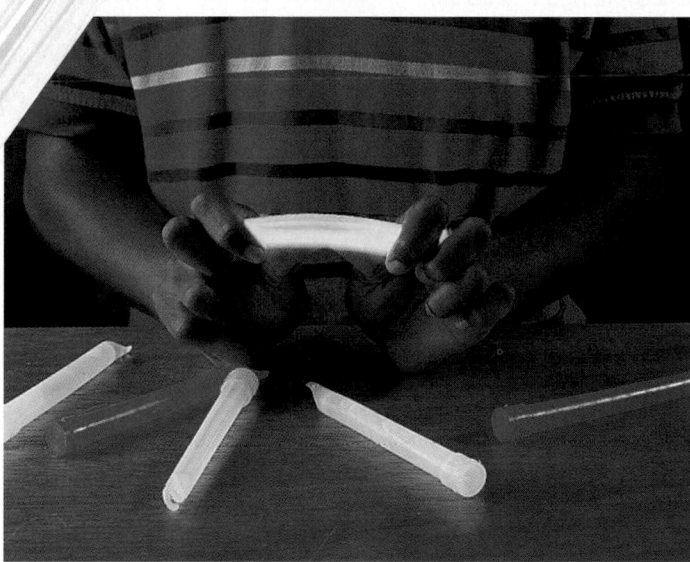

4. When the electrons drop back down to their original energy levels, the extra energy is given off as light. This light is sometimes referred to as cool light because no noticeable heat is given off in the reaction.

Thinking Critically

1. Explain why the chemical reaction in the light stick is not exothermic, even though it produces energy.

2. What are some advantages of light sticks over conventional light sources?

Thinking Critically

1. To be exothermic, a reaction must give off heat. Because the light stick does not get hot, the reaction observed does not give off heat.
2. Light sticks can be used when the electricity is cut off by a power outage, and because they do not get hot, they are safer to use in some situations such as under water or in wet areas.

Purpose

TECH PREP

Students will learn that emergency light sticks involve a reaction in which the energy generated is given off as light rather than heat.

Background

Cyalume light sticks have a solution containing a phenyl oxalate ester and a fluorescent dye in a plastic outer rod. Also in the plastic rod is a thin glass ampoule full of dilute hydrogen peroxide in a phthalic ester solvent. When the hydrogen peroxide and oxalate ester react, an intermediate is produced that transfers energy from the reaction to the fluorescent dye molecule. Light is emitted when the electrons return to the more stable, ground state. Some colored light sticks actually emit the wavelength of light observed; others appear to be different colors due to the color of the plastic outer rods.

Visual Learning

Darken the room and let students observe the light given off when a light stick is activated. Dissect a light stick to show students its parts.

Teaching Strategies

• Ask students to predict the effect of varying temperature on the light-emitting life span of a light stick. Have them compare active times observed for light sticks placed at room temperature, in a refrigerator, in a freezer, and in an incubator or hot-water bath. **L1** **LEP** **COOP LEARN**
• Ask students to think of other examples of chemical energy being converted to light energy. *Other examples include bioluminescent fireflies, bacteria, and many marine organisms.* **L1** **LEP**

Note that in the cake-baking word equation, the amounts of reactants were not specified; however, they are critical if the reaction is to proceed successfully. For example, 5 lbs of salt, 1 tsp sugar, 1 gal water, 4 doz eggs, etc. will result in a mess—not a cake. Emphasize that the amounts of reactants and products in a chemical reaction and equation are critical.

Quick Demo

Is mass conserved?

Pour approximately 5 mL of $0.1M$ lead(II) nitrate solution into one small test tube and 5 mL of $0.2M$ potassium iodide solution into another. Carefully lower both test tubes into a small flask so that the tubes remain upright, stopper the flask, and weigh the flask and contents. Tip the flask so that the test tubes spill their contents into the flask, causing the formation of the yellow, insoluble lead(II) iodide. Ask students to predict any change in mass. *Students might say that the mass increased because a dense solid was formed.* Then weigh the flask and contents to demonstrate the law of conservation of mass. Ask students to write the chemical equation for this reaction.

Disposal: Lead salts should be disposed of by a hazardous waste vendor. L1

Correcting Misconceptions

Students often have the incorrect idea that the coefficients in a chemical equation must balance. For example, in the reaction $2NH_3 \rightarrow N_2 + 3H_2$, the total reactant coefficients are 2 and the total product coefficients are 4; however, the atoms are correctly balanced.

Balancing Chemical Equations

What do you think happens to the atoms in reactants when they are converted into products? Some products, such as the CO_2 produced from baking powder when a cake is baked, seem to disappear into the air. What really happens to them?

The Law of Conservation of Mass

Recall from Chapter 2 that you can test a reaction to determine whether the same amount of matter is contained in the products and the reactants. That type of experiment was first carried out by the French scientist Antoine Lavoisier (1743-1794), **Figure 6.5.** His results indicated that the mass of the products is always the same as the mass of the reactants that react to form them. The law of conservation of mass summarizes these findings. Matter is neither created nor destroyed during a chemical reaction.

Conservation of Atoms

Remember that atoms don't change in a chemical reaction; they just rearrange. The number and kinds of atoms present in the reactants of a chemical reaction are the same as those present in the products. When stated this way, it becomes the law of conservation of atoms. For a chemical equation to accurately represent a reaction, the same number of each kind of atom must be on the left side of the arrow as are on the right side. If an equation follows the law of conservation of atoms, it is said to be balanced.

How can you count atoms in an equation? The easiest way to learn is to practice—first with a simple reaction and then with some that are more complex. For example, consider the equation that represents breaking down carbonic acid into water and carbon dioxide.

$$H_2CO_3(aq) \rightarrow H_2O(l) + CO_2(g)$$

Because a subscript after the symbol for an element represents how many atoms of that element are found in a compound, you can see that there are two hydrogen, one carbon, and three oxygen atoms on each side of the arrow. All of the atoms in the reactants are the same as those found in the products.

Figure 6.5
Lavoisier's Contribution One of the experiments that Lavoisier used to discover the law of conservation of mass was the decomposition of the red oxide of mercury to form mercury metal and oxygen gas. He weighed the amount of HgO that decomposed and found it to be the same as the total weight of Hg and O_2 produced.

Demonstration 6.3

A Model Science

Purpose: To represent a balanced chemical equation using models.

Materials: If you do not have a commercially available chemistry model kit, obtain 18 plastic foam balls and pipe cleaners or toothpicks. Use latex paint to color four balls black to represent C and four balls blue to represent H. Paint ten balls red to represent O.

Safety Precautions:

Disposal: Code F

Procedure: Use the balls and pipe cleaners or toothpicks to build two molecules of C_2H_2 and five molecules of O_2. Explain the color coding to students. Have a student rearrange the balls, using all of them, to form CO_2 and H_2O molecules. The student will quickly produce four CO_2 and two H_2O molecules. Point out that the student has balanced the equation for the reaction of acetylene and oxygen. L1 LEP

Examine the equation for the formation of sodium carbonate and water from the reaction between sodium hydroxide and carbon dioxide.

$$NaOH(aq) + CO_2(g) \rightarrow Na_2CO_3(s) + H_2O(l)$$

Do both sides of the equation have the same number of each type of atom? No. One carbon atom is on each side of the arrow, but the sodium, oxygen, and hydrogen atoms are not balanced. The equation, as written, does not truly represent the reaction because it does not show conservation of atoms.

A Balancing Act

To indicate more than one unit taking part or being formed in a reaction, a number called a **coefficient** is placed in front of it to indicate how many units are involved. Look at the previous equation with a coefficient of 2 in front of the sodium hydroxide formula.

$$2NaOH(aq) + CO_2(g) \rightarrow Na_2CO_3(s) + H_2O(l)$$

Is the equation balanced now? Two sodium atoms are on each side. How many oxygen atoms are on each side? You should be able to find four on each side. How about hydrogen atoms? Now two are on each side. Because one carbon atom is still on each side, the entire equation is balanced; it now represents what happens when sodium hydroxide and carbon dioxide react.

The balanced equation tells us that when sodium hydroxide and carbon dioxide react, two units of sodium hydroxide react with each molecule of carbon dioxide to form one unit of sodium carbonate and one molecule of water. Look at a different balanced equation in **Figure 6.6.**

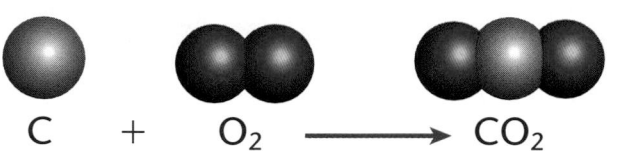

C + O₂ ⟶ CO₂

1 carbon atom   1 oxygen molecule   1 molecule of carbon dioxide

Why can't subscripts be changed when balancing an equation? Changing a subscript changes the identity of that substance. Look at the equation in Sample Problem 1. Changing the subscript of the oxygen in water to 2 changes water, H_2O, to hydrogen peroxide, H_2O_2, a different compound. Changing a coefficient simply means that you are changing the amount of that substance compared to the other substances in the reaction. Changing the coefficient of water to 2 means there are two molecules of water, $2H_2O$. The identity of water stays the same.

Figure 6.6

How many atoms? Examine the balanced equation that shows what happens when carbon reacts with oxygen to form carbon dioxide (left). If a piece of coal contains 10 billion C atoms, how many molecules of O_2 will it react with? How many molecules of CO_2 will be formed?

6.1   **Chemical Equations**   **199**

SECTION 6.1

Correcting Errors

Strongly emphasize the importance of writing correct formulas for all reactants and products first. Once correct formulas are written, they may not be changed during the balancing part of the process. Only the coefficients may be changed.

GLENCOE TECHNOLOGY

⊙ CD-Rom

Chemistry: Concepts and Applications
Balancing an Equation
Show the balancing of an equation by counting atoms on both sides and changing coefficients. [L1] [LEP]

Visual Learning

Figure 6.6: Ten billion C atoms will react with 10 billion O_2 molecules, forming 10 billion CO_2 molecules.

Transparency

Display **Problem Solving Transparency 6** to aid students in understanding balanced equations. [L1] [LEP]

Display **Problem Solving Transparency 7** to help teach students how to balance equations. [L1] [LEP]

Results: The balanced equation for the reaction is $2C_2H_2 + 5O_2 \rightarrow 4CO_2 + 2H_2O$.

Analysis: Ask these questions.

1. Why it is necessary to have the same number and kind of atom on both sides of the equation? *to obey the law of conservation of mass*

2. What do chemists call the number that is written in front of a chemical formula when balancing a chemical equation? *a coefficient*

✓ Assessment

Skill: Have students place a drawing of the models from this demonstration in their portfolios. [L1] [LEP] [P]

3 ASSESS

Check for Understanding

- Check for understanding of the section material by having students answer the Section Review questions.
- Provide toothpicks and gumdrops of various colors, and ask students to work in groups to construct models of the reactants and products in one or two simple chemical reactions. Check to see that their gumdrop equations are balanced. **L1** **LEP** **COOP LEARN**

Reteach

Discussion: Ask each student to share one of his or her Journal entries regarding a used or observed chemical reaction. **L1**

Extension

Emphasize that the balanced equation for a reaction specifies the ratios of atoms, ions, molecules, and formula units. Provide students with several chemical equations, and have them determine ratios of atoms for the equations. However, point out that in any practical situation, huge numbers (usually billions of billions) of these entities are involved in the reaction. **L1**

SAMPLE PROBLEM **1** **Writing a Simple Equation**

Write word and chemical equations for the reaction of hydrogen and oxygen gases to form gaseous water and release energy. This reaction powers the main stage of the space shuttle.

Analyze
- To write a word equation for the reaction, write the names of the reactants, draw an arrow, then write the name of any product. If there is more than one reactant or product, plus signs should separate them.

$$\text{hydrogen} + \text{oxygen} \rightarrow \text{water} + \text{energy}$$

Set Up
- To write the chemical equation, use chemical formulas to replace the names of the reactants and products in the word equation you wrote. Then add symbols to represent the physical state of each compound. Remember that hydrogen and oxygen occur as diatomic gases.

$$H_2(g) + O_2(g) \rightarrow H_2O(g) + \text{energy}$$

Solve
- To balance the atoms on each side of the arrow, count the number of atoms of each type on each side of the arrow. On the left are two hydrogen atoms and two oxygen atoms. On the right are also two hydrogen atoms but only one oxygen atom. Change the coefficient of the water to 2 so that the number of oxygen atoms will be balanced. Because that puts four hydrogen atoms on the right side of the arrow, the coefficient of hydrogen gas also must be changed to 2.

> **Problem-Solving HINT**
>
> Be sure to change only coefficients, not subscripts, when balancing equations.

$$2H_2(g) + O_2(g) \rightarrow 2H_2O(g) + \text{energy}$$

Check
- Make a final check of all atoms to make sure they are balanced.

SAMPLE PROBLEM **2** **Writing an Equation**

Write word and chemical equations for the reaction that takes place when an aqueous solution of magnesium chloride is added to a silver nitrate solution. Aqueous magnesium nitrate and solid silver chloride form.

Analyze
- To write a word equation for the reaction, write the names of the reactants, draw an arrow, then write the name of any product. If there is more than one reactant or product, plus signs should separate them.

magnesium chloride + silver nitrate →

magnesium nitrate + silver chloride

Set Up
- To write the chemical equation, use chemical formulas to replace the names of the reactants and products in the word equation you wrote. Remember to use the oxidation number of an element and the charge on a polyatomic ion to write a correct formula. Then add symbols to represent the physical state of each compound.

$$MgCl_2(aq) + AgNO_3(aq) \rightarrow Mg(NO_3)_2(aq) + AgCl(s)$$

PRACTICE

Guided Practice: Guide students through the sample problem, then have them work in class on Practice Problems 1-3. **L1**

Independent Practice: Your homework or classroom assignment can include the additional practice problems shown below. **L1**

Answers to Practice Problems

1. magnesium + water → magnesium hydroxide + hydrogen; Mg(s) +

$$2H_2O(l) \rightarrow Mg(OH)_2 + H_2(g)$$

2. hydrogen peroxide + lead(II) sulfide → lead(II) sulfate + water; $4H_2O_2(aq)$ + PbS(s) → $PbSO_4(s)$ + $4H_2O(l)$

3. manganese(II) sulfate heptahydrate + energy → water + manganese(II) sulfate monohydrate; $MnSO_4 \cdot 7H_2O(s)$ + energy → $6H_2O(l)$ + $MnSO_4 \cdot H_2O(s)$

4. potassium + water → potassium hydroxide + hydrogen gas; 2K(s) +

$$2H_2O(l) \rightarrow 2KOH(aq) + H_2(g)$$

Additional Practice Problem

Write word and chemical equations for the following reaction.

Water and carbon dioxide are produced when methane reacts with oxygen in a Bunsen burner. *methane + oxygen → water + carbon dioxide;* $CH_4(g) + 2O_2(g) \rightarrow 2H_2O(g) + CO_2(g)$

Solve
- To balance the atoms on each side of the arrow, count the number of atoms of each type on each side of the arrow. On the left are one magnesium atom and two chlorine atoms. On the right, there is also one magnesium atom but only one chlorine atom. Change the coefficient of the AgCl to 2 so that the number of chlorine atoms will be balanced. Because that puts two silver atoms on the right side of the arrow, the coefficient of $AgNO_3$ also must be changed to 2. This balances the oxygen and nitrogen atoms, as well.

$$MgCl_2(aq) + 2AgNO_3(aq) \rightarrow Mg(NO_3)_2(aq) + 2AgCl(s)$$

Check
- Make a final check of all atoms on both sides of the equation to make sure they are balanced.

PRACTICE PROBLEMS

Write word equations and chemical equations for the following reactions.

1. Magnesium metal and water combine to form solid magnesium hydroxide and hydrogen gas.

2. An aqueous solution of hydrogen peroxide (dihydrogen dioxide) and solid lead(II) sulfide combine to form solid lead(II) sulfate and liquid water.

3. When energy is added to solid manganese(II) sulfate heptahydrate crystals, they break down to form liquid water and solid manganese(II) sulfate monohydrate.

4. Solid potassium reacts with liquid water to produce aqueous potassium hydroxide and hydrogen gas.

4 CLOSE

Writing About Chemistry

Portfolio: Ask students to write short essays in which they discuss both the truth and the falsity of the statement, "We are rapidly using up our natural resources." Instruct them to include the essays in their portfolios. L1 P

SECTION REVIEW

Understanding Concepts

1. Explain how you can tell whether a chemical reaction has taken place.

2. Write balanced chemical equations for the reactions described.
   a) sodium metal + chlorine gas → sodium chloride crystals
   b) propane gas + oxygen → carbon dioxide + water vapor + energy
   c) zinc metal + hydrochloric acid → zinc chloride solution + hydrogen

3. Why is it important to balance a chemical equation?

Thinking Critically

4. **Applying Concepts** In the reaction of carbon with oxygen, one carbon atom reacts with one molecule of oxygen to form one molecule of carbon dioxide. If you react 20 billion atoms of carbon with 30 billion molecules of oxygen, will all the reactants be used up? What do you think the results will be?

Applying Chemistry

5. **Catalytic Converter** In the catalytic converter of a car, a reaction occurs when nitrogen monoxide gas, NO, reacts with hydrogen gas. Ammonia gas and water vapor are formed. Write a balanced equation for this reaction.

6.1 **Chemical Equations** **201**

SECTION REVIEW

1. Macroscopic changes that indicate a reaction has occurred are color changes, odor changes, energy absorption or release, gas release, and precipitation of solids.

2. **a.** $2Na(s) + Cl_2(g) \rightarrow 2NaCl(s)$ **b.** $C_3H_8(g) + 5O_2(g) \rightarrow 3CO_2(g) + 4H_2O(g) + energy$
   c. $Zn(s) + 2HCl(aq) \rightarrow ZnCl_2(aq) + H_2(g)$

3. The law of conservation of mass must be obeyed in a chemical reaction. For mass to be balanced on both sides of an equation that represents a reaction, the numbers and kinds of atoms present in the products must be the same as those present in the reactants.

4. If the carbon atoms and oxygen molecules react in a 1:1 ratio, then 20 billion molecules of oxygen would react with 20 billion atoms of carbon to produce 20 billion molecules of carbon dioxide. There would be 10 billion oxygen molecules in excess.

5. $2NO(g) + 5H_2(g) \rightarrow 2NH_3(g) + 2H_2O(g)$

SECTION
6.2

PREPARE

Key Concepts

The reasons for classifying chemical reactions are discussed, and the five general reaction types are described.

Planning Ahead

For the ChemLab Have ice available. Be sure pennies are newer than 1983. Prepare solutions as follows. $0.1M$ $CuSO_4$ may be prepared by adding 16 g of copper(II) sulfate to 900 mL of water, stirring until dissolved, and bringing the volume up to 1 L. $0.5M$ Na_2CO_3 may be prepared by adding 53 g of sodium carbonate to 900 mL of water, stirring until dissolved, and bringing the volume up to 1 L. $0.5M$ $CuCl_2$ may be prepared by adding 67 g of copper(II) chloride to 900 mL of water, stirring until dissolved, and bringing the volume up to 1 L. $6M$ HCl may be prepared by adding 300 mL of concentrated $(12M)$ hydrochloric acid to 300 mL of water. **CAUTION:** *Add the acid to the water.*

1 FOCUS

Focus Transparency

Display the **Section Focus Transparency** for Section 6.2 to introduce the five classes of reactions. L1 LEP

SECTION PREVIEW

Objectives

Distinguish among the five major types of chemical reactions. **Classify** a reaction as belonging to one of five major types.

Key Terms

synthesis
decomposition
single displacement
double displacement
combustion

Types of Reactions

Chemistry has a lot in common with cooking. What do you have to do to become a good cook? A lot of study and probably even more practice are necessary. The same is true of chemistry. If you combine the right amounts of the right reactants in the correct order under certain conditions, you will get the right products.

As you study chemistry, you will begin to recognize what types of reactants and conditions lead to certain products. In this section, you will learn to recognize five major classes of reactions. Just as an experienced chef can tell a sauté from a soufflé, you will soon be able to distinguish between combustion and decomposition.

Why Reactions Are Classified

Why are reactions grouped into classes? Think about why scientists classify plants and animals. If you were hiking in the Rocky Mountains for the first time, you might see animals you have never seen before, such as the ones in **Figure 6.7.** How would you decide whether or not to be cautious around them? You would probably decide what familiar animals they most resemble. If the animal looks like a cat, it is most likely a predator that you should be careful around. If it looks like a goat, you will probably realize that it is more likely to run from you than eat you.

Figure 6.7
Classifying
Around which animal would you be most cautious? Did you mentally classify the animals into categories of some type before answering?

202

Program Resources

Study Guide, pp. 23-24 L1
Laboratory Manual, 6.1, pp. 49-54 L1
Section Focus Transparency 12 and
   Master, p. 12 L1 LEP
ChemLab and MiniLab Worksheets,
   p. 45 L1
Basic Concepts Transparency 17 and
   Master, pp. 33-34 L1 LEP

Chemistry Journal

Using an Analogy

In their Chemistry Journals, have students compare reaction types with dance activities. *Synthesis—two persons come together to dance. Decomposition—the dance is over, and the two people part. Single-displacement—another person cuts in and displaces one of the two dancers. Double-displacement—two couples switch partners.* L1

Major Classes of Reactions

Just as there are thousands of species of animals, there are many different types of chemical reactions. Five types are common. If you can classify a reaction into one of the five major categories by recognizing patterns that occur, you already know a lot about the reaction.

In one type of reaction, two substances—either elements or compounds—combine to form a compound. Whenever two or more substances combine to form a single product, the reaction is called a **synthesis** reaction.

BIOLOGY CONNECTION

Air in Space

The concentration of carbon dioxide in Earth's atmosphere is regulated through a complex interplay of human, biological, and geological mechanisms. In a space vehicle, these mechanisms aren't available. If not controlled, carbon dioxide from the respiration of astronauts could become toxic to them. So how is air quality maintained within a space vehicle?

A 66-m³ atmosphere The volume of the crew compartment of the space shuttle Orbiter is about 66 m³. The air is maintained at a pressure of 100 kPa, which is similar to Earth's atmospheric pressure at sea level. During a flight, the composition of the air in the crew compartment is maintained at 79 percent nitrogen and 21 percent oxygen, which is almost identical to that of Earth's atmosphere. Oxygen is carried in the Orbiter as a liquid stored in two cryogenic tanks in the mid-fuselage. The gaseous oxygen from the tanks passes through pressurizing and heating nozzles and moves into the crew compartment. A five-member crew will normally consume about 4 kg of oxygen every day. Nitrogen is supplied from two systems, each made of two storage tanks, also in the mid-fuselage of the Orbiter. The atmosphere of the compartment is recycled about every seven minutes.

Filtering the air During recycling, odors are removed by filters containing activated charcoal granules, which absorb from the air the chemical substances that cause the odor. Carbon dioxide gas is removed from the air by reacting it with solid lithium hydroxide.

$$CO_2(g) + 2LiOH(s) \rightarrow Li_2CO_3(s) + H_2O(g)$$

The lithium hydroxide is stored in canisters that are changed every 12 hours. The used canisters are then stored for disposal when the Orbiter returns to Earth.

Connecting to Chemistry

1. **Applying** Identify by name the products of the reaction between carbon dioxide and lithium hydroxide.

2. **Comparing and Contrasting** How does the chemical removal of carbon dioxide from the atmosphere of the Orbiter compare with the geochemical removal of carbon dioxide from Earth's atmosphere?

SECTION 6.2

2 TEACH

BIOLOGY CONNECTION

Purpose
Students will learn how air quality is maintained in a space vehicle.

Teaching Strategies

- Point out that the part of Earth's atmosphere in which humans live has a volume of about 10^{17} m³ as compared to the 10^3-m³ atmosphere of the crew compartment of the *Orbiter*.

- Have students recall that the metabolism of a simple sugar such as glucose is represented by the following chemical equation: $6O_2 + C_6H_{12}O_6 \rightarrow 6CO_2 + 6H_2O$. Ask students for the word equation for this reaction. *One molecule of glucose and six molecules of diatomic oxygen react and form six molecules each of carbon dioxide and water.*

Extension

Have students investigate how human hygiene is maintained by the crew of the *Orbiter* during a flight and summarize their findings in their Journals. **L2**

Connecting to Chemistry

1. The products are solid lithium carbonate and gaseous water.
2. In the *Orbiter*, CO_2 is removed by reacting with solid lithium hydroxide. In Earth's geochemical cycle, CO_2 is removed from the atmosphere by reacting with water and being used in photosynthesis.

Integrating THE Sciences

Earth Science Of known minerals, the largest group is the silicates—minerals containing frameworks of silicon and oxygen linked by other elements. In mountains, sandy beaches, and soil, they make up the bulk of Earth's crust. Silicates form mainly in chemical reactions in which silicon tetroxide ions (SiO_4^{4-}) react to form chains, sheets, and three-dimensional networks. These structures are connected to one another with cations of metals. Which type of silicate mineral forms depends upon the environment during the crystallization process and the chemical composition of the molten rock in which the reactions occur. These reactions are synthesis reactions when two or more groups join together to form a larger one and single-displacement reactions when atoms of aluminum or other elements replace silicon atoms. Have students acquire information about the formulas of silicates and use model-building kits to make models of several different silicates. **L1** **LEP**

Concept Development

Point out these correlations to students. Synthesis reactions are usually exothermic. Decomposition reactions are usually endothermic. Combustion reactions are always exothermic. Single-displacement and double-displacement reactions may be exothermic or endothermic.

Figure 6.8

A Synthesis Reaction
When iron rusts, iron metal and oxygen gas combine to form one new substance, iron(III) oxide. The balanced equation for this synthesis reaction shows that there is more than one reactant but only one product.

$$4Fe(s) + 3O_2(g) \rightarrow 2Fe_2O_3(s)$$

WORD ORIGIN

synthesis:
syn (Gk) together
tithenai (Gk) to place
A synthesis reaction involves placing elements or compounds together to make a compound.

An example of a synthesis reaction involving elements as reactants is shown in **Figure 6.8.** A synthesis reaction also occurs when two compounds combine, such as when rainwater combines with carbon dioxide in the air to form carbonic acid, or when an element and a compound combine, as when carbon monoxide combines with oxygen to form carbon dioxide.

In a **decomposition** reaction, a compound breaks down into two or more simpler substances. The compound may break down into individual elements, such as when mercury(II) oxide decomposes into mercury and oxygen. The products may be an element and a compound, such as when hydrogen peroxide decomposes into water and oxygen, or the compound may break down into simpler compounds, as shown in **Figure 6.9.**

Figure 6.9

A Decomposition Reaction
When ammonium nitrate is heated to a high temperature, it explosively breaks down into dinitrogen monoxide and water. The decomposition reaction taking place is represented by a balanced equation that shows one reactant and more than one product.

$$NH_4NO_3(s) \rightarrow N_2O(g) + 2H_2O(g)$$

Demonstration 6.4

Reactions Involve Energy

Purpose: To demonstrate that some metals displace hydrogen in water.
Materials: 2 g of copper turnings, 2 g of calcium turnings, 5-cm-long piece of magnesium ribbon, 1 mL of phenolphthalein indicator, 3 test tubes, 40 mL of distilled water, 3 thermometers, test-tube rack, timer

Safety Precautions: 🐄 👓 🚫
Disposal: Code F
Procedure: Place a few drops of phenolphthalein in each of three test tubes, then nearly fill them with distilled water. Place a thermometer in each test tube, and after two minutes, record the temperatures. Place the three metals each in a different test tube. Have three students each watch

a test tube and report observations, including any temperature change. L1
Results: H_2 is liberated, and the solution turns pink for any metal more active than hydrogen. A slight temperature change is observed for the single-displacement reactions involving Ca and Mg. Copper does not react.

A Simple Exchange

Single-displacement reactions occur when one element replaces another in a compound. However, such a reaction does not automatically occur just because an element and a compound are mixed together. Whether or not such a reaction occurs depends on how reactive the element is compared to the element it is to displace.

miniLAB
2

Procedure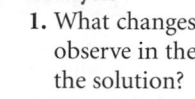

1. Pour $0.1M$ AgNO$_3$ solution into a test tube until it is half full. Clean a piece of copper wire or copper foil with steel wool.
2. Drop the copper foil or copper wire into the solution.
3. Keeping the test tube absolutely still, observe what happens over a half-hour period of time.

Analysis

1. What changes do you observe in the wire? In the solution?
2. Write a balanced equation for the reaction.
3. Does copper displace silver in silver nitrate? Does silver displace copper in copper(II) nitrate? How do you know?

In a **single-displacement** reaction, one element takes the place of another in a compound, as shown in **Figure 6.10**. The element can replace the first part of a compound, or it can replace the last part of a compound.

 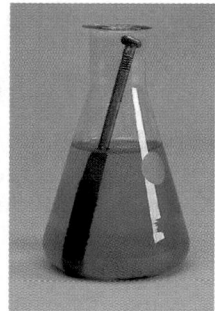

Figure 6.10

Single Displacement

◀ If an iron nail is placed into an aqueous solution of copper(II) sulfate, the iron displaces the copper ions in solution, and copper metal forms on the nail.

$$Fe(s) + CuSO_4(aq) \rightarrow FeSO_4(aq) + Cu(s)$$

When the chlorine gas in the flask on the left is bubbled through an aqueous solution of sodium bromide, the chlorine replaces the bromine in the compound. The reddish-brown bromine can be seen in the solution. ▶

$$Cl_2(g) + 2NaBr(aq) \rightarrow$$
$$2NaCl(aq) + Br_2(l)$$

Analysis: Ask these questions.
1. Which of these metals react with water? *Ca, Mg*
2. Write a balanced equation for one of the reacting metals. *Mg + 2H$_2$O → Mg(OH)$_2$ + H$_2$*
3. Was the reaction of the metal and water endothermic or exothermic? *exothermic*
4. What type of reaction occurred? *single-displacement*

✓ Assessment

Portfolio: Have students write a summary of this demonstration and add it to their portfolios.
L1 P

SECTION 6.2

miniLAB

A Simple Exchange

Purpose: Students will observe a single-displacement reaction and discover that the reactivities of different elements vary.

Process Skills: Observing, interpreting data, recognizing cause and effect

Safety Precautions: Caution students to keep the solution off their clothes and skin.

Disposal: Code B

Teaching Strategies

• This experiment can be run side by side with one in which a piece of silver is placed in a solution of copper(II) nitrate so the activities of the two metals can be readily compared.

Expected Results

The copper foil or wire will seem to disappear under the silver deposit as the more active copper displaces the silver in silver nitrate. The solution will turn blue as the copper(II) nitrate forms.

Analysis

1. The wire becomes smaller and covered with silver crystals. The colorless solution turns blue.
2. $2AgNO_3(aq) + Cu(s) \rightarrow Cu(NO_3)_2(aq) + 2Ag(s)$
3. Yes. No. Copper is a more active metal than silver.

✓ Assessment

Oral: Explain that some metals are more active than others. Ask students how they could carry out experiments to make a table of relative activities for all common metals. *More experiments like this one could be done using different metals and compounds of them in solution.* L1

ChemLab

Exploring Chemical Changes

Time Allotment: One class period.

Objectives

Review objectives with students before they begin the ChemLab.

Process Skills

Observing, inferring, interpreting data, communicating

Safety Precautions and Disposal

Students should wear goggles and aprons. They should be advised to avoid contact with the hydrochloric acid solution, which could damage their eyes, skin, or clothing. If contact occurs, they should rinse the area with tap water and notify the teacher. They should wash their hands thoroughly after the lab.

Disposal: Code E: The solid copper sulfide waste can be cooled and placed in a garbage can. Code A: Place the $CaCO_3$/CaO in a plastic or cardboard container, and bury it in an approved landfill site. Code C: Flush the limewater down the drain with excess water.

PREPARATION

See page 202 for preparation of solutions and other materials.

PROCEDURE

- Having the required materials and equipment available at each laboratory table or work area will greatly reduce student traffic, the potential for spillage or breakage, and the time required for the work. All students at a table can use a single hot plate.
- Set up reactions as stations to which groups of students rotate to minimize materials needed and the hazard presented when many students are using flames at the same time.

ANALYZE AND CONCLUDE

1. Synthesis: formation of a solid and color change; De-

ChemLab  SMALL SCALE

Exploring Chemical Changes

Most reactions can be classified into five major types. As you carry out this experiment, you'll observe examples of each of these types. In doing so, you will also learn to recognize many of the physical changes that accompany reactions.

Problem

What are some of the physical changes that indicate that a reaction has occurred?

Objectives

- **Observe** physical changes that take place during chemical reactions.
- **Compare** changes that take place during different types of chemical reactions.

PREPARATION

Materials

125-mL flasks (4)	large test tube and one-
balance	hole stopper with glass
hot plate	tube and rubber tubing
watch glass	attached
spatula	ring stand
stirring rod	test-tube clamp
lab burner	$0.1M$ $CuSO_4$
file	granular copper, Cu
new penny	powdered sulfur, S
250-mL flask	$CaCO_3$, finely ground
ice	saturated $Ca(OH)_2$
tongs	solution, limewater
100-mL graduated	$6M$ HCl
cylinder	$0.5M$ Na_2CO_3
	$0.5M$ $CuCl_2$

Safety Precautions

Wear an apron and goggles. Use care when handling hot objects. Dispose of the reaction mixture and products as instructed by your teacher.

PROCEDURE

For each of the following reactions, record in a data table all changes that you observe.

Synthesis Reaction

1. Place 50 mL $0.1M$ $CuSO_4$ in a 125-mL flask.
2. Place 1.6 g granular copper and 0.8 g powdered sulfur on a watch glass and mix together thoroughly with a spatula.
3. Heat the flask on a hot plate set at high until the solution begins to boil.
4. Stir the Cu/S mixture into the boiling $CuSO_4$ solution.
5. Continue boiling until a black solid forms.

Decomposition Reaction

1. Place 100 mL of saturated $Ca(OH)_2$ solution (limewater) in the 250-mL flask.
2. Add finely ground $CaCO_3$ to a large test tube until it is one-fourth full. Stopper the tube with the stopper/glass tube/rubber tubing assembly, and clamp the tube to the ring stand.
3. Light a laboratory burner, and begin to heat the test tube. Submerge the end of the rubber tubing into the limewater so that any gas produced in the tube will bubble through the limewater.
4. Continue heating the $CaCO_3$ until you observe a change in the limewater. The presence of CO_2 causes limewater to become cloudy.

Single-Displacement Reaction

1. Place 30 mL $6M$ HCl in a 125-mL flask.

---

composition: gas formation; Single Displacement: gas formation and change in mass and appearance of the penny; Double Displacement: formation of precipitate; Combustion: energy given off and water and black soot formed.

2. Physical changes could be observed during all of the reactions.

3. **a.** copper(II) sulfide, CuS **b.** carbon dioxide, CO_2 **c.** calcium oxide, CaO **d.** copper(II) carbonate, $CuCO_3$ **e.** water, H_2O

4. The zinc metal on the inside of the penny displaced hydrogen in hydrochloric acid to

form zinc chloride, leaving only the copper shell on the outside intact. A solid copper penny will not react because copper will not displace the hydrogen.

5. a product

APPLY AND ASSESS

1. No, all the major types of changes should have been observed.

2. Synthesis: Cu(s) + S(s) → CuS(s); Decomposition: $CaCO_3(s)$ → $CaO(s)$ + $CO_2(g)$; Single Displacement: 2HCl(aq) + Zn(s) →

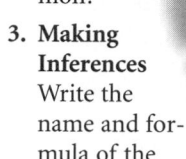

ChemLab

2. Using a file, cut six 0.2-cm notches evenly spaced around the perimeter of a new penny.

3. Place the penny in the flask of acid and leave it in a fume hood overnight.

Double-Displacement Reaction

1. Add 25 mL 0.5M Na$_2$CO$_3$ and 25 mL 0.5M CuCl$_2$ to a 125-mL flask.

2. Swirl the flask gently until you observe the formation of a precipitate.

Combustion Reaction

1. Light a laboratory burner and adjust the air and gas supplies until the flame is blue. Observe what happens.

2. Using tongs, hold a flask or beaker with ice in it about 10 cm over the flame for approximately one minute. Move the flask away from the flame and observe the bottom of the flask.

DATA AND OBSERVATIONS

Reaction	Observations
Synthesis	
Decomposition	
Single displacement	
Double displacement	
Combustion	

ANALYZE AND CONCLUDE

1. **Making Inferences** Which observations noted during each of the reactions indicated that a reaction had occurred?

2. **Comparing and Contrasting** What did all of the reactions have in common?

3. **Making Inferences** Write the name and formula of the

a) black solid formed in the synthesis reaction.

b) gaseous product of the decomposition reaction.

c) solid product of the decomposition reaction.

d) pale blue precipitate in the double-displacement reaction.

e) liquid product of the combustion reaction.

4. **Observing and Inferring** Explain how the penny changed during the single-displacement reaction. What would happen if a pre-1983 penny, which is solid copper, were used?

5. **Relating Concepts** Is energy a reactant or product of the combustion reaction?

APPLY AND ASSESS

1. Were there any physical changes that often occur during a reaction that you did not observe while doing this ChemLab? If so, what were they?

2. Write balanced chemical equations for all of the reactions carried out.

3. Why do you think pennies are no longer made from only copper metal?

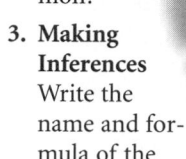

ZnCl$_2$(aq) + H$_2$(g); Double Displacement: Na$_2$CO$_3$(aq) + CuCl$_2$(aq) → 2NaCl(aq) + CuCO$_3$(s). Combustion: (if natural gas is used and methane is burned) CH$_4$(g) + 2O$_2$(g) → CO$_2$(g) + 2H$_2$O(g) or (if a propane tank is used) C$_3$H$_8$(g) + 5O$_2$(g) → 3CO$_2$(g) + 4H$_2$O(g).

3. Zinc is less expensive than copper.

✓ Assessment

Performance: Give some students pre-1983 pennies and others post-1983 pennies to use in the single-displacement reaction. Ask them to think up reasons why all results did not turn out the same. See whether the students can find out the difference between the floaters (post-1983 pennies) and sinkers (pre-1983 pennies) by examining the two groups of coins. L1 COOP LEARN

GLENCOE TECHNOLOGY

Videodisc

Chemistry: Concepts and Applications
Types of Chemical Reactions
Disc 1, Side 2, Ch. 3

Show this video to help you explain differences among the five major types of reactions.
L1 LEP

CD-Rom

Chemistry: Concepts and Applications
Types of Chemical Reactions
Have students use this video to help explain the characteristics of the different types of reactions. L1 LEP

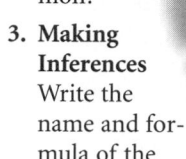

DATA AND OBSERVATIONS

The following observations should be made for each indicated reaction type. Synthesis—a black solid formed from a blue solution, a copper-colored solid, and a yellow solid. Decomposition—carbon dioxide gas bubbled from the tubing, turning the limewater milky white. Single Displacement—gas bubbles appear immediately; after one day, the middle of the penny is empty and only the copper outside remains. The penny shell floated on top of the acid solution. Double Displacement—a pale blue solid was formed when the colorless and blue solutions were mixed. Combustion—heat given off; black soot and drops of a clear liquid covered the bottom of the flask.

What is evidence of double displacement?

Place 0.16 g of calcium nitrate into a test tube containing 5 mL of H_2O. Stopper and shake until dissolved. Place 0.15 g of sodium sulfate in a second test tube containing 5 mL of H_2O. Stopper the test tube and shake until dissolved. Have students observe the tubes and record observations. Mix the contents of the two test tubes. The two clear, colorless solutions, when mixed, form a white precipitate. Ask students to name the products of the double-displacement reaction. *barium sulfate and sodium nitrate*

L1

GLENCOE TECHNOLOGY

Videodisc

Chemistry: Concepts and Applications
Five Types of Chemical Reactions
Disc 1, Side 2, Ch. 2

Use space-filled molecules to illustrate synthesis, decomposition, combustion, single-displacement, and double-displacement reactions. L1 LEP

CD-Rom

Chemistry: Concepts and Applications
Five Types of Chemical Reactions
Use space-filled molecules to illustrate synthesis, decomposition, combustion, single-displacement, and double-displacement reactions. L1 LEP

In **double-displacement** reactions, the positive portions of two ionic compounds are interchanged. For a double-displacement reaction to take place, at least one of the products must be a precipitate or water. An example of a double-displacement reaction is shown in **Figure 6.11**.

Figure 6.11
Double Displacement
When clear aqueous solutions of lead(II) nitrate and potassium iodide are mixed, a double-displacement reaction takes place and a yellow solid appears in the mixture. This solid is lead(II) iodide, and it precipitates out because it is insoluble in water, unlike the two reactants and the other product.

$$Pb(NO_3)_2(aq) + 2KI(aq) \rightarrow PbI_2(s) + 2KNO_3(aq)$$

Table 6.1 Types of Reactions

Reaction Type	General Equation
Synthesis	element/compound + element/compound → compound
Examples:	$2Na(s) + Cl_2(g) \rightarrow 2NaCl(s)$
	$CaO(s) + SiO_2(l) \rightarrow CaSiO_3(l)$
Decomposition	compound → two or more elements/compounds
Examples:	$PCl_5(s) \rightarrow PCl_3(s) + Cl_2(g)$
	$2Ag_2O(s) \rightarrow 4Ag(s) + O_2(g)$
Single displacement	*element a + compound bc → element b + compound ac
Example:	$2Al(s) + Fe_2O_3(s) \rightarrow 2Fe(s) + Al_2O_3(s)$
	element d + compound bc → element c + compound bd
Example:	$Cl_2(aq) + 2KBr(aq) \rightarrow 2KCl(aq) + Br_2(aq)$
Double displacement	compound ac + compound bd → compound ad + compound bc
Examples:	$PbCl_2(s) + Li_2SO_4(aq) \rightarrow PbSO_4(s) + 2LiCl(aq)$
	$BaCl_2(aq) + H_2SO_4(aq) \rightarrow 2HCl(aq) + BaSO_4(s)$
Combustion	element/compound + oxygen → oxide(s)
Examples:	$CH_4(g) + 2O_2(g) \rightarrow CO_2(g) + 2H_2O(g)$
	$C_6H_{12}O_6(s) + 6O_2(g) \rightarrow 6CO_2(g) + 6H_2O(l)$

*The letters a, b, c, and d each represent different elements or parts of compounds. For example, in compound ac, *a* represents the positive part of the compound, and *c* represents the negative part.

Figure 6.12
Combustion
When welding is done with an acetylene torch, acetylene combines with oxygen to form carbon dioxide and water. This combustion reaction is exothermic, and enough energy is released to melt metal.

$$2C_2H_2(g) + 5O_2(g) \rightarrow$$
$$4CO_2(g) + 2H_2O(g) + energy$$

WORD ORIGIN

combustion:
comburere (L) to burn up
When wood burns, it reacts with oxygen in a combustion reaction.

The fifth common type of reaction is a combustion reaction. A **combustion** reaction is one in which a substance rapidly combines with oxygen to form one or more oxides, as shown in **Figure 6.12.**

Although there are exceptions, many thousands of specific reactions fit into these five classes. One example is in the Chemistry and Technology feature on pages 216-217. **Table 6.1** summarizes important information about each of these reaction types.

SECTION REVIEW

Understanding Concepts

1. Explain why classifying reactions can be useful.

2. Using different symbols to represent different atoms, draw pictures to represent an example of each of the following kinds of reactions.

   a) synthesis
   b) decomposition
   c) single displacement
   d) double displacement
   e) combustion

3. Classify each of the following reactions.

   a) $N_2O_4(g) \rightarrow 2NO_2(g)$
   b) $2Fe(s) + O_2(g) \rightarrow 2FeO(s)$
   c) $2Al(s) + 3Cl_2(g) \rightarrow 2AlCl_3(s)$

   d) $BaCl_2(aq) + Na_2SO_4(aq) \rightarrow$
      $$BaSO_4(s) + 2NaCl(aq)$$
   e) $Mg(s) + CuSO_4(aq) \rightarrow Cu(s) + MgSO_4(aq)$

Thinking Critically

4. **Making Inferences** When a candle burns, wax undergoes a combustion reaction. Would a candle burn longer in the open or when covered with an inverted glass jar? Explain.

Applying Chemistry

5. **Decomposition** When a fungus breaks down the wood in a fallen tree, the biological process is called decomposition. What does that process have in common with the chemical decomposition reaction type?

SECTION REVIEW

1. Recognizing patterns of changes that take place during a reaction helps you predict changes that might occur in a new reaction that fits one of the patterns.

2. Sample answers: **a.** ⊗ + ⊕ → ⊗⊕
   b. ⊗⊕ → ⊗ + ⊕
   c. ⊗ + ⊕© → ⊗© + ⊕
   d. ⊗⊕ + ©® → ©⊕ + ⊗®
   e. ®⊕ + ⊗⊗ → ®⊗ + ⊕⊗

3. **a.** decomposition **b.** synthesis **c.** synthesis

d. double displacement **e.** single displacement

4. A candle would burn longer in the open. Burning wax is a combustion reaction, which requires oxygen as a reactant.

5. Both types of decomposition involve breaking down more complex materials into simpler ones. The fungus gets energy and nutrients by breaking down large carbohydrate and protein molecules in the wood.

3 ASSESS

Check for Understanding

• Check for understanding of the section material by having students answer the Section Review questions.

• Ask students to use a solubility table to predict which of these two double-displacement reactions will occur, yielding an insoluble product. **a.** AgCl + NaNO₃ → NaCl + AgNO₃ or **b.** NaCl + AgNO₃ → AgCl + NaNO₃. *reaction b, because AgCl is an insoluble product* ⎣L2⎦

Reteach

Discussion: Explain that tarnish may be removed from silver by reacting it with aluminum, yielding aluminum sulfide and silver. Ask students to write the balanced equation for the reaction and to state the reaction type. *3Ag₂S + 2Al → Al₂S₃ + 6Ag; single-displacement* ⎣L1⎦

Extension

For reactions in which polyatomic ions obviously remain intact, they may be considered as units to simplify the balancing process. For example: In the reaction 3Na₂SO₄ + 2AlCl₃ → 6NaCl + Al₂(SO₄)₃, three sulfate ions on each side of the equation balance sulfur and oxygen. Have students balance several equations containing intact polyatomic ions. ⎣L1⎦

4 CLOSE

Portfolio

Ask each student to go through all notes, assignments, and other activities for Sections 6.1 and 6.2 and compile a list of chemical equations, properly balanced and categorized by type. The list should be included in his or her portfolio. ⎣L1⎦ ⎣P⎦

Key Concepts

The reversibility of reactions is introduced and related to equilibrium. Le Châtelier's principle and the driving force concept for reactions are briefly discussed. Rates of chemical reactions, as well as rate-affecting factors such as activation energy, temperature, concentration, and catalysis are examined.

Planning Ahead

For the MiniLab Have ice available. The starch solution may be prepared by dissolving 4 g of starch in 1 L of boiling water and then adding 0.2 g of $Na_2S_2O_5$ (sodium metabisulfite) and 5 mL of $1M$ sulfuric acid. Alternatively, laundry starch in a can may be sprayed for about 5 seconds into 1 L of cold water before adding the other two reactants. The iodine solution may be prepared by adding 4.3 g of KIO_3 to 1 L of water.

For the Demonstrations For the Quick Demos, obtain effervescent antacid tablets from a pharmacy or grocery store and 12% hydrogen peroxide from a beauty salon or cosmetology supply store. Lycopodium powder for Demonstration 6.7 should be ordered, if not on hand.

1 FOCUS

Focus Transparency

Display the **Section Focus Transparency** for Section 6.3 to illustrate a reversible reaction. L1 LEP

SECTION 6.3

Nature of Reactions

You know that changes constantly occur, but some changes are not permanent. For example, liquid water freezes into ice, but then ice melts and becomes liquid water again. In other words, the freezing process is reversed. Can chemical reactions be reversed? Can the product of a reaction become a reactant?

Reversible Reactions

Many reactions can change direction. These reactions are called reversible reactions. Some reactions, such as the one shown in **Figure 6.13,** can be reversed based on the energy flow.

Not all chemical changes are reversible. Caves form, paint hardens, and fuel burns. These chemical changes result in new products, and the reactions are said to go to completion because at least one of the reactants is completely used up and the reaction stops. The reactions can't be reversed. But what happens when a reaction reverses?

Figure 6.13
Reversible Recharge
When an automobile battery releases energy when the car isn't running, the reaction below moves to the right. If you leave the lights on and need to recharge the battery with a jump start, the reaction moves to the left while the car engine runs.

$$Pb(s) + PbO_2(s) + 2H_2SO_4(aq) \rightleftarrows$$
$$2PbSO_4(s) + 2H_2O(l) + energy$$

Figure 6.14
No Net Change
For this train, the initial movement is all in one direction. Eventually, the number of passengers on the train will not be changing much because the number getting off the train will be about equal to the number getting on.

2 TEACH

Using an Analogy

Describe an automobile's movement as reversible. It may move in reverse or forward. If the reverse gear were broken, the situation would be like an irreversible reaction.

Quick Demo

How can you model equilibrium?

Put unequal amounts of water in two 5- or 10-gallon aquariums and measure and record both depths in centimeters. Give a student a 250-mL beaker to be used to scoop the water from tray 1 into tray 2, and give another a 600-mL beaker to simultaneously scoop the water from tray 2 into tray 1. Ask students what will happen as the exchanges occur. *They will think that the student having the 600-mL beaker will empty tray 2.* Allow the process to go on until an equilibrium state has been achieved. Tray 1 will have more water than tray 2; however, the 250-mL beaker can transfer only the same amount of water in each exchange as is accomplished by the less-than-full 600-mL beaker from tray 2. The amounts of reactants and products are usually not equal in an equilibrium situation. L1 LEP

Picture what happens in a subway station when the door of a train opens for the first time that day. Passengers rush in, streaming across the platform and through the doors. No passengers get out because this is the first train of the day. When the train stops at the next station, more passengers rush in. A few probably leave the train at this point, as well. With each successive stop, more passengers will be getting on, but more and more will also be getting off. Look at **Figure 6.14.** To describe a similar situation in chemical reactions in which a reaction automatically reverses and there is no net (overall) change, we use the term *equilibrium.*

Equilibrium

When no net change occurs in the amount of reactants and products, a system is said to be in **equilibrium.** In most situations, chemical reactions exist in equilibrium when products and reactants form at the same rate. Such a system, in which opposite actions are taking place at the same rate, is said to be in **dynamic equilibrium.**

Reactions at equilibrium have reactants and products changing places, much like the passengers getting on and off a subway train. In equilibrium, reactants are never fully used up because they are constantly being formed from products. Eventually, reactants and products form at the same rate.

One example of a reversible reaction that reaches equilibrium occurs when lime, CaO, which is used to make soils less acidic, is formed by decomposing limestone, $CaCO_3$.

$$CaCO_3(s) \rightleftarrows CaO(s) + CO_2(g)$$

Notice that the single arrow in the equation has been replaced with double arrows. Because an arrow shows what direction the reaction is going, the double arrow indicates that the reaction can go in either direction. In this case, $CaCO_3$ decomposes into CaO and CO_2. But, as those products form, CO_2 and CaO combine to form $CaCO_3$. The rate, or speed, of each reaction can be determined by how quickly a reactant disappears. Eventually, the rates of the two reactions are equal, and an equilibrium exists.

Meet Caroline Sutliff, Plant-Care Specialist

Teaching Strategies

• Ask students to speculate on the meaning of the terms *herbicides*, *fungicides*, and *pesticides*. Students should realize that the root *-cide* means "to kill." (Latin, *-cidere*, "to kill"). Therefore, herbicides, fungicides, and pesticides are agents that kill certain plants, fungi, and lower forms of animals, respectively. L1

• Have students investigate how herbicides and insecticides affect plant and animal physiology, respectively. Herbicides affect the plants in various ways such as inhibition of respiration, reduction of photosynthesis, and decreased production of hormones, which affects cell enlargement and division. Insecticides usually affect respiration and the nervous system. L1

• Have interested students contact the local office of the Agricultural Extension Service and find out what information and services it supplies to urban and suburban gardeners. L1

Background

Plant-care specialists care for plantings in private gardens; in public gardens and parks; and in shopping centers, golf courses, apartment and office building complexes, and cemeteries.

PEOPLE in CHEMISTRY

Meet Caroline Sutliff, Plant-care Specialist

An old gardening adage goes like this: A weed is just a flower that is growing where you don't want it. Landscape gardener Caroline Sutliff works to eradicate those hardy, invasive plants, such as *Taraxacum officinale* (better known as dandelions). In this interview, she tells about how she wages her weed battles with garden tools and chemicals.

On the Job

 Ms. Sutliff, what do you do on the job?

 I take care of flowers, shrubs, and trees. The company for which I work maintains the grounds of business properties, such as real estate offices and banks, as well as private homes. My job involves working with chemicals such as herbicides, fungicides, and pesticides.

Q **Are those chemicals hazardous?**

A They can be if you don't treat them properly. I have a spraying license that I earned through the Agricultural Service Extension, so I have learned all the safety precautions. Mixing the chemicals to get them ready for use is more hazardous than actually spraying them. For mixing, I wear a plastic-coated jacket and pants, heavy boots, rubber gloves, and a face mask.

 Do you and your coworkers know what to do in case of a mishap?

 We carry eyewash and first-aid kits in all of our trucks. We're trained to react quickly if we get a chemical on our clothes. In that case, we must remove the affected clothing and wash up quickly. When some chemicals come in contact with your skin, they are absorbed into the body and stored in fat cells. The cumulative effect could be harmful. Because I'm very careful and follow directions exactly, I haven't had any problems.

Q **Why are chemicals necessary in your work?**

A On large properties, it's a matter of economy and time—two hours of hand weeding and cultivating versus a few minutes of spraying. However, chemical use is just part of our program. Before we spray chemicals, we cultivate, do some hand weeding, and fertilize and aerate beds of plants. Healthy plants are more likely to have a fighting chance against weeds and insects.

Demonstration 6.5

Equilibrium Is Not 50:50

Purpose: To show that equilibrium may not indicate equal amounts of reactant and product.

Materials: two 100-mL graduated cylinders; two 35-cm lengths of glass tubing, 10-mm and 6-mm diameters

Safety Precautions:

Disposal: Pour down the drain.

Procedure: Fill a graduated cylinder, labeled *reactants*, with water. Leave the other cylinder, marked *products*, empty. Simultaneously lower the larger tube to the bottom of the reactant cylinder and the smaller tube to the bottom of the product cylinder. Cover each end with a finger and remove the tubes simultaneously. Transfer the water from the reactant cylinder to the product cylinder by draining the reactant tube. Mimic draining the product tube into the reactant cylinder. Repeat the process until the volumes of the cylinders remain constant during a transfer.

Q Do you use the information you learned in high school chemistry classes?

 A Yes. It's important to be familiar with the properties of chemicals—to know what can be safely mixed and what can't. The container used for mixing is important chemically, too. For example, stainless steel containers can erode and change the makeup of chemicals so they won't have the same effects. Heavy plastic is used instead.

Early Influences

 Q As a child, were you interested in plants?

 A I wasn't interested in planting things because I wanted immediate results. Plants don't work that way. Now I have more patience and love to tend to the flowers around my own home.

 Q Did you plan to enter the plant-care field later on?

 A No. My intentions were to get a degree in psychology and then go to law school. While I was looking for a job as a paralegal, I began working in plant care and discovered I love working outside. Now I can't imagine being inside all day at a desk job.

Personal Insights

 Q Would you recommend that students investigate a plant-care job like yours?

 A Only if they like hard, physical work. My workdays last ten or 12 hours, and I'm out in all kinds of weather. I certainly don't need to go to a gym to stay in shape! It's satisfying for me to see that more women are entering this field.

 Q How do you see the field of plant care changing in the years ahead?

 A I think people will become even more aware of the value of plants, particularly trees. Trees can do so much to improve a neighborhood. They clean the air, shade the sidewalks, and beautify an area. Here in Iowa City, an organization called Project Green helps plant trees in public places. I hope to get involved with that group as a volunteer.

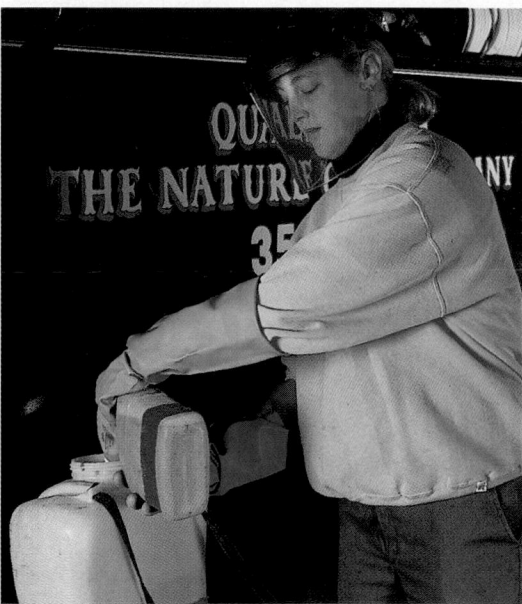

CAREER ▶ CONNECTION

The following careers are also associated with plant care.

Horticulturist Master's degree, research, and fieldwork
Landscape Architect Bachelor's degree, often followed by a licensing examination
Soil Conservation Technician Two-year college program

Results: The reverse reaction rate increases until equilibrium is reached.
Analysis: Ask these questions.
1. Initially, which rate is greater, forward or reverse? *forward*
2. As the rate of the forward reaction decreases, what happens to the rate of the reverse reaction? *It increases.*
3. How can you know that equilibrium has been reached? *The volumes in the two cylinders remain constant as water is transferred.*

✓ Assessment

Skill: Have students collect data and graph the volumes in the cylinders versus the number of transfers, representing time. **L2**

Describe the following dynamic equilibrium situation for students. A person is rowing a boat upstream at exactly the same speed as the current downstream. The boat has no net movement; however, a state of physical equilibrium exists.

GLENCOE TECHNOLOGY

 Videodisc

Chemistry: Concepts and Applications
Equilibrium
Disc 1, Side 2, Ch. 5

Use this animation to show a system at equilibrium. L1 LEP

 CD-Rom

Chemistry: Concepts and Applications
Equilibrium
Use this animation to show a system at equilibrium. L1 LEP

GLENCOE TECHNOLOGY

 Software

Mastering Concepts in Chemistry
Unit 13, Lesson 2
Chemical Equilibrium
Use this lesson to reinforce the concept of chemical equilibrium. L1 LEP

WORD ORIGIN

soluble:
solvere (L) capable of being dissolved in a liquid
Because sugar can be dissolved in water, it is water-soluble.

Did you ever have to make up your mind?

Saying that a reaction has reached equilibrium does not mean that equal amounts of reactants and products are present. Equilibrium just means that no net change is taking place—the amounts of reactants and products are not changing. Often, a reaction at equilibrium contains differing amounts of reactants and products.

Consider the following reaction in which phosphorus pentachloride decomposes into phosphorus trichloride and chlorine.

$$PCl_5(g) \rightleftharpoons PCl_3(g) + Cl_2(g)$$

Actual measurements of the amounts of reactant and products show that the system is in equilibrium. But the measurements also show that there is more PCl_5 than PCl_3 and Cl_2 present, even though the rates of the reactions are equal. The reversible reaction will favor the direction that produces the most stable products, which are those that are least likely to change. In this example, PCl_5 is less likely to decompose than the two products, PCl_3 and Cl_2, are likely to combine. Reversible reactions at equilibrium are in a balance that favors stability.

Changing Direction

If a reaction reaches equilibrium, how can you obtain large quantities of a product? Won't the product constantly become a reactant? Keep in mind that reactions in equilibrium are stable. The French scientist Henri Louis Le Chatelier proposed in 1884 that disturbing an equilibrium will make a system readjust to reduce the disturbance and regain equilibrium, **Figure 6.15.** This principle regarding changes in equilibrium is called Le Chatelier's principle.

Chemical engineers can apply this tendency of reactions to stay at equilibrium to find ways to increase the yield of products in a reaction. For example, if products are removed from a reaction at equilibrium, more reactants will go on to form products so that balance is regained. If this removal continues, most of the reactants can be converted into products. For example, recall the reaction in which limestone decomposes into lime and carbon dioxide.

$$CaCO_3(s) \rightleftharpoons CaO(s) + CO_2(g)$$

If the carbon dioxide is removed as it is produced, the reaction will favor the formation of more carbon dioxide to reestablish equilibrium. Thus, the reaction will shift in the direction that also produces more lime as a product.

Figure 6.15
A System in Stress
Before the dog started to drink, the water in this container reached a stable equilibrium level. As the dog drinks, this level is stressed and water comes from the bottle into the bowl to reestablish the equilibrium level.

Demonstration 6.6

A Reversible Reaction

Purpose: To demonstrate a reversible reaction.
Materials: 14 g of NaOH or KOH, 14 g of dextrose, 1-L Erlenmeyer flask with stopper, 1 mL of 1% methylene blue indicator solution
Safety Precautions:
Disposal: Code I
Procedure: Dissolve 14 g of NaOH or KOH in 500 mL of water in the flask. Add 14 g of dextrose and a few drops of 1% methylene blue indicator solution. (Resazurin can be used if you prefer a red color.) Stopper the flask tightly. Shake vigorously until the solution is blue in color. **CAUTION:** *This caustic solution damages skin and eyes. Be careful that the flask does not leak.* Allow the solution to sit until the color clears. Shake the solution again.

Have students note each color change.
Results: The solution turns blue when shaken but turns colorless upon standing. Shaking the solution dissolves oxygen, which then oxidizes the methylene blue indicator to its blue form. The oxidized indicator is then reduced back to its colorless or yellow form by the dextrose, a reducing sugar.

When one product is a gas and other products and reactants are not, as in the previous example and in **Figure 6.16,** it is easy to see how the gas can be removed from the reaction. How can products that are not isolated gases be removed from a reaction? They usually can't be picked out manually because of the mix of reactants and products.

A product that does not dissolve in water can be removed if all other products and the reactants dissolve in water. A compound is **soluble** in a liquid if it dissolves in it; it is **insoluble** if it does not. An insoluble product will form a solid precipitate that sinks to the bottom of a liquid solution, as shown in **Figure 6.17.** The precipitate cannot react easily because it is somewhat isolated from the other substances in the reaction.

Adding Reactants or Energy

Adding more reactants has basically the same effect as removing products. For the reaction to reestablish equilibrium amounts of reactants and products, more products must be made if more reactants are added.

Adding or removing energy, usually in the form of heat, can also influence the direction of a reaction. Because energy is a part of any reaction, it can be thought of as a reactant or a product. Just as adding more reactants to a reaction pushes it to the right, so does adding more energy to an endothermic reaction. For example, the equation for a reaction that produces aluminum metal from bauxite, an aluminum ore, shows that energy must be added for products to form.

$$3C + 2Al_2O_3(s) + energy \rightleftarrows 4Al(l) + 3CO_2(g)$$

If more energy is added, the reaction goes to the right, forming more aluminum and carbon dioxide.

For an exothermic reaction, adding more energy pushes the reaction to the left. In the Haber process for producing ammonia from hydrogen and nitrogen, for example, energy is produced.

$$N_2(g) + 3H_2(g) \rightleftarrows 2NH_3(g) + energy$$

Adding energy favors the formation of nitrogen and hydrogen. Thus, temperature must be carefully controlled in the Haber process so that the desired product, ammonia, is produced in large quantities.

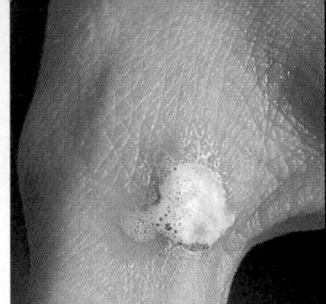

Figure 6.16

Removing a Gas
When hydrogen peroxide, H_2O_2, is poured onto a wound, the hydrogen peroxide decomposes to form water and oxygen. The gaseous oxygen bubbles away, preventing the re-formation of H_2O_2.

$$2H_2O_2(aq) \rightarrow 2H_2O(l) + O_2(g)$$

Notice that only a forward arrow is used because the removal of oxygen drives the reaction forward.

Figure 6.17

Forming a Precipitate
When potassium hydroxide is added to an aqueous solution of calcium chloride, calcium hydroxide and potassium chloride form. Calcium hydroxide is nearly insoluble in water, so it precipitates out as a solid.

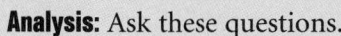

215

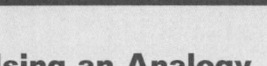
Using an Analogy

Describe this equilibrium situation for students. You withdraw $100 from your bank account on the first day of the month and deposit $100 on the last day of the month. Assuming that you earn no interest, the money you put into the bank equals the money you take from the bank. The bank has a great deal more money than you have; however, the equilibrium state only requires that the amount in equals the amount out.

Correcting Misconceptions

Students often think that any reaction that yields more stable products than reactants will occur quickly. Use the example of wood in a fireplace to illustrate that, until the activation energy is supplied in the form of a flame, the combustion reaction will not proceed.

Concept Development

Explain that molecules, in order to react, must collide, must have sufficient energy, and must collide with the correct orientation. Any factor, such as concentration, temperature, and catalysis, that influences those requirements will have an effect on the reaction rate.

Transparency

Display **Basic Concepts Transparency 18** to show that concentration and temperature affect reaction rate. L1 LEP

Analysis: Ask these questions.
1. Is the rate of appearance of the blue substance slower or faster than the rate of appearance of the yellow substance? *faster*
2. Does the chemical system in the flask reach equilibrium if the flask is not shaken? Explain. *Yes, the macroscopic properties do not change.*

3. If many of the biochemical reactions that take place in our bodies are reversible, why do we age? *Reactions are not completely reversible.*

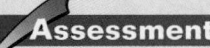

Assessment

Oral: Ask students to research and discuss a reversible reaction involving iron that occurs in the blood. *Hemoglobin contains*

iron in the 2+ state. It is easily oxidized to iron(III) in methemoglobin, which is not able to transport oxygen. An enzyme, diaphorase, causes the iron(III) to be reduced back to iron(II). L2

Purpose

Students will learn about the development of the Haber-Bosch process of ammonia production.

Background

Haber and Bosch were independently awarded the Nobel Prizes in Chemistry. Haber received the prize in 1918 and Bosch in 1931. There was much controversy over Haber's award. Haber had been dubbed the father of modern gas warfare because he was the first to use gaseous substances as weapons. In April 1915, he had canisters of deadly chlorine gas vented against Allied troops at Ypres, Belgium, in hopes of ending the war quickly.

GLENCOE TECHNOLOGY

Videodisc

Science and Technology Videodisc Series
Super Grit
Disc 2: Chemistry
Side 1, Ch. 9

Show this video to introduce the concept of activation energy. L1 LEP

Mining the Air

Diatomic nitrogen makes up about 79 percent of Earth's atmosphere. Nitrogen is an essential element for life, yet only a few organisms can use atmospheric nitrogen directly. A few species of soil bacteria can produce ammonia, NH₃, from atmospheric nitrogen. Other species of bacteria can then convert the ammonia into nitrite and nitrate ions, which can be absorbed and used by plants.

The Haber Process

Not all usable nitrogen compounds are produced naturally from atmospheric nitrogen. Ammonia can also be synthesized. The process of synthesizing large amounts of ammonia from nitrogen and hydrogen gases was invented by Fritz Haber, a German research chemist, and his English research assistant, Robert LeRossignol. The process was first demonstrated in 1909. Haber was granted the German patent for the process.

The Haber process involves a synthesis reaction. The two reactants are $H_2(g)$ and $N_2(g)$, and the products are $NH_3(g)$ and heat. The process produces high yields of ammonia by manipulating three factors that influence the reaction—pressure, temperature, and catalytic action.

Ammonia storage tanks

Fritz Haber

Pressure

In the ammonia synthesis reaction, four molecules of reactant, H_2 and N_2, produce two molecules of product, NH_3. According to Le Chatelier's principle, if pressure on the reaction or the system is increased, the forward reaction will speed up to lessen the stress because two molecules exert less pressure than four molecules. Increased pressure will also cause the reactants to collide more often, thus increasing the reaction rate. Haber's apparatus used a total pressure of 2×10^5 kPa, which was the highest pressure he could achieve in his laboratory.

Temperature

Two factors influence the temperature at which the process is carried out. A low temperature favors the forward reaction because this reduces the stress of the heat generated by the reaction. But high temperature increases the rate of reaction because of more collisions between the reactants. Haber carried out the process at a temperature of about 600°C.

216   Chapter 6   Chemical Reactions and Equations

Cultural Diversity

...and Call Me in the Morning

A German chemist named Felix Hofmann who worked at the Baeyer Company around the turn of the 20th century is usually given credit for being the first to make today's leading commercial analgesic (anti-pain), antipyretic (anti-fever), and anti-inflammatory medicine. However, doctors dating back to the ancient Greek healer Hippocrates knew that chewing willow bark relieved pain. This discovery was made independently in many different cultures including Native American and Chinese over thousands of years, but it wasn't until the 1800s that the active ingredient in the bark was isolated and identified as salicylic acid. The task worked on by Hofmann was how to modify salicylic acid to make it less acidic, yet still active. He found a solution in a displacement reaction in which a single hydrogen on salicylic acid is replaced with an acetyl group. The new compound formed is the more easily tolerated molecule acetylsalicylic acid, which was first sold as a powder called aspirin in envelopes and capsules in 1899. The tablets we are more familiar with were introduced in 1915.

appears. Either measurement will give an amount of substance changed per unit time, which is the rate of reaction.

Why is reaction rate important? The rate of a reaction is important to a chemical engineer designing a process to get a good yield of product. The faster the rate, the more product that can be made in a fixed amount of time. Rate of reaction is also important to food processors who hope to slow down reactions that cause food to spoil. Can the rate of a reaction be changed? Four major factors affect this rate.

Effect of Temperature

One factor that affects rate of reaction is temperature, **Figure 6.19.** Most reactions go faster at higher temperatures. Baking a cake speeds up the reactions that change the liquid batter into a spongy product. Lowering the temperature slows down most reactions. Photographic film and batteries stay useful longer if they are kept cool because the lower temperature slows the reactions that can ruin these products.

Figure 6.19
Effect of Temperature on Reaction Rate
Adding heat to reactants helps break bonds and increases the speed at which molecules and atoms are moving. The faster they move, the more likely it is that they will collide and react. Removing heat slows down reactions. That's why freezing food might help keep it from spoiling as quickly.

It's a Matter of Concentration

Changing the amounts of reactants present can also alter reaction rate. The amount of substance present in a certain volume is called the **concentration** of the substance. Raising the concentration of a reactant will speed up a reaction because there are more particles per volume. More particles result in more collisions, so the reaction rate increases, **Figure 6.20.** In most cases, concentration is increased by adding more reactant. If a fire is burning slowly, fanning the flames increases the amount of oxygen available and the fire burns faster. If a gas is involved, concentration may be increased by increasing pressure. Increasing pressure does not increase the number of particles, but it brings the particles closer together so collisions are more frequent. The Haber process, for example, uses high pressures to increase the rate of reaction of hydrogen and nitrogen to form ammonia. Lowering concentration decreases the rate of reaction. Many valuable historic documents are stored in sealed cases with most of the air removed to decrease the number of particles that might react with the paper.

Figure 6.20
Concentration and Reaction Rate
Adding more bumper boats increases the chance that a collision will occur. Taking some away decreases the odds of two boats meeting. When more particles are added to a reaction mixture, the chance that they will collide and react is increased.

2. Predict which would dissolve first, a sugar cube or the same mass of granulated sugar. *the granulated sugar*
3. What do you believe happens to the concentration of oxygen gas reacting with each dust particle as the surface area increases? *The concentration of oxygen effectively increases as surface area increases.*

 **Assessment**

Portfolio: Have interested students photograph the demonstration. Have students add the developed photograph to their portfolio with a written summary of the demonstration. L1
LEP P

SECTION 6.3

⚡ Quick Demo

Does temperature affect reaction rate? 🥽

Add an effervescent antacid tablet to each of three plastic sandwich bags, and break each tablet into four pieces. Place 50 mL of ice water into a 250-mL Erlenmeyer flask. Add 50 mL of hot water to a second Erlenmeyer flask. Add 50 mL of water at room temperature to a third flask. Secure a plastic bag to the neck of each flask using a rubber band. Have three students simultaneously empty the pieces of tablet from the bags into the flasks and note the time for the bags to inflate. Ask students to write a summary that reflects the usual relationship between reaction rate and temperature. *Increasing the temperature increases the reaction rate.* L1

GLENCOE TECHNOLOGY

 Software

Mastering Concepts in Chemistry
Unit 13, Lesson 1
Reaction Rates
Use this lesson to reinforce the concept of what factors affect reaction rates. L1 LEP

GLENCOE TECHNOLOGY

 Videodisc

Science and Technology Videodisc Series
Grain-Dust Explosions
Disc 2: Chemistry
Side 1, Ch. 5

Show this video to illustrate the relationship between reaction rate and surface area. L1 LEP

219

miniLAB

Starch-Iodine Clock Reaction

Purpose: Students will discover that changing concentrations and temperature will affect the rate of a reaction.
Process Skills: Observing, interpreting data, recognizing cause and effect, sequencing
Safety Precautions: Caution students that iodine is toxic.
Disposal: Code G

Teaching Strategies

• Ask students to name other factors that could be varied to alter the rate of a reaction. *pressure, presence of catalysts or inhibitors* L1

Expected Results

The solution in tube 5 turns blue faster than tube 1, which turns faster than tube 4. Tube 3 turns blue faster than tube 1, which turns faster than tube 2.

Analysis

1. Increasing concentration of a reactant speeds up a reaction because the more reactant particles there are, the more likely they are to collide and react. Decreasing concentration slows the rate of a reaction because that makes collisions less likely.
2. Lowering the temperature causes reactant particles to move more slowly, slowing down the reaction rate.
3. Raising the temperature causes reactant particles to move more quickly, increasing the reaction rate.

✔Assessment

Portfolio: Ask students to write thorough lab reports. L1 P

---

miniLAB 3

Starch-Iodine Clock Reaction

In a clock reaction, an observable change indicating that a reaction has taken place occurs at some point in time after two or more different reactants are mixed. How soon this change occurs depends on the speed or rate of the reaction. Altering any factors that change reaction rate will change how quickly the observable change is seen.

Starch is an organic compound that reacts immediately with iodine to form a dark blue compound. The starch-iodine clock reaction involves timing the rate of formation of this blue compound after mixing two solutions. One solution reacts to produce iodine, and the other solution contains starch.

Procedure

1. Use a wax pencil to label five large test tubes with the numbers 1 through 5, and place them in a test-tube rack.
2. Place 10 mL of starch-containing solution in each of the five test tubes.
3. Place tube 4 in an ice bath and tube 5 in a water bath at 35°C. Leave them there for at least ten minutes.
4. Add the following amounts of iodine-producing solution to the indicated test tubes, stir carefully with a clean stirring rod or a wire stirrer, and record the time it takes for the blue color to appear.

   a) 10 mL to tube 1
   b) 5 mL to tube 2
   c) 20 mL to tube 3
   d) 10 mL to tube 4
   e) 10 mL to tube 5

5. Summarize your results in a table.

Analysis

1. What effect does changing the amount, and therefore the concentration, of a reactant have on the rate of a reaction? Why?
2. What effect does lowering the temperature have on the rate of a reaction? Why?
3. What effect does raising the temperature have on the rate of a reaction? Why?

Keep in mind that it doesn't matter how much the concentration of one reactant is increased if another reactant is depleted. Consider the situation that exists in **Figure 6.21.** Sometimes, when two reactants are present in a reaction, more of one than the other is available for reacting. The one that there is not enough of is called the **limiting reactant.** When it is completely used up, the reaction must stop. The reaction is limited because the other reactant alone cannot form any product.

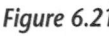

Figure 6.21
Limiting Reactants
How many s'mores can you make if you have six graham cracker halves, three marshmallows, and two pieces of chocolate? Chocolate is the limiting reactant. When it is completely used up, only two s'mores have been made, and the s'more making must stop.

220   **Chapter 6 Chemical Reactions and Equations**

Meeting Individual Needs

Learning Disabled To demonstrate the dynamic nature of the equilibrium state, ask two students to pull on opposite ends of a rope with just sufficient force to balance each other. (You don't want to create a real tug-of-war in the classroom.) Explain that this represents a state of physical equilibrium. Ask students whether it is correct to say that *nothing* is happening. *No. Two opposing forces are being exerted; however, they are in a state of balance.* L1

Everyday Chemistry

Stove in a Sleeve

That Napoleon stated "An army marches on its stomach" may be debated. However, few argue about the meaning of the statement. One of the more mundane necessities of military operations is supplying troops with daily meals, which are called *rations*. In the past, front-line troops often ate cold rations. Heating food by fire took too long, and the resulting smoke might have drawn enemy artillery fire. What was needed was a portable, smokeless, self-contained method of quickly heating rations with readily available materials. The flameless heater was developed to meet these needs.

Flameless ration heaters
The flameless ration heater uses a heat-releasing chemical reaction similar to that of magnesium and water in forming magnesium hydroxide, $Mg(OH)_2$, and hydrogen gas. But pure magnesium can't be used in the heaters. Magnesium reacts with oxygen in the air to form a coating of magnesium oxide, MgO, that prevents the magnesium from reacting with other materials. So other materials must be added to the magnesium to dissolve the MgO and promote the reaction between magnesium and water. A magnesium-iron alloy, composed of 95 percent magnesium and five percent iron, and a small amount of sodium chloride provide these desired effects. These materials are

mixed with a powdered plastic to form a porous pad. Then the heat-producing reaction is started by adding water to the pad.

Hot MREs The heater is used to warm an individual portion of rations called a ready-to-eat meal, or MRE (*M*eal, *R*eady-to-*E*at). Each MRE comes in a pouch inside a cardboard sleeve. The MRE pouch slips into a small, polyethylene bag containing the Mg-Fe pad. After the water is added, the plastic bag is placed inside the sleeve in which the MRE was packed. The reaction transfers heat to the MRE pouch. The cardboard sleeve acts as an insulator to prevent heat loss to the surrounding air. The reaction produces enough heat to raise the temperature of an 8-ounce ration serving by 60°C in about 12 minutes and keep the ration warm for about an hour. The heater and MRE can fit inside the utility pocket of a battle-dress uniform, so it can be carried easily by soldiers. A hot meal-to-go might just be the morale booster a field soldier needs.

Exploring Further

1. **Classifying** Is the chemical reaction in the heater classified as an endothermic or an exothermic reaction?

2. **Applying** What other applications might flameless heating have?

Everyday Chemistry

TECH PREP

Purpose
Students will learn how an exothermic reaction can be used as a heat source.

Teaching Strategies
- Have students write the balanced chemical equation for the following word equation. Magnesium and water form magnesium hydroxide and hydrogen gas. $Mg(s) + 2H_2O(l) \rightarrow Mg(OH)_2(s) + H_2(g)$ **L1**
- Point out that in the above reaction, 24 g of magnesium, when reacted with water, will release 351 kJ of energy. This is enough heat to raise the temperature of a liter (kilogram) of room-temperature water to its boiling point.
- Point out that a temperature increase of 60°C is about the same as a 100°F temperature increase.

Extension
Have students research the exothermic and endothermic processes that are used in instant hot and cold packs. **L2**

Exploring Further

1. The reaction is exothermic.
2. Accept reasonable answers such as flameless heaters might be used to warm an individual's food during camping and backpacking trips and to warm snack foods at outdoor sporting events.

Can a catalyst cause decomposition?

Hold up a small, clean beaker containing about 20 mL of 12% H_2O_2 and tell students that the liquid is made up of hydrogen and oxygen only. Most of them will think it is just water. Setting the beaker on the lab table next to a sink, stir in a small amount of liquid detergent and add 1 to 2 g of potassium iodide. As the liquid foams, producing oxygen-filled bubbles, test the bubbles with burning and glowing splints. The burning splints will flame up, and the glowing splints will often burst into flame, indicating the presence of oxygen. Ask students whether they know what the liquid was. Usually, a few students will guess that the liquid contains hydrogen peroxide, particularly if you tell them it is commonly available in pharmacies. Explain that the iodide ion from the potassium iodide was not consumed, but caused the decomposition reaction of hydrogen peroxide into water and oxygen to speed up. Ask students what the potassium iodide is called in this case. *a catalyst*
Disposal: Code C L1

GLENCOE TECHNOLOGY

 **Videotape**

MindJogger Videoquizzes for Chemistry
Chapter 6: Chemical Reactions and Equations
Have students work in groups as they play the videoquiz game to review key chapter concepts. L1 LEP COOP LEARN

Catalysts

Another way to change the rate of a reaction is to add or take away a catalyst. A **catalyst** is a substance that speeds up the rate of a reaction without being permanently changed or used up itself. Even though they increase the rate of reactions, catalysts do not change the position of equilibrium. Therefore, they do not affect how much product you can get from a reaction—just how fast a given amount of product will form.

How does a catalyst speed up a chemical reaction? You know that in chemical reactions, chemical bonds are broken and new ones are formed. The energy needed to break these bonds is the activation energy of the reaction. A catalyst speeds up the reaction by lowering the activation energy. This process can be compared to kicking a football over a goalpost. A player can kick the football over the goalpost at its regulation height, but if the height of the goalpost is lowered, the ball can be kicked over it using less energy.

Figure 6.22
Enzyme Action
Gelatin made with fresh pineapple will not harden properly, whereas gelatin made with canned pineapple will. Fresh pineapple contains active protease enzymes that break down protein molecules in the gelatin. Canned pineapple has been heated. Enzymes are heat-sensitive, so the proteases in the canned fruit are not active.

Many different compounds are able to act as catalysts. The most powerful catalysts are those found in nature. They are needed to speed up the reactions necessary for a cell to function efficiently. These biological catalysts are called **enzymes.** Enzymes help your body use food for fuel, build up your bones and muscles, and store extra energy as fat. Enzymes are involved in almost every process in a cell. For example, proteases are enzymes that break down proteins, as shown in **Figure 6.22.** These enzymes occur naturally in cells to help with recycling proteins so their parts can be used over and over. Proteases are also used in many common products, including contact lens-cleaning solution and meat tenderizer.

The importance of enzymes to our health can be seen when someone lacks a gene for a particular enzyme. For example, lactose intolerance results when an enzyme that breaks down lactose, the sugar in dairy products, is not produced in a person's digestive system, **Figure 6.23.** When lactose is not broken down, it accumulates in the intestine, causing bloating and diarrhea.

222   **Chapter 6   Chemical Reactions and Equations**

Chemistry Journal

Enzyme Deficiencies
Have students survey the class to find out how many students present are lactose intolerant. In their Journals, have them make a bar graph showing the number of those who are lactose intolerant and those who are not. L1

Slowing Down Reactions

Adding a catalyst speeds up a reaction. Would you ever want to slow down a reaction? Reactions that have undesirable products sometimes have to be slowed down. Food undergoes reactions that cause it to spoil. Medications decompose, destroying or limiting their effectiveness. Can any substances slow down these reactions?

A substance that slows down a reaction is called an **inhibitor.** Just as catalysts don't make reactions occur, inhibitors don't completely stop reactions. An inhibitor is placed in bottles of hydrogen peroxide to prevent it from decomposing too quickly into water and oxygen. If no inhibitor were added, the shelf life of hydrogen peroxide products would be much shorter because the molecules would decompose at a faster rate.

Figure 6.23
Enzyme Deficiency
Lactose tolerance or intolerance to dairy foods such as these is a trait you have inherited from your ancestors.

Connecting Ideas

Now you are familiar with the types of chemical reactions that occur around you. You know the importance of equilibrium and are aware of what factors can affect the rate of reaction. But why do some substances react with each other while others don't? In the next chapter, you will find out how the structure of an atom affects the way it reacts with other atoms.

SECTION REVIEW

Understanding Concepts

1. List and describe four factors that can influence the rate of a reaction.

2. Explain whether an exothermic reaction that is at equilibrium will shift to the left or to the right to readjust after each of the following procedures is followed.

   a) Products are removed.
   b) More reactants are added.
   c) More heat is added.
   d) Heat is removed.

3. Explain the difference between an inhibitor and a catalyst.

Thinking Critically

4. **Analyzing** Hydrogen gas can be produced by the reaction of magnesium and hydrochloric acid, as shown by this equation.

   $$Mg(s) + 2HCl(aq) \rightarrow MgCl_2(aq) + H_2(g)$$

   In a particular reaction, 6 billion molecules of HCl were mixed with 1 billion atoms of Mg.

   a) Which reactant is limiting?
   b) How many molecules of H_2 are formed when the reaction is complete?

Applying Chemistry

5. **Catalytic Converters** Catalytic converters on cars use the metals rhodium and platinum as catalysts to convert potentially dangerous exhaust gases to carbon dioxide, nitrogen, and water. Why don't cars need to have the rhodium and platinum replaced after they are used?

SECTION REVIEW

1. Answers may include temperature—raising the temperature increases reaction rate; concentration—increasing the concentration of one or more reactants increases reaction rate; catalysts—adding a catalyst increases reaction rate; inhibitors—adding an inhibitor decreases reaction rate.
2. **a.** to the right **b.** to the right **c.** to the left **d.** to the right
3. An inhibitor decreases the rate of a reaction, but a catalyst increases the rate of a reaction.
4. **a.** Mg **b.** 1 billion molecules
5. The metals provide a surface on which the reaction can take place, but the metals themselves are not used up in the reaction.

---

3 ASSESS

Check for Understanding

- Check for understanding of the section material by having students answer the Section Review questions.
- Ask students to use models to simulate the reaction $CH_4 + 2O_2 \rightarrow CO_2 + 2H_2O$, starting with $2CH_4$ molecules, and $3O_2$ molecules. Which is the limiting reactant? O_2 **L2**

Reteach

Quick Demo

What starts and what stops a reaction?

Hold an unlighted match and ask students why the match does not burn in a spontaneous, exothermic reaction. *Activation energy is required.* Provide activation energy by lighting the match. Light a wood splint with the match and put it up inside a test tube until the flame goes out. Ask students what the limiting reactant was. *oxygen*

Extension

Draw energy diagrams for uncatalyzed and catalyzed exothermic and endothermic reactions. Guide students to infer from the diagrams that there are lowered activation energies for the catalyzed reactions. **L2**

4 CLOSE

Portfolio

Ask each student to list common reactions that we normally wish to slow down and ones that we normally wish to speed up and the means by which we attempt to accomplish these desires. **L1** **P**

- Review the Reviewing Main Ideas statements and Key Terms with your students.
- Complete solutions to Chapter Review problems can be found in the *Problems and Solutions Manual*.

UNDERSTANDING CONCEPTS

1. **a.** odor and color changes **b.** odor changes, precipitation of a solid **c.** odor and color changes, energy release, gas given off **d.** light release **e.** color and odor changes in hot dog, energy release from burning gas
2. right; left
3. **a.** solid **b.** 2 **c.** 2 **d.** two
4. **a.** $2C_4H_{10}(g) + 13O_2(g) \rightarrow 10H_2O(g) + 8CO_2(g)$ **b.** $N_2O_4(g) \rightarrow 2NO_2(g)$ **c.** $4Li(s) + O_2(g) \rightarrow 2Li_2O(s)$ **d.** $H_2SO_4(aq) + 2KOH(aq) \rightarrow K_2SO_4(aq) + 2H_2O(l)$
5. **a.** silver nitrate + sodium bromide → silver bromide + sodium nitrate **b.** pentane + oxygen → carbon dioxide + water **c.** cobalt(II) carbonate + energy → cobalt(II) oxide + carbon dioxide **d.** barium carbonate + carbon + water → carbon monoxide + barium hydroxide
6. Changing subscripts changes the chemical identities of the substances involved.

REVIEWING MAIN IDEAS

6.1 Chemical Equations

- Changes in color or odor, production or absorption of heat or light, gas release, and formation of a precipitate are all observable macroscopic changes that indicate that a chemical reaction may have occurred.

- Chemical equations are used to represent reactions. They are written using symbols and formulas for elements and compounds. Once the symbols and formulas are written, the equation can be balanced only by changing coefficients.

- Examining a chemical equation can tell you how elements and compounds change during a reaction. It may also tell you whether a reaction is endothermic or exothermic.

- Chemical equations must be balanced according to the law of conservation of mass, which states that matter cannot be created or destroyed in a chemical reaction.

6.2 Types of Reactions

- Although thousands of individual chemical reactions are known, most can be classified into five major classes that are based on patterns of behavior of reactants and products.

- The five general classes of reactions are synthesis, decomposition, single-displacement, double-displacement, and combustion.

6.3 Nature of Reactions

- Reversible reactions are those in which the products can react to form the reactants.

- Equilibrium occurs when no net change is taking place in a reaction.

- The direction in which a reaction shifts can be influenced by disturbing the equilibrium of reactants and products.

- How fast a reaction occurs can be influenced by temperature, concentration of reactants, and the presence of catalysts or inhibitors.

Key Terms

For each of the following terms, write a sentence that shows your understanding of its meaning.

activation energy	equilibrium
catalyst	inhibitor
coefficient	insoluble
combustion	limiting reactant
concentration	product
decomposition	reactant
double displacement	single displacement
dynamic equilibrium	soluble
enzyme	synthesis

● UNDERSTANDING CONCEPTS

1. List one piece of evidence you expect to see that indicates that a chemical reaction is taking place in each of the following situations.

   a) A slice of bread gets stuck in the toaster and burns.
   b) An unused carton of milk is left on a kitchen counter for two weeks.
   c) Wood is burned in a campfire.
   d) A light stick is bent to activate it.
   e) Hot dogs are grilled on a gas stove.

2. On which side of the arrow are the products of a reaction usually found? On which side are the reactants found?

3. Use the equation below to answer the following questions.

$$2Sr(s) + O_2(g) \rightarrow 2SrO(s)$$

   a) What is the physical state of strontium?
   b) What is the coefficient of strontium oxide?
   c) What is the subscript of oxygen?
   d) How many reactants take part in this reaction?

✓ Assessment

Portfolio: Review the portfolio options that are provided throughout the chapter. Encourage students to select one product that demonstrates their best work for the chapter. Have students explain what they have learned and why they chose this example for placement into their portfolios. **P**

Additional portfolio options can be found in the following **Teacher Classroom Resources:**
Tech Prep Applications
Consumer Chemistry
Chemistry and Industry
Applying Scientific Methods in Chemistry

4. Change these descriptions of reactions to balanced chemical equations.

a) Butane vapors in a handheld lighter burn in oxygen to form water vapor and carbon dioxide.

b) Dinitrogen tetroxide gas decomposes into nitrogen dioxide gas.

c) Solid lithium metal reacts with oxygen gas to form solid lithium oxide.

d) Aqueous solutions of sulfuric acid and potassium hydroxide undergo a double-displacement reaction.

5. Use word equations to describe the chemical equations given.

a) $AgNO_3(aq) + NaBr(aq) \rightarrow$
$$AgBr(s) + NaNO_3(aq)$$

b) $C_5H_{12}(l) + 8O_2(g) \rightarrow 5CO_2(g) + 6H_2O(g)$

c) $CoCO_3(s) + energy \rightarrow CoO(s) + CO_2(g)$

d) $BaCO_3(s) + C(s) + H_2O(g) \rightarrow$
$$2CO(g) + Ba(OH)_2(s)$$

6. Explain why subscripts should never be changed when balancing an equation.

7. Balance the following equations.

a) $Fe(s) + O_2(g) \rightarrow Fe_3O_4(s)$

b) $NH_4NO_3(s) \rightarrow N_2O(g) + H_2O(g)$

c) $COCl_2(g) + H_2O(l) \rightarrow HCl(aq) + CO_2(g)$

d) $Sn(s) + NaOH(aq) \rightarrow Na_2SnO_2(aq) + H_2(g)$

8. Which of the five general types of reactions has only one reactant?

9. Classify each of these reactions as one of the five general types.

a) $2C_8H_{18}(l) + 25O_2(g) \rightarrow$
$$16CO_2(g) + 18H_2O(g)$$

b) $2Sb(s) + 3I_2(g) + energy \rightarrow 2SbI_3(s)$

c) $BaCl_2(aq) + 2KIO_3(aq) \rightarrow$
$$Ba(IO_3)_2(s) + 2KCl(aq)$$

d) $Cl_2(aq) + 2KBr(aq) \rightarrow 2KCl(aq) + Br_2(aq)$

e) $2NaCl(l) + energy \rightarrow 2Na(l) + Cl_2(g)$

10. Hydrogen gas and iodine gas combine in a reversible exothermic reaction to form hydrogen iodide gas.

a) Write a word equation for this reaction.

b) Write a balanced chemical equation for this reaction.

c) If more iodine gas is added after the reaction reaches equilibrium, will the reaction be shifted to the left or the right?

d) If heat is added after the reaction reaches equilibrium, will the reaction be shifted to the left or the right?

11. Chlorine dissolved in an aqueous solution of calcium iodide can react to replace the iodine in calcium iodide.

a) Write word and balanced chemical equations for this reaction.

b) Classify this reaction as one of the five major types.

● APPLYING CONCEPTS

12. List all the macroscopic changes that occur while a cake is made that indicate that chemical reactions have occurred.

13. One pollutant produced in automobile engines is nitrogen dioxide, NO_2. It changes to form nitrogen oxide, NO, and oxygen atoms when exposed to sunlight. Of what reaction type is this an example?

14. Few biochemical reactions occurring in our cells reach equilibrium. Most of the time, it is to our advantage to almost completely form products from reactants. Would our cells get more energy from the exothermic combustion of the sugar glucose if the reaction goes to completion or reaches an equilibrium?

Nerve cell

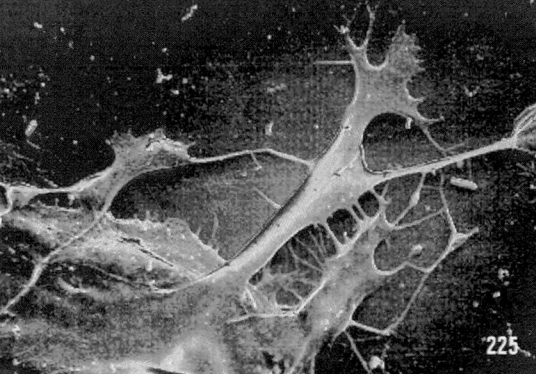

225

15. The acid is a catalyst.

16. The larger the surface area over which a catalyst is spread, the more reactants that can associate with the catalyst and undergo reaction at the faster rate.

17. Substances in the two aqueous solutions react when mixed to form a precipitate in a double-displacement reaction.

18. Taste change would indicate that a substance has changed identity, which indicates a chemical change.

19. Compounds that are safe alone can react together to form harmful compounds. Warning labels on compounds that are likely to react in this way could help prevent consumers from creating dangerous mixtures of chemicals.

20. The N_2 and H_2 molecules must be close enough (increased pressure) for there to be many collisions, and the molecules must be moving fast enough (increased temperature) that the collisions result in a reaction.

21. Student answers might include that spoilage of food and medicines could be prevented by lower temperatures.

22. Student answers might include that different colors could have different meanings. For example, if used as a distress signal, yellow could signify one type of problem, and green could signify another type.

23. Although some variation is tolerated, the human body is adapted to the proportions of gases breathed in air on Earth.

THINKING CRITICALLY

24. A tightly sealed bottle keeps reactive substances in air away from the chemicals so they can't react with them. Storing chemicals at low temperatures slows down the rate at which they react

15. Sucrose, which is table sugar, breaks down slowly in aqueous solutions to form smaller sugar molecules. If a small amount of acid is added to such a solution, the reaction goes much more quickly, even though the acid is not used up in the reaction. What role is the acid playing in this reaction?

16. Catalytic converters help break down pollutants that result from gasoline combustion and are required in automobiles in the United States. Most catalytic converters contain a platinum-rhodium catalyst that coats the extensive surface of a honeycomb structure in the converter. Why would having the catalyst spread out over a large surface area be useful?

17. In the book *The Chemical Catechism, with Notes, Illustrations, and Experiments,* published in 1808, the author (S. Parkes) describes experiments such as mixing clear solutions of magnesium sulfate and potassium hydroxide to obtain a solid as "chemical miracles." How would you describe what happens when the two solutions are mixed?

18. If you place a piece of a saltine cracker on your tongue for a few minutes, it begins to taste sweet. Do you think a chemical or physical change leads to this taste change? Why?

Everyday Chemistry

19. Explain why combining different cleaning products, such as those that contain ammonia and bleach, sometimes has deadly consequences. How could product manufacturers make the average person aware of hazards like this?

Chemistry and Technology

20. Why are high temperature and pressure needed for the Haber process?

Everyday Chemistry

21. What purposes might there be for a unit similar to an MRE that uses an endothermic reaction?

How It Works

22. When do you think it might be important that light sticks come in different colors?

Biology Connection

23. Why is it necessary that the gases breathed on the space shuttle are about the same composition as air on Earth?

● THINKING CRITICALLY

Relating Cause and Effect

24. Explain why it is best to store many chemicals in tightly sealed bottles in dark locations at cool or moderate temperatures.

Applying Concepts

25. **ChemLab** Why can't you find a reaction in the ChemLab that can be classified as two different reaction types?

Forming a Hypothesis

26. **MiniLab 1** Does the reaction taking place in MiniLab 1 go to completion, or does it reach an equilibrium? How do you know?

Making Predictions

27. **MiniLab 3** Predict the effect of each of the following on the speed of the starch-iodine clock reaction.

    a) adding an inhibitor to the reaction
    b) adding a catalyst to the reaction
    c) raising the temperature of the reaction

● CUMULATIVE REVIEW

28. Compare the charges, masses, and atomic locations of electrons, protons, and neutrons. (Chapter 2)

29. Compare the number of electrons in the outer energy levels of metals, nonmetals, and metalloids. (Chapter 3)

30. Explain why KCl is commonly used as a substitute for NaCl, rather than using another chloride such as $CaCl_2$ or CsCl. (Chapter 3)

31. Write a balanced equation for the formation of sodium chloride from sodium metal and chlorine gas. (Chapter 4)

32. Compare properties of ionic and covalent compounds. (Chapter 5)

to form other compounds. The dark bottle reduces light, which is a catalyst for some reactions.

25. Each type of reaction has its own patterns of reactants and products. The only time a reaction could be classified into more than one type is when some metals or carbon undergo combustion, in which case the reaction is also a synthesis.

26. It goes to completion because energy does not continue to be produced.

27. **a.** slows down reaction **b.** speeds up reaction **c.** speeds up reaction

CUMULATIVE REVIEW

28. Electrons: negative charge, light, move around nucleus; Protons: positive charge, heavier, found in nucleus; Neutrons: no charge, same mass as proton, found in nucleus

29. Metals—four or fewer outer-level electrons; Nonmetals—four or more outer-level electrons; Metalloids—three to seven outer-level electrons

Skill Review

1. **Relating Cause and Effect** Use what you have learned about factors that influence rates of chemical reactions to explain each of the following statements.

   a) Colors in the fabric of curtains in windows exposed to direct sunlight often fade.

   b) Meat is preserved longer when stored in a freezer rather than a refrigerator.

   c) Taking one or two aspirin will not harm most people, but taking a whole bottle at once can be fatal.

   d) BHA, or butylated hydroxyanisole, is an antioxidant often added to food, paints, plastics, and other products as a preservative.

2. **Modeling** Use toothpicks and Styrofoam balls of different colors to balance the following equations.

   a) $Cl_2O(g) + H_2O(l) \rightarrow HClO(aq)$

   b) $Fe_2O_3(s) + CO(g) \rightarrow Fe(s) + CO_2(g)$

   c) $H_2(g) + N_2(g) \rightarrow NH_3(g)$

   d) $ZnO(s) + HCl(aq) \rightarrow ZnCl_2(aq) + H_2O(l)$

3. **Interpreting Graphs** The graph shown represents the concentrations of two compounds, A and B, as they take part in a reaction that reaches equilibrium.

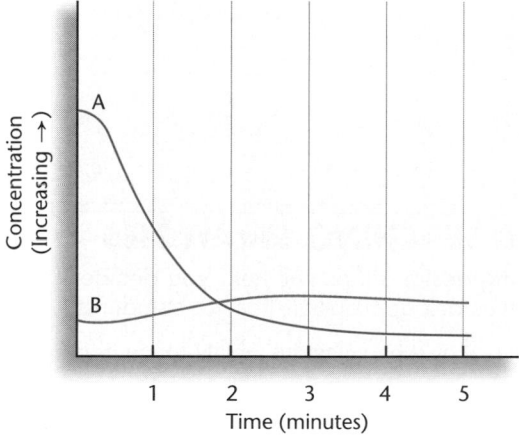

a) Which compound represents the reactant in this reaction? The product?

b) How long did it take for the reaction to reach equilibrium?

c) Explain how the graph will change if more product is added one minute after the reaction reaches equilibrium.

d) Summarize your answers in a report.

Writing in Chemistry

4. Write a short paper describing five chemical reactions that commonly occur in your home, school, or neighborhood every day. Describe what you can see, smell, feel, hear, and taste as a result of these reactions.

5. Investigate the label on a package of cold cuts. Research as many of the chemicals added as possible. Is anything added as an inhibitor? Prepare a written report on your findings.

Problem Solving

6. A chemist who is director of a project to synthesize a new medicine at a pharmaceutical company is trying to increase the yield of the medicine. Imagine that you are that chemist. What steps would you take to increase the yield of the reactions used to make the medicine?

Projects

7. Write the formulas for the active compounds in household bleach and ammonia. Predict what products will form when they react. Research household bleach and ammonia, and locate the equation for the reaction that occurs when they react. Was your prediction close? Also research the number of injuries and fatalities caused by careless mixing of these products each year. Present your results graphically on a poster.

8. Research common chemical reactions that occur safely in your home. Demonstrate one of these reactions. Tell what type of reaction it is and how it is used. Write a balanced equation for the reaction.

7. $NaClO$ and NH_3 react to form chloramines, NH_2Cl and $NHCl_2$, which form pungent fumes and can cause respiratory distress. Statistics will depend on student research.

8. Student answers will vary but may include reactions involved in cooking or using certain cleaners.

Program Resources

Chapter Review and Assessment,
   pp. 35-40 [L1]
Alternate Assessment in the Science Classroom [L1]
Computer Test Bank, Chapter 6 [L1]
Problems and Solutions Manual, Ch. 6 [L1]
Supplemental Practice Problems,
   Chapter 6 [L1]
Performance Assessment in the Science Classroom [L1]

---

30. Because potassium and sodium are in the same column of the periodic table and are just one row apart, the properties of KCl will be more similar to those of NaCl than will those of other chlorides.

31. $2Na(s) + Cl_2(g) \rightarrow 2NaCl(s)$

32. Ionic: hard, brittle, electrolytes, high melting points; Covalent: hard or soft, non-electrolytes, usually lower melting points

AUTHENTIC ASSESSMENT

1. **a.** Heat and light from the sunlight speed up a chemical reaction involving the dye compounds in the curtain fabric. **b.** Lowering the temperature of the meat decreases the rate of chemical reactions that spoil the meat. **c.** Increasing the concentration of the aspirin in someone increases the rate at which it undergoes potentially harmful reactions. **d.** BHA is an inhibitor that decreases the rate of chemical reactions that spoil foods and destroy other products.

2. **a.** $Cl_2O + H_2O \rightarrow 2HClO$ **b.** $Fe_2O_3 + 3CO \rightarrow 2Fe + 3CO_2$ **c.** $3H_2 + N_2 \rightarrow 2NH_3$ **d.** $ZnO + 2HCl \rightarrow ZnCl_2 + H_2O$

3. **a.** A, B **b.** about 4 min. **c.** The graph will at that point show a sharp rise in the B line, which will be followed by a new equilibrium being reached, with higher concentrations of both A and B.

4. Answer will depend on student research but may include reactions involved in cooking or those used in manufacturing of products.

5. Answers will vary but may include chemicals such as $NaNO_3$ as inhibitors.

6. Answer will depend on student research but should include using temperature, concentration, and catalysts to control the rate of reaction.

Chapter 7 Organizer

Section	Objectives	Activities/Features
7.1 **Expanding the Theory of the Atom** (3 days)	**1. Relate** emission spectra to the electron configurations of atoms. **2. Relate** energy sublevels and orbitals within the atom.	**Physics Connection:** Niels Bohr—Atomic Physicist and Humanitarian, p. 232 **MiniLab 1:** Colored Flames—A Window into the Atom, p. 234 **ChemLab:** Metals, Reaction Capacities, and Valence Electrons, pp. 236-237 **Chemistry and Technology:** Hi-Tech Microscopes, p. 240 **Discovery Demo:** 7.1 A Visible Periodic Table, pp. 228-229 **Demonstrations:** 7.2 The Uncertainty Principle, pp. 238-239
7.2 **The Periodic Table and Atomic Structure** (4 days)	**3. Distinguish** the *s, p, d,* and *f* blocks on the periodic table and relate them to an element's electron configuration. **4. Predict** the electron configurations of elements using the periodic table.	**MiniLab 2:** Electrons in Atoms, p. 245 **Everyday Chemistry:** Colors of Gems, p. 248 **Demonstrations:** 7.3 Emission Spectra of Elements, pp. 244-245; 7.4 Electrons and Reactivity, pp. 248-249

Activity Materials

ChemLab (pages 236-237)
thin-stemmed beral pipets, 50-mL graduated cylinder, water trough or plastic basin, hydrochloric acid, magnesium ribbon, aluminum foil, transparent waterproof tape, plastic wrap, forceps

MiniLab 1 (page 234)
wooden splints, Bunsen burner or propane torch, solutions of the chlorides of: lithium, sodium, potassium, calcium, strontium, barium

MiniLab 2 (page 245)
$8\frac{1}{2} \times 11$ inch white paper, pencils or markers, poster board, carbon paper, transparent tape, darts

Demonstrations For a complete list of materials for the demonstrations in this chapter, see pages 228, 238, 244, and 248.

KEY TO TEACHING STRATEGIES

The following designations will help you decide which activities are appropriate for your students.

L1 Level 1 activities should be within the ability range of all students.

L2 Level 2 activities should be within the ability range of the average to above-average student.

L3 Level 3 activities are designed for the ability range of above-average students.

LEP LEP activities should be within the ability range of Limited English Proficiency students.

COOP LEARN Cooperative Learning activities are designed for small group work.

P These strategies represent student products that can be placed into a best-work portfolio.

Completing the Model of the Atom

Teacher Classroom Resources

STUDENT MASTERS

Study Guide: pp. 27-28
Critical Thinking/Problem Solving: p. 8
Laboratory Manual: 7.1 Atomic Spectra, pp. 61-64
ChemLab and MiniLab Worksheets: pp. 47-50

TEACHING AIDS

Section Focus Transparency 14
Section Focus Transparency Master: p. 14
Basic Concepts Transparency 19, 20
Basic Concepts Transparency Masters: pp. 37-40
Problem Solving Transparency 8, 9
Problem Solving Transparency Masters: pp. 15-18
Lesson Plans: p. 14

Study Guide: pp. 29-30
Laboratory Manual: 7.2 Transition Metals, pp. 65-68
ChemLab and MiniLab Worksheets: p. 51

Section Focus Transparency 15
Section Focus Transparency Master: p. 15
Lesson Plans: p. 15

ASSESSMENT RESOURCES

Chapter Review and Assessment: pp. 41-46
Alternate Assessment in the Science Classroom
Computer Test Bank: Chapter 7
Performance Assessment in the Science Classroom
Problems and Solutions Manual: Chapter 7
Supplemental Practice Problems: Chapter 7

CHAPTER RESOURCES

Spanish Resources: Chapter 7
English/Spanish Audiocassettes: Chapter 7
Cooperative Learning in the Science Classroom
Lab and Safety Skills in the Science Classroom

GLENCOE TECHNOLOGY

Software
Mastering Concepts in Chemistry: *Unit 2* Lesson 1 Levels, Sublevels, Orbitals; Lesson 2 Basics of Electron Distribution; Lesson 3 Writing Electron Configurations and Dot Structures

Videotape
MindJogger Videoquizzes, Chapter 7

Videodisc
Chemistry: Concepts and Applications, Demonstrations, Videos, and Animations Videodisc and Bar Code Teacher Guide

CD-ROM Multimedia System
Chemistry: Concepts and Applications Same as for Videodisc, plus Interactive Exploration and Major Simulation, *The Periodic Table*

TECH PREP

The following Glencoe resources provide opportunities for integrating science and technology.

Student Edition: Chemistry and Technology, p. 240; Everyday Chemistry, p. 248

Teacher Wraparound Edition: Enrichment, p. 242

Teacher Classroom Resources: Tech Prep Applications, pp. 11-12; Chemistry and Industry, pp. 15-18; Consumer Chemistry, pp. 13-14

CHAPTER 7
Completing the Model of the Atom

Chapter Overview

Emission spectra are used to introduce concepts of energy sublevels and orbitals, and a new model of the atom emerges. Students learn how to use the periodic table to write electron configurations of the elements.

Theme Connection

Macro to Submicro The macroscopic phenomenon of emission spectra is explained by an expanded model of the atom. The periodic table becomes a macroscopic tool from which the submicroscopic structure of the atom can be predicted.

✓ Assessment Planner

Choose assessment strategies from the following pages to evaluate the progress of your students.
Assess, pp. 242, 244, 250
MiniLabs, pp. 234, 245
ChemLab, p. 236
Portfolio, pp. 240, 245, 252
Chapter Review, p. 252

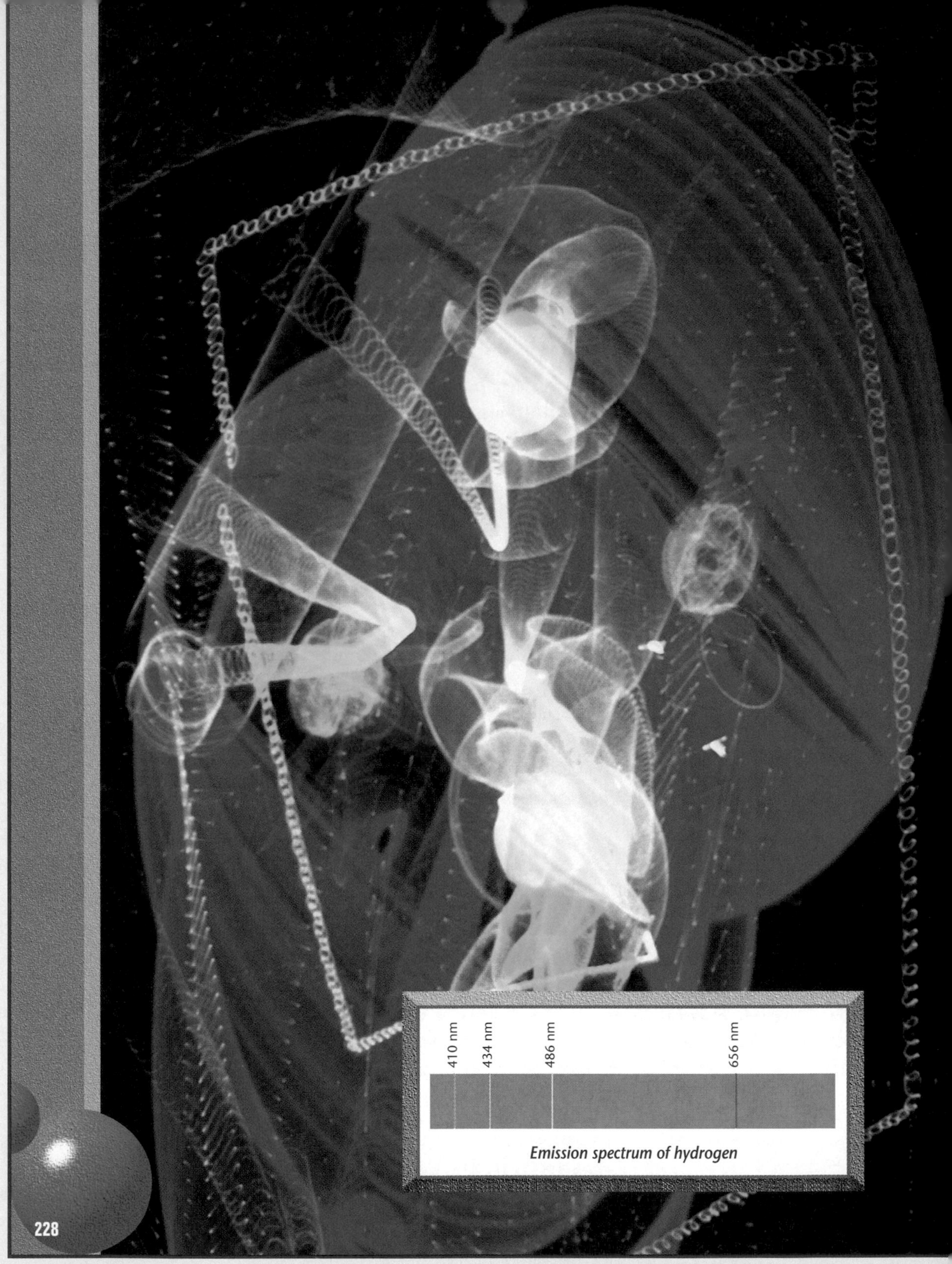

410 nm	434 nm	486 nm	656 nm

Emission spectrum of hydrogen

228

Discovery Demo 7.1

A Visible Periodic Table

Purpose: To observe properties of common elements.
Materials: 24 small, clear containers with plastic screw tops; elements from school's stockroom; kerosene; water; glue; paraffin; beaker; hot plate
Safety Precautions:

Disposal: Code F
Procedure: Place samples of the elements in individual, clear containers. Label each container with the element's symbol and name. Use elements such as aluminum, antimony, bismuth, bromine, cadmium, calcium, carbon, chromium, cobalt, copper, gallium, gold foil, iodine, iron, lead, lithium, magnesium, mercury, nickel, phosphorus (red or white), potassium,

selenium, silicon, silver, sodium, sulfur, tin, titanium, and zinc. **CAUTION:** *Handle these elements with care; some are carcinogenic. The Group 1 elements should be submerged in kerosene, and white phosphorus in water. Care should be exercised in preparing these exhibits. Once prepared and sealed, students can handle them safely if they are careful not to break the containers. Glue the lids onto the bottles, and dip the tops into*

CHAPTER 7

Completing the Model of the Atom

Turn on the lights and start the show! A dazzling display of pure colors dances through space when lasers entertain. Who are the stars of a laser show? None other than tiny electrons. Energize these atomic particles and they respond in predictable ways. They drop from a high energy level to a lower energy level and emit pure colored lights of definite frequencies.

Electrons star again as color blazes up when atoms are heated in a Bunsen flame. The characteristic color produced by an element in a flame test is sometimes used to identify that element. More information can be obtained if the colored light from the flame is put through a spectrophotometer that can separate it into the individual lines of an emission spectrum. You learned about the emission spectra of atoms in Chapter 2, and how spectra provide evidence about electrons in atoms.

Chemistry Around You

Laser light shows are fun to watch, and flame tests are valuable laboratory tests. But every day, you see evidence of electrons in atoms. The colors of gems, the brightness of neon signs and fireworks, and even lightning and the glow of fireflies all have their origins in the movement of electrons. In this chapter, you'll put together a more complete picture of the electron in the atom using information from the emission spectra of elements.

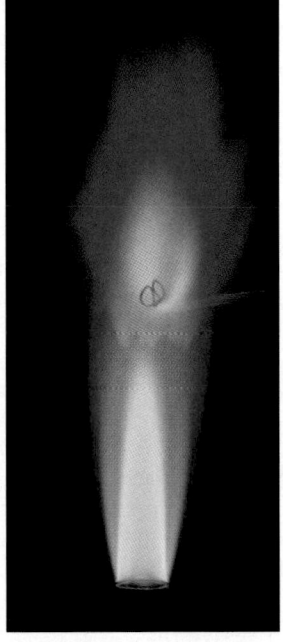

Strontium burns with a bright red flame.

229

Introducing the Chapter

Ask students to draw a diagram of a magnesium atom. Typical drawings will place 12 electrons in three orbits around the nucleus. Point out that the model of planetary electrons cannot explain the observations of emission spectra. But the idea of circular positions around the nucleus is a good model for energy levels. L1 LEP

Using the Illustrations The vivid colors of the laser show are light of definite frequencies and energies. When an element is energized—like the one pictured in the flame test—its electrons absorb energy and jump to a higher energy level. In a few seconds, the energy is emitted as the electrons drop back to a lower energy level.

Revealing Misconceptions

Students may think that energy levels are equally spaced. Data from emission spectra show that as the number of the energy level increases, the distance between energy levels decreases.

GLENCOE TECHNOLOGY

 Videodisc

Chemistry: Concepts and Applications
See the Videodisc Program Teacher Guide to access illustrations and other material available for this chapter.

---

melted paraffin to seal. Distribute the bottles around several lab stations, and have the students rotate among them recording their observations of each element. **CAUTION:** *Lithium, sodium, phosphorus, potassium, and bromine should be observed only with the teacher present.*
Results: Have students compare and contrast their observations of metals and nonmetals. Have them compare the properties of metals in a family. L1 COOP LEARN

Analysis: Ask these questions.
1. Describe the color and luster of most metals. *Most metals are gray and shiny.*
2. What color is the nonmetal sulfur? *yellow*
3. Are there more metals or nonmetals on the periodic table? *more metals*
4. The properties of elements reflect their electron arrangement. Do metals have similar properties because they have a similar number of outer electrons? *yes*

PREPARE

Key Concepts

The evidence that led to current theories about atomic structure is reviewed, then related to the distribution of electrons in energy levels, sublevels, and orbitals.

1 FOCUS

Focus Transparency

Display the **Section Focus Transparency** for Section 7.1 to build upon students' understanding of the material in the chapter opener. L1 LEP

2 TEACH

Visual Learning

The Gaming Wheel: Just as the outcome of a spin of the roulette wheel cannot be known in advance, so it is impossible to know the exact position of an electron in an atom. However, in both situations, probabilities can be calculated. The modern theory of the atom is based upon the probability of finding electrons in regions of space called orbitals. L1

Correcting Misconceptions

When students read about the Bohr theory or see diagrams in other books, they may think that electrons circle the atomic nucleus just as planets orbit the sun. Explain that much evidence indicates that this idea is incorrect.

Expanding the Theory of the Atom

SECTION PREVIEW

Objectives
Relate emission spectra to the electron configurations of atoms.
Relate energy sublevels and orbitals within the atom.

Key Terms
sublevel
Heisenberg uncertainty principle
orbital
electron configuration

Electrons are strange. They don't behave like everyday objects. They're so small, they move so fast, and they seem to be in perpetual motion. They form clouds around the nucleus of the atom, but you can never be sure exactly where they are. Truly, electrons are strange. Their world is described in terms of probability, and that is different from the macroscopic world.

Scientists have discovered that electrons occupy a complex world of energy levels. They've described this world in terms of uncertainty, probability, and orbitals. In this section, you'll learn how to describe the energy levels in an atom and you'll see how electrons are arranged in these levels. The ways in which electrons are distributed in the energy levels of an atom account for many of the physical and chemical properties of the element.

Developing a Model of Atomic Structure

You learned in Chapter 2 that in 1803, John Dalton proposed an atomic theory based on the law of conservation of matter. In his view, atoms were the smallest particles of matter. Then, in 1897, J.J. Thomson discovered electrons. The existence of electrons meant that the atom was made up of smaller particles. These particles include protons and neutrons, as well as electrons. At first, it wasn't clear how these subatomic particles were arranged. Scientists thought they were just mixed together like the ingredients in cookie dough. But in 1909, Ernest Rutherford performed an experiment in which he aimed atomic particles at a thin sheet of gold foil. He found that most particles went right through the foil, but some were deflected. These results suggested that most of the atom is empty space and that almost all of the mass of the atom is contained in a tiny nucleus. Rutherford proposed a nuclear model of the atom in which protons and neutrons make up the nucleus, and electrons move around in the space outside the nucleus.

Program Resources

Study Guide, pp. 27-28 L1
Critical Thinking/Problem Solving, p. 8 L2
English/Spanish Audiocassettes, Chapter 7 L1 LEP
Laboratory Manual, 7.1, pp. 61-64 L1
Section Focus Transparency 14 and **Master,** p. 14 L1 LEP
Basic Concepts Transparencies 19, 20 and **Masters,** pp. 37-40 L1 LEP

Problem Solving Transparencies 8, 9 and **Masters,** pp. 15-18 L1 LEP
Spanish Resources L1 LEP
ChemLab and MiniLab Worksheets, pp. 47-50 L1
Consumer Chemistry, pp. 13-14 L1
Tech Prep Applications, pp. 11-12 L1

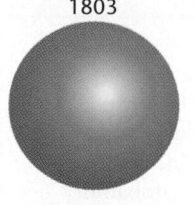

1803

Dalton's model

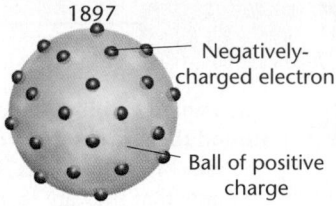

1897

Negatively-charged electron

Ball of positive charge

Thomson's model

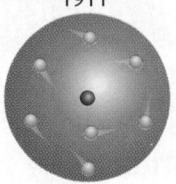

1911

Rutherford's model

1913

Bohr's atomic model

Today

Modern model

Figure 7.1
Building an Atomic Model
The atomic theory evolved over a period of 2000 years. But it's the experimental evidence of the last 200 years that reveals the complex nature of the submicroscopic world. Because electrons are responsible for an element's chemical properties, chemists need an atomic model that describes the arrangement of electrons.

⚡ **Quick Demo**

Where's the electron?
Have a fan rotating at high speed when students enter the classroom so they will not have seen the blade in a stopped position. Ask them to describe the blade. They will be able to tell the approximate length of the blade and little else. Explain that scientists experience the same situation in trying to describe electrons in atoms. The electrons move about the nucleus and appear to fill the entire volume, yet occupy little volume themselves. Due to the motion of the electrons and certain limitations in our ability to view them, we are unable to simultaneously describe exactly where the electrons are and where they are going.

In 1913, the Danish physicist Niels Bohr suggested that electrons revolve around the nucleus just as planets revolve around the sun. Bohr's model was consistent with the emission spectrum produced by the hydrogen atom, but the model couldn't be extended to more complicated atoms. **Figure 7.1** illustrates the evolution of the atomic theory.

By 1935, the current model of the atom had evolved. This model explains electron behavior by interpreting the emission spectra of all the elements. It pictures energy levels as regions of space where there is a high probability of finding electrons. Before going on to the modern atomic theory, take another look at what you already know about atoms and electrons.

Building on What You Know

In the present-day model of the atom, neutrons and protons form a nucleus at the center of the atom. Negatively-charged electrons are distributed in the space around the nucleus. The electrons with the most energy are farthest from the nucleus and occupy the outermost energy level. Recall from Chapter 2 that evidence for the existence of energy levels came from the interpretation of the emission spectra of atoms. It's important to know about energy levels in atoms because it helps explain how atoms form chemical bonds and why they form particular kinds of compounds, for example ionic or covalent compounds.

Valence Electrons and the Periodic Table

The periodic table reflects each element's electron arrangement. In Chapter 3, you learned that the number of valence electrons is equal to the group number for elements in Groups 1 and 2 and is equal to the second digit of the group number for Groups 13 through 18. The period number represents the outermost energy level in which valence electrons are found. For example, lithium—in Group 1, Period 2—has one valence electron in the second energy level; sulfur—in Group 16, Period 3—has six valence electrons in the third level.

WORD ORIGIN

spectrum:
spectrum (L) appearance, specter

An emission spectrum reveals what cannot be seen directly.

Meeting Individual Needs

Visually Impaired This activity will work well for all students. Give each pair of students a large glob of modeling clay with a marble embedded in it. The marbles you use should vary in size. Provide a plastic toothpick (or a long pin if the clay is too hard for the toothpick to penetrate easily) for students to poke into the clay to find out about the marble. Allow each pair of students 50 pokes, but caution them that they are not permitted to squeeze or deform the clay. As one student pokes the toothpick into the clay, the other student (not visually impaired) maps the pokes and records whether the marble was hit. After 50 pokes, each pair of students estimates the location and size of the marble. Explain that this process is similar to Rutherford's gold foil experiment. After students have turned in their poke map and estimations, you could allow them to remove and measure the marble to determine the accuracy of their estimate. L1 LEP COOP LEARN

Purpose

To show that Niels Bohr's work was important to both chemistry and physics, and that scientists are real people.

Teaching Strategies

- Show students a picture of Bohr's model and ask them to tell you what is wrong with it.
- Have students look up more information on Bohr. There are other interesting events in his life. For example, he dissolved Nobel medals in acid to keep the Nazis from getting them, and he fought the use of the atomic bomb he had helped to develop. L2

Extension

Have students research the life and work of Lise Meitner, a renowned physicist whose work in the field of nuclear fission helped make the atomic bomb possible. L2

Connecting to Chemistry

1. He was talking about valence electrons, which are involved in chemical bonding. The number of valence electrons an element has and how they combine with the valence electrons of other elements to form compounds is the subject matter of chemistry.
2. Bohr's work is important because the properties and behavior of each element are functions of its structure.
3. They were of many nationalities and had a wide variety of backgrounds. They were male and female, young and old. Some felt remorse because of their involvement. Neils Bohr and Richard Feynmann are two examples.

PHYSICS CONNECTION

Niels Bohr–Atomic Physicist and Humanitarian

Scientists are sometimes portrayed as strange and distant people. In fact, just like other people, they work, they interact with others, and they deal with everyday problems. They may differ from other people only in the way they use their intellects to be creative in areas of science. But they are like other people in that they may have strong ethical beliefs and the courage to fight for those beliefs. Niels Bohr's life exemplified these characteristics.

Bohr's atomic theory By the age of 28, Bohr had developed his atomic theory. Nine years later, in 1922, he received the Nobel Prize in Physics for this work. The basic ideas of Bohr's theory were that electrons move around the atom's nucleus in circular paths called orbits. These orbits are definite distances from the nucleus and represent energy levels that determine the energies of the electrons. Those electrons orbiting closest to the nucleus have the lowest energy; those farthest from the nucleus have the highest energy. If electrons absorb energy, they move to higher energy levels. When they drop down to lower energy levels, they release energy. Energy is absorbed and given off in definite amounts called quanta.

Theories change Parts of Bohr's theory are still accepted today, and parts are outdated. Electrons are thought to move around the nucleus, but not in definite paths. The idea of energy levels is correct, but now we know that they are regions of probability of finding electrons. An electron cannot be expected to be in any exact place. Electrons do jump to higher energy spaces as they gain energy and drop to lower ones when they lose energy.

Bohr helped develop the atomic bomb In 1939, Bohr attended a scientific conference in the United States, where he reported that Lise Meitner and Otto Hahn had discovered how to split uranium atoms—a process called fission. This dramatic announcement laid the groundwork for the development of the atomic bomb.

Bohr escaped from the Nazis In 1940, the Nazis invaded and occupied Niels Bohr's country, Denmark. Bohr was opposed to the Nazis, but he continued his position as director of the Copenhagen Institute for Theoretical Physics until 1943. When he learned that the Nazis planned to arrest him and force him to work in Germany on an atomic project, he and his family fled to Sweden under frightening conditions. In 1943, Bohr came to the United States and worked with scientists from all over the world on the Manhattan Project.

Connecting to Chemistry

1. **Interpreting** Bohr explained that the electrons in the outermost shell determine the chemical properties of an element. What did he mean?

2. **Thinking Critically** How important to chemistry is the physicist's work on atomic structure?

3. **Acquiring Information** Look up information on the Manhattan Project. Find out about the scientists involved, their nationalities, ages, genders, and their fields of expertise.

Meeting Individual Needs

Learning Disabled Demonstrate that energy changes accompany transitions of electrons from one energy level to another. Have students place a book on the floor. Tell them that the book represents an electron in the lowest energy level of an atom. Ask them to raise the book to a higher energy level (their chair). Ask if energy is required. *yes* Ask what happens when the book returns to the floor. *Energy is released.* Ask how the energy required to raise the book to the chair compares with the energy required to raise it to the desktop. *The energy required to raise the electron to the chair is less than that required to raise it to the desktop.* Explain that the energy needed to raise the electron is exactly the same as the energy released when the electron falls back down. L1

LEP

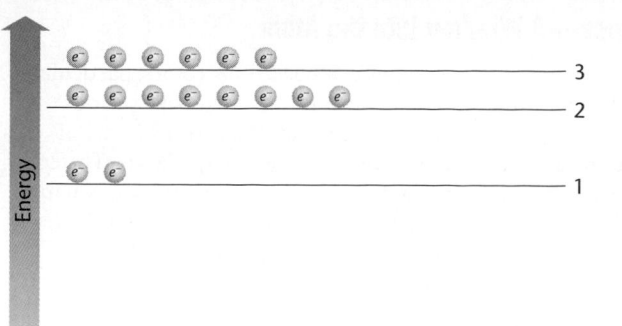

Energy

Energy Levels in Sulfur

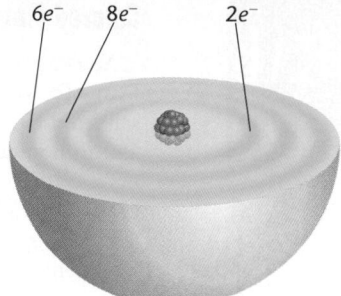

6e⁻   8e⁻   2e⁻

3

2

1

You can use the periodic table to determine a complete energy-level diagram for any element. The atomic number of sulfur is 16. This means that sulfur has a total of 16 electrons. Sulfur is in Group 16 and Period 3, so it has six valence electrons in the third energy level. The remaining ten electrons must be distributed over the first two energy levels. **Figure 7.2** shows energy levels in sulfur.

Electromagnetic Radiation and Energy

Light, or electromagnetic radiation, can be described as waves having a range of frequencies and wavelengths. The higher the frequency of a wave and the shorter the wavelength, the greater the energy of the radiation. The lower the frequency and the longer the wavelength, the lower the energy. These relationships are used to calculate the exact amount of energy released by the electrons in atoms.

In the emission spectrum of hydrogen, shown on page 228, you can see the four different colors of visible light released by hydrogen as its electron moves from higher energy levels to a lower energy level. In Chapter 2, the analogy of energy levels as rungs on a ladder showed that electrons can move from one energy level to another, but they can't land between energy levels. By absorbing a specific amount of energy, an electron can jump to a higher energy level. Then, when it falls back to the lower energy level, the electron releases the same amount of energy in the form of radiation with a definite frequency. **Figure 7.3** shows how electron transitions between energy levels are related to amounts of energy.

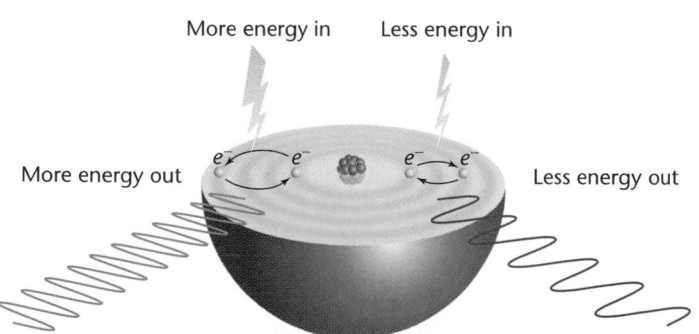

More energy in   Less energy in

More energy out   Less energy out

Figure 7.2

The Electron Distribution in Sulfur

The electrons in a sulfur atom, and in the atoms of all the other elements, are distributed over a range of energy levels shown in the diagram on the left. Relate the energies of the electrons in levels 1, 2, and 3 to the placement of the electrons in the model of the sulfur atom on the right.

Figure 7.3

Electron Transitions

The number of energy levels that an electron jumps depends on the amount of energy it absorbs. When an electron falls back to its original level, it emits energy in the form of light. The energy (color) of the light depends on how far the electron falls. The greater the energy given off, the more toward the violet end of the spectrum the color will be.

Chemistry Journal

Quantum Analogies

Have students find and write in their journals descriptions of everyday quantized events. Examples might be steps, elevator stops, digital clocks, and push-button radios. Ask students to explain how each example is an analogy for the energy levels in the atom. ☐L1

Tying to Previous Knowledge

Ask students what the atomic number 7 tells them about the nitrogen atom. *The atomic nucleus of nitrogen contains seven protons. Because atoms are neutral, the nitrogen atom must also contain seven electrons.* Remind students that the number of neutrons in the nucleus cannot be determined from the atomic number or from the atomic mass. ☐L1

Using an Analogy

When the people in a stadium transmit energy by making a stadium wave, the wave travels around the stadium as individual persons move up and down; however, the persons transmitting the wave remain in the same place.

Concept Development

Borrow a slinky from the physics department, and attach it securely to an object on one side of the room. Demonstrate the characteristics of waves—wavelength, frequency, and energy—by generating standing waves. Start with a half wave, showing the longest wavelength, and work up to two or two and a half standing waves. It will be obvious that you must use more energy as the number of standing waves increases. With each increase in the number of waves, ask students what is happening to frequency and wavelength, and how energy is changing. *Frequency is increasing, wavelength is decreasing, and energy is increasing.* ☐L1 ☐LEP

miniLAB

Colored Flames—A Window into the Atom

Purpose: To enable students to see and use the emission spectra of several metallic elements as flame tests.

Process Skills: Observing, comparing and contrasting, communicating

Safety Precautions: Remind students to be careful around open flames, to wear safety goggles, and to hold the soaked wood splints with forceps. Caution students to wash their hands thoroughly after the lab, as some of the heavy metal ions are toxic.

Preparation: Soak the wood splints overnight in saturated solutions of the chlorides of the metallic elements. If you do not have the chlorides, dissolve other compounds in 6*M* HCl to prepare the saturated chloride solutions. [For 6*M* HCl, add 1 part concentrated (12*M*) HCl to 1 part water.] **CAUTION:** *Add concentrated acid to water.*

Teaching Strategies

- Review with students how to adjust the flow of gas and the air supply on the Bunsen burner in order to obtain a blue flame.
- **Disposal:** Code D; neutralize solutions before washing them down the drain with copious amounts of water. All insoluble debris, including the extinguished stubs of the wood splints, should be disposed of in approved waste containers—not in sinks.

Expected Results

The flame colors are listed under Analysis question 1.

miniLAB 1

Colored Flames—A Window into the Atom

You've probably noticed that flames can be various colors, particularly in a fireworks display or when you burn logs treated with various salts in the fireplace. These colors are the result of electrons in metal atoms moving from higher energy levels to lower energy levels. The colors produced when compounds containing metals are heated in a flame can be used to identify the metals. The procedure is known as a flame test. In this MiniLab, you will study the flame tests of a few elements and identify an unknown element.

Procedure

1. Obtain from your teacher six wooden splints that are labeled and have been soaked in saturated solutions of the chlorides of lithium, sodium, potassium, calcium, strontium, and barium.

2. Light a laboratory burner or a propane torch held in a metal holder. Adjust the burner to give the hottest blue flame possible.

3. In turn, hold the soaked end of each of the splints in the flame for a short time. Observe and record the color of the flames. Extinguish the splint as soon as the flame is no longer colored.

4. Obtain a splint that has been soaked in a solution unknown to you. Perform the flame test and identify the unknown metallic element.

Analysis

1. What characteristic colors identify each of the metallic elements?

2. What is the identity of your unknown element?

3. Explain how you might test an unknown crystalline substance to determine whether it is table salt. A taste test is never recommended for an unknown compound.

Energy Levels and Sublevels

Just as the emission spectrum of hydrogen has four characteristic lines that identify it, so the emission spectrum for each element has a characteristic set of spectral lines. This means that the energy levels within the atom must also be characteristic of each element. But when scientists investigated multi-electron atoms, they found that their spectra were far more complex than would be anticipated by the simple set of energy levels predicted for hydrogen. **Figure 7.4** shows spectra for three elements.

Notice that these spectra have many more lines than the spectrum of hydrogen. Some lines are grouped close together, and there are big gaps between these groups of lines. The big gaps correspond to the energy released when an electron jumps from one energy level to another. The interpretation of the closely spaced lines is that they represent the movement of electrons from levels that are not very different in energy. This suggests that **sublevels**—divisions within a level—exist within a given energy level. If electrons are distributed over one or more sublevels within an energy level, then these electrons would have only slightly different energies. The energy sublevels are designated as *s, p, d,* or *f.*

Analysis

1. lithium, red; sodium, yellow; potassium, violet; calcium, orange (some red, some yellow); strontium, scarlet or red-orange; barium, green or lime green.

2. Student answers depend upon unknowns.

3. Test the substance for solubility in water. All sodium compounds are water soluble. If it is soluble, soak a splint in the solution and perform the flame test. A yellow flame indicates sodium.

✓ Assessment

Knowledge: Ask students which of the flame tests represents the electron energy level change of greatest energy. *Potassium's because violet light has the shortest wavelength, highest frequency, and highest energy of the colors observed in this MiniLab.* L1

Each energy level consists of sublevels that are close in energy. Each energy level has a specific number of sublevels, which is the same as the number of the energy level. For example, the first energy level has one sublevel. It's called the 1s sublevel. The second energy level has two sublevels, the 2s and 2p sublevels. The third energy level has three sublevels: the 3s, 3p, and 3d sublevels; and the fourth energy level has four sublevels: the 4s, 4p, 4d, and 4f sublevels. Within a given energy level, the energies of the sublevels, from lowest to highest, are s, p, d, and f. **Figure 7.5** shows a diagram of the first three energy levels and an inside view of the sublevels within them. Notice how the sublevels within an energy level are close together. This explains the groups of fine lines in an element's emission spectrum. For example, you might expect three spectral lines with slightly different frequencies because electrons fall from the 3s, 3p, and 3d sublevels to the 2s sublevel. Because each of these electrons initially has slightly different energy within the third energy level, each emits slightly different radiation.

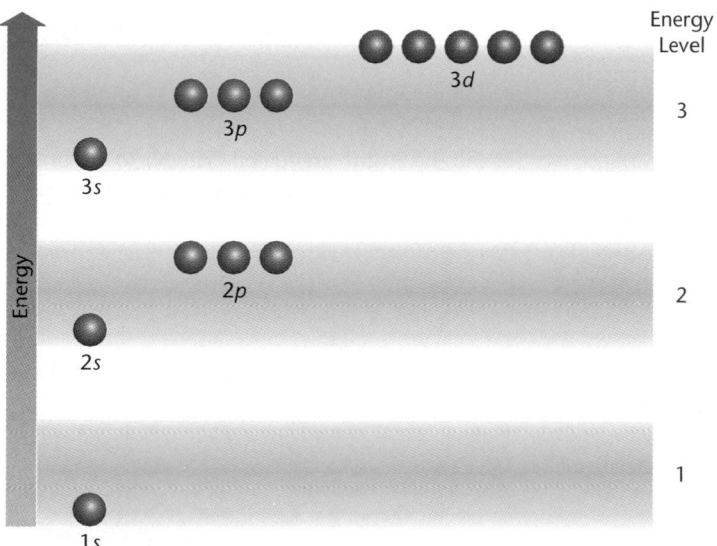

Figure 7.4

Comparison of Emission Spectra

The big gaps between spectral lines indicate that electrons are moving between energy levels that have a large difference in energy. The groups of fine lines indicate that electrons are moving between energy levels that are close in energy. The existence of sublevels within an energy level can explain the fine lines in the spectra of these elements.

Figure 7.5

Electron Distribution in an Atom

The diagram above shows the relative energies of the 1s, 2s, 2p, 3s, 3p, and 3d sublevels. Electrons in the 1s sublevel are closest to the nucleus. Electrons in the 3s, 3p, and 3d sublevels are farthest from the nucleus.

⚡ Quick Demo

Which is more energetic, red or blue?

You can easily demonstrate that red light has lower energy than blue light. Explain that you are making a solution of a fluorescent substance as you dissolve about 10 g of fluorescein in 100 mL of water. Put the solution into a 150-mL beaker and turn out the room lights. Shine the beam of a flashlight through a transparent red cellophane sheet into the fluorescein solution. When you turn out the flashlight, the solution will not fluoresce. Now, shine the flashlight through a blue cellophane sheet. The solution will fluoresce when you turn out the light. Ask students to explain the results. *The blue light waves have a higher frequency, shorter wavelength, and greater energy than the red light waves.* Explain that a certain amount of energy is needed to bring about the fluorescence, and red light does not provide enough. [L1]

Transparency

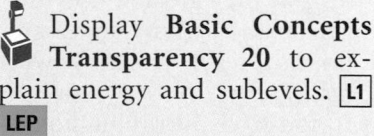 Display **Basic Concepts Transparency 20** to explain energy and sublevels. [L1] [LEP]

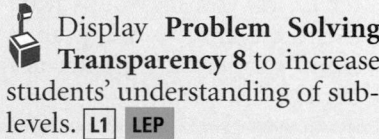 Display **Problem Solving Transparency 8** to increase students' understanding of sublevels. [L1] [LEP]

Across the Curriculum

Language

The names of famous scientists continue in the language of science long after the individual has died. Marie Curie, Albert Einstein, and Dimitri Mendeleev have chemical elements that bear their names. The Heisenberg uncertainty principle is so named to acknowledge its discoverer, Werner Heisenberg. Max Planck hypothesized that energy is emitted in discrete packets called photons. The energy of a photon is described by the equation $E = h\nu$, where h = Planck's constant. The Pauli exclusion principle is named after Wolfgang Pauli, who discovered that no two electrons in an atom can have the same four quantum numbers.

ChemLab

Metals, Reaction Capacities, and Valence Electrons

Time Allotment: One 45- to 60-min. period.

Objectives

Review objectives with students before they begin the ChemLab.

Process Skills

Observing, recognizing cause and effect, using numbers, measuring, interpreting data, communicating

Safety Precautions and Disposal

Students must wear safety goggles and aprons. Advise students to avoid contact with hydrochloric acid solutions because they could damage eyes, skin, or clothing. They should wash their hands thoroughly after the lab.
Disposal: Code D

PREPARATION

Prepare 3*M* hydrochloric acid by adding 10 mL of concentrated (12*M*) acid to 30 mL of water. Prepare 1*M* hydrochloric acid by adding 10 mL of 3*M* acid to 20 mL of water. **CAUTION:** *Add acid to water.*

Alternative Materials

• Other metals may be used as long as they are above hydrogen in the activity series and do not react violently with acid.
• If you have 50-mL eudiometer tubes, you may substitute them for the 50-mL graduated cylinders for greater accuracy.

PROCEDURE

• Magnesium ribbon and aluminum foil are usually uniform enough that you may determine the mass per unit length or mass per unit area and have students carefully cut the desired amounts.

Metals, Reaction Capacities, and Valence Electrons

Many metals react with acids, producing hydrogen gas. If a metal reacts with acids in this manner, the amount of hydrogen produced is related to the number of valence electrons in the atoms of the metal. In this ChemLab, you will react equal numbers of atoms of magnesium and aluminum with hydrochloric acid and compare the reaction capacities of the two metals. Each reaction proceeds only as long as there are metal atoms to react.

Problem

How do the reaction capacities of magnesium and aluminum compare, and how are these capacities related to the valence electrons in the atoms of the two elements?

Objectives

• **Compare** the reaction capacities of magnesium and aluminum.
• **Interpret** the results of the experiment in terms of the numbers of valence electrons in the atoms of the two elements.

 **PREPARATION**

Materials

thin-stemmed micropipets (2)	3*M* HCl
50-mL graduated cylinder	magnesium ribbon
	aluminum foil
water trough or plastic basin	transparent, waterproof tape
1*M* HCl	plastic wrap
	forceps (2)

Safety Precautions

Wear goggles and an apron. The bulb of the micropipet may become hot during the reaction. Hold the stem portion of the pipet with forceps.

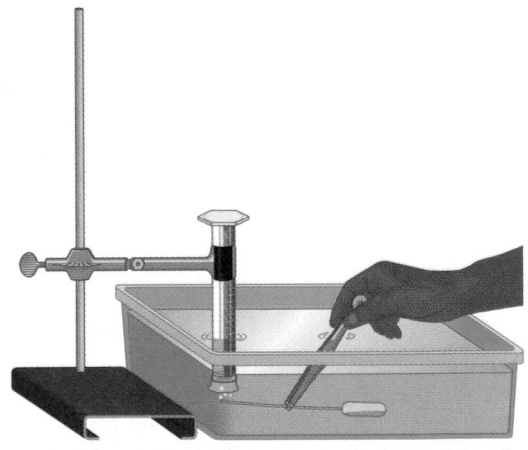 **PROCEDURE**

1. Cut small slits in two micropipets as shown. Obtain a 0.020-g sample of magnesium ribbon from your teacher and insert the magnesium into the bulb of one pipet. Obtain a 0.022-g sample of aluminum foil and insert it into the bulb of the other pipet. Seal the slits with transparent, waterproof tape and label the two pipets *Mg* and *Al*.

2. Fill the water trough or plastic basin nearly full of water.

3. Fill the 50-mL graduated cylinder to the brim with water and cover the top with plastic wrap.

Hold the wrap tightly so that no water can escape and invert the cylinder, placing the top beneath the surface of the water in the trough. Remove the plastic wrap. No appreciable amount of air should be in the cylinder. If this is not the case, repeat the procedure. The inverted cylinder may be clamped in place or held by hand.

4. Using the pipet containing the magnesium, squeeze most of the air from the pipet bulb and draw 3 mL of $1M$ HCl into the pipet. (**CAUTION:** *Handle the hydrochloric acid solution with care. It is harmful to eyes, skin, and clothing. If any acid contacts your skin or eyes or any spillage occurs, rinse immediately with water and notify your teacher.*)

5. Hold the pipet with forceps as shown, and quickly immerse it in the water in the trough so that the pipet tip is inside the open end of the graduated cylinder.

6. Collect the hydrogen gas in the graduated cylinder. Allow the reaction to proceed until the magnesium ribbon is completely gone.

7. While the graduated cylinder is still in place, read the volume of hydrogen gas produced and record it in a table like the one shown.

8. Remove the tape from the pipet and rinse the pipet with water.

9. Repeat steps 4 through 8 using the micropipet containing aluminum and $3M$ HCl. Read and record the volume of hydrogen gas produced.

DATA AND OBSERVATIONS

Hydrogen from magnesium (mL)	
Hydrogen from aluminum (mL)	

ANALYZE AND CONCLUDE

1. **Interpreting Data** The volume of the hydrogen gas you collected is proportional to the number of molecules of hydrogen produced in the two reactions. Which element, aluminum or magnesium, produced more hydrogen molecules?

2. **Comparing and Contrasting** You used approximately equal numbers of atoms of magnesium and aluminum and enough HCl to react with all the atoms. Which of the two elements produced more hydrogen molecules per atom?

3. **Drawing Conclusions** Which element has the greater reaction capacity per atom? Use your volume data to express the relative reaction capacities of the two elements as a ratio of small, whole numbers.

4. **Relating Concepts** In this experiment, the atoms of both metals react by losing electrons to form positive ions. Relate the ratio of reaction capacities to the number of valence electrons each element has.

APPLY AND ASSESS

1. Write the balanced equations for the two chemical reactions you performed.

2. Had you performed this experiment with sodium, using approximately the same number of atoms as you used of magnesium and aluminum, predict the volume of hydrogen gas that would have been produced. (**CAUTION:** *Because sodium reacts explosively with water, it cannot be used safely in this experiment.*)

DATA AND OBSERVATIONS

Volume of hydrogen from magnesium *19.8 mL*
Volume of hydrogen from aluminum *29.5 mL*

ANALYZE AND CONCLUDE

1. aluminum
2. aluminum
3. aluminum;
   Al/Mg = 29.5/19.8
   Al/Mg = 1.5/1 or 3/2
4. Aluminum has three valence electrons, whereas magnesium atoms have two valence electrons.

APPLY AND ASSESS

1. $Mg + 2HCl \rightarrow H_2 + MgCl_2$
   $2Al + 6HCl \rightarrow 3H_2 + 2AlCl_3$
2. 9-10 mL of hydrogen

 **Assessment**

Performance: Have students repeat the experiment with some transition metals that are above hydrogen in the activity series, and relate the results to the number of electrons the elements have in their *s* and *d* sublevels. L1

GLENCOE TECHNOLOGY

 Software

Mastering Concepts in Chemistry
Unit 2, Lesson 2
Basics of Electron Distribution
Use this lesson to develop students' understanding of how electrons are distributed in atoms. L1 LEP

Transparency

Display **Problem Solving Transparency 9** to show energy level diagrams. L1 LEP

Using an Analogy

You might describe the energy levels and sublevels of electrons in atoms as analogous to the floors (energy levels) and rooms on each floor (sublevels) of a multistory house. An *s* sublevel could be thought of as a small room, a *p* sublevel as a larger room, a *d* sublevel an even larger room, etc.

Content Background

Quantum Mechanics The mathematical description of energy levels in the atom is known as quantum mechanics. A central postulate of quantum mechanics is that matter has wave properties that are imperceptible at the macroscopic level but of considerable significance at the submicroscopic level. According to quantum mechanics, each electron in an atom can be described by four quantum numbers. Three of these (n, l, and m_l) are related to the probability of finding the electron at various points in space. The fourth (m_s) is related to the direction of electron spin—clockwise or counterclockwise. The principal quantum number, n, determines the energy level of an electron; l determines the energy sublevel and the shape of the orbital; m_l determines the orientation in space of an electron cloud; m_s determines the orientation of the electron's spin axis.

Transparency

Display **Basic Concepts Transparency 19** showing the electron cloud model. [L1]
[LEP]

The Distribution of Electrons in Energy Levels

A specific number of electrons can go into each sublevel. An *s* sublevel can have a maximum of two electrons, a *p* sublevel can have six electrons, a *d* sublevel can have ten electrons, and an *f* sublevel can have 14 electrons. **Table 7.1** shows how electrons are distributed in the sublevels of the first four energy levels. Notice that the first energy level has one sublevel, the 1*s*. The maximum number of electrons in an *s* sublevel is two, so the first energy level is filled when it reaches two electrons. The second energy level has two sublevels, the 2*s* and 2*p*. The maximum number of electrons in a *p* sublevel is six, so the second energy level is filled when it reaches eight electrons—two in the 2*s* sublevel and six in the 2*p* sublevel. Look at the third and fourth energy levels and you can see that the third energy level can hold ten more electrons than the second because there is a *d* sublevel. The fourth can hold 14 more electrons than the third because there is an *f* sublevel.

Table 7.1	Distribution of Electrons in the First Four Energy Levels		
Energy Level	**Sublevel**	**Electrons in Sublevel**	**Electrons in Level**
1	1*s*	2	2
2	2*s*	2	
2	2*p*	6	8
3	3*s*	2	
3	3*p*	6	
3	3*d*	10	18
4	4*s*	2	
4	4*p*	6	
4	4*d*	10	
4	4*f*	14	32

Orbitals

In the 1920s, Werner Heisenberg reached the conclusion that it's impossible to measure accurately both the position and energy of an electron at the same time. This principle is known as the **Heisenberg uncertainty principle.** In 1932, Heisenberg was awarded the Nobel Prize in Physics for this discovery, which led to the development of the electron cloud model to describe electrons in atoms.

The electron cloud model is based on the probability of finding an electron in a certain region of space at any given instant. Here's how it works. Pretend that you can photograph the single electron that is attracted to the nucleus of a hydrogen atom. Once every second, you click the shutter on your camera, forming images of the electron on the same piece of film. After a few hundred snapshots, you develop the film and find a scatter diagram.

Demonstration 7.2

The Uncertainty Principle

Purpose: To anticipate students' questions about Heisenberg's uncertainty principle.
Materials: half-sheet of paper, a ruler, rubber band
Safety Precautions:
Disposal: Code F
Procedure: Loosely crumple the paper into a ball and place it on a desk. Loop a rubber band over the end of a ruler and shoot it at the paper ball. **CAUTION:** *Avoid pointing the rubber band at students. Wear goggles.*
Results: When the rubber band strikes the paper ball, its energy is transferred to the ball, which moves. The rubber band simulates a photon and the ball simulates an electron.
Analysis: Ask these questions.
1. Does a radar beam reflecting from an airplane cause a deflection in the airplane's flight path? *No, the energy of the photon is too small.*

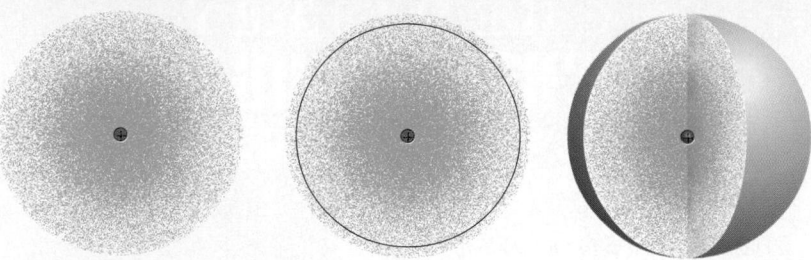

Sometimes, the electron is close to the nucleus. Other times, it's far away. Most of the time, it's in a small region of space that looks like a cloud. This electron cloud doesn't have a sharp boundary; its edges are fuzzy. If you ask someone to look at this picture, shown in **Figure 7.6,** and tell you where the electron is, he or she could say only that it's probably somewhere in the cloud. But this cloud of electron probability can be useful. If you draw a line around the outer edge enclosing about 95 percent of the cloud, within the region enclosed by the sphere, you can expect to find the electron about 95 percent of the time. The space in which there is this high probability of finding the electron is called an **orbital.**

Placing Electrons in Orbitals

Orbitals are regions of space located around the nucleus of an atom, each having the energy of the sublevel of which it is a part. Orbitals can have different sizes and shapes. There are four types of orbitals that accommodate the electrons for all the atoms of the known elements. Two simple rules apply to these four orbitals. First, an orbital can hold a maximum of two electrons. Second, an orbital has the same name as its sublevel. There is only one *s* orbital with, at most, 2 electrons. A *p* sublevel has, at most, 6 electrons, and, thus, there are three *p* orbitals, each with 2 electrons. **Figure 7.7** shows the shapes of *s* and *p* orbitals.

Figure 7.7

Models of *s* and *p* Orbitals
Orbitals have characteristic shapes that depend on the number of electrons in the energy sublevels. An *s* orbital is spherically symmetrical about the nucleus; a *p* orbital has a dumbbell shape. Three *p* orbitals are aligned along the *x, y,* or *z* axis at each energy level.

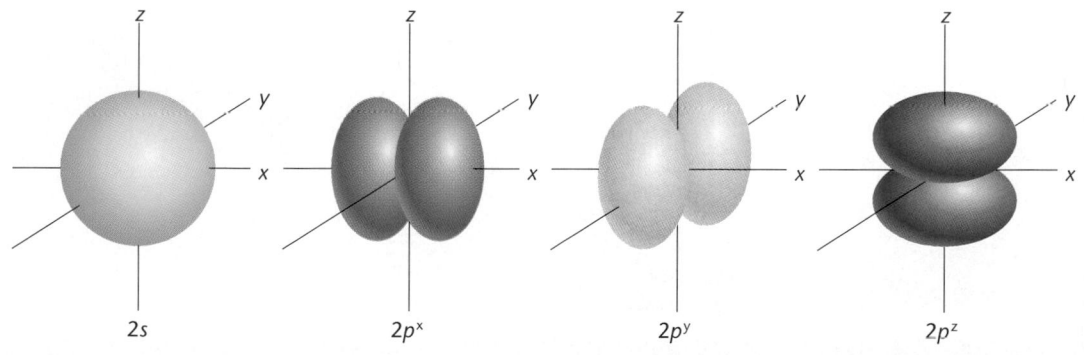

2*s*   2*p*ˣ   2*p*ʸ   2*p*ᶻ

Figure 7.6

Hydrogen's Electron Cloud
Most of the time, hydrogen's electron is within the fuzzy cloud in the two-dimensional drawing (left). A circle, with the nucleus at the center and enclosing 95 percent of the cloud, defines an orbital in two dimensions (center). The spherical model (right) represents hydrogen's 1*s* orbital in three dimensions.

7.1   Expanding the Theory of the Atom   **239**

2. What happens to the position of the paper ball (electron) when the photon (rubber band) reflects off it? *The paper ball changes position.*
3. Heisenberg stated that there is always uncertainty about specifying the position and energy of an electron. What is the name of that principle? *Heisenberg's uncertainty principle*

Purpose

To learn how SPMs, STMs, and AFMs work and what they are contributing to science and technology.

Background

Ordinary light microscopes can detect objects if they are 1000 times as large as an atom. Electron microscopes provide better resolution by using high-energy electrons rather than radiant light, but they penetrate below the surface of the material being examined. SPMs, STMs, and AFMs allow chemists to examine the surface of materials and can resolve details as small as 0.1 nm.

Teaching Strategies

- Have students determine which contributions of the microscopes are for pure science and which are technological or practical applications. L1
- Point out that the computer plays an important role in the rapid advancements of the new microscopes.
- Ernst Ruska invented the electron microscope in 1931 and received the Nobel prize for his work in 1986 along with Binning and Rohrer for their STM design. Discuss why deserving scientists are sometimes overlooked. For example, they do not work for prestigious establishments or they are not in favor. Rosalyn Yalow and Barbara McClintock are two scientists who were not recognized for their work for too long. Have students find out why, and write a short paper describing their findings and giving their own opinions about what could be done. Student papers could be placed in their portfolios. L2 P

Hi-Tech Microscopes

If you were a chemistry student in the 1960s or earlier, you would have been told that no one could see an atom. Now, with the aid of computers and revolutionary microscopes, it's possible to generate two- and three-dimensional images of atoms. It's even possible to move atoms around and observe their electron clouds. What kinds of instruments can accomplish feats that were not even thought of a decade or two ago? They include three kinds of microscopes: scanning probe, scanning tunneling, and atomic force.

Scanning Probe Microscope (SPM)

SPMs use probes to sense surfaces and produce three-dimensional pictures of a substance's exterior, as shown below. The precise arrangement of atoms can be viewed in three dimensions. This is accomplished by measuring changes in the current as the probe passes over a surface. SPMs can pick up atoms one at a time, move them around, and form letters, such as IBM shown below. Accomplishments like this suggest that in the future, all the information in the Library of Congress could be stored on an 8-inch silicon wafer. Scientists may

be able to use SPMs to learn why two surfaces stick together or how to shrink chip circuits to make computers work faster. They may even be able to put a thousand SPM tips on a 2-cm² silicon chip. Arrays of such chips could produce a working imitation of human sight. Could this mean that a small computer in the form of a visor similar to that worn by Geordi (above) on "Star Trek" will be available for the visually impaired in a few years?

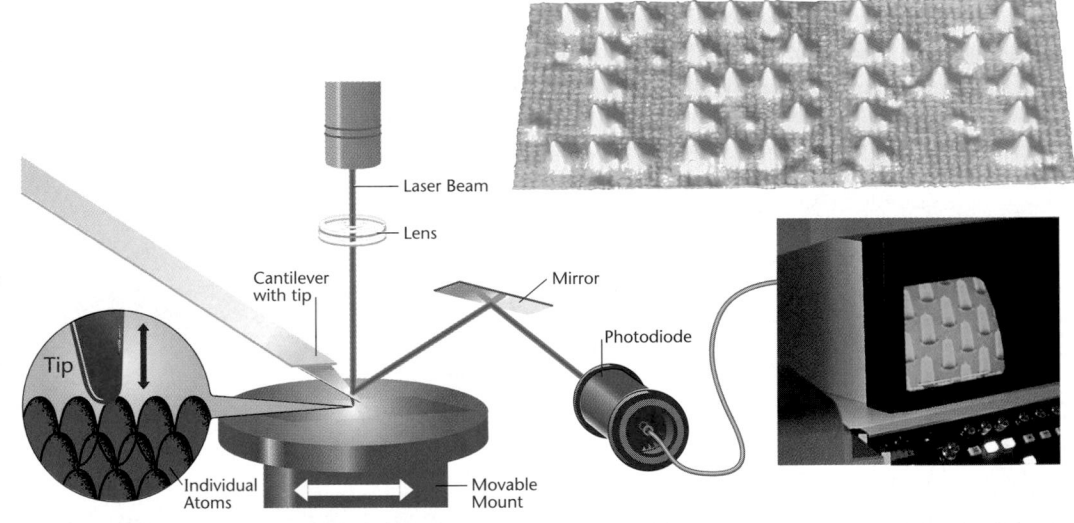

Laser Beam

Lens

Cantilever with tip

Mirror

Photodiode

Tip

Individual Atoms

Movable Mount

Scanning Tunneling Microscope (STM)

The scanning tunneling microscope also provides new technologies for chemists and physicists. The red areas in the photo below show the valence electrons of metal atoms that are free to move about in a metallic crystal. On the surface of the crystal, they can move in only two dimensions and behave like waves. Two imperfections on the surface of the crystal cause the electrons to produce concentric wave patterns.

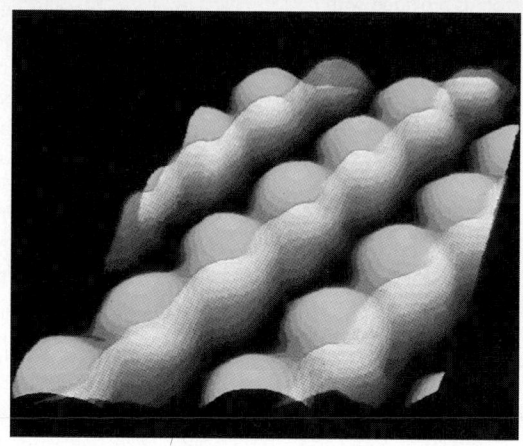

Wave patterns caused by electrons

Another current STM technology uses hydrogen to push aside surface atoms of semiconductors to reveal the atomic structure underneath. Using this method, atoms can be removed one layer at a time.

Recently, the STM has succeeded in forming images of electron clouds in atoms and molecules. The electron clouds of molecules can be calculated. An STM color-enhanced image of gallium arsenide is shown at right. The electron clouds of gallium appear blue. The electron clouds of arsenic appear red. Chemists have figured out the structure of atoms and molecules using indirect evidence from chemical reactions, physical experiments, and mathematical calculations. Now, they can view images of atoms while they experiment with them and base their work on both direct and indirect evidence.

Atomic Force Microscope (AFM)

The AFM, invented in 1985, uses repulsive forces between atoms on the probe's tip and those on the sample's surface to form images on the computer screen. An advantage of the AFM is that it does not have to work in a vacuum, so its samples require no special preparation. AFMs can provide images of molecules in living tissue and peel off membranes of living cells one layer at a time.

DISCUSSING THE TECHNOLOGY

1. **Comparing and Contrasting** Some scanning tunneling microscopes can remove layers of atoms. How does this capability compare with what the scanning probe microscope can do?

2. **Acquiring Information** Find information on the Near-Field Scanning Optical Microscope.

3. **Inferring** Compare the diagrams of the molecules with the computer-generated images of the molecules produced by the STM. How are the conclusions about the molecular structure obtained using conventional methods supported by the photos?

interNET CONNECTION

A gallery of Scanning Tunneling Microscope (STM) images of atoms is available on the Internet.

World Wide Web
http://almaden.ibm.com/vis/stm/gallery.html

Check for Understanding

- Check for understanding of the section material by having students answer the Section Review questions.
- Display a transparency of the energy-level diagram of hydrogen. Ask students how arrows should be drawn to represent transitions of the electron. Ask what colors should be used to draw the four lines. *a violet line from energy level 6 to energy level 2; a blue-violet line from 5 to 2; a blue-green line from 4 to 2; a red line from 3 to 2* L1 LEP

Reteach

Discussion Ask students which transition in each pair releases more energy: $2s$ to $1s$, or $3s$ to $1s$; $3p$ to $2p$, or $3p$ to $1s$; $3d$ to $3s$, or $3d$ to $2s$. *3s to 1s; 3p to 1s; 3d to 2s* L1

Extension

 Ask students what number of electrons could be contained in the fifth, sixth, and seventh energy levels. They could use the relationship: number of electrons = $2n^2$, where n is the number of the energy level. *50, 72, 98 electrons, respectively* L3

4 CLOSE

Enrichment

TECH PREP Have an astronomer come to class and explain the way in which the spectra of stars can be interpreted to find out what elements are present in the stars and their approximate ages, or ask an interested student to research the topic and present her or his findings to the class. L2

Electron Configurations

In any atom, electrons are distributed into sublevels and orbitals in the way that creates the most stable arrangement; that is, the one with lowest energy. This most stable arrangement of electrons in sublevels and orbitals is called an **electron configuration.** Electrons fill orbitals and sublevels in an orderly fashion beginning with the innermost sublevels and continuing to the outermost. At any sublevel, electrons fill the s orbital first, then the p. For example, the first energy level holds two electrons. These electrons pair up in the $1s$ orbital. The second energy level has four orbitals and can hold eight electrons. The first two electrons pair up in the $2s$ orbital, and the remaining six pair up in the three $2p$ orbitals. **Figure 7.8** shows that the overlap of the $2s$ and $2p$ orbitals results in a roughly spherical cloud. That's why the electrons in an atom can be represented as a series of fuzzy, concentric spheres. In the next section, you'll learn how the periodic table can be used to predict the electron configurations of the atoms.

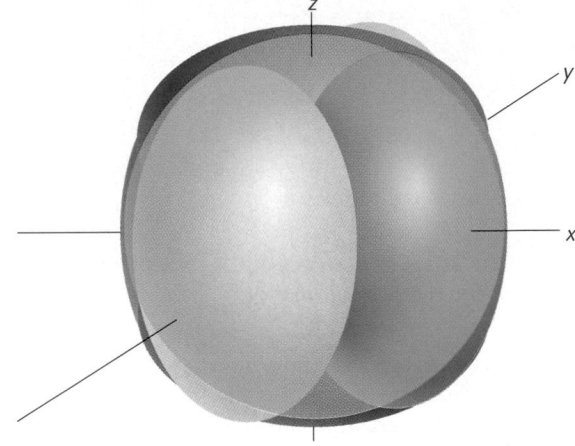

Figure 7.8
Overlapping Orbitals
Because orbitals are regions of space, they can be placed one on top of another. Imagine a pair of electrons moving around in an orbital. At any instant, most of the orbital is empty space, and this space can be used by another pair of electrons. That's how orbitals can overlap.

One 2s and three 2p orbitals

SECTION REVIEW

Understanding Concepts

1. How many s orbitals can the third energy level have? How many p orbitals? How many d orbitals?

2. How many electrons can each sublevel of the fourth energy level hold, and how many orbitals are required to accommodate them?

3. What is the shape of a p orbital? How do p orbitals at the same energy level differ from one another?

Thinking Critically

4. **Applying Concepts** What are two ways in which the elements with atomic numbers 12 and 15 are different?

Applying Chemistry

5. **Outdoor Lighting** Sodium vapor contained in a bulb or tube emits a brilliant yellow light when connected to a high-voltage source. Explain what is happening to the sodium atoms to produce this light.

SECTION REVIEW

1. The third energy level can have one s orbital, three p orbitals, and five d orbitals.
2. The sublevels of the fourth energy level are $4s$, $4p$, $4d$, and $4f$. The $4s$ sublevel can hold two electrons and needs only one orbital. The $4p$ sublevel can hold six electrons and needs three orbitals. The $4d$ sublevel can hold ten electrons and needs five orbitals. The $4f$ sublevel can hold 14 electrons and needs seven orbitals. A maximum of 32

electrons can be accommodated in the fourth energy level.
3. p orbitals are shaped like dumbbells. p orbitals at the same energy level differ in their spatial orientation. One is oriented along the x-axis, one along the y-axis, and the third along the z-axis.
4. Elements 12 and 15 have different numbers of protons and electrons. They have different numbers of valence electrons, and the

The Periodic Table and Atomic Structure

Suppose you were about to play your first game of Chinese checkers. How would you know where to place the marbles on the game board to start the game? The game board has many indentations to accommodate the marbles, but the marbles can be placed in only one way at the start of the game. Without the rules of the game or maybe a diagram of the starting game board, you wouldn't know what to do.

The same is true of placing electrons in orbitals around a nucleus. Imagine yourself with the bare phosphorus nucleus and a bag of electrons. So many energy levels and sublevels are available, each with one or more orbitals for electrons. How can you know where to put the electrons to produce the most stable electron configuration? You will learn some simple rules and discover that the periodic table provides a clear diagram for building electron configurations.

SECTION PREVIEW

Objectives
Distinguish the *s, p, d,* and *f* blocks on the periodic table and relate them to an element's electron configuration.
Predict the electron configurations of elements using the periodic table.

Key Terms
inner transition element

Patterns of Atomic Structure

Electrons occupy energy levels by filling the lowest level first and continuing to higher energy levels in numerical order. Valence electrons of the main group elements occupy the *s* and *p* orbitals of the outermost energy level. The position of any element in the periodic table shows which orbitals—*s, p, d,* and *f*—the valence electrons occupy.

Orbitals and the Periodic Table

The shape of the modern periodic table is a direct result of the order in which electrons fill energy sublevels and orbitals. The periodic table in **Figure 7.9** is divided into blocks that show the sublevels and orbitals occupied by the electrons of the atoms. Notice that Groups 1 and 2 (the active metals) have valence electrons in *s* orbitals, and Groups 13 to 18 (metals, metalloids, and nonmetals) have valence electrons in both *s* and *p* orbitals. Therefore, all the main group elements have their valence electrons in *s* or *p* orbitals. Groups 1 and 2 are designated as the *s* region of

7.2   The Periodic Table and Atomic Structure   **243**

---

orbital configuration of their valence electrons differs. Element 12 has both valence electrons in an *s* orbital; element 15 has two valence electrons in an *s* orbital and three in *p* orbitals.

5. The electrons in sodium absorb energy and move to higher energy levels. As they fall back to their original energy level, they emit radiation in the visible spectrum that has the frequency of yellow light.

Program Resources

Study Guide, pp. 29-30 [L1]
Laboratory Manual, 7.2, pp. 65-68 [L1]
Section Focus Transparency 15 and **Master,** p. 15 [L1] [LEP]
ChemLab and MiniLab Worksheets, p. 51 [L1]
Chemistry and Industry, pp. 15-18 [L1]

PREPARE

Key Concepts
The correlation between the electron configurations of the elements, the organization of the periodic table, and the periodicity of properties is emphasized. The characteristics of the noble gases and transition elements are shown to be directly related to their electron arrangements. Relative orbital sizes are discussed briefly.

Planning Ahead
For the MiniLab Carbon paper may be purchased from office supply stores. Solder a BB to the tips of darts or to the tips of large nails.

1 FOCUS

Focus Transparency
Display the **Section Focus Transparency** for Section 7.2 to introduce the relationship between electron configuration and the periodic table. [L1] [LEP]

2 TEACH

Concept Development
The sequence in which electrons are added to energy levels, sublevels, and orbitals in the electron configuration of an atom is known as the aufbau sequence. *Aufbau* is a German word meaning "building up."

GLENCOE TECHNOLOGY

CD-Rom
Chemistry: Concepts and Applications
The Periodic Table
Have students use this interactive periodic table to predict the electron configurations of elements. [L1] [LEP]

Assessment

Performance: Challenge interested students to create aufbau games. Examine their creations and consider using good ones in class in the *Assess* part of the section as a way of reinforcing the skill of writing electron configurations. L2 **COOP LEARN**

GLENCOE TECHNOLOGY

 Videodisc

Chemistry: Concepts and Applications
Emission Spectra of Elements
Disc 1, Side 2, Ch. 6

Show this video of Demonstration 7.3 to reinforce the concept of trends in reactivity of the elements of the halogen family. L1 **LEP**

 CD-Rom

Chemistry: Concepts and Applications
Emission Spectra of Elements
Have students use this video to help reinforce the concept of trends in reactivity within a group. L1 **LEP**

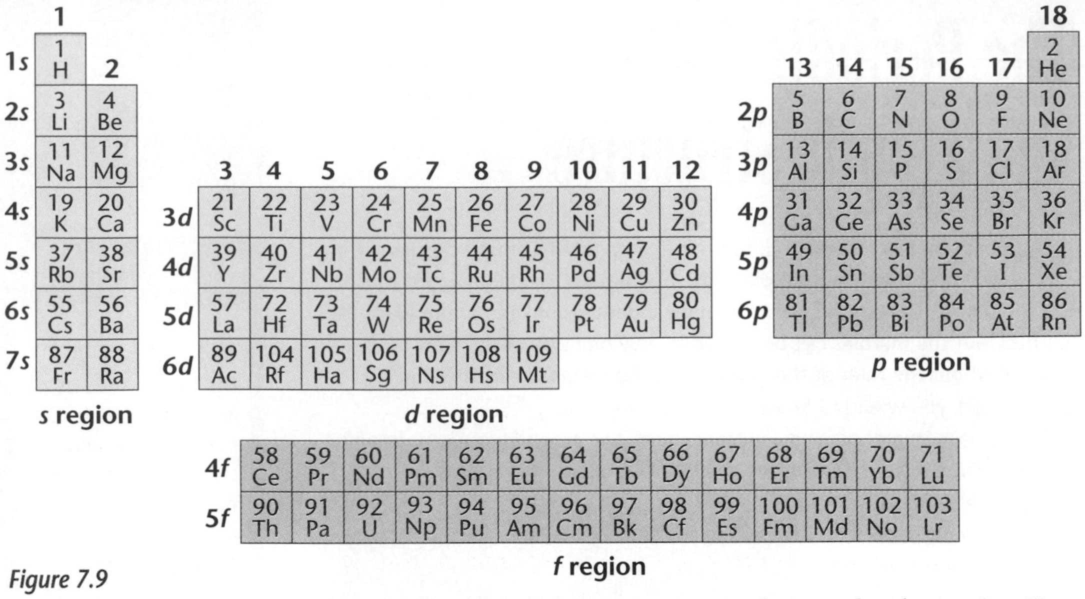

Figure 7.9

The Periodic Table: Key to Electron Configurations
It's not necessary to memorize electron configurations if you can interpret the *s, p, d,* and *f* blocks shown in this periodic table. When you write electron configurations, move from left to right through the periods, filling the orbitals that correspond to the *s, p, d,* and *f* blocks.

the periodic table, and Groups 13 to 18 are designated as the *p* region. Note that Groups 3 to 12 are designated as the *d* region and that each row of this region, except for that in Period 7, has ten elements. The block beneath the table is the *f* region, and each row in this region contains 14 elements.

Building Electron Configurations

Chemical properties repeat when elements are arranged by atomic number because electron configurations repeat in a certain pattern. As you move through the table, you'll notice how an element's position is related to its electron configuration. Hydrogen has a single electron in the first energy level. Its electron configuration is $1s^1$. This is standard notation for electron configurations. The number *1* refers to the energy level, the letter *s* refers to the sublevel, and the superscript refers to the number of electrons in the sublevel. Helium has two electrons in the $1s$ orbital. Its electron configuration is $1s^2$. Helium has a completely filled first energy level. When the first energy level is filled, additional electrons must go into the second energy level. Electrons enter the sublevel that will give the atom the most stable configuration, that with the lowest energy.

Lithium begins the second period. Its first two electrons fill the first energy level, so the third electron occupies the second level. Lithium's electron configuration is $1s^2 2s^1$. Beryllium has two electrons in the $2s$ orbital, so its electron configuration is $1s^2 2s^2$. As you continue to move across the second period, electrons begin to enter the *p* orbitals. Each successive element has one more electron in the $2p$ orbitals. Carbon, for example, has four electrons in the second energy level. Two of these are in the $2s$ orbital and two are in the $2p$ orbitals. The electron configuration for carbon is $1s^2 2s^2 2p^2$. At element number 10, neon, the *p* sublevel is filled with six electrons. The electron configuration for neon is $1s^2 2s^2 2p^6$. Neon has eight valence electrons; two are in an *s* orbital and six are in *p* orbitals.

Demonstration 7.3

Emission Spectra of Elements

Purpose: To illustrate the relationship between the electron configuration of nonmetals and their reactivity.

Materials: 20 mL 12*M* HCl, 6 g MnO₂, 3 small pieces sodium, 1 mL water, 1 g iodine, 1 mL bromine, dropper, stainless steel spoon, 3 widemouth glass bottles with watch glass covers, fume hood

Safety Precautions:

Disposal: Code H

Procedure: Halogen reactivity can be illustrated by reacting a small piece (1/8 the size of a pea) of sodium metal in Cl_2, Br_2, and I_2 vapors. **CAUTION:** *Do NOT use potassium as it will explode with bromine and iodine. Do ALL work in a fume hood while wearing goggles.* Cl_2 gas can be gen-

erated by adding 20 mL of 12*M* HCl to 6 g of MnO₂. **CAUTION:** *HCl is corrosive. Cl_2 gas is toxic. Collect the gas in a covered bottle in the fume hood.* Using a small stainless steel spoon, lower the small piece of sodium metal into the Cl_2 gas. Carefully put a drop or two of water on the sodium to begin the reaction. Cover the bottle. Repeat the procedure using Br_2 vapor. **CAUTION:** *Wear rubber gloves*

Electrons in Atoms

The modern theory of the atom cannot tell you exactly where the electrons in atoms are placed. However, it does define regions in space called orbitals, where there is a 95 percent probability of finding an electron. The lowest energy orbital in any atom is called the 1s orbital. In this MiniLab, you will simulate the probability distribution of the 1s orbital by noting the distribution of impacts or hits around a central target point.

Procedure

1. Obtain two pieces of blank, white, $8\frac{1}{2}'' \times 11''$ paper and draw a small but visible mark in the center of each of the papers. Hold the papers together toward a light and align the center marks exactly.

2. Around the center dot of one of the papers, which you will call the target paper, draw concentric circles having radii of 1 cm, 3 cm, 5 cm, 7 cm, and 9 cm. Number the areas of the target 1, 2, 3, 4, and 5, starting with number 1 at the center.

3. Place a piece of poster board on the floor, and lay the target paper face up on top of it. Cover the target paper with a piece of carbon paper, carbon side down. Then place the second piece of white paper on top with the center mark facing up. Use tape to fasten the three layers of paper in place on the poster board and to secure the poster board to the floor.

4. Stand over the target paper and drop a dart 100 times from chest height, attempting to hit the center mark.

5. Remove the tape from the papers. Separate the white papers and the carbon paper. Tabulate and record the number of hits in each area of the target paper.

Analysis

1. How many hits did you record in each of the target areas? What does each hit represent in the model of the atom?

2. Make a graph plotting the number of hits on the vertical axis and the target area on the horizontal axis.

3. Which of the target areas has the highest probability of a hit? Relate your findings to the model of the atom.

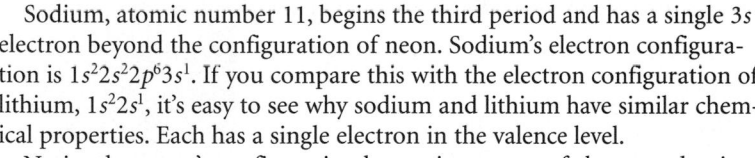

Sodium, atomic number 11, begins the third period and has a single 3s electron beyond the configuration of neon. Sodium's electron configuration is $1s^2 2s^2 2p^6 3s^1$. If you compare this with the electron configuration of lithium, $1s^2 2s^1$, it's easy to see why sodium and lithium have similar chemical properties. Each has a single electron in the valence level.

Notice that neon's configuration has an inner core of electrons that is identical to the electron configuration in helium ($1s^2$). This insight simplifies the way electron configurations are written. Neon's electron configuration can be abbreviated $[\text{He}]2s^2 2p^6$. In the abbreviated form, neon's electron configuration is represented by an inner core of electrons from

when working with Br_2 in the fume hood. Repeat using I_2 vapor in a separate bottle. Gently warm a few I_2 crystals until the bottle is filled with violet vapors.
Results: Water initiates an exothermic reaction. Sodium burns with a yellow flame and produces a salt vapor.
Analysis: Ask these questions.
1. Write the electron configurations of Cl, Br, and I. *[Ne]3s²3p⁵, [Ar]3d¹⁰4s²4p⁵, [Kr]4d¹⁰5s²5p⁵*

2. How do the outer electrons of each element in the halogen family compare? *Each element has seven outer electrons.*

✓ Assessment

Portfolio: Have students write a summary of this demonstration and add it to their portfolios. [L1] [P]

SECTION 7.2

miniLAB

Electrons in Atoms

Purpose: To simulate the probability distribution of an electron in the 1s orbital.
Process Skills: Observing, using space/time relationships, formulating models, communicating
Preparation: Solder a BB to the tip of each dart. Have available white paper, poster board, carbon paper, rulers, and tape.

Teaching Strategies

• Caution students that the darts are to be used only as described in the lab.
• Students should drop, not throw, the darts.

Expected Results

Typical data: area 1, 21 hits; area 2, 57 hits; area 3, 16 hits; area 4, 6 hits; area 5, 0 hits.

Analysis

1. See expected results. Each hit represents a location of the electron.

2.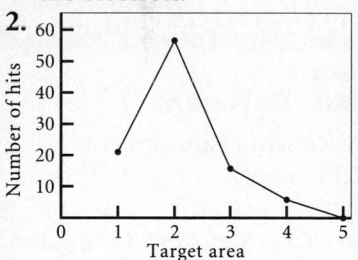

3. area 2; This area represents the region in the 1s orbital in which the probability of finding the electron is the highest.

✓ Assessment

Oral: Ask students if it is possible that the electron may be found even farther from the nucleus than the areas represented in the MiniLab. *Yes, it is possible but not probable.* [L1]

245

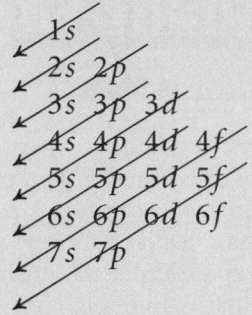

Draw this sublevel diagram on the chalkboard, then draw arrows through the diagram as shown. Relate the aufbau sequence to the arrows and to the periods and sublevel blocks on the periodic table.

Correcting Misconceptions

Students may think that because electrons are attracted to the positively charged protons in the nucleus of the atom, they will eventually fall into it. Point out that the electrons would lose energy if they did so and that this phenomenon has not been observed.

the noble gas in the preceding row (He), followed by the orbitals filled in the current period. The abbreviated electron configuration for sodium is [Ne]$3s^1$, where the neon core represents sodium's ten inner electrons. Its similarity to the [He]$2s^1$ configuration of lithium shows clearly that these Group 1 elements have the same number of valence electrons in the same type of orbital. **Table 7.2** shows electron configurations for all the elements in the second and third periods. Notice that elements in the same group have similar configurations. This is important because it shows that the periodic trends in properties, observed in the periodic table, are really the result of repeating patterns of electron configuration.

Table 7.2   Electron Configurations of Second and Third Period Elements

Second Period Elements	Configuration	Third Period Elements	Configuration
Lithium	[He] $2s^1$	Sodium	[Ne] $3s^1$
Beryllium	[He] $2s^2$	Magnesium	[Ne] $3s^2$
Boron	[He] $2s^22p^1$	Aluminum	[Ne] $3s^23p^1$
Carbon	[He] $2s^22p^2$	Silicon	[Ne] $3s^23p^2$
Nitrogen	[He] $2s^22p^3$	Phosphorus	[Ne] $3s^23p^3$
Oxygen	[He] $2s^22p^4$	Sulfur	[Ne] $3s^23p^4$
Fluorine	[He] $2s^22p^5$	Chlorine	[Ne] $3s^23p^5$
Neon	[He] $2s^22p^6$	Argon	[Ne] $3s^23p^6$

The Stable Configurations of the Noble Gases

Each period ends with a noble gas, so all the noble gases have filled energy levels and, therefore, stable electron configurations. The electron configurations for all the noble gases are shown in **Table 7.3.** All except helium have eight valence electrons. However, helium's two electrons fill its outermost energy level and are a stable configuration. These stable electron configurations explain the lack of reactivity of the noble gases. Noble gases don't need to form chemical bonds to acquire stability.

Table 7.3   The Electron Configurations of the Noble Gases

Noble Gas	Electron Configuration
Helium	$1s^2$
Neon	[He] $2s^22p^6$
Argon	[Ne] $3s^23p^6$
Krypton	[Ar] $4s^23d^{10}4p^6$
Xenon	[Kr] $5s^24d^{10}5p^6$
Radon	[Xe] $6s^24f^{14}5d^{10}6p^6$

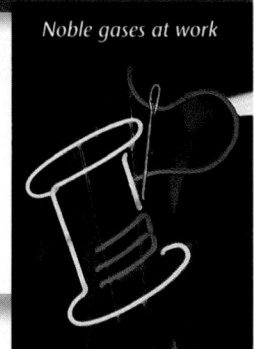

Noble gases at work

Integrating THE Sciences

Physics Since the time of Newton, the laws of physics were based on macroscopic observations. Now known as classical physics, Newton's laws accurately describe and predict the behavior of macroscopic objects. However, these laws failed to describe and predict behavior at the submicroscopic level. The branch of physics known as quantum mechanics was developed to explain submicroscopic behavior and has proved successful for more than 60 years. Today, scientists realize that the laws of classical mechanics do not disagree with the laws of quantum mechanics. The wave particle duality and energy considerations that are so significant at the submicroscopic level are still valid at the macroscopic level, but they are insignificant. The traditional laws of classical mechanics are valid approximations of the laws of quantum mechanics and work perfectly at the macroscopic level.

What happens in the fourth period?

You might expect that after the $3p$ orbitals are filled in argon, the next electron would occupy a $3d$ orbital, but this is not the case. Potassium follows argon and begins the fourth period. Its configuration is $[Ar]4s^1$. Compare potassium's configuration to the configurations of lithium and sodium in **Table 7.2,** and recall that potassium is chemically similar to the Group 1 elements. Experimental evidence indicates that the $4s$ and $3d$ sublevels are close in energy, with the $4s$ sublevel having a slightly lower energy. Thus, the $4s$ sublevel fills first because that order produces an atom with lower energy. The next element after potassium is calcium. Calcium completes the filling of the $4s$ orbital. It has the electron configuration $[Ar]4s^2$.

Transition Elements

Notice in the periodic table that calcium is followed by a group of ten elements beginning with scandium and ending with zinc. These are transition elements. Now the $3d$ sublevel begins to fill, producing atoms with the lowest possible energy. Ten electrons are added across the row and fill the $3d$ sublevel. Just as the s block at each energy level adds two electrons and fills one s orbital, and the p block at each energy level adds six electrons and fills three p orbitals, so the d block adds ten electrons. There are five d orbitals. The f block adds 14 electrons, and they are accommodated in seven orbitals. The electron configuration of scandium, the first transition element, is $[Ar]4s^23d^1$. Zinc, the last of the series, has the configuration $[Ar]4s^23d^{10}$. **Figure 7.10** shows the configurations of the $3d$ transition elements. The six elements following the $3d$ elements, gallium to krypton, fill the $4p$ orbitals and complete the fourth period.

	3	4	5	6	7	8	9	10	11	12
$3d$	Scandium 21 **Sc** $[Ar]4s^23d^1$	Titanium 22 **Ti** $[Ar]4s^23d^2$	Vanadium 23 **V** $[Ar]4s^23d^3$	Chromium 24 **Cr** $[Ar]4s^13d^5$	Manganese 25 **Mn** $[Ar]4s^23d^5$	Iron 26 **Fe** $[Ar]4s^23d^6$	Cobalt 27 **Co** $[Ar]4s^23d^7$	Nickel 28 **Ni** $[Ar]4s^23d^8$	Copper 29 **Cu** $[Ar]4s^13d^{10}$	Zinc 30 **Zn** $[Ar]4s^23d^{10}$

Figure 7.10

The Electron Configurations of 3d Transition Elements
Ten electrons are added across the d block, filling the d orbitals. Notice that chromium and copper have only one electron in the $4s$ orbital. Such unpredictable exceptions show that the energies of the $4s$ and $3d$ sublevels are close.

Like most metals, the transition elements lose electrons to attain a more stable configuration. Most have multiple oxidation numbers because their s and d orbitals are so close in energy that electrons can be lost from both orbitals. For example, cobalt (atomic number 27) forms two fluorides—one with the formula CoF_2 and another with the formula CoF_3. In the first case, cobalt gives up its two $4s$ electrons to fluorine. In the second case, cobalt gives up both of its $4s$ electrons and one $3d$ electron.

Chemistry Journal

An Electron's Story
Ask each student to assume he or she is an electron in the $3p$ sublevel of a phosphorus atom and write a short story summarizing his or her position in the atom, any feelings of attraction within the atom, and any feelings of repulsion. Have students include the stories in their journals. L1

WORD ORIGIN

orbital:
orbita (L) wheel, track, course, circuit

An orbital is the most probable location in atomic space for an electron. The name derives from the old idea that electrons orbit the nucleus of an atom.

Visual Learning

Figure 7.10: Point out that the electron configurations of chromium and copper are exceptions to the aufbau sequence of filling orbitals. One explanation for these anomalies is that a configuration in which a sublevel is filled or half-filled is lower in energy and, therefore, more stable. Other examples are molybdenum, palladium, and silver. Molybdenum's configuration is $[Kr]5s^14d^5$. Each of the six orbitals occupied by the valence electrons is half filled. Such arrangements avoid the repulsion of pairing electrons in an orbital and, in this way, lower energy. The ground state electron configuration of palladium is $[Kr]4d^{10}$ rather than the expected configuration, $[Kr]5s^24d^8$. The filled inner d orbital of palladium imparts stability (lower energy) compared to the expected configuration, which has a partially filled d sublevel. Palladium's neighbor, silver, with one more electron, retains the filled d sublevel, and its extra electron is placed in the $5s$ orbital. Silver's configuration is $[Kr]4d^{10}5s^1$.

Correcting Misconceptions

Electron configurations for the transition elements are often written with the sublevels in the same sequence as energy levels, for example, $[Kr]4d^25s^2$ for zirconium. Writing the configuration in this way does not mean that the filling of the orbitals doesn't follow the aufbau sequence. The $5s$ orbital is filled before electrons enter the $4d$ sublevel.

Everyday Chemistry

Purpose
To show students how transition elements give gemstones and glass their colors.

Teaching Strategies
- Explain how transition metals produce color. When white light shines on transition metals, *d* electrons of the metals absorb certain frequencies (colors) of the light and are excited to a higher energy state. The unabsorbed light is no longer white, but rather a color combination made up of the unabsorbed frequencies. These are transmitted or reflected and are what is perceived by the eye.
- Point out the role that chemistry plays in glass/ceramic art. Artists need to have a knowledge of chemistry.

Extension
Have an art teacher talk to your class about what chemicals are used in paints and/or glazes to produce the desired colors.

Exploring Further

1. Iron(II) and iron(III) thiocyanate are different colors because they have different oxidation numbers, but iron(III) sulfate differs from iron(III) thiocyanate because it has a different negative ion.
2. Amethyst: iron and manganese oxides; rose quartz: manganese and titanium oxides; citrine: Fe_2O_3. Most commercial citrine is heat-treated amethyst.
3. The chemical and physical properties are generally identical. Synthetics have superior quality. Naturals have inclusions that help identify them.

Colors of Gems

Have you ever wondered what produces the gorgeous colors in a stained-glass window or in the rubies, emeralds, or sapphires mounted on a ring? Compounds of transition elements are responsible for creating the entire spectrum of colors.

Transition elements color gems and glass Transition elements have many important uses, but one that is often overlooked is their role in giving colors to gemstones and glass. Although not all compounds of transition elements are colored, most inorganic colored compounds contain a transition element such as chromium, iron, cobalt, copper, manganese, nickel, cadmium, titanium, gold, or vanadium. The color of a compound is determined by three factors: (1) the identity of the metal, (2) its oxidation number, and (3) the negative ion combined with it.

Impurities give gemstones their color Crystals have fascinating properties. A clear, colorless quartz crystal is pure silicon dioxide (SiO_2). But a crystal that is colorless in its pure form may exist as a variety of colored gemstones when tiny amounts of transition element compounds, usually oxides, are present. Amethyst (purple), citrine

(yellow-brown), and rose quartz (pink) are quartz crystals with transition element impurities scattered throughout. Blue sapphires are composed of aluminum oxide (Al_2O_3) with the impurities iron(II) oxide (FeO) and titanium(IV) oxide (TiO_2). If trace amounts of chromium(III) oxide (Cr_2O_3) are present in the Al_2O_3, the resulting gem is a red ruby. A second kind of gemstone is one composed entirely of a colored compound. Most are transition element compounds, such as rose-red rhodochrosite ($MnCO_3$), black-grey hematite (Fe_2O_3), or green malachite ($CuCO_3 \cdot Cu(OH)_2$).

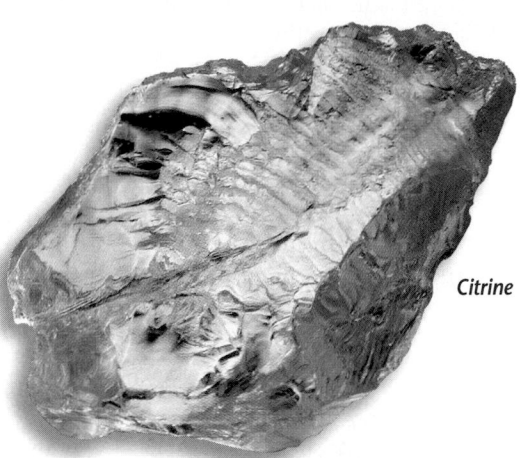
Citrine

Quartz

Amethyst

Demonstration 7.4

Electrons and Reactivity
Purpose: To show that electron configuration is related to chemical reactivity.
Materials: 1-L beaker with cover, 30 g of baking soda, 30 mL of vinegar, 2-cm piece of magnesium ribbon, wood splint, crucible tongs, laboratory burner, dark plastic film capable of screening out high-frequency light

Safety Precautions:

Disposal: Code A
Procedure: In the 1-L covered beaker, combine $NaHCO_3$ and vinegar to generate CO_2 gas. Using tongs, light a wood splint and hold it in the CO_2 atmosphere. Ignite the magnesium and hold it in the CO_2 gas. **CAUTION:** *Use a dark-colored plastic film to protect students' eyes from the bright light of the magnesium flame.*
Results: The burning splint goes out in a CO_2 atmosphere. The magnesium burns more vio-

How metal ions interact with light to produce color Why does the presence of Cr_2O_3 in Al_2O_3 make a ruby red? The Cr^{3+} ion absorbs yellow-green colors from white light striking the ruby, and the remaining red-blue light is transmitted, resulting in a deep red color. This same process occurs in all gems. Trace impurities absorb certain colors of light from white light striking or passing through the stone. The remaining colors of light that are reflected or transmitted produce the color of the gem.

Adding transition elements to molten glass for color Glass is colored by adding transition element compounds to the glass while it is molten. This is true for stained glass, glass used in glass blowing, and even glass in the form of ceramic glazes.

Rose quartz

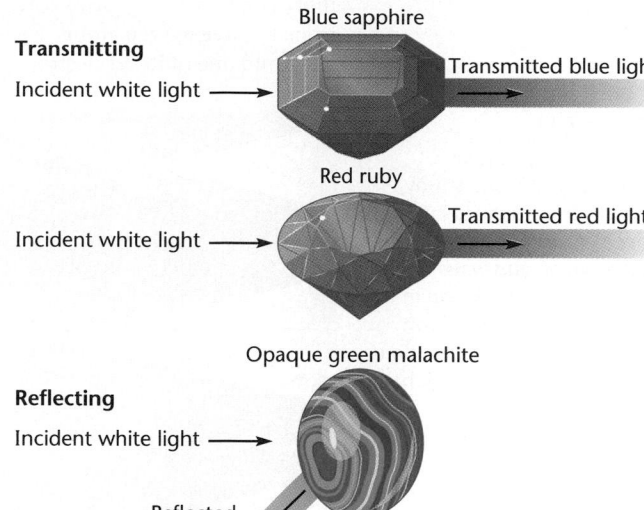

Transmitting

Incident white light → Blue sapphire → Transmitted blue light

Incident white light → Red ruby → Transmitted red light

Opaque green malachite

Reflecting

Incident white light →

Reflected green light

Gems are the color of the light they transmit.

Most of the coloring agents are oxides. When oxides of copper or cobalt are added to molten glass, the glass is blue; oxides of manganese produce purple glass; iron oxides, green; gold oxides, deep ruby red; copper or selenium oxides, red; and antimony oxides, yellow. Some coloring compounds are not oxides. Chromates, for example, produce green glass, and iron sulfide gives a brown color.

Exploring Further

1. **Applying** Explain why iron(III) sulfate is yellow, iron(II) thiocyanate is green, and iron(III) thiocyanate is red.

2. **Acquiring Information** Find out what impurities give amethyst, rose quartz, and citrine their colors.

3. **Comparing and Contrasting** Conduct library research to find the similarities and differences between synthetic and natural gemstones.

Correcting Misconceptions

Many students think that one proton in the nucleus of an atom attracts one electron. Emphasize that each electron is attracted to each and every proton in the nucleus.

Using an Analogy

To illustrate the previous concept about nuclear attraction for the electrons, ask students whether their weight automatically decreases when another person is born because Earth's gravity must now be divided more ways. *no; The gravitational attraction between Earth and each person depends only upon the mass of Earth, the mass of the person, and the distance between their centers of mass. Earth's gravity is not divided among the people on Earth.*

GLENCOE TECHNOLOGY

 CD-Rom

Chemistry: Concepts and Applications
Determining Electron Configurations
Have students use this Interactive Exploration to practice writing electron configurations and relating them to the periodic table. [L1] [LEP]

---

lently in CO_2 than in air.

$$Mg + CO_2 \rightarrow CO + MgO$$

Analysis: Ask these questions.
1. What are the electron configurations of carbon and magnesium? *C, $1s^2 2s^2 2p^4$; Mg, $1s^2 2s^2 2p^6 3s^2$*
2. In which element are the valence electrons closer to the nucleus? *carbon*
3. Classify magnesium and carbon as metals or nommetals. *Carbon is a non-*

metal; magnesium is a metal.

4. What reasons can you give for the difference in reactivity? *Carbon and magnesium are far from each other on the periodic table, so you would not expect them to have similar reactivities.*

✓ Assessment

• **Performance:** Ask students to write the electron configurations of calcium and

barium, two elements in the same group as magnesium. Ask them to decide which element is more reactive. [L1]

• **Oral:** Ask students whether they can think of any way to put out a fire in the fuselage of a jet airplane made of a magnesium alloy. *Sand is the only effective agent used to extinguish a magnesium fire.* [L1]

Check for Understanding

- Check for understanding of the section material by having students answer the Section Review questions.
- Use the best of the aufbau games created by students in the *Assessment* strategy on page 244.

Reteach

Cooperative Learning: Divide the class into groups of four students. They will act as a team in identifying elements by their electron configurations. Write the first electron configuration on the chalkboard, for example, [Kr]$5s^1$ for rubidium. The teams then work together to identify the element. The first group to agree on a correct identification can be awarded a credit point. Allow that winning team to write the next configuration. Continue the process as long as time permits, and interest and learning persist. **L1 COOP LEARN**

The rusting of the transition element iron shows that iron can have more than one oxidation number. In the process of rusting, **Figure 7.11,** iron first forms the compound FeO. This compound continues to react with oxygen and water to form the familiar orange-brown compound called rust, Fe_2O_3. Because oxygen requires two electrons to achieve a noble-gas configuration, it takes the two $4s$ electrons from iron to form the FeO compound. The more complex Fe_2O_3 forms when two iron atoms give up a total of six electrons to three oxygen atoms. Each iron atom must surrender its two $4s$ electrons and one of its $3d$ electrons.

Figure 7.11

The Rusting of Iron
Once beautiful and sturdy, mighty iron can turn into a mass of rust as a result of the action of air and water. First, iron reacts with oxygen in the presence of water to form FeO. In FeO, iron's oxidation number is 2+ because it has lost its two $4s$ electrons. Then, FeO continues to combine with oxygen to form the familiar orange-brown compound, Fe_2O_3. In this oxide, iron's oxidation number is 3+ because it has lost two $4s$ electrons and one $3d$ electron.

Inner Transition Elements

The two rows beneath the main body of the periodic table are the lanthanides (atomic numbers 58 to 71) and the actinides (atomic numbers 90 to 103). These two series are called **inner transition elements** because their last electron occupies inner-level $4f$ orbitals in the sixth period and the $5f$ orbitals in the seventh period. As with the d-level transition elements, the energies of sublevels in the inner transition elements are so close that electrons can move back and forth between them. This results in variable oxidation numbers, but the most common oxidation number for all of these elements is 3+.

The Size of Orbitals

Hydrogen and the Group 1 elements each have a single valence electron in an s orbital. Hydrogen's configuration is $1s^1$; the valence electron configuration of lithium is $2s^1$. For sodium, it's $3s^1$; for potassium, $4s^1$. Continuing down the column, rubidium, cesium, and francium have valence configurations of $5s^1$, $6s^1$, and $7s^1$, respectively. How do these s orbitals differ from one another? As you move down the column, the energy of the outermost sublevel increases. The higher the energy, the farther the outermost electrons are from the nucleus. The s orbitals, occupied by the

Cultural Diversity

Saving the Sperm Whale

The excellent qualities of sperm whale oil make it difficult for some nations to obey a 1986 ban on hunting this endangered species, but the efforts of Native Americans may have helped preserve the whale. Oil from the seeds of the jojoba plant, a native of the American southwest, contains compounds similar to those found in whale oil. William P. Miller, a Cherokee and an employee of the U.S. Bureau of Indian Affairs, saw jojoba as a way of providing jobs for Native Americans and also reducing pressure on the whale. Miller enlisted the San Carlos Apache tribe to collect wild jojoba seeds. The oil was extracted, and samples were sent to industries that normally use whale oil. The response was positive, so in the early 1980s, Miller encouraged Native Americans to invest in jojoba plantations. Jojoba oil is now considered one of the finest oils for use in cosmetics and personal care products.

valence electrons of hydrogen and the Group 1 elements, are described as spheres around the nucleus. As the valence electron gets farther from the nucleus, the *s* orbital it occupies gets larger and larger. **Figure 7.12** shows the relative sizes of the 1*s*, 2*s*, and 3*s* orbitals.

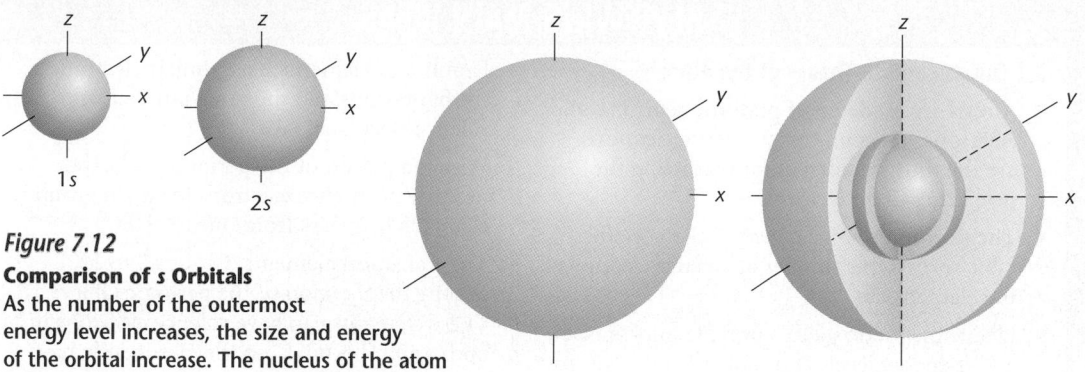

Figure 7.12
Comparison of *s* Orbitals
As the number of the outermost energy level increases, the size and energy of the orbital increase. The nucleus of the atom is at the intersection of the coordinate axes. The model on the right shows the overlap of the 1*s*, 2*s*, and 3*s* orbitals. This model could represent sodium.

3*s*

Overlapping 1*s*, 2*s*, and 3*s* orbitals

Connecting Ideas

An important skill that you learned in this chapter is how to use the periodic table to write electron configurations. It should be clear to you now that the organization of the table arises from the electron configurations of the elements. With this added insight, you are ready to learn in Chapter 8 about trends in properties and patterns of behavior of the elements. Knowing electron configurations and periodic trends will help you organize what may seem to be a vast amount of information.

SECTION REVIEW

Understanding Concepts

1. Use the periodic table to help you write electron configurations for the following atoms. Use the appropriate noble gas, inner-core abbreviations.

   a) Ca   d) Cl
   b) Mg   e) Ne
   c) Si

2. Identify the elements that have the following electron configurations:

   a) $1s^2 2s^2$ d) $1s^2 2s^2 2p^2$
   b) $1s^2$ e) $[Ne]3s^2 3p^4$
   c) $1s^2 2s^2 2p^5$ f) $[Ar]4s^1$

3. What is the difference between a filled and an unfilled orbital?

Thinking Critically

4. **Applying Concepts** What is the lowest energy level that can have an *s* orbital? What region of the periodic table is designated as the *s* region? Are the elements in this region mostly metals, metalloids, or nonmetals?

Applying Chemistry

5. **Periodic Table** Why do the fourth and fifth periods of the periodic table contain 18 elements rather than eight elements as in the second and third periods?

Extension

Ask students how many elements would be in period 8 of the periodic table if 200 elements were known and there were a *g* sublevel. *50* What sublevels would accommodate the electrons added in period 8? *The sublevels would be 8s, 5g, 6f, 7d, and 8p.* L2

4 CLOSE

Enrichment

Make and show a transparency of a scanning tunneling microscope photo showing the interference pattern (standing waves) formed by the interaction of an atom's electrons. Explain that much of the theory studied in this chapter is based upon the assumption that the electron has characteristics of a wave, as well as a particle.

GLENCOE TECHNOLOGY

Videotape

MindJogger Videoquizzes for Chemistry
Chapter 7: Completing the Model of the Atom
Have students work in groups as they play the videoquiz game to review key chapter concepts. L1 LEP COOP LEARN

SECTION REVIEW

1. a. $[Ar]4s^2$ **b.** $[Ne]3s^2$ **c.** $[Ne]3s^2 3p^2$
   d. $[Ne]3s^2 3p^5$ **e.** $[He]2s^2 2p^6$
2. a. Be **b.** He **c.** F **d.** C **e.** S **f.** K
3. A filled orbital has a pair of electrons. An unfilled orbital can be empty or contain a single electron.
4. The first energy level is the lowest energy level that can have an *s* orbital. Groups 1 and 2 of the periodic table are designated as the *s* region. The elements in the *s* region are metals.

5. The fourth and fifth periods contain 18 elements because ten elements with 3*d* orbitals appear in the fourth period, and ten elements with 4*d* orbitals appear in the fifth period.

CHAPTER REVIEW

- Review the Reviewing Main Ideas statements and Key Terms with your students.
- Complete solutions to Chapter Review problems can be found in the *Problems and Solutions Manual*.

UNDERSTANDING CONCEPTS

1. An electron cloud is a region of space around the nucleus of an atom where an electron is likely to be found.
2. An *s* orbital is a sphere and a *p* orbital has a dumbbell shape.
3. An octet is four pairs of electrons in an outer energy level. An octet of electrons is particularly stable.
4. two electrons
5. There can be five *d* orbitals in an energy level. The maximum number of *d* electrons in an energy level is ten. The third energy level is the lowest energy level that can have *d* orbitals.
6. The first energy level has a maximum of two electrons; the second, eight electrons; the third, 18 electrons; the fourth, 32 electrons.
7. An energy level can have three *p* orbitals. The second energy level is the lowest energy level that can have *p* orbitals.
8. 2*s* means second energy level, *s* sublevel; 4*d* means fourth energy level, *d* sublevel; 3*p* means third energy level, *p* sublevel; 5*f* means fifth energy level, *f* sublevel.
9. **a.** $[Kr]5s^2$ **b.** $[Kr]4d^{10}5s^25p^2$; **c.** $[Ar]4s^23d^6$; **d.** $[Kr]4d^{10}5s^2$
10. The maximum number of 2*s* electrons is two; of 3*p* electrons, six; of 4*d* electrons, ten; of 4*f* electrons, 14.
11. The *d* region is the central portion of the table including Groups 3 through 12, periods 4 to 7. The transition elements occupy this region.

REVIEWING MAIN IDEAS

7.1 Expanding the Theory of the Atom

- Atoms are made up of protons, neutrons, and electrons. The outermost valence electrons are the most important for predicting the properties of an element.
- The position of an element in the periodic table reveals the number of valence electrons the element has.
- Electromagnetic spectra provide information about energy levels and sublevels in an atom.
- Electrons are found only in levels of fixed energy in an atom. They cannot be located between energy levels.
- Energy levels have sublevels, which are partitioned into orbitals. An orbital is a region of space where there is a high probability of finding an electron. An orbital can hold a pair of electrons.

7.2 The Periodic Table and Atomic Structure

- The organization of the periodic table reflects the electron configurations of the elements.
- The active metals occupy the *s* region of the periodic table. Metals, metalloids, and nonmetals fill the *p* region.

- Families of elements have similar electron configurations and the same number of valence electrons.
- Within a period of the periodic table, the number of valence electrons for main group elements increases from one to eight.
- The transition elements, Groups 3 to 12, occupy the *d* region of the periodic table. These elements can have valence electrons in both *s* and *d* orbitals, so they frequently have multiple oxidation numbers.
- The lanthanides and actinides, called the inner transition elements, occupy the *f* region of the periodic table. Their valence electrons are in *s* and *f* orbitals. Inner transition elements exhibit multiple oxidation numbers.

Key Terms

For each of the following terms, write a sentence that shows your understanding of its meaning.

electron configuration
Heisenberg uncertainty principle
inner transition element
orbital
sublevel

● UNDERSTANDING CONCEPTS

1. Explain what an electron cloud is.
2. Describe the shapes of the *s* and *p* orbitals.
3. What is an octet of electrons? What is the significance of an octet?
4. How many electrons fill an orbital?
5. How many *d* orbitals can be in an energy level? What is the maximum number of *d* electrons in an energy level? What is the lowest energy level that can have *d* orbitals?
6. What is the maximum number of electrons in each of the first four energy levels?

7. How many *p* orbitals can an energy level have? What is the lowest energy level that can have *p* orbitals?
8. What does each of the following symbols represent: 2*s*, 4*d*, 3*p*, 5*f*?
9. Use the periodic table and write the electron configurations of the following elements:
   a) Sr
   b) the element with atomic number 50
   c) Fe
   d) Cd
10. What is the maximum number of 2*s*, 3*p*, 4*d*, and 4*f* electrons in any sublevel?

✓ Assessment

Portfolio: Review the portfolio options that are provided throughout the chapter. Encourage students to select one product that demonstrates their best work for the chapter. Have students explain what they have learned and why they chose this example for placement into their portfolios. **P**

Additional portfolio options can be found in the following **Teacher Classroom Resources:**
Tech Prep Applications
Chemistry and Industry
Consumer Chemistry

11. What region of the periodic table is designated as the *d* region? What series of elements occupies this region?

12. Where is the *f* region of the periodic table? Name the two series of elements that occupy the *f* region.

13. Why does the first period contain only two elements?

14. Gallium has the electron configuration $[Ar]3d^{10}4s^24p^1$ and is in Group 13, so gallium has three valence electrons. Why are gallium's 3*d* electrons not considered valence electrons?

15. What is the highest occupied sublevel in the structure of each of the following elements?

   a) H   c) Si
   b) Mg   d) Cl

APPLYING CONCEPTS

Physics Connection

16. What significance did Bohr's model of the atom have in the development of the modern theory of the atom?

Everyday Chemistry

17. Why do the transition metals impart color to gemstones such as emeralds and rubies?

Chemistry and Technology

18. Explain why the development of the scanning probe, the scanning tunneling, and the atomic force microscopes is important to modern chemists.

19. Astronomers gain information about the birth, life, and death of stars by examining the light they give off. How does analysis of starlight reveal what elements are present in the star?

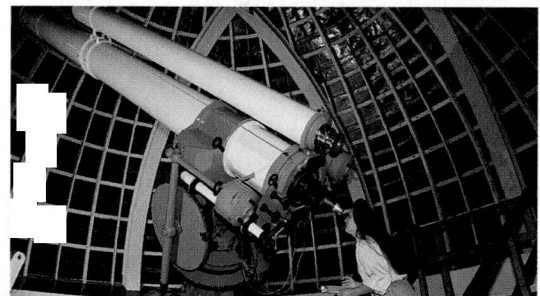

20. Write the electron configurations for each of the following atoms. Use the abbreviated form with the appropriate noble gas inner core.

   a) Be   c) Na   e) K
   b) O   d) P   f) Se

21. Given the colors red, green, and blue, which color has the highest frequency? The longest wavelength? The most energy?

22. How is the light given off by fireworks similar to an element's emission spectrum?

23. What are valence electrons? Where are they located in an atom? Why are they important?

24. Sodium and oxygen combine to form sodium oxide, which has the formula Na_2O. Use the periodic table to predict the formulas of the oxides of potassium, rubidium, and cesium. What periodic property of the elements are you using?

25. Where are the valence electrons represented in this electron configuration of sulfur, $[Ne]3s^23p^4$?

THINKING CRITICALLY

Observing and Inferring

26. **MiniLab 2** Why are electrons best described as electron clouds?

Applying Concepts

27. **ChemLab** In an experiment to find out the reaction capacities of magnesium and aluminum, different amounts of the two metals were used, but the amount of hydrochloric acid was kept the same. Would it be possible to obtain correct results in this way? Explain.

Relating Cause and Effect

28. Explain the following observation. A metal is exposed to infrared light for 3 seconds and no electrons are emitted. When the same metal is exposed to ultraviolet light, thousands of electrons are emitted.

12. The *f* region of the periodic table falls between La and Hf and between Ac and Rf. The lanthanides are elements 58 to 71 and the actinides are elements 90 to 103.

13. The first energy level is filled by two electrons.

14. Gallium's 3*d* electrons are not in the outer energy level.

15. **a.** 1*s*; **b.** 3*s*; **c.** 3*p*; **d.** 3*p*

APPLYING CONCEPTS

16. Bohr's interpretation of the hydrogen emission spectrum showed that electrons in hydrogen are restricted to certain energy values. His ideas led to the development of the present-day model of energy levels, sublevels, and orbitals.

17. The color arises because electrons in transition metals absorb certain frequencies of light, so we see the complementary color.

18. Images on the screen can add to or corroborate evidence gathered through experimentation.

19. Light coming from the stars can be analyzed to reveal the emission spectra of the elements present in the star.

20. **a.** $[He]2s^2$; **b.** $[He]2s^22p^4$; **c.** $[Ne]3s^1$; **d.** $[Ne]3s^23p^3$; **e.** $[Ar]4s^1$; **f.** $[Ar]3d^{10}4s^24p^4$

21. Blue has the highest frequency; red has the longest wavelength; blue has the most energy.

22. Electrons in the transition metal compounds used in fireworks release energy at their characteristic frequencies. This is the same process that occurs in the emission spectrum of an element.

23. Valence electrons are the electrons with the highest energy. They're in the outermost energy level. They are gained, lost, or shared in the formation of chemical bonds.

24. K_2O, Rb_2O, and Cs_2O; These elements are in the

same family as sodium and have the same number of valence electrons.

25. $3s^23p^4$

THINKING CRITICALLY

26. The electrons' positions cannot be known exactly.

27. Yes, the amount of each metal that could react would be determined by the amount of HCl that was used. The same reaction capacities would be obtained.

28. Electrons can absorb only certain frequencies of electromagnetic radiation—the frequencies that give them the exact amount of energy needed to jump to higher energy levels. Ultraviolet light provided enough energy for the electrons to jump high enough to escape the attractive force of the nucleus.

29. 521 nm, green; 461 nm, blue-green; 553 nm, orange; 397 nm, violet; 492 nm, green; 405 nm, violet; 519 nm, green; 589 nm, blue-violet; calcium's line with wavelength 397 nm is the most energetic.

30. The pair of lines results from electrons jumping from sublevels within a given level to a lower level.

31. Fluorine and chlorine have seven electrons in their outer energy levels. Fluorine has the helium core; its outer energy level is the second. Chlorine has the neon core; its outer energy level is the third. Oxygen and fluorine have different numbers of valence electrons. Both elements have the helium core, and the second energy level is their outermost energy level. Chlorine and argon have different numbers of valence electrons. Both elements have the neon core, and the third energy level is their outermost energy level. Argon has an octet in its valence level.

32. Electrons in the metal atoms absorb energy of a specific frequency and jump to a higher energy level. Then they fall back to the lower level and emit the energy as light of a specific frequency (color).

CUMULATIVE REVIEW

33. a. metal, 6; **b.** nonmetal, 7; **c.** metal, 1; **d.** metalloid, 4; **e.** nonmetal, 8

34. Atomic number is the number of protons in the nucleus. Mass number is the total number of protons and neutrons in the nucleus. Because the atomic number is also the number of electrons in the neutral atom, it determines the electron configuration of the atom and the number of valence electrons.

35. a. strontium, 2, 2+; **b.** barium, 2, 2+; **c.** lithium, 1, 1+; **d.** aluminum, 13, 3+; **e.** potassium, 1, 1+

36. The charge on the ion equals the number of valence electrons.

37. Dalton's model of the atom is a hard, homogeneous, indivisible sphere represent-

Using a Table

29. Each element emits light of a specific color in a flame test. The characteristic wavelengths emitted by some elements are shown in the following table. Using the electromagnetic spectrum shown on page 235, state the color associated with the wavelengths. Of the wavelengths listed, which represents the highest-energy light?

Element	Wavelength	Element	Wavelength
Ag	521 nm	Fe	492 nm
Au	461 nm	K	405 nm
Ba	553 nm	Mg	519 nm
Ca	397 nm	Na	589 nm

Interpreting Data

30. The emission spectrum of sodium shows a pair of lines close together near 500 nm. What might be the explanation for these lines?

Comparing and Contrasting

31. Refer to the periodic table and compare the similarities and differences in the electron configurations of the following pairs of elements: F and Cl, O and F, Cl and Ar.

Sequencing Events

32. MiniLab 1
Explain what happens to the electrons of a metal atom when a splint soaked with a chloride of the metal is placed in a flame and a color is produced.

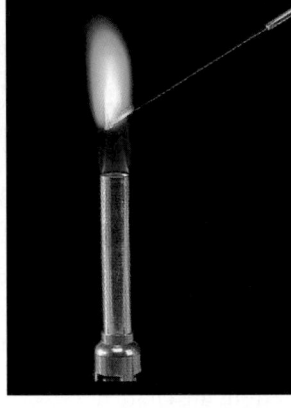

CUMULATIVE REVIEW

33. Classify the following elements as metal, nonmetal, or metalloid. How many valence electrons does each element have? (Chapter 3)

a) chromium   d) germanium
b) fluorine   e) argon
c) silver

34. Distinguish between atomic number and mass number. Which of the two numbers is important in determining the number of valence electrons an atom has? (Chapter 2)

35. Identify the element that is the positive ion in each of the following ionic compounds. What is the group number of the element, and what is the charge on the ion? (Chapter 5)

a) $Sr(NO_3)_2$ d) AlF_3
b) BaO e) K_2SO_4
c) $LiBr$

36. For elements that form positive ions, how does the charge on the ion relate to the number of valence electrons the element has? (Chapter 2)

37. Describe John Dalton's model of the atom, and compare and contrast it with the present-day atomic model. (Chapter 2)

38. Draw Lewis dot diagrams for these elements. (Chapter 2)

a) the element in Group 15, Period 4
b) the element in Group 2, Period 2
c) the element with atomic number 14
d) the halogen in Period 3
e) scandium

39. What is the significance of an octet of electrons in forming both ionic and covalent compounds? (Chapter 4)

40. Write the electron configuration of krypton (atomic number 36), and give the formula of a negative ion and a positive ion that have the same electron configuration. Draw the Lewis dot diagram for the three. (Chapter 4)

ing the smallest unit of matter. The present-day atom is composed of protons and neutrons in a tiny nucleus surrounded by electrons in energy levels outside the nucleus. The atom has no exterior boundary, but is a fuzzy sphere of electron probability.

38. a. $\cdot\ddot{A}s\cdot$ **b.** $\dot{B}e\cdot$ **c.** $\cdot\dot{S}i\cdot$ **d.** $:\ddot{C}l:$ **e.** $\cdot\dot{S}c\cdot$

39. In forming ionic compounds, elements either lose or gain electrons until they have an octet. In forming covalent compounds, elements share electrons to gain an octet.

40. $[Ar]3d^{10}4s^24p^6$; Br^-; Rb^+; $\left[:\ddot{R}b:\right]^+$ $:\ddot{K}r:$ $\left[:\ddot{B}r:\right]^-$

AUTHENTIC ASSESSMENT

1. Energy doubles when frequency doubles; energy is halved when frequency is halved.
2. Answers should be descriptions such as the first floor has only one room and one double bed, so only two electrons could stay there. The hotel will have an unusual shape because the higher levels (floors) have more rooms and bunk beds to accommodate the increasing number of electrons.

Skill Review

1. **Using a Graph** Using the graph of energy versus frequency, answer the following questions. What happens to energy when frequency is doubled? What happens to energy when frequency is halved?

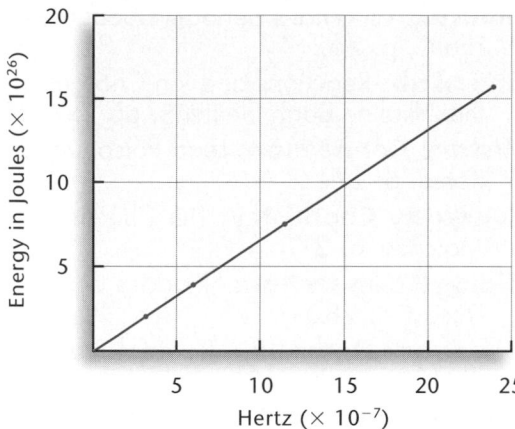

Energy Versus Frequency

Writing in Chemistry

2. You might think about the atom as an electron hotel with many floors. Specifications for the electron hotel are the relationships between energy levels, sublevels, and orbitals. Using the periodic table as a guide, write a description of how you would build an electron hotel. You can include a drawing.

3. The word *laser* is an acronym for *l*ight *a*mplification by *s*timulated *e*mission of *r*adiation. Lasers have a number of applications besides light shows. Find out how lasers are produced, what substances are used in lasers, and the ways laser light is used. Write a paper describing your findings.

Problem Solving

4. Use the periodic table to invent a system of street addresses that would locate every element and specify its electron configuration.

5. Use the diagram of the electromagnetic spectrum below to list the following types of radiation in order of increasing wavelength: microwaves that cook food, ultraviolet radiation from the sun, X rays used by dentists and doctors, the red light in a calculator display, and gamma rays. Which type of radiation has the highest energy? The lowest energy?

Projects

6. Find out about the wave-particle nature of electrons and the experiments that led to an understanding of electrons in atoms. Look for experiments on the photoelectric effect, X-ray diffraction of electrons, and mass spectrometry. Find out how evidence for particles is found in bubble chambers and cloud chambers. Prepare an oral report for your class, or present your findings in the form of a poster.

7. Wear safety goggles and use a sharp knife to cut a large Styrofoam ball in half. From one half of the ball, create a cutaway model of a 1*s* orbital. From the other half, make a cutaway model of an atom, for example, a phosphorus atom. Show the nucleus, the energy levels, and the electrons in the atom.

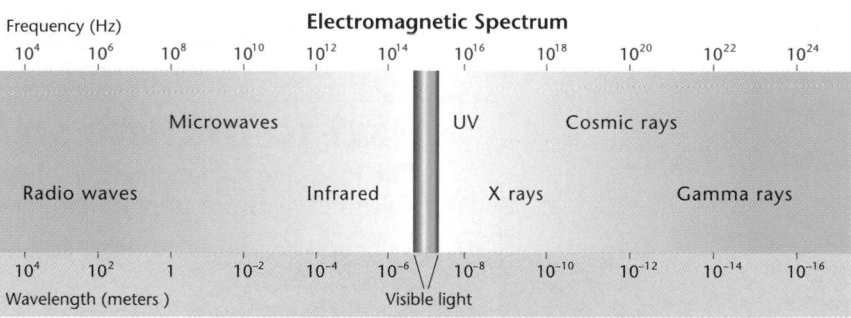

Electromagnetic Spectrum

Program Resources

Chapter Review and Assessment,
   pp. 41-46 [L1]
Alternate Assessment in the Science
   Classroom [L1]
Computer Test Bank, Chapter 7 [L1]
Problems and Solutions Manual,
   Chapter 7 [L1]

Supplemental Practice Problems,
   Chapter 7 [L1]
Performance Assessment in the Science
   Classroom [L1]

---

3. Students' papers will contain a range of information, which should include that laser materials are stimulated to produce intense radiation of a single, pure wavelength. Laser materials may be solids (aluminum oxide, neodymium), liquids (solutions of neodymium oxide or selenium oxychloride), or gases (helium, neon, carbon dioxide). Lasers are used in research, communications, and medicine.

4. Answers should contain the idea that the period number gives the horizontal location and the number of the highest energy level; the group number gives the vertical location and the number of electrons in the highest energy level. Look for interesting ways to specify electron configuration—atomic number or some connection to the preceding noble gas.

5. gamma rays, X rays, ultraviolet radiation, red light, microwaves. Of the types of radiation listed, gamma rays have the highest energy and microwaves the lowest.

6. Student presentations could include a range of information but should include a description of how the matter-energy interaction in the experiments gave information about the particle or wave nature of the electron.

7. Models of a 1*s* orbital should show, by dots or shading, how a spherical electron cloud might look in two dimensions. The model of phosphorus should show two electrons in the first energy level, eight electrons in the second energy level, and five electrons in the third energy level. The levels should be divided into sublevels. The third should be closer to the second than the second is to the first.

Chapter 8 Organizer

Section	Objectives	Activities/Features
8.1 **Main Group Elements** (5 days)	**1. Relate** the position of any main group element in the periodic table to its electron configuration. **2. Predict** chemical behavior of the main group elements. **3. Relate** chemical behavior to electron configuration and atomic size.	**MiniLab 1:** What's periodic about atomic radii?, p. 262 **ChemLab:** Reactions and Ion Charges of the Alkaline Earth Elements, pp. 266-267 **History Connection:** Lead Poisoning in Rome, p. 271 **Everyday Chemistry:** The Chemistry of Matches, p. 275 **Biology Connection:** Fluorides and Tooth Decay, p. 280 **Discovery Demo:** 8.1 Physical Properties of Hydrogen, pp. 256-257 **Demonstrations:** 8.2 A Quick Thaw, pp. 264-265; 8.3 Magnesium, pp. 268-269; 8.4 Group 2 Solubility Trend, pp. 270-271; 8.5 Chlorine, Bromine, and Iodine, pp. 278-279
8.2 **Transition Elements** (3 days)	**4. Relate** the chemical and physical properties of the transition elements to their electron configurations. **5. Predict** the chemical behavior of transition elements from their positions in the periodic table.	**How It Works:** Inert Gases in Lightbulbs, p. 284 **MiniLab 2:** The Ion Charges of a Transition Element, p. 285 **Chemistry and Technology:** Carbon and Alloy Steels, p. 288 **Demonstration:** 8.6 pp. 292-293

Activity Materials

ChemLab (pages 266-267)
96-well microplates (2), microtip beral pipets (4), black paper, toothpicks, marking pen, 0.1M calcium nitrate solution, 0.1M strontium nitrate solution, 0.1M barium nitrate solution, 0.1M sodium oxalate solution

MiniLab 1 (page 262)
96-well microplate, straws, scissors, ruler, graph paper

MiniLab 2 (page 285)
125-mL Erlenmeyer flasks (2), rubber stoppers (2), iron(III) chloride solution, iron(II) chloride solution, 1M sodium hydroxide solution

Demonstrations For a complete list of materials for the demonstrations in this chapter, see pages 256, 264, 268, 270, 278, and 292.

KEY TO TEACHING STRATEGIES

The following designations will help you decide which activities are appropriate for your students.

L1 Level 1 activities should be within the ability range of all students.

L2 Level 2 activities should be within the ability range of the average to above-average student.

L3 Level 3 activities are designed for the ability range of above-average students.

LEP LEP activities should be within the ability range of Limited English Proficiency students.

COOP LEARN Cooperative Learning activities are designed for small group work.

P These strategies represent student products that can be placed into a best-work portfolio.

Periodic Properties of the Elements

Teacher Classroom Resources

STUDENT MASTERS

Study Guide: pp. 31-32
Laboratory Manual: 8.1 Periodicity and Chemical Activity, pp. 69-72
Critical Thinking/Problem Solving: p. 9
ChemLab and MiniLab Worksheets: p. 57

TEACHING AIDS

Section Focus Transparency 16
Section Focus Transparency Master: p. 16
Basic Concepts Transparency 21, 22, 23
Basic Concepts Transparency Masters: pp. 41-46
Lesson Plans: p. 16

Study Guide: pp. 33-34
Laboratory Manual: 8.2 Comparing Activities of Selected Metals, pp. 73-76
ChemLab and MiniLab Worksheets: pp. 53-56, 58

Section Focus Transparency 17
Section Focus Transparency Master: p. 17
Problem Solving Transparency 10
Problem Solving Transparency Master: pp. 19-20
Lesson Plans: p. 17

ASSESSMENT RESOURCES
Chapter Review and Assessment: pp. 47-52
Computer Test Bank: Chapter 8
Performance Assessment in the Science Classroom
Supplemental Practice Problems: Chapter 8

CHAPTER RESOURCES
Spanish Resources: Chapter 8
English/Spanish Audiocassettes: Chapter 8
Cooperative Learning in the Science Classroom
Lab and Safety Skills in the Science Classroom

GLENCOE TECHNOLOGY

Software
Mastering Concepts in Chemistry: *Unit 8* Lesson 1
Videotape
MindJogger Videoquizzes: Chapter 8
Videodisc
Chemistry: Concepts and Applications, Demonstrations, Videos, and Animations
Science and Technology Videodisc Series Disc 2 Chemistry *Hydrogen Sponge; Advanced Composites; High Tech Ceramics*
CD-ROM Multimedia System
Chemistry: Concepts and Applications, Same as for Videodisc, plus Interactive Exploration and Major Simulation, *The Periodic Table*

TECH PREP

The following Glencoe resources provide opportunities for integrating science and technology.

Student Edition: Everyday Chemistry, p. 275; Biology Connection, p. 279; How It Works, p. 284; Chemistry and Technology, p. 288

Teacher Wraparound Edition: Applying Chemistry, pp. 263, 269, 272, 277, 278, 292; Demonstration 8.2, pp. 264-265; Extension, pp. 265-271

Teacher Classroom Resources: Tech Prep Applications, pp. 13-14; Consumer and Industry, pp. 19-22; Consumer Chemistry, pp. 15-16

CHAPTER 8 Periodic Properties of the Elements

Chapter Overview

The periodic nature of atomic and ionic radii is explained in terms of the force of attraction between the nucleus of the atom and its electrons. The chemical and physical properties of the main group elements and the transition elements are related to their electron configuration and atomic radius. Applications and uses of each family of elements illustrate patterns in chemical behavior.

Theme Connection

Macro to Submicro The behavior of submicroscopic particles is responsible for the periodicity in chemical and physical properties at the macroscopic level. These patterns in properties and behaviors are related to atomic size and electron configuration.

Assessment Planner

Choose assessment strategies from the following pages to evaluate the progress of your students.
Assess, pp. 260, 280, 294, 296
MiniLabs, pp. 262, 285
ChemLab, p. 266
Portfolio, pp. 271, 275, 280, 296
Chapter Review, p. 296

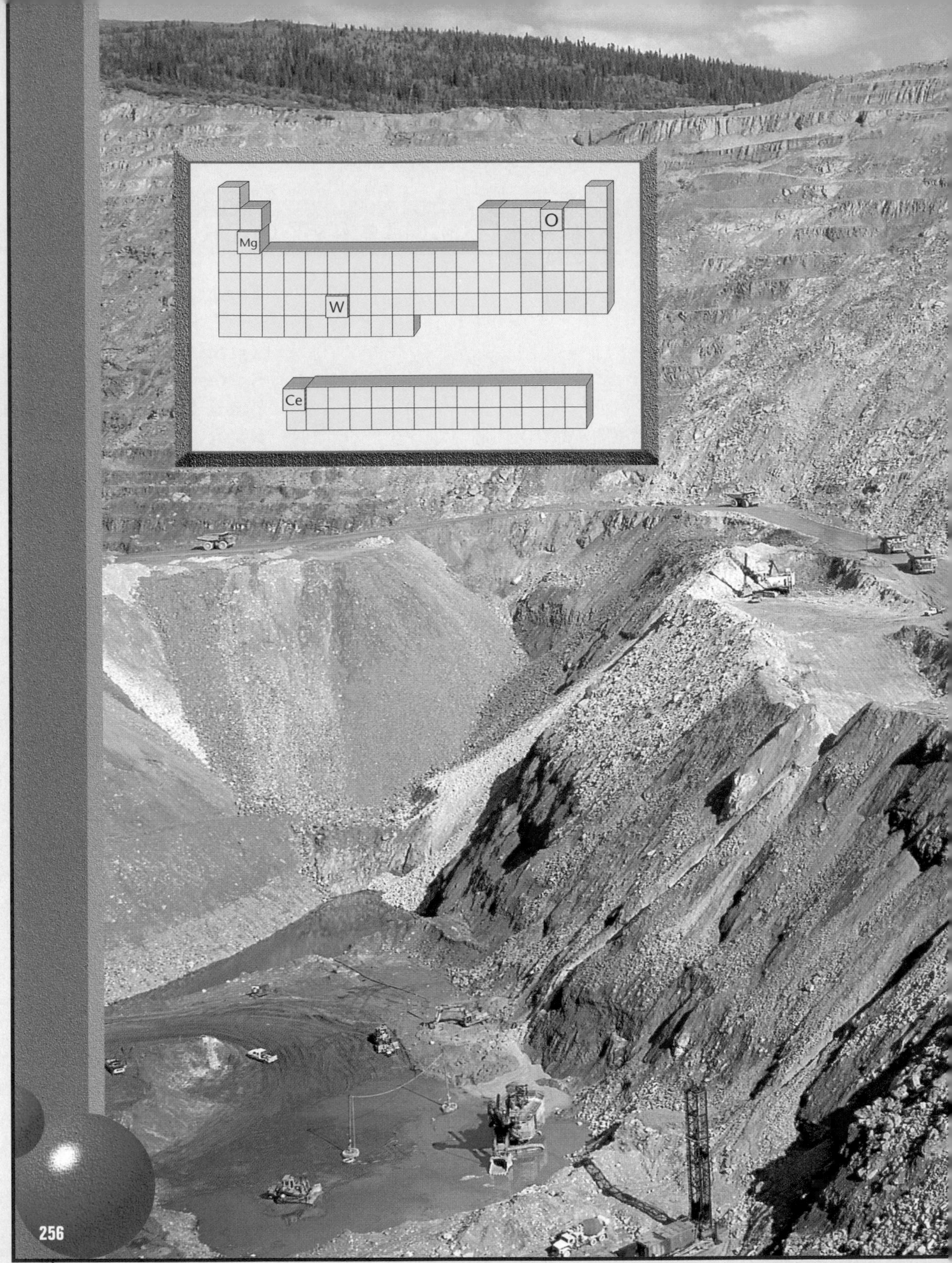

Discovery Demo 8.1

Physical Properties of Hydrogen

Purpose: To demonstrate the unique properties of hydrogen gas.

Materials: lecture cylinder of compressed hydrogen gas, regulator for cylinder, 9-inch balloon, 1-m length of string, a test tube

Safety Precautions:

Gas cylinders should be fastened to a wall or table. Use only the proper valves and regulators on the cylinders.

Disposal: Code F

Procedure: Open the regulator valve on a lecture cylinder of hydrogen gas. Inflate a 9-inch balloon with the hydrogen gas. **CAUTION:** *Hydrogen gas is explosive; keep the balloon away from flames and sparks.* Tie the mouth of the balloon closed. Use a length of string to tie the

CHAPTER 8

Periodic Properties of the Elements

arth's crust is a storehouse for almost all of the elements. Only a few elements are not obtained from the crust but from the atmosphere or from the oceans. Some are found free and uncombined in Earth's crust, but more often the elements are combined with each other in compounds, and the compounds are mixed together in rocks and soil. Over the ages, chemists have found ways to separate and purify each element and study its properties. What has been learned about the elements is shown in broad outline by the organization of the periodic table.

You have learned that an element's chemical and physical properties are closely related to its position in the periodic table and that certain properties repeat periodically. Both position in the periodic table and the properties of the elements arise from the electron configurations of their atoms. Understanding the relationship between electron configuration and position in the periodic table enables you to predict the properties of the elements and the outcome of many chemical reactions.

 Concept Check

Review the following concepts before studying this chapter.
Chapter 3: properties of metals, nonmetals, and metalloids and their positions on the periodic table

 Chapter Preview

8.1 Main Group Elements

8.2 Transition Elements

Laboratory Activities
ChemLab
Reactions and Ion Charges of the Alkaline Earth Elements
MiniLabs
1. What's periodic about atomic radii?
2. The Ion Charges of a Transition Element

Chemistry Around You

Having discovered, purified, and studied the elements, chemists found that they could use the unique properties of each element to create new materials to meet new needs. In the modern world, an awesome array of materials is available to fill a wide variety of purposes.

Oxygen
O
8

Tungsten
W
74

Magnesium
Mg
12

Cerium
Ce
58

257

balloon to the sink faucet on the demonstration table. Fill a test tube with hydrogen gas by the downward displacement of air. Have a student carefully sniff the test tube of hydrogen gas to see whether he or she can detect any odor.
Results: The balloon inflates when filled with hydrogen gas. The balloon rises toward the ceiling and is held down by the string attached

to the faucet. Hydrogen gas is odorless, colorless, and tasteless.
Analysis: Ask these questions.
1. Is hydrogen gas heavier or lighter than air? *lighter than air*
2. How would you describe the physical appearance and smell of hydrogen gas? *Hydrogen gas is odorless, colorless, and tasteless.*

SECTION 8.1

PREPARE

Key Concepts

The properties of the main group elements (Groups 1, 2, and 13 to 18) are discussed in detail and related to electron configuration and atomic size.

1 FOCUS

Focus Transparency

Display the **Section Focus Transparency** for Section 8.1 to introduce some useful compounds of the main group elements. L1 LEP

2 TEACH

Discussion

Explain that nuclear fission explosions in the atmosphere may release significant amounts of fallout containing the radioactive isotope strontium-90, which is then deposited on vegetation and may enter the food chain. Ask students why many scientists are concerned about potentially harmful physiological effects from fallout containing strontium-90. *Because strontium and calcium are in the same chemical group or family, they are chemically similar; therefore, some of the strontium from vegetables and dairy products may be incorporated into our bones, where the radioactivity may cause serious harm.*

SECTION PREVIEW

Objectives

Relate the position of any main group element in the periodic table to its electron configuration.

Predict chemical behavior of the main group elements.

Relate chemical behavior to electron configuration and atomic size.

Key Terms

alkali metal
alkaline earth metal
halogen

Main Group Elements

You have a chemical storehouse in your own home. Under the kitchen sink, you'll probably find dishwasher detergent, steel-wool soap pads, ammonia window cleaner, and a variety of other cleaning products. In your pantry, there could be vinegar, baking powder, and baking soda. Your medicine cabinet probably contains toothpaste, deodorants, and various medications. Read the labels on these products, and you'll find that they contain many simple compounds such as sodium chloride ($NaCl$), or table salt; sodium hydrogen carbonate ($NaHCO_3$), or baking soda; sodium hypochlorite ($NaOCl$), the active agent in bleach; or sodium hydroxide ($NaOH$), which is present in drain cleaners.

Patterns of Behavior of Main Group Elements

Recall from Chapter 7 that elements in the same group (vertical column) of the periodic table have the same number of valence electrons, and because of this, they have similar properties. But elements in a period (horizontal row) have properties different from one another. This is because the number of valence electrons increases from one to eight as you move from left to right in any row of the periodic table except the first. As a result, the character of the elements changes. **Figure 8.1** illustrates the main group elements and shows that each period begins with two or more metallic elements, which are followed by one or two metalloids. The metalloids are followed by nonmetallic elements, and every period ends with a noble gas.

Patterns in Atomic Size

Recall that the size of an atom increases in any group of elements as you go down the column because the valence electrons are found in energy levels farther and farther from the nucleus. But how does atomic radius change across a period from left to right? Take Period 2 as an example. You might expect the size of the second-period atoms to increase across the period from lithium on the left to fluorine on the right because both the atomic number and, therefore, the number of electrons increases. However, the opposite is true. The lithium atom, with only three

Program Resources

Study Guide, pp. 31-32 L1
Critical Thinking/Problem Solving, p. 9 L2
English/Spanish Audiocassettes, Chapter 8 L1 LEP
Laboratory Manual, 8.1, pp. 69-72 L1
Section Focus Transparency 16 and **Master,** p. 16 L1 LEP
Basic Concepts Transparencies 21, 22, 23

and **Masters,** pp. 41-46 L1 LEP
Spanish Resources L1 LEP
ChemLab and MiniLab Worksheets, p. 57 L1
Consumer Chemistry, pp. 15-16 L1
Chemistry and Industry, pp. 19-22 L1
Tech Prep Applications, pp. 13-14 L1

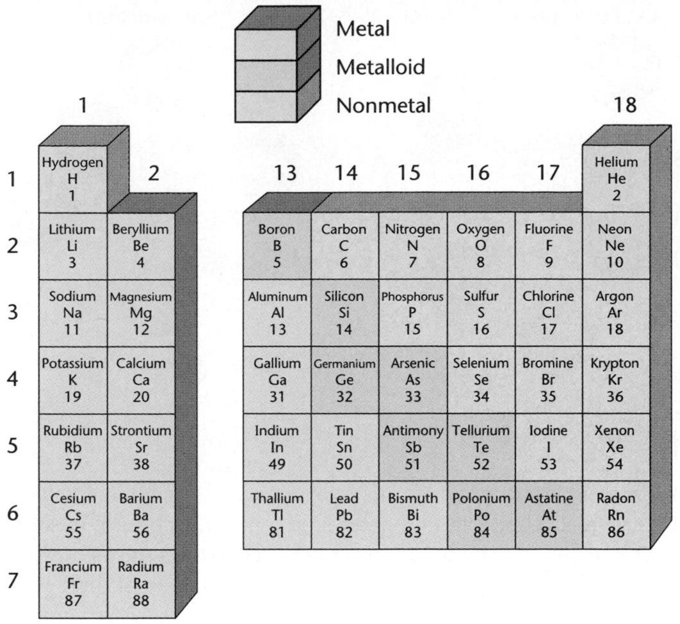

Figure 8.1

Trends in Metallic Properties

The pattern metal-metalloid-nonmetal-noble gas is typical for the main group elements in each period. Period 2 begins with a metal, lithium, and ends with a noble gas, neon. In between are the metal beryllium; the metalloid boron; and the nonmetals carbon, nitrogen, oxygen, and fluorine. Remember that the most active metals, Groups 1 and 2, are in the *s* region of the periodic table. The metalloids, nonmetals, and less active metals are in the *p* region of the periodic table.

Metal / Metalloid / Nonmetal

1	2		13	14	15	16	17	18
Hydrogen H 1								Helium He 2
Lithium Li 3	Beryllium Be 4		Boron B 5	Carbon C 6	Nitrogen N 7	Oxygen O 8	Fluorine F 9	Neon Ne 10
Sodium Na 11	Magnesium Mg 12		Aluminum Al 13	Silicon Si 14	Phosphorus P 15	Sulfur S 16	Chlorine Cl 17	Argon Ar 18
Potassium K 19	Calcium Ca 20		Gallium Ga 31	Germanium Ge 32	Arsenic As 33	Selenium Se 34	Bromine Br 35	Krypton Kr 36
Rubidium Rb 37	Strontium Sr 38		Indium In 49	Tin Sn 50	Antimony Sb 51	Tellurium Te 52	Iodine I 53	Xenon Xe 54
Cesium Cs 55	Barium Ba 56		Thallium Tl 81	Lead Pb 82	Bismuth Bi 83	Polonium Po 84	Astatine At 85	Radon Rn 86
Francium Fr 87	Radium Ra 88							

Figure 8.2: To begin the discussion of atomic size, refer students to this figure and point out how the sizes of the spheres change across each period and down each group. Make a transparency of the figure and display it throughout the discussion. These trends are important for developing an understanding of chemical bonding. **L1** **LEP**

Concept Development

Explain that the valence electrons in metal atoms surround the nuclei and the filled energy levels of the atoms in such a way that each electron is shared by a number of atoms. These delocalized electrons are responsible for characteristic properties of metals such as malleability, ductility, and electrical and thermal conductivity.

Transparency

Display **Basic Concepts Transparency 21** to introduce atomic and ionic radii. **L1** **LEP**

---

electrons, is actually larger than the fluorine atom, which has nine. **Figure 8.2** illustrates the relative sizes of atoms of the main group elements.

There's a simple explanation for the trend of decreasing atomic size across a period. Picture the valence electron on a lithium atom. The lithium nucleus has three protons, so there's an attractive force of 3+ acting on the electron. Lithium's valence electron is in the second energy level, and the attraction to the nucleus isn't too strong. Now, think about the valence electrons in a beryllium atom. Here, there is an attractive force of 4+ from the four protons in the beryllium nucleus. But the outer electrons are still in the second energy level, so the larger attractive force of the beryllium nucleus pulls these electrons a little closer to the nucleus, and the electron cloud gets a little smaller. With each increase in nuclear charge across the period, the outer electrons are attracted more strongly toward the nucleus, resulting in smaller size. In **Figure 8.2**, compare the size of fluorine, with a nuclear charge of 9+, to the size of lithium.

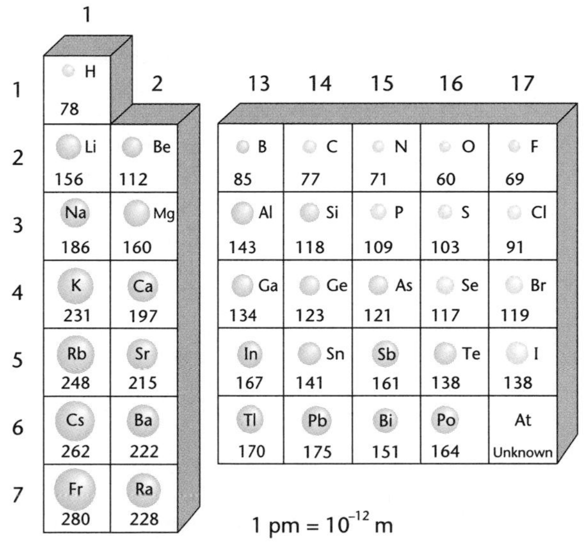

1	2		13	14	15	16	17
H 78							
Li 156	Be 112		B 85	C 77	N 71	O 60	F 69
Na 186	Mg 160		Al 143	Si 118	P 109	S 103	Cl 91
K 231	Ca 197		Ga 134	Ge 123	As 121	Se 117	Br 119
Rb 248	Sr 215		In 167	Sn 141	Sb 161	Te 138	I 138
Cs 262	Ba 222		Tl 170	Pb 175	Bi 151	Po 164	At Unknown
Fr 280	Ra 228						

$1 \text{ pm} = 10^{-12} \text{ m}$

Figure 8.2

Atomic Radii of Main Group Elements

Atomic radius (plural, *radii*) is a measure of the size of an atom. The spheres in the table represent the relative sizes of the atoms as measured by X-ray diffraction studies. The actual atomic radius is given in picometers beneath each sphere. Atomic size is a periodic property of the elements. Can you see the pattern in every period?

Across the Curriculum

History

Rachel Carson, a marine biologist and author, had a monumental impact on public policy in the United States when, in 1962, she published her fifth book, *Silent Spring*. The book warned about the dangers to the environment of the use of pesticides, and it generated a storm of international controversy. Because Carson's evidence was so thoroughly documented, and later substantiated by extensive government studies, DDT and a number of other deadly pesticides were banned in the United States.

Refer to the Chapter 3 MiniLab, *Trends in Reactivity Within Groups*, in which calcium is found to be more reactive than magnesium. Explain that the greater reactivity of calcium is related to its larger atomic radius.

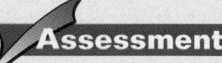

Performance: Give students blank periodic tables and ask them to fill in the Lewis electron dot diagrams for all the elements in Groups 1, 2, and 13 to 18. They will find that this seemingly laborious task is easy when they use the periodic pattern in the numbers of outer-level electrons. L1

Discussion

Ask students why hydrogen is placed in Group 1. *Hydrogen's valence electron configuration is s¹ like the elements in Group 1.* Then ask why hydrogen does not have the characteristics that we normally associate with Group 1 metals. *because it is a gas at normal temperatures and pressures* Explain that hydrogen at very low temperatures and very high pressures does have metallic characteristics. L1

Oral: You may wish to repeat Demonstration 3.1, *Activity of Alkali Metals*, on pages 84 and 85. As an assessment for the Demonstration, show students that alkali metals are soft and can be cut with a knife. Ask why the exposed silvery surface dulls quickly in air. *The metals are highly reactive and react with oxygen in the air.* L1

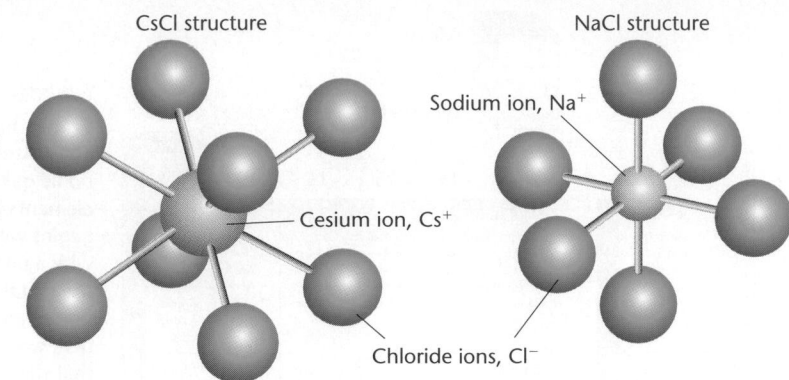

Figure 8.3

Ionic Size and Crystal Structure
The cesium ion is larger than the sodium ion so it's possible for eight chloride ions to fit around a single cesium ion in the CsCl crystal lattice. The smaller sodium ion can accommodate only six chloride ions in the NaCl structure.

Ionic Size

Atomic size is an important factor in the chemical reactivity of an element. Ionic size is also important in determining how ions behave in solution and the structure of solid ionic compounds. **Figure 8.3** shows how the structures of two ionic compounds differ because of the sizes of their positive ions. How does the size of an atom change when it becomes an ion? When metallic atoms lose one or more electrons to become positive ions, they acquire the configuration of the noble gas in the preceding period. This means that the outermost electrons of the ion are in a lower energy level than the valence electrons of the neutral atom. The electrons that are not lost by the atom experience a greater attraction to the nucleus and pull together in a tighter bundle with a smaller radius. The result is that all positive ions have smaller radii than their corresponding atoms. **Figure 8.4** shows a comparison of lithium and sodium with their positive ions.

When an atom gains electrons to become a negative ion, the atom acquires the electron configuration of the noble gas at the end of its period. But the nuclear charge doesn't increase with the number of electrons. In the case of fluorine, a nuclear charge of $9+$ must hold ten electrons in the F^- ion. The result is that all the electrons are held less tightly, and the radius of the ion is larger than the neutral atom. **Figure 8.4** shows how the sizes of the fluoride and chloride ions compare with the fluorine and chlorine atoms.

Figure 8.4

Sizes of Atoms and Their Ions
Lithium and sodium lose the single electron from their outermost energy level. The ions that form are smaller because the remaining electrons are at a lower energy level and are attracted more strongly to the nucleus. Fluorine and chlorine become negative ions by adding an electron. When electrons are added, the charge on the nucleus is not great enough to hold the increased number of electrons as closely as it holds the electrons in the neutral atom.

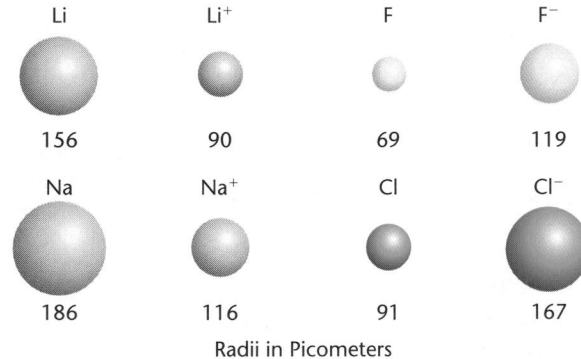

Radii in Picometers

interNET CONNECTION

Software for exploring the periodic table is available for download from the Internet.

Internet
gopher://ucsbuxa.ucsb.edu:3001/11/.Sciences/.Chemistry/.periodic.table

Patterns in Ionic Radii

In **Figure 8.4,** you can see that the sodium ion is larger than the lithium ion. This trend in increasing ionic size continues as you go down the periodic table in Group 1, as shown in **Figure 8.5.** Notice the same trend for the positive ions in Groups 2 and 3 and for the negative ions in Groups 15, 16, and 17. **Figure 8.5** also shows how the sizes of both positive and negative ions change across a period.

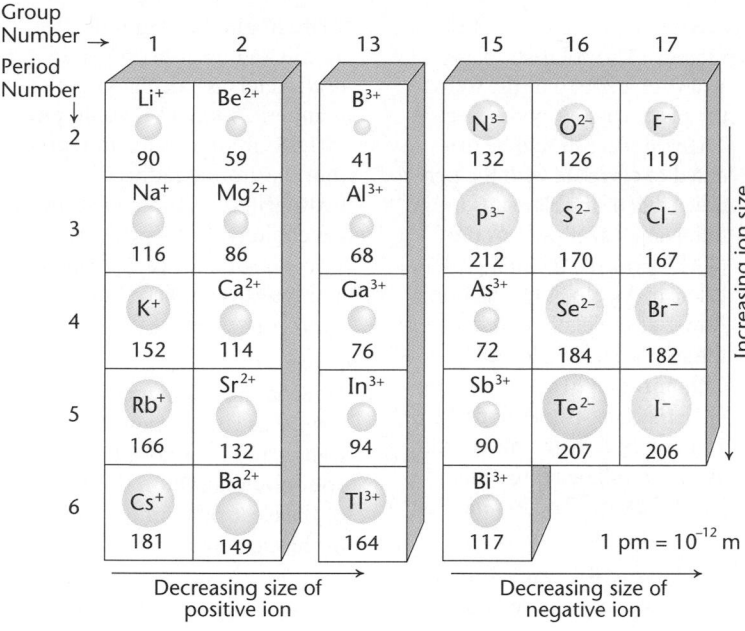

Figure 8.5

Trends in Ionic Radii

Ionic radii increase down the table in any group because of the increasing distance of the outermost electrons from the nuclear charge. Ions of atoms in the same period with 1+, 2+, and 3+ charges (Groups 1, 2, and 13) decrease in size from left to right. Although the ions have the same electron configuration, nuclear charge increases from left to right, resulting in a stronger attraction for electrons and smaller size. Negative ions in the same period with 3−, 2−, and 1− charges (Groups 15, 16, and 17) show the same trend in size. Ionic radii decrease because nuclear charge increases.

Patterns in Chemical Reactivity in Period 2

You've already noticed that the character of the Period 2 elements changes from metal to metalloid to nonmetal to noble gas as you move across the period. How are the electron configurations of these elements related to the tendency of the metals to lose electrons, the nonmetals to share or gain electrons, and the noble gases to be unreactive?

Lithium is the most active metal in the second period because it can attain the noble-gas configuration of helium by losing a single electron. If lithium loses one electron from its 2s sublevel, its electron configuration changes from $1s^2 2s^1$ to $1s^2$. The resulting lithium ion has a 1+ charge and the same electron configuration as a helium atom. While this is not an octet, it is a noble-gas configuration. Elements tend to react in ways that allow them to achieve the configuration of the nearest noble gas.

Beryllium, the next element in the second period, must lose a pair of 2s electrons to acquire the helium configuration. It's harder to lose two electrons than it is to lose one, so beryllium is slightly less reactive than lithium. Nevertheless, beryllium does react by losing both of its 2s electrons and forming a 2+ ion with the helium electron configuration.

8.1 **Main Group Elements**   **261**

Using an Analogy

Just as a building becomes higher as additional floors are added, the atoms of the elements in a group become larger as additional energy levels of electrons are added.

Visual Learning

Figure 8.5: This figure can help students predict oxidation numbers and write correct chemical formulas for many compounds. Emphasize that metallic elements, on the left side of the periodic table, tend to lose their valence electrons to become positive ions having charges equal to the number of their valence electrons. The ion formed has the electron configuration of the noble gas in the preceding period. Nonmetallic elements, on the right side of the table, tend to gain electrons to become negative ions. The number of electrons gained is equal to the number needed to achieve the electron configuration of the noble gas in their periods. Ask students what positive ion would be formed by cesium? Cs^+ What positive ion would be formed by barium? Ba^{2+} What negative ion would be formed by oxygen? O^{2-} What compounds would form between cesium and oxygen, and barium and oxygen? Cs_2O, BaO
L1

Transparency

Display **Basic Concepts Transparency 22** to reinforce the concept of size change in ion formation. L1 LEP

miniLAB

What's periodic about atomic radii?

Purpose: To study the periodic trends in the atomic radii of the first 36 elements (excluding the transition elements).

Process Skills: Measuring, using numbers, observing, interpreting data, communicating

Teaching Strategies

• To limit the time required, teams of two or three students can share the work in this MiniLab. **COOP LEARN**

•  You can save more class time if the calculation of picometers to centimeters is assigned as homework. **L1**

Expected Results

The straw lengths, calculated in centimeters to two significant digits, are as follows: H, 2.0; He, 3.0; Li, 3.9; Be, 2.8; B, 2.1; C, 1.9; N, 1.8; O, 1.5; F, 1.7; Ne, 3.3; Na, 4.6; Mg, 4.0; Al, 3.6; Si, 3.0; P, 2.7; S, 2.6; Cl, 2.3; Ar, 4.8; K, 5.8; Ca, 4.9; Ga, 3.4; Ge, 3.1; As, 3.0; Se, 2.9; Br, 3.0; Kr, 5.0.

Analysis

1. Atomic radii generally decrease from left to right. The outer-level electrons of the elements within a period are in the same energy level. Nuclear charge increases from left to right, and as it does, the electrons are attracted more strongly to the nucleus, thereby decreasing the size of the atom.

2. Atomic radii increase down a group. The elements within a group have the same outer-level electron configurations, but the number of energy levels increases down a group, and the atoms get larger.

---

miniLAB 1

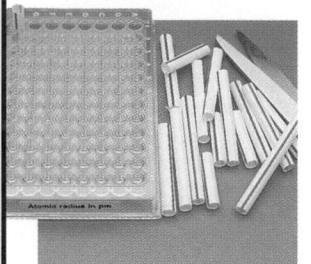

What's periodic about atomic radii?

Atomic radius is approximately the distance from the nucleus of an atom to the outside of the electron cloud where the valence electrons are found. The reactivity of the atom depends on how easily the valence electrons can be removed, and that depends on their distance from the attractive force of the nucleus. In this MiniLab, you will study the periodic trends in the atomic radii of the first 36 main group elements from hydrogen through barium.

Procedure

1. Obtain a 96-well microplate, straws of a size to fit the wells in the plate, scissors, and a ruler. The well plate should be oriented to correlate with the periodic table of the elements in the following way: Row 1 of the plate represents the first period, H1 as hydrogen, A1 as helium; Row 2 of the plate represents the second period, from H2 (lithium) to A2 (neon). Rows 3 to 7 correlate with Periods 3 to 7; however, only the main group elements will be represented. Label the well plate *Atomic radius in pm* (picometers).

2. Use **Figures 8.1** and **8.2** to help you make your model. Look up the atomic radius of each of the elements in Table 4 in Appendix C.

3. Convert atomic radius in picometers to an enlarged scale in centimeters by multiplying the atomic radius in picometers by the conversion factor, 1 cm/40 pm. For example, the atomic radius of hydrogen in centimeters is calculated in this manner: 78 pm × 1 cm/40 pm = 1.95 cm or 2.0 cm. To represent the atomic radius of hydrogen, cut a piece of straw 2.0 cm long. Cut a piece of straw to scale for each element, and insert each piece into the appropriate well of the plate.

Analysis

1. How do the atomic radii change as you go from left to right across a period? Explain your observation on the basis of the electron configurations of the elements.

2. How do atomic radii change as you go from top to bottom within a group or family? Explain your observation on the basis of the electron configurations of the elements.

3. Why is the atomic radius of the elements described as a periodic property?

If this pattern continued, you would expect boron to lose three electrons to attain the helium configuration. Sometimes, boron does react by losing electrons, but often it reacts by sharing electrons. Boron is the only metalloid in the period. That means boron sometimes behaves like a metal and loses electrons like its neighboring metals, lithium and beryllium. When it loses electrons, boron achieves the noble-gas configuration of helium. But more often, boron acts like a nonmetal and shares electrons. Boron is unusual because it has only three electrons to share and cannot acquire an octet of electrons by just sharing. Later, you'll learn more about boron's chemistry.

---

3. With increasing atomic number, atomic radii increase and decrease in a periodic manner.

 **Assessment**

Performance: Have students predict the approximate atomic radius for rubidium. *240-300 pm; actual radius is 248 pm* **L1**

Meeting Individual Needs

Visually Impaired Assign a visually impaired student to work with a sighted student in this MiniLab. Allow the visually impaired student to feel the pattern of atomic radii depicted by the straws. **L1** **LEP**

$$\text{Li} - 1\ e^- = \text{Li}^+$$
$$\text{Li} \cdot - 1\ e^- = \text{Li}^+$$
$$[\text{He}]2s^1 - 1\ e^- = [\text{He}]$$

$$\text{F} + 1\ e^- = \text{F}^-$$
$$:\ddot{\text{F}}\cdot + 1\ e^- = :\ddot{\text{F}}:^-$$
$$[\text{He}]2s^2\,2p^5 + 1\ e^- = [\text{He}]\,2s^2\,2p^6 = [\text{Ne}]$$

$$-1\ e^- = \qquad\qquad +1\ e^- =$$

Figure 8.6

Steps in the Formation of LiF

When an alkali metal such as lithium loses an electron, it attains the configuration of the noble gas in the preceding period. When a nonmetal such as fluorine gains an electron, it acquires the configuration of the noble gas at the end of its period. Ionic compounds such as LiF are combinations of a positive metal ion and a negative nonmetal ion, each having a noble-gas configuration.

Carbon, nitrogen, oxygen, and fluorine are nonmetals. Carbon, with the configuration $[\text{He}]2s^2 2p^2$, and nitrogen, with the configuration $[\text{He}]2s^2 2p^3$, share electrons to attain the noble-gas configuration of neon, $[\text{He}]2s^2 2p^6$. Oxygen, with the configuration $[\text{He}]2s^2 2p^4$, gains two electrons to form the oxide ion, O^{2-}. Fluorine, with the configuration $[\text{He}]2s^2 2p^5$, gains one electron to become the fluoride ion, F^-. **Figure 8.6** gives an example of the loss and gain of electrons by two elements in Period 2 and the ionic compound formed by these elements.

The Main Group Metals and Nonmetals

You can learn a lot about the elements in the products you use every day by studying the chemistry of the main group elements. To do so, recall that elements in a group are chemically similar to the first element in the group.

The Alkali Metals

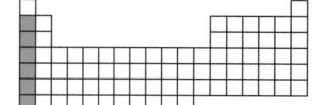

The Group 1 elements—lithium (Li), sodium (Na), potassium (K), rubidium (Rb), cesium (Cs), and francium (Fr)—are called the **alkali metals.** The alkali elements are soft, silvery-white metals and good conductors of heat and electricity. Their chemistry is relatively uncomplicated; they lose their *s* valence electron and form a 1+ ion with the stable electron configuration of the noble gas in the preceding period.

Because all of the alkali metals react by losing their single *s* valence electron, the most reactive alkali metal is the one that has the least attraction for this electron. Remember that the bigger the atom, the farther the valence electron is from the nucleus and the less tightly it's held. In the alkali metal family, francium is the largest atom and probably the most reactive, but francium has not been widely investigated because it is scarce and radioactive. Cesium (Cs) is usually considered the most active alkali metal—in fact, the most active of all the metals. Lithium, the smallest of the alkali metals, is the least reactive element in Group 1.

8.1 **Main Group Elements** **263**

Extension

The energy required to remove the first electron from an atom in the gas phase is called the first ionization energy. Similarly, the energy required to remove the second electron is called the second ionization energy. (Ionization energies are usually expressed in kilojoules per mole of atoms.) Interested students could look up the first ionization energies of the first 36 elements and graph them versus atomic number. Ask students to explain the periodic patterns. **L2**

Applying Chemistry

Ionization in Photocells The ionization energy of cesium is so low that this metal is often used in photocells such as those that control automatic doors. A light beam shines on the cesium, causing it to ionize. When an object or person blocks the light, ionization ceases to send the signal that causes the door to open.

GLENCOE TECHNOLOGY

 CD-Rom

Chemistry: Concepts and Applications
Tracing Periodic Patterns
Have students use this Interactive Exploration to gain practice in obtaining information from the periodic table. **L1** **LEP**

Applying Chemistry

Blue Jeans Require Sodium Sodium metal is used in the manufacture of indigo, the blue dye used to color denim, which is the material used to make blue jeans.

Meeting Individual Needs

Gifted Use Coulomb's law, $F = kq_1q_2/d^2$, to help gifted students understand the factors that affect atomic and ionic radii. These factors are the distance of the outer electrons from the nucleus, the magnitude of the nuclear charge (number of protons), and the repulsive or shielding effect of the inner-level electrons. In the equation, *F* is the force of attraction (unlike) or repulsion (like) between the two charges having magnitudes of q_1 and q_2, *d* is the distance between the charges, and *k* is the Coulomb's law constant. Ask students which has the greater effect on the force of attraction or repulsion: doubling one of the charges or doubling the distance between the charges? *Doubling the distance has the greater effect because it has an inverse squared relationship to the force.* **L3**

Sodium in the Diet Some dietary sodium is required to produce the carbonate buffer compounds Na_2CO_3 and $NaHCO_3$, which regulate blood pH. However, excessive dietary sodium has been linked with a number of health problems including high blood pressure. For this reason, many people must limit their intake of sodium. "Lite" salt (KCl) is often substituted for NaCl as a seasoning.

Extension

Ask students to list all the foods they eat in a typical day and to use information on the food packages or from a book about nutrition to estimate the amount of sodium in each food. Have them compare the amount of sodium they consume with the recommended daily intake of sodium and suggest dietary changes if the amounts differ significantly. L1 LEP

GLENCOE TECHNOLOGY

Videodisc

Chemistry: Concepts and Applications
Properties of Alkali and Alkaline Earth Metals
Disc 1, Side 2, Ch. 7

Show this video to reinforce the properties of elements in Groups 1 and 2. L1 LEP

CD-Rom

Chemistry: Concepts and Applications
Properties of Alkali and Alkaline Earth Metals
Have students use this video to reinforce the properties of alkali and alkaline earth metals. L1 LEP

Reactions and Uses of the Alkali Metals

Because of their chemical reactivity, the alkali metals don't exist as free elements in nature. Sodium, for example, is found mostly combined with chlorine in sodium chloride. Metallic sodium is obtained from NaCl through a process called electrolysis in which an electric current is passed through the molten salt.

1 Spontaneous Reactivity
Oil protects sodium metal from spontaneous reaction with oxygen or moisture in the air. The metal is soft enough to be cut with a knife, and inside you can see the shiny metallic surface. Sodium and the other Group 1 elements are among the most active of all the metals. All Group 1 metals react vigorously with water. When they do, they replace hydrogen and form a hydroxide, as shown in the following equation.

$$2K + 2H_2O \rightarrow H_2 + 2KOH$$

2 Alkali Metals Form Hydroxides
So much heat is generated in the rapid reaction of potassium and water that the hydrogen gas produced in the reaction bursts into flames. The pink color of the water is due to the presence of the indicator phenolphthalein, which turns pink when the solution is alkaline. The pink color of the flame is characteristic of potassium. Potassium hydroxide (KOH) formed in the reaction makes the solution alkaline. Hydroxides are important household and industrial chemicals.

Demonstration 8.2

TECH PREP

A Quick Thaw
Purpose: To demonstrate that soluble ionic compounds are de-icing agents.
Materials: 10-g samples of rock salt (NaCl), table salt, granular $CaCl_2$, and powdered $CaCl_2$; 1 m of cloth ribbon
Safety Precautions:
Disposal: Code C
Procedure: Freeze a large, shallow pan of water overnight in the freezer. Use ribbons to section

the surface of the pan of ice into four areas. Mass 10-g samples of rock salt, table salt, granular $CaCl_2$, and powdered $CaCl_2$. In the upper-left corner, sprinkle rock salt on the surface of the ice. In the lower-left corner, sprinkle granulated table salt on the ice. In the upper-right corner, add large granules of calcium chloride. Add powdered $CaCl_2$ to the lower right.
Results: The powdered $CaCl_2$ melted the most ice in the least amount of time. The powdered

3 **Household, Industrial, and Biological Uses**

Sodium hydroxide is used in the digestion of pulp in the process of making paper (left). It's also used in making soap, in petroleum refining, in the reclaiming of rubber, and in the manufacture of rayon (right). In your household chemical storehouse, you'll find sodium hydroxide (lye) in oven cleaners and in the granular material you use to unclog drains. It is sodium hydroxide's ability to convert fats to soap that makes it effective as a kitchen drain cleaner.

Compounds of sodium and potassium are important to the human body because they supply the positive ions that play a key role in transmitting nerve impulses that control muscle functions. Potassium is also an essential nutrient for plants. It's one of the three major components of fertilizers; the other two are also main group elements—nitrogen and phosphorus.

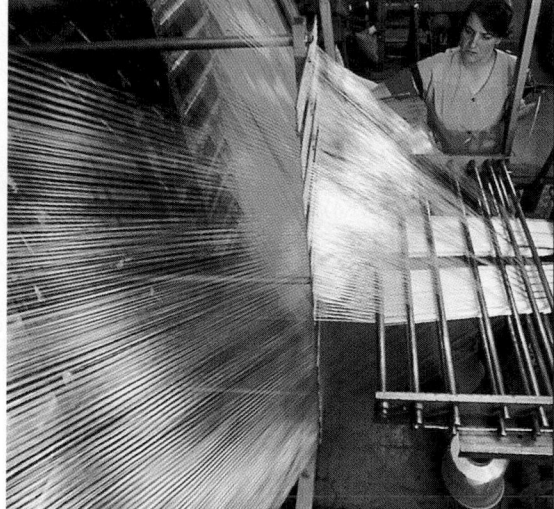

The Alkaline Earth Metals

The Group 2 elements—beryllium (Be), magnesium (Mg), calcium (Ca), strontium (Sr), barium (Ba), and radium (Ra)—are called the **alkaline earth metals.** Their properties are

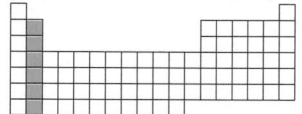

similar to those of the Group 1 elements. Like the alkali metals, they are too reactive to be found as free elements in nature. They lose both of their *s* valence electrons and form 2+ ions with the stable electron configuration of the noble gas in the preceding period. Because the Group 2 elements must lose two electrons rather than one, these metals are less reactive than the Group 1 elements. Each alkaline earth metal is denser and harder and has a higher melting point than the alkali metal that is its neighbor.

The most reactive element in the alkaline earth group is the one with the largest atomic radius and, therefore, the least attraction for its two valence electrons. Knowing this, you can predict that radium, the largest atom in the group, is the most reactive.

The trend to increasing reactivity with increasing size of atom for the alkaline earth metals is illustrated by the reaction of the elements with water, as shown in **Figure 8.7**. Beryllium does not react with water. Magnesium reacts with hot water. But calcium reacts with water to form calcium hydroxide [$Ca(OH)_2$], as shown by this equation.

$$Ca + 2H_2O \rightarrow H_2 + Ca(OH)_2$$

Figure 8.7

The Trend in Reactivity of the Alkaline Earth Metals
No reaction is visible when beryllium is placed in water, but bubbles of hydrogen are produced by the reaction of calcium with water. Strontium, barium, and radium react with water with increasing vigor.

SECTION 8.1

Discussion

People who exercise heavily and lose significant amounts of body fluids through perspiration are sometimes advised to drink fluids or to eat foods such as bananas that replenish electrolytes. Ask students to speculate about the identity of the electrolytes. *KCl and NaCl primarily* L1

Tying to Previous Knowledge

Ask students why compounds such as KCl and NaCl are called electrolytes. *because their aqueous solutions contain ions that conduct electricity*

***GLENCOE* TECHNOLOGY**

 Software

Mastering Concepts in Chemistry
Unit 8, Lesson 1
Periodic Properties
Use this lesson to reinforce the concept of periodicity in the properties of the elements when arranged by atomic number in the periodic table.

L1 LEP

---

form dissolved faster than the other forms of salt.

Analysis: Ask these questions.

1. Which chemical, and in which form, produced the best de-icing action? *powdered $CaCl_2$*

2. Use a chemical catalog to compare the costs of NaCl and $CaCl_2$. Which one do you think would be the most economical de-icing agent to buy? *Answers and*

costs *will vary with supplier. Typical prices for 500 g are the following: powdered $CaCl_2$, $18; flake $CaCl_2$, $8; rock salt, $3; NaCl crystals, $4.*

 Assessment

Knowledge: Ask students to compare the cost per gram of the same item in a grocery store when it is packaged in different

amounts and forms. An example might be sliced cheese compared with bulk cheese, or a large box of cereal compared with a smaller box. Point out that scientific knowledge can be used to make better consumer choices. L1 LEP

ChemLab

Reactions and Ion Charges of the Alkaline Earth Elements

Time Allotment: One 40- to 60-minute laboratory period.

Objectives
Review objectives with students before they begin the ChemLab.

Process Skills
Observing, inferring, predicting, communicating

Safety Precautions and Disposal
Advise students to wear safety goggles and aprons and not to touch or ingest the solutions because they are toxic and may irritate skin. Advise students to keep their goggles on and to be especially careful when rinsing the well plates because water may splash from the wells.
Disposal: Code J

PREPARATION
To make the $0.100M$ solutions, use the following masses with sufficient distilled water to make a total volume of 100 mL of each solution: 23.6 g of $Ca(NO_3)_2 \cdot 4H_2O$, 21.2 g of $Sr(NO_3)_2$, 26.1 g of $Ba(NO_3)_2$, 13.4 g of $Na_2C_2O_4$.

Alternative Materials
You may substitute barium chloride dihydrate for barium nitrate, calcium chloride dihydrate for calcium nitrate tetrahydrate, strontium chloride hexahydrate for strontium nitrate, and ammonium oxalate monohydrate for sodium oxalate.

ChemLab

Reactions and Ion Charges of the Alkaline Earth Elements

The alkaline earth elements, Group 2 on the periodic table, are beryllium (Be), magnesium (Mg), calcium (Ca), strontium (Sr), barium (Ba), and radium (Ra). The positive ions of most of these elements react with negative oxalate ions $(C_2O_4^{2-})$ to form compounds that are insoluble in water. In this ChemLab, you will study the reactions of calcium, strontium, and barium with oxalate ions and determine the formulas for the insoluble products.

Problem
In what ratios do the positive ions of calcium, strontium, and barium react with negative oxalate ions to produce compounds?

Objectives
- **Observe** the reactions of the calcium, strontium, and barium ions with oxalate ions.
- **Determine** the formulas for the insoluble products and the charges on the ions of the alkaline earth elements.
- **Relate** the ion charges of the alkaline earth elements to their electron configurations.

PREPARATION

Materials
96-well microplates (3)
microtip pipets (4)
black paper
toothpicks (3)
marking pen
$0.1M$ calcium nitrate solution
$0.1M$ strontium nitrate solution
$0.1M$ barium nitrate solution
$0.1M$ sodium oxalate solution

Safety Precautions

Wear an apron and goggles. The solutions are toxic if ingested and may irritate the skin. Do not touch or ingest the solutions. Dispose of the products as instructed by your teacher.

PROCEDURE

1. Obtain three 96-well microplates and four labeled micropipets containing the four solutions: calcium nitrate solution, strontium nitrate solution, barium nitrate solution, and sodium oxalate solution. Label the microplates *calcium, strontium,* and *barium.*

2. Lay the calcium microplate near the edge of the lab bench with row H aligned with the edge of the bench. You will use only wells H1 through H9. You will see the product best if the microplate rests on a piece of black paper.

3. Add one drop of the calcium solution to well H1, two drops to well H2, and three drops to well H3. Continue in this way until you get to well H9, which receives nine drops.

4. Add one drop of the sodium oxalate solution to well H9, two drops to well H8, and three drops to well H7. Continue in this way until you get to H1, which receives nine drops.

5. Use a toothpick to stir the mixtures.

6. It may take several minutes for the reactions to be complete. When the insoluble product has settled in the bottoms of the wells, stoop so that your eyes are level with the microplate.

7. Determine and record the identity of the well (or wells) that contains the greatest depth, and therefore the greatest amount, of the insoluble product. Record your observations in a data table like the one shown.

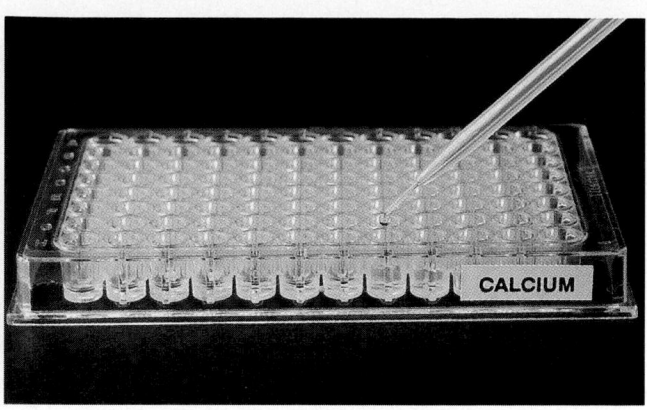

CALCIUM

8. Repeat procedures 2 through 7 with the strontium and oxalate solutions.

9. Repeat procedures 2 through 7 with the barium and oxalate solutions.

10. Dispose of the substances in all three well plates as directed by your teacher. Rinse the well plates with tap water and then with distilled water.

ANALYZE AND CONCLUDE

1. **Interpreting Data** What ratio of drops of the reactant solutions produced the maximum amount of insoluble product for each of the three reactions you performed?

DATA AND OBSERVATIONS

Sodium oxalate combined with	Number of well with maximum precipitate	Drops of oxalate solution	Drops of Group 2 ion solution
Calcium nitrate			
Strontium nitrate			
Barium nitrate			

2. **Interpreting Data** All the solutions you used were 0.1 molar, which means that they contained equal numbers of ions per drop. For example, if the maximum amount of product was produced from two drops of calcium solution and eight drops of oxalate solution, the insoluble product contained one calcium ion for each four oxalate ions. The formula for such a compound would be $Ca(C_2O_4)_4$. Interpret your results to determine the formulas for calcium oxalate, strontium oxalate, and barium oxalate.

3. **Drawing Conclusions** The oxalate ion has a charge of 2− and it combines with the positive alkaline earth ions in such a way that neutral compounds result. What do you think are the ion charges of calcium, strontium, and barium ions? Explain how you can infer these ion charges from your experimental results.

APPLY AND ASSESS

1. Write the electron configurations of calcium, strontium, and barium atoms. Relate the charges on calcium, strontium, and barium ions to the electron configurations of their respective atoms.

2. Use the same reasoning that you used to answer question 1 to predict the ion charges of potassium in Group 1 and gallium in Group 13. Explain your reasoning.

PROCEDURE
- Ask students, before the ChemLab, to make outlines or flowcharts of the procedure and prepare tables in their lab notebooks for recording data and observations. L1
- To thoroughly remove the precipitates from the well plates, students may have to soak them in hot, soapy water and use cotton swabs to scrub the wells.

DATA AND OBSERVATIONS
The greatest depth of insoluble product should occur in well 5 for all three reactions.

ANALYZE AND CONCLUDE
1. calcium and oxalate solutions, 5:5 or 1:1; strontium and oxalate solutions, 5:5 or 1:1; barium and oxalate solutions, 5:5 or 1:1
2. CaC_2O_4, SrC_2O_4, BaC_2O_4
3. Ca^{2+}, Sr^{2+}, Ba^{2+}

APPLY AND ASSESS
1. Calcium, $[Ar]4s^2$; strontium, $[Kr]5s^2$; barium, $[Xe]6s^2$; the atom of each alkaline earth element loses the two electrons in the outermost energy level to form the 2+ ion.
2. Potassium, $[Ar]4s^1$; the potassium atom would lose the single 4s electron to produce the 1+ ion. Gallium, $[Ar]4s^23d^{10}4p^1$; the gallium atom would probably lose the three electrons in the fourth energy level to produce the 3+ ion.

✓Assessment

Performance: Students could perform the same procedure with $0.100M$ solutions of aluminum nitrate and sodium oxalate, using the results to determine the formula for aluminum oxalate and the charge on an aluminum ion. L1

Is it alkaline?

Add a few drops of phenolph-thalein indicator solution to water, then drop in a small piece of calcium. The alkaline or basic nature of the result-ing solution is shown by the change of the colorless phe-nolphthalein to pink (ma-genta). Ask students what other metals would produce a pink color if they were added to water containing phe-nolphthalein. *All the Group 1 elements and the Group 2 ele-ments except beryllium.*
Disposal: Code C

Extension

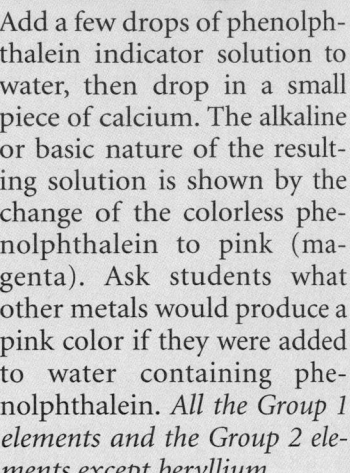

TECH PREP The presence of Ca^{2+} and Mg^{2+} in water is a major cause of hard-ness. Hard water re-sults in the formation of deposits of insoluble com-pounds in pipes and plumbing, and it cuts down on the effec-tiveness of soap in cleaning. Have interested students find out what can be done to elimi-nate the inconvenience of hard water. **L2**

Reactions and Uses of the Alkaline Earth Metals

Beryllium has become a strategically important metal in the nuclear and weapons industries. Magnesium and beryllium are valued for their individual properties, but they are especially important when alloyed with other metals.

❶ Important Properties of Magnesium
Alloys of magnesium are used where light weight and strength are important, as in this jet engine. Magnesium resists corrosion because it reacts with oxygen in the air to form a coating of magnesium oxide. The coating of MgO protects the metal under-neath from further reaction with oxygen.

❷ Reactions of Magnesium and Calcium
Magnesium oxide is also formed when magnesium is heated in air. It burns vigorously, producing a brilliant white light and magnesium oxide. In the process, magnesium loses two electrons to form the Mg^{2+} ion, and oxygen gains two electrons to form the O^{2-} ion. Together, they form the ionic com-pound MgO. The following equation shows what happens.

$$2Mg + O_2 \rightarrow 2MgO$$

Magnesium and calcium are essential elements for humans and plants. Plants need magnesium for photosynthesis because a magnesium atom is located at the center of every chlorophyll molecule.

Calcium ions are essential in your diet. They maintain heartbeat rate and help blood to clot. But the largest amount of dietary calcium ions is used to form and maintain bones and teeth. Bone is composed of protein fibers, water, and minerals, the most important of which is hydroxyapatite, $Ca_5(PO_4)_3OH$, a compound of calcium, phosphorus, oxygen, and hydrogen—all main group elements.

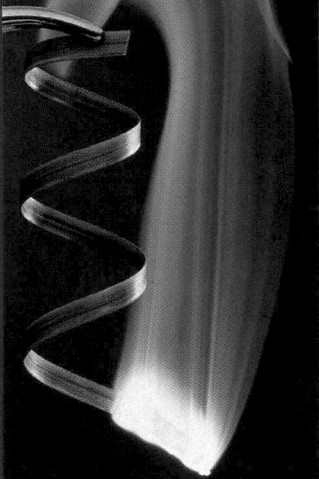

❸ Strontium Reveals Its Presence
Strontium is a less well-known element of Group 2, but it's important, nevertheless. Because of its chemical similarity to calcium, strontium can replace calcium in the hydroxyapatite of bones and form $Sr_5(PO_4)_3OH$. This could be a problem only if the strontium atoms are the radioactive isotope strontium-90, which is hazardous if it is incorporated into a person's bones.

Strontium makes its presence known by the brilliant red color of a fireworks display. The red color also identifies strontium in laboratory flame tests.

268   **Chapter 8   Periodic Properties of the Elements**

Demonstration 8.3

Magnesium
Purpose: To demonstrate the reactivity of an alkaline earth metal.
Materials: crucible tongs, 5-cm length of magnesium ribbon, laboratory burner, empty metal can

Safety Precautions: 🧤 🥽 👢
Caution students not to look directly at burning magnesium.

Disposal: Code A
Procedure: You can safely burn a short piece of magnesium ribbon for the class to see if you darken the room and use tongs to hold the burning ribbon inside a metal can or sink. The reflected light will not injure students' eyes. **CAUTION:** *Don't look directly at the burning magne-sium; doing so causes serious retinal burns.* Magnesium is used in fireworks, and in

parachute flares to illuminate large areas at night. These flares are also used to con-fuse heat-seeking antiaircraft missiles.
Results: The magnesium ribbon burns with a white-hot flame. The darkened room is illuminated by the reflected light.
Analysis: Ask these questions.
1. Magnesium is a member of which group or family? *Group 2, alkaline earth family*

Group 13 Elements

Only boron, the first element in Group 13, is a metalloid. The other Group 13 elements—aluminum (Al), gallium (Ga), indium (In), and thallium (Tl)—are metals. None of the metals are as active as the metals in Groups 1 and 2, but they're good conductors of heat and electricity. They are silvery in appearance and fairly soft. Group 13 metals tend to share electrons rather than form ionic compounds; in this respect, they resemble boron. Their valence configuration is s^2p^1, and they exhibit the 3+ oxidation number in most of their compounds.

Aluminum is the most abundant metallic element in Earth's crust and has so many desirable properties that it's becoming one of the world's most widely used metals. Aluminum's many uses result from its properties—low density, good electrical and thermal conductivity, malleability, ductility, and resistance to corrosion. Aluminum, like magnesium, is a self-protecting metal. On exposure to oxygen in the air, a protective layer of aluminum oxide (Al_2O_3) forms over the surface of the aluminum, preventing further reaction with oxygen.

$$4Al + 3O_2 \rightarrow 2Al_2O_3$$

Aluminum is obtained from its ore through a process that consumes 4.5 percent of the electricity produced in the United States. Recycling aluminum reduces costs by lowering the need for power.

The Uses of Group 13 Elements

Boron is a metalloid found in boric acid (H_3BO_3) and borax ($Na_2B_4O_7 \cdot 10H_2O$). Boric acid is one of the active ingredients in eyewash or contact lens-cleaning solution. Borax is the abrasive in some tough cleansing powders. It's also used as a water softener and is an important component in some types of glass.

1 The Importance of Aluminum
Think about how important aluminum is in your life. Aluminum foil and aluminum soda cans are everywhere. The antacid in your medicine cabinet may contain aluminum hydroxide, $Al(OH)_3$. Your antiperspirant or deodorant may contain an aluminum zirconium hydroxide or aluminum chlorohydrate.

Because aluminum is neither as hard nor as strong as steel, it is often alloyed with other metals to make structural materials. Aluminum alloys are used in automobile engines, airplanes, and truck bodies where high strength and light weight are important. The alloys used in the structure of airplanes are ten percent to 30 percent magnesium; the rest is aluminum. To save weight, some automobile engines use a magnesium alloy that is five percent to ten percent aluminum; the rest is magnesium.

Recycling aluminum from soda cans

269

Nearly 100 years ago, Charles Hall and Paul Héroult, working independently, both discovered the process that is used today to extract aluminum from its ore. Both men were born in 1863, and both developed the process to manufacture aluminum in 1886. Both died in 1914 at the age of 51. Prior to the development of what is now known as the Hall-Héroult process, aluminum metal was more expensive than gold, platinum, or silver because of the difficulty of obtaining it from its ore.

Applying Chemistry

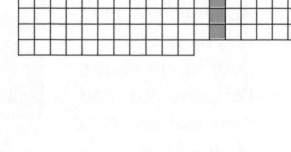

How Drain Cleaners Work
A kitchen drain usually becomes clogged because of accumulations of grease—a variety of fats. The sodium hydroxide in drain cleaners reacts with water in the drain and generates heat. The heat melts the grease, and the sodium hydroxide reacts with some of the fats and converts it to soap. The soap helps dissolve the rest of the grease. Some granular drain cleaners contain pieces of aluminum. These also react with NaOH to produce hydrogen gas that bubbles out of the clogged area and creates a stirring action that loosens the grease. Oven cleaners work according to the same principle, except that there's no hydrogen to stir things up.

Content Background

The Hall-Héroult Process Aluminum was recognized as an element well before it was possible to reduce it from its ore. It was not until 1854 that pure aluminum was produced by reducing it with sodium, a process so expensive that objects made of pure aluminum made similar artifacts made of gold and silver second best. However, in 1886, Charles Hall and Paul Héroult developed an economical electrolytic method for extracting aluminum from its ore. But the production of aluminum depends upon electricity. Ask students to predict what might happen to the cost of aluminum if there were a shortage of fuels for generating electricity. *The price would increase.* [L1]

2. Magnesium combines with oxygen to form magnesium oxide. Write and balance the chemical equation for this reaction. *$2Mg + O_2 \rightarrow 2MgO$*

3. Is the reaction of Mg and O_2 endothermic or exothermic? *exothermic*

4. Is the product of the reaction, magnesium oxide, more or less stable than the reactants, oxygen and magnesium metal? *more stable*

✓ Assessment

Oral: Ask students to contribute to a detailed list of observations of this demonstration. *Answers will vary but should include a description of the reactant, magnesium metal, and the product, MgO. The energy released in the form of heat and light should also be listed.* [L1]

Discussion

Ask students to speculate as to why a large part of chemistry, organic chemistry, is devoted to a single element, carbon. *Carbon atoms have four valence electrons in the second energy level, making them the best atoms at forming covalent bonds by sharing electrons. This characteristic accounts for the large number of carbon compounds.* L1

Discussion

Ask students if they have seen the James Bond movie *Diamonds Are Forever*. Explain that diamonds are formed at very high temperatures and pressures, either in volcanoes or by a synthetic process. At normal temperatures and pressures, graphite is the more stable form of carbon. Diamonds are, therefore, changing to graphite *very slowly*. Suggest that a chemically more accurate title for the movie might be *Graphite Is Forever*.

Transparency

Display **Basic Concepts Transparency 23** to focus on groups of elements. L1 LEP

2 Aluminum in Your Home

At home, you may find bicycles, outdoor furniture, ladders, and pots and pans that are made of aluminum or an aluminum alloy.

3 Aluminum as a Conductor

Even though aluminum doesn't conduct electricity as well as copper, it costs less to use aluminum than copper for transmission of electricity. Aluminum cables are much lighter than copper cables, so fewer support towers are needed to hold the miles and miles of cable that span the country. Fewer support towers means lower cost for all consumers of electricity.

4 Gallium's Low Melting Point

Gallium, indium, and thallium react much like aluminum. But gallium, shown here, has an unusually low melting point, 29.8°C. The heat of a hand is sufficient to liquefy the metal. For comparison, aluminum melts at 660°C.

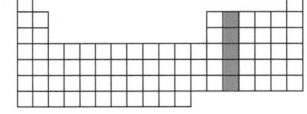

Group 14

The Group 14 elements—carbon (C), silicon (Si), germanium (Ge), tin (Sn), and lead (Pb)—exhibit a variety of properties. Carbon is a nonmetal, silicon and germanium are metalloids, and tin and lead are metals. Because the valence electron configuration for these elements is s^2p^2, a gain or loss of four electrons results in a noble-gas configuration. However, it's unusual for any element to gain or lose four electrons. Instead of gaining electrons to attain a noble-gas configuration, carbon, silicon, and germanium react by sharing electrons. But tin and lead, like the metals in the preceding groups, react by losing electrons. These larger elements at the bottom of the group lose electrons more easily than the smaller nonmetals at the top of the group because of the size of their atoms and the reduced attraction the nucleus has for the outermost electrons. As a group, the most common oxidation number is 4+.

Demonstration 8.4

Group 2 Solubility Trend

Purpose: To show the trend in solubility for Group 2 elements.

Materials: 1 g each of $MgSO_4$, $CaSO_4$, and $BaSO_4$; three 100-mL graduated cylinders with stoppers; 150 mL distilled water

Safety Precautions: 🚫 💪 🥽

Disposal: Code E

Procedure: Place 1 g of $MgSO_4$, $CaSO_4$, and $BaSO_4$ in three separate 100-mL gradu-

ated cylinders. The amount of each solid in the cylinders should be noted by the students before water is added. Add 50 mL of distilled water to each cylinder, stopper, and shake them. Allow the solutions to stand for a few minutes. Then have students observe each cylinder.

Results: $MgSO_4$ dissolved completely. $CaSO_4$ dissolved slightly, and $BaSO_4$ did not dissolve at all.

Analysis: Ask these questions.

1. Which metal sulfate is the most soluble, and which is the least soluble? *$MgSO_4$ is the most soluble; $BaSO_4$ is the least soluble.*

2. The relative solubilities of these three salts are $MgSO_4$, 1 g in 0.8 mL of water; $CaSO_4$, 1 g in 375 mL of water; $BaSO_4$, 1 g in 400 000 mL of water. From the trend noted, predict the solubility of the sulfates of the first five elements of

HISTORY ⬦ CONNECTION

Lead Poisoning in Rome

Could lead poisoning be partially responsible for the fall of the Roman Empire? Some researchers think so; they found that lead poisoning occurred in early history.

How lead poisoning happens Lead gets into the body when materials containing lead compounds are ingested with food and drink, or when lead-containing dust in the air is inhaled or absorbed through the skin. As lead intake increases, the body's ability to get rid of it decreases. Over a period of time, the accumulation of lead in the liver, kidneys, bones, and other body tissues becomes critical. Symptoms of abdominal pains, anemia, lethargy, and nerve paralysis of hands and feet develop.

An old and versatile metal Lead was highly valued in ancient times. The Egyptians refined and used lead as early as 3000 B.C. Deposits of galena, PbS, located near Athens, Greece, were processed and used by the Greeks in the sixth century B.C. But it was the Romans, in the first century B.C., who realized the full potential of lead. They processed it for a wide variety of purposes. Rome's famous system for supplying water to the populace was built with lead pipes. Beer and wine were stored in lead-glazed pottery and served in lead goblets. Cooking pots were made of lead.

Lead was everywhere The lead water pipes of the Roman plumbing system allowed lead to dissolve in the drinking water. Many types of food and drink were sweetened by a thick, sugary syrup called sapa. Sapa was made by boiling wine in a lead pot until much of the water and alcohol had evaporated. What remained was a tasty but poisonous confection. A small portion of sapa was lead(II) acetate, also known as sugar of lead. It was

impossible for anyone living in ancient Rome to avoid ingesting lead. Because lead poisoning causes listlessness and mental failure, some researchers think that it contributed to the breakdown of the Roman ruling class and hastened the fall of the Empire.

Lead emissions When Rome was at its peak in lead production, it produced 80 000 metric tons every year. In studies of changes in atmospheric composition throughout history, researchers measured residues of lead from ancient Rome and Greece found in British peat bogs and in Swedish lake sediments. Lead emissions from Roman smelters at the height of Roman power were nearly as great as they were during the years of the Industrial Revolution in England from 1760 to 1840.

Connecting to Chemistry

1. **Acquiring Information** Find out how lead can get into water supplies today.

2. **Comparing and Contrasting** Use the library to find out about the problems the United States has with lead pollution and lead poisoning.

3. **Interpreting** The formula for the acetate ion is $C_2H_3O_2{}^-$. What is the formula for lead(II) acetate?

---

the alkaline earth family by listing the elements in order of solubility with the most soluble first. *Be, Mg, Ca, Sr, Ba*

3. Even though barium compounds are poisonous, they are used in medicine to make clear X-ray photographs of the digestive tract. Why isn't the patient poisoned? *Barium compounds are insoluble in water and digestive juices. They are not absorbed by the body.*

✓ Assessment

Portfolio: Have students wrtie a summary of this demonstration and add it to their portfolios. L1 P

Applying Chemistry

Alumina The basic component of several gemstones, including rubies and sapphires, is aluminum oxide, Al_2O_3. Aluminum oxide, often called alumina, is used in many ceramic products, such as insulators for spark plugs, because it has a high melting point and does not conduct electricity.

GLENCOE TECHNOLOGY

 Videodisc

Science and Technology Videodisc Series
High-Tech Ceramics
Disc 2: Chemistry
Side 1, Ch. 13

Show this video to illustrate how new ceramic materials are being developed as substitutes for metals in automobile engines. **L1** **LEP**

The Uses of Group 14 Elements

You'll find Group 14 elements in many of the products in your household chemical storehouse—carbon in charcoal briquettes, lead pencils, diamond jewelry, and almost every food item in your home. You'll learn more about carbon and carbon compounds in Chapters 9 and 18.

1 From Sand to Many Uses
Silicon, like boron, is a metalloid. It occurs in sand as silicon dioxide, SiO_2—sometimes called silica. About 59 percent of Earth's crust is made up of silica. In its elemental form, silicon is a hard, gray solid with a relatively high melting point, 1410°C.

Silicon is in window glass and in the chips that run computers. Compounds of silicon are found in lubricants, caulking, and sealants.

2 Special Glasses from Silicon
Silicon is important in semiconductors. It's also important in making alloys and in ceramics, glass, and cement. The glass-ceramic shown here doesn't expand when heated so it won't break when exposed to large temperature changes.

3 "Tin" Cans and Alloys
Tin (Sn) is best known for its use as a protective coating for steel cans used for food storage. The coating protects the steel from corrosion. Tin is also a principal component in the alloys bronze, solder, and pewter. Tin is a soft metal that can be rolled into thin sheets of foil.

272   Chapter 8   Periodic Properties of the Elements

Cultural Diversity

Science Is Women's Work

Women from all cultures have made significant contributions to science, but their contributions have often gone unnoticed. Marguerite Perey discovered the radioactive element francium in 1939. Marie Curie was the first scientist to receive two Nobel prizes—only three other scientists have been so honored. Her first Nobel prize was awarded in 1903 for the discovery of radioactivity. The second was awarded in 1911 for the discovery of two elements, polonium and radium, and for isolating pure radium. Jewel Plummer Cobb, the granddaughter of a freed slave, graduated from pharmacy school and earned both master's and doctorate degrees from New York University. Her basic-cell research in cancer and chemotherapy brought her world recognition. She has published nearly 50 books, scholarly reports, and articles in this highly specialized field. Because of her pioneering work in environmental and sanitary engineering, Ellen Swallow Richards,

4 The Lead-Acid Storage Battery
Lead (Pb) has been known and used since ancient times. It's obtained from the ore galena (PbS). Lead is alloyed with tin in solder and cheaper grades of pewter. The most important use of lead is in the lead-acid storage batteries used in automobiles. The electrodes in this kind of battery are made of lead and lead(III) oxide (PbO₂).

Discussion

Ask students to explain the successive ionization energies of aluminum. The first is 577 kJ/mol, the second is 1815 kJ/mol, the third is 2740 kJ/mol, and the fourth is 11 600 kJ/mol. *After the first electron is removed, the second is harder to remove because it must be taken from a positive ion rather than a neutral atom. In addition, the electrons are pulled closer to the nucleus in the 1+ ion, making the second ionization more difficult. Therefore, it's not surprising that the second ionization energy is about three times larger than the first. The same factors hold true for the third ionization energy, which is about one and one-half times larger than the second. The fourth ionization energy is significantly larger than the third because the electron being removed is in the second energy level and part of a stable noble-gas configuration.*

L2

Group 15

The trend in metallic properties is obvious as you go from the top of Group 15 to the bottom—from nitrogen (N) to phosphorus (P) to arsenic (As) to antimony (Sb) and bismuth (Bi). Nitrogen and phosphorus are nonmetals. They form covalent bonds to complete their outer-level configuration. Arsenic and antimony are metalloids and either gain or share electrons to complete their octets. Bismuth is more metallic and often loses electrons.

Group 15 elements have five valence electrons. Their valence-electron configuration is s^2p^3. They need only three electrons to attain the configuration of the noble gas at the end of their period. Nitrogen, phosphorus, and arsenic have an oxidation number of 3− in some of their compounds, but they can also have oxidation numbers of 3+ and 5+. Nitrogen is a component of proteins, deoxyribonucleic acid (DNA), and ribonucleic acid (RNA) so it's essential to life. Phosphorus is equally important because the phosphate group (PO_4^{3-}) is a repeating link in the DNA chain. The DNA molecule carries the genetic code that controls the activities of cells for many living organisms. Another important biological molecule, adenosine triphosphate, ATP, also contains phosphate groups that store and release energy in living organisms.

Nitrogen, as the chemically unreactive molecule N_2, makes up 78 percent by volume of Earth's atmosphere. Plants and animals can't use nitrogen in this form. Lichens, soil bacteria, and bacteria in the root nodules of beans, clover, and other similar plants convert nitrogen to ammonia and nitrate compounds. Lightning also converts atmospheric nitrogen to nitrogen monoxide (NO). Plants use these simple nitrogen compounds to make proteins and other complex nitrogen compounds that become part of the food chain.

8.1   Main Group Elements   **273**

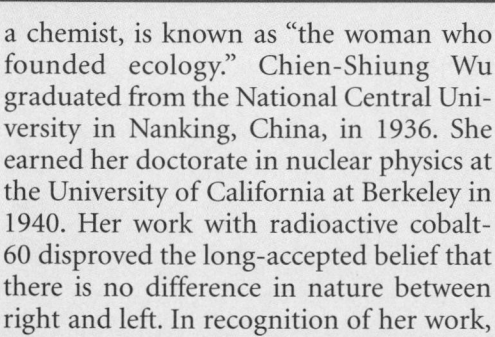

a chemist, is known as "the woman who founded ecology." Chien-Shiung Wu graduated from the National Central University in Nanking, China, in 1936. She earned her doctorate in nuclear physics at the University of California at Berkeley in 1940. Her work with radioactive cobalt-60 disproved the long-accepted belief that there is no difference in nature between right and left. In recognition of her work, she was named to the National Academy of Sciences (the seventh woman to be so honored) and, in 1976, received the National Medal of Science, the highest science award in the United States. Reatha Clark King was raised in rural Georgia and rose to become president of Metropolitan University in Minnesota. Dr. King received her Ph.D. from the University of Chicago at a time when there were few women enrolled in the graduate program. Her research focused on fluorine flame colorimetry. Today, Dr. King works to increase opportunities for women and minorities in higher education. Have students search out the names and achievements of other women who have contributed significantly to chemistry and write a paragraph about each. L1

GLENCOE TECHNOLOGY

 Software

Mastering Concepts in Chemistry
Unit 8, Lesson 2
Typical Elements
Use this lesson to reinforce information about the periodicity of elements. L1 LEP

GLENCOE TECHNOLOGY

 Videodisc

Science and Technology Videodisc Series
Advanced Composites
Disc 2: Chemistry
Side 1, Ch. 12

Show this video to illustrate how new materials, developed from nonmetals, have properties superior to those of metals. L1 LEP

Content Background

Arsenic Attacks Enzymes Arsenic is a very toxic element. Like many other heavy metals, arsenic is toxic because it reacts with sulfur atoms in enzymes and blocks their ability to carry out their necessary functions in cells.

The Uses of Group 15 Elements

Commercially, elemental nitrogen (N_2) is obtained from liquid air by fractional distillation. Much of it is converted to ammonia (NH_3), the familiar ingredient in some household cleaners.

1 Ammonia, the Essential Fertilizer
Ammonia is used as a liquid fertilizer applied directly to soil, or it can be converted to solid fertilizers such as ammonium nitrate, NH_4NO_3; ammonium sulfate, $(NH_4)_2SO_4$; or ammonium hydrogen phosphate, $(NH_4)_2HPO_4$. The bag of fertilizer shows the percentages of the essential main group elements—nitrogen, phosphorus, and potassium—that this fertilizer contains. Fertilizers are formulated in a variety of ways to provide the proper nutrients for different plant growth needs.

2 Two Allotropes of Phosphorus
White and red phosphorus are two common allotropes of phosphorus. Notice that the white phosphorus is photographed under a liquid because this form of phosphorus, which has the formula P_4, reacts spontaneously with oxygen in the air. Red phosphorus is used in making matches. You can read about it in *Everyday Chemistry*.

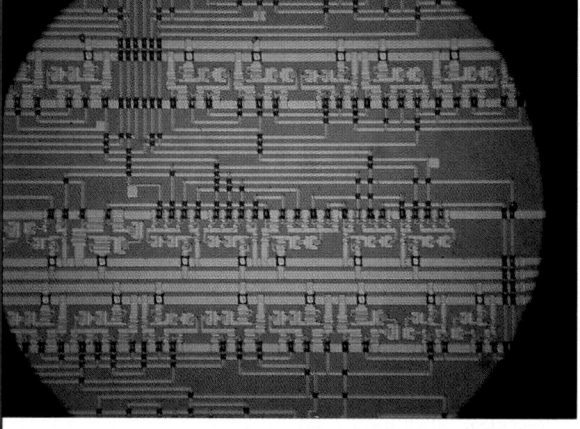

3 Gallium Arsenide Semiconductors
Arsenic is a metalloid found widely distributed in Earth's crust. An increasingly important use of the element is in the form of the binary compound gallium arsenide, GaAs. Because of its higher speed and performance, gallium arsenide is now replacing silicon in some of its semiconductor applications in electronic circuitry.

4 Antimony (Sb) is used primarily in alloys with other metals, particularly lead. Antimony improves the hardness and corrosion resistance of the metal.

274   Chapter 8   Periodic Properties of the Elements

Chemistry Journal

Arsenic

Have students find out about this toxic metalloid and write brief reports. The reports should include the products in which arsenic is used, the ways in which a person might ingest arsenic, the biological effects of arsenic, and the treatment for arsenic poisoning. L1

Everyday Chemistry

The Chemistry of Matches

Making and lighting matches involves a lot of chemistry. Two kinds of matches are currently used: friction matches and safety matches.

Making friction matches Pinewood matchsticks are cut and dipped into a solution of borax (sodium tetraborate, $Na_2B_4O_7 \cdot 10H_2O$) or ammonium phosphate [$(NH_4)_3PO_4$] to make the matches safer. Next, the match head end is dipped into paraffin and then into a mixture of glue, coloring, a combustible material, and an oxidizing agent. Sulfur or diantimony trisulfide (Sb_2S_3) is used as the combustible material. Potassium chlorate ($KClO_3$) and manganese dioxide (MnO_2) are common oxidizing agents. Adding the tip—a mixture of tetraphosphorus trisulfide (P_4S_3), powdered glass, and a binder—is the final step.

Chemistry of friction matches The striking surface on a box of friction matches is powdered glass and glue. The P_4S_3 tip has a low kindling temperature. When the tip is rubbed on the striking surface, the heat from the friction causes it to ignite.

$$P_4S_3(s) + 6O_2(g) \rightarrow P_4O_6(g) + 3SO_2(g) + \text{heat}$$

The heat produced causes the potassium chlorate to decompose.

$$2KClO_3(s) \rightarrow 2KCl(s) + 3O_2(g)$$

The oxygen given off, combined with the heat from the first reaction, causes the sulfur to catch fire, which ignites the paraffin.

$$S(s) + O_2(g) \rightarrow SO_2(g) + \text{heat}$$

The burning paraffin carries the flame from the head to the wooden stem.

How safety matches work The wooden or paperboard sticks of safety matches are treated in a similar manner. Their heads contain diantimony

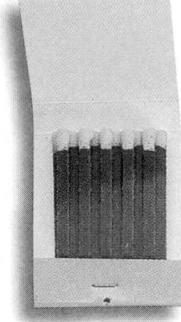
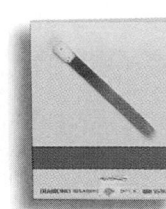

trisulfide or sulfur, potassium chlorate or some other oxidizing agent, ground glass, and glue with paraffin underneath. These matches are called safety matches because they generally ignite only when they are rubbed across the striking surface on their box or packet. The striking surface serves the same function as the tip on the friction match; it ignites the head. The striking surface is a layer of red phosphorus, powdered glass, and glue. The friction of the match on the striking surface changes red phosphorus to white phosphorus.

$$P(\text{red}) + \text{heat} \rightarrow P(\text{white})$$

A small amount of white phosphorus is formed; it ignites spontaneously in air and gives off enough heat to ignite the match head.

$$4P(\text{white})(s) + 5O_2(g) \rightarrow P_4O_{10}(s) + \text{heat}$$

Exploring Further

1. **Comparing and Contrasting** The first step in lighting a safety match is the conversion of red phosphorus to white phosphorus. Compare the chemical reactivities of these allotropes.

2. **Applying** Devise a method for making a match that would produce a colored flame.

Everyday Chemistry

Purpose
To show the composition and chemical reactions of friction and safety matches.

Teaching Strategies
- Display four matches—two different-colored friction matches, a wooden safety match, and a safety book match. Ask students to identify the types. Explain that friction matches are also called kitchen matches. Confusion arises when wooden safety matches are also referred to as kitchen matches.
- Have students research the history of the match, and work in groups to find out how the chemistry of the match developed and what problems were involved. **L1**
  COOP LEARN
- Paraffin is a mixture of hydrocarbons with these formulas: $C_{24}H_{50}$, $C_{25}H_{52}$, and $C_{26}H_{54}$. Paraffin could contain more than one compound having each of these formulas but different arrangements of the atoms (isomers).

Extension
Pose this problem: What types of surfaces can be used to ignite friction and safety matches other than those provided on their containers? Have students form hypotheses, plan experiments, and carry them out to find the answer. Have students put their hypotheses and plans in their portfolios. **L2** **P**

Exploring Further

1. White phosphorus is far more reactive than red phosphorus.
2. Students may suggest incorporating a salt of an element that produces a colored flame test.

Discussion

Point out to students that the oxides of sulfur produce acids that are the primary components of acid rain. Sulfur dioxide produces sulfurous acid (H_2SO_3) when it reacts with water in the atmosphere. Sulfur dioxide is also converted to sulfur trioxide (SO_3) by further reaction with atmospheric oxygen in the presence of sunlight. Ask students what the formula would be for the acid formed when sulfur trioxide reacts with water in the atmosphere. *H_2SO_4, sulfuric acid* Ask what happens when sulfurous and sulfuric acids dissolve in water and fall back to Earth. *acid rain* Explain that calcium oxide (limestone) can be added to soil and lakes to neutralize the acids that have accumulated over a period of years. L1

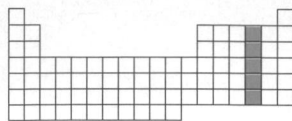

Group 16

The Group 16 elements—oxygen (O), sulfur (S), selenium (Se), and tellurium (Te)—are nonmetals, and polonium (Po) is a metalloid. Their valence-electron configuration is s^2p^4. With rare exceptions, oxygen gains two electrons and forms the oxide ion (O^{2-}) with the neon configuration. Oxygen reacts with both metals and nonmetals and, among the nonmetals, is second only to fluorine in chemical reactivity.

Oxygen is the most abundant element on Earth. It makes up 21 percent by volume of Earth's atmosphere and nearly 50 percent by mass of Earth's crust. Oxygen is present in the compound water and as oxides of other elements. Like nitrogen, oxygen gas (O_2) is obtained from fractional distillation of liquefied air.

There are two allotropes of oxygen—O_2, the most common, and O_3, called ozone. Ozone is a highly unstable and reactive gas that is considered a pollutant in the lower atmosphere. However, in the upper atmosphere, ozone protects Earth by absorbing harmful ultraviolet radiation from the sun. Ozone is responsible for the pungent odor you may notice during thunder and lightning storms or while operating your computer or other electronic equipment.

Both metals and nonmetals react directly with molecular oxygen to form oxides, as shown in the equations in **Table 8.1.**

Table 8.1 Reactions of Oxygen with Metals and Nonmetals	
Reaction of O_2 with Metals	**Reaction of O_2 with Nonmetals**
$4Na + O_2 \rightarrow 2Na_2O$	$C + O_2 \rightarrow CO_2$
$2Ca + O_2 \rightarrow 2CaO$	$S + O_2 \rightarrow SO_2$

Like oxygen, sulfur gains two electrons and forms the sulfide ion (S^{2-}) when it reacts with metals or with hydrogen. But in its reactions with nonmetals, sulfur can have other oxidation numbers. Much of the sulfur produced in the United States is taken from deposits of elemental sulfur by the Frasch process, shown in **Figure 8.8.**

Figure 8.8
Mining Sulfur
In the 1890s, the Frasch process (left) was invented by Herman Frasch. Hot water is pumped into an underground sulfur deposit where it melts the sulfur. Then, the liquid sulfur is brought to the surface by forcing air into the deposit. Liquid sulfur is shown solidifying after removal from a deposit (right).

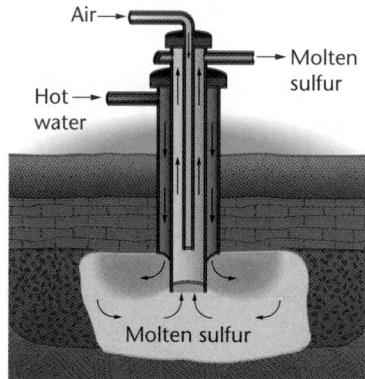

Integrating THE Sciences

Biology Ask students to research how it is that the nitrates and phosphates that are so useful for promoting growth in plants can cause the death of streams, lakes, and rivers. Ask them to prepare an oral report or a paper that shows clearly how agricultural runoff adversely affects aquatic life in waterways. L1

The Uses of Group 16 Elements

The largest industrial use of oxygen is in the production of steel, which is described later in Section 8.2. It's also used in the treatment of wastewater, as a part of H_2/O_2 rocket fuel, and in medicine to assist in respiration.

1 Unstable Hydrogen Peroxide
In your household chemical storehouse, you'll find oxygen in solutions of hydrogen peroxide (H_2O_2), shown in a brown bottle. Peroxides are unstable compounds that decompose to produce molecular oxygen. The brown container helps to slow the decomposition of hydrogen peroxide by excluding light. The reaction is described by this equation.

$$2H_2O_2 \rightarrow 2H_2O + O_2$$

2 Oxygen As an Antiseptic and Bleach
It's oxygen gas that produces the foam when you use hydrogen peroxide as an antiseptic to clean a cut or scrape. It's also oxygen that bleaches hair when a peroxide bleach is used. Some household cleansers use oxygen bleach rather than chlorine bleach.

3 One Use of Sulfuric Acid
Most elemental sulfur is converted to sulfuric acid (H_2SO_4), a key chemical in the production of a wide variety of products, such as fertilizers, automobile batteries, detergents, pigments, fibers, and synthetic rubber, as shown here.

4 An Application of Selenium's Photosensitivity
The chemistry of selenium and tellurium is similar to that of sulfur. Selenium has the property of increased electrical conductivity when exposed to light. This property has applications in security devices and mechanical opening and closing devices, where the interruption of a beam of light triggers an electrical response. But the most important application of selenium is in xerography, a process employed in modern photocopiers, as shown here.

Applying Chemistry

TECH PREP

Sulfur in Your Walls The walls in most homes are constructed of drywall, or they are plastered. Both drywall sheathing and plaster contain gypsum, a mineral with the formula $CaSO_4 \cdot 2H_2O$. Gypsum occurs commonly in Earth's crust, but the largest deposits are in the United States, Canada, France, Italy, Russia, and Great Britain. Gypsum is also used in fertilizers, as a filler in paper and textiles, and as a retarding agent in cement.

Content Background

The Beginning of Fluoridation In 1908, F. McKay, a dentist, noticed that patients with brown mottling on their teeth had a low number of cavities. He traced the cause of the mottling to a substance in their drinking water. In 1931, the substance was identified as a fluoride. On further checking, it was found that these people were drinking water with natural concentrations of 2-4 ppm of sodium fluoride. After careful experiments, scientists learned that water with 1 ppm of NaF has the same ability to reduce cavities but without the disfiguring brown mottling.

GLENCOE TECHNOLOGY

CD-Rom

Chemistry: Concepts and Applications
Classify the Element
Have students use this Interactive Experiment to determine the periodic properties of elements in the periodic table. **L1** **LEP**

Applying Chemistry

TECH PREP

Halocarbons All of the halogens react readily with hydrocarbons to form halocarbons. These are stable, non-flammable compounds with low toxicity and little tendency to react. Carbon tetrachloride is a well-known chlorocarbon, and polytetrafluoroethylene, commonly known as Teflon, is the best-known fluorocarbon. Halocarbons are used as refrigerants, lubricants, plastics, and in artificial blood. The use of halocarbons for some applications is being limited because these compounds have been associated with the destruction of the ozone layer in the atmosphere.

Content Background

Chlorine in Drinking Water The extensive use of chlorine as a disinfectant for drinking water has greatly reduced the incidence of many waterborne diseases in the United States; however, its use also causes the formation of chlorinated hydrocarbons, probable carcinogens, in many water supplies.

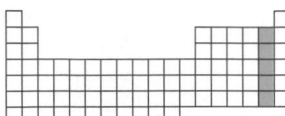

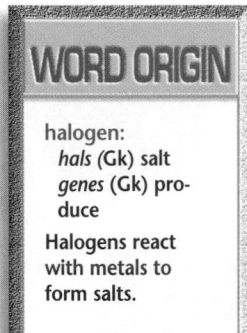

WORD ORIGIN

halogen:
hals (Gk) salt
genes (Gk) produce

Halogens react with metals to form salts.

Group 17

The **halogens**—fluorine (F), chlorine (Cl), bromine (Br), iodine (I), and astatine (At)—are active nonmetals. Because of their chemical reactivity, they don't exist as free elements in nature. Their chemical behavior is characterized by a tendency to gain one electron to complete their s^2p^5 valence-electron configuration and form a 1− ion with a noble-gas configuration. Chlorine, for example, has the configuration [Ne]$3s^23p^5$. When it gains an electron, the chloride ion Cl^- is formed with the argon configuration [Ne]$3s^23p^6$. The halogens can also achieve a noble-gas configuration by sharing electrons.

Because the halogens react by gaining an electron, the most reactive element in the group is the one with the strongest attraction for an electron. Fluorine is the smallest of the halogens so it has the greatest ability to attract and hold an electron. As the size of the atoms increases down the group, the ability of the nucleus to attract and hold outer-level electrons decreases. Consequently, the reactivity of the halogens decreases as you go from fluorine to iodine. Iodine is the least active halogen because of its large atomic radius. Astatine, the largest of the halogens, is probably even less active than iodine, but it is scarce and radioactive.

The elemental halogens exist as diatomic molecules that are both highly reactive and toxic. Many household products contain chlorine compounds that can generate chlorine gas if not handled properly.

Uses of the Halogens

Fluorine and chlorine are both abundant in nature, and both are present in biologically essential compounds. Table salt (NaCl), used to flavor foods, provides all the chloride ions needed for a healthy diet. The iodide ion is also an essential trace element in the diet.

❶ Biologically Important Iodine
The label on this box of table salt says that the salt is iodized. This means that the salt contains a small amount of the iodide ion, another biologically essential element. Iodine is absorbed by the thyroid gland, which regulates metabolism in the body. A swollen thyroid gland in the neck may indicate a lack of this trace element.

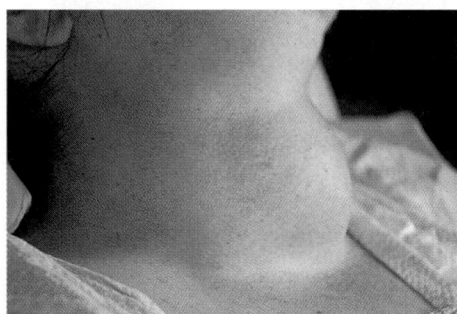

Demonstration 8.5

Chlorine, Bromine, and Iodine

Purpose: To observe the gaseous forms of chlorine, bromine, and iodine.
Materials: 20 mL of 6*M* HCl, 20 mL of chlorine bleach, three 500-mL glass bottles with stoppers, a few drops of liquid bromine, a few crystals of iodine
Safety Precautions:

Disposal: Code H, then C; G for iodine.
Procedure: Chlorine gas can be prepared by carefully adding a small amount of 6*M* HCl to 20 mL of chlorine bleach. (To make 6*M* HCl, add 10 mL of 12*M* HCl to 10 mL of water.) **CAUTION:** *Prepare the toxic chlorine gas in an operating fume hood. Add concentrated acid to water.* The gas will collect above the liquid. Stopper the 500-mL glass bottle after the bubbling

ceases so that students can observe the pale yellow-green chlorine gas. Add a few drops of liquid bromine to an empty glass bottle. Stopper it securely. **CAUTION:** *Use liquid bromine under a fume hood, and wear rubber gloves, goggles, and an apron.* Have students note the red-brown color of bromine vapor. To prepare iodine vapor, place a few crystals of solid iodine in an empty glass bottle and stopper it.

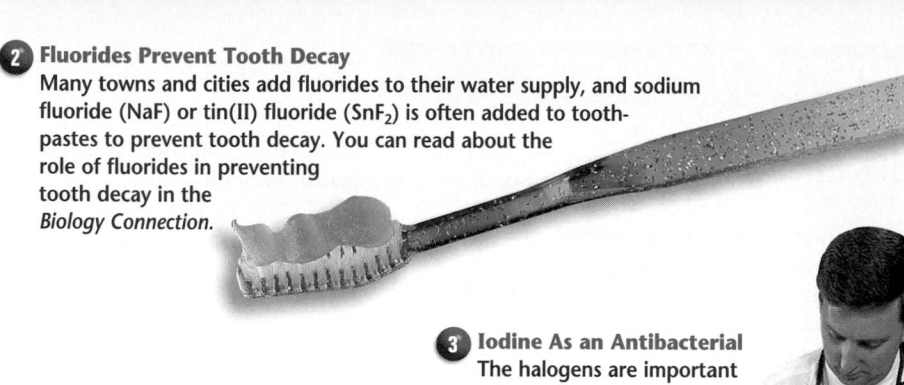

2 Fluorides Prevent Tooth Decay
Many towns and cities add fluorides to their water supply, and sodium fluoride (NaF) or tin(II) fluoride (SnF_2) is often added to tooth-pastes to prevent tooth decay. You can read about the role of fluorides in preventing tooth decay in the *Biology Connection.*

3 Iodine As an Antibacterial
The halogens are important as antibacterial agents. Doctors use an iodine solution to sterilize the skin before surgery.

4 Chlorine Makes Water Safe
Chlorine is used in the water supply of most cities and towns and in swim-ming pools to kill bacteria. Chlorine added to swimming pools makes the water slightly acidic, so if your eyes are irritated after swimming in a pool, it's probably because of the acid.

5 Silver Bromide Coats Photographic Film
The compounds of the halogens are more important than the free elements. Compounds of chlorine with carbon, such as carbon tetra-chloride and chloroform, are important sol-vents. Silver bromide (AgBr) is important in the light-sensitive coating on film.

279

Applying Chemistry

TECH PREP

Silver Bromide in Photog-raphy Black-and-white film is produced by coating an emulsion of 40% silver bromide, 50% gelatin, and 10% water on a base often made of cellulose acetate. The process of mixing the emulsion and coating it on the base must be done in com-plete darkness. The size of the crystals of silver bromide deter-mines the speed or resolving power of the film. Large parti-cles are more light sensitive, but photos taken with film coated with coarse-grained silver bro-mide cannot be greatly en-larged because a grainy photo is obtained. Interested students may do research to find out how the formulation of the emulsion of color film is differ-ent from that of black-and-white film. L2

The bottle can be placed in warm water to has-ten the sublimation process. Point out the faint violet color of iodine vapor.
Results: Students can observe the pale yellow-green chlorine gas, the red-brown color of bromine vapor, and the pale violet color of iodine vapor.
Analysis: Ask these questions.
1. What are the diatomic formulas of the three elements? Cl_2, Br_2, and I_2

2. How are the colors of the vapors of each substance described? Cl_2, *pale yellow-green;* Br_2, *red-brown;* I_2, *pale violet*

✓ Assessment

Knowledge: Ask students if the odors of any of the gases remind them of a common house-hold chemical. *Chlorine bleach is often men-tioned.* L1

Purpose
To explore the chemistry involved in the ability of fluorides to reduce tooth decay.

TECH PREP

Teaching Strategies

- Using information from question 1, have students debate whether fluorides should be added to city water supplies.
- Review reversible reactions with students.
- A good reference is "Toothpaste" by Brad Yohe, *ChemMatters,* Feb. 1986, pp. 12-13. Have students read the article and write a report for their portfolios. **L1** **P**

Extension

Have students design and carry out an experiment to see what happens to the pH of their mouths for an hour after drinking a sugary drink. Use narrow-range pH paper. **L2**

Connecting to Chemistry

1. Fluoridation is controversial. Pros: It is a safe preventive that cuts decay by 70%; it may protect against other dental diseases. Cons: It is forced medication; long-term effects need to be studied more.
2. NaF is in Crest and Colgate, SnF_2 in Oral-B Stop, MFP in Aquafresh, and there are no fluorides in Sensodyne.
3. $Ca_5(PO_4)_3OH(s) \rightleftarrows$ $5Ca^{2+}(aq) + OH^-(aq) +$ $3PO_4{}^{3-}(aq)$

3 ASSESS

Check for Understanding

- Check for understanding of the section material by having students answer the Section Review questions.

BIOLOGY CONNECTION

Fluorides and Tooth Decay

Do you worry that you might get a cavity? If so, it's not surprising because total prevention of cavities is not yet possible. But since the 1950s, the problem of tooth decay has been greatly reduced.

Fluoridation and fluorides in toothpaste
Drinking water with minute amounts of fluorides helps prevent tooth decay. Studies by the U.S. Public Health Service show that people who drink water containing fluorides have lower rates of tooth decay. In the 1950s, cities and towns began to add fluorides to their water supplies. The results show that when children drink water containing 1 part per million (ppm) of sodium fluoride, NaF, or sodium silicofluoride, Na_2SiF_4, they have 70 percent fewer cavities than those who drink non-fluoridated water. Protection from tooth decay can also be obtained by using a toothpaste containing fluorides in the form of NaF, SnF_2, and Na_2PO_3F (MFP).

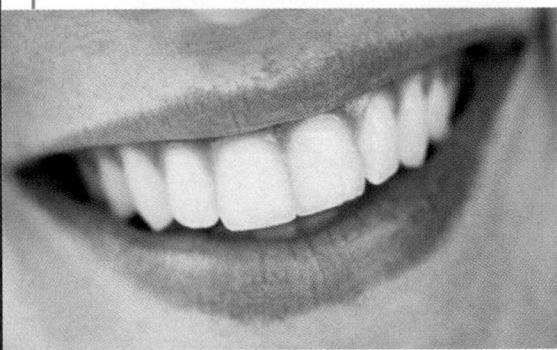

The decay process Tooth enamel is about 2 mm thick and 98 percent hydroxyapatite, $Ca_5(PO_4)_3OH$. Although hydroxyapatite is essentially insoluble in water, tiny amounts dissolve in the saliva in a process called demineralization.

$$Ca_5(PO_4)_3OH(s) \rightarrow$$
$$5Ca^{2+}(aq) + 3PO_4{}^{3-}(aq) + OH^-(aq)$$

The reverse process, remineralization, is the body's defense against bacterial acids.

$$5Ca^{2+}(aq) + 3PO_4{}^{3-}(aq) + OH^-(aq) \rightarrow$$
$$Ca_5(PO_4)_3OH(s)$$

The two equations show that this is a reversible reaction. In adults, the rates of demineralization and remineralization are equal, so equilibrium is established.

Bacteria and cavities Bacteria use sugar for energy and produce lactic acid. The acid causes the pH of saliva, which is normally 6.8, to drop below 6.0. When that happens, the rate of demineralization increases and tooth decay occurs.

Fluorides prevent cavities Fluoride compounds dissociate in water to form fluoride ions.

$$NaF(s) \rightarrow Na^+(aq) + F^-(aq)$$

$$SnF_2(s) \rightarrow Sn^{2+}(aq) + 2F^-(aq)$$

The fluoride ions replace the hydroxide ions in some of the $Ca_5(PO_4)_3OH$ and form fluorapatite, $Ca_5(PO_4)_3F$.

$$Ca_5(PO_4)_3OH(s) + F^-(aq) \rightarrow$$
$$Ca_5(PO_4)_3F(s) + OH^-(aq)$$

Fluorapatite is about 100 times less soluble than hydroxyapatite; it is also harder and denser, so tooth enamel is stronger and more resistant to bacterial attack.

Connecting to Chemistry

1. **Acquiring Information** Find out why only 50 percent of cities in the United States fluoridate their water supply.

2. **Applying** Check the ingredients on five different toothpastes to see whether they contain a fluoride compound, and if so, which one.

3. **Writing Equations** Write the reversible reaction of demineralization and remineralization of tooth enamel.

Group 18

Helium (He), neon (Ne), argon (Ar), krypton (Kr), xenon (Xe), and radon (Rn), the noble gases, were originally called the inert gases because chemists couldn't get them to react. Their lack of reactivity is understandable; all the noble gases have a full complement of valence electrons and, therefore, no tendency to gain or lose electrons. **Figure 8.9** shows one use of a noble gas that results from its lack of reactivity. In recent years, however, chemists have succeeded in making fluorine compounds of the heavier noble gases, krypton and xenon, but no reactions have been achieved for the lighter members of the group.

Trends in the chemical properties in each group of the main group elements are directly related to changes in atomic radii. For groups of elements that form compounds by losing electrons, the larger the atom, the more readily the atom gives up its electrons and the more reactive the atom is. For groups of elements that form compounds by gaining electrons, the larger the element, the less attraction it has for electrons and the less reactive the atom is. You will see similar trends among the transition elements.

Figure 8.9

Helium in Weather Balloons Because helium does not burn, it is used rather than the lighter gas, hydrogen, to carry weather instruments into upper levels of the atmosphere. The instruments gather information on weather and atmospheric conditions.

SECTION REVIEW

Understanding Concepts

1. Describe how the atomic radii of the main group elements change as you move across the third period. Give reasons for this trend.

2. Describe how the atomic radii of the elements in Group 2 change as you move down the group. Give reasons for this trend.

3. How does the size of a positive ion compare with the size of the neutral atom? How does the size of a negative ion compare with the size of the neutral atom? Give reasons for your answer.

Thinking Critically

4. **Comparing and Contrasting** Compare the way metals and nonmetals form ions and explain why they are different.

Applying Chemistry

5. **Hard Water** Soap scum forms when soap is used with hard water. This is because of the presence of magnesium and calcium ions. Which of the two elements, magnesium or calcium, dissolves more easily in water?

SECTION REVIEW

1. Atomic radii decrease as atomic number increases from 11 (Na) to 17 (Cl). The outer energy level remains the same, but nuclear charge increases; thus, the electrons are attracted more closely to the nucleus, and the size of the atom decreases.

2. Atomic radii increase as atomic number increases from 4 (Be) to 88 (Ra). Nuclear charge increases but the number of energy levels also increases and valence electrons are farther from the nucleus.

3. A positive ion is always smaller than the neutral atom; a negative ion is always larger than the neutral atom. When ions form, the nuclear charge remains the same. For positive ions, the remaining electrons are pulled in more closely around the nucleus. For negative ions, the nuclear charge is not sufficient to hold one more electron as closely as it could hold its own electrons.

4. Metals form positive ions by losing electrons. Nonmetals form negative ions by gaining electrons. Both metals and nonmetals form ions to attain a noble-gas configuration.

5. calcium

PREPARE

Key Concepts

The properties of several of the most important transition elements in Groups 3-12 are examined and related to their electron configurations. The inner transition elements are identified with *f* sublevels, and a few of the most important elements are discussed.

1 FOCUS

Focus Transparency

Display the **Section Focus Transparency** for Section 8.2 to introduce some uses of transition elements. `L1` `LEP`

2 TEACH

Correcting Misconceptions

Remind students that the terms *transition elements* and *transition metals* are used interchangeably because all have the physical and chemical properties of metals.

Transparency

Display **Problem Solving Transparency 10** to give students practice working with multiple oxidation numbers of transition elements. `L1` `LEP`

SECTION PREVIEW

Objectives

Relate the chemical and physical properties of the transition elements to their electron configurations.

Predict the chemical behavior of transition elements from their positions in the periodic table.

Transition Elements

Some of the most intriguing elements are those that are beautiful, rare, and expensive. The might and power of kings and queens have been displayed through ornaments, jewelry, and works of art made of precious platinum, silver, and gold. But important as these metals are, they are only three of the transition elements—elements with such a variety of physical and chemical properties that they provide materials to fill a tremendous range of purposes. Iron, for example, is the world's most important structural material. Copper is known for its electrical conductivity. Chromium prevents the corrosion of other metals, and molybdenum can be alloyed with iron to give added hardness, corrosion resistance, and solidity at high temperatures. These are some of the metals that occupy the *d* block in the periodic table.

Properties of the Transition Elements

Each of the transition elements has its own properties that result from its atomic structure. For example, iron is strong. It's used for the structural framework of bridges and skyscrapers. But iron can also be reduced to a pile of reddish-brown rust if it is exposed to water and oxygen. Other transition metals may not be as strong as iron, but they also may not disintegrate in air like iron does. Fortunately, transition elements can be used together in alloys, as shown in **Figure 8.10**.

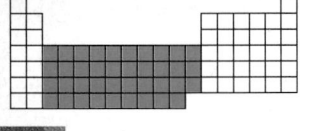

Figure 8.10

A Steel for All Purposes

When lightness, durability, and strength are needed for uses such as this racing wheelchair, a combination of transition elements, alloyed with iron, can provide the necessary properties. Here chromium, nickel, and magnesium (a main group element) are combined to produce a frame with the right properties.

282   Chapter 8   Periodic Properties of the Elements

Program Resources

Study Guide, pp. 33-34 `L1`
Laboratory Manual, 8.2, pp. 73-76 `L1`
Section Focus Transparency 17 and
    Master, p. 17 `L1` `LEP`
Problem Solving Transparency 10 and
    Master, pp. 19-20 `L1` `LEP`
ChemLab and MiniLab Worksheets,
    pp. 53-56, 58 `L1`

Trends in Properties of the Transition Elements

With the exception of the Group 12 elements (zinc, cadmium, and mercury), the transition metals have higher melting points and boiling points than those of almost all of the main group elements. For example, in the fourth period (scandium to copper), the melting points range from 1083°C for copper (Cu) to 1890°C for vanadium (V). When you compare these melting temperatures to the melting temperatures of the main group metals, you find that only beryllium (Be) melts above 1000°C. Most of the other main group elements melt well below this temperature.

In any period, the melting and boiling points of the transition metals increase from Group 3 and reach a maximum in Group 5 or 6. Then they decrease across the remainder of the period. Tungsten (W) in Period 6, Group 6 has a melting point of 3410°C, the highest of any metal. It's because of its high melting point that tungsten is used as the filament in lightbulbs, as you'll see in *How it Works*. Mercury (Hg) in Period 6, Group 12 melts at −38°C, the lowest melting point of any metal. Mercury's liquid state at room temperature and its high density make it an important liquid for use in thermometers and barometers, as shown in **Figure 8.11**.

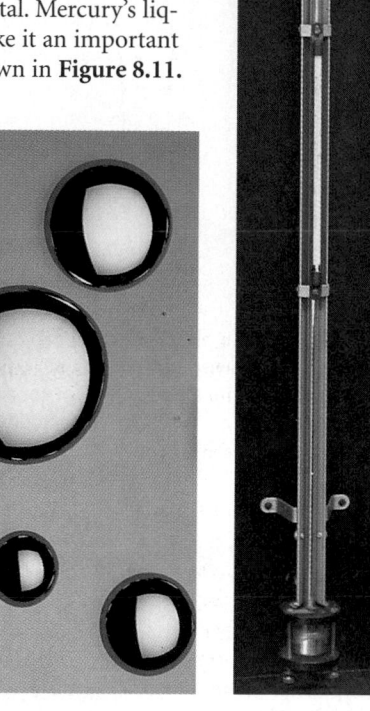

Figure 8.11

The Mercury Barometer
Any liquid could be used to make a barometer, but mercury is a good choice because a column of mercury only 76 cm high exerts a pressure approximately equal to the pressure of the atmosphere at sea level. This is because mercury has a high density for a liquid—13.6 g/mL. Other liquids, for example water with a density of 1.0 g/mL, require a column of liquid more than 30 feet high to equal the pressure of the atmosphere.

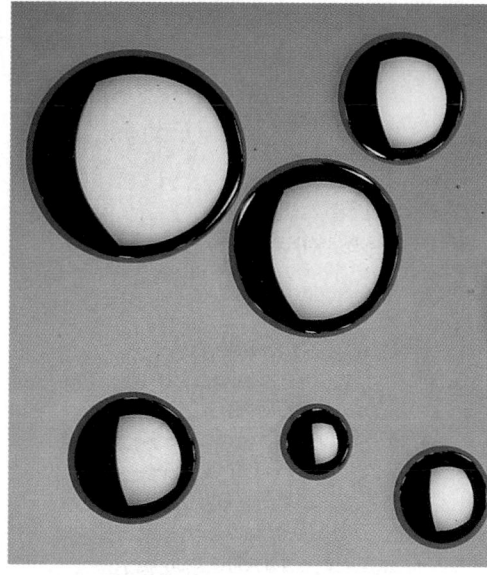

Multiple oxidation states are characteristic of the transition elements. Remember that iron gives up its two $4s$ electrons and forms the Fe^{2+} ion in its oxide, FeO. In another oxide, Fe_2O_3, iron gives up its two $4s$ electrons and one $3d$ electron to form the Fe^{3+} ion. Many of the transition elements can have multiple oxidation numbers ranging from 2+ to 7+. These oxidation numbers are due to involvement of the d electrons in chemical bonding. Recall that only some of the heavier main group elements such as tin, lead, and bismuth have multiple oxidation numbers. These elements also have d electrons that can be involved in bonding.

Content Background

One Transition Metal Silicate Zircon, $ZnSiO_4$, is found in both igneous and metamorphic rocks and in beach sands throughout the world, particularly in Australia, India, Brazil, and Florida. Gem varieties are found in streambeds and deposits of rock detritus. Although some crystals of zircon double for the colorless diamond, zircons are often colored. Red, orange, and yellow varieties are found. Sometimes the color is removed by heating under oxidizing conditions. It's also possible to add blue color by heating under reducing conditions. The gemlike quality of zircon is the result of high refractive indices.

Concept Development

Draw diagrams on the chalkboard that represent carbon steel and alloy steels. In the carbon steel, show the iron atoms as uniform spheres packed closely together, with carbon atoms fitting into the interstices or holes between the iron atoms. Show an alloy steel with atoms such as chromium, manganese, and molybdenum replacing some of the iron atoms. Explain that small amounts of these added elements make dramatic changes in the properties of iron.

GLENCOE TECHNOLOGY

 CD-Rom

Chemistry: Concepts and Applications
Organizing Elements in a Periodic Table
Have students use this Interactive Exploration to create a periodic arrangement of some elements. L1 LEP

Purpose

To relate the properties of incandescent light-bulbs to the elements inside the bulb.

Background

Metals conduct electricity, have high melting points, and radiate both heat and light under the influence of an electric current. These properties make them useful in generating light. Noble gases are chemically unreactive and will not react with metals even at the elevated temperatures needed to generate heat and light. These properties make them appropriate for use in lightbulbs because they help preserve the metal filament.

Visual Learning

Use the photos to point out the construction of the lightbulb and the difference between a burned-out bulb and a working bulb.

Teaching Strategies

• Ask students why the atmosphere inside the bulb is important. What would happen if the gases inside the bulb were pure oxygen, pure nitrogen, air, or a noble gas? L1
• Ask students to research the history of the incandescent lightbulb and present a written or oral report. L2

Thinking Critically

1. Look for substances that have high melting points and low reactivity with non-metals.
2. The cost of a lightbulb may be higher.

How it Works

Inert Gases in Lightbulbs

Tungsten (W) is used as a filament in incandescent lightbulbs because it has a high melting point, 3420°C, and boiling point, 5850°C. But, nothing lasts forever. Because electricity continually passes through the filament, the metal eventually breaks as it vaporizes, and the bulb burns out.

1. If a lightbulb were filled with air, the filament would react with oxygen, burn, break, and lose its ability to provide light. If there were no gas inside the bulb, the filament would quickly vaporize and no electricity would flow.

2. If you've ever looked closely at a burned-out bulb, you may have seen a black coating of condensed metal on the inside of the bulb from the evaporated filament.

3. To prevent the filament from reacting and to slow its evaporation, lightbulbs traditionally have been filled with a mixture of nitrogen and argon. These gases carry heat away from the metal filament so it doesn't overheat and boil away. Nitrogen and argon don't react with the filament, and traditional bulbs can last as long as 750 hours.

In the search for longer-lasting lightbulbs, manufacturers are experimenting with various combinations of inert gases to replace the nitrogen/argon mixture. The most promising combination is a mixture of argon, krypton, and xenon, which produces a lightbulb that lasts 7500 to 10 000 hours—ten times longer than an ordinary lightbulb.

Thinking Critically

1. How would you go about finding a substance other than tungsten to use as the filament of a lightbulb?

2. What might be a disadvantage to the consumer of lightbulbs containing the relatively rare elements argon, krypton, and xenon?

Chemistry Journal

Many Uses for Many Metals

Ask students to look around their homes and list all the objects that contain or are made of metal. Have them make a guess as to the type of metal used in each object and what property or properties of the metal make it suitable for use in the object. L1

The Ion Charges of a Transition Element

Iron, the most common transition element, has the electron configuration $[Ar]4s^23d^6$. The two electrons in the highest energy level, $4s^2$, are the ones most likely to be involved in the chemical reactions of iron. However, iron is a transition element and, like other transition elements, it has a partially filled d sublevel. Electrons in the d sublevel may also be involved in reactions. In this MiniLab, you will study some reactions of iron compounds and relate the results to the electron configuration of iron.

miniLAB

2

Procedure

1. Place about 20 mL of aqueous $FeCl_3$ solution into a 125-mL Erlenmeyer flask labeled *$FeCl_3$*. Add 2 drops of $1M$ NaOH solution. Describe the results in your data table.

2. Place about 20 mL of aqueous $FeCl_2$ solution into another 125-mL flask labeled *$FeCl_2$*. Add 2 drops of $1M$ NaOH solution. Describe the results in your data table.

3. Stopper the flask. Swirl and shake the mixture in the flask, labeled *$FeCl_2$*. Every 30 seconds, stop shaking the flask, and remove the stopper for a moment to admit more oxygen. Put the stopper back on the flask, and resume shaking until a change occurs.

4. Record your observations.

Analysis

1. Describe the colors of the two precipitates.

2. What are the charges on the iron ion in the two precipitates? Which electrons of the iron atom are probably lost to form the two ions?

3. Examine your results from procedure 3. Hypothesize what probably happened to the iron ions when oxygen entered the flask.

4. Iron(II) ions (Fe^{2+}) are useful as nutrients, whereas iron(III) ions (Fe^{3+}) are not. Using the results of this MiniLab, suggest a reason why elemental iron (Fe) is often added to breakfast cereals as a dietary supplement rather than an iron compound containing Fe^{2+} ions.

Trends in Atomic Size of Transition Elements

You learned that for the main group elements, atomic radius decreases from left to right in a period because of increasing nuclear charge. There is a similar trend for the transition elements, but the changes in atomic radii for the transition elements are not as great as the changes for the main group elements. You also learned that as you move from top to bottom in a main group, atomic radius increases. The same trend is seen among the transition elements. Atomic radius increases as you move from the fourth period to the fifth period in any group, but there is little change in atomic radius as you move from the fifth period to the sixth period. Because atomic size affects reactivity, you can expect the transition elements in Periods 5 and 6 to have similar chemical properties.

---

miniLAB

The Ion Charges of a Transition Element

Purpose: To study the reactions of two iron compounds and relate the results to the electron configuration of iron.

Process Skills: Observing, inferring, predicting

Preparation: Prepare the iron(II) chloride and iron(III) chloride solutions by dissolving about 1 g of solid in enough water to make 100 mL of solution. Prepare $1M$ sodium hydroxide solution by dissolving 40 g of solid NaOH in enough distilled water to make 1 L of aqueous solution. (Add the sodium hydroxide to the water.)

Teaching Strategies

• Have students make a flow-chart or outline of the procedure in their lab notebooks prior to the MiniLab. **L1** **LEP**

Expected Results

A rust-brown precipitate of iron(III) hydroxide will form in procedure step 2. A blue-green precipitate of iron(II) hydroxide will form in procedure step 3. It will change to rust-brown in procedure step 4.

Analysis

1. $Fe(OH)_2$ is blue-green and $Fe(OH)_3$ is rust-brown.
2. Fe^{2+} and Fe^{3+}; iron atoms lose either the $4s$ electrons to form the 2+ ion or the $4s$ electrons and one $3d$ electron to form the 3+ ion.
3. The iron 2+ ions are converted to (oxidized to) the 3+ ion by the oxygen in the air.

4. Elemental iron may be converted (oxidized) to the 2+ ion, a useful ion. If the iron already existed in the 2+ state, it might be oxidized to the nonuseful 3+ ion.

✔ Assessment

Knowledge: Ask students to use the same rationale to predict the ion charges or oxidation numbers for chromium. *2+ if only the 4s electrons are lost, 6+ if the 3d electrons are lost, as well. Both are correct, and 3+ occurs also.* **L1**

Point out to students that the separation of iron from its ore is the reverse of the process of rusting. Ask students whether they think rusting is a spontaneous reaction and why. *yes, because it just happens* Ask whether obtaining iron from its ore (essentially rust) in a blast furnace is a spontaneous reaction and why they think so. *No, it is not spontaneous. A high temperature is needed to start the process.* Ask whether removing iron from its ore is an exothermic or endothermic reaction and how they know. *endothermic, because high temperature is required* [L1]

GLENCOE TECHNOLOGY

Videodisc

Science and Technology Videodisc Series
Hydrogen Sponge
Disc 2: Chemistry
Side 1, Ch. 10

Show this video to illustrate how metals can be designed to fill special needs. [L1] [LEP]

Iron: First Among the Transition Elements

Iron, and many steel alloys that are made from it, have been known and used since ancient times. Iron is the fourth most abundant element in Earth's crust and the second most abundant metal after aluminum.

① Iron Is Essential in Human Life
Besides its importance as a structural metal, iron is an essential element in biological systems. It is the iron ion at the center of the heme molecule that binds oxygen. Iron in heme can be the Fe^{2+} or Fe^{3+} ion, but only the Fe^{2+} ion can bind oxygen. Heme is bound in the proteins hemoglobin and myoglobin. Hemoglobin supports life by transporting oxygen through the blood from the lungs to every cell of the body. Myoglobin stores oxygen for use in certain muscles.

Heme molecule

Blast furnace

② Separation of Iron from Its Ore
Iron is obtained from its ore (oxides) in a blast furnace. Iron oxide, Fe_2O_3, is mixed with carbon (coke) and limestone ($CaCO_3$) and fed continuously into the top of the furnace. Hot air is fed into the bottom of the furnace. Temperatures of 2000°C are reached as the mixture reacts while falling through the furnace. The process is complex, but the following equations are the important steps. First, coke is ignited in the presence of the hot air and converted to carbon dioxide.

$$C(s) + O_2(g) \rightarrow CO_2(g)$$

Then, CO_2 reacts with more coke to form carbon monoxide.

$$CO_2(g) + C(s) \rightarrow 2CO(g)$$

Carbon monoxide then converts iron ore (Fe_2O_3) to iron.

$$Fe_2O_3(s) + 3CO(g) \rightarrow 2Fe(l) + 3CO_2(g)$$

The molten iron, called pig iron, drops to the bottom of the furnace and is drawn off as a liquid. Impurities, called slag, form a layer on top of the iron and are drawn off separately. Pig iron, as it comes from the blast furnace, is a crude product containing impurities such as carbon, silicon, and manganese. The crude iron is refined, and then most of it is converted to steel.

Pig iron

3 Steelmaking

The first step in the production of steel is the removal of impurities from the pig iron. The second step is the addition of carbon, silicon, or any of a variety of transition metals in controlled amounts. Different elements give special properties to the final steel. Some steels are soft and pliable and are used to make things like fence wire. Others are harder and are used to make railroad tracks, girders, and beams. The hardest steels are used in surgical instruments, drills, and ordinary razor blades. These steels are made from iron mixed with small amounts of carbon.

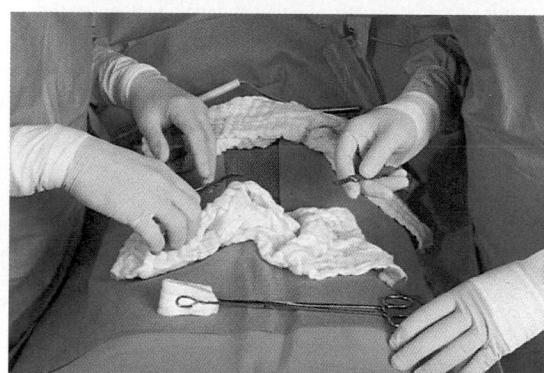

SECTION 8.2

GLENCOE TECHNOLOGY

 Videodisc

Chemistry: Concepts and Applications
Properties of Transition Metals
Disc 1, Side 2, Ch. 9

Show this video to help you demonstrate common properties and uses of transition metals. L1 LEP

 CD-Rom

Chemistry: Concepts and Applications
Properties of Transition Metals
Have students use this video to reinforce an appreciation of the properties and uses common to transition metals. L1 LEP

4 Heat-Treating

Heat-treating is a final step in the production of steel. When purified steel is heated to 500°C, the small amount of carbon contained in it combines with iron to form a carbide (Fe_3C), which dissolves in the steel. This makes the steel harder. Cooling the steel quickly in oil or water makes the hardness permanent.

8.2 **Transition Elements** **287**

Purpose
To discuss the different kinds of steel, their composition, properties, and uses.

Background
Nickel and cobalt have magnetic properties like iron. The three elements (Fe, Co, and Ni) are termed *ferromagnetic*. Two hundred million tons of steel are produced yearly in the United States. It is the most used alloy. When steels other than UHCSs are used, about 50% ends up as waste. A good article is "Scientists Rediscover Superplastic Steel," *Chemecology*, March 1992, p. 6.

Teaching Strategies
- Relate the magnetic properties of nickel and cobalt to their use in alnico.
- Review alloys. Ask students why the fact that tungsten steel stays hard when it is heated is important for high-speed cutting tools. Explain that a low coefficient of thermal expansion means that when heated, the steel does not expand greatly. Ask students why this characteristic is important for precision instruments and measuring devices. Ask why manufacturers might look for a steel that is ductile and malleable for use in food cans and automobile bodies.
- Ask a physics teacher or an electrician to visit your class and take off the cover of an ammeter or voltmeter. Have her or him show the permanent alnico magnet and explain how the magnetic field set up by the incoming current in the coil is affected by the permanent magnetic field and results in a deflection of the meter pointer.

Carbon and Alloy Steels

Steelmaking was already well advanced in ancient times. The earliest known steel objects were made between 1500 and 1200 B.C. About 1000 B.C., wootz steel was developed in India by heating a mixture of iron ore and wood in a sealed container.

Wootz steel later became known as Damascus steel because sword blades made from it had wavy surface patterns like Damask fabric. Damascus steel became famous because these swords kept their sharpness and strength after many battles. The knowledge of how to make Damascus steel was lost in the 1800s, but recently the process was redeveloped under the name superplastic steel. Collector hunting knives worth thousands of dollars are being made from superplastic steel.

What is steel and how is it classified?

Steel is an iron alloy containing small amounts of carbon (0.2 to 1.8 percent) and sometimes other elements such as chromium, manganese, nickel, tungsten, molybdenum, and vanadium. Steel cannot contain more than 1.8 percent carbon without becoming brittle, and the most common steels usually have closer to 0.2 percent carbon content. Commercial iron contains 2 to 4 percent carbon and is very brittle. Carbon steel contains only iron and carbon.

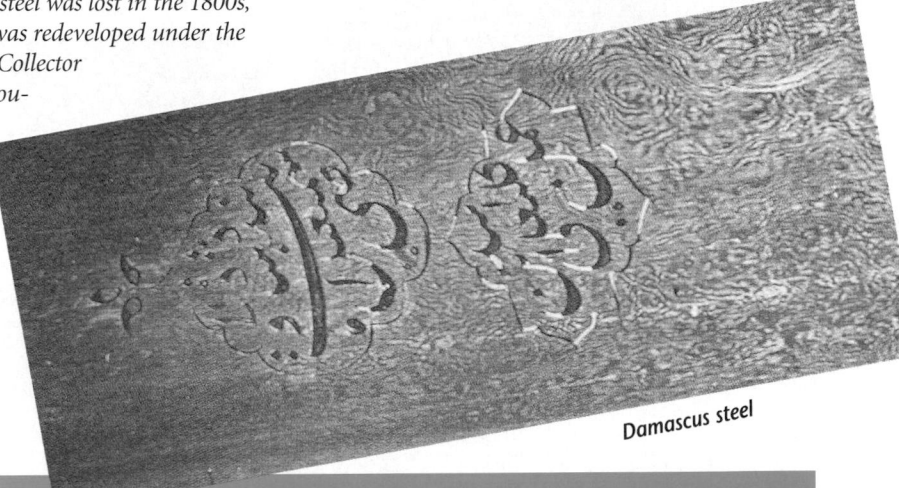

Damascus steel

Table 1   Carbon Steels

Name of Steel	Composition	Characteristics	Uses
Mild	Fe, less than 0.2% C	Malleable, ductile	Steel food cans, automobile bodies
Medium	Fe, 0.2–0.6% C	Less malleable/ductile	Structural purposes—beams, bridge supports
High	Fe, 0.7–1.5% C	Hard, brittle	Farm implements, drill bits, knives, springs, razor blades
UHCS, Superplastic*	Fe, 1.8% C	Corrosion- and wear-resistant, highly malleable	Engine components, earth movers, underside of tractors

*A few UHCSs are alloy steels.

Ninety Percent of Steel Is Carbon Steel

The properties of carbon steels depend upon the percent of carbon present. They are classed as mild, medium, and high on this basis. Because mild-carbon steels are ductile, sheets of it can be cold-formed to mold fenders and body parts for cars. Medium-carbon steels have more strength but are less ductile so they are used as structural materials. High-carbon steels are hard and brittle; they are used for wear-resistance purposes.

Superplastic Steel Is New Again

A new steel that can be formed into complex shapes called superplastic or ultrahigh-carbon steel (UHCS) has been developed. Damascus steels can be stretched up to 100 times their length without breaking. The limit for most steels is less than 1 1/2 times. At high temperatures, UHCS pulls like taffy. There is almost no waste from their use, thus conserving material and energy.

The process of separating a metal from its ore is known as reduction because the mass of the metal is obviously less than that of the ore. Tell students that they will learn a related meaning for the term *reduction* in a later chapter.

Content Background

Characteristics of Transition Elements All of the transition elements except for those in Group 12 are hard solids at room temperature and have high melting and boiling points. Those in Groups 6 and 7 are the hardest and have the highest melting and boiling points.

Table 2   Alloy Steels

Name of Steel	Composition in %*	Characteristics	Uses
Alnico I	Ni-20, Al-12, Co-5	Strongly magnetic	Loudspeakers, ammeters
Invar	Ni-36 to 50	Low coefficient of thermal expansion	Precision instruments, measuring tapes
Manganese steel	Mn-12 to 14	Holds hardness and strength	Safes
18-8 Stainless	Cr-18, Ni-8	Corrosion resistant	Surgical instruments, cooking utensils, jewelry
Tungsten steel	W-5	Stays hard when hot	High-speed cutting tools

* All alloys contain iron and 0.1-1.5% carbon.

Alloy Steels Contain Carbon and Other Elements

In alloy steels, iron is mixed with carbon and varying amounts of other elements, mainly metals. Added metals produce desired properties such as hardness and corrosion resistance (Cr), resistance to wear (Mn), toughness (Ni), heat resistance (W and Mo), and springiness (V). Stainless steel is a well-known, corrosion-resistant alloy steel. It contains ten to 30 percent chromium and sometimes nickel and/or silicon. Because of its outstanding magnetic properties, Alnico steel is used to make permanent magnets. Alnico magnets are used in voltmeters and ammeters to rotate the coil of wire connected to the pointer.

Working Steel

Two methods are used to shape steel into various shapes for specific purposes: hot-working and cold-working. In the hot-working process, steel is hammered, rolled, pressed, or extruded while it is very hot. Hammering and pressing are called forging. These processes were originally done by hand, in fact, hand forging continues as a craft in blacksmith shops. However, steam-powered hammers and hydraulic presses are used to forge most of the steel produced today. Extrusion involves forcing molten steel through a die that is cut to the desired

The development of methods of producing and working steel led to a revolution in construction in the 1880s. Concrete, reinforced by steel, became an important structural material. Steel beams made skyscrapers possible, and changed the shape of modern cities. Although the Sears tower is the highest building in the United States, it has so many tall neighbors that it no longer dominates the skyline of Chicago.

shape. Rolling is the most widely used method for shaping steel. The metal is passed between two rollers that move in opposite directions. The shape of the finished product determines the type of rollers used. Railroad tracks and I-beams are shaped in this way.

Cold-working includes rolling and extrusion, and in addition, drawing. Drawing involves pulling steel through a die rather than pushing it as in extrusion. Wires, tubes, sheets, and bars are shaped by drawing. Often cold-working processes follow hot-working to produce a more finished product.

DISCUSSING THE TECHNOLOGY

1. **Applying** The steel used in airplane motors is 43 percent Fe and 0.4 percent C; the other 56.6 percent is Cr, Ni, and Mo. What kind of steel is this? What are the functions of Cr, Ni, and Mo?

2. **Inferring** Using Table 2, how would you classify most of the metals used? (Hint: Look at their locations on the periodic table.) Draw a conclusion about most metals used in alloy steels.

3. **Acquiring Information** Find out how heat treatments, quenching, tempering, and annealing further alter the properties of different steels.

Integrating THE Sciences

Medicine High levels of the trace element zinc in victims of Alzheimer's disease have been linked to the onset of the disease. Test-tube experiments show that zinc ions can convert certain proteins to an insoluble plaque that forms deposits on and kills brain cells. This plaque has been found in the brains of people with Alzheimer's disease. At the same time, zinc plays an important role in learning and memory, the functioning of the nervous system, and a variety of other bodily processes. The amount of zinc consumed in food or daily vitamins is beneficial, but the research just mentioned suggests that megadoses of zinc may not be advisable. It also casts some doubt on ten-year-old studies that proposed that there is a connection between aluminum and Alzheimer's disease. Have students find out about the changes in brain tissue that occur in this disease and write a short paper summarizing their findings. L2

291

Applying Chemistry

TECH PREP

Colored Compounds The fact that many colored compounds contain transition metal ions suggests that partially filled *d* orbitals may play a role in producing the colors. Further evidence is found in the fact that compounds of Zn^{2+} are colorless. The zinc ion has no unfilled *d* orbitals. In addition, compounds of Sc^{3+} are colorless. Scandium(III) has no *d* electrons. Colored compounds of transition elements are thought to have some electrons that require energy of the frequencies of visible light to raise them to a higher energy state. When a particular frequency is absorbed from incident white light, the reflected or transmitted light is colored.

Applying Chemistry

TECH PREP

Platinum Mufflers Automobile mufflers contain small amounts of platinum and rhodium, which catalyze the decomposition of several harmful combustion products such as nitrogen monoxide.

The principal ore of zirconium, a Group 4 element, is a silicate crystal called zircon. A zircon is a colorless crystal that can substitute for a diamond in jewelry. Often used in costume jewelry, jewelers refer to them as cubic zirconium.

Other Transition Elements: A Variety of Uses

You've learned that some transition elements are important in the production of steel because they impart particular properties to the steel. In addition, most of the transition elements, because of their individual properties, have a variety of uses in the production of the infrastructure and consumer products of the modern world.

The Iron Triad, Platinum Group, and Coinage Metals

Iron (Fe), cobalt (Co), and nickel (Ni) have nearly identical atomic radii, so it isn't surprising that these three elements have similar chemical properties. Like iron, both cobalt and nickel are naturally magnetic. Because of their similarities, the three elements are called the iron triad (group of three). Notice the positions of iron, cobalt, and nickel on the periodic table, as shown in **Figure 8.12**. They are in Period 4, Groups 8, 9, and 10. The elements below the iron triad in Periods 5 and 6—ruthenium (Ru), rhodium (Rh), palladium (Pd), osmium (Os), iridium (Ir), and platinum (Pt)—all resemble platinum in their chemical behavior and are called the platinum group. The platinum group elements are used as catalysts to speed up chemical reactions. Copper (Cu), silver (Ag), and gold (Au) in Group 11 are the traditional metals used for coins because they are malleable, relatively unreactive, and in the case of silver and gold, rare. You may have predicted that these elements would have similar chemical properties because they are in the same group. Notice the positions of the platinum group and the coinage metals in **Figure 8.12**.

Figure 8.12

The Iron Triad, Platinum Group, and Coinage Metals
The nearly identical atomic radii of the iron triad—iron, cobalt, and nickel—help explain the similar chemistry of these three elements. The similarities among the platinum group elements in Periods 5 and 6 emphasize the fact that there is little difference between the atomic radii of the elements in these periods in which inner *d* orbitals are being filled. The coinage metals show the expected similarity among elements in the same group.

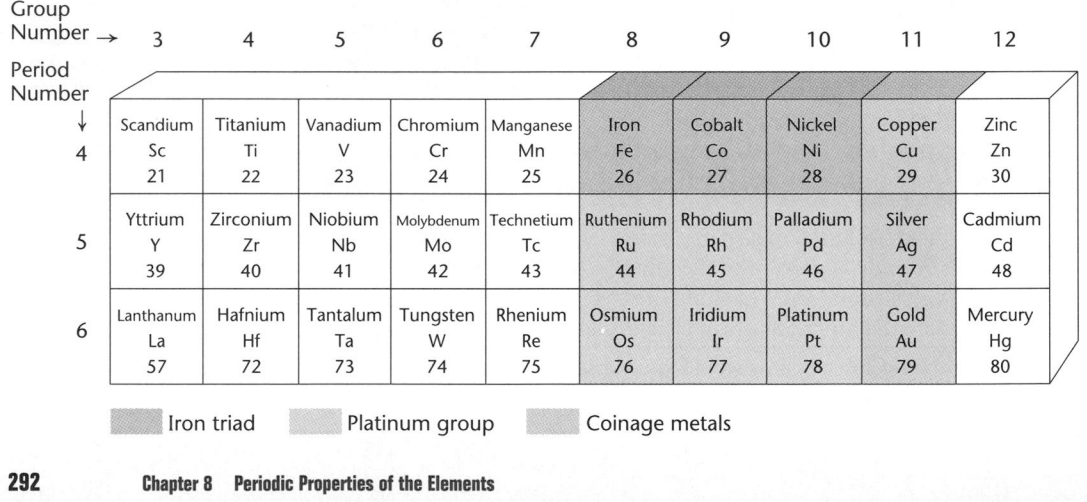

Iron triad   Platinum group   Coinage metals

Demonstration 8.6

Oxidation States of Vanadium

Purpose: To demonstrate the colorful ions formed as the oxidation number of vanadium changes from 5+ to 4+, 3+, and 2+.

Materials: granular zinc-mercury(II) chloride(aq) amalgam*, ammonium metavanadate (NH_4VO_3) solution*, 500-mL Erlenmeyer flask with stopper, powder funnel, 250-mL graduated cylinder
*This demonstration is available as a kit

from Flinn Scientific, Inc., catalog no. AP8472, by calling (800) 452-1261.

Safety Precautions:

Disposal: Code A; this demonstration will produce hazardous waste as described by EPA regulators and should be disposed of accordingly.

Procedure: A detailed set of instructions developed by Jim and Julie Ealy is avail-

able from Flinn Scientific as Chem Fax publication no. 8472, P.O. Box 219, Batavia, IL 60510. The following is used by permission of Flinn Scientific, Inc. The zinc-mercury amalgam is stored under water. After pouring off the water, pour 63 g of the zinc-mercury amalgam into a 500-mL Erlenmeyer flask through a powder funnel. Add 140 mL of ammonium metavanadate solution to the zinc-

Chromium

When chromium is alloyed with iron, tough, hard steels or steels that are corrosion-resistant are formed. Chromium is also alloyed with other transition metals to produce structural alloys for use in jet engines that must withstand high temperatures. A self-protective metal, chromium is often plated onto other materials to protect them from corrosion.

Chromium has the electron configuration $[Ar]4s^13d^5$ and exhibits oxidation numbers 2+, 3+, and 6+. When chromium loses two electrons, it forms the Cr^{2+} ion and has the configuration $[Ar]3d^4$. The Cr^{3+} ion results when chromium loses a second $3d$ electron. Chromium can lose six valence electrons and have an oxidation number of 6+. When it does, it loses all of its s and d electrons and assumes the electron configuration of argon. Potassium chromate (K_2CrO_4) and potassium dichromate ($K_2Cr_2O_7$) are two compounds in which chromium's oxidation number is 6+. **Figure 8.13** shows the brilliant colors that are typical of compounds of transition elements. Chromium gets its name from the Greek word for color, *chroma*, and many of its compounds are brightly colored—yellow, orange, blue, green, and violet.

Figure 8.13

Two Compounds of Chromium
The brilliant colors of these chromium compounds are typical of many transition metal compounds. In both the yellow potassium chromate (K_2CrO_4) and the orange potassium dichromate ($K_2Cr_2O_7$), chromium has a 6+ oxidation number.

Zinc

Like chromium, zinc is a corrosion-resistant metal. One of its principal uses is as a coating on iron and steel surfaces to prevent rusting. In the process called galvanizing, a surface coating of zinc is applied to iron by dipping the iron into molten zinc. Zinc is also important when alloyed with other metals. The most important of these alloys is the combination of zinc with copper in brass. Brass is used for making bright and useful objects like those shown in **Figure 8.14.**

Figure 8.14

The Importance of Brass
Brass can be worked into smooth shapes and drawn into the long, thin-walled tubes needed for musical instruments. Look around and see how many brass items you find that are both decorative and useful.

Correcting Misconceptions

Students may think that the inner transition elements, the lanthanides and actinides, are not part of the periodic table because they are set off at the bottom. Show students where they fit in the chart.

GLENCOE TECHNOLOGY

 Videodisc

Chemistry: Concepts and Applications
Oxidation States of Vanadium
Disc 1, Side 2, Ch. 8

Show this video of the Demonstration on this page to introduce the concept of multiple oxidation states. [L1] [LEP]

 CD-Rom

Chemistry: Concepts and Applications
Oxidation States of Vanadium
Have students use this video of the Demonstration to help reinforce the concept of oxidation numbers and electron configuration. [L1] [LEP]

---

mercury amalgam in the flask. **CAUTION:** *The solutions are toxic and corrosive.* Note the color change as vanadium with oxidation number 5+ (VO_2^+, yellow) is converted to vanadium with oxidation number 4+ (VO^{2+}, blue). The yellow and blue of the two ions mix to give a green color. Stopper the flask and shake gently until the color of the solution changes from green to blue to blue-green as vanadium with oxidation number 4+ (VO^{2+}, blue) is converted to V^{3+}, blue-green. Finally, shake the stoppered flask vigorously to change V^{3+} to V^{2+}, violet.

Results: Equilibrium reactions account for the mixtures of colors.

$$Zn + 2VO_2^+ + 4H^+ \rightleftarrows 2VO^{2+} + Zn^{2+} + 2H_2O$$

$$Zn + 2VO^{2+} + 4H^+ \rightleftarrows 2V^{3+} + Zn^{2+} + 2H_2O$$

$$2V^{3+} + Zn \rightleftarrows 2V^{2+} + Zn^{2+}$$

Analysis: Ask these questions.
1. Where is vanadium found on the periodic table? *Group 5, period 4, among the transition elements*
2. Which electrons are lost as vanadium's oxidation number changes from 0 to 5+? *the 4s and 3d electrons*

3 ASSESS

Check for Understanding

- Check for understanding of the section material by having students answer the Section Review questions.
- Ask students to draw a periodic table that includes the inner-transition elements in the body of the chart. *The inner-transition elements should be placed between Groups 3 and 13 in periods 6 and 7.* L1 LEP

Reteach

Discussion: Ask students which groups of the periodic table could be referred to as:
the *s* sublevel block, *Groups 1 and 2*
the *p* sublevel block, *Groups 13-18*
the *d* sublevel block, *Groups 3-12*
the *f* sublevel block, *the inner-transition elements* L1

The lanthanides were once called the rare earth elements. However, these elements are not so rare in the United States and Canada. More than half of the world's supply of the lanthanides comes from a single mine in California. Many of these elements are used to make ceramic superconductors.

Lanthanides and Actinides: The Inner Transition Elements

The inner transition elements are found in the *f* block of the periodic table. In the lanthanides, electrons of highest energy are in the 4*f* sublevel. The lanthanides were once called *rare earth elements* because all of these elements occurred in Earth's crust as *earths,* an older term for oxides, and seemed to be relatively rare. The highest-energy electrons in the actinides are in the 5*f* sublevel. You probably won't find these elements among your household chemicals. Their names are unfamiliar except for uranium and plutonium, which are the elements associated with nuclear reactors and weapons. However, many of these elements, especially lanthanides, have important practical uses.

Cerium: The Most Abundant Lanthanide

Cerium is the principal metal in the alloy called misch metal. Misch metal is 50 percent cerium combined with lanthanum, neodymium, and a small amount of iron. Misch metal is used to make the flints for lighters. Cerium is often included in alloys of iron and other metals such as magnesium. A high-temperature alloy of three percent cerium with magnesium is used for jet engines. Some of cerium's compounds—for example, cerium(IV) oxide (CeO_2)—are used to polish lenses, mirrors, and television screens; in glass manufacturing to decolorize glass; and to make porcelain coatings opaque.

Other Lanthanides

Other lanthanides are used in the glass industry. Neodymium (Nd) is used not only to decolorize glass but to add color to glass. When added to the glass used for welders' goggles, neodymium and praseodymium (Pr)

absorb the eye-damaging radiation from welding, as shown in **Figure 8.15.** They also decrease reflected glare when used in the glass of a television screen. A combination of the oxides of yttrium (Y), a transition element, and europium (Eu) produce a phosphor that glows a brilliant red when struck by a beam of electrons, such as in a TV picture tube. This phosphor is used with blue and green phosphors to produce realistic-looking television pictures. When rare earth phosphors are used in mercury-arc outdoor lighting, they change the bluish light of the mercury arc to a clear white light.

Figure 8.15

Lanthanides Provide Eye Protection
Intense light from a welding torch can harm eyes. Neodymium and praseodymium, incorporated into the lenses of goggles, absorb the damaging wavelengths.

Integrating THE Sciences

Geology By no means a rare element, uranium is found in several ores including pitchblende and uraninite. From these ores, uranium is obtained for use as a nuclear fuel. Uranium is a hard, silvery-white metal with the metallic properties of ductility and malleability, but it is not a good conductor of electricity. Have students find out what isotopes are present in naturally occurring uranium and in what proportion. *Uranium-238 comprises 99.27% of natural uranium. Uranium-235 makes up 0.72%, and uranium-234, 0.006%.* L1

Because europium, gadolinium (Gd), and dysprosium (Dy) are good absorbers of neutrons, they are used in control rods in nuclear reactors. Promethium (Pm) is the only synthetic element in the lanthanide series. It is obtained in small quantities from nuclear reactors and is used in specialized miniature batteries. Samarium (Sm) and gadolinium are used in electronics. Terbium (Tb) is used in solid-state devices and lasers.

Radioactivity and the Actinides

Uranium (U) is a naturally occurring, radioactive element used as a source of nuclear fuel and other radioactive elements. Plutonium (Pu) is one of the elements obtained from the use of uranium as a nuclear fuel. The isotope Pu-238 emits radiation that is easily absorbed by shielding. Pu-238 is used as a power source in heart pacemakers and navigation buoys. Other isotopes of plutonium are used as nuclear fuel and in nuclear weapons. Plutonium is the starting material for the synthetic production of the element americium, which is used in smoke detectors.

Some actinides have medical applications; for example, radioactive californium-252 (Cf) is used in cancer therapy. Better results in killing cancer cells have been achieved using this isotope of californium than by using the more traditional X-ray radiation.

Connecting Ideas

If you can locate an element on the periodic table, you can predict its properties. Each element has unique characteristics because of its unique electron configuration. Together, the elements, their alloys, and compounds provide a wide variety of materials for countless applications. Compounds of the elements range from ionic to covalent, from polar to nonpolar. They have size and shape. In Chapter 9, you'll learn more about the formation of compounds and how to predict their shape and polarity.

> **WORD ORIGIN**
>
> actinide:
> *aktis* (Gk) a ray
> The actinides are named for actinium, a radioactive element.

SECTION REVIEW

Understanding Concepts

1. What are transition elements? What are inner transition elements? Describe where all the transition elements are found on the periodic table.

2. How do the electron configurations of the transition metals and inner transition elements differ from those of the main group metals?

3. Iron, aluminum, and magnesium all have important uses as structural materials, yet iron corrodes (rusts), whereas aluminum and magnesium usually do not corrode. Explain what causes iron to rust.

Thinking Critically

4. **Applying Concepts** Why do the transition metals have multiple oxidation numbers whereas the alkali metals and the alkaline earth metals have only one oxidation number, 1+ and 2+, respectively?

Applying Chemistry

5. **Steelmaking** Name three transition metals that are added to iron and steel to improve their properties and help prevent corrosion. Explain how each additive works.

8.2 Transition Elements   **295**

SECTION REVIEW

1. The transition elements are those in the *d* block in the periodic table—Groups 3 through 12, periods 4 through 7. The inner-transition elements form the *f* block of the periodic table. They are the lanthanides and the actinides.
2. The transition metals fill *d* orbitals; the inner-transition elements fill *f* orbitals.
3. Iron rusts by reacting with oxygen and moisture in the air. This is a problem because the iron oxide flakes off and exposes the inner metal, which reacts with oxygen and water to form more rust.
4. The inner *d* electrons of the transition metals are involved in their chemical reactions.
5. Chromium and nickel help prevent corrosion and harden the steel. Manganese increases the hardness of steel.

SECTION 8.2

Extension

Ask students how many elements would be included in Period 8 of the periodic table if 200 elements existed. *50, according to the predicted continuation of the aufbau sequence* [L1]

4 CLOSE

Discussion

Ask students to draw the Lewis electron dot diagrams for arsenic, antimony, and bismuth and to explain why each of the elements has the common oxidation numbers 3+ and 5+. Ask them to explain why arsenic and antimony also have the oxidation number 3−. *The atoms of the three elements have five electrons in the highest energy level: two in the s sublevel and three in the p sublevel. If the atoms lose the three electrons in the p sublevel, they become 3+ ions. If they lose all five electrons, they become 5+ ions. If the more nonmetallic atoms gain three electrons to attain the noble-gas configuration, they become 3− ions.* [L1]

GLENCOE TECHNOLOGY

 Videotape

MindJogger Videoquizzes for Chemistry
Chapter 8: Periodic Properties of the Elements
Have students work in groups as they play the videoquiz game to review key chapter concepts. [L1] [LEP] [COOP LEARN]

- Review the Reviewing Main Ideas statements and Key Terms with your students.
- Complete solutions to Chapter Review problems can be found in the *Problems and Solutions Manual.*

UNDERSTANDING CONCEPTS

1. The atom with the larger atomic radius is **a.** K; **b.** Na; **c.** Ca; **d.** Cs; **e.** Sr; **f.** S

2. **a.** Ca^{2+} because the outer electrons are in a higher energy level
   b. S^{2-} because the number of electrons is the same, but Cl^- has a greater nuclear charge
   c. F^- because the number of electrons is the same, but Na^+ has a greater nuclear charge
   d. Ca^{2+} because its outer electrons are in a higher energy level, and Al^{3+} has a greater positive charge

3. As atomic radius increases, chemical reactivity decreases. Fluorine is the most active halogen; iodine is the least active.

4. As atomic radius increases, chemical reactivity increases. Cesium is the most active common alkali metal; lithium is the least active. Students may correctly answer that francium is the most reactive.

5. **a.** the potassium atom because the nuclear charge is the same, but the potassium atom has its $4s$ valence electron while the potassium ion has no electrons in the fourth energy level
   b. the bromide ion because the nuclear charge is the same, but the bromide ion has an additional electron

6. Elements in a group have similar chemical properties because they have similar valence-electron configurations.

7. Hydrogen gas and a hydroxide are formed from both.

REVIEWING MAIN IDEAS

8.1 Main Group Elements

- In a period of the periodic table, the number of valence electrons increases as atomic number increases. As a result, elements change from metal to metalloid to nonmetal to noble gas.

- Atomic size is a periodic property. As atomic number increases in a period, atomic radius decreases. As atomic number increases in a group, atomic radius increases.

- Positive ions have smaller atomic radii than the neutral atoms from which they derive. Negative ions have larger atomic radii than their neutral atoms.

- Positive ions in the same group increase in size down the group.

- In a group, each element has the same number of valence electrons. As a result, the elements in a group show similar chemical behavior.

- Metals react by losing electrons. The most reactive metals are those that give up electrons most easily. The metal with the biggest atom and smallest number of valence electrons is the most active metal. Cesium in the lower-left-hand corner of the periodic table is the most active metal.

- Nonmetals react by gaining or sharing electrons. The most reactive nonmetals are those that attract and hold electrons most strongly. The nonmetal with the smallest atom and greatest number of valence electrons is the most active nonmetal. Fluorine in the upper right-hand corner of the periodic table is the most active nonmetal.

8.2 Transition Elements

- Transition elements react by losing the valence electrons in their s orbitals. Many of the transition metals have more than one oxidation number because they also lose electrons from their d orbitals.

- The lanthanides and actinides react by losing the valence electrons in their s orbitals. Because these elements can also lose electrons from their d and f orbitals, they have multiple oxidation numbers.

Key Terms

For each of the following terms, write a sentence that shows your understanding of its meaning.

alkali metal
alkaline earth metal
halogen

UNDERSTANDING CONCEPTS

1. From each of the following pairs of atoms, select the one with the larger atomic radius.

   a) K, Ca   d) Rb, Cs
   b) F, Na   e) Ca, Sr
   c) Mg, Ca   f) S, C

2. From each of the following pairs of ions, choose the ion with the larger ionic radius and explain your choice.

   a) Mg^{2+}, Ca^{2+} c) F^-, Na^+
   b) S^{2-}, Cl^- d) Ca^{2+}, Al^{3+}

3. What is the effect of atomic radius on the chemical reactivity of the halogens? Which is the most active halogen? Which is the least active?

4. What is the effect of atomic radius on the chemical reactivity of the alkali metals? Which is the most active alkali metal? Which is the least active?

5. Choose the larger from the following pairs and explain your choice.

   a) a potassium atom or a potassium ion (K^+)
   b) a bromine atom or a bromide ion (Br^-)

✓ Assessment

Portfolio: Review the portfolio options that are provided throughout the chapter. Encourage students to select one product that demonstrates their best work for the chapter. Have students explain what they have learned and why they chose this example for placement into their portfolios. **P**

Additional portfolio options can be found in the following **Teacher Classroom Resources:**
Tech Prep Applications
Chemistry and Industry
Consumer Chemistry

6. Why do elements in a group have similar chemical properties?

7. What are the products of a reaction between an alkali metal and water? An alkaline earth metal and water?

8. Zinc is one of the few transition metals that has a single oxidation number. What is the oxidation number for zinc? How does zinc's electron configuration account for this oxidation number?

9. What is the electron configuration of the fluoride ion? What noble gas has that configuration? Why do elements tend to attain a noble-gas configuration?

10. Use the graph of atomic radius versus period number to answer these questions. How does the size of a lithium atom compare with that of a cesium atom? How does the size of a fluorine atom compare with that of an iodine atom? Which has the larger radius, lithium or fluorine? Based on their atomic radii, which is the most active alkali metal? Which is the most active halogen? Explain.

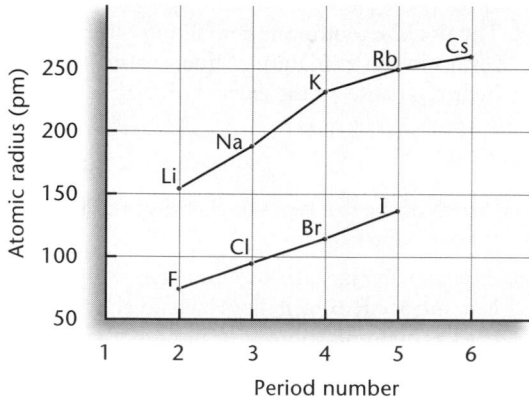

Chemistry and Technology

12. Explain how steel can fill a variety of purposes ranging from making farm equipment to food cans.

History Connection

13. Use the periodic table to determine the oxidation numbers of lead and sulfur in galena, PbS, the common ore of lead.

Everyday Chemistry

14. The friction of a safety match rubbed on the striking surface causes red phosphorus to change to white phosphorus. Is the reaction endothermic or exothermic?

Biology Connection

15. Hydroxyapatite, $Ca_5(PO_4)_3OH$, in tooth enamel undergoes demineralization and remineralization. Explain what happens in the two processes and how they are related.

16. Nitrogen, oxygen, and phosphorus are essential elements for maintaining life. What essential biological molecules contain these elements?

17. Write a balanced equation for the reaction of potassium with oxygen.

18. Describe one way in which nitrogen (N_2) is converted into a form that plants can use.

19. What household substances are likely to contain the following sodium compounds? What is each substance used for?

   a) $NaHCO_3$ c) $NaOH$
   b) $NaOCl$ d) $NaCl$

20. Magnesium and aluminum both react with oxygen to form oxides. Explain why they are considered corrosion-resistant metals.

● THINKING CRITICALLY

Comparing and Contrasting

21. Compare the chemical behavior of oxygen when it combines with an active metal such as calcium to its behavior when it combines with a nonmetal such as sulfur.

● APPLYING CONCEPTS

How It Works

11. Why are noble gases used in lightbulbs? What advantages do they offer over a vacuum?

---

8. 2+; zinc's $3d$ sublevel is filled and only the $4s$ electrons are available.

9. $1s^2 2s^2 2p^6$; neon; a noble-gas configuration is an octet of electrons in the outermost energy level. Noble-gas configurations are stable.

10. A lithium atom is smaller than a cesium atom. A fluorine atom is smaller than an iodine atom. A lithium atom is larger than a fluorine atom. The most active alkali metal is cesium because it has the largest atomic radius in its family and it gives up electrons most easily. Fluorine is the most active halogen because it has the smallest atomic radius in its family and it attracts and holds electrons most strongly.

APPLYING CONCEPTS

11. Noble gases carry heat away from the filament and prevent it from boiling away.

12. Steels are formulated with different amounts of carbon and other elements, especially metals. These additives impart specific characteristics to the steel that suit particular purposes.

13. lead, 2+; sulfur, 2−

14. endothermic

15. In the process of demineralization, $Ca_5(PO_4)_3OH$ breaks up into five Ca^{2+}, three PO_4^{3-}, and one OH^-. In remineralization, the ions recombine to form $Ca_5(PO_4)_3OH$.

16. proteins, RNA, and DNA

17. $4K(s) + O_2(g) \rightarrow 2K_2O(s)$

18. Nitrogen is converted into ammonia and nitrates by the action of soil bacteria and lightning.

19. a. baking soda; baking, cleaning, antacid
b. bleach; laundry bleach, disinfectant, cleaning
c. drain cleaner; open clogged drains
d. table salt; nutritional source of sodium

---

20. Magnesium and aluminum form self-protective oxide coatings that resist further reaction with oxygen.

THINKING CRITICALLY

21. When oxygen combines with calcium, it forms an ionic bond. When oxygen combines with sulfur, it forms covalent bonds.

22. sharing electrons; $\ddot{O}::C::\ddot{O}$ The shared electrons are counted in the octets of both elements involved in the bond.

23. ZnC_2O_4
24. atomic size; cesium
25. Fe, $[Ar]4s^2 3d^6$; Fe^{2+}, $[Ar]3d^6$; Fe^{3+}, $[Ar]3d^5$
26. Both produce a hydroxide and hydrogen gas.
   $2Na + 2H_2O \rightarrow H_2 + 2NaOH$
   $2Cs + 2H_2O \rightarrow H_2 + 2CsOH$
27. Ionization energy is a periodic property. Ionization energy generally increases across a period and decreases down a group.

CUMULATIVE REVIEW

28. Metals have metallic luster, conduct heat and electricity, can be shaped, and are all solids except mercury. Nonmetals may be solids, liquids, or gases; do not conduct electricity; and do not conduct heat well. Solid nonmetals may be hard or soft and are usually brittle, have no luster, and are various colors. Metalloids have properties similar to both metals and nonmetals.

29. a. $[He]2s^22p^5$; **b.** $[Ne]3s^23p^1$; **c.** $[Ar]4s^23d^2$; **d.** $[Ne]3s^23p^6$

30. Elements in Groups 1 and 2 lose electrons and become positive ions. Elements in Groups 16 and 17 gain electrons and become negative ions. Positive and negative ions then react to form compounds. KBr; Na_2O

31. a. $Al_2(SO_4)_3$, ionic; **b.** N_2O, covalent; **c.** $NaHCO_3$, ionic; **d.** $Pb(NO_3)_2$, ionic

32. single-displacement

33. Helium, neon, argon, krypton, xenon, and radon are unreactive noble gases.

AUTHENTIC ASSESSMENT

1. Atomic radius decreases across Periods 2 and 3.

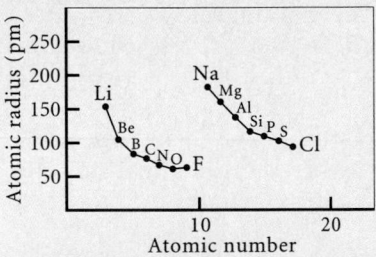

2. Atomic radius increases down Groups 2 and 17.

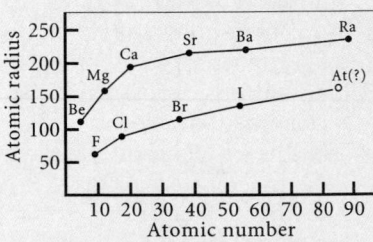

3. There are no abrupt changes in the properties of elements across a row or down a column. It is logical to expect

---

Relating Concepts

22. Is carbon dioxide, CO_2, produced by sharing or transferring electrons? Draw the Lewis dot diagram for CO_2 and explain how each element achieves an octet of electrons.

Identifying Patterns

23. ChemLab Zinc exhibits the same oxidation number as the alkaline earth metals. What is the chemical formula of the compound formed when zinc combines with an oxalate ion?

Relating Cause and Effect

24. MiniLab 1 What factor contributes to the difference in reactivity between potassium and cesium? Which element is more reactive?

Relating Concepts

25. MiniLab 2 Write the electron configurations of the neutral iron atom and the ions Fe^{2+} and Fe^{3+}.

Making Predictions

26. Both sodium and cesium react with water. Predict the products of the reactions of sodium and cesium with water. Write balanced equations for both reactions.

Using a Table

27. The energy needed to remove the first electron from a gaseous atom to produce a gaseous ion is called the first ionization energy. First ionization energies for the main group elements are shown in the table. Examine the data and decide whether ionization energy is a periodic property. Describe how ionization energies change within a period and within a group.

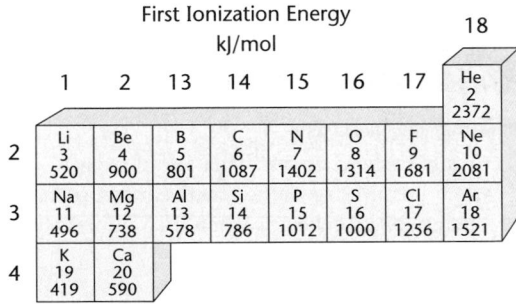

First Ionization Energy kJ/mol

---

one or two elements in each row or column to have properties intermediate between those of metals and nonmetals. B, Si, Ge, As, Sb, Te, Po, and At look metallic but are often brittle like nonmetals. They are also semiconductors.

4. Nitrogen is responsible for leaf production and greener leaves. Phosphorus hastens plant maturity and increases seed yields. Potassium helps in carbohydrate and protein synthesis. Fertilizers are designed to compensate for soil deficiencies or to supply nutrients for particular plant needs.

CUMULATIVE REVIEW

28. Describe the physical properties of metals, metalloids, and nonmetals. (Chapter 3)

29. Write the electron configurations for each of the following atoms. Use the appropriate noble-gas inner core abbreviations. (Chapter 7)

  a) fluorine   **c)** titanium
  b) aluminum   **d)** argon

30. Explain why many compounds are combinations of a Group 1 or a Group 2 element with an element from Group 16 or Group 17. Give the formula of a compound formed from an element in Group 1 and an element in Group 17. Give the formula of a compound formed from an element in Group 1 and an element in Group 16. (Chapter 4)

31. Write formulas for the following compounds. Are they ionic or covalent? (Chapter 4)

  a) aluminum sulfate
  b) dinitrogen monoxide
  c) sodium hydrogen carbonate
  d) lead(II) nitrate

32. The use of a toothpaste containing fluoride results in the formation of fluorapatite, from hydroxyapatite, in the enamel of teeth.

$$Ca_5(PO_4)_3OH(s) + F^-(aq) \rightarrow$$
$$Ca_5(PO_4)_3F(s) + OH^-(aq)$$

Which of the five types of chemical reactions is this? (Chapter 6)

33. Explain why the gaseous nonmetals—hydrogen, nitrogen, oxygen, fluorine and chorine—exist as diatomic molecules, but other gaseous nonmetals—helium, neon, argon, krypton, xenon and radon—exist as single atoms. (Chapter 4)

---

5. Within a period, ionization energy increases as atomic radius decreases. Within a group, ionization energy decreases as atomic radius increases.

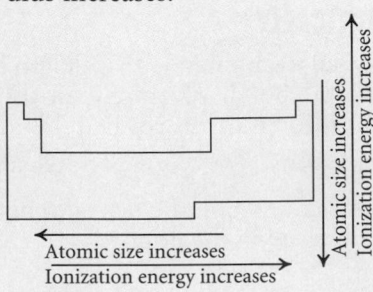

Skill Review

1. **Making and Using a Graph** Use the data in the figure below to draw a graph of atomic radius versus atomic number for the second and third period elements in Groups 1, 2, 13, 14, 15, 16, and 17. Describe any patterns you observe.

1	2		13	14	15	16	17
1 H 78							
2 Li 156	Be 112		B 85	C 77	N 71	O 60	F 69
3 Na 186	Mg 160		Al 143	Si 118	P 109	S 103	Cl 91
4 K 231	Ca 197		Ga 134	Ge 123	As 121	Se 117	Br 119
5 Rb 248	Sr 215		In 167	Sn 141	Sb 161	Te 138	I 138
6 Cs 262	Ba 222		Tl 170	Pb 175	Bi 151	Po 164	At Unknown
7 Fr 280	Ra 228						

$1 \text{ pm} = 10^{-12} \text{ m}$

2. Using the data in the above figure draw a graph of atomic radius versus atomic number for the elements in Groups 2 and 17. Describe any patterns you observe.

Writing in Chemistry

3. Write a short article explaining the properties of elements called metalloids. Use the concept of periodicity as the basis for your article. List the symbols of the metalloids and describe some of the properties that relate metalloids to both metals and nonmetals.

4. The numbers 5-10-5 on a fertilizer package refer to the percentages of nitrogen (N), phosphorus (P), and potassium (K), respectively, in the fertilizer. Write a short paper describing the role of each of these elements in plants. Research other fertilizer compositions, and explain the purpose of varying the proportions of these nutrients.

Problem Solving

5. Draw a sketch of the outline of a periodic table. Use the data from the diagram in question 1 of the Skill Review to help you draw arrows on the table showing changes in atomic radius and ionization energy in periods and groups as atomic number increases. Write a sentence describing the relationship between ionization energy and atomic radius in both periods and groups.

Projects

6. You rarely come in contact with most of the main group metals in the free state, but their salts have many uses in your daily life. Find out about the salts of the metals in Groups 1 and 2, and write a paragraph describing the use of one salt of each of the following elements: Li, Na, K, Be, Mg, and Ca.

7. Find out more about the structure of bone tissue and the role of calcium and other minerals present in bone tissue. Make a poster or prepare an oral report on your findings. Include information about artificial joints and how they are made to simulate natural bone tissue.

8. The graph of atomic radii below illustrates a phenomenon called the lanthanide contraction. Investigate the lanthanide contraction and write a report explaining what it is and the effect it has on the chemical properties of the transition elements. Use data from the graph to support your conclusions.

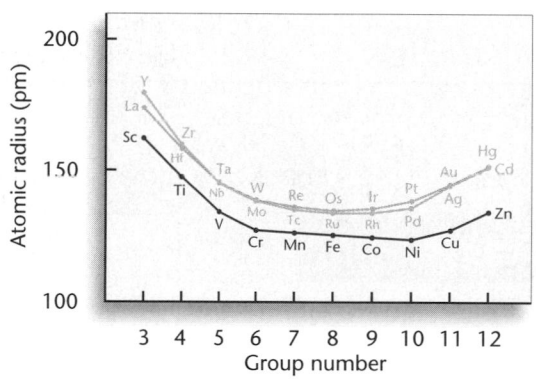

Program Resources

Chapter Review and Assessment,
  pp. 47-52 L1
Alternate Assessment in the Science
  Classroom L1
Computer Test Bank, Chapter 8 L1
Performance Assessment in the Science
  Classroom L1

Problems and Solutions Manual,
  Chapter 8 L1
Supplemental Practice Problems,
  Chapter 8 L1

6. Probable answers include lithium carbonate used to treat depression, sodium chloride used as table salt, potassium chloride used as a salt substitute and in fertilizers along with potassium nitrate and potassium sulfate, magnesium hydroxide used to treat indigestion, and calcium chloride used to melt ice on roads. Beryllium salts are toxic and play no role in daily life.

7. Answers should describe the role of calcium in the formation of bone tissue and include reasons why calcium is needed in the diet. A diagram could show the components of bone: an outer membrane that supplies blood vessels and nerves, and inner layers of compact and spongy bone that surround a core of bone marrow. An advanced essay may describe the movement of calcium ions into and out of bone tissue. Bioceramic coatings, being researched and developed for artificial joints, include calcium and phosphorus. Other artificial joints use titanium.

8. The graph shows the expected increase in the size of the $4d$ transition elements compared with the $3d$ elements. However, little difference is seen in the sizes of the $4d$ and $5d$ elements. This is because the lanthanide elements that come between the $4d$ and $5d$ elements add 14 electrons to the inner $4f$ sublevel. Because the $4f$ orbitals are inner orbitals, the added electrons do not increase the size of the atom. Instead, the increase in nuclear charge tends to decrease the size. The result of the contraction in size is that the chemistry of the $4d$ and $5d$ transition elements is remarkably similar.

Chapter 9 Organizer

Section	Objectives	Activities/Features
9.1 **Bonding of Atoms** (4 days)	1. **Predict** the type of bond that forms between atoms by using electronegativity values. 2. **Compare and contrast** characteristics of ionic, covalent, and polar covalent bonds. 3. **Interpret** the sea of electrons model of metallic bonding.	**History Connection:** Linus Pauling: An Advocate of Knowledge and Peace, p. 307 **MiniLab 1:** Coffee Filter Chromatography, p. 312 **Discovery Demo:** 9.1 Forming Ionic and Covalent Bonds, pp. 300-301 **Demonstrations:** 9.2 Combustion of Ethanol, pp. 304-305; 9.3 Paper Chromatography, pp. 310-311
9.2 **Molecular Shape and Polarity** (4 days)	4. **Diagram** electron dot structures for molecules. 5. **Formulate** three-dimensional geometry of molecules from electron dot structures. 6. **Predict** molecular polarity from three-dimensional geometry and bond polarity.	**People in Chemistry:** Dr. William Skawinski, Chemist, pp. 316-317 **Everyday Chemistry:** Jiggling Molecules, p. 320 **MiniLab 2:** Modeling Molecules, p. 325 **Chemistry and Technology:** Chromatography, pp. 326-327 **ChemLab:** What colors are in your candy?, pp. 328-329 **Demonstrations:** 9.4 Electron Cloud Geometries, pp. 318-319; 9.5 Like Charges Repel, pp. 322-323

Activity Materials

ChemLab (pages 328-329)
chromatography paper, colored candies, toothpicks, salt, jars, small cups, food coloring, ruler, pencils

MiniLab 1 (page 312)
coffee filters, washable markers, cups

MiniLab 2 (page 325)
model kits

Demonstrations For a complete list of materials for the demonstrations in this chapter, see pages 300, 304, 310, 318, and 322.

KEY TO TEACHING STRATEGIES

The following designations will help you decide which activities are appropriate for your students.

- **L1** Level 1 activities should be within the ability range of all students.
- **L2** Level 2 activities should be within the ability range of the average to above-average student.
- **L3** Level 3 activities are designed for the ability range of above-average students.
- **LEP** LEP activities should be within the ability range of Limited English Proficiency students.
- **COOP LEARN** Cooperative Learning activities are designed for small group work.
- **P** These strategies represent student products that can be placed into a best-work portfolio.

Chemical Bonding

Teacher Classroom Resources

STUDENT MASTERS

Study Guide: pp. 35-36
ChemLab and MiniLab Worksheets: pp. 59-61, 64-65
Laboratory Manual: 9.1 Diagnostic Properties of Bonds, pp. 77-82

Study Guide: pp. 37-38
Critical Thinking/Problem Solving: p. 10
ChemLab and MiniLab Worksheets: pp. 62-63
Laboratory Manual: 9.2 Electron Clouds, pp. 83-86

TEACHING AIDS

Section Focus Transparency 18
Section Focus Transparency Master: p. 18
Basic Concepts Transparency 24, 25
Basic Concepts Transparency Masters: pp. 47-50
Lesson Plans: p. 18

Section Focus Transparency 19
Section Focus Transparency Master: p. 19
Basic Concepts Transparency 26
Basic Concepts Transparency Master: pp. 51-52
Problem Solving Transparency 11
Problem Solving Transparency Master: pp. 21-22
Lesson Plans: p. 19

ASSESSMENT RESOURCES
Chapter Review and Assessment: pp. 53-58
Alternate Assessment in the Science Classroom
Computer Test Bank: Chapter 9
Performance Assessment in the Science Classroom
Problems and Solutions Manual: Chapter 9
Supplemental Practice Problems: Chapter 9

CHAPTER RESOURCES
Applying Scientific Methods in Chemistry: pp. 24-27
Spanish Resources: Chapter 9
English/Spanish Audiocassettes: Chapter 9
Cooperative Learning in the Science Classroom
Lab and Safety Skills in the Science Classroom

GLENCOE TECHNOLOGY

Software
Mastering Concepts in Chemistry: *Unit 9* Lesson 1 Electronegativity and Bond Type
Videotape
MindJogger Videoquizzes: Chapter 9
Videodisc
Chemistry: Concepts and Applications, Demonstrations, Videos, and Animations
Science and Technology Videodisc Series: Disc 2 Chemistry *Snowflakes*
CD-ROM Multimedia System
Chemistry: Concepts and Applications, Same as for Videodisc, plus Interactive Exploration and Major Simulation, *The Periodic Table*

TECH PREP

The following Glencoe resources provide opportunities for integrating science and technology.

Student Edition: Everyday Chemistry, p. 320; Chemistry and Technology, p. 326

Teacher Wraparound Edition: Integrating the Sciences, p. 308

Teacher Classroom Resources: Consumer Chemistry, pp. 17-18; Chemistry and Industry, pp. 23-26

9 Chemical Bonding

Chapter Overview

Chemical bonding is viewed as a continuum from the complete transfer of electrons and the formation of ions, to equal sharing of electrons in a purely covalent bond. Metallic bonding is also introduced. The chapter shows a method for predicting the three-dimensional geometry of molecules, and explains how the geometry of a molecule determines its polarity.

Theme Connection

Macro to Submicro Just as the properties of a building are determined by an underlying structure that cannot be seen, the properties of matter are determined at the submicroscopic level by the type of chemical bonds that hold the atoms together and by their spatial arrangement.

✔ Assessment Planner

Choose assessment strategies from the following pages to evaluate the progress of your students.
Assess, pp. 306, 308, 313, 332
MiniLabs, pp. 312, 325
ChemLab, p. 328
Portfolio, pp. 311, 320, 334
Chapter Review, p. 334

300

Discovery Demo 9.1

Forming Ionic and Covalent Bonds

Purpose: To demonstrate the formation of an ionic bond and a covalent bond.
Materials: 5 cm of magnesium ribbon, 2 g of roll sulfur, tongs, large metal can, deflagrating spoon, laboratory burner

Safety Precautions:
Disposal: Codes A and H

Procedure: Place a large can in the sink. Carefully ignite a 5-cm length of magnesium ribbon. **CAUTION:** *Wear goggles when burning magnesium. Do not look directly at the burning metal.* Hold the magnesium with tongs while it burns inside the can to prevent students from looking directly at the burning metal. If the room lights are turned off, students can safely observe the reflected light. In a darkened room, place a small piece of roll sulfur in a deflagrating spoon and briefly heat it in a burner flame. The sulfur will ignite and burn with a low blue flame. **CAUTION:** *Use a fume hood; SO$_2$ vapor is toxic.*

Results: Magnesium burns with a bright, white-hot flame. Sulfur burns with a low blue flame. The stability of MgO is evidenced by the large amount of heat and light produced when it is formed com-

CHAPTER
9
Chemical Bonding

You just finished playing basketball on a hot summer day. You're really thirsty, so you take a sip of water. That sip of water was about 600 sextillion (600 000 000 000 000 000 000 000) water molecules. If each water molecule were the size of a basketball, 600 sextillion basketballs would cover Earth to a height of more than 1500 km. Six hundred sextillion is a large number when you're talking about basketballs, but is it a lot of water? Not really. It's 6×10^{23} molecules—about 18 mL. That's not even a good gulp. So if you're thirsty, 600 000 000 000 000 000 000 000 water molecules won't quench your thirst. Water molecules are incredibly small.

What does a water molecule look like? It isn't possible to see a single molecule, but in Chapter 4, you used a covalent bonding model to explain how two hydrogen atoms and one oxygen atom bond together to form the water molecule. Does this model tell anything about the shape of the molecule? It's important to find out because the bonding patterns and the shapes of all molecules have a lot to do with their properties.

Concept Check

Review the following concepts before studying this chapter.
Chapter 2: the use of electron dot diagrams to depict valence electrons
Chapter 4: properties of ionic compounds and covalent compounds
Chapter 8: trends in periodic properties of the elements

Chapter Preview

9.1 Bonding of Atoms

9.2 Molecular Shape and Polarity

Laboratory Activities
ChemLab
What colors are in your candy?

MiniLabs
1. Coffee Filter Chromatography
2. Modeling Molecules

Chemistry Around You

Materials as simple as gumdrops and toothpicks can be used to model the bonding and shape of simple molecules like water. More complicated molecules may require more sophisticated modeling methods such as those created on supercomputers. Images of actual atoms and molecules can be viewed only by using advanced electron microscopes as in the image on the left made from 28 carbon monoxide molecules on a platinum surface.

301

Introducing the Chapter
Pour 18 mL of water (6×10^{23} molecules) into a small beaker and display it while you brainstorm the extraordinary properties of water. Let the small volume of the water sample make the point that a water molecule is extremely small.

Using the Illustrations The opening photo of a thirsty player drinking water sets the stage for moving from a macro to a submicro view of chemical bonding. Gumdrops and toothpicks are tools to model the submicroscopic world. The electron microscope image of the carbon monoxide/platinum surface provides a partial glimpse into the atomic realm.

Revealing Misconceptions

Students often view electron dot diagrams as pictures of molecules. Remind students that electron dot models are only convenient devices that show the valence electrons of atoms and how valence electrons combine with the electrons of other atoms in chemical bonds.

GLENCOE TECHNOLOGY

 Videodisc

Chemistry: Concepts and Applications
Forming Ionic and Covalent Bonds
Disc 1, Side 2, Ch. 10

Show this video of the Discovery Demo to introduce the concept of bonding. L1 LEP

 CD-Rom

Chemistry: Concepts and Applications
Forming Ionic and Covalent Bonds
Have students use this video to help reinforce the concept of bonding. L1 LEP

pared to the heat and light produced in the formation of SO_2. Students have not been introduced to ΔH_f, but they can appreciate the difference in the magnitudes of the energies given off in the two reactions.
$2Mg(s) + O_2(g) \rightarrow 2MgO(s)$
$\Delta H_f = -602$ kJ/mol
$S(s) + O_2(g) \rightarrow SO_2(g)$
$\Delta H_f = -297$ kJ/mol
Point out to students that the greater energy released in the formation of MgO indicates that the ionic bond in MgO is stronger (more stable) than the covalent bonds in SO_2.
Analysis: Ask these questions.
1. With what element do magnesium and sulfur react? *oxygen*
2. Write balanced chemical equations for the chemical reactions. *See results.*
3. Which reaction releases more heat and light? *magnesium*

Key Concepts

Electronegativity is used as the key to differentiating among covalent, polar covalent, and ionic bonds. Metallic bonding is presented as a different way of sharing electrons that explains the unique properties of metals.

Planning Ahead

For the Demonstrations If you do not have a 6-gallon plastic bottle for Demonstration 9.2, obtain one from a bottled-water dealer.

1 FOCUS

Focus Transparency

Display the **Section Focus Transparency** for Section 9.1 to show competition between atoms for electrons. L1 LEP

2 TEACH

Visual Learning

Figure 9.1: When atoms form bonds, the atom with the greater attraction for electrons acquires a greater share in the electron pair. This can mean either a complete transfer of an electron from one atom to the other or some partial transfer. Point out to students that the degree of sharing in all bonds falls somewhere on the continuum from pure covalent to ionic. L1 LEP

SECTION 9.1

Bonding of Atoms

SECTION PREVIEW

Objectives

Predict the type of bond that forms between atoms by using electronegativity values.

Compare and contrast characteristics of ionic, covalent, and polar covalent bonds.

Interpret the sea of electrons model of metallic bonding.

Key Terms

electronegativity
shielding effect
polar covalent bond
malleable
ductile
conductivity
metallic bond

You've probably used glue to repair a broken object or to make something new. The glue joins separate pieces together to form a stable whole. In Chapter 4, you learned that atoms bond with other atoms to form molecules and compounds. The glue that holds atoms together in a molecule is the sharing of electrons between the atoms.

A Model of Bonding

Atoms either transfer electrons and then form ionic compounds or they share electrons to form covalent compounds. In both cases, the bond forms because of an increase in stability. By forming bonds, atoms acquire an octet of electrons and the stable electron configuration of a noble gas. Atoms are often more stable when they're bonded in compounds than when they're free atoms.

Dividing compounds into two bonding types, ionic and covalent, is convenient. If you know the type of bonds in a compound, you can predict many of its physical properties. **Table 9.1** summarizes the physical properties of ionic and covalent compounds. You can also reason in the opposite direction. If you know the physical properties of an unknown compound,

Table 9.1 Physical Properties of Ionic and Covalent Compounds		
Property	**Ionic Compound**	**Covalent Compound**
State at room temperature	crystalline solid	liquid, gas, solid
Melting point	high	low
Conductivity in liquid state	yes	no
Water solubility	high	low
Conductivity of aqueous solution	yes	no

Program Resources

Study Guide, pp. 35-36 L1
English/Spanish Audiocassettes, Chapter 9 L1 LEP
Laboratory Manual, 9.1, pp. 77-82 L1
Section Focus Transparency 18 and **Master,** p. 18 L1 LEP

Basic Concepts Transparencies 24, 25 and **Masters,** pp. 47-50 L1 LEP
Spanish Resources L1 LEP
ChemLab and MiniLab Worksheets, pp. 62-63 L1
Consumer Chemistry, pp. 17-18 L1
Applying Scientific Methods in Chemistry, pp. 24-25 L1

you can predict its bond type. But predictions may not always be correct because there is no clear-cut division between ionic and covalent compounds. A compound may be partly covalent and partly ionic. A more realistic view of bonding is to consider that all chemical bonds involve a sharing of electrons. Electrons may be shared equally, but they also may be shared only slightly—almost not at all. The properties of any compound, particularly its physical properties, are related to how equally the electrons are shared. **Figure 9.1** shows a model of bonding as a sharing of electrons.

In Chapter 4, you explored the two extremes of this range of electron sharing: the ionic bond and the covalent bond. A purely ionic bond results when the sharing is so unequal that it is best described as a complete transfer of electrons from one bonding atom to another. A purely covalent bond results when electrons are shared equally. Most compounds fall somewhere in between these two extremes; they have some ionic characteristics and some covalent characteristics.

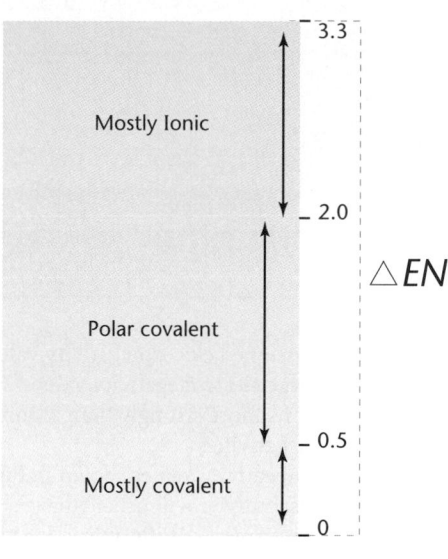

Figure 9.1
Electron Sharing Model of Bonding
◀ The bonding between atoms in compounds can be viewed as a range of electron sharing measured by electronegativity, *EN*. This range contains three main classes of bonds—ionic, polar covalent, and covalent. Differences in electronegativity, Δ*EN*, are explained on the next two pages.

▲ Bonding can be thought of as a tug-of-war between two atoms for shared electrons.

Electronegativity: An Attraction for Electrons

You can think of bonding atoms as being in a tug-of-war for the shared valence electrons, as shown in **Figure 9.1**. To use this model of electron sharing, you need some way of determining how much tug each atom exerts on the shared electrons. The measure of the tug is electronegativity. **Electronegativity** is a measure of the ability of an atom in a bond to attract electrons. How each atom fares in a tug-of-war for shared electrons is determined by comparing the electronegativities of the two bonded atoms.

Revealing Misconceptions

Students may think that all atoms attract each other and tend to bond. Explain that there are both attractive and repulsive forces between atoms. A bond forms only if the attractive forces are greater than the repulsive forces.

Concept Development

On the chalkboard, draw the potential energy diagram for the interaction of two hydrogen atoms at varying distances from each other.

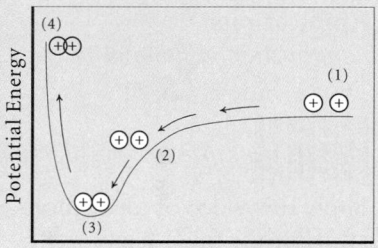

Distance between nuclei

Explain that at (1), the atoms are fully separated and have a particular potential energy. As the atoms move closer (2), their potential energy goes down. At (3), they are at the lowest energy. The distance between the nuclei at (3) is the bond length, which is the most stable distance. The hydrogen atoms have formed a bond, and the potential energy is lower than that of the separated atoms. Ask students what happens when the two hydrogen nuclei get even closer at (4). *The potential energy goes up because of nuclear repulsion.* Ask students why atoms form chemical bonds. *because by doing so they lower their energy and are more stable* Ask if a stable valence configuration is always an octet of electrons. *No, hydrogen has a stable duet.* [L1]

Meeting Individual Needs

Limited English Proficiency To demonstrate the bonding arrangement of two hydrogen atoms that has minimum potential energy, obtain two soft foam balls. Label each ball with a large *H*, and wrap the two with an elastic band that holds them together firmly. Ask a student to hold them in this bonding position while you point to the minimum potential energy position on the diagram in *Concept Development* above. As the student pushes the balls closer together, move your pointer up the potential energy curve to the left of the diagram, illustrating a less stable position because of repulsion. As the student pulls the balls further apart against the resistance of the elastic, move your pointer up the potential energy diagram to the right, illustrating first a less stable position and then a broken bond, resulting in two separated atoms. [L1] [LEP]

Concept Development

Despite the many differences in the properties of the elements, the common factor underlying the chemical bonding of the 90 naturally occurring elements is the tendency for an atom to attain a noble-gas electron configuration. This is a regularity that helps clarify the bonding in the millions of known compounds.

Figure 9.2

The Electronegativities of the Elements
Electronegativity is a periodic property. Notice how the heights of the columns change across each period and down each group. The heights are proportional to the electronegativities.

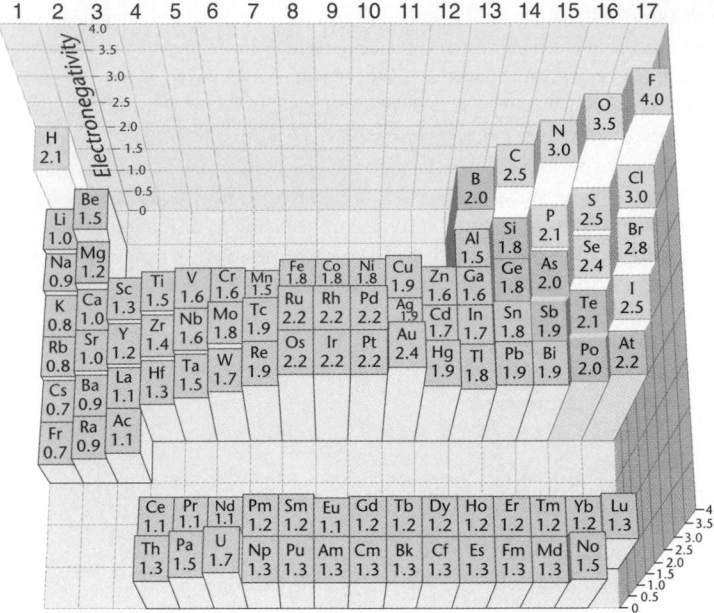

Atoms are assigned electronegativity values, as shown in **Figure 9.2**. Atoms with large electronegativity values, such as fluorine, attract shared valence electrons more strongly than atoms such as sodium that have small electronegativities.

The electronegativity values shown in **Figure 9.2** may not always be at your fingertips, but you will usually have a periodic table. Electronegativity is a periodic property. With only a few exceptions, electronegativity values increase as you move from left to right in any period of the periodic table. Within any group, electronegativity values decrease as you go down the group. That means that the most electronegative elements are in the upper-right corner of the table. Fluorine has the highest value of 4.0. It also follows that the elements having the lowest electronegativities are in the lower-left corner. The electronegativity of cesium is 0.7. The noble gases are considered to have electronegativity values of zero and do not follow the periodic trends.

The decrease in electronegativity as you move down a column occurs because the number of energy levels increases, and the valence electrons are farther from the positively charged nucleus. The nucleus has less attraction for its own valence electrons so they are held less tightly. Also, electrons in the inner energy levels tend to block the attraction of the nucleus for the valence electrons. This is known as the **shielding effect.** The shielding effect increases as you move down a column because the number of inner electrons increases. For example, magnesium has the electron configuration $1s^2 2s^2 2p^6 3s^2$ and an electronegativity of 1.2; calcium has the configuration $1s^2 2s^2 2p^6 3s^2 3p^6 4s^2$ and an electronegativity of 1.0. Although both have two valence electrons, calcium has eight more inner electrons than magnesium. These electrons shield the outer electrons from the attraction of the nucleus, as shown in **Figure 9.3**.

Demonstration 9.2

Combustion of Ethanol

Purpose: To show the combustion of vapors.
Materials: 6-gallon transparent plastic bottle, 40 mL ethanol, matches, tongs
Safety Precautions: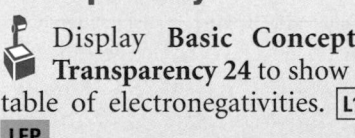
Use a safety shield.
Disposal: Code F
Procedure: Place 40 mL of ethanol in the empty bottle. Rotate the bottle on its side so that a mixture of ethanol vapor and air fills the bottle. Pour out any excess ethanol and wash it down the drain. Show students that the bottle is empty and set it upright. Turn off the room lights. Holding a burning match with tongs, ignite the vapors at the mouth of the bottle. **CAUTION:** *Keep bottle away from flammable materials because the burning vapors will roar out of the neck of the bottle.*

Results: The bottle will not be damaged by the burning vapors, but it will be warm to the touch. Condensed water vapor can be poured from the bottle.
Analysis: Ask these questions.
1. How does the rate of combustion of a vapor compare to the rate at which a liquid burns? *The vapor burns much faster.*

Figure 9.3

The Shielding Effect
The electrons in the first and second energy levels shield the two valence electrons in the magnesium atom from the full effect of 12 nuclear protons. ▼

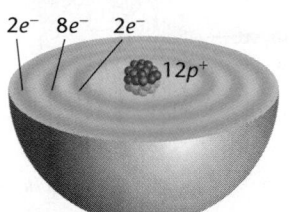

$2e^-$ $8e^-$ $2e^-$

$12p^+$

Magnesium atom

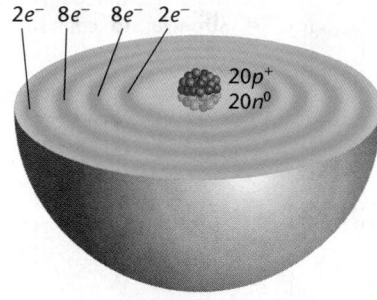

$2e^-$ $8e^-$ $8e^-$ $2e^-$

$20p^+$
$20n^0$

Calcium atom

◄ Calcium's two valence electrons are shielded from the full attraction of the 20+ nuclear charge by inner electrons in the first, second, and third energy levels. Because calcium has eight more inner electrons than magnesium, the valence electrons of calcium are held less tightly.

As you move across a row from left to right in the periodic table, the number of protons in the nucleus increases. With the increase in nuclear charge, the attraction of the nucleus for the valence electrons increases, and therefore electronegativity tends to increase across a row. In period 4, potassium in Group 1 has an electronegativity of 0.8, while bromine has a value of 2.8.

Because electronegativity varies in a periodic way, you can make predictions about differences in electronegativity by looking at the distance between bonding atoms on the table. In general, the farther the bonding atoms are from each other on the periodic table, the greater their electronegativity difference. **Figure 9.4** shows the trends in electronegativity.

The Ionic Extreme

The greater the difference between the electronegativities of the bonding atoms, the more unequally the electrons are shared. The electronegativity difference between two bonding atoms is often represented by the symbol ΔEN, where EN is an abbreviation for ElectroNegativity and Δ is the Greek letter *delta* meaning "difference." ΔEN is calculated by subtracting the smaller electronegativity from the larger, so ΔEN is always positive. For example, ΔEN for cesium and fluorine is $4.0 - 0.7 = 3.3$.

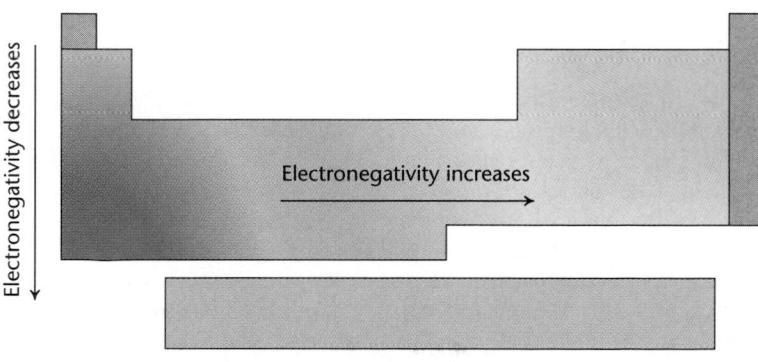

Electronegativity decreases

Electronegativity increases

Figure 9.4

Periodic Trends in Electronegativity
Electronegativity increases from left to right across a period and decreases from top to bottom down a group.

9.1   **Bonding of Atoms**   **305**

---

2. Did the vapor burn evenly throughout the bottle? *No, the flame was seen in different areas of the bottle.*

3. (Have students recall the definition of *volatile* from Chapter 1.) Is ethanol more or less volatile than water? *more volatile*

4. Is greater volatility associated with strong or weak intermolecular attractive forces? *weak*

Assessment

Knowledge: Ask students why engines must be turned off and cigarette smoking banned while a vehicle is being fueled. L1

SECTION 9.1

Correcting Misconceptions

When different symbols (dots, x's, circles, squares) are used in electron dot diagrams to represent electrons from different atoms, students may think that the electrons are different. Emphasize that different symbols are used only to keep track of the electrons of the various atoms. All electrons, as far as is known, are identical.

Using an Analogy

To help students understand the shielding effect of electrons, have them model two atoms, sodium and potassium, with different numbers of electrons but the same number of valence electrons. For the sodium atom, one student can represent the nucleus and ten students can form the $1s$, $2s$, and $2p$ sublevels by gathering around the nucleus in three circles. Another student can be sodium's valence electron in the $3s$ orbital. Ask students to observe the sodium model and then model potassium by adding seven more students to represent the completion of the $3s$ and $3p$ orbitals, and an eighth student to represent potassium's $4s$ valence electron. Ask students whether the nucleus could exert as much control over the valence electron in potassium as it could over the valence electron in sodium. *The intervening students prevented the nucleus from interacting with the valence electron. The effect was more noticeable for potassium.* L1

Knowledge: Ask students to determine which element of each pair—nitrogen or arsenic, and magnesium or phosphorus—would be expected to have the higher electronegativity and write a paragraph explaining their choices. *nitrogen, because nitrogen has fewer energy levels of electrons to shield the valence electrons from the nuclear charge; phosphorus, because although both atoms have three energy levels, the higher nuclear charge in phosphorus causes a greater attraction for the valence electrons* L1

GLENCOE TECHNOLOGY

Videodisc

Chemistry: Concepts and Applications
Electronegativity
Disc 1, Side 2, Ch. 12

Show this quick-time animation to demonstrate the submicroscopic process of bonding and how it is influenced by electronegativity. L1 LEP

CD-Rom

Chemistry: Concepts and Applications
Electronegativity
Have students use this animation to help reinforce their understanding of bonding and the role of electronegativity in determining the type of bond. L1 LEP

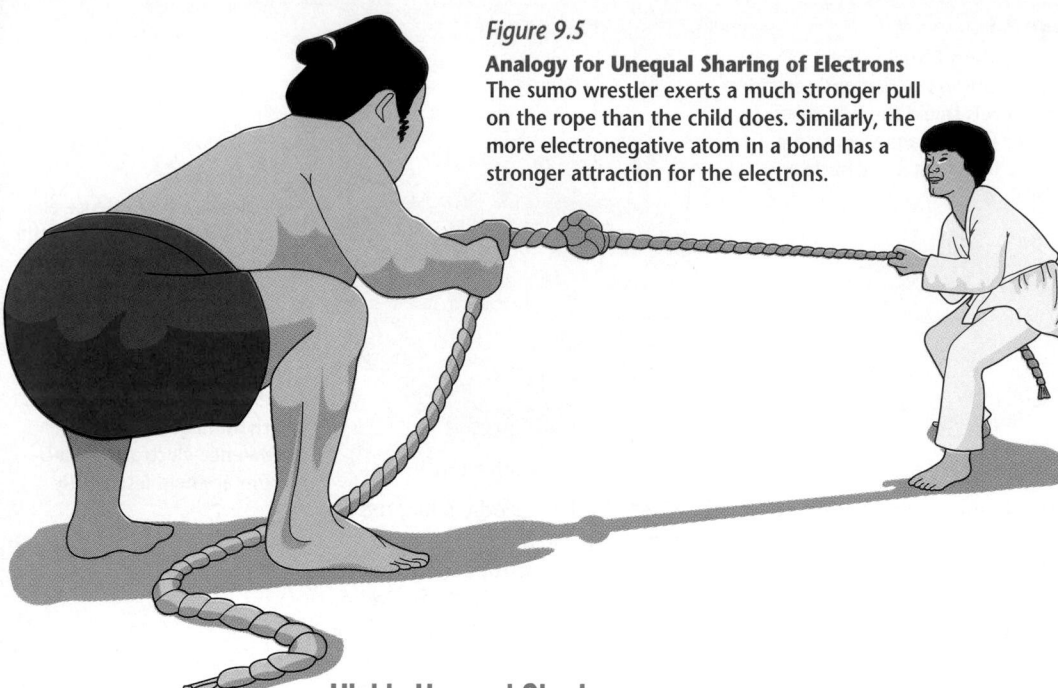

Figure 9.5
Analogy for Unequal Sharing of Electrons
The sumo wrestler exerts a much stronger pull on the rope than the child does. Similarly, the more electronegative atom in a bond has a stronger attraction for the electrons.

Highly Unequal Sharing

Using the tug-of-war analogy, consider the situation in which a sumo wrestler is at one end of a rope and a small child is at the other end, as shown in **Figure 9.5**. There's no contest. The same is true in a chemical bond when the ΔEN between bonding atoms is 2.0 or greater.

When the electronegativity difference in a bond is 2.0 or greater, the sharing of electrons is so unequal that you can assume that the electron on the less electronegative atom is transferred to the more electronegative atom. This electron transfer results in the formation of one positive ion and one negative ion. The bond formed by the two oppositely charged ions is classified as a mostly ionic bond. Many bonds are classified as ionic, but they have varying degrees of ionic character. The greater the difference in the electronegativities of the two atoms, the more ionic the bond. The greater the distance between the bonding atoms on the periodic table, the more ionic the bond between the atoms.

Ionic Bonding in Sodium Chloride

Electrons are transferred when sodium chloride forms. The electronegativity of sodium is 0.9, and the electronegativity of chlorine is 3.0, one of the highest values on the table. The ΔEN for this bond is 2.1. The sharing of a pair of electrons between sodium and chlorine is so unequal that the electrons are both essentially on the chlorine atom, creating a chloride ion, Cl^-. The sodium atom can't compete with chlorine for a share in the electron pair and becomes a sodium ion, Na^+. Sodium ions and chloride ions combine to form NaCl. Sodium chloride is best described as an ionic compound.

306   Chapter 9   Chemical Bonding

Meeting Individual Needs

Gifted/Visually Impaired Ask an interested and capable student to make a 3-D periodic table of electronegativities using a well plate and straws cut to lengths proportional to the elements' electronegativities. Visually impaired students can examine the patterns by touch. L2

HISTORY ▷ CONNECTION

Linus Pauling: An Advocate of Knowledge and Peace

Some proclaim him one of the 20 greatest scientists of all time. But Linus C. Pauling claimed merely to have been well prepared and to have been in the right place at the right time. The time was the mid-1920s at the beginning of quantum physics.

Quantum theory of chemistry In 1925, Pauling was awarded a Ph.D. in chemistry from the California Institute of Technology, where he studied the crystal structure of materials. A year later, he was granted a Guggenheim Fellowship and traveled to Europe to study the quantum theory of the atom. Returning to CalTech, he merged his knowledge of material structures and quantum theory into the concept of the chemical bond. Pauling's book *The Nature of the Chemical Bond* was influential in providing a framework for researchers to study and predict the structures and properties of inorganic, organic, and biochemical compounds. To acknowledge the importance of his work in understanding the chemical bond, Pauling was awarded the 1954 Nobel Prize for Chemistry.

Antinuclear armaments Pauling was an outspoken critic of the atmospheric testing of nuclear bombs. He was convinced that the

radioactive fallout from such testing would be hazardous to humans for many generations. Petitioning scientists worldwide, Pauling pleaded for an international ban on testing nuclear weapons. The announcement that Pauling had been awarded the 1962 Nobel Prize for Peace was made on the day that the world's first partial nuclear test ban went into effect.

Vitamin C In the early 1970s, Pauling became an advocate of the health benefits of taking megadoses (large doses) of vitamin C. His book *Vitamin C and the Common Cold* became a best-seller. Although his ideas are controversial, Pauling thought vitamin C might help eliminate minor ailments and be a possible cure for cancers.

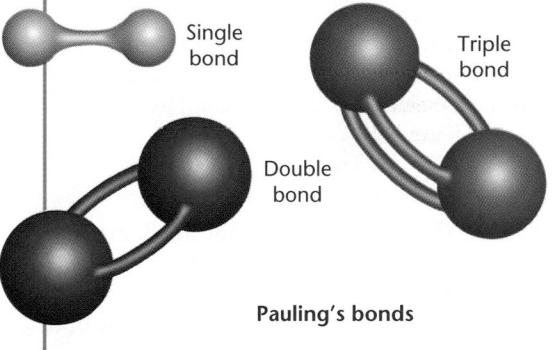

Single bond

Double bond

Triple bond

Pauling's bonds

Connecting to Chemistry

1. **Applying** Why is understanding the nature of chemical bonding important?

2. **Acquiring Information** Investigate the role Pauling played in the discovery of the structure of DNA.

HISTORY ▷ CONNECTION

Purpose
Students will learn about some aspects of the life of Linus C. Pauling.

Teaching Strategies
- Tell students that Pauling lived from Feb. 28, 1901, to August 1, 1994.
- Point out that Pauling is the only person to have been awarded, independently, two Nobel prizes.
- Have students identify the number of bonds shown in the illustrations of H_2, O_2, and N_2, as conceptualized by Pauling. *one, two, and three, respectively* L1

Extension
Have interested students investigate Pauling's work in nutrition. L2

Connecting to Chemistry
1. The chemical bond can account for the structure and properties of substances.
2. Among Pauling's contributions to the development of DNA are the helical structures of proteins and the concept of hydrogen bonding. Hydrogen bonding plays a large part in holding the two DNA strands together in the helix.

GLENCOE TECHNOLOGY

 Videodisc

Chemistry: Concepts and Applications
See the Videodisc Program Teacher Guide to access illustrations and other material available for this chapter.

Chemistry Journal

Seamless Changes
Ask students to describe at least five examples of changes that occur on a continuum and include them in their journals. L1

Knowledge: If students don't know the symbols for the elements, they may select incorrect electronegativities from the table. This would be a good time for a review quiz, particularly for pairs of elements that are often confused, such as sodium and sulfur, potassium and phosphorus, and magnesium and manganese. L1

Using an Analogy

To illustrate the meaning of a continuum, explain that most politicians are neither 100% conservative nor 100% liberal. The opinions and proposals of a politician generally vary from issue to issue on a continuum from 100% conservative to 100% liberal.

Correcting Misconceptions

Students may think that the breaks between covalent, polar covalent, and ionic bonding on the ΔEN scale are absolute. Explain that the dividing lines are arbitrary. They are established for convenience in talking about bonds. Dividing lines may vary from one text to another.

Tying to Previous Knowledge

HF is a molecular formula giving the number and kind of atoms in one molecule. KF is an empirical formula showing that there is a 1:1 ratio of potassium to fluorine ions in an array of ions in a KF crystal. It is not possible to differentiate between molecular formulas and empirical formulas unless the type of bonding is known.

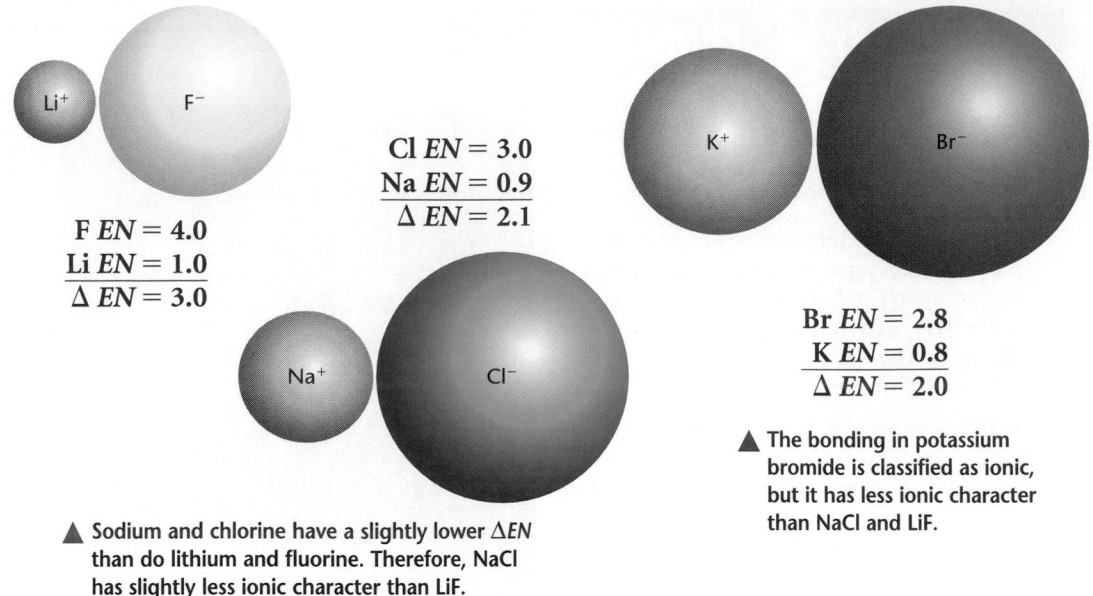

$$F\ EN = 4.0$$
$$\underline{Li\ EN = 1.0}$$
$$\Delta\ EN = 3.0$$

$$Cl\ EN = 3.0$$
$$\underline{Na\ EN = 0.9}$$
$$\Delta\ EN = 2.1$$

$$Br\ EN = 2.8$$
$$\underline{K\ EN = 0.8}$$
$$\Delta\ EN = 2.0$$

▲ The bonding in potassium bromide is classified as ionic, but it has less ionic character than NaCl and LiF.

▲ Sodium and chlorine have a slightly lower ΔEN than do lithium and fluorine. Therefore, NaCl has slightly less ionic character than LiF.

Figure 9.6

Three Ionic Compounds The electronegativity differences in lithium fluoride, sodium chloride, and potassium bromide show that they are best represented as ionic compounds.

Figure 9.6 compares the formation of sodium chloride with the formation of lithium fluoride and potassium bromide. For each of these salts, the ΔENs are equal to or greater than 2.0. Like sodium chloride, both lithium fluoride and potassium bromide are considered mostly ionic compounds. Notice that the two atoms in each bond are well separated from each other on the periodic table.

The Covalent Extreme

You have seen that when there is a large electronegativity difference between two atoms (2.0 or greater), the bond that forms between the two atoms is considered mostly ionic. What happens when there is no difference in electronegativity or if the electronegativity difference is small?

Equal Sharing

Now imagine a 300-pound sumo wrestler on one end of the rope, and another 300-pound sumo wrestler on the other end. This situation is the same as when two of the same atoms form a bond, for example, when two fluorine atoms form the fluorine molecule, F_2. Here is the electron dot structure for F_2.

Because two atoms of the same element are forming the bond, the difference in electronegativities is zero. In the fluorine molecule, a pair of valence electrons are shared equally. This type of bond is a pure covalent bond. All other diatomic molecules (Cl_2, Br_2, I_2, O_2, N_2, and H_2) have pure covalent bonds. In all these molecules, the electrons are shared equally.

Integrating THE Sciences

TECH PREP

Ecology New compounds are discovered every day. Some of these compounds are synthesized in laboratories, but the majority are naturally occurring substances, found mostly in plants. Chemists who study the compounds produced by nature are called natural products chemists. Aspirin is closely related to a chemical substance produced by willow trees, and taxol, a promising treatment for cancer, is produced by the Pacific yew tree. Ask students to find information about the work of natural products chemists and write a report describing their findings. An alternative might be to make a list of naturally occurring compounds and their sources, for example, antibiotics, and herbs and drugs used in folk remedies. L2

Sharing That's Close to Equal

Sometimes, the electronegativities of bonding atoms are close but not exactly the same. For example, carbon's electronegativity is 2.5 and hydrogen's is 2.1. A ΔEN of greater than zero always means unequal electron sharing. However, a bond in which the electronegativity difference is less than or equal to 0.5 is considered to be a covalent bond with electrons that are not shared exactly equally. All of the many compounds formed between carbon and hydrogen are considered covalent.

When the electronegativity difference is less than or equal to 0.5, the slightly unequal sharing of the electrons doesn't have a significant effect on the properties of the molecule. Low boiling points and melting points are typical of pure covalent compounds. Most of the elemental diatomic molecules are gases at room temperature—Cl_2, F_2, O_2, N_2, and H_2. Carbon disulfide, methane, and nitrogen dioxide, shown in **Figure 9.7,** are examples of covalent compounds in which the electron sharing is slightly unequal. These molecules are gases or low-boiling point liquids at room temperature.

Figure 9.7

Three Covalent Compounds
Carbon disulfide is an important solvent for waxes and greases. Methane is the principal component of natural gas. Nitrogen dioxide is used for making nitric acid and is also an atmospheric pollutant. All three compounds contain covalent bonds in which the sharing of electrons is more or less equal.

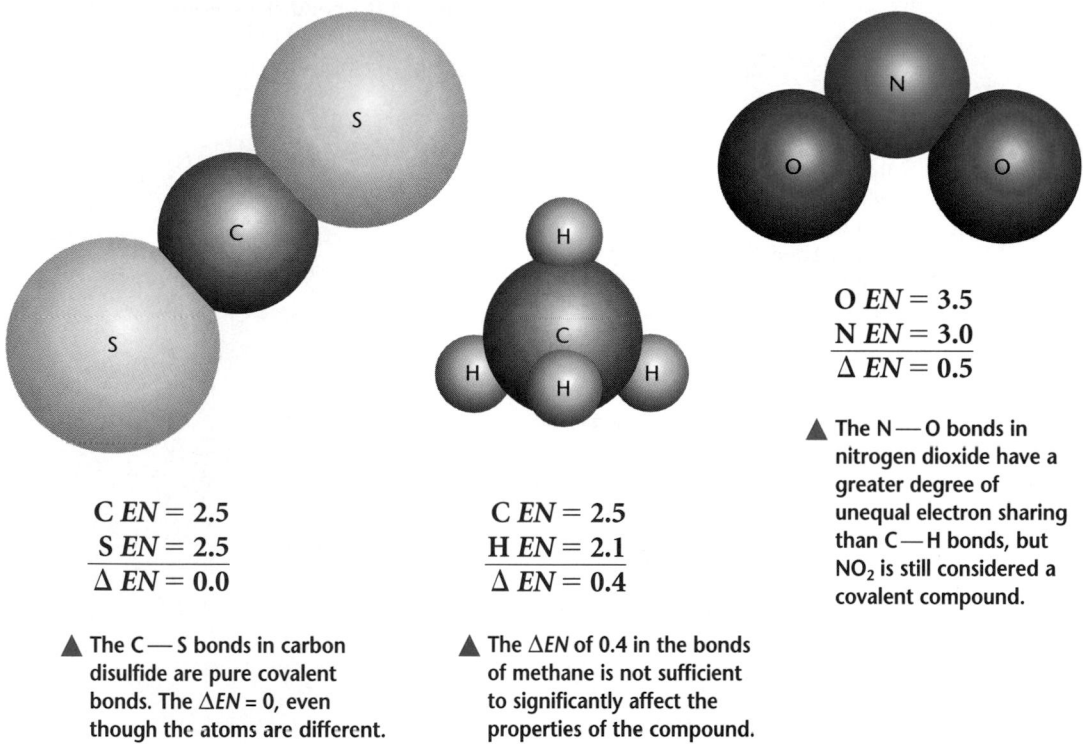

C *EN* = 2.5
S *EN* = 2.5
—————
Δ *EN* = 0.0

▲ The C—S bonds in carbon disulfide are pure covalent bonds. The ΔEN = 0, even though the atoms are different.

C *EN* = 2.5
H *EN* = 2.1
—————
Δ *EN* = 0.4

▲ The ΔEN of 0.4 in the bonds of methane is not sufficient to significantly affect the properties of the compound.

O *EN* = 3.5
N *EN* = 3.0
—————
Δ *EN* = 0.5

▲ The N—O bonds in nitrogen dioxide have a greater degree of unequal electron sharing than C—H bonds, but NO_2 is still considered a covalent compound.

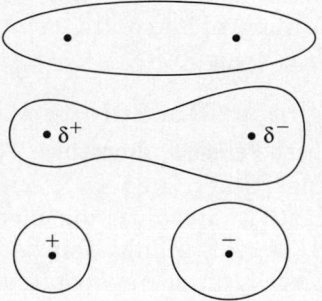
Across the Curriculum

History

In 1902, G.N. Lewis (1875-1946) discovered that he could represent the chemistry of many elements by a model in which electrons were arranged in atoms in concentric cubes. Each of the eight cube corners was a possible location for an electron. This was the birth of the octet rule. Lewis found that it was the number of electrons in the outermost cube that influenced the formation of bonds. His insight was astonishing. The existence of electrons had been discovered just five years earlier, and it would be nine more years before the nuclear model of the atom was proposed. In 1916, Lewis proposed the idea of the electron pair or covalent bond. At that time, it was believed that all bonds were ionic. Although deserving of a Nobel prize, Lewis never received one. However, he led the chemistry department at the University of California at Berkeley, where he attracted five future Nobel prize winners in chemistry.

Using an Analogy

Draw an analogy between the positive and negative poles of a polar molecule and the magnetic north and south poles of Earth. Point out that one meaning of the word *polar* is "directly opposite."

Content Background

Other Periodic Properties Two properties related to electronegativity are ionization energy and electron affinity. Ionization energy is the energy required to remove an electron from a gaseous atom to form a positive ion. Elements with the lowest ionization energies are at the lower left of the periodic table. Elements in the upper right (excluding the noble gases) require the most energy to remove an electron. Electron affinity is a measure of the ability of a neutral atom to accept an electron and become a negative ion. Not surprisingly, the elements with the greatest electron affinity are at the upper right of the table. Fluorine is the element with the highest electron affinity.

WORD ORIGIN

electronegative:
elektron (Gk)
amber (Rubbing amber produces electric charge.)
negare (L) to deny

When highly electronegative atoms attract other atoms' electrons, they become more negative.

Polar Covalent Bonds

Now consider a tug-of-war in which a 300-pound sumo wrestler is on one end of the rope and a 400-pound wrestler is on the other end. The unequal pull on the ends of the rope brings the middle of the rope closer to the 400-pound competitor. This is an analogy to a bond in which electrons are shared unequally. This type of bond falls between the two extremes of electron sharing: equal sharing (a covalent bond) and completely unequal sharing (an ionic bond). Bonds in which the pair of electrons is shared unequally have electronegativity differences between 0.5 and 2.0.

Unequal Sharing in Covalent Bonds

When the electronegativity difference between bonding atoms is between 0.5 and 2.0, the electron sharing is not so unequal that a complete transfer of electrons takes place. Instead, there is a partial transfer of the shared electrons to the more electronegative atom. The less electronegative atom still retains some attraction for the shared electrons. The bond that forms when electrons are shared unequally is called a **polar covalent bond.** A polar covalent bond has a significant degree of ionic character.

Polar covalent bonds are called *polar* because the unequal electron sharing creates two poles across the bond. Just as a car battery or a flashlight battery has separate positive and negative poles, so polar covalent bonds have poles, as shown in **Figure 9.8.** The negative pole is centered on the more electronegative atom in the bond. This atom has a share in an extra electron. The positive pole is centered on the less electronegative atom. This atom has lost a share in one of its electrons. Because there was

Figure 9.8

Bonds with Positive and Negative Ends
When the sharing of electrons in a bond isn't equal, the bond is polar, as in the H—Cl bond. Just like this battery, the bond has two poles, one positive and one negative. The symbols δ^+ and δ^- (delta plus and delta minus) are used to show the distribution of partial charges in a polar covalent bond. The arrow points in the direction of the negative end of the bond.

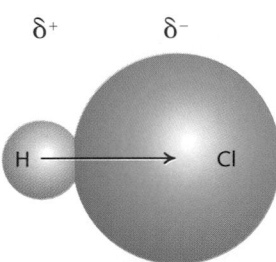

Paper Chromatography

Purpose: To demonstrate that the vegetable dye in food coloring is a mixture.
Materials: For each dye to be chromatographed: large test tube or gas collecting bottle with cork stopper, strip of filter or chromatography paper, thumbtack, 10 mL water, 5 mL isopropyl alcohol, food coloring, capillary tube

Safety Precautions:

Disposal: Code C; check any regulations in your area about pouring flammable liquid into drains.
Procedure: Use a thumbtack to attach a strip of filter or chromatography paper to the stopper. Place a small amount of a 2:1 mixture of water ρnd isopropyl alcohol into the bottle or test tube. Arrange the filter paper so that the bottom edge will be immersed in the water-alcohol solu-

tion when the cork is inserted. Draw a pencil line across the chromatography paper just above the solution level. Use a capillary tube to transfer a drop of food coloring to the center of the pencil line. Place the stopper and the attached chromatography paper in the bottle containing the solvent. Have students observe the contents of the bottle at the end of the class period, then remove the paper and

not a complete transfer of an electron, the charges on the poles are not 1+ and 1−, but δ+ and δ−. These symbols, delta plus and delta minus, represent a partial positive charge and a partial negative charge. This separation of charge, resulting in positively and negatively charged ends of the bond, gives the polar covalent bond a degree of ionic character.

Compounds with polar covalent bonds have different properties from compounds with pure covalent bonds. You saw that purely covalent compounds tend to have low melting points and boiling points. Carbon disulfide, CS_2, as shown in **Figure 9.7**, is a triatomic molecule, with a ΔEN equal to zero. Carbon disulfide boils at 46°C. Water is also a triatomic molecule, but the bonding in water is polar covalent. Even though water is a much lighter molecule than carbon disulfide, its boiling point is 100°C.

Water Has Polar Bonds

The ΔEN for the O—H bond is 1.4, so water molecules have polar bonds. When atoms of hydrogen and oxygen bond by sharing electrons, the shared pair of electrons is attracted toward the more electronegative oxygen. This unequal sharing causes an imbalance in the distribution of charge about the two atoms, as shown in **Figure 9.9**.

The effect of polar covalent bonds on boiling point is illustrated by comparing the boiling point of water with the other hydrides of Group 16 elements—sulfur, selenium, and tellurium. A hydride is a compound formed between any element and hydrogen. Water is the first hydride of Group 16. ΔEN for the O—H bond is 1.4. But ΔEN for both the S—H and the Se—H bonds is only 0.4, and ΔEN for the Te—H bond is 0.3. These electronegativity differences tell you that H_2O contains polar covalent bonds, but the bonds in H_2S, H_2Se, and H_2Te are essentially covalent. **Figure 9.10** shows that even though H_2O is the lightest of the four hydrides, its boiling point is significantly higher than the hydrides of the other members of Group 16.

Figure 9.9
Charge Distribution in an O—H Bond
Because oxygen is more electronegative than hydrogen, the electrons in an O—H bond spend more time near the oxygen atom than near the hydrogen atom. This distribution leads to a partial negative charge on oxygen and a partial positive charge on hydrogen.

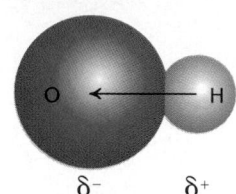

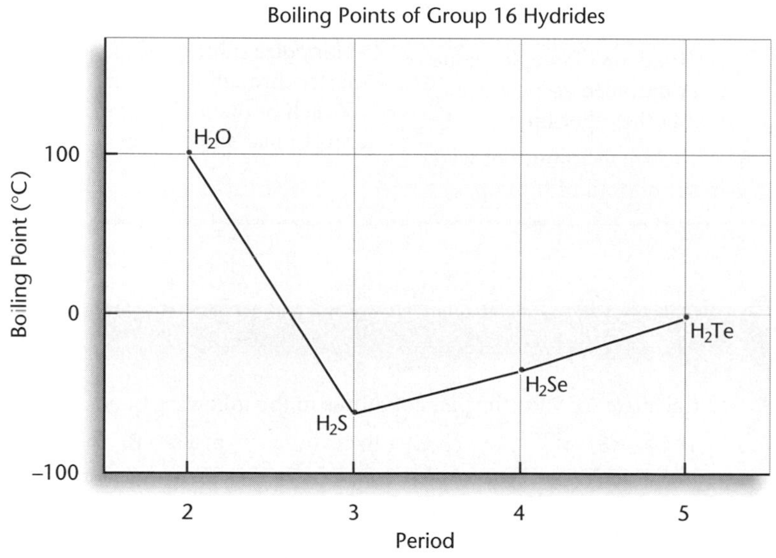

δ^- δ^+

SECTION 9.1

Concept Development
Point out that a pair of electrons in a covalent bond is often represented by a dash rather than two dots. Similarly, two dashes represent four electrons in a double bond, and three dashes represent six electrons in a triple bond.

GLENCOE TECHNOLOGY

Software

Mastering Concepts in Chemistry
Unit 9, Lesson 1
Electronegativity and Bond Type
Use this lesson to reinforce the concept of electronegativity as the key to bonding. L1 LEP

Boiling Points of Group 16 Hydrides

(graph: y-axis "Boiling Point (°C)" ranging −100 to 100, x-axis "Period" 2 to 5; data points labeled H_2O at ~100°C period 2, H_2S at ~−60°C period 3, H_2Se at ~−40°C period 4, H_2Te at ~−2°C period 5)

Figure 9.10
Boiling Points Reflect Bond Type
On the basis of the mass of the molecules, you might predict that H_2O would have the lowest boiling point of the four hydrides. Another factor, the polarity of the O—H bond, makes that prediction incorrect. Later, you'll learn how polar bonds often result in a polar molecule, as in the case of water.

let it air dry. Have students observe the paper strip the next day and discuss their observations.
Results: Some food colorings will show no separation because they are not mixtures. Other food colorings will produce colorful separations.

Analysis: Ask these questions.
1. Which food colorings are mixtures? *Answers will vary according to the manufacturer.*
2. Using one selected chromatogram, which component is the most polar? *the component that migrated the least distance up the paper because it has the most attraction for the paper*

✔ **Assessment**

Portfolio: Chemists use chromatography to test for drugs. Have students do research on drug testing and write a report to be placed in their portfolios. L2 P

miniLAB

Coffee Filter Chromatography

Purpose: This MiniLab is most appropriate if there is not time to do the ChemLab.

Process Skills: Observing, inferring, classifying

Teaching Strategies

- Have different brands of washable color markers on hand.
- Students may apply markers in the manner described, or more randomly if an artistic result is desired. If time permits, explore both methods.
- To explore the effect of the stationary phase, different filter or chromatography papers may be used.
- Set up the MiniLab as a crime lab. Have the class test a variety of pens, then write a message on paper with one of the pens. Students identify the pen by comparing the chromatograph of the message to the chromatographs of the known pens. L1 COOP LEARN

Expected Results

Some colors separate into a variety of colors. Black and brown markers show the largest number of colors.

Analysis

1. Some of the inks separate into different colors.
2. Answers will vary. Some inks separate into a number of colors.
3. The colors that are closest to the origin are the most polar—they have the most attraction for the paper and, therefore, move the slowest. The colors that travel the farthest are least polar—they have the lowest attraction for the paper

and, therefore, move the fastest.

 **Assessment**

Performance:
Have students predict how the separation would change if permanent (water-insoluble) markers were used. Have them chromatograph permanent markers using rubbing alcohol (isopropyl alcohol) as the solvent. L1

miniLAB 1

Coffee Filter Chromatography

Paper chromatography is a method of separating substances based upon their different attractions for the paper. The paper is called the stationary phase of the system, and the solvent is called the moving phase. In this experiment, the moving phase is water, a polar molecule. You will examine the composition of washable color marker inks by comparing the distances each component in the inks travels on a filter-paper chromatogram.

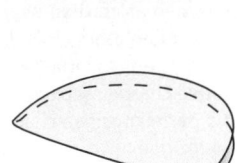

Fold

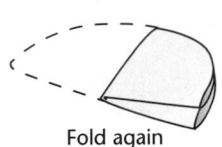

Fold again

Fold again

Procedure

1. Obtain a set of markers and two circular coffee filters.
2. Cover the bottom of a plastic cup with a small amount of water.
3. Lay the two coffee filters together on a dry surface. With the filters still together, fold the circles into eighths as shown.
4. Unfold the filters, but do not separate them. Find the center spot where all folds converge.
5. About 5 cm from the center, along a crease, make a dark mark with one of the pens. The mark should appear on both filters.
6. Continue for all eight creases, using a different color marker on each crease.
7. Separate the filters. Keep one filter unfolded as a control. Refold the other filter.
8. With a small amount of water in the bottom of the cup,

gently place the folded coffee filter, tip down, into the water.
9. When water reaches the top of the paper, gently remove the paper.
10. Gently open the paper. Compare each mark on the control filter with the corresponding mark on the chromatographed filter.

Analysis

1. Compare the marks on the control paper with the corresponding marks on the chromatographed filter. How are they different?
2. Which colors on the chromatographed filter are different from those on the control filter?
3. Polar substances tend to be attracted to other polar substances such as found in paper. Nonpolar substances move faster through the paper. Which of the inks contains the most polar substances?

PRACTICE PROBLEMS

1. Calculate ΔEN for the pairs of atoms in the following bonds.

   a) Ca—S   c) C—Br   e) H—Br

   b) Ba—O   d) Ca—F

2. Use ΔEN to classify the bonds in question 1 as covalent, polar covalent, or ionic.

Bonding in Metals

Bonding in metals doesn't result in the formation of compounds, but it is an interaction that holds metal atoms together and accounts for some of the typical properties of metals and alloys. What are some of these properties?

Properties That Reflect Metallic Bonding

Metals and alloys are malleable and ductile, and they conduct electricity. When a metal can be pounded or rolled into thin sheets, it is called **malleable.** Gold is an example of a malleable metal, as shown in **Figure 9.11.** A chunk of gold can be flattened and shaped by hammering until it is a thin sheet. **Ductile** metals can be drawn into wires. For example, copper can be pulled into thin strands of wire and used in electric circuits, as illustrated in **Figure 9.11.** Electrical **conductivity** is a measure of how easily electrons can flow through a material to produce an electric current. Metals such as silver are excellent conductors because there is low resistance to the movement of electrons in the metal. These properties—malleability, ductility, and electrical conductivity—are the result of the way that metal atoms bond with each other.

Sea of Valence Electrons

The valence electrons of metal atoms are loosely held by the positively charged nucleus. Sometimes, metal atoms form ionic bonds by losing one or more of their valence electrons and forming positive ions. However, in metallic bonding, metal atoms don't lose their valence electrons. Metal

Gold is malleable. An artisan practices the ancient art of making and using gold leaf. Gold leaf is gold metal that has been flattened until a sheet of gold foil, only a few hundred atoms thick, is obtained.

Figure 9.11

Malleability, Ductility, and Conductivity in Metals
Malleability, ductility, and electrical conductivity reflect the type of bonding in metals.

Copper is ductile and a good conductor of electricity. It is most commonly used in electrical circuits. ▼

SECTION 9.1

3 ASSESS

Check for Understanding

- Check for understanding of the section material by having the students answer the Section Review questions.
- To illustrate conductivity, have ten or more students stand in a line and represent metallic atoms. Distribute 3″ × 5″ cards with *e* written on them. The number of cards distributed should be one less than the number of students. Each card should be held jointly between two adjacent students, with the students at each end of the line holding only one card. Now give electron cards, one by one, to the student at one end. The student at the far end of the line should take the card he or she is sharing, transfer it to his or her other hand, and drop it into a container. The middle students should pass along the electron cards they are holding. This process mimics the movement of electrons in conduction. [L1] [LEP]

PRACTICE

Guided Practice: Have students work in class on Practice Problems 1-2. [L1]

Independent Practice: Your homework or classroom assignment can include the additional practice problems shown below. [L1]

Answers to Practice Problems

1. a. $\Delta EN = (S)2.5 - (Ca)1.0 = 1.5$
   b. $\Delta EN = (O)\ 3.5 - (Ba)\ 0.9 = 2.6$
   c. $\Delta EN = (Br)\ 2.8 - (C)\ 2.5 = 0.3$
   d. $\Delta EN = (F)\ 4.0 - (Ca)\ 1.0 = 3.0$
   e. $\Delta EN = (Br)\ 2.8 - (H)\ 2.1 = 0.7$
2. a. polar covalent b. ionic c. covalent d. ionic e. polar covalent

Additional Practice Problems

1. Calculate ΔEN for the pairs of atoms in the following bonds.

   a. Ca—O b. As—Cl c. Sn—Cl
   a. $\Delta EN = (O)\ 3.5 - (Ca)\ 1.0 = 2.5$
   b. $\Delta EN = (Cl)\ 3.2 - (As)\ 2.2 = 1.0$
   c. $\Delta EN = (Cl)\ 3.2 - (Sn)\ 2.0 = 1.2$

2. Are the bonds in question 1 covalent, polar covalent, or ionic?
   a. ionic b. polar covalent c. polar covalent

Reteach

Discussion: Ask a student to draw the Lewis electron dot notation for a nitrogen atom on the chalkboard. *The N has two dots on one side and one dot on each of the other three sides.* Then ask four students, in turn, to draw the Lewis electron dot notation for a hydrogen atom bonding to the nitrogen atom. *The fourth student cannot find a bonding site.* Suggest that the fourth student bond the hydrogen nucleus, H^+, to the nitrogen. Ask students the name and formula for the electron dot diagram that results. *the ammonium ion, NH_4^+* **L1**

Extension

The greater the bond energy and the shorter the bond, the stronger the bond. Write, on the chalkboard, the bond energies and bond lengths for the single and triple carbon-carbon bonds. The carbon-carbon triple bond has an energy of 839 kJ/mol and a length of 120 pm, while the carbon-carbon single bond has an energy of 347 kJ/mol and a length of 154 pm. Ask students which is the stronger bond? *the triple bond* Ask students to predict the bond energy and bond length of a carbon-carbon double bond. *Students will predict 573 kJ/mol and 137 pm. The actual values are 614 kJ/mol and 134 pm.* **L1**

4 CLOSE

Enrichment

Ask a chemist or technician who uses infrared spectroscopy to make a presentation illustrating the types of molecular motions that result in the characteristic absorptions observed in infrared spectra. Alternatively, ask an interested and capable student to visit a laboratory and make a class presentation. **L3**

atoms release their valence electrons into a sea of electrons shared by all of the metal atoms. The bond that results from this shared pool of valence electrons is called a **metallic bond. Figure 9.12** illustrates a model of metallic bonding. The bonding interaction at the submicroscopic level explains what you observe at the macroscopic level. Although the metal atoms are bonded together in a large network, they are not bonded to any single atom. This explains why metals are often malleable and ductile.

The conductivity of metals can also be explained by the sea of electrons model of metallic bonding, as shown in **Figure 9.12.** Because the valence electrons of all the metal atoms are not attached to any one metal atom, they can move through the metal when an external force, such as that provided by a battery, is applied.

Figure 9.12
Atomic View of Metallic Bonding
Each atom in this model of a Group 2 metal releases its two valence electrons into a pool of electrons to be shared by all of the metal atoms. ▼

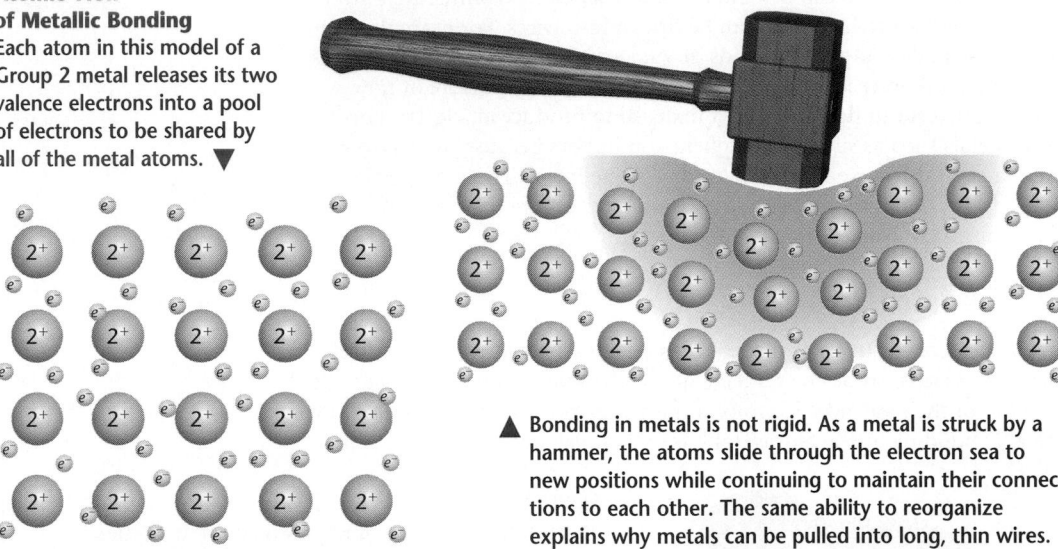

▲ Bonding in metals is not rigid. As a metal is struck by a hammer, the atoms slide through the electron sea to new positions while continuing to maintain their connections to each other. The same ability to reorganize explains why metals can be pulled into long, thin wires.

SECTION REVIEW

Understanding Concepts

1. Use ΔEN to classify the bonds in the following compounds as covalent, polar covalent, or ionic.

   a) H — S bond in H_2S
   b) S — O bond in SO_2
   c) Mg — Br bond in $MgBr_2$
   d) N — O bond in NO_2
   e) C — Cl bond in CCl_4

2. Using only a periodic table, rank these atoms from the least to most electronegative.

   Na, Br, I, F, Hg

3. Rank these bonds from the least to the most polar.

   a) C — F   **d)** Cl — F
   b) O — F   **e)** O — H
   c) Al — Br

Thinking Critically

4. **Properties of Metals** What aspect of metallic bonding is responsible for the malleability, ductility, and conductivity of metals?

Applying Chemistry

5. **Using Electronegativities** Do you expect LiF or LiCl to have a higher melting point? Explain.

SECTION REVIEW

1. **a.** covalent **b.** polar covalent **c.** polar covalent **d.** covalent **e.** covalent
2. Na < Hg < I < Br < F
3. O — F < Cl — F < Al — Br < O — H < C — F
4. Valence electrons are shared among more than two atoms; metal atoms can move within the structure.

5. The ΔEN for LiF is greater than for LiCl, so LiF has a greater degree of ionic character, a stronger ionic bond, and a higher melting point. (The melting point of LiF is 810°C and that of LiCl is 614°C.)

Molecular Shape and Polarity

SECTION 9.2

You've probably seen small-scale models of structures such as buildings or bridges. Designers find it helpful to work with small models before producing a detailed plan for a full-scale structure. Models can also be enlarged versions of objects that are so small that they can't be seen or handled. This is the case for atoms and molecules. So far, you have only seen molecules represented on paper by formulas or electron dot diagrams. These representations tell you the kinds of atoms contained in a molecule and how many there are of each. Electron dot diagrams also tell you how the valence electrons are distributed around each atom. But these models do not tell you how the atoms are arranged in space. Because many properties of molecules are determined by the structure of the molecule, it's important to develop models of molecules in three dimensions. In this section, you'll learn how models are used to represent the bonding between atoms and the shapes of molecules.

SECTION PREVIEW

Objectives
Diagram electron dot structures for molecules.
Formulate three-dimensional geometry of molecules from electron dot structures.
Predict molecular polarity from three-dimensional geometry and bond polarity.

Key Terms
double bond
triple bond
polar molecule

The Shapes of Molecules

Models can help you visualize the three-dimensional structures of molecules. Consider the simplest molecule that exists—hydrogen, H_2. Two hydrogen atoms share a pair of electrons in a nonpolar covalent bond, as shown in the electron dot structure.

$$H:H$$

A model of this molecule might be made by connecting two gumdrops of the same color by a toothpick. The gumdrops represent the hydrogen atoms, and the toothpick represents the shared pair of electrons that makes up the covalent bond.

Program Resources

Study Guide, pp. 37-38 L1
Critical Thinking/Problem Solving, p. 10 L2
Laboratory Manual, 9.2, pp. 83-86 L1
Section Focus Transparency 19 and **Master,** p. 19 L1 LEP
Basic Concepts Transparency 26 and **Master,** pp. 51-52 L1 LEP

Problem Solving Transparency 11 and **Master,** pp. 21-22 L1 LEP
ChemLab and MiniLab Worksheets, pp. 59-61, 64-65 L1
Chemistry and Industry, pp. 23-26 L1
Applying Scientific Methods in Chemistry, pp. 26-27 L1

SECTION 9.2

PREPARE

Key Concepts
Lewis electron dot diagrams of molecules are examined and electron-pair repulsions analyzed in order to determine molecular geometries and polarities. Properties of molecular compounds that are dependent upon polarities, such as melting points and boiling points, are discussed and compared with properties of ionic compounds.

Planning Ahead
For the ChemLab Obtain from a chemical supplier large sheets of filter paper or chromatography paper cut to size.

1 FOCUS

Focus Transparency
Display the **Section Focus Transparency** for Section 9.2 to introduce the concept of molecular shape. L1 LEP

2 TEACH

Concept Development
In this section, many models of molecules will be created and discussed. Emphasize these characteristics of models. (1) Models are human inventions and are based upon an incomplete understanding of nature. (2) Models are usually oversimplifications and can be wrong. (3) All models are based upon assumptions, and the assumptions for simple models are usually restrictive and probably inaccurate. (4) Models are modified as deficiencies are discovered. Sometimes, modifications make the model more complex and difficult to apply. (5) When a model is shown to be wrong, assumptions must be reexamined.

PEOPLE in CHEMISTRY

Meet Dr. William Skawinski, Chemist

An ordinary copy of *Mathematics for Physicists* is about two inches thick. However, the Braille version on Dr. Skawinski's bookshelf requires more than three feet of space. The chemist, who began losing his sight in childhood, consults this reference as he creates three-dimensional models of molecules. In the following interview, Dr. Skawinski talks about his innovative work and his love of chemistry.

On the Job

 Dr. Skawinski, can you tell us how you go about building molecular models?

 I start by putting information about molecular structures into a CAD (computer-aided design) program. The program uses the information to calculate a structure for the molecule. Then stereolithography is used to make a physical model from the calculations. This is basically how it works. A table covered by a polymer film sits at the top of a tank containing liquid plastic. A laser traces a cross section of the bottom of the molecule in the film, and a slice of the model hardens. Then the table moves down a little and the laser traces another cross section, and so on to the top of the molecular design. After eight to 12 hours, I get one solid piece of plastic that replicates a molecule of a particular compound.

 By what factor are these models bigger than the real thing?

 A carbon atom, for example, has a radius of about 0.2 nanometers. The model is about an inch in diameter—many billions of times larger.

 Why are these three-dimensional models useful?

A Dealing with a physical model of a mathematical distribution can give anyone—blind or sighted—a better insight into what it represents.

 Do you have a favorite of the molecular models you've built?

 Beta-cyclodextrin has doughnut-shaped molecules that are interesting. People have remarked that many of the molecular models look like pieces of artwork.

Early Influences

 Do you remember specific incidents from your childhood that influenced your interest in science?

 I can remember being two years old and watching the flames of a gas water heater. The bright blue lights on a perfectly black background fascinated me. I also remember at about the age of five hammering red, yellow, and gray rocks in the backyard, breaking them up into powder, and mixing them with water to make colored suspensions.

 What were things like for you in high school?

 Even though I was having serious problems with my vision by that time, I was involved in an amateur rocket group. These rockets weren't the little four-inch rockets you buy through the mail and send up a couple of hundred feet. One rocket for which I mixed the chemical propellant reached an altitude of 42 000 feet. I loved the idea of propelling a vehicle into space—and I still do. In fact, if anyone offered me a ticket to another planet, I'd only pause long enough to pack a toothbrush!

 Did your high school chemistry class interest you?

 My chemistry teacher made chemistry a part of the real world. Once he brought in a round, green glob about the size of a walnut. It was a flawed emerald. He explained that just a slight difference in chemical makeup differentiated that worthless stone from a priceless one.

Personal Insights

 Does the wonderment about science that you experienced as a child remain with you today?

 Of course. I think you really have to retain that trait to enter the field of science. Scientists are just a bunch of big kids who are fascinated with the world.

 How have you dealt with the challenge of your blindness?

I was able to see enough through college to learn a lot of critical information. But near the end of college, because my field of vision had narrowed so much, I could see just part of a word on a page at a time, and that was only with a strong magnifying glass. I managed to earn my master's degree and then my Ph.D. mostly through determination. My approach has always been to acknowledge obstacles and then concentrate on finding ways around them.

CAREER ▸ CONNECTION

The following job opportunities are also in the field of chemistry.

Chemical Engineer B.S. in chemical engineering
Chemical Laboratory Technician Two-year training program
Industrial Chemical Worker High school plus on-the-job training

Content Background

VSEPR and Hybridization The method used in this section to predict the shapes of molecules is called the Valence Shell Electron Pair Repulsion approach (VSEPR). VSEPR is simple and works well, but it says nothing about the actual bonds that hold atoms together. However, the atomic orbital hybridization approach, first proposed by Linus Pauling, predicts geometry and also explains how the bonding atoms share electrons. For example, the model for a water molecule derives from the hybridization (mixing) of oxygen's $2s$ and $2p$ orbitals. These four orbitals combine mathematically to form four hybrid orbitals designated sp^3. Each has a lobe extending toward the corner of a tetrahedron. Two of oxygen's sp^3 orbitals are filled; the other two are half-filled. When oxygen forms bonds with hydrogen to form water, the two half-filled hybrid orbitals overlap $1s$ orbitals on two hydrogen atoms. These two sp^3 orbitals are filled with bonding pairs of electrons; the remaining two are filled with nonbonding pairs. Therefore, the geometry of the molecule is bent—the same result obtained by the VSEPR approach.

Figure 9.13

The Geometry of Diatomic Molecules
Just as two gumdrops can be connected in only one way, all diatomic molecules are linear, whether they're composed of the same two atoms, as in H_2 and Cl_2, or different atoms, as in HCl.

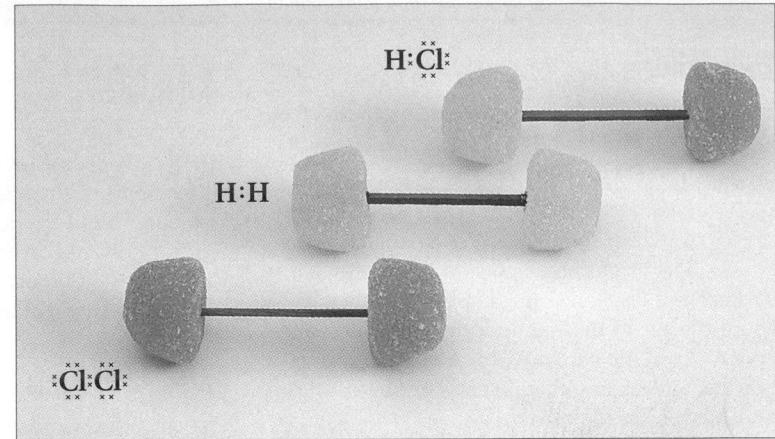

What does a gumdrop model tell about the shape of a hydrogen molecule? As **Figure 9.13** shows, there's only one way to put this model together. When the two gumdrops are connected with a toothpick, they lie in a straight line. Therefore, a hydrogen molecule is linear. You could model other diatomic molecules, such as oxygen, nitrogen, chlorine, iodine, fluorine, and even hydrogen chloride, HCl. The model always predicts the same geometry: linear.

Modeling Water

How would you build a gumdrop model of a water molecule? First, you would draw the electron dot diagram. Remember that the dot diagram models the arrangement of valence electrons in the molecule in two dimensions. The electron dot diagram for a water molecule shows that each hydrogen shares a pair of electrons with the oxygen.

The eight valence electrons are distributed in such a way that the oxygen atom has an octet of electrons and the stable electron configuration of the noble gas neon. Each hydrogen has two valence electrons and the stable configuration of helium. Two pairs of valence electrons are involved in the bonding. These electrons are called bonding pairs. The other two pairs of valence electrons are not involved in the bonding. These are called nonbonding pairs, or lone pairs.

From Electron Dot Diagram to Model

To make a model of the water molecule, you need two colors of gumdrops, such as red for oxygen and yellow for hydrogen. You also need two toothpicks to represent the two covalent bonds, and two additional toothpicks to represent the lone pairs of electrons. Even though the lone pairs of electrons are not part of any bonds, they play a major role in determining the shape of a molecule. They are present and they occupy space.

Demonstration 9.4

Electron Cloud Geometries

Purpose: To demonstrate the geometry of shared and unshared electron pairs.

Materials: 10 small and 6 large balloons, 16 20-cm lengths of string

Safety Precautions:

Disposal: Code A

Procedure: Inflate and tie off the ends of the balloons. Use a permanent felt marker to mark an *S* on each small balloon to represent a shared pair of electrons. Mark a *U* on each large balloon to represent an unshared pair of electrons. Tie a 20-cm length of string to each balloon. Form the following molecules by tying together the balloons: HF (3U, 1S); H_2O (2U, 2S); NH_3 (1U, 3S); CH_4 (4S). Discuss the geometries in terms of electron repulsion and the extra repulsion of unshared electron pairs.

Results: HF, linear; H_2O, bent, bond angle 105°; NH_3, triangular pyramid, bond angle 107°; CH_4, tetrahedral, bond angle 109.5°

Analysis: Ask these questions.

1. Describe the molecular shapes of the four models. *See Results section.*

2. Do you expect an unshared pair of electrons to exert equal, greater, or lesser repulsion compared to a shared pair? *greater repulsion*

How should the hydrogens be connected to the oxygen atom? Clearly, some rules are needed. Look at the dot diagram for water. Four pairs of electrons are on the oxygen. These electrons are all negatively charged, and because they all have the same charge, they repel each other. Therefore, they will form the three-dimensional arrangement around the oxygen atom that allows them to be as far from each other as possible. This arrangement is a geometric shape called a tetrahedron. In a tetrahedral arrangement, the repulsions between the electron pairs are minimized.

To create a model with tetrahedral geometry, place the four toothpicks in the red gumdrop in a three-dimensional arrangement with the largest possible angle between each adjacent toothpick. Two yellow gumdrops should then be placed at the ends of any two of the toothpicks. These represent the two O—H bonds. The other toothpicks represent the lone pairs. **Figure 9.14** shows the gumdrop model of a water molecule and visualizes the tetrahedral arrangement of four electron pairs.

From Model to Water Molecule

The models in **Figure 9.14** suggest that the three atoms in the water molecule are arranged in a bent structure. The angles in a perfect tetrahedron each measure 109.5°. The angle between the two bonds in the water molecule is 105°—a little less than the angle predicted by the gumdrop model. The difference between the predicted and experimental bond angle comes about because the lone pairs of electrons repel each other more than the shared pairs. In effect, the nonbonding electrons require more room, so they distort the tetrahedral arrangement by squeezing the bonding pairs closer together and decreasing the bond angle from 109.5° to 105°.

A space-filling model shows the electron clouds of each atom as spheres. The clouds overlap when two atoms form a bond. This model is a good representation of the water molecule. ▼

Figure 9.14

Visualizing a Water Molecule

In this gumdrop model of a water molecule, you can see that the three atoms form a bent structure. ▼

Four balloons, inflated equally and held together at a central point, arrange themselves in a tetrahedral shape. This is the most space-efficient arrangement for four things about a center point. In this model, the balloons represent the four electron pairs of the water molecule. ▶

Content Background

Isomers As the number of atoms in a molecule increases, the number of different ways those atoms may be connected also increases. An isomer is a compound with the same formula but with a different arrangement of atoms. Ethanol and dimethyl ether both have the formula C_2H_6O, but they differ in how the nine atoms are arranged and, as a result, they differ in chemical and physical properties.

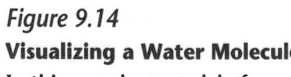

Ethanol

Dimethyl ether

Isomers illustrate how composition and structure determine the properties of a compound.

 Everyday Chemistry

Purpose

TECH PREP

Students will learn about the effect of microwave radiation on polar molecules.

Teaching Strategies

Model the effect of microwave radiation on polar molecules by using a small ceramic bar magnet and a magnetic compass. Lay the compass on a level surface. Securely tie a length of string to the center of the bar magnet so that the length of the suspended magnet is horizontal. Wind the string by spinning the magnet. As the string begins to unwind, place the spinning magnet next to the compass. Have students observe the motion of the compass needle in a changing magnetic field. Draw an analogy to the motion of polar molecules that absorb microwave energy of a specific frequency and increase their rotational, vibrational, and translational motion. **L1** **LEP**

Extension

Have students investigate and write a short paper about how microwave ovens differ from conventional ovens in producing energy and heating food. The paper could be placed in the students' portfolios. **L1** **P**

Exploring Further

1. $CHCl_3$ is a polar molecule and CCl_4 is a nonpolar molecule.
2. Heat must be transferred by conduction from the polar molecules that absorbed the microwave energy to the other molecules of food.

Jiggling Molecules

How often have you warmed a quick snack by microwaves—that is, in a microwave oven? Maybe you were too hungry to notice that the food was a lot hotter than the dish or container that you zapped it in. A clue to the cause of this sometimes-overlooked observation is the condensed steam you blew away from the food as you waited for it to cool.

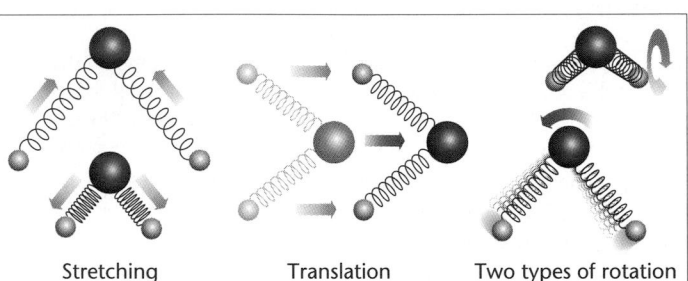

Stretching   Translation   Two types of rotation

Microwave heating Microwave radiation is a form of electromagnetic energy. The microwave radiation produced by a microwave oven has a wavelength of 11.8 cm and a frequency of 2.45 billion hertz. (A shorter way of writing a frequency of 2.45 billion hertz is 2.45 gigahertz, and a shorter way still is 2.45 GHz.) The radiation travels through space and materials in the form of a moving electromagnetic field; that is, a field having an electric-field component and a magnetic-field component. The frequency of the wave is the number of oscillations or waves that pass a point each second. The frequency is also a measure of the energy of the waves. Microwaves have little effect on most molecules. However, the microwave's oscillating electromagnetic field does interact with positively and negatively charged polar molecules. As a result, the charged molecules begin to oscillate (move back and forth). The increased motion means that the molecules have increased kinetic energy. (Kinetic energy is energy of motion and is directly related to temperature.) As

microwaves are absorbed by substances containing polar water molecules, the temperature rises rapidly. Heat is transferred by conduction from the water molecules to other parts of the substance.

Microwave decomposition If you ever jiggled a garden rake to dislodge a leaf, you know that violent motion can be used to separate things. Researchers are applying the same concept to decompose molecules of some toxic substances by using microwaves. Molecules of compounds such as dihydrogen sulfide, sulfur dioxide, and nitrogen dioxide, which contribute to air pollution, are polar and can be decomposed into their nontoxic elements by the oscillations caused by microwave radiation. The oscillations become so violent that the attractions between the atoms in the molecule are no longer great enough to hold them together, and the molecule decomposes or falls apart. Here, the energy of the microwaves has overcome the energy of the chemical bond. Similar research is being done to use microwaves to initiate reactions in which toxic polar organic compounds are reacted with other substances, such as oxygen, to form nontoxic products. Applications of the research may produce cost-effective methods of using microwaves to control air pollution and clean up hazardous wastes.

Exploring Further

1. **Applying** Why can microwave radiation be used to detoxify trichloromethane, $CHCl_3$, but not tetrachloromethane, CCl_4?
2. **Inferring** Why do the cooking instructions for most foods packaged for microwave preparation suggest that after microwaving, the food should stand for several minutes before being served?

*inter***NET**
CONNECTION
An animation of atomic orbitals can be viewed.

World Wide Web
http://www.chem.vt.edu/yip/organic/
compgraf.html

Modeling Carbon Dioxide

Are all triatomic molecules bent like the water molecule? To find out, you can model carbon dioxide, CO_2, as you did water. Begin by drawing the electron dot diagrams for the two atoms. Carbon has four valence electrons, and each oxygen atom has six.

$$\cdot \ddot{C} \cdot \qquad \cdot \ddot{O} \colon$$

To obtain a stable octet of electrons, the carbon atom needs four more electrons, and each oxygen atom needs two more electrons. Therefore, each oxygen atom must share two pairs of electrons with the carbon atom. A bond formed by sharing two pairs of electrons between two atoms is called a **double bond,** as illustrated by the electron dot structure for carbon dioxide.

$$\colon\ddot{O}\colon\colon C\colon\colon\ddot{O}\colon$$

If you count all the shared and unshared electrons around each of the three atoms, you'll see that each atom has an octet.

How can you determine the three-dimensional geometry of a carbon dioxide molecule? First, look at the arrangement of electrons around the central atom. Consider each double bond as one cloud of shared electrons. Two clouds are around the carbon. What geometry puts the two electron clouds as far apart as possible? Linear geometry, as illustrated in **Figure 9.15,** separates the clouds as much as possible. The model suggests that the three atoms in carbon dioxide are arranged in a straight line. This structure of CO_2 was proved correct by experiment.

Figure 9.15
Visualizing Carbon Dioxide
A gumdrop model of carbon dioxide predicts a linear structure with the C—O bonds pointed in opposite directions. ▼

The space-filling model of CO_2 is a good representation of the molecule. ▼

$$\colon\ddot{O}\colon\colon C\colon\colon\ddot{O}\colon$$

The balloon model represents the clouds of bonding electrons on either side of the carbon atom. A linear molecule results because of repulsion between the two clouds. ▶

9.2 **Molecular Shape and Polarity** **321**

Concept Development
Show that in Lewis electron dot diagrams for molecules such as ammonia, bonding pairs of electrons are often shown as dashes. This quickly differentiates them from nonbonding or lone pairs.

Integrating THE Sciences

Biology Ask students to list ways in which models are used in biology. Have them write a paragraph explaining how the models help in clarifying the concept they model. L1

Meeting Individual Needs

Hearing Impaired Make your own balloon models of some of the molecules discussed on this and the following pages, but on the balloons, print clearly the words *bonding pair* or *nonbonding pair.* Use balloons of one color for bonding pairs and balloons of another color for nonbonding pairs.

Construct models and draw chalkboard diagrams of tri-atomic molecules in order to differentiate linear molecules such as CO_2 and HCN from bent or angular ones such as H_2O and SO_2. Emphasize that a triatomic molecule will be linear only if the central atom has two bonding electron clouds and no lone pairs.

GLENCOE TECHNOLOGY

Videodisc

Chemistry: Concepts and Applications
Molecular Shapes
Disc 1, Side 2, Ch. 13

Show this video to help you explain how balloon models are made and what they represent. L1 LEP

CD-Rom

Chemistry: Concepts and Applications
Molecular Shapes
Have students use this video to help reinforce balloon models as representations of repelling electron clouds. L1 LEP

A Model for Ammonia

Ammonia, NH_3, has three bonds from a central nitrogen to hydrogen atoms. Nitrogen has five valence electrons, and each hydrogen has one electron. Each hydrogen shares a pair of electrons with the nitrogen. Nitrogen's remaining two electrons form a nonbonding pair. This arrangement gives nitrogen a complete octet of electrons.

$$H \overset{\cdot\cdot}{\underset{\overset{|}{H}}{N}} H$$

Count the number of electron pairs about the central nitrogen atom. There are four pairs—three bonding pairs and one lone pair. The four pairs avoid each other by arranging themselves in a tetrahedral arrangement, just as they did in the water molecule. But this time, three of the positions are N—H bonds, and the fourth position is the lone pair. The atoms in ammonia form a triangular pyramid structure. The three hydrogen atoms form the base of the pyramid with the nitrogen at the peak, as shown in **Figure 9.16.** Based on this tetrahedral arrangement, you would predict that the H—N—H bond angle is 109.5°. The experimentally determined structure is a triangular pyramid with a bond angle of 107°.

Figure 9.16

Looking at the Ammonia Molecule
Ammonia has one lone pair and three bonding electron pairs. When these are arranged around the central atom, the arrangement is tetrahedral, but the geometry of the four atoms is a triangular pyramid. The lone pair causes the H—N—H angles in ammonia to be 107°, slightly less than the predicted tetrahedral angle of 109.5°.

▲ The space-filling model of ammonia shows the overlap of the electron clouds of the nitrogen and hydrogen atoms.

Demonstration 9.5

Like Charges Repel

Purpose: To show that like charges repel.
Materials: 2 large inflated balloons; 1-m length of string; a piece of wool, silk, or fur
Safety Precautions: 👓
Disposal: Code F
Procedure: Tie a 50-cm length of string to each of two inflated balloons. Keeping the balloons separated, rub each balloon with a piece of cloth. Have a student help you by holding one of the balloons after you have rubbed it with the wool, silk, or fur. If your assistant has long, fine hair, you may want to bring the charged balloon near the hair to show that unlike charges attract. The hair will stick to the balloon's surface. Bring the strings together and observe how the balloons behave.
Results: The like-charged balloons will swing away from each other when the two strings are held together in one hand. (A dry day is best.)
Analysis: Ask these questions.
1. Draw the electron dot diagram of water.

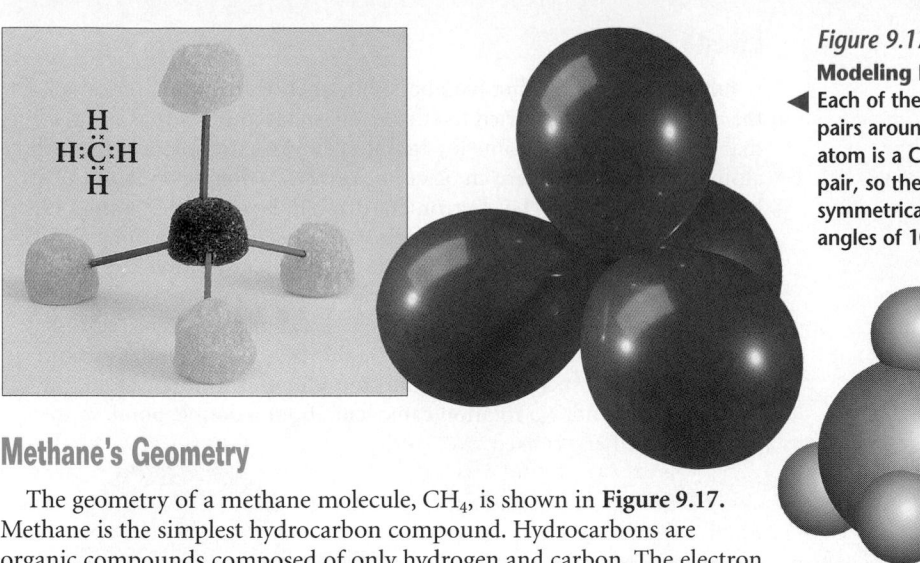

Figure 9.17
Modeling Methane
◀ Each of the four electron pairs around the carbon atom is a C—H bonding pair, so the structure is symmetrical with all bond angles of 109.5°.

▲ The space-filling model of methane displays the symmetry of the molecule.

Methane's Geometry

The geometry of a methane molecule, CH_4, is shown in **Figure 9.17.** Methane is the simplest hydrocarbon compound. Hydrocarbons are organic compounds composed of only hydrogen and carbon. The electron dot structure for methane consists of a central carbon atom with four C—H single bonds, as shown in **Figure 9.17.**

Four pairs of electrons are positioned in a tetrahedral arrangement around the carbon atom, just as they are in the water and ammonia molecules. But in methane, all four pairs are shared between the carbon atom and the four hydrogen atoms. Because there are no lone pairs requiring extra space, the structure of methane is a perfect tetrahedron with bond angles of 109.5°.

Ethane in 3-D

Ethane, C_2H_6, is the second member of the hydrocarbon series known as the alkanes. Methane is the first and simplest. Alkanes are hydrocarbons that contain only carbon and hydrogen atoms with single bonds between all the atoms. An ethane molecule has two carbon atoms that form a bond to each other and to three hydrogen atoms. The electron dot diagram of ethane is shown in **Figure 9.18.** Each carbon atom has four single bonds. Just as in methane discussed above, a tetrahedral arrangement of bonding electron pairs around each carbon atom provides the most space for the electrons.

The space-filling model of ethane shows two connected tetrahedral arrangements around the two carbon atoms. ▼

Figure 9.18
Modeling Ethane
The geometric arrangement of atoms around each carbon atom in ethane is tetrahedral. All bond angles are 109.5° as predicted from the geometry. ▶

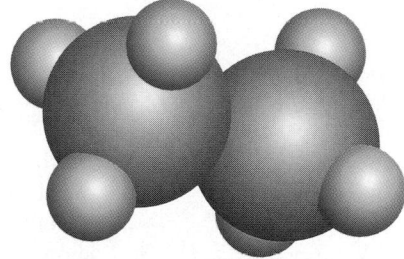

9.2 **Molecular Shape and Polarity** **323**

Using an Analogy

Borrow magnets from the physics department and let students handle them to experience the repulsion of two like poles when they come close together. Point out that electrons, having like charges, also repel each other when they are close together. Ask students whether it takes energy (effort) to push the repelling magnets together. *yes* Ask whether an arrangement of electrons in which two electrons occupy the same orbital is higher in energy than one in which the electrons occupy separate orbitals. *Yes, the paired electron arrangement has higher energy.* L1

Extension

Draw chalkboard diagrams such as those shown below that represent a bonding pair of electrons (a) and a lone pair of electrons (b) on the oxygen of a water molecule.

a.

b.

Because a lone pair is under the influence of only one nucleus, it tends to be drawn in closer and take up more space around the oxygen nucleus. A bonding pair is pulled between the oxygen and hydrogen nuclei. The region it occupies is elongated, and it takes up less space around the oxygen nucleus.

Transparency

Display **Basic Concepts Transparency 25** to show the geometry of the methane molecule. L1 LEP

2. Is the water molecule polar or nonpolar? Why? *Polar; two pairs of nonbonding electrons are arranged in an asymmetrical pattern.*
3. Because like charges repel, how do water molecules align when they are next to each other? *Molecules arrange themselves so that the positive end (hydrogen end) of one molecule is near the negative end (oxygen end) of the next molecule.*

Assessment

Skill: Give a team of students four balloons with strings attached. Ask them to rub the balloons with wool and demonstrate the 3-D arrangement of the four pairs of electrons around the oxygen atom. *The arrangement is tetrahedral.* L1 LEP COOP LEARN

How far can you stretch?
Students can demonstrate for themselves the geometry of a tetrahedron. Have them stand and put their feet as far apart as possible. Then have them twist at the waist until their shoulders are perpendicular to a line joining their feet. Now, have them raise their arms above their heads and spread them apart. Arms and legs are stretching toward the apices of a tetrahedron and can represent four bonds as in a molecule such as methane. Ask one or two students to don white gloves, and then have them remove one glove to model triangular pyramidal geometry. Have them remove the other glove or a shoe to model the bent geometry of a molecule such as water. Removal of two gloves and one shoe models linear geometry. L1

Correcting Misconceptions

When triple bonds are shown as two horizontal lines of three dots each, C⦙⦙⦙C, students often get the idea that each bond contains three electrons. A better representation of a triple bond is one in which two vertical lines of three dots each join the two atoms, C⦙⦙C. Point out to students that a bond is always one pair, two pairs, or three pairs of electrons.

Ethene in 3-D

Ethene, C_2H_4, is the first member of another hydrocarbon series called the alkenes. Ethene is related to ethane, but it has four hydrogens rather than six. The common name for ethene is ethylene. In order for the carbon atoms in ethene to acquire an octet of electrons, a double bond must exist between the carbons. The electron dot diagram is shown in **Figure 9.19**.

Each carbon atom has three bonds: two C—H bonds and one C=C double bond. The most space-efficient arrangement of three electron clouds about a central atom is a flat, triangular arrangement, as shown in **Figure 9.19**. The H—C—H and H—C—C bond angles are all 120°. The geometry is rigid because of the double bond between the carbon atoms. In alkanes such as ethane, the atoms are free to rotate about the single C—C bond. No rotation can occur about a double bond, so the structure of ethene is fixed.

Figure 9.19
Modeling Ethene
The arrangement of hydrogen atoms around each carbon atom in ethene is a flat, triangular arrangement. All six atoms lie in the same plane. ▼

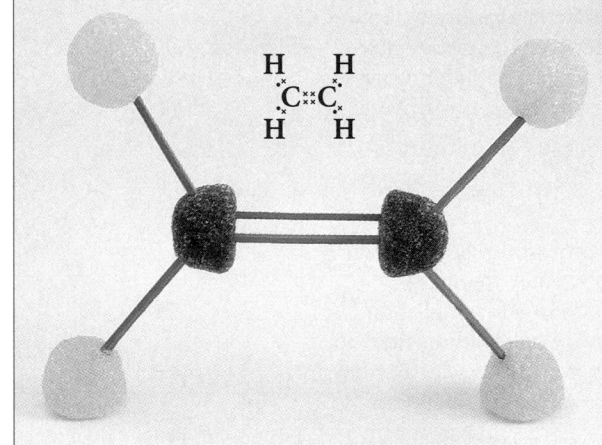

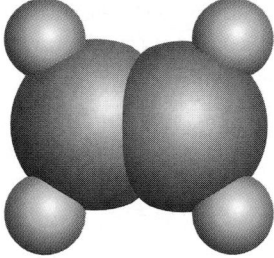

▲ Three equally inflated balloons, held at a central point, assume the triangular arrangement shown here. This is the most space-efficient arrangement of three things about a central point.

Like the hydrocarbon ethane, the space-filling model of ethene has two geometric centers. ▶

The Geometry of Ethyne

Ethyne, shown in **Figure 9.20**, is the first member of a hydrocarbon series called the alkynes. Its formula is C_2H_2. Notice that ethyne has two carbons like ethane and ethene, but only two hydrogens. Ethyne is more commonly known as acetylene. It is the fuel used in torches for cutting steel.

📓 Chemistry Journal

Hydrocarbon Series
Have students research the three hydrocarbon series that start with methane, ethene, and ethyne. Ask them to list, in their journals, the formulas for the first four members of each series. L2

Modeling Molecules

The ability to build and interpret models is an important skill in chemistry. This exercise will give you practice working with models.

Procedure

1. Obtain a model kit from your teacher.

2. Draw the electron dot structure for one of these molecules.

   H_2, HCl, H_2O, CO_2, NH_3, CH_4, C_2H_6, C_2H_4, C_2H_2

3. Build a three-dimensional model of the molecule.

4. Draw a sketch of the geometric shape you predict for the molecule.

5. Repeat steps 2-4 for each of the molecules in the list.

6. Make a table like the one shown here and fill in the boxes with your data.

Analysis

1. How did the electron dot structure of each molecule help you predict its geometry?

2. Pick one of your assigned models. How many lone pairs does it have? How many bonding pairs are there?

Formula	Electron Dot Structure	Sketch of Predicted Geometry

You learned that in ethene, the two carbon atoms share two pairs of electrons in a double bond. In ethyne, the carbon atoms share *three* pairs of electrons to obtain a stable octet. A bond formed by sharing three pairs of electrons between two atoms is called a **triple bond.** The electron dot diagram for ethyne is shown in **Figure 9.20.**

Each carbon has two bonds, a C—H single bond and a C≡C triple bond, so two electron clouds are around each carbon atom. Linear geometry places two electron clouds as far as possible from each other. In ethyne, the four atoms are arranged in a straight line, so ethyne is a linear molecule.

The space-filling model of ethyne shows the arrangement of the electron clouds. ▼

Figure 9.20
Looking at Ethyne
◄ Ethyne, when combined with excess oxygen, burns with a hot flame. It is used in welding torches.

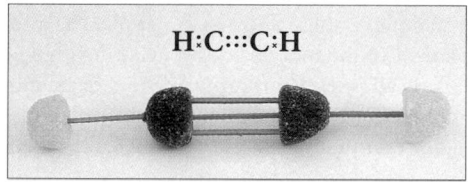

H:C:::C:H

◄ The model shows that the geometry of ethyne is linear. The triple bond makes the molecule rigid.

Formula	Dot Structure	Geometry
H_2	H:H	Linear
HCl	H:C̈l:	Linear
H_2O	H:Ö:H (H $\overset{..}{O}$ H)	Bent
CO_2	:Ö::C::Ö:	Linear
NH_3	H:N̈:H / H	Triangular pyramid

Formula	Dot Structure	Geometry
CH_4	H:C̈:H (with H above and below)	Tetrahedral
C_2H_6	H:C̈:C̈:H (with H H above and below)	Tetrahedral at each carbon
C_2H_4	H:C::C:H (with H above and below)	Planar with 120° bond angles
C_2H_2	H:C:::C:H	Linear

miniLAB

Modeling Molecules

Purpose: To introduce models and the three-dimensional shape of molecules.

Process Skills: Predicting, formulating models, using space/time relationships, communicating

Teaching Strategies

- Use commercially available model kits, or inexpensive kits may be assembled from a variety of materials including gumdrops, marshmallows, toothpicks, Styrofoam balls, soft clay, etc.
- Have students work in pairs. Students can help each other see the geometry. L1 LEP COOP LEARN

Expected Results

See tables below.

Analysis

1. The electron dot structure shows the distribution of electrons in the molecule. The geometry is determined by the number of bonding pairs and lone pairs on the central atom.

2. Answer depends on the molecule chosen.

✓ Assessment

Performance: Ask students to think about the flexibility of structures with multiple bonds. Have them build and compare models of CH_3CH_3 and CH_2CH_2 and write a paragraph describing how the structures are different in their geometry and in their ability to rotate. L1 LEP COOP LEARN

TECH PREP

Purpose

Students will learn about several types of chromatography used in analytical chemistry.

Background

Chromatography is a method of separating a mixture by virtue of differences in the mobilities of the components of the mixture. Once separated, the components can be identified in a variety of ways including comparison with known separations, mass spectroscopy, light spectroscopy, photoionization, and thermal conductivity.

Teaching Strategies

- Have students derive the word *chromatography* (*chromos*: Gk, "color;" *graphos*: Gk, "written").
- Explain that the migration rates of components are often compared by means of retention factors, R_f. The definition of the retention factor is

$$R_f = \frac{d_{solute}}{d_{solvent}}$$

where d_{solute} is the distance the solute (component) moved and $d_{solvent}$ is the distance the solvent moved.

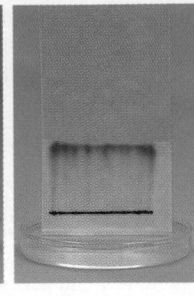
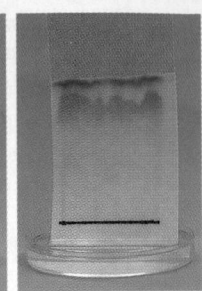

Chromatography

Much of the stuff that's interesting in this world is a mixture. Blood, dirt, air, pizza—to name a few—are mixtures. These substances are complex, and realizing that they are complex may make them more interesting. Scientists have devised ways to analyze complex things. One way is chromatography.

Going with the Flow

Believe it or not, this rained-upon poster is a good model of chromatography. Chromatography is a way of separating a mixture using differences in the abilities of the components to move through a material. All chromatography involves two phases—a stationary phase and a mobile phase. The movement of the mobile phase through the stationary phase allows separation to take place. Because the components of a mixture move at different rates, they eventually separate.

Paper Chromatography

One type of chromatography used to separate colored mixtures is paper chromatography. A porous paper is used as the stationary phase. Water or some other solvent is used as the mobile phase. A spot or a line of the mixture to be separated is placed on the paper. The solvent moves upward along the paper because of capillary action. As it reaches the spot, the mixture dissolves in the solvent. Now, the components of the mixture begin to migrate upward on the paper along with the solvent. Those components that have little attraction to the paper move almost as quickly as the solvent. Components of the mixture that have greater attraction for the paper migrate at a slower rate. The differences in the migration rates result in differences in the distances the separated components travel. Changing the solvent causes changes in the type of materials separated.

Thin-Layer Chromatography

In thin-layer chromatography, the stationary phase is a suspension made up of a material such as silica gel or cellulose in a solvent. The suspension is applied as a thin coating on a glass or metal plate, which is then dried. The

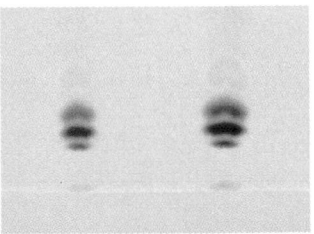

mixture being separated is applied to the bottom of the plate, and the plate is placed vertically in a solvent. Either the solvent vapor or the liquid itself acts as the mobile phase.

Gel Chromatography

Gel chromatography is often used to isolate the molecules of living, biological systems. The liquid phase carries the molecules through a stationary phase of a porous gel. The pore sizes can be manufactured precisely to affect the migration of molecules through the gel. Large molecules are blocked from moving through the gel. Because smaller molecules spend more time within the pores than midsize molecules, they take longer to migrate through the pores. As a result, gel chromatography can be used to isolate midsize molecules found in living materials.

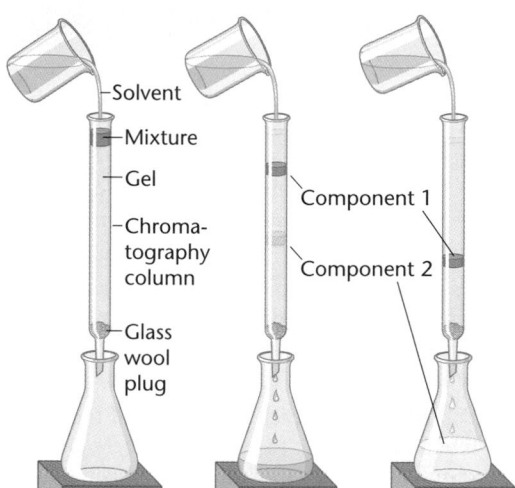

Gas Chromatography

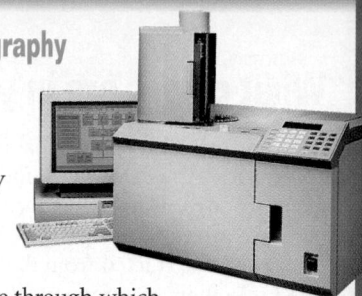

In gas chromatography, the mobile phase is a gas. The stationary phase is usually a liquid coating deposited on the interior of the tube through which the mixture, itself a mixture of gases, will migrate. Helium is often used as the mobile phase. The extent to which the components of a gaseous mixture interact with the coating of the tube determines their migration rates. The separated components of the mixture arrive at the end of the tube at different times, where they are analyzed and identified by a light spectrometer or mass spectrometer. Portable gas chromatographs, used in conjunction with mass spectrometers, can analyze trace amounts of gases that contribute to air pollution. Among other things, gas chromatography is used by researchers to analyze complex mixtures of compounds that constitute aromas and flavors.

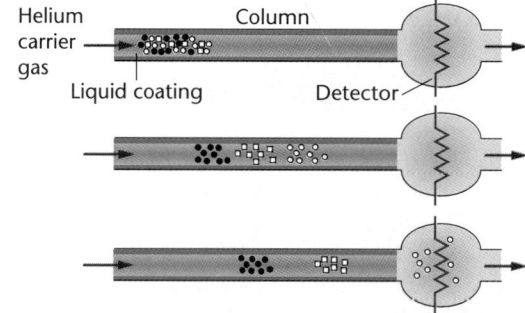

DISCUSSING THE TECHNOLOGY

1. **Analyzing** Identify the stationary and mobile phases of a rain-soaked poster.

2. **Designing** Design a laboratory method by which you could obtain pure samples of the components of a mixture that has been separated by using paper chromatography.

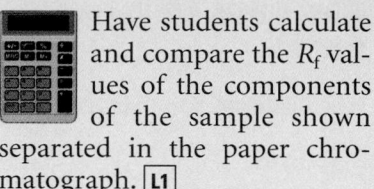
Have students calculate and compare the R_f values of the components of the sample shown separated in the paper chromatograph. [L1]

Visual Learning

Students can make several chromatographs of inks of water-soluble pens and markers. Working in groups, have them cut several strips from coffee-maker filters and draw a line across each strip with a different pen. The strips can be suspended in water just below the inked lines. After several minutes, have the students observe the separated colors of the inks and compare their results with the photo of the poster. [L1] **COOP LEARN**

Discussing the Technology

1. The poster board is the stationary phase and rainwater is the mobile phase.
2. Accept reasonable answers such as cutting the component spots from the dried chromatograph, placing each spot into a test tube containing a solvent, dissolving the component from the paper, removing the paper, and evaporating the solution.

ChemLab

ChemLab

What colors are in your candy?

Time Allotment: One class period.

Objectives

Review objectives with students before they begin the ChemLab.

Process Skills

Recognizing cause and effect, observing, inferring, sequencing, interpreting data, classifying, communicating

Safety Precautions and Disposal

Students should wear aprons and goggles.
Disposal: Chromatography paper may be discarded in wastebaskets.

PREPARATION

Obtain good-quality filter paper or chromatography paper cut to size. Paper is available from science suppliers as large sheets or precut. Obtain several bags of colored candies (M&Ms or Skittles, for example) and large jars with lids or chromatography chambers.

Alternative Materials

- Other colored candies may be used.
- Liquid drink mixes may also be used. These should be prepared as concentrated solutions and spotted with toothpicks.
- Coffee-filter paper may be used, but results will be less precise.
- An aquarium with a cover, large coffee cans with lids, or large beakers with plastic-wrap lids may be used as chromatography chambers.

328

What colors are in your candy?

Yellow dye #5 is an artificial food coloring approved by the FDA, but some people are allergic to this dye. Many candies contain Yellow #5 as part of a mixture to color the candies. Dye mixtures can be extracted from the candy and separated into their component colors using paper chromatography. The yellow food coloring that you buy in the grocery store contains Yellow #5 and can be used as a reference standard.

Separations by paper chromatography are possible because different substances have different amounts of attraction for the paper. The greater the attraction the substance has for the paper, the slower it will move up the paper with the solvent.

Problem

Are there any colored candies that a person with an allergy to Yellow #5 can safely eat?

Objectives

- **Observe** separation of colors in dye mixtures.
- **Interpret** data to determine which candies contain Yellow #5.

PREPARATION

Materials

10 cm × 10 cm piece of Whatman #1 filter paper
large jar with lid   water
colored candy   salt
yellow food coloring   toothpicks
small plastic cup   ruler

PROCEDURE

1. Make a data table like the one shown.
2. Make a fine line with a pencil about 3 cm from one edge of the piece of filter paper.

3. Place a small amount of water in a plastic cup.
4. Dip the tip of a toothpick into the water.
5. Dab the moistened tip of the toothpick onto a piece of colored candy to dissolve some of the colored coating.
6. Place the tip of the toothpick with dye onto the filter paper to form a spot along the pencil line.

7. Remoisten the tip of the toothpick, and dab the same piece of candy to dissolve additional coating. Place the tip of the toothpick onto the filter paper on the same spot made in step 6. Repeat this step until a concentrated spot is obtained.
8. Using a new toothpick and fresh water, repeat steps 4 to 7 with a different colored piece of candy. Make a new spot for each piece of candy, and keep a record in your data table.
9. Dip a fresh toothpick into a drop of the yellow food coloring to be used as a reference standard. Make a spot along the pencil line and mark the location of the reference spot.
10. Carefully roll the paper into a cylinder. The spots should be at one end of the cylinder. Staple the edges. Avoid touching the paper.
11. Add water to the jar to a level of about 1.5 cm from the bottom. Sprinkle in a pinch of salt. Close the lid and shake.

PROCEDURE

- Have students work in pairs.
- Require that students' laboratory notebooks contain the following before lab begins: Objectives; Safety Precautions; Outline of the Procedure; Data and Observations Table.
- Spots should be as concentrated as possible. Have students practice spotting on a piece of scrap paper. Several small applications are better than one large one.
- Spots should be evenly spaced on the pencil line. The line should be at least 2 cm from the edge of the paper. Spot patterns must be

recorded in lab notebooks.
- The yellow dye is a reference. If students have space, they may apply other colors of food colorings.
- The water will take 20 to 30 minutes to rise to the required level. Have students clean up while the chromatogram is developing, or use the time for related discussion or a MiniLab.
- If necessary, the development could be reserved for a second period.
- See the published experiment: Kandel, M. *J. Chem. Educ.*, 69, 1992, p. 989.

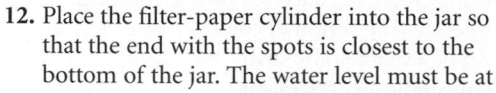

12. Place the filter-paper cylinder into the jar so that the end with the spots is closest to the bottom of the jar. The water level must be at least 1 cm below the pencil line. Adjust the amount of water if necessary and close the lid.

13. Allow the water to rise to about 1 cm from the top of the filter paper.

14. Carefully remove the filter paper, open it flat, and mark the solvent edge (the farthest point the water traveled) gently with a pencil. Lay the filter paper on a paper towel to dry.

15. For each piece of candy spotted, measure the distance from the original pencil line to the center of each separated spot. Record these data in your data table. Some candies may have more than one spot.

16. Measure and record the distance from the original pencil line to the marked solvent edge.

17. Record the distance from the original pencil line to the center of each spot separated from the reference spot of Yellow #5.

ANALYZE AND CONCLUDE

1. **Interpreting Observations** Do any of the candies contain Yellow #5? How can you tell?
2. **Comparing and Contrasting** Do any of the candies contain the same dyes? Explain.
3. **Inferring** Which candies would be safe to eat if you were allergic to Yellow #5?
4. **Designing an Experiment** Can you devise a better way to remove the dye from the candy and place a spot on the paper?

APPLY AND ASSESS

1. On what portion of the paper are the substances with the greater attraction for the paper? What conclusions can you draw about the molecular polarities of these dyes?
2. Why was it important to use a pencil instead of a pen to mark the paper?
3. Why was it important to do the experiment in a closed jar?
4. What makes the water move up the paper?
5. How did the rate of water movement up the paper change as the water got higher on the paper? Suggest reasons why it changed.

DATA AND OBSERVATIONS

Solvent distance : _____ (distance from first pencil mark to solvent edge)

Original Spot	Distance (color 1)	Distance (color 2)	Distance (color 3)
Yellow #5 Reference			
Candy 1			

APPLY AND ASSESS
1. The colors that stay close to the origin have the greatest attraction for the paper and are the most polar.
2. Pencil (graphite) is not soluble in water. Ink may be water soluble and would interfere.
3. This keeps the atmosphere saturated with water so that the paper does not dry out until the experiment is complete.
4. capillary action
5. As water moves higher, the rate of climbing slows because of competition between capillary action and the force of gravity.

✔ Assessment

Performance: Have students use chromatography to separate other colored substances. Encourage them to try solvents such as water with different salt concentrations and isopropyl alcohol (rubbing alcohol). L1

DATA AND OBSERVATIONS
These data are representative. Specific results will depend upon brand of candies used.

Solvent Front Distance: 9 cm

Candy Color	Colors and Distances After Development
Yellow	Yellow (3.8 cm)
Orange	Orange (3.2 cm)
Red	Red (1.9 cm)
Brown	Blue (6.1 cm) Yellow (3.8 cm) Orange (3.2 cm)
Tan	Blue (6.1 cm) Yellow (3.8 cm) Orange (3.2 cm)
Green	Blue (6.1 cm) Yellow (3.8 cm)
Yellow Food Coloring (Yellow #5)	Yellow (4.0 cm)

ANALYZE AND CONCLUDE
1. Answers will depend upon the candies used. In data shown, the yellow, brown, tan, and green candies contain yellow #5. If any separated spot matches the yellow spot of the reference (both in color and distance), a student could conclude that the candy contains yellow #5.
2. Answers will depend upon the candies used. In data shown, the brown, tan, and green all share similar colors. Tan and brown have the same colors but in different proportions.
3. Answers will depend upon the candies used. In data shown, the orange and red candies would be safe to eat.
4. Students may have a variety of suggestions, including soaking candies briefly in a small amount of water and using a moist candy to apply the spot.

How Polar Bonds and Geometry Affect Molecular Polarity

Have you ever pulled clothing from the dryer and found that it was stuck together because of static cling? Static cling results from electrostatic attraction between positive and negative charges. Positive and negative charges arise in other ways besides the action of the clothes dryer, and some physical properties can be explained by them. For example, water molecules tend to bead up on smooth surfaces, and raindrops take a spherical shape, as shown in **Figure 9.21.** Why should water molecules stick together to form drops? The reason is that water molecules have positive and negative ends. Oppositely charged ends of the molecules attract one another, and the molecules stick together like the clothes in the dryer. Water is an example of how polar bonds and molecular geometry act together to affect the properties of compounds.

Figure 9.21

Water Molecules Attract Each Other
Water molecules attract one another because they have positive and negative ends. The water on this leaf beads up because water molecules on the surface of the drop are attracted to other water molecules below them. These attractions also explain the typical shape of a falling drop of water.

Water: A Polar Molecule

You saw earlier that the water molecule has a bent structure. The ΔEN of the O—H bond is 1.4 so the two O—H bonds in a water molecule are polar bonds. The oxygen end of the bond has a partial negative charge, while the hydrogen end of the bond has a partial positive charge. Because of its bent shape, the water molecule as a whole has a negative pole and a positive pole, as shown in **Figure 9.22.**

Figure 9.22

The Polar Water Molecule
The O—H bonds in a water molecule are polar. Because of water's bent shape, the hydrogen side of the water molecule has a positive charge, and the oxygen side has a negative charge. The arrows indicate the directions in which electrons are pulled.

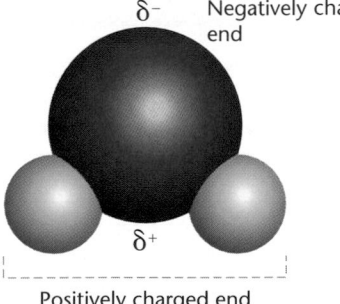

Meeting Individual Needs

Learning Disabled To demonstrate that polar bonds cancel under some circumstances and not under others, borrow spring scales from the physics department and attach strings from two scales to a movable object such as the base of a ring stand. While one student holds the ring stand in place, two other students use the spring scales to exert forces of the same magnitude in opposite directions. While both students are exerting the same pull, have the student release the ring stand.

The ring stand does not move because the forces cancel. To show that forces do not always cancel, repeat the demonstration with the students again exerting equal forces but at an angle of 90°. When the object is released, it moves in a direction midway between the directions of the forces. Draw chalkboard diagrams of carbon dioxide and selenium dichloride molecules and correlate them with the demonstrations. L1 LEP

The water molecule is an example of how polar bonds, arranged in certain geometries, can result in a **polar molecule,** that is, a molecule that has a positive and a negative pole. A polar molecule is also called a dipole.

Ammonia: Another Polar Molecule

Ammonia, NH_3, is another example of a molecule with polar bonds. The N—H bond has a $\Delta EN = 0.9$. The geometry of an ammonia molecule is a triangular pyramid, as shown in **Figure 9.23.** When the three polar N—H bonds are arranged in this geometry, a polar molecule results. A center of net positive charge is located at the base of the pyramid, while a center of negative charge is on the nitrogen atom.

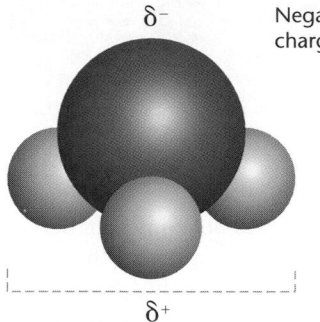

Negatively charged end

Positively charged end

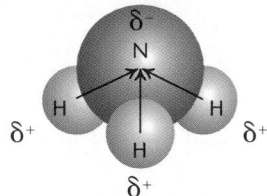

Figure 9.23

The Polar Ammonia Molecule
Like water, an ammonia molecule has two distinct sides. Because of polar bonds, the hydrogen side has a net positive charge and the nitrogen side has a net negative charge.

Carbon Dioxide: A Nonpolar Molecule

Carbon dioxide is another molecule with polar covalent bonds. The ΔEN for the C—O bond is 1.0, so the polarity of the C—O bond in carbon dioxide is similar to the polarity of the N—H bond in ammonia. But the geometry of CO_2 is different from the geometry of ammonia. The shape of a CO_2 molecule is linear. **Figure 9.24** shows that with the polar C=O bonds of a CO_2 molecule arranged in a straight line, the effects of the two polar bonds exactly cancel each other. As a result, there is no separation of positive and negative charge in the molecule. So, although carbon dioxide has relatively strong polar covalent bonds, CO_2 is a nonpolar molecule.

Figure 9.24

Nonpolar Carbon Dioxide
Both bonds in CO_2 have ΔEN equal to 1.0 and are polar, but the polar bonds oppose each other and therefore cancel each other's effects. Carbon dioxide is a nonpolar molecule. ▼

$$\delta^- \; O = C = O \; \delta^-$$

◀ Polar solvents such as water usually do not dissolve nonpolar substances. However, nonpolar CO_2 is slightly soluble in water; but under pressure, even more CO_2 dissolves. Bottling soda pop with CO_2 under pressure adds the fizz. When you release the pressure by opening the bottle, the CO_2 comes out of solution—sometimes too fast.

331

⚡ Quick Demo

Which disappears first?
Be sure your chalkboard composition is compatible with acetone, water, and ethanol before trying this demo. Draw circles on the board and label them C_2H_5OH (ethanol), H_2O, and C_3H_6O (acetone). Ask three students to put on safety goggles and aprons and help with the demo. Give each student a small beaker containing about 3 mL of one of the liquids and ask them to simultaneously splash the liquids on the appropriate circles. Ask all students to observe the rates of evaporation of the liquids and explain the results. *Acetone evaporates quickest because it has the lowest molecular polarity and intermolecular attractions. Water evaporates slowest because it has the greatest molecular polarity and intermolecular attractions.* L1

Transparency

Display **Basic Concepts Transparency 26** to show the structures of polar and nonpolar molecules. L1 **LEP**

Cultural Diversity

2500 Years of Research

Do you have an allergy? Make a tea of beetle legs, lichens, toadstools, and some mysterious colored powders. Drink a cup of this tea daily. Some proponents of herbal medicines say it works. The use of herbal remedies has been well documented in China over at least the past 2500 years. Western scientists and pharmaceutical companies are beginning to take a look at these remedies as sources of possible drugs that may have desirable properties in the treatment of cancer and AIDS. The National Cancer Institute recently funded the compilation of a database that catalogs the medicinal effects of Chinese herbs. In 1992, The National Institute of Health created an Office of Alternative Medicine to evaluate a variety of medical therapies, including five related to Chinese herbal medicine. Several U.S. pharmaceutical companies have formed over the past few years to develop drugs from Chinese plants.

3 ASSESS

Check for Understanding

- Check for understanding of the section material by having students answer the Section Review questions. L1
- Ask students to use the *Handbook of Chemistry and Physics* to look up the boiling points of these molecular compounds: methane, CH_4; ammonia, NH_3; and hydrogen fluoride, HF. Ask them to explain the large differences. *Methane is nonpolar and has the lowest boiling point, $-164°C$. Ammonia is somewhat polar and boils at $-33°C$. HF is very polar and boils at $19.5°C$.* L1

Reteach

Discussion: Ask students to name and write the formula for one molecular compound of each of the following types: a diatomic nonpolar molecule, a diatomic polar molecule, a tetrahedral nonpolar molecule that has polar bonds, and a tetrahedral polar molecule. *Typical answers might be: nitrogen, N_2; hydrogen chloride, HCl; tetrachloromethane or carbon tetrachloride, CCl_4; and difluoromethane, CH_2F_2.* L1

GLENCOE TECHNOLOGY

 Videotape

MindJogger Videoquizzes for Chemistry
Chapter 9: Chemical Bonding Have students work in groups as they play the videoquiz game to review key chapter concepts. L1 **LEP** **COOP LEARN**

Figure 9.25
Dipole Interactions in Liquids and Solids
The force between dipole molecules is an attraction of the positive end of one dipole for the negative end of another dipole. Shown here are representations of dipole-dipole attractions in liquids (left) and solids (right).

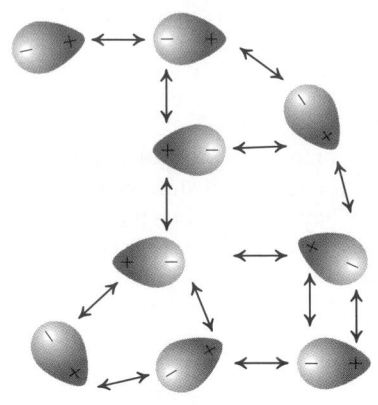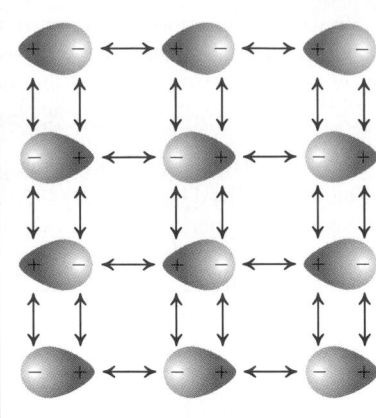

Methane and Water Compared

Polar molecules attract each other because they have positive and negative ends. The diagram in **Figure 9.25** shows how dipoles interact. Because of this attraction, the properties of polar molecules differ from those of nonpolar molecules. For example, melting points and boiling points of polar substances tend to be higher than those of nonpolar molecules of the same size.

When you compare the physical properties of the polar molecule water with the nonpolar molecule methane, you see major differences. Although water and methane are approximately the same size and both are covalently bonded, water is a liquid at room temperature, whereas methane is a gas. **Table 9.2** shows a comparison of the melting points and boiling points of water and methane. Notice that the boiling point of water is 264° higher than the boiling point of methane. This is macroscopic evidence for the submicroscopic attractions at work among water molecules.

Table 9.2 Comparison of Melting and Boiling Temperatures		
Compound	**Melting Point**	**Boiling Point**
methane (nonpolar)	$-182°C$	$-164°C$
water (polar)	$0°C$	$100°C$

Ions, Polar Molecules, and Physical Properties

Recall from Chapter 4 that the submicroscopic interactions between the particles of a substance determine many of its macroscopic physical and chemical properties. For ionic compounds, the strong attractive force that binds positive and negative ions into well-ordered crystals keeps the physical properties of ionic substances to a relatively narrow range of variability.

Chemistry Journal

High or Low—It Depends

Ask students to describe, in a paragraph in their journals, one or two situations in which liquids having low intermolecular attractions are desirable. In another paragraph, have them describe one or two situations in which liquids having high intermolecular attractions are desirable. L1

Figure 9.26
Ionic and Covalent Compounds
Ionic compounds exhibit a narrower range of physical properties than covalent compounds.

Ionic compounds, such as copper(II) nitrate (left) and potassium chloride (below), tend to be brittle, solid substances with high melting points. ▼

The ionic solids potassium chloride and copper(II) nitrate are shown in **Figure 9.26.** Both have high melting points because of the strong attractions among their oppositely charged ions. On the other hand, sugar is covalently bonded, but polar bonds create relatively strong interactions that hold the molecules together in a solid crystal structure at room temperature. Ethanol, C_2H_6O, has covalent bonds and is a liquid at room temperature. Ethanol boils at a temperature 44° higher than dimethyl ether, which has the same formula but a different arrangement of atoms. The difference in boiling point occurs because ethanol is a polar molecule whereas dimethyl ether is less polar. Chlorine, Cl_2, is the most nonpolar of the substances shown in **Figure 9.26** because the bond between the two chlorines is completely covalent and, therefore, nonpolar. Interactions between chlorine molecules are minimal, so chlorine is a gas at room temperature.

Connecting Ideas

Whether you use gumdrops, electron dot diagrams, or supercomputers, the ability to model bonding between atoms is useful. By determining the shape and polarity of a molecule, you can predict its behavior and properties. In Chapter 10, you'll learn more about the forces between particles and the effects they have on the physical states of substances.

▲ Covalent substances may be solids, liquids, or gases at room temperature. Sugar (top, right) is a solid, ethanol (top, left) is a liquid, and chlorine (bottom) is a gas, yet all are covalently bonded.

SECTION REVIEW

Understanding Concepts

1. Draw the electron dot diagrams for each of the following molecules.

   a) PH_3 c) HBr
   b) CCl_4 d) OCl_2

2. Describe the shape of each molecule in question 1.

3. What is the difference between a single bond, a double bond, and a triple bond?

Thinking Critically

4. **Molecular Polarity** Chloroform, $CHCl_3$, is a molecule similar to methane in structure. Is chloroform a polar or nonpolar molecule?

Applying Chemistry

5. **Properties and Covalent Compounds** Sugar, water, and ammonia are all covalent compounds. How do these compounds demonstrate the variety of types of covalent compounds?

9.2   Molecular Shape and Polarity   **333**

SECTION 9.2

Extension

On the chalkboard, write Lewis electron dot diagrams for these polyatomic ions: OH^-, NH_4^+, NO_3^-, SO_4^{2-}, and SiO_4^{4-}. Ask students whether the bonding pairs in each structure are covalent bonds. *yes* Ask students to predict the geometry of the ions. *linear, tetrahedral, triangular planar, tetrahedral, and tetrahedral* Ask students to predict the polarity of the ions. *polar, nonpolar, nonpolar, nonpolar, and nonpolar* Point out that although these ions have covalent bonds and may be polar or nonpolar, their net positive or negative charges indicate that they occur as part of ionic compounds. Ask students to write the formulas of the sodium compounds that the negative ions form. *NaOH, NaNO₃, Na₂SO₄, Na₄SiO₄* Ask students the formula of an ammonium compound. *NH₄Cl* L1

4 CLOSE

⚡ **Quick Demo**

Why does water condense?
Light a candle and hold a small beaker containing ice cubes or cold water about 5 cm above the flame. Water will condense on the beaker. Show that the condensate is water by testing with cobalt chloride paper. Ask students why water from the combustion reaction condenses on the beaker while the other product of combustion, carbon dioxide, does not. *Water molecules are polar, and carbon dioxide molecules are nonpolar. The intermolecular attractions in water are sufficient to cause it to condense at the temperature of cold water, but intermolecular attractions in carbon dioxide are not strong enough to cause it to condense.* L1

SECTION REVIEW

1.
   a. $H:\overset{..}{\underset{..}{P}}:H$ b. $\overset{xx}{\underset{xx}{Cl}}$
        $\overset{..}{H}$ $\overset{xx}{Cl}:\overset{xx}{C}:\overset{xx}{Cl}$
                                                   $\overset{xx}{\underset{xx}{Cl}}$

   c. $H:\overset{..}{\underset{..}{Br}}:$ d. $\overset{xx}{Cl}:\overset{..}{\underset{..}{O}}:$
                                                   $\overset{xx}{\underset{xx}{Cl}}$

2. **a.** triangular pyramid **b.** tetrahedral **c.** linear **d.** bent

3. In a single bond, one pair of electrons is shared between two atoms. In a double bond, two pairs of electrons are shared, and in a triple bond, three pairs are shared.

4. Chloroform is polar; the polarities of the three polar bonds do not cancel each other.

5. Answers may include that sugar is a solid, water is a liquid, and ammonia is a gas at room temperature.

- Review the Reviewing Main Ideas statements and Key Terms with your students.
- Complete solutions to Chapter Review problems can be found in the *Problems and Solutions Manual*.

UNDERSTANDING CONCEPTS

1. **a.** Mg **b.** F **c.** Ca **d.** O
2. **a.** ionic **b.** ionic **c.** polar covalent **d.** polar covalent
3. Ionic bonds are those with ΔEN of 2.0 or greater; covalent bonds have ΔEN of 0.5 or less; polar covalent bonds have ΔEN greater than 0.5 but less than 2.0.
4. Na — I
5. A polar covalent bond is one in which electrons are not shared equally between atoms. A nonpolar covalent bond is one in which electrons are shared relatively equally.
6. ΔEN is 1.0 so CO is polar.
7. H:N̈:H
   Ḧ
8. The shielding effect is the reduced attraction the nucleus of the atom has for its valence electrons when there are filled sublevels of electrons between the nucleus and its valence electrons. Shielding is more important for lead because lead has 78 electrons in filled sublevels whereas carbon has two.
9. :S̈:
   H × × H
10. In ionic bonding, electrons are transferred between atoms. In covalent bonding, electrons are shared between atoms. In metallic bonding, atoms donate electrons to a pool and all the atoms share in the pool. No compounds are formed, but the atoms are bonded into a network.
11. Metals conduct electricity and heat, indicating that the electrons are free to move. Metals are malleable and ductile, showing that atoms are not in fixed positions

---

REVIEWING MAIN IDEAS

9.1 Bonding of Atoms

- Bonding can be viewed as a sharing of electrons.
- Electronegativity is a measure of the attraction that an atom has for shared electrons.
- Electronegativity difference, ΔEN, is a measure of the degree of ionic character in a bond.
- Electronegativity can be estimated from the periodic table.
- Metal atoms bond by sharing in a sea of valence electrons.

9.2 Molecular Shape and Polarity

- Models help you visualize what you can't see.
- The first step in building a model of molecular shape is to draw the electron dot diagram.
- Electron pairs about the central atom are either lone (nonbonding) pairs or bonding pairs.
- Electron pairs, placed as far from each other as possible, determine the geometry of a molecule.
- The polarity of the bonds and the shape of the molecule determine whether a molecule is polar or nonpolar.
- Interparticle forces determine many of the physical properties of substances. Interparticle forces between dipoles such as water are especially strong.

Key Terms

For each of the following terms, write a sentence that shows your understanding of its meaning.

conductivity	metallic bond
double bond	polar covalent bond
ductile	polar molecule
electronegativity	shielding effect
malleable	triple bond

● UNDERSTANDING CONCEPTS

1. Using only the periodic table, predict which atom in each pair has the higher electronegativity.

   a) Mg, Na   **c)** Ca, Sr
   b) Cl, F   **d)** C, O

2. Classify the following bonds as ionic, covalent, or polar covalent.

   a) Mg — O   **c)** S — Cl
   b) B — F   **d)** Ti — Cl

3. How does the continuum model of bonding distinguish among ionic, covalent, and polar covalent bonding?

4. Which of these bonds has the most ionic character?

   a) Ti — Cl   **c)** Na — I
   b) O — H   **d)** Li — I

5. What is meant by a polar covalent bond? A nonpolar covalent bond?

6. Is carbon monoxide, CO, a polar or nonpolar molecule?

7. Ammonia, NH_3, is an important source of nitrogen for fertilizers. What is the electron dot structure of ammonia?

8. What is the shielding effect? Is shielding more important for carbon or for lead? Explain.

9. Dihydrogen sulfide is a smelly, poisonous gas produced by the decaying of organic matter. What is the electron dot structure of H_2S?

10. How does metallic bonding differ from ionic bonding? Covalent bonding?

11. What experimental evidence supports the model of metallic bonding?

✓ Assessment

Portfolio: Review the portfolio options that are provided throughout the chapter. Encourage students to select one product that demonstrates their best work for the chapter. Have students explain what they have learned and why they chose this example for placement into their portfolios. **P**

Additional portfolio options can be found in the following **Teacher Classroom Resources:**
Chemistry and Industry
Consumer Chemistry
Applying Scientific Methods in Chemistry

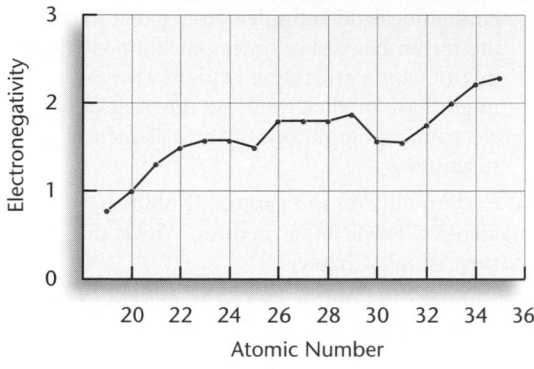

Electronegativities of Fourth Period Elements

12. Electronegativity versus atomic number is graphed for the fourth period elements. Describe in words the general trend in electronegativity in Period 4 as atomic number increases.

13. What is a lone pair of electrons? What is a bonding pair?

14. CO_2 and H_2O are both triatomic molecules. How are their structures different?

15. What is the most stable three-dimensional arrangement of four electron pairs (lone pairs or bonding pairs) attached to a central atom? Three electron pairs?

APPLYING CONCEPTS

16. From the periodic table, select three sets of elements that would form ionic compounds. Write the formulas of the compounds.

17. Methanol, CH_3OH, is the simplest of a series of alcohols. Methanol can be viewed as a derivative of water, in which one of the hydrogens in water is replaced by a CH_3 (methyl) group. Is methanol more polar or less polar than water?

18. Compounds containing the ammonium ion are often used as fertilizers. What is the geometry of an ammonium ion (NH_4^+)?

19. Draw the electron dot structure of propane, C_3H_8, the hydrocarbon fuel often used in gas grills. Describe the shape of the molecule.

20. Sulfur dioxide is an atmospheric pollutant resulting from the burning of sulfur-containing coal. What is the geometry of sulfur dioxide, SO_2? Is sulfur dioxide a polar or nonpolar molecule?

21. Metallic bonding is nondirectional. Using the model of metallic bonding shown in **Figure 9.12**, explain what is meant by the term *nondirectional*.

Chemistry and Technology

22. Which chromatographic technique would be appropriate for separating a mixture of gaseous hydrocarbons?

Everyday Chemistry

23. Explain how microwaves that cook food in a microwave oven might be used to decompose some pollutants in the atmosphere. What kinds of molecules could be decomposed in this way?

THINKING CRITICALLY

Comparing and Contrasting

24. Compare the molecules nitrogen trichloride (NCl_3) and carbon tetrachloride (CCl_4). How many pairs of electrons surround the central atom? How many pairs are bonding? Nonbonding? What are the shapes of the molecules?

Relating Cause and Effect

25. **MiniLab 1** In separating a mixture by paper chromatography, why is it important to keep the spot of colored mixture that you apply to the filter paper as small but as concentrated as possible?

Making Predictions

26. Ethanol, an alcohol, and dimethyl ether both have the molecular formula C_2H_6O. Look up the structural formulas and draw electron dot diagrams for the two compounds. Use your diagrams to decide whether there is a difference in the polarity of the two molecules. Explain.

and ductile, showing that atoms are not in fixed positions but can remain bonded even though they change their positions.

12. Electronegativity generally increases from left to right.

13. A lone pair is a pair of valence electrons that does not participate in bonding. A bonding pair is a pair of electrons that is shared between two atoms.

14. CO_2 is linear; H_2O is bent.

15. a tetrahedral arrangement; a triangular, planar arrangement

APPLYING CONCEPTS

16. Examples: LiF, LiCl, NaF, NaCl, KF, KCl, Li_2O, Na_2O

17. Methanol is less polar than water because it has only one O—H bond.

18. tetrahedral

19.
```
      H H H
H:C:C:C:H
      H H H
```
There are three carbon atoms in a chain connected by single bonds. Each carbon is a geometric center with four bonds arranged tetrahedrally around it. All bond angles are 109.5°.

20. bent; polar

21. The bonding electrons are not rigidly placed between two atoms as they are in covalent bonds, but are free to move about in all directions.

22. gas chromatography

23. Microwaves can cause molecules to vibrate and rotate so rapidly that they fall apart. Polar molecules respond to this treatment.

25. A large spot would allow the bottom part of the spot to dissolve and start to move before the top part. This would spread out each moving component and make the chromatogram less precise. A concentrated spot provides more of each component so that the results are clearer.

26.
```
  H H                    H   H
H:C:C:O:H           H:C:O:C:H
  H H                    H   H
```
Ethanol   Dimethyl ether

The two C—O bonds in dimethyl ether cancel each other, but the O—H bond in ethanol makes ethanol slightly polar.

27. Use paper chromatography to see whether the green spot separates into blue and yellow spots.

28. triangular planar geometry

29. Helium is unreactive so it will not affect the mixture being analyzed or be attracted to the stationary phase.

30. Like visible light, microwaves are electromagnetic energy, but they have longer wavelengths and lower frequencies.

THINKING CRITICALLY

24. NCl_3: There are four pairs of electrons around the central atom; three pairs are bonding, and one is nonbonding. The geometry is a triangular pyramid. CCl_4: There are four pairs of electrons around the central atom; all four pairs are bonding. The geometry is tetrahedral.

CUMULATIVE REVIEW

31. Ionic compound: NaCl
Submicroscopic: strong interparticle forces
Macroscopic: crystalline compound with a high melting point
Nonpolar covalent compound: CO_2
Submicroscopic: weak interparticle forces
Macroscopic: gas at room temperature
Polar compound: H_2O
Submicroscopic: strong interparticle forces
Macroscopic: liquid at room temperature, excellent solvent

32. Both volumes equal 1 cm³:
$1 \text{ cm}^3 = (l)(w)(t) = 15 \text{ cm} \times 15 \text{ cm} \times t$; $t = 1 \text{ cm}^3/(15 \text{ cm} \times 15 \text{ cm}) = 0.0044 \text{ cm}$

33. The lanthanides and actinides are filling inner $4f$ and $5f$ orbitals. Their outer valence electrons are shielded by large numbers of inner electrons so that there is little difference in how strongly these elements attract electrons.

34. Oxygen and sulfur are both in Group 16.

35. Group 15: Nitrogen has five valence electrons, so it needs three more electrons to acquire an octet of electrons. Therefore, nitrogen forms three covalent bonds. Group 16: Oxygen has six valence electrons so it needs two more electrons to acquire an octet of electrons. Therefore, oxygen forms two covalent bonds.

36. No, oxygen gains two electrons to form the O^{2-} ion. Potassium loses one electron to form the K^+ ion. The oxide ion would combine with two K^+ ions, not three.

37. Atoms will lose or gain electrons until they have eight electrons in their outer level. In $AlBr_3$, the bromine atoms have octets by each sharing an electron with aluminum, but aluminum has only three

Designing an Experiment

27. ChemLab A green dye could be produced by mixing a yellow dye with a blue dye. How could you determine whether a dye was a mixture of a yellow dye and a blue dye or whether it was a pure green dye?

Making Predictions

28. MiniLab 2 Below is the electron dot structure for $AlBr_3$. What geometric shape do you predict for this molecule?

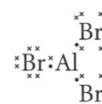

Relating Concepts

29. Chemistry and Technology Why do you think helium is often used as the mobile phase in gas chromatography?

Comparing and Contrasting

30. Everyday Chemistry How are microwaves like visible light? How are they different from visible light?

CUMULATIVE REVIEW

31. The relationship between submicroscopic structure and macroscopic properties is an important part of the study of chemistry. Use an ionic compound, a nonpolar covalent compound, and a polar covalent compound to illustrate this relationship. (Chapter 1)

32. Suppose you have a cube of gold measuring 1 cm on each side. You hammer it into a square measuring 15 cm on each side. What is the average thickness of the square? (Chapter 3)

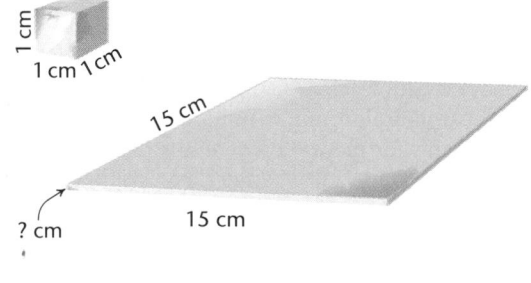

33. **Figure 9.2** gives electronegativity values for the lanthanides and actinides. Notice that the values for both series of inner transition elements do not vary a great deal. Explain why you might have predicted this on the basis of the electron configurations of these elements. (Chapter 7)

34. Carbon dioxide and carbon disulfide have identical Lewis dot structures. Why is this not surprising? (Chapter 7)

35. Elements in Groups 15 and 16 tend to form covalent bonds. Use the first element in each of these groups as an example and explain how the number of valence electrons the atom has determines the number of covalent bonds the element can form. (Chapter 2)

36. Would you expect K_3O to be a stable compound? Explain. (Chapter 4)

37. What is meant by the octet rule? How is the electron dot diagram in question 28 an example of the octet rule and an exception to it? (Chapter 4)

38. What is meant by the periodic law? Explain, in terms of periodicity, why you can determine the kind of bond two elements will form just by noting where the elements are on the periodic table. (Chapter 3)

39. Explain what happens to the nuclei of metal atoms and what happens to their valence electrons when a piece of metal is hammered or pulled. (Chapter 2)

40. The fizz of soda is caused by CO_2 gas. Identify the solvent and the solute in a can of soda. (Chapter 1)

41. Does the mass number of an atom change when the atom forms a chemical bond? Explain. (Chapter 2)

42. A semiconductor material such as silicon conducts electricity to a degree. Is the bonding in silicon represented by the sea of electrons model? Explain. (Chapter 3)

43. Why are no electronegativities recorded for the noble gases in **Figure 9.2**? (Chapter 3)

valence electrons and cannot achieve an octet by sharing.

38. When the elements are arranged in order of increasing atomic number, their properties repeat in a predictable way. Elements that are close together have similar electronegativities and, therefore, will form covalent bonds. Elements that are farther apart have larger differences between their electronegativities and, therefore, will form ionic or polar covalent bonds.

39. The nuclei of metals can move because they are not bound in a rigid bonding network. When hammered or pulled, the nu-

clei adjust their positions within the electron sea without disruption of the solid structure.

40. CO_2 is the solute and water (soda) is the solvent.

41. No, the mass number is the number of protons and neutrons in the nucleus. The nucleus is not involved in chemical reactions.

42. No, silicon, in Group 14, forms four covalent bonds.

43. The noble gases are stable atoms with no attraction for extra electrons.

Skill Review

1. Organizing Information The melting points for the sodium halide compounds are as follows.

NaF	992°C
NaCl	800°C
NaBr	755°C
NaI	651°C

Construct a graph of melting point versus ΔEN for these compounds. Explain the trend of the graph.

2. Organizing Information Complete the table below. Draw the best electron dot structure for each compound listed. In all cases, except H_2O, the first atom in the formula is the central atom. Predict the geometric arrangement of electron clouds around the central atom and use your prediction to determine the geometry of the molecule. From your predicted geometry, decide whether the molecule is polar or nonpolar. Water is given as an example.

Problem Solving

3. There are three different structures for the hydrocarbon pentane, C_5H_{12}. These structures are called geometric isomers because each has a different shape. Draw electron dot diagrams for the three possibilities. You can substitute a dash, like the stick in the gumdrop models, to represent each pair of electrons.

Writing in Chemistry

4. Pick one period in Linus Pauling's career and find out more about it. Locate a publication by Pauling from this period in his life and write a short article about his accomplishments.

5. Gilbert N. Lewis is known for inventing electron dot structures to describe the distribution of valence electrons in a molecule. But Lewis made many other important contributions to chemistry. Write a brief article that summarizes the work of G.N. Lewis.

Projects

6. Build balloon models like those in the chapter to show the preferred stable geometries when different numbers of balloons are attached at a central point. Prepare a display that describes the geometries, and give a live demonstration. You could also photograph different geometric arrangements of the balloons and display the photos or give a slide presentation.

Molecular Formula	Dot Structure	Geometry of Electron Pairs About the Central Atom	Molecular Geometry	Polar Bonds	Nonpolar Bonds	Polar Molecule
H_2O	H:Ö:H	tetrahedral	bent	O—H	none	yes
CCl_4						
$CHCl_3$						
CH_2Cl_2						
CH_3Cl						
CH_4						

AUTHENTIC ASSESSMENT

1.

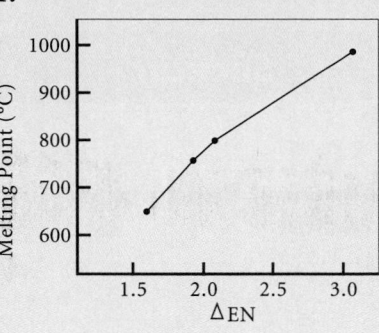

As ΔEN increases, melting point increases.

2. See table below.

3.

$$H-\underset{\underset{H}{|}}{\overset{\overset{H}{|}}{C}}-\underset{\underset{H}{|}}{\overset{\overset{H}{|}}{C}}-\underset{\underset{H}{|}}{\overset{\overset{H}{|}}{C}}-\underset{\underset{H}{|}}{\overset{\overset{H}{|}}{C}}-\underset{\underset{H}{|}}{\overset{\overset{H}{|}}{C}}-H$$

H H—C—H H
H—C————C————C—H
H H—C—H H
H

H H H H
H—C—C—C—C—H
H H—C—H H
H

Program Resources

Chapter Review and Assessment, pp. 53-58 [L1]

Alternate Assessment in the Science Classroom [L1]

Computer Test Bank, Ch. 9 [L1]

Performance Assessment in the Science Classroom [L1]

Supplemental Practice Problems, Chapter 9 [L1]

Molecular Formula	Dot Structure	Geometry of Electron Pairs	Molecular Geometry	Polar Bonds	Nonpolar Bonds	Polar Molecule?
H_2O	H:Ö:H	tetrahedral	bent	O-H	None	yes
CCl_4	:Cl:C:Cl: with :Cl: top and bottom	tetrahedral	tetrahedral	4 C-Cl	None	no
$CHCl_3$	H with :Cl:C:Cl: and :Cl:	tetrahedral	tetrahedral	3 C-Cl	C-H	yes

Molecular Formula	Dot Structure	Geometry of Electron Pairs	Molecular Geometry	Polar Bonds	Nonpolar Bonds	Polar Molecule?
CH_2Cl_2	H :Cl:C:H :Cl:	tetrahedral	tetrahedral	2 C-Cl	2 C-H	no
CH_3Cl	H H:C:H :Cl:	tetrahedral	tetrahedral	C-Cl	3 C-H	yes
CH_4	H H:C:H H	tetrahedral	tetrahedral	None	4 C-H	no

Chapter 10 Organizer

Section	Objectives	Activities/Features
10.1 **Physical Behavior of Matter** (2 days)	**1. Compare** characteristics of a solid, liquid, and gas. **2. Relate** the properties of a solid, liquid, and gas to the kinetic theory of matter. **3. Distinguish** among an amorphous material, liquid crystal, and plasma.	**MiniLab 1:** Molecular Race, p. 343 **Art Connection:** Glass Sculptures, p. 346 **Discovery Demo:** 10.1 Molecular Motion, pp. 338-339 **Demonstration:** 10.2 Glass Flows, pp. 344-345
10.2 **Kinetic Energy and Changes of State** (3 days)	**4. Interpret** changes in temperature and changes of state of a substance in terms of the kinetic theory of matter. **5. Relate** Kelvin-scale and Celsius-scale temperatures. **6. Analyze** the effects of temperature and pressure on changes of state.	**Everyday Chemistry:** Freeze Drying, p. 353 **Chemistry and Technology:** Fractionation of Air, pp. 354-355 **MiniLab 2:** Vaporization Rates, p. 357 **How It Works:** Pressure Cookers, p. 359 **ChemLab:** Molecules and Energy, pp. 362-363 **Demonstrations:** 10.3 Sublimation, p. 356; 10.4 Superheated Steam, pp. 358-359; 10.5 Heat of Vaporization, pp. 360-361

Activity Materials

ChemLab (pages 362-363)
hot plate, test tube (20 mm x 150 mm), 400-mL beakers, Celsius thermometer, beaker tongs, test-tube holder, clamp and ring stand, stearic acid, timer

MiniLab 1 (page 343)
apron, goggles, transparent straws, cellophane tape, black construction paper, cotton swabs, scissors, masking tape, concentrated ammonium hydroxide (NH_3OH), concentrated hydrochloric acid (HCl), meterstick

MiniLab 2 (page 357)
colored water, colored ethanol (C_2H_5OH), hexane (C_6H_{14}), hot plate, 250-mL beaker, Celsius thermometer, beaker tongs, thin-stem pipets, forceps, timer

Demonstrations For a complete list of materials for the demonstrations in this chapter, see pages 338, 344, 356, 358, and 360.

KEY TO TEACHING STRATEGIES

The following designations will help you decide which activities are appropriate for your students.

L1 Level 1 activities should be within the ability range of all students.

L2 Level 2 activities should be within the ability range of the average to above-average student.

L3 Level 3 activities are designed for the ability range of above-average students.

LEP LEP activities should be within the ability range of Limited English Proficiency students.

COOP LEARN Cooperative Learning activities are designed for small group work.

P These strategies represent student products that can be placed into a best-work portfolio.

The Kinetic Theory of Matter

Teacher Classroom Resources

STUDENT MASTERS

Study Guide: pp. 39-40
Laboratory Manual: 10.1 Crystal Shapes, pp. 87-90
ChemLab and MiniLab Worksheets: p. 70

Study Guide: pp. 41-42
Laboratory Manual: 10.2 Relating Gas Temperature and Pressure, pp. 91-96
Critical Thinking/Problem Solving: p. 11
ChemLab and MiniLab Worksheets: pp. 67-69, 71

TEACHING AIDS

Section Focus Transparency 20
Section Focus Transparency Master: p. 20
Basic Concepts Transparency 27
Basic Concepts Transparency Master: pp. 53-54
Problem Solving Transparency 12
Problem Solving Transparency Master: pp. 23-24
Lesson Plans: p. 20

Section Focus Transparency 21
Section Focus Transparency Master: p. 21
Basic Concepts Transparency 28, 29
Basic Concepts Transparency Masters: pp. 55-58
Lesson Plans: p. 21

ASSESSMENT RESOURCES
Chapter Review and Assessment: pp. 59-64
Alternate Assessment in the Science Classroom
Computer Test Bank: Chapter 10
Performance Assessment in the Science Classroom
Problems and Solutions Manual: Chapter 10
Supplemental Practice Problems: Chapter 10

CHAPTER RESOURCES
Applying Scientific Methods in Chemistry: pp. 29-30
Spanish Resources: Chapter 10
English/Spanish Audiocassettes: Chapter 10
Cooperative Learning in the Science Classroom
Lab and Safety Skills in the Science Classroom

GLENCOE TECHNOLOGY

Software
Mastering Concepts in Chemistry: *Unit 10* Lesson 1 Pressure; Lesson 2 Motion and Physical States
Videotape
MindJogger Videoquizzes, Chapter 10
Videodisc
Chemistry: Concepts and Applications, Demonstrations, Videos, and Animations
Science and Technology Videodisc Series: Disc 2 Chemistry *Sand Blasting with Dry Ice; Glass Making for Science*

CD-ROM Multimedia System
Chemistry: Concepts and Applications, Same as for Videodisc, plus Interactive Explorations and Major Simulation, *The Periodic Table*

TECH PREP

The following Glencoe resources provide opportunities for integrating science and technology.

Student Edition: Art Connection, p. 346; Everyday Chemistry, p. 353; Chemistry and Technology, pp. 354-355; How It Works, p. 359

Teacher Wraparound Edition: Discussion, p. 365

Teacher Classroom Resources: Consumer Chemistry, pp. 19-20

10 The Kinetic Theory of Matter

Chapter Overview

The kinetic model of matter is used to explain physical changes. Interparticle attractive forces explain structural differences among the states of matter. Students then learn about unique types of matter. Finally, students learn how to interpret graphs for heating and cooling processes that involve a change of state.

Theme Connection

Macro to Submicro/Energy This chapter emphasizes a *submicroscopic* model of matter in which particle motion accounts for matter's physical behavior and state changes. The theme of *energy* is discussed in relation to particle motion.

Assessment Planner

Choose assessment strategies from the following pages to evaluate the progress of your students.
Assess, pp. 341, 347, 360, 364
MiniLabs, pp. 343, 357
ChemLab, p. 362
Portfolio, pp. 345, 347, 366
Chapter Review, p. 366

GLENCOE TECHNOLOGY

 Videodisc

Chemistry: Concepts and Applications
Molecular Motion
Disc 1, Side 2, Ch. 14

Show this video of the Discovery Demo to introduce the concept of kinetic theory. L1
LEP

338

Discovery Demo 10.1

Molecular Motion

Purpose: To present a mental model of molecular motion.

Materials: 5 glass marbles, clear plastic lid from a card box, overhead projector, 5 mL 12M HCl, 5 mL conc. NH_3(aq), 2 widemouth gas-collecting bottles with stoppers

Safety Precautions: 🧤 💥 🥽 🔥

Disposal: Code F for marbles; combine HCl and NH_3(aq), check pH, and pour down drain with plenty of water.

Procedure: Place five glass marbles in the clear plastic lid on an overhead projector. Have students observe the collisions of the marbles with each other and the walls of the container. Simulate the effect of increased temperature by moving the box rapidly. Have students note the changes in the motion of the marbles. Place a small amount of 12M HCl in one of the widemouth bottles and stopper it. Place a

CHAPTER 10

The Kinetic Theory of Matter

Walking barefoot through the grass on an early summer morning makes you keenly aware that almost every blade is blanketed with a layer of cool moisture. If you live in a northern climate, wearing boots and walking through the same grass on an autumn morning might reveal a thin covering of ice on each blade. By noon, both dew and frost are usually gone.

The presence of dew and frost shows that water vapor in the atmosphere can change overnight into liquid water and then into solid water, better known as ice. Because dew and frost disappear with the increasing warmth of the day, it is clear that the changes in the states of matter—gas, liquid, or solid—are also reversible.

As you can see, there is a close relationship between the physical state of a substance and its temperature. Is the relationship always the same? Is a low temperature always related to the solid state of a substance? Is matter always a gas at high temperatures? You will discover that the state of any matter is determined by the motion of its particles, and the attractive forces between them.

 Concept Check

Review the following concepts before studying this chapter.
Chapter 1: physical change and physical properties
Chapter 2: the particle nature of matter
Chapter 9: structure of metallic solids

 Chapter Preview

10.1 Physical Behavior of Matter

10.2 Kinetic Energy and Changes of State

Laboratory Activities
ChemLab
Molecules and Energy

MiniLabs
1. Molecular Race
2. Vaporization Rates

Chemistry Around You

It's hard to believe that dew and frost are one substance, H_2O. The two substances look and feel so different. How is it possible for a snowflake to have such an intricate pattern and yet be made of the same molecules as the dew on these blades of grass? In general, the liquid, solid, and gaseous states of any substance are distinct. What causes matter to change its state? You will see that the positions and movements of atoms, ions, or molecules of a substance determine whether it is a gas, a liquid, or a solid.

Introducing the Chapter
Ask students to give examples of substances that are solids, liquids, and gases at room temperature. Have them describe the physical characteristics of each state of matter and the motion of atomic-level particles in each of these states. Continue the discussion by describing how water changes from solid to liquid to gas as the temperature rises. Point out that heat energy is needed to change state. Thus, energy and particle motion are related. L1

Using the Illustrations Point out that dew is a liquid when the temperature is above freezing and a solid when below freezing. Acknowledge that temperature plays a major role in determining the state of matter and that temperature is a measure of the kinetic energy of particles at the atomic level of matter.

Revealing Misconceptions

Students may think that matter ceases to exist when something is vaporized—in the explosion of an atomic bomb, for example—or if the temperature drops to absolute zero. Emphasize that conservation of matter means that matter can change form or appearance but can't be created or destroyed. Matter exists as a plasma at the very high temperatures in the sun. At temperatures near absolute zero, matter has unusual properties such as superconductivity.

similar small amount of concentrated $NH_3(aq)$ in the second collecting bottle and stopper it. **CAUTION:** *Add the HCl and $NH_3(aq)$ to the bottles in a fume hood. The vapors are harmful, and both liquids can damage tissue.* Open both bottles in the fume hood. Have students observe the formation of the NH_4Cl "smoke."

Results: Have students use the motion of the marbles to account for the formation of the

product, $NH_4Cl(cr)$. The reaction is as follows.
$$NH_3(g) + HCl(g) \rightarrow NH_4Cl(cr)$$
Analysis: Ask this question.
1. What happens to the speed of the molecules when the temperature increases? *The molecules move faster.*

GLENCOE TECHNOLOGY

 CD-Rom

Chemistry: Concepts and Applications
Molecular Motion
Show this video of the Discovery Demo to introduce the concept of the kinetic theory of matter. L1 LEP

Physical Behavior of Matter

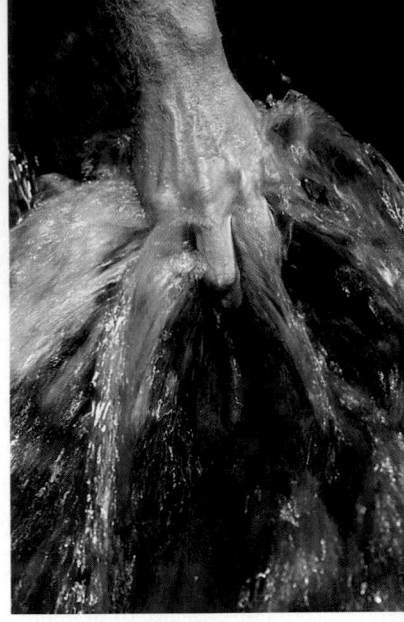

SECTION PREVIEW

Objectives
Compare characteristics of a solid, liquid, and gas.
Relate the properties of a solid, liquid, and gas to the kinetic theory of matter.
Distinguish among an amorphous material, liquid crystal, and plasma.

Key Terms
solid
liquid
gas
Brownian motion
kinetic theory
ideal gas
pressure
crystal lattice
liquid crystal
amorphous material
plasma

You know matter exists as gases, liquids, and solids because you can smell the perfume of flowers, pour fruit punch into a glass, and stack blocks of firewood. You know that air, water, and rocks all feel different, but can you describe their macroscopic properties? The properties that characterize these three states of matter—gas, liquid, and solid—reveal the organization of their submicroscopic particles.

States of Matter

A solid has a fixed volume that cannot be compressed into a smaller volume. Imagine trying to squeeze your textbook into a jelly jar. It can't be done. A **solid** is rigid with a definite shape, as shown in **Figure 10.1.** Solids are rigid because the atoms, ions, or molecules that make up a solid are fixed in place.

Figure 10.1
Properties of Solids
Whether its shape is natural, such as this crystal (left), or artificial, such as the silver coins (right) or platinum jewelry (bottom), a solid is rigid—it keeps its shape. Most elements are solids at room temperature.

340   Chapter 10   The Kinetic Theory of Matter

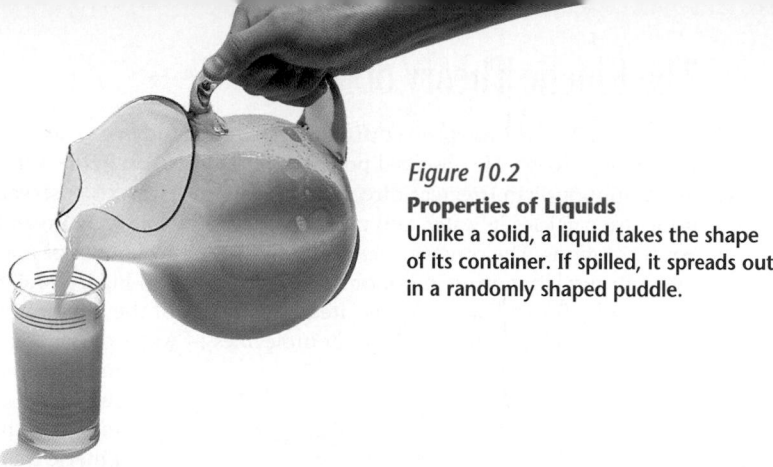

Figure 10.2
Properties of Liquids
Unlike a solid, a liquid takes the shape of its container. If spilled, it spreads out in a randomly shaped puddle.

The characteristics of a liquid differ dramatically from those of a solid. A **liquid** is flowing matter with a definite volume but an indefinite shape. Because its shape is indefinite, a liquid takes the shape of its container, as shown in **Figure 10.2.** If you have mopped up a spill, you have seen that the particles of a liquid can move and easily glide over each other. You can feel how a liquid flows by standing under a running shower.

Just as for a liquid, you can also feel the flow of gases that make up the atmosphere if you stand in a breeze. The flow of a gas feels different because the particles that make up a gas are farther apart than those of a liquid. If you blow up a balloon or a tire, you can observe some important properties of gases, as shown in **Figure 10.3.** From these observations, you see that a **gas** is flowing, compressible matter that has no definite volume or shape. The particles that make up a gas are much farther apart than they are in solids and liquids, and so, they can be easily pushed together.

Figure 10.3
Properties of Gases
The air completely fills the bag. Whatever the shape and volume of the bag, the air inside spreads out to fill it. If you squeeze the bag, you will observe that the air inside is compressible.

341

✓ Assessment

Performance: Have students work in small groups to look up in reference books the densities of the following substances at room temperature and pressure, and to explain the differences in the densities. hydrogen (8.99×10^{-5} g/mL), lead (11.4 g/mL), carbon dioxide (1.98×10^{-3} g/mL), water (1.00 g/mL), aluminum (2.70 g/mL), mercury (13.5 g/mL), ethanol (0.78 g/mL), gold (19.3 g/mL). *The substances with high densities, with the exception of mercury, are solids. The intermediate-density substances are liquids, and the very low-density substances are gases.* Then ask students what this implies about the distances between the atoms or molecules that make up the substances. *The atoms or molecules must be close together in solids and liquids and far apart in gases.* [L1]
COOP LEARN

⚡ Quick Demo

Is it Huey, Dewey, or Louie?

Breathe some helium gas from a balloon to demonstrate the effect of a low-density gas on the vibration of your vocal cords. Explain that the helium atoms are not only far apart, as with most gases, but their masses are also small and their attractions for each other are weak. **CAUTION:** *If you use a helium tank, be sure it is properly chained and correct valves are used.*

The Kinetic Theory of Matter

In 1827, Robert Brown, a Scottish botanist, was studying water samples with a microscope. He observed pollen grains suspended in the water moving continuously in irregular directions. Brown repeated his observations using dye particles in water, and noted that they too had random and erratic motions. The constant, random motion of tiny chunks of matter is called **Brownian motion** in honor of Robert Brown. In **Figure 10.4,** you can see why Brownian motion is cited as evidence of the random movement of particles of matter. Do only molecules of water display random motion? Is all matter in motion? The **kinetic theory** states that submicroscopic particles of all matter are in constant, random motion. The energy of moving objects is called kinetic energy.

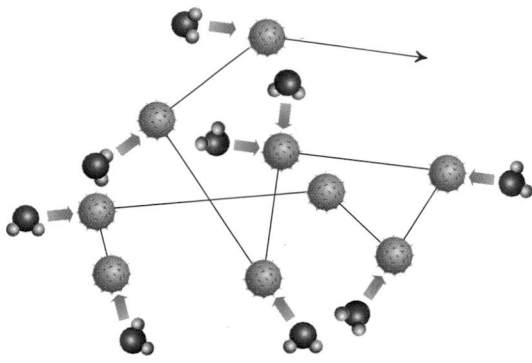

Figure 10.4

Brownian Motion
As viewed under a microscope, a pollen grain suspended in a water droplet traces an erratic path. Random movements of water molecules explain this path traced by the pollen grain. When a water molecule strikes the pollen grain, it pushes the grain in one direction. Then, another molecule strikes the grain and pushes it in a different direction.

Kinetic Model of Gases

In the kinetic theory, each particle of a gas moves like the air-hockey puck shown in **Figure 10.5.** The puck moves in a straight line until it strikes the side of the game board. Similarly, a gas particle can change direction only when it strikes the wall of its container or another gas particle.

Figure 10.5

Modeling the Motion of a Gas Particle
An air-hockey puck travels in a straight line until it strikes the side of the game board. Then, it rebounds in a straight line in a new direction. ▼

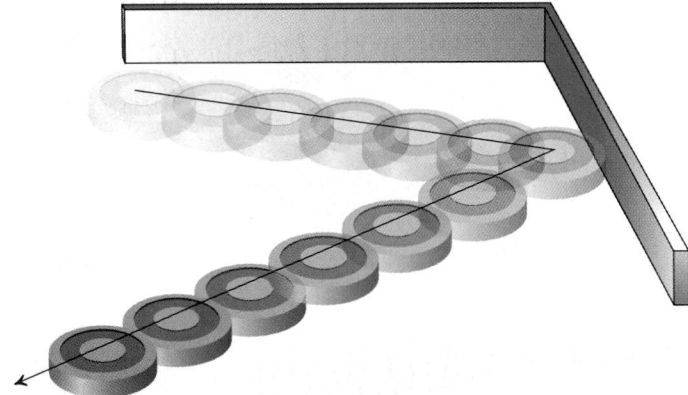

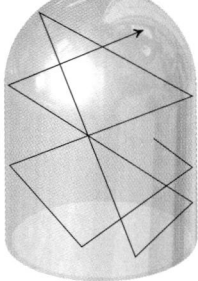

▲ Similarly, a gas particle moves through the space in its container in a straight line. The speed of the hockey puck is about 1 m per second, but the gas particle moves at a much faster rate, 10^2 to 10^3 m per second.

Meeting Individual Needs

Visually Impaired Put several drops of peppermint oil in one evaporating dish and several drops of oil of wintergreen in another. Place the peppermint oil dish on a hot plate set on low and the wintergreen dish in a beaker of crushed ice. Place the two dishes equidistant from the class. Ask students to explain their observations based upon the kinetic molecular theory of liquids. *The odor of peppermint oil reaches students first and is much stronger. Because they are at a higher temperature, more of the peppermint oil molecules have sufficient kinetic energy to escape the attractive forces of the liquid.*

Molecular Race

Odors may be evidence of the diffusion of gases. What inferences can you make about two molecular gases by observing their diffusion through air?

Procedure

1. Make a tube by taping two transparent straws together end to end.

2. Tape the tube onto a black, horizontal surface. Label one end of the tube "NH₃" and the other "HCl."

3. Cut a cotton swab in half with scissors and wrap each cut end with masking tape thick enough to seal the tube.

4. Put on an apron and goggles.

5. Obtain containers of concentrated solutions of ammonia, NH₃(aq), and hydrochloric acid, HCl, from your teacher. **CAUTION:** *Both solutions can damage eyes, skin, and clothing. Handle them with care. If any skin contact or spillage occurs, notify your teacher immediately.*

6. Dip one swab into the ammonia and the other into the hydrochloric acid. The cotton should be saturated but not dripping.

7. Simultaneously insert the swabs into the appropriate ends of the tube, pushing them far enough so that wrapped tape handles seal the ends of the tube as shown here.

8. Do not jostle or move the straws. After a few seconds, look closely for the white ring of ammonium chloride, the reaction product.

9. Measure and record the distances from the respective cotton swabs to the ammonium chloride ring.

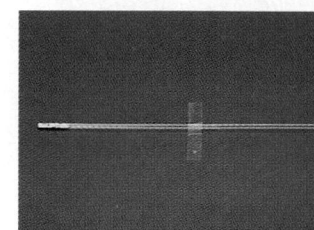

Analysis

1. Compare the diffusion rates of the two gases.

2. Propose an explanation for the difference in diffusion rates.

In air hockey, after each collision with the wall of the game board, the puck loses speed because some of its energy is transferred to the wall. When the puck has lost all its kinetic energy, it will stop gliding. But, unlike the puck, gas particles do not lose kinetic energy when they collide with the walls of their container or with another gas particle. Each gas particle rebounds without losing speed but in a new direction. Collisions of particles in a gas are called elastic collisions because no kinetic energy is lost.

A gas with particles that are in constant random motion but have no attraction for each other is called an ideal gas. The particles in an **ideal gas** undergo elastic collisions. Except at very low temperatures or very high pressures, nearly all real gases behave as ideal gases. The kinetic theory explains why gases fill their container and why they exert pressure on the walls of the container. The gas particles move randomly in all directions until they strike the walls or other particles and bounce back.

10.1   Physical Behavior of Matter   343

miniLAB

Molecular Race

Purpose: To compare and explain the relative diffusion rates of ammonia and hydrogen chloride gases.

Process Skills: Observing, measuring, interpreting data

Safety Precautions: Be sure that students wear safety goggles and aprons. The concentrated hydrochloric acid and ammonia solutions are hazardous. Collect the used cotton swabs in a beaker and thoroughly rinse them with water before disposing in containers for solid waste.

Teaching Strategies

- Demonstrate the formation of ammonium chloride by simultaneously opening containers of concentrated ammonia and concentrated hydrochloric acid and holding them side by side.

- To prevent students from touching the concentrated liquids or inhaling the fumes, bring covered containers of the liquids to the student work areas, and open the containers just long enough for the students to dip the swabs into the liquids.

Expected Results

The ammonia will diffuse about 1.5 times farther than the hydrogen chloride in the same period of time.

Analysis

1. NH₃ molecules diffuse faster than HCl molecules.

2. The NH₃ molecules have a lower molecular mass than the HCl molecules and, because both molecules have the same average kinetic energy at the same temperature, move at a higher average velocity.

Meeting Individual Needs

 **Gifted** Ask students who are capable mathematically to construct and display models of the seven crystal systems: cubic, tetragonal, hexagonal, rhombohedral, orthorhombic, monoclinic, triclinic. L3

✓ Assessment

Knowledge: Ask students to rank the diffusion rates of hydrogen, nitrogen, carbon monoxide, methane, and carbon dioxide from fastest to slowest. *H₂, CH₄, N₂, CO, and CO₂* L1

In the macroscopic world, blocking a volleyball is an example of how you can change the direction of motion. Remember the sting of the volleyball striking your hands? What you felt was pressure. **Pressure** is the force acting on a unit area of a surface; that is, for example, the force per square centimeter. Just as the volleyball exerted a force on a square centimeter of your skin, the particles in a gas exert a force on each square centimeter of the walls of the container when the walls deflect them. The outward pressure of the air inside a balloon is the force that keeps the balloon expanded. If the force is strong enough, the balloon will burst. You can feel this force by gently squeezing an inflated balloon or ball.

Earth's atmosphere, which is a mixture of gases, exerts a pressure, too. Atmospheric pressure is caused by the constant bombardment of the molecules and atoms in air. At sea level, this molecular bombardment exerts a pressure of about 14.7 pounds per square inch, as illustrated in **Figure 10.6.** Humans and other forms of life on Earth have adapted to atmospheric pressure. We are sensitive only to changes in pressure.

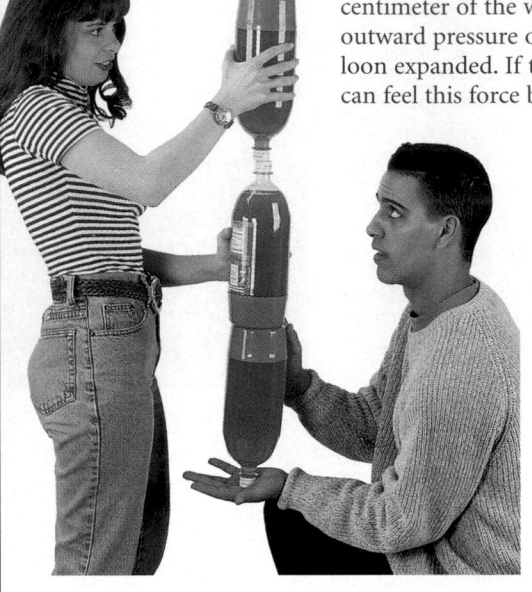

Figure 10.6

Experiencing the Pressure of Atmospheric Gases
The force of the bottle cap on the palm of the hand from three, 2-L bottles about equals the force that the gases of the atmosphere exert on the same area of skin. Of course, the pressure from these three bottles on this person's hand is in addition to normal atmospheric pressure.

Kinetic Model of Liquids

Just as a gliding air-hockey puck models a rebounding gas particle, **Figure 10.7** shows that marbles in a beaker model some behaviors of liquids. When liquids form puddles, interparticle forces maintain their volume but not their shape. The particles of a liquid can slide past each other, but they are so close together that they don't move as straight or as smoothly as an air-hockey puck. When you try to walk across a crowded sidewalk, you can't move quickly in a straight line, either.

Figure 10.7

Modeling Liquids
Magnetized marbles spread out evenly to fill the bottom of their container. The volume they occupy cannot be reduced. ▼

When the container is tipped, the magnetized marbles flow onto the table. ▼

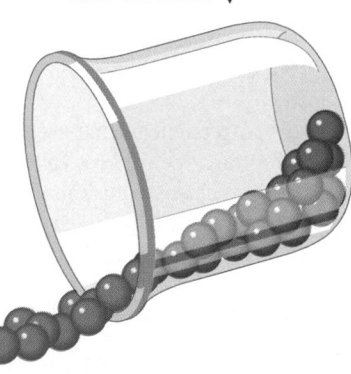

▲ When the container is swirled, the marbles flow with a swirling motion.

344   Chapter 10   The Kinetic Theory of Matter

Kinetic Model of Solids

According to the kinetic theory, strong forces between particles explain the rigid structure of solids. Particles of a solid cannot move past each other, but they are in constant motion, bouncing between neighboring particles. In a solid, the particles occupy fixed positions in a well-defined, three-dimensional arrangement. The arrangement, which is repeated throughout the solid, is called a **crystal lattice,** as shown in **Figure 10.8.**

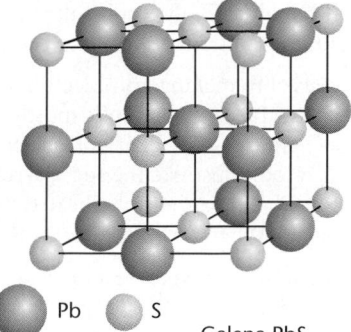

Pb S

Galena PbS

Figure 10.8
Crystal Lattices
Crystal lattices extend throughout solids. Shown here is the crystal lattice of lead(II) sulfide, which is found as the mineral galena.

Other Forms of Matter

Some forms of matter cannot readily be described as solid, liquid, or gas. Maybe they look like solids or gases but sometimes behave like a liquid. These forms of matter are liquid crystals, amorphous materials, and plasmas.

Liquid Crystals

When a solid melts, its crystal lattices disintegrate and its particles lose their three-dimensional pattern. However, when some materials called **liquid crystals** melt, they lose their rigid organization in only one or two dimensions. For example, the liquid crystal shown in **Figure 10.9** has rod-like molecules. The interparticle forces in a liquid crystal are relatively weak and their arrangement is easily disrupted. When the lattice is broken, the crystal can flow like a liquid. Liquid crystal displays (LCDs) are used in watches, thermometers, calculators, and laptop computers because liquid crystals change colors at specific temperatures.

In other liquid crystals, the parallel lines of molecules are arranged in layers. When these substances melt, the layers stay in place. ▼

Figure 10.9
Structures of Liquid Crystals
In some liquid crystals, the molecules are arranged in parallel lines. They keep the arrangement when the substances melt. ▶

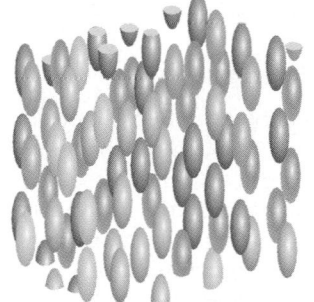

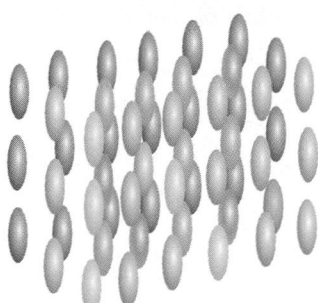

1. What evidence indicates that the particles in the ball move? *The bottom of the ball becomes flattened.*
2. Is the white solid an example of a crystalline solid? *no*
3. How would you classify the white ball? *an amorphous solid*

✔ Assessment

Portfolio: Have students write a summary of this demonstration and add it to their portfolios.
L1 P

Purpose
Students learn how art glass is made.

Teaching Strategies

- Ask students to use the story of Dale Chihuly's life to reflect on how unforeseen events can restructure one's life without destroying one's goals. [L1]
- Ask why most countries of the world have glass industries. *Students may suggest that the raw materials are readily available and that glass has many uses in most societies.*

Extension

Have students find out which countries are particularly well known for their art glass. *France, Italy, Sweden, and the Czech Republic. Northern Bohemia in the Czech Republic has been a leading center of the glass industry since the 1600s.* [L1]

Connecting to Chemistry

1. Because glass has a range of fusion points, it can be sculpted over a range of temperatures before hardening. Like other amorphous materials, glass can acquire a variety of textures by varying the conditions when it hardens.
2. Glass is amorphous and does not have a crystal lattice as do the crystals that characterize solids.

Correcting Misconceptions

Students may think that solids and liquids are easy to distinguish. Explain that liquids with high viscosities (resistance to flow) such as glass may be mistaken for solids; however, true solids are crystalline.

ART ▷ CONNECTION

Glass Sculptures

In 1976, an automobile accident cost Dale Chihuly the sight in his left eye. You might think this accident would have ended the career of the artist who founded the Pilchuck Glass Center, a school for training glass artisans, in Stanwood, Washington, just five years before. Although Chihuly lost his depth perception and so cannot blow glass safely, he continues to train and inspire others in the art of blowing glass.

Making glass The glass used in making art objects is soda-lead glass. It is commonly called crystal or lead glass. Its composition is mostly silica (sand), sodium oxide, and lead oxide. Soda-lead glass has excellent optical qualities.

When glass is molten, it looks just like the sugar syrup you might cook on your stove to make hard candy. The ingredients are heated in a furnace to 1370°C, turning them into a syrupy mass. When it cools, it forms the amorphous material glass.

Turning glass into art Chihuly begins by making a design. One of his students blows the glass to the specifications of the design. The glassblower sticks an iron blowpipe into the molten glass and blows a large glass bubble. When Chihuly is satisfied that the bubble is the right size, the blower rolls it in tinted glass dust, which adheres to the surface. The different colored glass layers are fused by another firing in the furnace.

More blowing and careful shaping by pulling and pushing followed by further heating continues to transform the bubble. When cooled, it becomes an artwork—an astonishing marvel of human creativity.

Connecting to Chemistry

1. **Applying** What properties of glass make it ideal for repeated blowing and forming?

2. **Thinking Critically** Why is it scientifically incorrect to call a glass goblet a crystal goblet?

Amorphous Materials

Is the peanut butter you spread on your bread a solid? What about the wax in candles? Although such materials have a definite shape and fixed volume, they are not classified as solids but rather as amorphous materials. An **amorphous material** has a haphazard, disjointed, and incomplete crystal lattice. Candles and cotton candy are everyday examples of amorphous materials. **Figure 10.10** compares the structure of a solid with that of an amorphous material.

Cultural Diversity

Obsidian Knives

Obsidian is a volcanic glass containing about 70% silica (silicon dioxide, SiO_2). Because of its fracture pattern, Native Americans fashioned it into knife blades and arrowheads. The Hopewell society of the southern Ohio and Illinois valleys flourished from 100 B.C. to A.D. 400. They carried on extensive trading from the northwest to the northeast coast.

One material they obtained from the northwest was obsidian mined in the Yellowstone area of Wyoming. The Hopewells struck the obsidian, like flint, into forms having sharp edges. However, obsidian can obtain a far sharper edge than flint. (Micrographs show that struck obsidian is sharper than the edges of metal razor blades.)

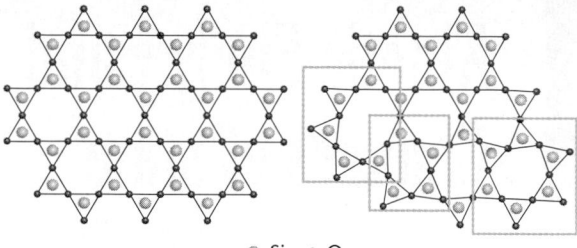

Figure 10.10

A Solid Becomes an Amorphous Material
The silicon dioxide crystal, SiO_2, has a regular honeycomb structure. ▶

◀ If melted and then quickly cooled, silicon dioxide loses its regularity and becomes an amorphous material.

○ Si   ● O

Plasmas

The most common form of matter in the universe but the least common on Earth is plasma. The sun and the other stars are composed of plasma. Plasmas can also be found in fluorescent lights, **Figure 10.11.** A **plasma** is an ionized gas. It can conduct electrical current but, like an ordinary conducting wire, it is electrically neutral because it contains equal numbers of free electrons and positive ions. Plasmas form at very high temperatures when matter absorbs energy and breaks apart into positive ions and electrons, or sometimes even into atomic nuclei and free electrons. In stars, the energy that ionizes the gases is produced by nuclear fusion reactions.

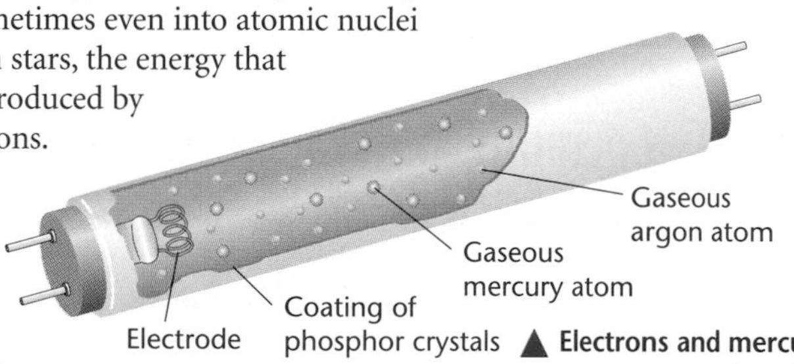

Gaseous argon atom

Gaseous mercury atom

Coating of phosphor crystals

Electrode

Figure 10.11

Fluorescent Light Tube
When a small electrical current heats the electrode, some electrons in the electrode material acquire enough energy to leave the surface and collide with molecules of argon gas, which become ionized. As more electrons are freed, they also ionize some of the mercury atoms, forming a plasma.

▲ Electrons and mercury ions collide with mercury atoms and excite their electrons to higher energy levels. When the excited electrons return to lower energy levels, they radiate energy in the form of invisible ultraviolet light. The fluorescent tube is coated with phosphor crystals, which absorb the ultraviolet light and radiate visible light.

SECTION REVIEW

Understanding Concepts

1. Compare liquids and gases in terms of particle spacing and particle motion.
2. How do the particles of an ideal gas behave?
3. Why is a plasma called a high-energy state of matter?

Thinking Critically

4. **Hypothesizing** Instead of the three states of matter described in this chapter, matter is sometimes classified into only two states. In what properties would these two states differ?

Applying Chemistry

5. **Plant Care** How could fluorescent light tubes be altered to supply indoor lighting in the blue and red region of the visible spectrum for raising plants?

10.1   **Physical Behavior of Matter**   **347**

SECTION REVIEW

1. Particles of gases are spread wide apart and move in straight lines until they collide with the walls of their container or with each other. Particles of liquids are close together because of interparticle forces. They are free to move past each other and, therefore, liquids flow.
2. The particles are in constant random motion and have no attraction for each other. They undergo perfectly elastic collisions.
3. High energy is needed to separate the atoms into ions and electrons and to maintain the separation.
4. Rigid materials and flowing materials. The two differ in rigidity and ability to flow.
5. alter the type of phosphor coating

SECTION REVIEW

---

3 ASSESS

Check for Understanding

- Check for understanding of the section material by having students answer the Section Review questions.
- Ask students to explain the following according to the kinetic model of gases. If the temperature of a gas is held constant and the volume of the container is decreased, the pressure inside the container increases. *The inside surface area of the container is decreased; therefore, more molecular collisions per unit time per unit area occur with the walls of the container.* [L1]

Reteach

⚡ Quick Demo

Why does the balloon inflate? 🧤 🥽 🥽

Crush some dry ice in a towel and put it through a funnel into a deflated balloon. **CAUTION:** *Wear gloves and do not allow the dry ice to come into contact with your skin.* Tie off the balloon. Ask students to explain the result. *The balloon inflates as the small amount of solid CO_2 sublimes and the gas fills the balloon.* [L1]

Extension

Ask students to draw diagrams of atoms or molecules in the solid, liquid, and gaseous states and to list the characteristics of each state. [L1] [LEP]

4 CLOSE

Portfolio

Ask students to make lists of common solids, liquids, and gases in their environment and to include the lists in their portfolios. [L1] [P]

SECTION 10.2

PREPARE

Key Concepts

Temperature is explained as a measure of the average kinetic energies of particles; then the Kelvin, Celsius, and Fahrenheit scales are compared. Changes in physical state are examined in terms of kinetic theory and related to energy changes.

Planning Ahead

For the Demonstrations Obtain the 45-cm length of copper tubing for Demonstration 10.4 from an appliance dealer or heating/cooling contractor.

1 FOCUS

Focus Transparency

Display the **Section Focus Transparency** for Section 10.2 to show the kinetic interpretation of boiling. L1 LEP

2 TEACH

Visual Learning

Figure 10.12: Show the change in the distribution of the velocities of gas particles as the temperature is increased.

Kinetic Energy and Changes of State

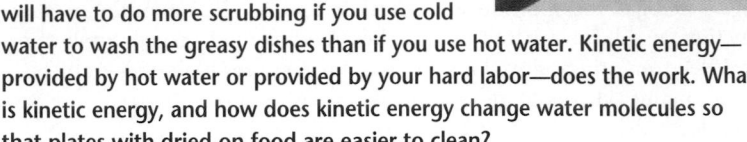

After a big evening meal with your family, you settle down to watch your favorite TV show. In the meantime, the leftover spaghetti and tomato sauce has dried and set hard on the plates. It's your turn to do the dishes. Will you use hot or cold water? You will have to do more scrubbing if you use cold water to wash the greasy dishes than if you use hot water. Kinetic energy—provided by hot water or provided by your hard labor—does the work. What is kinetic energy, and how does kinetic energy change water molecules so that plates with dried-on food are easier to clean?

SECTION PREVIEW

Objectives

Interpret changes in temperature and changes of state of a substance in terms of the kinetic theory of matter.
Relate Kelvin-scale and Celsius-scale temperatures.
Analyze the effects of temperature and pressure on changes of state.

Key Terms

temperature
absolute zero
Kelvin scale
kelvin (K)
diffusion
evaporation
sublimation
condensation
vapor pressure
boiling point
joule (J)
heat of vaporization
melting point
freezing point
heat of fusion

Temperature and Kinetic Energy

When you wash dishes with hot water, most of the water molecules are moving more rapidly than they do in cold water. They have more kinetic energy. Not all molecules of hot water in a sink have the same kinetic energy. They don't have the same speed. The same applies to a container of gas, such as an air-filled balloon. All the gas particles are moving randomly in the balloon at different rates.

Temperature and Particle Motion

The graph in **Figure 10.12** shows the distribution of speeds of the particles in a container of gas. The average speed (kinetic energy) of the particles is represented by the peak near the center of the graph. According to the kinetic theory, the **temperature** of a material is a measure of the average kinetic energy of the particles that make up the material. For example,

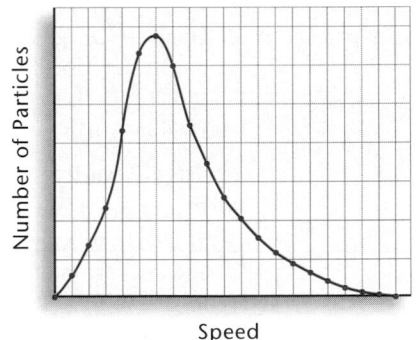

Figure 10.12
Distribution of Gas Particle Speeds
On a highway, some vehicles move slower and some faster than the flow of traffic. In a gas, some particles move slower or faster, but for most, the speed is near the average speed of the group. The peak of the graph represents the most common speed of the particles. The higher the temperature, the higher the average speed.

Program Resources

Study Guide, pp. 41-42 L1
Critical Thinking/Problem Solving, p. 11 L2
Laboratory Manual, 10.2, pp. 91-96 L1
Section Focus Transparency 21 and **Master,** p. 21 L1 LEP
Basic Concepts Transparency 28, 29 and **Masters,** pp. 55-58 L1 LEP

ChemLab and MiniLab Worksheets, pp. 67-69, 71 L1
Applying Scientific Methods in Chemistry, pp. 29-30 L1

Chemistry Journal

Water Molecules

Ask each student to imagine that he or she is a water molecule and to describe his/her positions and movements as the temperature changes from −10°C to +110°C. Ask students to include the accounts in their journals. L1

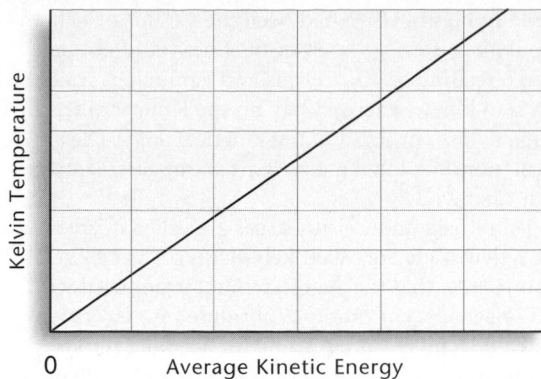

Figure 10.13

Relating Average Kinetic Energy and Temperature
Because the graph is a straight line, the temperature of a gas is directly proportional to the average kinetic energy of its particles.

Discussion
Give students the formula for kinetic energy, KE = (1/2) mv^2, and ask them for the physical meaning of molecules or atoms being in a state in which their kinetic energies are zero. *Because the atoms and particles cannot have zero mass, their velocities must be zero.*

as a gas is heated, the average kinetic energy of its particles increases. This increase in the average kinetic energy of the particles of the gas can be measured as an increase in the temperature of the gas, **Figure 10.13.** As a gas is cooled, the average kinetic energy and speed of its particles decreases.

The Kelvin Scale

According to the graph in **Figure 10.13,** as the temperature of a gas increases, the average kinetic energy of its particles increases and as the temperature decreases, its kinetic energy decreases. Could the kinetic energy be as low as zero—no motion at all? You know that, according to the kinetic theory of matter, the submicroscopic particles are in constant motion.

As a substance is cooled, it loses more of its kinetic energy. The temperature at which a substance would have zero kinetic energy is called **absolute zero.** At this temperature, none of the particles would be moving at all. Their speed and their kinetic energy would both be zero. The temperature of a substance can be lowered to values near zero, but absolute zero has never actually been reached.

In **Figure 10.13,** the scale used for temperature, shown on the vertical axis, is the **Kelvin scale.** This scale is defined so that the temperature of a substance is directly proportional to the average kinetic energy of the particles and so the zero on the Kelvin scale corresponds to zero kinetic energy. Therefore, absolute zero corresponds to the zero on the Kelvin scale. **Figure 10.14** shows how the Kelvin scale is related to the Celsius scale, used throughout the world, and to the Fahrenheit scale, the scale used by weather reporters and on household ovens.

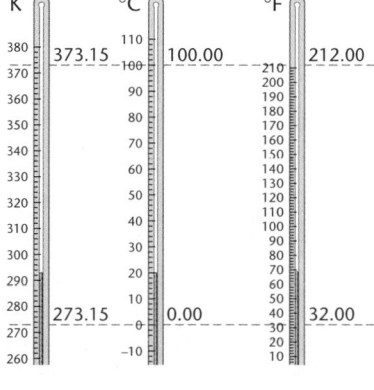

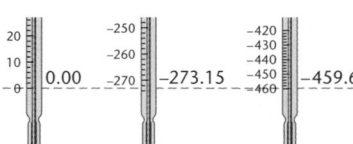

Figure 10.14

The Kelvin, Celsius, and Fahrenheit Scales
The Celsius scale is defined so that the temperature interval from the freezing point of water to the boiling point of water measures 100 degrees. Thus, on the Celsius scale, water freezes at 0°C and boils at 100°C. On the Fahrenheit scale, water freezes at 32°F and boils at 212°F. Note that on the Fahrenheit scale, the temperature interval between the freezing point and the boiling point of water is 180 degrees; Fahrenheit degrees are smaller than Celsius degrees.

Concept Development
Use the following argument to point out why there is no distribution of kinetic energies of the particles of a substance at a temperature of absolute zero. If particles of a substance at absolute zero have zero average kinetic energy, some particles must have more than zero kinetic energy and others must have less than zero kinetic energy. But no particle can have negative kinetic energy. Therefore, for the particles to have an average of zero kinetic energy, all particles must have zero kinetic energy.

Content Background
Absolute Zero Kinetic theory predicts that all molecular motion ceases at absolute zero, 0 K or −273°C. However, this temperature has never been reached, and it is probably the limiting temperature in the universe. The science of cryogenics is the study of the behavior of matter at temperatures close to absolute zero. Note that the kelvin is the SI unit of temperature and that the degree symbol that usually accompanies temperature units is not used with kelvin units.

10.2 Kinetic Energy and Changes of State **349**

Chemistry Journal

Temperature Scales
To help students better understand the meaning of everyday temperatures in Celsius and Kelvin, ask them to describe (by number of degrees) daily Fahrenheit temperatures ranging from extremely hot to bitter cold, then to convert the temperatures to Celsius and to Kelvin. Use °C = 5/9 (°F − 32°). Students should include the results in their journals. *Answers will be similar to this. A very hot day, 95°F: 35°C, 308 K. A moderate day, 75°F349¡ 24°C, 297 K. A cool day, 50°F: 10°C, 283 K. A cold day, 30°F: −1°C, 272 K. A very cold day, 0°F: −18°C, 255 K.* L1

Fact of the MATTER

Because the Kelvin scale measures temperatures above absolute zero and all readings are positive, the Kelvin scale is called the absolute temperature scale.

If the Kelvin scale used in **Figure 10.13** had been the Celsius or Fahrenheit scale, the graph would be a straight line but the line would not pass through the origin. A zero reading on the Celsius and Fahrenheit scales does not correspond to zero kinetic energy. Only on the Kelvin scale is the temperature reading directly proportional to the kinetic energy. The Kelvin scale makes calculations for solving problems about gases simple, as you will appreciate in Chapter 11.

The divisions of the Fahrenheit and Celsius scales are called degrees, but the divisions of the Kelvin scale are called **kelvins (K)**. The kelvin is the SI unit of temperature. Note that the degree symbol is not used with temperatures expressed in kelvins. For example, absolute zero is written as 0 K. On the Celsius scale, the temperature reading for absolute zero is $-273.15°C$, usually rounded to $-273°C$. A Celsius degree and a kelvin are the same size, so by shifting the Celsius scale up 273°, it will coincide with the Kelvin scale.

Temperature Conversions

Because the Celsius degree and the kelvin are the same size and Kelvin readings are 273 degrees higher than Celsius readings, any Celsius reading can be easily expressed as a Kelvin reading. Simply add 273 to the Celsius reading.

$$T_K = (T_C + 273) \text{ K}$$

For example, if the temperature of a room is 25°C, the Kelvin reading can be found as follows.

$$T_K = (25 + 273) \text{ K} = 298 \text{ K}$$

Similarly, a Kelvin reading can be expressed as a Celsius reading by subtracting 273.

$$T_C = (T_K - 273)°C$$

For example, human body temperature, 310 K, can be expressed on the Celsius scale.

$$T_C = (310 - 273) = 37°C$$

Use **Figure 10.14** to prove to yourself that 37°C is correct.

PRACTICE PROBLEMS

Complete the following table.

Temperature	Celsius, °C	Kelvin, K
1. Melting point of iron		1808
2. Household oven	175-205	
3. Food freezer		255
4. Sublimation point of dry ice, $CO_2(s)$	−78.5	
5. Boiling point of nitrogen, N_2		77.4

PRACTICE

 **Guided Practice:** Have students work in class on Practice Problems 1-5. L1
Independent Practice: Your homework or classroom assignment can include the additional practice problems shown below. L1

Answers to Practice Problems

1. 1535; $T_C = 1808 - 273 = 1535°C$
2. 448-478; $T_K = 175 + 273 = 448$; $205 + 273 = 478$

3. −18; $T_C = 255 - 273 = -18°C$
4. 194.5; $T_K = -78.5 + 273 = 194.5 \text{ K}$
5. −195.6; $T_C = 77.4 - 273 = -195.6°C$

Additional Practice Problems

1. Express 293 K on the Celsius scale. *20°C*
2. Express 25°C on the Kelvin scale. *298 K*

Mass and Speed of Particles

When a rolling bowling ball knocks over pins, you can see an effect of the kinetic energy of moving objects. You know that kinetic energy depends on speed, so you would expect that if the ball were rolling faster, it would have the kinetic energy to knock over more pins. You also know it's harder to move heavier objects than lightweight ones. In other words, it takes more work and more kinetic energy to move heavier objects. **Figure 10.15** shows that the kinetic energy of a moving object such as a wagon or a gas particle depends on its mass and speed.

Figure 10.15

Mass, Speed, and Kinetic Energy
Two wagons with loads of different masses move at the same speed. Their kinetic energies are different because their masses are different. The wagon with the greater mass has the greater kinetic energy. Kinetic energy depends on both the mass and the speed of the moving body. ▶

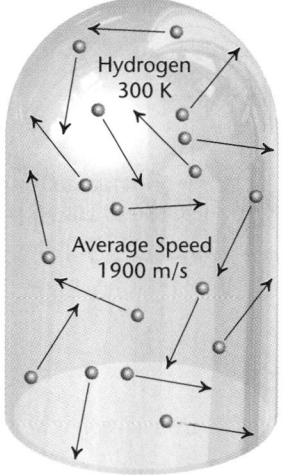
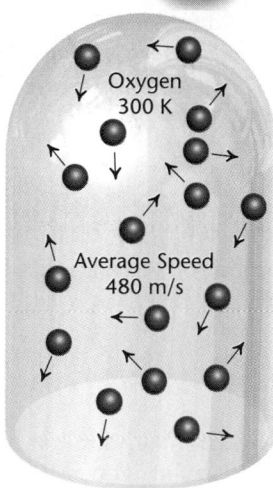
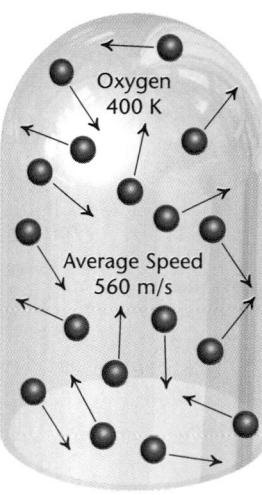

▲ The container of gas on the left contains hydrogen gas. Because the oxygen gas in the center and the hydrogen gas are at the same temperature, they have the same kinetic energy. Because the mass of an oxygen molecule is 32 u and the mass of a hydrogen molecule is only 2 u, the hydrogen molecules must have greater average speed.

▲ The two samples of oxygen on the right are at different temperatures. Because both gases are oxygen, the particles in the two containers have the same mass. The molecules of oxygen gas in the container at the higher temperature have greater kinetic energy because they are moving at a greater average speed.

10.2 **Kinetic Energy and Changes of State** 351

Across the Curriculum

Literature

Lavoisier published his *Traite elementaire de chimie* in 1789. In the preface, he acknowledged the importance of language on clear thinking. This book was instrumental in establishing chemistry as a science and was translated into five languages—German, English, Dutch, Italian, and Spanish. In the book, Lavoisier used the Arabic term *alcohol* to refer to the "spirit of wine" that resulted from fermentation. This word is still in use today.

Figure 10.16
Diffusion
Molecules of gas fill the left chamber. The right chamber is empty—a vacuum. ▼

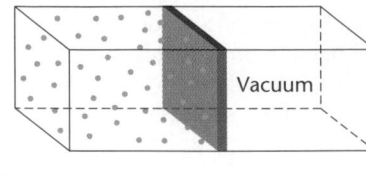

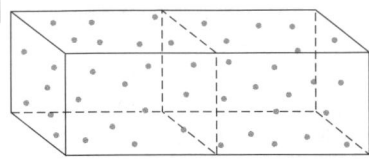

After the wall separating the two chambers has been removed, the moving gas particles pass into the right chamber. Gas flows to the right chamber until the number of molecules flowing to the left chamber is the same as the number flowing to the right. The gas diffuses between the two chambers until both have equal numbers of molecules. ▼

Diffusion

The motions of particles of a gas cause them to spread out to fill the container uniformly. **Diffusion** is the process by which particles of matter fill a space because of random motion, as shown in **Figure 10.16.** If you have seen dye such as food coloring spreading through a liquid, you have watched diffusion. Your sense of smell depends on diffusion and air currents for you to detect molecules of a gas that waft by your nose. Diffusion is slow, but in your lungs, oxygen reaches your blood rapidly enough by diffusion. Oxygen diffuses across the walls of tiny blood vessels called capillaries from the air sacs of your lungs that fill with air each time you inhale. The rate of diffusion of a gas depends upon its kinetic energy, that is, on the mass and speed of its molecules.

Changing State

You are very familiar with the changing states of water—from vapor to liquid and to ice. What environmental conditions are related to these changes of matter? When you remove ice from your freezer, it soon melts to water. When you boil vegetables in water, water vapor rises from the pot. From these observations, it is clear that temperature plays an important role in the changing of state of water, and indeed, of all matter.

Figure 10.17
Rate of Evaporation
The volume of water in the cup and puddle are the same. In the puddle, a larger surface area allows more molecules to escape.

Evaporation

You know that cans of turpentine must be kept tightly closed to prevent the liquid from evaporating. You may have noticed that wet laundry hung on a clothesline dries faster on a hot day and slower on a cold day. **Evaporation** is the process by which particles of a liquid form a gas by escaping from the surface. The area of the surface, as well as the temperature and humidity, affects the rate of evaporation, as shown in **Figure 10.17.**

Across the Curriculum

Language Arts
The heat that causes a solid to melt is called the heat of fusion because the word *fuse* means "to liquefy by heat or to melt."

Everyday Chemistry

Freeze Drying

Probably some of the foods you eat have been preserved by freeze drying. In this process, foods are frozen and the ice is removed from the food by sublimation at low pressures in a vacuum chamber. Then, the water vapor produced by sublimation is removed from the chamber by pumps or water-vapor ejectors.

Advantages of freeze drying foods Food products that would deteriorate during heat processing are often freeze dried. For example, orange juice, which loses its taste in heat processing, is freeze concentrated before freeze drying. First, the pulp is separated from the juice. Then some of the water is removed from the juice by partially freezing it and removing the ice crystals. The pulp is returned to the juice before freeze drying.

Freeze-dried foods are lightweight and take up little room. Hikers and persons in the armed forces often carry freeze-dried meals. Even foods such as chicken and potato salads have been freeze dried, packed in airtight cans, and eaten as field rations on military maneuvers.

Freeze drying biological specimens Freeze drying may be used to prepare specimens for scanning electron microscopy (SEM). Formerly, soft biological materials were chemically fixed and then air dried. Tissue structure collapsed and smaller appendages became plastered against the body during drying. Freeze-dried biological tissue does not shrink or become otherwise distorted. Specimens of organisms as small as amoebas have been preserved by freeze drying.

Exploring Further

1. **Comparing** Since the time of the Incas, people living in the Altiplano of Peru and Bolivia have eaten *chuño*, potatoes that have been preserved by drying in cold, dry air at 4000 m above sea level. Compare this practice with freeze drying.

2. **Applying** Name some foods that are processed by freeze drying.

Liquids that evaporate quickly, such as perfume and paint, are volatile liquids. Applying perfumes with a spray bottle or splashing them on your skin increases their volatility and scent by increasing the size of the surface where evaporation takes place. Like the particles in a gas, the particles in a liquid have a distribution of kinetic energies. **Figure 10.18** applies the kinetic model to evaporation.

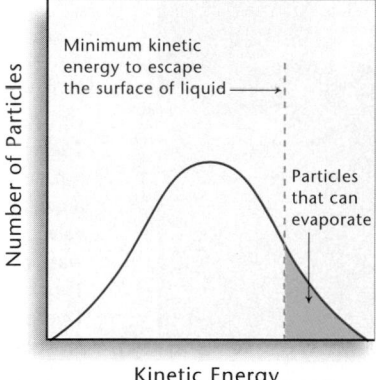

Figure 10.18
Evaporation of Liquids
Some particles in a liquid have enough kinetic energy to overcome interparticle forces. They leave the surface and become gas particles. Because the molecules that escape have higher kinetic energy, the average kinetic energy of the remaining particles decreases. Therefore, as liquids evaporate, they cool.

10.2 Kinetic Energy and Changes of State   **353**

Everyday Chemistry

Purpose
Students gain an understanding of freeze drying and some of its applications.

Teaching Strategies
- Mention that freeze drying removes 98% to 99% of the moisture from foods.
- Tell students that nonliving blood plasma, serum, and hormone solutions have been freeze dried, stored for years, and then used for transfusions. Freeze drying has been used to preserve bacteria; however, mammalian cells do not survive the process.

Extension
Ask students to investigate why freeze drying is not used to preserve more foods. *The main drawbacks of freeze drying are the high energy cost of long drying times and the cost of vacuum equipment. About 25% of the drying takes place in the first 5% of the drying time; 50% takes place in the first 15% of the drying time. After that, the rate slows down.* [L1]

Exploring Further

1. The potatoes of the Incas and their descendants froze at high altitude. Because the low pressure at 4000 m is similar to a partial vacuum, the frozen water in the potatoes sublimed. This process preserved the potatoes.
2. Answers will vary. Instant coffee is freeze dried.

Purpose

TECH PREP

Students learn how oxygen and nitrogen are recovered from air by the process of fractional distillation.

Background

Carl von Linde used the Joule-Thomson effect to make liquid air. He allowed compressed air to flow through valves into a series of larger and larger chambers. In each chamber, the air expanded more, and its temperature decreased. Finally, when the temperature of the air was lowered to −190°C, it liquefied.

Teaching Strategies

• Ask students to relate the formation of dry ice to the Joule-Thomson effect. *Dry ice is produced by forcing carbon dioxide under pressure through a small opening. As it expands, it cools and solidifies — an application of the Joule-Thomson effect.* **L2**

• Explain that whenever a mixture such as air contains several volatile components, it cannot be distilled in a single step because of the different boiling points. Briefly mention some products of fractional distillation of petroleum: light hydrocarbons (methane, ethane, propane, butane), naphtha or straight-run gasoline, kerosene, gas oil, paraffin, asphalt.

Fractionation of Air

In hospitals, at high altitudes, and on space walks, pure oxygen is used for life-support systems. Even greater supplies of oxygen are required by the steel and chemical manufacturing industries, where it is used in many reactions. Nitrogen, the most abundant gas in air, also has many industrial purposes. As a gas, it is used to insulate transformers and to exclude air from electric furnaces. Elemental nitrogen is used to manufacture fertilizers and explosives as well. Both oxygen and nitrogen are produced by the chemical industry by the fractional distillation of air.

Fractional Distillation

The components of dry air are nitrogen (78 percent), oxygen (21 percent), and smaller amounts of argon, carbon dioxide, neon, helium, krypton, hydrogen, xenon, and ozone (1 percent total). When a mixture such as air is fractionated, it is separated into its components. When the components are separated by differences in their boiling points, the method of separation is called fractional distillation.

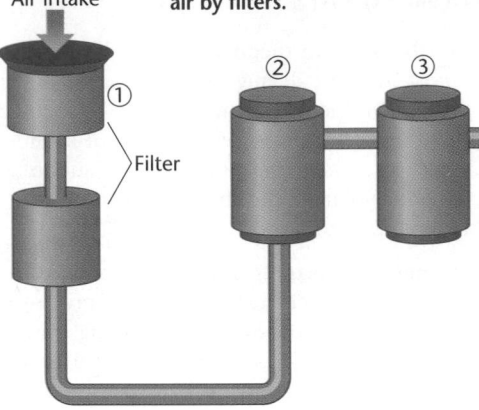

1. Soot and dirt are removed from air by filters.

Air intake

Filter

2. Water vapor is removed from air either by cooling and removing the liquid water produced or by absorbing the moisture with a silica gel sponge. Then carbon dioxide reacts with lime.

3. The air is compressed to more than 100 times atmospheric pressure. Compression raises the temperature of the air.

4. In a heat exchanger, the air releases heat to the cooler surrounding fluid.

5. The cooled, compressed air passes through a nozzle into a chamber of larger diameter called an expansion valve. As the air passes through the valve, it expands and cools. (This cooling effect was first described by James Joule and William Thomson and is called the Joule-Thomson effect.) The temperature difference is so great that the air liquefies.

6. As the liquefied air flows over several heated trays, it is warmed to the boiling point of nitrogen (−195.8°C). Most of the nitrogen, with a trace of oxygen, vaporizes.

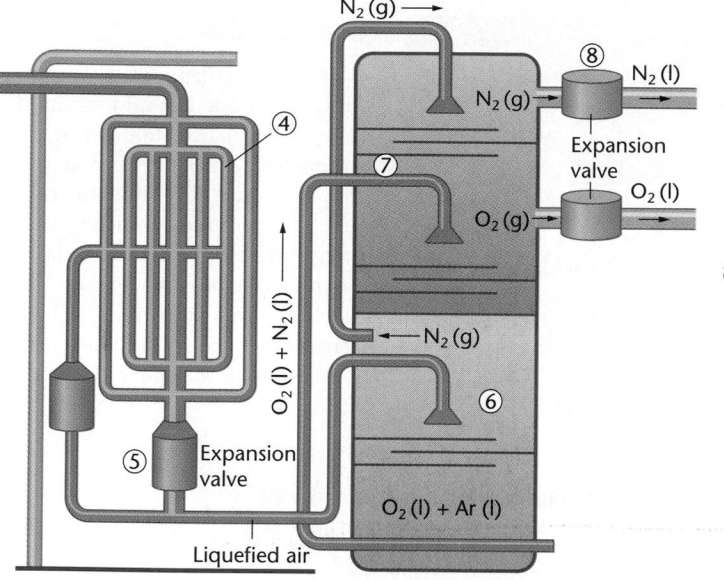

Liquefied air

7. The liquid oxygen and remaining liquid nitrogen are collected and passed into an upper chamber for another distillation at a higher temperature. Here the oxygen and nitrogen vaporize and separate because of their different densities.

8. After passing through expansion valves, the separated gases are liquefied again and bottled as liquid nitrogen and oxygen.

Liquid oxygen and liquid nitrogen are shipped in special insulated containers at temperatures slightly lower than their boiling points.

DISCUSSING THE TECHNOLOGY

1. **Analyzing** Why is it necessary to liquefy air to separate its components?

2. **Inferring** Why do you think liquid nitrogen is used to freeze food?

Discussing the Technology

1. Fractional distillation of air exploits the different boiling points of the volatile substances in a mixture. If air is liquefied, each component can be collected as a distillate when its boiling point is reached.
2. Liquid nitrogen is not chemically reactive, which makes it safe to use for preserving foods.

What's that odor?

After school the day before class, set a strong-smelling potpourri somewhere out of sight in the classroom. When students enter the room the following day, wait for them to ask about the odor. Then show them the potpourri and ask how the odor reached their noses. *Molecules of the aromatic substance leave the solid, enter the air of the room, circulate by Brownian motion and/or convection, and interact with the olfactory organs in the nose. Ask students what the process of molecules leaving the solid and going directly into the air is called.* sublimation

Transparency

Display **Basic Concepts Transparency 28** to show a phase diagram for water. L1

LEP

Sublimation

Some solid substances can change to the gaseous state directly, without melting first. The process by which particles of a solid escape from its surface and form a gas is called **sublimation.** For example, dry ice, solid carbon dioxide, does not melt but sublimes. Ice also sublimes. Some molecules in ice leave the surface and become water vapor. Sublimation of ice is the reason that food stored for a long time in a freezer becomes freezer burned.

Figure 10.19

Sublimation and Condensation of Iodine
When solid iodine is heated in a beaker, it forms a purplish gas, which is free molecules of I_2. The iodine rises to the top of the tube, where it cools. There, it condenses and forms solid iodine again.

Condensation

When dew forms, water vapor in the air condenses into its liquid state. Vapor is used to describe the gaseous state of a substance that is a liquid at room temperature. Condensation is the reverse of evaporation. In **condensation,** gaseous particles come closer together—that is, they condense—and form a liquid or sometimes, as in **Figure 10.19,** a solid.

Vapor Pressure and Boiling

When a rain puddle dries up, the water molecules leave the liquid state and become water vapor. Finally, the puddle is gone. However, if liquid water is in a closed container, only some of the water evaporates. **Figure 10.20** describes how a liquid in a closed container comes to equilibrium with its vapor.

Figure 10.20

Evaporation of a Liquid in a Closed Container
Initially, some particles have sufficient kinetic energy to escape the liquid and form vapor. ▼

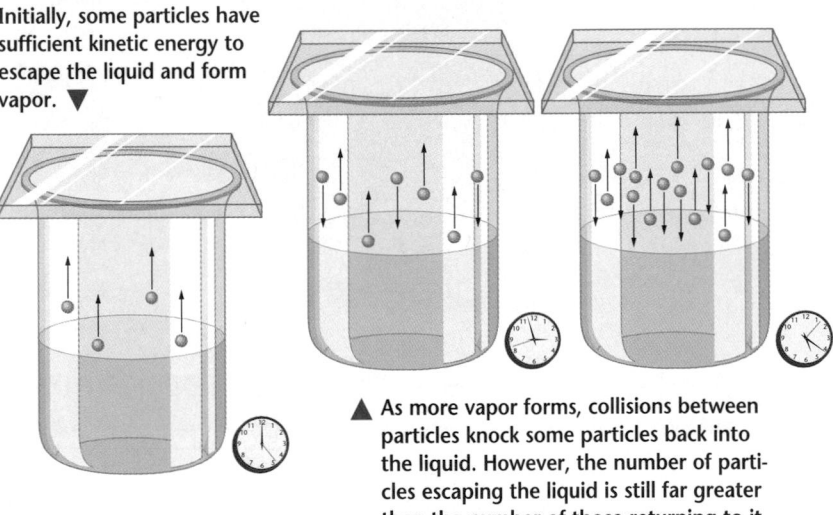

▲ As more vapor forms, collisions between particles knock some particles back into the liquid. However, the number of particles escaping the liquid is still far greater than the number of those returning to it.

◄ Eventually, the number of particles escaping from the liquid to the vapor will equal the number returning to the liquid from the vapor. Equilibrium has been reached. Although the liquid appears to have stopped evaporating, molecules are evaporating and condensing at equal rates.

Sublimation

Purpose: To demonstrate that some solids have a vapor pressure high enough to vaporize at room temperature.
Materials: a few iodine crystals, 250-mL glass flask, stopper, hot plate
Safety Precautions:

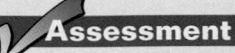

Disposal: Code G

Procedure: Place a few crystals of solid iodine in a stoppered glass flask. Have students observe the pale violet vapor that forms. The iodine will sublime faster if the flask is gently warmed on a hot plate.
Results: Pale violet vapors slowly fill the stoppered glass flask.
Analysis: Ask this question.
1. Because sublimation occurs at room temperature, would you describe the

intermolecular attractive forces as being strong or weak? *weak*

✔ **Assessment**

Performance: Have students place a small, wet cloth in a freezer, observe it after two hours, then every two days thereafter, and report their observations. *Ice will sublime, and the cloth will dry.* L1 LEP

Vaporization Rates

Not all molecules in a liquid move at the same speed. If the liquid is in an open container, the fastest molecules on the surface may have enough energy to escape from the liquid. What can you infer about liquids by observing their vaporization?

miniLAB 2

Procedure

1. Half fill a 250-mL beaker with water and set it on a hot plate. Put a thermometer in the water and monitor its temperature until it reaches about 60°C. Remove the beaker from the hot plate with beaker tongs and set it on the lab table.

2. Label three thin-stem pipets A, B, and C. Also, cut the bulbs of three other thin-stem pipets in half, creating three caps.

3. The teacher will provide small beakers of colored water, colored ethanol (C_2H_5OH), and hexane (C_6H_{14}). Draw water into the bulb of pipet A until it is about one-third full. Then invert it so that its stem remains filled with liquid and cap it as shown.

4. Grip the stem of the pipet with forceps, and immerse it bulb down in the hot water. Observe the liquid in the stem.

5. Measure and record the time it takes for the liquid to clear from the stem.

6. Repeat steps 3 to 5 for ethanol and hexane using pipets B and C, respectively.

Analysis

1. What can you infer about the rate of vaporization of these three liquids from the time they each take to clear from the pipet stem?

2. Rank the rate of vaporization of the three liquids from highest to lowest.

3. Considering how the molecules of the three substances vary in polarity, how do interparticle forces affect vaporization rates?

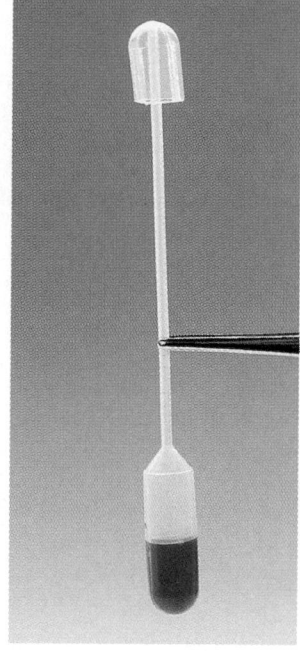

During evaporation of a liquid in a closed container, the amount of vapor and the pressure it exerts both increase. When equilibrium is reached, the pressure exerted by the vapor reaches its final, maximum value, and the volume of the liquid does not change. The pressure of a substance in equilibrium with its liquid is called its **vapor pressure.** The value of the vapor pressure of a substance indicates how easily the substance evaporates. When the pressure has reached this value, particles are evaporating and condensing at equal rates. For example, a volatile liquid such as ethanol has a high vapor pressure. The interparticle forces holding the ethanol molecules together as a liquid are weaker than those in water because ethanol molecules are less polar than water molecules. Therefore, it evaporates easily, and its vapor exerts more pressure in a closed cylinder. At the same temperature, a less volatile liquid such as water has a lower vapor pressure. Because the water molecule is polar, interparticle forces among water molecules are strong, and they tend to remain in the liquid state.

---

2. hexane, ethanol, water
3. The interparticle forces must be greatest in water, which consists of polar molecules and has hydrogen bonding; less in ethanol; and least in hexane, which consists of nonpolar molecules and has only London dispersion forces. Thus, strong interparticle forces slow down the rate of vaporization.

Assessment

Knowledge: Ask students to explain why gasoline blends that are used in low winter temperatures generally contain more nonpolar molecules than gasoline blends used in hot summer temperatures. *The nonpolar molecules vaporize more readily in cold temperatures, enabling automobiles to start and run better.* [L1]

---

miniLAB

Vaporization Rates
Purpose: To compare and explain the vaporization rates of ethanol and hexane.
Process Skills: Observing, inferring, interpreting data
Safety Precautions: Be sure students wear safety goggles and that they cap the pipets as instructed. The liquids may spurt from the tips of the pipets.
Disposal: Code J

Teaching Strategies

• Use the prelab discussion to go over safety precautions and to explain that heat from the hot water is absorbed by the liquid in the pipet bulb, causing some of the liquid to vaporize and clear the stem of liquid.

• Emphasize that students should note the time required for only the pipet stem (not the pipet bulb) to clear of liquid, and that they should remove the pipet from the hot water as soon as the stem clears.

• Ask students to pour the beakers of water, which contain small amounts of ethanol and hexane, into a large-surface-area dish. Place it in the fume hood or take it outside so the more volatile liquids may evaporate.

Expected Results

The times required for the pipet stems to clear should be about 5 to 6 seconds for water, 3 to 4 seconds for ethanol, and 1 to 2 seconds for hexane.

Analysis

1. The liquid that required the least time had the highest rate of vaporization, and the liquid that required the most time had the lowest rate.

SECTION 10.2

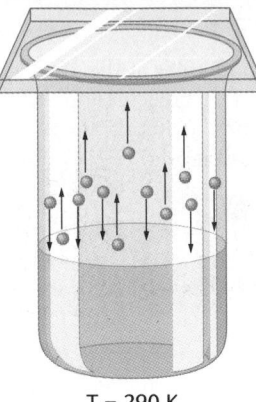 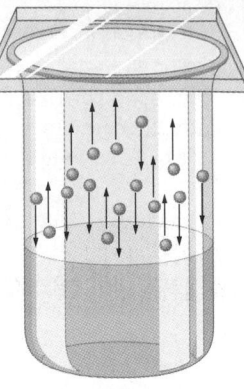

T = 290 K   T = 310 K

Figure 10.21
Vapor Pressure and Temperature
At the higher temperature, the vapor pressure is higher because more particles in the liquid have kinetic energy high enough to escape from the surface.

A liquid in a sealed container evaporates until its vapor pressure is high enough that rates of evaporation and condensation are equal. A liquid in an open container can never reach equilibrium with its vapor because the vapor is constantly escaping. But, exposed to the atmosphere, liquids having high vapor pressures will evaporate more quickly than those with low vapor pressures.

Temperature and vapor pressure are related, as shown in **Figure 10.21.** The rate of evaporation is higher at higher temperatures. If the temperature of a liquid such as isopropyl alcohol is raised high enough, not only will molecules of alcohol escape from its surface, but also, bubbles of the vapor will form below the surface, as shown in **Figure 10.22.** The **boiling point** of a substance is the temperature of the substance when its vapor pressure equals the pressure exerted on the surface of the liquid. For a liquid in an open container, the pressure exerted on its surface is atmospheric pressure. Atmospheric pressure is the force per unit area that the gases in the atmosphere exert on the surface of Earth. Normal boiling point is the temperature at which a liquid boils in an open container at normal atmospheric pressure. The normal boiling point of isopropyl alcohol is 82.3°C (355.5 K). The normal boiling point of mercury is 356.58°C (629.73 K) and that of ammonia is −33.35°C (239.80 K). Because mercury has such a low vapor pressure at room temperature, its temperature must be raised to a high level to boil under normal atmospheric pressure. On the other hand, the vapor pressure of ammonia at room temperature is so great that it boils well below room temperature at normal atmospheric pressure.

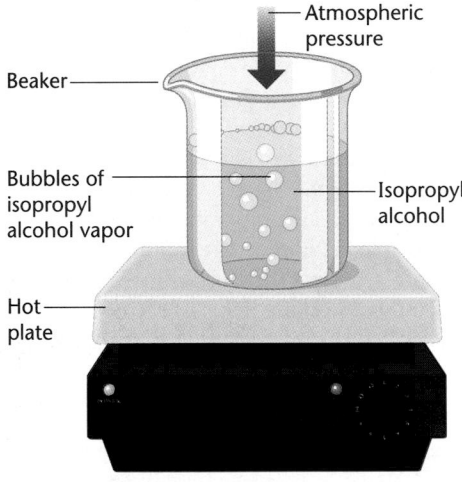

Atmospheric pressure

Beaker

Bubbles of isopropyl alcohol vapor

Isopropyl alcohol

Hot plate

Figure 10.22
Boiling
When a liquid is heated in an open container, the temperature rises until the vapor pressure equals the atmospheric pressure. Then the liquid boils. Small bubbles of vapor form below the surface and rise.

Demonstration 10.4

Superheated Steam

Purpose: To eliminate the misconception that the temperature of steam is always 100°C.

Materials: 2 laboratory burners, 45-cm length of small-diameter copper tubing such as that used to connect ice makers and furnace humidifiers, one-hole stopper, 250-mL Erlenmeyer flask, 75 mL H_2O, ring stand with utility clamps, piece of paper, tongs

Safety Precautions:

Disposal: Hardware can be reused.

Procedure: Make a coil of the tubing by wrapping it around a broom handle. Place one end of the coil in a one-hole stopper fitted to a 250-mL flask containing 75 mL H_2O. Heat the water to boiling. Use a large Meker burner to heat the coil of copper. **CAUTION:** *The steam will be-*

come invisible; take care not to burn yourself. Wear goggles. Using tongs, hold a piece of paper in front of the steam outlet.

Results: The water boils and "steam" is visible coming from the copper tubing. When the second burner heats the copper tubing, the "steam" becomes invisible and makes a noise as it escapes from the tubing. The paper in front of the steam outlet is scorched by the superheated steam.

Pressure Cookers

A pressure cooker is a heavy, covered pot designed to trap steam formed from boiling water. Because the pot is tightly sealed, the pressure of the steam builds up inside the pot, raising the boiling point of the water. Because both the water and steam are at higher temperatures, the food cooks faster. Some fast-food restaurants fry chicken in pressure cookers as well as in open pans so that they can offer customers chicken of both textures and flavors. A commercial pressure cooker reaches an internal temperature of 140°C (284°F) and cooks chicken in 12 minutes—much shorter than the hour or so a conventional oven would require, but no faster than open-pan frying, which requires a much higher temperature and therefore has greater energy costs.

1. The sturdy construction of the pot allows pressure to build up safely inside the cooker.

2. The interlocking cover forms an airtight seal that keeps steam safely inside.

5. The food rack supports food or jars above the water level. As steam fills the inside of the cooker, the pressure above the boiling water increases and the boiling temperature increases, as does the temperature of the steam. Food cooks more quickly because both steam and water are at a higher temperature than in an open pot.

3. The rocker is a small weight that rests on top of a tiny valve built into the cover of the pressure cooker. This weight regulates the internal pressure and allows excess steam to escape safely. When the rocker vibrates gently, the cooker is heating at the proper rate.

4. Some pressure cookers, such as those used for canning food, have a pressure gauge that indicates the internal pressure in pounds per square inch (psi) so that you can regulate the internal pressure by adjusting the external heat.

Thinking Critically

1. At higher altitudes, how does a pressure cooker offset the effects of the lower atmospheric pressure on the boiling of water?

2. How does the rocker regulate the pressure inside?

Analysis: Ask these questions.
1. Is water vapor (steam) visible or invisible? *invisible*
2. What constitutes a cloud of "steam"? *small water droplets*
3. What evidence do you have that indicates the temperature of steam is greater than 100°C? *The steam scorched paper.*

✓ Assessment

Oral: Ask students which would cause a more severe injury, steam at 100°C or boiling water at 100°C? Explain. *Steam contains the heat of vaporization and would cause a more severe injury.* L1

TECH PREP Purpose
Students will learn how increasing the pressure causes foods to cook faster.

Background

The total pressure of a mixture of gases is the sum of the partial pressures of the gases in the mixture. When water boils inside a pressure cooker, the steam cannot escape and its pressure adds to the pressure inside the vessel, which was equal to the atmospheric pressure. Because boiling point is the temperature at which the vapor pressure of a liquid equals the external pressure, the remaining water in the pressure cooker boils at a higher temperature. This higher temperature combined with the higher heat content of steam compared to that of boiling water causes atoms and molecules in the food to move faster. As the kinetic energy of the food particles increases, foods cook faster.

Visual Learning

Bring a pressure cooker into class, and show how its construction traps steam and lets the internal pressure increase safely.

Teaching Strategies

Ask students what effect any escaping steam might have on their skin. Explain that the escaping steam has a high heat content and could cause a serious burn. L1

Thinking Critically

1. A pressure cooker offsets the decrease in boiling point by raising the pressure and causing the water to boil above its normal boiling point.
2. When the pressure rises enough to lift the rocker, it also opens a valve that allows steam to escape, thus reducing the pressure.

Discussion

Ask students why foods cooked in boiling water at high altitudes must be cooked longer. *Water boils at a lower temperature because of the lower atmospheric pressure at the higher altitude. Therefore, the foods are not cooked as quickly.*

Concept Development

Explain that liquids have high heats of vaporization and high boiling points because of strong intermolecular attractions, the strongest of which is called hydrogen bonding.

Assessment

Performance: Before this assignment, be sure students understand the units of pressure, mm Hg. Show them a mercury barometer or draw a diagram of one on the chalkboard. Ask students to construct graphs with the approximate vapor pressures of methanol on the *y*-axis and the Celsius temperatures on the *x*-axis. *280 mm Hg at 40°C, 400 mm Hg at 50°C, 620 mm Hg at 60°C, 910 mm Hg at 70°C, 1310 mm Hg at 80°C* From the graphs, they should predict the approximate boiling point of methanol at standard atmospheric pressure, which is 760 mm Hg. *about 65°C* L1

Cloud physicists estimate that 2×10^{35} snowflakes have fallen on Earth in its 4.5-billion-year history. Water molecules bond in a hexagonal pattern when the liquid freezes. Because billions and billions of molecules make up a single snowflake and there are so many ways to arrange them, it is impossible to find two identical crystals.

When the pressure exerted on the surface of a liquid exceeds normal atmospheric pressure, vapor pressure must be higher than normal for the liquid to boil. To reach this higher vapor pressure, the liquid must be raised to a temperature higher than its normal boiling point. This is the principle behind the pressure cooker discussed in How It Works on page 559. Similarly, at pressures lower than normal atmospheric pressure, a liquid boils at a temperature below its normal boiling point. From these observations, you can conclude that the boiling point of a liquid increases when the pressure on the liquid increases and decreases when the pressure on the liquid decreases.

Heat of Vaporization

Probably, placing a pan of water on a burner, applying heat, and bringing the water to its boiling point is a familiar process to you. But you may be surprised to learn what the temperatures of steam and water are, as shown in **Figure 10.23.** You know that energy must be conserved. What happens to the energy supplied by the flame of the laboratory burner to the boiling water if it doesn't raise the temperature of the water? The rising bubbles of vapor as water changes state from liquid to gas within the boiling liquid are a clue.

The rising bubbles of steam are less dense than the liquid water because the molecules in the bubbles are farther apart than those in the liquid. To separate the particles and form a vapor requires overcoming the interparticle forces holding them together in the liquid. The energy to overcome these forces comes from the heat of the flame.

Figure 10.23

Steam Is No Hotter Than Boiling Water
The setup on the left at zero time shows water beginning to boil. After more than 10 minutes of boiling (right), the thermometers still have the same readings. Because liquid water and water vapor have the same Kelvin temperature during boiling, the molecules of steam must have the same average kinetic energy as the molecules of liquid water.

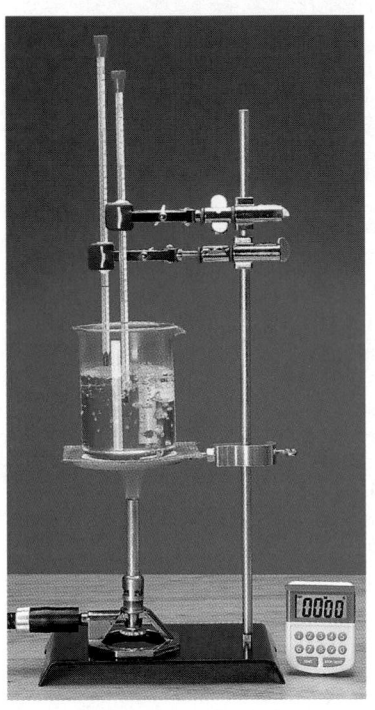
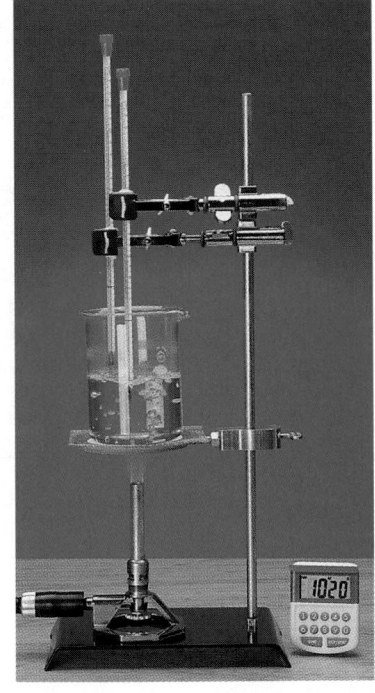

Demonstration 10.5

Heat of Vaporization

Purpose: To demonstrate the high specific heat and heat of vaporization of water.
Materials: 1 sheet of notebook paper, stapler, laboratory burner, wire screen, ring stand, 100 mL water
Safety Precautions: 🧤 ⚗️ 🔥 🥽
Disposal: none
Procedure: Fold a piece of notebook paper into an open-top box and staple the cor-

ners. Place it on a wire screen above a laboratory burner. Add water to the paper container to a depth of 1 cm. Light the burner and adjust it so it emits a bushy blue flame. Heat until the water begins to simmer. Ask a student to observe closely and report to the class what is happening to the water. After you pour out the steaming hot water, place the paper back on the wire gauze over the still-burning

flame. **CAUTION:** *The paper will dry and then burn in a few seconds.*
Results: The water will simmer and the edges of the paper box may brown a little, but the paper will not burn until the water is removed. Point out that the title of the book *Fahrenheit 451*, by Ray Bradbury, refers to the kindling temperature of paper.
Analysis: Ask these questions.

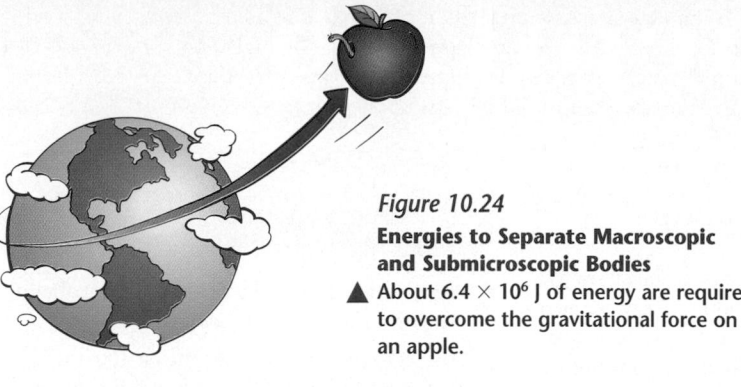

Figure 10.24
Energies to Separate Macroscopic and Submicroscopic Bodies
▲ About 6.4×10^6 J of energy are required to overcome the gravitational force on an apple.

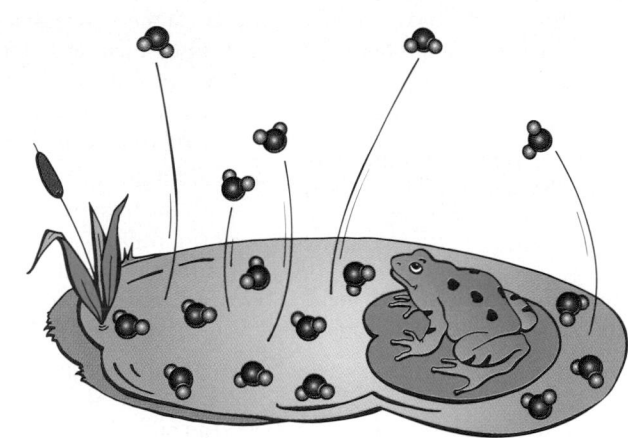

About 2.26×10^6 J of energy are required to overcome the interparticle forces on the molecules in 1 kg of liquid water. ▶

To overcome the force of gravity that keeps an apple on the ground requires energy. The energy that does the work of separating the apple and the ground by 1 m is about 1 joule. The **joule** is the SI unit of energy required to lift a 1-kg mass 1 m against the force of gravity. To move the apple so far that it escapes Earth's gravitational force would require 6.4×10^6 J. About one-third of that amount, 2.26×10^6 J, is the energy needed to move the molecules in 1 kg of water far enough apart that they form water vapor, as shown in **Figure 10.24.**

The energy absorbed when 1 kg of a liquid vaporizes at its normal boiling point is called its **heat of vaporization.** The heat of vaporization of water is 2.26×10^6 joules per kilogram, written as 2.26×10^6 J/kg, which is about 500 times larger than the energy required to raise the temperature of 1 kg of water by 1 Celsius degree, 4180 J.

Because energy is always conserved, energy is released when vapor changes to a liquid. As the vapor condenses, the particles move closer together. For example, when 1 kg of steam condenses at water's normal boiling point, 100.0°C, it releases as heat the same energy it gained when it vaporized, 2.26×10^6 J. Perhaps you know that burns from steam are usually more severe than burns received from hot water. This occurs because the steam transfers a great deal of heat to the skin when it condenses.

Using an Analogy

Use the following analogy to make the point that a hydrogen atom bonded to a fluorine, oxygen, chlorine, or nitrogen atom exerts a particularly strong intermolecular attraction (hydrogen bonding) for the negative pole of other molecules. If 1 dollar were stolen from a person who had only 1 dollar, that person would have a strong desire to get the dollar back. If 1 dollar were stolen from a person who had 25 dollars, the desire to get the dollar back would not be as strong. Similarly, a hydrogen atom assumes a partial positive charge when its single electron is shared with another element having a high electronegativity.

GLENCOE TECHNOLOGY

💿 CD-Rom

Chemistry: Concepts and Applications
How much heat does it take? Use this Interactive Experiment to give students practice determining heat of fusion and heat of vaporization. [L1]
[LEP]

1. Why doesn't the paper burn when the box contains water? *Because of its high specific heat and heat of vaporization, the water removes the energy from the paper box and prevents the paper from reaching its kindling temperature.*

2. If the energy from the flame wasn't used to raise the temperature, what did it do? *The energy from the flame is used to overcome the strong polar attractive forces between the water molecules and thus changes the water from a liquid to a vapor.*

✔ Assessment

Knowledge: Have students explain how the fact that large bodies of water act as storehouses for solar energy is related to the specific heat of water. [L2]

Molecules and Energy

Time Allotment: One class period.

Objectives

Review objectives with students before they begin the ChemLab.

Process Skills

Interpreting data, observing, measuring, formulating models, communicating

Safety Precautions

Be sure students use beaker tongs when handling the beaker of hot water and test-tube holders when handling the test tube of hot stearic acid.

PREPARATION

Alternative Materials

- If a fume hood is available for student use, other materials such as paradichlorobenzene or acetamide may be used.
- If you have a temperature probe and apparatus that may be connected to a computer input, you may be able to allow students to record and graph the temperature readings on the computer.

PROCEDURE

- If the water temperature is allowed to exceed 90°C, it may be cooled easily with ice cubes or water from a cooler.
- Rather than having students in subsequent periods measure out new stearic acid, the stearic acid and thermometer setups may be left in place and reused.

ANALYZE AND CONCLUDE

1.

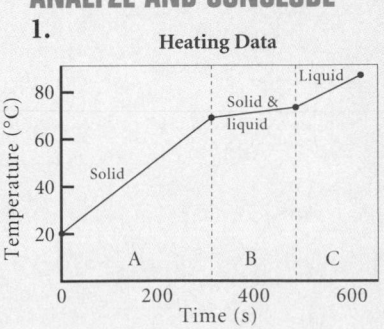

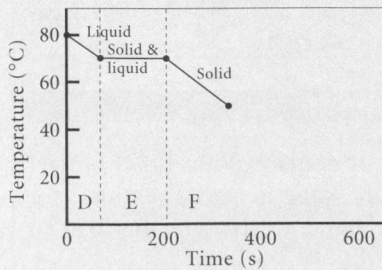

Molecules and Energy

As you break from a saunter to a full gallop to get to your next class, your kinetic energy changes. Your energy was increased by the muscles of your legs propelling you down the hallway. In terms of energy transfer, the muscles of your legs transferred energy obtained from the foods you have eaten. Can you observe changes in a substance as energy is transferred to it?

Problem

How does energy transferred to or from a molecular substance affect the average kinetic energy of its molecules?

Objectives

- **Observe** the temperature changes and changes of state when a molecular substance is heated and cooled.
- **Make and use graphs** to analyze temperature changes.
- **Interpret** temperature changes in terms of the changes in the average kinetic energy of a substance's molecules.

PREPARATION

Materials

timer	Celsius thermometers (2)
hot plate	beaker tongs
20-mm × 150-mm test tube	test-tube holder
	clamp and ring stand
400-mL beakers (2)	stearic acid

Safety Precautions

Use beaker tongs when handling the beaker of hot water and a test-tube holder when handling the hot test tube.

PROCEDURE

1. Prepare two tables like those shown. Label one *Heating* and the other *Cooling*.

2. Pour 300 mL of tap water into a 400-mL beaker and place the beaker on a hot plate.

3. Place a thermometer in the beaker of water. Turn on the heat and monitor the water temperature until it reaches 90°C. Maintain the water temperature at 90°C by using the heat control of the hot plate or by adding cold water.

4. Half fill the test tube with stearic acid. Gently push the bulb of the second thermometer down into the substance. After the temperature of the thermometer has adjusted to the stearic acid, record this temperature in the first line of the *Heating Data* table.

5. Attach the clamp to the test tube and immerse the tube in the beaker of hot water as shown. Read and record the temperature and the physical state or states of the stearic acid every 30 seconds until all of the material has melted and its temperature is about 80°C.

6. Pour 300 mL of cold tap water into the second 400-mL beaker.

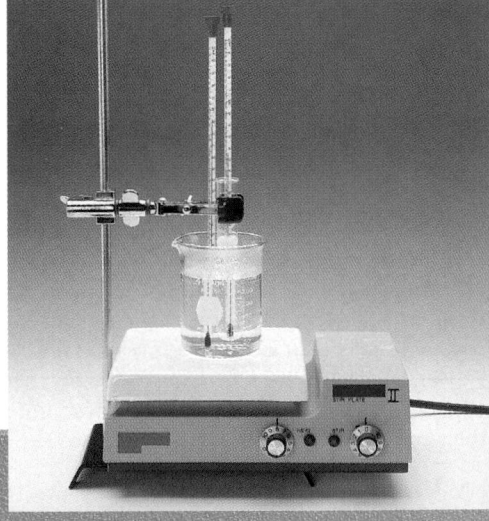

2. See graphs for question 1.

3. 69-72°C

4. Segment A: The kinetic energy of the molecules increased. Segment B: The kinetic energy of the molecules was unchanged. Segment C: The kinetic energy of the molecules increased. Segment D: The kinetic energy of the molecules decreased. Segment E: The kinetic energy of the molecules was unchanged. Segment F: The kinetic energy of the molecules decreased.

7. Remove the test tube and contents from the first beaker and immerse it in the cold water in the second beaker. Read and record in the *Cooling Data* table the temperature and physical state or states of the stearic acid every 30 seconds until the material has solidified.

ANALYZE AND CONCLUDE

1. Making Graphs Graph the heating data by plotting temperature readings on the vertical axis and time on the horizontal axis. Connect the data points with straight lines or smooth curves. Label the appropriate segments of the graph *solid, solid and liquid,* or *liquid*. Graph the cooling data in the same way.

2. Interpreting Data Divide each graph into three intervals by drawing two vertical lines at the points where the slope of the graph changes. Label the intervals of the first graph *A, B,* and *C* and those of the second graph *D, E,* and *F.*

3. Drawing Conclusions According to your data, what is the approximate melting point of stearic acid?

4. Relating Concepts Describe how the kinetic energy of the stearic acid molecules changed during each interval.

APPLY AND ASSESS

1. Describe how the molecular motion changed during each segment of the heating and cooling curves.

2. Suppose twice as much stearic acid were used. What would the graph look like? Make a sketch.

DATA AND OBSERVATIONS

Heating Data

Elapsed Time (s)	Temperature (°C)	Physical State
0		
30		
60		

Cooling Data

Elapsed Time (s)	Temperature (°C)	Physical State
0		
30		
60		

Assessment

Performance: Ask students to sketch a heating curve for carbon dioxide at normal atmospheric pressure from $-100°C$ to $-50°C$. Tell students that carbon dioxide sublimes at $-78.5°C$. L1

ChemLab

APPLY AND ASSESS

1. Segment A: The molecules speeded up. Segment B: The molecules separated. Segment C: The molecules speeded up. Segment D: The molecules slowed down. Segment E: The molecules came together. Segment F: The molecules slowed down.

2. The new graphs would have the same configurations as the original; however, each segment would be longer because more stearic acid is present to gain and lose heat energy.

DATA AND OBSERVATIONS

Heating Data

Time	Temperature (°C)	Physical State
0	20.5	solid
30	27.4	solid
60	33.2	solid
90	39.0	solid
120	44.9	solid
150	50.0	solid
180	54.4	solid
210	59.1	solid
240	64.0	solid
270	66.4	solid
300	69.2	solid + liquid
330	70.0	solid + liquid
360	70.6	solid + liquid
390	71.0	solid + liquid
420	71.7	solid + liquid
450	71.7	solid + liquid
480	72.0	solid + liquid
510	73.8	liquid
540	76.5	liquid
570	78.8	liquid
600	81.3	liquid

Cooling Data

Time	Temperature (°C)	Physical State
0	81.3	liquid
30	75.2	liquid
60	71.6	liquid + solid
90	70.0	liquid + solid
120	69.4	liquid + solid
150	68.6	liquid + solid
180	66.0	solid
210	61.2	solid
240	55.3	solid
270	49.2	solid
300	43.5	solid
330	39.0	solid

Figure 10.25

Changes in Temperature as a Solid Is Heated
A 0.100-kg sample of ice at −20°C is heated to water vapor at 120°C. The rising line shows the rise in temperature of the ice, water, and water vapor. Notice how the graph is flat during changes of state when heat is applied but that the temperature does not change.

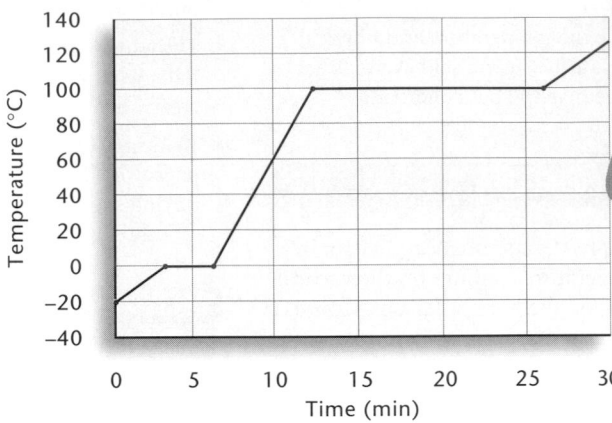

Fact of the MATTER

One definition of *energy* is "the ability to do work." The units of energy (joule, erg, foot-pound) are also the units of work.

Heat of Fusion

As with boiling and condensing, the kinetic energies of the particles of a substance do not change during melting or freezing. If enough heat is applied to a solid, its crystal lattice disintegrates and it becomes a liquid. The **melting point** is the temperature of the solid when its crystal lattice begins to disintegrate. If more heat is applied after the solid has reached its melting point, the additional energy is used to overcome the interparticle forces until the crystal lattice collapses and becomes a liquid. If a liquid substance is cooled, the temperature falls and the liquid becomes a solid. The temperature of a liquid when it begins to form a crystal lattice and becomes a solid is called its **freezing point.** The energy released as 1 kg of a substance solidifies at its freezing point is called its **heat of fusion.** The heat of fusion of water, for example, is 3.34×10^5 J/kg. The same energy is absorbed if 1 kg of a substance is heated until it melts.

As **Figures 10.25** and **10.26** show, the melting point and the freezing point of a substance are the same temperature, provided the pressure is the same. The rising lines and plateaus in the temperature graph as ice is heated to melting and then to boiling are characteristic of all solid substances being heated from a solid to a liquid and then to a gas. The plateaus represent periods when heat is being released or absorbed but the temperature is not changing. Similarly, when a gas is cooled to liquid and then to solid, the graph of the temperature changes has the same shape.

364   **Chapter 10   The Kinetic Theory of Matter**

Integrating THE Sciences

Earth Science Have students research and report on the mechanism and energy source of geysers, such as Old Faithful in Yellowstone National Park. Students may wish to expand their reports to include possible uses of geothermal energy. L2

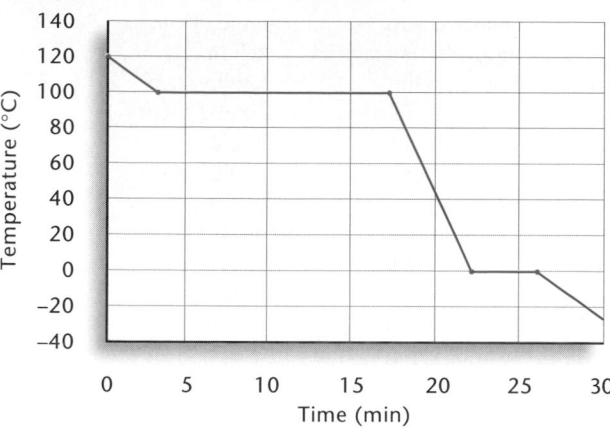

Temperature (°C) vs Time (min)

Figure 10.26
Changes in Temperature as a Gas Is Cooled
As a 0.100-kg sample of water vapor at 120°C is cooled, it liquefies to water, then freezes to ice, then cools to −20°C. The graph falls as the steam is cooled. It levels off while the steam is condensing, falls again after all the steam has condensed, levels off again while the liquid water freezes, and then falls after all the water has frozen.

4 CLOSE

Discussion

TECH PREP

Ask students why water is often sprayed on citrus fruit on the tree during cold weather, even though the water may freeze on the fruit. *During the freezing process, the water gives off the heat of fusion, which warms the fruit slightly and prevents it from freezing.*

GLENCOE TECHNOLOGY

 Videotape

MindJogger Videoquizzes for Chemistry
Chapter 10: The Kinetic Theory of Matter
Have students work in groups as they play the videoquiz game to review key chapter concepts. **L1** **LEP** **COOP LEARN**

Connecting Ideas

The kinetic theory of matter explains the properties of solids, liquids, and gases, and explains changes of state in terms of interparticle forces and energy. It also quantitatively relates the pressure, volume, and temperature of gases. By studying how gases behave under different conditions, you will soon begin to understand how all matter behaves.

SECTION REVIEW

Understanding Concepts

1. Rank the following temperature readings in increasing order.

    32.0°F, 32.0°C, 32.0 K, 102.1°C, 102.1 K

2. Why will water in a flask begin to boil at room temperature as air is pumped out of the flask?

3. In terms of changes in total energy, how does the melting of 1 kg of water at 0°C differ from the freezing of 1 kg of water at 0°C?

Thinking Critically

4. **Analyzing** Why are most elements solids at room temperature?

Applying Chemistry

5. **Aerosol Sprays** Explain why volatile liquids are often used as propellants in aerosol spray cans of such things as paints and deodorants.

SECTION REVIEW

1. 32.0 K, 102.1 K, 32.0°F, 32.0°C, 102.1°C
2. Decreasing air pressure decreases the boiling point.
3. There is no change in the average kinetic energy of the water molecules in either melting or freezing. The amount of energy absorbed by the ice as it melts equals the amount of energy released by the water as it freezes.
4. Most elements are metallic solids, which have large charge separation and, therefore, large interparticle forces holding them together as solids.
5. Volatile liquids quickly evaporate when the pressure is slightly lowered. They carry along the paint as the spray is expelled from the can.

- Review the Reviewing Main Ideas statements and Key Terms with your students.
- Complete solutions to Chapter Review problems can be found in the *Problems and Solutions Manual.*

UNDERSTANDING CONCEPTS

1. Boiling is a change of state in which the vapor pressure of the liquid equals the pressure exerted on the surface of the liquid, which is usually atmospheric pressure. In evaporation, some particles of a liquid acquire sufficient kinetic energy to leave the material. Boiling takes place at a precise temperature; evaporation takes place over a range of temperatures.

2. An amorphous material lacks the complete and uniform crystal lattice that is characteristic of a solid.

3. At absolute zero, the particles of a substance have zero kinetic energy.

4. Molecular solids are either polar covalent or nonpolar covalent substances. Because their molecules have little or no charge separation, interparticle forces are weak, unlike in ionic substances.

5. Increasing the temperature increases the average kinetic energy of the particles of a substance. More particles have sufficient kinetic energy to escape as a vapor. Equilibrium will be reached with a greater number of particles in the vapor state, which increases vapor pressure.

6. Liquids are less compressible than gases because the particles of a liquid are much closer together than those of a gas.

7. Liquids have no organization in any of the three dimensions. Liquid crystals are organized in one or two dimensions.

REVIEWING MAIN IDEAS

10.1 Physical Behavior of Matter

- The three ordinary states of matter are solid, liquid, and gas.
- The kinetic theory of matter postulates that particles of matter are in constant motion.
- Solids, liquids, and gases exist because of differences in interparticle forces.
- Amorphous materials lack the crystal lattice structures of solids.
- Plasma is the most common state of matter in the universe.

10.2 Kinetic Energy and Changes of State

- Changes in the temperature of a substance indicate changes in the average kinetic energy of its particles.
- The Kelvin-scale temperature of a substance is directly proportional to the average kinetic energy of its particles.
- Absolute zero, 0 K, is the temperature at which, theoretically, all particle motion stops.
- Changes of state occur without changing the temperature or kinetic energy of the particles.

Key Terms

For each of the following terms, write a sentence that shows your understanding of its meaning.

absolute zero	joule (J)
amorphous material	kelvin (K)
boiling point	Kelvin scale
Brownian motion	kinetic theory
condensation	liquid
crystal lattice	liquid crystal
diffusion	melting point
evaporation	plasma
freezing point	pressure
gas	solid
heat of fusion	sublimation
heat of vaporization	temperature
ideal gas	vapor pressure

UNDERSTANDING CONCEPTS

1. Compare and contrast evaporation and boiling.

2. What is the difference between an amorphous material and a solid?

3. Describe the physical interpretation of a temperature of absolute zero.

4. Why do most molecular solids have lower melting points than ionic solids?

5. Describe how vapor pressure is affected by increasing temperature.

6. Why are liquids less compressible than gases?

7. How do liquids and liquid crystals differ?

8. How is pressure exerted by a gas?

9. How does a real gas differ from an ideal gas?

APPLYING CONCEPTS

10. A bottle of vanilla extract was left uncovered in a cabinet overnight. When the cabinet door was opened, a strong odor of vanilla was detected. Explain this observation using the kinetic theory of matter.

11. Two persons, one burned by water at 100°C and the other by a similar amount of steam at 100°C over a similar area of skin, are taken to the emergency room of a hospital. Which person has the more severe burn? Why?

12. Explain how ice crystals form on the walls of a freezer even if the freezer door stays closed.

13. Steam sterilizers (autoclaves) reach temperatures of 121°C with boiling water. Explain how this is possible.

✔ Assessment

Portfolio: Review the portfolio options that are provided throughout the chapter. Encourage students to select one product that demonstrates their best work for the chapter. Have students explain what they have learned and why they chose this example for placement into their portfolios. **P**

Additional portfolio options can be found in the following **Teacher Classroom Resources: Consumer Chemistry Applying Scientific Methods in Chemistry Laboratory Manual**

14. Why do you blow on a spot of water on fabric to make the spot disappear faster?

15. Explain why you can detect the aroma of fried bacon more quickly than you can the aroma of cold bacon.

16. Explain why a car's radiator is likely to boil over on a trip through mountains, even though the air is colder.

Everyday Chemistry

17. Why is it necessary to place foods in low-pressure chambers during freeze drying?

Art Connection

18. How does glassblowing use the physical properties of gases?

Chemistry and Technology

19. Can fractional distillation be used to separate the substances in a compound? Explain.

● THINKING CRITICALLY

Comparing and Contrasting

20. A sample of $Cl_2(g)$ and $N_2(g)$ are both at 25°C. Compare the average kinetic energy and the most probable speed of the molecules of the gases.

Making Predictions

21. Predict whether the boiling point of water is greater or less than 100°C at the shoreline of the Dead Sea, which is 400 m below sea level. Explain your prediction.

Inferring

22. What shape would you expect an evaporating dish to have? Explain.

Comparing and Contrasting

23. Why does HCl have a higher boiling point than H_2?

Measuring in SI

24. How can both a Celsius-scale thermometer and a Kelvin-scale thermometer indicate the same temperature change but not the same final temperature reading?

Inferring

25. Explain why an amorphous material has a range of melting points rather than a fixed melting point.

Relating Cause and Effect

26. How does the vapor of CO_2 affect its sublimation in a closed container?

Observing and Inferring

27. Why can you cool a cup of hot water by swirling the cup?

APPLYING CONCEPTS (continued)

8. A gas exerts pressure because its particles collide with each other and with the walls of its container.
9. The particles of a real gas may lose energy when they collide with each other or with the walls of their container. Particles of an ideal gas always have elastic collisions.

APPLYING CONCEPTS

10. Vanilla extract is a volatile liquid. Its vapor diffused throughout the cabinet.
11. The person with steam burns has a more severe burn because a mass of steam at 100°C has more total energy than an equal mass of water at 100°C.
12. Ice cubes sublime, and then the vapor liquefies and freezes.
13. The pressure of the water vapor inside the autoclave is higher than atmospheric pressure.
14. Blowing on the spot removes vapor molecules that might return to the liquid state.
15. Increasing the temperature increases the evaporation of the volatile liquids in the bacon, so more vapor diffuses through the room.
16. At higher altitude, the pressure on the radiator is lower. Less energy is required for the water molecules to leave the liquid state, so the water boils at a lower temperature.
17. The low pressure increases the rate of sublimation of ice.
18. Gases exert pressure and fill their container. In glassblowing, gases expand a bubble of glass.
19. No. Substances in a compound are chemically bonded and, therefore, have lost unique physical properties, such as boiling points, by which they can be distinguished in the free or uncombined state.

THINKING CRITICALLY

20. Molecules of both gases have the same kinetic energy. Because the molecules of chlorine have greater mass, their average speed is less than that of the nitrogen molecules.
21. Because the air pressure is probably greater than 1 atm, the boiling point is higher than 100°C.
22. shallow, so that molecules of water are near the surface, and wide, so that a large surface area is open
23. The HCl molecule is more polar than the H_2 molecule.
24. Because a degree represents the same temperature change on each scale, the temperature change of a material will be the same on both scales. Because the zero on the two thermometers is different, the final temperature will be different.
25. Because there is no definite crystal lattice, disintegration of incomplete and haphazard lattices occurs over a range of temperatures.

CHAPTER REVIEW (continued)

26. As vapor pressure builds up in the container, sublimation decreases.

27. Because swirling the cup increases the surface area of the liquid, evaporation proceeds faster and the temperature of the liquid falls more rapidly.

28. a. propane, radon; radon
b. bromine, mercury; propane
c. silver; bromine, mercury, silver
d. radon
e. Yes, subtracting positive numbers is easier than subtracting positive and negative numbers.

29. Equilibrium is reached when the numbers of particles evaporating and particles condensing are equal, not when the numbers of particles in vapor and in liquid are equal.

30. decrease temperature and increase pressure

31. The material is a solid mixture of two substances.

32. The ring would be closer to the HCl swab.

33. Evaporation rate decreases as molecular mass increases.

CUMULATIVE REVIEW

34. The physical and chemical properties of the elements repeat in a regular pattern when the elements are arranged in order of increasing atomic number. Mendeleev said that the properties of the elements repeat in a regular pattern when the elements are arranged in order of increasing atomic mass.

35. a. metal b. nonmetal c. metal d. metalloid e. nonmetal f. nonmetal

36. a. $2K(s) + 2H_2O(l) \rightarrow H_2(g) + 2KOH(aq)$
b. balanced as written
c. $2C_4H_{10}(g) + 13O_2(g) \rightarrow 8CO_2(g) + 10H_2O(g)$
d. $2HCl(aq) + CaCO_3(s) \rightarrow CaCl_2(aq) + CO_2(g) + H_2O(l)$

37. The alkaline earth metals

Using a Table

28. Examine the table and answer the following questions.

Substance	Freezing Point, °C	Boiling Point, °C
Bromine	−7	58
Mercury	−39	357
Propane	−188	−42
Radon	−71	−62
Silver	961	2195

a) Which of the substances are gases at 50°C? At −50°C?
b) Which of the substances are liquids at 50°C? At −50°C?
c) Which of the substances are solids at 50°C? At −50°C?
d) Which substance has the smallest temperature range as a liquid?
e) Would it have been easier to answer part (d) if the Kelvin temperature scale had been used? Explain.

Observing and Inferring

29. How can a liquid and its vapor reach equilibrium in a closed container while the number of particles in the gaseous state and in the liquid state remain unequal?

Applying Concepts

30. Name two ways to liquefy a gas.

Interpreting Graphs

31. **ChemLab** What can you say about the material with a heating curve such as the one shown below?

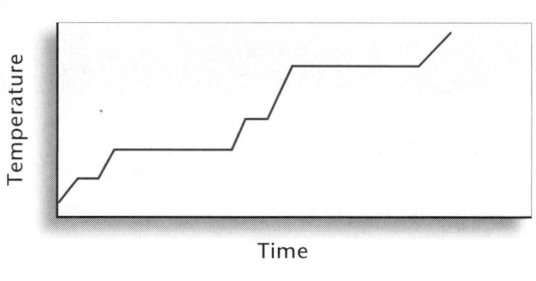

Making Predictions

32. **MiniLab 1** How would the result of the experiment change if the NH_3 were first heated?

Inferring

33. **MiniLab 2** How is the evaporation rate of nonpolar covalent molecules affected by molecular mass?

● CUMULATIVE REVIEW

34. What is the modern periodic law? How does it differ from the periodic law according to Mendeleev? (Chapter 3)

35. Classify the following elements as metal, nonmetal, or metalloid. (Chapter 3)
a) molybdenum   d) arsenic
b) bromine   e) neon
c) vanadium   f) phosphorus

36. Balance the following chemical equations. (Chapter 6)
a) $K(s) + H_2O(l) \rightarrow H_2(g) + KOH(aq)$
b) $Zn(s) + H_2SO_4(aq) \rightarrow ZnSO_4(aq) + H_2(g)$
c) $C_4H_{10}(g) + O_2(g) \rightarrow CO_2(g) + H_2O(g)$
d) $HCl(aq) + CaCO_3(s) \rightarrow CaCl_2(aq) + CO_2(g) + H_2O(l)$

37. Explain how atomic radii influence the chemical reactivity of the alkaline earth metals. (Chapter 8)

38. From the periodic table, predict which atom in each pair has the greater electronegativity. (Chapter 9)
a) Al, Si   c) Be, Mg   e) P, Ge
b) S, Rh   d) N, As

become more reactive as the atomic radius increases. As you move down their column in the periodic table, the energy level of the outer two electrons increases. Because these two electrons are held more loosely, they can be donated more easily.

38. a. Si b. S c. Be d. N e. P

AUTHENTIC ASSESSMENT

1. Interparticle forces determine the state of a substance and also affect the boiling point. Interparticle forces are stronger in water than they are in ammonia.

2.

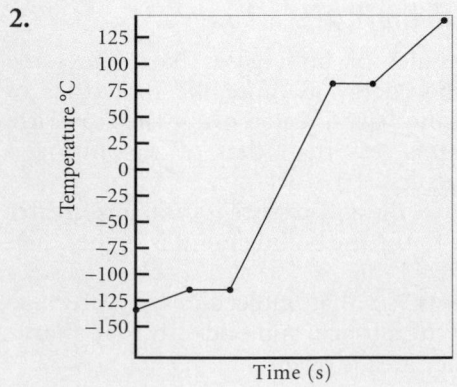

Skill Review

1. Water, H_2O, is a liquid at room temperature and ammonia, NH_3, is a gas. The normal boiling point of water is 100°C and that of ammonia is −33°C. Explain the differences in the states and boiling points of these compounds.

2. Ethanol boils at 79°C and melts at −114°C. A few grams of ethanol are heated from −130°C to +130°C. Graph the heating curve for ethanol. Show time on the horizontal axis and temperature on the vertical axis.

Writing in Chemistry

3. Microwave ovens cook most foods but have no effect on substances such as paper, microwave-safe glassware, and plastic. Write an article that explains how microwaves cook food and why they have no effect on certain other substances such as paper.

4. Compare the crystal structure of sodium chloride, diamond, and a metal such as copper in an essay. Illustrate each crystal with a drawing.

5. Research liquid crystals. Write a report about their most common characteristics and properties.

Problem Solving

6. A beaker of water is placed under a bell jar at room temperature. A vacuum pump removes all the air from the jar and the water begins to boil. Explain why the water boils at room temperature.

7. Ethanol melts at −114°C. Express the melting point of ethanol on the Kelvin scale.

Projects

8. Make your own illustration of the three states of matter and their physical properties. Show what the particles look like in each state. Show how the state changes as the substance is heated or cooled.

9. Collect pictures from magazines and newspapers that show products that use liquid crystals. Assemble the pictures into a collage.

10. Plasmas may be considered a fourth state of matter. Research plasmas. Find an example of a naturally occurring plasma.

Chapter 10   Review   **369**

Program Resources

Chapter Review and Assessment,
pp. 59-64 [L1]
Alternate Assessment in the Science Classroom [L1]
Computer Test Bank, Chapter 10 [L1]
Performance Assessment in the Science Classroom [L1]

Problems and Solutions Manual,
Chapter 10 [L1]
Supplemental Practice Problems,
Chapter 10 [L1]

---

3. Microwave radiation has enough energy to cause water molecules to reach their boiling point and change state. This energy doesn't affect paper and other microwave-safe cookware because these substances don't have water molecules.

4. The sodium chloride crystal lattice is a face-centered cubic unit. The diamond crystal is a covalent network solid. Each carbon atom is covalently bonded to four other carbon atoms in a tetrahedral unit. A metallic solid consists entirely of metal atoms, usually in a hexagonal closest-packed or cubic closest-packed structure. Each atom has eight to 12 adjacent atoms.

5. Liquid crystals are composed of long, rodlike molecules. Nematic and smectic phases are formed by parallel rodlike molecules, and they change orientation in an applied electric field.

6. At room temperature, when the air is evacuated from the jar, the pressure of the air on the water decreases until it equals the vapor pressure of the water. Then the water boils.

7. 159 K

8. Pictures should show the rigid shape of a solid, a liquid taking the shape of its container, and a gas filling all the available space. Particles should reflect these macroscopic differences.

9. Examples of pictures are laptop computers, calculators, digital watches, and fever strips.

10. Plasmas are composed of charged particles—electrons and positive ions. Plasmas exist only at extremely high temperatures, and they are affected by electric and magnetic fields. The sun is an example of a plasma.

Chapter 11 Organizer

Section	Objectives	Activities/Features
11.1 **Gas Pressure** (3 days)	**1. Model** the effects of changing the number of particles, mass, temperature, pressure, and volume on a gas using kinetic theory. **2. Measure** atmospheric pressure. **3. Demonstrate** the ability to use the factor label method to convert pressure units.	**MiniLab 1:** Relating Mass and Volume of a Gas, p. 375 **How It Works:** Tire-Pressure Gauge, p. 377 **Discovery Demo:** 11.1 Gas Against Fire, pp. 370-371 **Demonstrations:** 11.2 Air Bags, pp. 374-375; 11.3 Can Crushing, pp. 378-379
11.2 **The Gas Laws** (4 days)	**4. Analyze** data that relate temperature, pressure, and volume of a gas. **5. Model** Boyle's law and Charles's law using kinetic theory. **6. Predict** the effect of changes in pressure and temperature on the volume of a gas. **7. Relate** how volumes of gases react in terms of the kinetic theory of gases.	**ChemLab:** Boyle's Law, pp. 384-385 **Earth Science Connection:** Weather Balloons, p. 387 **MiniLab 2:** How Straws Function, p. 388 **Chemistry and Technology:** Hyperbaric Oxygen Chambers, pp. 390-391 **Everyday Chemistry:** Popping Corn, p. 397 **Demonstration:** 11.4 Demonstrating Boyle's Law, pp. 386-387

Activity Materials

ChemLab (pages 384-385)
thin-stem pipet, double-post screw clamp, fine-tip marker, food coloring, matches, metric ruler, scissors, small beaker, Bunsen burner

MiniLab 1 (page 375)
small zippered plastic bag, 1-quart zippered plastic bag, 1-gallon clear plastic bag, balance, gloves, tongs, dry ice

MiniLab 2 (page 388)
jar with screw-on lid, soda straws (2), hammer, nail, wax or clay

Demonstrations For a complete list of materials for the demonstrations in this chapter, see pages 370, 374, 378, and 386.

KEY TO TEACHING STRATEGIES

The following designations will help you decide which activities are appropriate for your students.

L1 Level 1 activities should be within the ability range of all students.

L2 Level 2 activities should be within the ability range of the average to above-average student.

L3 Level 3 activities are designed for the ability range of above-average students.

LEP LEP activities should be within the ability range of Limited English Proficiency students.

COOP LEARN Cooperative Learning activities are designed for small group work.

P These strategies represent student products that can be placed into a best-work portfolio.

Behavior of Gases

Teacher Classroom Resources

STUDENT MASTERS

Study Guide: pp. 43-44
Critical Thinking/Problem Solving: p. 12
ChemLab and MiniLab Worksheets: p. 76

TEACHING AIDS

Section Focus Transparency 22
Section Focus Transparency Master: p. 22
Lesson Plans: p. 22

Study Guide: pp. 45-46
Laboratory Manual: 11.1 Determining Absolute Zero, pp. 97-100; 11.2 Charles's Law, pp.101-104
ChemLab and MiniLab Worksheets: pp. 73-75, 77

Section Focus Transparency 23
Section Focus Transparency Master: p. 23
Basic Concepts Transparency 30, 31, 32
Basic Concepts Transparency Masters: pp. 59-64
Problem Solving Transparency 13
Problem Solving Transparency Master: pp. 25-26
Lesson Plans: p. 23

ASSESSMENT RESOURCES

Chapter Review and Assessment: pp. 65-70
Alternate Assessment in the Science Classroom
Computer Test Bank: Chapter 11
Problems and Solutions Manual: Chapter 11
Supplemental Practice Problems: Chapter 11
Performance Assessment in the Science Classroom

CHAPTER RESOURCES

Applying Scientific Methods in Chemistry: pp. 31-33
Spanish Resources
English/Spanish Audiocassettes: Chapter 11
Cooperative Learning in the Science Classroom
Lab and Safety Skills in the Science Classroom

GLENCOE TECHNOLOGY

Software
Mastering Concepts in Chemistry: *Unit 11* Lesson 1 General Concepts, Boyle's Law, Dalton's Law; *Unit 11* Lesson 2 Charles's Law, Combined Gas Law

Videotape
MindJogger Videoquizzes, Chapter 11

Videodisc
Chemistry: Concepts and Applications
Demonstrations, Videos, and Animations Videodisc and Bar Code Teacher Guide

CD-ROM Multimedia System
Chemistry: Concepts and Applications, Same as for Videodisc, plus Interactive Exploration and Major Simulation, *The Periodic Table*

TECH PREP

The following Glencoe resources provide opportunities for integrating science and technology.

Student Edition: How It Works, p. 377; Earth Science Connection, p. 387; Chemistry and Technology, p. 390; Everyday Chemistry, p. 397

Teacher Wraparound Edition: Reinforcement, p. 376; Extension, pp. 376, 398; Chemistry Journal, pp. 377, 383; Portfolio, p. 383; Discussion, p. 392

Teacher Classroom Resources: Tech Prep Applications, pp. 15-17; Consumer Chemistry, pp. 21-22

11 Behavior of Gases

Chapter Overview

The gas laws are developed using macroscopic observations and the kinetic theory of matter. Students are introduced to the factor label method of problem solving and use it in solving problems related to the gas laws. The law of combining gas volumes is introduced as a preamble to stoichiometry.

Theme Connection

Systems and Interactions/Macro to Submicro Macroscopic observations of gas behavior are evidence for the kinetic model of the ideal gas. A volume of confined gas is viewed as a system that interacts with its environment. As pressure and temperature change, the volume of the gas changes in a predictable way.

✔ Assessment Planner

Choose assessment strategies from the following pages to evaluate the progress of your students.
Assess, pp. 380, 383, 386, 398
MiniLabs, pp. 375, 388
ChemLab, p. 384
Portfolio, pp. 379, 381, 383, 397, 398, 399
Chapter Review, p. 399

GLENCOE TECHNOLOGY

Videodisc

Chemistry: Concepts and Applications
Gas Against Fire
Disc 2, Side 1, Ch. 1

Show this video of the Discovery Demo to introduce the properties of gases. L1 LEP

370

Discovery Demo 11.1

Gas Against Fire

Purpose: To show why carbon dioxide can be used to extinguish fires.
Materials: aquarium, 4 candles of different heights, 100 g baking soda, 200 mL vinegar
Safety Precautions: 🔥 🜂 🥽
Disposal: Code C
Procedure: Place four candles of different heights in an empty aquarium. Cover the bottom of the aquarium with baking soda. Light the candles. Have students predict what will happen when vinegar is poured into the aquarium. Lower the room lights and add the vinegar.

Results: Carbon dioxide gas, produced by the reaction of baking soda and vinegar, sinks to the bottom of the aquarium. As the gas layer becomes thicker and fills the aquarium, it extinguishes the candles one at a time, beginning with the shortest.

CHAPTER

11

Behavior of Gases

Scuba divers and firefighters know that their lives depend on breathing equipment. As divers descend, increased pressure affects their breathing, as well as the behavior of the oxygen-nitrogen mixture they breathe from their scuba gear. When toxic fumes are present, firefighters don masks attached to oxygen tanks. Breathing equipment is an application of our knowledge about how gases behave.

The properties of gases that are most easily observed are the relationships among pressure, volume, temperature, and mass. If you have ever inflated a balloon, baked a cake, or slept on an air mattress, you have observed how these properties are related. Because the laws of gases were developed from the study of their properties and behavior, it is now possible to predict the physical behavior of gases by the application of these laws.

 Concept Check

Review the following concepts before studying this chapter.
Chapter 10: properties of gases; the kinetic theory of gases

 Chapter Preview

11.1 Gas Pressure

11.2 The Gas Laws

Laboratory Activities
ChemLab
Boyle's Law

MiniLabs
1. Relating Mass and Volume of a Gas
2. How Straws Function

Chemistry Around You

An aerosol can doesn't look like an oxygen tank, but it is a pressurized container. The label warns against exposing the can to heat. By the study of the laws of gases, it becomes clear why warning labels on cans of spray paint or hair sprays are important.

371

Analysis: Ask these questions.
1. Why are the candles not extinguished by the gas at the same time? *The gas is more dense than air. It fills the aquarium from the bottom up, extinguishing the shortest candle first.*
2. How does a carbon dioxide fire extinguisher work? *It blankets the burning material with CO_2, displacing the lighter oxygen needed for combustion.*

Introducing the Chapter
Point out to students that as they breathe, the pressure of air entering the lungs is the same as the pressure the atmosphere exerts on the chest and that breathing is subject to the gas laws. Point out that a knowledge of the gas laws will enable them to predict the effect of changing conditions on the volume of a gas.

Using the Illustrations Scuba apparatus (Self-Contained Underwater Breathing Apparatus) allows the diver to increase the pressure of the gas that enters the lungs to compensate for the large pressure that water exerts on the chest. The firefighter needs a breathing apparatus because a fire raises the temperature of all the gases in the immediate vicinity and reduces the amount of oxygen available for breathing.

Revealing Misconceptions

Students may think that the ideal gas introduced in Chapter 10 is the same as a real gas. Emphasize that an ideal gas is a model of gas behavior. Over the range of pressures and temperatures for which the ideal gas law is valid, the kinetic model of gases both predicts and explains the gas laws. Remind students that the particles of a real gas experience attractive forces and that they have volume. These deviations from the ideal gas model cause the behavior of real gases to deviate from ideal behavior.

GLENCOE TECHNOLOGY

 CD-Rom

Chemistry: Concepts and Applications
Gas Against Fire
Show this video of the Discovery Demo to introduce the properties of gases. L1 LEP

PREPARE

Key Concepts

Gas pressure is related to volume, temperature, and the number of gas particles. Devices for measuring gas pressure are explained. The units of pressure are defined. Atmospheric pressure is explained, and the factor label method is used to convert units of pressure.

Planning Ahead

For the MiniLab Arrange with an ice-cream or frozen-food company to purchase dry ice for the MiniLab.

1 FOCUS

Focus Transparency

Display the **Section Focus Transparency** for Section 11.1 to introduce students to the barometer. L1 LEP

2 TEACH

Content Background

Real Gases The ideal gas law assumes that the particles of a gas do not attract one another. However, the pressure of a real gas is less than the pressure of the ideal gas because attractive forces do exist between molecules and slow down the molecules. Consequently, there are fewer collisions per unit time. The kinetic model also assumes that gas particles are point masses. However, real gas particles have volume, and the volume of a real gas is less than the volume of an ideal gas. The van der Waals gas equation, $(P + a/V^2)(V - b) = nRT$, corrects the ideal gas equation for these factors by adding the term a/V^2 to the pressure and subtracting the term b from the volume.

Gas Pressure

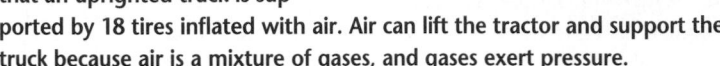

SECTION PREVIEW

Objectives

Model the effects of changing the number of particles, mass, temperature, pressure, and volume on a gas using kinetic theory.

Measure atmospheric pressure.

Demonstrate the ability to use the factor label method to convert pressure units.

Key Terms
barometer
standard
  atmosphere (atm)
pascal (Pa)
kilopascal (kPa)
factor label method

You might be surprised to see that an overturned tractor trailer can be uprighted by inflating several air bags placed beneath it. But, you're not at all surprised to see that an uprighted truck is supported by 18 tires inflated with air. Air can lift the tractor and support the truck because air is a mixture of gases, and gases exert pressure.

Defining Gas Pressure

Unless a ball has an obvious dent, you can't tell whether it is underinflated by looking at it. You have to squeeze it. If it's soft, you know it needs to be pumped up with more air. The springiness of a fully inflated ball is the pressure of the air inside. **Figure 11.1** shows how the pressure of air inside a soccer ball rises as air is added. How can changes in gas pressure be explained by the kinetic theory?

Figure 11.1

Pumping up a Soccer Ball
Molecules in air are in constant motion and exert pressure when they strike the walls of the ball. The pressure they exert counterbalances two other pressures—atmospheric pressure on the ball plus the pressure exerted by the tough rubber of the ball itself.

Pumping more air into the soft ball (left) increases the number of molecules inside. As a result, molecules strike the inner wall of the ball more often and the pressure increases (right). The increase in pressure is counterbalanced by increased pressure from the tough wall. As a result, the ball becomes firmer and bouncier. ▼

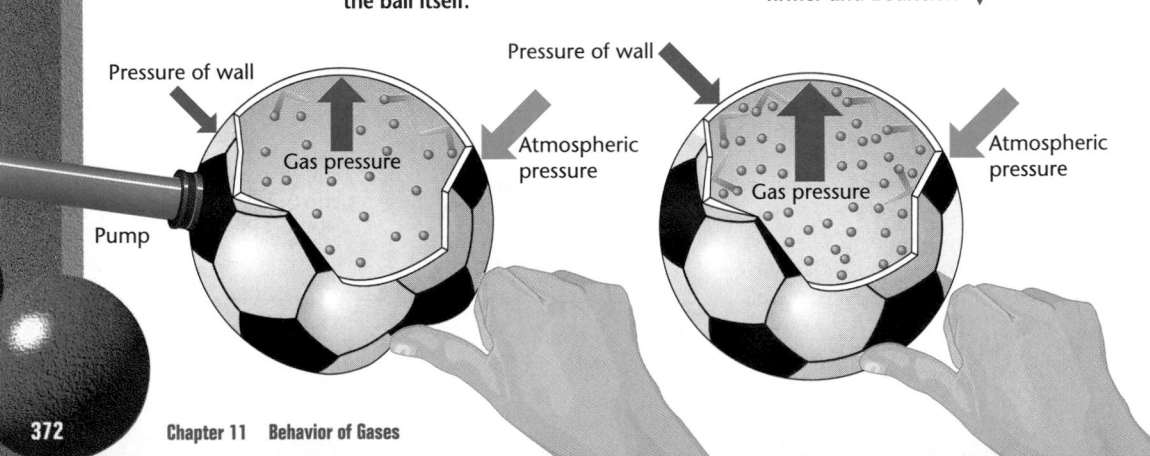

Pressure of wall

Gas pressure

Atmospheric pressure

Pump

Pressure of wall

Gas pressure

Atmospheric pressure

372   Chapter 11   Behavior of Gases

Program Resources

Study Guide, pp. 43-44 L1
Critical Thinking/Problem Solving, p. 12 L2
English/Spanish Audiocassettes, Chapter 11 L1 LEP
Section Focus Transparency 22 and **Master,** p. 22 L1 LEP
Spanish Resources L1 LEP

ChemLab and MiniLab Worksheets, p. 76 L1
Tech Prep Applications, pp. 15-17
Applying Scientific Methods in Chemistry, pp. 31-32 L1

How are number of particles and gas pressure related?

Recall from Chapter 10 that the pressure of a gas is the force per unit area that the particles in the gas exert on the walls of their container. As you would expect, more air particles inside the ball mean more mass inside. In **Figure 11.2,** one basketball was not fully inflated. The other identical ball was pumped up with more air. The ball that contains more air has higher pressure and the greater mass.

From similar observations and measurements, scientists from as long ago as the 18th century learned that the pressure of a gas is directly proportional to its mass. According to the kinetic theory, all matter is composed of particles in constant motion, and pressure is caused by the force of gas particles striking the walls of their container. The more often gas particles collide with the walls of their container, the greater the pressure. Therefore, the pressure is directly proportional to the number of particles. For example, doubling the number of gas particles in a basketball doubles the pressure.

Figure 11.2

Two Basketballs at Unequal Pressures
The mass of the ball on the left is greater than the mass of the one on the right. The ball on the left has greater mass because it has more air inside and, therefore, is at a higher pressure than the ball on the right.

Quick Demo

How do gas molecules do that?—Part 1
Use an EME Molecular Motion Demonstrator on an overhead projector to demonstrate the relationship between pressure and number of molecules. Put washable marks about 2 cm apart on one side of the moving chamber. Add several balls to the apparatus and set the speed at medium. Count and record the collisions between the lines for 30 seconds. Then double the number of balls in the apparatus and count collisions again. Ask students how the two results compare. *The number of collisions within the marked space approximately doubled when the number of balls was doubled.*

Reinforcement
Place a brick on the demonstration table in one of its three possible positions and state, for convenience, that the brick has a mass of approximately 1 kg and a weight of about 9.8 N. Place the brick, in turn, in the two other positions and ask in which position the brick exerts the most weight. *All positions are equal, 9.8 N.* Ask in which position the brick exerts the greatest pressure. *the upright position because pressure is force per unit area and the force or weight of 9.8 N is exerted on the smallest area* L1

Discussion
Ask students approximately how long a person can survive without food. *about one month* Without water? *several days* Then ask how long a person could survive without air. *only a few minutes* Explain that the average person breathes in approximately 14 kg of air per day. L1

Meeting Individual Needs

Gifted Refer to *Reinforcement* above and ask gifted students to calculate the pressure in N/m^2 (pascals) exerted by the brick in each of the three positions. Assume that the brick has dimensions of 18 cm × 9 cm × 6 cm. *9.8 N/ 0.016 m^2 = 610 N/m^2 or 610 Pa; 9.8 N/ 0.011 m^2 = 890 Pa; 9.8 N/0.0054 m^2 = 1800 Pa* L3

⚡ **Quick Demo**

Who likes to talk?

Blow up a balloon to demonstrate how easily it can be done. Then put an identical balloon into a volumetric flask or transparent bottle, and stretch the neck of the balloon over the mouth of the flask or bottle. Ask the class to identify the most talkative student and have this student blow up the balloon by blowing into the flask. Comment that it should be easy for a person with any lung power. Ask students to explain the result. *The student can inflate the balloon very little because of the pressure of the air trapped in the flask.* L1

⚡ **Quick Demo**

Does air really have weight?

Perform this demo without comment. Fill two equal-size balloons with air, tie off the openings, tie them to opposite ends of a meterstick, and balance the meterstick on a narrow edge or object. Use a sharp pin or needle to puncture one of the balloons. Ask students why the meterstick now tilts toward the end attached to the filled balloon. *When one balloon was broken, the compressed air was released from one end and the stick became unbalanced.*

To demonstrate how a gas at constant temperature can be used to do work, let's examine the action of a piston as shown in **Figure 11.3.** The piston inside the cylinder is like the cover of a jar because it makes an airtight seal, but it also acts like a movable wall. When gas is added, as shown in the top cylinder, the gas particles push out the piston until the pressure inside balances the atmospheric pressure outside.

If more gas is pumped into the cylinder, the number of particles increases, and the number of collisions on the walls of the container increases. Because the force on the inside face of the piston is now greater than the force on the outside face, the piston moves outward. As the gas spreads out into the larger volume, the number of collisions per unit area on the inside face falls. Then, the pressure of the gas inside the container falls until it equals the pressure of the atmosphere outside and the piston comes to rest in its new position, farther out. The atmospheric pressure remains constant while the piston moves and the gas expands.

Figure 11.3

Expanding a Gas at Constant Pressure
The constant bombardment of molecules and atoms of the atmosphere exerts a constant pressure on the outside face of the piston. This pressure balances the pressure caused by the confined gas bombarding the inside face of the piston (top). ▶

When gas is added to the cylinder, the piston is pushed out (bottom). When the pressure of the new volume inside the cylinder balances atmospheric pressure, the piston stops moving out. ▶

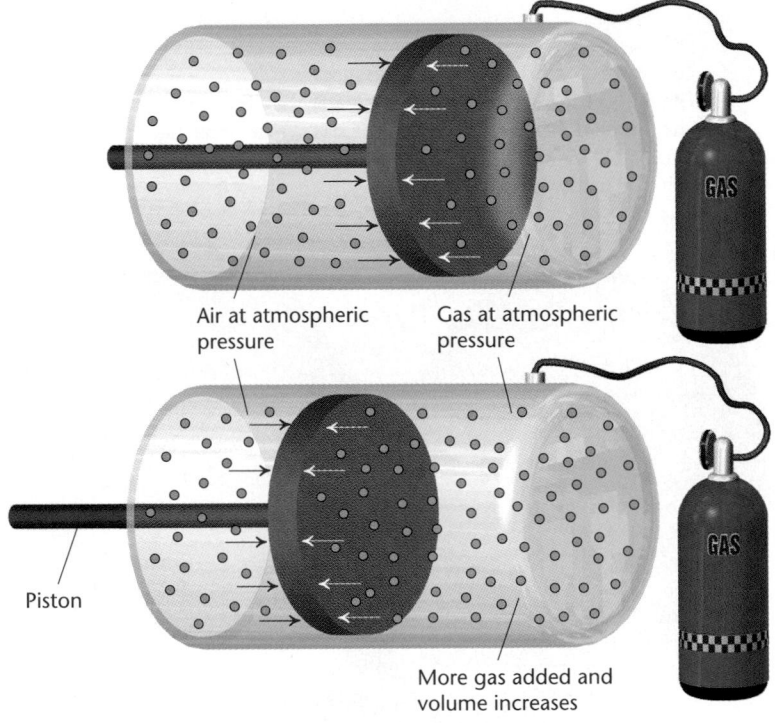

Air at atmospheric pressure   Gas at atmospheric pressure

Piston

More gas added and volume increases

How are temperature and gas pressure related?

How does temperature affect the behavior of a gas? You know from Chapter 10 that at higher temperatures, the particles in a gas have greater kinetic energy. They move faster and collide with the walls of the container more often and with greater force, so the pressure rises. If the volume of the container and the number of particles of gas are not changed, the pressure of a gas increases in direct proportion to the Kelvin temperature.

374   **Chapter 11**   **Behavior of Gases**

Demonstration 11.2

Air Bags

Purpose: To observe work done by an expanding gas.

Materials: 4 small, plastic sandwich bags; 4 one-hole rubber stoppers; 4 drinking straws; 4 rubber bands; cafeteria tray

Safety Precautions: 🥽

Disposal: Code F

Procedure: Use a rubber band to tightly at-tach a plastic sandwich bag to a one-hole stopper. Tightly fit the straw into the hole in the stopper. When air is blown into the straw, the plastic bag should inflate. Prepare four bags as described. Place the bags under the corners of a cafeteria tray. Ask four students to inflate the bags by blowing through the straws. Deflate the bags. Then, place a large mass on the tray and have the students inflate the bags again.

Results: The inflated bags lift the tray and the tray with the load.

Analysis: Ask these questions.

1. Ask the students who inflated the air bags whether high or low pressure was required to lift the tray with the load. *low pressure*

2. Was a long or short period of time required to inflate the bags and lift the load? *short time*

Relating Mass and Volume of a Gas

Recall from Chapter 4 that one property of carbon dioxide is that it changes directly from a solid (dry ice) to a gas. In other words, it sublimes. Can you determine some properties of gases from sublimed dry ice?

Procedure

1. Place a small, zipper-closure plastic bag on the pan of a zeroed balance. Wearing gloves and using tongs, insert a 20- to 30-g piece of dry ice into the bag.

2. Measure and record the mass of the bag and its contents. Quickly press the air out of the bag and zip it closed.

3. Immediately put the bag into a larger, clear plastic bag so that you can observe the sublimation process. After the inner bag is filled with gas, unzip the closure through the outer bag. Immediately remove the bag with dry ice from the larger bag. Press out the gas and zip it closed.

4. Measure and record the mass of the bag and its contents.

5. Use tongs and gloves to remove and dispose of the dry ice from the bag.

6. Determine the volume of the inner bag by filling it with water and pouring the water into a graduated cylinder. Record its volume.

7. Repeat steps 1-6 with a smaller zipper-closure bag and a smaller piece of dry ice.

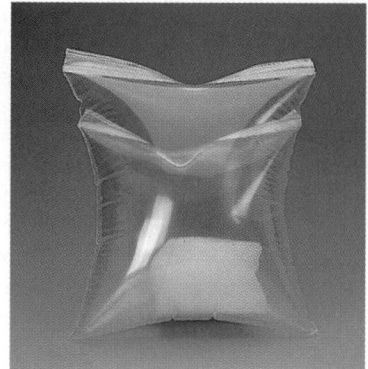

Analysis

1. Calculate the mass of carbon dioxide that has sublimed in each trial.

2. Calculate the mass ratio by dividing the mass of sublimed CO_2 in Trial 1 by the mass that sublimed in Trial 2. Calculate the volume ratio of CO_2.

3. How do the volume and mass ratios compare?

4. What can you infer about the relationship between the volume and mass of a gas?

What if the volume, such as in the piston in **Figure 11.3,** is not held constant? How would a change in temperature affect the volume and pressure of a gas? If the temperature of the gas inside the piston chamber is raised, the pressure will momentarily rise. But, if the gas is permitted to expand as the temperature is raised, the pressure remains equal to the atmospheric pressure and the volume expands. Thus, the volume of a gas at constant pressure is directly proportional to the Kelvin temperature.

When the cylinder is in an automobile engine, the other end of the piston rod, shown in **Figure 11.3,** is attached to the crankshaft. When the mixture of gasoline and air in the cylinder ignites and burns, gases are produced. The gas inside the cylinder expands because the heat of burning raises the temperature. The pressure of expanding gases drives the piston outward. As the piston rod moves out and then back in, it turns the crankshaft, delivering power to the wheels that propel the vehicle forward.

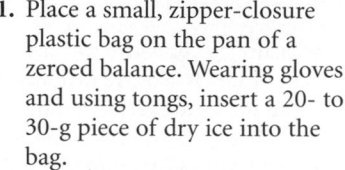

miniLAB

Relating Mass and Volume of a Gas

Purpose: To determine the relationship between the mass and volume of a gas.

Process Skills: Observing, measuring, inferring, communicating

Teaching Strategies

- If the reaction bag pops open because the pressure is too high, no harm is done. Tell students to immediately remove it and proceed.
- Set a small Styrofoam cooler near the balance and designate it for dry ice only.
- To save time in determining the volumes, use 500-mL or 1000-mL graduated cylinders.

Expected Results

One-Quart Bag
Initial mass of bag and dry ice: 26.18 g
Final mass of bag and dry ice: 23.38 g
Volume of bag: 1420 mL

Snack-Size Bag
Initial mass of bag and dry ice: 14.50 g
Final mass of bag and dry ice: 14.03 g
Volume of bag: 240 mL

Analysis

1. quart bag: 2.80 g CO_2; snack-size bag: 0.47 g CO_2

2. mass ratio: 6.0; volume ratio: 5.9

3. The mass and volume ratios are the same within the limits of experimental error.

4. A greater mass, and therefore a greater number of CO_2 molecules, occupies a proportionally greater volume.

✓ Assessment

Oral: A chemical reaction produces the gas that inflates automobile air bags. Ask students what factors are important in the design of air bags. *The chemical reaction must produce the gas almost instantly and in large enough quantities to produce a pressure in the bag that will support atmospheric pressure and the weight of a person.*
L1

✓ Assessment

Oral: Ask students how oxygen gas priced at $200 per 20-L cylinder by company #1 could be a better buy than oxygen priced at $100 per 20-L cylinder by company #2. *The pressure, and therefore the amount of oxygen per liter, must be more than twice as great in the cylinder from company #1.* **L1**

WORD ORIGIN

barometer:
baros (G) heavy
metron (G) to measure
A barometer measures the pressure (force per unit area) of the atmosphere.

Devices to Measure Pressure

Although you can compare a property such as gas pressure by touching partially and fully inflated basketballs, this method does not give you an accurate measure of the two pressures. What is needed is a measuring device.

The Barometer

One of the first instruments used to measure gas pressure was designed by the Italian scientist Evangelista Torricelli (1608-1647). He invented the **barometer**, an instrument that measures the pressure exerted by the atmosphere. His barometer was so sensitive that it showed the difference in atmospheric pressure between the top and bottom of a flight of stairs. **Figure 11.4** explains how Torricelli's barometer worked. The height of the mercury column measures the pressure exerted by the atmosphere. We live at the bottom of an ocean of air. The highest pressures occur at the lowest altitudes. If you go up a mountain, atmospheric pressure decreases because the depth of air above you is less.

One unit used to measure pressure is defined by using Torricelli's barometer. The **standard atmosphere (atm)** is defined as the pressure that supports a 760-mm column of mercury. This definition can be represented by the following equation.

$$1.00 \text{ atm} = 760 \text{ mm Hg}$$

Because atmospheric pressure is measured with a barometer, it is often called barometric pressure.

Figure 11.4
How Torricelli's Barometer Works
A barometer consists of a tube of mercury that stands in a dish of mercury. Because the mercury stands in a column in the closed tube, you can conclude that the atmosphere exerts pressure on the open surface of the mercury in the dish. This pressure, transmitted through the liquid in the dish, supports the column of mercury. ▶

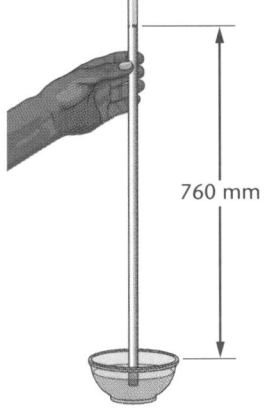

760 mm

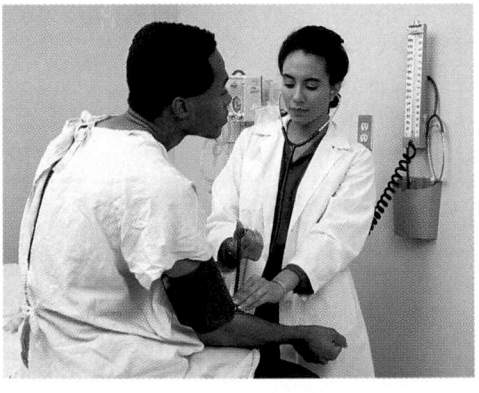

▲ The sphygmomanometer attached to the wall also uses a column of mercury to measure blood pressure.

The Pressure Gauge

Unfortunately, the barometer can measure only atmospheric pressure. It cannot measure the air pressure inside a bicycle tire or in an oxygen tank. You need a device that can be attached to the tire or tank. This pressure gauge must make some regular, observable response to pressure changes. If you have ever measured the pressure of an inflated bicycle tire, you're already familiar with such a device.

376   Chapter 11   Behavior of Gases

Integrating THE Sciences

Physics In 1663, Otto von Guericke, the German physicist who invented an air pump, performed a demonstration for the King of Prussia. Von Guericke placed two hemispheres together, pumped the air out of the resulting sphere, and demonstrated that even teams of horses were unable to pull the hemispheres apart. He then secretly did something to the sphere and pulled it apart himself. The king was so impressed that he awarded von Guericke a lifetime pension. Ask students to deduce what von Guericke did to the sphere. *He opened a valve, admitting air.* **L1**

Tire-Pressure Gauge

A tire-pressure gauge is a device that measures the pressure of the air inside an inflated tire or a basketball. Because an uninflated tire contains some air at atmospheric pressure, a tire-pressure gauge records the amount that the tire pressure exceeds atmospheric pressure.

The most familiar tire gauge is about the size and shape of a ballpoint pen. It is a convenient way to check tire pressure for proper inflation regularly. Proper inflation ensures tire maintenance and safety.

1. The pin in the head of the gauge pushes downward on the tire-valve inlet and allows air from the tire to flow into the gauge.

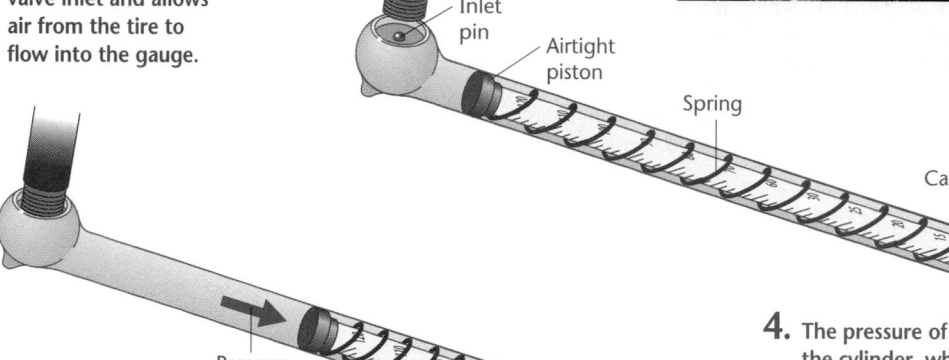

Tire valve

Inlet pin

Airtight piston

Spring

Calibrated scale

Pressure of tire

2. The air flowing into the gauge pushes against a movable piston that pushes a sliding calibrated scale.

3. The piston moves along the cylinder, compressing the spring until the force of air pressing on the surface of the piston is equal to the force of the compressed spring on the piston.

4. The pressure of the air in the cylinder, which is the same as the pressure of the air in the tire, is read from the scale.

Thinking Critically

1. What does the calibration of the scale indicate about the relationship between the compression of the spring and pressure?

2. Explain whether a tire-pressure gauge should be recalibrated for use at locations where atmospheric pressure is less than that at sea level.

Chemistry Journal

Atmospheric Pressure and Weather

TECH PREP Ask students to keep a daily or twice-daily record in their journals of barometric pressure for 30 days and to note daily weather and weather trends. Ask them to correlate barometric pressure and weather and summarize their correlations in their journals. [L1]

Purpose

TECH PREP Students will learn how pressure-volume relationships enable them to measure the air pressure in an automobile tire.

Background

Because the pressure of air inside a tire exceeds atmospheric pressure, the air rushes out when a tire gauge is connected to the valve of an inflated tire. The gauge contains an inflatable bladder that expands and moves a piston on a calibrated scale. The pressure of the air in the tire, in excess of atmospheric pressure, is read from the scale.

Visual Learning

Demonstrate how to use a tire gauge on a bicycle tire. If a tire gauge is available, have students measure the tire pressure of automobiles in the school parking lot. Use the illustrations in the text to supplement the measurement of tire pressure. [L1]

Teaching Strategies

Discuss the property of a confined gas that makes a tire gauge work. Explain that a confined gas will expand to fill the volume of its container. Molecular motion allows gas particles to move into regions of lower pressure and thus increases the volume of the gas.

Thinking Critically

1. The greater the compression of the spring, the greater the tire pressure.

2. Because the tire-pressure gauge measures pressures greater than 1 atm, it should be calibrated for use at locations where atmospheric pressure is much less than at sea level.

Content Background

SI Units Explain that the pascal unit is defined as a pressure of 1 N/m^2 and that a newton is the force needed to accelerate a 1-kg mass at a rate of 1 m/s^2.

Correcting Misconceptions

Students may think that because standard pressure is used in many calculations, atmospheric pressure is usually standard pressure. Explain that the standard pressure of 1 atm is merely a reference standard and that the pressure of the atmosphere in a given location is seldom exactly equal to standard pressure.

Visual Learning

Figure 11.5: When students learn that air pressure results from a column of air pressing on the area beneath it, they may think that air pressure is exerted only downward. Explain that air pressure is exerted in all directions. Use the following *Quick Demo* to enhance understanding of this concept.

Quick Demo

Can air pressure do that?
Fill a small glass having a smooth lip to the brim with water. Place a 3″ × 5″ card over the mouth of the glass, hold it firmly in place, and invert the glass. When you remove the hand supporting the card, the card will remain in place. Ask students to explain what keeps the card in place. *air pressure acting upward on the card*

Fact of the MATTER

The weight of a postage stamp exerts a pressure of about one pascal on the surface of an envelope.

When you measure tire pressure, you are measuring pressure above atmospheric pressure. The recommended tire inflation pressures listed by manufacturers are gauge pressures; that is, pressures read from a gauge. A barometer measures absolute pressure; that is, the total pressures exerted by all gases, including the atmosphere. To determine the absolute pressure of an inflated tire, you must add the barometric pressure to the gauge pressure.

Pressure Units

You have learned that atmospheric pressure is measured in mm Hg. Recall from Chapter 10, atmospheric pressure is the force per unit area that the gases in the atmosphere exert on the surface of Earth. **Figure 11.5** shows two additional units that are used to measure pressure of one standard atmosphere.

The SI unit for measuring pressure is the **pascal (Pa),** named after the French physicist Blaise Pascal (1623-1662). Because the pascal is a small pressure unit, it is more convenient to use the kilopascal. As you recall from Chapter 1, the prefix *kilo-* means 1000; so, 1 **kilopascal (kPa)** is equivalent to 1000 pascals. One standard atmosphere is equivalent to 101.3 kilopascals.

Figure 11.5

Atmospheric Pressure
The weight of the air in each column pushing on the area beneath it exerts a pressure of one standard atmosphere. Each column of air extends to the outer limits of the atmosphere.

If the unit of area is the square inch and the unit of force is the pound, then the unit of pressure is the pound per square inch (psi). Expressed in these units, one standard atmosphere is 14.7 pounds per square inch or 14.7 psi. ▼

Expressed in SI units, one standard atmosphere is 101 300 pascals. Note that SI units are based on the square meter area and not the square inch. ▼

14.7 psi   101 300 Pa
1 in.   1 in.   1 m
1 m   1 m

Demonstration 11.3

Can Crushing

Purpose: To demonstrate the relationships between temperature, pressure, and the number of particles of a confined gas.

Materials: empty soda can, hot plate, beaker tongs, 600-mL beaker filled with ice water

Safety Precautions: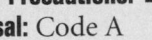

Disposal: Code A

Procedure: Place a small amount of water in an empty soda can. Heat the can on a hot plate until the water boils and steam comes from the opening in the top. Use beaker tongs to remove the can from the hot plate, and immediately plunge the open top into a large container of ice water.

Results: The can is crushed quickly and with some noise. The can was sealed when the opening was covered by the water. The water vapor in the can cooled and condensed, lowering the pressure inside the can. The pressure of the at-

Table 11.1 presents the standard atmosphere in equivalent units. Because there are so many different pressure units, the international community of scientists recommends that all pressure measurements be made using SI units, but pounds per square inch continues to be widely used in engineering and almost all nonscientific applications in the United States.

Table 11.1 Equivalent Pressures			
1.00 atm	760 mm Hg	14.7 psi	101.3 kPa

Pressure Conversions

You can use **Table 11.1** to convert pressure measurements to other units. For example, you can now find the absolute pressure of the air in a bicycle tire. Suppose the gauge pressure is 44 psi. To find the absolute pressure, add the atmospheric pressure to the gauge pressure. Because the gauge pressure is given in pounds per square inch, use the value of the standard atmosphere that is expressed in pounds per square inch. One standard atmosphere equals 14.7 psi.

$$44 \text{ psi} + 14.7 \text{ psi} = 59 \text{ psi}$$

The following Sample Problems show how to use the values in **Table 11.1** to express pressure in other units.

SAMPLE PROBLEM **1** **Converting Barometric Pressure Units**

In weather reports, barometric pressure is often expressed in inches of mercury. What is one standard atmosphere expressed in inches of mercury?

Analyze
- You know that one standard atmosphere is equivalent to 760 mm of Hg. What is that height expressed in inches? A length of 1.00 inch measures 25.4 mm on a meterstick.

Set Up
- Select the appropriate equivalent values and units given in **Table 11.1**. Multiply 760 mm by the number of inches in each millimeter to express the measurement in inches.

$$760 \text{ mm} \times \left(\frac{1.00 \text{ in.}}{25.4 \text{ mm}} \right)$$

The factor on the right of the expression above is the conversion factor.

Solve
- Notice that the units are arranged so that the unit *mm* will cancel properly and the answer will be in inches.

$$\frac{760 \text{ mm}}{} \left| \frac{1.00 \text{ in.}}{25.4 \text{ mm}} \right. = 29.9 \text{ in.}$$

Check
- Because 1 mm is much shorter than 1 in., the number of mm, 760, should be much larger than the equivalent number of inches.

11.1 **Gas Pressure** 379

---

mosphere was then sufficient to crush the can.

Analysis: Ask these questions.

1. What happens to the water vapor inside the can when the can is inverted in the ice water? *It condenses to a liquid.*

2. Why was the can crushed on all sides symmetrically? *Air molecules strike the can equally on all sides.*

3. Why was the can full of water at the conclusion of the demo? *A partial vacuum resulted from the condensation of the water vapor, and water entered to fill the volume.*

✔ Assessment

Portfolio: Have students write a description of the changes in temperature, pressure, and number of particles of gas in the soda can and how these changes resulted in the crushing of the can. L1 P

SECTION 11.1

Correcting Errors

Students may think that they can obtain the true barometric pressure by calling the local weather service rather than using the classroom barometer. Point out that meteorologists convert all barometric pressures to sea-level equivalents so that high-altitude cities do not always seem to be in low-pressure regions and low-altitude cities do not always seem to have high pressure. Explain that the barometric pressures reported by the local meteorologist are accurate only at sea level.

Correcting Misconceptions

Because the terms *air pressure* or *atmospheric pressure* are often used, students may think that an "air" molecule exists. Point out that air is a mixture of molecules and that its composition varies from time to time and place to place.

Discussion

Point out that the key to using the factor label method is knowing a conversion factor that links the unit you have to the one you want to obtain. One method is to memorize a few important conversion factors. Another method is to look up conversion factors in a text. Sometimes, calculators include a reference card listing some frequently used conversion factors, or they display a factor when you type in a unit. Ask students where they can find conversion factors in their text. *in tables, in the chapters, and in the Appendix*

3 ASSESS

Check for Understanding

- Check for understanding of the section material by having students answer the Section Review questions.
- Fill equal-sized balloons with helium (0.18 g/L), natural gas (0.55 g/L), air (1.0 g/L), and carbon dioxide (2.0 g/L). Don't reveal the identities of the gases. Have four students throw the balloons as far as possible. Ask students to explain the results. *The balloon that goes farthest must be filled with the gas with the greatest mass (CO_2). The balloon that goes the shortest horizontal distance (and up) must have the lowest mass (He).* L1

Reteach

⚡ Quick Demo

Why doesn't water pour out?

Punch a small hole in the side of an empty, 2-L, plastic soda bottle. Hold your finger over the hole and completely fill the bottle with water. Screw the cap on and remove your finger. Ask students why water doesn't come out the hole. *Air pressure holds it in.* Remove the cap to demonstrate that water will come out the hole if air is allowed into the top of the bottle. Ask students how tall the soda bottle would have to be before the water is no longer held in by atmospheric pressure. *about 33 feet* L2

The method used in the Sample Problem to convert measurement to other units is called the **factor label method.** Study the Sample Problem again. The following equation gives the conversion factor because it contains both the given unit and the desired unit.

$$1.00 \text{ in.} = 25.4 \text{ mm}$$

You know that you can divide both sides of an equation by the same value and maintain equality. For example, you can divide both sides of the equation 1.00 in. = 25.4 mm by 25.4 mm, as shown below.

$$\frac{1.00 \text{ in.}}{25.4 \text{ mm}} = \frac{25.4 \text{ mm}}{25.4 \text{ mm}}$$

Now, simplify the right side. The conversion factor is on the left.

$$\frac{1.00 \text{ in.}}{25.4 \text{ mm}} = 1$$

In the Sample Problem, the height of mercury in millimeters was being multiplied by this conversion factor. Because multiplying any quantity by 1 doesn't affect its value, the height of the mercury column isn't changed—only the units are. The factor label method changes the units of the measurement without affecting its value. You will use the factor label method to solve problems in this chapter and in several of the following chapters. You will find more information on the factor label method on pages 801-803 of Appendix A, the Skill Handbook.

SAMPLE PROBLEM ② Converting Pressure Units

The reading of a tire-pressure gauge is 35 psi. What is the equivalent pressure in kilopascals?

Analyze
- The given unit is pounds per square inch (psi), and the desired unit is kilopascals (kPa). According to **Table 11.1,** the relationship between these two units is 14.7 psi = 101.3 kPa.

Set Up
- Write the conversion factor with kPa units as the numerator and psi units as the denominator. Note that the psi units will cancel and only the kPa units will appear in the final answer.

> **Problem-Solving HINT**
>
> In the factor label method, terms are arranged so that units will cancel out.

$$35 \text{ psi} \times \left(\frac{101.3 \text{ kPa}}{14.7 \text{ psi}} \right)$$

Solve
- Multiply and divide the values and units.

$$\frac{35 \text{ psi} \mid 101.3 \text{ kPa}}{14.7 \text{ psi}} = \frac{35 \times 101.3 \text{ kPa}}{14.7} = 240 \text{ kPa}$$

Notice that the given units (psi) will cancel properly and the quantity will be expressed in the desired unit (kPa) in the answer.

Check
- Because 1 psi is a much greater pressure than 1 kPa, the number of psi, 35, should be much smaller than the equivalent number of kilopascals.

PRACTICE

Guided Practice: Guide students through the sample problems, then have them work in class on Practice Problems 1-5. See Appendix D for complete solutions. L1

Independent Practice: Your homework or classroom assignment can include the additional practice problems shown here. L1

Answers to Practice Problems

1. 29.4 psi

2. 3.80×10^2 mm Hg

3. 152 kPa

4. 131 kPa

5. 29.3 psi

Additional Practice Problems

1. 850 mm Hg to kPa

$$\frac{850 \text{ mm Hg} \mid 1.00 \text{ atm} \mid 101 \text{ kPa}}{760 \text{ mm Hg} \mid 1.00 \text{ atm}} = 113 \text{ kPa}$$

2. 2.50 atm to mm Hg

$$\frac{2.50 \text{ atm} \mid 760 \text{ mm Hg}}{1.00 \text{ atm}} = 1900 \text{ mm Hg}$$

3. 65.0 in. Hg to atm

$$\frac{65.0 \text{ in. Hg} \mid 25.4 \text{ mm Hg} \mid 1.00 \text{ atm}}{1.00 \text{ in. Hg} \mid 760 \text{ mm Hg}} = 2.17 \text{ atm}$$

Your skill in converting units will help you relate measurements of gas pressure made in different units.

Use Table 11.1 and the equation 1.00 in. = 25.4 mm to convert the following measurements.

1. 59.8 in. Hg to psi
2. 7.35 psi to mm
3. 1140 mm to kPa
4. 19.0 psi to kPa
5. 202 kPa to psi

SECTION REVIEW

Understanding Concepts

1. Compare and contrast how a barometer and a tire-pressure gauge measure gas pressure.

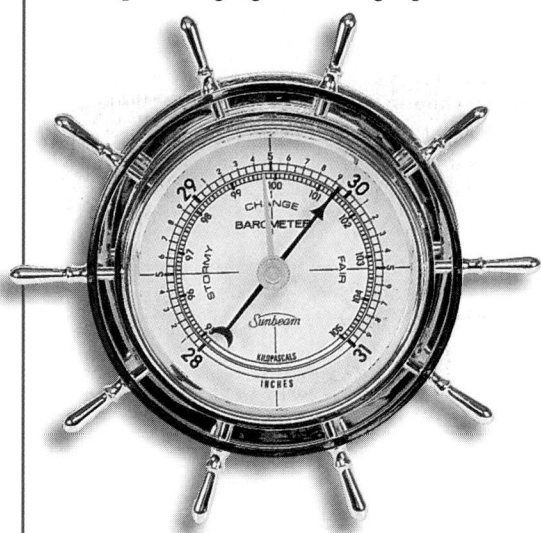

2. At atmospheric pressure, a balloon contains 2.00 L of nitrogen gas. How would the volume change if the Kelvin temperature were only 75 percent of its original value?

3. A cylinder containing 32 g of oxygen gas is placed on a balance. The valve is opened and 16 g of gas are allowed to escape. How will the pressure change?

Thinking Critically

4. **Interpreting Data** Sketch graphs of pressure versus time, volume versus time, and number of particles versus time for a large, plastic garbage bag being inflated until it ruptures.

Applying Chemistry

5. **Overloaded** Why do tire manufacturers recommend that tire pressure be increased if the recommended number of car passengers or load size is exceeded?

SECTION 11.1

Extension

Ask interested groups of students to devise a setup to demonstrate standard atmospheric pressure, 101 kPa or 14.7 lbs/in². [L2] **COOP LEARN**

4 CLOSE

Writing About Chemistry

Ask students to write a short story about a molecule of air confined in the tire of an automobile and the experiences of the molecule as the car travels from the northern part of the country, where temperatures are below freezing, to the south where temperatures are in the eighties. Students should include what happens when the tire pressure is measured, when air is added, and what the molecule experiences in the event of a blowout. The events in the story should be related to the kinetic model of gases. Students' stories may be placed in their portfolios. [L1] [P]

SECTION REVIEW

1. A barometer measures the height of a column of mercury supported by the pressure of the atmosphere; a tire-pressure gauge measures the compression of a spring resulting from the pressure of air in the tire.
2. 2.00 L × 0.75 = 1.50 L
3. The new mass is half the old mass, so there are half the original number of gas particles. Pressure is directly proportional to the number of particles, so the new pressure must be half the original pressure.
4. P versus time: constant slow rise and then rapid rise when the bag has reached its limit; V versus time: rapid rise until the bag reaches its limit, then leveling off; number of particles versus time: constant rise
5. The additional air exerts greater pressure and therefore gives greater rigidity to the tires so that they can support the added weight of the passengers or load.

PREPARE

Key Concepts

The behavior of gases is expressed and defined by Boyle's law, Charles's law, and the combined gas law. The law of combining gas volumes is related to Avogadro's principle.

Planning Ahead

For the ChemLab Be sure you have an adequate number of heavy, double-post screw clamps.

1 FOCUS

Focus Transparency

Display the **Section Focus Transparency** for Section 11.2 to show the relationship between temperature and gas volume. L1 LEP

2 TEACH

Discussion

Ask students to explain how a person's breathing in and out illustrates Boyle's law. *As a person breathes in, the diaphragm moves downward and outward, increasing the lungs' volume and decreasing the pressure in the lungs. External air pressure then forces air into the lungs. As a person breathes out, the diaphragm moves upward and inward, decreasing lung volume and increasing pressure, which expels air.* L2

The Gas Laws

SECTION PREVIEW

Objectives

Analyze data that relate temperature, pressure, and volume of a gas.
Model Boyle's law and Charles's law using kinetic theory.
Predict the effect of changes in pressure and temperature on the volume of a gas.
Relate how volumes of gases react in terms of the kinetic theory of gases.

Key Terms

Boyle's law
Charles's law
combined gas law
standard temperature and pressure, STP
law of combining gas volumes
Avogadro's principle

An uninflated air mattress doesn't make a comfortable bed because comfort depends upon the pressure of the air inside it. So, you inflate it by huffing and puffing or by using the exhaust feature of a vacuum cleaner. Recall that when more air is added to the mattress, the pressure of the air inside increases. When the mattress is filled and its air valve is closed, it is far more comfortable. The air inside supports the walls of the mattress and pushes against the force exerted by atmospheric pressure on the outside surface of the mattress. As you add your weight to the mattress, do you know how the air inside is now acting upon the mattress walls, against the force of the atmosphere on its surface, and on you?

Boyle's Law: Pressure and Volume

As you know from squeezing a balloon, a confined gas can be compressed into a smaller volume. Robert Boyle (1627-1691), an English scientist, used a simple apparatus like the one pictured in **Figure 11.6** to

Figure 11.6

Relation Between Pressure and Volume

1. The pressure of the trapped air in the J tube balances the atmospheric pressure, 1 atm or 760 mm Hg.

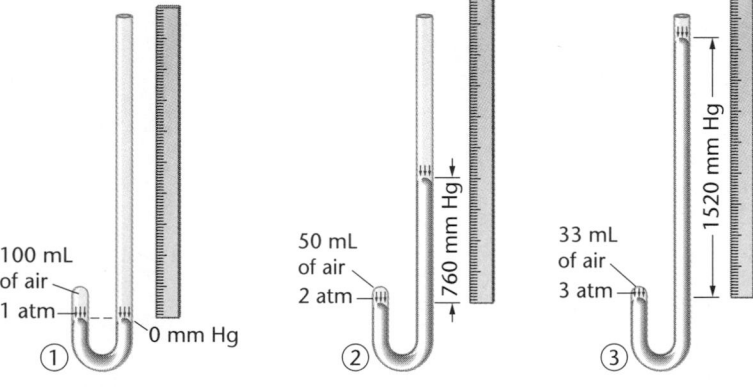

2. When mercury is added to a height of 760 mm above the height in the closed end, the volume of the trapped air is halved, and the pressure the air exerts is now 2 atm.

3. When an additional 760 mm of mercury is added to the column, the pressure of the trapped air is tripled.

Program Resources

Study Guide, pp. 45-46 L1
Laboratory Manual, 11.1, pp. 97-100; 11.2, pp. 101-104 L1
Section Focus Transparency 23 and **Master,** p. 23 L1 LEP
Basic Concepts Transparencies 30, 31, 32 and **Masters,** pp. 59-64 L1 LEP

Problem Solving Transparency 13 and **Master,** pp. 25-26 L1 LEP
ChemLab and MiniLab Worksheets, pp. 73-75, 77 L1
Applying Scientific Methods in Chemistry, p. 33 L1
Consumer Chemistry, pp. 21-22 L1

compress gases. The weight of the mercury in the open end of the tube compresses air trapped in the closed end.

After performing many experiments with gases at constant temperatures, Boyle had four findings.

a) If the pressure of a gas increases, its volume decreases proportionately.
b) If the pressure of a gas decreases, its volume increases proportionately.
c) If the volume of a gas increases, its pressure decreases proportionately.
d) If the volume of a gas decreases, its pressure increases proportionately.

Because the changes in pressure and volume are always opposite and proportional, the relationship between pressure and volume is an inverse proportion. By using inverse proportions, all four findings can be included in one statement called Boyle's law. **Boyle's law** states that the pressure and volume of a gas at constant temperature are inversely proportional.

Recall from Chapter 10 that the kinetic energy of an ideal gas is *directly* proportional to its temperature and that the graph of a direct proportion is a straight line. The graph of an *inverse* proportion is a curve like that shown in **Figure 11.7**. Just as a road runs both ways, you can think of a gas following the curve A-B-C or C-B-A. The path A-B-C represents the gas being compressed, which forces its pressure to rise. As the gas is compressed by half, from 1.0 L to 0.5 L, the pressure doubles from 100 kPa to 200 kPa. As it is compressed by half again, from 0.5 L to 0.25 L, the pressure again doubles from 200 kPa to 400 kPa. Look at **Figure 11.9** on page 386 to see another example of this relationship. The reverse path C-B-A represents what happens when the pressure of the gas decreases and the volume increases accordingly.

Figure 11.7

Boyle's Law: An Inverse Relationship
As you follow the curve from left to right, the pressure increases and the volume decreases. Look at the change from points A to C. The volume is reduced from 1.0 L to 0.25 L; that is, the gas is compressed to one-fourth of its volume, and the pressure rises from 100 kPa to 400 kPa, which is four times as high. The volume/pressure relationship is a two-way system.

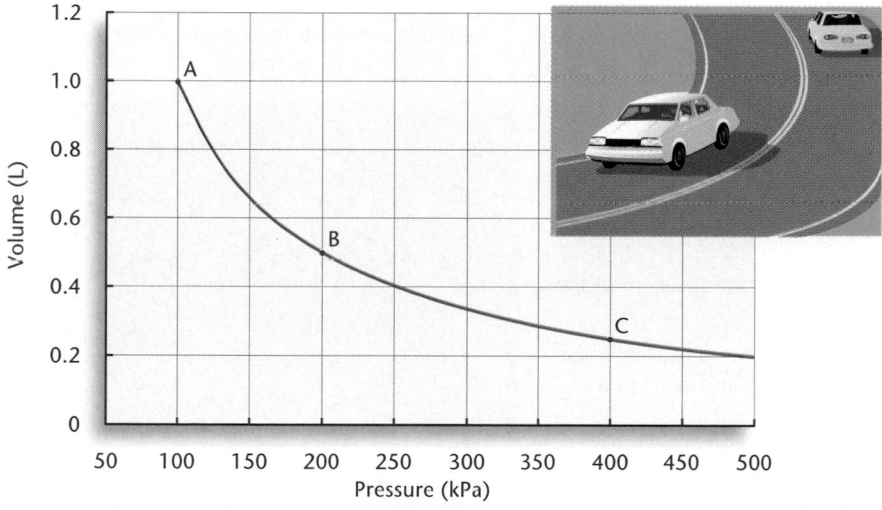

✔ **Assessment**

Portfolio: Have the students write the following questions in their portfolios. Ask them to write the answers to the questions as soon as they have learned enough from their study of Section 11.2.

1. Why are tennis balls sold in pressurized cans? *Tennis balls are filled with gas at higher pressures to give them bounce. If the balls are not kept in pressurized cans, the gas gradually escapes.*

2. Why are you advised to open windows slightly if a tornado approaches? *The sudden decrease in air pressure within a tornado could cause the building's windows to blow out or the whole structure to explode. Opening a window allows indoor and outdoor pressures to equalize.*

3. Why do hot-air balloons rise? *Heated air expands, causing it to be less dense.*

4. Why do breads and pastries rise when baked? *Carbon dioxide, created by yeast or baking powder, expands as it is heated.* L1 P

Transparency

Display **Basic Concepts Transparency 30** to illustrate Boyle's law. L1 LEP

Display **Basic Concepts Transparency 31** to illustrate Boyle's law graphically. L1 LEP

Meeting Individual Needs

Learning Disabled Many students will find solving gas law problems easier if they use simple formulas and solve for the unknown. For example, if Boyle's law is expressed mathematically as $P_1V_1 = P_2V_2$, those students can easily insert given pressures and volumes and solve for the unknown quantity. L1

Chemistry Journal

 The Vacuum Cleaner and the Gas Laws

TECH PREP Ask students to write a journal entry in which they evaluate the validity of the statement "A vacuum cleaner sucks up dirt." Entries should be supported by references to gas behavior. L1

ChemLab

Boyle's Law

Time Allotment: One laboratory period.

Objectives

Review objectives with students before they begin the ChemLab.

Process Skills

Observing, measuring, inferring, interpreting data, communicating

Safety Precautions

Advise students to be careful with open flames and hot objects when heating and sealing the pipet stems.

PREPARATION

To save time and conserve pipets, the sealed pipets may be prepared prior to the laboratory period.

PROCEDURE

- Require that students' notebooks contain the following items before they begin: Objectives, Safety Precautions, Outline or Flow Chart of the Procedure, Data and Observations Table.
- Emphasize that T is *total* number of turns, i.e., $T_{Trial\ 2} = T_{Trial\ 1} + 1$.

384

ChemLab

Boyle's Law

The quantitative relationship between the volume of a gas and the pressure of the gas, at constant temperature, is known as Boyle's Law. By measuring quantities directly related to the pressure and volume of a tiny amount of trapped air, you can deduce Boyle's law.

Problem

What is the relationship between the volume and the pressure of a gas at constant temperature?

Objectives

- **Observe** the length of a column of trapped air at different pressures.
- **Examine** the mathematical relationship between gas volume and gas pressure.

PREPARATION

Materials

thin-stem pipet	matches
double-post	metric ruler
screw clamp	scissors
fine-tip marker	small beaker
food coloring	water

Safety Precautions

Use care in lighting matches and melting the pipet stem.

PROCEDURE

1. Cut off the stepped portion of the stem of the pipet with the scissors.
2. Place about 20 mL of water in a beaker, add a few drops of food coloring, and swirl to mix.

---

3. Draw the water into the pipet, completely filling the bulb and allowing the water to extend about 5 mm into the stem of the pipet.

4. Heat the tip of the pipet gently above a flame until it is soft. (**CAUTION:** *If the stem accidentally begins to burn, blow it out.*) Use a metallic or glass object to flatten the tip against the countertop so that the water and air are sealed inside the pipet. Holding the pipet by the bulb, tap any water droplets in the stem down into the liquid. You should observe a cylindrical column of air trapped in the stem of the pipet.

5. Center the bulb in a double-post screw clamp, and tighten the clamp until the bulb is just held firmly. Mark the knob of the clamp with a fine-tip marker, and tighten the clamp three or four turns so that the length of the air column is 50 to 55 mm. Record this number of turns T under Trial 1 in a data table like the one shown. Measure and record the length L of the air column in millimeters for Trial 1.

6. Turn the clamp knob one complete turn, and record the trial number and the total number of turns. Measure and record the length of the air column.

7. Repeat step 6 until the air column is reduced to a length of 25 to 30 mm.

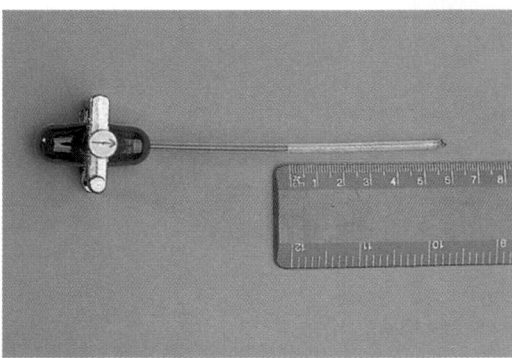

ChemLab

ANALYZE AND CONCLUDE

1. **Observing and Inferring** Explain whether the volume *V* of the air in the stem is directly proportional to the length *L* of the air column. What inferences can be made about the pressure *P* of the air in the column and the number of turns *T* of the clamp screw?

2. **Interpreting Data** Calculate the product *LT* and the quotient *L/T* for each trial. Which calculations are more consistent? If *L* and *T* are directly related, *L/T* will yield nearly constant values for each trial. On the other hand, if *L* and *T* are inversely related, *LT* will yield almost constant values for each trial. Are *L* and *T* directly or inversely related?

3. **Drawing Conclusions** Explain whether the data indicate that gas volume and gas pressure at constant temperature are directly related or inversely related.

APPLY AND ASSESS

1. For a mercury column barometer to measure atmospheric pressure as 760 mm Hg, the column containing the mercury must be completely evacuated. However, if the column is not completely evacuated, the barometer can still be used to correctly measure *changes* in barometric pressure. How is the second statement related to Boyle's law?

2. Using the kinetic theory, explain how a decrease in the volume of a gas causes an increase in the pressure of the gas.

From Robert Boyle's experiments, we know the relationship between the pressure and volume of a gas at constant temperature.

DATA AND OBSERVATIONS

Pressure and Length Data				
Trial	Turns, *T*	Length of Air Column, *L*, (mm)	Numerical Value, *LT*	Numerical Value, *L/T*

DATA AND OBSERVATIONS

Trial	Turns, *T*	Length of Air Column, *L* (mm)	Numerical Value, *LT*	Numerical Value, *L/T*
1	4	55	220	14
2	5	45	225	9
3	6	37	222	6
4	7	29	203	4

ANALYZE AND CONCLUDE

1. The volume of the air in the stem is directly proportional to the length of the column because the column is a cylinder of constant cross-sectional area. The pressure increases directly with the number of turns.

2. The calculations of *LT* are more consistent, which indicates that *L* and *T* are inversely related.

3. Because gas volume *V* is directly proportional to *L*, and gas pressure *P* is directly related to *T*, *V* and *P* must have the same mathematical relationship as *L* and *T*, which is an inverse relationship.

APPLY AND ASSESS

1. In an unevacuated column, trapped air exerts pressure on the mercury column equal to the pressure of the atmosphere on the mercury pool outside the column. Therefore, an increase or decrease in atmospheric pressure will cause the volume of the gas to decrease or increase because its pressure always balances atmospheric pressure. The increase or decrease in the volume causes the height of the mercury column to rise or fall directly as atmospheric pressure increases or decreases.

2. A decrease in the volume of a gas causes the molecules of gas to strike the walls of the container more frequently and, therefore, increases the pressure.

✓ Assessment

Performance: Ask students to make graphs of pressure vs. volume and pressure vs. 1/volume and to explain the meaning of the graphs. Students can use the analogous data they generated in the ChemLab. [L1]

Performance: If a bell jar and vacuum pump are available, have a student place a marshmallow in the bell jar and seal the jar. Ask students to predict what will happen when the jar is evacuated. *The marshmallow will expand as the pressure in the bell jar decreases.* Have the student evacuate the jar using a vacuum pump to test the prediction. Ask students what will happen if air is allowed to reenter the bell jar. *The marshmallow will shrink back to its normal size.* L1

GLENCOE TECHNOLOGY

Videodisc

Chemistry: Concepts and Applications
Demonstrating Boyle's Law
Disc 2, Side 1, Ch. 3

Show this video of the demonstration on these pages to reinforce the concepts embodied in Boyle's law. L1 LEP

CD-Rom

Chemistry: Concepts and Applications
Demonstrating Boyle's Law
Use this video to reinforce the concepts embodied in Boyle's law. L1 LEP

Figure 11.8
Weather Balloons and Boyle's Law
As the balloon rises, the weight of the column of air above it is shorter, and the pressure on the helium gas in the balloon decreases. At an altitude of 15 km, atmospheric pressure is only 1/10 as great as at sea level. When a balloon that was filled with helium at 1.00 atm reaches this altitude, its volume is about ten times as large as it was on the ground.

The weather balloon in **Figure 11.8** illustrates Boyle's law. When helium gas is pumped into the balloon, it inflates, just as a soccer ball does, until the pressure of gas inside equals the pressure of the air outside. Because helium is less dense than air, the mass of helium gas is less than the mass of the same volume of air at the same temperature and pressure. As a result, the balloon rises. As it climbs to higher altitudes, atmospheric pressure becomes less. According to Boyle's law, because the pressure on the helium decreases, its volume increases. The balloon continues to rise until the pressures inside and outside are equal, when it hovers and records weather data.

Kinetic Explanation of Boyle's Law

You know that by compressing air in a tire pump, the air's pressure is increased and its volume is reduced. According to the kinetic theory, if the temperature of a gas is constant and the gas is compressed, its pressure must rise. Boyle's law, based on volume and pressure measurements made on confined gases at constant temperature, quantified this relationship by stating that the volume and pressure are inversely proportional. **Figure 11.9** shows how the kinetic theory explains Boyle's observations of gas behavior.

Figure 11.9
Modeling Boyle's Law
1. When the piston of the bicycle tire pump is pulled all the way out, the air pressure inside balances the air pressure outside.

3. When the piston is forced down farther and the air is compressed into one-fourth the volume of the pump, the frequency of collisions with the walls is four times as great. The air pressure inside the pump is 4 atm.

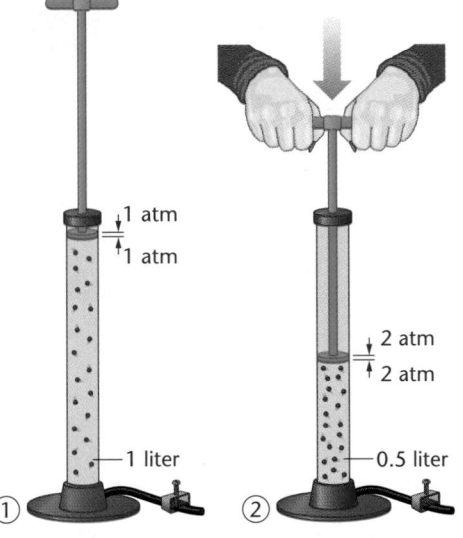

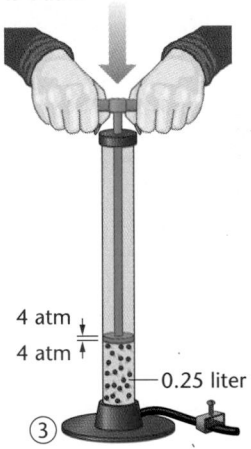

2. When the piston is forced halfway down, the average kinetic energy of the air particles is unchanged because the temperature is unchanged. They strike the wall with the same average force but, because they have been compressed into half the volume of the pump, the frequency of collisions with the walls doubles. The air pressure inside the pump is now 2 atm.

Demonstration 11.4

Demonstrating Boyle's Law
Purpose: To demonstrate Boyle's law.
Materials: empty, colorless, 2-L, soft-drink bottle with cap; dropper; food coloring

Safety Precautions: 🧤 🥽

Disposal: Code A; bottles and droppers can be reused, recycled, or disposed of in a landfill. Rinse water and dye down the drain.

Procedure: Remove the label from an empty, colorless, 2-L, soft-drink bottle. Add water up to 4 cm from the top. Fill a dropper halfway with water colored with vegetable dye. Put the dropper into the 2-L bottle and screw the cap on tight. Squeeze the sides of the bottle firmly until the dropper descends.
Results: When the bottle is squeezed, the air in the dropper is compressed and displaced by water. The level of the colored water in the

EARTH SCIENCE ▸ CONNECTION

Weather Balloons

If you watch the sky any day at 1 P.M. EST, you may see a weather balloon like the one in the photograph. Every 12 hours, an instrument-packed helium or hydrogen balloon is launched at each of 70 sites around the United States. The balloons provide the data used to make the country's weather forecasts.

Weather balloon system Since World War II, meteorologists have used weather balloons to provide profiles of temperature, pressure, relative humidity, and wind velocity in the upper atmosphere. Besides the 70 launching sites in this country, there are more than 700 other sites around the world. The balloons are released at the same time in all countries—at 6:00 and 18:00 hours Greenwich Mean Time (GMT). Data from weather stations all over the world reach the United States via a central computer in Maryland. From there, the data are sent to regional weather service stations throughout the country for release to newspapers and television and radio networks.

Sending the balloons aloft The rubber weather balloons are inflated with either helium or hydrogen to a diameter of about 2 m. The inflated balloons become buoyant because the density of either gas is less than the density of air. An instrument package is attached to each balloon by a 15-m cord. As the atmospheric pressure decreases, the gas inside the balloon expands, carrying the balloon higher. When it reaches a height of 30 km, the volume of the expanded gas is so great that the balloon bursts. Then a parachute carries the package containing the radio transmitter and measuring instruments safely to the ground.

Gathering data While aloft, devices in the instrument box record and transmit temperature, relative humidity, and barometric pressure.

Relative humidity readings are obtained from a device containing a polymer that swells when it becomes moist. The swelling causes an increase in the electrical resistance of a carbon layer in the polymer. Electrical measurements of the resistance of the material indirectly measure changes in relative humidity.

Wind velocity is determined by tracking the balloon by radar or by one of the radio location systems, known as LORAN-C, OMEGA, or VLF.

The instrument package transmits barometric, relative humidity, and temperature data to the ground by radio transmission using a multiplexing system. The system transmits the different kinds of data in rotation, for example, temperature first, followed by humidity, and then pressure.

Connecting to Chemistry

1. **Thinking Critically** Some people thought that when weather satellites were deployed, weather balloons would become obsolete. Why do you think this did not happen?

2. **Inferring** Why would a meteorologist have to understand the gas laws?

---

dropper rises, the dropper has more mass, and it descends.

Analysis: Ask these questions.

1. As the pressure is increased on the air in the dropper, does the volume increase or decrease? *The volume decreases.*

2. Is the relationship between gas volume and pressure an inverse or direct relationship? *inverse*

✔ Assessment

Knowledge: Have students write a statement of Boyle's law. *The pressure and volume of a confined gas at constant temperature are inversely related.* L1

EARTH SCIENCE ▸ CONNECTION

TECH PREP

Purpose
To relate the operation of weather balloons to the behavior of gases.

Teaching Strategies

- Discuss what is involved in forecasting the weather. *To make an accurate weather forecast for the East coast of the United States, a meteorologist must know what is happening to the weather over most of North America.*

- Point out that to forecast the weather three days in advance, a meteorologist must know what is happening over the entire northern hemisphere. For information about the weather five days in the future, forecasters need to study all of Earth's weather. Relate this to the importance of the worldwide weather balloon launchings that take place regularly.

Extension

Ask students to investigate zero-pressure balloons and how they can be regulated to float at a certain altitude. *As a zero-pressure balloon rises, the air pressure around the balloon decreases, causing it to expand. However, excess gas escapes through an open tube called the appendix, and this helps the balloon maintain a fixed altitude. The balloon can also be regulated by operators on the ground who release ballast in the form of steel grit until the desired altitude is reached.* L2

Connecting to Chemistry

1. The weather balloons can provide accurate weather information at a precise location. Weather satellites provide invaluable information, but it is less specific.

2. Because a meteorologist studies the atmosphere, he or she must understand how gases behave.

387

miniLAB

How Straws Function

Purpose: To determine how soda straws work.

Process Skills: Observing, inferring, communicating

Teaching Strategies

- Have a large number of soda straws on hand or have students bring in straws of their own.
- You could assign the Mini-Lab as homework.

Expected Results

Students will be unable to pull water into the straw.

Analysis

1. I was unable to lower the pressure in my mouth because I was drawing air in through one straw. Therefore, no air or water would rise up the other straw.
2. There was no air above the water in the jar to exert pressure on the surface of the liquid and force it up the straw.
3. Inhaling reduces the number of air molecules in the straw above the liquid, and the pressure in the straw goes down. Air pressure on the surface of the liquid pushes the liquid up the straw.

✓ Assessment

Oral: Ask students whether there is any limit to the height that water may be raised by a straw or similar devices. *Yes; normal air pressure can raise water only about 33 feet.* **L2**

miniLAB 2

How Straws Function

Each of us has used straws to sip soda, milk, or another liquid from a container. How do they function?

Procedure

1. Obtain a clean, empty food jar with a screw-on lid. Fill the jar halfway with tap water.
2. Put the ends of two soda straws into your mouth. Place the other end of one of the straws into the water in the jar and allow the end of the other straw to remain in the air. Try to draw water up the straw by sucking on both straws simultaneously. Record your observations.
3. Using a hammer and nail, punch a hole of about the same diameter as that of a straw in the lid of the jar. Push the straw 2 to 3 cm into the hole and seal the straw in position with wax or clay.
4. Fill the jar to the brim with water, and carefully screw the lid on so that no air enters the jar. Try to draw the water up the straw. Record your observations.

Analysis

1. Explain your observations in step 2.
2. Explain your observations in step 4.
3. Explain how a soda straw functions.

SAMPLE PROBLEM **3** **Boyle's Law: Determining Volume**

The maximum volume a weather balloon can reach without rupturing is 22 000 L. It is designed to reach an altitude of 30 km. At this altitude, the atmospheric pressure is 0.0125 atm. What maximum volume of helium gas should be used to inflate the balloon before it is launched?

Analyze
- At 30 km, the pressure exerted by the helium in the balloon equals the atmospheric pressure at that altitude, 0.0125 atm. What volume does this amount of helium occupy at 1 atm? The pressure is greater at high altitude by the factor 1.0 atm/0.0125 atm, which is greater than 1. By Boyle's law, the volume at launch is smaller than at high altitude by the factor 0.0125 atm/1.0 atm, which is less than 1.

Set Up
- Multiply the maximum volume by the factor given by Boyle's law.

$$V = 22\ 000\ \text{L} \times \left(\frac{0.0125\ \text{atm}}{1.0\ \text{atm}} \right)$$

Solve
- Multiply and divide the values and units.

$$V = \frac{22\ 000\ \text{L}}{} \left| \frac{0.0125\ \cancel{\text{atm}}}{1.0\ \cancel{\text{atm}}} \right. = \frac{22\ 000\ \text{L} \times 0.0125}{1.0} = 275\ \text{L}$$

Check
- As expected, the volume is less at sea level because the gas is compressed.

PRACTICE

 **Guided Practice:** Guide students through the sample problems, then have them work in class on Practice Problems 6-11. See Appendix D for complete solutions. **L1**

Independent Practice: Your homework or classroom assignment can include the Additional Practice Problems shown here. **L1**

Answers to Practice Problems

6. 55.3 mL
7. 200 L
8. 20 000 L
9. 0.087 L
10. 1210 L
11. 2.40 atm

Additional Practice Problems

1. Oxygen from the decomposition of hydrogen peroxide is collected in a 5.00-L glass bulb at a pressure of 740 mm Hg. It is then pumped into an evacuated container having a volume of 1.50 L. What is the pressure of the gas?

SAMPLE PROBLEM **4** **Boyle's Law: Determining Pressure**

Two liters of air at atmospheric pressure are compressed into the 0.45-L canister of a warning horn. If its temperature remains constant, what is the pressure of the compressed air?

Analyze • The initial pressure of the air that was forced into the canister was, of course, 1 atm. Because the volume of air is reduced, its pressure increases. Multiply the pressure by the factor with a value greater than 1.

Set Up • $P = 1.00 \text{ atm} \times \left(\dfrac{2.0 \text{ L}}{0.45 \text{ L}} \right)$

Solve • $P = \dfrac{1.00 \text{ atm}}{} \left| \dfrac{2.0 \cancel{L}}{0.45 \cancel{L}} \right. = \dfrac{1.00 \text{ atm} \times 2.0}{0.45} = 4.4 \text{ atm}$

Check • Estimate and reason: Because the volume of the air was reduced from 2 L to about half a liter, the volume changed by the factor $\frac{0.5}{2}$ or about $\frac{1}{4}$. Thus, the pressure changed by a factor of about $\frac{4}{1}$. The final pressure, 4.4 atm, is about four times as large as the initial pressure, 1.00 atm.

GLENCOE TECHNOLOGY

 CD-Rom

Chemistry: Concepts and Applications
Using the Gas Laws
Have students use this Interactive Exploration to enhance their understanding of Boyle's law and Charles's law. L1 LEP

PRACTICE PROBLEMS

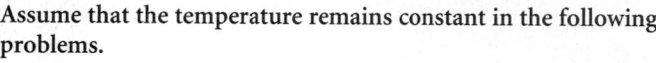

Assume that the temperature remains constant in the following problems.

6. Bacteria produce methane gas in sewage-treatment plants. This gas is often captured or burned. If a bacterial culture produces 60.0 mL of methane gas at 700.0 mm Hg, what volume would be produced at 760.0 mm Hg?

7. At one sewage-treatment plant, bacteria cultures produce 1000 L of methane gas per day at 1.0 atm pressure. What volume tank would be needed to store one day's production at 5.0 atm?

8. Hospitals buy 400-L cylinders of oxygen gas compressed at 150 atm. They administer oxygen to patients at 3.0 atm in a hyperbaric oxygen chamber. What volume of oxygen can a cylinder supply at this pressure?

9. If the valve in a tire pump with a volume of 0.78 L fails at a pressure of 9.00 atm, what would be the volume of air in the cylinder just before the valve fails?

10. The volume of a scuba tank is 10.0 L. It contains a mixture of nitrogen and oxygen at 290.0 atm. What volume of this mixture could the tank supply to a diver at 2.40 atm?

11. A 1.00-L balloon is filled with helium at 1.20 atm. If the balloon is squeezed into a 0.500-L beaker and doesn't burst, what is the pressure of the helium?

11.2 The Gas Laws **389**

$\dfrac{740 \text{ mm Hg}}{} \left| \dfrac{5.00 \cancel{L}}{1.50 \cancel{L}} \right. = 2470 \text{ mm Hg}$

2. An air-filled balloon has a volume of 0.500 L at sea level, where atmospheric pressure is 760 mm Hg. What is the volume of the balloon at the top of a mountain where atmospheric pressure is 745 mm Hg?

$\dfrac{0.500 \text{ L}}{} \left| \dfrac{760 \cancel{\text{mm Hg}}}{745 \cancel{\text{mm Hg}}} \right. = 0.510 \text{ L}$

Purpose
TECH PREP Students will learn various medical uses of hyperbaric oxygen chambers.

Background
Hyperbaric oxygen therapy, over periods of about two hours, is often used for patients with gangrene and tetanus. The microbes that cause these diseases are anaerobic—they cannot grow in an oxygen environment. The presence of concentrated oxygen during HBO slows or even stops the spread of the infection.

Teaching Strategies
- Ask students to brainstorm other uses of the hyperbaric oxygen chamber for healing. They may mention that it might be used successfully as a treatment for carbon monoxide poisoning because the combination of increased pressure and pure oxygen may force the carbon monoxide off the hemoglobin molecules to be replaced by oxygen. **L1**
- Discuss with students the information about the treatment of gangrene and tetanus found in *Background* above. Students may wonder how anaerobic bacteria can live in the human body, where the oxygen supply is constantly replenished. Point out that when body tissue is severely damaged, the tissue dies and is cut off from oxygen supplied by the circulatory system. The environment becomes anaerobic, and if the anaerobic pathogens that cause tetanus or gangrene are present, infection may occur.

Discussing the Technology
1. Healing occurs because the air pressure is three times higher than normal and pure oxygen is forced to the injured part of the body.

390

Hyperbaric Oxygen Chambers

Hyperbaric oxygen (HBO) chambers have been used since the 1940s. They were used then by the Navy to treat divers with decompression sickness. In the last 20 years, researchers have shown that HBO treatment has many more medical applications. One of these applications is healing bone and muscle injuries.

Healing Sports Injuries

The Dallas Cowboys, San Francisco 49ers, and New York Jets football teams have each recently acquired an HBO unit. The units have been dubbed *space capsules* because of their appearance. A player with a severe sprain, ordinarily dooming him to warm the bench for several

weeks, may instead spend three one-hour sessions in the HBO. The device compresses the air to 3 atm of pressure around the injured player, who is breathing pure oxygen through a mask. The higher air pressure in the HBO forces more oxygen to dissolve in the bloodstream, speeding up the flow of oxygen to 15 times the normal rate. The abnormal oxygen flow causes the blood vessels in the injured muscles to constrict, limiting swelling in that area. In two days, the injured player may walk without crutches. By the end of the week, the player could be back on the field.

After a Heart Attack

HBO therapy is also being used along with clot-dissolving drugs to save heart muscle and improve the quality of life for patients who have suffered heart attacks. In a study of heart-attack patients, half were given t-PA, a drug that prevents clots. The other half received the same drug, followed by two hours of HBO treatment. The patients who received the drug alone had lingering chest pains for an average of 10.75 hours. Those who had HBO therapy experienced pains for only 4.5 hours. In addition, normal electrical activity was restored sooner in the hearts of the patients who were in the hyperbaric chamber. Their electrocardiograms were normal in ten to 15 minutes—half the time it took for the other group to approach that goal.

Breath of Fresh Air for Premature Babies

One of the most important uses of HBO is to treat babies with hyaline-membrane disease. These babies suffer from respiratory distress soon after birth because the alveoli in their lungs fail to inflate. HBO therapy has increased the chances that the lungs will normalize so that these babies can survive.

DISCUSSING THE TECHNOLOGY

1. **Thinking Critically** What two factors cause the remarkable effects of HBO? Explain how.

2. **Hypothesizing** How might HBO be effective for treating second- and third-degree burns over large portions of the body?

Oxygen is needed for the growth of new cells, and HBO therapy delivers oxygen to injured parts at 15 times the normal rate.
2. HBO may be effective for treating second- and third-degree burns over large portions of the body because it facilitates rapid delivery of oxygen to the tissues of the body. It may stimulate the growth of new tissue to replace the damaged tissue.

Charles's Law: Temperature and Volume

You may have observed the beautiful patterns and graceful gliding of a hot-air balloon, but what happens to a balloon when it is cold? **Figure 11.10** shows some dramatic effects of cooling and warming a gas-filled balloon.

The French scientist Jacques Charles (1746-1823) didn't have liquid nitrogen, but he was a pioneer in hot-air ballooning. He investigated how changing the temperature of a fixed amount of gas at constant pressure affected its volume. The relationship Charles found can be demonstrated, as shown in **Figure 11.11**.

Figure 11.10

Cooling and Warming a Gas-Filled Balloon

At 77 K, the nitrogen in the beaker is so cold it is a liquid. It rapidly cools the air-filled balloons, shrinking them by reducing the volume of the balloons. When the balloons are removed from the beaker and the temperature of the air inside rises to room temperature, the balloons expand.

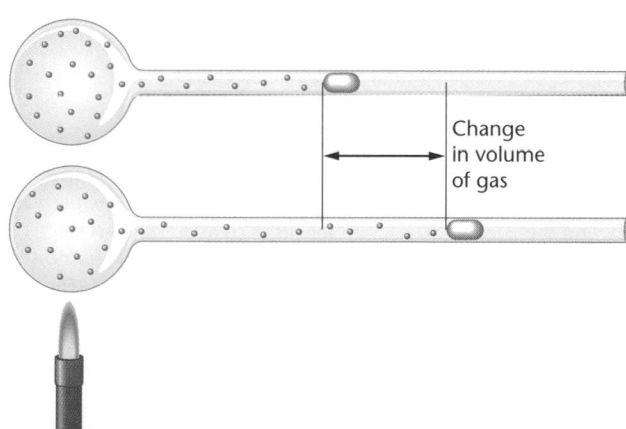

Change in volume of gas

Figure 11.11

Demonstrating Charles's Law

The mercury plug is free to move back and forth in the horizontal tube because the end of the glass tube is open to the atmosphere. The pressure of the gas inside the bulb is always equal to the atmospheric pressure. The distance the mercury plug moves right or left measures the increase or decrease in the volume of gas as it is heated or cooled.

11.2   The Gas Laws   **391**

Cultural Diversity

Inventor Garrett Morgan

Born in Kentucky in 1875, Garrett Morgan moved to Ohio at the age of 14, having received only six years of schooling. He worked as a sewing-machine adjuster and later opened his own sewing-machine repair business. In 1913, Morgan discovered a process for straightening hair. The discovery made him wealthy and allowed him to continue to develop new products. One of these, a gas mask, won grand prize at a New York Safety and Sanitation fair, but it was essentially ignored until 30 workers were trapped in a tunnel filled with poisonous gases below Lake Erie. Garrett used his breathing device to rescue the workers. Fire departments saw the value of Morgan's gas mask, but that did not mean ready acceptance. Because Morgan was African American, he had to have a white man represent him while he acted as a Native American assistant. As a result of his experiences dealing with the white business world, Morgan founded the *Cleveland Call* to meet the needs of African Americans.

Figure 11.12
Volume and Temperature for Three Gases
The three straight lines show that the volume of each gas is directly proportional to its Kelvin temperature. The solid part of each line repre-sents actual data. Part of each line is dashed because, as you recall from Chapter 10, when the temperature of a gas falls below its boiling point, a gas condenses to a liquid.

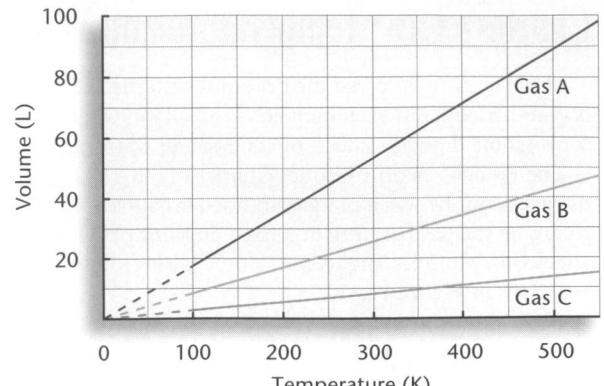

Charles's law states that at constant pressure, the volume of a gas is directly proportional to its Kelvin temperature, as shown in the graph in **Figure 11.12**. The straight lines for each gas indicate that volume and temperature are in direct proportions. For example, if the Kelvin tempera-ture doubles, the volume doubles, and if the Kelvin temperature is halved, the volume is halved.

Kinetic Explanation of Charles's Law

Why did the air in the balloons expand when heated and contract when cooled? Study **Figure 11.13** to learn how the kinetic theory of matter explains Charles's law.

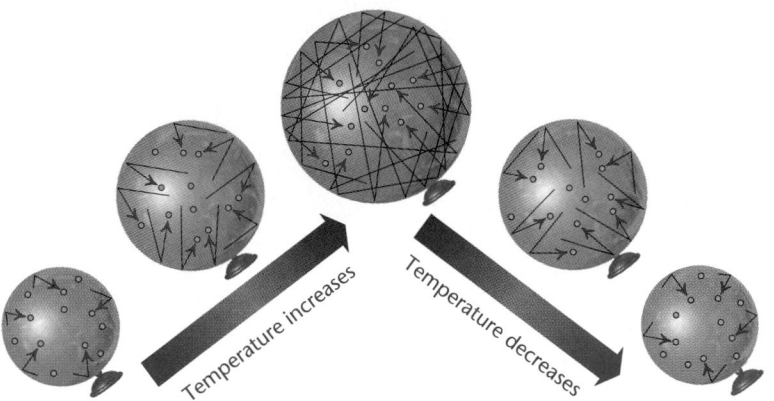

Figure 11.13
Modeling Charles's Law
▲ When the balloon is heated, the tempera-ture of the air inside increases, and the average kinetic energy of the particles in the air also increases. They exert more force on the balloon, but the pressure inside does not rise above the original pressure because the balloon expands.

▲ When the balloon cools, the tempera-ture of the air inside falls and the average kinetic energy of the particles in the air decreases. The particles move slower and strike the balloon less often and with less force. The bal-loon contracts and the pressure of the air inside the balloon continues to bal-ance the pressure of the atmosphere.

392   Chapter 11   **Behavior of Gases**

Additional Practice Problems

Assume that pressure remains constant in the following problems.
1. A sample of air in a piston at 25°C occupies 35 mL. What volume will it occupy if the temperature is raised to 250°C?

$$\frac{35 \ mL}{} \ \frac{523 \ K}{298 \ K} = 61 \ mL$$

SAMPLE PROBLEM 5 Charles's Law

A balloon is filled with 3.0 L of helium at 22°C and 760 mm Hg. It is then placed outdoors on a hot summer day when the temperature is 31°C. If the pressure remains constant, what will the volume of the balloon be?

Analyze
- Because the volume of a gas is proportional to its Kelvin temperature, you must first express the temperatures in this problem in kelvins. As in Chapter 10, add 273 to the Celsius temperature to obtain the Kelvin temperature.

> **Problem-Solving HINT**
>
> Remember that gas volumes and pressures are proportional to temperature only if the temperature is expressed in kelvins.

$$T_K = T_C + 273$$
$$T_K = 22 + 273 = 295 \text{ K}$$
$$T_K = 31 + 273 = 304 \text{ K}$$

Because the temperature of the helium increases from 295 K to 304 K, its volume increases in direct proportion. The temperature increases by the factor 304 K/295 K. Therefore, the volume increases by the same factor.

Set Up
- $V = 3.0 \text{ L} \times \left(\dfrac{304 \text{ K}}{295 \text{ K}} \right)$

Solve
- $V = \dfrac{3.0 \text{ L}}{} \dfrac{304 \text{ K}}{295 \text{ K}} = \dfrac{3.0 \text{ L} \times 304}{295} = 3.1 \text{ L}$

Check
- Does the answer have volume units? Does the volume increase as expected? Because the answer to both questions is yes, this solution is reasonable. Check your calculations to make sure your answer is correct.

PRACTICE PROBLEMS

Assume that the pressure remains constant in the following problems.

12. A balloon is filled with 3.0 L of helium at 310 K and 1 atm. The balloon is placed in an oven where the temperature reaches 340 K. What is the new volume of the balloon?

13. A 4.0-L sample of methane gas is collected at 30.0°C. Predict the volume of the sample at 0°C.

14. A 25-L sample of nitrogen is heated from 110°C to 260°C. What volume will the sample occupy at the higher temperature?

15. The volume of a 16-g sample of oxygen is 11.2 L at 273 K and 1.00 atm. Predict the volume of the sample at 409 K.

16. The volume of a sample of argon is 8.5 mL at 15°C and 101 kPa. What will its volume be at 0.00°C and 101 kPa?

11.2 The Gas Laws   393

2. A sample of hydrogen gas collected from the reaction of magnesium and HCl occupies a volume of 125 mL at 300 K. Predict its volume at standard temperature.

$$\dfrac{125 \text{ mL}}{} \dfrac{273 \text{ K}}{300 \text{ K}} = 114 \text{ mL}$$

Combined Gas Law

You know that according to Boyle's law, if you double the volume of a gas while keeping the temperature constant, the pressure falls to half its initial value. You also know that according to Charles's law, if you double the Kelvin temperature of a gas while keeping the pressure constant, the volume doubles. What do you think would happen to the pressure if you doubled the volume and doubled the temperature? Suppose, like Robert Boyle and Jacques Charles, you were to investigate how gases behave. You would conduct experiments with gases just as they did. You would measure temperature, pressure, and volume for a sample of gas; expand the gas to twice the volume; raise the temperature to twice as high; and then measure the pressure. Following accepted principles of scientific methods, you would make many such experiments. You might choose to triple the volume and the temperature or change the volume and temperature by half or less to get a wide range of data. You might then use a computer to help make a graph of your data and look for relationships among your three variables.

Another approach might be to first double the volume while keeping the temperature constant or to double the volume and temperature and then measure the pressure. In what other ways could you compare relationships among temperature, pressure, and volume? Any one of the variables could be kept constant while varying another and measuring the effect on the third. You would find that doubling the volume first and the temperature second has exactly the same result as doubling the temperature first and the volume second. These results should not be surprising when you remember that you can read the curve in the graph for Boyle's law in either direction. Whether the gas was expanding or contracting, if you knew its volume you could determine its pressure. If you knew its pressure, you could determine its volume. One application of these gas laws can be seen at hot-air balloon events, **Figure 11.14**.

Figure 11.14
Flying High
Hot-air ballooning is a popular sport all around the world. The height of the balloon over the ground is controlled by varying the temperature of the air within the balloon.

*inter*NET
CONNECTION
A software tutorial is available on the Internet for gas laws.
Anonymous FTP

ftp://truth.chem.sfu.ca/pub/chem1/mac/gasesdemo.sea

Because Boyle's law of gases and Charles's law of gases are equally valid, it is possible to determine one of the three variables, regardless of the order in which the other two are changed. If you doubled the volume and the temperature at the same time, you would get exactly the same result as if you had doubled one first, then the other. For example, suppose you had 3 L of gas at 200 K and 1 atm. If you double the volume, according to Boyle's law, the pressure falls to 0.5 atm. If you then double the temperature, according to Charles's law, the pressure is increased to twice as high again, or 1 atm. Note that doubling the volume and then doubling the temperature brings the sample back to its initial pressure. The effect on the pressure of doubling the volume and doubling the temperature offset each other because pressure is inversely proportional to volume but directly proportional to temperature.

The combination of Boyle's law and Charles's law is called the **combined gas law.** The factors are the same as in the previous sample problems, but you may have more than one factor in a problem because more than one quantity may vary. The set of conditions 0.00°C and 1 atm is so often used that it is called **standard temperature and pressure** or **STP.**

SAMPLE PROBLEM **6** | **Determining Volumes at STP**

A 154-mL sample of carbon dioxide gas is generated by burning graphite in pure oxygen. If the pressure of the generated gas is 121 kPa and its temperature is 117°C, what volume would the gas occupy at standard temperature and pressure, STP?

Analyze
- Reducing the pressure from 121 kPa to 101 kPa increases the volume of the carbon dioxide gas. Therefore, Boyle's law gives the factor 121 kPa/101 kPa. Because this factor is greater than 1, when it is multiplied by the volume, the volume increases. Cooling the gas from 117°C to 0.00°C reduces the volume of gas. To apply Charles's law, you must express both temperatures in kelvins.

$$T_K = T_C + 273 \qquad\qquad T_K = T_C + 273$$
$$= 117 + 273 \qquad\qquad\ = 0.00 + 273$$
$$= 390\ K \qquad\qquad\qquad = 273\ K$$

Because the temperature decreases, the volume decreases. Charles's law gives the factor 273 K/390 K, which is less than 1. Therefore, multiplying the volume by this factor decreases the volume.

Set Up
- Multiply the volume by these two factors.

$$V = 154\ mL \times \left(\frac{121\ kPa}{101\ kPa}\right) \times \left(\frac{273\ K}{390\ K}\right)$$

Solve
- The combined gas law equation is solved in the following steps.

$$V = \frac{154\ mL}{} \left| \frac{121\ kPa}{101\ kPa} \right| \frac{273\ K}{390\ K} =$$

$$\frac{154\ mL \times 121 \times 273}{101 \times 390} = 129\ mL$$

11.2 **The Gas Laws** **395**

Correcting Misconceptions

If students round their answers to combined gas law problems to the correct numbers of significant digits, slight differences in answers may result. To minimize rounding errors, advise students to carry several extra digits until the calculations are complete, then round the final answers to the correct number of digits.

Discussion

Ask students if each of the following descriptions is a satisfactory expression of the amount of matter: (1) 25.0 mL of water, (2) 20.0 L of carbon dioxide, and (3) a lead cube measuring 2.80 cm on an edge. *1 and 3 are satisfactory because the volumes of liquid water and solid lead are not changed significantly by changes in temperature and pressure. 2 is unsatisfactory because CO_2 is a gas and the volume of a gas varies with temperature and pressure. An adequate description of the amount of matter for gases must state the temperature and pressure.* L1

Correcting Misconceptions

Students may think that the volume of the gas molecules is the volume of the gas. Point out that the molecules themselves occupy very little of the volume occupied by the gas.

Across the Curriculum

History

Airships, also known as blimps and dirigibles, were a development of the hot-air balloon. These lighter-than-air craft consist of a balloon that contains the gas, usually helium, an engine that propels the ship, and one or more gondolas for the passengers and crew. The French engineer Henri Giffard constructed the first successful airship in 1852. Ferdinand von Zeppelin built a rigid-frame airship in 1900 that became the prototype for later models of airships. von Zeppelin's ship was driven by two 15- horsepower Daimler internal combustion engines. In its first flight, the ship carried five passengers, reached an altitude of 396 m, and traveled 6 km. In the period from 1900 to 1937, airships traveled from Europe to America and were routinely used in intercontinental travel. For example, the *Los Angeles*, an airship made for the U.S. Navy by the Zeppelin airship works in Germany, made nearly 250 flights in the United States. In 1937, the Hindenburg, after making ten transatlantic crossings, crashed and was destroyed by fire while landing at Lakehurst, New Jersey.

Concept Development

If the gases are available, put 1 L of argon, methane, hydrogen, and nitrogen in separate balloons and write the identifying symbols on the balloons. If the gases are unavailable, sketch the balloons on the chalkboard. Ask students whether the masses of the gases are equal. *no* The volumes? *yes* The numbers of molecules? *yes* Tell students that there are 2.5×10^{22} molecules if the pressure in the balloons is 1.0 atm and the temperature is 298 K. [L1]

Content Background

Molar Volume The converse of Avogadro's principle—equal numbers of gas particles at the same temperature and pressure have the same volume—allows you to reinforce the mole concept by showing how equal numbers of molecules of different gases can have the same volume and different masses. For example, 6.02×10^{23} molecules of oxygen at STP have the same volume as 6.02×10^{23} molecules of carbon dioxide, but the mass of 6.02×10^{23} molecules of oxygen is 32 g, while the mass of 6.02×10^{23} molecules of carbon dioxide is 44 g.

Check
- Estimate to see whether the answer is reasonable. The reduction in pressure would expand the volume by a factor of about 12/10. Cooling the gas would contract it by a factor of about 7/10. Both factors together would alter the volume by a factor of about 84/100, which is less than 1. The final volume should be less than the initial volume. The final volume, 129 mL, is less than 154 mL, the initial volume.

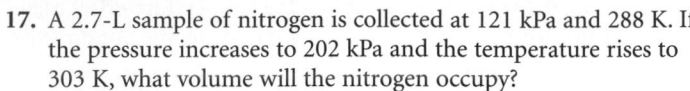

PRACTICE PROBLEMS

17. A 2.7-L sample of nitrogen is collected at 121 kPa and 288 K. If the pressure increases to 202 kPa and the temperature rises to 303 K, what volume will the nitrogen occupy?

18. A chunk of subliming carbon dioxide (dry ice) generates a 0.80-L sample of gaseous CO_2 at 22°C and 720 mm Hg. What volume will the carbon dioxide gas have at STP?

Fact of the MATTER

The law of combining volumes was first stated by the French chemist Joseph Louis Gay-Lussac (1778-1850) in 1809. Among his many pursuits, Gay-Lussac was an avid hot-air balloonist, an activity that was popular in early 19th-century France.

The Law of Combining Gas Volumes

When water decomposes into its elements—hydrogen and oxygen gas—the volume of hydrogen produced is always twice the volume of oxygen produced. Because matter is conserved, in the reverse synthesis reaction, the volume of hydrogen gas that reacts is always twice the volume of oxygen gas.

Experiments with many other gas reactions show that volumes of gases always react in ratios of small whole numbers. **Figure 11.15** presents the combining volumes for a synthesis reaction and a decomposition reaction. The observation that at the same temperature and pressure, volumes of gases combine or decompose in ratios of small whole numbers is called the **law of combining gas volumes.**

Figure 11.15

Comparing Volumes in Gas Reactions
When two liters of hydrogen chloride gas decompose to form hydrogen gas and chlorine gas, equal volumes of hydrogen gas and chlorine gas are formed—1 L of each. The ratio of hydrogen to chlorine is 1 to 1, and the ratio of volumes of hydrogen to hydrogen chloride is 1/2 to 1 or 1 to 2, the same as the ratio of chlorine to hydrogen chloride. Consider the reverse reaction—the composition of hydrogen gas with chlorine gas to form hydrogen chloride gas. What is the ratio of the reactants and the ratio of the product to each reactant?

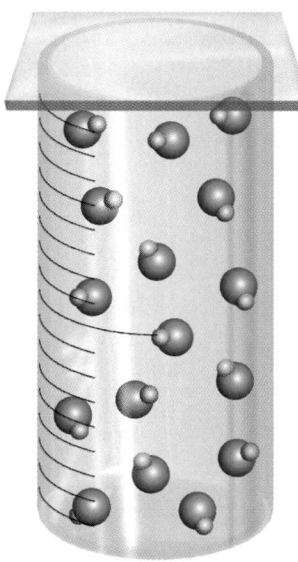

2 liters HCl

PRACTICE

 **Guided Practice:** Guide students through the sample problem, then have them work in class on Practice Problems 17-18. See Appendix D for complete solutions. [L1]

Independent Practice: Your homework or classroom assignment may include the additional practice problems shown here. [L1]

Answers to Practice Problems
17. 1.7 L
18. 0.70 L

Additional Practice Problems

1. A sample of gas was collected in a 2.5-L bulb at 295 K and 99 kPa. What volume would the gas occupy at 273 K and 101 kPa?

$$\frac{2.5\ L}{}\ \Big|\ \frac{273\ K}{295\ K}\ \Big|\ \frac{99\ kPa}{101\ kPa} = 2.3\ L$$

2. At −30°C and 200 kPa, a gas occupies a volume of 15.0 L. What volume would it occupy at STP?

$$\frac{15.0\ L}{}\ \Big|\ \frac{273\ K}{243\ K}\ \Big|\ \frac{200\ kPa}{101\ kPa} = 33.4\ L$$

Everyday Chemistry

Popping Corn

You easily recognize the smell, and that characteristic popping noise is a dead giveaway. Popcorn! What causes the kernels of popcorn to go through their rapid, explosive changes?

History of popcorn The ancestor of the popcorn you eat was cultivated in the New World by Native Americans more than 7000 years ago. Popcorn was used as food, as decoration, and in religious ceremonies. It was one of the foods shared with the Pilgrims during the first Thanksgiving.

Popcorn kernels Popcorn kernels are extremely small and hard. They are wrapped in a tough, shell-like covering called the hull. The hull protects the embryo and its food supply. This food supply is starch located within the endosperm. Each kernel also contains a small amount of water.

Exploding the kernel When heated to about 204°C, the water in the kernel turns to steam. The expansion of the steam rips the remarkably tough hull with an explosive force, and the popcorn bursts open to 30 to 40 times its original size. The

heat released by the steam bakes the starch into the fluffy product people like to eat.

Water content The amount of water in the kernel is an important aspect of popcorn popping. Food chemists have found that the kernel must contain about 13.5 percent water by mass to pop properly.

Exploring Further

1. **Hypothesizing** Suggest why too much or too little water present in the kernel greatly increases the number of unpopped kernels.

2. **Applying** Why is popcorn better stored in the freezer or refrigerator rather than on the shelf at room temperature?

3. **Acquiring Information** More than 1000 varieties of corn are grown in the world today. Write a report about as many uses for corn and corn products as you can.

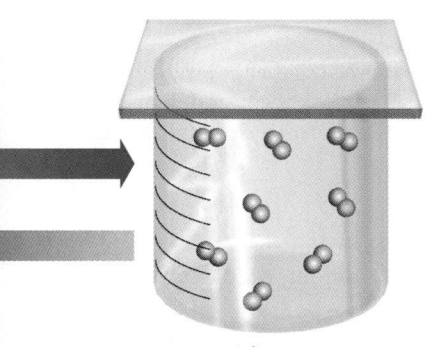

1 liter H_2

\+

1 liter Cl_2

11.2 The Gas Laws   **397**

Exploring Further

1. Too much water makes the tough endosperm flexible so that it does not hold back the steam. Too little water prevents adequate steam production.

2. Popcorn stored at room temperature has a greater tendency to dry out.

3. The germ provides oil cake for cattle feed; crude corn oil for making soap, glycerin, and plastic resins; and refined corn oil. The hull provides bran for cattle feed. The endosperm provides gluten for cattle feed and

raw starch for making corn syrup, corn sugar, industrial starch, dextrin, and edible starch. Reports may be placed into students' portfolios. **L1** **P**

Everyday Chemistry

Purpose

TECH PREP

Students will learn the characteristics of popcorn and the conditions under which it pops.

Teaching Strategies

- Discuss on the molecular level the difference between liquids and gases. Before reviewing the article in class, borrow a portable microwave oven. Have a student measure the size of an unpopped envelope of microwave popcorn and then pop the corn and re-measure the new size of the bag. Let students explain where the steam comes from, why the bag enlarges, and why the manufacturer put a hole in the bag. **L1**

- Discuss the large heat of vaporization of water ($\Delta H_{vap} = 2.26 \times 10^6$ J/kg). Compare it to the specific heat of water ($c = 4180$ J/kg·°C). Help students understand that only 4180 J are needed to raise the temperature of 1 kg of water from 99°C to 100°C, but that 540 times as much energy is needed to change 1 kg of liquid water to gas at 100°C.

Extension

Have teams of students perform consumer popcorn research using three to five different brands of popcorn and a hot-air popper. Have them make a table comparing the following information: (1) number of kernels used (keep constant), (2) mass before popping, (3) length of time needed to pop, (4) mass after popping, (5) number of unpopped kernels, and (6) percent of mass lost during popping. **L1** **COOP LEARN**

Check for Understanding

- Check for understanding of the section material by having students answer the Section Review questions.
- Use a pressure-temperature bulb in a large beaker of water at various temperatures to demonstrate the direct relationship between temperature and pressure at constant volume. Using data obtained from this demonstration, have students make a graph of pressure (*y*-axis) versus temperature in kelvins (*x*-axis). Have them extrapolate the data to zero pressure and explain the meaning of the result. L1

Reteach

Software: *Gas Law 542.* Lab simulation in which students control volume, temperature, and pressure. AP403, PC2602, Project SERAPHIM.

Extension

TECH PREP Ask a representative of a local scuba group to give a classroom presentation on the application of the gas laws to scuba diving.

4 CLOSE

Writing About Chemistry

Ask students to write a story in which they draw analogies between the atoms or molecules of solids and persons living in a city, liquids and persons living in the suburbs, and gases and persons living in the country. Stories may be placed into students' portfolios. L1 P

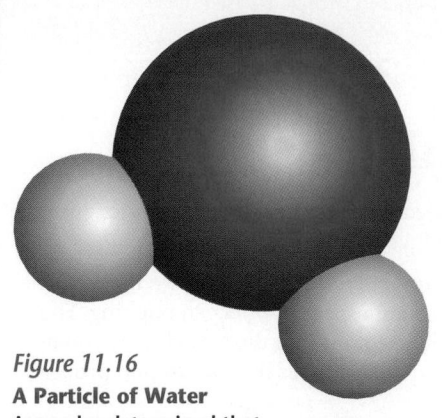

Figure 11.16
A Particle of Water
Avogadro determined that water is composed of particles.

Fact of the **MATTER**

Because of his observations of reactions involving oxygen, nitrogen, and hydrogen gases, Avogadro was the first to suggest that these gases were made up of diatomic molecules.

In the synthesis of water, two volumes of hydrogen react with one of oxygen, but only two volumes of water vapor are formed. You might expect three. In the reverse reaction, which is the decomposition of water, two volumes of water vapor yield one volume of oxygen gas and two volumes of hydrogen gas. That mysterious third volume appears again. How can we account for it? Amedeo Avogadro (1776-1856), an Italian physicist, made the same observations and asked the same question. He first noticed that when water forms oxygen and hydrogen, two volumes of gas become three volumes of gas, and he hypothesized that water vapor consisted of particles, as shown in **Figure 11.16.** Each particle in the water vapor broke up into two hydrogen parts and one oxygen part. Two volumes of hydrogen and one volume of oxygen formed because there were twice as many hydrogen particles as oxygen particles. Today, we can show this by doing something that Avogadro could not do—write the chemical equation for the formation of water.

$$2H_2(g) + O_2(g) \rightarrow 2H_2O(g)$$

Avogadro was the first to interpret the law of combining volumes in terms of interacting particles. He reasoned that the volume of a gas at a given temperature and pressure must depend on the number of gas particles. **Avogadro's principle** states that equal volumes of gases at the same temperature and pressure contain equal numbers of particles.

Connecting Ideas

Knowing that equal volumes of gases at the same temperature and pressure have the same number of particles is the first step toward knowing how many particles are present in a sample of a substance. Knowing the number of particles in a reactant and the ratio in which the particles interact enables us to predict the amount of a product. This important number has been estimated.

SECTION REVIEW

Understanding Concepts

1. The gas laws deal with the following variables—the number of gas particles, temperature, pressure, and volume. Which of these are held constant in Boyle's law, in Charles's law, and in the combined gas law?

2. Explain how an air mattress supports the weight of a person lying on it.

3. Why is it important to determine the volume of a gas at STP?

Thinking Critically

4. **Analyzing** If the volume of a helium-filled balloon increases by 20 percent at constant temperature, what will the percentage change in the pressure be?

Applying Chemistry

5. **Aerosol Cans** Use the kinetic theory to explain why pressurized cans carry the message "Do not incinerate."

SECTION REVIEW

1. Boyle's law, number of gas particles and temperature; Charles's law, number of gas particles and pressure; combined gas law, number of particles

2. Lying on a mattress decreases the volume of the mattress and increases the pressure. The increased pressure supports the weight added.

3. Determining the volume of a gas at STP allows it to be compared to other gases.

4. $$P = 100\% \frac{100\%}{120\%} = \frac{100\%}{1} \times \frac{100\%}{120\%}$$
$$= \frac{100\% \times 100}{120} = 83\%$$

Pressure changes by $100\% - 83\% = 17\%$.

5. Heating increases the average kinetic energy of the gas particles and the pressure. If the can continues to be heated, the pressure of the gas may exceed the strength of the container and explode.

REVIEWING MAIN IDEAS

11.1 Gas Pressure

- The pressure of a gas at constant temperature and volume is directly proportional to the number of gas particles.

- The volume of a gas at constant temperature and pressure is directly proportional to the number of gas particles.

- At sea level, the pressure exerted by gases of the atmosphere equals one standard atmosphere (1 atm).

- Gas pressure can be expressed in the following units: atmospheres (atm), millimeters of mercury (mm Hg), inches of mercury (in. Hg), pascals (Pa), and kilopascals (kPa).

- The factor label method is a simple tool for converting measurements from one unit to another.

11.2 The Gas Laws

- Boyle's law states that the pressure and volume of a confined gas are inversely proportional.

- Charles's law states that the volume of any sample of gas at constant pressure is directly proportional to its Kelvin temperature.

- According to the combined gas law, when pressure and temperature are both changing, both Boyle's and Charles's laws can be applied independently.

- The law of combining gas volumes states that in chemical reactions involving gases, the ratio of the gas volumes is a small whole number.

- Avogadro's principle states that equal volumes of gases at the same temperature and pressure contain equal numbers of particles.

Key Terms

For each of the following terms, write a sentence that shows your understanding of its meaning.

Avogadro's principle
barometer
Boyle's law
Charles's law
combined gas law
factor label method
kilopascal (kPa)
law of combining gas volumes
pascal (Pa)
standard atmosphere (atm)
standard temperature and pressure, STP

● UNDERSTANDING CONCEPTS

1. Name three ways to increase the pressure inside an oxygen tank.

2. What Celsius temperature corresponds to absolute zero?

3. In a chemical plant, xenon hexafluoride is stored at STP. Express these conditions using the Kelvin scale.

4. An oxygen storage tank contains 12.0 L at 25°C and 3 atm. After a 4-L canister of oxygen at 25°C and 3 atm is emptied into the storage tank, what do you expect the temperature and pressure of the tank to be?

5. At 1250 mm Hg and 75°C, the volume of a sample of ammonia gas is 6.28 L. What volume would the ammonia occupy at STP?

6. What physical conditions are specified by the term STP?

7. What factors affect gas pressure?

● APPLYING CONCEPTS

8. The passenger cabin of an airplane is pressurized. Explain what this term means and why it is done.

9. Explain why balloons filled with helium rise in air but balloons filled with carbon dioxide sink.

UNDERSTANDING CONCEPTS

- Review the Reviewing Main Ideas statements and Key Terms with your students.
- Complete solutions to Chapter Review problems can be found in the *Problems and Solutions Manual.*

UNDERSTANDING CONCEPTS

1. raise the temperature, decrease the volume, and add more particles of gas
2. −273°C
3. 273 K and 1 atm
4. 25°C; 3 atm × 16 L/12 L = 4 atm
5. 8.10 L
6. A gas is at 0°C and 1 atm pressure.
7. temperature, number of particles, and volume

APPLYING CONCEPTS

8. The pressure of the atmosphere decreases as the altitude increases. To maintain a comfortable environment, the passenger compartment must be maintained at a higher pressure than the outside air.
9. The density of helium is less than the density of air, and the density of carbon dioxide is greater than the density of air.
10. 1100 mL
11. 100 mL
12. 2.0 L
13. 37 mL
14. 1370 mL
15. 13 mL
16. 10 L
17. 15 times greater
18. Density is mass divided by volume. A popped kernel has a slightly smaller mass and a much larger volume than an unpopped kernel, so the popped kernel has a lower density.
19. No; chlorine is heavier than air and poisonous.

✔ Assessment

Portfolio: Review the portfolio options that are provided throughout the chapter. Encourage students to select one product that demonstrates their best work for the chapter. Have students explain what they have learned and why they chose this example for placement into their portfolios. **P**

Additional portfolio options may be found in the following **Teacher Classroom Resources:**
Tech Prep Applications
Consumer Chemistry
Applying Scientific Methods in Chemistry

THINKING CRITICALLY

20. Increasing pressure (turns of clamp or addition of mercury) results in decreasing volume (height of the gas in a tube).

21. The cold temperature made the closure brittle or the closure was not strong enough to withstand the pressure.

22. The pressure of the atmosphere holds the liquid in the pipet because the space above the liquid is essentially a vacuum and, therefore, there is little opposing pressure beyond the weight of the liquid.

23. Density increases because Boyle's law states that increasing pressure results in decreasing volume. The same mass is confined in a smaller volume.

CUMULATIVE REVIEW

24. Allotropes are forms of an element that differ in crystalline or molecular structure. Examples are carbon, sulfur, phosphorus, and oxygen.

25. Yes, the second energy sublevel or p sublevel holds a maximum of six electrons.

26. a. polar covalent **b.** ionic **c.** covalent **d.** polar covalent

27. a. increases the boiling point **b.** slightly decreases the melting point

GLENCOE TECHNOLOGY

 Videotape

MindJogger Videoquizzes for Chemistry
Chapter 11: Behavior of Gases
Have students work in groups as they play the videoquiz game to review key chapter concepts. L1 | LEP | COOP LEARN

10. The volume of a gas is 550 mL at 760 mm Hg. If the pressure is reduced to 380 mm Hg and the temperature is constant, what will be the new volume?

11. The volume of a gas at 101 kPa is 400 mL. If the pressure changes at constant temperature to 404 kPa, what will be the new volume?

12. The volume of a gas is 1.5 L at 27°C and 1 atm. What volume will the gas occupy if the temperature is raised to 127°C at constant pressure?

13. A 50-mL sample of krypton gas at STP is cooled to −73°C at constant pressure. What will be the volume?

14. A sample of carbon dioxide is collected at STP. Its volume is 500 L. What volume will it have if the pressure is 380 mm Hg and the temperature is 100°C?

15. A 120-mL sample of nitrogen is collected at 900 mm Hg and 0°C. What volume will the nitrogen have when its temperature rises to 100°C and the pressure increases to 15 atm?

16. How many liters of hydrogen will be needed to react completely with 10 L of oxygen in the composition reaction of hydrogen peroxide?

Chemistry and Technology

17. Compare the amount of oxygen you would receive in an HBO chamber with the amount you would receive breathing ordinary air.

Everyday Chemistry

18. Explain why the density of a popped kernel of popcorn is less than that of an unpopped kernel.

Earth Science Connection

19. Would chlorine gas be suitable for weather balloons? Explain.

● THINKING CRITICALLY

Comparing and Contrasting

20. ChemLab How are inferences made about the relationship between volume and pressure in

this chapter's ChemLab similar to inferences made from observing the apparatus described in **Figure 11.6**?

Observing and Inferring

21. MiniLab 1 How might you account for the possible popping of the bag containing the subliming dry ice?

Applying Concepts

22. MiniLab 2 Explain how air pressure helps you transfer a liquid with a pipet.

Making Predictions

23. Use Boyle's law to predict the change in density of a sample of air if its pressure is increased.

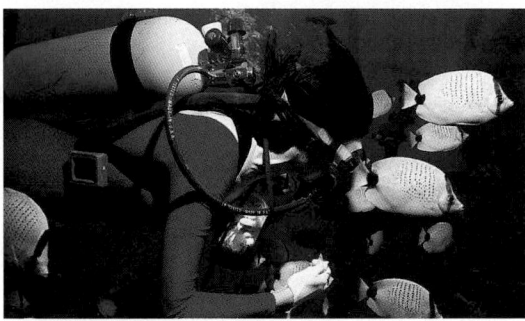

● CUMULATIVE REVIEW

24. What are allotropes? Name three elements that occur as allotropes. (Chapter 5)

25. Can an atom of an element have six electrons in the 2*p* sublevel? (Chapter 7)

26. Classify the following bonds as ionic, covalent, or polar covalent using electronegativity values. (Chapter 9)

   a) NH   **c)** BH
   b) BaCl   **d)** NaI

27. How does an increase in atmospheric pressure affect the following? (Chapter 10)

   a) the boiling point of water
   b) the melting point of ice

Skill Review

1. **Factor Label Method** In a laboratory, 400 L of methane are stored at 600 K and 1.25 atm. If the methane is forced into a 200-L tank and cooled to 300 K, how does the pressure change?

Projects

2. Ballooning today is primarily a sport. But balloons were once used for travel. As you learned in this chapter, Charles was a pioneer in ballooning, and this interest motivated him to study the relationship between temperature and volume of a gas. Investigate the early use of balloons or lighter-than-air vehicles for transportation, and make a presentation of one notable journey to the class.

Writing in Chemistry

3. Modern manufacturing applies the properties of gases in many ways. List three examples of things you see and use every day that apply gas properties. Include at least one household product. Write a short explanation of the property used in each example.

4. Research decompression sickness (also called the bends). Explain the causes and symptoms of decompression sickness. Prepare an instruction pamphlet for trainee scuba divers on how to avoid this problem.

5. Apply your knowledge of gas pressure to explain how a plunger can be used to open a clogged drainpipe.

Problem Solving

6. A sample of argon gas occupies 2.00 L at −33.0°C and 1.50 atm. What is its volume at 207°C and 2.00 atm?

7. A welding torch requires 4500 L of acetylene gas at 2 atm. If the acetylene is supplied by a 150-L tank, what is the pressure of the acetylene?

8. At a depth of 20 m, a 0.5-mL bubble of exhaled air is released from a scuba diver's mouthpiece. If the volume of the bubble just as it reaches the surface of the water is 1.5 mL, what was the pressure on the diver?

9. To produce 15.4 L of nitrogen dioxide at 300 K and 2 atm, how many liters of nitrogen gas and oxygen gas are required?

Program Resources

Chapter Review and Assessment,
 pp. 65-70 **L1**
Alternate Assessment in the Science
 Classroom **L1**
Computer Test Bank, Chapter 11 **L1**
Problems and Solutions Manual,
 Chapter 11 **L1**

Supplemental Practice Problems,
 Chapter 11 **L1**
Performance Assessment in the Science
 Classroom **L1**

AUTHENTIC ASSESSMENT

1. no change in pressure
2. Answers should include some of the following. In 1783, Jacques and Joseph Montgolfier sent up a balloon filled with hot air, Jacques Charles successfully launched a hydrogen-filled balloon, and Jean de Rozier made the first manned balloon flight near Paris. In 1785, Jean Blanchard and John Jeffries made the first balloon crossing of the English Channel. The first balloon ascent in America occurred in 1793. In 1836, the first balloon flight from London, England, to Weilburg, Germany, covered a distance of 800 km (500 miles) in 18 hours. In 1870, balloons were used in the Franco-Russian War for military observation. In 1914, a manned balloon distance record was set by the Berliner balloon traveling from Bitterfeld, Germany, to Perm, Russia. In 1931, Auguste Picard ascended 15 797 m into the stratosphere in a specially constructed, hydrogen-filled balloon. Since then, high-altitude ascents continued to set new records until 1961, when Malcolm Ross and Victor Prather set a record of 34 679 m on a flight from a U.S. Navy aircraft carrier. The first successful balloon flight across the Atlantic occurred in 1978. Also see *Across the Curriculum,* page 395.
3. Some examples are that helium under pressure in a tank is used to inflate balloons, air pressure in an automobile tire increases when more air is pumped into the tire, and a deflated basketball or soccer ball can be inflated by an air pump.
6. 3.00 L
7. 60 atm
8. 3 atm
9. 15.4 L of O_2 and 7.7 L of N_2

Chapter 12 Organizer

Section	Objectives	Activities/Features
12.1 **Counting Particles of Matter** (2 days)	1. **Compare and contrast** the mole as a number and the mole as a mass. 2. **Relate** counting particles to weighing samples of substances. 3. **Solve** stoichiometric problems using molar mass.	**MiniLab 1:** Determining Number Without Counting, p. 408 **How It Works:** Electronic Balances, p. 410 **Art Connection:** Asante Brass Weights, p. 411 **Discovery Demo:** 12.1 Percent Sugar in Bubble Gum, pp. 402-403 **Demonstrations:** 12.2 A Mole Ratio, pp. 406-407; 12.3 Limiting Reactants, pp. 410-411
12.2 **Using Moles** (6 days)	4. **Predict** quantities of reactants and products in chemical reactions. 5. **Determine** mole ratios from formulas for compounds. 6. **Identify** formulas of compounds by using mass ratios.	**Everyday Chemistry:** Air Bags, p. 417 **MiniLab 2:** Bagging the Gas, p. 420 **ChemLab:** Analyzing a Mixture, pp. 422-423 **Chemistry and Technology:** Improving Percent Yield in Chemical Synthesis, pp. 424-425 **Demonstrations:** 12.4 Molar Volume, p. 415; 12.5 Gas Volume and the Mole, pp. 426-427

Activity Materials

ChemLab (pages 422-423)
0.5-0.6 g sodium sulfate-sodium chloride mixture, 15 mL 0.500M strontium chloride solution, distilled water, weighing dish or paper, 250-mL beakers (2), 50-mL graduated cylinder, funnel, wash bottle, stirring rod, filter paper, spatula, iron ring, ring stand, balance

MiniLab 1 (page 408)
many small, uniform items (beans, beads, nails, grains of rice), balance, weighing dish or paper, self-sealing plastic bag, balance

MiniLab 2 (page 420)
5 g baking soda, 60 mL 1M acetic acid, 1-quart self-sealing plastic bag, 100-mL graduated cylinder or measuring cup, weighing paper, balance, plastic-coated twist tie, trash can or explosion screen

Demonstrations For a complete list of materials for the demonstrations in this chapter, see pages 402, 406, 410, 415, and 426.

KEY TO TEACHING STRATEGIES

The following designations will help you decide which activities are appropriate for your students.

L1 Level 1 activities should be within the ability range of all students.

L2 Level 2 activities should be within the ability range of the average to above-average student.

L3 Level 3 activities are designed for the ability range of above-average students.

LEP LEP activities should be within the ability range of Limited English Proficiency students.

COOP LEARN Cooperative Learning activities are designed for small group work.

P These strategies represent student products that can be placed into a best-work portfolio.

Chemical Quantities

Teacher Classroom Resources

STUDENT MASTERS

Study Guide: pp. 47-48
Critical Thinking/Problem Solving: p. 13
Laboratory Manual: 12.1 Determining the Avogadro Constant, pp. 105-110
ChemLab and MiniLab Worksheets: p. 81

TEACHING AIDS

Section Focus Transparency 24
Section Focus Transparency Master: p. 24
Basic Concepts Transparency 33
Basic Concepts Transparency Master: pp. 65-66
Lesson Plans: p. 24

Study Guide: pp. 49-50
Laboratory Manual: 12.2 Stoichiometry of Chemical Reactions, pp. 111-114; 12.3 Molar Volume of a Gas, pp. 115-118
ChemLab and MiniLab Worksheets: pp. 79-80, 82

Section Focus Transparency 25
Section Focus Transparency Master: p. 25
Basic Concepts Transparency 34
Basic Concepts Transparency Master: pp. 67-68
Problem Solving Transparency 14
Problem Solving Transparency Master: pp. 27-28
Lesson Plans: p. 25

ASSESSMENT RESOURCES
Chapter Review and Assessment: pp. 71-76
Alternate Assessment in the Science Classroom
Computer Test Bank: Chapter 12
Problems and Solutions Manual: Chapter 12
Supplemental Practice Problems: Chapter 12
Performance Assessment in the Science Classroom

CHAPTER RESOURCES
Applying Scientific Methods in Chemistry: pp. 34-38
Spanish Resources: Chapter 12
English/Spanish Audiocassettes: Chapter 12
Cooperative Learning in the Science Classroom
Lab and Safety Skills in the Science Classroom

GLENCOE TECHNOLOGY

Software
Mastering Concepts in Chemistry: *Unit 6,* Lesson 1, Lesson 3; *Unit 7,* Lesson 1, Lesson 2; *Unit 12,* Lesson 1, Lesson 2

Videotape
MindJogger Videoquizzes, Chapter 12

Videodisc
Chemistry: Concepts and Applications, Demonstrations, Videos, and Animations
Science and Technology Videodisc Series: Disc 2 Chemistry *Losing Weight by Design*

CD-ROM Multimedia System
Chemistry: Concepts and Applications, Same as for Videodisc, plus Interactive Exploration and Major Simulation, *The Periodic Table*

TECH PREP

The following Glencoe resources provide opportunities for integrating science and technology.

Student Edition: How It Works, p. 410; Everyday Chemistry, p. 417; Chemistry and Technology, pp. 424-425

Teacher Wraparound Edition: Extension, p. 417

Teacher Classroom Resources: Consumer Chemistry, pp. 23-24

12 Chemical Quantities

Chapter Overview

The mole concept is introduced and used in stoichiometric calculations. Students learn to convert the mass of a substance, or its volume at STP, to the number of particles and the number of moles of the substance. This connection is extended to all the substances in chemical reactions through calculations based upon the chemical equation.

Theme Connection

Macro to Submicro/Conservation
The mole concept is the link that relates the *macroscopic* measurement of mass or volume with the *submicroscopic* number of particles. Stoichiometric problems show that matter is always conserved.

✓ **Assessment Planner**

Choose assessment strategies from the following pages to evaluate the progress of your students.
Assess, pp. 406, 412, 429
MiniLabs, pp. 408, 420
ChemLab, p. 422
Portfolio, pp. 424, 428, 430
Chapter Review, p. 430

GLENCOE TECHNOLOGY

Videodisc

Chemistry: Concepts and Applications
Percent Sugar in Bubble Gum
Disc 2, Side 1, Ch. 5

Show this video of the Discovery Demo to introduce molar relationships and chemical analysis. L1 LEP

Discovery Demo 12.1

Percent Sugar in Bubble Gum

Purpose: To have students calculate the mass percentage of an ingredient in a commercial product.
Materials: 5 pieces of sugared bubble gum, paper cup, balance
Safety Precautions:
Disposal: Code A
Procedure: Have students work together in groups of five and follow this procedure. Use a balance to determine the mass of a clean paper cup. Unwrap five pieces of bubble gum containing sugar and place them in the cup. Determine the mass of the cup and the gum. Each person in the group should chew a piece of gum to remove the sugar. After about five minutes, collect the chewed gum in the massed cup and wash your hands. Determine the mass of the cup and gum. Calculate the mass of sugar dissolved from the gum. Calculate the percent-

CHAPTER 12
Chemical Quantities

Household ammonia, chlorine bleach, and oxygen bleach can be found in almost everyone's cleaning closet. But ammonia, chlorine, and oxygen have many other uses. They are some of the most important reactants and products of the chemical industry.

When ammonia gas is synthesized, nitrogen gas and hydrogen gas react by volume in the following way.

$$1 \text{ L } N_2(g) + 3 \text{ L } H_2(g) \rightarrow 2 \text{ L } NH_3(g)$$

According to the reaction above, 1 L of nitrogen and 3 L of hydrogen combine to form 2 L of ammonia.

According to Avogadro's principle, the volumes of gas combine in simple whole-number ratios because 1 L of nitrogen contains as many particles as 1 L of hydrogen. But how many particles is that?

The volumes of nitrogen, hydrogen, and ammonia gas are easily measured, but the particles in a gas are too small and too numerous to be counted easily. How can you determine the number of particles in a gas without counting?

Chemistry Around You

Is it possible to count the atoms, formula units, and molecules of substances? When a liter of methanol, for example, is formed from hydrogen gas and carbon dioxide gas, how many molecules of methanol are formed? How many molecules of hydrogen gas and carbon dioxide gas reacted?

METHANOL

403

Introducing the Chapter

Point out to students that even when dealing with macroscopic particles such as rice or beans, it is impractical to count a large quantity, and therefore rice and beans are sold by the pound. However, if the mass of a small, countable quantity of rice or beans is known, you can calculate the number of particles in a large mass. The same is true of invisible, submicroscopic particles such as atoms, ions, or molecules.

Using the Illustrations Ask students whether counting by weighing would work if the coins saved in the photo were a mixture of pennies, nickels, dimes, and quarters. *No, the coins must all be the same.* This is a good way to make the point that the mole concept can be applied only to a pure substance. L1

Revealing Misconceptions

Students may have difficulty grasping the difference between the mass of a substance and the number of particles in the substance. Stress the idea that the same number of different things must have different masses, and equal masses of different things contain different numbers of things. To illustrate equal numbers but unequal masses, count and mass 20 grains of rice and 20 beans. To illustrate equal masses and unequal numbers, measure enough rice and beans to make 20 g of each. Compare the numbers in each 20-g mass.

age of sugar in the gum by dividing the mass of the dissolved sugar by the mass of the unchewed gum and multiplying by 100. **Results:** Sample data: mass of cup, 7.94 g; mass of cup and gum before chewing, 48.31 g; mass of gum before chewing, 40.37 g; mass of cup and gum after chewing, 22.30 g; mass of dissolved sugar, 26.01 g; percentage of sugar, 64.43%

Analysis: Ask these questions.
1. What is the percentage of sugar? *64.4%*
2. What is the molar mass of the sugar, $C_{12}H_{22}O_{11}$? *342 g/mol*
3. Convert the mass of dissolved sugar to moles. *0.0761 mol*
4. How many molecules of sugar are in the dissolved sugar? *4.58 × 10²² molecules*

GLENCOE TECHNOLOGY

CD-Rom

Chemistry: Concepts and Applications
Percent Sugar in Bubble Gum
Have students use this video of the Discovery Demo to enhance understanding of molar relationships and chemical analysis. L1 LEP

PREPARE

Key Concepts

The concept of counting a large number of the same object by determining their mass introduces Avogadro's constant, the mole concept, and molar masses of elements and compounds.

1 FOCUS

Focus Transparency

Display the **Section Focus Transparency** for Section 12.1 to introduce the concept of the mole. L1 LEP

2 TEACH

Discussion

If you have not yet done so, discuss scientific calculators and the functions that are needed for this course and future science courses. Advise students to purchase suitable calculators soon because they will be needed throughout this chapter.

Concept Development

Explain that stoichiometry involves measuring or calculating the amounts of matter involved in chemical reactions. The mole ratios derived from the chemical equation allow calculations that predict the amount of each substance needed for, or produced in, a chemical reaction.

SECTION PREVIEW

Objectives

Compare and contrast the mole as a number and the mole as a mass.
Relate counting particles to weighing samples of substances.
Solve stoichiometric problems using molar mass.

Key Terms

stoichiometry
mole
Avogadro constant
molar mass
molecular mass
formula mass

Counting Particles of Matter

Louis Staffilino of Dillonvale, Ohio, saved pennies for 65 years. When he deposited them in a bank in 1994, he had 40 large drums of pennies. His wealth represented an enormous counting job for the bank teller.

Just as a bank teller counts coins and bills, a chemist counts atoms, molecules, and formula units of substances. Unlike a bank teller, a chemist cannot count individual items because the particles of matter are too small and just too numerous. How do you determine the number of particles in a sample of matter without counting?

Stoichiometry

In Chapter 11, you learned that volumes of gases always combine in definite ratios. This observation, called the law of combining volumes, is based on measurements of the gas volumes. When Avogadro suggested that gases combine in fixed ratios because equal volumes of gases at the same temperature and pressure contain equal numbers of particles, he may have been thinking of particles rearranging themselves. Individual gas particles are so small that their rearranging cannot be observed, but the volumes of gases can be measured directly. Avogadro's principle is one of the earliest attempts to relate the number of particles in a sample of a substance to a direct measurement made on the sample.

Today, by using the methods of **stoichiometry,** we can measure the amounts of substances involved in chemical reactions and relate them to one another. For example, a sample's mass or volume can be converted to a count of the number of its particles, such as atoms, ions, or molecules.

Pennies are not small, but counting a drum full of pennies, like counting the number of particles in a gas, is a formidable task. Can we measure the pennies by grouping them in some conveniently large quantity and then counting the number of groups to find the total number of pennies in the drum? Can methods of stoichiometry help? **Figure 12.1** presents two ways to count large numbers of pennies. To count the pennies directly, you would have to handle every one—all 24 216 pennies. But you

Program Resources

Study Guide, pp. 47-48 L1
Critical Thinking/Problem Solving, p. 13 L2
Laboratory Manual, 12.1, pp. 105-110 L1
English/Spanish Audiocassettes, Chapter 12 L1 LEP
Section Focus Transparency 24 and Master, p. 24 L1 LEP

Basic Concepts Transparency 33 and Master, pp. 65-66 L1 LEP
Spanish Resources L1 LEP
ChemLab and MiniLab Worksheets, p. 81 L1

could also just make three measurements: the weight of the drum of pennies, the weight of the empty drum, and the weight of a group of 1000 pennies. Now, if you want to count atoms, what size group is most suitable? You will need a larger group than you used for the pennies—much larger than a thousand or even a million.

Atoms are so tiny that an ordinary-sized sample of a substance contains so many of these submicroscopic particles that counting them by grouping them in thousands would be unmanageable. Even grouping them by millions would not help. The group or unit of measure used to count numbers of atoms, molecules, or formula units of substances is the **mole** (abbreviated *mol*). The number of things in one mole is 6.02×10^{23}. This big number has a short name: the **Avogadro constant,** as illustrated in **Figure 12.2.**

Fact of the MATTER

The most precise value of the Avogadro constant is 6.0221367×10^{23}. For most purposes, rounding to 6.02×10^{23} is sufficient.

Figure 12.1
Two Ways to Count Pennies
You could count the pennies directly. ▶

▲ You could calculate the number of pennies in a drum by using the mass of all the pennies, 70 000 g, and the mass of 1000 pennies. The mass of 1000 pennies is 2890.7 g. Use the ratio $\frac{1000 \text{ pennies}}{2890.7 \text{ g}}$ to find the total number of pennies.

$$70\ 000\ \cancel{g} \times \frac{1000 \text{ pennies}}{2890.7\ \cancel{g}} = 24\ 216 \text{ pennies}$$

The drum contains 24 216 pennies, worth $242.16.

Figure 12.2
One Mole Is a BIG Number.
If 6.02×10^{23} sheets of paper were placed in a stack, they would reach from Earth to the sun—more than a million times. The thickness of a sheet of paper is small, but an atom is much smaller. One mole of magnesium atoms is hardly a handful.

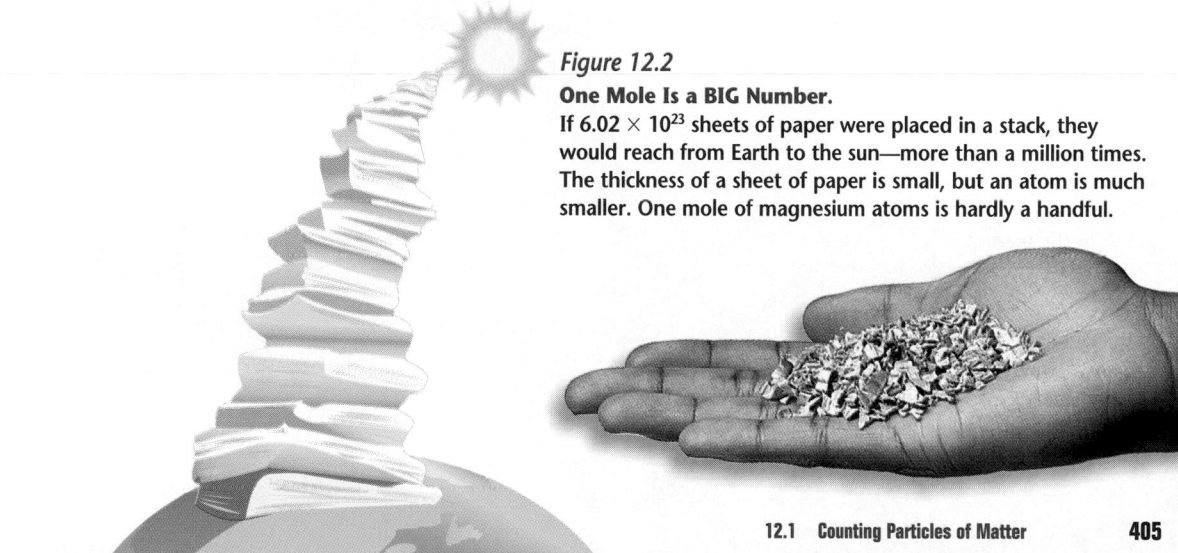

12.1 Counting Particles of Matter **405**

Across the Curriculum

History
John Dalton published the first set of atomic masses in 1808. Dalton's masses were all integers, with hydrogen assigned a mass of 1. Based on these values, English chemist William Proust suggested in 1815 that all elements were composed of hydrogen. This seemed reasonable until the ability to measure atomic masses became more refined and masses that were not integers were found. Oxygen was determined to have a mass of 15.9 on Dalton's scale. Because oxygen is a central element in the chemical reactions of almost every other element, it was chosen as a new reference. The mass of oxygen was defined as 16 amu. This change made only slight differences in the masses determined using hydrogen as the standard. In 1961, with the discovery of isotopes, the mass of ^{12}C, 12 amu, was adopted as the standard.

Visual Learning

Figure 12.2: Ask students how many inches are in one foot. *12* How many pounds in one ton? *2000* How many atomic mass units in 1 g? *1.00 g × (1 u/1.66 × 10⁻²⁴ g) = 6.02 × 10²³ u* Explain that Avogadro's number was not chosen arbitrarily to be 6.02×10^{23}. It is the number of atomic mass units per gram.

Concept Development
Chemistry teachers sometimes refer to the mass of a mole as molecular weight, gram molecular weight, formula weight, and molar mass. It helps to be consistent. The term *molar mass* is the most universal term and the one used in this text.

Using an Analogy
Give students this simplified recipe for vegetable soup that will serve four: 4 potatoes (0.3 lbs each), 2 onions (0.2 lb each), 8 carrots (0.1 lb each), 4 stalks of celery (0.05 lb each), 1.4 lbs water. Show that the mass of the soup is = 4.0 lbs. Ask students what they might do to the recipe if they were serving eight people. *double the recipe* Ask what they might do if they were making soup for an army of people. *multiply the recipe by a large factor* Suggest that they might use 1.2 tons of potatoes, 0.4 tons of onions, 0.8 tons of carrots, 0.2 tons of celery, and 1.4 tons of water to produce 4.0 tons of soup. Emphasize that when the units are changed, the proportion of one ingredient to another remains the same. Explain that the same is true when atomic masses are changed from atomic mass units to grams. Follow this discussion immediately with the *Concept Development* on page 406. L1

Concept Development

Ask students to write the balanced equation for the formation of carbon disulfide from carbon and sulfur and to write the meaning of each formula. *C (1 atom of carbon) + 2S (2 atoms of sulfur) → CS₂ (1 molecule of carbon disulfide)* Ask them to write another equation including the masses of the reactants and product. *12.0 u C + 64.0 u S → 76.0 u CS₂* Explain that because individual atoms of carbon and sulfur and molecules of carbon disulfide are far too small to be weighed, the recipe for making CS₂ could be greatly expanded by reacting one molar mass of carbon, 12.0 g, rather than one atomic mass, and two molar masses of sulfur, 64.0 g, rather than two atomic masses. Ask students how much carbon disulfide this recipe would yield. *one molar mass of carbon disulfide, 76.0 g* Ask students how many times larger this product is than the original recipe. *6.02×10^{23} times* L1

Oral: Ask students individually to use the periodic table to determine the mass of 6.02×10^{23} atoms of a chemical element chosen randomly. *The atomic mass in grams is the molar mass.* L1

All kinds of submicroscopic particles can be conveniently counted using the Avogadro constant. There are 6.02×10^{23} carbon atoms in a mole of carbon, and there are 6.02×10^{23} carbon dioxide molecules in a mole of carbon dioxide. There are 6.02×10^{23} sodium ions in a mole of sodium ions. For that matter, there are 6.02×10^{23} eggs in a mole of eggs, but eggs are so large compared with the particles that interact in chemistry that you would have no need to estimate how many there are in a typical sample. Can you see a relationship here? For counting many small things, use a large unit of measure; for counting fewer things or larger things, use a small unit of measure, as shown in **Figure 12.3**.

Molar Mass

How many moles of methanol are in 500 g of methanol? Methanol is formed from CO_2 gas and hydrogen gas according to the balanced chemical equation below.

$$CO_2(g) + 3H_2(g) \rightarrow CH_3OH(g) + H_2O(g)$$

Suppose you wanted to produce 500 g of methanol. How many grams of CO_2 gas and H_2 gas would you need? How many grams of water would be produced as a by-product? Those are questions about the masses of reactants and products. But the balanced chemical equation shows that three molecules of hydrogen gas react with one molecule of carbon dioxide gas. The equation relates molecules, not masses, of reactants and products.

Like Avogadro, you need to relate the macroscopic measurements—the masses of carbon dioxide and hydrogen—to the number of molecules of methanol. To find the mass of carbon dioxide and the mass of hydrogen needed to produce 500 g of methanol, you first need to know how many molecules of methanol are in 500 g of methanol.

WORD ORIGIN

stoichiometry
stoichen (GK)
element or part
metreon (GK)
measure
Stoichiometry allows you to determine the number of atoms in a substance by relating other measurable quantities of that substance.

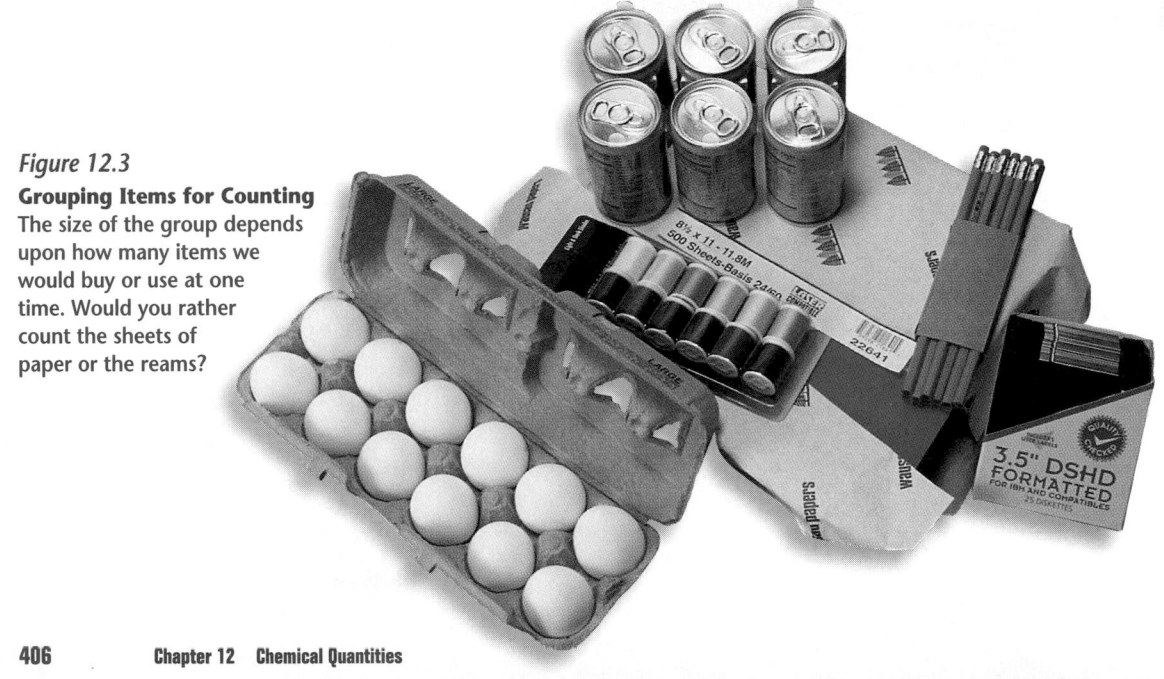

Figure 12.3
Grouping Items for Counting
The size of the group depends upon how many items we would buy or use at one time. Would you rather count the sheets of paper or the reams?

A Mole Ratio

Purpose: To prepare magnesium oxide and determine its empirical formula.

Materials: crucible, crucible tongs, laboratory burner, clay triangle, ring stand and ring, dropper, glass stirring rod, 70-cm length of magnesium ribbon, balance, 1 mL distilled water

Safety Precautions:
Disposal: Code A

Procedure: Record the mass of the crucible. Loosely roll the 70-cm length of magnesium ribbon into a ball and place it in the crucible. Mass the crucible and magnesium. Place the crucible on a clay triangle and heat it strongly with a laboratory burner. When the magnesium begins to burn, turn off the burner. Allow the magnesium to continue to burn. **CAUTION:** *Do not look at the burning magnesium.*

After the crucible has cooled, powder the contents with the tip of a glass stirring rod. Have a student note any odor as you slowly add 15 drops of distilled water. Heat the crucible strongly for three to five minutes to dry the residue. Cool it for five minutes. Record the mass of the crucible and contents.

Results: The magnesium ribbon burns in air to form MgO and Mg_3N_2. Three moles

Remember the drum of pennies? By knowing the mass of one group of 1000 pennies and knowing the mass of all the pennies, you can find the number of groups of 1000 pennies. Then, finding the total number of pennies is an easy matter. You have a suitable unit of measure, the mole, and you know the mass of methanol you want to produce. But, you still need to know the mass of 1 mol of methanol molecules.

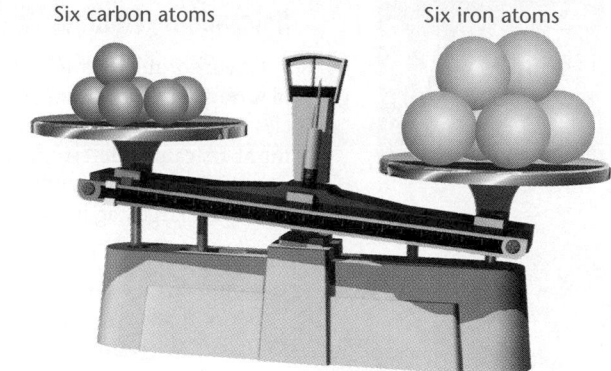
Six carbon atoms   Six iron atoms

Molar Mass of an Element

You know from Chapter 2 that average atomic masses of the elements are given on the periodic table. For example, the average mass of one iron atom is 55.8 u, where *u* means "atomic mass units." The atomic mass unit is defined so that the atomic mass of an atom of the most common carbon isotope is exactly 12 u, and the mass of 1 mol of the most common isotope of carbon atoms is exactly 12 g. The mass of 1 mol of a pure substance is called its **molar mass.** For example, the molar mass of iron is 55.847 g, and the molar mass of platinum is 195.08 g. Relative masses of elements are demonstrated in **Figure 12.4.** The molar mass is the mass in grams of the average atomic mass.

If an element exists as a molecule, remember that the particles in 1 mol of that element are themselves composed of atoms. For example, the element oxygen exists as molecules composed of two oxygen atoms, so a mole of oxygen molecules contains 2 mol of oxygen atoms. Therefore, the molar mass of oxygen molecules is twice the molar mass of oxygen atoms: 2×16.00 g $= 32.00$ g. **Figure 12.5** shows the molar masses of some elements.

Figure 12.4
Relative Masses of Iron and Carbon
An average iron atom is 4.65 times as heavy as an average carbon atom. Six iron atoms are 4.65 times as heavy as six carbon atoms. One mole of iron atoms is 4.65 times as heavy as 1 mol of carbon atoms.

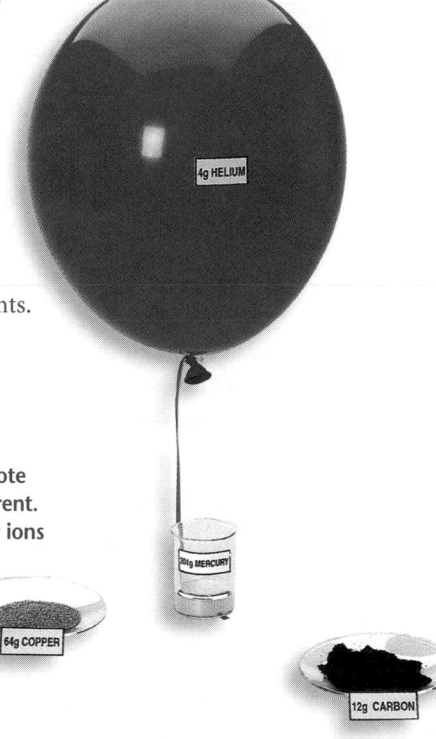

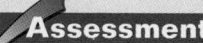

Figure 12.5
Molar Masses of Some Elements
Each sample contains 6.02×10^{23} atoms. Note that the masses of the samples are all different. Moles relate counts of atoms, molecules, or ions to mass because a molar mass always contains an Avogadro constant of particles.

Correcting Misconceptions

Students may think that Avogadro's constant, expressed with its eight significant figures, 6.022 136 7 $\times$ 10²³, is a very exact number—perhaps even accurate to a single atom. Point out that the constant has limited precision when one considers the magnitude of the number.

Tying to Previous Knowledge

The mass of a mole of carbon atoms is 12.01 u. This value is the mass percent average of the masses of the three naturally occurring carbon isotopes—carbon-12, carbon-13, and carbon-14. One molar mass of carbon, 12.01 g, can be thought of as containing 6.02×10^{23} identical atoms of carbon, each having a mass of 12.01 u. Ask students to judge by the mass percent average, 12.01 u, which of the three isotopes is the most abundant. *carbon-12* L1

GLENCOE TECHNOLOGY

⊙ **Software**

Mastering Concepts in Chemistry
Unit 6, Lesson 1
The Mole Concept
Use this lesson to reinforce the concept of the mole. L1 LEP

of water react with the Mg_3N_2 to form 3 mol of MgO and 2 mol of NH_3. Sample data: 0.57 g Mg ribbon, 0.96 g MgO, and 0.39 g oxygen.
Analysis: Ask these questions.
1. Calculate the number of moles of magnesium and the number of moles of oxygen in MgO. *0.024 mol Mg, 0.024 mol O*

2. What is the simplest whole-number mole ratio of Mg to O? *1:1*
3. Use this whole-number ratio to write the empirical formula of magnesium oxide. *MgO*

✓ Assessment

Performance: Have students repeat the procedure using Epsom salt, hydrated mag-

nesium sulfate. Have them determine the moles of water driven off, the moles of anhydrous $MgSO_4$, the mole ratio, and the formula of the hydrate. *1:7; the formula is $MgSO_4 \cdot 7H_2O$.* L1

miniLAB

Determining Number Without Counting

Purpose: To determine the approximate number of small, identical items by weighing.

Process Skills: Using numbers, measuring, communicating

Teaching Strategies

• Use small, uniform items such as beans, beads, nails, grains of rice, or jelly beans.

• Some students may need help with the calculations.

• Require that students use the factor label method.

Expected Results

Results depend upon the item(s) and the size of bag used.

Analysis

1. Answers depend on the items.

2. One possible determination uses the average mass of a group of one dozen items, m_{12}, to determine the number of groups in the bag. Number of groups = (mass of full bag – mass of bag)/m_{12}. To determine the total number of items in the bag, multiply the number of groups by 12.

✔ Assessment

Performance: Give students the following information and ask them to calculate how many nucleons (protons and neutrons) make up Earth. (1) The mass of all the electrons in atoms is insignificant. (2) The mass of Earth is approximately 5.98×10^{24} kg. (3) Protons and neutrons have approximately equal masses of 1.67×10^{-27} kg. 3.58×10^{51} nucleons [L1]

Determining Number Without Counting

Chemists and chemical engineers usually need to control the number of atoms, molecules, and ions in their reactions carefully. These particles are too small to be seen and too numerous to count, but their numbers can be determined by measuring their masses. Simulate this process by determining the approximate number of small, identical items in a large bag by weighing instead of counting.

Procedure

1. Count out a convenient number of items and weigh them. Record the number of items and the mass.

2. Weigh and record the mass of an empty, self-sealing plastic storage bag.

3. Completely fill the bag with items and seal it.

4. Weigh and record its mass.

5. Develop and carry out a procedure to determine the number of items in the bag without opening it.

Analysis

1. How many items are in the bag?

2. Explain how you determined that number.

SAMPLE PROBLEM **1** **Number of Atoms in a Sample of an Element**

The mass of an iron bar is 16.8 g. How many Fe atoms are in the sample?

Analyze
• Use the periodic table to find the molar mass of iron. The average mass of an iron atom is 55.8 u. Then the mass of 1 mol of iron atoms is 55.8 g.

Set Up
• To convert the mass of the iron bar to the number of moles of iron, use the mass of 1 mol of iron atoms as a conversion factor.

> **Problem-Solving HINT**
>
> Remember that the units of molar mass are grams per mole, which can be used as a conversion factor.

$$\frac{16.8 \text{ g Fe}}{} \left| \frac{1 \text{ mol Fe}}{55.8 \text{ g Fe}} \right.$$

Now, use the number of atoms in a mole to find the number of iron atoms in the bar.

$$\frac{16.8 \text{ g Fe}}{} \left| \frac{1 \text{ mol Fe}}{55.8 \text{ g Fe}} \right| \frac{6.02 \times 10^{23} \text{ Fe atoms}}{1 \text{ mol Fe}}$$

Solve
• Simplify the expression above.

$$\frac{16.8 \text{ g Fe}}{} \left| \frac{1 \text{ mol Fe}}{55.8 \text{ g Fe}} \right| \frac{6.02 \times 10^{23} \text{ Fe atoms}}{1 \text{ mol Fe}} =$$

$$\frac{16.8 \times 6.02 \times 10^{23} \text{ Fe atoms}}{55.8} = 1.81 \times 10^{23} \text{ Fe atoms}$$

Check
• Notice how the units of measure cancel to leave only Fe atoms.

PRACTICE

Guided Practice: Guide students through the sample problem, then have them work in class on Practice Problems 1 and 2 on page 412. See Appendix D for complete solutions. [L1]

Independent Practice: Your homework or classroom assignment can include Additional Practice Problem 1 shown on page 409. [L1]

Answers to Practice Problems (Page 412)

1. Because sulfur has a smaller molar mass, there will be a greater number of atoms in 50.0 g of sulfur than in 50 g of tin. 9.38×10^{23} S atoms; 2.53×10^{23} Sn atoms

2. **a.** 2.95×10^{23} Hg atoms
   b. 1.39×10^{23} Au atoms
   c. 9.28×10^{23} Li atoms
   d. 4.73×10^{23} W atoms

Molar Mass of a Compound

As you learned in Chapter 4, covalent compounds are composed of molecules, and ionic compounds are composed of formula units. The **molecular mass** of a covalent compound is the mass in atomic mass units of one molecule. Its molar mass is the mass in grams of 1 mol of its molecules. The **formula mass** of an ionic compound is the mass in atomic mass units of one formula unit. Its molar mass is the mass in grams of 1 mol of its formula units. How to calculate the molar mass for ethanol, a covalent compound, and for calcium chloride, an ionic compound, is shown below.

Ethanol, C_2H_6O, a covalent compound

2 C atoms	2×12.0 u =	24.0 u
6 H atoms	6×1.00 u =	6.00 u
1 O atom	1×16.0 u =	+16.0 u
molecular mass of C_2H_6O		46.0 u

mass of 1 mol C_2H_6O molecules	=	46.0 g
molar mass of ethanol		46.0 g/mol

Calcium chloride, $CaCl_2$, an ionic compound

1 Ca atom	1×40.1 u =	40.1 u
2 Cl atoms	2×35.5 u =	+71.0 u
formula mass of $CaCl_2$		111.1 u

mass of 1 mol $CaCl_2$ formula units	=	111.1 g
molar mass of calcium chloride		111.1 g/mol

By making calculations with molar masses, you can find the number of molecules of methanol in your 500 g; the number of molecules of carbon dioxide and hydrogen that reacted; and the masses of carbon dioxide, hydrogen, methanol, and water in grams. **Figure 12.6** shows the molar masses of some compounds.

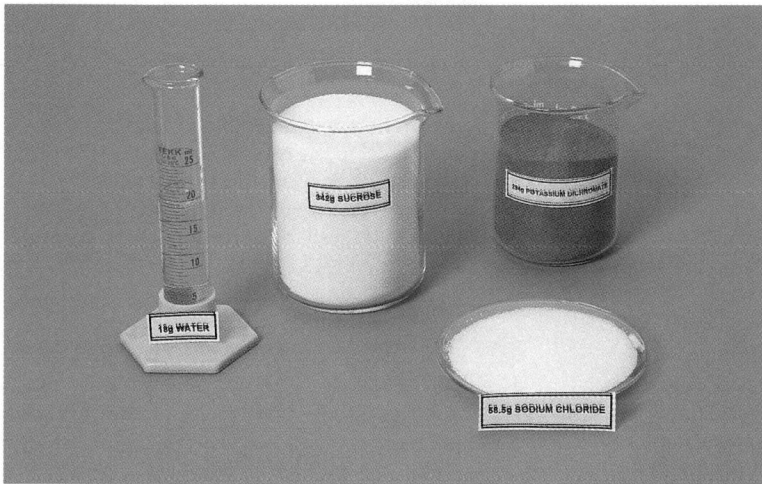

Figure 12.6
Molar Masses of Some Compounds
Each sample contains 6.02×10^{23} molecules of a covalent compound or 6.02×10^{23} formula units of an ionic compound. Each compound has its own molar mass.

Additional Practice Problem

1. Calculate the number of atoms in each sample.
   a. 25.0 g silicon, Si
   $$\frac{25.0 \text{ g Si}}{1} \left| \frac{1 \text{ mol Si}}{28.1 \text{ g Si}} \right| \frac{6.02 \times 10^{23} \text{ Si atoms}}{1 \text{ mol Si}}$$
   $= 5.36 \times 10^{23}$ atoms Si
   b. 1.29 g chromium, Cr
   $$\frac{1.29 \text{ g Cr}}{1} \left| \frac{1 \text{ mol Cr}}{52.0 \text{ g Cr}} \right| \frac{6.02 \times 10^{23} \text{ Cr atoms}}{1 \text{ mol Cr}}$$
   $= 1.49 \times 10^{22}$ atoms Cr

Meeting Individual Needs

Learning Disabled Display 12 grapefruit, 12 oranges, and 12 grapes and explain that the numbers of fruit per dozen are the same but the masses of the dozens differ. Demonstrate this by using a suitable balance to weigh the items. Explain that, in a similar way, 1-mol portions of different substances, each containing 6.02×10^{23} particles, have different masses. **L1** **LEP**

 **Quick Demo**

What's the same about moles?

 Display 1 mol each of several solid and liquid elements and compounds. Label each with the correct formula and 6.02×10^{23} atoms, molecules, ions, etc. Ask students to calculate the molar mass for each of the substances. Collect and compare answers for accuracy. **L1**

Reinforcement

Help students set up and use a general formula or algorithm for each of the following conversions: mass to moles; moles to mass; moles to atoms, molecules, or ions; and atoms, molecules, or ions to moles. **L1**

GLENCOE TECHNOLOGY

 **Videodisc**

Chemistry: Concepts and Applications
Molar Mass
Disc 2, Side 1, Ch. 6

Show this video to help you explain how different substances have different molar masses. **L1** **LEP**

 CD-Rom

Chemistry: Concepts and Applications
Molar Mass
Show this video to help reinforce the concept of molar mass. **L1** **LEP**

Transparency

Display **Basic Concepts Transparency 33** to help students visualize how atoms are counted. **L1** **LEP**

How it Works

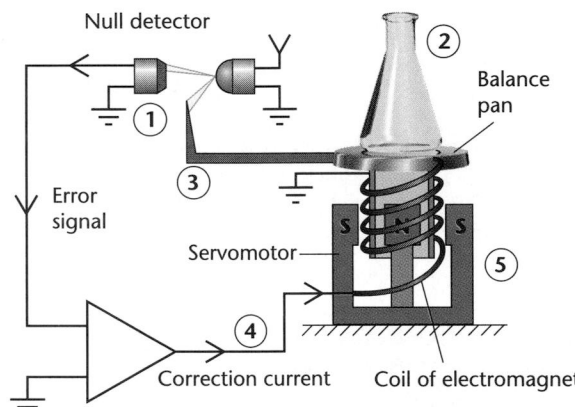

Purpose

Students are introduced to the basics of an electronic balance.

Background

If an electronic balance is not available, students will still be familiar with them from grocery-store checkouts.

Visual Learning

If possible, demonstrate the operation of an electronic balance. Show the utility of the automatic tare and rezero features of the balance. Compare its operation to a triple-beam or mechanical balance.

Teaching Strategies

• Have students consider that although an electronic balance may be more expensive than a mechanical balance, the speed of the electronic balance means that one electronic balance may replace two or more mechanical balances. L1

• Depending upon the balance, demonstrate other features, such as automatic calibration and alternative mass units.

Thinking Critically

1. They are fast and durable. They tend to be smaller than mechanical balances and so are more portable.

2. Electronic balances are often shared. To avoid errors, a balance must always read zero before massing.

Electronic Balances

Balances are devices used to measure mass. When a sample of a substance is placed on the balance pan, gravity on it pushes the pan down with a force proportional to the object's mass. A sensor uses an electromagnet to push the pan back to its original position. The electric current required to push the pan back to its original position is directly proportional to the gravitational force on the object on the pan. The mass of the object is displayed on a digital readout.

1. **The Null Position** The position of the balance is sensed by a light beam. When the beam is blocked, the balance is at the null position. The first step in using an electronic balance is to set the balance to zero so that the readout is zero when the pan is empty. This usually requires just the press of a button.

2. Most electronic balances allow the user to automatically tare the balance so that when an empty container is on the pan, the readout is zero.

3. **Weighing** When an object to be weighed is placed on the balance pan, the arm drops down and allows light to reach the detector, which sends an electrical signal to the control circuit.

4. The control circuit sends a correction current to an electromagnet under the balance pan. The electromagnet produces a magnetic force that pushes the pan up against gravity.

5. The control circuit sends enough current to make the magnetic force exactly balance the gravitational force. The pan returns to the null position.

6. The current required to null the balance is proportional to the mass of the sample. The control circuit converts this current into a numerical value, a number of grams or kilograms, which can be seen in the readout on the balance.

Thinking Critically

1. What are some advantages of electronic balances over mechanical balances?

2. Why is it good practice to always re-zero an electronic balance before using it?

Demonstration 21.3

Limiting Reactants

Purpose: To demonstrate the effects of limiting a reactant.

Materials: 10 test tubes, cafeteria tray, four 150-mL beakers, 20 rubber stoppers that fit the test tubes

Safety Precautions: 👓

Disposal: Code F

Procedure: Place ten empty test tubes on a cafeteria tray. In each of four small beakers, place five rubber stoppers. Show students a model of a synthesis reaction: reactant *A* (test tube) plus reactant *B* (stopper) react to form product *AB* (a stoppered test tube). Ask students how much product will form when the first beaker containing five stoppers is added to the cafeteria tray containing the ten test tubes. *five units of product* Add another beaker containing five stoppers and re-

peat the question. *ten units of product* Add the third beaker and then the fourth beaker, each time asking how much product will be produced. *ten each time* Ask students which reactant is in excess and which one is the limiting reactant. *Stoppers are in excess; test tubes are limiting.* L1

Results: Students observe that the model demonstrates the effect of limiting reactants.

ART ☐ CONNECTION

Asante Brass Weights

Imagine waking up in the morning after a rain and finding that gold has popped from the ground. Does it sound like a fairy tale? In the past, this was a common experience for people living in Ghana. Their land is an alluvial gold field. Besides picking up gold that has been bared by erosion and finding gold by panning streams, the Asante, one of about 35 language groups in Ghana, recovered gold by digging into sediment where smaller streams joined larger ones and by crushing gold-bearing quartz and rock and separating the gold by washing. Gold dust and small gold nuggets were their national currency.

Asante standard weights Although the Asante state was founded about 1700 A.D., as early as about 1400, the Asante created artistic, standardized weights. Their first weights were simple blocks, but later they made cubes, pyramids, triangular blocks, and rectangular blocks with carved designs. They cast natural objects like seeds, beetles, shells, and chicken feet. In the early 1700s, the carved designs on flattened shapes became more elaborate. While Europeans were colonizing America, the Asante were creating works of art with their weights.

Besides royal gold jewelry, they made artistic brass weights, spoons, and boxes. They also worked copper, zinc, tin, lead, and nickel. They cast objects in brass by the lost wax method. In this process, shapes were carved in wax, then encased in clay and baked. The hot wax was poured out, leaving a ceramic mold. Molten metal was poured into the mold. When the metal cooled, the mold was broken to remove the casting.

Today, Asante weights are collected as art forms—evidence of the advanced Asante culture of the past few centuries. Compared to the plain weights used in the United States, Asante weights add interest and beauty to everyday life. How enriching it must have been to use a 100-g porcupine or a 50-g headless fish to weigh gold dust for a purchase instead of paying with paper currency.

Connecting to Chemistry

1. **Hypothesizing** Do you think standard weights will soon be obsolete in scientific laboratories? Why?

2. **Applying** Why does gold form few compounds?

12.1 Counting Particles of Matter **411**

Analysis: Ask these questions.

1. For a school bake sale, you volunteer to bake several cakes. You have three boxes of cake mix and ten eggs. How many cakes can be baked if each mix requires two eggs? *three cakes*

2. Which of the items is the limiting reactant? *cake mixes*

3. How much of the excess reactant remains after the cakes are baked? *four eggs*

✔ Assessment

Skill: Ask groups of students to brainstorm additional examples of limiting reactants observed in everyday occurrences. L1 **COOP LEARN**

ART ☐ CONNECTION

Purpose

To show the importance of standardized measuring devices and how one culture combined need and aesthetics.

Teaching Strategies

- Invite a local artist to talk about casting. If possible, let students visit his or her studio for a demonstration.
- The steps in casting are: (1) making a wax model or selecting an organic object, (2) making a mold that encases the model or object, (3) burning the model out of the mold, (4) pouring molten metal into the mold, (5) breaking the mold, and (6) cleaning and polishing the piece.
- Switch roles with the art teacher for a day. You could teach art students about the chemistry of art. The art teacher can help your students sculpt 100-g standardized weights from a material of their choice such as clay, wax, wood, paper, or soap. L1 **LEP**
- Students can find out more about the Asante culture and its art. Discuss factors that indicate they were an advanced people. L1

Extension

Have students do more research about alluvial gold fields. L1

Connecting to Chemistry

1. Yes, electronic balances are becoming popular because they are quick and easy to use and their prices are competitive; or no, a standard weight will always be needed to check the accuracy of weighing devices.

2. Gold is an unreactive chemical element.

3 ASSESS

Check for Understanding

- Check for understanding of the section material by having students answer the Section Review questions.
- Celebrate *Mole Day* on any convenient day, or preferably on October 23 (10/23) from 6:02 A.M. to 6:02 P.M. Students might earn mole points for activities such as the following: (1) creating a "moley" food such as mole cake, molasses cookies, coca mola, or guacomole to share with the class; (2) making an item of mole apparel and wearing it; (3) composing a mole song or mole rap; (4) creating and solving mole riddles such as: Avogadro's favorite expression when startled—*Holy Moley,* night cream that Avogadro created—*Oil of Molay,* Avogadro's favorite board game—*Moleopoly.* This activity can reinforce the mole concept, bring out creativity in your students, and liven the student-teacher relationship. L1

GLENCOE TECHNOLOGY

Videodisc

Chemistry: Concepts and Applications
See the Videodisc Program Teacher Guide to access illustrations and other material available for this chapter.

SAMPLE PROBLEM 2 — Number of Formula Units in a Sample of a Compound

The mass of a quantity of iron(III) oxide is 16.8 g. How many formula units are in the sample?

Analyze
- Use the periodic table to calculate the mass of one formula unit of Fe_2O_3.

$$2 \text{ Fe atoms} \quad 2 \times 55.8 \text{ u} = \quad 111.6 \text{ u}$$
$$3 \text{ O atoms} \quad 3 \times 16.0 \text{ u} = \underline{+ \; 48.0 \text{ u}}$$
$$\text{formula mass of } Fe_2O_3 \quad\quad 159.6 \text{ u}$$

Therefore, the molar mass of Fe_2O_3 (rounded off) is 160 g.

Set Up
- $$\dfrac{16.8 \text{ g } Fe_2O_3}{} \bigg| \dfrac{1 \text{ mol } Fe_2O_3}{160 \text{ g } Fe_2O_3}$$

Now, multiply the number of moles of iron oxide by the number in a mole.

$$\dfrac{16.8 \text{ g } Fe_2O_3}{} \bigg| \dfrac{1 \text{ mol } Fe_2O_3}{160 \text{ g } Fe_2O_3} \bigg| \dfrac{6.02 \times 10^{23} \; Fe_2O_3 \text{ formula units}}{\text{mol } Fe_2O_3}$$

Solve
- $$\dfrac{16.8 \text{ g } Fe_2O_3}{} \bigg| \dfrac{1 \text{ mol } Fe_2O_3}{160 \text{ g } Fe_2O_3} \bigg| \dfrac{6.02 \times 10^{23} \; Fe_2O_3 \text{ formula units}}{1 \text{ mol } Fe_2O_3} =$$

$$\dfrac{16.8 \times 6.02 \times 10^{23} \; Fe_2O_3 \text{ formula units}}{160} =$$

$$6.32 \times 10^{22} \; Fe_2O_3 \text{ formula units}$$

Check
- The units of measure show that the calculation is set up correctly. The multiplication also checks.

PRACTICE PROBLEMS

1. Without calculating, decide whether 50.0 g of sulfur or 50.0 g of tin represents the greater number of atoms. Verify your answer by calculating.

2. Determine the number of atoms in each sample below.
   a) 98.3 g mercury, Hg
   b) 45.6 g gold, Au
   c) 10.7 g lithium, Li
   d) 144.6 g tungsten, W

3. Determine the number of moles in each sample below.
   a) 6.84 g sucrose, $C_{12}H_{22}O_{11}$
   b) 16.0 g sulfur dioxide, SO_2
   c) 68.0 g ammonia, NH_3
   d) 17.5 g copper(II) oxide, CuO

In the previous problems, you used the molar mass to convert a mass measurement to a number of moles. Now, you will learn to convert a number of moles to a mass measurement.

PRACTICE

Guided Practice: Guide students through the sample problems on pages 412 and 413, then have them work in class on Practice Problems 3 and 4. See Appendix D for complete solutions. L1

Independent Practice: Your homework or classroom assignment can include the Additional Practice Problem shown here. L1

Answers to Practice Problems

3. **a.** 2.00×10^{-2} mol $C_{12}H_{22}O_{11}$
   b. 0.250 mol SO_2 **d.** 0.220 mol CuO
   c. 4.00 mol NH_3

4. **a.** 98.9 g Si   **c.** 20.9 g F_2
   b. 225 g $C_9H_8O_4$ **d.** 919 g BaI_2

Additional Practice Problem

1. How many molecules are in each sample?

a. 89.0 g sodium oxide, Na_2O

$$2 \text{ Na atom} \quad 2 \times 23.0 \text{ u} = \quad 46.0 \text{ u}$$
$$1 \text{ O atom} \quad 1 \times 16.0 \text{ u} = \underline{+ 16.0 \text{ u}}$$
$$\text{molecular mass } Na_2O \quad\quad 62.0 \text{ u}$$

molar mass Na_2O = 62.0 g/mol

$$\dfrac{89.0 \text{ g } Na_2O}{1} \bigg| \dfrac{1 \text{ mol } Na_2O}{62.0 \text{ g } Na_2O} \bigg| \dots$$

$$\dfrac{6.02 \times 10^{23} \text{ molecules } Na_2O}{1 \text{ mol } Na_2O}$$

$$= 89.0 \times 1/62.0 \times 6.02 \times 10^{23}$$

SAMPLE PROBLEM ⓷ Mass of a Number of Moles of a Compound

SECTION 12.1

What mass of water must be weighed to obtain 7.50 mol of H_2O?

Analyze
- The molar mass of water is obtained from its molecular mass.

2 H atoms	$2 \times 1.00\ u =$	$2.00\ u$
1 O atom	$1 \times 16.0\ u =$	$16.0\ u$
molecular mass of H_2O		$18.0\ u$

The molar mass of water is 18.0 g/mol.

Set Up
- Use the molar mass to convert the number of moles to a mass measurement.

$$\frac{7.50\ mol\ H_2O}{} \Bigg| \frac{18.0\ g\ H_2O}{1\ mol\ H_2O}$$

Solve
$$\frac{7.5\ \cancel{mol\ H_2O}}{} \Bigg| \frac{18.0\ g\ H_2O}{1\ \cancel{mol\ H_2O}} = 7.50 \times 18.0\ g\ H_2O = 135\ g\ H_2O$$

Check
- The final units are grams of H_2O, as expected. The multiplication also checks.

PRACTICE PROBLEMS

4. Determine the mass of the following molar quantities.

   a) 3.52 mol Si
   b) 1.25 mol aspirin, $C_9H_8O_4$
   c) 0.550 mol F_2
   d) 2.35 mol barium iodide, BaI_2

The concept of molar mass makes it easy to determine the number of particles in a sample of a substance by simply measuring the mass of the sample. The concept is also useful in relating masses of reactants and products in chemical reactions.

SECTION REVIEW

Understanding Concepts

1. How is counting a truckload of pennies like counting the atoms in 10.0 g of aluminum? How is it different?

2. The average atomic mass of nitrogen is 14 times greater than the average atomic mass of hydrogen. Would you expect the molar mass of nitrogen gas, N_2, to be 14 times greater than the molar mass of hydrogen gas, H_2? Why?

3. Why is the following statement meaningless? An industrial process requires 2.15 mol of a sugar-salt mixture.

Thinking Critically

4. **Analyzing** Determine the mass in grams of one average atomic mass unit.

Applying Chemistry

5. **Balances** If the mole is truly central to quantitative work in chemistry, why do balances in chemistry labs measure mass and not moles?

Using Moles

PREPARE

Key Concepts

The mole concept is used in solving stoichiometry problems involving both molar mass and molar volume. The ideal gas law is used in solving problems involving gases at nonstandard conditions. Theoretical yield, actual yield, percent yield, the determination of empirical and molecular formulas, and percent composition are explained.

1 FOCUS

Focus Transparency

Display the **Section Focus Transparency** for Section 12.2 to introduce the use of moles as a measure of number of particles. L1 LEP

2 TEACH

Correcting Misconceptions

Students sometimes become confused about the meaning of the coefficients in a balanced chemical equation. Have them read a chemical equation from the chalkboard. For example, students should read the equation for the formation of water as *two moles of hydrogen plus one mole of oxygen react to form two moles of water.* Emphasize that the coefficients reveal the molar relationship between the reactants and the product and provide the conversion factors needed for solving stoichiometry problems.

SECTION PREVIEW

Objectives
Predict quantities of reactants and products in chemical reactions.
Determine mole ratios from formulas for compounds.
Identify formulas of compounds by using mass ratios.

Key Terms
molar volume
ideal gas law
empirical formula

The rose's distinctive scent is mostly the odor of one compound, geraniol. The balanced chemical equation for the formation of geraniol shows the ratios of carbon, hydrogen, and oxygen that interact. Although the particles are too small to be seen or weighed, a mole of carbon, hydrogen, or oxygen is large enough. Can you use the chemical equation to predict the masses of carbon, hydrogen, and oxygen that react and the mass of geraniol that is formed?

Using Molar Masses in Stoichiometric Problems

In Chapter 11, you saw how balanced chemical equations indicate the volume of gas required for a reaction or the volume of gas produced. Similarly, you can use balanced chemical equations and numbers of moles of each substance to predict the masses of reactants or products.

SAMPLE PROBLEM **4** **Predicting Mass of a Reactant**

As you review this problem, note that nitrogen and hydrogen are related in terms of moles, not mass. In chemical reactions, when substances react, their particles react. Ammonia gas is synthesized from nitrogen gas and hydrogen gas according to the balanced chemical equation below.

$$N_2(g) + 3 H_2(g) \rightarrow 2 NH_3(g)$$

How many grams of hydrogen gas are required for 3.75 g of nitrogen gas to react completely?

Analyze
- The amount of hydrogen needed depends upon the number of nitrogen molecules present in 3.75 g and the mole ratio of hydrogen gas to nitrogen gas in the balanced chemical equation.

Set Up
- Find the number of moles of N_2 molecules by using the molar mass of nitrogen.

$$\frac{3.75 \text{ g } N_2}{} \frac{1 \text{ mol } N_2}{28.0 \text{ g } N_2}$$

Program Resources

Study Guide, pp. 49-50 L1
Laboratory Manual, 12.2, pp. 111-114; 12.3, pp. 115-118 L1
Section Focus Transparency 25 and **Master,** p. 25 L1 LEP
Basic Concepts Transparency 34 and **Master,** pp. 67-68 L1 LEP

Problem Solving Transparency 14 and **Master,** pp. 27-28 L1 LEP
ChemLab and MiniLab Worksheets, pp. 79-80, 82 L1
Applying Scientific Methods in Chemistry, pp. 34-38 L1
Consumer Chemistry, pp. 23-24 L1

To find the mass of hydrogen needed, first find the number of moles of H_2 molecules needed to react with all the moles of N_2 molecules. The balanced chemical equation shows that 3 mol of H_2 molecules react with 1 mol of N_2 molecules. Multiply the number of moles of N_2 molecules, as shown in *Set Up*, by this ratio.

$$\frac{3.75 \text{ g } N_2}{} \left|\frac{1 \text{ mol } N_2}{28.0 \text{ g } N_2}\right| \frac{3 \text{ mol } H_2}{1 \text{ mol } N_2}$$

The units in the expression above simplify to moles of H_2 molecules. To find the mass of hydrogen, multiply the number of moles of hydrogen molecules by the mass of 1 mol of H_2 molecules, which is 2.00 g.

$$\frac{3.75 \text{ g } N_2}{} \left|\frac{1 \text{ mol } N_2}{28.0 \text{ g } N_2}\right| \frac{3 \text{ mol } H_2}{1 \text{ mol } N_2}\left|\frac{2.00 \text{ g } H_2}{1 \text{ mol } H_2}\right.$$

Solve • $\dfrac{3.75 \text{ g } \cancel{N_2}}{} \left|\dfrac{1 \cancel{\text{ mol } N_2}}{28.0 \text{ g } \cancel{N_2}}\right| \dfrac{3 \cancel{\text{ mol } H_2}}{1 \cancel{\text{ mol } N_2}}\left|\dfrac{2.00 \text{ g } H_2}{1 \cancel{\text{ mol } H_2}}\right. =$

$$\frac{3.75 \times 1 \times 3 \times 2.00 \text{ g } H_2}{28.0} = 0.804 \text{ g } H_2$$

Check • Check the multiplication to verify that the result is correct.

SAMPLE PROBLEM **5** **Predicting Mass of a Product**

What mass of ammonia is formed when 3.75 g of nitrogen gas react with hydrogen gas according to the balanced chemical equation below?

$$N_2(g) + 3 H_2(g) \rightarrow 2 NH_3(g)$$

Analyze • The amount of ammonia formed depends upon the number of nitrogen molecules present and the mole ratio of nitrogen and ammonia in the balanced chemical equation.

Set Up • As in Sample Problem 4, the number of moles of nitrogen molecules is given by the expression below.

> **Problem-Solving HINT**
>
> The balanced chemical equation is the source of conversion factors relating moles of one substance to moles of another substance.

$$\frac{3.75 \text{ g } N_2}{} \left|\frac{1 \text{ mol } N_2}{28.0 \text{ g } N_2}\right.$$

To find the mass of ammonia produced, first find the number of moles of ammonia molecules that form from 3.75 g of nitrogen. Use the mole ratio of ammonia molecules to nitrogen molecules to find the number of moles of ammonia formed.

$$\frac{3.75 \text{ g } N_2}{} \left|\frac{1 \text{ mol } N_2}{28.0 \text{ g } N_2}\right| \frac{2 \text{ mol } NH_3}{1 \text{ mol } N_2}$$

Use the molar mass of ammonia, 17.0 g, to find the mass of ammonia formed.

$$\frac{3.75 \text{ g } N_2}{} \left|\frac{1 \text{ mol } N_2}{28.0 \text{ g } N_2}\right| \frac{2 \text{ mol } NH_3}{1 \text{ mol } N_2}\left|\frac{17.0 \text{ g } NH_3}{1 \text{ mol } NH_3}\right.$$

Concept Development

Describe the four steps required to solve a stoichiometry problem and work with students to make a map or algorithm to represent the process. (1) Write the balanced equation for the reaction. (2) Convert the given mass or volume of a reactant or product to moles. (3) Use the coefficients in the balanced equation to set up the appropriate mole ratio and multiply to calculate moles of the desired reactant or product. (4) Convert moles of the desired reactant or product to mass or volume as required.

GLENCOE TECHNOLOGY

 **Software**

Mastering Concepts in Chemistry
Unit 7, Lesson 2
Mass Stoichiometry
Use this lesson to reinforce the significance of chemical equations in stoichiometry. L1 LEP

Assessment

Performance: As an assessment for the demonstration on this page, ask a student to fit the inflated bag into a cardboard box that has a volume of approximately 22.4 L. Ask students to calculate the length of the side of a cube having a volume of 22.4 L (22 400 cm³). *28.2 cm* L1

Demonstration 12.4

Molar Volume
Purpose: To demonstrate the molar volume of a gas.
Materials: 100 g dry ice, large plastic garbage bag with tie, hammer, towel, balance, tongs or thermal glove
Safety Precautions: 🧤 🏋 🥽
Disposal: Code A
Procedure: CAUTION: *Use tongs or ther-mal glove to handle dry ice.* Obtain a large block of dry ice. Break off a piece large enough to make 1 mol, 44 g, of crushed dry ice. Cover the dry ice with a towel and use a hammer to crush it. Place 44 g of the crushed CO_2 into a large plastic garbage bag. Tightly twist and seal the end of the bag. Continue class discussion as students observe.
Results: In ten to 30 minutes, the garbage bag inflates as the 1 mol of CO_2 sublimes.
Analysis: Ask these questions.
1. What is the name of the process in which a solid changes directly to vapor? *sublimation*
2. What is the mass of 1 mol of CO_2? *44 g*
3. What volume does 1 mol of the gas occupy at STP? *22.4 L*

Use the reaction $N_2 + 3H_2 \rightarrow 2NH_3$ to introduce the terms *stoichiometric quantities* and *limiting reactant*. Explain that 0.50 mol of N_2 and 1.50 mol of H_2 would react completely and are, therefore, stoichiometric quantities of the two reactants in this reaction; however, 1.00 mol of N_2 would react only partially with 1.50 mol of H_2, leaving 0.50 mol of N_2 in excess. Ask students which reactant is the limiting reactant. *hydrogen* Which reactant is in excess? *nitrogen* L1

Correcting Errors

Students sometimes confuse moles and masses. For example, in the reaction $N_2 + 3H_2 \rightarrow 2NH_3$, students may think that 10.0 g of N_2 would react stoichiometrically with 30.0 g of H_2. Emphasize that the coefficients are the number of moles.

Visual Learning

Figure 12.7: Ask students to represent the equation for the formation of ammonia, given in *Correcting Errors* above, in terms of molar volumes of reactant and product gases at STP. Have them draw and label representative 22.4-L cubes—one cube for nitrogen, three for hydrogen, and two for ammonia. Ask students to explain why the equation can be represented by these volumes of gases. *Each gas volume represents 1 mol of reactant or product at STP.* L1

Solve
$$\frac{3.75 \text{ g N}_2}{1} \cdot \frac{1 \text{ mol N}_2}{28.0 \text{ g N}_2} \cdot \frac{2 \text{ mol NH}_3}{1 \text{ mol N}_2} \cdot \frac{17.0 \text{ g NH}_3}{1 \text{ mol NH}_3} =$$

$$\frac{3.75 \times 1 \times 2 \times 17.0 \text{ g NH}_3}{28.0} = 4.55 \text{ g NH}_3$$

Check
• Convince yourself that the factors are set up correctly and that the multiplication is correct.

PRACTICE PROBLEMS

5. The combustion of propane, C_3H_8, a fuel used in backyard grills and camp stoves, produces carbon dioxide and water vapor.

$$C_3H_8(g) + 5O_2(g) \rightarrow 3CO_2(g) + 4H_2O(g)$$

What mass of carbon dioxide forms when 95.6 g of propane burns?

6. Solid xenon hexafluoride is prepared by allowing xenon gas and fluorine gas to react.

$$Xe(g) + 3F_2(g) \rightarrow XeF_6(s)$$

How many grams of fluorine are required to produce 10.0 g of XeF_6?

7. Using the reaction in Practice Problem 6, how many grams of xenon are required to produce 10.0 g of XeF_6?

Fact of the MATTER

The first compounds of noble gases were synthesized by Niel Bartlett in 1962.

Figure 12.7

Molar Volumes of Gases
One mole of any gas at STP occupies 22.4 L. How large is that? It is the volume of a cube that is 28.2 cm on each edge. Each such volume contains 6.02×10^{23} atoms of a gaseous element or 6.02×10^{23} molecules of a molecular element or compound, but the mass of 1 mol is different for each element. One mole of helium floats because its mass is less than the mass of 22.4 L of air.

Using Molar Volumes in Stoichiometric Problems

In Chapter 11, you used the law of combining volumes. Avogadro inferred from that law that equal volumes of gases contain equal numbers of particles. In terms of moles, Avogadro's principle states that equal volumes of gases at the same temperature and pressure contain equal numbers of moles of gases.

The **molar volume** of a gas is the volume that a mole of a gas occupies at a pressure of one atmosphere (equal to 101 kPa) and a temperature of 0.00°C. Under these conditions of STP, the volume of 1 mol of any gas is 22.4 L, as shown in **Figure 12.7.** Like the molar mass, the molar volume is used in stoichiometric calculations.

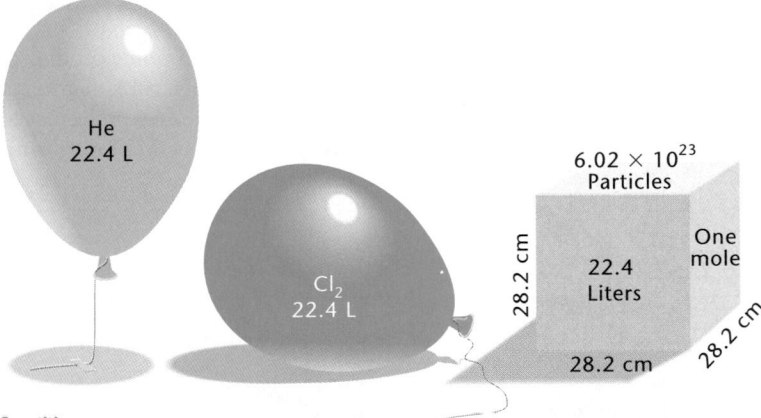

PRACTICE

Guided Practice: Guide students through the sample problems, then have them work in class on Practice Problems 5-7. See Appendix D for complete solutions. L1

Independent Practice: Your homework or classroom assignment can include the Additional Practice Problem shown here. L1

Answers to Practice Problems

5. 287 g CO_2
6. 4.65 g F_2
7. 5.35 g Xe

Additional Practice Problem

1. When potassium chlorate, $KClO_3$, is heated, it decomposes to form potassium chloride and oxygen gas.
$$2KClO_3(s) \rightarrow 2KCl(s) + 3O_2(g)$$

Everyday Chemistry

Air Bags

In 1990, on a Virginia hilltop, two cars collided in a head-on crash. Would you believe that both drivers walked away with only minor injuries? A chemical reaction, along with seat belts, saved their lives. This was the first recorded head-on collision between two cars having air bags.

How air bags function Although air bags seem to inflate instantaneously, the process occurs in steps.
1. When a car collides with a rigid barrier at a speed of 12 mph or greater, two or three sensors on the front of the car send an electric current to fire the control unit 0.01 s after impact.
2. After 0.05 s, a chemical reaction in the stored air bag creates a gaseous product that inflates it and pushes open the cover on the steering wheel or on the passenger-side dash.
3. The driver or passenger strikes the inflated bag.
4. The bag deflates in 0.045 s as the gas escapes through holes at the base of the bag.

The chemical reactions Sodium azide (NaN_3) is the chemical that produces nitrogen gas to inflate the air bag. Sodium azide pellets, an igniter, inflator, and a tightly folded nylon air bag are stored under a breakaway cover in the steering wheel or dash. The igniter provides a current to decompose the sodium azide into nitrogen gas and sodium.

$$2NaN_3(s) \xrightarrow{\text{electricity}} 3N_2(g) + 2Na(s)$$

The sodium immediately reacts with iron(III) oxide in the pellet to form sodium oxide and iron.

$$6Na(s) + Fe_2O_3(s) \rightarrow 3Na_2O(s) + 2Fe(s)$$

The sodium oxide reacts with carbon dioxide and water vapor in the air to form sodium hydrogen carbonate.

$$Na_2O(s) + 2CO_2(g) + H_2O(g) \rightarrow 2NaHCO_3(s)$$

Reliability Air bags are almost perfectly reliable. During a ten-year period, one car in a million *may* have an air-bag defect. Why are they so successful? There are no moving parts to wear out, all exposed components are tightly sealed, and the gold-plated electrical connectors corrode slowly.

Gas-volume relationships The driver's air bag requires 0.0650 m³ of nitrogen to inflate—no more, no less. The passenger's air bag needs 0.1340 m³. The pellet must have the exact amount of sodium azide needed to produce the correct amount of nitrogen. As in all applications of expanding gases, pressure and temperature affect the amount of sodium azide needed. For example, because the nitrogen gas is formed in an explosion, it has to be cooled before it goes into the air bag.

L28247

Exploring Further

1. **Applying** If 130 g of sodium azide are needed for the driver's air bag, how much is needed for the passenger's bag? Explain.

2. **Inferring** What effect does the heat from the sodium azide reaction have on the pressure and volume of the nitrogen gas formed?

Everyday Chemistry

Purpose
TECH PREP Students will learn how chemical reactions, gas laws, and stoichiometry are applied in air bags.

Teaching Strategies
- Discuss the structure of sodium azide, NaN_3, an ionic compound. One possible Lewis diagram is $Na^+[:\ddot{N}::N::\ddot{N}:]^-$. Ask students to write two other possible structures. [L1]
- A 1989 study showed that air bags used with seat belts reduced deaths of front-seat passengers by 43% to 46%.
- Have students investigate hybrid inflation air bags, which use argon. They have been in use since 1994 and have several advantages over the type discussed here.
- A good reference is Smith, J. "Caution: Air Bags at Work." *Popular Science*, June 1994, p. 36.

Extension

TECH PREP Assign a different car manufacturer to each of four groups of students. Have each group research the extent to which each company has incorporated air bags into their vehicles and the effectiveness of the bags. Groups may combine their findings and make a report to the class. [L1]
COOP LEARN

Exploring Further

1. 260 g; the volume of the passenger's air bag is twice that of the driver's; assuming other conditions are the same, twice as much NaN_3 is needed.
2. The heat increases the volume and pressure of the nitrogen gas according to the ideal gas law.

---

a. How many grams of KCl are formed when 28.0 g of $KClO_3$ decompose?

$$\frac{28.0 \text{ g } KClO_3}{1} \left| \frac{1 \text{ mol } KClO_3}{122.6 \text{ g } KClO_3} \right| \frac{1 \text{ mol } KCl}{1 \text{ mol } KClO_3}$$

$$\frac{74.6 \text{ g } KCl}{1 \text{ mol } KCl} = 17.0 \text{ g } KCl$$

b. Use the mass of KCl you determined in Part a to calculate the mass of oxygen gas produced in the reaction.

$$\frac{17.0 \text{ g } KCl}{} \left| \frac{1 \text{ mol } KCl}{74.6 \text{ g } KCl} \right| \frac{3 \text{ mol } O_2}{2 \text{ mol } KCl} \left| \frac{32 \text{ g } O_2}{1 \text{ mol } O_2} \right.$$

$$= 10.9 \text{ g } O_2$$

SAMPLE PROBLEM **6** **Using Molar Volume**

In the space shuttle, exhaled carbon dioxide gas is removed from the air by passing it through canisters of lithium hydroxide. The following reaction takes place.

$$CO_2(g) + 2LiOH(s) \rightarrow Li_2CO_3(s) + H_2O(g)$$

How many grams of lithium hydroxide are required to remove 500.0 L of carbon dioxide gas at 101 kPa pressure and 25.0°C?

Analyze • In this problem, you use the molar volume to find the number of moles.

Set Up • The volume of gas at 25°C must be converted to a volume at STP.

$$V = 500.0 \text{ L CO}_2 \left(\frac{273\text{K}}{298\text{K}} \right) = 458 \text{ L CO}_2$$

Now, find the number of moles of CO_2 gas as below.

$$\frac{458 \text{ L CO}_2 | 1 \text{ mol CO}_2}{| 22.4 \text{ L CO}_2}$$

The chemical equation shows that the ratio of moles of LiOH to CO_2 is 2 to 1. Therefore, the number of moles of lithium hydroxide is given by the expression below.

$$\frac{458 \text{ L CO}_2 | 1 \text{ mol CO}_2 | 2 \text{ mol LiOH}}{| 22.4 \text{ L CO}_2 | 1 \text{ mol CO}_2}$$

To convert the number of moles of LiOH to mass, use its molar mass, 23.9 g/mol.

$$\frac{458 \text{ L CO}_2 | 1 \text{ mol CO}_2 | 2 \text{ mol LiOH} | 23.9 \text{ g LiOH}}{| 22.4 \text{ L CO}_2 | 1 \text{ mol CO}_2 | 1 \text{ mol LiOH}}$$

Solve •
$$\frac{458 \text{ L CO}_2 | 1 \text{ mol CO}_2 | 2 \text{ mol LiOH} | 23.9 \text{ g LiOH}}{| 22.4 \text{ L CO}_2 | 1 \text{ mol CO}_2 | 1 \text{ mol LiOH}} =$$

$$\frac{458 \times 2 \times 23.9 \text{ g LiOH}}{22.4} = 977 \text{ g LiOH}$$

Check • A mass of 1000 g of lithium hydroxide is about 40 mol of the compound. According to the chemical equation, about half as much, or 20 mol of CO_2, will be removed from the air. At STP, 20 mol of any gas occupy about 450 L, which will expand to about 500 L at 25°C.

PRACTICE PROBLEMS

8. What mass of sulfur must burn to produce 3.42 L of SO_2 at 273°C and 101 kPa? The reaction is $S(s) + O_2(g) \rightarrow SO_2(g)$.

9. What volume of hydrogen gas can be produced by reacting 4.20 g of sodium in excess water at 50.0°C and 106 kPa? The reaction is $2 \text{ Na} + 2 \text{ H}_2\text{O} \rightarrow 2 \text{ NaOH} + \text{H}_2$.

PRACTICE

 **Guided Practice:** Guide students through the sample problem, then have them work in class on Practice Problems 8-9. See Appendix D for complete solutions. L1
Independent Practice: Your homework or classroom assignment can include the Additional Practice Problems shown. L1

Answers to Practice Problems

8. 2.45 g S
9. 2.31 L H_2

Additional Practice Problems

1. What mass of glucose, $C_6H_{12}O_6$, must be broken down in your body to produce 2.50 L of CO_2 at 273°C and 760 mm Hg? $C_6H_{12}O_6(aq) + 6O_2(g) \rightarrow 6H_2O(l) + 6CO_2(g)$

$$\frac{2.50 \text{ L CO}_2 | 1 \text{ mol CO}_2 | 1 \text{ mol } C_6H_{12}O_6}{1 | 22.4 \text{ L CO}_2 | 6 \text{ mol CO}_2} \cdots$$

$$\frac{180 \text{ g } C_6H_{12}O_6}{1 \text{ mol } C_6H_{12}O_6} = 3.35 \text{ g } C_6H_{12}O_6$$

2. When steel wool burns in air, this reaction occurs: $4\text{Fe}(s) + 3O_2(g) \rightarrow 2\text{Fe}_2O_3(s)$. What volume of oxygen, measured at 725 mm Hg and 25.0°C, is required to react with 100 g of iron?

Gases are much less dense than solids, as illustrated by **Figure 12.8.**

Ideal Gas Law

Exactly how the pressure P, volume V, temperature T, and number of particles n of gas are related is given by the **ideal gas law** shown here.

$$PV = nRT$$

The value of the constant R can be determined using the definition of molar volume. At STP, 1 mol of gas occupies 22.4 L. Therefore, when $P = 101.3$ kPa, $V = 22.4$ L, $n = 1$ mol, and $T = 273.15$ K, the equation for the ideal gas law can be shown as follows.

$$101.3 \text{ kPa} \times 22.4 \text{ L} = 1 \text{ mol} \times R \times 273.15 \text{ K}$$

Now, we can solve for R.

$$R = \frac{101.3 \text{ kPa}}{} \bigg| \frac{22.4 \text{ L}}{1 \text{ mol}} \bigg| \frac{1}{273.15 \text{ K}} = \frac{8.31 \text{ kPa} \cdot \text{L}}{\text{mol} \cdot \text{K}}$$

Now, you can find the volume, pressure, temperature, and number of moles of a gas.

Figure 12.8
Explosives
Gases produced in powerful chemical reactions of explosives expand with great energy. One mole of a gas occupies much more space than 1 mol of any solid.

SAMPLE PROBLEM **7** **Using the Ideal Gas Law**

How many moles are contained in a 2.44-L sample of gas at 25.0°C and 202 kPa?

Analyze • Solve the ideal gas law for n, the number of moles.

$$n = \frac{PV}{RT}$$

Set Up • $n = \dfrac{202 \text{ kPa} \times 2.44 \text{ L}}{\left(\dfrac{8.31 \text{ kPa} \cdot \text{L}}{\text{mol} \cdot \text{K}}\right) \times 298 \text{ K}}$

Solve • $n = \dfrac{202 \text{ kPa}}{} \bigg| \dfrac{2.44 \text{ L}}{} \bigg| \dfrac{1}{\dfrac{8.31 \text{ L} \cdot \text{kPa}}{\text{mol} \cdot \text{K}}} \bigg| \dfrac{1}{298 \text{ K}} =$

$$\frac{202 \text{ kPa} \times 2.44 \text{ L} \times 1 \text{ mol} \cdot \text{K}}{8.31 \text{ L} \cdot \text{kPa} \times 298 \text{ K}} = \frac{202 \text{ mol} \times 2.44}{8.31 \times 298} = 0.199 \text{ mol}$$

Check • First, find the volume that 2.44 L of a gas would occupy at STP.

$$V = \frac{2.44 \text{ L}}{} \bigg| \frac{273 \text{ K}}{298 \text{ K}} \bigg| \frac{202 \text{ kPa}}{101 \text{ kPa}} = 4.47 \text{ L}$$

Then, find the number of moles in this volume.

$$\frac{4.47 \text{ L}}{} \bigg| \frac{1 \text{ mol}}{22.4 \text{ L}} = 0.200 \text{ mol}$$

0.200 mol is close to the calculated value.

$$\frac{100 \text{ g Fe}}{1} \bigg| \frac{1 \text{ mol Fe}}{55.8 \text{ g Fe}} \bigg| \frac{3 \text{ mol O}_2}{4 \text{ mol Fe}} \bigg| \frac{22.4 \text{ L}}{1 \text{ mol O}_2} \bigg| \cdots$$

$$\frac{298 \text{ K}}{273 \text{ K}} \bigg| \frac{760 \text{ mm Hg}}{725 \text{ mm Hg}} = 34.5 \text{ L O}_2$$

Meeting Individual Needs

Learning Disabled Show students a ball-and-stick molecular model of glucose, $C_6H_{12}O_6$, and determine its mass. Then disassemble the model and form six CH_2O units to demonstrate that the empirical formula is CH_2O. Obtain the mass of a CH_2O unit and show that it has one-sixth the mass of the glucose molecule. **L1** **LEP**

Correcting Misconceptions

Explain to students that $PV = nRT$ is called the ideal gas law because its validity is generally adequate under ideal conditions—normal or lower-than-normal pressures and normal or higher-than-normal temperatures. At high pressures and low temperatures, the ideal gas law does not adequately predict the behavior of real gases.

Correcting Errors

Remind students that the value of the constant R in the ideal gas law depends upon the units of pressure used in the equation. Temperature is always in kelvins and volume is usually in liters, but pressure can be expressed in kilopascals, atmospheres, or millimeters of mercury. Ask students to calculate the value of R when pressure is in atmospheres. $0.08205 \text{ atm} \cdot L/(mol \cdot K)$ Ask them to calculate the value when pressure is in millimeters of mercury. $62.36 \text{ mm Hg} \cdot L/(mol \cdot K)$ You can decide whether you want students to use different values of R or whether you want them to consistently convert pressures to kilopascals. **L1**

Reinforcement

Review with students the algebraic steps required to solve the ideal gas equation for each of the variables P, V, n, or T. Use dimensional analysis to demonstrate that each variable has the appropriate units.

Tying to Previous Knowledge

Add the methods for converting a given volume to moles and moles to volume to the stoichiometry algorithms discussed in *Concept Development* on page 415.

miniLAB

miniLAB 2

Bagging the Gas

TECH PREP

Purpose: To determine the amount of baking soda that yields one quart of carbon dioxide when reacted with excess vinegar.

Process Skills: Measuring, predicting, interpreting data

Safety Precautions: Students should wear safety goggles and aprons. Explain that no harm is done if the closure of the bag pops open. It just indicates that too much CO_2 was produced.

Teaching Strategies

You may wish to conduct a contest to determine which group fills the bag best. Explain to students that they are unlikely to win the contest if the bag is substantially underfilled or if the bag pops open because it is overfilled.

L1 **LEP**

Expected Results

If students' calculations and laboratory procedures are carefully done, the bags will be nicely filled with CO_2.

Analysis

1. Reaction is $NaHCO_3 + HC_2H_3O_2 \rightarrow NaC_2H_3O_2 + H_2O + CO_2$
Sample student data and calculations follow.
Molar volume of CO_2 at 1.00 atm, 25 °C:
$$V = \frac{22.4 \text{ L}}{1} \cdot \frac{298 \, K}{273 \, K} = 24.4 \text{ L}$$
Mass of $NaHCO_3$ if volume of bag is 1.34 L:
$$\frac{1.34 \text{ L } CO_2}{1} \cdot \frac{1 \text{ mol } CO_2}{24.4 \text{ L } CO_2} \cdots$$
$$\frac{1 \text{ mol } NaHCO_3}{1 \text{ mol } CO_2} \cdots$$
$$\frac{84.0 \text{ g } NaHCO_3}{1 \text{ mol } NaHCO_3}$$
$$= 4.61 \text{ g } NaHCO_3$$

Bagging the Gas

Chemists and chemical engineers often need to determine amounts of reactants and products that will react efficiently and cost effectively. Use the molar volume to determine the amount of baking soda required to react with vinegar to yield just enough carbon dioxide to fill a one-quart, self-sealing plastic bag.

Procedure

1. Write the balanced equation for the reaction of baking soda (sodium hydrogen carbonate) and vinegar (acetic acid) that produces sodium acetate, water, and carbon dioxide.

2. Find the volume of a 1-quart, self-sealing plastic bag by filling it with water and then pouring the water into a graduated cylinder or measuring cup.

3. Calculate the mass of sodium hydrogen carbonate that will fill the bag with CO_2 gas when the compound reacts with excess acetic acid.

4. Put on an apron and goggles.

5. Weigh the calculated amount of sodium hydrogen carbonate and place it in a bottom corner of the bag. Use a plastic-coated twist tie to seal off this corner.

6. Pour about 60 mL of $1M$ acetic acid into the other bottom corner of the bag. Be careful not to allow the reactants to mix.

7. Place the bag in a large trash can or behind an explosion shield.

8. Release the twist tie, quickly mix the reactants, and allow the reaction to proceed to completion.

Squeeze the air from the bag and seal the zipper top.

Analysis

1. Show and explain the calculations that you used to determine the required mass of sodium hydrogen carbonate.

2. What mass of sodium hydrogen carbonate would be required to react with excess acetic acid to produce 20 000 L of carbon dioxide gas at STP for a water-treatment plant?

3. What would have happened in step 8 if the amount of acetic acid added were insufficient to react with all of the baking soda?

PRACTICE PROBLEMS

10. How many moles of helium are contained in a 5.00-L canister at 101 kPa and 30.0°C?

11. What is the volume of 0.020 mol Ne at 0.505 kPa and 27.0°C?

12. How much zinc must react in order to form 15.5 L of hydrogen, $H_2(g)$, at 32.0°C and 115 kPa?
$$Zn(s) + H_2SO_4 \rightarrow ZnSO_4(aq) + H_2(g)$$

2. Mass of $NaHCO_3$ by using molar volume:
$$\frac{20\ 000 \text{ L } CO_2}{1} \cdot \frac{1 \text{ mol } CO_2}{22.4 \text{ L } CO_2} \cdots$$
$$\frac{1 \text{ mol } NaHCO_3}{1 \text{ mol } CO_2} \cdots$$
$$\frac{84.0 \text{ g } NaHCO_3}{1 \text{ mol } NaHCO_3}$$
$$= 75.0 \text{ kg } NaHCO_3$$

3. If acetic acid were the limiting reactant, some baking soda would have been left over and less CO_2 than calculated would have been produced.

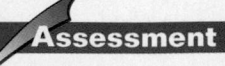

Assessment

Performance: Ask students to make an analysis of the error in this MiniLab. **L2**

Theoretical Yield and Actual Yield

The amount of product of a chemical reaction predicted by stoichiometry is called the theoretical yield. As shown earlier, if 3.75 g of nitrogen completely react, a theoretical yield of 4.55 g of ammonia would be produced. The actual yield of a chemical reaction is usually less than predicted. The collection techniques and apparatus used, time, and the skills of the chemist may affect the actual yield.

When actual yield is less than theoretical yield, you express the efficiency of the reaction as percent yield. The percent yield of a reaction is the ratio of the actual yield to the theoretical yield expressed as a percent. Suppose the actual yield was 3.86 g of ammonia. Then, the percent yield could be calculated by using the following equation.

$$\text{percent yield} = \frac{\text{actual yield}}{\text{theoretical yield}} \times 100\%$$

$$\frac{3.86 \text{ g NH}_3}{4.55 \text{ g NH}_3} \times 100\% = 84.8\%$$

This means that 84.8 percent of the possible yield of ammonia was obtained from the reaction. Calculating percentage yield is similar to calculating a baseball player's batting average, as shown in **Figure 12.9.**

A manufacturer is interested in producing chemicals as efficiently and inexpensively as possible. High yields make commercial manufacturing of substances possible. For example, in human experiments, taxol, a naturally occurring complex compound, is a strong agent against cancer. For ten years, chemists tried to synthesize this compound in the lab. In 1994, two independent academic research groups succeeded. However, the process is so complicated and time-consuming that the percent yield is probably not even one percent.

Figure 12.9
Batting Average and Percent Yield
In 1995, Alfredo Spinelli batted .352, which means:
$$\frac{\text{hits}}{\text{attempts}} = \frac{352}{1000}.$$
Just as batting averages measure a hitter's efficiency, percent yield measures a reaction's efficiency.

GLENCOE TECHNOLOGY

 Videodisc

Science and Technology Videodisc Series
Losing Weight by Design
Disc 2: Chemistry
Side 2, Ch. 19

Show this video to relate concepts of mass and weight. L1
LEP

GLENCOE TECHNOLOGY

 **Software**

Mastering Concepts in Chemistry
Unit 6, Lesson 3
Analytical Calculations Using Moles
Use this lesson to reinforce the use of moles in calculations.
L1 LEP

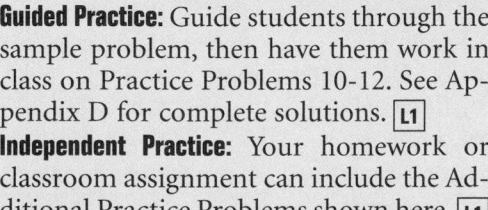

Guided Practice: Guide students through the sample problem, then have them work in class on Practice Problems 10-12. See Appendix D for complete solutions. L1
Independent Practice: Your homework or classroom assignment can include the Additional Practice Problems shown here. L1

Answers to Practice Problems

10. 0.201 mol
11. 99 L

12. 46.0 g Zn

Additional Practice Problems

1. What is the pressure of 50.0 mol of oxygen gas confined in a 1000-L cylinder at 25.0°C?
   $P = nRT/V = (50.0 \text{ mol}) (8.31 \text{ kPa} \cdot L / 1 \text{ mol} \cdot K) (298 K) (1/1000 L) = 124 \text{ kPa}$

2. Calcium fluoride reacts with sulfuric acid according to this equation.

$$CaF_2(s) + H_2SO_4(aq) \rightarrow CaSO_4(s) + 2HF(g)$$

What mass of CaF_2 is needed to produce 2.00 L of HF at 28.0°C and 98.0 kPa?

$$g\ CaF_2 = \frac{(98.0 \text{ kPa}) (2.00 L)}{\left(\frac{8.31 \text{ kPa} \cdot L}{1 \text{ mol} \cdot K}\right)(301 K)} \Big| \cdots$$

$$\frac{1 \text{ mol } CaF_2}{2 \text{ mol } HF} \Big| \frac{78.1 \text{ g } CaF_2}{1 \text{ mol } CaF_2} = 3.06 \text{ g } CaF_2$$

ChemLab

Analyzing a Mixture

Time Allotment: One and one-half laboratory periods.

Objectives

Review objectives with students before they begin the ChemLab.

Process Skills

Observing, inferring, using numbers, recognizing cause and effect, communicating

Safety Precautions and Disposal

Remind students to wear safety goggles and aprons while performing the ChemLab and to wash their hands thoroughly before leaving the laboratory. **Disposal:** Evaporate the filtrate on a hot plate. Disposal Code A for the residue and for the precipitate.

PREPARATION

- Prepare $0.500M$ $SrCl_2$ solution by dissolving 13.3 g of $SrCl_2 \cdot 6H_2O$ in sufficient distilled water to make 100. mL of solution.
- Carefully weigh amounts of anhydrous sodium sulfate and sodium chloride and mix them well so that student samples will be uniform. Record the masses.
- If you wish to eliminate any possible nonuniformity of student samples, you could weigh individual masses of sodium sulfate and sodium chloride into small bottles and assign them to specific students.

422

Analyzing a Mixture

Chemists often analyze mixtures to determine their compositions. Their analytical procedures may include gas chromatography, mass spectrometry, or infrared spectroscopy. In this ChemLab, you will use a double-displacement reaction between strontium chloride and sodium sulfate to analyze a mixture of sodium sulfate and sodium chloride.

Problem

What is the mass percent of sodium sulfate in a mixture of sodium sulfate and sodium chloride?

Objectives

- **Observe** the double-displacement reaction between strontium chloride and sodium sulfate.
- **Quantify** the amount of strontium sulfate produced.
- **Compare** the mass of strontium sulfate produced with the mass of sodium sulfate that reacted.

PREPARATION

Materials

funnel
wash bottle with distilled water
filter paper
250-mL beakers (2)
50-mL graduated cylinder
stirring rod
ring stand
iron ring
spatula
weighing dish or paper
balance

Safety Precautions

Advise students to wear safety goggles and an apron.

PROCEDURE

1. Weigh the weighing dish or weighing paper and record its mass in your data table where indicated.

2. Transfer a 0.500- to 0.600-g sample of the sodium sulfate-sodium chloride mixture to the dish or paper and weigh it. Record the total mass of the sample and container in your data table.

3. Transfer all of the mixture to a 250-mL beaker, add approximately 50 mL of distilled water, and stir slowly until the sample is completely dissolved.

4. Use a graduated cylinder to measure 15 mL of $0.500M$ strontium chloride solution, then pour it into the solution in the beaker. Stir slowly for 30 s to completely precipitate the strontium sulfate.

5. Obtain a piece of filter paper, fold it in half, and tear off one corner. Fold the paper in half again. Weigh the folded paper and record its mass in your data table where shown.

PROCEDURE

- Require that students' notebooks contain the following items before they begin: Objectives, Safety Precautions, Outline or Flow Chart of the Procedure, Data and Observations Table.
- Set aside an area, safe from disturbances and contamination, for the residues to dry overnight.
- If you prepare individual samples for students and if students are sufficiently competent in their laboratory techniques, you may wish to relate part of the lab grade to the accuracy of results.

DATA AND OBSERVATIONS

Mass of sample + dish	10.852 g
Mass of dish	0.300 g
Mass of sample	0.552 g
Mass of $SrSO_4$ + filter paper	1.274 g
Mass of filter paper	0.863 g
Mass of $SrSO_4$	0.411 g

6. Clamp a funnel support or an iron ring to a ring stand, put the funnel in the support or ring, insert the folded filter paper in the funnel, and wet the paper with a small amount of distilled water from your wash bottle. Set an empty 250-mL beaker under the funnel to receive the filtrate.

7. Carefully transfer the mixture, including all of the precipitate, from the beaker to the filter paper, using a stirring rod and a wash bottle as demonstrated by your teacher.

8. Rinse the precipitate by pouring about 10 mL of distilled water into the filter paper.

9. When liquid no longer drips from the funnel, carefully remove the filter paper and residue from the funnel. Unfold the paper and spread it on a paper towel to dry overnight. Write your name on the paper towel.

10. Dispose of the filtrate according to your teacher's instructions.

11. Weigh the dry filter paper the following day, and record the mass of filter paper with residue in your data table.

DATA AND OBSERVATIONS

Mass of sample + dish	
Subtract mass of dish	
= Mass of sample	
Mass of SrSO$_4$ + filter paper	
Subtract mass of filter paper	
= Mass of SrSO$_4$	

ANALYZE AND CONCLUDE

1. **Interpreting Data** What is the balanced equation for the reaction between sodium sulfate and strontium chloride in aqueous solution?

2. **Making Inferences** After the reaction, does the sodium chloride produced and the unreacted sodium chloride from the original sample appear in the precipitate or the filtrate?

3. **Interpreting Data** What mass of strontium sulfate is produced in the reaction? Subtract the mass of filter paper from the mass of strontium sulfate plus filter paper.

4. **Interpreting Data** What is the mass and mass percent of sodium sulfate in the original sample?

APPLY AND ASSESS

1. Could you precipitate the sulfate from the sodium sulfate in the sample by adding an excess of aluminum chloride solution? Explain.

2. If your strontium sulfate precipitate was not completely dry when you weighed it, how would it affect your value of the mass percent of sodium sulfate in the sample?

3. How could you have tested the precipitate to be sure it contained strontium?

ANALYZE AND CONCLUDE

1. $Na_2SO_4(aq) + SrCl_2(aq) \rightarrow SrSO_4(s) + 2NaCl(aq)$
2. in the filtrate
3. 0.411 g
4. 0.411 g SrSO$_4$ × 1 mol SrSO$_4$/187 g SrSO$_4$ × 1 mol Na$_2$SO$_4$/1 mol SrSO$_4$ × 142 g Na$_2$SO$_4$/1 mol Na$_2$SO$_4$ = 0.312 g Na$_2$SO$_4$
   0.312 g Na$_2$SO$_4$/0.552 g sample × 100 = 56.5% Na$_2$SO$_4$

APPLY AND ASSESS

1. No, aluminum sulfate is soluble so no precipitate would form.
2. It would be higher.
3. Do a flame test on the precipitate; strontium compounds produce a red flame.

Assessment

Performance: Ask students to write and submit a formal report for the ChemLab. L1

*inter*NET
CONNECTION
A hypercard stack for reviewing the concept of moles can be downloaded.
Anonymous FTP

ftp://archive.umich.edu/mac/misc/chemistry/moleconcept.sit.hqx

Purpose

Students will learn how chemists vary conditions in reactions to improve yields of industrial chemicals.

Background

- The sulfuric acid industry started in the mid-1700s with the lead-chamber process. The contact process was patented in England in 1831.
- A simplified set of reactions for the lead-chamber process follows.

$$S(s) + O_2(g) \rightarrow SO_2(g)$$
$$2NO(g) + O_2(g) \rightarrow 2NO_2(g)$$
$$NO_2(g) + SO_2(g) \rightarrow NO(g) + SO_3(g)$$
$$SO_3(g) + H_2O(l) \rightarrow H_2SO_4(aq)$$

- Recent research using antibodies as catalysts has produced a 98% yield of a desired product. Without the antibody, a complex mixture of by-products formed. Some enantiomeric (mirror-image isomers) pharmaceuticals that had low yields dramatically improved with antibody catalysts.

Teaching Strategies

- Discuss the reactions for the contact process. Point out that the reactions in step 2 are important because they can significantly affect yield.
- Discuss the reactions for the lead-chamber process. Ask students to compare and contrast the two processes. Ask students why the chamber process is less efficient. [L1]
- Have students do research and write a report on how environmentalists and chemists can work together to improve and preserve the environment. Students' reports may be placed in their portfolios. [L2] [P]

Improving Percent Yield in Chemical Synthesis

The single most important industrial chemical in the world is probably sulfuric acid. In the United States, its production exceeds 40 million tons per year. The strength of the U.S. economy can be gauged by the amount of sulfuric acid produced annually. Over the years, the manufacturing process of sulfuric acid has been improved to provide a higher and more economical yield.

Uses of Sulfuric Acid

Sixty percent of all manufactured sulfuric acid is used to make fertilizers. It is also used in manufacturing detergents, photographic film, synthetic fibers, pigments, paints, drugs, and other acids. It is the electrolyte in some batteries, acts as a catalyst and a dehydrating agent, and is a component in refining petroleum and metals.

The Lead-Chamber Process—A Low-Yield Process

The first industrial method, the lead-chamber process, is not commonly used now because its purity is low and its percent yield is only 60 to 80 percent. But it is much cheaper than the later and more productive contact process. The lead-chamber process is used for manufacturing sulfuric acid for applications that do not demand high purity.

The Contact Process—The High-Yield Process

The contact process is the most widely used commercial method. It is more expensive than the lead-chamber process, but it is simple and it produces high-purity sulfuric acid at a high percent yield—about 98 percent. In addition, it creates no by-products that pollute the atmosphere. The contact process has four steps.

Step 1

Sulfur is burned in air to produce sulfur dioxide, a stable compound. This reaction takes place quickly and readily.

$$S(s) + O_2(g) \rightarrow SO_2(g)$$

Impurities produced in combustion are removed from the sulfur dioxide so they will not react with and impede the catalyst in the next step.

Sulfur melter

Melted sulfur

O_2

Air

O_2

Water

H_2O

Maximizing Percent Yield

Reaction yields in the contact process were increased in several ways—by selecting an optimum temperature, using an efficient catalyst, removing a product from a reaction that does not go to completion, and by controlling the rate of reaction of SO_3 with water. Keeping operating pressures at the correct values also increases yield.

Using Scientific Methods to Attain the Best Yields

To develop higher yields that are closer to the theoretical values, chemists adjust temperatures, pressures, or other conditions in their industrial processes. They look for better catalysts or new ways to deal with undesirable side reactions. Unwanted by-products are serious concerns to the chemist. Because they may cause environmental damage or be expensive to dispose of, by-products may raise production costs.

Chemistry Journal

Percent Na and K in Additives

Ask students to look on the labels of foods or beverages, and note the names of at least two additive compounds that contain sodium or potassium. Have them look up the formulas in the *Merck Index* or another reference book, and calculate the mass percent of sodium or potassium in the additives. Ask students to include the data and calculations in their journals. [L1]

Step 3
Because sulfur trioxide reacts violently with water, it is bubbled through 98 percent concentrated sulfuric acid and forms pyrosulfuric acid, $H_2S_2O_7$.

$$SO_3(g) + H_2SO_4(l) \rightarrow H_2S_2O_7(l)$$

Step 4
When water is added to the pyrosulfuric acid, high-quality, 98 percent concentrated sulfuric acid is formed.

$$H_2S_2O_7(l) + H_2O(l) \rightarrow 2 H_2SO_4(l)$$

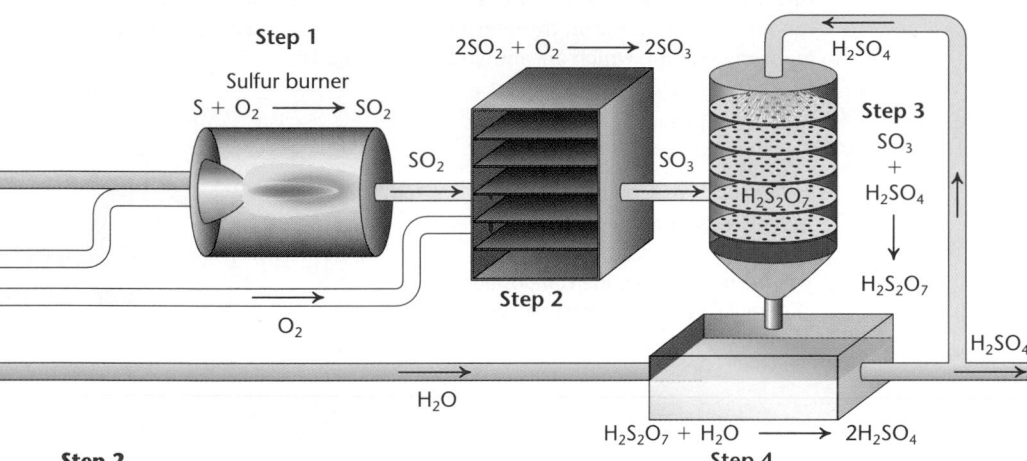

Step 1
Sulfur burner
$S + O_2 \rightarrow SO_2$

$2SO_2 + O_2 \rightarrow 2SO_3$

Step 2

SO_2

SO_3

O_2

H_2O

Step 3
SO_3
+
H_2SO_4
$\downarrow$
$H_2S_2O_7$

H_2SO_4

H_2SO_4

$H_2S_2O_7 + H_2O \rightarrow 2H_2SO_4$
Step 4

Step 2
Because sulfur dioxide reacts slowly with excess oxygen, a catalyst, either vanadium pentoxide (V_2O_5) or finely divided platinum at a temperature of 400°C, is used. The reaction produces sulfur trioxide.

$$2SO_2(g) + O_2(g) \xrightarrow{catalyst} 2SO_3(g)$$

Sulfur trioxide is quickly removed from the contact chamber because it tends to form sulfur dioxide and oxygen again. Removing sulfur trioxide promotes the production of more sulfur trioxide.

Green Chemistry

The contact process is economically sound because the reactants are abundant and inexpensive and because it produces no unwanted by-products, which may be expensive to store or dispose of and which may pollute the environment. Ecologically sound chemical manufacturing is called green chemistry or green technology.

DISCUSSING THE TECHNOLOGY

1. **Hypothesizing** Roasting iron pyrite, FeS_2, can replace burning sulfur in Step 1 to produce sulfur dioxide. Write an equation to show what you think the reaction with pyrite is.

2. **Acquiring Information** Find out how sulfuric acid is used to manufacture hydrochloric acid and write the equation(s) of the reactions.

- A good article for background information on this subject is Illman, D. "Environmentally Benign Chemistry Aims for Processes That Don't Pollute." *C&E News*, Sept. 5, 1994, pp. 22-27. Also see *C&E News*, June 13, 1994, pp. 43-44 and Sept. 19, 1994, pp. 38-43.

Discussing the Technology

1. Two reactions are possible.
$2FeS_2 + 5O_2 \rightarrow 4SO_2 + 2FeO$
$4FeS_2 + 11O_2 \rightarrow 8SO_2 + 2Fe_2O_3$

2. The laboratory preparation and 5% of the industrial preparation use the reaction of concentrated sulfuric acid and sodium chloride.
$H_2SO_4(aq) + NaCl(s) \rightarrow HCl(g) + NaHSO_4(aq)$
Hydrogen chloride gas is bubbled through water to form hydrochloric acid.
$HCl(g) + H_2O(l) \rightarrow H_3O^+(aq) + Cl^-(aq)$

GLENCOE TECHNOLOGY

 CD-Rom

Chemistry: Concepts and Applications
How much oxygen is available? Have students use this interactive experiment to practice determining yields of a product in a reaction. **L1** **LEP**

Cultural Diversity

Norbert Rillieux—Chemical Engineer

"...the most important chemical engineer in America." These words, quoted from the *1993 Blackfax Calendar*, describe Norbert Rillieux for his contributions to science and engineering. Born of mixed-race parents, Rillieux suffered racial discrimination throughout his life, but his creative genius and scientific achievements place him among the most

remarkable free blacks in early America. Having observed the labor-intensive refining of sugarcane, Rillieux perfected a vacuum-evaporating process that produced a finer grade of sugar at nearly half the cost. Mexican and Cuban sugar planters quickly adopted Rillieux's method, and soon American sugar plantations followed. In 1856, Rillieux discov-

ered that his process had been copied and applied unsuccessfully to the refining of beet sugar. It took him 19 years to prove his methods productive and receive new patents for refining beet sugar. Today, Rillieux's sugar-refining methods are being applied to the production of products such as condensed milk, gelatin, soap, and glue.

Determining Mass Percents

As you know, the chemical formula of a compound tells you the elements that comprise it. For example, the formula for geraniol (the main compound that gives a rose its scent) is $C_{10}H_{18}O$. The formula shows that geraniol is comprised of carbon, hydrogen, and oxygen. Because all these elements are nonmetals, geraniol is probably covalent and comprised of molecules.

In addition, the formula $C_{10}H_{18}O$ tells you that each molecule of gera-niol contains ten carbon atoms, 18 hydrogen atoms, and one oxygen atom. In terms of numbers of atoms, hydrogen is the major element in geraniol. How can you tell whether it is the major element by mass? You can answer this question by determining the mass percents of each ele-ment in geraniol, which are shown in **Figure 12.10**.

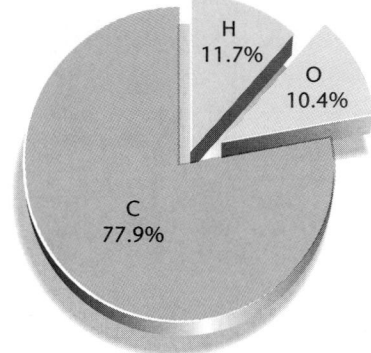

Figure 12.10

Mass Percents of Elements in Geraniol

This pie graph shows the composition of geraniol in terms of mass percents of the elements.

Suppose you have a mole of geraniol. Its molar mass is 154 g/mol. Of this mass, how many grams do the carbon atoms contribute? The formula shows that one molecule of geraniol includes ten atoms of carbon. Therefore, 1 mol of geraniol contains 10 mol of carbon. Multiply the mass of 1 mol of carbon by 10 to get the mass of carbon in 1 mol of geraniol.

$$\frac{10 \ \cancel{mol} \ | \ 12.0 \ g \ C}{\cancel{mol}} = 120 \ g \ C$$

Now, use this mass of carbon to find the mass percent of carbon in geraniol.

$$\% \ C = \frac{120 \ g \ C}{154 \ g \ C_{10}H_{18}O} \times 100\% = 77.9\%$$

The mass percents of the other elements are calculated below in a similar fashion.

Mass of hydrogen in 1 mol geraniol:

$$18 \ \cancel{mol} \times \frac{1.00 \ g \ H}{\cancel{mol}} = 18.0 \ g \ H$$

$$\text{Mass percent of H} = \frac{\text{mass of H}}{\text{mass of geraniol}} \times 100\%$$

$$= \frac{18.0 \ g \ H}{154 \ g \ C_{10}H_{18}O} \times 100\% = 11.7\% \ H$$

Mass of oxygen in 1 mol geraniol:

$$\frac{1 \; \cancel{mol} \;|\; 16.0 \text{ g O}}{|\; \cancel{mol}} = 16.0 \text{ g O}$$

$$\text{Mass percent of O} = \frac{\text{mass of O}}{\text{mass of geraniol}} \times 100\%$$

$$= \frac{16.0 \text{ g O}}{154 \text{ g C}_{10}\text{H}_{18}\text{O}} \times 100\% = 10.4\% \text{ O}$$

In the example above, you learned how to use the chemical formula and the molar masses to find the mass percent of a compound. You can also solve the reverse problem. You can use the mass percent to find the chemical formula of an unknown compound.

Determining Chemical Formulas

Suppose you analyzed an unknown compound and found that, by mass, it was 18.8 percent sodium, 29.0 percent chlorine, and 52.2 percent oxygen. Because this compound contains some metal and nonmetal elements, it may be an ionic compound. To determine its chemical formula, find the relative numbers of sodium, chlorine, and oxygen atoms in the formula unit of the compound.

Suppose you have 100.0 g of the unknown compound. Because you know the sample includes 18.8 g of sodium, 29.0 g of chlorine, and 52.2 g of oxygen, use the molar mass to find the number of moles of each element.

$$\frac{18.8 \; \cancel{\text{g Na}} \;|\; 1 \text{ mol Na}}{|\; 23.0 \; \cancel{\text{g Na}}} = 0.817 \text{ mol Na}$$

$$\frac{29.0 \; \cancel{\text{g Cl}} \;|\; 1 \text{ mol Cl}}{|\; 35.5 \; \cancel{\text{g Cl}}} = 0.817 \text{ mol Cl}$$

$$\frac{52.2 \; \cancel{\text{g O}} \;|\; 1 \text{ mol O}}{|\; 16.0 \; \cancel{\text{g O}}} = 3.26 \text{ mol O}$$

You know that atoms combine in ratios of small whole numbers to form compounds. To find the whole-number ratios, divide the mole numbers by the smallest one.

$$\frac{0.817 \text{ mol Na}}{0.817} = 1.00 \text{ mol Na}$$

$$\frac{0.817 \text{ mol Cl}}{0.817} = 1.00 \text{ mol Cl}$$

$$\frac{3.26 \text{ mol O}}{0.817} = 3.99 \text{ mol O}$$

These values are whole numbers or very close to whole numbers. Therefore, the mole ratio for this compound is 1 mol Na :1 mol Cl :4 mol O. Similarly, the ratio of atoms in this compound is 1 Na :1 Cl :4 O.

Concept Development

To begin discussion of how the formulas of compounds are determined, use this example. Fish in some lakes and rivers have been found to contain a mercury compound, possibly a contaminant from the processing of wood pulp. Analysis of this compound gives the following mass percentages. Carbon, 5.57%; hydrogen, 1.40%; and mercury, 93.03%. From this information, determine the empirical formula of the compound. *Carbon: 5.57 g/12.01 g/mol = 0.464 mol. Hydrogen 1.40 g/1.008 g/mol = 1.39 mol. Mercury: 93.03 g/200.59 g/mol = 0.464 mol. The mole ratio C:H:Hg is 1.00:3.00:1.00, and the empirical formula is CH₃Hg, methyl mercury.* L1

Transparency

Display **Problem Solving Transparency 14** so students can practice solving mass-mass problems. L1 LEP

Display **Basic Concepts Transparency 34** to show mole relationships. L1 LEP

---

sium is consumed before enough gas is produced to fill the cylinder.
2. What is the balanced chemical equation for the reaction? *See Results.*
3. What is the mole ratio of magnesium to hydrogen gas in the equation? *1:1*

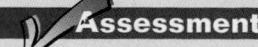
Assessment

Performance: Have students repeat this demon-

stration as a laboratory experiment. Measurements of volume, temperature, pressure, and mass of magnesium can be made and then used to introduce related calculations using *PV = nRT.* For greater accuracy, use a 50-mL gas-measuring tube instead of the graduated cylinder. L1

Correcting Errors

Point out that in determining the formula of a substance, the molar mass must be close to a whole-number multiple of the mass of the empirical formula. If it is not, there is probably an error in the data or the calculations.

Revealing Misconceptions

Students sometimes find it mysterious that when the composition of a substance is given in mass percents, the percentages are taken to be grams in the formula calculations. Point out that the easiest way to deal with percent composition data is to assume that you have a 100-g sample; then the percentage of each element is the same as its mass.

The formula of a compound having the smallest whole-number ratio of atoms in the compound is called the **empirical formula.** The empirical formula of this unknown compound is $NaClO_4$.

What is the chemical formula for this compound? You have learned that the formula for an ionic compound represents the simplest possible ratio of the ions present and is called a formula unit. Chemical formulas for most ionic compounds are the same as their empirical formulas. Because the unknown compound is ionic, the chemical formula for a formula unit of the compound is the same as its empirical formula, $NaClO_4$. The compound is called sodium perchlorate.

As another example, suppose the mass percents of a compound are 40.0 percent carbon, 6.70 percent hydrogen, and 53.3 percent oxygen. Because all its elements are nonmetals, the compound is covalent. Imagine you have 100 g of the compound. Then you have 40.0 g of carbon, 6.70 g of hydrogen, and 53.3 g of oxygen. Use the molar masses of these elements to find the number of moles of each element.

$$\frac{40.0 \text{ g C}}{} \cdot \frac{1 \text{ mol C}}{12.0 \text{ g C}} = 3.33 \text{ mol C}$$

$$\frac{6.70 \text{ g H}}{} \cdot \frac{1 \text{ mol H}}{1.00 \text{ g H}} = 6.70 \text{ mol H}$$

$$\frac{53.3 \text{ g O}}{} \cdot \frac{1 \text{ mol O}}{16.0 \text{ g O}} = 3.33 \text{ mol O}$$

Now, divide all the mole numbers by the smallest one.

$$\frac{3.33 \text{ mol C}}{3.33} = 1.00 \text{ mol C}$$

$$\frac{6.70 \text{ mol H}}{3.33} = 2.01 \text{ mol H}$$

$$\frac{3.33 \text{ mol O}}{3.33} = 1.00 \text{ mol O}$$

Because the ratio of moles is the same as the ratio of atoms, CH_2O is the empirical formula for this compound. But, the empirical formula is not always the chemical formula. Many different covalent compounds have the same empirical formula, as demonstrated in **Figure 12.11**, because atoms can share electrons in different ways.

WORD ORIGIN

empirical
empeirikos (GK)
working from experience

From experimental work, the empirical formula of an unknown compound can be derived.

Figure 12.11

Compounds Having the Empirical Formula CH_2O
For each of these compounds, the ratio of atoms is 1C : 2H : 1O. Because one molecule of each compound has a different number of each atom, it is a different compound with its own molecular formula.

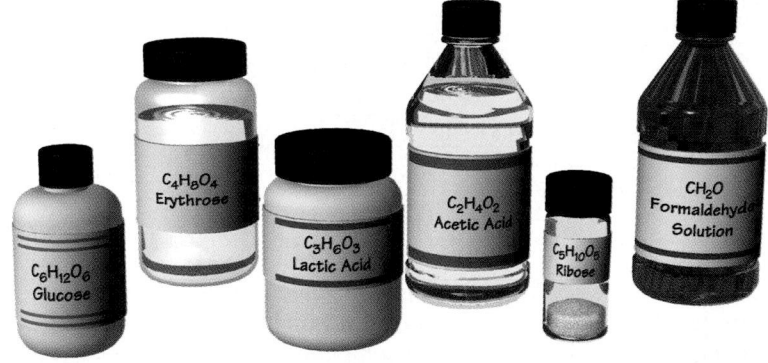

Integrating THE Sciences

Atmospheric Science In 1995, the Nobel Prize in Chemistry was awarded to the scientists who, in the early 1970s, proposed the mechanism for the depletion of the ozone layer. The ozone layer is a protective layer of ozone molecules, O_3, in the upper atmosphere. It protects Earth from high-energy solar radiation. One part of the mechanism for ozone destruction involves chlorine atoms that are produced by high-energy reactions between sunlight and compounds called chlorofluorocarbons (CFCs). In these reactions, one chlorine atom may cause the destruction of 100 000 ozone molecules. Have interested students investigate the mechanism for ozone destruction and write a paper in which they discuss the problems associated with banning CFCs. Student papers may be placed into their portfolios. L2 P

To decide which multiple of the empirical formula is the correct molecular formula, you need the molar mass of the compound. Suppose a separate analysis shows that the molar mass of the compound is 90.0 g/mol. Therefore, the molecular mass of the compound is 90.0 u. The molecular mass of a CH_2O molecule is 30.0 u. By dividing the molecular mass of the compound by 30.0 u, you find the multiple.

$$\frac{90.0\ u}{30.0\ u} = 3$$

The molecular formula of the compound contains three empirical formula units. The molecular formula is $C_3H_6O_3$. The compound is lactic acid, the sour-tasting substance in spoiled milk. Lactic acid is also produced by active muscles, causing them to feel sore after strenuous activity.

Connecting Ideas

In this chapter, you learned to solve several kinds of stoichiometric problems. For each kind, you used the mole concept because when substances react, their particles interact. The number of particles at this submicroscopic level controls what happens macroscopically. In the next chapter, you will use the mole concept and the particle nature of matter to study the mixtures of substances called solutions.

SECTION REVIEW

Understanding Concepts

1. Octane, C_8H_{18}, is one of the many components of gasoline. Write the balanced chemical equation for the combustion of octane to form carbon dioxide gas and water vapor. Identify as many mole ratios among the substances in the combustion as you can.

2. If 25.0 g of octane burn as in Problem 1, how many grams of water will be produced? How many grams of carbon dioxide?

3. A manufacturer advertises a new synthesis reaction for methane with a percent yield of 110 percent. Comment on this claim.

Thinking Critically

4. **Predicting** The reaction of iron(III) oxide and aluminum is called the thermite reaction because of its intense heat. The iron produced is molten and was formerly used to weld railroad tracks in remote areas. Its balanced chemical equation is shown below.

$$Fe_2O_3\ (s) + 2Al(s) \rightarrow Al_2O_3(s) + 2Fe(l)$$

If 20.0 g of each reactant are used, which is the limiting reactant?

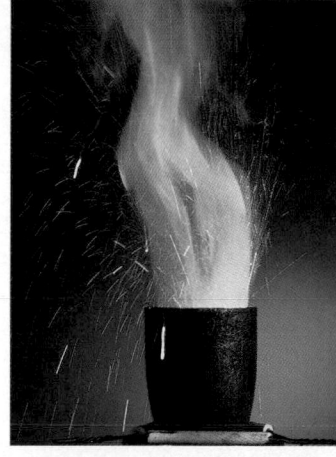

Applying Chemistry

5. **Blue Jeans** Indigo, the dye used to color blue jeans, is prepared using sodium amide. Sodium amide contains the following mass percents of elements: hydrogen, 5.17 percent; nitrogen, 35.9 percent; and sodium, 58.9 percent. Find an empirical formula for sodium amide.

SECTION REVIEW

1. $2C_8H_{18}(l) + 25O_2(g) \rightarrow 16CO_2(g) + 18H_2O(g)$
   16 mol CO_2/2 mol C_8H_{18}; 18 mol H_2O/2 mol C_8H_{18}; 25 mol O_2/2 mol C_8H_{18}; 16 mol CO_2/25 mol O_2; 16 mol CO_2/18 mol H_2O, etc.
2. 35.5 g H_2O, 77.2 g CO_2
3. Because a balanced chemical equation is based on the law of conservation of matter, a yield higher than 100% is impossible.

4. Fe_2O_3 is the limiting reactant.
5. 5.17 g H × 1 mol H/1.01 g H = 5.12 mol H
   35.9 g N × 1 mol N/14.0 g N = 2.56 mol N
   58.9 g Na × 1 mol Na/23.0 g Na = 2.56 mol Na
   Ratio H:N:Na = 2:1:1; empirical formula is $NaNH_2$

SECTION 12.2

3 ASSESS

Check for Understanding
- Check for understanding of the section material by having students answer the Section Review questions.
- Create a pretest consisting of one or more of each type of problem introduced in this section. Divide the class into groups and have each group work each problem. Compare the answers and have the groups resolve any discrepant answers by working those problems on the chalkboard. [L1] COOP LEARN

Reteach
Cooperative Learning: Divide the class into teams of four students. Have each team make up one problem similar to each sample problem in this section. Each problem should be solved by the team that created it; then teams can exchange their problems and challenge each other to solve them. [L1] COOP LEARN

Extension
Ask an interested student to make a poster of the stoichiometry problem-solving process or algorithm. Display the poster and review the process with the class. [L2]

4 CLOSE

Discussion
One cylinder at STP contains 2.50×10^{23} atoms of helium. Another cylinder, half the size of the first, also at STP, contains 2.50×10^{23} atoms of hydrogen. Ask students how it is possible that the two samples of gas can have the same number of atoms at the same temperature and pressure but different volumes. *The hydrogen atoms are combined into diatomic molecules so there are only 1.25×10^{23} gas particles in the smaller cylinder.* [L1]

CHAPTER 12 REVIEW

- Review the Reviewing Main Ideas statements and Key Terms with your students.
- Complete solutions to Chapter Review problems can be found in the *Problems and Solutions Manual.*

UNDERSTANDING CONCEPTS

1. 1345 lb × 454 g/lb × 50 nickels/232 g × \$1/20 nickels = \$6580

2. 20 000 bolts, 40 000 washers, 60 000 nuts

3. Weigh 1000 washers. Divide the total mass of the washers by the mass of 1000 washers and then multiply by 1000.

4. **a.** 160 g/mol **b.** 39.9 g/mol **c.** 331 g/mol **d.** 92.0 g/mol

5. 0.498 mol Cu_2SO_4 × 223 g Cu_2SO_4/1 mol Cu_2SO_4 = 111 g Cu_2SO_4

6. One sucrose molecule has the largest mass, 342 u.

7. One mole of sucrose has the largest mass, 342 g.

8. 352 u; 352 g/mol

9. 68.7 g Fe × 1 mol Fe/55.8 g Fe × 1 mol Cu/1 mol Fe × 63.5 g Cu/1 mol Cu = 78.2 g Cu

10. The mass of a Cu atom is greater than the mass of an Fe atom.

11. **a.** 157 g/mol **b.** 294 g/mol **c.** 149 g/mol **d.** 242 g/mol

12. 23.0 g + 14.0 g + 32.0 g = 69.0 g/mol
0.345 mol $NaNO_2$ × 69.0 g $NaNO_2$/1 mol $NaNO_2$ = 23.8 g $NaNO_2$

13. **a.** 34.5 g **b.** 3.01 × 10^{23} atoms
c. 1.10 mol
d. 10.8 g $MgBr_2$ × 1 mol $MgBr_2$/184 g $MgBr_2$ × 3 mol ions/1 mol $MgBr_2$ × 6.02 × 10^{23} ions/mol = 1.06 × 10^{23} ions

14. **a.** 28.0 g/mol **b.** 254 g/mol **c.** 32.0 g/mol **d.** 58.7 g/mol

15. $2NaNO_3 \rightarrow 2NaNO_2 + O_2$
128 g O_2 × 1 mol O_2/32.0 g O_2 × 2 mol $NaNO_3$/1 mol O_2 × 85.0 g $NaNO_3$/1 mol $NaNO_3$ = 680 g $NaNO_3$

REVIEWING MAIN IDEAS

12.1 Counting Particles of Matter

- The number of particles (atoms, molecules, or ions) in macroscopic matter controls the consumption and formation of substances in chemical reactions.

- One mole equals 6.02 × 10^{23}.

- Use the molar mass to convert mass to moles or moles to mass.

12.2 Using Moles

- A balanced chemical equation provides mole ratios of the substances in the reaction.

- The mole is a central concept in making chemical calculations.

- The ideal gas law is expressed in the following equation.

$$PV = nRT$$

- Percent yield measures the efficiency of a chemical reaction.

$$\text{Percent yield} = \frac{\text{actual yield}}{\text{theoretical yield}} \times 100\%$$

- Percent composition can be determined from the chemical formula of a compound.

- The empirical formula of a compound can be determined from its percent composition.

- The chemical formula of a compound can be determined if the molar mass and the empirical formula are known.

Key Terms

For each of the following terms, write a sentence that shows your understanding of its meaning.

Avogadro constant	molar volume
empirical formula	mole
formula mass	molecular mass
ideal gas law	stoichiometry
molar mass	

● UNDERSTANDING CONCEPTS

1. Your uncle leaves you a barrel of nickels. You determine that the nickels weigh 1345 pounds and that 50 nickels weigh 232 g. How many dollars is the barrel of nickels worth?

2. A manufacturer must supply 20 000 connector units, each consisting of a bolt, two washers, and three nuts. How many of each part are needed?

3. Explain how you would use weighing to count 40 000 washers.

4. What is the molar mass of the following substances?
   a) bromine   **c)** lead(II) nitrate
   b) argon   **d)** dinitrogen tetroxide

5. A reaction requires 0.498 mol of Cu_2SO_4. How many grams must you weigh to obtain this number of formula units?

6. Which has the largest mass?
   a) three atoms of magnesium
   b) one molecule of sucrose, $C_{12}H_{22}O_{11}$
   c) ten atoms of helium

7. Which has the largest mass?
   a) 3 mol of magnesium
   b) 1 mol of sucrose, $C_{12}H_{22}O_{11}$
   c) 10 mol of helium

✓ Assessment

Portfolio: Review the portfolio options that are provided throughout the chapter. Encourage students to select one product that demonstrates their best work for the chapter. Have students explain what they have learned and why they chose this example for placement into their portfolios. **P**

Additional portfolio options can be found in the following **Teacher Classroom Resources:**
Consumer Chemistry
Applying Scientific Methods in Chemistry
Laboratory Manual

8. What is the molecular mass of UF_6? What is the molar mass of UF_6?

9. What mass of copper contains the same number of atoms as 68.7 g of iron?

10. Explain why the mass of copper is not equal to the mass of iron in question 9.

11. Determine the molar mass of each of the following compounds.

   a) C_6H_5Br c) $(NH_4)_3PO_4$
   b) $K_2Cr_2O_7$ d) $Fe(NO_3)_3$

12. Calculate the mass of 0.345 mol of sodium nitrite, $NaNO_2$.

13. Calculate each of the following:

   a) the mass in grams of 0.254 mol of calcium sulfate
   b) the number of atoms in 2.0 g of helium
   c) the number of moles in 198 g of glucose, $C_6H_{12}O_6$
   d) the total number of ions in 10.8 g of magnesium bromide

14. What are the molar masses of the following elements?

   a) nitrogen   c) oxygen
   b) iodine   d) nickel

15. Sodium nitrate decomposes upon heating to form sodium nitrite and oxygen gas. This reaction is sometimes used to produce small quantities of oxygen in the lab. How many grams of sodium nitrate must be heated to produce 128 g of oxygen?

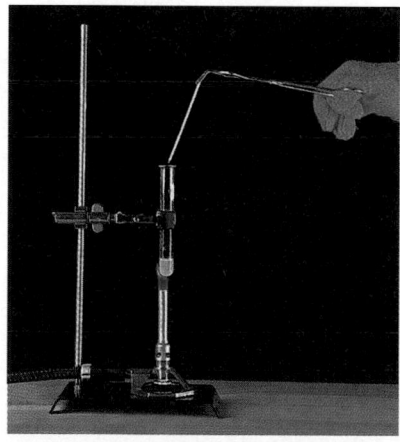

16. Calculate the mass of each product formed when 85.6 g of silver sulfide, Ag_2S, react with excess hydrochloric acid, HCl.

17. Aluminum nitrite and ammonium chloride react to form aluminum chloride, nitrogen, and water. What mass of aluminum chloride is present after 43.0 g of aluminum nitrite and 43.0 g of ammonium chloride have reacted completely?

18. A 10.0-g sample of magnesium reacted with excess hydrochloric acid to form magnesium chloride by the reaction below. When the reaction was complete, 30.8 g of magnesium chloride were recovered. What was the percent yield?

   $$Mg(s) + 2HCl(aq) \rightarrow MgCl_2(aq) + H_2(g)$$

19. What is the molecular formula for each of the following compounds?

   a) empirical formula: CH_2;
      molar mass: 42 g/mol
   b) empirical formula: CH;
      molar mass: 78 g/mol
   c) empirical formula: NO_2;
      molar mass: 92 g/mol

20. An oxide of nitrogen is 26 percent nitrogen by mass. The molar mass of the oxide is approximately 105 g/mol. What is the formula of the compound?

21. How many moles are in each of the following samples?

   a) 43.6 g NH_3
   b) 5.0 g of aspirin, $C_9H_8O_4$
   c) 15.0 g CuO

● APPLYING CONCEPTS

22. Hydrogen fuels are rated with respect to their hydrogen content. Determine the percent hydrogen for the following fuels.

   a) ethane, C_2H_6
   b) methane, CH_4
   c) whale oil, $C_{32}H_{64}O_2$

Chapter 12   Review   431

16. $Ag_2S + 2HCl \rightarrow 2AgCl + H_2S$
   85.6 g $Ag_2S \times$ 1 mol Ag_2S / 248 g $Ag_2S \times$ 2 mol AgCl/ 1 mol $Ag_2S \times$ 143 g AgCl/mol AgCl = 98.7 g AgCl
   85.6 g $Ag_2S \times$ 1 mol Ag_2S/248 g $Ag_2S \times$ 1 mol H_2S/1 mol $Ag_2S \times$ 34.1 g H_2S/mol H_2S = 11.8 g H_2S

17. $Al(NO_2)_3 + 3NH_4Cl \rightarrow AlCl_3 + 3N_2 + 6H_2O$
   43.0 g $Al(NO_2)_3 \times$ 1 mol $Al(NO_2)_3$/165 g $Al(NO_2)_3 \times$ 1 mol $AlCl_3$/1 mol $Al(NO_2)_3 \times$ 134 g $AlCl_3$/1 mol $AlCl_3$ = 34.9 g $AlCl_3$
   43.0 g $NH_4Cl \times$ 1 mol NH_4Cl/53.5 g $NH_4Cl \times$ 1 mol $AlCl_3$/3 mol $NH_4Cl \times$ 134 g $AlCl_3$/1 mol $AlCl_3$ = 35.9 g $AlCl_3$
   $Al(NO_2)_3$ is limiting. Reaction produces 34.9 g $AlCl_3$.

18. Theoretical yield:
   10.0 g Mg $\times$ 1 mol Mg/24.3 g Mg $\times$ 1 mol $MgCl_2$/1 mol Mg $\times$ 95.2 g $MgCl_2$/1 mol $MgCl_2$ = 39.2 g $MgCl_2$
   % yield = 30.8 g/39.2 g $\times$ 100 = 78.6%

19. **a.** C_3H_6 **b.** C_6H_6 **c.** N_2O_4

20. 26 g N $\times$ 1 mol N/14.0 g N = 1.9 mol N
   74 g O $\times$ 1 mol O/16.0 g O = 4.6 mol O
   The mole ratio of oxygen to nitrogen is 4.6:1.9 or 5:2. The empirical formula is N_2O_5; molar mass = 108 g/mol; formula is N_2O_5.

21. **a.** 2.56 mol NH_3 **b.** 0.028 mol $C_9H_8O_4$
   c. 0.189 mol CuO

APPLYING CONCEPTS

22. **a.** 6.0 g/30 g $\times$ 100 = 20%
   b. 4.0 g/16 g $\times$ 100 = 25%
   c. 64 g/480 g $\times$ 100 = 13%

23. 76.5 g C $\times$ 1 mol C/12.0 g C = 6.37 mol C
   12.1 g H $\times$ 1 mol H/1.01 g H = 12.0 mol H
   11.3 g O $\times$ 1 mol O/16.0 g O = 0.71 mol O
   The C:H:O mole ratio is 9:17:1; empirical formula is $C_9H_{17}O$; empirical molar mass = 141 g/mol

known molar mass/empirical molar mass = 274 g/141 g = 2; molecular formula is $C_{18}H_{34}O_2$

24. 68.2 g C $\times$ 1 mol C/12.0 g C = 5.68 mol C
   6.86 g H $\times$ 1 mol H/1.01 g H = 6.79 mol H
   15.9 g N $\times$ 1 mol N/14.0 g N = 1.14 mol N
   9.08 g O $\times$ 1 mol O/16.0 g O = 0.568 mol O
   The C:H:N:O mole ratio is 10:12:2:1; empirical formula = $C_{10}H_{12}N_2O$; empirical molar mass = 176 g/mol. The molecular formula is $C_{10}H_{12}N_2O$.

25. Since 1901, all units of measure have been set by the U.S. National Institute of Standards and Technology.

26. 0.0650 m³ $\times$ 1000 L/1 m³ = 65.0 L
   V (at STP) = 65.0 L $\times$ 2.00 atm/1.00 atm $\times$ 273 K/313 K = 113 L N_2
   113 L $N_2 \times$ 1 mol N_2/22.4 L $N_2 \times$ 2 mol NaN_3/3 mol $N_2 \times$ 65.0 g NaN_3/1 mol NaN_3 = 219 g NaN_3

27. Sulfuric acid is used in manufacturing fertilizers and detergents. It is also used as an electrolyte, as a catalyst, as a dehydrating agent, and as a component in refining petroleum and metals.

431

THINKING CRITICALLY

28. 18.0 g H_2O/1 mol H_2O × 1 mol H_2O /6.02 × 10^{23} molecules = 2.99 × 10^{-23} g

29. A mole can represent 6.02 × 10^{23} particles or it can represent the mass of 6.02 × 10^{23} particles.

30. Molar mass is the mass of 1 mol of a pure substance. Molecular mass is the mass of one molecule of a covalent substance. Formula mass is the mass of one formula unit of an ionic substance.

31. Add NaCl to the dissolved silver nitrate and sodium nitrate to precipitate insoluble AgCl. From the mass of AgCl, find the moles of AgCl formed. This number equals the moles of $AgNO_3$ in the sample. Use the mass of $AgNO_3$ and the mass of the sample to find the mass percent of $AgNO_3$.

32. The molar mass is used to convert the mass of a sample of atoms, ions, or molecules of a substance to the number of particles it contains.

33. Equation for reaction using baking soda:
$NaHCO_3 + HC_2H_3O_2 \rightarrow NaC_2H_3O_2 + H_2O + CO_2$
Equation for reaction using calcium carbonate:
$CaCO_3 + 2HC_2H_3O_2 \rightarrow Ca(C_2H_3O_2)_2 + H_2O + CO_2$
Twice as much acetic acid will be needed and a larger mass of calcium carbonate than sodium hydrogen carbonate will be needed because calcium carbonate has a larger molar mass.

CUMULATIVE REVIEW

34. Atomic number is the number of protons in the nucleus. Mass number is the number of protons and neutrons. Isotopes have the same atomic number but different mass numbers.

23. Oleic acid is a component in olive oil. It is 76.5 percent C, 12.1 percent H, and 11.3 percent O. The molar mass of the compound is approximately 282 g/mol. What is the molecular formula of oleic acid?

24. Serotonin is a compound that conducts nerve impulses in the brain. Serotonin is 68.2 percent C, 6.86 percent H, 15.9 percent N, and 9.08 percent O. The molar mass is 176 g/mol. What is the molecular formula of serotonin?

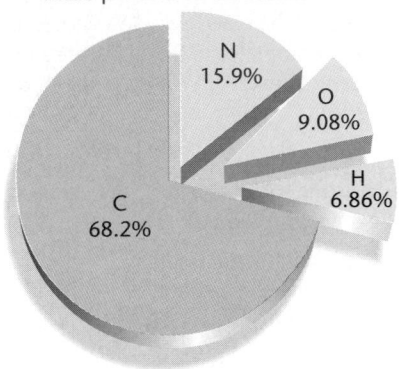

Mass percent in serotonin

N 15.9%
O 9.08%
H 6.86%
C 68.2%

Art Connection

25. How does the United States standardize measurements of mass and weight?

Everyday Chemistry

26. How many grams of sodium azide, NaN_3, are needed to fill a 0.0650-m^3 driver's air bag with nitrogen if the pressure required is 2.00 atmospheres and the gas temperature is 40.0°C? (Hint: 1 m^3 = 1000 L)

Chemistry and Technology

27. What are some uses of sulfuric acid?

THINKING CRITICALLY

Applying Concepts

28. What is the mass in grams of a single molecule of water?

29. Explain how a mole is used in chemistry as both a number and a mass.

Comparing and Contrasting

30. Distinguish between molar mass, formula mass, and molecular mass.

31. **ChemLab** Suggest a procedure to determine the mass percent of silver nitrate, $AgNO_3$, in a mixture of silver nitrate and sodium nitrate, $NaNO_3$.

Comparing and Contrasting

32. **MiniLab 1** Explain how your procedure in this activity is like using the molar mass to determine the number of particles in a sample of a known substance.

Relating Cause and Effect

33. **MiniLab 2** If you used calcium carbonate instead of baking soda, how would the amount of reactants needed change?

CUMULATIVE REVIEW

34. Distinguish between atomic number and mass number. How do these two numbers compare for isotopes of an element? (Chapter 2)

35. Answer the following questions about sugar and table salt. (Chapter 4)

   a) What type of compound is each substance?
   b) Which compound is an electrolyte?
   c) Which compound is made up of molecules?

36. Why does the second period of the periodic table contain eight elements? (Chapter 7)

37. Explain why the metals zinc, aluminum, and magnesium are resistant to corrosion, while iron is not. (Chapter 8)

38. The volume of a gas at STP is 8.50 L. What is its volume if the pressure is 1250 mm Hg and the temperature is 75.0°C? (Chapter 11)

Item	Formula	Molar Mass (g/mol)	Number of Particles	Rank
The formula units in 10.0 g of calcium fluoride	CaF_2	78.1	7.71 × 10^{22}	2
The sodium ions in 10.0 g of sodium chloride	NaCl	58.4	1.03 × 10^{23}	3
The water molecules in 10.0 g of water	H_2O	18.0	3.34 × 10^{23}	7
The hydrogen atoms in 10.0 g of water	H_2O	18.0	6.68 × 10^{23}	8
The carbon dioxide molecules in 10.0 g of carbon dioxide	CO_2	44.0	1.37 × 10^{23}	4

Skill Review

1. **Using a Table** Complete the table below. In the last column, rank the number of particles from smallest to largest. Did you use the number of moles or the number of individual particles for your ranking? Explain.

Writing in Chemistry

2. A mole is such a large number that it is hard to envision just how big 6.02×10^{23} is. Write three of your own examples of the number of things in a mole. One example must use time, one must use distance, and the third is up to you.

Problem Solving

3. A student carries out a sequence of chemical reactions on a 0.635-g sample of pure copper. The copper is converted to copper(II) nitrate, then to copper(II) hydroxide, then to copper(II) oxide, then to copper(II) sulfate, and finally back to elemental copper. The student isolates and weighs the product at the end of each step before proceeding to the next. What is the theoretical yield of each step?

Projects

4. As you read in *Chemistry and Technology* in this chapter, sulfuric acid is the number one industrial chemical produced in the United States each year. Research and prepare a poster or class presentation about another important industrial chemical. Your presentation should give last year's production (available in *Chemical and Engineering News*) and explain why so much of this chemical is sold. It should describe how this chemical is made and what its major uses are. Are some everyday items produced using this chemical?

Item	Formula	Molar Mass (g/mol)	Number of Particles	Rank
The formula units in 10.0 g of calcium fluoride				
The sodium ions in 10.0 g of sodium chloride				
The water molecules in 10.0 g of water				
The hydrogen atoms in 10.0 g of water				
The carbon dioxide molecules in 10.0 g of carbon dioxide				
The carbon monoxide molecules in 10.0 g of carbon monoxide				
The aspirin molecules, $C_9H_8O_4$, in 10.0 g of aspirin				
The carbon atoms contained within 10.0 g of aspirin				
The valence electrons in 10.0 g of aspirin				

Item	Formula	Molar Mass (g/mol)	Number of Particles	Rank
The carbon monoxide molecules in 10.0 g of carbon monoxide	CO	28.0	2.15×10^{23}	5
The aspirin molecules, $C_9H_8O_4$, in 10.0 g of aspirin	$C_9H_8O_4$	180	3.34×10^{22}	1
The carbon atoms in 10.0 g of aspirin	$C_9H_8O_4$	180	3.01×10^{23}	6
The valence electrons in 10.0 g of aspirin	$C_9H_8O_4$	180	2.27×10^{24}	9

CHAPTER REVIEW

35. **a.** Sugar is a covalent compound; table salt is an ionic compound. **b.** table salt **c.** sugar
36. In the second period, the second energy level is being filled. Eight electrons are needed to fill the $2s$ orbital and the three $2p$ orbitals, so eight elements are possible.
37. Zn, Al, and Mg form an oxide coating that protects the metal from further reaction. Iron oxide flakes off and exposes the metal to further corrosion.
38. 8.50 L $\times$ 760 mm Hg/1250 mm Hg $\times$ 348 K/273 K = 6.59 L

AUTHENTIC ASSESSMENT

1. See the tables below and on page 432. Most students will say it's easier to count moles than individual particles, but both methods give the same result.
2. Some possibilities follow. Time: A computer that counts 10 million per second would need almost 2 billion years to count to 1 mol. Distance: 1 mol of baseballs would cover Earth to a depth of 100 miles.
3. 1.88 g $Cu(NO_3)_2$; 0.976 g $Cu(OH)_2$; 0.795 g CuO; 1.60 g $CuSO_4$; 0.635 g Cu

Chapter 13 Organizer

Section	Objectives	Activities/Features
13.1 **Uniquely Water** (4 days)	1. **Demonstrate** the uniqueness of water as a chemical substance. 2. **Model** the three-dimensional geometry of a water molecule. 3. **Relate** the physical properties of water to the molecular model.	**MiniLab 1:** How many drops can you put on a penny?, p. 443 **Chemistry and Society:** p. 447 **People in Chemistry:** Alice Arellano, Wastewater Operator, pp. 448-449 **Discovery Demo:** 13.1, pp. 434-435 **Demonstrations:** 13.2, pp. 440-441; 13.3, pp. 444-445; 13.4, pp. 448-449
13.2 **Solutions and Their Properties** (6 days)	4. **Compare and contrast** the ability of water to dissolve ionic and covalent compounds. 5. **Distinguish** solutions from colloids. 6. **Compare and contrast** colligative properties.	**MiniLab 2:** Hard and Soft Water, p. 452 **Everyday Chemistry:** p. 455 **ChemLab:** Solution Identification, pp. 456-457 **Everyday Chemistry:** Antifreeze, p. 466 **How It Works:** p. 468 **Chemistry and Technology:** Versatile Colloids, pp. 470-471 **Demonstrations:** 13.5, pp. 454-455; 13.6, pp. 470-471

Activity Materials

ChemLab (pages 456-457)
96-well microplate; 10 mL each of 0.1M solutions of sodium carbonate, sodium iodide, copper(II) sulfate, copper(II) nitrate, lead(II) nitrate, barium nitrate; 6 microtip pipets; 15 toothpicks; distilled water; rinse bottle

MiniLab 1 (page 443)
penny, 2 microtip pipets, detergent solution

MiniLab 2 (page 452)
small test tubes (3), hard water, soft water, distilled water, droppers (2), 0.1M acetic acid solution, 0.1M sodium oxalate solution, large test tubes (3), soap solution

Demonstrations For a complete list of materials for the demonstrations in this chapter, see pages 434, 440, 444, 448, 454, and 470.

KEY TO TEACHING STRATEGIES

The following designations will help you decide which activities are appropriate for your students.

L1 Level 1 activities should be within the ability range of all students.

L2 Level 2 activities should be within the ability range of the average to above-average student.

L3 Level 3 activities are designed for the ability range of above-average students.

LEP LEP activities should be within the ability range of Limited English Proficiency students.

COOP LEARN Cooperative Learning activities are designed for small group work.

P These strategies represent student products that can be placed into a best-work portfolio.

Water and Its Solutions

Teacher Classroom Resources

STUDENT MASTERS

Study Guide: pp. 51-52
ChemLab and MiniLab Worksheets: p. 87

Study Guide: pp. 53-54
Critical Thinking/Problem Solving: p. 14
Laboratory Manual: 13.1 Relating Solubility and Temperature, pp. 119-122; 13.2 The Effect of a Solute on Freezing Point, pp. 123-126
ChemLab and MiniLab Worksheets: pp. 83-86, 88

TEACHING AIDS

Section Focus Transparency 26
Section Focus Transparency Master: p. 26
Basic Concepts Transparency 35
Basic Concepts Transparency Masters: pp. 69-70
Lesson Plans: p. 26

Section Focus Transparency 27
Section Focus Transparency Master: p. 27
Basic Concepts Transparency 36, 37
Basic Concepts Transparency Masters: pp. 71-74
Problem Solving Transparency 15
Problem Solving Transparency Master: pp. 29-30
Lesson Plans: p. 27

ASSESSMENT RESOURCES

Chapter Review and Assessment: pp. 77-82
Alternate Assessment in the Science Classroom
Computer Test Bank: Chapter 13
Supplemental Practice Problems: Chapter 13
Performance Assessment in the Science Classroom

CHAPTER RESOURCES

Applying Scientific Methods in Chemistry: pp. 39-40
Spanish Resources: Chapter 13
English/Spanish Audiocassettes: Chapter 13
Cooperative Learning in the Science Classroom
Lab and Safety Skills in the Science Classroom

GLENCOE TECHNOLOGY

Software
Mastering Concepts in Chemistry: *Unit 6* Lesson 1, The Mole Concept; Lesson 2, Molarity; Lesson 3, Analytical Calculations Using Moles

Videotape
MindJogger Videoquizzes, Chapter 13

Videodisc
Chemistry: Concepts and Applications, Demonstrations, Videos, and Animations
Videodisc and Bar Code Teacher Guide
Science and Technology Videodisc Series: Disc 2 Chemistry *Snowflakes, Cloud Chemistry*

CD-ROM Multimedia System
Chemistry: Concepts and Applications, Same as for Videodisc, plus Interactive Exploration and Major Simulation, *The Periodic Table*

TECH PREP

The following Glencoe resources provide opportunities for integrating science and technology.

Student Edition: Chemistry and Society, p. 447, People in Chemistry, p. 448; MiniLab 2, p. 452; Everyday Chemistry, p. 455; Everyday Chemistry, p. 466; How It Works, p. 468

Teacher Wraparound Edition: Integrating the Sciences, p. 438; Applying Chemistry, pp. 454, 461, 469; Extension, p. 467

Teacher Classroom Resources: Consumer Chemistry, pp. 25-26; Tech Prep Applications, pp. 19-20; Chemistry and Industry, pp. 27-30

13 Water and Its Solutions

Chapter Overview

The unique physical properties of water are introduced, and the relationship between these properties and the structure of water molecules is explored. The properties of water solutions are discussed within the context of their significance for life on Earth.

Theme Connection

Macro to Submicro The unique properties of water and water solutions are ultimately linked to the structure of water molecules and how they interact with each other and with dissolved solutes.

✔ Assessment Planner

Choose assessment strategies from the following pages to evaluate the progress of your students.

Assess, pp. 450, 472
MiniLabs, pp. 443, 452
ChemLab, p. 456
Portfolio, pp. 450, 465, 473, 474
Chapter Review, p. 474

GLENCOE TECHNOLOGY

 Videodisc

Chemistry: Concepts and Applications
Hydrogen Bonding
Disc 2, Side 1, Ch. 8

Show this video of the Discovery Demo to demonstrate the concept of hydrogen bonding.

L1 LEP

H₂O

434

◈ Discovery Demo 13.1

Hydrogen Bonding

Purpose: To demonstrate hydrogen bonding while reinforcing the concepts of equilibrium and Le Chatelier's principle.

Materials: 1.4 g $CoCl_2 \cdot 2H_2O$, 60 mL acetone, 100-mL graduated cylinder, 30 mL H_2O

Safety Precautions: 🔥 🧤 🥽 ☠️

Disposal: Code C

Procedure: Dissolve 1.4 g of $CoCl_2 \cdot 2H_2O$ in 30 mL of H_2O. Place this solution in a 100-mL graduated cylinder. Slowly pour 60 mL of acetone down the inside of the graduated cylinder. **CAUTION:** *The cobalt chloride solution is toxic and acetone is flammable.* Have students observe the blue layer on top.

Results: The acetone forms hydrogen bonds with water and decreases the effective concentration of water in the top layer of the solution. The equilibrium shifts to the right to form the blue chloro complex in the following reaction.

CHAPTER
13
Water and Its Solutions

Try to imagine your world without water. It's pretty tough to do. Water is all around you, and it has nearly endless uses. Think of the many different ways you use water every day: drinking, cooking, washing, and flushing. The average water use in the United States is a surprising 300 L per person per day. That's 300 kg—greater than the mass of four average persons. Water is inside you, too. Your body is about 60 percent water by weight. In fact, almost every chemical reaction that happens in your body happens in a water environment—an aqueous solution. Water is essential to your life. Without it, you wouldn't survive more than a week. It really is difficult to think of a world without water. As a matter of fact, without water, it's impossible to think at all: remember that your brain is mostly water, too.

Chemistry Around You

One thing that makes Earth unique among the planets in our solar system is the amount of liquid water. It covers almost three-fourths of Earth's surface. Water is what has earned Earth the nickname "the blue planet."

Concept Check

Review the following concepts before studying this chapter.
Chapter 4: the meaning of electron dot structures
Chapter 9: how to determine molecular geometry; the meaning of electronegativity; the basis of polarity
Chapter 10: structure of solids, liquids, and gases; kinds of changes of state; energy in state changes

Chapter Preview

13.1 Uniquely Water

13.2 Solutions and Their Properties

Laboratory Activities
ChemLab
Solution Identification

MiniLabs
1. How many drops can you put on a penny?
2. Hard and Soft Water

435

$$Co(H_2O)_6^{2+} + 4Cl^- \rightleftarrows CoCl_4^{2-} + 6H_2O$$
(pink)   (blue)
Analysis: Ask these questions.
1. Why does the blue layer form on top of the water? *Acetone is less dense than water.*
2. What do chemists call the formation of an attractive force between the hydrogen atoms of one water molecule and the oxygen atom of another water molecule? *hydrogen bonding*

3. Using the chemical equation that represents the system at equilibrium, which direction will be favored when the water is attracted to the acetone and the water's effective concentration is lowered? *The equilibrium will shift to the right, favoring the formation of the blue product, $CoCl_4^{2-}$.*
4. Is a hydrogen bond an attractive force or a full covalent chemical bond? *attractive force*

Introducing the Chapter
Divide students into cooperative groups and have them discuss and record the number of different ways that they use water. After a few minutes of brainstorming, have student groups share the results of their discussions. Challenge students to estimate the volume of water they use and consume daily. *approximately 300 L* Discuss that although Earth is covered with water, most of that water is not suitable for human use; it is salty, frozen, or inaccessible. **L1**
COOP LEARN

Using the Illustrations Draw attention to the photo and art on the facing page. Explain the point of each and the relationship between them—namely, that water is essential for life and that the structure of the water molecule is the key to understanding its properties and uses.

Revealing Misconceptions
Students often have problems distinguishing between the forces that hold atoms together in molecules (intramolecular forces) and those forces that exist between molecules (intermolecular forces). Emphasize that although the intermolecular forces are the result of the intramolecular organization, the two kinds of forces are different. **Figure 13.4** may help the students to distinguish the difference.

GLENCOE TECHNOLOGY

 CD-Rom
Chemistry: Concepts and Applications
Hydrogen Bonding
Show this video of the Discovery Demo to demonstrate hydrogen bonding. **L1** **LEP**

SECTION
13.1

Key Concepts

The unique physical characteristics of water are discussed, then related to its molecular structure and intermolecular attractions.

1 FOCUS

Focus Transparency

 Display the **Section Focus Transparency** for Section 13.1 to show students ways in which water is useful. L1 LEP

2 TEACH

GLENCOE TECHNOLOGY

 **Videodisc**

Chemistry: Concepts and Applications
The Unique Properties of Water
Disc 2, Side 1, Ch. 11

Show this video to illustrate the unique properties of water, including its expansion on freezing. L1 LEP

 CD-Rom

Chemistry: Concepts and Applications
The Unique Properties of Water
Show this video to illustrate the unique properties of water.
L1 LEP

SECTION PREVIEW

Objectives
Demonstrate the uniqueness of water as a chemical substance.
Model the three-dimensional geometry of a water molecule.
Relate the physical properties of water to the molecular model.

Key Terms
hydrogen bonding
surface tension
capillarity
specific heat

Uniquely Water

Water may be a common substance on Earth, but its properties are anything but common. Those properties make water essential for life. You may well ask how it is that a substance like water can have such vital properties. A water molecule, after all, is just two atoms of hydrogen and one atom of oxygen linked by covalent bonds. Molecules don't get much simpler than H_2O. However, the simplicity of the composition of a water molecule is deceiving. As you will see, the electron distribution and the three-dimensional arrangement of those three atoms in the molecule are the source of the unique properties of water—those properties that cause it to play a major role in so many aspects of your world.

Water: The Molecular View

Because water is so much a part of life, its properties are easy to take for granted. If you step back a bit and examine water scientifically, you will find that it is unusual among the compounds found on Earth. **Table 13.1** compares the physical properties of water against a group of molecules of similar mass. Look at the differences in physical state, melting point, and boiling point. Water definitely behaves differently from other small molecules. It is a unique substance.

When you use water for drinking or most other purposes, you use it in its liquid state. When you try to think of other common liquids, you probably think of water solutions. Water is most often thought of as a liquid. However, solid water, called ice, and gaseous water, called steam or water vapor, also exist in large quantities on Earth. Water is the only substance on Earth that exists in large quantities in all three states. Gaseous water is found around geysers and hot springs and in the atmosphere, where it plays a major role in determining weather. Ice occurs in glaciers and ice caps and acts as a huge reservoir of fresh water.

Program Resources

Study Guide, pp. 51-52 L1
English/Spanish Audiocassettes, Chapter 13
   L1 LEP
Section Focus Transparency 26 and **Master,**
   p. 26 L1 LEP
Basic Concepts Transparency 35 and **Master,**
   pp. 69-70 L1 LEP

Spanish Resources L1 LEP
ChemLab and MiniLab Worksheets, p. 87
   L1
Consumer Chemistry, pp. 25-26 L1
Tech Prep Applications, pp. 19-20 L1

Table 13.1 The Uniqueness of Water

Substance	Formula	Molecular Mass (u)	State at Room Temperature	Melting Point (°C)	Boiling Point (°C)
Methane	CH_4	16	gas	–183	–161
Ammonia	NH_3	17	gas	–78	–33
Water	H_2O	**18**	**liquid**	**0**	**100**
Nitrogen	N_2	28	gas	–210	–196
Oxygen	O_2	32	gas	–218	–183

Most substances tend to be more dense as solids than they are as liquids. Water is an important exception. It is less dense as a solid than as a liquid. For that reason, ice floats in liquid water rather than sinking. If ice were like most solids and denser than the liquid state, lakes and ponds would freeze from the bottom up. The fact that the ice forms and remains on the top of lakes and ponds allows aquatic life in the liquid water below it to survive the winter cold. Also, if ice were like most solids, popular sports such as ice hockey, figure skating, and ice fishing, **Figure 13.1,** could never have developed.

Figure 13.1
Ice Has Low Density
The fact that ice is less dense than liquid water is the reason that ice collects at the top of the liquid rather than at the bottom. If it were not for this fact, ice-skating and ice fishing would not be possible.

13.1 **Uniquely Water** 437

Discussion

Write the following molar masses and boiling points of substances on the chalkboard and ask students what general trend or correlation seems to exist: 2 g/mol, −253°C; 10 g/mol, −246°C; 28 g/mol, −192°C; 44 g/mol, −78.5°C (sublimes); 81 g/mol, −67°C; 154 g/mol, +76°C; 18 g/mol, +100°C; 332 g/mol, +189°C. *As the molar mass increases, the boiling point generally increases.* Then ask whether any substance seems out of place in terms of the trend. *The substance that has a molar mass of 18 g/mol and a boiling point of 100°C is out of place.* Ask students whether they know what substance it is and why the boiling point is so high for such a low molar mass. *The substance is water. The molecules must have great attraction for each other.* Then supply the identities of the eight other substances: H_2, Ne, CO, CO_2, HBr, CCl_4, H_2O, CBr_4.

Visual Learning

Table 13.1: The data presented in this table provide examples of the physical properties of water and their uniqueness. These properties are extremely significant because they provide the insight into the nature of interactions at the particle level. Keep in mind that the simplest physical property to observe is physical state: solid, liquid, or gas. The state of matter at room temperature provides an immediate indication of the strength of forces that hold the matter together. These forces are interparticle forces. The measurement of the temperatures at which matter changes states further refines the picture of the submicroscopic structure.

Chemistry Journal

Properties of Water
Ask students to describe in their journals how the properties of water would change if the molecules were linear. Ask them to add to the journal entry as various properties of water are discussed throughout Section 13.1.
L1

Tying to Previous Knowledge

Briefly review molecular shape and polarity from Chapter 9, and ask students to draw the Lewis electron dot structure for water. Ask them to describe the geometric arrangement or orientation of the four electron pairs around the central oxygen atom. *The arrangement is tetrahedral.* Then ask them to describe the shape of the molecule. *It is bent, or angular.* L1

Correcting Misconceptions

Because hydrogen bonding has such a significant effect on the properties of water, students may believe that hydrogen bonding is strong compared with covalent bonding. Point out that this is not true and that, in fact, the energy required to break the O—H covalent bonds in 1 mol of water is 934 kJ, compared with only 40.7 kJ/mol needed to overcome hydrogen bonding and vaporize water.

Quick Demo

How does water do it?

Ask each of five students to hold a molecular model of water, and demonstrate the way in which a central water molecule may form four hydrogen bonds with other molecules. L1

What factors will explain the unique physical properties of water in its various states? **Figure 13.2** reviews some of the things you have already learned about the composition of the water molecule, its electron distribution, and its three-dimensional structure.

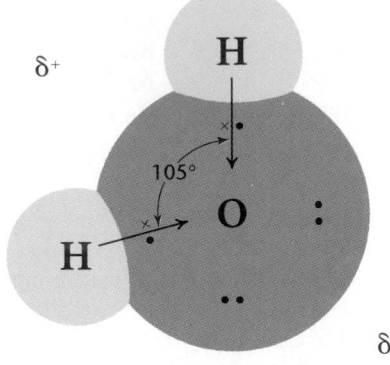

Figure 13.2
Geometry of the Water Molecule
The arrangement of electrons about the central oxygen in the water molecule relates to its three-dimensional geometry. There is a large electronegativity difference between the covalently bonded hydrogen and oxygen. Therefore, the electron pair is shared unequally. Because of the molecule's bent structure, the poles of positive and negative charge in the two bonds do not cancel, and the water molecule as a whole is polar.

Intermolecular Forces in Water

Recall from Chapter 9 that molecules that are dipoles, such as water, have interparticle attractive forces. You can model the behavior of a group of water molecules by imagining that the molecules act like little bar magnets, as shown in **Figure 13.3.**

Notice that if you pulled one of the magnets in **Figure 13.3,** the position of all of the magnets would change. The same thing would happen if you could somehow reach into a group of water molecules and pull just one. The model demonstrates that attractive forces between objects do not create interactions between just two objects. The forces can also combine to give organization to groups of objects, whether they are bar magnets or water molecules.

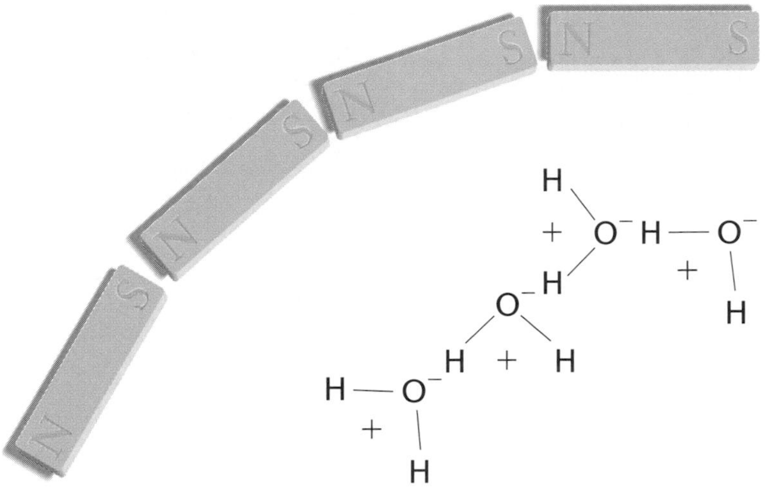

Figure 13.3
Attractions and Order
A group of magnets will tend to orient themselves with the opposite poles of the different magnets pulled toward one another. A group of water molecules will do the same kind of thing at the molecular scale because of electrical forces. The opposites attract and create order among molecules. This effect is especially great at low temperatures.

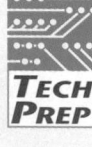

Integrating the Sciences

Meteorology TECH PREP Ice cores from the Arctic Circle and the South Pole provide scientists with information about the ancient climate that goes back more than 100 000 years. Every year, a new layer of ice and snow falls on these remote regions, trapping dust, pollen, and gases. Over the years, the ice piles up, preserving lower layers under an increasingly massive quantity of ice. By removing an ice core from this layered system and analyzing the material trapped at various depths, scientists can determine changes in temperature, precipitation, and atmospheric composition that occurred over the time the ice formed. The longest ice core to date (about 3000 m) spans a time range greater than 100 000 years. The ice record can even indicate specific global events such as major forest fires or volcanic eruptions. The cores may provide useful information regarding long-term climatic change and ozone depletion.

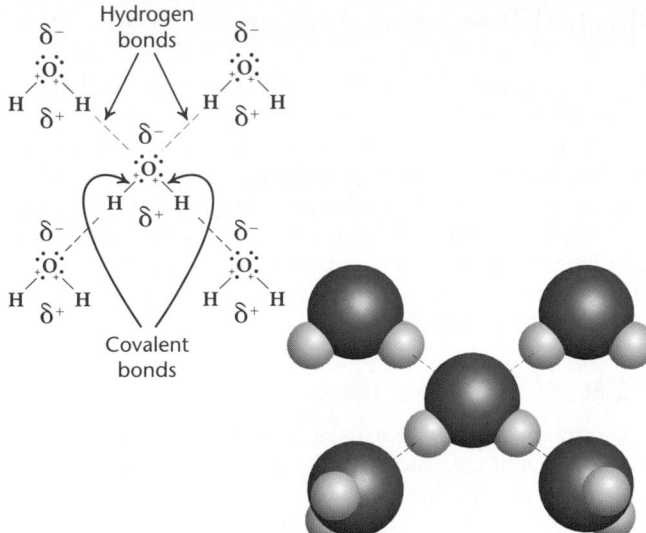

Hydrogen bonds

Covalent bonds

Figure 13.4

Hydrogen Bonding Versus Covalent Bonding in Water
Notice that water molecules have both intermolecular forces (hydrogen bonds, dashed lines) and intramolecular forces (covalent bonds, solid lines). The oxygens are highly electronegative and strongly attract hydrogen's electrons. As a result, the hydrogen nucleus, a proton, is partially exposed. The exposed hydrogen nucleus is attracted to the non-bonding pairs of electrons on the oxygen atom. The two hydrogens on the central water molecule are attracted at the same time to the oxygens of two other water molecules.

Modeling Water: Hydrogen Bonding

Although the intermolecular forces in water create order that involves large numbers of molecules, you can get a good picture of what's going on by modeling the interactions between just a few water molecules, as shown in **Figure 13.4.** Notice that the oxygen on one water molecule is attracted to the hydrogen atoms on other water molecules. The connections between the molecules are not full covalent bonds, but they are still fairly strong. The formation of such a connection between the hydrogen atoms on one molecule and a highly electronegative atom on another is called **hydrogen bonding.** Atoms that are electronegative enough to cause hydrogen bonding include oxygen, fluorine, and nitrogen.

Water: The Hydrogen Bonding Champion

Oxygen-hydrogen groups in molecules tend to promote intermolecular hydrogen bonding. In pure water, each water molecule may form hydrogen bonds with four other water molecules.

Any molecule that contains O—H bonds has the potential to form hydrogen bonds. Oxygen is sufficiently electronegative to attract hydrogen's single electron strongly so that the hydrogen atom almost becomes an exposed proton. Alcohols, organic compounds that contain O—H bonds, also form hydrogen bonds. Not surprisingly, some of the physical properties of alcohols are similar to those of water, as shown in **Figure 13.5.**

Because they have many O—H bonds, biological molecules such as proteins, nucleic acids, and carbohydrates also have the ability to form hydrogen-bond networks. These networks can be extensive, and their three-dimensional shape is important in determining their functions in living things.

Figure 13.5

Hydrogen Bonding in Methanol
Molecules other than water also form hydrogen bonds. This simplest alcohol, methanol, CH_3OH, has an O—H bond with which to form hydrogen bonds. This hydrogen bonding gives methanol some physical properties that are similar to those of water.

SECTION 13.1

Discussion

Show students a photograph or slide of liquid water, an iceberg, and clouds. Ask how many states of water appear in the photograph. *Three states appear: solid, liquid, and gas.* If you have photos of other planets and/or moons, show them to students and ask what major difference exists between them and Earth. *One obvious difference is the small amount of water on most other objects in the solar system.* Point out that life as we know it is dependent upon water, and that most of the rest of the solar system seems to contain relatively little water.

Quick Demo

Does the volume change when water freezes?
Place 25 mL of water in a 50-mL plastic graduated cylinder, and stopper the cylinder or cover it with plastic wrap. Place the cylinder in a freezer overnight, show it to the students, and ask them to explain the result. *The water expands when it freezes. Hydrogen bonds hold the molecules in an arrangement in which they are farther apart than in the liquid state.*

GLENCOE TECHNOLOGY

 Videodisc

Chemistry: Concepts and Applications
See the Videodisc Program Teacher Guide to access illustrations and other material available for this chapter.

Meeting Individual Needs

Learning Disabled In order to illustrate the significance of the attractions that a water molecule may have due to multiple hydrogen bonds, ask a student to hold a basketball or football with one hand. Demonstrate that the ball may easily be taken from the student. Now ask the student to hold the ball with both hands, and demonstrate that much more effort must be exerted to separate the ball from his or her hands. Explain that water molecules may each form hydrogen bonds with up to four other molecules. L1 LEP

Water: Physical Properties Revisited

Look back at the properties in **Table 13.1.** You can see that compared to similar-sized molecules, water's physical properties are different. The most obvious difference is that water is a liquid at room temperature, whereas the others are gases. Water also has a high melting point and high boiling point for such a small molecule. Many of these unique physical properties are a result of the hydrogen bonding that occurs between water molecules. This is just another example of the relationship between submicroscopic structure and the macroscopic, directly observable behavior of substances. Let's examine a few of these properties in more detail.

States of Water

Water occurs primarily in the liquid and solid states on Earth, rather than as a gas. The intermolecular hydrogen bonds hold the water molecules together strongly enough that they cannot readily escape into the gaseous state at ordinary temperatures. That is why water has such a high boiling point for such a small molecule, 100°C. In order for water to boil, the temperature must be increased, adding enough energy to overcome the hydrogen bonds that are holding the water molecules together and separating them so that they can enter the gaseous state.

Ice Floats

You know that if you drop an ice cube into a glass of water, the ice floats. You also know this means that the density of the solid water is less than that of liquid water. **Figure 13.6** shows how the density of water changes as its temperature changes. Like most substances, liquid water shrinks and becomes denser when it is cooled. As water cools from 60°C, its volume decreases and its density increases. The water molecules move less rapidly, and they are able to be drawn closer together by dipole-dipole attractions. The volume of the water decreases because the molecules pull together. Meanwhile, the mass of water stays the same, so density increases.

However, when the water is cooled to about 4°C, something unusual happens. The volume stops decreasing. The density has reached a maximum. At this point, the water molecules have been pulled as close together as they can get. As the temperature is lowered below 4°C, the volume of the water begins to expand and its density decreases.

Figure 13.6
Density of Water at Various Temperatures
Water achieves its minimum volume, and therefore its maximum density, at 3.98°C.

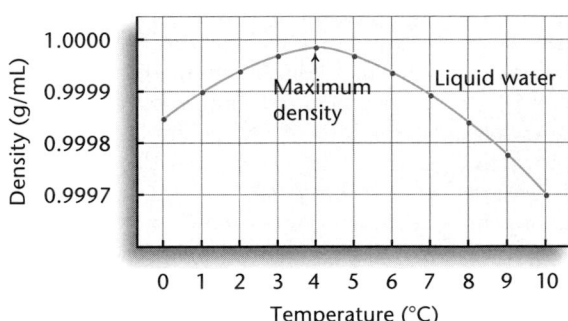

Demonstration 13.2

Surface Tension
Purpose: To demonstrate surface tension.
Materials: 1-pint canning jar with ring lid, wire screen, 1 L water with food coloring added, piece of card stock

Safety Precautions:
Disposal: Save hardware, pour solution down drain, and throw paper in trash.

Procedure: Cut a piece of wire window screen to fit inside the ring lid of a 1-pint canning jar. Screw on the lid with the wire-screen center. Pour the colored water through the screen to fill the canning jar to overflowing. Have students predict what will happen when the jar is inverted. Cover the top of the jar with a piece of card stock. Turn the jar upside down over a sink. Remove the card stock by sliding it off to

Figure 13.7
Freezing Water

Its Molecular Structure The six-pointed snowflakes reflect the open arrangement of bonds. In order to freeze, molecules of liquid water separate slightly from one another as a result of hydrogen bonding. The volume of cooling liquid increases as the molecules move apart and the density decreases. At 0°C, the liquid water freezes to solid water, ice. The volume rises and the density drops dramatically, from 1.000 g/mL to 0.917 g/mL, as the water molecules assume the open arrangement. ▶

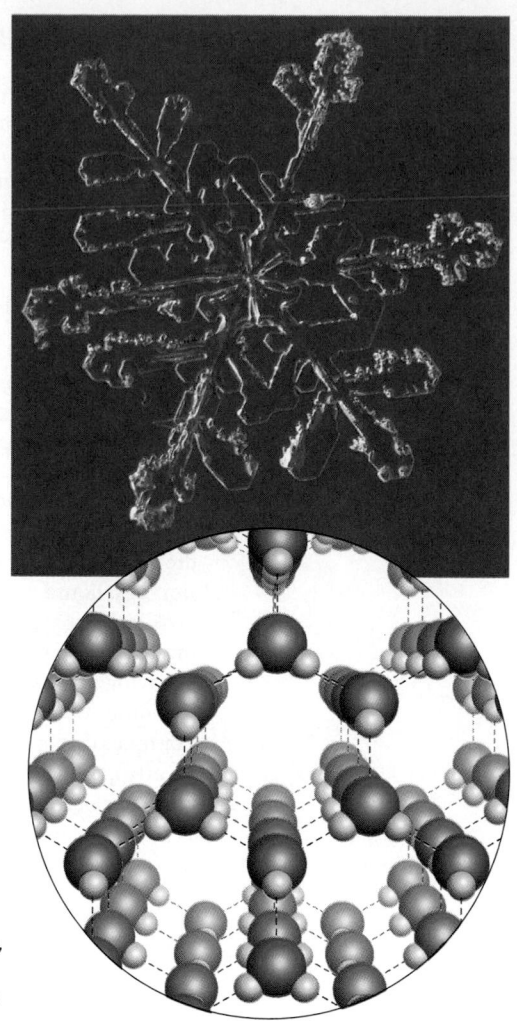

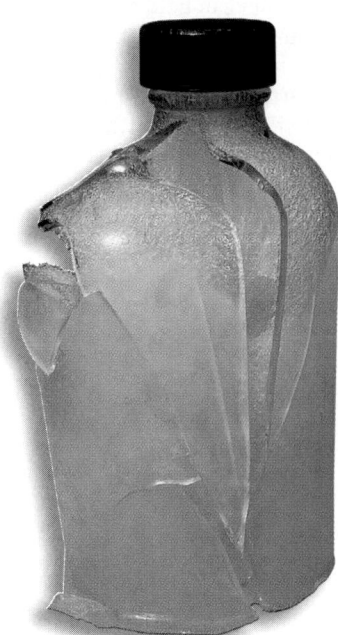

◀ **Its Expansion** The volume expansion that occurs when water freezes has some disadvantages, as you can see by looking at this burst water bottle.

You can account for this if you know what is happening to the molecular arrangement. Below 4°C, the water molecules are beginning to approach the solid state, which is highly organized. The water molecules begin to form the open arrangement shown in the model in **Figure 13.7.** The arrangement results from hydrogen bonding and is the most stable structure for the molecules in or near the solid state.

The fact that ice floats in liquid water has advantages for life that you read about earlier. However, it also has other consequences, as shown in **Figure 13.7.** Because liquid water expands significantly when it freezes, frozen water pipes can break and sidewalks can crack. The forces involved in the expansion of freezing water are surprisingly great—strong enough eventually to break boulders into small pieces. The processes of freezing and thawing contribute to soil formation and erosion and have been transforming the surface of Earth for many millions of years.

Using an Analogy

Ask two students to represent water molecules at 0°C by assuming positions used in formal dancing, at arms' length. Explain that water molecules in ice can move a little in place and are held in rigid positions by hydrogen bonds. Now, ask another two students to represent water molecules at 4°C by assuming a close, slow-dance position. Explain that water molecules at this temperature can move but are close together. Finally, ask two more students to represent water molecules at temperatures above 4°C by assuming disco-dance positions, fairly far apart. Explain that water molecules at higher temperatures move fast enough to remain farther apart. ⬜L1

GLENCOE TECHNOLOGY

 Videodisc

Science and Technology Videodisc Series
Snowflakes
Disc 2: Chemistry
Side 1, Ch. 3

Show this video to illustrate how snowflakes form and how the properties of water affect its transition to the solid state. ⬜L1 ⬜LEP

---

the side. Have a student look up into the jar and report what he or she observes. While holding the jar over the sink, slowly tip the inverted jar sideways until the water flows through the wire screen into the sink.
Results: The surface tension is sufficient to form a film that holds the water in the jar when inverted.

Analysis: Ask these questions.
1. Why does the water remain in the inverted jar? *The surface tension of water forms a film that covers the little holes in the window screen.*
2. Why does water fall from the jar when it is tilted? *Air is able to enter the jar, and it displaces the water and pushes it out of the jar.*

✔Assessment

Performance: Have students design and conduct experiments to determine whether the results of this demonstration depend on the size of the jar. Students can also investigate whether the results depend on the size of the holes in the wire screen. ⬜L1

Does anyone have the magic touch?

Before the demo, give a student a small plastic container containing some liquid detergent, and ask the student to keep the container in his or her pocket or purse, out of sight from the class. Instruct the student to secretly dip a finger into the detergent before touching the surface of water in a beaker that you will pass around the class. Sprinkle some powdered sulfur on the surface of some water in a large beaker, and ask students whether touching the surface will have any effect on the sulfur's floating. *It should not.* Demonstrate that it has no effect and ask several students to touch the surface also, saving your secret helper until last. *When the student touches the surface, the sulfur will sink because the detergent lowers the surface tension of the water.* **Disposal:** Filter out sulfur and dispose in solid waste. L1

Using an Analogy

Ask students to imagine that they are water molecules and that they are attracted toward one another. Choose a student in the middle of the class and ask whether there is a net attractive force on the student that will tend to pull the student in any direction. *No, the student is attracted equally from all sides.* Then choose a student in the front row and ask the same question. *Yes, there is a net attraction on the student that would pull him or her toward the center of the class.* Explain that surface water molecules experience a similar net attraction and that this attraction causes surface tension. L1

More Evidence for Water's Intermolecular Forces

The strong intermolecular forces that bind water molecules together show up in other properties of water. These observable properties provide more macroscopic evidence for the scientific model of the submicroscopic structure of water molecules. Keep in mind that the source of all of these properties is the unique combination of the water molecule's geometry and the electron distribution that makes the O—H bond highly polar.

Surface Tension

Have you ever watched water drip from a faucet? Each drop is composed of an enormous number of water molecules—roughly 2×10^{21}. The observation that this large number of molecules can hold together as a unit to form a single drop is more evidence for the presence of intermolecular forces.

A water molecule forms a drop because of **surface tension,** which is the force needed to overcome intermolecular attractions and break through the surface of a liquid or spread the liquid out. The higher the surface tension of a liquid is, the more resistant the liquid is to having its surface broken. **Figure 13.8** shows a molecular model of a water drop. The net inward force makes the surface of the drop contract and seem to toughen, behaving like a sort of skin.

Liquids other than water exhibit surface tension. Mercury is a good example of a liquid that has a high surface tension and strong interparticle attractive forces. When mercury is spilled, it forms droplets, much like the water beads that you see when it rains on a freshly waxed car.

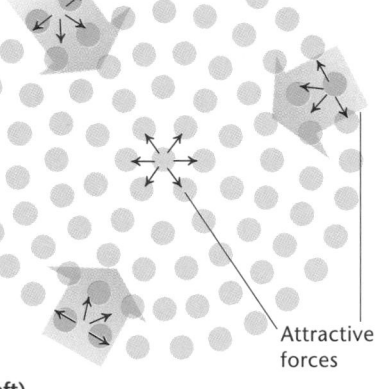

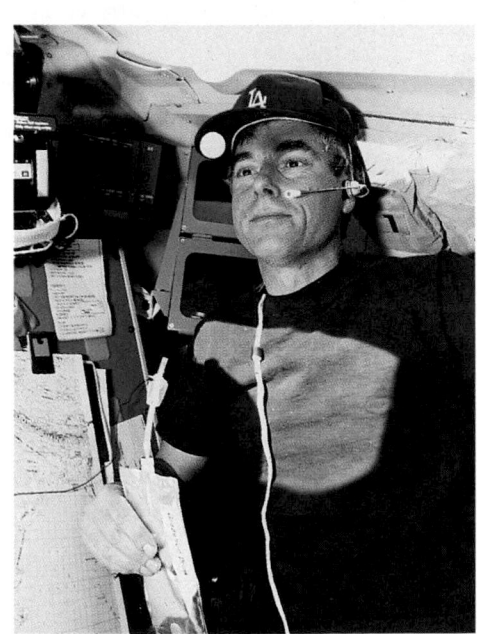

Figure 13.8
Surface Tension
Water drops in space, such as this one photographed on a space shuttle mission (right), show the perfectly spherical shape that is due to surface tension. In the drawing of a droplet of water (left), molecules in the middle of the drop are completely surrounded by other molecules and experience attractive forces (the small arrows) in all directions. Water molecules at the surface are not completely surrounded by other molecules. The forces acting on them are not equal in all directions, but they all pull inward with a net force (the large arrows). A water drop takes on its spherical shape because of this response to the inward force.

Attractive forces

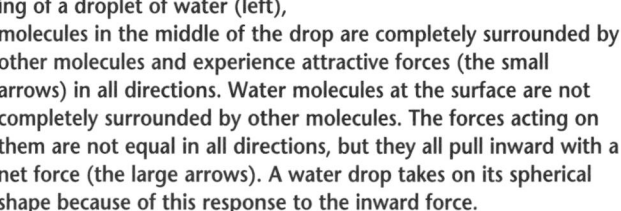

Chemistry Journal

Surface Tension

Have students observe everyday examples illustrating surface tension. They should record a description of each situation and provide an explanation of what is occurring. They should then write a summary that describes what is common to the different events, and tie this information to what they have learned about interparticle forces. L1

How many drops can you put on a penny?

The surface tension of water allows it to bead up on many surfaces. In this MiniLab, you will compete to see who can deposit the most drops of water and the most drops of an aqueous detergent solution on a penny.

miniLAB
①

Procedure

1. Lay a penny flat on your lab table.
2. Fill a microtip pipet with tap water, and count the number of drops you can deposit on the penny before water spills over the edge. Record the number of drops.
3. Fill another microtip pipet with a detergent solution prepared by your teacher, and repeat the process. Record the number of drops.

Analysis

1. How is surface tension demonstrated in this experiment?
2. Which has the lower surface tension: the water or the aqueous detergent solution? What accounts for this fact?

Capillarity

Have you ever had a small sample of blood drawn from a finger during a routine blood test as shown in **Figure 13.9?** After your finger is pricked, the blood is drawn by touching a thin glass tube, called a capillary tube, to the blood drop. Even without any suction being applied, the blood rises up into the tube. You have seen another version of this effect if you have ever observed the curved meniscus surface that results when you place water in a narrow glass tube or graduated cylinder. The water level next to the walls of the glass tube is higher than that at the center, giving a concave surface at the top of the liquid.

These examples of the rising of liquids in narrow tubes are illustrations of **capillarity,** or, as it is sometimes called, capillary action. Capillarity results from the competition between the interparticle attractive forces between the molecules of liquid and the attractive forces between the liquid and the tube that contains it.

WORD ORIGIN

capillarity:
capillus (L) hair
A liquid will rise by capillarity in a tube that is as narrow as a hair.

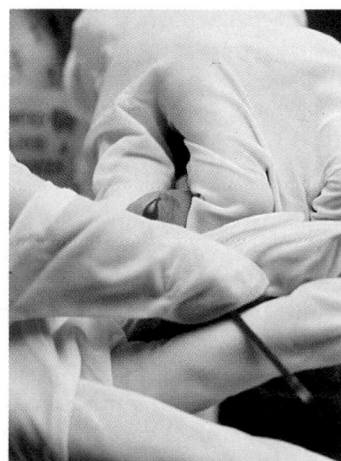

Figure 13.9
Capillarity
Capillarity draws blood into a thin tube after a finger is pricked.

13.1 Uniquely Water   443

SECTION 13.1

miniLAB

How many drops can you put on a penny?

Purpose: To determine the effect of a detergent solution on surface tension.
Process Skills: Observing, measuring, inferring, communicating

Teaching Strategies

• Ask students to predict how many drops can be deposited. *Usually, their predictions will be low.* [L1]
• Ask students to think about what property might account for water's ability to build up on a coin without spilling over. *The property is surface tension.*

Expected Results

Number of drops of water: about 43
Number of drops of detergent solution: about 14

Analysis

1. The beading and the ability of the dome of water to resist spilling over demonstrate surface tension.
2. The detergent solution has the lower surface tension because the nonpolar character of part of the detergent molecules interferes with the intermolecular attractions between water molecules.

✓ Assessment

Performance: Ask students to make a hypothesis about how many drops of a 70% isopropanol solution (rubbing alcohol) may be similarly deposited on a penny, relative to water and to the detergent solution. Students should test their hypotheses. *About 22–24 drops can be deposited.* [L1]

[LEP]

Chemistry Journal

Capillary Action

Ask students to write a journal entry describing capillary action from the point of view of a water molecule. The account should be imaginative and lively, but should be based on an accurate representation of the interplay between molecules. [L1]

Display **Basic Concepts Transparency 35** to reinforce the concepts of capillarity and surface tension. L1 LEP

Tying to Previous Knowledge

Ask students to recall from Chapter 10 that a heating curve for a substance in the solid, liquid, and gaseous states shows temperatures rising at three different rates. Explain that the specific heats for the three states of a single substance are different partially because the particles are at different average distances from each other. Explain that the specific heat for ice, for example, is 2.02 J/g°C, which is considerably lower than for liquid water.

Extension

Ask an interested student to conduct library research on the operation of a house or auto air conditioner, to prepare visual materials, and to make a class presentation that explains how endothermic and exothermic changes of state are used in the cooling of the house or auto. L2

Figure 13.10

Comparison of Water and Mercury in Glass Tubes
Water forms a concave meniscus, whereas mercury produces a convex meniscus. In the case of mercury, there is essentially no attractive force between the mercury atoms and the silicon dioxide to compete with the attractive forces between the mercury atoms themselves. The mercury forms a convex (high-centered) meniscus because the only force is the interparticle attractive force between mercury atoms, which produces surface tension.

In this case of a glass tube, water molecules can form hydrogen bonds to the oxygen atoms in the silicon dioxide that makes up the glass. This attractive force between the water and the SiO_2 draws the water up the walls of the tube. Because the water molecules are also attracted to each other, more water rises upward. If the tube is narrow, like the capillary tubes used in blood tests, the liquid will be drawn high into the tube because nearly all the molecules are close to the walls and are thus strongly attracted to them. If the glass tube is of larger diameter, like a graduated cylinder, you see the upward capillary effect near the tube walls, causing the liquid surface to have a meniscus shape. **Figure 13.10** contrasts the behavior of water in a glass tube and mercury in a glass tube.

Water: Earth's Thermostat

Have you ever jumped into a pool of cool water on a hot day early in summer? Despite the fact that the air temperature is high, the water temperature lags behind and tends to stay lower. On the other hand, if you return and jump into that same pool in the evening when the air temperature has dropped, the pool water will be warmer than the outside air. The water once again has lagged behind its surroundings in changing temperature. Once it has heated up, it does not cool down quickly. You may have noticed the same kind of effect if you have ever put a pot of water on to boil and had to wait a long time for it to heat up.

Demonstration 13.3

Capillary Action
Purpose: To demonstrate capillary rise.
Materials: 4 glass capillary tubes, 5 mL ethanol, 5 mL water, 5 mL glycerol, 5 mL saturated NaCl solution, food coloring

Safety Precautions:
Disposal: Code C

Procedure: Demonstrate capillary rise by using a number of capillary tubes in liquids that have different surface tensions, such as ethanol, water, glycerol, and saturated salt solution. Food coloring can be used to make the solutions more visible in the capillary tubes.
Results: The molecules of all of the various liquids are attracted to the SiO_2 molecules of the glass in the capillary tube and are drawn up the tube, but by different amounts.

What you were observing in these examples is the high specific heat of water. **Specific heat** measures the amount of heat, in joules, needed to raise the temperature of 1 g of substance by 1°C. The specific heats of water and some other common substances are shown in **Table 13.2.**

Table 13.2 Specific Heats of Some Substances

Substance	Specific Heat (J/g°C)
Gold, Au	0.129
Copper, Cu	0.385
Iron, Fe	0.450
Glass	0.84
Cement	0.88
Wood	1.76
Ethanol, $C_2H_5OH(l)$	2.46
Water, $H_2O(l)$	4.18

Notice that water has the highest specific heat of any of the substances listed. This means that water must absorb or release more heat for its temperature to change by one Celsius degree than any of the other substances. The specific heat of water is 4.18 J/g°C. (The unit is read "joules per gram per degree Celsius.") In order to raise the temperature of 1 g of water by 1°C, you must add 4.18 J of heat. On the other hand, you must remove 4.18 J of heat from a 1-g sample of water to lower its temperature by 1°C. That is why it takes time for water in a swimming pool to warm up in the early part of the year and why the water, once warmed, cools off slowly when the outdoor temperature drops, **Figure 13.11.**

Figure 13.11
Warm Water Cools Slowly
The water in this swimming pool heats up during the day. Even though the outside temperature drops at night, the water tends to remain warm. Because of its high specific heat, the warm water has stored a great deal of energy.

13.1 Uniquely Water   **445**

SECTION 13.1

Visual Learning

Table 13.2: How does the high specific heat of water give evidence for strong intermolecular forces in water? Have students compare the specific heats of Cu and H_2O in the table. They should remember that the temperature of a sample is a measure of the average kinetic energy (energy of motion) of the particles in the sample. As heat is added to the water sample, the water molecules move more rapidly but greatly resist the changes in their motion because of the strong forces that act between them. Therefore, it takes a large amount of heat energy to make a water molecule move more rapidly—an amount greater than is necessary for a copper atom in a sample of copper. Students should notice that ethanol, C_2H_5OH, also has a relatively large specific heat. Ethanol molecules undergo hydrogen bonding, but not as strongly as water.

Analysis: Ask these questions.
1. In a glass graduated cylinder containing water, does the edge of the meniscus turn up or down? Explain. *Up; the water molecules are attracted to the molecules of glass.*
2. In a plastic graduated cylinder, the meniscus of water is flat. Explain. *The molecules of water do not form hydrogen bonds with the molecules of plastic.*

3. Why did the different liquids rise to different heights in the capillary tubes? *The molecules differ in their polarities and abilities to form hydrogen bonds.*

✔ **Assessment**

Knowledge: Ask whether students think capillary action has any effect in helping to lift water from the roots to the top of a

tall tree. *Capillary rise plays a role in helping to lift water to the top of a tall tree. The dipole-dipole attraction and hydrogen bonding between water molecules tend to cause water molecules to pull on one another, and this cohesion helps move the water upward.* L1

445

Discussion

Ask students which process loses more energy: the cooling of 1 mol of water from its boiling point to its freezing point or the condensation of 1 mol of steam at 100°C to 1 mol of water at the same temperature. *Students will probably think more energy is lost in the cooling process because the temperature change is 100°C.* Demonstrate the calculation for the cooling: 18.0 g × 4.18 J/g°C × 100°C = 7520 J. Explain that the value is much less than that for the condensation of steam at 100°C, 40 700 J. L2

GLENCOE TECHNOLOGY

Videodisc

Science and Technology Videodisc Series
Cloud Chemistry
Disc 2: Chemistry
Side 1, Ch. 22

Show this video to illustrate the concepts of condensation and water pollution, tying it to the problem of water purification. Also, use it as a way to lead into the topic of water as a solvent.

L1 LEP

Figure 13.12
Low Heat Capacity of Dry Soil and Rock
A desert's temperature changes greatly at twilight. Without water, heat is given off quickly.

Water serves as a great heat reservoir that moderates the temperature at Earth's surface. During the day, heat from the sun is absorbed by the oceans. Because water has a high specific heat, and because there's a great deal of water, the temperature of the oceans does not increase detectably despite all the heat the water has absorbed. In the evening, the stored heat in the world's oceans is released, helping to maintain the air temperature in the absence of the sun's energy. If there were no water on Earth's surface to serve as a temperature moderator, the daytime temperature would soar. Rocks have a much lower specific heat than water does. In the evening, Earth would turn frigid because rocks have so little capacity to store heat and to release it slowly. You have noticed this effect if you have ever experienced the extreme temperature difference between day and night in a desert during clear weather, **Figure 13.12.**

Water Evaporation/Condensation: No Sweat?

Recall that vaporization is the change in state of liquid to gas, and that the amount of heat required to vaporize a quantity of a liquid is called the heat of vaporization. During the vaporization of a liquid, interparticle forces must be overcome. Therefore, vaporization of a liquid is an endothermic (energy-absorbing) process. Condensation, the formation of a liquid from a gas, is an exothermic (energy-releasing) process. Because of the strong intermolecular forces, water absorbs a great deal of heat when it is vaporized. Thus, it has a high heat of vaporization. Water also loses a great deal of heat when it condenses—thus the effectiveness of steam in heating a room or in causing burns.

Evaporation is vaporization from the surface of a liquid. Evaporation of water is one of the mechanisms by which your body regulates its temperature. On a hot day, particularly if you are as active as the person in **Figure 13.13,** you perspire and cool down.

Figure 13.13

Perspiration and Cooling
Perspiration is a mechanism by which your body passes water to the surface of your skin via sweat glands. The vaporization of water is an endothermic (energy-absorbing) process, and as the water evaporates, it removes heat from the surface of your skin.

CONNECTION
Software tools for exploring molecules and the periodic table are available for download.

World Wide Web
http://www.cchem.berkeley.edu

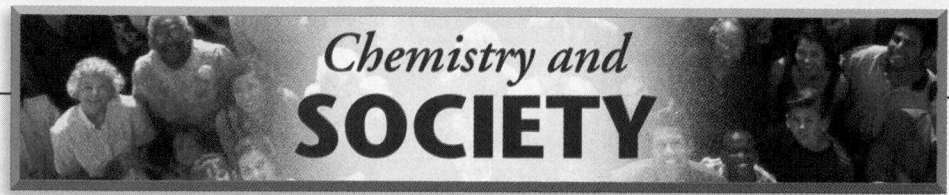

Water Treatment

With a twist of your wrist, you can make clean water flow from the faucet. Like most other people in industrialized countries, you may never stop to think how fortunate you are to have accessible, safe water. In many parts of the world, people are forced to carry water long distances. Then, after all their hard work, they have no assurance that the water does not carry disease. The water you use is probably safe because it is purified in a reliable water-treatment plant through the following series of steps.

1. Water is often conveyed over long distances through pipes from a lake or other freshwater source to the treatment plant. When the water to be treated enters the intake basin, it passes through bar screens that remove large, suspended solids and trash.

2. After the screening, pumps lift the water 6 m or more so that it can flow, by gravity, through the rest of the tanks.

3. Some particles present in water are so fine that they can be trapped only by coagulation. To do this, chemicals such as alum, chlorine, and lime are added by chemical applicators. The purpose of the alum—$Al_2(SO_4)_3 \cdot 18H_2O$—is to produce alum floc, a coagulant that causes fine, suspended solids to clump together. Chlorine kills bacteria. Lime causes precipitation of calcium carbonate and clears the water.

4. In the settling tank, bacteria, silt, and other impurities stick to the coagulants and settle to the bottom. The rest of the water moves on to a filtering tank.

5. The filtering tank allows the water to trickle through sand, gravel, and sometimes charcoal for a final cleansing.

6. When the water reaches the reservoir, it receives another treatment of chlorine. The purified water remains in the reservoir until it is pumped to homes, factories, and businesses.

Analyzing the Issue

1. **Acquiring Information** Find out and report on how biological treatment is used to clear water of organic waste from industries.

2. **Writing** In an essay, explain community activities that could be followed to ensure that local drinking water would not become polluted.

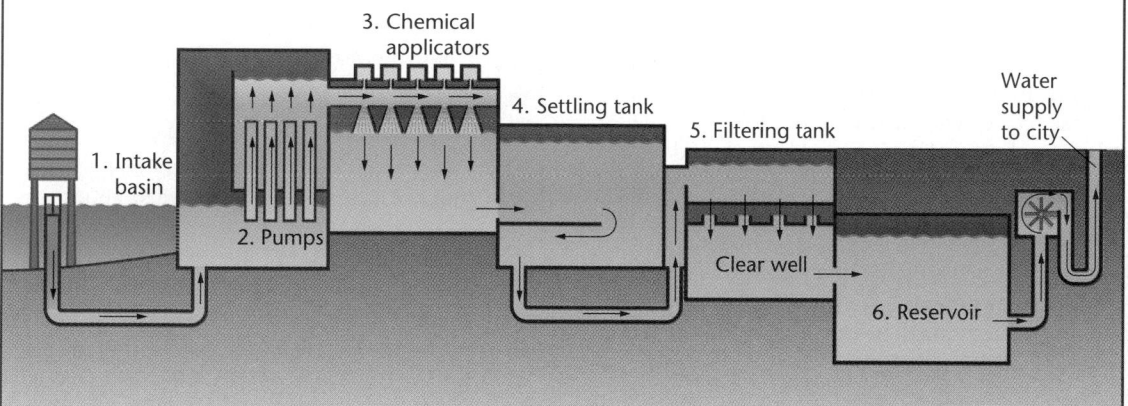

3. Chemical applicators

1. Intake basin

2. Pumps

4. Settling tank

5. Filtering tank

Clear well

6. Reservoir

Water supply to city

13.1 Uniquely Water 447

Cultural Diversity

Clean Water—A Global Perspective

In 1900, there were more than 35 000 deaths from typhoid in the United States. Typhoid is a disease caused by bacteria and spread by water. Today, as a result of chemical treatment, clean drinking water is taken for granted in the United States. However, waterborne diseases are still common in much of the rest of the world. World health scientists estimate that 80% of the world's sickness is related to contaminated water. It is estimated that less than 10% of the world's people have access to safe drinking water.

Meet Alice Arellano, Wastewater Operator

Teaching Strategies

- Ask students to account for the 200-400 L (roughly 50-100 gallons) of wastewater that each individual in the United States generates daily. *The water is used for washing the self, dishes, and cars; for laundry, and for flushing toilets.* L1

- Suggest that interested students videotape a field trip to a local water-treatment plant for classroom presentation. L1 LEP

- Have interested students research the treatment of wastewater on the space shuttle and compare it with municipal wastewater treatment. L2

Background

Wastewater is generally about 99.94% water and 0.06% dissolved or suspended solids. Wastewater may contain pathogens such as bacteria, viruses, and protozoans; pesticides; heavy metals; and nutrients.

PEOPLE in CHEMISTRY

Meet Alice Arellano, Wastewater Operator

When alarm clocks buzz on weekday mornings, millions of people head for the bathroom to prepare for the day. Showers run, toilets flush, and teeth are brushed, generating a tremendous flood of wastewater that flows toward treatment plants. Wastewater operators like Alice Arellano are on the job, making sure this water is properly treated and protecting the health of people and the environment.

On the Job

 Ms. Arellano, can you explain your job responsibilities to us?

 I'm a wastewater control-room operator. Our plant is responsible for cleaning wastewater from this part of Austin, Texas.

 How does the cleaning process work?

 First, the wastewater comes into our plant through a big line, about 54 inches in diameter. I've seen some pretty nasty stuff come through that pipe! The water goes through some bar screens, which remove the debris. The heavy solids, such as inorganic metals and sand, settle to the bottom of the chamber, where augers move them to a dump truck. The truck takes the inorganic waste to a landfill. The material that stays in suspension, the floatable stuff, goes into primary treatment. After primary treatment, then it goes to secondary treatment and some aeration

tanks. Microorganisms break down the harmful bacteria. As I like to explain it in nontechnical language, the microorganisms start at the top of the tank and eat and party all the way down.

 After the water has been cleaned, where does it go?

 Into the creeks and eventually into the Colorado River and Lake Powell. The water is filtered and used again. There's a finite amount of water on Earth, so we have to use it over and over.

 Does the weather have an effect on your work?

 When it rains, my work can get really tough. Ordinarily, the plant treats about 40 million gallons of water a day. Capacity is about 120 million gallons. But an unusually heavy storm can overwhelm the system with 400 million gallons. Then, all we can do is to add chlorine, open the valves, and let the water through.

Demonstration 13.4

Making Soap

Purpose: To prepare soap by performing a saponification reaction.

Materials: 25 g solid vegetable shortening, 250-mL beaker, 10 mL ethanol, 1.2 g NaOH, 40 g NaCl, hot plate, stirring rod, cheesecloth, evaporating dish

Safety Precautions:

Disposal: Code C

Procedure: Place 25 g of a solid vegetable shortening in a 250-mL beaker. Add 10 mL of ethanol and 5 mL of 6M NaOH solution (1.2 g of NaOH in 5 mL of water). Using a hot plate, heat the mixture over medium heat and stir constantly for 15 minutes. **CAUTION:** *Ethanol is flammable.* The soap can be colored by adding a small piece of crayon and can be scented with a drop of perfume oil. Cool the mixture in a cold-water bath, then add 25 mL of a saturated NaCl solution. The soap should now appear as curds. Collect the soap by filtering it through several layers of cheesecloth, and press it into an evaporating dish to shape it. **CAUTION:** *After carrying out the procedure, rinse your hands until they no longer feel slippery.* Air dry the soap for several days. **CAUTION:** *Do not use this harsh lye soap on the skin.*

How do people react when you talk about your profession?

 Some people are jealous because the job pays well and is steady work. Not everyone would like to do this job, though. The odor of ammonia can be strong. Once in a while, I get stomach viruses, even though I wear gloves and glasses and use all my safety equipment. However, I find the work fascinating. I'm trying to learn everything that goes on at the plant, so I ask everyone questions—the mechanics, the instrumentation people, and the electricians.

Early Influences

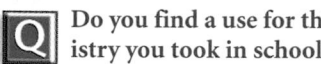 **Was this position your first one after you left school?**

 I got my GED late because I became a mother quite young. I had a friend, a teammate on my softball team, who worked in the plant and encouraged me to get certified as an operator. I took classes in wastewater management because I really wanted to better myself.

 Do you find a use for the math and chemistry you took in school?

 I have to know the formulas for the amount of chlorine to add, as well as for the sulfur dioxide. That chemical removes the chlorine so the water discharge won't harm fish in the creeks. More than my knowledge of chemistry, I use all my senses. It's often a matter of experience.

Personal Insights

 Your job isn't a typical one for a woman, is it?

 Men are in the majority here, but that's changing. My daughter came to Take-Our-Daughters-to-Work Day and became interested in what I do.

 Do you have any advice for kids who might not like school?

 Just to try hard, and stay in school. I encourage my own daughter to leave the adult behavior until the time she's ready for adult responsibilities. There are always about a dozen kids who make my home their gathering place these days. Because I had problems when I was young, they trust me to help them with their problems.

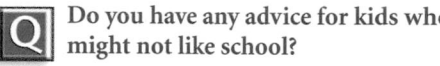

Other jobs vital for keeping water clean include the following.

Environmental Health Inspector College degree plus licensing

Environmental Technician Two-year accreditation program

Industrial Machinery Mechanic High school plus apprenticeship program

CAREER CONNECTION

- **Career Path** A student interested in a career as a wastewater operator should have mechanical aptitude. He or she should take high school courses in mathematics and chemistry. Two-year technical programs are offered, as well as on-the-job training.
- **Career Issue** Have students research how changes in the Safe Drinking Water Act will affect wastewater-treatment operations. **L2**

For More Information

Students interested in careers in wastewater operations can obtain information by writing to the National Environmental Training Association, 2930 E. Camelback Road, Phoenix, AZ 85016.

GLENCOE TECHNOLOGY

 Videodisc

Chemistry: Concepts and Applications
Making Soap
Disc 2, Side 1, Ch. 10

Show this video of the demonstration on these pages so students can visualize the process of soap making. **L1** **LEP**

 CD-Rom

Chemistry: Concepts and Applications
Making Soap
Show this video of the demonstration on these pages so students can see soap being made.
L1 **LEP**

Results: A small bar of soap is formed that dries hard after several days and can be removed from the dish.

Analysis: Ask these questions.

1. When fats react with a base, what products are formed? *Glycerol and the metallic salt of fatty acids (soap) are formed.*
2. How does soap clean? *It has a polar end and a nonpolar end, and can thus dissolve in greasy dirt and in water, and wash the dirt away.*
3. The soap you have just seen made is not safe to use on skin. Explain. *The sodium hydroxide (lye) used to make the soap is caustic, and some of it is probably still present.*

 **Assessment**

Skill: Have students test various soaps with pH paper and report their results on a chart to be displayed on the bulletin board. **L1**

3 ASSESS

Check for Understanding

- Check for understanding of the section material by having students answer the Section Review questions.
- Have students test various types of paper for adhesion/cohesion effects with water (properties related to capillarity) by dipping them in water. Ask students to explain the results. *Some of the papers carry the water more quickly because the water molecules may be more strongly attracted to the fibers or because some of the papers are more absorbent.* L1

Reteach

Software: Use the software program WAQUAL, which simulates a wastewater-treatment plant. AR 801, Project SERAPHIM L1 LEP

Extension

Allow students to use small drinking glasses that are filled to the rim with water and to determine how many pennies may be dropped carefully into the glasses before the water overflows. *Typically, about 50 to 60 pennies may be dropped.* L1 LEP

4 CLOSE

Portfolio

Ask students to imagine several relevant occurrences in a world in which water has been replaced by methanol, a substance that has a lower freezing point, does not expand when freezing, and has weaker hydrogen bonding than water. Students should write an essay describing their occurrences and consider putting it in their portfolios. L1 P

Water: The Super Solvent

Most of the water on Earth is not pure, but rather is present in solutions. Water is difficult to keep pure because it is an excellent solvent for a variety of solutes. Water is such a versatile solvent that it is sometimes called the universal solvent. Its ability to act as a solvent is one of its most important physical properties. As you will see, it is again the attraction of water molecules for other molecules, as well as for one another, that accounts for these solvent properties.

As **Figure 13.14** shows, nearly all of the water you consume is in aqueous solution. Aqueous solutions provide efficient means of transporting nutrients in plants, as well as the nutrients in your blood. Almost all of the life-supporting chemical reactions that occur take place in an aqueous environment. In the absence of water, these reactions would not occur.

Figure 13.14

Aqueous Beverages
Unless you buy distilled water or distill your own, everything you drink is an aqueous solution. Soft drinks, tea, coffee, spring water, and even tap water are all aqueous solutions.

SECTION REVIEW

Understanding Concepts

1. List five physical properties that distinguish water from most other compounds of similar molecular size.

2. What is hydrogen bonding?

3. What is surface tension? Use the concept of surface tension to explain why a drop of water forms a sphere.

Thinking Critically

4. **Comparing** Ethylene glycol is a molecule with the formula $C_2H_4(OH)_2$. Each molecule has two O—H bonds. Ethanol, C_2H_5OH, has one O—H bond. How would you expect the boiling points of ethanol and ethylene glycol to compare?

Applying Chemistry

5. **Humidity and Cooling** Hot weather is more uncomfortable if the humidity is high. Explain why.

SECTION REVIEW

1. Water is liquid at room temperature, has a high boiling point, exists in all three states on Earth, has a high specific heat, is a versatile solvent, and has a high heat of vaporization.
2. Hydrogen bonding is the formation of a connection between the hydrogen atoms on one molecule and a highly electronegative atom on another.
3. Surface tension is the force needed to break through the surface of a liquid or to spread a liquid out. A water drop is spherical be- cause the initially unbalanced forces that cause surface tension cause surface molecules to press inward. A sphere has a surface that is as close as possible to the center.
4. Ethylene glycol's boiling point should be higher because of greater hydrogen bonding.
5. The body cools itself by means of evaporating perspiration. On a dry day, evaporation and cooling take place efficiently, but on a humid day, atmospheric water recondenses on the skin, giving off heat.

Solutions and Their Properties

The polar water molecule is capable of dissolving a range of compounds, from ionic compounds, such as sodium chloride, to covalent compounds, such as sugars. What properties of the resulting aqueous solutions make them different from pure water? Because most of the water with which you come into contact contains dissolved materials, these aqueous solution properties play an important role in your everyday life.

The Dissolving Process

Just as with the other physical properties of water, you can relate water's solvent properties to its molecular structure. The submicroscopic interactions that occur between water molecules and various solute particles determine the extent to which water is able to dissolve the solutes.

Water Dissolves Many Ionic Substances

You have probably added table salt, sodium chloride, to water and noticed that the salt dissolved. If you made a careful measurement, you'd discover that you could dissolve about 36 g of sodium chloride in 100 mL of water at room temperature. Salt, like a great many ionic compounds, is soluble in water. The salt solution is also an excellent conductor of electricity. This high level of electrical conductivity is always observed when ionic compounds dissolve to a significant extent in water.

What's going on in such a situation? Your model of water and its interactions explains why salt and many other ionic compounds dissolve in water and why the solutions conduct electricity. Remember that ionic solids are composed of a three-dimensional network of positive and negative ions, which form strong ionic bonds. Look at **Figure 13.15** to see how a crystal of sodium chloride dissolves in water.

13.2 Solutions and Their Properties 451

SECTION PREVIEW

Objectives
Compare and contrast the ability of water to dissolve ionic and covalent compounds.
Distinguish solutions from colloids.
Compare and contrast colligative properties.

Key Terms
dissociation
unsaturated solution
saturated solution
supersaturated solution
heat of solution
osmosis
colloid
Tyndall effect

SECTION 13.2

PREPARE

Key Concepts
The formation and characteristics of aqueous solutions are examined. Solution concentration is explained, and the common expression of concentration—molarity—is defined and calculated. Several of the useful properties of solutions are then discussed.

Planning Ahead
For the MiniLab If you do not have a source of soft water, order or purchase a commercial softening cartridge.

1 FOCUS

Focus Transparency
Display the **Section Focus Transparency** for Section 13.2 to illustrate the solubility of sugar. [L1] [LEP]

2 TEACH

Concept Development
Point out that although one deals most often with liquid-solvent solutions, nine types of two-component solutions are possible: liquid in liquid, gas in liquid, solid in liquid, gas in gas, liquid in gas, solid in gas, solid in solid, gas in solid, and liquid in solid. Ask students to list one or more examples of each of the nine types of solutions. *Examples include air (gas in gas), metal alloys (solid in solid), rubbing alcohol (liquid in liquid), and club soda (gas in liquid).* [L1]

Program Resources

Study Guide, pp. 53-54 [L1]
Critical Thinking/Problem Solving, p. 14 [L2]
Laboratory Manual, 13.1, pp. 119-122; 13.2, pp. 123-126 [L1]
Section Focus Transparency 27 and **Master,** p. 27 [L1] [LEP]
Basic Concepts Transparencies 36, 37 and **Masters,** pp. 71-74 [L1] [LEP]

Problem Solving Transparency 15 and **Master,** pp. 29-30 [L1] [LEP]
ChemLab and MiniLab Worksheets, pp. 83-86, 88 [L1]
Applying Scientific Methods in Chemistry, pp. 39-40 [L1]
Chemistry and Industry, pp. 27-30

miniLAB

Hard and Soft Water

Purpose: To test hard water, soft water, and distilled water for Ca^{2+} and Mg^{2+} ions and to compare the ability of soap to form suds in the three types of water.

Process Skills: Observing, measuring, interpreting data
Disposal: Code C

Teaching Strategies

- Advise students to clean all three test tubes thoroughly and rinse them with distilled water before using them for testing.
- Advise students to measure the hard water with a small graduated cylinder, place it in the appropriate test tube, and then approximate the amounts of soft and distilled water by filling the tubes to the same level as the hard water.
- If your tap water is not hard enough to produce a precipitate in step 3, you may prepare your own hard water by adding sufficient $0.1M$ $CaCl_2$ to tap water to effect precipitation.
- If soft water is not available, you may pass tap water through a commercial softening cartridge.

Expected Results

Oxalate test: white, cloudy precipitate with hard water; no reaction with soft or distilled water
Height of suds: 1 cm or less with hard water; 8 cm or more with soft and distilled water

Analysis

1. No; the Ca^{2+} and/or Mg^{2+} ions may be present in such low concentrations that no discernible precipitate forms.

miniLAB 2

Hard and Soft Water

Water is said to be hard if it has significant concentrations of certain ions, usually Ca^{2+} and/or Mg^{2+}. Water in which calcium and magnesium ions are present in low concentrations only or in which they have been substantially replaced with Na^+ ions is called soft water. Distilled water contains only dissolved gases and exceedingly few ions. Test hard water, soft water, and distilled water with a sodium oxalate ($Na_2C_2O_4$) solution. This solution will cause Ca^{2+} and Mg^{2+} to precipitate as calcium oxalate (CaC_2O_4) and magnesium oxalate (MgC_2O_4). How well do the three types of water produce suds when they are mixed with a soap solution?

Procedure

1. Wear an apron and safety goggles.

2. Add about 1 mL of hard water to a small test tube, about 1 mL of soft water to another small test tube, and about 1 mL of distilled water to a third small test tube.

3. Add to each of the three solutions 2 drops of $0.1M$ acetic acid and 2 drops of $0.1M$ sodium oxalate. Mix the contents of each of the tubes by tapping them with your finger. Observe the solutions against a black background and record your observations.

4. Add about 2 mL of hard water to a large test tube, about 2 mL of soft water to another large test

tube, and about 2 mL of distilled water to a third large test tube.

5. Add to each of the three solutions about 1 mL of soap solution.

6. Shake each of the three solutions ten to 15 times by holding the test tube in your hand with your thumb over the top. Measure and record the height of the suds in each tube.

Analysis

1. If the oxalate hardness test is negative (that is, if no precipitate forms), is a complete absence of calcium and magnesium ions confirmed?

2. How does the hardness of water affect the ability to make soap suds?

Figure 13.15

A Model of the Dissolving of NaCl
When water is added, the polar water molecules surround the sodium and chloride ions, and the ionic compound dissociates. Because water molecules are polar and have a negative end and a positive end, they are attracted to both the positively charged sodium ions and the negatively charged chloride ions. Water molecules surround both types of ions. The opposite charges attract.

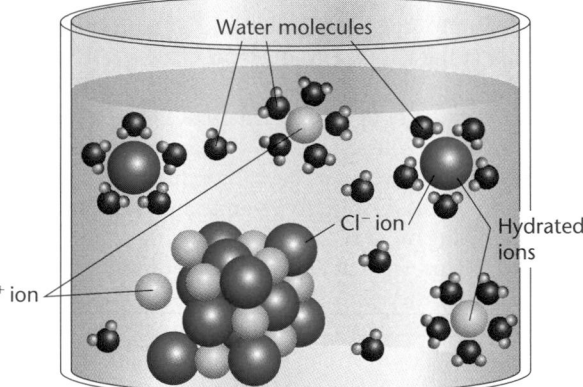

2. The hard water significantly diminishes the ability of the soap to form suds.

✓ Assessment

Performance: Ask students to write net ionic equations for the reactions between calcium ions and magnesium ions and the aqueous sodium oxalate solution. $Ca^{2+}(aq) + C_2O_4^{2-}(aq) \rightarrow CaC_2O_4(s)$ and $Mg^{2+}(aq) + C_2O_4^{2-}(aq) \rightarrow MgC_2O_4(s)$ **L1**

The process by which the charged particles in an ionic solid separate from one another is called **dissociation.** You can represent the process of dissolving and dissociation in shorthand fashion by the following equation.

$$NaCl(s) \xrightarrow{H_2O} Na^+(aq) + Cl^-(aq)$$

Water Dissolves Many Covalent Substances

Water is not only good at dissolving ionic substances. It also is a good solvent for many covalent compounds. Consider the covalent substance sucrose, commonly known as table sugar, as an example. You have probably observed that this substance, with the formula $C_{12}H_{22}O_{11}$, dissolves in water. In fact, it is highly soluble. It's possible to dissolve almost 200 g of sugar in 100 mL of water.

Take a look at **Figure 13.16.** It shows the molecular structure of a sucrose molecule. Notice that the structure has a number of O—H bonds. As you learned earlier, if a molecule contains O—H bonds, it will tend to be polar and it can form hydrogen bonds. One of the reasons that sugar is a solid rather than a liquid at room temperature is that sugar molecules have the ability to form hydrogen bonds with each other. The attractions hold the molecules in a crystal arrangement that ordinarily cannot be broken without an input of heat.

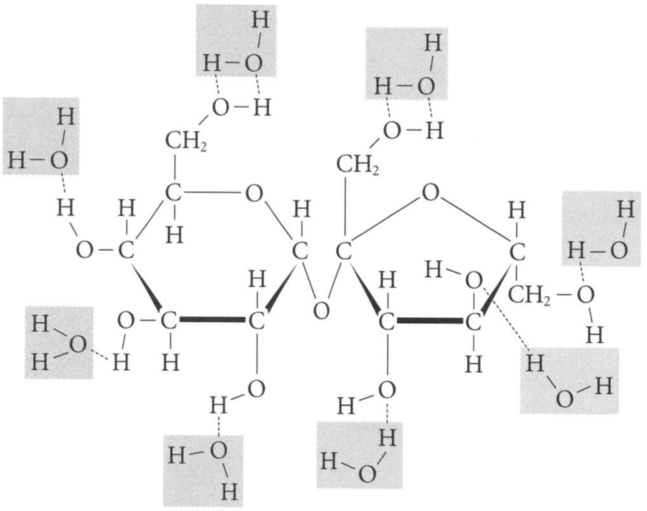

Figure 13.16
Sucrose

Sucrose, $C_{12}H_{22}O_{11}$, is a molecule that contains eight O—H bonds. When water is added to solid sucrose, each of the O—H bonds on the sucrose molecule is a possible site for hydrogen bonding with water. The intermolecular attractive forces between sucrose molecules are overcome and replaced by water-sucrose intermolecular attractive forces. This is why sugar is highly soluble.

This covalent sucrose model is similar to the ionic sodium chloride model in some ways. In both cases, interparticle attractive forces between the solvent and solute particles overcome attractive forces between solute particles. However, because sucrose is covalent, the sucrose molecules are simply separated from one another by water molecules. They do not dissociate into charged particles, but remain neutral molecules. Neutral molecules cannot conduct electricity because they have no charge. Therefore, an aqueous solution of sucrose is a nonconductor. The dissolving of sugar is represented by the following simple equation. Note that no ions are formed.

$$C_{12}H_{22}O_{11}(s) \xrightarrow{H_2O} C_{12}H_{22}O_{11}(aq)$$

Concept Development

Explain that the solvation mechanism can be thought of as the sum of three processes: (1) breaking up of the solute into separate component particles, (2) overcoming of interparticle forces in the solvent to make room for the solute (expanding the solvent), and (3) interaction of the solute and solvent particles to form the solution.

Correcting Misconceptions

Students may confuse the meaning of the designations *(l)* for *liquid* and *(aq)* for *aqueous.* Point out that *(l)* designates a pure liquid substance, whereas *(aq)* denotes a water solution of a substance.

GLENCOE TECHNOLOGY

CD-Rom

Chemistry: Concepts and Applications
Dissolving Table Salt
Have students use this animation to illustrate the dissociation of an ionic solid in water.
L1 LEP

Videodisc

Chemistry: Concepts and Applications
Dissolving Table Salt
Disc 2, Side 1, Ch. 12

Have students use this animation to illustrate the dissociation of an ionic solid in water.
L1 LEP

Transparency

Display **Basic Concepts Transparency 37** to reinforce the concept of dissociation. L1 LEP

Meeting Individual Needs

Learning Disabled Use colored markers to label some small steel balls as *Na⁺* and *Cl⁻*, and layer the balls alternately in the top of a box to represent ions in a crystal lattice. Allow students to use a clay ball-and-stick model of a water molecule, in which you have embedded small but powerful magnets near the surface of the hydrogen and oxygen balls, to pull the ions from the lattice. L1 LEP

Applying Chemistry

TECH PREP

Water in Gasoline Explain that water may be present in an automobile's fuel tank, either from contaminated fuel or from condensation of water from air in the tank. The water may interfere with the combustion process and may freeze in the fuel lines, plugging them. Point out that a dry-gas product, which is a methanol solution, may be added in such a case. It will solvate water in the fuel tank and burn during the combustion process in the engine, effectively eliminating the water by including it among the combustion products.

Transparency

Display **Problem Solving Transparency 15** to compare solubilities of two compounds. L1 LEP

Like Dissolves Like

Although water dissolves an enormous variety of substances, both ionic and covalent, it does not dissolve everything. The phrase that scientists often use when predicting solubility is "like dissolves like." The expression means that dissolving occurs when similarities exist between the solvent and the solute.

Consider the two examples of salt and sugar being dissolved in water. In the case of the ionic salt, water is "like" a charged ionic compound in the sense that it is polar, meaning that it has partially charged ends. The interactions that may occur between water molecules—dipole-dipole attractions and hydrogen bonds—are somewhat similar to the water-ion interactions that occur when salt dissolves. Thus, "like dissolves like." Water is polar and it tends to dissolve ionic substances.

In the case of sugar, water is like sucrose in that both compounds contain O—H bonds. More importantly, both substances are made up of polar molecules, with partially positive and negative ends. The molecular interactions between sucrose molecules are like the molecular interactions between water molecules. Again, like dissolves like. Water is polar and has hydrogen bonding, and it tends to dissolve substances that are polar or that form hydrogen bonds.

Oil and water are a classic example of two substances that do not mix; they do not form a solution. Oil is a mixture of nonpolar covalent compounds made up primarily of carbon and hydrogen. Given the composition of oil, you should not be surprised that oil does not dissolve in water. They are simply too *unlike*. When you try to place them in contact, there is little intermolecular force between solvent and solute that would allow them to mix. As shown in **Figure 13.17,** even after vigorous shaking, oil and water will rapidly separate into layers. However, different oils, which are nonpolar, are "like" enough to remain mixed.

Figure 13.17

Oil and Water Don't Mix, but Different Oils Do
When a bottle containing a mixture of baby oil (colorless) and water (dyed blue) is shaken, the oil and water seem at first to mix.
Upon sitting, the substances separate into two layers. ▶

In contrast, olive oil (greenish-gold) and safflower oil (nearly colorless) mix readily and remain mixed. ▶

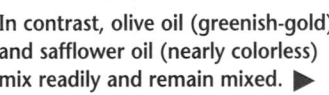  

Demonstration 13.5

Like Dissolves Like

Purpose: To demonstrate that nonpolar solutes are most readily dissolved by nonpolar solvents.

Materials: 1 g iodine crystals, 100-mL graduated cylinder, 30 mL water, 30 mL dichloromethane (methylene chloride) or TTE (trichlorotrifluoroethane), 30 mL hexane, deflagrating spoon

Safety Precautions: 🚫 ⚗️ 🥽 ☠️ 🧤

Disposal: B, G; hazardous material disposal may be required for TTE and dichloromethane. Check with local EPA.

Procedure: Place the dichloromethane or TTE in the graduated cylinder. Slowly pour 30 mL of water carefully down the side, avoiding mixing. Slowly add 30 mL of hexane to form the third distinct layer. Place several large I₂ crystals in a defla-

grating spoon, and slowly lower the spoon into the cylinder. **CAUTION:** *Avoid contact with the solvents. Do not breathe the vapors. Use a fume hood for the demonstration.* Begin at the top, and slowly move the spoon through each layer. Stir each layer with the spoon, allowing time for some of the I₂ to dissolve. Slowly remove the deflagrating spoon. You can evaporate the solvents in an operating fume hood, if

Everyday Chemistry

Soaps and Detergents

Has this ever happened to you? A slippery piece of pizza makes a greasy stain on your favorite jeans. You pour water on the telltale spot, but the water just runs off. You are learning the lesson that oil and water don't mix. The saying certainly applies—unless you add soap. Oil is normally insoluble in water, but soap molecules achieve the seemingly impossible by causing oil to mix with water.

Making soap Most people use soap for bathing and other detergents for washing clothes. However, there are important similarities between soaps and detergents.

People may first have learned how to make soap when ashes from a fire fell into a pot of boiling fat. The people soon found uses for the smooth white gel that floated to the top of the mixture. The ashes supplied lye, which is a strong base such as sodium hydroxide or potassium hydroxide. When fats or oils react chemically with lye, the end products are soap—often sodium stearate—and glycerin. Examine the chemical formula of sodium stearate. A negative carboxyl group ion ($-COO^-$) is tied to a positive sodium ion (Na^+) and a long chain of $-CH_2-$ groups.

$$CH_3-(CH_2)_{16}-COO^-Na^+$$

Like soaps, detergents have molecules with a polar end and a nonpolar end. Detergents typically contain sulfonates, which have SO_3 groups attached to a carbon chain or ring. The detergents are typically sodium salts of these sulfonates, and their molecules have an SO_3^- end rather than the COO^- end that is typical of soaps.

Soap forms a scumlike precipitate in the presence of the Mg^{2+} and Ca^{2+} ions that are present in hard water. Detergents, on the other hand, form soluble sulfonate salts in the presence of Mg^{2+} and Ca^{2+}. That makes detergents more effective than soaps in hard water.

How soaps and detergents work A soap molecule has two widely different parts: one end is hydrophilic, attracted to water, and the other end is hydrophobic, repelled by water. The hydrophobic part of the soap molecule consists of a long hydrocarbon chain that is structurally similar to oil and is therefore soluble in oil.

The hydrocarbon chains in soap are attracted to particles of oily grease and dirt. That part of the soap molecule forms a protective layer around the oily material. The hydrophilic end of the soap molecule (the $-COONa$ end) is attracted by polar water molecules. That causes the entire soap molecule, together with the oily material, to pull toward the wash water. The oil-soap complex is suspended in the water and rinsed away.

Exploring Further

1. **Inferring** How might rubbing and scrubbing with soap help to remove a greasy stain from clothing?

2. **Comparing** Some ancient peoples, such as the Egyptians and Romans, washed by rubbing themselves with oil, which was then scraped off. Compare the effectiveness of this method with the use of soap and water today.

---

legal, then collect the I_2 for reuse.
Results: The nonpolar iodine dissolves in the top and bottom layers, but not in the water layer.
Analysis: Ask these questions.
1. Which type of solvent, polar or nonpolar, dissolves more iodine? *nonpolar solvents*
2. What is meant by the phrase *like dissolves like? Nonpolar solutes are more soluble in nonpolar solvents, and polar solutes are more soluble in polar solvents.*

Assessment

Knowledge: Ask students to investigate why pretreating spots at a dry-cleaning store can be more effective if the nature of the spot is known. *Different solvents are used to remove different stains.* L1

Everyday Chemistry

Purpose

TECH PREP

To learn what soaps and detergents are and how they work.

Teaching Strategies

• Ask students how soaps make water and hydrocarbons compatible. *Soaps dissolve partly in hydrocarbons and partly in water, forming a link between these two dissimilar materials.*

• Have students carefully place a tack, point side up, on water. Because of the surface tension of the water, the tack should not sink. However, if students put detergent on the tip of a toothpick and poke it into the water near the floating tack, the tack will sink. Ask students why this happens. *The detergent lowers the surface tension of the water.* Relate this to a detergent's ability to clean. L1

Extension

Have students find out why soaps do not form a true solution. *The anions in soap are dispersed in water not individually, but rather as groups of ions. A soap-water mixture is a suspension of micelles. Students should relate this to the reason why soapy water looks cloudy.* L2

Exploring Further

1. Rubbing and scrubbing break down the grease into smaller droplets. The surfactant is then able to surround each grease particle with a thin layer, separating it from the clothing and making it easier to wash away.

2. Washing with oil is not much less effective than washing with soap. Oil and dirt on the body would dissolve in the washing oil. The main difficulty would be in removing the wash oil completely from the skin.

ChemLab

ChemLab

Solution Identification

Time Allotment: One laboratory period and two or three class periods.

Objectives

Review objectives with students before they begin the ChemLab.

Process Skills

Observing, classifying, communicating, interpreting data

Safety Precautions and Disposal

Advise students to wear safety goggles and aprons.
Disposal: Code J

PREPARATION

Make 100 mL each of 0.100M solutions of sodium carbonate, sodium iodide, copper(II) sulfate, copper(II) nitrate, lead(II) nitrate, and barium nitrate by adding the following masses to sufficient distilled or deionized water to make 100 mL of solution: 2.86 g of $Na_2CO_3 \cdot 10H_2O$, 1.60 g of NaI, 2.50 g of $CuSO_4 \cdot 5H_2O$, 2.42 g of $Cu(NO_3)_2 \cdot 3H_2O$, 3.31 g of $Pb(NO_3)_2$, 2.61 g of $Ba(NO_3)_2$.

Alternative Materials

You may substitute other solutions, as long as sufficient insoluble and differentiable products result.

PROCEDURE

- In order that students know what reactions to expect, have students complete *Analyze and Conclude* questions 1 and 2 prior to carrying out the laboratory work.
- You may wish to make and duplicate solubility charts and lists of colors of compounds. Allow students to use the lists to obtain the necessary information.

ANALYZE AND CONCLUDE

1. $Cu(NO_3)_2 + CuSO_4 \rightarrow$ $CuSO_4 + Cu(NO_3)_2$ (NR); $Cu(NO_3)_2 + Ba(NO_3)_2 \rightarrow$ $Cu(NO_3)_2 + Ba(NO_3)_2$ (NR);

456

Solution Identification

In an aqueous solution, ionic compounds are completely dissociated into ions. For example, an aqueous solution of barium nitrate, $Ba(NO_3)_2$, contains Ba^{2+} ions and NO_3^- ions. If aqueous solutions of ionic compounds are mixed, some ions may interact to form an insoluble product called a precipitate. For example, if aqueous solutions of barium nitrate and sodium sulfate are mixed, insoluble barium sulfate will precipitate. The complete formula equation for this reaction is written as follows:

$$Ba(NO_3)_2(aq) + Na_2SO_4(aq) \rightarrow$$
$$2NaNO_3(aq) + BaSO_4(s)$$

The barium sulfate is a white precipitate. Its formation as a result of mixing two unknown solutions could help you to identify the two reacting solutions. In this ChemLab, you will work with the aqueous solutions of six unknown ionic compounds. By observing the solutions and their interactions, you will be able to determine the identities of the compounds.

Problem

What are the identities of six unknown aqueous solutions?

Objectives

- **Observe** the interactions of the aqueous solutions of six compounds.
- **Interpret** the results of the interactions and use the results to identify the solutions.

PREPARATION

Materials

96-well microplate
10 mL each of 0.1M solutions of sodium carbonate, sodium iodide, copper(II) sulfate, copper(II) nitrate, lead(II) nitrate, barium nitrate—identified only as A through F
6 microtip pipets, labeled A through F
15 toothpicks
distilled water in a rinse bottle

Safety Precautions

Wear an apron and safety goggles. Wash your hands thoroughly after performing the lab. Some of the solutions are toxic.

PROCEDURE

1. Obtain a 96-well microplate. To produce all the combinations of solutions, you will be using only the 15 wells in the upper-right-hand corner.

2. Use the appropriately labeled microtip pipets to place three drops of solution A in each of five wells in the first horizontal row, three drops of B in each of four wells in the second row, three drops of C in each of three wells in the third row, three drops of D in each of two wells in the fourth row, and three drops of E in one well in the fifth row. The first letter in each box in the table in Data and Observations shows the arrangement.

3. For your second additions, use the appropriately labeled microtip pipets to add three drops of the solutions to the same wells to obtain the combinations shown in the Data and Observations table. Thus, you will place three drops of solution B in the single well in the first column, three drops of C in both of the wells in the second column, and so on up to the fifth column to which you will add three drops of solution F. You have now created all the possible combinations of pairs of solutions in the wells.

$2Cu(NO_3)_2 + 4NaI \rightarrow 2CuI + I_2 + 4NaNO_3$ (Note that this is unexpected; students will probably write $Cu(NO_3)_2 + 2NaI \rightarrow CuI_2 + 2NaNO_3$.);
$Cu(NO_3)_2 + Pb(NO_3)_2 \rightarrow Cu(NO_3)_2 + Pb(NO_3)_2$ (NR)
$Cu(NO_3)_2 + Na_2CO_3 \rightarrow CuCO_3 + 2NaNO_3$
$CuSO_4 + Ba(NO_3)_2 \rightarrow Cu(NO_3)_2 + BaSO_4$;
$2CuSO_4 + 4NaI \rightarrow 2CuI + I_2 + 2Na_2SO_4$;
$CuSO_4 + Pb(NO_3)_2 \rightarrow Cu(NO_3)_2 + PbSO_4$;

$CuSO_4 + Na_2CO_3 \rightarrow CuCO_3 + Na_2SO_4$;
$Ba(NO_3)_2 + 2NaI \rightarrow BaI_2 + 2NaNO_3$;
$Ba(NO_3)_2 + Pb(NO_3)_2 \rightarrow Ba(NO_3)_2 + Pb(NO_3)_2$ (NR);
$Ba(NO_3)_2 + Na_2CO_3 \rightarrow BaCO_3 + 2NaNO_3$;
$2NaI + Pb(NO_3)_2 \rightarrow 2NaNO_3 + PbI_2$;
$NaI + Na_2CO_3 \rightarrow Na_2CO_3 + NaI$ (NR);
$Pb(NO_3)_2 + Na_2CO_3 \rightarrow PbCO_3 + 2NaNO_3$

2. Reactants: copper(II) nitrate, blue; copper(II) sulfate, blue; barium nitrate, white (colorless in solution); sodium iodide, white; lead(II) nitrate, white; sodium carbonate, white.

4. Stir the solutions in each well, using a clean toothpick for each.

5. Place the microplate on a white surface and look down through each well to detect the presence of a precipitate. Cloudiness is evidence of a suspended precipitate. Repeat the procedure on a black surface. Record in your data table the color of any precipitates. If no precipitate is formed, write *NR*, meaning "no reaction."

2. **Classifying** Use a handbook of chemistry to find out the colors of each of the products and whether they are soluble or insoluble in water.

3. **Drawing Conclusions** Identify the six solutions by relating your data and observations to the predicted colors and interactions. Assume that the color of an aqueous solution is generally the same as that of the corresponding solute compound, except that solutions of white compounds are colorless.

DATA AND OBSERVATIONS

Combinations

A + B	A + C	A + D	A + E	A + F
	B + C	B + D	B + E	B + F
		C + D	C + E	C + F
			D + E	D + F
				E + F

ANALYZE AND CONCLUDE

1. **Formulating Models** Write balanced equations for all 15 possible double-replacement reactions, regardless of whether they actually occurred. Show both possible products in each case. If the two possible products are the same as the two reactants, you can rule out the possibility of reaction, so write *NR* (no reaction) after the equation.

APPLY AND ASSESS

1. Explain in general terms the reasoning you used in your identification process.

2. What are some other methods you might have used to identify the solutions?

Products: copper(I) iodide, brown-white, insoluble; iodine(I_2), brown in water, insoluble; copper(II) carbonate, blue-green, insoluble; barium sulfate, white, insoluble; barium iodide, colorless, soluble; barium carbonate, white, insoluble; sodium nitrate, white, soluble; sodium sulfate, white, soluble; lead(II) sulfate, white, insoluble; lead(II) iodide, yellow, insoluble; lead(II) carbonate, white, insoluble. (Copper(II) iodide, an expected product, does not exist. Instead, copper(I) iodide and elemental iodine are formed.)

3. A is copper(II) nitrate; B is copper(II) sulfate; C is barium nitrate; D is sodium iodide; E is lead(II) nitrate; F is sodium carbonate.

APPLY AND ASSESS

1. Students should explain that they made observations that allowed them to relate information in a logical way and to rule out possibilities so as to identify the various precipitates and solutions.

2. The original six solutions might have been evaporated to test for crystal form and color. Flame tests also might have been done.

✔ **Assessment**

Performance: Make up a set of eight solutions different from the ones already used, and allow interested students to carry out a similar testing and identification process. L1

DATA AND OBSERVATIONS

Assuming solution A is copper(II) nitrate; B is copper(II) sulfate; C is barium nitrate; D is sodium iodide; E is lead(II) nitrate; and F is sodium carbonate, the grid should look as shown at the right.

A + B NR	A + C NR	A + D brn-wh	A + E NR	A + F blu-grn
	B + C wh	B + D brn-wh	B + E wh	B + F blu-grn
		C + D NR	C + E NR	C + F NR
			D + E yel	D + F NR
				E + F wh

457

Explain that solution concentration is generally expressed as amount of the solute per amount of solution or as the amount of solute per amount of solvent.

Discussion

Ask students the following question. What is the concentration of oxygen in the stream? Then ask them which of the following responses is more appropriate. Response 1: The oxygen concentration in the stream is fairly high. Response 2: The concentration is 8.2 parts per million. *Students will probably correctly say that response 2 is better.* Explain that general or qualitative expressions of concentration such as *dilute* or *concentrated* are satisfactory and appropriate in many situations. However, more specific expressions of concentration, such as *8.2 ppm*, are required for purposes such as analysis.

Correcting Misconceptions

Students often believe that *saturated* and *concentrated* have the same meaning. Explain that *concentrated* means "containing a large amount of solute per amount of solvent or solution," and *saturated* means "containing all the solute that can be dissolved under the given conditions."

Concept Development

Explain that if positive and negative ions such as Ca^{2+} and $C_2O_4^{2-}$ are introduced separately into a solution in concentrations that exceed the solubility of the substance they form by combining, a solid called a precipitate (abbreviated *ppt*) will form—in this case, calcium oxalate, CaC_2O_4.

Solution Concentration

Suppose someone handed you a bottle and said, "This is an aqueous ammonia solution." You'd know that it consists of ammonia dissolved in water, but you wouldn't know how much. In other words, you wouldn't know the concentration of the solution, which is the relative amount of solute and solvent.

Concentrated Versus Dilute

If you were making tea, you might choose the approximate concentration you desire based upon personal taste. If you like strong tea, you make what a chemist might call a concentrated solution of tea; a relatively large amount of tea is dissolved in the water, so the concentration is high. On the other hand, if you like weak tea, a chemist might call the resulting mixture a dilute solution; relatively little tea is dissolved in the water, so the concentration is low. **Figure 13.18** summarizes this terminology. Chemists never apply the terms *strong* and *weak* to solution concentrations. As you'll see in the next chapter, these terms are used in chemistry to describe the chemical behavior of acids and bases. Instead, use the terms *concentrated* and *dilute*.

Figure 13.18

Qualitative Expressions of Solution Concentration
When your tea is strong, it's concentrated. If you like weak tea, you prefer a dilute solution.

Fact of the MATTER

The Dead Sea in Israel receives water from several streams but has no outflow. Instead, water evaporates from it, which concentrates salt. The Dead Sea has a salinity of 25 percent—as compared with only about five percent for ocean water—and is saturated with salt. The salt makes the water uninhabitable for fishes but so buoyant that it is nearly impossible for swimmers to sink in it.

Unsaturated Versus Saturated

Another way of providing information about solution composition is to express how much solute is present relative to the maximum amount the solution could hold. If the amount of solute dissolved is less than the maximum that could be dissolved, the solution is called an **unsaturated solution.** The oceans of Earth are examples of unsaturated saltwater solutions. They could hold a higher concentration of salt than they do now. The maximum concentration of salt water is approximately 36 g of salt dissolved per every 100 g of water, or 36 percent by mass. Such a solution, which holds the maximum amount of solute per amount of the solution under the given conditions, is called a **saturated solution.**

Figure 13.19
Fudge from a Supersaturated Solution
In making fudge, you heat a highly concentrated mixture of sugar, chocolate, and a water-based solvent such as milk to a high enough temperature to make a sugar solution that is supersaturated. Next, you slowly cool the mixture to room temperature with a great deal of stirring. If you've done everything right, your fudge will be soft and creamy because the sugar will crystallize out as very small crystals.

An interesting third category of solution is called a **supersaturated solution.** Such solutions contain more solute than the usual maximum amount and are unstable. They cannot permanently hold the excess solute in solution and may release it suddenly. Supersaturated solutions, as you might imagine, have to be prepared carefully. Generally, this is done by dissolving a solute in the solution at an elevated temperature, at which solubility is higher than at room temperature, and then slowly cooling the solution. Fudge making involves preparation of a supersaturated solution, as shown in **Figure 13.19.**

Effect of Temperature on Solubility

As you realized in the fudge example above, the solubility of sugar increases as the temperature increases. More and more solute is able to dissolve at higher and higher temperatures. Temperature has a significant effect on solubility for most solutes. **Figure 13.20** shows how the solubilities of six different solutes change with temperature.

Notice that each solute behaves differently with temperature. The solubilities of some solutes, such as sodium nitrate and potassium nitrate, increase dramatically with increasing temperature. Notice how steeply the curves climb upward in the figure. Other solutes, like NaCl and KCl, show only slight increases in solubility with increasing temperatures. A few solutes, like cerium(III) sulfate, $Ce_2(SO_4)_3$, decrease in solubility as temperature increases.

Figure 13.20
Solubility Versus Temperature
The amount of solute required to achieve a saturated solution in water depends upon the temperature, as this graph shows. Most solutes increase in solubility as temperature increases.

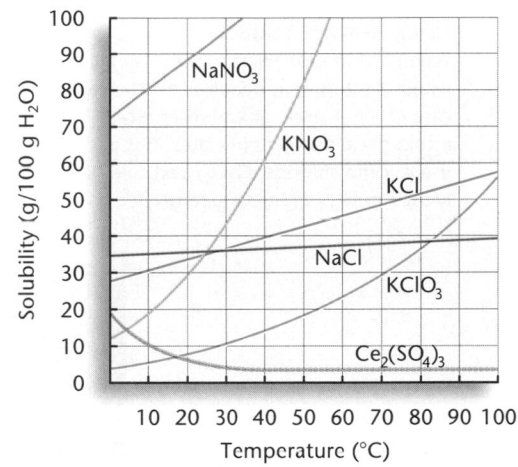

Chemistry Journal

Explain that substances that have endothermic heats of solution tend to be more soluble at higher temperatures, and substances that have exothermic heats of solution tend to be more soluble at lower temperatures.

Quick Demo

Which is exo, which is endo?

Place beakers of water that contain magnetic stirring bars and thermometers on two magnetic stirrers set on medium speed. Add several small scoops of ammonium chloride to one beaker and several small scoops of calcium chloride to the other. Ask students to observe the temperature changes and determine which compound has an exothermic heat of solution and which has an endothermic heat of solution. *Calcium chloride dissolves exothermically; ammonium chloride dissolves endothermically.* [L1]

As you can see, heat plays an important role in determining solubility. The heat taken in or released in a dissolving process is called the **heat of solution.** For most solutes, the process of dissolving in a solvent is an endothermic process. The dissolving of ammonium nitrate, for example, is endothermic. You can write the equation for the process, incorporating the heat term, as follows.

$$NH_4NO_3(s) + heat \xrightarrow{H_2O} NH_4^+(aq) + NO_3^-(aq)$$

Notice that the heat in an endothermic process is written as if it were a reactant because it must be added to the substances that will form the products. To increase the solubility of NH_4NO_3, add more heat, which increases the temperature. This forces the process toward the production of more aqueous ions. When working with a solute that dissolves endothermically, you may notice that the mixture becomes cooler during the dissolving. The heat required for the process is taken from the solution, which therefore cools off.

The dissolving of some solutes is exothermic. For these solutes, the dissolving process releases heat. Calcium chloride is a good example of such a solute. Note that in the equation for the dissolving process, heat shows up on the product side.

$$CaCl_2(s) \xrightarrow{H_2O} Ca^{2+}(aq) + 2Cl^-(aq) + heat$$

Hot and cold packs for sports injuries, shown in **Figure 13.21,** use solutes whose dissolving is highly exothermic or endothermic. Therefore, the solutes have large heats of solution.

Molarity

Suppose you're a worker in a hospital pharmacy lab and you must prepare a salt solution that matches the salt concentration of a patient's blood. You need to measure things carefully. Blood is a dilute salt solution, but the term *dilute* gives only qualitative information. In this case, you need a quantitative concentration unit.

Figure 13.21

Using Heats of Solution
When hot or cold packs for sports injuries are activated, a solute that dissolves exothermically in the case of a hot pack, or endothermically in the case of a cold pack, dissolves in water. Hot packs generally use calcium chloride, $CaCl_2$, and cold packs generally use ammonium nitrate, NH_4NO_3.

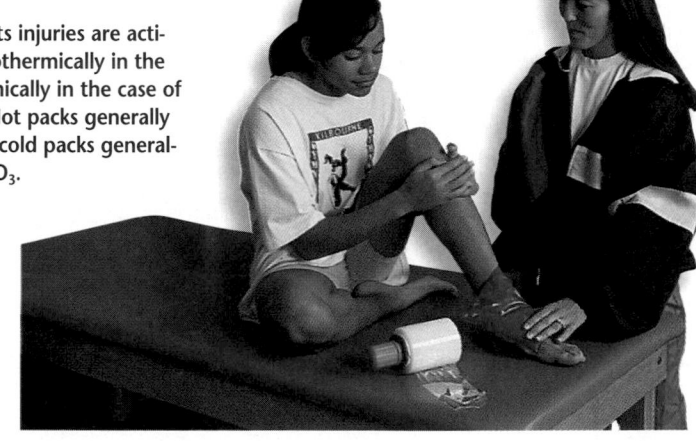

Chemistry Journal

Units of Concentration

Students should find a description of a mixing process, such as a recipe that involves combining of ingredients. They should calculate the relative proportions in terms of concentration units such as grams or moles per unit volume. Students should then rewrite the description, using the new units and recasting it in a more formal, scientific style. [L1]

Concentration units can vary greatly. They express a ratio that compares an amount of the solute with an amount of the solution or the solvent. For chemistry applications, the concentration term *molarity* is generally the most useful. Molarity is defined as the number of moles of solute per liter of solution.

Molarity = moles of solute/liter of solution

Note that the volume is the total solution volume that results, not the volume of solvent alone. Suppose you need 1.0 L of the salt solution mentioned above. In order to be at the same concentration as the salt in the patient's blood, it needs to have a concentration of 0.15 moles of sodium chloride per liter of solution. In other words, it must have a molarity of 0.15. To save space, you refer to the solution as 0.15*M* NaCl, where the *M* stands for "moles/liter" and is pronounced like the word *molar*. Thus, you need 1.0 L of a 0.15-molar solution of NaCl. How are you going to prepare it?

Assuming you're making an aqueous solution, you need to know only three things when working quantitatively: the concentration, the amount of solute, and the total volume of solution needed. The factor label approach to problem solving, as explained in Appendix A, is useful when you work problems such as the one on the next page. **Figure 13.22** shows the general solution-making steps in more detail using equipment that is available in labs where solutions are prepared quantitatively.

Step 1

Step 2

Step 3

Step 4

Figure 13.22

Quantitative Solution Preparation
Preparing solutions is important lab work. In step 1, the solute is weighed. In step 2, the solute is transferred to a volumetric flask, which holds a known volume. In step 3, enough water is added to dissolve the solute, and then more water is added to bring the solution volume up to the calibration mark on the flask. In step 4, the solution is shaken, stored in a stoppered container, and labeled.

Correcting Errors
Students often make mistakes in molarity calculations by neglecting water of hydration in determining formula mass. The difference in values can be dramatic. For example, the formula mass of $CuSO_4$ is 159.60 g/mol, but the formula mass of the common pentahydrate form, $CuSO_4 \cdot 5H_2O$, is 249.68 g/mol.

Concept Development
Explain that molar mass is expressed in g/mol and mg/mmol and that molarity is expressed in mol/L and mmol/mL. Point out that calculations may be simplified by using these relationships, particularly for solutions in which the mass of solute is known in milligrams and the volume of solution is known in milliliters.

Applying Chemistry

TECH PREP

Recyclable Heat Packs Explain that some of the heat packs used by athletic trainers contain supersaturated solutions of sodium acetate, a compound that has an endothermic heat of solution. Small packets of seed crystals of the compound are also present and are squeezed into the solution to initiate the exothermic precipitation. A pack can be reused by placing it in boiling water until the sodium acetate is redissolved, cooling it to room temperature, and squeezing out another seed crystal as needed.

SAMPLE PROBLEM **1** **Preparing 1 L of an NaCl Solution**

How would you prepare 1.0 L of a 0.15M sodium chloride solution?

Analyze • First, determine the mass of NaCl to add to a 1.0-L container. The 0.15M solution must contain 0.15 moles of NaCl per liter of solution.

Set Up • You will need to use the molarity of the solution (0.15 mol NaCl/L solution) as a conversion factor to get from molarity to number of moles of NaCl. You will then use the molar mass of NaCl as a conversion factor to change moles of NaCl to grams of NaCl. To find the molar mass of NaCl (58.5 g/mol), add the atomic masses of Na and Cl, and apply the unit *grams/mole* to the sum.

> **Problem-Solving HINT**
>
> Remember that a conversion factor is any relationship that compares two different units.

Solve • The proper setup, showing the conversion factors, is as follows.

$$\frac{1.0 \text{ L solution}}{} \left| \frac{0.15 \text{ mol NaCl}}{1 \text{ L solution}} \right| \frac{58.5 \text{ g NaCl}}{1 \text{ mol NaCl}}$$

Then carry out cancellations and calculate the answer.

$$\frac{1.0 \text{ L solution}}{} \left| \frac{0.15 \text{ mol NaCl}}{1 \text{ L solution}} \right| \frac{58.5 \text{ g NaCl}}{1 \text{ mol NaCl}} =$$

$$\frac{1.0}{} \left| \frac{0.15}{1} \right| \frac{58.5 \text{ g NaCl}}{1} = 8.8 \text{ g NaCl}$$

The result means you need to measure 8.8 g of NaCl, add some water to dissolve it, and then add enough additional water to bring the total volume of the solution to 1.0 L.

Check • Check to be sure that all units were correctly handled and that the calculation can be repeated with the same result.

SAMPLE PROBLEM **2** **Preparing a Different Volume of a Glucose Solution**

How would you prepare 5.0 L of a 1.5M solution of glucose, $C_6H_{12}O_6$?

Analyze • You need to determine the number of grams of glucose to add to a 5.0-L container. The 1.5M solution must contain 1.5 mol of glucose per liter of solution.

Set Up • Use the solution molarity as a conversion factor (1.5 mol glucose/L solution) to change from liters of solution to moles of glucose. Then, use the molar mass of glucose as another conversion factor to change from moles to grams of glucose. To find the molar mass of glucose (180 g/mol), add the atomic masses of 6 C, 12 H, and 6 O, and apply the unit *grams/mole* to the sum.

Solve • The proper setup, showing the conversion factors, is as follows.

$$\frac{5.0 \text{ L solution}}{} \left| \frac{1.5 \text{ mol glucose}}{1 \text{ L solution}} \right| \frac{180 \text{ g glucose}}{1 \text{ mol glucose}}$$

PRACTICE

Guided Practice: Guide students through Sample Problems 1 and 2, then have them work in class on Practice Problems 1-4. L1

Independent Practice: Your homework or classroom assignment can include the additional practice problems shown. L1

Answers to Practice Problems

1. Dissolve 63.8 g of CuSO$_4$ in 1.00 L of solution;

Molar mass CuSO$_4$ = 159.61 g/mol

$$\frac{1.00 \text{ L soln.}}{1} \left| \frac{0.400 \text{ mol CuSO}_4}{1 \text{ L soln.}} \right| \dots$$

$$\frac{159.61 \text{ g CuSO}_4}{1 \text{ mol CuSO}_4} = 63.8 \text{ g CuSO}_4$$

2. Dissolve 202 g of KNO$_3$ in 2.50 L of solution;
Molar mass KNO$_3$ = 101.10 g/mol

$$\frac{2.50 \text{ L soln.}}{1} \left| \frac{0.800 \text{ mol KNO}_3}{1 \text{ L soln.}} \right| \dots$$

$$\frac{101.10 \text{ g}}{1 \text{ mol}} = 202 \text{ g KNO}_3$$

3. 173 g of sucrose;
Molar mass C$_{12}$H$_{22}$O$_{11}$ = 342.30 g/mol

$$\frac{460 \text{ mL soln.}}{1} \left| \frac{1 \text{ L}}{10^3 \text{ mL}} \right| \dots$$

$$\frac{1.10 \text{ mol C}_{12}\text{H}_{22}\text{O}_{11}}{1 \text{ L soln.}} \left| \frac{342.30 \text{ g}}{1 \text{ mol}} \right| \dots$$

$$= 170 \text{ g C}_{12}\text{H}_{22}\text{O}_{11}$$

Cancel units and carry out the calculation.

$$\frac{5.0 \; \cancel{\text{L solution}}}{} \; \left|\frac{1.5 \; \cancel{\text{mol glucose}}}{1 \; \cancel{\text{L solution}}}\right| \frac{180 \; \text{g glucose}}{1 \; \cancel{\text{mol glucose}}} =$$

$$\frac{5.0}{} \; \left|\frac{1.5}{1}\right| \frac{180 \; \text{g glucose}}{1} = 1400 \; \text{g glucose}$$

The mass of glucose required is 1400 g. Weigh this mass, add it to a 5.0-L container, add enough water to dissolve the glucose, and fill with water to the 5.0-L mark.

Check • Check to be sure that all units were correctly handled and that the calculation can be repeated with the same result. All aspects of setup and calculation appear to be correct.

PRACTICE PROBLEMS

1. How would you prepare 1.00 L of a $0.400M$ solution of copper(II) sulfate, $CuSO_4$?

2. How would you prepare 2.50 L of a $0.800M$ solution of potassium nitrate, KNO_3?

3. What mass of sucrose, $C_{12}H_{22}O_{11}$, must be dissolved to make 460 mL of a $1.10M$ solution?

4. What mass of lithium chloride, LiCl, must be dissolved to make a $0.194M$ solution that has a volume of 1.00 L?

SAMPLE PROBLEM 3 Calculating Molarity

You add 32.0 g of potassium chloride to a container and add enough water to bring the total solution volume to 955 mL. What is the molarity of this solution?

Analyze • You have all the information you need about the preparation of the solution. You know the solute, the mass, and the total volume of the solution in milliliters. What you need to know is the molarity (moles KCl/L) of the solution.

Set Up • You are given that there are 32.0 g of solute per 955 mL of solution, so this relationship can be expressed in fraction form with the volume in the denominator. Therefore, the initial part of the setup is as follows.

$$\frac{32.0 \; \text{g KCl}}{955 \; \text{mL solution}}$$

Determine that the molar mass of KCl is 74.6 g/mol by adding the atomic masses of K and Cl and applying the unit *grams/mole* to the sum. The conversion factor that must be used to convert from grams to moles of KCl is 1 mol KCl/74.6 g KCl.

$$\frac{32.0 \; \text{g KCl}}{955 \; \text{mL solution}} \; \left|\frac{1 \; \text{mol KCl}}{74.6 \; \text{g KCl}}\right| \ldots$$

SECTION 13.2

GLENCOE TECHNOLOGY

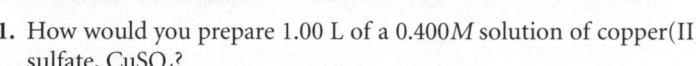 **Software**

Mastering Concepts in Chemistry
Unit 6, Lesson 3
Analytical Calculations Using Moles
Use this lesson to reinforce the concept of moles as used in solution calculations. **L1** **LEP**

---

4. 8.23 g of LiCl;
Molar mass LiCl = 42.40 g/mol

$$\frac{1.00 \; \cancel{\text{L soln.}}}{1} \; \left|\frac{0.194 \; \cancel{\text{mol}} \; \text{LiCl}}{1 \; \cancel{\text{L soln.}}}\right| \ldots$$

$$\frac{42.40 \; \text{g}}{1 \; \cancel{\text{mol}}} = 8.23 \; \text{g LiCl}$$

Additional Practice Problems

1. What mass of NaCl must be used to make 1.00 L of a $0.635M$ solution?

$$\frac{1.00 \; \cancel{\text{L soln.}}}{1} \; \left|\frac{0.635 \; \cancel{\text{mol}} \; \text{NaCl}}{1 \; \cancel{\text{L soln.}}}\right| \ldots$$

$$\frac{58.5 \; \text{g}}{1 \; \cancel{\text{mol}}} = 37.1 \; \text{g NaCl}$$

2. What mass of $NaNO_3$ must be dissolved to make 838 mL of a $1.25M$ solution?

$$\frac{838 \; \cancel{\text{mL soln.}}}{1} \; \left|\frac{1 \; \cancel{\text{L}}}{10^3 \; \cancel{\text{mL}}}\right| \ldots$$

$$\frac{1.25 \; \cancel{\text{mol}} \; \text{NaNO}_3}{1 \; \cancel{\text{L}} \; \text{soln.}} \; \left|\frac{85.0 \; \text{g NaNO}_3}{1 \; \cancel{\text{mol}}}\right| \ldots$$

$$= 89.0 \; \text{g NaNO}_3$$

Colligative Properties Colligative properties result mainly from the fact that a solution containing a nonvolatile solute has a lower vapor pressure than the corresponding pure solvent. Emphasize that values for this set of properties (freezing-point depression, boiling-point elevation, and osmotic pressure) depend only on the concentration of solute particles, not on the nature of the particles.

Quick Demo

How does salt melt ice?

Sprinkle some salt into crushed ice or snow in a beaker and ask students how salt melts ice. *Most students will incorrectly think that salt has a highly exothermic heat of solution and that the heat melts the ice.* Then put a thermometer into the mixture to demonstrate the decreasing temperature and explain that melting is due to freezing-point depression, a colligative property.

Next, to convert milliliters to liters, given that there are 1000 mL solution/L solution, use that conversion factor in the setup.

$$\frac{32.0 \text{ g KCl}}{955 \text{ mL solution}} \left| \frac{1 \text{ mol KCl}}{74.6 \text{ g KCl}} \right| \frac{1000 \text{ mL solution}}{1 \text{ L solution}} \cdots$$

Solve • Cancel units and carry out the calculation, using the setup just developed.

$$\frac{32 \text{ g KCl}}{955 \text{ mL solution}} \left| \frac{1 \text{ mol KCl}}{74.6 \text{ g KCl}} \right| \frac{1000 \text{ mL solution}}{1 \text{ L solution}} =$$

$$\frac{32.0}{955} \left| \frac{1 \text{ mol KCl}}{74.6} \right| \frac{1000}{1 \text{ L solution}} =$$

$$0.449 \text{ mol KCl/L solution} = 0.449M \text{ KCl}$$

Check • Check to be sure that all units were correctly handled and that the calculation can be repeated with the same result. All aspects of setup and calculation appear to be correct.

PRACTICE PROBLEMS

5. What is the molarity of a solution that contains 14 g of sodium sulfate, Na_2SO_4, dissolved in 1.6 L of solution?

6. Calculate the molarity of a solution, given that its volume is 820 mL and that it contains 7.4 g of ammonium chloride, NH_4Cl.

Solution Properties and Applications

Throwing salt onto an icy sidewalk and adding coolant to a car radiator—what do these familiar activities have in common? They both relate to a set of interesting and useful solution properties that depend only on the number and concentration of solute particles. These properties are freezing-point depression and boiling-point elevation.

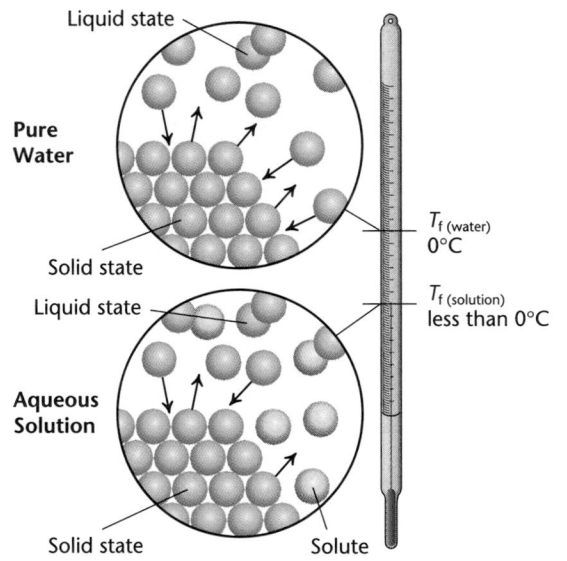

Liquid state
Pure Water
Solid state
$T_{f \text{ (water)}}$ 0°C
Liquid state
Aqueous Solution
$T_{f \text{ (solution)}}$ less than 0°C
Solid state
Solute

Figure 13.23
Disrupting Solvent Organization
A solute disrupts the high level of organization that is necessary if water is to be in the solid state. The freezing point of the solution is lower than the freezing point of the pure water. If the surrounding temperature is above the freezing point of the solution, the solution melts. The effect occurs because solute particles have replaced some of the solvent particles, and they interfere with the freezing process. The more solute particles there are, the more the freezing point is lowered.

PRACTICE

Guided Practice: Guide students through Sample Problem 3, then have them work in class on Practice Problems 5-6. **L1**

Independent Practice: Your homework or classroom assignment can include the additional practice problems shown below. **L1**

Answers to Practice Problems

5. $0.062M$ Na_2SO_4;
Molar mass $Na_2SO_4 = 142.06$ g/mol

$$\frac{14 \text{ g } Na_2SO_4}{1.6 \text{ L soln.}} \left| \frac{1 \text{ mol}}{142.06 \text{ g}} \right.$$

$= 0.062$ mol Na_2SO_4/L soln.

$= 0.062M$ Na_2SO_4

6. $0.17M$ NH_4Cl;
Molar mass $NH_4Cl = 53.50$ g/mol

$$\frac{7.4 \text{ g } NH_4Cl}{820 \text{ mL}} \left| \frac{1 \text{ mol}}{53.50 \text{ g}} \right| \frac{10^3 \text{ mL}}{1 \text{ L}}$$

$= 0.17$ mol NH_4Cl/L soln.

$= 0.17M$ NH_4Cl

Additional Practice Problems

1. What is the molarity of a solution that contains 25.1 g of KI dissolved in 4.3 L of solution?

Freezing-Point Depression

Figure 13.24

Applying Freezing-Point Depression Freezing-point depression has some practical uses, such as in deicing airplanes, a process that uses ethylene glycol as a solute.

A solution always has a lower freezing point than the corresponding pure solvent. If you are interested only in aqueous solutions, this means that any aqueous solution will have a freezing point lower than 0°C. The amount that the freezing point is depressed relative to 0°C depends only upon the concentration of the solute. You have observed this property if you have carried out the winter salt-throwing you just read about. Salt placed on an icy sidewalk causes the ice to melt. This melting is a result of freezing-point depression. The salt dissolves in the water that makes up the ice and forms a solution that has a lower freezing point than pure water, as illustrated in **Figure 13.23.**

An ionic solute produces greater depression of freezing point than a covalent one because it dissociates into ions. One mole of ionic NaCl thus produces 2 moles of solute particles. It produces twice the interference with the freezing process and twice the desired effect as 1 mol of sucrose. Likewise, 1 mol of $CaCl_2$ has three times the effect of 1 mol of sucrose because each formula unit of $CaCl_2$ dissociates into three ions: one Ca^{2+} and two Cl^-. **Figure 13.24** gives some additional examples of the application of freezing-point depression.

Boiling-Point Elevation

You have just learned that the freezing point of a solution is lower than the freezing point of the pure solvent. It turns out that the boiling point of a solution is higher than the boiling point of the pure solvent. For aqueous solutions, this means that the solution boiling point will be greater than 100°C, assuming standard atmospheric pressure. The solute must also be nonvolatile; that is, not able to evaporate readily.

Solute particles affect boiling point because they take up space at the surfaces where the liquid and gas meet, which interferes with the ability of the solvent particles to escape the liquid state. The solute particles lower the vapor pressure of the solvent, so a higher temperature is required to bring the vapor pressure up to atmospheric pressure and cause boiling to occur. The higher the concentration of solute particles, the greater is the boiling-point elevation. Read about how antifreeze works as a coolant in an automobile engine in *Everyday Chemistry.*

The world's largest water-purification plant that uses reverse osmosis is in Jubail, Saudia Arabia. The plant provides 50 percent of the country's drinking water by using reverse osmosis to remove salt from seawater taken from the Persian Gulf.

Reinforcement

Ask interested students to put tap water, a saturated sugar solution, and a saturated sodium chloride solution in three small plastic containers and to place the containers in a freezer overnight. They should write a report on their findings and consider placing it in their portfolios. [L1] [P]

Discussion

Ask students why either methanol or ethanol would make an effective antifreeze but not an effective coolant for an automobile, particularly during the summer. *These alcohols are volatile. Although they would lower the freezing point of water, they would tend to boil at high engine-operating temperatures and could possibly cause the cooling system to burst. They could also simply boil away, leaving no antifreeze protection.*

$$\frac{25.1 \, g\,KI}{4.3 \, L} \left| \frac{1 \, mol \, KI}{166 \, g\,KI} \right. = 0.035M \, KI$$

2. What is the molarity of a solution that has a volume of 705 mL and that contains 10.6 g of KNO_3?

$$\frac{10.6 \, g\,KNO_3}{705 \, mL} \left| \frac{10^3 \, mL}{1 \, L} \right| \frac{1 \, mol \, KNO_3}{101 \, g\,KNO_3}$$

$$= 0.15M \, KNO_3$$

Chemistry Journal

Everyday Situations

Have students write journal entries about everyday situations in which freezing-point depression and boiling-point elevation play a role. These might include processes such as addition of antifreeze to car radiators, spreading of salt on icy roads, and addition of salt to boiling water to increase temperature and shorten cooking time. [L1]

Everyday
Chemistry

Purpose

TECH PREP

To show how the addition of solutes to water in a radiator can keep a car from overheating in warm weather and freezing in cold weather.

Teaching Strategies

- Remind students that after a snowfall, salt is spread on an icy sidewalk. Ask students to explain how this helps to prevent ice from forming. *The salt is a solute that lowers the freezing point of water and melts ice or keeps it from forming.* L1

- Discuss the need for responsible disposal or recycling of coolants. Have students suggest special procedures that would be necessary when draining a car radiator and disposing of the used fluid. *Answers will vary. A possible answer would be to purify the liquid to make certain that lead is removed.* L1

Extension

Ask students to investigate the chemical explanation of why propylene glycol would be a safer coolant than ethylene glycol. *Unlike ethylene glycol, propylene glycol can be metabolized to lactic and pyruvic acids, which are relatively harmless products normally formed during processes taking place in the body.* L2

Exploring Further

1. Temperatures below 0°C are needed to make ice cream. People add rock salt to the ice in a hand-cranked ice-cream maker. This causes a lowering of the freezing point so that the temperature of the ice-water mixture decreases enough below 0°C to make ice cream. This is

Antifreeze

In 1885, Karl Benz of Germany invented and patented the first radiator for an automobile. This was a big change over simply cooling the engine by evaporation, which removed 4 L of water each hour and required constant replenishment. A radiator was able to recirculate the water used to cool the engine. After air-cooled water removed the heat created when the engine burned fuel, the water returned to the radiator to be recooled.

How chemical coolants work The first use of a chemical engine coolant, ethylene glycol, was tried in England in 1916 for high-performance military aircraft engines.

Ethylene glycol, $C_2H_4(OH)_2$, is still the major constituent of most automobile antifreeze solutions used in the United States. When it is added to water, the solution that is formed has a higher boiling point than plain water. This keeps the water in the radiator from boiling off, as it would without the coolant. Recall that the boiling point of a substance is the temperature at which the vapor pressure of the liquid phase equals the atmospheric pressure. When a solute is present, it crowds out some of the water molecules at the surface of the solution. This reduces the number of water molecules that can escape as water vapor. As a result, the solution has a lower vapor pressure than pure water. If the vapor pressure of the solution in the radiator is lowered, additional kinetic energy is needed to raise it to the same level as the atmospheric pressure. You can see that this would make the boiling point of the solution higher than that of water.

Ethylene glycol as an antifreeze The same process of adding a solute to water also lowers the temperature at which water freezes. The number of degrees the freezing point is lowered is proportional to the number of solute particles dissolved in the water. Any solute added to water decreases

its freezing point. A solution with a low freezing point, such as an ethylene glycol solution in a car radiator, is less likely to freeze in winter.

Properties of a good engine coolant Besides raising the boiling point and lowering the freezing point, a good coolant doesn't corrode the materials in the radiator and the pump. Over the years, material and design changes have caused problems of corrosion of metal parts in contact with coolants. Adding inhibitors to the coolant or changing the materials solves this problem.

A good coolant should also be easy to dispose of. Ethylene glycol is biodegradable. However, it is toxic to humans and other mammals if ingested. It has a sweet smell and taste, which seems to appeal to animals and sometimes children. Care must be taken in disposing of these chemicals until a safer product is found.

Exploring Further

1. **Acquiring Information** Investigate how ice cream is made at home. Relate the information gathered to the use of an antifreeze in an automobile during winter.

2. **Applying** If a coolant becomes oily, murky, or rusty, what would this suggest?

analogous to the freezing-point depression caused by antifreeze in a radiator.

2. It would suggest that the coolant has been contaminated with a foreign material or that excessive corrosion in the cooling system has occurred.

Osmosis

Have you ever seen water being sprayed onto vegetables in a super-market? Some of the water is absorbed by the vegetables, which makes them plumper, crisper, and fresher looking. The water is able to move inside because cell membranes on the outside of the vegetables are selectively permeable—that is, they allow certain materials, such as water, to pass through them.

As you may recall from Chapter 10, gas molecules have a tendency to diffuse from an area of high concentration to an area of low concentration. Particles in a liquid have a similar tendency to move. In the case of the vegetables, the water that is naturally inside them contains solutes, such as sugars and salts. Because of the presence of these solutes, there is less water per unit volume than in pure water. If pure water is sprayed onto the outside of the vegetables, the water will tend to diffuse into the vegetables. It moves from the area of higher water concentration outside to the area of lower water concentration inside the vegetables. This flow of solvent molecules through a selectively permeable membrane, driven by concentration difference, is called **osmosis.**

Figure 13.25 shows a system that illustrates osmosis. Pure water is on the left side of a selectively permeable membrane, and an aqueous solution of sucrose is on the right side. The membrane allows water molecules to pass, but not sucrose molecules.

WORD ORIGIN

osmosis:
  osmos (GK)
  impulse
  othein (GK) push
During osmosis, solvent molecules push through a selectively permeable membrane.

Figure 13.25
Osmosis
Note what takes place when pure water and a sucrose solution are separated by a selectively permeable membrane.

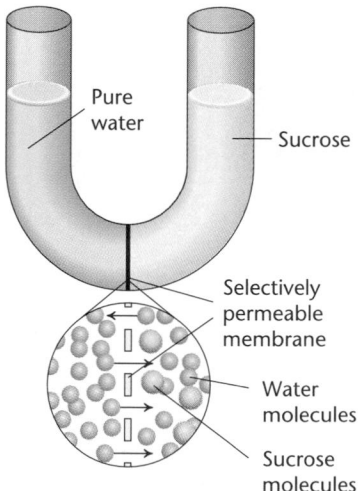

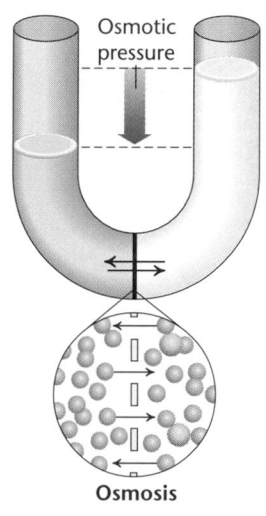

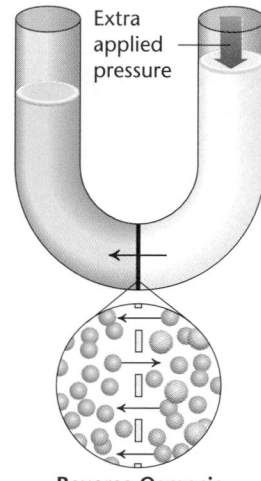

▲ As osmosis begins, water molecules diffuse more rapidly from the water into the sucrose solution than they diffuse from the sucrose solution into the water. As a result, the sucrose solution gains water, becomes more dilute, and the volume rises.

▲ The increasing height of the sucrose solution exerts a pressure that opposes the diffusion of water molecules from left to right. Eventually, the pressure becomes high enough that the rates of diffusion of water in both directions become the same.

▲ Adding extra pressure to the side with the sucrose solution can cause the water diffusion to move in the opposite direction, forcing water out of the solution and producing pure water. This process, called reverse osmosis, can be used to purify water, as discussed in *How It Works.*

13.2 Solutions and Their Properties **467**

Extension

TECH PREP Ask interested students to research the dialysis process that is used to purify blood in artificial kidney machines. The students might arrange to visit a hospital or clinic in which the procedure is carried out. Ask them to make a class presentation, using visual aids and explaining the role of osmosis in the process. **L2** **LEP**

Visual Learning

Figure 13.25: You may wish to introduce and explain the term *osmotic pressure,* which is the minimum pressure required to stop osmosis between two solutions of different concentration. The units for osmotic pressure are generally mm of Hg (torr) or atmospheres.

Across the Curriculum

Literature

The famous poem *The Rime of the Ancient Mariner* by Samuel Taylor Coleridge provides a great example of osmosis and osmotic pressure, as well as the difference between salt water and fresh water.

> *Water, water everywhere,*
> *And all the boards did shrink;*
> *Water, water, everywhere,*
> *Nor any drop to drink*

The boards shrank because of the movement of water out of the boards and into the salty seawater. The problem of there not being *any drop to drink* ties in with the feature on portable reverse osmosis systems for sailors.

How it Works

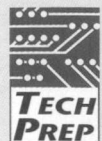

Purpose

TECH PREP

Students will learn about an important practical application of reverse osmosis.

Background

Reverse osmosis (RO) units depend upon the application of large amounts of pressure. Emphasize that the large pressures that are required to reverse the spontaneous direction of water movement (from seawater into pure water) are the major obstacle that prevents RO from being more widely used in large-scale water-purification systems.

Visual Learning

The details of the figure are less important than the fact that the design allows for an individual to generate the required pressure using a hand-pumped device. Emphasize that light weight, simplicity of operation, and portability are important features of the design.

Teaching Strategies

Have students discuss other situations in which portable RO units could be useful. *Backpacking, camping, and responses to natural disasters are some possibilities.* Have students look through sailing or backpacking catalogs for descriptions and prices of portable RO units. L1

Thinking Critically

1. Application of 27 atm of pressure on the solution side only balances the flow across the membrane, so no net movement occurs. A larger applied pressure is required to get a net drinkable quantity of water from the unit.
2. It is expensive, operates rather slowly, and can be done only on a small scale.

How it Works

A Portable Reverse Osmosis Unit

Reverse osmosis (RO) is the process by which water from a solution is forced through a selectively permeable membrane by the application of pressure to the solution side of the membrane. Portable RO units are commercially available that can purify seawater of its salts and make it drinkable. Approximately 27 atm of pressure needs to be applied to seawater in order to counteract the flow of solvent into the seawater through a selectively permeable membrane. In order to get a usable amount of water through the membrane, you need to apply about twice that pressure. An RO unit can apply these large pressures.

1. An operating handle is pushed down, and an attached piston puts pressure on seawater to drive it into a cylinder. Some of the water moves through a selectively permeable membrane, but the salts are left behind.

3. The remaining salty water that has not passed through the membrane is returned to the area behind the piston, where it provides some of the pressure needed to process the seawater.

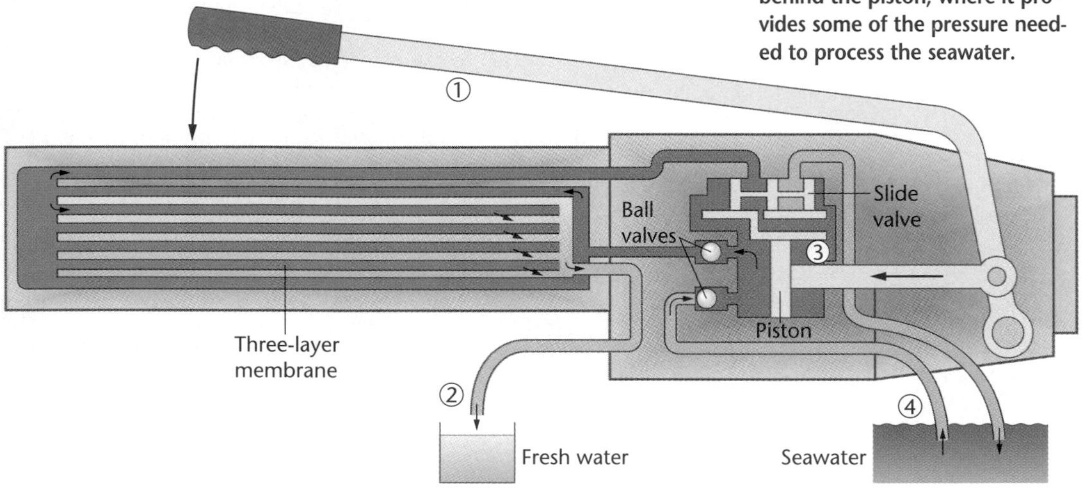

2. The pure water leaves the device and is collected.

4. The operating arm is lifted. The seawater behind the piston flows out as waste. New seawater enters and passes around movable ball joints, and the process is ready to begin again.

Thinking Critically

1. Why must the pressure required to operate a reverse osmosis unit be more than 27 atm?
2. Why isn't reverse osmosis more widely used as a purification method for water?

Solutions of Gases in Water

If you observe an unopened bottle of club soda, the liquid inside looks just like pure water. However, when you unscrew the cap, small bubbles of gas appear throughout the liquid and rise to the top. If the club soda is warm, the fizziness can be so intense that the liquid rises up and spurts out the top. **Figure 13.26** explains this sort of behavior.

The solubility of a gas in a liquid depends on the pressure of the gas pushing down on the liquid. The higher the pressure, the more soluble is the gas. For solutions of gases in liquids, gas solubility decreases as temperature increases. That's why club soda fizzes more vigorously when it is warm. The dependence of gas solubility on temperature and pressure is important in a number of areas, as shown in **Figure 13.26**.

Figure 13.26
Solutions of Gases in Water
The club soda (left) is a solution of carbon dioxide in water. In the unopened bottle, the gas is dissolved under pressure in the water. When the bottle is opened, the carbon dioxide gas trapped above the liquid escapes and the pressure drops. The solution is now supersaturated in carbon dioxide. Some of the dissolved carbon dioxide reenters the gaseous state and forms the bubbles that you see. ▼

▲ Deep under water, pressures are high. The nitrogen gas from the air in the lungs of divers gets dissolved at higher-than-normal concentrations in their blood. As they ascend from the depths, the pressure decreases and so does the blood solubility of the nitrogen. If divers come up too rapidly, the released nitrogen can form dangerous and painful bubbles in the blood vessels in a condition called the bends. To help prevent this, divers may use tanks of gas that contain helium mixed with oxygen instead of the usual nitrogen-oxygen mixture. The helium is less soluble in blood.

The fishes depend on dissolved oxygen in the water. If water temperature goes too high, the solubility of the oxygen may fall too low, and the fishes may die. That is one of the reasons thermal pollution from power plants can pose serious problems to aquatic life. ▶

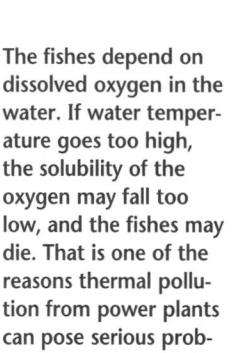

Applying Chemistry

Lake Nyos Tragedy On August 21, 1986, the water in Lake Nyos in the African nation of Cameroon "overturned," perhaps due to winds or cold monsoon rains. The colder water from the bottom of the lake, which was supersaturated with carbon dioxide due to the high pressures and low temperatures at that depth, came to the surface. The CO_2 was released, smothering approximately 2000 people.

⚡ Quick Demo

Where is the fat in homogenized milk?
Explain that homogenized milk is a colloid of small butterfat globules dispersed in water. Allow students to observe homogenized whole milk, low-fat milk, and skim milk at $10\times$ magnification under a microscope to observe the individual globules. **L1** **LEP**

Applying Chemistry

The Cottrell Precipitator Find a diagram of a Cottrell precipitator, which is used in *TECH PREP* smokestacks to attract charged colloidal particles and remove them from smoke emissions. Copy the diagram for students and discuss the operation of the device with them.

GLENCOE TECHNOLOGY

📼 Videotape

MindJogger Videoquizzes for Chemistry
Chapter 13: Water and Its Solutions
Have students work in groups as they play the videoquiz game to review key chapter concepts. **L1** **LEP** **COOP LEARN**

📓 Chemistry Journal

Deep-Sea Dive
Have students write an exciting fictional account of a deep-sea dive, a situation in which dissolved blood gases play a crucial role. For example, students might describe an escape from a shark or other threatening organism that might also expose the diver to the danger of acquiring the bends by a too-rapid ascent. Students may wish to add illustrations to their account. **L1**

Purpose

Students become familiar with various kinds of colloids and recognize the importance of these mixtures in many fields of science.

Background

The main factor that keeps the particles in a colloid suspended in a medium seems to be electrostatic repulsion. Although a colloid is electrically neutral overall, the dispersed particles all migrate to the same electrode when the colloid is placed in an electric field. This indicates that the dispersed particles carry a similar charge or partial charge. For this reason, the colloidal particles repel each other and do not easily aggregate to form clumps of particles large enough to precipitate.

Teaching Strategies

• Demonstrate the Tyndall effect, which shows that light is reflected in all directions from particles in colloids and suspensions but not in solutions. In a darkened room, focus a beam of light through a pinhole onto a glass containing a salt solution. The light will not reflect noticeably from the particles in solution. However, the same beam of light will be reflected and dispersed by the particles in a colloid or a suspension, such as colloidal sulfur. See the demonstration below.

Versatile Colloids

When you look at the photographs of the colloids on these pages, think about what these materials have in common. Recall that colloids are mixtures composed of tiny particles of one substance that are dispersed or evenly distributed in another substance. The particles of a colloid are intermediate in size between those of suspensions and true solutions. They range somewhere between the size of large molecules and a size great enough to be seen through a microscope.

Liquid Aerosols

Fog is an example of a liquid aerosol, which is formed when fine liquid droplets are dispersed in a gas. Fog appears when moist air near Earth's surface is cooled to the point at which the water vapor begins to condense. Other liquid aerosols with which you are familiar include spray deodorants and hair spray.

Solid Aerosols

At times, solid particles are dispersed in a gas. One example of a solid aerosol is the polluting soot particles that may be released into the air from an industrial smokestack. Such a problem could be remedied by placing a

470

Cottrell precipitator in the smokestack. The charged plates of that type of precipitator attract colloidal soot particles and remove them from the air. This is how air pollution has been reduced in many industrial cities.

Emulsions

Milk and mayonnaise are examples of emulsions. Emulsions are dispersions of fine droplets of liquid—generally a fat—in another liquid. Many emulsions are able to maintain their stability with the help of materials such as gums. Gums and other stabilizers thicken the liquid phase, making it less likely for the dispersed droplets to come together.

Sols

A sol is a fluid colloidal system in which fine solid particles are dispersed in a liquid medium. Most household paints are sols with finely ground pigments mixed with acrylic resins dissolved in water. Paint is applied to a surface as a liquid. With time, the water dries and the resins harden, leaving a thin solid film on the surface.

Demonstration 13.6

Tyndall Effect

Purpose: To demonstrate the Tyndall effect.
Materials: 1 g sodium thiosulfate ($Na_2S_2O_3$), 120 mL distilled water, 150-mL beaker, overhead projector, opaque paper, 1 mL concentrated ($12M$) HCl, dropper, stirring rod
Safety Precautions:
Disposal: Code C

Procedure: Dissolve the sodium thiosulfate in the distilled water in a beaker. Cover the top of an overhead projector with opaque paper. Cut a hole in the paper slightly smaller than the beaker. Turn on the projector, and turn off the room lights. Place the beaker containing the $Na_2S_2O_3$ over the hole in the paper. Focus the image on the screen. Add the $12M$ HCl drop by drop to the beaker, stirring

occasionally. Students should look at the beaker from the side and at the screen.
Results: The colloidal sulfur that is formed tends to scatter light that is mainly of shorter wavelengths (on the blue end of the spectrum). When viewed from the side, the liquid in the beaker appears blue as the light is scattered (Tyndall effect). The transmitted light on the screen, in contrast, becomes dimmer and red be-

Gels

The food industry uses many colloids that have the ability to thicken or gel liquid foods. Gels are dispersions of giant macromolecules in liquid. Many gels are natural gums found in seaweeds and land plants. The natural substance pectin, found in fruits, is responsible for the gel structure in jelly.

Pastes

The basis for a paste is a concentrated dispersion of solids in a limited amount of liquid. To make a beautiful porcelain vase like the one in the photo, ground quartz and feldspar are mixed with a white clay called kaolin in

a small portion of water. A paste is formed in which the water adheres to the surface of the clay, making the clay easy to work with.

Foam

You are familiar with the foam that forms when egg whites are beaten. Foam is a dispersion of gas bubbles in a liquid. Yeasts provide another kind of foam, as can be seen in bread dough. They do this by fermenting carbohydrates and giving off carbon dioxide gas, which produces tiny holes in the dough.

DISCUSSING THE TECHNOLOGY

1. **Classifying** Classify several health and beauty products according to the kinds of colloids described in this feature.

2. **Thinking Critically** Colloids do not pass through selectively permeable membranes. What can you conclude from this?

3. **Hypothesizing** Destruction of a colloid through a clumping process called coagulation can usually be achieved by heating. How would heat accomplish this?

Quick Demo

Where does the foam come from? 👓

Grind 1 g of laundry detergent and 7 g of $Al_2(SO_4)_3 \cdot 18H_2O$ into a powder, using a mortar and pestle. Dissolve the powder in 50 mL of water. Dissolve 5 g of $NaHCO_3$ in another 50 mL of water in a 250-mL beaker. Pour the first solution into the second, and quickly stir to produce a chemical foam. The foam is likely to be so stable that the beaker may be inverted over a sink without the foam spilling out.

3 ASSESS

Check for Understanding

- Check for understanding of the section material by having students answer the Section Review questions.
- Prepare a sodium silicate solution (also called water glass) by mixing one part of sodium silicate with four parts of water. Ask students to wear aprons and goggles, pour about 100 mL of the solution into 250-mL beakers, and take them to their lab tables. Ask them to drop one or two crystals each of iron(III) chloride, cobalt(II) nitrate, and zinc sulfate into the silicate solution and to observe the formation of colored columns from the crystals. Have them explain why their observations are due to osmosis. *The positive ions from the salts interact with silicate ions to form insoluble metal silicates that function as gelatinous, semipermeable membranes. Water molecules move through them into the internal salt solutions, increasing internal pressure until the membranes rupture to expose fresh solutions, which interact and repeat the process.* [L1] [LEP]

WORD ORIGIN

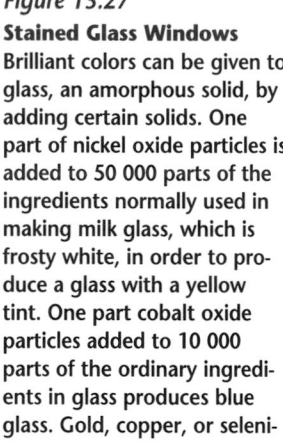

colloid:
kolla (GK) glue
eidos (GK) form

A colloid often contains particles that are stuck together in small clumps.

Colloids

Sometimes, mixtures are partway between true solutions and heterogeneous mixtures. Such mixtures, called **colloids,** contain particles that are evenly distributed through a dispersing medium, and remain distributed over time rather than settling out. The major difference between a colloid and a solution is the size of the solute particles. Colloid particles are generally clumps that are ten to 100 times larger than typical ions or molecules dissolved in solutions. Because of their relatively large particle size, colloids play important roles in a variety of processes. Sometimes solid particles are dispersed in another solid, as shown in **Figure 13.27.** Some biological molecules, such as proteins, are large enough that their behavior is often best understood using a colloid model.

Because colloid particles are evenly dispersed, it is sometimes hard to tell colloids from true solutions. However, the larger particle size gives colloids some unique properties that help in identifying them. **Figure 13.28** illustrates one of these properties. Notice how a beam of light looks different when it passes through a solution and through a colloid. In the solution, the beam's path is hardly visible. As the light moves into the colloid, the light is partially scattered and reflected by the dispersed particles, and the beam becomes visible and broadens. You can observe the same phenomenon when you see the path of sunbeams through dusty air or the path of headlight beams on a foggy night. This light-scattering effect is called the **Tyndall effect.** It occurs because the dispersed colloid particles are about the same size as the wavelength of visible light (400 to 700 nanometers). The solute particles in a true solution are too small to produce this effect.

Figure 13.27
Stained Glass Windows
Brilliant colors can be given to glass, an amorphous solid, by adding certain solids. One part of nickel oxide particles is added to 50 000 parts of the ingredients normally used in making milk glass, which is frosty white, in order to produce a glass with a yellow tint. One part cobalt oxide particles added to 10 000 parts of the ordinary ingredients in glass produces blue glass. Gold, copper, or selenium oxide particles are used to produce red glass.

Chemistry Journal

Colloids
Have students conduct library research on types of colloids and their uses. They should write a detailed description of each type, and include information about the sources they consulted. [L2]

Figure 13.28
Tyndall Effect
Notice how the light beam becomes easily visible in the colloid because of light scattering. The headlight beam is visible in fog for the same reason.

Connecting Ideas

Now that you have learned about some of the properties of water, solutes, and colloids, you can understand why it is sometimes said that the chemistry of water is the chemistry of solutions. As you have discovered in this chapter, the physical properties of water make it a unique and important substance. Its ability to form solutions gives it an essential role in every aspect of life, including the reactions of acids and bases. As you will see, in these reactions, water is a subtle but important partner.

SECTION REVIEW

Understanding Concepts

1. Explain how a water molecule can be attracted to both a positive ion and a negative ion when dissolving an ionic compound.

2. Write equations for the dissociation of the following ionic compounds when they dissolve in water.

   a) Na_2SO_4 b) $NaOH$ c) $CaCl_2$

3. Characterize each of the following solutions as unsaturated, saturated, or supersaturated before the addition of solute.

   a) a solution that will produce a large amount of crystalline solid if only a small additional amount of solute is added

   b) a solution in which additional solute can be dissolved and remain in solution

   c) a solution in which, if additional solute is added, that solute will remain undissolved at the bottom of the container

Thinking Critically

4. **Comparing** Determine which solution has the highest concentration. Then rank the solutions from the one with the lowest to the one with the highest freezing point, and explain your answer.

   a) 0.10 mol of KBr in 100.0 mL of solution

   b) 1.1 mol of NaOH in 1.00 L of solution

   c) 1.6 mol of $KMnO_4$ in 2.00 L of solution

Applying Chemistry

5. **A Kitchen Colloid** Oil and aqueous solutions normally do not mix. However, if oil and vinegar or lemon juice are beaten together with egg, a stable mixture is formed. Explain how this is possible.

13.2   Solutions and Their Properties   **473**

SECTION REVIEW

1. Because water is polar, it has both a positive pole and a negative pole. The positive pole is attracted to negative ions and repelled by positive ions, and the negative pole is attracted to positive ions and repelled by negative ions.

2. **a.** $Na_2SO_4(s) \rightarrow 2Na^+(aq) + SO_4^{2-}(aq)$
   b. $NaOH(s) \rightarrow Na^+(aq) + OH^-(aq)$
   c. $CaCl_2(s) \rightarrow Ca^{2+}(aq) + 2Cl^-(aq)$

3. **a.** The solution is supersaturated.
   b. The solution is unsaturated.
   c. The solution is saturated.

4. The molarities are $1.0M$ KBr, $1.1M$ NaOH, and $0.80M$ $KMnO_4$. The NaOH solution has the highest concentration. The $1.1M$ NaOH solution has the lowest (most depressed) freezing point, and $0.80M$ $KMnO_4$ has the highest. Freezing-point depression depends only upon concentration, and the NaOH solution is the most concentrated.

5. The egg acts as an emulsifier, allowing the oil and water-based droplets to mix together as a colloid.

473

- Review the Reviewing Main Ideas statements and Key Terms with your students.
- Complete solutions to Chapter Review problems can be found in the *Problems and Solutions Manual.*

UNDERSTANDING CONCEPTS

1. It is a liquid at ordinary temperatures, despite its small molecular size; it exists in all three states on Earth; its solid form is less dense than its liquid form; it has a high specific heat; it is an excellent solvent.

2. The molecule is bent, and the polar O—H bonds do not cancel one another's polarity.

3. As liquid water cools, the molecules move closer and the density thus increases. When the temperature goes below 4°C, the molecules begin to organize into the open ice structure, and the density begins to decrease.

4. Any molecule in which hydrogen is bonded to a highly electronegative atom, such as oxygen, is a candidate for hydrogen bonding.

5. Surface tension is the force needed to overcome intermolecular attractions and break through the surface of a liquid or spread a liquid out. It occurs because the particles at the surface experience an unbalanced attractive force that pulls them inward.

6. There are strong attractive forces between mercury atoms, producing high surface tension, but there is little attraction between the glass and the mercury.

7. Capillarity is the rising of a liquid in a narrow tube. It results from competition of the interparticle forces between the liquid molecules and the attractive forces between the liquid and the tube. The rising of blood in a thin tube and the formation of a concave meniscus by water are examples.

8. Bond b is covalent; bonds a and c are hydrogen bonds.

REVIEWING MAIN IDEAS

13.1 Uniquely Water

- Water has a number of unusual properties, such as high boiling point for its molecular size.

- On Earth, water exists primarily as liquid but also as solid and gas.

- The polarity of the water molecule is the source of many of water's unusual physical properties.

- Hydrogen bonds are formed by strong interactions between the hydrogen atoms on one molecule and a highly electronegative atom on another.

- Specific heat is the amount of heat needed to raise the temperature of 1 g of a substance by 1°C. Water has a high specific heat.

13.2 Solutions and Their Properties

- Like dissolves like. For example, polar solvents, such as water, tend to dissolve polar and ionic solutes, but not nonpolar ones.

- Interparticle forces between solvent and solute strongly influence solution formation.

- Ionic compounds dissociate when they dissolve in water.

- Hydrogen bonding plays an important role in the dissolving of many covalent compounds, such as sugar, in water.

- Solutions can be unsaturated, saturated, or supersaturated.

- Temperature affects solubility; generally, solubility of solids increases with increasing temperature.

- The molarity of a solution is equal to the moles of solute per liter of solution.

- Colligative properties of solutions, such as freezing-point depression and boiling-point elevation, are dependent only upon concentration of solute particles.

- Osmosis is movement of a solvent through a selectively permeable membrane under the influence of a concentration difference.

- Colloids are mixtures in which the dispersed particles are larger than those in solutions.

Key Terms

For each of the following terms, write a sentence that shows your understanding of its meaning.

capillarity	saturated solution
colloid	specific heat
dissociation	supersaturated solution
heat of solution	surface tension
hydrogen bonding	Tyndall effect
osmosis	unsaturated solution

⬤ UNDERSTANDING CONCEPTS

1. List several ways in which water is unusual as a chemical substance.

2. Explain why water is polar.

3. Describe how the density of water relates to temperature, and explain why it does so.

4. What types of molecules form hydrogen bonds?

5. What is surface tension? Explain why it occurs.

6. When mercury is placed in a glass cylinder, it forms a convex meniscus. Why does it do so?

7. Explain what capillarity is and give two examples of it.

8. In the figure to the right, which shows water molecules, identify the bonds as covalent or as hydrogen bonds.

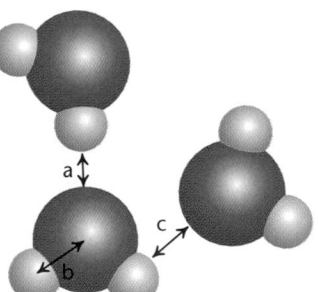

✔ Assessment

Portfolio: Review the portfolio options that are provided throughout the chapter. Encourage students to select one product that demonstrates their best work for the chapter. Have students explain what they have learned and why they chose this example for placement into their portfolios. **P**

Additional portfolio options can be found in the following **Teacher Classroom Resources: Consumer Chemistry Tech Prep Applications Chemistry and Industry Applying Scientific Methods in Chemistry**

9. Why does sucrose dissolve so well in water?

10. Explain how you would try to prepare a super-saturated solution.

11. Explain what is meant by the term *heat of solution*.

12. When two solutions of two different concentrations are placed on opposite sides of a selectively permeable membrane, in which direction is the net solvent flow?

13. Define *specific heat*. Why is the high specific heat of water important to Earth?

14. Why does ice float? What is the significance of this fact for aquatic life?

15. How does your body use perspiration to stay cool?

16. What is the molarity of a solution prepared by dissolving 0.217 mol of ethanol, C_2H_5OH, in enough water to make 100.0 mL of solution?

17. How many grams of sodium carbonate are needed to make 1.30 L of 0.890M Na_2CO_3?

● APPLYING CONCEPTS

18. What mass of iron(III) chloride, $FeCl_3$, is needed to prepare 1.00 L of a 0.255M solution?

19. What is the molarity of a solution that contains 4.13 g of magnesium bromide, $MgBr_2$, in 0.845 L of solution?

20. What mass of potassium iodide, KI, is needed to prepare 5.60 L of a 1.13M solution?

21. Explain how adding antifreeze to your car radiator protects it from freezing.

22. Explain why fresh water is a precious resource, despite the fact that Earth has plenty of water.

23. Certain nations have proposed harvesting icebergs from the oceans and transporting them back to their countries as a source of fresh water. Comment on the advantages and disadvantages of this proposal.

24. When you make pickles, the first step after washing the cucumbers is to soak them in a concentrated salt solution. What is the purpose of this step?

25. Calcium chloride is found in salt mixtures that are used to melt ice on roads in the winter. The dissolving of $CaCl_2$ is exothermic. Provide two reasons why $CaCl_2$ is a good choice for this application as compared to a salt such as NaCl or a salt that dissolves endothermically.

26. The specific heat of aluminum is 0.903 J/g°C; that of copper is 0.385 J/g°C. Suppose you have one cube each of Al and Cu; both cubes have a mass of 100 g and are at a temperature of 100°C. Which cube will release more heat when it cools to 20°C?

27. If you prepared a saturated aqueous solution of potassium chloride at 25°C, then heated the solution to 50°C, would you describe the resulting solution as saturated, unsaturated, or supersaturated? Explain.

28. An aqueous solution that contains 5.85 g of sodium chloride in 100 mL of solution has a slightly higher boiling point than a solution that contains 34.2 g of sucrose in 100 mL of solution. Determine the molarity of each, and explain why the boiling points are different.

---

9. Sucrose contains several OH groups and can form hydrogen bonds to water.

10. Using a solute whose solubility increases with temperature, saturate a solution of the solute at high temperature, then cool slowly.

11. Heat of solution is the heat absorbed or released during the dissolving of a solute in a solvent.

12. The net flow of solvent is from the dilute solution to the concentrated solution.

13. Specific heat is a measure of the amount of heat required to change the temperature of 1 g of a substance by 1°C. Water's high specific heat is important because it moderates temperatures on Earth.

14. Because of its open structure, ice is less dense than liquid water. Bodies of water freeze from the top. The ice provides an insulating layer that protects aquatic life from freezing.

15. The vaporization of the water in the perspiration absorbs heat, cooling the body.

16. 0.217 mol ethanol/0.1000 L solution = 2.17M ethanol

17. (1.30 L)(0.890 mol Na_2CO_3/L)(106 g Na_2CO_3/mol Na_2CO_3) = 123 g Na_2CO_3

APPLYING CONCEPTS

18. (1.00 L)(0.255 mol $FeCl_3$/L)(162 g $FeCl_3$/mol $FeCl_3$) = 41.3 g $FeCl_3$

19. (4.13 g $MgBr_2$/0.845 L)(1 mol $MgBr_2$/184 g $MgBr_2$) = 0.0266 mol $MgBr_2$/L = 0.0266M $MgBr_2$

20. (1.13 mol KI/L)(5.60 L)(166 g KI/mol KI) = 1.05 × 10³ g KI

21. Adding antifreeze to water lowers the freezing point. If the concentration of the antifreeze is high enough, the freezing point will be lower than the lowest winter temperature.

22. Fresh water is essential for most life on Earth. Most water on Earth is seawater, and most of the fresh water

---

is locked up in polar ice or is otherwise inaccessible.

23. The icebergs could provide a good deal of fresh water. The benefits would have to be weighed against considerations of ease, efficiency, time, and cost.

24. Soaking them in brine causes water to flow from them into the concentrated salt solution. This makes the pickles crisp, which is desirable. The process occurs because of osmosis.

25. $CaCl_2$ dissociates to give three ions, Ca^{2+} and $2Cl^-$, and thus produces more

solute particles than a salt such as NaCl, and it has a greater depressive effect on freezing point. Also, the exothermic dissolving process gives off heat, which contributes to further melting.

26. (0.903 J/g°C)(100 g)(80°C) = 7.2 × 10³ J
(0.385 J/g°C)(100 g)(80°C) = 3.1 × 10³ J
The aluminum cube releases more heat.

27. The solubility of KCl in water increases with temperature, as shown in **Figure 13.20,** so at 50°C the solution could hold more KCl and is unsaturated.

28. The molar mass of NaCl is 58.5 g/mol and that of sucrose is 342 g/mol, so both are 1.0M solutions. NaCl dissociates into two ions, whereas sucrose does not dissociate, so the increased number of particles creates a greater boiling-point elevation for NaCl.

29. The hydrophobic end of the soap molecule is attracted to the grease, and the hydrophilic end of the soap is attracted to water. The water pulls away the soap, which pulls the grease away with it.

30. The ethylene glycol crowds out some water molecules at the surface of the solution, lowering the vapor pressure. A higher temperature would thus be needed to raise the vapor pressure high enough to equal the atmospheric pressure and cause boiling.

31. Dead Sea water contains a higher concentration of salt and would therefore require application of greater pressure to counteract solvent flow and produce reverse osmosis.

32. Emulsions are colloidal dispersions of liquids in liquids. Sols are colloidal dispersions of solids in liquids. Milk is an emulsion. Paint is a sol.

33. The water that must be purified usually contains particles that are too small to settle as a result of gravity. They must be made to coagulate or be filtered out.

THINKING CRITICALLY

34. Water has a high surface tension and its surface is resistant to being broken, so a carefully placed razor blade can float.

35. The condensation of water releases heat to the glass, so the temperature of the glass of lemonade increases.

36. At 10°C, the solubility of NaCl is almost twice that of KNO$_3$. The solubility of

476

Everyday Chemistry

29. Explain why soap is effective in removing greasy dirt.

30. How is a solute such as ethylene glycol able to help prevent the water in a car radiator from boiling?

How It Works

31. Why would it be more difficult to use a reverse osmosis unit to purify water from the Dead Sea than water from the ocean?

Chemistry and Technology

32. Contrast emulsions and sols and give an example of each.

Chemistry and Society

33. Explain why gravity settling alone is generally not sufficient as a water-purification treatment.

● THINKING CRITICALLY

Relating Cause and Effect

34. MiniLab 1 If you carefully place a steel razor blade flat on the surface of water, the razor blade can be made to float. Explain this result, given that the density of steel is much greater than that of water.

Drawing Conclusions

35. On a humid day, you may notice that water droplets condense on the outside of a glass of cold lemonade. What effect does this condensation have on the temperature of the glass of lemonade?

Interpreting Graphs

36. Use **Figure 13.20** to compare the solubilities in water of sodium chloride and potassium nitrate over the temperature range from 10°C to 40°C.

Drawing Conclusions

37. ChemLab Suppose you have spilled some blue copper(II) sulfate solution on a shirt. In washing the shirt, why should you avoid using washing soda (sodium carbonate)?

Making Predictions

38. MiniLab 2 Suppose that your tap water does not tend to produce suds when soap is mixed with it. What would you expect to happen if you added sodium oxalate test solution to a sample of your water? Why would you expect this to happen?

● CUMULATIVE REVIEW

39. The formula for a certain compound is CHCl$_3$. What does this formula tell you about the compound? (Chapter 1)

40. Based on physical state at room temperature, which of these common substances has the highest melting point: carbon dioxide, mercury, or table sugar? (Chapter 1)

41. Explain how a potassium atom bonds to an atom of bromine. (Chapter 4)

42. What is the maximum number of electrons in the second energy level? In the third energy level? Explain. (Chapter 7)

43. Why does your skin feel cool if you dab it with rubbing alcohol? (Chapter 10)

44. How many liters of nitrogen are needed to react completely with 28 L of oxygen in the synthesis of nitrogen dioxide? (Chapter 11)

KNO$_3$ increases more rapidly with temperature so that the two solubilities are equal at about 22°C. At 40°C, the solubility of KNO$_3$ is almost twice that of NaCl.

37. The carbonate ion will form insoluble copper(II) carbonate, which may remain.

38. Magnesium and/or calcium oxalate would precipitate because ions of magnesium or calcium are present in hard water, and the water described behaves like hard water.

CUMULATIVE REVIEW

39. A unit of the compound contains one carbon atom, one hydrogen atom, and three chlorine atoms.

40. Table sugar has the highest melting point because it is the only one that is solid at room temperature.

41. The potassium atom transfers its single valence electron to the bromine atom. The resulting K$^+$ ion and Br$^-$ ion are attracted to each other and bond ionically.

42. In the second energy level, a maximum of eight electrons can fill the 2s orbital and the three 2p orbitals. In the third energy level, a maximum of 18 electrons can fill

Skill Review

1. **Written Summary of a Graph** Examine the graph below and write a summary of the information in it. In the course of your summary, answer the following questions.

   a) What is being graphed?

   b) What do the four different lines represent?

   c) Which compound has the lowest boiling point? Which has the highest?

   d) Moving down a group on the periodic table, what is the general trend in boiling point for a particular family of hydrogen compounds?

   e) Moving across a period, what is the general trend in boiling point for the hydrogen compounds shown?

   f) What does the dashed line show?

   g) Are boiling points of the second-period hydrogen compounds predicted from the behavior of the rest of the group? Explain any anomalies or unpredicted behavior.

   h) Which of the compounds are liquids at room temperature?

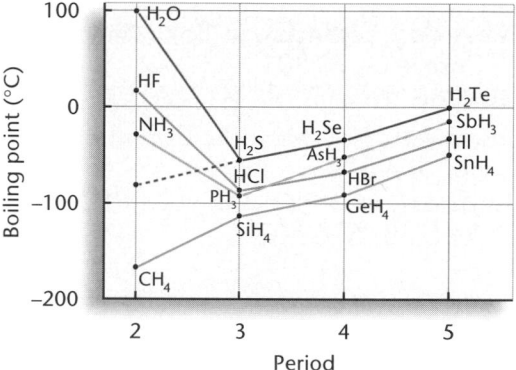

Writing in Chemistry

2. Drinking water is a valuable resource. Write an article that traces the route and treatment of water from your faucet, down the drain, to your local municipal sewage-treatment plant, and back, ultimately, to your faucet. Your paper should give a general overview of municipal sewage treatment and drinking-water treatment.

3. Industry and agriculture are the major users of fresh water. Write a brief paper that describes this water usage in more detail. Which kinds of industries use the most water, and why? What areas of the country rely most heavily on water for agricultural purposes? What benefits and problems are associated with this water usage? What are some ways in which water demands for these usages are being successfully decreased?

Problem Solving

4. Suppose you dipped a glass pipet tube into a beaker of water. Would the water level in the tube be higher or lower than the water level in the beaker? Explain. Suppose that you then immersed a clean, dry capillary tube into a beaker of molten wax and withdrew the tube in such a way that the inside walls of the tube became coated with a thin film of solid wax as the tube cooled. What would happen if you dipped this wax-coated tube into the beaker of water? Would the level of water inside the tube be higher or lower than it was for the non-wax-coated tube? Explain.

Projects

5. Monitor the different things for which you use water over the course of several days, other than drinking. Make a poster that illustrates each use and shows the property of water that is being relied upon in each case.

6. Under adult supervision, follow a recipe that involves the preparation and use of concentrated sugar solutions, as in the making of preserves, candy, or fudge. Bring your finished creations to class and explain the solution processes that took place during the preparation.

the $3s$ orbital, three $3p$ orbitals, and five $3d$ orbitals.

43. Rubbing alcohol evaporates quickly and therefore removes heat from the skin.

44. The balanced equation is $N_2 + 2O_2 \rightarrow 2NO_2$. The volume of nitrogen is one-half the volume of oxygen, or 28 L/2 = 14 L.

AUTHENTIC ASSESSMENT

1. **a.** Boiling point of hydrogen compounds versus period of the nonmetal bonded to hydrogen is being graphed.
   b. The lines represent Groups 14, 15, 16, and 17 of the periodic table.
   c. CH_4 has the lowest boiling point at $-161°C$. H_2O has the highest boiling point at 100°C.
   d. The boiling point generally increases moving down a group.
   e. The boiling point generally increases moving from left to right across the period.
   f. The dashed line shows the projected boiling point of water ($-100°C$), assuming it follows the trend of the other members of its family.
   g. Only CH_4 follows the trend of the other members of its group. The others—HF, H_2O, and NH_3—show higher boiling points than predicted on the basis of group trends, due to strong hydrogen bonding.
   h. Only water is a liquid at room temperature. (HF liquefies at 15°C.)

4. The water in the capillary tube would rise above the level in the beaker because of capillary action, in which the strong attractive forces between the polar molecules and polar glass would pull the water molecules up the tube. In the wax-coated tube, the water would not tend to rise above the water level in the beaker because of the lack of attraction between the polar water and the wax, which is nonpolar.

5. Examples may include:

washing (which makes use of the property of polarity that makes water a good solvent for many solutes), flushing (property of high boiling point, due to hydrogen bonding, which allows water to be a liquid under ordinary conditions), use of ice in cooling (high freezing point that allows water to be frozen at practical temperatures), hot-water heating (high specific heat), swimming in hot weather (high specific heat), use of ice for ice-skating (low density of solid compared to liquid), and use of steam in cooking and heating (high heat of vaporization).

Program Resources

Chapter Review and Assessment, pp. 77-82 [L1]

Alternate Assessment in the Science Classroom [L1]

Computer Test Bank, Chapter 13 [L1]

Problems and Solutions Manual, Ch. 13 [L1]

Supplemental Practice Problems, Ch. 13 [L1]

Performance Assessment in the Science Classroom [L1]

Chapter 14 Organizer

Section	Objectives	Activities/Features
14.1 **Acids and Bases** (4 days)	1. **Distinguish** acids from bases by their properties. 2. **Relate** acids and bases to their reactions in water. 3. **Evaluate** the central role of water in the chemistry of acids and bases.	**MiniLab 1:** What do acids do?, p. 482 **Chemistry and Technology:** Manufacturing Sulfuric Acid, p. 484 **Biology Connection:** Measurement of Blood Gases, p. 487 **People in Chemistry:** Fe Tayag, Cosmetic Bench Chemist, pp. 490-491 **Chemistry and Society:** Atmospheric Pollution, p. 495 **Discovery Demo:** 14.1 pp. 478-479 **Demonstrations:** 14.2 pp. 492-493; 14.3 pp. 494-495
14.2 **Strengths of Acids and Bases** (5 days)	4. **Relate** different electrical conductivities of acidic and basic solutions to their degree of dissociation or ionization. 5. **Distinguish** strong and weak acids or bases by their degree of dissociation or ionization. 6. **Compare and contrast** the composition of strong and weak solutions of acids or bases. 7. **Relate** pH to the strengths of acids and bases.	**Everyday Chemistry:** Balancing pH in Cosmetics, p. 501 **MiniLab 2:** Antacids, p. 503 **ChemLab:** Household Acids and Bases, pp. 504-505 **Demonstrations:** 14.4 pp. 498-499; 14.5 pp. 506-507

Activity Materials

ChemLab (pages 504-505)
red cabbage, hot plate, beaker tongs, 100-mL beakers (2), distilled water, microtip pipets (9), 96-well microplate, white paper, toothpicks, 100-mL graduated cylinder, solutions of: eyewash, lemon juice, white vinegar, table salt, baking soda, borax, drain cleaner

MiniLab 1 (page 482) microtip pipets (3), 24-well microplate, $3M$ HCl, $3M$ H_2SO_4, $3M$ $HC_2H_3O_2$ blue litmus paper, dropper, marble chips, bromothymol blue indicator, pieces of zinc, pieces of aluminum, egg whites, distilled water

MiniLab 2 (page 503)
plastic zipper bags (4), white vinegar, red cabbage, markers, antacid tablets, 10-mL graduated cylinder, dropper

Demonstrations For a complete list of materials for the demonstrations in this chapter, see pages 478, 492, 498, and 506.

KEY TO TEACHING STRATEGIES

The following designations will help you decide which activities are appropriate for your students.

L1 Level 1 activities should be within the ability range of all students.

L2 Level 2 activities should be within the ability range of the average to above-average student.

L3 Level 3 activities are designed for the ability range of above-average students.

LEP LEP activities should be within the ability range of Limited English Proficiency students.

COOP LEARN Cooperative Learning activities are designed for small group work.

P These strategies represent student products that can be placed into a best-work portfolio.

Acids, Bases, and pH

Teacher Classroom Resources

STUDENT MASTERS

Study Guide, pp. 55-56
Critical Thinking/Problem Solving, p. 15
Laboratory Manual: 14.1 Acidic and Basic Anhydrides, pp. 127-130
ChemLab and MiniLab Worksheets: p. 92

TEACHING AIDS

Section Focus Transparency 28
Section Focus Transparency Master: p. 28
Basic Concepts Transparency 38, 39
Basic Concepts Transparency Masters: pp. 75-78
Problem Solving Transparency 16
Problem Solving Transparency Master: pp. 31-32
Lesson Plans: p. 28

Study Guide: pp. 57-58
Laboratory Manual: 14.2 Using Indicators to Determine pH, pp. 131-134
ChemLab and MiniLab Worksheets: pp. 89-91, 93

Section Focus Transparency 29
Section Focus Transparency Master: p. 29
Basic Concepts Transparency 40
Basic Concepts Transparency Master: pp. 79-80
Problem Solving Transparency 17
Problem Solving Transparency Master: pp. 33-34
Lesson Plans: p. 29

ASSESSMENT RESOURCES

Chapter Review and Assessment: pp. 83-88
Computer Test Bank: Chapter 14
Performance Assessment in the Science Classroom
Problems and Solutions Manual: Chapter 14
Supplemental Practice Problems: Chapter 14

CHAPTER RESOURCES

Applying Scientific Methods in Chemistry: p. 41
Spanish Resources: Chapter 14
English/Spanish Audiocassettes: Chapter 14
Cooperative Learning in the Science Classroom
Lab and Safety Skills in the Science Classroom

GLENCOE TECHNOLOGY

Software
Mastering Concepts in Chemistry: *Unit 14* Lesson 1 Background and Theories; Lesson 2 Names and Formulas of Common Acids and Bases; *Unit 15* Lesson 1 Water Equilibria

Videotape
MindJogger Videoquizzes, Chapter 14

Videodisc
Chemistry: Concepts and Applications, Demonstrations, Videos, and Animations
Science and Technology Videodisc Series: Disc 2 Chemistry *Dealing with Hazardous Materials*

CD-ROM Multimedia System
Chemistry: Concepts and Applications, Same as for Videodisc, plus Interactive Explorations and Major Simulation, *The Periodic Table*

TECH PREP

The following Glencoe resources provide opportunities for integrating science and technology.

Student Edition: Chemistry and Technology, p. 484; Biology Connection, p. 487; Chemistry and Society, p. 495; Everyday Chemistry, p. 501; ChemLab, pp. 504-505

Teacher Wraparound Edition: Cultural Diversity, p. 486; Demonstration 14.5, pp. 506-507

Teacher Classroom Resources: Tech Prep Applications, pp. 21-22; Consumer Chemistry, pp. 27-28, Chemistry and Industry, pp. 31-34

CHAPTER
14 Acids, Bases, and pH

Chapter Overview

This chapter describes acids and bases in terms of their properties. The role of water in the chemistry of acids and bases is emphasized. The relative strengths of acids and bases are introduced. pH as an efficient way to express the range of strengths and concentrations of acids and bases is demonstrated.

Theme Connection

Systems and Interactions/Equilibrium and Change Aqueous solutions of acids and bases are two important chemical systems. How a dissolved substance interacts with water determines whether the substance is an acid, a base, or neutral. If the substance is an acid or a base, the interaction with water determines the relative strength. Equilibrium and change are discussed in relation to degree of ionization and dissociation of acids and bases.

✓ Assessment Planner

Choose assessment strategies from the following pages to evaluate the progress of your students.

Assess, pp. 489, 494, 507
MiniLabs, pp. 482, 503
ChemLab, p. 504
Portfolio, pp. 503, 507, 509
Chapter Review, p. 509

GLENCOE TECHNOLOGY

💿 CD-Rom

Chemistry: Concepts and Applications

Ionization of a Weak Acid
Use this Discovery Demo to observe a visible indicator of degree of ionization. L1 LEP

Formic acid

Discovery Demo 14.1

Ionization of a Weak Acid

Purpose: To provide students with an indicator of percent of ionization.
Materials: 100 mL glacial acetic acid, 100 mL distilled water, conductivity apparatus with a lightbulb (see Demonstration 14.4), 250-mL beaker
Safety Precautions:
Disposal: Code D
Procedure: CAUTION: *Acetic acid is corrosive.*

Use care with conductivity apparatus to avoid electrical shock. Carefully add acetic acid to the beaker to a depth of about 2 cm. Lower the conductivity apparatus into the beaker until about 1 cm of the electrodes is in the acetic acid. Test the conductivity of the acetic acid, and have students record the results. With the conductivity apparatus disconnected, slowly add about half of the distilled water to the apparatus. Stir and test the conductivity again,

CHAPTER

14

Acids, Bases, and pH

Have you ever tasted sour milk? Have you ever exercised so hard that your muscles were tired? Ever had an upset stomach after eating too much or been bitten by an ant? Cleaned a toilet with toilet-bowl cleaner? Taken a vitamin C tablet to prevent a cold?

Have you ever taken an antacid for an upset stomach? Have you cleaned out a clogged drain with drain cleaner? Or used lime to improve the soil in a potted plant or garden? Have you washed windows or scrubbed the floor with household cleaners?

You probably answered yes to many, if not all, of these questions. If you answered yes to any of the questions in the first paragraph, you were experiencing the chemistry of an acid. If you answered yes to any of the questions in the second paragraph, you were experiencing the chemistry of a base. Although the experiences are different, they all share a common chemistry—the chemistry of acids and bases.

 Concept Check

Review the following concepts before studying this chapter.
Chapter 5: names and formulas of common acids and bases
Chapter 13: dissociation of ionic compounds; hydrogen bonding

 Chapter Preview

14.1 Acids and Bases

14.2 Strengths of Acids and Bases

Laboratory Activities
ChemLab
Household Acids and Bases

MiniLabs
1. What do acids do?
2. Antacids

Chemistry Around You

Acid-base chemistry plays an important role in many processes that occur in your body. An example of the role acids play takes place during exercising. When you exercise, lactic acid is produced as a by-product of cell processes. If the exercise is vigorous, more lactic acid may be produced than can be removed by your circulatory system. This accumulation may cause muscle soreness until the body removes the acid.

479

---

recording the results. Add more distilled water to the beaker. Test the solution a third time after stirring. Record your observations of the diluted acid.

Results: The more dilute the acid becomes, the better it conducts.

Analysis: Ask these questions.

1. Write the equation for the ionization of acetic acid. $HC_2H_3O_2 + H_2O \rightleftarrows H_3O^+ + C_2H_3O_2^-$

2. The percents of ionization of the three acetic acid solutions are 17.4M, 0%; 10.0M, 0.132%; and 0.100M, 1.32%. Do these figures agree with the observations? Explain. *Yes. The greater number of ions permits conductivity to occur.*

3. Will continued dilution eventually reduce the conductivity of the solution? *yes*

Introducing the Chapter

The introduction emphasizes that many common substances are acids or bases. The Chem-Lab introduces students to common household acids and bases and the measurement of their pH using red cabbage as an acid-base indicator. Testing some of these materials, as well as others, with a simple litmus test would be an excellent experimental introduction to the topic of acids and bases.

Using the Illustrations The opening photo sequence focuses on a fire ant. Some ants inject formic acid, $HCHO_2$, a weak organic acid, into their victims when they bite. Asking how students treat an ant bite (baking soda, which is a base) may be a starting point for talking about acids and bases.

Revealing Misconceptions

In order to make it easy for students to recognize acids, formulas of acids are written with the acidic hydrogens as the first element. Acetic acid ($HC_2H_3O_2$) and citric acid ($H_3C_6H_5O_7$) are examples. This formula style is helpful in recognition and classification but may lead to a student misconception that acids are ionic compounds that contain hydrogen ions, H^+. Acids are always molecular (HCl) or polyatomic ions (HCO_3^-). They do not contain free H^+ ions. The H^+ ion is produced in the interaction with the water.

GLENCOE TECHNOLOGY

 Videodisc

Chemistry: Concepts and Applications
Ionization of a Weak Acid
Disc 2, Side 1, Ch. 13

Show this video of the Discovery Demo to introduce ionization of acids. L1 LEP

SECTION 14.1

PREPARE

Key Concepts

The properties of acids and bases are explored and related to their reactions in water. Some conceptual definitions of acids and bases are also examined. Last, the chemistry of acidic and basic anhydrides is discussed.

Planning Ahead

For the MiniLab Prepare 100 mL of $3M$ HCl by adding 25 mL of concentrated $(12M)$ HCl to 75 mL of water. Prepare 100 mL of $3M$ H_2SO_4 by adding 17 mL of concentrated $(18M)$ H_2SO_4 to 83 mL of water. Prepare 100 mL of $3M$ $HC_2H_3O_2$ by adding 5 mL of glacial acetic acid to 95 mL of water.
For the Demonstrations Make arrangements to obtain dry ice for Demonstration 14.3.

1 FOCUS

Focus Transparency

Display the **Section Focus Transparency** for Section 14.1 to introduce the properties of acids and bases. L1 LEP

2 TEACH

Correcting Misconceptions

Students may think that alkalis are something different from bases. Point out that they are the same.

Visual Learning

Figure 14.1: Explain that if aluminum is present in the drain cleaner, a reaction occurs that produces hydrogen gas bubbles, which then loosen the clogging material.

480

SECTION 14.1

SECTION PREVIEW

Objectives
Distinguish acids from bases by their properties.
Relate acids and bases to their reactions in water.
Evaluate the central role of water in the chemistry of acids and bases.

Key Terms
acid
hydronium ion
acidic hydrogen
ionization
base
acidic anhydride
basic anhydride

Acids and Bases

As you have discovered, classifying substances into broad categories simplifies the study of chemistry. Consider some of the chemistry classification schemes you have used. In each scheme, the categories have generally been opposites, such as metal versus nonmetal, ionic versus covalent, and soluble versus insoluble. The materials in each category do not share exactly the same properties, but they share similar properties. Likewise, substances classified as acids or bases can be considered opposites. Substances in each category share some general properties that make them different from other substances. In this section, you will examine acids and bases from both a macroscopic and a submicroscopic level.

Macroscopic Properties of Acids and Bases

Because they are present in so many everyday materials, acids and bases have been recognized as interesting substances since the time of alchemists. Simple, observable properties distinguish the two.

It's a Matter of Taste and Feel

Although taste is not a safe way to classify acids and bases, you probably are familiar with the sour taste of acids. Lemon juice and vinegar, for example, are both aqueous solutions of acids. Bases, on the other hand, taste bitter.

Bases have a slippery feel. Like taste, feel is not a safe chemical test for bases, but you are familiar with the feel of soap, a base, on the skin. Bases, such as soap, react with protein in your skin, and skin cells are removed. This reaction is part of what gives soaps a slippery feel, as well as a cleansing action. **Figure 14.1** shows how this reaction makes some bases excellent drain cleaners.

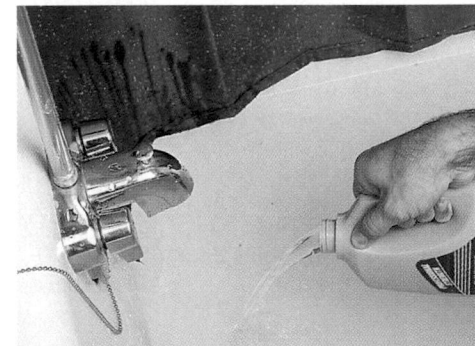

Figure 14.1
Bases and Protein
Certain bases are excellent at dissolving hair, which is often the source of clogged drains. Hair is composed of protein.

480   Chapter 14   Acids, Bases, and pH

Program Resources

Study Guide, pp. 55-56 L1
Critical Thinking/Problem Solving, p. 15 L2
English/Spanish Audiocassettes, Chapter 14 L1 LEP
Laboratory Manual, 14.1, pp. 127-130 L1
Section Focus Transparency 28 and **Master,** p. 28 L1 LEP
Basic Concepts Transparency 38, 40 and **Masters,** pp. 75-76, 79-80 L1 LEP

Problem Solving Transparency 16 and **Master,** pp. 31-32 L1 LEP
Spanish Resources L1 LEP
ChemLab and MiniLab Worksheets, p. 92 L1
Consumer Chemistry, pp. 27-28 L1
Chemistry and Industry, pp. 31-34 L1
Applying Scientific Methods in Chemistry, pp. 41-42 L1

Table 14.1 Top Ten Industrial Chemicals Produced in the United States in 1994

Chemical	Billions of Pounds	Acid/Base	Some Uses
Sulfuric acid, H_2SO_4	78.70	Acid	Car batteries; manufacture of chemicals, fertilizer, and paper
Nitrogen, N_2	67.54		
Oxygen, O_2	49.67		
Ethylene, C_2H_4	48.53		
Lime, CaO	38.35	Base	Neutralizes acidic soils
Ammonia, NH_3	37.93	Base	Fertilizer; cleaner; making rayon, nylon, and nitric acid
Propylene, C_3H_6	28.84		
Sodium hydroxide, NaOH	25.83	Base	Drain and oven cleaners; manufacture of soap and chemicals
Phosphoric acid, H_3PO_4	25.26	Acid	Making detergents and fertilizers; soft drinks
Chlorine, Cl_2	24.20		

Acids React with Bases

As you have learned, substances with opposite properties, such as acids and bases, tend to react with each other. You'll learn more about these acid-base reactions in Chapter 15.

The reactions of acids and bases are central to the chemistry of living systems, the environment, and many important industrial processes. **Table 14.1** shows the top ten industrial chemicals produced in the United States in 1994. Not surprisingly, acids and bases make up half of the top ten.

Litmus Test and Other Color Changes

Acids and bases cause certain colored dyes to change color. The most common of these dyes is litmus. When mixed with an acid, litmus is red. When added to a base, litmus is blue. Therefore, litmus is a reliable indicator of whether a substance is an acid or a base. **Figure 14.2** shows how vegetable dyes change color in the presence of an acid or a base. Dyes such as these are called acid-base indicators because they are often used to indicate whether substances are acids or bases.

Fact of the MATTER

In the chemistry lab, litmus test paper is used. Litmus paper is made by soaking paper in a solution of litmus and then drying it to remove the water. Litmus papers are usually available in a slightly basic form (blue) and a slightly acidic form (red).

Figure 14.2
Acid-Base Indicators
The ability of a substance to change the color of certain dyes is a good indication of whether the substance is an acid or a base. Common materials that act as acid-base indicators include litmus, red cabbage, radishes, tulips, and rose petals.

14.1 Acids and Bases **481**

Across the Curriculum

Etymology

Calcium hydroxide is formed when water is added to calcium oxide. Calcium oxide is commonly called lime, and calcium hydroxide is called *slaked lime*. To *slake* means "to quench," as in thirst.

Using an Analogy

Point out at the beginning of the chapter that acids and bases are described and defined in various ways, depending upon the circumstances, and that students should not be confused by this practice. In an analogous way, the same person might be called, in different situations, a student, a teacher, a juror, a golfer, a driver, a passenger, etc.

Applying Chemistry

The Litmus Test Explain that the test of acidic and basic solutions with litmus paper is well known. The same terminology is often used to describe any definitive test. For example, one might say, "The primary election will be a *litmus test* of his political ability."

Quick Demo

Who has just the right voice?

Put about 250 mL of 95% ethanol in a 500-mL flask. Add five or six drops of thymolphthalein indicator solution and enough dilute sodium hydroxide solution to produce a blue color. Stopper the flask until ready for the demo. Pass the flask around the class, and ask each student to remove the stopper, politely ask the solution to turn color, re-stopper the flask, and give the liquid a gentle swirl. At some point, enough CO_2 will have dissolved in the solution to lower the pH sufficiently that the thymolphthalein will change color from blue to yellow. Ask students if they know what caused the color to change. *Some of the exhaled CO_2 dissolved, producing a weak carbonic acid solution, which then changed the color of the indicator.* L1 LEP

miniLAB

What do acids do?

Purpose: To test the reactivities of three acids and to develop an operational definition for acidic solutions.

Process Skills: Observing, classifying, using space/time relationships, communicating, sequencing

Safety Precautions: Ask students to be particularly careful when rinsing the well plate.

Disposal: Code D; collect all acidic solutions in the fume hood.

Teaching Strategies

React marble chips, Zn, and Al with HCl in large test tubes and test the CO_2 (does not burn or support combustion) and H_2 (burns, does not support combustion) gases produced with glowing and burning splints.

Expected Results

See table below.

Analysis

1. Acids turn litmus red, turn bromothymol blue yellow, react with carbonates to produce carbon dioxide, react with zinc and aluminum to produce hydrogen gas, and coagulate egg whites.
2. Acetic acid ionizes less.

✓ Assessment

Performance: Ask students to write a balanced equation for the reaction between hydrochloric acid and marble chips. $2HCl + CaCO_3 \rightarrow H_2O + CO_2 + CaCl_2$ [L1]

miniLAB 1

What do acids do?

Most acids tend to be reactive substances. Test the reactivities of three acids with several common substances, and develop an operational definition for acidic solutions.

Procedure

1. Wear an apron and goggles.
2. Use a labeled microtip pipet to add 10 drops of $3M$ hydrochloric acid, HCl, to wells D1-D6 of a clean, 24-well microplate. In the same manner, add 10 drops of $3M$ sulfuric acid, H_2SO_4, to wells C1-C6 and 10 drops of $3M$ acetic acid, $HC_2H_3O_2$, to wells B1-B6.
3. Dip blue litmus paper into the solutions in wells D1, C1, and B1. Record your observations.
4. Add 2 drops of bromothymol blue indicator solution to wells D2, C2, and B2. This indicator turns from blue to yellow as the solutions become more acidic. Record your observations.
5. In a similar manner, add marble chips (calcium carbonate) to wells D3, C3, and B3; pieces of zinc to wells D4, C4, and B4; pieces of aluminum to wells D5, C5, and B5; and a small amount of egg white to wells D6, C6, and B6. Record your observations.
6. Dispose of all materials as directed by your teacher. Rinse the microplate with tap water, then distilled water.

Analysis

1. Summarize the reactions of the three acids with the substances you tested. This summary constitutes an operational definition of an acid.
2. Which acid, although it had the same molar concentration as the other acids, reacted less noticeably? Explain this behavior.

Fact of the MATTER

Except for Group I carbonates, carbonate-containing compounds are almost completely insoluble in water. This makes naturally occurring substances such as marble and limestone stable materials for sculpting and building.

Reactions with Metals and Carbonates

Another characteristic property of an acid is that it reacts with metals that are more active than hydrogen. **Figure 14.3** shows how iron metal rapidly reacts with hydrochloric acid, HCl, to form iron(II) chloride, $FeCl_2$, and hydrogen gas. However, if you were to add a piece of copper metal to the acid, you could see that the acid will not react with copper metal. This property explains why acids corrode most metals. Bases do not commonly react with metals.

Another simple test that distinguishes acids from bases is the reaction of acids with ionic compounds that contain the carbonate ion, CO_3^{2-}, to form carbon dioxide gas, water, and another compound, as shown in **Figure 14.3**. A similar reaction, also shown in **Figure 14.3,** is the source of the destructive action of acidic pollution on marble and limestone sculptures. Bases do not react with carbonates.

482 Chapter 14 Acids, Bases, and pH

Other Reactant	3M HCl	3M H₂SO₄	3M HC₂H₃O₂
blue litmus	red	red	red
bromothymol blue	yellow	yellow	yellow
marble chips	bubbles	bubbles	bubbles a little
zinc	bubbles rapidly	bubbles rapidly	bubbles slowly
aluminum	bubbles rapidly	bubbles rapidly	bubbles slowly
egg white	white, coagulates	white, coagulates	some white, coagulates some

Figure 14.3

Acids with Metals and Carbonates
Typical behavior in certain chemical reactions helps identify substances as acids.

lithium
potassium
calcium
sodium
magnesium
aluminum
zinc
chromium
iron
nickel
tin
lead
hydrogen
copper
mercury
silver
platinum
gold

Decreasing activity

▲ Vinegar, a solution of acetic acid, reacts with egg shell, which is primarily calcium carbonate, to produce carbon dioxide, calcium acetate, and water.

$$2HC_2H_3O_2(aq) + CaCO_3(s) \rightarrow$$
$$CO_2(g) + Ca(C_2H_3O_2)_2(aq) + H_2O(l)$$

▲ Acids react with metals that are more active than hydrogen to form both a compound of the metal and hydrogen gas.

$$Fe(s) + 2HCl(aq) \rightarrow FeCl_2(aq) + H_2(g)$$

Calcium carbonate is the major component in limestone and marble. In the presence of acids in the environment, marble and limestone sculptures and buildings can be damaged or destroyed. ▶

Defining Acids and Bases— A Submicroscopic Look

The description of acids and bases in terms of their physical and chemical properties is useful for classification purposes. However, to understand these properties, you need to know about the behavior of acids and bases at the submicroscopic level.

Submicroscopic Behavior of Acids

The submicroscopic behavior of acids when they dissolve in water can be described in several ways. The simplest definition is that an **acid** is a substance that produces hydronium ions when it dissolves in water. A **hydronium ion**, H_3O^+, consists of a hydrogen ion attached to a water molecule.

14.1 **Acids and Bases** 483

Integrating THE Sciences

Medicine A recent research study by the California Birth Defects Monitoring Program found that women who take folic acid prior to becoming pregnant and during the first two months of pregnancy may reduce the incidence of cleft lip and cleft palate in their offspring. The 1995 study surveyed 731 women whose children had this deformity and 734 women whose children had no birth defects. Folic acid also seems to prevent spina bifida, which is traced to a folic acid deficiency. Women who took multiple vitamins, which ordinarily contain folic acid, had fewer children with the deformities. This preliminary study did not rule out that another vitamin or mineral might be responsible for the difference. The current recommendation for folic acid for women of childbearing age is 0.4 mg/day. Folic acid is found in liver, mushrooms, yeast, and green leafy vegetables.

SECTION 14.1

Visual Learning

Figure 14.3: Ask students why limestone is sometimes put in acidic streams. *In the process of reacting with the limestone, the acid is rendered less corrosive and less harmful to the environment.*

Concept Development

Explain that descriptions of what acids and bases do are known as *operational definitions*. Molecular-level descriptions and theories are known as *conceptual definitions* of acids and bases.

Discussion

Ask students whether they would ever put acids in their mouths. *They will probably say no because they generally think of acids as highly corrosive substances. Then ask whether any foods taste sour to them. oranges, grapefruit, lemons, vinegar, apples, etc.* Explain that these foods contain acids that taste sour and that, in fact, the Latin word for "sour" is *acidus*.

GLENCOE TECHNOLOGY

 Videodisc

Chemistry: Concepts and Applications
The Actions of Acids and Bases
Disc 2, Side 1, Ch. 15

Show this video to help you explain chemical properties of acids and bases. L1 LEP

 CD-Rom

Chemistry: Concepts and Applications
The Actions of Acids and Bases
Use this video to help explain chemical properties of acids and bases. L1 LEP

CHEMISTRY
&TECHNOLOGY

Purpose

TECH PREP

Students learn how sulfuric acid is manufactured.

Background

Used sulfuric acid can be regenerated by thermal decomposition. High temperatures cause the acid to decompose to sulfur dioxide. All organic impurities can be burned away. The sulfur dioxide can then be subjected to the contact process.

Teaching Strategies

- Sulfuric acid is made commercially available in a number of concentrations, ranging from 77.7% to 100%. Mention that dilute sulfuric acid does not readily react with copper. Concentrated sulfuric acid is needed for reaction with copper.
- Mention that sulfuric acid is used in making paper, textiles, plastics, paints, petroleum, batteries, and explosives. Most industries use some of these products. L1

Visual Learning

Ask questions or have students write questions that are related to the chemical equations that show the contact process. Questions may relate to order and sequence, to kinds of reactants used, and to products produced. L1

Discussing the Technology

1. Students may hypothesize that sulfur trioxide reacts too violently with water.
2. Because sulfuric acid is used in steel production, petroleum refining, and in the production of organic and inorganic chemicals, increased production might indicate that important aspects of the economy are improving.

Manufacturing Sulfuric Acid

You might not expect that a simple acid would acquire worldwide status, but sulfuric acid has done just that. Most industrialized nations produce significant quantities of the chemical. The United States alone produces 40 million tons of sulfuric acid every year.

With such quantities being produced, this product must have many uses. Ninety percent of the sulfuric acid made in the United States is used in the production of liquid fertilizers and other inorganic chemicals. The rest is used in refining petroleum, in steel production, and in producing organic chemicals. Sulfuric acid is also useful in removing unwanted materials from ores.

The Manufacturing Process

The production of sulfuric acid is fairly simple. It starts with burning sulfur to produce sulfur dioxide.

$$S(s) + O_2(g) \rightarrow SO_2(g)$$

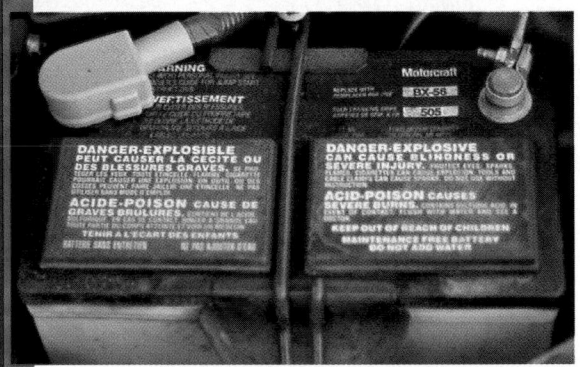

The next step in the process is called the contact method because the sulfur dioxide and oxygen molecules are in contact with a catalyst, usually vanadium pentoxide, V_2O_5. When the sulfur dioxide and oxygen gases pass through a heated tube that contains layers of the pellet-size catalyst, the sulfur dioxide is converted to sulfur trioxide. To make sure the reaction is complete, contact with the catalyst takes place twice.

$$2SO_2(g) + O_2(g) \xrightarrow{catalyst} 2SO_3(g)$$

Then the sulfur trioxide is bubbled through a solution of sulfuric acid to produce pyrosulfuric acid, $H_2S_2O_7$. Pyrosulfuric acid is then added to water to produce sulfuric acid.

$$SO_3(g) + H_2SO_4(l) \rightarrow H_2S_2O_7(l)$$
$$H_2S_2O_7(l) + H_2O(l) \rightarrow 2H_2SO_4(l)$$

DISCUSSING THE TECHNOLOGY

1. **Hypothesizing** Why do you suppose sulfur trioxide is bubbled through a solution of sulfuric acid instead of through water to produce sulfuric acid? Isn't this inefficient?

2. **Inferring** Some people use the quantity of sulfuric acid produced by an industrialized nation as an economic indicator. Why is sulfuric acid production useful in this regard?

Across the Curriculum

History

In approximately 1300, a medieval alchemist, probably Spanish, writing under the pseudonym of Geber, was the first person known to write about sulfuric acid. Geber also described the formation of nitric acid, another important mineral acid. Mineral acids, such as HCl, HNO_3, and H_2SO_4, are derived from nonliving things, such as rocks. These were the first acids that were not from a living source, such as acetic acid, which was derived from vinegar. The discovery and use of mineral acids is seen by science historians as perhaps the next-most-significant chemical advancement after the production of iron 3000 years earlier. However, the alchemists did not recognize the significance of mineral acids and continued to pursue the transmutation of base metals into gold.

For example, hydrochloric acid is produced by dissolving hydrogen chloride gas, HCl, in water. Remember from Chapter 13 that water is a polar molecule that is able to form strong hydrogen bonds with solutes that also form hydrogen bonds. When HCl dissolves in water, it produces hydronium ions by the reaction shown below. HCl is definitely an acid; it produces H_3O^+ when dissolved in water.

$$HCl(g) + H_2O(l) \rightarrow H_3O^+(aq) + Cl^-(aq)$$

Acetic acid, $HC_2H_3O_2$, undergoes a similar reaction when it dissolves in water to form a vinegar solution.

$$HC_2H_3O_2(aq) + H_2O(l) \rightarrow H_3O^+(aq) + C_2H_3O_2^-(aq)$$

Notice the similarities in these two reactions. In both cases, the dissolved substance reacts with water to form hydronium ions and a negatively charged ion.

Acidic Hydrogen Atoms

How and why are hydronium ions formed? At the submicroscopic level, the reaction of an acid with water is a transfer of a hydrogen ion, H^+, from an acid to a water molecule. This transfer forms the positively charged hydronium ion, H_3O^+, and a negatively charged ion. In an acid, any hydrogen atom that can be transferred to water is called an **acidic hydrogen.**

Take another look at the acetic acid example. Although a molecule of acetic acid, $HC_2H_3O_2$, contains four hydrogen atoms, only one is an acidic hydrogen that participates in the transfer. The other hydrogen atoms remain a part of the acetate ion. **Figure 14.4** shows that it is possible for acids to have more than one acidic hydrogen per molecule.

To help distinguish acids from other hydrogen-containing molecules, acidic hydrogens are written first in the formula. Any time hydrogen is the first element in a formula of a compound, the substance is an acid.

▲ Acids such as acetic acid, $HC_2H_3O_2$, and hydrochloric acid, HCl, are called monoprotic acids. Monoprotic acids contain only one acidic hydrogen.

H_2SO_4	$H_3C_6H_5O_7$
Sulfuric acid	Citric acid
a **di**protic acid	a **tri**protic acid

▲ All acids that have more than one acidic hydrogen per molecule are called polyprotic acids. Polyprotic acids with two acidic hydrogens are diprotic acids. Those with three acidic hydrogens are triprotic acids.

Figure 14.4

Acidic Hydrogen
If a hydrogen atom loses its electron, all that remains is a proton. Prefixes, used with the term *protic,* which refers to the remaining proton, indicate how many acidic hydrogens are present in an acid.

WORD ORIGIN

acid:
acidus (L) sour
One property of acids is that they taste sour.

Content Background

Acidic Hydrogens The acidic hydrogens in acids are all bonded to electronegative elements such as O, N, or Cl. In the structures of these molecules, these bonds are very polar because of the extremely unequal sharing of the electron pair. When these molecules are dissolved in water, water molecules are strongly attracted to these polar bonds by hydrogen bonds. These hydrogen bonds are strong enough to break the polar covalent bond, leaving behind the stable anion and forming a separate hydronium ion, H_3O^+.

Concept Development

Explain that the strength of the bonding of a hydrogen atom in a covalent compound determines whether the hydrogen is acidic. If the bonding is weak enough that the hydrogen ion is lost in a particular circumstance, the compound has reacted as an acid.

Transparency

Display **Basic Concepts Transparency 40** to introduce the concept of acids and bases as proton donors and acceptors. L1 LEP

GLENCOE TECHNOLOGY

 Videodisc

Chemistry: Concepts and Applications
See the Videodisc Program Teacher Guide to access illustrations and other material available for this chapter.

Meeting Individual Needs

Learning Disabled Have students make paper cutouts to represent the atoms of hydrogen, oxygen, and chlorine in the reaction between hydrochloric acid and water. Use thumbtacks to attach them to a piece of poster board or a bulletin board, then physically transfer the H^+ from HCl to H_2O to create H_3O^+ and Cl^-. L1 LEP

Chemistry Journal

Acids in Foods
Ask students to look on the labels of food items in their home and to list the acids that are present in the foods. Have students include the lists of foods and acids in their journals. L1 LEP

⚡ Quick Demo

Why is iron sometimes acid-dipped? 🥽 🧤 🖌

Wearing safety goggles and an apron, use forceps to dip a piece of rusty iron into a 6*M* HCl solution. Move the piece around until the rust begins to fall off as the iron reacts. Ask students why iron is often dipped into a vat of hydrochloric acid before using it to manufacture something such as an automobile. *to remove rust and other surface impurities* Ask students to write the balanced equation for the reaction between iron and hydrochloric acid. *2Fe + 6HCl → 3H₂ + 2FeCl₃ or Fe + 2HCl → H₂ + FeCl₂*
Disposal: Code D

Concept Development

Explain that acids and bases are often defined according to three different models. The **Arrhenius** concept is: *Acids produce hydrogen ions in aqueous solution, whereas bases produce hydroxide ions.* The **Brønsted-Lowry** model is: *Acids are proton (H^+) donors, and bases are proton acceptors.* The **Lewis** model is: *Acids are electron-pair acceptors, and bases are electron-pair donors.* To lessen confusion, be careful to distinguish the conceptual definitions when you use them to explain acid or base behavior.

Tying to Previous Knowledge

Ask students to draw Lewis electron dot structures for all reactants and products for the reaction of hydrogen chloride with water. Point out and emphasize the formation of the hydronium ion by the transfer of the H^+ ion from hydrogen chloride to water. L1

Figure 14.5
Steel-Making
Sulfuric acid, which is used to make steel, is an example of a diprotic acid.

Chemical Reaction Shorthand

You know that you can write an equation for the ionization of a specific acid. However, it is sometimes handy to represent the formation of hydronium ions when acids dissolve in water by a general equation. In this general equation, any monoprotic acid is represented by the general formula HA. Compare this general equation to the specific equation for the ionization of HCl.

$$HCl(g) + H_2O(l) \rightarrow H_3O^+(aq) + Cl^-(aq)$$

$$HA + H_2O(l) \rightarrow H_3O^+(aq) + A^-(aq)$$

Although the form of the general equation written above is the most complete, it is more convenient to use a shorthand form of the reaction. In the shorthand form, water is not shown as participating in the reaction, and the hydronium ion is represented as an aqueous hydrogen ion.

$$HA(aq) \rightarrow H^+(aq) + A^-(aq)$$

Similar equations apply to the transfer of hydrogen ions from polyprotic acids, such as sulfuric acid, which is used in **Figure 14.5**. Equations depicting the transfer are shown in **Figure 14.6**.

When using this convenient shorthand style, keep in mind that the water molecule is always an active participant in the reaction, even though it is not written in the equation.

Figure 14.6
Ionization of Polyprotic Acids
Polyprotic acids lose their acidic hydrogens one at a time. For a diprotic acid, there are two steps (A). For triprotic acids, there are three steps (B).

A	General:		Example:	
	$H_2A(aq) \rightarrow H^+(aq) + HA^-(aq)$		$H_2SO_4(aq) \rightarrow H^+(aq) + HSO_4^-(aq)$	
	$HA^-(aq) \rightarrow H^+(aq) + A^{2-}(aq)$		$HSO_4^-(aq) \rightarrow H^+(aq) + SO_4^{2-}$	

B				
	$H_3A(aq) \rightarrow H^+(aq) + H_2A^-(aq)$		$H_3PO_4(aq) \rightarrow H^+(aq) + H_2PO_4^-(aq)$	
	$H_2A^-(aq) \rightarrow H^+(aq) + HA^{2-}(aq)$		$H_2PO_4^-(aq) \rightarrow H^+(aq) + HPO_4^{2-}(aq)$	
	$HA^{2-}(aq) \rightarrow H^+(aq) + A^{3-}(aq)$		$HPO_4^{2-}(aq) \rightarrow H^+(aq) + PO_4^{3-}(aq)$	

486 Chapter 14 Acids, Bases, and pH

Cultural Diversity

TECH PREP

Pickling

Pickling is when foods are preserved in an acidic solution. The acidic solution inhibits the growth of microorganisms that cause spoilage. The process is done by placing slightly cooked foods in an acidic solution, such as vinegar, or in brine. In the latter, the food ferments and produces a solution of lactic acid that acts as the preservative in foods such as sauerkraut.

Many types of foods are pickled, including fruits, vegetables, and some fish and fowl. Pickled fish was an important foodstuff of the Maglemosian culture of Scandinavia, which flourished from 10 000-6000 B.C. Today, pickled fish is still an important part of the Scandinavian diet. Greek historians noted that the ancient Egyptians were fond of eating whole, small birds that were pickled.

BIOLOGY CONNECTION

Measurement of Blood Gases

There are only four hydrogen ions in blood for every 100 000 000 other ions and molecules in blood. But enzyme reactions in the body are sensitive to small changes in the concentration of hydrogen ions, so it is crucial. Hydrogen ions affect acid-base relationships in body fluids.

Normal Ranges of Some Blood Components	
Plasma Component	**Normal Range**
HCO_3^-	23–29 m Eq/liter*
P_{CO_2}	35–45 mm Hg
P_{O_2}	75–100 mm Hg
pH	7.35–7.45

*expressed in molar equivalents per liter

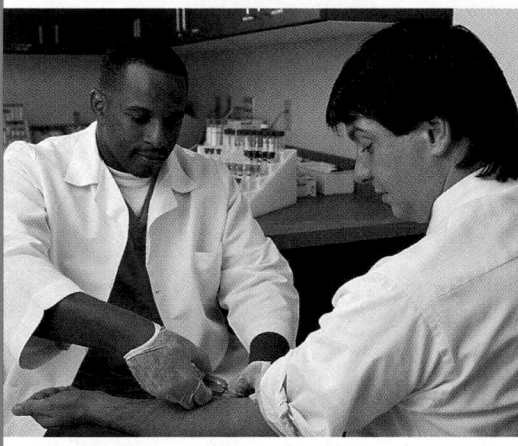

Interpreting acid-base status in blood
When a patient is ill, the doctor's role is to diagnose the patient's condition. Sometimes, this is difficult because different conditions may have similar symptoms. One helpful tool that the physician has is a blood test that provides information about the acid-base relationships in blood. This particular blood test will provide data on acidity (pH), pressure caused by dissolved carbon dioxide (P_{CO_2}), pressure caused by dissolved oxygen (P_{O_2}), and hydrogen carbonate concentration (HCO_3^-). The normal ranges of these components are shown in the table above.

Case histories To see how a physician uses acid-base relationships, it is interesting to consider case histories. In one case, after a pneumonia patient was put on a respirator, she failed to improve. Her blood test showed the following.

P_{CO_2}	17 mm Hg	HCO_3^-	18 m Eq/liter
P_{O_2}	75 mm Hg	pH	7.65

The low carbon dioxide level and the high pH while on the respirator were unexpected. These levels led the physician to check the settings on the respirator. The volume adjustment on the respirator had slipped. The patient was receiving twice the recommended quantity of air. This caused respiratory alkalosis, a condition of decreased acidity of the blood and tissues. When the respirator was adjusted, the blood levels returned to normal as the acid-base balance was reestablished, and the patient began to recover.

Connecting to Chemistry

1. **Hypothesizing** In a heart attack, blood flow to some parts of the heart may be stopped or greatly reduced. How might this affect the acid-base relationship in blood in the heart muscle?

2. **Applying** Blood gases may fail to show major abnormalities while the patient is at rest. Suggest a way to overcome this problem if the patient is not bedridden.

Purpose

TECH PREP Students gain an understanding of how the measurement of blood gases can lead to a diagnosis of pulmonary problems.

Teaching Strategies

- Discuss why analysis of arterial blood gases (ABGs) has become one of the most useful tests in pulmonary medicine. Determination of P_{O_2} in a sample of arterial blood gives a reliable estimate of the pulmonary O_2 transport function.

- Use the table to discuss that P_{CO_2} measures the pressure of the carbon dioxide in the blood. By definition, an elevated P_{CO_2} means too little oxygen is taken into the alveoli of the lungs. A low P_{CO_2} means that more oxygen than is needed is taken into the alveoli.

Extension

Mention that sodium hydrogen carbonate forms an important part of the body's acid-base balance. Have interested students investigate the role of the hydrogen carbonate ion in blood chemistry. **L2**

Connecting to Chemistry

1. Students may hypothesize that the blood components in the heart muscle would change. The heart would be receiving less oxygen, so the P_{O_2} component would decline. The pH and electrolytes would also change.

2. In selected cases, exercise can be performed while O_2 consumption is either measured or estimated. Blood can be obtained at various points during the exercise, at the peak, and during recovery.

Acid Ionization

In the reaction of an acid with water, ions are formed from a covalent compound. When ions form from a covalent compound, the process is called **ionization.** Specifically, acids form ions in a process called acid ionization.

Acids Are Electrolytes

Because acids ionize to form ions in water, acidic solutions conduct electricity. As you learned in Chapter 4, substances that dissolve in water to give conducting solutions are called electrolytes. **Figure 14.7** compares the electrical conductivities of water, a solution of an ionic compound, and solutions of two different acids.

Figure 14.7
Conductivity
The electrical conductivities of solutions are easy to compare by using a simple circuit. The brightness of the light indicates the relative electrical conductivity of the solutions.

▼ Distilled water is nonconducting.

◄ 1*M* NaCl is an excellent conductor.

▲ 1*M* $HC_2H_3O_2$ is a weak conductor.

As opposed to solutions of ionic compounds (such as table salt), which are always excellent conductors of electricity, acidic solutions have electrical conductivities ranging from strong to weak. The range of electrical conductivities exhibited by different acidic solutions distinguishes acid ionization from ionic dissociation. The range also indicates that acids vary in their ability to produce ions.

Submicroscopic Behavior of Bases

The behavior of bases is also described at the molecular level by the interaction of the base with water. A **base** is a substance that produces hydroxide ions, OH^-, when it dissolves in water. There are two mechanisms by which bases produce hydroxide ions when they dissolve in water.

Simple Bases: Metal Hydroxides

The simplest kind of base is a water-soluble ionic compound, such as sodium hydroxide, that contains the hydroxide ion as the negative ion. When NaOH dissolves in water, for example, it dissociates into aqueous sodium ions and hydroxide ions, as shown below.

$$NaOH(s) \xrightarrow{H_2O} Na^+(aq) + OH^-(aq)$$

NaOH is definitely a base because it produces hydroxide ions when it dissolves in water. You can predict that any water-soluble or slightly water-soluble metal hydroxide will be a base when added to water.

Water plays a different role here than in the formation of hydronium ions when acids ionize in water. Water molecules do not chemically react with this type of base. The hydroxide ion is formed by simple ionic dissociation, and no transfer occurs between the base and the water molecules to form the hydroxide ions.

Just as a polyprotic acid in water produces more than one hydronium ion, it is possible for a formula unit of a metal hydroxide to produce more than one hydroxide ion. Calcium hydroxide, $Ca(OH)_2$, and aluminum hydroxide, $Al(OH)_3$, are examples of such bases, as shown in **Figure 14.8.**

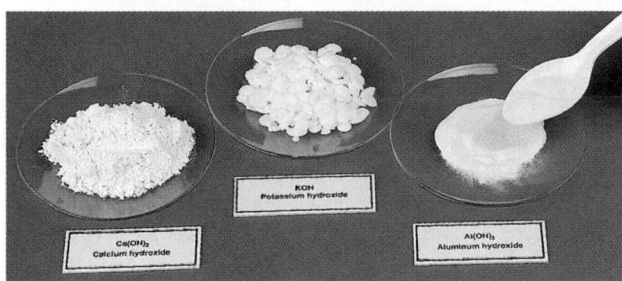

Figure 14.8
Dissociation of Some Metal Hydroxides
All of these compounds are bases because they produce hydroxide ions when they dissolve in water.

Reversing the Transfer: Bases That Accept H⁺

A few bases are covalent compounds that produce hydroxide ions by an ionization process when dissolved in water. The ionization involves the transfer of a hydrogen ion from water to the base. The most common example of this type of base is ammonia, NH_3.

When ammonia gas dissolves in water, some of the aqueous ammonia molecules react with water molecules to form ammonium ions and hydroxide ions, as shown in this reaction.

$$NH_3(g) + H_2O(l) \rightarrow NH_4^+(aq) + OH^-(aq)$$

Ammonia is a base. It produces hydroxide ions in water, but by a different mechanism than that in the NaOH example. In the reaction with ammonia, the water molecule is an active chemical reactant. Water molecules transfer hydrogen ions to ammonia molecules.

It is also helpful to have a general reaction for the ionization of a covalent base, which is represented by the letter *B*. Study the equation for the general reaction.

$$B + H_2O(l) \rightarrow BH^+(aq) + OH^-(aq)$$

14.1 **Acids and Bases** **489**

Concept Development

Explain that water and sometimes other substances are often involved in chemical reactions in such a way that their concentrations are essentially constant. In many of these cases, they are left out of the equations for the reactions, and, in most cases, their concentrations are incorporated as constants into calculations.

Reinforcement

Explain that this definition of a base, *a substance that produces hydroxide ions when it dissolves in water,* is the Arrhenius definition that was explained earlier in the section.

✓Assessment

Performance: Ask students to write the chemical equation for the reaction of water with $Al(OH)_6^{3+}$. $Al(OH)_6^{3+} + H_2O \rightarrow H_3O^+ + Al(OH)_5O^{2+}$ L1

Transparency

Display **Problem Solving Transparency 16** to teach the concept of acids and bases as electrolytes. L1 LEP

Display **Basic Concepts Transparency 38** to introduce the concept of ionization of acids and bases. L1 LEP

Cultural Diversity

Chemistry?

The origins of the English word *chemistry* are probably impossible to trace. Its earliest origins go back to a time when humans first began to manipulate their environment. According to one theory, the word derives from the Egyptian word for "Egypt" *(Kham),* and *khemeia* might be "the Egyptian art." It is also thought that the word might derive from the Greek *khumos,* which means "the juice of a plant." *Khemeia* might refer to "the art of extracting juices." The only thing for certain is that chemistry is deeply rooted in many diverse cultures, including those of China and India.

Meet Fe Tayag, Cosmetic Bench Chemist

Teaching Strategies

Explain to students that surfactants, such as sodium lauryl sulfate, $NaC_{21}H_{41}NO_4S$, decrease the surface tension of water so that stable bubbles form freely.

Background

Viscosity is a measure of a material's resistance to flow. The metric unit of viscosity is the poise (p). The SI viscosity unit is the pascal-second (Pa·s). A viscosity of 2000 to 3000 centipoise (cp) is 2-3 Pa·s. The table below is useful for comparing viscosities.

Substance	Viscosity (Pa·s)
water (20°C)	0.001
olive oil (20°C)	0.08
motor oil, heavy (20°C)	0.6
pitch (15°C)	5×10^9

PEOPLE in CHEMISTRY

Meet Fe Tayag, Cosmetic Bench Chemist

It pays to be a careful reader of labels. Here's a hint from Ms. Tayag, who has formulated cosmetics for more than 20 years. Many cosmetic companies find it is a good selling point to add sunscreen to their products. But unless the container specifies an SPF (sun protection factor) as a number, there probably isn't enough sunscreen to do much good. In this interview, Ms. Tayag shares other cosmetics savvy that can make you a wiser shopper.

On the Job

 Ms. Tayag, what do you do on your job?

 I'm in the Research and Development Department of a cosmetic laboratory. I formulate cosmetics, such as shampoos, lotions, and bubble baths, for cosmetic companies.

 Can you give us an idea of a typical formula for a shampoo?

 A simple shampoo consists of a mixture of water, sodium lauryl sulfate, and an amide to make it foam. I heat the mixture and then I have to adjust the acidity. Most shampoos are neutral. If it's more basic than that, I adjust it with a citric acid solution. Then I cool it and check the viscosity, or how it flows. I don't want it to be either water-thin or molasses-thick, so I'll adjust the viscosity by using a 20 percent sodium chloride solution. Perfume and color are added to make it smell and look good.

 The label on a cosmetic usually lists water as the first ingredient. Does that mean it's mostly water?

 Yes. Face creams have the lowest amount of water, about 60 percent. The amount of water used depends on the skin type.

 What colorings do you add to the cosmetics you make?

 I use colors approved by the Food and Drug Administration. They come in basic colors, but I can combine them to produce other colors. Green is the most popular coloring for shampoo. That color seems to be associated with freshness and cleanliness.

 What trends do you see developing in cosmetics?

 There are more cosmetics created especially for different ethnic groups. Sunscreen is being added to more and more cosmetics as people become increasingly aware of the damage that ultraviolet light can do to the skin. And, because a large part of the population is getting older, ingredients that are supposed to delay the aging effects, such as antioxidants, are becoming more popular. In addition, there's lots of interest in antiallergenic, natural cosmetics that are both unscented and uncolored.

Early Influences

 How did you get interested in becoming a chemist?

 My father was a warehouse man for a cosmetics company in the Philippines, where I grew up. He was fascinated by the process of producing cosmetics. Because he thought the job of a chemist was a very dignified one, he encouraged me to study chemistry. Our family, which included eight children, was poor, but he somehow found a way to send me to college.

 Did you enjoy the study of chemistry in college?

 Not at first. I had a tough time in college and kept asking myself, "Why did I take this course?" I cried at exam time, but I couldn't bear to disappoint my father. I prayed a lot, and I thought of my father struggling to find money to pay my tuition. That helped me find the courage to continue. My third year of college marked a turning point, and my studies became easier for me.

 Were there people besides your father who influenced you in your career?

 A friend of my father helped me get a job in the cosmetics lab while I was still in school. I worked days and attended school at nights. That way, I got to see what chemistry was all about in the real world. I felt I was ahead of students who lacked practical experience. For instance, in my colloid chemistry class, I brought materials to school and demonstrated how to make a cleansing cream.

Personal Insights

 Do you consider chemistry to be a little like cooking?

 Yes. When I cook, I rely heavily on my senses of sight and smell. In my opinion, a keen sense of smell and a good ability to make observations are very important for a chemist, too.

 Some people think that cosmetics are frivolous. Do you agree?

 No. I think it's important for people's self-confidence to look nice. Because people want to remain attractive and young-looking, this is an industry that will never die out.

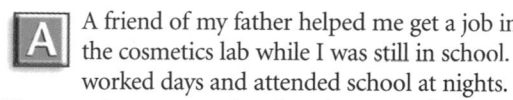

These career opportunities are related to cosmetic chemistry.

Food and Drug Inspector College degree plus written examinations
Manufacturers' Sales Representative High school diploma
Cosmetologist State-administered exam

CAREER CONNECTION

- **Career Path** Students interested in becoming cosmetic bench chemists should explore the following courses: high school courses in biology, chemistry, and physics; and college courses in general chemistry, biology, physics, organic chemistry, biochemistry, inorganic chemistry, physical chemistry, and human physiology.
- **Career Issue** Have interested students investigate the functioning of antioxidants in the aging process. L2

For More Information

Students can obtain information by writing to the American Chemical Society, 1155 16th Street, NW, Washington, DC 20036.

Ammonia Plus Water—What happens?

CAUTION: *Perform this demo in a well-ventilated area.* Wet a piece of red litmus paper with distilled water, and hold the litmus paper with forceps over the mouth of a bottle of concentrated ammonia solution. *The litmus color will change from red to blue.* Ask students what this indicates regarding the interaction of ammonia with the water in the paper. *A basic solution has been produced.*

Figure 14.9

Bases as Electrolytes
The electrical conductivity of $1M$ NaOH is greater than that of $1M$ NH_3. These differences show that aqueous solutions of bases may be strong or weak electrolytes, based on whether many or few ions are in solution.

Applying Chemistry

Why is ammonia slippery? Ask a student to feel a window-cleaning solution that contains ammonia. Explain that the basic solution reacts with oils and greases on the window, dissolving them, and that plain water would not do this well. L1 LEP

Bases Are Electrolytes

Because a base in water produces ions, you can predict that aqueous solutions of bases will conduct electricity. **Figure 14.9** compares the conductivity of a $1M$ NaOH solution with that of a $1M$ NH_3 solution. As with acids, the ability of basic solutions to conduct electricity varies, depending upon the base. This variability is evidence that differences exist in the ability of different bases to produce ions.

Why does water transfer H^+ to bases?

Think back to why acids transfer hydrogen ions to water. The same model can be used to explain why water molecules lose hydrogen ions to covalent bases when they dissolve in water. Consider the example of ammonia. Ammonia is a polar molecule because it contains polar covalent N — H bonds. The nitrogen end of the molecule has a slight negative charge, and the hydrogen atoms each have a slight positive charge. A lone pair of electrons is also on the central nitrogen. Look at **Figure 14.10** to see what happens when polar ammonia molecules dissolve in polar water molecules.

Other Acids and Bases: Anhydrides

Two related classes of compounds do not fit the previous models of acids and bases, but they still act as acids or bases. These compounds are both oxides, which are compounds containing oxygen bonded to just one other element. These oxides are called anhydrides, which means that they contain no water.

Anhydrides differ, depending upon whether the oxygen is bonded to a metal or a nonmetal. Nonmetal oxides form acids when they react with water and are called **acidic anhydrides.** Metal oxides, on the other hand, react with water to form bases and are called **basic anhydrides.** In both of these reactions, water is an active reactant. Now, examine some examples of anhydrides.

Demonstration 14.2

Forming a Basic Anhydride

Purpose: To demonstrate that metal oxides react with water to form a basic solution.

Materials: 10-cm length of magnesium ribbon, crucible, stirring rod, evaporating dish, mortar and pestle, 1 mL distilled water, red litmus paper, hot plate

Safety Precautions:

Disposal: Code F

Procedure: Before class, burn a few centimeters of magnesium ribbon in a crucible. **CAUTION:** *Do not look directly at the bright light.* Use a mortar and pestle to powder the white residue, MgO. Place the MgO in an evaporating dish and add 1 mL of distilled water. Heat to near boiling. Test the resulting $Mg(OH)_2$ with red litmus paper.

Figure 14.10
Forming Ammonium and Hydroxide Ions
Because an aqueous solution of ammonia contains ammonium ions and hydroxide ions, such a solution is commonly referred to as ammonium hydroxide.

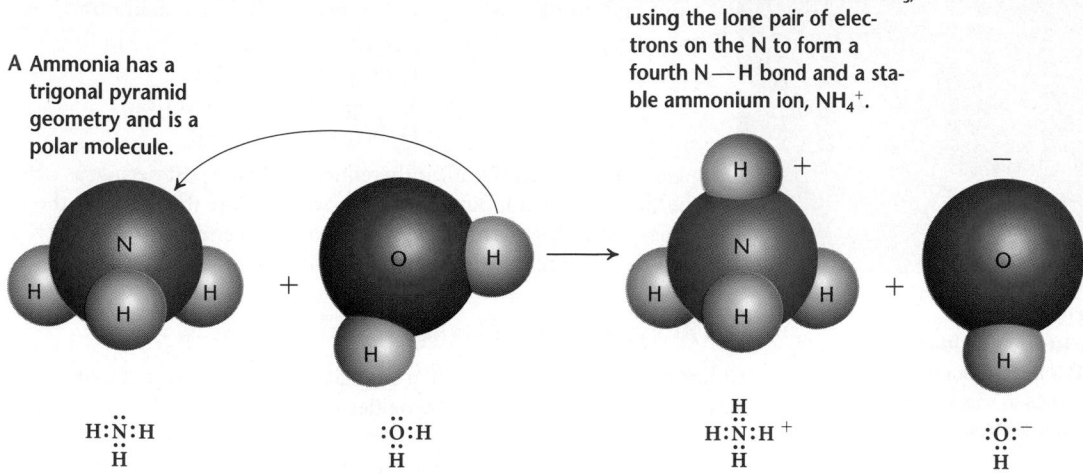

A Ammonia has a trigonal pyramid geometry and is a polar molecule.

C The H$^+$ bonds to the N in NH$_3$, using the lone pair of electrons on the N to form a fourth N—H bond and a stable ammonium ion, NH$_4^+$.

B The hydrogen bond that forms between the N end of NH$_3$ and the H end of H$_2$O is strong enough to pull an H$^+$ completely away from H$_2$O. The two electrons in the broken O—H bond remain as a lone pair on the O. The result is a stable hydroxide ion, OH$^-$.

D The electron dot structures show that these ions are stable. Each atom in the dot structures has a stable number of valence electrons.

Acidic Anhydrides and Acid Rain

Probably the most familiar acidic anhydride is carbon dioxide, CO$_2$. Water that has had carbon dioxide bubbled through it turns blue litmus to red, indicating that CO$_2$ and water form an acid, carbonic acid, H$_2$CO$_3$. A solution of CO$_2$ also has a slightly sour taste, which is one of the reasons that carbonated water is such a refreshing beverage.

Carbon dioxide is a minor component in Earth's atmosphere and an important component in the carbon cycle. Because CO$_2$ is always in the atmosphere, when it rains, CO$_2$ dissolves in rainwater, forming carbonic acid, H$_2$CO$_3$. The result is that rain is always slightly acidic. If rain is always acidic, why is increased acidity in rain such an environmental concern?

The acidity of normal rain does not damage the environment. However, other nonmetal oxides such as sulfur oxides are sometimes present in the atmosphere. Also, levels of carbon dioxide are sometimes higher than normal.

The major source of sulfur oxides in the atmosphere is the burning of sulfur-containing coal in power plants. As this type of coal burns in a furnace, sulfur dioxide gas, SO$_2$, is produced. The SO$_2$ escapes into the atmosphere, where it reacts with more oxygen to form sulfur trioxide, SO$_3$.

14.1 Acids and Bases 493

Visual Learning

Figure 14.10: Draw the Lewis electron dot structure for aminomethane (often called methyl amine) on the chalkboard, and show how amines, as well as ammonia, may react as bases by accepting hydrogen ions from water and other substances.

Quick Demo

What is the brown gas?

CAUTION: *Do this demo in a fume hood or in a well-ventilated room. Wear safety goggles and an apron.* Stir several drops of phenolphthalein indicator into a slightly basic water solution in a 600-mL beaker, which will produce a pink or magenta color. Generate some nitrogen dioxide by placing a piece of copper into concentrated nitric acid in a small beaker. Do not tell students the identity of the gas at this time. Pour the brown NO$_2$ gas, which is more dense than air, into the beaker containing the basic solution. Stir the solution with a glass stirring rod. *The gas will dissolve in the water, causing the pink color to disappear as the solution becomes more acidic.* Explain that the brown gas is the oxide of an element. Ask students if the element is a metal or a nonmetal. *A nonmetal; the gas is NO$_2$.*
Disposal: Code D

Results: Metal oxides react with water to form a basic solution. MgO is a basic anhydride and caused the red litmus paper to turn blue.
Analysis: Ask these questions.
1. What does the litmus test indicate when red litmus paper changes to the blue color? *The solution is basic.*
2. If a metal oxide is a base without water, what do you think a nonmetal oxide is? *an acidic anhydride (an acid without water)*

 **Assessment**

Skill: Ask students to design a similar experiment that would test an acidic anhydride, which is a nonmetal oxide. *Copper metal can be reacted with nitric acid and the nitrogen oxides collected and tested.* **L2**

Discussion

Ask students to write the chemical equation for the complete combustion of octane, C_8H_{18}, yielding carbon dioxide and water. $2C_8H_{18} + 25O_2 \rightarrow 16CO_2 + 18H_2O$ Then ask why automobile emissions are acidic. *Carbon dioxide is an acidic anhydride, and it can react with water to produce acidic solutions.* Explain that the high-temperature combustion that occurs in automobile engines causes nitrogen in the air to react with oxygen to produce nitrogen oxides, which are also acidic anhydrides. Catalytic conversion is not 100% efficient at removing these oxides of nitrogen. [L1]

3 ASSESS

Check for Understanding

- Check for understanding of the section material by having students answer the Section Review questions.
- Ask students to name the following acids: HF, HCN, H_2SO_3, $HBrO_3$, and $HBrO_4$. *hydrofluoric acid, hydrocyanic acid, sulfurous acid, bromic acid, perbromic acid* Briefly review the system for naming acids. **1.** If the anion does not contain oxygen, the acid is named with the prefix *hydro-*, the root name of the anion, and the suffix *-ic*. **2.** If the anion contains oxygen, the acid name is formed from the root name of the anion with a suffix of *-ic* or *-ric*, which replaces *-ate*; or *-ous* or *-rous*, which replaces *-ite*. [L1]

Reteach

Discussion: Ask students why many water-treatment plants bubble carbon dioxide through the water at some stage. *The carbon dioxide is used to adjust the acidity of the water.*

Figure 14.11
Nitrogen Oxides
The major source of nitrogen oxides in the atmosphere is automobiles.

At room temperature, the reaction between nitrogen, N_2, and O_2 is slow and insignificant. At the high temperatures in an automobile engine, the reaction between N_2 and O_2 goes quickly, and large amounts of nitrogen oxides are produced in exhaust, **Figure 14.11**.

When sulfur oxides, nitrogen oxides, and increased amounts of carbon dioxide dissolve in rain, they undergo acid-forming reactions and produce what is commonly referred to as acid rain.

Examples:
$$SO_2(g) + H_2O(l) \rightarrow H_2SO_3(aq)$$
$$SO_3(g) + H_2O(l) \rightarrow H_2SO_4(aq)$$
$$2NO_2(g) + H_2O(l) \rightarrow HNO_3(aq) + HNO_2(aq)$$
$$CO_2(g) + H_2O(l) \rightarrow H_2CO_3(aq)$$

Acid rain has been significantly reduced over the past decade as new mechanisms for trapping nonmetal oxides before they get into the atmosphere have been developed. Read the Chemistry and Society feature for more information about this type of air pollution.

Basic Anhydrides and Making Your Garden Grow

Unlike nonmetal oxides, which are covalent compounds, metal oxides are ionic compounds. When metal oxides react with water, they produce hydroxide ions.

Gardeners sometimes use lime to treat their soil, as shown in **Figure 14.12**. Lime is the common name for the chemical compound calcium oxide, CaO. When CaO is spread on soil, it reacts with water in the soil to form calcium hydroxide, $Ca(OH)_2$. This compound then forms calcium and hydroxide ions.

$$CaO(s) + H_2O(l) \rightarrow Ca(OH)_2(aq)$$
$$Ca(OH)_2(aq) \rightarrow Ca^{2+}(aq) + 2OH^-(aq)$$

A similar reaction was used historically to produce an important commodity, soap. Early soap makers used the basic properties of metal oxides. When wood burns, the metal atoms in the wood form solid metal oxides in the burning process. These metal oxides are predominantly those of sodium, potassium, and calcium. These metal oxides are ionic so they are solids, even at the high temperature of a roaring fire. They are the major component of the ash that is left when the fire burns out.

Figure 14.12

Making Soil More Basic
Adding lime to soil makes the soil less acidic and more favorable for growing many types of plants. Because $Ca(OH)_2$ is only slightly water soluble, it provides a longer-lasting source of base than provided by more soluble ionic hydroxides.

494 Chapter 14 Acids, Bases, and pH

Demonstration 14.3

An Acidic Fizz

Purpose: To demonstrate the formation of carbonic acid from an acidic anhydride, CO_2.

Materials: 1-L graduated cylinder, several walnut-size pieces of dry ice, 5 mL universal indicator, stirring rod, 5 mL household ammonia solution

Safety Precautions: 🧤 🥼 🥽

Disposal: Code C

Procedure: Fill a 1-L graduated cylinder with

water that has a few drops of universal indicator added to it. Add household ammonia solution dropwise, with stirring, until the color is in the basic color range for the indicator. Place pieces of dry ice in the cylinder. A light box will enable students to better see color changes.

Results: The solid CO_2 sinks to the bottom; bubbles rise to the top, producing a familiar-looking fog. The color of the universal indicator changes through the entire color range to

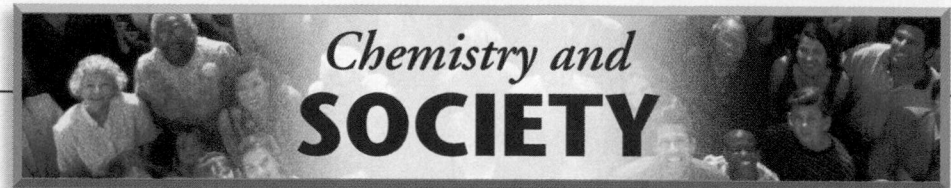

Chemistry and SOCIETY

Atmospheric Pollution

The air you breathe is literally a matter of life and death. Atmospheric oxygen is taken into your body and, during respiration, reacts with glucose to produce the energy required for all the life processes that keep you going.

Unfortunately, the same air, at times, may contain materials that cause respiratory diseases and bring about other harmful effects. Air is often polluted with chemicals produced by human activity. Even Earth itself coughs up some of the same air pollutants during volcanic eruptions.

Introducing the major air pollutants The major chemicals that pollute the air are carbon monoxide, CO; carbon dioxide, CO_2; sulfur dioxide, SO_2; nitrogen monoxide, NO; nitrogen dioxide, NO_2; hydrocarbons; and suspended particles.

In addition, pollutants form under the influence of sunlight when oxygen, nitrogen oxides, and hydrocarbons react. These reactions produce ozone, O_3, and aldehydes such as formaldehyde, CH_2O. Why are pollutants a problem?

Acid rain What do the salmon and the pine trees in the photos have in common? Both have succumbed to the acid environment in which they live. They are two of the many victims of acid rain.

Unpolluted rain is not harmful. However, many industrial and power plants burn coal and oil. The smoke produced may contain large quantities of sulfur oxides, suspended particles, and nitrogen oxides. Automobiles also contribute to the problem by emitting similar oxides. These chemicals react with water in the air to form acids, such as sulfuric acid. These acids reach the surface of Earth in fog, rain, snow, and dew. Acid rain can have a disastrous effect when it reaches bodies of water and waterways. But if a lake has a high limestone content, it is able to somewhat neutralize the acid.

Smog Large cities with many automobiles may have another problem with airborne pollutants. It is called smog, which is a haze or fog that is made harmful by the chemical fumes and suspended particles it contains.

A type of smog known as photochemical smog frequently occurs in large cities in sunny, dry climates. When the pollutants from automobile exhaust enter the air and are exposed to sunlight, they interact to produce photochemical smog. This type of smog is generally worse on hot days and between 11 A.M. and 4 P.M. when exhaust has accumulated in the air.

Analyzing the Issue

1. **Acquiring Information** Read the November 1993 issue of *Scientific American,* pages 50-57. How might the issue of free trade influence the problems of air pollution in this country?

2. **Thinking Critically** Older-model cars are responsible for the greatest amount of air pollutants being vented into the atmosphere. Hold a debate on whether these cars should be banished from the highways.

Purpose

TECH PREP

Students will gain an understanding of some causes of pollution.

Background

When warm air containing pollutants cannot rise, the amount of pollutants can reach dangerous levels.

Teaching Strategies

Discuss that although the countries of eastern Europe and China still experience industrial smog, most industrialized nations have overcome this problem. These nations take the polluting sulfur out of the coal that they burn as fuel. They do this by adding crushed limestone to powdered coal to be burned. Ask students how the limestone achieves its purpose. *Limestone will react with the sulfur dioxide, which will reduce acid rain.*

Analyzing the Issue

1. Industries may choose to produce goods in countries that have no restrictions on pollution because production costs will be cut. The result could be increased world air pollution.

2. Pro: increased air pollution caused by older cars; Con: frequent smog tests that older cars are subjected to in some areas make them clean up their car emissions; banishing these cars could put an unfair burden on those who cannot afford a new vehicle

---

indicate that the water is becoming acidic.
Analysis: Ask these questions.
1. Ask a student to read the label of a soda container to determine whether carbonated water is listed. *Carbonated water is listed.*
2. Write a chemical equation that shows how carbonated water is made. $H_2O(l) + CO_2(g) \rightarrow H_2CO_3(aq)$
3. Name each compound in the equation from question 2. *liquid water, carbon dioxide gas,*

carbonic acid (hydrogen carbonate)

✓ Assessment

Knowledge: While looking at the periodic table, ask students to write the name and formula of another acidic anhydride. *Possible answers: NO, nitrogen(II) oxide; NO_2, nitrogen(IV) oxide; SO_2, sulfur dioxide; SO_3, sulfur trioxide*

L1

Extension

Ask students to write the formula for acetic acid. $HC_2H_3O_2$ Ask why the single hydrogen atom is written separately from the other three. *It is bonded differently and can be more easily ionized.* Explain that acids that contain the carboxyl group, —COOH, are called carboxylic or organic acids and are named differently from inorganic acids. The acid name begins with the root name of the organic compound, then the *-e* ending is replaced with *-oic acid.* Draw the structural formulas and names for ethanoic acid and propanoic acid. Ask students how ethanoic acid differs from acetic acid. *Though their formulas are often written differently, they are the same acid.* L2

4 CLOSE

Figure 14.13
Making Soap
Lye was produced by collecting wood ashes and soaking them in water. After several days, the highly basic solution was separated from the undissolved ash and combined with animal fat. The lye reacted with the fat to make soap.

This early soap-making process is shown in **Figure 14.13**. The reaction of water and sodium oxide, Na_2O, one of the metal oxides in wood ash, is similar to that shown for lime and water.

$$Na_2O(s) + H_2O(l) \rightarrow 2NaOH(aq)$$

$$NaOH(aq) \rightarrow Na^+(aq) + OH^-(aq)$$

The Macroscopic-Submicroscopic Acid-Base Connection

As you have discovered, the properties of acids and bases are determined by the submicroscopic interactions between the acid or base and the solvent water. For example, HCl and $HC_2H_3O_2$ both interact with water to cause a transfer of hydrogen ions from the acid to water molecules to form hydronium ions. Both solutions turn blue litmus red. Even though these properties are the same for both acids, remember that the conductivity of a $1M$ HCl solution is much greater than that of a $1M$ $HC_2H_3O_2$ solution.

A base interacts with water molecules to form hydroxide ions either by ionic dissociation or by the transfer of a hydrogen ion from a water molecule to the base. Consider a $1M$ NaOH solution and a $1M$ NH_3 solution. Both solutions are basic. They each turn red litmus to blue. But the NaOH solution shows strong electrical conductivity, while the NH_3 solution is only weakly conducting.

Why do different acids have some properties in common and yet differ in other properties? Why is the same thing true for bases? Section 14.2 will explain these differences.

SECTION REVIEW

Understanding Concepts

1. Make a table that compares and contrasts the properties of acids and bases.

2. Use a chemical equation to show how aqueous HNO_3 fits the definition of an acid.

3. Consider the oxides MgO and CO_2. For each oxide, tell whether it is a basic anhydride or an acidic anhydride. Write an equation for each to demonstrate its acid-base chemistry.

Thinking Critically

4. **Applying Concepts** Chemists often call a hydrogen ion a proton. Explain why an acid is sometimes called a proton donor, and a base is sometimes called a proton acceptor.

Applying Chemistry

5. **Using Soap** After using soap to wash dishes by hand, it is sometimes difficult to keep your hands from remaining slick. Explain why rinsing your hands in lemon juice would make them less slick.

SECTION REVIEW

1. Tables should show that acids have a sour taste, turn blue litmus red, react with active metals to form H_2, react with carbonates to form CO_2, and react with bases. Bases have a bitter taste and a slippery feel, turn red litmus to blue, and react with acids.

2. $HNO_3(aq) + H_2O(l) \rightarrow H_3O^+(aq) + NO_3^-(aq)$; HNO_3 forms the hydronium ion when dissolved in water. $HNO_3(aq) \rightarrow H^+(aq) + NO_3^-(aq)$ could also be used.

3. $MgO(s) + H_2O(l) \rightarrow Mg(OH)_2(aq) \rightarrow Mg^{2+}(aq) + 2OH^-(aq)$, basic; $CO_2(aq) + H_2O(l) \rightarrow H_2CO_3(aq) \rightarrow 2H^+(aq) + CO_3^{2-}(aq)$ or $H^+(aq) + HCO_3^-(aq)$, acidic

4. An acid gives up a hydrogen ion, which is a proton, and a base will accept one.

5. The acid in the lemon juice would react with the base in the soap, so the slick property of the base would no longer be present.

Strengths of Acids and Bases

From reading Section 14.1, you know that all acids have a sour taste, but they may differ in how readily they react with another substance. You wouldn't hesitate to use the acetic acid in vinegar on a salad, but you certainly wouldn't use hydrochloric acid, which may be used to clean brick, on any type of food. All bases also share some properties but differ in others. You would readily use a dilute ammonia solution as a cleaner, but you certainly wouldn't let sodium hydroxide, which is used in drain cleaners, come in contact with your skin. These two bases differ greatly in how they react.

Strong Acids and Bases

What's going on? You know that when acids and bases are mixed with water, they form ions. Much of the behavior of acids and bases depends on how many ions are formed by a particular acid or base in water. The degree to which bases and acids produce ions depends on the nature of the acid or base.

Acids and bases are classified into one of two categories depending upon their strength, which is the degree to which they form ions. The strong category is reserved for those substances, such as NaOH and HCl, that completely dissociate or ionize and produce the maximum number of ions when dissolved in water. All other acids and bases are classified as weak because they produce few ions when dissolved in water.

Strong Bases

Sodium hydroxide, NaOH, is a **strong base** because when NaOH dissolves in water, all NaOH formula units dissociate into separate sodium and hydroxide ions. The dissociation of the base is complete. The strength of a base is based on the percent of units dissociated, not the number of OH^- ions produced. Some bases, such as $Mg(OH)_2$, are not very soluble in water, and they don't produce a large number of OH^- ions. However, they are still considered to be strong bases because all of the base that does dissolve completely dissociates.

14.2   Strengths of Acids and Bases   **497**

SECTION PREVIEW

Objectives

Relate different electrical conductivities of acidic and basic solutions to their degree of dissociation or ionization.
Distinguish strong and weak acids or bases by their degree of dissociation or ionization.
Compare and contrast the composition of strong and weak solutions of acids or bases.
Relate pH to the strengths of acids and bases.

Key Terms

strong base
strong acid
weak acid
weak base
pH

SECTION 14.2

PREPARE

Key Concepts
The differences between strong and weak acids and bases are explained and related to the degree of ionization or dissociation of the respective compounds. The pH scale is then introduced and related to the concentrations of hydronium and hydroxide ions.

Planning Ahead
For the ChemLab Have available purple cabbage, eyewash ($0.1M$ boric acid solution), lemon juice, white vinegar, table salt, laundry soap (not detergent), baking soda, borax, and drain cleaner.
For the MiniLab Obtain a variety of antacids.

1 FOCUS

Focus Transparency

Display the **Section Focus Transparency** for Section 14.2 to introduce some aspects of chemistry in the kitchen.
L1   LEP

Program Resources

Study Guide, pp. 57-58 L1
Laboratory Manual, 14.2, pp. 131-134 L1
Section Focus Transparency 29 and Master, p. 29 L1 LEP
Basic Concepts Transparency 39 and Master, pp. 77-78 L1 LEP

Problem Solving Transparency 17 and Master, pp. 33-34 L1 LEP
ChemLab and MiniLab Worksheets, pp. 89-91, 93 L1
Tech Prep Applications, pp. 21-22 L1

2 TEACH

Concept Development

Explain that the strengths of acids are determined by the strengths and polarities of the H—X bonds. They are dependent primarily upon the relative electronegativities of H, X, and other elements that may be bonded to X. Generally, the weaker the H—X bond, the stronger the acid.

Using an Analogy

Students often assume that an acid or a base is strong simply because it has numerous hydrogen atoms per molecule or hydroxide ions per formula unit. For example, H_3PO_4 and $Ti(OH)_4$ might be assumed to be strong, but they are not. Explain that one cannot determine the degree of generosity of a person based solely upon the amount of money the person has. For example, a billionaire might be greedy and stingy and give nothing to charity, while a person of modest wealth might give a relatively large amount to charity. Similarly, one cannot ascertain the strength of an acid or base by the amount of H^+ or OH^- contained in the substance. For example, HBr is a strong acid because virtually all of its molecules ionize, and H_3PO_4 is a weak acid because few of its molecules ionize.

Table 14.2 Common Strong Acids and Bases

Strong Acids	Strong Bases
Perchloric acid, $HClO_4$	Lithium hydroxide, LiOH
Sulfuric acid, H_2SO_4	Sodium hydroxide, NaOH
Hydriodic acid, HI	Potassium hydroxide, KOH
Hydrobromic acid, HBr	Calcium hydroxide, $Ca(OH)_2$
Hydrochloric acid, HCl	Strontium hydroxide, $Sr(OH)_2$
Nitric acid, HNO_3	Barium hydroxide, $Ba(OH)_2$
	Magnesium hydroxide, $Mg(OH)_2$

The strong bases shown in **Table 14.2** are all ionic compounds that contain hydroxide ions. NaOH and KOH are the most common strong bases you will encounter. A $1M$ solution of NaOH and a $1M$ solution of KOH will each contain $1M$ OH^- because both compounds completely dissociate.

Strong Acids

HCl is a **strong acid** because no HCl molecules are in a water solution of HCl. Because of the strong attraction between the water molecules and HCl molecules, every HCl molecule ionizes. A $1M$ HCl solution contains $1M$ H_3O^+ and $1M$ Cl^-. Similarly, a $1M$ HNO_3 solution contains $1M$ H_3O^+ and $1M$ NO_3^-. As before, the labeling of a solution as $1M$ HNO_3 is not particularly descriptive of the submicroscopic composition of the solution.

Table 14.2 lists common strong acids and bases. Because they are used so often, it is helpful to memorize their names and formulas. They all completely dissociate or ionize into ions when they dissolve in water. If an acid or base is not listed in this group, it is considered to be a weak acid or base. However, the terms *strong* and *weak* are not absolute. Strength of acids and bases covers a wide range from extremely strong to extremely weak. Notice that the strongest bases are all hydroxides of Group I, the alkali metals, and Group II, the alkaline earth metals. *Alkali* is a term frequently used to refer to materials that have noticeably basic properties.

Figure 14.14
Some Common Weak Acids
Some common weak acids vary in their structures.

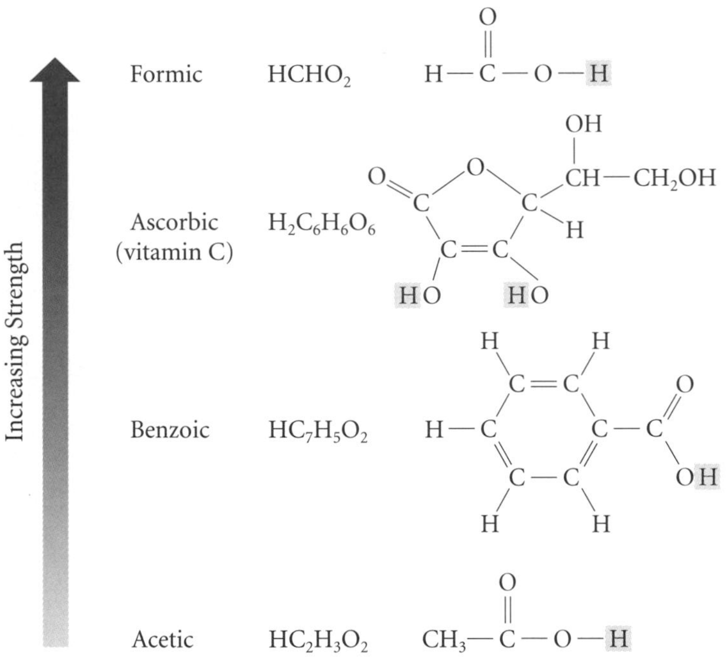

Demonstration 14.4

When is an acid strong?

Purpose: To demonstrate relative ion concentrations by conductivity.

Materials: 60 mL 12M HCl, 41 mL glacial acetic acid, eight 100-mL beakers, conductivity tester with a lightbulb, 1 L distilled water. A safe conductivity apparatus can be easily made using an AC adapter (battery eliminator). Remove the special plug from the end of the DC wire. Attach

alligator clips to the ends of the two wires. Wire in series a flashlight bulb socket into one side of the wire. Place a bulb in the socket. The alligator clips can be attached to straightened paper clips (electrodes) that have been pushed through cardboard. The conductivity apparatus can now be used with a small beaker.

Safety Precautions:

Disposal: Code D

Procedure: Prepare the following solutions as indicated. **CAUTION:** *Both acids are corrosive.* 6.0M HCl: Add 50 mL 12M HCl to 50 mL H_2O. 1.0M HCl: Add 8.3 mL 12M HCl to 91.7 mL H_2O. 0.1M HCl: Add 10 mL 1.0M HCl to 90 mL H_2O. 6.0M acetic acid: Add 34.5 mL glacial acetic acid to 65.5 mL H_2O. 1.0M acetic acid: Add 5.8 mL glacial acetic acid to 94.2 mL

Weak Acids and Bases

The weak category of acids and bases contains those with a wide range of strengths. This is the category into which most acids and bases fall. Instead of being completely ionized, weak acids and bases are only partially ionized.

Weak Acids

Acetic acid, $HC_2H_3O_2$, is a good example of a **weak acid.** In a $1M$ $HC_2H_3O_2$ solution, less than 0.5 percent of the acetic acid molecules ionize, and 99.5 percent of the acetic acid molecules remain as molecules. Another way to think of this is to consider 1000 acetic acid molecules in a water solution. On the average, only five of the 1000 molecules transfer their single hydrogen ion to a water molecule. The molarity of hydronium ion produced in a $1M$ $HC_2H_3O_2$ solution is much less than $1M$ due to this partial ionization. Other common weak acids include phosphoric acid, H_3PO_4, and carbonic acid, both of which are found in soft drinks.

The molecular structure of a weak acid determines the extent to which the acid ionizes in water. **Figure 14.14** shows the variety of structures of some common weak acids.

Figure 14.15 uses a graph format to show the dramatic difference in degree of ionization between a solution of a weak acid and one of a strong acid. A solution of weak acid contains a mixture of un-ionized acid molecules, hydronium ions, and the corresponding negative ions. The concentration of the un-ionized acid is always the greatest of the three concentrations.

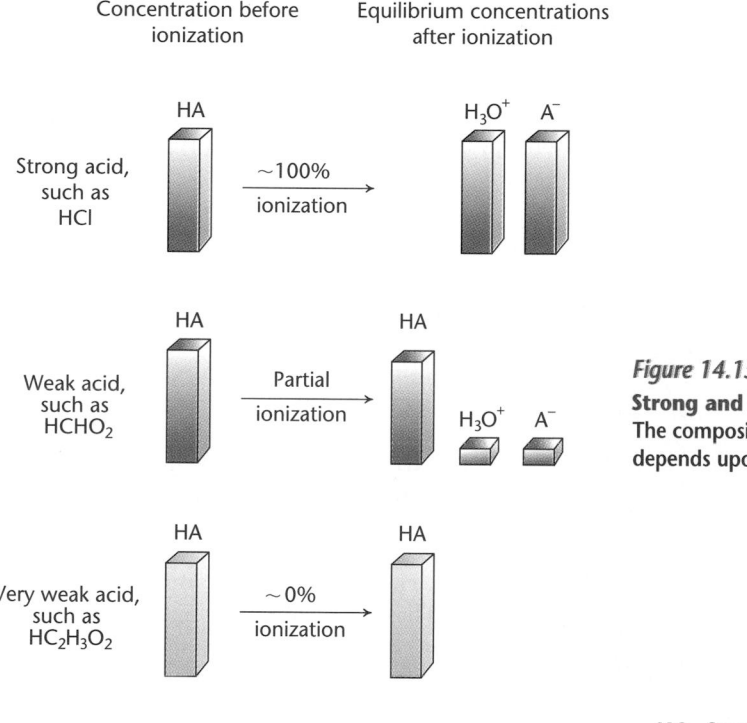

Figure 14.15
Strong and Weak Acids
The composition of an acidic solution depends upon the strength of the acid.

Applying Chemistry
Hydroxides as Antacids The low solubilities of aluminum hydroxide and magnesium hydroxide make them useful as antacids. If they were very soluble compounds, the high concentrations of hydroxide ions would harm the tissues of the mouth, esophagus, and stomach.

Visual Learning

Figure 14.15: Explain that the relative strengths of strong and weak acids may be shown in a rough way through the use of long and short arrows between reactants and products. For example, $H_2SO_4 \longrightarrow H^+ + HSO_4^-$ indicates that H_2SO_4 is a strong acid, and $HF \rightarrow H^+ + F^-$ indicates that HF is a weak acid. This notation also serves as an unspoken introduction to the concept of equilibrium and equilibrium mixtures of molecules and ions, which will be studied in later chapters.

Transparency

Display **Basic Concepts Transparency 39** to compare strong and weak acids. L1
LEP

H_2O. 0.1M acetic acid: Add 10 mL 1.0M acetic acid to 90 mL H_2O. 0.01M acetic acid: Add 10 mL 0.1M acetic acid to 90 mL H_2O. Dim the room lights as you test each solution with the conductivity tester. Have students note the brightness of the bulb for each test.
Results: The HCl solutions conduct electricity. However, the bulb becomes dimmer as the lower concentrations are

tested. The bulb does not light for any of the acetic acid solutions.
Analysis: Ask these questions.
1. As the concentration of the HCl decreased, what happened to the conductivity as measured by the lightbulb's brightness? *The conductivity decreased.*
2. Did the acetic acid conduct an electric current? *no*

Assessment

Skill: Have students list other common household acids and predict whether they are strong or weak acids. Under supervision, have them use the conductivity apparatus to check their predictions. Most examples, such as lemon juice or vinegar, are weak acids. L1

Weak Bases

Ammonia is a **weak base** because most of its molecules don't react with water to form ions. In a $1M$ aqueous solution of ammonia, only about 0.5 percent of the ammonia molecules react with water to form ammonium and hydroxide ions. About 99.5 percent of the ammonia molecules remain as intact molecules. The molarity of hydroxide ions in a $1M$ ammonia solution is much less than $1M$. The major dissolved component in a weak base solution is the un-ionized base. Other examples of bases that produce so few OH^- ions that they are considered to be weak bases are $Al(OH)_3$, and $Fe(OH)_3$.

Weak Is Not Insignificant

Although most acids and bases are classified as weak, their behavior is extremely significant. Most of the acid-base chemistry in living systems occurs through interactions between weak acids and bases. For example, amino acids, the small molecules that serve as the building blocks of proteins, have properties of both weak acids and weak bases. The amino portion of the molecule acts as a base when it comes into contact with a strong acid, and the acid part of the molecule acts as a weak acid when exposed to a base. The coiling of the DNA into a double helix is also due to the interactions between weak acids and bases. *Weak* does not mean "insignificant."

Strength Is Not Concentration

Figure 14.16
Developing the pH Scale
The Danish biochemist S.P.L. Sørenson developed the pH scale in 1909 while working on brewing beer. *pH* is an abbreviation in French for "pouvier d'hydrogene" or, in English, the "power of hydrogen."

Although the terms *weak* and *strong* are used to compare the strengths of acids and bases, *dilute* and *concentrated* are terms used to describe the concentration of solutions. The combination of strength and concentration ultimately determines the behavior of the solution. For example, it is possible to have a concentrated solution of a weak acid or weak base or a dilute solution of a weak acid or weak base. Similarly, you can have a concentrated solution of a strong acid or strong base, as well as a dilute solution of a strong acid or strong base.

The pH Scale

Because of the range of solution concentration, the range of possible concentrations of hydronium ions and hydroxide ions in solutions of acids or bases is huge. For example, a $6M$ solution of HCl has an H_3O^+ molarity of $6M$, but a $6M$ solution of $HC_2H_3O_2$ has an H_3O^+ molarity of $0.01M$.

In most applications, the observed range of possible hydronium or hydroxide ion concentrations spans $10^{-14}M$ to $1M$. This huge range of concentrations presents a problem when comparing different acids and bases. To make this range of possible concentrations easier to work with, the pH scale was developed by S.P.L. Sørenson, **Figure 14.16.**

Meeting Individual Needs

Gifted Ask gifted students to research and explain the relative strengths of the oxyacids of chlorine. *Hypochlorous acid, HOCl, is the weakest because the chlorine draws little of the electron density and strength from the H—O bond. Perchloric acid, $HClO_4$, is the strongest because the three oxygen atoms that are bonded only to the chlorine atom draw a great deal of the electron density and strength from the Cl—O and H—O bonds.* L3

Everyday Chemistry

Balancing pH in Cosmetics

With so many shampoos available, you may find it hard to know which type is best for your hair. Advertisements for each type tell you that their shampoo has more to offer than any other shampoo on the market. How can you know which one to choose? How can you believe product claims?

Cosmetic chemists have many tools for determining the effect of their products on different hair types. Using one technique, developed by NASA, they place a hair under a microscope connected to a TV screen that is hooked up to a computer. The computer evaluates the hair before and after treatment with the shampoo. Working with different types of hair allows these chemists to determine the best treatment for each type of hair.

Shampoos and pH balance The clear, outer layer of a strand of hair is the cuticle, which consists of the protein keratin. The cells of the cuticle are arranged like overlapping shingles. Shampoos that have a high pH make the entire hair shaft swell and push the cells of the cuticle away from the rest of the shaft. Harsh basic substances in solutions for permanents and hair coloring dissolve some of the cuticle, damaging the hair. Hair can also be damaged by the sun and excessive blow drying. Damaged hair is dull and dry.

In contrast, acidic substances in shampoos of low pH make the hair shaft tight and smooth by shrinking it and causing the cells of the cuticle to lie flat. Low-pH shampoos help restore damaged hair to its original condition and make it shine again. They also strengthen the keratin and increase the flexibility and elasticity of the hair. But don't get the idea that everyone will be helped by shampoos with low pH. People with coarse, curly hair can benefit from using alkaline or high-pH shampoos. These products soften and relax the hair, making it softer and less curly.

Why balance pH in skin products? The outer layer of skin has a keratin structure just as hair does. Products aimed at making the skin look brighter and clearer have a higher pH. Their purpose is to remove the top layer of keratin, which may consist of dead cells. The new cells underneath look fresh and vibrant. Occasional use of these basic skin products may be helpful, but regular use damages healthy skin by removing too many layers of cells.

Another problem with skin products that are bases is related to an acid mantle that bathes the top layer of the skin, the epidermis. This fluid—composed of oil, sweat, and other cell secretions—is a natural defense against bacterial infections. Strongly basic soaps can neutralize the protective acid mantle. People with acne or oily skin must be careful not to remove the acid mantle.

Exploring Further

1. **Inferring** If you live in an area where the water is hard, your hair may look dull. Why would rinsing with water to which lemon juice has been added help?

2. **Thinking Critically** Why would a person with acne use skin products that are pH-neutral or mildly acidic?

Everyday Chemistry

Purpose

TECH PREP

Students learn how the pH of shampoos and skin products affects hair and skin.

Teaching Strategies

• Ask students to bring in the shampoos they use in order to test the pH of these products. They can dip a separate piece of pH-sensitive paper into a sample of each shampoo and record the pH. Using this feature and further research, students can list the type of hair that would best be helped by each shampoo. **L1** **LEP**

• Ask students to work in groups to list the ingredients of several shampoos. They will probably find many of the same ingredients in different shampoos. Challenge them to find the ingredients that help achieve the proper pH. Have them brainstorm why each shampoo contains so many substances. **L1** **COOP LEARN**

Extension

Have students investigate what has caused cosmetic companies to rely more than ever before on findings in science and medicine to develop their products and why their products, when used properly, do what they claim. **L1**

Exploring Further

1. Students may infer that substances that cause water to be hard must be basic because they react with the weakly acidic lemon solution.

2. People with acne have to wash their faces frequently. Alkaline products react with the acid mantle on the skin, making the skin prone to additional bacterial infections.

Content Background

pH and Water Ionization The mathematical definition of pH is $pH = -\log[H^+]$, where $[H^+]$ symbolizes the hydrogen ion concentration. Remember that a logarithm is a power of ten; if $a = \log(b)$, then $b = 10^a$. pH is easy to determine with a scientific calculator. The mathematical calculation of pH may be introduced here, but it is not required. The student should have an idea for trends in pH—that pH > 7 is basic and pH < 7 is acidic—as well as that a unit change in pH is a significant change in $[H^+]$.

Water acts as both a weak acid and a weak base, even when there is no added acid or base.

$$H_2O + H_2O \rightleftharpoons$$
$$H_3O^+(aq) + OH^-(aq)$$

Equilibrium theory requires that the product of the hydronium and hydroxide concentration always be 10^{-14} at 25°C; $[H_3O^+][OH^-] = 10^{-14}$. This requirement means that as acid is added to water, the $[H_3O^+]$ increases, and the $[OH^-]$ decreases by the same factor.

GLENCOE TECHNOLOGY

Software

Mastering Concepts in Chemistry
Unit 15, Lesson 1
Water Equilibria
Use this lesson to reinforce equilibria that occur with water. L1 **LEP**

Transparency

Display **Problem Solving Transparency 17** to introduce the concept of the pH scale. L1 **LEP**

What is pH?

pH is a mathematical scale in which the concentration of hydronium ions in a solution is expressed as a number from 0 to 14. A scale of 0 to 14 is much easier to work with than a range from 1 to 10^{-14} (10^0 to 10^{-14}). The pH scale is a convenient way to describe the concentration of hydronium ions in acidic solutions, as well as the hydroxide ions in basic solutions.

Think about the pH numbers 0 to 14 and the hydronium ion concentration range. Notice that the pH value is the negative of the exponent of the hydronium ion concentration. For example, a solution with a hydronium ion concentration of $10^{-11}M$ has a pH of 11. A solution with a pH of 4 has a hydronium ion concentration of $10^{-4}M$.

How do these numbers relate to hydroxide ion concentrations? Experimental evidence shows that when the hydronium ion concentration and the hydroxide ion concentration in aqueous solution are multiplied together, the product is 10^{-14}. So, if the pH of a solution is 3, the hydronium ion concentration is $10^{-3}M$, and the hydroxide ion concentration is $10^{-14}/10^{-3}M$, which is $10^{-14-(-3)}M$, or $10^{-11}M$.

PRACTICE PROBLEMS

Find the pH of each of the following solutions.
1. The hydronium ion concentration equals:

    a) $10^{-5}M$ **b)** $10^{-12}M$ **c)** $10^{-2}M$

2. The hydroxide ion concentration equals:

    a) $10^{-4}M$ **b)** $10^{-11}M$ **c)** $10^{-8}M$

There are several easy ways to measure pH. Two common methods are shown in **Figure 14.17.**

▲ pH meters are instruments that measure the exact pH of a solution.

Figure 14.17

Measuring pH
pH is convenient because there are simple methods for measuring it in the lab or in the field.

Indicators register different colors at different pHs. A convenient way to store indicators is by soaking strips of paper in them and then drying the paper. Indicator paper is frequently sold because it is easy to use. ▶

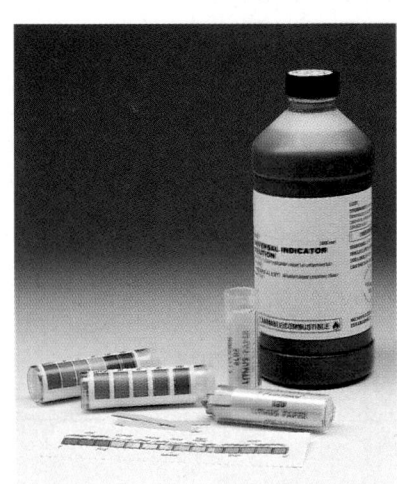

PRACTICE

Guided Practice: Have students work in class on Practice Problems 1-2. L1
Independent Practice: Your homework or classroom assignment can include the additional practice problems. L1

Answers to Practice Problems

1. a. pH = 5   **b.** pH = 12   **c.** pH = 2
2. a. pH = 10   **b.** pH = 3   **c.** pH = 6

Additional Practice Problems

1. What is the hydronium ion concentration if a solution has the following pH?
  a. pH = 6   **b.** pH = 12   **c.** pH = 2
  a. 10^{-6} *b.* 10^{-12} *c.* 10^{-2}

2. What is the hydroxide ion concentration in the solutions with the pH values in problem 1 above?
  a. 10^{-8} *b.* 10^{-2} *c.* 10^{-12}

Antacids

Many people experience a burning sensation known as heartburn after eating certain foods. Heartburn is caused by excess acid in the stomach and esophagus. Use your knowledge of acid chemistry to evaluate the effects of antacids that are commonly used to treat heartburn.

Procedure

1. Obtain four snack-size zipper-closure bags, and mark each with the name of an antacid to be tested.

2. To each of the four bags, add 5 mL of white vinegar, 10 mL of water, and enough cabbage juice indicator (probably 30 to 40 drops) to impart a distinct color.

3. Add the appropriate antacid tablet to each bag, squeeze out the excess air, and zip the bag closed. Be sure that the antacid tablet is immersed in the vinegar solution.

4. Squeeze the antacid tablets to break them into small pieces. Record your observations.

When the reactions have ceased or slowed markedly, note and record the colors and approximate pH values of the solutions. Consult the pH-color chart from the ChemLab on the next page.

Analysis

1. Describe the different ways in which the antacids reacted with the vinegar. Infer which of the antacids contain carbonates. Explain your answer.

2. Which of the antacids created the most basic final solution? Explain this answer in terms of how well the antacid works.

Interpreting the pH Scale

The pH scale is shown in **Figure 14.18.** The scale is divided into three areas. If a solution has a pH of exactly 7, the solution is said to be neutral. It is neither acidic nor basic.

Figure 14.18

The pH Scale
A pH of 7 is neutral. A pH less than 7 is acidic, and a pH greater than 7 is basic. As the pH drops from 7, the solution becomes more acidic. As pH increases from 7, the solution becomes more basic. The flowers shown are hydrangeas. Their blooms are blue when the plants are in acidic soil and pink when the soil is basic.

Neutral
↓

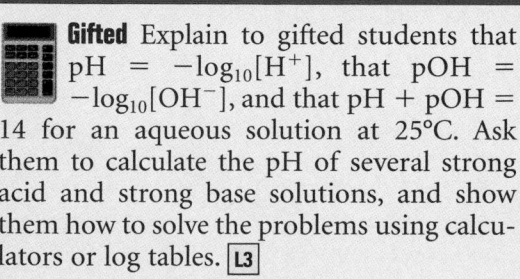

0 1 2 3 4 5 6 7 8 9 10 11 12 13 14

← More acidic ——— More basic →

14.2 **Strengths of Acids and Bases** **503**

---

miniLAB
2

miniLAB

Antacids

Purpose: To test and evaluate the effects of some antacids.
Process Skills: Observing, measuring, interpreting data

Teaching Strategies

• If the same brand of antacid is sold in different strengths, students could test all strengths to verify potency claims.
• See ChemLab steps 1 and 2 on page 504 of the Student Edition for preparation of the cabbage juice indicator.

Expected Results

Some tablets react quickly, creating a significant amount of gas. The solutions change from red-orange to a cloudy lavender. The final pH values are about 6 to 7. Others produce little reaction, although they began to dissolve, and the solutions changed from red-orange to a blue-green. The final pH values were about 9.

Analysis

1. See Expected Results. Those that reacted a lot probably contained carbonates because CO_2 gas was produced.
2. Those that produced less gas produced more basic solutions having pH values of about 9. These would react with more acid.

✓ Assessment

Portfolio: Ask students to write reports, including balanced equations for the reactions of hydrochloric acid with both magnesium carbonate and magnesium hydroxide, for the MiniLab. L1 P

Meeting Individual Needs

Gifted Explain to gifted students that pH = $-\log_{10}[H^+]$, that pOH = $-\log_{10}[OH^-]$, and that pH + pOH = 14 for an aqueous solution at 25°C. Ask them to calculate the pH of several strong acid and strong base solutions, and show them how to solve the problems using calculators or log tables. L3

Chemistry Journal

Acids in Foods

Ask students to refer to their journal entries from Section 14.1 that pertain to acids in foods. Have them evaluate the strengths of the acids and explain why strong acids are not used in foods. *Citric, phosphoric, acetic, and carbonic acids are weak. Strong acids would probably destroy the foods and would definitely make them unsafe for consumption.* L1

SMALL SCALE

Household Acids and Bases

Household Acids and Bases

TECH PREP

Time Allotment: One class period.

Objectives

Review objectives with students before they begin the Chem-Lab.

Process Skills

Observing, measuring, classifying, interpreting data, communicating, sequencing

Safety Precautions and Disposal

Caution students to use care when handling the drain cleaner. Warn students to not eat anything from the lab.
Disposal: Code J

PREPARATION

Prepare the eyewash solution by dissolving 0.6 g boric acid, H_3BO_3, per 100 mL of distilled water. Use a weak solution of laundry soap (not detergent) for the soap solution. Use approximately 1 g of drain cleaner per 100 mL of water for the drain cleaner solution.

Alternative Materials

- Encourage students to bring in other safe household liquids to be tested. **CAUTION:** *Students should bring all liquids to be tested to this classroom at the beginning of the school day.*
- If you have time, you may prepare standard pH solutions, add cabbage juice indicator to each, and let students compare their colors to the standards rather than using the chart. Prepare 1.0*M* HCl as pH 0 and pH values of 2, 4, and 6 as serial dilutions of pH 0. Use a commercial buffer for pH 7 or prepare one from a formula in the *Chemistry-Physics Handbook.* Prepare 1.0*M* NaOH and use serial dilutions for pH values of 12, 10, and 8. L1

Indicators often are used to determine the approximate pH of solutions. In this ChemLab, you will make an indicator from red cabbage and use the indicator to determine the approximate pH values of various household liquids. The cabbage juice indicator contains a molecule, anthocyanin, that accounts for the color changes.

Problem

What are the approximate pH values of various household liquids?

Objectives

- **Measure** and **compare** the pH values for various household liquids.
- **Compare** the functions of the liquids to their chemical makeup.

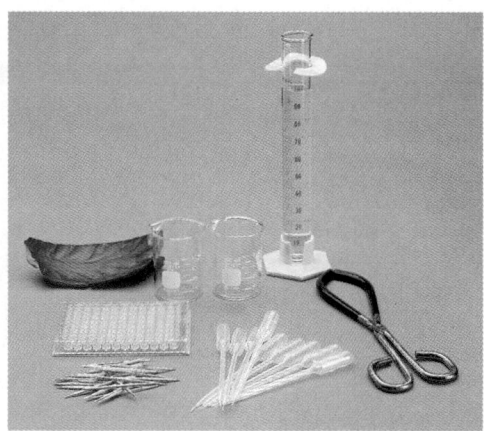

Indicator color	Relative pH
bright red	strong acid
red	medium acid
reddish purple	weak acid
purple	neutral
blue green	weak base
green	medium base
yellow	strong base

Materials

red cabbage	toothpicks
hot plate	solutions of:
beaker tongs	eyewash
100-mL beakers (2)	lemon juice
distilled water	white vinegar
microtip pipets (9)	table salt
96-well microplate	soap
piece of white paper	baking soda
100-mL graduated	borax
cylinder	drain cleaner

Safety Precautions

Use beaker tongs to handle hot beakers. Wear an apron and goggles. Some of the solutions to be tested are caustic, especially the drain cleaner. Avoid all contact with skin and eyes. If contact occurs, immediately wash with large amounts of water and notify the teacher.

PROCEDURE

1. Tear a red cabbage leaf into small pieces, and layer the pieces in a 100-mL beaker to a depth of about 2 cm. Add about 30 mL of distilled water.

2. Set the beaker on a hot plate, and heat until the water has boiled and become a deep purple color. Remove the beaker from the hot plate using beaker tongs, and allow it to cool. Pour off the cabbage juice indicator liquid into a clean beaker.

3. Set a clean microplate on a piece of white paper. Use the pipets to add 5 drops of eyewash to well H1, lemon juice to H2, white vinegar to H3, and solutions of table salt to H4, soap to H5, baking

PROCEDURE

- One hot plate per lab table should be sufficient to prepare the cabbage juice indicator for several groups.
- When students conduct the MiniLab in this section, they may wish to prepare enough indicator for both activities.

soda to H6, borax to H7, and drain cleaner to
H8. Use a clean pipet for each solution.

4. Draw the cabbage juice indicator solution into
a clean pipet, and add 5 drops to each of the
solutions in wells H1-H8. Stir the solution in
each well with a clean toothpick.

5. Looking down through the wells, note and
record the color of each solution in a data
table such as the one shown. Using the color
chart, record in the data table the approximate
pH of each of the solutions.

ANALYZE AND CONCLUDE

1. **Interpreting Data** Are food items such as
lemon juice or vinegar acidic or basic? These
solutions are either tart or sour, so what ion
probably accounts for this characteristic?

2. **Interpreting Data** Were the cleaning solutions acidic or basic? What ion is probably
involved in the cleaning process?

3. **Observing and Inferring** How can you
account for the great pH difference between
lemon juice (citric acid solution) and eyewash
(boric acid solution)?

4. **Using Variables, Constants, and Controls**
Suppose that, in addition to the solutions, you
tested a well containing pure distilled water.
What purpose would this test serve?

APPLY AND ASSESS

1. Would your indicator work well to determine
the pH of ketchup? Explain.

2. You may have noted that some shampoos are
described as pH-balanced. What do the manufacturers mean by this phrase? Why would
they do this to a soap or detergent?

3. Hypothesize about how other solutions at
home would react with the cabbage juice indicator. Explain your predictions.

DATA AND OBSERVATIONS

Solution	Color	Approximate pH
Eyewash		
Lemon juice		
White vinegar		
Table salt		
Soap		
Baking soda		
Borax		
Drain cleaner		

ANALYZE AND CONCLUDE

1. acidic; the hydronium ion or
hydrogen ion
2. basic; the hydroxide ion
3. Citric acid is a stronger acid
than boric acid.
4. Water would be a control.

APPLY AND ASSESS

1. No. The red color of the
ketchup would obscure the
indicator color.
2. The pH of the shampoo is
adjusted to be neutral or
slightly acidic, compatible
with the natural pH of skin
and hair. Soaps and detergents tend to be basic and
potentially harsh to skin and
hair.
3. Answers will vary with solutions chosen. Be sure students can justify choices.

 **Assessment**

Performance: Encourage interested students to bring other
colored vegetable, fruit, or flower
matter to school, process it as
the cabbage, and test its viabil-
ity as an acid-base indicator. [L1]

DATA AND OBSERVATIONS

Solution	Color	Approximate pH
eyewash	lavender	5-6
lemon juice	light red	2
white vinegar	red-pink	3
table salt	purple	7
soap	green	10-11
baking soda	green	8-9
borax	yellow-green	9-10
drain cleaner	yellow	13-14

Figure 14.19: Have students examine the figure to reinforce the concept that an acidic solution contains some hydroxide ions, a basic solution contains some hydronium ions, and a neutral solution contains equal concentrations of both ions. **L1**

Concept Development

Ask students how they would design a scale for the acidity or alkalinity of solutions. *They may suggest a scale from 1 to 10 with 5 being neutral, or 1 to 100 with 50 being neutral. Ask why Sörenson would pick a scale from 1 to 14 with 7 being neutral. The concentrations of both hydronium and hydroxide ions are $10^{-7}M$ at 25°C.* **L2**

GLENCOE TECHNOLOGY

 **Videotape**

MindJogger Videoquizzes for Chemistry
Acids, Bases, and pH
Have students work in groups as they play the videoquiz game to review key chapter concepts. **L1** **LEP** **COOP LEARN**

Figure 14.19
Comparison of Concentrations
Look at the hydronium ion and hydroxide ion concentrations for three common solutions. For each solution, what is the product of the two concentrations?

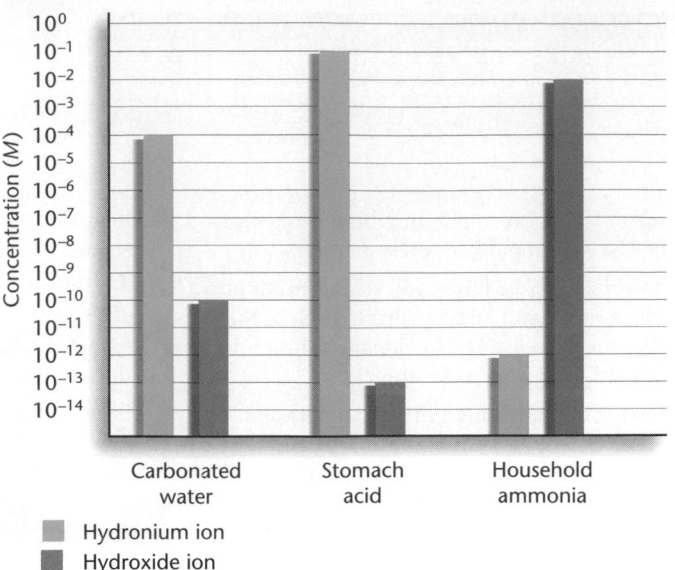

Carbonated water — Stomach acid — Household ammonia

■ Hydronium ion
■ Hydroxide ion

Figure 14.20
pH of Common Materials
The pH of an item may vary depending upon the solution concentration, the brand tested, and other variables.

As the pH decreases, the concentration of hydronium ions increases, and the concentration of hydroxide ions decreases. Every one unit decrease in the pH means a factor of 10 increase in the hydronium ion concentration. For example, a solution with a pH of 4 and a solution with a pH of 3 are both acidic because their pHs are less than 7. The solution with a pH of 3 has ten times the concentration of H_3O^+ of the solution with a pH of 4. Small changes in pH can mean big changes in hydronium ion concentration.

Similarly, as the pH increases above 7, the concentration of hydroxide ions increases, and the concentration of hydronium ions decreases. For example, suppose you have a solution with a pH of 10 and another solution with a pH of 11. The solution with a pH of 11 has ten times more hydroxide ions than the solution with a pH of 10. The solution with a pH of 11 has one-tenth the concentration of hydronium ions of the solution that has a pH of 10.

In a neutral solution, the concentration of hydroxide ions and the concentration of hydronium ions are equal. **Figure 14.19** compares hydroxide and hydronium ion concentrations for several solutions with different pHs.

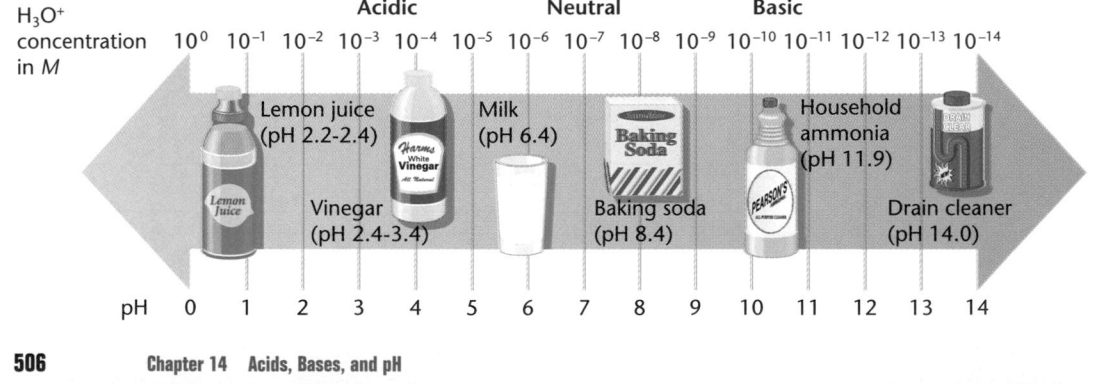

H_3O^+ concentration in M — **Acidic** — **Neutral** — **Basic**

Lemon juice (pH 2.2-2.4) · Vinegar (pH 2.4-3.4) · Milk (pH 6.4) · Baking soda (pH 8.4) · Household ammonia (pH 11.9) · Drain cleaner (pH 14.0)

Demonstration 14.5

At Home with the pH Scale

Purpose: To determine the pH values of common household products.
Materials: small samples of household products, such as pickle juice, toothpaste, liquid soap, aspirin, baking soda, vinegar, milk, salt water, antacid liquid or tablet, and baby shampoo; 2.1 mL 12*M* HCl; 1.05 g NaOH; 1 L distilled water; 4 petri dishes;

overhead projector; stirring rods; wash bottle; 14 test tubes; 5 mL universal indicator
Safety Precautions: 🧤 👓 🥼
Disposal: Code C
Procedure: In advance, prepare a set of solutions from pH 1 to pH 13. To prepare standard acid solutions, mix 2.1 mL of 12*M* HCl with 248 mL distilled H_2O to make a 0.1*M* HCl solution with a pH of 1. Add

5 mL of the 0.1*M* HCl to 45 mL of H_2O to prepare a solution with a pH of 2. Repeat using 5 mL of each previously diluted hydrochloric acid solution and 45 mL of water to prepare solutions with pHs of 3 to 6. Use boiled distilled H_2O for a material with a pH of 7. Add 1.05 g NaOH to 249 mL distilled H_2O to make a 0.1*M* NaOH solution with a pH of 13. To prepare solutions with pHs of 12 to 8, add 5 mL of

pH of Common Substances

Figure 14.20 gives the pHs of some common materials. Notice the range of pHs from the low pH of lemon juice to the high pH of drain cleaner.

Compare the pHs of vinegar and milk. Depending upon the brand and type of vinegar, the pH ranges from 2.4 to 3.4. If you have a bottle of vinegar with a pH of 3.4, the vinegar is definitely acidic. Milk, with a pH of 6.4, is also acidic, but much less so. The difference between 3.4 and 6.4 (3 pH units) may not seem like much, but remember that each unit of pH represents a power of 10. The hydronium ion concentration of the vinegar is 10^3, or 1000, times the hydronium concentration in the milk.

Using Indicators to Measure pH

Perhaps you've been to a swimming pool and have seen someone testing the water as in **Figure 14.21**. Among the tests done on swimming pool water is a test for pH by using indicators. The pH tells the condition of the water and its suitability for swimming.

The colored solutions that are used in this test are indicators that have different colors at different pHs. These dyes are not as precise as a pH meter, but they allow you to find approximate pH by comparing the color to a standard chart. **Figure 14.22** on the next page shows the colors of several indicators at different pHs. By choosing the right combination of these indicators, you can estimate pH across the entire pH range.

Figure 14.21

Checking Pool Water
A check of the pH of the water in a swimming pool is quick, easy, and inexpensive. Water that is not within a certain range of pH can harm skin or encourage the growth of bacteria.

14.2 Strengths of Acids and Bases   **507**

SECTION 14.2

Discussion

Explain that the enzymes in our bodies are proteins, and acids and bases can denature proteins. Yet the acids and bases we ingest do not destroy our enzymes. Ask how this can happen. *Our bodies contain substances called buffers that protect it against drastic changes in pH.* Don't go into detail about buffers at this time. Mention that they will be studied in the next chapter.

3 ASSESS

Check for Understanding

- Check for understanding of the section material by having students answer the Section Review questions. L1
- Ask students to calculate the pH values for each of the following solutions: 0.10*M* KOH, *13;* 0.000010*M* HI, *5.* L1

Reteach

Discussion: Explain that the bond between the hydrogen atom and the remainder of the molecule for the oxyacids of bromine is strongest in hypobromous acid, HBrO; weaker in bromous acid, $HBrO_2$; weaker yet in bromic acid, $HBrO_3$; and weakest in perbromic acid, $HBrO_4$. Ask students which of the acids is the strongest and why. *perbromic The weakest bond means that the hydrogen ions are most readily lost to form hydronium ions.*

---

the previous solution to 45 mL H_2O.
   CAUTION: *Both low and high pH solutions cause injury to skin and eyes.* Add 5 mL of each solution and 1 drop of universal indicator into appropriately labeled test tubes. Use this colored set as pH standards. Place several small petri dishes on an overhead projector. Add a small amount of distilled water to each. Place a small amount of a substance to be tested

in the dish and stir to mix. Add 1 drop of universal indicator to each of the petri dishes. Using the pH standards prepared earlier, have students group the products into acid and base categories and arrange the substances within each group in order of increasing pH.
Results: Student results will depend upon the samples tested.
Analysis: Ask these questions.

1. Are most household cleansers acidic or basic? *basic*
2. What is the pH range for acids? *<7*
3. What is the pH range for bases? *>7*

✔**Assessment**

Portfolio: Have students write a summary and add it to their portfolios. L1 P

508

Extension

Ask students what would happen if a weight-lifting contest were held among the five strongest persons in the world in order to determine who was strongest by using a single 100-pound weight. *Each of the contestants would lift the 100-pound weight with ease, and it would be impossible to determine who was strongest.* Point out that the use of only the minimal weight "leveled" the strengths of the contestants. Explain that water has a leveling effect on the strengths of strong acids and strong bases. The polarities of water molecules enable them to pull virtually every hydrogen ion or hydroxide ion from the dissolved acid or base, rendering them 100% ionized or dissociated and thus equally strong. Explain that in solvents other than water, this may not be the case. [L1]

4 CLOSE

Discussion

Ask students how they would respond to this comment. "The treatment of the acid rain problem in this area has been largely ineffective because the pH of the rain has been raised only from 4 to 6." *The comment is misleading. pH 6 rain is 100 times less acidic than pH 4 rain.* [L1]

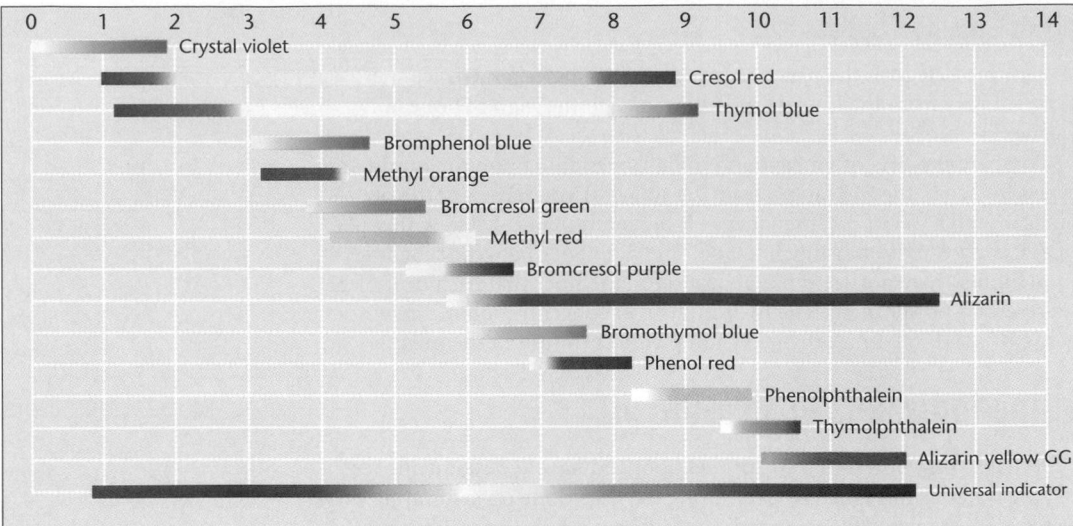

Figure 14.22

Indicators

Notice the colors of these different acid-base indicators over the pH scale. Some are better indicators at low pH, others at moderate pH, and yet others at high pH.

Connecting Ideas

You may have noticed that this chapter, which introduces the properties of acids and bases, follows a chapter that focuses on the properties of water. Although acids and bases have different properties, the common link between the two is the role that water plays in making the chemistry of acids and bases happen.

Except for those cases where ionic hydroxides dissolve in water to form free hydroxide ions, the behavior and strength of acids and bases are due to their ability to cause hydrogen ions to move to water molecules from an acid or from water molecules to a base. This movement of hydrogen ions between particles in solution can be used to demonstrate different types of acid-base reactions.

SECTION REVIEW

Understanding Concepts

1. Classify the following acids and bases as strong or weak: NH_3, KOH, HBr, $HCHO_2$, HNO_2, $Ca(OH)_2$.

2. What distinguishes a strong acid or base from a weak acid or base? Are there more strong acids and bases or weak acids and bases?

3. Other than water molecules, what particle has the highest concentration in an aqueous solution of ammonia? What has the lowest concentration? Why?

Thinking Critically

4. **Interpreting Data** The pH of normal rain is about 5.5 due to dissolved CO_2. Consider a sample of rainfall with a pH of 3.5. How does the hydronium ion concentration in these two rain samples compare?

Applying Chemistry

5. **Finding pH** A solution of unknown pH gives a pink color with phenolphthalein indicator and a blue-gray color with universal indicator. Use **Figure 14.22** to estimate the pH of the unknown solution.

SECTION REVIEW

1. NH_3 (weak base), KOH (strong base), HBr (strong acid), $HCHO_2$ (weak acid), HNO_2 (weak acid), $Ca(OH)_2$ (strong base)

2. The degree of dissociation or ionization distinguishes strong (completely forms ions) and weak (does not completely ionize or dissociate). Most acids and bases are weak.

3. NH_3 is in highest concentration. Because NH_3 is a weak base, there is a low concentration of OH^- and NH_4^+.

4. $5.5 - 3.5 = 2.0$ pH units difference. The more acidic solution has 100 times the hydronium ion concentration.

5. pH is approximately 9.

REVIEWING MAIN IDEAS

14.1 Acids and Bases

- Acids have a sour taste, turn litmus red, and react with active metals, carbonates, and bases.

- Bases have a bitter taste, a slippery feel, turn litmus blue, and react with acids.

- Acids ionize by losing hydrogen ions to water molecules to form hydronium ions, H_3O^+.

- Bases form hydroxide ions, OH^-, when dissolved in water. Ionic hydroxides are bases because their ions dissociate. Covalent bases ionize by hydrogen ion transfer from water molecules to the base.

- Acidic anhydrides are nonmetallic oxides that react with water to form acids. Basic anhydrides are metallic oxides that react with water to form bases.

- Water plays a central role in the chemistry of acids and bases.

14.2 Strengths of Acids and Bases

- Strong acids and bases completely dissociate or ionize.

- Most acids and bases are weak. Weak acids and bases form few ions.

- The structure of acids and covalent bases determines their strength.

- The pH scale is a convenient way to compare the acidity and basicity of solutions.

Key Terms

For each of the following terms, write a sentence that shows your understanding of its meaning.

acid	ionization
acidic anhydride	pH
acidic hydrogen	strong acid
base	strong base
basic anhydride	weak acid
hydronium ion	weak base

● UNDERSTANDING CONCEPTS

1. Show that solid magnesium hydroxide is a base when it is dissolved in water.

2. You test several solutions and find that they have pHs of 7.6, 9.8, 4.5, 2.3, 4.0, and 11.6. Which solution is the closest to being neutral? Assuming the concentrations are the same, which solution is the weakest base?

3. Classify each of the following as an acid, a base, or neither when mixed with water.

   a) NaOH   d) CO_2
   b) $H_2C_2O_4$ e) CH_4
   c) NH_3

4. Write the chemical equation that shows the ionization of the acid HBr in water.

5. Write balanced equations that show what happens when NH_3 and KOH are placed separately in water. How do these reactions differ?

6. Explain the role of water in the ionization of an acid. Use the general equation to illustrate your explanation.

7. Explain the role of water in the ionization of a covalent base. Use the general equation for the reaction to illustrate your explanation.

UNDERSTANDING CONCEPTS

1. $Mg(OH)_2(s) \rightarrow Mg^{2+}(aq) + 2OH^-(aq)$; it produces OH^- by dissociation when dissolved in water.

2. 7.6, 7.6

3. **a.** base **b.** acid **c.** base **d.** acid **e.** neither

4. $HBr(aq) + H_2O(l) \rightarrow H_3O^+(aq) + Br^-(aq)$, or $HBr(aq) \rightarrow H^+(aq) + Br^-(aq)$

5. $KOH(s) \rightarrow K^+(aq) + OH^-(aq)$; KOH dissociates. $NH_3(aq) + H_2O(l) \rightarrow NH_4^+(aq) + OH^-(aq)$; the process is ionization, not dissociation.

6. $HA(aq) + H_2O(l) \rightarrow H_3O^+(aq) + A^-(aq)$; the acid acts as an H^+ donor. Because of the strong H bonds between the acid and the water, an H^+ is transferred to a water molecule, forming H_3O^+.

7. $B(aq) + H_2O(l) \rightarrow BH^+(aq) + OH^-$; the base acts as an H^+ acceptor. Because of the strong H bonds between the base and the water, an H^+ is transferred to the base, B, from the H_2O, forming OH^-.

- Review the Reviewing Main Ideas statements and Key Terms with your students.
- Complete solutions to Chapter Review problems can be found in the *Problems and Solutions Manual.*

✔Assessment

Portfolio: Review the portfolio options that are provided throughout the chapter. Encourage students to select one product that demonstrates their best work for the chapter. Have students explain what they have learned and why they chose this example for placement into their portfolios. **P**

Additional portfolio options can be found in the following **Teacher Classroom Resources:**
Tech Prep Applications
Consumer Chemistry
Chemistry and Industry
Applying Scientific Methods in Chemistry

8. Polyprotic acids are acids that have more than one acidic hydrogen per acid molecule. *Poly-* means "many." HI is not polyprotic because it has only one acidic hydrogen.

9. $Zn(s) + 2HCl(aq) \rightarrow ZnCl_2(aq) + H_2(g)$

10. $Na_2CO_3(s) + 2HNO_3(aq) \rightarrow 2NaNO_3(aq) + CO_2(g) + H_2O(l)$

11. $HClO_4(aq) + KOH(aq) \rightarrow H_2O(l) + KClO_4(aq)$

12. $Mg(OH)_2(s) + 2HC_2H_3O_2(aq) \rightarrow 2H_2O(l) + Mg(C_2H_3O_2)_2(aq)$

13. KOH, a base

14. H_2SO_4, sulfuric acid

15. **a.** H_2SO_4, diprotic **b.** $HClO_4$, monoprotic **c.** H_3PO_4, triprotic **d.** HF, monoprotic **e.** $HC_2H_3O_2$, monoprotic

16. **a.** NH_3, weak **b.** LiOH, strong **c.** $Ca(OH)_2$, strong **d.** $Ba(OH)_2$, strong

17. weak base

18. More substance dissolved means that more units are dissociated, producing more OH^- ions.

19. The molecular structure of the acid determines its ability to interact with water. The stronger the interaction, the more complete the ionization and the stronger the acid.

20. **a.** acidic **b.** basic **c.** basic **d.** neutral

21. Indicators have different colors in solutions of different pH. If you know the color behavior of an indicator, you can use it to estimate pH.

22. The H_3O^+ concentration decreases by a factor of 10. The solution becomes less acidic.

8. What are polyprotic acids? From this definition, infer what the prefix *poly-* means. Is HI a polyprotic acid? Explain.

9. Write the balanced equation for the reaction when hydrochloric acid reacts with zinc.

10. Write the balanced equation for the reaction of nitric acid with sodium carbonate.

11. The reaction of perchloric acid with potassium hydroxide produces water and potassium perchlorate. Write an equation for this reaction.

12. You want to prepare magnesium acetate by an acid-base reaction. Write a reaction that will do this.

13. Give the formula of the product formed when potassium oxide reacts with water. Is the product an acid or a base?

14. What product forms when sulfur trioxide reacts with water? Is the product an acid or is it a base?

15. Write the formula of each of the following acids and identify each as a diprotic, a triprotic, or a monoprotic acid.
    a) sulfuric acid
    b) perchloric acid
    c) phosphoric acid
    d) hydrofluoric acid
    e) acetic acid

16. Give the formula for each of the following bases, and identify each as a strong base or a weak base.
    a) ammonia
    b) lithium hydroxide
    c) calcium hydroxide
    d) barium hydroxide

17. Methylamine, CH_3NH_2, is a gas that reacts with water to produce relatively few hydroxide ions and methylammonium ions, $CH_3NH_3^+$. Classify methylamine according to whether it is a strong acid, a weak acid, a strong base, or a weak base.

18. Why are soluble ionic hydroxides always strong bases?

19. What is the distinguishing factor that differentiates strong acids from weak acids?

20. Indicate whether each of the following pH values represents an acidic, a neutral, or a basic solution.
    a) 3.5 **c)** 8.8
    b) 11.5 **d)** 7.0

21. Explain how acid-base indicators are useful in determining pH.

22. If the pH of a solution increases from 5.0 to 6.0, what has happened to the hydronium ion concentration?

APPLYING CONCEPTS

23. Why is natural rainfall acidic?

24. Hydrogen sulfide, H_2S, reacts slightly with water to form relatively few hydronium ions and hydrogen sulfide, HS^-, ions. Classify hydrogen sulfide.

25. Blood has a pH of 7.4. Milk of magnesia, a common antacid for upset stomachs, has a pH of 10.4. What is the difference in pH between the two bases? Compare the hydroxide ion concentrations of the two solutions.

26. Lemon juice has a pH of about 2. Should you cut a lemon on a marble surface? Explain.

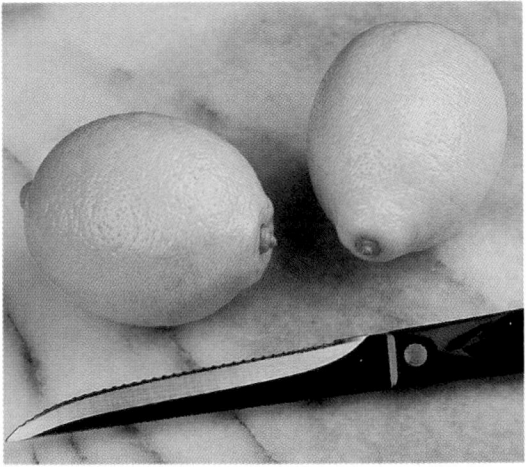

APPLYING CONCEPTS

23. CO_2 in the atmosphere dissolves in rainwater, forming dilute H_2CO_3.

24. a weak, diprotic acid

25. 3.0; milk of magnesia has 1000 times more OH^- ions.

26. Lemon juice is an acid. It will react with the marble, $CaCO_3$.

27. A substance that can act as either an acid or a base is called an amphoteric compound. Use a solution of nitric acid and a solution of ammonia as examples to show that water is an amphoteric compound. Include two balanced chemical equations in your explanation.

28. An aqueous ammonia solution is sometimes called ammonium hydroxide. Why?

29. Why does acid rain dissolve marble statues? Write the reaction that occurs when rain containing sulfuric acid reacts with a statue made of marble, $CaCO_3$. Calcium sulfate is one of the products formed.

30. One of the most acidic rainfalls ever recorded occurred in West Virginia. The pH of the rain was about 3. Approximately how much more concentrated is the hydronium ion in this acid rain than that in rain with a pH of 5?

31. Finland, a country that has been hit hard by acid rain, has used lime, CaO, to try to return life to acidified lakes. Why was lime used?

32. A sour taste is related to acids. A bitter taste is related to bases. How is a salty taste related to acids and bases?

Everyday Chemistry

33. Some people rinse their freshly shampooed hair in diluted lemon juice or vinegar. Why might doing this be beneficial to your hair?

Biology Connection

34. After exercise, muscles sometimes become sore because of a buildup of lactic acid. How might this buildup temporarily affect the acid-base balance in the blood?

Chemistry and Society

35. Why might city dwellers with respiratory diseases such as asthma or emphysema be advised to stay inside on hot days?

Chemistry and Technology

36. From what you know about reactions that acids will undergo, infer how sulfuric acid might be used to remove unwanted materials from ores.

● THINKING CRITICALLY

Applying Concepts

37. What is the molarity of HCl, H_3O^+, and Cl^- in a 0.50M solution of HCl?

Relating Cause and Effect

38. Estimate the molarity of $HC_2H_3O_2$, H_3O^+, and $C_2H_3O_2^-$ in 0.50M $HC_2H_3O_2$.

Observing and Inferring

39. Examine the formulas of several polyprotic acids and compare them to the formulas of several monoprotic acids. What element is always present in monoprotic acids? What two elements are usually present in polyprotic acids?

Relating Cause and Effect

40. Baking powder, when mixed with water, produces bubbles that make a cake rise. From what you know about the properties of acids, infer what types of compounds are contained in baking powder and what gas is contained in the bubbles released by the reaction.

Applying Concepts

41. Hard water deposits around sinks may be composed of calcium carbonate and magnesium carbonate. You can buy commercial cleaners to remove these insoluble compounds, or you could use something from your kitchen. What might you try?

Chapter 14   Review   **511**

Program Resources

Chapter Review and Assessment, pp. 83-88 [L1]

Alternate Assessment in the Science Classroom [L1]

Computer Test Bank, Chapter 14 [L1]

Problems and Solutions Manual, Chapter 14 [L1]

Supplemental Practice Problems, Chapter 14 [L1]

Performance Assessment in the Science Classroom [L1]

27. $HNO_3(aq) + H_2O(l) \rightarrow H_3O^+(aq) + NO_3^-(aq)$: Water accepts a hydrogen ion and acts as a base. $NH_3(g) + H_2O(l) \rightarrow NH_4^+(aq) + OH^-(aq)$: Water loses a hydrogen ion and acts as an acid.

28. The solution contains both ammonium and hydroxide ions.

29. Sulfuric acid converts insoluble calcium carbonate to more soluble calcium sulfate. This causes the statue to dissolve more easily. $CaCO_3(s) + H_2SO_4(aq) \rightarrow CaSO_4(aq) + CO_2(g) + H_2O(l)$

30. 100 times more concentrated

31. Lime is CaO, a basic anhydride.

32. Salt can be formed by an acid-base reaction.

33. Both are acidic and will tighten the cuticle, making hair shiny.

34. Blood carries away impurities from the muscles. Having to remove lactic acid might cause a temporary decrease in blood pH.

35. Photochemical smog is worse on sunny days. Also, cities have more industry and cars and therefore more pollution.

36. Student answers might include that the acid may react with any free metal in the ore, separating it from the other materials. If carbonates, such as limestone, are present, acid will react with them, separating them from the ore.

THINKING CRITICALLY

37. HCl, 0; H_3O^+, 0.50M; Cl^-, 0.50M

38. Because $HC_2H_3O_2$ ionizes so little, its concentration is approximately 0.50M. The concentrations of H_3O^+ and $C_2H_3O_2^-$ are equal, but much less than 0.50M.

39. H; H and O

40. an acid and a carbonate; carbon dioxide

CUMULATIVE REVIEW

41. an acid, such as lemon juice or vinegar, that would react with the carbonate

42. a. base **b.** acid **c.** acid **d.** acid

43. The acidic hydrogen(s) of an acid are those hydrogens that may be transferred to water. Three of the hydrogens in acetic acid are bonded to carbon. These bonds are nonpolar. The acidic hydrogen is connected to oxygen and is a polar covalent bond that H-bonds with water.

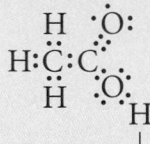

Acidic hydrogen

44. approximately 7

45. The surface area of powered zinc is so large, the reaction might occur too quickly to be safe.

46. Answers will vary, but common ingredients include pain relievers such as aspirin or acetaminophen and flavorings including sweeteners. Pain relievers might relieve symptoms without getting rid of the problem, so less antacid might be included. Other ingredients are sometimes bulky and take up room in the tablet.

CUMULATIVE REVIEW

47. a. CaO(s) + H₂O(l) → Ca(OH)₂(aq); synthesis **b.** H₂SO₄(aq) + Zn(s) → ZnSO₄(aq) + H₂(g); single displacement **c.** HNO₃(aq) + KOH(aq) → KNO₃(aq) + H₂O(l); double displacement

48. a. 100.1 g **b.** 70.9 g **c.** 134 g

49. 200 g NaOH

50. Solution *a* has more solute particles per volume of solution.

Interpreting Data

42. Identify the first compound in the following reactions as an acid or a base.

a) $C_5H_5N + H_2O \rightarrow C_5H_5NH^+ + OH^-$
b) $HClO_3 + H_2O \rightarrow H_3O^+ + ClO_3^-$
c) $HCHO + H_2O \rightarrow H_3O^+ + CHO^-$
d) $C_6H_5SH + H_2O \rightarrow H_3O^+ + C_6H_5S^-$

Interpreting Chemical Structures

43. What is an acidic hydrogen? Draw the dot structure for acetic acid, and use it to explain why only one of the four hydrogens is acidic.

Observing and Inferring

44. ChemLab What would you expect the pH of an aqueous solution of the salt calcium chloride to be?

Forming a Hypothesis

45. MiniLab 1 Small pieces of zinc must be added to the acids instead of powdered zinc. Hypothesize why powdered zinc can't be used safely.

Observing and Inferring

46. MiniLab 2 Examine antacid ingredient lists. Some antacid tablets contain ingredients other than the antacid itself. What are some of these ingredients? Infer why the presence of these ingredients might affect the amount of antacid contained in an antacid tablet.

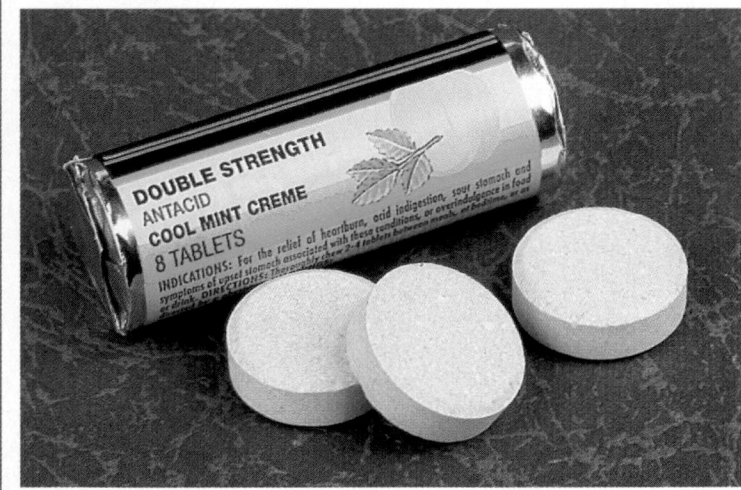

● CUMULATIVE REVIEW

47. Write a balanced equation for each of the following reactions involving acids and bases, and classify each reaction as one of the five general reaction types. (Chapter 6)

a) the reaction of water and lime to form calcium hydroxide
b) the reaction of sulfuric acid with zinc
c) the reaction of nitric acid with aqueous potassium hydroxide

48. What is the molar mass of each of the following substances? (Chapter 12)

a) calcium carbonate
b) chlorine gas
c) ammonium sulfate

49. How many grams of sodium hydroxide are needed to prepare 2.50 L of 2.00*M* NaOH? (Chapter 13)

50. Which solution has the lower freezing point? Explain. (Chapter 13)

a) 0.70 mol of NaCl dissolved in 3.00 L of solution
b) 0.15 mol of KNO₃ in 1.10 L of solution

1.

Name of Substance	Formula of Substance	Acid, Base, or Neutral	Strong or Weak Electrolyte	pH: >7, <7, or 7	Litmus Color	Major Particle(s) in Solution
Acetic acid	HC₂H₃O₂	Acid	Weak	<7	Red	HC₂H₃O₂
Hydrochloric acid	HCl	Acid	Strong	<7	Red	H⁺, Cl⁻
Nitric acid	HNO₃	Acid	Strong	<7	Red	H⁺, NO₃⁻
Ammonia	NH₃	Base	Weak	>7	Blue	NH₃
Lithium hydroxide	LiOH	Base	Strong	>7	Blue	Li⁺, OH⁻

Skill Review

1. **Making and Using Tables** Complete each row of the table below using the information provided. For some cells, there is more than one possible correct answer. Assume that all substances are dissolved in pure water, and the solutions all have a molarity of 0.10*M*.

2. **Designing an Experiment** Design an experiment to determine the best material to use to store lemon slices. Explain your conclusions.

3. **Sequencing** Create a bar graph that shows solutions with the following pHs in order from the most acidic to the most basic: 7.6, 9.8, 4.5, 2.3, 4.0, 11.6.

Writing in Chemistry

4. Find out about how sulfuric acid is made and its many commercial uses. Write an advertisement for sulfuric acid, including any research information you found out. Include graphics that illustrate things such as uses of the acid or a graph that indicates the amount of production by different countries.

Problem Solving

5. A student buys some distilled water. She measures the pH of the water and finds it to be 6.0.

She then boils the water, fills a container to the top with the hot water, and puts a lid on the container. When the water is at room temperature, the pH is 7.0. The student pours half the water into one container and stirs it for five minutes. She blows air through a straw into the other sample of water. She measures the pH of the samples of water and finds it to be 6.0 in both samples. Write an entry in your Science Journal that explains this series of pH measurements.

Projects

6. Check ingredient lists on an assortment of household products to find the names or formulas for several different acids and bases. For each, explain how the acid or base helps the product achieve its intended purpose. Prepare a presentation of your findings. Possible methods of presentation include a poster or a display that includes information about the acid or base along with the product that contains it. Your presentation should include the name, formula, and use of each acid or base.

7. Write a brief paper about citric acid, $H_3C_6H_5O_7$. Your paper should include information such as its discovery, its sources, and possible uses. Include the structure of the molecule, and indicate the acidic hydrogens.

Name of Substance	Formula of Substance	Acid, Base, or Neutral	Strong or Weak Electrolyte	pH: >7, <7, or 7	Litmus Color	Major Particle(s) in Solution
Acetic acid						
	HCl					
		Acid	Strong			
Ammonia						
	LiOH					
		Base	Strong			
			Weak	<7		
		Neutral				Na$^+$, Cl$^-$
						H$^+$, NO$_3^-$
Phosphoric acid						

Name of Substance	Formula of Substance	Acid, Base, or Neutral	Strong or Weak Electrolyte	pH: >7, <7, or 7	Litmus Color	Major Particle(s) in Solution
Sodium hydroxide	NaOH	Base	Strong	>7	Blue	Na$^+$, OH$^-$
Formic acid	HCHO$_2$	Acid	Weak	<7	Red	HCHO$_2$
Sodium chloride	NaCl	Neutral	Strong	7	Unchanged	Na$^+$, Cl$^-$
Nitric acid	HNO$_3$	Acid	Strong	<7	Red	H$^+$, NO$_3^-$
Phosphoric Acid	H$_3$PO$_4$	Acid	Weak	<7	Red	H$_3$PO$_4$

1. See the table at the bottom of pages 512 and 513. Those rows that are shaded contain sample answers. Those answers may vary.

2. Use different materials to wrap the lemon slices and observe what happens to the wrap. Students will probably notice that plastic works well; lemon juice is acidic and reacts with foil.

3. Graphs should reflect the following scale: acidic 2.3, 4.0, 4.5, 7.6, 9.8, 11.6 basic.

4. Students may get some ideas from the Chemistry and Technology feature. The ad should include its manufacture—both process and amount produced by different countries—and uses.

5. The initial pH is due to carbonic acid, formed when CO_2 from the air dissolves in the water. CO_2 is less soluble in hot water, so the gas is released, and the pH indicates a more neutral solution. The pH again drops when CO_2 is dissolved from the air or exhaled.

6. Students may also wish to compare results and find items that contain the same acid or base. From that information, they might be able to infer the purpose of the acid or base in the products.

7. Student information will vary.

Acidic hydrogens

513

Chapter 15 Organizer

Section	Objectives	Activities/Features

15.1
Acid and Base Reactions
(4 days)

1. **Distinguish** the overall, ionic, and net ionic equations for an acid-base reaction.
2. **Classify** acids and bases using the hydrogen transfer definition.
3. **Predict** and **explain** the final results of an acid-base reaction.

MiniLab 1: Acidic, Basic, or Neutral?, p. 518
How It Works: Taste, p. 519
Earth Science Connection: Cave Formation, p. 524
Discovery Demo: 15.1 Buffer Preparation, pp. 514-515
Demonstrations: 15.2 An Acid-Base Reaction, pp. 518-519; 15.3 Water Not Required, pp. 526-527

15.2
Applications of Acid-Base Reactions
(5 days)

4. **Evaluate** the importance of a buffer in controlling pH.
5. **Design** strategies for doing acid-base titrations, and calculate results from titration data.

MiniLab 2: What does a buffer do?, p. 532
Everyday Chemistry: Hiccups, p. 534
Chemistry and Society: The Development of Artificial Blood, p. 537
ChemLab: Titration of Vinegar, pp. 542-543
How It Works: Indicators, p. 545
Demonstrations: 15.4 The Buffering Effect, pp. 532-533; 15.5 A Salt, pp. 538-539; 15.6 Indicators in Nature, pp. 540-541

Activity Materials

ChemLab (pages 542-543)
24-well microplate, vinegar (several different brands), microtip pipets, marker, sodium hydroxide, distilled water, phenolphthalein indicator, toothpicks, wash bottle, white paper

MiniLab 1 (page 518)
microtip pipets (6), 24-well microplate, sodium acetate, potassium nitrate, ammonium chloride, sodium carbonate, sodium chloride, aluminum sulfate, dropper, bromothymol blue indicator, toothpicks, white paper, distilled water

MiniLab 2 (page 532)
test tubes (4), marker, sodium chloride, distilled water, 10-mL graduated cylinders (2), droppers (4), methyl orange indicator, hydrochloric acid, 0.100M sodium hydroxide, phenolphthalein indicator

Demonstrations For a complete list of materials for the demonstrations in this chapter, see pages 514, 518, 526, 532, 538, and 540.

KEY TO TEACHING STRATEGIES
The following designations will help you decide which activities are appropriate for your students.

L1 Level 1 activities should be within the ability range of all students.

L2 Level 2 activities should be within the ability range of the average to above-average student.

L3 Level 3 activities are designed for the ability range of above-average students.

LEP LEP activities should be within the ability range of Limited English Proficiency students.

COOP LEARN Cooperative Learning activities are designed for small group work.

P These strategies represent student products that can be placed into a best-work portfolio.

Acids and Bases React

Teacher Classroom Resources

STUDENT MASTERS

Study Guide: pp. 59-60
Laboratory Manual: 15.1 Hydrolysis of Salts, pp. 135-140
ChemLab and MiniLab Worksheets: p. 98

TEACHING AIDS

Section Focus Transparency 30
Section Focus Transparency Master: p. 30
Basic Concepts Transparency 41
Basic Concepts Transparency Master: pp. 81-82
Lesson Plans: p. 30

Study Guide: pp. 61-62
Critical Thinking/Problem Solving: pp. 16-17
Laboratory Manual: 15.2 Acid/Base Titration, pp. 141-146
ChemLab and MiniLab Worksheets: pp. 95-97, 99

Section Focus Transparency 31
Section Focus Transparency Master: p. 31
Basic Concepts Transparency 42
Basic Concepts Transparency Master: pp. 83-84
Problem Solving Transparency 18
Problem Solving Transparency Master: pp. 35-36
Lesson Plans: p. 31

ASSESSMENT RESOURCES

Chapter Review and Assessment: pp. 89-94
Alternate Assessment in the Science Classroom
Computer Test Bank: Chapter 15
Performance Assessment in the Science Classroom
Problems and Solutions Manual: Chapter 15
Supplemental Practice Problems: Chapter 15

CHAPTER RESOURCES

Spanish Resources: Chapter 15
English/Spanish Audiocassettes: Chapter 15
Cooperative Learning in the Science Classroom
Lab and Safety Skills in the Science Classroom

GLENCOE TECHNOLOGY

Software
Mastering Concepts in Chemistry: *Unit 14* Lesson 3 Salts, Neutralization; *Unit 15* Lesson 2 Titration

Videotape
MindJogger Videoquizzes, Chapter 15

Videodisc
Science and Technology Videodisc Series: Disc 2 Chemistry *Treating Acid Lakes*
Chemistry: Concepts and Applications, Demonstrations, Videos, and Animations

CD-ROM Multimedia System
Chemistry: Concepts and Applications, Same as for Videodisc, plus Interactive Exploration and Major Simulation, *The Periodic Table*

TECH PREP

The following Glencoe resources provide opportunities for integrating science and technology.

Student Edition: Earth Science Connection, p. 524; Chemistry and Society, p. 537

Teacher Wraparound Edition: Introducing the Chapter, p. 515; Chemistry Journal, p. 535; Integrating the Sciences, p. 536

Teacher Classroom Resources: Tech Prep Applications, pp. 23-24; Consumer Chemistry, pp. 29-30

15 Acids and Bases React

Chapter Overview

This chapter focuses on the reactions of acids and bases. Three categories of significant acid-base reactions are introduced. The applications of buffers and several common acid-base reactions are discussed.

Theme Connection

Equilibrium and Change This chapter shows how the *equilibrium* involved in acidic and basic solutions is changed when the acids and bases react. The actual amount of *change* is determined by the strength of the acid and the strength of the base.

✔ Assessment

Choose assessment strategies from the following pages to evaluate the progress of your students.

Assess, pp. 517, 528, 529, 546
MiniLabs, pp. 518, 532
ChemLab, p. 542
Portfolio, pp. 517, 527, 528
Chapter Review, p. 547

GLENCOE TECHNOLOGY

⊙ Videodisc

Chemistry: Concepts and Applications
Using a Buffer
Disc 2, Side 1, Ch. 16

Show this video to introduce the concept of buffer systems.
L1 LEP

⊙ CD-Rom

Chemistry: Concepts and Applications
Using a Buffer
Use this video to help introduce the concept of buffer systems. L1 LEP

$$HCl(aq) + NaHCO_3(aq) \rightarrow NaCl(aq) + H_2CO_3(aq)$$
$$H_2CO_3(aq) \rightarrow H_2O(l) + CO_2(g)$$

514

Discovery Demo 15.1

Using a Buffer

Purpose: To prepare a buffer system.
Materials: 1 mL glacial acetic acid, three 100-mL beakers, 60 mL distilled water, 4 g sodium acetate, 3 large test tubes, 10 drops universal indicator, 3 stirring rods, 50 mL 4.0M NaOH (16 g in 100 mL water), dropper

Safety Precautions:

Disposal: Codes C, D; check pH on all solutions. Adjust pH to between 3 and 10, dilute, and pour down the drain.
Procedure: A buffer system can absorb moderate amounts of acid or base without a significant change in pH. To prepare a system, add 1 mL of glacial acetic acid to 49 mL distilled H_2O. **CAUTION:** *Acid is corrosive.* Divide the acid between two 100-mL beakers. Dissolve 4 g of sodium acetate in the diluted acid in one of the

CHAPTER
15

Acids and Bases React

Suppose you've enjoyed too much good food. You have heartburn, acid indigestion, acid stomach—whatever you call it—and you need some relief. Your stomach produces hydrochloric acid along with enzymes to digest food. But sometimes, particularly after a big meal, your stomach produces a bit too much acid.

An acid-base reaction is what you need. You know that bases react with acids to diminish, or neutralize, their properties. In this case, the property of interest is the irritating action of too much acid on your stomach lining. The bases that compose commercial products used to treat acid stomachs are called anti-acids—*antacids* for short—just for this reason.

This problem, and many others, require you to use the reaction of an acid with a base. Whether you're trying to neutralize stomach acid or predict the effects of acid rain on a forest, lake, or sculpture, the primary reactions are acid-base reactions.

Chemistry Around You

Take a look at the counter where antacids are sold. Obviously, you're not the only person with an acid stomach who is looking for a base. Americans purchase more than $250 million worth of antacids every year. That's a lot of stomach acid being neutralized. In this chapter, you'll learn about acid-base reactions and their importance.

● **Concept Check**

Review the following concepts before studying this chapter.
Chapter 12: the mole concept; using the factor label method
Chapter 13: solution concentration using molarity
Chapter 14: properties and definitions of acids and bases

● **Chapter Preview**

15.1 Acid and Base Reactions

15.2 Applications of Acid-Base Reactions

Laboratory Activities
ChemLab
Titration of Vinegar

MiniLabs
1. Acidic, Basic, or Neutral?
2. What does a buffer do?

515

Introducing the Chapter

TECH PREP Have students examine the labels on a variety of commercial antacids. Using their knowledge from the last chapter, have them write down ingredients and identify the base(s) used by the different brands. Have students share their results and attempt to group the various antacids with respect to the type of base used. Although students may not have much experience with personal use of antacids, they will certainly recognize these products from their availability and advertisements. [L1]

COOP LEARN

Using the Illustrations The photo of the common fizzing antacid tablet serves as the introduction to the reactions of acids and bases. More than 8000 different brands of antacids are available in the United States to treat acid indigestion. It's no wonder that the student in the photo looks a bit confused. Although there are many different antacids, they all react with acids to neutralize their properties. Survey the class to determine what antacids are used by the students and their families and why they favor one brand over another.

Revealing Misconceptions

Students will confuse the fact that acids and bases react to neutralize each other's properties, but acids and bases do not necessarily react with each other to give neutral (pH = 7) solutions.

beakers. Stir and label it "buffer." Label three large test tubes *acid, buffer,* and *water*. Add equal amounts of the diluted acetic acid, the buffer, and distilled water separately to the respectively labeled test tubes. Add a few drops of universal indicator to each test tube. Place a stirring rod in each test tube. Add a filled dropper of 4*M* NaOH to each solution. **CAUTION:** *NaOH is caustic.* Stir and observe the color. Fill the dropper and repeat many times, and record your observations.
Results: The buffered test tube does not change color as rapidly as do the other two test tubes.
Analysis: Ask these questions.
1. Which of the three test tubes served as a control in this experiment? *The control was the distilled water.*
2. How was the definition of a buffer ver-

ified? *The indicator's color did not change as several measures of base were added.*
3. Do you think that buffered aspirin would be better for your stomach than unbuffered aspirin? *Buffered aspirin would compensate for the acidic nature of aspirin.*

515

Key Concepts

Strong acid-strong base, strong acid-weak base, weak acid-strong base, and weak acid-weak base neutralization reactions are examined and represented by molecular, ionic, and net ionic equations. The hydrogen ion-transfer definitions (Brønsted-Lowry) of acids and bases are also explained.

Planning Ahead

For the Demonstrations Check to be sure the electrical conductivity apparatus is available and in working order for Demonstration 15.2.

1 FOCUS

Focus Transparency

Display the **Section Focus Transparency** for Section 15.1 to show acid-base strengths as a tug of war. L1 LEP

2 TEACH

Visual Learning

Figure 15.1: Write the generic word and formula equations for an acid-base reaction so that students may see the pattern: Acid + Base → A Salt + Water; HX(aq) + YOH(aq) → YX(s or aq) + H₂O(l). Also explain that HX is called the *parent acid* of the salt YX, and YOH is called the *parent base*. Illustrate with the equation for the reaction between sodium hydroxide, the parent base, and hydrochloric acid, the parent acid, yielding sodium chloride, the salt, and water.

SECTION 15.1

Acid and Base Reactions

SECTION PREVIEW

Objectives
Distinguish the overall, ionic, and net ionic equations for an acid-base reaction.
Classify acids and bases using the hydrogen transfer definition.
Predict and explain the final results of an acid-base reaction.

Key Terms
neutralization reaction
salt
ionic equation
spectator ion
net ionic equation

As you learned in Chapter 14, acids and bases are opposites in most of their properties. But you also learned that there is a big difference between strong acids and weak acids and between strong bases and weak bases. Do all acids react with all bases? Do all acid-base reactions produce a neutral solution?

Types of Acid-Base Reactions

The reaction of an acid and a base is called a **neutralization reaction** because the properties of both the acid and base are diminished or neutralized when they react.

In most cases, the reaction of an acid with a base produces water and a salt. **Salt** is a general term used in chemistry to describe the ionic compound formed from the negative part of the acid and the positive part of the base. In the language of chemistry, sodium chloride, common table salt, is just one of a large number of ionic compounds that are called salts. KCl, NH₄NO₃, and Fe₃(PO₄)₂ are other examples of salts.

Consider the following neutralization reaction. Hydrochloric acid, HCl, is a common household and laboratory acid. Muriatic acid is the common household name of hydrochloric acid. It is often sold in hardware stores to be used in masonry work to remove excess mortar from brick. Sodium hydroxide, NaOH, is a common household and laboratory base. The common name of sodium hydroxide is lye. It is the primary component of many drain cleaners. **Figure 15.1** shows litmus tests before and after mixing these substances together.

Figure 15.1

Neutralization Reactions
A solution of hydrochloric acid, HCl, is added to exactly the amount of a solution of basic sodium hydroxide, NaOH, that will react with it. Litmus papers show that the resulting salt solution is neither acidic nor basic.
NaOH(aq) + HCl(aq) → NaCl(aq) + H₂O(l)

Basic   Acidic   No color changes

516   Chapter 15   Acids and Bases React

Program Resources

After the reaction, the mixture contains only the salt sodium chloride, NaCl, dissolved in water. The litmus test shows no acid or base present in the reaction products.

Because both acids and bases may be either strong or weak, four possible combinations of acid-base reactions may occur. **Table 15.1** summarizes the possibilities. As you will find out later in this section, only three of the types are significant in everyday chemistry.

As long as one of the reactants is strong, the acid-base reaction goes to completion. As you learned in Chapter 6, a reaction goes to completion when the limiting reactant is completely consumed.

Although all of the reactions in **Table 15.1** are acid-base reactions, the submicroscopic interactions in each are different. Examine each possible type of acid-base reaction and see how they compare.

Table 15.1 Types of Acid-Base Reactions

Acid	Base
Strong	Strong
Strong	Weak
Weak	Strong
Weak	Weak

Strong Acid + Strong Base

A typical type of acid-base reaction is one in which both the acid and base are strong. The reaction of aqueous solutions of hydrochloric acid and sodium hydroxide shown in **Figure 15.1** is a good example of this type of reaction.

A Macroscopic View

It is easy to write and balance equations for strong acid-strong base reactions. In the previous example, HCl is the acid; NaOH is the base. The products are NaCl, which is a salt, and water. Now take a closer look at these reactants and their products.

The Submicroscopic View: Ionic Reactions

Recall from Chapter 14 that HCl, when dissolved in water, completely ionizes into hydronium ions and chloride ions because HCl is a strong acid. As you learned in Chapter 14, the hydronium ion is more conveniently written in shorthand as H^+.

$$HCl(aq) \rightarrow H^+(aq) + Cl^-(aq)$$

You also know that sodium hydroxide in water completely dissociates into sodium ions and hydroxide ions because NaOH is a strong base.

$$NaOH(aq) \rightarrow Na^+(aq) + OH^-(aq)$$

An overall equation for the reaction between NaOH and HCl shows each substance involved in the reaction. An overall equation does not indicate whether these substances exist as ions. The best way for you to model the submicroscopic behavior of an acid-base reaction is to show reactants and products as they actually exist in solution. Instead of an overall equation, an **ionic equation,** in which substances that primarily exist as ions in solution are shown as ions, can be written.

$$H^+(aq) + Cl^-(aq) + Na^+(aq) + OH^-(aq) \rightarrow$$
$$Na^+(aq) + Cl^-(aq) + H_2O(l)$$

15.1   Acid and Base Reactions   517

SECTION 15.1

Concept Development
Throughout Chapters 14 and 15, acids and bases are defined in two different ways. These definitions are named after the people who developed them. The Arrhenius theory defines acids as substances that produce H^+ ions in solution and bases as substances that produce OH^- ions in solution. Although this name is not used in the student text, it is used sometimes for convenience in teacher information. The Brønsted (or Brønsted-Lowry) theory defines acids and bases in terms of proton transfer.

Tying to Previous Knowledge
Review with students the concept that strong acids and strong bases ionize nearly completely in aqueous solution and should be written as ions in equations. Weak acids and weak bases ionize little and should be written as molecules or, for ionic compounds, formula units in equations.

✔ **Assessment**

Portfolio: Ask students to copy the list of the common strong acids and bases as found in Chapter 14, Section 2, and to keep the list in their portfolios for reference. Remind students to consider all other acids and bases to be weak unless they are informed otherwise. L1 P

Using an Analogy
Ask students what it would mean if the manager of a professional sports team said a certain opposing player were neutralized. *The opposing player was rendered ineffective.* Explain that what acids do to bases and bases do to acids is similar.

Chemistry Journal

Neutral or Not?
Ask students to think of three situations other than acid-base reactions in which they might use the terms *neutral, neutralized,* or *neutralization* and to write the appropriate descriptions in their journals. L1

Chemistry Journal

Operational Definitions
Ask students to write an operational definition for the neutralization of a sodium hydroxide solution with a hydrochloric solution, using litmus paper as an indicator. *The hydrochloric acid solution should be added slowly to the sodium hydroxide solution just until the litmus paper color begins to change from blue to red.* L1

Acidic, Basic, or Neutral?

Purpose: To test several salt solutions with bromothymol blue indicator to determine whether the solutions are acidic, basic, or neutral.

Process Skills: Observing, classifying, inferring

Disposal: Code J

Teaching Strategies

- Any concentrations of salt solutions between $0.1M$ and $1M$ will work for the MiniLab. 100 mL of $0.100M$ aqueous solutions may be prepared using the following amounts of the anhydrous solid salts: KNO_3, 8.51 g; $NaC_2H_3O_2$, 8.20 g; NH_4Cl, 5.35 g; Na_2CO_3, 10.60 g; $NaCl$, 5.84 g; $Al_2(SO_4)_3$, 34.2 g. If all solute does not easily dissolve, a solution of lesser concentration may be used.

Expected Results

$NaC_2H_3O_2$, Na_2CO_3—blue
KNO_3, $NaCl$—green
NH_4Cl, $Al_2(SO_4)_3$—yellow

Analysis

1. $NaC_2H_3O_2$, Na_2CO_3: basic; KNO_3, $NaCl$: neutral; NH_4Cl, $Al_2(SO_4)_3$: acidic
2. strong acid-strong base, neutral; strong acid-weak base, acidic; weak acid-strong base, basic
3. the salt of a strong acid and a weak base

✓ Assessment

Knowledge: Ask students to predict whether solutions of KCl, $NaC_2H_3O_2$, and NH_4NO_3 would be acidic, basic, or neutral and to explain their predictions. Have students keep their predictions to be confirmed by the results of the *Quick Demo* on page 527. **P**

---

miniLAB

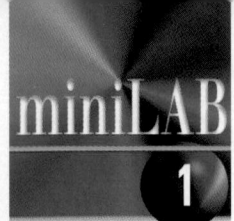

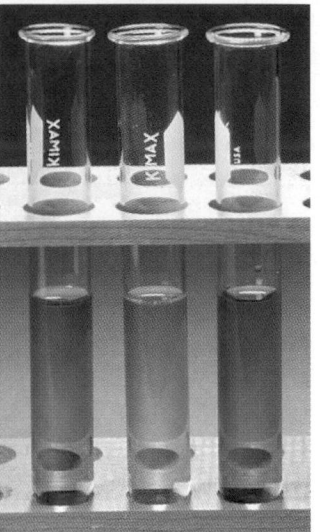

Acidic, Basic, or Neutral?

The salt potassium bromide forms in the acid-base neutralization reaction between hydrobromic acid and potassium hydroxide.

$$HBr(aq) + KOH(aq) \rightarrow H_2O(l) + KBr(aq)$$

Hydrobromic acid and potassium hydroxide are referred to as the parent acid and the parent base of potassium bromide. Test several aqueous salt solutions with bromothymol blue indicator to determine whether the solutions are acidic (yellow), basic (blue), or neutral (green).

Procedure

1. Wear laboratory aprons and safety goggles.
2. Use labeled microtip pipets to put six drops of sodium acetate solution in A1, potassium nitrate solution in A2, ammonium chloride solution in A3, sodium carbonate solution in A4, sodium chloride solution in A5, and aluminum sulfate solution in A6 of a 24-well microplate.
3. Add two drops of bromothymol blue indicator solution to each of the salt solutions. Stir each with a separate toothpick.
4. Set the microplate on a piece of white paper and look down through each well to determine the color of the solution. Record your results.

Analysis

1. According to the color of each solution that has bromothymol blue indicator added, is each of the solutions acidic, basic, or neutral?
2. Relate the relative strengths of the parent acids and bases to the results of using the indicator with the salt solutions.
3. What type of salt might be added to a product, such as shampoo, in order to make it slightly acidic so that it will not be harmful to skin and hair? Check the label of several brands of shampoo for the presence of such a salt.

Notice in the previous equation that in addition to showing the acid as completely ionized and the base as completely dissociated, the ionic compound NaCl is also dissociated. Water does not ionize much, so it is indicated as a molecule rather than H^+ and OH^- ions. The ionic aspects of this ionic equation are confirmed in **Figure 15.2**.

Figure 15.2

What ions are present?
When a conductivity apparatus is placed in solutions of each reactant and product of the reaction between a strong acid and a strong base, you can see that the acid, base, and salt exist as ions in solution.

Demonstration 15.2

An Acid-Base Reaction

Purpose: To demonstrate a typical acid-base neutralization reaction.

Materials: 0.6 mL of $18M$ H_2SO_4, 0.32 g $Ba(OH)_2 \cdot 8H_2O$, conductivity tester with light-bulb, two 250-mL beakers, magnetic stirrer

Safety Precautions: 🔥 🥽 🧤

Disposal: Code E

Procedure: Prepare 100 mL of $0.1M$ H_2SO_4 by adding 0.6 mL $18M$ H_2SO_4 to 99.4 mL water.

CAUTION: *Sulfuric acid is corrosive.* Add 10 mL of the $0.1M$ H_2SO_4 to 90 mL water to prepare 100 mL of $0.01M$ H_2SO_4. Separately, prepare 100 mL of $0.01M$ $Ba(OH)_2 \cdot 8H_2O$ by adding 0.32 g to 100 mL water. **CAUTION:** *The solution is toxic.* Add the $0.01M$ sulfuric acid to the beaker of a conductivity apparatus so that the beaker is filled halfway with acid. Place the electrodes in the beaker. While stirring, add the $0.01M$ $Ba(OH)_2$ solution by drops.

Taste

Although the total flavor of a food comes from the complex combination of taste, smell, touch, texture or consistency, and temperature sensations, taste is a major factor. Three of the four fundamental tastes are directly linked to acids and bases. Your tongue has four different types of taste buds—sweet, salty, bitter, and sour—that are located at different places on your tongue. Only certain molecules and ions can react with these specific buds to produce a signal that is sent to a certain region of your brain. When these signals are received, your brain processes them, and you sense taste.

2. The taste buds that detect a sour taste are along the sides of the tongue. The sour taste comes from acids in your food. Sour-tasting foods include vinegar and citrus fruits.

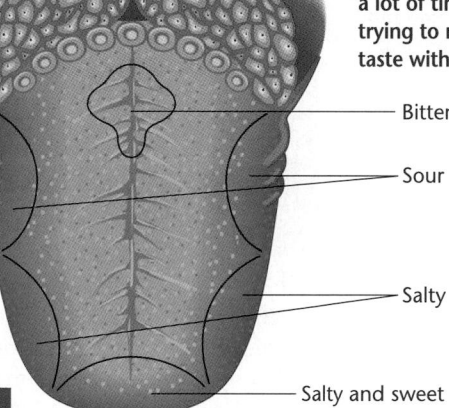

1. The taste buds that sense a bitter taste are located at the base of the tongue. Bases taste bitter. Many medications are basic, and pharmaceutical companies spend a lot of time and research trying to mask the bitter taste with other tastes.

— Bitter

— Sour

— Salty

— Salty and sweet

3. The taste buds that detect the salty and sweet tastes are located at the tip of your tongue. A salt is the product of an acid-base reaction. The sweet taste seems to depend a great deal on the properties of both acids and bases that are combined on a single molecule, but this taste is not as clear-cut as the other tastes.

Sweetness of Some Compounds

Compound	Relative Sweetness
Lactose	16
Glucose	74
Sucrose	100
Fructose	173
Aspartame	16 000
Saccharin	50 000

As the table shows, aspartame, the popular artificial sweetener in soft drinks, is 160 times sweeter than sucrose, common table sugar. The sweetening ability of aspartame comes from a molecular structure that creates an impact 160 times that of sucrose on your taste buds that detect the sweet taste.

Thinking Critically

1. How are three of the four fundamental tastes linked to the properties of acids and bases?

2. What types of companies do you think would support research on taste?

3. What benefits might artificial sweeteners, such as aspartame, have over natural sweeteners?

Have students observe both the formation of the salt, $BaSO_4$, and the conductivity change indicated by the lightbulb. Have students explain. [L1]

Results: $Ba^{2+} + 2OH^- + 2H^+ + SO_4^{2-} \rightarrow 2H_2O + BaSO_4$; the lightbulb dims as the number of ions decreases.

Analysis: Ask these questions.

1. What is the name and formula of the precipitate? *barium sulfate, $BaSO_4$*

2. As the amount of precipitate increases, what happens to the number of ions in solution? *The number decreases.*

✔ Assessment

Oral: Ask students for evidence that blood and other body fluids contain ions. *People get electric shocks.* [L1]

Purpose

Students will be introduced to the role that acids, bases, and salts play in the sensation of taste.

Background

Students have already been introduced in the previous chapter to the characteristic tastes of acids (sour) and bases (bitter).

Visual Learning

Refer to the diagram of the human tongue and note the areas that are most sensitive to the different tastes. Emphasize to students that taste buds responsible for a specific taste may be located anywhere on the tongue, but they are concentrated in the indicated areas.

Teaching Strategies

• Students should understand the table that compares the sweetness of different sweeteners. Ask this question: How many teaspoons of aspartame are needed to give the same sweet taste as a teaspoon of sucrose? *1/160 teaspoon* [L1]

• Ask students why inert ingredients are added to aspartame that is sold at the store. *Because it only takes so little aspartame to get the sweetness of one teaspoon of sugar, inert (no sweet taste) ingredients are added so that the aspartame mixture may be used like sugar from a sugar bowl.*

Thinking Critically

1. sour: acid; bitter: base; salt: acid and base

2. food and beverage companies

3. possibly useful in dieting and weight control, useful for people with diabetes who must carefully monitor sugar intake, prevention of tooth decay

-  Explain that the quantitative aspects of acid-base reactions are important. Ask students to calculate the mass of sodium hydroxide required to neutralize 10.0 g of hydrogen chloride in hydrochloric acid. *10.0 g HCl (1 mol HCl/36.5 g HCl)(1 mol NaOH/1 mol HCl) (40.0 g NaOH/1 mol NaOH) = 11.0 g NaOH* L2

- Ask students to write a short paragraph that explains the role that water plays in the ionization of HCl and the dissociation of NaOH. *Polar water molecules combine with hydrogen ions from HCl molecules to form hydronium ions. Polar water molecules exert forces that pull sodium and hydroxide ions apart.* L1

Reinforcement

 Review the process of adding algebraic equations with students with this example. Add the two equations $x + 5y = 20$ and $15 = 5y$. $x + 5y + 15 = 20 + 5y$. Ask students to subtract items common to both sides of the equation to derive the net equation. $x = 5$ L2

Concept Development

Even though spectator ions do not participate directly in the chemistry of acid-base reactions, it is important for students to realize that their presence is essential for charge balance and neutrality.

Spectator Ions and the Net Ionic Reaction

Note that the ionic equation gives more information about how a strong acid-strong base reaction occurs. When you examine the two sides of the ionic equation on page 517, you see that Na^+ and Cl^- are present both as reactants and as products. Although they are important components of an overall equation, they do not directly participate in the chemical reaction. They are called **spectator ions** because they are present in the solution but do not participate in the reaction, **Figure 15.3**.

Figure 15.3
Spectators
The presence of spectators at a sporting event is important, but spectators do not actually participate in the game and do not determine the final outcome.

Why show the spectator ions in an equation if they aren't really involved in the reaction? An ionic equation can be simplified to take care of that problem. Just as in a mathematical equation, items common to both sides of the equation can be subtracted. This process simplifies the equation so that the reactants and products that actually change can be seen more clearly.

$$H^+(aq) + Cl^-(aq) + Na^+(aq) + OH^-(aq) \rightarrow Na^+(aq) + Cl^-(aq) + H_2O(l)$$

When ions common to both sides of the equation are removed from the equation, the result is called the **net ionic equation** for the reaction of HCl with NaOH.

$$H^+(aq) + OH^-(aq) \rightarrow H_2O(l)$$

The net ionic equation describes what is really happening at the submicroscopic level. Although solutions of HCl and NaOH are mixed, the net ionic equation is hydrogen ions reacting with hydroxide ions to form water.

Even though the strong acid and strong base in Sample Problem 1 are different from those in the HCl reaction with NaOH, the net ionic equation is the same. Hydrogen ions from the acid react with hydroxide ions from the base to form water. This equation is always the net ionic equation for a strong acid-strong base reaction.

Cultural Diversity

Folk Remedy to Miracle Drug

A folk remedy for pain that was common to many societies was to chew willow bark. In 1860, salicylic acid was isolated from willow bark and demonstrated to be an active pain/fever reliever ingredient. Salicylic acid, $H_2C_7H_4O_3$, a weak acid, caused stomach upset in many people who used it. In the late 1800s, salicylic acid was chemically modified slightly to produce acetylsalicylic acid, $HC_8H_7O_4$, an

even weaker acid that maintained the medicinal properties of salicylic acid. In 1899, the German Bayer Company patented this compound, calling it aspirin. Today, more than 60 billion aspirin tablets, under a variety of brand names, are sold in the United States each year. In addition to pain and fever relief, aspirin appears to have other health benefits as a preventative in heart disease and stroke.

PRACTICE

Guided Practice: Guide students through the sample problem, then have them work in class on Practice Problems 1 4. L1

Independent Practice: Your homework o classroom assignment can include th additional practice problem on thi page. L1

Write the overall, ionic, and net ionic equations for the reaction of sulfuric acid with potassium hydroxide.

Analyze
- Decide whether the acid is a strong acid or a weak acid and whether the base is strong or weak. A list of strong acids and bases can be found in Chapter 14, **Table 14.2.** By looking at this table, you can see that sulfuric acid, H_2SO_4, is a strong acid. Potassium hydroxide, KOH, is a strong base.

Set Up
- Write an equation for the overall reaction. Because sulfuric acid is a diprotic acid, you need two moles of KOH for every one mole of H_2SO_4. Two moles of water and one mole of K_2SO_4 will be produced.

$$H_2SO_4(aq) + 2KOH(aq) \rightarrow K_2SO_4(aq) + 2H_2O(l)$$

- Write the ionic equation by showing H_2SO_4, KOH, and K_2SO_4 as ions. You must keep track of the coefficients from the overall equation and the formulas of the substances when writing the coefficients of the ions.

$$2H^+(aq) + SO_4^{2-}(aq) + 2K^+(aq) + 2OH^-(aq) \rightarrow$$
$$2K^+(aq) + SO_4^{2-}(aq) + 2H_2O(l)$$

Solve
- Look for spectator ions. In this reaction, K^+ and SO_4^{2-} are spectator ions. Subtract them from both sides of the equation to get the net ionic equation.

$$2H^+(aq) + \cancel{SO_4^{2-}(aq)} + \cancel{2K^+(aq)} + 2OH^-(aq) \rightarrow$$
$$\cancel{2K^+(aq)} + \cancel{SO_4^{2-}(aq)} + 2H_2O(l)$$

$$2H^+(aq) + 2OH^-(aq) \rightarrow 2H_2O(l)$$

- Simplify the balanced net reaction by dividing coefficients on both sides of the equation by the common factor of 2.

> **Problem-Solving HINT**
>
> Coefficients should be in the smallest whole-number ratio possible.

$$\frac{2H^+(aq) + 2OH^-(aq) \rightarrow 2H_2O(l)}{2} \text{ results in } H^+(aq) + OH^-(aq) \rightarrow H_2O(l)$$

Check
- Take a final look at the net ionic equation to make sure no ions are common to both sides of the equation.

PRACTICE PROBLEMS

Write overall, ionic, and net ionic equations for each of the following reactions.
1. hydroiodic acid, HI, and calcium hydroxide, $Ca(OH)_2$
2. hydrobromic acid, HBr, and lithium hydroxide, LiOH
3. sulfuric acid, H_2SO_4, and strontium hydroxide, $Sr(OH)_2$
4. perchloric acid, $HClO_4$, and barium hydroxide, $Ba(OH)_2$

15.1 **Acid and Base Reactions** **521**

Correcting Errors

Students often assume that all ions in chemical reactions are spectators. Point out that ions may be reactants in many reactions and cannot, in that case, be omitted. This generalization is also true for oxidation-reduction reactions, which will be studied in a later chapter.

Using an Analogy

Ask students to describe situations in which the phrase "The bottom line is . . ." is used. Ask them what is meant by the expression. *Eliminating all of the intermediate and/or superfluous information, this is what really matters or actually happens.* Explain that in business accounting, "The bottom line" refers to the process of subtracting expenses from gross receipts in order to determine the net profits or losses. In a similar way, the net ionic equation for a reaction focuses on the substances that actually change. L1

GLENCOE TECHNOLOGY

 Software

Mastering Concepts in Chemistry
Unit 14, Lesson 3
Salts, Neutralization
Use this lesson to reinforce the concept of neutralization reactions. L1 LEP

Answers to Practice Problems

1. $2HI(aq) + Ca(OH)_2(aq) \rightarrow CaI_2(aq) + 2H_2O(l)$; $2H^+(aq) + 2I^-(aq) + Ca^{2+}(aq) + 2OH^-(aq) \rightarrow Ca^{2+}(aq) + 2I^-(aq) + 2H_2O(l)$; $H^+(aq) + OH^-(aq) \rightarrow H_2O(l)$
2. $HBr(aq) + LiOH(aq) \rightarrow LiBr(aq) + H_2O(l)$; $H^+(aq) + Br^-(aq) + Li^+(aq) + OH^-(aq) \rightarrow Li^+(aq) + Br^-(aq) + H_2O(l)$; $H^+(aq) + OH^-(aq) \rightarrow H_2O(l)$

3. $H_2SO_4(aq) + Sr(OH)_2(aq) \rightarrow SrSO_4(aq) + 2H_2O(l)$; $2H^+(aq) + SO_4^{2-}(aq) + Sr^{2+}(aq) + 2OH^-(aq) \rightarrow Sr^{2+}(aq) + SO_4^{2-}(aq) + 2H_2O(l)$; $H^+(aq) + OH^-(aq) \rightarrow H_2O(l)$
4. $2HClO_4(aq) + Ba(OH)_2(aq) \rightarrow 2H_2O(l) + Ba(ClO_4)_2(aq)$; $2H^+(aq) + 2ClO_4^-(aq) + Ba^{2+}(aq) + 2OH^-(aq) \rightarrow 2H_2O(l) + Ba^{2+}(aq) + 2ClO_4^-(aq)$; $H^+(aq) + OH^-(aq) \rightarrow H_2O(l)$

Additional Practice Problem

Write the overall, ionic, and net ionic equations for the following reaction.
1. hydrochloric acid and calcium hydroxide $2HCl + Ca(OH)_2 \rightarrow 2H_2O + CaCl_2$; $2H^+ + 2Cl^- + Ca^{2+} + 2OH^- \rightarrow 2H_2O + Ca^{2+} + 2Cl^-$; $H^+ + OH^- \rightarrow H_2O$

Students may think that all neutralization reactions produce neutral solutions. Point out that this is not so. Explain that *neutralization* occurs when acids and bases react in stoichiometric amounts, whereas a *neutral solution* is one that has a pH of 7. Emphasize that complete acid-base reactions may result in acidic, basic, or neutral solutions.

Discussion

Ask students why HBr is described as a monoprotic acid, H_2SeO_4 as a diprotic acid, and H_3AsO_4 as a triprotic acid. *The HBr molecule is capable of donating one hydrogen ion, the H_2SeO_4 is capable of donating two protons, and the H_3AsO_4 molecule is capable of donating three protons.*

Transparency

 Display **Basic Concepts Transparency 41** to review and reinforce the concept of neutralization reactions. L1 LEP

GLENCOE TECHNOLOGY

 CD-Rom

Chemistry: Concepts and Applications
An Acid Base Reaction
Have students use this animation to help reinforce the process that occurs when an acid and a base are mixed. L1 LEP

Videodisc

Chemistry: Concepts and Applications
An Acid Base Reaction
Disc 2, Side 1, Ch. 17

Use this animation to help reinforce the concept of neutralization. L1 LEP

The pH Perspective

The net ionic equation also shows why this reaction is called a neutralization reaction. The hydrogen ion from the acid reacts with the hydroxide ion from the base to form water, which is neutral.

Figure 15.4 shows the reaction of 50.0 mL of 0.100M HCl with 50.0 mL of 0.100M NaOH, which is exactly the right amount of base to react with all of the acid. As the NaOH solution is added to the HCl solution, the pH of the solution increases. After all of the NaOH solution is added, the pH of the final solution is 7.

What happens if one of the reactants is strong and the other is weak? A similar approach can be used.

Figure 15.4

Final pH
The reaction of a strong acid and a strong base is definitely a neutralization. The pH of a 0.100M HCl solution (top) is 1. The pH of a 0.100M NaOH solution (bottom left) is close to 13. When the reaction is complete, the pH is 7, which is the pH of a neutral solution (bottom right).

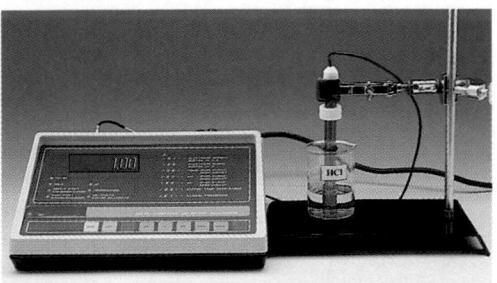

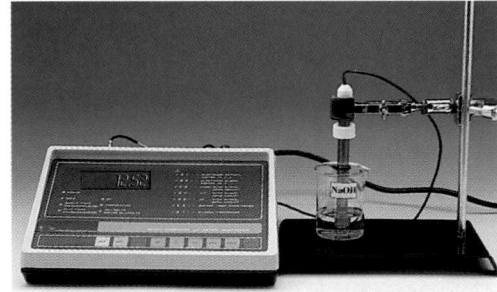

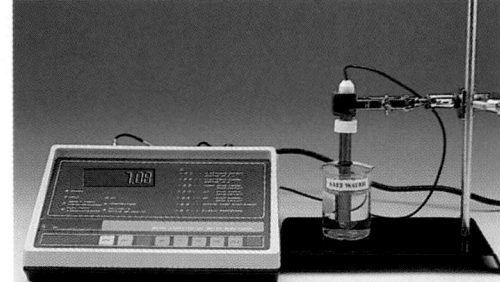

Strong Acid + Weak Base

Do the reactions change when the strength of an acid or base changes? Look at an example of what happens when a strong acid and a weak base are mixed together. Consider the reaction of hydrobromic acid and aluminum hydroxide. The overall equation shows the reactants and products.

$$3HBr(aq) + Al(OH)_3(s) \rightarrow AlBr_3(aq) + 3H_2O(l)$$

Hydrobromic acid, a strong acid, completely ionizes in water. All of the $Al(OH)_3$ that dissolves dissociates, so it is technically a strong base. However, because it is so insoluble, few OH^- ions are produced, and $Al(OH)_3$ acts as a weak base. Therefore, the ionic equation shows little dissociation of the base. The dissociated salt, $AlBr_3$, is also shown as ions.

$$3H^+(aq) + 3Br^-(aq) + Al(OH)_3(s) \rightarrow Al^{3+}(aq) + 3Br^- + 3H_2O(l)$$

522   **Chapter 15   Acids and Bases React**

Meeting Individual Needs

Gifted Explain to gifted students that any polyprotic acid ionizes in steps and, even if it is a strong acid, the second and any subsequent ionizations are weak. Ask students to write the complete ionic and the net ionic equations for the partial, first hydrogen neutralization of sulfuric acid with potassium hydroxide. $H^+ + HSO_4^- + K^+ + OH^- \rightarrow K^+ + HSO_4^- + H_2O$ and $H^+ + OH^- \rightarrow H_2O$ Then ask students to write the complete ionic and net ionic equations for the second hydrogen neutralization. $HSO_4^- + K^+ + OH^- \rightarrow K^+ + SO_4^{2-} + H_2O$. $HSO_4^- + OH^- \rightarrow SO_4^{2-} + H_2O$ Emphasize that the second net ionic equation is written differently from the first because HSO_4^- is a weak acid, whereas H_2SO_4 is a strong acid. L3

The spectator ions in this equation are bromide ions. They are removed from both sides of the equation to produce the net ionic equation.

$$3H^+(aq) + Al(OH)_3(s) \rightarrow Al^{3+}(aq) + 3H_2O(l)$$

Compare this equation to the net ionic equation for a strong acid-strong base reaction.

A Strong Acid + NH₃

Recall from Chapter 14 that the most common weak base does not contain the hydroxide ion. Consider an equation for the reaction between hydrochloric acid and ammonia.

$$HCl(aq) + NH_3(aq) \rightarrow NH_4Cl(aq)$$

Notice that although the product is a salt, NH_4Cl, no water is produced in this overall reaction. As before, use the net ionic equation to understand the submicroscopic processes for this reaction of a weak base with a strong acid.

The Ionic Reaction: What's in Solution?

Recall that when ammonia dissolves in water, some of the ammonia molecules react with water to form ammonium ions and hydroxide ions. However, most of the ammonia molecules remain as molecules. Other than water, the major particle present in an aqueous solution of ammonia is ammonia molecules, NH_3.

A solution of ammonia is best represented by $NH_3(aq)$. A solution of HCl, as in the last case, is best represented as $H^+(aq)$ and $Cl^-(aq)$. The ionic reaction is written, as in Sample Problem 1, by representing what is actually in the reactant and product solutions.

$$H^+(aq) + Cl^-(aq) + NH_3(aq) \rightarrow NH_4^+(aq) + Cl^-(aq)$$

The ionic salt NH_4Cl is written as dissociated ions. **Figure 15.5** shows which of these particles are actually involved in the reaction.

Figure 15.5

What actually reacts?
A quick look at the ionic reaction shows that the chloride ion is a spectator ion because it appears on both sides of the reaction. You get the net ionic equation by subtracting the spectator Cl⁻ from both sides of the ionic equation.

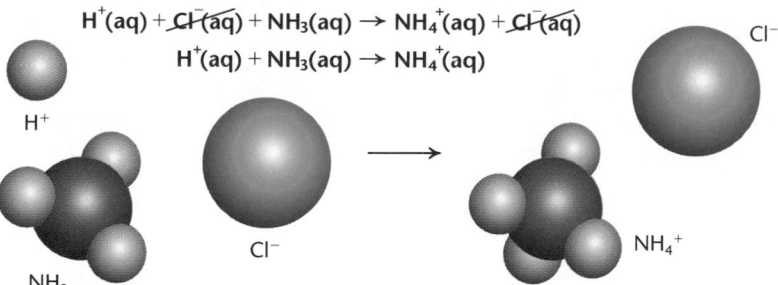

$$H^+(aq) + \cancel{Cl^-(aq)} + NH_3(aq) \rightarrow NH_4^+(aq) + \cancel{Cl^-(aq)}$$

$$H^+(aq) + NH_3(aq) \rightarrow NH_4^+(aq)$$

Concept Development

Explain that common ions, as opposed to spectator ions, affect the degree of ionization of weak acids, making them even weaker. Use this example. If sodium fluoride is added to an aqueous hydrofluoric acid solution, the added fluoride ions react with some of the hydrogen ions in solution to produce more HF molecules.

Fact of the **MATTER**

In medieval times, natural deposits of ammonium chloride, a compound derived from ammonia, were first mined in Egypt near a temple of the Egyptian god Ammon.

GLENCOE TECHNOLOGY

Videodisc

Chemistry: Concepts and Applications
See the Videodisc Program Teacher Guide to access illustrations and other material available for this chapter.

Meeting Individual Needs

Learning Disabled Use molecular models to illustrate the acid-base reaction between hydrogen chloride and ammonia, yielding the chloride ion and the ammonium ion. [L1]
[LEP]

*inter*NET
CONNECTION
Hypermedia text reviews of basic chemistry concepts are available for use on the Internet.

World Wide Web
http://www.scimedia.com

Purpose

To show how chemistry explains the formation of caves.

TECH PREP

Teaching Strategies

- Remind students that the amount of gas that dissolves in a liquid depends upon the partial pressure of the gas.
- A good resource is: Tanis, David. "Underground Sculpture." *ChemMatters*, Feb. 1994, pp. 10-11.
- A liter of cold, neutral water will dissolve 0.014 g of $CaCO_3$, which is about the size of a rice grain.

Extension

Design an Experiment Have student groups design and do an experiment to determine the difference in the amount of $CaCO_3$ that will dissolve in distilled water and acidic water found in nature. **L2**

COOP LEARN

Connecting to Chemistry

1. Physical: dissolving of $Ca(HCO_3)_2$, evaporation of water droplets; Chemical: acidic rainwater reacts with limestone, carbonic acid changes back into CO_2 and water
2. All equations would be the same as those for $CaCO_3$ except: $H_3O(aq) + MgCO_3(s) \rightarrow Mg^{2+}(aq) + HCO_3^-(aq) + H_2O(l)$.

EARTH SCIENCE ⊳ CONNECTION

Cave Formation

When you think of a cave, what images come to mind? A massive, damp, cool, underground chamber running deep into Earth or a fantasy underground world with stone icicles rising from the floor and ornate columns seeming to support the ceiling? Whatever visual pictures you may imagine, caves are one of nature's wonders.

How caves are formed Caves form in limestone regions throughout the world. Limestone is calcium carbonate, which is only slightly soluble in water. The caves that form within these rocks are called solution caves.

What causes natural water to be acidic? Most rainwater is slightly acidic because it contains carbon dioxide from the atmosphere. A small amount of the CO_2 dissolves in the water, but some of it reacts with the water to form carbonic acid.

$$CO_2(g) + H_2O(l) \rightarrow H_2CO_3(aq)$$

Carbonic acid forms a hydronium ion and a hydrogen carbonate ion.

$$H_2CO_3(aq) + H_2O(l) \rightarrow H_3O^+(aq) + HCO_3^-(aq)$$

How does this acid form caves? The hydronium ions react with limestone to produce soluble ions.

$$H_3O^+(aq) + CaCO_3(s) \rightarrow Ca^{2+}(aq) + HCO_3^-(aq) + H_2O(l)$$

The acidic water dissolves the limestone rocks, producing open spaces that contain water.

Stalactites and stalagmites During the second phase, clay, silt, sand, or gravel moves into the spaces. In the third phase, streams partially remove these materials and modify and enlarge the spaces. Stalactites and stalagmites now form by the reverse of the chemical and physical processes that formed the cave. The

water containing dissolved CO_2 and H_2CO_3 is saturated with $Ca(HCO_3)_2$. As it seeps through the roof of the cave, the water in each droplet slowly evaporates.

Some of the carbonic acid changes back into carbon dioxide and water. The pH of the water increases, and the solubility of $Ca(HCO_3)_2$ decreases. The $CaCO_3$ precipitates out slowly, forming stalactites over thousands of years. As the saturated water drops hit the floor, the same processes slowly form stalagmites. Sometimes, the two formations grow together, forming pillars.

Connecting to Chemistry

1. **Applying** Identify two physical changes and two chemical changes that occur in cave formation.

2. **Thinking Critically** Write the equations for cave formation from the $MgCO_3$ part of dolomite, $CaCO_3 \cdot MgCO_3$.

Across the Curriculum

Health

Cocaine, $C_{17}H_{21}NO_4$, is a widely abused drug. It is a weak base that is extracted from the coca plant by its reaction with a strong acid. The salt formed is a white powder called cocaine hydrochloride, $C_{17}H_{22}NO_4Cl$. In the middle 1980s, a new and much more addictive form of cocaine appeared on the street. This form, called crack, is easily produced. Because of its ease of preparation, potency, and highly addictive nature, crack has become a major drug of abuse and a major source of crime. Law-enforcement officials estimate that more than half of the crime in some urban areas is related to crack. Ask students to find more information about the societal impact of crack. **L1**

PRACTICE

Guided Practice: Guide students throug[h] the sample problem, then have the[m] work in class on Practice Problems 5[-] 7. **L1**

Independent Practice: Your homework o[r] classroom assignment can include th[e] additional practice problems show[n] here. **L1**

Write the overall, ionic, and net ionic equations for the reaction of nitric acid with ammonia.

Analyze • Decide whether the acid is strong or weak and whether the base is strong or weak. Nitric acid, HNO_3, is a strong acid. Ammonia is a weak base.

Set Up • Write an equation for the overall reaction.

$$HNO_3(aq) + NH_3(aq) \rightarrow NH_4NO_3(aq)$$

• Write the ionic equation. Because HNO_3 is a strong acid, you write it as completely ionized. NH_3 is a weak base, so you write it as NH_3. You dissociate the salt, ammonium nitrate, NH_4NO_3, into its component ions because it is an ionic compound.

$$H^+(aq) + NO_3^-(aq) + NH_3(aq) \rightarrow NH_4^+(aq) + NO_3^-(aq)$$

Solve • Look for spectator ions. In this reaction, only the nitrate ion, NO_3^-, is a spectator ion. Subtract NO_3^- from both sides of the equation to get the net ionic equation.

$$H^+(aq) + \cancel{NO_3^-(aq)} + NH_3(aq) \rightarrow NH_4^+(aq) + \cancel{NO_3^-(aq)}$$
$$H^+(aq) + NH_3(aq) \rightarrow NH_4^+(aq)$$

• Note that this is the same net equation as in the HCl and NH_3 example shown in **Figure 15.5**.

Check • Take a final look at the equation to make sure no ions are common to both sides of the equation.

Reinforcement

Ask students to write the Lewis electron dot structures for all reactants and products in the net ionic equation for the reaction between hydrochloric acid and ammonia. L1

PRACTICE PROBLEMS

Write overall, ionic, and net ionic equations for each of the following reactions.

5. perchloric acid, $HClO_4$, and ammonia, NH_3

6. hydrochloric acid, HCl, and aluminum hydroxide, $Al(OH)_3$

7. sulfuric acid, H_2SO_4, and iron(III) hydroxide, $Fe(OH)_3$

For all these examples of strong acid-weak base reactions, the net ionic equation differs from that for a strong acid and a strong base. The submicroscopic interactions in these strong acid-weak base reactions are between hydrogen ions and the bases.

Strong Acid-Weak Base and pH

Figure 15.6 shows that a solution of $0.100M$ ammonia is definitely a base. It has pH greater than 7. If you compare the pH of $0.100M$ NaOH with the pH of $0.100M$ NH_3, you see that ammonia is a weaker base because it has a lower pH.

Answers to Practice Problems

5. $HClO_4(aq) + NH_3(aq) \rightarrow$
$NH_4ClO_4(aq)$; $H^+(aq) + ClO_4^-(aq) +$
$NH_3(aq) \rightarrow NH_4^+(aq) + ClO_4^-(aq)$;
$H^+(aq) + NH_3(aq) \rightarrow NH_4^+(aq)$

6. $3HCl(aq) + Al(OH)_3(aq) \rightarrow AlCl_3(aq)$
$+ 3H_2O(l)$; $3H^+(aq) + 3Cl^-(aq) +$
$Al(OH)_3(aq) \rightarrow Al^{3+}(aq) + 3Cl^-(aq)$
$+ 3H_2O(l)$; $3H^+(aq) + Al(OH)_3(aq)$
$\rightarrow Al^{3+}(aq) + 3H_2O(l)$

7. $3H_2SO_4(aq) + 2Fe(OH)_3(aq) \rightarrow$
$Fe_2(SO_4)_3(aq) + 6H_2O(l)$; $6H^+(aq) +$
$3SO_4^{2-}(aq) + 2Fe(OH)_3(aq) \rightarrow$
$2Fe^{3+}(aq) + 3SO_4^{2-}(aq) + 6H_2O(l)$;
$3H^+(aq) + Fe(OH)_3(aq) \rightarrow Fe^{3+}(aq)$
$+ 3H_2O(l)$

Additional Practice Problems

Write the overall, ionic, and net ionic equations for the following reactions.

1. sulfuric acid and ammonia $H_2SO_4 +$
$2NH_3 \rightarrow (NH_4)_2SO_4$; $2H^+ + SO_4^{2-} +$
$2NH_3 \rightarrow 2NH_4^+ + SO_4^{2-}$; $H^+ + NH_3$
$\rightarrow NH_4^+$

2. hydrobromic acid and cobalt(II) hydroxide $2HBr + Co(OH)_2 \rightarrow 2H_2O +$
$CoBr_2$; $2H^+ + 2Br^- + Co(OH)_2 \rightarrow$
$2H_2O + Co^{2+} + 2Br^-$; $2H^+ + Co(OH)_2$
$\rightarrow 2H_2O + Co^{2+}$

Content Background

Brønsted Theory Explain that the hydrogen-ion donor or acceptor theory of acids and bases, which was developed independently by Johannes Brønsted and Thomas Lowry and presented in 1923, is most often referred to as the Brønsted-Lowry theory or as simply the Brønsted theory because Brønsted developed the theory more fully.

Discussion

Ask students why a hydrogen ion may be called a proton. *A hydrogen atom consists of a nucleus, which usually contains just one proton, and a single electron. If the electron is removed, one proton remains.*

Using an Analogy

Explain that, just as money is given by a *donor* to a charity, which is the *acceptor,* hydrogen ions or protons are given by the acid, which is the *donor,* and received by the base, which is the *acceptor.*

Tying to Previous Knowledge

To minimize student confusion as you are explaining the Brønsted-Lowry definition of acids and bases, review and summarize the operational and Arrhenius definitions from Chapter 14.

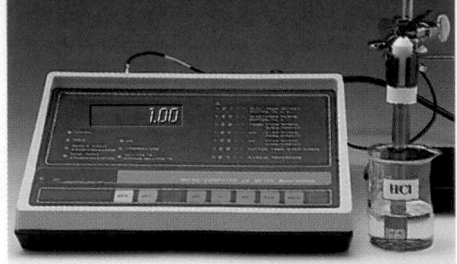

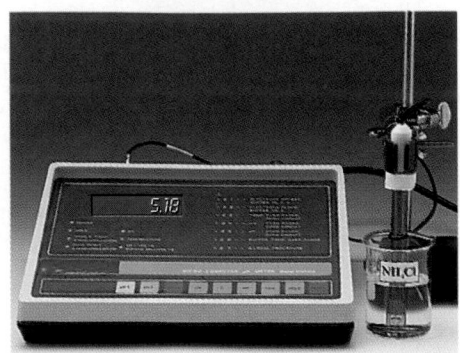

Figure 15.6
The Product Is Acidic
The reaction of a strong acid and weak base is not quite a neutralization. The pH of a 0.100M HCl solution is 1, and the pH of a 0.100M NH_3 solution is approximately 11. When equal volumes of the solutions are mixed and the reaction is complete, the pH is approximately 5.

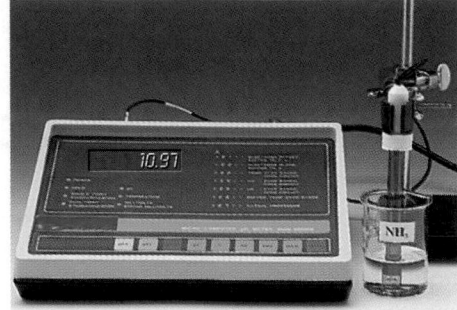

Figure 15.6 also shows that when equal volumes of solutions of ammonia and hydrochloric acid of equal molarity are mixed, the pH of the final mixture is less than 7. A pH less than 7 means that the final reaction mixture is acidic. Therefore, more hydronium than hydroxide ions must be present in the final reaction mixture.

How can equal moles of base and acid react to produce a neutral solution in one reaction and an acidic solution in the second? The result must have something to do with the relative strengths of the acid and the base.

A Broader Definition of Acids and Bases

The reaction of a strong acid with a weak base demonstrates the need for a slightly broader definition of acids and bases. As you learned in the last chapter, much of the behavior of acids and bases in water can be explained by a model that focuses on the hydrogen ion transfer from the acid to the base. This model will also help explain why every acid-base reaction does not result in a neutral solution.

Hydrogen-Ion Donor or Acceptor

You can use the ability to exchange a hydrogen ion as the basis of a broader definition of an acid or a base. In this definition, called the Brønsted-Lowry definition of acids and bases, an acid is defined as a substance that donates, or gives up, a hydrogen ion in a chemical reaction. A base, not surprisingly, is just the opposite. A base is a substance that accepts a hydrogen ion in a chemical reaction.

Demonstration 15.3

Water Not Required

Purpose: To demonstrate that the reaction of an acidic anhydride, CO_2, with a basic anhydride, CaO, is an example of an acid-base reaction.

Materials: 20 g powdered soda lime (a mixture of NaOH and CaO), 500-mL side-arm vacuum flask with stopper, 200 mL vinegar, 75 g $NaHCO_3$, empty litmus paper vial with stopper, 12-inch balloon, scissors, 250-mL beaker, rubber tubing

Safety Precautions:

Disposal: Code C

Procedure: Place 20 g freshly powdered soda lime, NaOH and CaO, into an empty litmus paper vial. **CAUTION:** *Soda lime is caustic.* Loosely stopper the vial and place it into a 250-mL beaker. Use vinegar and $NaHCO_3$ to generate CO_2 gas in a gas generator (stoppered side-arm vacuum flask). Fill the 250-mL beaker containing the vial with CO_2 by the upward displacement of air. Cut the bottom third off a large (12-inch) balloon. Use the top of the balloon to cover and seal the beaker containing the CO_2 and vial. Stretch the balloon as you reach through it to remove the stopper of the vial. Shake the balloon-covered beaker so as to spill the soda lime from the vial.

Figure 15.7

Defining Acids and Bases by H⁺ Transfer
In a reaction between aqueous HCl and aqueous NH₃, several H⁺ transfers occur.

A In the transfer of H⁺ from HCl to a water molecule, HCl acts as an acid, and water acts as a base.

A. $HCl(aq) + H_2O(l) \rightarrow H_3O^+(aq) + Cl^-(aq)$
   acid   base

B. $H_3O^+(aq) + NH_3(aq) \rightarrow NH_4^+(aq) + H_2O(l)$
   acid   base

C. $H_2O(l) + NH_3(aq) \rightarrow NH_4^+(aq) + OH^-(aq)$
   acid   base

B In the transfer of H⁺ from a hydronium ion to an ammonia molecule, the hydronium ion acts as an acid, and the ammonia molecule acts as a base.

C Water also reacts with ammonia molecules. Water acts as an acid, and ammonia acts as a base.

Take another look at the net ionic equation of HCl with NH₃ on page 523. If you adhere strictly to the definition of a base as a hydroxide-ion producer in water, none of these equations define ammonia as a base. Remember that hydroxide ions are produced, but the amount is so small that it is not shown in the equations.

$$H^+(aq) + NH_3(aq) \rightarrow NH_4^+(aq)$$

Using the new definition, you can definitely say that ammonia is acting as a base. It is accepting a hydrogen ion. **Figure 15.7** shows this reaction written in its most complete form and clearly shows the hydrogen ion transfer. The hydronium ion is acting as the acid because it donates the hydrogen ion.

It Takes Two to Transfer

Notice from **Figure 15.7** that the definitions of an acid as a hydronium-ion producer and a base as a hydroxide-ion producer are included in this H⁺-transfer definition. When HCl reacts with water, it acts as an H⁺ donor, so it is an acid. Water acts as an H⁺ acceptor, so it is a base. When ammonia reacts with water, the ammonia molecule accepts H⁺ from the water. Ammonia is the base and water is the acid. Remember that it takes two to transfer, so for every acid (an H⁺ donor), there must be a base (an H⁺ acceptor).

Although you generally think of water as neutral, a unique property of water can now be observed. Water can act as either an acid or a base, depending upon what else is in solution.

Water Not Required

Although most of the reactions that you will study occur in water, the H⁺-transfer definition does not require water to be present. For example, the reaction between HCl(g) and NH₃(g) is shown in **Figure 15.8.** This reaction occurs in the gas phase. It involves the transfer of a hydrogen ion from a gaseous HCl molecule to a gaseous ammonia molecule to form a solid product. This gas reaction can now be classified as an acid-base reaction.

15.1 Acid and Base Reactions   **527**

Tying to Previous Knowledge

Write Table 15.1 on page 517 on the chalkboard and, to the right of the table, make a column titled "Condition of solution after neutralization." Enter these descriptions where appropriate in the column: *Neutral, Acidic, Basic.* Whether each is acidic, basic, or neutral depends upon the relative strengths of the acid and the base. This will help students summarize many pages of text.

Quick Demo

Aren't all salt solutions neutral?

To convince students of the need for more than one definition for acids and bases, ask them how they would categorize potassium chloride, sodium acetate, and ammonium nitrate. *They are all salts.* Add several drops of bromothymol blue indicator solution to aqueous solutions of the three salts. *A green color in potassium chloride solution indicates that it is neutral. A blue color in sodium acetate solution indicates that it is basic. A yellow color in ammonium nitrate solution indicates that it is acidic.* Point out that ions in the sodium acetate and ammonium nitrate solutions react with water as Brønsted-Lowry bases and acids, respectively.
Disposal: Code C

Results: An exothermic reaction occurs as a solid reacts with a gas to produce a solid. Air pressure forces the balloon into the beaker in a few seconds.
Analysis: Ask these questions.
1. Use electron dot diagrams to show the reaction of the OH⁻ ion from NaOH with CO₂ to form the HCO₃⁻ ion.

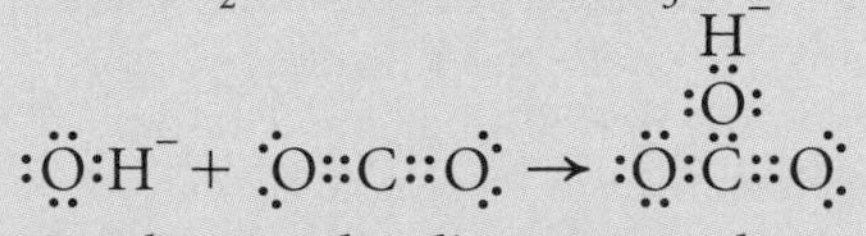

$$:\ddot{O}:H^- + :\ddot{O}::C::\ddot{O}: \rightarrow :\ddot{O}:\ddot{C}::\ddot{O}:$$

2. Use electron dot diagrams to show the reaction of the O²⁻ ion from CaO with CO₂ to form the CO₃²⁻ ion.

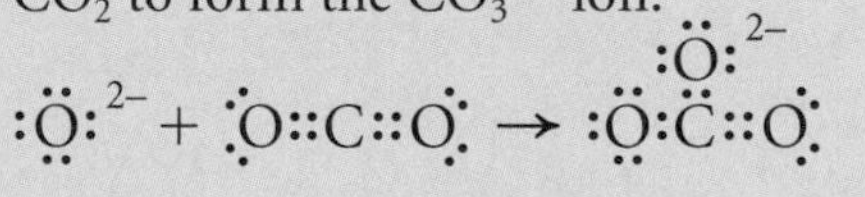

$$:\ddot{O}:^{2-} + :\ddot{O}::C::\ddot{O}: \rightarrow :\ddot{O}:\ddot{C}::\ddot{O}:$$

Assessment

Portfolio: Have students write a summary of this demonstration and add it to their portfolios. [L1] [P]

Figure 15.8

A Gas Phase Acid-Base Reaction
HCl(g) and NH_3(g) react to form NH_4Cl(s). Gases from the concentrated aqueous solutions react to form a smoke of solid ammonium chloride.

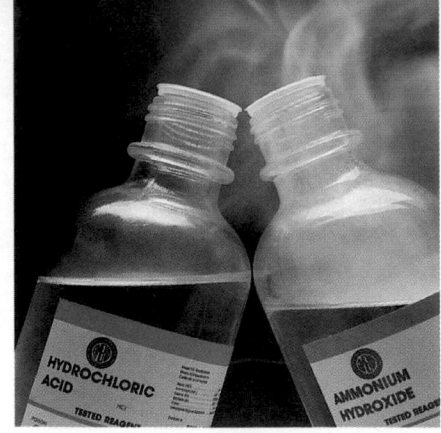

WORD ORIGIN

vinegar:
vinaigre (Fr) sour wine

When wine becomes sour, it has turned to vinegar.

Fact of the MATTER

Vinegar is a dilute solution of acetic acid. Pure acetic acid is often called glacial acetic acid. It was first purified in 1700 by the distillation of vinegar. At room temperature, pure acetic acid is a liquid, but it freezes at 17°C. The term *glacial* means "icelike." In poorly heated chemistry labs, pure acetic acid freezes. Acetic acid is 34th on the 1994 list of top industrial chemicals.

Weak Acid + Strong Base

Considering this new H^+-transfer acid-base definition, take a look at the type of acid-base reaction in which the acid is weak and the base is strong. An example is the reaction of acetic acid, $HC_2H_3O_2$, the weak acid present in vinegar, with sodium hydroxide. The equation of the overall reaction is similar to that of a strong acid-strong base reaction.

$$HC_2H_3O_2(aq) + NaOH(aq) \rightarrow NaC_2H_3O_2(aq) + H_2O(l)$$

The Ionic Reaction: What's in Solution?

As you know, acetic acid is a weak acid. In a solution of acetic acid, only a small fraction of the acetic acid molecules ionize. Other than water, the major particle present in an aqueous solution of acetic acid is $HC_2H_3O_2$. NaOH, as seen before, completely dissociates. The ionic equation shows what is present when the acid and base react. As in the previous cases, the salt is also written as dissociated ions.

$$HC_2H_3O_2(aq) + Na^+(aq) + OH^-(aq) \rightarrow$$
$$Na^+(aq) + C_2H_3O_2^-(aq) + H_2O(l)$$

Net Ionic Reaction: H^+ Transfer

The ionic equation contains the sodium ion as a spectator ion. Subtracting Na^+ from each side of the ionic equation gives the net ionic equation.

$$HC_2H_3O_2(aq) + \cancel{Na^+(aq)} + OH^-(aq) \rightarrow \cancel{Na^+(aq)} + C_2H_3O_2^-(aq) + H_2O(l)$$
$$HC_2H_3O_2(aq) + OH^-(aq) \rightarrow C_2H_3O_2^-(aq) + H_2O(l)$$

The net ionic equation shows that a weak acid and a strong base react by hydrogen ion transfer from the weak acid to the hydroxide ion. The acetic acid is the H^+ donor and serves as the acid. The hydroxide ion is the H^+ acceptor and serves as the base. Although this is an acid-base reaction, notice that H^+ is not involved as a reactant or product of the reaction. However, using the H^+-transfer definition, it is easy to include this reaction as an acid-base reaction.

528   **Chapter 15   Acids and Bases React**

PRACTICE

Answers to Practice Problems

8. H_2CO_3(aq) + 2NaOH(aq) $\rightarrow$
   Na_2CO_3(aq) + 2H_2O(l); H_2CO_3(aq) +
   2Na^+(aq) + 2OH^-(aq) $\rightarrow$ 2Na^+(aq) +
   CO_3^{2-}(aq) + 2H_2O(l); H_2CO_3(aq) +
   2OH^-(aq) $\rightarrow$ CO_3^{2-}(aq) + 2H_2O(l)

9. H_3BO_3(aq) + 3KOH(aq) $\rightarrow$ K_3BO_3(aq)
   + 3H_2O(l); H_3BO_3(aq) + 3K^+(aq) +
   3OH^-(aq) $\rightarrow$ 3K^+(aq) + BO_3^{3-}(aq) +
   3H_2O(l); H_3BO_3(aq) + 3OH^-(aq) $\rightarrow$
   BO_3^{3-}(aq) + 3H_2O(l)

10. 2$HC_2H_3O_2$(aq) + Ca(OH)$_2$(aq) $\rightarrow$
    Ca($C_2H_3O_2$)$_2$(aq) + 2H_2O(l);

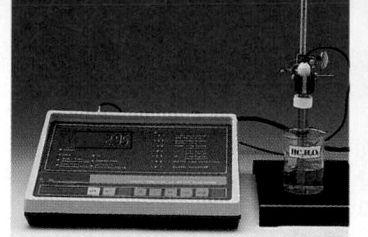 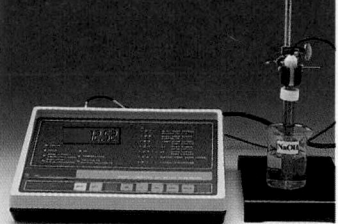 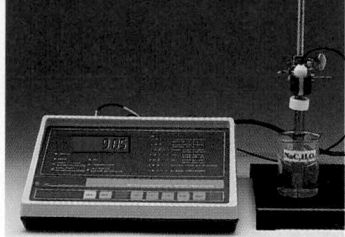

The photos in **Figure 15.9** show what happens when 0.100*M* solutions of sodium hydroxide and acetic acid are mixed. Notice that the pH of the 0.100*M* $HC_2H_3O_2$ is greater than the pH of the 0.100*M* HCl solution in **Figure 15.6.**

As the sodium hydroxide solution is added to the acetic acid solution, the pH increases as the acidic hydrogens from acetic acid molecules react with the hydroxide ions. When equal volumes of the two are mixed, the final pH is greater than 7. The final reaction mixture is basic.

As in the case of the weak base-strong acid reaction, mixing equal moles of acid and base does not produce a neutral solution. Because the final pH in this case is basic, more hydroxide than hydronium ions must be present in the final reaction mixture.

Figure 15.9
Weak Acid, Strong Base
The reaction of a weak acid and strong base does not result in a neutral solution. The pH of a 0.100*M* $HC_2H_3O_2$ solution is approximately 3, and the pH of a 0.100*M* NaOH solution is approximately 13. When the reaction is complete, the pH of the resulting solution is approximately 9.

SECTION 15.1

3 ASSESS

Check for Understanding

• Check for understanding of the section material by having students answer the Section Review questions.

• Ask students to hard-boil two eggs at home and bring them to school for this activity. Ask them to bring and prepare two egg-dying solutions, one according to the instructions on the packet and the other without vinegar, and set the eggs in the two solutions for 10 minutes. *The dye solution without the vinegar imparts much less intensity of color to the egg.* Tell students that the outer layer of the eggshell is mostly a protein cuticle that contains amine ($-NH_2$) groups and that the amine groups must react with an acid in order to be receptive to the dye. Ask students to write the net ionic equation for the reaction of acetic acid with the $-NH_2$ group.
$HC_2H_3O_2 + -NH_2 \rightarrow C_2H_3O_2^- + -NH_3^+$ $\boxed{\text{L1}}$

SAMPLE PROBLEM **3** **Equations—Weak Acid, Strong Base**

Write the overall, ionic, and net ionic equations for the reaction of phosphoric acid with lithium hydroxide.

Analyze
• Identify the acid and base as strong or weak. Phosphoric acid, H_3PO_4, is a weak acid. Lithium hydroxide, LiOH, is a strong base.

Set Up
• Write a balanced equation for the overall reaction. The salt is lithium phosphate, Li_3PO_4. Water is also a product. You'll need three moles of LiOH for each mole of H_3PO_4. Three moles of water will be produced.

$$H_3PO_4(aq) + 3LiOH(aq) \rightarrow Li_3PO_4(aq) + 3H_2O(l)$$

Solve
• Now, write the ionic equation. Because H_3PO_4 is a weak acid, it is only partially ionized, and you write it in the ionic equation as H_3PO_4. LiOH is completely dissociated, as is the salt lithium phosphate. Be careful to keep the ionic equation balanced.

$$H_3PO_4(aq) + 3Li^+(aq) + 3OH^-(aq) \rightarrow$$
$$3Li^+(aq) + PO_4^{3-}(aq) + 3H_2O(l)$$

• Check for spectator ions. Li^+ is a spectator ion. Subtract Li^+ from both sides of the equation to get the net ionic equation.

$$H_3PO_4(aq) + \cancel{3Li^+(aq)} + 3OH^-(aq) \rightarrow \cancel{3Li^+(aq)} + PO_4^{3-}(aq) + 3H_2O(l)$$
$$H_3PO_4(aq) + 3OH^-(aq) \rightarrow PO_4^{3-}(aq) + 3H_2O(l)$$

Check
• Check to see that the reaction occurs through hydrogen ion transfer from phosphoric acid molecules to hydroxide ions. Every weak acid-strong base reaction occurs by this type of H^+ transfer.

15.1 Acid and Base Reactions **529**

$2HC_2H_3O_2(aq) + Ca^{2+}(aq) + 2OH^-(aq) \rightarrow Ca^{2+}(aq) + 2C_2H_3O_2^-(aq) + 2H_2O(l);$
$HC_2H_3O_2(aq) + OH^-(aq) \rightarrow C_2H_3O_2^-(aq) + H_2O(l)$

Additional Practice Problems

Write the overall, ionic, and net ionic equations for the following reaction.

1. tartaric acid, $H_2C_4H_4O_6$, and lithium hydroxide $H_2C_4H_4O_6 + 2LiOH \rightarrow 2H_2O + Li_2C_4H_4O_6$; $H_2C_4H_4O_6 + 2Li^+ + 2OH^- \rightarrow 2H_2O + 2Li^+ + C_4H_4O_6^{2-}$; $H_2C_4H_4O_6 + 2OH^- \rightarrow 2H_2O + C_4H_4O_6^{2-}$

2. benzoic acid, $HC_7H_5O_2$, and calcium hydroxide $2HC_7H_5O_2 + Ca(OH)_2 \rightarrow 2H_2O + Ca(C_7H_5O_2)_2$; $2HC_7H_5O_2 + Ca^{2+} + 2OH^- \rightarrow 2H_2O + Ca^{2+} +$

$2C_7H_5O_2^-$; $HC_7H_5O_2 + OH^- \rightarrow H_2O + C_7H_5O_2^-$

Reteach

![software icon] **Software:** Use the software program Acid Strength, PC 2701, Project SERAPHIM, to give students practice working with acid dissociation. [L1] [LEP]

Extension

Ask students to work in small groups, construct ball-and-stick molecular models of acetic acid and an amine group, notify you when they are ready, and demonstrate how the amine group acts as a base in the reaction from *Check for Understanding.* [L1] [LEP]

[COOP LEARN]

4 CLOSE

Enrichment

Explain that organic acids are often described as electron-pair acceptors, and organic bases as electron-pair donors. To illustrate this type of reaction, have students draw electron dot structures for the reactants and products of the reaction between H_2O and NH_3, yielding NH_4^+ and OH^-. [L1]

PRACTICE PROBLEMS

Write overall, ionic, and net ionic equations for the following reactions.

8. carbonic acid, H_2CO_3, and sodium hydroxide, NaOH

9. boric acid, H_3BO_3, and potassium hydroxide, KOH

10. acetic acid, $HC_2H_3O_2$, and calcium hydroxide, $Ca(OH)_2$

Weak and Weak: It's Uncertain

The strong-strong reaction plus the two types of weak-strong reactions are the favorable acid-base reactions. Looking at an acid-base reaction as occurring by H^+ transfer helps you to understand why the weak-weak reaction is not considered a favorable reaction.

Because neither a weak acid nor a weak base has a strong tendency to transfer a hydrogen ion, transfer between the two may occur, but it is uncommon. Reactions between a weak acid and a weak base generally do not play an important role in acid-base chemistry, as shown in **Figure 15.10.**

Figure 15.10

Weak Acid-Weak Base The weak acetic acid does not react with the weak base aluminum hydroxide.

SECTION REVIEW

Understanding Concepts

1. Write the overall, ionic, and net ionic equations for the following reactions.

   a) perchloric acid and sodium hydroxide
   b) sulfuric acid and ammonia
   c) citric acid, $H_3C_6H_5O_7$, and potassium hydroxide

2. For each of the reactions in question 1, predict whether the pH of the product solution is acidic, basic, or neutral. Explain.

3. Identify the acid and the base in each of the following reactions.

   a) $HBr(aq) + H_2O(l) \rightarrow H_3O^+(aq) + Br^-(aq)$
   b) $NH_3(aq) + H_3PO_4(aq) \rightarrow NH_4^+(aq) + H_2PO_4^-(aq)$
   c) $HS^-(aq) + H_2O(l) \rightarrow H_2S(aq) + OH^-(aq)$

Thinking Critically

4. **Applying Concepts** Think about what acid and what base would react to form each of the following salts. When each of the salts is dissolved in water, will its solution be acidic, basic, or neutral? If not neutral, use an equation to explain why.

   a) NH_4Cl c) $LiC_2H_3O_2$
   b) $NaCl$ d) $NH_4C_2H_3O_2$

Applying Chemistry

5. **Lactic Acid Reaction** Lactic acid is produced in muscles during exercise. It is also produced in sour milk due to the action of lactic acid bacteria. The formula for lactic acid is $HC_3H_5O_3$. Write the overall, ionic, and net ionic equations for the reaction of lactic acid with sodium hydroxide. Will the pH of the product solution be greater than 7, exactly 7, or less than 7?

SECTION REVIEW

1. All reactants and products are (aq) unless otherwise indicated. **a.** $HClO_4 + NaOH \rightarrow NaClO_4 + H_2O(l)$; $H^+ + ClO_4^- + Na^+ + OH^- \rightarrow Na^+ + ClO_4^- + H_2O(l)$; $H^+ + OH^- \rightarrow H_2O(l)$ **b.** $H_2SO_4 + 2NH_3 \rightarrow (NH_4)_2SO_4$; $2H^+ + SO_4^{2-} + 2NH_3 \rightarrow 2NH_4^+ + SO_4^{2-}$; $H^+ + NH_3 \rightarrow NH_4^+$ **c.** $H_3C_6H_5O_7 + 3KOH \rightarrow K_3C_6H_5O_7 + 3H_2O(l)$; $H_3C_6H_5O_7 + 3K^+ + 3OH^- \rightarrow 3K^+ + C_6H_5O_7^{3-} + 3H_2O(l)$; $H_3C_6H_5O_7 + 3OH^- \rightarrow C_6H_5O_7^{3-} + 3H_2O(l)$

2. **a.** neutral, pH = 7, only water as product **b.** acidic, pH < 7, NH_4^+ is a weak acid **c.** basic, pH > 7, $C_6H_5O_7^{3-}$ is a weak base

3. **a.** acid, HBr; base, H_2O **b.** acid, H_3PO_4; base, NH_3 **c.** acid, H_2O; base, HS^-

4. **a.** acidic, $NH_4^+(aq) \rightarrow NH_3(aq) + H^+(aq)$ **b.** neutral **c.** basic, $C_2H_3O_2^-(aq) + H_2O(l) \rightarrow HC_2H_3O_2(aq) + OH^-(aq)$ **d.** neutral

5. $HC_3H_5O_3 + NaOH \rightarrow NaC_3H_5O_3 + H_2O$; $HC_3H_5O_3 + Na^+ + OH^- \rightarrow Na^+ + C_3H_5O_3^- + H_2O$; $HC_3H_5O_3 + OH^- \rightarrow C_3H_5O_3^- + H_2O$; pH > 7

Applications of Acid-Base Reactions

Acid-base reactions play an important role in many chemical systems, whether you're interested in blood chemistry, the chemistry of acid rain and lakes, or the chemistry of your favorite shampoo. How is the acid-base balance maintained in each of these situations? Look at several applications of acid-base reactions and how balance is maintained.

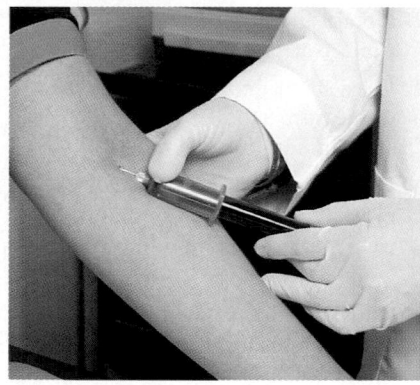

SECTION PREVIEW

Objectives
Evaluate the importance of a buffer in controlling pH. Design strategies for doing acid-base titrations, and calculate results from titration data.

Key Terms
buffer
titration
standard solution

Buffers to Regulate pH

Much blood chemistry depends on the acid-base balance in blood. The pH of your blood is slightly basic, about 7.4. If you're healthy, the pH of your blood does not vary by more than one-tenth of a pH unit. If you stop and think about all of the materials that go into and out of your blood, it is amazing that the pH remains so constant. Many of those materials are acidic or basic, which is why the constancy of the pH of blood is fascinating. Blood is an example of an effective buffer.

Buffers Defined

A **buffer** is a solution that resists changes in pH when moderate amounts of acids or bases are added. It contains ions or molecules that react with OH^- or H^+ if one of these ions is introduced into the solution.

Buffer solutions are prepared by using a weak acid with one of its salts or a weak base with one of its salts. For example, a buffer solution can be prepared by using the weak base ammonia, NH_3, and an ammonium salt, such as NH_4Cl. If an acid is added, NH_3 reacts with the H^+.

$$NH_3(aq) + H^+(aq) \rightarrow NH_4^+(aq)$$

If a base is added, the NH_4^+ ion from the salt reacts with the OH^-.

$$NH_4^+(aq) + OH^-(aq) \rightarrow NH_3(aq) + H_2O(l)$$

Look at another system that contains the weak acid acetic acid, $HC_2H_3O_2$, and the salt sodium acetate, $NaC_2H_3O_2$. If a strong base, OH^-, is added to the buffer system, the weak acid reacts to neutralize the addition.

$$HC_2H_3O_2(aq) + OH^-(aq) \rightarrow C_2H_3O_2^-(aq) + H_2O(l)$$

15.2 Applications of Acid-Base Reactions **531**

Program Resources

Study Guide, pp. 61-62 L1
Critical Thinking/Problem Solving, pp. 16-17 L2
Laboratory Manual, 15.2, pp. 141-146 L1
Section Focus Transparency 31 and **Master,** p. 31 L1 LEP
Basic Concepts Transparency 42 and **Master,** pp. 83-84 L1 LEP

Problem Solving Transparency 18 and **Master,** pp. 35-36 L1 LEP
ChemLab and MiniLab Worksheets, pp. 95-97, 99 L1
Tech Prep Applications, pp. 23-24 L1

SECTION 15.2

PREPARE

Key Concepts
The acid-base reactions involved in buffering and antacid use are examined. The stoichiometry of acid-base reactions is explained and related to titration.

Planning Ahead
For the ChemLab Order or purchase several brands of vinegar for the ChemLab.
For the MiniLab Purchase or prepare a pH 7 buffer solution.
For the Demonstrations Purchase plant material for Demonstration 15.6.

1 FOCUS

Focus Transparency

Display the **Section Focus Transparency** for Section 15.2 to show a soda-acid fire extinguisher. L1 LEP

2 TEACH

Content Background
Conjugate Acids and Bases A conjugate pair of acids and bases are related by a hydrogen ion: the acid has it; the base doesn't. Ammonium, NH_4^+, and ammonia, NH_3, are a conjugate pair. A buffer is always comprised of a solution that contains ideally equal quantities of an acid and its conjugate base or a base and its conjugate acid.

Using an Analogy
Ask students to explain the function of a buffer zone between two hostile countries. *It partially protects the countries and lessens the impact of any strategic change by one of them.* Point out that, in a similar manner, a chemical buffer lessens the impact on pH when any acid or base is added to a solution. L1

531

miniLAB

What does a buffer do?

Purpose: To compare the amounts of acidic and basic solutions required to effect similar changes in pH for a pH 7 sodium chloride solution and a pH 7 buffered solution.

Process Skills: Observing, measuring, interpreting data, using space/time relationships, communicating

Disposal: Code J

Teaching Strategies

• Remind students not to use the same graduated cylinders for the two solutions.

• 120 mL of 0.100M HCl may be prepared by adding 1.00 mL concentrated (12M) HCl to 119 mL distilled water. 0.100M NaOH solution may be prepared by adding 0.400 g NaOH to distilled water to make 100.0 mL of solution.

Expected Results

Drops of HCl to change pH of NaCl; of buffered solution *one; 10-20* Drops of NaOH to change pH of NaCl; of buffered solution *one; 10-20*

Analysis

1. A great deal more HCl is required to lower the pH of the buffered solution.
2. A great deal more NaOH is required to raise the pH of the buffered solution.
3. The buffer tends to resist any change in the pH of the solution.

Oral: Ask students to describe situations in which buffering might be beneficial. *Answers may include aspirin and lakes and streams.* [L1]

miniLAB 2

What does a buffer do?

Compare the amounts of acidic and basic solutions required to cause similar pH changes in two solutions: one a sodium chloride solution at a pH of 7 and the other a buffered solution with a pH of 7.

Procedure

1. Wear aprons and safety goggles.
2. Obtain and number four test tubes 1-4.
3. Use a small, graduated cylinder to measure 5.0 mL of NaCl solution into tubes 1 and 3.
4. Use another small, graduated cylinder to measure 5.0 mL of pH 7 buffer solution into tubes 2 and 4.
5. Add two drops of methyl orange indicator to tubes 1 and 2. The methyl orange is yellow if the pH is greater than 4.4 and will change to orange-red as the pH is lowered from 4.4 to 3.2.
6. Add 0.100M HCl solution drop by drop to tubes 1 and 2, stirring after each drop, just until the solution color changes to orange-red. Record the number of drops required for each solution.

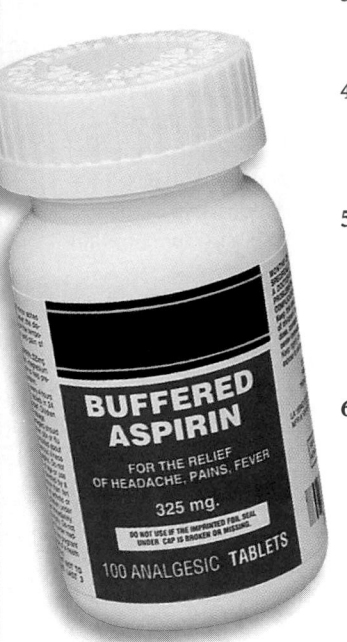

7. Add two drops of phenolphthalein indicator to tubes 3 and 4. The phenolphthalein is colorless if the pH is lower than 8.2 and will change to pink or magenta as the pH is raised from 8.2 to 10.0.
8. Add 0.1M NaOH solution drop by drop to tubes 3 and 4, stirring after each drop, just until the solution color changes to pink or magenta. Record the number of drops required for each solution.

Analysis

1. Compare the number of drops of acid required to lower the pH of the two solutions sufficiently that the methyl orange changed color.
2. Compare the number of drops of base required to raise the pH of the two solutions sufficiently that the phenolphthalein changed color.
3. How would you describe the effect of the buffer on the pH of the solution?

This reaction takes care of the added OH$^-$. If H$^+$ is added, the acetate ion from the NaC$_2$H$_3$O$_2$ is available to neutralize the added H$^+$.

$$C_2H_3O_2^-(aq) + H^+(aq) \rightarrow HC_2H_3O_2(aq)$$

This system is shown in **Figure 15.11**.

Notice in **Figure 15.11** that the pH does not remain constant in the buffer solution. It changes slightly in the direction of the pH of the added acid or base. These pH changes are insignificant when you compare them to the changes that occur in the unbuffered solution.

These two buffer systems are common ones used in many laboratories. Take a closer look at the specific buffer chemistry of your blood, which operates just like these laboratory buffers.

532   **Chapter 15 Acids and Bases React**

Demonstration 15.4

The Buffering Effect

Purpose: To demonstrate the effect of increasing the concentration of one ion involved in a reaction at equilibrium.

Materials: two 100-mL graduated cylinders, 40 g NaCl, 200 mL water, 2 effervescent antacid tablets, filter paper, funnel

Safety Precautions:

Disposal: Code C

Procedure: Add 40 g NaCl to 100 mL of water while stirring. Filter this solution into a 100-mL graduated cylinder. Fill a second cylinder with water. Simultaneously drop an effervescent antacid tablet into each cylinder.

Results: The tablet in the saturated salt solution did not fizz as much as the one in plain water. Discuss the students' observations in terms of Le Chatelier's principle.

Analysis: Ask these questions.

1. The tablet in which cylinder does not react

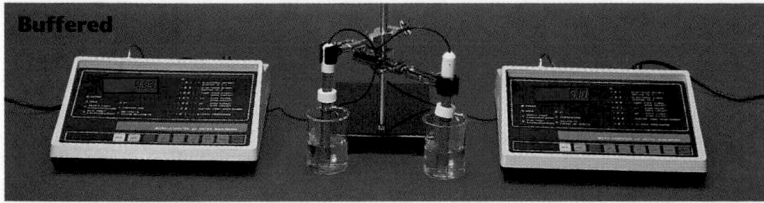

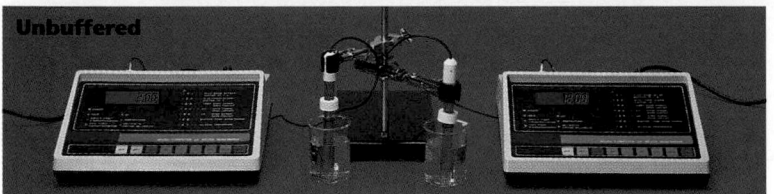

Figure 15.11
Buffers
A buffer maintains the pH of a solution at a fairly constant value. Compare what happens when acid and base are added to an acetic acid/sodium acetate buffer system at pH = 5 (top) to what happens when the same amount of acid and base are added to an unbuffered solution of pH = 5 (bottom).

⚡ Quick Demo

Does buffering aspirin really matter? 👓

Put a crushed, plain aspirin tablet in 100 mL of distilled water in a 250-mL beaker and a crushed, buffered aspirin tablet in another similar setup. Put magnetic stirring bars in both beakers and set them on magnetic stirrers. Turn the stirrers on low and set digital pH meter probes in both solutions. Record the initial pH of each solution and the pH after 5 minutes and after 10 minutes. *The buffered tablet will dissolve faster and will produce a solution of about pH 6. The unbuffered tablet will produce a solution of about pH 3.* Point out that persons who are prone to acid indigestion or who have ulcers should use only buffered aspirin.

Blood Buffer: Dissolved CO_2

The ability of blood to maintain a constant pH of 7.4 is due to several buffer systems. Dissolved carbon dioxide makes up one of the systems. Remember that when carbon dioxide dissolves in water, it produces carbonic acid, H_2CO_3.

$$CO_2(g) + H_2O(l) \rightarrow H_2CO_3(aq)$$

The other part of the blood buffer is the hydrogen carbonate ion, HCO_3^-. If something happens to increase OH^- in your blood, H_2CO_3 reacts to lower the OH^- concentration and keep the pH from increasing. If H^+ enters the blood, HCO_3^- reacts to keep the pH from decreasing.

added OH^-: $H_2CO_3(aq) + OH^-(aq) \rightarrow HCO_3^-(aq) + H_2O(l)$

added H^+: $HCO_3^-(aq) + H^+(aq) \rightarrow H_2CO_3(aq)$

Keep in mind that the level of CO_2 in the blood, and therefore the level of carbonic acid, is ultimately controlled by the lungs. If the amount of H^+ in the blood increases, a large amount of H_2CO_3 is produced, which lowers the H^+ concentration. In order to reduce the carbonic acid concentration produced, the lungs work to remove CO_2. Yawning is a mechanism that your body uses to get rid of extra CO_2.

On the other hand, rapid and deep breathing can cause a deficiency of carbon dioxide in the blood. This problem is called hyperventilation. It often occurs when a person is nervous or frightened. In this case, the air in the lungs is exchanged so rapidly that too much CO_2 is released. CO_2 blood levels drop, which causes the amount of carbonic acid in the blood to decrease. This causes the blood pH to increase and can be fatal if steps are not taken to stop this type of breathing.

When a person hyperventilates, he or she needs to become calm and breathe regularly. If the person breathes with a paper bag covering his or her nose and mouth, the concentration of CO_2 is increased in the air breathed. More CO_2 is forced into the blood through the lungs, and the pH of the blood drops to its normal level. **Figure 15.12** shows the narrow pH range of blood and summarizes the behavior of the buffer that controls the pH.

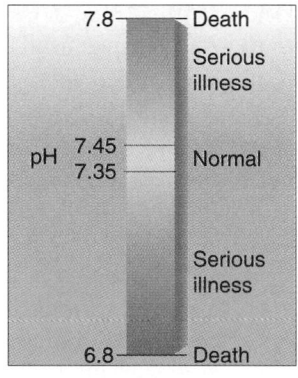

Figure 15.12
Blood pH
The pH of human blood is maintained within a narrow range by a mixture of buffers. The H_2CO_3/HCO_3^- system is one of the important parts of the blood buffer.

Extension

Ask interested students to prepare their own carbonic acid-hydrogen carbonate ion buffer system by blowing through a soda straw into 100 mL of $0.10M$ $NaHCO_3$ solution for 5 minutes. Then let them test 50 mL distilled water and 50 mL of the buffer solution for changes in pH, adding 5 drops of universal indicator to both, when $0.20M$ HCl is added drop-by-drop. Also have them do a similar experiment with $0.20M$ NaOH. ⬜L1

Correcting Misconceptions

Students may think that hyperventilating could cause serious physical damage. Point out that the body has a natural defense mechanism, which is fainting. Ask students what would occur to correct the problem if a person fainted. *The person's breathing and, consequently, the blood level of CO_2, would return to normal.*

as you expected? *The cylinder with the salt water didn't fizz as expected.*
2. The effervescent antacid tablet contains the sodium citrate-citric acid buffer system and $NaHCO_3$. What effect would increasing the concentration of the sodium ion have on the system? *The high Na^+ concentration keeps the sodium citrate and $NaHCO_3$ in the undissociated form.*

✓ Assessment

Performance: Have students predict the results of using a salt that does not contain sodium. They should then perform the experiment to test their hypothesis. ⬜L2

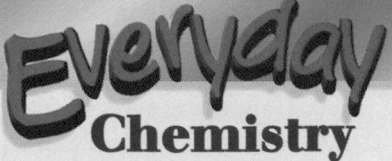

Purpose

To discuss the causes and cures of hiccups and how they relate to CO_2 in the blood.

Teaching Strategies

- Hiccups are classed as benign (those lasting no longer than 48 hours), persistent (longer than two days and less than a month), and intractable (longer than a month). The last two categories are rare but cause serious health problems—sometimes death.
- Two excellent references are: Rousseau, P. "Hiccups." *Southern Medical Journal*, Feb. 1995, pp. 175-179; and Tiermey et al. *Current Medical Diagnosis and Treatment*, 34th ed., 1995, pp. 479-480.
- Point out that a 6.8-7.35 blood pH range is known as acidosis and 7.45-7.8 as alkalosis. Any pH below 6.8 or above 7.8 results in death.
- Ask students how holding one's breath would affect the concentration of CO_2 and pH of the blood. *The CO_2 concentration would increase, and the pH would decrease.*

Extension

Have students do research to find problems involved in hiccups lasting longer than 48 hours and possible cures. L1

Exploring Further

1. Student answers may reflect cures that stop the rhythmic reflex and those that interrupt normal breathing.
2. Because the bag contains exhaled air, the CO_2 concentration is higher than that in atmospheric air. An increased CO_2 concentration helps control hiccups.

Hiccups

Have you ever wondered what causes hiccups? They are a reflex reaction. Hiccups afflict almost everyone, but remain a scientific puzzle. They serve no useful purpose.

What are hiccups? Most of the time, hiccups are a minor, harmless annoyance lasting a few minutes to several hours. They are repeated, involuntary spasms of your diaphragm, which is the dome-shaped breathing muscle separating the chest area from the abdomen. When hiccups start, the diaphragm jerks, and the air coming in is abruptly stopped when a small flap (the epiglottis) suddenly closes the opening to the windpipe (the glottis), resulting in the familiar "hic." No one is sure why this happens. It might start in a hiccup center in the brain stem by abnormal stimulation of nerves that control the diaphragm and the glottis.

How are they caused? Hiccups often occur when your stomach is distended due to gas, overeating, or drinking too many carbonated beverages. Other causes are sudden temperature changes, such as drinking hot or cold beverages or taking a cold shower, and sudden excitement or stress.

Possible cures An astounding number of common cures stop the rhythmic reflex. These methods include massaging the area with a cotton swab, gargling with water, sipping ice water, eating a spoonful of dry sugar, and biting on a lemon. Interruptions of normal breathing such as sneezing, coughing, and sudden pain or fright may stop hiccups.

Most cures that seem to work are related to an increase of CO_2 in the blood, which slightly lowers

the pH. The body has a buffer system to maintain a blood pH range of 7.35 to 7.45. Slight decreases in pH might turn off certain nervous controls that cause hiccups.

Normal Breathing and Hiccuping

A When you breathe in, the diaphragm contracts and flattens. The glottis and epiglottis open, and air rushes in.

B When you breathe out, the diaphragm relaxes and resumes its dome shape, pushing the air out.

Epiglottis
Glottis
Larynx

Position of ribs when inhaling

Position of ribs when exhaling

Position of diaphragm when exhaling

Lung when exhaling

Lung when inhaling

Position of diaphragm when inhaling

C When you hiccup, your diaphragm twitches. Air is forced past the glottis and epiglottis, which snap shut and cause a hiccup.

Exploring Further

1. **Comparing and Contrasting** Compare and contrast possible cures for hiccups.
2. **Hypothesizing** Hypothesize as to why breathing in and out of a paper bag or holding your breath might stop hiccups.

Acid Rain Versus Acid Lakes

In Chapter 14, you learned about the sources of acid rain and about its impact on plant and animal life, as well as on human-made objects such as monuments and buildings. When acid rain falls on lakes and streams, it might be expected that the acidity of the water increases and the pH decreases.

This effect is true in some cases. Many lakes in the northeastern United States, southern Canada, northern Europe, and the Scandinavian countries have pHs as low as 4.0. The pH of a healthy lake is about 6.5.

Other lakes, many of which are in the midwestern United States, appear to get the same amount of acid rain as these low-pH lakes, but they do not show a dramatic lowering of pH. The key to the ability of these lakes, one of which is shown in **Figure 15.13,** to resist the pH-lowering effects of acid rain is their local geology.

Figure 15.13
Buffered Lakes
If the rock and soil that compose and surround a lake bed are rich in limestone, the lake can neutralize acid rain by acid-base reactions. These lakes have a capacity to absorb acid rain without an appreciable change in pH. The water in these lakes behaves as a buffer.

Limestone is primarily calcium carbonate, $CaCO_3$. Calcium carbonate reacts with carbon dioxide and water to form calcium hydrogen carbonate, $Ca(HCO_3)_2$, a water-soluble compound. Lakes in areas rich in limestone have significant concentrations of hydrogen carbonate ions.

$$CaCO_3(s) + CO_2(g) + H_2O(l) \rightarrow Ca(HCO_3)_2(aq)$$

Just as in the regulation of pH in the blood, hydrogen carbonate ions produced from calcium hydrogen carbonate form a base that can neutralize acid in lakes.

$$HCO_3^-(aq) + H^+(aq) \rightarrow CO_2(g) + H_2O(l)$$

The Acid-Base Chemistry of Antacids

The acid-base chemistry of stomach upset is big business. Even if you never use commercial antacids, a lot of other people do. The terms *acid indigestion* or *acid stomach* are a bit misleading. You need an acid stomach in order to be healthy. Remember that the pH of stomach acid, which is mostly hydrochloric acid, is about 2.5.

Discussion

Explain to students that farmers provide soda (carbonated) water to chickens during hot summers to prevent problems with chicken eggshells, which may become so thin that they break easily. Also explain that chickens pant to keep cool and that the carbon to create $CaCO_3$, the major component of the shells, is derived primarily from CO_2 in the blood. Ask students why the problem occurs and how soda water helps solve it. *Panting by the chickens is comparable to hyperventilation and lowers the level of CO_2 in their blood. The soda water is saturated with CO_2, which subsequently increases the CO_2 level in the chickens' blood.*

Discussion

Ask students how they would respond to the statement "The buffering allowed only a slight change in acidity, from pH 5.0 to 4.5." *The pH scale is logarithmic; therefore, a change from 5.0 to 4.5 is appreciable.* Point out that it is, in fact, approximately a three-fold change in acidity.

GLENCOE TECHNOLOGY

 Videodisc

Science and Technology Videodisc Series
Treating Acid Lakes
Disc 2, Chemistry
Side 2, Ch. 2

Use this lesson to reinforce the concept of neutralization reactions in the environment.
L1 LEP

Chemistry Journal

TECH PREP

Measuring pH
Ask students to investigate the pH of soil, rain, lake water, and stream water in their local area and to explain the reasons for the pH levels. Provide opportunities for students to test samples with pH meters and/or indicators, and ask them to include the results in their journals. L1

Meeting Individual Needs

Gifted Ask gifted students to write the equations for the reactions of both added hydrogen ion and hydroxide ion with a carbonic acid-hydrogen carbonate ion buffer solution. $H^+ + HCO_3^- \rightarrow H_2CO_3$ and $OH^- + H_2CO_3 \rightarrow HCO_3^- + H_2O$ L3

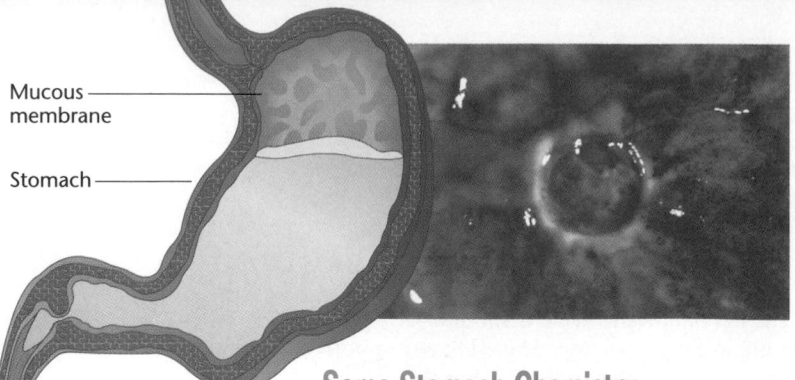

Mucous membrane

Stomach

Figure 15.14
Stomach Ulcer
A gastric ulcer forms when the membrane that protects the stomach lining breaks down and stomach acid attacks the stomach wall.

Some Stomach Chemistry

The combination of an acidic environment and enzymes in the stomach works together to break down the complex molecules that you eat into smaller molecules that can be transported through your blood and delivered to every cell in your body. These smaller molecules are the source of energy and structural material for the cells.

Your stomach is made of protein that is not much different from the protein in a hamburger. The inner walls of your stomach are coated by a basic mucous membrane that protects the stomach from the digestive power of the acid and enzymes. If the stomach contents become too acidic, this basic membrane breaks down by acid-base neutralization reactions. At the point where the barrier is neutralized, the gastric juices can begin to digest the protein that makes up the stomach wall, as shown in **Figure 15.14.** This causes the discomfort of acid indigestion and may cause more serious problems.

In most individuals, this breakdown of the mucous membrane corrects itself in a short time, and there is no long-term damage to the stomach. However, in some cases, the damage may be more long-term, and if not treated, a stomach ulcer may result. The correction process can be speeded up by using an antacid.

Fact of the MATTER

An alternative approach to controlling acid indigestion is now available in over-the-counter medication. These drugs work by decreasing the secretion of acid in the stomach. These drugs were previously available only by prescription for persons who suffer severe acid indigestion or who have a tendency for gastric ulcers.

Antacids = Anti-Acids = Bases

Although a variety of over-the-counter and prescription antacids is available, they all share common acid-base chemistry. **Table 15.2** gives a list of the bases that are most widely used in antacids. The compounds can be divided into two categories: the hydroxide-containing bases and the carbonate-containing bases.

Table 15.2   Compounds Used in Antacids

Insoluble Hydroxides	Carbonate-Based
Aluminum hydroxide, $Al(OH)_3$	Calcium carbonate, $CaCO_3$
Magnesium hydroxide, $Mg(OH)_2$	Magnesium carbonate, $MgCO_3$
	Sodium hydrogen carbonate, $NaHCO_3$
	Potassium hydrogen carbonate, $KHCO_3$

Integrating THE Sciences

TECH PREP

Earth Science The geology surrounding lakes plays a major role in determining how susceptible lakes are to acid rain. Lakes in the midwestern United States often have limestone ($CaCO_3$) lake beds. Because $CaCO_3$ is a weak base, these lakes have the ability to neutralize acid rain that falls on the lake. Mountain lakes in the northeastern United States, on the other hand, often have granite, SiO_2, lake beds. The silicon dioxide has no neutralizing power, and these lakes are often severely affected by acid rain. Have students research different areas of the country and relate geology to problems with acid lakes. L1

Chemistry and SOCIETY

The Development of Artificial Blood

You have a serious accident and need a blood transfusion immediately. There is no time to type your blood. Would you rather have someone else's blood or artificial blood? Right now, this choice is not available, but it may be in the near future. Under the described situation, artificial blood might be the better choice.

Why is artificial blood needed? The fear of blood transfusions is on the rise. Even though many safeguards are used, people are afraid of contracting diseases such as AIDS and hepatitis. Shortages of certain types of blood occur frequently, especially if a catastrophic disaster happens. Some people think the solution to these problems is artificial blood. Researchers have invested hundreds of millions of dollars to develop such a substance.

Types of artificial blood In real blood, hemoglobin in red blood cells carries oxygen. Many types of artificial blood focus on the hemoglobin aspect of the blood. One group is working on using the hemoglobin from outdated human blood from blood banks. Other scientists obtain it from genetically engineered bacteria, cattle, or pigs. Hemoglobin that has been removed or that was never in red blood cells causes problems. Its molecule falls apart. It does not carry oxygen well, and it sometimes clogs blood vessels. Scientists from different companies are working on these problems. Some are doing animal testing, some are already testing humans, and others have plans to do so in the near future.

Another approach to this problem is to use an oxygen-carrying perfluorocarbon emulsion that has been around since 1966. At that time, its oxygen-holding power was demonstrated by the fact that mice could breathe oxygen and survive when submerged in it. It is not considered a blood substitute by its manufacturer, but rather a drug-delivery system.

Artificial success? It will probably be several years before most oxygen-carrying blood substitutes will be approved by the FDA. However, one artificial blood component is currently in use. In the past, hemophiliacs have run a much higher risk of getting HIV because many units of blood were needed to get enough blood-clotting factor VIII. Now, the FDA has approved two genetically engineered factor VIIIs—Recombinate and Kogenate—so people who need factor VIII have a safe alternative.

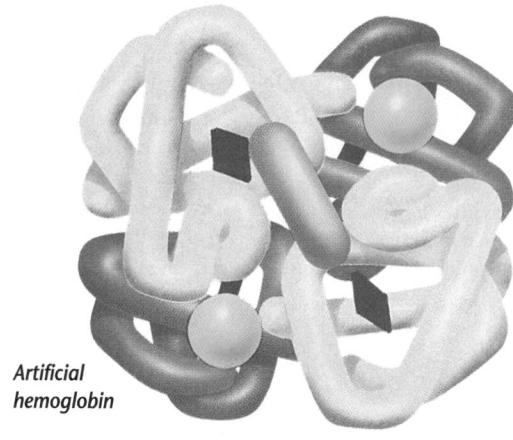

Artificial hemoglobin

Analyzing the Issue

1. **Acquiring Information** Research the structure of hemoglobin, and make a simplified model of hemoglobin.

2. **Inferring** Analyze why scientists working to develop artificial blood need an extensive knowledge of chemistry.

3. **Thinking Critically** What problems might arise with government control of artificial blood research? What advantages might there be?

15.2 **Applications of Acid-Base Reactions** 537

Analyzing the Issue

1. Hemoglobin has two long and two short polypeptide chains. Each chain is connected to an iron-containing heme group. The four chains and their groups are wound in such a way as to form a roughly spherical shape.

2. To understand the complex chemical structure of blood components and the way they work involves a tremendous understanding of chemistry. To duplicate or make a substance to replace these substances requires the same amount of knowledge.

3. The discussion might include ethical issues such as whether it is right to give a placebo in a life-death situation to establish experimental correctness with a control group.

Chemistry and SOCIETY

Purpose
To discuss alternatives to using homologous blood transfusions and the types of blood substitutes being developed.

Background
- The gene for human factor VIII is put into hamster ovary cells, causing them to make the recombinant form.
- Making money is one motivation for developing artificial blood. Potential manufacturers estimate making more than $2 billion a year.
- The iron in the four hemoglobin heme groups holds the oxygen. When one heme group takes up an oxygen molecule, it makes it easier for the next one to do the same, and so on.
- When hemoglobin is taken out of red blood cells, it gives up its oxygen more reluctantly. It also falls apart and clogs blood vessels in the kidneys.
- Two good articles are: Farley, D. "Alternatives to Regular Blood Transfusions." *FDA Consumer,* July/Aug. 1994; and Edelson, Edward. "Fear of Blood." *Popular Science,* June 1993, pp. 108-111+.

Teaching Strategies
- Divide the class into five groups, and have each group research a method for developing oxygen-carrying artificial blood. Group reports can be given. [L1] **COOP LEARN**
- Ask the local director of the Red Cross Bloodmobile to speak to your classes about how donated blood is handled, how is it stored, how and why it is separated, and the safety measures put in place to prevent contracting diseases such as AIDS and hepatitis. These questions could also serve as a guide for a student extra-credit research project. [L2]

537

Ask students why bases such as sodium, potassium, and barium hydroxides are not used as antacids. *They are highly reactive, corrosive, and toxic.*

Tying to Previous Knowledge

Ask students to write the net ionic equations for the reactions of calcium carbonate and magnesium hydroxide with hydrochloric acid. $CaCO_3 + 2H^+ \rightarrow Ca^{2+} + H_2CO_3$ *and* $Mg(OH)_2 + 2H^+ \rightarrow Mg^{2+} + 2H_2O$

GLENCOE TECHNOLOGY

Videodisc

Chemistry: Concepts and Applications
Forming a Salt
Disc 2, Side 1, Ch. 18

Show this video of the demonstration on these pages to introduce the concept of salt formation. L1 LEP

CD-Rom

Chemistry: Concepts and Applications
Forming a Salt
Have students use this video of the demonstration on these pages to help introduce the concept of salt formation. L1 LEP

Hydroxide Antacids

If you examine the ingredient list on an antacid, you probably recognize any hydroxide-containing compounds as bases. The hydroxides used in antacids have low solubility in water. Because the pH of saliva is neutral to slightly basic, these insoluble hydroxides do not dissolve and react until they get past your mouth and upper digestive tract and into the highly acidic environment of the stomach.

Milk of magnesia, a suspension of magnesium hydroxide in water, is a good example of this type of antacid. If you have ever used unflavored milk of magnesia, you may have noticed that it has a bitter taste, which is typical of a base.

Carbonate Antacids

You are familiar with the acid-neutralizing abilities of the carbonates and hydrogen carbonates from the discussion about blood buffering and acid rain. Carbonate and hydrogen carbonate antacids react with HCl to form carbonic acid, which decomposes into carbon dioxide gas and water.

$$CaCO_3(s) + 2HCl(aq) \rightarrow CaCl_2(aq) + H_2CO_3(aq)$$
$$NaHCO_3(aq) + HCl(aq) \rightarrow NaCl(aq) + H_2CO_3(aq)$$
$$H_2CO_3(aq) \rightarrow H_2O(l) + CO_2(g)$$

Many carbonates and hydrogen carbonates are insoluble in water and have great neutralizing power. They are the primary components of many over-the-counter antacid tablets. See an example of such an antacid in **Figure 15.15.**

As with any over-the-counter medication, antacids are designed for occasional use. Anyone who needs to use them frequently for attacks of indigestion may have a more serious health problem and should consult a physician.

Figure 15.15

Antacids
Antacids are hydroxide-containing bases or carbonate-containing compounds.

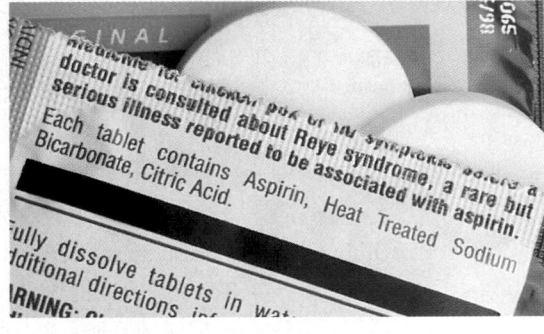

Some antacids contain a carbonate or hydrogen carbonate and a weak acid, such as citric acid. When added to water, these compounds react, producing a basic salt and carbonic acid, which decomposes into water and carbon dioxide. ▶

Demonstration 15.5

Forming a Salt

Purpose: To demonstrate that a salt is a product of an acid-base reaction.

Materials: 1 mL $6M$ HCl, 1 mL $6M$ NaOH, dropper, one 100-mL beaker, hot plate, microscope slide, dissecting microscope or magnifying hand lens

Safety Precautions:

Disposal: Codes C, D

Procedure: Use a dropper to place 20 drops of $6M$ HCl and 20 drops of $6M$ NaOH in a small beaker. Slowly evaporate the water from the beaker using a hot plate. Have students observe the NaCl salt crystals under a microscope or with a hand lens.

Results: Students observe small white cubic crystals of NaCl.

Stoichiometry Revisited: Acid-Base Titrations

Although acid-base reactions do not have to happen in aqueous solutions, many are water-based. In these reactions, it is frequently important to know the concentrations of the solutions involved.

The general process of determining the molarity of an acid or a base through the use of an acid-base reaction is called an acid-base **titration.** In a titration, the molarity of one of the reactants, acid or base, is known, but the other is unknown. The known reactant molarity is used to find the unknown molarity of the other solution. Solutions of known molarity that are used in this fashion are called **standard solutions.**

Titration of an Acid with a Base

Suppose you are in charge of cleaning up and doing inventory in the chemistry stockroom. Your first job is to catalog the variety of solutions that are on the shelf. One of the solutions you find is a tightly stoppered, well-labeled bottle of 0.100M NaOH.

On the next shelf, you find a tightly stoppered bottle that is not as well-labeled. The label has been damaged, and the only thing left of the label is M HCl. This tells that the solution contains HCl, but it does not tell the molarity. For the inventory form, you need to know the molarity of the solution. What can you do?

Do a Titration

You know that NaOH and HCl react completely.

$$HCl(aq) + NaOH(aq) \rightarrow NaCl(aq) + H_2O(l)$$

You know the concentration of the NaOH solution, so it is your standard solution. You can use the reaction, the volumes of acid and base used, plus the molarity of the base to determine the molarity of the unlabeled HCl. Follow the first part of this process in **Figure 15.16.**

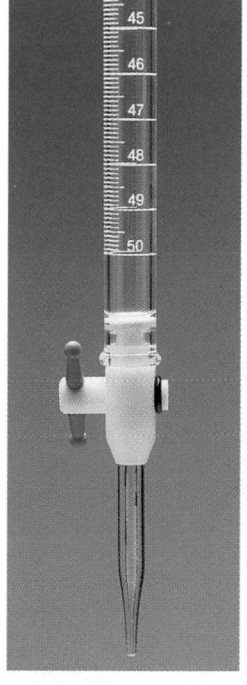

Figure 15.16
Set Up the Titration
A titration requires mixing measured volumes of the standard solution and the unknown solution.

To add the NaOH to the HCl and to measure the amount of NaOH solution needed, a burette is used. A burette is a long, calibrated tube with a valve at the bottom. ▶

◀ You need to carefully measure a volume of the unknown HCl solution into a flask. For careful volume measurements, you can use a pipet. For this titration, add 20.0 mL of the unknown acid to the flask using a 20.0-mL pipet.

Analysis: Ask these questions.
1. Did the positive metallic ion in the salt come from the acid or the base? *base*
2. Other than a salt, what product forms when an acid reacts with a base? *water*

Assessment

Oral: Have students review the classification of chemical reactions from Chapter 6. Ask students how chemists would classify a chemical reaction between an acid and a base. *double-displacement reaction* [L1]

SECTION 15.2

Concept Development
Explain that the success and accuracy of titration depend in large part upon carefully measuring the volumes of both reactant solutions and that titration is a form of volumetric analysis. Show students a buret, demonstrate proper titration technique, and demonstrate the correct method of reading delivered volume to 0.01 mL.

Correcting Misconceptions
Indicators have colorful and interesting names, and students tend to think of them as mysterious and magical substances. Draw structural formulas for the colorless and pink forms of phenolphthalein on the chalkboard, and ask students what happens to the colorless form that transforms it to the pink form. *It loses an H^+ ion and a water molecule.* Then ask how the colorless form is categorized as it reacts, yielding the pink form. *an acid because it loses a proton* Explain that this change occurs when the OH^- concentration in the solution has reached a certain level of about pH 9.

Transparency
 Display **Problem Solving Transparency 18** to give students practice using indicators. [L1] [LEP]

GLENCOE TECHNOLOGY

Videotape
MindJogger Videoquizzes for Chemistry
Chapter 15: Acids and Bases React
Have students work in groups as they play the videoquiz game to review key chapter concepts. [L1] [LEP] [COOP LEARN]

Using Indicators in a Titration

When the solution is neutral, you know that you have added exactly enough base to react with the amount of acid present, and you are at what is known as the endpoint of the titration. But how do you know when the endpoint is reached? Probably the best way to indicate the endpoint of a titration is to use an acid-base indicator. Different indicators change color at different pH values. Look at the indicators in **Figure 15.17.** For a titration of NaOH and HCl, the endpoint is reached when the solution reaches a pH of 7. Therefore, you need an indicator that changes color as close to pH = 7 as possible. The best indicator for this titration is bromothymol blue, which changes from yellow to blue close to pH = 7.

Figure 15.17
Indicators and Titration
The graphs show some of the more popular acid-base indicators used in the chemistry laboratory and the pH at which they change color.

Titration Process

Assume you have a burette containing an NaOH standard solution and a flask containing 20.0 mL of HCl of unknown concentration, as described in **Figure 15.16.** You know what indicator to use, according to **Figure 15.17.** To actually perform the NaOH-HCl titration, follow the process outlined in **Figure 15.18,** and record the data collected.

The reaction between a strong acid and a strong base results in a completely neutral solution. Bromothymol blue is an effective indicator for such reactions because it changes color at a pH of 7. ▶

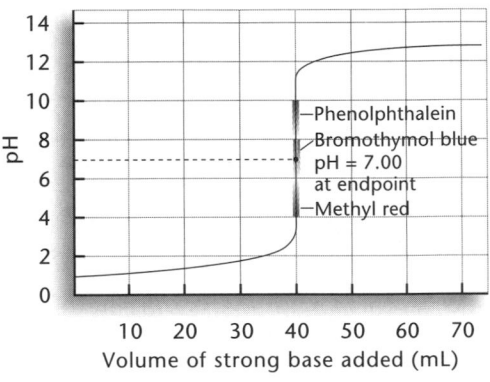

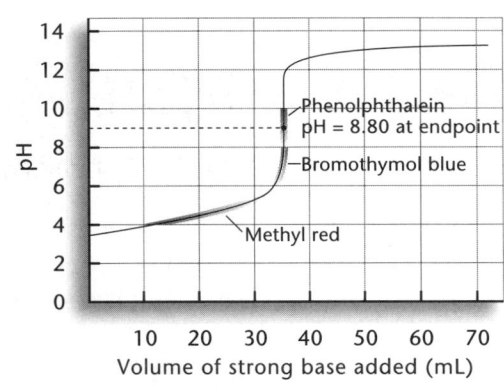

▲ Because the reaction between a weak acid and a strong base results in a slightly basic solution, the endpoint pH for a weak acid-strong base titration is greater than 7. For such a titration, phenolphthalein changes color at the endpoint.

The titration of a weak base with a strong acid has an endpoint pH that is less than 7. For this titration, methyl red is a good indicator because its pH range matches closely the endpoint pH. ▶

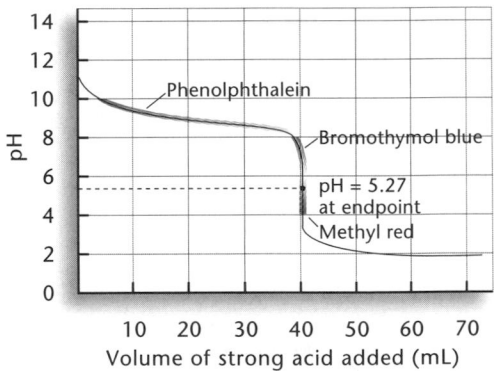

Demonstration 15.6

Indicators in Nature

Purpose: To demonstrate that natural indicators can be made from material such as blueberry juice, radishes, rose petals, carrots, and red beets. (Remind students that they used red cabbage as a natural indicator in the ChemLab in Chapter 14.)

Materials: 30 g of vegetable, 25 mL ethanol, vegetable shredder, 250-mL beaker, hot plate, series of solutions having a pH range of 1 to 13,

13 test tubes

Safety Precautions: 🚫 💪 🥽 🔥
Disposal: Code C
Procedure: Place 30 g of shredded vegetable in a small beaker with 100 mL H_2O and boil the mixture. After a few minutes, remove from heat, and add 25 mL ethanol. **CAUTION:** *Ethanol is flammable.* Prepare a series of solutions having a pH range of 1 to 13 by diluting 0.1M HCl and 0.1M NaOH. (See Demonstration 14.5.)

Figure 15.18
Titration Process
The following process shows how to perform and obtain data for an HCl-NaOH titration.

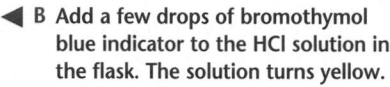

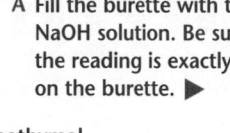

A Fill the burette with the NaOH solution. Be sure the reading is exactly 0 on the burette. ▶

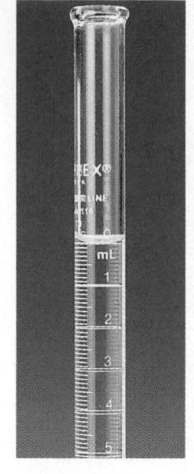

◀ **B** Add a few drops of bromothymol blue indicator to the HCl solution in the flask. The solution turns yellow.

C Titrate the HCl with the NaOH by slowly adding NaOH from the burette to the flask as the solution is constantly stirred. Continue to add the NaOH slowly until the color changes from yellow to blue. ▼

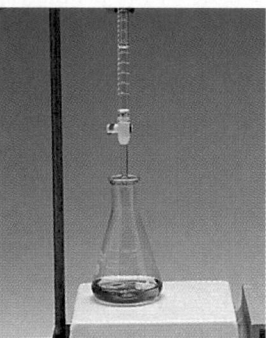

D Measure the final volume of NaOH solution, which is 19.9 mL in this example. This reading means that the volume of NaOH used in this titration is 19.9 mL. ▼

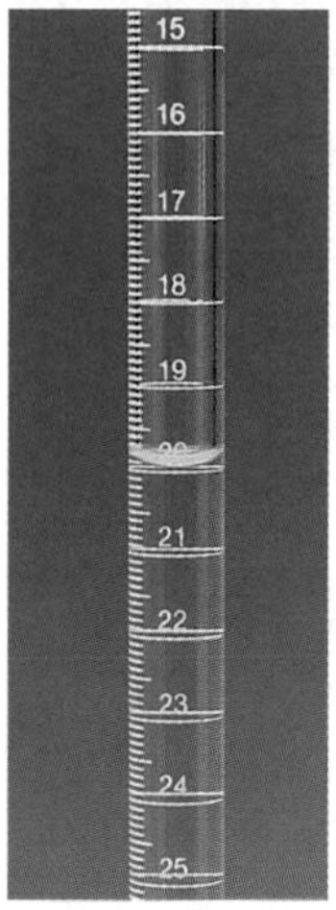

Determining Concentration: Using Stoichiometry

You now have all your experimental data. How do you combine these experimental data into an experimental result—the molarity of the HCl solution?

First, summarize what you know. You know that 20 mL of the HCl solution reacts with 19.9 mL of 0.100M NaOH solution to reach the endpoint. From the balanced equation for the reaction, you know that one mole of HCl reacts with one mole of NaOH. Therefore, the number of moles of HCl in 20.0 mL of the HCl solution equals the number of moles of NaOH in 19.9 mL of 0.100M NaOH solution.

Now, use the factor label method to solve this solution stoichiometry problem, just as you used it to solve other stoichiometry problems. Because you know the concentration of the NaOH solution, first find the number of moles of NaOH involved in the reaction.

$$\frac{19.9 \text{ mL soln}}{} \left| \frac{1 \text{ L}}{10^3 \text{ mL}} \right| \frac{0.100 \text{ mol NaOH}}{1.00 \text{ L soln}} = 1.99 \times 10^{-3} \text{ mol NaOH}$$

Next, examine the balanced equation for the reaction and determine that, because their coefficients are the same, equal numbers of moles of NaOH and HCl react.

Concept Development

Explain that, using an appropriate indicator, at the endpoint of a titration, mol H_3O^+ (from the acid) = mol OH^- (from the base). Thus, the stoichiometric calculations may be simplified by using the volumes of both solutions in milliliters, rather than converting them to liters. Point out that this is valid because M = mol/L = mmol/mL and because mmol H_3O^+ = mmol OH^- at the equivalence point.

GLENCOE TECHNOLOGY

Videodisc
Chemistry: Concepts and Applications
An Acid-Base Titration
Disc 2, Side 1, Ch. 19

Show this video to reinforce different measurements of pH in titrations. **L1** **LEP**

CD-Rom
Chemistry: Concepts and Applications
An Acid-Base Titration
Have students use this video to show different methods of determining pH in a titration. **L1** **LEP**

CAUTION: *Acids and bases are corrosive.* Place equal amounts of the pH standards in labeled test tubes. Add a few drops of the vegetable indicator to each tube. Have students observe the pH range where the indicator changes color. Emphasize that a good indicator changes color over a narrow pH range.
Results: The vegetable materials undergo distinct color changes.

Analysis: Ask this question.
1. Why is it important that a good indicator change color over a narrow pH range? *A narrow range will pinpoint a pH value.*

Assessment

Performance: Have students take a piece of chocolate laxative and grind it with ethanol. **CAUTION:** *Ethanol is flammable.* Place the mixture in a petri dish on an overhead projector and add a few drops of 1M NaOH (1 g NaOH in 25 mL H_2O). Ask students which indicator has this characteristic pink color. *Chocolate laxative contains phenolphthalein as the active ingredient.*

ChemLab

Titration of Vinegar

Time Allotment: One laboratory period if one to three types of vinegar are titrated.

Objectives

Review objectives with students before they begin the ChemLab.

Process Skills

Observing, measuring, using numbers, communicating

Safety Precautions

Remind students to wear safety goggles and aprons.

PREPARATION

The NaOH solution should be prepared to be as close to $0.100M$ as possible by adding 4.00 g of dry, solid NaOH to sufficient distilled water, which has been boiled and cooled to room temperature, to make 1.000 L of solution. The molarity may be determined to a greater degree of accuracy by titrating approximately 0.500-g samples of potassium hydrogen phthalate (204.44 g/mol), a primary standard, using phenolphthalein as an indicator. The titrations should require approximately 25 mL of NaOH solution.

Alternative Materials

• Potassium hydroxide may be used instead of sodium hydroxide, and other indicators, such as o-cresolphthalein and thymolphthalein, may replace phenolphthalein.

• More accurate titrations may be performed by students if they are allowed to use volumetric pipets to measure the vinegar samples and burets for the sodium hydroxide solution.

Titration of Vinegar

Vinegar is a solution of mostly acetic acid in water. It varies in concentration from about three percent to five percent by volume. The acetic acid in vinegar may be neutralized by adding sodium hydroxide.

$$HC_2H_3O_2(aq) + NaOH(aq) \rightarrow$$
$$H_2O(l) + NaC_2H_3O_2(aq)$$

You will titrate several brands of commercial vinegar with sodium hydroxide solution of a known concentration, and use your data to calculate the molarities and volume percentages of acetic acid in the vinegars.

Problem

What are the volume percentages of acetic acid in several brands of vinegar?

Objectives

• **Observe** acid-base titrations of several vinegars with a standard sodium hydroxide solution.

• **Calculate** the volume percentages of acetic acid in the vinegars.

• **Compare** the acetic acid concentrations of various brands of vinegar.

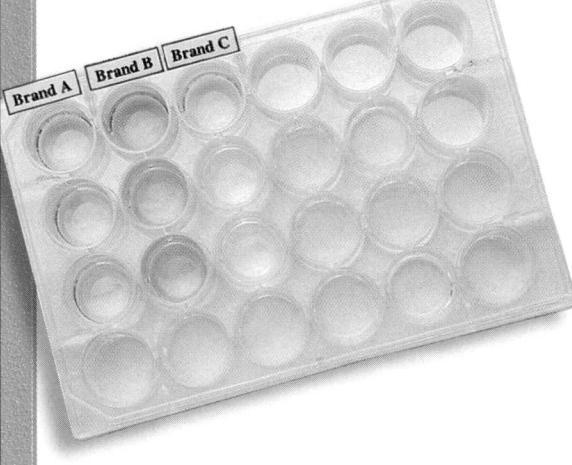

PREPARATION

Materials
24-well microplate
several brands of vinegar
labeled microtip pipets
standard NaOH solution
phenolphthalein solution
toothpicks
distilled water in a wash bottle

Safety Precautions

Wear aprons and goggles. Sodium hydroxide is caustic and can damage skin and eyes. If you come into contact with any of this solution, rinse the affected area with a large volume of water and notify the teacher. Wash hands thoroughly when you complete the lab.

PROCEDURE

1. Make a data table like the one shown.

2. Use a microtip pipet to add ten drops of the first type of vinegar to each of the wells—A1, B1, and C1—of the microplate. Record the brand of the first vinegar in your data table.

3. Use a clean microtip pipet to add one drop of phenolphthalein indicator solution to each of the three wells.

4. Set the microplate on a piece of white paper.

5. Use a clean pipet to carefully add a drop of the standard NaOH solution to the solution in well A1, and stir with a toothpick. Pause for about 30 seconds and look down through the well for evidence of a persistent, light pink, phenolphthalein color that indicates the endpoint of the titration. Repeat this process with each drop until the endpoint is reached. Record the num-

---

PROCEDURE

• Require that students' notebooks contain the following items before they begin: Objectives, Safety Precautions, Outline or Flow Chart of the Procedure, and Data and Observations Table.

• Endpoint phenolphthalein color fades with time, as CO_2 is absorbed from the air. Students may, by vigorous stirring for a period of time, reverse the endpoint. Point out that this is an invalid method of obtaining a good endpoint.

ANALYZE AND CONCLUDE

1. The sample calculation is shown assuming an NaOH molarity of 0.0982. Brand A: (0.0982 mol NaOH/1 L NaOH)(88 L NaOH/10 L acetic acid)(1 mol acetic acid/ 1 mol NaOH) = 0.864M acetic acid. A similar calculation for brand B yields a molarity of 0.555.

2. Brand A: 0.864M acetic acid $\times$ (1.00%/0.175M) = 4.94% acetic acid by volume. A similar calculation for brand B yields 3.17%.

3. brand A

ChemLab

ber of drops of sodium hydroxide solution required to titrate the vinegar to the endpoint.

6. Repeat procedure 5 with the second sample of this vinegar, which is in well B1. If the result differs by more than one drop from that of the first titration, repeat again with the sample in well C1.

7. Repeat procedures 2 through 6 with the other brands of vinegar, using other columns of wells of the microplate. Record your data after each titration.

ANALYZE AND CONCLUDE

1. **Interpreting Data** From the two closest trials, find the average number of drops of NaOH required to titrate each vinegar. Use this average number of drops and the given molarity of the NaOH to calculate the molarity of acetic acid in each of the brands of vinegar. Assuming identical volumes for drops of vinegar and drops of NaOH solution, the ratio of reacting volumes in liters is the same as the ratio of reacting volumes in drops.

2. **Interpreting Data** Use your results to calculate the volume percentage of acetic acid in each brand of vinegar according to the formula:
$M_{acetic\ acid} \times (1.00$ percent acetic acid$/0.175M$ acetic acid$) =$ percent acetic acid by volume.

3. **Comparing and Contrasting** Which of the brands of vinegar you tested contained the highest volume percent of acetic acid?

APPLY AND ASSESS

1. How could you have changed the experimental procedure in order to more accurately determine the concentrations of the vinegars?

2. If the cost and volume of each brand of vinegar are available, calculate the cost per percent of acetic acid per unit volume for each. Which is the best buy based upon this criterion?

DATA AND OBSERVATIONS

Type of Vinegar	Trial 1—Drops of NaOH	Trial 2—Drops of NaOH	Trial 3—Drops of NaOH
Brand A			

1. The volumes of the vinegar and sodium hydroxide solution could have been measured much more accurately if pipets and burets were available.

2. Assuming a cost of 79 cents/pint for brand A and 59 cents/pint for brand B, these are the calculations. Brand A: 79 cents/4.94%/1 pint = 16.0 cents/%/pint. Brand B: 59 cents/3.17%/1 pint = 18.6 cents/%/pint.

✓ Assessment

Knowledge: Ask students to formulate a similar procedure that might be used to determine the molarity of household ammonia solutions. Ask interested and capable students to carry out their procedures for testing household ammonia solutions. **L2**

GLENCOE TECHNOLOGY

CD-Rom

Chemistry: Concepts and Applications
Titrate to Determine Acid Concentration
Have students use this Interactive Experiment to practice doing acid-base titrations. **L1**

LEP

DATA AND OBSERVATIONS

Type of Vinegar	Trial 1—Drops of NaOH	Trial 2—Drops of NaOH	Trial 3—Drops of NaOH
Brand A	88	88	—
Brand B	59	56	57

Correcting Errors

- Remind students that unless an equation is correctly balanced, the mole ratios will not be correct when performing stoichiometric calculations.
- Students who are inexperienced in titration might bypass the endpoint. Demonstrate to students how they can return to the endpoint and how to include the amounts of solutions used in their calculations.

Applying Chemistry

Titration and Neutralization Ask students to brainstorm a list of practical applications of titration and neutralization reactions that are not listed in the student text. Applications may include control of certain industrial processes, such as the amount of acid present in soft drinks, or cleaning up spills of acids or bases. Students can present their ideas in a poster, collage, or bulletin board.

Transparency

Display **Basic Concepts Transparency 42** to reinforce the titration procedure.
L1 LEP

$$NaOH(aq) + HCl(aq) \rightarrow NaCl(aq) + H_2O(l)$$

Because 1.99×10^{-3} mol NaOH react, 1.99×10^{-3} mol HCl present in solution also react.

Finally, use the volume to find the molarity of the acid.

$$\frac{1.99 \times 10^{-3}\, \text{mol HCl}}{20.0\, \text{mL soln}} \bigg| \frac{10^3\, \text{mL}}{1\, \text{L}} = \frac{0.0995\, \text{mol HCl}}{1\, \text{L soln}} = 0.0995M\, \text{HCl}$$

Based on your single titration, the molarity of the HCl solution is $0.0995M$. However, before you put this value on the label, you probably would repeat the titration for several additional trials in order to verify your analysis and be more confident of the value on the label. Study the following Sample Problem that gives another example of a titration.

SAMPLE PROBLEM 4 Finding Molarity

A 15.0-mL sample of a solution of H_2SO_4 with an unknown molarity is titrated with 32.4 mL of $0.145M$ NaOH to the bromothymol blue endpoint. Based upon this titration, what is the molarity of the sulfuric acid solution?

Analyze
- Because the molarity of the NaOH solution is known, the number of moles of NaOH involved in the titration can be calculated. The corresponding number of moles of H_2SO_4 can then be determined, and this figure can be used to calculate the molarity of the acid.

Set Up
- Write the balanced equation for the reaction. Remember that sulfuric acid is a diprotic acid.

$$H_2SO_4(aq) + 2NaOH(aq) \rightarrow 2H_2O(l) + Na_2SO_4(aq)$$

Solve
- Because the concentration of the NaOH solution is known, find the number of moles of NaOH used in the titration.

$$\frac{32.4\, \text{mL soln}}{} \bigg| \frac{1\, \text{L}}{10^3\, \text{mL}} \bigg| \frac{0.145\, \text{mol NaOH}}{1.00\, \text{L soln}} = 4.70 \times 10^{-3}\, \text{mol NaOH}$$

- Using the balanced equation, find the number of moles of H_2SO_4 that react with 4.70×10^{-3} mol NaOH.

$$\frac{4.70 \times 10^{-3}\, \text{mol NaOH}}{} \bigg| \frac{1\, \text{mol H}_2\text{SO}_4}{2\, \text{mol NaOH}} = 2.35 \times 10^{-3}\, \text{mol H}_2\text{SO}_4$$

- Find the concentration of the H_2SO_4.

$$\frac{2.35 \times 10^{-3}\, \text{mol H}_2\text{SO}_4}{15.0\, \text{mL soln}} \bigg| \frac{10^3\, \text{mL}}{1\, \text{L}} = \frac{0.157\, \text{mol H}_2\text{SO}_4}{1\, \text{L soln}} = 0.157M\, \text{H}_2\text{SO}_4$$

Check
- Check to make sure that the final units are what they are supposed to be.

544　　Chapter 15　Acids and Bases React

PRACTICE

Guided Practice: Guide students through the sample problem on this page, then have them work in class on Practice Problems 11-13 on page 546. See Appendix D for complete solutions. L1

Independent Practice: Your homework or classroom assignment can include the additional practice problems shown on this page. L1

Answers to Practice Problems

11. $0.210M$ HBr　　12. $0.208M$ HNO$_3$
13. $0.227M$ Ca(OH)$_2$

Additional Practice Problems

1. A $0.125M$ NaOH solution is added to a buret. The initial buret reading is 1.5 mL. The solution is used to titrate 25.0 mL of H_2SO_4. The volume in the buret at the endpoint is 45.5 mL. What is the

molarity of the acid?

$$\frac{44.0\, \text{mL NaOH}}{1} \bigg| \frac{1\, \text{L}}{10^3\, \text{mL NaOH}} \cdots$$
$$\frac{0.125\, \text{mol NaOH}}{1\, \text{L}} \bigg| \frac{1\, \text{mol H}_2\text{SO}_4}{2\, \text{mol NaOH}} \cdots$$
$$\frac{1}{25.0\, \text{mL H}_2\text{SO}_4} \bigg| \frac{10^3\, \text{mL H}_2\text{SO}_4}{1\, \text{L}} \cdots$$
$$= 0.110M\, \text{H}_2\text{SO}_4$$

2. What volume of $0.100M$ HCl is required

Indicators

Indicators are acids and bases that have complicated structures and change color as they lose or gain hydrogen ions. The molecules must have acidic hydrogens, and they must have a relatively large number of carbon-carbon double bonds. The structure of the indicator thymol blue, $H_2C_{27}H_{28}O_5S$, satisfies both of these criteria.

1. At low pH, thymol blue contains two acidic hydrogens. In this form, the molecule is red.

3. At pH of 8.9, the thymol blue molecule loses its second acidic hydrogen. With no acidic hydrogens, the molecule is blue.

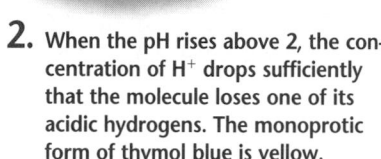

2. When the pH rises above 2, the concentration of H^+ drops sufficiently that the molecule loses one of its acidic hydrogens. The monoprotic form of thymol blue is yellow.

However, a single indicator can give only a general indication of pH. For example, if a solution is blue with thymol blue, it does not reveal the actual pH. It indicates only that the pH is above 8.9. By combining indicators that span the range of pH from 1 to 14, it is possible to create color changes that allow more precise measurement of pH. pH paper is such a mixture. When the color of the paper is compared to the color chart, the pH can be determined to within about one pH unit.

Thinking Critically

1. What are two characteristics that molecules of indicators must have?

2. What do you think is the relationship between the number of acidic hydrogens in an indicator molecule and the number of color changes?

3. What might be some advantages of using indicator papers instead of a pH meter?

15.2 Applications of Acid-Base Reactions   545

to titrate 25.0 mL of 0.150*M* $Ca(OH)_2$? (Hint: Step 1—Find the number of moles of $Ca(OH)_2$ involved in the titration. *0.00375 mol $Ca(OH)_2$* Step 2—Find the corresponding number of moles of HCl. *0.00750 mol HCl* Step 3—Use the number of moles and molarity of HCl to find volume.) *75.0 mL HCl solution*

Thinking Critically

1. It must have a series of double bonds that cause the molecule to absorb light. It must be able to act as either an acid or a base that changes color as acidic hydrogens are lost or gained.

2. In many indicators, a color change occurs for each acidic hydrogen lost or gained.

3. Indicator papers are quick and inexpensive. They also have applications in the field.

Purpose

This feature provides more background on indicators, which were introduced in Chapter 14 as a quick way to approximate the pH of a solution.

Background

Indicators must be molecules that are also able to act as both acids and bases. They must absorb light in the visible portion of the electromagnetic spectrum. The arrangement of double bonds in a molecule determines the color of light absorbed. The loss or gain of an acidic hydrogen alters the arrangement slightly and causes the molecule to change color. Molecules that impart color are often called dyes.

Visual Learning

Have students examine the thymol blue line of **Figure 14-22** on page 508. Ask them to determine whether thymol blue would best be used as an indicator for a strong acid-strong base, a strong acid-weak base, or a weak acid-strong base reaction. *The color change at pH = 2 is too low to indicate the product of a neutralization. The change at pH = 8.9 might be used for the titration of a weak acid and a strong base.*

Teaching Strategies

Use the thymol blue example to illustrate how indicators "indicate" pH. If a solution turns red when thymol blue is added, what is the pH? *The best you can indicate is that the pH of the solution is less than 2.* If the solution is yellow? *2 < pH < 9* If the solution turns blue when the indicator is added? *pH > 9* Stress that sometimes knowing the approximate pH is good enough.

3 ASSESS

Check for Understanding

- Check for understanding of the section material by having students answer the Section Review questions.
- Ask students to write a summary of the titration process in which they define or describe the role of each of the following terms or processes: *acid, base, neutralization, salt, titration, indicator, equivalence point, endpoint.* L1

Reteach

Discussion: Ask students which ion in an $H_2PO_4^-/HPO_4^{2-}$ buffer system would react with added H^+. *HPO_4^{2-}* With added OH^-? *$H_2PO_4^-$* Ask them to write the equations for the reactions and to specify the acid and base in each. *H^+ (acid) + HPO_4^{2-} (base) → $H_2PO_4^-$ and OH^- (base) + $H_2PO_4^-$ (acid) → $H_2O + HPO_4^-$* L1

Extension

Allow interested students to work in groups to conduct an acid-base titration with a pH meter, using the data to construct a pH curve, or to use a pH probe-computer interface setup and demonstrate the computer drawing of the curve for other students. L2
COOP LEARN

4 CLOSE

Enrichment

Ask students how bulimia, habitual overeating then vomiting, would affect the pH in a person's body. *It can lead to a severe loss of stomach acid and, thus, a high blood pH.* Explain that such a condition, called alkalosis, exists and can cause weak, irregular breathing and muscle contractions, convulsions, and even death. Have students summarize their findings in a report.

---

11. A 0.100M LiOH solution was used to titrate an HBr solution of unknown concentration. At the endpoint, 21.0 mL of LiOH solution had neutralized 10.0 mL of HBr. What is the molarity of the HBr solution?

12. A 0.150M KOH solution fills a burette to the 0 mark. The solution was used to titrate 25.0 mL of an HNO_3 solution of unknown concentration. At the endpoint, the burette reading was 34.6 mL. What was the molarity of the HNO_3 solution?

13. A $Ca(OH)_2$ solution of unknown concentration was used to titrate 15.0 mL of a 0.125M H_3PO_4 solution. If 12.4 mL of $Ca(OH)_2$ are used to reach the endpoint, what is the concentration of the $Ca(OH)_2$ solution?

Connecting Ideas

Acid-base reactions are usually double-displacement reactions. If you examine the oxidation number of each element involved in a double-displacement reaction, you can see that the oxidation numbers of the elements do not change. Are there types of reactions in which the oxidation numbers do change?

Many important reactions, such as the rusting of an old car or the burning of a fuel, involve chemical reactions in which oxidation numbers do change. It's important to find out what causes these changes and investigate this important type of reaction.

SECTION REVIEW

Understanding Concepts

1. How does a solution that contains dissolved ammonia and ammonium chloride act as a buffer? Use net ionic equations to show how this buffer responds to added H^+ and OH^-.

2. Use chemical equations to show that the buffer system in blood works like the buffer in question 1.

3. Sequence the steps in an acid-base titration.

Thinking Critically

4. **Applying Concepts** The following are the endpoint pHs for three titrations. From the endpoint pH, indicate whether the titration involves a weak acid-strong base, a weak base-strong acid, or a strong acid-strong base reaction. Use **Figure 15.17** to select the best indicator for the reaction. What color will you see at the endpoint?

   a) pH = 4.65   b) pH = 8.43   c) pH = 7.00

Applying Chemistry

5. **Antacids** You wish to compare two different antacid tablets, brand X and brand Y. You crush each tablet and add each to 100 mL of 1.00M HCl. After stirring, you titrate the leftover HCl in the resulting solution with 1.00M NaOH. The brand X tablet requires less NaOH than the brand Y tablet requires. Which antacid neutralizes more acid?

SECTION REVIEW

1. $NH_4^+ + OH^- \rightarrow NH_3 + H_2O$; $NH_3 + H^+ \rightarrow NH_4^+$
2. $H_2CO_3 + OH^- \rightarrow HCO_3^- + H_2O$; $HCO_3^- + H^+ \rightarrow H_2CO_3$
3. **a.** You must have an acid (or base) of unknown molarity. **b.** You must have a standard solution of either acid or base, opposite the unknown. **c.** Select an indicator based on the strengths of the acid and the base. **d.** Measure a precise volume of the unknown solution. **e.** Titrate to the endpoint with standard solution, measuring volume using a burette. **f.** Calculate the concentration of the unknown.

4. **a.** strong acid-weak base; methyl red; red **b.** weak acid-strong base; phenolphthalein; pink **c.** strong acid-strong base; bromothymol blue; green

5. Brand X requires less NaOH to completely finish neutralizing the HCl. Therefore, brand X has more neutralizing power.

REVIEWING MAIN IDEAS

15.1 Acid and Base Reactions

- Acid-base reactions are classified by the strength of the acid and base. Three reactions are of interest: strong acid-strong base, weak acid-strong base, and weak base-strong acid.

- Representing acid-base reactions by ionic and net ionic equations shows what is happening submicroscopically.

- Acids and bases in reactions can be identified using a hydrogen-ion transfer definition. An acid is an H^+ donor; a base is an H^+ acceptor.

- When acids and bases react, the pH of the final solution is dependent upon the nature of the reactants.

15.2 Applications of Acid-Base Reactions

- A buffer is a solution that maintains a relatively constant pH when H^+ ions or OH^- ions are added.

- The pH of blood is controlled in part by a buffer composed of carbonic acid, H_2CO_3, and the hydrogen carbonate ion, HCO_3^-.

- Antacids are bases that react with stomach acid.

- An acid-base titration uses an acid-base reaction to determine the molarity of an unknown acid or base.

Key Terms

For each of the following terms, write a sentence that shows your understanding of its meaning.

buffer
ionic equation
net ionic equation
neutralization reaction
salt
spectator ion
standard solution
titration

UNDERSTANDING CONCEPTS

1. What are the three types of acid-base reactions that always go to completion? Give an example of each by writing an overall equation.

2. Describe what it means for a reaction to go to completion.

3. Write and balance the overall equation for each of the following reactions. Identify the type of acid-base reaction represented by the equation.

   a) potassium hydroxide + phosphoric acid
   b) formic acid, $HCHO_2$ + calcium hydroxide
   c) barium hydroxide + sulfuric acid

4. For each of the reactions in question 3, write the ionic and net ionic equations.

5. For each of the reactions in question 4, what ions are spectators? Will the final reaction mixture be acidic, basic, or neutral? Explain.

6. In words, not equations, explain the use and differences in the overall, ionic, and net ionic equations.

7. Define a *buffer solution*.

8. Write the pH control reactions for the carbon dioxide-based buffer in blood.

CHAPTER REVIEW

- Review the Reviewing Main Ideas statements and Key Terms with your students.
- Complete solutions to Chapter Review problems can be found in the *Problems and Solutions Manual.*

UNDERSTANDING CONCEPTS

1. strong acid-strong base: $HCl + NaOH \rightarrow NaCl + H_2O$; weak acid-strong base: $HC_2H_3O_2 + NaOH \rightarrow NaC_2H_3O_2 + H_2O$; weak base-strong acid: $NH_3 + HCl \rightarrow NH_4Cl$

2. The reactants combine to form the maximum possible amount of product. When the reaction is complete, the limiting reactant is completely used up.

3. **a.** $3KOH + H_3PO_4 \rightarrow K_3PO_4 + 3H_2O$; weak acid-strong base **b.** $2HCHO_2 + Ca(OH)_2 \rightarrow Ca(CHO_2)_2 + 2H_2O$; weak acid-strong base **c.** $Ba(OH)_2 + H_2SO_4 \rightarrow BaSO_4 + 2H_2O$; strong acid-strong base

4. **a.** $3K^+ + 3OH^- + H_3PO_4 \rightarrow 3K^+ + PO_4^{3-} + 3H_2O$; $3OH^- + H_3PO_4 \rightarrow PO_4^{3-} + 3H_2O$ **b.** $2HCHO_2 + Ca^{2+} + 2OH^- \rightarrow Ca^{2+} + 2CHO_2^- + 2H_2O$; $HCHO_2 + OH^- \rightarrow CHO_2^- + 2H_2O$ **c.** $Ba^{2+} + 2OH^- + 2H^+ + SO_4^{2-} \rightarrow Ba^{2+} + SO_4^{2-} + 2H_2O$; $OH^- + H^+ \rightarrow H_2O$

5. **a.** K^+, basic, PO_4^{3-} is a weak base. **b.** Ca^{2+}, basic, CHO_2^- is a weak base. **c.** Ba^{2+} and SO_4^{2-}, neutral, only water is produced in the net reaction.

6. The overall reaction is a description of the solutions that are mixed and the products that are produced; ions are not shown. The ionic reaction shows all reactants and products in their submicroscopic form as molecules or ions. The net ionic equation focuses on the submicroscopic particles that are actually part of the reaction; spectator ions are deleted.

7. A buffer solution is one that maintains a relatively constant pH when H^+ and OH^- are added.

8. $HCO_3^- + H^+ \rightarrow H_2CO_3$; $H_2CO_3 + OH^- \rightarrow HCO_3^- + H_2O$

9. The pH of a buffer solution may change slightly; it remains relatively constant rather than completely constant.

10. From the reactions in question 8, H_2CO_3 increases and HCO_3^- decreases as the H^+ is consumed. H^+ is slightly greater than before the increase. The relative change is small.

11. As blood passes through the lungs, excess CO_2 is expelled to maintain the H_2CO_3 level at a constant value.

12. $2H^+ + CaCO_3 \rightarrow H_2O + CO_2 + Ca^{2+}$; $2H^+ + Ca(HCO_3)_2 \rightarrow Ca^{2+} + 2H_2O + 2CO_2$

13. Insoluble hydroxide bases: $Mg(OH)_2$; Carbonate/Hydrogen carbonate bases: $CaCO_3$; $NaHCO_3$

14. $Al(OH)_3(s) + 3HCl(aq) \rightarrow AlCl_3(aq) + 3H_2O(l)$

15. $0.494M$

16. $0.145M$

17. $0.0616M$

18. pH is higher for the weak acid-strong base titration.

19. pH is higher for the strong acid-strong base titration.

20. $0.364M$

21. Hyperventilation is when a person breathes rapidly and deeply, which causes the blood to lose more CO_2 than normal. This causes the pH of the blood to increase. $CO_2 + H_2O \rightarrow H_2CO_3 \rightarrow H^+ + HCO_3^-$; as CO_2 decreases, H_2CO_3 decreases, and H^+ decreases.

APPLYING CONCEPTS

22. **a.** HBr is acid; H_2O is base **b.** HS^- is base; H_2O is acid **c.** NH_3 is base; $HCHO_2$ is acid **d.** HCO_3^- is base; H_2O is acid **e.** HCO_3^- is acid; H_2O is base

9. Why is it incorrect to define a buffer as a solution that maintains a constant pH?

10. If the H^+ concentration in blood increases, what happens to the concentrations of H_2CO_3, HCO_3^-, and H^+?

11. What role do the lungs play in regulating blood pH?

12. Write two reactions that explain why lakes in limestone areas are capable of resisting pH decreases due to acid rain.

13. Antacids are classified into two types. What are they? Give an example of each type.

14. Consider an antacid that contains aluminum hydroxide. Write the overall equation that shows how this antacid reduces the acidity of stomach acid.

15. A student found that 53.2 mL of a $0.232M$ solution of NaOH was required to titrate 25.0 mL of an acetic acid solution of unknown molarity to the endpoint. What is the molarity of the acetic acid solution?

16. A student neutralizes 30.0 mL of a sample of sodium hydroxide with 28.9 mL of $0.150M$ HCl. What is the molarity of the sodium hydroxide?

17. A student finds that 23.1 mL of $0.200M$ potassium hydroxide are required to react completely with 25.0 mL of a phosphoric acid solution. What is the molarity of the H_3PO_4?

18. How does the endpoint pH of a strong acid-strong base titration compare with that of a weak acid-strong base titration?

19. How does the endpoint pH of a strong base-strong acid titration compare with that of a weak base-strong acid titration?

20. A 50.0-mL sample of an unknown monoprotic acid is titrated to the endpoint with 45.5 mL of $0.200M$ $Ca(OH)_2$. What is the molarity of the acid solution?

21. What is hyperventilation? How does it change the pH of blood?

APPLYING CONCEPTS

22. In the following reactions, identify whether each of the reactants is acting as an acid or a base according to the hydrogen-ion transfer definition.

    a) $HBr + H_2O \rightarrow H_3O^+ + Br^-$
    b) $HS^- + H_2O \rightarrow H_2S + OH^-$
    c) $NH_3 + HCHO_2 \rightarrow NH_4^+ + CHO_2^-$
    d) $HCO_3^- + H_2O \rightarrow H_2CO_3 + OH^-$
    e) $HCO_3^- + H_2O \rightarrow CO_3^{2-} + H_3O^+$

23. Methylamine, CH_3NH_2, is a weak base, as ammonia is. When methylamine completely reacts with hydrochloric acid, the final solution has a pH less than 7. Why are the products of this "neutralization" reaction not neutral? Use the net ionic equation to help in your explanation.

24. Vitamin C is also known as ascorbic acid, $HC_6H_7O_6$. A solution made from a vitamin C tablet is titrated to the endpoint with 12.3 mL of $0.225M$ NaOH. Assuming that vitamin C is the only acid present in the tablet, how many moles of vitamin C are in the tablet?

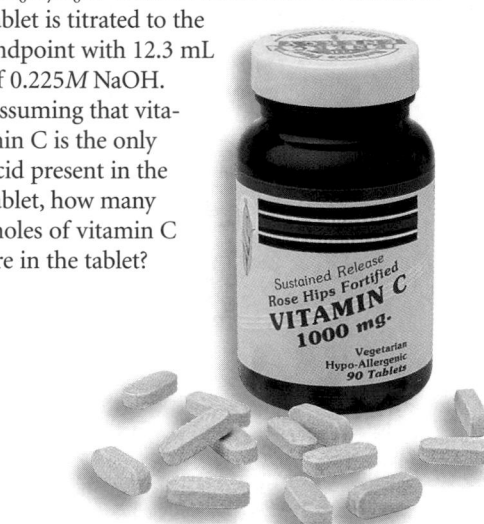

25. Why are magnesium hydroxide and aluminum hydroxide effective antacids, but sodium hydroxide is not?

26. How many milliliters of $0.100M$ NaOH are required to neutralize 25.0 mL of $0.150M$ HCl?

23. $CH_3NH_2 + H^+ \rightarrow CH_3NH_3^+$; $CH_3NH_3^+$ is a weak acid.

24. 0.00277 mol

25. $Mg(OH)_2$ and $Al(OH)_3$ are insoluble. Therefore, they do not dissolve and react until they get into the stomach. NaOH is soluble. It would dissolve immediately if taken internally, and the strong base would damage tissue in the mouth and esophagus.

26. 37.5 mL

27. Complete and balance the following overall equations.

a) $KOH(aq) + HNO_3(aq) \rightarrow$
b) $Ba(OH)_2(aq) + HCl(aq) \rightarrow$
c) $NaOH(aq) + H_3PO_4(aq) \rightarrow$
d) $Ca(OH)_2(aq) + H_3PO_4(aq) \rightarrow$

28. Dihydrogen phosphate and monohydrogen phosphate ions play an important role in maintaining the pH in intracellular fluid. Write equations that show how these ions maintain the pH.

29. The concentration of H_2CO_3 in blood is 1/20 of the concentration of HCO_3^-, yet the blood buffer is capable of buffering the pH against bases, as well as against acid. Explain.

30. Water is an example of an amphoteric substance. An amphoteric substance may act as either an acid or a base. There are other amphoteric substances. The hydrogen carbonate ion is also amphoteric. Write reactions that demonstrate this property of HCO_3^-.

31. Tartaric acid is often added to artificial fruit drinks to increase tartness. A sample of a certain beverage contains 1.00 g of tartaric acid, $H_2C_4H_4O_6$. The beverage is titrated with 0.100M NaOH. Assuming no other acids are present, how many milliliters of base are required to neutralize the tartaric acid?

32. An antacid tablet that contains sodium hydrogen carbonate is titrated with 0.200M HCl. If 0.400 g of the tablet requires 22.0 mL of HCl to reach the endpoint, what is the mass percent of $NaHCO_3$ in the tablet?

33. Stomach acid is approximately 0.0200M HCl. What volume of stomach acid does an antacid tablet that contains 45.5 percent $Mg(OH)_2$ and weighs 355 mg neutralize?

Chemistry and Society

34. What function of blood is most important when developing artificial blood?

Earth Science Connection

35. What acid most likely causes groundwater to be acidic? How does groundwater become acidic?

Everyday Chemistry

36. How is the CO_2 concentration in the blood related to hiccups?

How It Works

37. Explain how molecules and ions are related to taste.

How It Works

38. Can an indicator provide an exact pH? Explain.

● THINKING CRITICALLY

Observing and Inferring

39. A sample of rainwater turns blue litmus red. Fresh portions of the rainwater turn thymol blue indicator yellow, bromophenol blue indicator green, and methyl red indicator red. Estimate the pH of the rainwater.

27. a. $KOH(aq) + HNO_3(aq)$
$\rightarrow KNO_3(aq) + H_2O(l)$
b. $Ba(OH)_2(aq) +$
$2HCl(aq) \rightarrow BaCl_2(aq) +$
$2H_2O(l)$
c. $3NaOH(aq) +$
$H_3PO_4(aq) \rightarrow Na_3PO_4(aq)$
$+ 3H_2O(l)$
d. $3Ca(OH)_2(aq) +$
$2H_3PO_4(aq) \rightarrow$
$Ca_3(PO_4)_2(aq) + 6H_2O(l)$

28. $H_2PO_4^- + OH^- \rightarrow$
$HPO_4^{2-} + H_2O$; HPO_4^{2-}
$+ H^+ \rightarrow H_2PO_4^-$

29. As H_2CO_3 is depleted, dissolved CO_2 produces more.

30. $HCO_3^- + H_2O \rightarrow H_2CO_3$
$+ OH^-$; in this reaction,
HCO_3^- acts as a base.
$HCO_3^- + H_2O \rightarrow CO_3^{2-}$
$+ H_3O^+$; in this reaction,
HCO_3^- is acting as an acid.

31. 133 mL

32. 92.4%

33. 277 mL

34. Answers may vary but should include the ability to carry oxygen.

35. Carbonic acid, H_2CO_3; carbon dioxide in the air dissolves in rainwater, forming carbonic acid.

36. Possibly, an increase in CO_2 in the blood lowers the pH and may turn off nerves that control hiccups.

37. Only certain molecules and ions react with specific taste buds to produce signals to your brain, signaling taste.

38. No, an indicator can indicate only a range of pHs because it remains a certain color over a range of pHs, not at a specific pH.

THINKING CRITICALLY

39. Litmus to red means pH < 7; thymol blue as yellow means pH > 2.5; bromophenol blue as green means pH is about 4; methyl red as red means pH < 5. Best estimate of pH is about 4.

40. $HCHO_2 + OH^- \rightarrow$ $CHO_2^- + H_2O$; CHO_2^- is a weak base.

41. a. $NaOH + HCl \rightarrow NaCl + H_2O$ **b.** $Ca(OH)_2 + H_2SO_4 \rightarrow CaSO_4 + 2H_2O$ **c.** $Mg(OH)_2 + 2HCl \rightarrow MgCl_2 + 2H_2O$ **d.** $2NH_3 + H_2SO_4 \rightarrow (NH_4)_2SO_4$ **e.** $KOH + HBr \rightarrow KBr + H_2O$

42. Vinegar is within a range of concentrations. Different batches may vary in concentration.

43. acidic

44. Phenolphthalein changes color in a slightly basic solution, and methyl orange changes color in a slightly acidic solution.

CUMULATIVE REVIEW

45. a. physical **b.** physical **c.** chemical **d.** physical

46. manganese nitrate tetrahydrate; Mn, 3; N, 6; O, 30; H, 24

47. $C_8H_6O_4$

48. a. $CuSO_4(s) \rightarrow Cu^{2+}(aq) + SO_4^{2-}(aq)$ **b.** $Ca(NO_3)_2(s) \rightarrow Ca^{2+}(aq) + 2NO_3^-(aq)$ **c.** $Na_2CO_3(s) \rightarrow 2Na^+(aq) + CO_3^{2-}(aq)$

49. A monoprotic acid, such as $HC_2H_3O_2$, contains only one ionizable hydrogen atom. A triprotic acid, such as H_3PO_4, contains three ionizable hydrogen atoms.

Interpreting Data

40. When formic acid, $HCHO_2$, reacts completely with NaOH, the resulting solution has a pH greater than 7. Why are the products of this neutralization reaction not neutral? Use the net ionic equation to help in your explanation.

Applying Concepts

41. Write an overall equation for the acid-base reaction that would be required to produce each of the following salts.

a) NaCl   c) $MgCl_2$ e) KBr
b) $CaSO_4$ d) $(NH_4)_2SO_4$

Observing and Inferring

42. ChemLab Explain why different bottles of the same brand of vinegar might contain solutions that have different pHs.

Making Predictions

43. MiniLab 1 Would a solution of iron(III) bromide, $FeBr_3$, be acidic, basic, or neutral?

Relating Cause and Effect

44. MiniLab 2 Explain why phenolphthalein and methyl orange are used as indicators in Mini-Lab 2.

● CUMULATIVE REVIEW

45. Identify each of the following as either a chemical or a physical property of the substance mentioned. (Chapter 1)

a) The element hydrogen has a low density.
b) At room temperature, solid iodine changes to gaseous iodine without passing through the liquid state.
c) Aluminum corrodes when exposed to air.
d) Sodium nitrate is a white crystalline solid.

46. Give the name of the compound represented by the formula $Mn(NO_3)_2 \cdot 4H_2O$, and determine how many atoms of each element are present in three formula units of the compound. (Chapter 5)

47. Terephthalic acid is an organic compound used in the formation of polyesters. It contains 57.8 percent C, 3.64 percent H, and 38.5 percent O. The molar mass is known to be approximately 166 g/mol. What is the molecular formula of terephthalic acid? (Chapter 12)

48. Write equations for the dissociation of the following ionic compounds when they dissolve in water. (Chapter 13)

a) $CuSO_4$
b) $Ca(NO_3)_2$
c) Na_2CO_3

49. What is a monoprotic acid? A triprotic acid? Give an example of each. (Chapter 14)

Solution	pH	Bromphenol blue	Methyl red	Thymol blue
A	5.45	Blue	Red-yellow	Yellow
D	4.50	Blue	Red	Yellow
C	3.45	Green-yellow	Red	Yellow
E	2.36	Yellow	Red	Red-yellow
B	1.00	Yellow	Red	Red

Skill Review

1. **Data Table** Solutions of five different monoprotic acids are all 0.100M. The pH of each solution is given. Rank the acids in the following table from weakest to strongest. For each solution, use the indicator table in **Figure 14.22** to predict the color that each solution would produce with the given indicator.

Solution	pH
A	5.45
B	1.00
C	3.45
D	4.50
E	2.36

Solution (weakest to strongest acid)	pH	Color in Bromphenol Blue	Color in Methyl Red	Color in Thymol Blue

Writing in Chemistry

2. Write an article about the effect of acid rain on a specific aspect of a local environment such as a lake or a forest. Give some history of the problem and indicate when local residents first realized a problem exists. What, if any, corrective measures have been taken to correct the problem? Is the environmental damage reversible?

Problem Solving

3. Because antacids are frequently insoluble in water, they are often analyzed by dissolving them in a known volume of HCl with a known molarity. After the antacid has completely reacted, there is still HCl left in the solution. This excess HCl is then titrated with a standard NaOH solution. A 165-mg sample of an antacid tablet containing calcium carbonate is dissolved in 50.0 mL of 0.100M HCl. After complete reaction, the excess HCl is titrated with 15.8 mL of 0.150M NaOH. Sketch a flowchart that shows the steps in the analysis.

Projects

4. Acids and bases compose eight of the top 15 industrial chemicals produced in the United States in 1994. Select one of these acids or bases, and prepare a poster presentation about the compound. Your presentation should include a description of the chemical and its properties. Discuss how it is produced and who the major producers are. Why is the compound so important? What is it used for? Consider interviewing an individual associated with the production or use of this chemical. Indicate what, if any, everyday products might have been made using this chemical.

AUTHENTIC ASSESSMENT

1. See table at bottom of page 550.
2. Articles will vary. Have students discuss any differences in findings.
3. The flow chart should include the following steps: calculate the number of moles of NaOH involved in the titration and the total number of moles of HCl originally added to the antacid; subtract the moles of NaOH from the moles of HCl to get the number of moles of HCl that reacted with the antacid; calculate the number of moles of antacid that reacted; calculate the grams of CaCO₃ reacted; divide this number by 0.165 g to get the mass percent of CaCO₃. (80%)
4. Student work will vary. Be sure each of the acids and bases listed is researched by a student.

Program Resources

Chapter Review and Assessment, pp. 89-94
  L1

Alternate Assessment in the Science Classroom L1

Computer Test Bank, Chapter 15 L1

Performance Assessment in the Science Classroom L1

Problems and Solutions Manual, Chapter 15
  L1

Supplemental Practice Problems, Chapter 15 L1

Chapter 16 Organizer

Section	Objectives	Activities/Features
16.1 **The Nature of Oxidation-Reduction Reactions** (3 days)	**1. Analyze** the characteristics of an oxidation-reduction reaction. **2. Distinguish** between oxidation reactions and reduction reactions by definition. **3. Identify** the substances that are oxidized and those that are reduced in a redox reaction. **4. Distinguish** oxidizing and reducing agents in redox reactions.	**MiniLab 1:** Corrosion of Iron, p. 557 **ChemLab:** Copper Atoms and Ions: Oxidation and Reduction, pp. 560-561 **Discovery Demo:** 16.1 pp. 552-553 **Demonstrations:** 16.2 pp. 558-559
16.2 **Applications of Oxidation-Reduction Reactions** (2 days)	**5. Analyze** common redox processes to identify the oxidizing and reducing agents. **6. Identify** some redox reactions that take place in living cells.	**Physics Connection:** p. 566 **MiniLab 2:** Testing for Alcohol by Redox, p. 568 **How It Works:** Breathalyzer Test, p. 569 **Everyday Chemistry:** p. 571 **Chemistry and Technology:** Forensic Blood Detection, p. 573 **Demonstrations:** 16.3 pp. 564-565; 16.4 pp. 566-567; 16.5 pp. 574-575

Activity Materials

ChemLab (pages 560-561)
copper(II) oxide, powdered charcoal, weighing paper, balance, large Pyrex or Kimex test tubes (2), 1-hole rubber stopper, glass tube with 90° bend, 150-mL beakers (2), small graduated cylinder, glass stirring rod, Bunsen or Tirrill burner, ring stand, test-tube clamp, limewater, thermal glove

MiniLab 1 (page 557)
unflavored gelatin, 250-mL beaker, phenolph-thalein, potassium hexacyanoferrate(III), widemouth glass jar or petri dish, iron nail, aluminum nail, gal-vanized iron nail, painted iron nail, 250-mL and 10-mL graduated cylinders, glass stirring rod

MiniLab 2 (page 568)
small test tubes (5), 5 household products (some containing alcohol such as mouthwash), small grad-uated cylinder, potassium dichromate reagent, glass stirring rod, dropper

Demonstrations For a complete list of materials for the demonstrations in this chapter, see pages 552, 558, 564, 566, and 574.

KEY TO TEACHING STRATEGIES
The following designations will help you decide which activities are appropriate for your students.

L1 Level 1 activities should be within the ability range of all students.

L2 Level 2 activities should be within the ability range of the average to above-average student.

L3 Level 3 activities are designed for the ability range of above-average students.

LEP LEP activities should be within the ability range of Limited English Proficiency students.

COOP LEARN Cooperative Learning activities are designed for small group work.

P These strategies represent student products that can be placed into a best-work portfolio.

Oxidation-Reduction Reactions

Teacher Classroom Resources

STUDENT MASTERS

Study Guide: pp. 63-64

 **Laboratory Manual:** 16.1 Oxidation/Reduction of Vanadium, pp. 147-150; 16.2 Corrosion as an Electrochemical Process, pp. 151-154

Critical Thinking/Problem Solving: p. 18

ChemLab and MiniLab Worksheets: p. 105

Study Guide: pp. 65-66

ChemLab and MiniLab Worksheets, pp. 101-104, 106

TEACHING AIDS

Section Focus Transparency 32
Section Focus Transparency Master: p. 32
Basic Concepts Transparency 43
Basic Concepts Transparency Master: pp. 85-86
Problem Solving Transparency 19
Problem Solving Transparency Master: pp. 37-38
Lesson Plans: p. 32

Section Focus Transparency 33
Section Focus Transparency Master: p. 33
Basic Concepts Transparency 44, 45
Basic Concepts Transparency Masters: pp. 87-90
Lesson Plans: p. 33

ASSESSMENT RESOURCES

Chapter Review and Assessment: pp. 95-100
Alternate Assessment in the Science Classroom
Computer Test Bank: Chapter 16
Performance Assessment in the Science Classroom
Problems and Solutions Manual: Chapter 16
Supplemental Practice Problems: Chapter 16

CHAPTER RESOURCES

Applying Scientific Methods in Chemistry: p. 42
Spanish Resources: Chapter 16
English/Spanish Audiocassettes: Chapter 16
Cooperative Learning in the Science Classroom
Lab and Safety Skills in the Science Classroom

GLENCOE TECHNOLOGY

Videotape
MindJogger Videoquizzes: Chapter 16
Videodisc
Chemistry: Concepts and Applications,
 Demonstrations, Videos, and Animations
Videodisc and Bar Code Teacher Guide
CD-ROM Multimedia System
Chemistry: Concepts and Applications, Same as
 for Videodisc, plus Interactive Exploration and Major
 Simulation, *The Periodic Table*

TECH PREP

The following Glencoe resources provide opportunities for integrating science and technology.

Student Edition: Physics Connection, p. 566; How It Works, p. 569; Everyday Chemistry, p. 571; Chemistry and Technology, p. 573

Teacher Wraparound Edition: Chemistry Journal, p. 568; Across the Curriculum, p. 570

Teacher Classroom Resources: Chemistry and Industry, pp. 35-36; Consumer Chemistry, pp. 31-32; Tech Prep Applications, pp. 25-26

CHAPTER 16 Oxidation-Reduction Reactions

Chapter Overview

Students learn through examples what oxidation and reduction reactions are. After the oxidation-reduction process is defined, students learn to examine equations of redox reactions and identify which substance is oxidized, acting as the reducing agent, and which is reduced, functioning as the oxidizing agent. Several important processes that involve redox reactions are then described.

Theme Connection

Systems and Interactions/Conservation The *systems and interactions* theme is apparent in discussions of oxidation-reduction reactions, which are chemical systems in which oxidizing and reducing agents interact. The theme of *conservation* also is apparent in discussions of conservation of charge in redox reactions.

✔ Assessment Planner

Choose assessment strategies from the following pages to evaluate the progress of your students.
Assess, pp. 562, 575
MiniLabs, pp. 557, 568
ChemLab, p. 560
Portfolio, pp. 562, 576, 577
Chapter Review, p. 577

GLENCOE TECHNOLOGY

💿 CD-Rom

Chemistry: Concepts and Applications
Forming Crystals by Redox
Show this video to introduce the concept of redox reactions. L1 LEP

Oxygen
molecules

552

Discovery Demo 16.1

Forming Crystals by Redox

Purpose: To demonstrate a Cu/Ag redox reaction.

Materials: petri dish, overhead projector, 0.85 g AgNO₃, 50 mL distilled water, 8-cm length of copper wire, small piece of steel wool

Safety Precautions: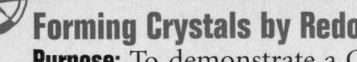

Disposal: Code A, or collect silver crystals and recycle back to AgNO₃.

Procedure: Place a petri dish on the stage of an overhead projector. Partially fill the dish with 0.1*M* AgNO₃ solution (0.85 g of AgNO₃ dissolved in 50 mL of distilled water). **CAUTION:** *AgNO₃ solution stains skin and clothes and is toxic. Fresh stains may be bleached using an aqueous solution of Na₂S₂O₃, hypo.* Clean a short piece of copper wire with steel wool, and place the wire in the silver nitrate solution. Have students record their observations in a

CHAPTER
16
Oxidation-Reduction Reactions

Remember the last time you were really hungry and bit into a fresh, juicy apple? Mmmmm—wasn't it delicious? Once you got past the skin, each bite into the crisp, white fruit was better than the next. But suppose you put the apple down for a few minutes and then went back to finish it. Would it still have looked as appealing? The mottled brown splotches that appeared on the fruit might not have been so appetizing.

What causes the brown color seen in fruits and some vegetables after the skin has been broken? Why doesn't the color change occur in unbroken fruit? Recall that color change is one of the signs that a chemical reaction has taken place. Apples, like all plant and animal tissues, contain lots of molecules that can undergo the many different reactions that occur in living things. What kind of reaction takes place in a bitten apple? An important clue to solving this puzzle is the observation that fruit turns brown only when its skin is broken.

Concept Check

Review the following concepts before studying this chapter.
Chapter 3: patterns of valence electrons
Chapter 5: predicting oxidation number from the periodic table
Chapter 6: types of reactions

Chapter Preview

16.1 The Nature of Oxidation-Reduction Reactions

16.2 Applications of Oxidation-Reduction Reactions

Laboratory Activities
ChemLab
Copper Atoms and Ions: Oxidation and Reduction

MiniLabs
1. Corrosion of Iron
2. Testing for Alcohol by Redox

Chemistry Around You

If you are a slow eater, an apple you're enjoying may begin to turn brown before you finish it. The reaction that causes the brown color involves the oxygen molecules shown opposite. Many of the reactions in familiar processes also involve oxygen. After you have studied these reactions, you will be able to explain not only why apples turn brown, but also why nails rust, why the Statue of Liberty is green, and even why fireflies glow.

553

diagram several minutes after the reaction begins.
Results: A beautiful, feathery deposit of metallic silver appears around the copper wire and continues to change for several hours. Allow the setup to remain overnight, and have students observe it the next day.

Analysis: Ask these questions.
1. What happened to the silver ions? *They were converted to silver atoms and were deposited around the wire.*
2. What happened to the copper atoms in the wire? *They became copper ions and went into solution.*
3. Write the balanced redox equation for the reaction. $Cu + 2Ag^+ \rightarrow Cu^{2+} + 2Ag$
4. Is the reaction spontaneous? *yes*

Introducing the Chapter
Ask students why food should be stored in tightly sealed containers. If students answer that air must be kept out to keep the food from spoiling, ask them what in the air causes foods to spoil. *They should be able to answer that chemical substances are involved, especially water and oxygen.* Find out whether students can list physical changes occurring during food spoilage that hint of chemical reactions taking place. *Odor, color, and texture changes take place.* Tell the class that a reaction in which oxygen takes part is called an oxidation reaction, and that the focus of this chapter is on oxidation reactions and the reduction reactions that accompany them.

Using the Illustrations Bring in two or three apples. Have students suggest hypotheses for why apples turn brown when cut, and then suggest ways to test the hypotheses. They should conclude that apples turn brown due to chemical reactions between substances in the apple and oxygen in the air.

Revealing Misconceptions
Make sure students know that the system for classifying reactions that they learned in Chapter 6 is still valid. Reactions can be classified in more than one way, and many reactions that they would label as synthesis or combustion are also redox reactions.

GLENCOE TECHNOLOGY

 Videodisc

Chemistry: Concepts and Applications
Forming Crystals by Redox
Disc 2, Side 2, Ch. 1

Show this video to introduce the concept of redox reactions. `L1` `LEP`

Key Concepts

Several oxidation-reduction reactions are described, then oxidation and reduction reactions are defined and shown as half-reactions. Finally, methods for identifying redox reactions, oxidizing agents, and reducing agents are explained and demonstrated.

Planning Ahead

For the MiniLab Order these nails from a hardware or builders' supply store if you do not have them on hand: aluminum, iron, galvanized iron, painted iron.

For the Demonstrations Be sure that you have fresh hydrogen peroxide for Demonstration 16.2.

1 FOCUS

Focus Transparency

Display the **Section Focus Transparency** for Section 16.1 to show students a way to prevent an oxidation-reduction reaction. L1 LEP

2 TEACH

Discussion

Take a photograph of a rusty car, preferably one that is owned by one of your students. If the car is in the school parking lot, you might see it out the window or take the class out to see it. Tell students only that this car illustrates the next topic of study. Have students tell you what that topic is. *oxidation and reduction, as represented by the oxidation or corrosion of the car* L1

SECTION 16.1

The Nature of Oxidation-Reduction Reactions

Objectives

Analyze the characteristics of an oxidation-reduction reaction.

Distinguish between oxidation reactions and reduction reactions by definition.

Identify the substances that are oxidized and those that are reduced in a redox reaction.

Distinguish oxidizing and reducing agents in redox reactions.

Key Terms
oxidation-reduction reaction
oxidation
reduction
oxidizing agent
reducing agent

Oxygen undergoes many reactions when it encounters other substances. One of these reactions is responsible for the browning of fruits. Another forms the rust that eats away at the metal parts of bikes and cars. In both of these cases, a type of reaction called oxidation is taking place. You can probably guess how this reaction got its name; oxygen is a reactant. But you will learn that not all oxidation reactions involve oxygen. And oxidation reactions are never lonely because they always have partners—reduction reactions. You will see what the characteristics of these reactions are and why they always take place together.

What is oxidation-reduction?

Oxygen is the most abundant element in Earth's crust. It is very reactive and can combine with almost every other element. An element that bonds to oxygen to form a new compound, called an oxide, usually loses electrons because oxygen is more electronegative. You will recall that an electronegative element has a strong attraction for electrons. Because of this strong attraction, oxygen is able to pull electrons away from other atoms. The reactions in which elements combine with oxygen to form oxides were among the first to be studied by early chemists, who grouped them together and called them *oxidation reactions*. Later, chemists realized that some other nonmetal elements can combine with substances in the same way as oxygen and that these reactions are similar to oxidation reactions. Modern chemists use the term *oxidation* to refer to any chemical reaction in which an element or compound loses electrons to another substance.

A common oxidation reaction occurs when iron metal loses electrons to oxygen. Each year in the United States, corrosion of metals—especially the iron in steel—costs billions of dollars as automobiles, ships, and bridges and other structures are slowly eaten away. **Figure 16.1** shows some of this damage and how it can be prevented.

Program Resources

Study Guide, pp. 63-64 L1
Critical Thinking/Problem Solving, p. 18 L2
English/Spanish Audiocassettes, Chapter 16 L1 LEP
Laboratory Manual, 16.1, pp. 147-150; 16.2, pp. 151-154 L1
Section Focus Transparency 32 and **Master,** p. 32 L1 LEP
Basic Concepts Transparency 43 and **Master,** pp. 85-86 L1 LEP

Problem Solving Transparency 19 and **Master,** pp. 37-38 L1 LEP
Spanish Resources L1 LEP
ChemLab and MiniLab Worksheets, p. 105 L1
Applying Scientific Methods in Chemistry, p. 42 L1

Figure 16.1
Corrosion of Iron

When iron corrodes, iron metal reacts with oxygen to form iron(III) oxide—rust. Corrosion of iron can be prevented by covering the surface of exposed steel with paint or other coatings such as plastic. If the protective coating is damaged or cracked, rust forms quickly. ▶

◀ Steel can be protected from oxidation if it is coated with a more active metal such as zinc. Zinc loses electrons to oxygen more readily than iron does, so the zinc is oxidized preferentially, forming a tough protective layer of zinc oxide. The coating of zinc and zinc oxide prevents the formation of rust by keeping oxygen from reaching the iron. Steel that has been coated with zinc is called galvanized steel. The bucket on the left has been galvanized.

Redox

What happens to the zinc in galvanized steel? It reacts with oxygen to form zinc oxide in the following reaction.

$$2Zn(s) + O_2(g) \rightarrow 2Zn^{2+}(s) + 2O^{2-}(s)$$

Does this type of reaction look familiar? You learned in Chapter 6 that this is classified as a synthesis reaction. You also know that early chemists called it an oxidation reaction because oxygen is a reactant. The formation of zinc oxide falls into another, broader class of reactions characterized by the transfer of electrons from one atom or ion to another. This type of reaction is called an **oxidation-reduction reaction,** commonly known as a redox reaction. Many important chemical reactions are redox reactions. Formation of rust is one example; combustion of fuels is another. In each redox reaction, one element loses electrons, and another element takes them.

How do atoms or ions lose electrons in a redox reaction? If you examine the equation for the reaction between zinc and oxygen more closely, you can see which atoms are gaining electrons and which are losing them. You also can determine where the electrons go during a redox reaction by comparing the oxidation number of each type of atom or ion before and after the reaction takes place. Recall from Chapter 5 that the oxidation number of an ion is equal to its charge. All elements, when in their free form, have a charge of zero and are assigned an oxidation number of zero. In the formation of zinc oxide, the zinc atom and the diatomic oxygen molecule that react each has an oxidation number of zero. In the ionic compound formed, each oxide ion has a 2− charge and an oxidation number of 2−. Because the compound must be neutral, the total positive charge must be 4+; thus, each zinc ion must have a charge and an oxidation number of 2+.

16.1   The Nature of Oxidation-Reduction Reactions   555

Oxidation

You have learned that a reaction in which an element loses electrons is called an **oxidation** reaction. The element that loses the electrons becomes more positively charged; that is, its oxidation number increases. That element is said to be oxidized during the reaction. Zinc is oxidized during the formation of zinc oxide because metallic zinc atoms each lose two electrons. The oxidation reaction can be written by itself to show how zinc changes during the redox reaction. Here's what happens to each atom of zinc.

$$Zn\colon \rightarrow [Zn]^{2+} + 2e^- \quad \text{(loss of electrons)}$$

Reduction

What happens to the electrons that are lost by the zinc atom? Electrons do not wander around by themselves; they must be transferred to another atom or ion. This is why oxidation reactions never occur alone. They are always paired with reduction reactions. A **reduction** reaction is one in which an element gains one or more electrons. The element that picks up the electrons and becomes more negatively charged during the reaction is said to be reduced. Its oxidation number decreases, or is reduced. Because oxidation and reduction reactions occur together, each is referred to as a half-reaction.

In every redox reaction, at least one element undergoes reduction while another undergoes oxidation. Just as a successful pass in football requires a quarterback to throw the ball and a receiver to catch it, a redox reaction must have one element that gives up electrons and one that accepts them. The electronic structure of both reactants changes during a redox reaction.

Figure 16.2 shows the movement of electrons in the formation of zinc oxide. Oxygen accepts the electrons that zinc loses. Oxygen is reduced during the reaction between zinc and oxygen because each oxygen atom gains two electrons. Like the oxidation reaction, the reduction reaction can be written by itself. Here's what happens to each atom of oxygen.

$$\cdot \ddot{O}\colon + 2e^- \rightarrow \colon\!\ddot{O}\colon^{2-} \quad \text{(gain of electrons)}$$

Figure 16.2

Formation of Zinc Oxide
In the formation of zinc oxide, the zinc atom loses two electrons during the reaction, becoming a zinc ion. Its oxidation number increases from zero to 2+. The oxygen atom gains the two electrons from zinc, becoming an oxide ion. Its oxidation number decreases from zero to 2−.

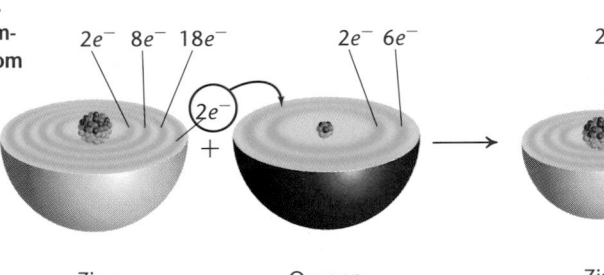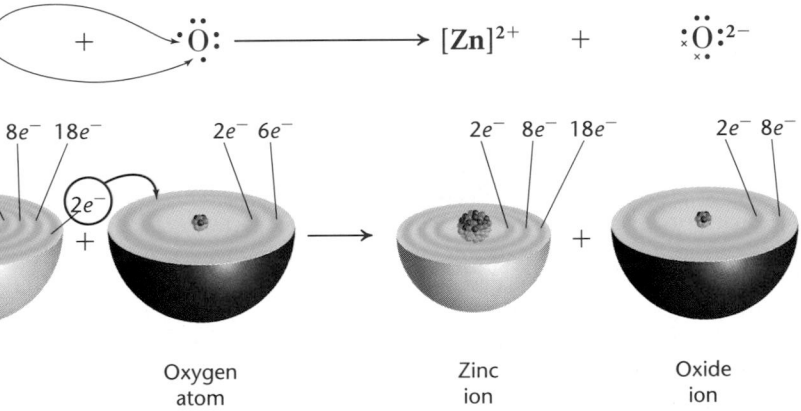

| Zinc atom | Oxygen atom | Zinc ion | Oxide ion |

Corrosion of Iron

Corrosion is the term generally used to describe the oxidation of a metal during its interaction with the environment. In this MiniLab, you will study the corrosion of a nail and determine the factors that affect this process.

Procedure

1. Dissolve a package of clear, unflavored gelatin in about 200 mL of warm water. Stir in 2 mL of phenolphthalein solution and 2 mL of potassium hexacyanoferrate(III) solution. Pour the prepared solution into a widemouth glass jar or petri dish to a depth of about 1 cm.

2. In the liquid gelatin, place a plain iron nail, an aluminum nail, a galvanized iron nail, and a painted iron nail of the type used for paneling. Space the nails far apart.

3. Label the jar or petri dish with your name, and leave it for several hours or overnight. Handle the jar or dish carefully until the gelatin has set.

4. Record your observations regarding any interactions of the nails with the substances in the gelatin.

Analysis

1. Which of the nails have reacted with the substances in the gelatin? What is the evidence of corrosion?

2. If any of the nails have not corroded in the solution, can you suggest a reason why they haven't? What methods are commonly used to prevent or minimize corrosion?

3. Any blue color in the gelatin is due to the formation of iron(II) ions and their interaction with the hexacyanoferrate(III) ion. Any pink or red color in the gelatin is due to the gaining of electrons by oxygen and water molecules, forming basic hydroxide ions that turn the phenolphthalein pink. Which of these reactions is oxidation, and where does it occur on the reacting nail?

Combining the Half-Reactions

The equation for the reduction half-reaction shows one atom of oxygen reacting. However, oxygen is not found in nature as single atoms; two atoms combine to form a diatomic molecule of O_2. The reduction equation must be multiplied by 2 to reflect this. Thus, the balanced equation for the reduction reaction is written as follows.

$$O_2 + 4e^- \rightarrow 2:\ddot{O}:^{2-}$$

Note that four electrons are gained by the oxygen molecule. To produce those four electrons, two atoms of zinc must take part in the reaction. Therefore, the balanced equation for the oxidation reaction must be written as follows.

$$2Zn: \rightarrow 2[Zn]^{2+} + 4e^-$$

---

preference to the iron. Galvanized iron is used in many applications, including automobiles. The painted nail does not corrode because the paint covers the iron, preventing it from contacting the reactive solution. Many metal objects are painted to protect them from corrosion.

3. The reaction of iron atoms to form the iron Fe^{2+} ions is oxidation. It seems to occur mostly at the head and point of the reacting nail.

Assessment

Performance: Ask students to write a report for the MiniLab and to include descriptive drawings of all observed reactions. [L1]

---

miniLAB

Corrosion of Iron

Purpose: Students will study the corrosion of an iron nail and determine some factors that affect this process.

Process Skills: Observing, inferring, interpreting data

Safety Precautions: Caution students that the gelatin prepared in this MiniLab is poisonous.

Preparation

- Purchase the phenolphthalein indicator solution or prepare it according to any standard formula.
- Prepare 100 mL of 0.5M potassium ferricyanide solution by adding 16.5 g of solid to sufficient water to yield 100 mL of solution.

Teaching Strategies

- Agar may be used in place of gelatin.
- You may wish to bend the plain iron nails with pliers. This will produce areas of stress in the nail, and corrosion will occur faster at these points.

Expected Results

See answer to Analysis question 1.

Analysis

1. Only the plain iron nail has reacted. A blue color appears at the head and tip of this nail, and a pink color appears around the middle.

2. The aluminum nail actually does corrode slightly on the surface, forming a protective coating of aluminum oxide. Aluminum objects often are used in situations where iron objects would be corroded. The galvanized nail does not corrode because the more reactive zinc on the surface reacts in

Correcting Misconceptions

Students often are confused when associating *reduction* with the *gain* of electrons. Draw a vertical number line covering the range from 6+ to 2−. Begin with the sulfur 6+ ion, and show how the oxidation number is reduced as two electrons, then four more, then two more are gained, changing the oxidation number of sulfur from 6+ to 4+ to 0 to 2−.

Correcting Errors

- To help students remember that **L**oss of **E**lectrons is **O**xidation and **G**ain of **E**lectrons is **R**eduction, tell them that **LEO** says **GER.**
- Remind students to consider both coefficients and subscripts when determining loss or gain of electrons in a half-reaction. For example, in the reaction $2Al + 3Cl_2 \rightarrow 2AlCl_3$, the reduction half-reaction is $3Cl_2 + 6e^- \rightarrow 6Cl^-$.

Concept Development

Point out to students that when they combine and add half-reactions, they should write the half-reaction equations one over the other, line up the arrows, and be sure that the numbers of electrons lost and gained in the two half-reactions are equal.

Reinforcement

To reinforce the idea that *reduction* refers to the loss of mass when a metal is refined from its ore, assign this problem. What mass of copper metal, in kilograms, may be derived from the reduction of the copper in 1000 kg of copper(II) oxide, according to the equation $2CuO \rightarrow 2Cu + O_2$? *(1000 kg CuO) (1000 g CuO/1 kg CuO) (1 mol CuO/79.55 g CuO) (2 mol Cu/2 mol CuO) (63.55 g Cu/1 mol Cu) (1 kg Cu/1000 g Cu) = 798.9 kg Cu* L2

Figure 16.3

Summarizing the Reaction
Two zinc atoms combine with one diatomic oxygen molecule to form two formula units of zinc oxide. Because zinc loses four electrons in the oxidation reaction and oxygen gains four electrons in the reduction reaction, all electrons are accounted for; the two half-reactions are balanced.

Each atom of oxygen accepts $2e^-$ from a zinc atom and is reduced.

$$2Zn^0 + O_2^0 \rightarrow 2Zn^{2+} + 2O^{2-}$$

with $+4e^-$ from Zn^0 above, and $-4e^-$ to O_2 below.

Each atom of zinc donates $2e^-$ to an oxygen atom and is oxidized.

The balanced overall equation for the reaction now can be written as shown in **Figure 16.3**. This equation is the same as the equation for the formation of zinc oxide that you read at the beginning of the discussion on redox reactions. Now you know that it represents the net oxidation-reduction reaction and is the sum of an oxidation half-reaction and a reduction half-reaction.

If an element is gaining electrons, why is this called a reduction reaction? After all, you don't gain weight when you reduce. You have learned that the reason is because there is a reduction in the charge or oxidation number of an atom of the substance that is reduced. An older, historic reason for the use of the term *reduction* is that the name was first applied to processes in furnaces in which metals are isolated from their ores at high temperatures, **Figure 16.4.** During these processes, oxygen is removed from the ores in which it is combined with the metal, so the ore is reduced to the free metal. There is a reduction in the amount of solid material and a considerable decrease in volume.

Figure 16.4

Furnaces: Old and Modern

For thousands of years, metals have been used by many different cultures for making jewelry, cookware, and weapons. Because metals are normally found combined with other elements as ores, furnaces operating at high temperatures are used to separate the free metal from other elements. The positively charged metal ions in the ore are reduced to the elemental state, while oxygen and other negatively charged elements in the ore are oxidized. Iron smelting in medieval England is shown here. ▼

▲ A modern industrial iron blast furnace is shown here.

558   Chapter 16   Oxidation-Reduction Reactions

Demonstration 16.2

Hydrogen Peroxide Can Do Both

Purpose: To demonstrate that H_2O_2 can be both an oxidizing agent and a reducing agent in different redox reactions.

Materials: 3 g KI, 2 drops 18*M* H_2SO_4, 50 mL 3% H_2O_2, dropper, 0.1 g $KMnO_4$, 150 mL distilled water, two 150-mL beakers

Safety Precautions:

Disposal: Code C

Procedure: In a beaker, dissolve 0.1 g of $KMnO_4$ in 75 mL of distilled water that has been acidified with a drop of 18*M* H_2SO_4. **CAUTION:** *H_2SO_4 is corrosive. $KMnO_4$ and H_2O_2 are skin irritants.* Add H_2O_2 by drops until the solution becomes colorless. In a second beaker, dissolve 3 g of KI in 75 mL of distilled water that has been acidified as above. Add H_2O_2 by drops until a red-brown color is visible. Have students note the color change.

Identifying a Redox Reaction

The oxidation of zinc is a redox reaction in which oxygen is a reactant. You have learned that elements other than oxygen can accept electrons and become reduced during redox reactions. You are already familiar with the explosive reaction in which sodium and chlorine combine to form table salt.

$$2Na(s) + Cl_2(g) \rightarrow 2NaCl(s)$$

Are electrons transferred during this reaction? Yes, because each sodium atom loses one electron to become a sodium ion with a charge of 1+. The oxidation number of sodium increases from 0 to 1+. Each chlorine atom gains one electron to form a chloride ion. The oxidation number of chlorine decreases from 0 to 1−. Therefore, this is another example of a redox reaction.

$$\overset{\overset{\displaystyle 2e^-}{\frown}}{2Na^0 + Cl_2^{\ 0}} \rightarrow 2Na^+ + 2Cl^-$$

Oxidizing and Reducing Agents

Another redox reaction that doesn't involve oxygen occurs when a strip of zinc metal is placed in a solution of copper(II) sulfate. The progress of this reaction can be followed easily because a readily observable change takes place. As shown in **Figure 16.5**, copper metal quickly begins to form on the zinc strip.

Cu^{2+} is the oxidizing agent, and Zn^0 is the reducing agent. ▶

$$Cu^{2+} \xrightarrow[\text{reduced to}]{2e^-} Cu^0 \qquad Zn^0 \xrightarrow[\text{oxidized to}]{2e^-} Zn^{2+}$$

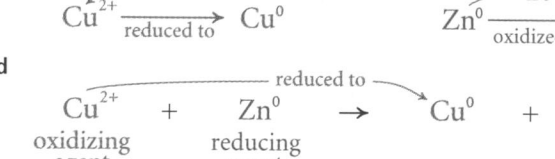

$$\underset{\substack{\text{oxidizing} \\ \text{agent}}}{Cu^{2+}} + \underset{\substack{\text{reducing} \\ \text{agent}}}{Zn^0} \rightarrow Cu^0 + Zn^{2+}$$

reduced to

oxidized to

Figure 16.5

The Reaction Between Zinc and Copper(II) Sulfate
A blue copper(II) sulfate solution gradually becomes colorless if a strip of zinc metal is placed in it. The zinc gives up electrons, becoming oxidized to zinc ions. The colorless Zn^{2+} ions that form go into solution. The Cu^{2+} ions pick up electrons from zinc and become reduced to copper metal atoms, which are deposited on the strip. ▼

Using an Analogy

To illustrate the concept that an oxidizing agent enables oxidation but is not itself oxidized, and a reducing agent enables reduction but is not itself reduced, ask students this question. If I go to a travel agent to obtain tickets for a trip, who does the traveling, and who does the enabling? *I travel, and the agent enables me to travel.*

⚡ Quick Demo

Manganese ions, what colors?

Add 2 mL of $1M$ NaOH solution to 20 mL of $0.01M$ $KMnO_4$ solution in a small beaker. **CAUTION:** *Wear goggles, plastic gloves, and an apron.* Stir the mixture, and ask students to note the color of the permanganate ion. *purple* Fold a heavy, white paper napkin three or four times, and pour the $KMnO_4$ solution on it. Ask students to note the color of the manganese 6+ ion that is formed when the permanganate ions react with cellulose in the paper napkin. *green* Ask students what has happened to manganese in the reaction. *It is reduced as its oxidation number changes from 7+ to 6+.*

Results: The permanganate ion is a stronger oxidizing agent than H_2O_2 and causes H_2O_2 to be oxidized (purple). $2MnO_4^- + 5H_2O_2 + 6H^+ \rightarrow 5O_2 + 8H_2O + 2Mn^{2+}$ (pale pink) Hydrogen peroxide is a strong oxidizing agent, and the colorless I^- is oxidized. $H_2O_2 + 2I^- + 2H^+ \rightarrow 2H_2O + I_2$ (red-brown)

Analysis: Ask this question.

1. Using the above equations, identify the reducing agent and the oxidizing agent in each reaction. *(1) oxidizing agent: MnO_4^-; reducing agent: H_2O_2 (2) oxidizing agent: H_2O_2; reducing agent: I^-*

✓ Assessment

Performance: Students are familiar with hydrogen peroxide as a first aid for cuts. Have students place some 3% H_2O_2 in a test tube and add several drops of blood taken from a meat-wrapping tray. Ask students to write a short summary describing what happens when H_2O_2 is poured onto a cut. *When H_2O_2 is oxidized, H_2O and O_2 gas form.* L1

Copper Atoms and Ions: Oxidation and Reduction

Time Allotment: One class period.

Objectives

Review objectives with students before they begin the ChemLab.

Process Skills

Observing, classifying, interpreting data, communicating

Safety Precautions and Disposal

The concentrated HNO_3 should be used with care. It can cause severe burns to skin and eyes. The fumes produced from reactions of HNO_3 are irritating and poisonous. The HNO_3 reactions must be performed in the fume hood or outside and by the teacher.

Disposal: Dispose of the solution by diluting it with large amounts of water and rinsing it down the drain, Code C.

PREPARATION

• Have stopper/glass tube combinations made ahead of time so that class time is conserved and the danger from breakage during assembly is eliminated.

PROCEDURE

• Require that students' notebooks contain the following items before they begin: Objectives; Safety Precautions; Outline or Flow Chart of the Procedure; Data and Observations Table.

• If the copper oxide pieces are large, better reaction will occur if they are pulverized with a mortal and pestle before use.

• Advise students to wait at least 10 minutes after completing step 4 before doing step 5.

Copper Atoms and Ions: Oxidation and Reduction

Copper atoms and ions often take part in reactions by losing or gaining electrons, which are oxidation and reduction, respectively. If copper atoms lose electrons to form positive ions, copper is oxidized. Other atoms or ions must gain the electrons that copper atoms lose. These atoms or ions are reduced and are called oxidizing agents. In this ChemLab, you will observe two reactions that involve the oxidation or reduction of copper.

Problem

What are some typical reactions that involve the oxidation or reduction of copper?

Objectives

• **Observe** reactions that involve the oxidation or reduction of copper.

• **Classify** the reactants as substance oxidized, reducing agent, substance reduced, and oxidizing agent.

PREPARATION

Materials

copper(II) oxide
powdered charcoal (carbon)
weighing paper
balance
large Pyrex or Kimax test tubes (2)
1-hole rubber stopper fitted with glass tube with bend as shown
150-mL beakers (2)
graduated cylinder, small
glass stirring rod
Bunsen or Tirrill burner
ring stand
test-tube clamp

---

thermal glove
limewater (calcium hydroxide solution)

Safety Precautions

Wear apron and goggles at all times. Care should be taken in handling hot objects and when working around open flames. Do not breathe in the fumes that are produced during the teacher demonstration in step 1.

PROCEDURE

1. **Teacher Demonstration** Your teacher will perform this reaction as a demonstration either in the fume hood or outside the building.
   CAUTION: *Do not perform this procedure by yourself.* A 1-cm square of copper foil will be placed in a porcelain evaporating dish. First, 5 mL of water, then 5 mL of concentrated HNO_3 will be added.

   Note the color of the evolved gas and the color of the resulting solution. Record your observations in a table similar to the one under Data and Observations.

2. On a piece of weighing paper, thoroughly mix approximately 1 g of copper(II) oxide with twice its volume of powdered charcoal. Place the mixture in a clean, dry Pyrex or Kimax test tube. Add about 10 mL of limewater to a second test tube, and stand it in a 150-mL beaker. Assemble the apparatus as shown here, with the copper-oxide test tube sloped slightly downward and the delivery tube extending into the limewater.

DATA AND OBSERVATIONS

Step 1: Gas: Brown and colorless gases are formed.
Solution: The solution turns blue.
Step 3: Limewater: The limewater turns cloudy and white.
Step 5: Product: Metallic copper is formed.

ANALYZE AND CONCLUDE

1. In the first reaction, a brown gas is formed and the solution turns from colorless to blue. In the second reaction, a colorless gas is formed that turns the limewater cloudy, and copper is produced.

2. In the first reaction, copper is oxidized and the nitrogen in nitric acid is reduced. In the second reaction, carbon is oxidized and the copper in CuO is reduced.

ChemLab

3. Heat the mixture in the test tube, gently at first and then strongly. As soon as you notice a change in the limewater, carefully remove the stopper and delivery tube from the reaction test tube. **CAUTION:** *Do not stop heating as long as the tube is in the limewater.* Record your observations of the limewater in your table.

4. Continue heating the reaction test tube until a glow spreads throughout the reactant mixture. Turn off the burner.

5. After the reaction test tube has cooled to nearly room temperature, empty the contents into a beaker that is about half full of water. In a sink, slowly stir the mixture while running water into the beaker until all the unreacted charcoal has washed away. Observe the product that remains in the beaker, and record your observations.

ANALYZE AND CONCLUDE

1. **Interpreting Data** What evidence of chemical change did you observe in each reaction?

2. **Interpreting Data** In the first reaction, a blue-colored solution indicates the presence of Cu^{2+} ions, a brown gas is NO_2, and a colorless

gas is NO. In the second reaction, if limewater becomes cloudy and white, carbon dioxide gas has reacted with the calcium hydroxide to form insoluble calcium carbonate. Using this information, analyze your data and observations. Determine which reactants (Cu and HNO_3 for the first reaction, CuO and C for the second reaction) were oxidized and which were reduced in each reaction.

3. **Classifying** Classify each of the four reactants as an oxidizing agent or a reducing agent.

APPLY AND ASSESS

1. The mass of copper produced in the second reaction is less than the mass of the reacting copper(II) oxide. Why, then, is the gain of electrons known as reduction?

2. What mass of copper may be produced from the reduction of 1.000 metric ton of copper(II) oxide? Hint: Determine the formula mass of copper(II) oxide.

3. If the chlorine gas used at a water-treatment plant reacts with organic materials in the water to yield chloride ions, how would you classify the chlorine gas in terms of oxidation and reduction?

DATA AND OBSERVATIONS

	Observations
Step 1: Gas	
Solution	
Step 3: Limewater	
Step 5: Product	

3. In the first reaction, copper is the reducing agent and nitric acid is the oxidizing agent. In the second reaction, carbon is the reducing agent and CuO is the oxidizing agent.

APPLY AND ASSESS

1. The oxidation number of copper is reduced from 2+ to 0 as copper gains electrons. A reduction in the mass of the oxide ore also occurs when it is refined to the elemental copper.

2. 79.89% of 1.000 metric ton is 0.7989 metric ton of copper.

3. If the neutral chlorine atoms change to 1− chloride ions, they gain electrons and are reduced. Therefore, chlorine is the oxidizing agent.

✔ Assessment

 Performance: Instruct interested and capable students to conduct the experiment as indicated, but to weigh the copper oxide reactant and the copper product and to calculate the theoretical and percent yields for the process. L2

Transparency

✍ Display **Basic Concepts Transparency 43** to show how oxidation or reduction occurs in different metals. L1 LEP

GLENCOE TECHNOLOGY

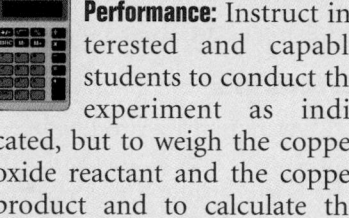 **Videodisc**

Chemistry: Concepts and Applications
See the Videodisc Program Teacher Guide to access illustrations and other material available for this chapter.

Check for Understanding

- Check for understanding of the section material by having students answer the Section Review questions.
- Have students determine the oxidation number for chlorine in each of the following species, and rank the species in sequence from the most oxidized state to the most reduced state for chlorine: ClO^-, Cl_2O_5, Cl_2, Cl^-, ClO_4^-, $NaClO_2$. ClO_4^- (7+); Cl_2O_5 (5+); $NaClO_2$ (3+); ClO^- (1+); Cl_2 (0); Cl^- (1−) [L1]

Reteach

Chemistry Journal: Ask students to examine each of the oxidation reactions that they listed in their journals earlier in the section and to determine and list the oxidizing agent, if possible. [L1]

Extension

Ask students to write balanced equations for a reaction in which the nitrate ion is a spectator ion and a reaction in which the nitrate ion is an oxidizing agent. Also, ask students to explain why it is not possible to describe a reaction in which the nitrogen in the nitrate ion acts as a reducing agent. *In nitrate, nitrogen is assigned a 5+ oxidation number, its most oxidized state. Therefore, it cannot lose additional electrons.* [L1]

4 CLOSE

Portfolio

Ask students to make tables for their portfolios in which they list all the polyatomic ions from Chapter 5 and the oxidation numbers for all the atoms contained in the polyatomic ions. [L1] [P]

What role do the copper ions play in the redox reaction? Each copper ion is reduced to uncharged copper metal when it accepts electrons from the zinc metal. Because the copper ion is the agent that oxidizes zinc metal to the zinc ion, Cu^{2+} is called an oxidizing agent. An **oxidizing agent** is the substance that gains electrons in a redox reaction. The oxidizing agent is the material that's reduced. Because oxidation and reduction go hand in hand, a reducing agent must be present. Zinc metal is the agent that supplies electrons and reduces the copper ion to copper metal; therefore, zinc is called the reducing agent. A **reducing agent** is the substance that loses electrons in a redox reaction. The reducing agent is the material that's oxidized. **Figure 16.6** summarizes the roles of oxidizing and reducing agents in redox reactions.

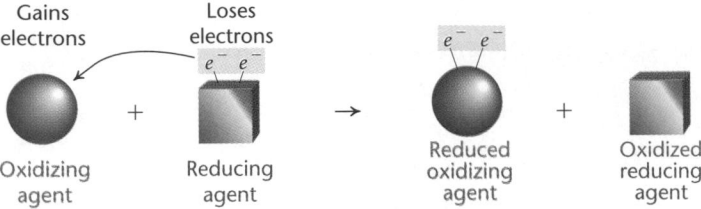

Figure 16.6
Oxidizing and Reducing Agents
When electrons are transferred from one element to another, a combination of an electron-gaining—or reduction—reaction and an electron-losing—or oxidation—reaction takes place. This combination is called a redox reaction. The element that is reduced oxidizes another element by attracting electrons from it, so it is called an oxidizing agent. The element that is oxidized reduces the first element by transferring electrons to it, so it is called a reducing agent.

SECTION REVIEW

Understanding Concepts

1. Name and define the two half-reactions that make up a redox reaction.

2. Identify which reactant is reduced and which is oxidized in each of the following reactions.

   a) $C_5H_{12}(l) + 8O_2(g) \rightarrow 5CO_2(g) + 6H_2O(g)$
   b) $2Al(s) + 3Cu^{2+}(aq) \rightarrow 2Al^{3+}(aq) + 3Cu(s)$
   c) $2Cr^{3+}(aq) + 3Zn(s) \rightarrow 2Cr(s) + 3Zn^{2+}(aq)$
   d) $2Au^{3+}(aq) + 3Cd(s) \rightarrow 2Au(s) + 3Cd^{2+}(aq)$

3. What is the oxidizing agent when iron rusts? What is the reducing agent?

Thinking Critically

4. **Applying Concepts** The following equation represents the reaction between an acid and a base to form a salt and water. Determine the oxidation number for each element. Is this a redox reaction? Explain.

   $2KOH(aq) + H_2SO_4(aq) \rightarrow$
   $$K_2SO_4(aq) + 2H_2O(l)$$

Applying Chemistry

5. **Antioxidants** Compounds that are easily oxidized can act as antioxidants to prevent other compounds from being oxidized. Vitamins C and E protect living cells from oxidative damage by acting as antioxidants. Why does adding lemon juice to fruit salad prevent browning of the fruit?

SECTION REVIEW

1. Redox reactions are combinations of oxidation and reduction half-reactions. An element gains one or more electrons during a reduction; during an oxidation, an element loses one or more electrons.
2. **a.** O_2 is reduced and C in C_5H_{12} is oxidized.
   b. Cu^{2+} is reduced and Al is oxidized.
   c. Cr^{3+} is reduced and Zn is oxidized.
   d. Au^{3+} is reduced and Cd is oxidized.
3. O_2 is the oxidizing agent, and Fe is the reducing agent.

4. $K(1+)$, $O(2-)$, $H(1+)$, $S(6+)$; this is not a redox reaction because electrons are not being transferred. Only ions are being exchanged.
5. The lemon juice contains vitamin C. Vitamin C is more easily oxidized than the compounds in the fruit, which turns brown. The vitamin C acts as an antioxidant when it is oxidized in preference to the other compounds in fruit.

Applications of Oxidation-Reduction Reactions

Natural redox reactions are going on around you every day, everywhere. This is partly due to the abundance of oxygen, which acts as the oxidizing agent as it is reduced in some redox reactions. Other oxidizing agents take part in different redox reactions, especially in environments where not much oxygen gas is found. Near the vents of volcanoes, where sulfur compounds explode out from deep within Earth, enormous deposits of solid yellow sulfur are found. The element sulfur acts both as an oxidizing agent and as a reducing agent in the reaction that forms the sulfur deposits. Can you tell which sulfur compound serves each function in this reaction?

$$2H_2S(g) + SO_2(g) \rightarrow 3S(s) + 2H_2(g) + O_2(g)$$

Note that more than one element in a reaction can be oxidized or reduced. The sulfur in hydrogen sulfide and the oxygen in sulfur dioxide both are oxidized. Sulfur in sulfur dioxide and hydrogen in hydrogen sulfide both are reduced. Each reactant acts as both a reducing agent and an oxidizing agent.

SECTION PREVIEW

Objectives

Analyze **common** redox processes to identify the oxidizing and reducing agents. Identify some redox reactions that take place in living cells.

Say Cheese: Redox in Photography

Understanding natural redox reactions such as the one that occurs in sulfur volcanoes has allowed chemists to develop many processes that make use of oxidation and reduction reactions. Without them, photographs or steel wouldn't exist, and stains would be much harder to remove from clothing.

16.2   Applications of Oxidation-Reduction Reactions   563

Program Resources

Study Guide, pp. 65-66 L1
Section Focus Transparency 33 and
   Master, p. 33 L1 LEP
Basic Concepts Transparencies 44, 45 and
   Masters, pp. 87-90 L1 LEP

ChemLab and MiniLab Worksheets,
   pp. 101-104, 106 L1
Consumer Chemistry, pp. 31-32 L1
Chemistry and Industry, pp. 35-36 L1
Tech Prep Applications, pp. 25-26 L1

PREPARE

Key Concepts
The oxidation-reduction reactions that occur in several common redox processes are discussed.

Planning Ahead
For the MiniLab Order or obtain several household hygiene, cosmetic, and cleaning products. Be sure that several, but not all, of the products contain ethanol.

1 FOCUS

Focus Transparency
Display the **Section Focus Transparency** for Section 16.2 to show students why aluminum should be recycled. L1 LEP

2 TEACH

Discussion
Show students that it is possible for a single element to be both oxidized and reduced in a reaction. This is called disproportionation. Ask students which elements are oxidized and reduced in this reaction: $6KCl + 3I_2 + 3H_2O \rightarrow 5KI + KIO_3 + 6HCl$. I_2 is oxidized to I^{5+} in KIO_3 and is reduced to I^- in KI. Further, ask students whether the loss and gain of electrons are equal. *yes, five lost and five gained*

Transparency
Display **Basic Concepts Transparency 44** to preview redox reactions in photography. L1 LEP

Extension

Ask a professional photographer, a photography lab manager, a staff member who teaches photography at your school, or a competent student to make a class presentation about photography with an emphasis on the role of chemical reactions in the exposure, developing, and printing processes. If possible, a few interested students or the entire class might take a field trip to a photo lab after the presentation to see these processes. L1 LEP

GLENCOE TECHNOLOGY

Videodisc

Chemistry: Concepts and Applications

Tollen's Silver Mirror Test

Disc 2, Side 2, Ch. 3

Show this video of the Demonstration on these pages to illustrate a redox reaction. L1 LEP

CD-Rom

Chemistry: Concepts and Applications

Tollen's Silver Mirror Test

Use this video to show a redox reaction. L1 LEP

Figure 16.7

Early Photos

In a daguerreotype, a redox reaction between silver and iodine fumes produced a layer of light-sensitive silver iodide on the surface of the polished photographic plate. Exposure to light caused decomposition of the silver iodide into elemental silver, which was then treated with the fumes of heated mercury to form bright amalgam areas. The image of Paris shown here was made by Daguerre himself.

WORD ORIGIN

photograph:
photos (GK) light
graphein (GK) to
write

Light is used to
record the image
of an object in a
photograph.

Leonardo da Vinci described a primitive "camera" before 1519, in which someone had to trace images focused on a glass plate inside a box. However, it wasn't until 1838 that the French inventor L.J.M. Daguerre successfully fixed the images in a camera on highly polished, silver-plated copper to make the first photographs. These early photographs were called daguerreotypes in his honor, **Figure 16.7.**

Modern photographic film is made of a plastic backing covered with a layer of gelatin, in which millions of grains of silver bromide are embedded. When light strikes a grain, silver and bromide ions are converted into their elemental forms through a redox reaction. The equation for this redox reaction is as follows.

$$2Ag^+ + 2Br^- \rightarrow 2Ag + Br_2$$

The reaction begins when the shutter on a camera is opened. Light from the scene being photographed passes through the camera's lens and shutter and strikes the light-sensitive silver bromide on the film. The light energy causes electrons to be ejected from a few of the bromide ions, oxidizing them to elemental bromine. The electrons are transferred to silver ions, reducing them to metallic silver atoms. These grains are now activated. The developing chemicals continue the redox reaction by causing the activated grains to be converted to metallic silver. In areas where the light is brightest, more grains are activated, and after developing, they become the darker areas. No silver atoms form in areas of the film that are not struck by light, and that part of the film remains transparent. The exposed film is then developed into a negative, during which time the remaining AgBr and Br_2 is washed away. **Figure 16.8** describes the developing and printing processes.

Demonstration 16.3

Tollen's Silver Mirror Test

Purpose: To demonstrate how a reducing sugar changes the Ag^+ ion into an Ag metal mirror.

Materials: 12.8 g AgNO$_3$, 3.4 g KOH, 2.3 g dextrose (D-glucose C$_6$H$_{12}$O$_6$), 15M NH$_3$ (aq), 50 mL concentrated nitric acid, 600-mL beaker, dropper, 1- to 2-L stoppered container to be silvered, graduated cylinder, stirring rod, 1.5 L distilled water

Safety Precautions: [icons]

Disposal: Code C; remove Ag from inside of container with 50 mL of 1M HNO$_3$; recover AgNO$_3$ by evaporation.

Procedure: Prepare solutions A, B, and C in advance. Store in separate bottles. (A): Dissolve 12.8 g AgNO$_3$ in 150 mL distilled water. **CAUTION:** *This solution is toxic.* (B): Dissolve 3.4 g KOH in 75 mL distilled water. (C): Dissolve 2.3 g dextrose in 50

mL distilled water. Place 150 mL of A into a 600-mL beaker. Add 75 mL of B to the beaker. Slowly add enough 15M NH$_3$(aq) with stirring to just dissolve the precipitate. The container to be silvered must be very clean. Rinse it with 50 mL conc. nitric acid and then rinse thoroughly with distilled water. Pour 50 mL of C and the entire contents of the beaker (AgNO$_3$ + KOH + NH$_3$) into the vessel to be sil-

Figure 16.8
Developing and Printing Pictures
Making photographic negatives by developing exposed film involves several steps. The process describes how black-and-white pictures are made. For color photos, light-sensitive dyes are combined with the silver bromide in layers on the film.

▲ 1. The exposed film is transferred to a canister, where it is developed using a solution of a reducing agent, or developer. The organic compound hydroquinone is usually used for this purpose. The developer reduces all the silver ions to silver atoms in any grain of silver bromide that was hit by light, but it does not react with silver ions in grains that were not exposed to light. Because metallic silver is dark and silver bromide is light, an image having light and dark areas is produced.

▲ 2. After the film has been developed, a solution of a fixer containing thiosulfate ions is added. Thiosulfate ions react with unreduced silver ions to form a soluble complex, which is washed away. This prevents unreduced silver ions from becoming reduced and darkening slowly over time. The reaction follows.
$$AgBr(s) + 2S_2O_3^{2-}(aq) \rightarrow [Ag(S_2O_3)_2]^{3-}(aq) + Br^-(aq)$$

◄ 3. The fixed film is washed to remove any remaining developer or fixer solution. The photographic negative is the reverse of the image photographed; that is, light areas in the scene are dark on the film, and vice versa.

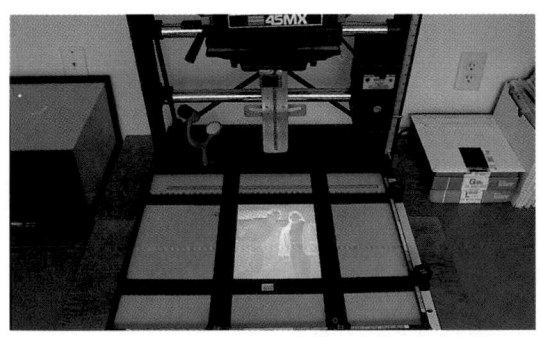

▲ 4. When light is shone through the negative onto light-sensitive photographic paper, a photographic print is made. The print is positive; light and dark areas are identical to those in the scene.

16.2   **Applications of Oxidation-Reduction Reactions**   565

⚡ Quick Demo

Is it the light? 👓 🧤
Use a brush to spread some $0.05M$ silver nitrate solution on a piece of white paper. Put a clearly defined object on part of the paper, and set it in the sunlight at the start of the class period. Near the end of the period, remove the object and show the paper to the class. The outline of the object will be clear and will demonstrate the interaction of light with the silver nitrate. **CAUTION:** *Wear disposable gloves when using silver nitrate and handling the paper.*

Reinforcement

Ask students to determine the substance oxidized, substance reduced, oxidizing agent, and reducing agent in the photographic reaction between silver ions and bromide ions. *Br^-, Ag^+, Ag^+, Br^-, respectively* L1

Transparency

Display **Basic Concepts Transparency 45** to introduce redox reactions in a blast furnace. L1 **LEP**

---

vered. Rotate and tilt the stoppered container to keep all of the surfaces wet. After a minute or two, a silver mirror will begin to form. Continue to rotate and tilt until the mirror completely covers the inside surface. Pour out the reaction mixture, and rinse the inside of the container two times with tap water. **CAUTION:** *Prepare fresh solutions for each demo. To dispose, dilute solutions with large quantities of* *water and flush immediately. The leftover solution forms an explosive precipitate if allowed to stand for several hours.*

Results: The aldehyde group from the dextrose is oxidized to a carboxylic acid as the Ag^+ is reduced to Ag^0, which is deposited on the inside of the reaction vessel.

Analysis: Ask these questions.
1. What is the process called when Ag^+ becomes Ag^0? *reduction*

2. When reduction occurs, what other process also occurs? *oxidation*

✔ Assessment

Skill: Have students do library research to answer these questions. What forms when an aldehyde is oxidized? *a carboxylic acid* What is the test called that identifies a reducing sugar? *Tollen's test* L1

Purpose
TECH PREP Students will gain an understanding of the reactions and resultant forces of the solid rocket fuel of the space shuttle.

Teaching Strategies

- Drop a small quantity of baking soda loosely wrapped in aluminum foil into a 2-L pop bottle containing 50 mL of vinegar. Plug loosely with a cork. **CAUTION:** *Students should stand back.* Discuss the evolution of gas pressure and the resulting force needed to pop the cork.
- Discuss how the popular pump-up water rocket toy uses compressed air to push water out of the rocket and provide the thrust necessary to launch the rocket.

Extension

Aluminum perchlorate is a member of the powerful oxidizing family of perchlorates. Have students use the library to find other members of the perchlorate family and list uses for each compound. ▣

Connecting to Chemistry

1. The aluminum is a reducing agent.
2. Answers will vary.

PHYSICS CONNECTION

Solid Rocket Booster Engines

If you have ever built and launched a model rocket, you probably noticed that the rocket engine was made of a solid, highly combustible material packed into a cardboard tube. After ignition, the expansion and expulsion of the gases produced enough downward force to launch the lightweight rocket quickly into the air. Space shuttles use a similar type of technology, but on a much larger scale.

Engine systems The space shuttle has two different engine systems. The three main engines attached directly to the shuttle operate on liquid hydrogen and liquid oxygen reservoirs carried in the large, centrally located disposable fuel tank. The two smaller, reusable, strap-on booster rockets on either side of the main fuel tank are loaded with a solid fuel, which undergoes a powerful, thrust-producing, oxidation-reduction reaction that helps boost the shuttle into orbit.

Solid rocket fuel The solid rocket fuel is a mixture containing 12 percent aluminum powder, 74 percent ammonium perchlorate, and 12 percent polymer binder. Once ignited, the engine cannot be extinguished. The extremely reactive ammonium perchlorate supplies oxygen to the easily oxidized aluminum powder, providing a greatly exothermic and fast reaction. The purpose of the polymer binder is to hold the mixture together and to help it burn evenly. The overall redox reaction is shown here.

$$\overset{\overbrace{\hspace{2em}24e^-\hspace{1em}}\downarrow}{8Al + 3NH_4ClO_4 \rightarrow 4Al_2O_3 + 3NH_4Cl}$$

Shuttle forces Each solid rocket booster weighs 591 000 kg at liftoff, produces 11.5 million N of force, and operates for about two minutes into the flight. For comparison, a 1000-kg car accelerating from 0 to 26.8 m/s (60 mph) in 7 seconds would require a force of only 3830 N. The tremendous release of chemical energy and expansion of hot gases due to the oxidation-reduction reaction through the engine of the solid rocket booster produces the tremendous thrust needed to get the 2 million-kg shuttle from 0 to almost 700 m/s (1500 mph) in just 132 seconds.

Connecting to Chemistry

1. **Applying** Powdered aluminum is used in another greatly exothermic reaction, the thermite reaction, which is used for welding metals. The reaction is as shown.

$$2Al + Fe_2O_3 \rightarrow Al_2O_3 + 2Fe$$

What role does the powdered aluminum play in this reaction?

2. **Acquiring Information** Investigate the lives and research of Robert Goddard and Werner Von Braun, who both experimented with rockets in the 1930s and helped guide the United States into the space age. Write a short report about these men.

Demonstration 16.4

Copper Nail

Purpose: To demonstrate that a metal replacement reaction is also a redox reaction.

Materials: 150-mL beaker, small piece of steel wool, 75 mL vinegar, 5 g NaCl, 10 dull darkened pennies, small iron nail, stirring rod

Safety Precautions:

Disposal: Code C

Procedure: Place ten dark pennies in the bottom of the beaker, and add vinegar. Use steel wool to clean the outside of the nail, then rinse it. Have students note the color of the nail. Add the 5 g of NaCl to the vinegar, and stir to mix. Place the iron nail in the vinegar solution so it is touching the copper pennies. Allow the beaker to stand until the end of the class period. Have students observe the nail. Leave the

nail in the solution until the next day, and again have students observe the nail.

Results: The nail becomes copper-colored. Point out that the copper ions from the pennies' oxide coating dissolve in the acid and then plate out on the iron nail. The iron also dissolves into the solution.

Having a Blast: Redox in a Blast Furnace

Iron is seldom found in the elemental form needed to make steel. Metallic iron must be separated and purified from iron ore—usually hematite, Fe_2O_3. This process takes place in a blast furnace in a series of redox reactions. The major reaction in which iron ore is reduced to iron metal uses carbon monoxide gas as a reducing agent.

First, a blast of hot air causes coke, a form of carbon, to burn, producing CO_2 and heat. Limestone, $CaCO_3$, which is mixed with the iron ore in the furnace, decomposes to form lime (CaO) and more carbon dioxide. The carbon dioxide then oxidizes the coke in a redox reaction to form carbon monoxide, which is used to reduce the iron ore to iron. The process is outlined here and illustrated in **Figure 16.9.**

$$CaCO_3(s) \rightarrow CaO(s) + CO_2(g)$$

$$CO_2(g) + C(s) \rightarrow 2CO(g)$$

$$2Fe^{3+}(s) + 3O^{2-}(s) + 3CO(g) \rightarrow 2Fe^0(l) + 3CO_2(g)$$
$$\overset{6e^-}{\longrightarrow}$$

Redox in Bleaching Processes

Bleaches can be used to remove stains from clothing. Where do the stains go? Bleach does not actually remove the chemicals in stains from the fabric; it reacts with them to form colorless compounds. In chlorine bleaches, an ionic chlorine compound in the bleach reacts with the compounds responsible for the stain. This ionic compound is sodium hypochlorite (NaOCl). The hypochlorite ions oxidize the molecules that cause dark stains.

$$OCl^-(aq) + \text{stain molecule(s)} \rightarrow Cl^-(aq) + \text{oxidized stain molecule(s)}$$
$$\text{(colored)} \qquad\qquad\qquad\qquad \text{(colorless)}$$

▲ Molten iron is drawn off at the bottom of the furnace. A combination of by-products known as slag is also removed at the bottom.

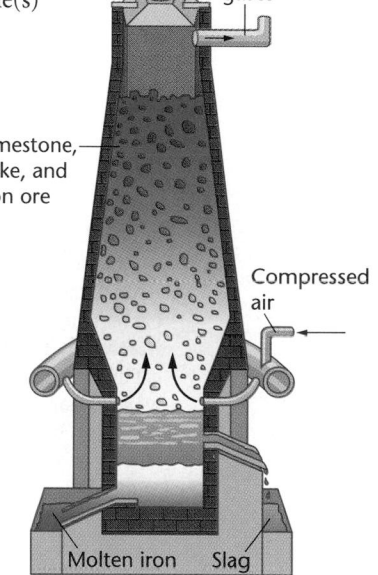

Exhaust gases

Limestone, coke, and iron ore

Compressed air

Molten iron   Slag

Figure 16.9

Blast Furnace
Iron ore (Fe_2O_3), coke (C), and limestone ($CaCO_3$) are added at the top of the furnace. Hot air at about 900°C, blasted into the bottom of the furnace, burns the coke in an exothermic reaction. This reaction causes temperatures in a blast furnace to reach about 2000°C. ▶

⚡ Quick Demo

Can silicon be removed from sand?

Mix 5 g of powdered magnesium and 7.5 g of dry silica sand. Place the mixture in a large Pyrex or Kimax test tube. Clamp the tube on a ring stand, and heat it with a lab burner until a red-orange glow indicates that the reaction has begun. Turn off the burner and let the reaction continue. White magnesium oxide will deposit near the top of the tube, and silvery-gray elemental silicon will deposit below it on the inside of the tube. **CAUTION:** *Wear safety goggles and do this demo behind a safety shield. Caution students not to look directly at the burning magnesium.* Ask students to write the balanced equation for the reaction and to identify the substance oxidized. $2Mg + SiO_2 \rightarrow Si + 2MgO$; *magnesium is oxidized.* L1

Fact of the MATTER

GLENCOE TECHNOLOGY

💿 Videodisc

Chemistry: Concepts and Applications
Iron Smelting
Disc 2, Side 2, Ch. 4

Show this video to help you explain the redox reactions in a blast furnace. L1 LEP

💿 CD-Rom

Chemistry: Concepts and Applications
Iron Smelting
Use this video to help reinforce the redox process in a blast furnace. L1 LEP

Analysis: Ask these questions.
1. Write the reduction half-reaction that shows the copper(II) ions depositing on the iron nail. $Cu^{2+} + 2e^- \rightarrow Cu$
2. Write the oxidation half-reaction that shows the iron nail dissolving into the acetic acid/salt solution. $Fe \rightarrow Fe^{2+} + 2e^-$
3. Write the balanced chemical equation for this redox reaction. $Cu^{2+} + Fe \rightarrow Cu + Fe^{2+}$

✔ Assessment

Knowledge: Ask students to determine the oxidizing agent and the reducing agent from the balanced redox equation. *oxidizing agent:* Cu^{2+}; *reducing agent: Fe* L1

miniLAB

Testing for Alcohol by Redox

Purpose: To use a redox reaction to test for the presence of alcohol in several household hygiene, cosmetic, and cleaning products.

Process Skills: Observing, classifying, communicating, comparing and contrasting

Safety Precautions: Students should wear disposable plastic gloves when doing the MiniLab.

Disposal: Code F or J

Teaching Strategies

- Use one product that is a blue-green color to obscure the test results.
- You may wish to use a product that contains methanol to illustrate that other alcohols react as well.
- Caution students that none of the products are to be tasted, even if the identities seem obvious.

Expected Results

Depending upon the products tested, several will give a positive test for alcohol.

Analysis

1. Answers will vary depending upon the products tested.
2. Alcohol is the reducing agent.

✓ Assessment

Performance: Ask students to bring in and test several other products from their homes. Advise them to leave the labeled products in the chemistry lab before school rather than carrying them to other classes, lunch, etc. L1

miniLAB 2

Testing for Alcohol by Redox

Organic alcohols react with orange dichromate ions, producing blue-green chromium(III) ions. This reaction is used in a Breathalyzer test to test for the presence of alcohol in a person's breath. In this MiniLab, you will use this reaction to test for the presence of alcohol in a number of household hygiene, cosmetic, and cleaning products.

Procedure

1. Label five small test tubes with the names of the products to be tested.

2. Place approximately 1 mL of each product in the appropriate tube.

3. Wearing apron and goggles, add three drops of dichromate reagent to each tube, and stir to mix the solutions.

CAUTION: *Do not allow dichromate reagent to come into contact with skin. Wash with large volumes of water if it does.*

4. Observe and record any color changes that occur within one minute.

Analysis

1. Which of the products that you tested contain alcohol? Was the presence of alcohol noted on the label of the products?

2. If the orange $Cr_2O_7^{2-}$ ion reacts with alcohol to produce the blue-green Cr^{3+} ion, what substance is the reducing agent in the reaction?

Bleaches containing hypochlorite should be used carefully because hypochlorite is a powerful oxidizing agent that can damage delicate fabrics. These bleaches usually have a warning label telling the user to test an inconspicuous part of the fabric before using the product. In addition to acting as a bleaching agent, hypochlorite ions are also used as disinfectants, as **Figure 16.10** shows.

Figure 16.10

Hypochlorite as a Disinfectant
Hypochlorite is used in disinfectants to kill bacteria in swimming pools and in drinking water. In both cases, the hypochlorite ions act as oxidizing agents. Bacteria are killed when important compounds in them are destroyed by oxidation. In this photo, the amount of chlorine in the water is being monitored. Chlorine reacts with the water to form hypochlorite ions.

Chemistry Journal

Bleach Labels

Ask students to read the label(s) on the products used to bleach and/or to remove stains in their homes to determine whether any of the products contain hypochlorites or other chlorine compounds. Students should include the results of their investigations in their journals. L1

Chemistry Journal

Water Treatment

TECH PREP

Ask a few interested students to investigate the use of disinfectants in the treatment of local drinking water, and perhaps arrange a visit to a water-treatment plant. Ask the students to write up their findings in their journals and to make a class presentation. L2

COOP LEARN

How it Works

Breathalyzer Test

The alcohol in beverages, hair spray, and mouthwashes is ethanol. Ethanol is a volatile liquid that evaporates rapidly at room temperature. Because of this volatility, drinking an alcoholic beverage results in a level of gaseous ethanol in the breath that is proportional to the level of alcohol in the bloodstream. About 50 percent of all automobile accidents that result in a fatality are caused by intoxicated drivers. Law officers can determine quickly whether a person is legally intoxicated by using an instrument called a breath analyzer, or Breathalyzer.

1. A simple Breathalyzer device has an inflatable plastic bag attached to a tube containing an orange solution of potassium dichromate and sulfuric acid.

2. During a Breathalyzer test, a person blows into the mouthpiece of the bag.

3. If alcohol vapors are present in the person's breath, ethanol undergoes a redox reaction with the dichromate. As ethanol is oxidized, the orange Cr^{6+} ions are reduced to blue-green Cr^{3+} ions.

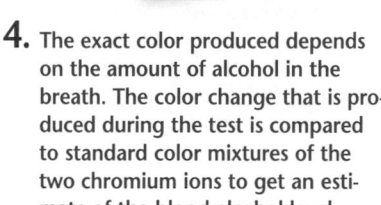

4. The exact color produced depends on the amount of alcohol in the breath. The color change that is produced during the test is compared to standard color mixtures of the two chromium ions to get an estimate of the blood alcohol level.

Thinking Critically

1. Suppose a person used mouthwash shortly before taking a Breathalyzer test. What might be the result?
2. How would the color produced in a Breathalyzer test change as the ethanol content of the blood increases?

16.2   Applications of Oxidation-Reduction Reactions   569

Thinking Critically

1. Many mouthwashes contain ethanol, so any mouthwash remaining in the mouth might cause the test to indicate incorrectly a high blood alcohol level.
2. As the ethanol content of the blood increases, the color in the test becomes more blue-green and less orange.

How it Works

Purpose

TECH PREP

To learn about the role a redox reaction plays in the operation of a Breathalyzer test.

Background

Redox reactions provide the basis for several tests used to detect the presence and amount of organic chemicals that contain specific functional groups. The test used in the Breathalyzer is also called the Jones test, and it indicates the presence of an alcohol functional group. In this test, the alcohol serves as a reducing agent that reduces chromium ions from the 6+ state, in which they have an orange color, to the 3+ state, in which they have a green color.

Visual Learning

Show students what an early alcohol breath analyzer from around 1960 looked like by using a clip from the video of the Disney movie *The Absent Minded Professor*. In the movie, a police officer administers a breath test to someone who says he has seen the professor flying in his automobile. The old breath analyzer tests consisted of blowing into a bottle containing a large volume of orange dichromate solution and watching for a color change. Today's Breathalyzer instruments are much more complex and give a quantitative measurement of alcohol level using spectroscopy.

Teaching Strategies

- Demonstrate how the Breathalyzer test works by mixing a sample of alcohol with a dichromate solution, as is done in MiniLab 2.
- Demonstrate how alcohol concentration in air, due to its volatility, is significant by running tubing from a container of alcohol with a narrow opening into a solution of dichromate.

Figure 16.11
The Green Lady
The green color of the Statue of Liberty in New York Harbor is due to a layer of patina, or protective coating, that covers the copper sheets making up the statue. The presence of the patina helps keep the statue from corroding further because oxygen cannot get through the patina to reach the copper layers underneath.

Corrosion of Metals

Did you know that the Statue of Liberty is made of copper sheets attached to a steel skeleton? Why does it appear green rather than the reddish-brown color of copper? When copper is exposed to humid air that contains sulfur compounds, it undergoes a slow oxidation process. Under these conditions, the copper metal atoms each lose two electrons to produce Cu^{2+} ions, which form the compounds $CuSO_4 \cdot 3Cu(OH)_2$ and $Cu_2(OH)_2CO_3$. These compounds are responsible for the green coat or patina found on the surface of copper objects that have been exposed to air for long periods of time, **Figure 16.11.**

You have learned that iron is oxidized by oxygen in the air to form rust. Aluminum is a more active metal than iron. As a result of its greater activity, aluminum is oxidized more quickly than iron. If this is true, why does an aluminum can degrade much more slowly than a tin can, which is made of iron-containing steel that is coated with a thin layer of tin? The reason is that, like copper, aluminum is oxidized to form a compound that coats the metal and protects it from further corrosion, as shown in **Figure 16.12.** Aluminum reacts with oxygen to form aluminum oxide in a redox reaction.

$$4Al(s) + 3O_2(g) \rightarrow 2Al_2O_3(s)$$

A coating of aluminum oxide is tough and does not flake off easily, as iron oxide rust does. When rust flakes fall off a surface, additional metal is exposed to air and becomes corroded.

Figure 16.12
Corrosion of Iron and Aluminum
Because iron rust is porous and flaky, it does not form a good protective coating for itself. ▶

A tin coating offers some protection to the iron. However, if a hole or crack develops in the thin tin coating, the underlying iron corrodes rapidly. A tin-coated steel can will degrade completely in about 100 years. The aluminum oxide coating on an aluminum can is tough and closely packed. It protects the underlying aluminum from further corrosion so that the can will take about 400 years to degrade. ▶

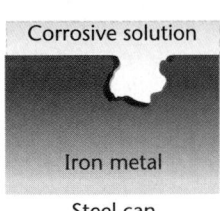
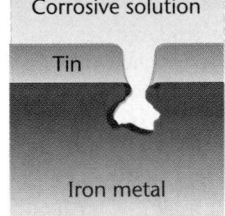

Corrosive solution	Corrosive solution	Corrosive solution
Iron metal	Tin / Iron metal	Aluminum oxide / Aluminum metal
Steel can	Tin-coated steel can	Aluminum can

ACROSS THE Curriculum

Art

TECH PREP

Since the 17th century, chemists have supplied artists with the colors used in paintings, sculptures, and other art forms. In 1809, Vauquelin, who discovered the element chromium in 1797, suggested that chromium compounds could be used in ceramic glazes. All chromium compounds are colored. Chromium oxide, Cr_2O_3, is the most stable green pigment ever developed. Because it is a dull-green shade that reflects infrared light in a manner similar to chlorophyll, the green pigment in plants, it is used in military camouflage paint. It is prepared by heating a mixture of a dichromate salt and excess sulfur to 1100°C. The equation for this redox reaction follows.

$$20Na_2Cr_2O_7 + 3S_8 \rightarrow$$
$$20Cr_2O_3 + 12Na_2SO_4 + 8Na_2O + 12SO_2$$

In this reaction, sulfur serves as a reducing agent that reduces chromium ions from the 6+ state to the green 3+ state.

Everyday Chemistry

Lightning-Produced Fertilizer

All plants need nutrients just as animals do, but did you know that one of the main nutrients needed is nitrogen? Even though the air surrounding Earth is almost 80 percent nitrogen, the nitrogen is in the form of N_2 molecules, a form that most plants and animals cannot use. Nitrogen from the air is converted to a form that plants can use by a process called nitrogen fixation. Plants can best use nitrogen when it is in the form of the ammonium ion, NH_4^+, where the nitrogen has an oxidation number of 3−, but they can also use the nitrate ion, NO_3^-, with nitrogen having an oxidation number of 5+.

Nitrogen fixation Nitrogen can be fixed for plants in three different ways: by lightning, by nitrogen-fixing bacteria living in the roots of certain plants or in the soil, and by commercial synthesis reactions such as the Haber ammonia process.

Nitrogen is a fairly inert gas because the triple bond of N_2 is strong and resists breaking. However, the exceptionally high energy and temperatures of lightning can easily break bonds and allow for recombination of gases in the atmosphere.

Lightning-driven reactions In the process of lightning-driven nitrogen fixation, nitrogen and oxygen combine to form nitrogen monoxide. Nitrogen monoxide then combines with more oxygen to form nitrogen dioxide. This nitrogen dioxide mixes with water in the air to form nitric acid and more nitrogen monoxide, which is available to continue the cycle.

$$N_2 + O_2 \rightarrow 2NO$$
$$2NO + O_2 \rightarrow 2NO_2$$
$$3NO_2 + H_2O \rightarrow 2HNO_3 + NO$$

Fertilizer production The pH of rainwater is naturally slightly acidic, and you can see that some of this acidity is due to the dissolved nitric acid, HNO_3, from nitrogen fixation. As the rain soaks into the soil, bacteria convert the nitrate ions into ammonium ions.

How does nature's manufacturing of fixed nitrogen compare with commercial production of fixed nitrogen? You may think that lightning isn't all that common, but it is estimated that there are approximately 10 000 lightning storms every day over the surface of Earth. Stated another way, lightning strikes 100 times a second on the planet as a whole. Approximately 10 billion kg of nitrogen are fixed yearly in the atmosphere. Biological agents such as bacteria fix about 100 billion kg of nitrogen yearly, and an amount equal to that is fixed through the manufacture of fertilizer and other industrial processes.

Exploring Further

1. **Classifying** Nitrogen fixation in the soil is accomplished by bacteria living in the roots of certain plants. Name some of these plants.

2. **Applying** In each of the three equations shown, what is oxidized and what is reduced?

3. **Acquiring Information** The process by which nitrogen is put back into the air is called denitrification. Find out what conditions are necessary for this process and what reaction occurs.

2. In the first two equations, nitrogen is oxidized and oxygen is reduced. In the third reaction, nitrogen is both oxidized and reduced.

3. Denitrification occurs under anaerobic conditions by bacteria and stresses the need for varied life processes to complete Earth's cycles.

Purpose
TECH PREP

Students will gain an understanding of the redox reactions that occur during the fixation of atmospheric nitrogen by lightning.

Teaching Strategies

- Ask students what purpose fertilizers serve and how they work. Discuss the need for certain ionic compounds as plant food. Have the students name natural fertilizers and list them on the board. [L1]

- Compare different fertilizer formulas such as 12-12-12 and 30-5-5, and discuss the formulations for different uses or needs of plants.

- Discuss the catalytic Haber process ($H_2 + N_2 \rightleftarrows NH_3$) for making ammonia using high pressure (300 atm) and high temperatures (400°-500°C). Discuss the advantages nature has by performing these same reactions at common Earth temperatures. Explain how anhydrous ammonia is placed directly into the soil during planting season by many farmers.

Extension

The fertilizer industry produces tremendous amounts of fertilizer to help in crop production. Have students research the main chemical compounds comprising synthetic fertilizer and determine which part of the plant each compound targets. [L1]

Exploring Further

1. Various bacteria are symbiotically associated with the following plants: alfalfa, clover, peas, beans, lupines, soybeans, alder tree, maize, sorghum, pearl millet, Douglas fir, white fir.

Silver Tarnish: A Redox Reaction

Imagine if, along with your usual chores of taking out the trash, washing dishes, feeding your pets, and taking care of your younger siblings, you also had to polish the silver—as people did back in your great-grandparents' days. How would you find time for any fun? Fortunately, other materials such as stainless steel have replaced most "silverware." Why do silver utensils have to be polished, but those made of stainless steel or aluminum don't? Silver becomes tarnished through a redox reaction that is a form of corrosion, as rusting is. Tarnish is formed on the surface of a silver object when silver reacts with H$_2$S in air. The product, black silver sulfide, forms the coating of tarnish on the silver.

$$O_2(g) + 4Ag(s) + 2H_2S(g) \rightarrow 2Ag_2S(s) + 2H_2O(l)$$

Many commercial silver polishes contain abrasives that help to remove tarnish. Unfortunately, they also remove some of the silver. A more gentle way to remove tarnish from the surface of a silver object involves another redox reaction. In this reaction, aluminum foil scraps act as a reducing agent.

$$2Al^0(s) + 6Ag^+(s) + 3S^{2-}(s) + 6H_2O(l) \rightarrow 6Ag^0(s) + 2Al^{3+}(aq) + 6OH^-(aq) + 3H_2S(g)$$

with $6e^-$ transfer noted above the equation.

This reaction is essentially the reverse of the reaction that forms tarnish. Here, silver ions in the Ag$_2$S tarnish are reduced to silver atoms, while aluminum atoms in the foil are oxidized to aluminum ions. The tarnish-removing solution usually includes baking soda (sodium hydrogen carbonate) to help remove any aluminum oxide coating that forms and to make the cleaning solution more conductive. **Figure 16.13** shows how this method of silver cleaning is done.

Figure 16.13
Removing Silver Tarnish
Even though corrosion is an unwanted redox reaction, removing the tarnish makes use of another redox reaction. A nest of crumpled aluminum foil scraps is made at the bottom of a large pot. The tarnished silver object is added, making sure the silver is in contact with the foil scraps. Baking soda is added, and the silver is covered with water. When the pot is heated on a stove, the silver sulfide tarnish is reduced to silver atoms, and the silver object becomes shiny and bright.

572   Chapter 16   Oxidation-Reduction Reactions

Cultural Diversity

Ancient African Steel

Researchers have found that carbon steel was manufactured in Africa on the western shore of Lake Victoria 2000 years ago. The method employed a forced-draft oven using a technology similar to that developed in Europe in the 19th century. The conical slag and mud ovens were built over a liner made from the base of a termite mound. Iron ore placed inside the oven was heated by burning charcoal. Technological sophistication was demonstrated by the use of air conduits around the base of the oven. These conduits allowed the air to be heated as it entered the oven, thus allowing the oven to reach the high internal temperatures needed to make carbon steel.

Forensic Blood Detection

The gas station at the corner was robbed, and the cashier was shot. On television, police announce that Suspect A has been taken into custody. They have confiscated a jacket, allegedly worn by the suspect. After preliminary examination by the police department, the jacket is sent to a forensic laboratory for scientific investigation. One of the first tests a technician at the laboratory will carry out determines whether or not there are blood stains on the jacket.

The Luminol Test

The technician may choose from several chemical tests for blood, all based on the fact that the hemoglobin in blood catalyzes the oxidation of a number of organic indicators to produce a colored product that emits light, or luminesces.

The technician on this case chooses the luminol test. Luminol has an organic double-ring structure, shown below. In 1928, German chemists first observed the blue-green luminescence when the compound was oxidized in alkaline solution. It was soon found that a number of oxidizing agents, such as hydrogen peroxide, bring about the luminescence. Later, workers noted that the luminescence was greatly enhanced by the presence of blood, which led to its current use in forensic investigations.

Luminol

The technician carefully mixes an alkaline solution of luminol with aqueous sodium peroxide and, in a darkened workplace, sprays the solution onto suspected spots on the jacket. Bingo! An intense, blue-green chemiluminescence is emitted from several spots. Because the glow will last for a few minutes, the technician photographs the spots and their telltale light.

Ruling Out with Luminol

You may wonder if this relatively simple procedure will serve to convict Suspect A. Certainly not. However, if the test had been negative, Suspect A might have been cleared from suspicion. A negative result ensures that a stain is *not* blood. But, because this is not the case with the stains on the jacket, the luminol test is preliminary and will be used with other tests.

The luminol test is especially useful because it works well with both fresh and dried blood. Luminol has one particularly useful feature. The same stains can be made luminescent over and over again if the spray is allowed to dry and the stains are resprayed.

A positive test should not be taken as absolute proof of blood because luminol reacts with copper and cobalt ions, as well as with the iron in hemoglobin. However, it reacts much more strongly with hemoglobin. A large number of forensic authorities believe that the luminol test has value as a preliminary sorting technique.

DISCUSSING THE TECHNOLOGY

1. **Applying** If a luminol test yields a positive reaction, what is the next logical step?

2. **Hypothesizing** Why can it almost never be assumed that stains are uncontaminated, although stain evidence is important in a criminal investigation?

Integrating THE Sciences

Health Chemicals that are so easily oxidized that they protect other chemicals from being oxidized are called antioxidants. Zinc functions as an antioxidant when it is used in galvanizing steel and iron to prevent corrosion. Because vitamin C (ascorbic acid) is so readily oxidized that it protects other food components from oxidation, it is used as a food preservative. Two other common antioxidant food preservatives are BHA (butylated hydroxyanisole) and BHT (butylated hydroxytoluene). Some scientific evidence indicates that the presence of antioxidants in the diet helps protect cells from oxidative damage that hastens aging and can lead to cancer.

Figure 16.14
Chemiluminescence

◀ When lightning is produced by an electrical discharge in the atmosphere, electrons in molecules of O_2 and N_2 gases are excited to higher energy levels. Energy from the electricity breaks the molecules into atoms. When the atoms recombine to form molecules and the electrons return to lower energy levels, light energy is released through chemiluminescence.

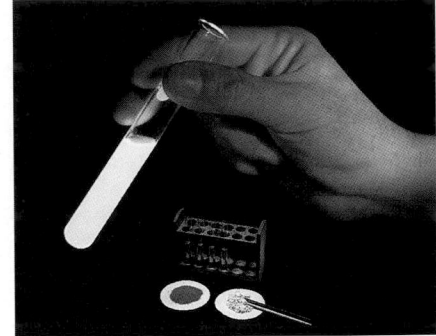

When luminol is oxidized and is observed in the dark, an eerie blue-green glow is produced through chemiluminescence. ▶

Chemiluminescence: It's Cool

Some redox reactions can release light energy at room temperature. The production of this kind of cool light by a chemical reaction is called chemiluminescence. The light from chemiluminescent reactions can be used in emergency light sticks that work without an external energy source. You may recall learning in Chapter 6 how these light sticks work. Now you know that the reaction that takes place when the two solutions in the light sticks are mixed involves an oxidation and a reduction.

Some chemiluminescent redox reactions occur naturally in the atmosphere as a result of lightning, **Figure 16.14.** Other chemiluminescent reactions involve luminol, an organic compound that emits cool light when it is oxidized. Luminol reactions are utilized by forensic chemists to analyze evidence in crime investigations. They spray luminol onto a location where the presence of blood is suspected. If blood is present, the iron(II) ions in the blood oxidize the luminol to form a chemiluminescent compound that glows in the dark. The iron is reduced by the luminol. **Figure 16.14** shows the glow from the oxidized form of luminol.

Biochemical Redox Processes

How are bears able to stay warm enough to keep from freezing during their winter hibernation? How do marathon runners get the energy to finish a race without stopping to eat? In both cases, fats stored in the body are oxidized. Oxygen molecules from the air are reduced as they gain electrons to form water. In a series of redox reactions called respiration,

Demonstration 16.5

Fresh Fruit
Purpose: To demonstrate that redox reactions occur in living cells.
Materials: 50 mL orange juice, lemon juice, or a vitamin C tablet; an apple

Safety Precautions:
Disposal: Code A

Procedure: Cut an apple in half and leave the cut side exposed to the air for a few minutes. Have students observe the apple. Take a second freshly cut piece of apple, and dip it in orange juice, lemon juice, or a solution made by crushing and dissolving a vitamin C tablet. Have students observe that the oxidation of the apple is much slower.

Figure 16.15
Keeping Warm
Although it is common to think that only mammals keep warm, in truth, all plants and animals maintain a temperature at which their enzymes function best. Plants keep from freezing because heat is produced as a by-product of respiration and photosynthesis. One of the first plants to poke through the snow in early spring is the heat-producing skunk cabbage. The heat it releases allows it to get a head start on other plants and also contributes to the unpleasant odor that gives it its name.

energy is released. **Figure 16.15** shows one effect of this heat in plants. Respiration will be discussed in Chapter 19. Many other redox reactions take place in living things. Electrons are transferred between molecules in redox reactions during photosynthesis and in the reactions that fireflies use to flash light signals to potential mates. You will study photosynthesis in Chapter 20.

Some organisms can use the energy released during redox reactions to convert chemical energy into light energy, a process called bioluminescence. You are probably familiar with the flashing lights given off by fireflies during courtship, but did you know that many different organisms—including some fish, at least one type of mushroom, and a caterpillar known as a glowworm—also are bioluminescent? **Figure 16.16** shows bioluminescence in fireflies.

Now that you have learned what redox reactions are and have read about some of the processes of which they are part, you can reexamine the redox reaction that makes cut fruit turn brown. The color is due to brown pigments that are formed by the oxidation of colorless compounds normally present in the cells of the fruit. Oxygen in the air is the oxidizing agent that reacts with the colorless compounds to produce the brown pigments. The oxygen is reduced when it accepts electrons from the pigments, so the pigments function as reducing agents. This combination of oxidation and reduction goes hand in hand in a redox reaction because electrons that are lost by one element must be gained by another.

WORD ORIGIN

bioluminescent:
bios (GK) life
lumen (L) light
escentis (L) beginning to be, have, or do

A bioluminescent substance undergoes a chemical reaction in living things in which potential energy in chemical bonds is converted into light energy.

Figure 16.16
Firefly Signals
Fireflies use flashes of light to attract mates. Light energy is released during an enzyme-catalyzed redox reaction. Luciferase is the name given to the enzyme that speeds up the reaction in which the organic molecule luciferin is oxidized.

16.2   Applications of Oxidation-Reduction Reactions   575

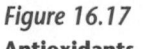

Figure 16.17
Antioxidants
Vitamin C owes its antioxidant properties to the fact that it reacts so readily with oxygen. When added to a food product, oxygen reacts preferentially with vitamin C, thereby sparing the food product from oxidation. Other anti-oxidant food additives include the synthetic compounds BHA and BHT and the natural antioxidant, vitamin E.

Fact of the MATTER

Myoglobin, found in muscle tissue, is an iron-containing protein that stores oxygen. Myoglobin in living muscle tissue is bound to oxygen and is a red color. It becomes pale purple after death when the oxygen is lost. Heating meat results in oxidation of the iron in myoglobin, which then has the brown color that tells you the meat is cooked.

The skin of a fruit keeps oxygen out, which is why unbroken fruit does not turn brown. Coating cut fruit with an antioxidant can prevent browning and keep a fruit salad looking fresh longer. The vitamin C in lemons is a good antioxidant. If lemon juice is squirted onto cut banana or apple slices, they will not brown as quickly because the vitamin C reacts with oxygen more readily than do the fruit-browning compounds, **Figure 16.17**.

Connecting Ideas

Most reactions involve electron transfer and thus are redox reactions. You have learned to identify which element is reduced and which is oxidized when you are given the equation for a redox reaction. You might wonder why one element accepts electrons from another and whether you can predict which element will be oxidized and which will be reduced. Learning to make those predictions is the next step in your study of electron-transfer processes in compounds and will help you understand how redox reactions in batteries produce electricity.

SECTION REVIEW

Understanding Concepts

1. What role does the reducing agent hydroquinone play in the production of a photographic negative?

2. How is most of the iron that is used for making steel purified from iron ores?

3. Why do aluminum cans degrade more slowly than cans made of iron?

Thinking Critically

4. **Applying Concepts** Oxygen is required for the production of light by fireflies. What role does the oxygen play in the reaction?

Applying Chemistry

5. **Bleaching** Why can't rust stains be removed with bleach?

SECTION REVIEW

1. Hydroquinone is a reducing agent that darkens the image on the negative by converting all Ag^+ ions to Ag atoms in grains of silver bromide that have been exposed to light.
2. Iron is obtained from its ore through a redox reaction with carbon monoxide. In this reaction, iron ions in the ore are reduced to iron metal, while carbon monoxide is oxidized to carbon dioxide.
3. Aluminum metal on the surface of a can is oxidized into a tough coating or patina of aluminum oxide that protects the metal underneath. Iron in the steel of a tin can is oxidized into flaky rust compounds that break off the surface easily and do not provide much protection for the iron metal underneath.
4. Oxygen serves as an oxidizing agent.
5. The iron(III) in rust is already in the oxidized form. It cannot be further oxidized by bleach.

REVIEWING MAIN IDEAS

16.1 The Nature of Oxidation-Reduction Reactions

- Oxidation occurs when an atom or ion loses one or more electrons and attains a more positive oxidation number. Reduction takes place when an atom or ion gains electrons and attains a more negative oxidation number.

- Oxidation and reduction reactions always occur together in a net process called a redox reaction.

- An oxidizing agent is the substance that gains electrons and is reduced during a redox reaction. A reducing agent is the substance that loses electrons and is oxidized during a redox reaction.

16.2 Applications of Oxidation-Reduction Reactions

- In photography, light triggers the reduction of silver ions to silver metal on photographic film.

- Bleach removes stains from clothing by oxidizing colored molecules to form colorless molecules.

- Metals such as copper and aluminum are resistant to corrosion even though they are easily oxidized because the products of their reactions with oxygen form protective coatings on the surface of the metal.

- Chemiluminescent reactions in emergency light sticks, lightning, and the luminol reaction convert the energy of chemical bonds into light energy.

- Some organisms use redox reactions to produce light, which they use in communication. This light production is called bioluminescence.

- Cut fruits turn brown because compounds in the fruit cells react with oxygen in a redox reaction to produce brown pigments. Coating the fruits with antioxidants can prevent this browning.

Key Terms

For each of the following terms, write a sentence that shows your understanding of its meaning.

oxidation
oxidation-reduction reaction
oxidizing agent
reducing agent
reduction

UNDERSTANDING CONCEPTS

1. What is the difference between an oxidizing agent and a reducing agent?

2. Which of the changes indicated are oxidations and which are reductions?

   a) Cu becomes Cu^{2+}
   b) Sn^{4+} becomes Sn^{2+}
   c) Cr^{3+} becomes Cr^{6+}
   d) Ag becomes Ag^+

3. Identify the oxidizing agent in each of the following reactions.

   a) $Cu^{2+}(aq) + Mg(s) \rightarrow Cu(s) + Mg^{2+}(aq)$
   b) $Fe_2O_3(s) + 3CO(g) \rightarrow 2Fe(l) + 3CO_2(g)$

4. What is the oxidizing agent in household bleach?

5. Why does a photographic negative need to be fixed?

6. In which direction do electrons move during a redox reaction: from oxidizing agent to reducing agent or vice versa?

7. Why is aluminum metal used to remove tarnish from silver?

8. What chemical process do hibernating animals use to stay warm?

9. Write the equation for the redox reaction that occurs when a piece of iron metal is dipped in a solution of copper(II) sulfate.

CHAPTER REVIEW

- Review the Reviewing Main Ideas statements and Key Terms with your students.
- Complete solutions to Chapter Review problems can be found in the *Problems and Solutions Manual*.

UNDERSTANDING CONCEPTS

1. An oxidizing agent is reduced as it causes another substance to be oxidized, whereas a reducing agent is oxidized as it causes another substance to be reduced.

2. **a.** oxidation **b.** reduction **c.** oxidation **d.** oxidation

3. **a.** Cu^{2+} **b.** Fe_2O_3

4. the hypochlorite ion, ClO^-

5. During the photographic fixing process, any remaining Ag^+ ions in the film are removed. If they were not removed, exposure to light would reduce them to silver atoms over time, and the entire negative would darken.

6. Electrons move from a reducing agent to an oxidizing agent.

7. Aluminum is a more active metal than silver, so it acts as a reducing agent to reduce the Ag^+ ions in the tarnish to silver atoms. The aluminum metal atoms are oxidized, losing electrons to become aluminum ions.

8. All animals produce heat as a by-product of respiration. Hibernating animals produce extra heat at the expense of energy. Because they are not active, they need less energy.

9. $Fe(s) + Cu^{2+}(aq) \rightarrow Fe^{2+}(aq) + Cu(s)$

✓ Assessment

Portfolio: Review the portfolio options that are provided throughout the chapter. Encourage students to select one product that demonstrates their best work for the chapter. Have students explain what they have learned and why they chose this example for placement into their portfolios. **P**

Additional portfolio options can be found in the following **Teacher Classroom Resources:**
Tech Prep Applications
Consumer Chemistry
Chemistry and Industry
Applying Scientific Methods in Chemistry

CHAPTER 16 REVIEW

10. oxidizing agent: H_2O_2; reducing agent: I^-

11. This is an oxidation reaction.

APPLYING CONCEPTS

12. The zinc that coats a galvanized nail undergoes a redox reaction with the iodine atoms. In this reaction, iodine molecules pull electrons away from zinc metal atoms to form Zn^{2+} ions. As the zinc is oxidized, the iodine is reduced to I^-. The brown color disappears when all the brown I_2 molecules have been reduced and returns when bleach oxidizes the I^- ions back into I_2.

13. a. $CH_4(g) + 2O_2(g) \rightarrow CO_2(g) + 2H_2O(g) +$ energy
b. Yes. CH_4 is oxidized (loses electrons); O_2 is reduced (gains electrons).
c. reduced

14. The steel could be painted, galvanized by coating it with zinc, or coated with plastic.

15. The red-brown color is due to I_2, which forms when the I^- ions in the potassium iodide solution are oxidized by the hydrogen peroxide.

16. Many double-displacement reactions are not redox reactions. One example is: $2AlCl_3(aq) + 3Na_2CO_3(aq) \rightarrow Al_2(CO_3)_3(s) + 6NaCl (aq)$; no electrons are transferred in this reaction.

17. The indigo dye is an organic compound that is yellow when reduced by the sodium hydrosulfite and blue when oxidized by oxygen in the air.

18. a. Cu^{2+} **b.** formaldehyde **c.** Cu^{2+} **d.** formaldehyde

19. a. MnO_4^- **b.** oxalic acid

10. Identify the oxidizing and reducing agents in the following reaction. The equation is not balanced.

$$H_2O_2(l) + I^-(aq) + H^+ \rightarrow I_2(aq) + H_2O(l)$$

11. Identify the following as an oxidation reaction or a reduction reaction.

$$Fe^{2+} \rightarrow Fe^{3+} + e^-$$

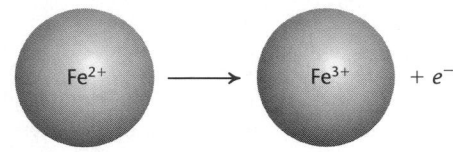

● APPLYING CONCEPTS

12. If galvanized nails, which have been coated with zinc, are placed in a brown solution containing I_2, the solution slowly turns colorless. Adding a few drops of bleach to the colorless solution results in a return of the brown color. Explain what makes these changes occur.

13. When methane in natural gas undergoes complete combustion, carbon dioxide and water are produced in an exothermic reaction.

a) Write a balanced equation that represents this reaction.
b) Is this a redox reaction? Explain.
c) Is the oxygen molecule oxidized or reduced during this reaction?

14. List several ways in which a steel chain-link fence could be treated to prevent corrosion.

15. When hydrogen peroxide is added to a colorless solution of potassium iodide, a red-brown color appears. What substance is responsible for the color?

16. Write the equation for a reaction that is not a redox reaction. Are electrons transferred in this reaction?

17. Indigo is one of the oldest known dyes. It has been detected in cloth used to wrap mummies that are more than 5000 years old. When cotton jeans are dyed with indigo, they are dipped into a yellow solution of indigo and sodium

hydrosulfite, which is a good reducing agent. Within minutes after being taken out of the solution, the jeans turn blue. How can you explain this?

18. A shiny copper mirror can be formed on the inside of a test tube in which the following reaction takes place.

$$H_2C{=}O(aq) + Cu^{2+}(aq) + 2OH^-(aq) \rightarrow$$
formaldehyde
$$Cu(s) + HCOOH(aq) + H_2O(l)$$
formic acid

a) Identify the substance that is reduced during this reaction.
b) Identify the substance that is oxidized during this reaction.
c) What is the oxidizing agent in this reaction?
d) What is the reducing agent in this reaction?

19. Potassium permanganate ($KMnO_4$) is a stain remover that can be used to clean most white fabrics. However, there is one problem with using it; it is a deep purple color that leaves behind its own stain. Fortunately, the purple stain can then be removed by treating the fabric with oxalic acid. The reaction is as follows.

$$5H_2C_2O_4 + 2MnO_4^- + 6H^+ \rightarrow$$
(oxalic acid) (purple)
$$10CO_2 + 2Mn^{2+} + 8H_2O$$
(colorless)

a) What is the oxidizing agent in this reaction?
b) What is the reducing agent in this reaction?

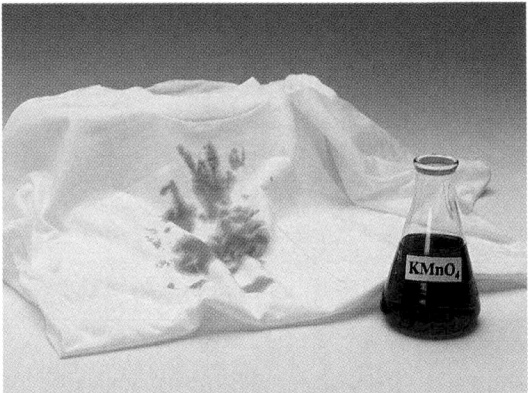

20. Is oxygen a necessary reactant for an oxidation reaction? Explain.

21. Sodium nitrite is often added to meat to inhibit the growth of microorganisms and to keep the meat from spoiling. Under the acidic conditions in our stomachs, nitrites can be converted into potentially cancer-causing substances. Vitamin C can convert nitrite ions into nitrogen monoxide gas and may help protect us from the effects of these ions.

$$NO_2^-(aq) \rightarrow NO(g)$$

a) Is the nitrite ion oxidized or reduced in this reaction?

b) Does vitamin C act as an oxidizing agent or a reducing agent?

22. A redox reaction involving silver is used in a chemical test to determine whether an unknown organic compound is an aldehyde. This test is called a Tollen's test. It is also sometimes called the silver mirror test because a spectacular shiny layer of elemental silver plates out on the inside of a test tube if an aldehyde is present. In this test, a silver nitrate solution is mixed with a solution of the unknown substance, and the mixture is observed to see whether the mirror forms.

a) Is silver reduced or oxidized when the mirror forms?

b) Does the aldehyde function as an oxidizing agent or a reducing agent?

23. Why is gold rather than copper used to coat electrical connections in expensive electronic equipment?

Everyday Chemistry

24. Just before World War I, a German chemist named Fritz Haber developed a process for fixing atmospheric nitrogen into ammonia. The ammonia produced this way can be converted into ammonium nitrate, an important fertilizer and explosive.

$$3H_2 + N_2 \rightarrow 2NH_3$$

a) What element is oxidized during this reaction? What is reduced?

b) What is the oxidizing agent? What is the reducing agent?

Physics Connection

25. By passing an electric current through water, the water can be separated into its component elements in the reverse of the reaction used to power the main stage of the space shuttle.

$$2H_2O(l) + energy \rightarrow 2H_2(g) + O_2(g)$$

a) Is this a redox reaction? If so, what element is oxidized?

b) Where does the energy for this endothermic reaction come from?

How It Works

26. If ethanol were less volatile, how might the usefulness of a Breathalyzer test be affected? Explain.

Chemistry and Technology

27. Why should a positive result from the luminol test not be taken as proof of the presence of blood?

20. Oxygen is not necessary for an oxidation reaction. Any substance that can gain electrons can serve as an oxidizing agent.

21. a. reduced **b.** reducing agent

22. a. reduced **b.** reducing agent

23. Gold is a less active metal than copper, which means it won't corrode as easily.

24. a. Hydrogen is oxidized; nitrogen is reduced. **b.** oxidizing agent: N_2; reducing agent: H_2

25. a. Yes; oxygen is oxidized. **b.** Energy comes from the electric current.

26. Less alcohol would be present in the breath, so it would be more difficult to determine accurately the degree of intoxication.

27. Luminol reacts with copper and cobalt ions, as well as with the iron ion in hemoglobin.

THINKING CRITICALLY

28. a. oxygen, hydrogen peroxide, potassium permanganate, chlorine, potassium dichromate, nitric acid, sodium hypochlorite, potassium chlorate

b. O_2: combustion; H_2O_2: restoration of paintings; $KMnO_4$: stain removal; Cl_2: water treatment; $K_2Cr_2O_7$: Breathalyzer; HNO_3: production of nitrate chemicals; NaClO: bleach; $KClO_3$: fireworks

c. some common reducing agents: H_2, carbon, hydroquinone, vitamin C, vitamin E, sodium hydrosulfite, zinc, hydrogen peroxide, sodium iodide, tin(II) chloride, sodium borohydride, lithium aluminum hydride; hydrogen peroxide is both an oxidizing agent and a reducing agent

29. Hydrogen peroxide could be used to remove tarnish from silver because hydrogen peroxide would react with silver sulfide to form silver sulfate. However, the silver sulfate residue would remain.

30. $Ca(OH)_2(aq) + CO_2(g) \rightarrow CaCO_3(s) + H_2O(l)$ This is not a redox reaction because no electrons are transferred between atoms.

31. The head and point of a nail are the areas where the metal is subjected to stress caused by the wearing away of the surface.

32. a. reducing agent: NH_4^+; oxidizing agent: $Cr_2O_7^{2-}$

b. In both this reaction and the Breathalyzer reaction, the dichromate ion acts as an oxidizing agent, turning from orange to green.

CUMULATIVE REVIEW

33. a. mixture **b.** mixture **c.** mixture **d.** pure substance **e.** mixture **f.** mixture

34. Metals have high melting points, are good conductors of heat and electricity, are hard and shiny, and

THINKING CRITICALLY

Using a Table

28. The table below lists some of the most common compounds that are used as oxidizing agents.

Common Oxidizing Agents	
O_2	$K_2Cr_2O_7$
H_2O_2	HNO_3
$KMnO_4$	NaClO
Cl_2	$KClO_3$

a) Name each of the compounds in the table.
b) List at least one practical application, mentioned in this chapter or from a reference book, of each of these oxidizing agents.
c) Make a similar table for common reducing agents. Should any compounds be listed in both tables?

Making Predictions

29. Hydrogen peroxide (H_2O_2) can be used to restore white areas of paintings that have darkened from the reaction of lead paint pigments with polluted air containing hydrogen sulfide gas.

$$PbS + 4H_2O_2 \rightarrow PbSO_4 + 4H_2O$$
$$\text{(black)} \qquad \text{(white)}$$

Could hydrogen peroxide be used to remove tarnish from silver objects? Would the reaction have any undesirable effects?

Interpreting Data

30. ChemLab Write the balanced equation for the reaction that caused the limewater to become cloudy. Is this a redox reaction? Explain.

31. MiniLab 1 Why do you think corrosion seems to occur mostly at the head and point of a nail?

Making Inferences

32. MiniLab 2 When a pile of orange ammonium dichromate is ignited, it decomposes in an exothermic reaction in which the green product and flames shoot upward like an erupting volcano. (**CAUTION:** *Do NOT perform this reaction.*)

$$(NH_4)_2Cr_2O_7(s) \rightarrow$$
$$Cr_2O_3(s) + N_2(g) + 4H_2O(g)$$

a) What is the reducing agent in this reaction? The oxidizing agent?
b) How is this reaction similar to the Breathalyzer reaction?

CUMULATIVE REVIEW

33. Identify each of the following as a pure substance or a mixture. (Chapter 1)

a) petroleum   d) diamond
b) fruit juice   e) milk
c) smog   f) iron ore

34. List some characteristic properties of metals. (Chapter 3)

35. Name each of the following ionic compounds. (Chapter 5)

a) NaF   d) $Na_2Cr_2O_7$
b) CaS   e) KCN
c) $Al(OH)_3$ f) NH_4Cl

36. How many grams of nitrogen are needed to react completely with 346 g of hydrogen to form ammonia by the Haber process? (Chapter 12)

$$N_2 + 3H_2 \rightarrow 2NH_3$$

37. Draw Lewis electron dot diagrams for each of the following covalent molecules. (Chapter 9)

a) $CHCl_3$ c) CH_3CH_3
b) CH_3CH_2OH

tend to lose electrons (become oxidized) during chemical reactions.

35. a. sodium fluoride **b.** calcium sulfide **c.** aluminum hydroxide **d.** sodium dichromate **e.** potassium cyanide **f.** ammonium chloride

36. 1615 g (1610 g in three significant figures)

37.

a.

b.

c.

Skill Review

1. **Designing an Experiment** Do you think silver will tarnish more quickly in clean air or in polluted air? Design an experiment to test your hypothesis.

Writing in Chemistry

2. Research the evidence that suggests that the antioxidant properties of vitamin C may help prevent cancer in people who take large doses of this vitamin. Write a summary of your findings in which you propose how you would do more tests to determine whether or not vitamin C has anticarcinogenic properties.

3. Solutions of good reducing agents, such as photographic developers, have a limited shelf life. Write a short paragraph explaining why this is true.

Problem Solving

4. A flask filled with acid-washed steel wool is fitted with a long, thin glass tube in a rubber stopper. When the flask is inverted so the tube opening is in a beaker of colored water, the water slowly begins to rise in the tube. Write a summary of this experiment, as if you had performed it. Explain what makes the water rise. Predict what portion of the flask will be filled with water at the end of the experiment.

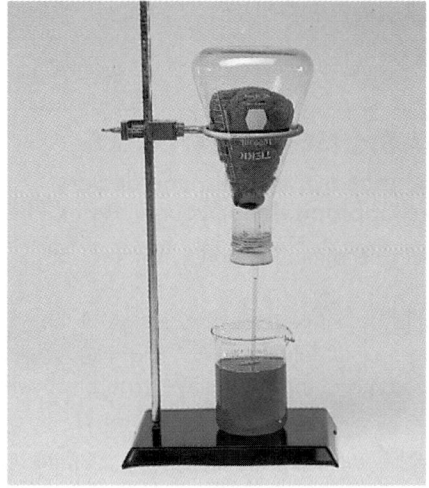

5. The patina coating on the Statue of Liberty has preserved most of the copper metal in the statue. Some damage does occur wherever steel rivets are in contact with copper and exposed to water. Do library research to determine why those sites are more susceptible to corrosion than the rest of the statue. Write up your findings in a short report. Include a diagram or make a poster showing the movement of electrons in the process.

Projects

6. Determine for yourself whether or not paint and other coatings will protect iron from rusting by designing and carrying out an experiment using iron nails.

7. Design and carry out a project to verify the purpose of the different steps of black-and-white photo developing.

AUTHENTIC ASSESSMENT

1. Silver will tarnish more quickly in polluted air that contains hydrogen sulfide. Silver spoons can be placed into identical containers such as self-sealing plastic bags. One bag can contain air, and hydrogen sulfide or some other pollutant can be introduced into the other bag.

3. Solutions of good reducing agents are oxidized by the oxygen in the air.

4. The iron in the steel wool undergoes a redox reaction with oxygen in the air in the flask. Oxygen gas forms oxide ions that make up the rusty iron oxide compounds. As O_2 gas is removed by this reaction, a vacuum is created, drawing liquid up through the tube. The flask will be about 20% ($\frac{1}{5}$) filled with water because this is the mole percentage of oxygen in air.

5. When two different metals are in contact with each other and are surrounded by an electrolyte, one metal loses electrons and is oxidized. The other metal serves as a conductor of electrons to the oxygen, which is reduced.

Program Resources

Chapter Review and Assessment, pp. 95-100 `L1`
Alternate Assessment in the Science Classroom `L1`
Computer Test Bank, Chapter 16 `L1`

Performance Assessment in the Science Classroom `L1`
Problems and Solutions Manual, Chapter 16 `L1`
Supplemental Practice Problems, Chapter 16 `L1`

Chapter 17 Organizer

Section	Objectives	Activities/Features
17.1 **Electrolysis: Chemistry from Electricity** (3 days)	1. **Explain** how a non-spontaneous redox reaction can be driven forward during electrolysis. 2. **Relate** the movement of charge through an electrolytic cell to the chemical reactions that occur. 3. **Apply** the principles of electrolysis to its applications such as chemical synthesis, refining, plating, and cleaning.	**MiniLab 1:** Electrolysis, p. 587 **Chemistry and Technology:** Copper Ore to Wire, pp. 590-592 **Everyday Chemistry:** Manufacturing a Hit CD, p. 594 **People in Chemistry:** pp. 596-597 **Discovery Demo:** 17.1 Electrolysis of KI, pp. 582-583 **Demonstration:** 17.2, pp. 590-591
17.2 **Galvanic Cells: Electricity from Chemistry** (3 days)	4. **Relate** the construction of a galvanic cell to how it functions to produce a voltage and an electrical current. 5. **Trace** the movement of electrons in a galvanic cell. 6. **Relate** chemistry in a redox reaction to separate reactions occurring at electrodes in a galvanic cell.	**MiniLab 2:** The Lemon with Potential, p. 600 **ChemLab:** Oxidation-Reduction and Electrochemical Cells, pp. 606-607 **Health Connection:** Lithium Batteries in Pacemakers, p. 610 **How It Works:** Nicad Rechargeable Batteries, p. 612 **How It Works:** p. 614 **Demonstrations:** 17.3, pp. 602-603; 17.4, pp. 604-605; 17.5, pp. 608-609

Activity Materials

ChemLab (pages 606-607)
craft stick support with V-cut and slit cut, dialysis tubing, magnesium ribbon, copper foil (1-cm and 2-mm widths), metric ruler, 250-mL beaker, 10 × 100 mm test tubes (2), wire leads with alligator clips (2), DC voltmeter with 2-V or 3-V scale, flash-light bulb for AAA batteries, 9-V transistor radio, 0.5M NaCl solution, 0.5M CuCl$_2$ solution, 0.1M MgCl$_2$ solution

MiniLab 1 (page 587)
0.5M CuSO$_4$ solution, 250-mL graduated cylinder, 250-mL beaker, wire leads with alligator clips, 4-inch pencil leads (2), 9-V battery

MiniLab 2 (page 600)
lemon, knife, zinc strip, lead strip, wire leads with alligator clips, voltmeter, aluminum strip, fine steel wool

Demonstrations For a complete list of materials for the demonstrations in this chapter, see pages 582, 590, 602, 604, and 608.

KEY TO TEACHING STRATEGIES

The following designations will help you decide which activities are appropriate for your students.

L1 Level 1 activities should be within the ability range of all students.

L2 Level 2 activities should be within the ability range of the average to above-average student.

L3 Level 3 activities are designed for the ability range of above-average students.

LEP LEP activities should be within the ability range of Limited English Proficiency students.

COOP LEARN Cooperative Learning activities are designed for small group work.

P These strategies represent student products that can be placed into a best-work portfolio.

Electrochemistry

Teacher Classroom Resources

STUDENT MASTERS

Study Guide: pp. 67-68
ChemLab and MiniLab Worksheets: p. 110

TEACHING AIDS

Section Focus Transparency 34
Section Focus Transparency Master: p. 34
Basic Concepts Transparency 46, 47
Basic Concepts Transparency Masters: pp. 91-94
Lesson Plans: p. 34

Study Guide: 69-70
ChemLab and MiniLab Worksheets: pp. 107-109, 111
Critical Thinking/Problem Solving: p. 19
Laboratory Manual: 17.1 Comparing the Abilities of Metals to Give Up Electrons, pp. 155-160; 17.2 The Six-Cent Battery, pp. 161-166

Section Focus Transparency 35
Section Focus Transparency Master: p. 35
Basic Concepts Transparency 48
Basic Concepts Transparency Master: pp. 95-96
Problem Solving Transparency 20
Problem Solving Transparency Master: pp. 39-40
Lesson Plans: p. 35

ASSESSMENT RESOURCES

Chapter Review and Assessment: pp. 101-106
Alternate Assessment in the Science Classroom
Computer Test Bank: Chapter 17
Problems and Solutions Manual: Chapter 17
Supplemental Practice Problems: Chapter 17
Performance Assessment in the Science Classroom

CHAPTER RESOURCES

Applying Scientific Methods in Chemistry: p. 43
Spanish Resources: Chapter 17
English/Spanish Audiocassettes: Chapter 17
Cooperative Learning in the Science Classroom
Lab and Safety Skills in the Science Classroom

GLENCOE TECHNOLOGY

Videotape
MindJogger Videoquizzes, Chapter 17
Videodisc
Chemistry: Concepts and Applications, Demonstrations, Videos, and Animations
Videodisc and Bar Code Teacher Guide
Science and Technology Videodisc Series: Disc 2
Chemistry *Hydroelectric Power, Mini-Hydroelectric Power Plants, Battery Science*
CD-ROM Multimedia Guide
Chemistry: Concepts and Applications, Same as for Videodisc, plus Interactive Exploration and Major Simulation, *The Periodic Table*

TECH PREP

The following Glencoe resources provide opportunities for integrating science and technology.

Student Edition: Everyday Chemistry, p. 594; People in Chemistry, pp. 596-597; Health Connection, p. 610; How It Works, pp. 612, 614

Teacher Wraparound Edition: Assessment, p. 593; Enrichment, p. 598; Integrating the Sciences, p. 610; Applying Chemistry, p. 613; Extension, p. 613

Teacher Classroom Resources: Tech Prep Applications, pp. 27-28; Chemistry and Industry, pp. 37-40; Consumer Chemistry, pp. 33-34

CHAPTER
17 Electro-chemistry

Chapter Overview

Redox reactions are briefly reviewed. The process of electrolysis is discussed, along with some of its applications. After learning how electricity is used to drive electrolysis, students learn that electrochemical reactions can be used to produce electricity. Students then learn the chemistry of several batteries.

Theme Connection

Energy/Conservation In this chapter, *energy* is discussed in the context of electrolysis to drive chemical reactions and in the harnessing of energy from reactions in batteries. The theme of *conservation* is evident in the discussion of recycling of aluminum.

GLENCOE TECHNOLOGY

CD-Rom

Chemistry: Concepts and Applications
Electrolysis of KI
Show this video of the Discovery Demo to introduce an electrolytic reaction. L1 LEP

✓ Assessment Planner

Choose assessment strategies from the following pages to evaluate the progress of your students.
Assess, pp. 589, 593, 598, 611, 615
MiniLabs, pp. 587, 600
ChemLab, p. 606
Portfolio, pp. 603, 607, 613, 615, 616
Chapter Review, p. 616

582

Aluminum foil
Amalgam filling
e⁻
Nerve

✦ Discovery Demo 17.1

Electrolysis of KI

Purpose: To demonstrate electrolysis.
Materials: 4 drops 1% phenolphthalein indicator, 2 g KI, 50 mL distilled H_2O, petri dish, 2 pieces copper wire or pencil lead, 0-9V variable DC power supply

Safety Precautions:
Disposal: Code C

Procedure: On the stage of an overhead projector, place 4 drops of 1% phenolphthalein indicator into a solution of 2 g of KI dissolved in 50 mL of distilled H_2O in a petri dish. Connect the two pieces of copper wire or pencil lead electrodes to the 0-9V variable DC power supply. Dip the electrodes into the KI solution. Have students observe the electrodes.

CHAPTER 17

Electrochemistry

You can imagine Luigi Galvani's reaction in the late 1700s when he discovered that a severed frog's leg jumped when it received an electric shock, just as though the frog were still attached. Galvani believed, erroneously, that the electricity came from the leg's muscles. Actually, the source of the electricity was a machine that Galvani's student was operating nearby. The machine generated static electricity, which produced a spark. When the frog's leg was touched with a metal knife, the leg jumped. Later, Galvani noticed, and Alessandro Volta proved, that the generator of the electricity could be replaced by two dissimilar metals. A chemical reaction was causing muscle contractions in the frog leg.

Although it would be many years before Galvani's and Volta's experiments were well understood, the long path to an explanation of electrochemical reactions began a few years later in 1800, when Volta constructed the first battery, shown below. Within weeks of the report of Volta's work, electricity was used to decompose water into hydrogen and oxygen. Thus, electrolysis was born.

Chemistry Around You

Have you ever bitten down on a piece of aluminum foil that was accidentally left on your sandwich? If you have fillings in your teeth, you won't easily forget the strange pain you felt as the electricity generated when the foil, a silver amalgam filling, and the electrolyte solution provided by the saliva in your mouth all came into contact and caused a current to flow through your tooth to the nerve hiding beneath that filling.

583

Results: Bubbling occurs at the cathode as $H_2(g)$ is produced, and the solution around the cathode turns pink as OH^- is generated in the presence of phenolphthalein. The solution around the anode turns brown as I_2 is generated and combines with I^- to form brown I_3^-. The overall reaction for the electrolysis is $2H_2O + 2I^- \rightarrow H_2 + I_2 + 2OH^-$. Reflecting the formation of I_3^-, the reaction is written as $2H_2O + 3I^- \rightarrow H_2 + I_3^- + 2OH^-$.

Analysis: Ask these questions.
1. What evidence demonstrates that the KI is being broken down by the electric current? *the brown color around the anode*
2. What evidence indicates that a chemical bond in water is being broken? *hydrogen gas bubbles and the pink color of the indicator from the OH^- ions*

PREPARE

Key Concepts

Oxidation and reduction are briefly reviewed and are related to electrolysis, the process in which electrical energy is used to cause a non-spontaneous chemical reaction to occur. The functioning of electrolytic cells and several applications of electrolysis are then examined.

Planning Ahead

For the MiniLab Purchase 4-inch pencil leads. Scripto Long Leads work well.

For the Demonstrations Check your low-voltage power supply for proper operation because it is required for several demonstrations.

1 FOCUS

Focus Transparency

Display the **Section Focus Transparency** for Section 17.1 to introduce the concept of electrolytic decomposition.
L1 LEP

2 TEACH

Reinforcement

Remind students to use the *LEO says GER* mnemonic from Chapter 16 to identify oxidation and reduction reactions.

Visual Learning

Ask students to examine the structure of cisplatin and determine what type of change platinum atoms undergo as they react to form cisplatin. *The bonding of platinum atoms in cisplatin indicates an oxidation number of 4+; therefore, platinum atoms have undergone oxidation.* L1

584

Electrolysis: Chemistry from Electricity

SECTION PREVIEW

Objectives

Explain how a non-spontaneous redox reaction can be driven forward during electrolysis.

Relate the movement of charge through an electrolytic cell to the chemical reactions that occur.

Apply the principles of electrolysis to its applications such as chemical synthesis, refining, plating, and cleaning.

Key Terms

electrical current
electrolysis
cathode
anode
electrolytic cell
cation
anion

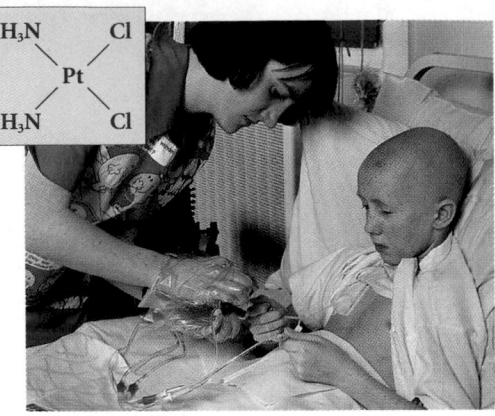

Sometimes, unexpected results in scientific research, such as Galvani's frog, lead to important discoveries. In 1964, a group of researchers studying how electricity affects the growth of bacteria found that bacterial cells stopped dividing if they were subjected to an electric current. This was an important finding because agents that inhibit cell division have the potential to work as cancer treatments. Cancer is a disease in which body cells divide uncontrollably. Upon doing more careful studies, the researchers found that it wasn't the electricity that was preventing the cells from dividing. Rather, a compound was made when platinum from the electrodes used to provide the current took part in a reaction caused by electrical energy. This compound was named cisplatin, and today it is a major medicine used in chemotherapy.

Redox Revisited

You have learned how oxidation and reduction always occur simultaneously. Think about the chemistry of corrosion you studied in Chapter 16. When iron metal reacts with oxygen, a redox reaction creates rust, iron oxide. Electrons are always transferred when a redox reaction occurs. In the rust reaction, electrons are transferred from the reducing agent, iron, to the oxidizing agent, oxygen.

Suppose you could separate the oxidation and reduction parts of a redox reaction and cause the electrons to flow through a wire. The flow of electrons in a particular direction is called an **electrical current.** In other words, you are using a redox reaction to produce an electrical current. This is what occurs in a battery—one form of an electrochemical cell in which chemical energy is converted to electrical energy. You can reverse the process and use a current to cause a redox reaction to occur.

In this section, you will examine **electrolysis,** the process in which electrical energy is used to cause a non-spontaneous chemical reaction to occur. In the second part of the chapter, you will learn about the process that is the reverse of electrolysis—electrochemical reactions that can be used to produce electricity.

Program Resources

Study Guide, pp. 67-68 L1
English/Spanish Audiocassettes,
   Chapter 17 L1 LEP
Section Focus Transparency 34 and **Master,**
   p. 34 L1 LEP
Basic Concepts Transparencies 46, 47 and
   Masters, pp. 91-94 L1 LEP

ChemLab and MiniLab Worksheets, p. 110
   L1
Chemistry and Industry, pp. 37-40 L1
Spanish Resources L1 LEP

Electrolysis

Not too long after Volta's invention of the first electrochemical cell, the British chemist Humphry Davy built a cell of his own and used it to pass electricity through molten salts. An electrochemical cell consists of two electrodes and a liquid electrolyte. One electrode, the **cathode,** brings electrons to the chemically reacting ions or atoms in the liquid; the other electrode, the **anode,** takes electrons away, **Figure 17.1.** The electrons act as chemical reagents at the electrode surface. The liquid electrolyte acts as the chemical reaction medium.

In Davy's electrolysis of molten NaCl, sodium ions were reduced to metallic sodium at the cathode. The oxidation of chloride ions to chlorine gas occurred at the second electrode, the anode. Each half-reaction in the electrolysis of molten sodium chloride is shown.

$$2Na^+(l) + 2e^- \rightarrow 2Na(l) \qquad 2Cl^-(l) \rightarrow Cl_2(g) + 2e^-$$

When the equations for the two half-reactions are combined, the equation for the overall reaction can be written as follows.

$$2Na^+(l) + 2Cl^-(l) \rightarrow 2Na(l) + Cl_2(g)$$

Davy discovered several elements in this way, beginning in 1807. After releasing purified potassium metal from potassium hydroxide, it took him only a year to produce magnesium, strontium, barium, and calcium. Fewer than 30 elements had been isolated by 1800, but by 1850, more than 50 were known. Most of these new elements were isolated using electrolysis. **Figure 17.2** shows the modern commercial electrolysis of molten rock salt. Rock salt is sodium chloride, NaCl. In this process, pure sodium metal and chlorine gas are produced.

Figure 17.1

Oxidation and Reduction
You can remember that reduction always occurs at the cathode and oxidation always occurs at the anode by studying this diagram.

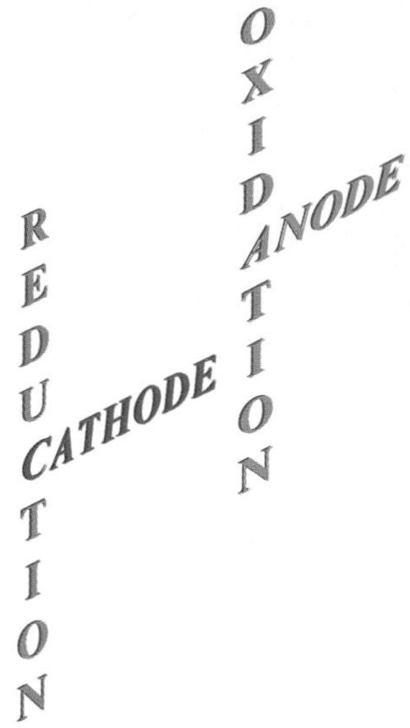

OXIDATION ANODE

REDUCTION CATHODE

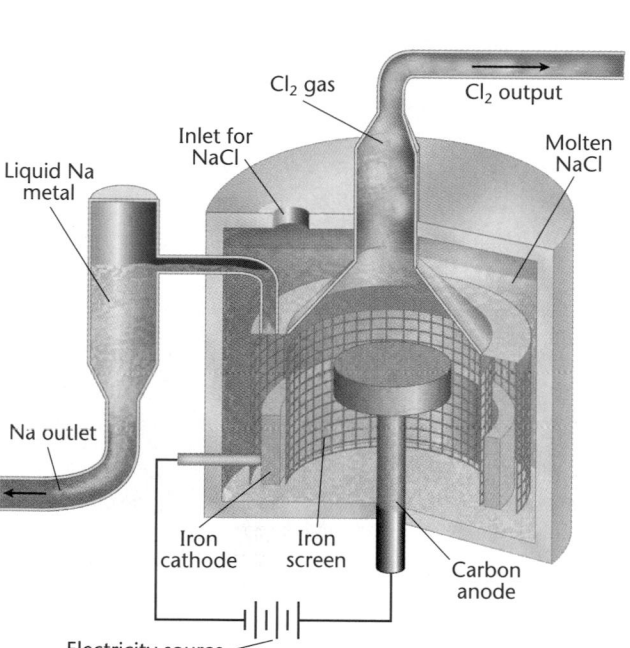

Figure 17.2

Production of Sodium and Chlorine
A device called a Downs cell is used for the electrolysis of molten sodium chloride. As an electrical current is passed through the cell, liquid sodium forms at the circular iron cathode. Because liquid sodium is less dense than molten sodium chloride, the sodium floats to the surface and is collected. Chlorine gas forms at the carbon anode and is collected at the top. An iron screen separates the two electrodes.

Cl₂ gas · Cl₂ output · Inlet for NaCl · Molten NaCl · Liquid Na metal · Na outlet · Iron cathode · Iron screen · Carbon anode · Electricity source

interNET

CONNECTION

Maps, water resource information, and other environmental data are available from the U.S. Geological Survey.

World Wide Web
http://internet.er.usgs.gov

Correcting Errors

Students are likely to become confused if they always attempt to relate oxidation and reduction with positive and negative electrodes in both electrolytic and galvanic cells. Explain that it is, however, always correct to relate oxidation with the anode and reduction with the cathode.

Visual Learning

Figure 17.2: Point out to students that calcium chloride is added to the sodium chloride in the Downs cell, although the calcium chloride does not react. Ask students whether they know what function the calcium chloride serves. *As a nonvolatile solute, calcium chloride lowers the melting point (same as the freezing point) of sodium chloride from about 800°C to about 600°C; therefore, much less energy is required to melt the sodium chloride and maintain it in the liquid state.* [L1]

Quick Demo

Why only the filament?

Turn on a light containing an unfrosted bulb, and ask students what part of the bulb conducts electricity. *the filament* Then, ask why the glass in the bulb does not conduct electricity whereas the metal filament does, although both are made up of atoms that contain electrons. *In order for conduction to occur, some of the atoms' electrons must be mobile; that is, they must be able to move from atom to atom.* [L1]

The Electrolysis Process

Electrolysis takes place in a type of electrochemical cell called an **electrolytic cell,** in which a source of electricity, such as a battery, is added to an external circuit connecting the electrodes. The electrolysis process occurs when the electrons are transferred between the electronic conductors—the metal electrodes—and the ions or atoms at the electrode surfaces. In the liquid, the charge is conducted by ions such as the Na^+ and Cl^- in the molten rock salt. Of course, ions contain electrons too, but these electrons are held tightly to individual ions. Ions can conduct current through a liquid only when they move through the liquid. This type of conduction is called electrolytic conduction. A positive ion is known as a **cation.** A negative ion is called an **anion.** Notice in **Figure 17.3** that the cations, Na^+ ions, move toward the cathode. The negative anions, Cl^- ions, move toward the anode.

What happens when the moving ions reach an electrode surface? If the electrodes are inert, which means that they don't react chemically with the ions in the solution, then only electron transfer will take place at the electrodes. Electrons are being pumped from the battery toward the cathode, where reduction will occur. At the cathode, the ion that reacts is the one that most readily reacts with electrons. In **Figure 17.3,** Na^+ and Cl^- are both present in the liquid near the surface of the cathode. Na^+ accepts electrons more readily, so each Na^+ cation gains one electron, being reduced to the metal.

At the anode, electrons are transferred from the ion that most easily gives them up to the anode. In this case, Cl^- holds onto its electrons more loosely, so each Cl^- anion loses an electron and is oxidized to a chlorine atom. Chlorine atoms then combine to form Cl_2 molecules. The electrons released by the chloride ions flow through the external circuit to the battery and are recycled to the cathode, where they continue the reduction reaction. Because only as many electrons are available at the cathode as are removed at the anode, the reduction process at the cathode must

Figure 17.3

Electrolytic Cell
Electrolysis, the splitting of compounds by electricity, occurs when two electrodes, an anode and a cathode, are inserted into a liquid electrolyte such as molten sodium chloride and connected to a source of electrical energy such as a battery. When electrical current flows into the electrolytic cell, chemical reactions occur. Anions and cations conduct the current by moving freely through the liquid. In the external circuit, electrons move out of the anode, through the battery, and into the cathode.

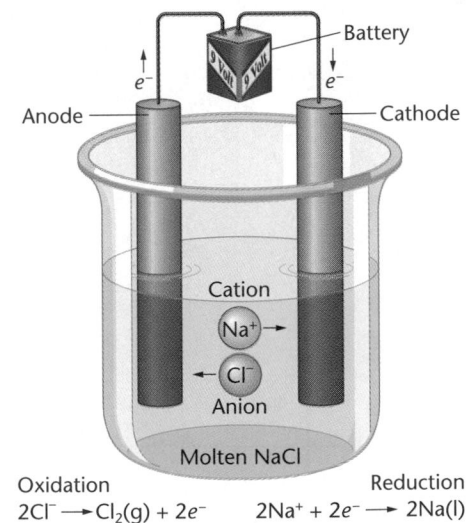

Oxidation
$2Cl^- \longrightarrow Cl_2(g) + 2e^-$

Reduction
$2Na^+ + 2e^- \longrightarrow 2Na(l)$

Meeting Individual Needs

Electrolysis

Electrolysis is the process in which an electric current is used to produce a chemical change. In this MiniLab, you will use the current from a 9-V battery to produce chemical changes in a simple electrolytic cell.

Procedure
1. Pour about 200 mL of 0.5M copper(II) sulfate into a 250-mL beaker.

2. Use wires with alligator clips to attach two 4-inch pencil leads (actually graphite, not lead) to the terminals of a 9-V battery.

3. Put the pencil leads into the copper(II) sulfate solution, keeping them as far from each other as possible.

4. Observe the reactions that occur at the two pencil leads for five minutes. Record your observations, including the

polarities (+ and −) of the leads.

Analysis
1. Describe the reaction that occurs at the positive electrode, the anode, of the electrolytic cell. Write the equation for the oxidation reaction.

2. Describe the reaction that occurs at the negative electrode, the cathode. Write the equation for the reduction reaction.

3. Describe how you might use an electrolytic cell to silver plate an iron spoon.

always occur together with the oxidation process at the anode. Charge transfer at the two electrodes must exactly balance because, just as in the redox reaction, the liquid and its contents must always remain electrically neutral. Therefore, in an electrolytic cell, the overall result of the two electrolysis processes is to carry out a balanced redox reaction, even though the two half-reactions take place at different locations.

Producing Chemicals by Electrolysis

In the electrolysis of molten sodium chloride, a redox reaction taking place in an electrolytic cell can be used to generate chemicals that are important commercially. Because electrolysis consumes large amounts of energy when it is carried out on a commercial scale, many companies that use this process have made their homes in locations where electric power is inexpensive. The abundant hydroelectric power available from Niagara Falls has made that area of New York State a prime location for companies that use electrolysis. One commercial electrolytic process that is carried out in this area is the electrolysis of rock salt solutions to produce chlorine, hydrogen, and sodium hydroxide. The overall change taking place in this process is a redox as well as a substitution reaction.

$$2NaCl(aq) + 2H_2O(l) \rightarrow Cl_2(g) + H_2(g) + 2NaOH(aq)$$

How is electrolysis of a sodium chloride *solution* different from electrolysis of *molten* sodium chloride? In molten NaCl, the only ions present are Na^+ and Cl^-. What ions are present in an aqueous solution of rock salt? Recall that water dissociates slightly to form H^+ and OH^- ions.

17.1 Electrolysis: Chemistry from Electricity   **587**

miniLAB

Electrolysis

Purpose: To use the current from a 9-V battery to produce chemical changes in a simple electrolytic cell.
Process Skills: Observing, classifying, interpreting data

Teaching Strategies
- Prepare 0.5M copper(II) sulfate solution by dissolving 125 g of $CuSO_4 \cdot 5H_2O$ in sufficient distilled water to make 1 L of solution.
- Carbon rods may be used for the electrodes in place of pencil leads. You can also use two wooden pencils by breaking off the erasers and sharpening both ends of each pencil. Test each pencil with a conductivity tester to be sure the lead isn't broken inside the pencil.
- If students ask you about the gas that is formed at the positive electrode, explain that if it were collected and tested, it would not burn but would support combustion.

Expected Results

Copper is deposited on the negative electrode. Bubbles of oxygen gas are formed at the positive electrode.

Analysis
1. Bubbles of a colorless gas are formed.
$$2H_2O \rightarrow 4H^+ + O_2 + 4e^-$$
2. Copper metal is deposited.
$$Cu^{2+} + 2e^- \rightarrow Cu$$
3. Make the iron spoon the cathode in an electrolytic cell, immersing it in a solution that contains silver ions. Make the anode a silver object.

✔ Assessment

Knowledge: Ask students to write equations for the cathodic reactions that would occur if silver, aluminum, and tin were plated from solutions of their ions (assume tin from an Sn^{4+} solution).
$$Ag^+ + e^- \rightarrow Ag,$$
$$Al^{3+} + 3e^- \rightarrow Al,$$
$$Sn^{4+} + 4e^- \rightarrow Sn \boxed{L1}$$

Where do those gases come from?

Have a Hoffman apparatus set up and operating as students arrive in class. **CAUTION:** *Wear safety goggles, and keep all flames and sparks away from the apparatus.* Ask students if they know what is happening. If no one knows, turn off the power to demonstrate that the bubbling stops. Ask students what gases are produced at the anode and cathode during the electrolysis of water. *oxygen and hydrogen, respectively* Also ask students to note the relative volumes of the gases produced at the two electrodes. Point out that electricity causes water to decompose into hydrogen and oxygen and that the 2:1 volume ratio occurs because two molecules of hydrogen are produced for each molecule of oxygen.

Visual Learning

Figure 17.4: Relate to students the story of Charles Martin Hall who, as a student at Oberlin College in Ohio, heard a comment by one of his professors that anyone who could find a way to manufacture aluminum cheaply would make a fortune. Tell students that Hall did just that, working out his process in a wooden shed near his home and using an iron frying pan as a reaction vessel, a blacksmith forge as a heat source, and fruit jars as galvanic cells. Point out that he founded Aluminum Corporation of America (ALCOA) and became a multimillionaire.

Therefore, a rock salt solution contains Na^+, H^+, Cl^-, and OH^- ions. At the anode, the chloride ions lose electrons more easily than the other ions present, just as they did in the electrolysis of molten rock salt. This oxidation forms chlorine gas that can be used for making PVC plastic and other consumer products. At the cathode, the reaction that occurs in molten salt doesn't occur. Instead, H^+ ions are easier to reduce than Na^+, Cl^-, or OH^- ions, so hydrogen ions pick up electrons from the cathode and are reduced to form hydrogen gas. Hydrogen is used in industrial processes such as the catalytic hydrogenation of vegetable oils to form margarine. The Na^+ and OH^- ions are left dissolved in water after the electrolysis process has removed H^+ and Cl^- ions. This is a solution of the base sodium hydroxide, NaOH. Sodium hydroxide is an important industrial and household chemical.

As you have seen, electrochemical transformation of a simple salt solution has produced three valuable products. Each can be sold to pay for the electrical energy that must be invested to make them, with a little left over for a profit to the manufacturer.

Applications of Electrolysis

Electrolysis has numerous useful applications in addition to the generation of chemical substances. The process can also be used to purify metals from ores, coat surfaces with metal, and purify contaminated water. The applications range from the world of art to the world of heavy industry.

Figure 17.4

Hall-Héroult Method of Producing Aluminum
A Hall-Héroult electrolytic cell is used to produce aluminum metal. It is made of a steel shell lined with carbon that forms the cathode. Anodes of carbon hang down into the solution of aluminum oxide dissolved in cryolite. ▶

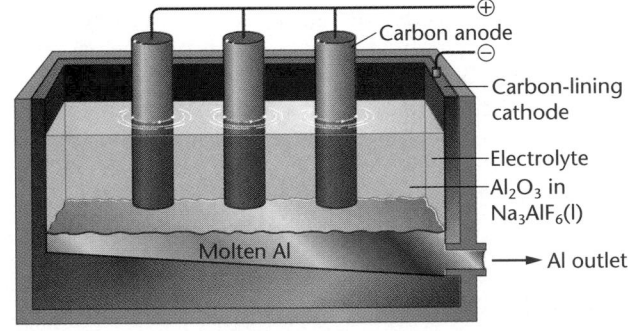

Molten aluminum is drawn off from the bottom of the electrolytic cell, where it accumulates during the electrolysis process. ▶

 Scrap aluminum is melted in a furnace and then re-formed into other products.

Refining Ores

Just as sodium can be produced from melted NaCl by electrolysis, many metals are separated from their ores, or refined, using electrolysis. Today, the metal produced in largest quantity by electrolysis is aluminum from aluminum oxide. First, bauxite ore is heated, driving off the water and leaving aluminum oxide, Al_2O_3. Pure aluminum oxide melts at about 2000°C, so cryolite, Na_3AlF_6, is added to lower the melting point to about 1000°C. The molten salt solution is placed in a large electrolytic cell lined with carbon, which acts as the cathode during electrolysis. Large carbon anodes are dipped into the molten salt to complete the cell, as shown in **Figure 17.4.**

The electron transfers cause complex chemical reactions during electrolysis of molten Al_2O_3 / Na_3AlF_6, but the net reactions are simple. At the cathode, electrons reduce aluminum ions to aluminum metal, which is molten at this temperature. At the anodes, oxide ions lose electrons to form oxygen. The oxygen then combines with the carbon anodes to produce carbon dioxide. The carbon anodes must be replaced periodically because they are gradually used up. The overall redox reaction follows.

$$2Al_2O_3(s) + 3C(s) \rightarrow 4Al(l) + 3CO_2(g)$$

This electrolytic method for producing aluminum is called the Hall-Héroult process because it was developed simultaneously by Charles Martin Hall of the United States and Paul Héroult of France in 1886. Aluminum was so rare before this process was developed that it was more expensive than silver or gold. Today, about 10 million metric tons of aluminum are produced per year worldwide using this process.

The Hall-Héroult process is expensive and consumes large amounts of electrical energy. Recycling aluminum metal prevents some of the expense of producing new aluminum by electrolysis. A lot of waste aluminum is available from discarded aluminum containers, and the energy invested in the original refining of aluminum can be saved if metallic aluminum can be melted down, **Figure 17.5.** It takes only about seven percent as much energy to make new aluminum cans from old ones as it does to make new cans from aluminum ore.

✓ Assessment

Knowledge: Ask students which elements lose and which gain electrons, as well as how many are lost and gained, in the reaction that occurs in the electrolysis of salt brine. *Two chloride ions lose a total of two electrons. Two hydrogen ions gain a total of two electrons.*

Extension

Ask interested students to research and report on the many uses of chlorine gas, hydrogen gas, and sodium hydroxide. L1

Discussion

Ask students why it is much more difficult to reduce aluminum in ore to aluminum metal than iron in iron ore to metallic iron. *Aluminum is a much more active element than iron, and aluminum atoms lose electrons more easily than iron atoms.*

GLENCOE TECHNOLOGY

 **Videodisc**

Chemistry: Concepts and Applications
Aluminum from Bauxite
Disc 2, Side 2, Ch. 7

Show this video to help you describe the production of aluminum from bauxite. L1
LEP

 CD-Rom

Chemistry: Concepts and Applications
Aluminum from Bauxite
Have students use this video to learn about the industrial production of aluminum from ore. L1 **LEP**

Purpose

To show how refined metal products are made from ores.

Background

Copper is found at a level of 70 ppm in Earth's crust in the form of more than 160 known copper-bearing minerals. Copper ore, like that found in the Lake Superior region, is of such high quality that it needs only to be mixed and heated with flux to reduce it to copper metal. (Flux can be made from pine rosins or borax.) Most copper ores contain less than 2% copper and must be enriched using processes like oil flotation before they can be profitably smelted. These processes capitalize on copper ore's inability to be wetted with water. The world's largest producer of copper currently is Chile, with other large producers being the United States, Canada, Zaire, Zambia, Peru, and many countries in the former Soviet Union. Yearly world production averages about 8 million metric tons. A new refining process called solvent extraction, which dissolves the copper ores in a weak acid solution and separates the copper by electrolysis, is proving successful because production costs are much lower than those for smelting.

Teaching Strategies

- Discuss the importance of copper and copper ores to the metal ages of history, particularly the Bronze Age (an alloy of copper and tin), the production of strong weapons, and the introduction of brass (an alloy of copper and zinc).
- Discuss the many properties of copper that make it a valuable, reasonably priced metal with many uses. Include its properties of low oxidation, strength, melting point, and ease of use.

590

Copper Ore to Wire

It would be hard to imagine life without many of the common metals that are used today. Copper, for example, is used in many of the pots and pans in your kitchen, the cooling coils in the air-conditioning system, and some of the pennies you carry in your pocket. More importantly, copper is the metal of choice for most of the electrical wiring in your appliances, homes, and cars because of its good conductivity and low cost. How does copper get from rocks to the finished wire?

Copper can be found in the ground as the free metal and was used about 5000 B.C. In Roman times, much of the copper was obtained from the island of Cyprus, whose name means "copper." Coppersmiths soon learned that copper could be shaped by exposing it to a slow, softening heat in a process called annealing. Annealing soon led to the development of the smelting process. By 3000 B.C., smiths were adept at the metallurgical processes of hammering, annealing, oxidation and reduction, melting, alloying, and removing impurities. Unfortunately, these processes could produce only small quantities of copper. Large-scale production was not mastered until modern furnaces and rolling techniques were developed.*

1. Mining

Copper can be found as the free metal in many parts of the world. However, most copper is found as $CuFeS_2$ (chalcopyrite), Cu_2S (chalcocite), and CuS (covellite). The ores are typically surface mined and ground into powders.

2. Ore Enrichment

Because each ore contains only from one percent to ten percent copper, the copper ore must be concentrated by the flotation process. A frothing agent such as pine oil is mixed with the powdered copper ore, and air is blown in to froth the mixture. Because copper ores are hydrophobic (not wetted by water), the copper and iron sulfides cling to the oil and float to the top, where they can be continuously removed.

Copper mine, Arizona

Ore enrichment

Demonstration 17.2

Plating Cu Metal

Purpose: To observe the functioning of an electrolytic cell and the anode and cathode products.

Materials: 2 soft lead pencils, 2 connecting leads with alligator clips attached to one end of each pencil, 0-9V variable DC power supply, 250-mL beaker, balance, 6.8 g $CuCl_2$, 100 mL distilled water, stirring rod, sharp knife, masking tape

Safety Precautions:

Disposal: Code A

Procedure: CAUTION: *Perform this demonstration in a fume hood.* Dissolve 6.8 g of $CuCl_2$ in 100 mL of distilled water in a 250-mL beaker. Carefully cut the eraser ends off two pencils. Use the pencil sharpener to sharpen both ends of each pencil. Tape the pencils to the inside wall of the beaker. Attach the alligator clips to the pencil tips, and attach the connection leads on the other end of the wires to the power supply. Set the supply at 4.5V, and operate the ap-

Roasting

SO_2. The blistered surface of the copper is due to the escaping gas. The blister copper can be drawn off the bottom and cast into large blocks.

5. Purification

The copper can be purified by electrolysis to 99.95 percent purity. The large blocks of blister copper are used as anodes suspended in a solution of aqueous copper(II) sulfate. Pure copper is used as the cathode. During electrolysis, copper is oxidized at the anode, moves through the solution as Cu^{2+} ions, and is deposited on the cathode. Waste products left after the dissolution of the anode produce a sludge on the bottom of the electrolysis vessel. The sludge, which is often rich in silver and gold, can be recovered for profit.

3. Roasting

The preparation of copper metal involves roasting the ore with oxygen to convert the metallic sulfides to metallic oxides. Usually, both iron and copper are present in the mix.

$$2Cu_2S(s) + 3O_2(g) \rightarrow 2Cu_2O(s) + 2SO_2(g)$$
$$2FeS(s) + 3O_2(g) \rightarrow 2FeO(s) + 2SO_2(g)$$

4. Smelting

The copper and iron oxides are smelted by mixing and heating them with SiO_2, air, and limestone. The result is dense blister copper, lightweight iron(II) calcium silicate slag, and gaseous

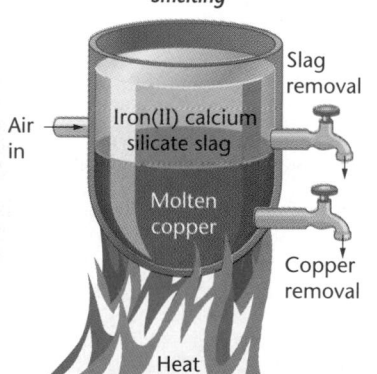

Smelting

Air in

Slag removal

Iron(II) calcium silicate slag

Molten copper

Copper removal

Heat

Electrolysis

Pure copper cathode

Blister copper anode

$CuSO_4$

Purification

paratus until the end of the laboratory period. Disconnect the power supply. Have students inspect the pencils and record their observations.
Results: The products of the reaction can be identified by sight and smell. Copper is plated onto one pencil lead, and the odor of chlorine is retained by the other pencil for a short time.

Analysis: Ask these questions.
1. The solution contains Cu^{2+} and Cl^- ions. Write the half-reaction that occurs at the anode. $2Cl^- \rightarrow Cl_2 + 2e^-$
2. Write the half-reaction that occurs at the cathode. $Cu^{2+} + 2e^- \rightarrow Cu$

Assessment

Knowledge: After copper has plated onto

one of the electrodes, have a student disconnect the power supply and measure the voltage between the electrodes. Then, have the student reconnect the power supply, reversing the polarity of the cell. Students should explain why the copper leaves one electrode and plates onto the other. *This is similar to what happens when a battery is recharged.* **L1**

Discussing the Technology

1. Lightweight slag allows for easy removal of the impurities from the more dense copper. Also, the slag can be skimmed off the top as the reaction continues, leaving more room for the copper.

2. Tough-pitch copper: downspouts, flashing, gutters, radiators, kettles
   Deoxidized copper: hydraulic fluid lines, oil coolers
   Oxygen-free copper: conductors, gasoline supply lines, refrigeration lines, water piping
   Silver-bearing copper: heavy-duty motor windings, brazing solders
   Arsenical copper: boilers, radiators, heat exchangers
   Free-cutting copper: soldering coppers, welding-torch tips

3. The sludge could be dissolved in a strong acid such as nitric acid or aqua regia. The gold and silver could then be removed from solution using electrolysis.

6. Wire Bar Production

The electrolytic copper is cast into wire bars ranging from 60 to 227 kg. The wire bars are heated to 700-850°C and rolled without further reheating into rods approximately 1 cm in diameter.

Wire bar production

7. Drawing Wire

The 1-cm rods of copper are drawn through successively smaller dies until the desired size of wire is reached. The dies must be made of exceptionally hard materials because of the tremendous amount of wear from drawing the wire. The dies typically are made of tungsten carbide or diamond.

Drawing wire

8. Coatings

The finished wire may be coated with plastic, enamel, or another metal to help protect it from moisture and oxidation, or it may be left bare.

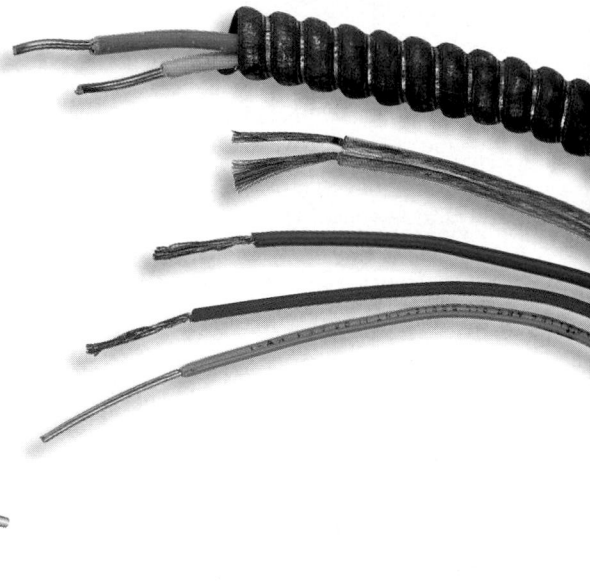

DISCUSSING THE TECHNOLOGY

1. **Thinking Critically** Why is it important that the slag formed during smelting be lightweight?

2. **Acquiring Information** Copper can be made into many different forms and alloys. Research and list several of the forms and uses for each.

3. **Hypothesizing** How might the gold and silver be removed from the sludge formed beneath the anode?

Cultural Diversity

Early Steel Production

Although specularite iron ore was mined for use as a colorful cosmetic in Swaziland for 28 000 years, it was the development of a direct-reduction method of smelting iron ores in furnaces at high temperatures that helped South Africans move out of the Stone Age and into the Iron Age as early as A.D. 250. Metalworking was attended by secrecy and ritual, and men did not allow women near the furnaces. A form of steel actually was produced during smelting because carbon particles were present in the ore. The steel was used to make spear points, axes, hoes, and ornamental bangles. The development of high-quality metal tools led to improved agricultural techniques, which resulted in the production of sufficient food for settled communities with growing populations. In the mid-1800s, the British army occupying Natal used local iron metal to make wagon fittings after they found it to be superior in quality to British steel. The Iron Age developed earlier in northern parts of the African continent, where iron working began as early as 1000 B.C.

Electroplating

You have learned that many metals can be protected from corrosion by plating them with other metals. Zinc coatings are often used to keep iron from rusting. Metal garbage cans are galvanized by dipping them into molten zinc. This process produces an uneven, lumpy coating both inside and outside of the can. That's OK for a garbage can, but the lumpy surface wouldn't look good under the snazzy red paint job on a new sports car. That's why automobile manufacturers electroplate zinc onto the steel used for car bodies. This process involves dragging a sheet of steel across the surface of an electrolyte in an electrolytic cell. The process produces a thin (cost-saving), uniform (smooth and clean) coating of zinc on only one side of the sheet of steel, saving half the coating cost. Because only one side of the car body is exposed to the corrosive effects of water and salt, it is not worth the cost of coating the inside. **Figure 17.6** shows other uses of electroplating.

In zinc electroplating, zinc ions are reduced to zinc atoms at the surface of the metal object to be coated, which becomes the cathode in an electrolytic cell. At the anode, which is made of zinc, atoms of zinc are oxidized to ions. The electrolyte solution contains dissolved zinc salt.

Figure 17.6
Electroplating
Chromium is often electroplated onto a softer metal to improve its hardness, stability, and appearance. Chrome bumpers and trim can be found on many vintage cars. ▼

Reduction of silver ions onto cheaper metals forms silverplate. The object to be plated is made the cathode. At the pure silver anode, oxidation of silver metal to silver ions replaces the silver ions removed from the solution by plating at the cathode. ▶

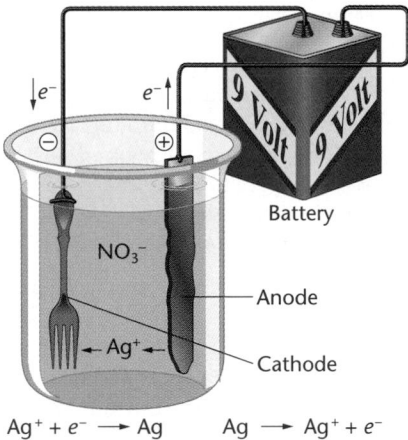

$e^-\downarrow$ $e^-\uparrow$

9 Volt   9 Volt

Battery

NO_3^-

Anode

Ag^+

Cathode

$Ag^+ + e^- \longrightarrow Ag$ $Ag \longrightarrow Ag^+ + e^-$

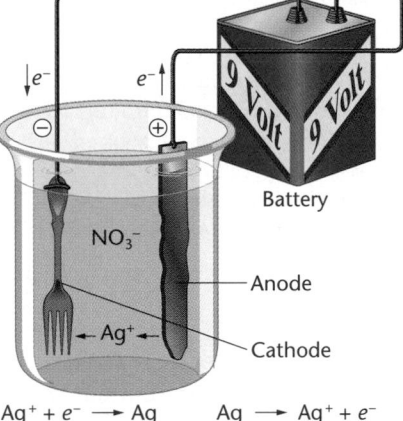

◀ In the early 1980s, because of high inflation of the U.S. currency and a worldwide shortage of copper, the cost of producing a copper penny became nearly equal to the value of the copper metal itself. The U.S. Mint was instructed to begin making pennies from a cheaper metal, such as zinc. Knowing that the American public would not accept dusty-gray zinc pennies, the Mint began to produce pennies by electroplating zinc disks with a copper coat. Many people think that pennies are still made of pure copper.

17.1   **Electrolysis: Chemistry from Electricity**   593

Everyday Chemistry

Purpose

To describe the chemical and electroplating processes necessary in the production of compact discs.

Teaching Strategies

- Discuss the addition of binary numbers to show students how information can be stored as a series of on-off signals. For example, 37 + 54 = 91 in binary would be 100101 + 110110 = 1011011.
- Discuss the need for direct current to be used rather than alternating current during the plating process. *This is so metallic ions can be plated onto the source of electrons.*
- Explain to students that the photo is a false color scanning electron micrograph, magnified 1200 times, of a compact disc, edge on, that has been cracked to show the musical layer (foreground) below the outer plastic covering (background). The disc is made of plastic, which is pressed with a series of fine depressions, shown, representing a digitized musical signal capable of being read by a laser. To reflect the laser light, the music layer is coated with a fine film of metal (unseen), which follows the depressions exactly. This layer is covered by another layer of transparent plastic. The music is sealed between two layers of plastic, which prevents dust and scratches from affecting the sound.

Extension

Compact disc information is transferred off the disc using a laser beam. Have students conduct library research to determine the type of laser used in CD readers, how it works, and why the particular color of laser light is produced. [L2]

594

Manufacturing a Hit CD

The sound from your new compact disc (CD) is so clear and crisp that it seems as if the musicians are in the same room. The exceptional clarity of the sound is possible thanks to the chemical process of metallic depositing and electroplating used in the manufacturing of CDs.

Data pickup A CD is a collection of binary data in a long, continuous spiral that runs from the inside edge to the outside. Nothing touches the information stored on the CD spiral except laser light, which is reflected from the CD and read by the CD player computer as a binary signal. This signal is then transformed by a tiny computer into audio signals, which are amplified to produce the sound heard from your speakers.

Master copy production Your CDs are stamped from a master copy called a stamper master. But how is this stamper made? The musical data are first transferred onto a glass disc using a high-powered laser that etches small pits in the glass as shown here. This glass master copy then contains all of the musical information in binary data form.

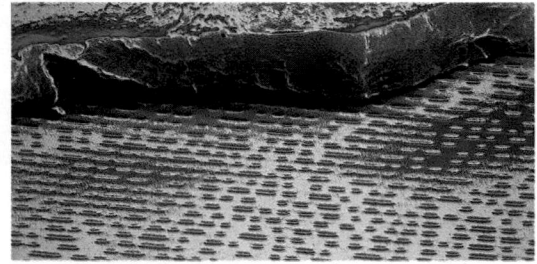

Etched pits, magnified 1200×

The master is coated with a dilute solution of silver diammine complex—$[Ag(NH_3)_2]^+$—followed by a solution of formaldehyde that acts as a reducing agent for the silver. The result is a redox reaction that deposits a thin silver mirror that completely

plates the etched disc. This mirrored master disc is produced in the following reaction.

$$2Ag^+ + HCHO + H_2O \rightarrow 2Ag + HCOOH + 2H^+$$

Electrodeposition The silver coating forms the surface onto which a thin layer of nickel is electrodeposited to make a nickel-coated disc called the mother disc.

$$Ni^{2+}(aq) + 2e^- \rightarrow Ni(s)$$

A second coating of nickel is then electrodeposited onto the mother disc. This nickel layer, the stamper master, is peeled off and used to stamp the data impression onto melted polycarbonate plastic discs. The polycarbonate disc now has all of the pits found on the original glass master disc etched by the laser. Because polycarbonate is clear, it is vacuum coated with a thin aluminum film to produce the reflective layer required by the laser. This delicate aluminum layer is covered with a protective layer of polycarbonate to prevent aluminum oxidation and marring of the data surface. The back of the CD can now be covered with information in the form of art and lettering.

Stamping characteristics Each nickel stamper master can make about 20 000 copies before it fails to produce good indentations. At this point, the nickel stamper can be recycled into an aqueous nickel solution and used to make more nickel stampers. A hit CD may go through as many as 50 nickel stampers plated onto the mother disc when making a platinum million seller. CD-ROMs for your computer are made in the same way.

Exploring Further

1. **Hypothesizing** Why do you think nickel is used for stamping the polycarbonate discs?
2. **Acquiring Information** Polycarbonate is the base material used for CDs. Research its properties and some of its other major uses.

Exploring Further

1. Nickel is a strong metal that can withstand multiple stampings, is fairly easy to electroplate, is low in cost, and is recyclable.
2. Polycarbonate is a synthetic thermoplastic that is transparent, noncorrosive, of high-impact strength, heat-resistant, excellent for molding, and easily fabricated. Uses include beverage bottles, molded parts, tubes and piping, prosthetic devices, nonbreakable windows, household appliances, and streetlight globes.

The thickness of the zinc metal coating can be controlled exactly by controlling the total charge (number of electrons) used to plate the object. Only that portion or side of the object that is immersed in the cell electrolyte receives a zinc coat.

Because coatings adhere best to a chemically clean foundation, objects to be coated are usually degreased, cleaned with soap, and then treated with a corrosive fluid to remove any dirt on their surface before electroplating. Then, the object is immersed in an electrolyte containing the salt of the metal to be deposited. Because the object acts as the cathode of the electrolytic cell, it must be a conductor. Metal objects are electroplated most often because metals usually are excellent conductors. The anode is made of the same metal that is being plated so that it will replenish the metal ions in the electrolyte that are removed as the plating proceeds. The net effect is that when an electric current is passed through the electroplating bath, metal is transferred from the anode and distributed over the cathode. Eventually, the entire object becomes coated with a thin film of the desired metal.

Electrolytic Cleaning

Electrolysis can be used to clean objects by pulling ionic dirt away from them. The process has been used to restore some of the many metal artifacts taken from the shipwrecked cruise ship *Titanic*, which sank in the northern Atlantic Ocean in 1912, **Figure 17.7.** Coatings of salts containing chloride ions, which came from the seawater, were removed by electrolysis. The electrolysis cell for this cleaning process includes a cathode that is the object itself, a stainless steel anode, and an alkaline electrolyte. When an electric current is run through the cell, the chloride ions are drawn out. Hydrogen gas forms and bubbles out, helping to loosen corrosion products. Among the objects that have been recovered are a porthole, a chandelier, and buttons from the uniforms of crew members.

WORD ORIGIN

corrosion:
com (L)
thoroughly
rodere (L) to
gnaw

Metals wear away gradually during corrosion, as if they were being eaten.

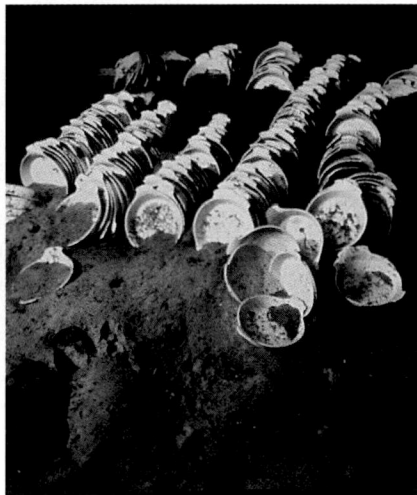

Figure 17.7
When That Great Ship Went Down
Chemistry played a key role in restoring many items taken from the wreck of the *Titanic*. Electrolysis was used to clean and stabilize many metal artifacts, and electrophoresis was used to remove corrosion from bank notes, leather, and objects such as these casserole dishes. Chemicals that attract and hold metal atoms or ions were used to remove iron stains from delicate objects made of organic materials such as newspapers, textiles, and letters. The study of objects from the ship may help scientists compile information for long-term storage and containment under seawater.

Chemistry Journal

Electrophoresis

Ask interested students to visit a medical testing laboratory and learn about electrophoresis as it is used there. Ask them to make a poster or video presentation for the entire class and to write a summary of their visit in their journal. L1 LEP

Extension

Ask an interested student to contact the Glidden Company through a local paint store to obtain information about their Rustmaster Pro paint. A poster or class presentation regarding the ability of the paint coating to prevent rust can be made. L1
LEP

Applying Chemistry

Cleaning Silver from Shipwrecks
Silver bars and coins that have been in wooden boxes in the ocean for long periods of time become coated with silver sulfide due to bacteria that convert sulfate ions to hydrogen sulfide gas. When the coins are connected to the cathode of an electrolytic cell that has an unreactive anode and a sodium hydroxide electrolyte, the silver in silver sulfide is reduced to its original metallic state. Two cathodic reactions occur:
$Ag_2S + 2e^- \rightarrow 2Ag + S^{2-}$ and $2H_2O + 2e^- \rightarrow H_2(g) + 2OH^-$. The anodic reaction is $2H_2O \rightarrow O_2 + 4H^+ + 4e^-$.

GLENCOE TECHNOLOGY

 Videodisc

Chemistry: Concepts and Applications
Chromium Electroplating
Disc 2, Side 2, Ch. 6

Use this video to show the electroplating of chromium. L1 LEP

 CD-Rom

Chemistry: Concepts and Applications
Chromium Electroplating
Use this video to show the electroplating of chromium. L1 LEP

Meet Harvey Morser, Metal Plater

TECH PREP

Teaching Strategies

- Be sure students understand that electroless nickel plating does not use electrical currents to reduce the plating metal. Interested students can research how the nickel is plated onto the base metal and report to the class.

- Have students compare and contrast standard electrolytic and electroless nickel plating. L1

Background

The object to be plated is placed in aqueous solution containing nickel chloride or nickel sulfate and sodium hypophosphite. In the presence of heat and a catalyst on the surface of the object to be plated, some hypophosphite ions are oxidized to orthophosphite with some hydrogen being absorbed by the catalytic surface.

$$H_2PO_2^- + H_2O \xrightarrow[\text{heat}]{\text{catalyst}}$$
$$H^+ + HPO_3^{2-} + 2H$$

Nickel ions are reduced by the absorbed hydrogen and plated onto the surface.

$$Ni^{2+} + 2H \rightarrow Ni + 2H^+$$

A small amount of absorbed hydrogen at the catalytic surface also reduces hypophosphite.

$$H_2PO_2^- + H \rightarrow H_2O + OH^- + P$$

As a result, some phosphorus is deposited onto the surface so that the plating is about 8% P and 92% Ni. Most hypophosphite is oxidized to orthophosphite and gaseous hydrogen independent of the Ni and P plating.

$$H_2PO_2^- + H_2O \rightarrow HPO_3^{2-} + H_2$$

Meet Harvey Morser, Metal Plater

A seasonal highlight in Reno, Nevada, is a celebration called Hot August Nights. On display are thousands of beautifully restored antique cars. "Muscle cars"—like a 1934 Ford sedan with a $12 000 custom paint job and immaculately plated chrome—draw admiring crowds. Harvey Morser watches with pride as the cars sporting his shop's work on grilles and door handles parade by.

On the Job

 Mr. Morser, will you tell us what you do on the job?

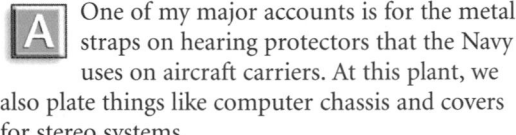 Although I'm now the owner of Western Metal Finishing, I still go out on the shop floor and do plating. All iron metal needs some type of protective coating to keep it from rusting. There are different types of applications that you can use. Cadmium plating is probably the best, but it's very toxic. Hardchrome plating also releases fumes into the atmosphere. So, because of environmental concerns, electroless nickel is the favored process. It's a chemically applied nickel plating done without electricity. In my opinion, this process is twice as effective because it goes on easily and consistently, unlike hard plating, which you have to apply and then grind back down. Chemists have developed an electroless nickel that has the same Rockwell factors (the hardness factors) as hardchroming. I know that a major heavy-equipment manufacturer converted from hardchrome over to electroless nickel and saved something like $3 million the first year.

 What kinds of metal products do you plate?

One of my major accounts is for the metal straps on hearing protectors that the Navy uses on aircraft carriers. At this plant, we also plate things like computer chassis and covers for stereo systems.

Why does the electroless nickel process produce a more even coating?

Visualize a flat, square plate. When you put a hook in one edge of it and hang it in the tank to hardplate it, the electrical current will reach the corners first, then travel down the side edges, and finally spread into the center of the plate. The plating will go on in the same way, building up probably twice as fast on the edges as in the center. So the edges might have eight ten-thousandths of an inch of plating, whereas the center might only have four ten-thousandths. With electroless nickel, a metal part will plate perfectly evenly because it's put on chemically. Electroless

nickel has the highest corrosion resistance next to cadmium plating, so these plated parts resist corrosion as well as being uniform.

Early Influences

 What training did you have in metal plating?

 All my training came on the job. I started out polishing the metal prior to plating. That's a tough and dirty job, but an important one. Plating duplicates a surface, so it has to be polished like a mirror. Otherwise, plating will magnify even a tiny pit. In those early days, I had to work a side job in a bowling alley as a pin chaser, unsticking the balls and unjamming the pins. Along the way, I picked up carpentry, welding, and electrical skills, which have come in very handy in doing the maintenance at the metal plant.

 How did you work your way up to owning the plant?

 It all boiled down to learning quickly, not whining, and working hard. Plating is tough and heavy work. The heat on the lines is terrible, with the humidity and steam off the tanks. All the tanks are running about 160°, so in the summertime it will be 104° on the line. The joke here is that we don't charge extra for the steam bath. Nine years ago, I became the owner of the business. To me, that's the Great American Dream.

Personal Insights

 If someone came to you just out of high school, would you give him or her a chance on the job?

 In a heartbeat. I believe people need to get their schooling, but they also need to have common sense. Every employee here gets on-the-job training, just like I did. What I look for in a prospective employee is honesty and dependability, plus an ability and desire to learn.

 Is this a stressful business to be in?

 Absolutely! I carry a pager all the time and even take a cellular phone out on the lake when I go fishing. There's always the possibility of an industrial accident. Earthquakes don't announce that they are on their way. If I see the numbers 1-5 on my pager, I know it's all clear. Seeing five zeros is what I dread.

 What appeals to you about the plating business?

 I was always turned on by the fact that I could take something that looked terrible, like an old car part, and make it look gorgeous. I want even the most modest plating job to look good, even when it's on a part that probably won't be visible after it's installed.

CAREER CONNECTION

These jobs also involve working with metals.

Metallurgical Technician Two-year training program
Mining Engineer Bachelor's degree in engineering
Scrap Metal Processing Worker On-the-job training after high school

PEOPLE in CHEMISTRY

CAREER CONNECTION

- **Career Path** High school students interested in various fields of metallurgy, such as metallurgical technology, would take algebra, chemistry, and physics courses. Two-year training programs in metallurgical technology are offered by local community colleges and technical institutes. These programs may include chemistry, physics, instrumentation, and spectroscopy courses.
- **Career Issue** Have students discuss some of the traits that one would need to be the owner of a small, technology-based business. L1

For More Information

Students interested in related fields of metallurgy may obtain information by writing The American Society of Metals, Metals Park, IL 44073, and the Metallurgical Society, 420 Commonwealth Drive, Warrendale, PA 15086.

Transparency

Display **Basic Concepts Transparency 47** to reinforce the process of electroplating. L1 LEP

Across the Curriculum

Music

Lip-vibrated aerophone instruments were common in most early civilizations. The earliest instruments were made of natural objects such as conch shells, tortoise shells, animal horns, or hollow wood or cane strips. Later, some were also made of ceramics or glass. The Romans were known to use these early instruments to send signals and make ceremonial sounds. It was in Egypt around 1500 B.C. that metal instruments were first developed. By 1000 B.C., Scandinavian metalworkers made tenor-sized trumpets to use in both military and civil ceremonies. Musical brasses were not developed in Europe until the late 14th and early 15th centuries, after which the famous Nuremberg brass-making dynasties were founded. Beginning in the middle of the 19th century, brass instruments were electroplated with nickel, silver, or gold to protect them from corrosion, but today most instruments made of metals that tarnish easily are kept shiny with a coating of lacquer.

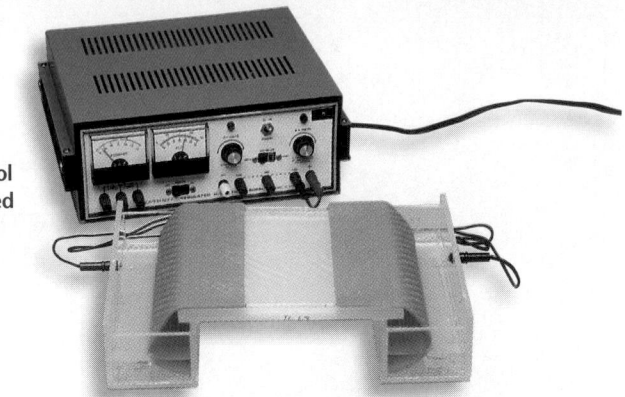

Figure 17.8
Electrophoresis
Electrophoresis is a valuable laboratory tool used to separate and identify large charged particles such as DNA and proteins. Negatively charged particles move toward the anode. Positively charged particles move toward the cathode.

Electrophoresis is another electrochemical process that was used to restore some of the ceramic and organic artifacts from the *Titanic*. Electrophoresis involves placing an artifact in an electrolyte solution between positive and negative electrodes and applying a current. The current breaks up salts, dirt, and other particles as their charged components migrate to the electrodes. Electrophoresis is also used in laboratories to separate and identify large molecules, **Figure 17.8.**

Electrolysis of Toxic Wastes

The plating baths used in the various applications of electrolysis often contain toxic materials or produce toxic by-products. After bath solutions have been used for a period of time, they must be changed and the toxic contents disposed of in a safe manner. Remarkably, electrolysis offers one of the safest and most thorough means of cleaning up toxic metal-containing wastes. When the bath solution is subjected to electrolysis, the toxic metal ions are reduced to free metal at the cathode. The metal can then be recycled or disposed of safely.

SECTION REVIEW

Understanding Concepts

1. Draw and label the parts of an electrolytic cell.
2. Of what value is electroplating? Why is the process used?
3. How is electrolysis used for cleaning objects?

Thinking Critically

4. **Applying Concepts** What effect would electroplating steel jewelry with gold have on the rate of corrosion of the jewelry?

Applying Chemistry

5. **Magnesium from the Sea** Magnesium in seawater is found mostly as $Mg(OH)_2$, which can be converted to $MgCl_2$ by reacting it with HCl. Magnesium metal can then be purified by electrolysis of molten $MgCl_2$.

   a) What reaction takes place at the cathode during electrolysis?
   b) What reaction takes place at the anode during electrolysis?
   c) Write an equation for the net reaction that occurs.

SECTION REVIEW

1. See **Figure 17.3.**
2. Electroplating is done to improve the hardness or appearance of an object or to protect it from corrosion.
3. Electrolytic cleaning loosens dirt and reduces corroded metals to their original metallic state.
4. It will protect the steel from corrosion because gold is more resistant to oxidation than is iron.
5. **a.** cathode reaction: $Mg^{2+} + 2e^- \rightarrow Mg(s)$
   b. anode reaction: $2Cl^- \rightarrow Cl_2(g) + 2e^-$
   c. $MgCl_2(aq) \rightarrow Mg(s) + Cl_2(g)$

Galvanic Cells: Electricity from Chemistry

Suppose it's nearing the end of half-time during the seventh game of the NBA championship finals, and your favorite team is leading by just two points. Suddenly . . . the electricity goes out. You might find that battery-powered radio you haven't used for months, but its batteries are dead. You can't go buy more batteries because the electricity is out all over town. Must you resign yourself to missing the end of this game?

Electrochemical Cells

You don't have to miss the end of the NBA game because you have at hand all the ingredients for making a battery that will power your radio. All you need are several lemons or pieces of fruit (or even glasses of fruit juice), two different kinds of metal (a penny and a steel nail will do), and some pieces of wire for connecting everything together. Once you have those items, you need only a little knowledge of electrochemistry, and you'll soon be listening to your team go all the way.

How can it be that simple? The energy in oxidation-reduction reactions can be harnessed to do useful work, if listening to an NBA game can be called work. A battery is the tool that makes this possible.

When Luigi Galvani used two dissimilar metals to produce an electrical current that stimulated the nerve in the frog leg, he didn't know that he had invented the first battery. Batteries are electrochemical cells. The battery made out of fruit, which allowed you to hear the end of the basketball game, is such a cell. In a battery, the two halves of a spontaneous redox reaction are separated and made to transfer electrons through a wire.

The Lemon Battery

How does the lemon battery produce electrical energy? The lemon itself is a container for a solution of electrolyte—the lemon juice. As you know, lemon juice is sour; that is, it is acidic. The hydrogen ions from partially dissociated citric acid give it a sour taste and also provide the ions for conduction of charge through the lemon battery. The two dissimilar metal strips are the electrodes at which an oxidation reaction and a reduction reaction take place to provide the battery's power source.

17.2 Galvanic Cells: Electricity from Chemistry   599

SECTION PREVIEW

Objectives
Relate the construction of a galvanic cell to how it functions to produce a voltage and an electrical current.
Trace the movement of electrons in a galvanic cell.
Relate chemistry in a redox reaction to separate reactions occurring at electrodes in a galvanic cell.

Key Terms
potential difference
voltage
galvanic cell

PREPARE

Key Concepts
The electrochemical processes that occur in galvanic cells and batteries are examined, then the design and operation of typical batteries are described.

Planning Ahead
For the ChemLab Obtain dialysis tubing from a laboratory supply company, a hospital, or a medical supply company.
For the Demonstrations Purchase flashbulbs or flashcubes for Demonstration 17.3.

1 FOCUS

Focus Transparency
Display the **Section Focus Transparency** for Section 17.2 to show a lemon battery.
L1 LEP

2 TEACH

Content Background
Cell Potential Electrical current is produced in a galvanic cell when electrons are pushed from the anode through an external wire to the cathode. The force with which electrons move through the cell is called the cell potential or electromotive force (emf), and it is measured in volts (V). The cell potential depends upon the concentrations of ions in the cell, the partial pressures of any gases involved, and the temperature. A potentiometer is a device that can be used to measure a cell's potential. A potential within the device opposes the potential generated by the cell, and when the two potentials are equal, no current flows. At this point, the cell's potential can be read from the potentiometer.

Program Resources

Study Guide, pp. 69-70 L1
Critical Thinking/Problem Solving, p. 19 L2
Laboratory Manual, 17.1, pp. 155-160; 17.2, pp. 161-166 L1
Section Focus Transparency 35 and **Master,** p. 35 L1 LEP
Basic Concepts Transparency 48 and **Master,** pp. 95-96 L1 LEP
Problem Solving Transparency 20 and **Master,** pp. 39-40 L1 LEP

ChemLab and MiniLab Worksheets, pp. 107-109, 111 L1
Applying Scientific Methods in Chemistry, p. 43 L1
Tech Prep Applications, pp. 27-28 L1
Consumer Chemistry, pp. 33-34 L1

miniLAB

The Lemon with Potential

Purpose: To investigate the interactions of zinc and aluminum with lead when the metals are placed in a lemon.
Process Skills: Observing, measuring, interpreting data
Safety Precautions: Remind students that the lemons are not to be tasted or eaten after use because they contain metal ions.

Teaching Strategies

• The lemon pulp and membranes act somewhat as a salt bridge and a barrier.
• Hydrogen ions will probably be reduced at the lead electrode unless sufficient time is allowed for some of the lead to dissolve in the lemon prior to measurement. If the lead strip is allowed to remain in the lemon before measurements are made, the measured potential differences will be closer to theoretical values.
• Be sure the metal strips are clean and free of corrosion. They should be rinsed and dried immediately after use.
• A vacuum tube voltmeter (VTVM) or other high-impedance voltmeter works best in this MiniLab.

Expected Results

The measured potential difference between zinc and lead will probably be between 0.5V and 0.6V, fairly close to the theoretical standard value of 0.63V. The potential difference between aluminum and lead will be close to the theoretical standard value of 1.55V if the oxide is removed from the aluminum.

miniLAB 2

The Lemon with Potential

Lemons are good for more than just making lemonade. By adding some metal strips, lemons have other "potential" uses. In this MiniLab, you will investigate the interactions of zinc and aluminum with lead when the metals are placed in a lemon.

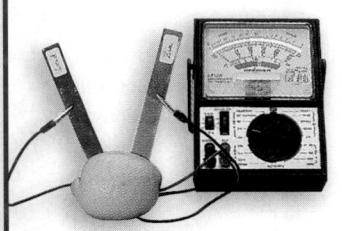

Procedure

1. Gently knead a lemon without breaking the skin. Make two slits about 1 cm in depth on opposite sides of the lemon.

2. Insert a strip of zinc in one of the slits and a strip of lead in the other slit.

3. Connect an alligator clip wire to each of the metal strips, touching or connecting the other end of each wire to the poles of a voltmeter. If the voltmeter gives no reading, reverse the wires.

4. Read and record the voltage.

5. Repeat steps 1 through 4 with strips of lead and aluminum, making a new slit for the aluminum and lightly buffing the metal with fine steel wool *immediately* before inserting to remove the oxide coating. **CAUTION:** *Discard the lemon. Do not use for food.*

Analysis

1. What causes the potential difference between the zinc and lead strips?

2. Why is the potential difference greater when aluminum is substituted for zinc?

3. If strips of zinc and magnesium, rather than zinc and lead, were used in the MiniLab, would the reaction that occurs at the zinc strip be the same? Explain.

In the lemon battery shown in **Figure 17.9,** a different chemical reaction occurs at each of the metal-strip electrodes. The electrode made of the metal that is more easily oxidized becomes the anode—the electrode at which the oxidation reaction occurs. The second electrode becomes the cathode, and a reduction reaction proceeds at its surface. The substance in a lemon that is most easily reduced is the abundant hydrogen ion of the electrolyte. When these two reactions occur together, in the same cell, they combine to produce a spontaneous redox reaction. This type of reaction is represented by the equation below, where M is the metal that is oxidized.

$$M + 2H^+ \rightarrow M^{2+} + H_2$$

This spontaneous reaction generates the cell voltage of the battery by producing a different electrical potential at each electrode.

Figure 17.9
Lemon Battery
A battery can be made by inserting iron and copper strips into a lemon and connecting them with a conducting wire in an external circuit. Electrons travel through the wire by metallic conduction and through the lemon by electrolytic conduction.

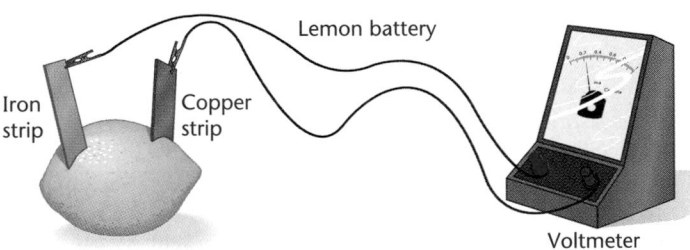

Iron strip   Copper strip   Lemon battery   Voltmeter

Analysis

1. The zinc metal has a greater tendency to lose electrons (oxidation potential) than lead; therefore, oxidation occurs at the zinc strip and reduction at the lead strip.
2. Aluminum has a greater oxidation potential than zinc, creating an even higher potential difference with lead.
3. No. Oxidation would occur at the magnesium strip and reduction at the zinc strip.

✓Assessment

Performance: Ask students to try other fruits in place of lemons and to report on their results.
L1

Potential Difference

Electrons in the metal electrodes of the lemon battery move through the external circuit as a current and can do useful work. The portable radio you connected to a lemon battery to listen to the game needs power to work, and it gets this power from the electrons flowing through the wires from the anode to the cathode of the battery. The chemical reaction at the anode gives off electrons, which enter the metal and then flow through the external part of the circuit connecting the anode to the cathode. At the cathode, the electrons are used up in a reduction reaction. Just as adding water to a container raises the level of the water, adding electrons builds up a negative potential at the anode. This electrical potential is often described as a force, or a pressure of electrons produced by raising the level of the electron sea, **Figure 17.10.**

Why do the electrons travel in one direction and not in the reverse? The electron pressure at the cathode is kept low by the reduction reaction, and the electrons flow from a region of high pressure (negative potential at the anode) to a region of low pressure (positive potential at the cathode). This **potential difference** between the electrodes in the lemon battery causes an electrical current to flow. If there is no potential difference between the electrodes, no current will flow. The size of the current depends upon the size of the potential difference. As the electrons move from a region of more negative potential to a region of more positive potential, they lose energy, so the discharge of a lemon battery is a spontaneous process. Potential energy stored in chemical bonds is released as electrical energy and, finally, as heat. An electrical potential difference is called **voltage** and is expressed in units of volts in honor of Alessandro Volta.

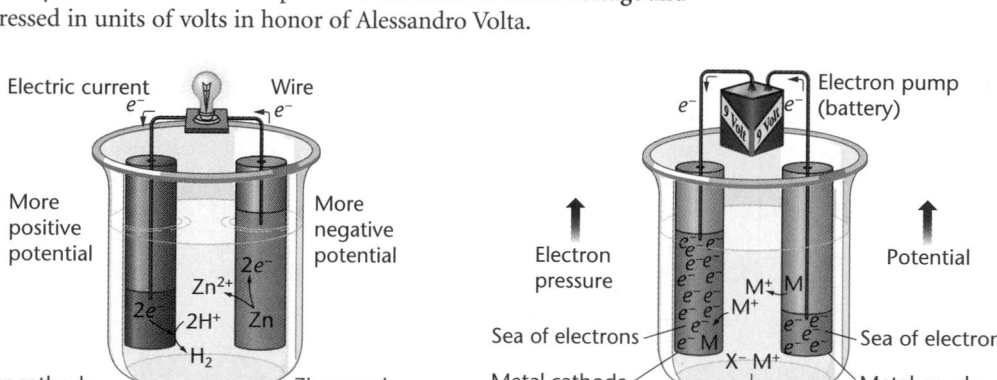

Figure 17.10
Potential Difference

▲ In this model of a lemon battery, the level of the electron sea is raised or lowered by the chemical reactions at the electrode surfaces, creating a potential difference across the battery. A spontaneous oxidation reaction raises the electron pressure (potential) at the anode, and a spontaneous reduction reaction reduces the pressure at the cathode. The "sea level" in the lemon juice is uniform throughout and is intermediate between the levels at the two electrodes.

▲ Because the redox reactions that take place during electrolysis are not spontaneous, a battery is needed to pump electrons from an area of low potential to one of high potential.

SECTION 17.2

Using an Analogy

Describe electrical potential as analogous to water pressure in a pipe, and amperage in a conductor as analogous to the rate of water flow in the pipe.

Correcting Errors

Students may think that cell potential, potential difference, electromotive force (emf), and voltage are different. Explain that they are often used interchangeably, although *cell potential* and *electromotive force* normally refer to galvanic cells and *voltage* refers to the unit, the volt.

Visual Learning

Figure 17.10: Explain that potential difference results from differences in the tendencies of different substances and/or reactions to lose and gain electrons.

GLENCOE TECHNOLOGY

 CD-Rom

Chemistry: Concepts and Applications
Build an Electrochemical Cell Have students use this Interactive Experiment to practice building electrochemical cells.

L1 LEP

Meeting Individual Needs

Gifted Work with gifted students for a short time to show them how to use a table of standard reduction potentials to determine potentials for several standard galvanic cells. Explain that, as long as one works with cells only under standard conditions such as $1M$ concentrations of ions and molecules, the process is simple. **1.** Reverse the half-reaction having the more negative reduction potential, thereby expressing it as an oxidation. **2.** Change the sign of the potential for the reversed reaction. **3.** Add the oxidation and reduction potentials to calculate the cell potential. **4.** To achieve and express the overall reaction, multiply the half-reactions (but not the potentials) by integers to balance the loss and gain of electrons, and add the half-reactions. Take, for example, an Al/Al^{3+} and Cd/Cd^{2+} cell.
1. $Al \rightarrow Al^{3+} + 3e^-$
2. $+1.67V$
3. $(+1.67V) + (-0.40V) = 1.27V$
4. $2(Al \rightarrow Al^{3+} + 3e^-) + 3(Cd^{2+} + 2e^- \rightarrow Cd) = 2Al + 3Cd^{2+} \rightarrow 2Al^{3+} + 3Cd$
L3

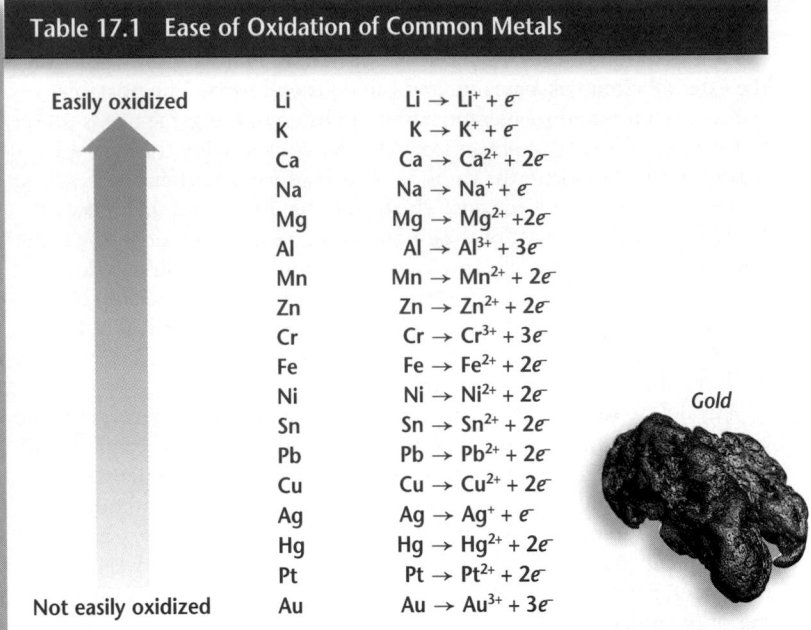

Table 17.1 Ease of Oxidation of Common Metals

Easily oxidized

Li	$Li \rightarrow Li^+ + e^-$
K	$K \rightarrow K^+ + e^-$
Ca	$Ca \rightarrow Ca^{2+} + 2e^-$
Na	$Na \rightarrow Na^+ + e^-$
Mg	$Mg \rightarrow Mg^{2+} + 2e^-$
Al	$Al \rightarrow Al^{3+} + 3e^-$
Mn	$Mn \rightarrow Mn^{2+} + 2e^-$
Zn	$Zn \rightarrow Zn^{2+} + 2e^-$
Cr	$Cr \rightarrow Cr^{3+} + 3e^-$
Fe	$Fe \rightarrow Fe^{2+} + 2e^-$
Ni	$Ni \rightarrow Ni^{2+} + 2e^-$
Sn	$Sn \rightarrow Sn^{2+} + 2e^-$
Pb	$Pb \rightarrow Pb^{2+} + 2e^-$
Cu	$Cu \rightarrow Cu^{2+} + 2e^-$
Ag	$Ag \rightarrow Ag^+ + e^-$
Hg	$Hg \rightarrow Hg^{2+} + 2e^-$
Pt	$Pt \rightarrow Pt^{2+} + 2e^-$
Au	$Au \rightarrow Au^{3+} + 3e^-$

Not easily oxidized

Gold

Chemists at the University of California at Irvine have made the world's smallest galvanic cell. It is too small to be seen without an electron microscope and much smaller than most human cells. The galvanic cell consists of two mounds each of copper and silver attached to a graphite surface. Although it probably will never be used as a practical battery, it may allow scientists to study redox reactions at the atomic level.

Iron is readily oxidized partly because the transfer of electrons from iron to an oxidizing agent releases a large amount of energy. You learned that other metals also are oxidized in corrosion reactions. However, different substances release different amounts of energy when they become oxidized, and this fact may be used to construct a table such as **Table 17.1.** It may be used as a general guide to the ease with which a substance will lose electrons. By examining this table, you can see why copper, gold, and silver are the metals most commonly used in jewelry. All three are hard to oxidize and are, thus, resistant to corrosion.

Galvanic Cells

In the lemon battery, a redox reaction occurs spontaneously to produce a separation of charge at the two electrodes. The reaction begins as soon as the two electrodes are connected by a conductor so that current can flow. An electrochemical cell in which an oxidation-reduction reaction occurs spontaneously to produce a potential difference is called a **galvanic cell.** In a galvanic cell, chemical energy is converted into electrical energy. *Galvanic cells* are sometimes called *voltaic cells;* both terms refer to the same device. A galvanic cell that has been packaged as a portable power source is often called a battery.

Sometimes, the chemical change taking place in a galvanic cell can be seen easily, such as in the simple magnesium-copper galvanic cell shown in **Figure 17.11.** Because magnesium is more easily oxidized than copper, the magnesium loses electrons and becomes oxidized, forming Mg^{2+} ions. The potential of the magnesium anode becomes more negative because of the increased electrical pressure from the released electrons. At the same time, the Cu^{2+} ions pick up electrons from the copper electrode and are reduced to copper metal. The potential of the copper electrode becomes more positive because electrical pressure is lowered as electrons are removed from the cathode. If a wire is connected between the electrodes, current flows from the magnesium electrode to the copper electrode, and the voltmeter in the external circuit reads a voltage of 2.696 V. The energy released during discharge of the cell can be used to power a device such as a radio by connecting the wire from the electrodes through the radio. The overall reaction in the copper-magnesium cell is a redox reaction.

$$Mg(s) \rightarrow Mg^{2+}(aq) + 2e^- \qquad Cu^{2+}(aq) + 2e^- \rightarrow Cu(s)$$
Oxidation half-reaction   **Reduction half-reaction**

$$Mg(s) + Cu^{2+}(aq) \rightarrow Mg^{2+}(aq) + Cu(s)$$
Net redox reaction

Figure 17.11

Magnesium-Copper Galvanic Cell
A piece of magnesium metal is placed in a beaker containing a solution of magnesium sulfate, and a piece of copper metal is placed in a beaker containing a solution of copper(II) sulfate. The two beakers are connected via a salt bridge, which is a porous barrier containing a salt solution; this prevents the two solutions from mixing but permits the movement of ions from one side of the cell to the other. An external circuit containing a voltmeter connects the two metal electrodes. ▶

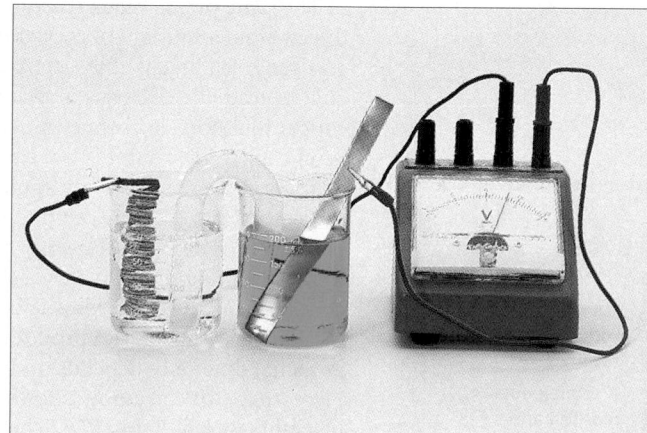

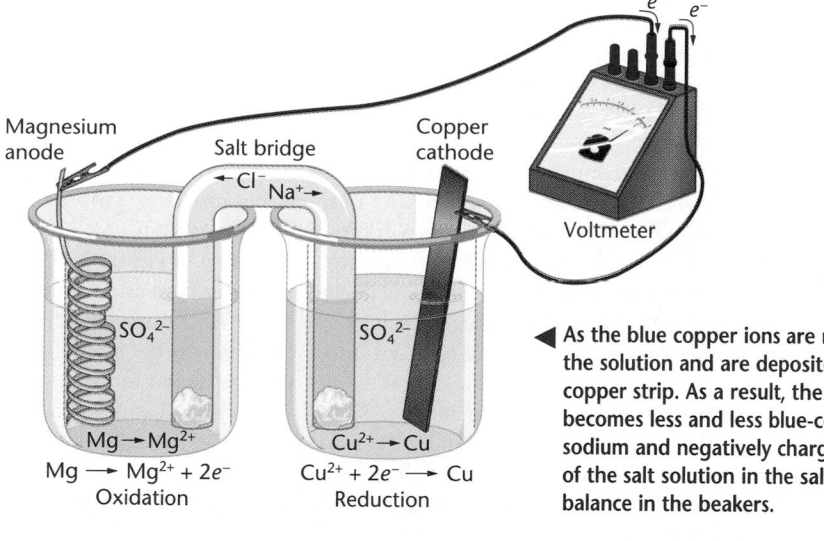

◀ As the blue copper ions are reduced, they move out of the solution and are deposited as copper metal on the copper strip. As a result, the blue copper solution becomes less and less blue-colored. Positively charged sodium and negatively charged chloride ions move out of the salt solution in the salt bridge to restore charge balance in the beakers.

17.2   **Galvanic Cells: Electricity from Chemistry**   **603**

Correcting Misconceptions

Students may think a salt bridge functions as a pathway for the transfer of ions between the anodic and cathodic half-cells, as an automobile bridge serves as a pathway for cars. Explain that the ions initially present in the salt bridge move out of the bridge into the half-cells to a greater extent than ions from the anode and cathode move through the bridge.

Quick Demo

Does the salt bridge really matter?

Remove the salt bridge from a galvanic cell or use two separate lemons for a lemon battery to show that if no salt bridge exists, no current registers on an ammeter.

GLENCOE TECHNOLOGY

 **Videodisc**

Science and Technology Videodisc Series
Mini-Hydroelectric Power Plants
Disc 2: Chemistry
Side 2, Ch. 16

Show this video to illustrate the development of new sources of power. **L1** **LEP**

Figure 17.12
Magnesium-Copper Redox Reaction
When magnesium metal is added to a blue solution of $CuSO_4$, both the magnesium metal and the blue color disappear. ▶

The more easily oxidized magnesium forms colorless Mg^{2+} ions, which dissolve in the solution. The blue copper(II) ions are reduced to the red-brown copper metal that can be seen at the bottom of the beaker. ▶

WORD ORIGIN

spontaneous:
sponte (L) of free will

A spontaneous reaction arises from the inherent qualities of the reactants and occurs with no external input of energy.

The same overall redox reaction occurs if the magnesium metal is placed directly into a solution of copper sulfate, **Figure 17.12.** However, this is not a galvanic cell because the electrons do not flow through an external circuit. Instead, the electrons move directly from the magnesium metal to the copper ions, forming copper metal. This is a way to make copper metal from copper ions, but it is not a way to make electrical power.

You can see that for every spontaneous redox reaction, you theoretically can construct a galvanic cell that can capture the energy released by the reaction. The amount of energy released depends upon two properties of the cell: the amount of material that is present and the potential difference between the electrodes. The more material there is in the electrode, the more electrons it can produce during the course of the reaction. The potential difference depends upon the nature of the reaction that takes place; that is, it corresponds to the relative positions of the two substances in a table such as **Table 17.1.** The farther apart the two substances are in the table, the greater the potential difference between the electrodes, and the greater the energy delivered by each electron that flows through the external wire.

How do you know which substance will be oxidized and which reduced in any cell? Look back at **Table 17.1.** Experimental chemists such as Humphry Davy and his student Michael Faraday did many experiments from which this type of table could be made. The table is used today to predict the outcome of new experiments. For example, in a Zn-Cu galvanic cell, zinc will be oxidized and copper reduced. Because zinc is more easily oxidized than copper, electrons will flow from zinc to copper.

A cell voltage should register on the voltmeter shown in **Figure 17.11** because a potential difference exists between the magnesium and copper electrodes. What function does the salt bridge serve? As the half-reactions continue, magnesium ions are released into the solution at the anode, and copper ions are removed at the cathode. Ions must be free to move between the electrodes to neutralize positive charge (Mg^{2+} cations) created at the anode and negative charge (anions) left over at the cathode. The solution of ions in the salt bridge allows ionic conduction to complete the electrical circuit and prevent a buildup of excess charge at the electrodes.

604 **Chapter 17 Electrochemistry**

Demonstration 17.4

Complete a Circuit

Purpose: To reinforce that electrolytes conduct an electric current.

Materials: lightbulb conductivity tester, 150 mL ethanol, 150 mL distilled water, 30 g NaCl, stirring rod

Safety Precautions:

Disposal: Code C

Procedure: Use a lightbulb conductivity tester to demonstrate that electrolytes conduct an electric current. **CAUTION:** *There is a risk of electrical shock.* Darken the room. Immerse the electrodes in ethanol to demonstrate that a nonelectrolyte does not conduct current. **CAUTION:** *Ethanol is flammable.* Repeat with distilled water. Add a small amount of NaCl slowly to the distilled water, and have students observe the lightbulb as the solution is stirred slowly.

Results: Ethanol is a nonelectrolyte. Distilled water does not conduct enough to light the bulb until the NaCl is added. The bulb brightens as more salt dissociates.

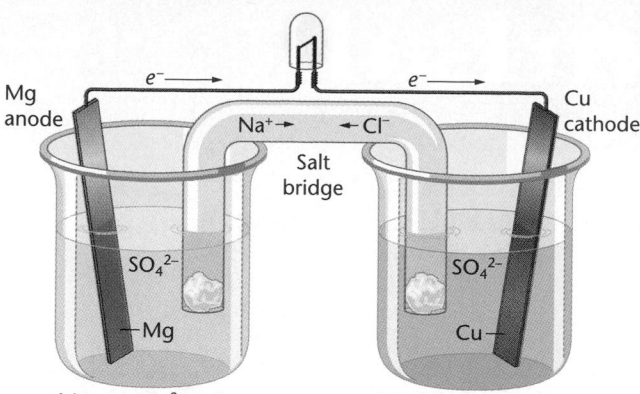

Mg anode — e^- → — Salt bridge — Na^+ → ← Cl^- — e^- → Cu cathode

SO_4^{2-} Mg

SO_4^{2-} Cu

$Mg \longrightarrow Mg^{2+} + 2e^-$

$Cu^{2+} + 2e^- \longrightarrow Cu$

Figure 17.13

Batteries Perform Work
When a simple galvanic cell does useful work, it is called a battery. If the external circuit is connected with a wire, electrons flow from the site of oxidation at the magnesium strip and through the LED to the surface of the copper strip, where reduction of Cu^{2+} ions takes place. The voltage pushes electrons through the LED, causing it to light up.

Useful work may be done if the voltmeter is replaced by wires connecting the galvanic cell to a lightbulb. Then, electrical energy will be converted into light energy, a useful process in a dark room. In **Figure 17.13,** wires connect the cell to a light source with a low-voltage requirement, called a light-emitting diode or LED. If the circuit to the cell is complete, the LED lights up, showing that the cell is doing useful work. With time, the light intensity will fade. Why doesn't it stay lit indefinitely? Eventually, all of the magnesium in the anode becomes oxidized. The capacity of the battery has been exceeded, the magnesium is gone and, if there is no electrode, there can be no cell.

Better and Better Batteries

Although the galvanic cell made from magnesium and copper can do useful work, it isn't something you'd want to bring along on a camping trip. The wet solutions could be sloppy, the glass could break easily, and the capacity is limited. Fortunately, scientists have developed much better batteries that are smaller, lighter, provide higher voltages, and last longer. **Figure 17.14** shows an assortment of commonly used batteries. Experimental batteries no thicker than a sheet of paper have already been developed. And, although you might think batteries always have to be made of metal and acids, some batteries of the future may be made of microorganisms that use the energy in sugar to make electricity. A living fuel cell has been developed that someday could be used to power an automobile for up to 15 miles on two pounds of sugar.

How are batteries designed? The farther apart two metals are in **Table 17.1,** the larger the voltage of a battery that can be constructed from them. If you wanted to make a high-voltage battery to power your radio, you would choose metals that are far apart in the table. A copper penny with an iron nail will yield a larger voltage than a penny with a piece of nickel because copper is farther away from iron in the table than it is from nickel.

Figure 17.14

Modern Batteries
Modern batteries come in a wide variety of sizes, shapes, and strengths. Each type of battery serves a different purpose.

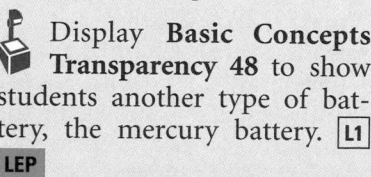

Concept Development

Explain that the concentrations of ions in the anodic and cathodic compartments can cause an increase or a decrease in the potential of a cell and that an equation (the Nernst equation) can be used to calculate cell potential if concentrations other than $1M$ are encountered.

⚡ Quick Demo

How does a pH meter work?

Show students a pH meter with both pH and millivolt scales, and demonstrate its use. Explain that the meter is able to measure pH due to the effect that the concentration of H^+ has on the potential difference between the electrodes of the meter.

GLENCOE TECHNOLOGY

💿 CD-Rom

Chemistry: Concepts and Applications
Building a Battery
Have students use this Interactive Exploration to practice building batteries with specific voltages. L1 LEP

Transparency

Display **Basic Concepts Transparency 48** to show students another type of battery, the mercury battery. L1
LEP

Analysis: Ask these questions.
1. When electricity is added to an electrochemical cell, what particles must be present for conductivity to occur? *ions*
2. What are the names used to describe the two electrodes present on the conductivity tester? *anode and cathode*

✓ Assessment

Performance: Have students bring in different solutions from the kitchen, garage, and bathroom to be tested for the presence of ions. L1
LEP

ChemLab

ChemLab

Oxidation-Reduction and Electrochemical Cells

Time Allotment: One laboratory period.

Objectives

Review objectives with students before they begin the ChemLab.

Process Skills

Observing, inferring, measuring, interpreting data, communicating

Safety Precautions and Disposal

Remind students to wear aprons and safety goggles and to wash their hands before leaving the laboratory.

Disposal: Codes B and C; Have students dispose of magnesium and copper pieces and strips in solid-waste containers, and rinse solutions down the drain with large amounts of tap water.

PREPARATION

- Prepare 0.5M CuCl$_2$ solution by dissolving 8.5 g of CuCl$_2$· 2H$_2$O in sufficient distilled water to make 100 mL of solution.
- Prepare 0.1M MgCl$_2$ solution by dissolving 2.03 g of MgCl$_2$·6H$_2$O in sufficient distilled water to make 100 mL of solution.
- Prepare 0.5M NaCl solution by dissolving 29.2 g of NaCl in sufficient distilled water to make 1 L of solution.
- Prior to the ChemLab, prepare craft sticks with V-cuts and slit cuts as shown in the diagram or use commercial dialysis bag holders.

Oxidation-Reduction and Electrochemical Cells

Redox reactions involve the loss and gain of electrons. By separating the oxidation process from the reduction process and connecting them electrically through an external circuit, many spontaneous redox reactions can be utilized to produce an electrical potential and an electrical current. Devices that perform these functions are called electrochemical cells. In this ChemLab, you will investigate a redox reaction and use it to construct an electrochemical cell.

Problem

How may a spontaneous redox reaction be used to construct an electrochemical cell?

Objectives

- **Observe** a simple oxidation-reduction reaction.
- **Relate** the reaction to the oxidation tendencies of the reactants.
- **Utilize** the reaction to construct an electrochemical cell that can operate electrical devices.

PREPARATION

Materials

craft-stick support with V-cut and slit cut
25-mm (flat diameter) dialysis tubing (15 cm in length)
magnesium ribbon (10 cm in length)
magnesium ribbon (1 cm in length)
copper foil (10 cm × 1 cm strip)
copper foil (1 cm × 2 mm piece)
metric ruler
250-mL beaker
10 × 100 mm test tubes (2)
wire leads with alligator clips (2)
DC voltmeter with a 2-V or 3-V scale

flashlight bulb for 2 AAA batteries
9-V transistor radio
0.5M sodium chloride solution
0.5M copper(II) chloride solution
0.1M magnesium chloride

Safety Precautions

Wear an apron and safety goggles. Rinse the solutions down the drain with large amounts of tap water. Wash your hands after performing the lab.

PROCEDURE

1. Soak the dialysis tubing in tap water for about ten minutes while you complete steps 2 and 3. Tie two knots near one end of the tubing, and open the other end by sliding the material between your fingers.

2. Pour a small amount of the copper(II) chloride solution into a 10 × 100 mm test tube, and drop a 1-cm length of magnesium ribbon into the solution. Observe the system for about one minute, and record your observations in a data table similar to the table in Data and Observations. Pour the solution down the drain and discard the piece of magnesium in a wastebasket.

3. Repeat step 2 using magnesium chloride solution and the small piece of copper foil.

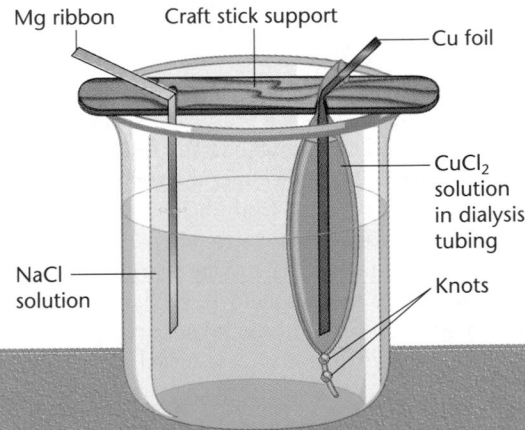

4. Pour copper(II) chloride solution into the open end of the tubing to a depth of 6 cm to 8 cm, and insert the strip of copper foil. Slide the top of the tubing and copper strip into the V-cut in the stick as shown below. Suspend the tubing in the beaker, as shown.

5. Slide the length of magnesium ribbon into the slit cut in the craft stick, as shown.

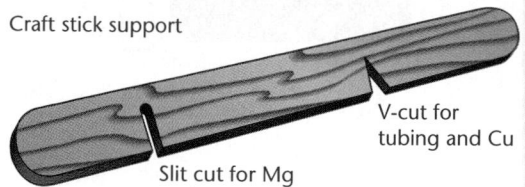

Craft stick support

V-cut for tubing and Cu

Slit cut for Mg

6. Pour about 200 mL of sodium chloride solution into the beaker.

7. Connect leads to the pieces of copper and magnesium, and touch the leads to a DC voltmeter—the lead from the Cu electrode to the + terminal and the lead from the Mg electrode to the − terminal. Read and record the potential difference, or voltage.

8. Cooperate with other lab groups in the following way to light the bulb and to operate the transistor radio. The flashlight bulb requires a voltage of about 3V, and the radio requires a voltage of about 9V. Connect your electrochemical cells in series (copper to magnesium) to provide the desired voltages. **CAUTION:** *Be sure to connect your cells to the battery terminals of the radio in the correct polarity.* Connected in series, the voltages are additive; for example, five 2-V cells in series yield a voltage of 10V. Such combinations of electrochemical cells are called batteries.

9. Disassemble your cell, observing the pieces of copper and magnesium and recording your observations. Rinse the pieces of copper and magnesium, and dispose of them according to the instructions of your teacher.

DATA AND OBSERVATIONS

	Data and Observations
$Mg - Cu^{2+}$	
$Cu - Mg^{2+}$	
Voltage	
Pieces of Cu and Mg	

ANALYZE AND CONCLUDE

1. **Interpreting Data** Write the balanced equation for the single-replacement reaction between magnesium and copper(II) chloride that occurred in step 2. Which metallic element, Cu or Mg, has the greater tendency (or oxidation potential) to lose electrons?

2. **Relating Concepts** In an electrochemical cell, oxidation occurs at the anode, and reduction occurs at the cathode. Which metal was the anode and which was the cathode? Write the equations for the half-reactions.

APPLY AND ASSESS

1. When an electrochemical cell is used to operate an electrical device, in which direction do the electrons move in the external circuit?

2. Is it possible to construct an electrochemical cell in which lead is the anode and lithium is the cathode? Explain.

Alternative Materials

- Copper(II) nitrate and magnesium nitrate solutions may be used in place of copper(II) chloride and magnesium chloride solutions.
- Separate cells with salt bridges or porous barriers connecting them may be used instead of dialysis tubing.
- Low-current hobby motors may be purchased and used in addition to or in place of lightbulbs and transistor radios.

PROCEDURE

- Require that students' notebooks contain the following items before they begin: Objectives; Safety Precautions; Outline or Flow Chart of the Procedure; Data and Observations Table.

ANALYZE AND CONCLUDE

1. $Mg + CuCl_2 \rightarrow MgCl_2 + Cu$. Magnesium has the greater tendency to lose electrons.
2. Magnesium is the anode. $Mg \rightarrow Mg^{2+} + 2e^-$ Copper is the cathode. $Cu^{2+} + 2e^- \rightarrow Cu$

APPLY AND ASSESS

1. The electrons move from the anode to the cathode in the external circuit.
2. No. Lithium has a much greater tendency (oxidation potential) to lose electrons than lead. Therefore, lithium would be the anode.

✓Assessment

Portfolio: Ask students to write formal reports for the ChemLab and to include them in their portfolios. L1 P

DATA AND OBSERVATIONS

$Mg-Cu^{2+}$	The magnesium ribbon turned black and somewhat copper colored. Bubbles were produced.
$Cu-Mg^{2+}$	No reaction
Voltage	1.5-1.6V
Pieces of Cu and Mg	The copper foil has more copper plated onto it. The magnesium ribbon partially dissolved and has some copper plated onto it.

Correcting Misconceptions

The term *battery* is often used incorrectly to refer to a single galvanic or electrochemical cell, such as a dry cell. Point out that a battery is usually considered to consist of two or more galvanic cells that are electrically connected.

Discussion

Ask students how dry cells are able to function if they are really dry. If students don't know, explain that dry cells contain moisture in a thick slurry (as opposed to the liquid in an automobile battery) and would become inoperable if they were really dry because no ion movement would occur.

Extension

Have students research the method of retarding the rusting of metal oceangoing ships by hanging large pieces of magnesium along the hull of the ship. The more active magnesium is oxidized instead of the iron. When depleted, the magnesium strips are replaced. L1

Fact of the MATTER

If you have metal fillings in your teeth, you may have received an electric shock after biting down on a bit of aluminum foil. The aluminum foil acts as an anode, and the silver-mercury amalgam, an alloy, in the filling acts as a cathode. The saliva in your mouth serves as an electrolyte. The galvanic cell in your mouth is short-circuited when the foil comes into contact with the filling, causing a weak current to flow between the electrodes. This current is detected as pain by the nerve of the tooth.

Although the term *battery* usually refers to a series of galvanic cells connected together, some batteries have only one such cell. Other batteries may have a dozen or more cells. When you put a battery into a flashlight, radio, or CD player, you complete the electrical circuit of the galvanic cell(s), providing a path for the electrons to flow through as they move from the reducing agent (the site of oxidation) to the oxidizing agent (the site of reduction). The most powerful batteries combine strong oxidizing agents and strong reducing agents to give the largest possible potential difference. But those agents aren't necessarily safe, convenient, or economical to use. To get a higher voltage from a cell type with a relatively small potential difference, several of the cells can be connected in series, as **Figure 17.15** shows.

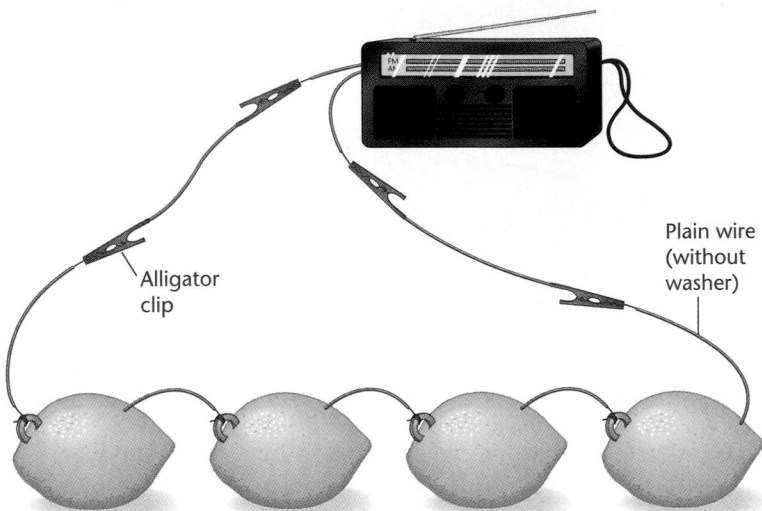

Alligator clip

Plain wire (without washer)

Figure 17.15

A More Powerful Lemon Battery
One lemon cell wouldn't provide enough voltage to power a transistor radio, but several connected together in series would. This means that the positive terminal of one cell is connected to the negative terminal of the next, and so on. The electrodes in this battery are steel washers and copper wire. The total voltage of the battery is the sum of all the voltages of the individual cells.

Carbon-Zinc Dry Cell

Whenever you put two or more common D batteries into a flashlight, you are connecting them in series. They have to be placed in the correct order so that electrons flow through both cells. These relatively inexpensive batteries are carbon-zinc galvanic cells, and they come in several types, including standard, heavy-duty, and alkaline. This type of battery is often called a dry cell because there is no aqueous electrolyte solution; a semisolid paste serves that role. Examine the cutaway view of the carbon-zinc battery in **Figure 17.16** to see if you can locate the parts of the galvanic cell it contains.

Demonstration 17.5

A 9-Volt Battery

Purpose: To demonstrate that the 9 volts in a battery are achieved by connecting six galvanic cells in series.

Materials: used 9-V battery, small screwdriver, pair of needle-nose pliers

Safety Precautions:

Disposal: Code F

Procedure: Obtain a used, 9-V alkaline battery. Use a small screwdriver and a pair of pliers to open the case. Show students that the battery consists of six galvanic cells connected in series. **CAUTION:** *Do not use a mercury cell. Mercury salts are difficult and expensive to dispose of. Warn students not to open the case of batteries at home.* Alkaline batteries contain a zinc anode and a manganese(IV) oxide, MnO_2, cathode. Mn^{4+} is reduced to Mn^{3+} during the cell reaction.

Results: Students will be able to see a separator that allows negative ions to pass while preventing mixing of the half-cell components. The insulator prevents the anode and cathode from touching and shorting out the battery.

Figure 17.16

Carbon-Zinc Dry Cell
A standard D battery is shown both whole and cut in half to reveal the structure of the carbon-zinc dry cell. Beneath the outside paper cover of the battery is a cylinder casing made of zinc. The zinc serves as the anode and will be oxidized in the redox reaction. ▼

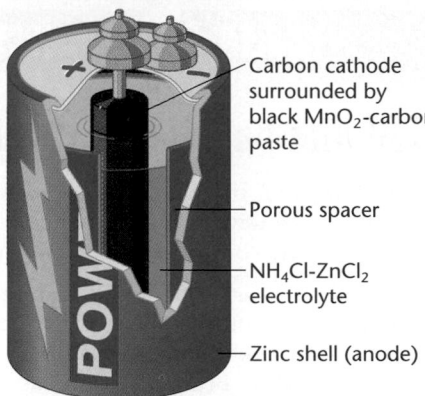

Carbon cathode surrounded by black MnO_2-carbon paste

Porous spacer

NH_4Cl-$ZnCl_2$ electrolyte

Zinc shell (anode)

▲ The carbon rod in the center of the cylinder—surrounded by a moist, black paste of manganese(IV) oxide (MnO_2) and carbon black—acts as a cathode. Ammonium chloride (NH_4Cl) and zinc chloride ($ZnCl_2$) serve as electrolytes. Alkaline batteries contain potassium hydroxide (KOH) in place of the ammonium chloride electrolyte, and they maintain a high voltage for a longer period of time.

What is missing in this galvanic cell? Notice that the circuit is not complete, so the electrons that are produced at the zinc cylinder have no external conductor through which to travel to the carbon. This is by design and is not a defect in the battery. The circuit will be complete when the battery is placed in something designed to be powered by it, such as a flashlight. When the flashlight is turned on, the redox reaction starts. Electrons travel out of the zinc casing into a piece of metal built into the flashlight. There, they travel through a bulb, causing it to light up. The electrons then reenter the battery at the top and move down through the carbon rod and into the black paste, where they take part in the reduction reaction.

The flow of electrons from the zinc cylinder through the electrical circuits of an appliance and back into the battery provides the electricity needed to power a flashlight, radio, CD player, toy, clock, or other item. When electrons leave the casing, zinc metal is oxidized.

$$Zn \rightarrow Zn^{2+} + 2e^-$$

The reactions in the carbon rod and the paste are much more complex, but one major reduction that takes place is that of manganese in manganese(IV) oxide. In this reaction, the oxidation number of manganese is reduced from 4+ to 3+.

$$2MnO_2 + H_2O + 2e^- \rightarrow Mn_2O_3 + 2OH^-$$

Adding the two half-reactions together gives the major redox reaction taking place in a carbon-zinc dry cell.

$$Zn + 2MnO_2 + H_2O \rightarrow Zn^{2+} + Mn_2O_3 + 2OH^-$$

17.2   Galvanic Cells: Electricity from Chemistry   609

Analysis: Ask these questions.
1. Draw a diagram of the dissected battery. *Students should be able to label the parts of the battery.*
2. Can a single cell produce 9 volts? *No, six 1.5-V cells are used to produce the 9 volts.*

Assessment

Performance: Students realize that these batteries can and should be recycled to avoid contamination of the environment. Have students contact battery manufacturers to determine the locations of recycling centers in your area that accept batteries. L1

Visual Learning

Figure 17.16: Cut open and display several batteries and cells, such as a 6-V lantern battery containing four 1.5-V carbon-zinc cells, a 1.5-V carbon-zinc dry cell, and a 9-V transistor radio battery consisting of six small 1.5-V carbon zinc cells. See Demonstration 17.5. Ask students to examine but not touch them and to deduce which ones consist of more than one cell and are, therefore, batteries. L1 LEP

Extension

Have students research and report on air-activated batteries that are used in hearing aids and other such devices. This type of battery is supplied with a removable seal. After the seal is removed, the battery becomes operational. Ask students to explain in their reports how this type of battery is activated by air and what advantage(s) this type of battery might have. *The battery becomes activated by absorbing water molecules from the air. The major advantage is that this type of battery has a long storage life until the seal is removed, as there is no water to aid in any deterioration of the chemicals in the battery.* L1

Extension

Ask students to find out why some batteries should not be recharged and to report their findings to the class. *Hydrogen gas may be produced from moisture in the battery, causing it to explode.* L1

HEALTH CONNECTION

Purpose

To demonstrate the use of lithium metal in commercial battery reactions.

Teaching Strategies

- Attach wires to small zinc and copper metal strips. Attach the wires to a galvanometer and insert the metal strips into a potato. Have students note the direction of deflection. Switch the wires on the poles and reinsert them into the potato. Discuss why the galvanometer deflected in the opposite direction. L1

- Obtain and demonstrate a two-potato clock sold by many science supply houses. Use a variety of substances as the electrolyte instead of potatoes, such as soft drinks, apples, oranges, or a student's hands. Also try dry substances such as bread, marshmallows, or a pile of salt. Discuss the properties required by the electrolyte.

Extension

Obtain at least six different types of metal strips or rods. Have students use a voltmeter to determine the voltage output of a simple battery made by inserting two of the metals into opposite ends of a potato. Have them make a table that lists the metals involved, which metal acted as the anode or cathode, and the voltage output for all 15 possible two-metal combinations. L1

Connecting to Chemistry

1. $4Li(s) + 2SOCl_2(l) \rightarrow 4LiCl(aq) + S(s) + SO_2(g)$
2. $2Li + 2H_2O \rightarrow 2LiOH + H_2$; lithium is the reducing agent and water is the oxidizing agent.

HEALTH CONNECTION

Lithium Batteries in Pacemakers

It's always frustrating to have batteries go dead just when it seems you need them most. However, imagine needing a battery upon which your life depends. What materials could it be made from? Would it last long enough?

Heart stimulation Consider for a moment that, over a period of time, you began to feel light-headed, dizzy, weak, or fatigued. It could be that the chambers of your heart are not beating rhythmically or fast enough. You might be a candidate for a heart pacemaker. This device, which is inserted inside the body, monitors the heart's activity. When necessary, the pacemaker supplies the electrical impulses needed to stimulate the heart. In order to be most effective, the batteries in a pacemaker need to be fully powered for long periods of time and survive in the hostile, saline environment of the human body without breaking down.

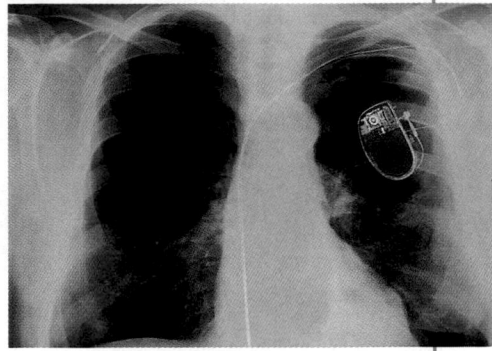

Powerful electrodes One type of battery used in pacemakers is a cell based on lithium and sulfuryl chloride. Lithium is a popular choice for battery anodes because of its strong tendency to be oxidized. Lithium is oxidized during the reaction, and the sulfur in sulfuryl chloride is reduced. The unbalanced half-reactions are given below.

$$Li(s) \rightarrow Li^+(aq) + e^- \qquad \text{oxidation}$$
$$SOCl_2(l) + 4e^- \rightarrow Cl^-(aq) + S(s) + SO_2(g) \quad \text{reduction}$$

Battery characteristics Lithium is the least dense of all nongaseous elements, with a density of only 0.534 g/mL. The lightweight lithium contributes little weight to the small, detachable battery pack, which is a circular disk only about 5 cm by 1 cm in size. The pacemaker battery pack is implanted under the patient's collarbone and has a life expectancy of about seven to ten years, depending upon how often it is needed to stimulate the heart.

Lithium hazards Lithium also presents some potential problems when used in a wet environment. The tremendous activity of lithium makes it dangerously reactive with a variety of compounds, including water. If the battery pack is not adequately sealed against the body's corrosive fluids, the resulting exothermic reaction of lithium in water accompanied by the production of H_2 gas could have serious consequences.

As battery technology continues to advance, medical applications requiring portable, long-lasting power supplies will become more and more common.

Connecting to Chemistry

1. **Interpreting** Write the balanced overall reaction of lithium with sulfuryl chloride.

2. **Interpreting** Write the balanced overall reaction of lithium with water. What is the oxidizing agent? The reducing agent?

Integrating THE Sciences

Medicine The electrical impulses that travel through the heart muscle as it contracts are conducted through body fluids to the surface of the body, where the current can be detected by electrodes placed on the skin and recorded as an electrocardiogram (EKG or ECG). One of the world's first electrocardiograms was made by Dr. Augustus Waller, who used his pet bulldog Jimmie as an experimental subject. This experience did not harm Jimmie, who stood in buckets of salty water. Electricity passed from Jimmie's heart to the surface of his skin, where the salt water carried the signals to a crude monitoring device. Today, EKG sensors can detect even faint abnormalities in the workings of the heart, and computers are used to analyze signals from the beating heart.

Automobile Lead Storage Battery

The most common type of battery used in cars is a lead-acid, 12-volt storage battery. It contains six 2-volt cells connected in series. Although much larger than carbon-zinc batteries and relatively heavy, this type of battery is durable, supplies a large current, and can be recharged. When you turn your key in the ignition, it is the battery that supplies electricity to start the car. It also provides energy for any demands not met by the car's alternator, such as running the radio or using the lights when the engine is off. Leaving on the lights or radio for too long with the engine off can make the battery go dead because it is the engine that recharges the battery as the car runs.

Each galvanic cell in a lead-acid battery has two electrodes—one made of a lead(IV) oxide (PbO_2) plate and the other of spongy lead metal, as **Figure 17.17** shows. In each cell, lead metal is oxidized as lead(IV) oxide is reduced. The lead metal is oxidized to Pb^{2+} ions as it releases two electrons at the anode. The Pb^{4+} ions in lead oxide gain two electrons, forming Pb^{2+} ions at the cathode. The Pb^{2+} ions combine with SO_4^{2-} ions from the dissociated sulfuric acid in the electrolyte solution to form lead(II) sulfate at each electrode. Thus, the net reaction that takes place when a lead-acid battery is discharged results in the formation of lead sulfate at both of the electrodes.

$$PbO_2 + Pb + 2H_2SO_4 \rightarrow 2PbSO_4 + 2H_2O$$

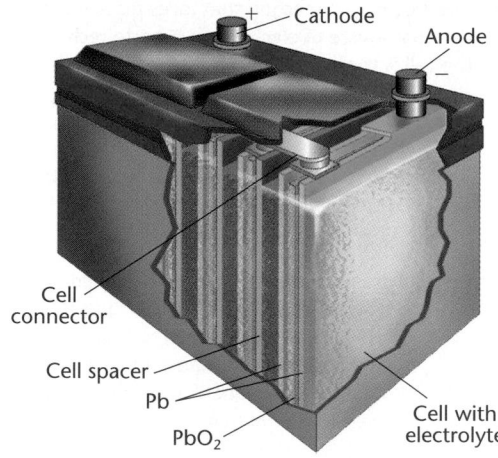

Figure 17.17
Lead Storage Battery
◄ A lead storage battery is not a dry cell because it contains several connected cells filled with an aqueous solution of sulfuric acid, which serves as the electrolyte. The electrodes are alternating plates of lead metal and lead(IV) oxide. The case surrounding the battery is hard plastic. It holds the cells in place and acts as an insulator because it does not conduct electricity itself. This helps keep the electricity inside the battery.

A car with a dead battery can still be started. Electricity from a second car is used to jump-start the car, bypassing the dead battery. ►

GLENCOE TECHNOLOGY

 Videodisc

Science and Technology Videodisc Series
Battery Science
Disc 2: Chemistry
Side 2, Ch. 8

Show this video to illustrate the use of new batteries. L1
LEP

Quick Demo

How does this thing test the battery?
Hold up or describe a battery tester of the type that uses a hygrometer to determine the density of the sulfuric acid in a lead storage battery. Ask students how it can determine the degree of charge of the battery. *When the battery is fully charged, the H_2SO_4 is at maximum concentration and density (about 1.28 g/mL). As the battery is discharged, the H_2SO_4 is consumed and the density of the solution decreases.*

Assessment

Knowledge: Ask students what is oxidized and what is reduced as a lead storage battery discharges. *Pb is oxidized, and Pb in PbO_2 is reduced.* L1

Tying to Previous Knowledge

People are sometimes advised to turn on their headlights for about five seconds during extremely cold weather to warm the battery before trying to start the car. What does this indicate about the discharging reaction? *It is exothermic.* The charging reaction? *It is endothermic.*

Integrating THE Sciences

Biology Although signals from the nervous system make skeletal muscle contract, cardiac muscle in the heart contracts spontaneously. Even if all nerves leading to the heart were destroyed, the heart would keep on beating due to signals produced and conducted by pace-making cells. Seventy or 80 electrical impulses per minute are conducted out from the sino-atrial node (SA, or cardiac pacemaker), spreading over both atria and causing them to contract. When the signals reach the atrio-ventricular (AV) node, they move more slowly, allowing the atria to finish contracting before the ventricles do. Artificial electrical signals also can keep the heart beating in a normal rhythm. Some artificial pacemakers implanted in patients are powered by nuclear energy from plutonium-238 rather than by more common batteries. This radioisotope is sealed in a case so its emissions won't damage tissues around the pacemaker.

TECH PREP

Purpose

Students will learn what redox reactions take place during discharge of a nickel-cadmium rechargeable battery, and how electricity can be used to reverse the reactions when the battery is recharged.

Background

The rechargeable nicad battery consists of cadmium and an oxide of nickel. The cadmium making up the anode undergoes oxidation, while the NiO_2 cathode undergoes reduction. The potential provided by nicad cells is about 1.4V. They have a longer life than rechargeable lead storage batteries.

Visual Learning

Use a battery tester to show students that a nicad battery is fully discharged, and then test it again after recharging the battery overnight.

Teaching Strategies

Ask students to name some advantages and disadvantages of using rechargeable rather than disposable batteries. *Rechargeable batteries are less expensive in the long run and produce less waste, but they are more expensive initially, need to be charged before the first use, can require recharging at inconvenient times, and the cadmium waste from expired nicad batteries is toxic.* L1

Thinking Critically

1. oxidation reaction: Ni → $Ni^{2+} + 2e^-$; reduction reaction: $Cd^{2+} + 2e^- →$ Cd
2. Using nicad batteries helps eliminate some of the solid waste generated by ordinary disposable batteries.

Nicad Rechargeable Batteries

The nickel-cadmium, or nicad, cell is a common storage battery that can usually be discharged and recharged more than 500 times. These batteries are used in calculators, cordless power tools and vacuum cleaners, and rechargeable electric toothbrushes and shavers. Once nicad batteries have been spent, disposal presents a problem because cadmium is toxic. Nicads can be recycled, but the process is expensive. Although rechargeable batteries containing less toxic metals are being developed, none have been found that can sustain a constant rate of discharge as well as the nicad.

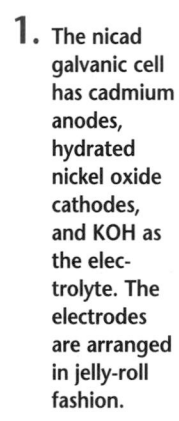

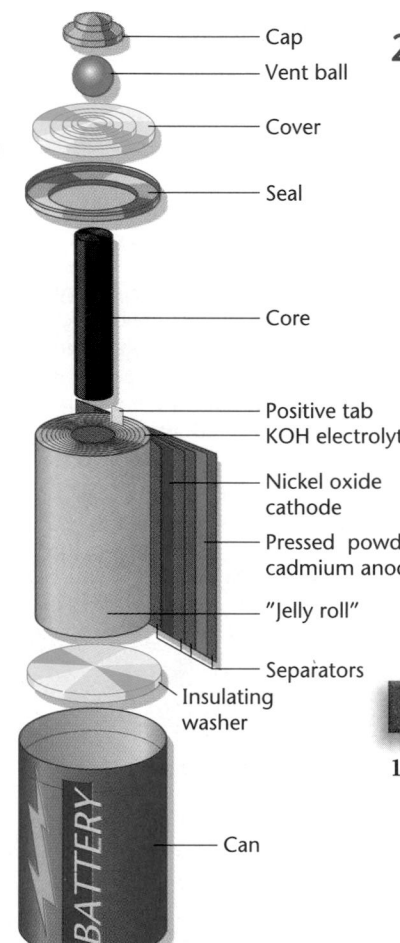

1. The nicad galvanic cell has cadmium anodes, hydrated nickel oxide cathodes, and KOH as the electrolyte. The electrodes are arranged in jelly-roll fashion.

- Cap
- Vent ball
- Cover
- Seal
- Core
- Positive tab
- KOH electrolyte
- Nickel oxide cathode
- Pressed powdered cadmium anode
- "Jelly roll"
- Separators
- Insulating washer
- Can

2. In the redox reaction that takes place during discharge, nickel oxide is reduced at the cathode, and cadmium is oxidized at the anode.

Cd + NiO → CdO + Ni

3. The electrolysis reaction that takes place when an external source of electricity is used to recharge the cell is the reverse of the discharge reaction.

CdO + Ni → Cd + NiO

4. Newly purchased nicad batteries must be charged before use.

5. Nicad batteries are not suitable for devices that are left idle for long stretches—such as smoke detectors, cameras, and flashlights—because they will lose about one percent of their charge daily even when not being used.

Thinking Critically

1. What are the equations for the oxidation and reduction half-reactions that occur during recharging of a nicad battery?

2. What might be an environmental advantage to using nicad batteries?

The reaction that occurs during discharge of a lead-acid battery is spontaneous and requires no energy input. The reverse reaction, which recharges the battery, is not spontaneous and requires an input of electricity from the car's alternator. Current enters the battery and provides energy for the reaction in which lead sulfate and water are converted into lead(IV) oxide, lead metal, and sulfuric acid.

$$2PbSO_4 + 2H_2O \rightarrow PbO_2 + Pb + 2H_2SO_4$$

Sulfuric acid is corrosive. It is important to be careful when working around a car battery, as well as disposing of it properly when it finally goes dead for good. These batteries can usually be discharged and recharged numerous times and last about three to five years.

Better Batteries for Electric Cars

At the end of the 19th century, most cars were powered by steam or by electric batteries, although today most cars have internal combustion engines that are powered by gasoline. Electric cars, **Figure 17.18,** could help reduce our dependence on fossil fuels, cause less pollution, and be more economical in the long run, but they have several disadvantages, such as high initial cost, limited driving range, low speed, and long recharge time. They also present a disposal problem because cadmium is a toxic metal.

These disadvantages would disappear if a battery that is cheap enough, powerful enough, and safe enough for running an electric car could be developed. Two new experimental types of batteries for use in electric cars show early promise as candidates. One is a rechargeable, nickel-metal hydride or NiMH battery. This type of battery is less toxic and has a higher storage capacity than the batteries now used in electric cars. Another experimental battery is a lithium battery with a water-based electrolyte. Lithium is more easily oxidized than any other metal but has a drawback that has limited its use in batteries: it explodes violently when it comes into contact with water. Lithium is used in some batteries

Figure 17.18

Electric Cars
Most of the nickel-cadmium batteries that are used in electric cars today power the car for only 50 to 100 miles before they run down and need to be recharged, a process that takes many hours. In Randers, Denmark, special parking spaces with electric hookups are available for battery-operated cars.

SECTION 17.2

Concept Development

As traditional or older car batteries were being charged, water was converted to hydrogen by electrolysis, creating a dangerous situation and requiring frequent addition of water to the battery. Newer batteries use an alloy of lead and calcium, do not consume water, and are completely sealed.

Applying Chemistry

TECH PREP

Better Batteries for Electric Cars BMW has created an electric car powered by a 584-pound battery developed by Asea Brown Boveri, a German battery manufacturer. The car has a top speed of 60 mph, a range of about 100 miles, and a battery life of more than 100 000 miles. The battery uses liquid sodium as the anode, liquid sulfur as the cathode, and a solid ceramic material as the electrolyte. It can store about four times as much energy per unit mass as the lead storage battery.

Extension

TECH PREP

Have interested students research the operation of fuel cells used on NASA missions. Posters and reports can be made and added to students' portfolios. L1 P

GLENCOE TECHNOLOGY

  Videotape

MindJogger Videoquizzes for Chemistry
Chapter 17: Electrochemistry
Have students work in groups as they play the videoquiz game to review key chapter concepts. L1 LEP COOP LEARN

Chemistry Journal

Everyday Electrochemistry
Ask students to list all the uses or occurrences of electrochemical reactions that they encounter in a typical day and to describe them in their journals. L1

Purpose

Students will learn how the hydrogen-oxygen fuel cell is used to produce electricity and water through a redox reaction.

Background

The hydrogen-oxygen fuel cell uses gaseous fuels to convert chemical energy into electrical energy. Oxygen gas undergoes reduction at one electrode, producing OH^- ions, while hydrogen gas is oxidized at the other electrode to form H^+ ions. These ions combine to form water vapor, which can be condensed and used as drinking water. The space shuttle uses fuel cells to power its orbiters and provide drinking water to the astronauts. Fuel cells can generate energy indefinitely as long as the external supply of hydrogen and oxygen fuels is maintained. Their efficiency is almost 75%, which is much higher than the 40% efficiency of generators powered by burning fossil fuels.

Visual Learning

You can demonstrate the power of the hydrogen-oxygen fuel cell in a dramatic way by blowing up an eggshell. **CAUTION:** *Carry out all steps in a fume hood, and do not allow students to do this.* Make a pinhole in each end of a raw egg, carefully blow out the contents, and let the egg dry. Fill the eggshell with hydrogen gas by mixing aluminum foil and $6M$ HCl in a flask with a stopper to which tubing fitted with a Pasteur pipet is attached. When a steady stream of hydrogen gas is produced, place the tip of the pipet in one of the holes in the eggshell, and allow H_2 to fill it for about 1 minute. Place the eggshell on top of an empty 2-L soft-drink bottle in which several airholes have been punched. Ignite the hydrogen

614

Hydrogen-Oxygen Fuel Cell

Recall that the combustion of a fuel is a redox reaction in which the fuel molecules are oxidized and oxygen is reduced to form an oxide. For years, scientists have worked to find a way to separate the oxidation and reduction reactions to make them produce an electric current. The simplest fuel cell involves the oxidation of the fuel hydrogen gas to form water. Today, hydrogen-oxygen fuel cells are used to supply electricity to the space shuttle orbiters. The fuel cells have a weight advantage over storage batteries, and the water produced during their operation can be used for drinking.

1. A simple hydrogen-oxygen fuel cell differs in two major ways from a galvanic cell: the electrodes are made of an inert material that doesn't react during the process, and hydrogen and oxygen gas are fed in continuously.

2. Hydrogen is fed onto an electrode on one side of the fuel cell, and oxygen is fed onto an electrode on the other side.

3. Concentrated KOH serves as the electrolyte in the fuel cell.

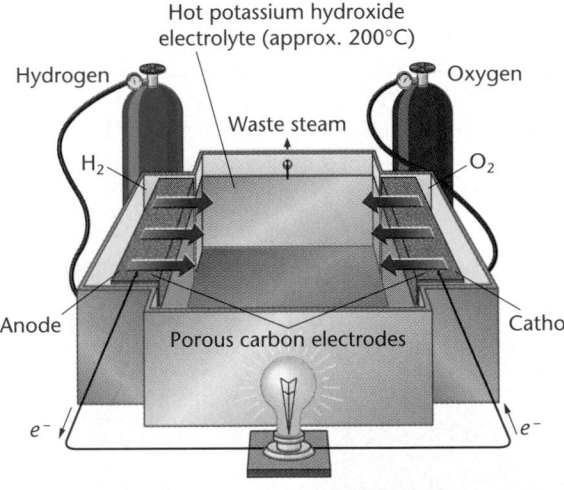

4. The electrons lost by hydrogen molecules, which are oxidized at the anode, flow out of the fuel cell, through a circuit, and then back into the fuel cell at the cathode, where oxygen is reduced.

5. Water vapor—steam—is produced in the fuel cell, as up to 75 percent of the chemical energy is converted into electricity. The steam can be condensed and used for drinking water.

$$2H_2(g) + O_2(g) \rightarrow 2H_2O(g) + energy$$

6. If more inexpensive and longer-lasting fuel cells can be developed, they may someday produce electricity in power plants.

Thinking Critically

1. What causes electrons to flow from hydrogen to oxygen in a fuel cell?

2. If fuel cells are about 75 percent efficient, what happens to the rest of the potential energy?

614   Chapter 17   Electrochemistry

seeping out of the top hole in the egg. You should hear a soft pop. Make sure the door to the fume hood is lowered, and wait for the explosion that will take place as soon as air moving into the egg through the bottom hole displaces enough hydrogen to reach the critical hydrogen-oxygen ratio.

Teaching Strategies

Ask students why fuel cells aren't used routinely to power automobiles. *Although fuel cells are an efficient, clean energy source, they are still expensive and require cylinders of hydrogen and oxygen gas, which are dangerous to transport. As their price goes down and safe ways of carrying the gaseous fuels are developed, they will probably be used more often.*

Thinking Critically

1. The potential difference that exists between hydrogen and oxygen causes electrons to flow. Oxygen has a greater tendency to attract electrons.

2. The rest of the potential energy is lost as heat.

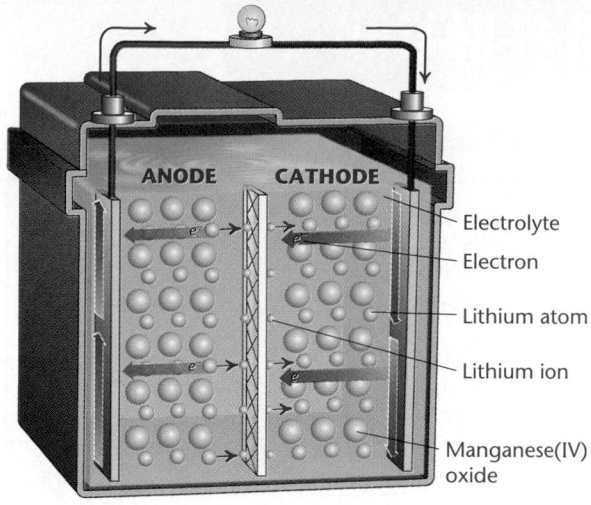

Figure 17.19
Aqueous Lithium Battery
How can a lithium battery have an aqueous electrolyte? Two facets of the construction of this new battery keep the lithium metal from reacting with water. First, the lithium is in the form of individual atoms embedded in a material such as manganese(IV) oxide, rather than as a solid metal. Second, the electrolyte is full of dissolved lithium salts, so the lithium ions that are produced travel to the site of reduction without reacting with water.

Labels on figure: ANODE, CATHODE, Electrolyte, Electron, Lithium atom, Lithium ion, Manganese(IV) oxide

to power camcorders, but they require an expensive, nonaqueous electrolyte. **Figure 17.19** shows the construction of the experimental aqueous lithium battery. This battery is less toxic and will probably be cheaper to manufacture than the nickel-cadmium batteries used in most electric cars in operation today.

Connecting Ideas

In your study of electrochemistry, you have seen how chemical reactions in batteries can be used to generate electricity. However, most of the electricity you use comes from another chemical source—the fossil fuels petroleum, natural gas, and coal. Although inorganic chemicals are usually used to fuel batteries, fossil fuels are a major source of a large group of chemicals, the carbon-containing organic compounds. In the next chapter, you'll learn that organic chemicals also provide us with most medicines, dyes, plastics, and textiles.

SECTION REVIEW

Understanding Concepts

1. Describe the movement of electrons in a galvanic cell.
2. Draw a diagram of a simple galvanic cell.
3. How are zinc-carbon and lead-acid batteries different?

Thinking Critically

4. **Using a Table** A piece of copper metal is placed in a 1M solution of silver nitrate (AgNO$_3$).

a) Use **Table 17.1** to predict which metal will be reduced and which will be oxidized.
b) Write an equation for the net redox reaction that occurs. HINT: Cu^{2+} is formed.
c) Is this system a galvanic cell? Explain.

Applying Chemistry

5. **Dry Cells** A dry cell cannot really be dry. Explain why.

SECTION REVIEW

1. The electrons move from the anode through the external circuit to the cathode, where they take part in the reduction reaction.
2. See **Figure 17.11**.
3. A zinc-carbon battery is a dry cell; is nonrechargeable; and has a small, low voltage. A lead-acid battery has a liquid acid electrolyte; is rechargeable; and has a large, higher voltage.

4. **a.** Ag$^+$ ions are reduced to Ag metal, and copper metal is oxidized to Cu^{2+} ions.
   b. 2AgNO$_3$(aq) + Cu(s) → Cu(NO$_3$)$_2$(aq) + 2Ag(s)
   c. No. Although this is a redox reaction, it is not a galvanic cell because the electrons do not flow through an external circuit.
5. If the cell were completely dry, ions would not be able to move freely through the electrolyte.

---

3 ASSESS

Check for Understanding
- Check for understanding of the section material by having students answer the Section Review questions.
- Divide students into small groups and assign each group to make a poster showing a diagram of a specific galvanic cell, such as a copper-zinc cell. Ask groups to label all parts of the cell, write all cell reactions, and show movement of ions within the cell and electrons in the external circuits. Briefly review and critique each of the cells with the class. **L1** **LEP** **COOP LEARN**

Reteach
Portfolio: Ask students to write definitions of the following terms in their portfolios: *galvanic cell, electrode, anode, cathode, salt bridge, porous barrier, electrolyte, cell potential* and *volt, current,* and *ampere.* Ask students to use all of the terms as they write a descriptive paragraph on the operation of a galvanic cell. **L1** **P**

Extension
Ask students to research and prepare posters about other types of cells or batteries, such as mercury cells that are often used in calculators, alkaline dry cells, or silver cells. **L2**

4 CLOSE

Discussion
Ask students to name electrical devices that operate on power from galvanic cells. Then discuss the characteristics—such as high voltage for a photographic flash, small size for a hearing aid, absolute reliability and long life for a heart pacemaker, and high current output for a car battery—that each type of cell should possess. **L1**

- Review the Reviewing Main Ideas statements and Key Terms with your students.
- Complete solutions to Chapter Review problems can be found in the *Problems and Solutions Manual*.

UNDERSTANDING CONCEPTS

1. The salt bridge allows the electrical circuit to be completed by ionic conduction and prevents a buildup of excess charge at the electrodes.
2. An electrolytic cell requires an external source of electricity to power a redox reaction that is not spontaneous. A galvanic cell requires no power source because the redox reaction is spontaneous.
3. A galvanic cell is a chemical system that proceeds spontaneously while producing an electric current.
4. The zinc metal in the case is oxidized to Zn^{2+} ions.
5. Energy in the form of electricity is applied to power the reaction.
6. The electrolyte completes the circuit for electric current to flow through.
7. This is a redox reaction because Ag^+ is reduced to silver metal atoms, and copper metal is oxidized to Cu^{2+} ions. No current is produced because the electrons move directly from the copper metal to the silver ions.
8. Cl_2, H_2, NaOH
9. The acid serves as an electrolyte.
10. Chlorine gas can be prepared by electrolysis of either molten sodium chloride or an aqueous solution of sodium chloride.

REVIEWING MAIN IDEAS

17.1 Electrolysis: Chemistry from Electricity

- An electrolytic cell is a chemical system that uses an electric current to drive a non-spontaneous redox reaction. Electrolysis is the process that takes place in such a cell.
- An electrical current is the flow of charged particles such as electrons.
- Reduction takes place at the cathode in an electrolytic cell.
- Oxidation takes place at the anode in an electrolytic cell.
- Electrolysis can be used to produce compounds, separate metals from ores, clean metal objects, and plate metal coatings onto objects.

17.2 Galvanic Cells: Electricity from Chemistry

- A potential difference between two substances is a measure of the tendency of electrons to flow from one to the other.

- A galvanic cell is a chemical system that produces an electric current through a spontaneous redox reaction.
- Batteries contain one or more galvanic cells.

Key Terms

For each of the following terms, write a sentence that shows your understanding of its meaning.

anion	electrolysis
anode	electrolytic cell
cathode	galvanic cell
cation	potential difference
electrical current	voltage

● UNDERSTANDING CONCEPTS

1. What is the function of the salt bridge in a galvanic cell?
2. What is the difference between an electrolytic cell and a galvanic cell?
3. What is a galvanic cell?
4. What happens to the case of a carbon-zinc dry cell as the cell is used to produce an electric current?
5. How can you make a non-spontaneous redox reaction take place in a cell?
6. Why is the electrolyte necessary in both galvanic and electrolytic cells?
7. If a strip of copper metal is placed in a solution of silver nitrate, will a redox reaction take place? Will a current be produced? Explain.

8. What products are formed from the electrolysis of an aqueous solution of rock salt?
9. What is the function of the acid in the lead-acid storage battery used in cars?
10. By what process can chlorine gas be prepared commercially?

● APPLYING CONCEPTS

11. What will happen if a rod made of aluminum is used to stir a solution of iron(II) nitrate?
12. Can a solution of copper(II) sulfate be stored in a container made of nickel metal? Explain.
13. How would gold-electroplated jewelry compare to jewelry made of solid gold in terms of price, appearance, and durability?
14. Tests conducted on different types of common commercial batteries involved measuring the

✓ Assessment

Portfolio: Review the portfolio options that are provided throughout the chapter. Encourage students to select one product that demonstrates their best work for the chapter. Have students explain what they have learned and why they chose this example for placement into their portfolios. **P**

Additional portfolio options can be found in the following **Teacher Classroom Resources:**
Chemistry and Industry
Tech Prep Applications
Consumer Chemistry
Applying Scientific Methods in Chemistry

voltage drop over time during simulated non-stop use of a motorized toy. Based on the following graph of data obtained from testing rechargeable, alkaline, and heavy-duty batteries, which battery type would be best to use if you wanted to run the toy for a long period of time? Which battery type goes dead abruptly?

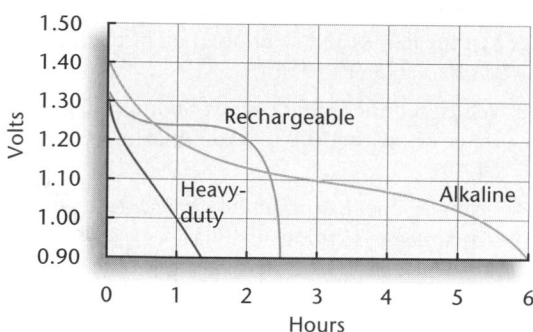

15. What is the purpose of the lemon used in making a lemon battery? If you didn't have any lemons available for constructing your battery during that NBA final, what else could you have used?

16. What would you expect to see if you placed

    a) a strip of copper metal into a solution of zinc sulfate?

    b) a strip of zinc metal into a solution of copper sulfate?

17. To plate a steel pendant with gold, the pendant can be placed in an electrolytic cell that has a gold electrode and an electrolyte solution containing gold ions. The pendant functions as the second electrode. A battery may function as the power source. Is the pendant the anode or cathode?

18. How could lead be removed from drinking water by electrolysis?

Everyday Chemistry

19. The information on a CD stamper master is the reversal of that on the original glass master cut by the recording laser. Explain why this reversal is necessary.

Chemistry and Technology

20. Should the anode or the cathode be made of pure copper in an electrolytic cell designed for refining copper metal? Explain.

Health Connection

21. What happens to lithium metal if it comes into contact with water?

How it Works

22. What are the advantages and disadvantages of using rechargeable batteries instead of conventional types?

How it Works

23. Write the equations for the two half-reactions that take place in a hydrogen-oxygen fuel cell.

● THINKING CRITICALLY

Making Predictions

24. **MiniLab 1** Would it be possible to plate a silver spoon or a gold spoon with copper?

18. An electrolytic cell can be set up using a lead cathode and an anode made from a metal more easily oxidized than lead, such as iron.

19. The stamping master must be the reversal of the original data so that the stamped CD will be identical to the original.

20. The cathode should be pure copper because that is where more copper will plate out as ions from the ore are reduced. The anode should be blister copper.

21. It explodes violently in a redox reaction in which water acts as an oxidizing agent.

22. Advantages: can be used many times, lower cost in the long run, cuts down on battery waste in landfills
Disadvantages: higher initial cost, must be charged initially, don't last as long as conventional batteries between recharges, cadmium waste is toxic

23. $2H_2 \rightarrow 4H^+ + 4e^-$
$O_2 + 4e^- \rightarrow 2O^{2-}$

APPLYING CONCEPTS

11. The aluminum metal would be oxidized to aluminum ions as the iron ions are reduced to iron metal. As a result, the rod would slowly dissolve and solid iron would precipitate out.

12. No. Because nickel is more easily oxidized than copper, the nickel in the container would be oxidized to nickel ions as the copper ions are reduced to copper metal.

13. Electroplated jewelry would be cheaper because an inexpensive metal could be plated with just a thin layer of gold. It would look almost identical to jewelry made from solid gold unless some of the plating is uneven or is scratched off. It wouldn't be as durable as pure gold because the plating could be scratched off fairly easily, and the cheap metal underneath could begin to corrode.

14. The alkaline battery lasts the longest, so it would work best for running a toy for a long period of time. The rechargeable battery goes dead abruptly.

15. The juice in the lemon is an electrolyte because it contains many different compounds dissolved in water, some of them ionic. Although lemons are a rich source of electrolytes, many others could be used—an orange or lime, vinegar, soft drinks, or a potato.

16. **a.** no reaction because copper is not as easily oxidized as zinc
b. The zinc metal would be oxidized to Zn^{2+} ions as the copper ions are reduced to copper metal. The color of the solution would change from blue to colorless or light blue.

17. the cathode

THINKING CRITICALLY

24. Yes. Although copper won't plate spontaneously onto gold or silver, electricity can be supplied to make this reaction go. The gold or silver spoon would be the cathode in the electrolytic cell, and the copper would be the anode. Copper ions would be dissolved in the electrolytic solution.

25. Cleaning them removes corrosive buildup and exposes the metal so it can react.

26. a. The potential difference, or voltage reading, would be lower because Zn has a lower oxidation potential than Mg.
b. The potental difference, or voltage reading, would be greater because Ag has a smaller oxidation potential than Cu.

27. Answers will vary. Some factors to be considered are cost, size, weight, useful life, and voltage.

28. Possible answers include K, Ca, and Na.

29. These artifacts are nonconductors and are not suitable to be used as the cathode in electrolytic cleaning.

CUMULATIVE REVIEW

30. a. $[Ca]^{2+}$ **b.** $[:\overset{..}{\underset{..}{Cl}}:]^-$

c. $[:\overset{..}{\underset{..}{O}}:H]^-$ **d.** $[:\overset{..}{\underset{..}{O}}:]^{2-}$

31. a. MnO_2 **b.** KI **c.** $CuSO_4$
d. $AlCl_3$ **e.** H_2SO_4 **f.** Fe_2O_3

32. helium, He; neon, Ne; argon, Ar; krypton, Kr; xenon, Xe; radon, Rn

33. The heat of vaporization of water (2260 J/g) is much greater than the heat of fusion (334 J/g), so boiling the water requires more energy than melting the ice. Boiling involves separating the water molecules, which requires the breaking of the hydrogen bonds between molecules.

Drawing Conclusions

25. MiniLab 2 Why should the metal pieces used as electrodes in the lemon battery be cleaned with steel wool?

Relating Cause and Effect

26. ChemLab How would your result in step 7 of the ChemLab have been different if

a) a piece of zinc were used instead of the piece of magnesium?
b) silver were used instead of copper?

Making Decisions

27. What factors must be considered in designing or selecting batteries for the following applications?

a) flashlight   **c)** pacemaker
b) hearing aid   **d)** toy car

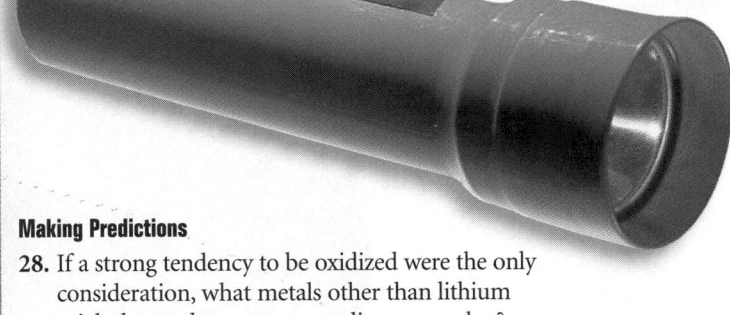

Making Predictions

28. If a strong tendency to be oxidized were the only consideration, what metals other than lithium might be used to power a cardiac pacemaker?

Forming a Hypothesis

29. Why was electrophoresis rather than electrolysis used to restore ceramic and organic artifacts from the *Titanic*?

● CUMULATIVE REVIEW

30. Draw Lewis electron dot structures for the ions listed. (Chapter 2)

a) Ca^{2+} **b)** Cl^- **c)** OH^- **d)** O^{2-}

31. Give the chemical formulas for the following compounds. (Chapter 5)

a) manganese(IV) dioxide
b) potassium iodide
c) copper(II) sulfate
d) aluminum chloride
e) sulfuric acid
f) iron(III) oxide

32. List the names and symbols of all of the noble gases. (Chapter 8)

33. Which requires more energy: boiling 100 g of water or melting 100 g of ice? Explain. (Chapter 10)

34. Compare the hydronium ion concentrations in two aqueous solutions that have pH values of 9 and 11. (Chapter 14)

35. You test several solutions and find that they have the following pH values: 8.6, 4.7, 10.4, 13.1, 2.6, 6.1. Determine the pH of each of the following. (Chapter 14)

a) the strongest acid
b) the strongest base
c) the weakest acid
d) the solution that is closest to being neutral

36. Which of the changes indicated are oxidations and which are reductions? (Chapter 16)

a) Cu^+ becomes Cu^{2+}
b) S becomes S^{2-}
c) Na becomes Na^+
d) Mn^{4+} becomes Mn^{2+}

37. Green plants make glucose and oxygen in photosynthesis, a redox process shown below. (Chapter 16)

$$6CO_2 + 6H_2O + energy \rightarrow C_6H_{12}O_6 + 6O_2$$

a) Is the carbon in carbon dioxide reduced or oxidized in this process?
b) What kind of energy do you think is used in this process?

34. 11 − 9 = 2 pH units difference. The less basic solution (pH 9) has a hydronium ion concentration 100 times greater than the more basic solution (pH 11).

35. a. 2.6 **b.** 13.1 **c.** 6.1 **d.** 6.1

36. a. oxidation
b. reduction
c. oxidation
d. reduction

37. a. reduced
b. light energy

Skill Review

1. **Interpreting Scientific Illustrations** A process called cathodic protection is sometimes used to protect a buried steel pipeline from corrosion. In this process, the pipeline is connected to a more active metal such as magnesium, which is corroded preferentially before the iron. The diagram below illustrates how the two metals are connected and shows the reactions that take place.

   a) What acts as the cathode in this process? What acts as the anode?

   b) What is the oxidizing agent?

   c) Write a short summary describing how the magnesium is preferentially corroded.

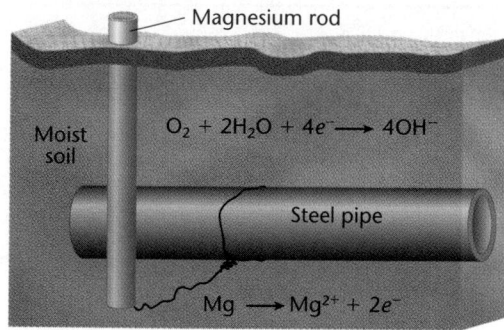

Magnesium rod

Moist soil

$O_2 + 2H_2O + 4e^- \longrightarrow 4OH^-$

Steel pipe

$Mg \longrightarrow Mg^{2+} + 2e^-$

2. **Making Scientific Illustrations** Draw a diagram of a galvanic cell in which the reaction is $Ni(s) + 2Ag^+(aq) \rightarrow Ni^{2+}(aq) + 2Ag(s)$. Label the cathode and the anode. Show the ions present in both compartments, and indicate the direction of electron flow in the external circuit.

Writing in Chemistry

3. Write an article about the development and uses for the Daniell cell, an early battery made in 1836 by John Frederic Daniell of Great Britain. Find out how it improved on the Volta cell and whether or not this type of battery is used much today.

4. Read "Leaf Jewelry" by Rachel Wilt, *ChemMatters* 5, no. 4 (December 1987): pp.14-15. Write a report on how biological materials such as leaves can be made into jewelry.

Problem Solving

5. When a test tube containing copper(II) oxide and powdered charcoal (carbon) is heated, a gas and copper metal are produced in a redox reaction. If the gas is bubbled into a test tube containing limewater (aqueous calcium hydroxide solution), a milky white precipitate forms. Write a lab report of this experiment as if you had performed it in the lab. Answer the following questions in your report.

   a) What is the gas that forms?

   b) Write an equation for the redox reaction that takes place.

   c) What is the identity of the white precipitate?

Projects

6. Conduct your own tests to compare different types and brands of batteries. Test them for lifetime, durability, and performance in both hot and cold weather conditions. Carefully log your procedures and collected data in your chemistry journal. Write a report presenting the results and using graphs to summarize your conclusions. Compare your results with other published studies and the advertising claims made by battery companies. Consider including the project in your portfolio.

AUTHENTIC ASSESSMENT

1. **a.** Magnesium is the anode and the steel pipe is the cathode.

   b. oxygen gas

   c. Magnesium is the more active metal and is more easily oxidized, so it is corroded preferentially.

5. **a.** CO_2

   b. $2CuO(s) + C(s) \rightarrow 2Cu(s) + CO_2(g)$

   c. $CaCO_3$, formed when limewater reacts with CO_2

Program Resources

Chapter Review and Assessment,
   pp. 101-106 [L1]
Alternate Assessment in the Science Classroom [L1]
Computer Test Bank, Chapter 17 [L1]
Problems and Solutions Manual,
   Chapter 17 [L1]

Supplemental Practice Problems,
   Chapter 17 [L1]
Performance Assessment in the Science Classroom [L1]

Chapter 18 Organizer

Section	Objectives	Activities/Features
18.1 **Hydrocarbons** (5 days)	1. **Write** and **interpret** structural formulas of linear, branched, and cyclic alkanes, alkenes, and alkynes. 2. **Distinguish** among isomers of a given hydrocarbon. 3. **Infer** the relationship between fossil fuels and organic chemicals.	**MiniLab 1:** p. 630 **Biology Connection:** p. 632 **People in Chemistry:** pp. 634-635 **Discovery Demo:** 18.1 pp. 620-621 **Demonstrations:** 18.2 pp. 624-625; 18.3 pp. 628-629; 18.4 pp. 636-637
18.2 **Substituted Hydrocarbons** (2 days)	4. **Compare and contrast** the structures of the major classes of substituted hydrocarbons. 5. **Summarize** properties and uses of each class of substituted hydrocarbons.	**MiniLab 2:** A Synthetic Aroma, p. 646
18.3 **Plastics and Other Polymers** (3 days)	6. **Identify** monomers that form specific polymers; draw structural formulas for polymers made from given monomers. 7. **Differentiate** between condensation and addition polymerization reactions. 8. **Summarize** the relationship between structure and properties of polymers.	**ChemLab:** pp. 650-652 **MiniLab 3:** p. 653 **Everyday Chemistry:** p. 657 **Chemistry and Society:** p. 659 **Demonstrations:** 18.5 pp. 648-649; 18.6 pp. 654-655; 18.7 pp. 656-657

Activity Materials

ChemLab (pages 650-652)
fabric samples (7), Bunsen burner, beaker, forceps, test-tube holder, test-tube rack, test tubes (4), balance, watch glass, red litmus paper, stirring rod, 100-mL beakers (2), 10-mL pipet, 25-mL cylinder, $Ca(OH)_2$, 3M $BaCl_2$, conc. H_2SO_4, iodine solution, 0.05M $CuSO_4$, 3M NaOH, acetone

MiniLab 1 (page 630)
peanut oil, canola oil, 2 small flasks, dropper, tincture of iodine, stirring rod, hot plate

MiniLab 2 (page 646) 250- or 450-mL beaker, hot plate, methyl alcohol, salicylic acid, large test tube, stirring rod, dropper, concentrated sulfuric acid, small beaker, distilled water, watch glass

MiniLab 3 (page 653) disposable diapers (2 different brands), 100-mL graduated cylinder

KEY TO TEACHING STRATEGIES

The following designations will help you decide which activities are appropriate for your students.

L1 Level 1 activities should be within the ability range of all students.

L2 Level 2 activities should be within the ability range of the average to above-average student.

L3 Level 3 activities are designed for the ability range of above-average students.

LEP LEP activities should be within the ability range of Limited English Proficiency students.

COOP LEARN Cooperative Learning activities are designed for small group work.

P These strategies represent student products that can be placed into a best-work portfolio.

Demonstrations For a complete list of materials for the demonstrations in this chapter, see pages 624, 628, 636, 648, 654, and 656.

Organic Chemistry

Teacher Classroom Resources

STUDENT MASTERS

Study Guide: pp. 71-72
Laboratory Manual: 18.1 Examples of Organic Reactions, pp. 167-172
Critical Thinking/Problem Solving: p. 20
ChemLab and MiniLab Worksheets: p. 118

TEACHING AIDS

Section Focus Transparency 36
Section Focus Transparency Master: p. 36
Basic Concepts Transparency 49, 50
Basic Concepts Transparency Masters: pp. 97-100
Lesson Plans: p. 36

Study Guide: pp. 73-74
Laboratory Manual: 18.2 Analysis of Aspirin, pp. 173-176
ChemLab and MiniLab Worksheets: p. 119

Section Focus Transparency 37
Section Focus Transparency Master: p. 37
Problem Solving Transparency 21
Problem Solving Transparency Master: pp. 41-42
Lesson Plans: p. 37

Study Guide: pp. 75-76
ChemLab and MiniLab Worksheets: pp. 113-116, 120

Section Focus Transparency 38
Section Focus Transparency Master: p. 38
Basic Concepts Transparency 51
Basic Concepts Transparency Master: pp.101-102
Lesson Plans: p. 38

ASSESSMENT RESOURCES

Chapter Review and Assessment: pp. 107-112
Alternate Assessment in the Science Classroom
Computer Test Bank: Chapter 18
Performance Assessment in the Science Classroom
Problems and Solutions Manual: Chapter 18
Supplemental Practice Problems: Chapter 18

CHAPTER RESOURCES

Applying Scientific Methods in Chemistry: pp. 44-46
Spanish Resources: Chapter 18
English/Spanish Audiocassettes: Chapter 18
Cooperative Learning in the Science Classroom
Lab and Safety Skills in the Science Classroom

GLENCOE TECHNOLOGY

Videotape
MindJogger Videoquizzes, Chapter 18
Videodisc
Chemistry: Concepts and Applications, Demonstrations, Videos, and Animations
Science and Technology Videodisc Series: Disc 2 Chemistry
CD-ROM Multimedia System
Chemistry: Concepts and Applications, Same as for Videodisc, plus Interactive Exploration and Major Simulation, *The Periodic Table*

TECH PREP

The following Glencoe resources provide opportunities for integrating science and technology.

Student Edition: MiniLab 1, p. 630; Biology Connection, p. 632; MiniLab 3, p. 653; Everyday Chemistry, p. 657; Chemistry and Society, p. 659

Teacher Wraparound Edition: Applying Chemistry, p. 633, 643; Chemistry Journal, p. 645; Check for Understanding, p. 647; Extension, p. 661

Teacher Classroom Resources: Consumer Chemistry, pp. 35-36; Chemistry and Industry, pp. 41-52; Tech Prep Applications, pp. 29-30

CHAPTER

18 Organic Chemistry

Chapter Overview

Students are introduced to principles and applications of organic chemistry. The chapter begins with hydrocarbons, covering their structure, nomenclature, properties, sources, and uses. Next, the topic turns to the major organic functional groups. Polymers are the focus of the last section.

Theme Connection

Macro to Submicro The theme of this chapter is how the *submicroscopic* structure of organic compounds affects their *macroscopic* chemical and physical properties. How structure determines properties and, hence, function is shown for the homologous series of hydrocarbons, as well as for several polymer examples.

✓ Assessment Planner

Choose assessment strategies from the following pages to evaluate the progress of your students.

Assess, pp. 638, 641, 647, 660
MiniLabs, pp. 630, 646, 653
ChemLab, p. 650
Portfolio, pp. 625, 641, 646, 653, 660, 662
Chapter Review, p. 662

GLENCOE TECHNOLOGY

Videodisc

Chemistry: Concepts and Applications
Modeling Hydrocarbons
Disc 2, Side 2, Ch. 10

Show this video of the Discovery Demo for this chapter to introduce the concept of molecular modeling. **L1** **LEP**

Discovery Demo 18.1

Modeling Hydrocarbons

Purpose: To reinforce visually the fact that a member of a homologous series differs from each other member by a specific structural unit.

Materials: modeling kit or fresh gumdrops and toothpicks

Safety Precautions: 🥽

Disposal: Code F

Procedure: Prepare models of methane, CH_4;

ethane, C_2H_6; and propane, C_3H_8. Arrange the bonds tetrahedrally. Ask students how many bonds each carbon atom forms. Note that because there are no double or triple bonds, the compound is said to be saturated. As students view the completed three-dimensional models, ask them how each larger molecule differs from the previous smaller one.

Results: Students gain a better understanding of the structure of alkanes.

CHAPTER 18

Organic Chemistry

"Sodium lauryl ether sulfate, d-limonene, cocamide DEA, isopropyl alcohol, panthenol, diazolidinyl urea, methylchloroisothiazolinone, FD & C yellow No. 5 . . ." Those are just a few of the ingredients in one brand of dog shampoo. Do many of the chemical names look familiar? You may have trouble pronouncing some of them, and even someone who has taken several chemistry courses wouldn't be able to draw structures for all of those names. What kinds of chemicals are these?

The dog-shampoo ingredients listed are all organic compounds, which are compounds that contain carbon. About 10 million different organic compounds are now known—far more than all known inorganic compounds. About 10 000 new organic compounds are discovered or synthesized each year. There must be a way to name and organize such a large number of compounds and to study their properties and reactions. In this chapter, you will examine the properties of carbon that set it apart from all other elements, and you will be introduced to some of the many organic compounds that are used in everyday life.

Chemistry Around You

Organic compounds are found in many products you use every day, such as clothing, gasoline, oil, paints, foods, plastics, medicines, and shampoos. The long molecule on the opposite page is a detergent—sodium lauryl sulfate. It consists of a long chain of carbon and hydrogen atoms. An oxygen atom connects the carbon chain to a sulfate ion, which is ionically bonded to a sodium ion. You may recognize the name of this compound as an ingredient in many shampoos and see it listed in the ingredients label shown here. Look around. The world is full of organic substances.

Concept Check

Review the following concepts before studying this chapter.
Chapter 4: bonding in covalent compounds
Chapter 5: naming of hydrocarbons
Chapter 6: combustion of hydrocarbons
Chapter 9: bonding in hydrocarbons

Chapter Preview

18.1 Hydrocarbons

18.2 Substituted Hydrocarbons

18.3 Plastics and Other Polymers

Laboratory Activities
ChemLab
Identification of Textile Polymers

MiniLabs
1. How unsaturated is your oil?
2. A Synthetic Aroma
3. When Polymers Meet Water

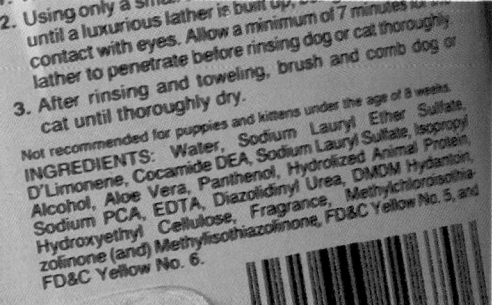

2. Using only a small [amount] until a luxurious lather is built up, [...] contact with eyes. Allow a minimum of 7 minutes for [the] lather to penetrate before rinsing dog or cat thoroughly.
3. After rinsing and toweling, brush and comb dog or cat until thoroughly dry.

Not recommended for puppies and kittens under the age of 8 weeks.
INGREDIENTS: Water, Sodium Lauryl Ether Sulfate, D'Limonene, Cocamide DEA, Sodium Lauryl Sulfate, Isopropyl Alcohol, Aloe Vera, Panthenol, Hydrolized Animal Protein, Sodium PCA, EDTA, Diazolidinyl Urea, DMDM Hydantion, Hydroxyethyl Cellulose, Fragrance, Methylchloroisothia-zolinone (and) Methylisothiazolinone, FD&C Yellow No. 5, and FD&C Yellow No. 6.

621

Analysis: Ask these questions.
1. Do the three models represent saturated hydrocarbons? *yes*
2. What family of hydrocarbons is represented by the models? *alkane family*
3. What makes each member of the family different in this homologous series? *the structural unit —CH₂—*

GLENCOE TECHNOLOGY

CD-Rom

Chemistry: Concepts and Applications
Modeling Hydrocarbons
Have students use this video of the Discovery Demo to help reinforce the process of molecular modeling. **L1** **LEP**

Introducing the Chapter
Ask students to name some "natural" and "artificial" objects or materials in the classroom. Choose one or two of the artificial objects named, and ask what they are made from. Explain that most organic chemicals used to make synthetic or artificial materials are derived from fossil materials such as petroleum. Because petroleum was made from the remains of living organisms, the line between natural and artificial substances is not clear.

Using the Illustrations Point out some of the long names of compounds listed on the dog shampoo bottle, and have students find others on products brought to class. Explain to students that not even experienced chemists can draw or name all compounds. More important to chemists than memorizing long names is understanding how structure and properties are related. If chemists can make predictions about the properties of a compound based on structure, they will be able to predict its applications.

Revealing Misconceptions
Because unbranched hydrocarbons are often referred to as being straight-chain, students may begin picturing that as their actual structure. Chains of saturated carbon atoms actually form a zigzag structure due to the 109.5° angles of the sp^3 bonds that link them. Having students make models of structures using organic model kits or gumdrops and toothpicks can help them better visualize what the shape of these compounds looks like. **L1** **LEP**

SECTION 18.1

PREPARE

Key Concepts

The section opens with a brief description of the bonding capabilities of carbon atoms, which are subsequently related to the great number of organic compounds. The structure, naming, and properties of each of the various classes of hydrocarbons—alkanes, alkenes, alkynes, and aromatics—are then presented.

Planning Ahead

For the Demonstrations Order calcium carbide, sometimes called miners' lamp, for Demonstration 18.4.
For the MiniLab Purchase peanut oil and canola oil.

1 FOCUS

Focus Transparency

Display the **Section Focus Transparency** for Section 18.1 to introduce uses of hydrocarbons. L1 LEP

2 TEACH

Discussion

Begin the study of organic compounds by asking the following questions.
1. What are the names of some commonly used organic fuels? *Butane is used in handheld lighters; propane in gas grills, camping stoves, and liquid-propane (LP) tanks; and methane in natural gas used for cooking, heating, and in lab Bunsen burners.*
2. Why is there an odor when a lab gas spigot or gas stove is turned on if methane is odorless? *Small amounts of smelly compounds are added to the natural gas so that leaks in gas lines can be detected.*

622

Hydrocarbons

Have you ever seen a spectrum of colors on the top of an otherwise-drab puddle of water? What causes those bright colors to appear? They form as a result of pollution, namely small amounts of gasoline or oil that have leaked out of automobiles and formed a miniature oil spill on the puddle.

Gasoline is composed mostly of hydrocarbons, the organic compounds with the names and structures you studied in Chapters 5 and 9. Hydrocarbons have properties that are different from those of water. They are insoluble in water, which is why they form a distinct layer on a puddle of water. Hydrocarbons are less dense than water; that's why the layer floats on top of the puddle. What causes the colorful effect? The spill forms an extremely thin layer of hydrocarbon molecules on the water, which reflects sunlight. In this section, you will learn more about the structures and names of hydrocarbons, as well as the sources of these useful compounds.

SECTION PREVIEW

Objectives
Write and interpret structural formulas of linear, branched, and cyclic alkanes, alkenes, and alkynes.
Distinguish among isomers of a given hydrocarbon.
Infer the relationship between fossil fuels and organic chemicals.

Key Terms
saturated hydrocarbon
alkane
isomer
unsaturated hydrocarbon
alkene
alkyne
aromatic hydrocarbon
fractional distillation
cracking
reforming

Millions and Millions of Organic Compounds

Carbon is unique among elements in that it can bond to other carbon atoms to form chains containing as many as several thousand atoms. Because a carbon atom can bond to as many as four other atoms at once, these chains can have branches and form closed-ring structures that make possible an almost endless variety of compounds. In addition, carbon can bond strongly to elements such as oxygen and nitrogen, and it can form double and triple bonds. Thus, carbon forms an enormous number of compounds with chains and rings of various sizes, each with a variety of bond types and atoms of other elements bonded to them. Fortunately, you don't need to study each of these millions of compounds to understand organic chemistry because they can be classified into groups of compounds that have similar structures and properties.

622   Chapter 18   Organic Chemistry

Program Resources

Study Guide, pp. 71-72 L1
Critical Thinking/Problem Solving, p. 20 L2
Laboratory Manual, 18.1, pp. 167-172 L1
English/Spanish Audiocassettes,
    Chapter 18 L1 LEP
Section Focus Transparency 36 and **Master,**
    p. 36 L1 LEP
Basic Concepts Transparencies 49, 50 and
    Masters, pp. 97-100 L1 LEP

Spanish Resources L1 LEP
ChemLab and MiniLab Worksheets,
    p. 118 L1
Applying Scientific Methods in Chemistry,
    p. 44 L1
Chemistry and Industry, pp. 45-48 L1
Tech Prep Applications, pp. 29-30 L1

Saturated Hydrocarbons

Gasoline is a mixture of organic compounds that is derived from petroleum. Most of the compounds in gasoline are hydrocarbons. You will recall learning that hydrocarbons are organic compounds containing only hydrogen and carbon atoms. A hydrocarbon in which all the carbon atoms are connected to each other by single bonds is called a **saturated hydrocarbon.** Another name for a saturated hydrocarbon is an **alkane.** Although burning alkanes for fuel is their most common use, they are also used as solvents in paint removers, glues, and other products.

Alkanes

Alkanes are the simplest hydrocarbons. The carbons in an alkane can be arranged in a chain or a ring, and both chains and rings can have branches of other carbon chains attached to them. Alkanes that have no branches are called straight-chain alkanes. Methane, CH_4; ethane, C_2H_6; propane, C_3H_8; and butane, C_4H_{10} are all common fuels. Their structural formulas show that each differs from the next by an increment of $—CH_2—$.

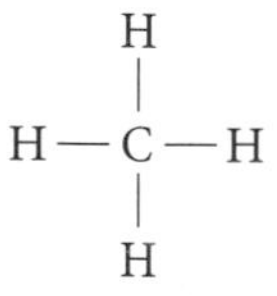

Methane 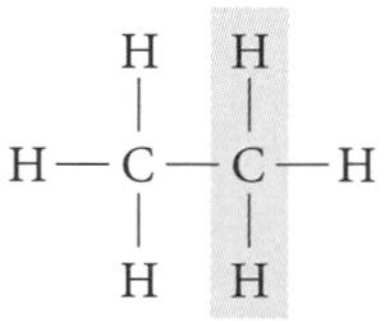   Ethane 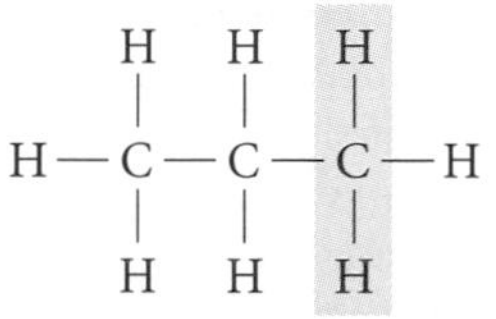   Propane 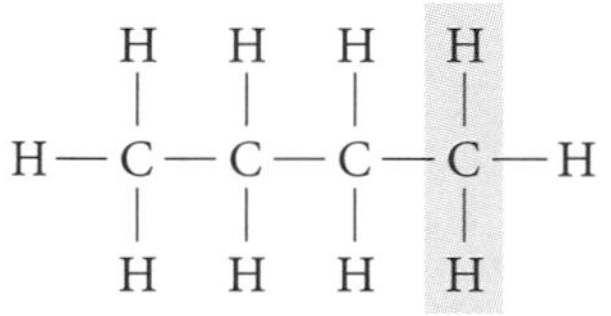   Butane

Some alkanes have a branched structure. In these compounds, a chain of one or more carbons is attached to a carbon in the longest continuous chain, which is called the parent chain. If a chain containing a single carbon is branched off the second carbon of a propane parent chain, a branched alkane with the following structure results.

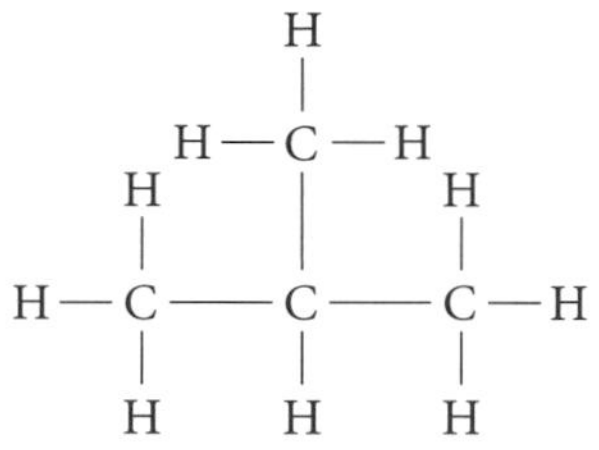

2-methylpropane

The carbon atoms in alkanes can also link up to form closed rings. The most common rings contain five or six carbons. The structures of these compounds can be drawn showing all carbon and hydrogen atoms.

Cyclopentane   Cyclohexane

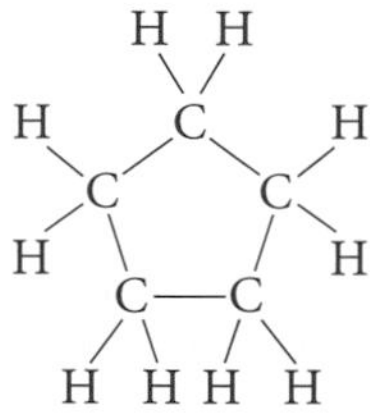

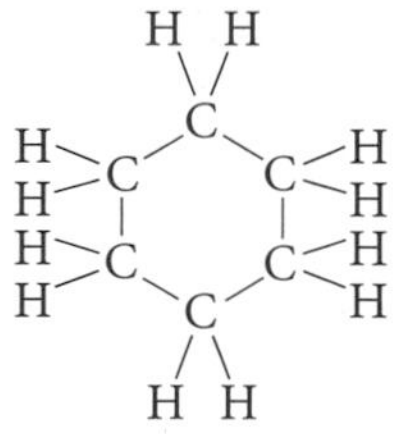

18.1 Hydrocarbons   623

Integrating THE Sciences

Ecology Cyclopropane rings are not found often in naturally occurring compounds. One reason cyclopropane rings are rare is that they are not stable due to high angle strain. The saturated carbon atoms in them are connected by angles of 120° rather than the ideal 109.5° geometry. Thus, they react readily to form more stable compounds. An exception is the group of cyclopropane-containing compounds known as pyrethrins that are found in chrysanthemums. They act as potent natural insecticides that protect the leaves from insect predators. Scientists are studying natural insecticides such as these in the hope that they can be safely used with other types of plants, and that this will cut down on the use of some of the more toxic synthetic insecticides currently marketed.

Quick Demo

Why so many organic compounds?

Hold up a chemistry handbook and show students the great disparity in the sizes of the sections about physical constants of inorganic compounds (about 80 pages) and physical constants of organic compounds (about 500 pages). Ask students to speculate as to why a single element, carbon, is able to form so many compounds compared with all the other chemical elements. *Carbon atoms are capable of bonding in several different ways, with atoms of many other elements, and with other carbon atoms.* L1

Tying to Previous Knowledge

Remind students that carbon atoms have four valence electrons. Demonstrate the bonding versatility of carbon by drawing carbon atom electron dot structures and having students use molecular models showing the common bonding patterns: four single bonds, two single bonds and one double bond, one single bond and one triple bond, and two double bonds. L1 LEP

Concept Development

Due to the scope of organic chemistry, you should consider telling students at the outset that they may use all their notes on tests and quizzes. This will encourage students to take careful and complete notes, ask questions when required, and focus their efforts on understanding rather than attempting to memorize great amounts of material.

Figure 18.1: The high compression ratio in modern internal combustion engines makes the engines efficient but can also lead to disruptions in the normally smooth explosions of the fuel-air mixture in the engine's cylinders. This is called knocking, and it leads to a reduction of power in the engine. Different hydrocarbons vary in knocking tendency, and the gasoline that is obtained by distillation of petroleum can be improved by adding compounds that have strong antiknock properties. An octane rating system that indicates the relative antiknock tendency of a hydrocarbon was developed based on a scale of zero for heptane, which has a strong tendency to knock, to 100 for 2,2,4-trimethylpentane (also called isooctane), which has only a small tendency to knock. Today, there are compounds with even stronger antiknock tendencies that have been given octane numbers greater than 100. Have interested students investigate some of these compounds and report to the class. **L3**

Cyclopentane

Cyclohexane

Figure 18.1

Components of Gasoline and Octane Rating
Gasolines are rated on a scale known as octane rating, which is based on the way they burn in an engine. The higher the octane rating, the greater the percentage of complex-structured hydrocarbons that are present in the mixture, the more uniformly the gasoline burns, and the less knocking there is in the automobile engine. Thus, a gasoline rated 92 octane will burn more smoothly than one rated 87 octane.

Structural diagrams can be simplified by using straight lines to represent the bonds between atoms in the rings. In these ring diagrams, each corner represents a carbon atom. Because carbon usually forms four bonds, it is understood that enough hydrogen atoms are bonded to each carbon to give it four bonds. Rings containing five and six carbons are drawn as a pentagon and a hexagon, respectively.

Structural diagrams of straight- and branched-chain hydrocarbons also can be written in a simplified way by leaving out some of the bonds. For example, the formula for propane can be written as CH_3—CH_2—CH_3. In this type of shorthand structure, the bonds between C and H are understood. In an even more simplified type of shorthand, the condensed structural formula for propane can be written as $CH_3CH_2CH_3$. Here, the bonds both between C and C and between C and H are understood. These condensed structural formulas will be used throughout this chapter.

Figure 18.1 shows the names and structures of some saturated hydrocarbons found in typical gasoline mixtures. The properties of a gasoline mixture—such as how well it burns in an engine—are determined by the relative amounts of various components present in the mixture. Hydrocarbons that contain rings and many branches burn at a more uniform rate than do straight-chain alkanes, which tend to explode prematurely, causing engine knock. The octane number on a gasoline pump is an index that indicates the relative amounts of branched and ring structures in that blend of gasoline.

Table 18.1	The First Ten Alkanes
Formula	**Name**
CH_4	methane
C_2H_6	ethane
C_3H_8	propane
C_4H_{10}	butane
C_5H_{12}	pentane
C_6H_{14}	hexane
C_7H_{16}	heptane
C_8H_{18}	octane
C_9H_{20}	nonane
$C_{10}H_{22}$	decane

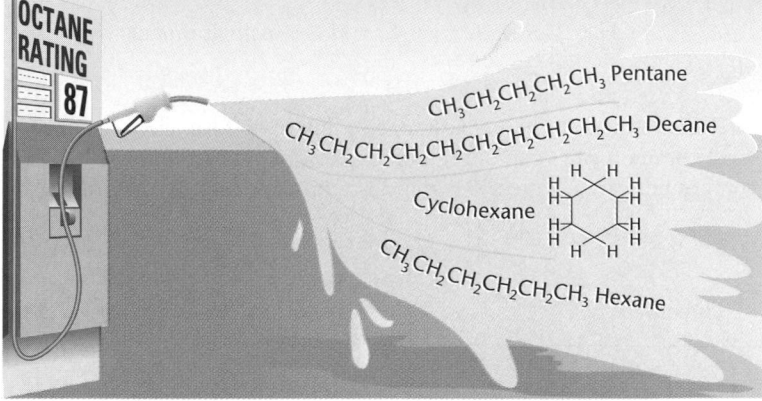

Demonstration 18.2

Octane's 18 Isomers
Purpose: To show the large number of isomers octane has, due to the ability of the carbon atoms to form long chains.
Materials: 144 fresh gumdrops, 1 box of toothpicks
Safety Precautions:
Disposal: Code F
Procedure: Copy the structural diagrams for each of the 18 isomers of octane, shown here, onto a piece of paper, and cut them apart. Give each of 18 students eight gumdrops, seven toothpicks, and one of the diagrams. Have each student put the gumdrops together according to his or her diagram. Have students place the isomers they constructed on the demonstration table and make diagrams of all 18 models. **CAUTION:** *Do not allow any of the gumdrops to be eaten.* **L1** **LEP**

Results: Students are quickly able to assemble the 18 isomers of octane, a hydrocarbon in gasoline.
Analysis: Ask these questions.
1. How many of octane's isomers have a parent chain that consists of eight carbon atoms? *one*
2. How many of octane's isomers have a parent chain that consists of four carbon atoms? *one*

Naming Alkanes

The names of the first ten straight-chain alkanes, shown in **Table 18.1**, are used as the basis for naming most organic compounds. To name a branched alkane, you must be able to answer three questions about its structure.

1. How many carbons are in the longest continuous chain of the molecule?

2. How many branches are on the longest chain and what is their size?

3. To which carbons in the longest chain are the branches attached?

For convenience, the carbon atoms in organic compounds are given position numbers. In straight-chain hydrocarbons, the numbering can begin at either end. It makes no difference. In branched hydrocarbons, the numbering begins at the end closest to the branch.

Examine the structure of this branched alkane.

$$\overset{\textstyle CH_3}{\underset{\textstyle }{|}}$$
$$^1CH_3\overset{2}{C}H\overset{3}{C}H_2{}^4CH_3$$

Four carbons are in the longest continuous chain, so butane is the parent chain and will be part of the compound's name. There is only one branch, and it contains one carbon. Instead of calling this a methane branch, change the *-ane* in methane to *-yl*. Thus, this is a methyl branch. Because the methyl branch is attached to the second carbon of the butane chain, this compound has the name 2-methylbutane.

Now, examine the structure of a different hydrocarbon.

$$\overset{\textstyle CH_3}{\underset{\textstyle }{|}}$$
$$^1CH_3\overset{2}{C}CH_3$$
$$\underset{\textstyle CH_3}{\overset{\textstyle }{|}}$$

Propane will be part of this compound's name because the longest continuous chain has three carbons. Two methyl branches are present, both on the second carbon. To indicate the presence of more than one branch of the same kind, use the same Greek prefixes presented in Chapter 5 for naming hydrates and molecules. The prefix to use when two of anything are present is *di-*. Thus, the name of this compound is 2,2-dimethylpropane. Note that the positions where the methyl groups are attached to the parent chain are written, separated by a comma.

Follow the three-step process to name the following alkane.

$$CH_3-CH_2-CH-CH-CH_2-CH_3$$
with branches: CH (position 3) has CH$_3$ above; CH (position 4) has CH$_2$—CH$_2$—CH$_3$ below

---

Fact of the MATTER

The terms *saturated* and *unsaturated* originated before chemists understood the structures of organic substances. They knew that some hydrocarbons would take up hydrogen in the presence of a catalyst. When the substance would react with no more hydrogen, it was said to be saturated. Hydrocarbons that would take up additional hydrogen were said to be unsaturated. Today, we know that unsaturated hydrocarbons contain double and triple bonds that will react with hydrogen to form single bonds.

SECTION 18.1

Concept Development

You may wish to begin your presentation about alkanes by constructing the molecular models and naming the alkanes from methane, CH_4, through pentane, C_5H_{12}. Simultaneously, draw the structural formulas on the board, emphasizing that the alkanes are not really *straight* chains. Follow this with the heptane isomer activity in the Visual Learning strategy on the next page.

Correcting Misconceptions

Students may think that the co-valent bonds shown in structural formulas written in the style $CH_3-CH_2-CH_3$ actually connect carbon atoms with two or three hydrogen atoms from an adjacent carbon atom. Be sure to construct molecular models for a few of these compounds to dispel this idea.

GLENCOE TECHNOLOGY

Videodisc

Chemistry: Concepts and Applications
See the Videodisc Program Teacher Guide to access illustrations and other material available for this chapter.

✔ Assessment

Portfolio: Have students add the drawings of the 18 isomers of octane to their portfolios. L1 P

Isomers of Octane

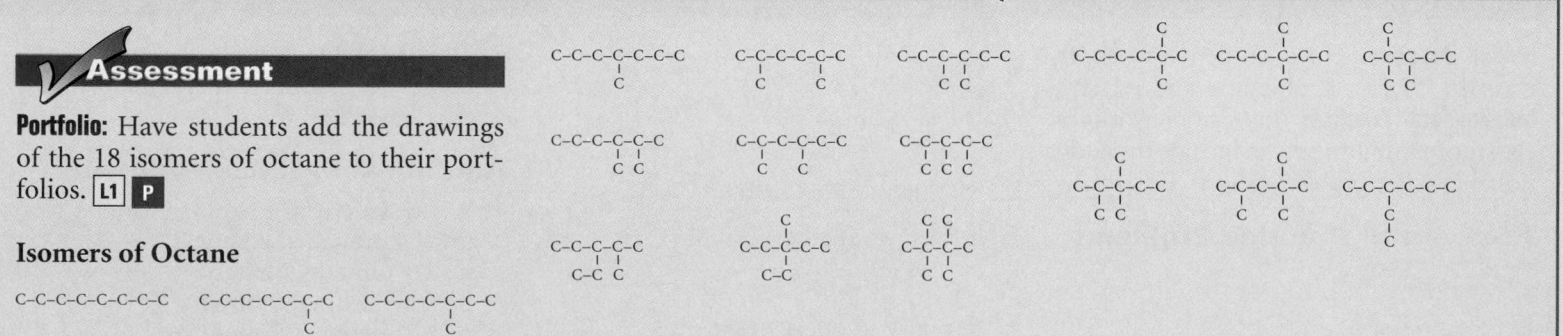

A group of isomers such as the hexanes includes compounds with names like 2-methylpentane and 2,3-dimethylbutane. Because compounds are named by the longest chain present in them and not by the total number of carbon atoms present, students sometimes have trouble categorizing compounds into the proper groups. Instead of asking them to draw structures of all of the isomeric hexanes, you might want to ask them to draw all of the hydrocarbons that contain six carbons. L1

Transparency

Display **Basic Concepts Transparency 49** to help students visualize the structures of hydrocarbons. L1 LEP

1. Count the number of carbons in the longest continuous chain, which may not always be written in a straight line. What looks like a branch may be part of the parent chain. Because this compound has a chain of seven carbons, heptane will be part of the name.

2. Note the size and number of branches present. The compound shown has one branch that is one carbon long and one branch that is two carbons long. *Ethyl* is the name given to a branch that has two carbons, and you know that the name for a one-carbon branch is *methyl*. If a compound has two different branches, they should be named in alphabetical order. Thus, part of the name will be ethylmethylheptane.

3. Number the carbons in the chain starting at the end nearest a branch. In the compound shown, position numbers will start on the right side.

$$CH_3 - CH_2 - \overset{4}{C}H - \overset{3}{C}H - \overset{2}{C}H_2 - \overset{1}{C}H_3$$

with branches: CH₃ (top, on carbon 3), and ⁵CH₂ — ⁶CH₂ — ⁷CH₃ (bottom, on carbon 4)

The position numbers of the carbons that have branches are 3 and 4. Now, put the name together. The position numbers of the branches come first, followed by a hyphen and then the name of the branch. The complete name for this compound is 4-ethyl-3-methylheptane.

Alkanes containing rings are named using the same rules, but the prefix *cyclo-* is placed before the name. Cyclohexane has six carbons connected in a ring, and cyclononane has a nine-carbon ring.

PRACTICE PROBLEMS

1. Write the structural formulas for the following branched alkanes.
   a) 2-methylbutane
   b) 1,3-dimethylcyclohexane (Hint: Begin numbering at any carbon in the ring, then attach the methyl groups.)
   c) 4-propyldecane
   d) 2,3,4-trimethylheptane

2. Name each of the following alkanes.
   a)

PRACTICE

Guided Practice: Have students work in class on Practice Problems 1 and 2. L1
Independent Practice: Your homework or classroom assignment can include the additional practice problems shown below. L1

Answers to Practice Problems

1a.
$$CH_3CHCH_2CH_3$$
(with CH₃ branch)

b.

c. CH₃CH₂CH₂CHCH₂CH₂CH₂CH₂CH₂CH₃
    |
    CH₂CH₂CH₃

d. CH₃CH−CH−CHCH₂CH₂CH₃
    | | |
    CH₃ CH₃ CH₃

2. a. 3-methylpentane b. 2,3-dimethylbutane c. 2,3-dimethylheptane d. 3-ethylhexane

Additional Practice Problems

1. Write structural formulas for the following branched alkanes.
   a. 2,3-dimethylpentane
   b. 3,4,5-trimethylnonane
   c. 1,2-diethylcyclobutane
   d. 5-butyldecane

b)

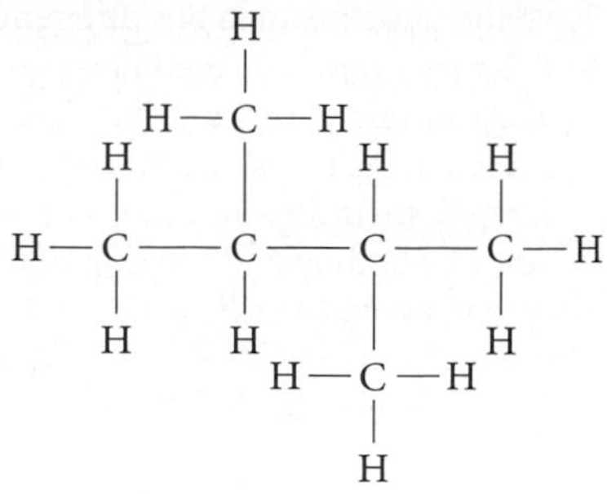

$$
\begin{array}{c}
\text{H} \\
| \\
\text{H}-\text{C}-\text{H} \\
\end{array}
$$

(structure b)

c)

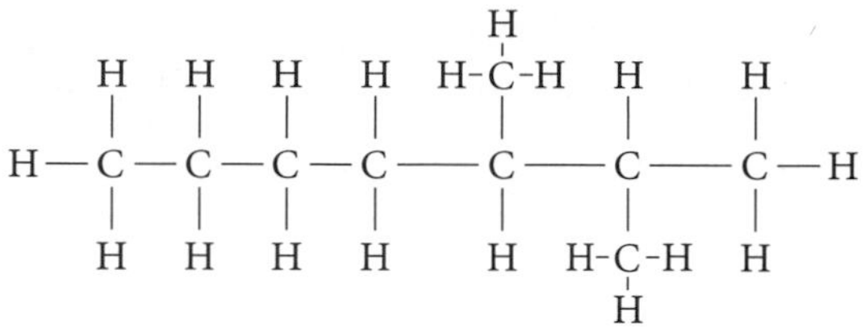

(structure c)

d)

(structure d)

Content Background

Isomers Two kinds of isomers are covered in this chapter: structural isomers of alkanes, in which branches on a hydrocarbon chain vary in size and position, and geometric isomers of alkenes, in which the spatial positions of groups held rigidly around the carbons in a double bond vary. There are many other types of isomers, as well. Another type of structural isomers includes compounds with the same formula but different functional groups, such as ethanol and dimethyl ether. Positional isomers contain the same functional groups attached to different locations in a molecule, such as 1-chlorobutane and 2-chlorobutane. One group of stereoisomers, in addition to the geometric isomers, is the mirror-image or enantiomer isomers, such as D-glucose and L-glucose. Mirror-image isomers differ only in how four different groups attached to single tetrahedral carbon atoms are oriented in space. The chemical and physical properties of a pair of isomers can vary greatly, as in the case of most structural isomers, or be nearly identical, as in the case of mirror-image isomers.

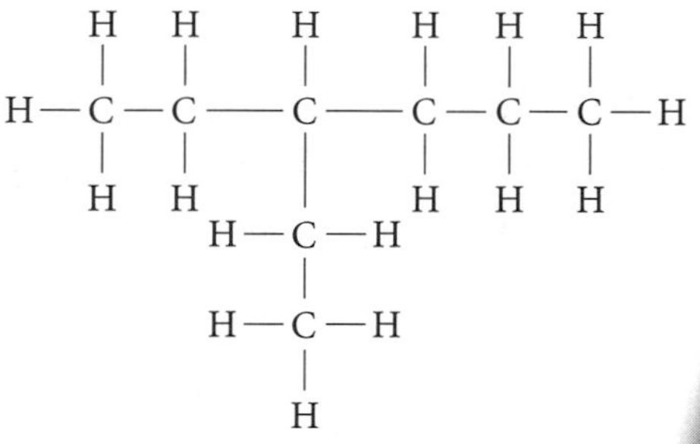

Isomers

Butane and 2-methylpropane are two alkanes with different names and different structures.

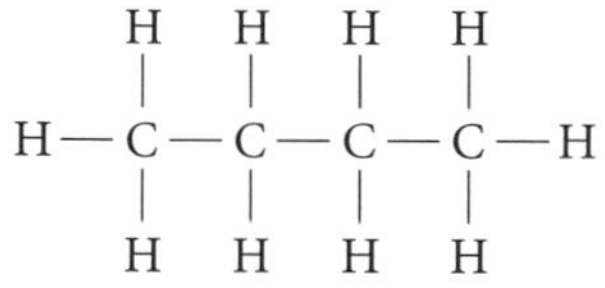

Butane   2-methylpropane

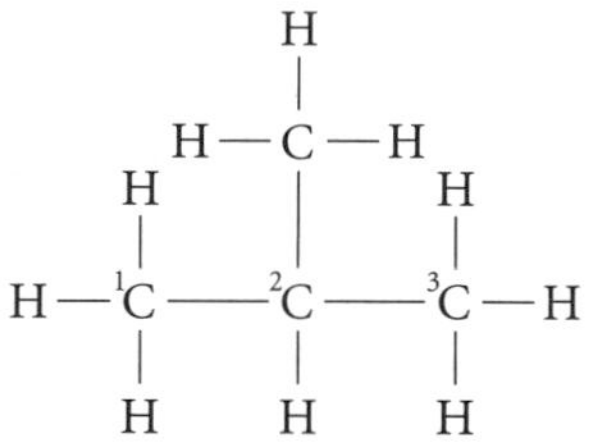

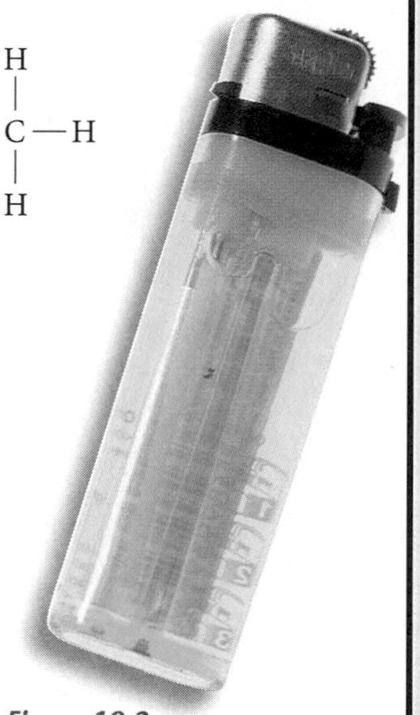

Figure 18.2

Disposable Lighter Fuel
Although butane and 2-methylpropane are gases at room temperature and atmospheric pressure, they can be liquefied under higher pressures in closed containers. Most disposable lighters contain one or both of these compounds, which are flammable enough to be ignited by a spark.

Both of these compounds are familiar fuels that burn to give off heat in handheld lighters, **Figure 18.2.** Is there any relationship between these two compounds? If you count the number of carbon and hydrogen atoms in each, you will find that they both have four carbons and ten hydrogens, which gives them the same molecular formula, C_4H_{10}. Compounds that

18.1 Hydrocarbons **627**

---

1. a. 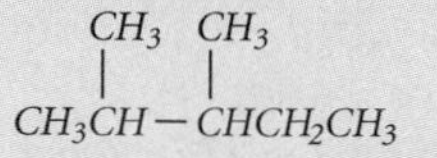
$$CH_3CH-CHCH_2CH_3$$
with CH_3 CH_3 groups

b. 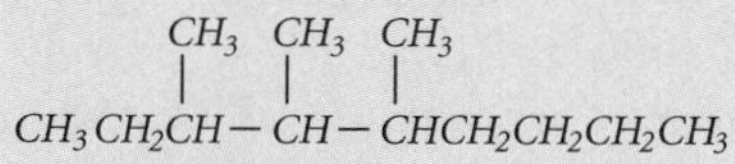
$$CH_3CH_2CH-CH-CHCH_2CH_2CH_2CH_3$$
with CH_3 CH_3 CH_3 groups

c. 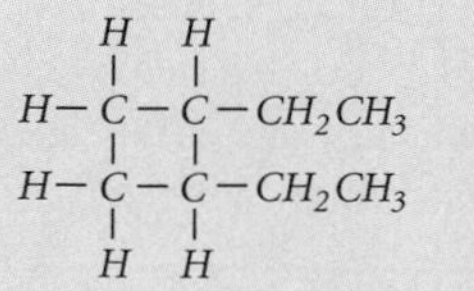
$$H-\overset{H}{\underset{H}{C}}-\overset{H}{\underset{H}{C}}-CH_2CH_3$$
$$H-\overset{H}{\underset{H}{C}}-\overset{H}{\underset{H}{C}}-CH_2CH_3$$

d. 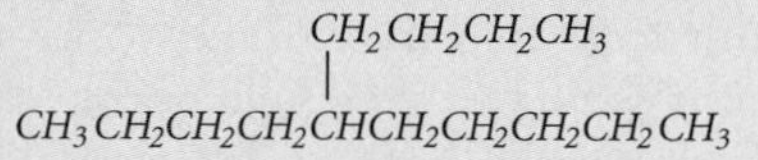
$$CH_3CH_2CH_2CH_2CHCH_2CH_2CH_2CH_3$$
with $CH_2CH_2CH_2CH_3$ group

2. Name each of the following alkanes.

a. 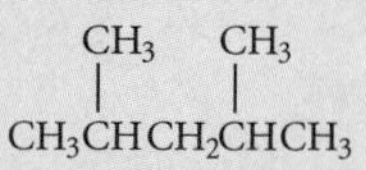
$$CH_3CHCH_2CHCH_3$$
with CH_3 CH_3 groups

b. 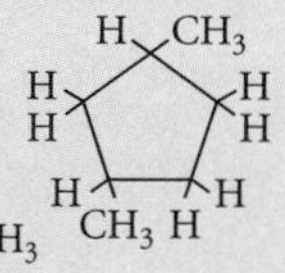 (cyclopentane structure)

c. 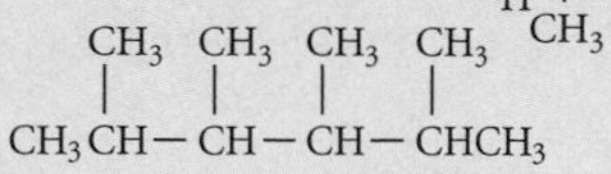
$$CH_3CH-CH-CH-CHCH_3$$
with CH_3 CH_3 CH_3 CH_3 groups and CH_3 H

d. 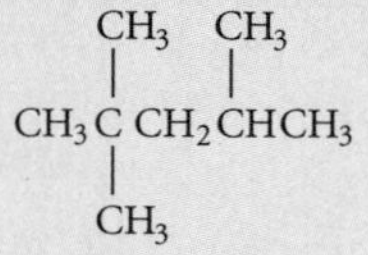
$$CH_3C\,CH_2CHCH_3$$
with CH_3 CH_3 groups and CH_3

2. a. *2,4-dimethylpentane*
b. *1,4-dimethylcyclopentane*
c. *2,3,4,5-tetramethylhexane*
d. *2,2,4-trimethylpentane*

Allow students to work in small groups with molecular model sets to construct all possible structural isomers of heptane, C_7H_{16}. Students should also draw the structural formulas on paper. Do not tell them the actual number of isomers (nine), but mention that hexane has five isomers and octane has 18.

L1 **LEP** **COOP LEARN**

Concept Development

Follow the heptane isomer activity above by putting the structural formulas for the nine isomers of heptane on the chalkboard and pointing out that a naming system is obviously required to specify each formula adequately. Refer to the IUPAC system in the text and name each of the isomers.

GLENCOE TECHNOLOGY

Videodisc

Chemistry: Concepts and Applications

Isomers

Disc 2, Side 2, Ch. 11

Show this quick-time animation to demonstrate the rearrangements of hexane to form its isomers. **L1** **LEP**

CD-Rom

Chemistry: Concepts and Applications

Isomers

Have students use this animation to help reinforce the process of molecular rearrangement to form hexane's isomers. **L1** **LEP**

Figure 18.3

Properties of Pentane Isomers

Structure and properties are closely related, as you can see by examining the isomers of pentane. Although all three compounds have the formula C_5H_{12}, differences in the amount of branching affect their properties. Note the differences in the shapes of the molecules.

have the same formula but different structures are called **isomers.** Butane and 2-methylpropane are known as structural isomers. Each has the molecular formula C_4H_{10}, but they have different structural formulas because the carbon chains have different shapes.

Despite their identical molecular formulas, isomers have different properties. The boiling and melting points of 2-methylpropane and butane are different, as are their densities and solubilities in water. In addition, their chemical reactivity is different. **Figure 18.3** shows some property differences in the isomers of pentane.

$$CH_3—CH_2—CH_2—CH_2—CH_3$$

Pentane
bp 36°C

$$CH_3—CH_2—\overset{\overset{\displaystyle CH_3}{|}}{CH}—CH_3$$

2-methylbutane
bp 28°C

$$CH_3—\overset{\overset{\displaystyle CH_3}{|}}{\underset{\underset{\displaystyle CH_3}{|}}{C}}—CH_3$$

2,2-dimethylpropane
bp 9.5°C

Although only two alkane isomers have four carbons, the number of possible isomers increases rapidly as carbon atoms are added to the parent chain. This is because longer chains provide more locations for branches to attach. A methyl branch on a six-carbon parent chain can be attached to either the second or the third carbon from the end of the chain. To make sure that these two compounds are really isomers, count the number of carbon and hydrogen atoms in the structures formed by placing the methyl group at those two positions, and write the molecular formulas.

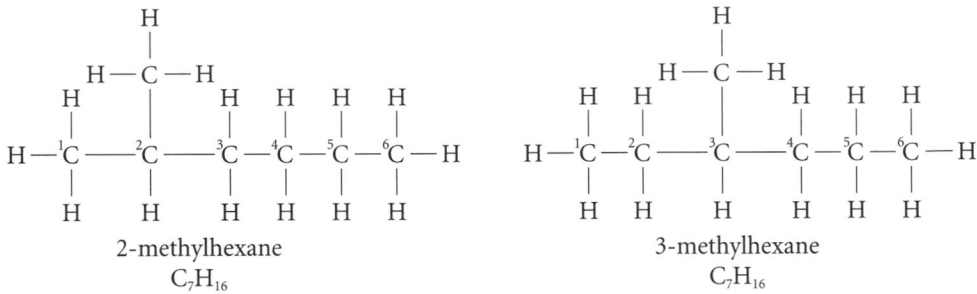

2-methylhexane
C_7H_{16}

3-methylhexane
C_7H_{16}

There are three isomers of pentane, five of hexane, and more than 4 billion isomers of the alkane with the formula $C_{30}H_{62}$.

Demonstration 18.3

Reactive Alkenes

Purpose: To demonstrate that the double bond is more reactive than the weaker single bond.

Materials: 5 mL cyclohexane, 5 mL cyclohexene, 10 mL bromine water, 3 test tubes with stoppers

Safety Precautions:

Disposal: Collect solvents for disposal via commercial recycler or waste-disposal vendor.

Procedure: Show the presence and reactivity of a double bond by adding 3 mL of bromine water to 5 mL of cyclohexene. Stopper the test tube and shake to mix. **CAUTION:** *Hydrocarbons are both flammable and toxic. Use a fume hood.* Repeat the procedure using 5 mL of cyclohexane. Have students record their observations.

Results: Bromine water becomes nearly colorless when shaken with cyclohexene. There is no evident change when it is shaken with cyclohexane.

Properties of Alkanes

Properties are affected by the structure or arrangement of atoms present in a molecule. Another factor that affects properties of alkanes is chain length. In general, the more carbons present in a straight-chain alkane, the higher its melting and boiling points. At room temperature, straight-chain alkanes that have from one to four carbon atoms are gases, those with from five to 16 carbon atoms are liquids, and those with more than 16 carbon atoms are solids.

A property shared by all alkanes is their relative unreactivity. You will recall from Chapter 9 that the carbon-carbon and carbon-hydrogen bonds found in alkanes are nonpolar. Because alkanes don't have any polar bonds, they undergo only a small number of reactions and will dissolve only those organic compounds that are nonpolar or that have low polarity, such as oils and waxes. The nonpolar and low-reactive nature of alkanes makes them good organic solvents. Paints, paint removers, and cleaning solutions often contain hexane or cyclohexane as solvents.

Four ways to represent molecules are shown here. At the top are ball-and-stick models, with springs representing double and triple bonds. At the bottom are space-filling models, which show the actual shape of the molecule. ▶

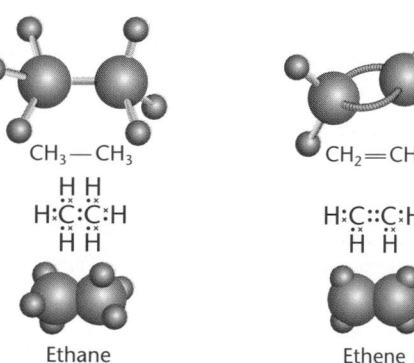

$$CH_3 - CH_3$$

H H
H:C:C:H
H H

Ethane

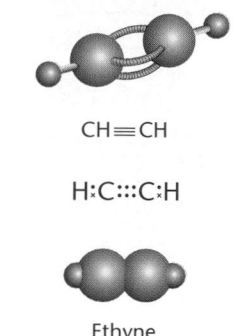

$$CH_2 = CH_2$$

H:C::C:H
H   H

Ethene

$$CH \equiv CH$$

H:C:::C:H

Ethyne

Unsaturated Hydrocarbons

You learned in Chapter 9 that carbon atoms in organic compounds can be connected by single, double, or triple bonds. A hydrocarbon that has one or more double or triple bonds between carbons is called an **unsaturated hydrocarbon.** Molecules with single, double, and triple bonds are compared in **Figure 18.4.**

Alkenes

Gasoline contains several hydrocarbons with double bonds. A hydrocarbon in which one or more double bonds link carbon atoms together is called an **alkene.** The most common alkenes found in gasoline are pictured in **Figure 18.5.**

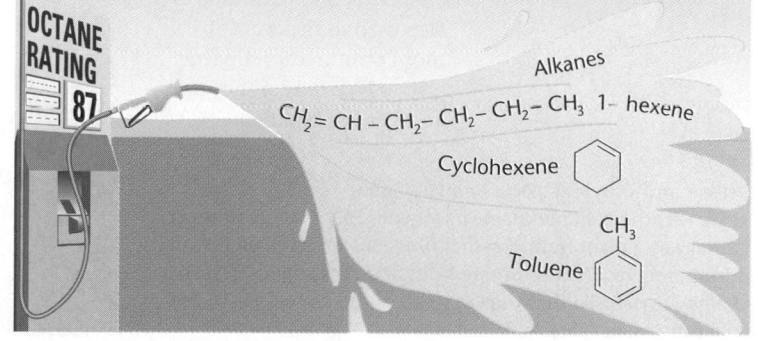

OCTANE RATING
87

Alkanes

$$CH_2 = CH - CH_2 - CH_2 - CH_2 - CH_3 \quad \text{1- hexene}$$

Cyclohexene

Toluene

Figure 18.4
Single, Double, and Triple Bonds
Two carbon atoms can share one, two, or three pairs of electrons. In saturated hydrocarbons, carbon atoms share only one pair, whereas in unsaturated hydrocarbons, the carbon atoms participating in double or triple bonds share two or three pairs. Because a carbon atom forms four bonds, unsaturated hydrocarbons contain fewer than the maximum number of hydrogen atoms.

Figure 18.5
Alkenes in Gasoline
Gasoline contains some alkenes as well as alkanes. Alkenes burn more uniformly, and their presence raises the octane rating of a gasoline.

Correcting Errors

The general formula for alkanes is C_nH_{2n+2}; however, cyclic alkanes are C_nH_{2n} because the two end carbon atoms are linked.

Discussion

Ask students to speculate on the meaning of polyunsaturated fats and oils. *Polyunsaturated* means "containing many double and/or triple bonds."

Transparency

Display **Basic Concepts Transparency 50** to help students visualize the structures of isomers. L1 LEP

18.1 **Hydrocarbons** 629

Analysis: Ask these questions.
1. What happens to the color of the bromine water when mixed with an alkene? *It discolors.*
2. Explain how the bromine, Br₂, might react with a double bond. *One of the bonds of the C-C double bond breaks. The bond between the Br₂ breaks. Each Br atom bonds to the carbon atoms of the former double bond.*

✔ **Assessment**

Performance: After students have observed the difference between an alkane and an alkene when mixed with bromine water, they can test for double bonds by adding bromine water to various oils and observing the color difference. This test can be used to determine the degree of unsaturation in cooking oils. L1

miniLAB

TECH PREP

How unsaturated is your oil?

Purpose: Students will test two cooking oils to determine the degree of unsaturation in each.
Process Skills: Observing, inferring, classifying, comparing and contrasting, communicating, predicting
Disposal: Codes A, G

Teaching Strategies

- Ask students before they begin the lab to predict which oil will be the more unsaturated. [L1]
- After completing the lab, have students check their results against the information on the product labels of the oils. [L1]
- For further comparisons, have students repeat the test using corn, coconut, olive, sunflower, and other oils. [L1]

Expected Results

The solution in the flask containing the canola oil should decolorize before the solution in the flask containing the peanut oil.

Analysis

1. The red-brown iodine reacts by adding across the double bonds in the unsaturated fat molecules to form colorless alkyl halides.
2. **a.** canola **b.** sunflower

Assessment

Knowledge: Ask students to draw structures of saturated and unsaturated hydrocarbons. Have them translate these structures into models using model kits or gumdrops and toothpicks. [L1]

miniLAB
1

How unsaturated is your oil?

Most animal fats are saturated hydrocarbons and are solids at room temperature, whereas most vegetable fats are unsaturated and are liquids at room temperature. Both types of fats are essential in our diets, but medical research has shown that eating high levels of saturated fats can contribute to health problems such as heart disease. Different amounts of unsaturation in fats can be compared by testing how quickly a red-brown iodine solution added to each fat is decolorized. Iodine adds to carbons that take part in multiple bonds, forming colorless organic halogen compounds when the double or triple bond is broken.

Procedure

1. Place 20 mL of peanut oil in one small flask and 20 mL of canola oil in another flask. Label the flasks.

2. Add five drops of tincture of iodine to each oil and swirl to mix well. Note the color of each solution.

3. Heat both flasks on a hot plate set on low.

4. Note which oil returns to its original color first. This oil is more unsaturated than the other.

5. Now read the food label on each bottle of oil, and determine whether your test results agree with how much unsaturated fat the labels say are in each oil.

Analysis

1. What happens to the red-brown iodine when it is added to a relatively unsaturated oil?

2. Examine food labels on bottles of the oils listed below, and predict which of each pair will decolorize faster.

   a) canola or corn oil
   b) coconut or sunflower oil

Alkenes are named using the root names of the alkanes, with the *-ane* ending changed to *-ene.* The simplest alkene is ethene, $CH_2 = CH_2$, which contains two carbons linked in a chain. Ethene, a gas at room temperature, is the most important organic compound used in the chemical industry. Almost half of the ethene used is converted into plastics. It is also used to make automobile antifreeze, ethylene glycol. **Figure 18.6** shows another use of ethene.

Figure 18.6

Ethene and Ethylene: One and the Same
Ethylene is the common name of ethene. This compound occurs naturally as a plant hormone that functions to speed up ripening of fruits and vegetables. Unripe fruits and vegetables can be treated with ethylene so that they all ripen at the same time, making harvesting more efficient.

Chemistry Journal

Geometric Isomers

Have students practice drawing geometric isomers by asking them to draw in their journals the *cis* and *trans* isomers for a hydrocarbon molecule with five carbon atoms and one double bond. *The molecule should have a methyl group on one carbon and an ethyl group on the other carbon involved in the double bond.* [L1]

The alkene with three carbons in a chain is called propene, CH_2=$CHCH_3$. When four or more carbons are present in a chain, the double bond can be located at more than one possible position. When naming these alkenes, a number must be added at the beginning of the name to indicate where the double bond is located. You should use the following steps in naming alkenes with long carbon chains.

1. Count the number of carbons in the longest continuous chain that contains the double bond, and assign the appropriate alkene name.

2. Number the carbons consecutively in the longest chain, starting at the end of the chain that will result in the lowest possible number for the first carbon to which the double bond is attached.

3. Write the number corresponding to the first carbon in the double bond, followed by a hyphen and then the alkene name.

What is the name of the compound that has the following structure?

$$CH_2$$=$$CHCH_2CH_3$$

This compound has four carbons in a chain with one double bond, so butene will be part of its name. Numbering the carbons starting on the left side of the compound gives the first carbon that is part of the double bond the position number of one. Thus, this compound is 1-butene.

$$\overset{1}{C}H_2=\overset{2}{C}H\overset{3}{C}H_2\overset{4}{C}H_3 \quad \text{1-butene}$$

An isomer of 1-butene is 2-butene, CH_3CH=$CHCH_3$. These two compounds have the same molecular formulas but different structures. They are called positional isomers because they differ only by the position of the double bond. Positional isomers have different properties, just as structural isomers do.

The formation of a double bond prevents the carbons on each side of the bond from rotating with respect to each other. If the two groups attached to either carbon are different, the alkene can have two different geometric structures. These structures are geometric isomers. Study the two geometric isomers of 2-butene that are modeled in **Figure 18.7**. In the isomer called *cis*-2-butene, the hydrogen atoms and —CH_3 groups are on the same side of the double bond. In *trans*-2-butene, the hydrogen atoms and —CH_3 groups are on opposite sides of the double bond.

The properties of a pair of geometric isomers vary. *Trans* isomers are more symmetrical than *cis* isomers, and they can pack together more closely when in the solid state. This close packing makes the molecules harder to pull apart. As a result, *trans* isomers generally have higher melting points than do *cis* isomers. ▼

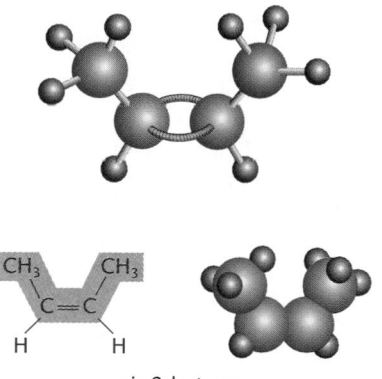
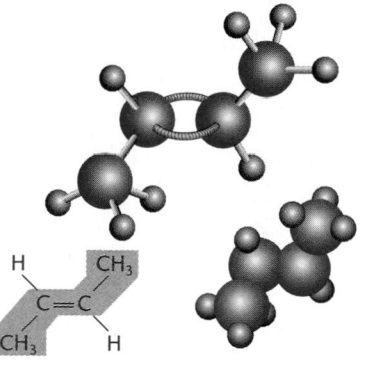

Figure 18.7
Geometric Isomers
Cis-2-butene and *trans*-2-butene are geometric isomers. Note their shapes in both ball-and-stick models and space-filling models. ▶

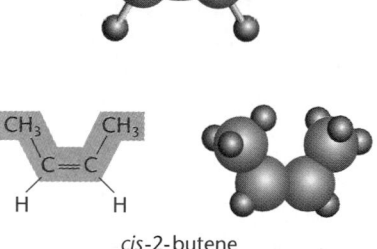

cis-2-butene

trans-2-butene

SECTION 18.1

Concept Development

Mention that the meaning of *trans* may be remembered by noting that *transcontinental* means "across the continent."

Concept Development

Ask two students to make models of butene, having prearranged with them that one will construct 1-butene and the other *trans*-2-butene. Note for the class that the names must differentiate the two compounds, and explain the naming, omitting the *trans* prefix. Continue with the development by asking the student who constructed 1-butene to convert it to 2-butene, having previously instructed him or her to construct the *cis* isomer. Ask the students to display the two molecules, asking the class if they are the same. *no* Explain *cis-trans* isomerism and the naming of the two compounds. L1

Cultural Diversity

Cosmetics

Although the origins of cosmetics are, no doubt, prehistoric, the culture of ancient Egypt was one of the first to make the use of cosmetics an important part of daily life among the privileged. One of the most prominent characteristics of Egyptian cosmetics as evidenced by sculptures and sarcophagi was the extensive application of eyeliner by both men and women. The eyeliner consisted of a black, gray, or colored powder called kohl. Kohl was made from carbon black, with the addition of materials such as powdered antimony, manganese oxide, lead, copper oxide, iron oxide, and the green-blue copper ore, chrysocolla. Often, the kohl was applied in several layers around the eyes. Used predominately for beautification, kohl may also have had the beneficial effect of protecting the eyes from desert sands and sun glare. Today, most cosmetics are made from inorganic dyes incorporated into organic mixtures of beeswax, lanolin, waxes, and hydrocarbons.

BIOLOGY CONNECTION

Purpose

Students will gain an understanding of the chemistry of vision and learn how night vision is enhanced by vitamin A.

Teaching Strategies

- Mention that at least three times during evolution, eyes have developed independently in mollusks, arthropods, and vertebrates. Although these eyes vary greatly in structure, they all have separately taken 11-*cis*-retinal as an important part of their visual pigments. Ask students to hypothesize why this might be so. *They may hypothesize that 11-*cis*-retinal is used by all three groups for vision because of its particularly suitable properties as a photoreceptor molecule.*
- Tell students that when someone reenters a bright room after being in the dark, retinal sensitivity decreases from 100 000 back down to 1 within another 10 minutes. Ask students to explain why. *The concentration of rhodopsin is diminished by the light.*

Extension

Have students investigate whether their diet supplies ample vitamin A. After preparing a list of sources of β-carotene, they should compare their diet with the list. Ask them to decide which sources of dietary vitamin A they enjoy eating. L1

Connecting to Chemistry

1. Students may hypothesize that because owls and bats are not exposed to much light, the amount of rhodopsin in the animals' rods would probably be high.
2. The change from 11-*cis*-retinal to the all-*trans*-iso-

BIOLOGY CONNECTION

Vision and Vitamin A

Vision is a process usually studied in biology class. That's where you may have learned that light rays pass through the eye to reach the retina, where the rods and cones are located. Rods and cones are nerve receptors that are excited by light. More than 120 million rods in each eye detect white light and provide sharpness of visual images. The 7 million cones in each eye detect color. The pigment molecules responsible for vision are attached to the ends of the rods and cones. One of these pigments is called rhodopsin. Rhodopsin has two parts—a protein called opsin and a small molecule called retinal.

Chemistry of vision Beta-carotene, the natural orange pigment found in carrots and other yellow or green vegetables, breaks down in your body to form vitamin A, which is then converted to 11-*cis*-retinal. In the retina of the eye, 11-*cis*-retinal attaches to the protein opsin to form rhodopsin. When light is absorbed by the 11-*cis* isomer portion of rhodopsin, the energy causes the pigment molecule to undergo a change in shape. The right end of the molecule rotates about a double bond to form all-*trans*-retinal, in which all the groups are in the *trans* position.

11-*trans*-retinal

11-*cis*-retinal

When the retinal portion of rhodopsin is isomerized to the all-*trans* form, it separates from the opsin. Thus, light energy causes the rhodopsin to break down into the substances from which it was formed. As the rhodopsin molecule splits, the rods become excited, probably due to ionic charges that develop on the splitting surfaces. These charges last only for a second, but they generate nerve signals that are transmitted to the optic nerve and then to the brain. After the rhodopsin is activated, the *trans*-retinal returns to the *cis* form and recombines with opsin. This process is relatively slow, which is why your eyes need time to adjust to dim light.

Night vision When large quantities of light energy strike the rods, large amounts of rhodopsin are broken down. As a consequence, the concentration of rhodopsin in the rods falls to a low level. If you leave a bright area and enter a darkened room, the quantity of rhodopsin in the rods at first is small. As a result, you experience temporary blindness. The concentration of rhodopsin gradually builds up until it becomes high enough for even a small amount of light to stimulate the rods. During dark adaptation, the sensitivity of the retina can increase as much as a thousandfold in only a few minutes and as much as 100 000 times in an hour or more.

Connecting to Chemistry

1. **Hypothesizing** Owls and bats avoid daylight and are at home in the dark. Their eyes contain only rods. Hypothesize as to the relative amount of rhodopsin in their rods compared with that in human eyes, and explain.

2. **Analyzing** What is the significance of the change from 11-*cis*-retinal to all-*trans*-retinal for the vision process?

mer causes rhodopsin to split. The rods become excited because of ionic charges on the splitting surfaces. These charges generate nerve signals that are transmitted to the optic nerve.

Figure 18.8
Saturated and Unsaturated Fats
▲ Most vegetable oils contain unsaturated alkenes, whereas butter and lard—both animal products—contain mostly saturated alkanes. Because alkanes have higher melting points than alkenes, fats from animals are solid at room temperature. Fats from plants generally are liquids.

◄ When corn oil is hydrogenated, it becomes solid. This is how margarine is produced and explains why margarine is solid at room temperature, whereas the vegetable oil from which it is derived is liquid.

Applying Chemistry

TECH PREP **Hydrogenation** Have interested students research the subject of catalytic hydrogenation of cooking oils in the formation of margarines, solid vegetable shortenings, and chocolate candy bars. Students should determine why food manufacturers use partially hydrogenated oils rather than completely hydrogenated oils. *A completely hydrogenated oil is hard and brittle, with the consistency of candle wax. Partially hydrogenated oils are soft and creamy.* L2

Alkenes are more reactive than alkanes because the two extra electrons in the double bond are not held as tightly to the carbons as are the electrons in a single bond. Alkenes readily undergo synthesis reactions in which smaller molecules or ions bond to the atoms on either side of the double bond. An unsaturated alkene can be converted into a saturated alkane by adding hydrogen to the double bond. This reaction is called hydrogenation. Its effects are shown in **Figure 18.8.**

$$CH_2{=}CH_2 + H_2 \rightarrow CH_3CH_3$$

Ethene   Ethane
(unsaturated)   (saturated)

Alkynes

Another type of unsaturated hydrocarbon, called an **alkyne,** contains a triple bond between two carbon atoms. Alkynes are named using the alkane root name for a given carbon chain length and changing the *-ane* ending to *-yne*. Ethyne, known more commonly as acetylene, is the most important commercial alkyne. Most acetylene produced in the United States is used to make vinyl and acrylic materials, although about ten percent is burned in oxyacetylene torches. These torches are used to cut and weld metals. Few alkynes are known to occur naturally because they are very reactive. However, they can be synthesized from other organic compounds. The names and structures of some small alkyne molecules are shown in **Table 18.2.**

Table 18.2   Simple Alkynes

Chemical Name	Common Name	Structure
ethyne	acetylene	$HC{\equiv}CH$
propyne	methylacetylene	$HC{\equiv}CCH_3$
1-butyne	—	$HC{\equiv}CCH_2CH_3$
2-butyne	—	$CH_3C{\equiv}CCH_3$

Meeting Individual Needs

Visually Impaired Have visually impaired students feel the consistency of fats of various degrees of saturation. The consistency and texture of vegetable oil, partially hydrogenated oil such as margarine and shortening, and candle wax can be compared. Be sure students understand the chemical process that created these differences in consistency. L1 LEP

Meet John Garcia, Pharmacist

TECH PREP

Teaching Strategies

- Alluding to Mr. Garcia's statement that his Mexican mother knew many home remedies, point out that the oldest medical book produced in the Americas was a manuscript of an Aztec herbal, a listing of medically beneficial plants and their uses, known as the *Bandianus Manuscript*. The manuscript, produced in 1552, is a Latin translation by the Aztec scholar, Juannes Bandianus, of an herbal written in the local Aztec language (Nahuatl) by Martinus de la Cruz, a prominent physician at the College of Santa Cruz.

Background

At one time, pharmacists prepared as well as dispensed prescription and over-the-counter drugs. Today, most drugs are manufactured.

PEOPLE in CHEMISTRY

Meet John Garcia, Pharmacist

"A treasure house going up in smoke"—that's how pharmacist John Garcia describes the fires set to clear land in the Amazon. He's concerned about the potential medicines that humans might be destroying as they make inroads into formerly wild areas. "However," he continues, "endangered sources for new medicines aren't only in jungles. Lincomycin, an antibiotic, was discovered in soil in Lincoln, Nebraska." In this interview, Mr. Garcia describes some of the changes he has seen in his four decades as a pharmacist.

On the Job

 Mr. Garcia, can you tell us what you'll be doing first today in your pharmacy?

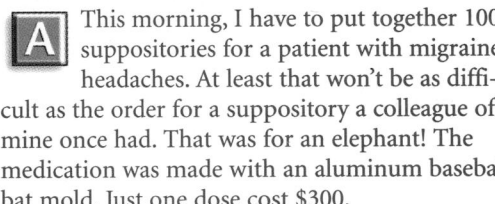 This morning, I have to put together 100 suppositories for a patient with migraine headaches. At least that won't be as difficult as the order for a suppository a colleague of mine once had. That was for an elephant! The medication was made with an aluminum baseball bat mold. Just one dose cost $300.

 Do you prepare prescriptions for other animals?

 I've made them for rabbits. I added raspberry flavor because rabbits enjoy the flavor.

 Do your human patients get a choice of flavors, too?

 I've got about 40 flavors—piña colada, bubble gum, peppermint. Tutti-frutti is my personal favorite. These flavorings aren't just frills. For instance, I prepare chloroquine, an antimalarial drug for children, and put it in a chocolate base. Because the molecules of chocolate are fairly large, they block the taste buds and cover up the bitterness of the medicine.

 You own and operate your own independent pharmacy. How does it differ from most pharmacies?

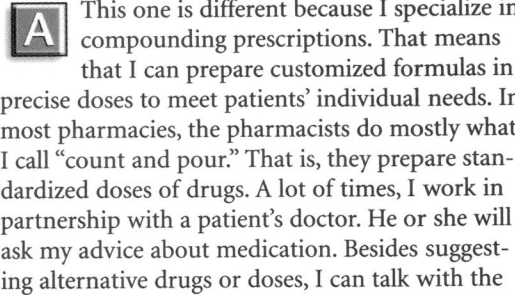 This one is different because I specialize in compounding prescriptions. That means that I can prepare customized formulas in precise doses to meet patients' individual needs. In most pharmacies, the pharmacists do mostly what I call "count and pour." That is, they prepare standardized doses of drugs. A lot of times, I work in partnership with a patient's doctor. He or she will ask my advice about medication. Besides suggesting alternative drugs or doses, I can talk with the doctor about methods of administering a drug.

 How has the practice of pharmacy changed in the 40 years you've been a pharmacist?

A New laws have given pharmacists the responsibility of explaining carefully to a patient the workings of a drug and its possible side effects. At the same time, changes in the insurance industry have added lots of paperwork and reduced our fees. So that means that pharmacists have to work more quickly while maintaining careful standards.

Q Seriously ill patients are living longer today. How has that change affected your practice?

A I work with several hospices, which care for terminally ill patients. By the time patients are admitted to a hospice, they typically are in the last stages of an illness and have about 40 days to live. Our priority is to make the patient comfortable by using a variety of medications to relieve pain and discomfort.

Early Influences

Q How did you come to be a pharmacist?

A I got an after-school job in a small pharmacy when I was in high school. I started out making deliveries and eventually helped fill prescriptions. I liked working with the public, and the pharmacist, Nathan Fisher, encouraged me to apply to pharmacy school. I feel honored that he was my mentor.

Q Were you, as a child, aware of pharmacies and medications?

A Not really. I don't think I even saw a doctor until I needed a physical exam to play high school sports. My parents are from Mexico, and my mother knew many home remedies. She gave me things like charcoal or mint for an upset stomach.

Personal Insights

Q What qualities do you think it takes to be a good pharmacist?

A It's obvious that you have to like people and be accurate in mathematical and chemical calculations. I also like to say that pharmacy is a lot like cooking—a little of this, a little of that. Just like a cook, I want my preparations to contain the right proportion of ingredients, to taste good, and also to look elegant.

Q What changes do you see coming in the field of pharmacy during the 21st century?

A I think people will be able to insert a prescription in something like an automated teller machine. They'll also insert medical cards encoded with their personal medical history and their age, height, and weight. The machine will dispense their medication.

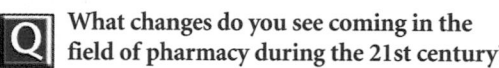

CAREER ▶ CONNECTION

These careers are all related to the field of pharmacy.

Pharmacologist Four to six years post-college study at a medical school or school of pharmacy
Pharmaceutical Technician Two years in a training program after high school
Pharmaceutical Production Worker High school diploma and on-the-job training

CAREER CONNECTION

- **Career Path** A career in pharmacy would include these high school courses: biology, Earth science, chemistry, and physics. Colleges offer five-year degree programs in pharmacy, which include courses in biology, general chemistry, organic chemistry, inorganic chemistry, physics, biochemistry, physiology, and pharmacology.
- **Career Issue** Have interested students research how the AIDS epidemic has affected the method by which drugs are approved by the U.S. Food and Drug Administration (FDA). **L2**

For More Information

Students may obtain more information about careers in pharmacy by writing the American Association of Colleges of Pharmacy, 1426 Prince Street, Alexandria, VA, 22314, and the American Council of Pharmaceutical Education, 311 West Superior Street, Chicago, IL 60610.

Fact of the MATTER

August Kekulé (1829-1896) was a German student of architecture who became interested in chemistry. One of the biggest puzzles of 19th-century chemistry involved the structure of benzene, which seemed to have too few hydrogen atoms to exist as a straight-chain molecule. Kekulé solved the puzzle after having a dream in which a dancing snake formed a ring and bit its own tail. When he awoke, he worked out the ring structure of benzene. Kekulé's work is considered to be among the most important in 19th-century theoretical organic chemistry.

Melting and boiling points of alkynes increase with increasing chain length, just as they did for alkanes and alkenes. Alkynes have physical and chemical properties similar to those of alkenes; their melting points are lower than those of alkanes, and they undergo synthesis reactions. For example, hydrogen molecules can be added to an alkyne in a stepwise fashion to form an alkene and then an alkane.

$$CH \equiv CH + H_2 \rightarrow CH_2 = CH_2$$
$$CH_2 = CH_2 + H_2 \rightarrow CH_3CH_3$$

Aromatic Hydrocarbons

Another group of unsaturated hydrocarbons has distinctive, six-carbon ring structures. The simplest compound in this group is benzene, with the molecular formula C_6H_6. Benzene contains six carbons joined together in a flat ring. Its structure was originally thought to contain alternating single and double bonds between the carbons, as seen in **Figure 18.9,** but this structure is now known to be incorrect.

Although it contains double bonds, benzene does not share most of the properties of alkenes. It is unusually resistant to hydrogenation, whereas most alkenes readily become hydrogenated. To account for this inertness, chemists have suggested that the extra electrons are shared equally by all six carbons in the ring rather than being located between specific carbon atoms. The currently accepted structure for the benzene molecule is shown in **Figure 18.9.**

This sharing of electrons among so many carbons gives benzene and similar compounds unique properties. The name **aromatic hydrocarbon** is used to describe a compound that has a benzene ring or the type of bonding exhibited by benzene. Aromatic hydrocarbons were originally named because most of them have distinctive aromas. Naphthalene, formerly used in mothballs to prevent moth damage to woolen clothes,

Figure 18.9

The Structure of Benzene
The structure of benzene can be represented in different ways.

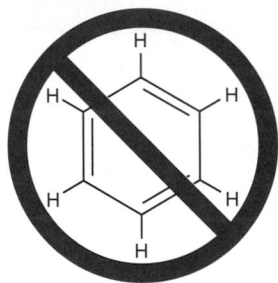

▲ This structure of benzene, showing alternating single and double bonds, is now known to be incorrect because it does not account for benzene's inertness.

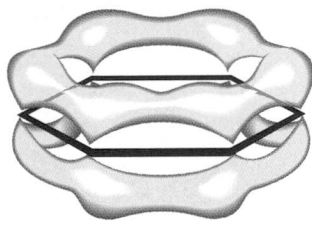

The flat benzene molecule is shown with clouds of shared electrons above and below the plane of the ring. ▼

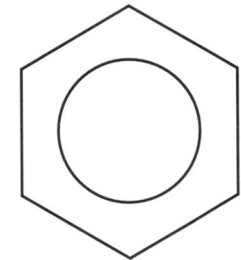

The hexagon diagram is a more accurate shorthand representation. In this hexagon, each corner represents a carbon atom. The circle in the middle of the structure represents the cloud of six electrons that are shared equally by the six carbon atoms in the molecule. ▼

636   Chapter 18   Organic Chemistry

Demonstration 18.4

An Active Alkyne

Purpose: To produce ethyne (acetylene), and to observe complete and incomplete combustion.

Materials: 3 chunks of calcium carbide, 3 test tubes with stoppers, three 400-mL beakers, 3 wood splints, laboratory burner

Safety Precautions:

Disposal: Be sure all solid CaC_2 has reacted, then dilute and flush down drain.

Procedure: Fill a 400-mL beaker with water. Completely fill a test tube with water. Place a chunk of calcium carbide in the water in the beaker, and cover with the inverted, water-filled test tube. Collect the gas by water displacement. Collect two additional tubes of gas, one containing 50% gas (start with a tube filled halfway

with water and halfway with air), and one containing 10% gas (start with a tube filled 10% with water, 90% with air). Keeping the mouths of the test tubes down, ignite the gas mixtures by placing a burning splint in the mouth of each test tube. **CAUTION:** *Fire risk.*

Results: The tube that had 100% ethyne gas burned at the mouth of the tube only and produced soot. The tube that had

consists of two benzene rings attached side-by-side. Aromatic hydrocarbons are unusually stable because the carbons are bonded tightly together by so many electrons.

Sources of Organic Compounds

You have examined structures and learned how to name basic organic compounds, but have you stopped to think about where these compounds come from? Most hydrocarbons come from fossil fuels, especially petroleum, but also natural gas and coal, **Figure 18.10.** Other important sources include wood and the fermentation products of plant materials.

Natural gas contains large quantities of methane, along with smaller amounts of alkanes up to about five carbons in length. Natural gas passes easily through pipes and is used mostly as a fuel, but it also serves as a raw material for making many small organic compounds.

Petroleum is a complex mixture, mostly of alkanes and cyclic alkanes. Products we get from petroleum include gasoline, jet fuel, kerosene, diesel fuel, fuel oil, asphalt, and lubricating oil. To use these organic products, they have to be separated from one another. What properties of hydrocarbons might allow them to be separated?

Figure 18.10

Sources of Aromatic Hydrocarbons
Soot is full of aromatic hydrocarbons produced by combustion of organic materials such as wood or coal. In 1775, a British physician investigating the high incidence of cancer among chimney sweeps theorized that soot in the chimneys was the cause. In the 1930s, it was shown that a number of large hydrocarbons with ring structures found in the coal tar in soot were carcinogenic, or cancer-causing. ▼

▲ Valuable deposits of petroleum and natural gas often are found by offshore oil drilling. These deposits are common under the oceans because microscopic marine organisms—including algae, bacteria, and plankton—died and settled on the seafloor where their remains were altered by high temperatures and pressure. Petroleum deposits also are found under dry land because at one time, this land was under a sea.

WORD ORIGIN

petroleum:
petra (L) a rock
oleum (L) oil
Petroleum is an oily fossil fuel found naturally in rock strata of certain geological formations.

SECTION 18.1

GLENCOE TECHNOLOGY

 Videodisc
Chemistry: Concepts and Applications
Petroleum Refinery
Disc 2, Side 2, Ch. 12

Show this video to help you demonstrate the refining of petroleum. L1 LEP

 CD-Rom
Chemistry: Concepts and Applications
Petroleum Refinery
Use this video to help reinforce the process of petroleum refining. L1 LEP

50% gas produced large amounts of soot. The 10% gas tube burned with no soot being produced. It made a sound as the hot gases left the tube.
Analysis: Ask these questions.
1. Why did the test tube that had 100% gas burn only at the mouth? *There was no oxygen inside the tube, and it is required for combustion.*
2. Why did the tube that had 50% gas

produce soot? *There was not enough air for efficient and complete combustion. The soot is unburned carbon.*
3. Which tube was the cleanest after combustion? Why? *In the 10% gas tube, the fuel was efficiently burned, releasing the energy that was stored in its chemical bonds.*

Knowledge: Have students research and discuss the pollution caused by an improperly tuned automobile engine. They should relate the results of this demonstration to the pollution caused by improperly tuned automobile engines. L1

3 ASSESS

Check for Understanding

• Check for understanding of the section material by having students answer the Section Review questions.

• Ask students why gasoline is formulated differently for different regions of the country and for different seasons. *If the gasoline is used in a hot climate, it must contain a greater proportion of larger molecular mass molecules in order to decrease its volatility. Conversely, if it is used in cold weather, it must contain a higher proportion of smaller molecules to ensure adequate vaporization and ignition.* L1

Reteach

On a handout, draw the structural formulas of several hydrocarbons. Ask students to determine whether each compound is an alkane, an alkene, or an alkyne. Then, have students name each compound. L1

Figure 18.11
Fractional Distillation of Petroleum
To separate important components of petroleum using fractional distillation, petroleum or hot crude oil is heated in a furnace. The liquid alkanes are vaporized and allowed to rise in the fractionating tower. Gases that were dissolved in the petroleum are removed at the top of the tower and condensed into liquids that are sold in cylinders. Gasoline is part of the next-lower group of materials drawn off. Below the gasoline fraction come mixtures of heavier hydrocarbons, such as kerosene, fuel and lubricating oils, and asphalt.

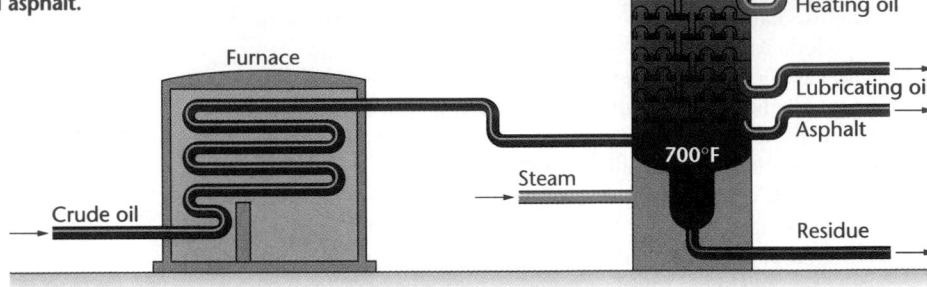

WORD ORIGIN

distillation:
de (L) down
stillare (L) to drop
Distillation is a process by which one compound is removed from others drop by drop as it evaporates out of a solution.

One property is boiling point. As you learned in Chapter 5, distillation is a technique used to separate substances that have different boiling points. In the petroleum industry, huge towers are used to distill petroleum into its component liquids. Inside the tower, many plates provide multiple surfaces on which repeated vaporization-condensation cycles take place. Repeated cycles provide for more efficient separations and allow fractions containing only one or a few different compounds to be isolated. This method of separation is called **fractional distillation.**

Figure 18.11 shows where different products are drawn off according to boiling points at different levels of the tower. The temperature of the tower is controlled so that it is hotter at the bottom and cooler at the top. Because boiling point increases along with molecular weight, the heavier, higher-boiling components condense near the bottom, and the lighter, lower-boiling molecules condense toward the top. Pipes connected at various points allow chemical engineers to draw off each fraction.

The process of fractional distillation does not produce quantities of hydrocarbons in the proportions needed in industry. For example, gasoline is usually the fraction of petroleum most in demand, but this fraction makes up less than half of petroleum. How can the quantities of useful hydrocarbons be increased? The hydrocarbons in the fractions coming off the lower parts of the tower are mostly large alkanes. These can be converted into smaller, more useful alkanes and alkenes by a process called **cracking.** Cracking uses catalysts or high temperatures in the absence of air to break down or rearrange large hydrocarbons. Cracking is also used to increase the yield of natural gas by producing small alkenes from larger molecules. The cracking of propane produces some methane and ethene, in addition to propene and hydrogen.

$$2CH_3CH_2CH_3 \xrightarrow{\quad\quad} CH_4 + CH_2{=}CH_2 + CH_3CH{=}CH_2 + H_2$$
$$500{-}700°C$$

Chemistry Journal

Uses of Petroleum

Ask students to write a short essay, noting their individual use of petroleum and its products, either for fuel or as plastics and other materials. They should also discuss ways in which they can conserve this resource by their actions. Instruct them to include this essay in their journals. L1

Underground coal mining can be difficult, dirty, and dangerous work. The coal seams are often found deep underground, and tunnels must be dug so that miners can reach them. A mining machine breaks up the coal and delivers it to a conveyor belt, which carries the coal chunks out of the mine. ▼

Figure 18.12
Sources of Coal
▲ Coal is our most plentiful fossil fuel. The most inexpensive way to obtain coal that is not deeply buried is by strip-mining, in which large areas of land are stripped bare of all vegetation. This causes environmental problems when the exposed soil washes away after the coal is removed. Laws in the United States now require that most stripped areas be restored.

Another process that uses heat, pressure, and catalysts to convert large alkanes into other compounds is called **reforming.** It is used to form aromatic hydrocarbons.

Like petroleum and natural gas, coal also is a fossil fuel. It is formed from the remains of plants that became buried underwater and were subjected to increasing pressures as layers of mud built up. Coal is composed primarily of carbon but also contains many mineral impurities. It is used mostly as a fuel and as a source of aromatic hydrocarbons. Coal must be obtained from underground or surface mines, as shown in **Figure 18.12.**

SECTION 18.1

Extension
Have interested students use a chemistry handbook to verify that straight-chain hydrocarbons have higher melting and boiling points than do their branched isomers. [L3]

4 CLOSE

Quick Demo

Why does napthalene smell?
Before the class period begins, set a small amount of napthalene in a beaker in an inconspicuous place in the classroom. Wait for students to notice the odor and ask about it. Then draw the molecular structure on the chalkboard and ask what the odor tells about the compound. *It has high volatility and, therefore, weak intermolecular bonds.*

SECTION REVIEW

Understanding Concepts

1. Name the organic compounds shown.

   a) $CH_3CH=CHCH_2CH_2CH_3$

   b) $CH_3CHCH_2CH_2CH_3$ with CH_3 branch

   c) $CH_3CH_2CH_2CH_2CH_2CH_2CH_3$

   d) $CH_3CH_2CHCH_2CHCH_3$ with CH_3 and CH_3 branches

2. Draw structures that correspond to the names of the hydrocarbons given.

   a) hexane
   b) 3-ethyloctane
   c) *trans*-2-pentene

   d) ethyne
   e) 1,2-dimethylcyclopropane
   f) 1-butyne

3. What are the major sources of hydrocarbons? Where are they found?

Thinking Critically

4. **Interpreting Chemical Structures** Halogen molecules such as Br_2 can be added to double bonds in a reaction similar to hydrogenation. Draw the structure of the product that forms when Br_2 is added to propene.

Applying Chemistry

5. **Storage of Hydrocarbons** Would it be more important to store octane or pentane in a tightly sealed bottle at a low temperature? Why?

SECTION REVIEW

1. **a.** 2-hexene
   b. 2-methylpentane
   c. heptane
   d. 2,4-dimethylhexane

2. **a.** $CH_3CH_2CH_2CH_2CH_2CH_3$
   b. $CH_3CH_2CHCH_2CH_2CH_2CH_2CH_3$ with CH_2CH_3 branch

   c. H and CH_2CH_3 on C=C; H_3C and H

   d. $HC\equiv CH$

   e. H CH_3 on C; H, H on C—C; H, CH_3

   f. $HC\equiv CCH_2CH_3$

3. fossil fuels such as coal, natural gas, and petroleum; deep in Earth, under the ocean

4. Br Br on CH_2CHCH_3

5. Pentane. Because it has a shorter carbon chain than octane, it has a lower boiling point and will be more likely to vaporize away.

Substituted Hydrocarbons

PREPARE

Key Concepts
In this section, organic compounds that contain atoms of elements other than carbon and hydrogen are introduced and discussed.

Planning Ahead
For the MiniLab Be sure you have methanol and salicylic acid.

1 FOCUS

Focus Transparency

Display the **Section Focus Transparency** for Section 18.2 to illustrate some uses of substituted hydrocarbons. L1
LEP

SECTION PREVIEW

Objectives
Compare and contrast the structures of the major classes of substituted hydrocarbons. Summarize properties and uses of each class of substituted hydrocarbon.

Key Terms
substituted hydrocarbon
functional group

Doesn't that apple pie look good, fresh out of the oven? Looking at the photo, you can almost smell the apples, cinnamon, nutmeg, and perhaps vanilla. What causes these ingredients to have aromas? Apples, cinnamon, nutmeg, vanilla, and many other fruits and spices all contain molecules with a distinctive group of atoms located at the end of the molecule. This atomic arrangement imparts a pleasant odor to the molecule.

Functional Groups

When chlorine and methane gases are mixed in the presence of heat or ultraviolet light, an explosive reaction takes place in which a mixture of the following four products is produced. Because structures of these compounds are the same as hydrocarbons except for the substitution of atoms of another element for part of the hydrocarbon, they are called **substituted hydrocarbons.** The part of a molecule having a specific arrangement of atoms that is largely responsible for the chemical behavior of the parent molecule is called a **functional group.** Functional groups can be atoms, groups of atoms, or bond arrangements. Notice that in the structures shown, one or more of the hydrogen atoms in methane is replaced by a chlorine atom.

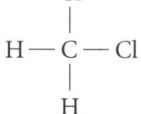

Chloromethane
(methyl chloride)

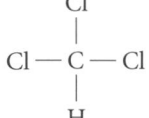

Dichloromethane
(methylene chloride)

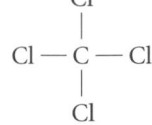

Trichloromethane
(chloroform)

Tetrachloromethane
(carbon tetrachloride)

Replacing part of a hydrocarbon with functional groups changes the structure, properties, and uses that we have for the compounds. For example, chloromethane, a gas, is used as a refrigerant, and dichloromethane, a liquid, is used as a solvent to decaffeinate coffee. You may know trichloromethane by its common name, chloroform. It was an early anesthetic, used to put people to sleep during surgery. Tetrachloromethane, commonly called carbon tetrachloride, is a solvent that was used in dry cleaning and in fire extinguishers.

Program Resources

Study Guide, pp. 73-74 L1
Laboratory Manual, 18.2, pp. 173-176 L1
Section Focus Transparency 37 and **Master,** p. 37 L1 LEP
Problem Solving Transparency 21 and **Master,** pp. 41-42 L1 LEP
ChemLab and MiniLab Worksheets, p. 119 L1

Applying Scientific Methods in Chemistry, pp. 45-46 L1
Chemistry and Industry, pp. 49-52 L1

Some functional groups are complex in structure and consist of a group of atoms rather than a single atom. These groups of atoms often contain oxygen or nitrogen, and some contain sulfur or phosphorus. Double and triple bonds also are considered to be functional groups. Many organic compounds contain more than one type of functional group. Simple organic compounds are grouped into categories depending on the functional group they contain. You will study structures and examples of the most important functional groups.

Functional Groups: Structure and Function

As you study these functional groups, notice that the similarity in properties among molecules containing a given functional group leads to the use of members of that group for similar purposes. In the structures shown below, R and R' each represent a hydrocarbon part of the molecule. R and R' may be the same, or they may be different. For example, in the formula for an ether, $R—O—R'$, where R represents $CH_3CH_2—$ and R' represents CH_3, the ether has the formula $CH_3CH_2OCH_3$.

1 Halogenated Compounds

structure: $R—X$, where X = F, Cl, Br, or I

functional group: halogen atoms
properties: high density
uses: refrigerants, solvents, pesticides, moth repellents, some plastics
biological functions: thyroid hormones
examples: chloroform, dichloromethane, thyroxin, Freon, DDT, PCBs, PVC

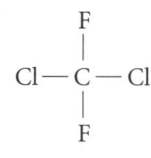

$$Cl—\underset{\underset{\displaystyle F}{|}}{\overset{\overset{\displaystyle F}{|}}{C}}—Cl$$

Freon

Chlorofluorocarbons (CFCs) are substituted compounds containing chlorine and fluorine atoms bonded to carbon. The most common CFC, Freon, has the formula CCl_2F_2. CFCs were once widely used as propellents in aerosol cans; as solvents; as the foaming agent in the manufacture of plastic foam materials; and as refrigerants in air conditioners, refrigerators, and freezers. However, in 1987, the major industrial nations of the world agreed to a gradual reduction in the use of CFCs because they cause a depletion of the valuable ozone in Earth's upper atmosphere. CFCs are being replaced by other halogenated compounds that are not as damaging to the atmosphere.

18.2 Substituted Hydrocarbons   641

Across the Curriculum

Etymology

Because the carboxylic acids were one of the first groups of organic chemicals studied, their common names were derived from their sources rather than their chemical structures, which were not known at the time. The acid with a single carbon, formic acid (Latin *formica,* ant), is found in ant secretions and causes the sting that results from ant bites. Butyric acid (Latin *butyrum,* butter) is the four-carbon acid that causes the smell of rancid butter. The *Valeriana* genus of old-world herbs gave the five-carbon acid found in those plants its name, valeric acid. The six, eight, and ten-carbon acids were all discovered in goat fat, so they are called caproic, caprylic, and capric acids (Latin *caper,* goat). Palmitic (Latin *palma,* palm) acid with its 16 carbons is found in the yellowish oil from the fruit of several varieties of palm, and the 18-carbon oleic (Latin *oleum,* oil) acid got its name because it is found in a great variety of oils.

Correcting Misconceptions

Students are often confused about the exact meaning of *R* when it is used to represent an alkyl group in the generic formula for a type of organic compound. For example, the *R* in the generic formula for an alcohol, *R*—OH, must contain at least one carbon atom. However, the *R* in the generic formula for a carboxylic acid, an aldehyde, and an ester may be, in the simplest case, just a hydrogen atom. It is useful to replace such *R* groups in the generic formulas with an asterisk (*) to lessen confusion.

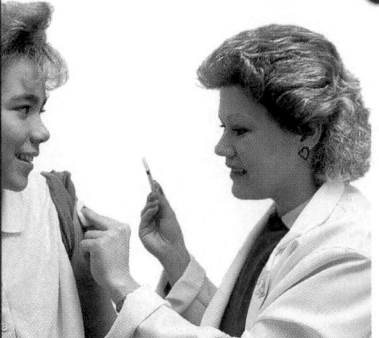

② **Alcohols**

structure: *R*—O—H

functional group: hydroxyl group (—OH); hydrogen atom bonded to an oxygen atom, which is bonded to the hydrocarbon part of the molecule

properties: polar, so water molecules are attracted to it; high boiling point; alcohols with low molecular weights are water-soluble

uses: solvents, disinfectants, mouthwash and hair-spray ingredient, antifreeze

biological functions: reactive groups in carbohydrates, product of fermentation

examples: methanol, ethanol, isopropanol (one type of rubbing alcohol), cholesterol, sugars

Ethanol

An organic compound that contains at least one hydroxyl group is called an alcohol. The name of an alcohol ends in *-ol*. Alcohols have many different uses. One of the most important is as a disinfectant for killing bacteria and other potentially harmful microorganisms. For this reason, ethanol is put into mouthwashes, and rubbing alcohol is used as a disinfectant. Antifreeze also is an alcohol.

③ **Carboxylic Acids**

structure: *R*—C—O—H (with O double-bonded to C)

functional group: carboxyl group (—COOH); oxygen atom double-bonded to a carbon, which is also bonded to a hydroxyl group and the hydrocarbon part of the molecule

properties: acidic, usually water-soluble, strong unpleasant odors, form metal salts in acid-base reactions

uses: vinegar, tart flavoring, skin-care products, production of soaps and detergents

biological functions: pheromones, ant-sting toxin; causes rancid-butter and smelly-feet odors

examples: acetic acid (in vinegar), formic acid, citric acid (in lemons), salicylic acid

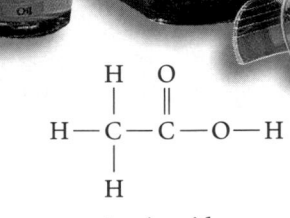

Acetic acid

A compound containing a carboxyl group is known as a carboxylic acid, or organic acid. Many pheromones contain carboxyl functional groups. Pheromones are organic compounds used by animals to communicate with each other. When an ant finds food, it leaves behind a pheromone trail that other ants in its colony can follow to get to the food source.

642   Chapter 18   Organic Chemistry

Integrating THE Sciences

Biology Pheromones are organic compounds produced by an organism to communicate with others of the same species. Some pheromones are used to attract members of the opposite sex, some to sound an alarm when danger is sensed, others to mark territories, and still others to mark trails. Most pheromones known so far are made by organisms such as insects that have few other means of communicating, but many mammalian pheromones are known, as well. This chemical communication can cover vast distances; a male gypsy moth can detect the pheromone of a female a mile away. Many pheromones are long-chain alkenes containing a small number of functional groups.

Reactions involving pheromones are isomer-specific. Sometimes, only the *cis* or the *trans* isomer of a given alkene pheromone is biologically active, and the presence of even a small amount of the other isomer can render some pheromones completely inactive. Pheromones have been used in agriculture to trap insect pests and to induce porcine mating.

4 **Esters**

structure: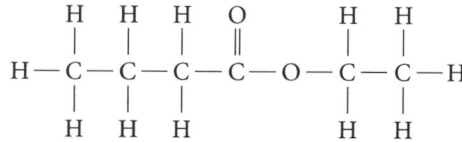

derived from carboxylic acids in which the — OH of the carboxyl group has
    been replaced with an — OR from an alcohol
properties: strong aromas, volatile
uses: artificial flavorings and fragrances, polyester fabric
biological functions: fat storage in cells, DNA phosphate-sugar backbone,
    natural flavors and fragrances, beeswax
examples: banana oil, oil of wintergreen, triglycerides (fats)

A compound formed from the reaction of an organic acid and an alcohol is called an ester. Some esters are used in the production of polyester fabrics. Many esters are used as flavorings in food products. Natural flavors are often complex mixtures of esters and other compounds, whereas artificial flavors usually contain fewer compounds and may not taste exactly the same as the natural flavor.

Ethyl butyrate
(pineapple flavoring)

Applying Chemistry

Polyesters Have interested students research and report to the class about polyesters. They should learn that polyester is not a single compound. Rather, the term refers to a class of synthetic polymers. Dacron, Fortran, and Kodel are some of the fabrics that are produced when these polymers are drawn into fibers and woven. [L2]

18.2   **Substituted Hydrocarbons**   **643**

Chemistry Journal

Esters

Have students research the various food aromas that are due to esters. Students should choose one or two aromas and write a journal entry for each, describing the aroma and giving the structure of the ester and the carboxylic acid and alcohol from which it is derived. [L1]

***inter*NET**
CONNECTION

Molecule models in MIME format can be downloaded.

World Wide Web
http://www.ch.ic.ac.uk/
chemical_mime.html

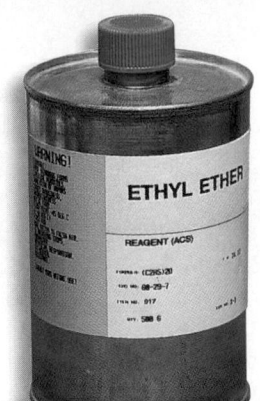

5 Ethers

structure: R—O—R'; oxygen atom bonded to two hydrocarbon groups

properties: mostly unreactive, insoluble in water, volatile
uses: anesthetics, solvents for fats and waxes
examples: diethyl ether

Ethyl ether

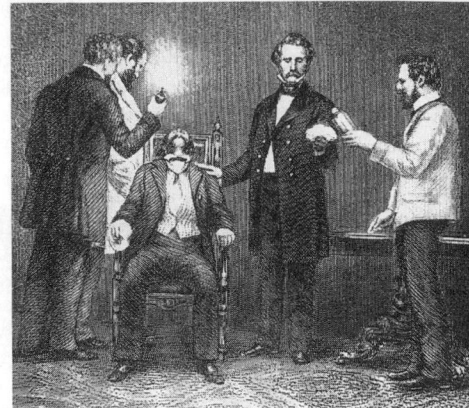

An ether is an organic compound in which an oxygen atom is bonded to two hydrocarbon parts of the molecule. Diethyl ether, often simply called ether, is an effective anesthetic. Because it is insoluble in water, it passes readily through the membranes surrounding cells. Ether is rarely used as an anesthetic today because it is highly flammable and because it causes nausea.

6 Ketones and Aldehydes

structures: ketones: R—C—R' aldehydes: R—C—H

functional group: carbonyl group (—CO); carbon atom double bonded to an oxygen atom
properties: very reactive, distinctive odors
uses: solvents, flavorings, manufacture of plastics and adhesives, embalming agent
examples: acetone; formaldehyde; cinnamon, vanilla, and almond flavorings

Vanillin

Acetone

Aldehydes and ketones are compounds that contain a carbonyl group. If the carbonyl group is on the end of the carbon chain, the compound is called an aldehyde. If the carbonyl group is not at the end of the chain, the compound is a ketone. Acetone is a ketone solvent commonly used in nail-polish remover. Nail polish is not soluble in water—if it were, it would come off when you wash your hands. It is soluble in many organic compounds such as acetone, which is used to remove it.

644

Meeting Individual Needs

Gifted Every common monosaccharide contains either an aldehyde or a ketone group. Have interested students examine the structures of several monosaccharides to determine the existence of one or the other of these two carbonyl groups. Students can then classify the monosaccharides as aldoses, those with an aldehyde group, or ketoses, those with a ketone group. Students might also be interested in learning about some of the different properties of these two classes of monosaccharides. **L3**

⑦ Amines and Amides

structure: amine: $R{-}NH_2$ amide: $R{-}\overset{\overset{\displaystyle O}{\|}}{C}{-}NH_2$

functional group:
amine: amino group ($-NH_2$); two hydrogen atoms bonded to a nitrogen atom, which is bonded to the hydrocarbon part of the molecule
amide: amino group bonded to a carbonyl group ($-CONH_2$)
properties: amines: basic, ammonia-like odor
        amides: neutral, most are solids
uses: solvents, synthetic peptide hormones, fertilizer, nylon synthesis
biological functions: in amino acids, peptide hormones, and proteins; distinctive odor of some cheeses
examples: urea, putrescine, cadaverine, Nutrasweet

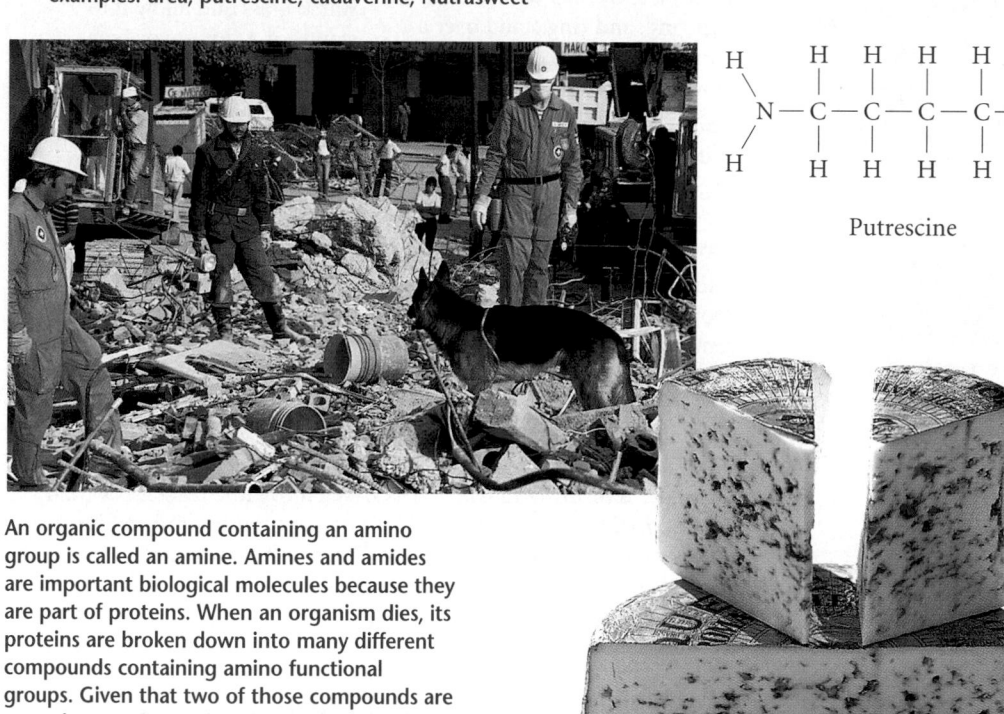

Putrescine

An organic compound containing an amino group is called an amine. Amines and amides are important biological molecules because they are part of proteins. When an organism dies, its proteins are broken down into many different compounds containing amino functional groups. Given that two of those compounds are named putrescine and cadaverine, what type of odor do you think they have? These compounds have a distinctive, unpleasant smell that specially trained sniffer dogs can use to locate human remains and to help in forensic investigations. Cadaverine also contributes to bad breath.

Discussion

Discuss with students the role of amines in the ecological nitrogen cycle. All decaying animal tissues release amines into their environment. In the oceans, the deaths of marine animals provide a constant supply of amines to ocean water and to the air above the oceans. These amines are taken up by living organisms, providing essential nitrogen. Amines are partly responsible for the characteristic odor of the ocean and sea organisms.

18.2   Substituted Hydrocarbons   **645**

Chemistry Journal

Synthetic Dyes

TECH PREP

Most synthetic dyes are organic molecules containing functional groups that act as chromophores, or color-producers. Among these chromophores are hydroxyl, amino, methyl, and ether groups, along with aromatic rings. Different numbers and arrangements of these groups cause compounds to selectively absorb and reflect different wavelengths of light. Have interested students research and write reports in their journals about some of these synthetic dyes. [L3]

miniLAB

A Synthetic Aroma

Purpose: To produce the ester methyl salicylate and associate it with its characteristic wintergreen odor.

Process Skills: Observing, measuring, interpreting data

Safety Precautions: Aprons and safety goggles must be worn. The concentrated sulfuric acid is corrosive and must be used carefully.

Disposal: Codes A, D

Teaching Strategies

- When students discover the wintergreen odor, tell them that the molecules that constitute many flavors and odors occur naturally, but often the same molecules can be made synthetically.

Expected Results

There will be no visible reaction, but the odor of wintergreen will be noticeable in step 7.

Analysis

1. wintergreen
2.

Salicylic acid   Methanol

Methyl salicylate   Water

3. ethanol and butyric acid

✓ Assessment

Portfolio: Ask students to look through their food and drug products at home. Have them identify synthetic colors and flavors and include the list in their portfolios along with the report for the MiniLab.

L1 P

miniLAB 2

A Synthetic Aroma

When brought together under the proper conditions, an organic acid and an alcohol will react to form an ester. Esters are generally volatile compounds having distinctive odors. In this MiniLab, you will combine methanol and salicylic acid to produce the ester, methyl salicylate.

Procedure

1. Put on an apron and goggles.
2. Set up a hot-water bath for the reaction by filling a 250- or 400-mL beaker about halfway with tap water and setting it on a hot plate or on a wire gauze, iron ring, and ring stand over a burner flame. Allow the water to warm but not boil.
3. Place 3 mL of methyl alcohol and 1 g of salicylic acid in a large (20 mm × 150 mm) test tube. Mix the reactants with a glass stirring rod.
4. Add about 0.5 mL (about ten drops from a dropper) of concentrated sulfuric acid to the reaction mixture and stir. Do not draw the acid into the rubber bulb of the dropper. (**CAUTION:** *Concentrated sulfuric acid is corrosive to skin, eyes, body tissues, and most other organic materials. If any contact occurs, notify your teacher immediately.*)

Plants are a common source of esters.

5. Place the test tube and contents in the hot-water bath and allow it to warm for five or six minutes.
6. Pour the contents of the test tube into about 50 mL of cold, distilled water in a small beaker. Cover the beaker with a watch glass and allow it to stand for a minute or two.
7. Remove the watch glass and waft the methyl salicylate odor toward your nose. Record your observations.

Analysis

1. What familiar odor is caused by methyl salicylate?
2. Look up the structural formulas for the reactants in a reference book, and write the equation for the reaction using structural formulas for all organic reactants and products.
3. What organic reactants would be required to produce the ester, ethyl butyrate?

The Sources of Functional Groups

How can different compounds containing functional groups be made? Alkanes are not very reactive, so it can be difficult to react them directly to form substituted molecules. You read earlier of one reaction used to substitute chlorine atoms for hydrogen atoms in methane. However, that reaction is not practical because a mixture of four different products is formed. Each must be separated and purified before it can be used.

Alcohols are a better choice than alkanes as the raw materials for synthesizing organic molecules. Alcohols can be converted into compounds containing almost every other type of functional group. For example, an acid and an alcohol react together to form an ester and water.

Across the Curriculum

History

From cave paintings to bright neon signs, people have surrounded themselves with color. Today, natural dyes that were used for thousands of years have largely been replaced with synthetics. Since 1600 B.C., one of the few natural permanent purple dyes has been extracted from small mollusks in the Mediterranean Sea. The mollusks are hard to collect, and it takes about 9000 of them to make 1 g of dye, so the dye has always been expensive. This is why Tyrian purple became associated with the royalty who could afford it. A British chemist named William Perkin accidentally made the first synthetic purple dye in 1856 in his home laboratory while he was trying to prepare the antimalarial drug quinine. Perkin made a successful business selling the dye to the French, who called it *mauve.*

Figure 18.13
Production of Ethanol
Ethanol can be produced industrially by the fermentation of sugars and starches by yeast cells. These sugars and starches come from plant material such as corn, molasses from sugarcane, and the grapes grown in vineyards like this one.

Where do all the alcohols needed by industry come from? Cracking of petroleum products produces alkenes, which can readily be converted into alcohols by reaction with water. This process is used to synthesize ethanol, the two-carbon compound that contains an alcohol functional group.

$$
\begin{array}{ccc}
\underset{\text{Ethene}}{\overset{\displaystyle H\diagdown \quad \diagup H}{\underset{\displaystyle H\diagup \quad \diagdown H}{C=C}}} & + & \underset{\text{Water}}{H_2O} & \rightarrow & \underset{\text{Ethanol}}{\overset{\displaystyle H \quad OH}{\underset{\displaystyle H \quad H}{H-C-C-H}}}
\end{array}
$$

A second important natural source of ethanol provides a convenient supply of this reactive molecule. Ethanol is produced by yeast cells when they ferment the sugars and starches in plant materials, as discussed in **Figure 18.13.** Fermentation produces much smaller quantities of ethanol than does the reaction between ethene and water; it is used in industry mainly to produce the ethanol in alcoholic beverages.

SECTION REVIEW

Understanding Concepts

1. Explain why organic compounds with similar structures often have similar uses.

2. Distinguish between an aldehyde and a ketone.

3. Redraw the following structure of thyroxine, a thyroid hormone. Circle and name each type of functional group present.

$$HO-\overset{I}{\underset{I}{\bigcirc}}-O-\overset{I}{\underset{I}{\bigcirc}}-CH_2CHCOOH \\ \qquad\qquad\qquad\qquad\qquad\qquad | \\ \qquad\qquad\qquad\qquad\qquad\qquad NH_2$$

Thinking Critically

4. **Inferring** Explain why alkenes and alkynes contain functional groups but are not substituted hydrocarbons.

Applying Chemistry

5. **Antifreeze** Ethylene glycol, used as an antifreeze in car radiators, has two hydroxyl groups, as shown below. What can you infer about the boiling point, freezing point, and solubility in water of this compound from its structure?

$$\overset{\displaystyle OH \quad OH}{\underset{\displaystyle H \quad H}{H-C-C-H}}$$

18.2   **Substituted Hydrocarbons**   **647**

SECTION REVIEW

1. Uses depend on properties, which depend on structures.
2. Ketones have a carbonyl group bonded between two hydrocarbon groups. In an aldehyde, one of the hydrocarbon groups is replaced by a hydrogen atom.
3. one hydroxyl group, four halogen (I) atoms, one ether group, one amino group, one carboxyl group, two benzene rings
4. Alkenes and alkynes contain multiple

bonds, which are functional groups. Substituted hydrocarbons contain groups of atoms in place of a part of the hydrocarbon. Alkenes and alkynes do not have any substitute groups.
5. The two hydroxyl groups make ethylene glycol very polar. Therefore, it could be expected to have a high boiling point and low freezing point, and be soluble in water.

PREPARE

Key Concepts

Synthetic and natural polymers are examined along with the processes of addition and condensation polymerization. Natural and synthetic rubbers and plastics are briefly discussed.

Planning Ahead

For the Demonstrations Obtain hydrolyzed polyvinyl alcohol for Demonstration 18.6. Check to be sure you have TTE, sebacoyl chloride, and 1,6-diaminohexane for Demonstration 18.7.

For the Quick Demo in *Close* If you do not have a set of "smart/stupid" balls, order them from Flinn Scientific.

For the ChemLab You will need seven types of fabrics: silk, wool, cotton, nylon, acrylic, acetate, and polyester. You might obtain small samples of each from the home economics department.

1 FOCUS

Focus Transparency

 Display the **Section Focus Transparency** for Section 18.3 to show students how polymers are formed. L1 LEP

Plastics and Other Polymers

SECTION PREVIEW

Objectives

Identify the monomers that form specific polymers, and draw structural formulas for polymers made from given monomers.
Differentiate between condensation and addition polymerization reactions.
Summarize the relationship between structure and properties of polymers.

Key Terms

polymer
monomer
addition reaction
condensation reaction
cross-linking
thermoplastic
thermosetting

On April 6, 1938, a young chemist working on the preparation of a compound used in refrigeration opened the valve on a tank of the reactant he planned to use in this process, tetrafluoroethene gas. Roy J. Plunkett was puzzled when no gas came out after he opened the valve because the tank was heavy enough to indicate that it was full of gas. He poked a wire through the opening of the valve on the tank and found that it wasn't clogged.

What do you think you would have done if you had been in Plunkett's shoes? Would you have discarded the tank in favor of a new one? Or would you have seen this problem as an intriguing puzzle to be investigated? Plunkett was curious, and when he cut open the tank, he discovered a white, waxy solid in place of the gas he expected. Today, this solid is known as Teflon, a large molecule with remarkable properties. You are probably familiar with Teflon as a nonstick coating on pots and pans, but it is also used for dentures, artificial joints and heart valves, space suits, and fuel tanks on space vehicles.

Monomers and Polymers

How did a white, waxy solid form from the colorless gas tetrafluoroethene (also called tetrafluoroethylene)? One clue to what type of reaction occurred can be found in the properties of those two substances. Most nonpolar molecules that are relatively small tend to be gases at room temperature, whereas larger molecules usually are solids. When the structure of Teflon was determined, the molecule was found to consist of chains of carbons with fluorine atoms attached. Somehow, the tetrafluoroethylene molecules in the gas had reacted with each other to form these long-chain molecules. A large molecule that is made up of many smaller, repeating

---

Program Resources

Study Guide, pp. 75-76 L1
Section Focus Transparency 38 and
  Master, p. 38 L1 LEP
Basic Concepts Transparency 51 and
  Master, pp. 101-102 L1 LEP
Consumer Chemistry, pp. 35-36 L1
ChemLab and MiniLab Worksheets,
  pp. 113-116, 120 L1
Chemistry and Industry, pp. 41-44 L1

Demonstration 18.5

Polymers

Purpose: To help students see the relationship between a polymer and a monomer.
Materials: 10 spring-type clothespins, 1 box of paper clips
Safety Precautions:
Disposal: Code F
Procedure: Clip one clothespin onto the leg of another one and continue this process to form a polymer chain. Each clothespin repre-

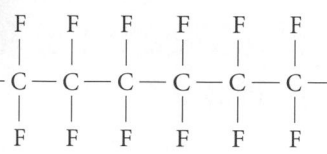

F₂C=CF₂ structure (Tetrafluoroethylene):

$$F_2C=CF_2$$

Tetrafluoroethylene

Figure 18.14
Teflon
Teflon provides a good coating for this skillet because it is unreactive and food will not stick to it.

Teflon polymer structure:

$$-C-C-C-C-C-C-$$

Teflon polymer

units is called a **polymer.** A polymer forms when hundreds or thousands of these small individual units, which are called **monomers,** bond together in chains. The monomers that bond together to form a polymer may all be alike, or they may be different. When Teflon formed in the tank, many small tetrafluoroethylene monomers combined to form long polymers of polytetrafluoroethylene. Examine the structures in **Figure 18.14** to help you visualize this reaction.

The properties of a polymer are different from those of the monomers that formed it. For example, the polyethylene plastic in milk jugs is made when molecules of gaseous ethylene react to form long chains. The unique properties of a given polymer, such as tensile strength, water-repellency, or flexibility, are related to the polymer's enormous size and the way its monomers join together.

Synthetic Polymers

Polymers are everywhere, making fabrics such as nylon and polyester, plastic wrap and bottles, rubber bands, and many more products you see every day. How many polymers can you identify in **Figure 18.15?** What do the polymers that make up such different substances have in common? All are large molecules made of smaller, repeating units.

Chemists have been synthesizing polymers in laboratories for only about 100 years. Can you imagine what life was like before people had these synthetic polymers? You might have gotten wet on the way to school without a nylon raincoat, eaten a stale sandwich for lunch without plastic wrap or a container to put it in, and worn a heavier cotton uniform in your sports activity instead of a lightweight, synthetic fabric. Many polymers have added conveniences that we take for granted in our lives.

WORD ORIGIN

polymer:
  poly (GK) many
  meros (GK) part
Polymers are large molecules made of many smaller parts.

Figure 18.15
Synthetic Polymers
Sports activities would be different today without synthetic polymers. Balls, uniforms, artificial turf, bandages used to wrap sprains, and nets used in hoops and goals are usually made of synthetic polymers.

SECTION 18.3

2 TEACH

Visual Learning

Figure 18.14: Construct two or three molecular models of tetrafluoroethylene, and then demonstrate their polymerization and the formation of Teflon by disrupting the double bond in each molecule and linking the molecules.

GLENCOE TECHNOLOGY

 **Videodisc**
Chemistry: Concepts and Applications
Polymers
Disc 2, Side 2, Ch. 13

Show this video of the demonstration on these pages to reinforce the concept of polymer formation from monomers. L1 LEP

 CD-Rom
Chemistry: Concepts and Applications
Polymers
Use this video to help reinforce the process of making polymers from monomers. L1 LEP

sents a monomer. You can also use paperclips to represent the monomers and form a polymer chain by linking them together. If you already have a paper-clip chain made, place it in a small box before class begins. In front of your students, add loose, individual paper clips to the box. Cover the box and shake it to simulate the reaction. Uncover the box and pull out the previously connected paper clips to the amazement of your students.
Results: The monomers link together to form a long-chain polymer.
Analysis: Ask these questions.
1. What are the small molecules that link together called? *monomers*
2. Are the properties of polymers different from those of monomers? *yes*

✓ Assessment

Knowledge: Ask students to write as many names as they can that begin with the prefix *poly.* Possible responses: *polypropylene, polyethylene, polyvinyl chloride, polystyrene, polyester, polyvinylacetate.* L1

ChemLab

Identification of Textile Polymers

Time Allotment: One class period.

Objectives

Review objectives with students before they begin the ChemLab.

Process Skills

Observing, recognizing cause and effect, interpreting data, communicating, inferring, comparing and contrasting, classifying

Safety Precautions and Disposal

Students should wear goggles and aprons. They should be advised to avoid contact with the concentrated sulfuric acid and sodium hydroxide solutions, which could damage their eyes, skin, or clothing. If contact occurs, they should rinse the area with tap water and call for help. Warn students not to inhale odors from burning plastics. They should wash their hands thoroughly after the lab.

Disposal: Neutralize the $3M$ NaOH with dilute hydrochloric acid solution or vinegar before pouring the liquid down the drain. The barium chloride waste should be converted to barium sulfate by adding 10% sodium sulfate to the barium chloride salt. Filter this solution to collect the fine barium sulfate precipitate, which should be stored until it can be disposed of properly. Other liquid wastes produced in this experiment can be diluted with water and poured down the drain.

PREPARATION

- Prepare $3M$ BaCl$_2$ by adding 62.4 g of barium chloride to 70 mL of water, stirring until dissolved, and then bringing the volume to 100 mL.

ChemLab

Identification of Textile Polymers

For centuries, polymers have been used for clothing in the form of cotton, wool, and silk. Chemists trying to improve these natural polymers to fit specific purposes have designed many synthetic polymers that also are used to make fabrics. Among these are nylon, rayon, acetate, and the polyesters.

Many synthetic fabrics are designed to look and feel like natural polymers but also have the superior properties for which they were designed, such as being wrinkle-resistant, water-repellent, or quick-drying. However, these imitators can't fool a chemist; simple tests that can be done in minutes will distinguish the real thing from a pretender. This is because differences in their structures lead to differences in properties. In this experiment, you will identify fabric samples by testing them for characteristic properties.

Problem

What tests can be performed that will differentiate among the polymers used to make fabrics?

Objectives

- **Analyze** changes in fabric samples that are subjected to flame and chemical tests.
- **Classify** fabrics by polymer type based on test results.

PREPARATION

Materials

fabric samples A-G (7 types; six 0.5 × 0.5 cm squares of each type)	beaker containing water
Bunsen burner	forceps
	test-tube holder
	test-tube rack

medium test tubes (4)   Ca(OH)$_2$
balance $3M$ BaCl$_2$
watch glass   concentrated H$_2$SO$_4$
red litmus paper   iodine solution
stirring rod $0.05M$ CuSO$_4$
100-mL beakers (2) $3M$ NaOH
10-mL pipet   acetone
25-mL graduated cylinder

Safety Precautions

Use care when working with an open flame and handling concentrated acids and bases. Do not inhale odors from plastics. Wear an apron and goggles at all times.

PROCEDURE

Flame Tests

1. Use forceps to hold a square of fabric A in a Bunsen burner flame for 2 seconds.

2. Remove the fabric from the flame, and blow out the flame if the fabric keeps burning.

3. Observe the odor by wafting smoke from the smoldering sample toward your nose. Be sure the fabric is no longer burning by immersing it in a beaker of water.

4. Make a data table similar to Table 1 in Data and Observations on page 652. Record your observations about the way the fabric burns in the flame, odor observed, and characteristics of the residue left after burning.

5. Repeat steps 1 to 4 six times, using fabric samples B-G.

6. Use the following table to make a preliminary identification of your samples.

- Prepare $0.05M$ CuSO$_4$ by adding 4 g of copper(II) sulfate to 450 mL of water, stirring until dissolved, and bringing the volume to 500 mL.
- Prepare $3M$ NaOH by adding 24 g of sodium hydroxide to 150 mL of water, stirring until dissolved, and then bringing the volume to 200 mL. **CAUTION:** *Add the base to the water carefully.*

Polymer Type	Type of Burning	Burning Odor	Residue Type
Silk or wool	Burns and chars	Hair	Crushable bead
Cotton	Burns and chars	Paper	Ash
Nylon, polyester, acetate, or acrylic	Burns and melts	Chemical	Plastic bead

Chemical Tests

Use your preliminary identifications to determine which tests are necessary to identify each fabric sample. You should not have to carry out every test on every sample. Make a data table similar to Table 2 on page 652, and record all your results.

7. **Nitrogen Test** Place a fabric square in a test tube and add 1 g of $Ca(OH)_2$. Using a test-tube holder, heat the tube gently in a Bunsen burner flame while holding a piece of red litmus paper with forceps over the mouth of the tube. If the litmus paper turns blue, nitrogen is present. Only silk, wool, nylon, and acrylic contain nitrogen.

8. **Sulfur Test** Add a fabric square to 10 mL of $3M$ NaOH in a test tube, and gently heat to boiling by holding the tube in a Bunsen burner flame. Be careful that the tube is not pointing toward anyone. Cool the solution, add ten drops of $BaCl_2$ solution, and observe whether or not a precipitate forms. Only wool contains enough sulfur to give a barium sulfide precipitate.

9. **Cellulose Test** Place a fabric square in a beaker, cover with approximately 2 mL of concentrated H_2SO_4, and then carefully transfer

the contents to another beaker containing ten drops of iodine solution in 25 mL of water. Rinse the empty beaker with large quantities of water. Cotton gives a dark blue color within 1 to 2 minutes, and acetate gives this color after 1 to 2 hours. (**CAUTION:** *Handle the beaker containing sulfuric acid with care.*)

10. **Protein Test** Place a fabric square on a watch glass, and add ten drops of $0.05M$ $CuSO_4$. Wait 5 minutes, and then use forceps to dip the fabric into $3M$ NaOH in a test tube for 5 seconds. Silk and wool are protein polymers, and a dark violet color will appear on those fabrics after doing this test.

DATA AND OBSERVATIONS
Flame Test Observations

Sample	Type of Burning	Burning Odor	Residue Type
silk, wool	burns & chars	hair	crushable bead
cotton	burns & chars	paper	ash
nylon, polyester, acetate, acrylic	burns & melts	chemical	plastic bead

- Require that students' laboratory notebooks contain the following items before they begin the lab: Objectives, Safety Precautions, Outline of the Procedure, Table for Observations.
- If fabric and sulfuric acid stick to the bottom of the beaker when being transferred to the beaker of iodine water in the cellulose test, have students use a wooden splint to facilitate the transfer. They should *not* use metal forceps. Tell them to discard the splint immediately after use because it will probably have concentrated acid on it.
- Have students make flow-charts to follow in order to identify the fabric samples most efficiently.
- Having the required materials and equipment available at each laboratory table or work area will greatly reduce student traffic, the potential for spillage or breakage, and the time required for the work.
- To save time and reagents, tell students to try to do the minimum number of tests needed to determine the identities of the fabrics. For example, a sample that gave off a chemical odor when burned and left a plastic bead residue is not wool or cotton, so the sulfur and cellulose tests are unlikely to give any more useful information.
- Textile samples that have been washed before being used in this lab often are easier to identify because various stainproof and wrinkle-resistant coatings sometimes put on fabrics by manufacturers can interfere with test results. It is also best to use white or other light-colored samples because the blue color that appears in two of the tests is hard to observe on dark fabrics.

ChemLab

ChemLab

ANALYZE AND CONCLUDE

1. The sodium hydroxide helps free the sulfur atoms that are part of the amino acid cysteine, which is abundant in the wool protein.
2. Synthetic polymers burn and melt, giving off an acrid chemical smell and leaving behind a hard plastic bead residue. Natural polymers give off different odors, similar to burning of natural substances such as hair or paper, and they leave behind a softer, ashy residue.
3. Student answers will vary depending on the samples tested.

APPLY AND ASSESS

1. The polymers that leave behind plastic beads have structures similar to those found in plastics.
2. Hair is a protein polymer similar to those that make up silk and wool.
3. A polyester-cotton blend would leave both ash and plastic bead residues after burning. The fabric would turn blue in a cellulose test.
4. yes

✔ Assessment

Skill: Ask students to predict what the flame and chemical test results would be for textile samples that are blends of two or more polymer types, such as cotton-polyester and polyester-nylon. Obtain some samples of blended fabrics, and let students attempt to identify the blends through experimentation. L1

11. **Formic Acid Test** Place any sample to be subjected to the formic acid test in a test tube and bring it to your teacher, who will perform this test in the fume hood. The teacher will add 1 mL of formic acid to the test tube and stir with a glass rod. Note whether or not the fabric dissolves in the solution. Silk, acetate, and nylon dissolve in formic acid.
12. **Acetone Test** Add a fabric square to 1 mL of acetone in a test tube, stir with a glass rod, and note whether or not the fabric dissolves. Only acetate dissolves in acetone. (**CAUTION:** *Be careful to do this test away from any open flames.*)
13. Dispose of all products of your tests as directed by your teacher.

ANALYZE AND CONCLUDE

1. **Thinking Critically** Why do you think it was necessary to add NaOH and heat before adding $BaCl_2$ in the sulfur test?

2. **Comparing and Contrasting** Can you make any conclusions about the way that synthetic and natural polymers burn?
3. **Classifying** Use your data from the flame and chemical tests to classify each fabric sample you tested as a silk, cotton, wool, nylon, acrylic, acetate, or polyester polymer.

APPLY AND ASSESS

1. What does the plastic, beadlike residue left by some polymers after burning tell about the polymers' structures?
2. If burning silk and wool smell like burning hair, what does this tell about the structure of hair?
3. What results of the flame and chemical tests would you expect to see if you were testing a fabric sample that was a polyester-cotton blend?
4. Cellulose is a major component of wood as well as of cotton. Would you expect wood to also give a dark blue color in the cellulose test?

DATA AND OBSERVATIONS

Table 1   Flame Test Observations

Sample	Type of Burning	Burning Odor	Residue Type
A			

Table 2   Chemical Test Observations

Sample	Nitrogen	Sulfur	Cellulose	Protein	Formic Acid	Acetone
A						

DATA AND OBSERVATIONS
Chemical Test Observations

Sample	Positive Tests
silk	nitrogen, protein, formic acid
wool	nitrogen, sulfur, protein
cotton	cellulose
nylon	nitrogen, formic acid
polyester	none
acetate	formic acid, acetone, cellulose (only after 1-2 hours)
acrylic	nitrogen

When Polymers Meet Water

Sometimes, the ability of certain polymers to repel water is useful—this property is what keeps you dry when you wear a raincoat. Other times, it is desirable for polymers to absorb water. This is why wool socks keep your feet warm by wicking water away from your skin and why a diaper helps keep a baby dry. Cloth diapers are made of cotton, a natural polymer that absorbs water well. Why do disposable diapers hold so much more water than cloth diapers? They contain a super-absorbent polymer that can hold hundreds of times its weight in water. In this MiniLab, you will determine how much water the polymers in two different brands of diapers can hold.

miniLAB
3

Procedure

1. Obtain two diapers of approximately the same size but of different brands.

2. Fill a 100-mL graduated cylinder with tap water, and slowly pour the water into the center of one of the diapers. Stop when the water begins leaking out of the diaper.

3. Record the volume of water the diaper holds.

4. Repeat steps 2 and 3 with the other diaper.

Analysis

1. What could be different about the two diapers that makes one of them hold more water than the other?

2. How might placing a superabsorbent polymer around the roots of a houseplant help it to grow?

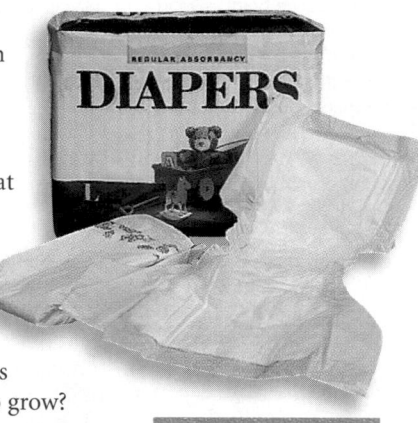

Natural Polymers

Laboratories are not the only place where polymers are synthesized. Living cells are efficient polymer factories. Proteins, DNA, the chitin exoskeletons of insects, wool, silky spiderwebs and moth cocoons, and the jellylike sacs that surround salamander eggs are polymers that are synthesized naturally. The strong cellulose fibers that give tree trunks enough strength and rigidity to grow hundreds of feet tall are formed from monomers of glucose, which is a sweet crystalline solid. Many synthetic polymers were developed by scientists trying to improve on nature. For example, nylon was developed as a possible silk substitute. The idea for the process by which synthetic threads are formed in factories was borrowed from spiders. Observe in **Figure 18.16** the similarities between the spinneret of a spider and an industrial spinneret.

Figure 18.16

Thread Spinners
Long, fine threads are spun when polymer molecules are forced through tiny holes in spinnerets, both natural and industrial.

2. The polymer will prevent water from quickly leaching through the soil and will store water near the roots for the plant to use.

✓ Assessment

Portfolio: Ask students to research and write a report comparing cloth and disposable diapers in regard to cost, quality, consumer preference, and long-term disposal. [L1] [P]

SECTION 18.3

miniLAB

When Polymers Meet Water

TECH PREP

Purpose: Students will compare the water-holding properties of polymers in two brands of disposable diapers.

Process Skills: Observing, interpreting data, recognizing cause and effect

Disposal: Code A

Teaching Strategies

• Ask students to predict how much water a diaper will hold before starting the experiment.

• You may wish to test a cloth diaper as well as disposable brands for comparison.

• One superabsorbent polymer used in some diapers is polymethacrylate. It is also found in products such as *Water Grabber,* which is sold in gardening stores. It is inserted in soil near plant roots to absorb water for plant use during extended periods. Demonstrate how much water this polymer can hold by weighing a sample before and after it soaks up water. Many superabsorbent polymers can hold as much as 600 to 800 times their weight in water. Ask students to come up with other applications for superabsorbent polymers. [L1]

Expected Results

Results will depend on the brand of diaper used. Many generic brands hold less water than name brands.

Analysis

1. How much water each brand of diaper will hold depends on what polymer each contains and how much polymer is used in each diaper.

Fact of the MATTER

When nylon was first discovered by Wallace Carothers and his colleagues at DuPont in the 1930s, it was not thought to have any useful properties and was set aside on a shelf without even being patented by the company. It wasn't until some chemists at the company were playing games to see how far they could stretch nylon drawn out into a string that the strength and silky appearance of the nylon threads were noticed. Nylon hosiery was introduced at the 1939 New York World's Fair in a product display advertised as "Nylon, the Synthetic Silk Made from Coal, Air, and Water!" Four million pairs of nylons were sold in the first few hours after they were offered for sale in New York City on May 15, 1940.

Structure of Polymers

If you examine the structure of a polymer, you can identify the repeating monomers that formed it. Because polymer molecules are large, they are commonly represented by showing just a piece of the chain. The piece shown must include at least one complete repeating unit. Look carefully at the structure of a segment of a cellulose molecule shown in **Figure 18.17.** Cellulose is a polymer found in the cell walls of plant cells such as those of wood, cotton, and leaves. It is responsible for giving plants their structural strength. Can you find the portions of the cellulose structure that repeat? Notice that the ring parts of the molecule are all identical. These are the monomer units that combine to form the polymer. Glucose is the name of the monomer found in cellulose. In **Figures 18.17** and **18.18,** the glucose units are shown in simplified form without carbon and hydrogen atoms. The complete structure of glucose is shown below.

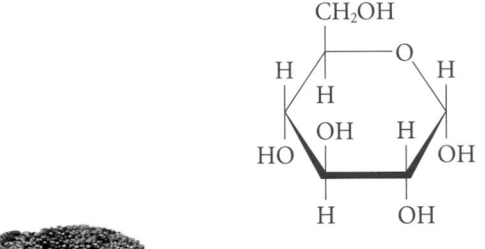

Glucose

Cellulose

Figure 18.17
Cellulose
Cellulose, the main component of plants, is probably the most abundant organic compound on Earth. This plant material, found in the cell walls of fruits and vegetables, cannot be digested by humans. It passes through the digestive tract unchanged, serving as dietary fiber that keeps the human digestive system healthy. Cellulose molecules range in length from several hundred to several thousand glucose units, depending on their source.

Demonstration 18.6

A Cross-Linked Polymer

Purpose: To prepare a sample of a cross-linked polymer.
Materials: for 50 students: 40 g of 98-100% hydrolyzed polyvinyl alcohol, 12 g sodium borate, magnetic stirrer-hot-plate, stirring rods, fifty 3-ounce disposable plastic cups, few drops of food coloring

Safety Precautions:

Disposal: Code A
Procedure: You will need to prepare 20 mL of 4% polyvinyl alcohol for each student. A magnetic stirrer-hot-plate is helpful. While stirring, slowly add 40 g of 98-100% hydrolyzed polyvinyl alcohol to 960 mL of tap water. Heat this suspension to 80°C while stirring. Add a few drops of food coloring. This will prepare enough for 50 students. Do not overheat the solu-

tion. Cool, and store it in a stoppered bottle. The cross-linker (borax) is also a 4% solution. For 50 students, dissolve 12 g of sodium borate in 288 mL of warm tap water. Into a small disposable plastic cup for each student, pour 20 mL of 4% polyvinyl alcohol. Have students observe the viscosity of the polymer as they stir it. They should add 6 mL of 4% sodium borate solution while stirring, and continue

Figure 18.18
Starch
Starch is digested by humans and serves as an important nutrient. During digestion, starch is broken down into glucose molecules that can be absorbed in the digestive tract. When starchy potatoes and cereal grains are cooked, they form a paste that thickens soups and stews. Starch molecules are usually several hundred glucose units in length. In addition to forming linear chains, glucose molecules form a branched polymer. Starches are a mixture of linear and branched molecules, both consisting only of glucose monomers.

Starch

Another natural plant polymer that is formed from glucose monomers is starch. Examine the structure of starch in **Figure 18.18** to see if you can find the difference between starch and cellulose. Both cellulose and starch are made from only glucose monomers. The difference between them is the way that these monomers are bonded to each other. In starch, the oxygen atom joining each pair of monomers is pointed downward; in cellulose, the oxygen atom is pointed upward. This appears to be only a minor difference, but it changes the properties of the two polymers dramatically.

Polymerization Reactions

How do you think the many different types of polymers are formed? Polymerization is a type of chemical reaction in which monomers are linked together one after another to make large chains. The two main types of polymerization reactions are addition polymerization and condensation polymerization. The type of reaction that a monomer undergoes depends upon its structure.

18.3 Plastics and Other Polymers **655**

to stir until the polymer is completely cross-linked. It will break into small chunks. Students can use their hands to scoop the material out of the cup and knead the polymer into a ball. It will flow when draped over the hand, forming long strands. If it is pulled quickly, it will break. **CAUTION:** *Do not place the polymer on finished wood or paper.*

Results: Students have formed "slime." The borax forms two bonds that link two polymer chains together, like rungs on a ladder.
Analysis: Ask this question.
1. What happens to the liquid when the polymer chains cross-link together? *The polymer becomes more rigid.*

Assessment

Performance: Students should allow the polymer to flow so as to form a square. Have them drape the square of polymer over a 2-L plastic bottle from which the label has been removed. The square will flow over the bottle and form a plastic sheet when dried. L1 LEP

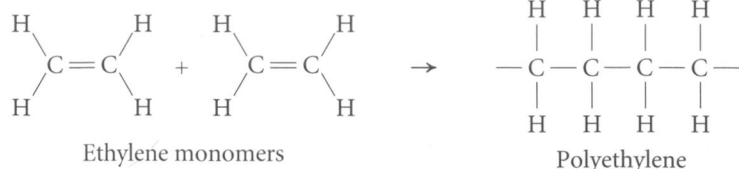

Ethylene monomers → Polyethylene

Monomers → Polymer + More monomers → Larger polymer

Figure 18.19

Addition Reactions
Ethylene monomers undergo an addition reaction to form the polyethylene that is used in plastic bags, food wrap, and bottles. The extra pair of electrons from the double bond of each ethylene monomer is used to form a new bond to another monomer.

Addition Reactions

The reaction by which Teflon is made from its monomer, tetra-fluoroethylene, is called an **addition reaction.** In this type of reaction, monomers that contain double bonds add onto each other, one after another, to form long chains. The product of an addition polymerization reaction contains all of the atoms of the starting monomers. Notice in **Figure 18.19** that the ethylene monomer contains a double bond, whereas there are none in polyethylene. When monomers are added onto each other in addition polymerization, the double bonds are broken. Thus, all of the carbons in the main chain of an addition polymer are connected by single bonds. Other polymers made by addition polymerization are illustrated in **Figure 18.20.**

Figure 18.20

Polymers Made by Addition Polymerization
Molten low-density polyethylene can be formed into a film. This tough plastic forms a barrier to food odors, which makes it useful for wrapping and storing foods. ▼

▲ Polyvinylacetate is another plastic made by addition polymerization. When mixed with sugar, flavoring, glycerol (for softening), and other ingredients, it becomes chewing gum.

656 **Chapter 18 Organic Chemistry**

Demonstration 18.7

Substitute Silk

Purpose: To prepare nylon.
Materials: 250-mL beaker, 50 mL 1,1,2-trichloro-1,2,2-trifluoroethane(TTE), 1 mL sebacoyl chloride, 100-mL beaker, 1.1 g 1,6-diaminohexane, 1 g anhydrous sodium carbonate, food coloring, forceps, graduated cylinder, paper towels

Safety Precautions:
Disposal: Code A

Procedure: CAUTION: *The vapors are hazardous. Use a fume hood for this demonstration. Goggles must be worn because the solution can easily splash into the eyes.* In a 250-mL beaker, place 50 mL of TTE. Add 1 mL of sebacoyl chloride to the TTE, and stir vigorously. It is important that the solution be thoroughly mixed. In a 100-mL beaker, dissolve 1.1 g of 1,6-diaminohexane and 1 g of anhydrous sodium car-

bonate in 25 mL of H_2O. Add a drop of food coloring to the H_2O to help students see the two layers. Carefully pour the water solution down the side of the 250-mL beaker onto the TTE solution without mixing the two. Using forceps, pick up the center of the film that forms at the interface of the two liquids. Slowly pull the nylon fiber from the beaker, and roll it up on a graduated cylinder that has a paper

Chemistry and Permanent Waves

A series of bad-hair days may drive you to the hair stylist in search of a permanent wave. Your hair has to be able to endure the chemical changes caused by a perm.

What happens in a permanent wave? Permanent-wave lotions are designed to penetrate the scales of the cuticle, the outer layer of the hair shaft. The lotions work because they affect the structure of the proteins that make up the hair. The amino acid cysteine, which contains an atom of sulfur, is found in human hair protein. Sulfur atoms in neighboring cysteine molecules within the hair protein form strong, covalent, disulfide (S—S) bonds. This cross-linking between cysteine molecules holds the strands of hair protein in place and affects its shape and strength.

When you have a permanent wave, the waving lotion that is first applied breaks the disulfide bonds. Then, after the hair is set in a new style, it is treated with a neutralizing lotion, which creates new cross-links between cysteine molecules to hold the hair in the new style. Thus, two chemical reactions occur. In the first, the waving lotion reduces each disulfide bond to two —SH groups. In the second, the neutralizing lotion oxidizes the —SH groups to form new disulfide bonds.

Kinds of permanent waves Two kinds of permanent-wave lotions are available to suit the needs of different types of hair. Alkaline lotions have as the reducing agent ammonium thioglycolic acid. The alkalinity causes the scales of the cuticle to swell and open, allowing the solution to penetrate rapidly. The advantages of the alkaline permanent are that it forms stronger, longer-lasting curls; takes a shorter time (usually 5 to 20 minutes); and occurs at room temperature. The alkaline permanent is used on hair capable of resisting damage.

Acid-balanced permanent-wave lotions contain monothioglycolic acid. Normally, acidic lotions penetrate the cuticle only slightly. That's why the process takes a longer time and requires heat in order for curls to develop. The advantages of the acid-balanced wave are that it forms softer waves, is more easily controlled, and can be used for delicate hair or hair that has been colored.

$$
\begin{array}{ccc}
\ \mid\ & & \ \mid\ \\
NH & & NH \\
\mid & & \mid \\
O{=}C & & O{=}C \\
\mid & & \mid \\
CH{-}CH_2{-}S{-}S{-}CH_2{-}CH \\
\mid & & \mid \\
NH & & NH \\
\mid & & \mid \\
O{=}C & \text{Linkage between} & O{=}C \\
\mid & \text{two cysteine} & \mid \\
& \text{molecules} &
\end{array}
$$

The next time you consider perming your hair, give some thought to the interesting chemistry that will take place on the top of your head.

Exploring Further

1. **Investigating** Find out more about the protein in hair. Where else is this protein found?

2. **Applying** Why might hair that has been colored or recently permed absorb lotion more rapidly than untreated hair?

Purpose

TECH PREP

Students will learn why permanent waves can change their hair.

Teaching Strategies

- Have students hold a small lock of hair at the crown of their head between the index finger and the thumb of the other hand, and slide these fingers toward the crown. If the hair is smooth, the cuticle is hard and the hair will not absorb liquids easily. If the hair is slightly rough, the cuticle is open and the hair is porous and will easily absorb liquids. **L1**

- Mention that the three-dimensional structure of a protein is necessary for it to function properly. Relate this to what happens when the disulfide links in cysteine are broken by a reducing agent in the wave lotion. The hair set in curlers then alters the three-dimensional structure of the protein. Ask students how the oxidizing agent in the neutralizer solves this problem.

Extension

Have students research the effect of breaking and re-forming S—S bonds in the formation of new textiles. Students will see how chemical changes affect the texture of wool fibers. **L2**

Exploring Further

1. Hair is composed of a hard protein called keratin, which is also found in fingernails, mammalian hooves, reptilian scales, and the feathers of birds.

2. The use of hair coloring or permanent lotions causes the scales of the cuticle to swell open, making them able to absorb liquids more readily.

---

towel taped around it.

Results: A continuous thread of nylon is pulled from the center of the beaker. Students can observe the formation of the fiber at the interface of the two solvents.

Analysis: Ask this question.

1. Nylon is hydrophobic. How is this property of use to clothing manufacturers? *Nylon is useful for waterproof clothing and umbrellas because it sheds water.*

 **Assessment**

Oral: Ask interested students to find the names of other synthetic fibers and report on the processes used to make them. **L1**

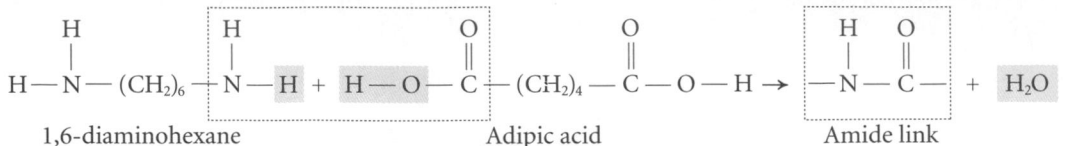

H—N—(CH₂)₆—N—H + H—O—C—(CH₂)₄—C—O—H → —N—C— + H₂O

1,6-diaminohexane   Adipic acid   Amide link

Figure 18.21

Condensation Reactions
The condensation of two different monomers—1,6-diaminohexane and adipic acid—is used to make the most common type of nylon. Nylons are named according to the number of carbon atoms in each monomer unit. There are six carbons in each monomer, so this type of nylon is called nylon 66.

Figure 18.22

Natural and Synthetic Rubber
Latex or natural rubber harvested from rubber trees is soft and tacky when hot. ▼

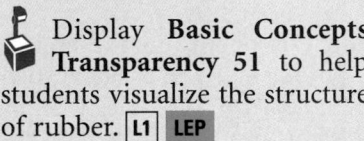

Condensation Reactions

In the second type of polymerization reactions, monomers add on one after another to form chains, as they do in an addition reaction. However, with every new bond that is formed, a small molecule—usually water—is also formed from atoms of the monomers. In such a reaction, each monomer must have two functional groups so it can add on at each end to another unit of the chain. This type of polymerization is called a **condensation reaction** because a portion of the monomer is not incorporated into the polymer but is split out—usually as water—as the monomers combine. A hydrogen atom from one end of a monomer combines with the —OH group from the other end of another monomer to form water. The condensation reaction used to make one type of nylon is shown in **Figure 18.21.** Other plastics made by condensation polymerization include Bakelite—which is the hard, ovenproof plastic used for handles of toasters and cooking utensils—and Dacron, which is used as fiber for clothing and carpets, backing on audiotapes and videotapes, and plastic wrap.

Rubber

Another process that often occurs in combination with addition or condensation reactions is the linking together of many polymer chains. This is called **cross-linking,** and it gives additional strength to a polymer, **Figure 18.22.** In 1844, Charles Goodyear discovered that heating the latex from rubber trees with sulfur can cross-link the hydrocarbon chains in the liquid latex. The solid rubber that is formed can be used in tires and rubber balls. The process is called vulcanizing, named for Vulcan, the Roman god of fire and metalworking.

Rubber has a white to brownish-yellow color. Automobile tires are black because the carbon allotrope carbon black is added to strengthen the polymer. ▶

When a piece of vulcanized, cross-linked rubber such as a rubber band is stretched and then released, the cross-links pull the polymer chains back to their original shape. Without vulcanization, the chains would slide past one another. ▼

Sulfur links

Unstretched rubber

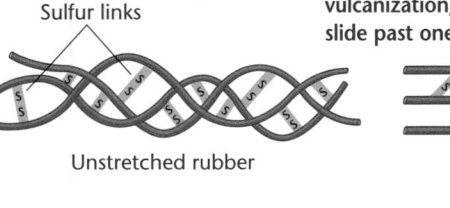

Stretched rubber

658   **Chapter 18   Organic Chemistry**

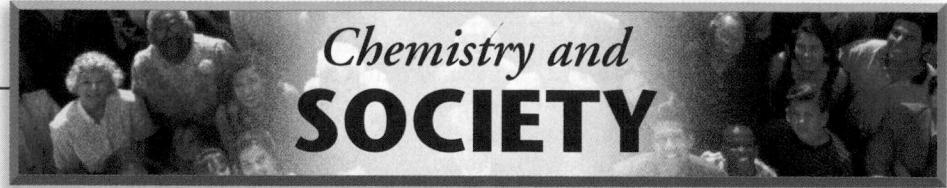

Chemistry and SOCIETY

Recycling Plastics

The growing mass of throwaway materials has caused more and more landfills to reach their capacity. In response, recycling has caught on in many parts of the country. People sort their trash into categories: garbage, paper, glass, and plastics. Because garbage and paper are biodegradable and glass can be reused, the focus is on plastics. Thirty percent of the volume of waste in the United States is composed of plastics. Unfortunately, recycling of plastics is more complex than for most other materials. Five plastics are commonly found in landfills. They are polyethylene—both high- and low-density, polyethylene terephthalate, polystyrene, polyvinyl chloride, and polypropylene.

How are plastics used? Polyethylene is the most widely used plastic. High-density polyethylene (HDPE) is used in rigid containers such as milk and water jugs and in household and motor-oil bottles. Low-density polyethylene (LDPE) is usually used for plastic film and bags. Polyethylene terephthalate (PET) is found in rigid containers, particularly carbonated-beverage bottles. Polystyrene (PS) is best known as a foam in the form of plates, cups, and food containers, although in its rigid form, it is used for plastic knives, forks, and spoons. Polyvinyl chloride (PVC) is a tough plastic that is used in plumbing and construction. It is also found as shampoo, oil, and household-product containers. Finally, polypropylene (PP) has a variety of uses, from snack-food packaging to battery cases to disposable-diaper linings.

The Society of the Plastics Industry has designated code numbers and acronyms to help people distinguish among plastics and to make communication about them uniform. The codes are useful in sorting plastics and in deciding what method will be used for recycling. In addition to having a distinct chemical composition, each kind of plastic has distinct physical properties that determine how it can be used.

Code	Material
♲1 PET	Polyethylene terephthalate
♲2 HDPE	High-density polyethylene
♲3 PVC	Polyvinyl chloride
♲4 LDPE	Low-density polyethylene
♲5 PP	Polypropylene
♲6 PS EPS	Polystyrene Foamed polystyrene

Recycling plastics PET beverage bottles and HDPE milk and water bottles receive the most attention because they are the easiest to collect and sort. PET bottles require a drawn-out recycling process because they are made of several materials. Only the body of the bottle is PET. The base is HDPE, the cap is another kind of plastic or aluminum, and the label has adhesives. The bottles are shredded and ground into chips for processing. The adhesives can be removed with a strong detergent. The lighter HDPE is separated from PET in water because one sinks and the other floats. The aluminum is removed electrostatically. What is left are plastic chips that may be sold to manufacturers who can use the chips in making other plastics. However, the FDA prohibits the use of recycled plastics in food containers, making one major market out of bounds for recycled plastics.

Analyzing the Issue

1. **Debating** Debate whether industries should be held responsible for recycling the packaging in which their products are sold.

2. **Writing** Prepare a letter or article for a newspaper in which you support the use of less packaging for food products.

Chemistry and SOCIETY

Purpose

TECH PREP

Students will become aware of what is involved in recycling plastics and should realize the importance of cooperative efforts.

Background

Several chemical processes are used to recover used plastics. Eastman Kodak employs a methanolysis reaction to reprocess tons of Kodak film scrap. Methanolysis begins with clean PET scrap to which a catalyst and methanol are added. This mixture is heated under pressure to force the PET to depolymerize. The end products are ethylene glycol and dimethyl terephthalate (DMT), which can be used as feedstocks for the synthesis of new polyesters.

Teaching Strategies

- Have students experiment to find the best way to remove labels from PET bottles. Have them test time, temperature, and amount of detergent required. **L1** **LEP**
- Mention that PET and HDPE can be processed together because they have different densities. PET and PVC cannot be processed together because they have similar densities.
- Discuss why the greatest return on recycled plastics is achieved when the resins are reused in their original application.
- Ask students what factors may influence whether there is an economic advantage in using reclaimed materials. *They would have to take into account the investment required for reprocessing equipment, the supply of recyclable materials available, and the cost of purchasing new resins in order to decide whether there is an economic advantage.*

Analyzing the Issue

1. Although it seems like an ideal situation to have the industries recycle their own packaging, this policy may backfire. Packaging is meant to safeguard the customer. If industries are forced to dispose of all packaging at great cost, they may cut corners at the expense of the health and safety of the consumer. Besides, the cost of recycling will certainly be passed to the consumer, who has already been taxed by having to return the packaging to the producer.

2. Students may suggest that they could bring their own carryalls to the supermarket in order to reduce the use of plastic bags. They may collect PET bottles and take them to a recycling center. They may choose products that use glass bottles instead of plastic because these can be reused.

3 ASSESS

Check for Understanding

- Check for understanding of the section material by having students answer the Section Review questions.
- Ask four students to help in this activity to illustrate the difference between addition and condensation polymerization. Two students should construct ball-and-stick models of four ethene molecules and show their addition polymerization to form polyethylene. Two other students should construct models of two 1,3-diaminopropane molecules and two ethandioic acid molecules and show their condensation polymerization to form a polyester and water molecules. `L1` `LEP`

Reteach

Ask students to list the ways in which naturally occurring and synthetic polymers are useful parts of their lives. Students might include the list in their portfolios. `L1` `P`

GLENCOE TECHNOLOGY

Videotape

MindJogger Videoquizzes for Chemistry
Chapter 18: Organic Chemistry
Have students work in groups as they play the videoquiz game to review key chapter concepts. `L1` `LEP` `COOP LEARN`

When World War II resulted in the cutting off of the Allies' supply of natural rubber, the polymer industry grew rapidly as chemists searched for rubber substitutes. Some of the most successful substitutes developed were gas- and oil-resistant neoprene, now used to make hoses for gas pumps, and styrene-butadiene rubber (SBR), which is now used along with natural rubber to make most automobile tires. Although synthetic substitutes for rubber have many desirable properties, no one synthetic has all the desirable properties of natural rubber.

Plastics

Although the terms *plastic* and *polymer* are often used synonymously, not all polymers are plastics. Plastics are polymers that can be molded into different shapes. What physical state has a fixed volume that you can pour into a mold and that will take the shape of its mold? Only liquids will. After a polymer has formed, it must be heated enough to become liquefied if it is to be poured into a mold. After pouring, the plastic will harden if it is allowed to cool, **Figure 18.23.**

Some plastics will soften and harden repeatedly as they are heated and cooled. This property is described as being **thermoplastic.** Thermoplastic materials are easy to recycle because each time they are heated, they can be poured into different molds to make new products. Polyethylene and polyvinylchloride are examples of this type of polymer.

Other plastics harden permanently when molded. Because they are set permanently in the shape they first form, they are called **thermosetting** polymers. Thermosetting plastics usually are rigid because they have many cross-links. No matter how much they are reheated, they won't soften enough to be remolded; instead, they get harder when heated because the heat causes more cross-links to form. Bakelite is this type of molecule. Even though thermosetting polymers are more difficult to reuse than thermoplastics, they are durable. **Figure 18.24** shows the relative amounts of various plastics produced for packaging in 1987.

Figure 18.23

Molded Plastics
Molded plastics are used for making objects with strength and durability qualities that are superior to those of earlier materials. Whereas a wood picnic table eventually will decay, plastic is so difficult to break down that it can create disposal problems as plastic materials pile up in landfills. Many picnic tables are already made of recycled plastic.

Chemistry Journal

Plastics
Have students choose one type of plastic and write a journal entry discussing the manufacture, structure, properties, and uses of this plastic. `L2`

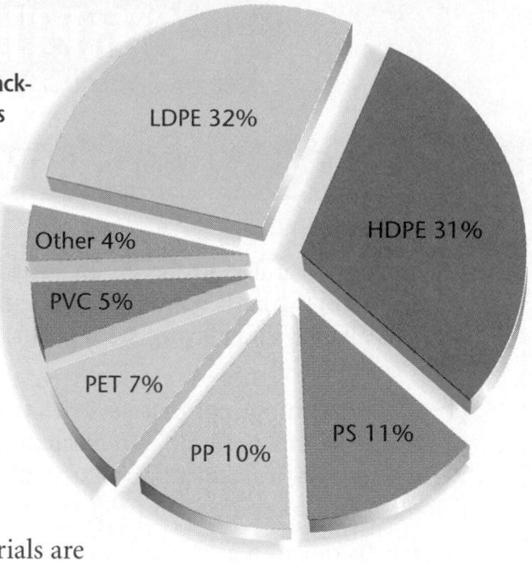

Figure 18.24
Recycling Plastics
The relative amounts of various plastics produced for packaging in 1987 are shown here. 87 percent of the plastics sold for packaging are thermoplastic. Polyethylenes and PVC are the most recyclable forms of plastic because they are so easy to remelt and reprocess.

LDPE 32%
HDPE 31%
Other 4%
PVC 5%
PET 7%
PP 10%
PS 11%

Connecting Ideas

Although polymers are large, complex molecules, their structures are easier to understand when you realize they are made from chains of simpler building blocks. Living things consist mostly of organic compounds, many of them extremely complex polymer structures. Keep in mind, however, that the chemical reactions through which living things get energy and building materials are exquisitely organized processes whereby a relatively small number of building blocks are used to make the many thousands of biochemicals that cells need. You will study these biomolecules in Chapter 19.

Extension

Lead a discussion on recycling plastic and paper, beginning by asking students the following question. When you are asked whether you want paper or plastic bags at the grocery, how do you respond and why? Relevant information is: plastic requires about 30% less energy to manufacture; paper is a renewable resource whereas plastic is not; paper is easier to recycle; if the two are incinerated as waste, plastic bags emit fewer atmospheric emissions and produce less ash; plastic bags occupy about 80% less volume in landfills. [L1]

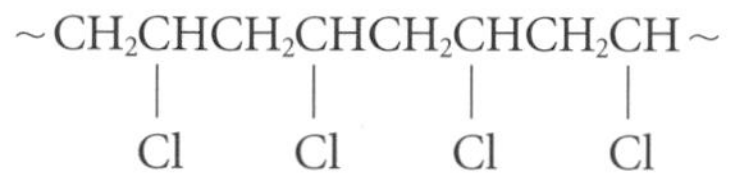

4 CLOSE

Quick Demo

Smart and stupid balls?
Show the difference in resiliency between "smart" and "stupid" balls, which contain different amounts of sulfur cross-linking and have somewhat different monomers. It is best not to describe the result of bouncing the two balls prior to the demo. Ask students to speculate as to what causes the difference in properties after you have bounced the balls.

SECTION REVIEW

Understanding Concepts

1. Identify the repeating unit that appears in each of the following polymers.

   a) polystyrene
   $$\sim CH_2CHCH_2CHCH_2CHCH_2CH\sim$$

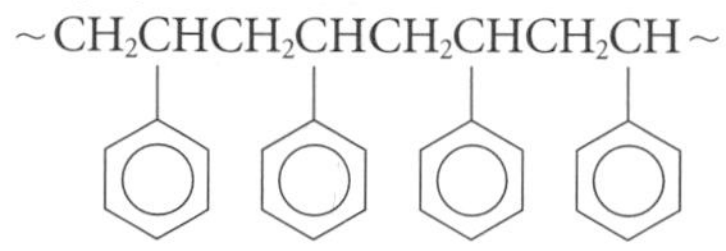

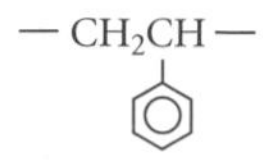

   b) 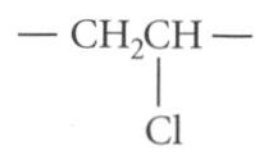 polyvinylchloride
   $$\sim CH_2CHCH_2CHCH_2CHCH_2CH\sim$$
   $$\qquad Cl \quad\ Cl \quad\ Cl \quad\ Cl$$

   c) 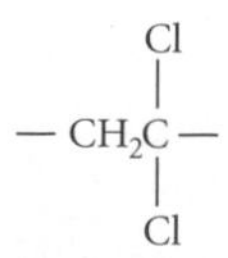 Saran
   $$\qquad Cl \ \ Cl \ \ Cl \ \ Cl$$
   $$\sim CH_2CCH_2CCH_2CCH_2C\sim$$
   $$\qquad Cl \ \ Cl \ \ Cl \ \ Cl$$

   d) rubber
   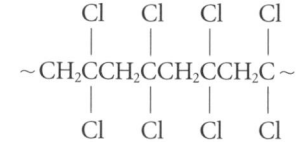

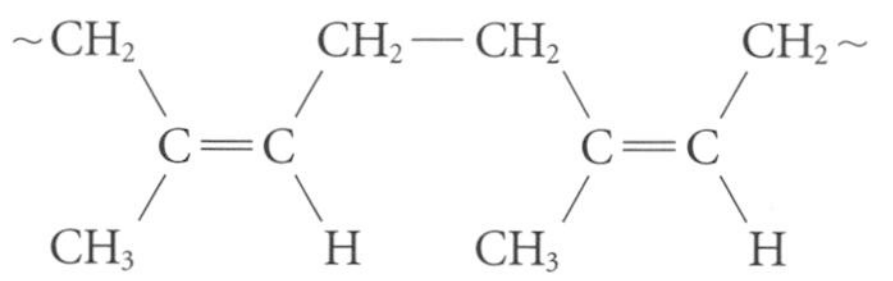

2. Draw the structure of the polymer that will be formed from each of the monomers shown.

   a) $CH_2{=}CHCl$
   b) NH_2CH_2COOH
   c) $CH_2{=}CHOH$

3. Compare and contrast addition and condensation polymerization reactions.

Thinking Critically

4. **Making Predictions** Can a polymer be made from alkane monomers by addition polymerization? Explain.

Applying Chemistry

5. **Paint Polymers** Paints usually contain three components: a binder that hardens to form a continuous film, a colored pigment, and a volatile solvent that evaporates. In latex paints, one of these components is a polymer. Use what you have learned about polymer properties to decide which of the three most likely is a polymer. Explain.

SECTION REVIEW

1a. $-CH_2CH-$

b. $-CH_2CH-$
$\qquad\ \ |$
$\qquad\ \ Cl$

c.
$\qquad\ Cl$
$\qquad\ |$
$-CH_2C-$
$\qquad\ |$
$\qquad\ Cl$

d.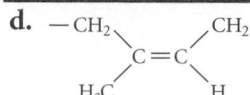

2a. 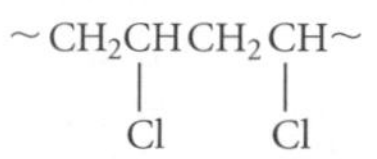 $\sim CH_2CHCH_2CH\sim$
$\qquad\ \ |\qquad\ \ |$
$\qquad\ \ Cl\quad\ Cl$

b.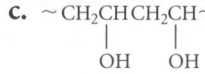

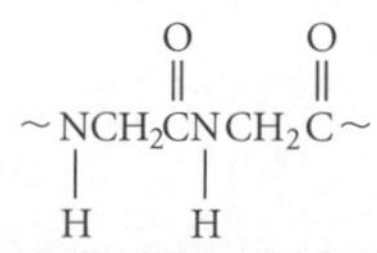

c. $\sim CH_2CHCH_2CH\sim$
$\qquad\ \ |\qquad\ \ |$
$\qquad\ \ OH\ \ OH$

3. Addition reactions involve adding monomers together by using electrons present in double bonds to make new single bonds without releasing water. Condensation reactions result in release of a small molecule, usually water, every time two monomers combine.

4. Polymers cannot be made from alkanes by addition polymerization because this process requires that the monomer have a double or triple bond. Alkanes have neither.

5. The binder will most likely be a polymer because it hardens into a film. Most polymers are not volatile, and most pigments are smaller molecules.

CHAPTER 18 REVIEW

CHAPTER REVIEW

- Review the Reviewing Main Ideas statements and Key Terms with your students.
- Complete solutions to Chapter Review problems can be found in the *Problems and Solutions Manual.*

UNDERSTANDING CONCEPTS

1. **a.** $CH_3CH_2CH_2CH_2CH_3$

   b. $CH_2=CHCH_3$

   c. $CH\equiv CCH_2CH_3$

   d. $CH_3(CH_2)_7CH_3$

   e. $CH_3CHCH_2CH_2CH_3$ (with CH_3 branch)

   f.
$$\underset{H}{\overset{H}{}}C=C\underset{CH_3}{\overset{CH_3}{}}$$

2. **a.** butane   **c.** propyne
   b. 1-hexene **d.** 1-heptene
3. **a.** alkyne   **c.** alkane
   b. alkyne   **d.** alkene
4. Examples could include the following.
   a. ethanol
   b. diethyl ether
   c. acetic acid
   d. Freon
   e. putrescine
   f. banana oil
   g. cinnamon and vanilla flavors
   h. Nutrasweet
5. Natural: cellulose, starch, DNA, protein
   Synthetic: nylon, Teflon, Bakelite, PVC
6. usually water
7. **a.** 3,5-diethylheptane
   b. 2-methylhexane
8. Boiling points increase as chain length increases in the alkanes.
9. IUPAC rules require that positional numbering begin at the end of the molecule that gives the lowest number for the added group. A compound named 4-methylhexane is identical to 3-methylhexane.
10. Thermosetting plastics become permanently molded the first time they set, whereas thermoplastics can be melted and re-formed repeatedly.

REVIEWING MAIN IDEAS

18.1 Hydrocarbons

- Alkanes are saturated hydrocarbons that contain only single bonds between carbon atoms, whereas alkenes and alkynes are unsaturated hydrocarbons.

- Isomers are compounds that have the same molecular formula but different structures. Because properties depend upon structure, isomers often have different properties.

- Benzene is one of a group of unusually stable cyclic hydrocarbons. This stability is due to the sharing of six electrons by all the carbon atoms in the molecule.

- Fossil fuels are our major sources of hydrocarbons. The many components in petroleum are separated by fractional distillation.

18.2 Substituted Hydrocarbons

- Special bond organizations, atoms, or groups of atoms that give predictable characteristics to molecules are called functional groups.

- Atoms and groups of atoms are substituted in carbon rings and chains to form substituted hydrocarbons.

- Compounds containing a specific functional group share characteristic properties.

18.3 Plastics and Other Polymers

- Polymers are large molecules made from many smaller units called monomers that repeat over and over again. Many synthetic polymers can be made from smaller monomers. Living cells also make many polymers.

- Addition and condensation are the main reactions by which polymers are made. In addition reactions, all atoms of monomers add end-to-end to form chains. In condensation reactions, a molecule of water is released as each new bond forms between two monomer units.

- Thermoplastic polymers can be recycled for different purposes because they can be repeatedly softened by heating and hardened by cooling. Thermosetting plastics become hardened permanently.

Key Terms

For each of the following terms, write a sentence that shows your understanding of its meaning.

addition reaction	monomer
alkane	polymer
alkene	reforming
alkyne	saturated hydrocarbon
aromatic hydrocarbon	
condensation reaction	substituted hydrocarbon
cracking	
cross-linking	thermoplastic
fractional distillation	thermosetting
functional group	unsaturated hydrocarbon
isomer	

● UNDERSTANDING CONCEPTS

1. Draw structures of the following hydrocarbons.

   a) pentane   **d)** nonane
   b) propene   **e)** 2-methylpentane
   c) 1-butyne   **f)** methylpropene

2. Name the following hydrocarbons.

   a) $CH_3CH_2CH_2CH_3$
   b) $CH_2=CHCH_2CH_2CH_2CH_3$
   c) $CH_3C\equiv CH$
   d) $CH_3CH_2CH_2CH_2CH_2CH=CH_2$

✓ Assessment

Portfolio: Review the portfolio options that are provided throughout the chapter. Encourage students to select one product that demonstrates their best work for the chapter. Have students explain what they have learned and why they chose this example for placement into their portfolios. **P**

Additional portfolio options can be found in the following **Teacher Classroom Resources:**
Consumer Chemistry
Chemistry and Industry
Applying Scientific Methods in Chemistry
Tech Prep Applications

3. Identify the following hydrocarbons as alkane, alkene, or alkyne.

   a) 1-hexyne
   b) acetylene
   c) 3-methyldecane
   d) *trans*-3-heptene

4. Name an example of each type of organic molecule listed.

   a) alcohol
   b) ether
   c) carboxylic acid
   d) halogenated compound
   e) amine
   f) ester
   g) aldehyde
   h) amide

5. List four natural and four synthetic polymers.

6. What molecule is usually released when monomers combine in a condensation reaction?

7. Name the branched alkanes shown.

   a)

   $$CH_3 \quad CH_3$$
   $$| \qquad |$$
   $$CH_2 \quad CH_2$$
   $$| \qquad |$$
   $$CH_3CH_2CHCH_2CHCH_2CH_3$$

   b)

   $$CH_2CH_2CH_3$$
   $$|$$
   $$CH_2$$
   $$|$$
   $$CH_3CHCH_3$$

8. Describe how the boiling points of alkanes change as chain length increases.

9. Why is there no compound named 4-methylhexane? What is the correct name for this compound?

10. Distinguish between thermoplastic and thermosetting materials.

11. Show the structure of the product of each of the following reactions.

    a) $CH_3CH{=}CHCH_2CH_3 + H_2 \rightarrow ?$
    b) $CH_3C{\equiv}CH + 2H_2 \rightarrow ?$

12. List two biological functions of esters.

● APPLYING CONCEPTS

13. Copy the following structure of penicillin G, an antibiotic. Circle all the functional groups in the molecule.

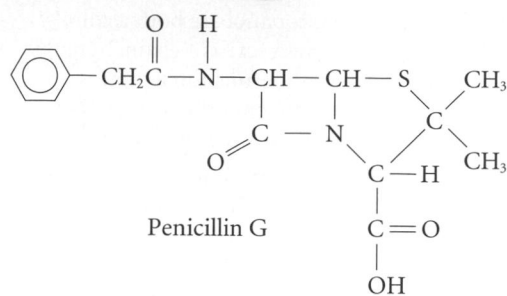

Penicillin G

14. Decide which substance in each pair listed will have the higher melting point, and briefly explain why.

    a) *cis*-2-pentene or *trans*-2-pentene
    b) butter or margarine
    c) vegetable oil or partially hydrogenated vegetable oil

15. Orlon is a polymer that is usually woven into fabric and used for clothing. Bakelite is a hard, rigid plastic that is used for handles of utensils such as this waffle iron. What can you infer about the amount of cross-linkage in these two polymers based on their properties and usage?

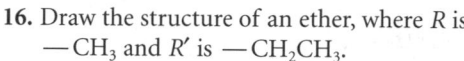

16. Draw the structure of an ether, where *R* is —CH₃ and *R'* is —CH₂CH₃.

17. Predict whether an amine will be more soluble in an acidic or in a basic aqueous solution. Explain.

18. Water is released in a reaction in which two different functional groups condense to form an ester. What two functional groups take part in this reaction?

19. Low-density polyethylene (LDPE) contains linear polymer chains, whereas high-density polyethylene (HDPE) contains highly branched chains. Predict which will be a harder polymer and explain why.

CHAPTER REVIEW

11. a. $CH_3CH_2CH_2CH_2CH_3$
    b. $CH_3CH_2CH_3$
12. Esters are present in DNA (genetic material) and in triglycerides (energy storage). They also are found in natural flavors and fragrances and in beeswax.

APPLYING CONCEPTS

13. aromatic ring, two amide groups, carboxylic acid group, thioether (—S—) group
14. a. *trans*-2-pentene; molecules pack together tighter
    b. butter; more saturated
    c. partially hydrogenated vegetable oil; more saturated
15. Orlon is not a cross-linked polymer; Bakelite is heavily cross-linked.
16. $CH_3{-}O{-}CH_2CH_3$
17. An amine is basic so it will be more soluble in an acidic solution. Bases react with acids to form water-soluble salts.
18. Hydroxyl and carboxylic acid groups condense to form esters.
19. HDPE is harder than LDPE because linear chains pack together tightly to form an orderly structure that is hard to pull apart. The structure of LDPE is a tangled, loose network of chains.

20. Glycerol will be more soluble in water than isopropanol because it can form more hydrogen bonds.

21. Fabric B contains many alcohol functional groups, which are hydrophilic and will stick to water, so it would not be good to use for an umbrella. Fabric A has fewer polar groups that can bind to water, so it is more hydrophobic and will repel water as an umbrella should.

22. Vitamin A has five double bonds, so five moles of H_2 molecules are required to fully hydrogenate one mole of vitamin A.

23. Alkaline permanent wave lotions penetrate the hair cuticle rapidly, allowing the formation of stronger, longer-lasting curls. The permanent process takes less time, occurs at room temperature, and is used on hair capable of resisting damage. Acidic lotions penetrate the cuticle only slightly so the process takes longer and requires heat. An acidic lotion can be used on delicate hair, and it allows softer waves to form.

24. polyethylene terephthalate or PET (1); high-density polyethylene or HDPE (2)

THINKING CRITICALLY

25. alkene pattern: C_nH_{2n}
alkyne pattern: C_nH_{2n-2}

26. Each fabric has to be tested using only the iodine test for cellulose. The polyester will give a negative result (no blue color), and the cotton-polyester blend will give a positive result (blue color).

27. Iodine will be decolorized more quickly by margarine because it is more unsaturated than butter.

28. ethyl acetate

29. Polymer A should be on the outside because it is hydrophobic and won't allow water to leak. Polymer B should be on the inside

20. Glycerol and isopropanol are both alcohols. Each contains a three-carbon chain, but glycerol has three alcohol functional groups, whereas isopropanol has only one. Predict which alcohol will be more soluble in water.

21. Compare the structures of the following two polymers, and decide which would produce a better fabric to use for making umbrellas. Explain.

a)

b) ~CH_2CHCH_2CH~
      | |
     OH   OH

Biology Connection

22. How many moles of H_2 molecules would be required to fully hydrogenate one mole of vitamin A, the structure shown here?

Everyday Chemistry

23. Compare and contrast the two different kinds of permanent-wave lotions.

Chemistry and Society

24. What are the full names and the corresponding codes of the two most commonly recycled plastics?

● THINKING CRITICALLY

Identifying Patterns

25. The molecular formulas of the noncyclic alkanes follow the pattern C_nH_{2n+2}, where n is the number of carbon atoms. What pattern is followed by the noncyclic alkenes with one double bond? What pattern is followed by the noncyclic alkynes with one triple bond?

Designing an Experiment

26. ChemLab Design an experiment that shows how you could distinguish between two fabric samples if one is polyester and the other is a cotton-polyester blend.

Making Predictions

27. MiniLab 1 Predict whether iodine will be decolorized more quickly by melted butter or by melted margarine, assuming that both are at the same temperature.

Applying

28. MiniLab 2 What is the name of the ester formed by reacting ethanol and acetic acid?

Making Predictions

29. MiniLab 3 Disposable diapers contain two different polymers. Examine the structures of the polymers shown, and determine which would best function on the inside and which on the outside of a diaper.

a) ~$CH_2CH_2CH_2CH_2$~

b) ~CH_2—CH——CH_2—CH~
           | |
          C=O   C=O
           | |
          OH   OH

● CUMULATIVE REVIEW

30. Explain why different compounds have different properties. (Chapter 4)

31. Write the formulas for the following compounds. (Chapter 5)
   a) acetic acid
   b) cesium chloride
   c) carbon disulfide

32. What shape is the methane molecule? (Chapter 9)

33. Compare the laws of Boyle and Charles. (Chapter 11)

because it is hydrophilic and will soak up urine.

CUMULATIVE REVIEW

30. Different compounds have different properties because they have different structures, which determine properties.

31. a. $HC_2H_3O_2$
   b. CsCl
   c. CS_2

32. tetrahedral with a central carbon atom and bond angles of 109.5°

33. Boyle's law states that the pressure of a gas is inversely proportional to volume at a constant temperature. Charles's law states that the volume of a gas is directly proportional to temperature at constant pressure.

Skill Review

1. **Interpreting a Graph** Examine the following graph and answer the questions.

   a) How does the number of carbon atoms in an alkene affect its boiling point?

   b) Is there a relationship between the amount of branching and the boiling point of an alkene? If so, what is the relationship?

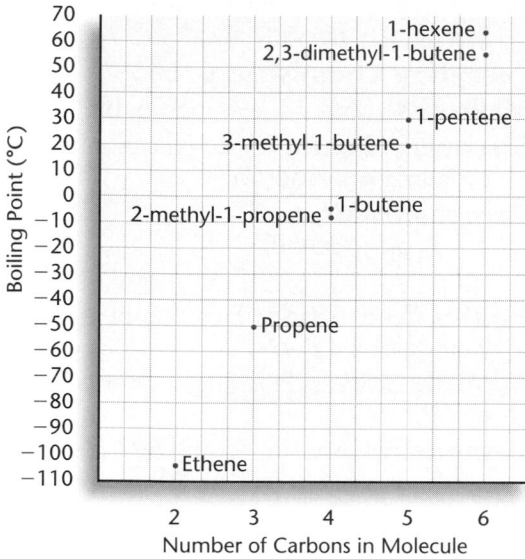

Boiling Point As a Function of Alkene Branching and Molecule Size

2. **Hypothesizing** The gasoline blend sold in hot climates consists of hydrocarbons of larger molecular mass than the gasoline blend sold in cold climates. Write a report suggesting a reason why refiners might vary the blends in this way.

3. **Designing an Experiment** Imagine you are a chemist at a textile company and are in charge of a research project aimed at developing a synthetic cotton substitute. What kind of information would you have to collect or questions would you have to ask to solve the problem? Design an experiment for beginning this research.

Writing in Chemistry

4. Do library research to find information about government regulations controlling levels of sulfur in coal-burning emissions. Write a report giving arguments for or against these regulations. If you can think of a better way to reduce sulfur emissions, give your ideas in your report.

Problem Solving

5. How might the greenhouse effect be affected by the development of new techniques that use atmospheric carbon dioxide as a carbon source for making organic compounds? Make a poster showing such possible effects.

Projects

6. Investigate fossil fuel use in the United States. Which type of fossil fuel do Americans use the most? The least? Which do Americans use the most for home heating? What are current predictions about how long it will be before the United States runs out of each? Make posters showing these statistics graphically.

7. Celluloid was the first polymer to be created by modification of a natural polymer. Research the history of celluloid, especially its discovery and uses in the early part of the 20th century.

8. Use gumdrops and toothpicks to construct models of the two structural isomers of butane. Study the models carefully to see how the shapes of the molecules are different.

9. Use gumdrops and toothpicks to make a model of 1-butene, CH_2=$CHCH_2CH_3$. Study your model to prove to yourself that this compound has no geometric isomers because one of the carbons in the double bond has two identical hydrogen atoms bonded to it.

1. **a.** Alkene boiling points increase with increasing numbers of carbon atoms.
   b. Alkene boiling points decrease as branching increases, as long as the double bond is in the same location (in this case, on carbon 1).

2. In a cold climate, hydrocarbons of low molecular mass vaporize more readily than do larger hydrocarbons, allowing an engine to start almost as easily as in a hot climate.

3. You could determine the functional groups present in natural cotton polymers, and then determine the structures of these polymers. Then you could plan a synthesis for a polymer with a similar structure and test it to determine how closely its properties approach those of cotton.

5. If more atmospheric CO_2 is used in synthetic processes, the influence of the greenhouse effect should slow down because CO_2 is the major molecule responsible for the greenhouse effect.

Chapter 19 Organizer

Section	Objectives	Activities/Features
19.1 **Molecules of Life** (3 days)	1. **Compare and contrast** the structures and functions of proteins, carbohydrates, lipids, and nucleic acids. 2. **Analyze** the relationship between the three-dimensional shape of a protein and its function.	**ChemLab:** Catalytic Decomposition—It's in the Cells, pp. 674-675 **People in Chemistry:** pp. 678-679 **Everyday Chemistry:** Clues to Sweetness, p. 683 **Everyday Chemistry:** p. 685 **MiniLab 1:** DNA—The Thread of Life, p. 688 **Discovery Demo:** 19.1 What's in a protein?, pp. 666-667 **Demonstrations:** 19.2, pp. 680-681; 19.3, pp. 686-687
19.2 **Reactions of Life** (2 days)	3. **Distinguish** between the reactions that cells use in the presence and in the absence of oxygen to extract energy from fuel molecules. 4. **Explain** how a small number of biochemical building blocks can be used to make the extraordinary variety of molecules needed to perform life's chemical functions.	**Health Connection:** Function of Hemoglobin, p. 693 **MiniLab 2:** Yeast Plus Sugar—Let It Rise, p. 699 **Demonstration:** 19.4, pp. 694-695

Activity Materials

ChemLab (pages 674-675)
hot plate, 8 small test tubes (13 × 100 mm), test-tube rack, 4 small beakers, carrot-cell slurry, 3% H_2O_2, 10-mL graduated cylinder, crushed ice, thermometer, glass stirring rod, timer or clock, metric ruler, marking pencil, thermal glove

MiniLab 1 (page 688)
mortar and pestle, uncooked wheat germ, cell lysis solution (citric acid, dish detergent, and table salt) cheesecloth or kitchen strainer, 250- or 400-mL beaker, funnel, 90% isopropanol, glass stirring rod, microscope slide, methylene blue dye, microscope

MiniLab 2 (page 699)
2 zipper-close sandwich bags, plastic dishpan, thermometer dry yeast, sucrose (table sugar), flour, 100-mL graduated cylinder or measuring cup, timer or clock

Demonstrations For a complete list of materials for the demonstrations in this chapter, see pages 666, 680, 686, and 694.

KEY TO TEACHING STRATEGIES

The following designations will help you decide which activities are appropriate for your students.

L1 Level 1 activities should be within the ability range of all students.

L2 Level 2 activities should be within the ability range of the average to above-average student.

L3 Level 3 activities are designed for the ability range of above-average students.

LEP LEP activities should be within the ability range of Limited English Proficiency students.

COOP LEARN Cooperative Learning activities are designed for small group work.

P These strategies represent student products that can be placed into a best-work portfolio.

The Chemistry of Life

Teacher Classroom Resources

STUDENT MASTERS

Study Guide: pp. 77-78
 **Laboratory Manual:** 19.1 Biochemical Reactions, pp. 177-182
ChemLab and MiniLab Worksheets: pp. 121-124

Study Guide: pp. 79-80
Critical Thinking/Problem Solving: p. 21
Laboratory Manual: 19.2 Qualitative Analysis of Food, pp. 183-186
ChemLab and MiniLab Worksheets: p. 125

ASSESSMENT RESOURCES
Chapter Review and Assessment: pp. 113-118
Alternate Assessment in the Science Classroom
Computer Test Bank: Chapter 19
Problems and Solutions Manual: Chapter 19
Supplemental Practice Problems: Chapter 19
Performance Assessment in the Science Classroom

TEACHING AIDS

Section Focus Transparency 39
Section Focus Transparency Master: p. 39
Basic Concepts Transparency 52, 53
Basic Concepts Transparency Masters: pp. 103-106
Problem Solving Transparency 22
Problem Solving Transparency Master: pp. 43-44
Lesson Plans: p. 39

Section Focus Transparency 40
Section Focus Transparency Master: p. 40
Basic Concepts Transparency 54
Basic Concepts Transparency Master: pp. 107-108
Lesson Plans: p. 40

CHAPTER RESOURCES
Applying Scientific Methods in Chemistry: pp. 47-48
Spanish Resources: Chapter 19
English/Spanish Audiocassettes: Chapter 19
Cooperative Learning in the Science Classroom
Lab and Safety Skills in the Science Classroom

GLENCOE TECHNOLOGY

Videotape
MindJogger Videoquizzes: Chapter 19
Videodisc
Chemistry: Concepts and Applications, Demonstrations, Videos, and Animations
Videodisc and Bar Code Teacher Guide
Science and Technology Videodisc Series: Disc 2 Chemistry *Straw as Feed; Energy-Integrated Farms*
CD-ROM Multimedia System
Chemistry: Concepts and Applications, Same as for Videodisc, plus Interactive Exploration and Major Simulation, *The Periodic Table*

TECH PREP

The following Glencoe resources provide opportunities for integrating science and technology.

Student Edition: People in Chemistry, p. 678; Everyday Chemistry, p. 685; Health Connection, p. 693

Teacher Wraparound Edition: Extension, p. 691; Chemistry Journal, p. 699; Discussion, p. 700

Teacher Classroom Resources: Tech Prep Applications, pp. 31-33; Consumer Chemistry, pp. 37-38

19 The Chemistry of Life

Chapter Overview

Students are first introduced to the structures and functions of the four major groups of bio-molecules. Emphasis is placed on learning how a relatively small number of building blocks is used to build up the vast number of larger molecules in each group. The structures and functions of vitamins A, C, and D are then presented. Finally, the fate of a molecule of glucose is followed, first through aerobic respiration and then through fermentation. The role of ATP as the energy currency of cells is discussed.

Theme Connection

Systems and Interactions/Energy

The *systems and interactions* and *energy* themes are apparent in discussions of metabolism, in which many different molecules interact to convert the chemical energy of food into ATP.

✓ Assessment Planner

Choose assessment strategies from the following pages to evaluate the progress of your students.

Assess, pp. 670, 690, 700
MiniLabs, pp. 688, 699
ChemLab, p. 674
Portfolio, pp. 670, 681, 688, 690, 701
Chapter Review, p. 701

Vitamin C

666

⊕ Discovery Demo 19.1

What's in a protein?

Purpose: To perform a destructive distillation of a protein.

Materials: a few peanuts, test tube, red litmus paper, lead acetate test paper, cobalt chloride test paper

Safety Precautions: 🐦 🧤 🔥 🥽

Disposal: Code A; all products and residue can be placed in local solid waste.

Procedure: Place a few peanuts in a test tube and heat strongly. **CAUTION:** *Do this in a hood or with good ventilation.* Place a piece of moist red litmus paper over the mouth of the tube while heating to test for ammonia. In a similar way, use lead acetate paper to test for hydrogen sulfide. Use cobalt chloride test paper to indicate the presence of water.

CHAPTER

19

The Chemistry of Life

When ships set sail from Europe in the 15th century to explore remote parts of the world, they had to carry provisions that would last for the many months or years of the voyage. Grains and salted and cured meat made up the bulk of the sailors' diets while aboard ship because fresh fruits and vegetables could not be stored without spoiling. As the long voyage progressed, the sailors developed a group of symptoms. Their gums began to bleed, then their teeth loosened and fell out. They bruised easily, and they became more susceptible to other diseases. These symptoms were caused by the disease called *scurvy,* a well-known and dreaded disease in those early seafaring days. The Crusaders who traveled to the Middle East in the 1300s suffered and died from the effects of scurvy. Vasco da Gama, the Portuguese explorer who sailed around the southern tip of Africa in 1497, lost more than half his crew to scurvy.

Eventually, ships' doctors came to realize that these symptoms could be prevented if the sailors ate fresh fruit, especially lemons and limes. However, it wasn't until 1795 that the British navy required ships to carry lemons or limes when sailing on long expeditions. British sailors, who ate the fruit on board ship, became known as limeys.

Chemistry Around You

Today, scurvy is a rare disease in developed countries because nutritionists know that the vitamin C in lemons and other fresh fruits and vegetables is an essential nutrient. People are aware of the importance of eating a variety of foods that contain the major groups of biomolecules—proteins, carbohydrates, and lipids—as well as essential vitamins and minerals. In this chapter, you will examine the structure and functions of biomolecules, such as the vitamin C shown opposite, to learn why each is an essential part of the dynamic processes of living things.

 Concept Check

Review the following concepts before studying this chapter.
Chapter 9: polar molecules; the geometry of hydrocarbons
Chapter 13: hydrogen bonding
Chapter 16: oxidation-reduction reactions
Chapter 18: functional groups, natural polymers

Chapter Preview

19.1 Molecules of Life

19.2 Reactions of Life

Laboratory Activities
ChemLab
Catalytic Decomposition—It's in the Cells

MiniLabs
1. DNA—The Thread of Life
2. Yeast Plus Sugar—Let It Rise

667

Introducing the Chapter
Show your class a copy of the food pyramid and ask them to name the molecules represented at each level. Ask students whether they know what functions these different nutrients carry out in their bodies. Explain that these are all organic molecules, and that their functions are different because the structures of their carbon chains and rings and the functional groups each type contains are different.

Using the Illustrations Direct students to the structure of vitamin C, and ask whether it is an organic or inorganic molecule. Ask them to name its functional groups. *alkene, four hydroxyl groups, one ester group* Point out that although it is found naturally in citrus fruits and some vegetables, most vitamin C in tablets is synthetic. While sailors needed to take fresh fruit along on their voyages, getting enough vitamin C on a long trip at sea today is much easier.

Revealing Misconceptions

People often think of lipids, especially cholesterol, as "bad for us." Explain that lipids are essential in the diet. Lipids make up most of the membranes of cells, are needed to synthesize many hormones and messenger molecules, and form the important myelin sheaths surrounding nerve cells that make nerve transmission possible.

GLENCOE TECHNOLOGY

 Videodisc

Chemistry: Concepts and Applications
See the Videodisc Program Teacher Guide to access illustrations and other material available for this chapter.

Results: The protein will blacken when heated, indicating carbon. The ammonia will turn the red litmus paper to blue. The lead acetate paper turns black for a positive test for H_2S. The cobalt chloride test paper indicates that water is present.

Analysis: Ask these questions.
1. What does the darkening of the protein indicate is in a protein? *carbon*
2. What other elements are shown to be present by the test papers? *S, H, O, N*

SECTION
19.1

PREPARE

Key Concepts

Biochemistry is defined and discussed, then the structures and functions of proteins, carbohydrates, lipids, nucleic acids, and vitamins are examined.

1 FOCUS

Focus Transparency

Display the **Section Focus Transparency** for Section 19.1 to introduce students to some sources of biomolecules.

L1 LEP

2 TEACH

Discussion

Display photos of these food items: an egg, bread, and butter. Ask students to relate each of the foods to the classes of proteins, carbohydrates, and lipids. *egg, mostly protein; bread, mostly carbohydrate; butter, mostly lipid* Ask students why we eat these classes of foods.

Concept Development

As you lead students through the study of proteins, carbohydrates, lipids, and nucleic acids, emphasize the structural similarities and differences among examples of each class of molecules.

Discussion

Write the formula $H_{5000}O_{2050}$ $C_{750}N_{100}Ca_{20}P_{20}K_5Cl_3S_3Na_2Mg$ on the chalkboard and ask students what it represents. *Excluding many trace elements, this is the formula for a human.* Ask how the atoms get into our bodies. *from our foods* You may want to extend the discussion to include the types of foods that provide the essential elements.

Objectives
Compare and contrast the structures and functions of proteins, carbohydrates, lipids, and nucleic acids.
Analyze the relationship between the three-dimensional shape of a protein and its function.

Key Terms
biochemistry
protein
amino acid
denaturation
substrate
active site
carbohydrate
lipid
fatty acid
steroid
nucleic acid
DNA
RNA
nucleotide
vitamin
coenzyme

Molecules of Life

Many of the most important molecules in your body are polymers. Proteins, carbohydrates, and nucleic acids, all extremely large molecules, are formed from small monomer subunits. Although lipids are usually not considered to be polymers, they, too, are formed from smaller molecules that have been linked together.

You need relatively large amounts of proteins, carbohydrates, and lipids in your diet. Complex reactions in your cells use some of these molecules and a few others to make a fourth group of biomolecules, the nucleic acids, which are needed in smaller amounts. Vitamins and minerals, too, are required by your body. Cells need all of these compounds to form the many structural materials of which they are made. They also use and store some of these compounds as a source of energy.

Biochemistry

All living things, no matter how different they appear, are composed of a relatively few kinds of chemicals. Some of these chemicals are molecular and some are ionic. They are arranged into complex and highly organized materials that provide support, carry out energy transformations, or duplicate themselves. The study of the chemistry of living things is called **biochemistry.** This science explores the substances involved in life processes and the reactions they undergo. Other than water, which can account for 80 percent or more of the weight of an organism, most of the molecules of life—the biomolecules—are organic.

Program Resources

Study Guide, pp. 77-78 L1
English/Spanish Audiocassettes, Chapter 19 L1 LEP
Consumer Chemistry, pp. 37-38 L1
Laboratory Manual, 19.1, pp. 177-182 L1
Section Focus Transparency 39 and **Master,** p. 39 L1 LEP
Basic Concepts Transparencies 52, 53 and **Masters,** pp. 103-106 L1 LEP

Problem Solving Transparency 22 and **Master,** pp. 43-44 L1 LEP
Spanish Resources L1 LEP
ChemLab and MiniLab Worksheets, pp. 121-124 L1
Applying Scientific Methods in Chemistry, pp. 47-48 L1
Tech Prep Applications, pp. 31-34 L1

Are these biomolecules any different from the other organic chemicals that you have studied? Scientists once thought that these molecules possessed a vital force and did not follow the same chemical and physical principles that govern all other matter. Until the early 1800s, no biomolecule had been synthesized outside of a living cell, although many had been identified. But in 1828, a German physician and chemist named Friedrich Wöhler reported that he could "make urea without needing a kidney."

The organic molecule urea is normally made in your kidneys and excreted in urine to dispose of excess nitrogen. Wöhler had produced urea in the laboratory from ammonia and cyanic acid. He showed that it is possible to take lifeless molecules and produce one of the molecules of life in the laboratory. Today, it is possible to synthesize artificially many thousands of complex biomolecules. However, living cells still are the most efficient laboratories, and it can take months or years for chemists to synthesize a large molecule that a cell can make in seconds or minutes.

The elemental composition of living things is different from the relative abundance of elements in Earth's crust. As shown in **Figure 19.1,** oxygen, silicon, aluminum, and iron are the most abundant atoms in Earth's crust. However, more than 95 percent of the atoms in your body are hydrogen, oxygen, carbon, and nitrogen. All four of these elements can form the strong covalent bonds found in organic molecules. Along with two other elements, sulfur and phosphorus, they are the only elements needed to make most of the proteins, carbohydrates, lipids, and nucleic acids found in every cell.

$$\begin{array}{c} O \\ \| \\ C \\ H_2N \quad \quad NH_2 \end{array}$$

Urea

Figure 19.1

Composition of Earth's Crust and the Human Body
The composition of Earth's crust and that of the human body are significantly different. The numbers represent percentages by mass in each. The elements oxygen and hydrogen are found in both, but carbon is concentrated in living things.

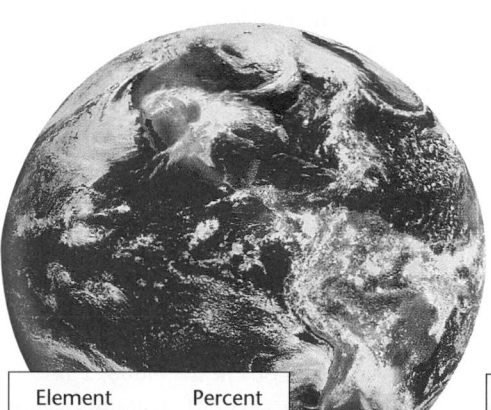

Element	Percent
oxygen	46.0
silicon	28.0
aluminum	8.0
iron	6.0
magnesium	4.0
calcium	2.4
potassium	2.3
sodium	2.1
hydrogen	0.9
other	0.3

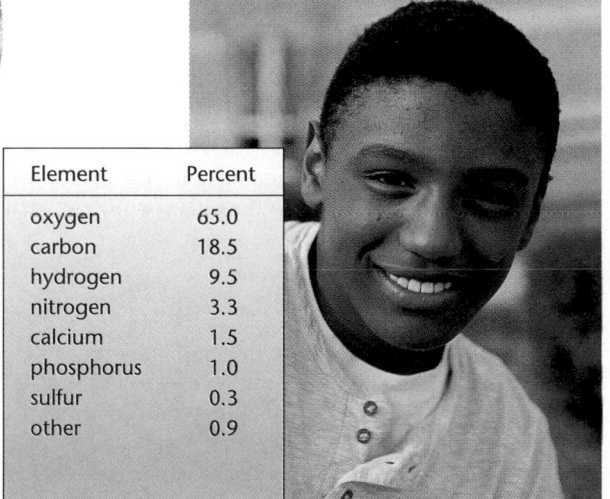

Element	Percent
oxygen	65.0
carbon	18.5
hydrogen	9.5
nitrogen	3.3
calcium	1.5
phosphorus	1.0
sulfur	0.3
other	0.9

19.1 Molecules of Life   **669**

Using an Analogy

Emphasize that biomolecules are no different from other types of molecules except that they can be arranged by plants and animals into complex and highly organized materials. Use the following analogy. Display a pile of miscellaneous nuts, bolts, screws, electrical parts, and plastic pieces. Also display the kit and assembly instructions for a small, simple radio. Ask students what the difference is. *The kit for the radio contains the parts that, if assembled according to the instructions, may be used to create a working mechanism. The pile of parts probably cannot be used, lacking the correct specifications and assembly instructions.* Draw the analogy between biomolecules and the radio kit. *Foods provide the parts, and cells provide the instructions.*

Visual Learning

Figure 19.1: Distribute copies of the periodic table. Ask students to color with one color marker the 11 essential elements that are most abundant in living systems—those listed in the formula on page 668. Students should color the 19 essential trace elements—Li, B, F, Si, V, Cr, Mn, Fe, Co, Ni, Cu, Zn, As, Se, Mo, Cd, Sn, I, and Pb—with another color. Point out to students that the table in **Figure 19.1** lists the most common elements *by mass*, whereas the formula lists the elements *by number of molecules.* L1 LEP

- **Portfolio:** Ask each student in the class to research the biochemical role of one of the 30 essential elements, write a short summary of the research, include the summary in his or her portfolio, and make a copy of the summary for each classmate. L1 P

- **Knowledge:** Distribute copies of the previously described element reports to all class members and briefly discuss them, asking relevant questions to the preparers of the reports. Consider giving a quiz on the information. L1

Tying to Previous Knowledge

Distribute a molecular model kit to each pair of students. Ask them to construct the models of 2-aminoethanoic acid and 2-aminopropanoic acid. Then give the biochemical names for the compounds glycine and alanine, respectively, and show why they are called amino acids. Also explain that the building blocks of all proteins are alpha (α) amino acids, those in which the amino group is bonded to the carbon adjacent to the carboxyl group. Ask each pair of students to create the dipeptide glycylalanine by removing the hydroxyl group from the glycine and one hydrogen from the amine group in alanine, bonding the two molecules at those sites and forming water in the process. Ask students to include the structural formula equation and word equation for the reaction in their journals. L2 COOP LEARN

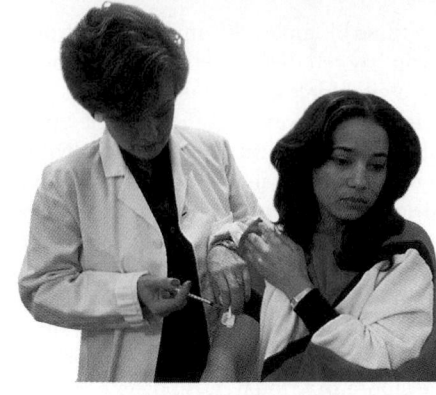

▲ The antibodies that fight disease organisms in your body are proteins. When you are vaccinated against a disease, your body produces protein antibodies that can attack the specific organism or virus that causes that disease.

Figure 19.2
The Functions of Proteins
▲ Keratin, the protein in hair and nails, is also the protein in the soft fur and tough hooves and horns of animals such as these bighorn sheep.

Fact of the MATTER

Some arctic fishes make unique antifreeze proteins that prevent ice crystals from forming in their blood. Scientists have transferred the genes for these proteins to tomato plants to make the plants tolerant of cold weather.

Proteins

What makes foods like chicken, beans, and fish so nutritious? They are composed largely of proteins. Proteins are one of the most important classes of biomolecules needed by living cells. When proteins were first discovered and named in the 19th century, their name was taken from the Greek word *proteios,* meaning "first" or "primary," because they were believed to be the chemical essence of life.

The Role of Proteins

Although about half of the proteins in your body function as catalysts in cell reactions, proteins play other important roles in living systems, as well. **Figure 19.2** shows some of these many functions. Structural proteins include collagen and keratin. Collagen is found in ligaments, tendons, and cartilage, and it provides the structural glue that holds your cells and tissues together. Keratin is the protein in hair, and it is also found in fur, hooves, skin, and fingernails. Other proteins make up your muscles and function to transport substances through your body in the blood. The hemoglobin that makes your blood red and carries oxygen from the lungs to the cells is a protein. Structural proteins provide strength to your body and give it shape.

Structure of Proteins

Proteins contain the elements carbon, hydrogen, oxygen, nitrogen, and sulfur. How are atoms of these elements arranged? Small monomer molecules link together by amide groups to form a polymer called a **protein.** The monomers that form proteins are organic compounds called **amino acids.** Although many different amino acids exist, only 20 are used by human cells to make proteins.

Meeting Individual Needs

Gifted Ask gifted students who have studied permutations and combinations in mathematics to calculate how many polypeptides composed of 20 different amino acids may exist. Most calculators have a factorial key (!). *20 factorial, which is 2.43 × 10^{18} or 2.43 quintillion* L3

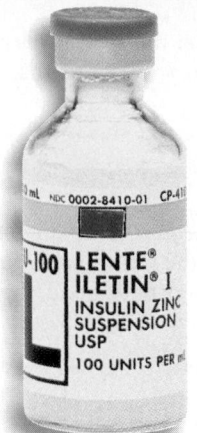

◀ Some proteins help cells communicate with each other. Insulin is a protein made in the pancreas. After a meal, insulin signals the body's cells to take in and use glucose from the blood. It does this by binding to receptor proteins embedded in the membranes surrounding the cells.

SECTION 19.1

Extension

Explain that more than 2000 natural and synthetic amino acids exist; however, only 20 occur in the proteins of organisms. Eight of these 20 (isoleucine, leucine, lysine, methionine, phenylalanine, threonine, tryptophan, and valine) cannot be synthesized in the human body and must be taken in at the same time in order for our bodies to manufacture the proteins that are necessary for life. Point out this example of the need for balanced diets, particularly for young persons who are still growing and for vegetarians who exclude certain items from their diets. Have students find out why vegetarians must be especially careful to eat a balanced diet. L1

All amino acids have certain parts of their structure in common. Look at this general structure of an amino acid and see if you can recognize the two functional groups present. The name *amino acid* should give you a clue.

$$\text{Amino group} \quad NH_2 - \underset{\underset{R}{|}}{\overset{\overset{H}{|}}{C}} - COOH \quad \text{Carboxyl group}$$

H   Hydrogen atom

R   Variable side chain

On the left side of the structure is an amino group, —NH₂, and on the right side is a carboxyl group, —COOH. Both of these groups are attached to a central carbon atom that also has a hydrogen atom and a side chain attached. The side chain, represented in the structure by the letter *R*, may consist of a number of different atoms or groups of atoms and is what makes each of the 20 amino acids unique with its own particular properties. The structures of several different amino acids are shown here. Glycine is the simplest amino acid. Its side chain consists simply of a hydrogen atom. Alanine is the next simplest with a methyl-group side chain. Phenylalanine contains a benzene ring as part of its side chain, and cysteine contains a polar —SH group.

Glycine   Alanine   Phenylalanine   Cysteine

The polymerization reactions that form proteins are condensation reactions, similar to those that are used to make some of the plastics you studied in Chapter 18. When two amino acids bond together, a hydrogen (—H) from the amino group of one amino acid combines with the hydroxyl (—OH) part of the carboxyl group of the other amino acid to

19.1   Molecules of Life   671

Integrating THE Sciences

Health Of the 20 amino acids in proteins, human cells can make only about half the nonessential amino acids. The others, which we must get from food, are called essential amino acids. Because we can't easily store amino acids, it is important to eat protein-rich foods that contain all the essential amino acids in ratios similar to those in the proteins in our bodies. Complete or high-quality proteins such as those in eggs, fish, meat, and milk contain all the essential amino acids in the desired proportions. Incomplete or low-quality proteins such as those found in corn, beans, and wheat are deficient in one or more of the essential amino acids. Although eating single foods with high-quality proteins is the most efficient way to make sure you get enough of all the essential amino acids in your diet, eating a combination of low-quality proteins from several foods can also provide all the essential amino acids. Vegetarians who don't eat animal products must eat a variety of plant materials that contain complementary proteins, combinations of incomplete protein foods that, when taken together, provide the necessary ratio of essential amino acids.

Discussion

To make the point that a great number of proteins are possible because the amino acids may be arranged in various sequences, ask students how many tripeptides containing tyrosine (tyr), cysteine (cys), and histidine (his) may exist. *six: tyr-cys-his; tyr-his-cys; cys-tyr-his; cys-his-tyr; his-tyr-cys; his-cys-tyr* **L2**

Visual Learning

Figure 19.3: Explain that the great diversity of proteins is explained not only by the sequence of amino acids, which is called the primary structure, but also by the secondary, folded structure; the tertiary, twisted structure; and the quaternary, clustering of polypeptides structure. Point out that much of the secondary, tertiary, and quaternary structure of proteins is due to the polar, or hydrophilic (water-loving), nature of some of the amino acid side chains and the nonpolar, or hydrophobic (water-fearing), nature of others.

---

form a molecule of water (H_2O). When the water molecule is released, an amide group has formed, linking the two amino acids. Recall that an amide group has the following structure.

$$
\begin{array}{c}
\overset{\displaystyle O}{\underset{\displaystyle}{}} \quad \overset{\displaystyle H}{\underset{\displaystyle}{}} \\
\| \qquad | \\
-C-N- \\
\end{array}
$$

Biochemists call an amide group by the name *peptide bond* when it occurs in a protein. Although the names are different, the functional groups are identical. When two amino acids are linked by a peptide bond, the resulting chain of two amino acids is called a dipeptide.

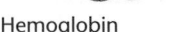

Glycine   +   Lysine   →   A dipeptide   +   Water

Additional amino acids can be added to a dipeptide by the same reaction to form a long chain. The chain is called a polypeptide because it is a polymer of amino acids held together by peptide bonds. A protein can consist of a single polypeptide chain, although most proteins contain two or more different polypeptide chains.

Three-Dimensional Protein Structure

Proteins have a three-dimensional structure because the polypeptide chains of which they are composed fold up like those shown in **Figure 19.3.** The polypeptide chains are held together by hydrogen bonds, ionic bonds, and disulfide cross-links between the side chains of neighboring amino acids.

Figure 19.3

Three-Dimensional Structure of Proteins

Proteins can fold into either round, globular structures or long, fibrous structures. The amino acid chains are held in place in three-dimensional structures by attractive forces between the side chains of different amino acids, which have been brought close together by the bending and folding of the polypeptide chains.

Hemoglobin   Collagen

Cultural Diversity

Sickle-Cell Anemia

Sickle-cell anemia is an inherited disorder that affects the blood's capacity to carry oxygen. The red blood cells of people who have two damaged copies of the hemoglobin gene have a tendency to assume a crescent shape, due to the production of an abnormal hemoglobin protein that forms large, stiff filaments. The sickled cells are rigid and have trouble passing through narrow capillaries, resulting in blocked blood flow that causes pain, tissue damage, and death. This disorder is prevalent among people of equatorial African or Mediterranean descent. In these geographic areas, malaria also is prevalent. Individuals who carry only one damaged gene tend to be resistant to malaria, which is why the sickle-cell gene has not been eliminated from the population. Have interested students research and compare the amino acid sequences of normal and sickle-cell hemoglobins. **L2**

The amino acids in a folded protein may be twisted into a helix or held tightly in flat sheets by hydrogen bonds that form between the $>$N—H and $>$C$=$O groups of neighboring amino acids. You may remember from Chapter 18 that disulfide bonds in the protein keratin give hair its strength and shape. These bonds form between sulfur atoms in the side chains of two molecules of the amino acid cysteine.

A molecule's geometrical shape is important in determining how chemical interactions take place. For example, the antibodies in your blood are folded into shapes so specific that each will bond to only one type of molecule specific to an invader. These antibodies protect your body from invasion by foreign organisms such as bacteria by bonding to molecules on the surface of the invader. Once bound, the invaders are destroyed by your immune system. Because bonding with invaders is so specific, your body must make a different kind of antibody for each kind of invader.

What happens when the three-dimensional structure of a protein is disrupted? Think of the difference between the consistency of a raw egg white and that of a hard-boiled egg. When the forces holding a polypeptide chain in its three-dimensional shape are broken, the protein is unfolded in a process called **denaturation.** High temperatures can denature proteins, which is why cooking foods that contain proteins results in the proteins' denaturation. Denatured proteins form the solid white of a hard-boiled egg. In addition, proteins can be denatured by extremes in pH, mechanical agitation, and chemical treatments. When egg whites are beaten, the proteins are denatured, as shown in **Figure 19.4.** Because the folded shape of a protein is essential for its function, denaturation of a protein results in loss of its function. This is one reason why organisms can live only in a narrow temperature and pH range.

The proteins of organisms that live in extremely cold or hot surroundings have three-dimensional structures held together by covalent bonds such as disulfide linkages. These bonds are stronger than the non-covalent hydrogen bonds that hold together the proteins of organisms living in more moderate locations. This strong bonding prevents denaturation and loss of protein function. The proteins of bacteria that live in hot springs such as this one in Yellowstone National Park are resistant to heat denaturation. ▼

Figure 19.4
Protein Denaturation
▲ The meringue on a lemon meringue pie owes its structure to denaturation of the proteins in egg whites. Beating the whites incorporates air and denatures some of the proteins, which provide the firm structure of the meringue.

Visual Learning

Figure 19.3: To introduce the concept of three-dimensional structure of proteins, have students twist three different-colored ropes together, then coil the structure into a three-dimensional super helix. The ropes represent the three polypeptide strands in collagen, which is found in tendons. Explain that the interwoven strands lend strength to the tendons. Illustrate the denaturation of the protein by having students unravel and unfold the ropes. L1 LEP

GLENCOE **TECHNOLOGY**

  Videodisc

Science and Technology Videodisc Series
Drugs from Snake Venom
Disc 2: Chemistry
Side 1, Ch. 16

Show this video to illustrate a use of proteins in medicine. L1 LEP

*inter***NET**
C O N N E C T I O N
A library of amino acid molecule art and formulas can be viewed on the Internet.

World Wide Web
http://www.chemie.fu.berlin.de/
chemistry/bio/amino-acids.html

Chemistry Journal

Cooking
Have students research various ways that proteins are denatured in food preparation. For example, vinegar, which denatures protein by lowering pH, is used to pickle fish and meats. Students might consider including recipes in their journal reports. L1

ChemLab  SMALL SCALE

Catalytic Decomposition— It's in the Cells

Time Allotment: One 45- to 60-minute laboratory period.

Objectives

Review objectives with students before they begin the ChemLab.

Process Skills

Observing, comparing and contrasting, using space/time relationships, communicating, using numbers, interpreting data

Safety Precautions and Disposal

Remind students to wear safety goggles to protect their eyes and aprons to protect their clothing.

PREPARATION

Prepare the carrot-cell slurry by grinding raw carrots in a food processor or blender with a small amount of water. Keep refrigerated.

Alternative Materials

- Potatoes or turnips may be substituted for carrots.
- You may wish to place the test tubes containing the hydrogen peroxide and carrot-cell slurry in Styrofoam cups rather than beakers in order to keep the temperature more constant.

PROCEDURE

- Cooperative lab groups should consist of four teams of two persons each. Each team can then be responsible for a trial at one of the four temperatures.
- Remind students that the test tubes must be very clean and completely free of detergent residue.
- You may want to demonstrate the splint test for oxygen with some hydrogen peroxide and carrot-cell slurry prior to the lab.

674

Catalytic Decomposition— It's in the Cells

You may have used hydrogen peroxide, H_2O_2, as an antiseptic when you cut or scratched your skin. If you have, you will have noticed that it seems to fizz and bubble when it contacts your skin. Hydrogen peroxide decomposes to form water and oxygen. The decomposition reaction is more rapid when the hydrogen peroxide comes into contact with your skin. Catalase, an enzyme found in the cells of your skin as well as in most other cells, is responsible for this rapid decomposition.

The activity of an enzyme is affected by environmental factors such as pH and temperature. Every enzyme has optimum conditions at which its reaction rate is fastest. In this ChemLab, you will study the decomposition of hydrogen peroxide as catalyzed by the catalase in carrot cells, and you will determine the optimum temperatures under which this enzyme works.

Problem

How does temperature affect the catalytic decomposition of hydrogen peroxide by catalase from carrot cells?

Objectives

- **Observe** the action of catalase on the decomposition of hydrogen peroxide.
- **Compare** the rate of reaction at various temperatures.
- **Make and use graphs** to interpret results.

Materials

hot plate	10-mL graduated
small (13 mm ×	cylinder
100 mm), clean,	crushed ice
unscratched test	Celsius thermometer
tubes (8)	glass stirring rod
test-tube rack	timer or clock
small beakers (4)	metric ruler
carrot-cell slurry	marking pencil
3% hydrogen	thermal glove
peroxide solution	

Safety Precautions

Hydrogen peroxide can harm the eyes. Wear safety goggles when using it.

PROCEDURE

1. Place carrot-cell slurry into each of four test tubes to a depth of about 2 cm.

2. Carefully measure 3.0 mL of 3% H_2O_2 solution into each of another four test tubes. If large quantities of bubbles are produced, obtain a new tube or thoroughly clean the original one.

3. Set one carrot tube and one hydrogen peroxide tube in each of four small beakers. Label the beakers *A, B, C,* and *D*. Add crushed ice and tap water to one beaker, hot tap water to a second beaker, and room-temperature tap water to the remaining two beakers.

ANALYZE AND CONCLUDE

1.

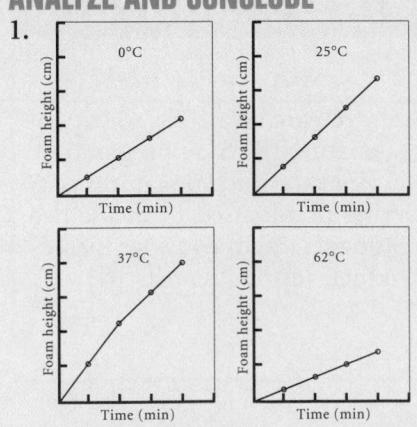

2.

3. 37°C. This is the same as human body temperature.

4. Set the hot-water beaker and one of the room-temperature-water beakers on the hot plate, and heat until the water temperatures reach 60-65°C and 37-38°C, respectively.

5. Make a data table similar to the one shown under Data and Observations. Take the temperature of the water in each of the four beakers, and record the temperature and label of each beaker. Let the test tubes sit in the beakers of water at 0°C, room temperature (probably about 18-25°C), 37-38°C, and 60-65°C for five to ten minutes.

6. Pour the tubes of H_2O_2 into the tubes of carrot-cell slurry at the same temperature. Stir each tube quickly with a glass stirring rod, and start the timer at the same moment. Measure the height of the foam, from the top of the liquid mixture to the top of the foam, at one-minute intervals for four minutes or until the foam reaches the top of the test tube. Record your data in the table.

ANALYZE AND CONCLUDE

1. **Interpreting Data** Use your data to construct and label a graph for each of your temperature trials. Plot foam height on the vertical axis and time on the horizontal axis.

2. **Interpreting Data** Construct another graph, plotting foam height at three minutes on the vertical axis and temperature on the horizontal axis.

3. **Comparing and Contrasting** Which of the temperatures appears to be optimum for the catalyzed decomposition of hydrogen peroxide by catalase? How does this temperature compare with human body temperature?

APPLY AND ASSESS

1. Peroxide ions are produced in plants and animals as a result of cellular reactions. Given that these peroxides can oxidize and damage cell structures, why is the presence of catalase in cells beneficial?

2. Suggest a way in which the gas in the foam might have been tested to verify that it was oxygen.

APPLY AND ASSESS

1. The catalase probably acts as an antioxidant, protecting the cells.

2. If a glowing splint were thrust into the bubbles, it would burst into flame.

✓ Assessment

Performance: Have students test other vegetables or fruits, such as those suggested under Alternative Materials, and compare the results with those for carrots. L1

DATA AND OBSERVATIONS

Beaker Number	Temperature of Reactants	Height of Foam After 1 Minute	Height of Foam After 2 Minutes	Height of Foam After 3 Minutes	Height of Foam After 4 Minutes
A					
B					
C					
D					

DATA AND OBSERVATIONS

Temperature of reactants	Height of Foam After 1 Minute	Height of Foam After 2 Minutes	Height of Foam After 3 Minutes	Height of Foam After 4 Minutes
0°C	0.5 cm	1.1 cm	1.6 cm	2.1 cm
25°C	0.8 cm	1.7 cm	2.6 cm	3.2 cm
37°C	1.0 cm	2.1 cm	3.0 cm	3.9 cm
62°C	0.4 cm	0.7 cm	0.9 cm	1.2 cm

The Role of Proteins as Enzymes

Although proteins are important for building strong muscles, that is only one of their many functions in living things. One of the most important roles they play is as the biological catalysts called enzymes. You may recall that enzymes help speed up reactions without being changed themselves during the reaction. A spontaneous chemical reaction may occur so slowly that it seems not to occur at all. Enzymes greatly speed up a reaction that would occur eventually, but they cannot make a reaction go if it would never occur spontaneously. How are enzymes able to speed up reactions? Enzymes provide specific sites, usually a pocket or groove in their three-dimensional structure, where each reactant, called a **substrate**, can bind. The pocket that can bind a substrate taking part in the reaction is called the **active site**. After a specific substrate binds to its active site, the active site changes shape to fit the substrate. This process of recognition is called induced fit. The shape of the substrate matches the shape of the active site, much as the shape of a key matches the hole in the lock it opens. Thus, only substrates that have a shape allowing them to bind to the active site can take part in a particular enzyme-catalyzed reaction. **Figure 19.5** shows how enzymes work.

Almost all reactions that take place in the body are catalyzed by enzymes. During digestion, enzymes speed up the breakdown of foods into molecules small enough to be absorbed by cells. Enzymes also make possible the reactions required for cells to extract energy from these nutrients. Enzymes are even involved in the production of other enzymes in cells.

Enzymes are now used in medicine to treat disorders that result from their deficiency. For example, lactase is an enzyme that breaks down milk sugar, which is called lactose, so that the two smaller sugar molecules of which it is composed can be absorbed by the digestive tract. Certain people who suffer from lactose intolerance gradually lose the ability to make the enzyme lactase as they mature. When this enzyme isn't present in the body, eating dairy products causes lactose to accumulate in the digestive system, resulting in gas,

Figure 19.5

Enzyme Action
Substrates are brought close together in the active sites of an enzyme, which lowers the activation energy of the reaction by facilitating the bonding together of the substrates to form a product. After the substrates have reacted, the product is released. The enzyme is then able to bind more substrate molecules and continue catalyzing the reaction. ▼

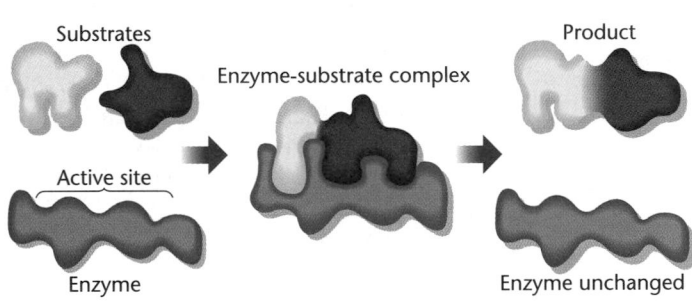

Substrates — Active site — Enzyme — Enzyme-substrate complex — Product — Enzyme unchanged

Substrates with shapes that don't match the shape of the active site won't undergo the catalyzed reaction, just as a house key can't be used to start a car. ▶

Figure 19.6
Uses of Enzymes
A group of enzymes found in many common household products is the proteases. These enzymes hydrolyze proteins into free amino acids. The proteases in meat tenderizer make steak more tender by breaking down the proteins. ▼

▲ Proteases in contact lens-cleaning solutions help remove the grimy buildup on lenses that results from proteins secreted by cells around the eyes.

bloating, and diarrhea. If lactase pills are taken along with the dairy products, the enzyme in the medicine breaks down the lactose just as the naturally produced enzyme would. **Figure 19.6** shows other everyday applications of enzymes.

Carbohydrates

Why do marathon runners eat large amounts of pasta the day before a big race? Foods such as pasta, bread, and fruit are rich in carbohydrates. A **carbohydrate** is an organic molecule that contains the elements carbon, hydrogen, and oxygen in a ratio of about two hydrogen atoms and one oxygen atom for every carbon atom. Early chemists thought that carbohydrates were chains of carbon with water attached—hydrated carbon—which is how carbohydrates were named. Chemists now know that carbohydrates are not hydrated carbon chains, but the name persists.

The Role of Carbohydrates

When the carbohydrates in pasta and other foods are broken down in the body, the sugar glucose is formed. Glucose is also called blood sugar because it is the sugar that is used in most body processes. The oxidation of glucose in cells provides most of the energy needed for life. When excess glucose is produced, as when athletes eat large quantities of pasta, the energy that can't be used right away is stored in the cells. Animals store glucose in their livers and muscles as glycogen, a polymer of glucose. In plants, glucose is stored as starch, the polymer you studied in Chapter 18.

19.1   Molecules of Life   677

⚡ **Quick Demo**

Why doesn't it gel?

At the end of one class period, mix and dissolve some gelatin in a sample of fresh or frozen pineapple juice and in a separate sample of canned pineapple juice. Label the two samples and leave them until the next day, having students observe the results. *The fresh juice contains an enzyme, bromelin, which breaks down protein in the gelatin, so the juice does not gel.* Ask students to explain why the canned juice has gelled but the frozen juice has not. *The canned juice has been heated, denaturing the enzyme and enabling the gelatin to gel.*

Concept Development

Point out that the general empirical formula for carbohydrates, such as glucose, $C_6H_{12}O_6$, is CH_2O, or carbon plus water; hence, the name *carbohydrate*. Stress also that carbohydrates are simple sugars or polymers of sugars.

Tying to Previous Knowledge

Draw the structural formulas for glucose and for fructose. Ask students to identify the functional groups. *The hydroxyl and carbonyl groups make these compounds alcohols and ketones, or alcohols and aldehydes.* If you draw the carbon skeletons in linear fashion, it will be more obvious to students that the position of the carbonyl group determines whether the sugar is an aldehyde or a ketone. Draw the structure of sucrose, and point out to students that the link between monomers is an ether linkage.

Meet Dr. Lynda Jordan, Biochemist

Teaching Strategies

- Have students refer to this feature during their study of enzymes.
- Obtain a copy of the video *Jewels in a Test Tube,* program no. 102, of the PBS documentary series *Discovering Women,* for a classroom presentation. The video features the teaching and research career of Dr. Jordan. **L1 LEP**

Background

Diabetes is a disease in which glucose cannot enter the cells where it is metabolized, causing excess levels of glucose to build up in the blood. Glucose cannot enter the cells either because there is a lack of functional insulin or because the cells develop resistance to insulin.

PEOPLE in CHEMISTRY

Meet Dr. Lynda Jordan, Biochemist

The door to Dr. Lynda Jordan's office at the North Carolina Agricultural and Technical State University is nearly always open. Students and younger faculty members continually stop by for help with lab problems and personal problems. Dr. Jordan comments, "I know what my life would have been like if there hadn't been people saying, 'You can do it.' That's why I decided to 'put my business on the street.' Students need to know that being a good, hard-working person does pay off."

On the Job

 A Public Broadcasting Service TV special has honored you and your work with enzymes. Can you tell us about this project?

 I'm studying an enzyme involved in diseases such as asthma, arthritis, bronchial disorders, gastrointestinal disorders, preterm labor, and diabetes. I want to find out about this enzyme's structural and functional relationship and how it behaves in human cells. This enzyme is isolated from cells of the human placenta. This is not an easy task. My colleagues and I have classified it by its molecular weight, pH, and calcium dependencies.

 Why is this research so important to you, personally?

 I first started working on the enzyme, which is called phospholipase A, as part of my post-doctoral research at the Institute Pasteur in Paris, France, in 1985. During that time, we discovered that phospholipase was present in human placentas, and we were one of the first laboratory groups to isolate this calcium-independent protein. It was a very exciting time for me. I know many people that have the diseases associated with this enzyme. If I could understand how this enzyme functions, then perhaps I could make a contribution to understanding what happens in the diseased state. This information would help us in the scientific community to devise strategies that could circumvent these biomedical problems.

 Which are the diseases that most interest you?

 Recent scientific evidence has shown a link between the enzyme phospholipase A_2 and diabetes. I am most interested in investigating the relationship between PLA_2 and glucose metabolism. Many of my family members, including myself, have diabetes. Understanding this disease on a molecular level would be rewarding for me on a personal as well as scientific level.

Early Influences

 What were you like as a high school student?

 I was a kid from an inner-city high school, with no hope for going to college. Although I was a good student, I used to skip classes. One day, when I was 15 years old, I was skipping class with a girlfriend in the rest room. The hall monitor came in, and I ran into the first room I came to, which happened to be an auditorium where an Upward Bound orientation was taking place. The speaker really got my attention. He asked me, 'What are you going to do the rest of your life—stand on the street corner and watch the world go by?' That's about all I was doing, and I was lucky not to end up a sad statistic. I signed up for the program right away.

Can you describe the program?

It was a rigorous six-week, college-level program at Brandeis University. We lived in college dormitories and were not allowed to have contact with the inner-city environment. The objective was to change our environment, so we were scheduled from 6 A.M. to 11 P.M. every day. We were expected to give our very best, and we were pushed and pushed.

Personal Insights

Mentors were crucial to your development. Are you yourself a mentor now?

Of all the things I do, the main thing is developing people. It's the most taxing, especially when I'm dealing with people who are first-generation college students, as I was. But it's also the most rewarding. I try to help students and younger faculty members with both their scientific problems and their personal problems, so they'll have the confidence to survive in the big, bad world. Character has a lot to do with success. Honesty, trustworthiness, dependability, and acceptance of responsibility sometimes have to be learned.

 Can you draw a parallel between solving problems in the lab and solving them in everyday life?

Both are step-by-step processes. First, you have to collect information. Then, you devise a strategy for reaching a goal. In the meantime, you need to think of backup strategies to utilize in case things don't work out as you anticipate. While you're progressing, you need to monitor how far you've come. If you're investigating careers, for example, and are considering being a doctor, the first thing to do is visit a hospital and maybe an emergency room. All along the way to solving any problem, you need to take little baby steps and monitor your progress.

CAREER ▶ CONNECTION

These fields are associated with biochemistry.

Medical Technologist Three years of college followed by a one-year program in medical technology

Medical Laboratory Technician High school plus a two-year training program

Medical Records Technician High school plus a two-year training program

CAREER ▶ CONNECTION

- **Career Path** A career in biochemistry would include these high school courses: biology, Earth science, chemistry, and physics. College courses would include biology, general chemistry, microbiology, organic chemistry, inorganic chemistry, physics, and biochemistry. Graduate programs would include advanced biochemistry and cellular physiology courses, among other courses.

- **Career Issue** Many university professors both teach and do research. Have students discuss some of the conflicts that may arise between the roles of teacher and researcher. `L1`

For More Information

Have interested students contact specific colleges or universities to find out what courses are required for a bachelor's degree or advanced degree in biochemistry and what career opportunities are available for those who hold biochemistry degrees.

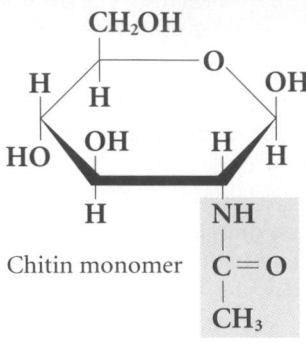

Figure 19.7

Chitin

Chitin is a structural carbohydrate that makes up the protective exoskeleton of arthropods such as beetles and lobsters. Chitin is one of the most abundant natural polymers on Earth due to the enormous number of insects on our planet. The monomer in chitin is similar to glucose, but it has a side chain containing an amide group attached to one carbon in the ring.

Chitin monomer

Carbohydrates also play a structural role in living things. You have learned that cellulose, also a glucose polymer, provides the strength and support needed in plants. Chitin, a natural polymer made from a sugar similar to glucose, forms the hard exteriors of insects and other arthropods, as shown in **Figure 19.7.** Carbohydrates are also found in cotton and rayon fabrics, wood, and paper.

Structures of Carbohydrates

Carbohydrates are a diverse group of molecules that include many simple sugar molecules, as well as larger molecules made by linking together different combinations of these sugars. Most simple sugars contain five, six, or seven carbon atoms arranged in a ring structure. An oxygen atom forms one corner of the ring, and hydroxyl groups are attached to each carbon. The most common simple sugars are glucose, fructose, and ribose.

Glucose   Fructose   Ribose

Simple sugar molecules are called monosaccharides. Two monosaccharides can link together in a condensation reaction to form a disaccharide, such as the most common carbohydrate in cakes and candies, sucrose. The structure of sucrose, which is also called table sugar, is made up of a glucose and a fructose monomer. **Figure 19.8** shows two sources of simple and complex carbohydrates in the diet.

Demonstration 19.2

Luminescence in Sugar

Purpose: To demonstrate triboluminescence.
Materials: wintergreen Lifesaver, pair of pliers, sugar cube

Safety Precautions:
Disposal: Code A

Procedure: If light is emitted when a crystal is crushed, the crystal is said to be triboluminescent. Sucrose can be used to demonstrate this very faint light. In a totally dark room, allow a few minutes for everyone's eyes to adjust to the dark. Rapidly crush a sugar cube in the jaws of a pair of pliers held at the students' eye level. Repeat the procedure, crushing a wintergreen Lifesaver.

Sucrose   + Water →   Glucose   +   Fructose

In order to be classified as a nutrient, a molecule must be able to enter cells. Disaccharides are too large to pass through cell membranes. However, enzymes in the digestive system catalyze the above reaction, breaking sucrose into glucose and fructose molecules that are small enough to be absorbed and used by cells as nutrients. Other common disaccharides in foods are milk sugar (lactose) and malt sugar (maltose).

Polysaccharides

What happens when you toast a slice of bread? Have you ever noticed that toast has a slightly sweet taste? The sweetness results from maltose, a disaccharide formed from two glucose molecules. Untoasted bread does not seem to taste sweet, and it contains no maltose. Where does the maltose come from when bread is toasted? It is the product of a reaction in which the large carbohydrates in bread—starch and cellulose—are broken down by heat. As you recall, these large carbohydrates are polymers that are formed from many glucose molecules linked together. Another common polymer consisting only of glucose is glycogen, also called animal starch. Starch, cellulose, and glycogen are examples of polysaccharides. Polysaccharides can contain hundreds or even thousands of monosaccharide molecules bonded together. A polysaccharide may contain only one type of monosaccharide, or it may contain more than one type.

Figure 19.8

Carbohydrates
Bread and jelly are both rich in carbohydrates. The sweet taste of jelly is a result of the disaccharide sucrose, whereas the structure and starchy taste of bread come from the polysaccharides cellulose and starch. Wood also contains cellulose. Check the ingredient label on your bread; sometimes, finely ground wood is added to increase the amount of fiber in the bread. This will be listed as *cellulose.*

SECTION 19.1

Discussion
Ask students if they have been eating much cellulose, the polymer of glucose that cannot be digested by humans. Point out that the dietary fiber that is present in many of our foods and is beneficial for our digestive systems is actually cellulose. Celery is an example of a food that contains so few nutrients and so much cellulose that more energy is expended in eating it than is derived from digesting it.

Transparency
Display **Basic Concepts Transparency 52** to introduce students to the structures of amylose and cellulose. L1
LEP

Results: A very faint flash is visible.
Analysis: Ask this question.
1. What is the name of another crystal effect that is observed when a crystal is squeezed and it produces electricity? *piezoelectric effect*

Assessment

Portfolio: Have students write a summary of this demonstration and add it to their portfolios.
L1 P

Figure 19.9
Starch, Cellulose, and Glycogen
The starch in bread, the cellulose in cotton, and the glycogen in meat are all polymers of glucose. The glucose units in cellulose are bonded together somewhat like a chain-link fence. Starch molecules can be either branched or unbranched, and glycogen is highly branched.

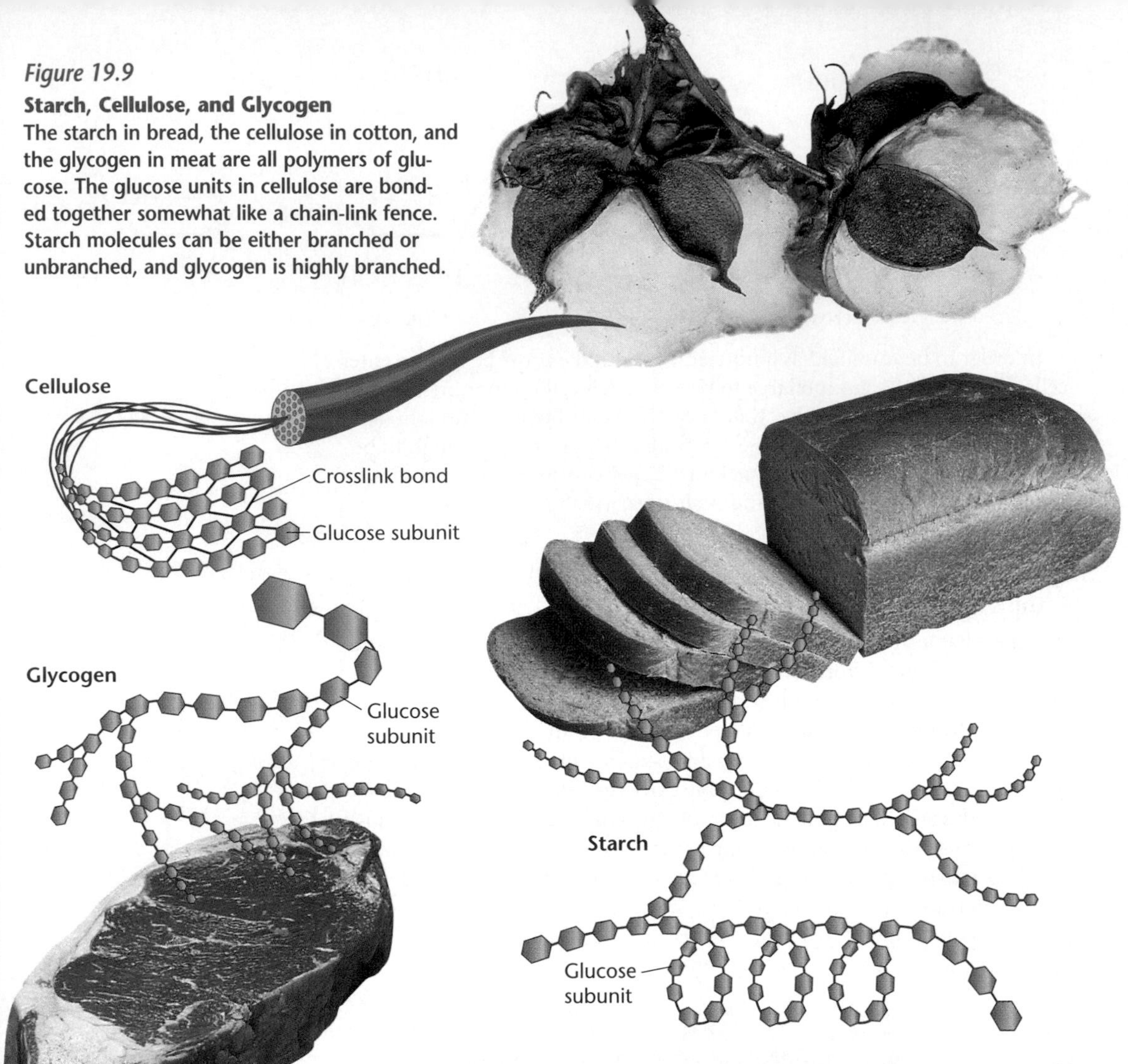

Cellulose
— Crosslink bond
— Glucose subunit

Glycogen
— Glucose subunit

Starch
Glucose — subunit
Glucose subunit

The repeating glucose units in starch, cellulose, and glycogen are the same in the starch of a potato as in the cellulose in the wood of a pencil. So, how can the structures of these three polymers be different? The major differences are the way the glucose molecules are bonded to each other, the number of glucose units in the polymer, and the amount of branching. These small differences in structure lead to large differences in properties and function. Their structures are compared in **Figure 19.9.**

Lipids

A biological compound that contains a large proportion of C—H bonds and less oxygen than in carbohydrates is called a **lipid.** Fats, oils, and waxes are all lipids. Lipids are insoluble in water and soluble in non-polar organic solvents. In general, lipids that are derived from animals are called fats, and plant lipids are called oils. Lipid molecules are found in

Meeting Individual Needs

Gifted Use molecular models to show gifted students the assymetric or chiral carbon atom that causes glyceraldehyde to have optical isomers. Explain that in fructose, for example, the three chiral carbon atoms lead to a total of 2^3 or 8 isomers, which differ in their ability to rotate polarized light. The optical isomers that rotate the plane of polarized light to the right are called the dextrorotatory or D forms, and those that rotate the plane to the left are called the levorotatory or L forms. Explain that many biomolecules, because they are chiral, interact or bind specifically with other molecules. To reinforce this concept, ask two students to perform a handshake using the right hand of one student and the left hand of the other. *It does not work, as the two hands, which seem to be identical, are mirror images of each other.* L3

Clues to Sweetness

What makes ripe strawberries taste so deliciously sweet? It has something to do with the sweetness message delivered to your brain when certain molecules from the strawberry fit into appropriate receptor sites in the taste buds on your tongue. The molecules lock onto the receptors in a specific way, which is determined by their structure. When this happens, chemical processes that produce the sweet response are stimulated. Even molecules of artificial sweeteners, many of which are not carbohydrates and are not metabolized by the body, are capable of matching the sweet receptor sites of the taste buds. This is interesting, but there's more to the secret of sweetness than meets the tongue.

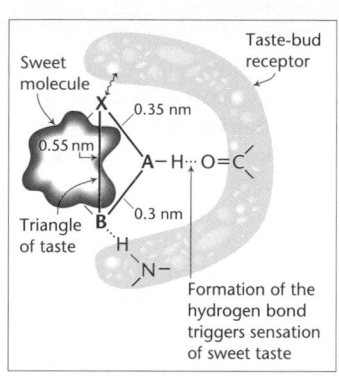

Formation of the hydrogen bond triggers sensation of sweet taste

Breaking the code for sweetness Hydrogen bonding is the most common interaction between molecules. When a hydrogen atom is bonded to a highly electronegative atom, its electron is pulled toward the more electronegative atom, and the hydrogen develops a partial positive charge. Consequently, it is capable of attracting other atoms that have partial negative charges. Nitrogen, oxygen, fluorine, and other nonmetals are all highly electronegative atoms that are attracted to hydrogen.

What does hydrogen bonding have to do with sweetness? Decades ago, chemists hypothesized that sweet-tasting molecules have two hydrogen-bonding sites—A and B—located close to each other. They postulated that the sites are separated by a distance of 0.3 nm and that site A has a hydrogen atom covalently bonded to atom A. They called these two sites AH and B, and named the structure the AH,B system. The AH,B system interacts with a similar system, which consists of

an $>$NH group and a $>$C$=$O group, located on the sweet receptors of the tongue.

Protein molecules are ideally suited for forming pairs of hydrogen bonds because they have both $>$C$=$O and $>$NH sites. The more electronegative oxygen in the $>$C$=$O group seeks a hydrogen atom on another molecule, and the hydrogen in the $>$NH group seeks a more electronegative atom such as nitrogen or oxygen. Interestingly, the $>$C$=$O and $>$NH sites are strategically placed about 0.3 nm apart. A molecule that has complementary sites, also 0.3 nm apart, can bind to the protein. The AH,B system of a sweet molecule, with sites AH and B, fits this description. If the sweet molecule bonds to the protein that forms part of the sweet taste receptors at the tip of the tongue, the brain receives a signal of sweetness.

Refining the sweetness model Chemists also noticed that sweet molecules shared a third common factor. Part of a sweet molecule is hydrophobic, which means it repels water. This part was called the X site. They found that X must be further from site B than from site AH. The whole molecular arrangement forms a triangle of sweetness.

Exploring Further

1. **Analyzing** What causes something to taste sweet?

2. **Acquiring Information** In a reference book, find a table comparing the relative sweetness of various sugars and artificial sweeteners. How do the following artificial sweeteners compare in sweetness with sucrose (table sugar): sucralose, aspartame, saccharin, and acesulfame-K?

Exploring Further

1. Certain characteristics of the structure of the molecule make it compatible with the sweet receptor on the tongue. These characteristics include a common feature of two electronegative atoms, A and B, separated by 0.3 nm, with a hydrogen atom covalently linked to A. This configuration forms an AH,B system that interacts with a similar system containing NH and CO groups on the receptors of the tongue.

2. times sweeter than 2% sucrose: sucralose, 750; aspartame, 200; saccharin, 500; acesulfame-K, 250

Purpose

Students will learn that certain structural features make a molecule more likely to elicit a sweet taste.

Teaching Strategies

- Ask students how someone might prove that the model presented in this feature is accurate. They may mention that if a chemist could make an artificial molecule to meet the specifications laid out by the model and find that it causes a sweet response, the chemist would have achieved the first step in showing the model to be correct. Then, the chemist would have to control each part of the sweetness triangle to see whether the resulting molecule is still sweet. `L2`

- Have interested students construct a three-dimensional model based on the diagram shown. They will be better able to visualize the conformation of the sweet molecule and taste receptor. `L2`

Extension

Research has shown that L-phenylalanine tastes bitter, whereas D-phenylalanine is seven times sweeter than sucrose. These two molecules differ only in that one is the mirror image of the other, which changes the location of X. Have interested students investigate the D- and L-designations of amino acids and build models of these two isomers of phenylalanine to show how the model discussed in the feature is supported. `L3`

Figure 19.10: Explain that waxes are large-molecule esters derived from monohydroxy alcohols rather than from glycerol. Beeswax, for example, has the formula $CH_3(CH_2)_{14}COO(CH_2)_{29}CH_3$ and is the ester formed from palmitic or hexadecanoic acid and the alcohol 1-tricontanol. Guide interested and capable students through the drawing of the structural formulas for these compounds. L3

Discussion

Ask students to draw the structural formula for glycerol given the IUPAC name: 1,2,3-propantriol. Draw the structure on the chalkboard, and show how the molecule may combine with three fatty acid molecules in a condensation reaction that yields a triglyceride and three water molecules. L1

Transparency

Display **Basic Concepts Transparency 53** so students can compare the structures of proteins, carbohydrates, and lipids. L1 LEP

Figure 19.10
Lipids
Honeybees make lipids that they use to form the honeycomb structure of their hive. The walls of the honeycomb are made of a mixture of lipids known as beeswax. ▼

▲ Lipstick is a mixture mostly of oils and waxes. These lipids dissolve the dyes and cause the lipstick to flow smoothly over the lips. They also help keep the skin of the lips soft and moist.

The avocados in this guacamole contain large amounts of oils. ▼

the fatty foods you eat such as butter, margarine, peanut butter, and the oil your french fries are cooked in. The waxes you use to polish the family car or to grease the bottom of a pair of skis also are lipids. Waxes are produced by both plants and animals. **Figure 19.10** shows some other common lipids.

The Structure of Lipids

Most of the oils and fats in your diet consist of long-chain carboxylic acids called **fatty acids** that are bonded to a glycerol molecule. Glycerol is a small carbon chain with three hydroxyl functional groups. Three fatty acid molecules combine with a single glycerol in a condensation reaction to form three molecules of water and a lipid molecule with three ester functional groups. Each fatty acid contributes the hydroxyl part of its carboxyl (—COOH) group, and each hydroxyl group on the glycerol contributes the hydrogen atom to form the water molecules. The lipid formed is called a triglyceride.

$$
\begin{array}{l}
CH_2OH \\
| \\
CHOH \\
| \\
CH_2OH
\end{array}
\quad + \quad
\begin{array}{l}
HO\overset{O}{\overset{||}{C}}(CH_2)_{14}CH_3 \\
HO\overset{O}{\overset{||}{C}}(CH_2)_{16}CH_3 \\
HO\overset{O}{\overset{||}{C}}(CH_2)_{18}CH_3
\end{array}
\quad
\begin{array}{l}
CH_2-O-\overset{O}{\overset{||}{C}}-(CH_2)_{14}-CH_3 \\
| \\
CH-O-\overset{O}{\overset{||}{C}}-(CH_2)_{16}-CH_3 \\
| \\
CH_2-O-\overset{O}{\overset{||}{C}}-(CH_2)_{18}-CH_3
\end{array}
\quad + \quad 3\ H_2O
$$

Glycerol + 3 fatty acids → Triglyceride + Water

Integrating THE Sciences

Biology Perfumes consist of three main components: a blend of fragrant essential oils, a solvent that dilutes the other components and allows them to be spread thinly, and a fixative that slows evaporation and stabilizes the scent. Most essential oils are complex mixtures of compounds with many different functional groups, including alcohols, ethers, aldehydes, ketones, and esters. The sense of smell is believed to result from the shape of molecules that bind to seven different kinds of receptor sites in the olfactory cells of the nose. Molecules in each of the seven primary odor families (floral, musky, pepperminty, ethereal, camphorous, pungent, and putrid) share similar shapes even though their molecular formulas are often quite different. The shapes are complementary to the shapes of the binding sites for the odor molecules in the receptors, and nerve impulses are initiated when a molecule of the correct size and shape fits into a binding site.

Everyday Chemistry

Fake Fats and Designer Fats

Fats have acquired a nasty reputation for causing numerous health problems. At the same time, fats are an essential nutrient and are the reason why so many foods taste, look, and feel so good as you eat them. Food scientists around the world are working on the problem of how to provide food that promotes good health, while at the same time satisfying consumers with the pleasing taste, appearance, and texture of fats.

Carbohydrates as fake fats Two kinds of carbohydrates—starches and cellulose—are being used to replace fats in foods. From the structures of these molecules in the chapter, you might wonder how starch and cellulose molecules can mimic fat's properties. However, when a starch is mixed with water, it forms gels that have the texture and bulk of fat. The gels can replace the fat in some foods, but they can't be used for frying.

Usually, carbohydrates contribute only four Calories per gram as opposed to the nine Calories per gram that come from fats. But cellulose, a carbohydrate in the cell walls of plants, contributes no Calories at all because the body is not able to metabolize it. Avicel is a form of natural cellulose that, when mixed with water, produces a texture similar to fat. It can replace fats in frozen desserts and bakery products.

Proteins for fats In order for a protein to mimic a fat, it has to be cut into tiny particles that measure only from 0.1 to 3.0 μm in size. Simplesse is the only protein based fat replacer available in the United States. In one version, egg white and milk protein are subjected to high heat and stress to form small, spherical particles. The small size of these protein particles causes them to be perceived by the mouth as smooth and creamy. Like carbohydrates, protein replacers provide only four Calories per gram. However, they are not suitable for frying, and they lack the fat flavor.

Chemically altered fats Searching for the perfect no-Calorie, health-promoting fat, food chemists have come up with chemically altered fats. They alter the size, shape, or structure of real fat molecules so that the human body digests and uses them to a lesser extent or not at all. One chemically altered fat is Olestra, a sucrose polyester. Food technologists claim that Olestra will satisfy your taste buds without providing saturated fats and Calories. It can be substituted for butter and grease, and can even be used for frying. A fat consists of three fatty acids attached to glycerol. The altered fat Olestra has six to eight fatty acids, all derived from vegetable oils, attached to sucrose. Being larger than a typical fat molecule, it is not absorbed by the cells in the digestive system. Also, enzymes fail to break the sucrose-fatty acid bond, so the molecule passes through the body undigested.

Exploring Further

1. **Analyzing** What are the dangers of eating only foods in which all the fats have been replaced?

2. **Thinking Critically** Products made with nondigestible fake fats could trap fat-soluble vitamins or medicines in the intestines. What effect would this have on the body? How could this problem be solved?

Exploring Further

1. In the worst-case scenario, a person eating only foods that supply no Calories from fats would be deprived of energy and would certainly waste away.

2. The body would be prevented from absorbing the vitamins and would be deprived of their beneficial effects. Fat-soluble vitamins or medicines could be taken in capsule form, in which they are dissolved in lipid.

Everyday Chemistry

Purpose

TECH PREP

Students will learn about the advantages of fake and designer fats and some of the problems involved with using them in foods.

Teaching Strategies

- Discuss one example of an already-existing fat in which the Calories and cholesterol have been changed. In the altered fat SALATRIM, stearic acid, which the body does not absorb easily, is substituted for another fatty acid. The number of Calories the altered fat contributes drops to 5 Calories per gram instead of 9 from ordinary fat. Stearic acid also contributes to the body's ability to remove cholesterol.

- Have students examine food labels of foods in which fats have been replaced or altered. Ask them to classify the products according to what they have learned in this feature about the kinds of fat replacers. **L1** **LEP**

Extension

Ask students to research the product carrageenan, currently used by meat processors to formulate low-fat ground beef products. Have students find out what type of molecule it is, its source, and its function in meat products. *It is a polysaccharide that is found in red seaweed. It makes the meat juicy.* **L1**

Figure 19.11: Have students examine the diagram of oleic acid for the kink that develops in the carbon chain at the point of the double bond. Mention that some polyunsaturated fatty acids have as many as four double bonds. This reduces their Caloric content, as fewer hydrogen atoms are available for oxidation in cells. [L1]

How saturated is coconut oil?

Pour boiling water over shredded fresh coconut at the beginning of a class period, strain the mixture, and allow the liquid to sit in a cool place. Coconut oil, a saturated lipid, will rise to the top of the liquid where it may be easily seen. Encourage students to examine foods, particularly snack foods, for the presence of tropical oils such as coconut, palm, or palm kernel oils, which are highly saturated. [L1]

Tying to Previous Knowledge

Ask students to recall the relative reactivity of unsaturated compounds compared with saturated compounds. *The double and/or triple bonds of the unsaturated compounds make them more reactive.* Explain that foods containing unsaturated fats are particularly likely to react with oxygen at the double bonds and become rancid. Ask students to look on packages of crackers or cookies for the presence of antioxidants or alpha-tocopherol, which tend to react with oxygen before it can react with the unsaturated fats. [L1]

The presence of a double bond introduces a bend in the molecule that prevents oleic acid molecules from packing together as tightly as do stearic acid molecules. As a result, oleic acid—like most plant lipids—is a liquid at room temperature, whereas stearic acid—like most animal lipids—is a solid. ▼

Figure 19.11

Saturated and Unsaturated Fatty Acids

▲ Structures of two 18-carbon fatty acids, one saturated and the other monounsaturated, are shown. Stearic acid is found in pork and beef tissue; oleic acid is a major component of olive oil.

The most common fatty acids are chains of 12 to 26 carbon atoms with a carboxylic acid group at one end. They usually have an even number of carbon atoms because they are made from smaller molecules with two carbons. Saturated fatty acids, like saturated hydrocarbons, have only single bonds connecting the carbon atoms. Monounsaturated fatty acids have one double bond between two of the carbon atoms, as you can see in **Figure 19.11.** Fatty acids that are polyunsaturated have two or more double bonds. In general, animal lipids are more saturated than are plant lipids.

Steroids

Another class of lipids is the steroids. A **steroid** is a lipid with a distinctive four-ring structure. Important steroids include cholesterol, some sex hormones, vitamin D, and the bile salts that are produced in the liver to aid in digestion of fats.

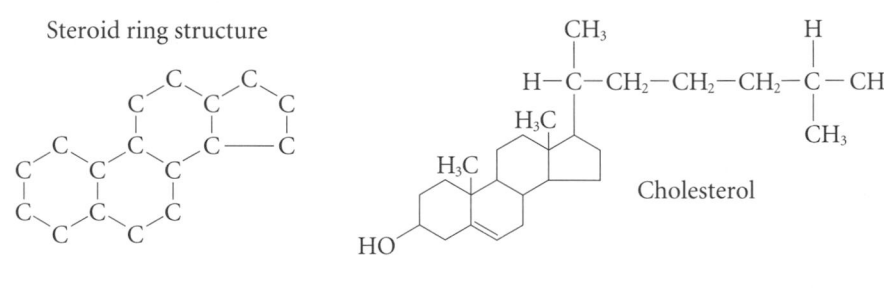

Steroid ring structure

Cholesterol

Demonstration 19.3

How much fat?

Purpose: To determine the percent of fat in peanuts.

Materials: 20-g sample of peanuts, balance, mortar and pestle, 30 mL acetone, stirring rod, 250-mL beaker, 100-mL beaker, funnel with filter paper, hot plate

Safety Precautions:

Disposal: Code A

Procedure: Place a 5-g sample of peanuts in a mortar, and grind it with a pestle. Repeat this procedure with additional 5-g samples until you have collected about 15 g of ground peanuts. Determine the mass of the ground peanuts. Use a stirring rod to mix the ground peanuts thoroughly with 30 mL of acetone in a 250-mL beaker. Decant the acetone into a filter setup that drains into a clean, massed 100-mL beaker. In a fume hood, place the 100-mL beaker containing

Eating a diet high in saturated fats has been linked to the development of cardiovascular problems such as heart disease. The reason for this is not fully understood, but it may be due partly to the liver's ability to break down fatty acids into small pieces used to make cholesterol. High levels of cholesterol in the blood are associated with a stiffening and thickening of the artery walls known as atherosclerosis, illustrated in **Figure 19.12**. This condition can result in high blood pressure and heart disease. Reducing the amount of saturated fats and cholesterol in the diet—especially animal fats found in eggs, cheese, and red meats—is one way to lower blood cholesterol levels. However, changing the diet alone does not work for everyone because genetics, exercise, stress, and other factors also affect cholesterol level.

The Functions of Lipids

Lipids have two major biochemical roles in the body. When an organism takes in and processes more food than it needs, excess energy is produced. The organism stores this excess energy for future use by using it to bond atoms together in lipid molecules. Later, when energy is needed, enzymes break these same bonds, releasing the energy used to form them. You have learned that carbohydrates also store energy; however, the process is not as efficient as in lipids. Therefore, long-term storage of energy is usually in the form of lipids.

Lipids also form the membranes that surround cells and many of their parts. The lipids found in membranes include cholesterol and phospholipids. Phospholipids are molecules in which a phosphate group and two fatty acids, rather than three fatty acids, are bonded to the three carbon atoms of glycerol.

$$CH_2O - PO_3^{2-}$$
$$|$$
$$CHO - \text{fatty acid}$$
$$|$$
$$CH_2O - \text{fatty acid}$$

Phospholipid

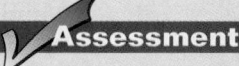

Fact of the **MATTER**

Many vegetable oils for cooking are promoted as being cholesterol-free. Because cholesterol occurs only in animal tissues, all plant products—fruits, vegetables, and all vegetable oils—are cholesterol-free.

Content Background

Cholesterol A cholesterol-free diet doesn't mean your cells lack cholesterol because your liver cells can make it from acetyl CoA. The rate of cholesterol synthesis in the cells is regulated by the amount of dietary cholesterol taken in; the more cholesterol in the diet, the less made by the cells and vice versa. Cholesterol and other lipids are transported in the blood by a number of lipid-binding proteins called lipoproteins. Low-density lipoproteins (LDLs) carry esters of cholesterol to liver cells and other types of cells. The LDLs bind to membrane receptors and deliver cholesterol into the cells. Absence of the LDL receptor in a disease called familial hypercholesterolemia leads to high plasma cholesterol levels, which cause atherosclerosis and heart disease at a young age. High-density lipoproteins (HDLs) pick up cholesterol that is released into plasma by dying cells and membranes being recycled and convert it into esters, which are then transferred to LDLs. The HDLs are sometimes called "good cholesterol" because evidence shows that the HDLs can remove some cholesterol from deposits on artery walls. The LDLs are sometimes called "bad cholesterol" because they deposit cholesterol on artery walls.

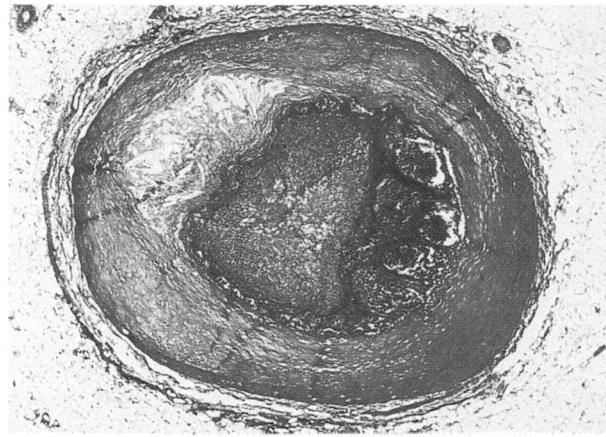

Figure 19.12
Atherosclerosis
Cholesterol forms part of the plaque material clogging this human artery. People with high levels of cholesterol in the blood have an increased risk of developing heart disease due to blocked arteries. Although cholesterol has a bad reputation, cells require it for making membranes, steroid hormones, and bile salts. The human body synthesizes about 1 g of cholesterol each day.

the acetone on a hot plate, using the low heat setting. **CAUTION:** *Acetone is flammable.* Evaporate the acetone until the pale yellow oil extracted from the peanuts thickens and the odor of acetone is absent. Cool the beaker, and determine the mass of the oil. Calculate the percentage of fat extracted from the peanuts.
Results: The peanut oil dissolves in the acetone. As the acetone evaporates, the pale yellow oil thickens. Peanuts contain about 40% fat, but this percentage is not obtained unless vacuum filtration is used. Using gravity filtration, you can expect about 20% fat.
Analysis: Ask these questions.
1. What is the percentage of fat in the sample of peanuts? *Answers will range between 15% and 45%, depending on the method of filtration used.*

2. Is peanut oil a saturated or unsaturated fat? *unsaturated*

■ **Assessment**

Knowledge: Have students look at a food label and determine the number of Calories per gram of fat. *9-10 Calories per gram of fat* L1

DNA — The Thread of Life

Purpose: To extract and examine the DNA from wheat-germ cells.

Process Skills: Observing, communicating, interpreting data

Safety Precautions: Caution students to wear aprons and safety goggles and to avoid inhalation of isopropanol fumes. Collect the residual isopropanol from students and allow it to evaporate in the fume hood. If a fume hood is not available, it may be evaporated outdoors.

Teaching Strategies

- Prepare 1 L of cell lysis solution by mixing 10 g of citric acid, 8 g of sodium chloride, and 300 mL of liquid dishwashing detergent with 500 mL of water. Stir until thoroughly mixed, then add water to bring the total volume to 1 L, and stir again.
- If you have no fume hood, you may wish to minimize the isopropanol fumes and the disposal of materials by performing steps 1-4 for the entire class, then distributing samples of the DNA to individuals for step 5.

Expected Results

A large amount of DNA will be extracted and will be visible as long strands.

Analysis

1. The large amount would suggest that it is compacted in cells.
2. The DNA evidently exists in long strands.

miniLAB 1

DNA—The Thread of Life

You have learned that DNA is a two-stranded molecule that consists of many polymerized nucleotides. DNA acts as the master blueprint for the cell's activity because it contains the coded instructions for every protein made by the cell. In addition, DNA enables a cell to pass on these instructions to the next generation because it is duplicated before the cell divides. Newly formed cells receive exact copies of the parent cell's DNA. In this MiniLab, you will extract and examine the DNA from wheat-germ cells.

Procedure

1. Use a clean mortar and pestle to grind approximately 5 g of wheat germ in 50 mL of cell lysis solution for one minute. Your teacher will supply the cell lysis solution. It contains chemicals that break open the wheat cells and remove unwanted cellular material.

2. Pour the mixture through a piece of cheesecloth or a kitchen strainer into a 250-mL or 400-mL beaker. Discard the solid.

3. Add 100 mL of 91 percent isopropanol to the liquid, and stir briefly.

4. Carefully wind the strands of DNA onto a glass rod.

5. Spread a thin piece of the DNA on a microscope slide, and add two drops of methylene blue dye. Examine the DNA under a microscope.

Analysis

1. Does the large amount of DNA that you extracted suggest that it is loosely arranged or compacted inside the cells?

2. Does the physical appearance of the DNA enable you to describe its physical structure?

Nucleic Acids

The fourth major class of biomolecules—nucleic acids—is not listed on food-product labels, although members of this group are found in every plant or animal cell that humans use for food. A **nucleic acid** is a large polymer containing carbon, hydrogen, and oxygen, as well as nitrogen and phosphorus. Nucleic acids are present in cells only in tiny amounts and are not essential in the human diet because they can be made by the body from amino acids and carbohydrates. Nucleic acids contain the coded genetic information that cells need to reproduce themselves, and they regulate the cell by controlling synthesis of the proteins that carry out so many functions in cells. The two kinds of nucleic acids found in cells are **DNA** (deoxyribonucleic acid) and **RNA** (ribonucleic acid). Nucleic acids are so named because they were first discovered in the nucleus of cells.

688   Chapter 19   The Chemistry of Life

✓ Assessment

Portfolio: Ask students to write reports for the MiniLab. Reports should include a drawing of the DNA. Students should relate the appearance of DNA to its structure. Students should consider adding the reports to their portfolios. [L1] [P]

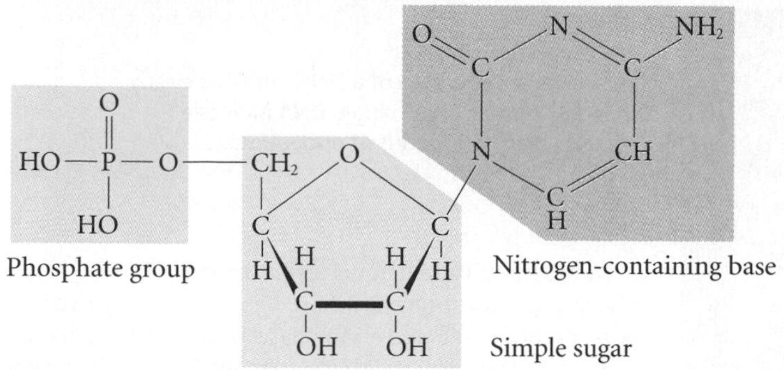

Figure 19.13
Structure of a Nucleotide
Each nucleotide is made of three smaller molecules bonded together: a phosphate group, a simple sugar, and a nitrogen-containing base. The phosphate group consists of phosphorus, oxygen, and hydrogen atoms linked as illustrated. One of the oxygen atoms attaches the phosphate group to the sugar.

The Structure of Nucleic Acids

Nucleic acids are polymers made from building blocks that look complex at first glance. Understanding the structure of these building blocks, called **nucleotides,** is easier if you break them down into three parts, as shown in **Figure 19.13.**

Nucleotides can contain either of two similar sugars. RNA nucleotides contain a five-carbon sugar called ribose, whereas DNA nucleotides contain deoxyribose, which has the same general structure as ribose but with a single hydrogen atom in place of one of the hydroxyl groups.

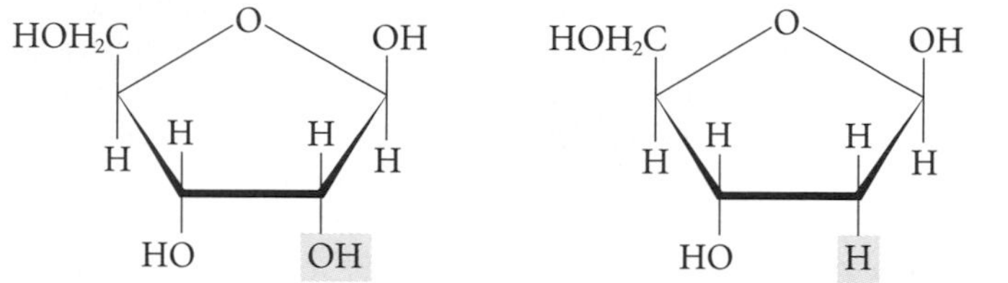

Ribose   Deoxyribose

Five different nitrogen-containing bases are found in nucleotides. Thus, there are five different nucleotide building blocks. The names of these bases are often abbreviated by a single letter—*A* for adenine, *C* for cytosine, *G* for guanine, *T* for thymine, and *U* for uracil. DNA contains A, C, G, and T, but never U. In RNA, U is found instead of T. A nucleic acid polymer is made of chains in which the sugar of one nucleotide is linked to the phosphate of another. The chain of alternating sugars and phosphates is often called the backbone of the polymer. Attached to each sugar is one of the five bases.

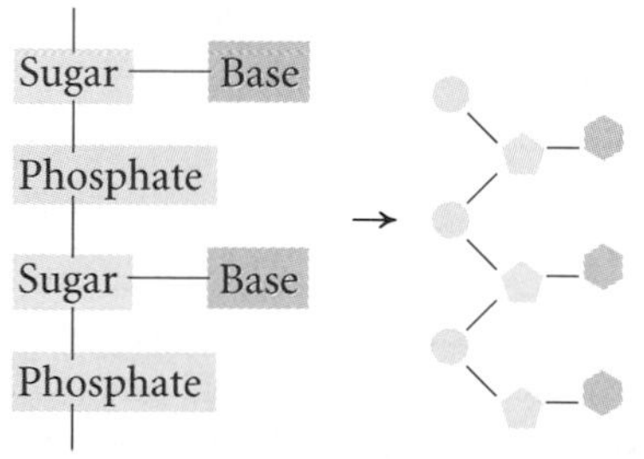

Using an Analogy

Compare DNA to a zipper, and illustrate the unzipping of the two strands that occurs as the first step of replication. Explain that each of the two unzipped strands then forms a complementary strand on itself. The two double strands are each identical to the original.

Quick Demo

What does a helix look like?

If students are unable to envision a helix, show them a large screw (which is normally a right-handed helical convolute). Wrap colored wire around the screw in a helical pattern, then carefully remove the wire from the screw so students can observe the helix.

Concept Development

Have students draw structural formulas for thymine and adenine on the chalkboard and examine how their shapes enable hydrogen bonds to exist between them. L1

Meeting Individual Needs

Visually Impaired Make models of ribose and deoxyribose with a model-building kit so that visually impaired students can understand the difference in structure between the two sugars. L1 LEP

3 ASSESS

Check for Understanding

- Check for understanding of the section material by having students answer the Section Review questions.

- Ask students to write down all the food items they consume for one complete day and bring the list to class. Assemble information from food labels, restaurant menus, and dietary tables that will enable students to calculate the number of grams of proteins, carbohydrates, and fats consumed. Ask students to then calculate the number of Calories of each food type using 4 Cal/g for proteins and carbohydrates and 9 Cal/g for fats. Obtain information from health or biology teachers that will enable students to compare their results with the recommended caloric requirements for persons of their age, sex, weight, height, and level of activity. [L1]

Reteach

Portfolio: Ask students to make a summary chart listing the molecular structure, properties, and occurrence for proteins, carbohydrates, lipids, and nucleic acids. The charts should be included in their portfolios and used as a study guide. [L1]

[P]

Figure 19.14
The Structure of DNA
This model of a portion of a DNA molecule clearly shows its complexity. A single DNA molecule contains many thousands of nucleotides.

DNA

In the three-dimensional structure of DNA, two sugar-phosphate backbone chains are held together by hydrogen bonds that form between the bases attached to each chain. Each base has a specific shape that allows it to hydrogen-bond to only one other base. In DNA, adenine bonds only to thymine, and cytosine bonds only to guanine. When the bases attached to the two DNA backbones bond together, the DNA forms a structure similar to a ladder. The bases form the rungs of the ladder, and the backbone chains form its sides. Now, imagine winding this ladder around a pole. The DNA ladder twists into a spiral structure known as a double helix, as **Figure 19.14** shows.

The specific sequence of the bases in an organism's DNA forms its genetic code. This master code controls all the characteristics of the organism because it contains the instructions for the structure of every protein made by that organism. The code is passed from one generation to the next because offspring receive copies of their parents' DNA. Francis Crick and James Watson were awarded the 1963 Nobel Prize in Chemistry for determining the structure of DNA. They correctly predicted how this structure permits DNA to be readily duplicated by cells so the genetic information can be transmitted to the next generation of cells. This discovery opened the way to the modern science of genetics and its genetic engineering applications.

RNA

RNA also is a polymer of nucleotides, but some important differences can be found when its structure is compared with that of DNA. You already learned that the sugars found in the two nucleic acids are different and that the base uracil replaces the thymine found in DNA. Uracil binds to adenine in the same way that thymine does. The three-dimensional structure of RNA is different from that of DNA, also. RNA has only a single chain of nucleotides, which is twisted into a helix. RNA functions in the cell to carry the genetic information from the DNA to the site of protein synthesis, where it directs the order of amino acids in the proteins.

Vitamins

Have you ever noticed that the labels on containers of milk all indicate that vitamin D has been added? Why do you think dairies do this? Cells require one more major group of organic molecules to carry out their functions. **Vitamins** are organic molecules that are required in small

WORD ORIGIN

vitamin:
  vita (L) life
Vitamins are required in small quantities for life.

Chemistry Journal

Vitamins
Have each student choose a vitamin to research. Students should discover the structure and function of their vitamin and write a report in their journals. They might also give oral reports to the class. [L1]

amounts in the diet. Vitamin D aids in the absorption of calcium and phosphorus through the wall of the intestine and into the bloodstream. These minerals are essential for the formation of healthy bones. If children do not get enough vitamin D in their diets, they develop rickets, a disease that causes bowed legs and other severe bone malformations. Adding vitamin D to milk assures that children will get enough vitamin D, as well as the calcium and phosphorus that are plentiful in milk.

Vitamins fall into two major classes: the water-soluble vitamins that dissolve in water and the fat-soluble vitamins that dissolve in nonpolar organic solvents. Vitamin D is a fat-soluble vitamin. When added to milk, it dissolves in the milk's fats.

Unlike the functions of proteins, carbohydrates, and lipids, vitamins are not used directly for energy or as building blocks for making structural materials in the body. Many serve as coenzymes in cell reactions. A **coenzyme** is an organic molecule that assists enzymes in catalyzing reactions. Vitamin C is a coenzyme for the reaction that modifies the protein collagen to make its structure stable enough to hold your tissues together.

Deficiencies of vitamins result in specific diseases. Recall the devastating effects of scurvy among British sailors caused by a deficiency of vitamin C. Too much of a fat-soluble vitamin also can result in disease because, unlike water-soluble vitamins, excess fat-soluble vitamins cannot be dissolved in urine and excreted. **Figure 19.15** describes vitamin A, another fat-soluble vitamin.

Figure 19.15
Vitamin A
Vitamin A is needed to maintain healthy eyes, skin, and mucous membranes. It is stored in the body's fat cells, especially in the liver. Polar bears store huge quantities of vitamin A in their livers. In the 19th century, many Arctic explorers died from vitamin A toxicity after eating large amounts of polar bear liver.

 **Extension**

TECH PREP

Ask interested and capable students to investigate the process of DNA fingerprinting, perhaps visiting a forensic laboratory of the local police department or the state police. Students can prepare a report that utilizes visual aids for the class. **L3** **COOP LEARN**

4 CLOSE

Enrichment

Oxytocin, a hormone that triggers contraction of the uterus and milk production, and vasopressin, which regulates blood pressure and kidney function, are both nine-amino-acid polypeptides. They differ in only two amino acids. Have students research the structures of both hormones and determine that this is a striking example of structure determining function. *Oxytocin is cys-tyr-ile-gln-asn-cys-pro-leu-gly, and vasopressin is cys-tyr-phe-gln-asn-cys-pro-arg-gly.* **L2**

SECTION REVIEW

Understanding Concepts

1. What groups of atoms are common to all amino acids?

2. Describe the general structure of a triglyceride.

3. How do the structures of DNA and RNA differ?

Thinking Critically

4. **Applying Concepts** Why is it possible to form two different dipeptides from two given amino acids? Using the structural formulas for glycine and alanine in this section, draw the two possible dipeptides.

Applying Chemistry

5. **Polyunsaturated Fats** Nutritionists believe that polyunsaturated fats in the diet are more healthful than saturated fats. What are polyunsaturated fats, and what is their source in the human diet?

SECTION REVIEW

1. Amino acids contain amino, carboxyl, and hydrogen atom groups attached to a central carbon atom.
2. A triglyceride is made up of a glycerol molecule to which three fatty acids are attached by ester linkages.
3. DNA is a double helix that contains the sugar deoxyribose and the bases adenine, thymine, cytosine, and guanine. RNA is a single chain of nucleotides that contains the sugar ribose and the bases adenine, uracil, cytosine, and guanine.
4. The amino group from glycine can combine with the carboxyl group from alanine, or the amino group from alanine can combine with the carboxyl group from glycine. The two dipeptides formed are different.

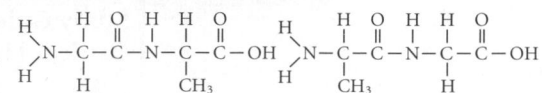

Glycylalanine dipeptide   Alanylglycine dipeptide

5. Polyunsaturated fats are lipids in which the fatty acids contain more than one double or triple bond. Vegetable oils are the best sources of dietary polyunsaturated fats.

SECTION
19.2

Reactions of Life

Key Concepts

Section 19.2 briefly examines two major metabolic processes—respiration and fermentation—as well as discusses the role that energy plays in these reactions.

Planning Ahead

Obtain a copy of the "Carbon Cycle Game" from a biology teacher or from the Massachusetts Audubon Society, South Great Road, Lincoln, MA 01773. This game can be used in *Check for Understanding* under ASSESS.

1 FOCUS

Focus Transparency

Display the **Section Focus Transparency** for Section 19.2 so students can recognize the relationship between energy and metabolism. L1 LEP

2 TEACH

Concept Development

Advise students to think of living cells as the laboratories in which the reactions and functions that sustain life occur.

SECTION PREVIEW

Objectives

Distinguish between the reactions that cells use in the presence and in the absence of oxygen to extract energy from fuel molecules. **Explain how** a small number of biochemical building blocks can be used to make the extraordinary variety of molecules needed to perform life's chemical functions.

Key Terms

metabolism
hormone
respiration
aerobic
ATP
electron transport chain
anaerobic
fermentation

Have you ever bitten into a banana that looked ready to eat but wasn't really ripe yet? How did it taste? Probably not very sweet because the sweet taste of a banana results from sugars that are produced when starch is broken down as the banana ripens. The reaction that breaks starch down into sugar is just one of many that occur in cells. Living cells are dynamic arenas in which molecules and elements are rarely on the sidelines. To uncover the action going on, start with the processes cells use to harness energy.

Metabolism

What happens to the sugars and starch in a banana after you eat it? The sugars are small enough to be transported into the cells lining your digestive system. From there, they are moved into your bloodstream and sent to other cells. The starch molecules are too large to be transported into cells. As soon as you bite into the fruit, enzymes in your saliva begin to hydrolyze the starch into glucose. In the same way, the amazing variety of large and complex molecules in foods must be broken down into smaller units before they can be absorbed into cells.

Enzymes—which have been coded by DNA and synthesized with the help of RNA—catalyze the reactions in which proteins, carbohydrates, and lipids are broken down; this process is called digestion. Only small building blocks are able to enter cells to be used in the many reactions that are involved in metabolism. **Metabolism** is the sum of all the chemical reactions necessary for the life of an organism. These numerous intertwined cellular reactions transform the chemical energy stored in the bonds of nutrients into other forms of energy and synthesize the biomolecules needed to provide structure and carry out the functions of living things. Where does the chemical energy that is stored in nutrient molecules come from? This energy was converted from light energy by plants during photosynthesis. Ultimately, all the energy we need to live comes from sunlight. You will learn about photosynthesis and energy conversion in Chapter 20.

Program Resources

Study Guide, pp. 79-80 L1
Critical Thinking/Problem Solving, p. 21 L2
Laboratory Manual, 19.2, pp. 183-186 L1
Section Focus Transparency 40 and Master, p. 40 L1 LEP
Basic Concepts Transparency 54 and Master, pp. 107-108 L1 LEP
ChemLab and MiniLab Worksheets, p. 125 L1

Function of Hemoglobin

In 1864, a British physicist discovered that hemoglobin, the pigment of the blood, binds and releases oxygen. As it does so, it changes color from red to bluish-red. Hemoglobin binds oxygen to its iron atoms in the alveoli of the lungs and transports it to the tissues of the body. The shape of the hemoglobin molecule changes as it moves from the lungs through the blood vessels to the capillaries near the cells, and that change in shape is what causes it to drop off the oxygen where it is needed.

Structure and function of hemoglobin The physiological properties of hemoglobin can be explained by studying the structure of the molecule. The hemoglobin molecule (Hb) consists of two copies of each of two slightly different polypeptide chains—alpha and beta. Each of the four polypeptide chains has a heme group near its center. Notice that a heme group consists of an iron atom associated with four nitrogen atoms, each of which is part of a ring structure. Thus, the structure of heme is a circle of nitrogenous rings with iron in the center.

Each iron atom in the hemoglobin molecule can bind with one molecule of diatomic oxygen, which contains two oxygen atoms. Thus, every hemoglobin molecule can hold eight oxygen atoms when saturated.

Carbon monoxide poisoning Because it is similar in size to oxygen, the poisonous gas carbon monoxide combines with hemoglobin in almost the same way that oxygen does. Unfortunately, carbon monoxide has an affinity for hemoglobin that is about 210 times that of oxygen. To make matters worse, both oxygen and carbon monoxide combine with hemoglobin at the same point on the molecule. Thus, they can't both be bound to hemoglobin at the same time. Even at concentrations of 0.1 percent, carbon monoxide is dangerous. At that concentration, half of the hemoglobin in the blood will combine with the carbon monoxide, leaving only half of the hemoglobin to combine with oxygen. When the concentration of carbon monoxide rises to about 0.2 percent, the quantity of hemoglobin free to transport oxygen is too small to support life, and death occurs.

Connecting to Chemistry

1. **Hypothesizing** Anemia occurs when the number of red blood cells falls below normal. The patient feels weak and tired. Hypothesize the cause of these symptoms.

2. **Analyzing** When a person is in danger of dying from carbon monoxide poisoning, pure oxygen is administered at six times the normal alveolar oxygen pressure. How might this help?

Connecting to Chemistry

1. Because the number of red blood cells is below normal, the amount of hemoglobin is insufficient to provide the oxygen needed for body functions. With less oxygen, the body produces less energy from food.

2. The amount of oxygen delivered in pure oxygen, which is five times the amount in ordinary air, and the force with which oxygen can combine with hemoglobin is increased by six times; hopefully, this will force the carbon monoxide off the hemoglobin molecule much more rapidly than would occur without treatment.

Purpose

Students will learn that the structure of hemoglobin explains its physiological function.

Teaching Strategies

- Point out that the heart of the hemoglobin molecule is the heme group, an iron atom surrounded by a porphyrin ring. The central porphyrin ring has various side chains: methyl ($-CH_3$), vinyl ($-CH=CH_2$), and propionic acid ($-CH_2-CH_2-COOH$).

- Mention that every milliliter of blood has approximately 5 billion red blood cells, and each red blood cell is packed with 280 million molecules of hemoglobin. Capable students can look up the volume of blood in the body and calculate the total number of hemoglobin molecules. **L3**

- Students may recall that tennis player Vitas Gerulaitis died from carbon monoxide poisoning caused by a faulty connection of a pipe from a swimming-pool heater to the guest house in which he was staying.

Extension

Ask students to investigate whether oxygen is transported in ways other than by hemoglobin. *Some oxygen dissolves directly in the plasma, but the solubility of oxygen gas in blood plasma is directly proportional to the partial pressure of oxygen in contact with the plasma. At the 100-mm oxygen pressure in the lungs, only 0.3 mL of oxygen can dissolve per 100 mL of plasma, which is far too little to keep a human alive.* **L2**

Concept Development
Take a moment to discuss the way in which insulin molecules, which are polypeptides, fit into shape-specific receptor sites in liver and muscle cells, enabling glucose to enter the cells.

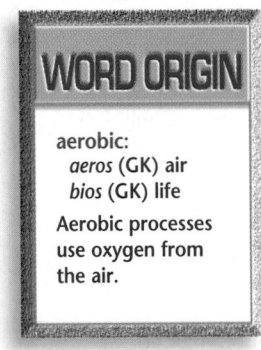

WORD ORIGIN

aerobic:
aeros (GK) air
bios (GK) life

Aerobic processes use oxygen from the air.

Look at the metabolic map in **Figure 19.17,** which gives you an idea of how the varied metabolic reactions are interconnected. Some important reactions involve the trapping of energy and its release and use. Other reactions involve the formation and processing of essential molecules, such as protein enzymes, antibodies, and hemoglobin. Cells perform only the reactions they need at any given time, and this allows them to conserve both energy and building block molecules. How do cells switch the reactions off and on? They use an intricate series of control processes. Many of these can be triggered by signal molecules called **hormones** that are made in specific organs of the body and travel through the bloodstream to communicate with cells in other locations. Insulin is one example of a hormone. After a meal, insulin is released by the pancreas to signal to cells that glucose is available. One of the metabolic processes that insulin triggers in cells is the energy-releasing process called respiration.

Respiration

When cells need energy, they oxidize fuels such as carbohydrates and fats. This process results in the formation of carbon dioxide and water, the same products that are formed when a fuel such as gasoline is burned in an engine. Energy is released in this reaction.

$$fuel\ molecule + oxygen \rightarrow carbon\ dioxide + water + energy$$

Most of the energy used by cells comes from the oxidation of carbohydrates. In oxidation, as in most chemical reactions, energy is needed initially to break bonds. This energy usually is supplied in the form of heat. In an automobile engine, hydrocarbons undergo combustion after the spark plugs heat the gasoline-oxygen mixture to the point where the molecules can react. Only then can the explosive reaction occur that drives the pistons in the engine. Heat speeds up a reaction, but high temperatures kill living cells. How can oxidation of carbohydrates take place at normal body temperatures? The answer can be found in the catalytic power of enzymes. The complex series of enzyme-catalyzed reactions used to extract the chemical energy from glucose is called **respiration.** Respiration is an **aerobic** process, which means that it takes place only in the presence of oxygen. The athlete shown in **Figure 19.16** could not perform without oxygen.

To harness the energy from glucose, cells must break the bonds in which the energy is stored. The net exothermic reaction that takes place when the bonds in glucose are broken is similar to the combustion of hydrocarbons.

$$C_6H_{12}O_6 + 6O_2 \rightarrow 6CO_2 + 6H_2O + energy$$

Figure 19.16

Respiration and Breathing
Although many people might describe respiration as the heavy breathing that occurs when they compete in an athletic event, a biochemist's definition of respiration explains what happens to molecules in the cells when oxygen is available as a result of breathing. Glucose reacts with oxygen to form water and carbon dioxide, and energy is released.

694   Chapter 19   The Chemistry of Life

Metabolism and Cold Light
Purpose: To demonstrate a redox reaction that emits energy in the form of visible light.
Materials: 0.2 g luminol, 0.5 g NaOH, 500 mL distilled water, 20 mL 3% hydrogen peroxide, 0.5 g potassium ferricyanide ($K_3Fe(CN)_6$), two 400-mL beakers, balance, funnel, ring stand, length of glass tubing, short piece of rubber tubing,

600-mL beaker
Safety Precautions:
Disposal: Code C
Procedure: Prepare the luminol solution within a few days of intended use because it has a short shelf life. Prepare Solution A by placing 0.2 g of luminol and 0.5 g of NaOH in 250 mL of distilled water. Prepare Solution B by placing 20 mL of 3% hydrogen peroxide and 0.5 g of

$K_3Fe(CN)_6$ in 250 mL of distilled water.
CAUTION: *Both solutions are toxic.* Use a ring stand to set up a funnel that has been attached to a long piece of glass tubing with a short piece of rubber tubing. Place the end of the glass tubing in a large beaker or pan. In a totally darkened room, simultaneously pour the two solutions together into the funnel.

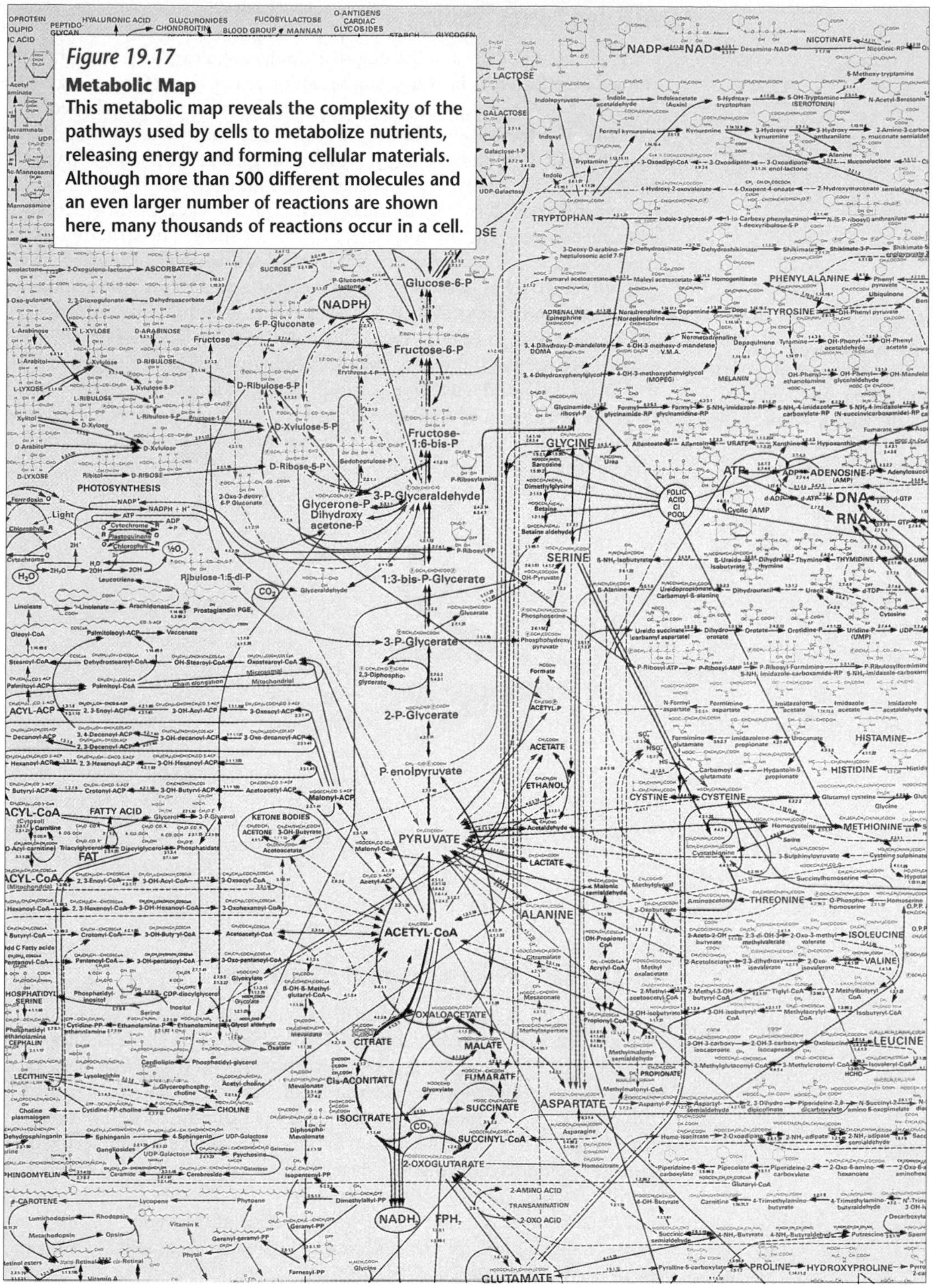

Figure 19.17

Metabolic Map

This metabolic map reveals the complexity of the pathways used by cells to metabolize nutrients, releasing energy and forming cellular materials. Although more than 500 different molecules and an even larger number of reactions are shown here, many thousands of reactions occur in a cell.

19.2 **Reactions of Life**   695

Visual Learning

Figure 19.17: The metabolic map shown in this figure can be ordered as a 16 × 28-inch chart or a 33 × 50-inch poster for a nominal fee from Sigma Chemical Company (toll-free telephone 800-325-3010).

Transparency

Display **Basic Concepts Transparency 54** to introduce the concept of ATP as an energy-storage molecule. L1
LEP

Results: As the mixture flows through the glass tubing into the beaker, it emits a blue light.

Analysis: Ask these questions.

1. Explain whether the redox reaction is endothermic or exothermic. *Exothermic; light is emitted.*

2. Do any of the reactants contain oxygen? Could the reaction simulate respiration? *hydrogen peroxide; yes*

Assessment

Knowledge: Have students recall seeing lightning bugs or fireflies twinkling on a summer evening just after dark. Point out that the insects emit light from their abdomens through a redox reaction. Ask students how the firefly reaction is different from the luminol reaction. *The firefly reaction is a metabolic process that involves enzymes. The luminol reaction is not.* L1

Is this oxidation reaction exothermic?

Illustrate the amount of energy available in a piece of candy by this demo. **CAUTION:** *Wear safety goggles and an apron, and do this demo in a fume hood or behind a safety shield.* Turn out the room lights. Fill a 20 × 150-mm Pyrex or Kimax test tube that is very clean and free from any organic matter about one-third full of reagent-grade potassium chlorate. Heat the tube over a lab burner until the $KClO_3$ is completely molten. Use forceps to drop an M&M candy into the $KClO_3$, and immediately cover the mouth of the tube with a folded section of wire gauze to prevent the M&M from jumping out. *The oxidation of the candy will occur and produce an amazing amount of heat and light energy.* Explain that the potassium chlorate decomposes, producing oxygen that then reacts with the sugar and fat in the candy. After cooling, the tube may be cleaned with a brush and water. Ask students to write the balanced equation for the decomposition of potassium chlorate. $2KClO_3 \rightarrow 2KCl + 3O_2$ L1

ATP and Energy Storage

When gasoline burns in the engine cylinders of a car, large amounts of energy are released as heat in a single explosive reaction. If the energy from the oxidation of glucose and other nutrients were to be released in the same way, the cells would not be able to use it all. Metabolic reactions require small amounts of energy. How can the energy in food be "packaged" so the cell can use it as needed? When nutrients are broken down, the energy is transferred from the broken bonds to energy-storage molecules called adenosine diphosphate, or *ADP* for short. The structure of ADP is identical to one of the building-block nucleotides of nucleic acids, except that ADP has two phosphate groups attached to the ribose molecule, as shown in **Figure 19.18.** Cells store energy by bonding a third phosphate group to ADP to form adenosine triphosphate—**ATP.** When the cell needs energy, the phosphate-phosphate bond in ATP is broken, producing ADP and a phosphate group and releasing stored energy. In cells, the conversion of ADP to ATP and vice versa occurs over and over as metabolic reactions release and use energy. Many of these reactions take place during respiration.

Figure 19.18
ATP and Energy
Energy is stored when ATP is made from ADP and phosphate. Energy is released when ATP breaks down into ADP and phosphate.

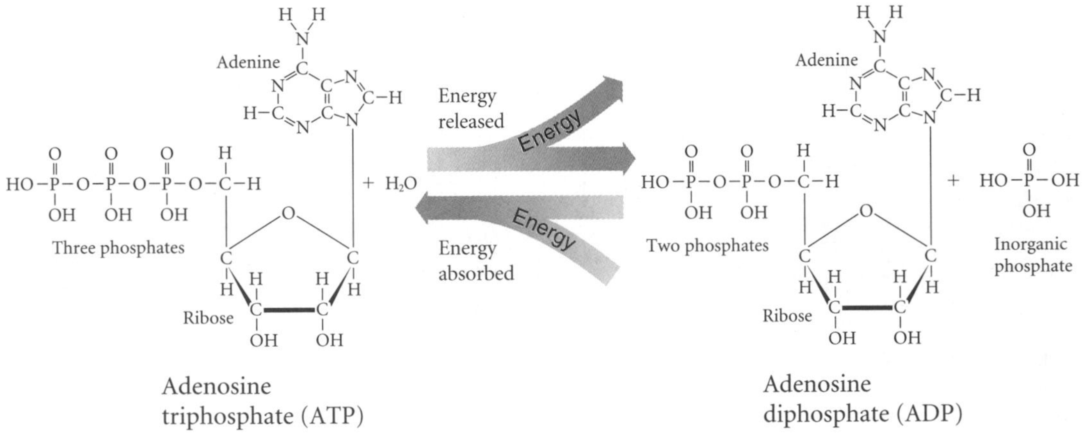

Adenosine triphosphate (ATP)

Adenosine diphosphate (ADP)

Glycolysis—The First Stage of Respiration

The process of respiration consists of many reactions that can be grouped into three stages. **Figure 19.19** summarizes the process, simplifying the reactions by showing only the carbon atoms. In the first stage of respiration, a series of nine reactions breaks down glucose into a pair of three-carbon compounds. Most of the energy from the broken bonds is transferred to ATP. A net yield of two molecules of ATP per molecule of glucose is produced during this stage. This series of reactions is called *glycolysis,* which means "splitting glucose."

Why do cells need ATP? Energy in the bonds of this molecule can be accessed much more quickly and in more manageable amounts than energy in triglycerides, starch, or glycogen.

Meeting Individual Needs

Gifted Ask able students to calculate the amount of energy, in Calories, that could be derived from the oxidation of a 4.00-ounce package of candy that is entirely glucose. Give students the following information: 1275 kJ of energy is released from the combustion of 1 mol of glucose, 1 oz = 28.35 g, 1 Calorie (actually 1 kilocalorie) = 4.186 kJ. *(4.00 oz)(28.35 g/1 oz)(1 mol/180 g)(1275 kJ/mol) (1 Cal/4.186 kJ) = 192 Cal* L3

Figure 19.19
Respiration

In the first stage of respiration, a six-carbon glucose molecule is split into a pair of three-carbon molecules. Hydrogen ions and electrons also are produced in glycolysis. These combine with electron carrier ions called nicotinamide adenine dinucleotide (NAD^+) to form NADH. NADH is a coenzyme that is made from vitamin B_4, which is also called niacin or nicotinic acid. ATP and NADH serve as temporary storage sites for energy and electrons, respectively, during respiration. Two molecules of ATP are used, and four molecules are produced. ▼

1. Glycolysis

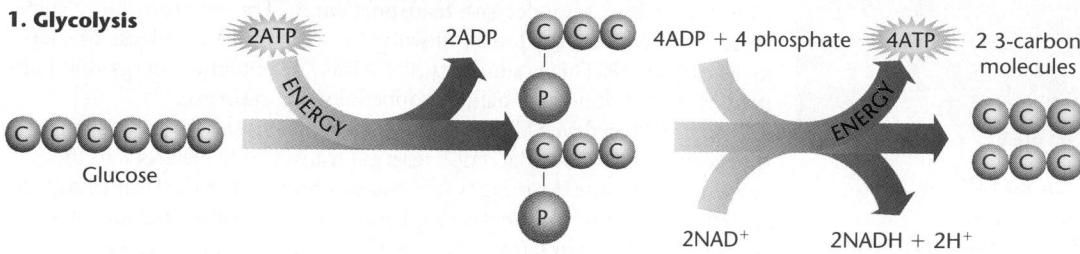

Glucose

2. Tricarboxylic Acid Cycle

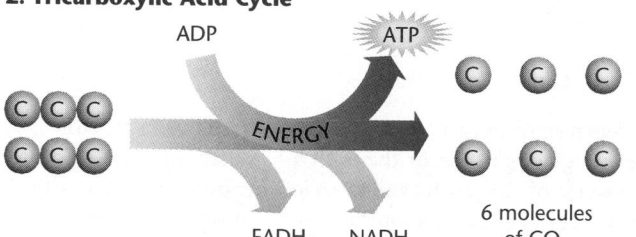

FADH$_2$ NADH

6 molecules of CO_2

◄ In the second stage, the pair of three-carbon molecules that was made from glucose is converted into six molecules of carbon dioxide. More ATP and NADH are produced, along with another coenzyme molecule called flavin adenine dinucleotide (FADH$_2$).

3. Electron Transport Chain

In the third stage, NADH and FADH$_2$ bring the electrons and hydrogen atoms from glucose to a series of carrier molecules, the electron transport chain. A chain-like series of redox reactions takes place. In the final reaction, the electrons and hydrogen are transferred to oxygen to produce water. At the top of the chain, the electrons have high energy. As they pass down the chain, the energy given off is captured in molecules of ATP. The energy available from the breakdown of each glucose molecule can be used to make as many as 38 ATP molecules, a net of 2 during glycolysis and 36 from the electron transport chain. ▶

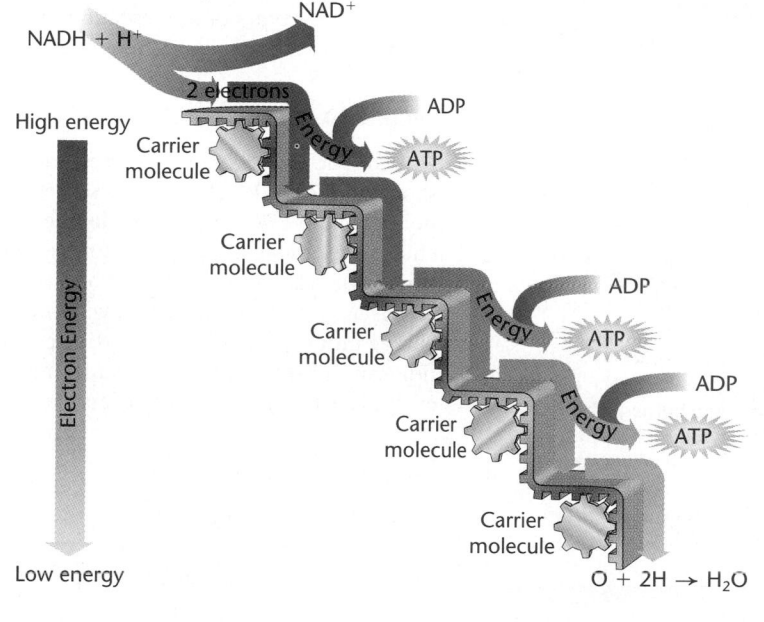

Visual Learning

Figure 19.19: When teaching the stages of respiration, stress the breakdown of the six-carbon chain of glucose and the production of energy, rather than have students memorize the individual reactions in the process. You may wish to have students build models of glucose and follow the model through the three stages of respiration. L1 LEP

Correcting Misconceptions

The bonds that are broken when ATP is converted to ADP are often referred to as high-energy bonds. Explain that the actual amount of free energy change is about −34 kJ/mol, which is not an exceedingly high amount but a suitable amount for biological reactions.

GLENCOE TECHNOLOGY

Videotape
MindJogger Videoquizzes for Chemistry
Chapter 19: The Chemistry of Life
Have students work in groups as they play the videoquiz game to review key chapter concepts. L1 LEP COOP LEARN

Concept Development

Ask a student who is interested in automobiles, particularly in engines, to research and make a class report on the functioning of engines with ethanol as the fuel or as a component of the fuel. In conjunction with this, ask another student to research and make a report about the process and economics of ethanol production by the fermentation of corn. L2

Figure 19.20
Energy Release
When a marshmallow is burned, the combustion reaction releases carbon dioxide, water, and a lot of energy all at once. In respiration, the same overall reaction occurs, but the energy is released step-by-step.

In the 1600s, distilleries distilled the results of fermentation to concentrate the ethanol. The distillates were tested for flammability by burning them on small piles of gunpowder. If the distillate contained at least 50 percent alcohol, the gunpowder would ignite after the alcohol had burned. This result was proof of the alcohol's quality and is the reason why 50 percent alcohol is called 100 proof.

The Tricarboxylic Acid Cycle—The Second Stage

In the second stage of respiration, carbon dioxide is produced in a series of reactions. These reactions are called the tricarboxylic acid cycle because some of the molecules formed in the intermediate reactions have three carboxyl groups.

The Electron Transport Chain—The Third Stage

Most of the energy from glucose is released in the final stage of respiration, referred to as an **electron transport chain.** The electrons move step-by-step to lower energy levels, allowing for the controlled release of energy, **Figure 19.20.** This is similar to the release of potential energy that happens at every bounce if a ball is dropped down a staircase.

Energy is stored as ATP during these steps, as phosphate groups are transferred to ADP. A final redox reaction transfers the electrons to the oxygen you breathe, forming water. You can see that the electron transport chain can't operate without oxygen because oxygen is the final acceptor of the electrons that came from glucose. If oxygen is not present, the entire transport chain stops because the electrons have nowhere to go.

Fermentation

If cells are deprived of oxygen, they must use alternative routes for extracting the energy of glucose, or they will die. Metabolic processes that occur in the absence of oxygen are called **anaerobic** processes. Cells can generate energy from glucose anaerobically by the process of **fermentation.** Because the first stage of respiration can take place without oxygen, glucose is broken down through the reactions of glycolysis into the two smaller, three-carbon units before fermentation reactions proceed. There are two major types of fermentation. In one, ethanol and carbon dioxide are produced. In the other, the product is lactic acid.

Alcoholic Fermentation

Alcoholic fermentation occurs in some bacteria and in yeast cells. As shown in **Figure 19.21,** the process yields much less energy than respiration; much of the energy from glucose remains in the bonds of the ethanol. Although it is relatively inefficient, fermentation provides enough energy for cells to carry on their basic functions. Fermentation by yeast cells is used in the brewing and wine-making industries, and also in the baking industry. The carbon dioxide produced during alcoholic fermentation causes bread dough to rise, making the dough light and spongy. The ethanol that is also produced evaporates during baking.

Lactic Acid Fermentation

Rapidly contracting muscle cells sometimes use up oxygen faster than it can be supplied by the blood. When the cells run out of oxygen, respiration cannot continue. Instead, a type of fermentation called lactic acid

Across the Curriculum

History

Although fermentation processes such as leavening of dough and making wine are recorded in the Bible, it wasn't until the late 1800s that chemists began to determine the identities of the molecules involved and what reactions were taking place. A German chemist named Eduard Buchner investigated whether fermentation was inseparable from life. In 1896, he ground up yeast with sand to produce a cell-free extract, and then he observed what happened when he mixed the extract with sugar. To his surprise, bubbles of carbon dioxide soon began forming, indicating that fermentation could take place outside of living cells. Buchner showed that the chemical processes within cells were not catalyzed by a "vital force" present only in living things, but rather by other molecules—proteins that were soon called *enzymes.* He was awarded the Nobel Prize in Chemistry for his work on fermentation in 1907.

Yeast Plus Sugar—Let It Rise

You may have used yeast to make bread dough or pizza dough rise. Dry yeast is the dormant form of a single-celled fungus that, when given favorable living conditions and food in the form of a carbohydrate, begins to break down the carbohydrate. One of the products of respiration is carbon dioxide. In this MiniLab, you will mix yeast with a disaccharide, sucrose, and with a polysaccharide, the starch in flour, and compare the rates at which carbon dioxide is produced.

Procedure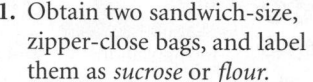

1. Obtain two sandwich-size, zipper-close bags, and label them as *sucrose* or *flour*.

2. Fill a water trough or plastic dishpan about ⅔ full of hot water, and adjust the temperature to between 40°C and 50°C.

3. Put one package of dry yeast and one tablespoon of the appropriate carbohydrate into each of the bags. Mix well.

4. Working quickly, measure and add 50 mL (¼ cup) of the warm water from the trough to each of the bags, and thoroughly mix the contents. Remove all the air you

can and seal the bags. Start timing when both bags are sealed.

5. Put both bags into the trough of warm water. Determine and record the time required for sufficient carbon dioxide to fill each bag completely. If either of the bags is not completely filled in 30 minutes, estimate the fraction of the bag that is filled.

Analysis

1. Based upon your data, rank the rates at which yeast breaks down each type of carbohydrate.

2. Why might the carbohydrates be broken down at different rates?

Figure 19.21

Alcoholic Fermentation
In alcoholic fermentation, each three-carbon molecule that is produced during glycolysis is split to form a two-carbon molecule—the alcohol ethanol—and a one-carbon molecule, carbon dioxide. Two molecules of ATP are produced during glycolysis. ▼

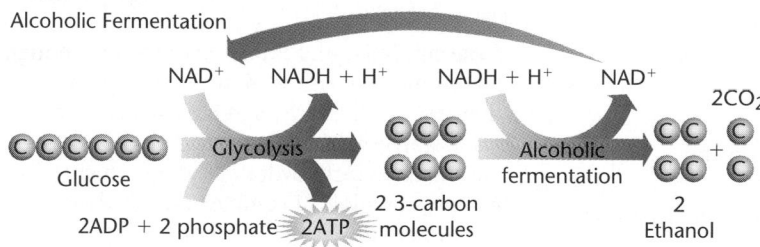

Alcoholic Fermentation

NAD⁺ NADH + H⁺ NADH + H⁺ NAD⁺

Glucose — Glycolysis → 2 3-carbon molecules — Alcoholic fermentation → $2CO_2$ + 2 Ethanol

2ADP + 2 phosphate → 2ATP

Yeast cells ferment the sugars in fruits to produce ethanol and carbon dioxide. The anaerobic process must be carried out in tightly sealed containers such as these large tanks. One-way valves allow carbon dioxide to escape from the fermenting liquid without letting air in. ▶

Chemistry Journal

Uses of Yeast

TECH PREP

Have interested students research the use of yeast in the brewing, winemaking, or baking industries and write reports in their journals. They might investigate how much alcohol is produced and how the alcohol concentration limits the growth of the yeast.

L2

---

miniLAB

Yeast Plus Sugar — Let it Rise

Purpose: To observe and compare the rates at which carbon dioxide is produced as yeast metabolizes a disaccharide and a polysaccharide.

Process Skills: Observing, comparing and contrasting, recognizing cause and effect

Teaching Strategies

• Use snack bags with a capacity of about 1 cup or 1/2 pint.

• If many trials are being done by students, yeast may be purchased bulk from natural food stores or in jars from supermarkets at less expense than the small packets. The jar label specifies how much is equivalent to a packet.

Expected Result:

The sucrose bag will fill completely with CO_2 in 20-30 minutes, and the flour bag will be only partially filled in that time.

Analysis

1. The sucrose-yeast reaction is much faster than the flour-yeast reaction.

2. Flour contains a polysaccharide, which requires significant time to be broken down into the simple sugars that react quickly.

✓ Assessment

Knowledge: Ask students why most yeast bread recipes specify a small amount of white or brown sugar, molasses, or honey in addition to flour. *Without sugar, a great deal of time would be required for the yeast to metabolize the flour and produce sufficient CO_2 to make the bread rise.* **L1**

Check for Understanding

- Check for understanding of the section material by having students answer the Section Review questions.
- Use a "Carbon Cycle Game" to teach the reactions of a carbon atom through cycles of photosynthesis and respiration for 500 million years. See *Planning Ahead* under PREPARE for ordering this game. L1 LEP

Reteach

Have students make posters or charts comparing the energy output of the metabolic processes discussed in this section. L1 LEP

Extension

 Have interested students calculate how many kilojoules of energy can be stored from the breakdown of 113 g of glucose (about 4 ounces) if 38 molecules of ADP are converted to ATP for each molecule of glucose. Students should assume that 29 kJ of energy are stored per mole of ATP. $(113 \, g \, C_6H_{12}O_6)(1 \, mol \, C_6H_{12}O_6 / 180 \, g \, C_6H_{12}O_6)(38 \, mol \, ATP / 1 \, mol \, C_6H_{12}O_6)(29 \, kJ / 1 \, mol \, ATP) = 690 \, kJ$ L3

4 CLOSE

Discussion

 **TECH PREP** Discuss the importance of lactic acid fermentation in the food industry. Ask students to research the production of particular foods such as buttermilk, yogurt, and sauerkraut prior to the discussion. L1

Figure 19.22

Lactic Acid Fermentation Strenuous exercise has caused lactic acid to build up inside this racer's muscle cells, resulting in muscle fatigue. When oxygen becomes available, the body will break down the lactic acid. ▼

fermentation begins. **Figure 19.22** shows the process of lactic acid fermentation. If the anaerobic state continues for too long, lactic acid can build up in the muscles, causing a stiffness called muscle fatigue. Because some of the energy of glucose remains trapped in lactic acid, which cannot be broken down anaerobically, this kind of fermentation does not release as much energy as respiration does.

Lactic Acid Fermentation

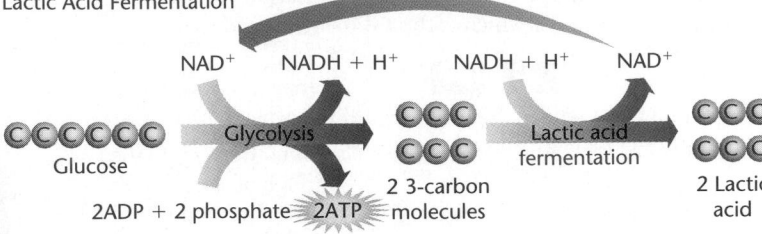

▲ In an anaerobic state, glucose is converted into a pair of three-carbon molecules called lactic acid through glycolysis followed by lactic acid fermentation. Two molecules of ATP are produced during glycolysis. Lactic acid fermentation occurs in some bacteria, in fungi, and in most animals including humans. The dairy industry uses bacteria to make yogurt and buttermilk by this process.

Connecting Ideas

Respiration involves the oxidation of fuel molecules such as glucose in the cells of the body. Animals must take in fuel molecules in the form of food, but how do plants get them? In Chapter 20, you will study photosynthesis, the process in which plants trap energy from sunlight and use this trapped energy to make food molecules. Energy in the form of heat and light commonly is involved in chemical processes. It is important to be able to recognize, as well as measure, the energy changes that accompany chemical reactions. Gaining those skills is the focus of the next chapter.

SECTION REVIEW

Understanding Concepts

1. Explain why respiration provides cells with more energy per molecule of glucose than does fermentation.

2. How do alcoholic fermentation and lactic acid fermentation differ?

3. How many moles of CO_2 are produced from one mole of glucose during respiration?

Thinking Critically

4. **Relating Cause and Effect** Marathon runners go through a crisis called hitting the wall when they use up all their stored glycogen and start using stored lipids as fuel. What would be the advantage to carbohydrate loading, which means eating lots of complex carbohydrates the day before running a marathon?

Applying Chemistry

5. **Aerobic Exercise** Explain what aerobic exercise has in common with aerobic processes in cells.

SECTION REVIEW

1. Respiration results in the complete breakdown of glucose, so all of the chemical energy in glucose's bonds is extracted. Fermentation is an incomplete oxidation in which some chemical energy remains trapped in the ethanol or lactic acid.

2. The products of alcoholic fermentation are ethanol and carbon dioxide. The product of lactic acid fermentation is lactic acid.

3. 6 mols of CO_2

4. Carbohydrate loading might increase the amount of glycogen stored in the liver and muscles so that more energy would be available to a runner.

5. Both aerobic exercise and aerobic respiration require large amounts of oxygen.

REVIEWING MAIN IDEAS

19.1 Molecules of Life

- Proteins, carbohydrates, lipids, and nucleic acids are the four major classes of biomolecules.

- Proteins are polymers of amino acids. These polymers are chains, each containing hundreds or thousands of amino acids, that fold up into a three-dimensional structure.

- Many proteins function as enzymes that speed up reactions. Substrates bind to pockets called active sites on the enzymes, where they are converted into products. The substrate fits tightly and specifically in the active site, much like a key fits in a lock.

- Carbohydrates contain carbon, hydrogen, and oxygen. Simple sugar units combine to form carbohydrate polymers. Simple sugars are used by cells to produce energy, whereas polymers are used to store energy.

- Lipids are molecules that are insoluble in water and soluble in nonpolar solvents. Lipids function to store energy and form cell membranes.

- Nucleic acids are polymers of repeating units called nucleotides, which consist of a sugar, one of five different bases, and a phosphate group. RNA functions in making proteins, and DNA transmits genetic information from one generation to the next.

- Vitamins are complex organic molecules necessary in small amounts for good health. Many vitamins function as coenzymes.

19.2 Reactions of Life

- Metabolism consists of a complex series of interconnected pathways through which fuel molecules are used to trap energy and to make all the different molecules and cell parts needed for the functioning of living cells.

- In the process of respiration, energy is extracted from fuel molecules through a series of enzymatic reactions in which oxygen is used and carbon dioxide and water are produced.

- Fermentation is a process in which energy is extracted from fuel molecules in the absence of oxygen, resulting in the production of ethanol or lactic acid. Less energy is obtained through fermentation than through respiration because some chemical energy is left behind in the products.

Key Terms

For each of the following terms, write a sentence that shows your understanding of its meaning.

active site	fermentation
aerobic	hormone
amino acid	lipid
anaerobic	metabolism
ATP	nucleic acid
biochemistry	nucleotide
carbohydrate	protein
coenzyme	respiration
denaturation	RNA
DNA	steroid
electron transport chain	substrate
fatty acid	vitamin

- Review the Reviewing Main Ideas statements and Key Terms with your students.
- Complete solutions to Chapter Review problems can be found in the *Problems and Solutions Manual*.

UNDERSTANDING CONCEPTS

1. Amino and carboxyl groups combine to form a peptide bond.
2. Six different tripeptides can be formed.
3. DNA is a double helix in which two nucleic acid polymer backbones are made up of alternating phosphate groups and deoxyribose sugar molecules. The two polymer chains are linked together through hydrogen bonds between the bases that extend inward from the sugar-phosphate backbone.
4. Unsaturated fats contain double and/or triple bonds, which form kinks in the long, straight molecule. These kinks prevent the fatty acids from stacking together tightly, so they don't form a rigid solid at room temperature.
5. A disaccharide contains two simple sugar molecules bonded together, whereas a polysaccharide contains many simple sugar molecules bonded together in straight or branched chains.

✓ Assessment

Portfolio: Review the portfolio options that are provided throughout the chapter. Encourage students to select one product that demonstrates their best work for the chapter. Have students explain what they have learned and why they chose this example for placement into their portfolios. **P**

Additional portfolio options can be found in the following **Teacher Classroom Resources:**
Consumer Chemistry
Tech Prep Applications
Applying Scientific Methods in Chemistry

CHAPTER REVIEW

6. ester groups

7. A triglyceride consists of three fatty acids bonded to a glycerol molecule. A phospholipid consists of two fatty acids and a phosphate group bonded to glycerol.

8. A hydroxyl group from one sugar molecule and a hydrogen from another sugar molecule combine to form water. The sugars bond together to form an ether group.

9. Enzymes speed up a reaction that otherwise would occur at a slow rate at body temperature.

10. a phosphate, a sugar, and a nitrogen-containing base

11. DNA stores genetic information and serves as a blueprint for making proteins. RNA is involved in protein synthesis.

12. Hydrogen bonds are found in the rungs of the DNA ladder, linking the bases A and T and C and G together. They hold the two backbone chains together in a double helix.

13. A coenzyme is a molecule, often derived from a vitamin, that aids an enzyme in catalyzing a reaction. NADH and $FADH_2$ are coenzymes involved in respiration.

14. hydrogen bonds, disulfide bonds

15. Lipids are not soluble in water because they are nonpolar and water is polar.

APPLYING CONCEPTS

16. As we age and lose the function of digestive enzymes, it becomes harder to get enough nutrients. Malnutrition and digestive problems can result.

17. Plant seeds contain large quantities of starch that is used for food by the developing plant embryo in the seed.

18. **a.** lipid **b.** amino acid **c.** carbohydrate **d.** amino acid

● UNDERSTANDING CONCEPTS

1. Name the two functional groups in an amino acid that become linked when peptide bonds are formed.

2. Calculate the number of tripeptides (chains of three amino acids) that can be formed from three amino acids—A, B, and C—if each amino acid can be used only once.

3. Describe the three-dimensional structure of the DNA double helix.

4. Why do unsaturated fats tend to be liquids rather than solids at room temperature?

5. Distinguish between a disaccharide and a polysaccharide.

6. What kind of functional group is formed when fatty acids combine with glycerol?

7. Describe the structures of a triglyceride and a phospholipid.

8. Describe the process by which a disaccharide forms from two simple sugar molecules.

9. What is the function of an enzyme?

10. What three structural units make up a nucleotide?

11. Compare the functions of DNA and RNA.

12. Where are hydrogen bonds found in the structure of DNA? What function do they have?

13. What is a coenzyme and what is its function? Name a coenzyme that takes part in respiration.

14. List the types of bonds involved in holding a protein in its folded, three-dimensional shape.

15. Lipids are defined by their solubility properties. Why are lipids usually not soluble in water?

● APPLYING CONCEPTS

16. Enzymes in the mouth, stomach, and intestine break down protein and carbohydrate polymers into amino acids and simple sugar units, respectively. Humans often produce smaller quantities of these enzymes as we age. How does this affect the nutrients we get from food?

17. Plant seeds contain large amounts of starch. What do you think is the function of this starch?

18. Examine the structures of the molecules shown and decide whether each is a carbohydrate, a lipid, or an amino acid.

a) $CH_3-(CH_2)_7-CH=CH-(CH_2)_7-\overset{\overset{\displaystyle O}{\|}}{C}-OH$

b)

c)

d)

19. Digestion of carbohydrates and proteins cannot take place in the absence of water. Explain.

20. Explain how changes in temperature and pH of a cell affect protein structure.

21. Which of the following amino acids would you expect to be the most soluble in water: alanine, cysteine, or phenylalanine? Explain.

22. Vitamins are classified by their solubility properties into two groups: water-soluble and fat-soluble. Use the structures of the vitamins shown to predict which group each falls into.

Vitamin A

Vitamin C

Vitamin D

23. Soap is made by reacting a triglyceride with a strong base such as sodium hydroxide. The products of this reaction are glycerol and soap, which is the sodium salt of fatty acids. Draw the general structure of the sodium salt of a fatty acid.

Health Connection

24. Crocodiles can hold their breath under water for up to 90 minutes, which allows them time to kill their prey by drowning. When an animal holds its breath, hydrogen carbonate ions (HCO_3^-) build up in the blood. Normally, only a small portion of the oxygen bonded to hemoglobin is released to cells for use in respiration, but when hydrogen carbonate ions bind to a crocodile's hemoglobin, the hemoglobin delivers most of its oxygen to cells. Suggest a mechanism by which HCO_3^- ions might have this effect on crocodile hemoglobin.

Everyday Chemistry

25. Why are protein molecules ideally suited for forming pairs of hydrogen bonds?

Everyday Chemistry

26. How is the structure of a fat molecule different from the structure of the chemically altered fat Olestra?

⬤ THINKING CRITICALLY

Making Inferences

27. Burning hair has a characteristic odor due to the presence of large amounts of the amino acid cysteine in keratin, the major protein in hair. What element do you think is responsible for this smell?

Forming a Hypothesis

28. When fresh pineapple is added to a gelatin solution, the gelatin fails to gel, or solidify. When cooked or canned pineapple is used, the gelatin gels. Explain.

19. Carbohydrates and proteins are broken down by the addition of water molecules. This reaction cannot take place in the absence of water.

20. Changes in temperature and pH both result in the disruption of bonds that hold a protein in its unique, three-dimensional structure, causing denaturation.

21. Of the three amino acids listed, cysteine will be the most soluble in water because it has a polar —SH functional group in its side chain that can form hydrogen bonds with water. The side chains in alanine and phenylalanine are nonpolar and do not bond well with water.

22. Vitamin C is water-soluble, and vitamins A and D are fat-soluble.

23. $CH_3(CH_2)_nCOO^-Na^+$

24. The hydrogen carbonate ion binds to the hemoglobin protein, changing its three-dimensional shape so it cannot bind as strongly to oxygen.

25. Protein molecules have both carbonyl and amino sites. The negative oxygen in the carbonyl group seeks a positive hydrogen atom on another molecule, and the positive hydrogen in the amino group seeks a more negative atom.

26. A fat molecule consists of three fatty acids attached to glycerol. Olestra has six to eight fatty acids attached to sucrose. The Olestra molecule is too large to be metabolized in the digestive system.

THINKING CRITICALLY

27. sulfur

28. An enzyme in pineapple breaks down gelatin. This enzyme is denatured by the heat of cooking and canning and cannot act on the gelatin.

29. a. pH 7 **b.** Enzyme activity decreases as pH gets lower or higher than pH 7. **c.** The pH affects the activity of the enzyme because it affects the bonds that hold the enzyme in its active 3-D shape. Both high and low pH values result in denaturation or unfolding of the enzyme as bonds break.

30. The DNA formed using each solvent can be quantified by weighing stirring rods before and after winding up the DNA.

31. This is the optimum temperature for yeast enzymes.

32. Addition polymerization involves the saturation of double and triple bonds. There are no such bonds in the carbon chains of the monomers that form proteins, polysaccharides, and nucleic acids.

CUMULATIVE REVIEW

33. a. H:Ö:H   **b.** Ö::C::Ö

c. H:C̈:H   **d.** H:C̈:C̈:Ö:H (with H above and below)

34. carbon dioxide < ammonia < water. The differences are due to the amount of hydrogen bonding present.

35. a. **b.**

c.

Interpreting Data

29. ChemLab The following table contains data obtained by measuring the activity of the enzyme salivary amylase in solutions buffered at different pH levels.

  a) At which pH is the enzyme most active?
  b) Can you identify any trend in activity as pH moves lower than the optimum pH? Higher?
  c) Why does pH affect enzyme activity?

pH	Time for Color to Disappear
4	10 minutes
5	8 minutes
6	1 minute
7	20 seconds
8	40 seconds
9	4 minutes

Designing an Experiment

30. MiniLab 1 Design an experiment that would show whether ethanol or isopropanol is more effective at precipitating DNA.

Relating Cause and Effect

31. MiniLab 2 Why was the experiment conducted in a 40-50°C water bath?

Identifying Patterns

32. Explain why the polymerization reactions that form proteins, polysaccharides, and nucleic acids are condensation and not addition reactions.

● CUMULATIVE REVIEW

33. Draw Lewis electron dot structures for the following molecules. (Chapter 2)

  a) water   **c)** methane
  b) carbon dioxide   **d)** ethanol

34. Compare the boiling points of water, carbon dioxide, and ammonia, and explain their differences. (Chapter 10)

35. Draw the following molecules, and show the locations of hydrogen bonds that form between the molecules. (Chapter 9)

  a) two water molecules
  b) two ammonia molecules
  c) one water molecule and one ammonia molecule

36. Explain what a buffer is and why buffers are found in body fluids. (Chapter 15)

37. Name the hydrocarbons shown. (Chapter 18)

  a) $CH_3CH_2CH_2CH_3$

  b)
$$CH_3CH_2\underset{\underset{CH_3}{|}}{CH}CH_2CH_2CH_3$$

  c)
$$\begin{array}{c} CH_2 \\ CH_2 \quad CH_2 \\ CH_2 - CH_2 \end{array}$$

  d)
$$CH_3 - \underset{\underset{CH_3}{|}}{\overset{\overset{CH_3}{|}}{C}} - CH_3$$

  e) $CH_3(CH_2)_5CH_3$

  f)
$$CH_3CH\underset{\underset{CH_3}{|}}{\overset{\overset{H_3C \quad CH_3}{|\quad|}}{C}} - CH_2CH_3$$

36. A buffer is a solution containing a mixture of an acid and its conjugate base that is able to resist changes in pH when acid or base is added. We have buffers in our body fluids to maintain a constant pH because the reactions that take place there are sensitive to changes in pH. This is due mainly to the loss of protein activity that occurs because of changes in protein structures (denaturation) when the pH is changed.

37. a. butane **b.** 3-methylhexane **c.** cyclopentane **d.** 2,2-dimethylpropane **e.** heptane **f.** 2,3,3-trimethylpentane

Skill Review

1. **Interpreting Chemical Structures** Use library references to make a poster showing the structures of the 20 common amino acids. Use one color to highlight the side chains that are different in each amino acid. Use other colors to highlight the functional groups that are common to all amino acids.

2. **Making and Using Graphs** Every enzyme has an optimum pH at which it is most active. The following graph shows activity ranges for two digestive enzymes, pepsin and trypsin. Pepsin is found in the stomach, and trypsin is found in the intestine. From the graph, determine the optimum pH for each enzyme. In a paragraph, describe the pH in the organ where each enzyme is found. What would happen in the stomach if pepsin had an optimum pH of 8?

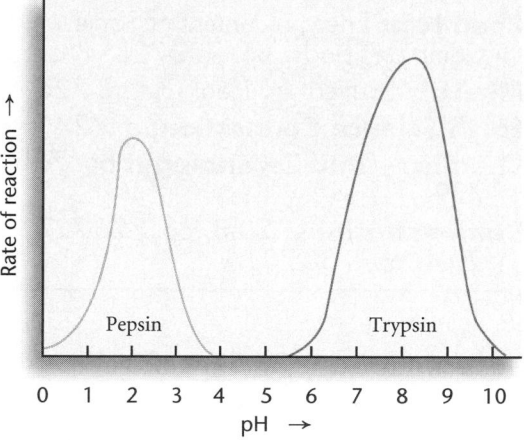

Writing in Chemistry

3. Read *In Search of the Double Helix* by John Gribbin, and write a report discussing whether Watson and Crick or Rosalind Franklin really made the discovery of the DNA structure. Include a discussion of how serendipity played a part in the discovery of the structure.

4. Investigate and write a report about claims that ingesting large amounts of certain vitamins—especially C, A, and E—can help prevent cancer.

Problem Solving

5. Using what you have learned about the change in melting points of alkanes as carbon chain length increases, predict how an increase in the number of $-CH_2-$ groups will affect the melting point of a saturated fatty acid. Give a short oral report to your class explaining your reasoning.

6. One serving of the cereal represented by the food label shown here provides 26 g of carbohydrates, which is nine percent of the recommended daily value for someone consuming 2000 Calories per day. Calculate how many total grams of carbohydrate that person should eat per day.

Nutrition Facts
Serving Size: 1 Cup (30g)
Servings Per Package: About 14

Amount Per Serving	1 Cup Cereal	Cereal With 1/2 Cup Skim Milk
Calories	110	160
Calories from Fat	10	10

		% Daily Value*	
Total Fat 1g		1%	2%
Saturated Fat 0g		0%	2%
Cholesterol 0mg		0%	1%
Sodium 130mg		5%	8%
Potassium 35mg		1%	7%
Total			
Carbohydrate 26g		9%	11%
Sugars 14g			
Protein 2g			

*Percent Daily Values are based on a 2000 calorie diet. Your daily values may be higher or lower depending on your calorie needs.

7. Why aren't the electrons from glucose transferred directly to oxygen during respiration? Why is the electron-transport chain needed? Write a short report in which you predict what would happen to a cell if the electrons were transferred directly to oxygen.

Projects

8. Cooking an egg denatures the egg white proteins, changing them from liquid to solid. Design and carry out an experiment to determine whether changing the pH of the egg white can cause the same kind of denaturation.

9. Keep a detailed record of your food intake for one week. Consult a nutrition book to calculate how closely you came to the recommended daily values of Calories and nutrients each day. Based on your calculations, plan ways to better balance your diet, if necessary.

Chapter 19 Review   **705**

AUTHENTIC ASSESSMENT

2. Optimum pH for pepsin is 2; optimum pH for trypsin is 8. The stomach, where pepsin is found, is highly acidic with a pH of 2. The intestine has a basic environment with a pH of 8. If pepsin had an optimum pH of 8, it would be denatured in the stomach.

5. Melting point will increase as more $-CH_2-$ groups are added.

6. 289 g

7. If the electrons were transferred directly to oxygen, an exothermic reaction would release all the energy at once as heat. The heat would kill the cell. The electron transport chain captures and stores this energy in the bonds of ATP molecules, which can release it more slowly.

Program Resources

Chapter Review and Assessment,
pp. 113-118 [L1]
Alternate Assessment in the Science Classroom [L1]
Computer Test Bank, Chapter 19 [L1]

Performance Assessment in the Science Classroom [L1]
Problems and Solutions Manual,
Chapter 19 [L1]
Supplemental Practice Problems,
Chapter 19 [L1]

Chapter 20 Organizer

Section	Objectives	Activities/Features
20.1 **Energy Changes in Chemical Reactions** (4 days)	1. **Compare and contrast** exothermic and endothermic chemical reactions. 2. **Analyze** the energetics of typical chemical reactions. 3. **Illustrate** the meaning of *entropy,* and trace its role in various processes.	**How It Works:** Hot and Cold Packs, p. 710 **MiniLab 1:** Dissolving—Exothermic or Endothermic?, p. 712 **Everyday Chemistry:** p. 715 **Discovery Demo:** 20.1, pp. 706-707 **Demonstration:** 20.2, pp. 716-717
20.2 **Measuring Energy Changes** (4 days)	4. **Sequence** the technique of calorimetry and illustrate its use. 5. **Compare** the heat generated by some common fuels and by some foods. 6. **Analyze** the efficiency of industrial processes and the need to conserve resources.	**ChemLab:** Energy Content of Some Common Foods, pp. 722-723 **MiniLab 2:** Heat In, Heat Out, p. 726 **Earth Science Connection:** p. 727 **Chemistry and Technology:** pp. 728-729 **Demonstrations:** 20.3, pp. 726-727; 20.4, pp. 730-731
20.3 **Photosynthesis** (4 days)	7. **Analyze** the process and importance of photosynthesis. 8. **Compare** the energy efficiency of photosynthesis and processes that produce electricity. 9. **Trace** how energy from the sun passes through a food web.	**Demonstration:** 20.5, pp. 734-735

Activity Materials

ChemLab (pages 722-723) oven mitt; bottle opener; empty soft-drink can with tab closure; empty can, minus the top and bottom; lids; ring stand; iron ring; beaker tongs; thermometer; 100-mL graduated cylinder; stirring rod; balance; paper clip; matches; pecan half; 2 small marshmallows; food sample

MiniLab 1 (page 712) Epsom salt ($MgSO_4 \cdot 7H_2O$), anhydrous borax (fused borax) ($Na_2B_4O_7$), anhydrous calcium chloride ($CaCl_2$), and ammonium chloride (NH_4Cl); goggles, apron, 250-mL beaker, thermometer, glass stirring rod

MiniLab 2 (page 726) goggles, apron, commercial bleach solution, 100-mL graduated cylinder, 150-mL beaker, 0.5M sodium sulfite solution

KEY TO TEACHING STRATEGIES

The following designations will help you decide which activities are appropriate for your students.

L1 Level 1 activities should be within the ability range of all students.

L2 Level 2 activities should be within the ability range of the average to above-average student.

L3 Level 3 activities are designed for the ability range of above-average students.

LEP LEP activities should be within the ability range of Limited English Proficiency students.

COOP LEARN Cooperative Learning activities are designed for small group work.

P These strategies represent student products that can be placed into a best-work portfolio.

Demonstrations For a complete list of materials for the demonstrations in this chapter, see pages 706, 716, 726, 730, and 734.

Chemical Reactions and Energy

Teacher Classroom Resources

STUDENT MASTERS

Study Guide: pp. 81-82
Laboratory Manual: 20.1 Energy Changes in Physical and Chemical Processes, pp. 187-190
ChemLab and MiniLab Worksheets: p. 130

Study Guide: pp. 83-84
Laboratory Manual: 20.2 Measuring the Heat of Reaction, pp. 191-194; 20.3 Heat of Hydration, pp. 195-200
Critical Thinking/Problem Solving: pp. 22-23
ChemLab and MiniLab Worksheets: pp. 127-129, 131

Study Guide: pp. 85-86

ASSESSMENT RESOURCES
Chapter Review and Assessment: pp. 119-124
Computer Test Bank: Chapter 20
Supplemental Practice Problems: Chapter 20
Performance Assessment in the Science Classroom

TEACHING AIDS

Section Focus Transparency 41
Section Focus Transparency Master: p. 41
Basic Concepts Transparency 55
Basic Concepts Transparency Master: pp. 109-110
Problem Solving Transparency 23
Problem Solving Transparency Master: pp. 45-46
Lesson Plans: p. 41

Section Focus Transparency 42
Section Focus Transparency Master: p. 42
Basic Concepts Transparency 56
Basic Concepts Transparency Master: pp. 111-112
Lesson Plans: p. 42

Section Focus Transparency 43
Section Focus Transparency Master: p. 43
Basic Concepts Transparency 57
Basic Concepts Transparency Master: pp. 113-114
Lesson Plans: p. 43

CHAPTER RESOURCES
Applying Scientific Methods in Chemistry: p. 49
Spanish Resources: Chapter 20
English/Spanish Audiocassettes: Chapter 20
Lab and Safety Skills in the Science Classroom

GLENCOE TECHNOLOGY

Videotape
MindJogger Videoquizzes, Chapter 20
Videodisc
Chemistry: Concepts and Applications, Demonstrations, Videos, and Animations
Science and Technology Videodisc Series: Disc 2 Chemistry: *Solar House, Solar Food Dryer, Solar Tower, Wind Power, Geothermal Wells*

CD-ROM Multimedia System
Chemistry: Concepts and Applications, Same as for Videodisc, plus Interactive Exploration and Major Simulation, *The Periodic Table*

TECH PREP

The following Glencoe resources provide opportunities for integrating science and technology.

Student Edition: How It Works, p. 710; Everyday Chemistry, p. 715; Earth Science Connection, p. 727; Chemistry and Technology, pp. 728-729

Teacher Wraparound Edition: Extensions, pp. 721, 725; Chemistry Journal, p. 725

Teacher Classroom Resources: Chemistry and Industry, pp. 53-54; Consumer Chemistry, pp. 39-40

CHAPTER 20 Chemical Reactions and Energy

$$KClO_4(s) + 2S(s) \rightarrow KCl(s) + 2SO_2(g) + \text{Energy}$$

Chapter Overview

Students learn how to recognize and describe exothermic and endothermic chemical reactions and how to use calorimetry to measure heat change. Finally, they learn about photosynthesis.

Theme Connection

Energy/Conservation The theme of *energy* is discussed in relation to heat changes during chemical reactions. The theme of *conservation* occurs in the idea that energy can change form but is never destroyed.

GLENCOE TECHNOLOGY

CD-Rom

Chemistry: Concepts and Applications
A Thermite Reaction
Use this video of the Discovery Demo to show that reactions can release great amounts of energy. L1 LEP

✔ Assessment Planner

Choose assessment strategies from the following pages to evaluate the progress of your students.
Assess, pp. 718, 724, 730, 736
MiniLabs, pp. 712, 726
ChemLab, p. 722
Portfolio, pp. 713, 718, 723, 731, 735, 738
Chapter Review, p. 738

706

Discovery Demo 20.1

A Thermite Reaction

Purpose: To demonstrate that an exothermic chemical reaction can release tremendous amounts of energy.

Materials: thermite mixture, thermite starter, 11-cm round piece of filter paper, bucket of dry sand, crucible tongs, 8-cm length of Mg ribbon, laboratory burner, hammer

Safety Precautions: 🔥 🧪 💧 🥽
A safety shield must be used with this demo.

Do not allow students to come closer than 10 feet during the reaction.

Disposal: Code A

Procedure: Commercially prepared thermite mixture and thermite starter are available from chemical-supply houses. Fill a large bucket with dry sand and place it in the sink of a demonstration table. Form a large, funnel-shaped depression in the sand. Use 11-cm filter paper to form a

paper cone. Place the paper cone in the sand near the center of the bucket at the bottom of the depression. There must be at least 10 cm of sand all around and under the cone. Place enough thermite mixture in the paper cone to almost fill it, and mix in a teaspoonful of the starter. Make a slight depression in the center of the thermite mixture and fill with thermite starter. Insert 4 cm of an 8-cm length

CHAPTER
20
Chemical Reactions and Energy

You use energy every day. It occurs in so many different forms that you are probably often unaware that you are using it. When you turn on a light or turn a page in this book, you are using energy. When you ride on a bus or walk up a flight of stairs, you are using energy. When you cook food or store ice cream in the freezer, you are using energy. The whole world runs on energy, and a great deal of this energy comes from chemical reactions.

In this chapter, you will learn about the energy changes that go along with chemical reactions. You will see how energy changes are measured and how industrial and biological processes can use energy produced by chemical reactions.

Chemistry Around You

The reaction of potassium perchlorate and sulfur in these highway flares produces a spontaneous release of energy as heat and light. The bright red color comes from the emission of light by strontium in the form of strontium nitrate mixed with the other substances. Chemists have learned how to control and manage the energy from chemical reactions and make it do useful work.

Concept Check

Review the following concepts before studying this chapter.
Chapter 1: energy in chemical changes
Chapter 6: reasons why reactions go forward; equilibrium; and reaction rates
Chapter 10: temperature and particle motion

Chapter Preview

20.1 Energy Changes in Chemical Reactions

20.2 Measuring Energy Changes

20.3 Photosynthesis

Laboratory Activities
ChemLab
Energy Content of Some Common Foods

MiniLabs
1. Dissolving—Exothermic or Endothermic?
2. Heat In, Heat Out

Introducing the Chapter
Ask students to give examples of everyday processes that take place without an evident input of energy and compare these to processes that require an input of energy to get started.
Using the Illustrations Point out that flares release heat and light only after the flare has received an input of energy, usually in the form of heat from friction.

Revealing Misconceptions

Students may think that all chemical reactions give off heat or light or that a reaction that does not give off noticeable energy involves no energy change at all. Nearly all chemical reactions result in some net energy change.

GLENCOE TECHNOLOGY

Videodisc
Chemistry: Concepts and Applications
A Thermite Reaction
Disc 3, Side 1, Ch. 1

Show this video of the Discovery Demo to show that chemical reactions can release great amounts of energy. [L1]
[LEP]

---

of Mg ribbon into the starter and thermite. Allow the remaining 4 cm of ribbon to stick out so it can be used as a fuse. Place a safety shield between the class and the sink. Darken the room. Bend the end of the Mg ribbon over to form a fuse and light it with a burner. **CAUTION:** *Wear safety goggles. Move the students to the back of the room. Move away quickly because the heat generated will produce white-* *hot molten iron and sparks.* After the reaction stops, have students come and look at the molten sand and metal in the pan while the room is still dark. Allow the sand and metal to cool for about 15 minutes, then use crucible tongs to hold the metal in a stream of water. When the metal is completely cooled, use a hammer to break off the melted sand around the iron core. Have students observe the pieces and test them with a magnet.
Results: Aluminum reacts with iron(III) oxide to produce iron metal.
Analysis: Ask these questions.
1. Write the balanced equation for the reaction. $Fe_2O_3 + 2Al \rightarrow Al_2O_3 + 2Fe + energy$ ($\Delta H = -204$ kJ/mol)
2. What was used to supply the high activation energy required for the reaction? *the heat of the burning Mg ribbon*

SECTION 20.1

Energy Changes in Chemical Reactions

SECTION PREVIEW

Objectives
Compare and contrast exothermic and endothermic chemical reactions.
Analyze the energetics of typical chemical reactions.
Illustrate the meaning of *entropy*, and trace its role in various processes.

Key Terms
heat
law of conservation of energy
fossil fuel
entropy

"Smile," says the photographer, as she pushes a button. The camera shutter opens and an electric current from a small lithium battery sparks across a gap in the flash unit. This spark ionizes xenon gas, creating a bright flash of light. The energy from the chemical reaction in the lithium battery has been successfully put to use.

Exothermic and Endothermic Reactions

As you learned in Chapter 1, chemical reactions can be exothermic or endothermic. Recall that an exothermic reaction releases heat, and an endothermic reaction absorbs heat. **Figure 20.1** pictures an exothermic process as a reaction that's going downhill energetically and an endothermic process as a reaction that's going uphill.

Exothermic Reactions

If you have ever started a campfire or built a fire in a fireplace, you know that the burning of wood is an example of an exothermic process. Once you have ignited the wood, the reaction generates enough heat to keep itself going. A net release of heat occurs, which is what makes the reaction exothermic.

Figure 20.1
Exothermic and Endothermic Reactions
The exothermic reaction (left) gives off heat because the products are at a lower energy level than the reactants. The endothermic reaction (right) absorbs heat because the products are at a higher energy level than the reactants.

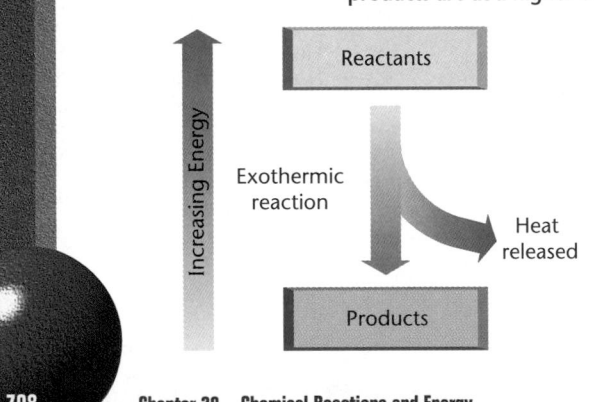

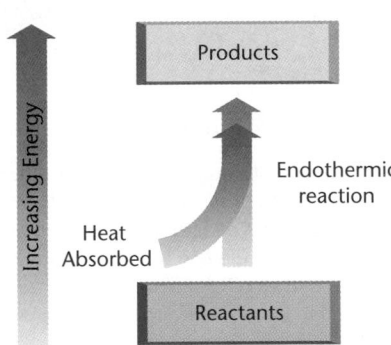

Program Resources

Figure 20.2

The Exothermic Formation of Water
It takes only a small amount of energy to start the reaction between hydrogen and oxygen to form water. The energy released by the reaction is much greater, so the reaction is exothermic. The energy from this reaction has been used to power car and truck engines.

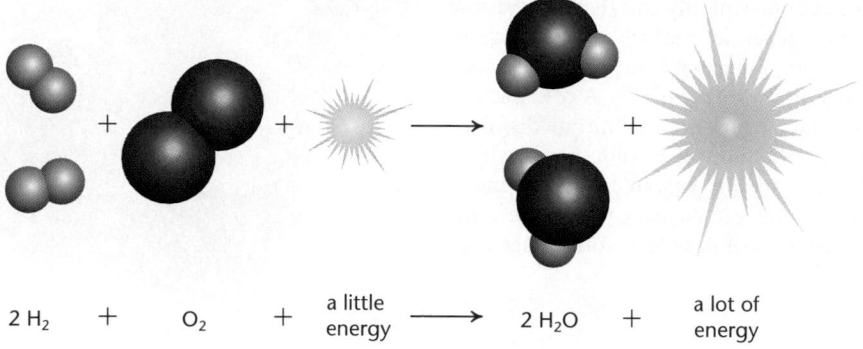

| 2 H₂ | + | O₂ | + | a little energy | ⟶ | 2 H₂O | + | a lot of energy |

The reaction between hydrogen gas and oxygen gas to form water, shown in **Figure 20.2,** is another example of an exothermic reaction. Once a small amount of energy—often just a spark—is added to the mixture of gases, the reaction continues to completion, usually explosively. No additional input of energy from outside is needed to keep it going. Once energy has been supplied to break the covalent bonds in the first few molecules of hydrogen and oxygen, the atoms combine to form water and release enough energy to break the bonds in additional hydrogen and oxygen molecules. The net energy is released as heat.

Endothermic Reactions

Consider the reverse of the reaction just discussed. Just as water can be formed from hydrogen and oxygen, it can also be decomposed to re-form hydrogen and oxygen. In the process of electrolysis, electrical energy is used to break the covalent bonds that unite the hydrogen atoms and the oxygen atoms in the water molecules. The hydrogen atoms pair up to form hydrogen molecules, and the oxygen atoms pair up to form oxygen molecules. The formation of the new bonds releases energy, but not as much as the amount required during the bond breaking. Additional energy must be added continuously during the electrolysis. The reaction absorbs energy and is, therefore, endothermic.

All endothermic reactions are characterized by a net absorption of energy. In the History Connection on page 58, you read about another example of an endothermic reaction: the decomposition of orange mercuric oxide into the elements mercury and oxygen. As long as heat is applied, the compound continues to decompose, but if the heat source is removed, the reaction stops. The net absorption of energy that is required is what makes the reaction endothermic.

*inter*NET
CONNECTION
A hypertext stack for reviewing basic concepts of thermodynamics is available for download.
Anonymous FTP

ftp://archive.umich.edu/mac/misc/chemistry/thermochemistry/sit.hqx

SECTION 20.1

Discussion

Ask students whether they know of any chemical reactions where it is necessary to continue to supply energy to keep the reaction going. This type of reaction is always endothermic. Typical responses may be that energy is needed throughout a decomposition reaction. For example, a continuous electric current is needed to decompose water into hydrogen and oxygen, and continuous heat is needed to decompose mercuric oxide into mercury and oxygen. L1

Transparency

 Display **Basic Concepts Transparency 55**. Use it to contrast exothermic and endothermic reactions. L1 LEP

GLENCOE TECHNOLOGY

 **Videodisc**

Chemistry: Concepts and Applications
Exothermic and Endothermic Reactions
Disc 3, Side 1, Ch. 2

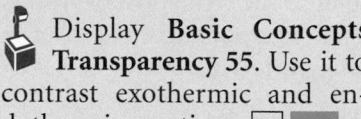

Show this animation to portray endothermic and exothermic reactions. L1 LEP

 CD-Rom

Chemistry: Concepts and Applications
Exothermic and Endothermic Reactions
Show this animation to portray endothermic and exothermic reactions. L1 LEP

Purpose

Students gain an understanding of how an exothermic or an endothermic process can be used to change the temperature of an object.

Background

Most hot and cold packs work on the principle of heat of solution. Chemicals such as calcium chloride dissolve in water exothermically, whereas chemicals such as ammonium nitrate dissolve endothermically. In some cases, a chemical reaction such as the oxidation of iron is used to generate heat over a long period of time.

Visual Learning

• Refer students to the diagram of the cold pack to emphasize how it works. **L1**
• Ask students whether they think that this kind of pack is reusable. *It is not reusable because the salt has dissolved in water. There is another type of hot pack that is reusable. It gives off heat when a supersaturated salt solution crystallizes. Heating the pack restores the contents to the liquid state.*

Teaching Strategies

• Discuss the conditions under which a hot or cold pack would be essential. Answers may include when an athlete is injured during a game and when traditional forms of heat or ice may not be available.
• Ask students to do research on the heat of solution of chemicals other than calcium chloride and ammonium nitrate to determine which ones would be suitable for use in hot or cold packs. **L1**

Hot and Cold Packs

Instant hot and cold packs create aqueous solutions that form exothermically or endothermically and therefore release or absorb heat. A hot pack generates heat when a salt such as calcium chloride dissolves in water that is stored in the pack. The calcium chloride dissolves exothermically. A cold pack absorbs heat when a salt such as ammonium nitrate dissolves in water. The ammonium nitrate dissolves endothermically. In both cases, the salt and water are separated by a thin membrane. All you have to do is squeeze the pack to mix the components and you have instant heat or cold at your fingertips.

1. The outer casing is strong and flexible. It resists puncture and can be shaped to fit the area that you want to heat or cool.

4. Salt is stored in the outer compartment. When the inner membrane breaks, the salt and water mix. The salt dissolves in the water and releases or absorbs energy.

Outer casing

2. Water is stored in an inner compartment separate from the solid salt.

Soluble salt

Water

Membrane of water pack

3. The inner membrane breaks easily when you knead or squeeze the pack or strike it sharply.

Thinking Critically

1. Another type of hand warmer contains fine iron powder and chemicals that cause the iron to rust. The rusting of iron lets the hand warmer maintain temperatures above 60°C for several hours. Explain how that is possible.

2. When a solid in a cold pack dissolves in water, the process takes place spontaneously. What causes the solution process to be spontaneous in spite of the fact that it is endothermic?

Thinking Critically

1. The oxidation of iron is exothermic and can heat a pack when the reaction occurs rapidly enough.
2. The solution process is spontaneous because, as the solid dissolves, entropy increases enough to offset the unfavorable energy change.

Heat

The energy that is involved in exothermic and endothermic reactions is usually in the form of heat. **Heat** is defined as the energy transferred from an object at high temperature to an object at lower temperature. Recall that energy is measured in joules; the symbol for joules is *J*. The symbol for a kilojoule, which is equal to 1000 J, is *kJ*.

Using Symbols to Show Energy Changes

Energy changes are frequently included in the equation for a chemical reaction. The amount of heat absorbed or evolved during a reaction is a measure of the energy change that accompanies the reaction. When 1 mol (18.0 g) of liquid water is produced from hydrogen and oxygen gas, 286 kJ of energy are given off. This means that the energy of the uncombined hydrogen gas and oxygen is greater than the energy of the water. When 1 mol (18.0 g) of liquid water decomposes to form hydrogen gas and oxygen gas, 286 kJ of energy are absorbed. This also shows that the energy of the uncombined hydrogen and oxygen is greater than the energy of the water. The graphs in **Figure 20.3** illustrate this relationship.

Scientists have observed that the energy released in the formation of a compound from its elements is always identical to the energy required to decompose that compound into its elements. This observation is an illustration of an important scientific principle known as the **law of conservation of energy.** That law states that energy is neither created nor destroyed in a chemical change, but is simply changed from one form to another. In an exothermic reaction, the heat released comes from the change from reactants at higher energy to products at lower energy. In an endothermic reaction, the heat absorbed comes from the opposite change.

WORD ORIGIN

energy:
en (GK) in
ergon (GK) work
A person who has a great deal of energy can work hard all day.

Figure 20.3

Formation and Decomposition of Water

As the graphs show, the energy produced when 1 mol of liquid water forms from the elements hydrogen and oxygen is equal in magnitude to the energy absorbed when the water decomposes.

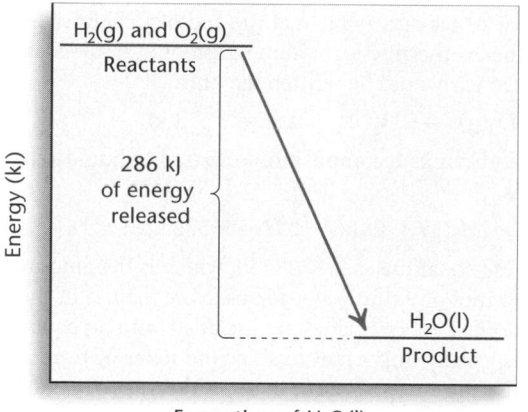

Formation of $H_2O(l)$

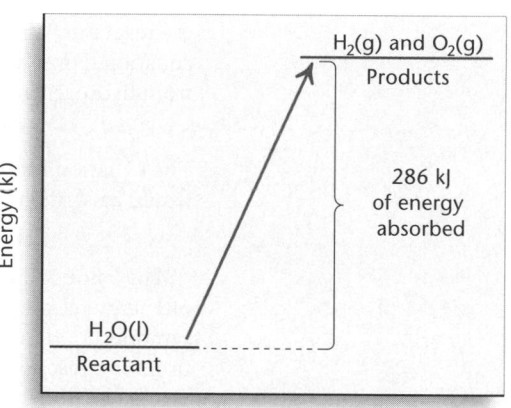

Decomposition of $H_2O(l)$

SECTION 20.1

Correcting Misconceptions

Students often confuse heat and temperature. Point out that temperature is a measure of the average kinetic energy of the particles of a substance and that heat is energy that is transferred from an object of higher temperature to an object of a lower temperature. Adding heat to an object may raise its temperature, but heat and temperature are not the same.

Extension

Ask interested student groups to make large posters or charts that depict energy use in the United States from approximately 1800 to the present. Categorize the sources as: wood, coal, petroleum, natural gas, nuclear, and hydroelectric. Discuss the reasons for and consequences of the evident trends.
`L1` `LEP` `COOP LEARN`

Visual Learning

Figure 20.3: Relate the negative ΔH sign for an exothermic reaction and the positive ΔH sign for an endothermic reaction to the changes in energy represented in **Figure 20.3** and the definition,
$\Delta H = H_{products} - H_{reactants}$.

GLENCOE TECHNOLOGY

Videodisc

Chemistry: Concepts and Applications
See the Videodisc Program Teacher Guide to access illustrations and other material available for this chapter.

Meeting Individual Needs

Gifted Explain that the ΔH value of -286 kJ for the formation of 1 mol of liquid water from its constituent elements in their standard states is known as the heat of formation of liquid water. Heats of formation may be used to calculate ΔH values for many reactions by subtracting the sum of the heats of formation of the reactants from the sum of the heats of formation of the products:
$\Delta H_{(reaction)} = \Sigma \Delta H_{f\ (products)} - \Sigma \Delta H_{f\ (reactants)}$.

Have students use a table of standard heats of formation to determine $\Delta H_{reaction}$ for the following reactions. Point out that the table values are molar heats of formation, so they must be multiplied by the number of moles in the reaction.
a. $2NO(g) + O_2(g) \rightarrow 2NO_2$ $(-144.14\ kJ)$
b. $Fe_2O_3(s) + 3CO(g) \rightarrow 3CO_2(g) + 2Fe(s)$ $(-24.8\ kJ)$ `L3`

miniLAB

Dissolving—Exothermic or Endothermic?

Purpose: To determine whether the dissolution of each of several common solids in water is exothermic or endothermic.

Process Skills: Observing, measuring, interpreting data, communicating

Disposal: Advise students to rinse the solutions down the drain with large amounts of water.

Teaching Strategies

- Require that students' notebooks contain the following items before they begin: Objectives; Safety Precautions; Outline or Flow Chart of the Procedure; Data and Observations Table.
- Suggested solids to be dissolved are magnesium sulfate heptahydrate (Epsom salts), anhydrous sodium borate (fused borax), anhydrous calcium chloride, and ammonium chloride. Do not use plain borax, which is a decahydrate. If you do not have fused borax, substitute anhydrous sodium carbonate, which also dissolves exothermically.

Expected Results

Epsom salts—temperature decreases
anhydrous borax—temperature increases
calcium chloride—temperature increases
ammonium chloride—temperature decreases

Analysis

1. Fused borax (anhydrous sodium borate) and anhydrous calcium chloride dissolve exothermically. Epsom salts (magnesium sulfate heptahydrate) and

miniLAB 1

Dissolving—Exothermic or Endothermic?

The dissolving of a solid in water, like most other processes, may liberate energy or absorb energy. If the dissolving process is exothermic, the liberated energy raises the temperature of the solution. If the process is endothermic, it absorbs energy from the solution, lowering the temperature. Examine the dissolution of several common solids in water.

Procedure

1. Measure 100 mL of water into a 250-mL beaker. Set a Celsius thermometer in the water and allow it to come to the water's temperature. Record that temperature. Remove the thermometer from the water.

2. Add to the water approximately 1 tablespoon of the solid to be tested, and stir with a stirring rod for 20 seconds. Put the thermometer back into the solution. Record the temperature.

3. Pour the solution down the drain with large amounts of tap water.

4. Repeat steps 1 through 3 for each of the solids to be tested.

Analysis

1. Which of the solids that you tested dissolve exothermically? Which dissolve endothermically?

2. Which of the solids that you tested could be mixed with water inside a flexible plastic container to produce a cold pack that might be used by medical personnel?

The difference in energy between products and reactants in a chemical change is symbolized ΔH (delta H), where the symbol Δ means a difference or change and the letter H represents the energy. The energy absorbed or released in a reaction ($\Delta H_{reaction}$) is related to the energy of the products and the reactants by the following equation.

$$\Delta H_{reaction} = H_{products} - H_{reactants}$$

For exothermic reactions, ΔH is negative because the energy stored in the products is less than that in the reactants. For endothermic reactions, ΔH is positive because the energy of the products is greater than that of the reactants. The value of ΔH is often shown at the end of a chemical equation. For example, the exothermic formation of 2 mol of liquid water from hydrogen and oxygen gas would be written like this.

$$2H_2(g) + O_2(g) \rightarrow 2H_2O(l) \quad \Delta H = -572 \text{ kJ}$$

The equation for the endothermic decomposition of 2 mol of liquid water would be written like this.

$$2H_2O(l) \rightarrow 2H_2(g) + O_2(g) \quad \Delta H = +572 \text{ kJ}$$

The value 572 kJ in these equations is 2×286 kJ, which is the amount of energy released when 1 mol of liquid water forms. Note the use of the symbols (s), (l), and (g). When energy values are included with an equation for a reaction, it is especially important to show the states of reactants and products because the energy change in a reaction can depend greatly upon physical states.

---

ammonium chloride dissolve endothermically.

2. Epsom salts or ammonium chloride

✓ Assessment

Knowledge: Ask students to explain how a cold pack might be designed. *The solid could be enclosed in a breakable pouch inside a larger packet or pouch of water. When the inner plastic pouch is broken, the solid will dissolve, making the solution colder.* **L1**

Activation Energy

The hydrocarbons found in petroleum and natural gas are the remains of plants and other organisms that lived millions of years ago. Oil and natural gas are called **fossil fuels** for this reason. Fossil fuels are a rich source of energy because when they react with oxygen to produce carbon dioxide and water, a great deal of energy is released in the form of heat.

However, fossil fuels do not burn automatically. Energy, usually in the form of heat or light, is needed to get the chemical reaction started. The combustion of hydrocarbon fuels requires this input of energy, called activation energy, to begin the reaction. For example, the butane gas in a disposable lighter requires a spark to start the combustion of the gas.

In Chapter 6, you learned that activation energy is needed to cause particles to collide with enough force to make them react. Activation energy is required in both exothermic and endothermic reactions. The fact that a fuel requires an input of energy—such as from a spark—to begin burning does not mean that the combustion reaction is endothermic. The reaction releases a net amount of heat, and so it is exothermic.

Look more closely at activation energy and heat of reaction, using as an example the burning of methane, which is the main component in natural gas, to yield carbon dioxide and water vapor. The equation for this reaction is as follows.

$$CH_4(g) + 2O_2(g) \rightarrow CO_2(g) + 2H_2O(g) \quad \Delta H = -802 \text{ kJ}$$

A graph of the energy change during the progress of this reaction is shown in **Figure 20.4.** Notice how the energy curve rises, then falls. The rise represents the activation energy, which is the energy difference between the reactants and the maximum energy stage in the reaction. The fall represents the energy liberated by the formation of new chemical substances. When 1 mol of methane burns, 802 kJ of heat are given off. The reaction is exothermic, as is shown by the negative sign of ΔH. The energy stored in the products is less than that stored in the reactants, so a net amount of energy is released. Some of the released energy provides the activation energy needed to keep the reaction going.

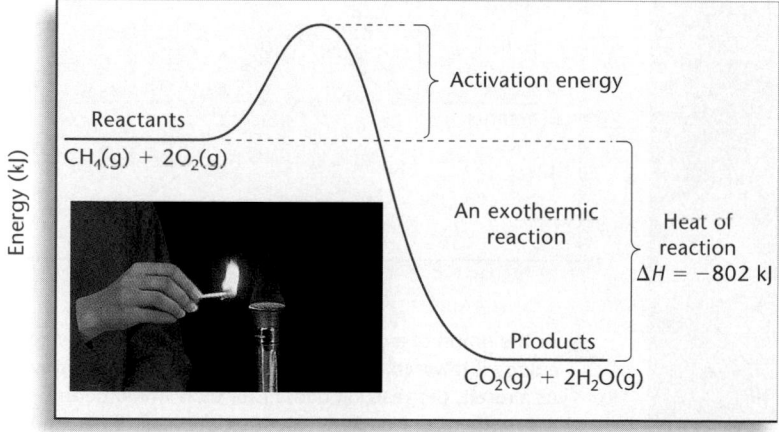

WORD ORIGIN

combustion:
combustus (L)
burned

The internal combustion engine in a car burns the hydrocarbons in gasoline.

Figure 20.4

Energy in an Exothermic Reaction

In order to occur, the combustion of methane, illustrated in the photo at the left, requires an input of activation energy—in this case, provided by a match or sparking device in the stove. Overall, the reaction releases 802 kJ of energy per mole of methane. Notice from the graph that the products are in a lower energy state than the reactants. The negative value of ΔH reflects this fact.

Using an Analogy

Explain that we usually measure the height of a mountain relative to sea level. In chemical changes, energy changes are measured relative to the energy of the reactants, which may be unknown and different for any set of reactants. This can be compared to measuring the height of a mountain relative to the lowest point in a nearby valley, although we may not know the absolute altitude of the valley. In chemistry, it is the change in energy and its direction that is important.

Discussion

Ask students to decide whether each of the following changes is exothermic or endothermic. (a) Tea is warmed in a microwave oven. *endothermic* (b) Droplets of water condense on a cold can or glass of soda. *exothermic (in reference to water)* (c) Natural gas is burned in the furnace of a home. *exothermic* (d) Iron is melted in a steel-manufacturing plant. *endothermic* L1

Concept Development

Explain to students that any chemical equation that includes energy or that displays a ΔH value is known as a thermochemical equation.

Correcting Misconceptions

Point out that the value +572 kJ included with the equation on page 712 is positive because it reflects the heat of decomposition of liquid water rather than the heat of formation. The value is twice the heat of decomposition of liquid water (286 kJ) because 2 mol of water are decomposed. ΔH values are *molar* heats.

Integrating THE Sciences

Biology Animals are classified as endotherms (or homeotherms) if they control their body temperature and ectotherms (poikilotherms) if they do not. Have interested students look up these terms and create a poster or display depicting animals that fall into each category. Both types release heat from cellular respiration. Students could research to find out what mechanisms enable endotherms to control body temperature and the adaptive advantages of each mechanism. Students should consider including this work in their portfolios. L1 P

Using an Analogy

Ask students to compare energy changes in a chemical reaction to profits and losses ($\Delta\$$) in a business. Each month, the business has receipts, which are positive dollar amounts, and expenses, which are negative dollar amounts. If receipts exceed expenditures, a positive dollar amount (profit) occurs and $\Delta\$$ is positive. If expenditures are greater than receipts, there is a loss and $\Delta\$$ has a negative value. L1

Discussion

Ask students to describe in their own words what **Figure 20.6** shows. Ask them to tell how the use of a catalyst affects ΔH and the energy values of reactants and products. *It has no effect.* Then ask them what the catalyst accomplishes. *It provides an easier pathway for the reaction by reducing the activation energy.* L1

Transparency

 Display **Problem Solving Transparency 23.** Use it to reinforce the concept of activation energy. L1 LEP

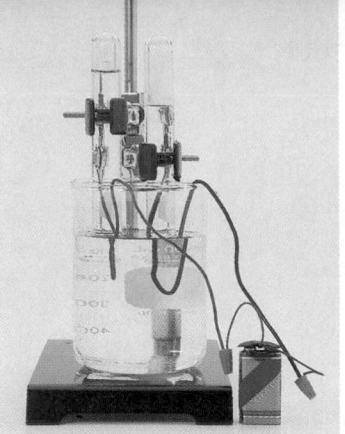

Figure 20.5
Energy in an Endothermic Reaction
The decomposition of water requires a continuous input of energy, such as electrical energy from a battery. Overall, the reaction absorbs 572 kJ of energy for every 2 mol of liquid water. The products are in a higher energy state than the reactant. The positive value of ΔH reflects this fact.

Figure 20.6
Effect of a Catalyst
It is easier for cars to go through the tunnel than for them to climb the mountain. ▼

Now, consider an example of an endothermic reaction—one that you have already read about—the decomposition of water. The equation for the reaction is as follows.

$$2H_2O(l) \rightarrow 2H_2(g) + O_2(g) \quad \Delta H = +572 \text{ kJ}$$

A graph of the energy changes during the progress of this reaction is shown in **Figure 20.5**. Notice how the energy curve rises above the level of the reactant, then falls only slightly when the products form. Therefore, the products have more stored energy than the reactants, and the graph clearly shows a net gain of energy. Because of this net gain, ΔH is positive. Energy must be added continuously to keep the reaction going.

Recall from Chapter 6 that a catalyst can be used to speed up a reaction. The catalyst provides a different reaction path—one for which the activation energy is lower. The catalyst thus creates a shortcut or tunnel through the energy hill between the reactants and products. More of the collisions will now be effective because less energy is required. The energy diagram in **Figure 20.6** shows the effect of a catalyst on activation energy.

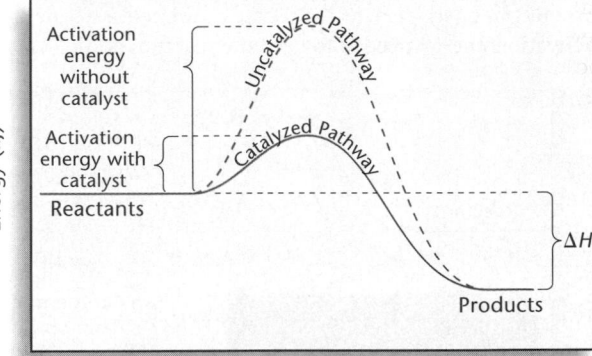

▲ As the graph of a catalyzed reaction shows, the activation energy is lowered, providing an easier reaction pathway. As a result, the reaction goes faster than would be the case without the catalyst. Note that the heat of reaction, ΔH, is the same in both cases.

714   Chapter 20   Chemical Reactions and Energy

Meeting Individual Needs

Learning Disabled Have students depict the reaction $2H_2O(l) \rightarrow 2H_2(g) + O_2(g)$ with ball-and-stick molecular models. Simultaneously draw the reactant molecules on an energy diagram—the reactant atoms at the higher energy state attained after the activation energy has broken the H—O bonds, and the product molecules at an energy state lower than the activation energy but higher than the reactants. L1 LEP

Chemistry Journal

Losses Equal Gains
Ask students to describe five common changes that involve loss or gain of energy; to define the system and the surroundings in each change; to identify which loses and which gains energy; and to include the changes, definitions, and identifications in their journals. L1

Everyday Chemistry

Catalytic Converters

Since 1975, every new car sold in the United States has a catalytic converter installed in the exhaust system. This device contains porous, heat-resistant material coated with a catalyst. The purpose of the catalytic converter is to reduce air pollution.

How a catalytic converter works A typical catalytic converter consists of particles of platinum and rhodium deposited on a ceramic structure that is like a honeycomb. The platinum and rhodium catalyze reactions that remove pollutants such as nitrogen monoxide (NO), carbon monoxide (CO), and unburned hydrocarbons. When nitrogen monoxide binds to the rhodium surface, it breaks down to oxygen and nitrogen. The bound oxygen reacts with carbon monoxide, which has also become bound to the rhodium surface. The reaction produces carbon dioxide. The oxidation of unburned hydrocarbons produces carbon dioxide and water.

The honeycomb arrangement of catalytic materials in the converter provides a large surface area for the reactions to take place. Such an arrangement increases the rate at which the pollutants are removed. Catalytic converters have reduced the pollutants released into the air by cars by as much as 90 percent.

Improving the catalytic converter The operating temperature of a catalytic converter is between 316° and 649°C, which is in the range of standard exhaust-gas temperatures for a vehicle being driven on the road. Below 316°C, the device does nothing. What about the pollutants emitted during the beginning of the driving period, when the converter is just heating up? They slip right through the catalytic converter without being changed.

In an effort to stop the emission of air pollutants during the warm-up period, some catalytic converters have heaters that raise their tempera-

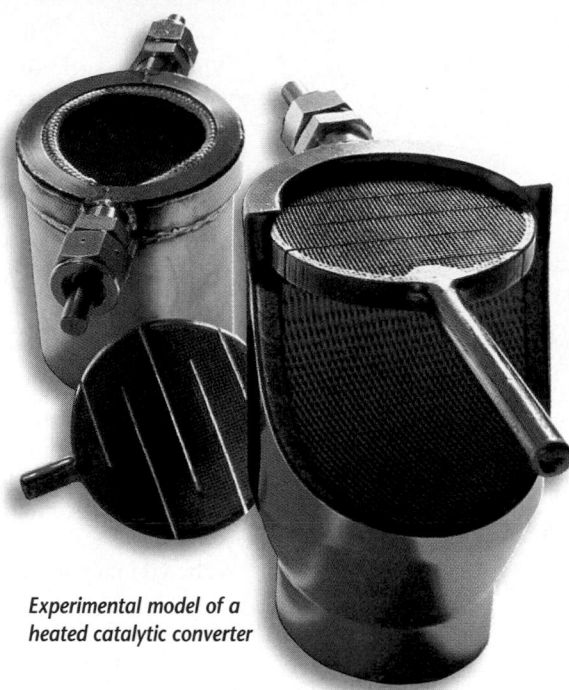

Experimental model of a heated catalytic converter

ture to 400°C within 5 seconds. The result is that almost as soon as the car is started, air pollutants in the exhaust are being broken down. Heated catalytic converters are proving to be highly efficient. Someday, perhaps even those 5 seconds of dirty emissions before the converter is heated will be eliminated.

Exploring Further

1. **Acquiring Information** Find out why cars with catalytic converters cannot use leaded gasoline.

2. **Applying** Instead of the honeycomb structure of the catalytic converter described above, some converters contain pellets coated with a platinum-palladium mixture. Why is the use of pellets effective?

Exploring Further

1. Cars with catalytic converters cannot use leaded gasoline because the lead will deactivate the catalyst.
2. The pellets coated with the catalyst would provide ample surface area on which the pollutants can be efficiently converted to nonpollutants.

Everyday Chemistry

Purpose
Students gain an understanding of how a catalytic converter helps remove pollutants from the exhaust system of automobiles.

Teaching Strategies
- Ask students to explain the function of a catalyst in a chemical reaction. *A catalyst is a substance that increases the speed of a chemical reaction without being consumed by the reaction.* [L1]
- Emphasize that many chemical reactions would occur so slowly without the aid of catalysts that they would not be effective. Typical examples are biochemical reactions in living things.
- Many of the catalyzed oxidations of pollutants in the converter are exothermic and release a great deal of heat. As a result, the catalytic converter can become hot. A quick look under a car will reveal metal shielding plates between the catalytic converter and the body of the car. Flammable objects such as papers, rags, or dry leaves can catch on fire when in contact with a hot catalytic converter.

Extension
Have students read "Catalysis on Surfaces" in *Scientific American,* April 1993, pp. 77-78 to find out how the rhodium catalyst in the catalytic converter works at the molecular level. *Nitric oxide interacts with irregular particles of rhodium and binds to a regular patterned surface of rhodium.* [L2]

What does the KI do?

🧤 🥽 🧪

Illustrate the effect of potassium iodide, which acts as a catalyst in the decomposition of hydrogen peroxide. **CAUTION:** *Wear safety goggles, an apron, and rubber gloves. 30% hydrogen peroxide is a strong oxidizing agent and can cause severe burns.* Set a 100-mL graduated cylinder in a trough or sink and add 5 mL of 30% H_2O_2 and 5 mL of dishwashing detergent. You may add a few drops of food coloring to make the liquid and resulting suds more visible. Use a scoop to add about 1 g of KI to the mixture. *The H_2O_2 decomposes rapidly, forming a great amount of oxygen-containing suds and releasing much heat.* Explain that the KI acts as a catalyst for the decomposition reaction.

Is disorder inevitable?

Show students a deck of cards in which all the cards are arranged in numerical order and by suit. Explain that this arrangement will represent the lowest entropy state for the deck of cards. Shuffle the cards thoroughly and show students the resulting arrangement that represents a much higher entropy state. Explain that the higher entropy state is simply much more probable. Ask students to speculate on the number of shuffles that would be necessary to restore the lowest entropy state. *Though perfect numerical order by suit is theoretically possible, the chance of shuffling and attaining this lowest entropy state is less than 1 out of 8×10^{67} arrangements.*

Forces That Drive Chemical Reactions

When pure aluminum metal is exposed to chlorine gas, aluminum chloride is formed. This chemical change is spontaneous; that is, it just happens on its own. The reaction is highly exothermic. The following equation summarizes the process.

$$2Al(s) + 3Cl_2(g) \rightarrow 2AlCl_3(s) + 1408 \text{ kJ}$$

What allows such reactions to occur spontaneously? The answer has to do with the general forces that drive all reactions.

Order, Disorder, and Entropy

Scientists have observed two tendencies in nature that explain why chemical reactions occur. The first tendency is for systems to go from a state of high energy to a state of low energy. The state of low energy is more stable. Thus, for example, exothermic reactions are more likely to occur than endothermic ones, all other things being equal. The second tendency is for systems to become more disordered. Spontaneous reactions tend to occur if energy decreases and if disorder increases.

Figure 20.7 shows an everyday spontaneous change that increases disorder. The first photo shows a full glass of milk on the breakfast table. The second photo shows the same glass of milk after someone has bumped into it and caused it to fall to the floor. Notice the disorder in the second picture. The glass is broken into seemingly random shards, and the milk is scattered into pools and droplets.

You have already learned a good deal about energy changes, but the concept of disorder in chemical changes may not be as familiar. Scientists use the term **entropy** to describe and measure the degree of disorder. Unlike energy, which is conserved in chemical changes and in the universe as a whole, entropy is not conserved. The natural tendency of entropy is to increase. The fallen glass illustrates how entropy increases in natural, spontaneous changes because when disorder increases, entropy increases. In most spontaneous, naturally occurring processes, entropy increases.

Figure 20.7
Disorder and Spontaneity
When the glass full of milk falls off a table, it breaks and the contents spill. The broken glass is in a state of greater disorder. This breaking and scattering is a spontaneous change. The reverse change—the glass reassembling itself and refilling with the milk—is extremely unlikely to occur. Changes that are spontaneous in one direction are not spontaneous in the other direction under the same conditions.

Demonstration 20.2

The Direction of a Chemical Reaction
Purpose: To observe a spontaneous, endothermic reaction.
Materials: 10 cm × 10 cm piece of corrugated cardboard, 250-mL Erlenmeyer flask, 32 g $Ba(OH)_2 \cdot 8H_2O$, 15 g NH_4SCN

Safety Precautions: 🧤 🥽 ☠️
Disposal: Code A
Procedure: Place the 32 g of barium hydroxide in the Erlenmeyer flask. Add the ammonium

thiocyanate crystals to the flask. **CAUTION:** *The substances used are poisonous.* Wet the cardboard with water and place the Erlenmeyer flask on it. (A piece of 2 × 4 or a square of plywood will be even more impressive. You should have a puddle of water on the surface of the wood.) Stopper the flask and agitate it until the two chemicals are mixed. Have students record their observations.
Results: When the two crystalline solids are

Some kinds of changes are apt to increase entropy, **Figure 20.8.** These changes include melting and evaporation. Increases in numbers of molecules also tend to increase entropy. Entropy increases during reactions that result in increases in the number of molecules. Such changes result in greater disorder at the submicroscopic level.

Figure 20.8
Increase in Entropy
Humpty-Dumpty's problem was too great an increase in entropy. An input of work can sometimes counter entropy increases. You know that living things require a constant supply of energy for life functions. One of the ways they use that energy is to maintain the strict organization of molecules required for life. The energy is used to do work to overcome entropy increases at the molecular level.

The Direction of a Chemical Reaction

At room temperature, most exothermic reactions tend to proceed spontaneously forward. In other words, they favor the formation of products. In an exothermic reaction, the released energy, usually in the form of heat, raises the temperature of the products and many more atoms and molecules in the surroundings. The energy is distributed more randomly than it was before the reaction. The motion of a greater number of atoms and molecules increases. Therefore, disorder increases.

You have read that both heat and entropy play a role in determining spontaneity. In general, the direction of a chemical reaction is determined by the magnitude and direction of the heat energy and entropy changes. For example, a reaction will proceed in the forward direction, toward formation of products, if that direction results in both a release of heat and an increase in entropy. As an example, consider the combustion of butane, C_4H_{10}.

$$2C_4H_{10}(g) + 13O_2(g) \rightarrow 8CO_2(g) + 10H_2O(g) + \text{heat}$$

This reaction occurs spontaneously in the forward direction because both energy and entropy changes are favorable. The energy decreases because the reaction is exothermic—a favorable change. The entropy increases partially because the total number of molecules increases, from 15 to 18—also a favorable change. The release of heat and increase in entropy combine to drive this reaction forward.

Look at another example of a reaction favored by both energy and entropy changes—the one between calcium carbonate and hydrochloric acid.

Concept Development

Explain that the tendencies of reactions to proceed from states of high to low energy and from low to high entropy are reflections of the second law of thermodynamics, which states that in any spontaneous process, the entropy of the universe always increases. The tendency from high to low energy exists simply because the energy lost by the system is gained by the surroundings, thereby increasing the thermal entropy of the surroundings. As we know, it is highly improbable that the dispersed thermal energy will reappear in any useful way. This would require a spot in the universe to suddenly become hot at the expense of the surroundings.

Extension

Ask student groups to make posters that depict the particle arrangements and motions that result during the following changes, all resulting in higher entropy states: (a) a solid melting; (b) a solid dissolving in a liquid; (c) a liquid evaporating; and (d) a gas at a lower temperature being heated to a higher temperature at constant volume. Discuss why the entropy increases in each case. [L1] [LEP]
[COOP LEARN]

---

mixed, a cold slushy liquid solution forms. The flask freezes to the cardboard. The odor of ammonia gas can be detected if the stopper is removed.
$Ba(OH)_2 \cdot 8H_2O(s) + 2NH_4SCN(s) \rightarrow 2NH_3(g) + Ba^{2+}(aq) + 2SCN^-(aq) + 10H_2O(l)$
Analysis: Ask these questions.
1. Is the entropy inside the flask increased or decreased when the two solids are

mixed? *increased*
2. The ΔH for the reaction is +90 kJ. Does that agree with the observation? *Yes, the flask feels cold because the reaction is endothermic.*
3. Based on the above answers, which absolute value is larger, the change in entropy or the change in energy? *The change in entropy is larger.*

Assessment

Knowledge: Ask students whether the reaction they observed is spontaneous? *yes* What evidence can they cite that the reaction was spontaneous? *Once begun, the reaction proceeded without outside intervention.* [L1]

3 ASSESS

Check for Understanding

- Check for understanding of the section material by having students answer the Section Review questions.
- Ask students to work in pairs and to roll two dice, adding the numbers on the two and recording the total after each roll. Ask them to conduct sufficient trials to determine which total is the most probable. *Seven is most probable.* Ask them to explain why 7 is the most probable result. *Seven can be achieved by the greatest number of combinations.* Explain that states of higher entropy are also more probable because the atoms or molecules have more possible arrangements or distributions of energy. L1 LEP

Reteach

Discussion: Burn a wood splint and ask students how the entropies of the system and surroundings change. *Both increase. The reaction is exothermic, increasing the thermal entropy of the surroundings, and the ashes have more entropy than the wood.*

Extension

Ask students to make a chart depicting the changes in the entropies of the system, surroundings, and universe for each of the four reaction situations in **Table 20.1.** L2

4 CLOSE

Portfolio

Ask students to write a short story about "Life Without Petroleum" set in the year 2030. Students should consider placing their stories in their portfolios. L1 P

$$CaCO_3(s) + 2HCl(aq) \rightarrow CaCl_2(aq) + H_2O(l) + CO_2(g) + heat$$

This reaction favors formation of products because it produces gases and liquids. They are more disordered than the solid $CaCO_3$, so entropy increases. Heat is given off, so the products are at a lower energy state than the reactants. That is also favorable. Once again, the decreased energy and increased entropy both drive the reaction in the forward direction.

What happens if only one of the changes—either energy or entropy—is favorable? If the favorable change is great enough to outweigh the unfavorable one, the reaction will still be spontaneous. Thus, even some endothermic reactions are spontaneous if disorder increases greatly. Also, some reactions that increase order are spontaneous if they are exothermic enough. Spontaneity depends on the balance between energy and entropy factors.

Table 20.1 summarizes these factors. Note that temperature plays a role in determining spontaneity when one factor is favorable and the other is not.

Table 20.1   Predicting Whether a Reaction Is Spontaneous in the Forward Direction

Energy Change	Entropy Change	Spontaneous?
decrease (exothermic)	increase	yes
decrease (exothermic)	decrease	yes at low temperature; no at high temperature
increase (endothermic)	increase	no at low temperature; yes at high temperature
increase (endothermic)	decrease	no

SECTION REVIEW

Understanding Concepts

1. Are reactions that occur spontaneously at room temperature generally exothermic or endothermic? Explain.

2. Describe the energy changes that occur when a match lights.

3. ΔH for a reaction is negative; compare the energy of the products and the reactants. Is the reaction endothermic or exothermic?

Thinking Critically

4. **Relating Cause and Effect** Describe each of the following processes as involving an increase or decrease in entropy.

   a) Water in an ice-cube tray freezes.

   b) You pick up scattered trash along a highway and pack it into a bag.

   c) Your campfire burns, leaving gray ashes.

Applying Chemistry

5. **Fermentation** Yeast can spontaneously ferment the sugar in grapes or apples and form ethanol. When the process occurs in grapes, it results in wine. In apples, it produces a beverage called hard cider. The first stage in the fermentation process is exothermic, and the balanced equation for the reaction is as follows.

   $$C_6H_{12}O_6(aq) \rightarrow 2C_2H_5OH(aq) + 2CO_2(g)$$

   Determine whether entropy increases in this reaction, and relate the change in heat energy and entropy to the spontaneity of the process.

SECTION REVIEW

1. These reactions are generally exothermic. The tendency is to change from a state of high energy to a state of low energy. In an energy-rich environment (at high temperatures), an exothermic reaction is not necessarily spontaneous.

2. When the match is struck on a hard surface, friction generates heat. The heat provides the activation energy that causes the chemicals in the match to react, releasing heat and sustaining the reaction.

3. The energy of the products is less than that of the reactants. The reaction is exothermic.

4. a. decrease
   b. decrease
   c. increase

5. Entropy increases in this reaction because four molecules are produced from one and because a gas is produced. If a reaction is exothermic and involves an increase in entropy, it is invariably spontaneous.

Measuring Energy Changes

SECTION
20.2

These days, many people are trying to avoid fattening foods and looking for things to eat that taste good and provide adequate nutrition without increasing body weight. Have you ever been on a diet and had to count Calories? If so, you may have wondered how food manufacturers determine the amount of energy available in a certain amount of food.

Calorimetry

The heat generated in chemical reactions can be measured by a technique called calorimetry and a device called a calorimeter. The device and its use are illustrated in **Figure 20.9,** which deals with an exothermic reaction. A calorimeter can also be used to study the heat absorbed in endothermic reactions. In the case of endothermic reactions, the surrounding water in the calorimeter supplies the heat and decreases in temperature.

In measurements involving a calorimeter, you calculate the heat lost or gained by the surrounding water. This is done by means of the following equation.

$$q_w = (m)(\Delta T)(C_w)$$

In this equation, the symbol q_w stands for heat absorbed by water, m is the mass of the water, ΔT is the temperature change of the water, and C_w is the specific heat of water, which equals 4.184 J/g•°C.

Thermometer

Stirrer
Ignition terminals

Water
Insulation

Sealed reaction chamber containing substance and oxygen

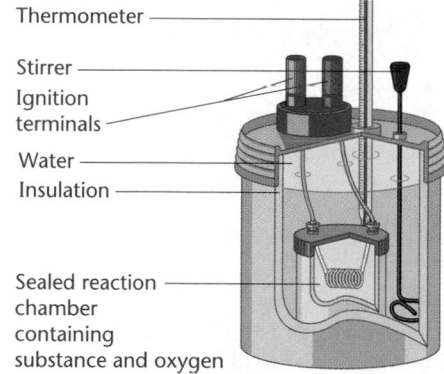

Figure 20.9
Calorimetry
A weighed sample of a substance is burned in pure oxygen inside a container called a reaction chamber. The heat that is released flows into the surrounding water and raises its temperature. Given the temperature change and mass of the water, you can calculate the heat given off in the reaction.

SECTION PREVIEW

Objectives
Sequence the technique of calorimetry and illustrate its use.
Compare the heat generated by some common fuels and by some foods.
Analyze the efficiency of industrial processes and the need to conserve resources.

Key Terms
calorie
kilocalorie
Calorie

PREPARE

Key Concepts
Calorimetry and its use in determining heats of reaction and the caloric content of foods are explained. The use of energy, energy efficiency, recycling, and use of catalysts as a means of saving energy are also examined.

1 FOCUS

Focus Transparency

Display the **Section Focus Transparency** for Section 20.2. Use it to draw attention to the fact that energy is stored in the bonds of certain molecules found in food. L1 LEP

2 TEACH

Concept Development
Explain that energy may be expressed in several interconvertible units. The joule is the SI unit; however, its small magnitude usually makes the kilojoule more practical. The calorie is equal to 4.184 J and was once defined as the energy required to change the temperature of 1 g of liquid water 1°C. What we often refer to as a Calorie of energy from food is actually 1000 calories or a kilocalorie. Most furnaces and air conditioners in the United States are rated in British Thermal Units, BTUs. 1 BTU = 1055 J and is the energy required to raise the temperature of one pound of water 1°F, from 63°F to 64°F.

Program Resources

Study Guide, pp. 83-84 L1
Critical Thinking/Problem Solving,
   pp. 22-23 L2
Laboratory Manual, 20.2, pp. 191-194;
   20.3, pp. 195-200 L1
Section Focus Transparency 42 and **Master,**
   p. 42 L1 LEP
Basic Concepts Transparency 56 and **Master,**
   pp. 111-112 L1 LEP

ChemLab and MiniLab Worksheets,
   pp. 127-129, 131 L1
Applying Scientific Methods in Chemistry,
   p. 49 L1
Chemistry and Industry, pp. 53-54 L1

By the law of conservation of energy, there is no net creation or destruction of energy. Any heat absorbed or released by the water has been released or absorbed by the reaction being studied. The symbol q_{reaction} stands for the heat change of the reaction. Heat loss by the reaction means heat gain for the water, and heat gain for the reaction means heat loss for the water. Therefore, the heat of reaction is equal to the negative of the heat change of the water.

$$q_{\text{reaction}} = -q_w$$

The following sample problem shows how calorimetry can be used to measure energy change in a chemical reaction.

SAMPLE PROBLEM **1** **Calculating Heat of Reaction for Combustion**

The burning of 1.60 g of methane in oxygen, to yield carbon dioxide gas and liquid water, causes the surrounding 1.52 kg of water in a calorimeter to change in temperature from 20.0°C to 34.0°C. What is the heat of reaction for this combustion of 1 mol of methane?

Analyze
- You have all the information you need to find the heat change for the water, q_w. This can be set equal to $-q_{\text{reaction}}$ for the burning of the 1.60 g of methane. What you then need to do is convert that value to find q_{reaction} for 1.0 mol of methane. That will equal ΔH.

Set Up
- Find the temperature change of the water, using the equation $\Delta T = T_{\text{final}} - T_{\text{initial}}$. Then, use the equation $q_w = (m)(\Delta T)(C_w)$. The negative of the value of q_w will equal q_{reaction} for the burning of 1.60 g of methane. Use the molar mass of methane, 16.0 g methane/1.0 mol methane, as a conversion factor to find the q_{reaction} for 1.0 mol of methane.

Solve
- First, calculate ΔT.
$$\Delta T = T_{\text{final}} - T_{\text{initial}}$$
$$= 34.0°C - 20.0°C$$
$$= 14.0°C$$
Then, calculate q_w.
$$q_w = (m)(\Delta T)(C_w)$$

> **Problem-Solving HINT**
>
> Be careful with signs in the calculation of ΔT. If the reaction is endothermic, the sign of ΔT will be negative.

$$= \frac{1.52 \times 10^3 \, \text{g} \mid 14.0°C \mid 4.184 \, \text{J}}{\text{g} \cdot °C} = 8.90 \times 10^4 \, \text{J} = 89.0 \, \text{kJ}$$

Now, find q_{reaction} for the 1.60 g of methane.
$$q_{\text{reaction}} = -q_w = -89.0 \, \text{kJ} = \text{heat released in burning 1.60 g methane}$$
Now, use a molar-mass conversion factor to find q_{reaction}, or ΔH, for 1 mol of methane.

$$q_{\text{reaction}} = \frac{-89.0 \, \text{kJ} \mid 16.0 \, \text{g methane}}{1.60 \, \text{g methane} \mid 1 \, \text{mol methane}} = -890 \, \text{kJ/mol methane}$$

$$q_{\text{reaction}} = \Delta H = -890 \, \text{kJ/mol methane}$$

Check
- Check to be sure that all units were correctly handled, and review the calculation to be sure that it was sound. The result does check out.

720   Chapter 20   Chemical Reactions and Energy

PRACTICE

 Guided Practice: Guide students through the sample problem, then have them work in class on Practice Problems 1-3. See Appendix D for complete solutions. L1

Independent Practice: Your homework or classroom assignment can include the additional practice problems shown here. L1

Answers to Practice Problems

1. 2.30 kJ
2. 12.6 kJ
3. 33.7 kJ for 1 g, 2220 kJ for 1 mol

1. How much heat is absorbed by a reaction that lowers the temperature of 500.0 g of water in a calorimeter by 1.10°C?

2. Aluminum reacts with iron(III) oxide to yield aluminum oxide and iron. Calculate the heat given off in the reaction if the temperature of the 1.00 kg of water in the calorimeter increases by 3.00°C.

3. When 1.00 g of a certain fuel gas is burned in a calorimeter, the temperature of the surrounding 1.000 kg of water increases from 20.00°C to 28.05°C. All products and reactants in the process are gases. Calculate the heat given off in this reaction. How much heat would 1.00 mol of the fuel give off, assuming a molar mass of 65.8 g/mol?

Energy Value of Food

Many years ago, chemists measured heat in calories instead of joules. A **calorie** is the heat required to raise the temperature of 1 g of liquid water by 1°C. A **kilocalorie** is a unit equal to 1000 calories. One calorie is equal to 4.184 J. One joule is equal to 0.239 calorie. Some nutritionists and dietitians still use Calories but are switching to kilojoules. The energy value of foods is measured in units called Calories. Note the capital *C*. One food **Calorie** is the same as 1 kilocalorie. One food Calorie is also equal to 4.184 kJ.

Chemical compounds in the food you eat provide you with energy. The compounds undergo slow combustion, combining with oxygen to produce the waste products carbon dioxide and water, along with compounds needed for body growth and development. Fats provide about 9 Calories per gram. In contrast, 1 g of carbohydrate or protein provides about 4 Calories. **Figure 20.10** illustrates some types of nutrients. **Table 20.2** shows how many Calories you get from typical servings of some foods.

Figure 20.10

Types of Nutrients
Foods such as those shown here contain a variety of chemical compounds. They can be grouped into three general types—protein, carbohydrate, and fat—that have different energy contents. Your body processes transform some of the energy released when these foods are broken down into work and heat. However, when you take in more food than you need, your body stores the extra energy by producing fat for later use.

Table 20.2 The Caloric Value of Some Foods

Food	Quantity	Kilojoules	Calories
butter	1 tbsp = 14 g	418	100
peanut butter	1 tbsp = 16 g	418	100
spaghetti	0.5 cup = 55 g	836	200
apple	1	283	70
chicken (broiled)	3 oz = 84 g	502	120
beef (broiled)	3 oz = 84 g	1000	241

20.2 Measuring Energy Changes **721**

Extension

TECH PREP

Ask interested students to look up the energy values of several food ingredients in addition to those listed in **Table 20.2**. Have them include ingredients such as mayonnaise, bread, meats, cheese, rice, potatoes, and so on. Students should collect enough information to determine the caloric value of a typical meal such as a lunch. Students should make a poster displaying their results for classroom use and discussion. Compare the calculated values with values obtained from a nutrition reference for typical servings. L1

GLENCOE TECHNOLOGY

Videodisc

Science and Technology Videodisc Series
Solar Food Dryer
Disc 2: Chemistry
Side 2, Ch. 11

Show this video to illustrate another use of solar energy.
L1 LEP

Videodisc

Science and Technology Videodisc Series
Solar Tower
Disc 2: Chemistry
Side 2, Ch. 13

Show this video to illustrate how solar power plants can be designed. L1 LEP

Additional Practice Problems

1. Burning a small quantity of hexane raises the temperature of 424 g of water in a calorimeter from 18.4°C to 32.7°C. How much heat was released in the reaction?

$q = (m)(\Delta T)(C_w)$
$\Delta T = 32.7°C - 18.4°C = 14.3°C$

$$q_w = \frac{424 \text{ g} \mid 14.3°C \mid 4.184 \text{ J}}{\text{g} \cdot °C}$$

$q_w = 25\ 400 \text{ J} = 25.4 \text{ kJ}$

2. A calorimeter contains 1.50 kg of water. What temperature change would result if a reaction released 22 500 J of heat?

$q = (m)(\Delta T)(C_w)$
$\Delta T = q/mC_w$

$$= \frac{22\ 500 \text{ J} \mid 1 \text{ kg} \mid \text{g} \cdot °C}{1.50 \text{ kg} \mid 1000 \text{ g} \mid 4.184 \text{ J}}$$

$\Delta T = 3.59°C$

Energy Content of Some Common Foods

Time Allotment: One laboratory period.

Objectives

Review objectives with students before they begin the ChemLab.

Process Skills

Observing, measuring, using numbers, interpreting data, communicating

Safety Precautions

Instruct students to wear safety goggles and aprons and to be careful when using matches and handling cans after the experiment as they may be hot.

PREPARATION

- Obtain and prepare combustion cans in advance.
- The combustion can should be a steel can such as a vegetable or fruit can, not a soft-drink can.

PROCEDURE

- Require that students' notebooks contain the following items before they begin: Objectives; Safety Precautions; Outline or Flow Chart of the Procedure; and Data and Observations Table.
- Instruct students to leave a small gap between the combustion can and the soft-drink can so that sufficient air flow for combustion occurs while heat loss is minimized.
- Be sure to check on students' food choices for procedure step 8 so that hazardous or noxious items are not used.

Energy Content of Some Common Foods

Foods supply the energy and nutrients you require to build and sustain your body and maintain your levels of activity. The energy is released within cells during the process of respiration. In that process, oxygen combines with energy-storage substances, such as glucose, to produce carbon dioxide, water, and heat. The reaction is essentially a slow combustion process. In this ChemLab, you will compare the amounts of energy liberated from the combustion of three food items.

Problem

How much energy is released during the combustion of some common food items?

Objectives

- **Interpret data** to calculate the energy released during the combustion of pecans, marshmallows, and a food item of your choice.
- **Compare** the energies obtained from the food items.
- **Infer,** based upon the chemical compositions of the foods and upon the amounts of energy obtained, which types of foods contain the greatest amount of energy.

Materials

oven mitt
bottle opener with can-piercing end
empty, clean soft-drink can with a tab closure
empty clean can, minus the top and bottom lids, of approximately the same diameter as the soft-drink can
ring stand

small iron ring
beaker tongs
Celsius thermometer
100-mL graduated cylinder
glass stirring rod
balance
large paper clip
matches
pecan half
2 small marshmallows
food sample of your choice

Safety Precautions

Be careful in using the match and in handling the cans after each experiment because they may be hot. Perform the experiment in a well-ventilated room because acrid fumes may be produced during the combustion process. Wear an apron and goggles.

PROCEDURE

1. Use a can opener to open holes near the bottom of the open-ended can. Bend a large paper clip to fashion the food support as shown in the figure.

DATA AND OBSERVATIONS

Food Item	Mass (g)	Initial H$_2$O Temp (°C)	Final H$_2$O Temp (°C)
pecan	1.61	19.8	65.4
marshmallows	1.15	22.6	26.0
other			

2. Use the graduated cylinder to measure 100 mL of tap water, and pour the water into the soft-drink can. Support this can by the ring stand as shown in the figure. Put the thermometer through the opening in the top of the can and allow it to reach the temperature of the water. Record the initial water temperature to the nearest 0.1°C.

3. Weigh a pecan half, record its mass, and impale it on the paper-clip food support. Hang the support on the edge of the open-ended can with the food sample inside, and rest the can on the ring stand.

4. Use a match to light the pecan. Quickly swing the water-filled soft-drink can assembly directly over the combustion can, allowing a small separation between the two cans.

5. Allow the pecan to burn as completely as possible. If the flame goes out during the process and the pecan is still mostly unburned, you must empty the water in the can and start over with fresh water and a new pecan half. After the pecan has burned out, allow the thermometer to reach the highest temperature. Record this temperature.

6. Carefully disassemble the apparatus, empty the water from the soft-drink can, and dispose of the burned food item.

7. Repeat steps 2-6 with two small marshmallows.

8. Repeat steps 2-6 with a small food item of your choice.

ANALYZE AND CONCLUDE

1. **Interpreting Data** Use your data to calculate the number of kilojoules of energy liberated per gram of each food item. Assume that the water in the soft-drink can has a density of 1.00 g/mL and that the specific heat of the water is 4.19 J/g°C.

2. **Comparing and Contrasting** Compare and rank the food items according to the amounts of energy liberated per gram.

3. **Inferring** Based upon the chemical compositions of the foods you tested (protein, carbohydrate, fat, etc.), what type of food provides the greatest amount of energy per gram?

APPLY AND ASSESS

1. Would it be desirable to eat mainly the type of food that provides the greatest amount of energy per unit mass? Explain.

2. What aspects of your procedure may have caused error in your calculated results? How would each of these sources of error affect your result?

DATA AND OBSERVATIONS

Food Item	Mass (g)	Initial H$_2$O Temperature (°C)	Final H$_2$O Temperature (°C)
Pecan			
Marshmallow			
Other food			

Sample Calculation
The following sample calculation assumes a temperature increase of 45.6° from the combustion of 1.6 g of pecans.

$$\frac{100 \text{ mL H}_2\text{O}}{} \left| \frac{1.00 \text{ g H}_2\text{O}}{1.00 \text{ mL H}_2\text{O}} \right| \frac{4.19 \text{ J}}{\text{g} \cdot {}^\circ\text{C}} \left| 45.6 {}^\circ\text{C} \right| \frac{1 \text{kJ}}{1000 \text{ J}} \left| \frac{1}{1.61 \text{ g pecan}} \right. = 11.9 \text{ kJ/g pecan}$$

ANALYZE AND CONCLUDE
1. pecans: 11.9 kJ/g
   marshmallows: 1.2 kJ/g
   See sample calculation below.
2. Pecans liberated more energy per gram than marshmallows. Other foods will vary.
3. fat

APPLY AND ASSESS
1. No. A diet consisting of predominantly fats would not have a balance of nutrients and would be harmful to the cardiovascular system. We take in food for reasons other than to obtain energy.
2. The food items were not completely burned; therefore, more energy should have been liberated. Energy was lost between the two cans and to heat parts of the apparatus; therefore, more energy was liberated than was measured.

 **Assessment**

Portfolio: Ask students to write formal reports for the ChemLab and to include the reports in their portfolios. L1 P

GLENCOE TECHNOLOGY

 CD-Rom

Chemistry: Concepts and Applications
Burning Calories
Use this Exploration to give students practice in using a calorimeter. L1 LEP

 CD-Rom

Chemistry: Concepts and Applications
Have a Blast
Use this Interactive Experiment to give students practice determining the heat of combustion of various alkanes. L1 LEP

Ask students to scan a list of specific heats in a reference source and to select substances that would be most suitable to store energy in a home that is heated by solar energy. *Substances of high specific heat, such as water, 4.184 J/g°C, and paraffin, 2.9 J/g°C, could store large amounts of heat. Ask students what other factors should be considered. cost, chemical properties, and ease of handling* [L1]

Correcting Errors

Instruct students to be careful when looking up and using specific heats to determine whether they are specific heat capacities expressed in J/g·°C or J/g·K or molar heat capacities expressed in J/mol·°C or J/mol·K.

Concept Development

Explain that when a substance is burned in a calorimeter, the energy evolved per gram of each substance is called the heat of combustion. If 1 mol of a substance is completely burned, the energy evolved is known as the molar heat of combustion.

 Assessment

Performance: Given the molar heat of combustion of sucrose ($C_{12}H_{22}O_{11}$), -5641 kJ/mol, ask students to write the thermochemical equation for the complete combustion of a mole of sucrose. $C_{12}H_{22}O_{11} + 12O_2 \rightarrow 12CO_2 + 11H_2O$ $\Delta H = -5641$ kJ [L1]

You can use calorimetry to measure the energy content of food. The food is burned rapidly in oxygen, rather than slowly, as in the body, but the amount of energy released is the same. The following sample problem shows you how to calculate food Calories from calorimetry data.

SAMPLE PROBLEM **Measuring Food Calories**

A 1.00-g sample of nuts reacts with excess oxygen in a calorimeter. The calorimeter contains 1.00 kg of water that has an initial temperature of 15.40°C and a final temperature of 20.20°C. Find the energy content of the nuts. Express your answer in kJ/g and Calories/g.

Analyze
- You first need to find the heat change for the water, q_w, which is equal to $-q_{reaction}$ for the burning of the 1.00 g of nuts.

Set Up
- Find the temperature change of the water. Then, use the equation $q_w = (m)(\Delta T)(C_w)$ after finding the temperature change of the water. The negative of the value of q_w will equal $q_{reaction}$ in kilojoules. A conversion factor can then be used to convert to Calories.

Solve
- $\Delta T = T_{final} - T_{initial}$
  $= 20.20°C - 15.40°C = 4.80°C$
  $q_w = (m)(\Delta T)(C_w)$

  $$= \frac{1.00 \times 10^3 \cancel{g} \mid 4.80\cancel{°C} \mid 4.184 J}{\cancel{g} \cdot \cancel{°C}}$$

  $= 2.01 \times 10^4$ J $= 20.1$ kJ
  Now, apply a conversion factor to convert to Calories.

  $$= \frac{20.1 \text{ kJ} \mid 1 \text{ Calorie}}{4.184 \text{ kJ}} = 4.80 \text{ Calories}$$

Check
- Check to be sure that all units were correctly handled, and repeat the calculation to be sure that it was sound. The result does check out.

PRACTICE PROBLEMS

4. A group of students decides to measure the energy content of certain foods. They heat 50.0 g of water in an aluminum can by burning a sample of the food beneath the can. When they use 1.00 g of popcorn as their test food, the temperature of the water rises by 24°C. Calculate the heat released by the popcorn, and express your answer in both kilojoules and Calories per gram of popcorn.

5. Another student comes along and tells the group in problem 4 that she has read the label on a popcorn bag that states that 30 g of popcorn yields 110 Calories. What is that value in Calories/gram? How can you account for the difference?

6. A 3.00-g sample of a new snack food is burned in a calorimeter. The 2.00 kg of surrounding water change in temperature from 25.0°C to 32.4°C. What is the food value in Calories per gram?

PRACTICE

 Guided Practice: Guide students through the sample problem, then have them work in class on Practice Problems 4-6. See Appendix D for complete solutions. [L1]
Independent Practice: Your homework or classroom assignment can include the additional problems shown below. [L1]

Answers to Practice Problems

4. 5.0 kJ/g, 1.2 Cal/g
5. 3.7 Cal/g; 1.2 Cal/g, the experimental value, is much lower than the value given on the popcorn bag because of loss of heat to the air in the student experiment. Also, the combustion of popcorn in the student experiment may be incomplete.
6. 4.93 Cal/g

Additional Practice Problems

1. A 1.00-g sample of vegetable oil is burned in a calorimeter. The calorimeter contains 2.00 kg of water. The water temperature changes from 16.8°C to 21.1°C. Calculate the heat released in the reaction. Convert to Calories.
   $q = (m)(\Delta T)(C_w)$
   $\Delta T = 21.1°C - 16.8°C = 4.3°C$

Energy Economics

Have you ever wondered why recycling has become so important? Part of the answer has to do with energy. For many materials, the energy required for recycling used objects is less than the energy of simply throwing the used objects away and making new ones from fresh raw materials.

Aluminum Production and Recycling

Aluminum is the metal of choice for soft-drink cans, as shown in **Figure 20.11**. It is lightweight, has a low specific heat (0.902 J/g°C), and conducts heat rapidly. As a result of its low specific heat, the aluminum itself absorbs little heat, so the beverage inside the container can be cooled quickly. In Chapter 17, you learned that aluminum is produced by electrolysis of its principal ore, bauxite. The production of aluminum requires large amounts of energy, particularly electrical energy. Therefore, one reason to recycle aluminum is to conserve that energy and reduce the cost of producing aluminum products.

Figure 20.11

Aluminum: A Useful Metal
The soft-drink cans and other objects shown here are all made from aluminum.

A comparison of the energy required to produce new aluminum cans and the energy required to produce cans from recycled aluminum reveals that 15-20 cans can be made from recycled aluminum for every one that can be made from new aluminum ore. Thus, the cost of using recycled aluminum is much less than that of using new aluminum. In addition to saving money on energy costs, recycling aluminum conserves valuable raw materials and uses less fossil fuel, which is the source of most of the energy required to produce the aluminum. To understand where the energy savings come from in reducing use of fossil fuels, it is important to look at the process used to change chemical energy into electrical energy.

Fact of the MATTER

Recycling aluminum consumes only about seven percent of the energy required to make the same amount of new aluminum from ore. Recycling 25 aluminum cans saves an amount of energy equivalent to that available from burning a gallon of gasoline.

SECTION 20.2

Extension

TECH PREP

Try to arrange a field trip to the nearest electrical power plant. As an alternative, a small group of students could tour on their own and record their experiences with a video camera. If that is not possible, ask a representative of the power company to make a class presentation on the energy processes involved or ask an interested student to interview the representative and make a class presentation. L1 LEP

GLENCOE TECHNOLOGY

 **Videodisc**

Science and Technology Videodisc Series
Wind Power
Disc 2: Chemistry
Side 2, Ch. 14

Show this video to illustrate the use of wind turbines to provide energy. L1 LEP

 **Videodisc**

Science and Technology Videodisc Series
Geothermal Wells
Disc 2: Chemistry
Side 2, Ch. 17

Show this video to illustrate a way to harness geothermal energy. L1 LEP

$$q_w = \frac{2.00 \text{ kg}}{} \left| \frac{1000 \text{ g}}{1 \text{ kg}} \right| \frac{4.3°C}{} \left| \frac{4.184 \text{ J}}{g \cdot °C} \right.$$

$q_w = 36\,000 \text{ J} = 36 \text{ kJ}$
   $= 36 \text{ kJ} \times (1 \text{ Cal}/ 4.184 \text{ kJ}) = 8.6 \text{ Cal}$

2. A mass of 3.22 g of a bread product advertised as "nonfat" is burned in a calorimeter containing 1325 g of water. The temperature of the water increases from 23.6°C to 29.7°C. Calculate the number of calories per gram of this product.

$q = (m)(\Delta T)(C_w)$
$\Delta T = 29.7°C - 23.6°C = 6.1°C$

$$q_w = \frac{1325 \text{ g}}{} \left| \frac{6.1°C}{} \right| \frac{4.184 \text{ J}}{g \cdot °C}$$

$q_w = 33\,817 \text{ J} = 34 \text{ kJ}$
   $= 34 \text{ kJ} \times (1 \text{ Cal}/ 4.184 \text{ kJ}) = 8.1 \text{ Cal}$
Calories per gram $= 8.1 \text{ Cal}/3.22 \text{ g} = 2.5 \text{ Cal/g}$

miniLAB

Heat In, Heat Out

Purpose: To determine the type of energy change that occurs in the reaction between sulfite ion and hypochlorite ion.

Process Skills: Observing, measuring, interpreting data

Safety Precautions: Advise students to wear safety goggles and aprons and to avoid touching the chlorine bleach.

Disposal: Code C

Teaching Strategies

• To save time, the MiniLab may be done easily as a demonstration.

• Remind students that the system in the reaction consists of sulfite ions and hypochlorite ions, while the surroundings are the water, beaker, and remainder of the universe.

Expected Results

Initial temperature: 21.5°C
Final temperature: 42.0°C

Analysis

1. exothermic; endothermic
2. $Na_2SO_3(aq) + 2NaOCl^-(aq) \rightarrow Na_2SO_4(aq) + 2NaCl(aq)$
   The oxidizing agent is NaOCl, and the reducing agent is Na_2SO_3.

✓ Assessment

Skill: Ask students to write the equation for this reaction in both full and net ionic form and then identify the spectator ions and the oxidizing and reducing ions. *The net ionic equation is* $SO_3^{2-}(aq) + 2OCl^-(aq) \rightarrow SO_4^{2-}(aq) + 2Cl^-(aq)$. *$Na^+(aq)$ ions are spectator ions. The oxidizing agent is OCl^- and the reducing agent is SO_3^{2-}.* [L1]

---

miniLAB 2

Heat In, Heat Out

If a chemical reaction is endothermic, the reverse reaction is exothermic. Examine the reaction between sulfite ion (SO_3^{2-}) and hypochlorite ion (OCl^-), which yields sulfate ion (SO_4^{2-}) and chloride ion (Cl^-).

Procedure

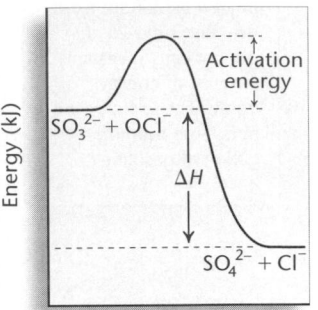

1. Pour approximately 40 mL of a commercial chlorine bleach solution into a 150-mL beaker. Put a thermometer into the solution and record the temperature. Be careful handling the bleach and avoid inhaling the fumes. Work in a well-ventilated room.

2. Pour approximately 40 mL of a 0.5M sodium sulfite solution into the beaker that contains the bleach, and stir gently for a few seconds.

3. Record the final temperature of the mixture.

Analysis

1. Based on the temperature change, was the reaction that occurred exothermic or endothermic? Would the reverse reaction be exothermic or endothermic? Refer to the graph for help.

2. Write the balanced equation for the reaction. Identify the oxidizing agent and the reducing agent in the reaction.

Figure 20.12

Energy Processes in a Coal-Fired Power Plant
In a typical electrical power plant, coal burns, producing heat that boils liquid water into steam. The kinetic energy of the moving steam drives a turbine, which drives a generator that produces the electrical energy.

Converting Chemical Energy to Electricity

The most convenient form of energy available at the present time is electricity. It provides light, heat, hot water, and the ability to run machines of all kinds. Televisions, air conditioners, home appliances, and computers are but a few of the conveniences you get to enjoy because of electricity. In the United States, most electricity must be produced by burning fossil fuels, typically coal. **Figure 20.12** illustrates the production of electricity in a coal-fired power plant.

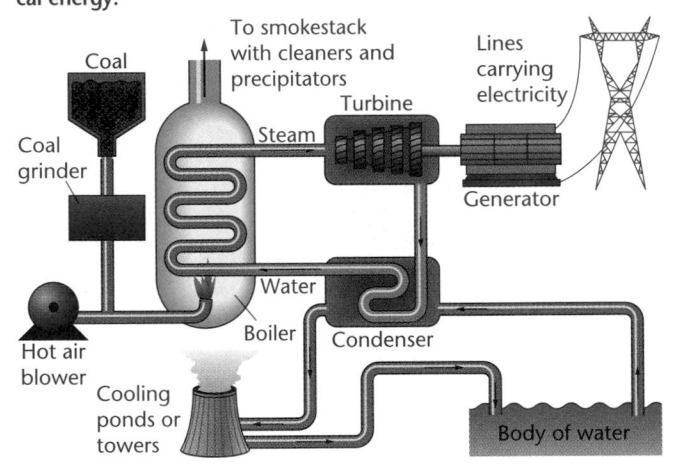

Demonstration 20.3

High-Energy Food

Purpose: To demonstrate that the chemical compounds in food contain energy.

Materials: large potato chip, tongs, book of matches

Safety Precautions:

Disposal: Code A

Procedure: Hold a potato chip with tongs over a pan of water and ignite it using a burning match. **CAUTION:** *Fire risk; hot oil will drip from the burning chip.* Darken the room as the students observe the large amounts of heat and flame produced by the burning carbohydrate and oil. A particularly greasy potato chip will produce a nice show of dripping, burning oil.

Results: The potato chip burns, releasing a large amount of heat, light, and some smoke.

Bacterial Refining of Ores

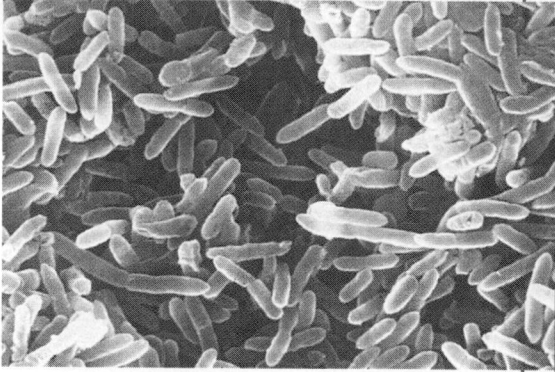

Roughly 2000 years ago, Roman miners began to investigate a blue liquid that they often noticed around piles of discarded, low-grade copper ore. They associated the blue color with copper, probably based on their experience with blue minerals, such as turquoise, which were often found in the vicinity of that metal. When they heated those minerals in a charcoal fire, they obtained copper.

The blue color of the liquid made them suspect that copper was present in it. They heated the liquid to see whether it really did contain copper. Their efforts yielded success. Pure copper was produced from copper salts in the liquid. What the Romans did not realize was that the copper in the liquid had been leached from the low-grade copper ore by bacteria.

Why bacteria are used to extract metals
After crude ores are mined, they are crushed and then the metals are extracted by chemical treatment or by heating to high temperatures. Another way to extract metals from ore is to use bacteria to do part of the job. This method is less damaging to the environment, costs less in terms of energy input, and offers a way to use low-grade ores productively. Today, 24 percent of the copper produced is the result of bioprocessing by bacteria.

The bacterium *Thiobacillus ferrooxidans* obtains its energy by means of reactions involving various minerals. As a result, it produces acid and an oxidizing solution of Fe^{3+} ions, which can react with metals in crude ore.

The process is simple and requires little energy. Low-grade ore is first treated with sulfuric acid to stimulate the growth of bacteria. The microbes process the ore and release copper ions into solution. The metal is then extracted from the solution.

There's gold in them thar' bacteria Gold is another prospect for biorefining. As the sources of high-grade gold ore are disappearing, miners are being forced to mine low-grade ores. The gold in these lesser ores is present in the form of sulfides. To burn off the sulfur, the ores are traditionally treated by roasting or pressure oxidation. Only then can the gold be extracted with cyanide ions.

In one bio-oxidation process, low-grade gold ore is mixed with a brew of bacteria in huge, stirred reactors. A newer process places open piles of gold ore on an impermeable base. Bacterial cultures and the fertilizers needed to nurture them are poured onto the gold ore. In both methods, *T. ferrooxidans* treats the gold sulfides at a lower cost and with an efficiency of recovery increased from 70 to 95 percent.

Connecting to Chemistry

1. **Acquiring Information** Thermophilic bacteria, which thrive at temperatures around 100°C or higher, are being considered as possible candidates for biorefining. Find out what chemical advantage these bacteria have over other bacteria.

2. **Thinking Critically** What are the energy advantages of biorefining?

Purpose
TECH PREP
Students learn that certain bacteria naturally leach out metals from ores, and this process is being used in mining.

Teaching Strategies
- Have students brainstorm about the kinds of problems that must be addressed during bioprocessing. Students may suggest that the size of the piles might have to be controlled so that most of the bacteria have access to sufficient oxygen. They may also suggest that excess heat would have to be avoided because that would kill bacteria. L1

Extension
Have students read the article "Bug in a Gilded Cage" in *Scientific American*, September 1992. Ask them to relate this to the topic of bio-mining. One scientist, John Watterson of the U.S. Geological Survey, used an electron microscope to observe gold particles. Among the particles, he found the cell walls of *Pedomicrobium*. Some gold particles seemed to be solid masses of gilded cells. L1

Connecting to Chemistry
1. Bio-oxidation on an industrial scale releases large amounts of heat. Because most bacteria cannot tolerate such heat, they die. Thermophilic bacteria thrive in such conditions.
2. Smelting ores uses up large amounts of energy. Biorefining by bacteria eliminates much of this energy cost. It also reduces the environmental pollution.

Analysis: Ask these questions.
1. Was the combustion reaction endothermic or exothermic? *exothermic*
2. What fraction of the available energy is released as the chip burns? *nearly 100%*
3. The potato chip package states that 28 g (about 19 chips) release 150 Calories, of which 110 come from fat. What percent of the energy is due to fat? *73%*

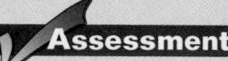
Assessment

Performance: Have students prepare a table similar to **Table 20.2** listing their favorite foods with the Calorie content per serving. Ask students to calculate and list the number of kilojoules per serving. L1

Purpose

TECH PREP

Students gain an understanding of how three forms of alternative energy work, as well as the pros and cons of each.

Background

A proposed geothermal source of energy is the hot, dry rock system, which does not require that magma chambers be found. Instead, this system could use Earth's heat to generate power anyplace in the world. Two deep wells are drilled. Cold water is pumped into one well to be heated by the hot rocks below. In between the wells, the rock would be broken so that the heated water has a way to reach the second well, in which it rises to the surface.

Teaching Strategies

- Discuss each of the alternative sources of energy. Invite students to debate the pros and cons of each source. **L1**
  COOP LEARN
- Ask students to do research on the hot, dry rock system mentioned in the *Background,* which would make it possible to obtain geothermal energy anyplace in the world. **L2**

Alternative Energy Sources

Energy can be neither created nor destroyed. Does that mean that people can keep using the same energy sources at the present rate forever? Fossil fuels are a finite source of energy that should be conserved. Using alternative, renewable sources of energy can prolong the supply of fossil fuels.

Solar Energy

One alternative source of energy is solar energy. The sun will be dispensing energy for the next 5 billion years. More than 300 000 homes in the United States are using solar energy to heat their living quarters. The solar home in the illustration was built to capture sunlight and convert it to heat that warms the rooms. Adobe walls, clay tile floors, triple windows, and heavily insulated walls and ceilings store the sun's warmth in winter. An overhang keeps the high-angle solar rays from entering the house in summer but does not block the lower-angle sun rays in winter.

Photovoltaic cells use solar energy to produce electricity. These solar cells consist of layers of silicon with trace amounts of gallium or phosphorus and are capable of emitting electrons when struck by sunlight. Two main problems must be overcome before solar cells are a reliable source of electricity for everyday use. The first is the efficiency of conversion of solar energy to electrical energy. The second problem is finding efficient ways to store the electricity for use during the dark hours or during periods of cloudy weather.

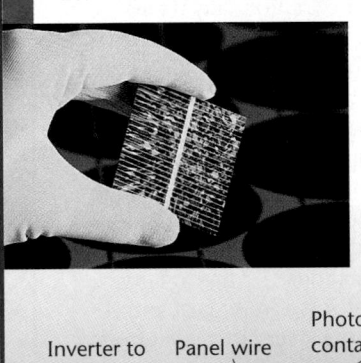

A single photovoltaic cell

Geothermal Energy

Magma, which is molten rock, can heat solid rock surrounding the magma chamber. When this occurs, water in the porous rock above the heated solid rock turns to steam. If there are cracks in the solid rock above, the steam escapes to the surface in the form of a geyser or hot spring. This is a natural source of geothermal energy. Many natural geothermal regions lie in the earthquake and volcano belts along Earth's crustal plate edges.

Geysers in Yellowstone National Park are an indication of the heat that is stored beneath the ground. This energy can be tapped and used to run power plants to produce electricity. The largest geothermal power plant in the world is at The Geysers in California. It generates 1000 megawatts of electric power—enough to serve the needs of 1 million people.

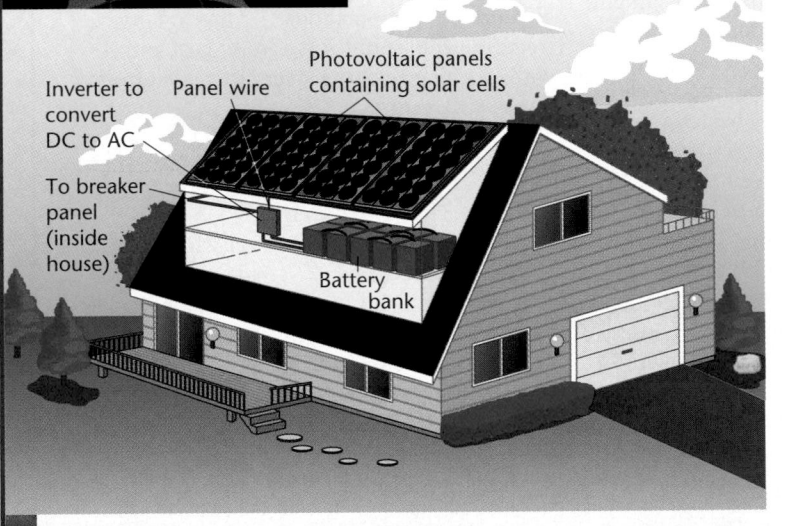

Inverter to convert DC to AC

Panel wire

Photovoltaic panels containing solar cells

To breaker panel (inside house)

Battery bank

Cultural Diversity

Louis H. Latimer (1848-1928)

An African American engineer, inventor, and draftsman named Louis Latimer was one of the pioneers in the search for more durable and long-lasting filaments in electric lightbulbs. He had a vision of the great possibilities of electric light. But, in 1880, a lightbulb had a carbon filament that was short lived. To solve this problem, Latimer invented and patented a carbon filament that was much more durable and longer lasting than the one it replaced. He also wrote and published a book that explained, in layman's terms, the workings of electric lights. He worked for the Edison Company and was one of a group known as the Edison Pioneers—those who worked with Thomas A. Edison before 1885. Apart from his talents as an inventor and engineer, Latimer was a poet, artist, musician, and teacher. He frequently found time to teach English to newly arrived immigrants. At the present time, his home in New York City is preserved as a museum in his honor.

Natural geothermal energy is available in only a few spots around the world. To create new sites requires deep drilling. Although a geothermal power plant generates electricity at one-fourth the cost of power from a new nuclear plant and one-half the cost of power from a new coal plant, it is expensive to drill down to a magma chamber. Another disadvantage of these power plants is that they produce air pollution in the form of hydrogen sulfide, ammonia, and radioactive materials released from deep within Earth.

Power plant at The Geysers

Wind Energy

In some parts of the country, wind power is an almost unlimited source of energy. Wind can turn turbines that produce mechanical energy. A generator converts the mechanical energy into electrical energy. Usually, hundreds of wind turbines are needed to run a power plant.

Wind power emits no air pollutants. It requires no water for cooling. However, wind power can be used only in areas that have reliable winds. When the winds die down, a backup system for producing electricity must also be available. At present, wind energy is sometimes more expensive than other energy sources. Cheaper wind turbines could change this.

DISCUSSING THE TECHNOLOGY

1. **Hypothesizing** Which form of alternative energy might be feasible in your state? Explain.

2. **Acquiring Information** Investigate and describe how a solar furnace works.

3. **Thinking Critically** The three kinds of alternative energy sources discussed in this feature are not effective in all places. Despite that fact, how can they help solve the nation's energy problems?

Chemistry Journal

Energy from Recycling

TECH PREP

Ask students to write down all items that they dispose of in a typical week that are recycled or could be recycled. Ask them to include the list in their journals. At the end of the week, have the class review the types of items listed and the materials from which they are made. Consider how the material is produced and discuss whether recycling a particular item would save energy or resources or both. L1

Visual Learning

• Ask students to discuss the purpose of the battery bank and the inverter that converts DC current to AC. Ask why these devices are needed. *Nearly all electrical devices run on alternating current. Batteries supply energy at night or on cloudy days and then are recharged during sunny times.* L1

• Have students study the photo of the windmill array and ask them to speculate on environmental hazards that might result from these devices. *The devices are noisy and kill birds that fly into them.* L1

Discussing the Technology

1. Student choices will vary according to the climatic conditions in their state. Some kinds of solar heating can be achieved by using proper building materials and design in nearly any location. Wind and geothermal energy depend on specific conditions that are found only in a few places.

2. Students should find that a solar furnace has a field of curved mirrors driven by computers to track the sun during the day. The mirrors focus the heat from the sun on a central heat-collection point located on a high tower. The concentrated heat is utilized to raise the temperature high enough to melt metals and to make steam for producing electricity.

3. If these energy sources are tapped in those parts of the country where they work efficiently, that will reduce the use of fossil fuels in those areas. This will prolong the supply of fossil fuels for the future.

Discussion

Ask students to compare the sources of energy used in the following activities in the year 1900 and the present. (a) Public transportation: 1900—trains burning coal, streetcars using electricity from coal; present—buses and airplanes burning petroleum products, subways using electricity from coal, hydro, and nuclear sources (b) Private transportation: 1900—horses, buggies, and wagons; present—cars burning petroleum products; (c) Home heating: 1900—coal and wood; present—natural gas, propane, oil, electricity from coal, hydro, nuclear sources; (d) Cooking: 1900—wood, coal; present—natural gas, propane, electricity from coal, hydro, nuclear sources; (e) Lighting: 1900—oil, gas, some electricity; present—electricity from coal, hydro, nuclear sources. Point out that the major change has been from burning fuels locally to burning them at more remote electrical-generating plants.

3 ASSESS

Check for Understanding

- Check for understanding of the section material by having students answer the Section Review questions.

Saving Energy with Catalysts

You have learned in Chapter 18 that polyethylene is a plastic used in many household products. It is made from ethylene, a simple hydrocarbon whose formula is C_2H_4. Ethylene is produced industrially by removing hydrogen from ethane (C_2H_6), another hydrocarbon, according to the following equation.

$$C_2H_6(g) + heat \rightarrow C_2H_4(g) + H_2(g)$$

In theory, 1 mol of ethylene could be made with the addition of only 137 kJ of heat. In practice, the reaction above uses nearly four times that amount of energy to produce 1 mol of ethylene because of inefficiency in transferring energy. Over many years, chemists have introduced refinements into the production process. The development of better catalysts has lowered the energy requirements for this and other chemical processes. Research on energy-saving catalysts continues as energy becomes more expensive.

Entropy: One of the Costs of Using Energy

Any time energy is produced or is converted from one form to another, as it is in power plants and industrial processes, some of the energy is lost. This may seem surprising because of the law of conservation of energy. However, in this case, the energy is not destroyed but is simply not usable to do work. What happens to it? It is wasted as heat. In many processes, the amount of energy lost as heat exceeds the amount of energy that can be harnessed to do useful work.

The waste heat generated in systems that use or produce energy cannot be reclaimed, reused, or recycled because it has increased the random motion, and therefore the disorder, of molecules in the environment. As a result of this increased disorder, entropy increases as shown in **Figure 20.13**. The environment cannot spontaneously reorganize itself to give back the energy that increased its entropy. This increased disorganization of the environment is part of the price of using energy.

Figure 20.13

Using Energy Increases Entropy
Every time energy is used to do useful work such as spinning turbines to generate electricity, some energy is lost to the environment, usually as heat. This heat increases the entropy of the environment and cannot be reclaimed. ▼

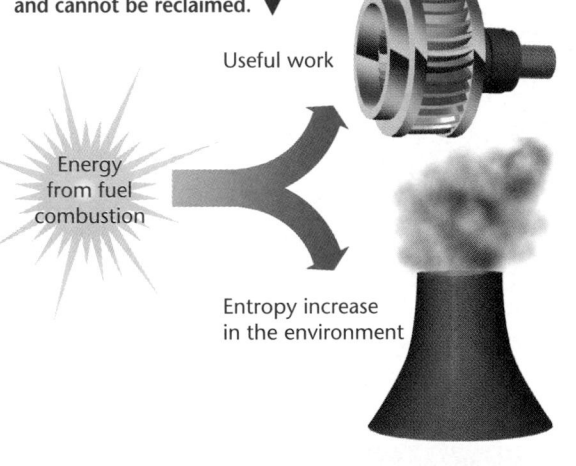

Useful work

Energy from fuel combustion

Entropy increase in the environment

In some cases, the heat is useful in situations where an increase in temperature is desirable, such as heating greenhouses in cold climates. ▼

730   Chapter 20   Chemical Reactions and Energy

Demonstration 20.4

A Living Catalyst

Purpose: To show that yeast is a living organism whose enzymes speed up the rate of a chemical reaction by providing an easier reaction pathway.
Materials: colorless, glass, disposable soda bottle; balloon; 1 tablespoon sugar; 2 tablespoons flour; 1 package yeast; 150 mL water at 50°C; 150-mL beaker; stirring rod

Safety Precautions: 🧤 💥 🥽
Disposal: Code A
Procedure: Mix the sugar, flour, and 100 mL of warm water together in the bottle. Cover the top of the bottle with a deflated balloon and observe for a few minutes. Add the yeast to the remaining 50 mL of warm water in a small beaker, stirring to mix. Remove the balloon and add the yeast-water mixture to the bottle with mixing. Replace

the balloon, sealing the bottle. **CAUTION:** *Do not use a screw-on bottle cap to seal the bottle. It will explode.* Have students observe the balloon every 10 minutes for the remainder of the class period.
Results: The sugar does not react with the flour or water. The balloon remains uninflated. After adding the yeast, the yeast enzymes oxidize the sugar, releasing CO_2 gas, which slowly inflates the balloon.

A process, such as recycling, that saves energy by conserving electricity does more than reduce the total consumption of electricity. It also saves the chemical energy of original fossil fuel that would otherwise be lost forever as waste heat and increased entropy. This sometimes amounts to a savings of 70 percent of the fuel, making its energy available for some other use.

Energy and Efficiency

Because of entropy increases, no power or industrial plant, no matter how well designed, can completely convert heat from chemical reactions into useful work. Waste heat is inevitable. This kind of loss is easiest to understand in terms of efficiency, which is the amount of work that can be obtained from a process compared to the amount that must go into it.

For example, a modern fossil-fuel-burning plant has a maximum theoretical efficiency of only 63 percent. This means that, at most, 63 percent of the chemical energy from the original fuel can be converted to useful work. The remainder is lost as waste heat.

Other inefficiencies further reduce the percentage of energy that is converted to useful work. **Table 20.3** illustrates some of these factors in regard to a modern electrical power plant. The data in the table allow you to compute the plant's overall efficiency, which is simply the product of the efficiencies of all the individual steps multiplied by the maximum theoretical efficiency. If the maximum theoretical efficiency is 63 percent, then the efficiency of generating electricity by the process detailed in **Table 20.3** can be calculated as follows.

$$(0.63)(0.90)(0.75)(0.95)(0.90) = 0.36 = 36\%$$

This low value is typical of the energy efficiency of power and industrial plants.

Table 20.3 Typical Efficiencies in Power Production

Maximum theoretical efficiency	63%
Efficiency of boiler	90%
Mechanical efficiency of turbine	75%
Efficiency of electrical generator	95%
Efficiency of power transmission	90%
Overall actual efficiency	36%

In a coal-fired power plant, a maximum of 36 percent of the energy available from coal is converted to electrical energy. ▶

• Ask students to set three 250-mL beakers on a single hot plate. One beaker should contain 200 mL of water, another 100 mL of water, and the third 100 mL of ethylene glycol. Students should put thermometers in all three beakers and record the initial temperatures of all three liquids, then turn the hot plate on low and note the time required to increase the temperature of the three liquids by 20°C. Ask them to explain the results. *200 mL of water require more time and more heat energy than 100 mL of water because of the greater mass. 100 mL of water require more time and more heat energy than 100 mL of ethylene glycol because water has a higher specific heat.* [L1]
COOP LEARN

Reteach

Portfolio: Ask students to consult with other family members to estimate the number of gallons of gasoline consumed during a typical week by the entire family in their private automobiles and the cost of the gasoline. Ask students to then calculate the savings in both gallons of gasoline and dollars if the efficiency of automobiles were increased from approximately 25% to 40%. Ask students to include their data and calculations in their portfolios. [L1] [P]

Analysis: Ask these questions.
1. What happens when the sugar, flour, and water are mixed and sealed in the bottle? *No reaction is observed.*
2. What is observed after the yeast cells are added to the mixture? *The balloon slowly inflates, indicating that a chemical reaction occurs.*
3. Yeast cells contain enzymes. Why are these enzymes considered to be cat-alytic? *The enzymes increase the rate of a chemical reaction.*

✓ Assessment

Performance: Have interested students get permission to bake bread at home. Bring in a cookbook and go over a recipe with the class. Contrast the cookbook recipe with the procedure followed in the demonstration. Ask students to describe in chemical terms the process of alcoholic fermentation that the yeast carry out. As an experiment, students could decide on a common recipe and then vary the amount of yeast to see the results. [L1]

4 CLOSE

Figure 20.14
Cooling Tower
A cooling tower is a common way to get rid of waste heat. Inside the tower, hot water is sprayed into the air while large fans draw air through the droplets. Although this process only transfers the heat to the atmosphere, it is an improvement over pumping the hot water back into streams, an action that killed organisms in the stream.

Improving the Efficiency of Industrial Processes

The efficiency of industrial processes can be increased and valuable energy resources can be conserved if plants that use improved energy-saving devices are designed and built. Also, if uses are found for waste heat—such as heating buildings and growing food in cold climates—energy can be conserved. Cooling towers, **Figure 20.14,** are one way of transferring waste heat. The development of more energy-efficient refrigerators, air conditioners, and lightbulbs can also improve the efficiency of electricity use.

Finally, the energy conversion processes of organisms can be studied and perhaps eventually applied to industrial methods. Biological processes convert energy from one form to another and at the same time create highly ordered—that is, low-entropy—systems. These natural processes are far more efficient than most industrial processes.

SECTION REVIEW

Understanding Concepts

1. Explain why it is impossible to convert 100 percent of the chemical energy in a fossil fuel to electrical energy.

2. The specific heat of aluminum is 0.902 J/g°C. The specific heat of copper is 0.389 J/g°C. If the same amount of heat is applied to equal masses of the two metals, which metal will increase more in temperature? Explain.

3. How much heat, in kilojoules, is given off by a chemical reaction that raises the temperature of 700 g of water in a calorimeter by 1.40°C?

Thinking Critically

4. **Using a Table** Using **Table 20.3** as a guide, determine the overall efficiency of a power plant that has a maximum theoretical efficiency of 50 percent, given that the other efficiencies are the same as those in the table.

Applying Chemistry

5. **Calorie Counting** Using **Table 20.2** as a guide, compare the Caloric content of 1 g of butter to 1 g of spaghetti. Which food gives you more Calories per gram?

SECTION REVIEW

1. It is impossible to convert 100% of the chemical energy in a fossil fuel to electrical energy because any conversion process is accompanied by a loss of energy as waste heat.

2. Copper's temperature will increase more because its specific heat is lower than that of aluminum. The copper requires fewer joules of heat for each degree of temperature change.

3. 4 kJ (Note that the answer must be rounded to one significant digit because 700 g has only one significant digit.)

4. 29%

5. 1 g of butter, 7.1 Calories; 1 g of spaghetti, 3.6 Calories; butter provides more Calories per gram.

Photosynthesis

All organisms need energy to survive, and the main source of energy for Earth is the sun. Somehow, the sun's energy must be captured as chemical energy so that living things can use it. This process is called photosynthesis.

The Basis of Photosynthesis

Plants and other photosynthetic organisms have at least one obvious characteristic in common: they are green, the color of chlorophyll. The chlorophyll in cells of plants and algae is found in organelles called chloroplasts, as shown in **Figure 20.15.** The chlorophyll in photosynthetic bacteria is found on membranes spread throughout the cells. However, in both cases, the process works essentially the same way. Chlorophyll absorbs light and changes it to chemical energy, triggering a complex series of changes that together make up photosynthesis. Let's examine the chemical process in more detail. Note, as you read, that most of the changes absorb energy.

SECTION PREVIEW

Objectives
Analyze the process and importance of photosynthesis.
Compare the energy efficiency of photosynthesis and processes that produce electricity.
Trace how energy from the sun passes through a food web.

Key Terms
photosynthesis

Figure 20.15
Chloroplasts
The photo (left) shows some plant cells as seen through a microscope. The green structures inside are chloroplasts, which contain the light-absorbing green pigment chlorophyll.
The diagram (right) shows a single chloroplast. Notice the grana, which are stacks of membranes on which the chlorophyll is located. The liquid surrounding the membranes is called stroma and is the place where sugar molecules are synthesized.

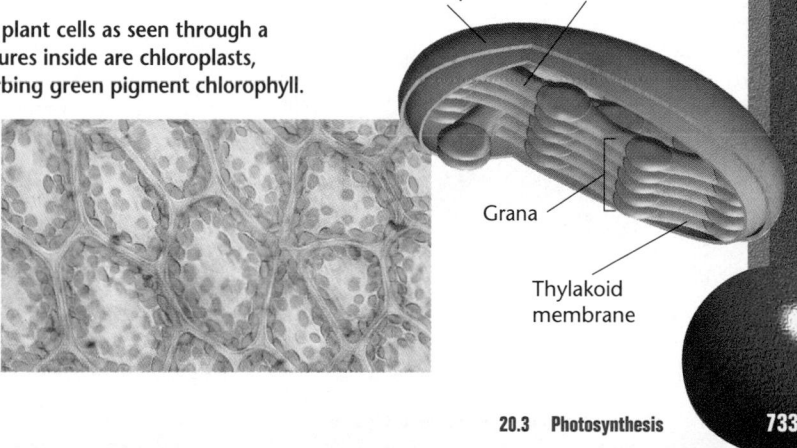

Chloroplast   Stroma

Grana

Thylakoid membrane

20.3   Photosynthesis   **733**

Program Resources

Study Guide, pp. 85-86 L1
Section Focus Transparency 43 and
   Master, p. 43 L1 LEP
Basic Concepts Transparency 57 and
   Master, pp. 113-114 L1 LEP

Meeting Individual Needs

Gifted Have students look up and study the molecular structure of the chlorophyll-*a* molecule. They should find out how the molecule functions to trap light energy. Point out the series of alternating single and double bonds that enable it to capture light. Have them explain why plants need magnesium as a nutrient. L3

PREPARE

Key Concepts
Photosynthesis is examined in terms of its chemistry, energy efficiency, and role in transforming energy.

1 FOCUS

Focus Transparency

Display the **Section Focus Transparency** for Section 20.3. Use it to focus students' attention on the fact that energy conversion by photosynthetic organisms is vital to life on Earth. L1 LEP

2 TEACH

Concept Development
Ask students to trace the radiant energy from the sun through processes leading to the formation of our food items and common fuels. Make chalkboard diagrams of the sequences of processes, and emphasize the role of photosynthesis in each. L1 LEP

Point out that the formation of glucose and oxygen during photosynthesis is an endothermic reaction consuming about 2870 kJ/mol of energy supplied by the sun. Explain that the oxidation of glucose, the reverse reaction, is exothermic as shown in the following Quick Demo.

⚡ Quick Demo

Does sugar burn?
Demonstrate that sugars do indeed burn exothermically (our bodies use them for fuel) by burning a sugar cube. Dab some fine ashes from a wood splint on the sugar cube first. Explain that the ashes act as a catalyst for the combustion of sucrose.

Visual Learning

Figure 20.16: Ask students to write the equation for the formation of oxygen, hydrogen ions, and electrons by the breakdown of water.
$2H_2O \rightarrow O_2 + 4H^+ + 4e^-$

WORD ORIGIN

chlorophyll:
chloros (GR)
green
phyllon (GR) leaf
Chlorophyll is the substance that gives a green color to the leaves of a plant.

The Chemistry of Photosynthesis

The process of **photosynthesis** involves a series of reactions in which green plants and some other organisms manufacture carbohydrates from carbon dioxide and water using energy from sunlight. The net reaction that is carried out in photosynthesis is one that produces simple sugars and oxygen gas from carbon dioxide and water. The balanced equation is generally written as follows.

$$6CO_2 + 6H_2O \rightarrow C_6H_{12}O_6 + 6O_2$$

The formula $C_6H_{12}O_6$ represents glucose, a simple sugar.

Photosynthesis is divided into two basic series of reactions—the light reactions and the Calvin cycle.

The Light Reactions

When light reaches a chlorophyll molecule, the energy of the light is absorbed. The energy excites electrons in the molecule, as described in **Figure 20.16**. A high-energy electron is released from the chlorophyll and transfers some of its energy to a molecule of ADP, adenosine diphosphate. That causes the ADP to bond to a third phosphate to form ATP, adenosine triphosphate. As you may recall from Chapter 19, ATP is the main energy storehouse in cells and also plays a critical role in respiration.

As you can see by following the process shown in **Figure 20.16**, electrons lost by the chlorophyll in the first step are restored to it by a reaction in which water is split into elemental oxygen and hydrogen ions. The oxygen is released into the atmosphere and can later be used by organisms, including you, for respiration.

Figure 20.16
The Light Reactions

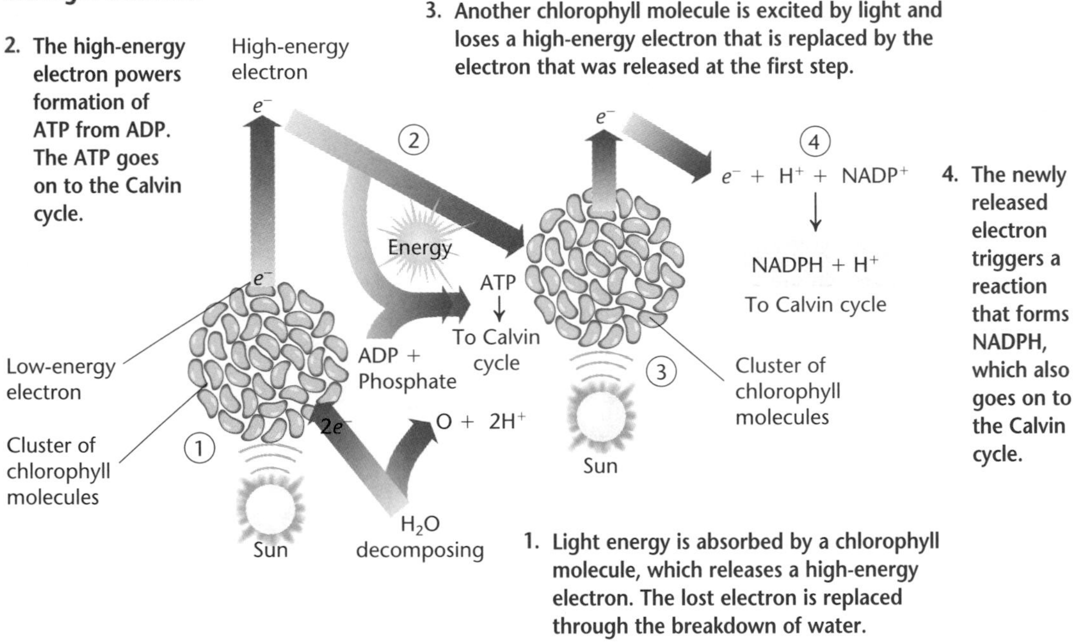

2. The high-energy electron powers formation of ATP from ADP. The ATP goes on to the Calvin cycle.

3. Another chlorophyll molecule is excited by light and loses a high-energy electron that is replaced by the electron that was released at the first step.

4. The newly released electron triggers a reaction that forms NADPH, which also goes on to the Calvin cycle.

1. Light energy is absorbed by a chlorophyll molecule, which releases a high-energy electron. The lost electron is replaced through the breakdown of water.

Demonstration 20.5

Photochemistry

Purpose: To demonstrate that the photochromism of methylene blue can be used to demonstrate that chemical changes are accompanied by energy changes.

Materials: 2 g $FeSO_4$, 96 mL 0.1M H_2SO_4 (add 1 mL concentrated (18M) H_2SO_4 to 95 mL H_2O), 5 small crystals of methylene blue, two 100-mL beakers, file folder, overhead projector

Safety Precautions: 🧤 🥽 🔥
Disposal: Code D
Procedure: The day of the demonstration, dissolve 2 g of $FeSO_4$ in 96 mL of 0.1M H_2SO_4. **CAUTION:** *Acid is corrosive.* Add four or five small crystals of methylene blue and stir to dissolve. Divide the solution between two 100-mL beakers. Place the opaque file folder on the stage of the overhead projector so that only half of the

screen is illuminated. Place one of the beakers on the illuminated area of the stage. If at first the solution does not go colorless, dilute the solution. Remove the beaker from the light. Have students observe that its blue color returns. Place the second beaker so that half of it is in the illuminated area of the stage.
Results: Students will observe that only the half that is exposed to light turns colorless.

Figure 20.17
The Calvin Cycle

1. In the Calvin cycle, atmospheric carbon dioxide reacts with a five-carbon molecule to form an unstable six-carbon molecule. That six-carbon molecule breaks down into two three-carbon molecules.

2. The two three-carbon molecules interact with ATP, NADPH, and H^+ from the light reactions, which provide energy and hydrogen atoms. As a result, the two three-carbon molecules are converted to two PGAL molecules, which are three-carbon sugars.

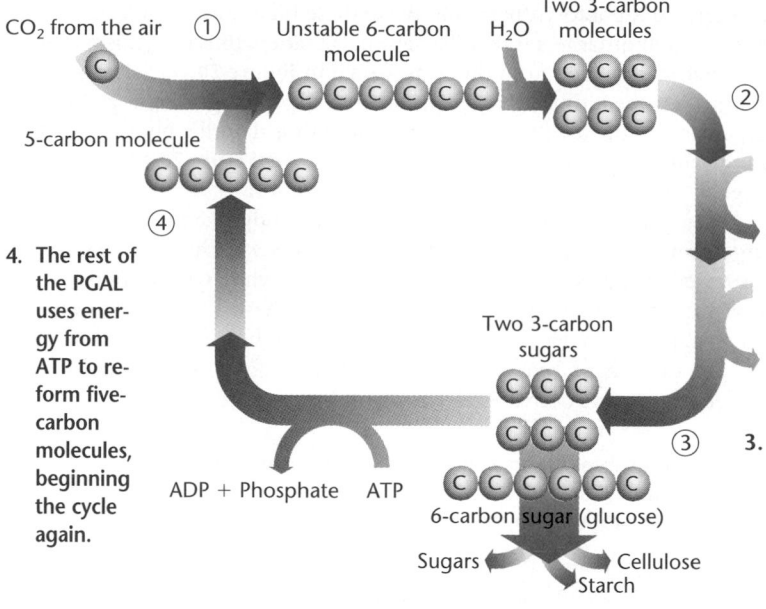

4. The rest of the PGAL uses energy from ATP to re-form five-carbon molecules, beginning the cycle again.

3. Some of the PGAL that is made in this way can combine to form the six-carbon sugar glucose. It can also be used directly as an energy source or to make fats or nucleic acids.

The electron from the original chlorophyll has now lost its energy and moves on to join a different chlorophyll molecule—one that also absorbs light energy but is involved in a different reaction. The transferred electron replaces an electron that is energized by light absorption and released. This newly released electron is involved in another reaction, which forms an important coenzyme called NADPH. The NADPH is formed from NADP and from hydrogen ions that were freed in the earlier breakdown of water. Both the NADPH and the ATP formed as a result of the first absorption of energy continue to the second series of photosynthesis reactions, the Calvin cycle.

The Calvin Cycle

Once the light reactions have produced ATP and NADPH, the second series of photosynthesis reactions can occur. As shown in **Figure 20.17,** the Calvin cycle, which takes place in the stroma of chloroplasts, makes use of NADPH from the light reactions, as well as carbon dioxide taken in from the surroundings. Simple sugars, such as glucose, result.

In the Calvin cycle, ATP turns back into ADP and provides energy for a series of reactions. In the first of these, NADPH loses a hydrogen ion and becomes NADP. The hydrogen ion from the NADPH, together with other H^+ ions produced during the light reactions, provides hydrogen for the synthesis of the sugars. The carbon and oxygen for the sugars are provided mainly by the carbon dioxide.

20.3 Photosynthesis **735**

When the folder is removed, the remainder will also become colorless. The process can be repeated because it contains a reversible redox pair. Fe^{2+} is oxidized to Fe^{3+} while the blue form of methylene blue is reduced to the colorless form.

Analysis: Ask these questions.
1. Is the reaction endothermic or exothermic? *endothermic*
2. What indication is there that the activation energy of this reaction is high? *Room light is not sufficient to start the reaction.*

✓ Assessment

Portfolio: The synthetic dye industry develops dyes that are colorfast and resist fading in sunlight. Have students do research reports on dyes and on William H. Perkin.

In 1856, at the age of 18 years, Perkin made the first synthetic dye accidentally from coal tar. More than 8000 synthetic dyes are used today. Students should consider including the reports in their portfolios. L1 P

3 ASSESS

Check for Understanding

- Check for understanding of the section material by having students answer the Section Review questions.
- Work with students to draw a large concept map linking the light reactions and the Calvin cycle of photosynthesis with the reactions of cellular respiration. Students should discover that respiration and photosynthesis are nearly the reverse of each other. L1 LEP

Reteach

Discussion: Ask students why rain forests have such a significant positive impact on the ecological balance of Earth. *They remove carbon dioxide from the air during photosynthesis, thus moderating the greenhouse effect. They produce oxygen that animals need for respiration and glucose that is used for food by both plants and animals.*

Extension

Ask students to research and explain the role of cytochromes in the electron transport chain. L2

Energy and the Role of Photosynthesis

As you read earlier, most of the reactions that occur during photosynthesis are endothermic. The absorbed energy is stored by using it to synthesize high-energy molecules. Only chlorophyll-containing organisms can carry out photosynthesis and make these high-energy molecules that they need in order to survive. The animals that eat those organisms get the energy they need from them, as shown in **Figure 20.18.** So, photosynthesis is the ultimate process for chemically storing the energy that all organisms in the food web need. You could thus describe photosynthesis as the entry point of energy for life on Earth. See **Figure 20.19** on the next page.

The organization and maintenance of your body, like that of every organism, depend upon energy that is used to power chemical changes. However, energy is not the only factor involved when systems undergo changes. There is also entropy. Natural systems have a tendency toward increased disorder. However, living things are highly ordered, and many of the processes that take place in them, such as building more complex molecules, involve decreases in entropy rather than increases.

When meat-eating animals, such as some kinds of birds, consume the plant eaters, they make use of the energy that was stored in their prey. ▼

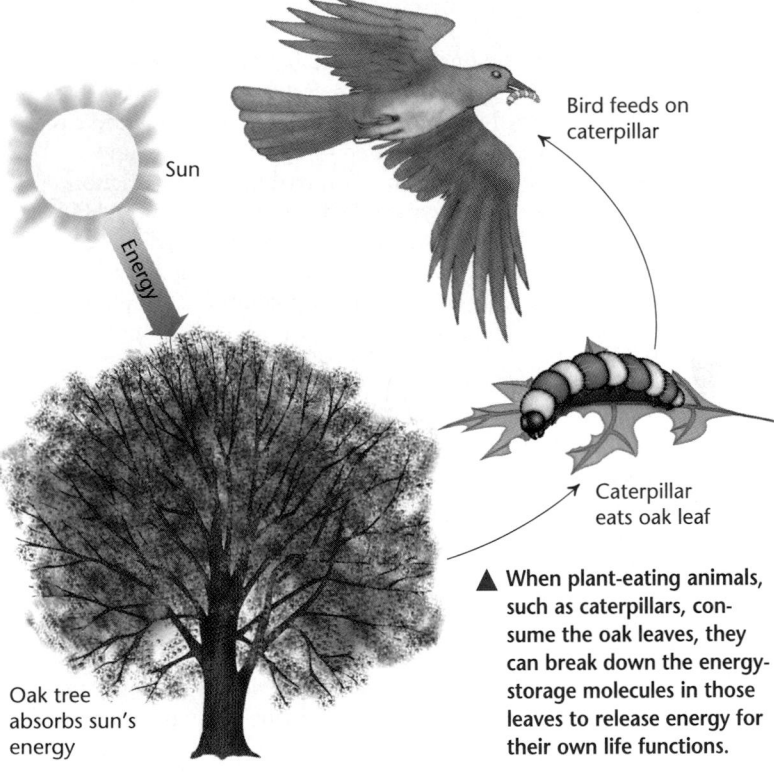

Figure 20.18
Energy and the Food Web
Ultimately, the sun is the source of the energy, and photosynthesis is the process that captures it and makes all life possible.

Sun

Energy

The oak tree absorbs energy from the sun and stores it in the molecules it makes during photosynthesis. The stored energy allows it to carry out essential functions. ▶

Oak tree absorbs sun's energy

Bird feeds on caterpillar

Caterpillar eats oak leaf

▲ When plant-eating animals, such as caterpillars, consume the oak leaves, they can break down the energy-storage molecules in those leaves to release energy for their own life functions.

Chemistry Journal

Changes in Vegetation

Obtain aerial photographs of the local area over a time period of at least the last 40 years. Ask students to study the photographs and to describe the changes in vegetation patterns in their specific neighborhoods, cities, or localities and to estimate any positive or negative effects on the amount of photosynthetic activity. Ask them to include their data and conclusions in their journals. L1 LEP

How is it possible, then, for organisms to overcome the tendency to disorder and maintain their complex structures and life functions? The secret of their success is their ability to absorb energy from outside sources, such as light and food. These outside sources continue pumping in the energy needed to keep the otherwise-nonspontaneous processes going.

When the body converts stored chemical energy to other uses during respiration, heat is produced. The heat, if it built up, could lead to destructive entropy increases and a breakdown of structures and functions. Some of the heat produced is radiated to the environment. The heat given off raises the entropy of the surroundings rather than that of the body. The low entropy of living things is thus maintained partially at the expense of the entropy of the surroundings.

In warm-blooded organisms, such as humans, some of the heat produced during respiration is retained by the body, which uses it to maintain a constant temperature. This temperature allows various biochemical processes to take place at the proper rates.

Figure 20.19
Photosynthesis and Respiration in Balance
This British chemist spent 15 days in a sealed chamber where his oxygen was entirely supplied by 30 000 wheat plants carrying out photosynthesis. All carbon dioxide for photosynthesis came from the man's respiration. This 1995 experiment was the first in a series to determine whether humans could live in a similar but larger chamber on the moon or another planet.

SECTION REVIEW

Understanding Concepts

1. Plants appear green because their leaves reflect green and yellow light. Explain what happens to the energy in the wavelengths of light that are not reflected.

2. How does chlorophyll function in photosynthesis?

3. Explain why a meat-eating animal such as an eagle depends on photosynthesis for its energy.

Thinking Critically

4. **Inferring** What would be the effect on Earth's atmosphere if all plants were temporarily unable to carry out the process of photosynthesis?

Applying Chemistry

5. **Opposite Processes** Compare the process of photosynthesis with the process of respiration in Chapter 19. Explain why they are often considered opposite processes.

SECTION REVIEW

1. The energy in the wavelengths of light that are not reflected is absorbed by electrons and converted to chemical energy.
2. Chlorophyll absorbs light energy and passes along the energy, which is used to produce sugars.
3. Animals that eat plants obtain energy from molecules that were synthesized using energy from sunlight. The eagle eats either these animals or other animals that have eaten these animals.
4. Oxygen in the atmosphere would decrease because it would no longer be produced during photosynthesis and would still be taken up for respiration by animals. Carbon dioxide would increase because the plants would not take it up and animals would continue to produce it.
5. The products of respiration (CO_2 and H_2O) are the reactants for photosynthesis, and the products of photosynthesis (O_2 and $C_6H_{12}O_6$) are the reactants for respiration. Respiration is exothermic and photosynthesis is endothermic.

- Review the Reviewing Main Ideas statements and Key Terms with your students.
- Complete solutions to Chapter Review problems can be found in the *Problems and Solutions Manual*.

UNDERSTANDING CONCEPTS

1. Electrical energy provides the spark that initiates a combustion reaction. Chemical energy is converted to mechanical energy when the combustion gases expand and force a piston to move. Some of the mechanical energy of the motor is used to charge a battery and generate electrical energy. Excess heat is lost to the environment.
2. Plants make sugars from simpler materials through photosynthesis. Most other organisms obtain sugars by eating food.
3. Most naturally occurring processes proceed in the direction of greater disorder.
4. Dispersal of waste heat from the engine is improved. In addition, some of the energy that would have been lost is put to use.
5. Endothermic reactions require a constant input of energy because they absorb heat and therefore require a continuing outside source of energy.
6. Entropy is a measure of disorder.
7. **a.** The reaction is exothermic.
   b. 121 kJ ($2H_2$, a mass of 4.00 g yields 484 kJ, so 1.00 g would yield 121 kJ.)
8. A positive ΔH means that products are at a higher energy than reactants, so the reaction involves a gain of energy and is endothermic. Conversely, a negative ΔH means that a reaction involves a release of energy and is exothermic.
9. Most exothermic reactions require an initial energy input to provide activation energy but, once started, they give off enough energy to sustain the reaction.

738

REVIEWING MAIN IDEAS

20.1 Energy Changes in Chemical Reactions

- Exothermic reactions give off energy; endothermic reactions absorb energy.
- Energy can be converted from one form to another, but cannot be created or destroyed.
- Activation energy is the energy needed to get a reaction started. A catalyst lowers the activation energy for a given reaction.
- Entropy is a measure of disorder. Spontaneous processes tend to proceed from a state of high energy to a state of low energy. They also tend to proceed from a state of less disorder to a state of greater disorder.

20.2 Measuring Energy Changes

- Chemical reactions are a source of energy because the energy of chemical bonds can be converted to heat, light, or electrical energy. Heat changes can be measured by calorimetry.
- Whenever energy is converted from one form to another, some energy is lost as heat. The lost energy generally cannot be used again or converted back to a useful form of energy.

20.3 Photosynthesis

- Photosynthesis is a process by which plants use light energy to manufacture carbohydrates from carbon dioxide and water. Because such substances are the primary source of energy for all organisms, the results of photosynthesis provide energy for life on Earth.
- The reactions that occur during photosynthesis include the light reactions and the Calvin cycle. Most of the reactions that occur during this process are endothermic.

Key Terms

For each of the following terms, write a sentence that shows your understanding of its meaning.

calorie	kilocalorie
Calorie	law of conservation
entropy	of energy
fossil fuel	photosynthesis
heat	

● UNDERSTANDING CONCEPTS

1. Explain how your car converts energy from one form to another.

2. How does the process by which plants obtain sugars differ from that by which most other organisms obtain sugars?

3. In terms of order and disorder, in what direction do most naturally occurring processes proceed?

4. The efficiency of an automobile engine is improved by heating the interior of the car. Why is this the case?

5. Why is energy needed to sustain an endothermic reaction?

6. What is meant by *entropy*?

7. Hydrogen is used as a fuel for the space shuttle because it provides more energy per gram than many other fuels. The combustion of hydrogen is described by the following equation.

$$2H_2(g) + O_2(g) \rightarrow 2H_2O(g) \quad \Delta H = -484 \text{ kJ}$$

   a) Is the reaction exothermic or endothermic?
   b) How much energy does complete combustion of 1.00 g of hydrogen provide?

8. What is the difference between a positive ΔH and a negative ΔH in terms of energy and the kind of reaction involved?

9. Why is outside energy needed to start most exothermic reactions but not to sustain them?

10. Why does it take more energy to decompose water than it does to boil water?

✓ Assessment

Portfolio: Review the portfolio options that are provided throughout the chapter. Encourage students to select one product that demonstrates their best work for the chapter. Have students explain what they have learned and why they chose this example for placement into their portfolios. **P**

Additional portfolio options can be found in the following **Teacher Classroom Resources:**
Consumer Chemistry
Chemistry and Industry
Applying Scientific Methods in Chemistry

11. Substances like nitroglycerine are powerful explosives partially because they are chemically unstable. Use the following equation and concepts of energy, entropy, and spontaneity to explain why nitroglycerine is unstable.

$$4C_3H_5(NO_3)_3(l) \rightarrow$$
$$6N_2(g) + O_2(g) + 12CO_2(g) + 10H_2O(g)$$
$$\Delta H = -5700 \text{ kJ}$$

12. Nitrogen gas reacts with hydrogen gas to form ammonia gas (NH_3). The heat of reaction is -46 kJ/mol NH_3. Write a balanced chemical equation that includes the reaction heat. Is the reaction exothermic or endothermic?

13. Describe the general trends of spontaneous reactions in terms of energy and entropy change.

● APPLYING CONCEPTS

Chemistry and Technology

14. Explain why garbage could be an excellent fuel for the production of electricity.

How It Works

15. Another type of pack that can be used to supply heat to injuries is reusable. It contains a supersaturated solution of a salt and a disc of metal. When the metal disc is bent, the solute begins to crystallize and releases heat. The pack can be reset by heating it in boiling water, which causes the salt to dissolve again. How can you account for the heat given off by this kind of pack?

Everyday Chemistry

16. Write balanced equations for the following reactions, which occur in catalytic converters: a) the breakdown of nitrogen monoxide into oxygen and nitrogen, and b) the reaction of carbon monoxide with oxygen to produce carbon dioxide.

Earth Science Connection

17. When copper ore is roasted during refining, the solid compound $CuFeS_2$ is combined with oxygen gas to yield solid copper(II) sulfide, iron(II) oxide, and gaseous sulfur dioxide. Write a balanced equation for this reaction.

18. What is the heat given off when a substance is burned completely in a calorimeter, given that the temperature of the 2.00 kg of water in the calorimeter rose from 25.0°C to 43.9°C?

19. List two factors that would make an endothermic reaction more likely to occur.

20. Why will the butane gas in a disposable lighter fail to ignite if the flint is worn out and does not generate a spark?

21. Cite three common examples that illustrate the natural tendency for entropy to increase.

22. Describe the change in entropy in each of the following chemical and physical changes, using the information on pages 716 and 717. a) Wood burns and forms carbon dioxide and water vapor. b) Dry ice sublimes at room temperature and forms carbon dioxide gas. c) Liquid oxygen freezes.

23. At a comfortable indoor temperature of 22°C, your body is at a higher temperature than its surroundings, so it is constantly radiating heat to the environment. However, your body needs to maintain an internal temperature of 37°C (98.6° F). How does it replace the heat that it is constantly losing to the environment?

24. A horseshoe can be shaped from an iron bar heated to temperatures near 1500°C. The hot iron is then dropped into a bucket of water and cooled. If an iron bar is heated from 1500°C and then cooled in 1.00 kg of water that was initially at 20.0°C, how much heat energy does the water absorb if its final temperature is 65.0°C?

739

10. It takes more energy to break chemical bonds than it does to overcome interparticle attractive forces, which are much weaker than chemical bonds.

11. Nitroglycerin is unstable because its decomposition is highly exothermic and also results in a great increase in entropy. (4 mol of liquid generate 29 mol of gas.) Therefore, the decomposition is highly spontaneous (highly favored to occur).

12. $N_2(g) + 3H_2(g) \rightarrow 2NH_3(g)$; $\Delta H = -92$ kJ. Note that 2 mol of $NH_3(g)$ are formed in the reaction. The reaction is exothermic because ΔH is negative.

13. For most spontaneous reactions, energy tends to decrease and entropy tends to increase. Variations from this rule of thumb occur at high and low temperatures.

APPLYING CONCEPTS

14. Garbage is abundant and costs nothing. It can ferment and produce methane gas, which is an excellent fuel, or the garbage itself could be burned to generate heat to produce electricity.

15. The heat given off is the heat of solution of the salt. Crystallization is generally exothermic because the solute gives up its heat of solution.

16. **a.** $2NO(g) \rightarrow N_2(g) + O_2(g)$
    b. $2CO(g) + O_2(g) \rightarrow 2CO_2(g)$

17. $2CuFeS_2(s) + 3O_2(g) \rightarrow 2CuS(s) + 2FeO(s) + 2SO_2(g)$

18. 158 kJ

19. high temperature; that the reaction results in a large increase in entropy

20. Activation energy is needed to initiate the reaction, and without the spark, this energy is not available.

21. Answers will vary but may include events such as the

decomposition of biological waste, the dispersal of a gaseous substance throughout a large volume, the dissolving and dispersal of a cube of sugar dropped into water, cards becoming randomized on shuffling, and so on.

22. **a.** Entropy increases.
    b. Entropy increases.
    c. Entropy decreases.

23. The heat from exothermic reactions during cellular respiration maintains the body temperature.

24. The temperature change of the water is 45.0°C. The water absorbs 188 kJ. Note

that the problem is simplified by assuming that water changed to steam remains in the system and the system equilibrates by recondensing the steam.

25. The lighted match can provide the activation energy required to energize collisions between freely moving and mixing gas particles so that chemical bonds break. The match can provide the activation energy to burn the molecules on the surface, but because of the limited surface area, the reaction cannot sustain itself.

26. The reaction is exothermic because it gives off energy in the form of heat and light. It

needs activation energy from an outside source simply to get started.

THINKING CRITICALLY

27. Water is a gas in the first reaction and a liquid in the second reaction. The energy difference in the reactions is due to the value of the heat of vaporization of water, which accounts for the difference in ΔH.

28. According to the kinetic model, particles of the dye are in rapid motion and move randomly throughout the water. The disorder of the system increases when the particles of dye spread out and become more mixed and randomized than they were in the original drop of dye. This increase in entropy favors the spontaneous dispersal of the dye.

29. HF is the most stable compound because its formation reaction is the most exothermic; therefore, its decomposition would require the largest input of heat, which makes it the least likely to occur spontaneously at room temperature. HI is the least stable because its formation requires an input of energy, and therefore its decomposition would release heat, which is favorable in terms of spontaneity.

30. Sample 1: 25 Cal/6.0 g = 4.2 Cal/g
Sample 2: 9.0 Cal/2.1 g = 4.3 Cal/g
Sample 2 releases more heat per gram.

31. Both the increase in entropy and the added heat promote dissolution of the solid that dissolves endothermically. In this case, adding heat raises the temperature, which facilitates an endothermic process. Only the increase in entropy promotes dissolution of the solid that dissolves exothermically. In this case,

740

---

25. A lighted match can instantly ignite the gas in a Bunsen burner, but it usually cannot cause a large piece of wood to catch fire. The wood tends simply to scorch. Explain the difference in terms of collisions of gas particles.

26. Heat from a burning match is necessary to ignite a candle. Why is it incorrect to say that the burning of a candle is, therefore, an endothermic reaction?

● THINKING CRITICALLY

Interpreting Data

27. Account for the fact that there is an energy difference between the following reactions.

$$CH_4(g) + 2O_2(g) \rightarrow CO_2(g) + 2H_2O(g)$$
$$\Delta H = -802 \text{ kJ}$$

$$CH_4(g) + 2O_2(g) \rightarrow CO_2(g) + 2H_2O(l)$$
$$\Delta H = -890 \text{ kJ}$$

Applying Concepts

28. The kinetic model of matter and the concept of entropy both explain the process of diffusion, which is the spontaneous spreading of liquid or gas particles throughout a volume. Explain, in terms of both kinetics and disorder, why a drop of a dye placed into a beaker of water will eventually color the entire volume of water.

Making Predictions

29. The heats of reaction for the formation of the hydrogen halides from their elements are listed in the table below. On the basis of this information, predict which compound is most stable, in terms of not breaking down into its elements. Which is least stable? Explain.

Compound	ΔH (kJ/mol)
HF	−268
HCl	−92
HBr	−36
HI	+25

Interpreting Data

30. ChemLab Two different foods are burned in a calorimeter. Sample 1 has a mass of 6.0 g and releases 25 Calories of heat. Sample 2 has a mass of 2.1 g and releases 9.0 Calories of heat. Which food releases more heat per gram?

Relating Cause and Effect

31. MiniLab 1 Which factor—entropy, added heat, or both—promotes the dissolution in water of a solid that spontaneously dissolves endothermically? Which factor promotes dissolution in water of a solid that spontaneously dissolves exothermically?

Inferring

32. MiniLab 2 A certain reaction proceeds in the forward direction with a ΔH of −16 kJ. Will the reverse reaction be endothermic or exothermic? What will be the value of ΔH for the reverse reaction?

● CUMULATIVE REVIEW

33. Identify each of the following as either chemical or physical properties of the substance mentioned. (Chapter 1)

a) The element mercury has a high density.
b) At room temperature, solid carbon dioxide changes to gaseous carbon dioxide without passing through the liquid state.
c) Iron rusts when exposed to moist air.
d) Sucrose is a white crystalline solid.

34. Distinguish between atomic number and mass number. How do each of these two numbers compare for isotopes of an element? (Chapter 2)

35. Classify each of the following elements as a metal, nonmetal, or metalloid. (Chapter 3)

a) molybdenum   c) arsenic
b) bromine   d) neon

36. Why does the second period of the periodic table contain eight elements? (Chapter 7)

37. Explain why the metals zinc, aluminum, and magnesium are resistant to corrosion, whereas iron is not. (Chapter 8)

---

heat given off must be dispersed, so added heat should hinder the process.

32. The reverse reaction will be endothermic, with a ΔH of +16 kJ.

CUMULATIVE REVIEW

33. a. physical
b. physical
c. chemical
d. physical

34. Atomic number is the number of protons, whereas mass number is the total number of protons and neutrons of a given isotope.

Isotopes of an element have the same atomic number but different mass numbers.

35. a. metal
b. nonmetal
c. metalloid
d. nonmetal

36. Eight electrons are needed to fill the $2s$ and $2p$ orbitals in the second energy level. The elements in the second period have from one to eight electrons in the second energy level.

37. Zinc, aluminum, and magnesium react with oxygen to form an unreactive oxide

Skill Review

1. Using a Graph The graph below projects the world production of coal (solid line) until reserves are depleted and also projects the demand for coal (dashed line). Use the graph to answer the following questions.

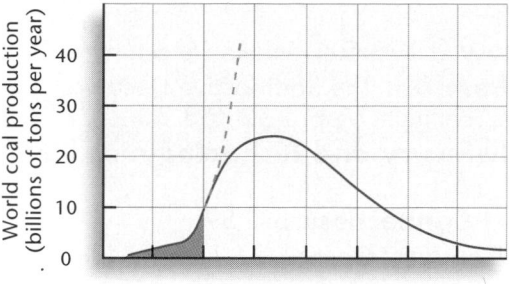

a) How much longer can the present growth rate in coal consumption continue?

b) In what year will maximum coal production be reached, and how much coal will be produced that year?

c) How does the amount of coal produced in 1900 compare to the production in the peak year?

2. Using a Data Table The chemical formulas, heats of combustion, and formula masses of three hydrocarbons are given in the following table.

Hydrocarbon	Heat of Combustion (kJ/mol)	Formula Mass (g/mol)
Methane, CH_4	−890	16
Butane, C_4H_{10}	−2859	58
Hexane, C_6H_{14}	−4163	86

a) According to the data, which hydrocarbon yields the greatest amount of energy per gram?

b) Using the data table, draw a graph of heat of combustion in kilojoules per mole versus formula weight. Use your graph to estimate the heat of combustion of propane (C_3H_8).

c) Conduct library research to find uses for these four hydrocarbons. Search out advertisements or magazine photos illustrating the uses. Prepare a display or bulletin-board presentation to illustrate your findings.

Problem Solving

3. The compound B_5H_9 was once proposed as a rocket fuel because it has a high heat of combustion. In a combustion experiment, a scientist observes that 1.00 g of B_5H_9 burned in excess oxygen in a calorimeter raises the temperature of 800 g of water from 24.0°C to 44.3°C. How much heat is generated in the reaction?

Writing in Chemistry

4. Research the meaning of the term *thermal pollution,* and write a news article that traces the possible sources of such pollution in the operation of a coal-fired power plant. Also, explain the impact of thermal pollution. Refer to **Figure 20.12** for assistance.

5. Scientists have discovered two laws that govern all use of energy—namely, that the energy of the universe is constant and that the entropy of the universe is increasing. In view of these laws, research the ways that we can better use our natural energy resources so as not to squander energy sources and to limit unnecessary entropy increases. Write a paper summarizing the results of your research.

Projects

6. Draw a large diagram or make a model of a fossil-fuel power plant. Write a description of the processes that take place in such a plant.

---

5. Responses will vary. Possible ideas may include: design of more efficient processes and industrial plants; use of renewable resources such as biomass, geothermal, solar, and wind energy; use of more efficient appliances and lightbulbs; capture and use of "waste" heat for home heating, etc.

6. Diagrams, models, and descriptions will vary but should correctly represent the structures and processes involved in burning the fuel and harnessing the resulting energy.

---

coating, which protects the metal from further reaction. Iron oxide flakes off, exposing fresh metal to further corrosion.

AUTHENTIC ASSESSMENT

1. a. The present growth rate in coal consumption can continue until approximately 2025, where the two curves separate, and production will probably no longer be able to keep up with demand.

b. Maximum coal production will be reached by about 2140; approximately 24 billion tons of coal will be produced.

c. The amount of coal produced in 1900 is roughly 1/12 the amount that will be produced in the peak year.

2. a. CH_4 yields the greatest amount of energy per gram (56 kJ).

b. The graph should show a nearly straight line of negative slope. Based on the graph, the heat of combustion of propane (C_3H_8) should be approximately 2000 kJ.

c. Students should find information and illustrations on uses such as methane in gas ranges, propane in home and industrial heating, butane in lighters, and hexane in synthesizing other chemicals.

3. 67.9 kJ were generated in the reaction.

4. Undesirable heat loss to the environment is thermal pollution. Thermal pollution in streams and other bodies of water usually is the major problem because increasing the temperature of water reduces the solubility of oxygen in it. Lower oxygen levels kill fish and other aquatic life. Plankton, which provide the food base for many other species, cannot generally survive when water temperature exceeds 24°C. They are then replaced by cyanobacteria, and the balance in the food web is severely damaged.

Chapter 21 Organizer

Section	Objectives	Activities/Features
21.1 **Types of Radioactivity** (4 days)	**1. Analyze** common sources of background radiation. **2. Compare and contrast** alpha, beta, and gamma radiation. **3. Apply** the concept of half-life of a radioactive element.	**How It Works:** Smoke Detectors, p. 748 **ChemLab:** The Radioactive Decay of "Pennium", pp. 752-753 **Chemistry and Technology:** pp. 754-755 **Art Connection:** p. 759 **Discovery Demo:** 21.1, pp. 742-743 **Demonstrations:** 21.2, pp. 746-747; 21.3, pp. 756-757
21.2 **Nuclear Reactions and Energy** (4 days)	**4. Compare and contrast** nuclear fission and nuclear fusion. **5. Demonstrate** equations that represent the changes that occur during radioactive decay. **6. Trace** the operation and structure of a fission reactor.	**MiniLab 1:** A Nuclear Fission Chain Reaction, p. 763 **Demonstration:** 21.4, pp. 764-765
21.3 **Nuclear Tools** (4 days)	**7. Distinguish** the biological effects of radiation and the units used to measure levels of exposure. **8. Illustrate** medical and nonmedical uses for radioactivity.	**Biology Connection:** A Biological Mystery Solved with Tracers, p. 772 **MiniLab 2:** Radon, p. 775 **Everyday Chemistry:** Radon—An Invisible Killer, p. 777 **Demonstration:** 21.5, pp. 774-775

Activity Materials

ChemLab (pages 752-753)
shoe box, 120 pennies, stopwatch or watch with second-hand display or second hand

MiniLab 1 (page 763)
domino tile sets

MiniLab 2 (page 775)
radon detection kits

Demonstrations For a complete list of materials for the demonstrations in this chapter, see pages 742, 746, 756, 764, and 774.

KEY TO TEACHING STRATEGIES

The following designations will help you decide which activities are appropriate for your students.

L1 Level 1 activities should be within the ability range of all students.

L2 Level 2 activities should be within the ability range of the average to above-average student.

L3 Level 3 activities are designed for the ability range of above-average students.

LEP LEP activities should be within the ability range of Limited English Proficiency students.

COOP LEARN Cooperative Learning activities are designed for small group work.

P These strategies represent student products that can be placed into a best-work portfolio.

Nuclear Chemistry

Teacher Classroom Resources

STUDENT MASTERS

Study Guide: pp. 87-88
Critical Thinking/Problem Solving: p. 24
Laboratory Manual: 21.1 Radioactive Dating—A Model, pp. 201-204
ChemLab and MiniLab Worksheets: pp. 133-136

Study Guide: pp. 89-90
ChemLab and MiniLab Worksheets: p. 137

Study Guide: pp. 91-92
Laboratory Manual: 21.2 How Does Microwave Radiation Affect Seeds? , pp. 205-208

ASSESSMENT RESOURCES
Chapter Review and Assessment: pp. 125-130
Alternate Assessment in the Science Classroom
Computer Test Bank: Chapter 21
Supplemental Practice Problems: Chapter 21
Performance Assessment in the Science Classroom

TEACHING AIDS

Section Focus Transparency 44
Section Focus Transparency Master: p. 44
Basic Concepts Transparency 58
Basic Concepts Transparency Master: pp. 115-116
Problem Solving Transparency 24
Problem Solving Transparency Master: pp. 47-48
Lesson Plans: p. 44

Section Focus Transparency 45
Section Focus Transparency Master: p. 45
Basic Concepts Transparency 59
Basic Concepts Transparency Master: pp. 117-118
Lesson Plans: p. 45

Section Focus Transparency 46
Section Focus Transparency Master: p. 46
Basic Concepts Transparency 60
Basic Concepts Transparency Master: pp. 119-120
Lesson Plans: p. 46

CHAPTER RESOURCES
Applying Scientific Methods in Chemistry: p. 50
Spanish Resources: Chapter 21
English/Spanish Audiocassettes: Chapter 21
Cooperative Learning in the Science Classroom
Lab and Safety Skills in the Science Classroom

GLENCOE TECHNOLOGY

Videotape
MindJogger Videoquizzes, Chapter 21
Videodisc
Chemistry: Concepts and Applications, Demonstrations, Videos, and Animations Videodisc and Bar Code Teacher Guide
Science and Technology Videodisc Series: Disc 2 Chemistry *Dating by Thermoluminescence, Dating of the Shroud of Turin, Neutron Activation Analysis of Paintings, Radon Danger, Carbon - 14 Dating*
CD-ROM Multimedia System
Chemistry: Concepts and Applications, Same as for Videodisc, plus Interactive Exploration and Major Simulation, *The Periodic Table*

TECH PREP

The following Glencoe resources provide opportunities for integrating science and technology.

Student Edition: How It Works, p. 748; Chemistry and Technology, pp. 754-755; Art Connection, p. 759; Biology Connection, p. 772; MiniLab 2, p. 775; Everyday Chemistry, p. 777

Teacher Wraparound Edition: Content Background, p. 762; Extension, p. 769; Chemistry Journal, p. 778

Teacher Classroom Resources: Tech Prep Applications, pp. 35-36; Consumer Chemistry, pp. 41-42

Chapter Overview

Students learn that unstable atomic nuclei can decay, and they use symbols to balance nuclear equations. After the concept of half-life is given, students learn about dating techniques, and they practice solving half-life problems. Fission and fusion reactions are then presented. Finally, practical uses of radioactivity and disposal concerns are discussed.

Theme Connection

Conservation/Energy This chapter includes an explanation of why the law of *conservation* of mass must be expanded to include *energy* when nuclear fission and fusion are involved. In such processes, small amounts of mass are converted into large amounts of energy.

✔ Assessment Planner

Choose assessment strategies from the following pages to evaluate the progress of your students.
Assess, pp. 756, 760, 762, 766, 771, 779
MiniLabs, pp. 763, 775
ChemLab, p. 752
Portfolio, pp. 757, 771, 780
Chapter Review, p. 780

GLENCOE TECHNOLOGY

Videodisc

Chemistry: Concepts and Applications
A Chain Reaction
Disc 3, Side 1, Ch. 4

Show this video of the Discovery Demo to introduce the concept of a chain reaction.

742

Discovery Demo 21.1

A Chain Reaction

Purpose: To demonstrate a chemical chain reaction in which one event causes other events to occur in rapid succession.

Materials: 3-4 books of paper matches, masking tape, ring stand and clamp, 50-cm-long metal rod, 1-m length of rubber tubing

Safety Precautions: 🔥 🔥 🥽

Disposal: Discard burned matchbooks in trash.

Procedure: Perform this demonstration in a well-ventilated room. Tape books of paper matches to a metal rod with the match heads exposed and extending outward. Use a ring stand to clamp the metal rod upright. Stand the rod in a sink containing some water. Darken the room, and then light the bottom match. The others will ignite in turn, demonstrating the concept of a chemical chain reaction. Use rubber tubing connected to a water tap to extinguish the flames. **CAUTION:** *This demonstration presents a fire risk.*

21 Nuclear Chemistry

Have you ever imagined what riches would be yours if you found a way to turn garbage into gold? Although you may have had that kind of idea on your own, so have a great many other people. For example, a goal of early philosopher-scientists called alchemists was to change common metals into precious ones. For centuries, they toiled away in their crude laboratories, hoping to find the elixir, a miraculous agent that would transmute one element into another.

Was the goal of transmuting elements unreasonable? After all, changing one element into another is a little more complicated than ordinary physical and chemical changes that you see every day. That's because a different part of the atom is involved: the nucleus.

Reactions that involve the nucleus can occur. In fact, elements can be transformed into one another through these nuclear reactions. In the 20th century, people have found uses for nuclear chemistry that the alchemists never imagined. Now, you are about to explore how the tiny core of the atom can serve these useful functions.

Chemistry Around You

In the second century B.C., a Chinese alchemist named Shaajun believed that learning to make gold would unlock the secret of immortality. Because gold never tarnished and reacted with almost no other substance, he thought it held the secret of eternal life. The promise of achieving extended youthfulness continues to draw many people today. Over-the-counter tonics and cosmetics that claim to reduce the effects of aging are bought in large quantities. Although that kind of search for a fountain of youth generally leads to disappointment, the dreams of the alchemists were not complete fantasies. Through nuclear chemistry, elements can be transformed into one another, as if by magic.

 Concept Check

Review the following concepts before studying this chapter.
Chapter 2: structure of the nucleus
Chapter 6: how to balance a chemical equation
Chapter 8: the actinides
Chapter 19: DNA structure

 Chapter Preview

21.1 Types of Radioactivity

21.2 Nuclear Reactions and Energy

21.3 Nuclear Tools

Laboratory Activities
ChemLab
The Radioactive Decay of "Pennium"

MiniLabs
1. A Nuclear Fission Chain Reaction
2. Radon—A problem in your home?

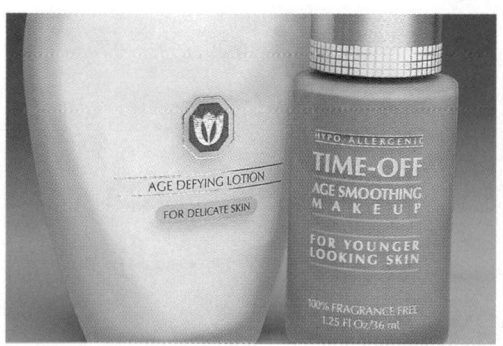

743

Results: The matchbooks rapidly ignite in succession. Point out that the Demonstration, which involves a chemical reaction, serves as an analogy to a nuclear fission chain reaction. Each match represents an atom that undergoes fission and emits neutrons, causing the next atom to undergo fission. Energy is released as an unstable, fissionable element changes into a more stable element.

Analysis: Ask these questions.
1. How does a series of matchbooks simulate a chain reaction? *As one matchbook reacts, it causes the next to react.*
2. What could be used in this demonstration to simulate a control rod? *Anything that would absorb the heat and prevent the next matchbook from igniting could be used.*

SECTION
21.1

PREPARE

Key Concepts

The discovery and early study of radioactivity by Becquerel and the Curies are discussed. Nuclear notation is introduced and used in the explanation of alpha, beta, and gamma decay, then characteristics and detection of the three types of radiation are described. Last, half-life and its use in radioactive dating are explained.

Planning Ahead

For the Demonstrations Check your source of dry ice for Demonstration 21.2, and be sure that your alpha source is adequate for Demonstrations 21.2 and 21.3. Prepare the photograph and photographic film for the *Quick Demo* on page 745.

1 FOCUS

Focus Transparency

Display the **Section Focus Transparency** for Section 21.1 to introduce students to the concept of radioactivity. L1 LEP

2 TEACH

Correcting Misconceptions

Students may believe that radiation and radioactivity are the same thing and that objects become radioactive when subjected to radiation. Point out that nuclear radiation is made up of matter or energy that has been released by a substance. Radioactivity is the ability of a substance to give off radiation. Thus, a radioactive substance is one that gives off radiation. Explain that objects do not become radioactive when subjected to radiation unless they actually absorb radioactive elements.

744

Types of Radioactivity

S ome elements have nuclei that are naturally unstable. Such nuclei disintegrate and form more stable nuclei. Marie Curie was one of the scientists who first studied this phenomenon and helped move the field of chemistry into the nuclear age. She and her husband, Pierre, carried out research together in a laboratory in France. They studied the changes that occur in atomic nuclei and discovered several elements that have unstable nuclei. In the days of the Curies, no one understood the dangers of dealing with unstable elements. Today, scientists know that they must be used with extreme care.

SECTION PREVIEW

Objectives
Analyze common sources of background radiation. Compare and contrast alpha, beta, and gamma radiation. Apply the concept of half-life of a radioactive element.

Key Terms
radioactivity
alpha particle
beta particle
gamma ray
half-life

Discovery of Radioactivity

You are probably familiar with novelty items that glow in the dark. Maybe you even have one of those ceiling constellation maps with stars that give off an eerie, green glow long after the room lights are turned out. Have you ever wondered what causes the stars in such maps to glow? The material in them is a phosphorescent form of the compound zinc sulfide, which means that after it is exposed to light, it continues to give off the light it has absorbed, as shown in **Figure 21.1.**

Figure 21.1
Phosphorescence
When light shines on phosphorescent materials, such as these glow-in-the-dark decorations, the absorbed energy raises the electrons in the atoms to higher energy levels. ▼

▲ When the electrons move back to lower energy levels, the extra energy is given off as light, as shown by the eerie, green glow of these stars. The glow disappears gradually, as the electrons return to their more stable configurations.

744   **Chapter 21   Nuclear Chemistry**

Program Resources

Study Guide, pp. 87-88 L1
English/Spanish Audiocassettes, Chapter 21 L1 LEP
Critical Thinking/Problem Solving, p. 24 L2
Laboratory Manual, 21.1, pp. 201-204 L1
Section Focus Transparency 44 and **Master,** p. 44 L1 LEP
Basic Concepts Transparency 58 and **Master,** pp. 115-116 L1 LEP

Problem Solving Transparency 24 and **Master,** pp. 47-48 L1 LEP
Spanish Resources L1 LEP
ChemLab and MiniLab Worksheets, pp. 133-136 L1
Tech Prep Applications, pp. 35-36 L1
Applying Scientific Methods in Chemistry, p. 50 L1

A French scientist named Henri Becquerel was studying a phosphorescent uranium compound in 1896 when he made an accidental discovery that changed the course of history. Becquerel was testing the compound to see whether the phosphorescence was the cause of a recently discovered type of electromagnetic radiation called X rays. After exposing the sample to sunlight, he placed paper-wrapped photographic film near it. X rays would be able to pass through the paper and expose the film, producing an image of the sample on it, just as they make an image of your teeth on film when you have an X ray taken. Becquerel found the image he expected. He believed that X rays were the cause and that the uranium gave off these rays because of absorbed sunlight. However, later he developed a piece of paper-wrapped film that had been in a dark drawer along with a crystal of the uranium. He did not expect to see an image on the film because the uranium sample had not been exposed to sunlight or any other source of energy. To his surprise, the film bore the image of the crystal, as shown in **Figure 21.2.**

What made the uranium give off radiation? It hadn't been exposed to light, so the radiation wasn't caused by phosphorescence. Was a chemical reaction taking place? No known chemical reactions had ever produced this effect. Becquerel suspected that the uranium was spontaneously emitting some type of radiation. Was the radiation the same as X rays? No one could tell for sure. If so, it was odd that the rays could be produced without an input of energy, which was generally needed to produce X rays.

Figure 21.2

Becquerel's Uranium Experiment
Becquerel discovered that uranium compounds spontaneously give off radiation that can pass through paper wrapping and produce an image on film. He also found that a coin prevents the radiation from reaching the film, producing a dark, unexposed circle in the image.

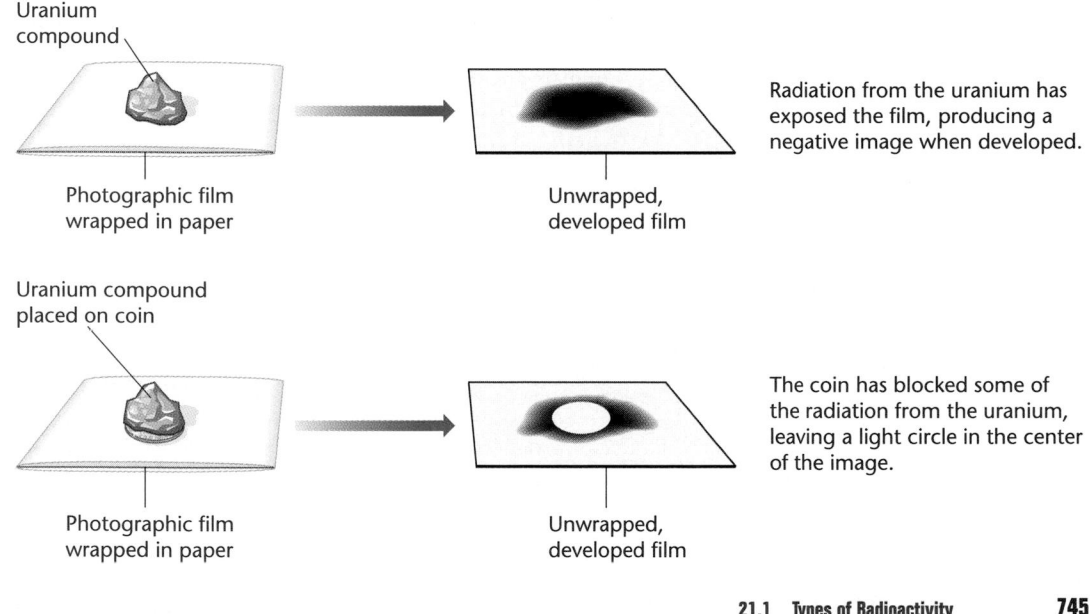

Uranium compound

Photographic film wrapped in paper

Unwrapped, developed film

Radiation from the uranium has exposed the film, producing a negative image when developed.

Uranium compound placed on coin

Photographic film wrapped in paper

Unwrapped, developed film

The coin has blocked some of the radiation from the uranium, leaving a light circle in the center of the image.

21.1   Types of Radioactivity   **745**

Marie and Pierre Curie's daughter, Irene, also became a scientist and Nobel prize winner. She and her husband, Frederick Joliot-Curie, discovered the first artificially created radioisotope and conducted nuclear fission with uranium. They also hampered Germany's use of nuclear installations during World War II by smuggling "heavy" water (D_2O) out of France after that country was invaded.

Becquerel presented the problem to Marie and Pierre Curie for further study. Their conclusion was that a nuclear reaction was taking place within the uranium atoms. Marie Curie named this spontaneous emission of radiation by an unstable atomic nucleus **radioactivity.**

For the first time, the effects of changes in the nucleus of an atom had been observed and correctly interpreted. The Curies went on to study the properties of radioactivity and discovered elements other than uranium that also give off radiation. Henri Becquerel and the Curies were awarded the 1903 Nobel Prize in Physics for their important work.

Nuclear Notation

Whereas chemical reactions involve changes in the number or configuration of electrons, nuclear reactions involve the protons and neutrons found in the nucleus. During nuclear reactions, a nucleus can lose or gain protons and neutrons. Recall that the number of protons is the same as the atomic number, which identifies the kind of element. Therefore, adding or taking away protons changes the identity of an element. In nuclear reactions, the dream of the alchemists becomes reality as atoms of one element are converted into atoms of another element.

You can represent nuclear changes using equations, just as you do for chemical reactions. The equations will look similar in most ways to those you already know how to write and balance. The reactants will be on the left and products on the right, following an arrow. However, now you will be considering the nuclei of elements as reactants and products.

When you write nuclear equations, it is important to indicate which isotopes of the given elements you are dealing with. For example, the element carbon has natural isotopes with three different atomic masses: carbon-12, carbon-13, and carbon-14. Two of these, carbon-12 and carbon-13, are stable isotopes and are not radioactive. However, carbon-14 is unstable and radioactive. There must be some way to distinguish between these isotopes in nuclear equations. To do this, you write the mass number as a superscript and atomic number as a subscript in the symbol for an isotope. These numbers are both placed to the left of the symbol for the element. So, the three isotopes of carbon are represented as $^{12}_{6}C$, $^{13}_{6}C$, and $^{14}_{6}C$, respectively. Each has six protons, but carbon-12 has a total of 12 protons and neutrons, carbon-13 has a total of 13, and carbon-14 has 14. **Figure 21.3** illustrates the isotopes of carbon and hydrogen.

Figure 21.3
Isotope Notation
Numerical superscripts and subscripts are used to show the mass numbers and atomic numbers of isotopes. The isotopes of carbon and of hydrogen are represented here.

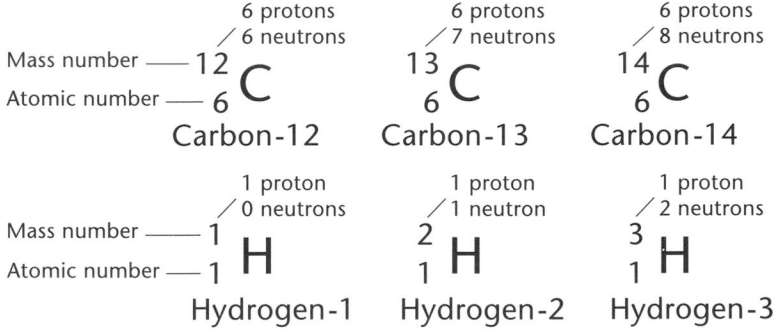

Radioactive Decay

The release of radiation by radioactive isotopes—radioisotopes, for short—is called decay. The nuclei of such radioisotopes are unstable. However, not all unstable nuclei decay in the same way. Some give off more powerful radiation than others or different kinds of radiation. Between 1896 and 1903, scientists had discovered three types of nuclear radiation. Each type changes the nucleus in its own way. These three types were named after the first three letters of the Greek alphabet: alpha (α), beta (β), and gamma (γ).

Alpha Decay

Alpha radiation consists of streams of **alpha particles,** which are helium nuclei consisting of two protons and two neutrons. Alpha particles can be represented as $^4_2He^{2+}$ or simply as α. Their relatively large size and charge, compared with other forms of radiation, cause them to collide frequently with atoms in air and objects they encounter. Alpha radiation does not deeply penetrate into matter and is easily stopped by a thin layer of material, such as paper and clothing, or even by air. Whenever alpha decay occurs, the decaying nucleus loses $2p^+$ and $2n^0$, and turns into the nucleus of an element with an atomic number that is 2 less and a mass number that is 4 less than that of the original.

Radioactive uranium-238 decays spontaneously in a long series of steps, the first of which is an alpha decay. When an alpha particle is emitted from a nucleus of uranium-238, the uranium atom loses two protons and two neutrons. The loss of protons makes the product another element, thorium, as shown in the following equation and in **Figure 21.4.**

$$^{238}_{92}U \rightarrow\ ^{234}_{90}Th + ^4_2He$$

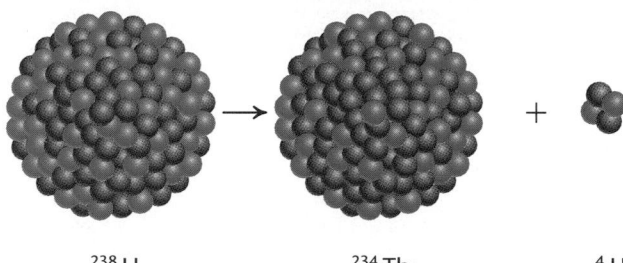

$$^{238}_{92}U \qquad\qquad ^{234}_{90}Th \qquad\qquad ^4_2He$$

Figure 21.4
Alpha Decay of Uranium-238
When a nucleus of uranium-238 undergoes alpha decay, a nucleus of thorium-234 and a nucleus of helium-4 are produced. The atomic numbers balance on the two sides of the equation, as do the mass numbers.

Is this equation balanced? Check to see whether the sum of the mass numbers on the right side of the equation is 238, which is the mass number of the uranium isotope on the left side. It is, because the mass number of the thorium atom is 234 and that of the alpha particle is 4. Also make sure that the sums of the atomic numbers on both sides of the equation are the same. Both sums equal 92, so the atomic number balances, and the nuclear equation as a whole is balanced. Note that a balanced nuclear equation differs from balanced chemical equations because the kinds of atoms involved need not remain the same.

Concept Development

Draw the following diagram on the chalkboard to illustrate the behavior of alpha (α), beta (β), and gamma (γ) rays in an electric field. Note that the light, negatively charged beta particles are strongly deflected toward the positive plate. The heavier, positively charged alpha particles are slightly deflected toward the negative plate. The uncharged gamma rays are undeflected.

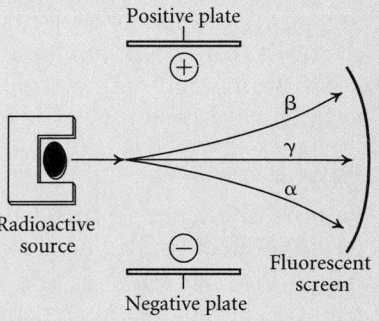

Reinforcement

Ask students to write equations for the alpha decay of thorium-230 and radium-226, and for the beta decay of magnesium-27 and sulfur-35.

$$^{230}_{90}Th \rightarrow\ ^4_2He + ^{226}_{88}Ra;$$
$$^{226}_{88}Ra \rightarrow\ ^4_2He + ^{222}_{86}Rn;$$
$$^{27}_{12}Mg \rightarrow\ ^{0}_{-1}e + ^{27}_{13}Al;$$
$$^{35}_{16}S \rightarrow\ ^{0}_{-1}e + ^{35}_{17}Cl\ \boxed{L1}$$

---

outward from the source, leaving momentary trails of condensed alcohol vapor.
Analysis: Ask these questions.
1. How big are the particles that caused the alcohol vapor to condense in a vapor trail? *submicroscopic (about 0.0012 picometers)*
2. The alpha particles that produced the tracks are positively charged helium nuclei. How might you show that they have an electrical charge? *put a strong magnet or charged ob-*

ject near the cloud chamber and observe that the tracks curve

✔ Assessment

Knowledge: Ask students whether the rate of emission of the particles producing the vapor trails is constant. *The rate appears to be more or less the same during the observation period, but it actually decreases slowly.* $\boxed{L1}$

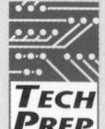

How it Works

Smoke Detectors

A smoke detector is a device that sounds an alarm in the presence of smoke particles from a smoldering or burning object. An ionizing smoke detector, like the one illustrated here, contains an electrical sensor that detects smoke particles that interrupt an electrical current. Ionizing detectors usually contain the unstable isotope americium-241 ($^{241}_{95}Am$), which decays to form $^{237}_{93}Np$ and $^{4}_{2}He$.

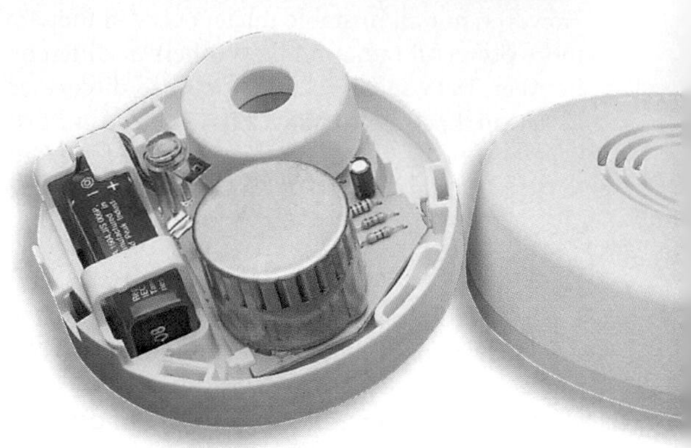

1. A battery or other power source provides electricity.

2. Americium-241 provides alpha particles that ionize air molecules.

3. In the absence of smoke, the ions carry current between a positive and a negative electrode.

4. A microchip monitors the flow of current between the electrodes.

5. Under normal conditions, the microchip permits no current to flow through the alarm. The alarm does not sound.

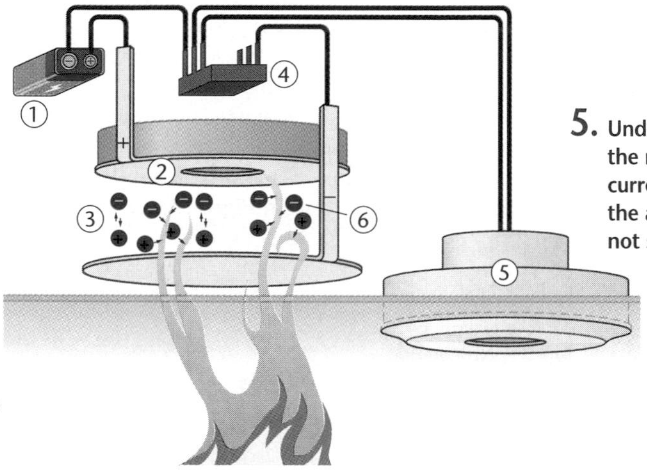

6. If a fire occurs, smoke particles rise up into the detector and interfere with the flow of ions between the electrodes. The microchip senses the drop in current and allows electricity to flow through the alarm circuit, sounding the alarm.

Thinking Critically

1. What kind of radioactive decay does americium-241 undergo?
2. Explain why the air that has been exposed to the americium-241 is able to carry a current between charged electrodes.

How it Works

Purpose

TECH PREP Students will learn about the role of a radioactive element in the operation of a smoke detector.

Background

The radioactive source is often americium-241, which undergoes alpha decay. The emitted alpha particles have sufficient energy to remove electrons from atoms they contact, creating positive ions. The ejected electrons can then join with other atoms and/or molecules to form negative ions.

Visual Learning

Remove the cover from a battery-operated smoke detector. If you have access to a Geiger counter, point out the general location of the Am-241 in the detector and demonstrate the ionizing radiation.

Teaching Strategies

- Ask students the following question. If the radiation from the smoke detector has sufficient energy to ionize some atoms in the air, what effect might it have on your skin if you came in contact with the americium-241? *The radiation could also ionize and damage the atoms in your skin. However, the small amount of radioactive material is sealed inside the smoke detector and presents no danger.*
- Show students the universal radiation symbol inside the smoke detector to illustrate an everyday use of radioactivity.

Thinking Critically

1. Americium-241 undergoes alpha decay, given that the process releases helium nuclei, which are alpha particles.
2. The alpha particles produced by the radioisotope produce positively and negatively charged particles when they strike the molecules in the air. Thus, the ions can be attracted by the oppositely charged electrodes and move toward them. Because the ions are moving electric charges, they can carry a current.

Integrating THE Sciences

Biology Ask students to research the composition of cigarette smoke, which causes smoke detectors to sound. They should find out why these substances affect the detector, and also should research the effects of the substances on the smoker. L1

Beta Decay

The second type of radiation from unstable atomic nuclei results in beta decay and produces beta radiation. Beta radiation is made up of particles that are much smaller and lighter than alpha particles, so they move much faster and have greater penetrating power. Each **beta particle** is a high-energy electron with a 1− charge and is written as $_{-1}^{0}e$ or β^-. Beta particles pass through matter more easily than do alpha particles. They can be stopped only by thick materials such as stacked sheets of metal, blocks of wood, or heavy clothing.

The electron produced in beta decay is not one of the original electrons from outside the nucleus. It is produced by the change of a neutron into a proton and an electron. Whenever beta decay occurs, transmutation of elements occurs. The decaying nucleus turns into the nucleus of an element with an atomic number that is 1 greater than that of the original and a mass number that is the same.

Carbon-14 is an example of an isotope that decays by beta emission.

$$_{6}^{14}\text{C} \rightarrow _{7}^{14}\text{N} + _{-1}^{0}e$$

Notice that transmutation of elements has occurred. An atom of carbon-14 is converted into an atom of nitrogen-14. Because the beta particle has only the mass of an electron, the mass number remains 14. Because the beta decay changes a neutron to a proton, the atomic number of the nucleus increases by one unit, from 6 to 7. The atomic number on the left side of the equation, 6, is the same as the sum on the right side of the equation ($6 = 7 + (-1)$). The nuclear equation as a whole is balanced.

Gamma Decay

In the third type of decay, gamma radiation is produced. A **gamma ray** is a high-energy form of electromagnetic radiation without mass or charge. Because of its high energy and penetrating ability, gamma radiation can cause great harm to living cells. Gamma rays are written as γ. Such rays are much harder to stop than alpha or beta particles, as shown in **Figure 21.5.** They can pass easily through most types of material, and stopping them requires thick blocks of lead or even thicker blocks of concrete.

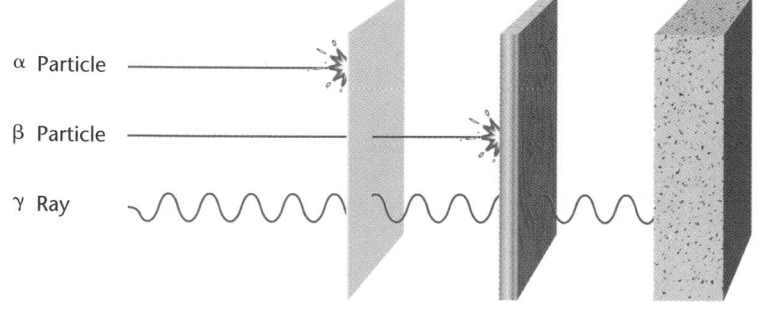

α Particle

β Particle

γ Ray

Sheet of paper   Dense wood   Thick blocks of lead or concrete

Figure 21.5

Blocking Radiation
The three types of radiation vary in their penetrating power. Alpha particles are stopped by even a thin sheet of paper, but beta particles require thicker shields, such as a dense piece of wood. The energetic gamma rays are stopped only by thick blocks of lead or concrete.

21.1   **Types of Radioactivity**   **749**

SECTION 21.1

Concept Development

Explain that transmutation in a nuclear reaction always involves a change of identity of one or more elements.

Reinforcement

Ask students to make a table in which they summarize the names, symbols, characteristics, and accompanying changes in mass number and atomic number for the three common types of radioactive decay. **L1**

Transparency

Display **Basic Concepts Transparency 58** to reinforce the different kinds of radioactivity. **L1** **LEP**

Meeting Individual Needs

Gifted Ask gifted students to research more specific correlations related to nuclear instability with respect to radioactive decay. For example, they might obtain information on the maximum number of protons possible in a stable nucleus. *83* They might also do research on the topic of neutron-to-proton ratios and discover that light nuclides are stable when the neutron/proton ratio is 1:1, but that heavier nuclides are stable when the neutron/proton ratio is greater than 1:1. They may also find out that nuclides that have even numbers of neutrons and protons tend to be more stable than those with odd numbers, and that nuclides having 2, 8, 20, 28, 50, 82, or 126 neutrons or protons are apt to be especially stable. **L3**

Students may assume that damage to living tissue is related only to penetrating power, not to the amount of ionization produced. Actually, both play an important role. Explain that alpha radiation may cause considerable damage at short distances because it has great ionizing ability. However, protecting against it is easy because its penetrating power is not great.

Transparency

Display **Problem Solving Transparency 24** to give students practice in balancing nuclear equations. L1 LEP

During gamma decay, only energy is given off. Gamma radiation is often omitted from nuclear equations because it does not affect mass number or atomic number. Gamma decay generally does not occur alone but accompanies other modes of decay. For example, the alpha decay of uranium to thorium involves simultaneous gamma decay.

$$^{238}_{92}\text{U} \rightarrow {}^{234}_{90}\text{Th} + {}^4_2\text{He} + \gamma$$

Once you know what kind of nucleus is produced by a simple nuclear reaction, you can determine what type of decay has taken place. If both the atomic number and mass number decrease, alpha decay has occurred. If the mass number stays the same but the atomic number increases, beta decay has occurred. If neither atomic number nor mass number changes, only gamma radiation has been emitted.

In more complex reactions, more than one type of decay can take place at the same time. In some cases, one kind of radioactive nucleus decays to form another, which decays to form a third kind, and so on. The decay series ends with the production of a stable nucleus. For example, uranium-238 undergoes a 14-step process involving alpha, beta, and gamma decay, and eventually forms stable lead-206.

SAMPLE PROBLEM 1 Writing and Balancing a Nuclear Equation

Potassium-40 can decay to form calcium-40. Write a balanced equation for this nuclear reaction and determine the type of decay.

Analyze
- Write the symbols for the two isotopes involved. Each symbol must include the abbreviation of the element represented, the mass number as a superscript, and the atomic number as a subscript.

Potassium-40 is $^{40}_{19}\text{K}$ and calcium-40 is $^{40}_{20}\text{Ca}$.

Set Up
- Set up the equation for the reaction. Potassium is the reactant and goes on the left. The product, calcium, goes on the right.

$$^{40}_{19}\text{K} \rightarrow {}^{40}_{20}\text{Ca} + ?$$

Now you must balance the nuclear equation. Balance the sums of both the mass numbers and the atomic numbers.

> **Problem-Solving HINT**
>
> Do not be tempted to change mass number or atomic numbers in solving this kind of problem. Consider only the mass number and atomic number of the product to produce a balanced equation.

Solve
- Because the mass numbers are already balanced (40 = 40), you have to balance only the atomic numbers. Potassium has an atomic number of 19, and the atomic number of calcium is 20. Because a beta particle ($_{-1}^{0}e$) has a negative charge and almost no mass, adding such a particle to the right side will balance the equation for atomic number.

$$^{40}_{19}\text{K} \rightarrow {}^{40}_{20}\text{Ca} + {}^0_{-1}e$$

Check
- The mass number on the left equals the sum (40 + 0) on the right. The atomic number on the left is 19, which equals the sum (20 + (−1)) on the right. The equation is balanced. Because an electron (a beta particle) is given off, beta decay has taken place.

Chemistry Journal

Decay of U-238

Ask students to make a table in their journals showing all 14 steps and intermediate nuclides in the decay process from U-238 to Pb-206. They should also provide a descriptive account of the decay process. Tell them the successive decays are as follows: α, ß, ß, α, α, α, α, α, ß, ß, α, ß, ß, α. L1

1. Write the balanced nuclear equation for the radioactive decay of radium-226 to give radon-222, and determine the type of decay.

2. Write a balanced equation for the nuclear reaction in which neon-23 decays to form sodium-23, and determine the type of decay.

Correcting Errors

Students have learned that, when writing balanced chemical equations, atoms of elements are conserved. They may attempt to impose the same rule for nuclear equations. Point out that kinds of atoms are not necessarily conserved in nuclear equations; however, mass numbers and atomic numbers are conserved.

Extension

Ask an interested student to research and prepare a brief report on the *chemical* toxicity of plutonium compared with that of other elements, such as arsenic, mercury, and lead. [L2]

Detecting Radioactivity

Radioactivity cannot be seen, heard, or touched, and it has no taste or smell. Other means of detection must be used to tell whether radioactive materials are nearby. Photographic film can serve as a simple detection device. However, for film to be effective as a detector, the source of radioactivity has to be close. Instruments that measure radiation are shown in **Figure 21.6.**

Figure 21.6

Uses of Radiation Detectors
People who work around radiation sources are required to wear film badges that show their levels of radiation exposure. The film in these badges is developed regularly to show how much exposure to radiation each person has received. ▶

A scintillation counter registers the intensity of radiation by detecting light. Radioactive samples to be measured are mixed with compounds that emit a flash of light, called scintillation, when exposed to radiation. The level of radiation is measured by the number of flashes of light recorded by the device. ▼

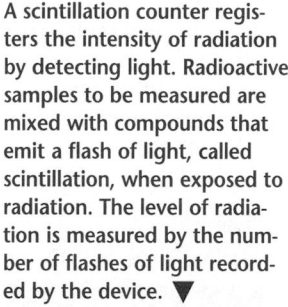

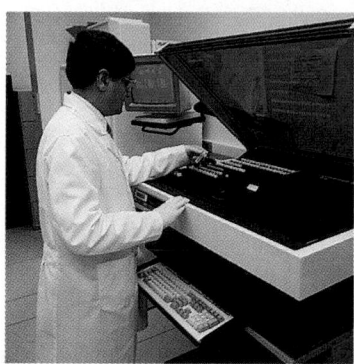

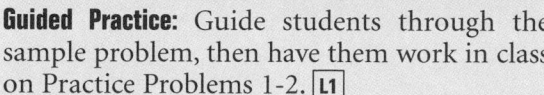

◀ Radioactive elements can be detected by a Geiger counter. This portable instrument maintains an electric circuit between its battery and a handheld, gas-filled cylinder. There is normally a gap in the circuit. If radiation is present, ions are formed in the gas. The charged particles complete the circuit, which activates a sounding device and a meter. The number of clicks per unit time and the meter reading reveal how much radiation is present.

21.1 Types of Radioactivity   **751**

PRACTICE

Guided Practice: Guide students through the sample problem, then have them work in class on Practice Problems 1-2. [L1]

Independent Practice: Your homework or classroom assignment can include the additional practice problems shown below. [L1]

Answers to Practice Problems

1. $^{226}_{88}Ra \rightarrow ^{222}_{86}Rn + ^{4}_{2}He$; alpha decay
2. $^{23}_{10}Ne \rightarrow ^{23}_{11}Na + ^{0}_{-1}e$; beta decay

Additional Practice Problems

1. Write a balanced nuclear equation for the radioactive decay of iron-60 to give cobalt-60, and determine the type of decay.
$^{60}_{26}Fe \rightarrow ^{60}_{27}Co + ^{0}_{-1}e$; *beta decay*

2. Write a balanced equation for the nuclear reaction in which uranium-234 decays to form thorium-230, and determine the type of decay.
$^{234}_{92}U \rightarrow ^{230}_{90}Th + ^{4}_{2}He$; *alpha decay*

ChemLab

The Radioactive Decay of "Pennium"

Time Allotment: One laboratory period.

Objectives

Review objectives with students before they begin the ChemLab.

Process Skills

Observing, using numbers, interpreting data, communicating, inferring, formulating models

PREPARATION

Ask students to bring in shoe boxes and pennies, but have some on hand in case some students fail to do so.

PROCEDURE

- Require that students' notebooks contain the following items before they begin: Objectives, Outline or Flow Chart of the Procedure, Data and Observations Table.
- Consider the proximity of other classes and ask students to avoid shaking the boxes so vigorously that they generate excessive noise.

ANALYZE AND CONCLUDE

1. The graph should show a descending curve that approaches but does not touch the *x*-axis.
2. The half-life is the time at which only half the original number of pennies remains. In this case, it is the shaking time, which is probably about 20 s.

The Radioactive Decay of "Pennium"

The nuclei of unstable atoms disintegrate or decay spontaneously, emitting alpha or beta particles and gamma radiation. Types of atoms that undergo this process are called radioactive isotopes. A decaying reactant isotope is referred to as a parent atom, and the atom produced is a daughter atom. In this ChemLab, heads-up pennies represent individual parent atoms of the fictitious element pennium, and tails-up pennies represent the daughter atoms of the decay. You will study the decay characteristics of pennium and will determine its half-life, which is the time required for one-half of the atoms to decay.

Problem

What is the half-life of the fictitious radioisotope pennium?

Objectives

- **Infer** the decay characteristics of pennium.
- **Analyze** your data to determine the half-life of pennium.
- **Make** and **use graphs** to interpret data.

PREPARATION

Materials

1 shoe box
120 pennies
stopwatch or watch with a second hand

PROCEDURE

1. Count 120 pennies and lay them heads-up in the bottom of the shoe box. In a table like the one in Data and Observations, record this as 0

daughter atoms and 120 parent pennium atoms at time 0.

2. Put the lid on the shoe box and, holding the lid on tightly, shake the box with moderate force up and down 20 times while your lab partner times the decay process to the nearest second. Assume that all the subsequent decay trial times are identical.

3. Open the box and count the tails-up pennies, which represent the daughter atoms, and remove them from the box. Subtract this number from the original number of heads-up pennies to determine the number of

remaining parent pennium atoms. Record the time taken for the shaking, the number of tails-up atoms removed, and the number of remaining parent atoms.

4. Repeat steps 2 and 3 four more times, recording your data after each trial. Simply add on the original shaking time repeatedly to arrive at the total elapsed time.

1. **Making a Graph** Construct a graph of your data, plotting the number of remaining parent pennies on the vertical axis (y-axis) and the total elapsed time on the horizontal axis (x-axis).

2. **Interpreting Data** What is the half-life of pennium in your experiment? Explain how you arrived at this number.

APPLY AND ASSESS

1. Does exactly the same fraction of pennium atoms decay during each half-life? What does this suggest about half-life? Why are such variations not likely to be obvious when actual atoms are involved?

2. If you had started with one mole of pennies (assuming you could actually take the time to count them!), how many would remain after ten half-lives? Is this a large number?

3. If you took a longer time to shake the box in this case, how would half-life be affected? Are the half-lives of atoms also controllable by a change in conditions?

DATA AND OBSERVATIONS

Total Elapsed Time	Number of Tails-Up Daughter Atoms Removed	Number of Heads-Up Parent Atoms Remaining

DATA AND OBSERVATIONS

Total Elapsed Time	Number of Tails-Up "Daughter" Atoms Removed	Number of Heads-Up "Parent" Atoms Removed
0 s	0	120
19 s	61	59
38 s	32	27
57 s	13	14
76 s	6	8
95 s	4	4

ChemLab

APPLY AND ASSESS

1. No; approximately but not exactly half the number of atoms that remain after the preceding half-life will decay. This suggests that half-life is not a precise quantity, in terms of exact numbers of atoms, but has to do with probability. Variations are not likely to be obvious because of the enormous numbers of atoms and also the difficulty of counting atoms.

2. $(6.02 \times 10^{23})(1/2^{10}) = 5.88 \times 10^{20}$, which is still a huge number

3. The half-life would be longer. This is not like the situation for atoms, the half-lives of which are not controllable by any change in conditions.

✓ Assessment

Performance: Ask students to carry out a variation of the ChemLab in which they shake the box of pennies only once each trial, each time removing the tails-up pennies. Ask them how they can use this sort of procedure to calculate a half-life on the basis of shakes rather than time. *They can determine the number of shakes required to allow them to remove half the pennies. That number is analogous to half-life but is expressed in numbers of shakes rather than in seconds.* Alternatively, allow interested students to carry out a similar investigation using more or fewer pennies. The half-life determined should remain essentially the same. **L1**

Purpose

TECH PREP

Students learn how archaeologists use the half-lives of radioactive isotopes to date materials and artifacts.

Background

Dating based on radioisotopes makes use of the fact that such isotopes decay with a known half-life to form stable isotopes. The ratio of radioactive to stable isotopes thus reveals the approximate age of the object being studied. Carbon-14 dating was first developed by Willard Libby, who received the 1960 Nobel Prize in Chemistry for his work.

Teaching Strategies

- Remind students that all isotopes of the same element react chemically in the same way. Thus, carbon-14 forms compounds such as carbon dioxide in the same way as carbon-12.
- To help students understand how carbon-14 cycles through living things, ask them how flesh-eating animals get their carbon-14. *They eat plant eaters, which have consumed plants that contain carbon-14.*
- Ask students where the proton in the nitrogen comes from in the decay of carbon-14. *A neutron decays to form the proton and an electron, or beta particle.*

Archaeological Radiochemistry

What do a piece of wood from an Egyptian pyramid, a bone tool from an archaeological dig, and the Dead Sea Scrolls have in common? They all contain carbon from dead organisms. Thus, archaeologists can determine the objects' ages by the carbon-14 method.

Development of Carbon-14 Dating

Archaeologists need accurate dating techniques for their studies of the artifacts and other remains of early humans. Until about 40 years ago, archaeological dating had to use indirect, time-consuming, and inaccurate methods. In 1946, Willard Libby, working at the University of Chicago, developed the carbon-14 method for dating objects that contain carbon. The first artifact to be dated using carbon-14 was a beam of cypress wood taken from a pharaoh's tomb. Since that time, carbon-14 dating has been widely used to date plant and animal fossils.

Carbon-14 is formed in the upper atmosphere. When cosmic rays collide with atoms in the upper atmosphere, the atoms break up and release subatomic particles. The carbon-14 is made when a neutron collides with a nitrogen atom, causing it to lose a proton.

$$^{14}_{7}N + ^{1}_{0}n \rightarrow ^{14}_{6}C + ^{1}_{1}p$$

When the ratio of radioactive carbon-14 to radioactive carbon-12 in a once-living thing is determined and compared with the ratio that must have existed while the organism was living, the age of the material can be determined.

The Ice Man

In 1991, a hiker found a man who had been frozen in a glacier for many centuries. Using carbon-14 dating, scientists established that the man lived about 5300 years ago, making the human remains the oldest ever found intact. The ancient ice man had many secrets to reveal about himself and his way of life, for he was completely outfitted and carried various tools and devices with him. Scientists discovered that people at that time had developed wooden-handled daggers. Also, the people cut their hair, had tattoos, and had knowledge of healthful properties of plants. Parts of Neolithic history could now be rewritten, and what made it all possible was being able to find out the age of the remains.

TAMS: An Advanced Dating Method

In the standard lab method of carbon-14 dating, a piece of the object being tested has to be burned. The carbon-14 atoms are then

Tandem Accelerator Mass Spectrometer

depends upon what radioactive substances are present in the material. The methods include rubidium-strontium dating and potassium-argon dating.

Archaeologists are also using a new method to date pottery, called the thermoluminescence (TL) method. As small amounts of radioactive materials such as uranium and thorium decay in clay, they excite other atoms in the clay to a higher energy state. When the clay is heated above 400°C, light called thermoluminescent glow is emitted as electrons fall back to their stable levels. Measuring this glow reveals roughly how long ago the artifact was made.

This painted amphora from Greece was dated at 500 B.C.

counted using a scintillator that counts the light flashes from beta radiation. This method requires a fairly large sample because of the small number of carbon-14 atoms, and is effective only for objects up to 60 000 years old. The counting also takes days or weeks because of the slow decay of carbon-14. With the recent development of a device called a tandem accelerator mass spectrometer (TAMS), carbon-14 dating may be effective for objects up to 100 000 years old. A TAMS needs only a few milligrams of sample, and counting can be done in less than an hour.

Other Radiodating Systems Used by Archaeologists

Sometimes an artifact contains little or no carbon or is too old to be dated with carbon-14. So an alternative method has to be used. The method

DISCUSSING THE TECHNOLOGY

1. **Acquiring Information** Carry out library research to obtain information on a radiodating technique not explained in this feature.

2. **Inferring** Iron is not a once-living material. How, then, can iron tools be dated using the carbon-14 method?

3. **Applying** The thermoluminescence method is not highly accurate. However, it is useful in exposing new forgeries in pottery. Explain how this could be so.

Unlike chemical reaction rates, which are sensitive to factors such as temperature, pressure, and concentration, the rate of spontaneous nuclear decay cannot be changed. Because the decay of an individual nucleus is a random event, it is impossible to predict when a specific nucleus in a sample of a radioactive material will undergo decay. However, the overall rate of decay is constant, which allows you to predict when a given fraction of the sample will have decayed.

Using an Analogy

To illustrate the concept of half-life, ask a student volunteer to stand 20 feet from the front wall of the classroom and to move half the *remaining* distance to the wall each minute. Ask students how much time will be required for the volunteer to reach the wall. *Virtually an infinite amount of time will be required.* Explain the analogy to the half-life of a radioisotope and the time required to attain decay of all the nuclei. L1 LEP

Rate of Decay

The **half-life** is the time it takes for half of a given amount of a radioactive isotope to undergo decay. Half-life, which is symbolized $t_{1/2}$, is easy to measure and has been determined for many different radioisotopes, some of which have important uses. Some half-lives are only fractions of a second, whereas others are billions of years. Half-lives for some of the most commonly used radioisotopes are listed in **Table 21.1**. The concept of half-life is illustrated in the graph in **Figure 21.7**.

With its predictable and unchanging rates, radioactive decay has provided scientists with a technique for determining the age of fossils, geological formations, and human artifacts. Using a knowledge of the half-life of a given radioisotope, one can estimate the age of an object in which the isotope is found. Four different isotopes are commonly used for dating objects: carbon-14, uranium-238, rubidium-87, and potassium-40. Now look at one of these techniques in more detail.

Assessment

Oral: Explain that Co-60 is a radioisotope used in hospitals to kill rapidly dividing cancer cells and that it has a half-life of 5.3 years. Ask students how much Co-60 will have to be purchased after 15.9 years to replenish an original 10.0-g sample. *It will require 8.75 g.* Ask how much time will be required for the original 10.0-g sample to decay to less than 0.1 g. *Somewhat less than seven half-lives, or about 36 years, will be required.* Ask how much time will be required for the entire 10.0-g sample to decay. *Virtually an infinite amount of time will be required.* L1

Table 21.1 Half-Life Values for Commonly Used Radioactive Isotopes

Isotope	Half-Life
$^{3}_{1}H$	12.26 years
$^{14}_{6}C$	5730 years
$^{32}_{15}P$	14.282 days
$^{40}_{19}K$	1.25 billion years
$^{60}_{27}Co$	5.271 years
$^{85}_{36}Kr$	10.76 years
$^{93}_{36}Kr$	1.3 seconds
$^{87}_{37}Rb$	48 billion years
$^{99m}_{43}Tc$	6.0 hours*
$^{131}_{53}I$	8.07 days
$^{131}_{56}Ba$	12 days
$^{153}_{64}Gd$	242 days
$^{201}_{81}Tl$	73 hours
$^{226}_{88}Ra$	1600 years
$^{235}_{92}U$	710 million years
$^{238}_{92}U$	4.51 billion years
$^{241}_{95}Am$	432.7 years

*The *m* in the symbol tells you that this is a metastable element, which is one form of an unstable isotope. Technetium 99m gives off a gamma ray to become a more stable form of the same isotope, with no change in either atomic or mass number.

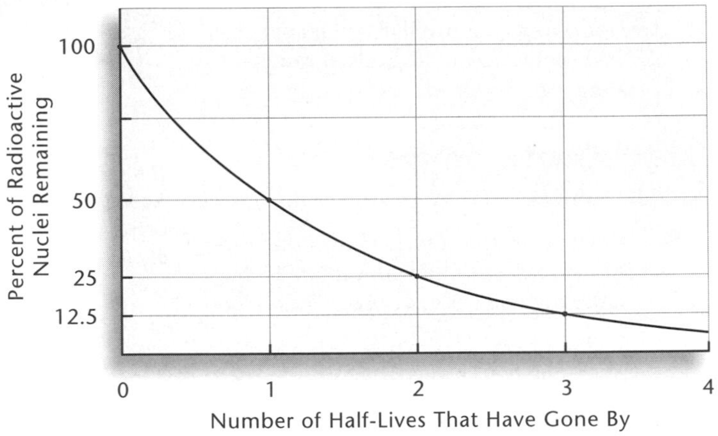

Figure 21.7

Graph Illustrating Half-Life
During each half-life period, half of the radioactive nuclei in a sample decay. After one half-life, 50 percent of the sample remains. After two half-lives, 25 percent of the sample remains. After three half-lives, 12.5 percent remains, and so on.

(Graph: y-axis "Percent of Radioactive Nuclei Remaining" with values 100, 50, 25, 12.5; x-axis "Number of Half-Lives That Have Gone By" with values 0, 1, 2, 3, 4)

Demonstration 21.3

Detecting Radioactivity

Purpose: To demonstrate how an electroscope can be used to show the presence of radioactivity.

Materials: electroscope, rubber or plastic rod, piece of fur or wool cloth, packaged alpha-radioactive source, tongs

Safety Precautions: ☢ 💪 👓

Disposal: Code F

Procedure: Rub the rod with the fur, then touch the ball of the electroscope. The two foil leaves will move apart. Use tongs to handle a packaged alpha-radioactive source. Bring the source close to the electroscope. The leaves will fall back together.

Results: The two foil leaves will spread apart as each is charged with electrons and they mutually repel one another. The positively charged alpha particles remove electrons and discharge the electroscope.

Carbon-14 Dating

Carbon-14 dating has been widely used to determine the ages of fossils. All organisms take in carbon during their lifetime—plants from carbon dioxide gas and animals from the plants and animals they eat. Most carbon taken in by organisms is in the form of the stable isotopes $^{12}_{6}C$ and $^{13}_{6}C$, but about one carbon atom in every million is the radioisotope $^{14}_{6}C$. As long as an organism is alive, the $^{14}_{6}C$ in its cells remains relatively constant at about one in every million carbon atoms. As $^{14}_{6}C$ atoms decay to form nitrogen and a beta particle, new ones are taken in from the surroundings. After death, the organism no longer takes in carbon, so the proportion of $^{14}_{6}C$ slowly decreases.

In carbon-14 dating, the fraction of $^{14}_{6}C$ that remains in material such as bone or skin is measured and compared with how much was in the material when it was alive. In this way, the age of the object can be estimated. For example, if half of the $^{14}_{6}C$ remains, one half-life period (5730 years) has passed, and the object is about 5730 years old. If only one-fourth of the $^{14}_{6}C$ remains, two half-life periods (11 460 years) have passed. **Figure 21.8** shows some examples in which carbon-14 dating proved useful.

The half-life of carbon-14, 5730 years, is short compared with the age of many fossils and geological formations. As a result, objects more than about 60 000 years old cannot be dated using this technique because the amount of carbon-14 left in them is too small to be measured accurately.

Figure 21.8

Determining the Age of Artifacts
A cypress beam from the cliff dwellings of Mesa Verde in southern Colorado was determined by carbon-14 dating to be about 700 years old. This date was confirmed by counting the annual tree rings in samples of old trees that came from the same area. The Anasazi Indians had moved to Mesa Verde and lived there from around 1100 to 1300 A.D. ▶

◀ This sandal, found in a cave in Oregon, was analyzed by carbon-14 dating and found to be about 9000 years old.

Analysis: Ask these questions.
1. Why do the leaves on the electroscope move apart? *They are both given a negative charge and repel each other.*
2. Why does an alpha-radioactive source discharge the electroscope? *The alpha particles that are emitted from the source are positively charged. They remove electrons from the air around the ball, resulting in a movement of electrons from the leaves.*

✓ Assessment

Portfolio: Have interested students write a report for their portfolios on the work of Becquerel and Curie in using the electroscope to detect radioactive sources. L1 P

SECTION 21.1

Concept Development

Point out that some half-lives are extremely short. Polonium-212, for example, has a half-life of approximately 3×10^{-7} s.

Concept Development

Explain that the amount of C-14 in the atmosphere remains approximately constant because the isotope is produced by collisions between N-14 atoms and high-energy neutrons from space at the same rate at which it undergoes beta decay. The equation for the formation is $^{14}_{7}N + ^{1}_{0}n \rightarrow ^{14}_{6}C + ^{1}_{1}H$.

GLENCOE TECHNOLOGY

 Videodisc

Science and Technology Videodisc Series
Carbon-14 Dating
Disc 2: Chemistry
Side 2, Ch. 4

Show this video to illustrate how a tandem particle accelerator mass spectrometer is used to date skeletal remains of prehistoric humans. L1
LEP

 Videodisc

Science and Technology Videodisc Series
Carbon-14 Dating of the Shroud of Turin
Disc 2: Chemistry
Side 2, Ch. 5

Show this video to illustrate how the age of the Shroud of Turin was determined through radiodating techniques. L1 LEP

SAMPLE PROBLEM **2** **Determining the Age of a Fossil**

A fossilized tree killed by a volcano was studied. It had 6.25 percent of the amount of carbon-14 found in a sample of the same size from a tree that is alive today. When did the volcanic eruption take place?

Analyze • The half-life of carbon-14 is 5730 years, and the fraction of carbon-14 remaining is 6.25 percent (0.0625).

Set Up • Calculate the number of half-lives that have passed for the carbon-14 in the sample. During each half-life of a radioactive isotope, one-half of the nuclei decay. The fossil sample has 6.25 percent (0.0625) of its original amount of carbon-14 left, so you need to find out how many times $1/2$ (0.5) must be used as a factor to produce 0.0625 as a result. The answer is *four times* because $1/2 \times 1/2 \times 1/2 \times 1/2 = 0.5 \times 0.5 \times 0.5 \times 0.5 = 0.0625$ (or, carrying out the calculation another way, 100 percent $\times 1/2 \times 1/2 \times 1/2 \times 1/2 = 6.25$ percent). Therefore, four half-lives have gone by.

> **Problem-Solving HINT**
>
> Repeatedly multiply ½ by itself until the number of factors produces the proper result.

Solve • Because four half-lives have gone by and each is 5730 years, multiply 5730 by 4.

$$5730 \times 4 = 22\ 920$$

The tree from which the sample was taken must have been killed by the volcano about 22 920 years ago.

Check • Check to be sure that the calculation can be repeated with the same result. All aspects of setup and calculation appear to be correct.

PRACTICE PROBLEMS

3. A rock was analyzed using potassium-40. The half-life of potassium-40 is 1.25 billion years. If the rock had only 25 percent of the potassium-40 that would be found in a similar rock formed today, calculate how long ago the rock was formed.

4. Ash from an early fire pit was found to have 12.5 percent as much carbon-14 as would be found in a similar sample of ash today. How long ago was the ash formed?

758   Chapter 21   Nuclear Chemistry

PRACTICE

Answers to Practice Problems

3. Because $(0.5)(0.5) = 0.25$, two half-lives have gone by. 2 half-lives $\times$ 1.25 billion years/half-life $= 2.50$ billion years

4. Because $(0.5)(0.5)(0.5) = 0.125$, three half-lives have gone by. 3 half-lives $\times$ 5730 years/half-life $= 17\ 190$ years

Additional Practice Problems

1. A rock sample has 12.5% of the potassium-40 that would be present in a similar rock formed today. How old is the rock sample?
3 half-lives $\times$ 1.25 billion years/half-life = 3.75 billion years old

2. How old is a piece of wood in which the carbon-14 is 3.12% of that in wood formed today?
5 half-lives $\times$ 5730 years/half-life = 27 250 years old

ART ▷ CONNECTION

Art Forger, van Meegeren– Villain or Hero?

Hans van Meegeren, a skilled Dutch artist, painted forgeries in the style of the great 17th-century Dutch painter Vermeer. In 1937, van Meegeren sold his best forged Vermeer, *Christ and His Disciples at Emmaus,* to a Dutch museum for $280,520. He continued to forge and sell Vermeers and other Dutch masters until 1945.

Wartime forgery During the Nazi occupation of Holland in the early 1940s, van Meegeren sold his forged Vermeer paintings to members of the Nazi government. Part of the price Meegeren exacted for his paintings was the return of 200 works of Dutch art that had been looted earlier in the war by the Nazis. At the end of the war, in 1945, a forged Vermeer was found with the Nazis' treasures and traced to van Meegeren. When he was charged as a Nazi collaborator for selling Dutch national treasures, he confessed to the forgeries. He pointed out that by exchanging the forgeries, he had saved many Dutch art treasures from the Nazis. However, the authorities did not believe that he was a forger. He was sentenced to a year in prison, but died before serving it.

Radioactive dating resolves the problem It was not until 1968 that an American scientist named Keisch presented irrefutable evidence that van Meegeren was an art forger. Keisch's work made use of uranium-lead dating. White lead, a combination of $PbCO_3$ and $Pb(OH)_2$, is an important pigment in oil painting. The parent source of the stable lead-206 in white lead is uranium-238. The uranium-238 has a half-life of 4.5 billion years, decaying into a series of unstable isotopes, including radium-226 and polonium-210, until it finally forms lead-206. During the processing of white lead, most of the isotopes preceding polonium-210 in the uranium-decay series are removed.

However, minute amounts of radium-226 remain. Because the radioactive elements that would be needed to make the polonium-210 are mostly removed, the amount of polonium-210 decreases as the pigment ages. The ratio between the amounts of the radium-226 and the polonium-210 is used to establish the age of paintings. A low ratio of radium to polonium indicates a recent painting.

Proof of a fake Keisch tested a white lead sample from the alleged fake "*Emmaus*" painting, shown in the photo below, and found a low Ra:Po ratio. The ratio indicated that the painting was only about 30 years old, not 260, and therefore a forgery.

Connecting to Chemistry

1. **Acquiring Information** Belgian chemist Coremans also exposed the fraudulent Vermeers. Find out how.

2. **Interpreting** You have two paintings. One is several hundred years old, the other one year old. Painting A has a Po concentration of 12 and a Ra concentra-tion of 0.3. Painting B has a Po concentration of 5 and a Ra concentration of 6. Which painting is which?

3. **Acquiring Information** What are some modern methods of analysis for revealing art forgeries? Write a brief paper tracing this history.

Meeting Individual Needs

Gifted Explain that radioactive decay is a first-order process; that is, one that has a rate directly dependent upon the amount of radioactive material. Perceptive students are likely to ask how one calculates the age of a substance by carbon-14 dating if anything other than a multiple of 5730 years has elapsed. Explain that living organisms contain an amount of carbon-14 that undergoes 15.3 decays per minute per gram of carbon. This is the value of N_0 in the equation $\ln(N/N_0) = -kt$, where t represents the time since the death of the organism, N represents the decays per minute per gram of carbon at time t, and k is the decay constant and is related to $t_{1/2}$ by the equation $k = 0.693/t_{1/2}$. Illustrate with this sample problem. What is the age of a fossil in which the carbon now undergoes 2.92 decays per minute per gram? $k = 0.693/5730 = 1.21 \times 10^{-4}$; $\ln(2.92/15.3) = -(1.21 \times 10^{-4})t$; $t = 13\ 700\ years$ **L3**

Purpose

Students learn how art forgeries can be detected using radioactive dating of white lead pigment.

Teaching Strategies

- Show students photocopies of Keisch's reports of his work, and ask them to analyze and comment on the work. You can find these reports in issues of *Science,* March 1967, pp. 1238-1241, and April 1968, pp. 413-415. **L1**
- A good reference on van Meegeren and chemical detection methods used by Coremans is H. Koningsberger, *The World of Vermeer,* New York: Time Inc., 1967, pp. 174-185. You may wish to assign selected students to read and report on this work. **L1**

Extension

Have students do research on forgeries of written and printed works, such as literary manuscripts. **L1**

Connecting to Chemistry

1. He used radiographs to show earlier painted figures on the canvas, found modern-type cobalt blue crystals in paint that should have contained only lapis lazuli blue, and found five layers of white lead on the canvas—not the three that would be expected in a genuine Vermeer.

2. The Ra:Po ratio in A is 0.3:12, which equals 0.025. The Ra:Po ratio in B is 6:5, which equals 1.2. The high ratio in B compared to A shows that B is the old painting and A is the new one.

3. Answers will vary, depending on the particular areas researched.

3 ASSESS

Check for Understanding

- Check for understanding of the section material by having students answer the Section Review questions.
- Ask students to work in small groups to create analogies for half-life. Consult with each group and present the best analogies to the entire class. **L1** **COOP LEARN**

Reteach

Discussion: Ask students what difference, if any, exists between an alpha particle and an ordinary helium atom. *The alpha particle, which has two protons and two neutrons but no electrons, has a 2+ charge. An atom of helium is neutral.* The difference between a beta particle and an electron? *The beta particle is an electron that originates in the nucleus of the radioactive atom as a neutron and usually has considerable kinetic energy when it is emitted.*

Extension

Have students make models of alpha particles and of a nucleus that undergoes alpha decay. They can use the models to construct a display of a decay process. **L1** **LEP**

4 CLOSE

Writing About Chemistry

Ask students to write a response to the comment, "All radiation is harmful." *Responses should allude to the fact that although nuclear radiation is dangerous, the danger depends somewhat on the kind of radiation. Also, radiation can have useful applications.* **L1**

To date objects that are more than 60 000 years old, techniques involving other isotopes must be used. Rocks and minerals that are up to billions of years old can be dated on the basis of the decay of radioisotopes with long half-lives, such as potassium-40 ($t_{1/2}$ = 1.25 billion years), uranium-238 ($t_{1/2}$ = 4.5 billion years), and rubidium-87 ($t_{1/2}$ = 48 billion years). Such techniques are illustrated in **Figure 21.9.**

Figure 21.9

Using Radioactive Dating
Radioactive dating techniques were used to determine that the cyanobacteria that left imprints in these rocks found in a shallow bay in Australia lived about 3.5 billion years ago. ▼

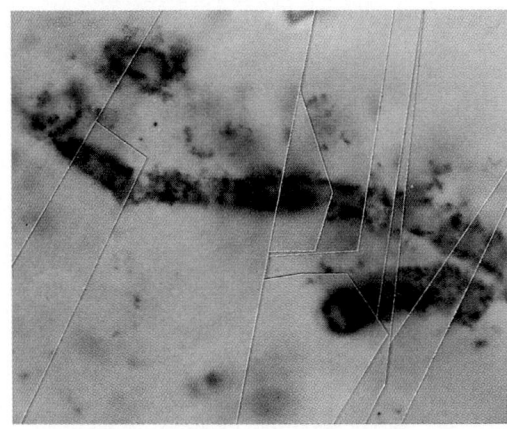

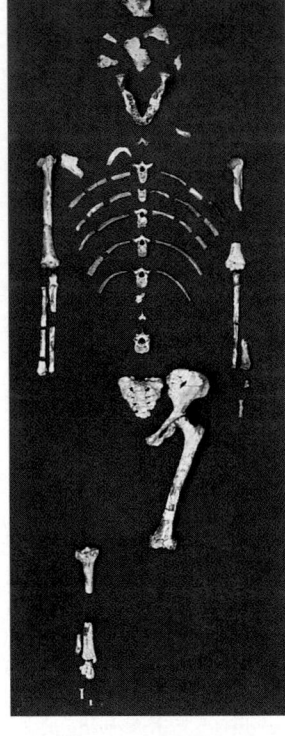

◀ Dating techniques established the age of a fossilized skeleton of *Australopithecus afarensis* found in Ethiopia to be 3.5 million years. The paleontologists who found it named this skeleton of an early humanlike primate "Lucy" after a song by the popular British singing group *The Beatles*. The scientists played the song around a campfire while celebrating their discovery.

SECTION REVIEW

Understanding Concepts

1. Explain what is meant by *radioactivity,* and provide an example of a radioactive isotope.

2. What is a Geiger counter used for? Describe its operation.

3. If half the atoms in a sample decay in one year, why is it incorrect to conclude that the other half will decay in one more year?

Thinking Critically

4. **Inferring** The oldest rocks dated so far on Earth are about 3.8 billion years old. Do you think that this fact was determined by means of carbon-14 dating? Explain.

Applying Chemistry

5. **Using Carbon Dating** Suppose you were given an ancient wooden box. If you analyze the box for carbon-14 activity and find that it is 50 percent of that of a new piece of wood of the same size, how old is the wood in the box?

SECTION REVIEW

1. Radioactivity is the spontaneous emission of radiation by an unstable atomic nucleus. Examples of radioactive isotopes include uranium-238 and carbon-14.
2. A Geiger counter is used to detect and measure radiation. The radiation forms in the handheld, gas-filled cylinder of the device. The ions complete a circuit, which causes a sounding device to be activated and a needle register on a meter to move.
3. Half-life is the time it takes for exactly half of any given quantity of a radioisotope to decay. In the second year, only half of the remaining nuclei will decay, not all of them.
4. It could not have been determined that way because carbon-14 dating can be used only for objects up to about 60 000 years old, given the relatively short half-life of carbon-14.
5. 1 half-life $\times$ 5730 years/half-life = 5730 years

Nuclear Reactions and Energy

B oth the United States and the former Soviet Union have sent nuclear reactors into space to power satellites. However, another giant nuclear reactor was there first—the sun, our local star. In the giant nuclear furnace of a star, nuclei react with one another and release enormous amounts of energy. You count on this energy to warm your air and water and to power your home, calculators, and other devices that run on solar power. Without this energy, life on Earth as you know it would not be possible. Is it possible to copy the reaction that takes place in stars and make a powerful generator that will provide electricity?

SECTION PREVIEW

Objectives
Compare and contrast nuclear fission and nuclear fusion.
Demonstrate equations that represent the changes that occur during radioactive decay.
Trace the operation and structure of a fission reactor.

Key Terms
nuclear fission
nuclear reactor
nuclear fusion
deuterium
tritium

The Power of the Nucleus

Compared to chemical reactions, nuclear reactions involve enormous energy changes. The energy change is due in part to the conversion of small amounts of the mass of nuclear particles into energy. Albert Einstein was the first scientist to realize the enormous amount of potential energy available in matter. He realized that mass and energy are equivalent and related by the following equation.

$$E = mc^2$$

In this equation, E represents energy, m is mass, and c is the speed of light $(3.00 \times 10^8$ m/s$)$. Because the value of the speed of light is so high, you can see that a small amount of mass can be converted into an enormous amount of energy. That explains why a nuclear weapon, such as an atomic bomb, is so much more powerful than a weapon that involves only chemical reactions, such as a bomb made with dynamite.

21.2   Nuclear Reactions and Energy   761

Program Resources

Study Guide, pp. 89-90 L1
Section Focus Transparency 45 and
Master, p. 45 L1 LEP
Basic Concepts Transparency 59 and
Master, pp. 117-118 L1 LEP
ChemLab and MiniLab Worksheets,
p. 137 L1

PREPARE

Key Concepts
The energy that accompanies nuclear reactions is described. Then, nuclear fission and fission reactors and nuclear fusion and fusion reactors are examined.

Planning Ahead
For the MiniLab Arrange to purchase, make, or borrow many sets of domino tiles for the MiniLab.
For the Demonstrations If you have no Geiger counter for Demonstration 21.4, ask the physics staff whether some of their apparatus may be assembled to operate as one. Alternatively, contact the local section of the American Nuclear Society or Civil Defense authority about the possibility of borrowing one.

1 FOCUS

Focus Transparency
Display the **Section Focus Transparency** for Section 21.2 to show some sources of nuclear energy. L1 LEP

2 TEACH

Content Background
Mass Defect The mass of an atomic nucleus is less than the mass of the constituent particles by an amount known as the mass defect. The difference for the oxygen-16 atom, for example, is 2.269×10^{-25} g/nucleus or 0.1366 g/mol of nuclei. The energy equivalent, which may be calculated using $E = mc^2$, is called binding energy and is a measure of the stability of the nucleus. For the stable O-16 nucleus, it is 1.23×10^{13} J/mol.

761

✓ Assessment

Knowledge: Ask students to look up the meaning of the word *fission* in a dictionary and to list its specific meaning in as many contexts as possible. [L1] [LEP]

Concept Development

Explain that more than 200 radio-isotopes, having half-lives up to millions of years, have been produced by fission reactions, making it difficult to dispose of waste from nuclear reactors.

Concept Development

Explain that in order for a fission reaction to be self-sustaining, at least one neutron from each fission event, on average, must go on to split another nucleus. If less than one neutron per fission event causes another fission event, the quantity is subcritical. If exactly one neutron per event causes another event, the quantity is critical. If more than one neutron per event causes another event, as in an atomic bomb, the quantity is supercritical.

Content Background

U-235 Uranium-235 is much more fissionable than uranium-238, which makes up most naturally occurring uranium. Therefore, it is desirable to concentrate the uranium-235. This uranium enrichment is often carried out by converting the naturally occurring uranium to the gas UF_6. The $^{235}UF_6$ diffuses more rapidly than the heavier $^{238}UF_6$, a fact that allows for separation of the two isotopes.

WORD ORIGIN

fission
findere (L) to split
When a nucleus undergoes fission, it is split into smaller pieces.

Nuclear Fission

Until 1938, all nuclear reactions observed involved the movement of radiation, such as alpha or beta particles, into or out of a decaying nucleus. Shortly before World War II, Otto Hahn, a German chemist, discovered that bombarding uranium with neutrons resulted in the formation of an isotope of barium that is only about 60 percent of the mass of the uranium isotope. This result was so unusual that Hahn asked another physicist, Lise Meitner, shown in **Figure 21.10,** to help him interpret his findings. She determined that the bombardment of uranium with neutrons had resulted in the splitting of the nucleus into two pieces. She called this transformation nuclear fission. The word *fission* means "splitting." When **nuclear fission** takes place, an atomic nucleus is split into two or more large fragments.

The Fission Process

Why did bombarding uranium atoms with neutrons result in fission? Remember that changing the ratio of neutrons to protons can make a nucleus less stable. Adding a neutron to a uranium-235 nucleus causes the nucleus to split. Atoms of two different elements are created along with more neutrons. Here is a typical reaction for uranium fission.

$$^{235}_{92}U + ^{1}_{0}n \rightarrow ^{140}_{56}Ba + ^{93}_{36}Kr + 3^{1}_{0}n$$

Notice that the sums of the mass numbers on the left and right sides are equal, and so are the sums of the atomic numbers. Therefore, the nuclear equation is balanced.

As World War II started, the race was on to find a way to sustain fission in a chain reaction. A chain reaction is a continuing series of reactions in which each produces a product that can react again. Each neutron produced during the fission of one atom of uranium has the potential to cause the fission of another nucleus. Notice that each fission reaction produces three neutrons—more than enough to keep the process going. The fission reactions can continue in a chain, as shown in **Figure 21.11.** For a chain reaction to take place, there must be enough fissionable material present for some of the neutrons produced to be able to collide with other nuclei. However, if the chain reaction takes place too quickly, an explosion will result and release an enormous amount of energy all at once. This is what happens when an atomic bomb explodes. If the rate of the fission chain can be controlled to release energy slowly, the energy can be used to heat materials such as water and then do useful work.

Figure 21.10
Lise Meitner
Lise Meitner was a physicist who was born in Austria but had to flee after the Nazis gained control of her country. She first described nuclear fission. Fission can release amounts of energy that are far greater than those released during radioactive decay.

Across the Curriculum

Music

The Polish composer Krzysztof Penderecki attempted to commemorate the agony of nuclear war in his 1960 composition *Threnody in Memory of the Victims of Hiroshima.* The horrors of one of the bombings that marked the end of World War II are expressed in the composition by violent torrents of dissonant, percussive sound, some produced by the beating of the bodies of the 52 stringed instruments for which the piece was scored. A 10-minute song of lamentation for the dead begins with a long, screaming tone produced by playing the highest pitches possible on violins.

A Nuclear Fission Chain Reaction

Nuclear fission provides the energy in nuclear power plants. During the fission process, a heavy atomic nucleus generally absorbs a neutron, then splits into two smaller atomic nuclei and several more neutrons. If sufficient fissionable material, known as the critical mass, is present, the neutrons produced from one fission initiate several more fissions, and a chain reaction is sustained. Your group will simulate a nuclear chain reaction with dominoes.

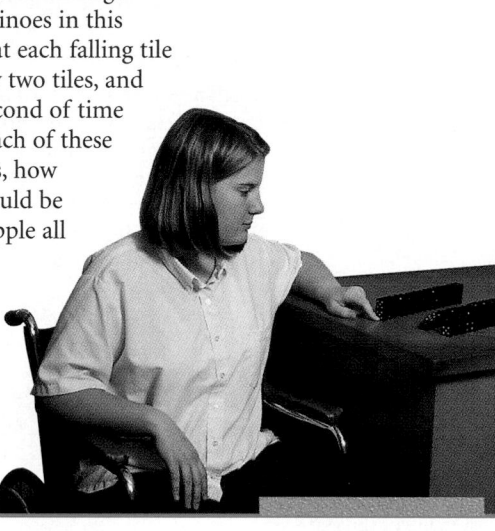

Procedure
1. Obtain two sets of domino tiles.
2. Set the individual tiles on end in patterns such that if you first topple a single tile, it will strike and topple two more tiles, each of which will strike and topple two others, and so on.
3. Practice with possible arrangements until you decide on the best one; that is, one that will topple all the tiles in the quickest, most efficient way. When you are ready, call the teacher to observe your domino chain reaction. You may perform a second trial if your first attempt is unsuccessful. The teacher will compare your best result with those of the class.

Analysis
1. How many tiles were involved in your best chain reaction? In the best of the entire class?
2. If you were able to arrange 1 million dominoes in this manner, so that each falling tile topples exactly two tiles, and assuming 1 second of time required for each of these double topples, how much time would be required to topple all the tiles?
3. Explain how this domino exercise simulates a nuclear fission chain reaction.

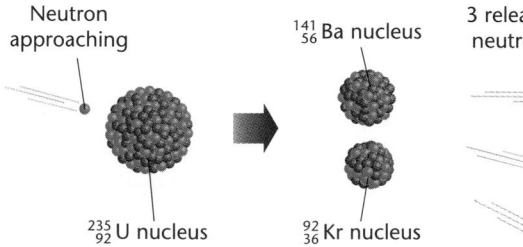

Figure 21.11
A Fission Chain Reaction
During a chain reaction, each step produces a reactant for the next step. In the fission of uranium, a neutron enters a uranium-235 nucleus. As a result, the nucleus breaks into two smaller nuclei. At the same time, it releases more neutrons, which are absorbed by other uranium-235 nuclei, and the reaction continues.

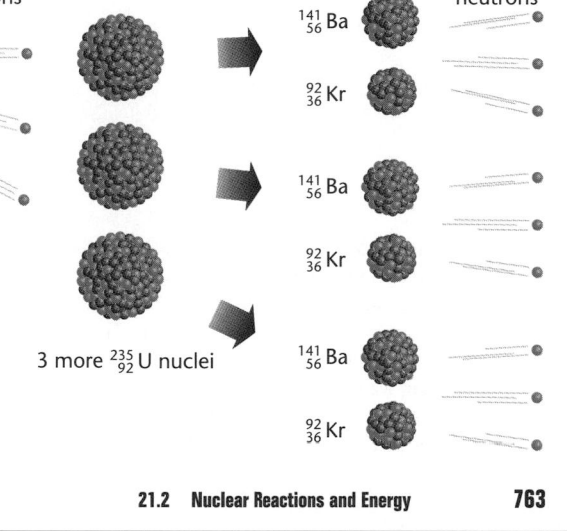

21.2   Nuclear Reactions and Energy   763

---

miniLAB
1

miniLAB

A Nuclear Fission Chain Reaction

Purpose: To simulate a nuclear chain reaction by using dominoes.

Process Skills: Observing, using numbers, formulating models, communicating

Teaching Strategies
- Emphasize the underlying concept of the MiniLab as well as the competitive aspect.
- Consider giving a reward or prize for the best result.

Expected Results
Students are likely to involve up to 32 tiles.

Analysis
1. Students are likely to involve up to 32 tiles.
2. It would take 19 seconds to involve a total of $2^{20} - 1 = 1\ 048\ 575$ tiles.
3. In a nuclear fission chain reaction, each fission releases neutrons that strike and initiate additional fissions, resulting in a cascade effect. Each individual domino tile was able to strike and topple two additional tiles, also resulting in a sort of chain reaction and cascade effect.

✓ Assessment

Knowledge: Ask students what would happen to the chain reaction if certain tiles were unable to contact other tiles and knock them over. *The overall chain reaction would be slowed or stopped.* Explain that control rods perform an analogous function in nuclear fission reactors. L1

Correcting Misconceptions

Students may think that fuel rods are made mostly of U-235. Explain that only about 3 percent of the uranium in the fuel rods or pellets is U-235—enough to sustain a chain reaction but far less than that required for a nuclear explosion. Point out that the other uranium present is nonfissionable U-238.

Correcting Misconceptions

Students may be confused about the role of water in nuclear reactors and power plants. Point out that although water does function as a coolant and as a heat-transfer medium, it also absorbs neutrons and slows fission reactions in some reactors. Explain that it has a high ability to capture neutrons.

GLENCOE TECHNOLOGY

 Videodisc

Chemistry: Concepts and Applications
Detecting Radiation
Disc 3, Side 1, Ch. 5

Use this video of the demonstration on these pages to show how some materials block radiation. L1 LEP

 CD-Rom

Chemistry: Concepts and Applications
Detecting Radiation
Use this video of the demonstration on these pages to show the effectiveness of shielding materials. L1 LEP

Fact of the MATTER

Operation Trinity was the name of the effort that led to the first atomic bomb explosion. The explosion took place in the desert on the Alamogordo Air Base, 120 miles southeast of Albuquerque, New Mexico, on July 16, 1945.

Fission Reactors

The first major step in developing usable, controllable nuclear power took place on December 2, 1942, in a squash court underneath an unused football field at the University of Chicago. It was there that Enrico Fermi, an Italian physicist, successfully carried out a sustained and controlled fission chain reaction.

Today, through a similar procedure, electrical energy is generated in nuclear power plants through the controlled fission of uranium. The device that is used to extract energy from a radioactive fuel is called a **nuclear reactor.** In the most common type of reactor used today, the fission of a sample of uranium enriched in the isotope uranium-235 releases energy that is used to generate electricity. It is, therefore, desirable to concentrate the uranium-235. Refer to the diagram of the nuclear reactor shown in **Figure 21.12** to learn how a fission reactor works.

Figure 21.12

Operation of a Fission Nuclear Power Plant

1. Uranium-235 in fuel rods produces fast-moving neutrons and heat in a fission chain reaction. The neutrons are slowed down by a moderator such as water or graphite so that they are not moving too quickly to be absorbed by other uranium-235 nuclei. The rate of the reaction is maintained using control rods that absorb some of the neutrons. These rods can be raised or lowered in the reaction chamber to slow or speed the reaction, respectively.

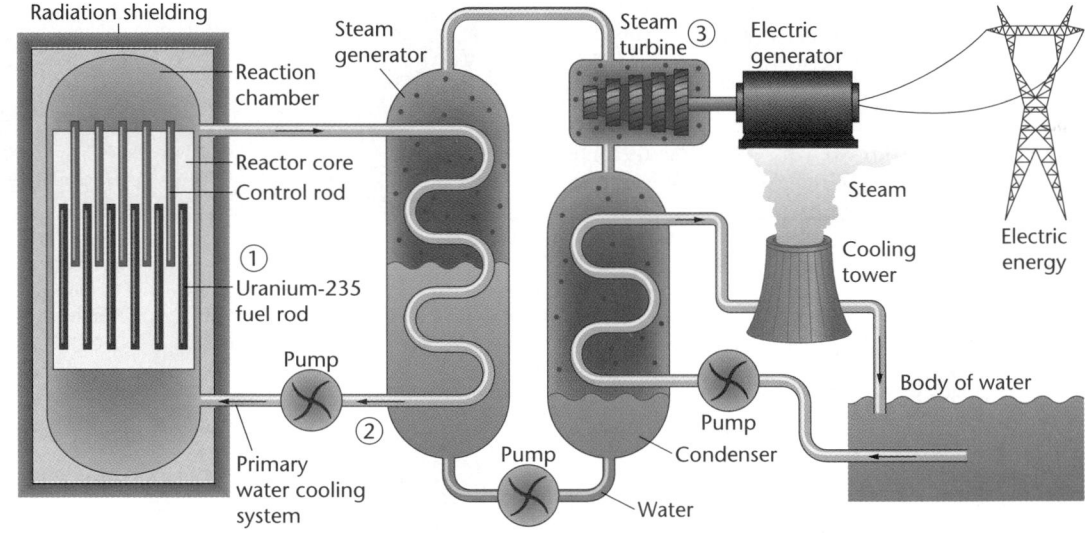

2. The neutrons then enter other fuel rods, where they initiate more fission reactions. The energy produced during these reactions heats water that is pumped among the fuel rods. The water functions as a coolant that keeps the rods from melting and also as a medium to transfer heat from the reaction chamber to a separate basin of water. That water is converted to steam.

3. The steam turns a turbine, which rotates the shaft of an electric generator. When the shaft of the generator turns, electricity is produced. The steam is then condensed back into water, using still other water from an outside source as a coolant. The newly condensed water is cycled back into the process.

Demonstration 21.4

Detecting Radiation

Purpose: To demonstrate shielding materials and their effectiveness, and to demonstrate the inverse square law.

Materials: Geiger counter with meter; packaged radioactive source; small pieces of cloth, aluminum foil, lead foil, iron sheeting, glass plates, roofing shingles, paper, and wood; meterstick; tongs

Safety Precautions: 🥽

Disposal: Code F

Procedure: Mount the Geiger tube about 5 cm from the packaged radioactive source. Record the meter reading. Place pieces of the shielding material (cloth, foil, etc.) being tested between the source and the Geiger tube, and take readings. You may want to graph the data. You can also have the students bring in materials that they want tested. Use tongs to move the radioactive source to different positions along a

More than 100 nuclear power plants are now in operation in the United States, providing almost 20 percent of all the electricity used. Thirty-three states have nuclear plants. Some of these states, such as Vermont, get most of their electricity from nuclear power. Many more plants are operating in other countries. The country that gets the highest percentage of its electrical power from nuclear reactors is France. There, more than 70 percent of the electricity produced comes from nuclear fission reactors.

Plutonium can also be used to power a nuclear fission reactor. Plutonium-239 undergoes a chain reaction that eventually produces more plutonium-239 from uranium-238, as well as heat that is used to generate electricity. This type of fission reactor, which produces fissionable material as it operates, is called a breeder reactor. A breeder reactor can actually produce more fissionable material than it uses. The construction of this type of reactor has been discouraged in many countries because of the health hazard that plutonium presents, as well as the fact that the plutonium-239 generated can be used to make powerful fission bombs.

Unlike burning fossil fuels, nuclear reactions do not produce pollutants such as carbon dioxide and acidic sulfur and nitrogen compounds. However, the nuclear reactions do form highly radioactive waste that is hard to dispose of safely. Other serious problems include the potential release of radioactive materials into the environment when fires or explosions take place, and also the limited supply of fissionable fuel and the higher cost of producing electricity using nuclear fuels rather than fossil fuels. Nuclear reactors that have experienced serious accidents are shown in **Figure 21.13**.

Figure 21.13

Nuclear Accident Sites
The gas that you see coming from the towers of the Three Mile Island nuclear power plant in Pennsylvania is all water vapor. Few chemical pollutants are released during the normal operation of a nuclear plant. Both equipment failure and human error resulted in overheating of the reaction chamber and a partial meltdown of fuel rods at this power plant in 1979. As a result, the building surrounding the reactor became flooded with water contaminated with radioactive material, and radioactive gas was released into the atmosphere. ▶

◀ Both human and mechanical errors led to overheating of the reaction chamber at the Chernobyl nuclear plant in the former Soviet Union in 1986. Water used to cool the chamber decomposed into hydrogen and oxygen gas, which exploded and blew the roof off the building that housed the reactor. A large amount of radioactive debris was released and traveled as far away as Scandinavia and England. Even now, almost 1000 square miles around the plant are considered too radioactive for permanent habitation.

21.2 Nuclear Reactions and Energy **765**

SECTION 21.2

Discussion
Ask students why there is such great resistance to nuclear power plants given the following facts. First, the environmental and health impacts of nuclear power plants in normal operation are minimal when compared with coal-fired power plants, which are probably responsible for the deaths of thousands of people per year due to air pollution. Second, a 1000-megawatt nuclear reactor uses 25 tons of fuel per year and produces 25 tons of radioactive waste. An equivalent coal-fired plant uses 2.5 million tons of fuel per year and produces 6.5 million tons of CO_2, 9000 tons of SO_2 (which contributes to acid rain), 4500 tons of NO_x (which also contributes to acid rain), and large amounts of ash. *Nuclear plants are exceedingly expensive to build. They produce radioactive waste that may cause long-term health problems, lasts virtually forever, and requires disposal that is difficult and controversial. The potential for a nuclear accident is small; however, such an accident could cause catastrophic loss of life and property damage.*

Visual Learning

Figure 21.13: Ask interested students to research and make detailed classroom presentations on the nuclear accidents at Three Mile Island and Chernobyl. L1

meterstick. Record the Geiger counter's meter reading as the distance between the tube and the source increases.
Results: The relative effectiveness of equal thicknesses of shielding materials (paper = 1) is: masonry, 16; iron, 50; and lead, 100. Other figures will vary with the type of material. As the distance between the radioactive source and the Geiger tube is doubled, the amount of radiation re-

ceived is reduced to one-fourth ($E = 1/d^2$).
Analysis: Ask these questions.
1. Which material provides the best shielding? *lead*
2. What happens to the radiation level as the distance from the source increases? *It falls rapidly.*

✔ Assessment

Performance: Have students simulate the inverse square relationship by varying the distance of a thermometer from a heat lamp and recording the thermometer's reading. The bulb of the thermometer should be wrapped with black paper. L1
LEP

GLENCOE TECHNOLOGY

Videodisc

Chemistry: Concepts and Applications
Inside a Nuclear Power Plant
Disc 3, Side 1, Ch. 7

Show this video to illustrate the design of a nuclear power plant. L1 LEP

CD-Rom

Chemistry: Concepts and Applications
Inside a Nuclear Power Plant
Show this video to illustrate the design of a nuclear power plant. L1 LEP

Transparency

Display **Basic Concepts Transparency 59** to compare fission and fusion. L1 LEP

3 ASSESS

Check for Understanding

- Check for understanding of the section material by having students answer the Section Review questions.
- Describe a hypothetical situation in which a power company proposes to build a nuclear power plant near your home. Assign each student to one of the following: town council, power-company board of directors, chamber of commerce, taxpayer association, and environmental-concerns committee. Ask the town council to prepare a format for a town meeting to discuss the issue. Give each of the other groups time to research the issues involved in the construction and operation of the power plant and to prepare a position report. Ask

Nuclear Fusion

Energy is also released in a type of nuclear reaction that is the opposite of fission. During **nuclear fusion,** two or more nuclei combine to form a larger nucleus. Fusion is the nuclear process that produces energy in stars like our sun. In the most typical fusion reaction in the sun, hydrogen nuclei fuse to form helium. The enormous amount of energy that is generated sustains all life on Earth.

The Fusion Process

Some of the fusion processes in which hydrogen forms helium have been carried out and studied in laboratories. In one common fusion reaction, two different isotopes of hydrogen combine to form helium and a neutron, as shown in **Figure 21.14.**

$$_{1}^{2}\text{H} + _{1}^{3}\text{H} \rightarrow _{2}^{4}\text{He} + _{0}^{1}n$$

Notice that the nuclear equation is balanced. The isotope of hydrogen with a mass number of 2 that is one of the reactants in the equation above is called **deuterium** (D). The hydrogen isotope with a mass number of 3 is called **tritium** (T). Nuclei such as these, which have a small mass, can combine to produce heavier, more stable nuclei. The helium produced has a stable neutron:proton ratio of 1:1. Two deuterium nuclei can also fuse to form either tritium or helium, as shown in the following two equations.

$$_{1}^{2}\text{H} + _{1}^{2}\text{H} \rightarrow _{1}^{3}\text{H} + _{1}^{1}p$$
$$_{1}^{2}\text{H} + _{1}^{2}\text{H} \rightarrow _{2}^{3}\text{He} + _{0}^{1}n$$

In the first case, tritium and a proton are produced; in the second, $_{2}^{3}\text{He}$ and a neutron are produced. The fusion of hydrogen produces 20 times the energy produced by the fission of an equal mass of uranium. That large quantity of energy makes the idea of fusion reactors an appealing one. Also, deuterium, the required fuel in fusion, is abundant on Earth. Energy production through nuclear fusion has some other potential advantages over production through fission. No radioactive products are produced, so waste disposal would not be as difficult. A fusion reaction is also easier to control, which would help prevent fires and explosions.

However, there is one major problem with creating a fusion reactor. Compared to uranium nuclei, hydrogen nuclei have much less tendency to react. A large initial input of energy must be provided to start the fusion. In the sun, enormous pressures and temperatures trigger fusion. To initiate a fusion reaction on Earth, a temperature of 200 million kelvins would be required. Any material used to contain the reaction

Figure 21.14
An Example of Nuclear Fusion
When a nucleus of deuterium (hydrogen-2) and a nucleus of tritium (hydrogen-3) undergo nuclear fusion, a nucleus of helium-4 and a neutron are produced.

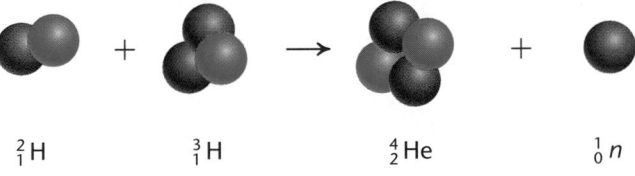

$$_{1}^{2}\text{H} \qquad _{1}^{3}\text{H} \qquad _{2}^{4}\text{He} \qquad _{0}^{1}n$$

Chemistry Journal

Energy Conservation
Point out that society's energy demands might be significantly reduced if more people exercised conservation measures. Ask students to keep a log of energy-consuming activities they perform during a typical day. Have them suggest ways in which they might conserve energy. They should include the log and suggestions in their journals. L1

Chemistry Journal

Integral Fast Reactor
Ask a student to research and present a class report on the new Integral Fast Reactor being investigated at Argonne National Laboratory. This reactor uses liquid sodium rather than water as a coolant, breeds more fuel, uses fuel efficiently, and produces less waste. Class members can take notes in their journals during the report. L3

would melt at that temperature. The difficulty in initiating and containing nuclear fusion has prevented its use as a practical energy source.

Scientists in a number of countries are researching ways to make commercially feasible fusion reactors. One promising reactor type is called a tokamak, which is shown in **Figure 21.15.** In this reactor, hydrogen nuclei are trapped by powerful magnetic fields produced by huge electromagnets. The nuclei are heated with radio waves to initiate fusion. Because a magnetic field confines the nuclei, no other container is needed. Such a reactor has been operated at the break-even point, where it produces as much energy as is required to run it, but because it produces no net energy, it is not useful. Better electromagnets and structural materials must be developed to correct the engineering problems encountered in the tokamak.

Figure 21.15

A Tokamak Fusion Reactor
The walls inside the vacuum chamber of the experimental DIII tokamak fusion reactor in San Diego are covered with carbon tiles. They can withstand the radiation and particles produced by the hydrogen undergoing fusion.

the town council to conduct the town meeting in which each of the groups presents their position. Then ask the town council to vote on the proposal and to justify their vote. L1 **COOP LEARN**

Reteach

Discussion: Ask how both of the following statements can be true. (1) Nuclear fusion has not been used as an energy source on Earth. (2) Nuclear fusion is the primary energy source for Earth. *Statement 1 is true because, although an attempt is being made to develop fusion power reactors, only the break-even point has been reached. Statement 2 is true because the fusion reactions in the sun provide essentially all of Earth's energy, either directly or indirectly.*

Extension

Ask interested students to research and make classroom reports on the latest information about so-called "cold fusion" experiments. L1

4 CLOSE

Extension

Ask students to research and prepare reports on stellar nucleosynthesis, the process in which elements are produced in stars. L2

SECTION REVIEW

Understanding Concepts

1. What is the difference between nuclear fission and nuclear fusion? List one current use for fission.

2. How is fission controlled and maintained in a nuclear reactor?

3. What advantages would a fusion reactor have over the fission-based reactors currently in use?

Thinking Critically

4. **Comparing and Contrasting** What are the advantages and disadvantages of using energy generated by nuclear fission reactors instead of energy from fossil fuels such as petroleum, natural gas, and coal?

Applying Chemistry

5. **Stellar Fusion** The sun's energy source was a mystery to people for a long time. One 19th-century astronomer calculated that if the sun were made entirely of coal, it would burn for only about 10 000 years. Today, it is known that the sun's fuel is hydrogen, consumed in fusion reactions in which helium is formed. After the sun has used up nearly all of its hydrogen, it is expected to begin a fusion reaction in which helium-4 nuclei combine to form carbon-12. Write a nuclear equation for this process, and verify that it is balanced.

21.2   **Nuclear Reactions and Energy**   **767**

SECTION REVIEW

1. Nuclear fission is the splitting of an atomic nucleus into two or more large fragments. Nuclear fusion is the combining of two or more small nuclei to form a single large nucleus. Fission is used in nuclear reactors.

2. Uranium-235 in fuel rods is used to produce neutrons that are absorbed by other uranium-235 nuclei, which then split and release more neutrons. The rate of the process is controlled by rods that absorb neutrons. At the proper rate, the reaction will sustain itself.

3. A fusion reactor would use readily available hydrogen fuel rather than the more limited supplies of uranium or plutonium. It would also produce much less radioactive waste and would be easier to control.

4. Nuclear power does not create the large amounts of pollutants that fossil fuels do. However, it creates hazardous radioactive waste that is hard to transport and store safely. In addition, nuclear power is expensive.

5. $3^4_2\text{He} \rightarrow {}^{12}_6\text{C}$. The total mass numbers and the atomic numbers are equal on the left and right, so the equation is balanced.

SECTION 21.3

Nuclear Tools

PREPARE

Key Concepts

Several applications of radiation and radioisotopes are investigated, and the production of radioisotopes in accelerators and particle generators is discussed. Natural sources of radioactivity, radiation exposure and health effects, and disposal of radioactive wastes are also examined.

Planning Ahead

For the MiniLab Initiate contacts with manufacturers of commercial radon-testing kits or local retailers to obtain the kits for a reduced cost for the Mini-Lab.

1 FOCUS

Focus Transparency

Display the **Section Focus Transparency** for Section 21.3 to show some uses of radioisotopes. L1 LEP

2 TEACH

Concept Development

Acquire or borrow a copy of the *Handbook of Radioactive Nuclides*, published by the Chemical Rubber Company (CRC). This handbook contains two sections (VI: "Radionuclides for Medical Applications," and VII: "Radionuclides for Industrial Applications") that are especially useful.

SECTION PREVIEW

Objectives

Distinguish the biological effects of radiation and the units used to measure levels of exposure.

Illustrate medical and nonmedical uses for radioactivity.

Key Terms

gray
sievert

Medical practice, as envisioned in many science fiction works, will differ from what it is today. Surgery, so common today, will supposedly become obsolete, replaced by non-invasive techniques. Only time will tell whether such changes take place. However, thanks to nuclear chemistry, the first steps in the new direction are already being taken. Although traditional surgery is still necessary, diagnosis has become less invasive. As you will see, the inside of the human body can even be explored in detail without the use of a scalpel.

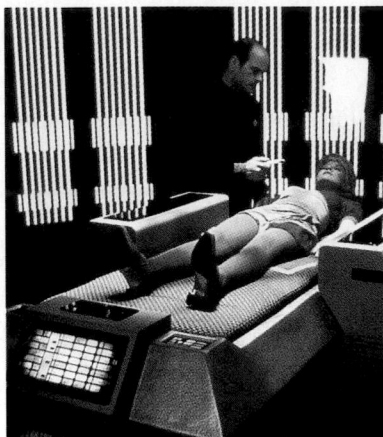

Sick bay on Star Trek's Voyager

Medical Uses of Radioisotopes

You are much more likely to be saved by some form of nuclear medicine than to be killed by the effects of radiation. Radioisotopes are widely used today in diagnosis to generate images of organs and glands, and in treatments for conditions such as cancer. Radioisotopes are also used as tracers to find out where certain chemicals move in the body and to identify abnormalities, as shown in **Figure 21.16**.

Figure 21.16

Use of Tracers in Medicine

A radioactive isotope of any element will undergo the same chemical reactions as a stable isotope. For example, if a molecule of glucose that contains a radioactive atom is injected into a patient, the radiolabeled glucose will be metabolized in a cell in the same way as nonradioactive glucose. A detector set to pick up only the type of radiation given off by the radioisotope in the glucose can be used to trace the molecule—that is, to tell where it moves or concentrates in the body. Because cancer cells take in and use fuels at a much higher rate than do more slowly dividing, normal cells, the radioisotope can be used to locate and identify a tumor.

Unlabeled and radiolabeled glucose being taken up by the tumor

Radiolabeled glucose

Cancerous tumor

Detection device

Program Resources

Study Guide, pp. 91-92 L1
Section Focus Transparency 46 and **Master,** p. 46 L1 LEP
Consumer Chemistry, pp. 41-42 L1
Basic Concepts Transparency 60 and **Master,** pp. 119-120 L1 LEP
Laboratory Manual, 21.2, pp. 205-208 L1

Medical Uses of Radioisotopes in Diagnosis and Treatment

Radioisotopes have become extremely useful tools in medical diagnosis and treatment. Using them allows doctors to detect many diseases early and to treat them more successfully than ever before.

A Positron emission tomography (PET) scans make use of a short-lived positron-emitting radioisotope. A positron is a particle that has the mass of an electron but that is positively charged. When a positron collides with an electron, two gamma rays that move apart at a 180° angle are produced. In a PET scan, rings of highly sensitive detectors surround the patient in the cylindrical enclosure and allow measurement of the distribution of radioactivity in the body. With the aid of a computer, a PET scan can produce a series of two-dimensional imaging slices through an organ. Different radioisotope-containing compounds can be used to measure rates of various processes in the body, such as glucose metabolism or blood flow. The patient shown here is undergoing a PET scan of the brain.

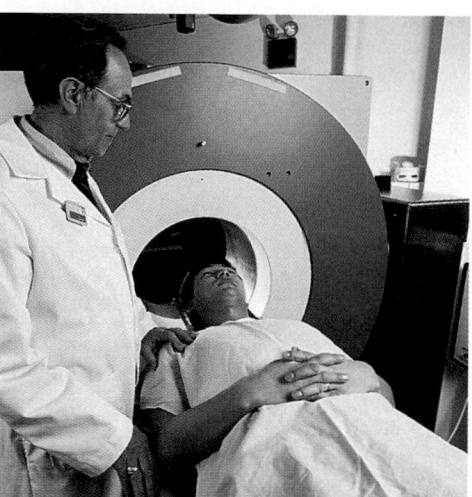

B Most of the iodine that enters your body ends up in the thyroid gland, where it is incorporated into hormones that regulate your growth and metabolism. If you ingest radioactive iodine-131, an image showing the size, shape, and activity of your thyroid gland can be obtained. The image is useful in diagnosing metabolic problems, such as hyperthyroidism. The photo shows a normal thyroid gland.

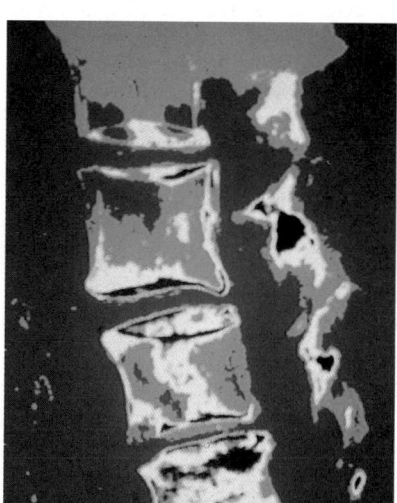

C Gadolinium-153 is widely used in medicine to detect osteoporosis, the reduction in bone mass that often accompanies aging. Bone density is determined by a scanning device that compares how much of the X rays and gamma rays produced by the decay of gadolinium-153 is absorbed by a bone. The scans shown compare a dense bone with one that has lost a great deal of mineral matter.

21.3 **Nuclear Tools** 769

Correcting Misconceptions

Students may think that only radioactive atoms are useful as tracers. Point out that some stable isotopes, such as oxygen-18, can be used and detected with a mass spectrometer.

Extension

TECH PREP

Ask an interested student to arrange a visit with a radiation therapist at a local clinic or hospital and to prepare a report for the class. Alternatively, you may wish to contact a radiation therapist and ask him or her to make a classroom presentation. The presentation might be videotaped for future use or use by other classes. L1

GLENCOE TECHNOLOGY

 Videodisc

Chemistry: Concepts and Applications
Nuclear Medicine
Disc 3, Side 1, Ch. 8

Show this video to illustrate the use of a PET scan in diagnosis. L1 LEP

 CD-Rom

Chemistry: Concepts and Applications
Nuclear Medicine
Show this video to illustrate the use of a PET scan in diagnosis. L1 LEP

SECTION 21.3

Concept Development

Explain that Tc-99m concentrates in areas in the body in which cell growth is most rapid. It is often used to locate tumors in the brain, thyroid gland, and kidneys.

Discussion

Ask students whether they would want to be subjected to radioisotopes for medical diagnosis or treatment under certain circumstances. Steer the discussion toward benefit-risk assessment.

Using an Analogy

Ask students what forms of travel they often use. *Automobile, foot (walking), bicycle, airplane, and bus are likely answers.* Ask whether any of these forms are without risk. Explain that if the benefit is perceived as outweighing the risk, the mode of travel is seen as justified. Point out that, analogously, one might consider allowing radioisotopes to be used to diagnose or treat a medical problem, particularly a serious problem.

Discussion

Ask students why radioisotopes that have short half-lives are preferred as medical tracers. *A radioisotope that has a short half-life is easily detectable and therefore usable in small amounts. It decays quickly to a virtually harmless level of radioactivity.*

D Technetium-99m is a metastable isotope that gives off gamma rays to become a more stable version of the same isotope, with no change in either atomic or mass number. Technetium-99m is one of the most commonly used isotopes in medicine because it produces no alpha or beta particles that could cause unnecessary damage to cells and because it has a half-life of only about six hours. It can be incorporated into different compounds for imaging bones or blood-flow patterns in the heart. The image on the right shows a damaged heart; the dark regions are not getting enough blood flow.

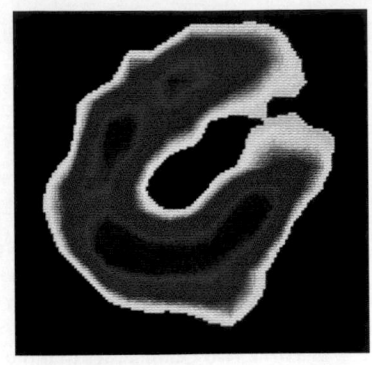

Fact of the MATTER

The first radioactive isotope used in medicine was phosphorus-32, which was administered in 1936 to a woman who had chronic leukemia. The phosphorus-32 was artificially produced in a machine called a cyclotron, in which elements become radioactive after they are bombarded with neutrons and deuterium nuclei.

E Scintigraphy is a bone-scanning technique used to search for stress fractures in racehorses. A radioisotope is injected into bone tissue, and the bone is scanned using a gamma-ray detector.

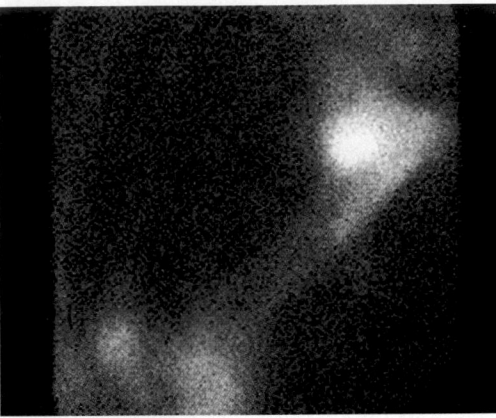

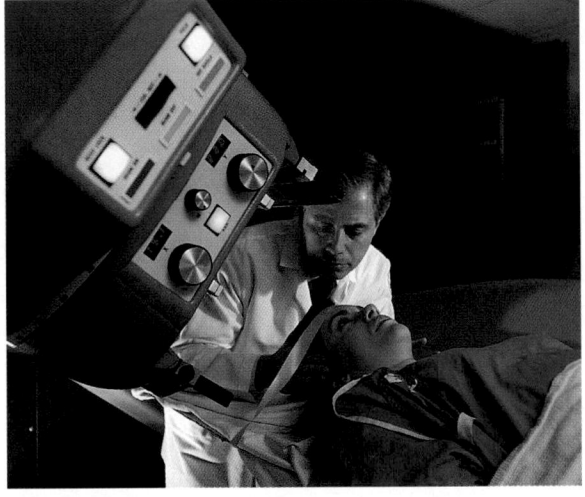

F The goal of radiation therapy is to destroy cancerous cells while minimizing damage to normal cells. Radiation produced by cobalt-60 is carefully aimed at tumor cells, which are more sensitive to the damaging effects of the radiation on DNA because they divide more rapidly than do normal cells. The patient shown in the photo is receiving radiation therapy for cancer. To minimize destruction of healthy cells, the cobalt source is moved around the outside of the patient's body in a circle, with the beam of rays sharply focused on the tumor at all times. This allows any given area of normal tissue to receive only a small dose of radiation while the tumor gets a large total dose.

770   Chapter 21   Nuclear Chemistry

Chemistry Journal

Benefit/Risk Assessment

Ask students to make a benefit/risk assessment for some of their major activities and travels during a typical day and to make a judgment as to whether the benefit seems to outweigh the risk in each instance. Ask them to include these assessments in their journals. L1

Nonmedical Uses of Radioisotopes

Hospitals are not the only locations where you'll find radioisotopes in use. You have already read about several nonmedical uses for radioactivity: nuclear reactors, nuclear weapons, and radiodating techniques. As you will see, radioisotopes are also used in research and in the food industry.

Practical Uses of Tracers

If a radioisotope is substituted for a nonradioactive isotope of the same element in a chemical reaction, all compounds formed from that element in a series of steps will also be radioactive. That makes it possible to follow the reaction pathway using instruments that can detect radiation. In this way, the series of steps involved in many important reactions has been studied, as shown in **Figure 21.17**. Tracers containing radioactive phosphorus-32 have also been used in biochemical research to help clarify complicated metabolic pathways. Tracers are also used to test structural weaknesses in mechanical equipment and to follow the pathways taken by pollutants.

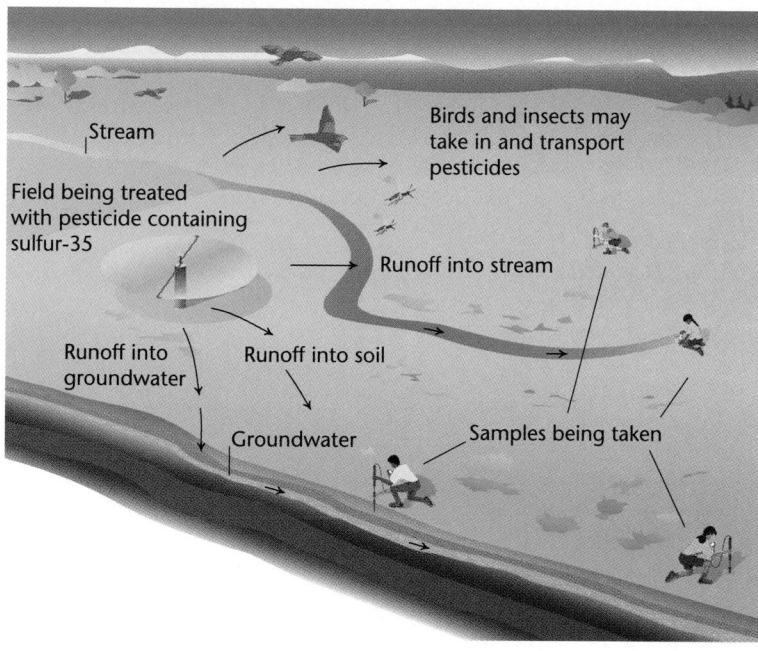

Figure 21.17

Using a Tracer to Study Pesticide Movement
A pesticide that is sprayed onto a field may be transported to other areas by runoff into soil, streams, and groundwater or by movement of animals that have taken in the pesticide. If a pesticide is "labeled" with a radioisotope, such as sulfur-35, its movement can be traced. Samples of soil or water can be taken at a number of points, and the amount of radiolabeled pesticide at each point can be measured.

Food Irradiation

Gamma radiation disrupts metabolism in cells, sometimes enough to kill them or at least keep them from multiplying. This property makes it useful for sterilization of food and surgical instruments. Exposing food to gamma radiation produced by the decay of cobalt-60 nuclei can keep the food from spoiling. The radiation destroys microorganisms and larger organisms such as insects. The food itself does not become radioactive.

21.3   Nuclear Tools   **771**

✔Assessment

Portfolio: Ask students to design an experiment using a radioactive tracer to evaluate the amount of wear of steel auto engine cylinders using different types of lubricating oils. They should consider placing their work in their portfolios. *Construct the steel cylinder using a percentage of a radioactive isotope of iron or some other constituent metal such as chromium. Run the engine for a period of time with particular lubricating oils, then measure the amount of the radioisotope in the oils as a measure of the wear.* [L2] [P]

Extension

Ask interested students to conduct library research and make a body chart showing the locations in which naturally occurring and artificial radioisotopes tend to enter or accumulate in the human body. *The chart might show the following. Lungs: radon-222, uranium-233, plutonium-239, krypton-85; thyroid: iodine-131; liver: cobalt-60; muscles: potassium-40, cesium-137; bones: radium-226, strontium-90, phosphorus-32, carbon-14. Explain that all living things contain some radioisotopes.* [L2]

Teaching Strategies

- Using Question 2 of *Connecting to Chemistry*, help students differentiate between results and conclusions by pointing out that conclusions are interpretations of results.

- Pose the following question. Does the oxygen gas given off as the final product in photosynthesis come from the H_2O or from the CO_2 taken in by the plant? Using the equations below, have students determine the answer. The red oxygens represent a tracer. **L1**

$$6CO_2 + 6H_2O \rightarrow C_6H_{12}O_6 + 6O_2$$

$$6CO_2 + 6H_2O \rightarrow C_6H_{12}O_6 + 6O_2$$

- Have students do research to find other ways that tracers are used. **L1**

Extension

Have students research the work of Melvin Calvin, who investigated some of the reactions of photosynthesis using $^{14}_{6}C$ as the tracer. **L3**

Connecting to Chemistry

1. Proteins are chains of amino acids. DNA is a double-helix composed of sugar, phosphate, and nitrogen bases.

2. Because the phages with the labeled DNA made new phages with unlabeled protein coats and labeled DNA cores, DNA is injected into the cell by the phages, and the DNA has the genetic information to make the entire phage. Also, because the phages with labeled coats made new phages with unlabeled coats, the phages do not inject their coats into the cell.

BIOLOGY CONNECTION

A Biological Mystery Solved with Tracers

For many years, biologists searched for the identity of the genetic material of life. Some scientists thought that the genetic material was protein. Others believed it was nucleic acids. By 1944, strong evidence pointed to DNA. In 1952, Alfred Day Hershey and Martha Chase published the results of their experiments, which confirmed that DNA, not protein, determines heredity.

Radioisotopes used as tracers solve scientific mysteries Radioactive and nonradioactive isotopes of the same element act the same way in a chemical reaction. When scientists want to put tags on a compound, they substitute a radioactive isotope for a nonradioactive one in the compound. Then they can use radiation detectors to track and locate the radioisotope tracer.

Hershey and Chase use tracers Bacteriophages (phages) are simple viruses that attack bacteria. They are composed of two parts: a protein coat and a DNA core, as shown here. The protein is composed of carbon, hydrogen, oxygen, nitrogen, and sulfur. The DNA is composed of carbon, hydrogen, oxygen, nitrogen, and phosphorus. Because sulfur is found only in protein and phosphorus only in DNA, Hershey and Chase chose radioactive ^{35}S and ^{32}P as the tracers for these substances to compare the inheritance of protein and DNA.

The question Hershey and Chase asked was: "When a phage infects a bacterial cell, does the phage inject its DNA core, its protein coat, or both?" They designed experiments using tracers to track the protein coat and DNA core separately. They grew bacteria in a culture containing ^{35}S. When they added phages to this ^{35}S-labeled culture, the new phages produced had protein coats containing ^{35}S. When they infected bacteria having no radioactive materials with the phages containing ^{35}S, they found the ^{35}S only in protein coats of phages outside the new bacteria. The phages made inside the new bacteria had no ^{35}S in their coats. This showed that the protein coat was not injected.

DNA is the genetic material Hershey and Chase did a similar experiment with bacteria grown in a culture containing ^{32}P, to which they added phages. The new phages produced had DNA containing ^{32}P. This time, the DNA with ^{32}P entered unlabeled bacterial cells and produced new phages that contained a significant amount of ^{32}P. The results showed that the phage's DNA core was injected. The phage's DNA alone was able to direct production of an entire phage, with coat. Therefore, Hershey and Chase concluded that DNA, not the protein, was the genetic material. In 1969, Hershey received the Nobel Prize in Medicine for his work.

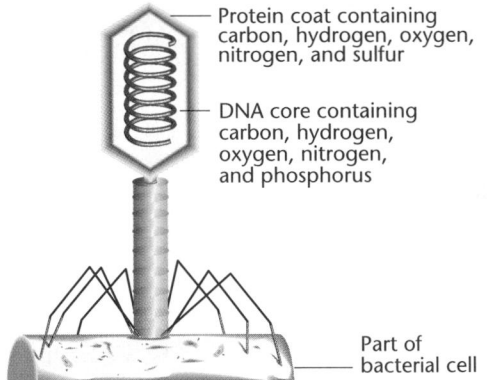

Protein coat containing carbon, hydrogen, oxygen, nitrogen, and sulfur

DNA core containing carbon, hydrogen, oxygen, nitrogen, and phosphorus

Part of bacterial cell

Connecting to Chemistry

1. **Comparing** Compare the structure of a protein with that of DNA.

2. **Interpreting** Explain how Hershey and Chase came to their conclusion.

Irradiation can extend the shelf life of food so that it can be stored for long periods of time without refrigeration, as shown in **Figure 21.18.** However, people who oppose food irradiation are concerned about what are called unique radiolytic products—URPs, for short. These products result from chemical changes caused by the ionizing effects of radiation. Whether URPs present a hazard has not yet been determined for sure. The U.S. Food and Drug Administration (FDA) has approved the use of irradiation for most fruits and vegetables.

Figure 21.18

Food Irradiation
Exposure to gamma radiation can extend the shelf life of food by preventing spoilage. Which group of mushrooms do you think was irradiated before storing?

Sources of Radioisotopes

Using nuclear chemistry, scientists today can change one element into another and even produce elements artificially. How are elements made artificially? Some are produced as by-products in nuclear reactors. However, most are made by bombarding nuclei with small particles that have been accelerated to high speeds. This is done mainly in three instruments, shown in **Figure 21.19.**

Figure 21.19

Accelerators and Particle Generators
◄ The large, expensive, and heavily shielded biomedical cyclotron is used to produce most radioisotopes used in PET scans.

A technetium-99m generator contains radioactive molybdenum-99 that decays to form technetium-99m. Hospitals can use the generator to produce and extract the short-lived technetium-99m just before it is needed in diagnostic techniques. ►

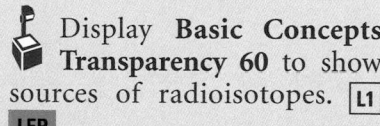

◄ The Tandem Cascade Accelerator (TCA) is smaller and cheaper to purchase and run than a cyclotron.

Discussion

Ask students to collect any irradiated foods they can obtain from local supermarkets or specialty food stores. Display the food items, and lead the class in a discussion regarding the benefits of and potential concerns about irradiated foods. L1

Correcting Misconceptions

Students may believe that all radioactive isotopes are synthetic. Point out that many occur naturally.

Concept Development

Explain that Rutherford accomplished the first synthetic nuclear transformation in 1919 by bombarding nitrogen-14 with alpha particles, producing oxygen-17 and a proton. Ask students to write the nuclear equation for the following reaction. $^{14}_{7}N + ^{4}_{2}He \rightarrow ^{17}_{8}O + ^{1}_{1}H$ L2

Tying to Previous Knowledge

Point out that many hospitals that use radioisotopes must generate them on site using small cyclotrons, accelerators, or reactors because of the short half-lives of the isotopes.

Transparency

Display **Basic Concepts Transparency 60** to show sources of radioisotopes. L1
LEP

Chemistry Journal

Accelerator
Have students carry out library research to obtain more information about a type of accelerator or particle generator. They should write a journal account on their findings and include a diagram of the device. L2

Where does radiation come from?

If you are able to obtain a Geiger counter with a scaler, a device that totals the number of ionizations, have it set up and turned on, with accompanying sound, as students enter class. They will be naturally curious as to what is happening and where the radiation is originating. You may detect varying amounts of background radiation at different locations in the classroom, particularly near a window or a brick wall. Use the results to lead the class in a discussion of the sources of this pervasive and prevalent radiation, such as cosmic rays, radon, and radioisotopes in building materials.

Concept Development

Ultraviolet radiation can cause damage to living things, even though it is nonionizing. Although cells can be damaged by natural and synthetic radiation, natural cellular processes normally repair or replace the damaged cells. Point out that this makes evaluation of the effects of small amounts of radiation difficult.

Problems Associated with Radioactivity

The understanding of radioactivity has grown rapidly in the 100 years since its discovery. When the Curies worked with radioisotopes, they did not realize how harmful such materials could be. Marie Curie died of leukemia that was probably caused by her years of contact with radioisotopes. Many more radioactive isotopes exist than the few studied by the Curies. In fact, most of the roughly 2000 known isotopes of all elements are unstable and undergo nuclear decay. Fortunately, most of those do not occur naturally, but are produced synthetically. Your surroundings contain mostly stable isotopes of the common elements, so you are not normally exposed to enough radiation to do you much harm.

However, you may be surprised to learn that you are constantly being bombarded with low levels of radiation. This radiation comes from many sources and is referred to as background radiation. Some of it is in the form of cosmic rays, which are particles that reach Earth from outer space. Small amounts of radioactive elements are found almost everywhere on Earth, as well—in wood and bricks used to make buildings, in the fabrics used in clothing, in the foods you eat, and even inside your body. Traces of uranium in rock layers beneath houses may also produce radioactive radon gas that enters the houses and can present health risks. Various sources of radiation are illustrated in **Figure 21.20.**

Exposure to radioactive elements can be hazardous to your health. That's because the radiation they give off is powerful enough to knock electrons loose from atoms and generate ions when it collides with neutral matter. Because it can result in the formation of ions, nuclear radiation is also known as ionizing radiation. In contrast, most electromagnetic radiation—ordinary visible light, for example—does not have enough energy to knock electrons out of neutral atoms and is, therefore, nonionizing. Only radiation such as cosmic rays and X rays has sufficient energy to generate ions. It is mainly ionizing power that makes radioactive elements dangerous.

Figure 21.20

Sources of Radiation
As this pie diagram shows, natural radioactivity accounts for about 82 percent of the radiation to which you are exposed. Most of the natural radiation is from radon produced by the decay of uranium present in rocks and soil. The remaining 18 percent of the radiation comes from sources produced by humans in the last century, including nuclear medical tools such as X rays and nuclear fuel used in reactors.

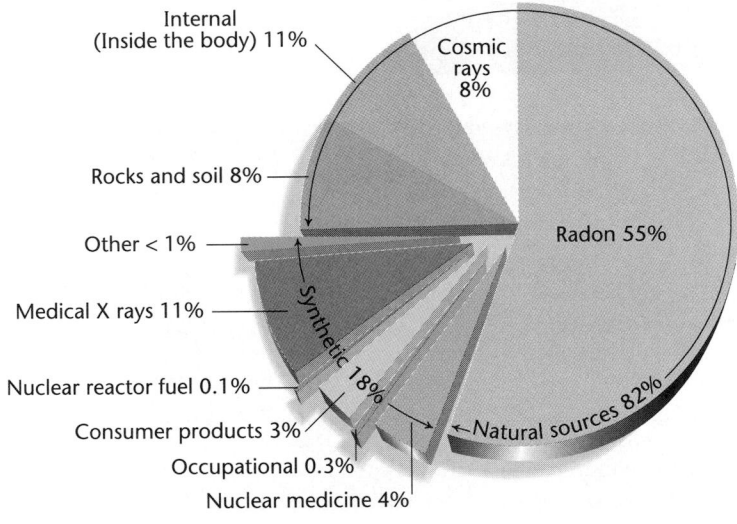

Internal (Inside the body) 11%
Cosmic rays 8%
Rocks and soil 8%
Other < 1%
Medical X rays 11%
Nuclear reactor fuel 0.1%
Consumer products 3%
Occupational 0.3%
Nuclear medicine 4%
Synthetic 18%
Radon 55%
Natural sources 82%

Too Close for Comfort
Purpose: To allow students to explore the inverse square law.
Materials: flashlight with batteries, dark paper or tempera paint, newspaper, felt-tip marker, meterstick

Safety Precautions: 🧤 🥽
Disposal: Code F
Procedure: Investigate what happens to the intensity of radiation you absorb when you move farther from the source. Cover a tabletop with newspaper. Cover the reflector of the flashlight with dark paper or tempera paint. In a darkened room, shine the flashlight on the desk from a height of 10 cm. Mark the limits of the lighted area. Repeat from 20 cm and 30 cm.
Results: The lighted area gets larger and the intensity of the light decreases as the distance increases. The inverse square law is represented by $E = 1/d^2$, where E is energy and d is distance. Point out that the intensity of nuclear or X radiation also varies as the inverse square of the distance from the source. Also point out that color televisions emit X rays.
Analysis: Ask these questions.
1. How does the intensity of the lighted area change as the light moves farther from the paper's surface? *It is less intense.*

Radon—A problem in your home?

A high level of radon in a home is correlated to a greater risk of lung cancer for the occupants. Radon-222 is an eventual product of the decay of uranium-238, which occurs naturally in many rocks and soils. In this MiniLab, you will study the results of radon tests in several homes.

Procedure

1. Your teacher will supply the class with several commercial radon-detection kits—one for each group of students. The four-day exposure canisters will probably be the most practical to use; however, the 30-day units will usually give more accurate results.

2. Familiarize yourself with the types of homes, geological and geographic features, and factors such as industries in your area. As a class, decide where each group will place its detector. Obtain permission to test for radon from those who own or live in the home you wish to test.

3. Follow the directions on the radon-testing kits, and expose the detectors in the selected home. If there are a sufficient number of kits for the class, some groups can expose their detectors in upper levels of homes, and other groups can do so in basements. Send the detectors to be evaluated according to the instructions.

4. Examine your data and collect and tabulate data from the rest of the class.

Analysis

1. Did any of the test results exceed the U.S. Environmental Protection Agency's recommended maximum, which is 4 picocuries per liter? How many did so, and by how much?

2. Carefully analyze the class data, looking for correlations between the radon levels and factors such as geological and geographic features, proximity to industrial sites, style of home, and materials used in home construction.

3. Suggest several ways in which radon levels in homes might be reduced.

Because of the danger, elaborate and expensive precautions must be taken to protect people who work with radioisotopes. Highly radioactive waste products that can take many thousands of years to decay must be stored carefully. Choosing storage sites is difficult, and so is transport of the wastes to the storage sites. Many people are concerned about nuclear processes as energy sources. The threat of nuclear war and concern over radioactive fallout have also caused some people to be reluctant to use any form of radioactivity.

21.3 Nuclear Tools   **775**

---

2. Calculate the area illuminated during each of the three trials. Use $A = \pi r^2$.
3. Compute a ratio of areas to the nearest whole number. *The area increases by a factor of 4 and then by a factor of 9.*
4. The same amount of energy continues to fall on the paper as the light moves farther away, but the energy is spread over a larger area. If the intensity was 1 unit at 10 cm, what was it at 20 cm? At 30 cm? *1/4; 1/9*

Assessment

Performance: Have students predict how much X-ray energy they would receive if they were 4 m from a TV instead of 1 m. Have them test their predictions using the flashlight 40 cm above the table. *X-ray radiation at 4 m is 1/16 that at 1 m.* **L1**

---

miniLAB

Radon—A problem in your home?

TECH PREP

Purpose: To perform radon tests in several homes and study the results.
Process Skills: Classifying, interpreting data, communicating

Teaching Strategies

- Work with student groups to determine the best and most diverse locations for the radon tests.
- Make sure that some detectors are placed in the homes of smokers.

Expected Results

Results will vary widely. However, radon levels are apt to be higher in basements, in homes with smokers, in brick or stone homes, in homes with attic fans, in homes with sump pumps, near certain industrial sites, and in geographic areas that contain uranium ores.

Analysis

1. Answers will vary.
2. Answers will vary. However, note the correlations listed under *Expected Results,* above.
3. Suggestions may include: stop smoking in the home, air out the home regularly, don't use an attic fan unless the basement is sealed from the rest of the house, and seal the sump and any cracks in the basement.

Assessment

Knowledge: Ask students to find out why newer homes may contain more radon than older homes. *Newer homes are generally built to be more airtight.* **L1**

Content Background

Radiation Units The amount of radiation emitted by a material is referred to as its activity, which is the number of nuclear changes per second. The SI unit of activity is the becquerel (Bk), which is defined as one nuclear disintegration per second. The curie (Ci) is an older unit. One curie is equal to the number of disintegrations per second in 1 g of radium, or 3.7×10^{10}. So, $1 \text{ Ci} = 3.7 \times 10^{10}$ Bk. Specific activity refers to the number of disintegrations per second per gram of sample. In SI units, specific activity is Bk/g. Often, it is useful to know not only how much radiation is coming from a sample, but also how much of that radiation is absorbed by something in its path. In SI units, this absorbed dose is expressed as the amount of energy in joules absorbed per kilogram of absorbing material. This unit is the gray (Gy) after British radiologist Harold Gray; $1 \text{ Gy} = 1 \text{ J/kg}$. An older unit for absorbed dose still used in the United States is the rad (radiation absorbed dose), which equals 10^{-5} J of energy absorbed per gram of absorbing material; $1 \text{ Gy} = 100$ rad. Neither the gray nor the rad unit takes into account how deeply a particular type of radiation penetrates a living thing. The unit that does this is the "radiation equivalent for man," or rem. Appropriate factors must be used for converting rad doses into rem, and different factors must be used to take into account the difference in penetrating power of alpha, beta, and gamma radiation.

Radiation Exposure

As you have read, radiation can do severe damage to parts of a single cell and can cause its death. Damage to DNA can be especially harmful because that substance is the blueprint from which genetic material for future generations is made. When a cell with damaged DNA divides, all cells made from it also have damaged DNA. When this damage occurs in egg or sperm cells of individuals, damaged or mutated DNA can be passed to offspring.

The more energy radiation has, the more dangerous it tends to be. Several different units can be used to measure the energy in a given amount of radiation. A unit used to measure the received dose of radiation is called the **gray.** One gray is equivalent to the transfer of 1 joule of energy in the form of radiation to 1 kg of living cells. However, not all of the radiation that reaches living tissue gets absorbed by the tissue. That fact is not taken into account when radiation is measured in grays. The biological damage caused by radiation is indicated best in terms of how much is actually absorbed, which is measured by a unit called a **sievert** (Sv). One sievert is equal to 1 gray, multiplied by a factor that takes into account how much of the radiation hitting the tissue is absorbed by it. Likely harmful biological effects of single doses of radiation of various strengths are listed in **Table 21.2.**

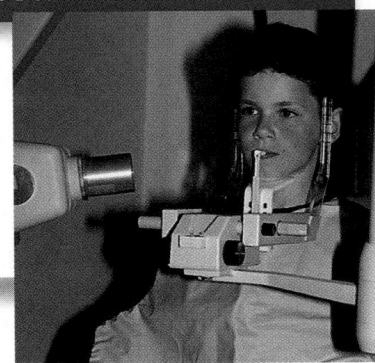

Table 21.2 Likely Physiological Effects on a Human from a Single Dose of Radiation	
Dose (Sv)	**Effect**
0-0.25	No immediate effect
0.25-0.50	Small temporary decrease in white blood cell count
0.50-1.0	Large decrease in white blood cell count, lesions
1.0-2.0	Nausea, hair loss
2.0-5.0	Hemorrhaging, possible death
>5.0	50 percent chance of death within 30 days of exposure

Background ionizing radiation from natural sources results in a dose of about 0.003 Sv per year for the average person. The U.S. government recommends that your total exposure to sources other than background radiation should be limited to 0.005 Sv per year. Workers in nuclear plants are permitted exposure to 0.05 Sv per year. Both these doses are far below the 2- to 5-Sv single dose that can be lethal. Besides the natural background level, the only significant source of radiation for most people who do not work around radioactive materials is medical radiation, mainly X rays. The dose equivalent of a chest X ray is about 0.0005 Sv, and that of a dental X ray is about 0.0002 Sv, presenting risks that most people would consider slight when compared to the potential diagnostic benefits of the X rays.

Chemistry Journal

Radiation Exposure

Have interested students write an account of the sources of their typical yearly radiation exposure, including background radiation and dental and medical X rays. They should also compute the total and compare it to the suggested yearly maximum. **L3**

Everyday Chemistry

Radon–An Invisible Killer

If you do not smoke or breathe in cigarette smoke produced by other people, do you still have a chance of getting lung cancer? The answer is yes. Each year, as many as 15 000 to 20 000 deaths due to lung cancer are the result of radon pollution. Radon, element 86, is the densest noble gas. All of its isotopes are radioactive. Radon-222 has the longest half-life, 3.823 days. Radon is formed in uranium deposits in Earth's crust. Because it is a gas, it can seep through the rocks and soil to the surface.

How does radon cause lung cancer? Even though radon is radioactive, it is not highly dangerous. Most of it is inhaled and rapidly exhaled, like any other gas. However, radon has a short half-life and quickly changes into radioactive isotopes of polonium and lead.

$$^{222}_{86}Rn \rightarrow \,^{218}_{84}Po + \,^{4}_{2}He$$
$$^{218}_{84}Po \rightarrow \,^{214}_{82}Pb + \,^{4}_{2}He$$

These radioactive isotopes are solids that can collect on dust. When the dust is inhaled, the radioactive solids remain in the lungs. High-energy alpha particles from the polonium and lead damage the DNA of lung cells, sometimes causing cancer.

Can radon get into your home? Radon can enter homes and other buildings through cracks in concrete foundations and slab floors; porous cinderblock walls; unsealed, poorly ventilated crawl spaces; and small areas around water and sewer pipes. In the early 1970s, radon started to cause more extensive problems. People had begun to make their homes airtight to save energy and money. That helped to seal in the unwanted radon.

Is your home safe? Because radon is a dense gas, it accumulates near the bottoms of structures. It is simple to test for its presence using detection kits purchased at a hardware store. The kits may be either short-term, which require two to seven days to test, or long-term, which may require 30 or more days. The Environmental Protection Agency (EPA) considers a result of more than 4 picocuries per liter (pCi/L) to be unsafe.

How can radon pollution be corrected? Fixing radon pollution problems usually requires only minor repairs or alterations. Sometimes, installing an exhaust fan in the polluted area or within a foundation will solve the problem. Sealing large cracks or spaces around pipes can also help prevent the entrance of radon.

Exploring Further

1. **Comparing and Contrasting** Compare the use, operation, and effectiveness of short- and long-term radon test devices.

2. **Thinking Critically** Some of the atoms of radon gas in the lungs can turn out to be dangerous. Explain.

Everyday Chemistry

Purpose
Students investigate the problem, detection, and elimination of radon pollution in the home.

Teaching Strategies
- Ask a local contractor who specializes in radon remediation to speak to your classes.
- A good reference is the article "Radon, Worth Learning About" in *Consumer Reports*, July 1995, pp. 464-465.
- Point out that chemical activity and radioactivity are different. A radioactive element has atoms with unstable nuclei that can produce different elements. A chemically reactive element has stable nuclei but can readily give off or take up electrons to form compounds.

Extension
Have students check the chemistry room for radon using different brands of short- and long-term radon kits. They should compare results from the detectors. L1

Exploring Further

1. Short-term devices work by trapping radon on charcoal granules. In long-term devices, alpha particles from radon decay cause imprints on a piece of specially formulated plastic. The two types are about equal in price and setup convenience. The long-term type requires more time to carry out and analyze the test, but it is more accurate.

2. Some of the radon atoms may decay while still in the lungs. They will turn into solid polonium and lead and will not be exhaled, presenting an ongoing danger.

Lifestyle and environment can result in more than the normal background exposure to radiation as shown in **Figure 21.21.** Cigarette smoke contains significant amounts of radioactive material that can contribute to lung cancer. Smoking two packs of cigarettes a day results in exposure to 0.1 Sv per year. Living at a high altitude or taking frequent trips in airplanes increases your exposure to cosmic rays from outer space. That's because the higher up you are, the less atmosphere there is to block the incoming radiation.

Figure 21.21
Two Sources of Radiation
Medical and dental X rays present only a slight risk due to radiation and are of great benefit in diagnosis. ▶

Other sources such as cigarettes present more serious radiation hazards and provide no benefits. ▼

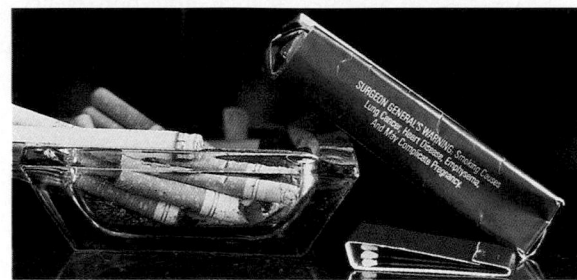

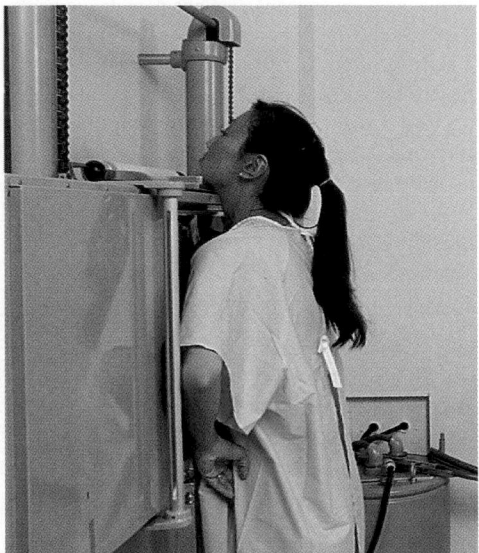

Waste Disposal

You probably associate radioactive waste with nuclear reactors, but more than 80 percent of all such waste is generated in hospitals. How can this hazardous waste be dealt with so that it cannot harm living things? Most of the radioactive waste material produced at hospitals contains isotopes with short half-lives. That kind of waste can simply be stored until the isotopes have decayed to a safe level.

Although waste from nuclear reactors may not be produced in quantities that are large compared to radioactive hospital waste, it is far more dangerous. For example, spent fuel rods from nuclear reactors contain both short-lived and long-lived radioisotopes produced by collisions of fast-moving particles with atoms in the nuclear fuel and in the walls of the reactor itself. These rods are usually stored at the reactor site for several decades until the isotopes with short half-lives have decayed.

What can be done after that point to isolate the radioisotopes with long half-lives? The plan that currently holds the most promise is to incorporate the unstable nuclei into stable material such as glass, which is then surrounded by canisters made of layers of steel and concrete. The canisters can then be buried deep underground in stable rock formations, as shown in **Figure 21.22.** The storage sites would be located in a dry, remote area.

Figure 21.22
Underground Storage of Radioactive Waste
Tunnels are being dug 400 m (about 1300 feet) beneath Yucca Mountain, Nevada, at a proposed site for storage of radioactive waste. It will be the largest radioactive storage facility in the country, capable of holding up to 63 500 tons of waste.

An alternative proposal being tested for nuclear waste disposal is burial in clay sediments deep in the ocean. Recycling of uranium for reuse in reactors is also being considered. The uranium can be extracted from used rods that have been chemically dissolved. This uranium can then be made into new rods. However, this recycling process is currently too expensive to be viable. Whether or not recycling and disposal technologies can keep pace with the use of nuclear materials will influence the development of more technologies using these materials.

Decisions regarding nuclear materials are not simply the business of government officials and scientists. You, too, will have to make decisions about such materials as you help decide how your community supplies energy, whether your country makes nuclear weapons, and, on a more immediate level, whether you undergo medical procedures that employ radioisotopes. How well you understand nuclear chemistry will influence your decisions.

SECTION REVIEW

Understanding Concepts

1. What is the effect of radiation on living cells?
2. What disease is radiation therapy most often used to treat?
3. Explain what a PET scan does.

Thinking Critically

4. **Illustrating** How could a radioactive tracer be used to study the way glucose is metabolized in cells?

Applying Chemistry

5. **Bone Scans** What radioactive isotopes might be useful in scanning techniques to examine bones for fractures and other abnormalities? Include at least two elements not discussed in this chapter, and explain your answer.

21.3 **Nuclear Tools**   779

SECTION REVIEW

1. Radiation can damage or kill living cells, and it can cause changes in DNA.
2. Radiation therapy is used most often to treat cancer.
3. A PET scan produces an image of a process or organ in the body using radioactive tracers and a computerized scanning instrument.
4. A stable element in glucose could be replaced with a radioisotope. Scanning instruments could then be used to detect the radiation given off by glucose molecules as they move through the body and are metabolized. Compounds to which glucose is converted during metabolism could also be isolated and any radiation in them detected.
5. The radioisotopes Gd-153 and Tc-99m can both be used in bone scans. Others that could be useful are isotopes of calcium, magnesium, and phosphorus because those elements concentrate in bones.

Check for Understanding

- Check for understanding of the section material by having students answer the Section Review questions.
- Ask students to explain why the procurement and storage of medical tracer radioisotopes pose special problems when those substances are compared with nonradioactive chemicals. *In addition to shielding and safety considerations, the short half-life of most of these isotopes requires that they be generated in the hospital or clinic where they are used or that they be shipped by rapid and reliable delivery systems.* L1

Reteach

Discussion: Ask students how they would respond if a proposal were made to dispose of radioactive hospital wastes in a nearby landfill. Point out that most of the radioisotopes included would have half-lives of less than 30 days. L1

Extension

Ask interested students to research and report on how each of the transuranium elements was created and named. L2

4 CLOSE

Enrichment

Ask interested students to investigate the latest information on the threshold and linear models for radiation damage. The threshold model assumes that radiation below a certain level causes little or no harm, and the linear model assumes that radiation damage is always proportional to dosage. L2

- Review the Reviewing Main Ideas statements and Key Terms with your students.
- Complete solutions to Chapter Review problems can be found in the *Problems and Solutions Manual.*

UNDERSTANDING CONCEPTS

1. The nucleus produces radioactivity.
2. Ionizing radiation is radiation that is capable of converting neutral matter into ions.
3. **a.** A helium nucleus, ^4_2He, is given off. **b.** An electron, $^0_{-1}e$, is given off. **c.** Gamma radiation is given off.
4. Alpha particles can be stopped easily by thin layers of paper or air; beta particles can be stopped by a dense block of wood, heavy clothing, or a thin sheet of metal; gamma radiation can be stopped only by thick blocks of materials such as lead or concrete.
5. **a.** $^{222}_{86}Rn$, **b.** $^0_{-1}e$, **c.** $^{218}_{84}Po$, **d.** $^{222}_{86}Rn$
6. Dinosaurs became extinct about 65 million years ago. Because of its relatively short half-life of 5730 years, carbon-14 can be used to date objects only up to about 60 000 years old.
7. The chain reaction is started when a neutron collides with a U-235 nucleus. The reaction keeps going because every such collision produces more neutrons that can collide with another nucleus. The rate is controlled by means of control rods that are lowered into the reaction chamber, as needed, to absorb neutrons and slow the reaction.
8. Nuclear fusion takes place in stars.
9. Radioisotopes are used as tracers, in radiation therapy, and in imaging processes such as PET scans, among other uses.

REVIEWING MAIN IDEAS

21.1 Types of Radioactivity

- Nuclear reactions involve protons and neutrons rather than electrons, and therefore differ from chemical reactions.
- Radioactivity involves the spontaneous emission of radiation by an unstable nucleus.
- Radioactivity can be detected and measured using film and instruments, including Geiger and scintillation counters.
- Unstable isotopes are called radioisotopes. Symbols for isotopes show the mass number as a superscript and the atomic number as a subscript to the left of the symbol for the element.
- When nuclei decay, they usually emit one or more of the three common types of radiation: alpha particles, beta particles, and gamma rays. During alpha decay, a ^4_2He particle is emitted. During beta decay, an electron is emitted. During gamma decay, high-energy electromagnetic gamma radiation is emitted.
- A nuclear decay process can be represented by an equation in which the sum of the mass numbers and the sum of the atomic numbers must each be equal on both sides of the arrow.
- The half-life of a radioactive element is the time it takes for one-half of the nuclei in a sample to decay. The rate of nuclear decay can be used to date fossils and artifacts.

21.2 Nuclear Reactions and Energy

- A small mass is converted into a large amount of energy during a nuclear reaction, according to Einstein's equation $E = mc^2$.
- Nuclear fission is the breaking apart of a nucleus into two or more smaller nuclei that are similar in size.

- Uranium-235 undergoes a fission chain reaction in a nuclear reactor, which converts some of the released energy into electricity.
- Nuclear fusion is the joining together of two smaller nuclei to form a larger one. Fusion in the sun provides energy for life on Earth. Fusion reactors have the potential to provide energy more safely than fission reactors.

21.3 Nuclear Tools

- Radioactive tracers are used for tracking compounds in an organism and for charting the movement of pollutants. Many different isotopes function as medical tracers for imaging organs and processes inside an organism.
- Radiation therapy involves selectively killing rapidly dividing cancer cells by targeting them with radiation generated by the decay of cobalt-60.
- Gamma radiation from cobalt-60 can be used to irradiate food to keep it from spoiling.
- Mostly natural sources contribute to the ionizing radiation exposure that people undergo all the time. The received dose of radiation is measured in grays. The absorbed dose of radiation that results in damage done to organisms is measured in sieverts.

Key Terms

For each of the following terms, write a sentence that shows your understanding of its meaning.

alpha particle	nuclear fission
beta particle	nuclear fusion
deuterium	nuclear reactor
gamma ray	radioactivity
gray	sievert
half-life	tritium

✔ Assessment

Portfolio: Review the portfolio options that are provided throughout the chapter. Encourage students to select one product that demonstrates their best work for the chapter. Have students explain what they have learned and why they chose this example for placement into their portfolios. **P**

Additional portfolio options can be found in the following **Teacher Classroom Resources: Tech Prep Applications Applying Scientific Methods in Chemistry Consumer Chemistry**

UNDERSTANDING CONCEPTS

1. What part of the atom produces radioactivity?

2. What is ionizing radiation?

3. What is given off by a nucleus undergoing each of the following types of decay?

   a) alpha   b) beta   c) gamma

4. What types of material will block alpha particles? Beta particles? Gamma rays?

5. Balance the following equations, which represent nuclear reactions in the uranium-238 decay series.

   a) $? \rightarrow {}^{218}_{84}Po + {}^{4}_{2}He$ c) $? \rightarrow {}^{214}_{82}Pb + {}^{4}_{2}He$
   b) ${}^{234}_{90}Th \rightarrow {}^{234}_{91}Pa + ?$ d) ${}^{226}_{88}Ra \rightarrow {}^{4}_{2}He + ?$

6. Explain why carbon-14 dating cannot be used to determine how old a dinosaur fossil is.

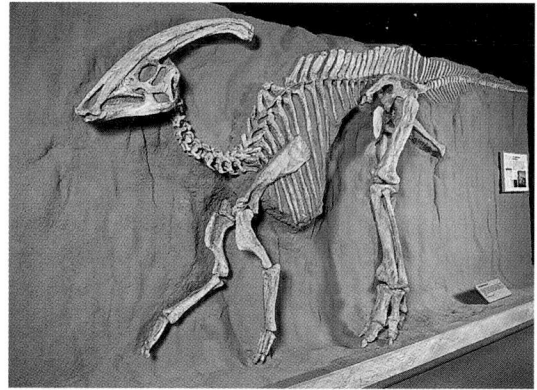

7. What starts the chain reaction that occurs in nuclear fission reactors used today? What keeps it going? How is its rate controlled?

8. What type of nuclear reaction takes place in stars?

9. List three ways that radioisotopes are used in medicine.

10. How is cobalt-60 used to treat cancer?

11. Why are foods irradiated? Does this make them radioactive?

12. Why is nuclear waste so difficult to dispose of safely?

13. Write the symbols for deuterium and tritium.

14. When each of the following is ejected from a nucleus, what happens to the atomic number and mass number of the atom?

    a) an alpha particle
    b) a beta particle
    c) gamma radiation

15. Where does the energy that is produced during nuclear reactions come from?

APPLYING CONCEPTS

16. The natural abundance of the uranium-235 needed for weapons and reactors is only about 0.7 percent of uranium found naturally; uranium-238 makes up most of the rest. During World War II, scientists separated out the uranium-235 by a process in which all the uranium was converted into uranium hexafluoride gas (UF_6). The separation was possible because the gas molecules of one of the isotopes diffuse faster than those of the other. Which isotope's molecules will diffuse faster? Explain.

17. What radioisotope is generally used in smoke detectors? What function does it serve?

18. Explain why radioisotopes with long half-lives are not administered internally in medical procedures. What are the half-lives of iodine-131, technetium-99m, and gadolinium-153, which are used in many medical procedures?

19. Tritium has a half-life of about 12.3 years, which makes it useful for dating objects up to about 100 years old. Tritium dating is sometimes used to verify the dates of aged alcoholic beverages like wines. How old is a bottle of wine if its tritium activity is 12.5 percent as high as that of new wine?

20. Krypton-81m is inhaled during studies to measure lung ventilation. Its half-life is only 13 seconds. How much time has passed before a sample of krypton-81m with an original activity of 260 decays per minute has decreased to an activity of 8 decays per minute?

lives are 8.07 days, 6.0 hours, and 242 days, respectively.

19. 12.5% = 100% × 1/2 × 1/2 × 1/2, so 3 half-lives have elapsed; 3 × 12.3 years = 36.9 years old

20. 260 × 1/2 × 1/2 × 1/2 × 1/2 × 1/2 = approximately 8, so 5 half-lives have elapsed; 5 × 13 s = 65 s

21. Sodium-24 is formed.
    ${}^{24}_{12}Mg + {}^{1}_{0}n \rightarrow {}^{24}_{11}Na + {}^{1}_{1}p$

22. ${}^{239}_{94}Pu \rightarrow {}^{235}_{92}U + {}^{4}_{2}He$

23. They would have observed
    a. no radioactivity inside cells, and ${}^{35}S$ outside;
    b. ${}^{32}P$ radioactivity inside cells;
    c. ${}^{18}O$ activity both inside and outside cells.

24. The proton that is formed along with the carbon-14 is a hydrogen nucleus.

25. The painting really is an old one. If it is a forgery, it was not done recently.

10. The gamma radiation given off by cobalt-60 is aimed at cancer cells during radiation therapy. Therapy is carried out from many angles to allow strong radiation to strike and kill the cancer cells, while healthy cells receive only a small dose of radiation.

11. Food is irradiated to keep it from spoiling. Cells of microorganisms and insects on the food are killed. This process does not leave traces of radioactive material in the food.

12. Nuclear waste can contain highly radioactive materials that have long half-lives, so the sites must remain isolated for thousands of years.

13. Deuterium is ${}^{2}_{1}H$ and tritium is ${}^{3}_{1}H$.

14. a. Atomic number decreases by 2; mass number decreases by 4.
    b. Mass number does not change; atomic number increases by 1.
    c. Both numbers remain the same.

15. Mass is converted into the energy produced during nuclear reactions.

APPLYING CONCEPTS

16. UF_6 molecules that contain U-235 will diffuse faster because they are lighter than those that contain U-238. The rate of diffusion of a gas decreases as the mass increases.

17. The americium-241 radioisotope is used; it gives off ionizing radiation that ionizes the air in the chamber of a smoke detector, allowing a current to flow. When smoke enters the chamber, it impedes the current, which causes an alarm to sound.

18. Radioisotopes with long half-lives would not decay quickly, so they would give off dangerous radiation in the body for a long time, damaging cells. The half-

THINKING CRITICALLY

26. Argon is a noble gas, which means that it will not form compounds that would remain in rock. It can escape through cracks and pores in a rock.

27. The box could be shaken for repeated short periods and the contents examined to find out the shaking time required to turn over half the pennies.

28. Reactive particles that are produced and that are needed to keep the chain reaction going can collide with each other, the walls of the reaction vessel, or substances other than the reactant and lose so much energy that they cannot keep the reaction going. When the quantity falls below a certain level, the chain reaction stops.

29. Either of two reasonable hypotheses could be developed. It might be easier for radon to move from deep inside Earth into a basement through a dirt floor, raising radon levels. Alternatively, rocks in the concrete itself might contain traces of uranium from which radon is formed. Also, concrete might keep the basement more tightly sealed and prevent radon from escaping. Either of the latter two possibilities would tend to raise radon levels.

CUMULATIVE REVIEW

30. It contains 64 electrons, 64 protons, and 89 neutrons.

31. a. liquid to gas **b.** solid to liquid **c.** solid to gas **d.** gas to liquid **e.** liquid to solid

32. They are converted into water and carbon dioxide, and energy is released.

33. Four factors that influence rate are temperature, pressure, concentration, and the presence of inhibitors or catalysts.

34. During oxidation, an atom

---

21. What new element is formed if magnesium-24 is bombarded with a neutron and then ejects a proton? Write a balanced nuclear equation for this transmutation.

22. Some types of heart pacemakers contain plutonium-239 as a power source. Write a balanced nuclear equation for the alpha decay of plutonium-239.

Biology Connection

23. What results would Hershey and Chase have observed if they had

  a) radiolabeled only protein?
  b) radiolabeled only DNA?
  c) labeled both the protein and DNA with oxygen-18?

Chemistry and Technology

24. Carbon-14 formation in the atmosphere also results in formation of hydrogen. Explain.

Art Connection

25. If the white lead in a sample of paint taken from an alleged forgery of a famous old painting turns out to have a high ratio of radium-226 to polonium-210, what can be concluded?

● THINKING CRITICALLY

Relating Concepts

26. Potassium-40 dating techniques can be tricky because of the properties of the decay product argon. What about argon makes it hard to measure?

Designing an Experiment

27. ChemLab Design an experiment to determine how long the half-life of pennium is, given standardized shaking of the box.

Drawing Conclusions

28. MiniLab 1 Why do most chain reactions stop before all of the reactant elements have been used up?

Forming a Hypothesis

29. MiniLab 2 Would you expect to find higher levels of radon in basements of homes that have floors of dirt or floors of thick concrete? Explain.

● CUMULATIVE REVIEW

30. Suppose that a neutral atom of gadolinium has an atomic number of 64 and a mass number of 153. How many electrons, protons, and neutrons does it contain? (Chapter 2)

31. List the change of state that takes place during each of the following. (Chapter 10)

  a) boiling   **d)** condensation
  b) melting   **e)** freezing
  c) sublimation

32. What happens to carbohydrates such as glucose as they are oxidized in your cells during metabolism? (Chapter 19)

33. What are four factors that influence the rate of a chemical reaction? (Chapter 6)

34. What happens during oxidation? During reduction? (Chapter 16)

35. What are allotropes? Give an example. (Chapter 5)

36. Define the term *strong acid* and give two examples of such acids. What type of reaction takes place when an acid is mixed with a base? (Chapter 14)

37. List two properties of bases that are different from those of acids. (Chapter 14)

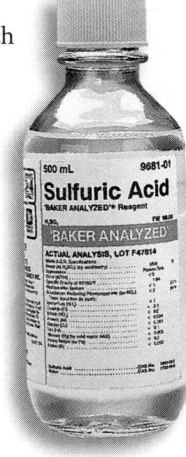

---

or ion loses electrons and/or becomes more positively charged. During reduction, an atom or ion gains electrons and/or becomes more negatively charged.

35. Allotropes are forms of an element that have different structures. Ordinary diatomic oxygen gas and ozone are oxygen allotropes; diamond and graphite are carbon allotropes.

36. A strong acid dissociates to a large extent into ions when dissolved in water. Hydrochloric acid and nitric acid are two examples. Neutralization occurs when an

acid and a base react.

37. Bases taste bitter and turn red litmus blue; acids taste sour and turn blue litmus red.

Skill Review

1. Using a Table Examine the table below, which lists information about nuclear power use in selected countries at the end of 1989. Then write a one-paragraph summary of the information, incorporating answers to the questions that follow the table.

Country	Percent Electricity from Nuclear Reactors
Argentina	11.4
Belgium	60.8
Brazil	0.7
Canada	15.6
France	74.6
Hungary	49.8
Japan	27.8
Netherlands	5.4
South Korea	50.2
Former Soviet Union	12.3
Spain	38.4
Sweden	45.1
Switzerland	41.6
United Kingdom	21.7
United States	19.1
Former West Germany	34.3

a) Which country listed gets the highest percentage of its electricity from nuclear power?

b) How many countries listed get more than 40 percent of their electricity from nuclear power?

c) Is the percentage of electricity from nuclear power in the United States greater or less than the average percentage for all countries listed?

Writing in Chemistry

2. Write a history of the development of the tokamak fusion reactor. Research the progress being made in getting experimental tokamak fusion reactors to operate efficiently and economically. When, if ever, do you expect them to exceed the break-even point in terms of energy produced compared to energy input? Are reactors of different designs likely to replace the tokamak before it is ever used commercially?

3. *Fat Man and Little Boy* is a 1989 film about the people involved in the Manhattan Project, the operation that developed the first atomic bomb during World War II. Watch a videotape of this movie and use library resources to read more about the Manhattan Project. Write a critical review of the film, including a discussion of how accurately you think the science and history are portrayed.

Problem Solving

4. When plutonium-239 is produced from uranium-238 in a breeder reactor, the process occurs in three steps. In the first step, uranium-238 absorbs a neutron. In the second step, a short-lived intermediate element is made through beta decay. In the third step, the plutonium-239 is produced through beta decay. Write a balanced nuclear equation for each of the three steps.

Projects

5. The light from many camping lanterns is produced when hydrocarbon fuels burn and heat up a small bag called a mantle, which is made of rayon mesh impregnated with salts of metals, usually cerium and thorium or yttrium. Those metal salts are converted into oxides when burned, and these oxides give off incandescent light when heated to high temperatures. The most common isotope of thorium, $^{232}_{90}\text{Th}$, is a radioactive alpha-particle emitter, so packages of mantles made with thorium have a warning telling you not to hold them close to your body for long periods of time. What product results when thorium-232 decays by emitting an alpha particle? Write an equation for the process.

AUTHENTIC ASSESSMENT

1. Student paragraphs should include the following answers: **a.** France; **b.** 6; **c.** less (the average is 31.8%).

2. Answers will vary but should provide a brief, accurate history and address the various points suggested.

3. Student reviews will vary but should discuss their reactions to the film and their assessments of its accuracy.

4. The equations are:
$$^{238}_{92}\text{U} + ^{1}_{0}n \rightarrow ^{239}_{92}\text{U}$$
$$^{239}_{92}\text{U} \rightarrow ^{239}_{93}\text{Np} + ^{0}_{-1}e$$
$$^{239}_{93}\text{Np} \rightarrow ^{239}_{94}\text{Pu} + ^{0}_{-1}e$$

5. $^{232}_{90}\text{Th} \rightarrow ^{228}_{88}\text{Ra} + ^{4}_{2}\text{He}$

Program Resources

Chapter Review and Assessment,
pp. 125-130 L1
Alternate Assessment in the Science
Classroom L1
Computer Test Bank, Chapter 21 L1
Problems and Solutions Manual,
Chapter 21 L1

Supplemental Practice Problems,
Chapter 21 L1
Performance Assessment in the Science
Classroom L1

APPENDICES
CONTENTS

APPENDIX A
Chemistry Skill Handbook

Measurement in Science

It's easier to determine if a runner wins a race than to determine if the runner broke a world's record for the race. The first determination requires that you sequence the runners passing the finish line—first, second, third. . . . The second determination requires that you carefully measure and compare the amount of time that passed between the start and finish of the race for each contestant. Because time can be expressed as an amount made by measuring, it is called a quantity. One second, three minutes, and two hours are quantities of time. Other familiar quantities include length, volume, and mass.

The International System of Units

In 1960, the metric system was standardized in the form of Le Système International d'Unités (SI), which is French for the "International System of Units." These SI units were accepted by the international scientific community as the system for measuring all quantities.

SI Base Units   The foundation of SI is seven independent quantities and their SI base units, which are listed in **Table A.1**.

Table A.1   SI Base Units

Quantity	Unit	Unit Symbol
Length	meter	m
Mass	kilogram	kg
Time	second	s
Temperature	kelvin	K
Amount of substance	mole	mol
Electric current	ampere	A
Luminous intensity	candela	cd

SI Derived Units   You can see that quantities such as area and volume are missing from the table. The quantities are omitted because they are derived—that is, computed—from one or more of the SI base units. For example, the unit of area is computed from the product of two perpendicular length units. Because the SI base unit of length is the meter, the SI derived unit of area is the square meter, m^2. Similarly, the unit of volume is derived from three mutually perpendicular length units, each represented by the meter. Therefore, the SI derived unit of volume is the cubic meter, m^3. The SI derived units used in this text are listed in **Table A.2**.

GLENCOE TECHNOLOGY

 Software

Mastering Concepts in Chemistry

Unit 5, Lesson 1
Review of the Metric System
Use this lesson to review the basic units of the metric system plus other units used in chemistry.

Table A.2   SI Derived Units

Quantity	Unit	Unit Symbol
Area	square meter	m^2
Volume	cubic meter	m^3
Mass density	kilogram per cubic meter	kg/m^3
Energy	joule	J
Heat of fusion	joule per kilogram	J/kg
Heat of vaporization	joule per kilogram	J/kg
Specific heat	joule per kilogram-kelvin	J/kg • K
Pressure	pascal	Pa
Electric potential	volt	V
Amount of radiation	gray	Gy
Absorbed dose of radiation	sievert	Sv

Other Useful Measurements

Metric Units   As previously noted, the metric system is a forerunner of SI. In the metric system, as in SI, units of the same quantity are related to each other by orders of magnitude. However, some derived quantities in the metric system have units that differ from those in SI. Because these units are familiar and equipment is often calibrated in these units, they are still used today. **Table A.3** lists several metric units that you might use.

Table A.3   Metric Units

Quantity	Unit	Unit Symbol
Volume	liter (0.001 m^3)	L
Temperature	Celsius degree	°C
Specific heat	joule per kilogram-degree Celsius	J/kg • °C
Pressure	millimeter of mercury	mm Hg
Energy	calorie	cal

SI Prefixes

When you express a quantity, such as ten meters, you are comparing the distance to the length of one meter. Ten meters indicates that the distance is a length ten times as great as the length of one meter. Even though you can express any quantity in terms of the base unit, it may not be convenient. For example, the distance between two towns might be 25 000 m. Here, the meter seems too small to describe that distance. Just as you would use 16 miles, not 82 000 feet, to express that distance, you would use a larger unit of length, the kilometer, km. Because the kilometer represents a length of 1000 m, the distance between the towns is 25 km.

In SI, units that represent the same quantity are related to each other by some factor of ten such as 10, 100, 1000, 1/10, 1/100, and 1/1000. In the example above, the kilometer is related to the meter by a factor of 1000; namely, 1 km = 1000 m. As you see, 25 000 m and 25 km differ only in zeros and the units.

To change the size of the SI unit used to measure a quantity, you add a prefix to the base unit or derived unit of that quantity. For example, the prefix *centi-* designates one one-hundredth (0.01). Therefore, a centimeter, cm, is a unit one one-hundredth the length of a meter and a centijoule, cJ, is a unit of energy one one-hundredth that of a joule. The exception to the rule is in the measurement of mass in which the base unit, kg, already has a prefix. To express a different size mass unit, you replace the prefix to the gram unit. Thus, a centigram, cg, represents a unit having one one-hundredth the mass of a gram. **Table A.4** lists the most commonly used SI prefixes.

Table A.4   SI Prefixes

Prefix	Symbol	Meaning	Multiplier	
			Numerical value	Expressed as scientific notation
Greater than 1				
giga-	G	billion	1 000 000 000	1×10^9
mega-	M	million	1 000 000	1×10^6
kilo-	k	thousand	1 000	1×10^3
Less than 1				
deci-	d	tenth	0.1	1×10^{-1}
centi-	c	hundredth	0.01	1×10^{-2}
milli-	m	thousandth	0.001	1×10^{-3}
micro-	μ	millionth	0.000 001	1×10^{-6}
nano-	n	billionth	0.000 000 001	1×10^{-9}
pico-	p	trillionth	0.000 000 000 001	1×10^{-12}

Practice Problems

Use Tables A.1, A.2, and A.3 to answer the following questions.

1. Name the following quantities using SI prefixes. Then write the symbols for each.
   a) 0.1 m
   b) 1 000 000 000 J
   c) 10^{-12} m
   d) 0.000 000 001 m
   e) 10^{-3} g
   f) 10^6 J

2. For each of the following, identify the quantity being expressed and rank the units in increasing order of size.
   a) cm, μm, dm
   b) Pa, MPa, kPa
   c) kV, cV, V
   d) pg, cg, mg
   e) mA, MA, μA
   f) dGy, mGy, nGy

1. a. decimeter, dm
    b. gigajoule, GJ
    c. picometer, pm
    d. nanometer, nm
    e. milligram, mg
    f. megajoule, MJ
2. a. centimeter, micrometer, decimeter; micrometer, centimeter, decimeter
    b. pascal, megapascal, kilopascal; pascal, kilopascal, megapascal
    c. kilovolt, centivolt, volt; centivolt, volt, kilovolt
    d. picogram, centigram, milligram; picogram, milligram, centigram
    e. milliampere, megaampere, microampere; microampere, milliampere, megaampere
    f. decigray, milligray, nanogray; nanogray, milligray, decigray

Relating SI, Metric, and English Measurements

Any measurement tells you how much because it is a statement of quantity. However, how you express the measurement depends on the purpose for which you are going to use the quantity. For example, if you look at a room and say its dimensions are 9 ft by 12 ft, you are estimating these measurements from past experience. As you become more familiar with SI, you will be able to estimate the room size as 3 m by 4 m. On the other hand, if you are going to buy carpeting for that room, you are going to make sure of its dimensions by measuring it with a tape measure no matter which system you are using.

Estimating and using any system of measurement requires familiarity with the names and sizes of the units and practice. As you will see in **Figure 1**, you are already familiar with the names and sizes of some common units, which will help you become familiar with SI and metric units.

Figure 1
Relating Measurements

Length
A paper clip and your hand are useful approximations for SI units. For approximating, a meter and a yard are similar lengths. To convert length measurements from one system to another, you can use the relationships shown here.

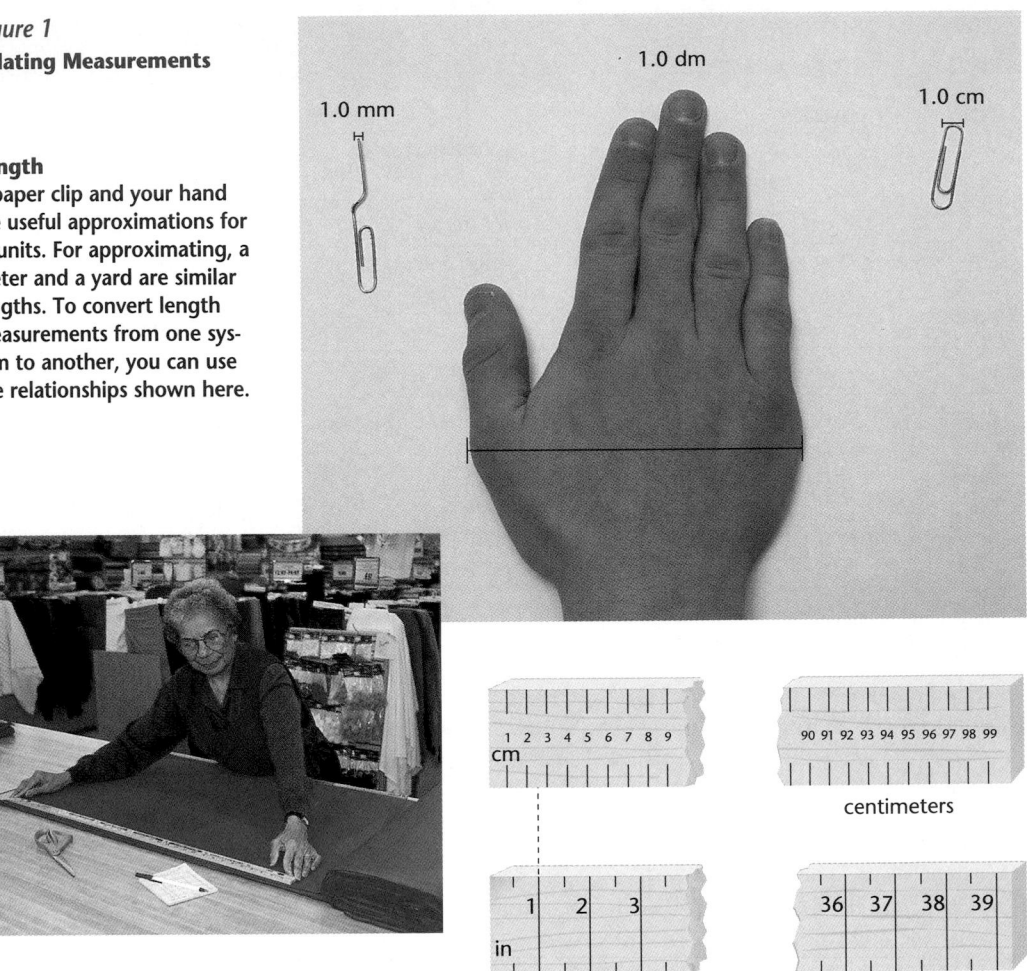

2.54 cm = 1.00 in.

1.00 m = 39.37 in.

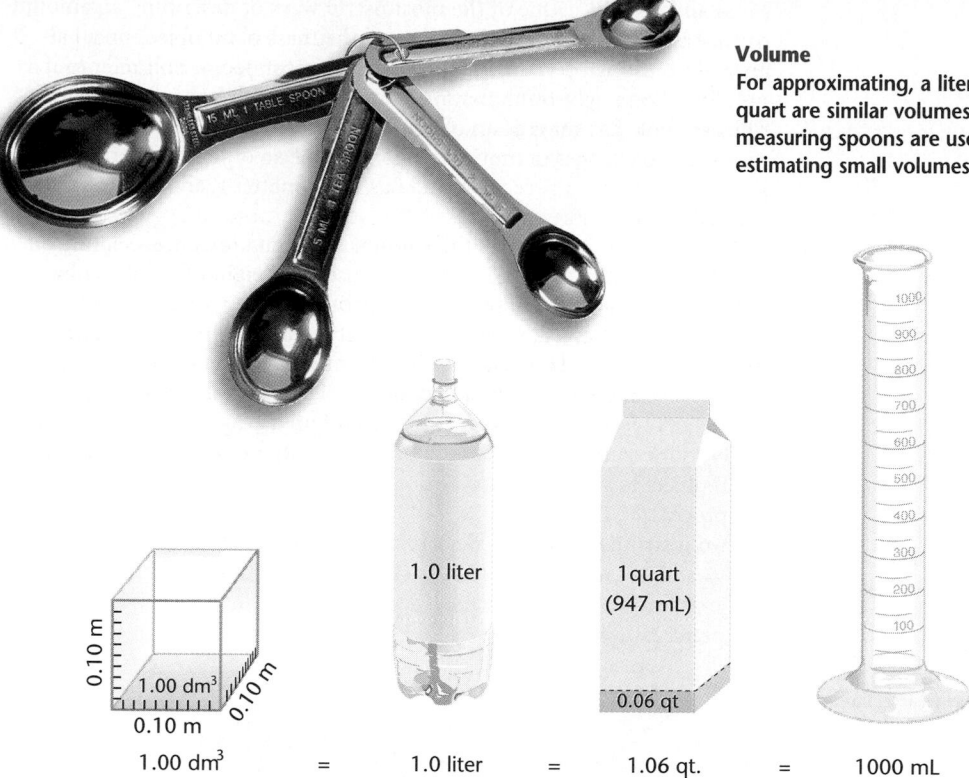

Volume

For approximating, a liter and a quart are similar volumes. Kitchen measuring spoons are used in estimating small volumes.

0.10 m
0.10 m
0.10 m
1.00 dm³

1.0 liter

1quart
(947 mL)

0.06 qt

1.00 dm³ = 1.0 liter = 1.06 qt. = 1000 mL

Temperature

Having a body temperature of 37° or 310 sounds unhealthy if you don't include the proper units. In fact, 37°C and 310 K are normal body temperatures in the metric system and SI, respectively.

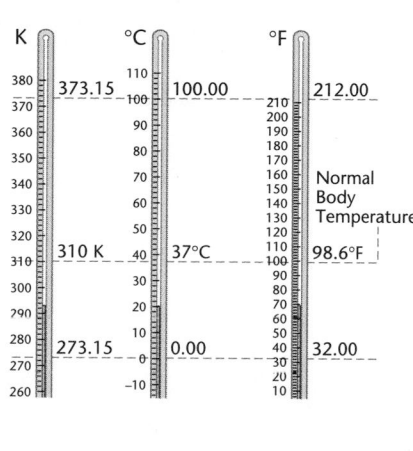

K   °C   °F
373.15   100.00   212.00

310 K   37°C   98.6°F   Normal Body Temperature

273.15   0.00   32.00

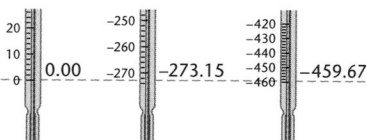

0.00   −273.15   −459.67

Mass and Weight   One of the most useful ways of describing an amount of stuff is to state its mass. You measure the mass of an object on a balance. Even though you are measuring mass, most people still refer to it as *weighing*. You might think that mass and weight are the same quantity. They are not. The mass of an object is a measure of its inertia; that is, its resistance to changes in motion. The inertia of an object is determined by its quantity of mass. A pint of sand has more matter than a pint of water; therefore, it has more mass.

The weight of an object is the amount of gravitational force acting on the mass of the object. On Earth, you sense the weight of an object by holding it and feeling the pull of gravity on it.

An important aspect of weight is that it is directly proportional to the mass of the object. This relationship means that the weight of a 2-kg object is twice the weight of a 1-kg object. Therefore, the pull of gravity on a 2-kg object is twice as great as the pull on a 1-kg object. Because it's easier to measure the effect of the pull of gravity rather than the resistance of an object to changes in its motion, mass can be determined by a weighing process.

Two instruments used to determine mass are the double-pan balance and a triple-beam balance. How each functions is illustrated in **Figure 2.** You can read how an electronic balance functions in *How it Works* in Chapter 12.

Figure 2

Measuring Mass
In a double-pan balance, the pull of gravity on the mass of an object placed in one pan causes the pan to rotate downward. To balance the rotation, an equal downward pull must be exerted on the opposite pan, as in a seesaw. You can produce this pull by placing calibrated masses on the opposite pan. When the two pulls are balanced, the masses in each pan are equal. ▶

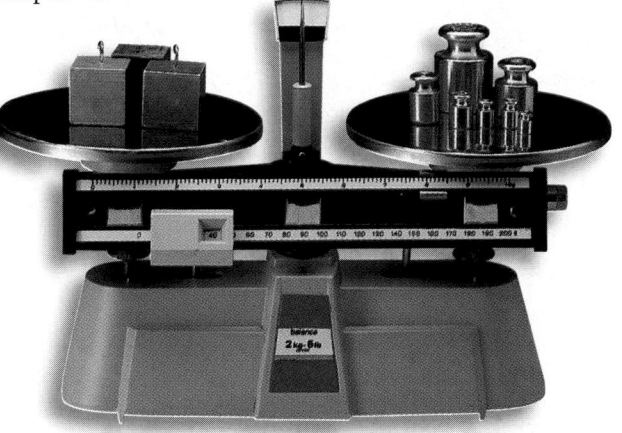

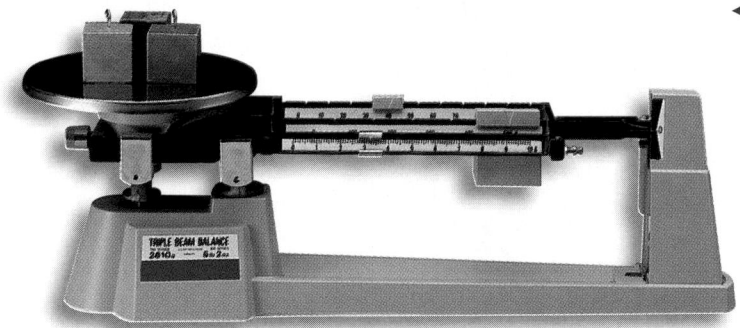

◀ In a triple-beam balance, weights of different sizes are placed at notched locations along each of two beams, and a third is slid along an arm to balance the object being weighed. To determine the mass of the object, you add the numbered positions of the three weights. Using a triple-beam balance is usually a much quicker way to determine the mass of an object than using a double-pan balance.

Figure 3 illustrates several familiar objects and their masses.

Figure 3
Mass
Because weight and mass are so closely related, the contents of cans and boxes are labeled both in pounds and ounces, the British/American weight units, and grams or kilograms, the metric mass units.

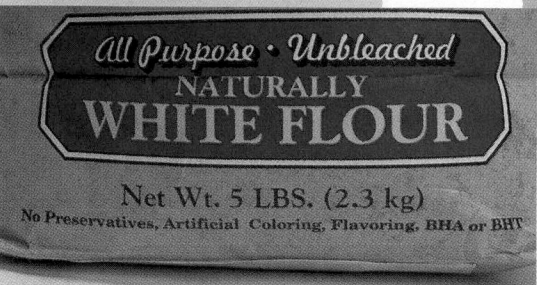

Making and Interpreting Measurements

Using measurements in science is different from manipulating numbers in math class. The important difference is that numbers in science are almost always measurements, which are made with instruments of varying accuracy. As you will see, the degree of accuracy of measured quantities must always be taken into account when expressing, multiplying, dividing, adding, or subtracting them. In making or interpreting a measurement, you should consider two points. The first is how well the instrument measures the quantity you're interested in. This point is illustrated in **Figure 4.**

Figure 4

Determining Length

Using the calibrations on the top ruler, you know that the length of the strip is between 4 cm and 5 cm. Because there are no finer calibrations between 4 cm and 5 cm, you have to estimate the length beyond the last calibration. An estimate would be 0.3 cm. Of course, someone might estimate it as 0.2 cm; others as 0.4 cm. Even though you record the measurement of the strip's length as 4.3 cm, you and others reading the measurement should interpret the measurement as 4.3 ± 0.1 cm.

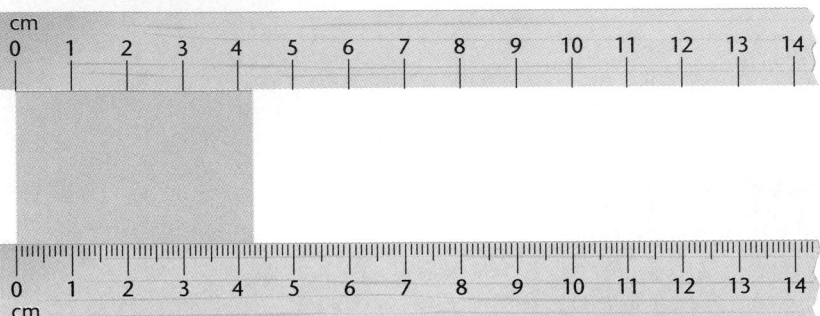

In the bottom ruler, you can see that the edge of the strip lies between 4.2 cm and 4.3 cm. Because there are no finer calibrations between 4.2 cm and 4.3 cm, you estimate the length beyond the last calibration. In this case, you might estimate it as 0.07 cm. You would record this measurement as 4.27 cm. You and others should interpret the measurement as 4.27 ± 0.01 cm.

Because the bottom ruler in **Figure 4** is calibrated to smaller divisions than the top ruler, the lower ruler has more precision than the upper. Any measurement you make with it will be more precise than one made on the top ruler because it will contain a smaller estimated value.

The second point to consider in making or interpreting a measurement is how well the measurement represents the quantity you're interested in. Looking at **Figure 4,** you can see how well the edge of the paper strip aligns with the value 4.27 cm. From sight, you know that 4.27 cm is a better representation of the strip's length than 4.3 cm. Because 4.27 cm better represents the length of the strip (the quantity you're interested in) than does 4.3 cm, it is a more accurate measurement of the strip's length.

In **Figure 5,** you will see that a more precise measurement might not be a more accurate measurement of a quantity.

GLENCOE TECHNOLOGY

 Software

Mastering Concepts in Chemistry
Unit 5, Lesson 2
Using Significant Digits
Use this lesson to introduce the concept of significant digits in expressing and calculating with measurements.

Figure 5
Precise and Accurate Measurements
Describing the width of the index card as 10.16 cm indicates that the ruler has 0.1-cm calibrations and the 0.06 cm is an estimation. Similarly, the uniform alignment of the edge of the index card with the ruler indicates that the measurement 10.16 cm is also an accurate measurement of width. ▶

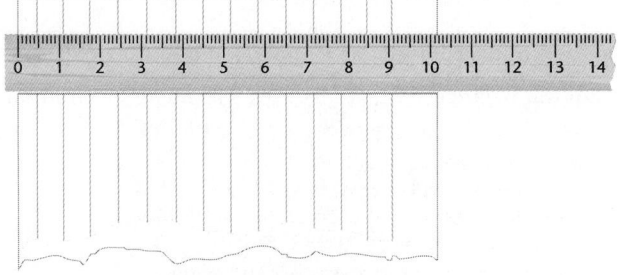

Describing the width of a brick as 10.16 cm indicates that this measurement is as precise as the measurement of the index card. However, 10.16 cm isn't a good representation of the width of the ragged- and jagged-edged brick. ▶

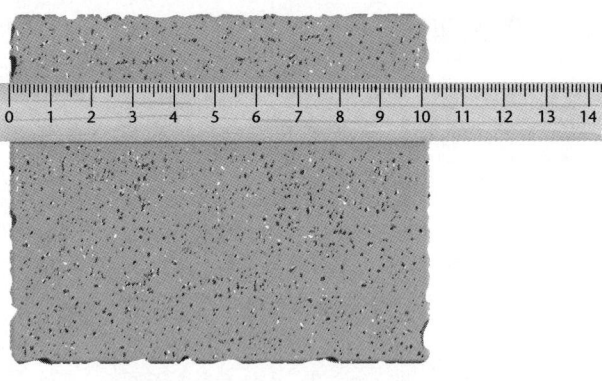

A better representation of the width of the brick is made using a ruler with less precision. As you can see, 10.2 cm is a better representation, and therefore a more accurate measurement, of the brick's width than 10.16 cm. ▶

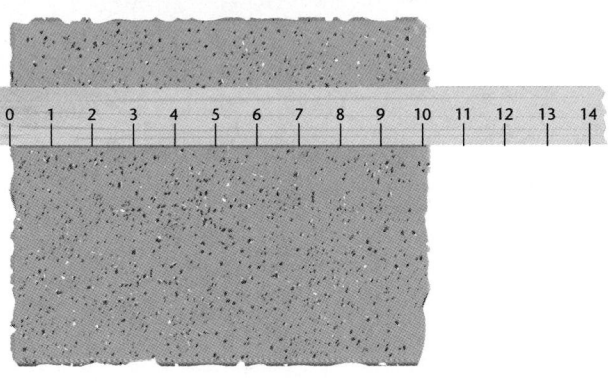

Expressing the Accuracy of Measurements

In measuring the length of the strip as 4.3 cm and 4.27 cm in **Figure 4,** you were aware of the difference in the calibration of the two rulers. This difference appeared in the way each measurement was recorded. In one measurement, the digits 4 and 3 were meaningful. In the second, the digits 4, 2, and 7 were meaningful. In any measurement, meaningful digits are called *significant digits.* The significant digits in a measured quantity include all the digits you know for sure, plus the final estimated digit.

Appendix A / Chemistry Skill Handbook   **793**

Significant Digits   Several rules can help you express or interpret which digits of a measurement are significant. Notice how those rules apply to readings from a digital balance (left) and a graduated cylinder (right).

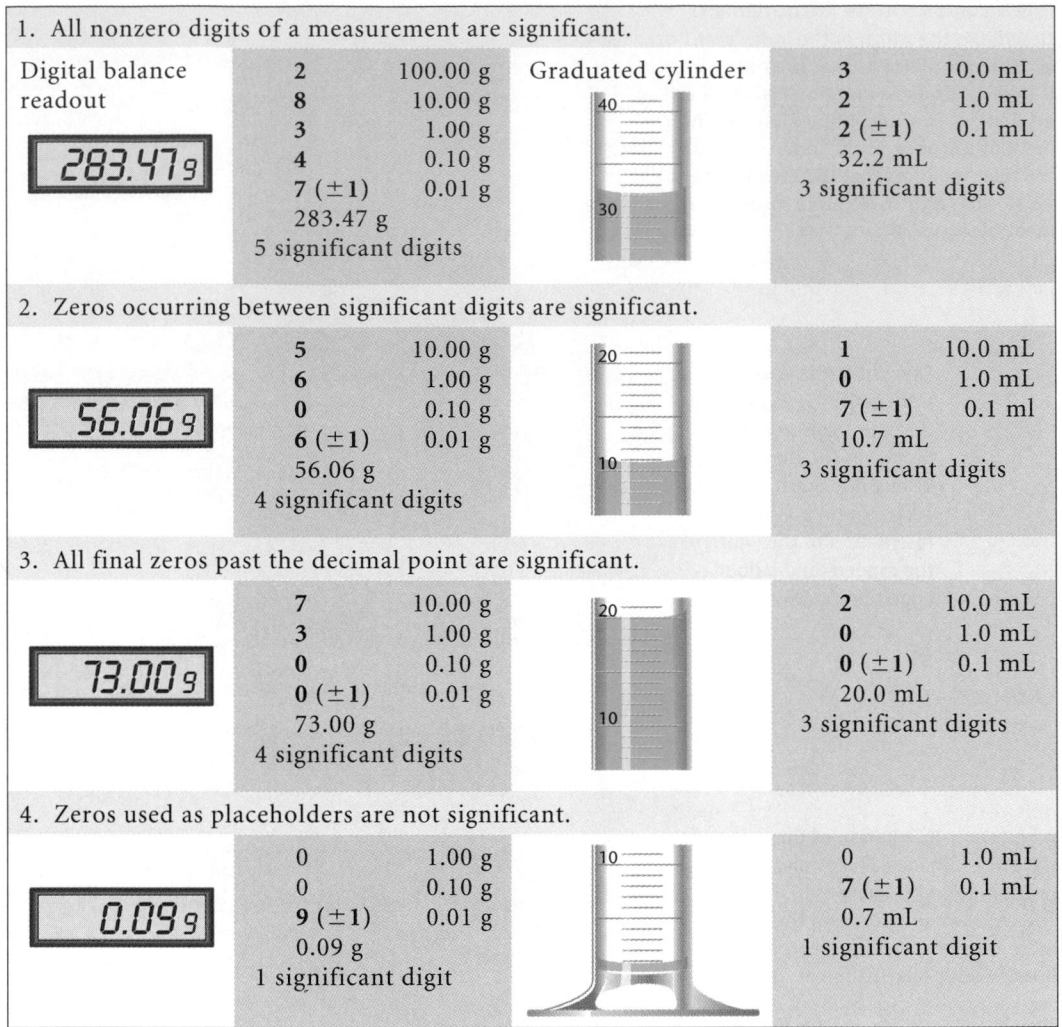

1. **All nonzero digits of a measurement are significant.**

Digital balance readout

283.47 g

2	100.00 g
8	10.00 g
3	1.00 g
4	0.10 g
7 (±1)	0.01 g

283.47 g
5 significant digits

Graduated cylinder

3	10.0 mL
2	1.0 mL
2 (±1)	0.1 mL

32.2 mL
3 significant digits

2. **Zeros occurring between significant digits are significant.**

56.06 g

5	10.00 g
6	1.00 g
0	0.10 g
6 (±1)	0.01 g

56.06 g
4 significant digits

1	10.0 mL
0	1.0 mL
7 (±1)	0.1 ml

10.7 mL
3 significant digits

3. **All final zeros past the decimal point are significant.**

73.00 g

7	10.00 g
3	1.00 g
0	0.10 g
0 (±1)	0.01 g

73.00 g
4 significant digits

2	10.0 mL
0	1.0 mL
0 (±1)	0.1 mL

20.0 mL
3 significant digits

4. **Zeros used as placeholders are not significant.**

0.09 g

0	1.00 g
0	0.10 g
9 (±1)	0.01 g

0.09 g
1 significant digit

| 0 | 1.0 mL |
| 7 (±1) | 0.1 mL |

0.7 mL
1 significant digit

The fourth rule sometimes causes difficulty in expressing such measurements as 20 L. Because the zero is a placeholder in the measurement, it is not significant and 20 L has one significant digit. You should interpret the measurement 20 L as 20 L plus or minus the value of the least significant digit, which is 10 L. Thus, a volume measurement of 20 L indicates 20 L ± 10 L, a range of 10-30 L. However, suppose you made the measurement with a device that is accurate to the nearest 1 L. You would want to indicate that both the 2 and the 0 are significant. How would you do this? You can't just add a decimal point after the 20 because it could be mistaken for a period. Adding a decimal point followed by a zero would indicate that the final zero past the decimal is significant (Rule 3), and the measurement would then be 20.0 ± 0.1 L. To solve the dilemma, you have to express 20 L as 2.0×10^1 L. Now the zero is a significant digit because it is a final zero past the decimal (Rule 3). The measurement has two significant digits and signifies $(2.0 \pm 0.1) \times 10^1$ L.

Expressing Quantities with Scientific Notation

The most common use of scientific notation is in expressing measurements of very large and very small dimensions. Using scientific notation is sometimes referred to as using *powers of ten* because it expresses quantities by using a small number between one and ten, which is then multiplied by ten to a power to give the quantity its proper magnitude. Suppose you went on a trip of 9000 km. You know that $10^3 = 1000$, so you could express the distance of your trip as 9×10^3 km. In this example, it may seem that scientific notation wouldn't be terribly useful. However, consider that an often-used quantity in chemistry is 602 000 000 000 000 000 000 000, the number of atoms or molecules in a mole of a substance. Recall that the mole is the SI unit of amount of a substance. Rather than writing out this huge number every time it is used, it's much easier to express it in scientific notation.

Determining Powers of 10   To determine the exponent of ten, count as you move the decimal point left until it falls just after the first nonzero digit—in this case, 6. If you try this on the number above, you'll find that you've moved the decimal point 23 places. Therefore, the number expressed in scientific notation is 6.02×10^{23}.

Expressing small measurements in scientific notation is done in a similar way. The diameter of a carbon atom is 0.000 000 000 000 154 m. In this case, you move the decimal point right until it is just past the first nonzero digit—in this case, 1. The number of places you move the decimal point right is expressed as a *negative* exponent of ten. The diameter of a carbon atom is 1.54×10^{-13} m. You always move the decimal point until the coefficient of ten is between one and less than ten. Thus, scientific notation always has the form, $M \times 10^n$ where $1 \le M < 10$.

Notice how the following examples are converted to scientific notation.

Quantities greater than 1

17.16 g   17 . 16 $\rightarrow$ 1.716×10^1 g   Decimal point moved 1 place left.

152.6 L   152 . 6 $\rightarrow$ 1.526×10^2 L   Decimal point moved 2 places left.

73 621 kg   73 621. $\rightarrow$ 7.3621×10^4 kg   Decimal point moved 4 places left.

Quantities between 0 and 1

0.29 mL   0 . 29 $\rightarrow$ 2.9×10^{-1} mL   Decimal point moved 1 place right.

0.0672 m   0 . 0672 $\rightarrow$ 6.72×10^{-2} m   Decimal point moved 2 places right.

0.0008 g   0 . 0008 $\rightarrow$ 8×10^{-4} g   Decimal point moved 4 places right.

Calculations with Measurements You often must use measurements to calculate other quantities. Remember that the accuracy of measurement depends on the instrument used and that accuracy is expressed as a certain number of significant digits. Therefore, you must indicate which digits in the result of any mathematical operation with measurements are significant. The rule of thumb is that no result can be more accurate than the least accurate quantity used to calculate that result. Therefore, the quantity with the least number of significant digits determines the number of significant digits in the result.

The method used to indicate significant digits depends on the mathematical operation.

Addition and Subtraction
The answer has only as many decimal places as the measurement having the least number of decimal places.

190.2 g	
65.291 g	
12.38 g	

190.2 $\ $ g
65.291 g
12.38 $\ $ g
267.871 g

The answer is rounded to the nearest tenth, which is the accuracy of the least accurate measurement.
267.9 g

Because the masses were measured on balances differing in accuracy, the least accurate measurement limits the number of digits past the decimal point.

Multiplication and Division
The answer has only as many significant digits as the measurement with the least number of significant digits.

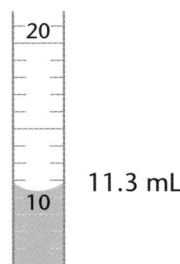

11.3 mL

13.78 g

$$density = \frac{mass}{volume}$$

$$D = \frac{m}{V} =$$

$$\frac{13.78 \text{ g}}{11.3 \text{ mL}} = 1.219469 \text{ g/mL}$$

The answer is rounded to three significant digits, 1.22 g/mL, because the least accurate measurement, 11.3 mL, has three significant digits.

Multiplying or dividing measured quantities results in a derived quantity. For example, the mass of a substance divided by its volume is its density. But mass and volume are measured with different tools, which may have different accuracies. Therefore, the derived quantity can have no more significant digits than the least accurate measurement used to compute it.

Practice Problems

3. Determine the number of significant digits in each of the following measurements.
   - **a)** 64 mL
   - **b)** 0.650 g
   - **c)** 30 cg
   - **d)** 724.56 mm
   - **e)** 47 080 km
   - **f)** 0.072 040 g
   - **g)** 1.03 mm
   - **h)** 0.001 mm

4. Write each of the following measurements in scientific notation.
   - **a)** 76.0°C
   - **b)** 212 mm
   - **c)** 56.021 g
   - **d)** 0.78 L
   - **e)** 0.076 12 m
   - **f)** 763.01 g
   - **g)** 10 301 980 nm
   - **h)** 0.001 mm

5. Write each of the following measurements in scientific notation.
   - **a)** 73 000 ± 1 mL
   - **b)** 4000 ± 1000 kg
   - **c)** 100 ± 10 cm
   - **d)** 100 000 ± 1000 km

6. Solve the following problems and express the answer in the correct number of significant digits.

   a)
   $$\begin{array}{r} 45.761 \text{ g} \\ -\ 42.65\ \text{ g} \\ \hline \end{array}$$

   c)
   $$\begin{array}{r} 0.340 \text{ cg} \\ 1.20\ \text{ cg} \\ +\ 1.018 \text{ cg} \\ \hline \end{array}$$

   b)
   $$\begin{array}{r} 1.6\ \text{ km} \\ +\ 0.62 \text{ km} \\ \hline \end{array}$$

   d)
   $$\begin{array}{r} 6\ 000\ \mu\text{m} \\ -\ 202\ \mu\text{m} \\ \hline \end{array}$$

7. Solve the following problems and express the answer in the correct number of significant digits.
   - **a)** 5.761 cm × 6.20 cm
   - **b)** $\dfrac{23.5 \text{ kg}}{4.615 \text{ m}^3}$
   - **c)** $\dfrac{0.2 \text{ km}}{5.4 \text{ s}}$
   - **d)** 11.00 m × 12.10 m × 3.53 m
   - **e)** $\dfrac{4.500 \text{ kg}}{1.500 \text{ m}^2}$
   - **f)** $\dfrac{18.21 \text{ g}}{4.4 \text{ cm}^3}$

Adding and Subtracting Measurements in Scientific Notation   Adding and subtracting measurements in scientific notation requires that for any problem, the measurements must be expressed as the same power of ten. For example, in the following problem, the three length measurements must be expressed in the same power of ten.

$$\begin{array}{r} 1.1012 \times 10^4 \text{ mm} \\ 2.31\ \ \times 10^3 \text{ mm} \\ +\ 4.573\ \ \times 10^2 \text{ mm} \\ \hline \end{array}$$

d. 11.00 m × 12.10 m × 3.53 m = 469.843 m³
rounds to 470 m³ but must be expressed as 4.70×10^2 to indicate three significant digits

e. $\dfrac{4.500 \text{ kg}}{1.500 \text{ m}^2} = 3.000 \text{ kg/m}^2$
Note that the result must be expressed as 3.000 kg/m² to show four significant digits.

f. $\dfrac{18.21 \text{ g}}{4.4 \text{ cm}^3} = 4.1386 \text{ g/cm}^3$
rounds to 4.1 g/cm³

3. **a.** 2
   b. 3
   c. 1
   d. 5
   e. 4
   f. 5
   g. 3
   h. 1

4. **a.** 7.60×10^1 °C
   b. 2.12×10^2 mm
   c. 5.6021×10^1 g
   d. 7.8×10^{-1} L
   e. 7.612×10^{-2} m
   f. 7.6301×10^2 g
   g. $1.030\ 198\ 0 \times 10^7$ nm
   h. 1×10^{-3} mm

5. **a.** 7.3000×10^4 mL
   b. 4×10^3 kg
   c. 1.0×10^1 cm
   d. 1.00×10^5 km

6. **a.**
   $$\begin{array}{r} 45.761 \text{ g} \\ -\ 42.65\ \text{ g} \\ \hline 3.111 \text{ g} \end{array}$$
   rounds to 3.11 g

   b.
   $$\begin{array}{r} 1.6\ \text{ km} \\ +\ 0.62 \text{ km} \\ \hline 2.22 \text{ km} \end{array}$$
   rounds to 2.2 km

   c.
   $$\begin{array}{r} 0.340 \text{ cg} \\ 1.20\ \text{ cg} \\ +\ 1.018 \text{ cg} \\ \hline 2.558 \text{ cg} \end{array}$$
   rounds to 2.56 cg

   d.
   $$\begin{array}{r} 6000\ \mu\text{m} \\ -\ 202\ \mu\text{m} \\ \hline 5798\ \mu\text{m} \end{array}$$
   rounds to 6000 μm

Notice that the value 6000 μm has only one significant digit. In this case, significant digits take precedence over decimal places. Use this example to point out to students why expressing quantities in scientific notation makes working with measurements less ambiguous.

7. **a.** 5.761 cm × 6.20 cm = 35.7182 cm²
   rounds to 35.7 cm²

   b. $\dfrac{23.5 \text{ kg}}{4.615 \text{ m}^3} = 5.09209 \text{ kg/m}^3$
   rounds to 5.09 kg/m³

   c. $\dfrac{0.2 \text{ km}}{5.4 \text{ s}} = 0.037037 \text{ km/s}$
   rounds to 0.04 km/s

In adding and subtracting measurements in scientific notation, all measurements are expressed in the same order of magnitude as the measurement with the greatest power of ten. When converting a quantity, the decimal point is moved one place to the left for each increase in power of ten.

$$2.31 \times 10^3 \qquad 2.31 \times 10^3 \rightarrow 0.231 \times 10^4$$

$$4.573 \times 10^2 \qquad 4.573 \times 10^2 \rightarrow 0.4573 \times 10^3 \rightarrow 0.04573 \times 10^4$$

$$
\begin{array}{r}
1.1012 \times 10^4 \text{ mm} \\
0.231 \times 10^4 \text{ mm} \\
+ \ 0.04573 \times 10^4 \text{ mm} \\
\hline
1.37793 \times 10^4 \text{ mm} = 1.378 \times 10^4 \text{ mm (rounded)}
\end{array}
$$

Multiplying and Dividing Measurements in Scientific Notation   Multiplying and dividing measurements in scientific notation requires that similar operations are done to the numerical values, the powers of ten, and the units of the measurements.

a) The numerical coefficients are multiplied or divided and the resulting value is expressed in the same number of significant digits as the measurement with the least number of significant digits.

b) The exponents of ten are algebraically added in multiplication and subtracted in division.

c) The units are multiplied or divided.

The following problems illustrate these procedures.

Sample Problem 1

$$(3.6 \times 10^3 \text{ m})(9.4 \times 10^3 \text{ m})(5.35 \times 10^{-1} \text{ m})$$
$$= (3.6 \times 9.4 \times 5.35) \times (10^3 \times 10^3 \times 10^{-1}) \ (\text{m} \times \text{m} \times \text{m})$$
$$= (3.6 \times 9.4 \times 5.35) \times 10^{(3+3+(-1))} \ (\text{m} \times \text{m} \times \text{m})$$
$$= (181.044) \times 10^5 \text{ m}^3$$
$$= 1.8 \times 10^2 \times 10^5 \text{ m}^3$$
$$= 1.8 \times 10^7 \text{ m}^3$$

Sample Problem 2

$$\frac{6.762 \times 10^2 \text{ m}^3}{(1.231 \times 10^1 \text{ m})(2.80 \times 10^{-2} \text{ m})}$$

$$= \frac{6.762}{1.231 \times 2.80} \times \frac{10^2}{10^1 \times 10^{-2}} \times \frac{\text{m}^3}{\text{m} \times \text{m}}$$

$$= 1.961819659 \times 10^{(2-(+1-2))} \text{ m}^{(3-(+2))}$$
$$= 1.96 \times 10^{(2-(-1))} \text{ m}^{(1)}$$
$$= 1.96 \times 10^3 \text{ m}$$

Practice Problems

8. Solve the following addition and subtraction problems.
   a) 1.013×10^3 g $+ 8.62 \times 10^2$ g $+ 1.1 \times 10^1$ g
   b) 2.82×10^6 m $- 4.9 \times 10^4$ m

9. Solve the following multiplication and division problems.
   a) 1.18×10^{-3} m $\times 4.00 \times 10^2$ m $\times 6.22 \times 10^2$ m
   b) 3.2×10^2 g $\div 1.04 \times 10^2$ cm^2 $\div 6.22 \times 10^{-1}$ cm

Computations with the Calculator

Working problems in chemistry will require you to have a good understanding of some of the advanced functions of your calculator. When using a calculator to solve a problem involving measured quantities, you should remember that the calculator does not take significant digits into account. It is up to you to round off the answer to the correct number of significant digits at the end of a calculation. In a multistep calculation, you should not round off after each step. Instead, you should complete the calculation and then round off. **Figure 6** shows how to use a calculator to solve a subtraction problem involving quantities in scientific notation.

Figure 6

Subtracting Numbers in Scientific Notation with a Calculator

A quantity in scientific notation is entered by keying in the coefficient and then striking the [EXP] or [EE] key followed by the value of the exponent of ten. At the end of the calculation, the calculator readout must be corrected to the appropriate number of decimal places. The answer would be rounded to the second decimal place and expressed as 2.56×10^4 kg.

To solve the problem

$$2.61 \times 10^4$$
$$- 5.2 \times 10^2$$

Keystrokes

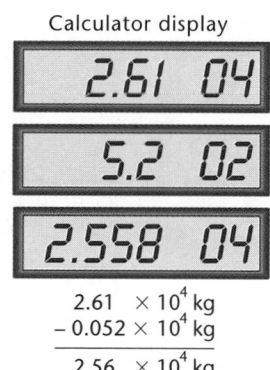

Calculator display

$$
\begin{array}{r}
2.61 \ \times 10^4 \text{ kg} \\
- 0.052 \times 10^4 \text{ kg} \\
\hline
2.56 \ \times 10^4 \text{ kg}
\end{array}
$$

8. **a.**
$$
\begin{array}{r}
1.013 \times 10^3 \text{ g} \\
8.62 \times 10^2 = 0.862 \times 10^3 \text{ g} \\
1.1 \times 10^1 = 0.011 \times 10^3 \text{ g} \\
\hline
1.886 \times 10^3 \text{ g}
\end{array}
$$

b.
$$
\begin{array}{r}
2.82 \ \times 10^6 \text{ m} \\
4.9 \times 10^4 = 0.049 \times 10^6 \text{ m} \\
\hline
2.771 \times 10^6 \text{ m}
\end{array}
$$

rounds to 2.77×10^6 m

9. **a.** $(1.18 \times 10^{-3}$ m$)(4.00 \times 10^2$ m$)(6.22 \times 10^2$ m$)$
$= (1.18 \times 4.00 \times 6.22) \times (10^{-3} \times 10^2 \times 10^2)$ (m $\times$ m $\times$ m)
$= (1.18 \times 4.00 \times 6.22) \times 10^{((-3)+2+2)}$ (m $\times$ m $\times$ m)
$= (29.3584) \times 10^1$ m^3
$= 2.94 \times 10^2 \times 10^1$ m^3
$= 2.94 \times 10^2$ m^3

b.
$$
\frac{3.2 \times 10^2 \text{ g}}{(1.04 \times 10^2 \text{ cm}^2) \times (6.22 \times 10^{-1} \text{ cm})}
$$
$$
= \frac{3.2}{1.04 \times 6.22} \times \frac{10^2}{10^2 \times 10^{-1}} \times \frac{\text{g}}{\text{cm}^2 \times \text{cm}}
$$

$= 0.494682166 \times 10^{(2-(2+(-1)))}$ g/cm$^{(2-1)}$
$= 0.49 \times 10^{(2-1)}$ g/cm^3
$= 0.49 \times 10^1$ g/cm^3
$= 4.9 \times 10^0$ g/cm^3 or 4.9 g/cm^3

Look at **Figure 7** to see how to solve multiplication and division problems involving scientific notation. The problems are the same as the two previous sample problems.

Figure 7

Multiplying and Dividing Measurements Expressed in Scientific Notation
A negative power of ten is usually entered by striking the [EXP] or [EE], entering the positive value of the exponent, and then striking the [±] key. ▼

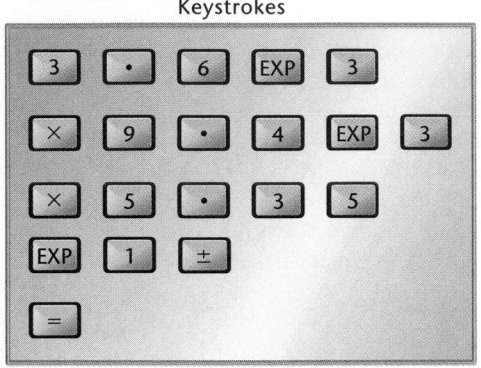

Keystrokes

3	.	6	EXP	3
×	9	.	4	EXP	3
×	5	.	3	5
EXP	1	±
=

Calculator display

3.6   03

9.4   03

5.35   -01

1.8104   07

Rounded off to 2 significant digits: 1.8×10^7

The numerical value of the answer can have no more significant digits than the measurement that has the least number of significant digits. ▼

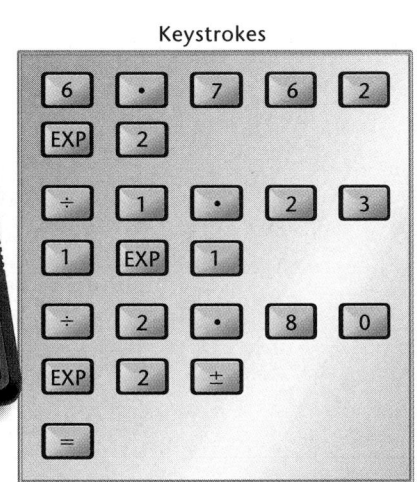

Keystrokes

6	.	7	6	2
EXP	2
÷	1	.	2	3
1	EXP	1
÷	2	.	8	0
EXP	2	±
=

Calculator display

6.762   02

1.231   01

2.80   -02

1.9618   03

Rounded off to 3 significant digits: 1.96×10^3

Practice Problems

10. Solve the following problems and express the answers in scientific notation with the proper number of significant digits.

a)
$$2.01 \times 10^2 \text{ mL}$$
$$3.1 \ \times 10^1 \text{ mL}$$
$$+ \ 2.712 \times 10^3 \text{ mL}$$

b)
$$7.40 \times 10^2 \text{ mm}$$
$$- \ 4.0 \ \times 10^1 \text{ mm}$$

c)
$$2.10 \times 10^1 \text{ g}$$
$$- \ 1.6 \ \times 10^{-1} \text{ g}$$

d)
$$5.131 \times 10^2 \text{ J}$$
$$2.341 \times 10^1 \text{ J}$$
$$+ \ 3.781 \times 10^3 \text{ J}$$

11. Solve the following problems and express the answers in scientific notation with the proper number of significant digits.

a) $(2.00 \times 10^1 \text{ cm})(2.05 \times 10^1 \text{ cm})$

b) $\dfrac{5.6 \times 10^3 \text{ kg}}{1.20 \times 10^4 \text{ m}^3}$

c) $(2.51 \times 10^1 \text{ m})(3.52 \times 10^1 \text{ m})(1.2 \times 10^{-1} \text{ m})$

d) $\dfrac{1.692 \times 10^4 \text{ dm}^3}{(2.7 \times 10^{-2} \text{ dm})(4.201 \times 10^1 \text{ dm})}$

Using the Factor Label Method

The factor label method is used to express a physical quantity such as the length of a pen in any other unit that measures that quantity. For example, you can measure the pen with a metric ruler calibrated in centimeters and then express the length in meters.

If the length of a pen is measured as 14.90 cm, you can express that length in meters by using the numerical relationship between a centimeter and a meter. This relationship is given by the following equation.

$$100 \text{ cm} = 1 \text{ m}$$

If both sides of the equation are divided by 100 cm, the following relationship is obtained.

$$1 = \frac{1 \text{ m}}{100 \text{ cm}}$$

To express 14.90 cm as a measurement in meters, you multiply the quantity by the relationship, which eliminates the cm unit.

$$14.90 \ \text{cm} \times \frac{1 \text{ m}}{100 \ \text{cm}} = \frac{14.90 \ \text{cm}}{1} \times \frac{1 \text{ m}}{100 \ \text{cm}} = \frac{14.90}{100} \text{ m} = 0.1490 \text{ m}$$

Students should work out problems 10 and 11 on a scientific calculator.

10. When using a calculator to do an addition or subtraction problem containing scientific notation, students must inspect the problem to determine where to round off. Often, this will be determined by the number of decimal places in the measurement with the largest power of ten. This rule of thumb holds in the following four sample problems. You might want to have students consider the following addition problem to see an exception. 4.1449×10^1 g $+ \ 3.2 \times 10^{-1}$ g $= 41.77$ g Here, the value 3.2×10^{-1} g expresses accuracy only to ± 0.01 g, which is less accurate than the measurement 4.1449×10^1 g (41.449 g ± 0.001 g).

a. 2.944×10^3 mL

b. 7.00×10^2 mm

c. 2.08×10^1 g

d. 4.318×10^3 J

11. a. 4.10×10^2 cm^2

b. 4.666666 rounds to 4.7×10^{-1} kg/m^3

c. 1.060224 rounds to 1.1×10^2 m^3

d. 1.491708 rounds to 1.5×10^4 dm

Copper weathervane

The factor label method doesn't change the value of the physical quantity because you are multiplying that value by a factor that equals 1. You choose the factor so that when the unit you want to eliminate is multiplied by the factor, that unit and the similar unit in the factor cancel. If the unit you want to eliminate is in the numerator, choose the factor which has that unit in the denominator. Conversely, if the unit you want to eliminate is in the denominator, choose the factor which has that unit in the numerator. For example, in a chemistry lab activity, a student measured the mass and volume of a chunk of copper and calculated its density as 8.80 g/cm³. Knowing that 1000 g = 1 kg and 100 cm = 1 m, the student could then use the following factor label method to express this value in the SI unit of density, kg/m³.

$$8.80\ \frac{g}{cm^3} \times \frac{1\ kg}{1000\ g} \times \left[\frac{100\ cm}{1\ m}\right]^3 =$$

$$8.80\ \frac{g}{cm \times cm \times cm} \times \frac{1\ kg}{1000\ g} \times \frac{100\ cm}{1\ m} \times \frac{100\ cm}{1\ m} \times \frac{100\ cm}{1\ m} =$$

$$\frac{8.80 \times (10^2 \times 10^2 \times 10^2)}{1000}\ \frac{kg}{m^3} = \frac{8.80 \times 10^{(2+2+2)}}{10^3}\ \frac{kg}{m^3} =$$

$$8.80 \times 10^{(6-3)}\ kg/m^3 = 8.80 \times 10^3\ kg/m^3$$

The factor label method can be extended to other types of calculations in chemistry. To use this method, you first examine the data that you have. Next, you determine the quantity you want to find and look at the units you will need. Finally, you apply a series of factors to the data in order to convert it to the units you need.

Sample Problem 1

The density of silver sulfide (Ag_2S) is 7.234 g/mL. What is the volume of a lump of silver sulfide that has a mass of 6.84 kg?

First, you must apply a factor that will convert kg of Ag_2S to g Ag_2S.

$$\frac{6.84 \text{ kg } Ag_2S}{} \left| \frac{1000 \text{ g } Ag_2S}{1 \text{ kg } Ag_2S} \cdots \right.$$

Next, you use the density of Ag_2S to convert mass to volume.

$$\frac{6.84 \text{ kg } Ag_2S}{} \left| \frac{1000 \text{ g } Ag_2S}{1 \text{ kg } Ag_2S} \right| \frac{1 \text{ mL } Ag_2S}{7.234 \text{ g } Ag_2S} = 946 \text{ mL } Ag_2S$$

Notice that the new factor must have grams in the denominator so that grams will cancel out, leaving mL.

Sample Problem 2

What mass of lead can be obtained from 47.2 g of $Pb(NO_3)_2$?

Because mass is involved, you will need to know the molar mass of $Pb(NO_3)_2$.

$$\begin{aligned} Pb &= 207.2 \text{ g} \\ 2N &= 28.014 \text{ g} \\ 6O &= 95.994 \text{ g} \end{aligned}$$

Molar mass of $Pb(NO_3)_2$ = 331.208 g.
Rounding off according to the rules for significant digits, the molar mass of $Pb(NO_3)_2$ = 331.2 g.

You can see that the mass of lead in $Pb(NO_3)_2$ is 207.2/331.2 of the total mass of $Pb(NO_3)_2$.

Now you can set up a relationship to determine the mass of lead in the 47.2-g sample.

$$\frac{47.2 \text{ g } Pb(NO_3)_2}{} \left| \frac{207.2 \text{ g Pb}}{331.2 \text{ g } Pb(NO_3)_2} \right. = 29.52850242 \text{ g Pb}$$
$$= 29.5 \text{ g Pb rounded to } 3 \text{ significant digits.}$$

Notice that the equation is arranged so that the unit *g $Pb(NO_3)_2$* cancels out, leaving only *g Pb*, the quantity asked for in the problem.

Practice Problems in the Factor Label Method

12. Express each quantity in the unit listed to its right.
    a) 3.01 g   cg   e) 0.2 L dm^3
    b) 6200 m   km   f) 0.13 cal/g   J/kg
    c) 6.24×10^{-7} g   μg   g) 5 ft, 1 in. m
    d) 3.21 L   mL   h) 1.2 qt   L

probably more realistic to express the metric conversion to the nearest centimeter than to the nearest 10 cm.

h.

$$\frac{1.2 \text{ qt}}{} \left| \frac{0.947 \text{ L}}{1 \text{ qt}} \right. = 1.1364 \text{ L}$$

= 1.1 L (rounded)

In this case, a quart and a liter are close in magnitude, so the number of significant digits should be the same as that in the quart measure.

12. a.
$$\frac{3.01 \text{ g}}{} \left| \frac{10 \text{ cg}}{1 \text{ g}} \right. = 30.1 \text{ cg}$$

b.
$$\frac{6200 \text{ m}}{} \left| \frac{1 \text{ km}}{1000 \text{ m}} \right. = 6.2 \text{ km}$$

c.
$$\frac{6.24 \times 10^{-7} \text{ g}}{} \left| \frac{1 \times 10^6 \text{ μg}}{1 \text{ g}} \right.$$
$$= 6.24 \times 10^{-1} \text{ μg}$$

d.
$$\frac{3.21 \text{ L}}{} \left| \frac{1000 \text{ mL}}{1 \text{ L}} \right.$$
$$= 3210 \text{ mL}$$

e.
$$\frac{0.2 \text{ L}}{} \left| \frac{0.001 \text{ m}^3}{1 \text{ L}} \right| \frac{1000 \text{ dm}^3}{1 \text{ m}^3}$$
$$= 0.2 \text{ dm}^3$$

f.
$$\frac{0.13 \text{ cal}}{} \left| \frac{1000 \text{ g}}{1 \text{ kg}} \right| \frac{4.184 \text{ J}}{\text{cal}}$$
$$= 0.54392 \text{ J/kg}$$

$$= 0.54 \text{ J/kg (rounded)}$$

g. first convert feet to inches
$$\frac{5 \text{ ft}}{} \left| \frac{12 \text{ in.}}{1 \text{ ft}} \right. = 60 \text{ in.}$$

then add inches
60 in. + 1 in. = 61 in.

$$\frac{61 \text{ in.}}{} \left| \frac{1 \text{ m}}{39.37 \text{ in.}} \right.$$
$$= 1.549403099 \text{ m}$$
$$= 1.55 \text{ m (rounded)}$$

Rules for handling significant digits sometimes must be regarded only as guidelines. You must make a judgment when converting from one system of measurement to another. You have to decide how to express 5 ft, 1 in. as a metric measure to the same degree of accuracy as the English measure. Practical knowledge of both systems is needed. If a measurement is made to the nearest inch, it is

It is often necessary to compare and sequence observations and measurements. Two of the most useful ways are to organize the observations and measurements as tables and graphs. If you browse through your textbook, you'll see many tables and graphs. They arrange information in a way that makes it easier to understand.

Making and Using Tables

Most tables have a title telling you what information is being presented. The table itself is divided into columns and rows. The column titles list items to be compared. The row headings list the specific characteristics being compared among those items. Within the grid of the table, the information is recorded. Any table you prepare to organize data taken in a laboratory activity should have these characteristics. Consider, for example, that in a laboratory experiment, you are going to perform a flame test on various solutions. In the test, you place drops of the solution containing a metal ion in a flame, and the color of the flame is observed, as shown in **Figure 8.** Before doing the experiment, you might set up a data table like the one below.

Flame Test Results		
Solution	Metal ion	Color of Flame
KNO_3	K^+	violet-pink

Figure 8
◀ **Flame Test**
A drop of solution containing the potassium ion, K^+, causes the flame to burn with violet color.

While performing the experiment, you would record the name of the solution and then the observation of the flame color. If you weren't sure of the metal ion as you were doing the experiment, you could enter it into the table by checking oxidation numbers afterward. Not only does the table organize your observations, it could also be used as a reference to determine whether a solution of some unknown composition contains one of the metal ions listed in the table.

Making and Using Graphs

After organizing data in tables, scientists usually want to display the data in a more visual way. Using graphs is a common way to accomplish that. There are three common types of graphs—bar graphs, pie graphs, and line graphs.

Bar Graphs   Bar graphs are useful when you want to compare or display data that do not continuously change. Suppose you measure the rate of electrolysis of water by determining the volume of hydrogen gas formed. In addition, you decide to test how the number of batteries affects the rate of electrolysis. You could graph the results using a bar graph as shown in **Figure 9.** Note that you could construct a line graph, but the bar graph is better because there is no way you could use 0.4 or 2.6 batteries.

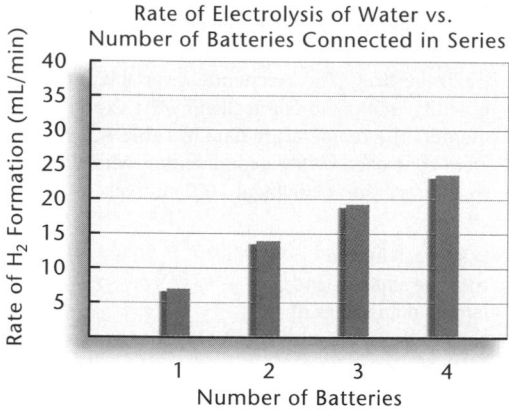

Figure 9
A Sample Bar Graph

Pie Graphs   Pie graphs are especially useful in comparing the parts of a whole. You could use a pie graph to display the percent composition of a compound such as sodium dihydrogen phosphate, NaH_2PO_4, as shown in **Figure 10.**

In constructing a pie graph, recall that a circle has 360°. Therefore, each fraction of the whole is that fraction of 360°. Suppose you did a census of your school and determined that 252 students out of a total of 845 were 17 years old. You would compute the angle of that section of the graph by multiplying $(252/845) \times 360° = 107°$.

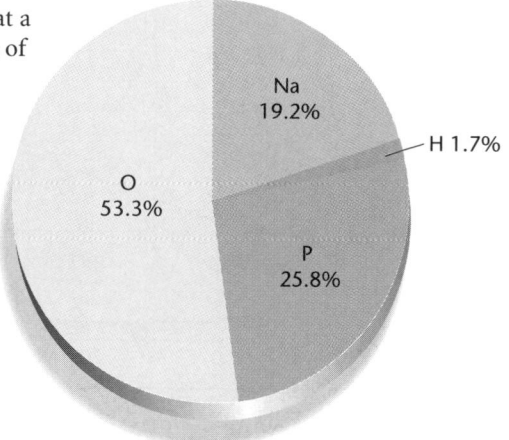

Composition of NaH_2PO_4

Figure 10
A Sample Pie Graph

Line Graphs   Line graphs have the ability to show a trend in one variable as another one changes. In addition, they can suggest possible mathematical relationships between variables.

Table A.5 shows the data collected during an experiment to determine whether temperature affects the mass of potassium bromide that dissolves in 100 g of water. If you read the data as you slowly run your fingers down both columns of the table, you will see that solubility increases as the temperature increases. This is the first clue that the two quantities may be related.

To see that the two quantities are related, you should construct a line graph as shown in **Figure 11**.

Table A.5 Effect of Temperature on Solubility of KBr	
Temperature (°C)	Solubility (g of KBr/100 g H₂O)
10.0	60.2
20.0	64.3
30.0	67.7
40.0	71.6
50.0	75.3
60.0	80.1
70.0	82.6
80.0	86.8
90.0	90.2

Figure 11

Constructing a Line Graph

1. Plot the independent variable on the *x*-axis (horizontal axis) and the dependent variable on the *y*-axis (vertical). The independent variable is the quantity changed or controlled by the experimenter. The temperature data in **Table A.5** were controlled by the experimenter, who chose to measure the solubility at 10°C intervals.

2. Scale each axis so that the smallest and largest data values of each quantity can be plotted. Use divisions such as ones, fives, or tens or decimal values such as hundredths or thousandths.

3. Label each axis with the appropriate quantity and unit.

4. Plot each pair of data from the table as follows.
   • Place a straightedge vertically at the value of the independent variable on the *x*-axis.
   • Place a straightedge horizontally at the value of the dependent variable on the *y*-axis.
   • Mark the point at which the straightedges intersect.

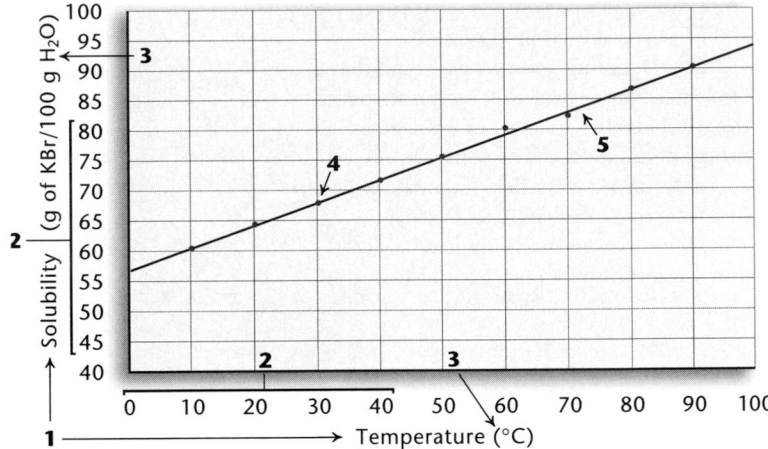

5. Fit the best straight line or curved line through the data points.

One use of a line graph is to predict values of the independent or dependent variables. For example, from **Figure 11** you can predict the solubility of KBr at a temperature of 65°C by the following method:

- Place a straightedge vertically at the approximate value of 65°C on the *x*-axis.

- Mark the point at which the straightedge intersects the line of the graph.

- Place a straightedge horizontally at this point and approximate the value of the dependent variable on the *y*-axis as 82 g/100 g H₂O.

To predict the temperature for a given solubility, you would reverse the above procedure.

Practice Problems

Use Figure 11 to answer these questions.

13. Predict the solubility of KBr at each of the following temperatures.
    a) 25.0°C   **c)** 6.0°C
    b) 52.0°C   **d)** 96.0°C

14. Predict the temperature at which KBr has each of the following solubilities.
    a) 70.0 g/100 g H₂O
    b) 88.0 g/100 g H₂O

Graphs of Direct and Inverse Relationships   Graphs can be used to determine quantitative relationships between the independent and dependent variables. Two of the most useful relationships are relationships in which the two quantities are directly proportional or inversely proportional.

When two quantities are directly proportional, an increase in one quantity produces a proportionate increase in the other. A graph of two quantities that are directly proportional is a straight line, as shown in **Figure 12.**

Figure 12

Graph of Quantities That Are Directly Proportional
As you can see, doubling the mass of the carbon burned from 2.00 g to 4.00 g doubles the amount of energy released from 66 kJ to 132 kJ. Such a relationship indicates that mass of carbon burned and the amount of energy released are directly proportional.

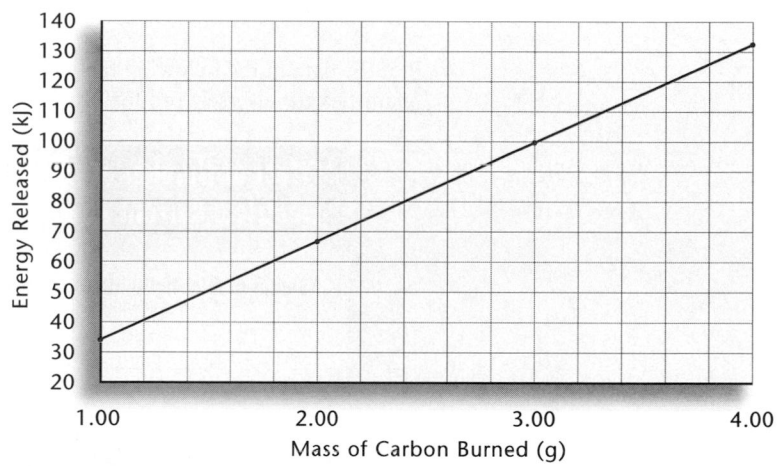

Mass of Carbon Burned (g)

13. **a.** 66 g KBr/100 g H₂O
    b. 76 g KBr/100 g H₂O
    c. 58 g KBr/100 g H₂O
    d. 93 g KBr/100 g H₂O
14. **a.** 36°C
    b. 87°C

15. See below, left. The quantities are directly proportional.

16. See below, right. The quantities are inversely proportional.

When two quantities are inversely proportional, an increase in one quantity produces a proportionate decrease in the other. A graph of two quantities that are inversely proportional is shown in **Figure 13.**

Figure 13

Graph of Quantities That Are Inversely Proportional
As you can see, doubling the pressure of the gas reduces the volume of the gas by one-half. Such a relationship indicates that the volume and pressure of a gas are inversely proportional.

Practice Problems

15. Plot the data in the following table and determine whether the two quantities are directly proportional.

Effect of Temperature on Gas Pressure

Temperature (K)	Pressure (kPa)
300.0	195
320.0	208
340.0	221
360.0	234
380.0	247
400.0	261

16. Plot the data in the following table and determine whether the two quantities are inversely proportional.

Effect of Number of Mini-Lightbulbs on Electrical Current in a Circuit

Number of mini-lightbulbs	Current (mA)
2	3.94
4	1.98
6	1.31
9	0.88

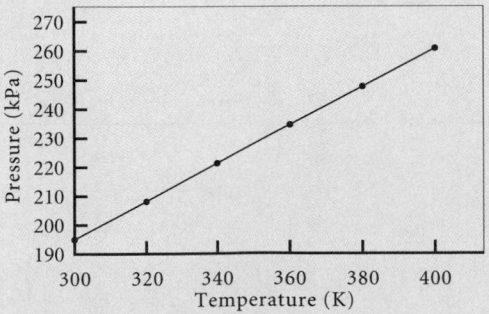

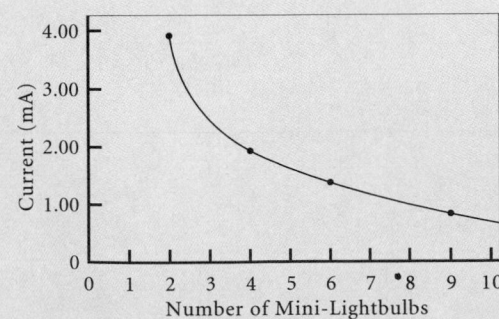

APPENDIX B
Safety Handbook

Safety Guidelines in the Chemistry Laboratory

The chemistry laboratory is a safe place to work if you are aware of important safety rules and if you are careful. You must be responsible for your own safety and for the safety of others. The safety rules given here will protect you and others from harm in the lab. While carrying out procedures in any of the **ChemLabs,** notice the safety symbols and caution statements. The safety symbols are explained in the chart on the next page.

1. Always obtain your teacher's permission to begin a lab.
2. Study the procedure. If you have questions, ask your teacher. Be sure you understand all safety symbols shown.
3. Use the safety equipment provided for you. Goggles and a safety apron should be worn when any lab calls for using chemicals.
4. When you are heating a test tube, always slant it so the mouth points away from you and others.
5. Never eat or drink in the lab. Never inhale chemicals. Do not taste any substance or draw any material into your mouth.
6. If you spill any chemical, wash it off immediately with water. Report the spill immediately to your teacher.
7. Know the location and proper use of the fire extinguisher, safety shower, fire blanket, first aid kit, and fire alarm.

8. Keep all materials away from open flames. Tie back long hair.
9. If a fire should break out in the classroom, or if your clothing should catch fire, smother it with the fire blanket or a coat, or get under a safety shower. **NEVER RUN.**
10. Report any accident or injury, no matter how small, to your teacher.

Follow these procedures as you clean up your work area.

1. Turn off the water and gas. Disconnect electrical devices.
2. Return materials to their places.
3. Dispose of chemicals and other materials as directed by your teacher. Place broken glass and solid substances in the proper containers. Never discard materials in the sink.
4. Clean your work area.
5. Wash your hands thoroughly after working in the laboratory.

First Aid in the Laboratory

Injury	Safe Response
Burns	Apply cold water. Call your teacher immediately.
Cuts and bruises	Stop any bleeding by applying direct pressure. Cover cuts with a clean dressing. Apply cold compresses to bruises. Call your teacher immediately.
Fainting	Leave the person lying down. Loosen any tight clothing and keep crowds away. Call your teacher immediately.
Foreign matter in eye	Flush with plenty of water. Use eyewash bottle or fountain.
Poisoning	Note the suspected poisoning agent and call your teacher immediately.
Any spills on skin	Flush with large amounts of water or use safety shower. Call your teacher immediately.

Safety Symbols

Disposal Alert

This symbol appears when care must be taken to dispose of materials properly.

Biological Hazard

This symbol appears when there is danger involving bacteria, fungi, or protists.

Open Flame Alert

This symbol appears when use of an open flame could cause a fire or an explosion.

Thermal Safety

This symbol appears as a reminder to use caution when handling hot objects.

Sharp Object Safety

This symbol appears when a danger of cuts or punctures caused by the use of sharp objects exists.

Fume Safety

This symbol appears when chemicals or chemical reactions could cause dangerous fumes.

Electrical Safety

This symbol appears when care should be taken when using electrical equipment.

Plant Safety

This symbol appears when poisonous plants or plants with thorns are handled.

Animal Safety

This symbol appears whenever live animals are studied and the safety of the animals and the students must be ensured.

Radioactive Safety

This symbol appears when radioactive materials are used.

Clothing Protection Safety

This symbol appears when substances used could stain or burn clothing.

Fire Safety

This symbol appears when care should be taken around open flames.

Explosion Safety

This symbol appears when the misuse of chemicals could cause an explosion.

Eye Safety

This symbol appears when a danger to the eyes exists. Safety goggles should be worn when this symbol appears.

Poison Safety

This symbol appears when poisonous substances are used.

Chemical Safety

This symbol appears when chemicals used can cause burns or are poisonous if absorbed through the skin.

APPENDIX C
Chemistry Data Handbook

Table C.1   Symbols and Abbreviations

α	=	rays from radioactive materials, helium nuclei	P	=	pressure, power
β	=	rays from radioactive materials, electrons	Pa	=	pascal (*pressure*)
γ	=	rays from radioactive materials, high-energy quanta	$\boldsymbol{p}$	=	momentum
Δ	=	change in	q	=	heat
λ	=	wavelength	R	=	gas constant
ν	=	frequency	S	=	entropy
Π	=	osmotic pressure	s	=	second (*time*)
A	=	ampere (*electric current*)	Sv	=	sievert (*absorbed radiation*)
Bq	=	becquerel (*nuclear disintegration*)	T	=	temperature
°C	=	Celsius degree (*temperature*)	U	=	internal energy
C	=	coulomb (*quantity of electricity*)	u	=	atomic mass unit
c	=	speed of light	V	=	volume
cd	=	candela (*luminous intensity*)	V	=	volt (*electromotive force*)
C_p	=	specific heat	v	=	velocity
D	=	density	W	=	watt (*power*)
E	=	energy, electromotive force	w	=	work
F	=	force, Faraday	x	=	mole fraction
G	=	free energy			
g	=	gram (*mass*)			
Gy	=	gray (*radiation*)			
H	=	enthalpy			
Hz	=	hertz (*frequency*)			
h	=	Planck's constant			
h	=	hour (*time*)			
J	=	joule (*energy*)			
K	=	kelvin (*temperature*)			
K_a	=	ionization constant (acid)			
K_b	=	ionization constant (base)			
K_{eq}	=	equilibrium constant			
K_{sp}	=	solubility product constant			
kg	=	kilogram			
M	=	molarity			
m	=	mass, molality			
m	=	meter (*length*)			
mol	=	mole (*amount*)			
min	=	minute (*time*)			
N	=	newton (*force*)			
N_A	=	Avogadro's number			
n	=	number of moles			

Table C.2 The Modern Periodic Table

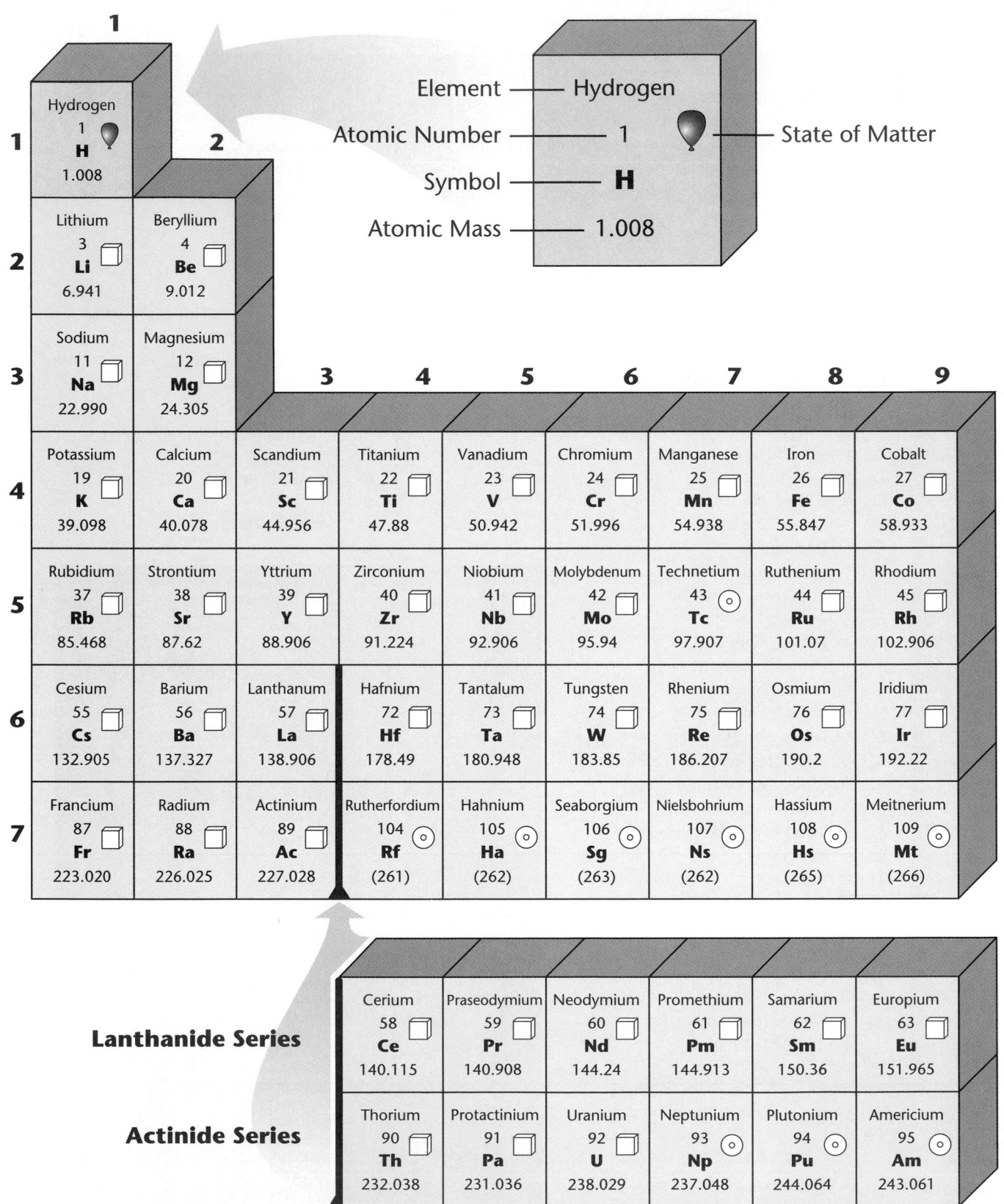

Lanthanide Series

Actinide Series

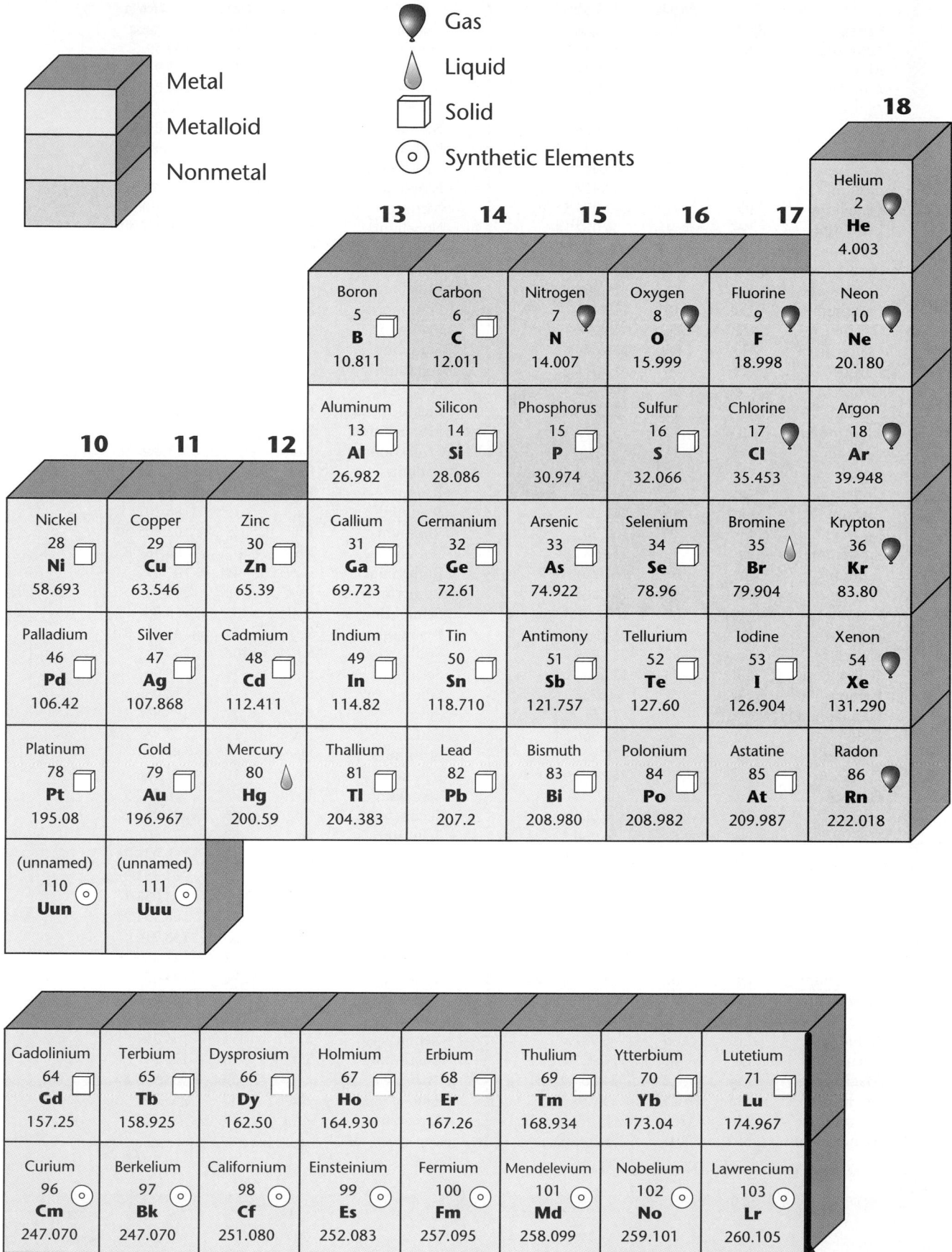

Metal
Metalloid
Nonmetal

Gas
Liquid
Solid
Synthetic Elements

13	14	15	16	17	18
					Helium 2 He 4.003
Boron 5 B 10.811	Carbon 6 C 12.011	Nitrogen 7 N 14.007	Oxygen 8 O 15.999	Fluorine 9 F 18.998	Neon 10 Ne 20.180
Aluminum 13 Al 26.982	Silicon 14 Si 28.086	Phosphorus 15 P 30.974	Sulfur 16 S 32.066	Chlorine 17 Cl 35.453	Argon 18 Ar 39.948

10	11	12						
Nickel 28 Ni 58.693	Copper 29 Cu 63.546	Zinc 30 Zn 65.39	Gallium 31 Ga 69.723	Germanium 32 Ge 72.61	Arsenic 33 As 74.922	Selenium 34 Se 78.96	Bromine 35 Br 79.904	Krypton 36 Kr 83.80
Palladium 46 Pd 106.42	Silver 47 Ag 107.868	Cadmium 48 Cd 112.411	Indium 49 In 114.82	Tin 50 Sn 118.710	Antimony 51 Sb 121.757	Tellurium 52 Te 127.60	Iodine 53 I 126.904	Xenon 54 Xe 131.290
Platinum 78 Pt 195.08	Gold 79 Au 196.967	Mercury 80 Hg 200.59	Thallium 81 Tl 204.383	Lead 82 Pb 207.2	Bismuth 83 Bi 208.980	Polonium 84 Po 208.982	Astatine 85 At 209.987	Radon 86 Rn 222.018
(unnamed) 110 Uun	(unnamed) 111 Uuu							

Gadolinium 64 Gd 157.25	Terbium 65 Tb 158.925	Dysprosium 66 Dy 162.50	Holmium 67 Ho 164.930	Erbium 68 Er 167.26	Thulium 69 Tm 168.934	Ytterbium 70 Yb 173.04	Lutetium 71 Lu 174.967
Curium 96 Cm 247.070	Berkelium 97 Bk 247.070	Californium 98 Cf 251.080	Einsteinium 99 Es 252.083	Fermium 100 Fm 257.095	Mendelevium 101 Md 258.099	Nobelium 102 No 259.101	Lawrencium 103 Lr 260.105

Table C.3 Alphabetical Table of the Elements

Element	Symbol	Atomic number	Atomic mass	Element	Symbol	Atomic number	Atomic mass
Actinium	Ac	89	227.027 8*	Mercury	Hg	80	200.59
Aluminum	Al	13	26.981 539	Molybdenum	Mo	42	95.94
Americium	Am	95	243.061 4*	Neodymium	Nd	60	144.24
Antimony	Sb	51	121.757	Neon	Ne	10	20.179 7
Argon	Ar	18	39.948	Neptunium	Np	93	237.048 2
Arsenic	As	33	74.921 59	Nickel	Ni	28	58.6934
Astatine	At	85	209.987 1*	Nielsbohrium	Ns	107	262*
Barium	Ba	56	137.327	Niobium	Nb	41	92 .906 38
Berkelium	Bk	97	247.070 3*	Nitrogen	N	7	14. 006 74
Beryllium	Be	4	9.012 182	Nobelium	No	102	259.100 9*
Bismuth	Bi	83	208.980 37	Osmium	Os	76	190.2
Boron	B	5	10.811	Oxygen	O	8	15.999 4
Bromine	Br	35	79.904	Palladium	Pd	46	106.42
Cadmium	Cd	48	112.411	Phosphorus	P	15	30.973 762
Calcium	Ca	20	40.078	Platinum	Pt	78	195.08
Californium	Cf	98	251.079 6*	Plutonium	Pu	94	244.064 2*
Carbon	C	6	12.011	Polonium	Po	84	208.982 4*
Cerium	Ce	58	140.115	Potassium	K	19	39.098 3
Cesium	Cs	55	132.905 43	Praseodymium	Pr	59	140.907 65
Chlorine	Cl	17	35.452 7	Promethium	Pm	61	144.912 8*
Chromium	Cr	24	51.996 1	Protactinium	Pa	91	231.035 88
Cobalt	Co	27	58.933 20	Radium	Ra	88	226.025 4
Copper	Cu	29	63.546	Radon	Rn	86	222.017 6*
Curium	Cm	96	247.070 3*	Rhenium	Re	75	186.207
Dysprosium	Dy	66	162.50	Rhodium	Rh	45	102.905 50
Einsteinium	Es	99	252.082 8*	Rubidium	Rb	37	85.467 8
Erbium	Er	68	167.26	Ruthenium	Ru	44	101.07
Europium	Eu	63	151.965	Rutherfordium	Rf	104	261*
Fermium	Fm	100	257.095 1*	Samarium	Sm	62	150.36
Fluorine	F	9	18.998 403 2	Scandium	Sc	21	44.955 910
Francium	Fr	87	223.019 7*	Seaborgium	Sg	106	263*
Gadolinium	Gd	64	157.25	Selenium	Se	34	78.96
Gallium	Ga	31	69.723	Silicon	Si	14	28.085 5
Germanium	Ge	32	72.61	Silver	Ag	47	107.868 2
Gold	Au	79	196.966 54	Sodium	Na	11	22.989 768
Hafnium	Hf	72	178.49	Strontium	Sr	38	87.62
Hahnium	Ha	105	262*	Sulfur	S	16	32.066
Hassium	Hs	108	265*	Tantalum	Ta	73	180.947 9
Helium	He	2	4.002 602	Technetium	Tc	43	97.907 2*
Holmium	Ho	67	164.930 32	Tellurium	Te	52	127.60
Hydrogen	H	1	1.007 94	Terbium	Tb	65	158.925 34
Indium	In	49	114.82	Thallium	Tl	81	204.383 3
Iodine	I	53	126.904 47	Thorium	Th	90	232.038 1
Iridium	Ir	77	192.22	Thulium	Tm	69	168.934 21
Iron	Fe	26	55.847	Tin	Sn	50	118.710
Krypton	Kr	36	83.80	Titanium	Ti	22	47.88
Lanthanum	La	57	138.905 5	Tungsten	W	74	183.85
Lawrencium	Lr	103	260.105 4*	Uranium	U	92	238.028 9
Lead	Pb	82	207.2	Vanadium	V	23	50.941 5
Lithium	Li	3	6.941	Xenon	Xe	54	131.29
Lutetium	Lu	71	174.967	Ytterbium	Yb	70	173.04
Magnesium	Mg	12	24.305 0	Yttrium	Y	39	88.905 85
Manganese	Mn	25	54.938 05	Zinc	Zn	30	65.39
Meitnerium	Mt	109	266*	Zirconium	Zr	40	91.224
Mendelevium	Md	101	258.098 6*				

* The mass of the isotope with the longest known half-life.

Table C.4 — Properties of Elements

Table C.4 Properties of Elements

Element	Symbol	Atomic Number (Z)	Atomic Mass* [u]	Melting Point [°C]	Boiling Point [°C]	Density (g/cm³) (gases measured at STP)	Atomic Radius (pm)	First Ionization Energy (kJ/mol)	Standard Reduction Potential (V) (for elements from or to oxidation state indicated)	Enthalpy of Fusion (kJ/mol)	Specific Heat [J/(g·°C)]	Enthalpy of Vaporization (kJ/mol)	Abundance in Earth's Crust (%)	Major Oxidation States
Actinium	Ac	89	[227.0278]	1050	3300	10.07	203	666	$(3+)-2.13$	14.3	0.120	293	trace	3+
Aluminum	Al	13	26.981539	660.37	2517.6	2.699	143	577.5	$(3+)-1.67$	10.71	0.9025	290.8	8.1	3+
Americium	Am	95	[243.0614]	994	2600	13.67	183	579	$(3+)-2.07$	10	—	238.5	2×10^{-5}	2+, 3+, 4+
Antimony	Sb	51	121.757	630.7	1635	6.697	161	834	$(3+)+0.15$	19.5	0.2072	193	2×10^{-4}	3+, 5+
Argon	Ar	18	39.948	−189.37	−185.86	0.001784	191	1521	—	1.18	0.52033	6.52	4×10^{-6}	—
Arsenic	As	33	74.92159	816 (2840 kPa)	615 (sublimes)	5.778	121	947	$(3+)+0.24$	27.7	0.3289	(sublimes)	1.9×10^{-4}	3+, 5+
Astatine	At	85	[209.98037]	300	350	—	203	916	$(1-)+0.2$	23.8	—	90.3	trace	1−, 5+
Barium	Ba	56	137.327	726.9	1845	3.62	222	502.9	$(2+)-2.92$	8.012	0.2044	140	0.039	2+
Berkelium	Bk	97	[247.0703]	986	—	14.78	170	601	$(3+)-2.01$	—	—	—	—	3+, 4+
Beryllium	Be	4	9.012182	1287	2468	1.848	112	899.5	$(2+)-1.97$	7.895	1.824	297.6	2×10^{-4}	2+
Bismuth	Bi	83	208.98037	271.4	1564	9.808	151	703	$(3+)+0.317$	10.9	0.1221	179	8×10^{-7}	3+, 5+
Boron	B	5	10.811	2080	3865	2.46	85	800.6	$(3+)-0.89$	50.2	1.026	504.5	9×10^{-4}	3+
Bromine	Br	35	79.904	−7.25	59.35	3.1028	119	1139.9	$(1-)+1.065$	10.571	0.47362	29.56	2.5×10^{-4}	1−, 1+, 3+, 5+
Cadmium	Cd	48	112.411	320.8	770	8.65	151	867.7	$(2+)-0.4025$	6.19	0.2311	100	1.6×10^{-5}	2+
Calcium	Ca	20	40.078	841.5	1500.5	1.55	197	589.8	$(2+)-2.84$	8.54	0.6315	155	4.66	2+
Californium	Cf	98	[251.0796]	900	—	—	186	608	$(3+)-2$	—	—	—	—	3+, 4+
Carbon	C	6	12.011	3620	4200	2.266	77	1086.5	$(4-)+0.132$	104.6	0.7099	711	0.018	4−, 2+, 4+
Cerium	Ce	58	140.115	804	3470	6.773	181.8	541	$(3+)-2.34$	5.2	0.1923	313	0.007	3+, 4+
Cesium	Cs	55	132.90543	28.4	674.8	1.9	262	375.7	$(1+)-2.923$	2.087	0.2421	67	2.6×10^{-4}	1+
Chlorine	Cl	17	35.4527	−101	−34	0.003214	91	1255.5	$(1-)+1.3583$	6.41	0.47820	20.41	0.013	1−, 1+, 3+, 5+
Chromium	Cr	24	51.9961	1860	2679	7.2	128	652.8	$(3+)-0.74$	20.5	0.4491	339	0.01	2+, 3+, 6+
Cobalt	Co	27	58.9332	1495	2912	8.9	125	758.8	$(2+)-0.277$	16.192	0.4210	382	0.0028	2+, 3+
Copper	Cu	29	63.546	1085	2570	8.92	128	745.5	$(2+)+0.34$	13.38	0.38452	304	0.0058	1+, 2+
Curium	Cm	96	[247.0703]	1340	3540	13.51	174	581	$(3+)-2.06$	—	—	—	—	3+, 4+
Dysprosium	Dy	66	162.5	1407	2600	8.536	178.1	572	$(3+)-2.29$	10.4	0.1733	250	6×10^{-4}	2+, 3+
Einsteinium	Es	99	[252.0828]	860	—	—	186	619	$(3+)-2$	—	—	—	—	3+
Erbium	Er	68	167.25	1497	2900	9.045	176.1	589	$(3+)-2.32$	17.2	0.1681	293	3.5×10^{-4}	3+
Europium	Eu	63	151.955	826	1439	5.245	208.4	547	$(3+)-1.99$	10.5	0.1820	176	2.1×10^{-3}	2+, 3+
Fermium	Fm	100	[257.0951]	—	—	—	—	627	$(3+)-1.96$	—	—	—	—	2+, 3+
Fluorine	F	9	18.9984032	−219.7	−188.2	0.001696	69	1681	$(1-)+2.87$	0.51	0.8238	6.54	0.0544	1−
Francium	Fr	87	[223.0197]	24	650	—	280	375	—	2	—	63.6	trace	1+
Gadolinium	Gd	64	157.25	1312	3000	7.886	180.4	592	$(3+)-2.29$	15.5	0.2355	311.7	6.3×10^{-4}	3+
Gallium	Ga	31	69.723	29.77	2203	5.904	134	578.8	$(3+)-0.529$	5.59	0.3709	256	0.0018	1+, 3+
Germanium	Ge	32	72.61	945	2850	5.323	123	761.2	$(4+)+0.124$	31.8	0.3215	334.3	1.5×10^{-4}	2+, 4+

* [ ] indicates mass of longest-lived isotope.

Table C.4 Properties of Elements (continued)

Element	Symbol	Atomic Number (Z)	Atomic Mass* (u)	Melting Point (°C)	Boiling Point (°C)	Density (g/cm³) (gases measured at STP)	Atomic Radius (pm)	First Ionization Energy (kJ/mol)	Standard Reduction Potential (V) (for elements from or to oxidation state indicated)	Enthalpy of Fusion (kJ/mol)	Specific Heat (J/g·°C)	Enthalpy of Vaporization (kJ/mol)	Abundance in Earth's Crust (%)	Major Oxidation States
Gold	Au	79	196.96654	1064	2808	19.32	144	889.9	(3+)+1.52	12.4	0.12905	324.4	3×10^{-7}	1+, 3+
Hafnium	Hf	72	178.49	2227	4691	13.28	159	654.4	(4+)-1.56	29.288	0.1442	661	3×10^{-4}	4+
Helium	He	2	4.002602	-269.7 (2536 kPa)	-268.93	0.00017847	122	2372	—	0.02	5.1931	0.084	—	
Holmium	Ho	67	164.9032	1461	2600	8.78	176.2	581	(3+)-2.33	17.1	0.1646	251	1.5×10^{-4}	3+
Hydrogen	H	1	1.00794	-259.19	-252.76	0.0000899	78	1312	(1+)0.0000	0.117	14.298	0.904		1-, 1+
Indium	In	49	114.82	156.61	2080	7.29	167	558.2	(3+)-0.3382	3.26	0.2407	231.8	2×10^{-5}	1+, 3+
Iodine	I	53	126.90447	113.6	184.5	4.93	138	1008.4	(1-)+0.5355	15.517	0.21448	41.95	4.6×10^{-5}	1-, 1+, 5+, 7+
Iridium	Ir	77	192.22	2447	4550	22.65	135.5	880	(4+)+0.926	26.4	0.1306	563.6	1×10^{-7}	3+, 4+, 5+
Iron	Fe	26	55.847	1536	2860	7.874	126	759.4	(3+)-0.4	13.807	0.4494	350	5.8	2+, 3+
Krypton	Kr	36	83.8	-157.2	-153.35	0.0037493	201	1351	—	1.64	0.2480	9.03		
Lanthanum	La	57	138.9055	920	3420	6.17	187	538	(3+)-2.37	8.5	0.1952	402	0.0035	3+
Lawrencium	Lr	103	[260.1054]	—		—			(3+)-2.06		—	—	—	3+
Lead	Pb	82	207.2	327	1746	11.342	175	715.6	(2+)-0.1251	4.77	0.1276	178	0.0013	2+, 4+
Lithium	Li	3	6.941	180.5	1347	0.534	156	520.2	(1+)-3.045	3	3.569	148	0.002	1+
Lutetium	Lu	71	174.967	1652	3327	9.84	173.8	524	(3+)-2.3	11.9	0.1535	414	8×10^{-5}	3+
Magnesium	Mg	12	24.305	650	1105	1.738	160	737.8	(2+)-2.356	8.477	1.024	127.4	2.76	2+
Manganese	Mn	25	54.93805	1246	2061	7.43	127	717.5	(2+)-1.18	12.058	0.4791	219.7	0.1	2+, 3+, 4+, 6+, 7+
Mendelevium	Md	101	[258.0986]	—	—	—		635	—		—	—	—	2+, 3+
Mercury	Hg	80	200.59	-38.9	357	13.534	151	1007	(2+)+0.8535	2.2953	0.13950	59.1	2×10^{-6}	1+, 2+
Molybdenum	Mo	42	95.94	2623	4679	10.28	139	685	(6+)0.114	36	0.2508	590	1.2×10^{-4}	4+, 5+, 6+
Neodymium	Nd	60	144.24	1024	3111	7.003	181.4	530	(3+)-2.32	7.13	0.1903	283.7	0.004	2+, 3+
Neon	Ne	10	20.1797	-248.61	-246.05	0.0008999	131	2081	—	0.34	1.0301	1.77	—	
Neptunium	Np	93	237.0482	640	3900	20.45	155	597	(5+)-0.91	9.46	0.4442	336	—	2+, 3+, 4+, 5+, 6+
Nickel	Ni	28	58.6934	1455	2883	8.908	124	736.7	(2+)-0.257	17.15	0.4442	375	0.0075	2+, 3+, 4+
Niobium	Nb	41	92.90638	2477	4858	8.57	146	664.1	(5+)-0.65	26.9	0.2648	690	0.002	4+, 5+
Nitrogen	N	7	14.00674	-210	-195.8	0.0012409	71	1402	(3-)-0.092	0.72	1.0397	5.58	0.002	3-, 2-, 1-, 1+, 2+, 3+, 4+, 5+
Nobelium	No	102	[259.1009]	—		—		642	(2+)-2.5		—	—	—	2+, 3+
Osmium	Os	76	190.2	3045	5025	22.57	135	840	(4+)+0.687	31.7	0.130	627.6	2×10^{-7}	4+, 6+, 8+
Oxygen	O	8	15.9994	-218.8	-183	0.001429	60	1313.9	(2-)0.815	0.44	0.91738	6.82	45.5	2-, 1-
Palladium	Pd	46	106.42	1552	2940	11.99	137	805	(2+)0.915	17.6	0.2441	362	3×10^{-7}	2+, 4+
Phosphorus	P	15	30.973762	44.2	280.5	1.823	109	1012	(3-)-0.063	0.659	0.76968	49.8	0.11	3-, 3+, 5+
Platinum	Pt	78	195.08	1769	3824	21.41	138.5	868	(4+)+1.15	19.7	0.1326	510.4	1×10^{-6}	2+, 4+
Plutonium	Pu	94	[244.0642]	640	3230	19.86	162	585	(4+)-1.25	2.8	0.138	343.5	—	3+, 4+, 5+, 6+
Polonium	Po	84	[208.9824]	254	962	9.4	164	813	(4+)+0.73	3.81	0.125	103	—	2-, 2+, 4+, 6+
Potassium	K	19	39.0983	63.2	766.4	0.862	231	418.8	(1+)-2.925	2.334	0.7566	76.9	1.84	1+
Praseodymium	Pr	59	140.90765	935	3343	6.782	182.4	522	(3+)-2.35	11.3	0.1930	332.6	9.1×10^{-4}	3+, 4+
Promethium	Pm	61	[144.9128]	1168	2460	7.2	183.4	536	(3+)-2.29	8.17	—	293	trace	3+

* [ ] indicates mass of longest-lived isotope.

Table C.4 Properties of Elements (continued)

Element	Symbol	Atomic Number (Z)	Atomic Mass* (u)	Melting Point (°C)	Boiling Point (°C)	Density (g/cm³) (gases measured at STP)	Atomic Radius (pm)	First Ionization Energy (kJ/mol)	Standard Reduction potential (V) [for elements from or to oxidation state indicated]	Enthalpy of Fusion (kJ/mol)	Specific Heat (J/g·°C)	Enthalpy of Vaporization (kJ/mol)	Abundance in Earth's Crust (%)	Major Oxidation States
Protactinium	Pa	91	231.03588	1552	4227	15.37	163	568	(5+) −1.19	14.6	—	481	trace	3+, 4+, 5+
Radium	Ra	88	226.0254	700	1630	5	228	509.1	(2+) −2.916	8.36	—	136.8	—	2+
Radon	Rn	86	[222.0176]	−71	−62	0.00973	232	1037	—	16.4	—	16.4	—	—
Rhenium	Re	75	186.207	3180	5650	21.232	137	760	(7+) +0.34	33.4	0.1368	707	1×10^{-7}	3+, 4+, 6+, 7+
Rhodium	Rh	45	102.9055	1960	3727	12.39	134	720	(3+) +0.76	21.6	0.2427	494	1×10^{-7}	3+, 4+, 5+
Rubidium	Rb	37	85.4678	39.5	697	1.532	248	403	(1+) −2.925	2.19	0.36344	69.2	0.0078	1+
Ruthenium	Ru	44	101.07	2310	4119	12.41	134	711	(4+) +0.68	25.5	0.2381	567.8	—	2+, 3+, 4+, 5+
Samarium	Sm	62	150.36	1072	1800	7.536	180.4	542	(3+) −2.3	8.9	0.1965	191	7×10^{-4}	2+, 3+
Scandium	Sc	21	44.95591	1539	2831	3	162	631	(3+) −2.03	15.77	0.5677	304.8	0.0022	3+
Selenium	Se	34	78.96	221	685	4.79	117	940.7	(2−) −0.67	5.43	0.3212	26.3	5×10^{-6}	2−, 2+, 4+, 6+
Silicon	Si	14	28.0855	1411	3231	2.336	118	786.5	(4−) −0.143	50.2	0.7121	359	27.2	2+, 4+
Silver	Ag	47	107.8682	961	2195	10.49	144	730.8	(1+) +0.7991	11.65	0.23502	255	8×10^{-6}	1+
Sodium	Na	11	22.989768	97.83	897.4	0.968	186	495.9	(1+) −2.714	2.602	1.228	97.4	2.27	1+
Strontium	Sr	38	87.62	776.9	1412	2.6	215	549.5	(2+) −2.89	7.4308	0.301	137	0.0384	2+
Sulfur	S	16	32.066	115.2	444.7	2.08	103	999.6	(2−) −0.45	1.7272	0.7060	9.62	0.03	2−, 4+, 6+
Tantalum	Ta	73	180.9479	2980	5505	16.65	146	760.8	(5+) −0.81	36.57	0.1402	737	2×10^{-4}	4+, 5+
Technetium	Tc	43	97.9072	2200	4567	11.5	136	702	(6+) +0.83	23.0	—	577	—	2+, 4+, 6+, 7+
Tellurium	Te	52	127.6	450	990	6.25	138	869	(2−) −1.14	17.4	0.2016	50.6	2×10^{-7}	2−, 2+, 4+, 6+
Terbium	Tb	65	158.92534	1356	2800	8.272	177.3	564	(3+) −2.31	10.3	0.1819	293	1×10^{-4}	3+, 4+
Thallium	Tl	81	204.3833	303.5	1457	11.85	170	589.1	(1+) −0.3363	4.27	0.1288	162	7×10^{-5}	1+, 3+
Thorium	Th	90	232.0381	1750	4787	11.78	179	587	(4+) −1.83	16.11	0.1177	543.9	8.1×10^{-4}	4+
Thulium	Tm	69	168.93421	1545	1727	9.318	175.9	596	(3+) −2.32	18.4	0.1600	213	5×10^{-5}	2+, 3+
Tin	Sn	50	118.71	232	2623	7.265	141	708.4	(4+) +0.064	7.07	0.2274	296	2.1×10^{-4}	2+, 4+
Titanium	Ti	22	47.88	1666	3358	4.5	147	658.1	(4+) −0.86	14.146	0.5226	425	0.63	2+, 3+, 4+
Tungsten	W	74	183.85	3680	6000	19.3	139	770.4	(6+) −0.09	35.4	0.1320	806	1.2×10^{-4}	4+, 5+, 6+
Unnilennium	Une	109	[266]	—	—	—	—	—	—	—	—	—	—	—
Unnilhexium	Unh	106	[263]	—	—	—	—	—	—	—	—	—	—	—
Unniloctium	Uno	108	[265]	—	—	—	—	—	—	—	—	—	—	—
Unnilpentium	Unp	105	[262]	—	—	—	—	—	—	—	—	—	—	—
Unnilquadium	Unq	104	[261]	—	—	—	—	—	—	—	—	—	—	—
Unnilseptium	Uns	107	[262]	—	—	—	—	—	—	—	—	—	—	—
Uranium	U	92	238.0289	1130	3930	19.05	156	584	(6+) −0.83	12.6	0.11618	423	2.3×10^{-4}	3+, 4+, 5+, 6+
Vanadium	V	23	50.9415	1917	3417	6.11	134	650.3	(4+) −0.54	22.84	0.4886	459.7	0.0136	2+, 3+, 4+, 5+
Xenon	Xe	54	131.29	−111.8	−108.09	0.0058971	218	1170	—	2.29	0.15832	12.64	—	2+, 4+, 6+
Ytterbium	Yb	70	173.04	824	1427	6.973	193.3	603	(3+) −2.22	7.66	0.1545	155	3.4×10^{-4}	2+, 3+
Yttrium	Y	39	88.50585	1530	3264	4.5	180	616	(3+) −2.37	17.15	0.2984	393	0.0035	3+
Zinc	Zn	30	65.39	419.6	907	7.14	134	906.4	(2+) −0.7626	7.322	0.3884	115	0.0076	2+
Zirconium	Zr	40	91.224	1852	4400	6.51	160	659.7	(4+) −1.7	20.92	0.2780	590.5	0.0162	4+

* [ ] indicates mass of longest-lived isotope.

Table C.5 Electron Configurations of the Elements

	Elements	1s	2s	2p	3s	3p	3d	4s	4p	4d	4f	5s	5p	5d	5f	6s	6p	6d	6f	7s
1	Hydrogen	1																		
2	Helium	2																		
3	Lithium	2	1																	
4	Beryllium	2	2																	
5	Boron	2	2	1																
6	Carbon	2	2	2																
7	Nitrogen	2	2	3																
8	Oxygen	2	2	4																
9	Fluorine	2	2	5																
10	Neon	2	2	6																
11	Sodium	2	2	6	1															
12	Magnesium	2	2	6	2															
13	Aluminum	2	2	6	2	1														
14	Silicon	2	2	6	2	2														
15	Phosphorus	2	2	6	2	3														
16	Sulfur	2	2	6	2	4														
17	Chlorine	2	2	6	2	5														
18	Argon	2	2	6	2	6														
19	Potassium	2	2	6	2	6		1												
20	Calcium	2	2	6	2	6		2												
21	Scandium	2	2	6	2	6	1	2												
22	Titanium	2	2	6	2	6	2	2												
23	Vanadium	2	2	6	2	6	3	2												
24	Chromium	2	2	6	2	6	5	1												
25	Manganese	2	2	6	2	6	5	2												
26	Iron	2	2	6	2	6	6	2												
27	Cobalt	2	2	6	2	6	7	2												
28	Nickel	2	2	6	2	6	8	2												
29	Copper	2	2	6	2	6	10	1												
30	Zinc	2	2	6	2	6	10	2												
31	Gallium	2	2	6	2	6	10	2	1											
32	Germanium	2	2	6	2	6	10	2	2											
33	Arsenic	2	2	6	2	6	10	2	3											
34	Selenium	2	2	6	2	6	10	2	4											
35	Bromine	2	2	6	2	6	10	2	5											
36	Krypton	2	2	6	2	6	10	2	6											
37	Rubidium	2	2	6	2	6	10	2	6			1								
38	Strontium	2	2	6	2	6	10	2	6			2								
39	Yttrium	2	2	6	2	6	10	2	6	1		2								
40	Zirconium	2	2	6	2	6	10	2	6	2		2								
41	Niobium	2	2	6	2	6	10	2	6	4		1								
42	Molybdenum	2	2	6	2	6	10	2	6	5		1								
43	Technetium	2	2	6	2	6	10	2	6	5		2								
44	Ruthenium	2	2	6	2	6	10	2	6	7		1								
45	Rhodium	2	2	6	2	6	10	2	6	8		1								
46	Palladium	2	2	6	2	6	10	2	6	10										
47	Silver	2	2	6	2	6	10	2	6	10		1								
48	Cadmium	2	2	6	2	6	10	2	6	10		2								
49	Indium	2	2	6	2	6	10	2	6	10		2	1							
50	Tin	2	2	6	2	6	10	2	6	10		2	2							
51	Antimony	2	2	6	2	6	10	2	6	10		2	3							
52	Tellurium	2	2	6	2	6	10	2	6	10		2	4							
53	Iodine	2	2	6	2	6	10	2	6	10		2	5							
54	Xenon	2	2	6	2	6	10	2	6	10		2	6							

Table C.5   Electron Configurations of the Elements (continued)

Elements		1s	2s	2p	3s	3p	3d	4s	4p	4d	4f	5s	5p	5d	5f	6s	6p	6d	6f	7s
55	Cesium	2	2	6	2	6	10	2	6	10		2	6			1				
56	Barium	2	2	6	2	6	10	2	6	10		2	6			2				
57	Lanthanum	2	2	6	2	6	10	2	6	10		2	6	1		2				
58	Cerium	2	2	6	2	6	10	2	6	10	2	2	6			2				
59	Praseodymium	2	2	6	2	6	10	2	6	10	3	2	6			2				
60	Neodymium	2	2	6	2	6	10	2	6	10	4	2	6			2				
61	Promethium	2	2	6	2	6	10	2	6	10	5	2	6			2				
62	Samarium	2	2	6	2	6	10	2	6	10	6	2	6			2				
63	Europium	2	2	6	2	6	10	2	6	10	7	2	6			2				
64	Gadolinium	2	2	6	2	6	10	2	6	10	7	2	6	1		2				
65	Terbium	2	2	6	2	6	10	2	6	10	9	2	6			2				
66	Dysprosium	2	2	6	2	6	10	2	6	10	10	2	6			2				
67	Holmium	2	2	6	2	6	10	2	6	10	11	2	6			2				
68	Erbium	2	2	6	2	6	10	2	6	10	12	2	6			2				
69	Thulium	2	2	6	2	6	10	2	6	10	13	2	6			2				
70	Ytterbium	2	2	6	2	6	10	2	6	10	14	2	6			2				
71	Lutetium	2	2	6	2	6	10	2	6	10	14	2	6	1		2				
72	Hafnium	2	2	6	2	6	10	2	6	10	14	2	6	2		2				
73	Tantalum	2	2	6	2	6	10	2	6	10	14	2	6	3		2				
74	Tungsten	2	2	6	2	6	10	2	6	10	14	2	6	4		2				
75	Rhenium	2	2	6	2	6	10	2	6	10	14	2	6	5		2				
76	Osmium	2	2	6	2	6	10	2	6	10	14	2	6	6		2				
77	Iridium	2	2	6	2	6	10	2	6	10	14	2	6	7		2				
78	Platinum	2	2	6	2	6	10	2	6	10	14	2	6	9		1				
79	Gold	2	2	6	2	6	10	2	6	10	14	2	6	10		1				
80	Mercury	2	2	6	2	6	10	2	6	10	14	2	6	10		2				
81	Thallium	2	2	6	2	6	10	2	6	10	14	2	6	10		2	1			
82	Lead	2	2	6	2	6	10	2	6	10	14	2	6	10		2	2			
83	Bismuth	2	2	6	2	6	10	2	6	10	14	2	6	10		2	3			
84	Polonium	2	2	6	2	6	10	2	6	10	14	2	6	10		2	4			
85	Astatine	2	2	6	2	6	10	2	6	10	14	2	6	10		2	5			
86	Radon	2	2	6	2	6	10	2	6	10	14	2	6	10		2	6			
87	Francium	2	2	6	2	6	10	2	6	10	14	2	6	10		2	6			1
88	Radium	2	2	6	2	6	10	2	6	10	14	2	6	10		2	6			2
89	Actinium	2	2	8	2	6	10	2	6	10	14	2	6	10		2	6	1		2
90	Thorium	2	2	6	2	6	10	2	6	10	14	2	6	10		2	6	2		2
91	Protactinium	2	2	6	2	6	10	2	6	10	14	2	6	10	2	2	6	1		2
92	Uranium	2	2	6	2	6	10	2	6	10	14	2	6	10	3	2	6	1		2
93	Neptunium	2	2	6	2	6	10	2	6	10	14	2	6	10	4	2	6	1		2
94	Plutonium	2	2	6	2	6	10	2	6	10	14	2	6	10	6	2	6			2
95	Americium	2	2	6	2	6	10	2	6	10	14	2	6	10	7	2	6			2
96	Curium	2	2	6	2	6	10	2	6	10	14	2	6	10	7	2	6	1		2
97	Berkelium	2	2	6	2	6	10	2	6	10	14	2	6	10	9	2	6			2
98	Californium	2	2	6	2	6	10	2	6	10	14	2	6	10	10	2	6			2
99	Einsteinium	2	2	6	2	6	10	2	6	10	14	2	6	10	11	2	6			2
100	Fermium	2	2	6	2	6	10	2	6	10	14	2	6	10	12	2	6			2
101	Mendelevium	2	2	6	2	6	10	2	6	10	14	2	6	10	13	2	6			2
102	Nobelium	2	2	6	2	6	10	2	6	10	14	2	6	10	14	2	6			2
103	Lawrencium	2	2	6	2	6	10	2	6	10	14	2	6	10	14	2	6	1		2
104	Rutherfordium	2	2	6	2	6	10	2	6	10	14	2	6	10	14	2	6	2		2?
105	Hahnium	2	2	6	2	6	10	2	6	10	14	2	6	10	14	2	6	3		2?
106	Seaborgium	2	2	6	2	6	10	2	6	10	14	2	6	10	14	2	6	4		2?
107	Nielsbohrium	2	2	6	2	6	10	2	6	10	14	2	6	10	14	2	6	5		2?
108	Hassium	2	2	6	2	6	10	2	6	10	14	2	6	10	14	2	6	6		2?
109	Meitnerium	2	2	6	2	6	10	2	6	10	14	2	6	10	14	2	6	7		2?

Table C.6   Useful Physical Constants

1 ampere is the constant current which, if maintained in two straight parallel conductors of infinite length, of negligible circular cross-section, and placed 1 meter apart in a vacuum, would produce a force of 2×10^{-7} newtons per meter of length between these conductors.

1 candela is the luminous intensity, in the perpendicular direction, of a surface of $1/600\ 000\ m^2$ of a blackbody at the temperature of freezing platinum at a pressure of 101 325 pascals.

1 cubic decimeter is equal to 1 liter.

1 kelvin is 1/273.16 of the thermodynamic temperature of the triple point of water.

1 kilogram is the mass of the international prototype kilogram.

1 meter is the distance light travels in 1/299 792 458 of a second.

1 mole is the amount of substance containing as many elementary entities as there are atoms in 0.012 kilogram of carbon-12.

1 second is equal to 9 192 631 770 periods of the natural electromagnetic oscillation during that transition of ground state $^2S_{1/2}$ of cesium-133, which is designated $(F = 4, M = 0) \leftrightarrow (F = 3, M = 0)$.

Avogadro constant = $6.022\ 136\ 7 \times 10^{23}$

1 electronvolt = $1.602\ 177\ 33 \times 10^{-19}$ J

Faraday constant = 96 485.309 C/mole e^-

Ideal gas constant = 8.314 471 J/mol·K = 8.314 471 dm³·kPa/mol·K

Molar gas volume at STP = 22.414 10 dm³

Planck's constant = $6.626\ 075 \times 10^{-34}$ J·s

Speed of light = $2.997\ 924\ 58 \times 10^8$ m/s

Table C.7   Names and Charges of Polyatomic Ions

1−	2−	3−	4−
Acetate, CH_3COO^-	Carbonate, CO_3^{2-}	Arsenate, AsO_4^{3-}	Hexacyanoferrate(II), $Fe(CN)_6^{4-}$
Amide, NH_2^-	Chromate, CrO_4^{2-}	Arsenite, AsO_3^{3-}	Orthosilicate, SiO_4^{4-}
Astatate, AtO_3^-	Dichromate, $Cr_2O_7^{2-}$	Borate, BO_3^{3-}	Diphosphate, $P_2O_7^{4-}$
Azide, N_3^-	Hexachloroplatinate, $PtCl_6^{2-}$	Citrate, $C_6H_5O_7^{3-}$	
Benzoate, $C_6H_5COO^-$	Hexafluorosilicate, SiF_6^{2-}	Hexacyanoferrate(III), $Fe(CN)_6^{3-}$	
Bismuthate, BiO_3^-	Molybdate, MoO_4^{2-}	Phosphate, PO_4^{3-}	
Bromate, BrO_3^-	Oxalate, $C_2O_4^{2-}$		

1−	2−	1+	2+
Chlorate, ClO_3^-	Peroxide, O_2^{2-}	Ammonium, NH_4^+	Mercury(I), Hg_2^{2+}
Chlorite, ClO_2^-	Peroxydisulfate, $S_2O_8^{2-}$	Neptunyl(V), NpO_2^+	Neptunyl(VI), NpO_2^{2+}
Cyanide, CN^-	Phosphite, HPO_3^{2-}	Plutonyl(V), PuO_2^+	Plutonyl(VI), PuO_2^{2+}
Formate, $HCOO^-$	Ruthenate, RuO_4^{2-}	Uranyl(V), UO_2^+	Uranyl(VI), UO_2^{2+}
Hydroxide, OH^-	Selenate, SeO_4^{2-}	Vanadyl(V), VO_2^+	Vanadyl(IV), VO^{2+}
Hypobromite, BrO^-	Selenite, SeO_3^{2-}		
Hypochlorite, ClO^-	Silicate, SiO_3^{2-}		
Hypophosphite, $H_2PO_2^-$	Sulfate, SO_4^{2-}		
Iodate, IO_3^-	Sulfite, SO_3^{2-}		
Nitrate, NO_3^-	Tartrate, $C_4H_4O_6^{2-}$		
Nitrite, NO_2^-	Tellurate, TeO_4^{2-}		
Perbromate, BrO_4^-	Tellurite, TeO_3^{2-}		
Perchlorate, ClO_4^-	Tetraborate, $B_4O_7^{2-}$		
Periodate, IO_4^-	Thiosulfate, $S_2O_3^{2-}$		
Permanganate, MnO_4^-	Tungstate, WO_4^{2-}		
Perrhenate, ReO_4^-			
Thiocyanate, SCN^-			
Vanadate, VO_3^-			

Table C.8　Solubility Guidelines

You will be working with water solutions, and it is helpful to have a few guidelines concerning what substances are soluble in water. A substance is considered soluble if more than 3 grams of the substance dissolve in 100 mL of water. The more common rules are listed below.

1. All common salts of the Group 1(IA) elements and ammonium ions are soluble.
2. All common acetates and nitrates are soluble.
3. All binary compounds of Group 17(VIIA) elements (other than F) with metals are soluble except those of silver, mercury(I), and lead.
4. All sulfates are soluble except those of barium, strontium, lead, calcium, silver, and mercury(I).
5. Except for those in Rule 1, carbonates, hydroxides, oxides, sulfides, and phosphates are insoluble.

Table C.9　Solubility Product Constants (at 25°C)

Substance	K_{sp}	Substance	K_{sp}	Substance	K_{sp}
AgBr	5.01×10^{-13}	$BaSO_4$	1.10×10^{-10}	Li_2CO_3	2.51×10^{-2}
$AgBrO_3$	5.25×10^{-5}	$CaCO_3$	2.88×10^{-9}	$MgCO_3$	3.47×10^{-8}
Ag_2CO_3	8.13×10^{-12}	$CaSO_4$	9.12×10^{-6}	$MnCO_3$	1.82×10^{-11}
AgCl	1.78×10^{-10}	CdS	7.94×10^{-27}	$NiCO_3$	6.61×10^{-9}
Ag_2CrO_4	1.12×10^{-12}	$Cu(IO_3)_2$	7.41×10^{-8}	$PbCl_2$	1.62×10^{-5}
$Ag_2Cr_2O_7$	2.00×10^{-7}	CuC_2O_4	2.29×10^{-8}	PbI_2	7.08×10^{-9}
AgI	8.32×10^{-17}	$Cu(OH)_2$	2.19×10^{-20}	$Pb(IO_3)_2$	3.24×10^{-13}
AgSCN	1.00×10^{-12}	CuS	6.31×10^{-36}	$SrCO_3$	1.10×10^{-10}
$Al(OH)_3$	1.26×10^{-33}	FeC_2O_4	3.16×10^{-7}	$SrSO_4$	3.24×10^{-7}
Al_2S_3	2.00×10^{-7}	$Fe(OH)_3$	3.98×10^{-38}	TlBr	3.39×10^{-6}
$BaCO_3$	5.13×10^{-9}	FeS	6.31×10^{-18}	$ZnCO_3$	1.45×10^{-11}
$BaCrO_4$	1.17×10^{-10}	Hg_2SO_4	7.41×10^{-7}	ZnS	1.58×10^{-24}

Table C.10   Acid-Base Indicators

Indicator	Lower Color	Range	Upper Color
Methyl violet	yellow-green	0.0–2.5	violet
Malachite green HCl	yellow	0.5–2.0	blue
Thymol blue	red	1.0–2.8	yellow
Naphthol yellow S	colorless	1.5–2.6	yellow
p-Phenylazoaniline	orange	2.1–2.8	yellow
Methyl orange	red	2.5–4.4	yellow
Bromophenol blue	orange-yellow	3.0–4.7	violet
Gallein	orange	3.5–6.3	red
2,5-Dinitrophenol	colorless	4.0–5.8	yellow
Ethyl orange	salmon	4.2–4.6	orange
Propyl red	pink	5.1–6.5	yellow
Bromocresol purple	green-yellow	5.4–6.8	violet
Bromoxylenol blue	orange-yellow	6.0–7.6	blue
Phenol red	yellow	6.4–8.2	red-violet
Cresol red	yellow	7.1–8.8	violet
m-Cresol purple	yellow	7.5–9.0	violet
Thymol blue	yellow	8.1–9.5	blue
Phenolphthalein	colorless	8.3–10.0	dark pink
o-Cresolphthalein	colorless	8.6–9.8	pink
Thymolphthalein	colorless	9.5–10.4	blue
Alizarin yellow R	yellow	9.9–11.8	dark orange
Methyl blue	blue	10.6–13.4	pale violet
Acid fuchsin	red	11.1–12.8	colorless
2,4,6-Trinitrotoluene	colorless	11.7–12.8	orange

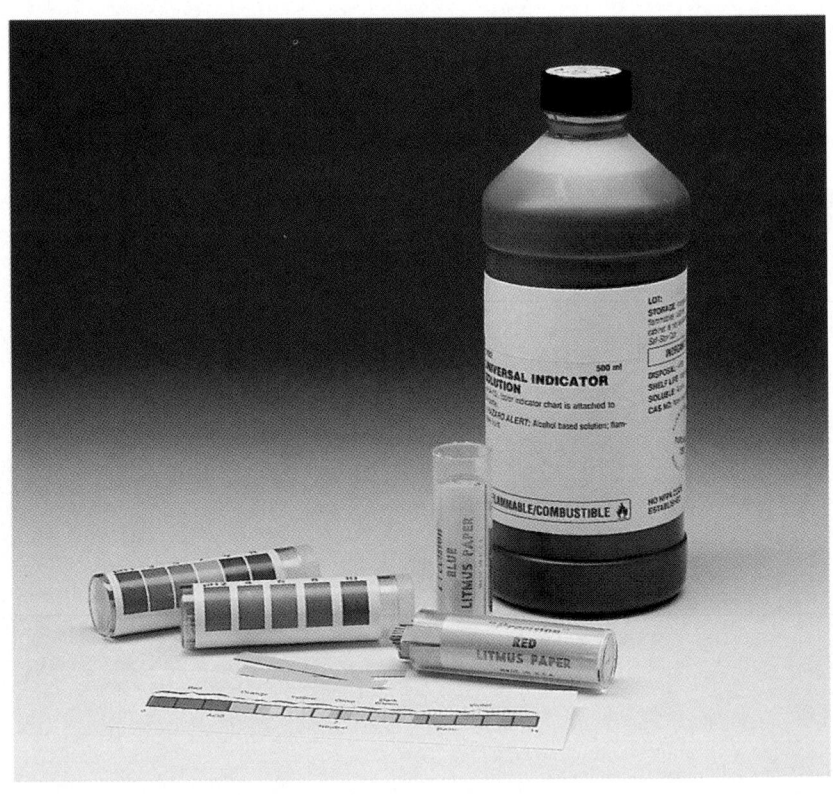

CHAPTER 5

1. Write the formula for each of the following compounds.
   a. lithium oxide
      Li_2O
   b. calcium bromide
      $CaBr_2$
   c. sodium oxide
      Na_2O
   d. aluminum sulfide
      Al_2S_3

2. Write the formula for the compound formed from each of the following pairs of elements.
   a. barium and oxygen
      BaO
   b. strontium and iodine
      SrI_2
   c. lithium and chlorine
      $LiCl$
   d. radium and chlorine
      $RaCl_2$

3. Write the formula for the compound made from each of the following ions.
   a. ammonium and sulfite ions
      $(NH_4)_2SO_3$
   b. calcium and monohydrogen phosphate ions
      $CaHPO_4$
   c. ammonium and dichromate ions
      $(NH_4)_2Cr_2O_7$
   d. barium and nitrate ions
      $Ba(NO_3)_2$

4. Write the formula for each of the following compounds.
   a. sodium phosphate
      Na_3PO_4
   b. magnesium hydroxide
      $Mg(OH)_2$
   c. ammonium phosphate
      $(NH_4)_3PO_4$
   d. potassium dichromate
      $K_2Cr_2O_7$

5. Write the formula for the compound made from each of the following pairs of ions.
   a. copper(I) and sulfite
      Cu_2SO_3
   b. tin(IV) and fluoride
      SnF_4
   c. gold(III) and cyanide
      $Au(CN)_3$
   d. lead(II) and sulfide
      PbS

6. Write the names of the following compounds.
   a. $Pb(NO_3)_2$
      lead(II) nitrate
   b. Mn_2O_3
      manganese(III) oxide
   c. $Ni(C_2H_3O_2)_2$
      nickel(II) acetate
   d. HgF_2
      mercury(II) fluoride

7. Name the following molecular compounds.
   a. S_2Cl_2
      disulfur dichloride
   b. CS_2
      carbon disulfide
   c. SO_3
      sulfur trioxide
   d. P_4O_{10}
      tetraphosphorus decoxide

8. Write the formulas for the following molecular compounds.
   a. carbon tetrachloride
      CCl_4
   b. iodine heptafluoride
      IF_7
   c. dinitrogen monoxide
      N_2O
   d. sulfur dioxide
      SO_2

CHAPTER 6

Write word equations and chemical equations for the following reactions.

1. Magnesium metal and water combine to form solid magnesium hydroxide and hydrogen gas.
   magnesium + water
   $\rightarrow$ magnesium hydroxide + hydrogen
   $Mg(s) + 2H_2O(1) \rightarrow Mg(OH)_2(s) + H_2(g)$

2. An aqueous solution of hydrogen peroxide (dihydrogen dioxide) and solid lead(II) sulfide combine to form solid lead(II) sulfate and liquid water.
   hydrogen peroxide + lead(II) sulfide
   $\rightarrow$ lead(II) sulfate + water
   $4H_2O_2(aq) + PbS(s) \rightarrow PbSO_4(s) + 4H_2O(1)$

3. When energy is added to solid manganese(II) sulfate heptahydrate crystals, they break down to form liquid water and solid manganese(II) sulfate monohydrate.
   manganese(II) sulfate heptahydrate + energy
   $\rightarrow$ water + manganese(II) sulfate monohydrate
   $MnSO_4 \cdot 7H_2O(s) + energy$
   $\rightarrow 6H_2O(1) + MnSO_4 \cdot H_2O(s)$

4. Solid potassium reacts with liquid water to produce aqueous potassium hydroxide and hydrogen gas.
   potassium + water
   $\rightarrow$ potassium hydroxide + hydrogen gas
   $2K(s) + 2H_2O(l) \rightarrow 2 KOH(aq) + H_2(g)$

1. Calculate ΔEN for the pairs of atoms in the following bonds.
   a. Ca—S
      1.5
   b. Ba—O
      2.6
   c. C—Br
      0.3
   d. Ca—F
      3.0
   e. H—Br
      0.7

2. Use ΔEN to classify the bonds in problem 1 as covalent, polar covalent, or ionic.
   a. polar covalent
   b. ionic
   c. covalent
   d. ionic
   e. polar covalent

CHAPTER 10

Complete the following table.
1. Melting point of iron
   1535; $T_C = 1808 - 273 = 1535°C$
2. Household oven
   448 - 478; $T_K = 175 + 273 = 448$; $205 + 273 = 478$
3. Food freezer
   -18; $T_C = 255 - 273 = -18°C$
4. Sublimation point of dry ice, $CO_2(s)$
   194.5; $T_K = -78.5 + 273 = 194.5$ K
5. Boiling point of nitrogen, N_2
   -195.6; $T_C = 77.4 - 273 = -195.6$

Use Table 11.1 and the equation 1.00 in. = 25.4 mm to convert the following measurements.
1. 59.8 in. Hg to psi

$$\frac{59.8 \text{ in. Hg}}{1} \left| \frac{25.4 \text{ mm Hg}}{1.00 \text{ in. Hg}} \right| \frac{14.7 \text{ psi}}{760 \text{ mm Hg}}$$

$$= \frac{59.8 \text{ in. Hg}}{1} \left| \frac{25.4 \text{ mm Hg}}{1.00 \text{ in. Hg}} \right| \frac{14.7 \text{ psi}}{760 \text{ mm Hg}}$$

$$= \frac{59.8 \times 25.4 \times 14.7 \text{ psi}}{760} = 29.4 \text{ psi}$$

2. 7.35 psi to mm Hg

$$\frac{7.35 \text{ psi}}{1} \left| \frac{1 \text{ atm}}{14.7 \text{ psi}} \right| \frac{760 \text{ mm Hg}}{1 \text{ atm}}$$

$$= \frac{7.35 \text{ psi}}{1} \left| \frac{1 \text{ atm}}{14.7 \text{ psi}} \right| \frac{760 \text{ mm Hg}}{1 \text{ atm}}$$

$$= \frac{7.35 \times 760 \text{ mm Hg}}{14.7} = 3.80 \times 10^2 \text{ mm Hg}$$

3. 1140 mm Hg to kPa

$$\frac{1140 \text{ mm Hg}}{1} \left| \frac{1.00 \text{ atm}}{760 \text{ mm Hg}} \right| \frac{101.3 \text{ kPa}}{1.00 \text{ atm}}$$

$$= \frac{1140 \text{ mm Hg}}{1} \left| \frac{1.00 \text{ atm}}{760 \text{ mm Hg}} \right| \frac{101.3 \text{ kPa}}{1.00 \text{ atm}}$$

$$= \frac{1140 \times 101.3 \text{ kPa}}{760} = 152 \text{ kPa}$$

4. 19.0 psi to kPa

$$\frac{19.0 \text{ psi}}{1} \left| \frac{1.00 \text{ atm}}{14.7 \text{ psi}} \right| \frac{101.3 \text{ kPa}}{1.00 \text{ atm}}$$

$$= \frac{19.0 \text{ psi}}{1} \left| \frac{1.00 \text{ atm}}{14.7 \text{ psi}} \right| \frac{101.3 \text{ kPa}}{1.00 \text{ atm}}$$

$$= \frac{19.0 \times 101.3 \text{ kPa}}{14.7} = 131 \text{ kPa}$$

5. 202 kPa to psi

$$\frac{202 \text{ kPa}}{1} \left| \frac{1.00 \text{ atm}}{101.3 \text{ kPa}} \right| \frac{14.7 \text{ psi}}{1.00 \text{ atm}}$$

$$= \frac{202 \text{ kPa}}{1} \left| \frac{1.00 \text{ atm}}{101.3 \text{ kPa}} \right| \frac{14.7 \text{ psi}}{1.00 \text{ atm}}$$

$$= \frac{202 \times 14.7 \text{ psi}}{101.3} = 29.3 \text{ psi}$$

Assume that the temperature remains constant in the following problems.
6. Bacteria produce methane gas in sewage-treatment plants. This gas is often captured or burned. If a bacterial culture produces 60.0 mL of methane gas at 700.0 mm Hg, what volume would be produced at 760.0 mm Hg?

$$\frac{60.0 \text{ mL}}{1} \left| \frac{700.0 \text{ mm Hg}}{760.0 \text{ mm Hg}} \right.$$

$$= \frac{60.0 \text{ mL}}{1} \left| \frac{700.0 \text{ mm Hg}}{760.0 \text{ mm Hg}} \right.$$

$$= \frac{60.0 \text{ mL} \times 700.0}{760.0} = 55.3 \text{ mL}$$

7. At one sewage-treatment plant, bacteria cultures produce 1000 L of methane gas per day at 1.0 atm pressure. What volume tank would be needed to store one day's production at 5.0 atm?

$$\frac{1000 \text{ L}}{1} \left| \frac{1 \text{ atm}}{5 \text{ atm}} \right. = \frac{1000 \text{ L}}{1} \left| \frac{1 \text{ atm}}{5 \text{ atm}} \right.$$

$$= \frac{1000 \text{ L}}{5} = 200 \text{ L}$$

8. Hospitals buy 400-L cylinders of oxygen gas compressed at 150 atm. They administer oxygen to patients at 3.0 atm in a hyperbaric oxygen chamber. What volume of oxygen can a cylinder supply at this pressure?

$$\frac{400 \text{ L}}{1} \left| \frac{150 \text{ atm}}{3 \text{ atm}} \right. = \frac{400 \text{ L}}{1} \left| \frac{150 \text{ atm}}{3 \text{ atm}} \right.$$

$$= \frac{400 \text{ L} \times 150}{3} = 20\ 000 \text{ L}$$

9. If the valve in a tire pump with a volume of 0.78 L fails at a pressure of 9.0 atm, what would be the volume of air in the cylinder just before the valve fails?

$$\frac{0.78\ L}{1}\ \left|\ \frac{1.0\ atm}{9.0\ atm}\ =\ \frac{0.78\ L}{1}\ \right|\ \frac{1.0\ \cancel{atm}}{9.0\ \cancel{atm}}$$

$$=\ \frac{0.78\ L}{9.0}\ =\ 0.087\ L$$

10. The volume of a scuba tank is 10.0 L. It contains a mixture of nitrogen and oxygen at 290.0 atm. What volume of this mixture could the tank supply to a diver at 2.40 atm?

$$\frac{10.0\ L}{1}\ \left|\ \frac{290.0\ atm}{2.40\ atm}\ =\ \frac{10.0\ L}{1}\ \right|\ \frac{290.0\ \cancel{atm}}{2.40\ \cancel{atm}}$$

$$=\ \frac{10.0\ L\ \times\ 290.0}{2.40}\ =\ 1210\ L$$

11. A 1.00-L balloon is filled with helium at 1.20 atm. If the balloon is squeezed into a 0.500-L beaker and doesn't burst, what is the pressure of the helium?

$$\frac{1.20\ atm}{1}\ \left|\ \frac{1.00\ L}{0.500\ L}\ =\ \frac{1.20\ atm}{1}\ \right|\ \frac{1.00\ \cancel{L}}{0.500\ \cancel{L}}$$

$$=\ \frac{1.20\ atm}{0.500}\ =\ 2.40\ atm$$

Assume that the pressure remains constant in problems 12 to 16.

12. A balloon is filled with 3.0 L of helium at 310 K and 1 atm. The balloon is placed in an oven where the temperature reaches 340 K. What is the new volume of the balloon?

$$\frac{3.0\ L}{1}\ \left|\ \frac{340\ K}{310\ K}\ =\ \frac{3.0\ L}{1}\ \right|\ \frac{340\ \cancel{K}}{310\ \cancel{K}}$$

$$=\ \frac{3.0\ L\ \times\ 340}{310}\ =\ 3.3\ L$$

13. A 4.0-L sample of methane gas is collected at 30.0°C. Predict the volume of the sample at 0°C.

$$\frac{4.0\ L}{1}\ \left|\ \frac{(0.0\ +\ 273)K}{(30.0\ +\ 273)K}\right.$$

$$=\ \frac{4.0\ L}{1}\ \left|\ \frac{273\ \cancel{K}}{303\ \cancel{K}}\right.$$

$$=\ \frac{4.0\ L\ \times\ 273}{303}\ =\ 3.6\ L$$

14. A 25-L sample of nitrogen is heated from 110°C to 260°C. What volume will the sample occupy at the higher temperature?

$$\frac{25\ L}{1}\ \left|\ \frac{(260\ +\ 273)K}{(110\ +\ 273)K}\right.$$

$$=\ \frac{25\ L}{1}\ \left|\ \frac{533\ \cancel{K}}{383\ \cancel{K}}\right.$$

$$=\ \frac{25\ L\ \times\ 533}{383}\ =\ 35\ L$$

15. The volume of a 16-g sample of oxygen is 11.2 L at 273 K and 1.00 atm. Predict the volume of the sample at 409 K.

$$\frac{11.2\ L}{1}\ \left|\ \frac{409\ K}{273\ K}\ =\ \frac{11.2\ L}{1}\ \right|\ \frac{409\ \cancel{K}}{273\ \cancel{K}}$$

$$=\ \frac{11.2\ L\ \times\ 409}{273}\ =\ 16.8\ L$$

16. The volume of a sample of argon is 8.5 mL at 15°C and 101 kPa. What will its volume be at 0.00°C and 101 kPa?

$$\frac{8.5\ mL}{1}\ \left|\ \frac{273\ K}{(15\ +\ 273)K}\right.$$

$$=\ \frac{8.5\ mL}{1}\ \left|\ \frac{273\ \cancel{K}}{288\ \cancel{K}}\right.$$

$$=\ \frac{8.5\ mL\ \times\ 273}{288}\ =\ 8.1\ mL$$

17. A 2.7-L sample of nitrogen is collected at 121 kPa and 288 K. If the pressure increases to 202 kPa and the temperature rises to 303 K, what volume will the nitrogen occupy?

$$\frac{2.7\ L}{1}\ \left|\ \frac{121\ kPa}{202\ kPa}\ \right|\ \frac{303\ K}{288\ K}$$

$$=\ \frac{2.7\ L}{1}\ \left|\ \frac{121\ \cancel{kPa}}{202\ \cancel{kPa}}\ \right|\ \frac{303\ \cancel{K}}{288\ \cancel{K}}$$

$$=\ \frac{2.7\ L\ \times\ 121\ \times\ 303}{202\ \times\ 288}\ =\ 1.7\ L$$

18. A chunk of subliming carbon dioxide (dry ice) generates a 0.80-L sample of gaseous CO_2 at 22°C and 720 mm Hg. What volume will the carbon dioxide gas have at STP?

$$\frac{0.80\ L}{1}\ \left|\ \frac{720\ mm\ Hg}{760\ mm\ Hg}\ \right|\ \frac{273\ K}{(22\ +\ 273)K}$$

$$=\ \frac{0.80\ L}{1}\ \left|\ \frac{720\ \cancel{mm\ Hg}}{760\ \cancel{mm\ Hg}}\ \right|\ \frac{273\ \cancel{K}}{295\ \cancel{K}}$$

$$=\ \frac{0.80\ L\ \times\ 720\ \times\ 273}{760\ \times\ 295}\ =\ 0.70\ L$$

CHAPTER 12

1. Without calculating, decide whether 50.0 g of sulfur or 50.0 g of tin represents the greater number of atoms. Verify your answer by calculating.

Because sulfur has a smaller molar mass, there will be a greater number of atoms in 50.0 g of sulfur than in 50.0 g of tin.

$$\frac{50.0\ \cancel{g\ S}}{1}\ \left|\ \frac{1\ \cancel{mol\ S}}{32.1\ \cancel{g\ S}}\ \right|\ \frac{6.02\ \times\ 10^{23}\ S\ atoms}{1\ \cancel{mol\ S}}$$

$$=\ \frac{50.0}{1}\ \left|\ \frac{1}{32.1}\ \right|\ \frac{6.02\ \times\ 10^{23}\ S\ atoms}{1}$$

$$=\ \frac{50.0\ \times\ 6.02\ \times\ 10^{23}\ S\ atoms}{32.1}$$

$$=\ 9.38\ \times\ 10^{23}\ S\ atoms$$

$$\frac{50.0\ \cancel{g\ Sn}}{1}\ \left|\ \frac{1\ \cancel{mol\ Sn}}{119\ \cancel{g\ Sn}}\ \right|\ \frac{6.02\ \times\ 10^{23}\ Sn\ atoms}{1\ \cancel{mol\ Sn}}$$

$$=\ \frac{50.0}{1}\ \left|\ \frac{1}{119}\ \right|\ \frac{6.02\ \times\ 10^{23}\ Sn\ atoms}{1}$$

$$=\ \frac{50.0\ \times\ 6.02\ \times\ 10^{23}\ Sn\ atoms}{119}$$

$$=\ 2.53\ \times\ 10^{23}\ Sn\ atoms$$

2. Determine the number of atoms in each sample below.
   a. **98.3 g mercury, Hg**

   $$\frac{98.3 \text{ g Hg}}{1} \left| \frac{\text{mol Hg}}{200.59 \text{ g Hg}} \right| \frac{6.02 \times 10^{23} \text{ Hg atoms}}{\text{mol Hg}}$$

   $= 2.95 \times 10^{23}$ Hg atoms

   b. **45.6 g gold, Au**

   $$\frac{45.6 \text{ g Au}}{1} \left| \frac{\text{mol Au}}{196.967 \text{ g Au}} \right| \frac{6.02 \times 10^{23} \text{ Au atoms}}{\text{mol Au}}$$

   $= 1.39 \times 10^{23}$ Au atoms

   c. **10.7 g lithium, Li**

   $$\frac{10.7 \text{ g Li}}{1} \left| \frac{\text{mol Li}}{6.941 \text{ g Li}} \right| \frac{6.02 \times 10^{23} \text{ Li atoms}}{\text{mol Li}}$$

   $= 9.28 \times 10^{23}$ Li atoms

   d. **144.6 g tungsten, W**

   $$\frac{144.6 \text{ g W}}{1} \left| \frac{\text{mol W}}{183.85 \text{ g W}} \right| \frac{6.02 \times 10^{23} \text{ W atoms}}{\text{mol W}}$$

   $= 4.73 \times 10^{23}$ W atoms

3. Determine the number of moles in each sample below.
   a) **6.84 g sucrose, $C_{12}H_{22}O_{11}$**

12 C atoms	12×12.0 u =	144.0 u
22 H atoms	22×1.0 u =	22.0 u
11 O atoms	11×16.0 u =	+176 u
molecular mass $C_{12}H_{22}O_{11}$		342 u
molar mass $C_{12}H_{22}O_{11}$		342 g/mol

   $$\frac{6.84 \text{ g } C_{12}H_{22}O_{11}}{1} \left| \frac{1 \text{ mol } C_{12}H_{22}O_{11}}{342 \text{ g } C_{12}H_{22}O_{11}} \right.$$

   $$= \frac{6.84 \text{ g } C_{12}H_{22}O_{11}}{1} \left| \frac{1 \text{ mol } C_{12}H_{22}O_{11}}{342 \text{ g } C_{12}H_{22}O_{11}} \right.$$

   $$= \frac{6.84}{1} \left| \frac{1 \text{ mol } C_{12}H_{22}O_{11}}{342} \right.$$

   $= 2.00 \times 10^{-2}$ mol $C_{12}H_{22}O_{11}$

   b) **16.0 g sulfur dioxide, SO_2**

   SO_2

1 S atom	1×32.1 u =	32.1 u
2 O atoms	2×16.0 u =	+32.0 u
molecular mass SO_2		64.1 u
molar mass SO_2		64.1 g/mol

   $$\frac{16.0 \text{ g } SO_2}{1} \left| \frac{1 \text{ mol } SO_2}{64.1 \text{ g } SO_2} \right.$$

   $$= \frac{16.0 \text{ g } SO_2}{1} \left| \frac{1 \text{ mol } SO_2}{64.1 \text{ g } SO_2} \right.$$

   $$= \frac{16.0}{1} \left| \frac{1 \text{ mol } SO_2}{64.1} \right.$$

   $= 0.250$ mol SO_2

   c) **68.0 g ammonia, NH_3**

   NH_3

1 N atom	1×14.0 u =	14.0 u
3 H atoms	3×1.0 u =	+ 3.0 u
molecular mass NH_3		17.0 u
molar mass NH_3		17.0 g/mol

   $$\frac{68.0 \text{ g } NH_3}{1} \left| \frac{1 \text{ mol } NH_3}{17.0 \text{ g } NH_3} \right.$$

   $$= \frac{68.0 \text{ g } NH_3}{1} \left| \frac{1 \text{ mol } NH_3}{17.0 \text{ g } NH_3} \right.$$

   $$= \frac{68.0}{1} \left| \frac{1 \text{ mol } NH_3}{17.0} \right.$$

   $= 4.00$ mol NH_3

   d) **17.5 g copper(II) oxide, CuO**

1 Cu atom	1×63.5 u =	63.5 u
1 O atom	1×16.0 u =	+16.0 u
formula mass CuO		79.5 u
molar mass CuO		79.5 g/mol

   $$\frac{17.5 \text{ g CuO}}{1} \left| \frac{1 \text{ mol CuO}}{79.5 \text{ g CuO}} \right.$$

   $$= \frac{17.5 \text{ g CuO}}{1} \left| \frac{1 \text{ mol CuO}}{79.5 \text{ g CuO}} \right.$$

   $$= \frac{17.5}{1} \left| \frac{1 \text{ mol CuO}}{79.5} \right.$$

   $= 0.220$ mol CuO

4. Determine the mass of the following molar quantities.
   a) **3.52 mol Si**

   $$\frac{3.52 \text{ mol Si}}{1} \left| \frac{28.1 \text{ g Si}}{1 \text{ mol Si}} \right.$$

   $$= \frac{3.52 \text{ mol Si}}{1} \left| \frac{28.1 \text{ g Si}}{1 \text{ mol Si}} \right.$$

   $= 3.52 \times 28.1$ g Si

   $= 98.9$ g Si

   b) **1.25 mol aspirin, $C_9H_8O_4$**

   $C_9H_8O_4$

9 C atoms	9×12.0 u =	108 u
8 H atoms	8×1.0 u =	+8.0 u
4 O atoms	4×16.0 u =	+64 u
molecular mass $C_9H_8O_4$		180.0 u
molar mass $C_9H_8O_4$		180.0 g/mol

   $$\frac{1.25 \text{ mol } C_9H_8O_4}{1} \left| \frac{180.0 \text{ g } C_9H_8O_4}{1 \text{ mol } C_9H_8O_4} \right.$$

   $$= \frac{1.25 \text{ mol } C_9H_8O_4}{1} \left| \frac{180.0 \text{ g } C_9H_8O_4}{1 \text{ mol } C_9H_8O_4} \right.$$

   $= 1.25 \times 180$ g $C_9H_8O_4$

   $= 225$ g $C_9H_8O_4$

   c) **0.550 mol F_2**

   F_2

2 F atoms	2×19.0 u =	38.0 u
molecular mass F_2		38.0 u
molar mass F_2		38.0 g/mol

   $(0.550 \text{ mol } F_2)(38.0 \text{ g } F_2/\text{mol } F_2)$

   $$= \frac{0.550 \text{ mol } F_2}{1} \left| \frac{38.0 \text{ g } F_2}{1 \text{ mol } F_2} \right.$$

   $= 0.550 \times 38.0$ g $F_2 = 20.9$ g F_2

   d) **2.35 mol barium iodide, BaI_2**

   BaI_2

1 Ba atom	1×137 u =	137 u
2 I atoms	2×127 u =	+254 u
formula mass BaI_2		391 u
molar mass BaI_2		391 g/mol

$$\frac{2.35 \text{ mol BaI}_2}{1} \Big| \frac{391 \text{ g BaI}_2}{1 \text{ mol BaI}_2}$$

$$= \frac{2.35 \text{ mol BaI}_2}{1} \Big| \frac{391 \text{ g BaI}_2}{1 \text{ mol BaI}_2}$$

$$= 2.35 \times 391 \text{ g BaI}_2 = 919 \text{ g BaI}_2$$

5. The combustion of propane, C_3H_8, a fuel used in back-yard grills and camp stoves, produces carbon dioxide and water vapor.

$$C_3H_8(g) + 5O_2(g) \rightarrow 3CO_2(g) + 4H_2O(g)$$

What mass of carbon dioxide forms when 95.6 g of propane burns?

$$\frac{95.6 \text{ g C}_3\text{H}_8}{1} \Big| \frac{1 \text{ mol C}_3\text{H}_8}{44.0 \text{ g C}_3\text{H}_8} \cdots$$

$$\frac{3 \text{ mol CO}_2}{1 \text{ mol C}_3\text{H}_8} \Big| \frac{44.0 \text{ g CO}_2}{1 \text{ mol CO}_3}$$

$$= \frac{95.6 \text{ g C}_3\text{H}_8}{1} \Big| \frac{1 \text{ mol C}_3\text{H}_8}{44.0 \text{ g C}_3\text{H}_8} \cdots$$

$$\frac{3 \text{ mol CO}_2}{1 \text{ mol C}_3\text{H}_8} \Big| \frac{44.0 \text{ g CO}_2}{1 \text{ mol CO}_2}$$

$$= 95.6 \times 3 \times \frac{44.0 \text{ g CO}_2}{44.0} = 287 \text{ g CO}_2$$

6. Solid xenon hexafluoride is prepared by allowing xenon gas and fluorine gas to react.

$$Xe(g) + 3F_2(g) \rightarrow XeF_6(s)$$

How many grams of fluorine are required to produce 10.0 g of XeF_6?

$$\frac{10 \text{ g XeF}_6}{1} \Big| \frac{1 \text{ mol XeF}_6}{245.3 \text{ g XeF}_6} \cdots$$

$$\frac{3 \text{ mol F}_2}{1 \text{ mol XeF}_6} \Big| \frac{38.0 \text{ g F}_2}{1 \text{ mol F}_2}$$

$$= \frac{10.0 \text{ g XeF}_6}{1} \Big| \frac{1 \text{ mol XeF}_6}{245.3 \text{ g XeF}_6} \cdots$$

$$\frac{3 \text{ mol F}_2}{1 \text{ mol XeF}_6} \Big| \frac{38.0 \text{ g F}_2}{1 \text{ mol F}_2}$$

$$= \frac{10.0 \times 3 \times 38.0 \text{ g F}_2}{245.3} = 4.65 \text{ g F}_2$$

7. Using the reaction in Practice Problem 6, how many grams of xenon are required to produce 10.0 g of XeF_6?

$$\frac{10.0 \text{ g XeF}_2}{1} \Big| \frac{1 \text{ mol XeF}_6}{245.3 \text{ g XeF}_6} \cdots$$

$$\frac{1 \text{ mol Xe}}{1 \text{ mol XeF}_6} \Big| \frac{131.3 \text{ g Xe}}{1 \text{ mol Xe}}$$

$$= \frac{10.0 \text{ g XeF}_6}{1} \Big| \frac{1 \text{ mol XeF}_6}{245.3 \text{ g XeF}_6} \cdots$$

$$\frac{1 \text{ mol Xe}}{1 \text{ mol XeF}_6} \Big| \frac{131.3 \text{ g Xe}}{1 \text{ mol Xe}}$$

$$= 10.0 \times 1 \times \frac{131.3 \text{ g Xe}}{245.3} = 5.35 \text{ g Xe}$$

8. What mass of sulfur must burn to produce 3.42 L of SO_2 at 273°C and 101 kPa? The reaction is
$$S(s) + O_2(g) \rightarrow SO_2(g).$$
Note that the volume must be corrected to standard temperature (0°C) by applying the factor

$$\frac{273 \text{ K}}{(273 + 273)\text{K}} = \frac{273 \text{ K}}{546 \text{ K}}$$

$$\frac{3.42 \text{ L SO}_2}{1} \Big| \frac{273 \text{ K}}{546 \text{ K}} \Big| \frac{1 \text{ mol SO}_2}{22.4 \text{ L SO}_2} \cdots$$

$$\frac{1 \text{ mol S}}{1 \text{ mol SO}_2} \Big| \frac{32.1 \text{ g S}}{1 \text{ mol S}}$$

$$= \frac{3.42 \text{ L SO}_2}{1} \Big| \frac{273 \text{ K}}{546 \text{ K}} \Big| \frac{1 \text{ mol SO}_2}{22.4 \text{ L SO}_2} \cdots$$

$$\frac{1 \text{ mol S}}{1 \text{ mol SO}_2} \Big| \frac{32.1 \text{ g S}}{1 \text{ mol S}}$$

$$= 3.42 \times 273 \times \frac{32.1 \text{ g S}}{546 \times 22.4} = 2.45 \text{ g S}$$

9. What volume of hydrogen gas can be produced by reacting 4.20 g sodium in excess water at 50.0°C and 106 kPa? The reaction is
$$2Na + 2H_2O \rightarrow 2NaOH + H_2$$

$$\frac{4.20 \text{ g Na}}{1} \Big| \frac{1 \text{ mol Na}}{23.0 \text{ g Na}} \Big| \frac{1 \text{ mol H}_2}{2 \text{ mol Na}} \cdots$$

$$\frac{22.4 \text{ L H}_2}{1 \text{ mol H}_2} \Big| \frac{323 \text{ K}}{273 \text{ K}} \Big| \frac{101 \text{ kPa}}{106 \text{ kPa}}$$

$$= \frac{4.20 \text{ g Na}}{1} \Big| \frac{1 \text{ mol Na}}{23.0 \text{ g Na}} \Big| \frac{1 \text{ mol H}_2}{2 \text{ mol Na}} \cdots$$

$$\frac{22.4 \text{ L H}_2}{1 \text{ mol H}_2} \Big| \frac{323 \text{ K}}{273 \text{ K}} \Big| \frac{101 \text{ kPa}}{106 \text{ kPa}}$$

$$= \frac{4.20 \times 22.4 \text{ L H}_2 \times 323 \times 101}{23.0 \times 2 \times 273 \times 106} = 2.31 \text{ L H}_2$$

10. How many moles of helium are contained in a 5.00-L canister at 101 kPa and 30.0°C?

$$n = \frac{VP}{RT} = \frac{(5.00 \text{ L})(101 \text{ kPa})}{\left(\dfrac{8.31 \text{ kPa} \cdot \text{L}}{1 \text{ mol} \cdot \text{K}}\right)303 \text{ K}}$$

$$= \frac{5.00 \text{ L} \times 101 \text{ kPa} \times \text{mol} \cdot \text{K}}{8.31 \text{ kPa} \cdot \text{L} \times 303 \text{ K}}$$

$$= \frac{5.00 \times 101 \times 1 \text{ mol}}{8.31 \times 303} = 0.201 \text{ mol}$$

11. What is the volume of 0.020 mol Ne at 0.505 kPa and 27.0°C?

$$V = \frac{nRT}{P} = \frac{(0.020 \text{ mol Ne})\left(\dfrac{8.31 \text{ kPa} \cdot \text{L}}{1 \text{ mol} \cdot \text{K}}\right)(300 \text{ K})}{0.505 \text{ kPa}}$$

$$= 99 \text{ L}$$

12. How much zinc must react in order to form 15.5 L of hydrogen, $H_2(g)$, at 32.0°C and 115 kPa?

$$Zn(s) + H_2SO_4(aq) \rightarrow ZnSO_4(aq) + H_2(g)$$

First, use the ideal gas law to determine moles of H_2.

$$PV = nRT \qquad n = \frac{PV}{RT}$$

$$n = \frac{(115 \text{ kPa})(15.5 \text{ L})}{\left(\dfrac{8.31 \text{ kPa} \cdot \text{L}}{1 \text{ mol} \cdot \text{K}}\right)(305 \text{ K})}$$

$n = 0.703 \text{ mol } H_2$

Now, determine the mass of zinc.

$$\frac{0.703 \text{ mol } H_2}{1} \left| \frac{1 \text{ mol } Zn}{1 \text{ mol } H_2} \right| \frac{65.39 \text{ g } Zn}{1 \text{ mol } Zn}$$

$= 46.0 \text{ g } Zn$

CHAPTER 13

1. How would you prepare 1.00 L of a 0.400M solution of copper(II) sulfate, $CuSO_4$?

Dissolve 63.8 g of $CuSO_4$ in 1.00 L of solution;
Molar mass $CuSO_4$ = 159.61 g/mol

$$\frac{1.00 \text{ L soln}}{1} \left| \frac{0.400 \text{ mol } CuSO_4}{1 \text{ L soln}} \right| \frac{159.61 \text{ g } CuSO_4}{1 \text{ mol } CuSO_4}$$

$= 63.8 \text{ g } CuSO_4$

2. How would you prepare 2.50 L of a 0.800M solution of potassium nitrate, KNO_3?

Dissolve 202 g of KNO_3 in 2.50 L of solution;
Molar mass KNO_3 = 101.10 g/mol

$$\frac{2.50 \text{ L soln}}{1} \left| \frac{0.800 \text{ mol } KNO_3}{1 \text{ L soln}} \right| \frac{101.10 \text{ g}}{1 \text{ mol}}$$

$= 202 \text{ g } KNO_3$

3. What mass of sucrose, $C_{12}H_{22}O_{11}$, must be dissolved to make 460 mL of a 1.10M solution?

173 g of sucrose;
Molar mass $C_{12}H_{22}O_{11}$ = 342.30 g/mol

$$\frac{460 \text{ mL soln}}{1} \left| \frac{1 \text{ L}}{10^3 \text{ mL}} \right| \frac{1.10 \text{ mol } C_{12}H_{22}O_{11}}{1 \text{ L soln}} \cdots$$

$$\left| \frac{342.30 \text{ g}}{1 \text{ mol}} = 170 \text{ g } C_{12}H_{22}O_{11} \right.$$

4. What mass of lithium chloride, LiCl, must be dissolved to make a 0.194M solution that has a volume of 1.00 L?

8.23 g of LiCl;
Molar mass LiCl = 42.40 g/mol

$$\frac{1.00 \text{ L soln}}{1} \left| \frac{0.194 \text{ mol } LiCl}{1 \text{ L soln}} \right| \frac{42.40 \text{ g}}{1 \text{ mol}}$$

$= 8.23 \text{ g } LiCl$

5. What is the molarity of a solution that contains 14 g of sodium sulfate, Na_2SO_4, dissolved in 1.6 L of solution?

0.062M Na_2SO_4;
Molar mass Na_2SO_4 = 142.06 g/mol

$$\frac{14 \text{ g } Na_2SO_4}{1.6 \text{ L soln}} \left| \frac{1 \text{ mol}}{142.06 \text{ g}} \right.$$

$= 0.062 \text{ mol } Na_2SO_4/\text{L soln}$
$= 0.062M \text{ } Na_2SO_4$

6. Calculate the molarity of a solution, given that its volume is 820 mL and that it contains 7.4 g of ammonium chloride, NH_4Cl.

0.17M NH_4Cl;
Molar mass NH_4Cl = 53.50 g/mol

$$\frac{7.4 \text{ g } NH_4Cl}{820 \text{ mL}} \left| \frac{1 \text{ mol}}{53.50 \text{ g}} \right| \frac{10^3 \text{ mL}}{1 \text{ L}}$$

$= 0.17 \text{ mol } NH_4Cl/\text{L soln}$
$= 0.17M \text{ } NH_4Cl$

CHAPTER 14

Find the pH of each of the following solutions.
1. The hydronium ion concentration equals:
  a) $10^{-5}M$
    5
  b) $10^{-12}M$
    12
  c) $10^{-2}M$
    2
2. The hydroxide ion concentration equals:
  a) $10^{-4}M$
    10
  b) $10^{-11}M$
    3
  c) $10^{-8}M$
    6

CHAPTER 15

Write overall, ionic, and net ionic equations for each of the following reactions.
1. hydroiodic acid, HI, and calcium hydroxide, $Ca(OH)_2$
$$2HI(aq) + Ca(OH)_2(aq) \rightarrow CaI_2(aq) + 2H_2O(l)$$
$$2H^+(aq) + 2I^-(aq) + Ca^{2+}(aq) + 2OH^-(aq)$$
$$\rightarrow Ca^{2+}(aq) + 2I^-(aq) + 2H_2O(l)$$
$$H^+(aq) + OH^-(aq) \rightarrow H_2O(l)$$
2. hydrobromic acid, HBr, and lithium hydroxide, LiOH
$$HBr(aq) + LiOH(aq) \rightarrow LiBr(aq) + H_2O(l)$$
$$H^+(aq) + Br^-(aq) + Li^+(aq) + OH^-(aq)$$
$$\rightarrow Li^+(aq) + Br^-(aq) + H_2O(l)$$
$$H^+(aq) + OH^-(aq) \rightarrow H_2O(l)$$
3. sulfuric acid, H_2SO_4, and strontium hydroxide, $Sr(OH)_2$
$$H_2SO_4(aq) + Sr(OH)_2(aq) \rightarrow SrSO_4(aq) + 2H_2O(l)$$
$$2H^+(aq) + SO_4^{2-}(aq) + Sr^{2+}(aq) + 2OH^-(aq)$$
$$\rightarrow Sr^{2+}(aq) + SO_4^{2-}(aq) + 2H_2O(l)$$
$$H^+(aq) + OH^-(aq) \rightarrow H_2O(l)$$
4. perchloric acid, $HClO_4$, and barium hydroxide, $Ba(OH)_2$
$$2HClO_4(aq) + Ba(OH)_2(aq) \rightarrow 2H_2O(l) + Ba(ClO_4)_2(aq)$$
$$2H^+(aq) + 2ClO_4^-(aq) + Ba^{2+}(aq) + 2OH^-(aq)$$
$$\rightarrow 2H_2O(l) + Ba^{2+}(aq) + 2ClO_4^-(aq)$$
$$H^+(aq) + OH^-(aq) \rightarrow H_2O(l)$$
5. perchloric acid, $HClO_4$, and ammonia, NH_3
$$HClO_4(aq) + NH_3(aq) \rightarrow NH_4ClO_4(aq)$$
$$H^+(aq) + ClO_4^-(aq) + NH_3(aq)$$
$$\rightarrow NH_4^+(aq) + ClO_4^-(aq)$$
$$H^+(aq) + NH_3(aq) \rightarrow NH_4^+(aq)$$

6. hydrochloric acid, HCl, and aluminum hydroxide, $Al(OH)_3$

$3HCl(aq) + Al(OH)_3(aq) \rightarrow AlCl_3(aq) + 3H_2O(l)$

$3H^+(aq) + 3Cl^-(aq) + Al(OH)_3(aq)$
$\rightarrow Al^{3+}(aq) + 3Cl^-(aq) + 3H_2O(l)$

$3H^+(aq) + Al(OH)_3(aq) \rightarrow Al^{3+}(aq) + 3H_2O(l)$

7. sulfuric acid, H_2SO_4, and iron(III) hydroxide, $Fe(OH)_3$

$3H_2SO_4(aq) + 2Fe(OH)_3(aq)$
$\rightarrow Fe_2(SO_4)_3(aq) + 6H_2O(l)$

$6H^+(aq) + 3SO_4^{2-}(aq) + 2Fe(OH)_3(aq)$
$\rightarrow 2Fe^{3+}(aq) + 3SO_4^{2-}(aq) + 6H_2O(l)$

$3H^+(aq) + Fe(OH)_3(aq) \rightarrow Fe^{3+}(aq) + 3H_2O(l)$

8. carbonic acid, H_2CO_3, and sodium hydroxide, NaOH

$H_2CO_3(aq) + 2NaOH(aq) \rightarrow Na_2CO_3(aq) + 2H_2O(l)$

$H_2CO_3(aq) + 2Na^+(aq) + 2OH^-(aq)$
$\rightarrow 2Na^+(aq) + CO_3^{2-}(aq) + 2H_2O(l)$

$H_2CO_3(aq) + 2OH^-(aq) \rightarrow CO_3^{2-}(aq) + 2H_2O(l)$

9. boric acid, H_3BO_3, and potassium hydroxide, KOH

$H_3BO_3(aq) + 3KOH(aq) \rightarrow K_3BO_3(aq) + 3H_2O(l)$

$H_3BO_3(aq) + 3K^+(aq) + 3OH^-(aq)$
$\rightarrow 3K^+(aq) + BO_3^{3-}(aq) + 3H_2O(l)$

$H_3BO_3(aq) + 3OH^-(aq) \rightarrow BO_3^{3-}(aq) + 3H_2O(l)$

10. acetic acid, $HC_2H_3O_2$, and calcium hydroxide, $Ca(OH)_2$

$2HC_2H_3O_2(aq) + Ca(OH)_2(aq)$
$\rightarrow Ca(C_2H_3O_2)_2(aq) + 2H_2O(l)$

$2HC_2H_3O_2(aq) + Ca^{2+}(aq) + 2OH^-(aq)$
$\rightarrow Ca^{2+}(aq) + 2C_2H_3O_2^-(aq) + 2H_2O(l)$

$HC_2H_3O_2(aq) + OH^-(aq) \rightarrow C_2H_3O_2^-(aq) + H_2O(l)$

Finding Molarity

11. A $0.100M$ LiOH solution was used to titrate an HBr solution of unknown concentration. At the endpoint, 21.0 mL of LiOH solution had neutralized 10.0 mL of HBr. What is the molarity of the HBr solution?

$0.210M$;

$$\frac{21.0 \text{ mL LiOH}}{1} \, \bigg| \, \frac{1 \text{ L}}{10^3 \text{ mL LiOH}} \, \bigg| \, \frac{0.100 \text{ mol LiOH}}{1 \text{ L}} \cdots$$

$$\frac{1 \text{ mol HBr}}{1 \text{ mol LiOH}} \, \bigg| \, \frac{1}{10.0 \text{ mL HBr}} \, \bigg| \, \frac{10^3 \text{ mL HBr}}{1 \text{ L}}$$

$= 0.210M$ HBr

12. A $0.150M$ KOH solution fills a burette to the 0 mark. The solution was used to titrate 25.0 mL of an HNO_3 solution of unknown concentration. At the endpoint, the burette reading was 34.6 mL. What was the molarity of the HNO_3 solution?

$0.208M$;

$$\frac{34.6 \text{ mL KOH}}{1} \, \bigg| \, \frac{1 \text{ L}}{10^3 \text{ mL KOH}} \cdots$$

$$\frac{0.150 \text{ mol KOH}}{1 \text{ L}} \, \bigg| \, \frac{1 \text{ mol HNO}_3}{1 \text{ mol KOH}} \cdots$$

$$\frac{1}{25.0 \text{ mL HNO}_3} \, \bigg| \, \frac{10^3 \text{ mL HNO}_3}{1 \text{ L}}$$

$= 0.208M$ HNO_3

13. A $Ca(OH)_2$ solution of unknown concentration was used to titrate 15.0 mL of a $0.125M$ H_3PO_4 solution. If 12.4 mL of $Ca(OH)_2$ are used to reach the endpoint, what is the concentration of the $Ca(OH)_2$ solution?

$0.227M$;

$$\frac{15.0 \text{ mL H}_3\text{PO}_4}{1} \, \bigg| \, \frac{1 \text{ L}}{10^3 \text{ mL H}_3\text{PO}_4} \, \bigg| \, \frac{0.125 \text{ mol H}_3\text{PO}_4}{1 \text{ L}} \cdots$$

$$\frac{3 \text{ mol Ca(OH)}_2}{2 \text{ mol H}_3\text{PO}_4} \, \bigg| \, \frac{1}{12.4 \text{ mL}} \, \bigg| \, \frac{10^3 \text{ mL}}{1 \text{ L}}$$

$= 0.227M$ $Ca(OH)_2$

CHAPTER 18

1. Write the structural formulas for the following branched alkanes.

a) 2-methylbutane

b) 1, 3-dimethylcyclohexane (Hint: Begin numbering at any carbon in the ring, then attach the methyl groups.)

c) 4-propyldecane

d) 2, 3, 4-trimethylheptane

2. Name each of the following alkanes.

a)

3-methylpentane

b)

2, 3-dimethylbutane

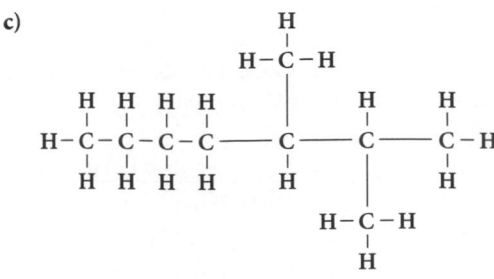

c)

2, 3-dimethylheptane

d)

3-ethylhexane

CHAPTER 20

1. How much heat is absorbed by a reaction that lowers the temperature of 500.0 g of water in a calorimeter by 1.10°C?

$q_w = (m)(\Delta T)(C_w)$
$= (500.0 \text{ g})(1.10°C)(4.184 \text{ J/g°C})$
$= 2300 \text{ J}$
$= 2.30 \text{ kJ}$

2. Aluminum reacts with iron(III) oxide to yield aluminum oxide and iron. Calculate the heat given off in the reaction if the temperature of the 1.00 kg of water in the calorimeter increases by 3.00°C.

$q_w = (m)(\Delta T)(C_w)$
$= (1.00 \times 10^3 \text{ g})(3.00°C)(4.184 \text{ J/g°C})$
$= 12600 \text{ J}$
$= 12.6 \text{ kJ}$

3. When 1.00 g of a certain fuel gas is burned in a calorimeter, the temperature of the surrounding 1.000 kg of water increases from 20.00°C to 28.05°C. All products and reactants in the process are gases. Calculate the heat given off in this reaction. How much heat would 1.00 mol of the fuel give off, assuming a molar mass of 65.8 g/mol?

$\Delta T = 28.05°C - 20.00°C = 8.05°C$

For 1 g:
$q_w = (m)(\Delta T)(C_w)$
$= (1.000 \times 10^3 \text{ g})(8.05°C)(4.184/g°C)$
$= 33700 \text{ J}$
$= 33.7 \text{ kJ}$

For 1 mol:
$$q_w = \frac{33.7 \text{ kJ}}{1 \text{ g}} \quad \frac{65.8 \text{ g}}{1 \text{ mol}}$$
$= 2220 \text{ kJ/mol}$

4. A group of students decides to measure the energy content of certain foods. They heat 50.0 g of water in an aluminum can by burning a sample of the food beneath the can. When they use 1.00 g of popcorn as their test food, the temperature of the water rises by 24°C. Calculate the heat released by the popcorn, and express your answer in both kilojoules and Calories per gram of popcorn.

$q_w = (m)(\Delta T)(C_w)$
$= (50.0 \text{ g})(24°C)(4.184 \text{ J/g°C}) \text{ for 1 g}$
$= 5000 \text{ J/g}$
$= 5.0 \text{ kJ/g}$

$$= \frac{5.0 \text{ kJ}}{1 \text{ g}} \quad \frac{1 \text{ Cal}}{4.184 \text{ kJ}} = 1.2 \text{ Cal/g}$$

5. Another student comes along and tells the group in problem 4 that she has read the label on a popcorn bag that states that 30 g of popcorn yields 110 Calories. What is that value in Calories/gram? How can you account for the difference?

$$\frac{110 \text{ Cal}}{30.0 \text{ g}} = 3.7 \text{ Cal/g}$$

1.2 Cal/g, the experimental value, is much lower than the value given on the popcorn bag because of loss of heat to the air in the student experiment. Also, the combustion of popcorn in the student experiment may be incomplete.

6. A 3.00-g sample of a new snack food is burned in a calorimeter. The 2.00 kg of surrounding water change in temperature from 25.0°C to 32.4°C. What is the food value in Calories per gram?

$\Delta T = 32.4°C - 25.0°C = 7.4°C$
$q_w = (m)(\Delta T)(C_w)$
$$= \frac{(2.00 \times 10^3 \text{ g})(7.4°C)(4.184 \text{ J/g°C}) \text{ for 3 g}}{3.00 \text{ g}}$$
$= 20.6 \text{ kJ/g}$

$$= \frac{20.6 \text{ kJ}}{1 \text{ g}} \quad \frac{1 \text{ Cal}}{4.184 \text{ kJ}} = 4.93 \text{ Cal/g}$$

CHAPTER 21

1. Write the balanced nuclear equation for the radioactive decay of radium-226 to give radon-222, and determine the type of decay.

$^{226}_{88}\text{Ra} \rightarrow {}^{222}_{86}\text{Rn} + {}^4_2\text{He}$; alpha decay

2. Write a balanced equation for the nuclear reaction in which neon-23 decays to form sodium-23, and determine the type of decay.

$^{23}_{10}\text{Ne} \rightarrow {}^{23}_{11}\text{Na} + {}^0_{-1}e$; beta decay

3. A rock was analyzed using potassium-40. The half-life of potassium-40 is 1.25 billion years. If the rock had only 25 percent of the potassium-40 that would be found in a similar rock formed today, calculate how long ago the rock was formed.

Because $(0.5)(0.5) = 0.25$, two half-lives have gone by.
2 half-lives × 1.25 billion years/half-life = 2.50 billion years

4. Ash from an early fire pit was found to have 12.5 percent as much carbon-14 as would be found in a similar sample of ash today. How long ago was the ash formed?

Because $(0.5)(0.5)(0.5) = 0.125$, three half-lives have gone by.
3 half-lives × 5730 years/half-life = 17 190 years

absolute zero: the temperature below which a substance would have zero kinetic energy. (Chap. 10, p. 349)

acid: a substance that produces hydronium ions when dissolved in water. (Chap. 14, p. 483)

acidic anhydride: a nonmetal oxide that reacts with water to form an acid. (Chap. 14, p. 492)

acidic hydrogen: in an acid, any hydrogen that can be transferred to water. (Chap. 14, p. 485)

actinide: any of the second series of inner transition elements with atomic numbers from 90 to 103; all are radioactive. (Chap. 3, p. 104)

activation energy: the amount of energy the particles in a reaction must have when they collide for the reaction to occur. (Chap. 6, p. 218)

active site: on an enzyme, the pocket or groove that can bind a substrate taking part in a reaction. (Chap. 19, p. 676)

addition reaction: a reaction where monomers that contain double bonds add onto each other to form long chains; the product contains all the atoms of the starting monomers. (Chap. 18, p. 656)

aerobic: a metabolic process that takes place only in the presence of oxygen. (Chap. 19, p. 694)

alkali metal: any element from Group 1: lithium, sodium, potassium, rubidium, cesium, francium. (Chap. 8, p. 263)

alkaline earth metal: any element from Group 2: beryllium, magnesium, calcium, strontium, barium. (Chap. 8, p. 265)

alkane: a saturated hydrocarbon that consists of only carbon and hydrogen atoms with single bonds between all the atoms. (Chap. 18, p. 623)

alkene: a hydrocarbon in which one or more double bonds link carbon atoms. (Chap. 18, p. 629)

alkyne: an unsaturated hydrocarbon that contains a triple bond between two carbon atoms. (Chap. 18, p. 633)

allotrope: any of two or more molecules of a single element that have different crystalline or molecular structures. (Chap. 5, p. 175)

alloy: a solid solution containing different metals, and sometimes nonmetallic substances. (Chap. 1, p. 23)

alpha particle: a helium nucleus consisting of two protons and two neutrons. (Chap. 21, p. 747)

amino acid: an organic compound; a monomer that forms proteins. (Chap. 19, p. 670)

amorphous material: a substance with a haphazard, disjointed, and incomplete crystal lattice. (Chap. 10, p. 346)

anaerobic: a metabolic process that takes place in the absence of oxygen. (Chap. 19, p. 698)

anhydrous: a compound in which all water has been removed, usually by heating. (Chap. 5, p. 168)

anion: a negative ion. (Chap. 17, p. 586)

anode: the electrode that takes electrons away from the reacting ions or atoms in solution. (Chap. 17, p. 585)

aqueous solution: a solution in which the solvent is water. (Chap. 1, p. 23)

aromatic hydrocarbon: a compound that has a benzene ring or the type of bonding exhibited by benzene; most have distinct odors. (Chap. 18, p. 636)

atom: the smallest particle of a given type of matter. (Chap. 2, p. 53)

atomic number: the number of protons in the nucleus of an atom of an element. (Chap. 2, p. 66)

atomic theory: the idea that matter is made up of fundamental particles called atoms. (Chap. 2, p. 53)

ATP: adenosine triphosphate, the energy storage molecule in cells. (Chap. 19, p. 696)

Avogadro constant: the number of things in one mole of a substance, specifically 6.02×10^{23}. (Chap. 12, p. 405)

Avogadro's principle: statement that at the same temperature and pressure, equal volumes of gases contain equal numbers of particles. (Chap. 11, p. 398)

barometer: an instrument that measures the pressure exerted by the atmosphere. (Chap. 11, p. 376)

base: a substance that produces hydroxide ions when it dissolves in water. (Chap. 14, p. 488)

basic anhydride: a metal oxide that reacts with water to form a base. (Chap. 14, p. 492)

beta particle: a high-energy electron with a 1− charge. (Chap. 21, p. 749)

binary compound: a compound that contains only two elements. (Chap. 5, p. 155)

biochemistry: the study of the chemistry of living things. (Chap. 19, p. 668)

boiling point: the temperature of a liquid where its vapor pressure equals the pressure exerted on its surface. (Chap. 10, p. 358)

Boyle's law: at a constant temperature, the volume and pressure of a gas are inversely proportional. (Chap. 11, p. 383)

Brownian motion: the constant, random motion of tiny chunks of matter. (Chap. 10, p. 342)

buffer: a solution that resists changes in pH when moderate amounts of acids or bases are added to it. (Chap. 15, p. 531)

calorie: the heat required to raise the temperature of 1 gram of liquid water by 1°C. (Chap. 20, p. 721)

Glossary

Calorie: a food Calorie, equal to 1 kilocalorie, used to measure the energy value of foods. (Chap. 20, p. 721)

capillarity: the rising of a liquid in a narrow tube, sometimes called capillary action. (Chap. 13, p. 443)

carbohydrate: an organic molecule that contains the elements carbon, hydrogen, and oxygen in a ratio of about two hydrogen atoms and one oxygen atom for each carbon atom. (Chap. 19, p. 677)

catalyst: a substance that speeds up the rate of a reaction without being used up itself or permanently changed. (Chap. 6, p. 222)

cathode: the electrode that brings electrons to the reacting ions or atoms in solution. (Chap. 17, p. 585)

cation: a positive ion. (Chap. 17, p. 586)

Charles's law: at constant pressure, the volume of a gas is directly proportional to its Kelvin temperature. (Chap. 11, p. 392)

chemical change: the change of one or more substances into other substances. (Chap. 1, p. 40)

chemical property: a property that can be observed only when there is a change in the composition of a substance. (Chap. 1, p. 40)

chemical reaction: another term for chemical change. (Chap. 1, p. 40)

chemistry: the science that investigates and explains the structure and properties of matter. (Chap. 1, p. 4)

coefficient: a number placed in front of the parts of a chemical equation to indicate how many are involved; always a positive whole number. (Chap. 6, p. 199)

coenzyme: an organic molecule that assists an enzyme in catalyzing a reaction. (Chap. 19, p. 691)

colloid: a mixture that contains particles that are evenly distributed through a dispersing medium and do not settle out over time. (Chap. 13, p. 472)

combined gas law: the combination of Boyle's law and Charles's law. (Chap. 11, p. 395)

combustion: term for a reaction in which a substance rapidly combines with oxygen to form one or more oxides. (Chap. 6, p. 209)

compound: a chemical combination of two or more different elements joined together in a fixed proportion. (Chap. 1, p. 30)

concentration: the amount of a substance present in a unit volume. (Chap. 6, p. 219)

condensation: the process where gaseous particles come together, that is, condense, to form a liquid or sometimes a solid. (Chap. 10, p. 356)

condensation reaction: reaction to form a polymer where a small molecule, usually water, is given off as each new bond is formed. (Chap. 18, p. 658)

conductivity: a measure of how easily electrons can flow through a material to produce an electrical current. (Chap. 9, p. 313)

covalent bond: the attraction of two atoms for a shared pair of electrons. (Chap. 4, p. 140)

covalent compound: a compound whose atoms are held together by covalent bonds. (Chap. 4, p. 140)

cracking: the use of a catalyst or high temperatures in the absence of air to break down or rearrange large hydrocarbons. (Chap. 18, p. 638)

cross-linking: the linking together of many polymer chains, giving the polymer increased strength. (Chap. 18, p. 658)

crystal: a regular, repeating arrangement of atoms, ions, or molecules in three dimensions. (Chap. 4, p. 134)

crystal lattice: the three-dimensional arrangement repeated throughout a solid. (Chap. 10, p. 345)

D

decomposition: the name applied to a reaction where a compound breaks down into two or more simpler substances. (Chap. 6, p. 204)

deliquescent: a substance that takes up enough water from the air that it dissolves completely to a liquid solution. (Chap. 5, p. 166)

denaturation: the name given to the process of unfolding of a protein when the forces holding the polypeptide chain in shape are broken. (Chap. 19, p. 673)

density: the amount of matter (mass) in a given unit volume. (Chap. 1, p. 36)

deuterium: the hydrogen isotope with a mass number of 2. (Chap. 21, p. 766)

diffusion: the process by which a gas enters a container and fills it, or when the particles of two gases or liquids mix together. (Chap. 10, p. 352)

dissociation: the process by which the charged particles in an ionic solid separate from one another, primarily when going into solution. (Chap. 13, p. 453)

distillation: the method of separating substances in a mixture by evaporation of a liquid and subsequent condensation of its vapor. (Chap. 5, p. 171)

DNA: deoxyribonucleic acid. (Chap. 19, p. 688)

double bond: a bond formed by the sharing of two pairs of electrons between two atoms. (Chap. 9, p. 321)

double displacement: a type of reaction where the positive and negative portions of two ionic compounds are interchanged; at least one product must be water or a precipitate. (Chap. 6, p. 208)

ductile: property of a metal that means it can easily be drawn into a wire. (Chap. 9, p. 313)

dynamic equilibrium: term describing a system in which opposite reactions are taking place at the same rate. (Chap. 6, p. 211)

electrical current: the flow of electrons in a particular direction. (Chap. 17, p. 584)

electrolysis: the process in which electrical energy causes a non-spontaneous chemical reaction to occur. (Chap. 17, p. 584)

electrolyte: any compound that conducts electricity when melted or dissolved in water. (Chap. 4, p. 144)

electrolytic cell: the electrochemical cell in which electrolysis takes place. (Chap. 17, p. 586)

electromagnetic spectrum: the whole range of electromagnetic radiation. (Chap. 2, p. 71)

electron: negatively-charged particle. (Chap. 2, p. 61)

electron cloud: the space around the nucleus of an atom where the atom's electrons are found. (Chap. 2, p. 77)

electron configuration: the most stable arrangement of electrons in sublevels and orbitals. (Chap. 7, p. 242)

electron transport chain: the controlled release of energy from glucose by the step-by-step movement of electrons to lower energy levels. (Chap. 19, p. 698)

electronegativity: the measure of the ability of an atom in a bond to attract electrons. (Chap. 9, p. 303)

element: a substance that cannot be broken down into simpler substances. (Chap. 1, p. 24)

emission spectrum: the spectrum of light released from excited atoms of an element. (Chap. 2, p. 74)

empirical formula: the formula of a compound having the smallest whole-number ratio of atoms in the compound. (Chap. 12, p. 428)

endothermic: chemical reaction that absorbs energy. (Chap. 1, p. 43)

energy: the capacity to do work. (Chap. 1, p. 42)

energy level: the regions of space in which electrons can move about the nucleus of an atom. (Chap. 2, p. 75)

entropy: term used to describe and measure the degree of disorder in a process. (Chap. 20, p. 716)

enzyme: a biological catalyst. (Chap. 6, p. 222)

equilibrium: term for a system where no net change occurs in the amount of reactants or products. (Chap. 6, p. 211)

evaporation: the process by which particles of a liquid form a gas by escaping from the liquid surface. (Chap. 10, p. 352)

exothermic: chemical reaction that gives off energy. (Chap. 1, p. 42)

factor label method: The problem-solving method in chemistry that uses mathematical relationships to convert one quantity to another. (Chap. 11, p. 380)

family: see group. (Chap. 3, p. 96)

fatty acid: a long-chain carboxylic acid. (Chap. 19, p. 684)

fermentation: the anaerobic process of generating energy from glucose. (Chap. 19, p. 698)

formula: a combination of chemical symbols that show what elements make up a compound and the number of atoms of each element. (Chap. 1, p. 31)

formula mass: the mass in atomic mass units of one formula unit of an ionic compound. (Chap. 12, p. 409)

formula unit: the simplest ratio of ions in a compound. (Chap. 5, p. 156)

fossil fuel: a fuel such as oil or natural gas, comprised of hydrocarbons that are the remains of plants and other organisms that lived millions of years ago. (Chap. 20, p. 713)

fractional distillation: the distillation of a mixture by the use of repeated vaporization-condensation cycles to increase the efficiency of separation. (Chap. 18, p. 638)

freezing point: the temperature of a liquid when it becomes a crystal lattice. (Chap. 10, p. 364)

functional group: the part of a molecule that is largely responsible for the chemical behavior of the molecule. (Chap. 18, p. 640)

galvanic cell: an electrochemical cell in which an oxidation-reduction reaction occurs spontaneously to produce a potential difference. (Chap. 17, p. 602)

gamma ray: a high-energy form of electromagnetic radiation with no charge and no mass. (Chap. 21, p. 749)

gas: a flowing, compressible substance with no definite volume or shape. (Chap. 10, p. 341)

gray: the unit used to measure a received dose of radiation. (Chap. 21, p. 776)

group: the elements in a vertical column of the periodic table. (Chap. 3, p. 96)

half life: the time it takes for half of a given radioactive isotope to decay (into a different isotope or element). (Chap. 21, p. 756)

halogen: any element from Group 17: fluorine, chlorine, bromine, iodine, astatine. (Chap. 8, p. 278)

heat: the energy transferred from an object at high temperature to an object at lower temperature. (Chap. 20, p. 711)

Glossary

heat of fusion: the energy released as one kilogram of a substance solidifies at its freezing point. (Chap. 10, p. 364)

heat of solution: the heat taken in or released in the dissolving process. (Chap. 13, p. 460)

heat of vaporization: the energy absorbed when one kilogram of a liquid vaporizes at its normal boiling point. (Chap. 10, p. 361)

Heisenberg uncertainty principle: the principle that it is impossible to accurately measure both the position and energy of an electron at the same time. (Chap. 7, p. 238)

hormone: a signal molecule that tells cells whether to start or stop a reaction. (Chap. 19, p. 694)

hydrate: a compound in which there is a specific ratio of water to ionic compound. (Chap. 5, p. 165)

hydrocarbon: an organic compound that consists of only hydrogen and carbon. (Chap. 5, p. 183)

hydrogen bonding: a connection between the hydrogen atoms on one molecule and a highly electronegative atom on another molecule, but not a full covalent bond. (Chap. 13, p. 439)

hydronium ion: a hydrogen ion attached to a water molecule. (Chap. 14, p. 483)

hygroscopic: a substance that absorbs water molecules from the air to become a hydrate. (Chap. 5, p. 166)

hypothesis: a prediction that can be tested to explain observations. (Chap. 2, p. 59)

ideal gas: a gas in which the particles undergo elastic collisions. (Chap. 10, p. 343)

ideal gas law: the equation that expresses exactly how pressure P, volume V, temperature T, and the number of particles n of a gas are related. $PV = nRT$. (Chap. 12, p. 419)

inhibitor: a substance that slows down a reaction. (Chap. 6, p. 223)

inner transition element: one of the elements in the two rows of elements below the main body of the periodic table; the lanthanides and the actinides. (Chap. 7, p. 250)

inorganic compound: a compound that does not contain carbon. (Chap. 5, p. 180)

insoluble: term describing a compound that does not dissolve in a liquid. (Chap. 6, p. 215)

interparticle forces: the forces between the particles that make up a substance. (Chap. 4, p. 144)

ion: an atom or group of combined atoms that has a charge because of the loss or gain of electrons. (Chap. 4, p. 134)

ionic bond: the strong attractive force between ions of opposite charge. (Chap. 4, p. 134)

ionic compound: a compound comprised of ions. (Chap. 4, p. 134)

ionic equation: an equation in which substances that primarily exist as ions in solution are shown as ions. (Chap. 15, p. 517)

ionization: the process where ions form from a covalent compound. (Chap. 14, p. 488)

isomer: a compound with a structure different from another compound with the same formula. (Chap. 18, p. 628)

isotope: any of two or more atoms of an element that are chemically alike but have different masses. (Chap. 2, p. 62)

joule: the SI unit of energy; the energy required to lift a one-newton weight one meter against the force of gravity. (Chap. 10, p. 361)

kelvin (K): a division on the Kelvin scale; the SI unit of temperature. (Chap. 10, p. 350)

Kelvin scale: the temperature scale defined so that temperature of a substance is directly proportional to the average kinetic energy of the particles and so that zero on the scale corresponds to zero kinetic energy. (Chap. 10, p. 349)

kilocalorie: a unit equal to 1000 calories. (Chap. 20, p. 721)

kilopascal (kPa): 1000 pascals. (Chap. 11, p. 378)

kinetic theory: the theory that states that submicroscopic particles of all matter are in constant, random motion. (Chap. 10, p. 342)

lanthanide: one of the first series of inner transition elements with atomic numbers 58 to 71. (Chap. 3, p. 104)

law of combining gas volumes: the observation that at the same temperature and pressure, volumes of gases combine or decompose in ratios of small whole numbers. (Chap. 11, p. 396)

law of conservation of energy: statement that energy is neither created nor destroyed in a chemical change, but is simply changed from one form to another. (Chap. 20, p. 711)

law of conservation of mass: in a chemical change, matter is neither created nor destroyed. (Chap. 1, p. 42)

law of definite proportions: the principle that the elements that comprise a compound are always in a certain proportion by mass. (Chap. 2, p. 54)

Lewis dot diagram: a diagram where dots or other small symbols are placed around the chemical symbol of an element to illustrate the valence electrons. (Chap. 2, p. 79)

limiting reactant: the reactant of which there is not enough; when it is used up, the reaction stops and no new product is formed. (Chap. 6, p. 220)

lipid: a biological compound that contains a large proportion of C—H bonds and less oxygen than in a carbohydrate; commonly called fats and oils. (Chap. 19, p. 682)

liquid: a flowing substance with a definite volume but an indefinite shape. (Chap. 10, p. 341)

liquid crystal: a material that loses its rigid organization in only one or two dimensions when it melts. (Chap. 10, p. 345)

malleable: property of a metal meaning it can be pounded or rolled into thin sheets. (Chap. 9, p. 313)

mass: the measure of the amount of matter an object contains. (Chap. 1, p. 4)

mass number: the sum of the neutrons and protons in the nucleus of an atom. (Chap. 2, p. 66)

matter: anything that takes up space and has mass. (Chap. 1, p. 4)

melting point: the temperature of a solid when its crystal lattice disintegrates. (Chap. 10, p. 364)

metabolism: name given to the sum of all the chemical reactions necessary for the life of an organism. (Chap. 19, p. 692)

metal: an element that has luster, conducts heat and electricity, and usually bends without breaking. (Chap. 3, p. 103)

metallic bond: the bond that results when metal atoms release their valence electrons to a pool of electrons shared by all the metal atoms. (Chap. 9, p. 314)

metalloid: an element with some physical and chemical properties of metals and other properties of nonmetals. (Chap. 3, p. 105)

mixture: a combination of two or more substances in which the basic identity of each substance is not changed. (Chap. 1, p. 18)

molar mass: the mass of one mole of a pure substance. (Chap. 12, p. 407)

molar volume: the volume that a mole of gas occupies at a pressure of one atmosphere and a temperature of 0.00°C. (Chap. 12, p. 416)

mole: the unit used to count numbers of atoms, molecules, or formula units of substances. (Chap. 12, p. 405)

molecular element: a molecule formed when atoms of the same element bond together. (Chap. 5, p. 174)

molecular mass: the mass in atomic mass units of one molecule of a covalent compound. (Chap. 12, p. 409)

molecular substance: a substance that has atoms held together by covalent rather than ionic bonds. (Chap. 5, p. 170)

molecule: an uncharged group of two or more atoms held together by covalent bonds. (Chap. 4, p. 140)

monomer: the individual, small units that make up a polymer. (Chap. 18, p. 649)

net ionic equation: the equation that results when ions common to both sides of the equation are removed, usually from an ionic equation. (Chap. 15, p. 520)

neutralization reaction: the reaction of an acid with a base, so called because the properties of both the acid and base are diminished or neutralized. (Chap. 15, p. 516)

neutron: a subatomic particle with a mass equal to a proton but with no electrical charge. (Chap. 2, p. 62)

noble gas: an element from Group 18 that has a full compliment of valence electrons and as such is unreactive. (Chap. 3, p. 98)

noble gas configuration: the state of an atom achieved by having the same valence electron configuration as a noble gas atom; the most stable configuration. (Chap. 4, p. 132)

nonmetal: an element that in general does not conduct electricity, is a poor conductor of heat, and is brittle when solid. Many are gases at room temperature. (Chap. 3, p. 104)

nuclear fission: the process in which an atomic nucleus splits into two or more large fragments. (Chap. 21, p. 762)

nuclear fusion: the process in which two or more nuclei combine to form a larger nucleus. (Chap. 21, p. 766)

nuclear reactor: the device used to extract energy from a radioactive fuel. (Chap. 21, p. 764)

nucleic acid: a large polymer containing carbon, hydrogen, and oxygen, as well as nitrogen and phosphorus; found in all plant and animal cells. (Chap. 19, p. 688)

nucleotide: the building blocks of nucleic acids; each consists of a simple sugar, a phosphate group, and a nitrogen-containing base. (Chap. 19, p. 689)

Glossary

nucleus: the small, dense, positively charged central core of an atom. (Chap. 2, p. 65)

O

octet rule: the model of chemical stability that states that atoms become stable by having eight electrons in their outer energy level except for some of the smallest atoms, which have only two electrons. (Chap. 4, p. 132)

orbital: the space in which there is a high probability of finding an electron. (Chap. 7, p. 239)

organic compound: a compound that contains carbon; a few exceptions exist. (Chap. 5, p. 180)

osmosis: the flow of molecules through a selectively permeable membrane driven by concentration difference. (Chap. 13, p. 467)

oxidation: a reaction in which an element loses electrons. (Chap. 16, p. 556)

oxidation number: the charge on an ion or an element; can be positive or negative. (Chap. 5, p. 157)

oxidation-reduction reaction: a reaction characterized by the transfer of electrons from one atom or ion to another. Also known as a redox reaction. (Chap. 16, p. 555)

oxidizing agent: the substance that gains electrons in a redox reaction. It is the substance that is reduced. (Chap. 16, p. 562)

P

pascal (Pa): the SI unit for measuring pressure. (Chap. 11, p. 378)

period: a horizontal row in the periodic table. (Chap. 3, p. 96)

periodic law: the statement that the physical and chemical properties of the elements repeat in a regular pattern when they are arranged in order of increasing atomic number. (Chap. 3, p. 94)

periodicity: the tendency to recur at regular intervals. (Chap. 3, p. 90)

pH: a mathematical scale in which the concentration of hydronium ions in a solution is expressed as a number from 0 to 14. (Chap. 14, p. 502)

photosynthesis: the process used by certain organisms to capture energy from the sun. (Chap. 20, p. 734)

physical change: a change in matter where its identity does not change. (Chap. 1, p. 20)

physical property: a characteristic of matter that is exhibited without a change of identity. (Chap. 1, p. 20)

plasma: an ionized gas. (Chap. 10, p. 347)

polar covalent bond: a bond where the electrons are shared unequally; there is some degree of ionic character to this type of bond. (Chap. 9, p. 310)

polar molecule: a molecule that has a positive pole and a negative pole because of the arrangement of the polar bonds; also called a dipole. (Chap. 9, p. 331)

polyatomic ion: an ion that consists of two or more different elements. (Chap. 5, p. 158)

polymer: a large molecule that is made up of many smaller repeating units. (Chap. 18, p. 649)

potential difference: the difference in electron pressure at the cathode (low) and at the anode (high) in an electrochemical cell. (Chap. 17, p. 601)

pressure: the force acting on a unit area of a surface. (Chap. 10, p. 344)

product: a new substance formed when reactants undergo chemical change. (Chap. 6, p. 192)

property: the characteristics of matter; how it behaves. (Chap. 1, p. 4)

protein: a polymer formed from small monomer molecules linked together by amide groups. (Chap. 19, p. 670)

proton: a positively charged subatomic particle. (Chap. 2, p. 62)

Q

qualitative: an observation made without measurement. (Chap. 1, p. 14)

quantitative: an observation made with measurement. (Chap. 1, p. 14)

R

radioactivity: the spontaneous emission of radiation by an unstable atomic nucleus. (Chap. 21, p. 746)

reactant: a substance that undergoes a reaction. (Chap. 6, p. 192)

reducing agent: the substance that loses electrons in a redox reaction. It is the substance that is oxidized. (Chap. 16, p. 562)

reduction: a reaction in which an element gains one or more electrons. (Chap. 16, p. 556)

reforming: the use of heat, pressure, and catalysts to convert large alkanes into other compounds, often aromatic hydrocarbons. (Chap. 18, p. 639)

respiration: the complex series of enzyme-catalyzed reactions that are used to extract chemical energy from glucose. (Chap. 19, p. 694)

RNA: ribonucleic acid. (Chap. 19, p. 688)

S

salt: the general term used in chemistry to describe the ionic compound formed from the negative part of an acid and the positive part of a base. (Chap. 15, p. 516)

saturated hydrocarbon: a hydrocarbon with all the carbon atoms connected to each other by single bonds. (Chap. 18, p. 623)

saturated solution: a solution that holds the maximum amount of solute under the given conditions. (Chap. 13, p. 458)

scientific law: a fact of nature that is observed so often that it is accepted as the truth. (Chap. 2, p. 59)

scientific model: a thinking device, built on experimentation, that helps us understand and explain macroscopic observations. (Chap. 1, p. 11)

semiconductor: an element that does not conduct electricity as well as a metal, but that does conduct slightly better than a nonmetal. (Chap. 3, p. 105)

shielding effect: the tendency for the electrons in the inner energy levels to block the attraction of the nucleus for the valence electrons. (Chap. 9, p. 304)

sievert: the unit of radiation equal to one gray multiplied by a factor that assesses how much of the radiation striking tissue is actually absorbed by the tissue and so is a measure of how much biological damage is caused. (Chap. 21, p. 776)

single displacement: a type of reaction where one element takes the place of another in a compound. (Chap. 6, p. 205)

solid: a substance in which the particles occupy fixed positions in a well-defined, three-dimensional arrangement. (Chap. 10, p. 340)

soluble: term describing a substance that dissolves in a liquid. (Chap. 6, p. 215)

solute: the substance that is being dissolved when making a solution. (Chap. 1, p. 23)

solution: a mixture that is the same throughout, or homogeneous. (Chap. 1, p. 22)

solvent: the substance that dissolves the solute when making a solution. (Chap. 1, p. 23)

specific heat: a measure of the amount of heat needed to raise the temperature of 1 gram of a substance 1°C. (Chap. 13, p. 445)

spectator ion: an ion that is present in solution but does not participate in the reaction. (Chap. 15, p. 520)

standard atmosphere (atm): the pressure that supports a column of mercury 760 millimeters in height. (Chap. 11, p. 376)

standard solution: a solution of known molarity used in a titration. (Chap. 15, p. 539)

standard temperature and pressure, STP: the set of conditions 0.00°C and 1 atmosphere. (Chap. 11, p. 395)

steroid: a lipid with a distinctive four-ring structure. (Chap. 19, p. 686)

stoichiometry: the study of relationships between measurable quantities, such as mass and volume, and the number of atoms in chemical reactions. (Chap. 12, p. 404)

strong acid: an acid that is completely ionized in water; no molecules exist in the water solution. (Chap. 14, p. 498)

strong base: a base that is completely dissociated into separate ions when dissolved in water. (Chap. 14, p. 497)

sublevel: the small energy divisions in a given energy level. (Chap. 7, p. 234)

sublimation: the process by which particles of a solid escape from its surface and form a gas. (Chap. 10, p. 356)

substance: matter with the same fixed composition and properties. (Chap. 1 p. 15)

substituted hydrocarbon: a compound that has the same structure as a hydrocarbon, except that other atoms are substituted for part of the hydrocarbon. (Chap. 18, p. 640)

substrate: the name given to a reactant in an enzyme-catalyzed reaction. (Chap. 19, p. 676)

supersaturated solution: a solution containing more solute than the usual maximum; they are unstable. (Chap. 13, p. 459)

surface tension: the force needed to overcome intermolecular attractions and break through the surface of a liquid or spread the liquid out. (Chap. 13, p. 442)

synthesis: the name applied to a reaction in which two or more substances combine to form a single product. (Chap. 6, p. 203)

T

temperature: the measure of the average kinetic energy of the particles that make up a material. (Chap. 10, p. 348)

theory: an explanation based on many observations and supported by the results of many experiments. (Chap. 2, p. 59)

thermoplastic: a plastic that will soften and harden repeatedly when heated and cooled. (Chap. 18, p. 660)

thermosetting: a plastic that hardens permanently when first formed. (Chap. 18, p. 660)

titration: the process of determining the molarity of an acid or base by using an acid-base reaction where one reactant is of known molarity. (Chap. 15, p. 539)

transition element: any of the elements in Groups 3 through 12 of the periodic table, all of which are metals. (Chap. 3, p. 103)

Glossary

triple bond: a bond formed by sharing three pairs of electrons between two atoms. (Chap. 9, p. 325)

tritium: the hydrogen isotope with a mass number of 3. (Chap. 21, p. 766)

Tyndall effect: the scattering effect caused when light passes through a colloid. (Chap. 13, p. 472)

unsaturated hydrocarbon: a hydrocarbon that has one or more double or triple bonds between carbon atoms. (Chap. 18, p. 629)

unsaturated solution: a solution in which the amount of solute dissolved is less than the maximum that could be dissolved. (Chap. 13, p. 458)

valence electron: an electron in the outermost energy level of an atom. (Chap. 2, p. 78)

vapor pressure: the pressure of a substance in equilibrium with its liquid. (Chap. 10, p. 357)

vitamin: an organic molecule required in small amounts; there are fat-soluble and water-soluble types. (Chap. 19, p. 690)

volatile: description of a substance that easily changes to a gas at room temperature. (Chap. 1, p. 35)

voltage: an electrical potential difference, expressed in units of volts. (Chap. 17, p. 601)

weak acid: an acid in which almost all the molecules remain as molecules when placed into a water solution. (Chap. 14, p. 499)

weak base: a base in which most of the molecules do not react with water to form ions. (Chap. 14, p. 500)

This index will help you locate important information in the text quickly and easily. Page numbers given in bold type indicate the location of a definition for that entry.

Index

B

Index

Index

Index

Index

Index

history of, 86-94; long version, 104 illus.; metal/nonmetal property patterns, 100-101; physical states/classes of elements, 102-105; pictorial version, 28-29 illus.; and valence electrons, 231, 232

Periodicity, 90, 94, 257

Permeable membranes, 468

Perms (hair), chemistry of, 657

Peroxides, 277 illus.

Perspiration, 446 illus.

Pesticide movement, 771 illus.

Petroleum, 145 illus.; deposits, 637 illus.; processing, 638 illus., 638-639

pH, 502; blood, 531, 533; common substances, 506 illus., 507; cosmetics, 501; measurement, 502 illus., 504-505

pH indicators, 502 illus., 504, 507-508, 508 illus.; as acid-base indicators, 481 illus., 482, 496; requirements for, 545; titration endpoint determination with, 540 illus.

pH scale, 500 illus.; 503 illus.; interpreting, 503, 506

Phages, 772

Pharmaceutical production worker, 635

Pharmaceutical technician, 635

Pharmacist, 634-635

Pharmacologist, 635

Phenolphthalein, 540 illus.

Phenylalanine, 671

Pheromones, 642 illus.

Phosphate group, 273, 689

Phospholipids, 687

Phosphorescence, 744 illus.

Phosphoric acid, 499

Phosphors, 294

Phosphorus: allotropes, 179 illus., 274 illus.; importance of, 273

Phosphorus pentachloride, 214

Photography, 563-564, 565 illus.

Photosynthesis, 43, 44 illus., 124-125, 733, **734**-735; and energy, 736-737

Photovoltaic cells, 728

Physical changes, 20, 20 illus., 35, *See also* Changes of state

Physical properties, 20, 34-35

Pie graphs, 805

Piston, expansion work done on, 374 illus., 375

Plant-care specialist, 212-213

Plasmas, 347

Plaster of paris, 167, 168

Plastics, 659, 660-661; recycling, 659, 661 illus.; thermosetting vs. thermoplastic, 660

Platen, 110

Platinum, 292

Platinum group, 292 illus.

Plutonium, 102, 295; in nuclear reactors, 765

Plutonium-238, 295

pnp-junctions, 112

Polar covalent bonds, 310-311

Polar molecules, 331-333

Polar solvents, 331 illus.

Polonium, 276

Polyatomic ionic compounds, 158-162

Polyatomic ions, 158, 159 illus.; common, 159 table

Polyester, 643 illus.

Polyethylene, 649; high-density, 659; low-density, 656 illus., 659

Polyethylene terephthalate, 659

Polymerization reactions, 655-658, 660

Polymers, 649; cross-linking, 658, 660; natural, 653; structure, 654-655; synthetic, 649 illus., 649-653; water affinity, 653

Polypeptides, 672, *See also* Proteins

Polypropylene, 659

Polyprotic acids, 485 illus.; ionization, 486 illus.

Polysaccharides, 681-682, 682 illus.

Polystyrene, 659

Polyunsaturated fatty acids, 686

Polyvinyl alcohol, 40

Polyvinyl chloride, 659

Polyvinylacetate, 656 illus.

Pool water, pH testing, 507 illus.

Popcorn, 396

Porcelain, 163

Positional isomers, 631

Positron emission tomography, 769 illus.

Potassium, 86; electron configuration, 247; as plant essential nutrient, 265 illus.; in water, 85 illus.

Potassium-argon dating, 755

Potassium bromide: formation, 518; ionic bonding, 308 illus.

Potassium chloride, 143 illus., 154; solubility-temperature relationship, 459

Potassium chromate, 293 illus.

Potassium dichromate, 293 illus.

Potassium hydroxide, 264 illus.; reaction with calcium chloride solution, 215 illus.; as strong base, 498

Potassium iodide, 154

Potassium nitrate, 459

Potassium perchlorate, 76, 707

Potential difference, 601, 601 illus.

Pottery, 163

Power plants, 726 illus.; energy efficiency, 731 table; nuclear, 732 illus., 764-765 illus.

Powers of ten, 795-796

Praseodymium: use in welding goggles, 294 illus.

Precipitates, 191 illus., 215 illus.

Pressure, 344; gases, 372-375; measurement, 376; standard atmosphere, 376; standard temperature and pressure (STP), 395; unit interconversion, 379-381; units of, 378-379, 379 table

Pressure cookers, 359, 360

Pressure gauges, 376-377

Private investigator, 13

Products, 192; physical state, 195; predicting mass of, 415-416

Professional opportunities, *see* Careers in chemistry

Promethium, 295

Propane, 145 illus., 182 illus., 623; combustion, 416; cracking, 638

Propene, 631

Properties, 4, 6 illus., *See also* Chemical properties; Physical properties

Proteases, 222, 677 illus.

Proteins, 670, 670 illus.; amino acid structure, 670-672; denaturation, 673; as enzymes, 676 illus., 676-677; as fat replacement, 685; hydrogen bonding in, 439, 672-673; three-dimensional structure, 672 illus., 672-673

Protons, 62, 66, 230; properties, 67 table

Proust, Joseph, 53-54

Pure substances, 15

Putrescine, 644

Pyrite, 165 illus.

Index

Qualitative observations, 14
Quantitative observations, 14
Quantities, 785
Quartz, 248 illus.
Quinine, 146

R **(ideal gas law constant),** 419
Radiation, *see* Electromagnetic radiation
Radiation exposure, 776 illus.
Radio waves, 72
Radioactive decay, 747-750; half-life, 756
Radioactivity, 295, **746;** detection, 751 illus.; discovery, 744-746; problems with, 775-779; waste disposal, 778-779
Radioisotope tracers, 768-770 illus., 771
Radioisotopes: medical applications, 768-770; nonmedical applications, 771, 773; sources of, 773
Radiometry, 754-755, 757 illus., 757-758, 760 illus.
Radium, 265
Radon, 132, 281, 775, 777
Rain forest, pharmaceuticals from, 146
Rare earth elements, *see* Lanthanides
Ration heaters, 221
Reactants, 192; limiting, 220; physical state, 195; predicting mass of, 414-415
Reaction capacity, 236-237
Reaction rate, *see* Chemical reaction rate
Reactions, *see* Chemical reactions
Rechargeable batteries, 210 illus., 612, 613 illus., 613-614
Recommended Daily Allowance (RDA), 128
Recycling, 55; aluminum, 589 illus., 725; glass, 60; plastics, 659, 661 illus.
Red phosphorus, 179 illus., 274 illus.

Redox reactions, *see* Oxidation-reduction reactions
Reducing agents, 559, **562**
Reduction reactions, 556, *See also* Oxidation-reduction reactions
Reforming, 639
Respiration, 124, 574-575, **694,** 694, 696-698, 737
Retinal, 632
Reverse osmosis, 465, 467 illus., 468
Rhodium, 292
Rhodopsin, 632
Ribonucleic acid, *see* DNA
Ribose, 689
RNA, 273, **688;** structure of, 690
RNA nucleotides, 689
Robotic arm, 109
Rock salt, *see* Sodium chloride
Rocket fuel, 566
Roman Empire, lead poisoning in, 271
Rose quartz, 249 illus.
Rubber, 658 illus., 660
Rubbing alcohol, 140 illus., 642
Rubidium-strontium dating, 755
Ruby, 249 illus.
Rust, 41 illus., 122, 204 illus., 250 illus., 555 illus., 557, 570 illus.
Ruthenium, 292
Rutherford, Ernest, 64, 230
Rutherford's atomic model, 65 illus., 230, 231 illus.

s **orbitals,** 239 illus., 242; and periodic table, 243-247; size, 250-251, 251 illus.
s **sublevels,** 235, 238
Saccharin, 5 illus.
Salt bridge, 602 illus., 604
Salt (table salt), *see* Sodium chloride
Salts, 516
Samarium, 295
Sapphire, 249 illus.
Saturated fats, 630, 633 illus.
Saturated fatty acids, 686 illus., 687
Saturated hydrocarbons, 623-629
Saturated solutions, 458
Scandium, 247

Scanning probe microscope (SPM), 240
Scanning tunneling microscope (STM), 10, 240-241
Scientific laws, 59
Scientific method, 59 illus.
Scientific models, 11
Scientific notation, 795-799
Scintigraphy, 770 illus.
Scintillation counters, 751 illus.
Scrap metal processing worker, 597
Scurvy, 667, 691
Sea, 18 illus.
Seaborgium, 102 illus.
Selenium, 276; electron configuration, 99 illus.; photosensitivity, 277 illus.
Semiconductor diodes, 112-113
Semiconductors, 105; electrical conduction in, 111-112
Serotonin, 432
Shampoos, pH, 501
Shape-filling molecular models, 319 illus.
Shape-memory metals, 108-109
Shielding effect, 304, 305 illus.
SI derived units, 785-786, 786 table
SI prefixes, 786 table, 786-787
SI units, 785 table
Sievert (radiation unit), 776
Significant digits, 793-795
Silicon, 270; doping, 111 illus., 112 illus.; uses, 272 illus.; valence electrons, 111 illus.
Silicon chips, 113 illus.
Silicon dioxide, 347 illus.
Silk, 653
Silver, 23, 106 illus.; as coinage metal, 86, 110, 292; conductivity, 313; electroplating, 593 illus.; tarnish, 572 illus.; in water, 85 illus.
Silver bromide, 30 illus.; in photographic film, 564, 565 illus.; use in films, 279 illus.
Silver nitrate, 200-201
Silver sulfide, 572
Simple sugars, *see* Sugars
Single bonds, 629 illus.
Single-displacement reactions, 205, 206-207, 208 table
Skawinski, Dr. William, 316-317
Skunk cabbage, 575 illus.
Slime, 40

Index

Index

Merken/Tony Stone Images, (r)Crown Studios; **258 262** Matt Meadows; **264** (tl)Yoav Levy/PhotoTake NYC, (tr)Chip Clark, (b)Stephen Frisch/Stock Boston; **265** (tl)William Rivelli/The Image Bank, (tr)Charles Gupton/Stock Boston, (b)Matt Meadows; **267** Matt Meadows; **268** (t)Joe Towers/The Stock Market, (c)Richard Megna/Fundamental Photographs, (b)Mark Burnett; **269** Will McIntyre/Photo Researchers; **270** (t)Glencoe photo, (c)David Parker/Scienc Photo Library/Photo Researchers, (b)Stephen Frisch/Stock Boston; **271** Scala/Art Resource, NY; **272** (t)H.P. Merten/The Stock Market, (c)Matt Meadows, (b)Ron Chapple/FPG International; **273** Mark Steinmetz; **274** (tl)Matt Meadows, (tr)Tony Stone Images, (c)Tim Courlas, (b)GTE Labs/Peter Arnold, Inc.; **275** (t)Matt Meadows, (b)Matt Meadows; **276** Farrell Grehan/Photo Researchers; **277** (t)Yoav Levy/PhotoTake NYC, (cl)John Maher/The Stock Market, (cr)Matt Meadows, (b)SuperStock; **278** (l)Barry Slaven/Medical Images, Inc., (r)Matt Meadows; **279** (t)Skip Comer, (cl)Yoav Levy/PhotoTake NYC, (cr)Gary Hansen/PhotoTake NYC, (b)Matt Meadows; **280** Jerome Tisne/Tony Stone Images; **281** Glencoe photo; **282** (t)Matt Meadows, (b)Jim Cummins/FPG International; **283** (l)Stephen Frisch/Stock Boston, (r)Tom Pantages; **284** Matt Meadows; **285** Keith Ledbury/Uniphoto; **286** (t)Richard Megna/Fundamental Photographs, (c)Gerard Fritz/FPG International, (b)Dawson Jones, Inc./The Stock Market; **287** (t)Alan Schein/The Stock Market, (cl)H.P. Merten/The Stock Market, (cr)David Joel/Tony Stone Images, (b)B. Roland/Image Works; **288** The Robert Elgood Collection, London; **289** (t)Terry Farmer/Tony Stone Images, (bl)Robert Sherbow/Uniphoto, (br)Matt Meadows; **290** (t)Matt Meadows, (b)Ken Kay/Fundamental Photographs; **291** (l)Jim Pickerell/The Image Works, (r)Aaron Haupt; **293** (t)Matt Meadows, (b)Mark E. Gibson/The Stock Market; **294** Stephen Simpson/FPG International; **300** Matt Meadows, (inset)IBM Research; **301** Matt Meadows; **302** Steven L. Alexander/Uniphoto; **307** Joe McNally/Sygma; **310** Mark Steinmetz; **312** Matt Meadows; **313** (l)Matt Meadows, (r)Mark Burnett; **315** Richard Pasley/Stock Boston; **316 317** Brian Buckley; **318** Matt Meadows; **319 through 324** Matt Meadows; **325** (t)Matt Meadows, (bl)Frank Siteman/Stock Boston, (br)Matt Meadows; **326** (c)Matt Meadows, (bl)Mark Steinmetz, (br)Sinclair Stammers/Science Photo Library/Photo Researchers, (others)E.R. Degginger/Color-Pic; **327** Perkin-Elmer Corporation; **328** Matt Meadows; **329** Matt Meadows; **330** (l)Stan Osolinski/Tony Stone Images, (r)Zefa-N.Y. Gold/The Stock Market; **331** Kristen Brochmann/Fundamental Photographs; **333** (tl, tc, tr)Matt Meadows, (b)Charles D. Winters; **337** Matt Meadows; **338** SuperStock; **339** Bill Auth/Uniphoto; **340** (t)Peter Menzel/Stock Boston, (cl)Tino Hammid, (cr)Will Crocker/The Image Bank, (b)James L. Amos; **341** (t)Mark Burnett, (bl br)Matt Meadows; **343 344** Matt Meadows; **345** Doug Martin; **346** James L. Amos; **348** Uniphoto; **351** Matt Meadows; **353** Mark Steinmetz; **354** Joseph Nettis/Stock Boston; **355** Doug Martin; **356 357** Matt Meadows; **359** Eric Leigh Simmons/The Image Bank; **360 362** Matt Meadows; **365** (l)Doug Martin, (r)David R. Frazier; **367** (t)Diane Schiumo/Fundamental Photographs, (b)Matt Meadows; **369** Julian Baum/Science Photo Library/Photo Researchers; **370** David R. Frazier Photolibrary; **371** Mark Steinmetz; **372** Joseph A. Dichello Jr.; **373 375** Matt Meadows; **376** Michael Krasowitz/FPG International; **377** David R. Frazier Photolibrary; **378** John Lemker/Earth Scenes; **381** (l)Matt Meadows, (r)Mark Burnett; **382** Matt Meadows; **384** Matt Meadows; **385** The Bettmann Archive; **386** Myrleen Ferguson Cate/PhotoEdit; **387** Kim Westerskov/Ford Scientific Films/Earth Scenes; **388** Bob Mullenix; **389** Mark Steinmetz; **390** Glenn Baglo/The Vancouver Sun; **391** Matt Meadows; **394** Vince Streano/Tony Stone Images; **397** Yoav Levy/PhotoTake NYC; **400** Bob Abraham/The Stock Market; **401** James Randklev/Tony Stone Images; **402** Matt Meadows, (inset)Obremski/The Image Bank; **403** Matt Meadows; **404** John Urban/Stock Boston; **405** Matt Meadows; **406** Mark Burnett; **407 through 410** Matt Meadows; **411** (t c)courtesy Rosenthal Art Slides, (b)Lakeview Museum of Arts & Sciences, Richard K. Meyer Collection; **414** Pete Salautos/The Stock Market; **417** Joe Caputo/Gamma Liaison; **419** Jim Cambon/Tony Stone Images; **420** Matt Meadows; **421** Frank Cezus/Tony Stone Images; **422** Matt Meadows; **429** Richard Megna/Fundamental Photos; **430** Mark Steinmetz; **431** Matt Meadows; **434** Jeanne Drake/Tony Stone Images; **435** Kevin Kelley/Tony Stone Images; **436** Tim Lynch/Stock Boston; **437** (l)Ryan-Beyer/Tony Stone Images, (r)Eric Neurath/Stock Boston; **441** (l)E.R. Degginger/Color-Pic, (r)Larry West/FPG International; **442** NASA; **443** (t)Geoff Butler, (b)Custom Medical Stock Photo; **444** Richard Megna/Fundamental Photographs; **445** (t)Aaron Haupt, (b)Robert Reiff/FPG International; **446** (t)E. Nagele/FPG International, (b)Bob Daemmrich/The Image Works; **448 449** Daniel Schaefer; **450** Matt Meadows; **451** Geoff Butler; **452** Phil Degginger/Color-Pic; **454** Matt Meadows; **455** Mark Steinmetz; **457** Matt Meadows; **458** Mark Steinmetz; **459** Geoff Butler; **460** Mark Burnett; **461** Matt Meadows; **465** Richard Kavlin/Tony Stone Images; **466** StudiOhio; **469** (t)Stephen Frink/The Stock Market, (bl)Ken Frick, (br)Vince McGuire; **470** (tl)David R. Frazier Photolibrary Photolibrary, (tr)Mark Steinmetz, (bl)PhotoTake NYC, (br)Skip Comer; **471** (t)Peter Johansky/FPG International, (bl)Giraudon/Art Resource, NY, (br)KS Studios; **472** Larry Hamill; **473** (l)Kip & Pat Peticolas/Fundamental Photographs, (r)George Disario/The Stock Market; **475** Kevin Kelley/Tony Stone Images; **476** StudiOhio; **478** Animals Animals/Richard Shiell; **479** Richard Hutchings/PhotoEdit; **480** Mark Steinmetz; **481 482** Matt Meadows; **483** (tl)E.R. Degginger/Color-Pic, (tr)Matt Meadows, (b)Martin Rogers/Stock Boston; **484** (t)Julie Houck/Stock Boston, (b)Chuck Holmes; **486** Tony Stone Images; **487** Charles Gupton/Stock Boston; **488 489** Matt Meadows; **490 491** Mark Tuschman; **492** Matt Meadows; **494** (t)Kent Knudson/Uniphoto, (b)Stephen R. Swinburne/Stock Boston; **495** (t)Shaun Van Steyn/Uniphoto, (b)Ray Pfortner/Peter Arnold, Inc.; **496** The Bettmann Archive; **497** Mark Burnett; **500** The Bettmann Archive; **501** Mark Steinmetz; **502** (l)Phil Degginger/Color-Pic, (r)Matt Meadows; **503** (t)Mark Steinmetz, (bl)Glencoe photo, (br)Tony Freeman/PhotoEdit; **504** Matt Meadows; **507** Bob Daemmrich/Tony Stone Images; **509** Bob Mullenix; **510** Mark Steinmetz; **511** Mark Burnett; **512** Mark Steinmetz; **514** Paul Silverman/Fundamental Photographs; **515** Doug Martin; **516** (t)Mark Steinmetz, (b)Matt Meadows; **517 518** Matt Meadows; **520** Bob Daemmrich/Stock Boston; **522** Matt Meadows; **524** Jeff Gnass/The Stock Market; **526** Matt Meadows; **528** Charles D. Winters; **529 530** Matt Meadows; **531** James Prince/Photo Researchers; **532** Mark Steinmetz; **533** Matt Meadows; **535** Mark Burnett; **536** (t)CNRI/Science Photo Library/Photo Researchers, (b)Mark Steinmetz; **537** Somatogen, Inc.; **538** Mark Steinmetz; **539 through 545** Matt Meadows; **547** Dennis Kunkel/PhotoTake NYC; **548** Mark Steinmetz; **549** John Terence Turner/FPG International; **550** Matt Meadows; **551** Rosenfeld Images/The Stock Market; **552** Charles Krebs/The Stock Market; **553** Hickson-Bender; **554** Kuhn, Inc./The Image Bank; **555** (l)Mark Burnett, (r)Michael Dalton/Fundamental Photographs; **557** Mark Burnett; **558** (l)Ann Ronan Picture Library/Image Select, (r)Nikolay Zurek/FPG International; **559 560 561** Matt Meadows; **563** Vulcain/Explorer/Photo Researchers; **564** Gernsheim Collection, Harry Ransom Humanities Research Center, University of TX Austin; **565**, (t)Mark Steinmetz, (others)Geoff Butler; **566** NASA; **567** Bryan F. Peterson/The Stock Market; **568** (t)Matt Meadows, (b)C.J. Allen/Stock Boston; **569** (t)Stacy Pick/Stock Boston, (b)Doug Martin; **570** (l)Kunio Owaki/The Stock Market, (r)Bill Bachman/Photo Researchers; **571** Zefa/The Stock Market; **572** Tim Courlas; **573** PhotoEdit; **574** (l)Telegraph Colour Library/FPG International, (r)Hickson Associates; **575** (t)Leonard Lee Rue III/Earth Scenes, (bl)Phil Degginger/Color-Pic, (br)Keith Kent/Science Photo Library/Photo Researchers; **576** Geoff Butler; **578** Matt Meadows; **579** E.R. Degginger/Color-Pic; **581** (l)Matt Meadows, (r)Rohan/Tony Stone Images; **582** Stephen Dalton/Photo Researchers; **583** Van Bucher/Photo Researchers; **584** Simon Fraser/Royal Victoria Infirmary, Newcastle/Science Photo Library/Photo Researchers; **587** Mark Steinmetz; **588** ALCOA; **589** (l)ALCOA, (r)M.E. Warren/Photo Researchers; **590** (t)Rick Raymond/Nawrocki Stock Photo, (b)courtesy Kennecott Utah Copper Corporation; **591** (t)John Zoiner/International Stock, (c)Donald L. Miller/International Stock, (b)Peter Arnold, Inc.; **592** (t)Pasquale Caprile/International Stock, (bl)Tom Pantages, (br)Mark Steinmetz; **593** (t)SuperStock, (b)Rick Gayle/The Stock Market; **594** Dr. Jeremy Burgess/Science Photo Library/Photo Researchers; **595** Capital Features/The Image Works; **596 597** Phillip DeManczuk; **598** Matt Meadows; **599** Jerry Wachter/Photo Researchers; **600** Matt Meadows; **602** Otto Rogge/The Stock Market; **603 604** Matt Meadows; **605** Mark Steinmetz; **609** StudiOhio; **610** Dept. of Clinical Radiology, Salisbury District Hospital/Science Photo Library/Photo Researchers; **611** Coco McCoy from Rainbow; **612** StudiOhio; **613** Russell D. Curtis/Photo Researchers; **617** Richard Laird/FPG International; **618** StudiOhio; **620** Ed Taylor/FPG International; **621** Mark Steinmetz; **622** Anne Sager/Photo Researchers; **627** Matt Meadows; **630** (t)Morton & White, (b)John Lund/PhotoTake NYC; **633** (l)Matt Meadows, (r)Matt Meadows; **634 635** Mark Tuschman; **637** (l)Saussier-Vanderstockt/Gamma Liaison, (r)Mark A. Leman/Tony Stone Images; **639** (l)Richard W. Brooks/Photo Researchers, (r)Mike Abrahams/Tony Stone Images; **640** Susanne Buckler/Gamma Liaison; **641** (l)Bob Daemmrich/Stock Boston, (r)Mark Steinmetz; **642** (t)Daniel A. Erickson, (c)Matt Meadows, (bl)Roger K. Burnard, (br)Matt Meadows; **643** Matt Meadows; **644** (tl c)Mark Steinmetz, (tr)The Bettmann Archive, (b)Matt Meadows; **645** (l)Owen Franken/Stock Boston, (r)John Sims/Tony Stone Images; **646** Alvin E. Staffan; **647** Spencer Grant/Photo Researchers; **648** NASA; **649** (t)Stephen Frisch/Stock Boston, (b)Robert J. Hack/International Sports Photo Agency; **651** Matt Meadows; **653** (t)Matt Meadows, (bl)courtesy Amoco Fabric and Fibers Company, (br)Nuridsany et Perennou/Photo Researchers; **654** Glencoe photo; **655** Charles D. Winters/Photo Researchers; **656** (l)Chris Springman/The Stock Market, (r)Mark Steinmetz; **657** Vince Streano/Tony Stone Images; **658** (l)Richard

Teacher's Notes

Teacher's Notes

Teacher's Notes

Teacher's Notes

Teacher's Notes